D1476162

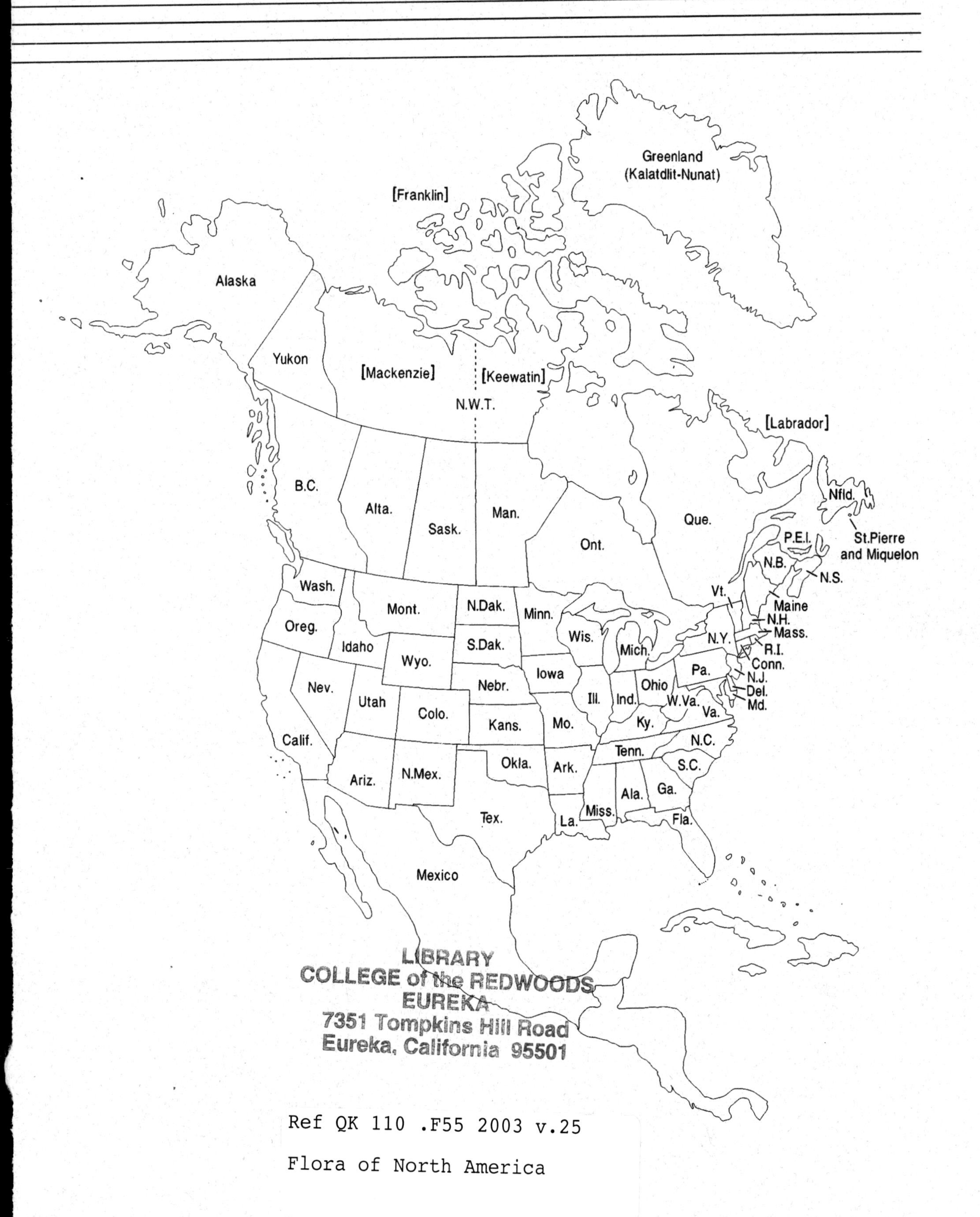
Greenland
(Kalatdlit-Nunat)
[Franklin]
Alaska
Yukon
[Mackenzie]
[Keewatin]
N.W.T.
[Labrador]
B.C.
Alta.
Sask.
Man.
Ont.
Que.
Nfld.
St.Pierre
and Miquelon
P.E.I.
N.B.
N.S.
Vt.
Maine
N.H.
Mass.
R.I.
Conn.
N.Y.
Wash.
Mont.
N.Dak.
Minn.
Oreg.
Idaho
Wyo.
S.Dak.
Wis.
Mich.
Pa.
N.J.
Del.
Md.
Iowa
Nev.
Utah
Colo.
Nebr.
Ill.
Ind.
Ohio
W.Va.
Va.
Calif.
Kans.
Mo.
Ky.
N.C.
Tenn.
Okla.
Ark.
S.C.
Ariz.
N.Mex.
Ala.
Ga.
Miss.
Tex.
La.
Fla.
Mexico

Flora of North America

Flora of North America

North of Mexico

Edited by FLORA OF NORTH AMERICA EDITORIAL COMMITTEE

VOLUME 25

Magnoliophyta: *Commelinidae* (in part): *Poaceae*, part 2

Edited by Mary E. Barkworth, Kathleen M. Capels, Sandy Long, and Michael B. Piep

Illustrated by Linda A. Vorobik

NEW YORK OXFORD • OXFORD UNIVERSITY PRESS • 2003

Oxford University Press
Oxford New York
Auckland Bangkok Buenos Aires Cape Town Chennai
Dar es Salaam Delhi Hong Kong Istanbul Karachi Kolkata
Kuala Lumpur Madrid Melbourne Mexico City Mumbai Nairobi
São Paulo Shanghai Singapore Taipei Tokyo Toronto

Published by Oxford University Press, Inc.
198 Madison Avenue, New York, New York 10016
www.oup-usa.com

Library of Congress Cataloging-in-Publication Data
(Revised for volume 25)
Flora of North America north of Mexico
edited by Flora of North America Editorial Committee.
Includes bibliographic references and indexes.
Contents: v.1. Introduction—v. 2. Pteridophytes and gymnosperms—
v. 3. Magnoliophyta: Magnoliidae and Hamamelidae—
v. 22. Magnoliophyta: Alismatidae, Arecidae, Commelinidae (in part), and Zingiberidae—
v. 26. Magnoliophyta: Liliidae: Liliales and Orchidales—
v. 23. Magnoliophyta: Commelinidae (in part): Cyperaceae—
v. 25. Magnoliophyta: Commelinideae (in part): Poaceae, part 2

ISBN-13 978-0-19-516748-1
ISBN 0-19-516748-1 (v. 25)
1. Botany —North America.
2. Botany—United States.
3. Botany—Canada.
4. Botany—Grasses.
1. Mary E. Barkworth, Kathleen M. Capels, Sandy Long, and Michael B. Piep.
QK110.F55 2003 581.97 92-30459

9 8 7 6 5 4 3 2
Printed in the United States of America
on acid-free paper

Contents

Acknowledgments

This volume, like others in the *Flora of North America*, has benefited from the efforts of numerous individuals and the resources of many institutions. We have been delighted by the willingness of taxonomists throughout the world to respond to our questions and extend our heartfelt thanks to all who have helped us.

Authors

Preparation of taxonomic treatments requires extensive research and observation. Old treatments must be examined and their ability to portray the observed variation assessed before the task of preparing descriptions and keys can be started. The descriptions in this volume are original. In some instances, existing descriptions were modified and amplified to meet the needs of the volume; other descriptions are essentially new. Most of the identification keys are new because this volume represents the first attempt to prepare a grass treatment for all grasses for North America north of Mexico.

We thank the authors of the treatments for so generously contributing their knowledge and time. They are listed below in alphabetical order and, in the body of the work, immediately after the name of the treatment(s) they authored. The Grass Phylogeny Working Group comprised 13 individuals who asked that their contributions be attributed to the group as a whole with the members being listed, if at all, alphabetically.

Charles M. Allen
Center for Environmental Management of Military Lands
Fort Polk, Louisiana

Kelly W. Allred
New Mexico State University
Las Cruces, New Mexico

Sharon J. Anderson
Florissant Fossil Beds National Monument
Lafayette, Colorado

Carol R. Annable
Kaneohe, Hawaii

Cynthia Aulbach
Botanical Services of South Carolina
Lexington, South Carolina

Nigel P. Barker
Rhodes University
Grahamstown, South Africa

Mary E. Barkworth
Utah State University
Logan, Utah

Christine M. Bern
Colorado State University
Fort Collins, Colorado

Christopher S. Campbell
University of Maine
Orono, Maine

Lynn G. Clark
Iowa State University
Ames, Iowa

Henry E. Connor
University of Canterbury
Canterbury, New Zealand

Stephen J. Darbyshire
Agriculture and Agri-Food Canada
Ottawa, Ontario

Gerrit Davidse
Missouri Botanical Garden
St. Louis, Missouri

Patricia Dávila Aranda
Escuela Nacional de Estudios Profesionales, Iztacala
Tlalnepantla, México

Jerrold I. Davis
Cornell University
Ithaca, New York

Melvin Duvall
Northern Illinois University
DeKalb, Illinois

Robert W. Freckmann
University of Wisconsin-Stevens Point
Stevens Point, Wisconsin

Mark L. Gabel
Black Hills State University
Spearfish, South Dakota

Grass Phylogeny Working Group
Nigel P. Barker
Lynn G. Clark
Jerrold I. Davis
Melvin Duvall
Gerald F. Guala
Catherine Hsiao
Elizabeth A. Kellogg
H. Peter Linder
Roberta J. Mason-Gamer
Sarah Y. Mathews
Mark P. Simmons
Robert J. Soreng
Russell E. Spangler

Gerald F. Guala
Fairchild Tropical Garden
Miami, Florida

David W. Hall
David W. Hall Consultant, Inc.
Gainesville, Florida

Barry E. Hammel
Missouri Botanical Garden
St. Louis, Missouri and
Instituto Nacional de Biodiversidad
Santo Domingo de Heredia, Costa Rica

LeRoy H. Harvey†
Smithsonian Institution
Washington, D.C.

Stephan J. Hatch
Texas A&M University
College Station, Texas

Khidir W. Hilu
Virginia Polytechnic Institute and State University
Blacksburg, Virginia

Ken M. Hiser
University of Missouri-St. Louis
St. Louis, Missouri

Catherine Hsiao
U.S. Department of Agriculture-Agricultural Research Service
Logan, Utah

Hugh H. Iltis
University of Wisconsin-Madison
Madison, Wisconsin

Elizabeth A. Kellogg
University of Missouri-St. Louis
St. Louis, Missouri

Michel G. Lelong
University of South Alabama
Mobile, Alabama

H. Peter Linder
University of Cape Town
Rondebosch, South Africa

Robert I. Lonard
University of Texas-Pan American
Edinburg, Texas

Roberta J. Mason-Gamer
University of Illinois
Chicago, Illinois

Sarah Y. Mathews
University of Missouri
Columbia, Missouri

Peter W. Michael
University of Sydney and *National Herbarium of New South Wales*
Sydney, Australia

Paul M. Peterson
Smithsonian Institution
Washington, D.C.

Sylvia M. Phillips
Royal Botanic Gardens
Kew, England

Charlotte G. Reeder
University of Arizona
Tucson, Arizona

John R. Reeder
University of Arizona
Tucson, Arizona

James M. Rominger
Northern Arizona University
Flagstaff, Arizona

Evangelina A. Sánchez
Museo Argentino de Ciencias Naturales Bernardino Rivadavia
Buenos Aires, Argentina

J. Gabriel Sánchez-Ken
Iowa State University
Ames, Iowa

Robert B. Shaw
Colorado State University
Fort Collins, Colorado

Mark P. Simmons
Ohio State University
Columbus, Ohio

James P. Smith, Jr.
Humboldt State University
Arcata, California

Neil W. Snow
University of Northern Colorado
Greeley, Colorado

Robert J. Soreng
Smithsonian Institution
Washington, D.C.

Russell E. Spangler
University of Minnesota
St. Paul, Minnesota

John W. Thieret
Northern Kentucky University
Highland Heights, Kentucky

Rahmona A. Thompson
East Central University
Ada, Oklahoma

Jesús Valdés-Reyna
Universidad Autonoma Agraria Antonio Narro
Saltillo, México

Alan S. Weakley
University of North Carolina-Chapel Hill
Chapel Hill, North Carolina

Robert D. Webster
U.S. Department of Agriculture-Agricultural Research Service
Beltsville, Maryland

J. K. Wipff
Pure Seed Testing, Inc.
Hubbard, Oregon

H. Oliver Yates
David Lipscomb University
Nashville, Tennessee

Data Contributors

Ideally, treatment authors would prepare the distribution maps for each species, based on a combination of fieldwork and examination of specimens in numerous herbaria. This is a daunting task for a region of approximately 21.5×10^6 km^2 and over 700 herbaria, an undertaking for which no author had the time or resources needed. To address this problem, a geographic database has been developed. The information in this database is obtained from a wide range of resources, including authors, reviewers, herbarium databases, unpublished regional treatments, and various published resources. We list here the individuals (other than authors) and herbaria that provided us with the data for the database. The data were donated free of charge, the only request being that we acknowledge our sources. We do so gladly and gratefully. A list of the publications consulted can be found in the Geographic Bibliography. The maps can also be viewed on the Web at http://herbarium.usu.edu/webmanual/. These maps are larger, provide some information on the source of the data for each record, and are updated at irregular intervals.

Individuals

Ken Allison
Canadian Food Inspection Agency
Ottawa, Ontario

Brock Benson
Natural Resources Conservation Service
Tremonton, Utah

David Biek
Oregon State University
Corvallis, Oregon

Curtis Randall Bjork
Washington State University
Pullman, Washington

Pam Brunsfeld
University of Idaho
Moscow, Idaho

Tom Cady
University of Iowa
Iowa City, Iowa

Jacques Cayouette
Agriculture and Agri-Food Canada
Ottawa, Ontario

Adolf Ceska
British Columbia Ministry of Environment
Victoria, British Columbia

William J. Cody
Agriculture and Agri-Food Canada
Ottawa, Ontario

Charles A. Davis
Ecological Consultant
Lutherville, Maryland

Myrna Fleming
Flathead Valley Community College
Flathead, Montana

Ben Franklin
Utah Natural Heritage Program
Salt Lake City, Utah

Alton M. Harvill, Jr.
Longwood College
Farmville, Virginia

Stuart G. Hay
Université de Montréal
Montréal, Québec

Douglass Henderson†
University of Idaho
Moscow, Idaho

Harold Hinds†
University of New Brunswick
Fredericton, New Brunswick

Ron Hise
Idaho Department of Parks and Recreation
Plummer, Idaho

Judy Hoy
Stevensville, Montana

John T. Kartesz
Biota of North America Program
Chapel Hill, North Carolina

Helen Kennedy
University of British Columbia
Vancouver, British Columbia

Roger Q. Landers, Jr.
Extension Specialist
San Angelo, Texas

Jeanette C. Oliver
Flathead Valley Community College
Flathead, Montana

Elizabeth L. Painter
Santa Barbara Botanic Garden
Santa Barbara, California

Steve J. Popovich
U.S. Department of Agriculture-Forest Service
Fort Collins, Colorado

Hans Roemer
University of British Columbia
Vancouver, British Columbia

Ronald J. Tyrl
Oklahoma State University
Stillwater, Oklahoma

Gordon C. Tucker
Eastern Illinois University
Charleston, Illinois

Margriet Wetherwax
University of California-Berkeley
Berkeley, California

Vernon Yadon
Pacific Grove Museum of Natural History
Pacific Grove, California

Herbaria

Some herbaria provided copies of information in their database for our use; others provided information on individual taxa; still others make their data available via the Web. The herbaria that provided data are listed below, alphabetized by country and state or province. The letters in parentheses are internationally recognized codes for the herbaria concerned. A complete list of these codes and herbarium addresses is available at http://www.nybg.org/bsci/ih/ih.html. If a code is not given, it means that the herbarium concerned does not have an official code.

Provincial Museum of Alberta (PMAE)
Edmonton, Alberta

Royal British Columbia Museum (V)
Victoria, British Columbia

University of British Columbia (UBC)
Vancouver, British Columbia

University of Manitoba (WIN)
Winnipeg, Manitoba

Mount Allison University
Sackville, New Brunswick

Auburn University (AUA)
Auburn, Alabama

Tongass National Forest Herbarium (TNFS)
Sitka, Alaska

University of Alaska Museum (ALA)
Fairbanks, Alaska

California State University-Chico (CHSC)
Chico, California

Humboldt State University (HSC)
Arcata, California

Jepson Herbarium (JEPS)
University of California-Berkeley
Berkeley, California

San Jose State University (SJSU)
San Jose, California

University of California-Berkeley (UC)
Berkeley, California

Colorado State University (CS)
Fort Collins, Colorado

University of Colorado (COLO)
Boulder, Colorado

United States National Herbarium (US)
Washington, D.C.

Fairchild Tropical Garden (FTG)
Miami, Florida

Florida Museum of Natural History (FLAS)
Gainesville, Florida

Florida State University (FSU)
Tallahassee, Florida

University of Florida (FLAS)
Gainesville, Florida

University of South Florida (USF)
Tampa, Florida

University of Georgia (GA)
Athens, Georgia

Valdosta State University (VSC)
Valdosta, Georgia

University of Idaho (ID)
Moscow, Idaho

Illinois Natural History Survey (ILLS)
Champaign, Illinois

Ada Hayden Herbarium (ISC)
Iowa State University
Ames, Iowa

University of Iowa (IA)
Iowa City, Iowa

R.L. McGregor Herbarium (KANU)
University of Kansas
Lawrence, Kansas

Institute for Botanical Exploration (IBE)
Mississippi State University
Mississippi State, Mississippi

University of Mississippi (MISS)
University, Mississippi

Missouri Botanical Garden (MO)
St. Louis, Missouri

University of Nebraska-Omaha (OMA)
Omaha, Nebraska

Nevada State Museum (NSMC)
Carson City, Nevada

University of Nevada-Las Vegas (UNLV)
Las Vegas, Nevada

Northern Prairie Wildlife Research Center (NPWRC)
Jamestown, North Dakota

Cleveland Museum of Natural History (CLM)
Cleveland, Ohio

Kent State University (KE)
Kent, Ohio

Miami University (MU)
Oxford, Ohio

Ohio State University (OS)
Columbus, Ohio

Ohio University (BHO)
Athens, Ohio

Oklahoma State University (OKLA)
Stillwater, Oklahoma

Robert Bebb Herbarium (OKL)
University of Oklahoma
Norman, Oklahoma

Malheur, Umatilla, and Wallowa-Whitman National Forests Herbarium
Baker, Oregon

Morton E. Peck Herbarium (WILLU)
Oregon State University
Corvallis, Oregon

Oregon State University (OSC)
Corvallis, Oregon

Botanical Research Institute of Texas (BRIT)
Fort Worth, Texas

C.L. Lundell Herbarium (LL)
University of Texas-Austin
Austin, Texas

Sam Houston State University (SHST)
Huntsville, Texas

S.M. Tracy Herbarium (TAES)
Texas A&M University
College Station, Texas

Spring Branch Science Center (SBSC)
Houston, Texas

Stephen F. Austin State University (ASTC)
Nacogdoches, Texas

Texas A&M University (TAMU)
College Station, Texas

University of Texas-Austin (TEX)
Austin, Texas

Welder Wildlife Foundation Herbarium (WWF)
Sinton, Texas

Intermountain Herbarium (UTC)
Utah State University
Logan, Utah

Virginia Polytechnic Institute and State University (VPI)
Blacksburg, Virginia

Pacific Lutheran University
Tacoma, Washington

University of Washington (WTU)
Seattle, Washington

Marion Ownbey Herbarium (WS)
Washington State University
Pullman, Washington

Western Washington University (WWB)
Bellingham, Washington

West Virginia University (WVA)
Morgantown, West Virginia

University of Wisconsin (WIS)
Madison, Wisconsin

Reviewers

Before being finalized, each treatment was reviewed by regional specialists, i.e., taxonomists with particular knowledge of the flora in some part of the *Flora* region. These individuals provided a wide range of advice, including provision of additional distributional information, suggestions for improving the keys, and observations on the ecology of taxa in their region. We join the contributors in thanking them for their assistance in improving the quality of the treatments.

Charles M. Allen
Center for the Environmental Management of Military Lands
Fort Polk, Louisiana

Kelly W. Allred
New Mexico State University
Las Cruces, New Mexico

Loran C. Anderson
Florida State University
Tallahassee, Florida

Ray Angelo
New England Botanical Club
Cambridge, Massachusetts

Alan R. Batten
University of Alaska-Fairbanks
Fairbanks, Alaska

Bruce Bennett
Yukon Department of the Environment
Whitehorse, Yukon

David E. Boufford
Harvard University
Cambridge, Massachusetts

Edwin L. Bridges
Botanical and Ecological Consultant
Bremerton, Washington

Luc Brouillet
Université de Montréal
Montréal, Québec

Julian J.N. Campbell
The Nature Conservancy
Lexington, Kentucky

Jacques Cayouette
Agriculture and Agri-Food Canada
Ottawa, Ontario

Adolf Ceska
British Columbia Ministry of the Environment
Victoria, British Columbia

Kenton L. Chambers
Oregon State University
Corvallis, Oregon

Edward W. Chester
Austin Peay State University
Clarksville, Tennessee

Anita F. Cholewa
University of Minnesota
St. Paul, Minnesota

Steven E. Clemants
Brooklyn Botanic Garden
Brooklyn, New York

William J. Cody
Agriculture and Agri-Food Canada
Ottawa, Ontario

Tom S. Cooperrider
Kent State University
Kent, Ohio

William Crins
Ontario Ministry of Natural Resources
Peterborough, Ontario

Amy Denton
University of Alaska-Fairbanks
Fairbanks, Alaska

H. R. DeSelm
University of Tennessee-Knoxville
Knoxville, Tennessee

George W. Douglas
British Columbia Ministry of Sustainable Resource Management
Victoria, British Columbia

Barbara Ertter
University of California-Berkeley
Berkeley, California

Kasia Fogarsi
Morris Arboretum
Philadelphia, Pennsylvania

Bruce A. Ford
University of Manitoba
Winnipeg, Manitoba

Donna Ford-Werntz
West Virginia University
Morgantown, West Virginia

Robert W. Freckmann
University of Wisconsin-Stevens Point
Stevens Point, Wisconsin

Craig C. Freeman
University of Kansas
Lawrence, Kansas

Arthur Haines
New England Wild Flower Society
Framingham, Maine

Richard Halse
Oregon State University
Corvallis, Oregon

H. David Hammond
Northern Arizona University
Flagstaff, Arizona

Bruce F. Hansen
University of South Florida
Tampa, Florida

Vernon L Harms
University of Saskatchewan
Saskatoon, Saskatchewan

Stephan L. Hatch
Texas A&M University
College Station, Texas

Robert R. Haynes
University of Alabama
Tuscaloosa, Alabama

Richard J. Hebda
Royal British Columbia Museum
Victoria, British Columbia

G. Frederic Hrusa
California Department of Food and Agriculture
Courtland, California

Charles G. Johnson
U.S. Department of Agriculture-Forest Service
Baker, Oregon

Walter S. Judd
University of Florida
Gainesville, Florida

Walter A. Kelly
Mesa College
Grand Junction, Colorado

Robert Kral
Botanical Research Institute of Texas
Fort Worth, Texas

Gordon Leppig
California Department of Fish and Game
Eureka, California

Aaron Liston
Oregon State University
Corvallis, Oregon

Charles T. Mason, Jr.
University of Arizona
Tucson, Arizona

Michelle McMahon
Washington State University
Pullman, Washington

James D. Morefield
Nevada Natural Heritage Program
Carson City, Nevada

Marian Zinck Munro
Nova Scotia Museum of Natural History
Halifax, Nova Scotia

John B. Nelson
University of South Carolina
Columbia, South Carolina

Francis E. Northam
Kansas State University
Manhattan, Kansas

Michael J. Oldham
Ontario Natural Heritage Information Centre
Peterborough, Ontario

Richard R. Olds
XID Services, Inc.
Pullman, Washington

Harold Ornes
Southern Utah University
Cedar City, Utah

John G. Packer
University of Alberta
Edmonton, Alberta

Donald Pinkava
Arizona State University
Phoenix, Arizona

Jim Pojar
British Columbia Forest Service
Smithers, British Columbia

Thomas Ranker
University of Colorado
Boulder, Colorado

Cindy Roché
Consultant
Medford, Oregon

Bruce A. Sorrie
Longleaf Ecological
Whispering Pines, North Carolina

Russell Spangler
University of Minnesota
St. Paul, Minnesota

Mary C. Stensvold
U.S. Forest Service
Sitka, Alaska

Peter F. Stickney
U.S. Forest Service
Missoula, Montana

John L. Strother
University of California-Berkeley
Berkeley, California

Scott Sundberg
Oregon State University
Corvallis, Oregon

David M. Sutherland
University of Nebraska-Omaha
Omaha, Nebraska

John W. Thieret
Northern Kentucky University
Highland Heights, Kentucky

R. Dale Thomas
University of Louisiana-Monroe
Monroe, Louisiana

Lowell E. Urbatsch
Louisiana State University
Baton Rouge, Louisiana

Edward G. Voss
University of Michigan
Ann Arbor, Michigan

Alan S. Weakley
University of North Carolina-Chapel Hill
Chapel Hill, North Carolina

Thomas F. Wieboldt
Virginia Polytechnic Institute and State University
Blacksburg, Virginia

Richard P. Wunderlin
University of South Florida
Tampa, Florida

James L. Zarucchi
Missouri Botanical Garden
St. Louis, Missouri

Peter F. Zika
University of Washington
Seattle, Washington

Specimen Loans

We were able to utilize the illustration skills of Linda Vorobik and her assistants because the herbaria of the University of California, Berkeley and the University of Washington were willing to provide staff assistance for processing the loans involved, cabinets for storing the specimens, and workspace. We thank the herbaria of both institutions for their support.

Loaning specimens creates considerable work for both the loaning and the borrowing institutions for each must keep track of the specimens as they are loaned, received, and returned. We thank the herbaria listed below for making specimens available for use in preparing the illustrations.

Crop Evolution Laboratory Herbarium (CEL)
University of Illinois
Urbana, Illinois

Florida Natural History Museum (FLAS)
Gainesville, Florida

Gray Herbarium (GH)
Harvard University
Cambridge, Massachusetts

Herbarium Senckenbergianum (FR)
Frankfurt, Germany

Intermountain Herbarium (UTC)
Utah State University
Logan, Utah

Jepson Herbarium (JEPS)
University of California-Berkeley
Berkeley, California

Missouri Botanical Garden (MO)
St. Louis, Missouri

Museo Argentino de Ciencias Naturales Bernardino Rivadavia (BA)
Buenos Aires, Argentina

National Botanical Institute (PRE)
Pretoria, South Africa

National Herbarium of New South Wales (NSW)
Sydney, Australia

Range Science Herbarium (NMCR)
New Mexico State University
Las Cruces, New Mexico

S.M. Tracy Herbarium (TAES)
Texas A&M University
College Station, Texas

United States National Herbarium (US)
Washington, D.C.

University of California-Berkeley (UC)
Berkeley, California

University of Louisiana-Monroe (NLU)
Monroe, Lousiana

University of South Florida (USF)
Tampa, Florida

Virginia Polytechnic Institute and State University (VPI)
Blacksburg, Virginia

In addition, J.K. Wipff and Hugh H. Iltis donated several specimens from their personal collections for use in preparing the illustrations and subsequent deposition in the Intermountain Herbarium. We thank both for their assistance and for enriching the Intermountain Herbarium.

Nomenclatural Consultants

One of the most difficult aspects of a flora for authors at institutions lacking extensive holdings of older taxonomic literature is nomenclature. Determining the correct scientific name for a taxon frequently requires consulting and interpreting literature that is available at few institutions. The task is particularly onerous for the grass volumes, as our nomenclatural file, which will be published separately, accounts for many more synonyms than are mentioned in other volumes of the *Flora of North America*. We sent all our nomenclatural questions to Dan H. Nicolson (U.S. National Herbarium) until his retirement in 1999, after which we sent them to K. Gandhi (Harvard University Herbaria). We thank both for their prompt and clear answers to our questions. We particularly thank Gandhi, whom we deluged with questions immediately prior to publication. We also thank members of the Committee for Spermatophyta of the International Association for Plant Taxonomy, whose assistance was sought by Nicolson and Gandhi on particularly difficult questions.

We also extend our thanks to Steve Cafferty (The Natural History Museum, England) for answering numerous questions concerning the typification of Linnaean names, and J.F. Veldkamp (Nationaal Herbarium Nederland, Leiden branch, The Netherlands) for responding to a variety of questions, many relating to the nomenclature of southeast Asian species that are now established in North America, and for *Grass Literature*, his intermittently published summaries of taxonomic literature on grasses.

Financial Supporters

We take pleasure in thanking our financial supporters.

Anonymous

Thad and Jenny Box

Flora of North America Association

Idaho Botanical Center

U.S. Department of Agriculture:
Agricultural Research Service
Animal and Plant Health Inspection Service
Forest Service
Natural Resources Conservation Service

U.S. Department of the Interior:
Bureau of Land Management
National Park Service

Utah Agricultural Experiment Station

Utah State University

The Agricultural Research Service of the U.S. Department of Agriculture provided the initial funding for preparation of a grass identification manual for North America. When it became apparent that additional funding was needed, the other U.S. federal agencies shown agreed to assist the project.

Utah State University and the Utah Agricultural Experiment Station have supported the project since its inception. The University has waived indirect costs and paid for most office supplies, mailing costs, and computer services. The Experiment Station has provided direct financial support. In addition, these two organizations support the Intermountain Herbarium, a resource that is essential to the project, and pay Barkworth's salary.

The Flora of North America Association helped defray the cost of modifying the original manuscripts for publication in the *Flora of North America*. Thad and Jenny Box, the Idaho Botanical Center, and two anonymous donors provided financial support to the project when the financial situation was particularly difficult. This non-governmental support has also been important in providing the matching funds needed to show that the project is of value to the public as a whole, not just government agencies.

Introduction

Two volumes of *Flora of North America* will be devoted to the *Poaceae* (Grass Family), volumes 24 and 25. The current volume treats six subfamilies of the *Poaceae*: the *Panicoideae, Arundinoideae, Chloridoideae*, *Centothecoideae*, *Aristidoideae*, and *Danthonioideae*. Together these six subfamilies constitute the PACCAD group of grasses (see p. 3). The other subfamilies present in the *Flora* region, the *Pharoideae*, *Bambusoideae*, *Ehrhartoideae*, and *Pooideae,* will be treated in volume 24.

Volume 25 is being published before volume 24, because volume 24 will contain a key to all the grass tribes in the *Flora* region and an artificial key to the genera. Because keys must be checked against the final descriptions for the taxa they contain, volume 24 cannot be completed until the tribal and generic treatments in both volumes have been finalized. The current volume contains a key to the tribes treated, but not one to the genera.

There is little or no disagreement over which genera belong in the *Poaceae*, but current perception concerning placement of the family differs from that adapted by Cronquist (1981). Cronquist placed the *Poaceae* next to the *Cyperaceae* in the order *Cyperales* and the *Juncaceae* in a separate, but closely related, order *Juncales*. Such a treatment was not universally supported (see, for example, Dahlgren and Clifford 1982) and is now generally rejected. Today, the more widely accepted perception is that the *Poaceae*, together with the *Joinvilleaceae*, *Restionaceae*, *Flagellariaceae*, and some other small families, belong to one lineage, and the *Cyperaceae* and *Juncaceae* to another, both lineages being included within a broadly defined *Poales*, one that encompasses such families as the *Typhaceae*, *Mayaceae*, *Bromeliaceae*, *Xyridaceae*, and *Eriocaulaceae* (Angiosperm Phylogeny Working Group 1998; Judd et al. 2002; Michelangeli et al. 2003).

The order of taxa in this volume is intended to reflect their relationships, except that infraspecific taxa are listed alphabetically under their species. A linear order cannot exactly mirror these relationships, even if all of them were known, because of the prevalence of polyploidy, both ancient and modern (Wendel 2000), among angiosperms. Nevertheless, placing closely related taxa together usually places morphologically similar taxa together partly because, for most taxa, the only indicator of relationships currently available is their appearance. The advantage of being reminded which taxa are considered closely related every time one opens a flora is that one unconsciously absorbs much information that an artificial ordering, such as an alphabetical ordering, conceals. It also means that, if following the key leads to a species that does not seem correct, it will be worth examining the descriptions and illustrations of the adjacent taxa before rekeying the specimen. The disadvantage is that, until familiar with the ordering, one must consult the index to locate the treatment of a particular taxon.

The subfamilial classification adopted in the two grass volumes is based on the comprehensive examination of molecular and morphological data conducted by the Grass Phylogeny Working Group (2001). The tribal treatment is similar to that adopted by Clayton and Renvoize (1986), with two exceptions: the *Cynodonteae* as recognized here includes the *Eragrostideae*, and *Gynerium* is placed in a tribe of its own, the *Gynerieae*, which is included in the *Panicoideae*; Clayton and Renvoize included *Gynerium* in the *Arundineae* of the *Arundinoideae*. Reasons for these exceptions are given after the description of the two tribes. The ordering of the tribes reflects the findings of the Grass Phylogeny Working Group.

In most instances, the authors of the generic treatments determined the limits of their genera, but the decision to recognize both *Panicum* and *Dichanthelium* was Barkworth's. The order of genera generally follows that of Clayton and Renvoize (1986), unless more recent study suggests an alternative arrangement. Controversial decisions are discussed following the description of the taxon concerned. References to alternative treatments are cited in appropriate places throughout the volume.

History

The two grass volumes differ somewhat in design, format, and content from other volumes of *Flora of North America* because their history is different. Their contents were originally intended for publication in a single volume identification manual similar in concept and design to Hitchcock's *Manual of Grasses of the United States* (1935a, 1951). Work on a new *Manual* began in 1986 when Hugh H. Iltis persuaded Terry B. Kinney, of the U.S.D.A.'s Agricultural Research Service, that a successor to Hitchcock's *Manual* was needed, one that would reflect current taxonomic knowledge, include Alaska and Canada, and provide more detailed distribution maps. Iltis reported on his discussions to grass taxonomists attending the International Symposium on Grass Systematics and Evolution (see Soderstrom et al. 1987) and invited expressions of interest in both contributing to and leading the project. Most of those present expressed interest in contributing to the project; Barkworth expressed interest in leading it.

Shortly afterwards, an editorial committee consisting of Susan G. Aiken, Mary E. Barkworth (Chair), Christopher S. Campbell, Hugh H. Iltis, and David W. Hall was formed. This committee prepared a *Guide for Contributors*, formed a list of the genera to be included, and identified potential contributors. Most invitations to contribute were accepted and several taxonomists volunteered to prepare treatments of "orphan" genera, those for which no one else had volunteered. In 1990, Kathleen M. Capels joined the project, taking over responsibility for general editing, maintenance of the nomenclatural and bibliographic files, and expansion of the geographic database.

A major focus of the scientific editing at this time was reducing each treatment to an average of 175 words per taxon. This limit was imposed in order to meet the goal of producing a single volume manual. In reducing the length of a treatment, the criterion was that: "After using the keys, reading the descriptions, studying the illustrations, and examining the distribution maps, users should either be confident that they have correctly identified their plant—or confident that they have not done so."

The illustrations in Hitchcock's *Manual* (1935a, 1951) were available copyright free, but many species now found in the *Flora* region were not illustrated in either edition. Moreover, many of the existing illustrations did not clearly portray the diagnostic features of the taxon. Attempts to obtain permission to use copyrighted illustrations from other publications were unsuccessful; some publishers were willing to grant permission for use of an illustration in a single print edition, but none of those contacted was willing to give permission for displaying the illustration on a Web site. This meant that the project had to prepare many more original illustrations than originally envisioned, because making the contents of the *Manual* available via the Web had become an integral part of the *Manual* project. Furthermore, all the illustrations from Hitchcock's *Manual* had to be redrawn, because the originals bore many annotations and other marks. The existing funding, however, precluded such extensive work. Linda A. Vorobik agreed to be responsible for the project's illustration needs once adequate funding was obtained, but no significant progress could be made in the absence of such funding.

In 1999, Barkworth and representatives of the Flora of North America Association agreed that everyone would benefit if treatments submitted to the *Manual* project were modified for initial publication in *Flora of North America.* This agreement opened up the Association's network of regional reviewers to the *Manual* project, eliminated the redundancy that had existed between the two projects, and probably helped persuade the U.S. Forest Service, Animal and Plant Health Inspection Service, Natural Conservation Service, Bureau of Land Management, and Park Service to provide the additional financial support needed by the *Manual* project.

In modifying the treatments that had been submitted to the *Manual* project, the emphasis has been on providing more detailed descriptions of the taxa and more extensive comments than on compatibility in format with other volumes of *Flora of North America.* The most obvious difference between the two grass volumes and other volumes in the *Flora* is the page layout, but there are other, less immediately obvious differences. These are discussed in the following paragraphs.

Comparison with Other FNA Volumes

The grass volumes, like other volumes of *Flora of North America* (*FNA*), provide full descriptions for all native species and established introduced species. Unlike other *FNA* volumes, they also provide full descriptions for many species that are known, within the *Flora* region, only as agricultural crops, ornamentals, research taxa, or waifs (introduced species that have failed to become established), as well as a few species that have never been found in the region but have been identified as being serious threats to American agriculture by the U.S. Department of Agriculture. The decision to include ornamental species is based partly on Hitchcock's (1935a, 1951) decision to do so, but also on the growing interest in grasses as ornamental species. The provision of full descriptions and illustrations for waifs was designed to benefit those working at small herbaria. The provision of such information for species grown only in connection with research programs was initially designed to benefit those interested in the origin of crop species; its extension to several forage species will, it is hoped, prove beneficial to those working with such species in North America. There are, of course, many other forage species that might have been included. The list was limited to taxa mentioned by Hitchcock (1935a, 1951) or appearing on a list of taxa occurring in one of the states or provinces without a clear indication that it was known only from a research planting.

The inclusion of potential noxious weeds is intended to help ensure that, should they be found in the *Flora* region, they will be correctly identified while eradication or control measures are still feasible. The species included are those on the U.S. Department of Agriculture's Noxious Weed List.

Text Material

The format of the text material differs slightly from that of other volumes. For instance, a bipartite number, e.g., 17.10, is used to indicate the position of the genera within the volumes. The number before the period indicates its tribal placement; the number after the period, its placement within the tribe. These bipartite numbers, rather than page numbers, are used to indicate the location of a treatment within the volume. Alphabetic and numerical listings of the genera, with page numbers, will be found on the last two pages of each volume.

Most scientific names are *italicized* in the treatments, including those above the generic level. This is recommended, but not required, by the most recent edition of the International Code of Botanical Nomenclature (Greuter et al. 2000). The exception is the use of **boldface** for the heading of the principal description of a taxon or, in a comment section, to indicate that a taxon is accepted even though no description is provided. This happens, for instance, when comparing two taxa, one of which does not occur in the *Flora* region. Both taxa are accepted, and hence their names are **boldfaced**, but only the taxon found in the *Flora* region is described.

Measurements, unless otherwise stated, refer to lengths. They may be preceded by the number of structures present, but are placed before comments on shape and other aspects of the structure concerned, e.g., "anthers 3, 1.4–5 mm, cylindrical." The provision of the number of anthers rather than the number of stamens is a carry-over from the concern with length that dominated editing when the manuscripts were being prepared for a single volume publication; "stamens 3, anthers 1.4–5 mm" takes more space than "anthers 3, 1.4–5 mm."

In the treatments, accent marks are shown only for people's names or the vernacular names of plants, not for names of places (e.g., "Mexico" not "México"). Similarly, English-language spelling of place names is adopted (e.g., "Greenland" not "Grønland"). In citing references, the language of the original article and source is used unless a non-Roman alphabet is involved; words in non-Roman alphabets are transliterated according to internationally recognized conventions. Journals cited in references are abbreviated according to *Botanico-Periodicum-Huntianum* (Lawrence et al. 1968) and its supplement (Bridson and Smith 1991).

Some of the information that other volumes of the *Flora* provide in a standardized format is presented in the grass volumes as part of the comments on a taxon. For instance, in the grass volumes there are no boxed codes indicating rarity or horticultural value, but this information is often in the comment section. Flowering time is rarely provided, even for species with a narrow distribution, because authors were not asked to provide this information. A summary of the number of species in each genus, and their global distribution, follows the generic description. If a species is rare or endangered, this is usually included in the comments that follow the species description, but the legal status is not always stated. Such information can be found on Web sites maintained by the agencies responsible for monitoring species of particular interest.

Illustrations

In the grass volumes, almost every species will be illustrated. The abundance of illustrations reflects the focus of the *Manual* project on identification.

The illustration plates were originally designed in one of four formats: full page, single-column, half-page horizontal, and quarter page. In most instances this layout has been retained, even if it means a single-column plate occupies a whole page. The layout of a few plates was modified in order to improve the appearance of a page.

Nomenclature

The nomenclatural information presented in the body of the work is limited to accepted names and an abbreviated citation of the author(s) of such names. The abbreviations follow Brummitt and Powell (1992). These abbreviated citations are appropriate for use in scientific journals and other publications. The placement of the citation is also shown for tautonymic infraspecific taxa (those in which the specific and infraspecific epithets are identical). Full publication information for the names mentioned in this volume can be found at http://www.ipni.org/index.html, the Web site of the International Plant Names Index, or at http://mobot.mobot.org/W3T/Search/vast.html, the Missouri Botanical Garden's nomenclatural Web site. The full name of the authors of plant names can be found at http://www.ipni.org/index.html or in Brummitt and Powell.

Frequently encountered synonyms of the names accepted in this volume are placed in the list of Names and Synonyms (pp. 731), together with the name used in this volume.

Vernacular names, often called common names, are a problem. Some can truly be described as common, but most have been created, often by translating scientific names, often scientific names based on old taxonomic treatments. Vernacular names of frequently encountered species usually remain unchanged, even when all taxonomists agree that they reflect an obsolete taxonomic treatment. Such names are, therefore, more stable than the corresponding scientific name. Unlike scientific names, however, they may be both regionally and linguistically restricted. Moreover, some vernacular names are used for two or more different taxa, and some taxa have more than one vernacular name, even within a single language.

The vernacular names included in this work came from various sources, including the U.S. Department of Agriculture's PLANTS Database (http://plants.usda.gov/) and, for French names, Centre ARICO's information for noxious weeds in Québec (http://www.agr.gouv.qc.ca/dgpar/arico/index3.htm), La Flore du Québec (http://www3.sympatico.ca/arold/famille2.html), and information from the federal Terminology Office, Montreal, and the Université de Montréal. Spanish language names are included if they were provided by the author of a treatment.

Some of the English names were created by the editors. In doing so, they followed the existing pattern for the genus, with preference being given to names that include the generic name and allude to the plants' morphology or ecology. All names in languages other than English were obtained from published sources or knowledgeable individuals.

Distribution Maps

The distribution maps in the grass volumes are prepared from a geographic database developed at the Intermountain Herbarium, the editorial center for the two volumes. This database was initially developed almost entirely from publications such as state, regional, or provincial atlases, county-level checklists, or, particularly for Canada, unpublished distribution maps. Around 1990, many herbaria began databasing their collections; several of these institutions have allowed us to incorporate their data into the database (see Acknowledgements above).

The maps in the grass volumes show where, in the *Flora* region, a species has been found. This is not always the same as either its current distribution or its native distribution. Many native species can no longer be found where they were once abundant; others have spread beyond their earlier range. Records known to be based on cultivated specimens are not shown in the maps, but some of the records shown may reflect escapes from cultivation that have not persisted in a particular area. We encourage those interested in the status of a species in a particular area to consult a regional expert.

In the grass volumes, the basic mapping unit used for the contiguous United States is the county, because most herbarium specimens show in which county they were collected. In addition, there are many county checklists available, and numerous atlases that indicate distribution by county. For Canada, Alaska, and Greenland, the basic mapping unit is the locality. Localities in these areas are plotted as circles with a diameter of 60 km. Locality data for the contiguous United States are shown as dots whose diameter does not change with the scale of the map.

Geographic coordinates for the collection locality are given on an increasing number of specimens, but most of those shown in this volume were estimated from the locality information on a label or from published or unpublished maps. The primary reference for Canadian localities was the Web site of the Geographical Names Board of Canada (http://geonames.nrcan.gc.ca/); for localities in the United States, the Web site of the Geographic Names Information Survey (http://geonames.usgs.gov/gnishome.html). None of the species in this volume grows in Greenland.

During the first half of 2002, all the authors of the treatments in this volume received a copy of the map(s) for the taxa that they had treated. Their comments led us to seek verification of several records and, in some instances, to delete records that appeared to reflect an error in data entry or identification.

Information about the distribution of infraspecific taxa is stored in the database but, because many sources do not provide such information, there are no maps for infraspecific taxa in this volume. Instead, there is a verbal synopsis of their distribution.

The maps provide an overview of the distribution of individual species within the *Flora* region, but much additional information resides in herbaria, primarily herbaria within the region. We hope that publication of the maps in this volume will lead to more research, more databasing, and the development of mechanisms for easier sharing of data. In the meantime, we continue to update the maps. "Updating" may involve replacing a record based on a print publication or other source with data from a voucher specimen, adding a completely new record, deleting an old record, or modifying an identification based on reexamination of a specimen. All new data will be reviewed for consistency with the approved maps before integration into the database. If a record suggests a major range extension, we shall ask the donor to double-check the accuracy of the identification and the mapping data. Updated maps are displayed at http://herbarium.usu.edu/webmanual/. This site also provides a limited amount of information on the basis for each record.

Bibliographic Indices

The *Geographic Bibliography* contains a listing of all the published sources that were consulted in preparing the distribution maps. The *General Bibliography* contains all references cited in this volume. Names of journals are abbreviated according to *Botanico-Periodicum-Huntianum* (Lawrence et al. 1968) and its supplement (Bridson and Smith 1991), and the names of books according to *Taxonomic Literature*, edition 2 and its supplements (Stafleu and Cowan 1976–1988, Stafleu and Mennega 1992+).

Numerical Summary

This volume treats 733 species plus 6 named interspecific hybrids. Of the 733 species, 489 are native to the *Flora* region and 237 are introduced. The other seven, all of which grow in the southern United States and in Mexico, may have been introduced since the arrival of Europeans or may be natives. Arrival after European settlement does not necessarily mean that human activity caused the establishment of the species within the *Flora* region, but it has certainly been a major factor affecting species distributions in the Americas during the last 500 or so years. Tables 1 and 2 present a summary of the above information by tribe. Obviously, adoption of different species limits from those presented in this volume would require modification of the tables.

Of the 489 grass species native to the *Flora* region, 305 ("South" + "South & Disjunct") have a range that extends to the south of the *Flora* region. Some of these extend only into Mexico; others extend as far south as Chile and Argentina. Twenty-one of these species have a disjunct distribution ("South & Disjunct"), seventeen of the disjunctions being from Mexico to South America and four from the southern

United States to southern Mexico. Thirteen species have a pan-tropical distribution and one, *Phragmites australis*, is global in distribution ("Other"). Only one species, *Danthonia intermedia*, extends beyond the *Flora* region in the north ("North"), being found also in Kamchatka. None of the species in this volume grows in Greenland. Seven species may be native to the *Flora* region (Table 2, "Unclear"), but their distribution lies primarily to the south of the United States, suggesting that they may be recent introductions rather than recently discovered natives.

Table 1. *Numerical summary of native species in this volume and their distribution outside the* Flora *region*. See text for further explanation of the column headings.

	NATIVE SPECIES						TOTAL NATIVE
	South	South & Disjunct	North	Other	*Flora* Endemic	US Endemic	
Arundinoideae							
Arundineae	0	0	0	1	0	0	1
Chloridoideae							
Cynodonteae	140	10	0	7	16	53	226
Pappophoreae	1	2	0	1	0	0	4
Orcuttieae	1	0	0	0	0	7	8
Danthonioideae							
Danthonieae	1	1	1	0	2	2	7
Aristidoideae							
Aristideae	14	0	0	0	3	12	29
Centothecoideae							
Centotheceae	0	0	0	0	0	5	5
Thysanolaeneae	0	0	0	0	0	0	0
Panicoideae							
Gynerieae	0	0	0	0	0	0	0
Paniceae	98	5	0	5	13	37	158
Andropogoneae	29	3	0	0	0	19	51
PACCAD Grasses	284	21	1	14	34	135	489

There are 135 native species that are known only from the United States ("US Endemics"); an additional 34 grow only in Canada and the United States ("*Flora* Endemics"). None of the species in this volume is restricted to Canada.

Of the 237 introduced species, 137 appear to be established (Table 2). This means that they are reproducing from plants that are not the result of cultivation or occasional seed contaminants. Sixty-eight of the species described are not established. Some of these are waifs, species that have been collected in the *Flora* region but have failed to become established here. Classic examples of such taxa are those associated with ballast dumps. Another category of introduced but not established species includes those known as "escapes", species that have been found growing outside of cultivation but appear not to persist. Many escapes are species that are cultivated as ornamentals, lawn grasses, or agricultural crops.

The status of 32 introduced species is questionable. They have been collected in the region, but whether or not they are reproducing in the region is not clear. It is probable that most such species are not established. What the tables do not reveal is the marked difference in distribution between the *Chloridoideae*, which is primarily southwestern, the *Panicoideae* and *Centothecoideae*, both of which are primarily southeastern, and the *Danthonioideae* which, in North America, tends to have western species and eastern species. The *Arundinoideae* are represented by only one native species, *Phragmites australis*. Differences in distribution of the various tribes were discussed by Barkworth and Capels (2000).

Table 2. *Numerical summary of introduced species in this volume by tribe, plus grand total of all species in each tribe.* See text for explanation of column headings.

	INTRODUCED SPECIES				STATUS UNCLEAR	NATIVE + INTRODUCED + UNCLEAR
	Established	Not Established	Question	Total Introduced		
Arundinoideae						
Arundineae	2	1	0	3	0	4
Chloridoideae						
Cynodonteae	39	22	10	71	3	300
Pappophoreae	1	0	0	1	0	5
Orcuttieae	0	0	0	0	0	8
Danthonioideae						
Danthonieae	9	1	0	10	0	17
Aristidoideae						
Aristideae	0	0	0	0	0	29
Centothecoideae						
Centotheceae	0	0	0	0	0	5
Thysanolaeneae	0	1	0	1	0	1
Panicoideae						
Gynerieae	0	1	0	1	0	1
Paniceae	64	20	16	100	4	262
Andropogoneae	22	22	6	50	0	101
PACCAD Grasses	137	68	32	237	7	733

Copyright and Citation

With the exception of the text of a few treatments, copyright for all the work in both grass volumes belongs to Utah State University. The exceptions are treatments prepared by employees of the U.S. and Canadian federal governments. United States law does not permit copyrighting of work conducted by its federal employees, so the text of treatments prepared by Paul M. Peterson (*Eragrostis*, *Muhlenbergia*, and *Sporobolus*) and Robert D. Webster (*Saccharum*) are free of copyright. Under Canadian law, the Crown owns the copyright on all material prepared by Canadian federal employees. Consequently, inquiries concerning the text of the treatments of *Danthonia*, *Karroochloa*, and *Rytidosperma* should be directed to the Minister of Public Works and Government Services, Canada.

If cited individually, this volume should be cited as:

Barkworth, M.E., K.M. Capels, S. Long, and M.B. Piep, eds. 2003. *Magnoliophyta: Commelinidae* (in part): *Poaceae*, part 2. *Flora of North America North of Mexico*, volume 25. Oxford University Press, New York.

If cited as one of several volumes in the *Flora*, it should be included in a general citation:

Flora North America Editorial Committee, eds. 1993+. *Flora of North America North of Mexico* 7+ vols. Oxford University Press, New York and Oxford.

It should, of course, be cataloged as a volume in the *Flora of North America.*

Volume Preparation

In fall 1999, Sandy Long and Michael B. Piep joined Barkworth and Capels in working on the grass volumes. The treatments for volume 25 were sent for review to the Flora of North America Association's (FNAA) regional reviewers and to John L. Strother (University of California-Berkeley), a member of the FNAA editorial committee. Many of the comments received reflected the difference in expectation between the *Manual* project and the *Flora.* After reading the comments, Long, Piep, and Barkworth referred some questions back to the authors but addressed others by consulting specimens in the Intermountain Herbarium and/or descriptions in other publications.

All treatment authors were sent a copy of their modified manuscript(s) for approval, together with a letter explaining that changes had been made and why. If they requested emendations in the modified manuscript, the changes were made unless they concerned matters of style rather than substance. Despite these efforts, the descriptions presented here are not as fully parallel, nor as long, as in other volumes of the *Flora.*

Christine M. Garrard wrote the programs that automated the map generation procedures and, in 2002, redesigned the geographic database. The new design is more robust than its precursor, and includes a module for importing data from other databases that first checks for their conformity with the nomenclature used in the grass volumes and for internal consistency of their state and county information, and another module that generates a listing of all counties in which a species has been found or, alternatively, all species that have been found in a county or set of counties. Garrard also designed and implemented the Web site (http://herbarium.usu.edu/webmanual/) where updated versions of the maps are posted at irregular intervals.

Linda A. Vorobik designed and prepared all the taxon illustrations. In selecting the features to be illustrated, she consulted the keys and descriptions provided by the authors. She was assisted in completing the illustrations by six individuals: Karen Klitz, Annaliese Miller, Linda Bea Miller, Hana Pazdírková, Cindy Roché, and Andy Sudkamp. The initials of the assistant illustrator for a given genus appear on at least one plate for the genus and are also shown to the right of this paragraph.

LAV Linda Ann Vorobik

K Karen Klitz

AM Annaliese Miller

LBM Linda Bea Miller

HP Hana Paźdirková

CTR Cindy Roché

AS Andy Sudkamp

The illustrations were labeled electronically, based on Vorobik's work, because this enabled us to use the same typeface as for the text material, even on those illustrations that had been prepared before the *Manual* project combined forces with the *Flora* project. It also permits greater versatility in using the illustrations. Long and Barkworth prepared the diagrams used to explain inflorescence structures in the *Andropogoneae.*

Michael Spooner, Director of Utah State University Press, supervised preparation of publication-ready copy in consultation with Barkworth. Ian Hatch completed most of the page composition; Laurel Anderton assisted in the final stages. Having page composition completed at Utah State University made it easier to address questions about page design arising from differences between the grass volumes and others in the *Flora of North America.*

Draft pages were shown to the FNAA's Management and Editorial Committees in October 2002 for comment and James L. Zarucchi, Managing Editor of *Flora of North America*, has reviewed the page proofs. Final decisions concerning the content and appearance of this volume were, however, made by Mary E. Barkworth. All comments and corrections should be sent to her.

Other

As lead editor, I thank all those who have helped bring this volume to fruition but particularly those at Utah State University who have been directly involved: Kathleen Capels, Sandy Long, and Michael Piep for assistance in editing, Michael Spooner for overseeing the page preparation process, and Russell Price for negotiating the agreement with the Flora of North America Association. I also extend my heartfelt thanks to my sister, Delia Rossi, for the encouragement, support, and sense of reality that she has provided throughout the duration of the project.

Mary Barkworth

SELECTED REFERENCES **Angiosperm Phylogeny Working Group.** 1998. An ordinal classification for the families of flowering plants. Ann. Missouri Bot. Gard. 85:531–553; **Barkworth, M.E.** and **K.M. Capels.** 2000. The Poaceae in North America: A geographic perspective. Pp. 327–346 *in* S.W.L. Jacobs and J. Everett (eds.). Grasses: Systematics and Evolution. International Symposium on Grass Systematics and Evolution (3rd:1998). CSIRO Publishing, Collingwood, Victoria, Australia. 408 pp.; **Bridson, G.D.R.** and **E.R. Smith** (eds.). 1991. B–P–H/S: Botanico–Periodicum–Huntianum/Supplementum. Hunt Institute for Botanical Documentation, Carnegie Mellon University, Pittsburgh, Pennsylvania, U.S.A. 1068 pp.; **Clayton, W.D.** and **S.A. Renvoize.** 1986. Genera Graminum: Grasses of the World. Kew Bull., Addit. Ser. 13. Her Majesty's Stationery Office, London, England. 389 pp.; **Cronquist, A.** 1981. An Integrated System of Classification of Flowering Plants. Columbia University Press, New York. 1262 pp.; **Dahlgren, R.M.T.** and **H.T. Clifford.** 1982. The Monocotyledons: A Comparative Study. Botanical Systematics: An Occasional Series of Monographs (series ed. V.H. Heywood). Academic Press, London and New York. 378 pp.; **Grass Phylogeny Working Group.** 2001. Phylogeny and subfamilial classification of the grasses (Poaceae). Ann. Missouri Bot. Gard. 88:373–457; **Hitchcock, A.S.** 1935a. Manual of the Grasses of the United States. U.S. Government Printing Office, Washington, D.C., U.S.A. 1040 pp.; **Hitchcock, A.S.** 1951 [title page 1950]. Manual of the Grasses of the United States, ed. 2, rev. A. Chase. U.S.D.A. Miscellaneous Publication No. 200. U.S. Government Printing Office, Washington, D.C., U.S.A. 1051 pp.; **Judd, W.S., C.S. Campbell, E.A. Kellogg, P.F. Stevens,** and **M.J. Donoghue.** 2002. Plant Systematics: A Phylogenetic Approach, ed. 2. Sinauer Associates, Sunderland, Massachusetts, U.S.A. 576 pp.; **Lawrence, G.H.M., A.F.G. Buchheim, G.S. Daniels,** and **H. Dolezal** (eds.). 1968. B–P–H: Botanico–Periodicum–Huntianum. Hunt Botanical Library, Pittsburgh, Pennsylvania, U.S.A. 1063 pp.; **Michelangeli, F.A., J.I. Davis,** and **D.W. Stevenson.** 2003. Phylogenetic relationships among Poaceae and related families as inferred from morphology, inversions in the plastid genome, and sequence data from the mitochondrial and plastid genomes. Amer. J. Bot. 90:93–106; **Stafleu, F.A.** and **R.S. Cowan.** 1976–1988. Taxonomic Literature: A Selective Guide to Botanical Publications and Collections with Dates, Commentaries and Types. 7 vols., ed. 2. Regnum Vegetabile [series], vols. 94, 98, 105, 110, 112, 115–116. Bohn, Scheltema and Holkema, Utrecht, The Netherlands; **Stafleu, F.A.** and **E.A. Mennega.** 1992+. Taxonomic Literature: A Selective Guide to Botanical Publications and Collections with Dates, Commentaries and Types. Supplement. 6+ vols. Regnum Vegetabile [series], vols. 125, 130, 132, 134–135, 137+. Koeltz Scientific Books, Königstein, Germany; **Wendel, J.F.** 2000. Genome evolution in polyploids. Pl. Molec. Biol. 42:225–249.

Flora of North America Association

The Flora of North America Association is an organization dedicated to the goal of completing a synoptic floristic account of all the plants of North America north of Mexico. Its governing body is the Flora of North America Editorial Committee, current members of which are listed below. The Editorial Management Committee, a subset of the Editorial Committee, is responsible for timely publication of the volumes. The names of its members are followed by an asterisk.

George W. Argus
Canadian Museum of Nature

Theodore M. Barkley*
Botanical Research Institute of Texas

Mary E. Barkworth*
Utah State University

Sharon Bartholomew-Began
West Chester University

David E. Boufford*
Harvard University Herbaria

Luc Brouillet*
Université de Montréal

William R. Buck
The New York Botanical Garden

Marshall R. Crosby
Missouri Botanical Garden

Paul G. Davison
University of North Alabama

Craig C. Freeman*
University of Kansas

K. Gandhi
Harvard University Herbaria

Ronald L. Hartman
University of Wyoming

Diana G. Horton
University of Iowa

Marshall C. Johnston
Austin, Texas

John T. Kartesz
Biota of North America Program

Robert W. Kiger*
Carnegie Mellon University

Aaron Liston
Oregon State University

John McNeill
Royal Botanic Garden, Edinburgh and *Royal Ontario Museum*

Nancy R. Morin*
The Arboretum at Flagstaff

David F. Murray
University of Alaska Museum

Norton G. Miller
New York State Museum

John G. Packer
University of Alberta

J. Scott Peterson
U.S. Department of Agriculture-Natural Resources Conservation Service

James B. Phipps
University of Western Ontario

Richard K. Rabeler
University of Michigan

Leila M. Shultz
Utah State University

Alan R. Smith
University of California-Berkeley

Richard W. Spellenberg
New Mexico State University

Lloyd R. Stark
University of Nevada

Raymond E. Stotler
Southern Illinois University

John L. Strother
University of California-Berkeley

John W. Thieret
Northern Kentucky University

Barbara M. Thiers
The New York Botanical Garden

Frederick H. Utech
Carnegie Mellon University

Dale H. Vitt
Southern Illinois University

Grady L. Webster
University of California-Davis

Richard P. Wunderlin
University of South Florida

Richard H. Zander*
Missouri Botanical Garden

James L. Zarucchi*
Missouri Botanical Garden

Flora of North America

PACCAD Grasses

E.A. Kellogg

The PACCAD group of grasses consists of the *Panicoideae*, *Arundinoideae sensu stricto*, *Chloridoideae*, *Centothecoideae*, *Aristidoideae*, and *Danthonioideae*. It includes more than half the species in the *Poaceae* as a whole and, according to every molecular study, constitutes a monophyletic group. There is not, however, any obvious morphological character that distinguishes it from other taxa or groups of taxa. Two characteristics that all its members share, so far as is known, are embryos with an elongated mesocotyl internode and lodicules that lack a distal membranous portion, but neither of these features is useful for routine identification.

Despite the lack of an obvious morphological character for recognizing PACCAD grasses, some general statements can be made. All C_4 grasses are in the PACCAD group, but so are many C_3 grasses. Perhaps because such a high proportion of the PACCAD grasses employ a C_4 photosynthetic pathway, the group is associated with warm climates and/or late summer blooming. Solid culms are common, but not universal, among PACCAD species; consequently a solid culm is a good indication that a grass belongs to one of the PACCAD subfamilies, but a hollow culm provides no information. Many PACCAD species have punctate hila, but so do many of the non-PACCAD grasses. It is generally easier to place a grass in its tribe or subfamily and from that determine whether it is a PACCAD grass than to work in the other direction.

Within the PACCAD grasses, molecular data indicate that *Panicoideae* and *Centothecoideae* are sister taxa, as are the two pairs *Arundinoideae* + *Chloridoideae* and *Aristidoideae* + *Danthonioideae*. The last four subfamilies form a group that members of the Grass Phylogeny Working Group (2000, 2001) refer to as the "Ligule of Hairs clade" because many (but not all) of its members have ligules made up of a fringe of hairs. Other, more cryptic characteristics shared by members of the four subfamilies are the presence of compound starch grains in the endosperm and embryonic leaf margins that meet rather than overlap.

Despite the strong evidence supporting the monophyly of the PACCAD grasses, the Grass Phylogeny Working Group (2000, 2001) decided not to give the group formal nomenclatural recognition. There were two reasons for this decision. One was the difficulty of identifying the group morphologically. The second was that it would lead to a drastic change in the meaning of the name *Panicoideae*, the name that would, according the International Code of Botanical Nomenclature, become the group name. "PACCAD" is simply an acronym, a pronounceable listing of the initials of the subfamilies in the group.

The following key to tribes includes only tribes in the PACCAD subfamilies. It is based primarily on morphological characters that can be observed with a hand lens, but many of the characters that delimit the tribes are microscopic or molecular. For this reason, several tribes have had to be keyed out more than once. A key to all tribes of grasses represented in the *Flora* region will be presented in Volume 24 of the *Flora*, as will as an artificial key to the genera.

SELECTED REFERENCES **Grass Phylogeny Working Group**. 2000. A phylogeny of the grass family (Poaceae), as inferred from eight character sets. Pp. 3–7 *in* S.W.L. Jacobs and J. Everett (eds.). Grasses: Systematics and Evolution. International Symposium on Grass Systematics and Evolution (3rd:1998). CSIRO Publishing, Collingwood, Victoria, Australia. 408 pp.; **Grass Phylogeny Working Group**. 2001. Phylogeny and subfamilial classification of the grasses (Poaceae). Ann. Missouri Bot. Gard. 88:373–457.

1. Spikelets with 1 floret; lemmas terminating in a 3-branched awn (the lateral branches sometimes greatly reduced); callus well-developed (*Aristidoideae*) .... 21. *Aristideae*
1. Spikelets with more than 1 floret or, if only 1, the lemma not terminating in a 3-branched awn; callus development various.
   2. Spikelets usually dorsally compressed, sometimes terete, rarely laterally compressed; sexually functional spikelets with 2 dimorphic florets (rarely only 1 floret), the lower floret

sterile or staminate, the upper floret bisexual or unisexual; upper glumes markedly different in texture from the upper lemmas, the thicker of the 2 structures coriaceous to indurate, rarely merely membranous, the thinner structure hyaline to membranous (*Panicoideae*, in part).

3. Glumes membranous, thinner than the upper lemmas; lower glumes usually shorter than the upper glumes, sometimes missing; upper glumes varying from slightly shorter to slightly longer than the upper lemmas; upper lemmas well-developed, usually indurate to coriaceous (sometimes membranous); lower lemmas similar in texture to the upper glumes; all spikelets pedicellate, often shortly so; inflorescence branches remaining intact at maturity . . . . 25. *Paniceae*
3. Glumes indurate, thicker than the lemmas, often subequal, at least 1 and usually both exceeding the florets; all lemmas hyaline; spikelets usually in pairs, sometimes in triplets, sometimes apparently solitary and sessile, 1 spikelet in each pair or triplet usually sessile, sometimes all spikelets pedicellate; inflorescence branches often breaking up at maturity . . . . 26. *Andropogoneae*

2. Spikelets usually laterally compressed, sometimes terete; spikelets with other than 2 florets or, if with 2, the sex distribution usually not as in the *Paniceae* or *Andropogoneae*; upper lemmas (if more than 1 floret present) usually as thick as or thicker than the upper glumes, usually membranous to coriaceous, rarely indurate.
4. Ligules absent; plants viscid annuals . . . . 19. *Orcuttieae*
4. Ligules present; plants annual or perennial but, if annual, not viscid.
5. Adaxial ligules composed of hairs or a membranous base and ciliate fringe, the fringe longer than the base, a cartilaginous ridge also sometimes present; hairs often present on either side of the collar region.
6. Lemmas flabellate or with (5)7–15 veins extending into teeth or awns (*Chloridoideae*, in part) . . . . 18. *Pappophoreae*
6. Lemmas lanceolate, rectangular, or ovate, with fewer than 5 veins or not all the veins extending into well-developed teeth or awns.
7. Lemmas usually awned, occasionally merely mucronate; awns once-geniculate, the basal portion flattened in cross section but strongly twisted; lemma apices bifid or bilobed (*Danthonioideae*, in part) . . . . 20. *Danthonieae*
7. Lemmas unawned, mucronate, or awned; awns, when present, usually not geniculate, neither flattened nor strongly twisted in the basal portion.
8. Culms 200–700 cm tall, terminating in a plumose panicle; leaves mostly basal (*Danthonioideae*, in part) . . . . 20. *Danthonieae*
8. Culms 1–350 cm tall; inflorescences various but never a terminal plumose panicle; leaves basal or cauline.
9. Lemmas 1–3 or (5)7–13-veined, the lateral veins prominent; glumes often more than ¾ as long as the spikelets, sometimes exceeding the distal florets; blades with Kranz anatomy (*Chloridoideae*, in part) . . . . . . 17. *Cynodonteae*
9. Lemmas 3-veined, the lateral veins not prominent; glumes up to ¾ as long as the spikelets; blades lacking Kranz anatomy (*Arundinoideae*, in part) . . . . 16. *Arundineae*
5. Adaxial ligules membranous, sometimes ciliolate or ciliate, but the cilia never as long as the membranous portion; hairs sometimes present at the sides of the collar region.
10. Culms 1–700 cm tall, usually less than 1 cm thick; leaves often mostly basal, sometimes mostly cauline, a few species with culms shorter than 80 cm having strongly distichous, mostly cauline leaves; inflorescences various, often with spikelike branches.
11. Lemmas with 1, 3, or 7–13 veins, the veins usually conspicuous; blades with Kranz leaf anatomy (*Chloridoideae*, in part) . . . . 17. *Cynodonteae*
11. Lemmas with 1–15 veins, the lateral veins usually inconspicuous; blades with non-Kranz leaf anatomy.
12. Lowest 1–4 florets in each spikelet sterile; paleas of fertile florets with gibbous bases and winged keels (*Centothecoideae*, in part) . . . . 22. *Centotheceae*

12. Lowest florets in each spikelet fertile; paleas of fertile florets not gibbous at the base, keels not winged (*Danthonioideae*, in part) .... 20. *Danthonieae*

10. Culms 80–1500 cm tall, 1–5 cm thick; leaves mainly cauline, evidently distichous and more or less evenly distributed; inflorescence a single terminal panicle, the branches not spikelike.
    13. Lemmas awned; leaves with a line of hairs across the collar (*Arundinoideae*, in part) .... 16. *Arundineae*
    13. Lemmas unawned; leaves without a line of hairs across the collar.
        14. Midvein of the leaf blades 5–15 mm wide near the base of the blades; blades 2–10 cm wide, those of the lower cauline leaves disarticulating; florets and plants unisexual; plants cultivated, not established, not reaching reproductive maturity when grown outside in the *Flora* region (*Panicoideae*, in part) .... 24. *Gynerieae*
        14. Midvein of the leaf blades 0.5–3 mm wide near the base of the blades; blades 0.2–10 cm wide, those of the lower cauline leaves sometimes disarticulating; florets bisexual; plants cultivated, not established in the *Flora* region, usually reaching reproductive maturity in at least some part of the region .... 23. *Thysanolaeneae*

5. ARUNDINOIDEAE Burmeist.

Grass Phylogeny Working Group
Kelly W. Allred

Plants usually perennial; cespitose or not, sometimes rhizomatous, sometimes stoloniferous. **Culms** 15–1000 cm, annual, herbaceous to somewhat woody, internodes usually hollow. **Leaves** usually mostly cauline, often conspicuously distichous; **sheaths** usually open; **auricles** usually absent; **abaxial ligules** usually absent (of hairs in *Hakonechloa*); **adaxial ligules** membranous or of hairs, if membranous, often ciliate; **blades** without pseudopetioles, sometimes deciduous at maturity; **mesophyll** usually non-radiate (radiate in *Arundo*); **adaxial palisade layer** absent; **fusoid cells** absent; **arm cells** usually absent (present in *Phragmites*); **Kranz anatomy** absent; **midribs** simple; **adaxial bulliform cells** present; **stomatal subsidiary cells** low dome-shaped or triangular; **bicellular microhairs** usually present, usually with long, narrow terminal cells; **papillae** usually absent. **Inflorescences** usually terminal, ebracteate, usually paniculate, occasionally spicate or racemose. **Spikelets** laterally compressed, with 1–several bisexual florets or all florets unisexual and the species dioecious; **florets** 1–several, terete or laterally compressed, distal florets often reduced; **disarticulation** above the glumes. **Glumes** 2, from shorter than the adjacent lemmas to exceeding the distal florets; **lemmas** (3)5–7-veined, lanceolate to elliptic, acute to acuminate, sometimes awned; **awns** 1 or 3, if 3 not fused into a single basal column; **paleas** subequal to the lemmas; **lodicules** 2, usually free, occasionally joined at the base, fleshy, usually glabrous, not, scarcely, or heavily vascularized; **anthers** (1)2–3; **ovaries** glabrous; **styles** 2, usually free, bases close together. **Caryopses** usually punctate (long-linear in *Molinia*); **endosperm** hard, without lipid; **starch grains** compound; **haustorial synergids** absent; **embryos** usually large compared to the caryopses, waisted or not; **epiblasts** absent; **scutellar cleft** present; **mesocotyl internode** elongate; **embryonic leaf margins** usually meeting (overlapping in *Hakonechloa*). x = 6, 9, 10, 12.

The *Arundinoideae* are interpreted here as including only one tribe, the *Arundineae*. The tribe used to be interpreted more broadly (e.g., Watson et al. 1985; Clayton and Renvoize 1986; Kellogg and Campbell 1987), but the broader interpretation was generally acknowledged to be somewhat artificial. Hsiao et al. (1998) showed support for inclusion of the *Danthonieae*, *Aristideae*, and *Arundineae* in a more broadly interpreted *Arundinoideae*, but other studies (e.g., Hilu et al. 1990; Barker et al. 1995, 1998; Grass Phylogeny Working Group 2001) have failed to support such a treatment.

SELECTED REFERENCES **Barker, N.P., H.P. Linder**, and **E.H. Harley**. 1998. Sequences of the grass-specific insert in the chloroplast *rpo*C2 gene elucidate generic relationships of the Arundinoideae (Poaceae). Syst. Bot. 23:327–350; **Barker, N.P., H.P. Linder**, and **E.H. Harley**. 1995. Polyphyly of Arundinoideae (Poaceae): Evidence from *rbcL* sequence data. Syst. Bot. 20:423–435; **Clark, L.G., W. Zhang**, and **J.F. Wendel**. 1995. A phylogeny of the grass family (Poaceae) based on *ndhF* sequence data. Syst. Bot. 20:436–460; **Clayton, W.D.** and **S.A. Renvoize**. 1986. Genera Graminum: Grasses of the World. Kew Bull., Addit. Ser. 13. Her Majesty's Stationery Office, London, England. 389 pp.; **Conert, H.J.** 1987. Current concepts in the systematics of the Arundinoideae. Pp. 239–250 *in* T.R. Soderstrom, K.W. Hilu, C.S. Campbell, and M.E. Barkworth (eds.). Grass Systematics and Evolution. Smithsonian Institution Press, Washington, D.C., U.S.A. 473 pp.; **Grass Phylogeny Working Group**. 2000. A phylogeny of the grass family (Poaceae), as inferred from eight character sets. Pp. 3–7 *in* S.W.L. Jacobs and J. Everett (eds.). Grasses: Systematics and Evolution. International Symposium on Grass Systematics and Evolution (3rd:1998). CSIRO Publishing, Collingwood, Victoria, Australia. 408 pp.; **Grass Phylogeny Working Group**. 2001. Phylogeny and subfamilial classification of the grasses (Poaceae). Ann. Missouri Bot. Gard. 88:373–457; **Hilu, K.W.** and **A. Esen**. 1990. Prolamin and immunological studies in the Poaceae. I. Subfamily Arundinoideae. Pl. Syst. Evol. 173:57–70; **Hsaio, C., S.W.L. Jacobs, N.P. Barker**, and **N.J. Chatterton**. 1998. A molecular phylogeny of the subfamily Arundinoideae (Poaceae) based on sequences of rDNA (ITS). Austral. Syst. Bot. 11:41–52; **Kellogg, E.A.** and **C.S. Campbell**. 1987. Phylogenetic analyses of the Gramineae. Pp. 310–322 *in* T.R. Soderstrom, K.W. Hilu, C.S. Campbell and M.E. Barkworth (eds.) Grass Systematics and Evolution. Smithsonian Institution Press, Washington, D.C., U.S.A. 473 pp.; **Linder, H.P., G.A. Verboom**, and **N.P. Barker**. 1997. Phylogeny and evolution in the *Crinipes* group of grasses (*Arundinoideae*: *Poaceae*). Kew Bull. 52:91–110; **Watson, L., H.T. Clifford**, and **M.J. Dallwitz**. 1985. The classification of the Poaceae: Subfamilies and supertribes. Austral. J. Bot. 33:433–484; **Watson, L.** and **M.J. Dallwitz**. 1992. The Grass Genera of the World. C.A.B. International, Wallingford, England. 1038 pp.

16. ARUNDINEAE Dumort.

Kelly W. Allred

See subfamily description.

There are still questions about the circumscription of the *Arundineae*, but it clearly includes the genera in this treatment. Its morphological circumscription is also difficult. The most abundant genera in North America, *Phragmites* and *Arundo*, have tall culms bearing numerous, conspicuously distichous, broad leaves and large, plumose panicles, a habit frequently described as "reedlike", but not all members of the tribe have this habit. Linder et al. (1997) noted that *Arundo*, *Phragmites*, and *Molinia* have hollow culm internodes, punctate hila, and convex sides to the adaxial ribs in the leaf blades, but these characters have not been examined in all genera of the tribe.

Members of the *Arundineae* are found in tropical and temperate areas around the world. The reedlike species are found in marshy to damp soils, but some of the other species grow in xeric habitats.

SELECTED REFERENCE **Linder, H.P., G.A. Verboom, and N.P. Barker.** 1997. Phylogeny and evolution in the *Crinipes* group of grasses (*Arundinoideae*: *Poaceae*). Kew Bull. 52:91–110.

1. Plants cespitose, not rhizomatous; rachillas and lemmas glabrous . . . . 16.01 *Molinia*
1. Plants rhizomatous or stoloniferous, sometimes also loosely cespitose; rachillas or lemmas hairy.
   2. Lemmas glabrous . . . . 16.03 *Phragmites*
   2. Lemmas hairy.
      3. Rachilla segments hairy; lemmas with papillose-based hairs on the margins . . . . 16.02 *Hakonechloa*
      3. Rachilla segments glabrous; lemmas pilose, the hairs not papillose-based . . . . 16.04 *Arundo*

16.01 MOLINIA Schrank

Mary E. Barkworth

Plants perennial; cespitose, not rhizomatous. **Culms** 15–250 cm, often disarticulating at the first node, basal internodes persistent, often swollen and clavate. **Leaves** mostly basal; **ligules** of hairs; **blades** flat or convolute, eventually disarticulating from the sheaths. **Inflorescences** terminal, panicles, not plumose. **Spikelets** laterally compressed, with (1)2–5 florets; **rachilla** prolonged beyond the distal florets, terminating in a rudimentary floret, internodes ⅓–½ as long as the florets, glabrous; **disarticulation** beneath the florets. **Glumes** exceeded by the florets, 1- or 3-veined; **calluses** 0.1–0.3 mm, blunt, glabrous or sparsely strigose, hairs to 0.5 mm; **lemmas** glabrous, inconspicuously 3(5)-veined, rounded over the back, acute to obtuse, unawned; **paleas** subequal to the lemmas; **anthers** 3. **Caryopses** falling free from the lemmas and paleas; **pericarps** loosely adherent. $x = 9$. Named for Juan Ignazio Molina (1740–1829), a Jesuit missionary and botanist, and the author of the first comprehensive summary of Chilean plants.

Molinia is a genus of two to five species, all of which are native to temperate Eurasia. One species is established in the *Flora* region.

SELECTED REFERENCE **Dix, W.L.** 1945. Will the stowaway, *Molinia caerulea*, become naturalized? Bartonia 23:41–42.

1. **Molinia caerulea** (L.) Moench [p. 9]
PURPLE MOORGRASS, MOLINIE BLEUE

Culms 15–150(250) cm, erect, rigid; **basal internodes** 2–6 cm, usually swollen and clavate. **Collars** marked with a line or narrow ridge; **sheaths** smooth, mostly glabrous, margins sparsely hairy distally; **blades** 0.2–1 cm wide. **Panicles** 5–40 cm, usually contracted; **branches** short and erect or long and lax. **Spikelets** 4–9 mm. **Lower glumes** 1.5–2.5 mm, hyaline, obtuse, slightly erose; **upper glumes** 2–3 mm, acute; **calluses** completely or almost glabrous; **lemmas** 2.5–4.5 mm, ovate, acute to obtuse; **anthers** 1.5–3 mm, purple. $2n$ = 18, 36, 90.

Molinia caerulea is established at scattered locations in the *Flora* region, but not at all the locations where it has been found. For instance, the record for Pennsylvania reflects a collection made in 1945 from an abandoned field; there are no extant populations known in the area. Most records are from southeastern Canada and the northeastern United States, but it has also been reported as being established in western Oregon.

Plants with long, lax panicle branches have been called **Molinia caerulea** subsp. **arundinacea** (Schrank) H. Paul rather than **Molinia caerulea** (L.) Moench subsp. **caerulea**, but there are many intermediates.

16.02 HAKONECHLOA Makino *ex* Honda

Mary E. Barkworth

Plants perennial; loosely cespitose, rhizomatous and stoloniferous. **Culms** 30–90 cm, erect or geniculate at the base. **Sheaths** open; **auricles** absent; **abaxial ligules** present, composed of a line of hairs across the collar; **adaxial ligules** membranous and sparsely ciliate, sometimes lacerate, cilia subequal to the base; **blades** flat, linear-lanceolate, resupinate, in living plants the glaucous-green adaxial surface facing downwards and the bright green abaxial surface facing upwards. **Panicles** not plumose. **Spikelets** pedicellate, somewhat laterally compressed, with 5–10 florets; **rachilla segments** conspicuously pilose; **disarticulation** at the base of the rachilla segment and below each spikelet. **Glumes** unequal, lanceolate, unawned; **calluses** 1.5–2 mm, strigose, hairs 1–1.5 mm; **lemmas** chartaceous, 3-veined, margins with papillose-based hairs near the base, apices inconspicuously bidentate, awned from between the teeth; **awns** 3–5 mm, straight; **paleas** 2-keeled. **Caryopses** glabrous. x = 10. Name from Hakone, a city on the island of Honshu, Japan, and the Greek *chloa*, 'grass'.

Hakonechloa is a monotypic genus, endemic to Japan, but grown as an ornamental in the *Flora* region. The resupination of the blades is not evident on herbarium specimens.

SELECTED REFERENCE Koyama, T. 1987. Grasses of Japan and Its Neighboring Regions: An Identification Manual. Kodansha, Ltd., Tokyo, Japan. 370 pp.

1. **Hakonechloa macra** (Munro) Makino [p. 9]
JAPANESE FOREST GRASS, HAKONE GRASS

Rhizomes and **stolons** covered with pale, coriaceous scales. **Culms** 30–90 cm tall, 1–1.5 mm thick, glabrous. **Adaxial ligules** 0.2-0.3 mm; **blades** 8–25 cm long, 0.4–1.2 cm wide, glabrous, abaxial surfaces green, adaxial surfaces often paler, turning orange-bronze in the fall. **Panicles** 6–12 cm long, 5–7 cm wide, open, nodding, with 15–30 spikelets; **branches** paired, somewhat stiff, scabrous. **Spikelets** 1–2 cm, yellowish-green, with 5–10 florets. **Glumes** broadly lanceolate; **lower glumes** 3–4 mm, 1–3-veined; **upper glumes** 3.8–5 mm, 3-veined; **calluses** 1.5–2 mm, strigose, hairs 1–1.5 mm; **lemmas** 6–7 mm long, 1.8–2.2 mm wide, chartaceous, 3-veined, margins sparsely pilose with long papillose-based hairs near the base, awned; **awns** 3–5 mm; **anthers** 2–3 mm. **Caryopses** about 2 mm. $2n$ = 50.

In Japan, *Hakonechloa macra* grows on rocks along rivers. Although rhizomatous, it is not an invasive species and is recommended for mass planting. Three forms are cultivated: forma *alboaurea* Makino *ex* Ohwi, with white- and yellow-striped leaves; forma *albovariegata* Makino *ex* Ohwi, with white-striped leaves; and forma *aureola* Makino *ex* Ohwi, with yellow leaves having narrow green stripes. This last form is the one most commonly available in the *Flora* region.

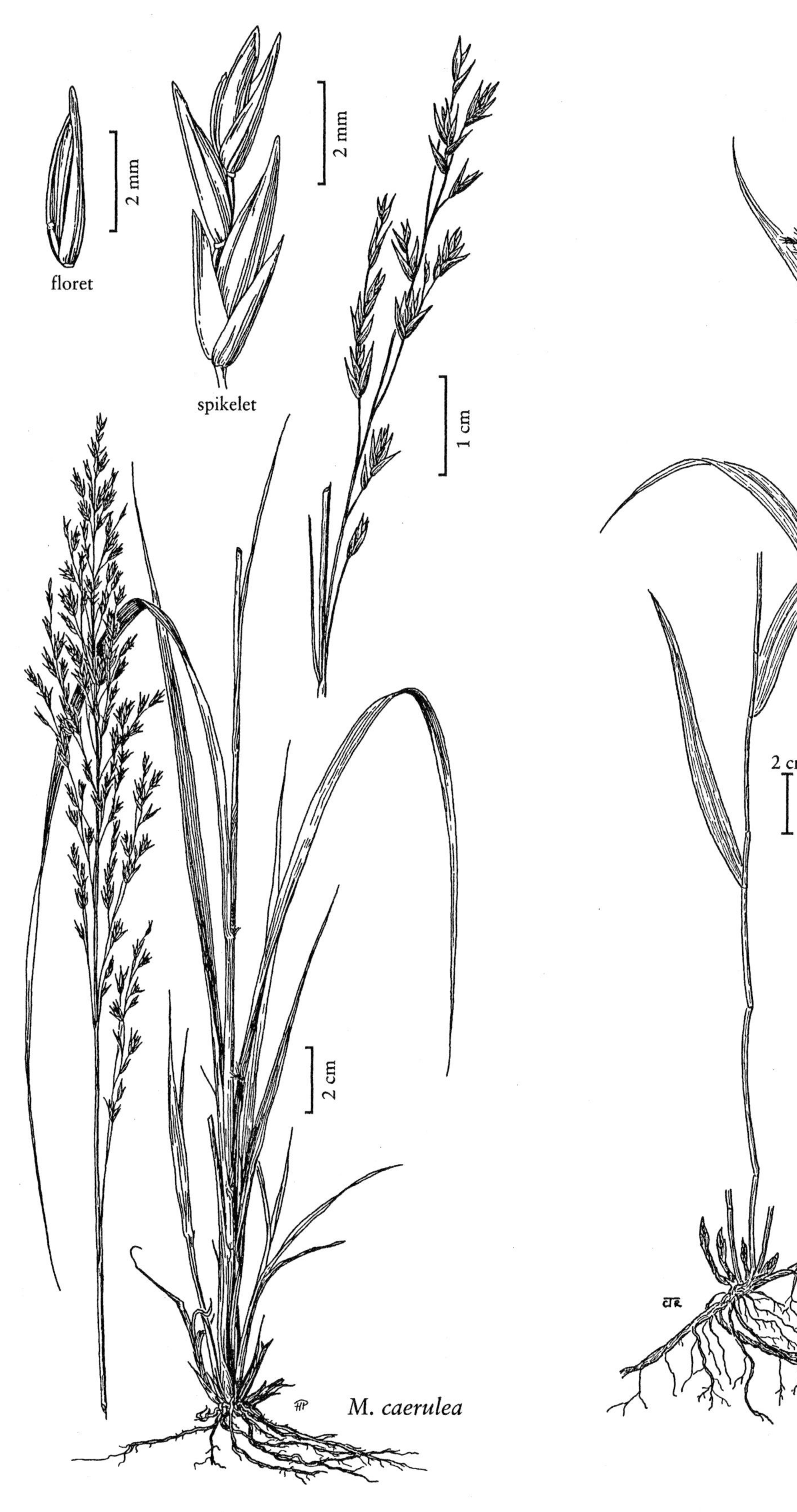

MOLINIA

2 cm
floret
2 mm
glume
glume
spikelet
rhizome

H. macra

HAKONECHLOA

16.03 PHRAGMITES Adans.

Kelly W. Allred

Plants perennial; rhizomatous or stoloniferous, often forming dense stands. **Culms** 1–4 m tall, 0.5–1.5 cm thick, leafy; **internodes** hollow. **Leaves** cauline, mostly glabrous; **sheaths** open; **ligules** membranous, ciliate; **blades** flat or folded. **Inflorescences** terminal, plumose panicles. **Spikelets** with 2–8 florets, weakly laterally compressed, lower 1–2 florets staminate, distal 1–2 florets rudimentary, remaining florets bisexual; **rachilla segments** sericeous; **disarticulation** above the glumes and below the florets. **Glumes** unequal, shorter than the florets, 1–3-veined, glabrous; **lower glumes** much shorter than the upper glumes; **calluses** pilose, hairs 6–12 mm; **lemmas** 3-veined, glabrous, unawned; **anthers** 1–3. **Caryopses** rarely maturing. $x = 12$. Name from the Greek *phragma*, 'fence', alluding to its fencelike growth.

Phragmites is interpreted here as a monotypic genus that has a worldwide distribution. Some taxonomists (e.g., Clayton 1970; Koyama 1987; Scholz and Bohling 2000) recognize 3–4 segregate species. Recent work has identified two different genotypes in North America (Saltonstall 2002) that preliminary data suggest may be morphologicaly distinct (see http://www.invasiveplants.net/). How these genotypes relate to the various segregate species that have been recognized is not yet known.

Plants of *Phragmites* are similar in overall appearance to *Arundo*, but the latter has subequal glumes, a glabrous rachilla, and hairy lemmas. Vegetatively, plants of *Arundo*, but not those of *Phragmites*, have a wedge-shaped, light to dark brown area at the base of the blades. They also tend to have thicker rhizomes, thicker and taller culms, and wider leaves than *Phragmites*, but there is some overlap. *Phragmites* is much more widely distributed than *Arundo* in North America.

SELECTED REFERENCES **Clayton, W.D.** 1967. Studies in the Gramineae. XIV. Kew Bull. 21:111–117; **Clayton, W.D.** 1970. Flora of Tropical East Africa, Gramineae (Part 1). Crown Agents for Oversea Governments and Administrations, London, England. 176 pp.; **Hocking, P.C., C.M. Finlayson,** and **A.J. Chick.** 1983. The biology of Australian weeds. 12. *Phragmites australis* Trin. *ex* Steud. J. Austral. Inst. Agric. Sci. 49:123–132; **Koyama, T.** 1987. Grasses of Japan and Its Neighboring Regions: An Identification Manual. Kodansha, Ltd., Tokyo, Japan. 370 pp.; **Saltonstall, K.** 2002. Cryptic invasion by a non-native genotype of the common reed, *Phragmites australis*, into North America. Proc. Natl. Acad. Sci. U.S.A. 99:2445–2449; **Scholz, H.** and **N. Böhling.** 2000. *Phragmites frutescens* (Gramineae) re-visited: The discovery of an overlooked, woody grass in Greece, especially Crete. Willdenowia 30:251–261.

1. **Phragmites australis** (Cav.) Trin. *ex* Steud. [p. 12]
COMMON REED, PHRAGMITE COMMUN, ROSEAU COMMUN

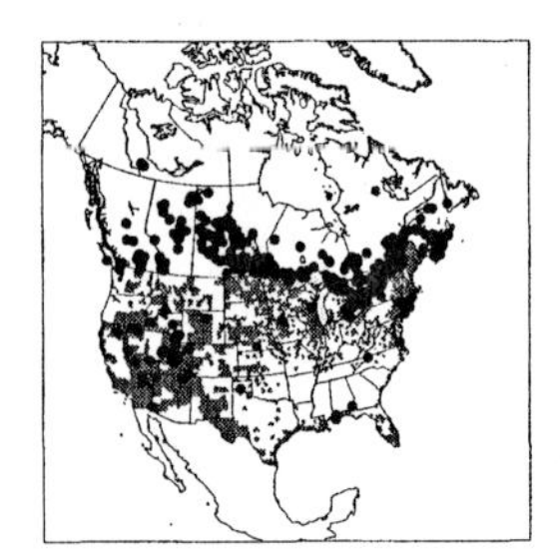

Culms 1–4 m tall, 0.5–1.5 cm thick, erect. **Ligules** about 1 mm; **blades** 15–40 cm long, 2–4 cm wide, long-acuminate, disarticulating from the sheath at maturity. **Panicles** 15–35 cm long, 8–20 cm wide, ovoid to lanceoloid, often purplish when young, straw-colored at maturity; **rachilla hairs** (4)6–10 mm. **Spikelets** with 3–10 florets. **Lower glumes** 3–7 mm; **upper glumes** (4)5–10 mm; **lemmas** 8–15 mm, glabrous, linear, margins somewhat inrolled, apices long-acuminate; **paleas** 3–4 mm, membranous; **anthers** 1.5–2 mm, purplish; **styles** persistent. **Caryopses** 2–3 mm, rarely maturing. $2n$ = 36, 42, 44, 46, 48, 49–52, 54, 72, 84, 96.

Phragmites australis grows in wet or muddy ground along waterways, in saline or freshwater marshes, and in sloughs throughout North America. Its tall, leafy, often persistent culms and plumose panicles make it one of our easier species to recognize. In Florida, *Neyraudia reynaudiana* is sometimes mistaken for *P. australis*, but the former has glabrous internodes and pilose lemmas.

It is also one of the most widely distributed flowering plants, growing in most temperate and tropical regions of the world, spreading quickly by rhizomes. Once established, it is difficult to eradicate. Its uses include thatching, lattices, arrow shafts, construction boards, mats, and erosion control, and it was used in the past to make cigarettes and superior pen quills.

Phragmites karka (Retz.) Trin. *ex* Steud. is sometimes attributed to the *Flora* region. It supposedly differs from *P. australis* as shown below, but all the characters intergrade. For this reason, they are treated here as components of a single species.

1. Blades smooth on the abaxial surfaces, the apices filiform, flexible; rachilla hairs 6–10 mm long; upper glumes 5–10 mm long . . . . . *P. australis*
1. Blades scabrous on the abaxial surface, the apices attenuate, stiff; rachilla hairs 4–7mm long; upper glumes 4–6 mm long . . . . *P. karka*

16.04 ARUNDO L.

Kelly W. Allred

Plants perennial; rhizomatous, rhizomes short, usually more than 1 cm thick. **Culms** 2–10 m tall, 1–3.5 cm thick, usually erect, occasionally pendant from cliffs; **nodes** glabrous; **internodes** hollow. **Leaves** cauline, conspicuously distichous, glabrous; **sheaths** open, longer than the internodes; **ligules** membranous, shortly ciliate; **blades** flat or folded, margins scabrous. **Panicles** terminal, plumose, silvery to purplish. **Spikelets** laterally compressed, with 1–several florets; **rachilla segments** glabrous; **disarticulation** above the glumes and between the florets. **Glumes** longer than the florets, 3–5-veined; **lemmas** pilose, hairs not papillose-based, 3–7-veined, apices entire or minutely awned; **paleas** shorter than the lemmas, 2-veined; **anthers** 3. x = 12. Name from the Latin *arundo*, 'reed'.

Arundo, a genus of three species, grows throughout the tropical and warm-temperate regions of the world. Only one species has been introduced to the Western Hemisphere.

Arundo is similar to, but usually larger than, *Phragmites*, a much more common genus in North America. In addition, *Arundo*, but not *Phragmites*, has a wedge-shaped, light to dark brown area at the base of its blades.

1. **Arundo donax** L. [p. 12]
GIANT REED

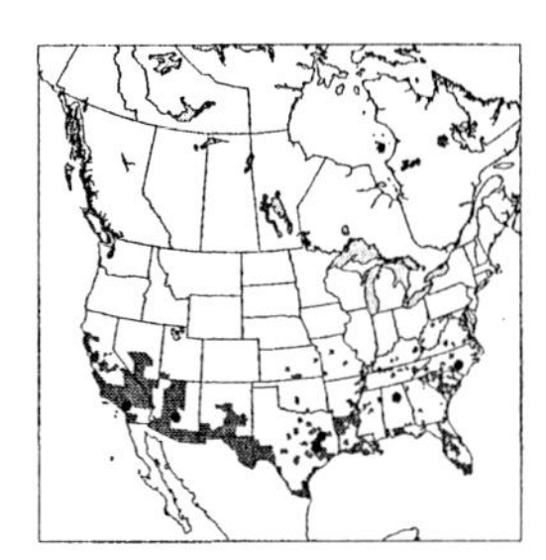

Culms (2)3–10 m, in large tussocks or hedges. **Leaves** distichous; **ligules** 0.4–1 mm; **blades** 30–100 cm long, 2–7(9) cm wide, with a wedge-shaped, light to dark brown area at the base. **Panicles** 30–60 cm long, to 30 cm wide. **Spikelets** 10–15 mm, with 2–4 florets. **Glumes** subequal, as long as the spikelets, thin, brownish or purplish, 3-veined, long-acuminate; **lemmas** 8–12 mm, 3–5-veined, pilose, hairs 4–9 mm, apices bifid, midvein ending into a delicate awn; **paleas** 3–5 mm, pilose at the base; **anthers** 2–3 mm. **Caryopses** 3–4 mm, oblong, light brown. $2n$ = 24, 100, 110.

Within the *Flora* region, *Arundo donax* grows in the southern half of the contiguous United States, being found along ditches, culverts, and roadsides where water accumulates. It has been used extensively as a windbreak, and planted for erosion control on wet dunes. It is also grown for the ornamental value of its tall, leafy culms and large panicles, but its tendency to spread is sometimes a disadvantage. Cultivars with striped or unusually wide leaves, e.g., 'Variegata' and 'Macrophylla', are of horticultural interest but do not merit taxonomic recognition.

Arundo donax has been used for thousands of years in making musical instruments, the stems being used for pipes and the tough inner rind for reeds in a wide variety of woodwind instruments. It is one of the species referred to as 'reed' in the Bible. It is still used in many parts of the world for house construction, lattice-work, mats, screens, stakes, walking sticks, and fishing poles.

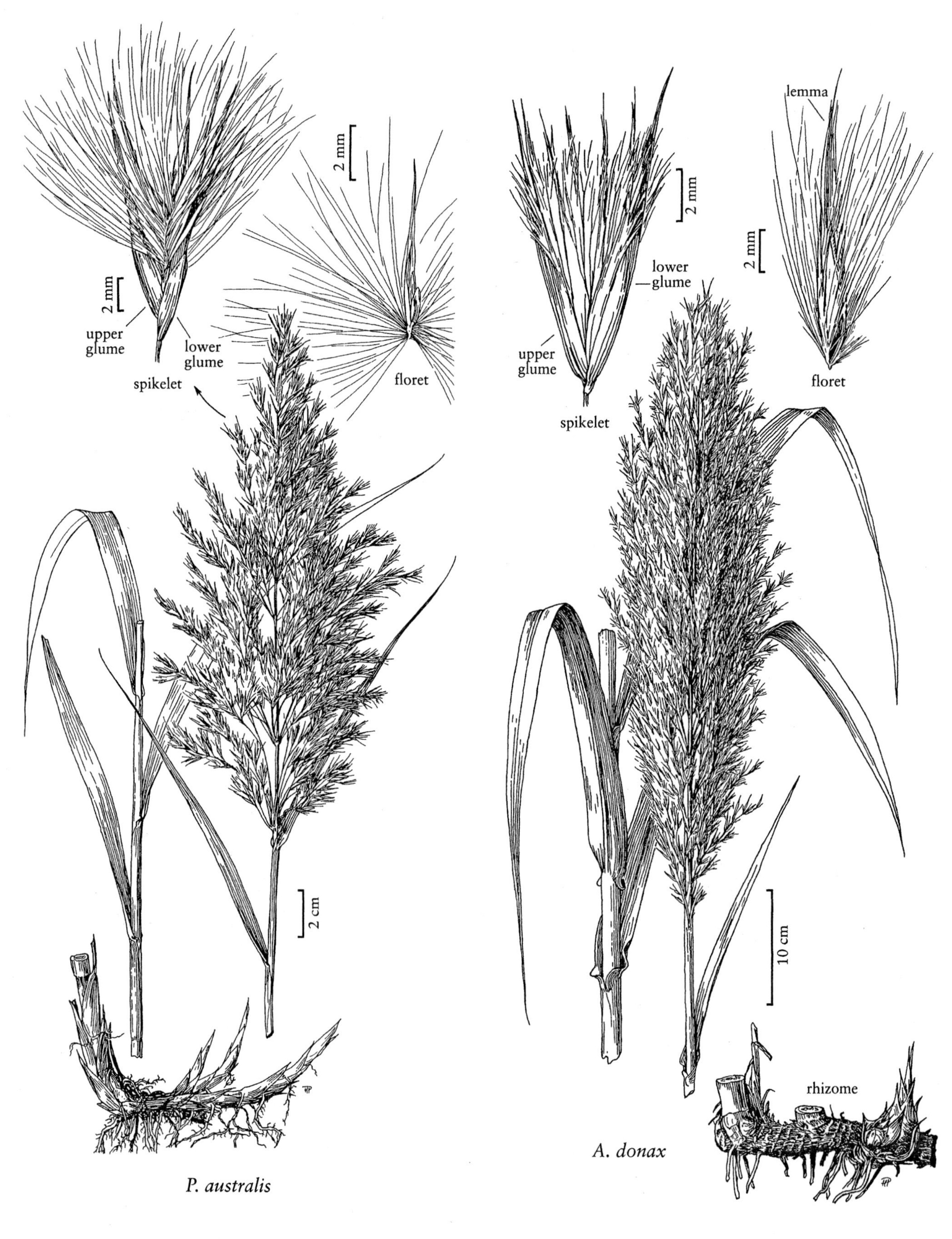

PHRAGMITES

ARUNDO

6. CHLORIDOIDEAE Kunth *ex* Beilschm.

Grass Phylogeny Working Group

Plants annual or perennial; usually synoecious, sometimes monoecious or dioecious; habit varied. **Culms** usually annual, sometimes becoming somewhat woody, internodes solid or hollow. **Leaves** sometimes conspicuously distichous; **sheaths** usually open; **auricles** absent; **abaxial ligules** usually absent, sometimes present as a line of hairs; **adaxial ligules** membranous, often ciliate with cilia longer than the membranous base, sometimes not ciliate; **blades** not pseudopetiolate; **mesophyll** usually radiate; **adaxial palisade layer** not present; **fusoid cells** absent; **arm cells** absent; **Kranz anatomy** present; **midrib** simple; **adaxial bulliform cells** present; **stomatal subsidary cells** dome-shaped or triangular; **bicellular microhairs** present, usually with a short, wide apical cell; **papillae** sometimes present. **Inflorescences** ebracteate, paniculate, racemose, or spicate (occasionally a single spikelet), if paniculate, often with spikelike branches; **disarticulation** usually beneath the florets, sometimes at the base of the panicle branches. **Spikelets** usually bisexual, usually laterally compressed, with 1–60 florets, distal florets often reduced. **Glumes** usually 2, shorter or longer than the lemmas, sometimes exceeding the distal florets, lower or both glumes occasionally missing; **lemmas** lacking uncinate hairs, sometimes awned, awns single or, if multiple, the bases not fused into a single column; **anthers** 1–3; **ovaries** glabrous; **styles** 2, separate throughout, bases close. **Caryopses** often with a free or loose pericarp; **hila** short; **endosperm** hard, without lipid; **starch grains** simple or compound; **haustorial synergids** absent; **embryos** usually large relative to the endosperm, not waisted; **epiblasts** usually present; **scutellar cleft** present; **mesocotyl internode** elongate; **embryonic leaf margins** usually meeting, rarely overlapping. x = (7, 8,) 9, 10 (12).

The subfamily *Chloridoideae* is most abundant in dry, tropical and subtropical regions. In the *Flora* region, it reaches its greatest diversity in the southwestern United States (Barkworth and Capels 2000). Almost all its members, and all those in the *Flora* region, have C_4 photosynthesis. Most employ the NAD-ME or PCK pathways, but *Pappophorum* utilizes the NADP-ME pathway.

The subfamily has been recognized, with essentially the same limits as here, for some time, although reservations have been expressed concerning its monophyly (Campbell 1985; Jacobs 1987; Kellogg and Campbell 1987). More recent studies, both morphological (Van den Borre and Watson 1997, 2000) and molecular (Soreng and Davis 1998; Hilu et al. 1999; Hsaio et al. 1999; Grass Phylogeny Working Group 2001; Hilu and Alice 2001) support its recognition as a monophyletic unit. There is less agreement concerning the subfamily's closest relative, some studies pointing to the *Arundinoideae* (Grass Phylogeny Working Group 2001) and some to the *Danthonioideae* (Barker et al. 1995; Hilu and Esen 1993; Hilu and Alice 2001).

There is considerable disagreement concerning the tribal treatment within the *Chloridoideae*, the number of tribes recognized varying from two (Prat 1936) to eight (Gould and Shaw 1983). Hilu and Wright (1982, p. 28) concluded, on the basis of their morphological study, that "... the boundaries between most of the tribes in this subfamily are not pronounced." They noted that Savile (1979) reached the same conclusion from considering the host specificity of various pathogenic fungi.

More recent work supports Hilu and Wright's conclusion. Van den Borre and Watson (1997, 2001) recognized eight informal groups within the subfamily. Five of the groups were large, the smallest including around 133 species and the largest around 380. The other three groups, which correspond to the *Orcuttieae*, *Pappophoreae*, and subtribe *Triodiinae*, include 9, 42, and 54 species, respectively. The difference in size is of no concern; the fact that all three of the small groups are embedded within one of the five large groups, the *Pappophoreae* and *Triodiinae* in a group than includes *Eragrostis* subg. *Eragrostis* and the *Orcuttieae* in the group that includes *Muhlenbergia*, is disturbing. Van den Borre and Watson noted that part of the problem was that that *Eragrostis*, and probably some of the other large genera, are not monophyletic.

Hilu and Alice (2001) recognized four clades within the *Chloridoideae*. Like Van den Borre and Watson, they found the *Orcuttieae* and *Triodiinae* to be monophyletic, although their placement within the subfamily was not clear. Unlike Van den Borre and Watson, Hilu and Alice found *Pappophorum*, and hence the *Pappophoreae*, to be polyphyletic.

The treatment presented here is conservative in recognizing the *Orcuttieae* and *Pappophoreae* as distinct tribes. It departs from most other treatments in merging all other North American taxa into a single tribe, the *Cynodonteae*. Consensus on how the *Cynodonteae sensu lato* should be broken up is unlikely to be reached until the generic limits of its members have been more thoroughly examined.

SELECTED REFERENCES **Barker, N.P., H.P. Linder,** and **E.H. Harley.** 1995. Polyphyly of Arundinoideae (Poaceae): Evidence from *rbcL* sequence data. Syst. Bot. 20:423–435; **Barkworth, M.E.** and **K.M. Capels.** 2000. The Poaceae in North America: A geographic perspective. Pp. 327–346 *in* S.W.L. Jacobs and J. Everett (eds.). Grasses: Systematics and Evolution. International Symposium on Grass Systematics and Evolution (3rd:1998). CSIRO Publishing, Collingwood, Victoria, Australia. 408 pp.; **Campbell, C.S.** 1985. The subfamilies and tribes of Gramineae (Poaceae) in the southeastern United States. J. Arnold Arbor. 66:123–199; **Gould, F.W.** and **R.B. Shaw.** 1983. Grass Systematics, ed. 2. Texas A&M University Press, College Station, Texas, U.S.A. 397 pp.; **Grass Phylogeny Working Group.** 2001. Phylogeny and subfamilial classification of the grasses (Poaceae). Ann. Missouri Bot. Gard. 88:373–457; **Hilu, K.W.** and **L.A. Alice.** 2001. A phylogeny of the Chloridoideae (Poaceae) based on *matK* sequences. Syst. Bot. 26:386–405; **Hilu, K.W., L.A. Alice,** and **H. Liang.** 1999. Phylogeny of the Poaceae inferred from *matK* sequences. Ann. Missouri Bot. Gard. 86:835–851; **Hilu, K.W.** and **A. Esen.** 1993. Prolamin and immunological studies in the Poacae: III. Subfamily Chloridoideae. Amer. J. Bot. 80:104–113; **Hilu, K.W.** and **K. Wright.** 1982. Systematics of Gramineae: A cluster analysis study. Taxon 31:9–36; **Hsiao, C., S.W.L. Jacobs, N.J. Chatterton,** and **K.H. Asay.** 1999. A molecular phylogeny of the grass family (Poaceae) based on the sequences of nuclear ribosomal DNA (ITS). Austral. Syst. Bot. 11:667–688; **Jacobs, S.W.L.** 1987. Systematics of the Chloridoid grasses. Pp. 871–903 *in* T.R. Soderstrom, K.W. Hilu, C.S. Campbell, and M.E. Barkworth (eds.) Grass Systematics and Evolution. Smithsonian Institution Press, Washington, D.C., U.S.A. 473 pp.; **Kellogg, E.A.** and **C.S. Campbell.** 1987. Phylogenetic analysis of the Gramineae. Pp. 310–322 *in* T.R. Soderstrom, K.W. Hilu, C.S. Campbell, and M.E. Barkworth (eds.) Grass Systematics and Evolution. Smithsonian Institution Press, Washington, D.C., U.S.A. 473 pp.; **Prat, H.** 1936. La systématique des Graminées. Ann. Sci. Nat., Bot., Ser. 10, 18:165–258; **Savile, D.B.O.** 1979. Fungi as aids in higher plant classification. Bot. Rev. 45:377–503; **Soreng, R.J.** and **J.I. Davis.** 1998. Phylogenetics and character evolution in the grass family (Poaceae): Simultaneous analysis of morphological and chloroplast DNA restriction site character sets. Bot. Rev. (Lancaster) 64:1–85; **Van den Borre, A.** and **L. Watson.** 1997. On the classification of the Chloridoideae (Poaceae). Austral. Syst. Bot. 10:491–531; **Van den Borre, A.** and **L. Watson.** 2000. On the classification of the Chloridoideae: Results from morphological and leaf anatomical data analyses. Pp. 180–183 *in* S.W.L. Jacobs and J. Everett (eds.). Grasses: Systematics and Evolution. International Symposium on Grass Systematics and Evolution (3rd:1998). CSIRO Publishing, Collingwood, Victoria, Australia. 408 pp.

1. Leaves with little or no distinction between the sheath and blade; ligules not present; plants annual, viscid . . . . 19. *Orcuttieae*
1. Leaves clearly differentiated into sheath and blade; ligules present; plants annual or perennial, not viscid.
   2. Lemmas 5–13-veined, all the veins extending into awns, often alternating with hyaline lobes or teeth . . . . 18. *Pappophoreae*
   2. Lemmas 1–11-veined, unawned or with 1 or 3 awns, sometimes with hyaline lobes on either side of the central awns . . . . 17. *Cynodonteae*

17. CYNODONTEAE Dumort.

Mary E. Barkworth

Plants annual or perennial. **Culms** 1–500 cm, not woody, usually not branched above the base. **Sheaths** usually open, often with coarse hairs at the top; **auricles** rarely present; **ligules** of hairs or membranous, if membranous, often ciliate, cilia sometimes longer than the membranous base; **blades** often with stiff, coarse marginal hairs adjacent to the ligules, glabrous or variously pubescent elsewhere. **Inflorescences** terminal, sometimes also axillary, simple panicles, panicles of 1–many spikelike branches, spikelike racemes, spikes, or, in 1 genus, a solitary spikelet, in dioecious taxa the staminate and pistillate inflorescences sometimes morphologically distinct; **disarticulation** usually beneath the fertile florets or the glumes but, particularly if the panicle branches are short, sometimes at the base of the branches. **Spikelets** usually laterally compressed, with 1–60 florets, sterile or reduced florets, if present, usually distal to the bisexual florets. **Glumes** from shorter than the adjacent florets to exceeding the distal florets; **lemmas** 1–3-veined or 7–13-veined, rarely 5-veined, if with 7–13 veins, the veins often in 3 groups; **lodicules** 2, or absent. x = 7, 8, 9, 10, 12.

Most members of the *Cynodonteae* in the *Flora* region can be recognized by their possession of two or more of the following characteristics: 1–3- or 7–13-veined lemmas, laterally compressed spikelets, spikelike inflorescence branches, and the presence of coarse hairs near the junction of the sheath and blade. All employ the NAD-ME or PCK C_4 photosynthetic pathways, have Kranz blade anatomy, and tend to grow in hot, dry areas. Having said this, it must be acknowledged that each of these characteristics can be found in other tribes and, within the *Cynodonteae*, there are genera that lack one or more of them.

The tribe *Cynodonteae*, as interpreted here, includes genera that are normally placed in two tribes, the *Cynodonteae sensu stricto*, and the *Eragrostideae* Stapf (Clayton and Renvoize 1986; Peterson et al. 2001; Grass Phylogeny Working Group 2001, but see Campbell 1985). Genera 17.01 to 17.34 correspond to the *Eragrostideae* and genera 17.35 to 17.53 to their *Cynodonteae*. The two are treated as one here because recent morphological, anatomical, and molecular studies (Van den Borre 1994; Van den Borre and Watson 1997; Hilu and Alice 2001) indicate that the distinction between the two is artificial. There is, however, no agreement on an alternative treatment as is indicated by Peterson et al. (2001) who listed 23 of the 53 genera treated here as being of uncertain position within the subfamily *Chloridoideae*. Part of the problem is that several of the genera (e.g., *Eragrostis*, *Chloris*, *Muhlenbergia*, and *Sporobolus*) are polythetic (Van den Borre and Watson 1997; Hilu and Alice 2001).

SELECTED REFERENCES **Campbell, C.S.** 1985. The subfamilies and tribes of Gramineae (Poaceae) in the southeastern United States. J. Arnold Arbor. 66:123–199; **Clark, L.G., W. Zhang, and J.F. Wendel.** 1995. A phylogeny of the grass family (Poaceae) based on *ndhF* sequence data. Syst. Bot. 20:436–460; **Clayton, W.D. and S.A. Renvoize.** 1986. Genera Graminum: Grasses of the World. Kew Bull., Addit. Ser. 13. Her Majesty's Stationery Office, London, England. 389 pp.; **Columbus, J.T., M.S. Kinney, R. Pant and M.E. Siqueiros Delgado.** 1998. Cladistic parsimony analysis of internal transcribed spacer region (nrDNA) sequences of *Bouteloua* and relatives (Gramineae: Chloridoideae). Aliso 17:99–130; **Grass Phylogeny Working Group.** 2001. Phylogeny and subfamilial classification of the grasses (Poaceae). Ann. Missouri Bot. Gard. 88:373–457; **Hilu, K.W. and L.A. Alice.** 2001. A phylogeny of the Chloridoideae (Poaceae) based on *matK* sequences. Syst. Bot. 26:386–405; **Peterson, P.M., R.J. Soreng, G. Davidse, T.S. Filgueras, F.O. Zuloaga, and E. Judziewicz.** 2001. Catalogue of New World Grasses (Poaceae): II. Subfamily Chloridoideae. Contr. U.S. Natl. Herb. 41:1–255; **Peterson, P.M., R.D. Webster, and J. Valdés-Reyna.** 1995. Subtribal classification of the New World Eragrostideae (Poaceae: Chloridoideae). Sida 16:529–544; **Van den Borre, A..** 1994. A taxonomy of the Chloridoideae (Poaceae), with special reference to the genus *Eragrostis*. Ph.D. dissertation, Australian National University, Canberra, New South Wales, Australia. 313 pp.; **Van den Borre, A. and L. Watson.** 1997. On the classification of the Chloridoideae (Poaceae). Austral. Syst. Bot. 10:491–531.

1. Inflorescences clearly exceeded by the upper leaves, often completely or almost completely enclosed in the upper leaf sheaths; culms 1–30(75) cm tall.
   2. Lemmas 3-lobed, the lobes ciliate; spikelets with 4 florets .... 17.17 *Blepharidachne*
   2. Lemmas not 3-lobed or the lobes not ciliate; spikelets with 1–60 florets.
      3. Spikelets (and often the plants) unisexual.
         4. Leaves strongly distichous; lemmas 9–11-veined; plants unisexual, growing in saline and alkaline soils.
            5. Pistillate and staminate inflorescences consisting of a single spikelet .... 17.07 *Monanthochloë*
            5. Pistillate and staminate inflorescences consisting of more than 1 spikelet .... 17.04 *Distichlis*
         4. Leaves not strongly distichous; lemmas 1–5-veined; plants unisexual or, if bisexual, with separate pistillate and staminate inflorescences, growing in a variety of soils.
            6. Spikelets 5–26 mm long; pistillate and staminate inflorescences similar, simple panicles; glumes and lemmas unawned, mucronate, or 1-awned .... 17.23 *Eragrostis*

6. Spikelets 2.5–7 mm long; pistillate and staminate inflorescences strongly dimorphic; staminate inflorescences with pectinate, spikelike branches; pistillate spikelets with conspicuously 3-awned glumes or distal florets.
  7. Upper glumes of the pistillate spikelets white, rigid, globose structures terminating in 3 awnlike teeth; pistillate spikelets with 1 floret, without rudimentary florets, the floret unawned or shortly 3-awned; staminate spikelets 4–6 mm long; anthers 2.5–3 mm long, brownish, red, or orange; widespread species of the central plains . . . . 17.48 *Buchloë*
  7. Upper glumes of the pistillate spikelets membranous, unawned; pistillate spikelets with one 3-awned floret and a distal 3-awned rudiment; staminate spikelets 3–4 mm long; anthers 2–2.5 mm long, pale; known, within the *Flora* region, only from Florida . . . . 17.47 *Opizia*

3. Spikelets bisexual, usually at least the lowest floret in each spikelet bisexual, in *Dasyochloa* the third floret in each spikelet bisexual or pistillate, if pistillate, the lowest 2 florets staminate.
  8. Lemma margins with a tuft of hairs at midlength, glabrous elsewhere; blades with white, thickened margins and sharply pointed . . . . 17.18 *Munroa*
  8. Lemma margins glabrous, or with hairs but the hairs not forming a tuft at midlength; blades without white cartilaginous margins, not sharply pointed.
    9. Lemmas awned, the awns 1–11 mm long.
      10. Plants stoloniferous; inflorescences 1–2.5 cm long, dense panicles; lemmas bilobed, the lobes about ½ as long as the lemmas; ligules of hairs . . . . 17.15 *Dasyochloa*
      10. Plants not stoloniferous; inflorescences 1.5–76 cm long, not dense; lemmas entire or minutely bilobed; ligules membranous, sometimes ciliate . . . . 17.19 *Leptochloa*
    9. Lemmas unawned, sometimes mucronate, with mucros less than 1 mm long.
      11. Spikelets with 2–20 florets; inflorescences panicles of 2–120 spikelike branches . . . . 17.19 *Leptochloa*
      11. Spikelets with 1(3) florets; inflorescences simple panicles, often highly contracted, without spikelike branches.
        12. Inflorescences 0.3–7.5 cm long, dense, spikelike or capitate panicles 1–8 times longer than wide; glumes strongly keeled; plants annual . . . . 17.31 *Crypsis*
        12. Inflorescences 1–60 cm long, sometimes dense and spikelike but, if less than 8 cm long, more than 8 times longer than wide; glumes rounded or weakly keeled; plants annual or perennial . . . . 17.30 *Sporobolus*

1. Inflorescences usually equaling or exceeding the upper leaves; culms 1–500 cm tall.
  13. Inflorescences with disarticulating branches, disarticulation at the base of the branches or (in *Lycurus*) the fused pedicels; branches 0.04–7 cm long, often globose or spikelike (fused to the rachis and not evident in *Lycurus*), usually with fewer than 15 spikelets per branch.
    14. Upper glumes with straight or uncinate spinelike projections; spikelets crowded, the branches condensed into burs . . . . 17.52 *Tragus*
    14. Upper glumes without spinelike projections; spikelets sometimes crowded, but not forming burlike clusters.
      15. Branches not fused to the rachises; spikelets more than 3 per branch, usually all alike, sometimes the proximal spikelet sterile or replaced with short secondary branches.
        16. Both glumes much longer than the florets, usually exceeding the distal florets; plants annual . . . . 17.22 *Dinebra*
        16. Lower glumes shorter than or subequal to the lower florets, 1 or both glumes usually exceeded by the distal florets; plants usually perennial.
          17. Spikelets with 1–2(3) florets; branches disarticulating promptly, before the spikelets disarticulate . . . . 17.46 *Bouteloua*
          17. Spikelets with 2–8 florets; branches disarticulating tardily, initially the spikelets disarticulating above the glumes . . . . 17.25 *Pogonarthria*

15. Branches sometimes fused to the rachises; spikelets 1–3 per branch, usually some spikelets on each branch sterile or staminate and 1 pistillate or bisexual.
   18. Axes of the branches extending beyond the base of the distal florets .... 17.46 *Bouteloua*
   18. Axes of the branches terminating at the base of the distal spikelet.
      19. Spikelets in pairs; glumes awned, the lower glumes (1)2(3)-awned, the upper glumes 1-awned; panicle branches often fused to the rachises ...... 17.34 *Lycurus*
      19. Spikelets in triplets; glumes unawned, 1-awned, or 3-awned; panicle branches sometimes appressed, but not fused, to the rachises.
         20. Branches straight at the base; central spikelets sessile .... 17.51 *Hilaria*
         20. Branches sharply curved at the base; central spikelets sessile or pedicellate.
            21. Central spikelets with 1 bisexual floret; lateral spikelets with 1 floret, varying to rudimentary .... 17.50 *Aegopogon*
            21. Central spikelets with 3–4 florets, the lowest floret pistillate, bisexual, or staminate, the distal florets staminate or sterile; lateral spikelets usually with 2 florets .... 17.49 *Cathestecum*

13. Inflorescences without disarticulating branches; branches, if present, often more than 4.5 cm long, variously shaped, including spikelike but not globose, often with more than 16 spikelets per branch.
   22. Inflorescences spikes or racemes.
      23. Spikelets with 1 bisexual or staminate floret and no additional florets.
         24. Rachises falcate or curved; both glumes exceeding the florets .... 17.43 *Microchloa*
         24. Rachises straight; lower glumes exceeded by the florets, sometimes absent.
            25. Spikelets solitary at each node; disarticulation below the glumes or the spikelets not disarticulating .... 17.53 *Zoysia*
            25. Spikelets paired, terminal on branches that are fused to the rachises; disarticulation at the base of the fused pedicel pairs .... 17.34 *Lycurus*
      23. Spikelets with more than 1 floret but sometimes only 1 floret bisexual, the additional florets sterile or staminate.
         26. All spikelets unisexual, the functional florets either staminate or pistillate; plants either unisexual or with both pistillate and staminate spikelets.
            27. Lemmas 9–11-veined; plants of saline habitats .... 17.04 *Distichlis*
            27. Lemmas 3-veined; plants of various habitats.
               28. Lemmas of the pistillate florets with awns 3.4–6.8 mm long; branches of staminate inflorescences pectinate; staminate spikelets with 1 floret .... 17.47 *Opizia*
               28. Lemmas of the pistillate florets with awns 30–150 mm long; branches of staminate inflorescences not pectinate; staminate spikelets with 5–20 florets .... 17.13 *Scleropogon*
         26. Some or all spikelets bisexual, the florets bisexual or unisexual, but both staminate and pistillate florets present within an individual spikelet.
            29. Lemmas of the pistillate or bisexual florets with awns 30–150 mm long; bisexual florets rarely found .... 17.13 *Scleropogon*
            29. Lemmas of the bisexual florets unawned or with awns less than 10 mm long; pistillate florets not present.
               30. Inflorescences 5–15 cm long, apparently a pectinate spike, actually a solitary, pectinate, spikelike branch .... 17.42 *Ctenium*
               30. Inflorescences 1.5–10 cm long, spikes, spikelike racemes, or panicles, linear or densely cylindrical to ovoid.
                  31. Inflorescences linear, (1.5)4–10 cm long; rachises not concealed by the spikelets .... 17.20 *Tripogon*
                  31. Inflorescences cylindrical to ovoid, 1.5–5 cm long; rachises concealed by the spikelets .... 17.01 *Fingerhuthia*
   22. Inflorescences simple panicles (sometimes highly condensed) or panicles of 1–120 spikelike branches.

32. Inflorescences simple panicles, sometimes highly contracted, even spikelike in appearance; spikelike branches not evident [for opposite lead, see p. 19].
    33. Spikelets usually with only 1 floret, occasionally with 2–3 florets.
        34. Ligules membranous, hyaline, or coriaceous, sometimes ciliate; lemmas 3-veined (occasionally appearing 5-veined), usually awned, sometimes unawned or mucronate.
            35. Lemmas and paleas densely sericeous over the veins and margins, glabrous between the veins . . . . 17.16 *Blepharoneuron*
            35. Lemmas and paleas glabrous to variously hairy but not densely sericeous over the veins and margins.
                36. Lemmas usually awned or mucronate; spikelets usually with 1 floret . . . . 17.33 *Muhlenbergia*
                36. Lemmas unawned or mucronate; spikelets frequently with 2–3 florets . . . . 17.23 *Eragrostis*
        34. Ligules of hairs; lemmas 1(3)-veined, unawned, sometimes mucronate.
            37. Panicles 0.3–4(7.5) cm long, 3–15 mm wide, spikelike or capitate, 1–8 times longer than wide; plants annual . . . . 17.31 *Crypsis*
            37. Panicles 1–80 cm long, 2–600 mm wide, dense to open, if less than 8 cm long, often 10 or more times longer than wide; plants annual or perennial.
                38. Calluses usually glabrous or almost so; paleas glabrous; fruits falling free of the lemma and palea . . . . 17.30 *Sporobolus*
                38. Calluses evidently hairy, the hairs ¼–⅞ as long as the lemmas; paleas hairy; fruits falling with the lemma and palea . . . . 17.32 *Calamovilfa*
    33. Spikelets with more than 1 floret.
        39. Lemmas with (5)9–11 veins (the lateral veins obscure in *Allolepis*).
            40. Spikelets unisexual; plants almost always unisexual, occasionally bisexual.
                41. Lemmas 9–11-veined; glumes 2–7 veined; plants rhizomatous and/or stoloniferous, found in saline or alkaline soils . . . . 17.04 *Distichlis*
                41. Lemmas 1–6-veined; lower glumes of the staminate spikelets 1-veined, those of the pistillate spikelets 1–5-veined; plants stoloniferous or rooting at the lower nodes, not rhizomatous, not found in saline or alkaline soils.
                    42. Plants perennial, stoloniferous; paleas of the pistillate florets completely surrounding the ovaries, the intercostal region coriaceous . . . . 17.06 *Allolepis*
                    42. Plants annual, rooting at the lower nodes; paleas of the pistillate florets not completely surrounding the ovaries, the intercostal region membranous or hyaline . . . . 17.23 *Eragrostis*
            40. All spikelets with at least 1 bisexual floret.
                43. Glumes longer than the adjacent lemmas.
                    44. Glumes exceeded by the distal florets; spikelets with 3–7 bisexual florets plus reduced distal florets; lemmas of the bisexual florets 5–7-veined throughout . . . . 17.03 *Swallenia*
                    44. Glumes exceeding the distal florets; spikelets with 2–4 florets, only the lowest floret bisexual; lemmas of the bisexual florets 3-veined basally, 5–7-veined distally . . . . 17.01 *Fingerhuthia*
                43. Glumes shorter than the adjacent lemmas.
                    45. Spikelets ovate-elliptical to ovate-triangular, 15–50 mm long, 6–16 mm wide; lower florets sterile, without paleas . . . . . 17.02 *Uniola*
                    45. Spikelets usually elliptical to lanceolate, 1–26 mm long, 0.6–9 mm wide; lower florets in each spikelet bisexual, with paleas.
                        46. Calluses glabrous or sparsely pubescent; lemmas (1)3(5)-veined; spikelets 1–26 mm long, 0.6–9 mm wide . . . 17.23 *Eragrostis*

46. Calluses densely pubescent; lemmas 5-, 7-, or 9-veined; spikelets 10–16 mm long, 2.5–5 mm wide . . . . 17.26 *Vaseyochloa*

39. Lemmas with 1–3 veins (occasionally with scabrous lines that may be mistaken for additional veins).
    47. Florets unisexual.
        48. Staminate and pistillate florets strongly dimorphic; plants unisexual or bisexual, bisexual plants with unisexual or bisexual spikelets; pistillate spikelets with 3–5 functional florets and lemma awns (30)50–150 mm long; staminate spikelets with 5–10(20) florets and unawned or shortly awned (to 3 mm) lemmas . . . . 17.13 *Scleropogon*
        48. Staminate and pistillate florets similar; plants unisexual; spikelets with 4–60 florets, all or almost all functional; lemmas 1.5–10.5 mm, unawned, sometimes mucronate.
            49. Plants perennial, stoloniferous; paleas of the pistillate florets completely surrounding the ovaries, the intercostal region coriaceous . . . . 17.06 *Allolepis*
            49. Plants annual, rooting at the lower nodes; paleas of the pistillate florets not completely surrounding the ovaries, the intercostal region membranous or hyaline . . . . 17.23 *Eragrostis*
    47. At least 1 floret in each spikelet bisexual.
        50. Lemmas, including the calluses, glabrous or inconspicuously hairy; lemma apices usually entire, sometimes minutely toothed.
            51. Spikelets with (1)2–60 florets; lemmas unawned, sometimes mucronate; ligules usually membranous and ciliate or ciliolate, sometimes of hairs . . . . 17.23 *Eragrostis*
            51. Spikelets with 1(2–3) florets; lemmas often awned, sometimes unawned or mucronate; ligules membranous, sometimes ciliolate, not ciliate . . . . 17.33 *Muhlenbergia*
        50. Lemma bodies conspicuously hairy over the veins and/or calluses conspicuously hairy; lemma apices usually with emarginate, bilobed, or trilobed apices, sometimes entire.
            52. Leaf margins evidently cartilaginous . . . . 17.14 *Erioneuron*
            52. Leaf margins not cartilaginous.
                53. Palea keels long hairy distally, the distal hairs 0.5–2 mm long . . . . 17.12 *Triplasis*
                53. Palea keels glabrous or with hairs less than 0.5 mm long.
                    54. Lemmas unawned, the midveins sometimes excurrent up to 0.5 mm.
                        55. Lemmas rounded to truncate, emarginate to bilobed; all 3 lemma veins often pilose basally . . . . 17.10 *Tridens*
                        55. Lemmas acute, entire or with 3 minute teeth, glabrous or shortly pubescent on the distal ⅔, the pubescence not confined to the veins . . . . . . 17.11 *Redfieldia*
                    54. Lemmas awned, the awns 1–7 mm long.
                        56. Plants 80–500 cm tall; panicles 35–73 cm long, plumose; lemma margins pilose; lemma apices bifid, awned from between the teeth; awns about 3 mm long . . . . 17.08 *Neyraudia*
                        56. Plants to 2–90 cm tall; panicles 6–30 cm long, not plumose; lemma margins sparsely pilose; lemma apices 3–4-lobed or -toothed and 3-awned; central awns 5–7 mm long, lateral awns 6–7 mm long . . . . 17.09 *Triraphis*

32. Inflorescences panicles of spikelike branches, the branches digitately or racemosely arranged on the rachises [for opposite lead, see p. 18].

57. Inflorescence branches 1 or more, if more than 1, arranged in terminal, digitate clusters, sometimes with additional branches or whorls below the terminal cluster.
    58. Inflorescences with 1(2) falcate branches.
        59. Spikelets with 2 well-developed sterile or staminate florets below the bisexual florets; additional sterile or staminate florets present distal to the bisexual floret . . . . 17.42 *Ctenium*
        59. Spikelets usually with 1, rarely 2, florets, the lowest or only floret bisexual . . . . 17.43 *Microchloa*
    58. Inflorescences with more than 1 branch or, if only 1, the branch not strongly falcate.
        60. Plants unisexual.
            61. Staminate spikelets 4–6 mm long, with 2 florets; upper glumes of the pistillate spikelets indurate, white . . . . 17.48 *Buchloë*
            61. Staminate spikelets 3–4 mm long, with 2 florets; upper glumes of the pistillate spikelets membranous . . . . 17.47 *Opizia*
        60. Plants bisexual, all spikelets with at least 1 bisexual floret.
            62. Spikelets with more than 1 bisexual floret.
                63. Panicle branches 0.4–7 cm long, terminating in a point . . . 17.29 *Dactyloctenium*
                63. Panicle branches 1–22 cm long, terminating in a functional or rudimentary spikelet.
                    64. Disarticulation eventually below the glumes, initially below the lemmas and caryopses, the paleas persistent; panicle branches terminating in a rudimentary spikelet . . . . 17.28 *Acrachne*
                    64. Disarticulation above the glumes, usually also below the florets; panicle branches terminating in a functional spikelet.
                        65. Lemmas 3-awned, the central awns 8–12 mm long . . . 17.38 *Trichloris*
                        65. Lemmas not 3-awned.
                            66. Lemmas usually with hairs over the veins, at least basally, the apices often toothed, some times mucronate or awned . . . . 17.19 *Leptochloa*
                            66. Lemmas glabrous, the apices entire, neither mucronate nor awned . . . . 17.27 *Eleusine*
            62. Spikelets usually with only 1 bisexual floret (occasionally 2 in some genera), often with additional staminate, sterile, or modified florets.
                67. Spikelets usually without sterile or modified florets; lemmas unawned . . . . 17.44 *Cynodon*
                67. Spikelets with 1 or more sterile florets distal to the bisexual floret; lemmas of the bisexual florets often awned.
                    68. Lowest lemmas in the spikelets 3-awned, the central awns 8–12 mm long, the lateral awns 0.5–12 mm long . . . 17.38 *Trichloris*
                    68. Lowest lemmas in the spikelets usually unawned or with a single awn, if 3-awned, the lateral awns less than 0.5 mm long.
                        69. Spikelets dorsally compressed . . . . 17.37 *Enteropogon*
                        69. Spikelets laterally compressed or terete.
                            70. Upper glumes truncate or bilobed; lowest lemmas unawned or with an awn to 1.2 mm long . . . . 17.36 *Eustachys*
                            70. Upper glumes acute to acuminate; lowest lemmas usually awned, the awns to 37 mm long . . . 17.35 *Chloris*
57. Inflorescence branches more than 1, racemosely arranged on the rachises.
    71. All spikelets unisexual.

72. Staminate spikelets 4–6 mm long, the anthers 2.5–3 mm long; pistillate spikelets with 1 unawned or shortly awned floret; widespread species of the central plains . . . . 17.48 *Buchloë*
72. Staminate spikelets 3–4 mm long, the anthers 2–2.5 mm long; pistillate lemmas with awns 3.4–6.8 mm long; in the *Flora* region, known only as an occasional escape from lawns and experimental plots in Florida . . . . 17.47 *Opizia*
71. All spikelets with at least 1 bisexual floret.
73. Inflorescence branches woody, terminating in hard, sharp points . . . 17.24 *Cladoraphis*
73. Inflorescence branches not particularly stiff or rigid, terminating in spikelets or points.
74. Spikelets with more than 1 bisexual floret, sometimes also with reduced florets.
75. Lemmas 7–11-veined, mucronate, the mucros 0.1–0.3 mm long; not established in the *Flora* region . . . . 17.05 *Aeluropus*
75. Lemmas 3-veined, unawned, mucronate, or awned, the awns often much more than 1 mm long; established in the *Flora* region.
76. Lower glumes exceeding the lowest lemmas, sometimes exceeding the distal lemmas . . . . 17.21 *Trichoneura*
76. Lower glumes not or only slightly exceeding the lowest lemmas.
77. Inflorescences with 50 or more closely spaced, arcuate, tardily deciduous branches . . . . 17.25 *Pogonarthria*
77. Inflorescences with 2–120 straight, nondisarticulating branches.
78. Spikelets with (2)3–12(20) bisexual florets . . . . 17.19 *Leptochloa*
78. Spikelets with 2–4 florets, but only the lowest 1(2) florets bisexual . . . . 17.41 *Gymnopogon*
74. Spikelets with 1 bisexual floret, sometimes with sterile, rudimentary, or modified florets distal to the bisexual floret.
79. Functional spikelets with sterile, rudimentary, or modified florets distal to the bisexual floret.
80. Spikelets widely spaced to slightly imbricate, appressed to the branch axes . . . . 17.41 *Gymnopogon*
80. Spikelets densely imbricate, varying from appressed to strongly divergent.
81. Inflorescence branches usually solitary at each node (sometimes only 1 per panicle); spikelets laterally compressed or terete . . . . 17.46 *Bouteloua*
81. Inflorescence branches usually more than 1 at the lower nodes; spikelets dorsally compressed . . . . . . 17.37 *Enteropogon*
79. Functional spikelets with only 1 floret, lacking sterile, rudimentary, or modified florets.
82. Spikelets distant to slightly imbricate, appressed to the branches; branches strongly divergent.
83. Blades with thick, white margins and a well-developed midrib . . . . 17.40 *Schedonnardus*
83. Blades lacking both thick, white margins and well-developed midribs . . . . 17.41 *Gymnopogon*
82. Spikelets clearly imbricate, appressed to strongly divergent; branches appressed to strongly divergent.
84. Spikelets laterally compressed, appressed to divergent . . . . 17.45 *Spartina*
84. Spikelets dorsally compressed, appressed . . . . 17.39 *Willkommia*

17.01 FINGERHUTHIA Nees

Mary E. Barkworth

Plants usually perennial, occasionally annual in desert areas; cespitose and shortly rhizomatous. **Culms** 5–120 cm, unbranched. **Leaves** mostly basal; **ligules** of hairs; **blades** 2–5 mm wide. **Inflorescences** terminal, exceeding the upper leaves, dense, cylindrical to ovoid panicles, occasionally reduced to racemes; **branches** short, non-disarticulating; **rachises** concealed by the spikelets; **disarticulation** beneath the glumes. **Spikelets** laterally compressed, with 2–4 florets, only the basal florets bisexual, the next 2 florets usually staminate, the fourth floret, if present, sterile. **Glumes** subequal, clearly exceeding the florets, awned or unawned; **lemmas** firmly membranous, 3-veined basally, 5–7-veined distally, mucronate to shortly awned, awns shorter than 10 mm; **anthers** 3; **ovaries** glabrous. x = 10. Named for Carl Anton Fingerhuth (1798–1876), a German botanist and physician.

Fingerhuthia is a genus of two species, one native to southern Africa and western Asia, the other endemic to southern Africa. One species has been grown in the *Flora* region.

1. **Fingerhuthia africana** Lehm. [p. 23]
THIMBLEGRASS, ZULU FESCUE

Plants cespitose and shortly rhizomatous. **Culms** 10–95 cm. **Blades** 2.5–40 cm long, 2–4 mm wide, scabrous distally. **Panicles** 1.5–5 cm, cylindrical or ovoid, cuneate below. **Spikelets** 4–5.5 mm. **Glumes** narrowly elliptic, ciliate on the keel and the distal portion of the margins; **lowest lemmas** 3.5–4 mm, glabrous, sometimes scabridulous, margins ciliate, apices obtuse and abruptly mucronate. $2n$ = 40.

Fingerhuthia africana is native to southern Africa and western Asia. It has been grown at the Santa Rita Experimental Range in Pima County, Arizona, but is not established in the *Flora* region.

17.02 UNIOLA L.

H. Oliver Yates

Plants perennial; rhizomatous or stoloniferous. **Culms** to 2.5 m, erect, glabrous, unbranched. **Ligules** of hairs; **blades** flat, becoming involute when dry, margins scabrous, tapering to an attenuate apex. **Inflorescences** terminal, simple panicles, exceeding the leaves; **disarticulation** below the glumes. **Spikelets** 8–50 mm long, 6–16 mm wide, ovate-elliptical to ovate-triangular, strongly laterally compressed, with 3–34 florets, lowest 2–8 florets sterile, remaining floret(s) bisexual. **Glumes** subequal, shorter than the adjacent lemmas, midveins keeled, serrate to serrulate, apices unawned; **lemmas** 3–9-veined, midveins keeled, serrate to serrulate, apices somewhat blunt to acute or mucronate, unawned; **paleas**, if present, from slightly shorter than to exceeding the lemmas, 2-keeled, keels winged, serrulate or ciliate; **anthers** 3; **ovary** glabrous; **styles** 1, with 2 style branches. **Caryopses** linear; **embryos** less than ½ as long as the caryopses. x = 10. Name from the Latin *unione glumarum*, 'united bracts', apparently a reference to the spikelets.

Uniola has two species, both of which grow on coastal sand dunes. There is one species native to the *Flora* region; the second, *U. pittieri* Hack., extends from northern Mexico to Ecuador, primarily along the Pacific coast. The genus used to be interpreted as including *Chasmanthium*, a genus that is now included in the *Centothecoideae*.

SELECTED REFERENCES **Grass Phylogeny Working Group.** 2001. Phylogeny and subfamilial classification of the grasses (Poaceae). Ann. Missouri Bot. Gard. 88:373-457; **Yates, H.O.** 1966a. Morphology and cytology of *Uniola* (Gramineae). SouthW. Naturalist 11:145–189; **Yates, H.O.** 1966b. Revision of grasses traditionally referred to *Uniola*, I. *Uniola* and *Leptochloöpsis*. SouthW. Naturalist 11:372–394; **Yates, H.O.** 1966c. Revision of grasses traditionally referred to *Uniola*, II. *Chasmanthium*. SouthW. Naturalist 11:415–455.

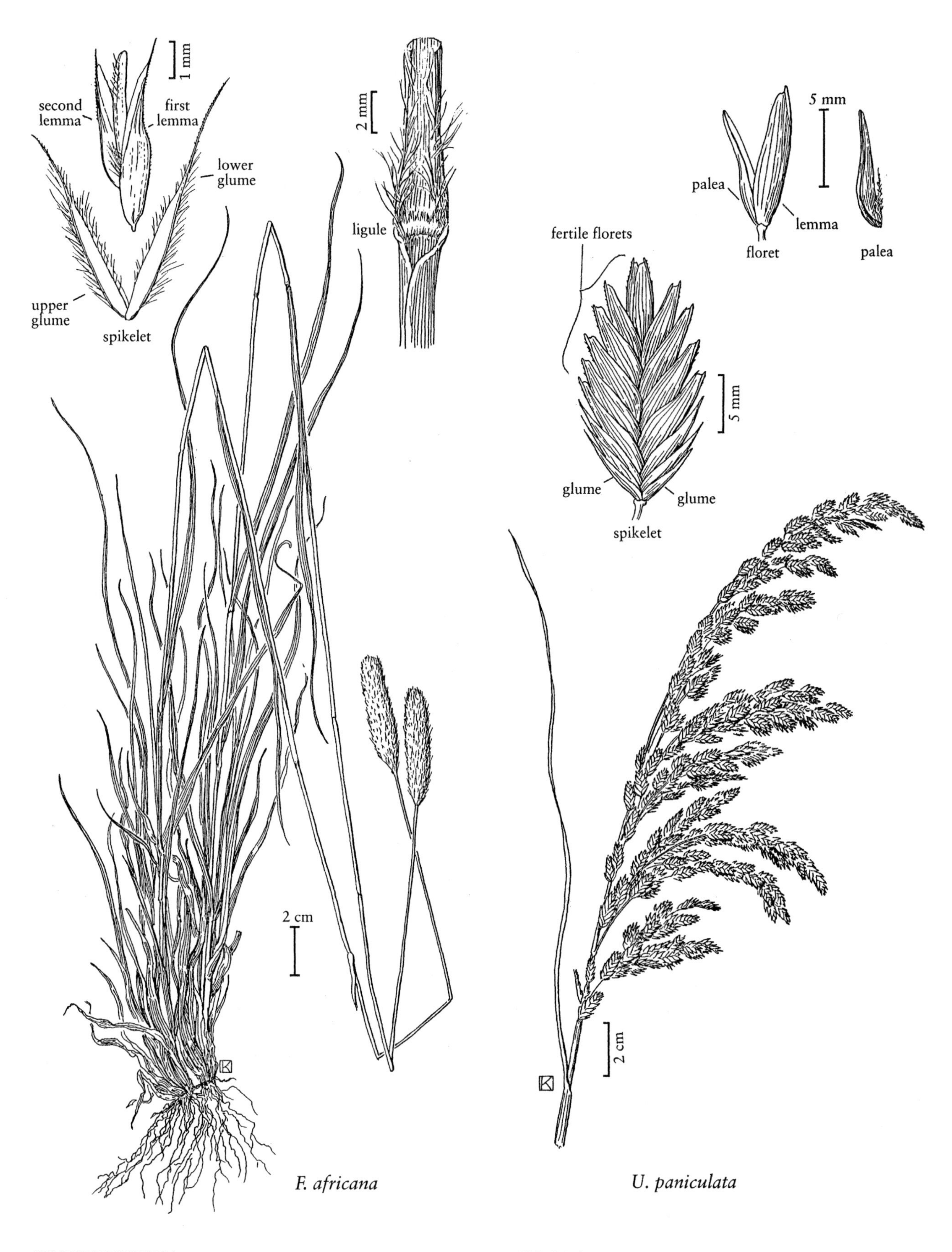

FINGERHUTHIA

UNIOLA

1. **Uniola paniculata** L. [p. 23]
SEA OATS

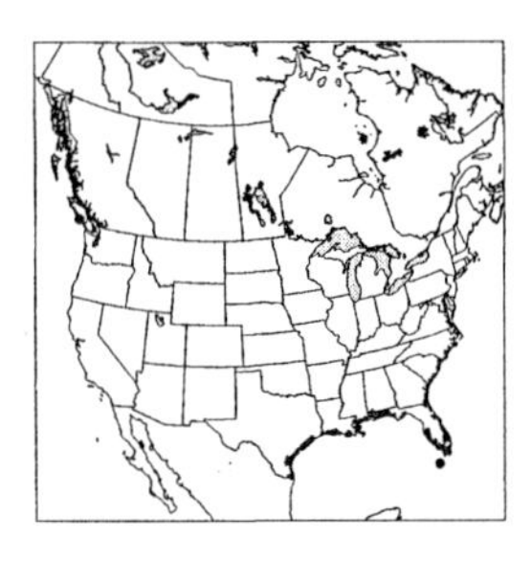

Plants perennial; rhizomatous. **Culms** to 2.5 m. **Sheaths** glabrate, mostly glabrous at maturity, with tufts of hairs near the collar; **collars** ciliate to pubescent; **blades** to 67 cm long, about 1 cm wide. **Panicles** 27–67 cm, open; **branches** drooping or nodding at maturity. **Spikelets** 15–30(50) mm long, 6–16 mm wide, ovate-elliptical, with (3)5–34 florets, the lower (3)4–5(8) sterile. **Glumes** 5–12 mm, acute; **lemmas** essentially glabrous, (7)9(13)-veined; **paleas** present only in the functional florets; **anthers** 4–6 mm. **Caryopses** 3–5 mm long, 1–1.5 mm wide. $2n = 40$.

Uniola paniculata grows on the beaches and sand dunes of the Atlantic and Gulf coastal plains from Maryland to Veracruz, Mexico, and on the Florida Keys, the Bahama Islands, and Cuba. Seed production is generally poor; the reason is not known.

17.03 SWALLENIA Soderstr. & H.F. Decker

James P. Smith, Jr.

Plants perennial; clumped, rhizomes woody. **Culms** 10–60 cm, branched above the base. **Leaves** mostly basal; **auricles** absent; **ligules** of hairs; **blades** flat, strongly veined, sharply pointed. **Inflorescences** terminal, usually exceeding the upper leaves, contracted panicles; **branches** ascending to erect; **disarticulation** beneath the caryopses. **Spikelets** laterally compressed, unawned, with 3–7 bisexual florets, distal florets reduced. **Glumes** subequal, longer than the adjacent lemmas but exceeded by the distal florets, acuminate; **lower glumes** 5–7-veined; **upper glumes** 7–11-veined; **calluses** hairy; **lemmas** membranous to papery, 5–7-veined, densely villous on the margins, sometimes also between the veins, unawned to mucronate; **paleas** equaling or exceeding the lemmas; **anthers** 3. **Caryopses** falling free from the lemma and palea. $x = 10$. Named for Jason Richard Swallen (1903–1991), a U.S. Department of Agriculture botanist and a former head of the Department of Botany at the Smithsonian Institution.

Swallenia is a monotypic genus, endemic to California. It is unusual in that only its caryopses break off the plant.

SELECTED REFERENCES **Gómez-Sánchez, M., P. Dávila-Aranda,** and **J. Valdés-Reyna.** Estudio anatómica de *Swallenia* (Poaceae: Eragroistideae: Monanthochloinae), un género monotípico de Norte América. Madroño 48:152–161 (2001) [publication date 2002]; **Henry, M.A.** 1979. A rare grass on the Eureka dunes. Fremontia 7:3–6; **Pavlik, B.M.** and **M.G. Barbour.** 1988. Demographic monitoring of endemic sand dune plants, Eureka Valley, California. Biol. Conservation 46:217–242.

1. **Swallenia alexandrae** (Swallen) Soderstr. & H.F. Decker [p. 26]
EUREKA VALLEY DUNEGRASS

Culms 10–40(60) cm, stiff, erect; **nodes** villous. **Sheaths** villous on the upper margins; **blades** 5–14 cm long, 3–8 mm wide. **Panicles** 4–10 cm; **branches** to 35 mm, with 1–3 spikelets. **Spikelets** 10–15 mm, persistent. **Glumes** 9–14 mm; **lemmas** 7–9 mm. **Caryopses** about 4 mm long, about 2 mm in diameter. $2n = 20$.

Swallenia alexandrae grows on sand dunes in Inyo County, California. It is only known from four sites, all between 900–1200 m, in the Eureka Valley of northern Inyo County. At these sites, it forms dense colonies 1–2 m across. It is state-listed as rare and federally-listed as endangered because of off-road vehicle activity.

17.04 DISTICHLIS Raf.

Mary E. Barkworth

Plants perennial; usually unisexual, occasionally bisexual; strongly rhizomatous and/or stoloniferous. **Culms** to 60 cm, usually erect, glabrous. **Leaves** conspicuously distichous; **lower leaves** reduced to scalelike sheaths; **upper leaf sheaths** strongly overlapping; **ligules** shorter than 1 mm, membranous, serrate; **upper blades** stiff, glabrous, ascending to spreading, usually equaling or exceeding the pistillate panicles. **Inflorescences** terminal, contracted panicles or racemes, sometimes exceeding the upper leaves. **Spikelets** laterally compressed, with 2–20 florets; **disarticulation** of the pistillate spikelets above the glumes and below the florets, staminate spikelets not disarticulating. **Glumes** 3–7-veined; **lemmas** coriaceous, staminate lemmas thinner than the pistillate lemmas, 9–11-veined, unawned; **paleas** 2-keeled, keels narrowly to broadly winged, serrate to toothed, sometimes with excurrent veins; **anthers** 3. **Caryopses** glabrous, free from the palea at maturity, brown. x = 10. Name from the Greek *distichos*, 'two-rowed', referring to the conspicuously distichous blades.

Distichlis, a genus of about five species, grows in saline soils of the coasts and interior deserts of the Western Hemisphere and Australia. All the species grow in South America, but only one, *Distichlis spicata*, is found in North America.

SELECTED REFERENCE Beetle, A.A. 1943. The North American variations of *Distichlis spicata*. Bull. Torrey Bot. Club 70:638–650.

1. **Distichlis spicata** (L.) Greene [p. 26]
SALTGRASS

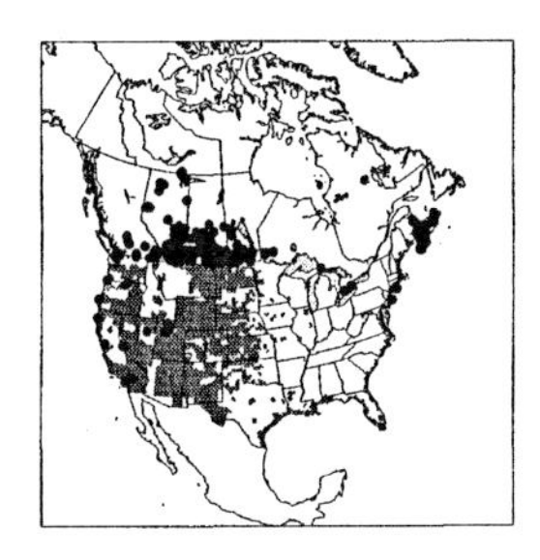

Plants rhizomatous and sometimes stoloniferous. **Culms** 10–60 cm, usually erect, sometimes decumbent or prostrate. **Blades of upper leaves** 1–8(20) cm, rigid and divaricate to lax and ascending, usually equaling or exceeding the pistillate panicles, varying with respect to the staminate panicles. **Pistillate panicles** 1–7 cm, often congested, with 2–20 spikelets. **Pistillate spikelets** 5–20 mm long, 4–7 mm wide, with 5–20 florets; **lower glumes** 2–3 mm; **upper glumes** 3–4 mm; **lemmas** 3.5–6 mm; **paleas** with serrate keels. **Caryopses** 2–5 mm, tapered or truncate. **Staminate panicles** and **spikelets** similar to the pistillate panicles and spikelets, but the lemmas somewhat thinner in texture and the paleas not bowed-out. **Anthers** 3–4 mm. $2n$ = 40.

Distichlis spicata grows in saline soils of the Western Hemisphere and Australia. Numerous infraspecific taxa have been recognized in the past, but none appears to be justified. Recent North American accounts of *Distichlis* have usually recognized plants from maritime coasts as distinct from those growing inland, supposedly having more congested inflorescences, but the range of variation is similar in the two habitats.

17.05 AELUROPUS Trin.

Mary E. Barkworth

Plants perennial; usually strongly rhizomatous or stoloniferous, rhizomes and stolons with persistent sheaths. **Culms** 5–40 cm, prostrate to erect, solitary or not; **internodes** numerous, short. **Sheaths** overlapping; **ligules** of hairs; **blades** usually stiffly spreading, flat or convolute below, folded distally, apices often cartilaginous and sharp. **Inflorescences** terminal, dense panicles of non-disarticulating spikelike branches racemosely arranged on elongate rachises, exceeding the upper leaves; **branches** 2.5–5 mm wide, axes triquetrous, spikelets closely packed, subsessile, in 2 rows. **Spikelets** bisexual, dorsally compressed, with 2–14 florets. **Glumes** unequal, exceeded by the florets, keeled; **lower glumes** 1–5(7)-veined; **upper glumes** 3–9-veined; **lemmas** 7–11-veined, mucronate, mucros 0.1–0.3 mm; **lodicules** 2; **anthers** 3, 0.8–1.6 mm; **style branches** 2. **Caryopses** ovoid-ellipsoid; **hila**

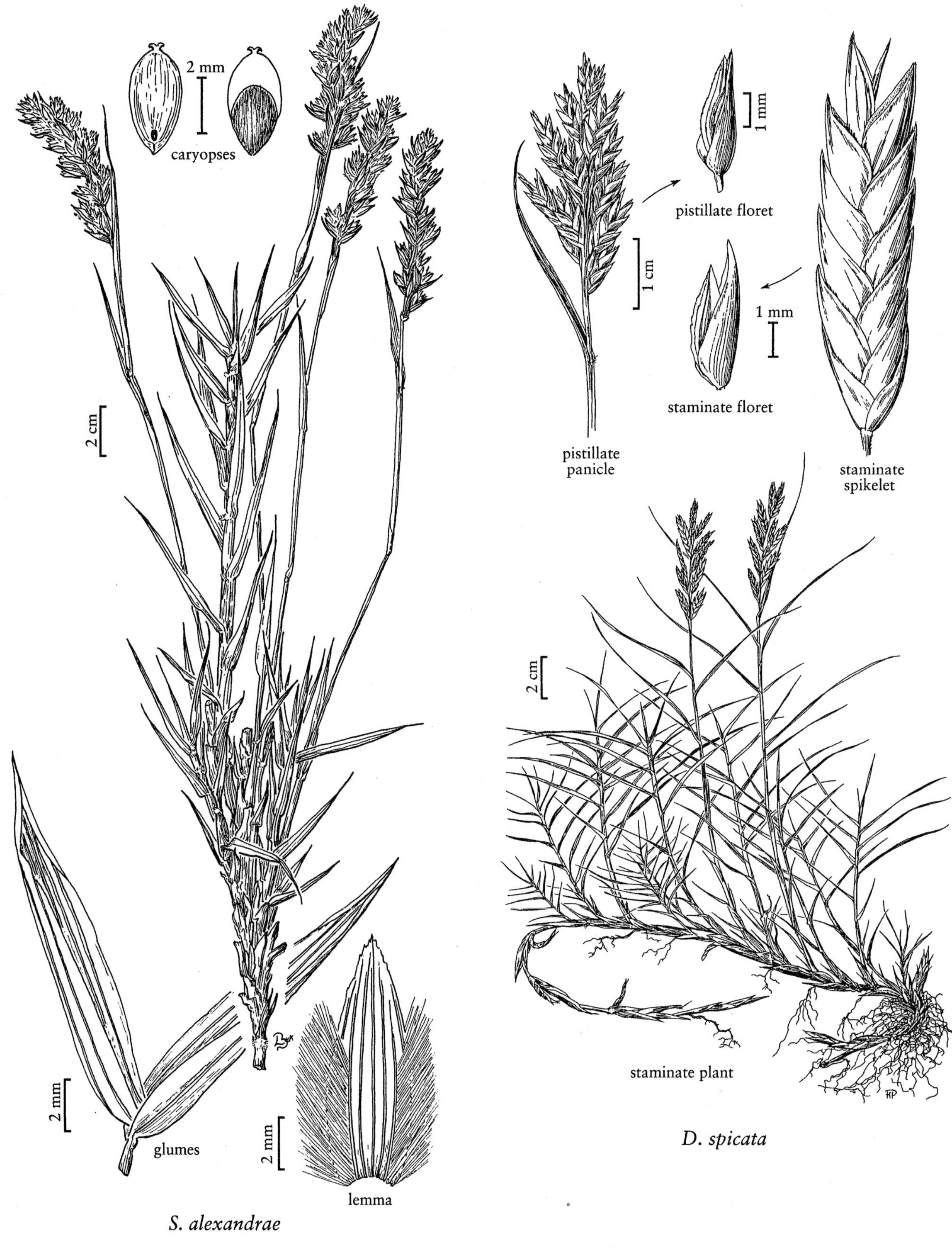

SWALLENIA

DISTICHLIS

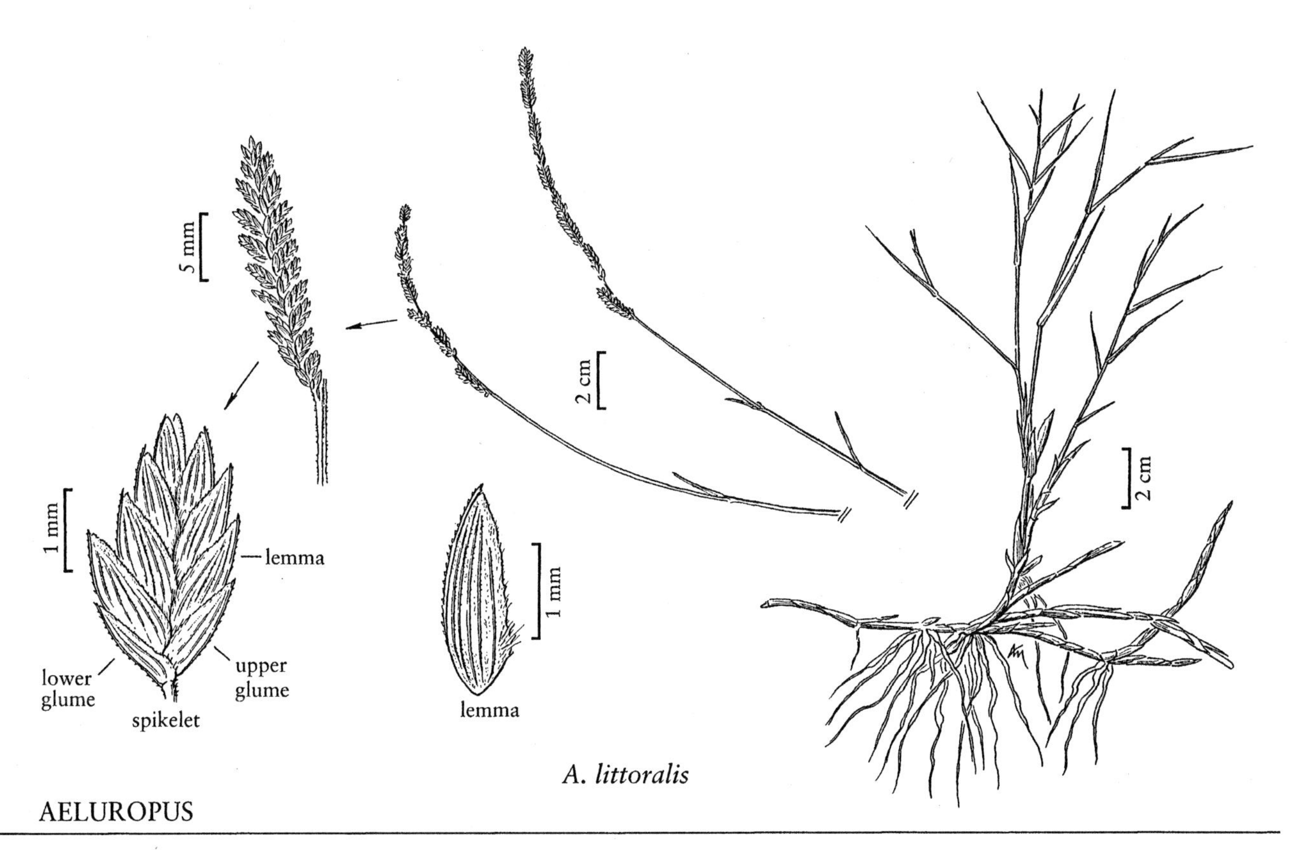

punctate, basal; embryos about ½ as long as the caryopses. x = 10. Name from the Greek *aeluros*, 'cat', and *pus*, 'leg', because the shape of the culms and panicles supposedly suggests a cat's leg.

Aeluropus is a genus of five species that extends from Portugal to China and northern India. The species are a good source of fodder and hay in Russia (Tzvelev 1976). One species has been cultivated in the *Flora* region.

SELECTED REFERENCE Tzvelev, N.N. 1976. Zlaki SSSR. Nauka, Leningrad [St. Petersburg], Russia. 788 pp. [In Russian].

1. **Aeluropus littoralis** (Gouan) Parl. [p. 27]

Plants both rhizomatous and stoloniferous. **Culms** 9–25 cm, erect or decumbent. **Sheaths** strongly ridged, glabrous; **blades** 1.5–7 cm, adaxial surfaces glabrous, scabridulous, or sparsely hairy, strongly and closely ridged. **Panicles** 1–7 cm long, 4–10 mm wide, with 4–13 branches, each bearing 12 or more spikelets; **lower branches** usually not overlapping. **Spikelets** 3–4.5 mm, ovoid, with 6–9 florets. **Glumes** unequal, veins scabrous, margins glabrous or hairy; **lower glumes** 1–2.5 mm, 1–4-veined; **upper glumes** 1.5–3 mm, 3–6-veined; **lemmas** 1.5–4 mm, 9–11-veined, veins scabrous, margins glabrous or hairy, apices mucronate, mucros 0.1–0.3 mm; **paleas** hyaline, keels ciliolate; **anthers** 1–1.5 mm. **Caryopses** strongly compressed. $2n$ = 20.

Aeluropus littoralis is native to Eurasia, where it grows in sandy, often saline habitats. It has been cultivated experimentally in Pima County, Arizona; it is not established in the *Flora* region.

17.06 **ALLOLEPIS** Soderstr. & H.F. Decker

J.K. Wipff

Plants perennial; dioecious, staminate and pistillate plants similar; stoloniferous. **Culms** 10–70 cm. **Leaves** mostly cauline; **ligules** membranous, ciliate. **Inflorescences** terminal panicles, exceeding the upper leaves; **branches** appressed or ascending, usually spikelet-bearing to the base. **Spikelets** not disarticulating. **Pistillate spikelets** with 5–10 florets; **lower glumes** 1–5-veined, only the midveins conspicuous; **upper glumes** with 3 conspicuous veins, sometimes with 2–4 inconspicuous lateral veins; **lemmas** 3–5(6)-veined, unawned; **paleas** shorter than the lemmas, completely surrounding the ovaries, saccate basally, narrowing distally and surrounding the styles, intercostal region coriaceous, veins keeled, scabrous and puberulent, margins scarious. **Staminate spikelets** with 4–21 florets; **lower glumes** 1-veined; **upper glumes** 1–3-veined; **lemmas** conspicuously 3-veined, unawned; **paleas** equaling or slightly longer than the lemmas, linear, 2-veined, veins ciliolate. **Lodicules** 2, cuneate. $x = 10$. Name from the Greek *allo*, 'different', and *lepis*, 'scale' or 'lemma', referring to the differing lemmas of the sexes.

Allolepis is a monotypic North American genus. It used to be included in *Distichlis*, but it differs in its paleal morphology, in never producing rhizomes, and in its soil preferences.

SELECTED REFERENCE **Soderstrom, T.R. and H.F. Decker.** 1965. *Allolepis*: A new segregate of *Distichlis* (Gramineae). Madroño 18:33–64.

1. **Allolepis texana** (Vasey) Soderstr. & H.F. Decker [p. 29]

FALSE SALTGRASS

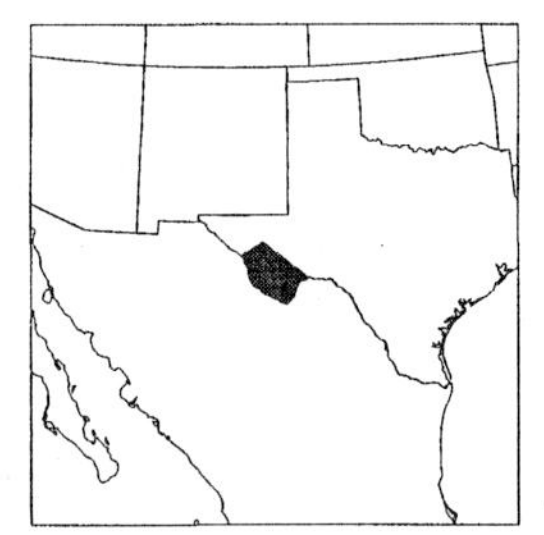

Culms (10) 25–70 cm, ascending, glabrous; **stolons** to 15 m long, 1–4 mm thick, strongly ribbed. **Sheaths** mostly glabrous, the apices with a tuft of hairs; **lower sheaths** shorter than the internodes; **ligules** 0.5–1.4 mm; **blades** (6)10–43 cm long, (2.5)4–6 mm wide, usually flat, glabrous, scabridulous near the apices, margins scabrous. **Panicles** (3)10–23 cm; **branches** 3–6 cm. **Pistillate spikelets** 10–30 mm. **Glumes** subcoriaceous, glabrous, midveins scabridulous; **lower glumes** 5.5–9 mm; **upper glumes** 6.8–10.3 mm; **lower lemmas** 6.5–10.5 mm, broadly ovate, coriaceous, glabrous, midveins scabrous distally, margins clasping the bases of the florets above. **Staminate spikelets** 9–25 mm long, 3–8 mm wide; **lower glumes** 3.5–5 mm, glabrous except for the scabrous midvein; **upper glumes** 4.5–6 mm, glabrous; **lowest lemma** 5–6.3 mm, glabrous; **anthers** 3, 3–3.7 mm, yellow. $2n = 40$.

Allolepis texana grows in the Big Bend region of Texas and northern Mexico. It is not common, but it may be locally abundant, growing on sandy or silty, but not alkaline, soils.

17.07 MONANTHOCHLOË Engelm.

John W. Thieret

Plants perennial; dioecious; mat-forming, with long, wiry stolons. **Culms** 5–20 cm, erect, with numerous short, leafy, lateral branches. **Leaves** distinctly distichous, clustered on distant to closely-spaced, short, lateral shoots; **ligules** thickly membranous ciliate rims; **blades** stiff, subulate. **Inflorescences** terminal, composed of a single glabrous spikelet, this enclosed, and almost concealed, by the uppermost leaf sheaths. **Pistillate spikelets** subterete, with 3–5 florets, distal florets rudimentary; **disarticulation** tardy, below the lowest floret; **glumes** absent; **lemmas** coriaceous, glabrous, 9-veined, acute; **paleas** coriaceous, keels prominently winged, wings overlapping and enclosing the caryopses. **Staminate spikelets** similar to the pistillate spikelets, but smaller and the glumes and lemmas thinner. $x = 10$. Name from the Greek *monos*, 'single', *anthos*, 'flower', and *chloë*, 'grass', alluding to the solitary spikelets.

Monanthochloë is a genus of two species, one growing along the southern coastlines of North America, the other on inland salt pans in Argentina. It is probably related to *Distichlis*, but differs from that genus, and all others in the *Flora* region, in its highly reduced inflorescence.

SELECTED REFERENCE **Villamil, C.B.** 1969. El género *Monanthochloë* (Gramineae). Estudios morfológicos y taxonómicos con especial referencia a la especie argentina. Kurtziana 5:369–391.

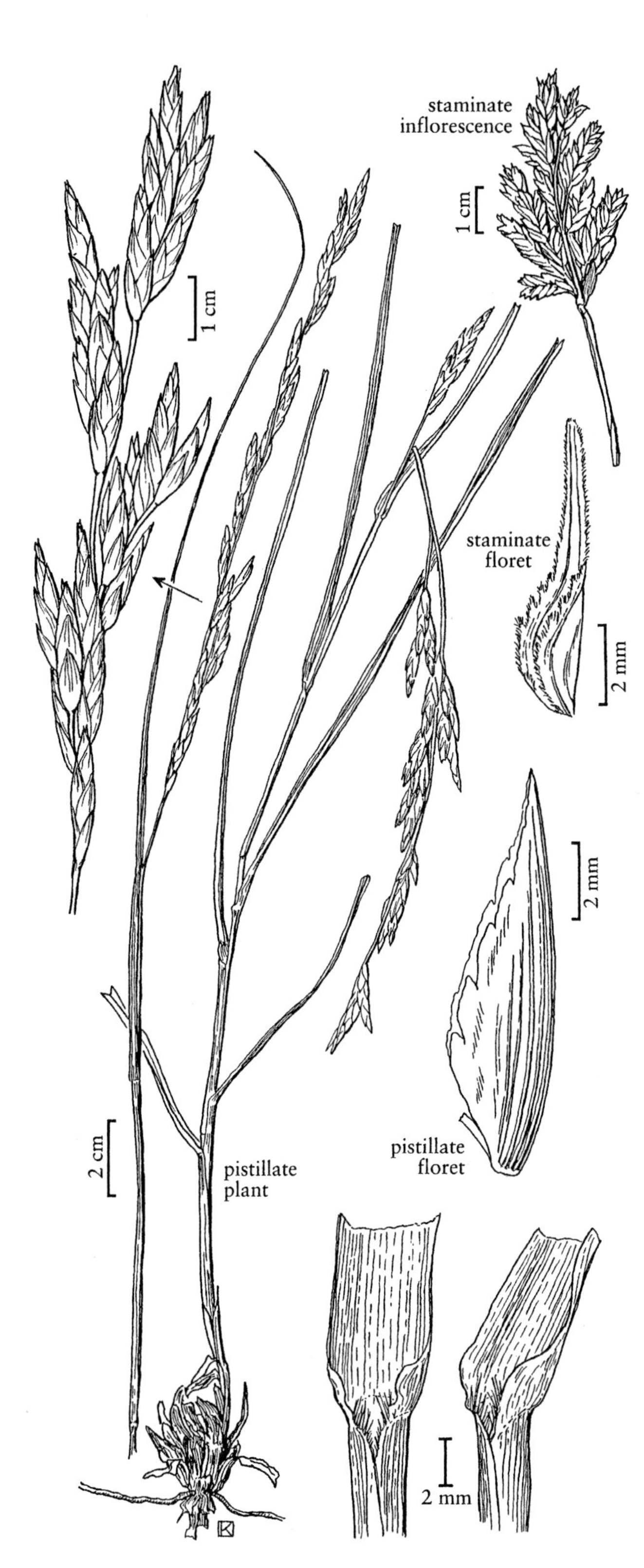

A. texana

ALLOLEPIS

1. **Monanthochloë littoralis** Engelm. [p. 30]
SHOREGRASS

Plants with extensively creeping stolons. **Culms** 8–15 cm, clustered, erect. **Sheaths** 4–6 mm, rounded, smooth, shiny, glabrous or puberulent at the base; **blades** 0.5–1.5 cm long, 1–2(3) mm wide, uniformly many-veined. $2n = 40$.

Monanthochloë littoralis grows in moist, sandy, saline soils along the coast of southern California and the southeastern United States, northeastern Mexico, and the Caribbean islands.

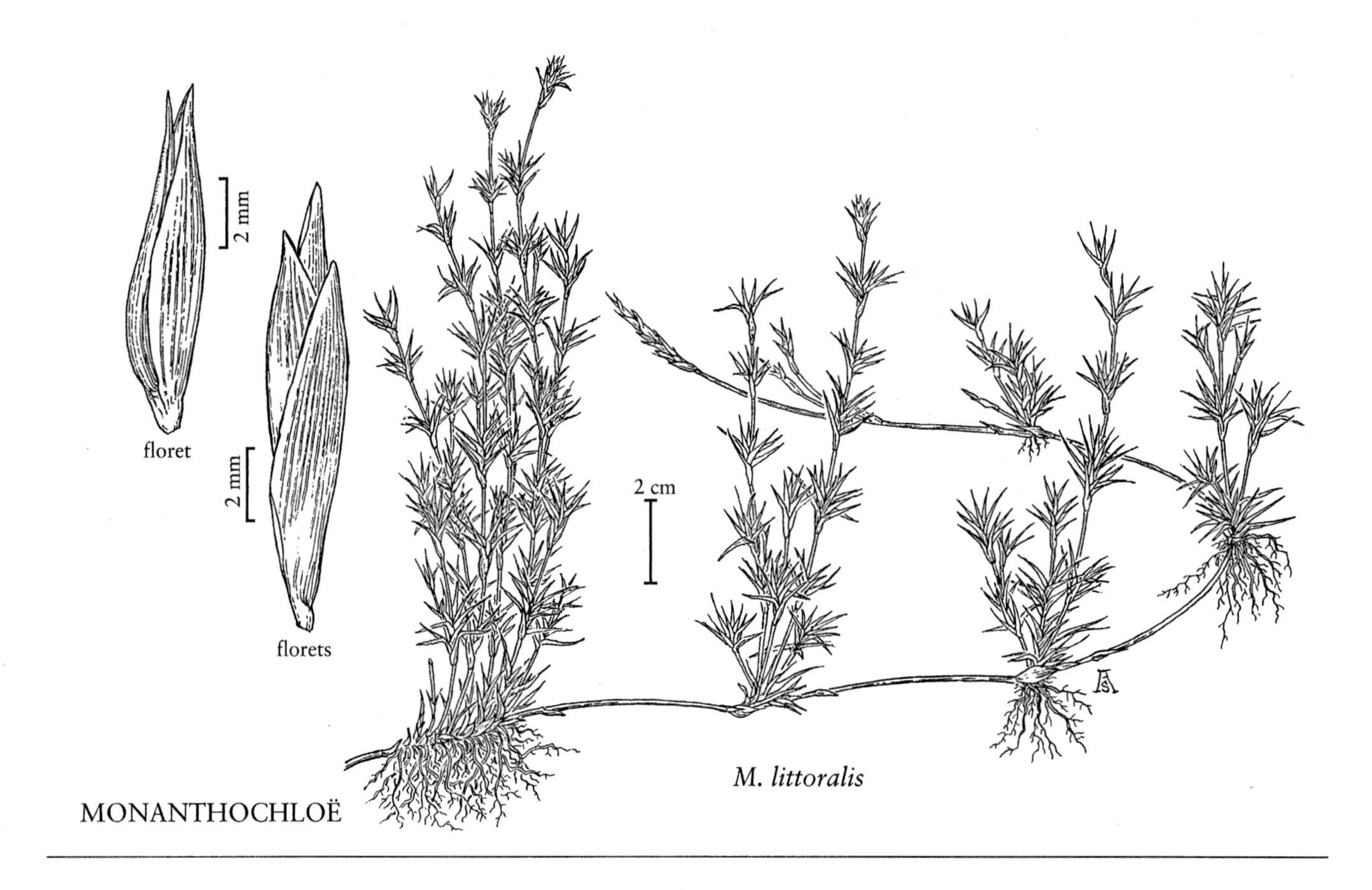

17.08 NEYRAUDIA Hook. f.

Gerald F. Guala

Plants perennial; cespitose, with short, thick, scaly rhizomes. **Culms** 80–500 cm tall, to 1.5 cm thick, reedlike, almost woody; **internodes** solid. **Leaves** cauline, evenly distributed, distichous; **sheaths** with tightly overlapping margins; **ligules** a cartilaginous ridge subtending a line of hairs; **blades** broad, flat, deciduous at maturity, margins not cartilaginous. **Inflorescences** terminal, plumose panicles, 30–80 cm, exceeding the upper leaves; **disarticulation** above the glumes and between the florets. **Spikelets** laterally compressed; **florets** (2)4–10, bisexual, lowest florets sometimes only an empty lemma; **rachilla internodes** glabrous. **Glumes** 2, shorter than the lemmas, subequal, narrowly ovate, glabrous, membranous, 1-veined, acute; **calluses** obtuse, pilose; **lemmas** narrowly ovate, thinly membranous to hyaline, 3-veined, midveins glabrous, margins and lateral veins long pilose on the distal ⅔, apices bifid, shortly awned from between the teeth; **paleas** shorter than the lemmas, hyaline, 2-keeled. x = unknown, frequently reported as 10. Named after A.A. Reynaud, a French surgeon and botanist, who was already honored by *Reynaudia*, a genus of grasses from Cuba; hence the anagram *Neyraudia*.

Neyraudia is a genus of 2–4 species from the tropics of the Eastern Hemisphere. One Asian species is established in Florida. It differs from the other species in the genus in having a sterile lemma at the base of the spikelets. Reports of *Neyraudia arundinacea* occurring in the *Flora* region have not been substantiated.

Neyraudia is often mistaken for *Phragmites*, but that genus has villous rachilla internodes and glabrous lemmas.

1. **Neyraudia reynaudiana** (Kunth) Keng *ex* Hitchc. [p. 32]
BURMA REED, SILK REED

Culms 0.8–3.8(5) m tall, 0.2–1.2 cm thick, often branching above the base. **Sheaths** glabrous, indurate at maturity; **ligules** both adaxial and abaxial; **blades** (7)25–89 cm long, 0.3–2.3 cm wide, auriculate, glabrous abaxially, sparsely hirsute adaxially, adaxial collar area lanate, hairs 2.2–7.2 mm, forming a conspicuous white tuft. **Panicles** 35–73 cm, longest branches 10–28 cm. **Spikelets** 4.3–9.5 mm; **florets** (3)5–10, basal floret in each spikelet reduced to an empty lemma. **Glumes** 0.5–0.8 mm wide; **lower glumes** (1.3)1.5–3.1 mm; **upper glumes** (1.6)2–3 mm; **basal lemmas** 2.5–4.2 mm; **fertile lemmas** 3–4.4 mm long, 0.9–1 mm wide, pilose on the lateral veins, otherwise glabrous or minutely papillate; **awns** about 3 mm, exceeding the bifid lemma apices by up to 1.5 mm; **paleas** 2.4–3 mm; **anthers** 3, 1.2–1.7 mm. **Caryopses** 1.5–3 mm, brown. $2n$ = 40.

Neyraudia reynaudia is an Asian species that was introduced at Chapman Field, U.S. Department of Agriculture, Coral Gables, Florida, probably from PI #39681, in 1915. It is now a troublesome weed in that state, growing in a variety of habitats from marshy areas to dry pinelands.

17.09 TRIRAPHIS R. Br.

J.K. Wipff

Plants annual or perennial; cespitose. **Culms** (1)4–140 cm, erect or geniculate at the lower nodes. **Leaves** cauline; **auricles** absent; **ligules** of hairs or membranous and long-ciliate; **blades** narrow, often involute. **Inflorescences** terminal, open or contracted (occasionally spikelike) simple panicles, exceeding the leaves. **Spikelets** laterally compressed, with 3–9 bisexual florets, reduced florets (if present) distal; **disarticulation** above the glumes and beneath the florets. **Glumes** subequal, exceeded by the florets, 1-veined; **calluses** short, bearded; **lemmas** 3-veined, 3–4-lobed or toothed, lateral veins pilose or ciliate, midveins glabrous or sparsely pubescent, all 3 veins extending into awns; **paleas** shorter than the lemmas, 2-veined; **lodicules** 2; **anthers** 3. **Caryopses** trigonous, falling free of the lemmas and paleas; **embryos** large relative to the caryopses. x = 10. Name from the Greek *treis*, 'three' and *rhaphis*, 'needle', alluding to the three awns of the lemma.

Triraphis is a genus of seven species that are often found in dry, open habitats in sandy or rocky soil. Most of its species are native to Africa and Arabia, but one species is native to Australia. The Australian species is established in the *Flora* region.

1. **Triraphis mollis** R. Br. [p. 32]
PURPLE NEEDLEGRASS

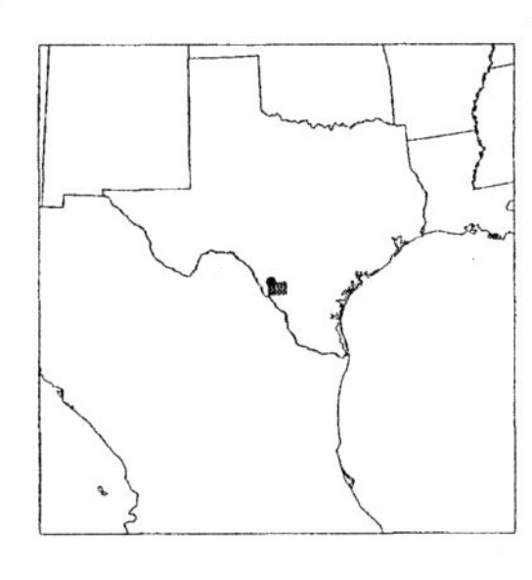

Plants short-lived perennials. **Culms** 2–90 cm, sometimes rooting at the nodes; **nodes** glabrous. **Ligules** membranous, long-ciliate, central cilia 0.5–1.6 mm, those at the sides to 3 mm, stiff; **blades** 7–24 cm long, 2–5 mm wide, usually involute, rarely flat. **Panicles** 6–30 cm long, 1.5–3 cm wide; **primary branches** to 4.5 cm, appressed or ascending. **Lower glumes** 3–4 mm, mucronate; **upper glumes** 4–5 mm, 2-toothed; **calluses** about 0.5 mm, with hairs; **lemmas** 3–5 mm, hairs on the veins to 2 mm, lobes about 1 mm, central awns 5–7 mm, lateral awns 6–7 mm. $2n$ = unknown.

Triraphis mollis usually flowers in response to rain. It is common on sandy soils in New South Wales, Australia. In the *Flora* region it is currently known only from Dimmit County, Texas, but it will probably spread.

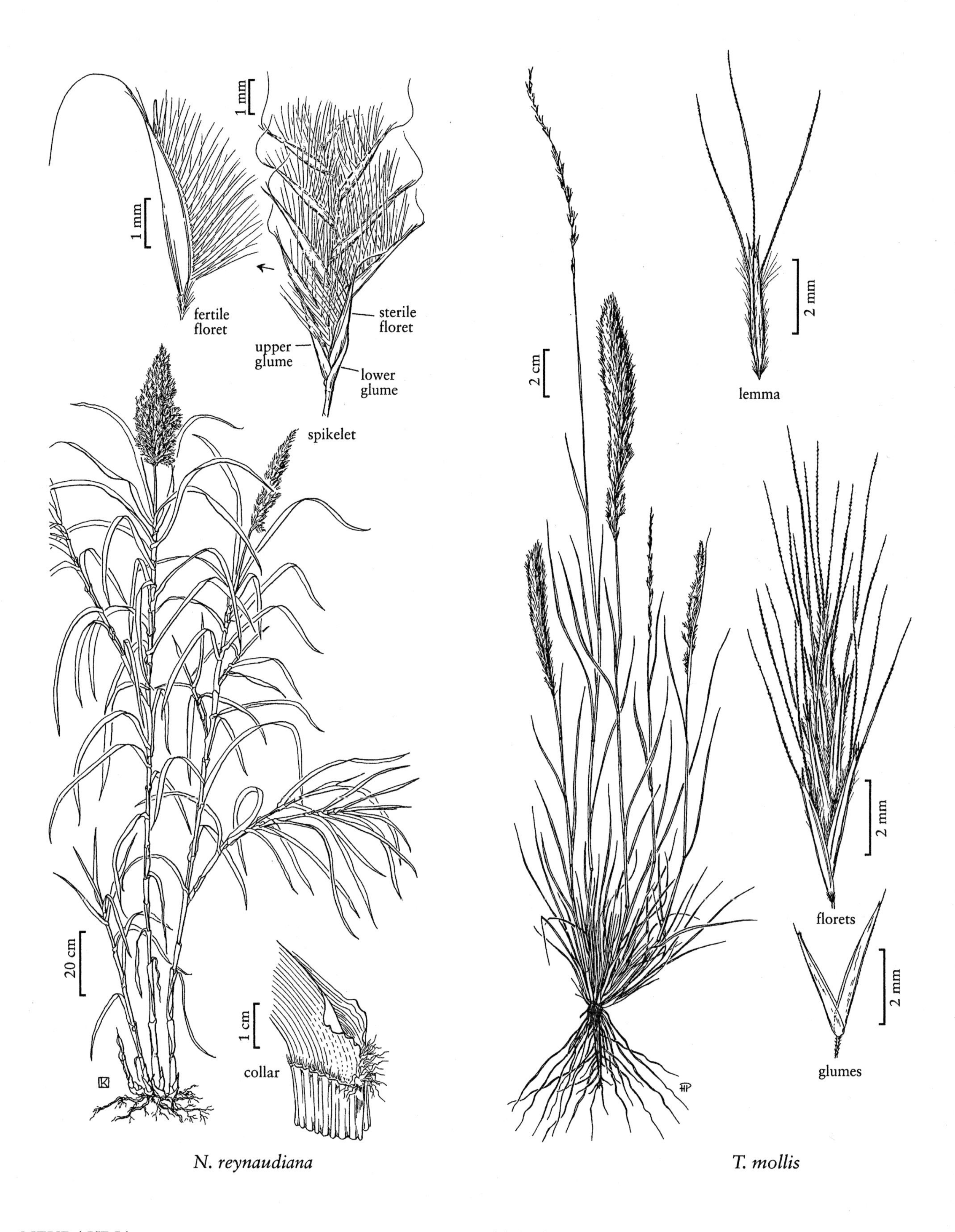

N. reynaudiana

T. mollis

NEYRAUDIA

TRIRAPHIS

17.10 TRIDENS Roem. & Schult.

Jesús Valdés-Reyna

Plants perennial; usually cespitose, often with short, knotty rhizomes, occasionally with elongate rhizomes, never stoloniferous. **Culms** 5–180 cm, erect, mostly glabrous, lower nodes sometimes with hairs. **Sheaths** shorter than the internodes, open; **ligules** membranous and ciliate or of hairs; **blades** 6–25 cm long, 1–8 mm wide, flat or involute, margins not thick and cartilaginous. **Inflorescences** terminal, usually panicles (sometimes reduced to racemes), 5–40 cm, exceeding the upper leaves, exserted. **Spikelets** 4–10(13) mm, laterally compressed, with 4–11(16) florets, more than 1 floret bisexual; **sterile florets** distal to the fertile spikelets; **disarticulation** above the glumes. **Glumes** from shorter than to equaling the distal florets; **lower glumes** 1(3)-veined; **upper glumes** shorter than or about equal to the lower glumes, 1–3(9)-veined, unawned; **calluses** usually glabrous, sometimes pilose; **lemmas** hyaline or membranous, 3-veined, veins usually shortly hairy below, apices rounded to truncate, emarginate to bilobed, midvein often excurrent to 0.5 mm, lateral veins not or more shortly excurrent; **paleas** glabrous or shortly pubescent on the lower back and margins, veins glabrous or ciliolate; **lodicules** 2, free or adnate to the palea; **anthers** 3, reddish-purple. **Caryopses** dorsiventrally compressed and reniform in cross section, dark brown; **embryos** about ⅖ as long as the caryopses. $x = 10$. Name from the Latin *tres*, 'three', and *dens*, 'tooth', referring to the three shortly excurrent veins of *Tridens flavus*, the type species.

Tridens, a genus of 14 species, is native to the Americas; all ten species described here are native to the the *Flora* region. Hitchcock (1951) included both *Erioneuron* and *Dasyochloa* in *Tridens*; Tateoka (1961) demonstrated that they should be excluded. One of the differences between *Tridens* and the other two genera lies in their chromosome bases numbers, 10 in *Tridens* and 8 in *Erioneuron* and *Dasyochloa*. *Tridens albescens* is exceptional within *Tridens* in having chromosome numbers that suggest two base numbers, 10 and 8.

SELECTED REFERENCES **Burbidge, N.T.** 1953. The genus *Triodia* R. Br. (Gramineae). Austral. J. Bot. 1:121–184; **Gould, F.W.** 1975. The grasses of Texas. Texas A&M University Press, College Station, Texas, U.S.A. 653 pp.; **Hitchcock, A.S.** 1951 [title page 1950]. Manual of the Grasses of the United States, ed. 2, rev. A. Chase. U.S.D.A. Miscellaneous Publication No. 200. U.S. Government Printing Office, Washington, D.C., U.S.A. 1051 pp.; **Tateoka, T.** 1961. A biosystematic study of *Tridens* (Gramineae). Amer. J. Bot. 48:565–573.

1. Primary panicle branches appressed to strongly ascending; panicles 0.3–4 cm wide, dense and spikelike.
   2. Lateral veins of the lemmas glabrous or pubescent only at the base . . . . 1. *T. albescens*
   2. Lateral veins of the lemmas pilose to well above the base.
      3. Glumes evidently longer than the adjacent lemmas, often twice as long, usually equaling or exceeding the distal florets . . . . 2. *T. strictus*
      3. Glumes from shorter than to equaling the adjacent lemmas, often exceeded by the distal florets.
         4. All 3 lemma veins shortly excurrent; calluses pilose . . . . 3. *T. carolinianus*
         4. Lateral lemma veins not excurrent, often terminating before the distal margin, the midvein sometimes excurrent; calluses glabrous or shortly pilose.
            5. Panicles 7–25 cm long, 0.3–0.8 cm wide; lemma midveins rarely excurrent . . . . 4. *T. muticus*
            5. Panicles 5–8(10) cm long, 1.2–2.5 cm wide; lemma midveins always shortly excurrent . . . 5. *T. congestus*
1. Primary panicle branches ascending to reflexed or drooping; panicles 1–20 cm wide, open, not spikelike.
   6. All pedicels shorter than 1 mm . . . . 6. *T. ambiguus*
   6. Some pedicels longer than 1 mm.
      7. Lateral veins of the lemmas rarely excurrent.
         8. Lemmas 4–6 mm; ligules 0.4–1 mm . . . . 7. *T. buckleyanus*

8. Lemmas 2–3.2 mm; ligules 1.2–3 mm . . . . 8. *T. eragrostoides*
7. Lateral veins of the lemmas commonly excurrent as short points.
9. Blades 1–5 mm wide; panicles 5–16 cm long . . . . 9. *T. texanus*
9. Blades mostly 3–10 mm wide; panicles 15–40 cm long . . . . 10. *T. flavus*

1. **Tridens albescens** (Vasey) Wooton & Standl. [p. 35]
White Tridens

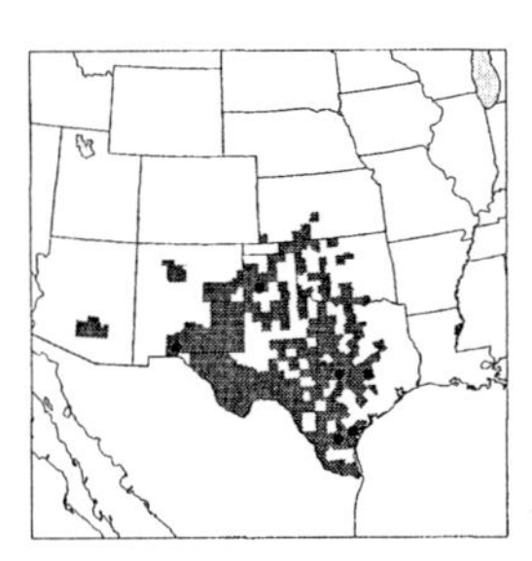

Plants cespitose, often with hard, knotty, shortly rhizomatous bases. **Culms** 30–100 cm; **lower nodes** sometimes sparsely bearded. **Sheaths** glabrous, not or obscurely keeled; **ligules** to 0.5 mm, membranous, ciliate; **blades** 1–4 mm wide, folded or involute, glabrous, apices sharp. **Panicles** 8–25 cm long, 0.5–1.3 cm wide, dense; **branches** appressed, lowest branches 2–6 cm; **pedicels** 1–2 mm. **Spikelets** 4–10 mm, with 4–11 florets. **Glumes** about as long as the adjacent lemmas, thin, 1-veined, acute or apiculate; **lower glumes** 4–4.5 mm; **upper glumes** 4–4.5 mm; **lemmas** 3–4(5) mm, thin, papery, mostly white, often purple distally, glabrous or the lateral veins with a few short hairs towards the base, all veins ending before the distal margin; **paleas** 3–3.5 mm, glabrous, bowed-out at the base; **anthers** 1–1.5 mm. **Caryopses** 1.5–1.8 mm. $2n$ = 60, 64, 72.

Tridens albescens grows in plains and open woods, often in clay soils that periodically receive an abundance of water. Its range extends into northern Mexico.

2. **Tridens strictus** (Nutt.) Nash [p. 35]
Longspike Tridens

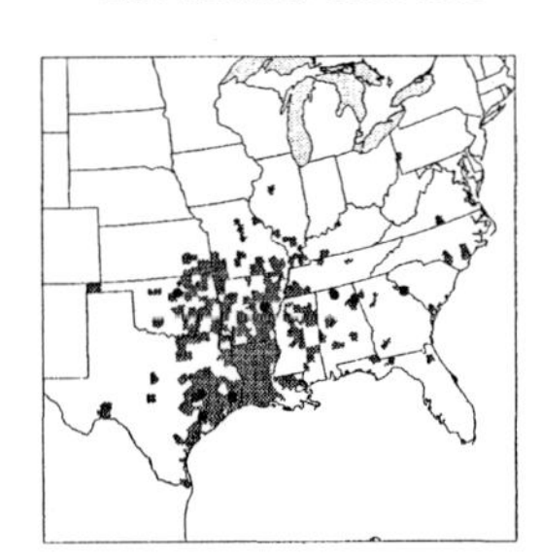

Plants with hard, knotty, shortly rhizomatous bases. **Culms** 50–170 cm, stiffly erect. **Sheaths** rounded, glabrous except for a few hairs on either side of the collar; **ligules** about 0.5 mm, membranous, ciliate; **blades** 2–8 mm wide, flat or loosely infolded, glabrous, tapering to the apices. **Panicles** 10–30(36) cm long, 1–2 cm wide; **branches** to 6 cm, erect or appressed; **pedicels** 1–1.5 mm, glabrous. **Spikelets** 4–7 mm, with 5–11 florets. **Glumes** 4–7 mm, always conspicuously exceeding and often twice as long as the adjacent lemmas, usually equaling or exceeding the distal florets, glabrous, 1-veined, tapering to acuminate apices; **calluses** pilose; **lemmas** (2)3–3.5 mm, veins pilose to well above midlength, lateral veins often excurrent; **paleas** 2–3 mm, bases not bowed-out; **anthers** 1–1.5 mm. **Caryopses** 1–1.5 mm. $2n$ = 40.

Tridens strictus grows in open woods, old fields, right of ways, and coastal grasslands. It is endemic to the United States.

3. **Tridens carolinianus** (Steud.) Henrard [p. 35]
Creeping Tridens

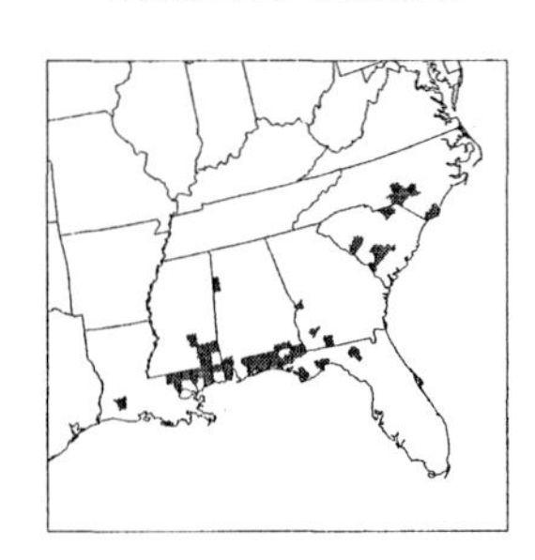

Plants rhizomatous; **rhizomes** elongate, 2.5–5 mm thick, scaly. **Culms** 80–120 cm. **Lower sheaths** pilose; **ligules** about 0.5 mm, membranous, ciliate; **blades** 2–7 mm wide, flat, both surfaces sparsely pilose basally, margins smooth or scabridulous; **upper leaves** with glabrous sheaths and blades. **Panicles** 9–15 cm long, 1–4 cm wide, nodding, purplish; **branches** appressed or narrowly ascending; **pedicels** 2–3(3.5) mm. **Spikelets** 7–10 mm, with 3–5 florets. **Glumes** glabrous, 1-veined; **lower glumes** 3.5–4.5 mm; **upper glumes** 4–5 mm; **calluses** sparsely pilose; **lemmas** 4–5 mm, veins pilose at least to midlength, all 3 veins excurrent as short points; **paleas** 3–3.5 mm, glabrous, bases bowed-out; **anthers** 1–2(2.5) mm. **Caryopses** 2–2.5 mm. $2n$ = unknown.

Tridens carolinianus grows in pinelands and open sandy woods along the coastal plain from North Carolina to Louisiana.

4. **Tridens muticus** (Torr.) Nash [p. 37]
Slim Tridens

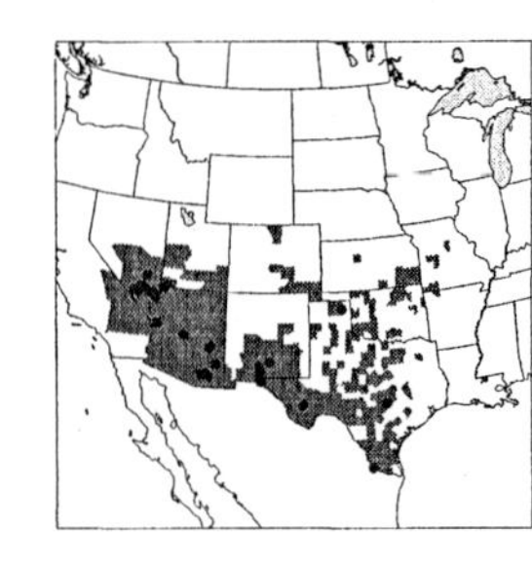

Plants cespitose, with knotty, shortly rhizomatous bases. **Culms** 20–80 cm; **nodes** often with soft, 1–2 mm hairs. **Sheaths** rounded, lower sheaths often strigose or pilose, upper sheaths glabrous or scabrous; **ligules** 0.5–1 mm, membranous, ciliate; **blades** 1–4 mm wide, usually involute or loosely infolded, glabrous, scabrous, or sparsely pilose, attenuate distally. **Panicles** 7–20(25) cm long, 0.3–0.8 cm wide; **branches** erect, spikelets imbricate but usually not crowded; **pedicels** 1–2 mm. **Spikelets** 8–13 mm, with 5–11 florets. **Glumes** glabrous, usually purple-tinged; **lower glumes** 3–8(10) mm, 1–3-veined; **upper glumes** 4–10 mm, 1–7-veined; **lemmas** 3.5–7 mm, usually purple-tinged, midveins pilose on the basal ⅓–½, rarely excurrent, lateral veins pilose to well

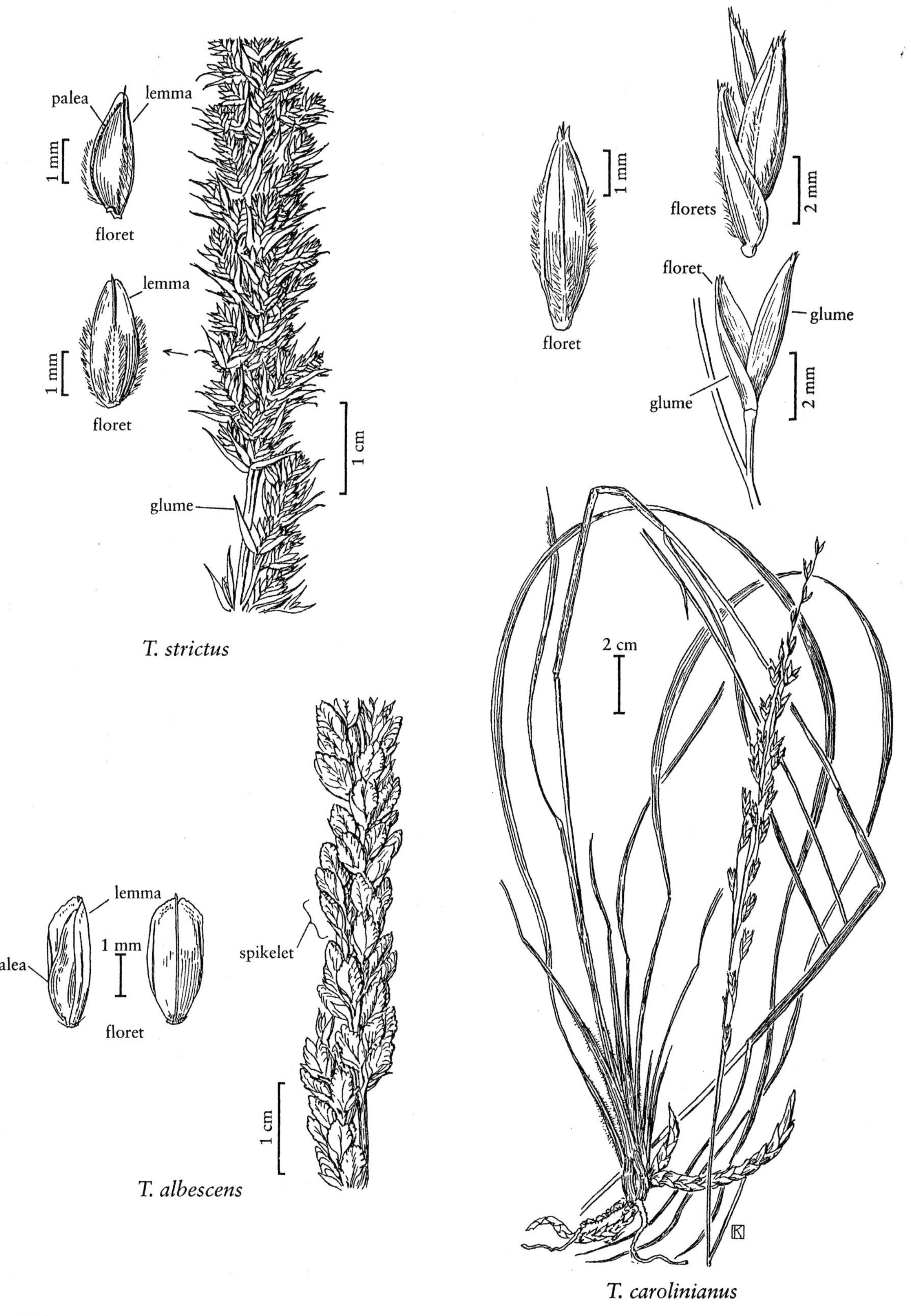
palea
lemma
1 mm
floret
lemma
1 mm
floret
1 cm
glume
T. strictus
1 mm
floret
florets
2 mm
floret
glume
glume
2 mm
2 cm
lemma
1 mm
palea
floret
spikelet
1 cm
T. albescens
T. carolinianus

TRIDENS

above midlength, never excurrent; **paleas** 1–2 mm shorter than the lemmas, margins pubescent; **anthers** 1–1.5 mm. **Caryopses** 1.5–2.3 mm. 2*n* = 40.

1. Upper glumes 4–5(6) mm long, 1-veined . . . .var. *muticus*
1. Upper glumes usually 5.5–10 mm long, 3–7-veined . . . . .var. *elongatus*

Tridens muticus var. **elongatus** (Buckley) Shinners

Culms usually 40–80 cm. **Blades** often 3–4 mm wide. **Upper glumes** usually 5.5–10 mm, 3-7-veined.

Tridens muticus var. *elongatus* grows on well-drained, clayey and sandy soils from Colorado to Missouri and from Arizona to Louisiana. It is not known from Mexico.

Tridens muticus (Torr.) Nash var. **muticus**

Culms usually 20–50 cm. **Blades** 1–2 mm wide. **Upper glumes** 4–5(6) mm, 1-veined.

Tridens muticus var. *muticus* is a common species on dry, sandy or clay soils in the arid southwestern United States and adjacent Mexico.

5. **Tridens congestus** (L.H. Dewey) Nash [p. 37]
PINK TRIDENS

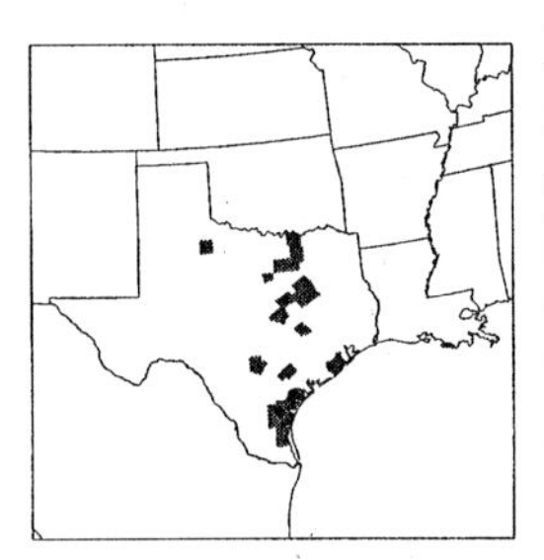

Plants cespitose, with shortly rhizomatous bases. **Culms** 30–75 cm. **Sheaths** glabrous, rounded; **ligules** to 0.5 mm, membranous, ciliate; **blades** 1.5–5 mm wide, tapering and involute distally. **Panicles** 5–8(10) cm long, 1.2–2.5 cm wide, dense; **branches** 0.5–3 cm, erect to ascending; **pedicels** 1–3 mm. **Spikelets** 5–10 mm, with 5–12 florets. **Glumes** glabrous, usually tinged with pink, 1-veined; **lower glumes** 3–4.5 mm; **upper glumes** 4–4.5(5) mm; **lemmas** 3–5 mm, usually tinged with pink, midveins and margins pubescent on the basal ⅓–½, midveins shortly excurrent, lateral veins pilose to midlength, terminating before the distal margins; **paleas** 0.5–1 mm, shorter than the lemmas, scabrous on the veins, broadened below, bases bowed-out at maturity; **anthers** 1–1.5 mm. **Caryopses** (1)1.5–2 mm. 2*n* = unknown.

Tridens congestus grows in moist depressions, ditches, and low flats of otherwise dry hills in Texas. It resembles *T. albescens*, but usually has shorter panicles, spikelets that are more or less evenly pink rather than purple-tipped, and more deeply cleft lemma apices.

6. **Tridens ambiguus** (Elliott) Schult. [p. 38]
PINE-BARREN TRIDENS

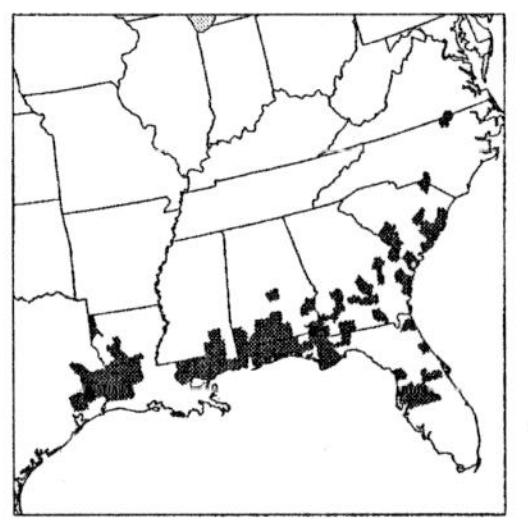

Plants cespitose, with knotty, shortly rhizomatous bases. **Culms** 60–125 cm. **Sheaths** rounded or the basal sheaths keeled, glabrous, except for a few hairs on either side of the collar; **ligules** 1–2 mm, membranous, ciliate; **blades** 2–5 mm wide, elongate, usually involute distally. **Panicles** 8–16(20) cm long, 1.5–4 cm wide, not dense; **branches** to 8(10) cm, erect to divergent, stiff; **pedicels** shorter than 1 mm. **Spikelets** 4–6 mm long, pale to dark purple, with 4–6 florets. **Glumes** 1-veined; **lower glumes** 4–4.5 mm; **upper glumes** about 5 mm; **lemmas** 3–4 mm, veins pubescent to midlength or beyond, midveins excurrent, lateral veins often excurrent; **paleas** 3–3.5 mm, veins ciliolate, bases bowed-out; **anthers** 1–1.5 mm. **Caryopses** 1.5–1.8 mm. 2*n* = 40.

Tridens ambiguus grows on the southeastern coastal plain, from North Carolina to Texas. It is usually found in mesic to perennially moist soils of pine flatwoods and pine-oak savannahs, in seasonally inundated depressions, and at the margins of pitcher plant bogs, often in disturbed sites.

7. **Tridens buckleyanus** (L.H. Dewey) Nash [p. 38]
BUCKLEY'S TRIDENS

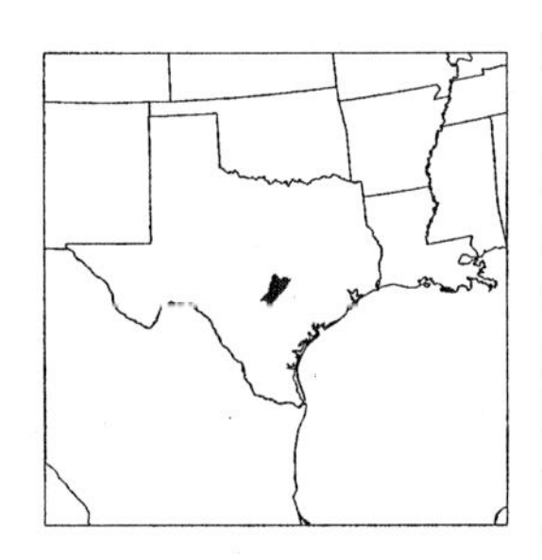

Plants cespitose. **Culms** 30–60 cm, erect; **lower nodes** sometimes hispid. **Sheaths** scabridulous, rounded; **ligules** 0.4–1 mm, membranous, ciliate; **blades** 1–4 mm wide, flat, apices attenuate. **Panicles** 10–28 cm long, 1–6 cm wide; **branches** (2)4–13 cm, widely-spaced, ascending to spreading; **pedicels** 1–2 mm. **Spikelets** 7–10 mm, pale to dark purple, with 2–5 florets. **Glumes** 1-veined; **lower glumes** 4–5(6) mm; **upper glumes** slightly shorter; **lemmas** 4–6 mm, midveins and back pubescent below midlength, lateral veins pubescent to well above midlength, midveins occasionally excurrent, lateral veins usually ending before the distal margins, rarely excurrent; **paleas** 3.5–4 mm, veins pubescent basally; **anthers** 1.5–2 mm. **Caryopses** about 2.5 mm. 2*n* = 32, "but this may be erroneous" (Gould 1975).

Tridens buckleyanus is endemic to the southeastern portion of the Edwards Plateau, Texas. It grows on rocky slopes along shaded stream banks and the borders of woodlands.

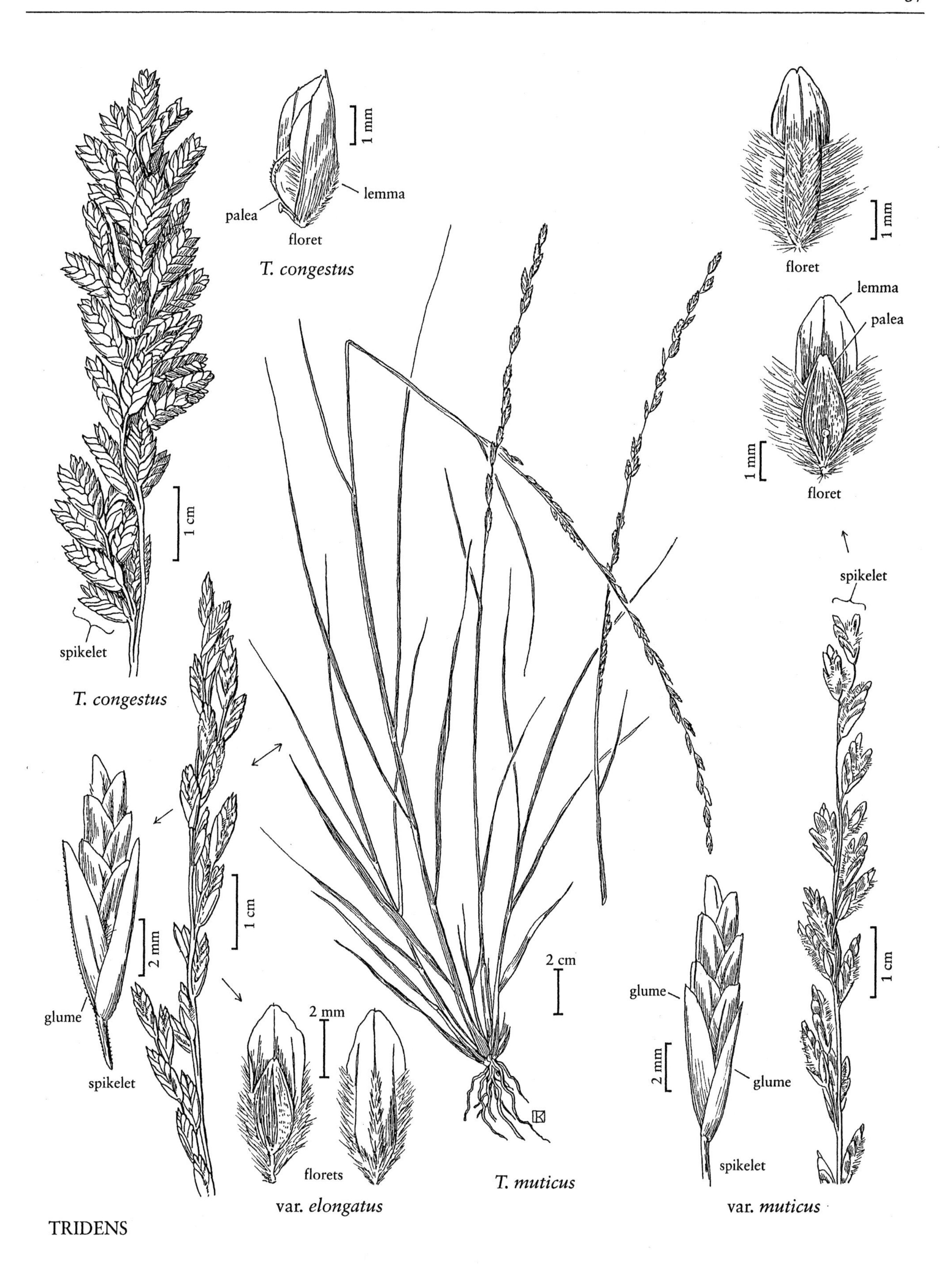

1 mm
palea
lemma
floret
T. congestus
1 mm
floret
lemma
palea
1 mm
floret
spikelet
1 cm
spikelet
T. congestus
2 mm
glume
spikelet
1 cm
2 mm
florets
var. elongatus
2 cm
T. muticus
glume
2 mm
glume
spikelet
1 cm
var. muticus

TRIDENS

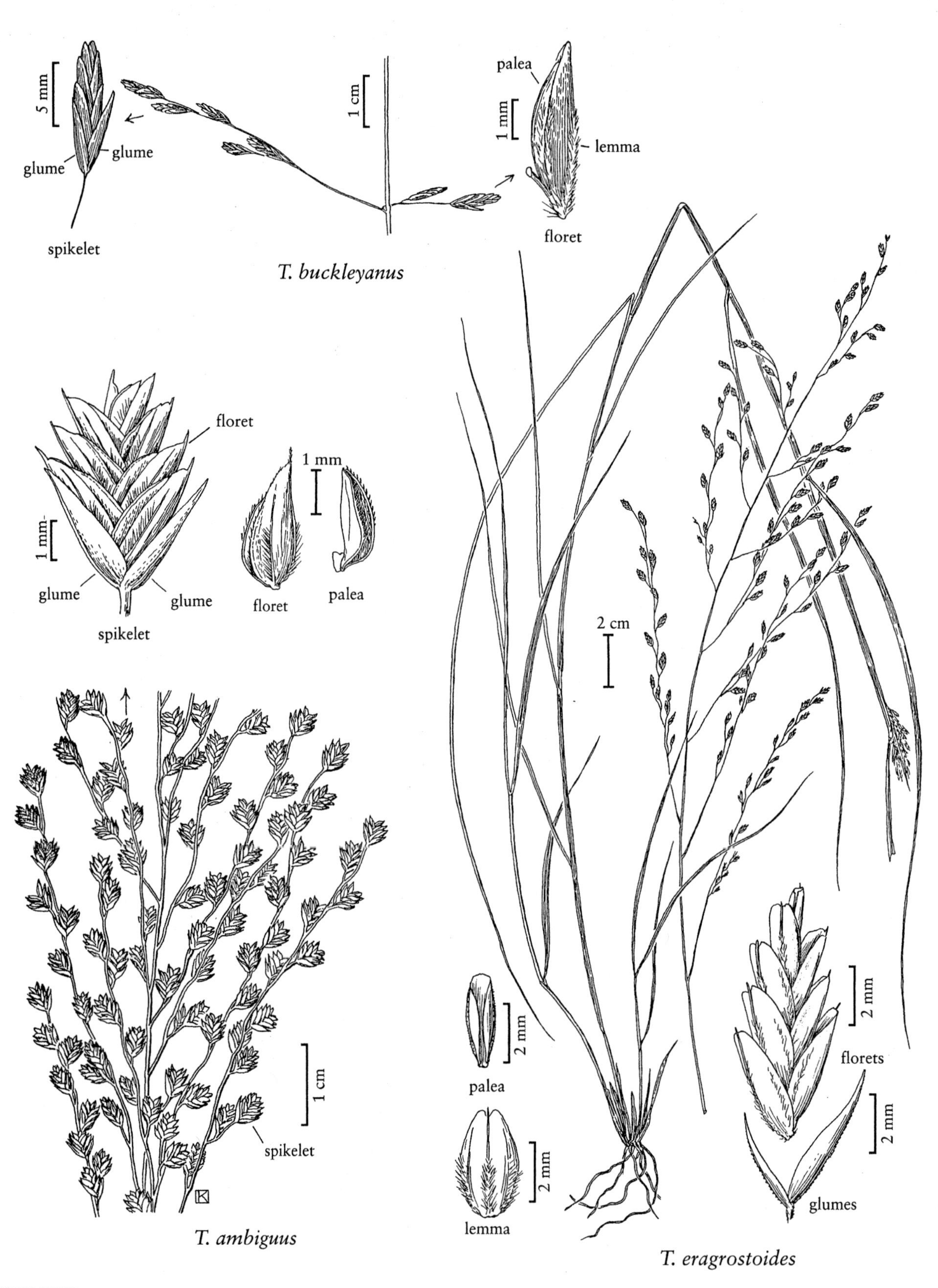
5 mm
glume
glume
spikelet
1 cm
palea
1 mm
lemma
floret
T. buckleyanus
floret
1 mm
1 mm
glume
glume
spikelet
floret
palea
2 cm
spikelet
1 cm
T. ambiguus
2 mm
palea
2 mm
lemma
2 mm
florets
2 mm
glumes
T. eragrostoides

TRIDENS

8. **Tridens eragrostoides** (Vasey & Scribn.) Nash [p. 38]
LOVEGRASS TRIDENS

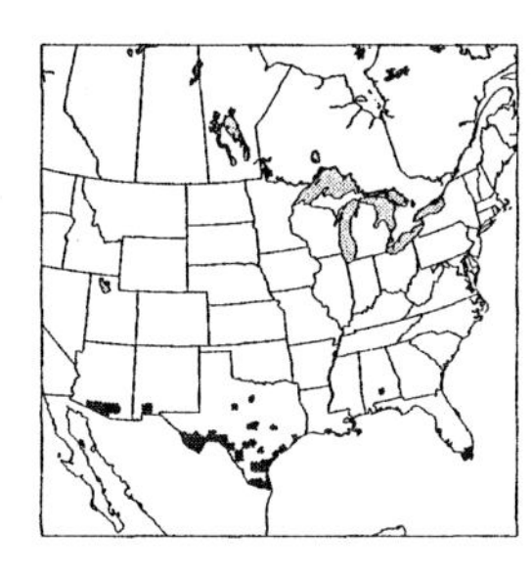

Plants cespitose, with knotty, shortly rhizomatous bases. **Culms** 50–100 cm; **nodes** sometimes sparsely bearded. **Sheaths** glabrous, scabrous, or sparsely pilose, rounded; **ligules** 1.2–3 mm, glabrous, membranous, usually lacerate; **blades** 10–15 cm long, 1.5–5 mm wide, scabrous (occasionally sparsely pilose), apices long-attenuate. **Panicles** 10–30 cm long, to 20 cm wide, open; **branches** 5–10(12) cm, lax, ascending to reflexed at maturity, proximal internodes longer than the distal internodes; **pedicels** (1.5)3–5 mm. **Spikelets** 3–7 mm, with 5–12 florets. **Glumes** glabrous, 1-veined, purple; **lower glumes** 2–2.5 mm; **upper glumes** 2–3.5 mm; **lemmas** 2–3.2 mm, veins puberulent to well above midlength, midveins sometimes excurrent, lateral veins rarely reaching the distal margins; **paleas** 1.5–2 mm, glabrous or scabrous basally, neither enlarged nor bowed-out; **anthers** 1–1.5 mm. **Caryopses** 1–1.3 mm. $2n = 40$.

Tridens eragrostoides grows in brush grasslands, generally in partial shade. Its range extends from the southern United States into Mexico and Cuba.

9. **Tridens texanus** (S. Watson) Nash [p. 40]
TEXAS TRIDENS

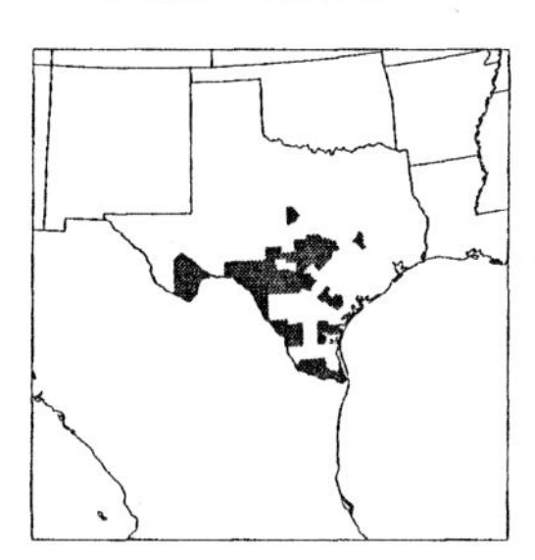

Plants cespitose, with knotty, shortly rhizomatous bases. **Culms** 20–75 cm, slender, strictly erect; **nodes** glabrous; **internodes** often pilose. **Sheaths** mostly glabrous or pilose throughout, collar and distal portion of the margins densely pilose; **ligules** to 0.5 mm, membranous, ciliate; **blades** 1–3(5) mm wide, flat or becoming inrolled, hispid, with long hairs on the adaxial surface just above the ligule, apices attenuate. **Panicles** 5–16 cm long, 2–9 cm wide, open or loosely contracted; **branches** (2)4–7 cm, slender, lax, strongly divergent to drooping, basal portion naked, spikelets confined to the distal portion; **pedicels** (2)3–6 mm. **Spikelets** 6–13 mm, with 6–12 florets. **Glumes** glabrous, 1-veined; **lower glumes** 3 mm; **upper glumes** 3.5–4 mm, veins bright green; **lemmas** 3–4.5 mm, usually purple or rosy-purple at maturity, veins pilose to midlength, lateral veins often excurrent as short points; **paleas** 3–3.5 mm, glabrous, abruptly broadened and bowed-out below; **anthers** 1–1.5 mm. **Caryopses** 1.5–2 mm. $2n = 40$.

Tridens texanus grows in clayey and sandy loam soils, often in the protection of shrubs and along fenced road right of ways. Its range extends from southern Texas into northern Mexico.

10. **Tridens flavus** (L.) Hitchc. [p. 40]
PURPLETOP TRIDENS

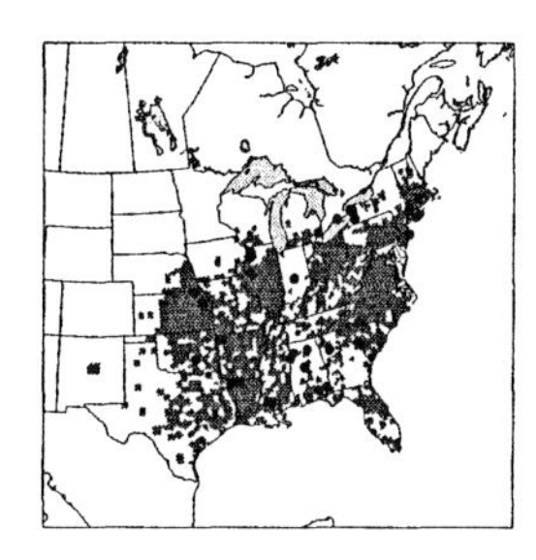

Plants with firm, knotty, shortly rhizomatous bases. **Culms** 60–180 cm. **Sheaths** keeled, mostly glabrous, collars pubescent; **ligules** to 0.5 mm, membranous, ciliate; **blades** 3–10 mm wide, glabrous or sparsely hispid, apices attenuate and involute. **Panicles** 15–40 cm long, 3–5 cm wide, erect or nodding; **branches** 10–15(25) cm, strongly divergent to drooping, stiff or lax, lower branches naked for ⅓–½ of their length; **pedicels** 3–8 mm. **Spikelets** 5–10 mm long, with 4–8 florets. **Lower glumes** 2.5–3 mm, often mucronate; **upper glumes** 3.5–4 mm; **lemmas** 3–5 mm, lateral veins puberulent or ciliate to well above midlength, midveins and lateral veins usually excurrent, midveins extending to 0.5 mm, lateral vein extensions shorter; **paleas** as long as the lemmas, widened below; **anthers** 1–1.5 mm. **Caryopses** 1.8–2 mm. $2n = 40$.

1. Panicles nodding; pulvini inconspicuously or conspicuously hairy, the hairs confined to the adaxial side of the branches .... var. *flavus*
1. Panicles erect throughout; pulvini always conspicuously hairy, the hairs extending around the base of the branches .... var. *chapmanii*

Tridens flavus var. **chapmanii** (Small) Shinners

Panicles usually erect throughout; **branches** stiff; **pulvini** conspicuously hairy, hairs extending around the base of the branches.

Tridens flavus var. *chapmanii* grows in pine and oak woods of the southeastern United States from Missouri to Virginia and south from eastern Texas to Florida.

Tridens flavus (L.) Hitchc. var. **flavus**

Panicles nodding distally; **branches** flexible; **pulvini** inconspicuously or conspicuously hairy, hairs confined to the adaxial side of the branches.

Tridens flavus var. *flavus* grows in old fields and open woods. Its range extends to Nuevo Léon, Mexico. It was discovered for the first time in Canada in 1976, growing along a railway track in southern Ontario.

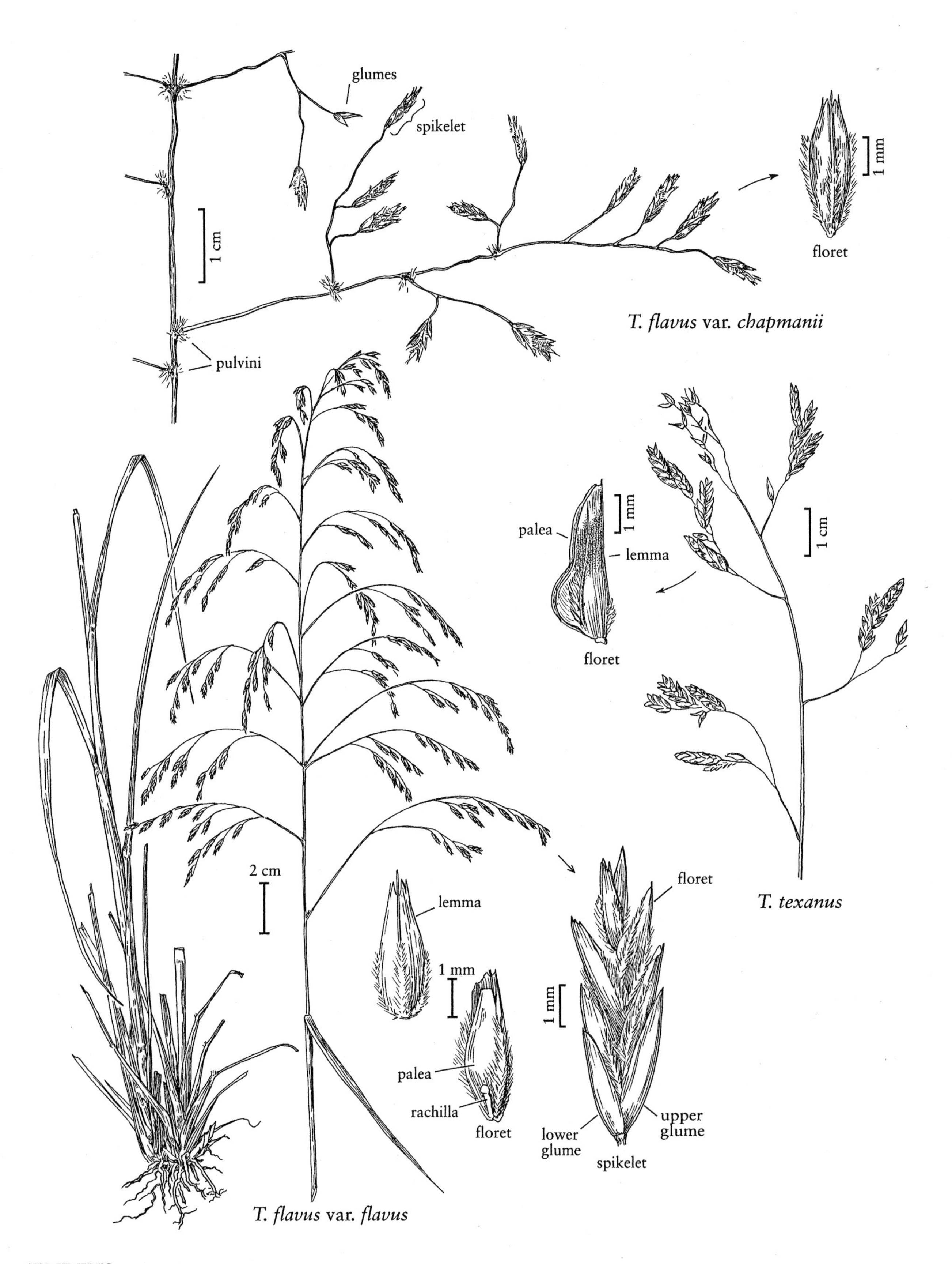
glumes
spikelet
1 mm
floret
1 cm
T. flavus var. chapmanii
pulvini
palea
1 mm
lemma
1 cm
floret
2 cm
floret
T. texanus
lemma
1 mm
1 mm
palea
rachilla
floret
lower glume
upper glume
spikelet
T. flavus var. flavus

TRIDENS

17.11 REDFIELDIA Vasey

Stephan L. Hatch

Plants perennial; with extensive, often deep, horizontal or vertical rhizomes. **Culms** 50–130 cm, erect, bases usually buried and rooting at the nodes. **Leaves** cauline; **sheaths** shorter than the internodes, open, ribbed; **ligules** membranous, ciliate; **auricles** absent; **blades** loosely involute, sometimes scabridulous, apices attenuate. **Inflorescences** terminal, conical to oblong panicles, open to diffuse, exceeding the upper leaves; **branches** slender, widely spreading. **Pedicels** longer than the spikelets, capillary, flexible. **Spikelets** ovate to obovate, olive-green to brownish, with (1)2–6 florets; **sterile florets** distal to the bisexual florets; **disarticulation** above the glumes and below the florets. **Glumes** unequal, usually exceeded by the florets, glabrous, acute; **lower glumes** 1-veined; **upper glumes** 1- or 3-veined; **calluses** with a tuft of soft hairs; **lemmas** lanceolate to falcate, glabrous or shortly pubescent, at least on the distal ⅔, 3-veined, lateral veins converging distally, apices acute to awn-tipped, entire or with 3 minute teeth; **anthers** 3. x = unknown. Named for John Howard Redfield (1815–1895), a Philadelphia businessman associated with the Philadelphia Academy of Natural Sciences.

Redfieldia is a monotypic genus that is endemic to the *Flora* region.

SELECTED REFERENCE Reeder, J.R. 1976. Systematic position of *Redfieldia* (Gramineae). Madroño 23:434–438.

1. **Redfieldia flexuosa** (Thurb. *ex* A. Gray) Vasey [p. 43]
BLOWOUT-GRASS

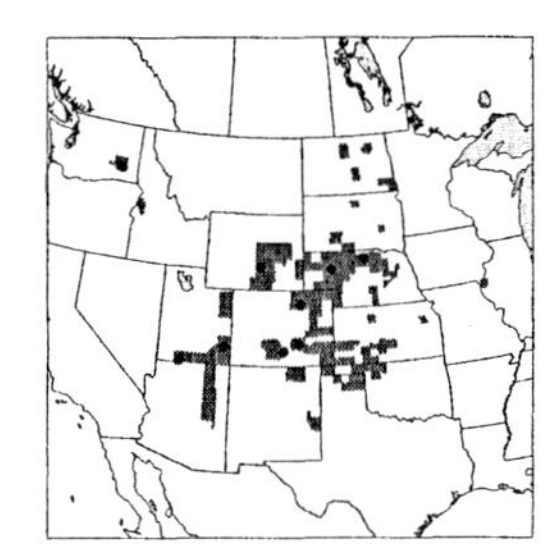

Culms 50–130 cm. **Ligules** to 1.5 mm; **blades** 15–45 cm long, 2–8 mm wide. **Panicles** 20–50 cm long, 8–25 cm wide. **Spikelets** (3)5–8 mm long, 3–5 mm wide. **Lower glumes** 3–4 mm; **upper glumes** 3.5–4.5 mm; **callus hairs** to 1.5 mm; **lemmas** 4.5–6 mm, glabrous or shortly pubescent, veins glabrous, entire or with 3 minute teeth; **paleas** glabrous; **anthers** 2–3.6 mm, yellow to reddish-purple; **lodicules** 2, truncate. **Caryopses** oblong, terete. $2n$ = 25.

Redfieldia flexuosa grows on sandhills and dunes. It is a common and important soil binder in blowout areas. It is only fair livestock forage but, because it grows in areas subject to blowout, this should not be of concern. The only Arizona collection was made in 1896; the Washington population was introduced for erosion control.

17.12 TRIPLASIS P. Beauv.

Stephan L. Hatch

Plants annual or perennial; cespitose, occasionally rhizomatous. **Culms** 14–100 cm, ascending to erect; **nodes** pubescent to hirsute. **Sheaths** open; **auricles** absent; **ligules** of hairs or membranous and ciliate; **blades** 1–5 mm wide, flat or involute. **Primary inflorescences** terminal, open panicles, with few spikelets, exerted or partially included in the upper sheath, apices exceeding the upper leaf blades, axillary panicles sometimes also present; **cleistogamous inflorescences** also present in the upper sheaths. **Spikelets** laterally compressed, purplish, with 2–5 florets; **sterile florets** above the fertile florets; **rachillas** prolonged; **disarticulation** above the glumes and beneath the florets and, subsequently, at the cauline nodes. **Glumes** equal or unequal, shorter than the first lemma, 1-veined, keeled; **calluses** hairy; **lemmas** 3-veined, veins villous, apices bilobed to incised, midveins sometimes extending into an awn, awns to 11 mm; **paleas** bowed-out, keels hairy, distal hairs 0.5–2 mm, longer than those below; **lodicules** 2, truncate; **anthers** 3, yellow or reddish-purple; **stigmas** pink to purple. **Caryopses** dorsiventrally compressed. x = 10. Name from the Greek *triplasios*, 'triple', alluding to the awn and long lobes of the type species, *Triplasis americana*.

Triplasis is an American genus of two species that is probably related to *Tridens*. The disarticulating culm, which helps disperse the cleistogenes, aids in distinguishing *Triplasis* from other genera.

1. Lemmas with lobes 4.5–8 mm long, tapering to acute tips; lemma awns 5–11 mm long; culm internodes puberulent to pilose . . . . . 1. *T. americana*
1. Lemmas with lobes about 1 mm long, rounded; lemma awns less than 2 mm long; culm internodes glabrous . . . . . 2. *T. purpurea*

1. **Triplasis americana** P. Beauv. [p. 43]
PERENNIAL SANDGRASS

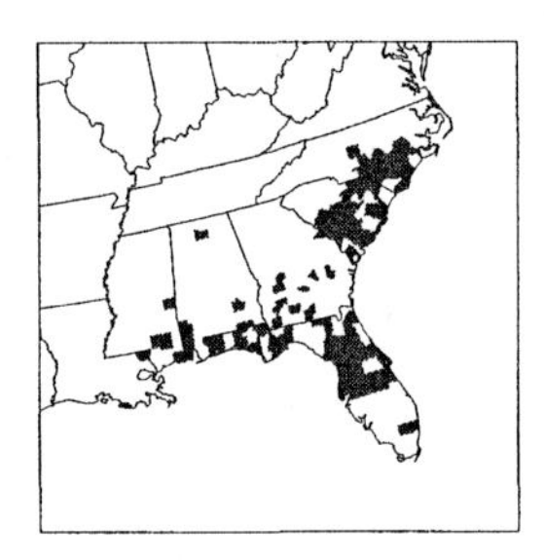

Plants perennial; cespitose. **Culms** 30–80 cm, usually erect; **nodes** and **internodes** appressed pubescent. **Sheaths** glabrous or pilose, margins ciliate; **ligules** to 2 mm, membranous, ciliate; **blades** to 20 cm long, usually less than 2 mm wide, filiform, scabrous adaxially. **Panicles** 1–5 cm long, 1–3 cm wide, occasionally reduced to a raceme. **Spikelets** 9–12 mm, with 2–5 florets. **Glumes** subequal, 3.4–4.5 mm, acuminate; **lemmas** 4–8 mm, lobes 4.5–8 mm, tapering to the acute apices; **awns** 8–11 mm, divergent; **paleas** 2–3 mm, keels ciliate; **anthers** 1.5–2 mm, yellow. **Caryopses** 1.5–2.5 mm, ovoid, tan. $2n$ = unknown.

Triplasis americana is endemic to the southeastern United States. It grows on sandy soils in prairies and woods, being less common in maritime dunes than *Triplasis purpurea*.

2. **Triplasis purpurea** (Walter) Chapm. [p. 43]
PURPLE SANDGRASS

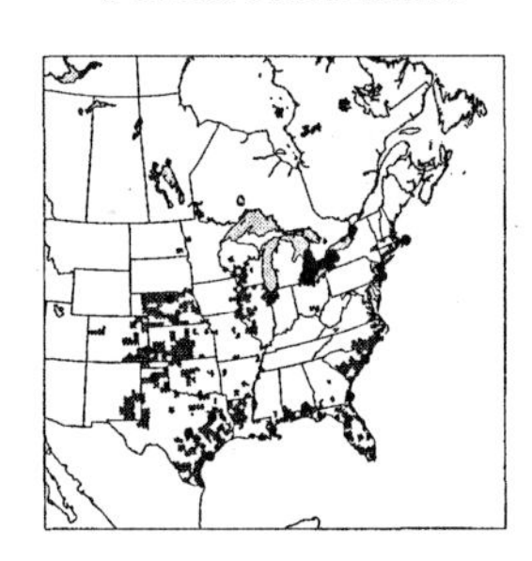

Plants annual and tufted or perennial and occasionally rhizomatous. **Culms** 14–100 cm, usually ascending; **internodes** glabrous. **Ligules** to 1 mm, of hairs; **blades** 1–5 mm wide, flat or involute, hispid or with papillose-based hairs. **Panicles** 3–7 cm long, 1–6 cm wide. **Spikelets** 6.5–9 mm, with 3–4 florets. **Glumes** about 2 mm, glabrous or scabrous, apices erose; **lemmas** 3–4 mm, lobes shorter than 1 mm, rounded; **awns** shorter than 2 mm, straight; **paleas** about 2.5 mm, keels ciliate; **anthers** about 2 mm, reddish-purple. **Caryopses** about 2 mm long, 0.6 mm wide, tapering distally, tan. $2n = 40$.

Triplasis purpurea grows in sandy soils throughout the eastern and central portion of the *Flora* region, extending southward through Mexico to Costa Rica. It is far more common in maritime dunes than *T. americana*. Plants in the *Flora* region belong to **Triplasis purpurea** (Walter) Chapm. var. **purpurea**.

17.13 SCLEROPOGON Phil.

John R. Reeder

Plants perennial; usually monoecious, less frequently dioecious, occasionally synoecious; bearing wiry, often arching, stolons with 5–15 cm internodes, sometimes also weakly rhizomatous. **Leaves** mostly basal; **sheaths** short, strongly veined, basal leaves commonly hispid or villous; **ligules** of hairs; **blades** firm, flat or folded. **Inflorescences** terminal, usually exceeding the upper leaves, spikelike racemes or contracted panicles with few spikelets, in bisexual plants staminate and pistillate florets in the same spikelet with the staminate florets below the pistillate florets or in separate spikelets, bisexual florets occasionally produced; **branches** not pectinate; **disarticulation** above the glumes and below the lowest pistillate floret in a spikelet, florets falling together, lowest floret with a bearded, sharp-pointed callus. **Staminate spikelets** with 5–10(20) florets; **glumes** membranous, pale, 1–3-veined, acuminate; **lemmas** 3-veined, similar to the glumes, unawned or awned, awns to 3 mm; **paleas** shorter than the lemmas, often conspicuously so. **Pistillate spikelets** appressed to the branch axes, usually the 3–5 lower florets functional,

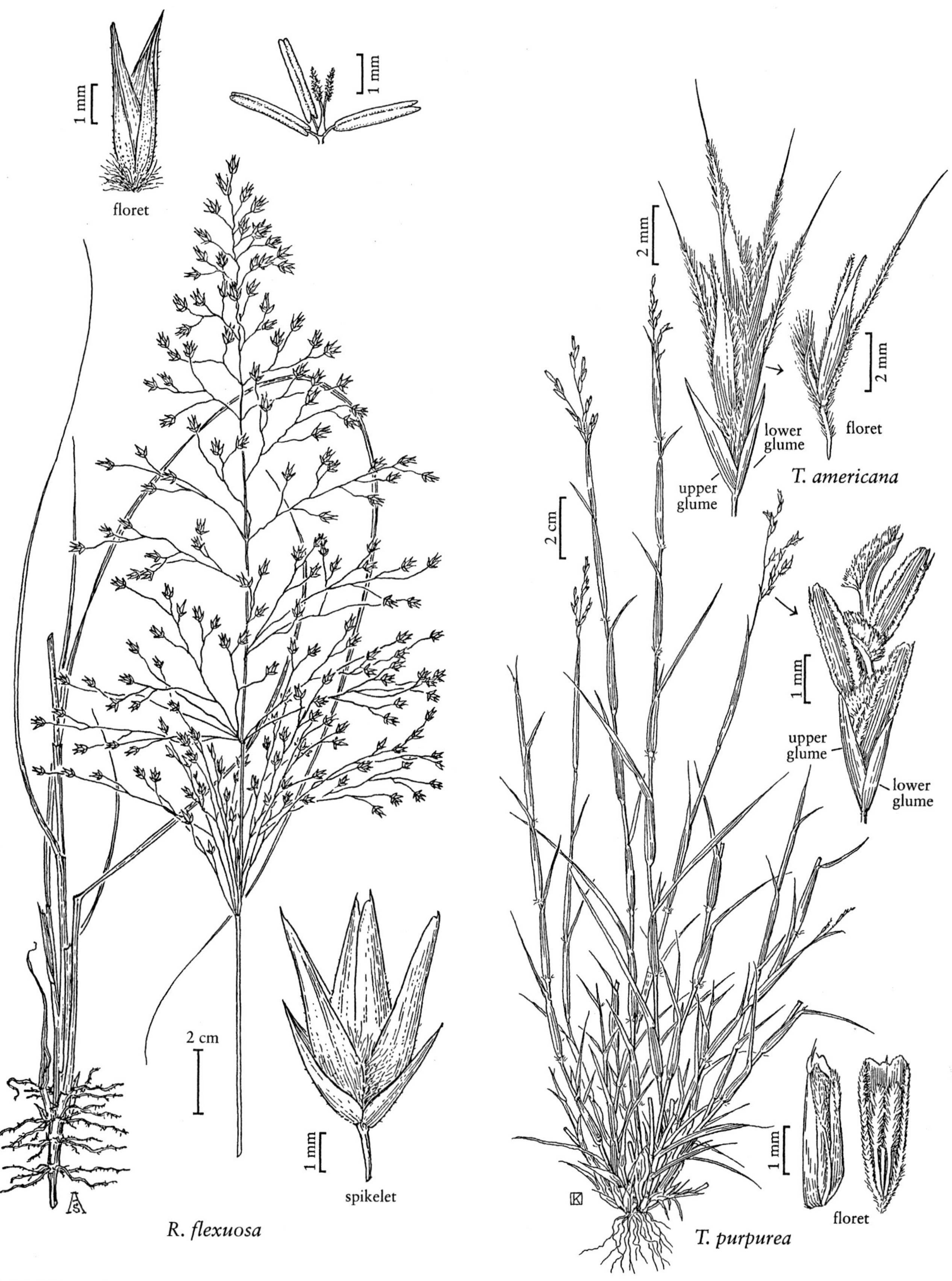

REDFIELDIA

TRIPLASIS

upper florets reduced to awns; **glumes** acuminate, strongly 3-veined, occasionally with a few fine accessory veins; **lemmas** narrow, 3-veined, veins extending into awns, awns (30)50–100(150) mm, spreading or reflexed at maturity. x = 10. Name from the Greek *skleros*, 'hard', and *pogon*, 'beard', in reference to the hard callus.

Scleropogon is a monotypic American genus with a disjunct distribution.

SELECTED REFERENCE **Reeder, J.R. and L.J. Toolin.** 1987. *Scleropogon* (Gramineae), a monotypic genus with disjunct distribution. Phytologia 62:267–275.

1. **Scleropogon brevifolius** Phil. [p. 46]
BURROGRASS

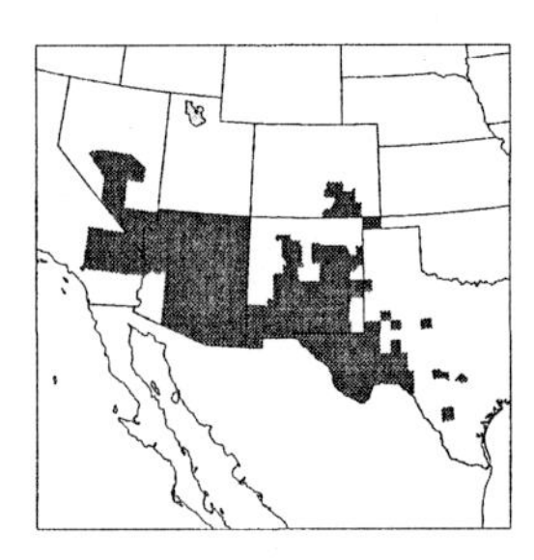

Stolons to 50 cm, wiry, internodes 5–15 cm. **Culms** (5)10–20 cm, erect. **Ligules** about 1 mm; **blades** 2–8(12) cm long, 1–2 mm wide. **Bisexual spikelets** 2-4 cm, staminate florets below the pistillate florets. **Staminate spikelets** 2–3 cm. **Pistillate spikelets** subtended by a glumelike bract; **lemma bodies** 2.5–3 cm. $2n$ = 40.

Scleropogon brevifolius grows on grassy plains and flats, generally being most abundant on disturbed or overgrazed land. Its North American range extends from the southwestern United States to central Mexico; its South American range is from Chile to northwestern Argentina.

17.14 **ERIONEURON** Nash

Jesús Valdés-Reyna

Plants perennial; usually cespitose, occasionally stoloniferous. **Culms** 6–65 cm, erect. **Leaves** mostly basal; **sheaths** smooth, glabrous, striate, margins hyaline, collars with tufts of 1–3 mm hairs; **blades** usually folded, pilose basally, margins white, cartilaginous, apices acute but not sharp. **Inflorescences** terminal, simple panicles (racemes in depauperate specimens), exserted well above the leaves. **Spikelets** laterally compressed, with 3–20 florets, distal florets staminate or sterile; **disarticulation** above the glumes and between the florets. **Glumes** thin, membranous, 1-veined, acute to acuminate; **calluses** with hairs; **lemmas** rounded on the back, 3-veined, veins conspicuously pilose, at least basally, apices toothed or obtusely 2-lobed, midveins often extended into awns, awns to 4 mm, lateral veins sometimes extended as small mucros; **paleas** shorter than the lemmas, keels ciliate, intercostal regions pilose basally; **lodicules** 2, adnate to the bases of the paleas; **anthers** 1 or 3. **Caryopses** glossy, translucent; **embryos** more than ½ as long as the caryopses. x = 8. Name from the Greek *erion*, 'wool', and *neuron*, 'nerve', a reference to the hairy veins of the lemmas.

Erioneuron is an American genus of three species. Its seedlings appear to have a shaggy, white-villous indumentum, but this is composed of a myriad of small, water-soluble crystals.

Stoloniferous plants are unusual in the region covered by the *Flora*, but they are quite common in populations of *Erioneuron nealley* and *E. avenaceum* from central Mexico.

SELECTED REFERENCES **Sánchez, E.** 1979. Anatomía foliar de las especies y variedades argentinas de los géneros *Tridens* Roem. et Schult. y *Erioneuron* Nash (Gramineae–Eragrostoideae–Eragrosteae). Darwiniana 22:159–175; **Valdés-Reyna, J. and S.L. Hatch.** 1997. A revision of *Erioneuron* and *Dasyochloa* (Poaceae: Eragrostideae). Sida 17:645–666.

1. Lemmas entire or with teeth to 0.5 mm long, the awns 0.5–2.5 mm long; both glumes shorter than the lowest floret . . . . 1. *E. pilosum*
1. Lemmas 2-lobed, the lobes 1–2.5 mm long, the awns 1–4 mm long; upper glumes equaling or exceeding the lowest floret.

2. Lemma lobes obtuse to broadly acute, 1–2 mm long; lateral veins not forming mucros; plants 7–40 cm tall . . . . 2. *E. avenaceum*
2. Lemma lobes rounded to truncate, 1.5–2.5 mm long; lateral veins forming mucros to 1 mm long; plants 15–65 cm tall . . . . 3. *E. nealleyi*

1. **Erioneuron pilosum** (Buckley) Nash [p. 46]
HAIRY TRIDENS, HAIRY WOOLYGRASS

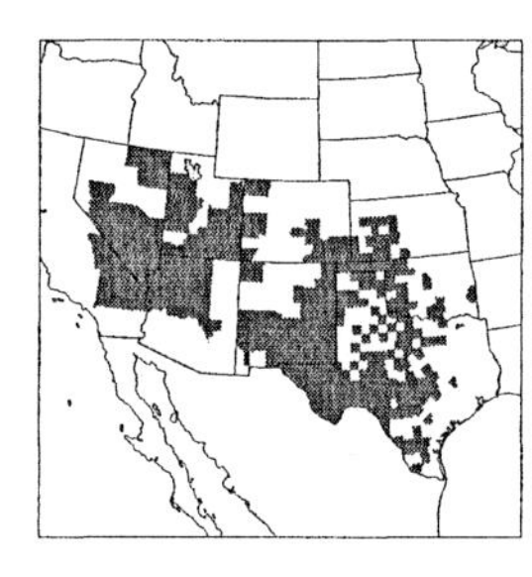

Culms (6)10–30(40) cm tall, (0.3)0.6–1(2.5) mm thick, glabrous or hispidulous. **Ligules** 2–3.5 mm; **blades** (1)3–6(9) cm long, (0.5)1–1.5(2.5) mm wide, both surfaces sparsely pilose or glabrous, grayish-green. **Panicles** 1–4(6) cm; **branches** with 3–9 shortly pedicellate spikelets. **Spikelets** 6–12(15) mm, with (5)6–12(20) florets. **Glumes** exceeded by the lowest florets, pale; **lower glumes** 4–7 mm; **upper glumes** 4–7 mm; **lemmas** 3–6 mm, green or purplish-green when young, becoming stramineous at maturity, awned, awns 0.5–2.5 mm, apices acute, entire or bidentate, teeth 0.3–0.5 mm; **anthers** usually 3, 0.3–1 mm. **Caryopses** 1–1.5 mm. $2n$ = 16.

Erioneuron pilosum grows on dry, rocky hills and mesas, often in oak and pinyon-juniper woodlands. In North America, it is represented by **E. pilosum** var. **pilosum**. This variety differs from the other two varieties, both of which are restricted to Argentina, in its longer, less equal glumes and shorter awns.

2. **Erioneuron avenaceum** (Kunth) Tateoka [p. 46]
LARGE-FLOWERED TRIDENS, SHORTLEAF WOOLYGRASS

Culms (7)10–30(40) cm tall, (0.4)0.7–1 mm thick, glabrous; **nodes** glabrous or villous. **Ligules** to 0.5 mm; **blades** (1.5)3–5(8) cm long, (0.5)1–1.5(2.5) mm wide, both surfaces sparsely pilose. **Panicles** 2–8(10) cm; **branches** with 2–10(16) shortly pedicellate spikelets. **Spikelets** 6–8(10) mm, purplish, with (4)6–12(20) florets; **lower glumes** 4–7 mm; **upper glumes** 6–9 mm, equaling or exceeding the lowest florets; **lemmas** 4–7 mm, purplish-green, awned from between the lobes, awns 2–4 mm, apices bilobed, lobes 1–2 mm, obtuse to acute; **anthers** 0.4–1 mm or (when monandrous) to 1.3 mm. **Caryopses** 1–1.4 mm. $2n$ = 16, 32.

Erioneuron avenaceum is common in rocky areas from the southwestern United States to central Mexico; it also grows in Bolivia and Argentina. North American plants belong to **E. avenaceum** (Kunth) Tateoka var. **avenaceum**. Stoloniferous plants occur in the *Flora* region, but they are most common in central Mexico.

3. **Erioneuron nealleyi** (Vasey) Tateoka [p. 46]
NEALLEY'S ERIONEURON, NEALLEY'S WOOLYGRASS

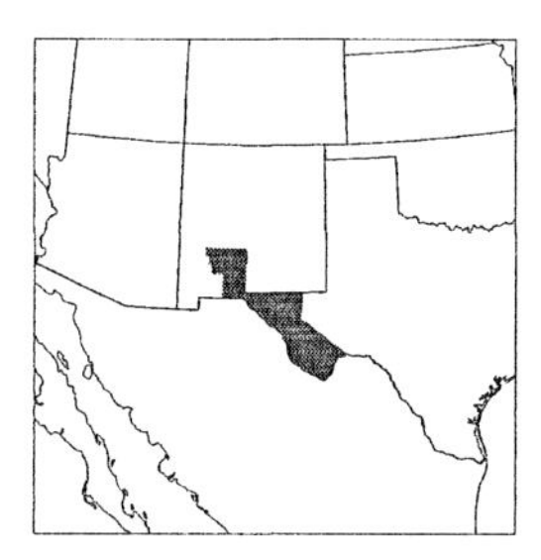

Culms (15)30–65 cm tall, 0.8–2 mm thick, glabrous or hispidulous; **nodes** glabrous or densely villous. **Ligules** 0.2–0.6 mm; **blades** 5–10 cm long, 2–2.5 mm wide, flat in moist conditions, both surfaces pilose to villous, green. **Panicles** compact (rarely open), 5–10 cm long, 1–3 cm wide, usually 2–4(6) times longer than wide, occasionally interrupted in the lower ½; **branches** with 5–17 shortly pedicellate spikelets. **Spikelets** 7–11 mm, purplish to pale, with 3–15 florets; **lower glumes** 5–7 mm; **upper glumes** 6–9 mm, generally equaling or exceeding the lowest florets; **lemmas** 4–6 mm, awned from between the lobes, awns 1–3.5 mm, apices bilobed, lobes 1.5–2.5 mm, rounded to truncate, lateral veins forming a mucro to 1 mm; **anthers** 1, 1–1.5 mm. **Caryopses** 1.3–1.5 mm. $2n$ = 16.

Erioneuron nealleyi is found on rocky slopes in the southwestern United States and central Mexico. Stoloniferous plants are known only from central Mexico.

17.15 DASYOCHLOA Willd. *ex* Rydb.

Jesús Valdés-Reyna

Plants perennial; stoloniferous, sometimes mat-forming. **Culms** (1)4–15 cm, initially erect, eventually bending and rooting at the base of the inflorescence. **Leaves** not basally aggregated on the primary culms; **sheaths** with a tuft of hairs to 2 mm at the throat; **ligules** of hairs; **blades** involute. **Inflorescences** terminal, short, dense panicles of spikelike branches, each subtended by leafy bracts and exceeded by the upper leaves; **branches** with 2–4 subsessile to shortly pedicellate

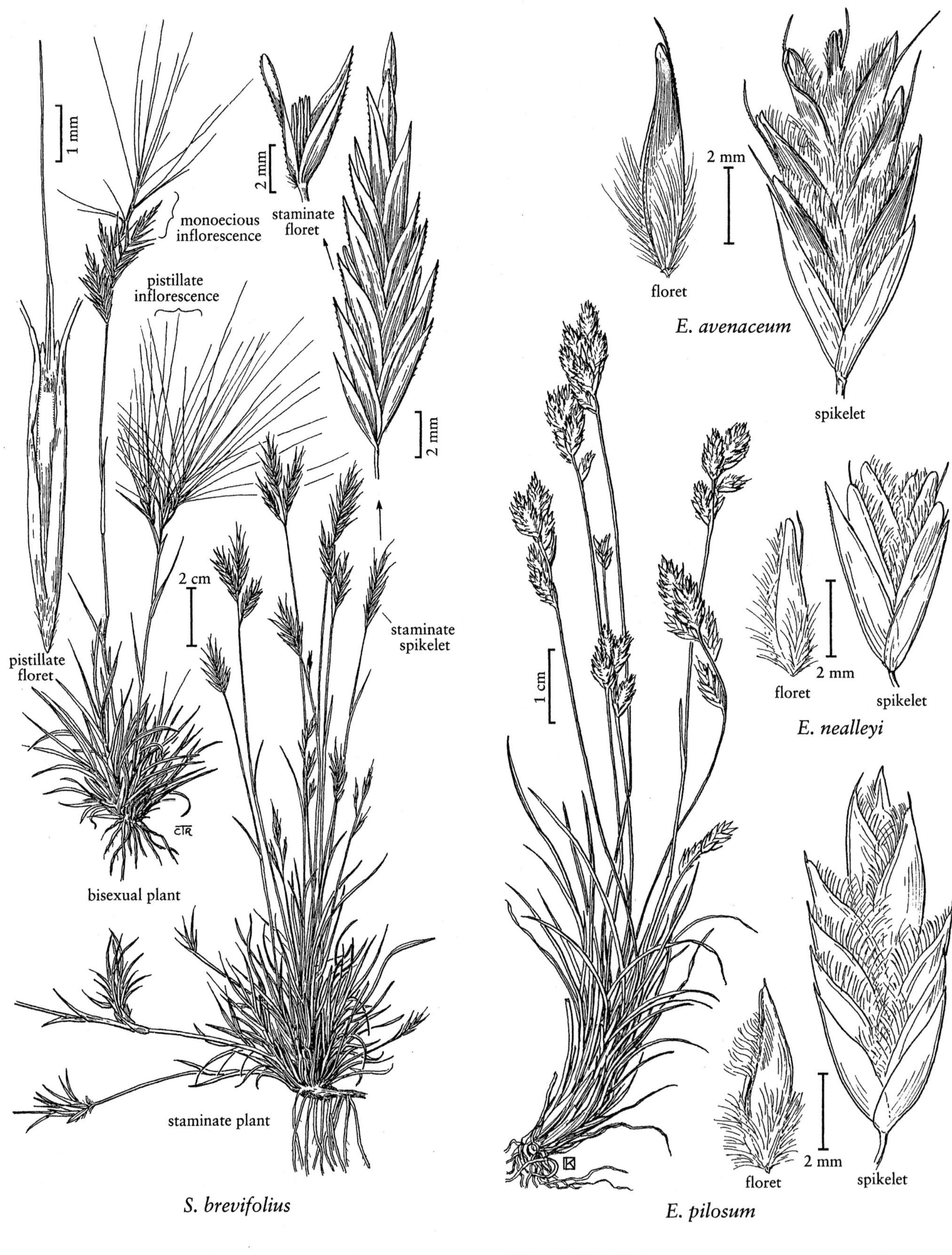

SCLEROPOGON

ERIONEURON

spikelets. **Spikelets** laterally compressed, with 4–10 florets; **disarticulation** above the glumes. **Glumes** subequal to the adjacent lemmas, glabrous, 1-veined, rounded or weakly keeled, shortly awned to mucronate; **florets** bisexual; **lemmas** rounded or weakly keeled, densely pilose on the lower ½ and on the margins, thinly membranous, 3-veined, 2-lobed, lobes about ½ as long as the lemmas and obtuse, midveins extending into awns as long as or longer than the lobes, lateral veins not excurrent; **paleas** about as long as the lemmas; **anthers** 3. **Caryopses** oval in cross section, translucent; **embryos** more than ½ as long as the caryopses. $x = 8$. Name from the Greek *dasys*, 'thick with hair' and *chloë*, 'grass'.

Dasyochloa is a monotypic genus that is restricted to the United States and Mexico. It has been included in the past in each of the following: *Triodia*, *Tridens*, and *Erioneuron*. *Dasyochloa* differs from all three of these genera, but resembles *Munroa*, in its leafy-bracteate inflorescence (Caro 1981). Seedlings of *Dasyochloa*, like those of *Erioneuron*, are shaggy-white-villous. This indumentum is composed of myriads of hairlike, water soluble crystals that wash off in water. They are the product of transpiration and evaporation.

SELECTED REFERENCES **Caro, J.A.** 1981. Rehabilitación del género *Dasyochloa* (Gramineae). Dominguezia 2:1–17; **Sánchez, E.** 1983. *Dasyochloa* Willdenow *ex* Rydberg (Poaceae). Lilloa 36:131–138; **Valdés-Reyna, J.** and **S.L. Hatch.** 1997. A revision of *Erioneuron* and *Dasyochloa* (Poaceae: Eragrostideae). Sida 17:645–666.

1. **Dasyochloa pulchella** (Kunth) Willd. *ex* Rydb. [p. 48]
FLUFFGRASS

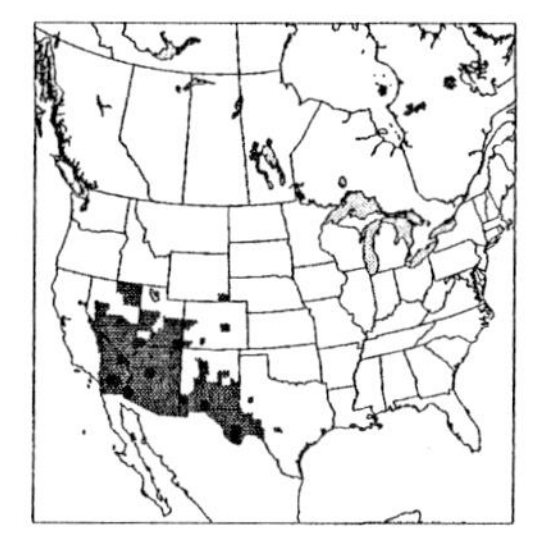

Culms (1)4–15 cm, scabrous or puberulent; **peduncles** (internode below the panicles) 3–7(11) cm. **Sheaths** striate, margins scarious; **ligules** 3–5 mm; **blades** (1)2–6 cm, abaxial surfaces scabrous, adaxial surfaces scabridulous. **Panicles** 1–2.5 cm long, 1–1.5 cm wide, densely white-pubescent, light green or purple-tinged. **Spikelets** (5)6–9(10) mm, with (4)6–10 florets. **Lower glumes** 6–8.5 mm; **upper glumes** 6.5–9 mm, as long as or longer than the florets; **lemmas** 3–5.5 mm, lobes (1)3–3.2 mm, midveins extending into straight (1.5)2.5–4 mm awns; **paleas** 2–3.5 mm, keels long pilose proximally, ciliate distally; **anthers** 0.2–0.5 mm. **Caryopses** 1–1.5 mm, translucent. $2n = 16$.

Dasyochloa pulchella grows in rocky soils of arid regions. It extends from the United States to central Mexico and is the most common grass in the *Larrea-Flourensia* scrub of the southwestern United States and adjacent Mexico.

17.16 BLEPHARONEURON Nash

Paul M. Peterson
Carol R. Annable

Plants annual or perennial. **Culms** 10–70 cm. **Sheaths** open, glabrous, usually longer than the internodes; **ligules** membranous or hyaline, truncate to obtuse, often decurrent; **blades** flat to involute, abaxial surfaces glabrous, sometimes scabrous, adaxial surfaces shortly pubescent. **Inflorescences** terminal, panicles, exceeding the leaves; **branches** spreading to ascending; **pedicels** capillary, lax, minutely glandular just below the spikelets. **Spikelets** with 1 floret, slightly laterally compressed, grayish-green; **disarticulation** above the glumes. **Glumes** subequal, ovate to obtuse, faintly 1-veined, glabrous; **lemmas** slightly longer and firmer than the glumes, 3-veined, veins and margins densely sericeous, hairs 0.1–0.7(1) mm, apices acute to obtuse, occasionally mucronate; **paleas** 2-veined, densely villous between the veins; **anthers** 3, purplish. $x = 8$. Name from the Greek *blepharis*, 'eyelash', and *neuron*, 'nerve', a reference to the sericeous veins of the lemmas.

Blepharoneuron is a genus of two species: a slender annual, *B. shepherdii* (Vasey) P.M. Peterson & Annable, which is known only from northern Mexico, and *B. tricholepis*, which is native to the *Flora* region.

SELECTED REFERENCE **Peterson, P.M.** and **C.R. Annable.** 1990. A revision of *Blepharoneuron* (Poaceae: Eragrostideae). Syst. Bot. 15:515–525.

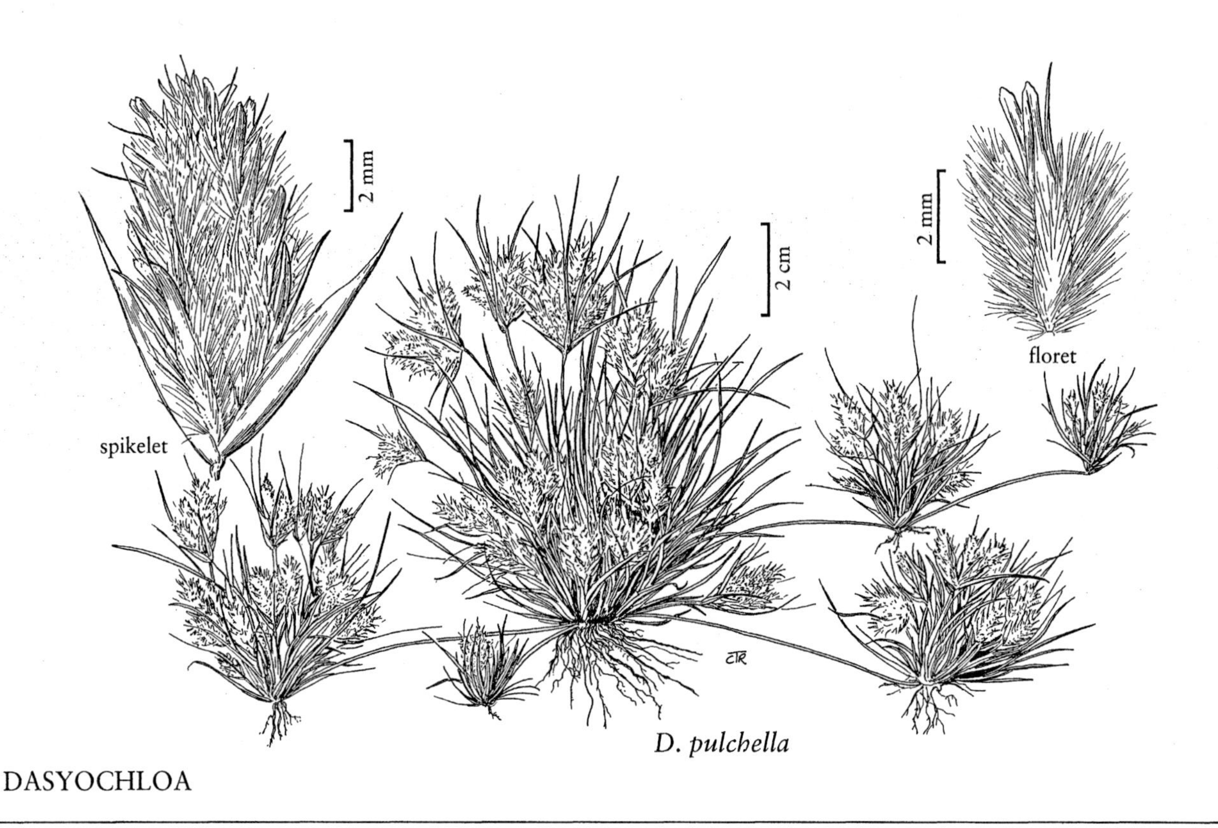

DASYOCHLOA

1. **Blepharoneuron tricholepis** (Torr.) Nash [p. 50]
HAIRY DROPSEED

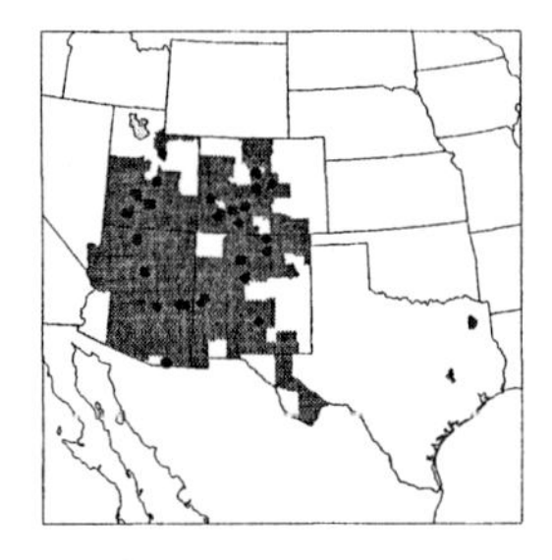

Plants perennial; densely cespitose. **Culms** 10–70 cm, erect, glabrous and smooth or scabrous just below the nodes. **Ligules** (0.3)0.7–2(2.7) mm, hyaline to membranous, entire; **blades** 1–15 cm long, 0.6–2.5 mm wide, scabrous. **Panicles** 3–25 cm long, 1–10 cm wide; **branches** ascending; **pedicels** 2–9 mm, straight or flexuous. **Glumes** (1.5)1.8–2.6(3) mm, often appearing 3-veined because of the characteristic infolding of the margins; **lemmas** (2)2.3–3.5(3.9) mm; **anthers** 1.2–2.1 mm, brownish. **Caryopses** 1.2–1.4 mm. $2n = 16$.

Blepharoneuron tricholepis grows in dry, rocky to sandy slopes, dry meadows, and open woods in pine-oak-madrone forests from Utah and Colorado to the state of Puebla, Mexico, at elevations of 700–3660 m. It flowers from mid-June through November.

17.17 BLEPHARIDACHNE Hack.

Jesús Valdés-Reyna

Plants perennial (rarely annual); cespitose, from a knotty base, often mat-forming. **Culms** 3–8(20) cm, often decumbent and rooting at the lower nodes, frequently branched above the bases, forming short spur shoots at the ends of long internodes; **internodes** minutely pubescent. **Leaves** clustered at the bases of the primary and spur shoots; **basal sheaths** shorter than the internodes; **ligules** of hairs or absent; **blades** linear to triangular, convolute to conduplicate, or flat to plicate, sharp, those of the upper leaves usually exceeding the inflorescences. **Inflorescences** terminal, compact panicles, exserted or partially included in the upper sheath(s). **Spikelets** laterally compressed, subsessile or pedicellate, with 4 florets per spikelet; **disarticulation** above the glumes but not between the florets. **Glumes** subequal to each other and the lowest lemma, rounded or weakly keeled, 1-veined, awn-tipped or unawned; **lowest 2 florets** in each spikelet staminate or sterile; **third floret** pistillate or bisexual; **lemmas** rounded on the back, 3-veined, mostly glabrous but pilose across the bases and on the margins, strongly 3-lobed, lateral lobes wider than the central lobes, all lobes ciliate on 1 or both margins, **lower lemmas** with the lateral lobes rounded or mucronate to awned, central lobes awned; **third lemmas** 3-lobed, lobes awned; **paleas** from slightly shorter to slightly longer than the lemmas; **lodicules** absent; **anthers** 2 or 3 (rarely 1); **style branches** 2. **Distal florets** rudimentary, 3-awned, plumose, or hairy. **Caryopses** laterally compressed. x = 7. Name from the Greek *blepharis*, 'eyelash', and *achne*, 'scale' or 'chaff', an allusion to the ciliate lemmas.

The four species comprising *Blepharidachne* are restricted to the Americas, growing in arid and semi-arid regions of the United States, Mexico, and Argentina. *Blepharidachne bigelovii* and *B. kingii* are endemic to North America, whereas *B. benthamiana* (Hack.) Hitchc. and *B. hitchcockii* Lahitte are native to Argentina. *Blepharidachne* differs from all other genera in the tribe in having four florets per spikelet, with the first two florets being sterile or staminate, the third bisexual or pistillate, and the fourth a rudimentary 3-awned structure.

SELECTED REFERENCE **Hunziker, A.T.** and **A.M. Anton.** 1979. A synoptical revision of *Blepharidachne* (Poaceae). Brittonia 31:446–453.

1. Glumes subacute, exceeded by the distal florets . . . . 1. *B. bigelovii*
1. Glumes acuminate or awn-tipped, exceeding the florets . . . . 2. *B. kingii*

1. **Blepharidachne bigelovii** (S. Watson) Hack. [p. 50]
BIGELOW'S DESERTGRASS

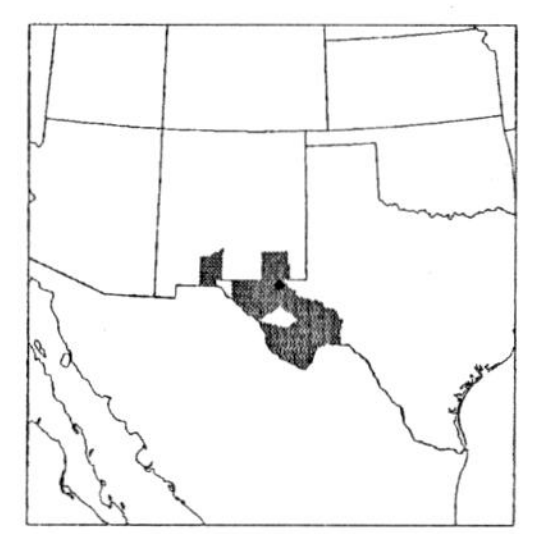

Plants cespitose, from a firm, often knotty base. **Culms** 6–20 cm, freely branched at and above the bases. **Sheaths** usually with a tuft of hairs on either side of the collars, often puberulent on the backs; **ligules** to 0.3 mm; **blades** 1–2 cm long, less than 1 mm wide, involute, convolute to conduplicate, firm, harshly puberulent, stiffly arcuate, lower blades deciduous. **Panicles** 15–30 mm long, 10–15 mm wide, exserted or partially included in the upper 2, subopposite, subtending sheaths; **rachises** and **branches** puberulent. **Spikelets** 5–7 mm. **Glumes** nearly equal to the lowest lemmas in the spikelets, exceeded by the distal florets, thin, translucent, smooth or the vein scabridulous, subacute; **lower glumes** 5–6 mm; **upper glumes** about 6 mm; **lowest florets** staminate, 4–5 mm; **second florets** sterile, about 5 mm, lateral lobes 1–3 mm; **paleas of lowest 2 florets** reduced, membranous; **third florets** pistillate, **third lemmas** 5.5–6 mm, lateral lobes 3–4.5 mm, awned, central awns 2.5–3 mm; **paleas of third florets** slightly longer than the lemmas; **anthers** 2(1), 1.2–1.5 mm. **Caryopses** 1.5–2 mm. $2n$ = 14.

Blepharidachne bieglovii grows on rocky slopes in western Texas and adjacent areas of New Mexico, and Coahuila and Zacatecas, Mexico.

2. **Blepharidachne kingii** (S. Watson) Hack. [p. 50]
KING'S EYELASH GRASS

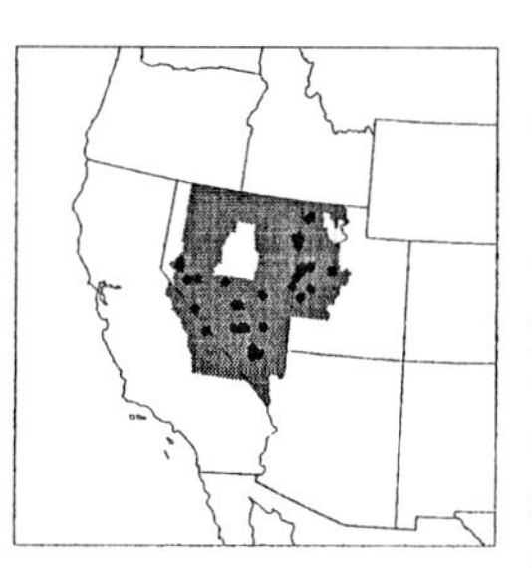

Plants cespitose. **Culms** 3–8(14) cm. **Sheaths** often with a tuft of hairs at the throat, sheaths immediately below the inflorescences often spathelike; **ligules** to 0.5 mm; **blades** 0.7–3 cm long, less than 1 mm wide, strongly convolute to conduplicate, stiffly arcuate, often deciduous. **Panicles** 10–25 mm, subcapitate, usually partially included in the 2 spathelike upper leaf sheaths. **Spikelets** 6–9 mm. **Glumes** exceeding the florets, papery and translucent, scabridulous toward the base, acuminate or awn-tipped, awns to about 1.3 mm; **lower glumes** 6–7.5 mm; **upper glumes** 6.8–8.5 mm; **lowest 2 florets** usually sterile; **lemmas of sterile florets** 3.4–5.8 mm, lateral lobes 2.2–3 mm; **paleas of sterile florets** linear, plumose; **third florets** bisexual; **third lemmas** with lateral lobes 0.5–1.5 mm, awned, central awns 3–5 mm; **paleas of third florets** subequal to slightly longer than the lemmas; **anthers** 2(1), about 1.5 mm. **Caryopses** about 2 mm, compressed. $2n$ = 14.

Blepharidachne kingii grows at scattered locations in arid regions of the Great Basin, sometimes being locally abundant.

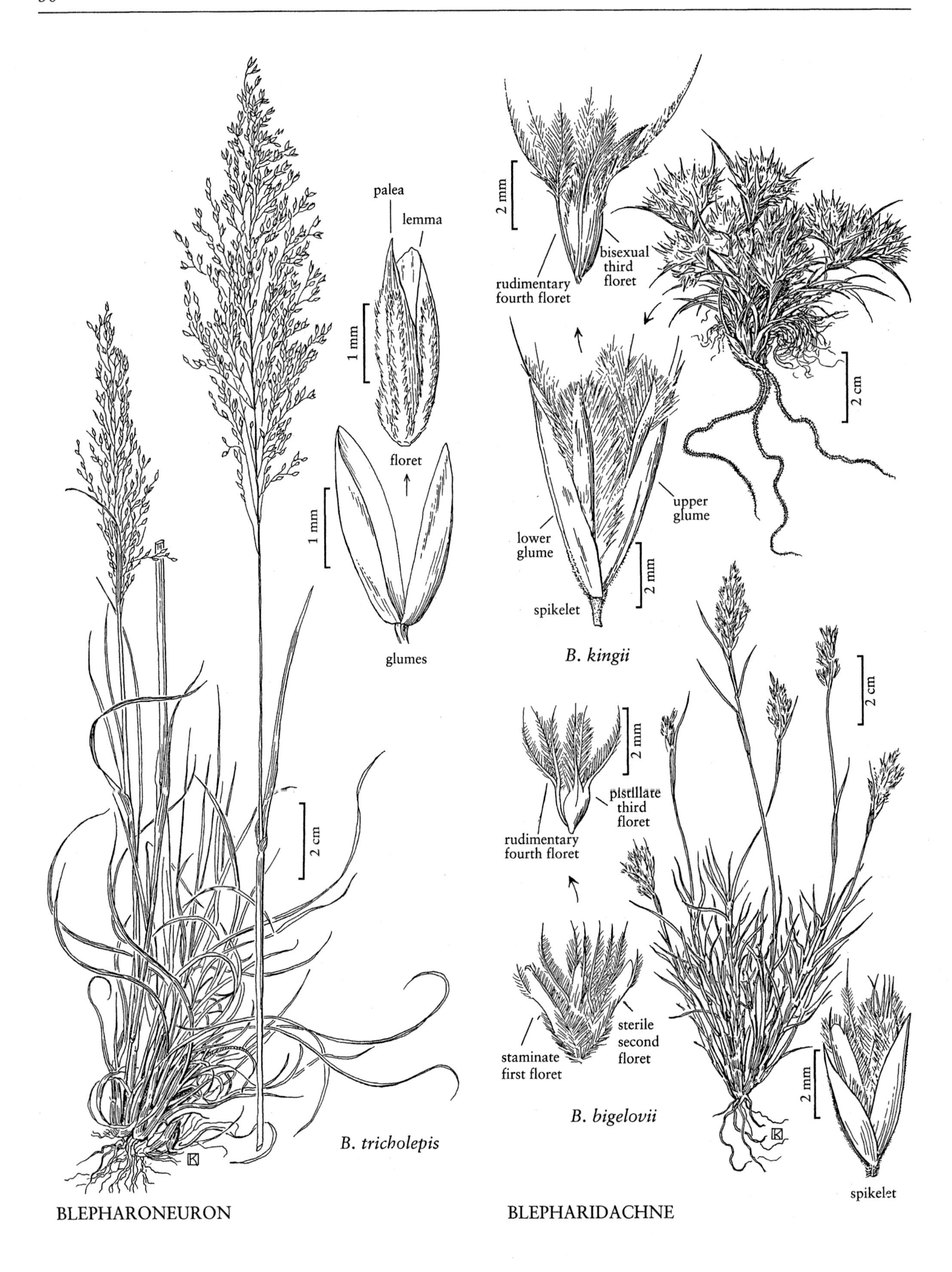
palea
lemma
1 mm
floret
1 mm
glumes
2 cm
B. tricholepis
BLEPHARONEURON
2 mm
bisexual third floret
rudimentary fourth floret
2 cm
upper glume
lower glume
2 mm
spikelet
B. kingii
2 mm
pistillate third floret
rudimentary fourth floret
2 cm
sterile second floret
staminate first floret
B. bigelovii
2 mm
spikelet
BLEPHARIDACHNE

17.18 MUNROA Torr.

Jesús Valdés-Reyna

Plants annual; stoloniferous, mat-forming; **stolons** 2–8 cm, terminating in fascicles of leaves from which new culms arise. **Culms** 3–15(30) cm. **Leaves** mostly basal, sometimes with a purple tint; **sheaths** with a tuft of hairs at the throat; **auricles** absent; **ligules** of hairs; **blades** linear, usually involute, sometimes flat or folded, with white, thickened margins, apices sharply pointed. **Inflorescences** terminal, capitate panicles of spikelike branches; **branches** almost completely hidden in a subtending leafy bract, bearing 2–4 subsessile or pedicellate spikelets. **Spikelets** laterally compressed, with 2–10 florets; **lower florets** bisexual or pistillate; **terminal florets** sterile; **disarticulation** above the glumes or beneath the leaves subtending the branches. **Glumes** shorter than the spikelets, keeled, 1-veined, unawned; **lower glumes** usually present on all spikelets (absent from all spikelets in *M. mendocina*); **upper glumes** absent or reduced on the terminal spikelet; **lemmas** with a pilose tuft of hairs along the margins at midlength, membranous or coriaceous, 3-veined, lateral veins occasionally shortly excurrent, apices emarginate or 2-lobed; **paleas** glabrous, smooth; **lodicules** present or absent, truncate; **anthers** 2 or 3, yellow; **style branches** elongate, 2(3), barbellate. **Caryopses** dorsally compressed. x = 7 or 8. Named for Sir William Munro (1818–1880), a British botanist who collected in Barbados, the Crimea, and India.

Munroa, a genus of five species, is endemic to the Western Hemisphere. One species occurs in the *Flora* region, the remainder being confined to South America. Its closest relatives are thought to be *Blepharidachne* and *Dasyochloa*, both of which are stoloniferous, mat-forming species with leafy-bracteate panicles. *Munroa* differs from both in its annual habit.

SELECTED REFERENCES **Anton, A.M.** and **A.T. Hunziker.** 1978. El género Munroa (Poaceae): Sinopsis morgológica y taxonómica. Bol. Acad. Nac. Ci. 52:229–252. **Parodi, L.R.** 1934. Contribución al estudio de las gramíneas del género *Munroa*. Revista Mus. La Plata 34:171–193; **Sánchez, E.** 1984. Estudios anatómicos en el género *Munroa* (Poaceae, Chloridoideae, Eragrostideae). Darwiniana 25:43–57.

1. **Munroa squarrosa** (Nutt.) Torr. [p. 52]
FALSE BUFFALOGRASS

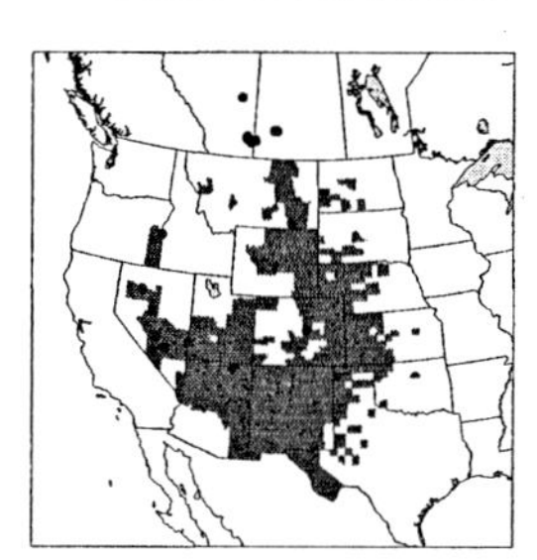

Stolons slender. **Culms** 3–15(30) cm, highly branched, scabrous, often minutely puberulent. **Ligules** 0.5(1) mm; **blades** 1–5 cm long, 1–2.5 mm wide. **Spikelets** 6–8(10) mm, with 3–5 florets. **Glumes of first 1–2 spikelets** subequal, 2.5–4.2 mm, narrow, 1-veined, acute; **glumes of upper spikelets** unequal, lower glumes reduced or even absent in the terminal spikelets; **lemmas in lower spikelets** scabridulous, lanceolate, midvein excurrent, forming a stout, scabrous 0.5–2 mm awn, lateral veins with a tuft of hairs on the margins near midlength, excurrent; **anthers** 1–1.5 mm. $2n$ = 16.

Munroa squarrosa grows in dry, open areas, usually in sandy soil or disturbed sites, from Saskatchewan and Alberta south to Chihuahua, Mexico.

17.19 LEPTOCHLOA P. Beauv.

Neil Snow

Plants annual or perennial; cespitose. **Culms** (3)10–250(300) cm, usually ascending to erect, often geniculate at the lower nodes, occasionally prostrate and rooting at the lower nodes, often branching at the aerial nodes; **nodes** usually glabrous; **internodes** usually hollow. **Leaves** usually primarily cauline, occasionally in basal rosettes; **sheaths** open; **ligules** 0.2–10(15) mm, obtuse to attenuate, usually membranous, sometimes ciliate; **blades** flat, involute when dry, usually ascending to erect, apices attenuate. **Primary inflorescences** terminal, panicles of 2–150 non-disarticulating, spikelike branches, usually exceeding the leaves; **branches** 1–22 cm, digitate,

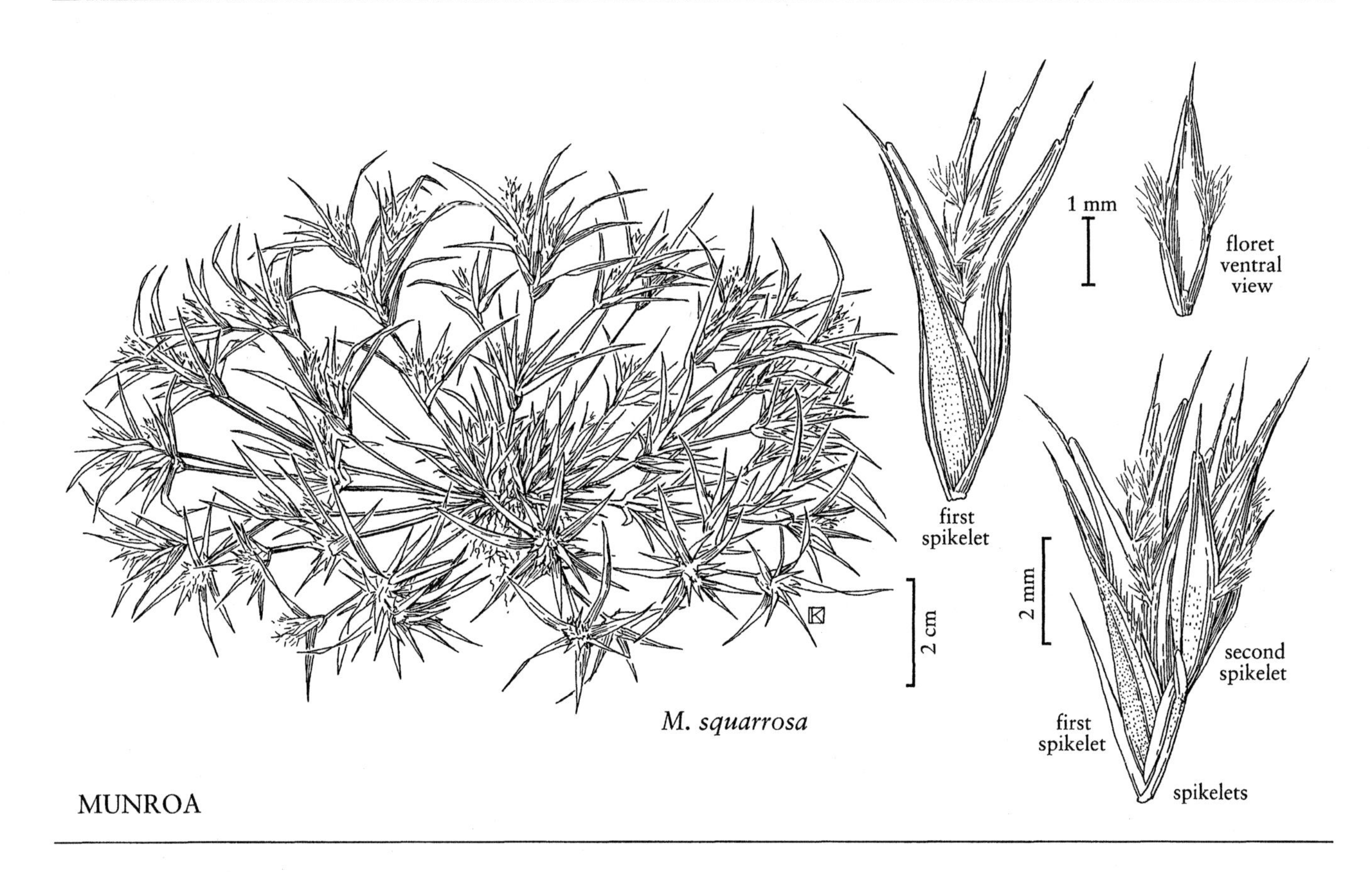

subdigitate, or racemose on the rachises, 1-sided, usually spikelet-bearing throughout their length, spikelets in 2 rows, axes terminating in a functional spikelet, lower branches occasionally with secondary branching; **secondary panicles** sometimes present, axillary to and concealed by the lower sheaths, their florets not disarticulating; **disarticulation** in the primary panicles beneath the florets. **Spikelets** rounded to slightly keeled, distant to tightly imbricate, not conspicuously pubescent, with (2)3–12(20) bisexual florets; **rachillas** rarely prolonged. **Glumes** usually unequal, sometimes subequal, exceeded by the florets, membranous, rounded to weakly keeled, 1-veined, veins scabrous, apices unawned (rarely mucronate); **lower glumes** 0.5–4.9 mm; **upper glumes** 0.9–6 mm; **florets** usually bisexual; **calluses** distinct or poorly developed, glabrous or pubescent; **lemmas** membranous, usually pubescent at least over the lower portion of the veins, 3(5)-veined, apices entire or minutely bilobed, unawned, mucronate, or awned; **paleas** usually subequal to the lemmas, membranous or hyaline; **anthers** 1–3, 0.1–2.7 mm. **Caryopses** obovate to elliptic, falling free of the lemmas and paleas. $x = 10$. Name from the Greek *leptos*, 'slender', in reference to the panicle branches, and *chloa*, 'grass'.

Leptochloa is a pantropical, warm-temperate genus of 32 species. Eight of the ten species in this treatment are native to the *Flora* region. Of the other two, *L. chloridiformis* was introduced over 60 years ago but has not become established. *Leptochloa chinensis* is not yet known from the region; it is included here because of its potential threat as an invasive weed. Cladistic studies (Snow 1997) do not support recognition of the segregate genus *Diplachne*.

Leptochloa tends to grow in somewhat basic soils. Many of the species, particularly the annual species, are poor ecological competitors and grow in relatively open, seasonally inundated soils, such as are found along rivers. In disturbed areas, they are associated with roadside ditches, the margins of reservoirs, and mesic agricultural lands. A few species, primarily perennial, grow on well-drained soils. The vegetative vigor of all species is greatly influenced by soil moisture availability.

SELECTED REFERENCES Nicora, E.G. 1995. Los géneros *Diplachne* y *Leptochloa* (Gramineae, Eragrosteae) de la Argentina y países limítrofes. Darwiniana 33:233–256; Snow, N. 1997. Phylogeny and systematics of *Leptochloa* P. Beauv. *sensu lato* (Poaceae, Chloridoideae). Ph.D. dissertation, Washington University, St. Louis, Missouri, U.S.A. 506 pp.; Snow, N. 1998. Nomenclatural changes in *Leptochloa* P. Beauvois *sensu lato* (Poaceae, Chloridoideae). Novon 8:77–80.

1. Panicle branches digitate or subdigitate; plants perennial.
  2. Lemma apices obtuse to truncate and often emarginate; lemmas membranous; plants often with secondary panicles concealed in the lower leaf sheaths . . . . 1. *L. dubia*
  2. Lemma apices usually acute; lemmas chartaceous; plants without secondary panicles concealed in the lower leaf sheaths.
    3. Panicle branches always digitate, 7–17 cm long; lemmas mucronate but not awned; in the *Flora* region, known only from a few old collections in Cameron County, Texas . . . . 2. *L. chloridiformis*
    3. Panicle branches subdigitate, 1.5–18 cm long; lemmas unawned, mucronate, or awned; native in much of the southeastern United States, including parts of Texas . . . . 3. *L. virgata*
1. Panicle branches racemose; plants annual or perennial.
  4. Ligules 2–8 mm long, attenuate, becoming lacerate at maturity . . . . 4. *L. fusca*
  4. Ligules 0.3–5.4 mm long, truncate to obtuse, erose, ciliate, or lacerate.
    5. Sheaths sparsely to densely hairy, the hairs papillose-based . . . . 5. *L. panicea*
    5. Sheaths lacking hairs or with hairs that are not papillose-based.
      6. Panicles with 25–150 branches.
        7. Lemmas 2.4–3 mm long; anthers 0.6–0.8 mm long; spikelets 4–5 mm long . . . . . . 10. *L. panicoides*
        7. Lemmas 1.2–2.4 mm long; anthers 0.2–0.6 mm long; spikelets 2.5–4.5 mm long.
          8. Leaf blades 8–16 mm wide; lemmas 2.1–2.4 mm long . . . . 7. *L. scabra*
          8. Leaf blades 4–8 mm wide; lemmas 1–2 mm long.
            9. Lower glumes 0.7–0.8 mm long; anthers 0.2–0.4 mm long; plants native to the southern United States . . . . 6. *L. nealleyi*
            9. Lower glumes 1.1–1.7 mm long; anthers 0.4–0.6 mm long; plants aggressive weeds, currently not known from the *Flora* region . . . . 8. *L. chinensis*
      6. Panicles with 2–25 branches.
        10. Plants perennial.
          11. Lemmas 4–5 mm long, membranous, their apices broadly acute, obtuse, or truncate, unawned; panicles with 2–15 branches; caryopses 1.9–2.3 mm long, strongly dorsally compressed; secondary panicles often present in, and concealed by, the lower leaf sheaths . . . . 1. *L. dubia*
          11. Lemmas 1.5–3.6 mm long, chartaceous, their apices usually acute, rarely obtuse, unawned, mucronate, or awned, the awns to 11 mm long; panicles with 9–25 branches; caryopses 1.3–1.8 mm long, somewhat laterally compressed; secondary panicles not present in the lower leaf sheaths . . . . 3. *L. virgata*
        10. Plants annual.
          12. Panicles 2–17 cm long, with 5–23 branches; anthers 0.4–0.5 mm long; caryopses 0.4–0.5 mm wide . . . . 9. *L. viscida*
          12. Panicles 20–35 cm long, with 20–90 branches; anthers 0.6–0.8 mm long; caryopses about 0.7 mm wide . . . . 10. *L. panicoides*

1. **Leptochloa dubia** (Kunth) Nees [p. 55]
GREEN SPRANGLETOP

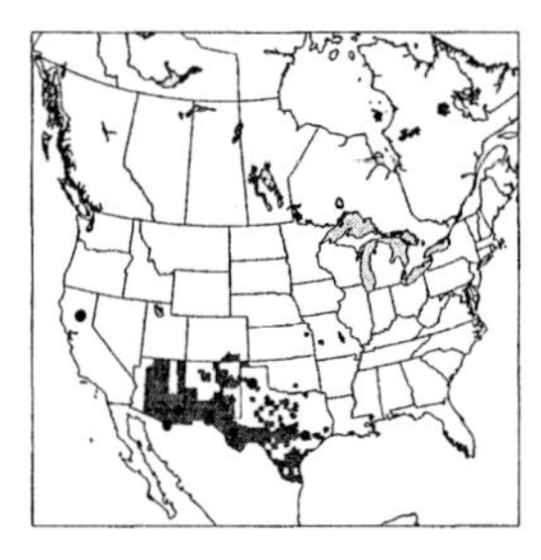

Plants perennial. **Culms** (10) 30–110 cm, round or basally compressed, tillering from the basal nodes, not branching from the aerial nodes, mostly glabrous, sometimes pilose basally; **internodes** solid. **Sheaths** sometimes with a pilose collar; **ligules** 1–2 mm, truncate, erose; **blades** (2)8–35 cm long, 2–8 mm wide, glabrous, strigose, or pilose. **Panicles** 8–20 cm, with 2–15 subdigitate or racemose branches; **secondary panicles** often hidden in the lowest leaf sheaths; **branches** 2–19 cm, ascending to spreading at maturity. **Spikelets** 4–12 mm, light brown to dark olive green, with 4–13 florets, often widely diverging at anthesis. **Glumes** narrowly triangular to ovate, acute; **lower glumes** 2.3–4.8 mm; **upper glumes** 3.3–6 mm; **lemmas** 3.5–5 mm, membranous, ovate to obovate, lateral veins glabrous or sericeous, hairs often restricted to the basal portion, sometimes also sericeous on the midvein and between the veins, apices obtuse to truncate, usually emarginate, unawned, sometimes mucronate; **paleas** ciliate on the margins; **anthers** 3, 0.3–1.6 mm. **Caryopses** 1.5–2.3 mm long, 0.9–1 mm wide, strongly dorsally compressed. $2n$ = 40, 60, 80.

Leptochloa dubia grows from the southwestern United States and Florida through Mexico to Argentina, often in well-drained, sandy or rocky soils. It provides fair to good forage, but is seldom abundant.

2. **Leptochloa chloridiformis** (Hack.) Parodi [p. 55]
ARGENTINE SPRANGLETOP

Plants perennial. **Culms** 60–200 cm, erect, compressed, unbranched; **internodes** hollow. **Sheaths** pubescent; **ligules** 0.5–1 mm, shortly ciliate; **blades** 25–50 cm long, 2.8–5.5 mm wide, ascending to reflexed, adaxial surfaces usually sparsely pilose behind the ligules, otherwise both surfaces glabrous. **Panicles** to 53 cm, with 5–20 digitate branches; **secondary panicles** not present in the lower sheaths; **branches** 7–17 cm, steeply ascending but drooping at the apices. **Spikelets** 4.5–5 mm, imbricate, green to straw-colored, with 3–4 florets. **Glumes** ovate, acute; **lower glumes** 1.3–2.6 mm; **upper glumes** 2.2–3.7 mm; **lemmas** 2.8–3.8 mm, lanceolate to ovate, chartaceous, lateral veins sericeous, apices acute to slightly obtuse, minutely emarginate, mucronate; **paleas** ciliolate over the veins; **anthers** 1, 0.4–0.6 mm. **Caryopses** 0.9–1.8 mm long, 0.3–0.4 mm wide, triangular in cross section. $2n$ = unknown.

Leptochloa chloridiformis is native to Uruguay, southern Paraguay, and northern Argentina. It was introduced in the early part of the twentieth century but has not become established in the *Flora* region. The only known collections are from Cameron County, Texas, the most recent having been made in the 1940s.

3. **Leptochloa virgata** (L.) P. Beauv. [p. 55]
TROPICAL SPRANGLETOP

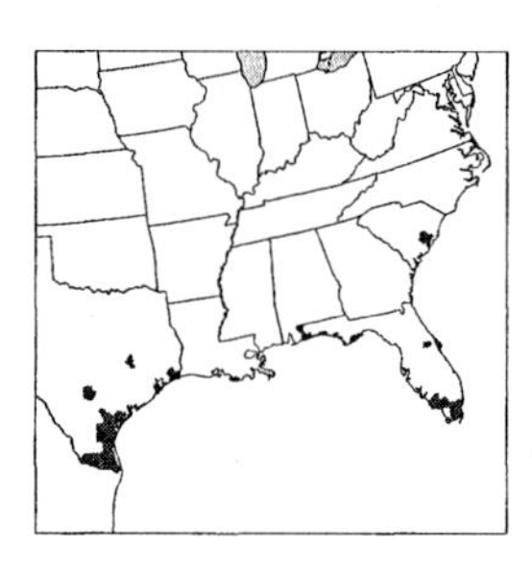

Plants perennial. **Culms** 30–200 cm, erect, occasionally geniculate below, compressed, branched; **internodes** solid. **Sheaths** glabrous, scabrous, or sparsely pilose below, hairs never papillose-based; **ligules** 0.3–1 mm, truncate, firmly membranous, fimbriate or erose, sometimes pilose at the sides; **blades** 5–45 cm long, 4–10 mm wide, glabrous abaxially, glabrous or sparsely pilose adaxially. **Panicles** 5–60 cm, with 9–25 usually racemose (rarely subdigitate) branches; **secondary panicles** not present in the lower sheaths; **branches** (1.5)3–18 cm, erect to spreading, somewhat flexible. **Spikelets** 2.5–4 mm, imbricate to distant, with 3–6(8) florets. **Lower glumes** 1.7–2.9 mm, lanceolate, acute; **upper glumes** 1.7–3.8 mm, lanceolate to ovate, acute, acuminate, rarely mucronate; **lemmas** (1.5)2.3–3.6 mm, ovate, chartaceous, veins sericeous, apices usually acute, rarely obtuse, unawned, mucronate, or awned, awns to 11 mm; **paleas** glabrous; **anthers** 2, 0.2–0.5 mm. **Caryopses** 1–1.8 mm long, about 0.5 mm wide, narrowly elliptic to ovate, somewhat laterally compressed. $2n$ = 40.

Leptochloa virgata is a common neotropical species that extends from the southeastern United States through the West Indies to Argentina. Awn length and lemma pubescence vary continuously and independently, precluding their use in recognizing additional taxa.

4. **Leptochloa fusca** (L.) Kunth [p. 57]

Plants annual or weakly perennial. **Culms** 5–170 cm, prostrate to erect; compressed, often branching; **internodes** hollow. **Sheaths** glabrous or scabrous; **ligules** 2–8 mm, membranous, attenuate, becoming lacerate at maturity; **blades** 3–50 cm long, 2–7 mm wide, glabrous or scabrous, those of the flag leaves sometimes exceeding the panicles. **Panicles** (1.5)10–105 cm long, 0.5–22 cm wide, with 3–35 racemose branches,

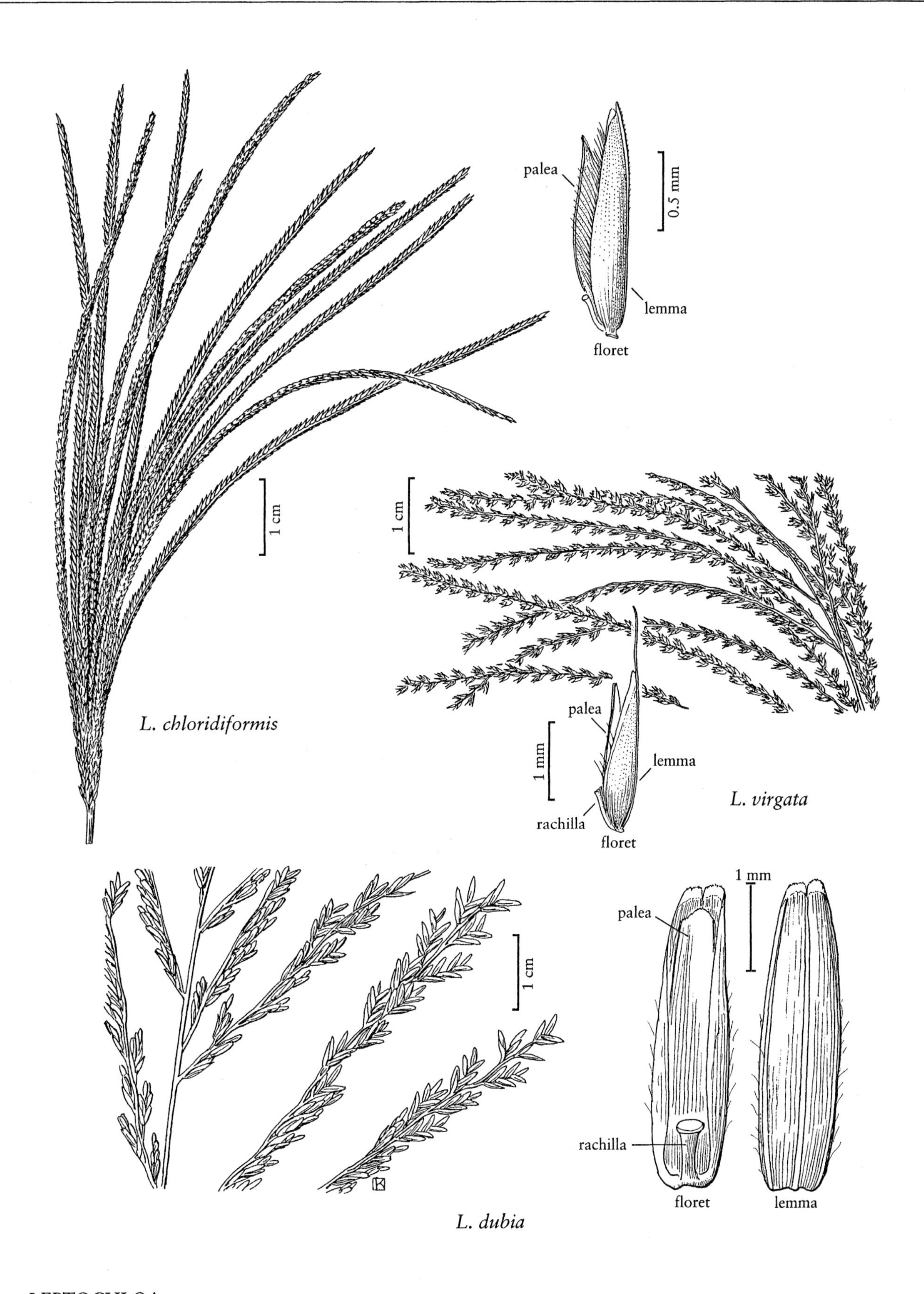
palea
0.5 mm
lemma
floret
1 cm
1 cm
L. chloridiformis
palea
1 mm
lemma
L. virgata
rachilla
floret
1 mm
palea
1 cm
rachilla
floret
lemma
L. dubia

LEPTOCHLOA

bases of the panicles sometimes remaining enclosed in the upper leaf sheaths at maturity; **branches** 1.5–20(22) cm, ascending to reflexed. **Spikelets** 5–12(14) mm, with 6–20 florets. **Lower glumes** 1–3(4.9) mm; **upper glumes** 1.8–5.5 mm; **lemmas** 2–6 mm, sometimes with a dark spot near the base, apices acute to truncate, sometimes emarginate to bifid, unawned, mucronate, or awned; **paleas** somewhat sericeous along the veins; **anthers** 1–3, 0.2–2.7 mm. **Caryopses** 0.8–2.4 mm, elliptic to ovate or obovate. $2n = 20$.

Leptochloa fusca grows in warm areas throughout the world. The two American subspecies, subsp. *uninervia* and subsp. *fascicularis*, are usually distinct, but they intergrade repeatedly with *L. fusca* subsp. *fusca*.

1. Uppermost leaf blades exceeding the panicles; panicles usually partially enclosed in the uppermost leaf sheaths; mature lemmas often smoky white with a dark spot in the basal ½ . . . . subsp. *fascicularis*
1. Uppermost leaf blades exceeded by the panicles; panicles usually completely exserted; mature lemmas usually lacking a dark spot.
   2. Anthers 0.5–2.7 mm long; spikelets 6–14 mm long; lemmas obtuse, acute, or acuminate, sometimes bifid, light brown to dark green . . . . subsp. *fusca*
   2. Anthers 0.2–0.6(1) mm long; spikelets 5–10 mm long; lemmas obtuse to truncate, usually notched and mucronate, often dark green or lead-colored . . . . subsp. *uninervia*

Leptochloa fusca subsp. **fascicularis** (Lam.) N. Snow
BEARDED SPRANGLETOP

Culms 5–110 cm, prostrate (in small circular clumps) to erect. **Blades** glabrous or scabrous, uppermost blades often exceeding the panicles. **Panicles** (1.5)10–72 cm long, 4–22 cm wide, with 3–35 branches, usually partially enclosed in the uppermost leaf sheaths; **branches** 3–12(22) cm, often spreading. **Spikelets** 5–12 mm. **Lower glumes** 2–3 mm, lanceolate, sometimes asymmetric; **upper glumes** 2.5–5 mm, elliptic to ovate; **lemmas** lanceolate, smoky white at maturity, often with a dark spot on the basal ½, apices acute, mucronate, or awned, awns to 3.5 mm; **anthers** 1–3, 0.2–0.5 mm. **Caryopses** 0.8–2 mm.

Leptochloa fusca subsp. *fascicularis* extends from southern British Columbia and Ontario to Argentina, although it has not yet been reported from Georgia. Coastal populations from Massachusetts to Florida with long lemma awns have been called *L. fascicularis* var. *maritima* (E.P. Bicknell) Gleason. They do not merit taxonomic recognition because long awns and salinity tolerance are common throughout the species.

Leptochla fusca subsp. *fascicularis* differs from *L. viscida*, which grows in the same region, in its longer panicles, frequently unawned or mucronate lemmas, and whitish florets.

Leptochloa fusca (L.) Kunth subsp. **fusca**
BEETLEGRASS SPRANGLETOP

Culms 40–170 cm, usually ascending to erect, sometimes decumbent and rooting at the lower nodes, often branching at the upper nodes. **Blades** glabrous or scabrous, not exceeding the panicles. **Panicles** 15–105 cm long, 2–20 cm wide, sometimes partially enclosed by the upper leaf sheaths; **branches** (1.5)4–20 cm, ascending to erect. **Spikelets** 6–14 mm. **Lower glumes** 1.9–3(4.9) mm, ovate, obtuse to acute, rarely bifid; **upper glumes** 3–4.7(5.5) mm, obtuse to acute; **lemmas** 3–4.7(6) mm, light brown to dark green at maturity, without a basal dark spot, apices obtuse, acute, or acuminate, sometimes bifid; **anthers** 3, 0.5–2.7 mm. **Caryopses** 1.6–2.3 mm.

Leptochloa fusca subsp. *fusca* is the most variable of the subspecies. In North America, it is known only from a few specimens collected at scattered locations in California; it may no longer be in the *Flora* region.

Leptochloa fusca subsp. **uninervia** (J. Presl) N. Snow
MEXICAN SPRANGLETOP

Culms (15)25–110 cm, more or less erect, often branching from the aerial nodes. **Blades** usually densely scabrous on both surfaces, not exceeding the panicle. **Panicles** 10–57 cm long, (0.5)3–18 cm wide, often ellipsoidal, usually completely exserted from the uppermost leaf sheaths; **branches** 2–11 cm, mostly ascending. **Spikelets** 5–10 mm. **Lower glumes** 1–2.6 mm, narrowly triangular to ovate; **upper glumes** 1.8–2.8 mm, obovate to widely obovate; **lemmas** 2–3.6 mm, light brown, dark green, or lead-colored, usually without a basal dark spot, apices usually truncate or obtuse, rarely broadly acute, sometimes bifid, sometimes mucronate; **anthers** 3, 0.2–0.6(1) mm. **Caryopses** 1–1.5 mm.

Leptochloa fusca subsp. *uninervia* is native from the southern United States to Argentina. It may be confused with *L. scabra*, from which it usually differs in its truncate or obtuse lemmas.

5. **Leptochloa panicea** (Retz.) Ohwi [p. 57]

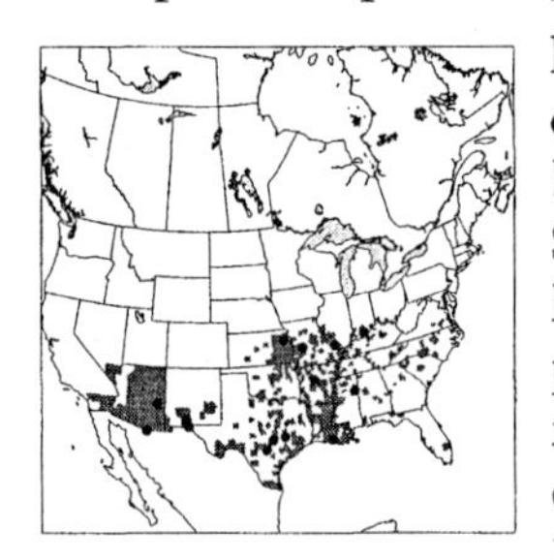

Plants annual. **Culms** (5)13–150 cm, usually erect, compressed, branching; **internodes** hollow. **Sheaths** sparsely or densely hairy, particularly distally, hairs papillose-based; **ligules** 0.6–3.2 mm, membranous, truncate, erose; **blades** 6–25 cm long, 2–21 mm wide, glabrous or sparsely pilose on both surfaces. **Panicles** 8–30 cm, with 3–100 racemose branches; **branches** 1–19 cm, ascending to reflexed. **Spikelets** 2–4 mm, distant to imbricate, green, magenta, or maroon, with 2–5(6) florets. **Glumes**

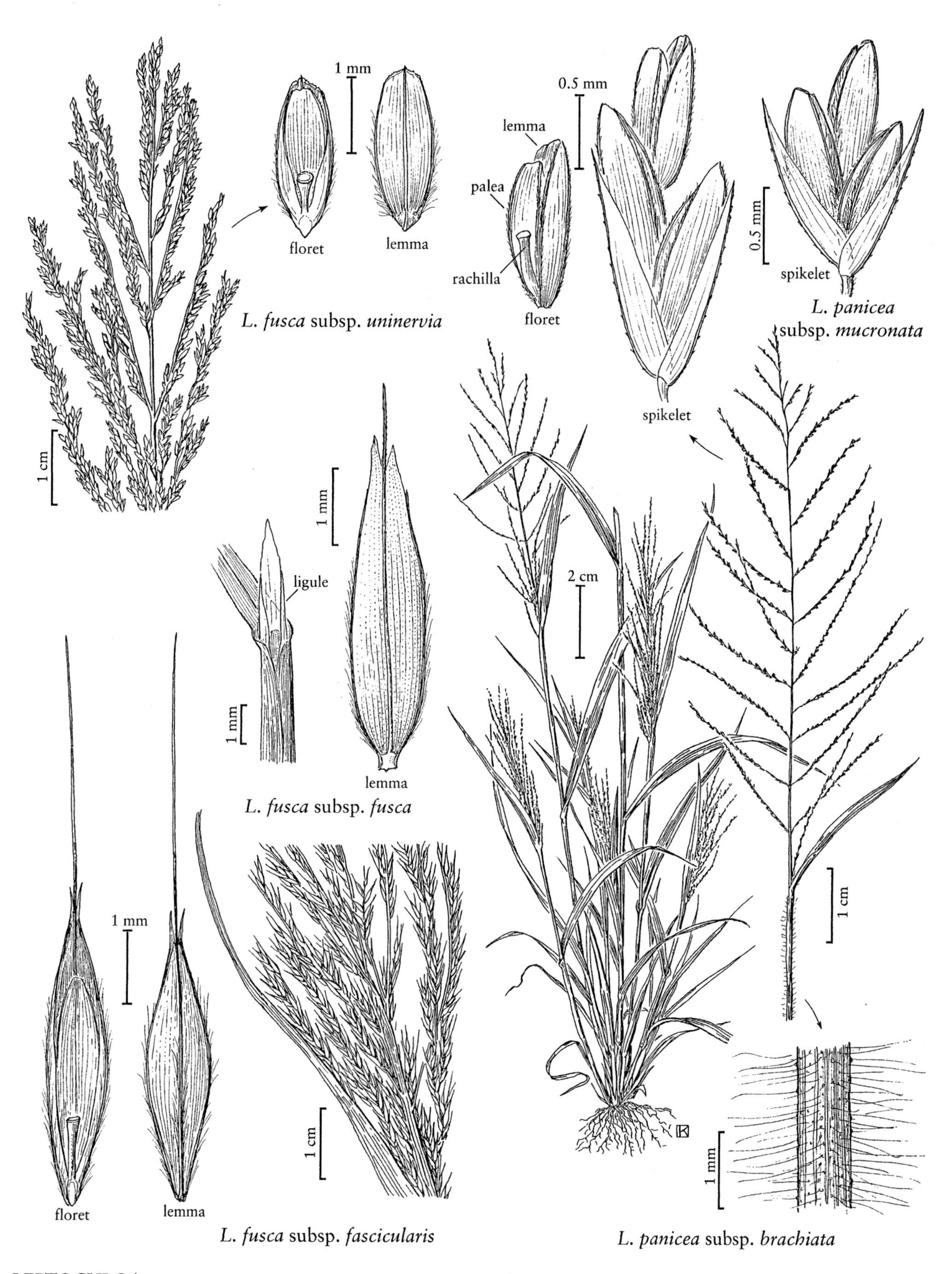
1 mm
floret
lemma
L. fusca subsp. uninervia
1 cm
0.5 mm
lemma
palea
rachilla
floret
spikelet
0.5 mm
spikelet
L. panicea subsp. mucronata
1 mm
ligule
1 mm
lemma
L. fusca subsp. fusca
2 cm
1 cm
1 mm
floret
lemma
1 cm
L. fusca subsp. fascicularis
1 mm
L. panicea subsp. brachiata

LEPTOCHLOA

sometimes exceeding the florets, linear to narrowly elliptic, acute, attenuate, or aristate; **lower glumes** 1.6–4 mm, linear to lanceolate; **upper glumes** 1.6–3.6 mm, lanceolate; **lemmas** 0.9–1.7 mm, glabrous or somewhat sericeous, acute to obtuse; **paleas** glabrous or sericeous; **anthers** 3, 0.2–0.3 mm. **Caryopses** 0.8–1.2 mm long, 0.5–0.6 mm wide, nearly round in cross section, with or without a ventral groove, apices acute to broadly obtuse.

Leptochloa panicea is a cosmopolitan species that somewhat resembles *L. chinensis*, an aggressive weed that has not yet been found in the *Flora* region. It differs in its sparsely to densely hairy, rather than glabrous or almost glabrous, sheaths and blades. Two of its three subspecies grow in the *Flora* region.

1. Glumes linear to narrowly lanceolate, exceeding the florets; lemmas 0.9–1.2 mm long; caryopses without a ventral groove, often somewhat coarsely rugose, the apices broadly obtuse . . . . subsp. *mucronata*
1. Glumes lanceolate to narrowly elliptic, not or only slightly exceeding the florets; lemmas 1.3–1.7 mm long; caryopses usually with a narrow, shallow ventral groove, smooth, the apices broadly obtuse to acute . . . . subsp. *brachiata*

Leptochloa panicea subsp. **brachiata** (Steud.) N. Snow
RED SPRANGLETOP

Culms to 150 cm. **Ligules** 0.9–3.2 mm; **blades** 2–21 mm wide. **Glumes** usually not exceeding the florets, lanceolate to narrowly elliptic; **lemmas** 1.3–1.7 mm, shortly sericeous along the veins. **Caryopses** 0.9–1.2 mm, widely depressed obovate or obdeltate in cross section, usually with a narrow, shallow ventral groove, apices broadly obtuse to acute. $2n$ = 20.

Leptochloa panicea subsp. *brachiata* extends from the southern half of the United States to Argentina. It is common in disturbed and mesic agricultural sites, and is considered a noxious weed by the U.S. Department of Agriculture.

Leptochloa panicea subsp. **mucronata** (Michx.) Nowack
MISSISSIPPI SPRANGLETOP

Culms to 110 cm. **Ligules** 0.6–1 mm; **blades** 5–9 mm wide. **Glumes** exceeding the florets, linear to narrowly lanceolate; **lemmas** 0.9–1.2 mm, glabrous or sparsely sericeous along the veins. **Caryopses** 0.8–0.9 mm, without a ventral groove, sometimes coarsely rugose, apices broadly obtuse. $2n$ = unknown.

Leptochloa panicea subsp. *mucronata* grows in the southern portion of the United States, primarily from Kansas and Missouri through Texas and Louisiana.

6. **Leptochloa nealleyi** Vasey [p. 60]
NEALLEY'S SPRANGLETOP

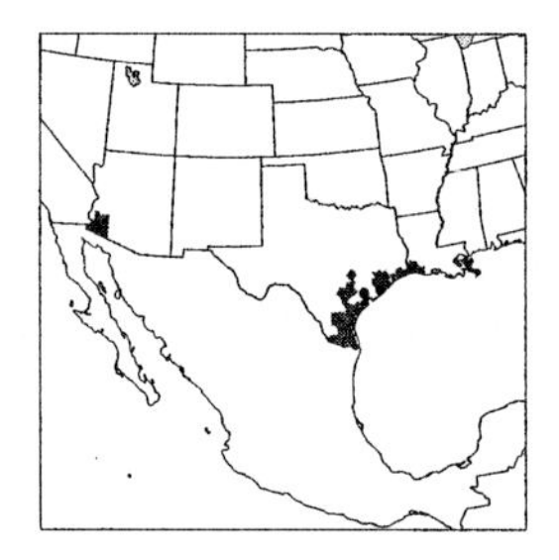

Plants annual. **Culms** (30)60–250 cm, mostly erect, compressed, sometimes branching from the lower nodes; **internodes** hollow. **Sheaths** glabrous, smooth or minutely scabrous; **ligules** 1.5–3 mm, membranous, truncate, erose, sometimes appearing ciliate because of the hairs at the base of the blades; **blades** 10–75 cm long, 4–7 mm wide, sometimes with stiff hairs behind the ligules, both surfaces scabridulous elsewhere. **Panicles** 30–76 cm, with 25–75 racemose branches; **branches** mostly 1–5(9) cm, steeply ascending to erect, stiff; **lower branches** sometimes included in the upper leaf sheaths. **Spikelets** 2.8–3.4 mm, imbricate, with 3–4 florets. **Glumes** lanceolate, **lower glumes** 0.7–0.8 mm, acute to narrowly obtuse; **upper glumes** 0.9–1.3 mm, obtuse; **lemmas** 1–2 mm, broadly lanceolate, membranous, veins sericeous basally, apices obtuse to acute or apiculate; **paleas** sericeous along the veins; **anthers** 3, 0.2–0.4 mm. **Caryopses** 0.5–1 mm long, 0.4–0.5 mm wide, elliptic to obovate, nearly round in cross section. $2n$ = 40.

Leptochloa nealleyi is native to coastal Louisiana, Texas, and Mexico; it also grows, but rarely, in Cuba. The species is not established in Arizona, but it was collected once from a farm in the Wellton area (NCU 303513). It is not clear whether the plants were being cultivated or growing as weeds.

The numerous, short, stiffly ascending or erect panicle branches make *Leptochloa nealleyi* easy to identify.

7. **Leptochloa scabra** Nees [p. 60]
ROUGH SPRANGLETOP

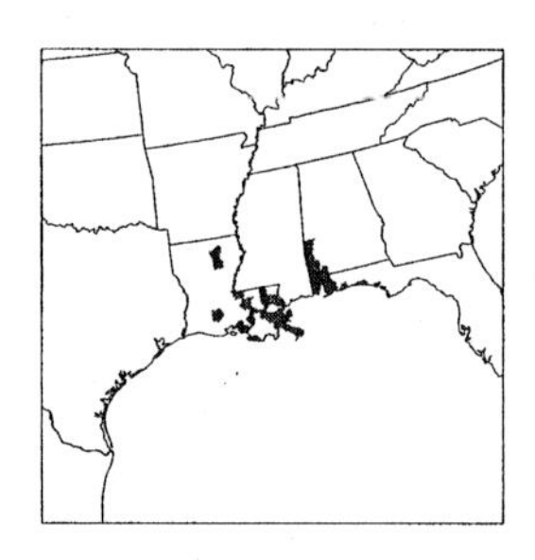

Plants annual. **Culms** (12)20–125 cm, mostly erect, often strongly compressed, branching; **internodes** hollow. **Sheaths** glabrous, smooth to scabrous; **ligules** 1.5–2 mm, membranous, truncate, erose; **blades** 25–35(50) cm long, 8–16 mm wide, scabrous on both surfaces. **Panicles** 8–35 cm, with 50–150 racemose branches; **branches** (2)5–12 cm, lax, sometimes arcuate, lower branches often remaining enclosed in the upper leaf sheaths. **Spikelets** 3–4.5 mm, usually tightly imbricate, green but straw-colored when dry, with 2–6 florets. **Glumes** sometimes mucronate; **lower glumes** 0.8–1.6 mm, narrowly triangular to lanceolate; **upper glumes** 1.1–2.1 mm, ovate; **rachilla segments** not visible between the florets; **lemmas** 2.1–2.4 mm, lanceolate to narrowly ovate, membranous, sparsely sericeous along the lateral veins, apices acute, unawned; **anthers** 0.2–0.4 mm. **Caryopses** 0.8–1.3 mm long, 0.3–0.5 mm wide, elliptic to obovate, depressed obovate in cross section. $2n = 60$.

Leptochloa scabra is a neotropical species that extends into Louisiana and southwestern Alabama. It is often confused with *L. panicoides*, but it has more, flexuous to arcuate panicle branches, shorter spikelets, and less prominent lemma veins. It may also be confused with *L. fusca* subsp. *uninervia*, from which it differs in its acute lemmas, and with *L. virgata*, from which it differs in its hollow, flattened culms and the complete lack of lemma awns.

8. **Leptochloa chinensis** (L.) Nees [p. 60]
ASIAN SPRANGLETOP

Plants annual or perennial. **Culms** 15–100 cm, round, glabrous or appressed-pubescent; **internodes** hollow. **Sheaths** sometimes flattened below, usually glabrous, sometimes sparsely pilose at the apices; **ligules** 1.8–5.4 mm, truncate, erose or ciliate; **blades** (1)5–30 cm long, 4–8 mm wide, glabrous, sometimes scabridulous. **Panicles** 20–50 cm, with 25–60 racemose branches; **branches** (1)4–8(14) cm, erect to slightly reflexed, usually straight. **Spikelets** 2.5–3.7 (4.2) mm, imbricate to somewhat distant, green to tan, with 4–6 florets. **Glumes** triangular to lanceolate; **lower glumes** 1.1–1.7 mm; **upper glumes** 1.6–2 mm; **rachilla segments** usually visible between the florets; **lemmas** 1.2–1.7 mm, ovate to elliptic, glabrous or hairy along and between the veins, apices broadly acute or obtuse; **paleas** glabrous or hairy; **anthers** 3, 0.4–0.6 mm. **Caryopses** 0.9–1.9 mm long, 0.6–0.8 mm wide. $2n = 40$.

Leptochloa chinensis is not yet known from the *Flora* region but, if introduced, it could become an aggressive weed because it competes well in undisturbed mesic sites. Although it resembles *L. panicea*, *L. chinensis* differs in its glabrous, or nearly glabrous, sheaths and blades.

9. **Leptochloa viscida** (Scribn.) Beal [p. 60]
SONORAN SPRANGLETOP

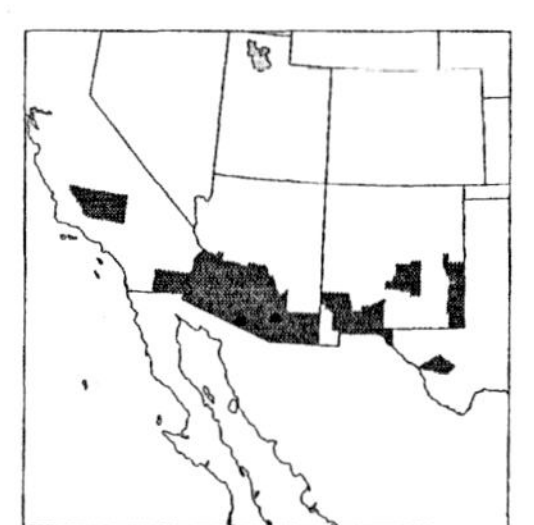

Plants annual. **Culms** (3)10–60 cm, prostrate or erect, round or somewhat compressed, often highly branched; **internodes** hollow. **Sheaths** glabrous (rarely sparsely pilose near the base), sometimes with a sticky exudate; **ligules** 1.2–2.5 mm, truncate, erose to lacerate; **blades** 1–15 cm long, 1.2–5.5 mm wide, glabrous abaxially and adaxially. **Panicles** 2–17 cm, with 5–23 racemose branches; **branches** 1–2.5(3.5) cm, stiff, often included in the upper leaf sheaths. **Spikelets** 4.5–7.5 mm, more or less imbricate, magenta or green, with 2–6 florets. **Glumes** triangular, acute; **lower glumes** 1.6–2 mm, acute; **upper glumes** 2–2.9 mm; **lemmas** 2.4–3.5 mm, ovate, membranous, sericeous along the lower veins, lateral veins pronounced, apices acute, obtuse, or truncate, awned, awns 0.5–1.5 mm; **paleas** minutely scabrous along the veins; **anthers** 3, 0.4–0.5 mm. **Caryopses** 1.2–1.6 mm long, 0.4–0.5 mm wide, narrowly elliptic to obovate, transversely elliptic in cross section. $2n = 40$.

Leptochloa viscida is a Sonoran Desert species that occurs from southern California to southwestern New Mexico and south into adjacent Mexico. It differs from *L. fusca* subsp. *fascicularis*, which grows in the same region, in its consistently short-awned lemmas, smaller panicles, often prostrate and much-branched growth habit, and often reddish florets.

10. **Leptochloa panicoides** (J. Presl) Hitchc. [p. 60]
AMAZON SPRANGLETOP

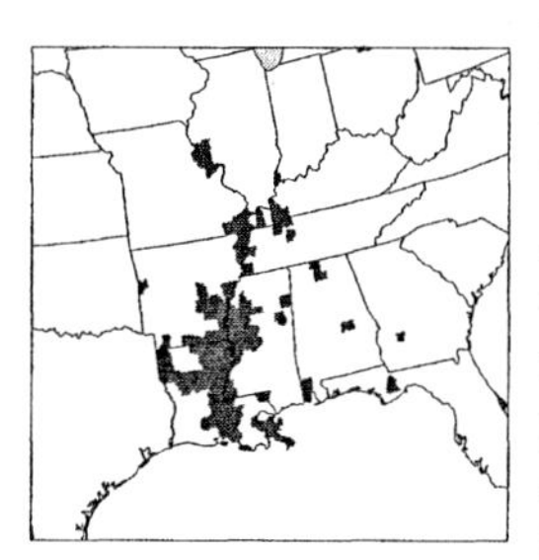

Plants annual. **Culms** (7)45–110 cm, often geniculate below, usually ascending to erect above, rarely branching at the base, often branching distally; **internodes** hollow. **Sheaths** glabrous, margins occasionally sparsely ciliate on the basal ½; **ligules** 2.2–3.8 mm, membranous, truncate, somewhat erose; **blades** 4–20 cm long, 4–8 mm wide, both surfaces smooth or scabridulous. **Panicles** 20–35 cm, with 20–30(90) racemose branches; **branches** 2.5–7 cm, ascending, mostly stiff. **Spikelets** 4–5 mm, usually somewhat imbricate, with 4–6(7) florets. **Lower glumes** 0.9–1.9 mm, usually lanceolate, sometimes falcate, acute; **upper glumes** 1.8–2.3 mm, ovate, acute to obtuse; **lemmas** 2.4–3 mm, narrowly elliptic to ovate, membranous, midveins and lateral veins sericeous basally, lateral veins prominent, excurrent, apices acute to broadly acute, unawned, sometimes mucronate; **paleas** glabrous; **anthers** 3, 0.6–0.8 mm. **Caryopses** 1.1–1.4 mm long, 0.7 mm wide, elliptic, depressed obovate in cross section. $2n = 20$.

Leptochloa panicoides is native from the central Mississippi and Ohio river drainages south through Mesoamerica to Brazil. It usually grows in somewhat mesic habitats. It has been reported from two counties in Texas, but no specimens documenting the reports have been found so they are not shown.

Nicora (1995) merged *Leptochloa panicoides* with *L. scabra*, but the two differ consistently in the number of panicle branches, spikelet length, and prominence of the lemma veins.

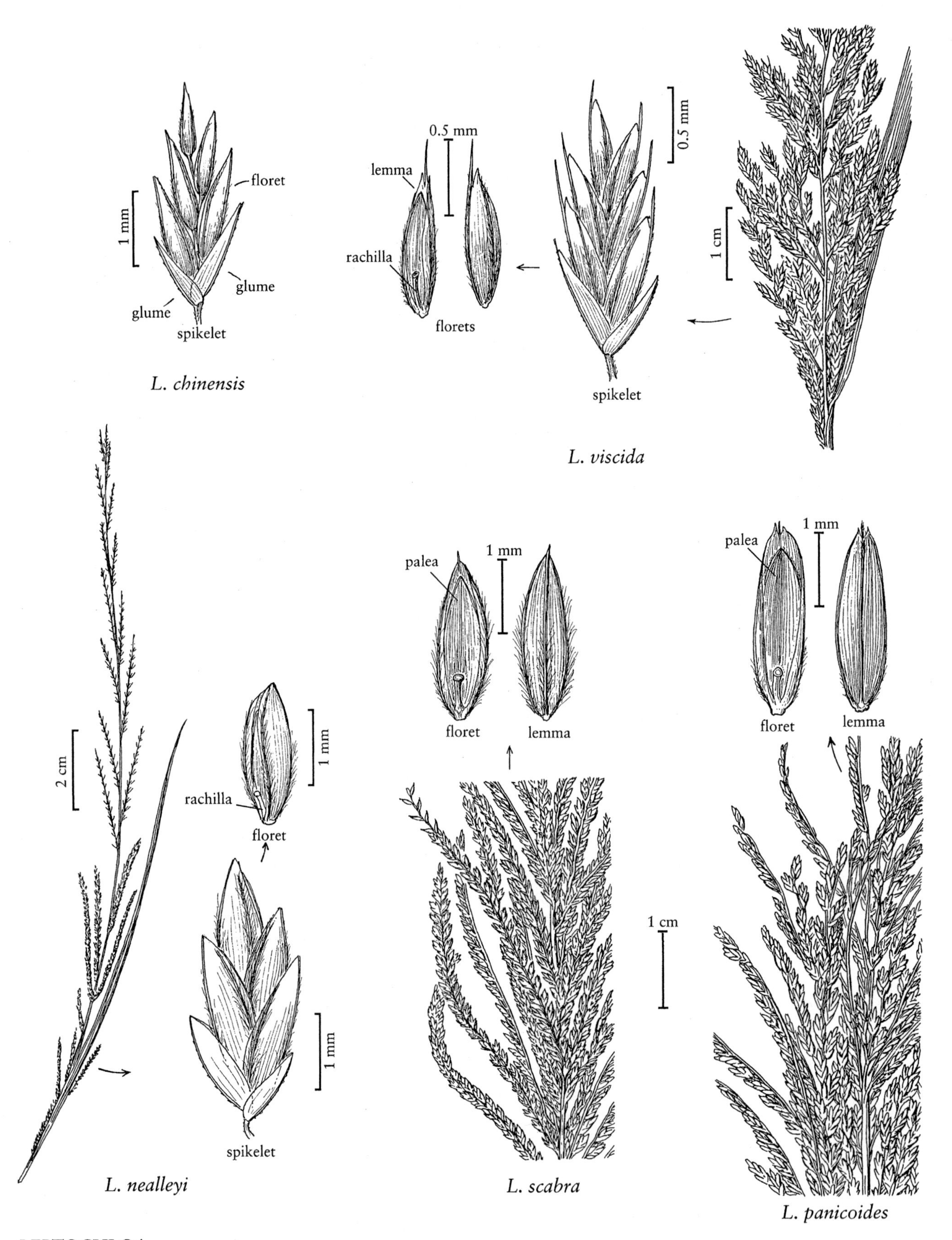
floret
1 mm
glume
glume
spikelet
L. chinensis
0.5 mm
lemma
rachilla
florets
0.5 mm
spikelet
1 cm
L. viscida
2 cm
1 mm
rachilla
floret
1 mm
spikelet
L. nealleyi
palea
1 mm
floret
lemma
1 cm
L. scabra
1 mm
palea
floret
lemma
L. panicoides

LEPTOCHLOA

17.20 TRIPOGON Roem. & Schult.

J.K. Wipff

Plants perennial or annual; cespitose or tufted. **Culms** 4–65 cm, erect, slender. **Leaves** linear, flat, usually becoming folded and filiform; **ligules** membranous, ciliate. **Inflorescences** terminal, unilateral linear spikes or spikelike racemes, with 1 spikelet per node, exceeding the leaves; **rachises** visible, not concealed by the spikelets. **Spikelets** appressed, in 2 rows along 1 side of the rachises, with 3–20 bisexual florets, distal florets sterile or staminate; **disarticulation** above the glumes and between the florets. **Glumes** unequal, 1(3)-veined; **lemmas** 1–3-veined, backs slightly keeled or rounded, apices lobed or bifid, mucronate or awned from between the lobes, lateral veins sometimes also excurrent, awns usually straight; **anthers** 1–3. $x = 10$. Name from the Greek *treis*, 'three', and *pogon*, 'beard', alluding to the hairs at the bases of the three lemma veins found in many of its species.

Tripogon is a genus of approximately 30 species, most of which are native to the tropics of the Eastern Hemisphere, especially Africa and India, but with one, *Tripogon spicatus*, native to the Western Hemisphere.

1. **Tripogon spicatus** (Nees) Ekman [p. 62]
American Tripogon

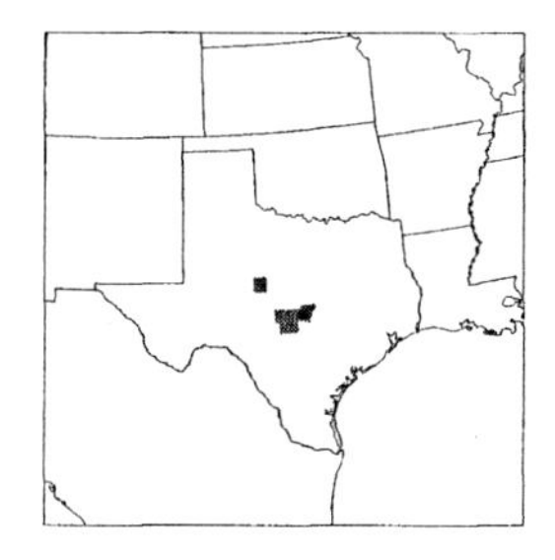

Plants perennial; cespitose. **Culms** (4.5)6–34 cm; **nodes** 2–3, glabrous. **Leaves** mostly basal; **sheaths** mostly glabrous, but with tufts of hairs flanking the collar; **ligules** 0.2–0.3 mm, truncate; **blades** 1.9–10 cm long, 0.2–1.1 mm wide, glabrous or the adaxial surfaces and margins sparsely pubescent. **Inflorescences** (1.5)4–10 cm long, 1.5–3.5 mm wide, with (6)13–22 spikelets; **pedicels** 0–0.5 mm, glabrous. **Spikelets** 4.5–12 mm long, 1–1.3 mm wide, with 5–14 florets; **rachilla segments** glabrous except for an apical tuft of hairs. **Glumes** unequal, exceeded by the basal florets; **lower glumes** (1.2)1.5–2.4 mm, glabrous, 1-veined, scabridulous over the veins; **upper glumes** 1.9–2.6 mm, glabrous, 1-veined; **lowest lemmas** 2.3–3.1 mm, 3-veined, apical sinuses 0.1–0.3 mm deep; **awns** 0.2–0.9 mm, straight; **paleas** 1.6–2.4 mm, glabrous on the back and minutely pubescent on the margins; **anthers** 3, 0.3–0.4 mm, yellow to purple. **Caryopses** 1–1.5 mm, reddish-brown. $2n = 20$.

Tripogon spicatus grows in shallow rocky soils, usually on granite outcroppings, occasionally on limestone. The flowering period, April–July(October, November), apparently depends on rainfall. Its range includes the West Indies, Mexico, and South America, in addition to central Texas.

17.21 TRICHONEURA Andersson

J.K. Wipff

Plants annual or perennial. **Culms** 12–155 cm, nodes glabrous, internodes solid. **Ligules** membranous; **blades** linear, narrow, usually flat. **Inflorescences** terminal, panicles of 5–40 racemosely-arranged, spikelike branches, exceeding the leaves; **branches** spreading to appressed, persistent, unilateral, with 1 spikelet per node. **Spikelets** 5.3–14 mm long, with 2 or more florets, typically with 2–8 bisexual florets, sterile or staminate florets sometimes present distal to the bisexual florets; **rachilla internodes** pilose basally, apices oblique; **disarticulation** above the glumes and below the florets. **Glumes** from shorter than to greatly exceeding the florets, equal or subequal to each other, narrow, apices acuminate and mucronate, awnlike, or awned; **calluses** well-developed, strigose; **lemmas** 3-veined, conspicuously hairy adjacent to and on the lateral veins, apices cleft, midveins excurrent from the sinuses, sometimes forming awns. $x = 10$. Name from the Greek *thrix*, 'hair', and *neuron*, 'nerve', alluding to the hairs on the lateral lemma veins.

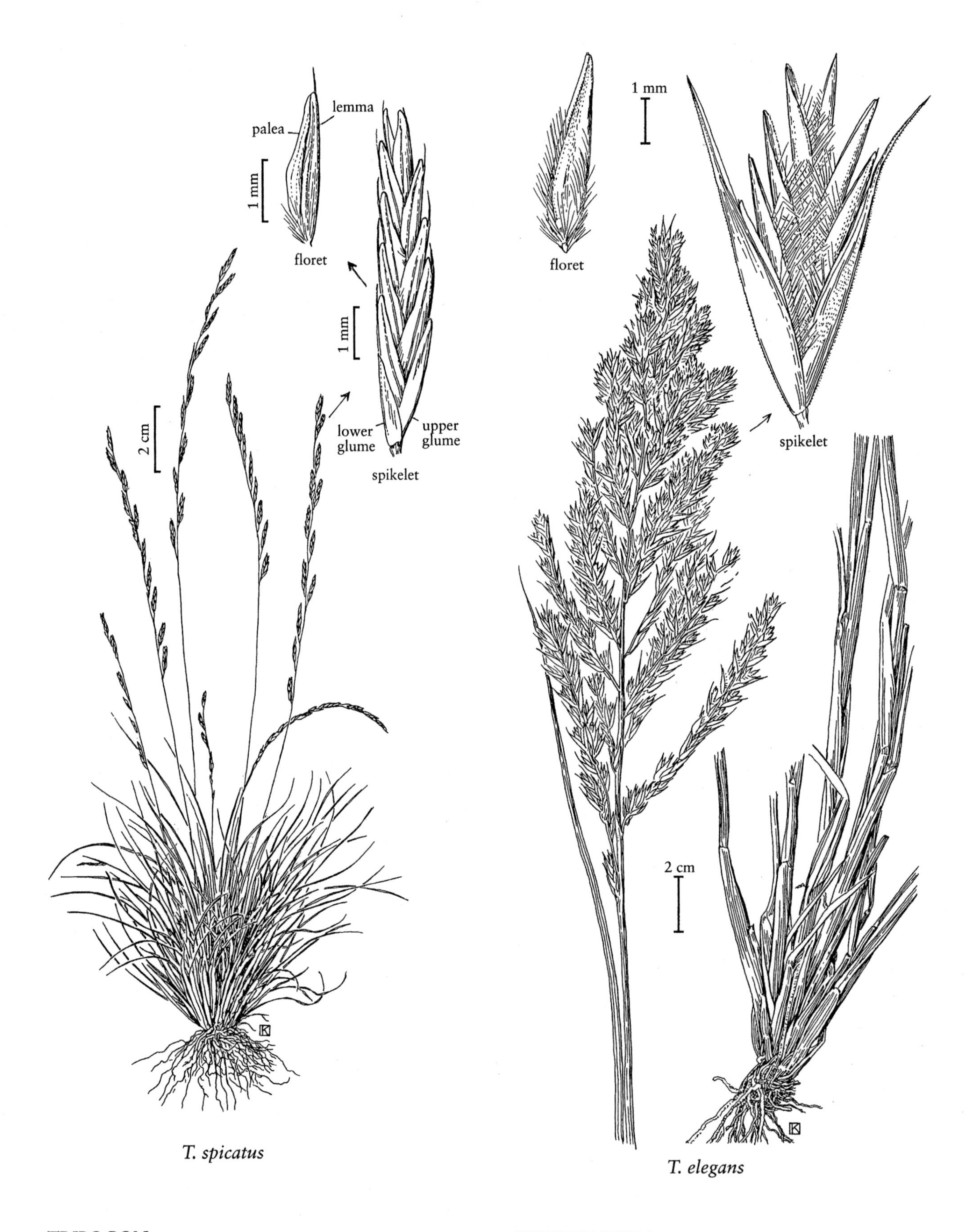

T. spicatus

T. elegans

TRIPOGON

TRICHONEURA

Trichoneura is a genus of seven species that grow in dry, sandy, or stony soils. Five species are native to the Eastern Hemisphere and two to the Western Hemisphere, one of which is native to the *Flora* region.

1. **Trichoneura elegans** Swallen [p. 62]
SILVEUS GRASS

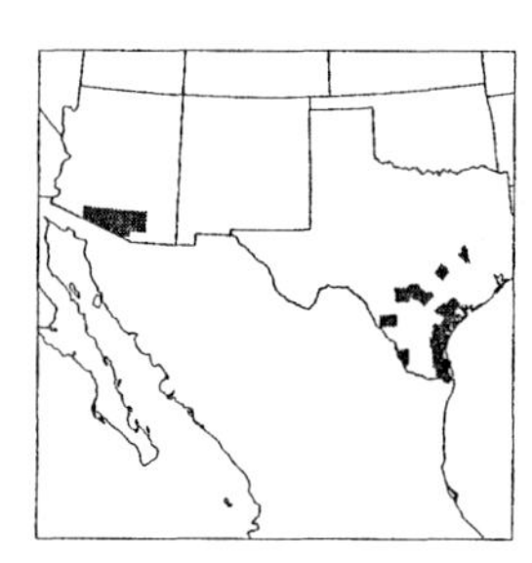

Plants annual; cespitose. **Culms** (12)30–115 cm, erect to decumbent, usually rooting at the lower nodes, mostly glabrous, pubescent beneath the panicles. **Lower sheaths** longer than the internodes, antrorsely scabridulous; **ligules** 1.2–2.7 mm, truncate, erose; **blades** (2.5)5–35.2 cm long, (1)2–9.3 mm wide, antrorsely scabridulous, usually flat, becoming involute or convolute when dry. **Panicles** (3)4–23 cm long, 1.5–8 cm wide, rachises pubescent; **branches** 5–24, ascending, axes triquetrous; **lower branches** (0.8)1–9.5 cm. **Spikelets** 8–11.5 mm, with 5–10 florets, distal 1–4 florets sterile or staminate. **Glumes** with 1 green to purplish vein, antrorsely scabridulous, apices acuminate, awnlike; **lower glumes** 5.9–11.5 mm; **upper glumes** 5.8–10.5 mm; **lowest lemmas** 4.5–5.7(6.4) mm, veins green, midveins excurrent to 1.2 mm, midsections hairy adjacent to and on the lateral veins, with 0.5–1.6 mm hairs, antrorsely scabridulous elsewhere; **paleas** 4–5 mm, antrorsely scabridulous on the apices and veins; **anthers** 3, 1.3–1.7 mm, yellow to purplish. **Caryopses** 2.6–3 mm long, 0.7–0.9 mm wide, flattened. $2n = 40$.

Trichoneura elegans usually grows in dry, deep, sandy soil. Its range extends from south central Texas to northern Tamaulipas, Mexico.

17.22 DINEBRA Jacq.

Mary E. Barkworth

Plants annual. **Culms** 13–120 cm, not woody. **Ligules** membranous, truncate, lacerate, sometimes ciliate; **blades** linear, flat. **Inflorescences** terminal, panicles of 1–70, 1-sided, spikelike branches, irregularly disposed on elongate rachises, clearly exceeding the upper leaves; **branches** with 2 rows of 1 or more closely imbricate, sessile spikelets, proximal spikelets sometimes replaced by short, tardily deciduous, secondary branches; **disarticulation** at the base of the branches or at the base of the secondary branches and (eventually) beneath the florets. **Spikelets** laterally compressed, cuneate, with 1–3 florets. **Glumes** subequal, much longer than the florets, usually exceeding the distal florets, coriaceous or membranous, strongly keeled, acuminate-aristate; **lemmas** thinly membranous, weakly keeled, 3-veined, pilose over the veins, apices acute to 2-lobed, central veins excurrent, forming mucros. **Caryopses** elliptic-oblong, trigonous. $x = 10$. Name a corruption of the Arabic *danaiba*, 'little tail', an allusion to the prolonged apices of the glumes.

Dinebra, a genus of three species, is native from Africa to Madagascar and India. One species has been reported from the *Flora* region.

SELECTED REFERENCE Reed, C.F. 1964. A flora of the chrome and manganese ore piles at Canton, in the Port of Baltimore, Maryland and at Newport News, Virginia, with descriptions of genera and species new to the flora of the eastern United States. Phytologia 10:321–405.

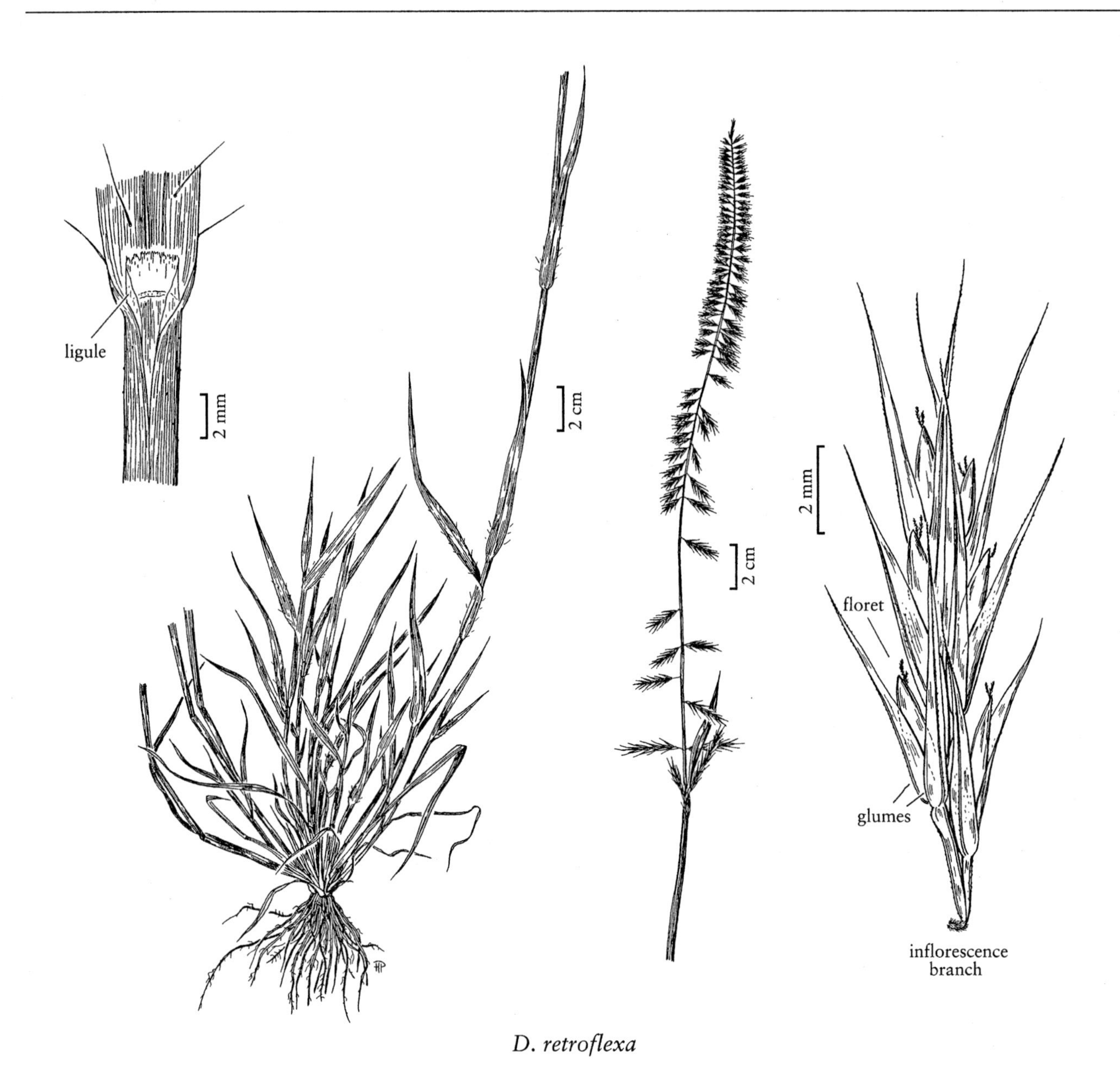

D. retroflexa

DINEBRA

1. **Dinebra retroflexa** (Vahl) Panz. [p. 64]
VIPER GRASS

Plants loosely tufted. **Culms** 13–120 cm, decumbent, straggling, often rooting at the lower nodes. **Leaves** sometimes glandular, particularly on the sheaths; **blades** 4.5–28 cm long, 4–8 mm wide, finely pointed. **Panicles** 8–34 cm; **branches** 0.6–5(7) cm, stiff, initially ascending, reflexed at maturity; **disarticulation** at the base of the branches. **Spikelets** 5.7–9 mm, with 1–3 florets. **Glumes** 5.7–9 mm, asymmetric, coriaceous, keels glandular, apices caudate-curving; **lemmas** 2.1–2.9 mm, narrowly ovate, appressed pubescent on the lateral veins and adjacent to the lower ½ of the central vein; **paleas** appressed pubescent on the flaps adjacent to the keels. $2n = 20$.

Dinebra retroflexa is native from southern Africa through tropical Africa to Egypt, Iraq, Pakistan, and India. It has reportedly been found on chrome ore piles in Canton, Maryland, a temporary unloading ground for ores in the Port of Baltimore (Reed 1964), and in Mecklenberg County, North Carolina. It is a common weed of rich soils in moist, tropical regions.

17.23 ERAGROSTIS Wolf

Paul M. Peterson

Plants annual or perennial; usually synoecious, sometimes dioecious; cespitose, stoloniferous, or rhizomatous. **Culms** 2–160 cm, not woody, erect, decumbent, or geniculate, sometimes rooting at the lower nodes, simple or branched; **internodes** solid or hollow. **Leaves** not strongly distichous; **sheaths** open, often with tufts of hairs at the apices, hairs 0.3–8 mm; **ligules** usually membranous and ciliolate or ciliate, cilia sometimes longer than the membranous base, occasionally of hairs or membranous and non-ciliate; **blades** flat, folded, or involute. **Inflorescences** terminal, sometimes also axillary, simple panicles, open to contracted or spike-like, terminal panicles usually exceeding the upper leaves; **pulvini** in the axils of the primary branches glabrous or not; **branches** not spikelike, not disarticulating. **Spikelets** 1–27 mm long, 0.5–9 mm wide, laterally compressed, with (1)2–60 florets; **disarticulation** below the fertile florets, sometimes also below the glumes, acropetal with deciduous glumes and lemmas but persistent paleas, or basipetal with the glumes often persistent and the florets usually falling intact. **Glumes** usually shorter than the adjacent lemmas, 1(3)-veined, not lobed, apices obtuse to acute, unawned; **calluses** glabrous or sparsely pubescent; **lemmas** usually glabrous, obtuse to acute, (1)3(5)-veined, usually keeled, unawned or mucronate; **paleas** shorter than the lemmas, longitudinally bowed-out by the caryopses, 2-keeled, keels usually ciliate, intercostal region membranous or hyaline; **anthers** 2–3; **ovaries** glabrous; **styles** free to the bases. **Cleistogamous spikelets** occasionally present, sometimes on the axillary panicles, sometimes on the terminal panicles. **Caryopses** variously shaped. $x = 10$. The origin of the name is obscure.

Eragrostis, a genus of approximately 350 species, grows in tropical and subtropical regions throughout the world. About 110 species are native or adventive in the Western Hemisphere; 25 species are native in the *Flora* region, 24 are introduced. In most taxa native to the Western Hemisphere, disarticulation is acropetal and the lemmas fall with the caryopses, leaving the paleas attached to the rachilla.

Nathaniel Wolf (1776), the person who first named *Eragrostis*, made no statement concerning the origin of its name. Clifford (1996) provides three possible derivations: from *eros*, 'love', and *Agrostis*, the Greek name for an indeterminate herb; from the Greek *er*, 'early' and *agrostis*, 'wild', referring to the fact that some species of *Eragrostis* are early invaders of arable land; or the Greek *eri-*, a prefix meaning 'very' or 'much', suggesting that the name means many-flowered *Agrostis*. Many authors have stated that the first portion of the name is derived from *eros*, but none has explained the connection between *Eragrostis* and passionate expressions of love, the kind of love to which *eros* applies.

SELECTED REFERENCES **Clifford, H.T.** 1996. Etymological Dictionary of Grasses, Version 1.0 (CD-ROM). Expert Center for Taxonomic Identification, Amsterdam, The Netherlands; **Harvey, L.H.** 1948. *Eragrostis* in North and Middle America. Ph.D. dissertation, University of Michigan, Ann Arbor, Michigan, U.S.A. 269 pp.; **Harvey, L.H.** 1975. *Eragrostis*. Pp. 177–201 *in* F.W. Gould. The Grasses of Texas. Texas A&M University Press, College Station, Texas, U.S.A. 635 pp.; **Koch, S.D.** 1974. The *Eragrostis pectinacea–pilosa* complex in North and Central America. Illinois Biol. Monogr. 48:1–74; **Sánchez Vega, I.** and **S.D. Koch**. 1988. Estudio biosistemático de *Eragrostis mexicana*, *E. neomexicana*, *E. orcuttiana*, y *E. virescens* (Gramineae: Chloridoideae). Bol. Soc. Bot. 48:95–112; **Van den Borre, A.** and **L. Watson**. 2000. On the classification of the Chloridoideae: Results from morphological and leaf anatomical data analyses. Pp. 180–183 *in* S.W.L. Jacobs and J. Everett (eds.). Grasses: Systematics and Evolution. International Symposium on Grass Systematics and Evolution (3rd:1998). CSIRO Publishing, Collingwood, Victoria, Australia. 408 pp.; **Wolf, N.M.** 1776. Genera Plantarum. [publisher unknown, Danzig, Germany]. 177 pp.

1. Plants annual, tufted or mat-forming, without innovations [for opposite lead, see p. 67].
   2. Palea keels prominently ciliate, the cilia 0.2–0.8 mm long.
      3. Spikelets 1–3.6 mm long, 0.9–2 mm wide, with 4–12 florets; lemmas 0.7–1.3 mm long.
         4. Anthers 2; pedicels 0.1–1 mm long, mostly shorter than the spikelets, straight . . . . 1. *E. ciliaris*

4. Anthers 3; pedicels 1–4(7) mm long, as long as or longer than the spikelets, mostly curved . . . . 3. *E. amabilis*
3. Spikelets 5–20 mm long, 1.4–4 mm wide, with 10–42 florets; lemmas 1.3–2.8 mm long.
5. Lemmas and culms without glands; anthers 0.1–0.2 mm long, purplish . . . . 2. *E. cumingii*
5. Lemmas with 1–3 crateriform glands on the keels, similar glands also often present below the cauline nodes; anthers 0.2–0.5 mm long, yellow . . . . 18. *E. cilianensis*
2. Palea keels smooth or scabrous, the scabridities less than 0.2 mm long.
6. Plants mat-forming; panicles 1–3.5 cm long; erect portion of culms (2)5–20 cm, the basal portion prostrate and rooting at the nodes.
7. Spikelets bisexual; anthers 2, 0.2–0.3 mm long . . . . 4. *E. hypnoides*
7. Spikelets and plants unisexual; anthers 3, 1.4–2.2 mm long . . . . 5. *E. reptans*
6. Plants usually not forming mats; panicles 3–55 cm long; culms (2)6–130 cm tall, not prostrate or rooting at the lower nodes.
8. Ligules membranous, neither ciliolate nor ciliate . . . . 6. *E. japonica*
8. Ligules membranous and ciliolate to ciliate, the cilia often longer than the basal membrane.
9. Caryopses with a shallow or deep ventral groove, ovoid to rectangular-prismatic or dorsally compressed, if dorsally compressed, the surface striate or smooth.
10. Bases of the caryopses greenish; caryopses dorsally compressed, the distal ⅔ translucent, the surface smooth; leaf sheaths with oblong glands; in the *Flora* region, known from a single collection at Canton, Maryland . . . . 7. *E. cylindriflora*
10. Bases of the caryopses reddish-brown or brownish; caryopses laterally compressed or rectangular-prismatic to ovoid, the distal ⅔ opaque, the surface striate; sheaths without oblong glands, sometimes with glandular pits; plants found at many locations in the *Flora* region.
11. Spikelets 4–11 mm long, with 5–15 florets; pedicels somewhat divergent to almost appressed . . . . 12. *E. mexicana*
11. Spikelets 1.4–5 mm long, with 2–7 florets; pedicels divergent.
12. Panicles 4–20 cm long, less than ½ the height of the plant; pedicels 1.5–5 mm long; glandular pits often present below the cauline nodes, on the rachises, and on the panicle branches . . . . 13. *E. frankii*
12. Panicles 10–45(55) cm long, ⅔ or more the height of the plant; pedicels 4–25 mm long; plants without glandular pits . . . . 14. *E. capillaris*
9. Caryopses without a ventral groove, usually globose, rarely flattened, pyriform, obovoid, ellipsoid, or rectangular-prismatic, the surface smooth to faintly striate.
13. Plants with glandular pits or bands somewhere, the location(s) various, including any or all of the following: below the cauline nodes, on the sheaths, blades, rachises, panicle branches, or pedicels, or on the keels of the lemmas and paleas.
14. Panicles 0.5–2 cm wide, contracted; primary panicle branches usually appressed, occasionally diverging up to 30° from the rachises; spikelets light yellowish, occasionally with reddish-purple markings . . . . 15. *E. lutescens*
14. Panicles 2–18 cm wide, open to somewhat contracted; primary panicle branches diverging 20–110° from the rachises; spikelets plumbeous, greenish, or reddish-purple.
15. Spikelets 1.7–5.6 mm long, with 3–6 florets . . . . 13. *E. frankii*
15. Spikelets (2)3.5–20 mm long, with (3)5–40 florets.
16. Spikelets 0.6–1.4 mm wide; pedicels 1–10 mm long, lax, appressed or divergent . . . . 16. *E. pilosa*
16. Spikelets 1.1–4 mm wide; pedicels 0.2–4 mm long, stiff, straight, usually divergent.
17. Lemmas 2–2.8 mm long, with 1–3 crateriform glands along the keels; spikelets 6–20 mm long, 2–4 mm wide, with 10–40 florets; disarticulation below the florets, the rachillas persistent; anthers yellow . . . . 18. *E. cilianensis*

17. Lemmas 1.4–1.8 mm long, rarely with 1 or 2 crateriform glands along the keels; spikelets 4–7(11) mm long, 1.1–2.2 mm wide, with 7–12(20) florets; disarticulation below the lemmas, both the paleas and rachillas usually persistent; anthers reddish-brown.
    18. Panicles with glandular regions below the nodes, the glandular tissue forming a ring or band, often shiny or yellowish; anthers 3; blade margins without crateriform glands; pedicels without glandular bands . . . . 19. *E. barrelieri*
    18. Panicles sometimes with areas, but rarely rings, of glandular spots or crateriform pits below the nodes, the glands usually dull greenish-gray to stramineous; anthers 2; blade margins sometimes with crateriform glands; pedicels usually with glandular bands . . . . 20. *E. minor*

13. Plants without glandular pits or bands.
    19. Spikelets (1.6)2–4 mm wide; florets disarticulating intact from the persistent rachillas . . . . 21. *E. unioloides*
    19. Spikelets 0.6–2.5 mm wide; lemmas disarticulating separately from the paleas, sometimes both the paleas and the rachillas persistent.
        20. Spikelets with 3–6 florets; plants of the central and northeastern United States and southern Ontario, Canada . . . . 13. *E. frankii*
        20. Spikelets with (3)5–42 florets; plants from throughout the contiguous United States and southern Ontario, Canada.
            21. Lemmas 1.6–3 mm long; caryopses 0.7–1.3 mm long, obovoid, smooth, light brown to white; plants cultivated, occasionally escaping . . . . 22. *E. tef*
            21. Lemmas 1–2.2 mm long; caryopses 0.3–1.1 mm long, subglobose, pyriform, or obovoid to prism-shaped, smooth or faintly striate, brownish; plants native species or established introductions, variously distributed.
                22. Lemmas with conspicuous, often greenish lateral veins; caryopses 0.3–0.6 mm long, ovoid, subglobose to obovoid.
                    23. Spikelets 5–12(18) mm long, with 12–42 florets; primary branches 6–10 per culm; lemmas 1.3–2 mm long; anthers 3 . . . . 2. *E. cumingii*
                    23. Spikelets 2–4.6 mm long, with 5–15 florets; primary branches (12)15–20 per culm; lemmas 1–1.3 mm long; anthers 2 . . . . 23. *E. gangetica*
                22. Lemmas with inconspicuous to moderately conspicuous lateral veins, the veins usually not greenish; caryopses 0.5–1.1 mm long, pyriform or obovoid to prism-shaped.
                    24. Lower glumes 0.5–1.5 mm long, at least ½ as long as the lowest lemmas; spikelets 1.2–2.5 mm wide; panicle branches solitary or paired at the lowest 2 nodes; lemmas with moderately conspicuous lateral veins . . . . 17. *E. pectinacea*
                    24. Lower glumes 0.3–0.6(0.8) mm long, usually less than ½ as long as the lowest lemmas; spikelets 0.6–1.4 mm wide; panicle branches usually whorled at the lowest 2 nodes; lemmas with inconspicuous lateral veins . . . . 16. *E. pilosa*

1. Plants perennial, sometimes rhizomatous, forming innovations at the basal nodes [for opposite lead, see p. 65].
    25. Paleas with a broad lower portion forming a wing or tooth on each side, these often projecting beyond the lemmas.
        26. Spikelets 5.5–16 mm long, 2.7–9 mm wide; lemmas 3–5 mm long, the keels without crateriform glands; pedicels with a narrow band or abscission line just below the apices; anthers 1.4–2.8 mm long . . . . 24. *E. superba*

26. Spikelets 2–5 mm long, 2–3.5 mm wide; lemmas 1.8–2.3 mm long, the keels with a few crateriform glands; pedicels without a narrow band or abscission line just below the apices; anthers 0.5–0.9 mm long . . . . 25. *E. echinochloidea*

25. Paleas without a broad lower portion forming a wing or tooth, the bases never projecting beyond the lemmas.

27. Plants rhizomatous; disarticulation always below the florets, the paleas falling with the lemmas and caryopses.

28. Plants with long, scaly rhizomes, 4–8 mm thick; spikelets 8–14 mm long; lemmas 3.8–4.5 mm long, 3-5-veined, the apices acute to obtuse, usually erose; caryopses 1.6–2 mm long . . . . 26. *E. obtusiflora*

28. Plants with short, knotty rhizomes less than 4 mm thick, often stout but never elongated; spikelets 2.5–7.6 mm long; lemmas 1–2.5 mm long, 3-veined, the apices acute, usually entire; caryopses 0.5–0.8 mm long.

29. Sheaths, blades, and culms not viscid or glandular; caryopses strongly flattened, the ventral surface with 2 prominent ridges separated by a groove; anthers 0.3–0.5 mm long; lemmas leathery . . . . 27. *E. spectabilis*

29. Sheaths, blades, and/or culms often viscid, sometimes glandular; caryopses terete, the ventral surfaces without 2 ridges separated by a groove; anthers 0.2–0.4 mm long; lemmas membranous.

30. Pedicels 0.2–1.2 mm long, appressed; lemmas 1.5–2.2 mm long; caryopses 0.6–0.8 mm long . . . . 28. *E. curtipedicellata*

30. Pedicels (1)1.5–12 mm long, divergent or appressed; lemmas 1.1–1.4 mm long; caryopses 0.5–0.6 mm long . . . . 29. *E. silveana*

27. Plants not rhizomatous; disarticulation often below the lemmas, the paleas persistent, sometimes below the florets and the the paleas falling with the lemmas and caryopses.

31. Panicles 0.3–0.6 cm wide, spicate, dense; spikelets with 2–3 florets . . . . 30. *E. spicata*

31. Panicles 1–45 cm wide, ovate to obovate or elliptic, open to somewhat condensed and glomerate; spikelets with 1–45 florets.

32. Caryopses with shallowly to deeply grooved adaxial surfaces, rectangular-prismatic to ellipsoid, ovoid, or obovoid in overall shape [for opposite lead, see p. 69].

33. Caryopses strongly dorsally compressed, translucent, mostly light brown, bases sometimes greenish.

34. Lemmas 1.8–3 mm long; panicles 16–35(40) cm long, (4)8–24 cm wide; blades 12–50(65) cm long; caryopses 1–1.7 mm long; ligules 0.6–1.3 mm long . . . . 9. *E. curvula*

34. Lemmas 1.4–1.7 long; panicles 6–18 cm long, 2–8 cm wide; blades 2–12 cm long; caryopses 0.4–0.8 mm long; ligules 0.3–0.5 mm long.

35. Plants without woolly hairs at the base; glumes unequal; lateral lemma veins not green, inconspicuous throughout; spikelets 0.8–1.2 mm; naturalized in the southwestern United States . . . . 10. *E. lehmanniana*

35. Plants with conspicuous, woolly hairs at the base; glumes subequal; lateral lemma veins green, conspicuous basally, obscure near the lemma apices; spikelets 1.3–2 mm wide; in the *Flora* region, known only from waste areas near a woolen mill in South Carolina . . . . 11. *E. setifolia*

33. Caryopses laterally compressed, terete, or slightly dorsally compressed, usually opaque, usually reddish-brown.

36. Lateral veins of the lemmas conspicuous, often greenish, the lemmas strongly keeled.

37. Panicles 2–8 cm wide, contracted to somewhat open, narrowly oblong to narrowly lanceolate; primary branches appressed or diverging up to 30° from the rachises; lemmas with punctate glands along the keels; pedicels 1–7 mm long, appressed; plants native to Africa, in the *Flora* region, known only from waste areas near sheep and cattle lots in South Carolina and Alabama . . . . 31. *E. plana*

37. Panicles 4–30 cm wide, open, ovate to oblong; primary branches diverging 10–90° from the rachises; lemmas without punctate glands on the keels; pedicels 0.4–22 mm long, usually diverging, occasionally appressed; plants native to the southern United States.
    38. Pedicels with a glandular band; culms with a glandular band below the nodes; anthers 0.3–0.5 mm long; restricted to southern Texas . . . . 32. *E. swallenii*
    38. Pedicels and culms without glandular bands; anthers 0.6–1.6 mm long; often found outside southern Texas.
        39. Glumes 1.8–4 mm long, the upper glumes generally equaling or exceeding the lower lemmas; spikelets 1.5–3.6 mm wide, greenish-yellow with a reddish-purple tinge; lemmas 2.2–3.5 mm long; caryopses 0.8–1.3 mm long . . . . 33. *E. trichodes*
        39. Glumes 1.1–2.2 mm long, the upper glumes exceeded by the lower lemmas; spikelets 1–2 mm wide, plumbeous; lemmas 2–2.6 mm long; caryopses 0.6–0.8 mm long . . . . 34. *E. palmeri*

36. Lateral veins of the lemmas inconspicuous, the lemmas sometimes only weakly keeled.
    40. Lemmas 1.2–1.8 mm long; culms 30–70 cm tall.
        41. Culms with a glandular ring below the nodes; bases of primary panicle branches with a glandular band; panicles 2–7 cm wide; pedicels glandular; known, in the *Flora* region, only from a few collections at Canton, Maryland . . . . 8. *E. trichophora*
        41. Culms without a glandular ring below the nodes; bases of primary panicle branches without a glandular band; pedicels not glandular at the base; panicles 5–27 cm wide; plants known from many parts of the southern United States.
            42. Spikelets 1.1–1.6 mm wide, uniformly plumbeous; sheaths sometimes densely pilose dorsally and on the collars; distal margins of the lemmas not hyaline . . . . . . 35. *E. polytricha*
            42. Spikelets 0.5–1(1.3) mm wide, plumbeous to reddish-purple; sheaths usually glabrous dorsally and on the collars; distal margins of the lemmas hyaline . . . . 36. *E. lugens*
    40. Lemmas 1.6–3 mm long; culms (30)40–110(120) cm tall.
        43. Spikelets greenish with a purplish tinge, with 2–6 florets; blades 25–60 cm long, 3–11 mm wide, flat to loosely involute; sheaths densely hirsute with papillose-based hairs on the collar, back, and base . . . . 37. *E. hirsuta*
        43. Spikelets olivaceous to plumbeous, with (3)5–12 florets; blades (4)10–35 cm long, 1–3.8 mm wide, involute or flat; sheaths never with papillose-based hairs, sometimes villous over the back.
            44. Lemmas 1.6–2.2 mm long; anthers 0.5–0.8 mm long, purplish . . . . 38. *E. intermedia*
            44. Lemmas 2–3 mm long; anthers 0.6–1.7 mm long, purplish to yellowish.
                45. Caryopses 0.8–1.6 mm long; lemmas 2.4–3 mm long . . . . 39. *E. erosa*
                45. Caryopses 0.6–0.8 mm long; lemmas 2–2.6 mm long . . . . 34. *E. palmeri*

32. Caryopses not grooved on the adaxial surfaces, ellipsoid, subellipsoid, ovoid, obovoid, globose, to pyriform in overall shape [for opposite lead, see p. 68].

46. Anthers 2.
  47. Panicles 15–45 cm wide, open, diffuse, broadly ovate to obovate; primary branches lax; pedicels 0.5–35(50) mm long, the lower pedicels longer or shorter than the spikelets.
    48. Spikelets with appressed pedicels; only the terminal pedicels of each branch longer than the spikelets; disarticulation usually in the rachilla beneath the florets . . . . 40. *E. refracta*
    48. Spikelets with divergent pedicels; all pedicels usually longer than the spikelets; disarticulation below the lemmas, the paleas persistent . . . . 41. *E. elliottii*
  47. Panicles (1)2–17 cm wide, contracted to open, narrowly ovate to oblong; primary branches stiff; pedicels absent or 0.3–6 mm long, always shorter than the spikelets.
    49. Spikelets 2.4–5 mm wide; glumes 1.7–4 mm long; lemmas 2–6 mm long, the apices usually acuminate or attenuate . . . . 42. *E. secundiflora*
    49. Spikelets 1–2.4 mm wide; glumes 1–2.2 mm long; lemmas 1.1–2.5 mm long, the apices usually acute, occasionally acuminate.
      50. Spikelets 0.7–1.4 mm wide; anthers 0.2–0.3 mm long; caryopses flattened ventrally . . . . 43. *E. prolifera*
      50. Spikelets 1.3–2.4 mm wide; anthers (0.2)0.3–0.7 mm long; caryopses rounded, not flattened ventrally.
        51. Terminal panicles 1–3.5 cm wide, contracted, condensed into glomerate lobes; primary branches 0.8–3 cm long . . . . 44. *E. elongata*
        51. Terminal panicles (1)2–17 cm wide, open to contracted; primary branches 1–15 cm long.
          52. Plants without axillary panicles; terminal panicles 15–45 cm long; blades (8)12–40 cm long, 2–5 mm wide, flat to involute; caryopses 0.6–0.8 mm long, striate, obovoid to ellipsoid . . . . 45. *E. bahiensis*
          52. Plants usually with axillary panicles, these contracted and partially to completely enclosed by the subtending sheaths; terminal panicles 5–15 cm long; blades 4–8(18) cm long, 1–2 mm wide, usually involute; caryopses 0.5–0.6 mm long, smooth, globose . . . . 46. *E. scaligera*
46. Anthers 3.
  53. Primary panicle branches not rebranched; proximal spikelets on each branch sessile or subsessile, the pedicels shorter than 0.4 mm . . . . 47. *E. sessilispica*
  53. Primary panicle branches usually with secondary branches; proximal spikelets on each branch usually pedicellate, the pedicels longer than 0.4 mm.
    54. Spikelets 1.3–2 mm long, with 1–3 florets; lemmas 0.8–1.2 mm long . . . . 48. *E. airoides*
    54 Spikelets 2–19 mm long, with 2–22 florets; lemmas 1.2–2.4 mm long.
      55. Spikelets 2–4.5(5) mm long.
        56. Blades 25–60 cm long, 3–11 mm wide; lemmas 1.6–2.4 mm long; spikelets 1–1.7 mm wide; sheaths densely hirsute, with papillose-based hairs on the base, back, and collar . . . . 37. *E. hirsuta*
        56. Blades 4–22 cm long, 1–3.5 mm wide; lemmas 1.2–1.8 mm long; spikelets 0.5–1.3 mm wide; sheaths

sometimes hirsute, at least partially, but the hairs never papillose-based . . . . 36. *E. lugens*

55. Spikelets 4–19 mm long.
    57. Spikelets with 10–22 florets; caryopses terete to laterally compressed, opaque, uniformly reddish brown . . . . 49. *E. atrovirens*
    57. Spikelets with 3–12(14) florets; caryopses dorsally compressed, translucent, greenish over the embryo.
        58. Lemmas 1.8–3 mm long; panicles 16–35(40) cm long, (4)8–24 cm wide; blades 12–50(65) cm long; caryopses 1–1.7 mm long; ligules 0.6–1.3 mm long . . . . 9. *E. curvula*
        58. Lemmas 1.4–1.7 long; panicles 6–18 cm long, 2–8 cm wide; blades 2–12 cm long; caryopses 0.4–0.8 mm long; ligules 0.3–0.5 mm long.
            59. Plants without woolly hairs on the base; glumes unequal; lateral lemma veins not green, inconspicuous throughout; spikelets 0.8–1.2 mm; naturalized in the southwestern United States . . . . 10. *E. lehmanniana*
            59. Plants with conspicuous, woolly hairs on the base; glumes subequal; lateral lemma veins green, conspicuous basally, obscure near the lemma apices; spikelets 1.3–2 mm wide; in the *Flora* region, known only from waste areas near a woolen mill in South Carolina . . . . 11. *E. setifolia*

1. **Eragrostis ciliaris** (L.) R. Br. [p. 73]
GOPHERTAIL LOVEGRASS

Plants annual; tufted, without innovations, without glands. **Culms** (3)9–75 cm, erect or geniculate in the lower portion, not rooting at the lower nodes, glabrous. **Sheaths** hairy on the margins and at the apices, hairs to 4 mm; **ligules** 0.2–0.5 mm; **blades** 1.8–12(15) cm long, 2–5 mm wide, usually flat, occasionally involute, glabrous or ciliate basally. **Panicles** 1.7–15 cm long, 0.2–5 cm wide, cylindrical, contracted or open, branches usually forming glomerate lobes, sometimes more open, often interrupted in the lower portion; **primary branches** 0.4–4 cm, appressed or diverging to 50° from the rachises; **pulvini** usually glabrous, occasionally sparsely pilose; **pedicels** 0.1–1 mm, erect, shorter than the spikelets, glabrous. **Spikelets** 1.8–3.2 mm long, 1–2 mm wide, elliptical-ovate to ovate-lanceolate, yellowish-brown, sometimes with a purple tinge, with 6–11 florets; **disarticulation** basipetal, glumes persistent. **Glumes** ovate to lanceolate, keels scabridulous, veins commonly green, apices acute; **lower glumes** 0.7–1.2 mm; **upper glumes** 1–1.6 mm; **lemmas** 0.8–1.3 mm, elliptical-ovate to lanceolate, membranous, keels scabridulous, lateral veins evident, apices obtuse to acute; **paleas** 0.8–1.3 mm, membranous, keels prominently ciliate, cilia 0.2–0.8 mm, apices obtuse to acute; **anthers** 2, 0.1–0.3 mm, purplish. **Caryopses** 0.4–0.5 mm, ovoid, reddish-brown. $2n$ = 20, 40.

Eragrostis ciliaris is native to the paleotropics. It is naturalized in parts of the United States, growing along roadsides, on waste sites, in xerothermic vegetation, and sometimes in saline habitats, at 0–200 m. It may be more widespread than indicated.

1. Panicles 0.2–1.5 cm wide, contracted, the branches mostly appressed to the rachises, congested, forming glomerate lobes; spikelets densely packed . . . . .var. *ciliaris*
1. Panicles 1.5–5 cm wide, open, the branches spreading 20–50° from the rachises; spikelets widely separated from each other . . . . var. *laxa*

Eragrostis ciliaris (L.) R. Br. var. **ciliaris**

Panicles 0.2–1.5 cm wide, contracted; **branches** mostly appressed to the rachises, forming glomerate lobes. **Spikelets** densely packed.

Eragrostis ciliaris var. *ciliaris* is more common than *E. ciliaris* var. *laxa* in the *Flora* region.

Eragrostis ciliaris var. **laxa** Kuntze

Panicles 1.5–5 cm wide, open; **branches** spreading 20–50° from the rachises. **Spikelets** widely separated from each other.

Eragrostis ciliaris var. *laxa* grows in five counties of Florida, the Caribbean Islands, and the Yucatan Peninsula, Mexico.

2. **Eragrostis cumingii** Steud. [p. 73]
CUMING'S LOVEGRASS

Plants annual; cespitose, without innovations, without glands. **Culms** 10–40(50) cm, erect to prostrate, sometimes geniculate, branching profusely from near the base, glabrous below the nodes. **Sheaths** sparsely hairy at the apices, hairs to 2.5 mm; **ligules** 0.1–0.2 mm; **blades** 3–10(12) cm long, 1–3 mm wide, flat to involute, sparsely pilose on the basal ½, scabridulous distally. **Panicles** 5–20 cm long, 2–8 cm wide, narrowly ovate, open, with 6–10 primary branches; **primary branches** 1–6 cm, widely spaced, axes trigonous, diverging to 90° from the rachises, densely spikelet-bearing to the base; **pulvini** sparsely pilose; **pedicels** 0.4–1(2) mm, stout, straight, flattened. **Spikelets** 5–12(18) mm long, 1.4–2.4 mm wide, linear-lanceolate, chartaceous, stramineous to greenish with reddish-purple tinges, with 12–42 florets; **disarticulation** acropetal. **Glumes** subequal in length, 1.2–1.9 mm, narrowly lanceolate to lanceolate, membranous; **lower glumes** narrower than the upper glumes; **lemmas** 1.3–2 mm, lanceolate to ovate, chartaceous, lateral veins conspicuous, greenish, apices acute; **paleas** 1–1.6 mm, hyaline, keels ciliate, cilia 0.1–0.2 mm, apices obtuse to acute; **anthers** 3, 0.1–0.2 mm, purplish. **Caryopses** 0.4–0.6 mm, ovoid, laterally compressed, finely striate, light brown. $2n = 40$.

Eragrostis cumingii is native to southeast Asia and Australia. Within the *Flora* region, it has become established in Florida, growing in waste places and along roadsides in sandy or gravelly soils, at 0–150 m.

3. **Eragrostis amabilis** (L.) Wight & Arn. *ex* Nees [p. 73]
JAPANESE LOVEGRASS

Plants annual; cespitose, without innovations, without glands. **Culms** 5–40 cm, erect, glabrous, occasionally with oblong glandular areas below the nodes. **Sheaths** hairy on the distal margins and at the apices, hairs to 4 mm, stiff; **ligules** 0.2–0.3 mm; **blades** 2–8 cm long, 2–4 mm wide, flat to involute, abaxial surfaces smooth, adaxial surfaces scabridulous, bases occasionally with papillose-based hairs. **Panicles** 4–15 cm long, 1–7 cm wide, cylindrical to narrowly ovate, open, rachises sometimes glandular below the nodes; **primary branches** 0.5–4 cm, diverging 20–100° from the rachises; **pulvini** sparsely pilose; **pedicels** 1–4(7) mm, as long as or longer than the spikelets, mostly pendent, lax, terete. **Spikelets** (1)1.5–2.5 mm long, 0.9–1.4 mm wide, ovate to oblong, reddish-purple to greenish, with 4–8 florets; **disarticulation** basipetal, glumes persistent. **Glumes** ovate, hyaline, keeled, veins commonly green; **lower glumes** 0.4–0.7 mm; **upper glumes** 0.7–1 mm; **lemmas** 0.7–1.1 mm, ovate to broadly oblong, membranous, lateral veins usually greenish, apices truncate to obtuse; **paleas** 0.6–1.1 mm, hyaline, keels ciliate, cilia 0.3–0.5 mm, apices obtuse to truncate; **anthers** 3, about 0.2 mm, purplish. **Caryopses** 0.3–0.5 mm, ellipsoid, translucent, light brown. $2n = 20$.

Eragrostis amabilis is native to the Eastern Hemisphere. It is now naturalized in the southeastern United States, growing in open areas such as cultivated fields, forest margins, and roadsides at 0–200 m.

4. **Eragrostis hypnoides** (Lam.) Britton, Sterns & Poggenb. [p. 75]
TEEL LOVEGRASS, ÉRAGROSTIDE HYPNOÏDE

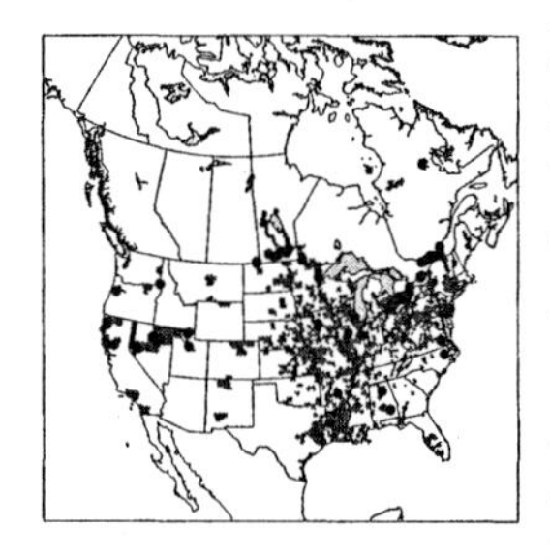

Plants annual; stoloniferous, mat-forming, without innovations, without glands. **Culms** decumbent and rooting at the lower nodes, erect portion (2)5–12(20) cm, often branched, glabrous or hairy on the lower internodes. **Sheaths** pilose on the margins, collars, and at the apices, hairs 0.1–0.6 mm; **ligules** 0.3–0.6 mm; **blades** 0.5–2.5 cm long, 1–2 mm wide, flat to involute, abaxial surfaces glabrous, adaxial surfaces appressed pubescent, hairs about 0.2 mm. **Panicles** terminal and axillary, 1–3.5 cm long, 0.7–2.5 cm wide, ovate, open to somewhat congested; **primary branches** 0.1–0.5 cm, appressed to strongly divergent, glabrous; **pulvini** sparsely pilose or glabrous; **pedicels** 0.2–1 mm, ciliate. **Spikelets** 4–13 mm long, 1–1.5 mm wide, linear-oblong, often arcuate, loosely imbricate, greenish-yellow to purplish, with 12–35 florets; **disarticulation** acropetal, paleas persistent. **Glumes** linear-lanceolate to lanceolate, hyaline; **lower glumes** 0.4–0.7 mm; **upper glumes** 0.8–1.2 mm; **lemmas** 1.4–2 mm, ovate, strongly 3-veined, veins greenish, apices acuminate; **paleas** 0.7–1.2 mm, hyaline, keels scabridulous, apices acute to obtuse; **anthers** 2, 0.2–0.3 mm, brownish. **Caryopses** 0.3–0.5 mm, ellipsoid, somewhat translucent, light brown. $2n = 20$.

Eragrostis hypnoides grows along muddy or sandy shores of lakes and rivers and in moist, disturbed sites, at 10–1600 m. It is native to the Americas, extending from southern Canada to Argentina.

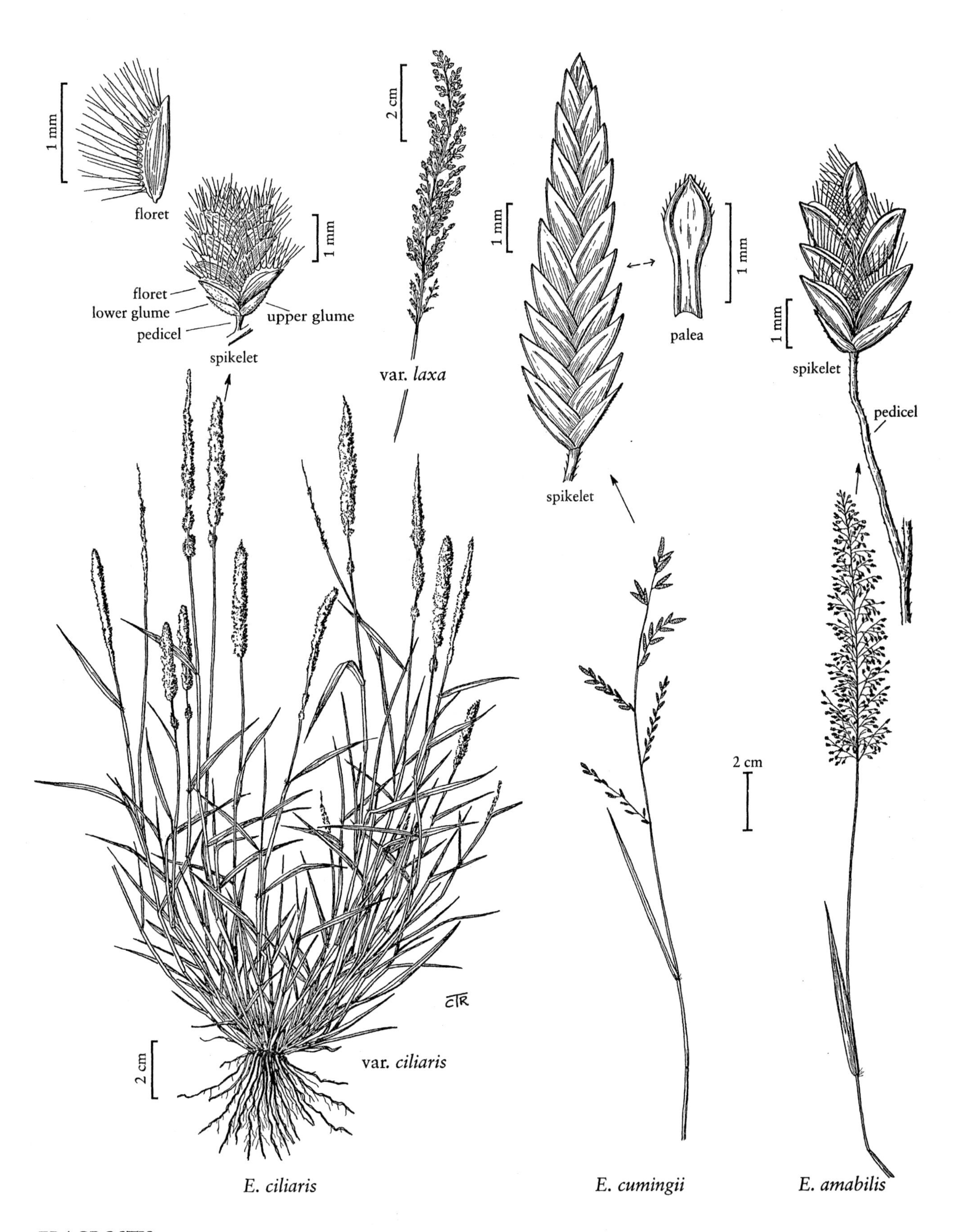
1 mm
floret
1 mm
floret
lower glume
pedicel
upper glume
spikelet
2 cm
var. laxa
1 mm
palea
1 mm
spikelet
1 mm
spikelet
pedicel
2 cm
CTR
2 cm
var. ciliaris
E. ciliaris
E. cumingii
E. amabilis

5. Eragrostis reptans (Michx.) Nees [p. 75]
CREEPING LOVEGRASS

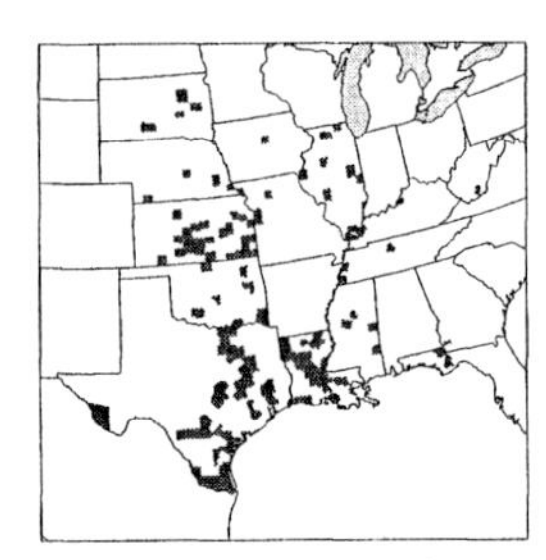

Plants annual; unisexual, pistillate and staminate plants morphologically similar; mat-forming, without innovations, without glands. **Culms** rooting at the lower nodes, erect portion 5–20 cm, glabrous, pilose, or villous, particularly below the panicles. **Sheaths** mostly scabrous, margins sometimes with 0.1–0.4 mm hairs; **ligules** 0.1–0.6 mm; **blades** 1–4 cm long, 1–4.5 mm wide, flat or conduplicate, abaxial surfaces glabrous, adaxial surfaces appressed pubescent, hairs about 0.2 mm. **Panicles** terminal, 1–3 cm long, 0.6–2.5 cm wide, ovate, contracted, exerted or partially included in the upper leaf sheaths, rachises somewhat viscid, pilose or glabrous; **primary branches** 0.5–1.5 cm, appressed to the rachises, each terminating in a spikelet; **pulvini** sparsely pilose or glabrous; **pedicels** 0.2–2 mm, shorter than the spikelets, glabrous or hairy. **Spikelets** 5–26 mm long, 1.5–4.7 mm wide, linear to ovate, greenish to stramineous, with 16–60 florets; **disarticulation** in the pistillate florets basipetal, the lemmas falling separately, staminate spikelets not or tardily disarticulating. **Glumes** unequal, ovate, hyaline, glabrous or sparsely hirsute; **lower glumes** 0.8–1.6 mm, 1-veined; **upper glumes** 1.5–2.5 mm, 1–3-veined; **lemmas** (1.5)1.8–4 mm, ovate, hyaline to membranous, lateral veins conspicuous, greenish, apices acute to acuminate, sometimes prolonged into a mucro, mucros to 0.4 mm; **paleas** 0.7–3.8 mm, hyaline, about ½ as long as the lemmas in pistillate florets, as long as the lemmas in staminate florets, keels scabridulous; **anthers** 3, 1.4–2.2 mm, reddish to yellowish. **Caryopses** 0.4–0.6 mm, ellipsoid, somewhat laterally compressed, smooth, light reddish-brown. $2n = 60$.

Eragrostis reptans grows in wet sand, gravel, and clay soils along rivers and lake margins from the United States to northern Mexico, at 0–400 m, frequently with *Cynodon dactylon* and *Heliotropium*. It flowers from April through November.

6. Eragrostis japonica (Thunb.) Trin. [p. 75]
POND LOVEGRASS

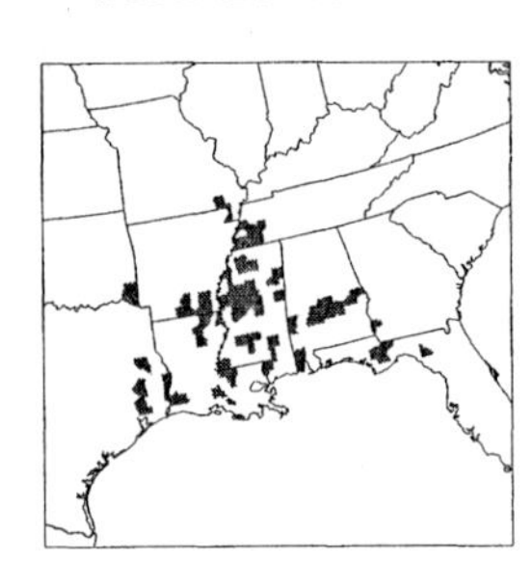

Plants annual; cespitose, without innovations, without glands. **Culms** 25–100(115) cm, erect or geniculate, lower portion glabrous and shiny. **Sheaths** glabrous at the apices and on the upper margins; **ligules** 0.4–0.6 mm, scarious, glabrous; **blades** (4)15–20(25) cm long, 1.5–6 mm wide, flat, sometimes auriculate, abaxial surfaces glabrous, smooth, adaxial surfaces scabridulous, **Panicles** 15–40 cm long, 0.8–5 cm wide, lanceoloid, contracted, interrupted near the base; **primary branches** 2–10 cm, appressed or diverging to 30° from the rachises, spikelet-bearing to near the base; **pulvini** glabrous; **pedicels** 0.5–1.5 mm, sinuous. **Spikelets** 2.2–3.8 mm long, 0.8–1.3 mm wide, oblong to narrowly lanceolate, yellowish-brown to whitish and hyaline, with 4–12 florets; **disarticulation** basipetal, rachillas and glumes persistent. **Glumes** subequal, 0.6–1 mm, ovate to ovate-lanceolate, hyaline; **upper glumes** without a midvein; **lemmas** 0.9–1.2 mm, ovate, hyaline, lateral veins conspicuous basally, greenish, apices acute; **paleas** 0.6–0.8 mm, hyaline, keels smooth basally, scabridulous distally, apices acute, often bifid; **anthers** 2, 0.1–0.2 mm, whitish to light brown. **Caryopses** 0.3–0.4 mm, obovoid, smooth, reddish-brown. $2n = 20$.

Eragrostis japonica is native to the tropics of the Eastern Hemisphere; it is now established in moist areas along rivers and streams in the southern portion of the contiguous United States, usually in sandy soils, at 0–200 m.

7. Eragrostis cylindriflora Hochst. [p. 77]

Plants annual; cespitose, without innovations. **Culms** 20–70 cm, erect or decumbent, with a ring of glands below the nodes, glabrous. **Sheaths** often with oblong glands below the collar, usually glabrous, rarely pilose, hairs to 1.5 mm, papillose-based, sometimes both glandular and with papillose-based hairs; **ligules** 0.4–0.6 mm; **blades** 3–15 cm long, 2–4 mm wide, flat to involute, abaxial surfaces glabrous or hairy, hairs to 3 mm, papillose-based. **Panicles** 5–22 cm long, 2.6–10 cm wide, open, oblong, usually diffuse; **primary branches** 2–9 cm, appressed or diverging to 80° from the rachises, lowest branches whorled, naked below; **pulvini** glabrous or scattered pilose; **pedicels** 0.8–7 mm, divergent. **Spikelets** 2–7 mm long, 1–1.5 mm wide, oblong, plumbeous to greenish-gray, with 3–14 florets; **disarticulation** acropetal, paleas persistent. **Glumes** subequal, 1.4–2 mm, narrowly ovate, hyaline; **lemmas** 1.4–1.7 mm, ovate, membranous, lateral veins inconspicuous, apices obtuse to acute; **paleas** 1.2–1.6 mm, hyaline, keels scabridulous, apices obtuse; **anthers** 3, 0.7–1 mm, yellowish. **Caryopses** 0.5–1.1 mm, ellipsoid to obovoid, dorsally compressed, with a shallow, broad adaxial groove, smooth, mostly translucent, mostly light brown, bases greenish. $2n$ = unknown.

Eragrstis cylindriflora is native to Africa. It is not established in the *Flora* region, but has been collected from a disturbed site in Canton, Maryland.

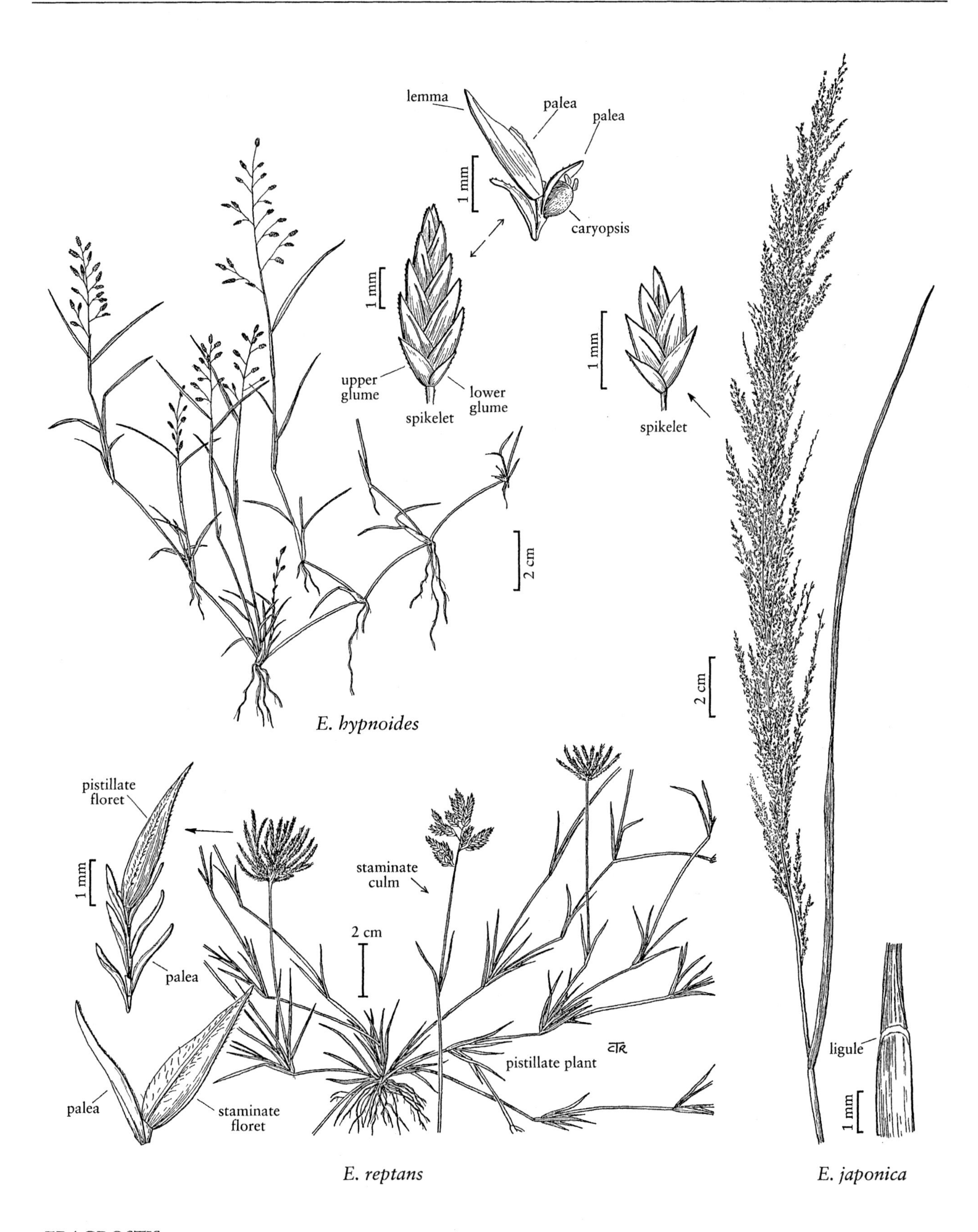
lemma
palea
palea
1 mm
caryopsis
1 mm
upper glume
lower glume
spikelet
1 mm
spikelet
2 cm
2 cm
E. hypnoides
pistillate floret
1 mm
palea
staminate culm
2 cm
pistillate plant
palea
staminate floret
ligule
1 mm
E. reptans
E. japonica

8. Eragrostis trichophora Coss. & Durieu [p. 77]

Plants perennial; cespitose, stoloniferous, forming innovations near the base. **Culms** 30–70 cm, erect, geniculate, or prostrate, often rooting at the lower nodes, glabrous, with a ring of glands below the nodes. **Sheaths** glabrous or with scattered papillose-based hairs over most of the surface, apices pilose, hairs 1–4 mm, a ring of oblong glands sometimes present below the collar; **ligules** 0.2–0.5 mm; **blades** 1.4–10 cm long, 2–3 mm wide, flat to involute, sparsely hairy with papillose-based hairs, abaxial surfaces often with glandular dots. **Panicles** 5–20 cm long, 2–7 cm wide, narrowly ovate, open; **primary branches** 2–7 cm, diverging 10–70° from the rachises, lowest branches whorled, naked proximally, bases with a glandular band; **pulvini** hairy; **pedicels** 0.3–3.3 mm, glandular. **Spikelets** 4–5.4 mm long, 1–1.5 mm wide, linear-lanceolate, plumbeous to greenish-gray, with 3–5 florets; **disarticulation** acropetal, paleas persistent. **Glumes** subequal, 1.4–1.8 mm, ovate-lanceolate, membranous; **lemmas** 1.5–1.8 mm, ovate, membranous, often hyaline, lateral veins inconspicuous, apices obtuse to acute; **paleas** 1.3–1.7 mm, hyaline, bases not projecting beyond the lemmas, apices obtuse; **anthers** 3, 0.7–1 mm, purplish. **Caryopses** 0.6–0.8 mm, ovoid, terete to dorsally compressed, shallowly grooved adaxially, translucent, mostly whitish to light brown, bases often greenish. 2*n* = unknown.

Eragrostis trichophora is native to Africa, where it often grows in moist, disturbed or overgrazed sites. It has been collected from disturbed sites at Canton, Maryland.

9. Eragrostis curvula (Schrad.) Nees [p. 77]
WEEPING LOVEGRASS

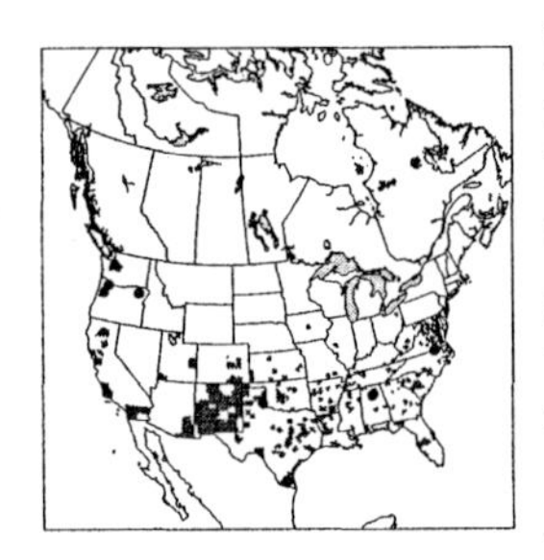

Plants perennial; cespitose, forming innovations at the basal nodes, without glands. **Culms** (45)60–150 cm, erect, glabrous or glandular. **Sheaths** with scattered hairs, hairs to 9 mm; **ligules** 0.6–1.3 mm; **blades** 12–50(65) cm long, 1–3 mm wide, flat to involute, abaxial surfaces glabrous, sometimes scabridulous, adaxial surfaces with scattered hairs basally, hairs to 7 mm. **Panicles** 16–35(40) cm long, (4)8–24 cm wide, ovate to oblong, open; **primary branches** 3–14 cm, diverging 10–80° from the rachises; **pulvini** glabrous or not; **pedicels** 0.5–5 mm, appressed, flexible. **Spikelets** 4–8.2(10) mm long, 1.2–2 mm wide, linear-lanceolate, plumbeous to yellowish, with 3–10 florets; **disarticulation** irregular to acropetal, proximal rachilla segments persistent. **Glumes** lanceolate, hyaline; **lower glumes** 1.2–2.6 mm; **upper glumes** 2–3 mm; **lemmas** 1.8–3 mm, ovate, membranous, lateral veins conspicuous, apices acute; **paleas** 1.8–3 mm, hyaline to membranous, apices obtuse; **anthers** 3, 0.6–1.2 mm, reddish-brown. **Caryopses** 1–1.7 mm, ellipsoid to obovoid, dorsally compressed, adaxial surfaces with a shallow, broad groove or ungrooved, smooth, mostly translucent, light brown, bases often greenish. 2*n* = 40, 50.

Eragrostis curvula is native to southern Africa. It is often used for reclamation because it provides good ground cover but, once introduced, it easily escapes. In the *Flora* region, it grows on rocky slopes, at the margins of woods, along roadsides, and in waste ground, at 20–2400 m, usually in pine-oak woodlands, and yellow pine and mixed hardwood forests.

10. Eragrostis lehmanniana Nees [p. 80]
LEHMANN'S LOVEGRASS

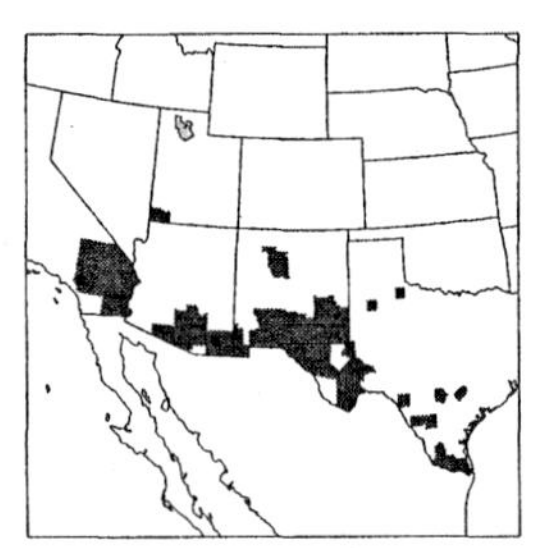

Plants perennial; cespitose, forming innovations at the basal nodes, without glands. **Culms** (20)40–80 cm, erect, commonly geniculate, sometimes rooting at the lower nodes, glabrous, lower portions sometimes scabridulous. **Sheaths** sometimes shortly silky pilose basally, hairs less than 2 mm, apices sparsely hairy, hairs to 3 mm; **ligules** 0.3–0.5 mm, ciliate; **blades** 2–12 cm long, 1–3 mm wide, flat to involute, glabrous, abaxial surfaces sometimes scabridulous, adaxial surfaces scabridulous. **Panicles** 7–18 cm long, 2–8 cm wide, oblong, open; **primary branches** 1–8 cm, appressed or diverging to 40° from the rachises; **pulvini** glabrous; **pedicels** 0.5–4 mm, diverging or appressed, flexible. **Spikelets** 5–12(14) mm long, 0.8–1.2 mm wide, linear-lanceolate, plumbeous to stramineous, with 4–12(14) florets; **disarticulation** irregular to basipetal, paleas usually persistent. **Glumes** oblong to lanceolate, membranous; **lower glumes** 1–1.5 mm; **upper glumes** 1.3–2 mm; **lemmas** 1.5–1.7 mm, ovate, membranous, lateral veins inconspicuous, apices acute to obtuse; **paleas** 1.4–1.7 mm, obtuse; **anthers** 3, 0.6–0.9 mm, yellowish. **Caryopses** 0.6–0.8 mm, ellipsoid to obovoid, dorsally compressed, sometimes with a shallow adaxial groove, smooth, translucent, mostly light brown, embryo region dark brown with a greenish ring. 2*n* = 40, 60.

Eragrostis lehmanniana is native to southern Africa, where it grows in sandy, savannah habitats. It was introduced for erosion control in the southern United States, where it often displaces native species. In the *Flora* region, it grows in sandy flats, along roadsides, on calcareous slopes, and in disturbed areas, at 200–1830 m. It is commonly found in association with *Larrea tridentata*, *Opuntia*, *Quercus*, *Juniperus*, and *Bouteloua gracilis*.

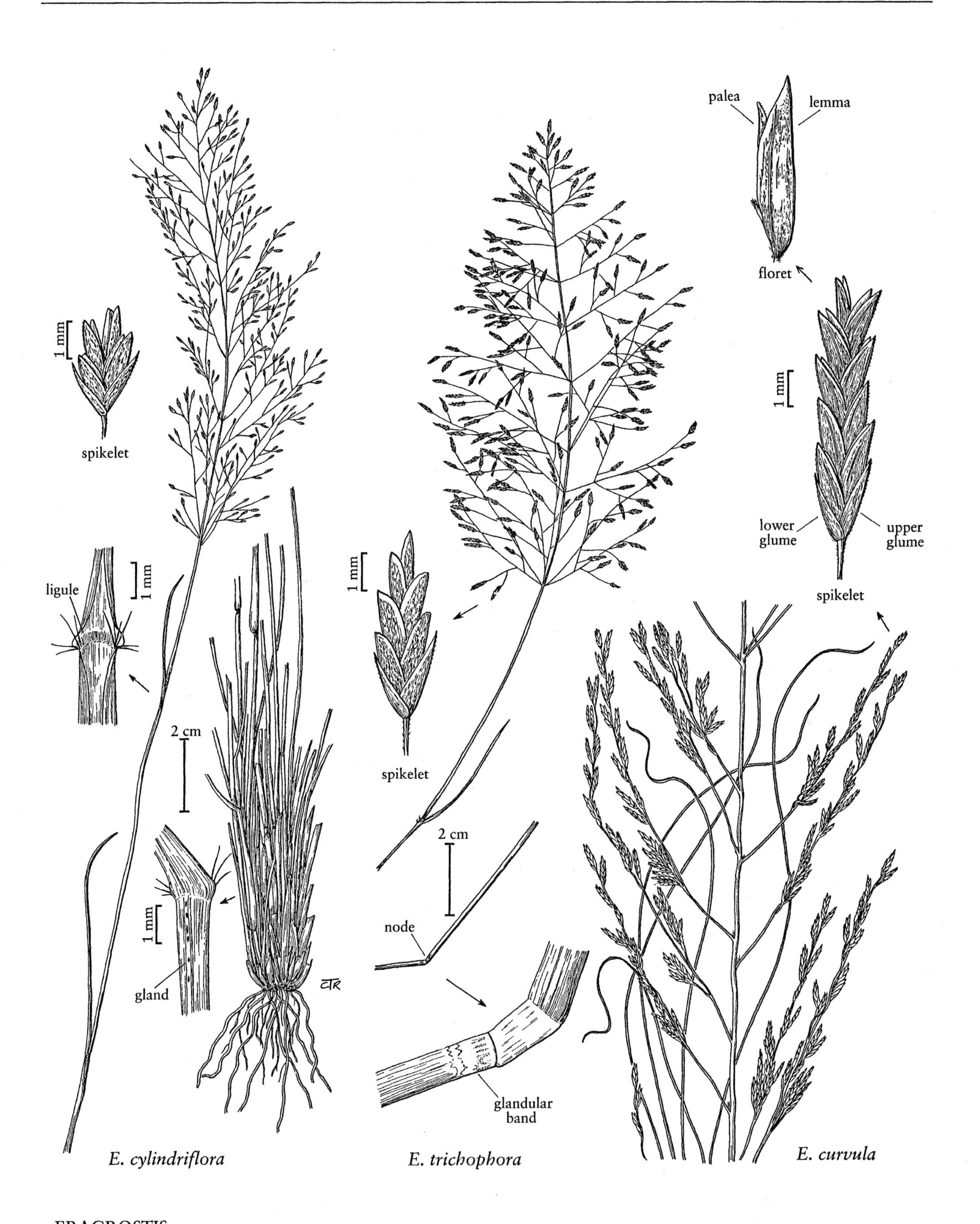

palea
lemma
floret
1 mm
spikelet
1 mm
spikelet
ligule
1 mm
2 cm
1 mm
gland
1 mm
spikelet
2 cm
node
glandular
band
1 mm
lower glume
upper glume
spikelet
E. cylindriflora
E. trichophora
E. curvula

11. **Eragrostis setifolia** Nees
NEVERFAIL LOVEGRASS

Plants perennial; cespitose, with innovations and thickened or knotty plant bases; plant bases with woolly hairs, hairs coarse, to 2 cm. **Culms** 12–60 cm, erect, glabrous or hairy near the base, often shiny below the nodes. **Sheaths** glabrous, summits shortly pilose, hairs to 0.5 mm; **ligules** 0.3–0.5 mm, ciliate; **blades** 3–11 cm long, 0.7–2 mm wide, involute or flat, glabrous abaxially, scabridulous adaxially. **Panicles** 6–11 cm long, 2–4 cm wide, narrowly ovate, loosely contracted; **primary branches** 1–3 cm, compact, appressed or diverging to 30° from the rachises, sometimes naked near the base; **pulvini** glabrous; **pedicels** 0.2–3 mm, diverging, scabridulous. **Spikelets** 3–6(15) mm long, 1.3–2 mm wide, linear lanceolate, stramineous, with 9–30 florets; **disarticulation** irregular or basipetal, paleas persistent. **Glumes** subequal, 1–1.3 mm, ovate, membranous, apices obtuse to acute; **lemmas** 1.4–1.6 mm, ovate, membranous, glabrous, lateral veins conspicuous, green, sometimes obscure towards the apices, apices obtuse; **paleas** 1.4–1.6 mm, hyaline, apices truncate, ciliolate; **anthers** 3, 0.5–0.8 mm, yellowish. **Caryopses** 0.4–0.5 mm, oblong-ellipsoid, strongly dorsally compressed, usually with a shallow dorsal groove, smooth to finely striate, mostly light brown, bases often greenish. $2n$ = unknown.

Eragrostis setifolia is an Australian species that was collected around the Santee Wool Combing Mill, Jamestown, Berkeley County, South Carolina, in 1958. It is not known to have spread from that location. There is no illustration of the species because it was a late addition to the treatment. For digital images, see http://herbarium.usu.edu/webmanual/.

12. **Eragrostis mexicana** (Hornem.) Link [p. 80]
MEXICAN LOVEGRASS

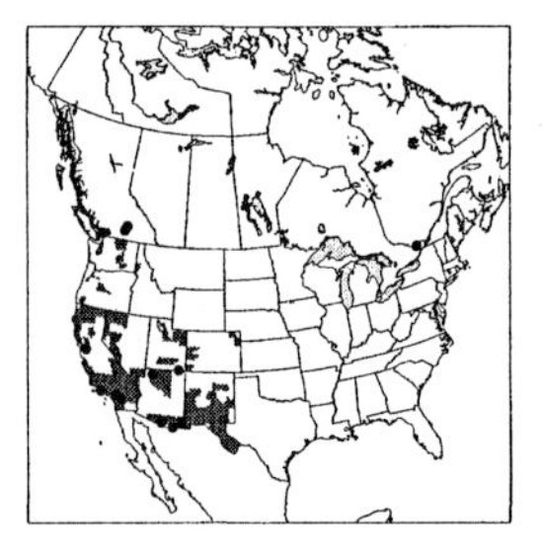

Plants annual; cespitose, without innovations. **Culms** 10–130 cm, erect, sometimes geniculate, glabrous, sometimes with a ring of glandular depressions below the nodes. **Sheaths** sometimes with glandular pits, pilose near the apices and on the collars, hairs to 4 mm, papillose-based; **ligules** 0.2–0.5 mm, ciliate; **blades** 5–25 cm long, 2–7(9) mm wide, flat, abaxial surfaces glabrous, adaxial surfaces scabridulous, occasionally pubescent near the base. **Panicles** (5)10–40 cm long, (2)4–18 cm wide, ovate, rachises angled and channeled; **primary branches** 3–12(15) cm, solitary to whorled, appressed or diverging to 80° from the rachises; **secondary branches** somewhat appressed; **pulvini** glabrous; **pedicels** 1–6(7) mm, almost appressed to narrowly divergent, stiff. **Spikelets** (4)5–10(11) mm long, 0.7–2.4 mm wide, ovate to linear-lanceolate, gray-green to purplish, with 5–11(15) florets; **disarticulation** acropetal. **Glumes** subequal, 0.7–2(2.3) mm, ovate to lanceolate, membranous; **lemmas** 1.2–2.4 mm, ovate, membranous, glabrous or with a few hairs, gray-green, lateral veins evident, often greenish, apices acute; **paleas** 1–2.2 mm, hyaline, keels scabrous, apices obtuse to truncate; **anthers** 3, 0.2–0.5 mm, purplish. **Caryopses** 0.5–0.8(1) mm, ovoid to rectangular-prismatic, laterally compressed, shallowly to deeply grooved on the adaxial surface, striate, reddish-brown, distal ⅔ opaque. $2n = 60$.

Eragrostis mexicana grows along roadsides, near cultivated fields, and in disturbed open areas, at 100–3000 m. It is native to the Americas, its native range extending from the southwestern United States through Mexico, Central and northern South America, to Argentina. Within the *Flora* region, it has been introduced beyond its native range, often becoming an established part of the flora.

1. Spikelets ovate to oblong in outline, 1.5–2.4mm wide; lower glumes 1.2–2.3 mm long; sum of the spikelet width and lower glume length 2.7–4.7 mm; culms and sheaths sometimes with glandular depressions . . . subsp. *mexicana*
1. Spikelets linear to linear-lanceolate, 0.7–1.4 wide; lower glumes 0.7–1.7 mm long; sum of the spikelet width and lower glume length 1.5–3.1 mm; culms and sheaths without glandular depressions . . . . subsp. *virescens*

Eragrostis mexicana (Hornem.) Link subsp. **mexicana**

Culms and **sheaths** sometimes with glandular depressions. **Spikelets** 1.5–2.4 mm wide, ovate to oblong in outline. **Lower glumes** 1.2–2.3 mm; **sum of spikelet width and lower glume length** 2.7–4.7 mm.

Eragrostis mexicana subsp. *mexicana* grows from Ontario through the midwestern United States to California, South Carolina, and Texas and southwards to Mexico.

Eragrostis mexicana subsp. **virescens** (J. Presl) S.D. Koch & Sánchez Vega

Culms and **sheaths** without glandular depressions. **Spikelets** 0.7–1.4 mm wide, linear to linear-lanceolate. **Lower glumes** 0.7–1.7 mm; **sum of spikelet width and lower glume length** 1.5–3.1 mm.

Eragrostis mexicana subsp. *virescens* has a disjunct distribution, growing in California and western Nevada and, in South America, from Ecuador to Chile, southern Brazil, and northern Argentina. It has also been found, as an introduction, at various other locations in North America, including eastern North America.

13. **Eragrostis frankii** C.A. Mey. *ex* Steud. [p.80]
SANDBAR LOVEGRASS, ÉRAGROSTIDE DE FRANK

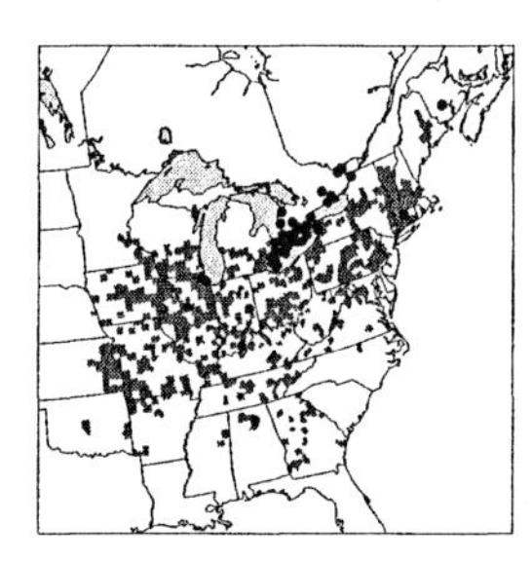

Plants annual; cespitose, without innovations. **Culms** 10–50 cm, erect to geniculate, glabrous, often with glandular pits below the nodes. **Sheaths** mostly glabrous, apices hirsute, hairs to 4 mm, often also with glandular pits; **ligules** 0.2–0.5 mm, ciliate; **blades** (2)4–10(21) cm long, 1–4 mm wide, flat to involute, glabrous abaxially, scabridulous adaxially. **Panicles** 4–20 cm long, less than ½ the height of the plants, 2–10(14) cm wide, narrowly elliptic, open; **primary branches** 2–6 cm, compact, diverging 20–70° from the rachises, capillary, sometimes with glandular pits, naked basally; **pulvini** glabrous; **pedicels** 1.5–5 mm, divergent. **Spikelets** (1.7)2–4(5.6) mm long, 1–2(2.5) mm wide, broadly ovate to lanceolate, plumbeous to reddish-purple, with 3–6 florets; **disarticulation** acropetal, paleas persistent. **Glumes** narrowly lanceolate to lanceolate, hyaline; **lower glumes** 1–1.5 mm; **upper glumes** 1–1.8 mm; **lemmas** 1.1–1.6 mm, broadly ovate, membranous, lateral veins inconspicuous, apices acute; **paleas** 1–1.5 mm, hyaline, keels scabridulous, apices obtuse; **anthers** 2 or 3, 0.2–0.3 mm, purplish. **Caryopses** 0.4–0.7 mm, ovoid to rectangular-prismatic, striate, reddish-brown, adaxial surfaces flat or shallowly grooved, distal ⅔ opaque. $2n$ = 40, 80.

Eragrostis frankii is native in the central and eastern United States, but it has been found, as an introduction, in southern Ontario, and appears to be increasingly common in the northeastern United States. It grows in moist meadows, along streams and sand bars, in forest openings, and along roadsides, at 5–1500 m, usually in association with *Pinus*, *Quercus*, *Acer*, and *Fagus grandiflora*. The record from Santa Fe County, New Mexico, is based on a specimen collected by Fendler in 1847; there are no other collections from the state. Fendler's specimens seem to represent either an accidental introduction that did not become established or a labeling error.

Eragrostis frankii is similar to *E. capillaris*, but differs in its frequent possession of glandular pits, its flat or more shallowly grooved caryopses, shorter pedicels, and glabrous sheath margins, and in having panicles that are usually less than half as long as the culms.

14. **Eragrostis capillaris** (L.) Nees [p. 82]
LACEGRASS

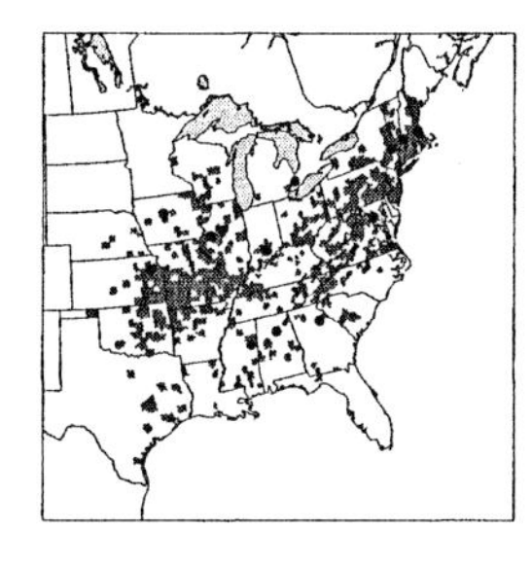

Plants annual; tufted, without innovations, without glands. **Culms** (15)20–50(60) cm, erect, glabrous, often shiny below the nodes. **Sheaths** pilose along the margins, apices hirsute, hairs to 7 mm; **ligules** 0.2–0.5 mm, ciliate; **blades** (6)8–20(30) cm long, (1)2–5 mm wide, flat, abaxial surfaces smooth, glabrous, adaxial surfaces scabridulous, with long scattered hairs. **Panicles** (10)15–45(55) cm long, (7)10–25 cm wide, to ⅔ the height of the plants, elliptic to ovate, open, rachises without glandular pits; **primary branches** (2)5–15 cm, diverging 20–90° from the rachises, capillary, naked basally; **pulvini** glabrous; **pedicels** (4)5–25 mm, divergent, scabridulous. **Spikelets** (1.4)2–5 mm long, 1–1.3(1.4) mm wide, ovate to lanceolate, plumbeous, occasionally reddish-purple, with 2–5(7) florets; **disarticulation** acropetal, paleas persistent. **Glumes** narrowly lanceolate to lanceolate, hyaline; **lower glumes** 1–1.2 mm, narrower than the upper glumes; **upper glumes** 1.2–1.4 mm; **lemmas** 1.2–1.7 mm, broadly ovate, membranous, keels scabridulous, lateral veins inconspicuous, apices acute; **paleas** 1.2–1.6 mm, hyaline, keels almost smooth to scabrous, scabridities to 0.1 mm, apices acute to obtuse; **anthers** 3, 0.2–0.3 mm, reddish-brown. **Caryopses** 0.4–0.7 mm, ovoid to rectangular-prismatic, adaxial surfaces deeply grooved, striate, bases reddish-brown, distal ⅔ opaque. $2n$ = 50, 100.

Eragrostis capillaris is native to the eastern portion of the *Flora* region. It grows in open, dry, sandy riverbanks, floodplains, rocky roadsides, and gravel pits, at 150–1500 m, usually in association with *Pinus*, *Quercus*, *Carrya*, and *Liquidambar styraciflua*. Its range extends into northeastern Mexico.

Eragrostis capillaris resembles *E. frankii*, but differs in its lack of glandular pits, deeply grooved caryopses, longer pedicels, pilose sheath margins, and larger panicles. The two species are sympatric over much of the eastern United States.

15. **Eragrostis lutescens** Scribn. [p. 82]
SIXWEEKS LOVEGRASS

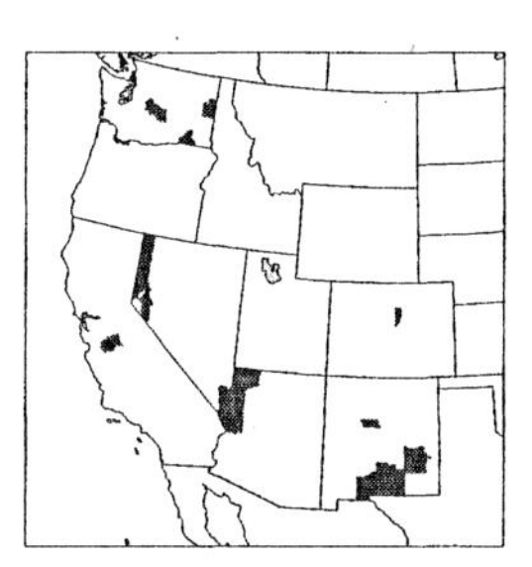

Plants annual; tufted, without innovations. **Culms** (2)6–25 cm, usually erect, sometimes decumbent, glabrous, with elliptical, yellowish, glandular pits below the nodes. **Sheaths** with elliptical glandular pits, sparsely hairy at the throat, hairs to 2 mm; **ligules** 0.2–0.5

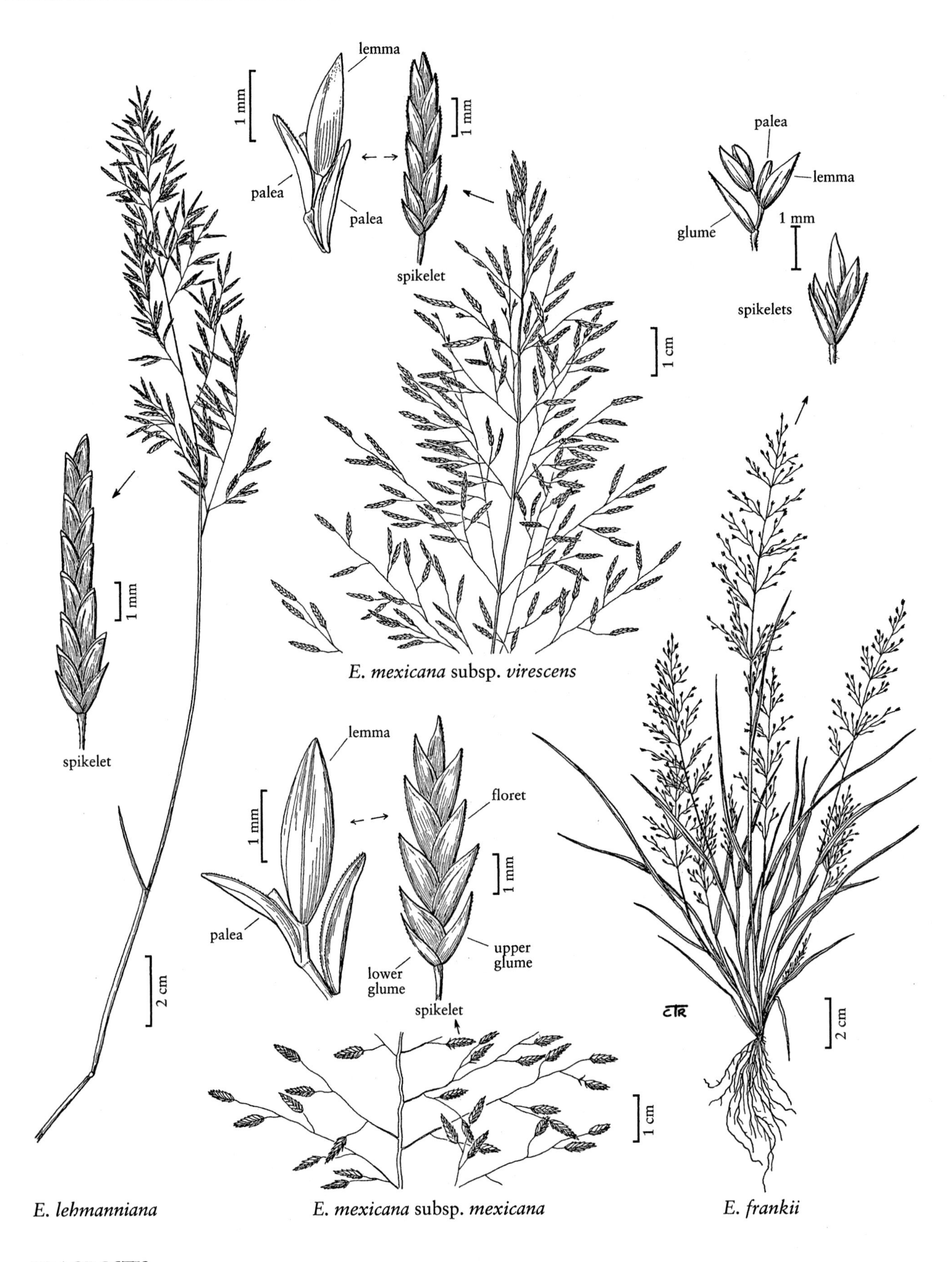

E. lehmanniana *E. mexicana* subsp. *mexicana* *E. frankii*

mm, ciliate; **blades** 2–12 cm long, 1–3 mm wide, flat to involute, abaxial surfaces scabridulous, bases with glandular pits. **Panicles** terminal, 4–10(15) cm long, 0.5–2 cm wide, narrowly elliptic, contracted, dense; **primary branches** alternate, usually appressed, occasionally diverging to 30° from the rachises, rachises and branches with glandular pits; **pulvini** glabrous; **pedicels** 1.4–10 mm, appressed or divergent. **Spikelets** 3.6–7.5 mm long, 1.2–2 mm wide, narrowly ovate, light yellowish, occasionally mottled with reddish-purple, with 6–11(14) florets; **disarticulation** acropetal, paleas persistent. **Glumes** subequal, ovate to lanceolate, hyaline; **lower glumes** (0.7)0.9–1.4 mm; **upper glumes** 1.2–1.8 mm; **lemmas** 1.5–2.2 mm, ovate, subhyaline, stramineous, veins greenish and conspicuous, apices acute; **paleas** 1.2–2 mm, hyaline, keels scabridulous, apices obtuse; **anthers** 3, 0.2–0.3 mm, purplish. **Caryopses** 0.5–0.8 mm, pyriform except slightly flattened adaxially, smooth, light brown. 2*n* = unknown.

Eragrostis lutescens grows on the sandy banks of streams and lakes and in moist alkaline flats of the western United States at 300–2000 m. It has not been reported from Mexico.

16. **Eragrostis pilosa** (L.) P. Beauv. [p. 82]
INDIA LOVEGRASS, ÉRAGROSTIDE POILUE

Plants annual; tufted, without innovations. **Culms** 8–45(70) cm, erect or geniculate, glabrous, occasionally with a few glandular depressions. **Sheaths** mostly glabrous, occasionally glandular, apices hirsute, hairs to 3 mm; **ligules** 0.1–0.3 mm, ciliate; **blades** 2–15(20) cm long, 1–2.5(4) mm wide, flat, abaxial surfaces glabrous, occasionally with glandular pits along the midrib, adaxial surfaces scabridulous. **Panicles** 4–20(28) cm long, 2–15(18) cm wide, ellipsoid to ovoid, diffuse; **primary branches** 1–10 cm, diverging 10–80°(110°) from the rachises, capillary, whorled on the lowest 2 nodes, rarely glandular; **pulvini** glabrous or hairy; **pedicels** 1–10 mm, flexible, appressed or divergent. **Spikelets** (2)3.5–6(10) mm long, 0.6–1.4 mm wide, linear-oblong to narrowly ovate, plumbeous, with (3)5–17 florets; **disarticulation** acropetal, paleas tardily deciduous, rachillas persisting longer than the paleas. **Glumes** narrowly ovate to lanceolate, hyaline; **lower glumes** 0.3–0.6(0.8) mm; **upper glumes** 0.7–1.2(1.4) mm; **lemmas** 1.2–1.8(2) mm, ovate-lanceolate, membranous to hyaline, grayish-green proximally, reddish-purple distally, lateral veins inconspicuous, apices acute; **paleas** 1–1.6 mm, membranous to hyaline, keels scabridulous to scabrous, apices obtuse; **anthers** 3, 0.2–0.3 mm, purplish. **Caryopses** 0.5–1 mm, obovoid to prism-shaped, adaxial surfaces flat, smooth to faintly striate, light brown. 2*n* = 40.

Eragrostis pilosa is native to Eurasia but has become naturalized in many parts of the world. In the *Flora* region, it grows in forest margins and disturbed sites such as roadsides, railroad embankments, gardens, and cultivated fields, at 0–2500 m.

1. Plants with numerous glandular pits scattered over the whole plant, especially on the midribs of the sheaths and blades; lemmas 1.8–2 mm long . . . . var. *perplexa*
1. Plants with a few glandular pits scattered on the culms or without any glandular pits; lemmas 1.2–1.8 mm long . . . . var. *pilosa*

Eragrostis pilosa var. **perplexa** (L.H. Harv.) S.D. Koch

Culms with numerous glandular pits. **Sheaths** with glandular pits; **blades** with glandular pits. **Spikelets** 0.6–1.4 mm wide. **Upper glumes** 1–1.4 mm; **lemmas** 1.8–2 mm. **Caryopses** 0.8–1 mm.

Eragrostis pilosa var. *perplexa* is known from widely scattered locations in Wyoming, North Dakota, Nebraska, Colorado, and northwestern Texas.

Eragrostis pilosa (L.) P. Beauv. var. **pilosa**

Culms with few or no glandular pits; **sheaths** and **blades** without glandular pits. **Spikelets** 0.6–1.3 mm wide. **Upper glumes** 0.7–1.2 mm; **lemmas** 1.2–1.8 mm; **caryopses** 0.5–0.9 mm.

Eragrostis pilosa var. *pilosa* is more common than *E. pilosa* var. *perplexa* in the *Flora* region. Most of the records shown on the map are for this variety.

17. **Eragrostis pectinacea** (Michx.) Nees [p. 84]
TUFTED LOVEGRASS, ÉRAGROSTIDE PECTINÉE

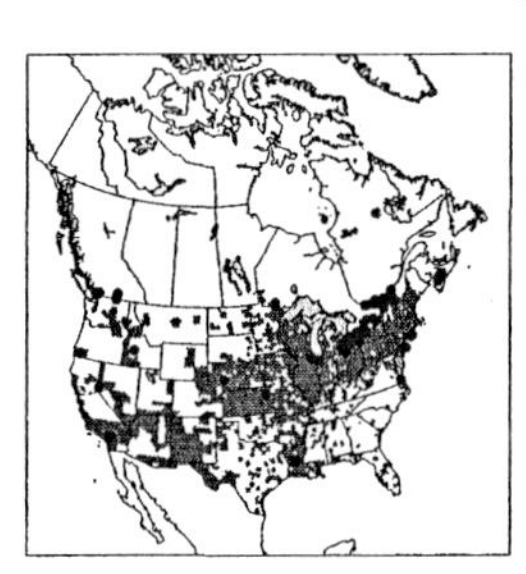

Plants annual; tufted, without innovations, without glandular pits. **Culms** 10–80 cm, erect to geniculate or decumbent below, glabrous. **Sheaths** hirsute at the apices, hairs to 4 mm; **ligules** 0.2–0.5 mm; **blades** 2–20 cm long, 1–4.5 mm wide, flat to involute, abaxial surfaces glabrous and smooth, adaxial surfaces scabridulous. **Panicles** 5–25 cm long, 3–12(15) cm wide, ovoid to pyramidal, usually open, sometimes contracted; **primary branches** 0.6–8.5 cm, appressed or diverging to 80° from the rachises, solitary or paired at the lowest 2 nodes; **pulvini** glabrous or sparsely hairy; **pedicels** 1–7 mm, flexible, appressed to widely divergent, sometimes capillary. **Spikelets** 3.5–11 mm long, 1.2–2.5 mm wide, linear-oblong to narrowly lanceolate, plumbeous, yellowish-brown, or dark reddish-purple, with 6–22 florets; **disarticulation** acropetal, paleas persistent. **Glumes** subulate to ovate-lanceolate, hyaline; **lower glumes**

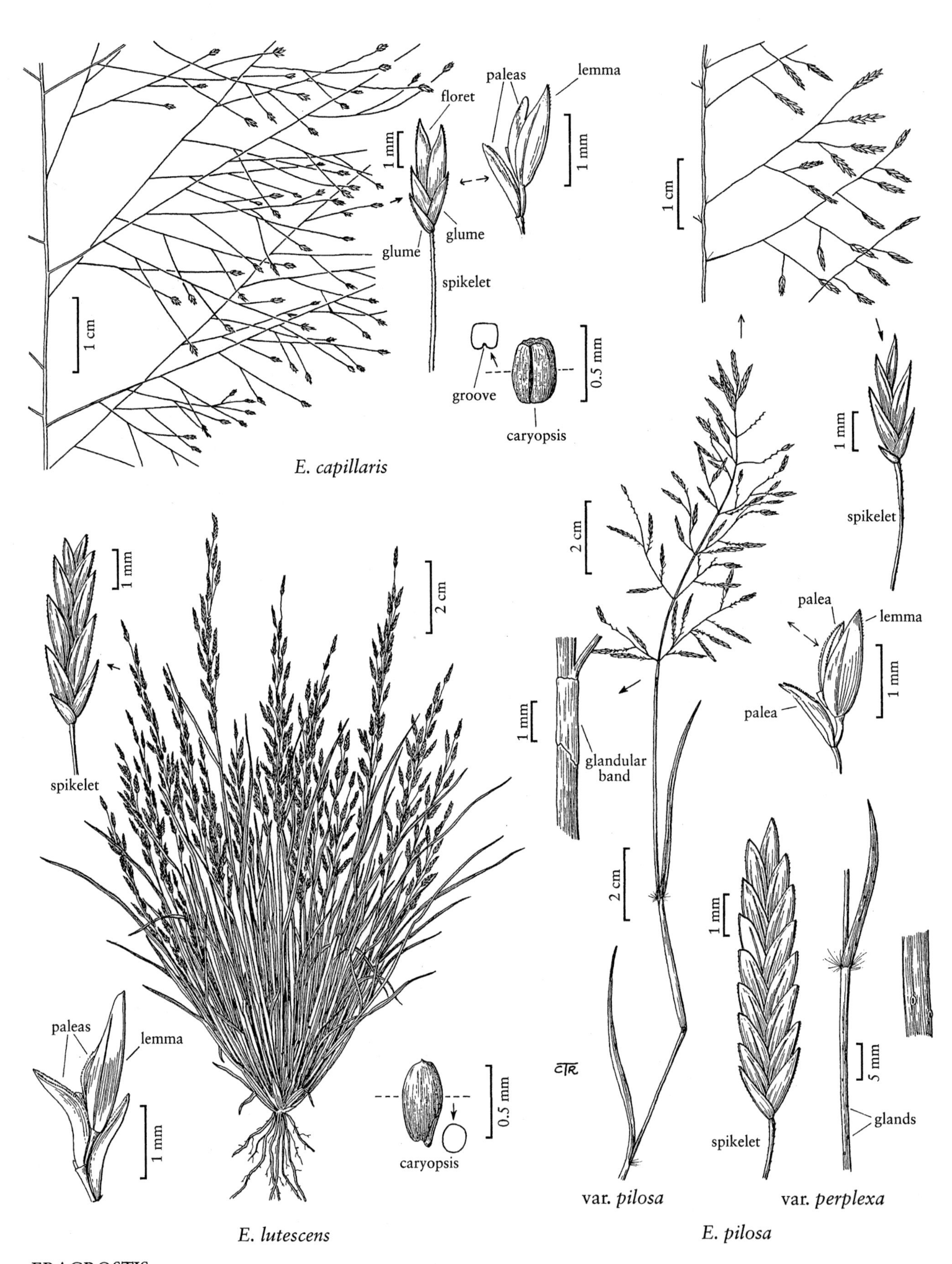
paleas
lemma
floret
1 mm
1 mm
glume
glume
spikelet
1 cm
1 cm
groove
0.5 mm
caryopsis
E. capillaris
1 mm
2 cm
spikelet
2 cm
1 mm
glandular band
spikelet
palea
lemma
1 mm
palea
2 cm
1 mm
5 mm
glands
paleas
lemma
1 mm
0.5 mm
caryopsis
spikelet
var. pilosa
var. perplexa
E. lutescens
E. pilosa

0.5–1.5 mm, at least ½ as long as the adjacent lemmas; **upper glumes** 1–1.7 mm, usually broader than the lower glumes; **lemmas** 1–2.2 mm, ovate-lanceolate, hyaline to membranous, grayish-green proximally, reddish-purple distally, lateral veins moderately conspicuous, apices acute; **paleas** 1–2 mm, hyaline to membranous, keels scabridulous, apices obtuse; **anthers** 3, 0.2–0.7 mm, purplish. **Caryopses** 0.5–1.1 mm, pyriform, slightly laterally compressed, smooth, faintly striate, brownish. $2n = 60$.

Eragrostis pectinacea is native from southern Canada to Argentina. In the *Flora* region, it grows in disturbed sites such as roadsides, railroad embankments, gardens, and cultivated fields, at 0–1200 m.

1. Anthers 0.5–0.7 mm long . . . . var. *tracyi*
1. Anthers 0.2–0.4 mm long.
   2. Pedicels appressed, rarely diverging to 20° from the branches . . . . var. *pectinacea*
   2. Pedicels widely divergent, usually diverging 20–60° from the branches . . . . var. *miserrima*

Eragrostis pectinacea var. **miserrima** (E. Fourn.) Reeder

Pedicels widely divergent, usually spreading 20–60° from the branches. **Anthers** 0.2–0.4 mm.

Eragrostis pectinacea var. *miserrima* grows in the southern United States, from Texas to Florida, and south through the lowland tropics to Brazil. It usually flowers from July–November in the *Flora* region.

Eragrostis pectinacea (Michx.) Nees var. **pectinacea**

Pedicels appressed or diverging to 20° from the branch axes. **Anthers** 0.2–0.4 mm.

Eragrostis pectinacea var. *pectinacea* grows throughout the range of the species, including most of the contiguous United States. Within the *Flora* region, it is most common in the eastern states and usually flowers from July–November.

Eragrostis pectinacea var. **tracyi** (Hitchc.) P.M. Peterson

Pedicels divergent, usually diverging 20–70° from the branches. **Anthers** 0.5–0.7 mm.

Eragrostis pectinacea var. *tracyi* is known from only Lee, Mantee, and Sarasota counties, Florida. It flowers from March–May and August–December in the *Flora* region.

18. **Eragrostis cilianensis** (All.) Vignolo *ex* Janch. [p. 84]
STINKGRASS, ÉRAGROSTIDE FÉTIDE

Plants annual; tufted, without innovations. **Culms** 15–45(65) cm, erect or decumbent, sometimes with crateriform glands below the nodes. **Sheaths** glabrous, occasionally glandular, apices hairy, hairs to 5 mm; **ligules** 0.4–0.8 mm, ciliate; **blades** (1)5–20 cm long, (1)3–5(10) mm wide, flat to involute, abaxial surfaces glabrous, sometimes glandular, adaxial surfaces scabridulous, occasionally also hairy. **Panicles** (3)5–16(20) cm long, 2–8.5 cm wide, oblong to ovate, condensed to open; **primary branches** 0.4–5 cm, appressed or diverging 20–80° from the rachises; **pulvini** glabrous or hairy; **pedicels** 0.2–3 mm, stout, straight, stiff, usually divergent, occasionally appressed. **Spikelets** 6–20 mm long, 2–4 mm wide, ovate-lanceolate, plumbeous, greenish, with 10–40 florets; **disarticulation** below the florets, each floret falling as a unit, rachillas persistent. **Glumes** broadly ovate to lanceolate, membranous, usually glandular; **lower glumes** 1.2–2 mm, usually 1-veined; **upper glumes** 1.2–2.6 mm, often 3-veined; **lemmas** 2–2.8 mm, broadly ovate, membranous, keels with 1–3 crateriform glands, apices obtuse to acute; **paleas** 1.2–2.1 mm, hyaline, keels scabrous, sometimes also ciliate, cilia to 0.3 mm, apices obtuse to acute; **anthers** 3, 0.2–0.5 mm, yellow. **Caryopses** 0.5–0.7 mm, globose to broadly ellipsoid, smooth to faintly striate, not grooved, reddish-brown or translucent. $2n = 20$.

Eragrostis cilianensis is an introduced European species that now grows in disturbed sites such as pastures and roadsides, at 0–2300 m, through most of the contiguous United States and southern Canada. The English name refers to the odor of fresh plants.

19. **Eragrostis barrelieri** Daveau [p. 86]
MEDITERRANEAN LOVEGRASS

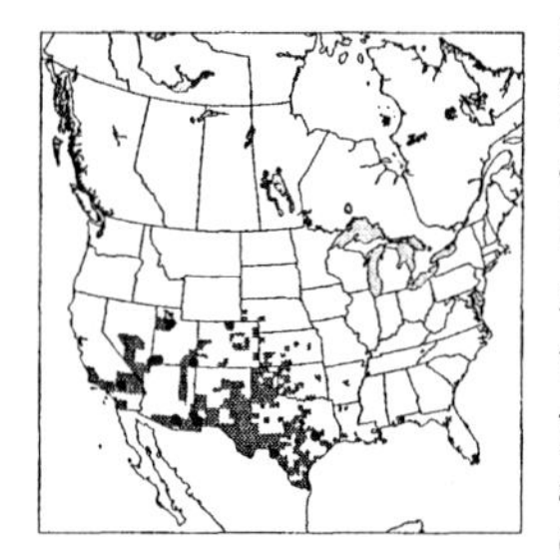

Plants annual; tufted, without innovations. **Culms** (5)10–60 cm, erect or decumbent, much-branched near the base, with a ring of glandular tissue below the nodes, rings often shiny or yellowish. **Sheaths** hairy at the apices, hairs to 4 mm; **ligules** 0.2–0.5 mm, ciliate; **blades** 1.5–10 cm long, 1–3(5) mm wide, flat, abaxial surfaces glabrous, adaxial surfaces glabrous, sometimes scabridulous, occasionally with white hairs to 3 mm, margins without crateriform glands. **Panicles** 4–20 cm long, 2.2–8(10) cm wide, ovate, open to contracted, rachises with shiny or yellowish glandular spots or rings below the nodes; **primary branches** 0.5–6 cm, diverging 20–100° from the rachises; **pulvini** glabrous; **pedicels** 1–4 mm, stout, stiff, divergent, without glandular bands. **Spikelets** 4–7(11) mm long, 1.1–2.2 mm wide, narrowly ovate, reddish-purple to greenish, occasionally grayish, with 7–12(20) florets; **disarticulation** acropetal, paleas persistent. **Glumes** broadly ovate, membranous, 1-veined; **lower glumes** 0.9–1.4 mm; **upper glumes** 1.2–1.6 mm; **lemmas** 1.4–1.8 mm, broadly ovate, membranous, apices acute to obtuse; **paleas** 1.3–1.7 mm, hyaline, keels scabrous, scabridities to 0.1 mm,

anther
paleas
1 mm
floret
lemma
spikelet
1 cm
var. *miserrima*
gland
palea
1 mm
palea
lemma
floret
1 mm
spikelet
2 cm
1 mm
floret
1 mm
spikelet
1 cm
spikelet
anther
1 mm
caryopsis
floret
CR
var. *pectinacea*
var. *tracyi*
E. pectinacea
E. cilianensis

apices obtuse to acute; **anthers** 3, 0.1–0.2 mm, reddish-brown. **Caryopses** 0.4–0.7 mm, ellipsoid, not grooved, smooth to faintly striate, light brown. 2*n* = 40.

Eragrostis barrelieri is a European species that is now naturalized in the *Flora* region, primarily in the southwestern United States. It grows on gravelly roadsides, in gardens, and other disturbed, sandy sites, especially near railroad yards, at 10–2000 m. The ring of glandular tissue is most conspicuous below the upper cauline nodes.

20. **Eragrostis minor** Host [p. 86]
LITTLE LOVEGRASS, ÉRAGROSTIDE FAUX-PÂTURIN

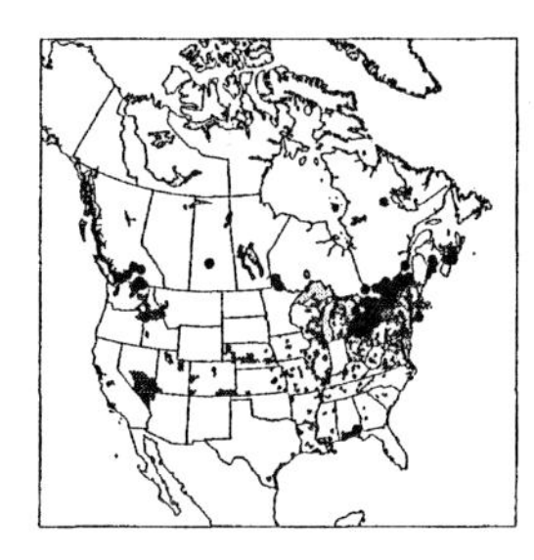

Plants annual; tufted, without innovations. **Culms** 10–45 cm, erect to decumbent, sometimes with a ring of glandular tissue below the nodes. **Sheaths** sometimes glandular on the midveins, hairy at the apices, hairs to 4 mm; **ligules** 0.2–0.5 mm, ciliate; **blades** 1.5–10 cm long, 1–3(4) mm wide, flat, glabrous or sparsely white-hairy, margins sometimes with crateriform glands. **Panicles** 4–20 cm long, 2.2–8(10) cm wide, ovate, open to contracted, rachises sometimes with glandular spots or pits below the nodes, rarely with a glandular ring, glands usually dull, greenish-gray to stramineous; **primary branches** 0.5–6 cm, diverging 20–100° from the rachises; **pulvini** glabrous or hairy; **pedicels** 1–4 mm, stiff, straight, divergent, usually with a distal ring of crateriform glands. **Spikelets** 4–7(11) mm long, 1.1–2.2 mm wide, narrowly ovate, mostly reddish-purple to greenish, occasionally grayish, with 7–12(20) florets; **disarticulation** acropetal, paleas persistent. **Glumes** broadly ovate, membranous; **lower glumes** 0.9–1.4 mm; **upper glumes** 1.2–1.6 mm; **lemmas** 1.4–1.8 mm, broadly ovate, membranous, keels occasionally with 1–2 crateriform glands, apices acute to obtuse; **paleas** 1.3–1.7 mm, hyaline, keels smooth or scabridulous, scabridities to 0.1 mm, apices obtuse to acute; **anthers** 2, 0.2–0.3 mm, reddish-brown. **Caryopses** 0.4–0.7 mm, ellipsoid, not grooved, striate, light brown. 2*n* = 40.

Eragrostis minor is a European species that now grows in gravelly roadsides and disturbed sites, especially near railroad yards, at 20–1600 m in southern Canada and the contiguous United States.

21. **Eragrostis unioloides** (Retz.) Nees *ex* Steud. [p. 86]
CHINESE LOVEGRASS

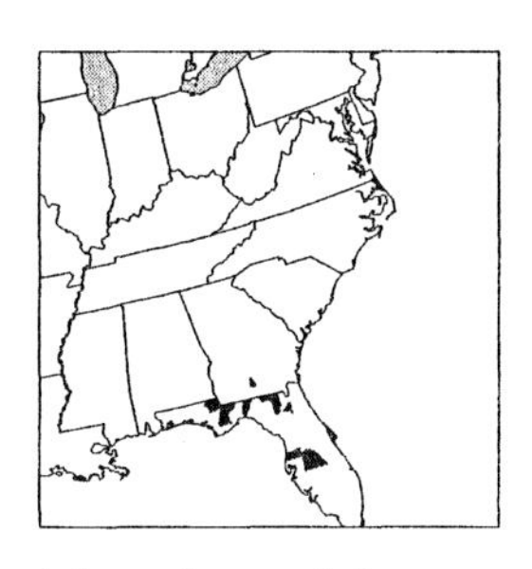

Plants annual; tufted, without innovations, without glands. **Culms** 10–50 cm, erect, glabrous below the nodes. **Leaves** mostly basal; **sheaths** mostly glabrous, apices pilose, hairs 0.4–3 mm; **ligules** 0.1–0.2 mm, ciliate; **blades** (1.8)5–12 cm long, 2–6 mm wide, flat to involute, abaxial surfaces glabrous, adaxial surfaces scabridulous and glabrous or sparsely hairy, hairs appressed. **Panicles** 6–15 cm long, 0.5–7 cm wide, ovate, open to contracted; **primary branches** 0.2–5 cm, appressed or diverging up to 70° from the rachises, glabrous; **pulvini** glabrous; **pedicels** 2–8 mm, glabrous. **Spikelets** 4–8(10) mm long, (1.6)2–4 mm wide, ovate-lanceolate, loosely imbricate, straw-colored to purplish, with 12–42 florets; **disarticulation** acropetal, paleas not persistent, rachillas persistent. **Glumes** ovate-lanceolate to lanceolate, hyaline to membranous; **lower glumes** 1–1.8 mm; **upper glumes** 1–2.2 mm; **lemmas** 1.5–1.9 mm, broadly ovate, membranous, lateral veins raised, apices obtuse to acute; **paleas** 1.4–1.9 mm, hyaline, keels scabridulous, apices acute to obtuse; **anthers** 2, 0.2–0.4 mm, purplish. **Caryopses** 0.6–0.9 mm, ellipsoid, laterally compressed, not grooved, smooth, light brown. 2*n* = 20, ca. 30.

Eragrostis unioloides is an Asian species that is now established in the southeastern United States, growing along roadsides and in disturbed ground, at 20–150 m.

22. **Eragrostis tef** (Zucc.) Trotter [p. 86]
TEFF

Plants annual; loosely tufted, without innovations, without glands. **Culms** 25–60 cm, erect, glabrous and shiny. **Sheaths** mostly glabrous, apices hairy, hairs to 5 mm; **ligules** 0.2–0.4 mm, ciliate; **blades** 10–30 cm long, 2–5.5 mm wide, flat to involute, glabrous abaxially, scabridulous adaxially. **Panicles** 10–45 cm long, 2.5–22 cm wide, ovate, open to contracted; **primary branches** 4–17 cm, appressed or diverging up to 50° from the rachises, flexible, naked below; **pulvini** glabrous or hairy, hairs to 5 mm; **pedicels** 2.5–17 mm, appressed or divergent. **Spikelets** 4–11 mm long, 1.3–2.5 mm wide, linear-lanceolate to ovate, stramineous, grayish-green to purplish, with 4–16 florets; **disarticulation** tardy, acropetal, caryopses falling before the glumes and lemmas, paleas persistent. **Glumes** lanceolate, membranous to hyaline; **lower glumes** 1–2 mm; **upper glumes** 1.5–2.8 mm; **lemmas** 1.6–3 mm, lanceolate, membranous, apices acute; **paleas** 1.4–2.2 mm, hyaline, keels scabridulous, apices obtuse; **anthers** 3, 0.2–0.5 mm, purplish. **Caryopses** 0.7–1.3 mm, obovoid, not grooved, smooth, light brown to whitish. 2*n* = 40.

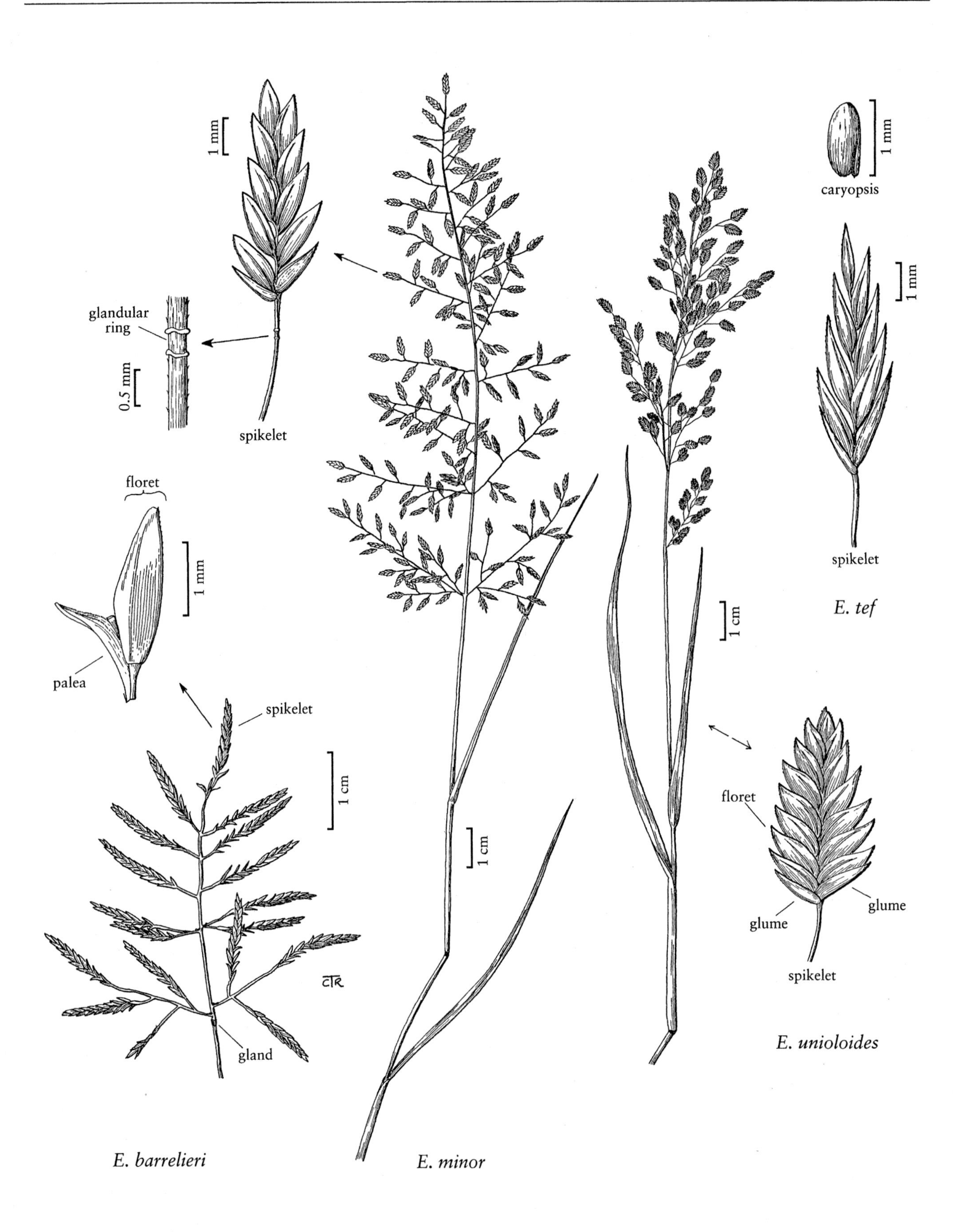
1 mm
glandular ring
0.5 mm
spikelet
floret
1 mm
palea
spikelet
1 cm
gland
E. barrelieri
1 cm
E. minor
1 cm
1 mm
caryopsis
1 mm
spikelet
E. tef
floret
glume
glume
spikelet
E. unioloides

Eragrostis tef is native to northern Africa. In Ethiopia, it is used both as a grain and as fodder for cattle. It is also grown, but not commonly, for these purposes in the *Flora* region and is occasionally found as an escape from cultivation.

23. **Eragrostis gangetica** (Roxb.) Steud. [p. 88]
SLIMFLOWER LOVEGRASS

Plants annual; tufted, without innovations, without glands. **Culms** (12)25–75 cm, usually erect, sometimes geniculate and branched below, glabrous, reddish. **Sheaths** glabrous, apices usually with 0.3–2.2 mm hairs; **ligules** 0.2–0.4 mm, ciliate; **blades** (5)7–17 cm long, 1–3 mm wide, flat to folded basally, involute apically, abaxial surfaces glabrous, adaxial surfaces scabridulous, sometimes with scattered hairs near the base. **Panicles** (6)11–21 cm long, 1–13 cm wide, ovate to somewhat contracted, open; **primary branches** 0.5–12 cm, (12)15–20 per culm, appressed or diverging up to 60° from the rachises, often capillary, naked near the base; **pulvini** glabrous; **pedicels** 0.3–5 mm, mostly appressed. **Spikelets** 2–4.6 mm long, 0.9–2 mm wide, narrowly ovate, greenish-yellow to plumbeous and with a reddish-purple tinge, with 5–15 florets; **disarticulation** acropetal, paleas persistent. **Glumes** lanceolate to ovate, membranous; **lower glumes** 0.4–0.9 mm; **upper glumes** 1–1.3 mm, occasionally 3-veined; **lemmas** 1–1.3 mm, broadly ovate, membranous, often reddish-purple, lateral veins conspicuous, often greenish, apices acute; **paleas** 0.9–1.1 mm, hyaline, keels scabridulous, apices obtuse; **anthers** 2, 0.1–0.2 mm, reddish-purple. **Caryopses** 0.3–0.5 mm, subglobose to obovoid, not grooved, translucent, faintly striate, reddish-brown. $2n = 80$.

Eragrostis gangetica is an Asian species that now grows in the southeastern United States. It can be found in the sandy margins of ponds, roadsides, and ditches, at 0–100 m, usually in association with *Pinus*, *Taxodium distichum*, *Rynchospora*, and *Steinchisma hians*. *Eragrostis gangetica* is similar to *E. bahiensis*, but differs from that species in its annual habit and shorter spikelets, lemmas, anthers, and caryopses.

24. **Eragrostis superba** Peyr. [p. 88]
SAWTOOTH LOVEGRASS

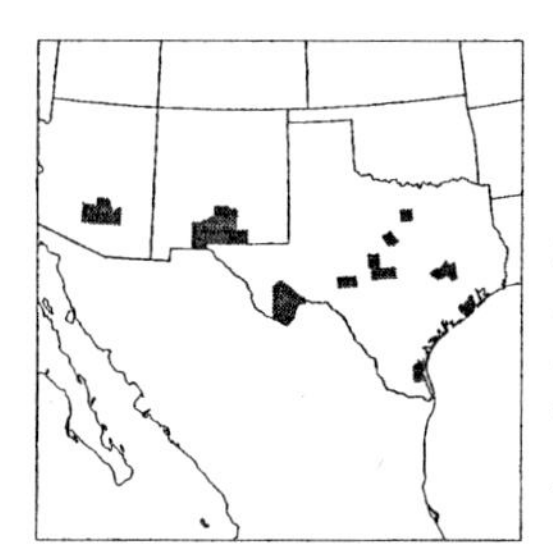

Plants perennial; cespitose, with innovations, without glands. **Culms** 45–95 cm, erect, glabrous. **Sheaths** hairy at the apices and on the margins, hairs to 6 mm; **ligules** 0.5–1.2 mm; **blades** 7–30 cm long, 2.5–7 mm wide, flat to loosely involute, glabrous abaxially, scabrous adaxially, margins sharply scabrous. **Panicles** 10–30 cm long, 1–6 cm wide, oblong, condensed, interrupted below; **primary branches** 1–11 cm, appressed or diverging to 40° from the rachises, naked basally; **pulvini** glabrous; **pedicels** 0.5–25 mm, with a narrow band or abscission line below the apices. **Spikelets** 5.5–16 mm long, 2.7–9 mm wide, ovate, flattened, greenish to stramineous, sometimes with a reddish-purple tinge, with 4–22 florets; **disarticulation** below the glumes, spikelets falling intact. **Glumes** equal, 3–4.5 mm, ovate, chartaceous; **lemmas** 3–5 mm, broadly lanceolate, chartaceous to leathery, lateral veins green, apices acute; **paleas** 3–5 mm, chartaceous to hyaline, keels broadly winged below, forming a wing or tooth on each side that often projects beyond the lemma bases, apices acuminate; **anthers** 3, 1.4–2.8 mm, golden-yellow. **Caryopses** 1–2 mm, ellipsoid, adaxial surfaces flattened, reddish-brown. $2n = 40$.

Eragrostis superba is native to Africa, where it is grown for hay, being fairly palatable and drought resistant. It is also used for erosion control and revegetation. In the *Flora* region, it grows on rocky slopes, in sandy flats, and along roadsides, at 480–1650 m, often with *Acacia*, *Prosopsis*, *Fouquieria splendens*, *Juniperus*, and *Quercus*. The English name is an appropriate description of the leaf blades.

25. **Eragrostis echinochloidea** Stapf [p. 88]
TICKGRASS

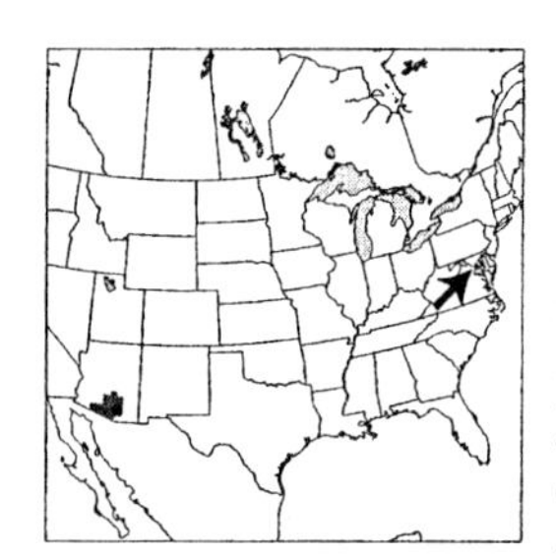

Plants perennial; cespitose, with innovations. **Culms** 30–100 cm, erect to geniculate, with narrow, sunken glandular bands. **Sheaths** sometimes glandular, apices hairy, hairs to 5 mm; **ligules** 0.4–1 mm; **blades** 5–21 cm long, 2–6(7) mm wide, flat to involute, with small crateriform glands on the keels and veins, sparsely pilose adaxially. **Panicles** 4–19 cm long, 0.8–7 cm wide, oblong to ovate, glomerate, spikelets clustered in 1-sided groups; **primary branches** 0.5–7.5 cm, diverging 10–90° from the rachises, angled, sinuous, glandular; **pulvini** hairy, hairs to 2 mm; **pedicels** 0.2–2 mm, stout, erect, without a narrow band or abscission line near the apices. **Spikelets** 2–5 mm long, 2–3.5 mm wide, broadly ovate, greenish, stramineous to plumbeous, with 7–14 florets; **disarticulation** basipetal, glumes persistent. **Glumes** subequal, 1.7–2.2 mm, ovate, membranous, keels with small crateriform glands, apices acute to acuminate; **lemmas** 1.8–2.3 mm, broadly ovate to orbicular, chartaceous, keels with small crateriform glands, apices acute to obtuse; **paleas** 1.7–2.2 mm, chartaceous, each side with a broad wing at the base, wings often projecting beyond the lemma bases, apices acute; **anthers** 3, 0.5–0.9 mm, yellowish. **Caryopses** 0.8–1.1 mm, ellipsoid, reddish-brown. $2n = 30$.

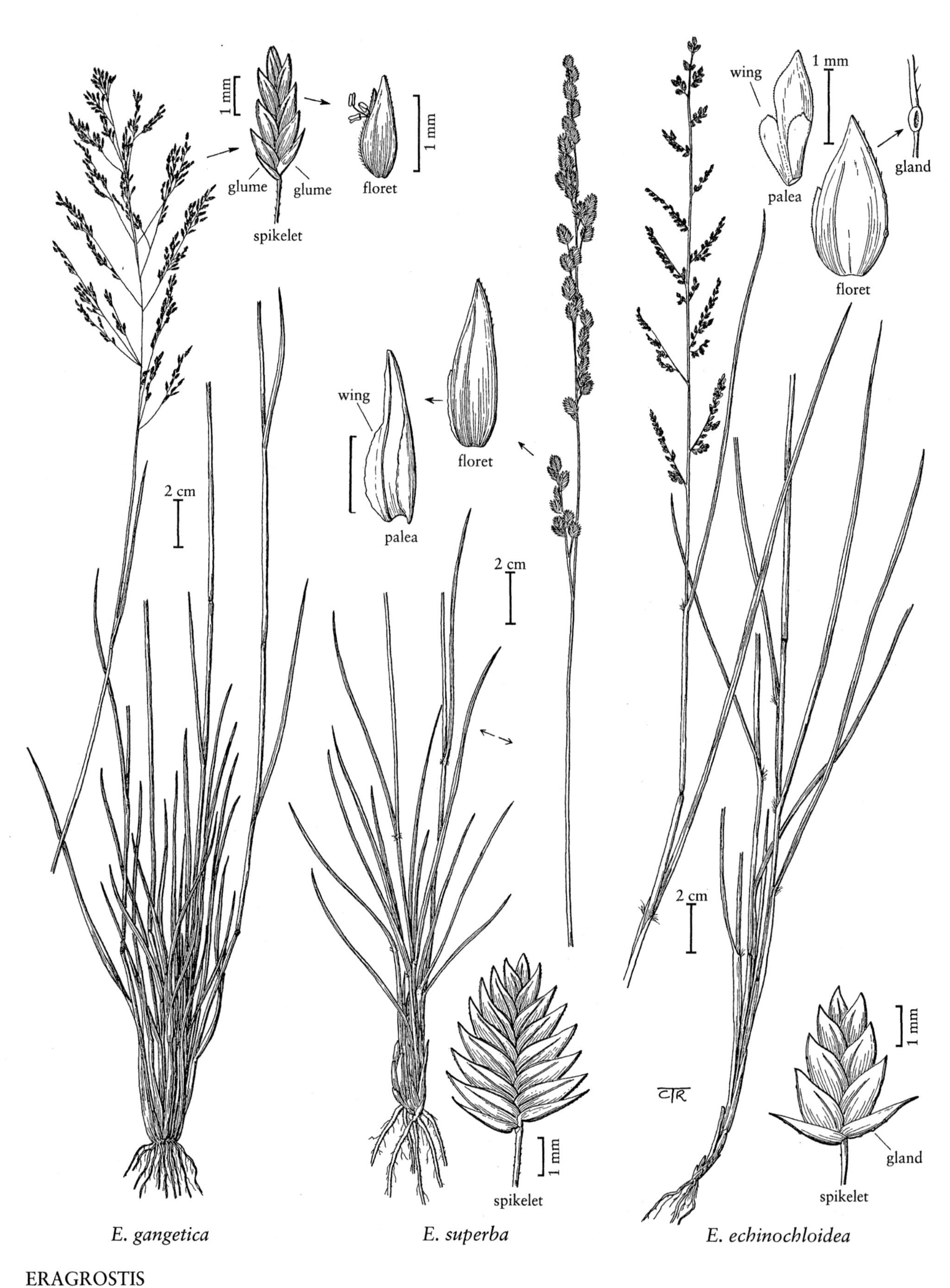

E. gangetica *E. superba* *E. echinochloidea*

ERAGROSTIS

Eragrostis echinochloidea is native to southern Africa. It is now established in Arizona, growing in gravel soils, often along roadsides and in sidewalks, from 700–1000 m. It has also been found in Prince George's County, Maryland.

26. **Eragrostis obtusiflora** (E. Fourn.) Scribn. [p. 90]
ALKALI LOVEGRASS

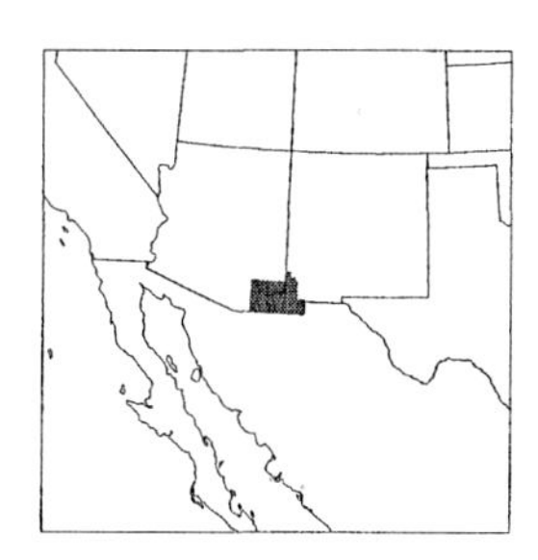

Plants perennial; rhizomatous, sometimes also stoloniferous, with many innovations and scaly, sharp-tipped rhizomes 4–8 mm thick. **Culms** 15–40(50) cm, erect, stiff, hard. **Sheaths** hairy at the apices, hairs to 2 mm; **ligules** 0.2–0.4 mm, membranous, ciliate; **blades** 2–15 cm long, (1)2–4 mm wide, involute, arcuate, glabrous abaxially, scabrous adaxially, apices sharply pointed. **Panicles** terminal, 6–20(24) cm long, 2–8(12) cm wide, ovate, open or contracted; **primary branches** 1–8(15) cm, appressed or diverging up to 50° from the rachises; **pulvini** glabrous or not; **pedicels** 0–8 mm, appressed, lower pedicels on each branch shorter than 1 mm. **Spikelets** 8–14 mm long, 1.4–3 mm wide, ovate to lanceolate, stramineous with a reddish-purple tinge, with 5–10 florets; **disarticulation** basipetal, glumes persistent. **Glumes** unequal, chartaceous; **lower glumes** 2.4–3.6 mm; **upper glumes** 3–4.5 mm, sometimes 3-veined; **lemmas** 3.8–4.5 mm, ovate, leathery, 3(4, 5)-veined, lateral veins evident, greenish, upper margins hyaline, apices acute to obtuse, usually erose; **paleas** 3.8–4.5 mm, membranous, keels scabridulous, apices obtuse to truncate; **anthers** 3, 2–2.4 mm, purplish to yellowish. **Caryopses** 1.6–2 mm, ellipsoid, dorsally flattened, with a shallow adaxial groove, striate, reddish-brown. $2n$ = 40.

Eragrostis obtusiflora is native to the southwestern United States and Mexico. It grows in dry or wet alkali flats, often in association with *Distichlis* and *Sarcobatus*, at 900–1400 m.

27. **Eragrostis spectabilis** (Pursh) Steud. [p. 90]
PURPLE LOVEGRASS, ÉRAGROSTIDE BRILLANTE

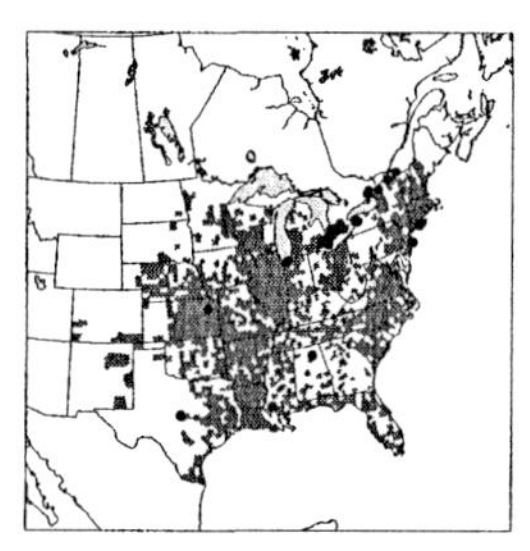

Plants perennial; cespitose, with innovations and short, knotty rhizomes less than 4 mm thick. **Culms** 30–70(85) cm, erect, glabrous. **Sheaths** hairy on the margins and at the apices, hairs to 7 mm; **ligules** 0.1–0.2 mm; **blades** 10–32 cm long, 3–8 mm wide, flat to involute, both surfaces usually pilose, sometimes glabrous on both surfaces or glabrous abaxially and sparsely pilose adaxially, often with a line of hairs behind the ligules, hairs to 8 mm. **Panicles** (15)25–45(60) cm long, 15–35 cm wide, broadly ovate to oblong, open, basal portions sometimes included in the uppermost leaf sheaths; **primary branches** (6)12–20 cm long, diverging 20–90° from the rachises, capillary, naked below; **pulvini** hairy, hairs to 6 mm; **pedicels** 1.5–17 mm, divergent or appressed. **Spikelets** 3–7.5 mm long, 1–2 mm wide, linear-lanceolate, reddish-purple, sometimes olivaceous, with (4)6–12 florets; **disarticulation** basipetal, glumes persistent. **Glumes** subequal to equal, (1)1.3–2.3 mm, lanceolate, membranous to chartaceous; **lemmas** (1)1.3–2.5 mm, ovate to lanceolate, leathery, 3-veined, apices acute; **paleas** (1)1.2–2.4 mm, membranous, keels sometimes shortly ciliate, apices obtuse to truncate; **anthers** 3, 0.3–0.5 mm, purplish. **Caryopses** 0.6–0.8 mm, ellipsoid, strongly flattened, adaxial surfaces with 2 prominent ridges separated by a groove, reddish-brown. $2n$ = 20, 40, 42.

Eragrostis spectabilis is native in the eastern portion of the *Flora* region, extending from southern Canada through the United States, Mexico, and Central America to Belize. It grows in fields and on the margins of woods, along roadsides, and in other disturbed sites, usually in sandy to clay loam soils, at 0–1830 m, and is associated with hardwood forests, *Prosopsis-Acacia* grasslands, and shortgrass prairies. A showy species, *E. spectabilis* is available commercially for planting as an ornamental.

28. **Eragrostis curtipedicellata** Buckley [p. 90]
GUMMY LOVEGRASS

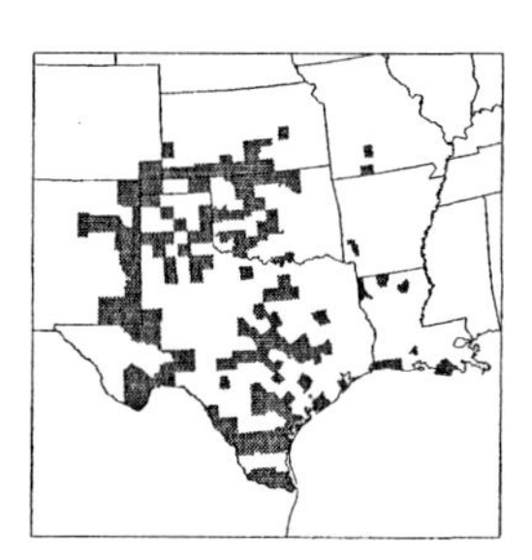

Plants perennial; cespitose, with innovations and short, knotty rhizomes less than 4 mm thick. **Culms** 20–65 cm, erect, viscid. **Sheaths** usually viscid, hairy at the apices and on the collars and margins, hairs to 6 mm; **ligules** 0.1–0.3 mm; **blades** 5–18 cm long, 2–4(5) mm wide, flat to involute, sometimes viscid, densely hairy behind the ligules, hairs to 8 mm. **Panicles** 18–35 cm long, 10–30 cm wide, broadly ovate, open; **primary branches** 3–18 cm, diverging 10–90° from the rachises, stiff, viscid, naked basally; **pulvini** hairy, hairs to 6 mm; **pedicels** 0.2–1.2 mm, appressed. **Spikelets** 3.5–6(7.6) mm long, 1–1.5 mm wide, linear-lanceolate, stramineous to reddish-purple, with 4–10 florets; **disarticulation** basipetal, glumes persistent. **Glumes** lanceolate, membranous; **lower glumes** 0.9–1.8 mm; **upper glumes** 1.2–2 mm, 1–3-veined; **lemmas** 1.5–2.2 mm, ovate to lanceolate, membranous, 3-veined, lateral veins evident, apices acute; **paleas** 1.2–2 mm, hyaline, not wider than the lemmas, apices obtuse; **anthers** 3, 0.2–0.4 mm,

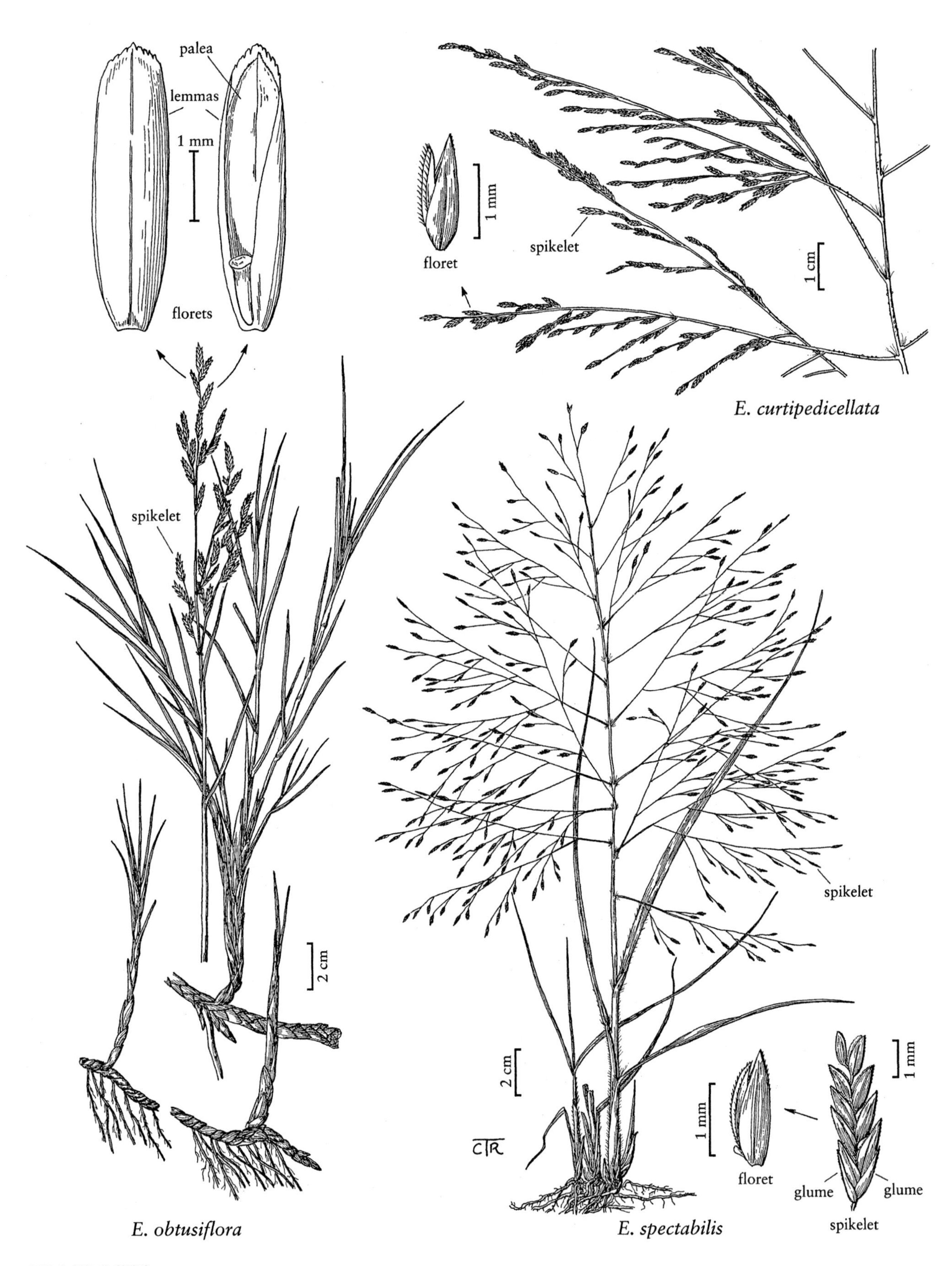
palea
lemmas
1 mm
florets
spikelet
E. obtusiflora
floret
1 mm
spikelet
1 cm
E. curtipedicellata
spikelet
2 cm
2 cm
CR
1 mm
floret
1 mm
glume
glume
spikelet
E. spectabilis

purplish. **Caryopses** 0.6–0.8 mm, ellipsoid, terete in cross section, neither ridged nor grooved, faintly striate, reddish-brown. $2n = 40$.

The range of *Eragrostis curtipedicellata* extends from southern Colorado, Kansas, and Missouri to northeastern Mexico. It grows near fields, along roadsides, and in the margins of woods, at 10–1525 m.

29. **Eragrostis silveana** Swallen [p. 92]
SILVEUS' LOVEGRASS

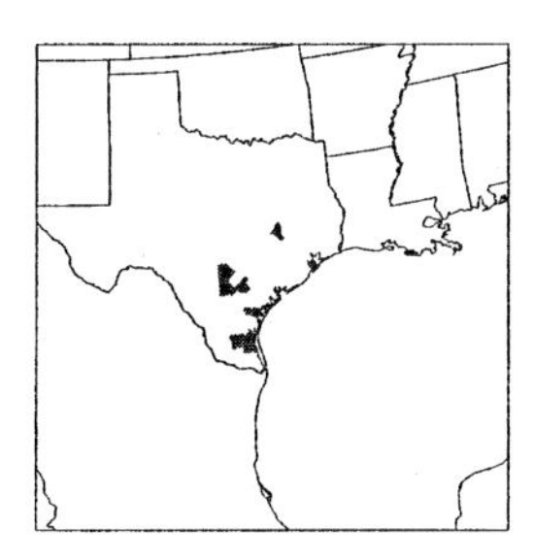

Plants perennial; cespitose, with innovations and short, knotty rhizomes less than 4 mm thick. **Culms** 45–60 cm, erect, often glandular below the nodes, sometimes viscid. **Sheaths** often viscid, sometimes sparsely pilose, hairy at the apices, hairs to 6 mm; **ligules** 0.2–0.3 mm; **blades** 8–25 cm long, 2–4 mm wide, flat to involute, glabrous, sometimes viscid. **Panicles** 20–35(42) cm long, 10–22 cm wide, broadly ovate, open, bases included in the uppermost leaf sheaths; **primary branches** 5–14 cm, diverging 20–90° from the rachises, capillary, sometimes viscid, naked basally; **pulvini** hairy, hairs to 6 mm; **pedicels** (1)1.5–12 mm, diverging or appressed. **Spikelets** (2.5)3–4.8 mm long, 0.9–1.4 mm wide, linear-lanceolate, reddish-purple, with 4–9 florets; **disarticulation** basipetal, glumes persistent. **Glumes** lanceolate, membranous; **lower glumes** 0.9–1.2 mm; **upper glumes** 1–1.3 mm; **lemmas** 1.1–1.4 mm, ovate to lanceolate, membranous, 3-veined, apices acute; **paleas** 1–1.4 mm, hyaline, not wider than the lemmas, apices obtuse; **anthers** 3, 0.2–0.3 mm, purplish. **Caryopses** 0.5–0.6 mm, ellipsoid, terete in cross section, neither ridged nor grooved, faintly striate, reddish-brown. $2n$ = unknown.

Eragrostis silveana grows in various open habitats, from sandy prairies to clay loam flats, near roadsides, railroads, and fields at 0–100 m. Its range is limited to the coastal plain of Texas and northern Mexico. Morphologically, *E. silveana* is somewhat intermediate between *E. spectabilis* and *E. curtipedicellata*, and grows where the distribution of these two species overlaps.

30. **Eragrostis spicata** Vasey [p. 92]
SPIKE LOVEGRASS

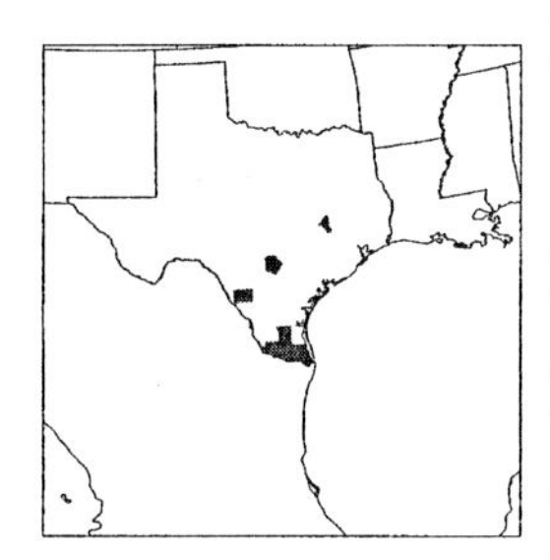

Plants perennial; cespitose, with innovations, without rhizomes. **Culms** 75–100 cm, erect, glabrous. **Sheaths** hirtellous on the margins when immature, apices glabrous or hairy shorter than 0.5 mm; **ligules** 0.2–0.3 mm; **blades** 20–40 cm long, 2–5(6) mm wide, flat to involute, glabrous abaxially, scabrous adaxially. **Panicles** 22–40 cm long, 0.3–0.6 cm wide, spikelike, dense; **primary branches** shorter than 1.2 cm, closely appressed, spikelet-bearing to the base; **pulvini** glabrous; **pedicels** 0.1–0.6 mm, mostly appressed, hirtellous. **Spikelets** 1.4–2.2 mm long, 0.9–1.2 mm wide, ovate, stramineous to light greenish, with 2–3 florets; **disarticulation** basipetal, in the rachilla below the individual florets or at the base of the florets, glumes persistent. **Glumes** elliptic to ovate, hyaline, keels ciliolate; **lower glumes** 0.7–1 mm; **upper glumes** 0.9–1.3 mm, apices obtuse; **lemmas** 1.5–2.1 mm, ovate, membranous to hyaline, apices acute to obtuse; **paleas** 1.1–1.6 mm, hyaline, not wider than the lemmas, apices obtuse; **anthers** 2, 0.3–0.4 mm, reddish-brown. **Caryopses** 0.7–1 mm, ellipsoid, somewhat ventrally flattened, smooth to faintly striate, reddish-brown. $2n = 40$.

Eragrostis spicata grows in moist areas in prairies, usually in deep, sandy, clay loam soils, at 0–70 m. It is native from southern Texas to Mexico and in Paraguay and Argentina. In North America, it grows with *Andropogon*, *Quercus stellata*, *Prosopsis glandulosa*, and *Acacia*.

31. **Eragrostis plana** Nees [p. 92]

Plants perennial; cespitose, with innovations, without rhizomes. **Culms** 65–100 cm, erect, glabrous. **Sheaths** flattened, smooth, shiny, glabrous or puberulent; **ligules** 0.2–0.4 mm; **blades** 15–50(70) cm long, 2–4 mm wide, folded, margins involute or revolute, abaxial surfaces glabrous or sparsely hairy, adaxial surfaces scabridulous. **Panicles** 13–28 cm long, 2–8 cm wide, narrowly oblong to narrowly lanceolate, contracted to open; **primary branches** 1–8 cm, appressed or diverging up to 30° from the rachises; **pulvini** glabrous or hairy; **pedicels** 1–7 mm, appressed, glabrous. **Spikelets** 6–14 mm long, 1.3–2.5 mm wide, linear-oblong, greenish to plumbeous, with 9–14 florets; **disarticulation** acropetal, paleas persistent. **Glumes** narrowly ovate to lanceolate, membranous to hyaline; **lower glumes** 0.4–1.2 mm, scalelike; **upper glumes** 1–1.5 mm; **lemmas** 1.8–3 mm, ovate, membranous, strongly keeled, keels with minute punctate glands, lateral veins conspicuous, apices acute to obtuse; **paleas** 1.8–3 mm, hyaline to membranous, bases not projecting beyond the lemmas, apices obtuse to truncate; **anthers** 3, 1.2–1.8 mm, reddish-purple. **Caryopses** 1–1.6 mm, rectangular-prismatic to ovoid, laterally compressed, adaxial surfaces deeply grooved, smooth, opaque, reddish-brown. $2n = 20$.

Eragrostis plana is native to southern Africa. It is known from two locations in the *Flora* region, both waste areas near sheep and cattle lots in Florence County, South Carolina.

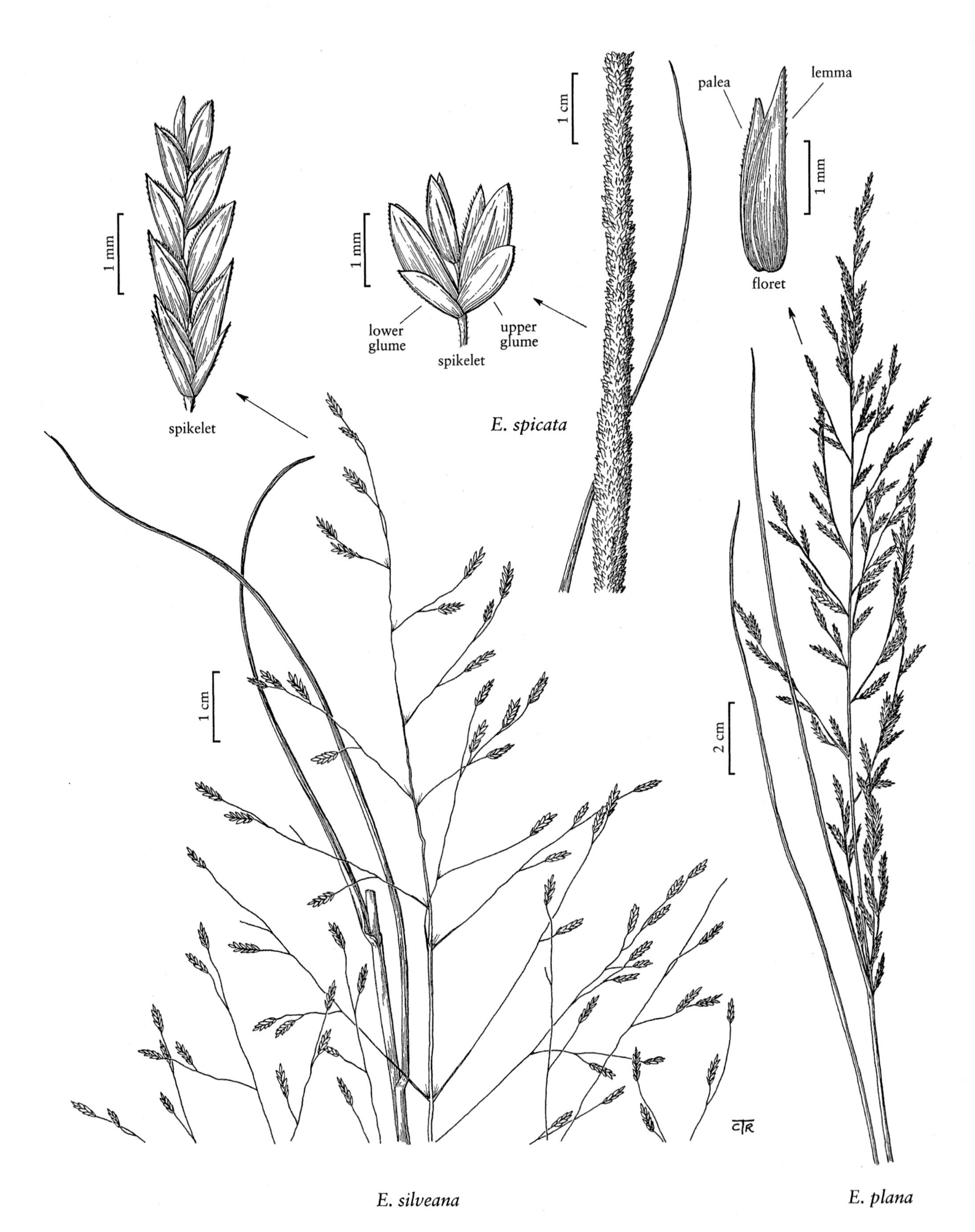

spikelet
1 mm
1 mm
lower glume
upper glume
spikelet
1 cm
E. spicata
palea
lemma
1 mm
floret
1 cm
2 cm
CTR
E. silveana
E. plana

32. **Eragrostis swallenii** Hitchc. [p. 94]
SWALLEN'S LOVEGRASS

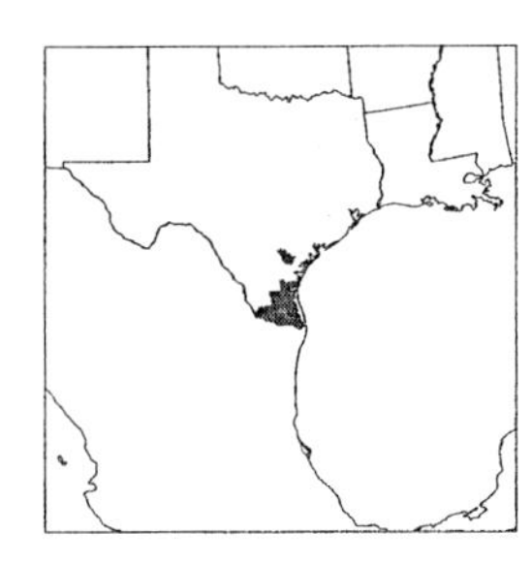

Plants perennial; cespitose, with innovations, without rhizomes. **Culms** 35–70 cm, erect, with glandular bands below the nodes. **Sheaths** hairy on the margins and at the apices, hairs to 4 mm; **ligules** 0.2–0.5 mm; **blades** (8)10–25(30) cm long, 1.5–4 mm wide, flat to involute, abaxial surfaces glabrous, adaxial surfaces scabridulous, sometimes also sparsely hairy, hairs to 4 mm. **Panicles** 12–30 cm long, 5–16 cm wide, ovate, open, an oblique glandular ring present below the lowest rachis node; **primary branches** 2–10 cm, diverging 10–70° from the rachises, flexible; **pulvini** glabrous; **pedicels** 1.5–14 mm, divergent, with a glandular band. **Spikelets** 5–16 mm long, 1.2–2.3 mm wide, linear-lanceolate, plumbeous to dark reddish-purple, with 5–25 florets; **disarticulation** acropetal, paleas persistent. **Glumes** ovate, membranous to hyaline; **lower glumes** 1.1–1.5 mm; **upper glumes** 1.4–2 mm; **lemmas** 1.5–2.5 mm, ovate, membranous, strongly keeled, keels without glands, lateral veins conspicuous, apices acute; **paleas** 1.2–2.1 mm, hyaline, narrower than the lemmas, apices obtuse to truncate; **anthers** 3, 0.3–0.5 mm, purplish. **Caryopses** 0.8–1.1 mm, rectangular-prismatic to ellipsoid, somewhat laterally compressed, with a well-developed adaxial groove, smooth, faintly striate, mostly opaque, light reddish-brown. $2n$ = 84.

Eragrostis swallenii grows in sandy sites along coastal grasslands and roadsides, often with *Andropogon* and *Spartina*, at 30–150 m. Its range extends around the Gulf coast from Texas to Mexico.

33. **Eragrostis trichodes** (Nutt.) Alph. Wood [p. 94]
SAND LOVEGRASS

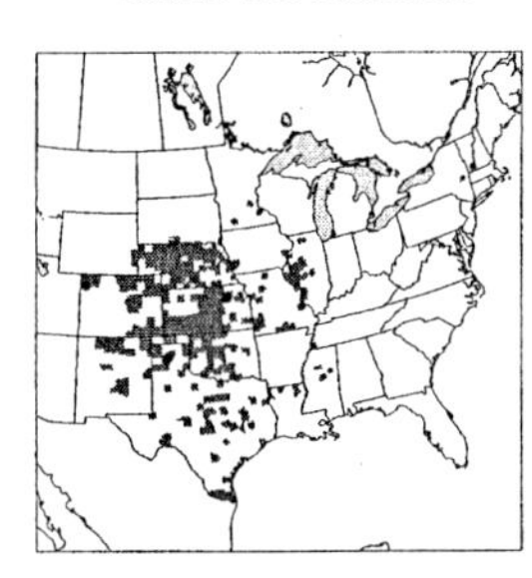

Plants perennial; cespitose, with innovations, without rhizomes, not glandular. **Culms** 30–120 (160) cm, erect, glabrous and non-glandular below the nodes. **Sheaths** sometimes villous along the margins, apices hairy, hairs to 5 mm; **ligules** 0.3–0.5 mm; **blades** 15–46(65) cm long, 1.5–8 mm wide, flat to involute, abaxial surfaces glabrous, adaxial surfaces scabridulous, sometimes also pilose on the basal ¼, hairs to 4 mm. **Panicles** 30–80 cm long, 6–30 cm wide, oblong to ovoid, diffuse; **primary branches** 2–35 cm, diverging 20–90° from the rachises, naked basally; **pulvini** hairy or glabrous; **pedicels** 2–22 cm, diverging, capillary. **Spikelets** 3–15 mm long, 1.5–3.6 mm wide, ovate to lanceolate, greenish-yellow with a reddish-purple tinge, with (2)4–18 florets; **disarticulation** acropetal, paleas persistent. **Glumes** subequal, 1.8–4 mm, narrowly ovate to linear-lanceolate, membranous, apices acuminate; **upper glumes** as long as or longer than the basal lemmas; **lemmas** 2.2–3.5 mm, broadly ovate to lanceolate, membranous, strongly keeled, keels not glandular, lateral veins conspicuous, apices acute; **paleas** 1.8–2.8 mm, hyaline, narrower than the lemmas, keels ciliolate, apices obtuse to truncate; **anthers** 3, 1–1.6 mm, purplish. **Caryopses** 0.8–1.3 mm, rectangular-prismatic, somewhat laterally compressed, with a wide, deep adaxial groove, faintly striate, opaque, dark reddish-brown. $2n$ = 40.

Eragrostis trichodes grows in sandy to gravelly prairies, open sandy woods, rocky slopes, and roadsides, at 100–2150 m, often in associations with *Quercus marilandica*, *Q. stellata*, *Juniperus*, and *Redfieldia flexuosa*. It is endemic to the contiguous United States, and is available commercially as an ornamental. Records from outside the primary range probably reflect introductions.

34. **Eragrostis palmeri** S. Watson [p. 94]
RIO GRANDE LOVEGRASS

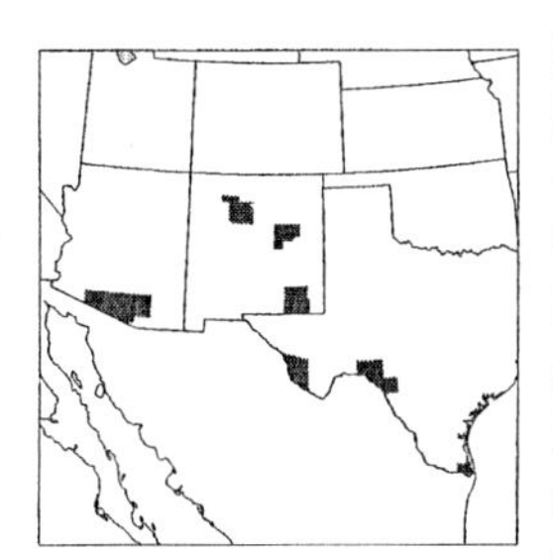

Plants perennial; cespitose, with innovations and knotty bases, without rhizomes, not glandular. **Culms** 50–90(120) cm, glabrous below the nodes. **Sheaths** villous and the hairs not papillose-based, or mostly glabrous, apices hairy, hairs to 5 mm, not papillose-based; **ligules** 0.2–0.4 mm; **blades** (14)20–35 cm long, 1–2.4 mm wide, involute, abaxial surfaces glabrous, adaxial surfaces scabridulous, sometimes sparsely hairy. **Panicles** 12–40 cm long, 4–20 cm wide, oblong, open; **primary branches** 2–20 cm, diverging 20–70° from the rachises, capillary; **pulvini** glabrous or sparsely hairy; **pedicels** (0.4)1–4(14) mm, appressed or diverging, only the terminal pedicels on each branch longer than 4 mm. **Spikelets** 4–6(7.3) mm long, 1–2 mm wide, linear-lanceolate, plumbeous, with 5–12 florets; **disarticulation** acropetal, paleas persistent. **Glumes** lanceolate to ovate, hyaline; **lower glumes** 1.1–1.8 mm; **upper glumes** 1.2–2.2 mm, exceeded by the basal lemmas; **lemmas** 2–2.6 mm, ovate, membranous, hyaline towards the apices and margins, keels weak or strong, without glands, lateral veins from inconspicuous to conspicuous, apices acute; **paleas** 1.7–2.4 mm, hyaline, bases not projecting beyond the lemmas, apices truncate, often notched; **anthers** 3, 0.6–1.3 mm, yellowish to purplish. **Caryopses** 0.6–0.8 mm, rectangular-prismatic to subellipsoid, laterally compressed, with a well-developed adaxial groove,

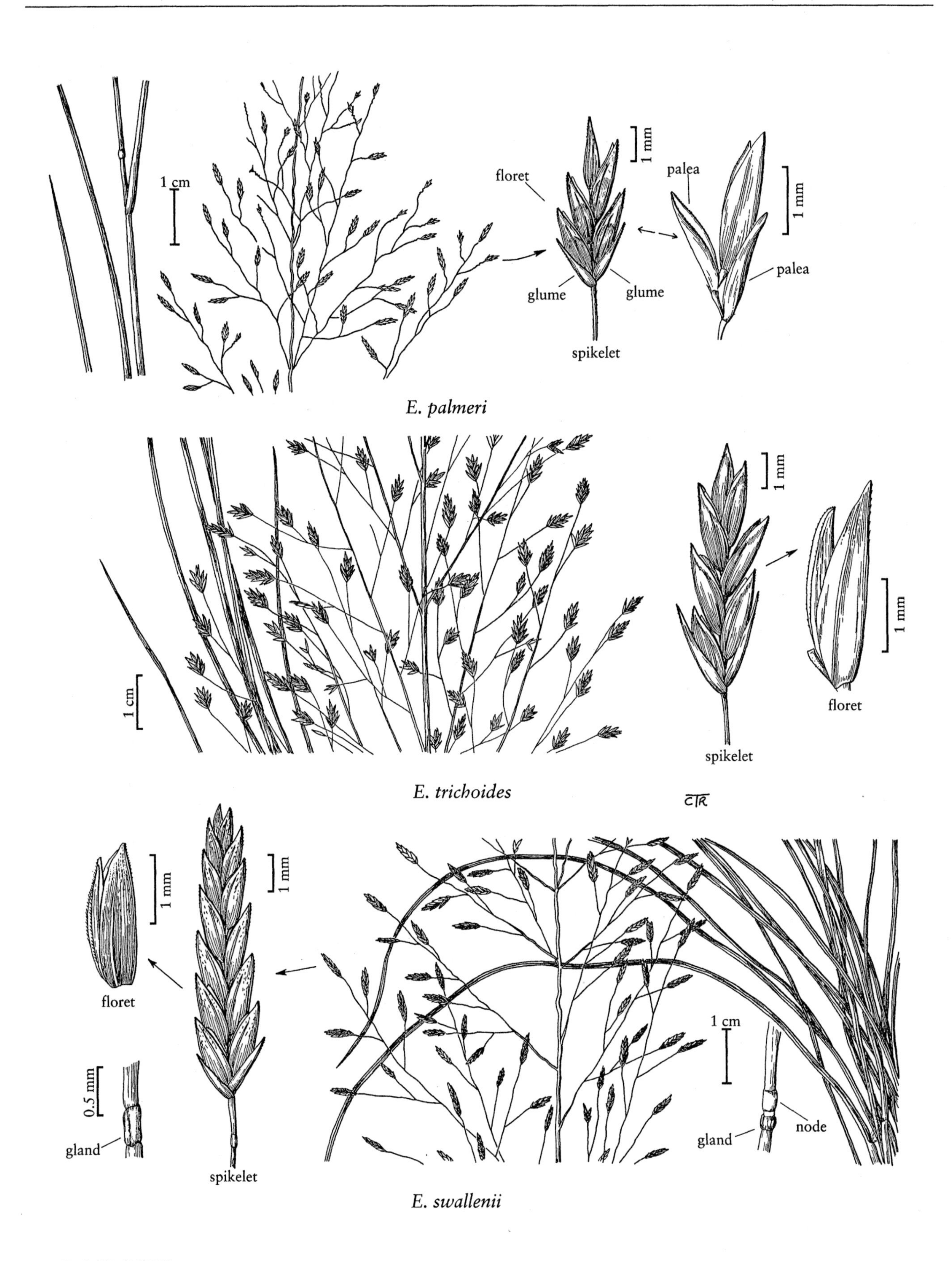
1 cm
floret
1 mm
palea
1 mm
palea
glume
glume
spikelet
E. palmeri
1 mm
1 mm
1 cm
floret
spikelet
E. trichoides
CTR
1 mm
1 mm
floret
1 cm
0.5 mm
gland
spikelet
gland
node
E. swallenii

faintly striate, opaque, reddish-brown. $2n = 40$.

Eragrostis palmeri grows on rocky slopes and hills between 300–2150 m, generally in association with *Pinus edulis*, *Juniperus monosperma*, *Bouteloua gracilis*, and *Prosopis*. Its range extends from the southwestern United States into Mexico. It resembles *E. erosa*, but differs in its shorter lemmas and caryopses.

35. **Eragrostis polytricha** Nees [p. 96]
HAIRYSHEATH LOVEGRASS

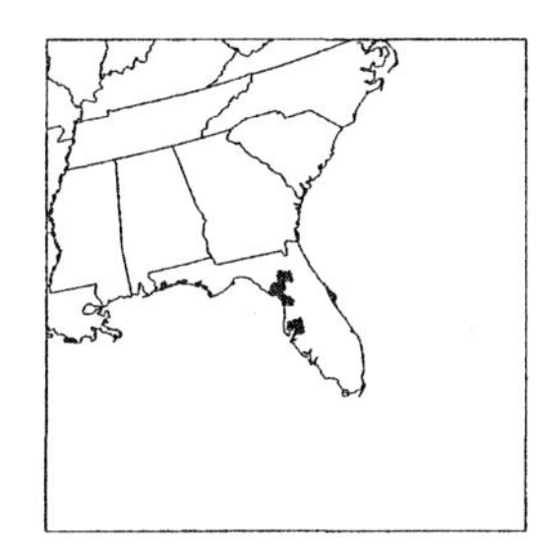

Plants perennial; cespitose, with innovations, without rhizomes, not glandular. **Culms** 30–62 cm, erect, glabrous and shiny below the nodes. **Sheaths** sometimes densely pilose dorsally and on the collars, margins and apices hairy, hairs to 5 mm; **ligules** 0.2–0.4 mm; **blades** 5–20(33) cm long, 1–3.5 mm wide, involute to flat, both surfaces with scattered hairs, adaxial surfaces densely hairy behind the ligules, hairs to 7 mm. **Panicles** 15–25 cm long, 5–27 cm wide, ovate, open; **primary branches** 0.6–15 cm, diverging up to 90° from the rachises, capillary, naked basally; **pulvini** hairy, hairs to 8 mm; **pedicels** 1.4–10(16) mm, divergent. **Spikelets** (2.5)3–5 mm long, 1.1–1.6 mm wide, narrowly lanceolate to linear-oblong, plumbeous, with 4–9 florets; **disarticulation** acropetal, paleas persistent. **Glumes** broadly ovate to narrowly lanceolate, hyaline to membranous; **lower glumes** 1.1–1.6 mm; **upper glumes** 1.2–1.8 mm; **lemmas** 1.2–1.8 mm, broadly ovate, membranous throughout, lateral veins inconspicuous, apices acute; **paleas** 1.1–1.7 mm, membranous to hyaline, narrower than the lemmas, apices obtuse; **anthers** 3, 0.3–0.5 mm, reddish-purple. **Caryopses** 0.5–0.8 mm, obovoid to somewhat prism-shaped, laterally compressed, with a well-developed adaxial groove, finely striate, opaque to translucent, reddish-brown. $2n = 60, 80$.

Eragrostis polytricha grows in sandy and rocky areas, at 0–30 m, usually in open pinelands. It is native to Florida but its primary range lies to the south of the *Flora region*, from southern Mexico through Central America to Venezuela, Chile, and Argentina.

36. **Eragrostis lugens** Nees [p. 96]
MOURNING LOVEGRASS

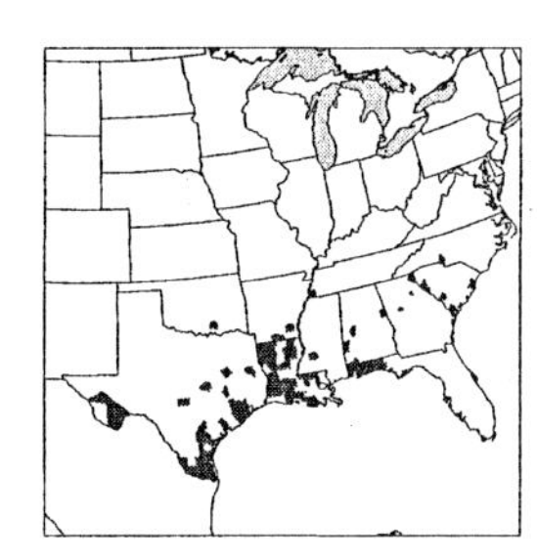

Plants perennial; cespitose, with innovations, without rhizomes, not glandular. **Culms** 30–70 cm, erect, sometimes geniculate, glabrous below the nodes. **Sheaths** usually mostly glabrous, summits hairy, hairs 2–5 mm, never papillose-based; **ligules** 0.2–0.3 mm; **blades** (4)8–22 cm long, 1–3.5 mm wide, involute to flat, both surfaces glabrous, margins sometimes with scattered hairs, hairs to 7 mm. **Panicles** 16–28 cm long, 10–21 cm wide, ovate, open; **primary branches** 0.6–15 cm, diverging up to 100° from the rachises, naked basally; **pulvini** hairy; **pedicels** 1.4–5(7) mm, diverging, wiry, present on all spikelets. **Spikelets** 2–4.5(5) mm long, 0.5–1(1.3) mm wide, narrowly lanceolate, plumbeous to reddish-purple, with 2–7 florets; **disarticulation** acropetal, paleas persistent. **Glumes** broadly ovate to narrowly lanceolate, hyaline, sometimes reddish-purple; **lower glumes** 0.6–1 mm; **upper glumes** 1.1–1.4 mm, usually broader than the lower glumes; **lemmas** 1.2–1.8 mm, broadly ovate, mostly membranous but the distal margins hyaline, lateral veins inconspicuous, apices acute; **paleas** 1.1–1.7 mm, membranous to hyaline, apices obtuse; **anthers** 3, 0.2–0.7 mm, reddish-purple. **Caryopses** 0.5–0.6 mm, obovoid to somewhat prism-shaped, terete to somewhat laterally compressed, sometimes with a weak adaxial groove, finely striate, usually opaque, faintly reddish-brown to whitish. $2n = 40, 80$, ca. 108.

Eragrostis lugens grows on sandy dunes and along river banks, at 1–300 m. Its range extends from the southern United States to Peru and Argentina.

37. **Eragrostis hirsuta** (Michx.) Nees [p. 96]
BIGTOP LOVEGRASS

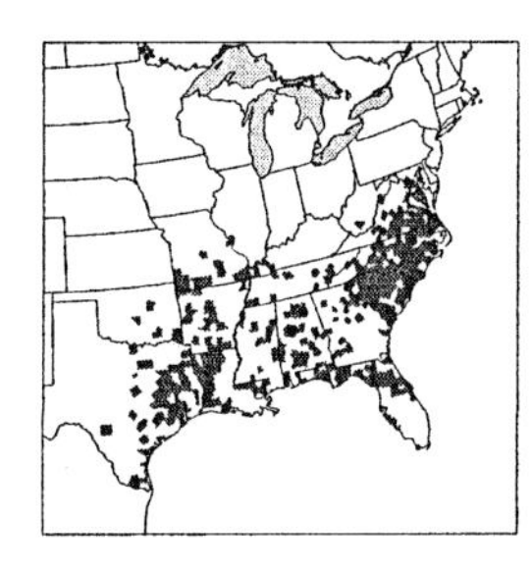

Plants perennial; cespitose, with innovations and hardened bases, without rhizomes, not glandular. **Culms** (30)45–100 cm, erect, glabrous below the nodes. **Sheaths** rarely glabrous, apices and distal margins usually hairy, sometimes also densely hairy basally, dorsally, and on the collars, hairs to 6 mm, papillose-based; **ligules** 0.2–0.4 mm; **blades** 25–60 cm long, 3–8(11) mm wide, flat to loosely involute, usually glabrous, adaxial surfaces sometimes hairy basally. **Panicles** 25–85 cm long, 15–40 cm wide, broadly ovate, open; **primary branches** mostly 4–35(45) cm, diverging 20–90° from the rachises, capillary; **pulvini** glabrous or hairy; **pedicels** 2–28 mm, divergent. **Spikelets** 2–4(5) mm long, 1–1.7 mm wide, lanceolate, greenish with purplish tinges, with 2–6 florets; **disarticulation** acropetal, paleas persistent. **Glumes** lanceolate, hyaline to membranous; **lower glumes** 1.1–2 mm; **upper glumes** 1.5–2.8 mm, apices acuminate to acute; **lemmas** 1.6–2.4 mm, ovate, membranous, hyaline near the margins, lateral veins inconspicuous, apices acute; **paleas** 1.2–2.2 mm, hyaline, bases not projecting beyond the lemmas, apices acute to obtuse; **anthers** 3, 0.3–0.8 mm, purplish. **Caryopses** 0.8–1 mm, rectangular-prismatic, somewhat laterally compressed,

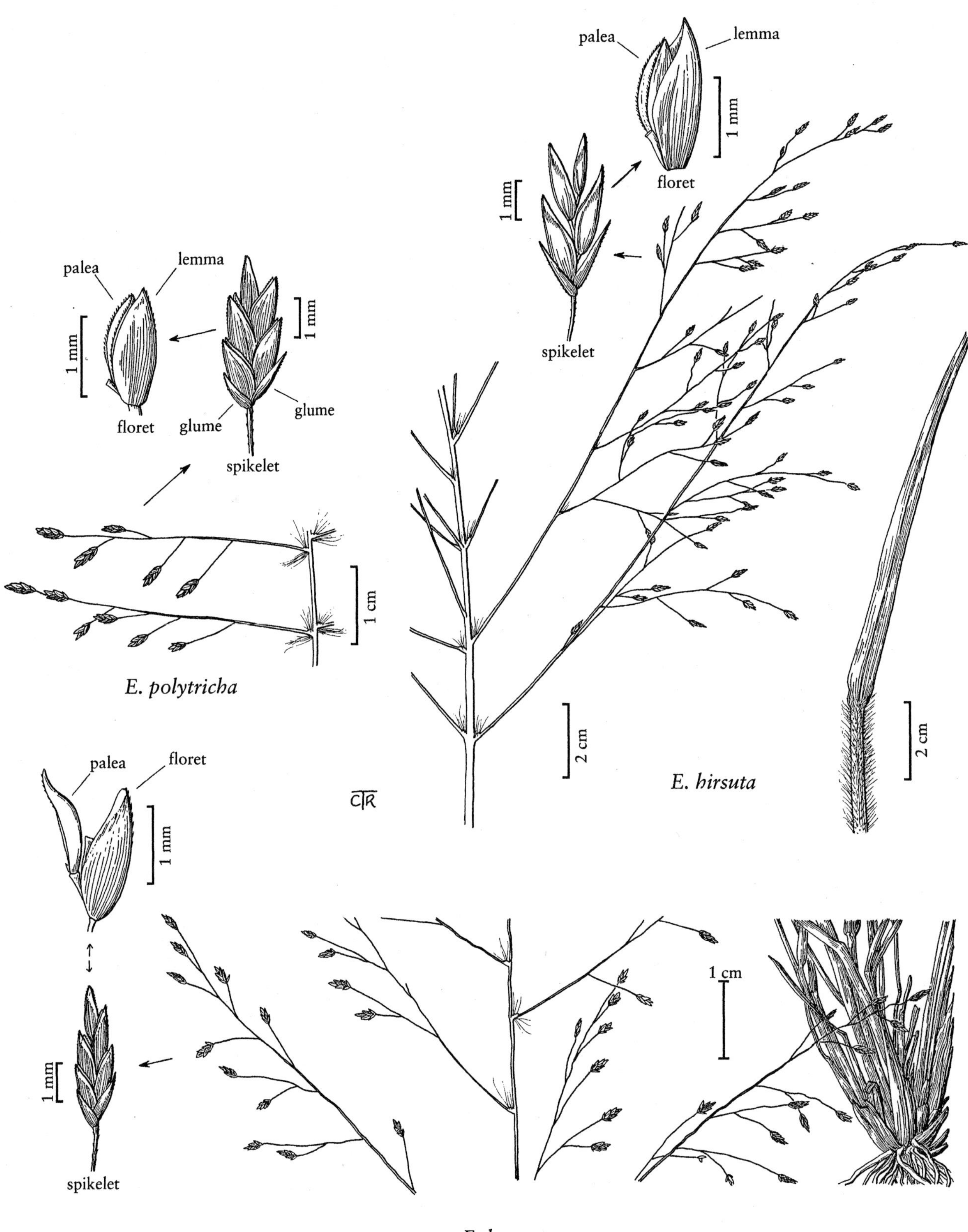
palea
lemma
1 mm
floret
1 mm
spikelet
E. polytricha
palea
lemma
1 mm
floret
glume
glume
spikelet
1 mm
1 cm
2 cm
E. hirsuta
2 cm
palea
floret
1 mm
1 mm
spikelet
1 cm
E. lugens

with or without a well-developed adaxial groove, striate, opaque, reddish-brown. $2n$ = 100.

Eragrostis hirsuta grows in sandy clay loams on the coastal plain and along roadsides, at 0–150 m, usually in association with *Pinus palustris* and *Quercus*. Its range extends from the southeastern United States through eastern Mexico to Guatemala and Belize.

38. **Eragrostis intermedia** Hitchc. [p. 98]
PLAINS LOVEGRASS

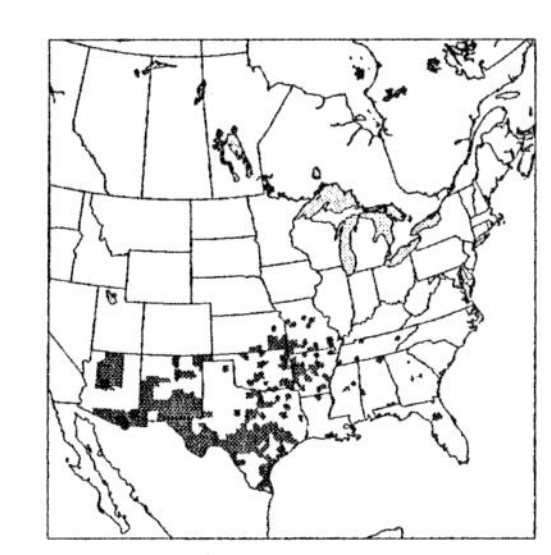

Plants perennial; cespitose, with innovations, without rhizomes, not glandular. **Culms** (30)40–90(110) cm, erect, glabrous below the nodes. **Sheaths** sparsely pilose on the margins, apices hairy, hairs to 8 mm, not papillose-based; **ligules** 0.2–0.4 mm; **blades** (4)10–20(30) cm long, 1–3 mm wide, flat or involute, abaxial surfaces glabrous, adaxial surfaces densely hairy behind the ligules, elsewhere usually glabrous, occasionally sparsely hairy. **Panicles** 15–40 cm long, (8.5)15–30 cm wide, ovate, open; **primary branches** 4–25 cm, diverging 20–90° from the rachises, capillary; **pulvini** hairy or glabrous; **pedicels** 2–14 mm, divergent. **Spikelets** 3–7 mm long, 1–1.8 mm wide, narrowly lanceolate, olivaceous to purplish, with (3)5–11 florets; **disarticulation** acropetal, paleas persistent. **Glumes** lanceolate to ovate, hyaline to membranous; **lower glumes** 1.1–1.7 mm, narrower than the upper glumes; **upper glumes** 1.3–2 mm, apices acuminate to acute; **lemmas** 1.6–2.2 mm, ovate, membranous, hyaline near the margins, lateral veins inconspicuous, apices acute; **paleas** 1.4–2.1 mm, hyaline, narrower than the lemmas, apices obtuse to acute; **anthers** 3, 0.5–0.8 mm, purplish. **Caryopses** 0.5–0.9 mm, rectangular-prismatic, somewhat laterally compressed, with a well-developed adaxial groove, striate, opaque, reddish-brown. $2n$ = ca. 54, 60, 72, ca. 74, 80, 100, 120.

Eragrostis intermedia grows in clay, sandy, and rocky soils, often in disturbed sites, at 0–1850 m. Its range extends from the United States through Mexico and Central America to South America. *Eragrostis intermedia* is similar to the more widespread *E. lugens*, but differs from that species in having wider spikelets, longer lemmas, and caryopses with a prominent adaxial groove.

39. **Eragrostis erosa** Scribn. *ex* Beal [p. 98]
CHIHUAHUA LOVEGRASS

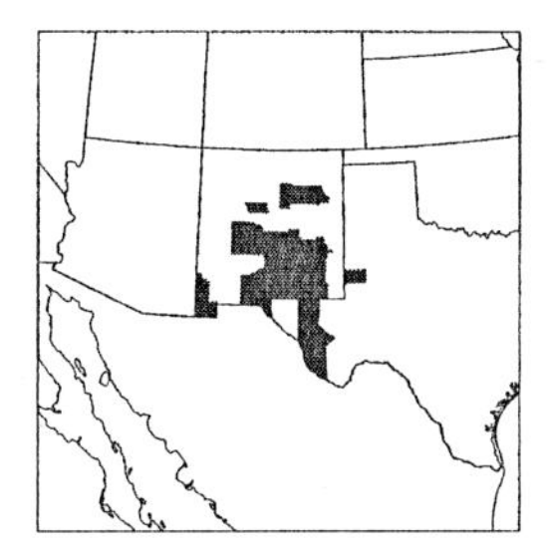

Plants perennial; cespitose, with innovations, without rhizomes, not glandular. **Culms** 70–110 cm, erect, glabrous below the nodes. **Sheaths** hairy at the apices and sometimes on the upper margins, hairs to 4 mm, not papillose-based; **ligules** 0.2–0.4 mm; **blades** (8)12–30 cm long, 1.5–3.8 mm wide, flat to involute, abaxial surfaces glabrous, adaxial surfaces scabridulous, glabrous or sparsely hairy, hairs to 4 mm. **Panicles** 25–45 cm long, (5)12–30 cm wide, broadly ovate, open; **primary branches** mostly 4–20 cm, diverging 20–90° from the rachises, capillary, sinuous; **pulvini** glabrous or hairy; **pedicels** 1–18 mm, appressed or divergent, proximal spikelets on each branch usually with pedicels shorter than 5 mm. **Spikelets** 5–9 mm long, 1–3 mm wide, lanceolate, plumbeous, with 5–12 florets; **disarticulation** acropetal, glumes first, then the lemmas, paleas persistent. **Glumes** lanceolate to ovate, membranous; **lower glumes** 1.3–2.4 mm; **upper glumes** 1.6–2.6 mm; **lemmas** 2.4–3 mm, ovate, mostly membranous, hyaline near the margins and apices, lateral veins inconspicuous, apices acute; **paleas** 1.5–3 mm, hyaline, narrower than the lemmas, apices obtuse to truncate; **anthers** 3, 0.6–1.7 mm, purplish. **Caryopses** 0.8–1.6 mm, subellipsoid, terete to somewhat laterally compressed, with a well-developed adaxial groove, faintly striate, opaque, reddish-brown. $2n$ = unknown.

Eragrostis erosa grows on rocky slopes and hills, at 1200–2300 m, often in association with *Pinus edulis*, *Juniperus monosperma*, and *Bouteloua gracilis*. Its range extends from New Mexico and western Texas to northern Mexico.

40. **Eragrostis refracta** (Muhl.) Scribn. [p. 98]
COASTAL LOVEGRASS

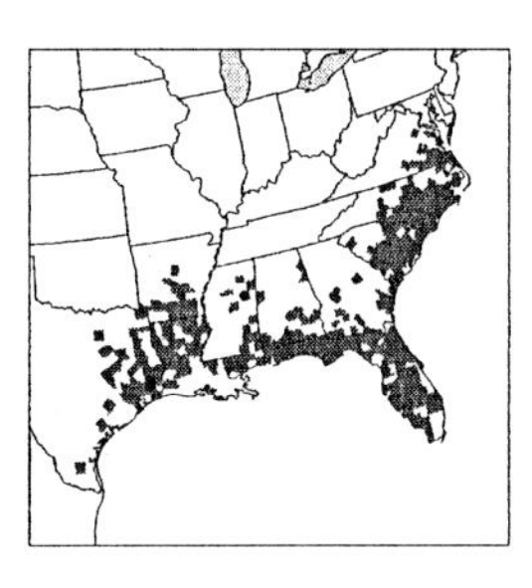

Plants perennial; cespitose, with innovations, without rhizomes, not glandular. **Culms** 30–85(110) cm, glabrous and shiny below the nodes. **Sheaths** sparsely hairy at the apices, hairs to 6 mm; **ligules** 0.1–0.4 mm; **blades** 10–35 cm long, 2–5 mm wide, flat to involute, glabrous abaxially, scabridulous and sparsely pilose adaxially, hairs to 7 mm. **Panicles** (25)30–60 cm long, 25–40 cm wide, broadly ovate to obovate, open, diffuse; **primary branches** 5–25 cm, diverging 20–90° from the rachises, capillary; **pulvini** hairy or glabrous; **pedicels** 0.5–25 mm, appressed, only the terminal pedicels on each branch longer than the

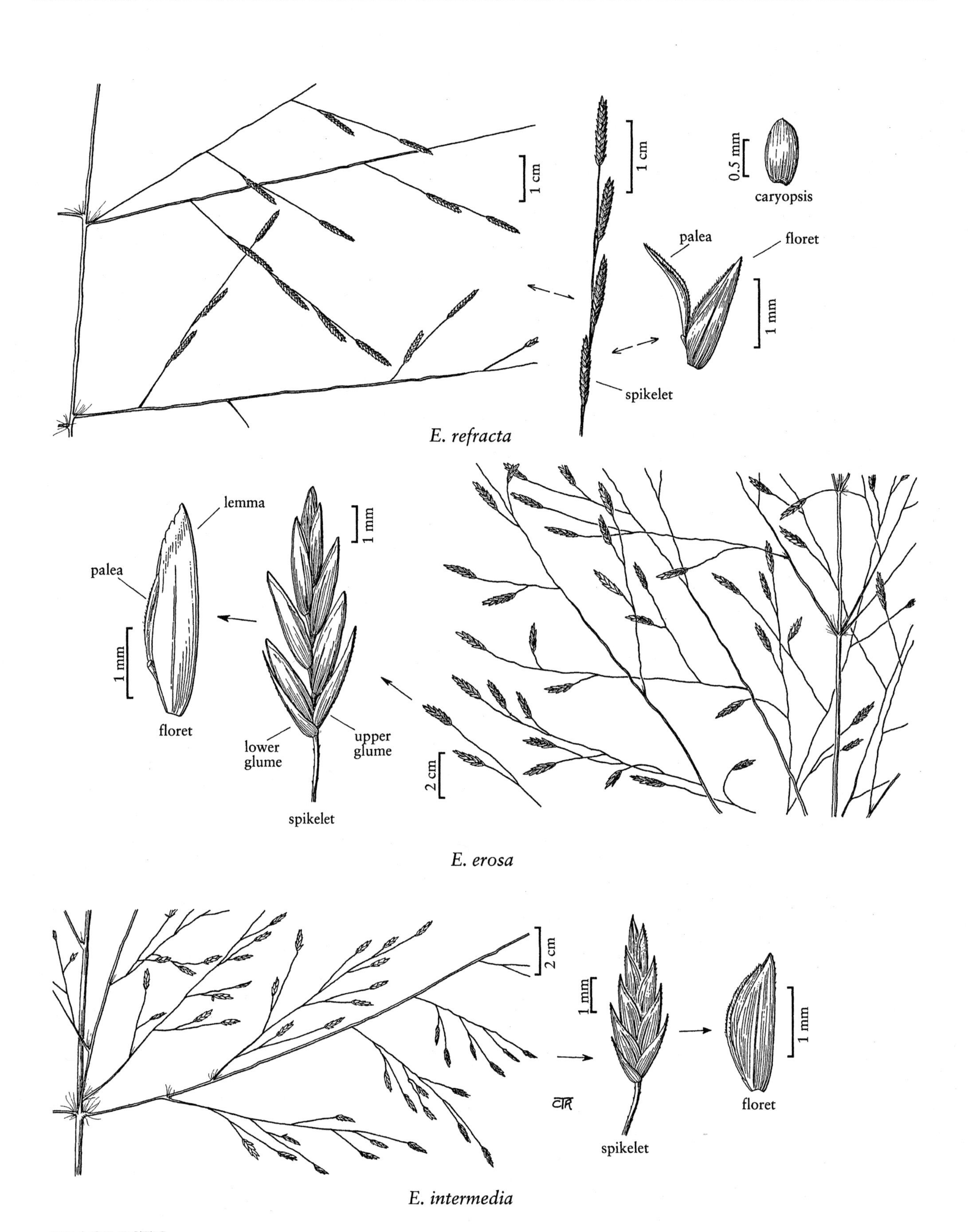

E. refracta

E. erosa

E. intermedia

spikelets. **Spikelets** 4–18(23) mm long, 1.4–3.4 mm wide, linear-lanceolate, grayish-green or stramineous to purplish, with (4)9–30 florets; **disarticulation** tardy, basipetal, in the rachillas below the florets, glumes persistent. **Glumes** narrowly lanceolate, membranous; **lower glumes** 0.8–2.4 mm; **upper glumes** 1.5–2.6 mm; **lemmas** 1.4–2.8 mm, lanceolate, membranous, apices acute to acuminate; **paleas** 1–2.6 mm, hyaline to membranous, narrower than the lemmas, apices obtuse to acute; **anthers** 2, 0.3–0.5 mm, purplish or brownish. **Caryopses** 0.5–0.9 mm, ovoid to ellipsoid, finely striate, reddish-brown. $2n = 28$.

Eragrostis refracta grows in sandy pinelands, savannahs, marshes, and woodlands on the coastal plain of the southeastern United States, at 0–150 m. It is not known from Mexico.

41. **Eragrostis elliottii** S. Watson [p. 100]
ELLIOTT'S LOVEGRASS

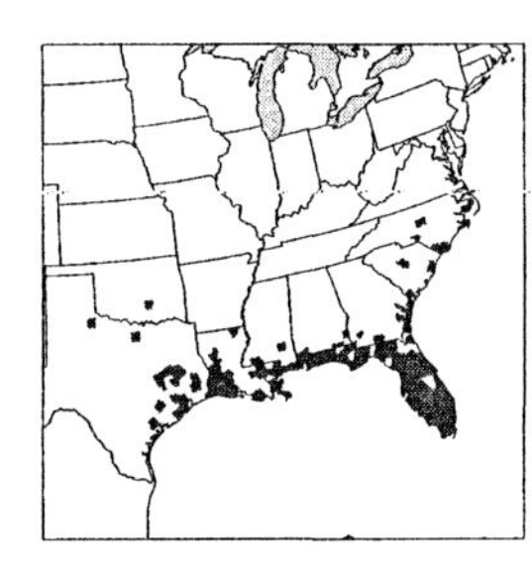

Plants perennial; cespitose, with innovations, without rhizomes, not glandular. **Culms** 25–80 cm, erect, glabrous and shiny below the basal nodes. **Sheaths** sparsely hairy at the apices, hairs to 6 mm; **ligules** 0.2–0.4 mm; **blades** 6–30(52) cm long, 2–4.5 mm wide, flat, abaxial surfaces glabrous, adaxial surfaces scabridulous, sometimes with a few scattered hairs near the base. **Panicles** (25)30–60 cm long, 15–45 cm wide, broadly ovate to obovate, open, diffuse; **primary branches** mostly 5–25(32) cm, diverging 20–90° from the rachises, capillary; **pulvini** hairy; **pedicels** (4)10–35(50) mm, widely diverging, capillary, all the pedicels on each branch longer than the spikelets. **Spikelets** 4–18 mm long, 1.4–3 mm wide, linear-lanceolate, grayish-green or stramineous to purplish, with (6)9–30 florets; **disarticulation** acropetal, below the lemmas, paleas persistent. **Glumes** narrowly lanceolate, membranous; **lower glumes** 1.1–3.4 mm; **upper glumes** 1.6–3.4 mm, apices acuminate; **lemmas** 1.8–4.4 mm, lanceolate, membranous, lateral veins evident to inconspicuous, sometimes greenish, apices acute to acuminate; **paleas** 1.1–3.5 mm, hyaline to membranous, narrower than the lemmas, apices obtuse; **anthers** 2, 0.3–0.8 mm, purplish or brownish. **Caryopses** 0.6–0.8 mm, ovoid to ellipsoid, finely striate, reddish-brown. $2n$ = unknown.

Eragrostis elliottii grows in sandy pinelands and live-oak woodlands on the coastal plain, at 0–150 m. Its range extends from the southeastern United States through the West Indies and Gulf coast of Mexico to Central and South America.

42. **Eragrostis secundiflora** J. Presl [p. 100]
RED LOVEGRASS

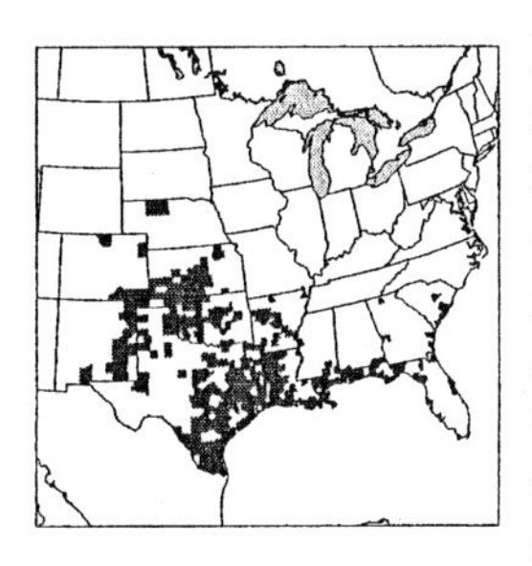

Plants perennial; cespitose, with innovations, without rhizomes, not glandular. **Culms** 30–75 cm, erect, glabrous below. **Sheaths** mostly glabrous, hairy at the apices, hairs to 4 mm; **ligules** 0.2–0.3 mm; **blades** 10–25(40) cm long, 1–5 mm wide, involute, glabrous abaxially, scabridulous adaxially, sometimes also sparsely pilose. **Panicles** (3)5–30 cm long, 1–15 cm wide, from narrowly oblong, glomerate, and interrupted below to ovate and open; **primary branches** 0.5–12(16) cm, appressed or diverging up to 40° from the rachises, stiff; **pulvini** glabrous or sparsely hairy; **pedicels** 0–1(3) mm, appressed, flattened. **Spikelets** 6–16(23) cm long, 2.4–5 mm wide, ovate to linear-elliptic, flattened, stramineous, with reddish-purple margins or completely reddish-purple, with 10–45 florets; **disarticulation** basipetal, florets falling intact and before the glumes. **Glumes** ovate-lanceolate to lanceolate, membranous; **lower glumes** 1.7–3 mm; **upper glumes** 2.2–4 mm, apices acuminate; **lemmas** 2–6 mm, ovate, membranous to leathery, apices usually acuminate or attenuate, sometimes acute; **paleas** 1.5–3 mm, membranous to leathery, narrower than the lemmas, apices obtuse, sometimes bifid; **anthers** 2, 0.2–0.5 mm, brownish. **Caryopses** 0.8–1.3 mm, ellipsoid, somewhat laterally flattened, smooth, reddish-brown. $2n = 40$.

There are two subspecies of *E. secundiflora*; plants from the *Flora* region belong to **E. secundiflora** subsp. **oxylepis** (Torrey) S.D. Koch. They grow in sandy soils, dunes, grasslands, beaches, and roadsides of the southern United States and northern Mexico, at 0–1000 m. **Eragrostis secundiflora** J. Presl subsp. **secundiflora** grows in Mexico and Central and South America.

43. **Eragrostis prolifera** (Sw.) Steud. [p. 100]
DOMINICAN LOVEGRASS

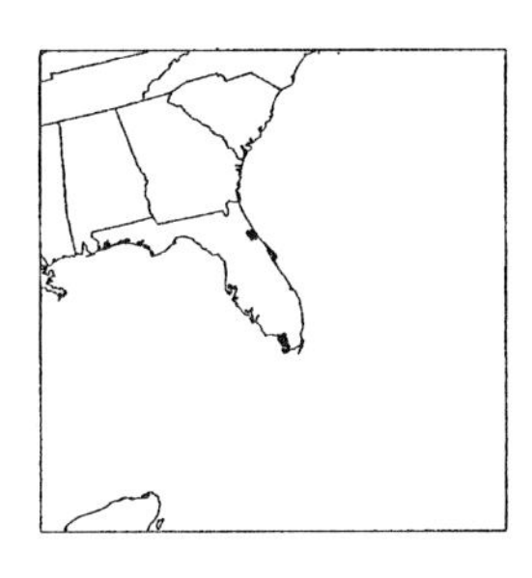

Plants perennial; cespitose, with innovations, without rhizomes, not glandular. **Culms** 85–130(150) cm, stiffly erect, glabrous below the nodes. **Sheaths** glabrous or hairy at the apices, hairs to 4 mm; **ligules** 0.1–0.2 mm; **blades** 25–50 cm long, 1.5–6 mm wide, flat to involute, glabrous abaxially, scabridulous adaxially, sometimes also with a few scattered hairs near the base. **Panicles** (10)20–50(60) cm long, 2–8(10) cm wide, narrowly ovate, contracted to open; **primary branches**

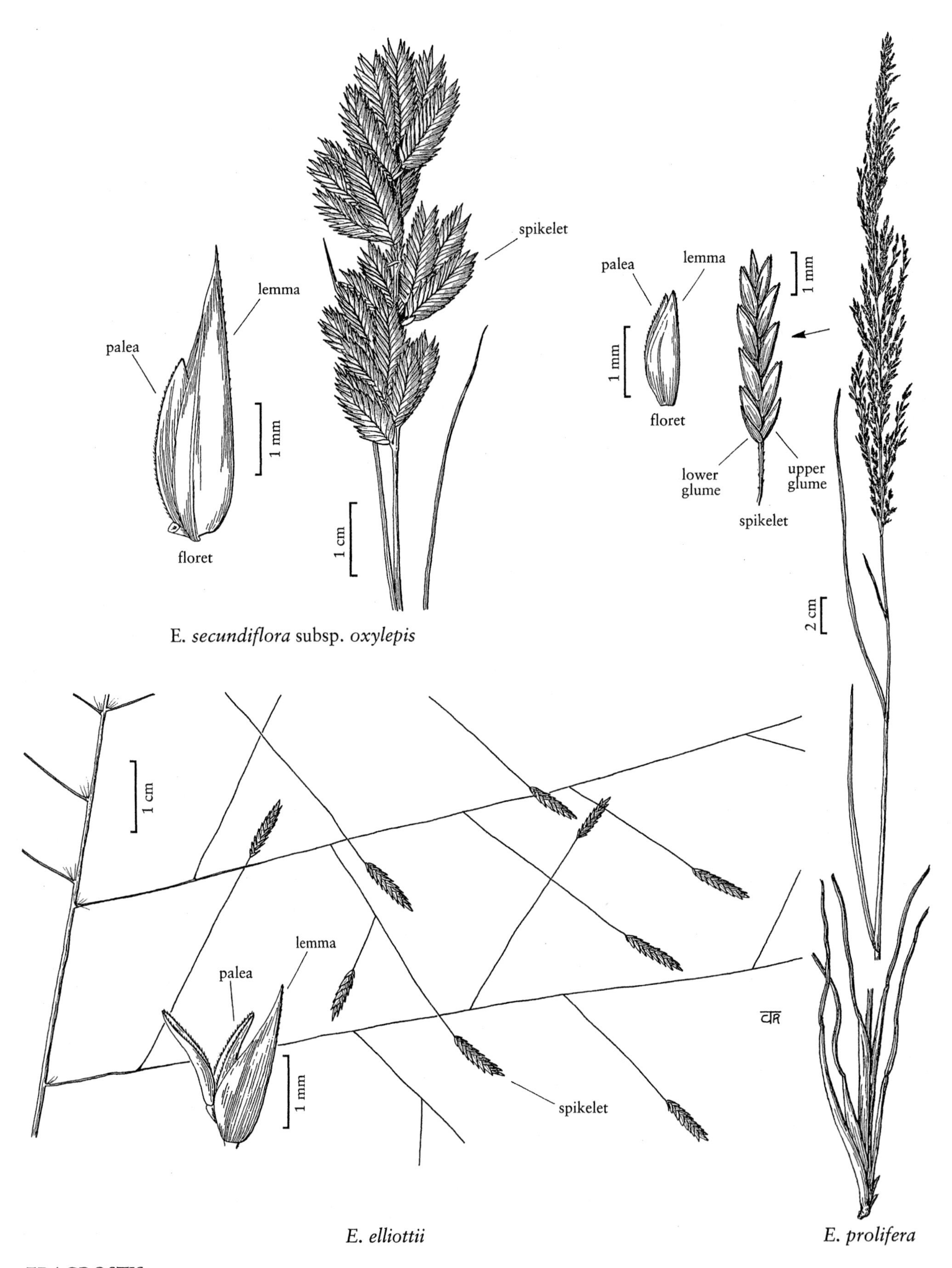
spikelet
lemma
palea
1 mm
floret
1 cm
E. secundiflora subsp. oxylepis
palea
lemma
1 mm
1 mm
floret
lower glume
upper glume
spikelet
2 cm
1 cm
lemma
palea
1 mm
spikelet
E. elliottii
E. prolifera

mostly 2–14 cm, appressed or diverging up to 50°(90°) from the rachises, spikelets congested near the base of the branches; **pulvini** glabrous; **pedicels** 0.3–2.4 mm, appressed, always shorter than the spikelets. **Spikelets** 3.2–10(12) mm long, 0.7–1.4 mm wide, linear-lanceolate, stramineous to plumbeous, sometimes with a reddish tinge, with (5)8–25 florets; **disarticulation** acropetal, glumes first, then the lemmas, paleas persistent. **Glumes** subequal, ovate to lanceolate, hyaline; **lower glumes** 1–1.5 mm; **upper glumes** 1.1–1.6 mm; **lemmas** 1.1–1.8(2) mm, ovate, membranous, apices acute; **paleas** 0.8–1.7 mm, hyaline, narrower than the lemmas, apices obtuse to truncate; **anthers** 2, 0.2–0.3 mm, purplish. **Caryopses** 0.6–0.9 mm, ovoid, flattened ventrally, finely striate, reddish-brown. $2n$ = 40.

Eragrostis prolifera grows on beaches, in brackish water, and along roadsides, at elevations below 5 m in Florida. Its range extends southward from Florida through Mexico and Central America to Colombia.

44. **Eragrostis elongata** (Willd.) Jacq. [p. 102]
LONG LOVEGRASS

Plants perennial; cespitose, with innovations, without rhizomes, not glandular. **Culms** 28–60 cm, erect to decumbent, glabrous below the nodes. **Sheaths** glabrous, apices sparsely hairy, hairs to 2 mm; **ligules** 0.3–0.4 mm; **blades** 5–20 cm long, 0.8–3 mm wide, flat to involute, abaxial surfaces glabrous, adaxial surfaces scabridulous, occasionally hairy near the base. **Panicles** terminal, (5)8–20(22) cm long, 1–3.5 cm wide, spicate to narrowly ovate, branches condensed into glomerate lobes; **primary branches** 0.8–3 cm, appressed or diverging up to 90° from the rachises, spikelet-bearing to the base; **pulvini** glabrous; **pedicels** 0.2–1.3 mm, flattened, mostly appressed, all shorter than the spikelets. **Spikelets** 3–7 mm long, 1.8–2.4 mm wide, linear-lanceolate, stramineous with a reddish-purple tinge, with 8–18 florets; **disarticulation** acropetal, glumes first, then the lemmas, paleas persistent. **Glumes** subequal, 1.2–2 mm, narrowly lanceolate to lanceolate, membranous; **lemmas** 1.5–2.2 mm, lanceolate to ovate, leathery, greenish, lateral veins conspicuous, apices acute; **paleas** 1.1–1.7 mm, hyaline, narrower than the lemmas, keels ciliate, cilia to 0.2 mm, apices obtuse to acute; **anthers** 2, 0.2–0.3 mm, purplish. **Caryopses** 0.4–0.5 mm, ovoid, not grooved, smooth, light brown. $2n$ = unknown.

Eragrostis elongata is native to southeastern Asia and Australia, where it grows in disturbed, sandy soils at 0–50 m. It was collected once near Washington, D.C., probably as an escape from the U.S. Department of Agriculture's experimental grass garden; it has not become established in the *Flora* region.

45. **Eragrostis bahiensis** (Schrad.) Schult. [p. 102]
BAHIA LOVEGRASS

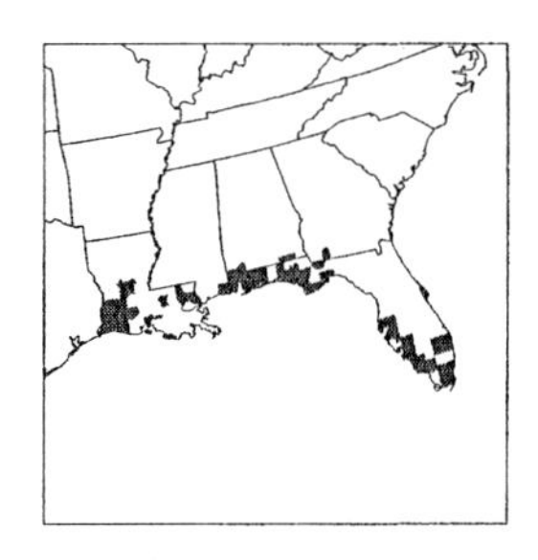

Plants perennial; cespitose, with innovations, without rhizomes, not glandular. **Culms** 25–95(110) cm, erect, glabrous. **Sheaths** glabrous, summits hairy, hairs 1–3 mm; **ligules** 0.2–0.4 mm; **blades** (8)12–40 cm long, 2–5 mm wide, flat to involute, abaxial surfaces glabrous, adaxial surfaces scabridulous and glabrous or long ciliate basally. **Panicles** terminal, 15–30(45) cm long, (4)8–17 cm wide, narrowly ovate, open to contracted; **primary branches** 5–15 cm, diverging 20–90° from the rachises, often capillary, usually naked basally; **pulvini** glabrous; **pedicels** 0.3–6 mm, mostly appressed, scabridulous, always shorter than the spikelets. **Spikelets** 6–15(18) mm long, 1.3–2(2.2) mm wide, narrowly lanceolate, plumbeous, occasionally with a reddish-purple tinge, with 8–30(40) florets; **disarticulation** usually in the rachilla below the florets, occasionally the lemmas falling separately, leaving the paleas on the rachilla. **Glumes** lanceolate to ovate, membranous to subhyaline, keeled; **lower glumes** 1–1.4 mm; **upper glumes** 1.4–1.7 mm; **lemmas** 1.5–2.2 mm, broadly ovate, leathery, scabridulous, lateral veins evident, apices acute; **paleas** 1.4–2.1 mm, hyaline, bases not projecting beyond the lemmas, keels scabridulous, apices acute to obtuse; **anthers** 2, 0.4–0.6 mm, reddish-purple. **Caryopses** 0.6–0.8 mm, obovoid to ellipsoid, terete, somewhat striate, reddish-brown. $2n$ = unknown.

Eragrostis bahiensis grows in sandy soils near river banks, lake shores, and roadsides, at 0–200 m. Its range extends south from the Gulf Coast of the United States through Mexico to Peru, Bolivia, Paraguay, and Argentina.

46. **Eragrostis scaligera** Salzm. *ex* Steud. [p. 102]

Plants perennial; cespitose, with innovations, without rhizomes, not glandular. **Culms** 25–75 cm, erect, glabrous below the nodes. **Sheaths** glabrous or with hairy apices, hairs to 4 mm; **ligules** 0.1–0.3 mm, ciliate; **blades** 4–8(18) cm long, 1–2 mm wide, involute, frequently deciduous, adaxial surfaces mostly glabrous, sometimes pilose near the base. **Panicles** terminal and axillary; **terminal panicles** 5–15 cm long, (1)2–10(12) cm wide, narrowly ovate, open; **axillary panicles** 2–5 cm long, 0.3–0.6 cm wide, usually contracted and partially to completely enclosed by the subtending sheath; **primary branches** 1–10 cm, appressed or diverging up to 90° from the

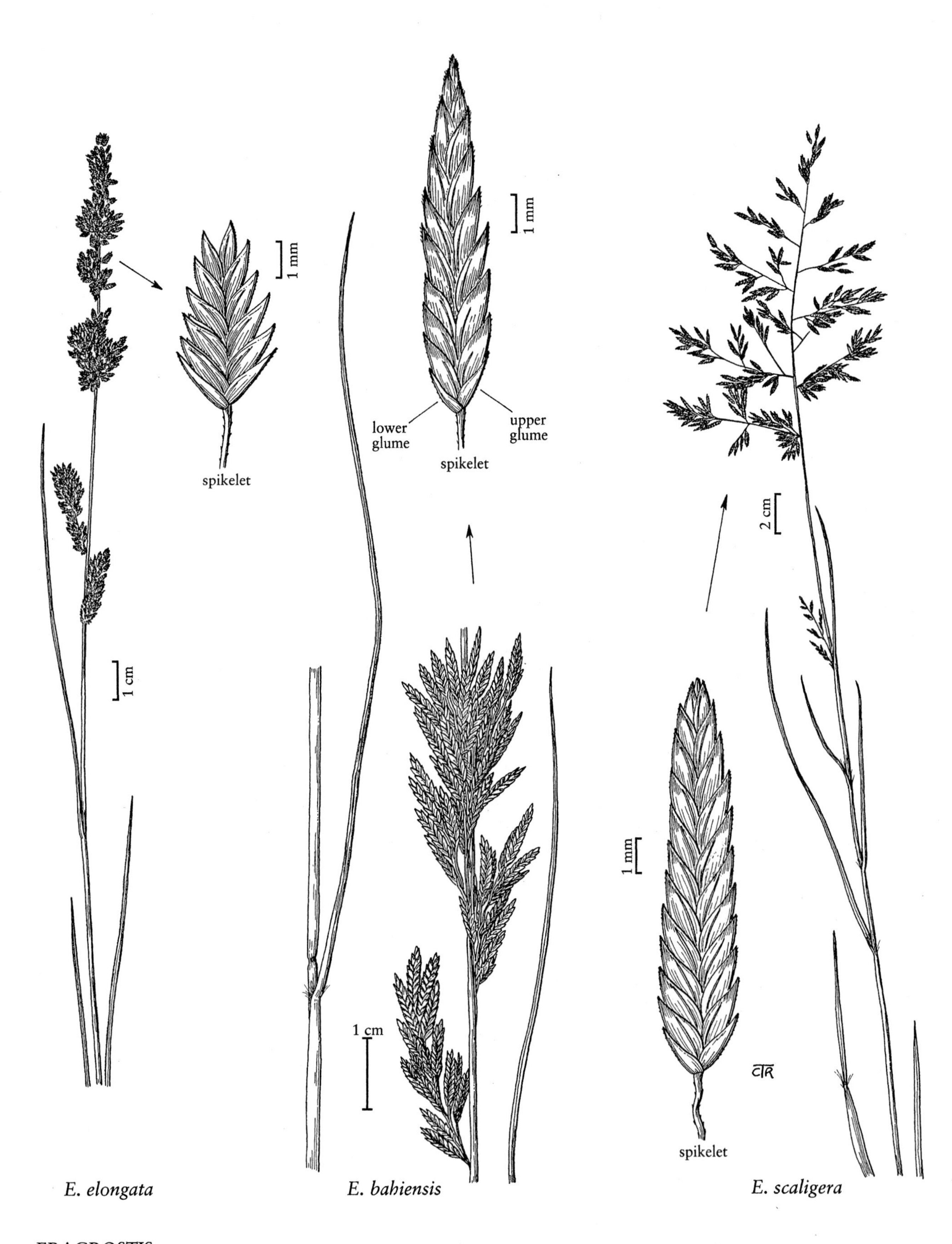
spikelet
1 mm
1 cm
lower glume
upper glume
spikelet
1 mm
1 cm
2 cm
1 mm
spikelet
CTR
E. elongata
E. bahiensis
E. scaligera

rachises, wiry; **pulvini** glabrous or hairy; **pedicels** 0.3–5 mm, appressed, flattened. **Spikelets** 6–15(27) mm long, 1.6–2.4 mm wide, ovate-lanceolate, plumbeous to greenish, often with a reddish-purple tinge, with 10–35(45) florets; **disarticulation** acropetal, glumes first, then the lemmas, paleas persistent. **Glumes** lanceolate to ovate, membranous; **lower glumes** 1.4–2.1 mm; **upper glumes** 1.6–2.2 mm; **lemmas** (1.7)2–2.5 mm, broadly ovate, leathery, apices acute to acuminate; **paleas** 1.1–1.7 mm, hyaline, apices obtuse to truncate; **anthers** 2, 0.3–0.7 mm, reddish-purple. **Caryopses** 0.5–0.6 mm, globose, not grooved, smooth, light reddish-brown. $2n = 40$.

Eragrostis scaligera is known from Lee and Collier counties, Florida, where it grows in sandy areas in the coastal scrub zone and along adjacent roadsides, at 0–10 m. It is native to French Guiana and Brazil.

47. **Eragrostis sessilispica** Buckley [p. 104]
TUMBLE LOVEGRASS

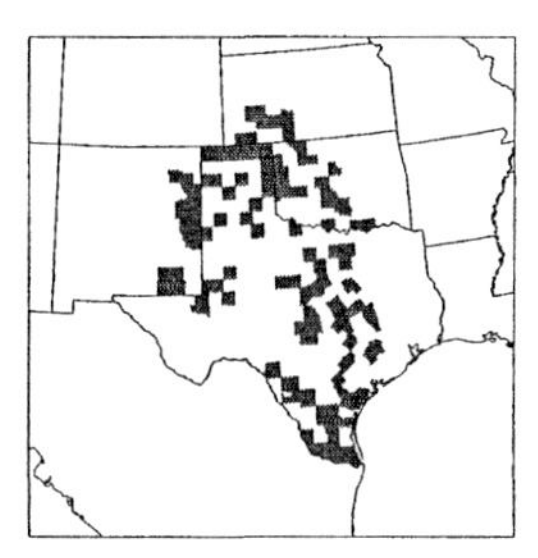

Plants perennial; cespitose, with innovations, without rhizomes, not glandular. **Culms** 30–90 cm, erect or decumbent, glabrous below the nodes. **Sheaths** hairy at the apices and on the collars, sometimes also on the distal portion of the margins, hairs to 5 mm; **ligules** 0.4–0.5 mm; **blades** 5–30 cm long, 1–3 mm wide, usually involute, sometimes flat, abaxial surfaces glabrous or sparsely pilose, hairs to 5 mm, adaxial surfaces scabridulous. **Panicles** 20–65 cm long, 10–35 cm wide, ovate, open; **primary branches** 2–20(24) cm, widely spaced, diverging 20–100° from the rachises, not rebranched, naked basally; **pulvini** hairy; **pedicels** 0–12 mm, appressed, proximal spikelets on each branch sessile or subsessile, the pedicels shorter than 0.4 mm. **Spikelets** 5–13 mm long, 1.4–3 mm wide, oblong to oblanceolate, stramineous to reddish-purple, with 3–12 florets; **disarticulation** tardy, basipetal, in the rachilla below the florets, glumes persistent. **Glumes** lanceolate, broad basally, indurate; **lower glumes** 2.5–6 mm; **upper glumes** 3–6 mm, apices acuminate; **lemmas** 3–5 mm, narrowly ovate to lanceolate, indurate, apices acuminate; **paleas** 2.4–4.6 mm, indurate, gibbous basally but the sides not projecting beyond the lemmas, keels ciliolate, apices obtuse; **anthers** 3, 0.3–0.5 mm, reddish-brown. **Caryopses** 0.9–1.5 mm, ovoid to pyriform, laterally flattened, tapering distally, smooth to faintly striate, brownish. $2n = 40$.

Eragrostis sessilispica grows in prairies, limestone mesas, partial forest openings, and grasslands, generally in sandy soils, at 0–1220 m, often in association with *Prosopsis* and *Quercus*. Its range extends into northern Mexico.

48. **Eragrostis airoides** Nees [p. 104]

Plants perennial; cespitose, with innovations, without rhizomes, not glandular. **Culms** 30–110 cm, erect, glabrous below the nodes. **Sheaths** glabrous or pilose, hairs to 5 mm; **ligules** 0.1–0.2 mm; **blades** 8–22 mm long, (1)2–4(5) mm wide, flat to folded, glabrous abaxially, scabridulous adaxially. **Panicles** 18–70 cm long, 3–25 cm wide, diffuse, ovate; **primary branches** 4–20 cm, appressed or diverging 10–70° from the rachises, naked basally; **pulvini** glabrous; **pedicels** 2.4–11 mm, divergent. **Spikelets** 1.3–2 mm long, 0.8–1.8 mm wide, ovate to lanceolate, plumbeous, with 1–3 florets; **disarticulation** acropetal, in the rachilla below the florets, glumes deciduous; **rachilla** prolonged above the terminal floret. **Glumes** lanceolate to ovate, membranous; **lower glumes** 0.8–1 mm; **upper glumes** 1.1–1.4 mm; **lemmas** 0.8–1.2 mm, ovate, membranous, plumbeous, keels and lateral veins inconspicuous, apices obtuse; **paleas** 0.8–1.2 mm, membranous, bases not projecting beyond the lemmas, apices obtuse; **anthers** 3, 0.3–0.5 mm, purplish. **Caryopses** 0.4–0.5 mm, ovoid, reticulate, reddish-brown. $2n = 36$ (Davidse, pers. comm.).

Eragrostis airoides is a South American species that, in the *Flora* region, is known only from roadsides and disturbed sites in Brazos County, Texas. It is an enigmatic species, often treated as *Sporobolus brasiliensis* (Raddi) Hack., which it resembles in its chromosome base number of $x = 9$ and caryopsis morphology, but its frequent possession of spikelets with more than 1 floret and its mode of spikelet disarticulation argue for its retention in *Eragrostis*.

49. **Eragrostis atrovirens** (Desf.) Trin. *ex* Steud. [p. 104]
THALIA LOVEGRASS

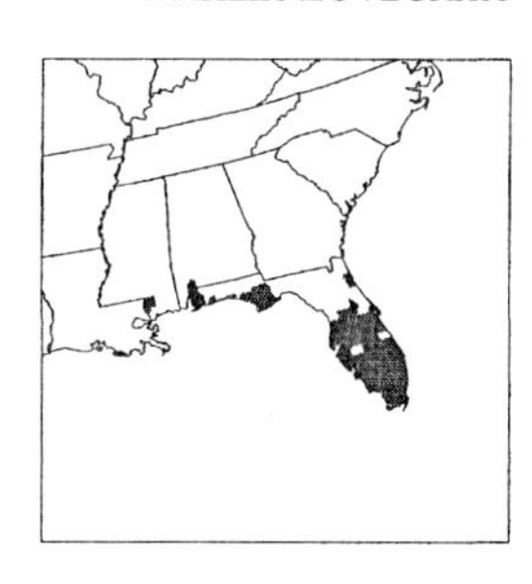

Plants perennial; cespitose, with innovations, without rhizomes, not glandular. **Culms** (60)75–130 cm, erect. **Sheaths** glabrous, apices hairy, hairs to 4 mm; **ligules** 0.1–0.3 mm; **blades** (5)8–20 cm long, (1)2–3(4) mm wide, flat to involute, abaxial surfaces glabrous, adaxial surfaces mostly scabridulous, long ciliate basally. **Panicles** (7)10–20(28) cm long, (2.5)4–15 cm wide, ovate, open; **primary branches** (3)5–10(13) cm, diverging 20–60° from the rachises, wiry, somewhat capillary, naked basally; **pulvini** glabrous or sparingly hairy, hairs shorter than 2 mm; **pedicels** 1–10 mm, appressed, scabridulous. **Spikelets** 5–10(19) mm long, 1.4–2.4 mm wide,

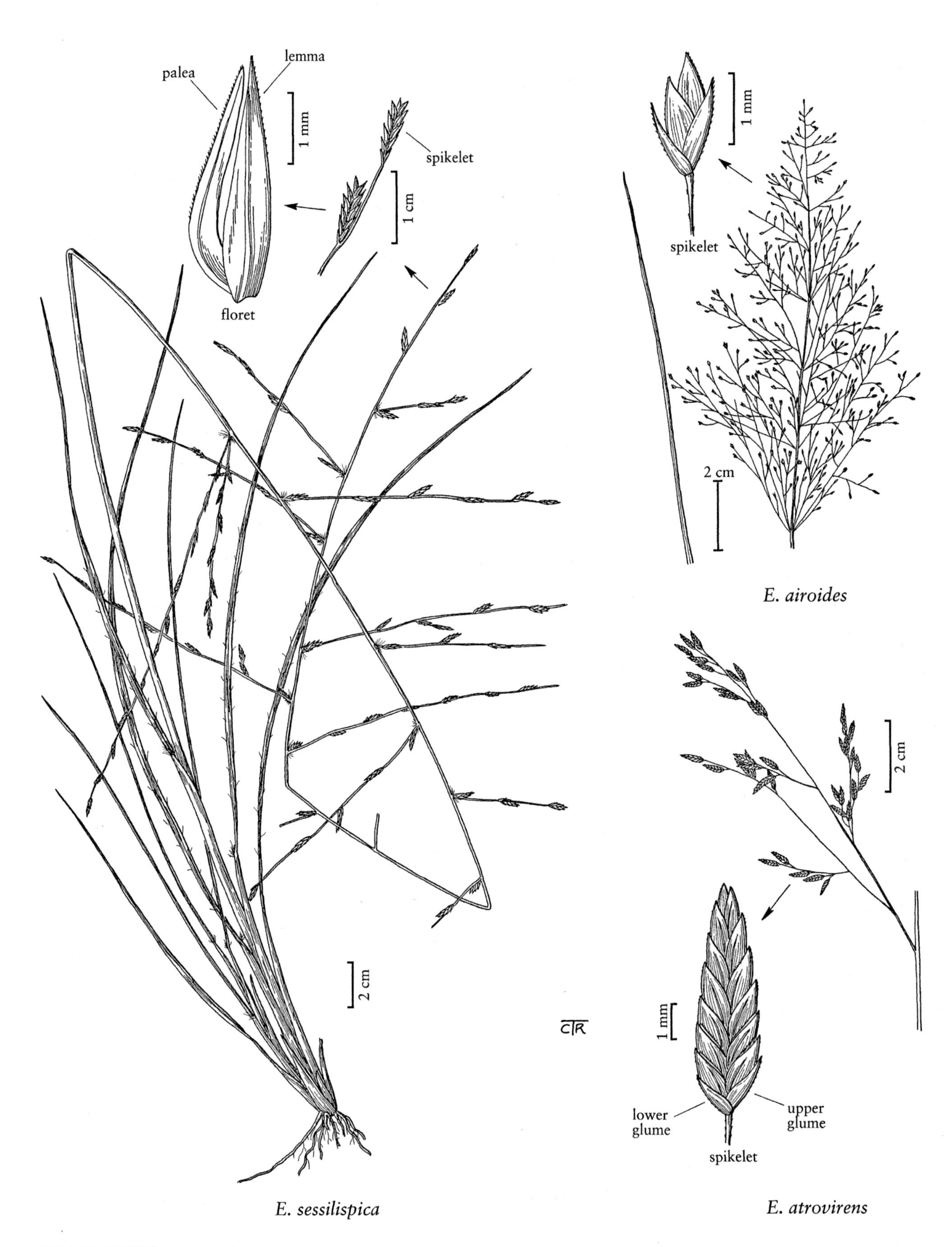

E. airoides

E. sessilispica

E. atrovirens

ovate-lanceolate, plumbeous to purplish, with 10–22 florets; **disarticulation** acropetal, glumes first, then the florets. **Glumes** subequal, lanceolate to ovate, membranous; **lower glumes** 1.2–1.4 mm, narrower than the upper glumes; **upper glumes** 1.4–1.7 mm; **lemmas** (1.5)1.7–2 mm, broadly ovate, leathery, keels scabridulous, lateral veins evident, apices acute; **paleas** 1.4–1.9 mm, hyaline, bases not projecting beyond the lemmas, keels scabridulous, apices acute to obtuse; **anthers** 3, 0.7–0.9 mm, reddish-purple. **Caryopses** 0.6–0.9 mm, obovoid to ellipsoid, terete, opaque, somewhat striate, reddish-brown. $2n$ = 60.

Eragrostis atrovirens is native to northern Africa, but it is now established in southeastern United States, where it grows along railways and roads, on beaches and in ditches, often in wet sandy soils and in association with *Pinus*, *Taxodium*, and *Sabal*.

EXCLUDED SPECIES

The following species have been reported from the *Flora* region, but no specimens supporting their presence, other than in experimental plots, have been found: *Eragrostis acutiflora* (Kunth) Nees, *Eragrostis leptostachya* (R. Br.) Steud., and *Eragrostis suaveolens* Becker *ex* Claus.

17.24 CLADORAPHIS Franch.

Mary E. Barkworth

Plants perennial; synoecious; rhizomatous, occasionally also stoloniferous. **Culms** 2–80 cm, hard, persistent, branched above the base. **Ligules** membranous, ciliate, cilia as long as or longer than the basal membrane; **blades** linear-lanceolate, becoming rolled, hard, and sharp-pointed. **Inflorescences** terminal, exceeding the upper leaves, panicles of racemosely arranged spikelike primary branches; **primary branches** woody, not disarticulating, apices hard, sharp; **secondary branches** shorter than 1 cm, otherwise similar to the primary branches, sometimes clustered. **Spikelets** 7–16 mm, laterally compressed, with 3–16(20) florets; **florets** bisexual; **disarticulation** above the glumes and beneath the florets. **Glumes** more or less equal, markedly exceeded by the florets; **lemmas** 3-veined, unawned; **lodicules** 2; **anthers** 3. **Caryopses** glabrous. x = unknown. Name from the Greek *klados*, 'branch', 'twig', or 'stem', and *rhaphis*, 'needle', alluding to the sharp-pointed inflorescence branches.

Cladoraphis is a southern African genus of two species, both of which grow in open, xeric, sandy habitats.

1. **Cladoraphis cyperoides** (Thunb.) S.M. Phillips [p. 107]
BRISTLY LOVEGRASS

Culms 2–80 cm. **Blades** 2–11 cm long, 4–9 mm wide, margins ciliate basally. **Primary branches** to 8 cm, widely spaced, often separated by more than their own length. **Spikelets** with 4–9(20) florets, densely clustered, appressed to and concealing the branch axes; **lemmas** about 3.5 mm. $2n$ = unknown.

Cladoraphis cyperoides was once collected on a ballast dump at Linnton (near Portland, Oregon). It is not known to have persisted in North America. The spinelike leaves and panicle branches would probably make an encounter memorable.

17.25 POGONARTHRIA Stapf

John R. Reeder

Plants annual or perennial; cespitose. **Culms** 13–100(250) cm, not woody. **Sheaths** open; **ligules** of hairs or membranous and ciliate; **blades** flat or loosely involute. **Inflorescences** terminal, panicles of numerous spikelike branches on elongate rachises. **Spikelets** in 2 rows on 1 side of the flat or trigonous branch axes, with 2–8 florets, additional reduced florets sometimes present distal to the functional florets; **rachilla internodes** tipped with a few short hairs; **disarticulation** initially above the glumes and between the florets or the lemmas falling and the paleas persistent, subsequently at the bases of the panicle branches. **Glumes** unequal, shorter than the spikelets, keeled, acute to acuminate, unawned; **lemmas** 3-veined, keeled, membranous, acute, acuminate, or shortly awned; **paleas** shorter than the lemmas. **Caryopses** ellipsoid to fusiform. $x = 10$. Name from the Greek *pogon*, 'beard', and *arthria*, 'joint', an allusion to the hairs on the rachilla joints.

Pogonarthria includes four species, all of which are native to tropical and southern Africa. One species has become established in Arizona.

1. **Pogonarthria squarrosa** (Licht.) Pilg. [p. 107]
HERRINGBONE GRASS, SEKELGRAS

Plants perennial; densely cespitose. **Culms** 27–100(140) cm, stiffly erect, unbranched, glabrous. **Sheaths** mostly shorter than the internodes, glabrous; **ligules** 0.5–1 mm, of hairs; **blades** 4–30 cm long, 2–5.5 mm wide, adaxial surfaces smooth to slightly scabrous. **Panicles** 20–30 cm, with 50+ branches; **rachises** more or less scabrous; **branches** 2–3 cm, arcuate, axes more or less scabrous. **Spikelets** usually with 4–8 florets. **Lower glumes** 0.8–1.5 mm, 1-veined; **upper glumes** about 2.5 mm, 1–3-veined; **lemmas** about 3 mm, glabrous, mostly smooth, keels scabridulous; **paleas** about 2 mm, keels scabrous; **anthers** about 1 mm. **Caryopses** 1.2–1.3 mm, fusiform, light brown; **embryos** less than ½ as long as the caryopses. $2n = 120$.

Pogonarthria squarrosa is native to eastern and southern Africa, where it is said to be common. In the *Flora* region, *P. squarrosa* grows spontaneously only in a small area in the foothills of the Huachuca Mountains, Cochise County, Arizona, at an elevation of about 1450 m, where it seems to be competing well with native grasses and *Eragrostis lehmanniana*, another African introduction. The plants tend to grow in rather dense colonies of a few square meters, scattered through the area. It is a handsome species that turns reddish-brown as it matures, causing it to stand out among its associates.

17.26 **VASEYOCHLOA** Hitchc.

Robert I. Lonard

Plants perennial; cespitose, occasionally rhizomatous. **Culms** 60–110 cm, erect, glabrous, unbranched. **Sheaths** glabrous; **ligules** membranous, ciliate, cilia subequal to or longer than the membranous base. **Inflorescences** terminal, lanceolate or lance-ovate panicles, exceeding the leaves; **branches** 1 per node. **Spikelets** bisexual, lanceolate, with 5–10 florets; **disarticulation** above the glumes and between the florets. **Glumes** shorter than the adjacent lemmas, about ¼ as long as the spikelets, lanceolate, glabrous, unawned; **calluses** densely pubescent; **lemmas** 5-, 7-, or 9-veined, lanceolate, densely pubescent below and glabrous above, apices truncate to rounded or obtuse, sometimes retuse, unawned; **paleas** shorter than the lemmas, glabrous, splitting down the midline as the caryopsis matures. **Caryopses** to 3 mm, suborbicular, concave, glabrous, amber, with 2 persistent hornlike style branches. x = unknown. Named for George Vasey (1822–1893), curator of the United States National Herbarium, and the Greek *chloa*, 'grass'.

Vaseyochloa is a monotypic genus endemic to the coastal zone of southern Texas. *Coelachyrum* Hochst. & Nees, a small genus of Africa and southwestern Asia, has been suggested as its nearest relative.

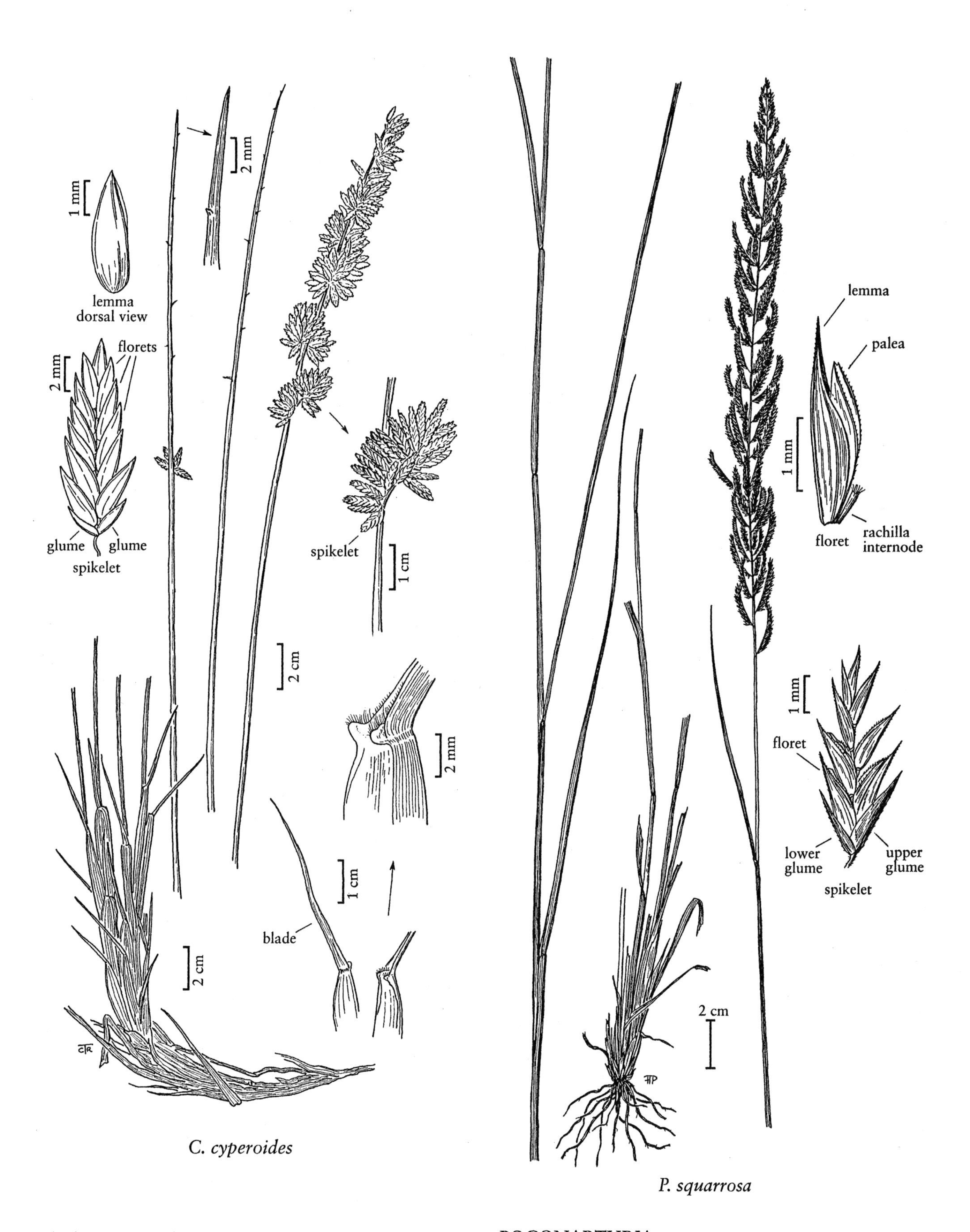

C. cyperoides

P. squarrosa

CLADORAPHIS

POGONARTHRIA

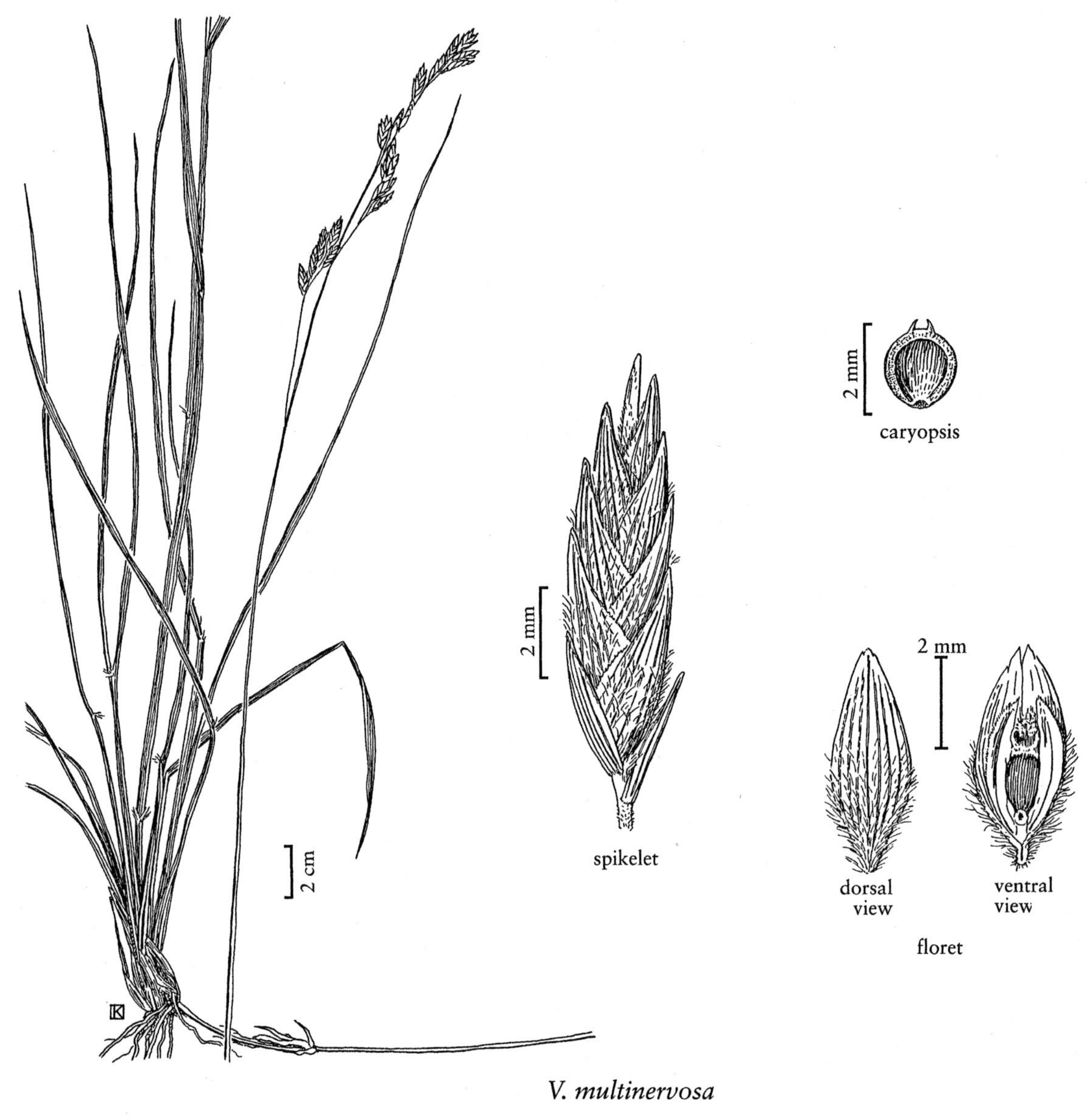

V. multinervosa

VASEYOCHLOA

1. **Vaseyochloa multinervosa** (Vasey) Hitchc. [p. 108]
TEXASGRASS

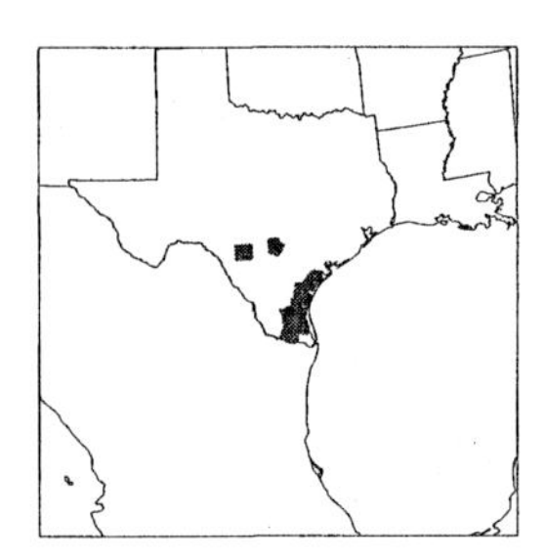

Culms 60–110 cm. **Ligules** 1–3 mm; **blades** 16–35 cm long, 1–6 mm wide, glabrous, lower blades usually folded, upper blades flat. **Panicles** 10–30 cm long, 1–3 cm wide; **branches** 4–13 cm, ascending or the lower branches occasionally spreading, each axil with a tuft of hairs. **Spikelets** 10–16 mm long, 2.5–5 mm wide. **Lower glumes** 2.5–4 mm, 1–7-veined; **upper glumes** 4–5.5 mm, 5–9-veined; **lemmas** 5–7 mm; **anthers** 3, 0.5–2 mm. $2n$ = 56, 60, 68.

Vaseyochloa multinervosa grows in islands of live oaks within rolling sand dunes on the Texas mainland, North Padre Island, and on naturally occurring islands in the Laguna Madre of Texas.

17.27 ELEUSINE Gaertn.

Khidir W. Hilu

Plants annual or perennial; cespitose. **Culms** 10–150 cm, herbaceous, glabrous, branching both at and above the base. **Sheaths** open; **ligules** membranous, ciliate. **Inflorescences** terminal, panicles of (1)2–20 non-disarticulating, spikelike branches, exceeding the upper leaves; **branches** 1–17 cm, all or most in a digitate cluster, sometimes 1(2) branch(es) attached immediately below the terminal whorl, axes flattened, terminating in a functional spikelet. **Spikelets** 3.5–11 mm, laterally compressed, with 2–15 bisexual florets; **disarticulation** above the glumes and between the florets (*E. coracana* not disarticulating). **Glumes** unequal, shorter than the lower lemmas; **lower glumes** 1–3-veined; **upper glumes** 3–5(7)-veined; **lemmas** 3-veined, glabrous, keeled, apices entire, neither mucronate nor awned; **paleas** sometimes with winged keels; **anthers** 3, 0.5–1 mm; **ovaries** glabrous. **Fruits** modified caryopses, pericarp thin, separating from the seed at an early stage in its development; **seeds** usually obtusely trigonous, the surfaces ornamented. x = 8, 9, 10. Name from Eleusis, a Greek town where Demeter, the goddess of harvests, was worshipped.

Eight of the nine species of *Eleusine* are native to Africa, where they grow in mesic to xeric habitats; the exception, *E. tristachya*, is native to South America. Three species have become established in the *Flora* region. When moistened, the seeds of all species are easily freed from the thin pericarp.

Eleusine coracana subsp. *africana*, *E. indica*, and *E. tristachya* are widely distributed weeds. *Eleusine coracana* subsp. *coracana* was domesticated in East Africa and subsequently introduced to India and China. It is frequently grown for grain in India and Africa.

SELECTED REFERENCES **Hilu, K.W.** 1980. Noteworthy collections: *Eleusine tristachya*. Madroño 27:177–178; **Hilu, K.W.** and **J.L. Johnson.** 1997. Systematics of *Eleusine* Gaertn. (Poaceae, Chloridoideae): Chloroplast DNA and total evidence. Ann. Missouri Bot. Gard. 84:841–847; **Phillips, S.M.** 1972. A survey of the genus *Eleusine* Gaertner (Gramineae) in Africa. Kew Bull. 27:251–70.

1. Panicles with 1–3 oblong branches 1–6(8) cm long, attached in a single digitate cluster . . . . 3. *E. tristachya*
1. Panicles with 4–20 linear branches 3.5–17 cm long, 1(2) of the branches attached below the terminal, digitate cluster.
   2. Lower glumes 1-veined; panicle branches 3–5.5 mm wide; surface of the seeds striate . . . . 1. *E. indica*
   2. Lower glumes 2- or 3-veined; panicle branches 5–15 mm wide; surface of the seeds granular . . . . 2. *E. coracana*

1. Eleusine indica (L.) Gaertn. [p. 111]
GOOSEGRASS, ELEUSINE DES INDES, ELEUSINE D'INDE

Plants annual. **Culms** 30–90 cm, erect or ascending, somewhat compressed; **lower internodes** 1.5–2 mm thick. **Sheaths** conspicuously keeled, margins often with long, papillose-based hairs, particularly near the throat; **ligules** 0.2–1 mm, truncate, erose; **blades** 15–40 cm long, 3–7 mm wide, with prominent, white midveins, margins and/or adaxial surfaces often with basal papillose-based hairs. **Panicles** with 4–10(17) branches, often with 1 branch attached as much as 3 cm below the terminal cluster; **branches** (3.5)7–16 cm long, 3–5.5 mm wide, linear. **Spikelets** 4–7 mm long, 2–3 mm wide, with 5–7 florets, obliquely attached to the branch axes. **Lower glumes** 1.1–2.3 mm, 1-veined; **upper glumes** 2–2.9 mm; **lemmas** 2.4–4 mm; **paleas** with narrowly winged keels. **Seeds** ovoid, rugulose and obliquely striate, usually not exposed at maturity. $2n$ = 18.

Eleusine indica is a common weed in the warmer regions of the world. In the *Flora* region, it usually grows in disturbed areas and lawns, and has been found in most states of the contiguous United States.

2. Eleusine coracana (L.) Gaertn. [p. 111]

Plants annual. **Culms** to 62 cm, often branching; **lower internodes** 6–10 mm thick. **Sheaths** glabrous; **ligules** 1–2 mm, ciliate, with 1–2 mm hairs; **blades** 10–60 cm long, 6–12 mm wide, sometimes longer than the culms, adaxial surfaces scabrous or pubescent. **Panicles** subdigitate, with 4–20 branches, 1(2) of the branches attached below the terminal cluster; **branches** 4–17 cm long, 5–15 mm wide, spreading at maturity. **Spikelets** 5–9 mm long, 3–6 mm wide, with 2–9 florets, sometimes not disarticulating at maturity. **Lower glumes** 1.2–3 mm, 2- or 3-veined; **upper glumes** 2.2–6.5 mm; **lemmas** 2.2–5 mm; **anthers** about 1 mm. **Seeds** oblong-globose, granular, usually exposed at maturity. $2n$ = 36.

Eleusine coracana is an allotetraploid, one of its genomes being derived from *E. indica.*

Two subspecies are recognized; only subsp. *coracana* is known from North America.

1. Seeds almost globose, the surface granular to smooth; florets not disarticulating ..... subsp. *coracana*
1. Seeds oblong, the surface shallowly ridged and uniformly granular; florets disarticulating at maturity .... subsp. *africana*

Eleusine coracana subsp. **africana** (Kenn.-O'Byrne) Hilu & de Wet
AFRICAN FINGER MILLET

Culms 21–62 cm. **Blades** 22–50 cm long, 6–10 mm wide. **Branches** slim, 4–17 cm long, 5–7 mm wide. **Spikelets** 5–8 mm long, 3–4 mm wide, with 2–6 florets, disarticulating at maturity. **Seeds** oblong, surfaces shallowly ridged, uniformly granular.

This weedy subspecies hybridizes freely with the cultivated subsp. *coracana.* It tends to have more slender branches than subsp. *coracana* (5–7 mm wide rather than 7–15 mm), which led to its previous inclusion in *E. indica.*

Eleusine coracana (L.) Gaertn. subsp. **coracana**
FINGER MILLET, RAGI

Culms to 17 cm. **Blades** 30–60 cm long, 6–12 mm wide. **Branches** 4–14 cm long, 7–15 mm wide, spikelets closely imbricate. **Spikelets** 5–9 mm long, brown, with 6–9 florets, florets not disarticulating at maturity. **Seeds** almost globose, brownish, surfaces granular to smooth.

Eleusine coracana subsp. *coracana* is the domesticated variant of *E. coracana.* Biochemical data suggest that it evolved from a few populations of the very variable subsp. *africana.* It is cultivated at various agricultural experiment stations and occasionally escapes.

Eleusine coracana subsp. *coracana* has a long historical record dating back at least 5000 years in Africa, and 3000 years in India. Five races, based on inflorescence morphology, are recognized in East Africa where it is widely cultivated for food and drink.

3. **Eleusine tristachya** (Lam.) Lam. [p. 111]
THREESPIKE GOOSEGRASS

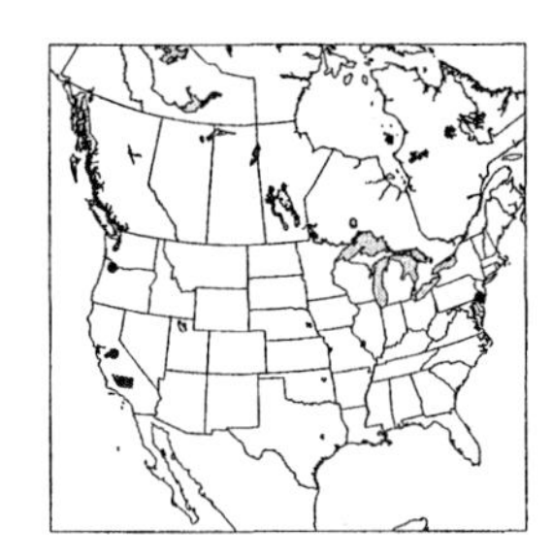

Plants annual. **Culms** 10–45 cm, compressed. **Blades** 6–25 cm long. **Panicles** digitate, with (1)2–3 branches; **branches** 1–6(8) cm long, 5–14 mm wide, oblong. **Spikelets** 8–10 mm, with 5–9(11) florets. **Glumes** unequal; **lower glumes** 2–3 mm; **upper glumes** 3–4 mm; **lemmas** 4–5 mm. $2n$ = 18.

In the 1800s and early 1900s, *Eleusine tristachya* was found on ballast dumps at various ports and transportation centers in the United States. It has since been found as a weed in the Imperial Valley of California (Hilu 1980), but records of collections outside of California appear to be historical, with no populations persisting. The species was originally thought to be native to tropical Africa and introduced into tropical America, but it occurs in Africa only as a rare adventive. It is now considered to be native to tropical America.

17.28 ACRACHNE Wight & Arn. *ex* Chiov.

Sylvia M. Phillips

Plants annual; tufted. **Culms** to approximately 50 cm, erect or geniculate, not woody. **Sheaths** open; **ligules** membranous, ciliate; **blades** broadly linear. **Inflorescences** terminal, panicles of spike-like branches, exceeding the upper leaves; **branches** 1.5–10 cm, subdigitate or in whorls along elongate rachises, axes flattened, with imbricate, subsessile spikelets, terminating in a rudimentary spikelet. **Spikelets** laterally compressed, with 3–25 florets; **disarticulation** of the spikelets below the glumes, of the lemmas within the spikelets acropetal, spikelets falling wholly or in part after only a few lemmas have fallen, paleas persistent. **Glumes** 1-veined, keeled, exceeded by the florets; **lemmas** 3-veined, strongly keeled, firmly membranous to cartilaginous, glabrous, cuspidate or awn-tipped. **Fruits** modified caryopses, pericarp hyaline, rupturing at maturity; **seeds** deeply sulcate, ornamented. x = 9. Name from the Greek *akros*, 'at the tip' and *achne*, 'scale', referring to the inflorescence branches which terminate in an aborted spikelet (Clifford 1996).

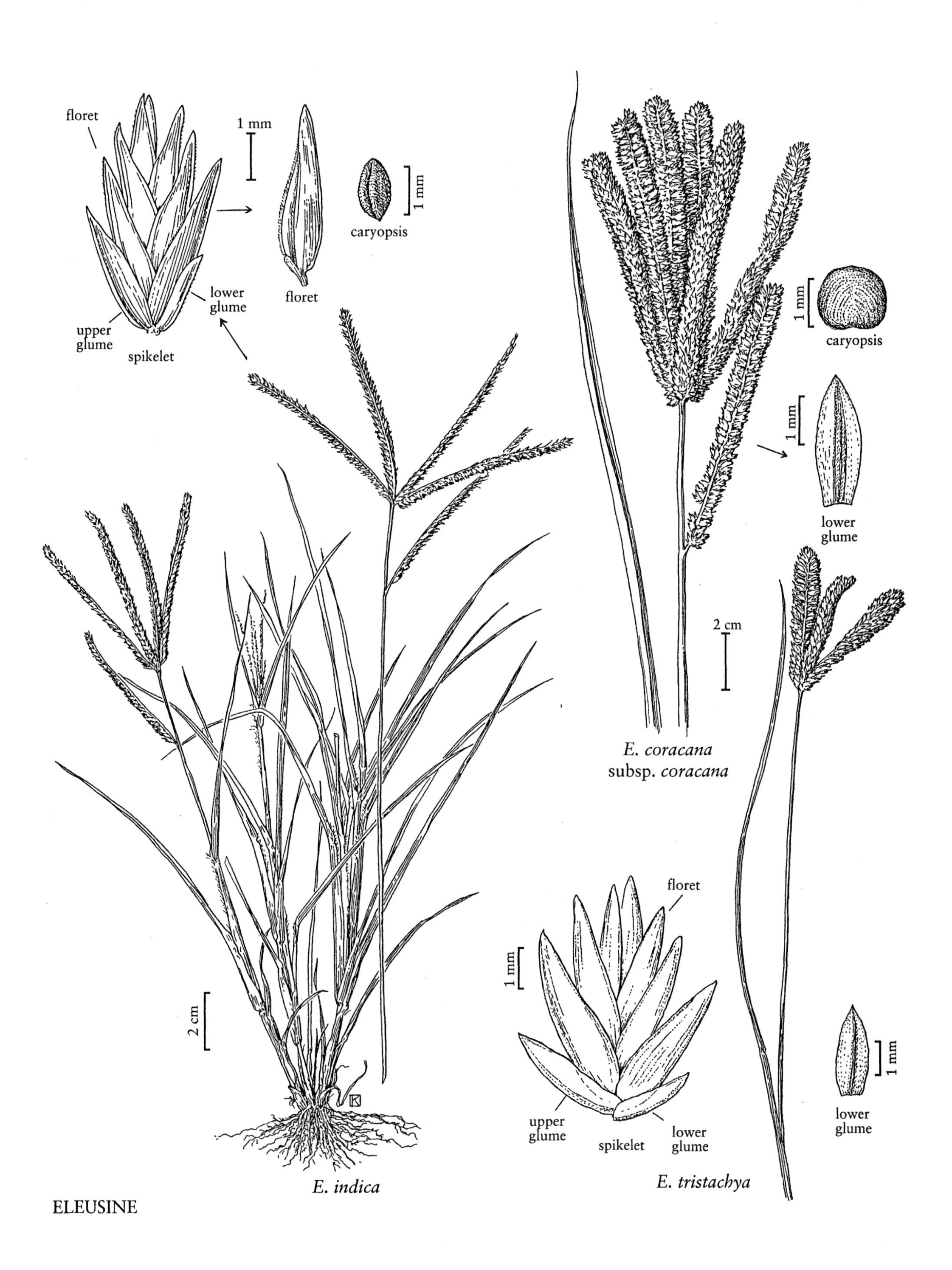
floret
1 mm
floret
1 mm
caryopsis
lower glume
upper glume
spikelet
1 mm
caryopsis
1 mm
lower glume
2 cm
E. coracana subsp. coracana
floret
1 mm
2 cm
upper glume
spikelet
lower glume
1 mm
lower glume
E. indica
E. tristachya

ELEUSINE

Acrachne has four species, all of which are native to the Eastern Hemisphere. One species, *Acrachne racemosa*, which is widely distributed in the tropics, was recently found in southern California. The genus resembles *Eleusine* and *Dactyloctenium* in its fruits and ornamented seeds, but differs from both in its mode of disarticulation.

SELECTED REFERENCES **Clifford, H.T.** 1996. Etymological Dictionary of Grasses, Version 1.0 (CD-ROM). Expert Center for Taxonomic Identification, Amsterdam, The Netherlands; **Phillips, S.M.** 1995. Flora of Ethiopia and Eritrea, vol. 7 (I. Hedberg and S. Edwards, eds.). National Herbarium, Biology Department, Science Faculty, Addis Ababa University, Addis Ababa, Ethiopia and Department of Systematic Botany, Uppsala University, Uppsala, Sweden. 420 pp.

1. **Acrachne racemosa** (B. Heyne *ex* Roem. & Schult.) Ohwi [p. 114]

Culms 17–50 cm. **Blades** 7–20 cm long, 5–11.5 mm wide, bases rounded, setaceous apically. **Inflorescence branches** 1.5-10 cm, in whorls of 2–5, either subdigitate or on rachises up to 15 cm long. **Spikelets** 5.5–13 mm, oblong, serrate in outline, with 6–25 florets. **Lower glumes** 1.2–2.9 mm, linear-oblong, acute, mucronate; **upper glumes** 1.5–3 mm, lanceolate, acuminate, awn-tipped; **lemmas** 2–2.8 mm, narrowly ovate, keels scabrous, shallowly concave distally, midveins excurrent, forming stout, 0.3–0.9 mm awns; **lateral veins** also slightly excurrent. **Fruits** ellipsoid; **seeds** blackish, rugose, finely granular. $2n = 36$.

Acrachne racemosa grows in areas of seasonal rainfall in tropical regions of Africa, Asia, and Australia. It has been found in Riverside County, California and may become established there.

17.29 DACTYLOCTENIUM Willd.

Stephan L. Hatch

Plants annual or perennial; tufted, stoloniferous, or rhizomatous. **Culms** 5–115(160) cm, erect or decumbent, often rooting at the lower nodes, not branching above the base. **Sheaths** not overlapping, open, keeled; **auricles** absent; **ligules** membranous, membranous and ciliate, or of hairs; **blades** flat or involute. **Inflorescences** terminal, panicles of 2–11, digitately arranged spicate branches; **branches** with axes 0.8–11 cm long, extending beyond the spikelets, terminating in a point, the spikelets imbricate in 2 rows on the lower sides. **Spikelets** with 3–7 bisexual florets, additional sterile florets distally; **disarticulation** usually above the glumes, the florets falling as a unit. **Glumes** unequal, shorter than the adjacent lemmas, 1-veined, keeled; **lower glumes** acute, mucronate; **upper glumes** subapically awned, awns curved; **calluses** glabrous; **lemmas** membranous, glabrous, 3-veined (lateral veins sometimes indistinct), strongly keeled, apices entire, mucronate, or awned; **paleas** glabrous; **anthers** 3, yellow; **ovaries** glabrous; **styles** fused. **Fruit** utricles; **seeds** falling free of the hyaline pericarp, transversely rugose or granular. $x = 10$. Name from the Greek *daktylos*, 'finger', and *ktenion*, 'a little comb', describing the comblike inflorescence branches.

Dactyloctenium is primarily an African and Australian genus of 10–13 species. Three species have been introduced in the *Flora* region, two of which have become established. *Dactyloctenium aegyptium* is widespread throughout the warmer areas of the world.

SELECTED REFERENCES **Black, J.M.** 1978. Gramineae. Pp. 88–249 *in* J.M. Black. Flora of South Australia, Part I, ed. 3 (rev. J.P. Jessop). D.J. Woolman, Government Printer, Adelaide, South Australia, Australia. 466 pp.; **Clayton, W.D., S.M. Phillips,** and **S.A. Renvoize.** 1974. Flora of Tropical East Africa. Gramineae (Part 2) (ed. R.M. Pohill). Whitefriars Press, Ltd., London, England. 373 pp.; **Jacobs, S.W.L.** and **S.M. Hastings.** 1993. *Dactyloctenium.* Pp. 527–529 *in* G.J. Harden (ed.). Flora of New South Wales, vol. 4. New South Wales University Press, Kensington, New South Wales, Australia. 775 pp.; **Koekemoer, M.** 1991. *Dactylocteniium* Willd. Pp. 99–101 *in* G.E. Gibbs Russell, L. Watson, M. Koekemoer, L. Smook, N.P. Barker, H.M. Anderson, and M.J. Dallwitz. Grasses of Southern Africa (ed. O.A. Leistner). National Botanic Gardens, Botanical Research Institute, Pretoria, Republic of South Africa. 437 pp.

1. Panicle branches 0.4–1.5 cm long; most spikelets touching those of an adjacent branch .... 2. *D. radulans*
1. Panicle branches 1.5–7 cm long; only the first few proximal spikelets on each branch in contact with those on an adjacent branch.
   2. Anthers 0.5–0.9 mm long; upper glume awns 1–2.5 mm long .... 1. *D. aegyptium*
   2. Anthers 1.1–1.7 mm long; upper glume awns 4.5–10 mm long .... 3. *D. geminatum*

1. **Dactyloctenium aegyptium** (L.) Willd. [p. 114]
DURBAN CROWFOOT

Plants tufted annuals or short-lived, shortly stoloniferous perennials. **Culms** 10–35(100) cm, usually geniculately ascending and rooting at the lower nodes. **Sheaths** keeled, with papillose-based hairs distally; **ligules** 0.5–1.5 mm, membranous, ciliate; **blades** 5–22 cm long, 2–8(12) mm wide, with papillose-based hairs. **Panicle branches** (1)2–6(8), 1.5–6 cm, only the first few spikelets in contact with the spikelets of adjacent branches; **branch axes** extending beyond the spikelets for 1–6 mm. **Spikelets** 3–4.5 mm long, about 3 mm wide. **Glumes** 1.5–2 mm; **lower glumes** ovate, acute; **upper glumes** oblong elliptic, obtuse, awned, awns 1–2.5 mm; **lemmas** 2.5–3.5 mm, ovate, midveins extended into curved, 0.5–1 mm awns; **paleas** about as long as the lemmas; **anthers** 0.5–0.8 mm, pale yellow. **Seeds** cuboid, about 1 mm long and wide, transversely rugose, light tan to reddish-brown. $2n$ = 20, 36, 40, 45, 48.

Dactyloctenium aegyptium is a widely distributed weed of disturbed sites in the *Flora* region. It is also considered a weed in southern Africa, but the seeds have been used for food and drink in times of famine. In addition, bruised young seeds have been used as a fish poison, and extracts are reputed to help kidney ailments and coughing (Koekemoer 1991). In Australia, it is planted as a sand stabilizer along the coast (Jacobs and Hastings 1993).

2. **Dactyloctenium radulans** (R. Br.) P. Beauv. [p. 114]
BUTTONGRASS

Plants annuals or short-lived perennials. **Culms** 5–20(50) cm, decumbent or ascending, rarely erect, usually branched. **Sheaths** glabrous or with papillose-based hairs, slightly keeled; **ligules** to 1 mm, membranous, truncate, ciliate; **blades** flat, bases with papillose-based hairs. **Panicle branches** 2–11, 0.4–1.5 cm, almost globose, most of the spikelets in contact with the spikelets of adjacent branches; **branch axes** extending beyond the distal spikelets as 1–1.5 mm points. **Spikelets** 3.5–5 mm, with 2–5 florets. **Glumes** strongly keeled; **lower glumes** 1–2 mm, ovate, acute; **upper glumes** 1.5–3 mm, oblong elliptic, acuminate, awned, awns 0.5–2.5 mm; **lemmas** 3–4.3 mm, ovate, keels scabrous, 1-veined, veins excurrent about 0.5 mm, apices acuminate to mucronate; **paleas** shorter than the lemmas; **anthers** 0.2–0.8 mm, pale yellow. **Seeds** about 1.2 mm long, about 0.7 mm wide, transversely rugose, brown. $2n$ = unknown.

Dactyloctenium radulans has been found at few locations in the *Flora* region, most of which were associated with wool waste. It is native to Australia, where it is regarded as a valuable ephemeral pasture grass in the drier inland areas but also as a garden weed. It resembles **Dactyloctenium aristatum** Link of tropical eastern Africa, differing primarily in having transversely rugose, rather than granular, caryopses.

3. **Dactyloctenium geminatum** Hack. [p. 114]
DOUBLE COMBGRASS

Plants perennial; stoloniferous, mat-forming. **Culms** 35–112 cm, ascending. **Blades** 4–25 cm long, 3–6 mm wide, flat, more or less glabrous. **Panicle branches** (1)2(3), 2.5–7 cm, often slightly falcate, only the first few spikelets in contact with the spikelets of adjacent branches. **Spikelets** 3–5.3 mm, with 3–6 florets. **Glumes** subequal, 1.3–1.8 mm, widely elliptic to ovate or obovate in profile, awned, awns 4.5–10 mm; **lemmas** 3–3.8 mm, lanceolate, keels smooth or finely scabridulous towards the acute or mucronate apices; **palea keels** not winged; **anthers** 1.1–1.7 mm. **Seeds** about 1 mm long, transversely rugose. $2n$ = unknown.

Dactyloctenium geminatum is native to tropical eastern Africa. It was found at one time on ballast dumps in Maryland, but has not survived in North America.

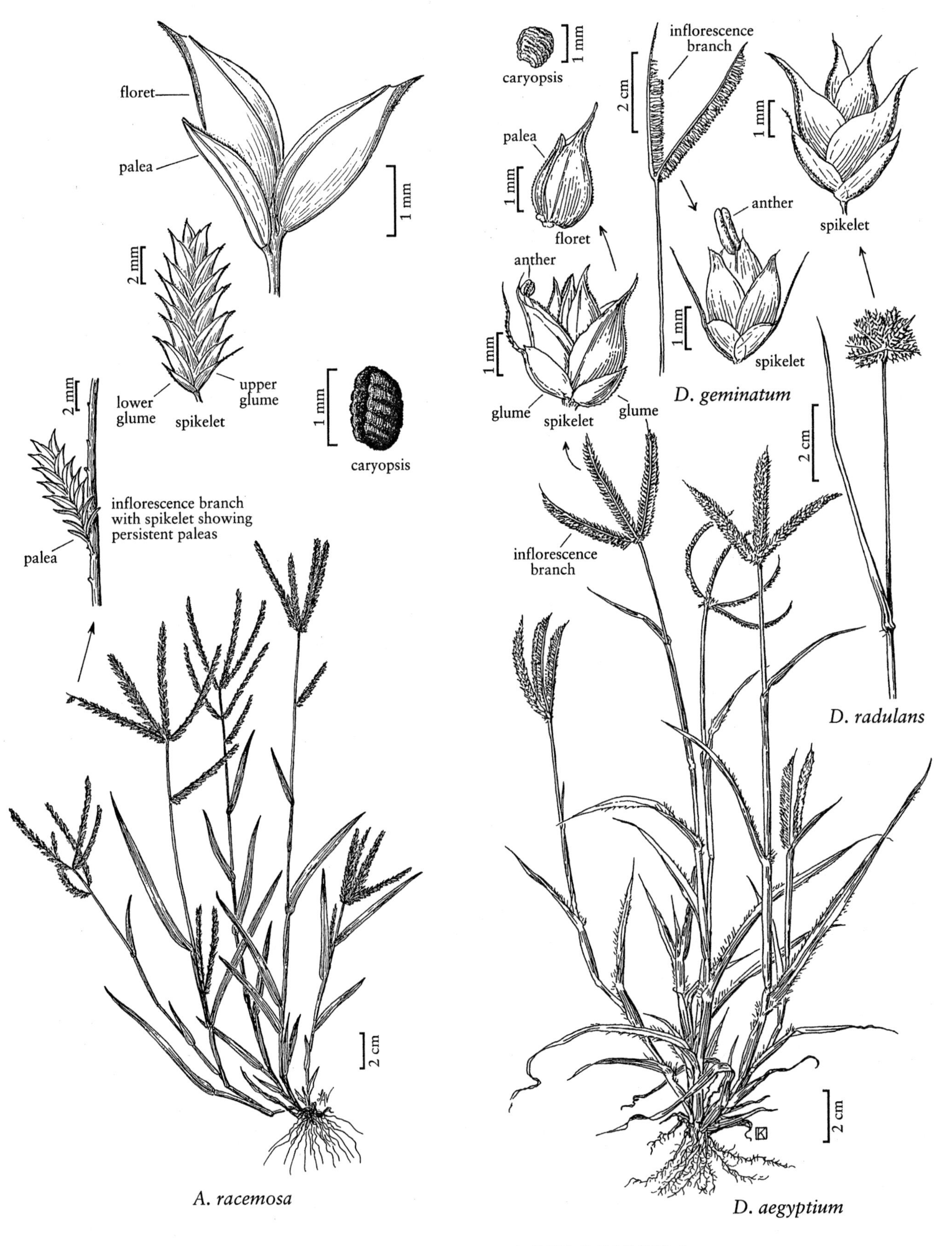

ACRACHNE

DACTYLOCTENIUM

17.30 SPOROBOLUS R. Br.

Paul M. Peterson
Stephan L. Hatch
Alan S. Weakley

Plants annual or perennial; usually cespitose, sometimes rhizomatous, rarely stoloniferous. **Culms** 10–250 cm, usually erect, rarely prostrate, glabrous. **Sheaths** open, usually glabrous, often ciliate at the apices; **ligules** of hairs; **blades** flat, folded, involute, sometimes terete. **Inflorescences** terminal, open or contracted panicles, sometimes partially included in the uppermost sheath. **Spikelets** rounded to laterally compressed, with 1(–3) floret(s) per spikelet; **disarticulation** above the glumes. **Glumes** 0–1-veined; **calluses** poorly developed, usually glabrous; **lemmas** membranous or chartaceous, 1(3)-veined, unawned; **paleas** glabrous, 2-veined, often splitting between the veins at maturity; **anthers** (2)3. **Fruits** utricles or achenes, ellipsoid, obovoid, fusiform, or quadrangular, pericarp free from the seed, becoming mucilaginous when moist in most species, remaining dry and partially adherent to the seed in *S. heterolepis* and *S. clandestinus*. **Cleistogamous spikelets** occasionally present in the lower leaf sheaths. $x = 9$. Name from the Greek *sporos*, 'seed', and *bolos*, 'a throw', referring to the free seeds, which are sometimes forcibly ejected when the mucilaginous pericarp dries.

Sporobolus is a cosmopolitan genus of more than 160 species that grow in tropical, subtropical, and warm-temperate regions throughout the world. Seventy-three species are native to the Western Hemisphere; 27 are native to the *Flora* region, three are established introductions, and one was introduced but has not persisted. Two genera of the Western Hemisphere, *Calamovilfa* and *Crypsis*, resemble *Sporobolus* in having hairy ligules, spikelets with 1 floret, 1-veined lemmas, and fruits with a free pericarp (Peterson et al. 1997).

SELECTED REFERENCES **Baaijens, G.J.** and **J.F. Veldkamp.** 1991. *Sporobolus* (Gramineae) in Malesia. Blumea 35:393–458; **Lægaard, S.** and **P.M. Peterson.** 2001. Flora of Ecuador 68 (ed. G. Harling & L. Andersson). 214(2). Gramineae (part 2): Subfam. Chloridoideae. Botanical Institute, University of Göteborg, Göteborg, Sweden and Section for Botany, Riksmuseum, Stockholm, Sweden. 131 pp.; **McGregor, R.L.** 1990. Seed dormancy and germination in the annual cleistogamous species of *Sporobolus* (Poaceae). Trans. Kansas Acad. Sci. 93:8–11; **Peterson, P.M., R.D. Webster** and **J. Valdés-Reyna.** 1997. Genera of New World Eragrostideae (Poaceae: Chloridoideae). Smithsonian Contr. Bot. 87:1–50; **Riggins, R.** 1977. A biosystematic study of the *Sporobolus asper* complex (Gramineae). Iowa State J. Res. 51:287–321; **Simon, B.K.** and **S.W.L. Jacobs.** 1999. Revision of the genus *Sporobolus* (Poaceae, Chloridoideae) in Australia. Austral. J. Bot. 12:375–448; **Weakley, A.S.** and **P.M. Peterson.** 1998. Taxonomy of the *Sporobolus floridanus* complex (Poaceae: Sporobolinae). Sida 18:247–270.

1. Plants annuals or short-lived perennials flowering in the first year.
   2. Lower panicle nodes with 7–15 branches; mature panicles pyramidal . . . . 2. *S. pyramidatus*
   2. Lower panicle nodes with 1–3 branches; mature panicles cylindrical to pyramidal.
      3. Spikelets 0.7–1.1 mm long; anthers 0.2–0.3 mm long . . . . 1. *S. tenuissimus*
      3. Spikelets 1.6–6 mm long; anthers 0.3–3.2 mm long.
         4. Mature panicles 10–35 cm long, 4.5–30 cm wide, open; secondary branches spreading; pedicels usually 6–25 mm long, spreading . . . . 19. *S. texanus*
         4. Mature panicles 1–5 cm long, 0.2–0.5 cm wide, contracted; secondary branches appressed; pedicels usually 0.1–4 mm long, appressed.
            5. Lemmas strigose; spikelets 2.3–6 mm long; mature fruits (1.1)1.8–2.7 mm long . . . 3. *S. vaginiflorus*
            5. Lemmas glabrous; spikelets 1.6–3 mm long; mature fruits 1.2–1.8 mm long . . . . 4. *S. neglectus*
1. Plants perennial.
   6. Plants with rhizomes.
      7. Spikelets 1.4–3.2 mm long.
         8. Panicles 0.4–1.6 cm wide, spikelike, blades usually conspicuously distichous . . . . 5. *S. virginicus*
         8. Panicles 2.4–8 cm wide, somewhat contracted to lax and open, blades not obviously distichous . . . . 12. *S. fimbriatus*
      7. Spikelets 4–10 mm long.

9. Panicles (0.6)1–8 cm wide, open to somewhat contracted, narrowly pyramidal, well-exerted from the uppermost sheath; branches without spikelets on the lower ⅓ ..... 25. *S. interruptus*

9. Panicles 0.04–1.6 cm wide, narrow or spikelike, partially to wholly included in the uppermost sheath; branches spikelet-bearing to the base.

10. Fruits 1–2 mm long; pericarp gelatinous, slipping from the seed when wet; panicles 5–30 cm long, 0.4–1.6 cm wide; lemmas glabrous, smooth ..... 6. *S. compositus*

10. Fruits (1.5)2.4–3.5 mm long; pericarp loose but neither gelatinous nor slipping from the seed when wet; panicles 5–11 cm long, 0.04–0.3 cm wide; lemmas minutely pubescent or scabridulous ..... 7. *S. clandestinus*

6. Plants without rhizomes.

11. Upper glumes usually less than ⅔ as long as the florets.

12. Lower panicle branches much shorter than the adjacent internodes, appressed to strongly ascending ..... 10. *S. creber*

12. Lower panicle branches usually as long as or longer than the adjacent internodes, appressed or ascending.

13. Spikelets 2–2.7 mm long; upper glumes usually ½–⅔ as long as the florets, acute to obtuse, entire ..... 8. *S. indicus*

13. Spikelets 1.3–1.8(2) mm long; upper glumes usually less than ½ as long as the florets, rarely longer; truncate, erose to denticulate.

14. Anthers 0.9–1.1 mm long, usually 3, rarely 2; branches spikelet-bearing to the base ..... 9. *S. jacquemontii*

14. Anthers 0.5–0.8 mm long, usually 2, rarely 3; branches without spikelets on the lower ¼ ..... 11. *S. diandrus*

11. Upper glumes at least ⅔ as long as the florets, often longer.

15. Spikelets 1–2.5(2.8) mm long [for opposite lead, see p. 117].

16. Lower sheaths keeled and flattened below ..... 13. *S. buckleyi*

16. Lower sheaths rounded below.

17. Panicles 12–35 cm wide, open.

18. Sheath apices with a conspicuous tuft of white hairs; flag blades nearly perpendicular to the culms ..... 17. *S. cryptandrus*

18. Sheath apices glabrous or with a few scattered hairs; flag blades ascending.

19. Secondary panicle branches spikelet-bearing to the base; pedicels mostly appressed, mostly 0.2–0.5 mm long; panicles 20–60 cm long ..... 15. *S. wrightii*

19. Secondary panicle branches without spikelets on the lower ¼–½; pedicels mostly spreading, mostly 0.5–25 mm long; panicles 10–45 cm long.

20. Pedicels 0.5–2 mm long; anthers 1.1–1.8 mm long ..... 16. *S. airoides*

20. Pedicels 6–25 mm long; anthers 0.3–1 mm long ..... 19. *S. texanus*

17. Panicles 0.2–12(14) cm wide, contracted to open.

21. Mature panicles 0.2–5 cm wide, contracted, often spikelike, the panicle branches appressed or diverging no more than 30° from the rachises.

22. Primary panicle branches without spikelets on the lower ⅛–½ of their length.

23. Leaf blades 1–1.5 mm wide; ligules 0.2–0.4 mm long ..... 20. *S. nealleyi*

23. Leaf blades 2–6 mm wide; ligules 0.3–1 mm long.

24. Lower panicle nodes with 7–12(15) branches; anthers 0.2–0.4 mm long ..... 2. *S. pyramidatus*

24. Lower panicle nodes with 1–3 branches; anthers 0.5–1 mm long ..... 17. *S. cryptandrus*

22. Primary panicle branches spikelet-bearing to the base.

25. Lower glumes usually 1-veined; mature panicles 0.2–0.8(1) cm wide; lemmas 2–3.2 mm long, linear-lanceolate; upper glumes 2–3.2 mm long; anthers 3, 0.3–0.5 mm long; plants primarily from west of the Mississippi River ..... 18. *S. contractus*

25. Lower glumes usually without veins; mature panicles 1–5 cm wide; lemmas 1.1–2 mm long, ovate; upper glumes 1.1–2 mm long; anthers 2 or 3, 0.5–1 mm long; plants primarily from east of the Mississippi River .... 14. *S. domingensis*

21. Mature panicles 4.5–30 cm wide, open, pyramidal to subovate or oblong, the panicle branches diverging more than 10° from the rachises, sometimes reflexed.

26. Lower panicle nodes with 7–12(15) branches; anthers 0.2–0.4 mm long .... 2. *S. pyramidatus*

26. Lower panicle nodes with 1–2(3) branches; anthers 0.4–1 mm long.

27. Pedicels 6–25 mm long, spreading; panicles 4.5–30 cm wide, about as long as wide, diffuse .... 19. *S. texanus*

27. Pedicels 0.1–3 mm long, appressed or spreading; panicles 0.3–14 cm wide, longer than wide, open and/or drooping.

28. Culms 10–50(60) cm tall, 0.7–1.2 mm thick near the base; plants with hard, knotty bases; blades (0.6)1.5–6(7) cm long, 1–1.5 mm wide, involute, spreading at right angles to the culms .... 20. *S. nealleyi*

28. Culms 30–120 cm tall, 1–3.5 mm thick near the base; plant bases not hard and knotty; blades (2)5–26 cm long, 2–6 mm wide, flat to involute, ascending or at right angles to the culms.

29. Pedicels appressed to the secondary branches; primary branches appressed, spreading, or reflexed; pulvini glabrous; rachises straight, erect; mature panicles narrowly pyramidal, lower branches longer than the middle branches .... 17. *S. cryptandrus*

29. Pedicels spreading from the secondary branches; primary branches reflexed; pulvini pubescent; rachises drooping or nodding; mature panicles subovate to oblong, lower branches no longer than those in the middle .... 21. *S. flexuosus*

15. Spikelets 2.5–10 mm long [for opposite lead, see p. 116].

30. Lower panicle nodes with 3 or more branches.

31. Mature panicles 2–6 cm wide, pyramidal; panicle branches diverging 20–100° from the rachises; blades 0.8–2 mm wide; fruits 1.4–1.8 mm long .... 23. *S. junceus*

31. Mature panicles 0.4–1.6 cm wide, narrow, contracted; panicle branches appressed or diverging to 20° from the rachises; blades 2–5 mm wide; fruits 1.8–2.3 mm long .... 24. *S. purpurascens*

30. Lower panicle nodes with 1–2(3) branches.

32. Mature panicles 0.04–4 cm wide, spikelike; panicle branches appressed.

33. Spikelets 4–6(10) mm long, stramineous to purplish-tinged; panicles terminal and axillary; sheaths without a conspicuous apical tuft of hairs.

34. Lemmas minutely pubescent or scabridulous, chartaceous and opaque; pericarps loose but neither gelatinous nor slipping off the seeds when wet; fruits (1.5)2.4–3.5 mm long .... 7. *S. clandestinus*

34. Lemmas usually glabrous and smooth, membranous to chartaceous and hyaline; pericarps gelatinous, slipping off the seeds when wet; fruits 1–2 mm long .... 6. *S. compositus*

33. Spikelets 1.7–3.5(4) mm long, whitish to plumbeous; panicles all terminal; sheaths with a conspicuous apical tuft of hairs.

35. Culms 40–100(120) cm tall, 2–4(5) mm thick near the base; mature panicles 0.2–0.8(1) cm wide; anthers 0.3–0.5 mm long ... 18. *S. contractus*

35. Culms 100–200 cm tall, (3)4–10 mm thick near the base; mature panicles 1–4 cm wide; anthers 0.6–1 mm long .... 22. *S. giganteus*

32. Mature panicles (0.6)1–30 cm wide, usually open, narrowly pyramidal to pyramidal or ovate; panicle branches appressed or spreading.
    36. Spikelets 2.3–3 mm long; panicles 4.5–30 cm wide, diffuse, about as long as wide; branches capillary; anthers 0.3–1 mm long . . . . 19. *S. texanus*
    36. Spikelets 3–7.2 mm long; panicles 0.6–15 cm wide, longer than wide, not diffuse; branches not capillary; anthers 1.5–5 mm long.
        37. Mature spikelets plumbeous; sheath bases dull, fibrous.
            38. Anthers 3–4.2 mm long; ligules 0.2–0.7 mm long; plants from Arizona . . . . 25. *S. interruptus*
            38. Anthers 1.7–3 mm long; ligules 0.1–0.3 mm long; plants not known from Arizona . . . . 26. *S. heterolepis*
        37. Mature spikelets purplish-brown to purplish; sheath bases shiny, indurate.
            39. Blades 0.5–1.2 mm wide, subterete to terete in cross section, at least at the base, sometimes channeled for portions of their length, sometimes becoming tightly involute distally, senescing or turning tan in late fall, the margins smooth; pedicels with scattered ascending hairs . . . . 27. *S. teretifolius*
            39. Blades 0.8–10 mm wide, flat or V-shaped in cross section, flat, folded, or involute when dry, remaining green well into winter or yellowing at maturity, the margins usually scabridulous, occasionally smooth; pedicels glabrous, sometimes scabridulous or scabrous.
                40. Lower glumes from 0.9 times as long as to longer than the upper glumes; culms 30–80(90) cm tall; panicles 10–25 cm long; pedicels 0.5–4(8) mm long, usually shorter than the spikelets, appressed . . . . 28. *S. curtissii*
                40. Lower glumes from 0.6–0.9 (0.94) times as long as the upper glumes; culms (30)45–250 cm tall; panicles 15–50 cm long; pedicels 2–22 mm long, spreading or appressed.
                    41. Pedicels appressed; lemmas 4.4–6.5 mm long; anthers 3.5–5 mm long; spikelets purplish . . . . 29. *S. silveanus*
                    41. Pedicels spreading; lemmas 3–4.3 mm long; anthers 2–3.4 mm long; spikelets purplish-brown.
                        42. Blades (2)3–10 mm wide, pale bluish-green, yellowing at maturity; panicles (18)30–50 cm long, 4–15 cm wide; lower glumes (0.6)0.75–0.94 times as long as the upper glumes . . . . 30. *S. floridanus*
                        42. Blades 1.2–2(3) mm wide, dark green, remaining green well into winter; panicles 15–30 cm long, 2–6 cm wide; lower glumes 0.6–0.83 times as long as the upper glumes . . . . 31. *S. pinetorum*

1. **Sporobolus tenuissimus** (Mart. *ex* Schrank) Kuntze [p. 120]
TROPICAL DROPSEED

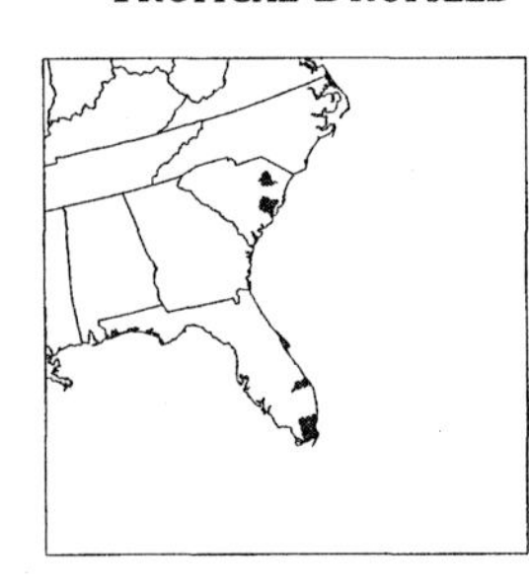

Plants annual; tufted. **Culms** 30–100 cm. **Sheaths** glabrous, including the apices; **ligules** 0.2–0.3 mm; **blades** 5–23 cm long, 2–4 mm wide, flat or folded, glabrous on both surfaces, margins glabrous. **Panicles** (8)15–30 cm long, 3.5–8 cm wide, open, diffuse, cylindrical; **lower nodes** with 1–2(3) branches; **primary branches** 0.6–5 cm, capillary, spreading 30–70° from the rachises, without spikelets on the lower ½; **secondary branches** spreading; **pedicels** 0.5–5 mm. **Spikelets** 0.7–1.1 mm, plumbeous to purplish. **Glumes** unequal, obovate to ovate, membranous; **lower glumes** 0.1–0.4 mm, occasionally absent; **upper glumes** 0.2–0.5 mm; **lemmas** 0.7–1.1 mm, elliptic, membranous, glabrous, acute to obtuse; **paleas** 0.7–1.1 mm, elliptic, membranous; **anthers** 0.2–0.3 mm, yellowish. **Fruits** 0.4–0.7 mm, pyriform or quadroid, somewhat laterally flattened, light brownish to whitish. $2n = 12$.

Sporobolus tenuissimus is native to the Western

Hemisphere, and introduced to Africa and Asia. Its native distribution in the Americas is tropical, extending from southern Mexico to Brazil and Paraguay. It has been found at a few locations in the southeastern United States, at 0–100 m. It grows in disturbed areas, often occurring as a weed in gardens and cultivated fields.

2. **Sporobolus pyramidatus** (Lam.) Hitchc. [p. 120]
WHORLED DROPSEED

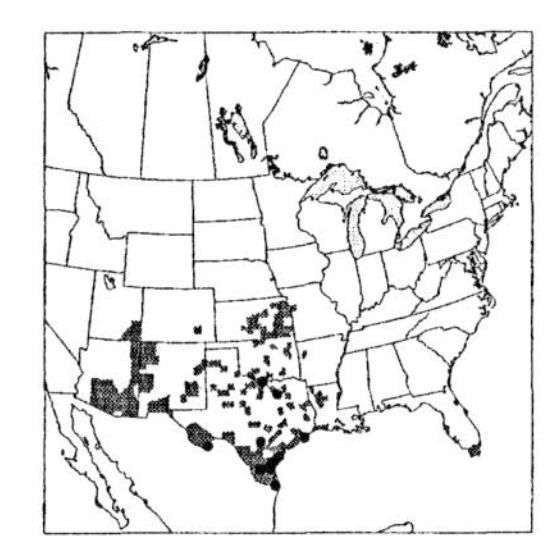

Plants annual, or short-lived perennials flowering in the first year; cespitose, not rhizomatous. **Culms** 7–35(60) cm, erect or decumbent. **Sheaths** rounded below, margins and apices hairy, hairs to 3 mm; **ligules** 0.3–1 mm; **blades** 2–12(20) cm long, 2–6 mm wide, flat, abaxial surface glabrous, adaxial surface scabridulous, sometimes sparsely hispid, margins ciliate-pectinate. **Panicles** 4–15(18) cm long, 0.3–6 cm wide, open (contracted when immature), pyramidal; **lower nodes** with 7–12(15) branches; **primary branches** 0.5–4.5 cm, spreading 30–90° from the rachis, with elongated glands, without spikelets on the lower ⅓–½; **secondary branches** appressed; **pedicels** 0.1–0.5 mm, appressed. **Spikelets** 1.2–1.8 mm, plumbeous or brownish, often secund along the branch. **Glumes** unequal, ovate to obovate, membranous; **lower glumes** 0.3–0.7 mm, without midveins; **upper glumes** 1.2–1.8 mm, at least ⅔ as long as the florets, often longer; **lemmas** 1.2–1.7 mm, ovate to elliptic, membranous, glabrous, acute; **paleas** 1.1–1.6 mm, ovate to elliptic, membranous, glabrous; **anthers** 0.2–0.4 mm, yellowish or purplish. **Fruits** 0.6–1 mm, obovoid, faintly striate, light brownish. $2n$ = 24, 36, 54.

Sporobolus pyramidatus is native to the Americas, extending from the southern United States to Argentina. It grows in disturbed soils, roadsides, railways, coastal sands, and alluvial slopes in many plant communities, at elevations from 0–1500 m. Morphologically, it is very similar to the Eastern Hemisphere **S. coromandelianus** (Retz.) Kunth, suggesting that they are closely related.

3. **Sporobolus vaginiflorus** (Torr. *ex* A. Gray) Alph. Wood [p. 120]
POVERTY GRASS, SPOROBOLE ENGAINÉ

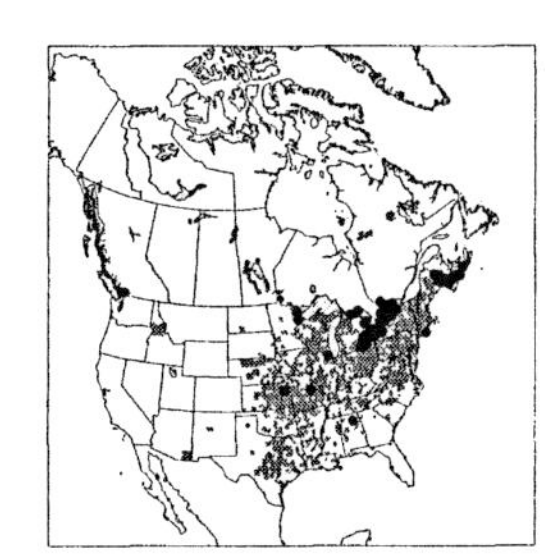

Plants annual; tufted, delicate. **Culms** 15–60(70) cm, erect to decumbent, wiry. **Sheaths** often inflated, sometimes with sparse hairs basally, hairs papillose-based, glabrous or the apices with small tufts of hairs, hairs to 3 mm; **ligules** 0.1–0.3 mm; **blades** 2–12(25) cm long, 0.6–2 mm wide, flat to loosely involute, glabrous abaxially, scabridulous adaxially, bases of both surfaces sometimes with a few papillose-based hairs, margins smooth or scabridulous. **Panicles** terminal and axillary, 1–5 cm long, 0.2–0.5 cm wide, contracted, cylindrical, enclosed in the uppermost sheath; **lower nodes** with 1–2(3) branches; **primary branches** 0.4–1.8 cm, appressed, spikelet-bearing to the base; **secondary branches** appressed; **pedicels** 0.2–4 mm, appressed, scabridulous. **Spikelets** 2.3–6 mm, yellowish to purplish- or grayish-mottled. **Glumes** subequal, linear-lanceolate to lanceolate-triangular or ovate, membranous to chartaceous, glabrous; **lower glumes** (2.2)2.8–4.7 mm; **upper glumes** (2.4)3–5 mm; **lemmas** (2.1)3–5.4 mm, lanceolate to lanceolate-triangular, 1–3-veined, chartaceous, often mottled with purplish or grayish areas, strigose, hairs less than 0.5 mm, apices acuminate or acute; **paleas** (2.1)3–6 mm, as long as or longer than the lemmas, sometimes tapering into a beak, lanceolate to lanceolate-triangular, chartaceous, strigose; **anthers** 3, 1.2–3.2 mm, yellowish or purplish. **Fruits** (1.1)1.8–2.7 mm, obovoid, laterally flattened, light brownish, translucent. $2n$ = 54.

Sporobolus vaginiflorus is a North American species, native to the eastern portion of the *Flora* region and probably introduced in the west. It grows in disturbed sites within many plant communities, commonly in sandy to sandy-clay soils, these often derived from calcareous parent materials. Its elevational range is 1–1250 m.

1. Sheath bases sparsely hairy; glumes usually longer than the florets; lemmas always faintly 3-veined . . . . var. *ozarkanus*
1. Sheath bases usually glabrous; glumes usually shorter than the florets; lemmas usually 1-veined . . . . var. *vaginiflorus*

Sporobolus vaginiflorus var. **ozarkanus** (Fernald) Shinners
OZARK DROPSEED

Sheath bases sparsely hairy, hairs papillose-based. **Spikelets** 2.3–4.6 mm. **Glumes** usually longer than the florets; **lemmas** 2.1–3.9 mm, faintly 3-veined; **paleas** 2.1–4 mm. **Fruits** 1.1–2 mm.

Sporobolus vaginiflorus var. *ozarkanus* grows primarily in the central and southeastern United States.

Sporobolus vaginiflorus (Torr. *ex* A. Gray) Alph. Wood var. **vaginiflorus**

Sheath bases usually glabrous. **Spikelets** mostly 3–6 mm. **Glumes** usually shorter than the florets; **lemmas** mostly 3–5.4 mm, usually 1-veined, sometimes faintly 3-veined; **paleas** mostly 3–5.4 mm. **Fruits** 1.8–2.7 mm.

Sporobolus vaginiflorus var. *vaginiflorus* is the most wide-ranging of the two varieties, extending north into Canada.

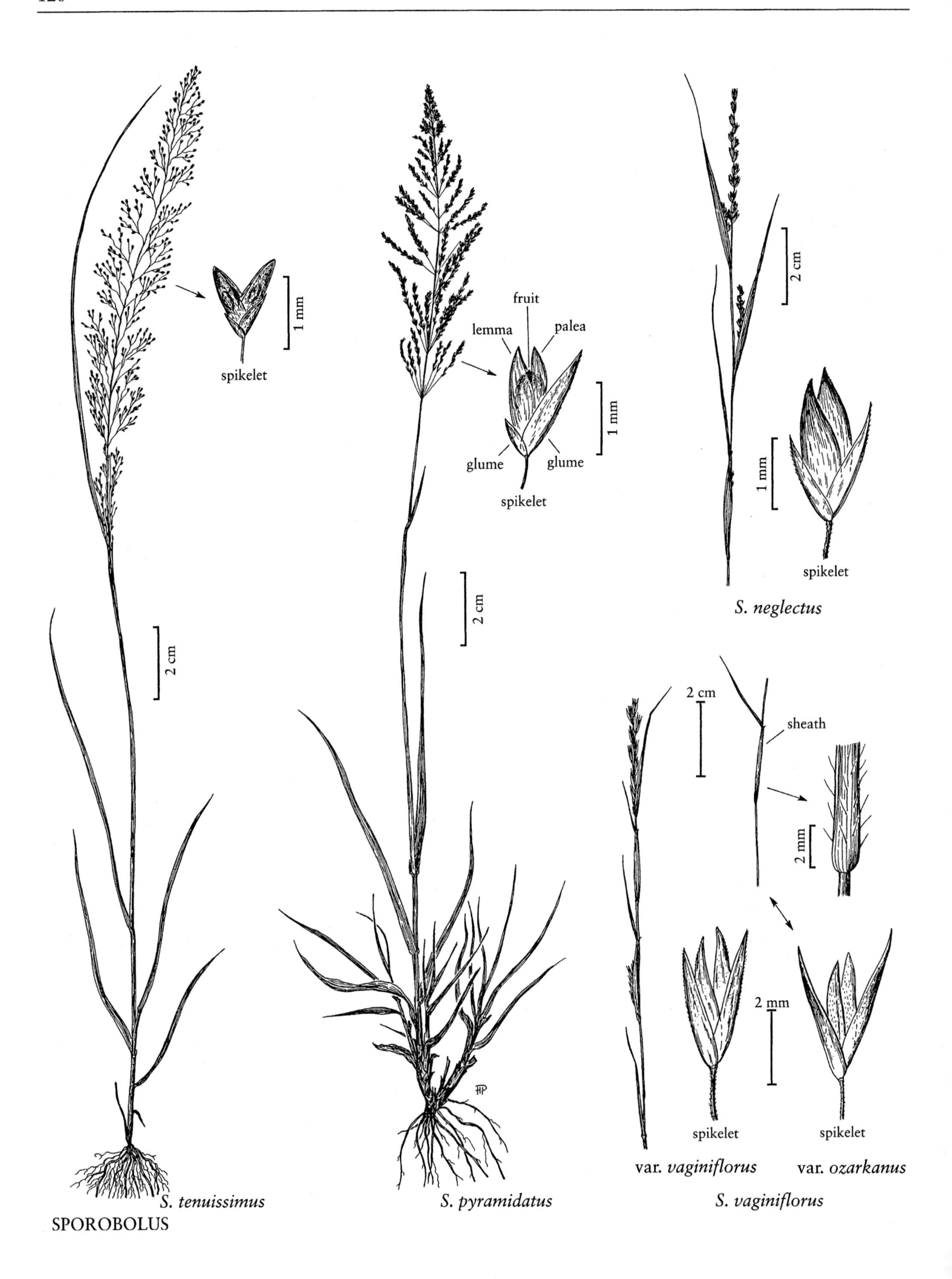
spikelet
1 mm
2 cm
S. tenuissimus
fruit
lemma
palea
glume
glume
spikelet
1 mm
2 cm
S. pyramidatus
2 cm
1 mm
spikelet
S. neglectus
2 cm
sheath
2 mm
2 mm
spikelet
spikelet
var. vaginiflorus
var. ozarkanus
S. vaginiflorus

SPOROBOLUS

4. **Sporobolus neglectus** Nash [p. 120]
PUFFSHEATH DROPSEED, SPOROBOLE NÉGLIGÉ

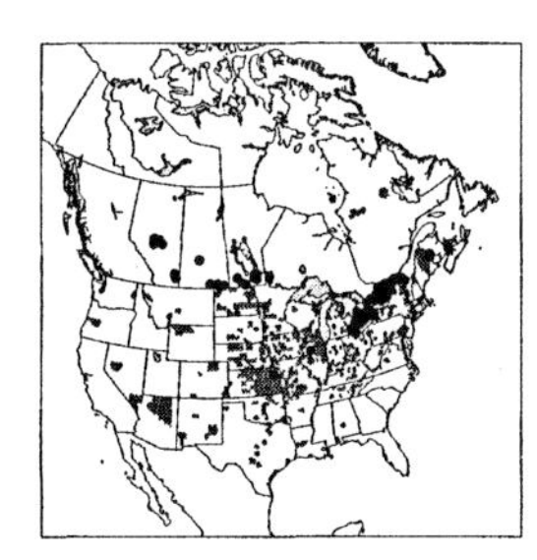

Plants annual; tufted, delicate, slender. **Culms** 10–45 cm, wiry, erect to decumbent. **Sheaths** inflated, mostly glabrous but the apices with small tufts of hairs, hairs to 3 mm; **ligules** 0.1–0.3 mm; **blades** 1–12 cm long, 0.6–2 mm wide, flat to loosely involute, abaxial surface glabrous, adaxial surface scabridulous, bases of both surfaces sometimes with papillose-based hairs, margins smooth or scabridulous. **Panicles** terminal and axillary, 2–5 cm long, 0.2–0.5 cm wide, contracted, cylindrical, included in the uppermost sheath; **lower nodes** with 1–2(3) branches; **primary branches** 0.4–1.8 cm, appressed, spikelet-bearing to the base; **secondary branches** appressed; **pedicels** 0.1–2.5 mm, appressed, scabridulous. **Spikelets** 1.6–3 mm, yellowish to cream-colored, sometimes purple-tinged. **Glumes** subequal, shorter than the florets, lanceolate to ovate, membranous to chartaceous, glabrous; **lower glumes** 1.5–2.4 mm, midveins often greenish; **upper glumes** 1.7–2.7 mm; **lemmas** 1.6–2.9 mm, ovate, chartaceous, glabrous, acute; **paleas** 1.6–3 mm, ovate, chartaceous, glabrous; **anthers** 3, 1.1–1.6 mm, purplish. **Fruits** 1.2–1.8 mm, obovoid, laterally flattened, light brownish or orangish-brown, translucent, finely striate. $2n$ = 36.

Sporobolus neglectus is native to the *Flora* region, and grows at 0–1300 m in sandy soils, on river shores, and in dry, open areas within many plant communities, often in disturbed sites. It appears to have been extirpated from Maine and Maryland and is considered endangered or of special concern in Connecticut, Massachusetts, New Hampshire, and New Jersey.

Sporobolus vaginiflorus is very similar to *S. neglectus*, but it differs in having strigose lemmas, sheaths that are sparsely hairy towards the base and, usually, longer spikelets.

5. **Sporobolus virginicus** (L.) Kunth [p. 123]
SEASHORE DROPSEED

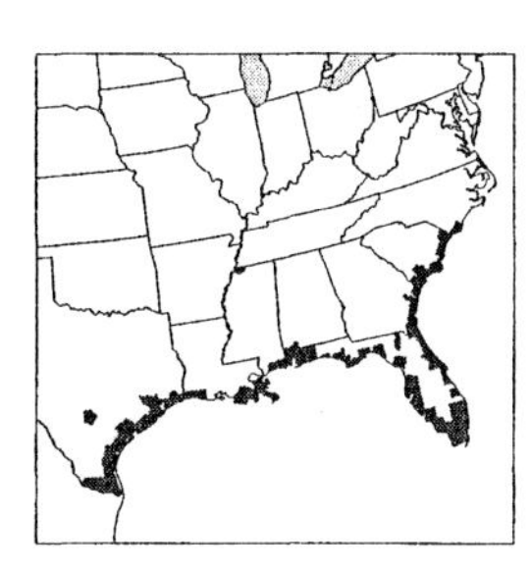

Plants perennial; rhizomatous, stoloniferous. **Culms** 10–65 cm, erect to decumbent. **Sheaths** overlapping, margins ciliate, apices with tufts of hairs, hairs to 2 mm; **ligules** 0.1–0.4 mm; **blades** usually conspicuously distichous, 4–16 cm long, 2–5 mm wide, flat to loosely involute, glabrous abaxially, scabridulous adaxially, margins scabridulous. **Panicles** 3–10 cm long, 0.4–1.6 cm wide, contracted, spikelike, dense; **primary branches** 0.5–2 cm, appressed, spikelet-bearing to the base; **pedicels** 0.2–1.4 mm, appressed. **Spikelets** (1.8)2–3.2 mm, yellowish-white to purplish-tinged, sometimes grayish. **Glumes** subequal, ovate-oblong, membranous; **lower glumes** 1.5–2.4 mm; **upper glumes** 1.8–3(3.2) mm; **lemmas** 2.1–3 mm, ovate to lanceolate, membranous, glabrous, acute; **paleas** 2.1–3 mm, ovate, membranous; **anthers** 3, 1–1.7 mm, yellowish. **Fruits** not known. $2n$ = 20, 30.

Sporobolus virginicus grows on sandy beaches, sand dunes, and in saline habitats, primarily along the southeastern coast, occasionally inland. Its range extends through Mexico and Central America to Peru, Chile, and Brazil. No fruits of this species have been found despite examination of several natural populations and over 200 herbarium specimens.

6. **Sporobolus compositus** (Poir.) Merr. [p. 123]
ROUGH DROPSEED, SPOROBOLE RUDE

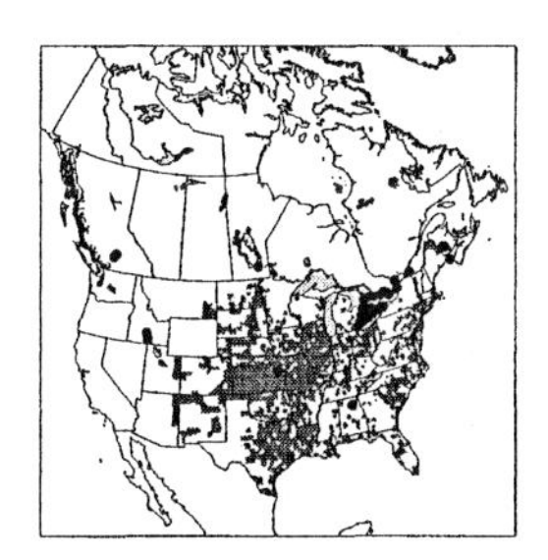

Plants perennial; cespitose, sometimes rhizomatous. **Culms** (20)30–130(150) cm. **Sheaths** with sparsely hairy apices, hairs to 3 mm; **ligules** 0.1–0.5 mm; **blades** not conspicuously distichous, 5–70 cm long, 1.5–10 mm wide, flat, folded, or involute, abaxial surface glabrous or pilose, adaxial surface glabrous or scabridulous, margins glabrous. **Panicles** terminal and axillary, 5–30 cm long, 0.4–1.6 cm wide, usually spikelike, partially included in the uppermost sheath, with 15–90 spikelets per cm^2 (exposed portion, when pressed); **lower nodes** with 1–2(3) branches; **primary branches** 0.4–6 cm, appressed, spikelet-bearing to the base; **secondary branches** appressed; **pulvini** glabrous; **pedicels** 0.3–3.5 mm, appressed, glabrous or scabridulous. **Spikelets** 4–6(10) mm, stramineous to purplish-tinged. **Glumes** subequal, lanceolate, membranous to chartaceous, midveins usually greenish; **lower glumes** (1.2)2–4 mm; **upper glumes** (2)2.5–5(6) mm, slightly shorter or longer than the lemmas; **lemmas** (2.2)3–6(10) mm, lanceolate, membranous to chartaceous and hyaline, glabrous, smooth, occasionally 2- or 3-veined, acute to obtuse; **paleas** (2.2)3–6(10) mm, ovate to lanceolate, membranous; **anthers** 0.2–3.2 mm, yellow to orangish. **Fruits** 1–2 mm, ellipsoid, laterally flattened, often striate, reddish-brown; **pericarps** gelatinous, slipping from the seeds when wet. $2n$ = 54, 88, 108.

Sporobolus compositus grows along roadsides and railroad right of ways, on beaches, and in cedar glades, pine woods, live oak-pine forests, prairies, and other partially disturbed, semi-open sites at 0–1600 m. Its range lies entirely within the *Flora* region.

The *Sporobolus compositus* complex is a difficult assemblage of forms, perhaps affected by their primarily autogamous breeding system (Riggins 1977). Asexual proliferation via rhizomes adds to the species' ability to maintain local population structure and to perpetuate unique character combinations.

1. Rhizomes present . . . . var. *macer*
1. Rhizomes absent.
   2. Culms slender, 1–2(2.5) mm thick; upper sheaths usually less than 2.5 mm wide; panicles with 16–36 spikelets per cm^2 when pressed . . . . var. *drummondii*
   2. Culms stout, 2–5 mm thick; upper sheaths usually 2.6–6 mm wide; panicles with 30–90 spikelets per cm^2 when pressed . . . . var. *compositus*

Sporobolus compositus (Poir.) Merr. var. **compositus**

Plants not rhizomatous. **Culms** stout, 2–5 mm thick. **Uppermost sheaths** usually 2.6–6 mm wide. **Panicles** with 30–90 spikelets per cm^2 when pressed. **Fruits** 1–1.8 mm.

Sporobolus compositus var. *compositus* is the most widespread of the three varieties, being found throughout most of the range shown for the species, but not in South Carolina or Florida.

Sporobolus compositus var. **drummondii** (Trin.) Kartesz & Gandhi

Plants not rhizomatous. **Culms** slender, 1–2(2.5) mm thick at the base. **Uppermost sheaths** 0.8–2.5 mm wide. **Panicles** with 16–36 spikelets per cm^2 when pressed. **Fruits** 1.1–1.7 mm.

Sporobolus compositus var. *drummondii* is most abundant in Kansas, Oklahoma, and Texas.

Sporobolus compositus var. **macer** (Trin.) Kartesz & Gandhi

Plants rhizomatous. **Culms** slender, 0.7–2.2 mm thick at the base. **Uppermost sheath** 0.7–2.5 mm wide. **Panicles** with 15–42 spikelets per cm^2 when pressed. **Fruits** 1.3–1.6 mm.

Sporobolus compositus var. *macer* is known only from the south central United States.

7. **Sporobolus clandestinus** (Biehler) Hitchc. [p. 123]
HIDDEN DROPSEED

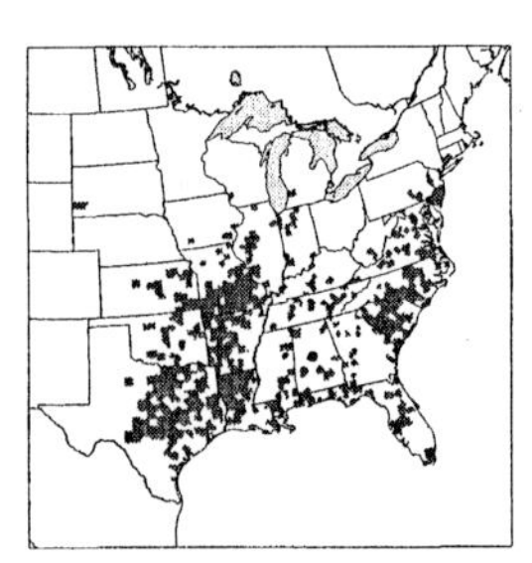

Plants perennial; cespitose, occasionally rhizomatous. **Culms** 40–130(150) cm tall, 1–4 mm thick, frequently glaucous. **Sheaths** with sparsely hairy apices, hairs to 3 mm, not conspicuously tufted; **uppermost sheaths** 0.5–3 mm wide; **ligules** 0.1–0.4 mm; **blades** 4–23 cm long, 1.5–4 mm wide, flat or involute, abaxial surface glabrous or pilose, adaxial surface glabrous or scabridulous, margins glabrous. **Panicles** terminal and axillary, 5–11 cm long, 0.04–0.2(0.3) cm wide, with 10–40 spikelets per cm^2, narrow, sometimes spikelike, included in the uppermost sheath; **lower nodes** with 1–2(3) branches; **primary branches** 0.4–5 cm, appressed, spikelet-bearing to the base; **secondary branches** appressed; **pulvini** glabrous; **pedicels** 0.3–3.5 mm, appressed, glabrous or scabridulous. **Spikelets** 4–9(10) mm, stramineous to purplish-tinged. **Glumes** subequal, lanceolate, membranous to chartaceous, midveins usually greenish; **lower glumes** 1.5–6.2 mm; **upper glumes** (2)2.5–5(6.5) mm, slightly shorter or longer than the lemmas; **lemmas** (2.2)3–7(7.4) mm, lanceolate, chartaceous and opaque, minutely appressed pubescent or scabridulous, occasionally 2- or 3-veined, acute to obtuse; **paleas** (2.2)3–9(10) mm, ovate to lanceolate, chartaceous; **anthers** 2.2–3.2 mm, yellow to orangish. **Fruits** (1.5)2.4–3.5 mm, ellipsoid, laterally flattened, often striate, reddish-brown; **pericarps** loose, but neither gelatinous nor slipping from the seeds when wet. $2n$ = unknown.

Sporobolus clandestinus grows primarily in sandy soils along the coast and, inland, along roadsides. In the southeastern United States, it is found in dry to mesic longleaf pine-oak-grass communities and cedar glades. Its range lies entirely within the *Flora* region.

8. **Sporobolus indicus** (L.) R. Br. [p. 125]
SMUTGRASS

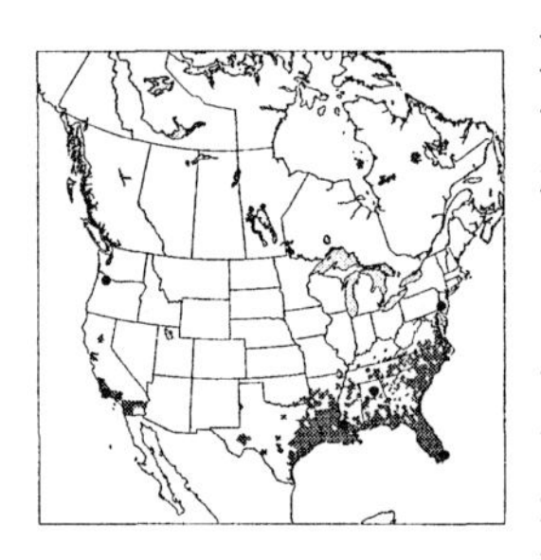

Plants perennial; cespitose, with tough fibrous roots, not rhizomatous. **Culms** 30–100 (120) cm. **Sheaths** usually keeled below, glabrous; **ligules** 0.2–0.5 mm; **blades** (6)10–30(50) cm long, 1–5 mm wide, flat, glabrous on both surfaces. **Panicles** 20–35(50) cm long, 0.3–2.2(3) cm wide, contracted, narrow, sometimes included in the uppermost sheath; **primary branches** 0.4–2.5(5) cm, appressed or spreading to 40° from the rachis, as long or longer than the adjacent internodes; **secondary branches** appressed, spikelet-bearing to near the base; **pulvini** glabrous; **pedicels** 0.1–1.8 mm, appressed. **Spikelets** 2–2.6(2.7) mm, plumbeous to light brownish. **Glumes** subequal, ovate or obovate, membranous; **lower glumes** 0.5–1 mm, often without midveins; **upper glumes** 0.8–1.6 mm, ½–⅔ as long as the florets, acute to obtuse, entire; **lemmas** 1.8–2.6(2.7) mm, ovate, membranous, glabrous, acute or obtuse; **paleas** 1.9–2.4 mm, ovate, membranous, glabrous; **anthers** 3, 0.5–1.1 mm, white, sometimes purple-tinged. **Fruits** 1–1.2 mm, quadrangular, laterally compressed, reddish-brown, truncate. $2n$ = 18, 24, 36.

Sporobolus indicus is a pantropical species. It commonly grows in disturbed places and open areas

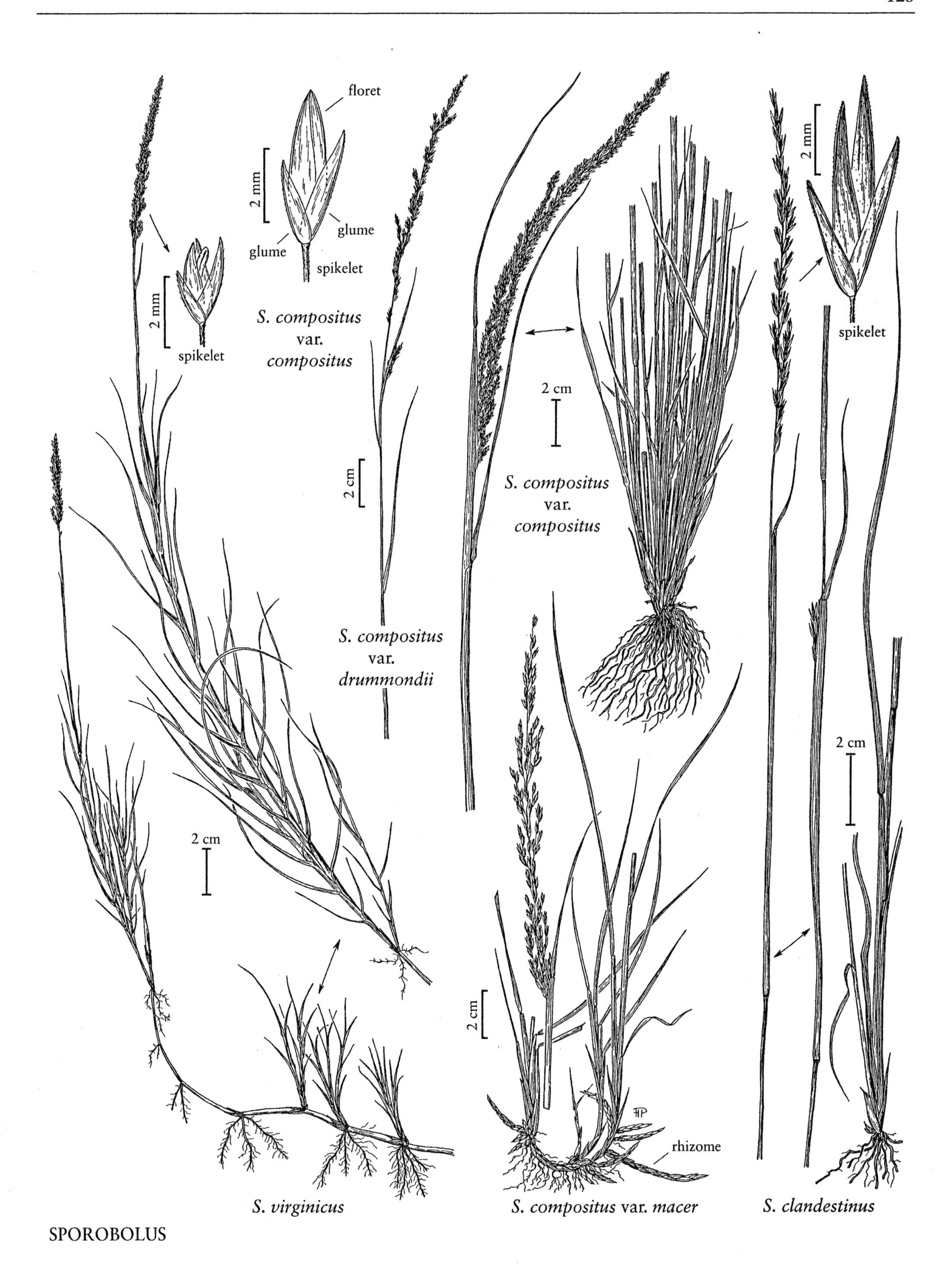
floret
2 mm
glume
glume
spikelet
S. compositus
var.
compositus
2 mm
spikelet
2 cm
S. compositus
var.
drummondii
2 cm
S. compositus
var.
compositus
2 mm
spikelet
2 cm
2 cm
2 cm
rhizome
S. virginicus
S. compositus var. macer
S. clandestinus

such as roadsides, pastures, and lake shores. In the *Flora* region, it is found on sandy or clay soils and is associated with many plant communities. The spikelets and upper leaves are often covered with hyphomycetous fungi (*Bipolaris* spp.); hence the common name of "smutgrass".

9. **Sporobolus jacquemontii** Kunth [p. 125]
RATSTAIL

Plants perennial; densely cespitose, not rhizomatous. **Culms** 40–100 cm. **Sheaths** keeled or rounded, glabrous, apices ciliate; **ligules** 0.2–0.4 mm; **blades** 10–40 cm long, 2–4 mm wide, flat but soon becoming involute, tapering to a fine point. **Panicles** 14–35 cm long, 0.4–3 cm wide, contracted, interrupted, and rather lax; **primary branches** appressed to strongly ascending, spikelet-bearing to the base, lower branches 1.5–5 cm, much longer than the adjacent internodes; **pedicels** 0.1–1.2(1.8) mm. **Spikelets** 1.4–1.8(2) mm, plumbeous to greenish. **Lower glumes** 0.3–0.5 mm, obtuse; **upper glumes** 0.4–0.7 mm, usually less than ½ as long as the florets, faintly 1-veined, truncate, erose to denticulate; **lemmas** 1.4–2 mm, elliptic, glabrous, 1-veined, acute; **paleas** 1.4–2 mm, elliptic; **anthers** 3(2), 0.9–1.1 mm. **Fruits** 0.7–1 mm, quadrangular, laterally compressed, reddish-brown, truncate. $2n$ = 24.

Sporobolus jacquemontii, like *S. indicus*, is native to North America. It is not a common species in the *Flora* region, being known only from coastal and low elevation sites in Florida. It is sometimes included in *S. indicus* (Baaijens and Veldkamp 1991) or *S. pyramidalis* P. Beauv. (Laegaard and Peterson 2001), but is retained here pending more definitive study.

10. **Sporobolus creber** De Nardi [p. 125]

Plants perennial; densely cespitose, not rhizomatous. **Culms** 6.5–100 cm. **Sheaths** rounded, margins ciliate, apices with tufts of hairs to 2 mm; **ligules** about 0.5 mm, the sides with a few hairs to 1.5 mm; **blades** 7–30 cm long, 1–2(3) mm wide, flat, becoming involute, tapering to a fine point. **Panicles** 20–40 cm long, 0.4–1 cm wide, narrowly contracted, sometimes spikelike; **primary branches** appressed to strongly ascending, spikelet-bearing to the base, lower branches 1.5–3 cm, much shorter than the adjacent internodes, usually appressed; **pedicels** 0.1–0.5 mm. **Spikelets** 1.1–1.5 mm, dark green. **Glumes** obtuse, often erose; **lower glumes** 0.4–0.5 mm; **upper glumes** 0.5–0.6 mm, less than ⅔ as long as the florets; **lemmas** 1.1–1.5 mm, glabrous, 1-veined, obtuse; **paleas** similar to the lemmas or slightly longer; **anthers** 2, 0.4–0.6 mm. **Fruits** 0.7–0.8 mm, often adhering to the floret at maturity, quadrangular to somewhat turbinate, red-brown, apices truncate and concave. $2n$ = unknown.

Sporobolus creber is an Australian species that was recently found growing spontaneously on a ranch in Glenn County, California. It is related to *S. indicus*, but differs in its widely spaced, closely appressed, and densely spikeleted branches.

11. **Sporobolus diandrus** (Retz.) P. Beauv. [p. 127]

Plants perennial; cespitose, not rhizomatous. **Culms** 30–80 cm. **Sheaths** keeled or rounded; **ligules** 0.2–0.5 mm; **blades** 10–30 cm long, 2–4 mm wide, flat, becoming folded. **Panicles** 15–35 cm long, 0.4–4 cm wide, contracted to rather lax and open; **primary branches** appressed to strongly ascending, without spikelets on the lower ¼, lower branches much longer than the internodes; **pedicels** 0.1–3 mm. **Spikelets** 1.3–1.8 mm, plumbeous to greenish. **Lower glumes** 0.4–0.8 mm, acuminate to truncate; **upper glumes** 0.7–1 mm, usually less than ½ as long as the florets, rarely longer, faintly 1-veined, truncate, erose to denticulate; **lemmas** 1.2–1.6(1.8) mm, elliptic, glabrous, 1-veined, acute to obtuse; **paleas** 1.4–1.8 mm, elliptic; **anthers** 2(3), 0.5–0.8 mm. **Fruits** 0.7–0.9 mm, quadrangular, laterally compressed, reddish-brown, truncate. $2n$ = 12.

Sporobolus diandrus is native from India to southeast Asia and Australia. It is not common in North America, being known only from a few counties in Florida, Mississippi, and Texas.

12. **Sporobolus fimbriatus** (Trin.) Nees [p. 127]

Plants perennial; rhizomatous. **Culms** 30–120(160) cm. **Sheaths** rounded and papery below; **ligules** 0.1–0.3 mm; **blades** 10–30 cm long, 2–4(5) mm wide, not obviously distichous, flat, becoming folded, pilose abaxially, tapering to the slender apices. **Panicles** 15–50 cm long, 2.4–8 cm wide, somewhat contracted to rather lax and open; **primary branches** appressed or ascending, spreading to 60°from the rachis; spikelet-bearing to the base or without spikelets on the lower ¼, lower branches mostly 2–9 cm, longer than the internodes; **pedicels** 0.7–3 mm. **Spikelets** 1.4–2.2 mm, plumbeous to greenish. **Glumes** unequal, linear-lanceolate to ovate, membranous; **lower glumes** (0.4)0.6–1.2(1.5) mm, without midveins, acuminate; **upper glumes** (0.9)1.4–2 mm, ⅔ to as long as the spikelet, faintly 1-veined, acute; **lemmas** (1.4)1.8–2.2 mm, narrowly ovate, glabrous, 1-veined, acute; **paleas**

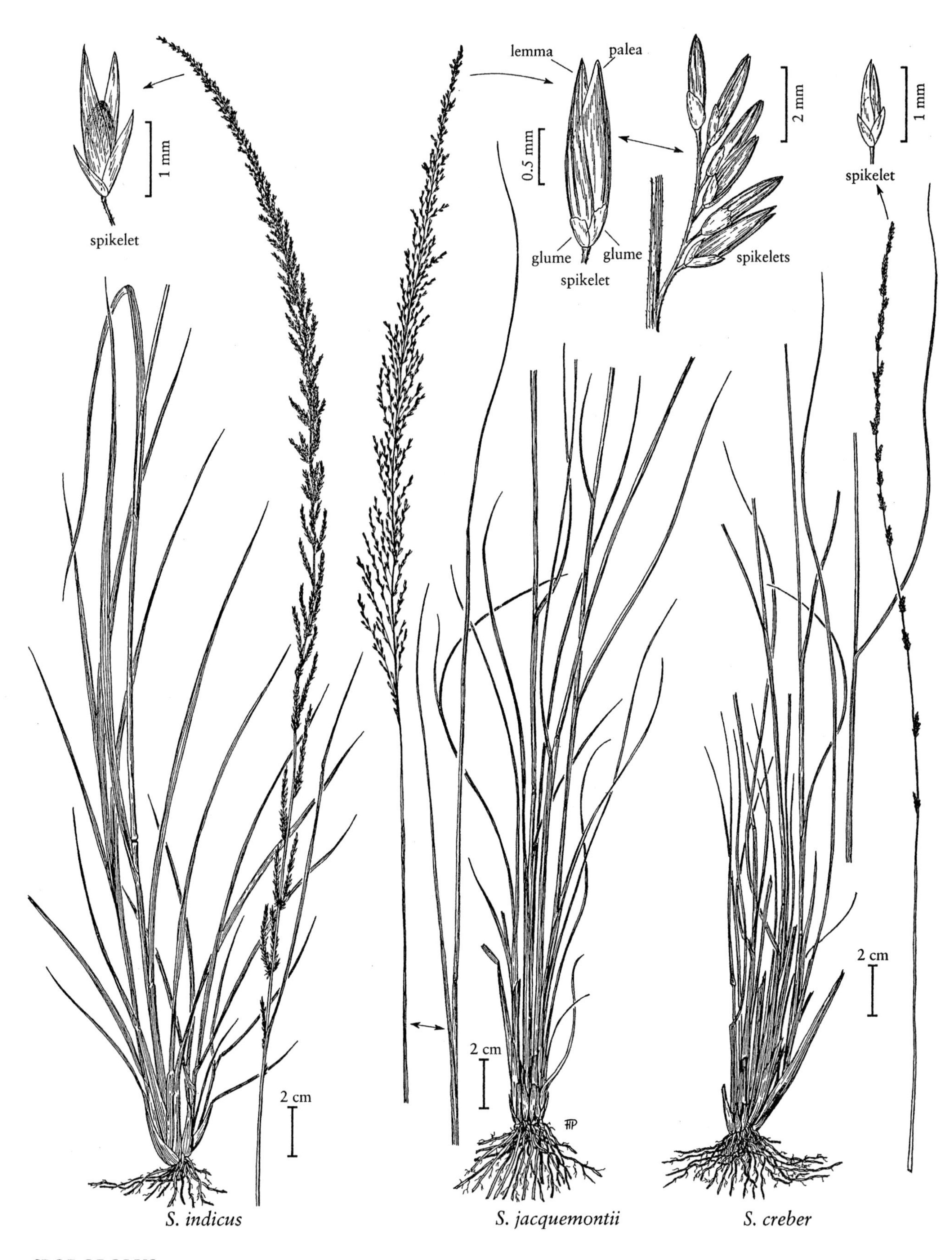
spikelet
1 mm
lemma
palea
0.5 mm
glume
glume
spikelet
2 mm
spikelets
1 mm
spikelet
2 cm
2 cm
2 cm
S. indicus
S. jacquemontii
S. creber

SPOROBOLUS

(1.2)1.6–2 mm, ovate; **anthers** 3, 09–1.2 mm. **Fruits** 0.6–1 mm, quadrangular, laterally compressed, whitish-brown, truncate. 2*n* = unknown.

Sporobolus fimbriatus is an African species that has only been found in waste areas near the sites of old wool mills in Berkeley and Florence counties, South Carolina.

13. **Sporobolus buckleyi** Vasey [p. 127]
BUCKLEY'S DROPSEED

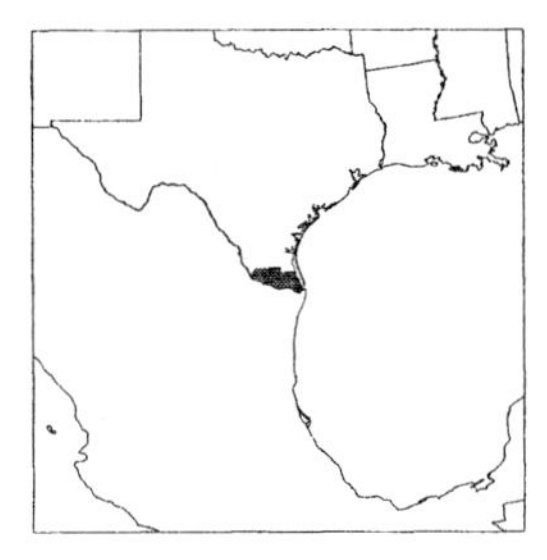

Plants perennial; cespitose, not rhizomatous. **Culms** 40–100 cm. **Sheaths** keeled and flattened below, margins occasionally hairy distally, hairs to 1.2 mm; **ligules** 0.2–0.4 mm; **blades** 12–35 cm long, 4–12 mm wide, flat, glabrous abaxially, scabridulous adaxially, margins smooth or scabridulous. **Panicles** 15–50 cm long, 7–22(30) cm wide, diffuse, ovate; **primary branches** 2–17 cm, spreading 40–100° from the rachis; **secondary branches** appressed to loosely spreading, without spikelets on the lower ¼–½; **pulvini** glabrous; **pedicels** 0.2–1.2 mm, appressed, scabridulous. **Spikelets** 1–2 mm, purplish to brownish. **Glumes** unequal, narrowly lanceolate to lanceolate, membranous, prominently keeled; **lower glumes** 0.6–1 mm; **upper glumes** 1.1–1.8 mm, slightly shorter than or subequal to the lemmas; **lemmas** 1.2–2 mm, lanceolate, membranous, glabrous, acute; **paleas** 1.2–2 mm, ovate, membranous, glabrous, often splitting in two between the veins at maturity; **anthers** 0.2–0.4 mm, purplish. **Fruits** 0.6–1 mm, ovoid, slightly flattened, reddish-brown. 2*n* = 40.

Sporobolus buckleyi grows between 0–150 m, in loamy soils near the margins of woods or thorn scrub, sometimes in partial sunlight. Its range extends from southeastern Texas to Belize.

14. **Sporobolus domingensis** (Trin.) Kunth [p. 128]
CORAL DROPSEED

Plants perennial; cespitose, not rhizomatous. **Culms** 20–100 cm. **Sheaths** rounded below, distal margins and apices hairy, hairs to 3 mm; **ligules** 0.2–1.2 mm; **blades** 5–20 cm long, 3–8 mm wide, flat to loosely involute, glabrous abaxially, scabridulous adaxially, margins scabridulous. **Panicles** 10–25(35) cm long, 1–5 cm wide, usually somewhat contracted, sometimes spikelike, often interrupted below; **primary branches** 0.7–7 cm, appressed or spreading to 30° from the rachis, spikelet-bearing to the base; **secondary branches** appressed; **pedicels** 0.2–1.4 mm, appressed. **Spikelets** 1.6–2 mm, yellowish-green to grayish. **Glumes** unequal, linear-lanceolate to ovate, membranous; **lower glumes** 0.5–1.1 mm, usually without veins; **upper glumes** 1.1–2 mm, subequal to the lemmas; **lemmas** 1.1–2 mm, ovate, membranous, glabrous (occasionally minutely pubescent), acute; **paleas** 1–2 mm, ovate, membranous; **anthers** 2 or 3, 0.5–1 mm, yellowish or purplish. **Fruits** 0.7–1.1 mm, ellipsoid, laterally flattened, light brownish. 2*n* = unknown.

Sporobolus domingensis grows in sandy, rocky, or alkaline soils, often in disturbed sites adjacent to the coast and below 20 m. Its range extends to the Antilles and the Yucatan Peninsula, Mexico.

15. **Sporobolus wrightii** Munro *ex* Scribn. [p. 128]
BIG ALKALI SACATON

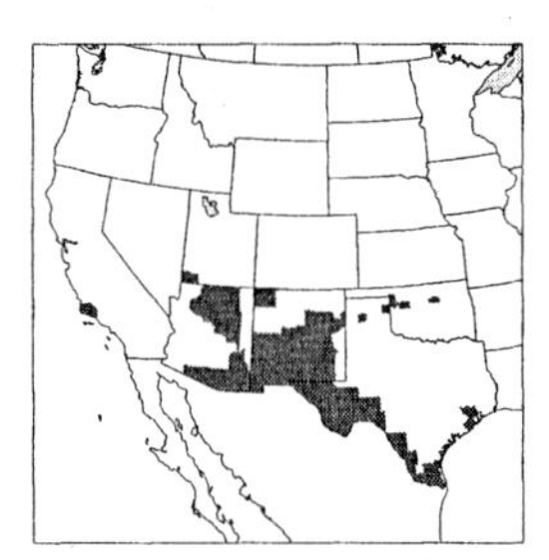

Plants perennial; cespitose, not rhizomatous. **Culms** 90–250 cm, stout. **Sheaths** rounded below, shiny, glabrous, rarely sparsely hairy apically, hairs to 6 mm; **ligules** 1–2 mm; **blades** 20–70 cm long, 3–10 mm wide, flat (rarely involute), glabrous abaxially, scabrous adaxially, margins scabrous; **flag blades** ascending. **Panicles** 20–60 cm long, 12–26 cm wide, open, broadly lanceolate, exserted; **primary branches** 1.5–10 cm, spreading 20–70° from the rachis; **secondary branches** appressed, spikelet-bearing to the base; **pulvini** glabrous; **pedicels** 0.2–0.5 mm, mostly appressed. **Spikelets** 1.5–2.5 mm, crowded, purplish or greenish. **Glumes** unequal, lanceolate to ovate, membranous; **lower glumes** 0.5–1 mm, often appearing veinless; **upper glumes** 0.8–2 mm, ⅔ or more as long as the florets; **lemmas** 1.2–2.5 mm, ovate, membranous, glabrous, acute to obtuse; **paleas** 1.1–2.5 mm, ovate, membranous, glabrous; **anthers** 1.1–1.3 mm, yellowish to purplish. **Fruits** 1–1.4 mm, ellipsoid, reddish-brown or blackish, striate. 2*n* = 36.

Sporobolus wrightii grows in moist clay flats and on rocky slopes near saline habitats, at elevations of 5–1800 m. Its range extends to central Mexico.

16. **Sporobolus airoides** (Torr.) Torr. [p. 128]
ALKALI SACATON

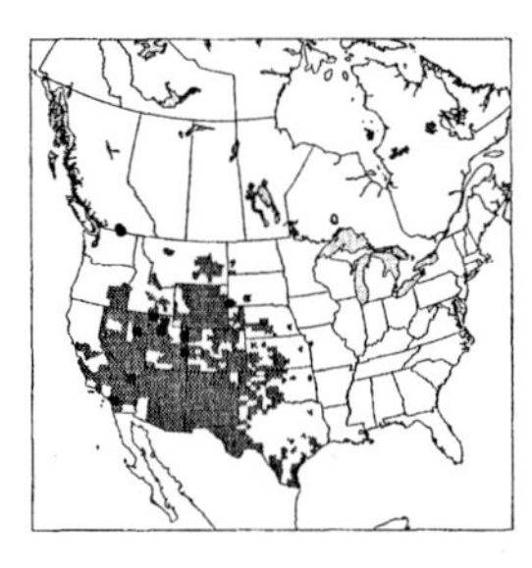

Plants perennial; cespitose, not rhizomatous. **Culms** 35–120 (150) cm, stout. **Sheaths** rounded below, shiny, apices glabrous or sparsely hairy, hairs to 6 mm; **ligules** 0.1–0.3 mm; **blades** (3)10–45(60) cm long, (1)2–5(6) mm wide, flat to

1 mm
spikelet
panicle branch
2 cm
S. diandrus
lemma
palea
glume
glume
1 mm
spikelet
2 cm
S. fimbriatus
1 mm
spikelet
2 cm
S. buckleyi

SPOROBOLUS

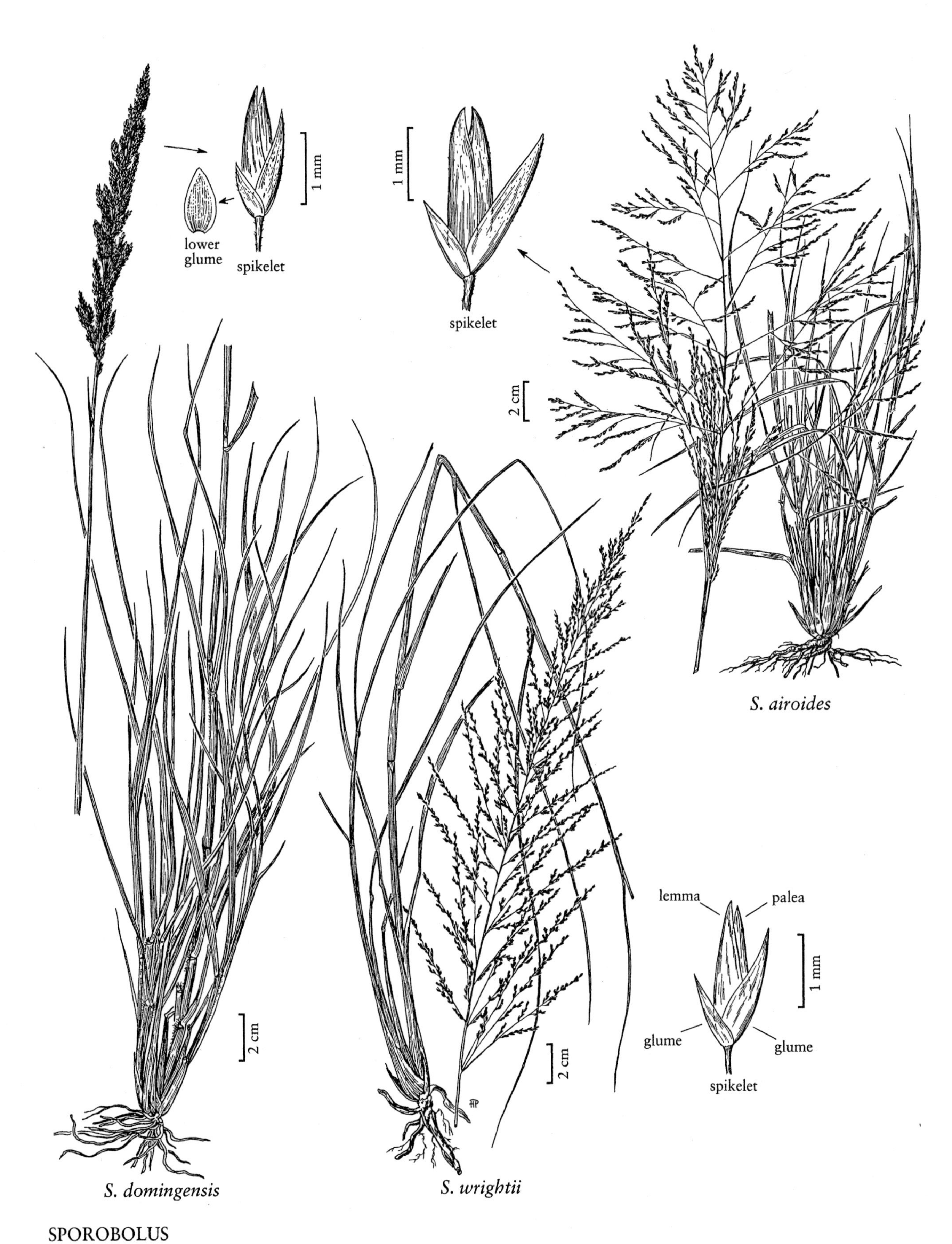
lower glume
spikelet
1 mm
1 mm
spikelet
2 cm
S. airoides
lemma
palea
1 mm
glume
glume
spikelet
2 cm
2 cm
S. domingensis
S. wrightii

SPOROBOLUS

involute, glabrous abaxially, scabridulous adaxially, margins smooth or scabridulous; **flag blades** ascending. **Panicles** (10)15–45 cm long, 15–25 cm wide, diffuse, subpyramidal, often included in the uppermost sheath; **primary branches** 1.5–13 cm, spreading 30–90° from the rachis; **secondary branches** spreading, without spikelets on the lower ¼–⅓; **pulvini** glabrous; **pedicels** 0.5–2 mm, spreading, glabrous or scabrous. **Spikelets** 1.3–2.8 mm, purplish or greenish. **Glumes** unequal, lanceolate to ovate, membranous; **lower glumes** 0.5–1.8 mm, often without midveins; **upper glumes** 1.1–2.4(2.8) mm, at least ⅔ as long as the florets; **lemmas** 1.2–2.5 mm, ovate, membranous, glabrous, acute; **paleas** 1.1–2.4 mm, ovate, membranous, glabrous; **anthers** 1.1–1.8 mm, yellowish to purplish. **Fruits** 1–1.4 mm, ellipsoid, reddish-brown, striate. $2n$ = 80, 90, 108, 126.

Sporobolus airoides grows on dry, sandy to gravelly flats or slopes, at elevations from 50–2350 m. It is usually associated with alkaline soils. Its range extends into northern Mexico.

17. **Sporobolus cryptandrus** (Torr.) A. Gray [p. 130]
SAND DROPSEED, SPOROBOLE À FLEURS CACHÉES

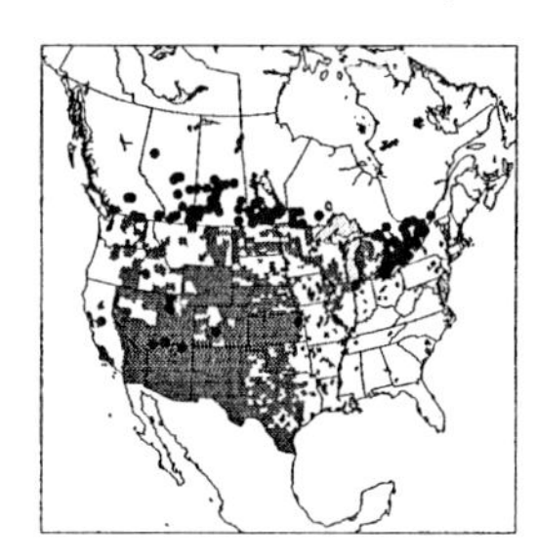

Plants perennial; cespitose, not rhizomatous, bases not hard and knotty. **Culms** 30–100(120) cm tall, 1–3.5 mm thick, erect to decumbent. **Sheaths** rounded below, glabrous or scabridulous, margins sometimes ciliate distally, apices with conspicuous tufts of hairs, hairs to 4 mm; **ligules** 0.5–1 mm; **blades** (2)5–26 cm long, 2–6 mm wide, flat to involute, glabrous abaxially, scabridulous to scabrous adaxially, margins scabridulous; **flag blades** nearly perpendicular to the culms. **Panicles** 15–40 cm long, 2–12(14) cm wide, longer than wide, initially contracted and spikelike, ultimately open and narrowly pyramidal; **rachises** straight, erect; **lower nodes** with 1–2(3) branches; **primary branches** 0.6–6 cm, appressed, spreading, or reflexed to 130° from the rachis, without spikelets on the lower ⅙–¼; **lower branches** longest, included in the uppermost sheath; **secondary branches** appressed; **pulvini** glabrous; **pedicels** 0.1–1.3 mm, appressed, glabrous or scabridulous. **Spikelets** 1.5–2.5(2.7) mm, brownish, plumbeous, or purplish-tinged. **Glumes** unequal, linear-lanceolate to ovate, membranous; **lower glumes** 0.6–1.1 mm; **upper glumes** 1.5–2.7 mm, at least ⅔ as long as the florets; **lemmas** 1.4–2.5(2.7) mm, ovate to lanceolate, membranous, glabrous, acute; **paleas** 1.2–2.4 mm, lanceolate, membranous; **anthers** 0.5–1 mm, yellowish to purplish. **Fruits** 0.7–1.1 mm, ellipsoid, light brownish to reddish-orange. $2n$ = 36, 38, 72.

Sporobolus cryptandrus is a widespread North American species, extending from Canada into Mexico. It grows in sandy soils and washes, on rocky slopes and calcareous ridges, and along roadsides in salt-desert scrub, pinyon-juniper woodlands, yellow pine forests, and desert grasslands. Its elevational range is 0–2900 m.

18. **Sporobolus contractus** Hitchc. [p. 130]
SPIKE DROPSEED

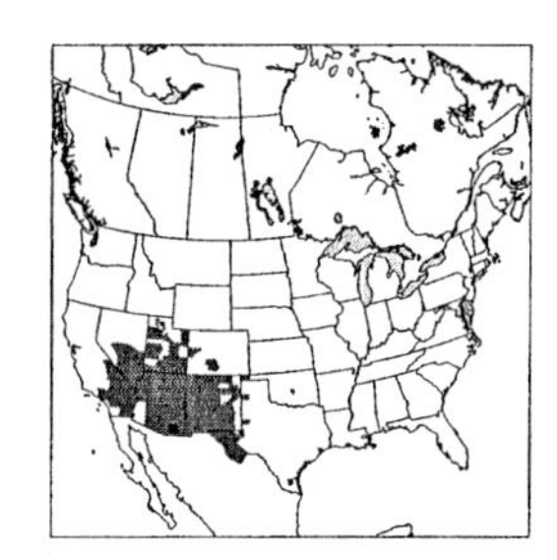

Plants perennial; cespitose, not rhizomatous. **Culms** 40–100 (120) cm tall, 2–4(5) mm thick near the base. **Sheaths** rounded below, margins hairy, particularly distally, hairs to 3 mm, apices with conspicuous tufts of hair; **ligules** 0.4–1 mm; **blades** (2)4–35 cm long, 3–8 mm wide, flat to involute, glabrous on both surfaces, margins whitish, somewhat scabridulous. **Panicles** all terminal, (10)15–45(50) cm long, 0.2–0.8(1) cm wide, contracted, spikelike, dense, usually included in the uppermost sheath; **lower nodes** with 1–2(3) branches; **primary branches** 0.3–1.5 cm, appressed, spikelet-bearing to the base; **secondary branches** appressed; **pulvini** glabrous; **pedicels** 0.2–2 mm, appressed, scabridulous. **Spikelets** 1.7–3.2 mm, whitish to plumbeous. **Glumes** unequal, narrowly lanceolate, membranous, prominently keeled; **lower glumes** 0.7–1.7 mm, usually 1-veined, acute to acuminate; **upper glumes** 2–3.2 mm, at least ⅔ as long as the florets; **lemmas** 2–3.2 mm, linear-lanceolate, membranous, glabrous, acute; **paleas** 1.8–3 mm, linear-lanceolate, membranous, glabrous; **anthers** 3, 0.3–0.5 mm, light yellowish. **Fruits** 0.8–1.2 mm, ellipsoid, laterally flattened, light brownish or translucent. $2n$ = 36.

Sporobolus contractus grows in dry to moist, sandy soils, at elevations from 300–2300 m. It is found occasionally in salt-desert scrub, desert grasslands, and pinyon-juniper woodlands. Its range extends to the states of Baja California and Sonora in Mexico.

19. **Sporobolus texanus** Vasey [p. 130]
TEXAS DROPSEED

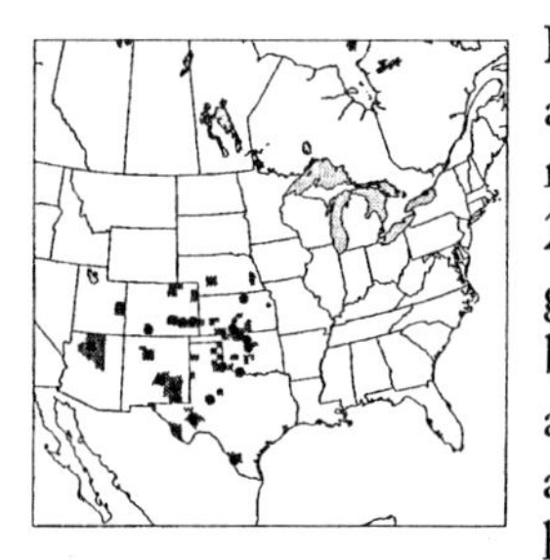

Plants perennial (often appearing annual); cespitose, with fibrous roots, not rhizomatous. **Culms** 20–70 cm, erect to decumbent, glabrous or scurfy roughened below. **Sheaths** rounded basally, apices glabrous or with scattered, appressed, papillose-based hairs, hairs to 4 mm; **ligules** 0.2–0.6 mm; **blades** 2.5–13(18) cm long, 1–4.2 mm wide, flat to involute, glabrous abaxially, scabrous adaxially, margins

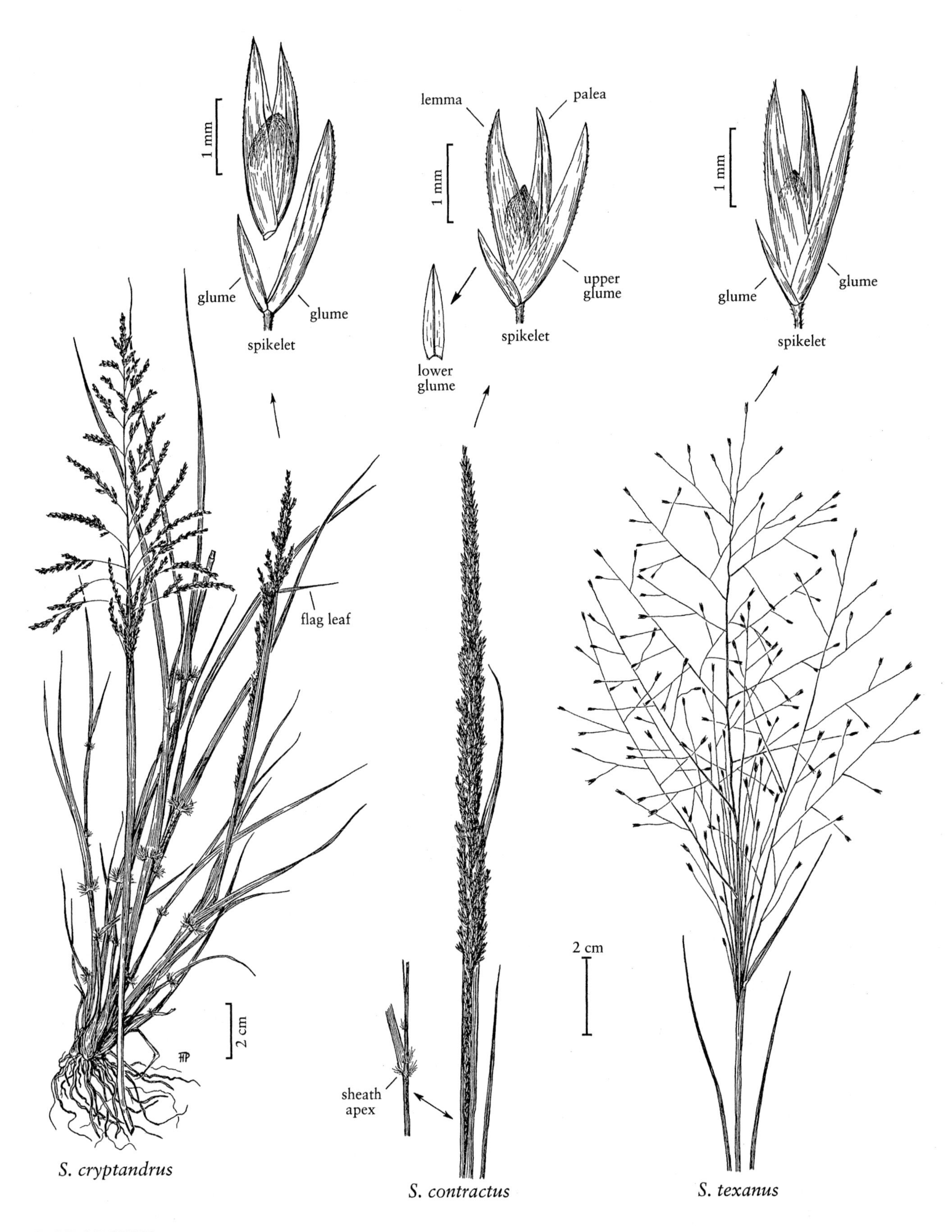
1 mm
glume
glume
spikelet
lemma
palea
1 mm
upper glume
spikelet
lower glume
1 mm
glume
glume
spikelet
flag leaf
2 cm
2 cm
sheath apex
S. cryptandrus
S. contractus
S. texanus

SPOROBOLUS

scabridulous, often also with a few papillose-based hairs; **flag blades** ascending. **Panicles** 10–35 cm long, 4.5–30 cm wide, open, diffuse, subpyramidal, about as long as wide, partially included in the uppermost leaf sheath; **lower nodes** with 1–2 branches; **primary branches** 4–14 cm, capillary, spreading 10–80° from the rachis; **secondary branches** spreading, without spikelets on the lower ⅓–½; **pedicels** 6–25 mm, spreading. **Spikelets** 2.3–3 mm, purplish-tinged. **Glumes** unequal, linear-lanceolate to lanceolate, membranous; **lower glumes** 0.5–1.7 mm, often without midveins; **upper glumes** 1.7–3 mm, at least ⅔ as long as the florets, often longer; **lemmas** 1.8–3 mm, lanceolate to ovate, membranous, glabrous, acute; **paleas** 1.7–2.9 mm, ovate, membranous, glabrous, often splitting as the fruit matures; **anthers** 0.3–1 mm, yellowish. **Fruits**1.1–1.5 mm, obovoid, light brown, translucent, occasionally rugulose. $2n$ = unknown.

Sporobolus texanus grows along rivers, ponds, and in wet alkaline habitats, at 100–3300 m. It is known only from the United States.

20. **Sporobolus nealleyi** Vasey [p. 132]
GYPGRASS

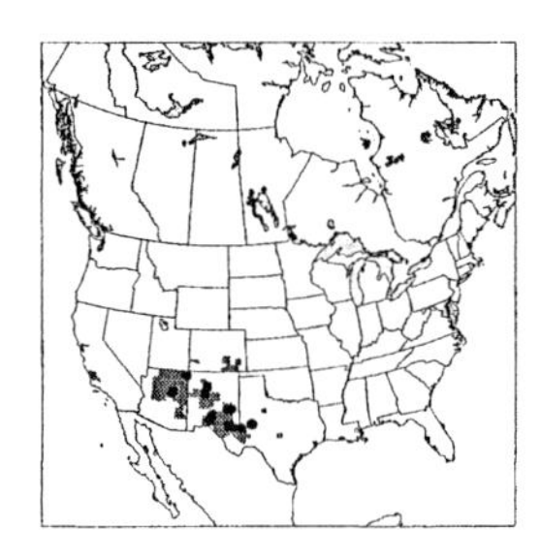

Plants perennial; cespitose, bases hard and knotty, not rhizomatous. **Culms** 10–50(60) cm tall, 0.7–1.2 mm thick. **Sheaths** rounded below, occasionally glabrous, usually villous to tomentose along the margins and back, with soft, kinky hairs to 4 mm; **ligules** 0.2–0.4 mm; **blades** (0.6)1.5–6(7) cm long, 1–1.5 mm wide, involute, stiff, spreading at right angles to the culm, glabrous abaxially, scabridulous adaxially, margins smooth. **Panicles** 3–10 cm long, (0.3)1–5(6) cm wide, longer than wide, ultimately open, subovate, lower portion often included in the uppermost sheath; **lower nodes** with 1–2 branches; **primary branches** 0.5–5 cm, appressed or spreading to 90° from the rachis; **secondary branches** appressed or spreading, without spikelets on the lower ⅛–¼; **pedicels** 0.2–2 mm, appressed or spreading. **Spikelets** 1.4–2.1 mm, purplish. **Glumes** unequal, linear-lanceolate to ovate, membranous; **lower glumes** 0.5–1.1 mm; **upper glumes** 1.3–2 mm, from slightly shorter than to subequal to the florets; **lemmas** 1.4–2.1 mm, ovate, membranous, glabrous, acute; **paleas** 1.4–2.1 mm, ovate, membranous; **anthers** 0.7–1 mm, purplish. **Fruits** 0.7–1 mm, orangish to whitish. $2n$ = 40.

Sporobolus nealleyi grows in sandy and gravelly soils, usually in those derived from gypsum, or near alkaline habitats associated with desert grasslands. It is known only from the southwestern United States, where it grows at 700–3000 m.

21. **Sporobolus flexuosus** (Thurb. *ex* Vasey) Rydb. [p. 132]
MESA DROPSEED

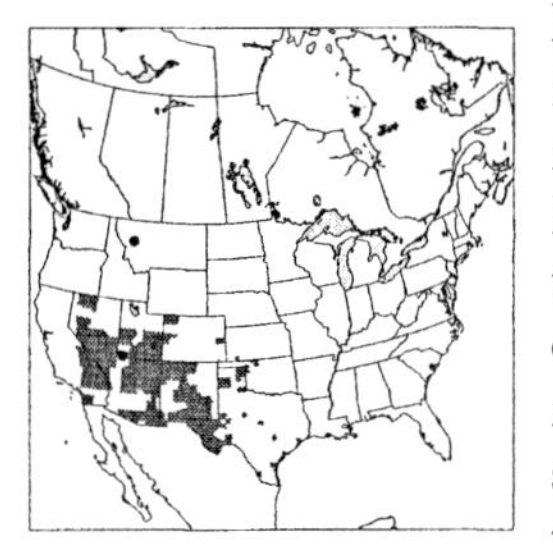

Plants perennial (rarely appearing annual); cespitose, not rhizomatous, bases not hard and knotty. **Culms** 30–100(120) cm tall, 1–3 mm thick near the base, erect to decumbent. **Sheaths** rounded below, smooth or scabridulous, margins sometimes ciliate distally, apices with tufts of hairs, hairs to 4 mm; **ligules** 0.5–1 mm; **blades** (2)5–24 cm long, 2–4(6) mm wide, ascending or strongly divergent, flat to involute, glabrous abaxially, scabridulous adaxially, margins scabridulous. **Panicles** 10–30 cm long, 4–12 cm wide, longer than wide, open, subovate to oblong; **rachises** drooping or nodding; **lower nodes** with 1–2 branches; **primary branches** 1–8(12) cm, flexible, diverging at least 70° from the rachis, often strongly reflexed to 130°, tangled with each other and with branches from adjacent panicles; **lower branches** no longer than those in the middle, usually included in the uppermost sheath; **secondary branches** widely spreading, without spikelets on the lower ⅛–½; **pulvini** pubescent; **pedicels** 0.3–3 mm, spreading, scabridulous. **Spikelets** 1.8–2.5 mm, plumbeous. **Glumes** unequal, ovate, membranous; **lower glumes** 0.9–1.5 mm; **upper glumes** 1.4–2.5 mm, subequal to the florets; **lemmas** 1.4–2.5 mm, lanceolate to ovate, membranous, glabrous, acute; **paleas** 1.4–2.4 mm, ovate, membranous; **anthers** 0.4–0.7 mm, yellow. **Fruits** 0.6–1 mm, ellipsoid, light brownish to reddish-orange. $2n$ = 36, 38.

Sporobolus flexuosus grows on sandy to gravelly slopes, flats, and roadsides in the southwestern United States and northern Mexico. It is associated with desert scrub, pinyon-juniper woodlands, and yellow pine forests. Its elevational range is 800–2100 m.

22. **Sporobolus giganteus** Nash [p. 132]
GIANT DROPSEED

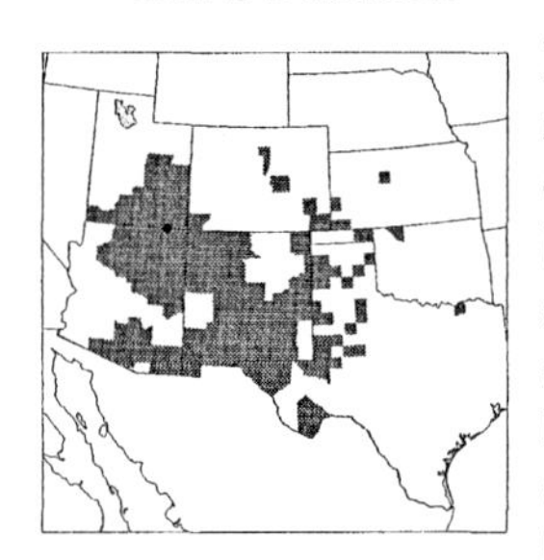

Plants perennial; cespitose, not rhizomatous. **Culms** 100–200 cm, (3)4–10 mm thick near the base. **Sheaths** rounded below, striate, margins hairy distally, apices with conspicuous tufts of hairs, hairs to 2 mm; **ligules** 0.5–1.5 mm; **blades** 10–50 cm long, (3)4–10(13) mm wide, flat, glabrous on both surfaces, margins whitish, scabridulous. **Panicles** all terminal, 25–75 cm long, 1–4 cm wide, spikelike, dense, usually included in the uppermost sheath; **lower nodes** with 1–2(3) branches;

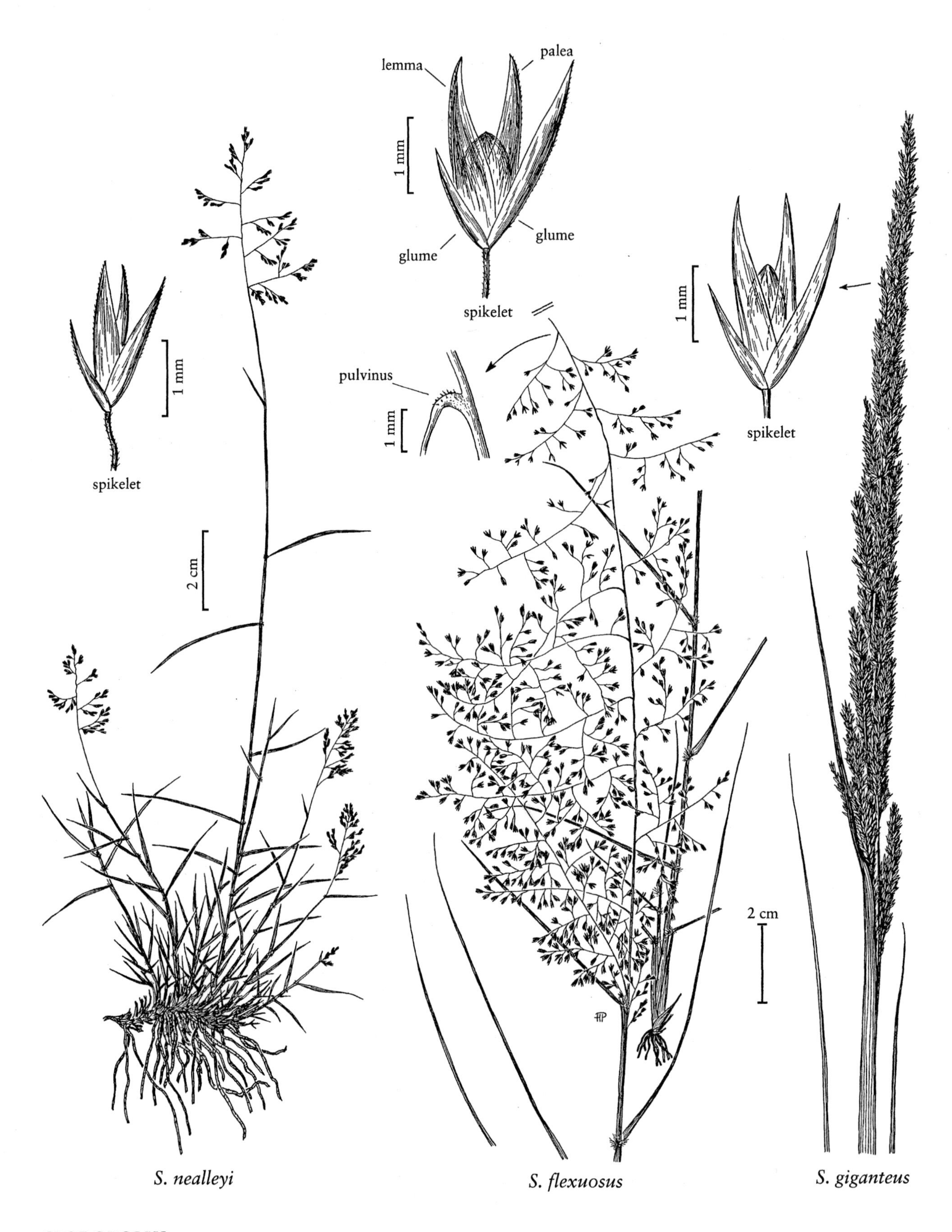
lemma
palea
1 mm
glume
glume
spikelet
1 mm
spikelet
1 mm
spikelet
pulvinus
1 mm
2 cm
2 cm
S. nealleyi
S. flexuosus
S. giganteus

primary branches mostly 0.5–6 cm, appressed or spreading to 30° from the rachis, spikelet-bearing to the base; **secondary branches** appressed; **pulvini** glabrous; **pedicels** 0.5–2 mm, appressed. **Spikelets** 2.5–3.5(4) mm, whitish to plumbeous. **Glumes** unequal, narrowly lanceolate, membranous, prominently keeled; **lower glumes** 0.6–2 mm; **upper glumes** 2–3.5(4) mm, subequal to the lemmas; **lemmas** 2.5–3.5(4) mm, linear-lanceolate, membranous, glabrous, acute; **paleas** 2.4–3.4(3.8) mm, linear-lanceolate, membranous, glabrous; **anthers** 0.6–1 mm, yellowish. **Fruits** 0.8–1.7 mm, ellipsoid, light yellowish-brown, sometimes translucent. 2*n* = 36.

Sporobolus giganteus grows in sand dunes and sandy areas along rivers and roadsides, at elevations from 100–1830 m. Its range extends from the southwestern United States into northern Mexico.

23. **Sporobolus junceus** (P. Beauv.) Kunth [p. 134]
PINEY WOODS DROPSEED

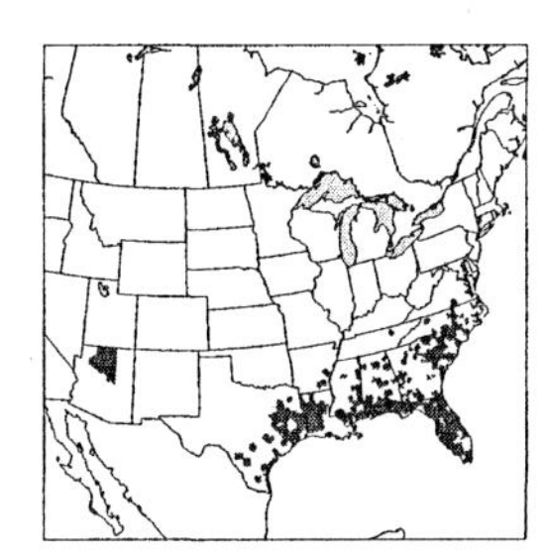

Plants perennial; cespitose, not rhizomatous. **Culms** (30)40–100 cm. **Sheaths** rounded below, margins and apices sometimes sparsely ciliate; **ligules** 0.1–0.2 mm; **blades** (6)10–30 cm long, 0.8–2 mm wide, flat to tightly involute, glabrous abaxially, scabridulous adaxially, margins scabrous, apices pungent. **Panicles** 7–28 cm long, 2–6 cm wide, open, pyramidal; **lower nodes** with 3 or more branches; **primary branches** 0.7–4.5 cm, spreading 20–100° from the rachis, whorled or verticellate, without spikelets on the lower ⅛–½; **secondary branches** appressed; **pedicels** 0.4–2.5 mm, appressed, scabridulous. **Spikelets** 2.6–3.8 mm, purplish-red. **Glumes** unequal, linear-lanceolate to lanceolate or ovate, hyaline to membranous; **lower glumes** 0.9–3 mm; **upper glumes** 2.6–3.8 mm, as long as or longer than the florets; **lemmas** 2–3.6 mm, ovate, membranous, glabrous, acute; **paleas** 2–3.6 mm, ovate, membranous; **anthers** 1.4–2 mm, purplish. **Fruits** 1.4–1.8 mm, ellipsoid, somewhat laterally flattened, somewhat rugulose, reddish-brown. 2*n* = unknown.

Sporobolus junceus grows in openings in pine and hardwood forests, coastal prairies, and pine barrens, usually in sandy to loamy soils, at 2-400 m. Its range lies entirely within the southern United States.

24. **Sporobolus purpurascens** (Sw.) Ham. [p. 134]
PURPLE DROPSEED

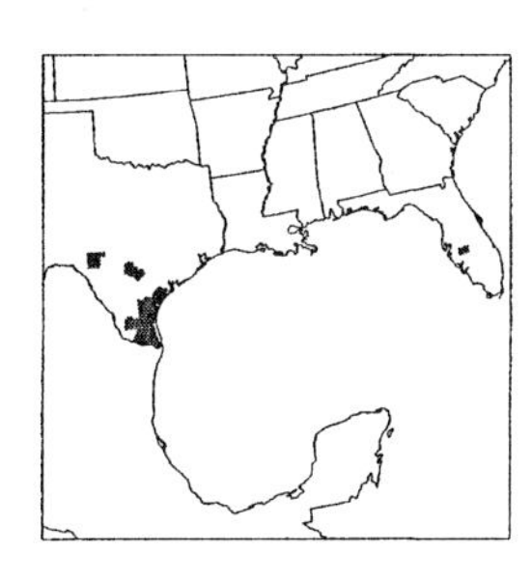

Plants perennial; cespitose, not rhizomatous. **Culms** 25–95 cm. **Sheaths** rounded below, upper margins sometimes sparsely hispid-ciliate, apices hairy, hairs to 5 mm; **ligules** 0.2–0.3 mm; **blades** 8–22 cm long, 2–5 mm wide, flat or involute, glabrous abaxially, scabridulous adaxially, margins scabrous, sometimes sparsely hispid. **Panicles** 5–30 cm long, 0.4–1.6 cm wide, narrow, contracted; **lower nodes** with 3 or 5 branches; **primary branches** 0.3–2 cm, appressed or spreading to 20° from the rachis, spikelet-bearing to near the base; **secondary branches** appressed; **pedicels** 0.2–2.5 mm, appressed, scabridulous. **Spikelets** 2.8–3.8 mm, purplish-red. **Glumes** unequal, linear-lanceolate to lanceolate or ovate, hyaline to membranous; **lower glumes** 0.9–3 mm; **upper glumes** 2.9–3.8 mm, subequal to the florets; **lemmas** 2.9–3.8 mm, ovate, membranous, glabrous, acute; **paleas** 2.9–3.8 mm, ovate, membranous; **anthers** 1.5–2 mm, yellowish to purplish. **Fruits** 1.8–2.3 mm, ellipsoid, somewhat laterally flattened, rugulose, reddish-brown. 2*n* = 60.

Sporobolus purpurascens grows in oak scrub, prairie grasslands, and sandy sites near railroad crossings and roadsides, at elevations from 2–300 m. It extends from southern Texas through eastern Mexico, the West Indies, and Central America to Brazil.

25. **Sporobolus interruptus** Vasey [p. 134]
BLACK DROPSEED

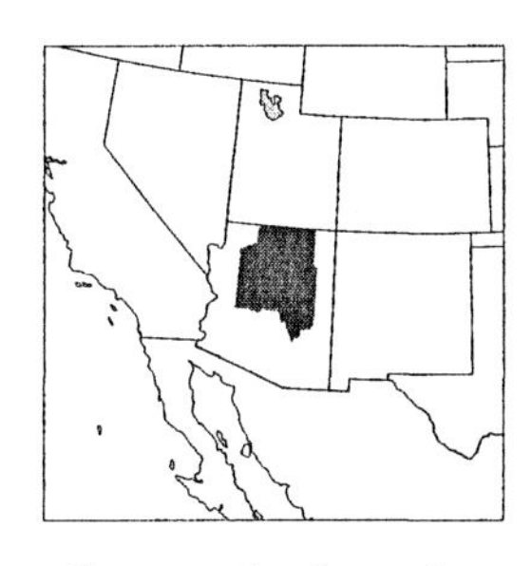

Plants perennial; cespitose but shortly rhizomatous, with tough, fibrous roots. **Culms** 25–60 cm. **Sheaths** dull and fibrous basally, with scattered, contorted hairs to 5 mm, margins glabrous; **ligules** 0.2–0.7 mm; **blades** (5)8–20 cm long, 1–2.5 mm wide, flat to folded, glabrous or scattered-pilose on both surfaces, margins glabrous. **Panicles** 5–20 cm long, (0.6)1–8 cm wide, longer than wide, narrowly pyramidal, open to somewhat contracted, not diffuse, well-exerted from the upper leaf sheath; **lower nodes** with 1–2(3) branches; **primary branches** 0.6–7 cm, appressed or spreading to 70° from the rachis, not capillary, without spikelets on the lower ⅓; **pedicels** 0.8–5.5 mm, appressed to spreading. **Spikelets** 4.5–6.6 mm, plumbeous. **Glumes** unequal, lanceolate, membranous; **lower glumes** (2)2.5–4.2 mm; **upper glumes** 3.8–6.5 mm, at least ⅔ as long as the florets; **lemmas** 5–6.5 mm, ovate, membranous, glabrous, acute; **paleas** 4.8–6.5 mm, ovate,

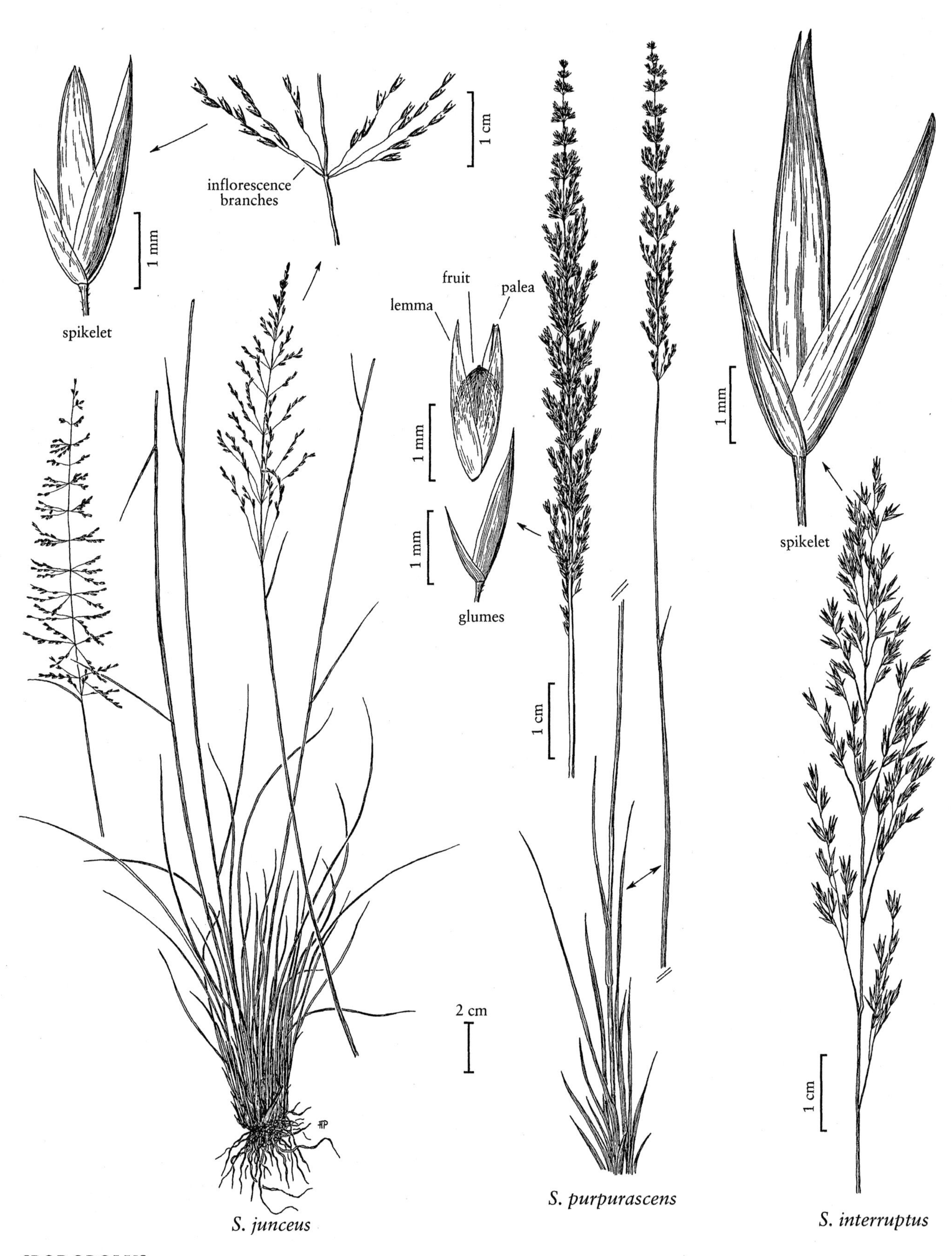
spikelet
1 mm
inflorescence branches
1 cm
fruit
palea
lemma
1 mm
1 mm
glumes
1 cm
1 mm
spikelet
2 cm
1 cm
S. junceus
S. purpurascens
S. interruptus

SPOROBOLUS

membranous; **anthers** 3–4.2 mm, yellow to purplish. **Fruits** about 3 mm long, 1.5–1.7 mm thick, pyriform-globose; **embryo** dark brown to blackish; **endosperm** reddish-brown. 2*n* = 30.

Sporobolus interruptus grows on rocky slopes and in dry meadows of open yellow pine and oak-pine forests and pinyon-juniper woodlands, at elevations from 1500–2300 m. It is an Arizonan endemic that is morphologically similar to *S. heterolepis*, but the two species are separated geographically, the range of the latter lying to the north and east of Arizona. The only reliable morphological difference between them is anther length (3–4.2 mm long in *S. interruptus*, 1.7–3 mm long in *S. heterolepis*). Cytologically, *S. interruptus* appears to be triploid, while *S. heterolepis* appears to be an octoploid (2*n* = 72).

26. **Sporobolus heterolepis** (A. Gray) A. Gray [p. 136]
PRAIRIE DROPSEED, SPOROBOLE À GLUMES INÉGALES

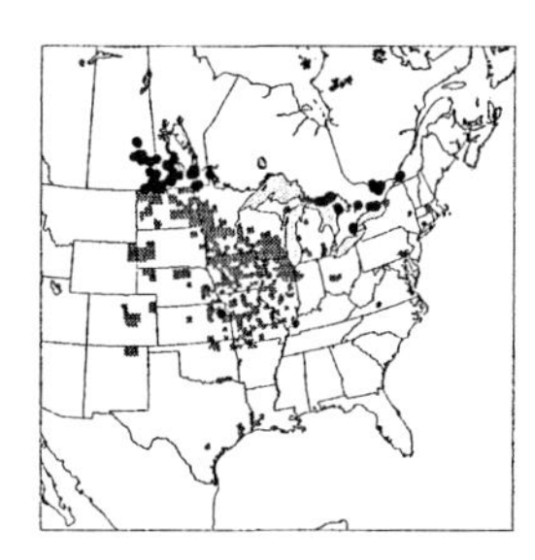

Plants perennial; cespitose, not rhizomatous. **Culms** 30–80(90) cm. **Sheaths** dull and fibrous basally, glabrous or sparcely pilose below, hairs to 4 mm, contorted; **ligules** 0.1–0.3 mm; **blades** 7–31 cm long, 1.2–2.5 mm wide, flat to folded, glabrous abaxially, scabridulous adaxially, margins scabrous. **Panicles** 5–22(25) cm long, (0.6)1–11 cm wide, open to somewhat contracted, longer than wide, narrowly pyramidal, not diffuse; **lower nodes** with 1–2(3) branches; **primary branches** 0.6–8(11) cm, appressed or spreading to 70° from the rachis, not capillary, without spikelets on the lower ⅓; **pedicels** 0.8–6 mm, appressed, occasionally spreading. **Spikelets** 3–6 mm, plumbeous. **Glumes** unequal, lanceolate, membranous; **lower glumes** (1.2)1.8–4.5 mm; **upper glumes** 2.4–6 mm, at least ⅔ as long as the florets, occasionally 3-veined; **lemmas** (2.7)3–4.3 mm, ovate, membranous, glabrous, acute; **paleas** 3.1–4.5 mm, slightly longer than the lemmas, ovate, membranous, glabrous; **anthers** 1.7–3 mm, yellowish to purplish. **Fruits** 1.4–2.1 mm, pyriform to globose, indurate, without a loose pericarp, smooth, shining, light brown. 2*n* = 72.

Sporobolus heterolepis grows at elevations of 40–2250 m, in lowland and upland prairies, along the borders of woods, roadsides, and swamps, and in north-facing swales. It is associated with many plant communities, and is also available commercially as an ornamental. It is restricted to the *Flora* region.

27. **Sporobolus teretifolius** R.M. Harper [p. 136]
WIRELEAF DROPSEED

Plants perennial; cespitose, not rhizomatous. **Culms** (20)35–80 (100) cm, wiry. **Sheaths** shiny and indurate basally, glabrous or appressed hairy elsewhere, hairs to 4 mm; **ligules** 0.2–0.4 mm; **blades** (10)25–54 cm long, 0.5–1.2 mm wide, terete or subterete at least basally, sometimes channeled for portions of their length, green to yellowish-green, senescing or turning tan in late fall, glabrous on both surfaces or the adaxial surface sparsely hairy basally, margins smooth. **Panicles** 10–26 cm long, 1–9 cm wide, open (contracted when immature), not diffuse, narrowly pyramidal to ovate; **lower nodes** with 1–2(3) branches; **primary branches** 1–8 cm, ascending or spreading to 40° from the rachis, not capillary, without spikelets on the lower ⅓; **secondary branches** spreading; **pulvini** hairy; **pedicels** 3–18 mm, longer than the spikelets, spreading, with scattered ascending hairs. **Spikelets** 4–5.6 mm, purplish-brown. **Glumes** unequal, linear-lanceolate, membranous; **lower glumes** 2–3.8 mm, 0.5–0.8 times as long as the upper glumes; **upper glumes** 4–5.6 mm, usually longer than the florets; **lemmas** 3.4–4.4 mm, ovate, membranous, glabrous, acute; **paleas** 3.3–4.4 mm, ovate, membranous, glabrous; **anthers** 1.5–2.6 mm, purplish. **Fruits** not seen. 2*n* = unknown.

Sporobolus teretifolius is restricted to the southeastern United States, where it grows in wet to moist flatwoods and savannahs, at elevations of 10–150 m.

28. **Sporobolus curtissii** Small *ex* Kearney [p. 136]
CURTISS' DROPSEED

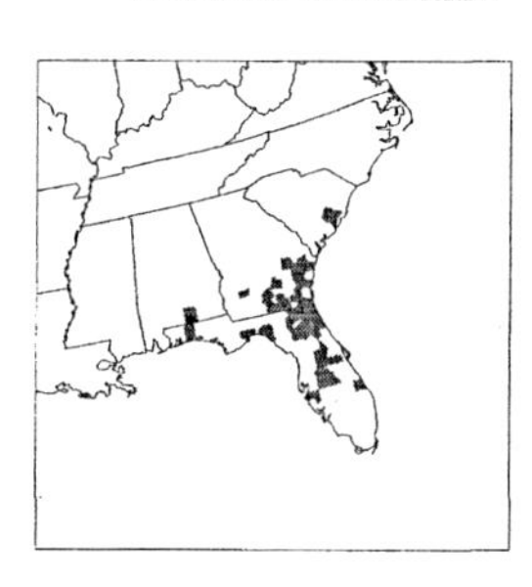

Plants perennial; cespitose, not rhizomatous. **Culms** 30–80(90) cm. **Sheaths** shiny and indurate basally, glabrous or appressed hairy elsewhere, hairs to 4 mm; **ligules** 0.2–0.6 mm; **blades** 5–22(28) cm long, 0.8–2(2.2) mm wide, flat to v-shaped in cross section, folded or involute when dry, green, remaining so well into winter, glabrous abaxially, mostly glabrous adaxially but densely pilose basally, margins glabrous or scabridulous. **Panicles** 10–25 cm long, 2–10(13) cm wide, longer than wide, open (contracted when immature), not diffuse, pyramidal to ovate; **lower nodes** with 1–2(3) branches; **primary branches** 2–9(10) cm, ascending or spreading to 80° from the rachis, not capillary, without spikelets on the lower ⅓; **secondary branches** mostly appressed;

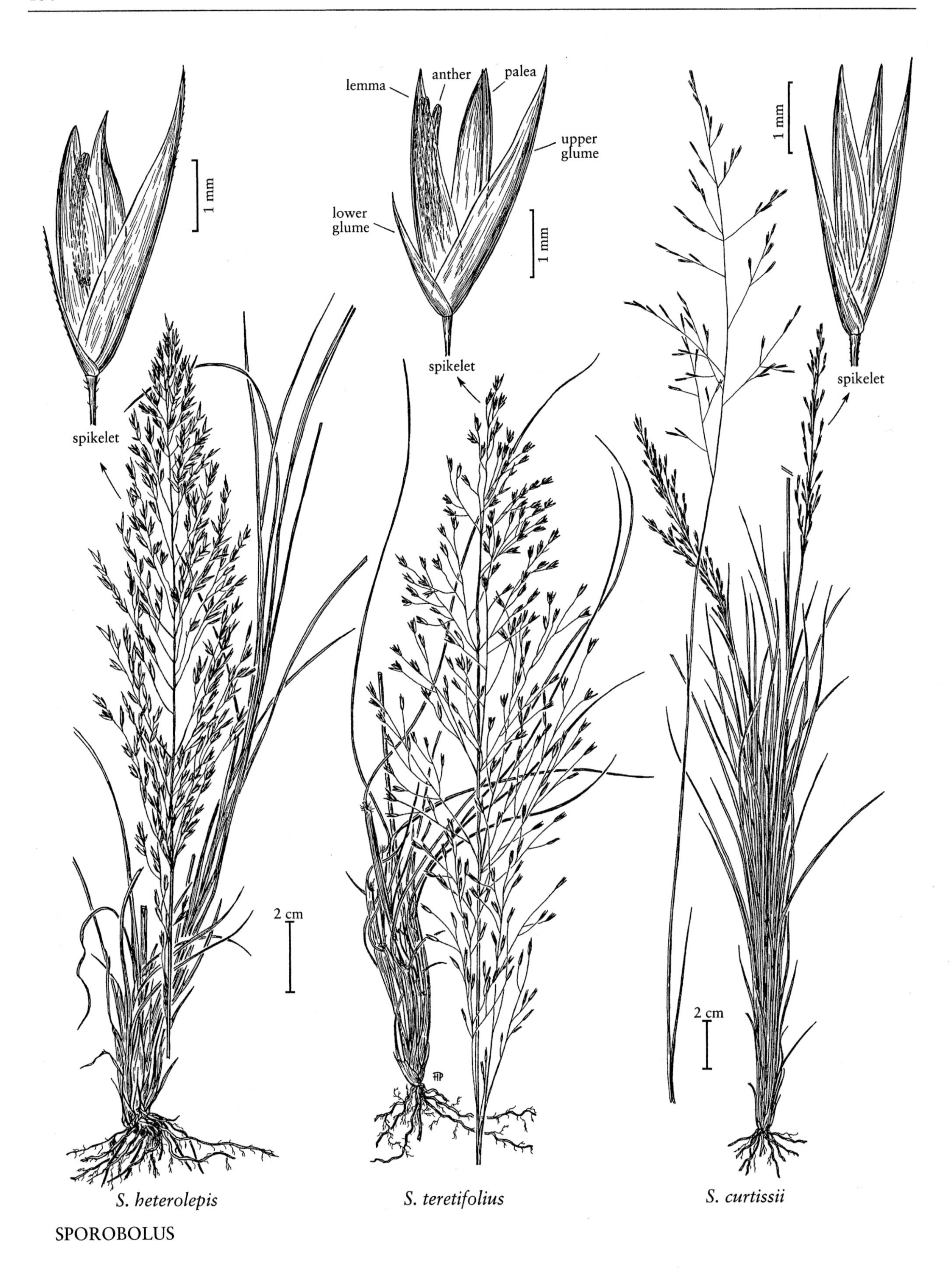
1 mm
spikelet
lemma
anther
palea
upper glume
lower glume
1 mm
spikelet
1 mm
spikelet
2 cm
2 cm
S. heterolepis
S. teretifolius
S. curtissii

SPOROBOLUS

pulvini glabrous or hairy; **pedicels** 0.5–4(8) mm, usually shorter than the spikelets, appressed, glabrous. **Spikelets** 3.5–6.6 mm, purplish-brown. **Glumes** equal to subequal, linear-lanceolate, membranous; **lower glumes** (2.9)3.5–6.2 mm, 0.9–1.2(1.3) times as long as the upper glumes; **upper glumes** 3.2–6.6 mm, from almost as long as to longer than the florets; **lemmas** 3.4–4.5 mm, ovate to lanceolate, membranous, glabrous, acute; **paleas** 3.4–4.5 mm, ovate, membranous, glabrous; **anthers** 1.5–2.8 mm, yellow to purplish. **Fruits** 1.1–1.4 mm, fusiform, reddish-brown. $2n$ = unknown.

Sporobolus curtissii is restricted to the southeastern United States, where it grows in dry-mesic to moist flatwoods, in soils seasonally saturated at the surface or rather well-drained throughout the year. Its elevational range is 0–100 m.

29. **Sporobolus silveanus** Swallen [p. 138]
SILVEUS' DROPSEED

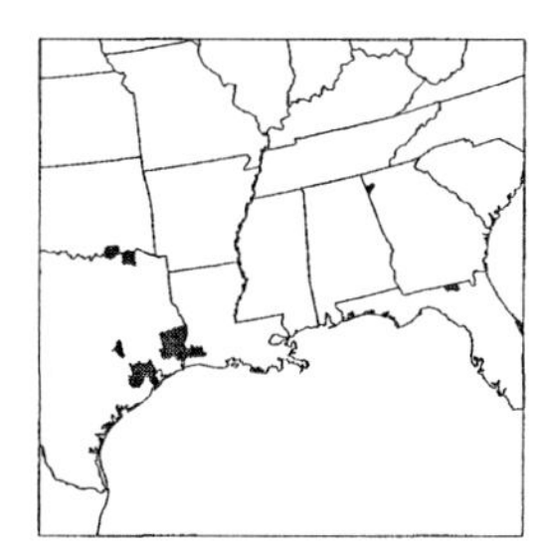

Plants perennial; cespitose, not rhizomatous. **Culms** 70–120 cm. **Sheaths** shiny and indurate basally, glabrous or appressed hairy elsewhere, hairs to 4 mm; **ligules** 0.2–0.8 mm; **blades** 15–52 cm long, 1–2.5 mm wide, flat, folded or involute, bluish-green, remaining so well into winter, glabrous on both surfaces, margins scabridulous. **Panicles** 21–50 cm long, 5–12(15) cm wide, longer than wide, open, not diffuse, pyramidal to ovate, with few spikelets; **lower nodes** with 1–2(3) branches; **primary branches** 6–20 cm, ascending, spreading 20–50° from the rachis, not capillary, without spikelets on the lower ¼–½; **secondary branches** appressed to loosely spreading; **pulvini** glabrous; **pedicels** 3–8(14) mm, longer or shorter than the spikelets, appressed, glabrous, scabrous. **Spikelets** 4.5–7(7.2) mm, purplish. **Glumes** subequal to unequal, linear-lanceolate to lanceolate, membranous; **lower glumes** 3–4.6 mm, 0.6–0.9 times as long as the upper glumes; **upper glumes** 4–7.2 mm, from slightly shorter to longer than the spikelets; **lemmas** 4.4–6.5 mm, lanceolate, membranous, glabrous, acuminate to acute; **paleas** 4.5–6.7 mm, lanceolate, membranous, glabrous; **anthers** 3, 3.5–5 mm, purplish. **Fruits** 1.8–2.2 mm, obovoid, laterally flattened, light brownish. $2n$ = unknown.

Sporobolus silveanus is restricted to the southeastern United States. It grows in wet to mesic pine woodlands and adjoining glades and barren openings, and in black-land prairies, at elevations of 5–200 m.

30. **Sporobolus floridanus** Chapm. [p. 138]
FLORIDA DROPSEED

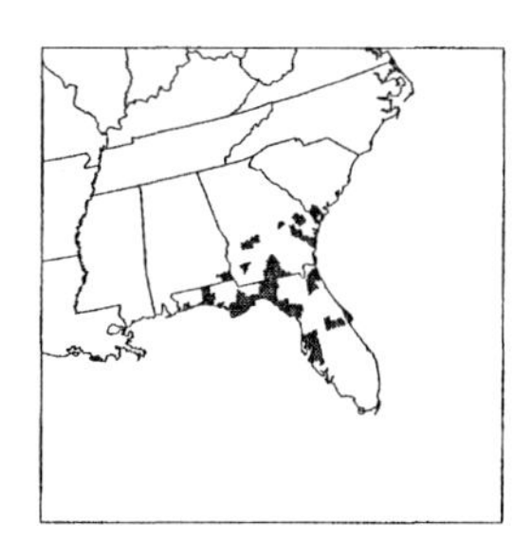

Plants perennial; cespitose, not rhizomatous. **Culms** (40)100–200(250) cm. **Sheaths** shiny and indurate basally, glabrous or appressed hairy elsewhere, hairs to 5 mm; **ligules** 0.2–0.7 mm; **blades** (10)25–50 cm long, (2)3–10 mm wide, flat to folded, pale bluish-green, yellowing at maturity, glabrous on both surfaces or the adaxial surface sparsely hairy basally, margins scabridulous. **Panicles** (18)30–50 cm long, 4–15 cm wide, open (contracted when immature), longer than wide, not diffuse, pyramidal to ovate; **lower nodes** with 1–2(3) branches; **primary branches** 4–15 cm, spreading 10–90° from the rachis, not capillary, without spikelets on the lower ⅓; **secondary branches** spreading; **pulvini** hairy or glabrous; **pedicels** 2–14 mm, longer than the spikelets, spreading, glabrous, sometimes scabridulous. **Spikelets** (3.7)4–6 mm, purplish-brown. **Glumes** linear-lanceolate, membranous; **lower glumes** 2.5–5.1 mm, (0.6)0.75–0.9(0.94) times as long as the upper glumes; **upper glumes** 3.7–5.7 mm, longer than the florets; **lemmas** 3–4 mm, ovate to lanceolate, membranous, glabrous, acute; **paleas** 3–4 mm, ovate, membranous, glabrous; **anthers** 2–3.1 mm, purplish. **Fruits** 1.7–2 mm, fusiform, reddish-brown. $2n$ = unknown.

Sporobolus floridanus grows in wet to mesic pine woodlands, seepage bogs, and treeless swales, in soils semi-permanently to seasonally saturated at the surface, and in places where water may pond for weeks, at elevations of 0–100 m. It is endemic to the southeastern United States.

31. **Sporobolus pinetorum** Weakley & P.M. Peterson [p. 138]
CAROLINA DROPSEED

Plants perennial; cespitose, not rhizomatous. **Culms** (30)45–120 (180) cm. **Sheaths** shiny and indurate basally, glabrous or appressed hairy elsewhere, hairs to 5 mm, margins hyaline; **ligules** 0.2–0.6 mm; **blades** 20–50 cm long, 1.2–2(3) mm wide, flat, folded or involute, dark green, remaining so well into winter, glabrous on both surfaces or the adaxial surface sparsely hairy basally, margins scabridulous. **Panicles** 15–30 cm long, 2–6 cm wide, open (contracted when immature), longer than wide, pyramidal to ovate, not diffuse; **lower nodes** with 1–2(3) branches; **primary branches** 2–8 cm, appressed or

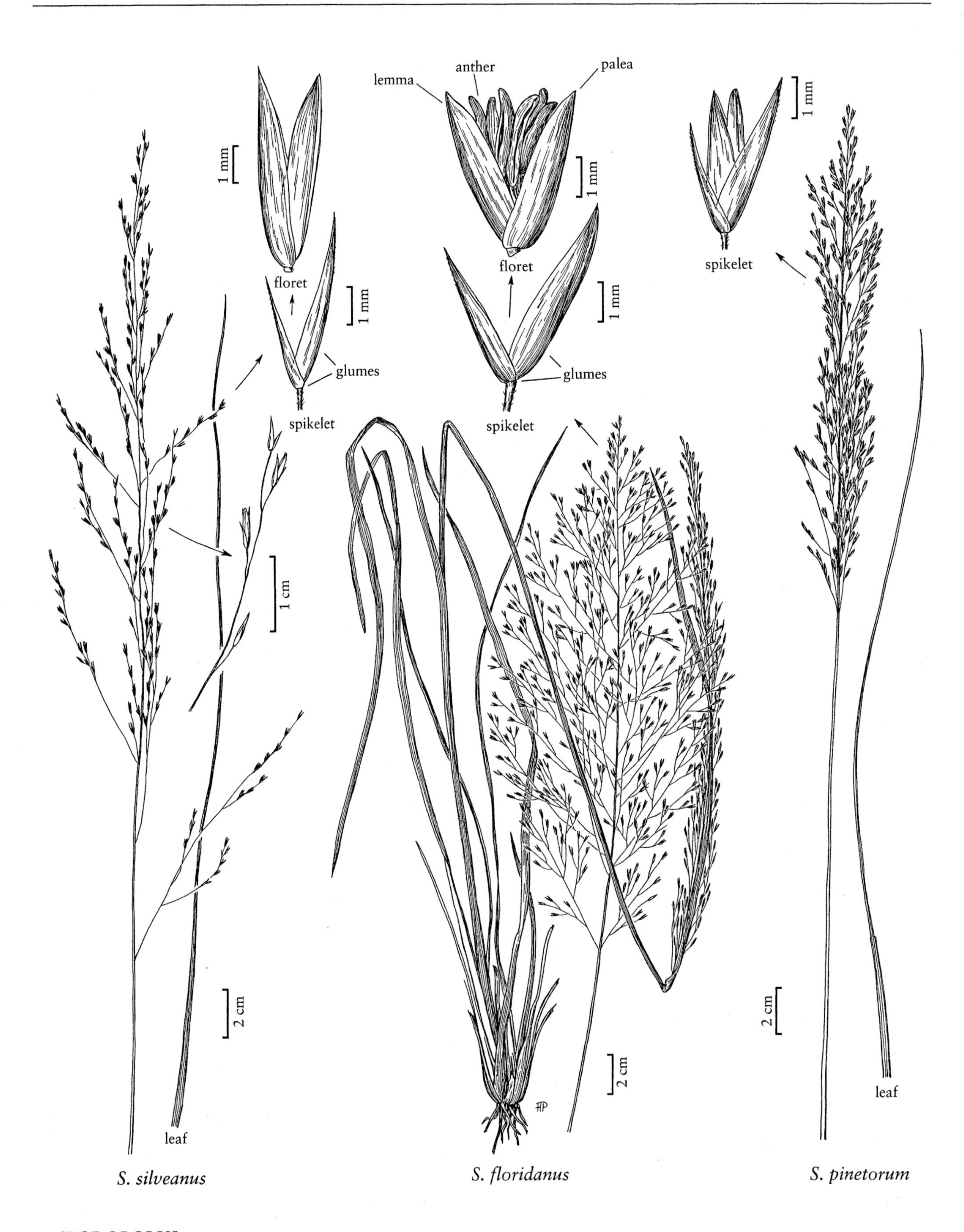
1 mm
floret
1 mm
glumes
spikelet
1 cm
2 cm
leaf
lemma
anther
palea
1 mm
floret
1 mm
glumes
spikelet
2 cm
1 mm
spikelet
2 cm
leaf
S. silveanus
S. floridanus
S. pinetorum

SPOROBOLUS

spreading to 50° from the rachis, not capillary, without spikelets on the lower ⅓; **secondary branches** spreading; **pulvini** hairy or glabrous; **pedicels** 2–22 mm, longer than the spikelets, spreading, glabrous, scabridulous. **Spikelets** 3.5–6.5 mm, purplish-brown. **Glumes** linear-lanceolate, membranous; **lower glumes** 2.4–4.5 mm, 0.6–0.83 times as long as the upper glumes; **upper glumes** (3.5)4–6(6.5) mm, as long as or longer than the florets; **lemmas** 3.4–4.3 mm, ovate to lanceolate, membranous, glabrous, acute; **paleas** 3.4–4.4 mm, ovate, membranous, glabrous; **anthers** 2.5–3.4 mm, purplish. **Fruits** 1.8–2.2 mm, fusiform, brown. $2n$ = unknown.

Sporobolus pinetorum grows in wet to moist pine woodlands, in soils seasonally to semi-permanently saturated, at elevations of 0–160 m. It is endemic to the southeastern United States.

17.31 **CRYPSIS** Aiton

Barry E. Hammel
John R. Reeder

Plants annual; synoecious. **Culms** 1–75 cm, erect to geniculately ascending, sometimes branching above the base; **nodes** usually exposed. **Sheaths** open, often becoming inflated, junction with the blades evident; **ligules** of hairs; **auricles** absent; **blades** often disarticulating. **Inflorescences** terminal or terminal and axillary, spikelike or capitate panicles subtended by, and often partially enclosed in, 1 or more of the uppermost leaf sheaths, additional panicles often present in the axils of the leaves below. **Spikelets** 2–6 mm, strongly laterally compressed, with 1 floret; **florets** bisexua; **disarticulation** above or below the glumes. **Glumes** 1-veined, strongly keeled; **lemmas** membranous, glabrous, 1-veined, strongly keeled, not lobed, unawned, sometimes mucronate; **paleas** hyaline, 1–2-veined; **lodicules** absent; **anthers** 2 or 3; **ovaries** glabrous. **Fruits** oblong, pericarp loosely enclosing the seed and easily removed when wet; **hila** punctate. x = 8. Name from the Greek *krupsis*, 'concealment', alluding to the partially concealed inflorescence.

Crypsis, a genus of eight species, is native from the Mediterranean region to northern China. Its species tend to occur in moist soils, often in areas subject to winter flooding. The three species found in the *Flora* region are very plastic in the lengths of their culms and leaves, e.g., the culms of *C. schoenoides* vary from 2 cm in dry sites to 75 cm under optimal conditions.

SELECTED REFERENCE Hammel, B.E. and J.R. Reeder. 1979. The genus *Crypsis* (Gramineae) in the United States. Syst. Bot. 4:267–280.

1. Spikelets 1.5–2.8 mm long; panicles 7–8 times longer than wide, usually completely exserted from the uppermost sheath at maturity . . . . 1. *C. alopecuroides*
1. Spikelets 2.5–4 mm long; panicles 1–5 times longer than wide, the bases usually enclosed in the uppermost sheath at maturity.
   2. Collars glabrous; glumes unequal in length, the margins glabrous; anthers 0.7–1.1 mm long . . . . 2. *C. schoenoides*
   2. Collars pilose; glumes subequal, at least the lower glumes pilose on the margin; anthers 0.5–0.9 mm long . . . . 3. *C. vaginiflora*

1. Crypsis alopecuroides (Piller & Mitterp.) Schrad. [p. 141]
FOXTAIL PRICKLEGRASS

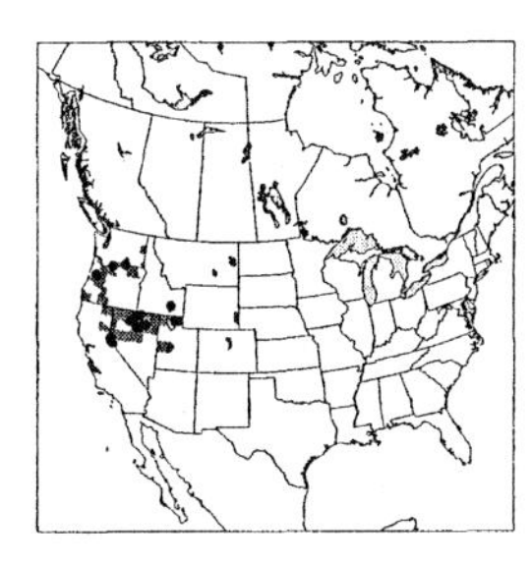

Culms (3)5–75 cm, rarely branched above the base. **Sheaths** glabrous; **collars** glabrous; **ligules** 0.2–1 mm; **blades** 5–12 cm long, 1.2–2.5 mm wide, not disarticulating. **Panicles** 1.5–6.5 cm long, 4–6 mm wide, 7–8 times longer than wide, often purplish, completely exserted from the uppermost sheath at maturity on peduncles at least 1 cm long. **Spikelets** 1.8–2.8 mm, remaining lightly attached until late in the season. **Lower glumes** 1.2–2 mm; **upper glumes** 1.4–2.4 mm; **lemmas** 1.7–2.8 mm; **paleas** faintly 2-veined; **anthers** 3, 0.5–0.6 mm. **Caryopses** 0.9–1.1 mm. $2n$ = 16.

Crypsis alopecuroides is common to abundant in sandy soils around drying lake margins in Oregon and southern Washington, and within the last forty years has become widespread in northern California; it is also known from several other western states. It was first collected in the Western Hemisphere in the late 1800s from shipyard areas in and around Philadelphia, but has not been collected in the eastern United States since. In the Eastern Hemisphere, it extends from France and northern Africa to the Urals and Iraq.

2. **Crypsis schoenoides** (L.) Lam. [p. 141]
SWAMP PRICKLEGRASS

Culms 2–75 cm, prostrate to erect, sometimes geniculate, usually not branching above the base, but some plants profusely branched. **Sheaths** glabrous or ciliate at the throat, often inflated; **collars** glabrous; **ligules** 0.5–1 mm; **blades** 2–10 cm long, 2–6 mm wide, not disarticulating. **Panicles** 0.3–4(7.5) cm long, 5–6(15) mm wide, 1–5 times longer than wide, bases usually enclosed in the uppermost leaf sheaths at maturity. **Spikelets** 2.7–3.2 mm, tardily disarticulating. **Lower glumes** 1.8–2.3 mm; **upper glumes** 2.2–2.7 mm; **lemmas** 2.4–3 mm; **paleas** 2-veined; **anthers** 3, 0.7–1.1 mm. **Caryopses** about 1.3 mm. $2n = 32$.

Crypsis schoenoides is common to abundant in clay or sandy clay soils around drying lake margins and vernal pools. In the *Flora* region, it is most abundant in California, but also appears to be established in a few other western states. It is known from a few collections in several eastern states (where it was first introduced in the late 1800s), though apparently none more recently than 1955. Its native range extends from southern Europe and northern Africa through western Asia to India.

3. **Crypsis vaginiflora** (Forssk.) Opiz [p. 141]
MODEST PRICKLEGRASS

Culms 1–30 cm, often profusely branching above the base, with 10–25 panicles per culm. **Sheaths** pilose on the margins; **collars** pilose; **blades** 1–5 cm long, 1–3 mm wide, soon disarticulating, thus many leaves on mature plants are bladeless. **Panicles** 0.3–1.5(3.5) cm long, 3–6(10) mm wide, 1–5 times longer than wide, sessile or almost so, mostly included in the sheaths of the upper 2 leaves. **Spikelets** 2.5–3.2 mm, readily disarticulating when disturbed, otherwise retained within the upper sheaths. **Glumes** about 3 mm, subequal; **lower glumes** pilose on the margins; **lemmas** subequal to the glumes; **paleas** minutely 2-veined; **anthers** 3, 0.5–0.9 mm. **Caryopses** 1.3–1.7 mm. $2n = 48$.

Crypsis vaginiflora is common to abundant in clay or sandy clay soil in California, where it was first introduced in the late 1800s. It has since been found at a few locations in Washington, Idaho, and Nevada, and will probably spread to additional sites with suitable habitat in the future. It is native to Egypt and southwestern Asia.

17.32 CALAMOVILFA Hack.

John W. Thieret

Plants perennial; synoecious; rhizomatous, rhizomes short or elongate. **Culms** 50–250 cm, solitary or few. **Leaves** cauline; **sheaths** open; **ligules** of hairs, dense, short; **blades** elongate, long tapering. **Inflorescences** terminal, simple panicles, usually exserted and exceeding the upper leaves, open to contracted; **panicles** 8–80 cm long, to 60 cm wide, simple, flexible, branches not spikelike, not disarticulating. **Spikelets** with 1 floret, laterally compressed, unawned; **rachillas** not prolonged; **disarticulation** above the glumes, achenes falling with the lemmas and paleas. **Glumes** subequal to unequal, 1-veined, acute; **calluses** evidently hairy, hairs ¼–⅞ as long as the lemmas; **lemmas** similar to the glumes, from shorter than to longer than the upper glumes, 1-veined, acute, unawned; **paleas** longitudinally grooved; **anthers** 3, 2.4–5.5 mm; **ovaries** glabrous. **Fruit** an achene, pericarp free from the seed. $x = 10$. Name from the Greek *kalamos*, 'reed', and *Vilfa*, a genus of grasses.

Calamovilfa is a genus of five species, all of which are endemic to temperate portions of the *Flora* region.

SELECTED REFERENCE Thieret, J.W. 1956. Synopsis of the genus *Calamovilfa* (Gramineae). Castanea 31:145–152.

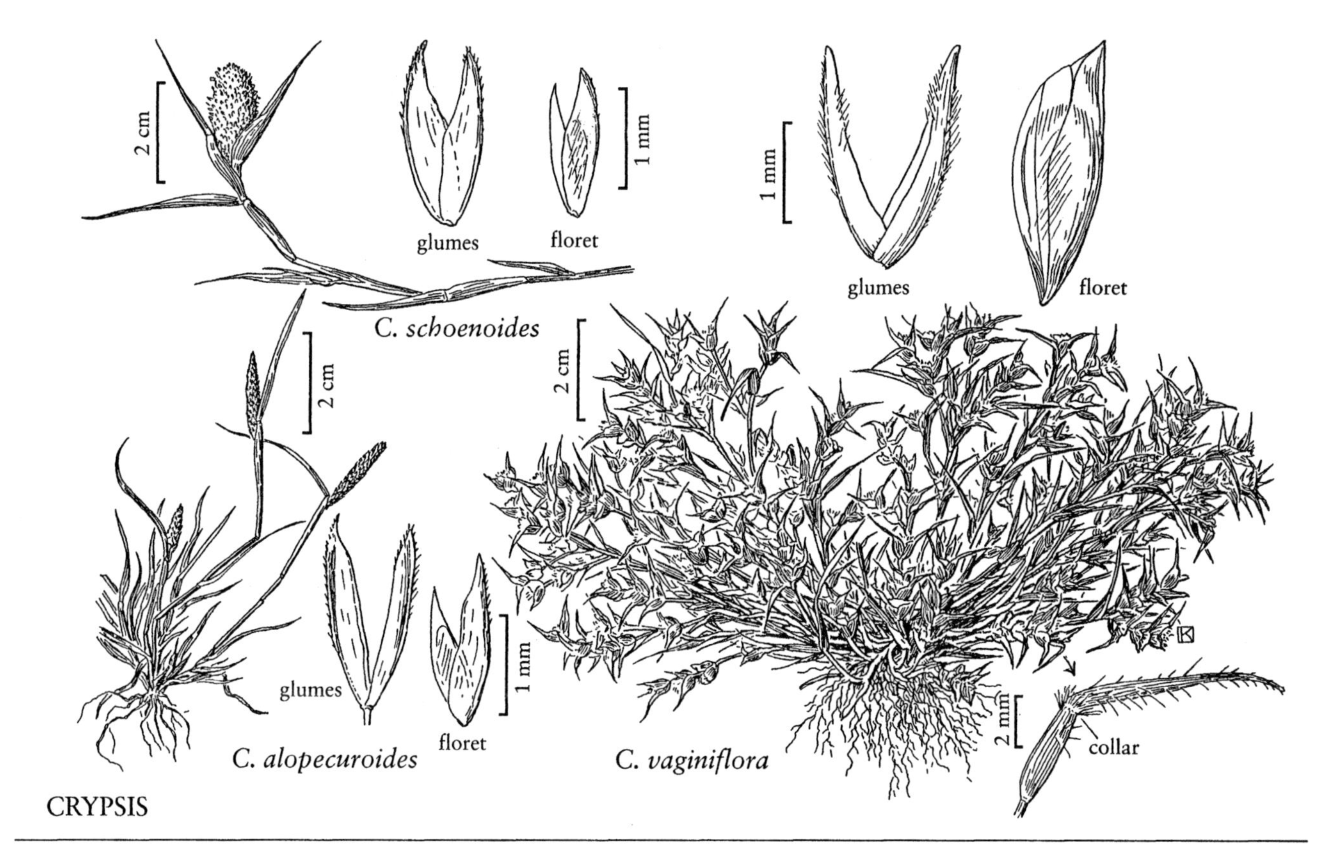

CRYPSIS

1. Rhizomes elongate; ligules 0.7–2.5 mm (sect. *Interior*).
   2. Lemmas or paleas (or both) pubescent, although sometimes sparsely so; spikelets 7–10.8 mm long . . . . 1. *C. gigantea*
   2. Lemmas and paleas glabrous; spikelets 5–8.5 mm long . . . . 2. *C. longifolia*
1. Rhizomes short; ligules to 0.7 mm (sect. *Calamovilfa*).
   3. Panicles contracted, to 3.5 cm wide . . . . 3. *C. curtissii*
   3. Panicles open, 4–40 cm wide.
      4. Spikelets 6–7.4 mm; glumes acute to acuminate, usually arcuate; lemmas 5.5–7 mm, arcuate, attenuate . . . . 4. *C. arcuata*
      4. Spikelets 4–5.8 mm; glumes acute, straight; lemmas 4–5.4 mm, straight, acuminate . . . . . . 5. *C. brevipilis*

1. **Calamovilfa gigantea** (Nutt.) Scribn. & Merr. [p. 143]
GIANT SANDREED

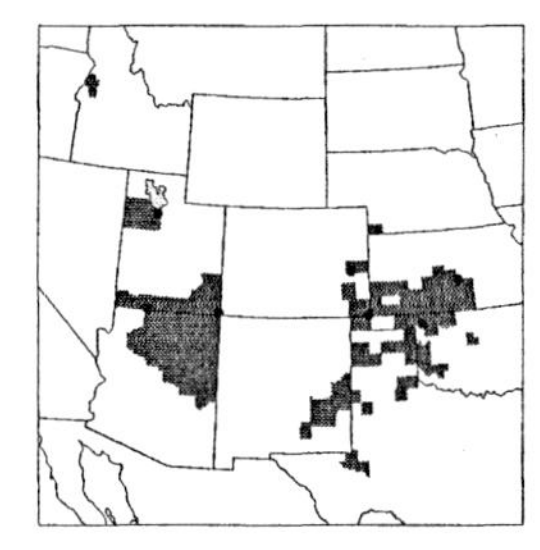

Rhizomes elongate, covered with shiny, coriaceous, scalelike leaves. **Culms** to 2.5 m. **Sheaths** entirely glabrous or pubescent at the throat; **ligules** 0.7–2 mm; **blades** to 90 cm long, about 12 mm wide. **Panicles** 20–80 cm long, 20–60 cm wide; **branches** to 35 cm, ascending to strongly divergent, lowermost branches sometimes reflexed. **Spikelets** 7–10.8 mm. **Glumes** straight; **lower glumes** 4.5–10.5 mm; **upper glumes** 6.4–10.1 mm; **callus hairs** ¼–¾ as long as the lemma; **lemmas** 6–10 mm, straight, pubescent, sometimes sparsely so, very rarely glabrous; **paleas** 6–8.3 mm, pubescent or glabrous; **anthers** 3–5.5 mm. $2n = 60$.

Calamovilfa gigantea grows on sand dunes, prairies, river banks, and flood plains in the Rocky Mountains and central plains from Utah and Nebraska to Arizona and Texas.

2. **Calamovilfa longifolia** (Hook.) Scribn. [p. 143]
PRAIRIE SANDREED

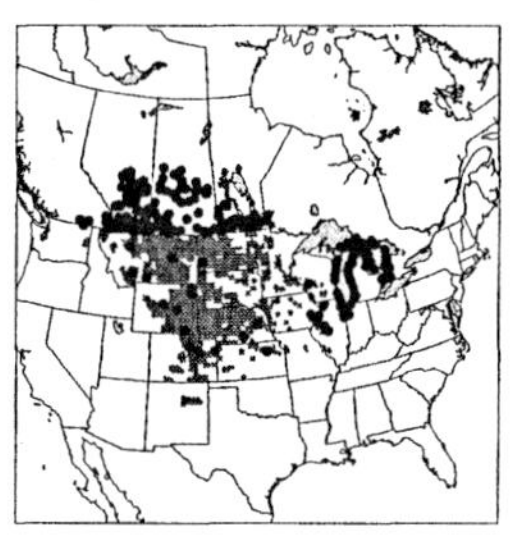

Rhizomes elongate, covered with shiny, coriaceous, scalelike leaves. **Culms** to 2.4 m. **Sheaths** glabrous to densely pubescent; **ligules** 0.7–2.5 mm; **blades** to 64 cm long, about 12 mm wide. **Panicles** 15–78 cm long, 1.7–26.4 cm wide; **branches** to 33 cm, erect to strongly divergent, lowermost branches sometimes reflexed.

Spikelets 5–8.5 mm. **Glumes** straight; **lower glumes** 3.5–6.5 mm; **upper glumes** 5–8.2 mm; **lemmas** 4.5–7.1 mm, straight, glabrous; **paleas** 4.4–6.9 mm, glabrous. $2n$ = 40, ca. 60.

Calamovilfa longifolia usually grows in sand or sandy soils, but is occasionally found in clay soils or loess. Two geographically contiguous varieties exist. They differ as shown in the following key; the differences between the two are more striking in the field.

1. Most spikelets overlapping no more than 1 other spikelet, usually with a brownish cast . . . var. *magna*
1. Most spikelets overlapping 2–3 other spikelets, usually without a brownish cast . . . . var. *longifolia*

Calamovilfa longifolia (Hook.) Scribn. var. **longifolia**

Sheaths usually glabrous, sometimes sparsely pubescent, rarely densely pubescent. **Panicles** to 55.5 cm long, usually 6.7–13.9 times as long as wide; **branches** to 23 cm long, erect or ascending, at least in the upper ⅓ of the panicle. **Spikelets** usually without a brownish cast, relatively closely imbricate.

Calamovilfa longifolia var. *longifolia* is a characteristic grass on the drier prairies of the interior plains, from southern Canada to northern New Mexico, with reports from southern Arizona. It also grows, as an adventive, in Washington, Wisconsin, Michigan, and Missouri.

Calamovilfa longifolia var. **magna** Scribn. & Merr.

Sheaths usually pubescent, often densely so, rarely glabrous. **Panicles** to 77.5 cm long, usually 1.3–5.5 times as long as wide; **branches** to 33 cm long, typically widely divergent, sometimes reflexed in the lower ⅓ of the panicle. **Spikelets** commonly with a brownish cast, relatively loosely imbricate.

Calamovilfa longifolia var. *magna* grows on dunes and sandy shores around lakes Superior, Michigan and Huron, with outlying stations in sand or sandy soils.

3. **Calamovilfa curtissii** (Vasey) Scribn. [p. 144]
FLORIDA SANDREED

Rhizomes short, covered with the persistent bases of the foliage leaves. **Culms** to 2 m. **Sheaths** to 30 cm; **ligules** to 0.5 mm; **blades** to 50 cm long, 2–5 mm wide. **Panicles** to 50 cm long, about 3.5 cm wide, contracted; **branches** to 20 cm, erect. **Spikelets** 3.5–5.4 mm. **Glumes** straight; **lower glumes** 2–4.7 mm; **upper glumes** 3.5–5.4 mm; **callus hairs** ¼–½ as long as the lemmas; **lemmas** 3.5–5.2 mm, straight, lightly to markedly pubescent; **paleas** 3.5–5 mm, slightly to markedly pubescent; **anthers** 2.5–3.1 mm. $2n$ = unknown.

Calamovilfa curtissii is a rare species, although sometimes locally common. It is restricted to two disjunct regions in Florida. Most Gulf coast populations grow in moist flatwoods or adjacent to wet cypress depressions; Atlantic coast populations occur in interdune swales.

4. **Calamovilfa arcuata** K.E. Rogers [p. 144]
CUMBERLAND SANDREED

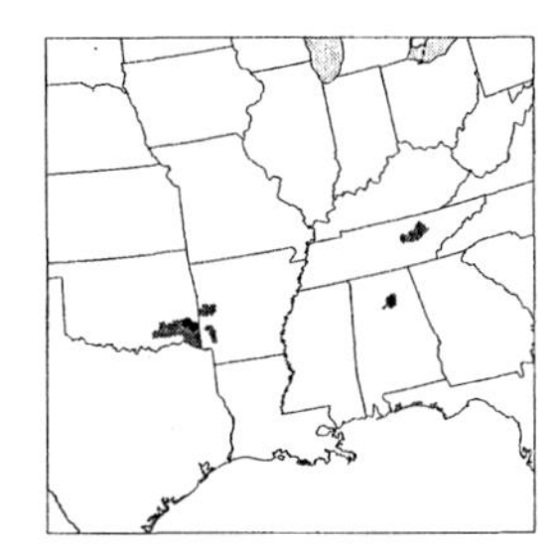

Rhizomes short, covered with the persistent bases of the foliage leaves. **Culms** to 1.5+ m. **Sheaths** to 22 cm; **ligules** to 0.7 mm; **blades** to 85 cm long, 1.5–6.5 mm wide. **Panicles** 15–45 cm long, 8–40 cm wide, open; **branches** to 22 cm, ascending to spreading. **Spikelets** 6–7.4 mm. **Glumes** usually arcuate, acute to acuminate; **lower glumes** 2.7–4.1 mm; **upper glumes** 4.2–5.4 mm; **callus hairs** ⅓–½ as long as the lemmas; **lemmas** 5.5–7 mm, arcuate, pubescent, attenuate; **paleas** 5.4–6.2 mm, pubescent; **anthers** 2.8–3.2 mm. $2n$ = unknown.

Calamovilfa arcuata is known only from a few scattered locations in the south central United States. It grows along streams and rivers.

5. **Calamovilfa brevipilis** (Torr.) Scribn. [p. 144]
PINE-BARREN SANDREED

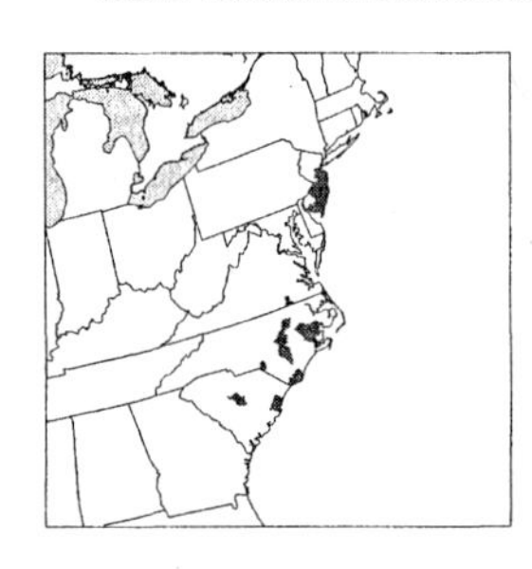

Rhizomes short, covered with the persisent bases of the foliage leaves. **Culms** to 1.5 m. **Sheaths** to 30 cm; **ligules** to 0.5 mm; **blades** to 50 cm long, 2–5 mm wide. **Panicles** 8–40 cm long, 4–20 cm wide, open; **branches** to 17 cm long, ascending to spreading. **Spikelets** 4–5.8 mm. **Glumes** straight, acute; **lower glumes** 1.7–4.1 mm; **upper glumes** 3.3–5 mm; **callus hairs** ¼–½ as long as the lemmas; **lemmas** 4–5.4 mm, straight, slightly to markedly pubescent, acuminate; **paleas** 3.8–5.3 mm, slightly to markedly pubescent; **anthers** 2.4–3.2 mm. $2n$ = unknown.

Calamovilfa brevipilis grows in moist to dry pine barrens, savannahs, bogs, swamp edges, and pocosins. It is a common grass on the New Jersey pine barrens and locally common across the coastal plain of North Carolina, but rare at present in Virginia and South Carolina. The length of the ligule hairs tends to increase from 0.3 mm or less in the north to 0.5 mm at the southern end of its range.

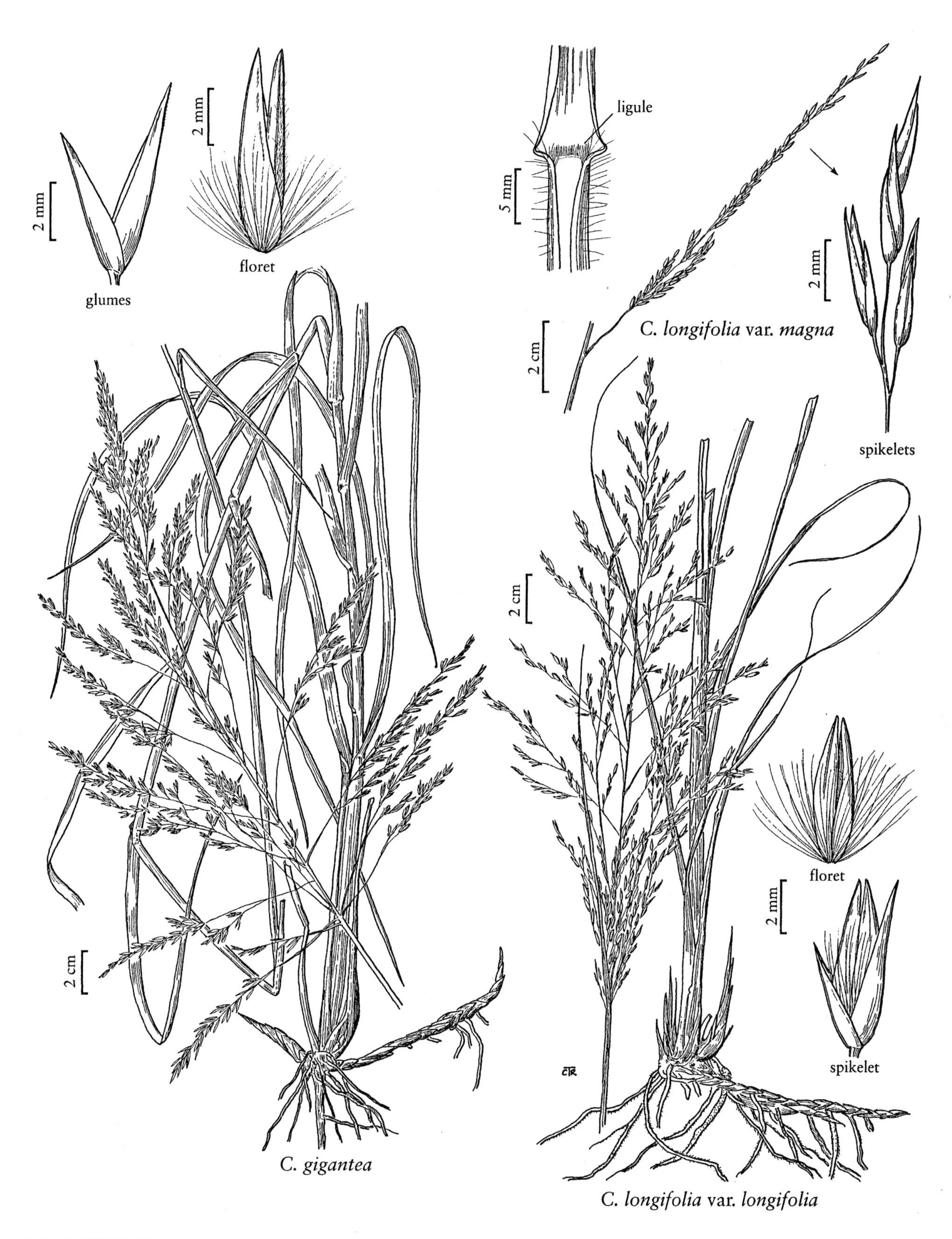

CALAMOVILFA

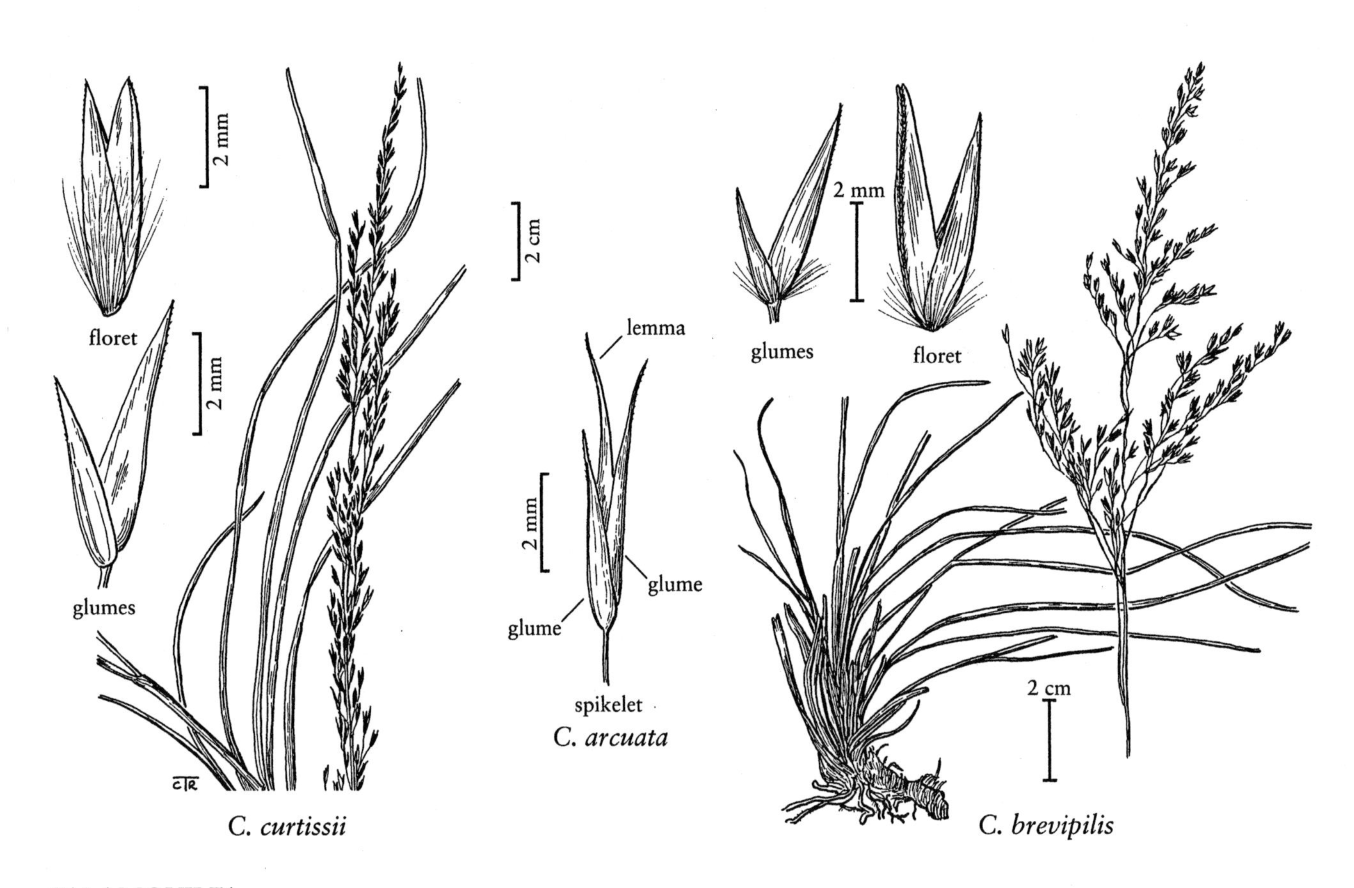

CALAMOVILFA

17.33 MUHLENBERGIA Schreb.

Paul M. Peterson

Plants annual or perennial; usually rhizomatous, often cespitose, sometimes mat-forming, rarely stoloniferous. **Culms** 2–300 cm, erect, geniculate, or decumbent, usually herbaceous, sometimes becoming woody. **Sheaths** open; **ligules** membranous or hyaline (rarely firm or coriaceous), acuminate to truncate, sometimes minutely ciliolate, sometimes with lateral lobes longer than the central portion; **blades** narrow, flat, folded, or involute, sometimes arcuate. **Inflorescences** terminal, sometimes also axillary, open to contracted or spikelike panicles; **disarticulation** usually above the glumes, occasionally below the pedicels. **Spikelets** with 1(2–3) florets. **Glumes** usually (0)1(2–3)-veined, apices entire, erose, or toothed, truncate to acuminate, sometimes mucronate or awned; **lower glumes** sometimes rudimentary or absent, occasionally bifid; **upper glumes** shorter than to longer than the florets; **calluses** poorly developed, glabrous or with hairs; **lemmas** glabrous, scabrous, or with short hairs, 3-veined (occasionally appearing 5-veined), apices awned, mucronate, or unawned; **awns**, if present, straight, flexuous, sinuous, or curled, sometimes borne between 2 minute teeth; **paleas** shorter than or equal to the lemmas, 2-veined; **anthers** (1–2)3, purple, orange, yellow, or olivaceous. **Caryopses** elongate, fusiform or elliptic, slightly dorsally compressed. **Cleistogamous panicles** sometimes present in the axils of the lower cauline leaves, enclosed by a tightly rolled, somewhat indurate sheath. $x = 10$. Named for Gotthilf Henry Ernest Muhlenberg (1753–1815), a Lutheran minister and pioneer botanist of Pennsylvania.

Muhlenbergia is primarily a genus of the Western Hemisphere. It has approximately 155 species. Sixty-nine of the 70 species treated here are native to the *Flora* region. The other species, *M. diversiglumis*, is included because it was recently reported from Texas, but no specimens supporting the report have been found. It may be based on a misidentification. Within the *Flora* region, *Muhlenbergia* is represented best in the southwestern United States. *Muhlenbergia montana* is an important range grass in the southwestern United States and northern Mexico.

In the key and descriptions, "puberulent" refers to having hairs so small that they can only be seen with a 10× hand lens.

SELECTED REFERENCES **Douglas, G.W., D. Meldinger, and J. Pojar.** 2002. Illustrated Flora of British Columbia, vol. 8. British Columbia Ministry of Sustainable Resource Management and British Columbia Ministry of Forests, Victoria, British Columbia, Canada. 457 pp.; **Herrera-Arrieta, Y.** 1998. A revision of the *Muhlenbergia montana* (Nutt.) Hitchc. complex (Poaceae: Chloridoideae). Brittonia 50:23–50; **Hitchcock, A.S.** 1935b. *Muhlenbergia* Schreb. Pp. 431–476 *in* M.A. Howe, H.A. Gleason, and J.H. Barnhart (eds.). North American Flora, vol. 17, part 6. New York Botanical Garden, New York, New York, U.S.A. 64 pp.; **Hitchcock, A.S.** 1951 [title page 1950]. Manual of the Grasses of the United States, ed. 2, rev. A. Chase. U.S.D.A. Miscellaneous Publication No. 200. U.S. Government Printing Office, Washington, D.C., U.S.A. 1051 pp.; **Kartesz, J.** and **C.A. Meacham.** 1999. Synthesis of the North American Flora, Version 1.0 (CD-ROM). North Carolina Botanical Garden, Chapel Hill, North Carolina, U.S.A.; **Kearney, T.H.** and **R.H. Peebles.** 1951. Arizona Flora. University of California Press, Berkeley and Los Angeles, California, U.S.A. 1032 pp.; **McGregor, R.L., T.M. Barkley, R.E. Brooks,** and **E.K. Schofield.** 1986. Flora of the Great Plains. Contribution No. 84-135-B, Division of Biology, Kansas Agricultural Experiment Station, Kansas State University. Contribution No. 1254, Department of Botany, North Dakota Agricultural Experiment Station, North Dakota State University. University Press of Kansas, Lawrence, Kansas, U.S.A. 1392 pp.; **Morden, C.W.** and **S.L. Hatch.** 1984. Cleistogamy in *Muhlenbergia cuspidata* (Poaceae). Sida 10:254–255; **Morden, C.W.** and **S.L. Hatch.** 1996. Morphological variation and synopsis of the *Muhlenbergia repens* complex (Poaceae). Sida 17:349–365; **Peterson, P.M.** 2000. Systematics of the Muhlenbergiinae (Chloridoideae: Eragrostideae). Pp. 195–212 *in* S.W.L. Jacobs and J. Everett (eds.). Grasses: Systematics and Evolution. International Symposium on Grass Systematics and Evolution (3rd:1998). CSIRO Publishing, Collingwood, Victoria, Australia. 408 pp.; **Peterson, P.M.** and **C.R. Annable.** 1991. Systematics of the annual species of *Muhlenbergia* (Poaceae-Eragrostideae). Syst. Bot. Monogr. 31:1–109; **Peterson, P.M.** and **Y. Herrera-Arrieta.** 2001. A leaf blade anatomical survey of *Muhlenbergia* (Poaceae: Muhlenbergiinae). Sida 19:469–506; **Soderstrom, T.R.** 1967. Taxonomic study of subgenus *Podosemum* and section *Epicampes* of *Muhlenbergia* (Gramineae). Contr. U.S. Natl. Herb. 34:75–189; **Swallen, J.R.** 1932. Five new grasses from Texas. Amer. J. Bot. 19:436–442; **Welsh, S.L., N.D. Atwood, S. Goodrich,** and **L.C. Higgins** (eds.). 1993. A Utah Flora, ed. 2, revised. Monte L. Bean Life Science Museum, Brigham Young University, Provo, Utah, U.S.A. 986 pp.

1. Plants annual [for opposite lead, see p. 147].
    2. Lemmas unawned or mucronate, the mucros to 1 mm long.
        3. Glumes strigulose, at least on the margins or towards the apices, the hairs 0.1–0.3 mm long.
            4. Pedicels of most spikelets strongly curved below the spikelets, often through 90° or more; anthers olivaceous, 0.6–1.2 mm long . . . . 65. *M. sinuosa*
            4. Pedicels of most spikelets straight to somewhat curved below the spikelets, rarely curved through 90°; anthers purplish, 0.2–0.7 mm long.
                5. Sheaths and culm internodes strigulose; lemmas 1.3–2 mm long, the apices acute to acuminate, mucronate or shortly awned . . . . 67. *M. texana*
                5. Sheaths and culm internodes glabrous, sometimes scabridulous; lemmas 0.8–1.5 mm long, the apices obtuse to subacute, unawned . . . . 66. *M. minutissima*
        3. Glumes glabrous.
            6. Panicles contracted, less than 0.5 cm wide; branches closely appressed at maturity; culms often rooting at the lower nodes . . . . 38. *M. filiformis*
            6. Panicles open or diffuse, 0.6–11 cm wide; branches spreading at maturity; culms not rooting at the lower nodes.
                7. Primary panicle branches 0.5–3.2 cm long; pedicels stout, 1–3 mm long, 0.5–1.5 mm thick; lemmas mottled, with greenish-black and greenish-white areas . . . . 70. *M. ramulosa*
                7. Primary panicle branches 0.4–6.2 cm long; pedicels delicate, 0.2–10 mm long, about 0.02 mm thick; lemmas not mottled, purplish, plumbeous, or brownish.
                    8. Primary panicle branches diverging 80–100° from the rachises; branches not developing below the lower leaf nodes; ligules with lateral lobes (vertical extensions of the sheath margins); plants truly annual . . . . 69. *M. fragilis*
                    8. Primary panicle branches diverging less than 80° from the rachises; branches developing below the lower leaf nodes; ligules without lateral lobes; plants perennial but often appearing annual . . . . 42. *M. uniflora*
    2. Lemmas awned, the awns 1–32 mm long.
        9. Upper glumes 2- or 3-veined, apices 2- or 3-toothed, not awned.
            10. Lemma awns olive-green, sinuous to curled; lemmas widest near the middle, 1.7–2.2 mm long . . . . 47. *M. crispiseta*
            10. Lemma awns purplish, flexuous; lemmas widest near the base, 1.4–4.2 mm long . . . . 48. *M. peruviana*
        9. Upper glumes l-veined or veinless, entire, erose, or awned.
            11. Lower glumes 2-veined, minutely to deeply bifid, the teeth aristate or with awns to 1.8 mm long; spikelets often in sessile-pedicellate pairs; disarticulation at the base of the pedicels.
                12. Glumes subequal to the lemmas; lemmas 2.5–4.5 mm long, those of the upper spikelets with awns 6–15 mm long . . . . 63. *M. depauperata*
                12. Glumes up to ⅔ as long as the lemmas; lemmas 3.5–6 mm long, those of the upper spikelets with awns 10–20 mm long . . . . 64. *M. brevis*
            11. Lower glumes, if present, veinless or l-veined, not bifid, unawned or with a single awn; spikelets borne singly; disarticulation above the glumes.
                13. Lemma awns 1–5 mm long.
                    14. Glumes 0.1–0.3 mm long, the lower glumes often almost absent; ligules 0.2–0.5 mm long . . . . 13. *M. schreberi*
                    14. Glumes 0.8–1.8 mm long; lower glumes 0.8–1.6 mm long; ligules 0.9–2.5 mm long.
                        15. Pedicels 2–7 mm long, usually longer than the florets, usually divergent; lemmas 1.3–2 mm long; lemma awns 0.1–1(2) mm long; caryopses 0.8–1 mm long; paleas minutely appressed-pubescent on the lower ½, 1.3–2 mm long . . . . 67. *M. texana*
                        15. Pedicels 1–2(3) mm long, usually shorter than the florets and appressed to the branches; lemmas (1.7)1.9–2.5 mm long; lemma awns 1.2–3.5 mm long; caryopses 1.3–2.3 mm long; paleas glabrous, 1.8–2.4 mm long . . . . . . 68. *M. eludens*
                13. Lemma awns 10–32 mm long.

16. Panicles secund; primary branches with 2–5 spikelets; secondary branches not developed; spikelets dimorphic with respect to the glumes, the glumes of the proximal spikelet on each branch subequal, to 0.7 mm long, orbicular and unawned, those of the distal spikelets evidently unequal, the lower glumes to 8 mm long and usually awned, the upper glumes orbicular, sometimes awned . . . . 18. *M. diversiglumis*
16. Panicles not secund; primary branches always with more than 2 spikelets, usually with more than 5; secondary branches well-developed; spikelets monomorphic with respect to the glumes.
17. Ligules 0.3–0.9 mm long, membranous and ciliate; distal portion of the sheath margins with hairs to 1 mm long; lemmas subulate to lanceolate, with a scabrous line between the midvein and lateral veins, giving the lemmas a 5-veined appearance . . . . 17. *M. pectinata*
17. Ligules 1–3(5) mm long, hyaline to membranous, often lacerate; sheath margins glabrous, even distally; lemmas lanceolate, smooth over most of their length, scabridulous to scabrous distally, not appearing 5-veined.
18. Cleistogamous panicles not present in the axils of the lower cauline leaves; upper glumes 1.5–2.8 mm long, acute . . . . 14. *M. tenuifolia*
18. Cleistogamous panicles of 1–3 florets present in the axils of the lower cauline leaves; upper glumes 0.6–2 mm long, obtuse to subacute.
19. Lemmas 2.5–3.8(5.3) mm long; glumes 0.4–1.3 mm long; ligules 1–2 mm long . . . . 15. *M. microsperma*
19. Lemmas 4–6.2(7.5) mm long; glumes 1–2 mm long; ligules 1.5–3 mm long . . . . 16. *M. appressa*

1. Plants perennial [for opposite lead, see p. 146].
20. Plants rhizomatous, usually not cespitose; rhizomes scaly and creeping [for opposite lead, see p. 149].
21. Panicles open, (2)4–20 cm wide; panicle branches capillary (0.05–0.1 mm thick), diverging 30–100° from the rachises at maturity.
22. Lemmas awned, awns 1–12(20) mm long.
23. Blades stiff and pungent; lemma awns 1–1.5(2) mm long, straight; primary branches of the panicles appearing fascicled in immature plants . . . . 32. *M. pungens*
23. Blades not stiff and pungent; lemma awns 4–12(20) mm long, flexuous; primary branches of the panicles not appearing fascicled . . . . 24. *M. arsenei*
22. Lemmas unawned, sometimes mucronate with mucros to 0.3 mm long.
24. Culms compressed-keeled; blades conduplicate; panicles cylindrical, 4–8 cm wide . . . . 39. *M. torreyana*
24. Culms terete to somewhat compressed-keeled near the base; blades usually flat, occasionally conduplicate; panicles ovoid, 4–16 cm wide.
25. Ligules 0.5–2 mm long, hyaline, with well-developed lateral lobes; blade margins and midveins prominent, white, thick . . . . 41. *M. arenacea*
25. Ligules 0.2–1 mm long, ciliate, without lateral lobes; blade margins and midveins not prominent, greenish, not particularly thick . . . . 40. *M. asperifolia*
21. Panicles contracted, 0.1–2(3) cm wide; panicle branches more than 0.1 mm thick, appressed or diverging up to 30(40)° from the rachises at maturity.
26. Culms 100–300 cm tall, 3–6 mm thick, woody and bamboolike . . . . 33. *M. dumosa*
26. Culms 4–100(140) cm tall, 0.5–2(3) mm thick, herbaceous, not bamboolike.
27. Blades 0.2–2(2.6) mm wide, flat, involute, or folded at maturity.
28. Lemmas awned, the awns 1–25 mm long.
29. Lemmas pubescent for ¾ their length, the hairs about 1.5 mm long . . . 21. *M. curtifolia*
29. Lemmas scabridulous or pubescent for no more than ½ their length, the hairs often less than 1.5 mm long.
30. Lemma awns generally less than 4(6) mm long . . . . 19. *M. glauca*
30. Lemma awns 4–25 mm long.

31. Lemmas and paleas mostly glabrous, the calluses with a few short hairs; ligules with lateral lobes, the lobes 1.5–3 mm longer than the central portion; culms erect; plants tightly cespitose at the base; sheaths and blades commonly with dark brown necrotic spots . . . . 22. *M. pauciflora*
31. Lemma midveins and margins and paleas appressed-pubescent on the lower ⅓–⅔; ligules without lateral lobes or with lobes less than 1.5 mm longer than the central portion; culms decumbent; plants loosely cespitose at the base; sheaths and blades without necrotic spots.
    32. Anthers orange, 1.5–2 mm long; lemmas elliptic, 2–3.5 mm long, the awns 10–20(25) mm long; panicles 0.3–1.8 cm wide . . . . 23. *M. polycaulis*
    32. Anthers purple, 1.3–3 mm long; lemmas lanceolate, 3.5–5 mm long, the awns 4–12(20) mm long; panicles 1–3(5) cm wide . . . . 24. *M. arsenei*

28. Lemmas unawned, mucronate, or shortly awned, the awns to 1 mm long.
  33. Lemmas and paleas pubescent, the hairs 0.4–1.2 mm long.
    34. Glumes acuminate, usually awned, the awns to 1.5 mm long; anthers orange; blades flat to involute distally, never arcuate . . . . 19. *M. glauca*
    34. Glumes acute, neither mucronate nor awned; anthers yellow, dark green, or purple; blades tightly involute, often arcuate.
      35. Lemmas 2.6–4 mm long; glumes nearly as long as the lemmas; anthers 2.1–2.3 mm long . . . . 20. *M. thurberi*
      35. Lemmas 1.4–2.5 mm long; glumes ½–⅔ as long as the lemmas; anthers 0.9–1.4 mm long . . . . 34. *M. villiflora*
  33. Lemmas and paleas scabrous, glabrous, or with hairs less than 0.3 mm long.
    36. Glumes more than ½ as long as the lemmas; lemmas 2.6–4.2 mm long, attenuate . . . . 35. *M. repens*
    36. Glumes ½ as long as the lemmas or less; lemmas 1.3–3.1 mm long; not attenuate.
      37. Ligules 0.2–0.8 mm long; panicles usually partially included, the rachises usually visible between the branches . . . . 36. *M. utilis*
      37. Ligules 0.8–3 mm long; panicles exserted, the rachises usually hidden by the branches . . . . 37. *M. richardsonis*

27. Blades (1.5)2–15 mm wide, flat at maturity.
  38. Glumes awn-tipped, 3–8 mm long (including the awns), about 1.3–2 times longer than the lemmas.
    39. Internodes dull, puberulent, usually terete, rarely keeled; culms seldom branched above the base; ligules 0.2–0.6 mm long; anthers 0.8–1.5 mm long . . . . 2. *M. glomerata*
    39. Internodes smooth and polished for most of their length, elliptic in cross section and strongly keeled; culms much branched above the base; ligules 0.6–1.7 mm long; anthers 0.4–0.8 mm long . . . . 1. *M. racemosa*
  38. Glumes unawned or awn-tipped, 0.4–4 mm long (including the awns), from shorter than to about 1.2 times longer than the lemmas.
    40. Lemmas usually completely glabrous, rarely with a few appressed hairs.
      41. Culms 30–100 cm tall, bushy and much branched above; axillary panicles common, partly included or exserted from the sheaths; lemmas shiny, stramineous or purplish, not mottled; anthers 0.3–0.5 mm long . . . . 5. *M. glabrifloris*

41. Culms 5–30 cm tall, often mat-forming; axillary panicles not present; lemmas dark greenish or plumbeous, sometimes mottled; anthers 0.9–1.6 mm long . . . . 37. *M. richardsonis*

40. Lemmas with hairs, these sometimes restricted to the callus.

42. Lemma bases with hairs about as long as the florets, usually 2–3.5 mm long . . . . 6. *M. andina*

42. Lemma bases glabrous or with hairs shorter than the florets, usually shorter than 1.5 mm.

43. Glumes unequal in length, the lower glumes 0.4–1.5 mm long, the upper glumes 0.8–1.9 mm long . . . . 7. *M. ×curtisetosa*

43. Glumes subequal in length, 1–4 mm long.

44. Axillary panicles often present, always partly included in the sheaths; internodes smooth, shiny.

45. Ligules 0.2–0.6 mm long; leaves of the lateral branches often shorter and narrower than those of the main branches; glumes 1.4–2 mm long, ⅓–⅔ as long as the lemmas . . . . 8. *M. bushii*

45. Ligules 0.7–1.7 mm long; leaves of the lateral branches similar in length and width to those of the main branches; glumes 2–4 mm long, about as long as the lemmas . . . . 9. *M. frondosa*

44. Axillary panicles, if present, exserted on elongated peduncles; internodes smooth, scabrous, or pubescent, sometimes smooth and shiny.

46. Glumes much shorter than the lemmas, acute.

47. Anthers 0.4–1 mm long; lemmas unawned, or with a short awn less than 1 mm long; internodes and sheaths glabrous . . . . 10. *M. sobolifera*

47. Anthers 1.1–2.2 mm long; lemmas awned (occasionally unawned), the awns to 12 mm long; internodes (and usually the base of the sheaths) pubescent . . . . 11. *M. tenuiflora*

46. Glumes nearly as long as or longer than the lemmas, acuminate.

48. Anthers 1–1.7 mm long; sheaths scabrous; lemma awns 0.2–2.2 mm long; restricted to the Transverse Ranges of southern California . . . . . . 4. *M. californica*

48. Anthers 0.3–0.8 mm long, sheaths smooth for most of their length; lemma awns to 18 mm long; widespread species, but not known from southern California.

49. Ligules 0.4–1 mm long; panicles dense; pedicels up to 2 mm long; anthers 0.3–0.5 mm long . . . . 3. *M. mexicana*

49. Ligules 1–2.5 mm long; panicles not dense, pedicels 0.8–3.5 mm long; anthers 0.4–0.8 mm long . . . . 12. *M. sylvatica*

20. Rhizomes absent; plants cespitose or bushy [for opposite lead, see p. 147].

50. Upper glumes usually 3-veined and 3-toothed; old sheaths flattened, ribbonlike or papery, sometimes spirally coiled.

51. Lemmas unawned, mucronate, or with awns to 5 mm long.

52. Lower glumes awned, the awns to 1.6 mm long; blades 2–6 cm long, filiform, tightly involute, sharp-tipped . . . . 43. *M. filiculmis*

52. Lower glumes unawned; blades (5)6–12 cm long, flat or loosely involute to subfiliform, not sharp-tipped . . . . 44. *M. jonesii*
51. Lemmas awned, the awns (2)6–27 mm long.
53. Upper glumes as long as or longer than the lemmas, the apices acuminate to acute, occasionally minutely 3-toothed; old sheaths conspicuously spirally coiled . . . . 45. *M. straminea*
53. Upper glumes ⅓–⅔ as long as the lemmas, the apices truncate to acute, 3-toothed; old sheaths occasionally spirally coiled . . . . 46. *M. montana*
50. Upper glumes usually 1-veined, (rarely 2- or 3-veined), rounded, obtuse, acute, or acuminate, entire or erose; old sheaths not flattened or papery, never spirally coiled.
54. Panicles loosely contracted, open, or diffuse, (1)2–40 cm wide; panicle branches usually not appressed at maturity, often naked basally [for opposite lead, see p. 152].
55. Culms arising from the bases of old depressed culms; plants loosely matted, delicate; lemmas 1.2–2 mm long, unawned . . . . 42. *M. uniflora*
55. Culms not arising from the bases of old depressed culms; plants cespitose, not delicate; lemmas 2–5 mm long, awned or unawned.
56. Lemmas unawned or with awns to 4(6) mm long.
57. Basal sheaths laterally compressed, usually keeled.
58. Glumes longer than the lemmas; ligules 10–25 mm long, membranous throughout; lemmas 2–3 mm long, usually awned, the awns to 15 mm long, occasionally unawned . . . . . . 49. *M. emersleyi*
58. Glumes shorter than the lemmas; ligules 3–12 mm long, firm and brown near the base, membranous distally; lemmas 3–4.2 mm long, awned, the awns 0.5–4 mm long . . . . 50. *M. ×involuta*
57. Basal sheaths rounded to somewhat flattened, not keeled.
59. Culms 10–60(70) cm tall, somewhat decumbent; blades 0.5–10(16) cm long.
60. Blade margins and midveins conspicuous, thick, white, and cartilaginous; ligules 1–2 mm long . . . . 29. *M. arizonica*
60. Blade margins not conspicuously thickened, greenish; ligules 2–9 mm long.
61. Blades arcuate, 0.3–0.9 mm wide, 1–3(5) cm long; usually no culm nodes exposed; most leaf blades reaching no more than ⅕ of the plant height . . . . 30. *M. torreyi*
61. Blades not arcuate, 1–2.2 mm wide, 4–10(16) cm long; 1 or more culm nodes exposed; leaf blades reaching ¼–½ of the plant height . . . . 31. *M. arenicola*
59. Culms 40–160 cm tall, erect from the base; blades (8)10–100 cm long.
62. Pedicels 0.1–2.5 mm long, usually shorter than the florets; glumes usually longer than the florets; ligules 10–30 mm long; panicle branches not capillary, appressed or spreading up to 60° from the rachises, the lower branches with 30–60 spikelets . . . . 51. *M. longiligula*
62. Pedicels 3–50 mm long, longer than the florets; glumes usually shorter than the florets; ligules 1.8–10 mm long; panicle branches capillary, spreading 30–100° from the rachises, the lower branches with 5–20 spikelets.
63. Panicles about as long as wide, not diffuse, 10–20(30) cm long; branches and pedicels stiff or flexible . . . . 55. *M. reverchonii*
63. Panicles longer than wide, diffuse, 15–60 cm long; branches and pedicels flexible.
64. Glumes more than ½ as long as the lemmas; lemmas usually unawned, if awned, the awns no more than 3 mm long . . . . 52. *M. expansa*

64. Glumes less than ½ as long as the lemmas; lemmas usually awned, the awns to 18 mm long .... 53. *M. capillaris*

56. Lemma awns 6–35 mm long.
 65. Plants conspicuously branched, bushy in appearance; culms wiry, with geniculate, stiff, widely divergent branches .... 26. *M. porteri*
 65. Plants not conspicuously branched, not bushy in appearance, usually typical bunchgrasses; culms not wiry, when branched, the branches not both geniculate and widely divergent.
  66. Basal sheaths laterally compressed, commonly keeled; glumes (excluding any awns) exceeding the florets; culms (50)80–150 cm tall .... 49. *M. emersleyi*
  66. Basal sheaths rounded; glumes (excluding any awns) exceeded by the florets; culms 30–160 cm tall.
   67. Glumes apices puberulent or scabridulous, acuminate to acute; lemmas long-acuminate, the demarcation of the lemma body and awn not evident .... 57. *M. elongata*
   67. Glume apices glabrous or sparsely hirtellous, obtuse to acute, sometimes awned; lemmas acute to acuminate, the demarcation of the lemma body and awn evident.
    68. Glumes neither awned nor mucronate; spikelets dark purple; lemmas scabrous distally.
     69. Panicles 8–41 cm wide, open, diffuse, the branches strongly divergent; pedicels 10–40(50) mm long; in the *Flora* region, restricted to the eastern United States, growing from Connecticut south to Kansas, Oklahoma, eastern Texas, and Florida .... 53. *M. capillaris*
     69. Panicles 2–5 (12) cm wide, loosely contracted to open, most branches appressed to ascending, occasionally a few diverging up to 80°; pedicels 1–10 mm long; in the *Flora* region, restricted to the southwestern United States, growing from Arizona to western Texas .... 56. *M. rigida*
    68. Glumes usually awn-tipped or mucronate; spikelets stramineous, brown, or purplish; lemmas smooth or scabrous distally.
     70. Lemmas shiny, smooth; blades tightly involute, 0.2–1.2 mm wide; panicles 2–5 cm wide, loosely contracted; branches diverging up to 70° from the rachises .... 58. *M. setifolia*
     70. Lemmas not shiny, usually scabrous, at least distally; blades flat or involute, 1–4 mm wide; panicles 5–40 cm wide, loosely contracted to open; branches diverging 30°–100° from the rachises.
      71. Panicles 10–20(30) cm long, about equally wide, not diffuse; branches spreading up to 80° from the rachises at maturity; spikelets stramineous, brownish or purple-tinged; lemmas awned, the awns 0.5–6 mm long .... 55. *M. reverchonii*
      71. Panicles 15–50(60) cm long, narrower than long, diffuse; branches spreading 30–100° from the rachises at maturity; spikelets usually purplish; lemmas unawned or, if awned, the awns to 35 mm long.

72. Upper glumes unawned or with awns to 5 mm long; lemmas without setaceous teeth or the teeth no more than 1 mm long; lemma awns 2–13(18) mm long . . . . 53. *M. capillaris*
72. Upper glumes awned, the awns 2–25 mm long; lemmas with setaceous teeth 1–5 mm long; lemma awns 8–35 mm long . . . . 54. *M. sericea*

54. Panicles narrow, 0.2–3(5) cm wide; branches appressed to ascending at maturity, usually spikelet-bearing their whole length [for opposite lead, see p. 150].
 73. Panicles spikelike, 0.1–1.2 cm wide, sometimes interrupted near the base; branches appressed, 0.3–1.2(4) cm long.
  74. Culms 3–20(35) cm tall, often decumbent and rooting at the lower nodes; blades 1–4(6) cm long . . . . 38. *M. filiformis*
  74. Culms (15)20–150 cm tall, stiffly erect, not rooting at the lower nodes; blades 1.4–50 cm long, at least some more than 6 cm long.
   75. Lemmas awned, awns 3–10 mm long; lemmas scabridulous, the veins and margins glabrous . . . . 59. *M. palmeri*
   75. Lemmas unawned, sometimes mucronate, the mucros up to 1 mm long; lemmas with hairs on the lower ⅙, the hairs sometimes extending to ¾ of the lemma length over the midvein and margins.
    76. Basal leaf sheaths rounded on the back; panicles 15-60 cm long . . . . 62. *M. rigens*
    76. Basal leaf sheaths compressed, the backs keeled; panicles 4-16 cm long.
     77. Ligules 0.2–0.8 mm long; paleas glabrous; glume apices gradually acute to acuminate, mucronate, the mucros to 0.3 mm long . . . . 27. *M. cuspidata*
     77. Ligules 1–3(5) mm long; paleas pubescent between the veins; glumes abruptly narrowed, acute or obtuse, awned, the awns 0.5–1 mm long . . . . 28. *M. wrightii*
 73. Panicles not spikelike, 0.6–5 cm wide; branches appressed, ascending, or diverging up to 70°, 0.2–13 cm long.
  78. Ligules 10–35 mm long, firm and brown basally, membranous distally; glumes as long as or longer than the florets.
   79. Basal sheaths rounded; lemmas unawned or with awns to 2 mm long; plants of Arizona and New Mexico . . . . 51. *M. longiligula*
   79. Basal sheaths compressed-keeled, at least basally; lemmas unawned or with awns to 4 mm long; endemic to south central Texas, sometimes grown as an ornamental . . . . 60. *M. lindheimeri*
  78. Ligules 0.2–10 mm long, usually membranous throughout, sometimes firmer basally, never brownish; glumes shorter than the florets.
   80. Lemma awns 0.5–6 mm long.
    81. Blades 10–60 cm long.
     82. Glumes acute, unawned; ligules 4–10 mm long, acute, lacerate; spikelets grayish-green . . . . 61. *M. dubia*
     82. Glumes acute to acuminate, awned, the awns to 1.5 mm long; ligules 1–3 mm long, truncate, ciliolate; spikelets yellowish-brown to purplish . . . . 59. *M. palmeri*
    81. Blades to 10 cm long.
     83. Glumes 2–4 mm long; lemmas 3.5–5 mm long, the awns flexuous; ligules 1–2 mm long, with lateral lobes less than 1.5 mm longer than the central portion; anthers 1.3–3 mm long, purple . . . . 24. *M. arsenei*
     83. Glumes less than 2 mm long; lemmas 1.8–3(3.4) mm long, the awns straight; ligules 0.2–1.1 mm long, without lateral lobes; anthers 0.2–0.9 mm long, yellow.

84. Upper glumes veinless, 0.1–0.3 mm long; lower glumes rudimentary or lacking, veinless . . . . 13. *M. schreberi*
84. Upper glumes 1(2)-veined, 0.8–1.9 mm long; lower glumes 0.4–1.5 mm long, veinless or 1-veined . . . . 7. *M.* ×*curtisetosa*

80. Lemma awns 6–40 mm long.
85. Glumes 0.3–1 mm long, obtuse to acute, sometimes erose, not awned; anthers 0.9–1.6 mm long . . . . 25. *M. spiciformis*
85. Glumes (1)1.2–3.5 mm long, acute to acuminate, usually awn-tipped; anthers 0.9–3 mm long.
86. Blades 15–50 cm long; lemma awns 3–10 mm long; plants of southern Arizona and Chihuahua, Mexico . . . . 59. *M. palmeri*
86. Blades 1–15 cm long; lemma awns 4–30 mm long; plants of the southwestern United States, including southern Arizona.
87. Lemmas and paleas almost glabrous, with only a few short hairs on the calluses; ligules with lateral lobes, the lobes 1.5–3 mm longer than the central portion; culms erect and plants tightly cespitose at the base; sheaths and blades usually with dark brown necrotic spots . . . . . 22. *M. pauciflora*
87. Lemmas and paleas pubescent on the lower ⅓–⅔ of the midveins and margins; ligules without lateral lobes or with lobes less than 1.5 mm longer than the central portion; culms decumbent and plants loosely cespitose; sheaths and blades without necrotic spots.
88. Anthers 0.9–1.5 mm long, yellowish; panicles usually 7–20 cm long . . . . 14. *M. tenuifolia*
88. Anthers 1.3–2 mm long, purplish or orange; panicles 2–13 cm long.
89. Anthers orange, 1.5–2 mm long; lemmas elliptic, 2–3.5 mm long, with awns 10–20(25) mm long; panicles 0.3–1.8 cm wide . . . . 23. *M. polycaulis*
89. Anthers purple, 1.3–3 mm long; lemmas lanceolate, 3.5–5 mm long, with awns 4–12(20) mm long; panicles 1–5 cm wide . . . . 24. *M. arsenei*

1. **Muhlenbergia racemosa** (Michx.) Britton, Sterns & Poggenb. [p. 155]
MARSH MUHLY

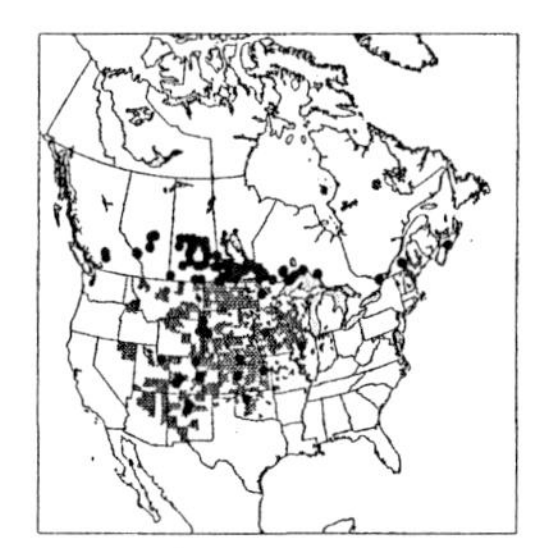

Plants perennial; rhizomatous, not cespitose. **Culms** 30–110 cm tall, 1–1.5 mm thick, stiffly erect, much branched above the middle, **internodes** mostly smooth, polished, glabrous or puberulent immediately below the nodes, elliptic in cross section and strongly keeled. **Sheaths** scabridulous, slightly keeled; **ligules** 0.6–1.5(1.7) mm, membranous, truncate, lacerate-ciliolate; **blades** 2–17 cm long, 2–5 mm wide, flat, usually scabrous or scabridulous, occasionally smooth. **Panicles** 0.8–16 cm long, 0.3–1.8 cm wide, lobed, dense; **branches** 0.2–2.5 cm, appressed; **pedicels** absent or to 1 mm, strigose. **Spikelets** 3–8 mm. **Glumes** subequal, 3–8 mm (including the awns), about 1.3–2 times longer than the lemmas, smooth or scabridulous distally, 1-veined, acuminate and awned, awns to 5 mm; **lemmas** 2.2–3.8 mm, lanceolate, pilose on the calluses, lower ½ of the midveins, and margins, hairs to 1.2 mm, apices scabridulous, acuminate, unawned or awned, awns to 1 mm; **paleas** 2.2–3.8(4.5) mm, lanceolate, intercostal region loosely pilose on the lower ½, apices acuminate; **anthers** 0.4–0.8 mm, yellowish. **Caryopses** (1.2)1.4–2.3 mm, fusiform, brown. $2n = 40$.

Muhlenbergia racemosa grows on rocky slopes, beside irrigation ditches, in seasonally wet meadows, on the margins of cultivated fields, railways and roadsides, in prairies, on sandstone outcrops, on stream banks, and in forest ecotones at elevations of 30–3400 m. It is most common in the north central United States, but can be found at scattered locations throughout the western United States, and extends into northern Mexico.

2. **Muhlenbergia glomerata** (Willd.) Trin. [p. 155]
SPIKE MUHLY, MUHLENBERGIE AGGLOMÉRÉE

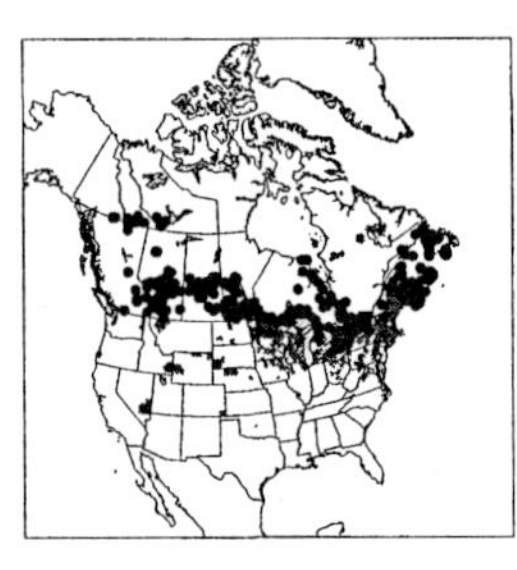

Plants perennial; rhizomatous, not cespitose. **Culms** 30–120 cm tall, 0.8–2.5 mm thick, erect, seldom branched above the base; **internodes** dull, mostly puberulent (sometimes sparsely so), terete, rarely keeled, strigose immediately below the nodes. **Sheaths** scabridulous, slightly keeled; **ligules** 0.2–0.6 mm, membranous, truncate, lacerate-ciliolate; **blades** 2–15 cm long, 2–6 mm wide, flat, usually scabrous or scabridulous, occasionally smooth. **Panicles** 1.5–12 cm long, 0.3–1.8 cm wide, lobed, dense; **primary branches** 0.2–2.5 cm, appressed; **pedicels** absent or to 1 mm, strigose. **Spikelets** 3–8 mm. **Glumes** subequal, 3–8 mm (including the awn), about 1.3–2 times longer than the lemmas, smooth or scabridulous distally, 1-veined, acuminate, awned, awns to 5 mm; **lemmas** 1.9–3.1 mm, lanceolate, pubescent on the calluses, midveins, and margins, hairs to 1.2 mm, apices scabridulous, acuminate, unawned or awned, awns to 1 mm; **paleas** 1.9–3.1 mm, lanceolate, loosely pilose between the veins, apices acuminate; **anthers** 0.8–1.5 mm, yellowish. **Caryopses** 1–1.6 mm, fusiform, brown. $2n = 20$.

Muhlenbergia glomerata grows in meadows, marshes, bogs, alkaline fens, lake margins, stream banks, beside irrigation ditches and hot springs, and on gravelly slopes, in many different plant communities, at elevations of 30–2300 m. It is most common in southern Canada and the northeastern United States, but grows sporadically throughout the western United States. It is not known from Mexico.

3. **Muhlenbergia mexicana** (L.) Trin. [p. 155]
WIRESTEM MUHLY, MUHLENBERGIE DU MEXIQUE, MUHLENBERGIE MEXICAINE

Plants perennial; rhizomatous, not cespitose. **Culms** 30–90 cm tall, 0.5–2 mm thick, erect, much branched above the base; **internodes** dull, puberulent or glabrous for most of their length, sometimes strigose immediately below the nodes. **Sheaths** smooth or scabridulous, somewhat keeled; **ligules** 0.4–1 mm, membranous, truncate, lacerate-ciliolate; **blades** 2–20 cm long, 2–6 mm wide, flat, scabrous or smooth, those of the secondary branches similar in length and width to those of the main branches. **Panicles** terminal and axillary, 2–21 cm long, 0.3–3 cm wide, dense; **primary branches** 0.3–5.5 cm, appressed or diverging up to 30° from rachises; **pedicels** to 2 mm, strigose; **axillary panicles** exserted on long peduncles. **Spikelets** 1.5–3.8 mm, often purple-tinged. **Glumes** subequal, 1.5–3.7 mm, equaling or slightly shorter than the lemmas, 1-veined, tapering from the bases to the acuminate apices, unawned or awned, awns to 2 mm; **lemmas** 1.5–3.8 mm, lanceolate, pubescent on the calluses, lower portion of the midveins, and margins, hairs shorter than 0.7 mm, apices scabridulous, acuminate, unawned or awned, awns to 10 mm; **paleas** 1.5–3.8 mm, narrowly lanceolate, apices acuminate; **anthers** 0.3–0.5 mm, yellow to purplish. **Caryopses** 1.1–1.6 mm, fusiform, brown. $2n = 40$.

Muhlenbergia mexicana usually grows in mesic to wet areas such as moist prairies and woodlands, stream banks, roadsides, ditch banks, lake margins, swamps, bogs, and hot springs, at elevations 50–3300 m, and is found in many different plant communities. Despite its name, *M. mexicana* grows only in Canada and the United States.

Plants with awns 3–10 mm long belong to **Muhlenbergia mexicana** var. **filiformis** (Torr.) Scribn., and those without an awn or with awns less than 3 mm long to **Muhlenbergia mexicana** (L.) Trin. var. **mexicana**. Early in the flowering season, *M. mexicana* may be confused with plants of *M. bushii* in which the axillary panicles are poorly developed, but they differ in their dull internodes and the fact that the blades on the secondary branches are usually similar in length and width to those of the main branches.

4. **Muhlenbergia californica** Vasey [p. 157]
CALIFORNIA MUHLY

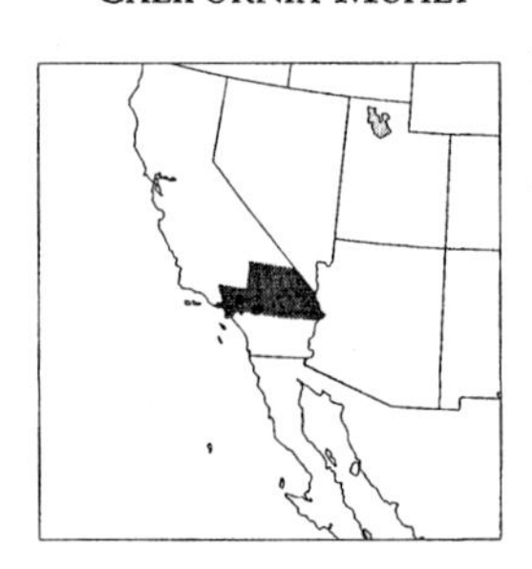

Plants perennial; rhizomatous, not cespitose. **Culms** 30–70 cm tall, 0.7–1.5 mm thick, decumbent; **internodes** dull, smooth, and glabrous for most of their length, sometimes strigose immediately below the nodes. **Sheaths** shorter than the internodes, scabrous, margins whitish; **ligules** 0.8–2 mm, membranous, truncate, ciliolate, irregularly toothed; **blades** 4–16 cm long, 2–6 mm wide, flat, scabridulous abaxially, scabrous to strigose adaxially. **Panicles** terminal, 5–13 cm long, 0.5–2.2 cm wide, dense; **branches** 0.5–3.2 cm, ascending, appressed or diverging up to 20° from the rachises; **pedicels** to 1.5 mm, stout, strigose; **axillary panicles** not present. **Spikelets** 2.8–4 mm. **Glumes** subequal, 2.5–4 mm, nearly as long as or slightly longer than the lemmas, scabrous (especially on the veins), 1-veined, tapering from the base to the acuminate apices, usually unawned, awns, if present, to 1.2 mm; **lemmas** 2.8–4 mm, narrowly lanceolate, with soft hairs on the calluses and lower portion of the lemma bodies, hairs to 1 mm, apices scabridulous, acuminate, awned, awns 0.2–2.2

1 mm
glumes
1 mm
floret
floret
1 mm
1 cm
glumes
M. glomerata
1 cm
floret
1 mm
node
glumes
M. racemosa
2 cm
M. mexicana

mm; **paleas** 2.8–4 mm, subequal to the lemmas, narrowly lanceolate, with short (less than 1.5 mm), soft hairs on the lower ½, apices scabridulous, acuminate; **anthers** 1–1.7 mm, yellow. **Caryopses** 1.7–2 mm, fusiform, brown. 2*n* = 80.

Muhlenbergia californica grows in canyons, along moist ditches, and on sandy slopes, at elevations of 100–2150 m. It is endemic to the Transverse Ranges of southern California.

5. **Muhlenbergia glabrifloris** Scribn. [p. 157]
INLAND MUHLY

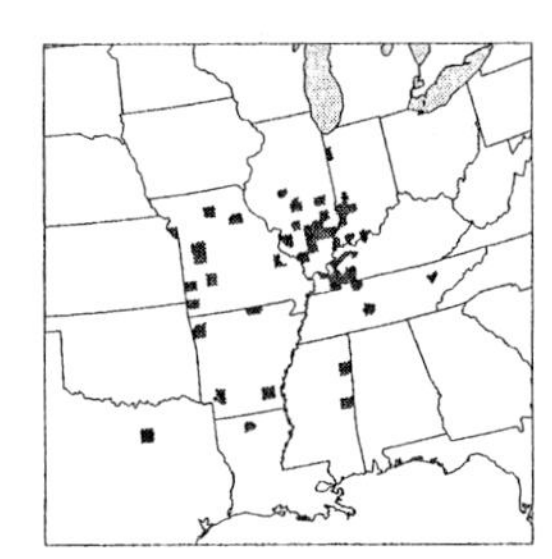

Plants perennial; rhizomatous, not cespitose. **Culms** 30–100 cm tall, 0.5–2 mm thick, herbaceous, ascending or decumbent, bushy and much branched above; **internodes** smooth and shiny for most of their length, scabridulous or strigulose below the nodes. **Sheaths** glabrous, margins hyaline; **ligules** 0.5–1.5 mm, membranous, truncate, lacerate-ciliolate; **blades** 3–8 cm long, (1.5)2–4 mm wide, flat, glabrous. **Panicles** 2–6.5 cm long, 0.2–0.8 cm wide, lobed, dense; **branches** 0.3–2.5 cm, ascending, closely appressed; **axillary panicles** common, partly included in or exserted from the subtending sheaths; **pedicels** 0.3–2.2 mm, scabrous to strigulose. **Spikelets** 2.2–3.5 mm. **Glumes** subequal, 1.5–3.5 mm, from ¾ as long as to longer than the lemmas, smooth or scabridulous near the apices, 1-veined; **upper glumes** acuminate, acute, unawned or awned, awns to 1.2 mm; **lemmas** 2.2–3.1 mm, narrowly lanceolate, shiny, stramineous or purplish, usually completely glabrous (rarely with a few appressed hairs), apices scabridulous, acuminate, unawned; **paleas** 2.2–3.1 mm, narrowly lanceolate; **anthers** 0.3–0.5 mm, yellow to purplish. **Caryopses** 1.2–1.4 mm, fusiform, brown. 2*n* = 40.

Muhlenbergia glabrifloris grows at the edge of dry forests, in prairies, thickets, and along roadsides in pine and oak associations, at elevations of 20–400 m. It is restricted to the southern portion of the central contiguous United States. It resembles *M. frondosa*, but differs from that species in its glabrous lemmas and shorter caryopses (1.2–1.4 mm rather than 1.6–1.9 mm).

6. **Muhlenbergia andina** (Nutt.) Hitchc. [p. 157]
FOXTAIL MUHLY

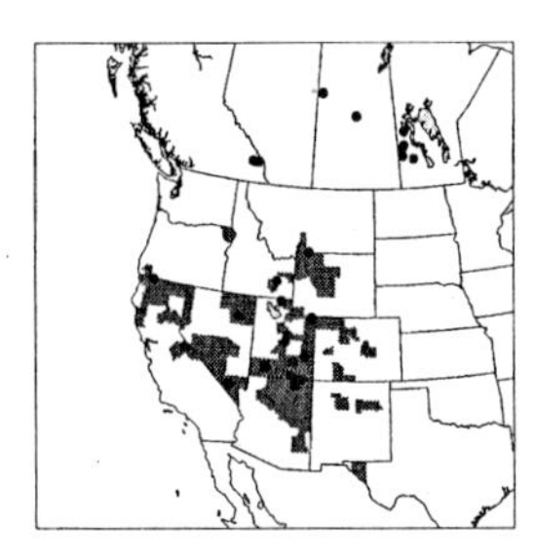

Plants perennial; rhizomatous, not cespitose. **Culms** 25–85 cm tall, 0.9–1.7 mm thick, ascending; **internodes** glabrous for most of their length, scabrous to strigose below the nodes. **Sheaths** scabridulous, especially basally; **ligules** 0.5–1.5 mm, membranous, truncate, lacerate to ciliate; **blades** 4–16 cm long, 2–4(5) mm wide, flat, scabrous abaxially, pubescent adaxially. **Panicles** 2–15 cm long, 0.5–2.8 cm wide, contracted, dense; **primary branches** 0.5–5 cm, appressed to strongly ascending; **pedicels** 0.5–1.5 mm, appressed, strigose. **Spikelets** 2–4 mm. **Glumes** equal to subequal, 2–4 mm, subequal to or longer than the florets, 1-veined, veins scabridulous, apices acuminate to awn-tipped; **lemmas** 2–3.5 mm, lanceolate, grayish-green, hairy on the calluses and lemma bases, hairs 2–3.5 mm, apices acuminate, awned, awns 1–10 mm; **paleas** 2–3.5 mm, lanceolate, bases with silky hairs between the veins, apices acuminate; **anthers** 0.4–1.5 mm, yellow. **Caryopses** 0.9–1.1 mm, cylindrical, yellowish-brown. 2*n* = 20.

Muhlenbergia andina grows in damp places such as stream banks, gravel bars, marshes, lake margins, damp meadows, around springs, and in canyons, at elevations of 700–3000 m. It grows only in the western part of southern Canada and the contiguous United States.

7. **Muhlenbergia ×curtisetosa** (Scribn.) Bush [p. 159]

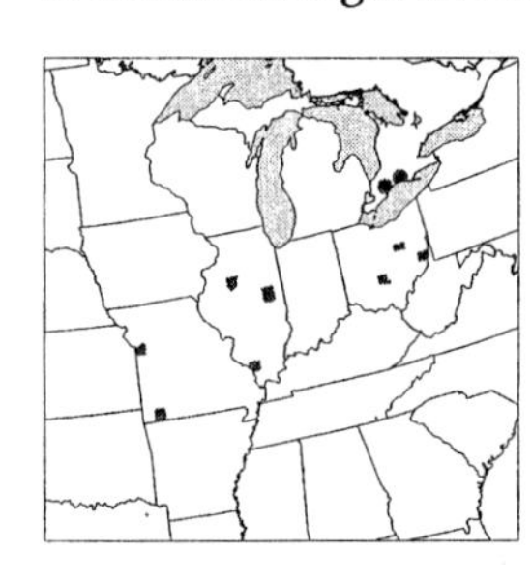

Plants perennial; occasionally rhizomatous. **Culms** 20–70 cm tall, less than 3 mm thick, erect, branched above; **internodes** smooth, shiny for most of their length, scabridulous or glabrous below the nodes. **Sheaths** glabrous, margins hyaline, old sheaths not flattened, papery, or spirally coiled; **ligules** 0.2–1.1 mm, membranous, truncate, sometimes ciliolate; **blades** 2–8.5 cm long, 2–5 mm wide, flat, smooth or scabridulous. **Panicles** 4.2–16.5 cm long, 0.2–1.5 cm wide, mostly exserted from the sheath; **primary branches** 0.8–7.2 cm, ascending to appressed; **pedicels** 0.6–3 mm, strigose. **Spikelets** 2.2–3.4 mm. **Glumes** shorter than the florets, veins scabridulous, unawned or awned, awns to 0.5 mm; **lower glumes** 0.4–1.5 mm, veinless (rarely 1-veined), usually truncate to rounded, occasionally acute, sometimes notched; **upper glumes** 0.8–1.9 mm, 1(2)-veined, acute to acuminate; **lemmas** 2.2–3(3.4) mm, lanceolate, hairy on the calluses and lower portion

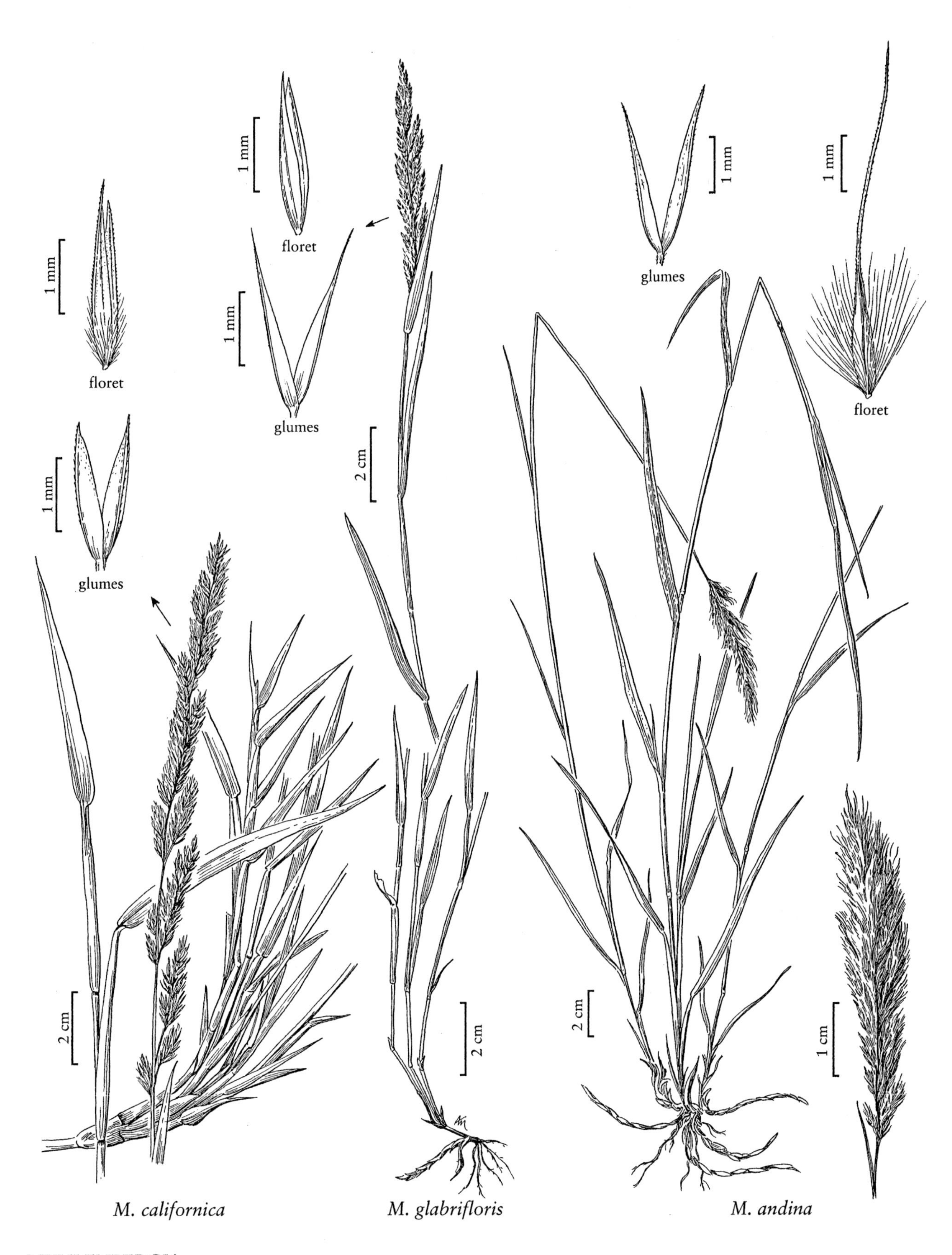

MUHLENBERGIA

of the margins and midveins, hairs shorter than 1.5 mm, apices scabridulous, acuminate, awned, awns 0.5–4 mm, straight; **paleas** 2.2–3.1(3.4) mm, lanceolate, intercostal region shortly pilose on the lower ½, apices acuminate; **anthers** usually not developed, occasionally 1 or 2 present, 0.3–0.9 mm, yellow. **Caryopses** 1.4–1.6 mm, fusiform, brown. 2*n* = unknown.

Muhlenbergia ×*curtisetosa* grows in abandoned fields and forest openings, often near bogs, at elevations of 20–300 m. It may be a hybrid between *M. schreberi* (which contributes the short glumes) and either of two rhizomatous species, *M. frondosa* and *M. tenuiflora*.

8. **Muhlenbergia bushii** R.W. Pohl [p. 159]
NODDING MUHLY

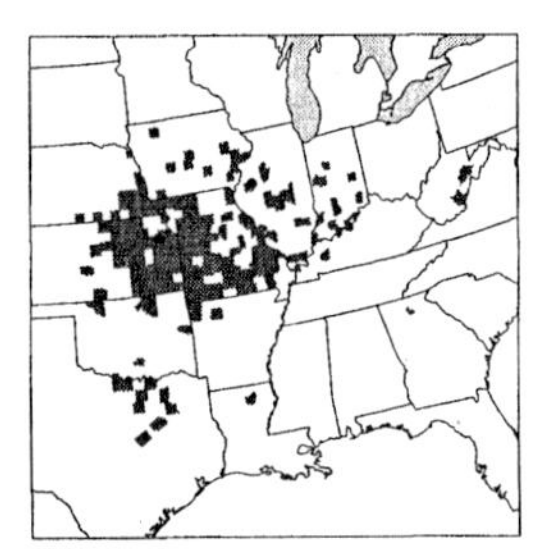

Plants perennial; rhizomatous, not cespitose. **Culms** 30–100 cm tall, 1–2 mm thick, herbaceous; **internodes** smooth, shiny, usually glabrous throughout. **Sheaths** glabrous, usually smooth, occasionally scabridulous distally; **ligules** 0.2–0.6 mm, membranous, truncate, ciliolate; **blades** 5–15 cm long, 2–7 mm wide, flat, glabrous abaxially, glabrous, smooth or scabridulous adaxially, those of the lateral branches often shorter and narrower than those of the primary branches. **Panicles** terminal and axillary, 4–15 cm long, 0.1–0.7 cm wide, narrow, not dense; **axillary panicles** common, partly included in the sheaths; **branches** 0.5–4 cm, ascending to appressed; **pedicels** 1–4 mm, glabrous. **Spikelets** 2.6–3.3 mm, not imbricate along the branches. **Glumes** subequal, 1.4–2 mm, about ⅓–⅔ as long as the lemmas, smooth or scabridulous distally, 1(2)-veined, acuminate or acute, unawned or awned, awns to 1 mm; **lemmas** 2.6–3.3 mm, narrowly lanceolate, usually hairy on the calluses and lower ⅓ of the margins and midveins (rarely glabrous), hairs 0.5–1 mm, apices scabridulous, acuminate, unawned or awned, awns 0.5–7 mm; **paleas** 2.2–3.1 mm, shorter than the lemmas, narrowly lanceolate, intercostal region shortly pilose on the basal ⅓, apices acuminate; **anthers** 0.3–0.6 mm, yellow. **Caryopses** 1.5–1.8 mm, fusiform, brown. 2*n* = 40.

Muhlenbergii bushii grows in sandy alluvium, open thickets, dry woodlands, and flood plains, at elevations of 10–250 m in the central portion of the contiguous United States.

Early season plants, in which the axillary panicles are poorly developed, can be distinguished from those of *M. mexicana* by their shiny internodes and the tendency of the blades on the secondary branches to be shorter and narrower than those on the main branches.

9. **Muhlenbergia frondosa** (Poir.) Fernald [p. 159]
WIRESTEM MUHLY, MUHLENBERGIE FEUILLÉE

Plants perennial; rhizomatous, not cespitose. **Culms** 50–110 cm tall, 0.7–1.8 mm thick, ascending, geniculate, or decumbent, bushy and much branched above; **internodes** smooth and shiny for most of their length, glabrous throughout. **Sheaths** glabrous, margins hyaline; **ligules** 0.7–1.7 mm, membranous, truncate, lacerate-ciliolate; **blades** 4–18 cm long, 2–7 mm wide, flat, glabrous, smooth or scabridulous, those of the lateral branches similar to those of the primary culms. **Panicles** 2–15 cm long, 0.3–2 cm wide, sometimes dense; **axillary panicles** common, partly included in the sheaths; **primary branches** 0.3–4.2 cm, ascending, appressed, occasionally diverging up to 30° from the rachises; **pedicels** 0.4–5 mm, smooth or scabrous. **Spikelets** 2.2–4 mm, imbricate along the branches. **Glumes** subequal, 2–4 mm, ¾ as long as to longer than the lemmas, 1-veined, unawned or awned, awns to 2 mm; **upper glumes** acuminate to acute; **lemmas** 2.2–4 mm, narrowly lanceolate, hairy on the calluses, lower ⅓ of the midveins, and margins, hairs 0.3–0.8 mm, apices scabridulous, acuminate, unawned or awned, awns 0.1–13 mm; **paleas** 2.3–4 mm, narrowly lanceolate, acuminate; **anthers** 0.3–0.6 mm, usually yellow, occasionally purplish. **Caryopses** 1.6–1.9 mm, fusiform, brown. 2*n* = 40.

Muhlenbergia frondosa grows in forest borders, thickets, clearings, alluvial plains, and disturbed sites within deciduous forests, at elevations of 20–1000 m. It grows only in southern Canada and the contiguous United States. The collection from Oregon was made in an artificial cranberry bog and was probably introduced with the plants.

Plants with unawned or shortly (less than 4 mm) awned lemmas can be called *M. frondosa* forma *frondosa*, and those with lemma awns 4–13 mm long, *M. frondosa* forma *commutata* (Scribn.) Fernald.

10. **Muhlenbergia sobolifera** (Muhl. *ex* Willd.) Trin. [p. 161]
ROCK MUHLY

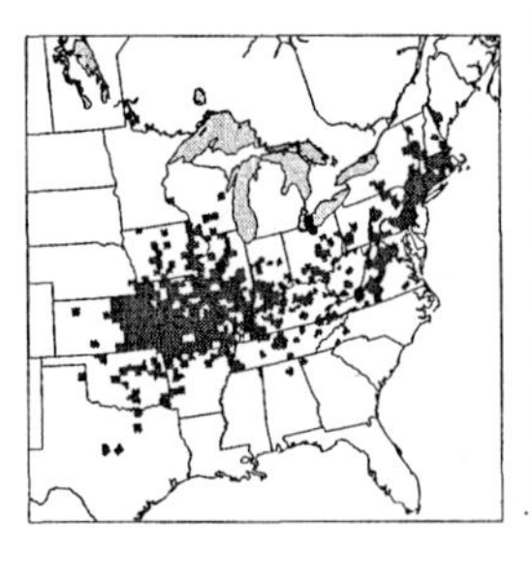

Plants perennial; rhizomatous, usually not cespitose. **Culms** 25–95 cm tall, 0.8–1.5 mm thick, erect or ascending; **internodes** smooth, shiny, and glabrous for most of their length, scabridulous immediately below the nodes. **Sheaths** glabrous, margins

MUHLENBERGIA

hyaline; **ligules** 0.3–1 mm, membranous, truncate, ciliolate; **blades** 4–16 cm long, 2–7 mm wide, flat, glabrous, usually smooth, occasionally scabridulous. **Panicles** 4–18 cm long, 0.2–0.8 cm wide, narrow, usually exserted; **axillary panicles** usually exserted, sometimes partially included in the subtending sheath; **primary branches** 0.6–4 cm, ascending to appressed; **pedicels** 0.3–1.6 mm, strigose. **Spikelets** 1.6–3 mm, erect, overlapping the next spikelet on the branch by ½ its length. **Glumes** equal to subequal, 1–2.5 mm, much shorter than the florets, scabridulous (particularly over the veins), 1-veined, narrowing from above the broad, overlapping bases to the acute apices, unawned or awned, awns to 1 mm; **lemmas** 1.6–2.8 mm, lanceolate, hairy on the calluses, lower ½ of the midveins, and margins, hairs 0.3–0.5 mm, apices acuminate, unawned or awned, awns to 1 mm; **paleas** 1.6–2.9 mm, lanceolate, basal ½ with hairs shorter than 1. 5 mm, apices scabridulous, acuminate; **anthers** 0.4–1 mm, yellow. **Caryopses** 1–1.5 mm, fusiform, brown. $2n = 40$.

Muhlenbergia sobolifera grows in dry upland forests, oak woodlands, and on rock outcrops of sandstone, chert, or limestone formations, at elevations of 0–1200 m. It is restricted to the *Flora* region.

11. **Muhlenbergia tenuiflora** (Willd.) Britton, Sterns & Poggenb. [p. 161]

SLIMFLOWERED MUHLY, MUHLENBERGIE TÉNUE

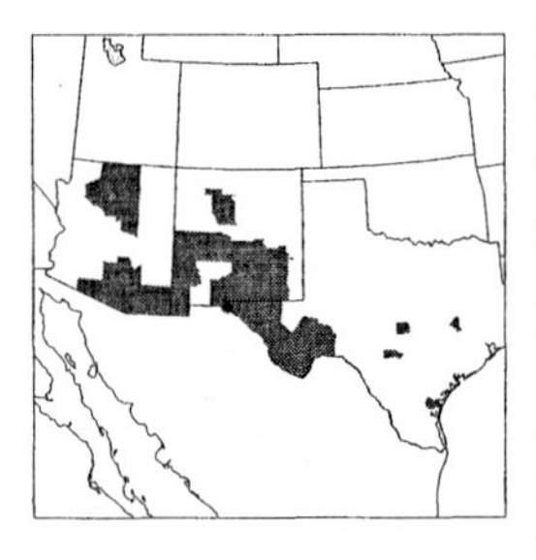

Plants perennial; rhizomatous, not cespitose. **Culms** 40–120 cm tall, less than 2 mm thick, erect; **internodes** mostly pubescent, retrorsely hirsute to strigose below the nodes. **Sheaths** mostly glabrous, usually pubescent near the base, scabridulous distally; **ligules** 0.4–1.2 mm, membranous, truncate, ciliolate; **blades** 6–20 cm long, 4–10(15) mm wide, flat, glabrous and smooth abaxially, occasionally scabridulous adaxially. **Panicles** usually terminal, 10–33 cm long, 0.2–0.8 cm wide, exserted; **branches** 1–10 cm, ascending to appressed; **pedicels** 1–6 mm, strigose. **Spikelets** 2.6–4.5 mm, overlapping the next spikelet on the branch by ¼ of its length. **Glumes** subequal, 1.3–3 mm, shorter than the lemmas, 1-veined (lower glumes rarely 2- or 3-veined), tapering from near the base, bases overlapping, apices scabridulous, acute, unawned or awned, awns to 1 mm; **lemmas** 2.6–4.5 mm, lanceolate to narrowly lanceolate, usually pubescent on the calluses, lower ⅓ of the midveins, and margins (hairs sometimes restricted to the callus), hairs shorter than 1.2 mm, apices acute or acuminate, usually awned, awns to 12 mm; **paleas** 2.6–4.5 mm, lanceolate to narrowly lanceolate, shortly pilose on the lower portion, apices acuminate; **anthers** 1.1–2.2 mm, yellowish. **Caryopses** 2–2.3 mm, fusiform, brown. $2n = 40$.

Muhlenbergia tenuiflora grows only in the *Flora* region, usually being found on sandy or rocky slopes derived from sandstone, chert, or limestone formations, in mixed hardwood and oak-hickory forests, at elevations of 40–1500 m. It resembles the Asiatic species **M. curviaristata** (Ohwi) Ohwi.

12. **Muhlenbergia sylvatica** (Torr.) Torr. *ex* A. Gray [p. 161]

WOODLAND MUHLY, MUHLENBERGIE DES BOIS

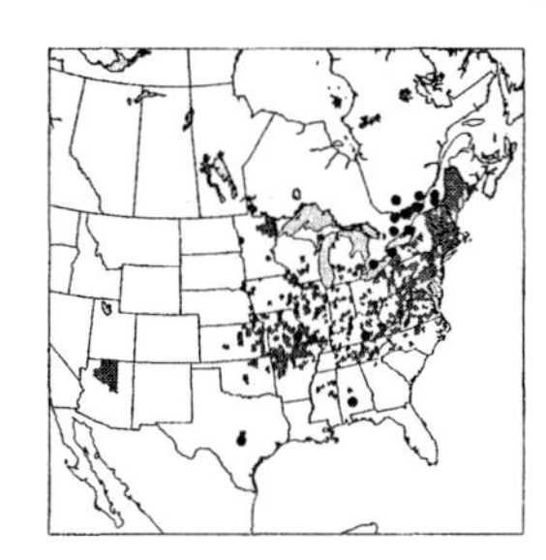

Plants perennial; rhizomatous. **Culms** 40–110 cm tall, 1–2 mm thick, erect; **internodes** puberulent for most of their length, strigose below the nodes. **Sheaths** glabrous and smooth for most of their length, scabridulous distally, margins hyaline; **ligules** 1–2.5 mm, membranous, truncate, lacerate-ciliolate; **blades** 5–18 cm long, 3–7 mm wide, flat, scabrous to scabridulous, occasionally smooth. **Panicles** terminal and axillary, 6–21 cm long, 0.2–1 cm wide, narrow, not dense; **axillary panicles** usually exserted at maturity; **branches** 0.8–6 cm, ascending to closely appressed; **pedicels** 0.8–3.5 mm, strigose. **Spikelets** 2.2–3.7 mm. **Glumes** subequal, 1.8–3 mm, nearly as long as the lemmas, 1-veined, tapering from near the base, apices scabridulous, acuminate, unawned or awned, awns to 1 mm; **lemmas** 2.2–3.7 mm, lanceolate to narrowly lanceolate, hairy on the calluses, lower ½ of the midveins, and margins, hairs 0.2–0.5 mm, apices scabridulous, acuminate, awned, awns 5–18 mm, purplish; **paleas** 2–3.5 mm, lanceolate, proximal ½ shortly pilose, apices scabridulous, acuminate; **anthers** 0.4–0.8 mm, yellow. **Caryopses** 1.4–2 mm, fusiform, brown. $2n = 40$.

Muhlenbergia sylvatica grows in upland forests, along creeks and hollows, on rocky ledges derived from sandstone, shale, or calcareous parent materials, moist prairies, and swamps, at elevations from 30–1500 m. It is restricted to the *Flora* region, its primary range being southeastern Canada and the midwestern and eastern United States. Reports from British Columbia were based on a misidentification (Douglas et al. 2002). The record from Arizona is based on the report in Kearney and Peebles (1951) of a collection made by Toumey at Grapevine Creek in the Grand Canyon.

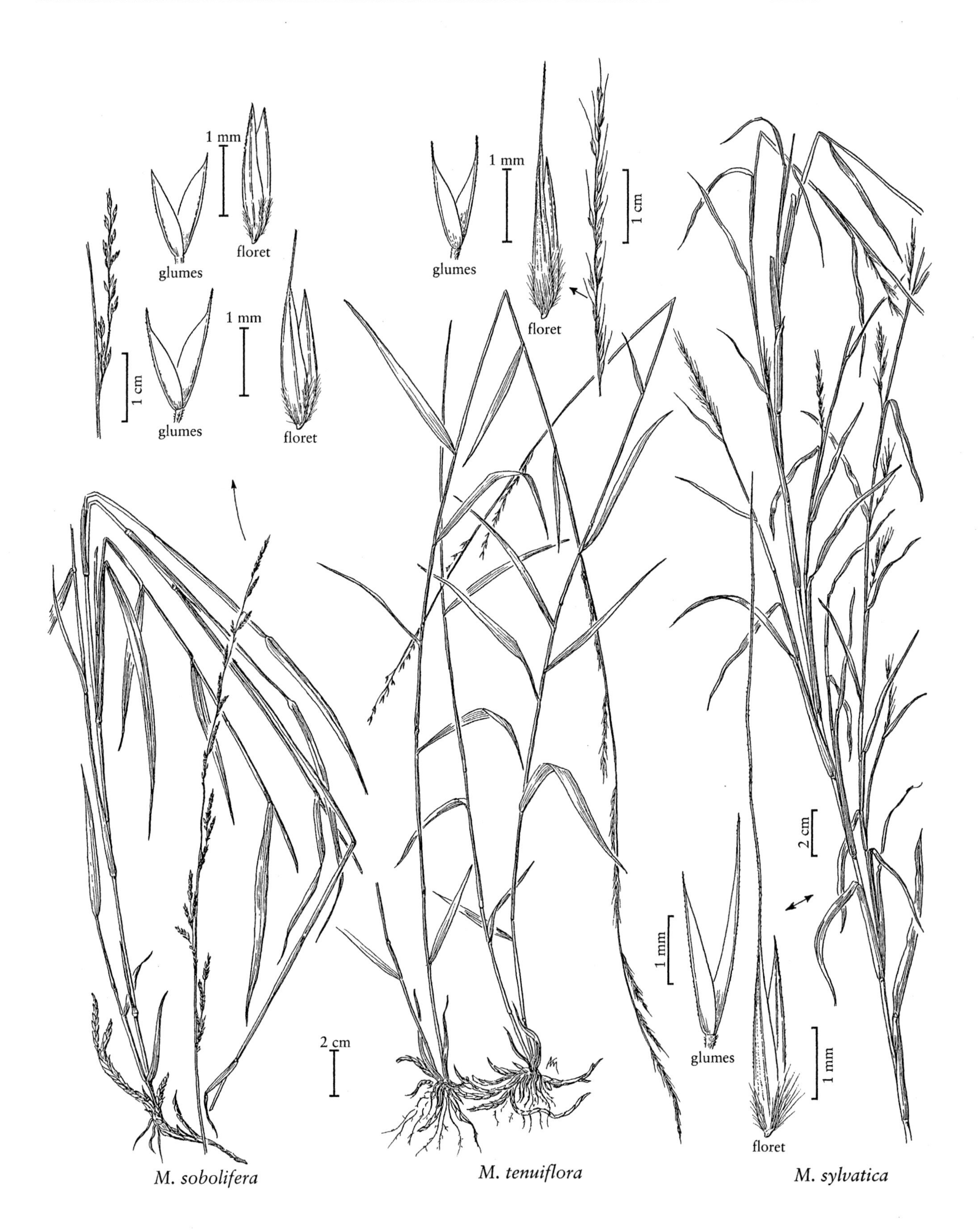

M. sobolifera *M. tenuiflora* *M. sylvatica*

13. Muhlenbergia schreberi J.F. Gmel. [p. 163]
NIMBLEWILL

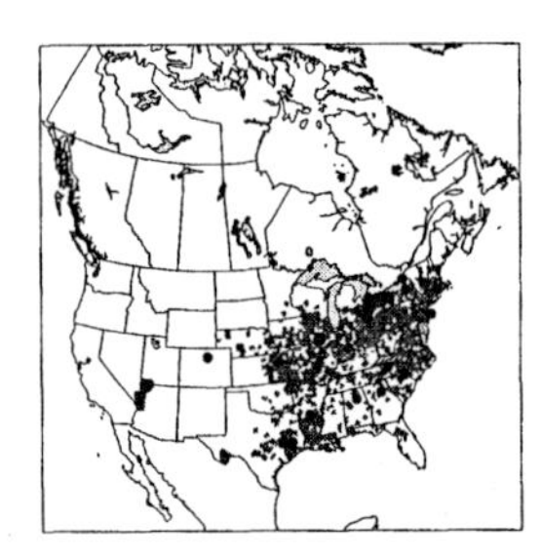

Plants perennial (appearing annual); usually cespitose, not rhizomatous, sometimes stoloniferous. **Culms** 10–45(70) cm, geniculate, often rooting at the lower nodes, glabrous or puberulent below the nodes; **internodes** often smooth, shiny, glabrous. **Sheaths** shorter than the internodes, glabrous for most of their length, margins shortly (0.3–1.2 mm) pubescent distally, not becoming spirally coiled when old; **ligules** 0.2–0.5 mm, truncate, erose, ciliate; **blades** (1)3–10 cm long, 1–4.5 mm wide, flat, smooth or scabridulous. **Panicles** 3–15 cm long, 1–1.6 cm wide, contracted, often interrupted below; **branches** 0.4–5.5 cm, appressed or diverging up to 30° from the rachises, spikelet-bearing to the base; **pedicels** 0.1–4 mm, scabrous to hirsute; **disarticulation** above the glumes. **Spikelets** 1.8–2.8 mm, borne singly. **Glumes** unequal, shorter than the florets, thin and membranous throughout, unawned; **lower glumes** lacking or rudimentary, veinless, rounded and often erose; **upper glumes** 0.1–0.3 mm, veinless; **lemmas** 1.8–2.8 mm, oblong-elliptic, mostly scabrous, calluses hairy, hairs to 0.8 mm, veins greenish, lower ¼ of the midveins with a few appressed hairs, apices acute to acuminate, awned, awns 1.5–5 mm, straight; **paleas** 1.8–2.8 mm, oblong-elliptic, acute to acuminate; **anthers** 0.2–0.5 mm, yellow. **Caryopses** 1–1.4 mm, fusiform, brownish. $2n$ = 40, 42.

Muhlenbergia schreberi grows in moist to dry woods and prairies on rocky slopes, in ravines, and along sandy riverbanks, at elevations of 60–1600 m. It is also common in disturbed sites near cultivated fields, pastures, and roads at these elevations. Its geographic range includes central, but not northern, Mexico. Records from the western United States probably reflect receent introductions. The species is considered a noxious, invasive weed in California.

14. Muhlenbergia tenuifolia (Kunth) Trin. [p. 163]
SLENDER MUHLY

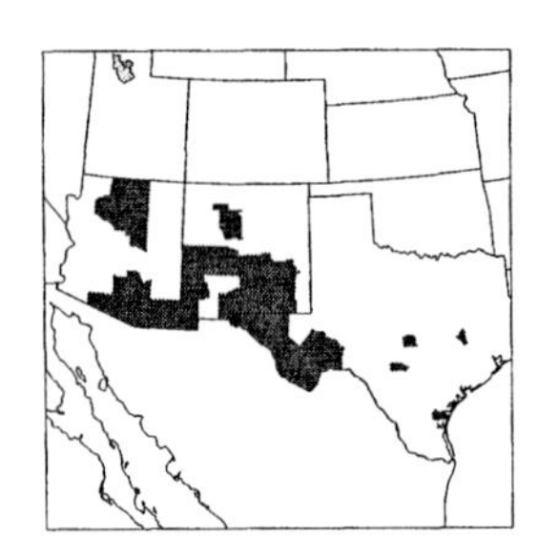

Plants annual or short-lived perennials; cespitose, not rhizomatous. **Culms** 20–70 cm, erect or decumbent; **internodes** mostly scabridulous or smooth, always scabridulous below the nodes. **Sheaths** usually shorter than the internodes, glabrous, smooth or scabridulous, usually without necrotic spots, not becoming spirally coiled when old; **ligules** 1.2–3(5) mm, membranous throughout, acute, often lacerate; **blades** 2–13 cm long, 1.2–2.5 mm wide, flat or loosely involute, scabridulous or glabrous abaxially, scabrous adaxially, usually without necrotic spots. **Panicles** numerous, terminal and axillary, 7–20 cm long, 0.3–1.4(3) cm wide, contracted, often lax, nodding, interrupted below; **primary branches** 3.5–7.5 cm, ascending or diverging up to 70° from the rachises, spikelet-bearing to the base; **pedicels** 1–3 mm, antrorsely scabrous; **disarticulation** above the glumes. **Spikelets** 2–4 mm, often purplish, borne singly. **Glumes** 1.2–2.8 mm, shorter than the florets, 1-veined, veins scabrous, apices often erose, unawned or awned, awns to 0.5 mm; **lower glumes** 1.2–2 mm, acute to acuminate; **upper glumes** 1.5–2.8 mm, acute; **lemmas** 2–3.5(4) mm, lanceolate, mostly smooth, scabridulous distally, pubescent on the calluses, lower ½ of the midveins, and margins, hairs 0.5–1.5 mm, apices acuminate to acute, awned, awns 10–30 mm, scabrous, sinuous to flexuous; **paleas** 1.8–3.4(3.8) mm, lanceolate, sparsely villous basally, apices acuminate to acute; **anthers** 0.9–1.5 mm, yellowish. **Caryopses** 1–2.2 mm, narrowly fusiform, brownish. **Cleistogamous panicles** not present. $2n$ = 20, 40.

Muhlenbergia tenuifolia grows in gramma grasslands and pine-oak woodlands on rocky slopes, limestone rock outcrops, gravelly roadsides, and in sandy drainages, at elevations of 1200–2200 m. Its range extends through Mexico to northern South America.

15. Muhlenbergia microsperma (DC.) Trin. [p. 163]
LITTLESEED MUHLY

Plants annual, sometimes appearing as short-lived perennials; tufted. **Culms** 10–80 cm, often geniculate at the base, much branched near the base; **internodes** mostly scabridulous or smooth, always scabridulous below the nodes. **Sheaths** often shorter than the internodes, glabrous, smooth or scabridulous; **ligules** 1–2 mm, membranous to hyaline, truncate to obtuse; **blades** 3–8.5(10) cm long, 1–2.5 mm wide, flat or loosely involute, scabrous abaxially, strigulose adaxially. **Panicles** 6.5–13.5 cm long, 1–6.5 cm wide, not dense, often purplish; **branches** 1.6–4 cm, ascending or diverging up to 80° from the rachises, spikelet-bearing to the base; **pedicels** 2–6 mm, appressed to divaricate, antrorsely scabrous; **disarticulation** above the glumes. **Spikelets** 2.5–5.5 mm, borne singly. **Glumes** 0.4–1.3 mm, exceeded by the florets, 1-veined, obtuse, often minutely erose; **lower glumes** 0.4–1 mm; **upper glumes** 0.6–1.3 mm; **lemmas** 2.5–3.8(5.3) mm, narrowly lanceolate, mostly smooth, scabridulous distally, hairy on the calluses, lower ½ of the margins, and midveins, hairs 0.2–0.5 mm, apices

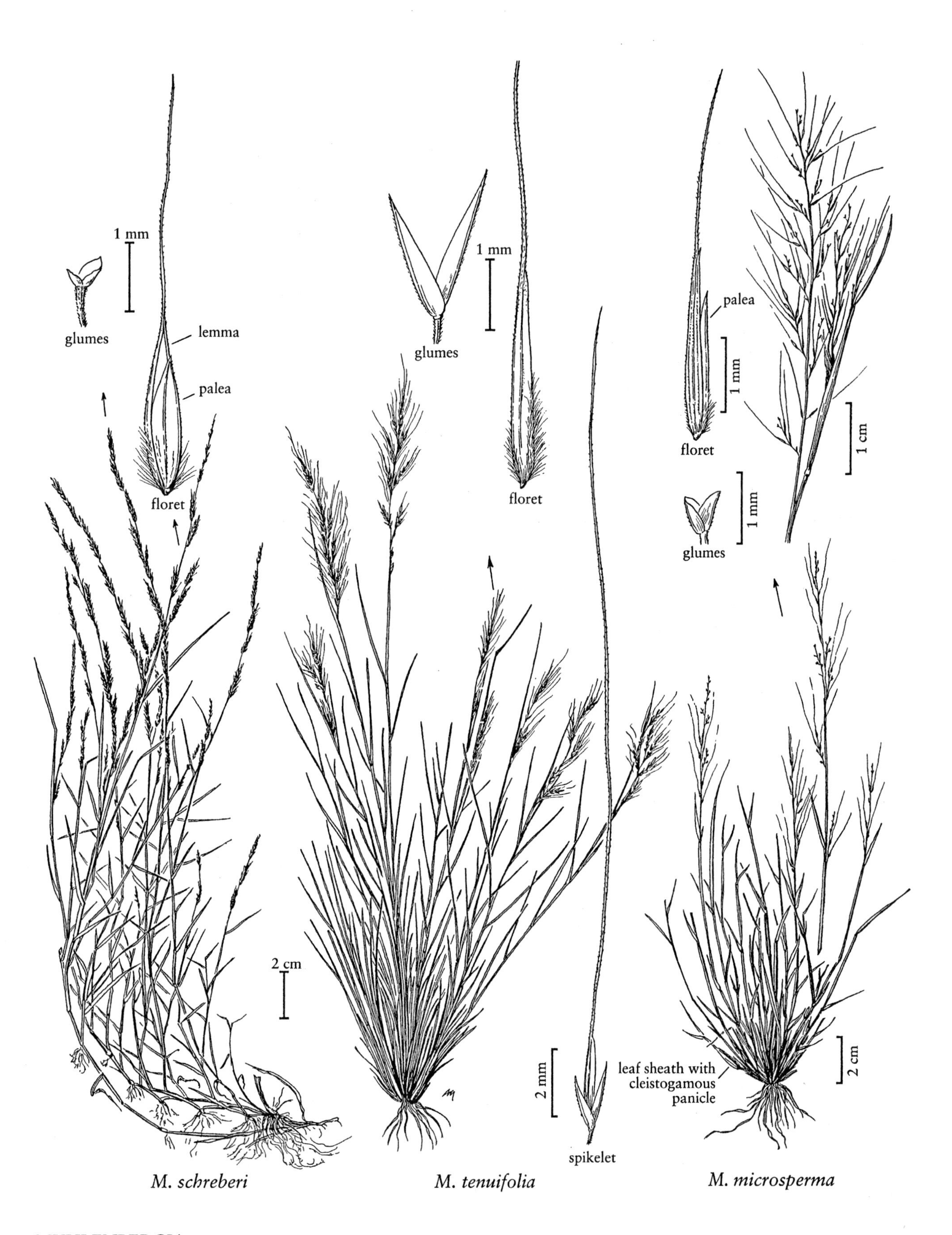

M. schreberi *M. tenuifolia* *M. microsperma*

acuminate, awned, awns 10–30 mm, straight to flexuous; **paleas** 2.2–4.8 mm, narrowly lanceolate, acuminate; **anthers** 0.3–1.2 mm, purplish. **Caryopses** 1.7–2.5 mm, fusiform, reddish-brown. **Cleistogamous panicles** with 1–3 spikelets present in the axils of the lower leaves. $2n$ = 20, 40, 60.

Muhlenbergia microsperma grows on sandy slopes, drainages, cliffs, rock outcrops, and disturbed roadsides, at elevations of 0–2400 m. It is usually found in creosote scrub, thorn-scrub forest, sarcocaulescent desert, and oak-pinyon woodland associations. Its range extends from the southwestern United States through Central America to Peru and Venezuela. Morphological variation among and within its populations is marked.

16. **Muhlenbergia appressa** C.O. Goodd. [p. 166]
DEVIL'S-CANYON MUHLY

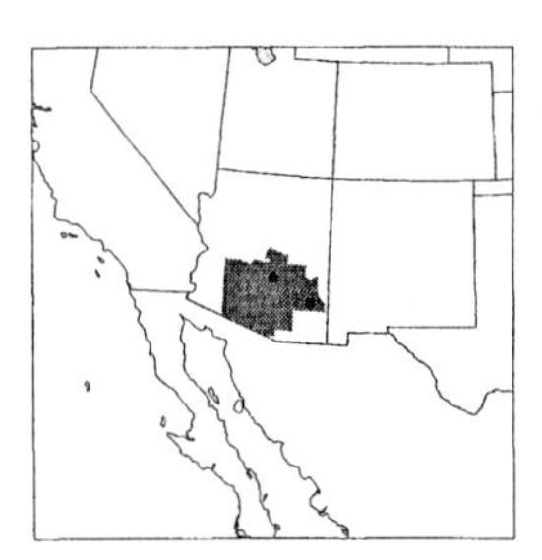

Plants annual. **Culms** 10–42 cm, erect or decumbent, much branched below; **internodes** mostly scabrous, scabrous or hispidulous below the nodes. **Sheaths** usually shorter than the internodes, flattened below, glabrous, scabridulous or smooth, striate; **ligules** 1.5–3 mm, hyaline, acute, lacerate, with lateral lobes; **blades** 1–5(7) cm long, 0.8–1.6(2) mm wide, flat or involute, scabrous abaxially, pubescent adaxially. **Panicles** 4–23 cm long, 0.5–2 cm wide, usually partially included in the subtending sheaths at maturity; **primary branches** 20–45 mm, ascending, closely appressed; **pedicels** 1–5 mm, appressed, scabrous; **disarticulation** above the glumes. **Spikelets** 4–7.5 mm, borne singly. **Glumes** 1–2 mm, 1-veined, veins conspicuous, scabrous, and greenish basally, apices obtuse to subacute, often erose; **lemmas** 4–6.2(7.5) mm, narrowly lanceolate, terete, mostly smooth, scabrous distally, hairy on the calluses and lower ¼ of the margins, hairs 0.2–0.3 mm, whitish, lemma bodies not appearing 5-veined, apices acuminate, awned, awns 10–30 mm, straight, scabrous; **paleas** 3.8–5.7(7) mm, narrowly lanceolate; **anthers** 0.3–1.1 mm, purplish. **Caryopses** 1.6–3 mm, narrowly fusiform, brownish to pinkish. **Cleistogamous spikelets** usually present in the axils of the lower leaves. $2n$ = unknown.

Muhlenbergia appressa grows in sandy drainages, canyon bottoms, rocky road cuts, and sandy slopes, at elevations of 20–1750 m. Its range extends from Arizona to Baja California, Mexico. It grows in gramma grasslands, oak-juniper woodlands, and chaparral associations.

17. **Muhlenbergia pectinata** C.O. Goodd. [p. 166]
COMBTOP MUHLY

Plants annual. **Culms** 10–33 cm, erect or decumbent, sometimes rooting at the lower nodes; **internodes** glabrous. **Sheaths** usually longer than the internodes, glabrous, margins sparsely hairy distally, hairs to 1 mm, coarse; **ligules** 0.3–0.9 mm, membranous, truncate, ciliate; **blades** 1–6 cm long, 0.6–2 mm wide, flat to loosely involute, appressed-pubescent to sparsely pilose. **Panicles** 4–12 cm long, 0.5–2.6(4) cm wide; **primary branches** 2–3.5 cm, appressed or diverging up to 70° from the rachises; **pedicels** 1–3 mm, glabrous, appressed; **disarticulation** above the glumes. **Spikelets** 2.5–5 mm, borne singly. **Glumes** glabrous, 1-veined, acute to acuminate, unawned or awned, awns to 1.5 mm; **lower glumes** 0.8–1.7 mm, entire; **upper glumes** 1–2(3) mm; **lemmas** (2.5)3–4.5(5) mm, subulate to lanceolate, calluses appressed-pubescent, lemma bodies appearing 5-veined (the intermediate "veins" are actually rows of short barbs on top of folded epidermal ridges), lateral veins occasionally ciliate, apices acuminate, awned, awns 10–32 mm, slender, flexuous; **paleas** 2.4–4.4 mm, narrowly lanceolate, glabrous, acuminate; **anthers** 0.3–0.6 mm, yellowish. **Caryopses** 0.6–3.1 mm, narrowly fusiform, light brownish. $2n$ = 20.

Muhlenbergia pectinata grows on rock outcrops, rocky cliffs, canyon walls, steep slopes, and road cuts, at elevations of 45–2400 m in thorn-scrub forests, gramma grasslands, and pine-oak woodlands. It is almost entirely restricted to vertical surfaces that are seasonally wet. Its range extends from southeastern Arizona to Oaxaca, Mexico.

18. **Muhlenbergia diversiglumis** Trin. [p. 166]
MIXEDGLUME MUHLY

Plants annual; sprawling. **Culms** 16–50 cm, decumbent, rooting at the lower nodes; **nodes** retrorsely pilose; **internodes** smooth or scabridulous. **Sheaths** 1.5–8.5 cm, sparsely or densely pilose, hairs to 3 mm, papillose-based; **ligules** 0.5–0.8 mm, membranous, truncate, erose; **blades** 2–6 cm long, 1.5–4 mm wide, flat, bases distinctly narrowed to the junction with the sheath, surfaces scabridulous and sparsely pilose, hairs papillose-based. **Panicles** 6–10.5 cm long, 2.0–4.5 cm wide, secund, open; **primary branches** 0.8–3.5 cm, secund, spreading at right angles or somewhat reflexed, with 2–5 spikelets;

secondary branches not developed; **pedicels** 1–5 mm, scabrous or shortly pilose, hairs papillose-based; **disarticulation** at the base of the primary branches. **Spikelets** 4–8 mm, dimorphic with respect to the glumes, proximal spikelets on each branch almost sessile. **Glumes of proximal spikelets on each branch** subequal, 0.2–0.7 mm, orbicular, truncate, often erose, unawned; **glumes of distal spikelets on each branch** markedly unequal; **lower glumes** to 8 mm, 1-veined, acute, usually awned, awns 0.5–3 mm; **upper glumes** orbicular, acute, sometimes awn-tipped; **lemmas** 4.0–7.6 mm, linear to broadly lanceolate, light greenish, smooth or scabrous, usually with greenish veins, apices acuminate, awned, awns 6–19 mm, usually straight, scabrous; **paleas** 3.7–6.8 mm, narrowly lanceolate, coarsely papillate or almost smooth, acuminate; **anthers** 0.4–0.8 mm, yellowish. **Caryopses** 1.8–3 mm, oblong-ovoid, flattened, brownish. $2n$ = 20.

Muhlenbergia diversiglumis has been collected from Galveston County, Texas. The species is native from Mexico to Peru and Venezuela, where it grows on moist cliffs, along water courses, sandy slopes, and road cuts, primarily in moist shaded environments of broadleaf evergreen forests and pine-oak forests, at elevations of 600–2500 m. The collection from Texas may represent a recent introduction.

19. **Muhlenbergia glauca** (Nees) B.D. Jacks. [p. 166]
DESERT MUHLY

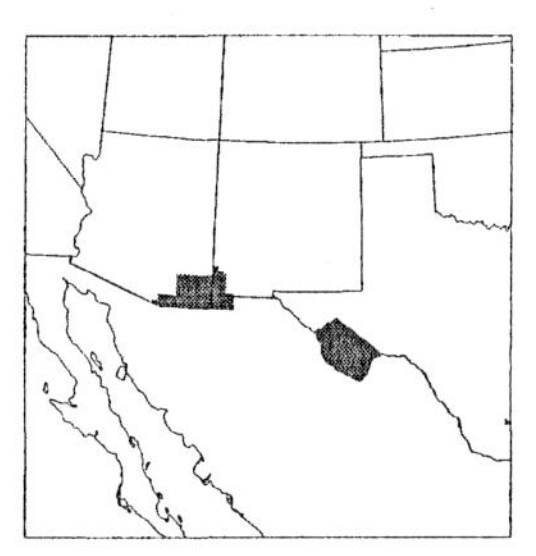

Plants perennial; rhizomatous, rhizomes slender, well-developed. **Culms** 25–60 cm tall, 1–2 mm thick, often decumbent, sometimes erect; **internodes** mostly scabrous, retrorsely hispidulous below the nodes. **Sheaths** longer than the internodes, scabridulous; **ligules** 0.5–2 mm, truncate to obtuse, erose or lacerate; **blades** 4–12 cm long, 1–2.6 mm wide, flat to involute distally, not arcuate, scabrous abaxially, hirsute or scabrous adaxially. **Panicles** 4–12(17) cm long, 0.3–2.4 cm wide, contracted, interrupted below; **branches** 0.3–3 cm, usually appressed, occasionally diverging up to 30° from the rachises; **pedicels** 0.1–1.2 mm, scabrous to hirsute. **Spikelets** 2.4–3.5 mm. **Glumes** equal, 1.5–3.5 mm, 1-veined, veins scabrous, apices acute or acuminate, usually awned, awns, if present, to 1.5 mm; **lemmas** 2.4–3.4 mm, elliptic, pubescent on the lower the ½ of the midveins and margins, hairs to 0.6 mm, tawny, apices acuminate to acute, awned, awns 0.1–3(5) mm; **paleas** 2.2–3.4 mm, elliptic, intercostal region pubescent on the lower ½, apices acuminate to acute; **anthers** 1.8–2.4 mm, orange. **Caryopses** 1.7–2 mm, fusiform, brownish. $2n$ = 60.

Muhlenbergia glauca grows on calcareous rocky slopes, cliffs, canyon walls, table rocks, and volcanic rock outcrops, at elevations of 1200–2780 m. Its range extends from the southwestern United States to central Mexico. *M. glauca* resembles *M. polycaulis*, but differs in its shorter lemma awns and strongly rhizomatous habit.

20. **Muhlenbergia thurberi** (Scribn.) Rydb. [p. 168]
THURBER'S MUHLY

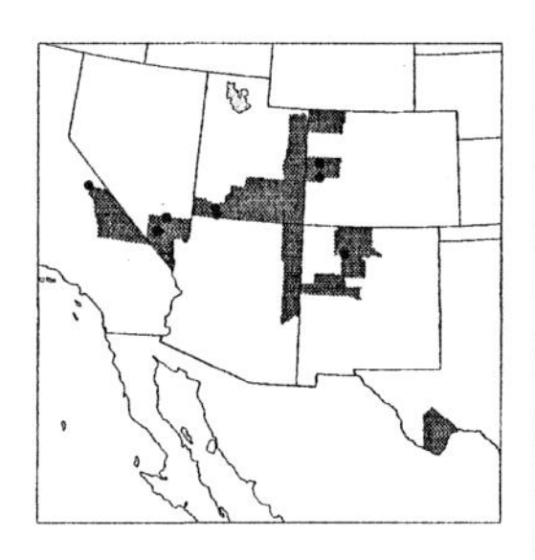

Plants perennial; rhizomatous, not cespitose. **Culms** 12–36 cm tall, 0.3–0.5 mm thick, often clumped, usually erect, somewhat decumbent at the base; **internodes** mostly or completely glabrous, sometimes strigose below the nodes. **Sheaths** shorter than the internodes, hirtellous near the margins; **ligules** 0.9–1.2 mm, membranous, truncate to obtuse, erose; **blades** 0.2–3.7 cm long, 0.2–1 mm wide, tightly involute, straight to arcuate, smooth or scabridulous abaxially, hirtellous adaxially. **Panicles** 0.7–5.5 cm long, 0.2–0.7 cm wide, contracted, not dense, sometimes interrupted below; **primary branches** 0.4–1.8 cm, appressed; **pedicels** 0.1–4 mm. **Spikelets** 2.6–4 mm. **Glumes** subequal, 1.6–3 mm, 1-veined, scabridulous on the veins, acute, unawned; **lemmas** 2.6–4 mm, lanceolate, hairy on the lower ¾, hairs to 1.2 mm, tawny, apices acuminate, unawned or awned, awns to 1 mm; **paleas** 2.6–4 mm, lanceolate, intercostal region tawny pubescent; **anthers** 2.1–2.3 mm, yellowish-purple. **Caryopses** 2–2.2 mm, fusiform, light brownish. $2n$ = unknown.

Muhlenbergia thurberi usually grows in moist soil in seeps near canyon cliffs, sandstone slopes, and rocky ledges, at elevations of 1350–2300 m. The species appears to be restricted to the southwestern United States. It flowers from July to September.

Muhlenbergia thurberi resembles *M. curtifolia*, but differs in its tightly involute blades, and longer anthers and ligules. The two species have been found growing within 50 m of each other in Apache County, Arizona, but in different habitats, *M. curtifolia* growing in a damp drainage areas whereas *M. thurberi* grew near a moist but dryer canyon cliff.

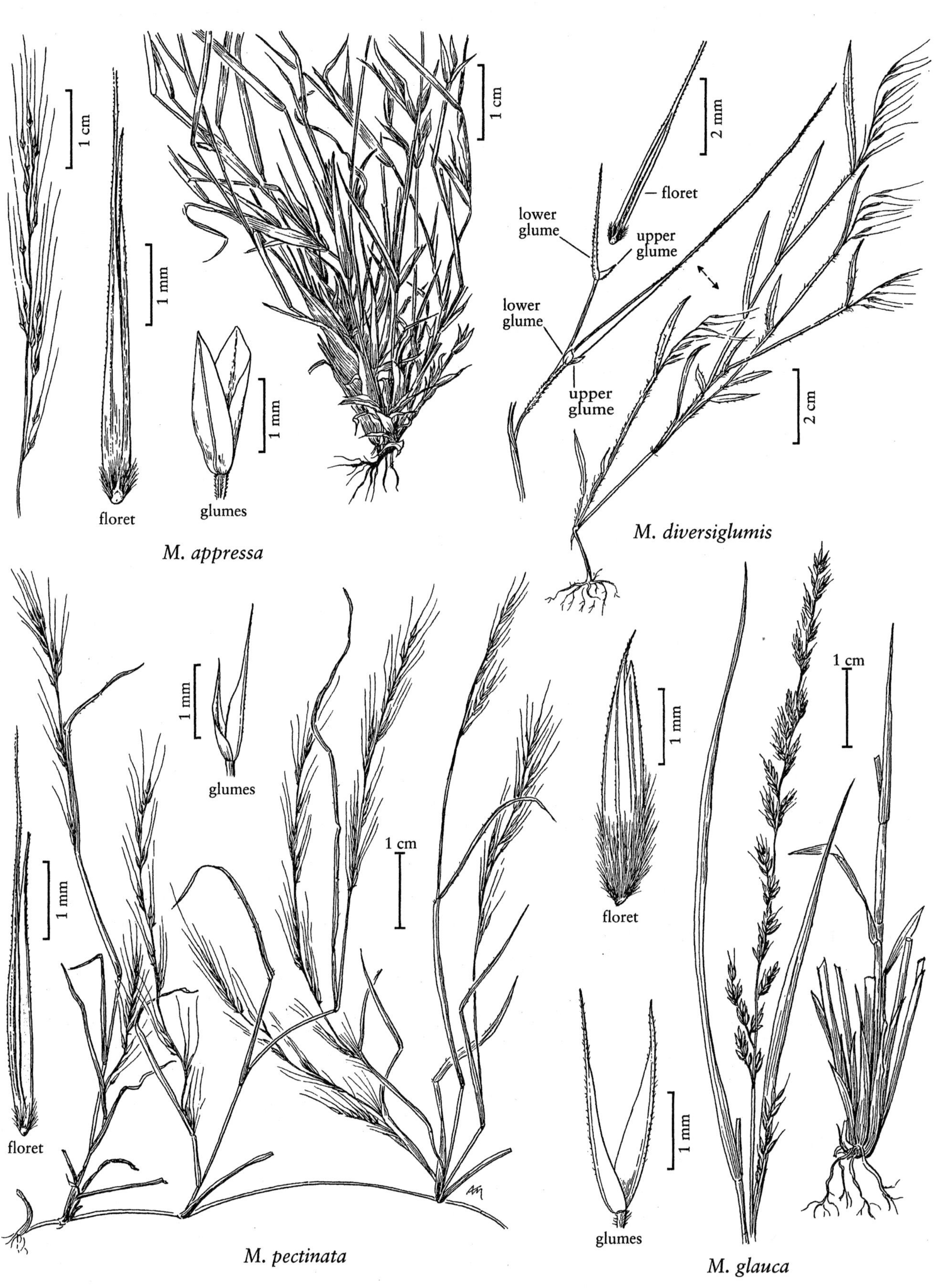
1 cm
1 mm
floret
1 mm
glumes
1 cm
M. appressa
2 mm
floret
lower glume
upper glume
lower glume
upper glume
2 cm
M. diversiglumis
1 mm
glumes
1 cm
1 mm
floret
M. pectinata
1 mm
floret
1 cm
1 mm
glumes
M. glauca

21. **Muhlenbergia curtifolia** Scribn. [p. 168]
UTAH MUHLY

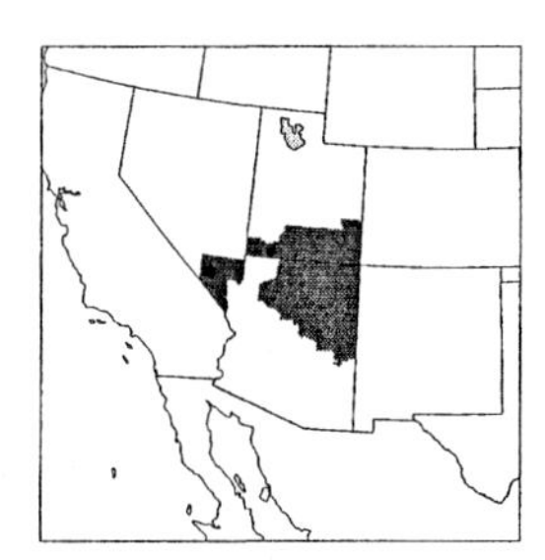

Plants perennial; rhizomatous, not cespitose. **Culms** 10–45 cm tall, 0.3–0.5 mm thick, erect or decumbent; **internodes** hirsute and strigose below the nodes. **Sheaths** shorter than the internodes, hirsute; **ligules** 0.2–0.6 mm, margins hyaline, apices truncate to obtuse, erose, ciliate; **blades** 0.5–4(5) cm long, 0.6–1.5(2.2) mm wide, awl-shaped, flat near the base, becoming slightly involute distally, strigose to scabrous abaxially, hirsute adaxially. **Panicles** 2–10 cm long, 0.2–1 cm wide, contracted, not dense; **primary branches** 0.3–2.5 cm, ascending to appressed; **pedicels** 0.1–3 mm. **Spikelets** 2.8–4 mm. **Glumes** equal, 2.5–4 mm, 1-veined, veins scabridulous, apices acute; **lemmas** 2.8–4 mm, lanceolate, hairy on the lower ¾, hairs to 1.5 mm, tawny, midveins scabridulous distally, apices acuminate, awned, awns usually 1–6(12) mm, delicate, straight; **paleas** 2.8–4 mm, lanceolate, intercostal region tawny pubescent on the basal ¾; **anthers** 1.2–1.6 mm, yellowish-purple. **Caryopses** 2–2.4 mm, fusiform, brownish. $2n$ = unknown.

Muhlenbergia curtifolia grows on damp ledges and in rock crevices of vertical cliffs, and beneath large calcareous boulders above the canyon floor, at elevations of 1600–2750 m, in the southwestern United States. It resembles *M. thurberi*, differing in its flatter leaf blades and shorter ligules and anthers. It also tends to grow in more mesic habitats than *M. thurberi.*

22. **Muhlenbergia pauciflora** Buckley [p. 168]
NEW MEXICAN MUHLY

Plants perennial; tightly cespitose, sometimes rhizomatous. **Culms** 30–70 cm tall, 0.7–1 mm thick, erect, geniculate and rooting at the lower nodes; **internodes** mostly glabrous, sometimes glaucous, often striate below the nodes. **Sheaths** usually shorter than the internodes, smooth or scabridulous, usually with dark brown necrotic spots, flat and spreading at maturity; **ligules** 1–2.5(5) mm, membranous, obtuse, with lateral lobes 1.5–3 mm longer than the central portion; **blades** (1)5–12(15) cm long, 0.5–1.5 mm wide, flat to involute, glabrous and smooth abaxially, scabridulous adaxially, often with dark brown necrotic spots. **Panicles** (2)5–15 cm long, 0.5–2.8 cm wide, contracted, interrupted below; **primary branches** 0.5–4(6) cm, appressed or diverging up to 30° from the rachises, spikelet-bearing to the base; **pedicels** 0.1–3 mm, scabrous. **Spikelets** 3.5–5.5 mm, occasionally with 2 florets. **Glumes** equal, 1.5–3.5 mm, shorter than the lemmas, 1-veined, acuminate to acute, often erose, unawned or awned, awns to 2.2 mm; **lemmas** 3–5.5 mm, lanceolate, often purplish, mostly glabrous, calluses with a few short hairs, apices acuminate, scabridulous, awned, awns (5)7–25 mm, flexuous; **paleas** 3–5.5 mm, lanceolate, glabrous, acuminate; **anthers** 1.5–2.1 mm, yellowish or purplish. **Caryopses** 2–2.5 mm, fusiform, brownish. $2n$ = unknown.

Muhlenbergia pauciflora grows in open or closed forests on rocky slopes, cliffs, canyons, and rock outcrops of granitic or calcareous origin, at elevations of 1200–2500 m. Its range extends from the southwestern United States to central Mexico.

23. **Muhlenbergia polycaulis** Scribn. [p. 170]
CLIFF MUHLY

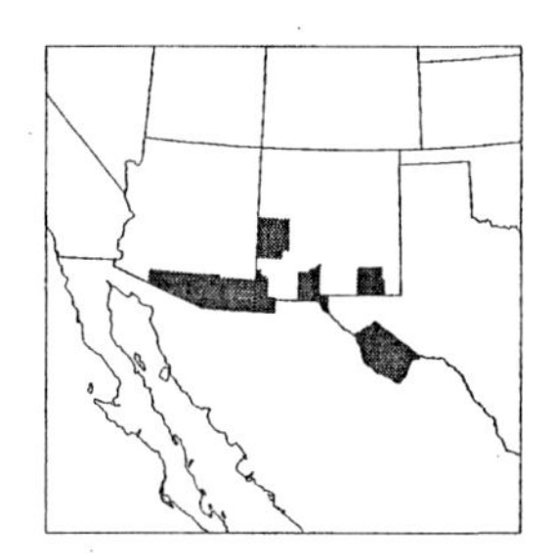

Plants perennial; loosely cespitose, sometimes shortly rhizomatous. **Culms** 15–40(50) cm tall, 0.8–1.5 mm thick, erect or decumbent; **internodes** glabrous or strigulose for most of their length, including below the nodes. **Sheaths** usually shorter than the internodes, smooth or scabridulous, without necrotic spots, not becoming spirally coiled when old; **ligules** 0.5–2.5 mm, membranous, obtuse to acute, without lateral lobes; **blades** 3–10 mm long, 0.5–2 mm wide, flat or involute (occasionally folded), smooth or scabridulous abaxially, hirsute or scabridulous adaxially, without necrotic spots. **Panicles** 2–12 cm long, 0.3–1.8 cm wide, contracted, interrupted below; **primary branches** 0.5–4 cm, appressed or diverging up to 30° from the rachises, spikelet-bearing to the base; **pedicels** 0.1–1.5 mm, scabrous. **Spikelets** 2.5–4 mm, plump near the middle. **Glumes** subequal, (1)1.5–2.6 mm, exceeded by the florets, 1-veined, acute or acuminate, unawned or awned, awns to 1.4 mm; **lemmas** 2–3.5 mm, elliptic, widest near the middle, appressed-pubescent on the lower ½–⅔ of the midveins and margins, hairs to 0.5 mm, apices scabridulous, acuminate, awned, awns 10–20(25) mm, flexuous; **paleas** 2–3.5 mm, elliptic, intercostal region appressed-pubescent on the basal ½, apices acuminate; **anthers** 1.5–2 mm, orange. **Caryopses** 1.5–2 mm, fusiform, brownish. $2n$ = 20, 40.

Muhlenbergia polycaulis grows in open vegetation on steep rocky slopes, canyon walls, cliffs, table rocks, and volcanic rock outcrops, at elevations of 1200–2400 m. Its range extends from the southwestern United States to central Mexico. It differs from *M. glauca*, with which it may be confused, in its longer lemma awns, shorter rhizomes, and loosely tufted habit.

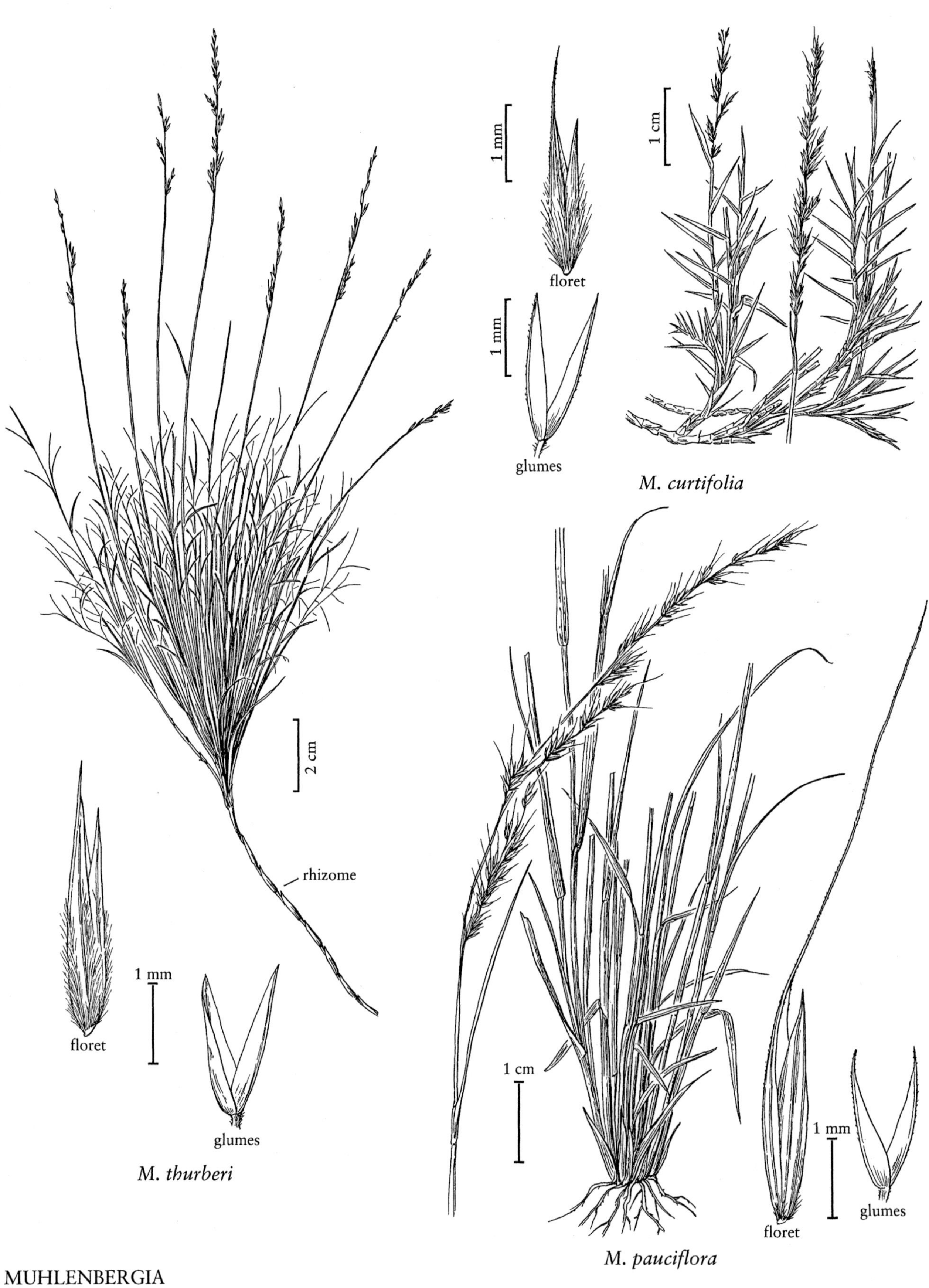
1 mm
floret
1 mm
glumes
1 cm
M. curtifolia
2 cm
rhizome
1 mm
floret
glumes
M. thurberi
1 cm
1 mm
floret
glumes
M. pauciflora

24. **Muhlenbergia arsenei** Hitchc. [p. 170]
NAVAJO MUHLY

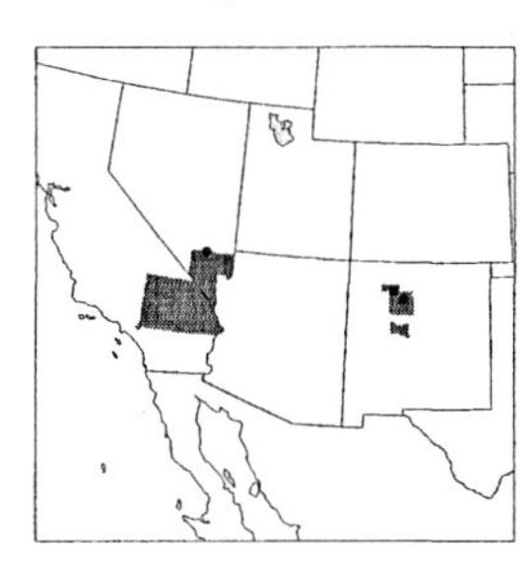

Plants perennial; rhizomatous, rhizomes sometimes short and the plants loosely cespitose. **Culms** 10–50 cm tall, 0.4–1 mm thick, decumbent; **internodes** glabrous or strigulose. **Sheaths** shorter than the internodes, strigulose or glabrous, usually without necrotic spots, not becoming spirally coiled when old; **ligules** 1–2 mm, membranous throughout, obtuse, strigulose or glabrous, erose or toothed, with lateral lobes, lobes less than 1.5 mm longer than the central portion; **blades** 1–6 cm long, 1–2 mm wide, flat to involute, smooth or scabridulous abaxially, hirsute adaxially, usually without necrotic spots. **Panicles** 4–13 cm long, 1–3(5) cm wide, not dense; **primary branches** 0.5–4 cm, appressed or diverging up to 30° from the rachises, spikelet-bearing to the base; **pedicels** 0.1–3 mm. **Spikelets** 3.5–5 mm. **Glumes** subequal, 2–4 mm, exceeded by the florets, 1-veined, scabrous on the veins and near the apices, apices acuminate, unawned or awned, awns to 1.2 mm; **lemmas** 3.5–5 mm, lanceolate, mostly purplish, veins conspicuously green, pubescent on the lower ⅓–½ of the midveins and margins, hairs to 1.5 mm, apices scabridulous, acuminate, awned, awns 4–12(20) mm, flexuous; **paleas** 3.5–5 mm, narrow-lanceolate, intercostal region pubescent, apices acuminate, veins sometimes extending into awns to 0.5 mm; **anthers** 1.3–3 mm, purple. **Caryopses** 2–2.3 mm, fusiform, brownish. 2*n* = unknown.

Muhlenbergia arsenei grows among granitic boulders, on rocky slopes, limestone rock outcrops, and in arroyos, at elevations of 1400–2850 m. Its range extends from the southwestern United States into Baja California, Mexico. It flowers from August to September.

25. **Muhlenbergia spiciformis** Trin. [p. 170]
LONGAWN MUHLY

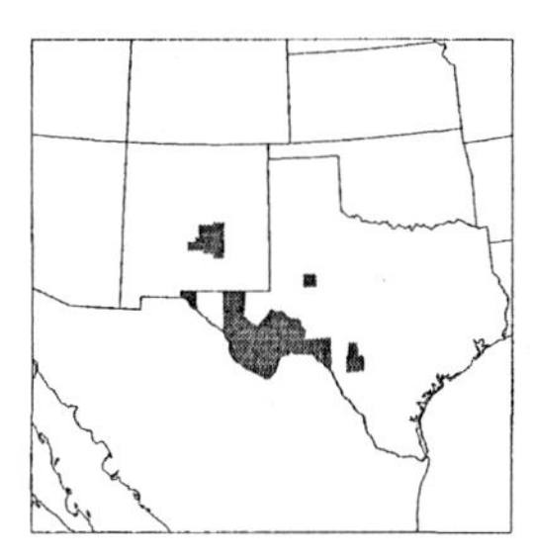

Plants perennial; cespitose, not rhizomatous. **Culms** 25–80 cm, erect, wiry; **internodes** mostly glabrous, strigose or glabrous below the nodes. **Sheaths** scabridulous, not becoming spirally coiled when old; **ligules** 1–3 mm, membranous, acuminate, deeply lacerate; **blades** 2–12 cm long, 1–3 mm wide, flat to involute, scabridulous abaxially, hirtellous or scabrous adaxially. **Panicles** 4–18(20) cm long, 0.2–2.8 cm wide, contracted, not dense, sometimes interrupted below; **panicle branches** 0.6–5 cm, appressed or diverging up to 30° from the rachises, spikelet-bearing to the base; **pedicels** 0.1–3 mm. **Spikelets** 3–4 mm. **Glumes** unequal, 0.3–1 mm, shorter than the florets, 1-veined, unawned; **lower glumes** shorter than the upper glumes, obtuse to acute, sometimes erose; **lemmas** 2.8–4 mm, narrowly lanceolate, purplish, scabrous, sparsely appressed-pubescent on the calluses and lower ¼ of the midveins and margins, hairs shorter than 0.3 mm, apices acuminate, awned, awns (10)20–40 mm; **paleas** 2.6–3.9 mm, narrowly lanceolate, intercostal region sparsely pubescent on the basal ⅓, apices acuminate, scabrous; **anthers** 0.9–1.6 mm, purplish. **Caryopses** 2–2.6 mm, fusiform, brownish. 2*n* = 40.

Muhlenbergia spiciformis grows on rocky slopes, cliffs, and calcareous rock outcrops, often in thorn-scrub and open woodland communities. Its elevational range is 450–2800 m; its geographic range extends from the southwestern United States to northern Mexico.

26. **Muhlenbergia porteri** Scribn. *ex* Beal [p. 172]
BUSH MUHLY

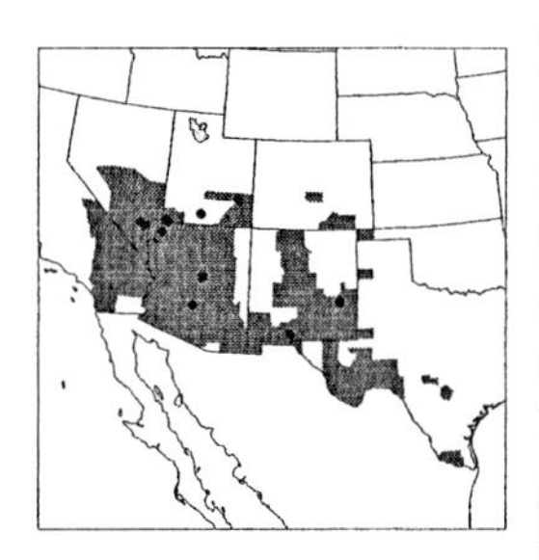

Plants perennial; loosely cespitose from a knotty base, not rhizomatous, distinctly bushy in appearance. **Culms** 25–100 cm tall, 0.5–1.5 mm thick, erect, geniculate, wiry, freely branched, branches stiff, geniculate, widely divergent; **internodes** scabridulous throughout. **Sheaths** shorter than the internodes, glabrous; **ligules** 1–2.5(4) mm, truncate, lacerate, with lateral lobes; **blades** 2–8 cm long, 0.5–2 mm wide, flat or folded, smooth or scabridulous abaxially, scabridulous adaxially. **Panicles** 4–14 cm long, 6–15 cm wide, open, not dense, usually purple; **primary branches** 1–7.5 cm, diverging 30–90° from the rachises, stiff, naked basally; **pedicels** 2–13(20) mm, scabrous. **Spikelets** 3–4.5 mm. **Glumes** 2–3 mm, shorter than the lemmas, 1-veined, veins scabrous, apices gradually acute to acuminate, occasionally mucronate, mucros to 0.4 mm; **lemmas** 3–4.5 mm, lanceolate, purplish, appressed-pubescent on the lower ½–¾ of the margins and midveins, apices acuminate, awned, awns 2–13 mm, straight; **paleas** 3–4.5 mm, lanceolate, acuminate; **anthers** 1.5–2.3 mm, purple to yellow. **Caryopses** 2–2.4 mm, oblong, compressed, yellowish-brown. 2*n* = 20, 23, 24, 40.

Muhlenbergia porteri grows among boulders on rocky slopes and on cliffs, and in dry arroyos, desert flats, and grasslands, frequently in the protection of shrubs, at elevations of 600–1700 m. Its geographic range extends from the southwestern United States to northern Mexico.

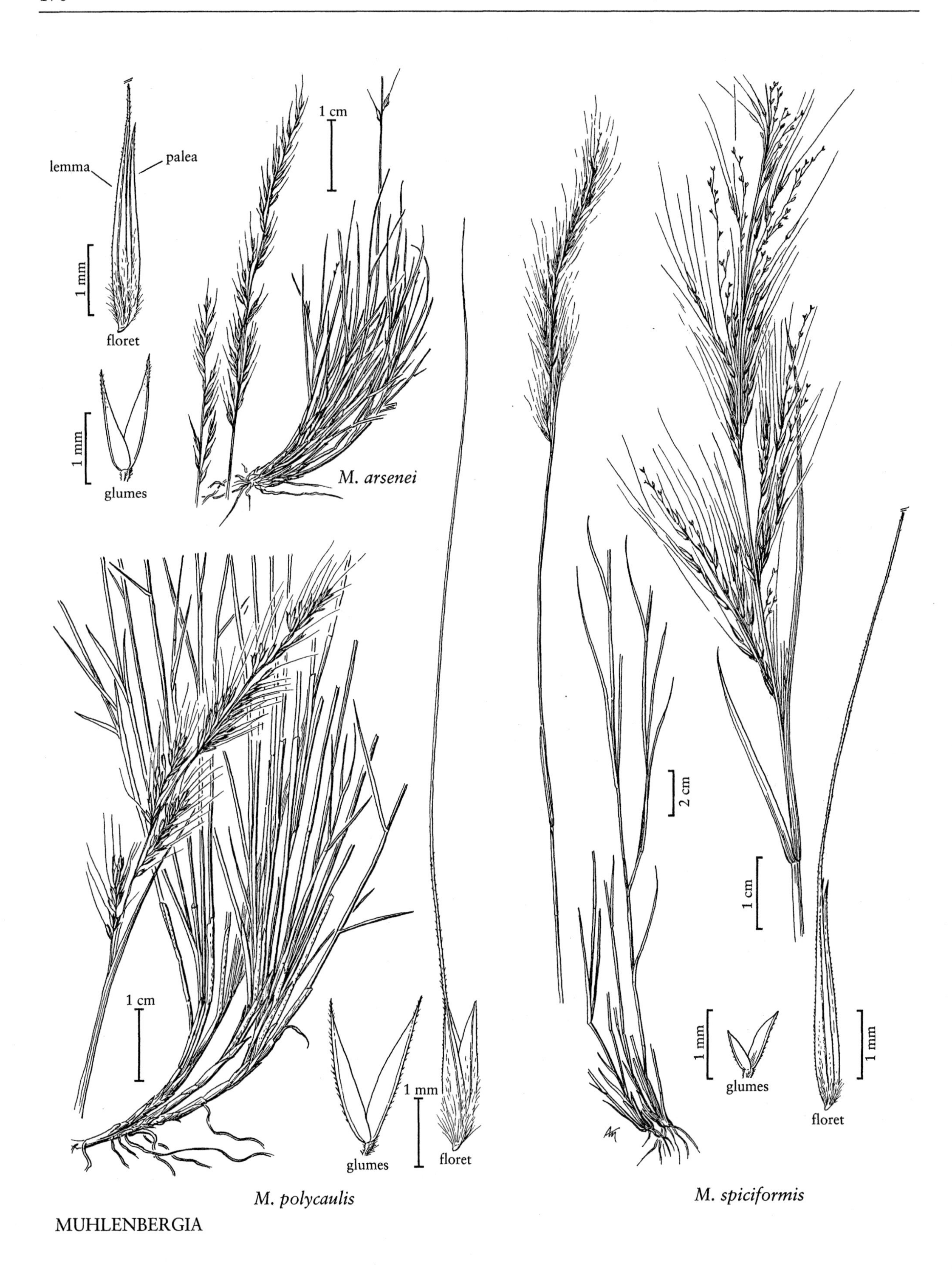
lemma
palea
1 mm
floret
1 mm
glumes
1 cm
M. arsenei
1 cm
1 mm
glumes
floret
M. polycaulis
2 cm
1 cm
1 mm
glumes
1 mm
floret
M. spiciformis

Muhlenbergia porteri is highly palatable to all classes of livestock, but is never abundant at any particular location.

27. **Muhlenbergia cuspidata** (Torr. *ex* Hook.) Rydb. [p. 172]
PLAINS MUHLY

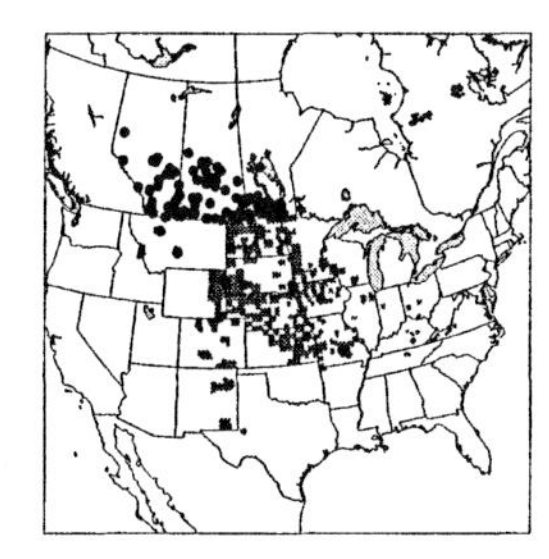

Plants perennial; cespitose, with knotty bases, not rhizomatous. **Culms** 20–60 cm tall, 1–2 mm thick, erect, not rooting at the lower nodes; **internodes** mostly glabrous or sparsely hispidulous, always hispidulous below the nodes. **Sheaths** shorter than the internodes, laterally compressed, keeled, smooth or scabridulous, not becoming spirally coiled when old; **ligules** 0.2–0.8 mm, membranous, truncate; **blades** 2–22 cm long, 0.5–2.7(3.5) mm wide, flat to folded, smooth or scabridulous abaxially, strigose adaxially. **Panicles** 4–14 cm long, 0.1–0.8 cm wide, spikelike, not dense; **primary branches** 0.4–3 cm, appressed; **pedicels** 0.1–1.2(4) mm. **Spikelets** 2.5–3.6 mm, dark green or plumbeous, occasionally with 2 florets. **Glumes** subequal, 1.2–3 mm, about ¾ as long as the spikelets, 1-veined, gradually acute to acuminate, occasionally mucronate, mucros to 0.3 mm; **lemmas** 2.5–3.6 mm, lanceolate, sometimes mottled with greenish-black areas, with short, appressed hairs on the basal ½–¾ of the midveins and margins, apices acuminate, sometimes mucronate, mucros to 0.6 mm; **paleas** 2.4–3.6 mm, lanceolate, glabrous, acuminate; **anthers** 1.2–1.8 mm, greenish. **Caryopses** 1.6–2.3 mm, fusiform, brownish. $2n$ = 20.

Muhlenbergia cuspidata grows in dry, gravelly prairies, on gentle rocky slopes, rocky limestone outcrops, and in sandy drainages, at elevations of 300–1400 m, primarily in the central portion of the *Flora* region. It flowers from June to October.

Muhlenbergia cuspidata is often confused with *M. richardsonis*, but that species has rhizomes and longer ligules. Morden and Hatch (1984) found cleistogamous panicles in the lower sheaths of a plant from Benton County, Oklahoma.

28. **Muhlenbergia wrightii** Vasey *ex* J.M. Coult. [p. 172]
SPIKE MUHLY

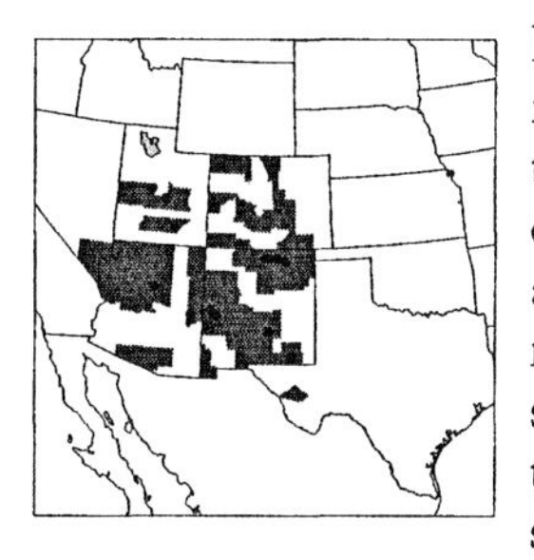

Plants perennial; cespitose, not rhizomatous. **Culms** 15–60 cm tall, 1.5–2.5 mm thick, compressed, erect, not rooting at the lower nodes; **internodes** mostly hispidulous or glabrous, strigose to hispidulous below the nodes. **Sheaths** usually shorter than the internodes, smooth or scabridulous, compressed-keeled, not becoming spirally coiled when old; **ligules** 1–3(5) mm, membranous, truncate; **blades** 1.4–12 cm long, 1–3 mm wide, flat to folded, smooth or scabridulous abaxially, strigose adaxially. **Panicles** 5–16 cm long, 0.2–1.2 cm wide, spikelike, dense; **primary branches** 0.3–2 cm, appressed; **pedicels** 0.1–1.4 mm. **Spikelets** 2–3 mm, dark green or plumbeous. **Glumes** equal, 0.5–1.6 mm, usually ½–¾ as long as the lemmas, 1-veined, scabridulous on the veins, acute or obtuse, abruptly narrowed to a short (0.5–1 mm) awn; **lemmas** 2–3 mm, lanceolate, appressed-pubescent on the basal ½–¾ of the midveins and margins, hairs about 0.5 mm, apices scabridulous, acute to acuminate, mucronate, mucros 0.3–1 mm; **paleas** 1.9–3 mm, lanceolate, intercostal region pubescent, apices acute to acuminate; **anthers** 1.3–1.8 mm, greenish. **Caryopses** 1.2–2 mm, fusiform, brownish. $2n$ = unknown.

Muhlenbergia wrightii grows in gravelly prairies, on rocky slopes, and in meadows on granitic, sandstone, or limestone-derived soils, at elevations of 1100–3000 m. Its range extends from the southwestern United States to northern Mexico.

29. **Muhlenbergia arizonica** Scribn. [p. 174]
ARIZONA MUHLY

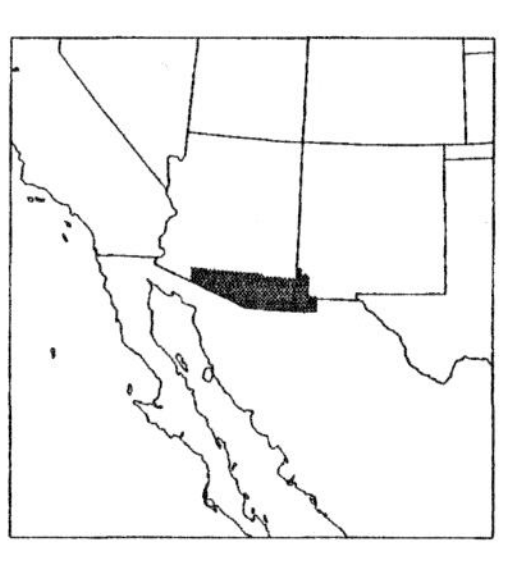

Plants perennial; cespitose, not rhizomatous. **Culms** 15–50 cm, erect to decumbent; **internodes** hispidulous or glabrous below the nodes. **Sheaths** from slightly shorter to slightly longer than the internodes, rounded to somewhat flattened but not keeled, hispidulous basally, glabrous distally, margins hyaline, not becoming spirally coiled when old; **ligules** 1–2 mm, hyaline, obtuse, minutely erose; **blades** 4–7 cm long, 0.8–2 mm wide, flat or folded, glabrous abaxially, scabridulous or hispidulous adaxially, midveins and margins conspicuous, thickened, white, and cartilaginous. **Panicles** 4–20 cm long, 4–15 cm wide, diffuse; **primary branches** 0.5–7.5 cm, capillary, diverging 40–90° from the

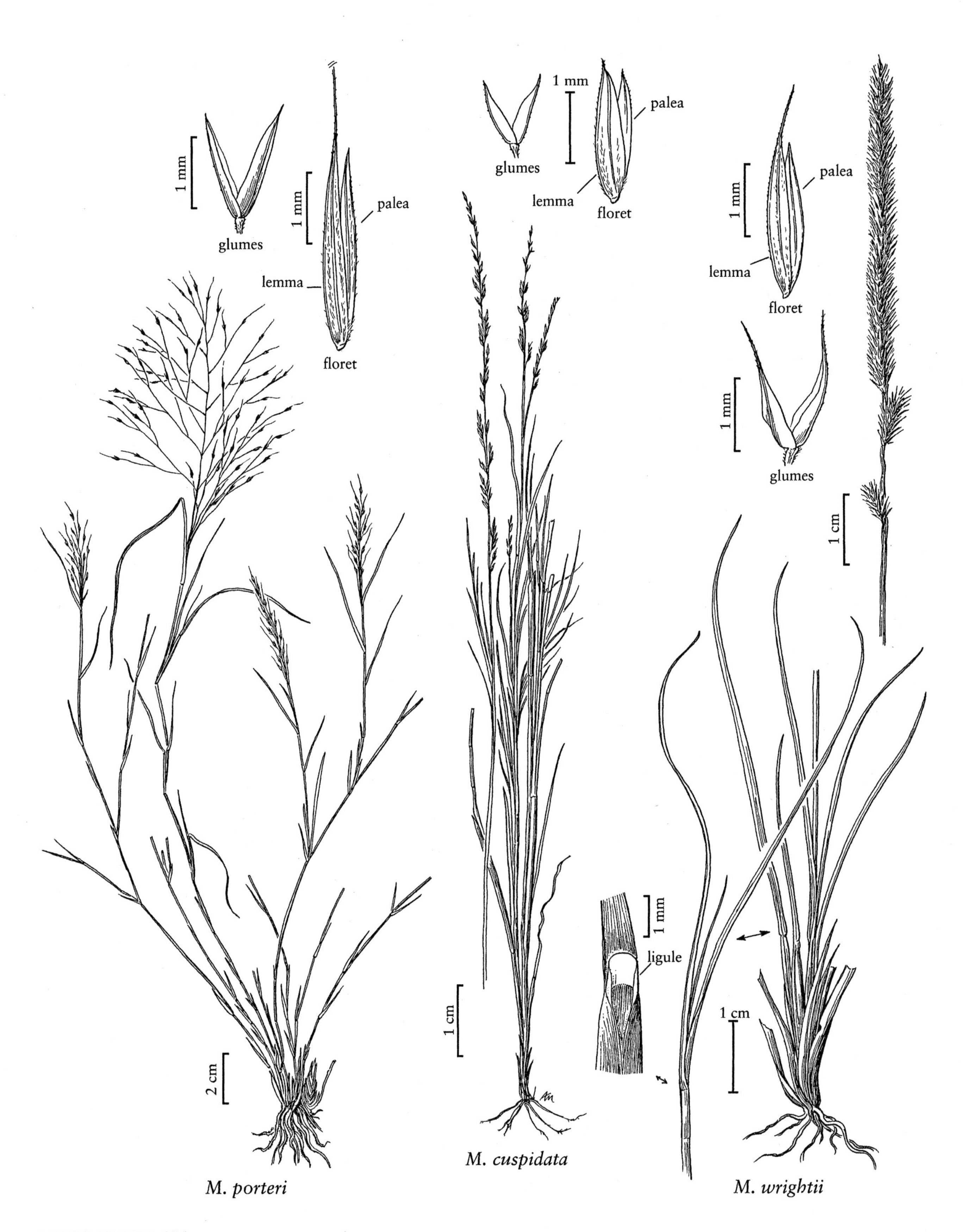
1 mm
glumes
1 mm
palea
lemma
floret
1 mm
glumes
1 mm
palea
lemma
floret
1 mm
palea
lemma
floret
1 mm
glumes
1 cm
1 mm
ligule
1 cm
1 cm
2 cm
M. porteri
M. cuspidata
M. wrightii

rachises, naked basally; **pedicels** 2–16 mm, flexuous. **Spikelets** 2.1–3.1 mm. **Glumes** equal, 1–1.5 mm, 1-veined, apices scabridulous, obtuse to acute, sometimes minutely erose, unawned; **lemmas** 2–3.1 mm, elliptic, purplish, appressed-pubescent on the lower ¾ of the midveins and margins, hairs to 0.6 mm, apices scabrous, acute, minutely bifid, awned, awns 0.5–1.1 mm; **paleas** 2.1–3.2 mm, elliptic, glabrous, acute; **anthers** 1.6–2.1 mm, purplish. **Caryopses** 1.3–1.7 mm, fusiform, sulcate dorsally, brownish. $2n$ = 20.

Muhlenbergia arizonica grows in sandy drainages and gravelly canyons, and on plateaus and rocky slopes in open desert grasslands, at elevations of 1220–2230 m. Its range extends from the southwestern United States into northwestern Mexico. Flowering is from August to October.

30. **Muhlenbergia torreyi** (Kunth) Hitchc. *ex* Bush [p. 174]
RING MUHLY

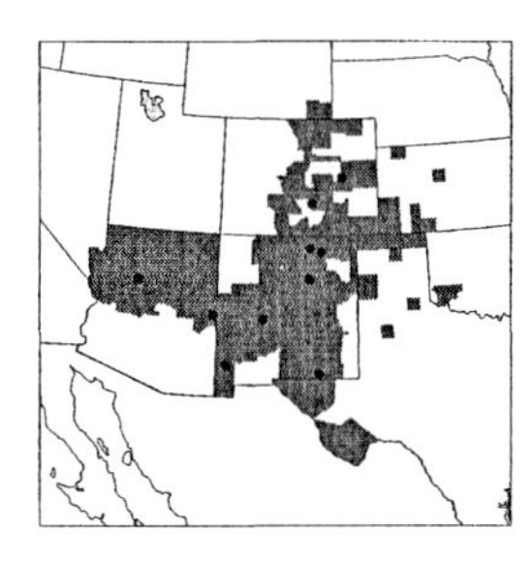

Plants perennial; cespitose, not rhizomatous. **Culms** 10–40(50) cm, decumbent, usually all the nodes concealed by the sheaths; **internodes** mostly scabrous or smooth, hispidulous below the nodes. **Leaves** strongly basally concentrated, most blades not reaching more than ⅕ of the plant height; **sheaths** shorter than the internodes, rounded, not keeled, scabridulous or smooth, not becoming spirally coiled when old; **ligules** 2–5(7) mm, hyaline, acuminate, lacerate, often with lateral lobes; **blades** 1–3(5) cm long, 0.3–0.9 mm wide, tightly involute or folded, arcuate, scabridulous, midveins and margins not thickened, green, apices somewhat sharp-pointed. **Panicles** 7–21 cm long, 3–15 cm wide, diffuse; **primary branches** 1–8 cm, diverging 30–90° from the rachises, stiff, naked basally; **pedicels** 1–8 mm, sometimes appressed to the branches. **Spikelets** 2–3.5 mm. **Glumes** equal, 1.3–2.5 mm, scabridulous, 1-veined, apices acute to acuminate, minutely erose, unawned or awned, awns to 1.1 mm; **lemmas** 2–3.2(3.5) mm, narrowly elliptic to lanceolate, appressed-pubescent on the basal ½–¾ of the margins and midveins, apices scabrous, acuminate, awned, awns 0.5–4 mm; **paleas** 2–3.2(3.5) mm, narrowly elliptic, intercostal region sparsely pubescent, apices acuminate; **anthers** 1.2–2.1 mm, greenish. **Caryopses** 1.7–2 mm, fusiform, brownish. $2n$ = 20, 21.

Muhlenbergia torreyi grows in desert grasslands and open woodlands on sandy mesas, calcareous rock outcrops, and rocky slopes, at elevations of 1000–2450 m. Its range extends from the southwestern United States to northern Mexico. It also grows, as a disjunct, in northwestern Argentina.

31. **Muhlenbergia arenicola** Buckley [p. 174]
SAND MUHLY

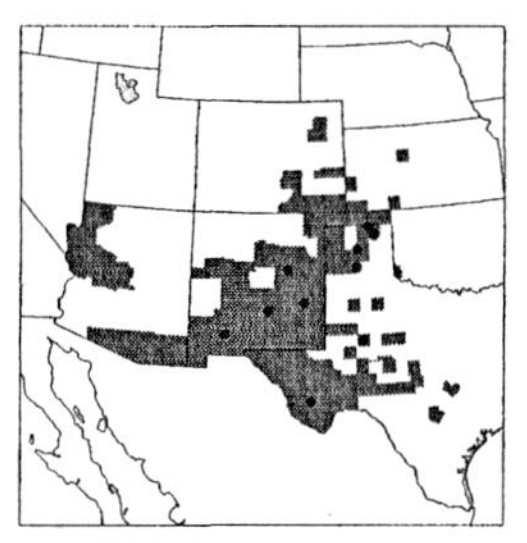

Plants perennial; cespitose, not rhizomatous. **Culms** (15)20–60 (70) cm, somewhat decumbent, 1 or more nodes exposed; **internodes** hispidulous below the nodes. **Leaves** somewhat basally concentrated, most blades not reaching more than (¼)½ of the plant height; **sheaths** usually a little shorter than the internodes, not keeled, scabridulous, margins hyaline, basal sheaths rounded, not becoming spirally coiled when old; **ligules** 2–9 mm, hyaline, acute, lacerate, often with lateral lobes; **blades** 4–10(16) cm long, 1–2.2 mm wide, not arcuate, flat, folded, or involute, scabrous, often glaucous, midveins and margins not thickened, green. **Panicles** 12–30 cm long, 5–20 cm wide, diffuse; **primary branches** 1–10 cm, diverging 30–80° from the rachises, naked basally; **pedicels** 1–4(6) mm. **Spikelets** 2.5–4.2 mm. **Glumes** equal, 1.4–2.5 mm, 1-veined, apices scabridulous, acute to acuminate, minutely erose, unawned or awned, awns to 1 mm; **lemmas** 2.5–4.2 mm, narrowly elliptic, usually purplish, scabrous distally, appressed-pubescent on the lower ½–¾ of the margins and midveins, apices acuminate, awned, awns 0.5–4.2 mm; **paleas** 2.5–3.5 mm, narrowly elliptic, intercostal region sparsely pubescent, apices acuminate, with 2 short (0.1–0.2 mm) awns; **anthers** 1.5–2.1 mm, greenish. **Caryopses** 1.9–2.3 mm, fusiform, brownish. $2n$ = 80, 82.

Muhlenbergia arenicola grows on sandy mesas, limestone benches, and in valleys and open desert grasslands, at elevations of 600–2135 m. Its range extends from the southwestern United States to central Mexico. It also grows, as a disjunct, in northwestern Argentina.

32. **Muhlenbergia pungens** Thurb. *ex* A. Gray [p. 176]
SANDHILL MUHLY

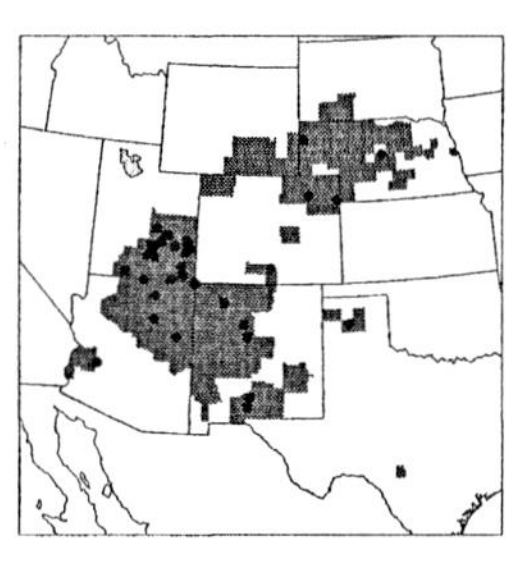

Plants perennial; rhizomatous, not cespitose. **Culms** 10–70 cm, decumbent below; **internodes** cinereous-lanate, glabrous, or scabrous for most of their length, always cinereous-lanate below the nodes. **Sheaths** longer than the internodes, cinereous-lanate below, glabrous and smooth or scabridulous distally; **ligules** 0.2–1 mm, densely ciliate, obtuse, with lateral lobes; **blades** 2–8 cm long, 1–2.2 mm wide, flat to tightly involute, scabrous abaxially, hirsute adaxially, stiff, pungent. **Panicles** (7)8–16(19) cm long, (2)4–14 cm wide, open; **primary branches** 1.5–8 cm, capillary, straight, lower branches

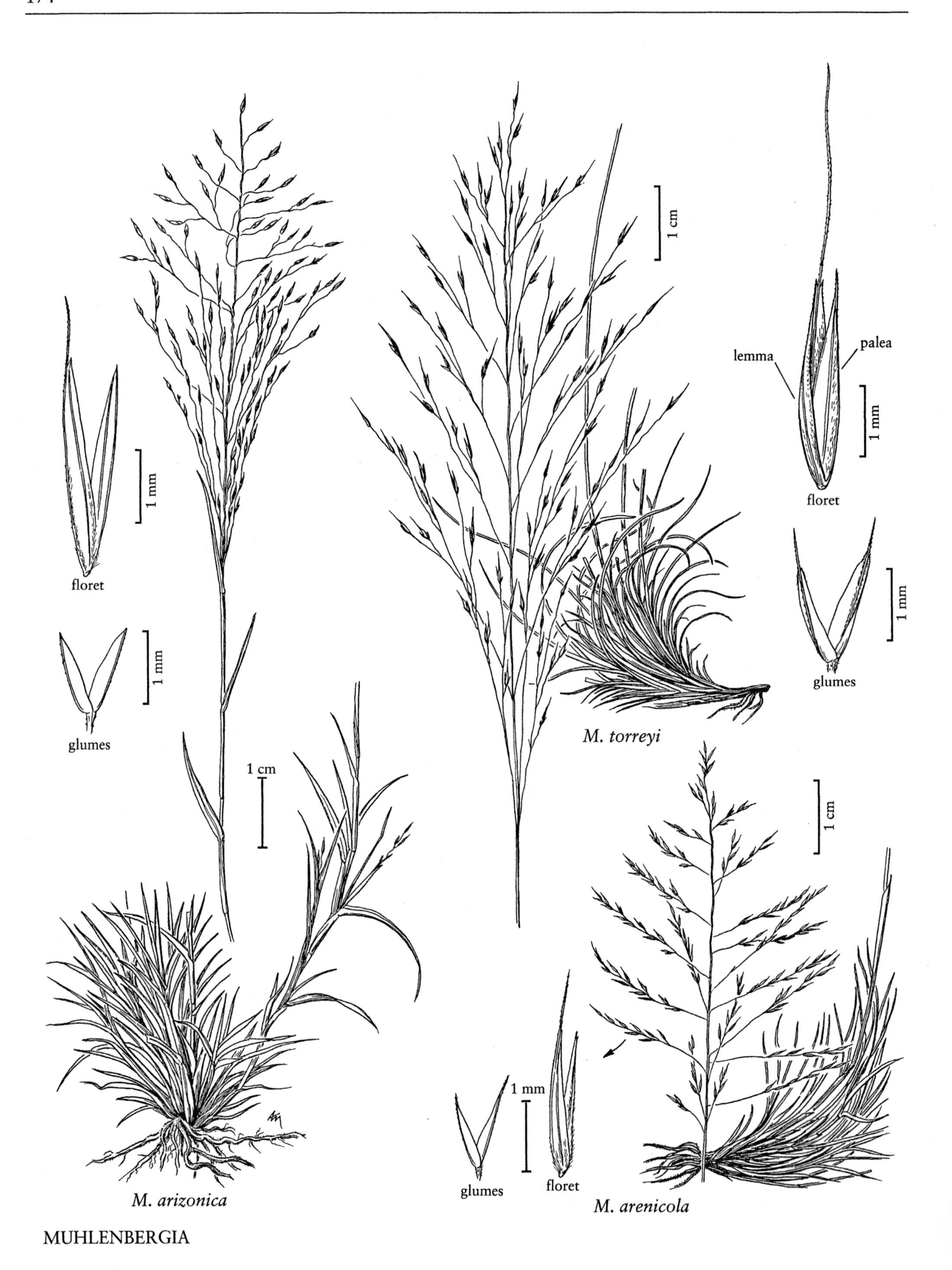

MUHLENBERGIA

diverging 70°–90° from the rachises in mature plants, often appearing fascicled in immature plants; **pedicels** 10–25 mm. **Spikelets** 2.6–4.5 mm. **Glumes** equal, 1.2–3 mm, purplish near the base, smooth or scabridulous distally, 1-veined, acuminate or acute, unawned or awned, awns to 1 mm; **lemmas** 2.6–4.5 mm, lanceolate, purplish, scabridulous distally and on the margins, apices acuminate, awned, awns 1–1.5(2) mm, straight; **paleas** 2.6–4.5 mm, lanceolate, glabrous, acuminate, 2-awned, awns to 1 mm; **anthers** 1.8–2.6 mm, purplish. **Caryopses** 1.8–2.5 mm, fusiform, brownish. $2n$ = 26, 42, 60.

Muhlenbergia pungens grows in loose sandy soils near sand dunes to sandy clay loam slopes and flats in desert shrub and open woodlands, at elevations of 600–2500 m. It is known only from the western and central contiguous United States.

33. **Muhlenbergia dumosa** Scribn. *ex* Vasey [p. 176]
BAMBOO MUHLY

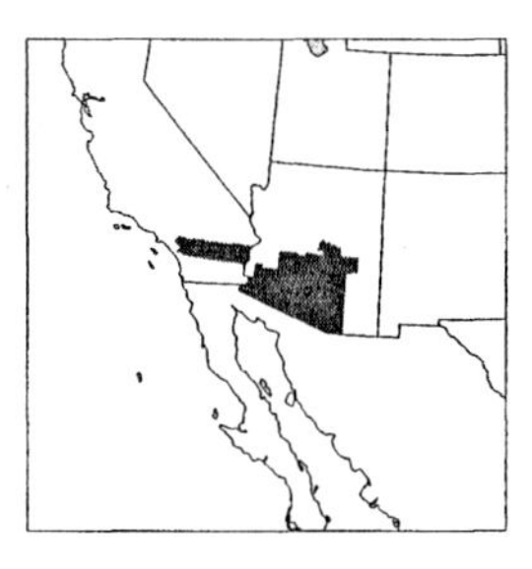

Plants perennial; rhizomatous, loosely cespitose. **Culms** 100–300 cm tall, 3–6 mm thick, erect or ascending, woody below, branching at the middle and upper nodes, branches numerous, fascicled, and spreading; **internodes** glabrous for most of their length, puberulent or glaucous below the nodes. **Sheaths** glabrous, enlarged and flattened basally, somewhat chartaceous; **ligules** 0.2–0.6 mm, membranous, truncate; **blades of the branch leaves** 1.2–8(12) cm long, 0.7–2.2 mm wide, flat or involute, glabrous abaxially, hirtellous adaxially; **blades of the cauline leaves**, particularly the lower cauline leaves, absent or greatly reduced. **Panicles** numerous, terminal on the main culms and the branches, 1–4 cm long, 0.3–1.4 cm wide, lax, inconspicuous; **primary panicle branches** appressed or loosely spreading up to 40° from the rachises; **pedicels** 0.1–1.5 mm. **Spikelets** 2.2–3.1 mm, green or purplish. **Glumes** subequal, 1–1.7 mm, glabrous, 1-veined, acute to acuminate, occasionally mucronate, mucros to 0.5 mm; **lemmas** 2.3–3.1 mm, lanceolate, appressed-pubescent on the calluses and lower portion of the margins, apices acuminate, awned, awns 1–5 mm, flexuous; **paleas** 2.3–3.1 mm, narrowly lanceolate, appressed-pubescent basally, acuminate; **anthers** 1.5–2 mm, purplish. **Caryopses** 1.2–1.6 mm, fusiform, reddish-brown. $2n$ = 40.

Muhlenbergia dumosa grows on rocky slopes, canyon ledges, and cliffs, in areas protected from grazing animals in oak-pine and thorn-scrub forests and oak-gramma savannahs, at elevations of 600–1800 m, from Arizona to southern Mexico.

The bladeless cauline leaves and abundant branching from the middle and upper nodes make 'Bamboo Muhly' a very apt English name. North American Indians used it, after boiling, for chest and bowel ailments. It is native from southern Arizona to southern Mexico, but is also grown as an ornamental in the southwestern United States.

34. **Muhlenbergia villiflora** Hitchc. [p. 176]
HAIRY MUHLY

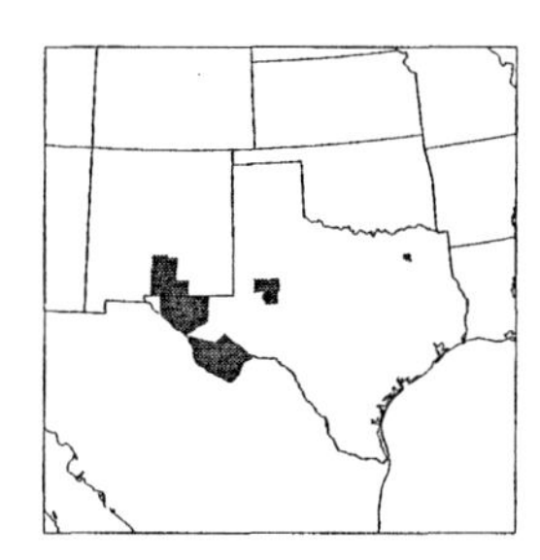

Plants perennial; rhizomatous, not cespitose. **Culms** 4–30 cm tall, to 2 mm thick, erect; **internodes** smooth or nodulose. **Sheaths** shorter than the internodes, smooth to nodulose; **ligules** 0.4–1.5 mm, membranous, acute, erose; **blades** 0.7–3 cm long, 0.2–1.2 mm wide, arcuate-spreading, tightly involute, glabrous abaxially, hirtellous adaxially. **Panicles** 1–5 cm long, 0.1–0.5 cm wide, contracted, not dense, usually completely exserted; **branches** 0.2–1.1 cm, appressed to ascending; **pedicels** 0.1–1.2 mm, setulose. **Spikelets** 1.4–2.5 mm. **Glumes** equal, 0.6–1.8 mm, ½–⅔ as long as the lemmas, glabrous, 1(2, 3)-veined, acute, unawned; **lemmas** 1.4–2.5 mm, lanceolate, green or purplish, midveins and margins densely villous for most of their length, hairs 0.4–1 mm, apices acute, unawned, sometimes mucronate, mucros to 0.5 mm; **paleas** 1.4–2.3 mm, lanceolate, intercostal region densely villous, apices acute; **anthers** 0.9–1.4 mm, yellow, dark green, or purple. **Caryopses** 1–1.4 mm, ellipsoid to fusiform, dark brown. $2n$ = 20, 40.

In the United States, *Muhlenbergia villiflora* grows in open ground with alkaline to calcareous soils and on gypsum rock flats, at elevations of 600–1200 m. It usually forms small, isolated populations.

Plants in the United States belong to **Muhlenbergia villiflora** var. **villosa** (Swallen) Morden. This variety differs from **M. villiflora** Hitchc. var. **villiflora,** which grows in Mexico, in its longer spikelets (1.8–2.5 mm versus 1.4–2.3 mm) and preference for calcareous, rather than gypsiferous, soils.

35. **Muhlenbergia repens** (J. Presl) Hitchc. [p. 178]
CREEPING MUHLY

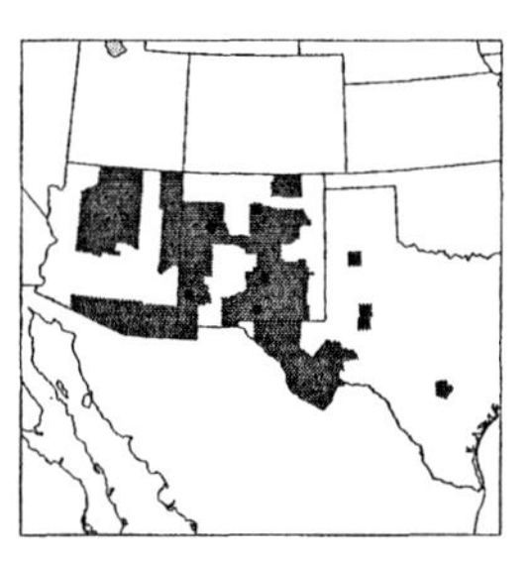

Plants perennial; rhizomatous, not cespitose. **Culms** 5–42 cm tall, 0.5–1 mm thick, decumbent near the base, forming dense mats; **internodes** glabrous, slightly nodulose. **Sheaths** shorter or longer than the

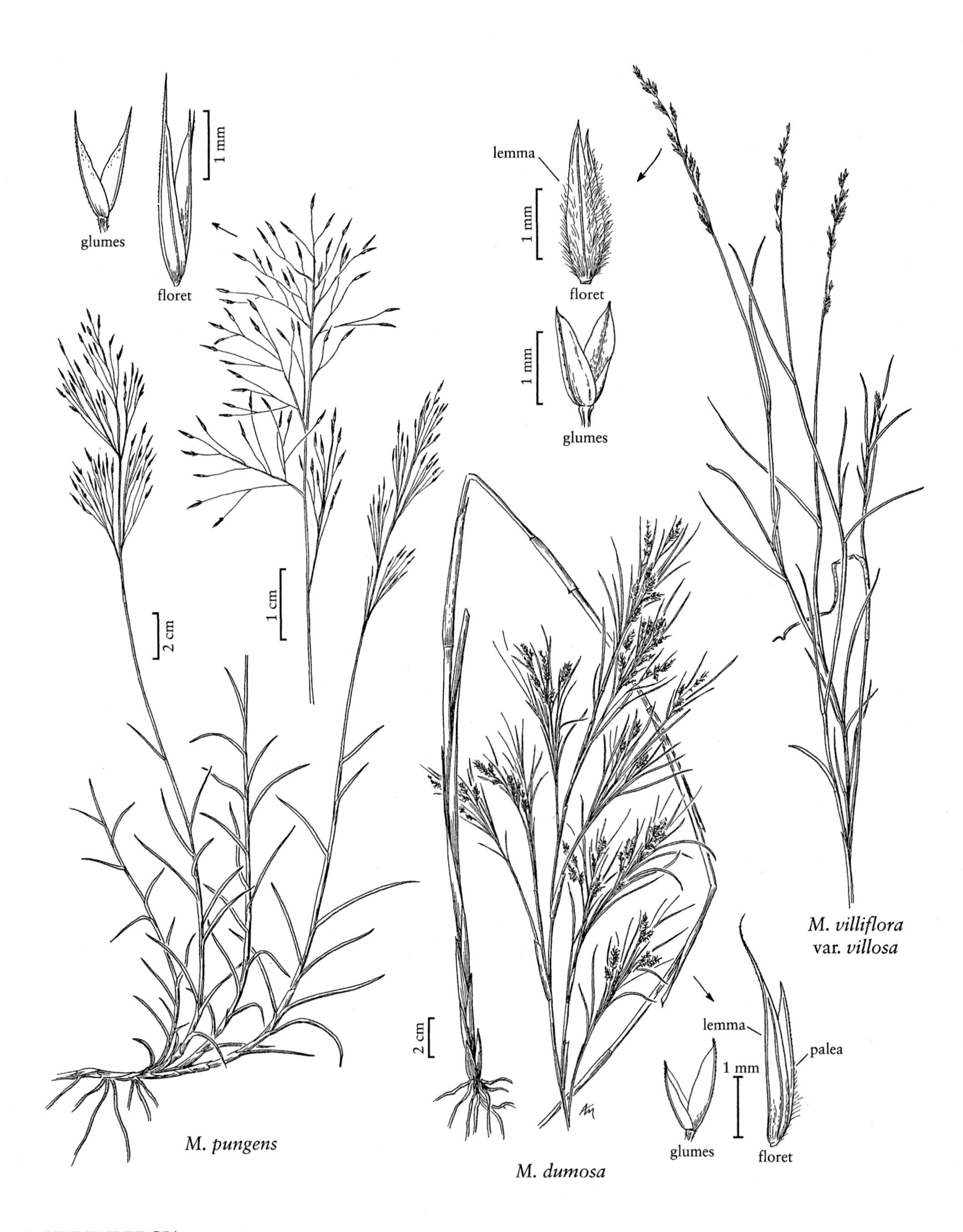

glumes
floret
1 mm
2 cm
1 cm
M. pungens
lemma
1 mm
floret
1 mm
glumes
2 cm
M. dumosa
lemma
palea
1 mm
glumes
floret
M. villiflora
var. villosa

internodes, glabrous; **ligules** 0.1–1(1.8) mm, membranous, truncate, occasionally lacerate; **blades** 0.4–6 cm long, 0.5–1.4 mm wide, involute, glabrous, smooth or scabridulous adaxially. **Panicles** 1–9 cm long, 0.1–0.6 cm wide, contracted, not dense, usually partially included in the upper leaf sheaths; **primary branches** 0.2–3 cm, usually closely appressed at maturity, rarely diverging up to 40° from the rachises; **pedicels** 0.2–3.6 mm, setulose. **Spikelets** 2.6–4.2 mm, occasionally with 2 florets. **Glumes** subequal, 1.1–3.6 mm, from ½ as long as to equaling the lemmas, light green, 1(2–3)-veined, acute, unawned; **lemmas** 2.6–3.2(4.2) mm, lanceolate, dark greenish or mottled, glabrous or the calluses and margins appressed-pubescent, hairs to 0.3 mm, apices scabridulous, attenuate, usually mucronate, mucros 0.1–0.3 mm; **paleas** 2.1–3.3 mm, lanceolate, smooth or scabridulous, acute; **anthers** 0.7–1.4 mm, yellow to purplish. **Caryopses** 1.1–1.5 mm, ellipsoid to ovoid, brownish. $2n$ = 60, 70–72.

Muhlenbergia repens grows in open, sandy meadows, canyon bottoms, calcareous rocky flats, gypsum flats, and on rolling slopes and roadsides, at elevations of 100–3120 m. Its range extends from the southwestern United States to southern Mexico.

36. **Muhlenbergia utilis** (Torr.) Hitchc. [p. 178]

APAREJOGRASS

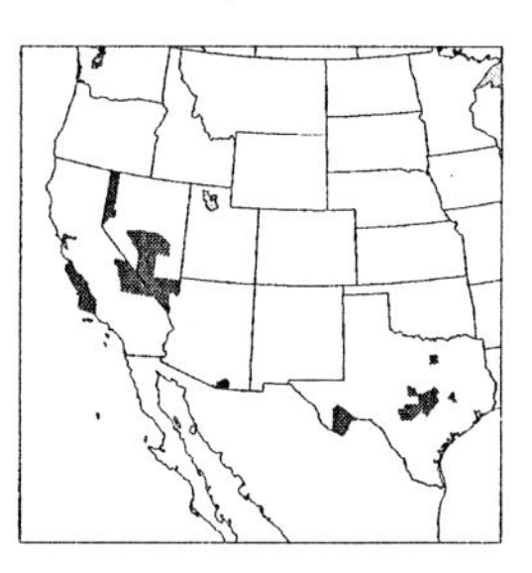

Plants perennial; rhizomatous, not cespitose. **Culms** 7–30 cm tall, 0.5–1 mm thick, erect to decumbent; **internodes** mostly smooth to slightly nodulose, minutely pubescent or glabrous below the nodes. **Sheaths** shorter or longer than the internodes, glabrous; **ligules** 0.2–0.8 mm, membranous, truncate; **blades** 0.5–4.7 cm long, 0.2–1.8 mm wide, usually involute, sometimes flat, often at right angles to the culm, glabrous abaxially, hirtellous adaxially. **Panicles** 1–5 cm long, 0.1–0.4 cm wide, contracted, usually partially included in the upper sheaths, rachises usually visible between the branches; **primary branches** 0.2–1.2 cm, usually closely appressed at maturity, rarely diverging up to 30° from the rachises; **pedicels** 0.1–1.1 mm, glabrous. **Spikelets** 1.4–2.4 mm. **Glumes** subequal, 0.5–1.4 mm, ⅓–½ as long as the lemmas, yellowish to light green, glabrous, 1(2–3)-veined, acute, unawned; **lemmas** 1.3–2.4 mm, lanceolate, green or purplish, glabrous or the calluses and margins appressed-pubescent, hairs shorter than 0.3 mm, apices acute, unawned; **paleas** 1–2 mm, lanceolate, glabrous, acute; **anthers** 0.7–1.4 mm, yellow to purplish. **Caryopses** 0.7–1.2 mm, ellipsoid to ovoid, brown. $2n$ = 20.

Muhlenbergia utilis grows in wet soils along streams, ponds, and depressions in grasslands and alkaline or gypsiferous plains, at elevations of 200–1800 m. Its range extends from the southern United States through Mexico to Costa Rica.

37. **Muhlenbergia richardsonis** (Trin.) Rydb. [p. 178]

MAT MUHLY, MUHLENBERGIE DE RICHARDSON

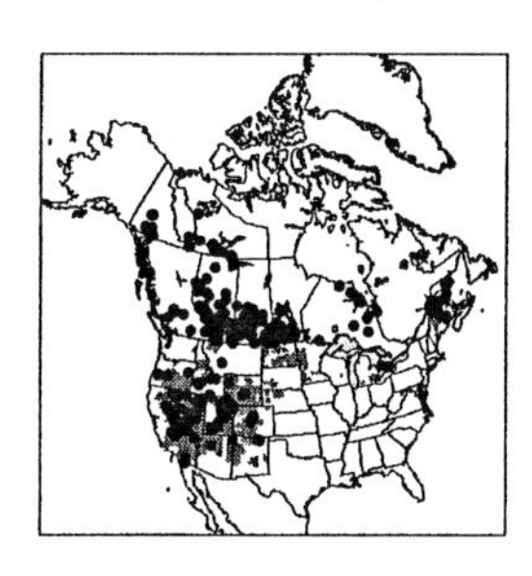

Plants perennial; rhizomatous, not cespitose, often mat-forming. **Culms** 5–30 cm tall, 0.4–1 mm thick, decumbent, geniculate, or erect; **internodes** usually nodulose (occasionally smooth) for most of their length, puberulent or nodulose below the nodes. **Sheaths** shorter or longer than the internodes, glabrous; **ligules** 0.8–3 mm, membranous, acute to truncate, erose; **blades** 0.4–6.5 cm long, 0.5–4.2 mm wide, flat or involute, straight or arcuate-spreading, glabrous abaxially, hirtellous adaxially. **Panicles** 1–15 cm long, 0.1–1.7 cm wide, exserted, narrow or spikelike, rachises usually concealed by the branches; **primary branches** 0.4–5 cm, usually closely appressed at maturity, rarely diverging up to 20° from the rachises; **pedicels** 0.2–2 mm, setulose. **Spikelets** 1.7–3.1 mm, occasionally with 2 florets. **Glumes** subequal, 0.6–2 mm, ⅓–½ as long as the lemmas, green, 1(2)-veined, acute, sometimes mucronate, mucros less than 0.2 mm; **lemmas** 1.7–2.6(3.1) mm, lanceolate, dark greenish, plumbeous, or mottled, glabrous, apices scabridulous, acute to acuminate, sometimes mucronate, mucros to 0.5 mm; **paleas** 1.2–2.4(2.9) mm, lanceolate, acute; **anthers** 0.9–1.6 mm, yellow to purplish. **Caryopses** 0.9–1.6 mm, narrowly ellipsoid, brown. $2n$ = 40.

Muhlenbergia richardsonis grows in open sites in alkaline meadows, prairies, sandy arroyo bottoms, talus slopes, rocky flats and the shores of rivers, at elevations of 60–3300 m. It is the most widespread species of *Muhlenbergia* in the *Flora* region, extending from the Yukon Territory to Quebec in the north and to northern Baja California, Mexico, in the south. Morden and Hatch (1996) reported that it also grows in Alaska, but no voucher specimen has been located.

Muhlenbergia richardsonis is often confused with *M. cuspidata*, which differs in lacking rhizomes and having shorter ligules, and sometimes with *M. filiformis*, which differs in being a weak annual with glabrous internodes and obtuse, erose glumes.

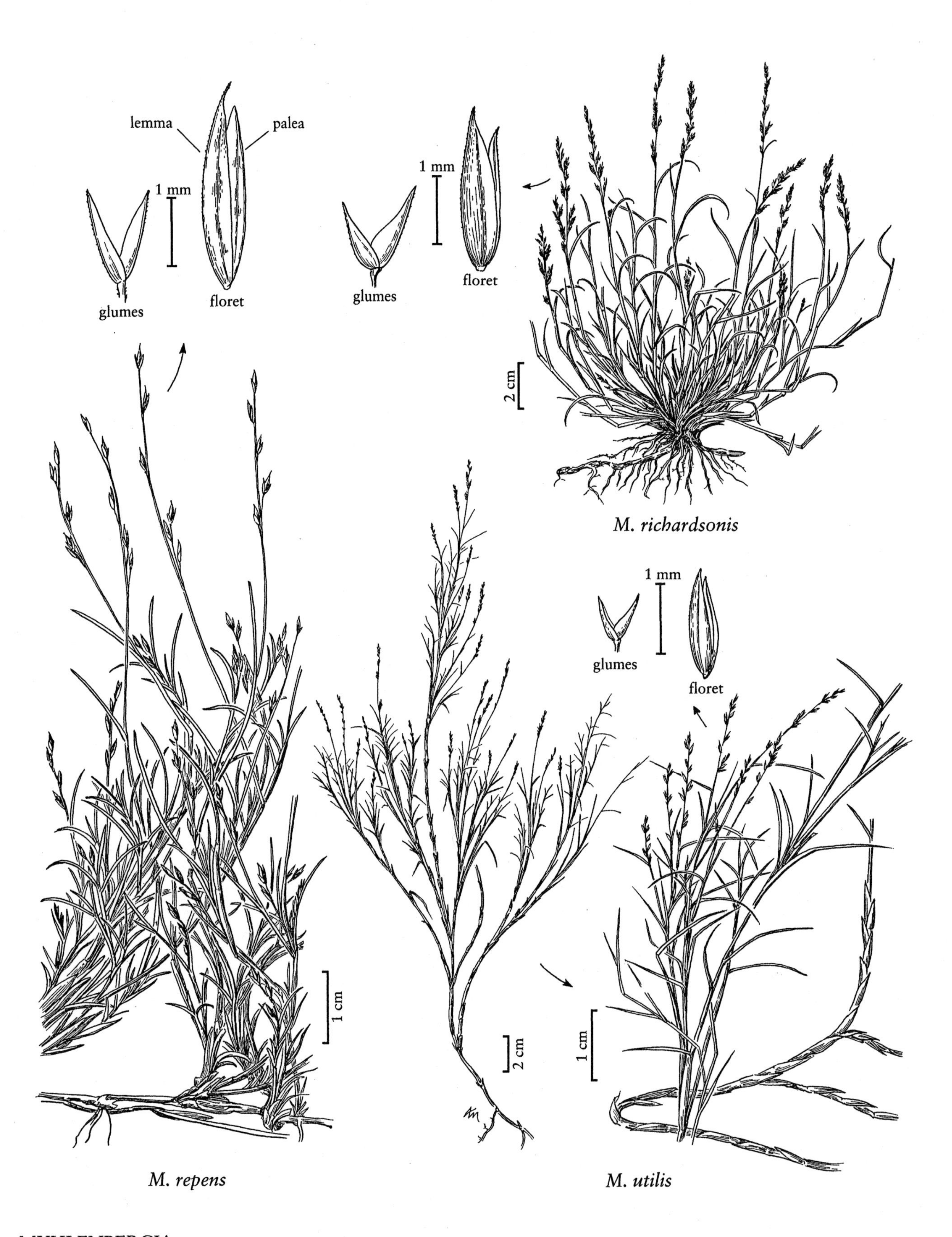

MUHLENBERGIA

38. **Muhlenbergia filiformis** (Thurb. *ex* S. Watson) Rydb. [p. 180]
Pull-Up Muhly

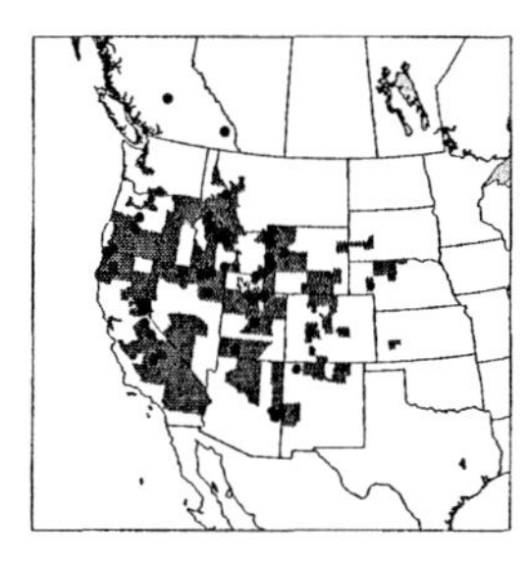

Plants annual (often appearing perennial), tufted. **Culms** (3)5–20(35) cm, erect or geniculate, often rooting at the lower nodes; **internodes** glabrous. **Sheaths** shorter or longer than the internodes, glabrous, smooth or scabridulous; **ligules** 1–3.5 mm, hyaline to membranous, rounded to acute; **blades** 1–4(6) cm long, 0.6–1.6 mm wide, flat or involute, smooth or scabridulous abaxially, scabrous or pubescent adaxially. **Panicles** 1.6–6 cm long, 0.2–0.5 cm wide, spikelike, interrupted near the base, long-exserted; **primary branches** 0.9–1.2 cm, closely appressed at maturity; **pedicels** 1–3 mm, scabrous. **Spikelets** 1.5–3.2 mm. **Glumes** greenish-gray, glabrous, 1-veined, rounded to subacute; **lower glumes** 0.6–1.4 mm; **upper glumes** 0.7–1.7 mm; **lemmas** (1.5)1.8–2.5(3.2) mm, lanceolate, dark greenish, appressed-pubescent on the margins and midveins, hairs shorter than 0.3 mm, apices scabridulous, acute to acuminate, unawned, sometimes mucronate, mucros shorter than 1 mm; **paleas** 1.6–2.6(3.1) mm, lanceolate, scabridulous distally; **anthers** 0.5–1.2 mm, purplish. **Caryopses** 0.9–1.5 mm, fusiform, reddish-brown. $2n$ = 18.

Muhlenbergia filiformis grows in open, moist meadows, on gravelly lake shores, along stream banks, and in moist humus near thermal springs, at elevations of 1060–3050 m. It is usually associated with yellow pine forests, but also grows in many other plant communities. Its range extends into northern Mexico.

Muhlenbergia filiformis resembles *M. richardsonis*, but differs in having glabrous internodes and subacute apices. Large, robust specimens have been referred to *M. simplex* Scribn. or *M. filiformis* var. *fortis* E.H. Kelso but, until there is more evidence to the contrary, it seems best to treat such plants as representing an extreme of the variation within *M. filiformis*.

39. **Muhlenbergia torreyana** (Schult.) Hitchc. [p. 180]
New Jersey Muhly

Plants perennial; rhizomatous, not cespitose. **Culms** 30–75 cm, compressed and keeled; **internodes** mostly glabrous, strigose on the keels and below the nodes. **Sheaths** strigose on the keels, basal sheaths much shorter than those above; **ligules** 0.3–1 mm, firm, truncate, ciliate; **blades** 6–20 cm long, 1–3.5 mm wide, conduplicate, scabrous on both surfaces, tapering to a fine sharp point. **Panicles** 10–28 cm long, 4–8 cm wide, cylindrical, open; **primary branches** 3–10 cm long, 0.05–0.1 mm thick, capillary, diverging 30–40° from the rachises, never appearing fascicled; **pedicels** 1.5–9 mm, usually longer than the spikelets. **Spikelets** 1.1–2.2 mm, occasionally with 2 florets. **Glumes** equal, 1–2 mm, purplish, scabridulous, 1-veined, acute, unawned; **lemmas** 1.1–2.2 mm, lanceolate, plumbeous, scabridulous, apices acute, unawned; **paleas** 1–2.1 mm, lanceolate, scabridulous, acute; **anthers** 1–1.4 mm, orange-yellow, turning purple at maturity. **Caryopses** about 1 mm, fusiform, brownish. $2n$ = unknown.

Muhlenbergia torreyana grows in perennially wet or moist, usually seasonally inundated habitats such as the sphagnous margins of shallow ponds and seasonally wet depressions, often within pine-oak or oak barrens and at elevations of 0–150 m. Kartesz and Meacham (1999) report that it appears to have been eliminated from New York, Delaware, and Georgia but that, in addition to the locations shown on the map, it grows in Maryland and New Jersey. Hitchcock (1951) reported that it grew in Kentucky, but no specimens documenting its presence there have been located. It is rare even in those states where it is still growing.

Morphologically, *Muhlenbergia torreyana* resembles the western *M. asperifolia* but differs in its strigose, strongly compressed, keeled culms and less strongly divergent panicle branches.

40. **Muhlenbergia asperifolia** (Nees & Meyen *ex* Trin.) Parodi [p. 180]
Scratchgrass

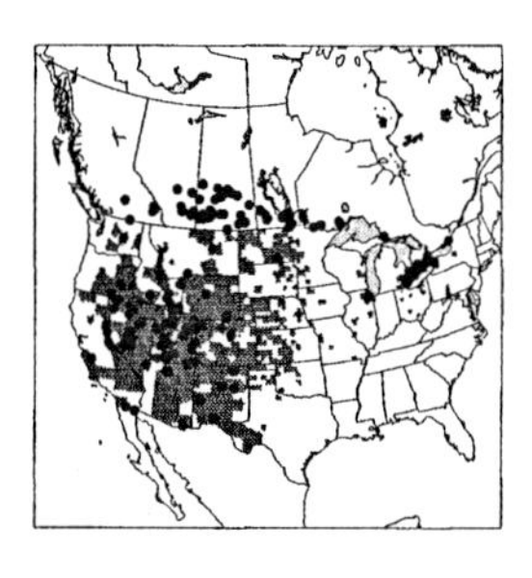

Plants perennial; rhizomatous, not cespitose, occasionally stoloniferous. **Culms** 10–60(100) cm, decumbent-ascending, bases somewhat compressed-keeled; **internodes** glabrous, shiny below the nodes. **Sheaths** glabrous, margins hyaline; **ligules** 0.2–1 mm, firm, truncate, ciliate, without lateral lobes; **blades** 2–7(11) cm long, 1–2.8(4) mm wide, flat, occasionally conduplicate, smooth or scabridulous abaxially, scabridulous adaxially, margins and midveins not conspicuously thickened, greenish, apices acute, not sharp. **Panicles** 6–21 cm long, 4–16 cm wide, broadly ovoid, open; **primary branches** 3–12 cm, capillary, lower branches spreading 30–90° from the rachises, never appearing fascicled; **pedicels** 3–14 mm, longer than the spikelets. **Spikelets** 1.2–2.1 mm, occasionally with 2 or 3 florets. **Glumes** equal, 0.6–1.7 mm, purplish, scabridulous, particularly on the veins,

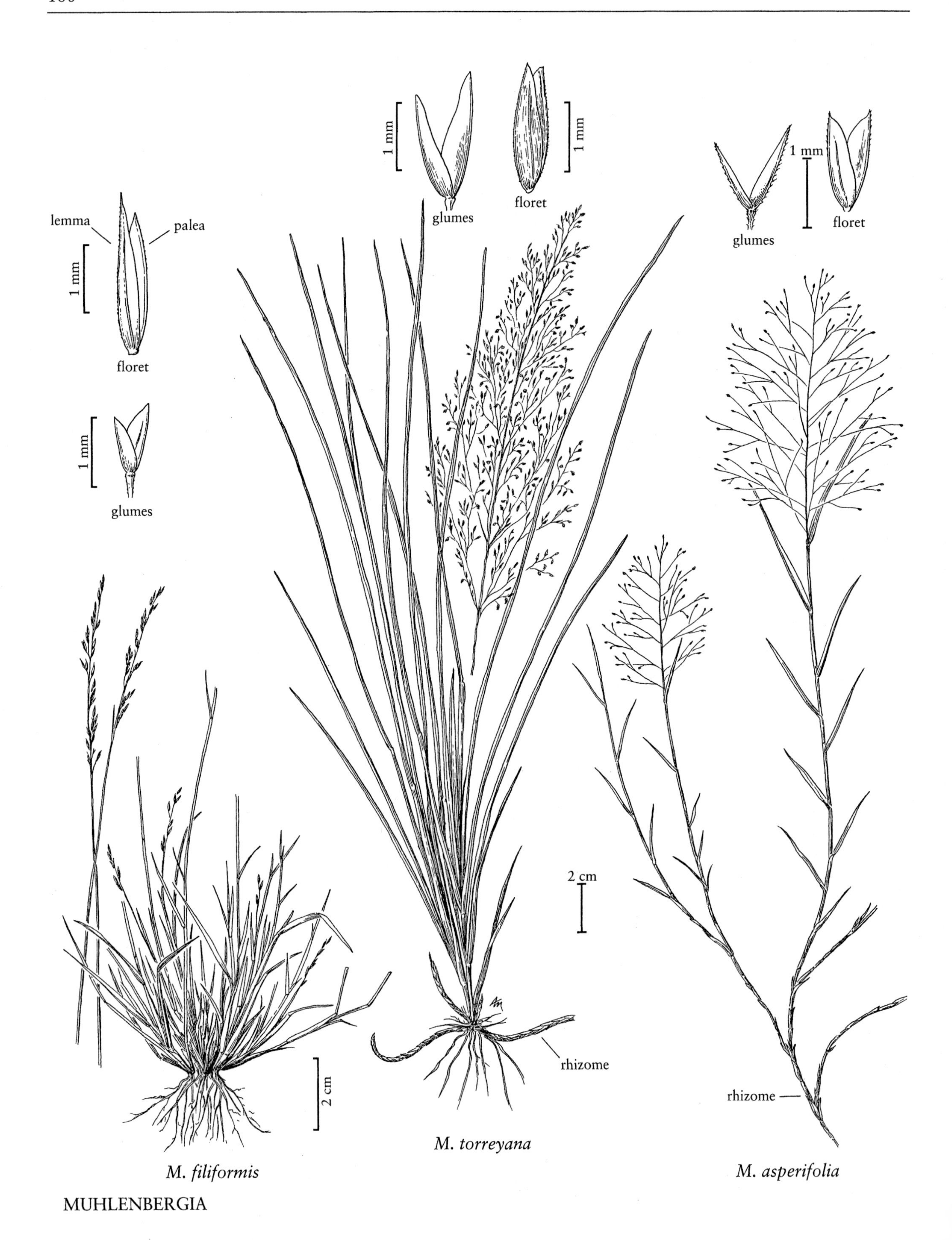

MUHLENBERGIA

1-veined, apices acute; **lemmas** 1.2–2.1 mm, lanceolate to oblong-elliptic, somewhat plumbeous, glabrous, usually smooth, occasionally scabridulous near the apices, apices acute, unawned or mucronate, mucros to 0.3 mm; **paleas** 1.2–2.1 mm, lanceolate, glabrous, acute; **anthers** 1–1.3 mm, greenish-yellow to purplish at maturity. **Caryopses** 0.8–1 mm, fusiform, brownish. $2n$ = 20, 22, 28.

Muhlenbergia asperifolia grows in moist, often alkaline meadows, playa margins, and sandy washes, on grassy slopes, and around seeps and hot springs, at elevations of 55–3000 m. Its geographic range includes northern Mexico. *Muhlenbergia asperifolia* is morphologically similar to the southeastern *M. torreyana*, but differs in having glabrous, weakly compressed culms and more widely divergent panicle branches.

The caryopses of *Muhlenbergia asperifolia* are frequently infected by a smut, *Tilletia asperifolia* Ellis & Everhart, which produces a globose body filled with blackish-brown spores.

41. **Muhlenbergia arenacea** (Buckley) Hitchc. [p. 182]
EAR MUHLY

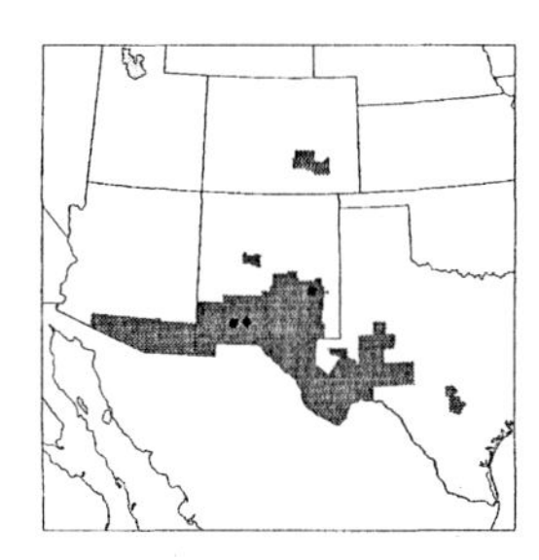

Plants perennial; rhizomatous, not cespitose. **Culms** 10–30(40) cm, decumbent, terete to somewhat compressed-keeled near the base; **internodes** scabridulous below the nodes. **Sheaths** about ½ as long as the internodes, margins hyaline; **ligules** 0.5–2 mm, hyaline, with lateral, 1–2 mm lobes; **blades** 0.7–4(6) cm long, 0.5–1.7 mm wide, flat, occasionally folded, tapering, scabrous abaxially, strigulose adaxially, margins and midveins thickened, whitish, apices narrow, often sharp. **Panicles** 5–15 cm long, 4–14 cm wide, broadly ovoid, open; **primary branches** 2–8 cm, capillary, straight to slightly flexuous, diverging 45–80(100)° from the rachises, never appearing fascicled, naked proximally; **pedicels** 1–11 mm, usually longer than the spikelets. **Spikelets** 1.5–2.6 mm, occasionally with 2 florets. **Glumes** equal, 0.9–2 mm, 1-veined, usually acute to acuminate, occasionally erose and mucronate, mucros to 0.2 mm; **lemmas** 1.5–2.5 mm, lanceolate to oblong-elliptic, plumbeous to purplish, sparsely appressed-pubescent on the lower ½ of the margins and midveins, hairs to 0.3 mm, apices acute to obtuse, sometimes shallowly bilobed, mucronate, mucros to 0.3 mm; **paleas** 1.5–2.6 mm, lanceolate, glabrous, obtuse to acute; **anthers** 1–1.5 mm, yellowish to purplish. **Caryopses** 1–1.3 mm, elliptic, brownish. $2n$ = unknown.

Muhlenbergia arenacea grows in sandy flats, plains, alluvial fans, washes, depressions, and alkaline mesas in open grasslands, at elevations of 1000–2200 m. Its range extends from the southwestern United States into northern Mexico.

42. **Muhlenbergia uniflora** (Muhl.) Fernald [p. 182]
BOG MUHLY, MUHLENBERGIE UNIFLORE

Plants perennial; loosely matted. **Culms** 5–45 cm, arising from the bases of old depressed culms, compressed-keeled, developing branches below the lower leaf nodes; **internodes** mostly glabrous, sometimes minutely puberulent below the nodes. **Sheaths** longer than the internodes, keeled, keels scabridulous, not becoming papery or spirally coiled when old; **ligules** 0.5–1.5 mm, membranous, truncate to obtuse, erose, without lateral lobes; **blades** 1–15 cm long, 0.8–2 mm wide, flat to conduplicate, smooth or scabridulous abaxially, hirtellous adaxially, midveins thickened and whitish proximally. **Panicles** 2–20 cm long, (0.2)2.5–6 cm wide, diffuse; **primary branches** 0.4–6 cm long, about 0.1 mm thick, ascending, diverging 10–60° from the rachises, naked basally; **pedicels** 0.2–7 mm, glabrous. **Spikelets** 1.3–2.1 mm, dark purplish to plumbeous, occasionally with 2 florets. **Glumes** equal, 0.4–1.3 mm, glabrous, 1-veined, apices scabrous, acute to obtuse, sometimes erose or notched, unawned; **lemmas** 1.2–2 mm, oblong-elliptic, dark purplish to plumbeous, glabrous, faintly 3-veined, apices acute to obtuse, unawned; **paleas** 1.3–2.1 mm, oblong-elliptic, glabrous, acute to obtuse; **anthers** 0.6–0.9 mm, dark purple. **Caryopses** 0.6–0.8 mm, ovoid, brownish. $2n$ = 42.

Muhlenbergia uniflora grows in bogs, wet meadows, and lake shores in sandy or peaty, often acidic, soils, at elevations of 0–650 m. It is native to eastern North America, but was collected once in British Columbia, probably having been introduced from ship ballast, and was recently collected from a commercial cranberry bog in Oregon. The collection from Texas may also be an introduction.

43. **Muhlenbergia filiculmis** Vasey [p. 182]
SLIMSTEM MUHLY

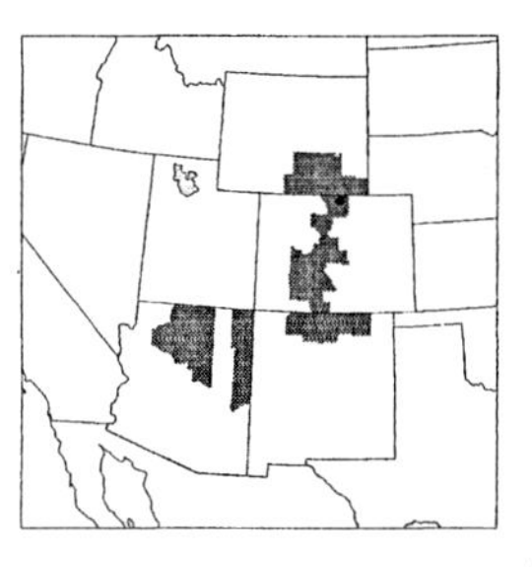

Plants perennial; cespitose, not rhizomatous. **Culms** 5–30(40) cm, erect, rounded near the base; **internodes** smooth or scabridulous. **Sheaths** longer than the lower internodes, glabrous, becoming flattened and papery or ribbonlike at maturity; **ligules** 2–4(5) mm, membranous, acute; **blades** 2–6 cm long, 0.4–1.6 mm wide, tightly involute, filiform, stiff, scabrous abaxially, sparsely hirtellous adaxially, apices sharp. **Panicles** 1.5–7 cm long, 0.4–2 cm wide, contracted, sometimes

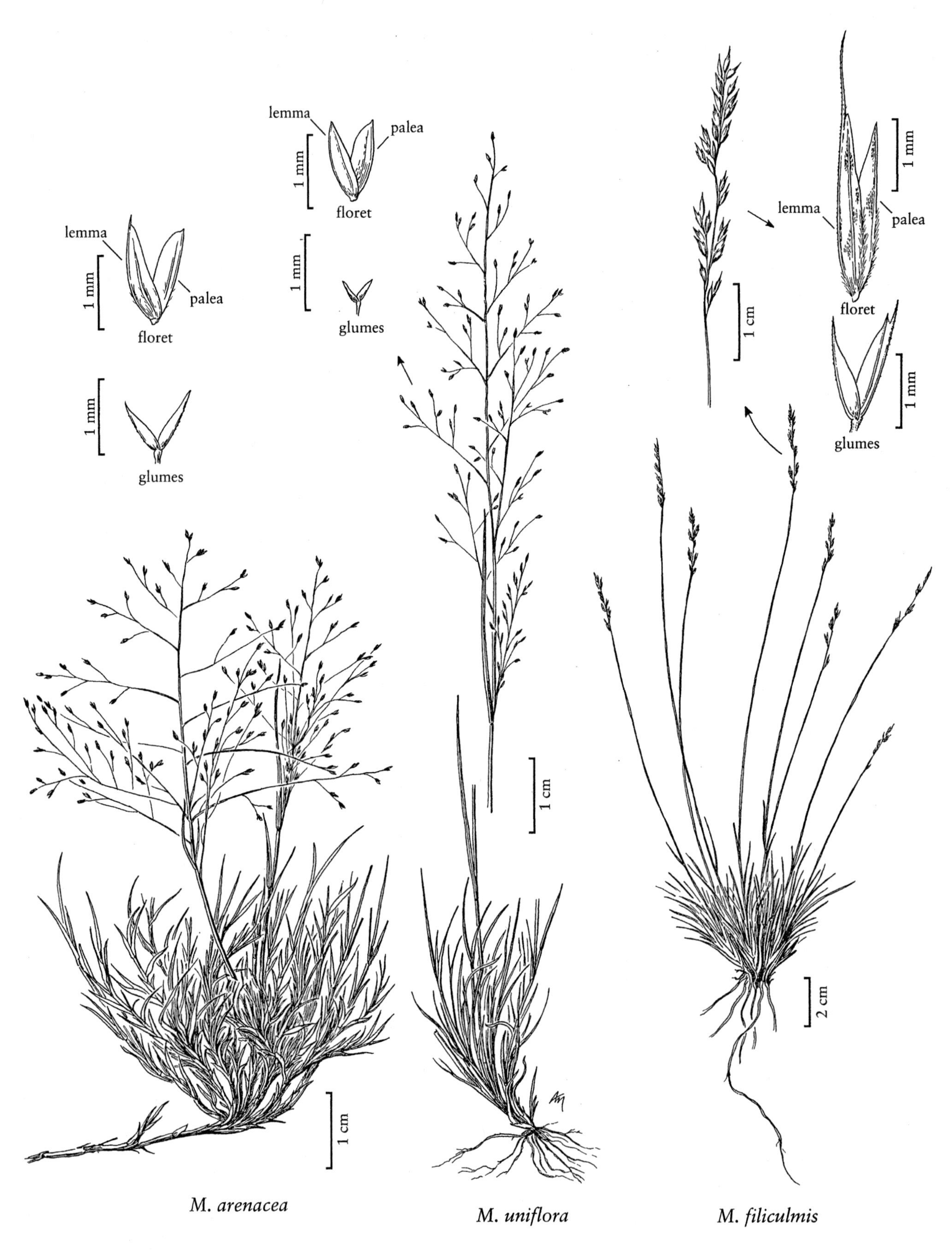
lemma
palea
1 mm
floret
1 mm
glumes
lemma
palea
1 mm
floret
1 mm
glumes
1 cm
1 cm
lemma
palea
1 mm
floret
1 mm
glumes
1 cm
2 cm
1 cm
M. arenacea
M. uniflora
M. filiculmis

dense; **primary branches** 0.1–2.8 cm, mostly tightly appressed or diverging up to 30° from the rachises; **pedicels** 0.1–2 mm, scabrous. **Spikelets** 2.2–3.5 mm. **Glumes** subequal, 0.8–2.5 mm, scabridulous distally; **lower glumes** 1-veined, awned, awns to 1.6 mm; **upper glumes** usually 3-veined, apices truncate to acute, 3-toothed, teeth sometimes shortly awned, awns to 0.6 mm; **lemmas** 2.2–3.5 mm, lanceolate, yellowish mottled with dark green, sparsely appressed-pubescent on the lower portion of the midveins and margins, hairs to 0.4 mm, apices scabridulous, acute to acuminate, awned, awns 1–5 mm, straight or flexuous; **paleas** 2.2–3.5 mm, lanceolate, acute to acuminate; **anthers** 1.5–2 mm, purplish. **Caryopses** 1.3–1.6 mm, fusiform, brownish. 2*n* = unknown.

Muhlenbergia filiculmis grows on rocky slopes, dry meadows, and dry gravelly flats in forest openings and grasslands, at elevations of 2500–3500 m in the southern Rocky Mountains and northern Arizona. Kartesz and Meacham (1999) report it as occurring in North Dakota and Utah, but neither of the sources cited (McGregor et al. 1986; Welsh et al. 1993) supports its occurrence in these states.

It is sometimes difficult to distinguish *Muhlenbergia filiculmis* from *M. montana*, but that species has longer spikelets and lemma awns, and leaf blades that are flatter and not sharply tipped.

44. **Muhlenbergia jonesii** (Vasey) Hitchc. [p. 184]
MODOC MUHLY

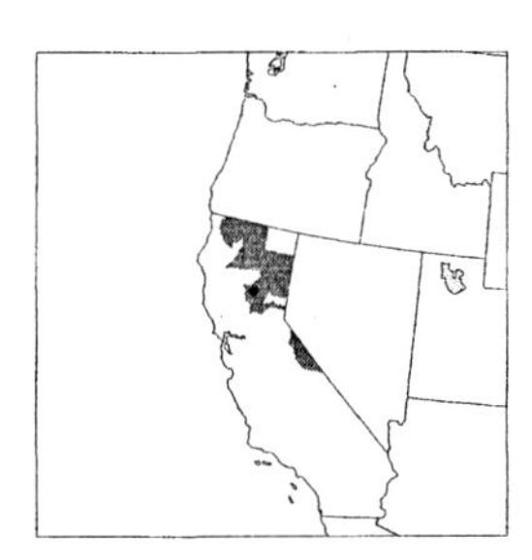

Plants perennial; tightly cespitose, not rhizomatous. **Culms** 18–50 cm, erect, rounded near the base; **internodes** glabrous. **Sheaths** glabrous, bases becoming flattened and papery, lower sheaths longer than the internodes; **ligules** 2–5 mm, membranous, acute to acuminate; **blades** (5)6–12 cm long, 1–2.5 mm wide, flat, becoming loosely involute to subfiliform, scabrous abaxially, hirsute adaxially, apices not sharp. **Panicles** 4–15 cm long, 1.5–4 cm wide, not dense; **primary branches** 0.5–5 cm, appressed or diverging up to 40° from the rachises; **pedicels** 0.5–6 mm, flattened, scabrous. **Spikelets** 2.8–3.5 mm. **Glumes** subequal, 0.6–1.8 mm, scabridulous distally, truncate to obtuse, unawned; **lower glumes** 1-veined; **upper glumes** 3-veined, 3-toothed, often erose; **lemmas** 2.8–3.5 mm, lanceolate, loosely pubescent on the basal ⅓ of the midveins and margins, hairs to 0.6 mm, apices scabridulous, acute, mucronate, mucros shorter than 1 mm; **paleas** 2.8–3.5 mm, lanceolate, acute; **anthers** 1.4–2.3 mm, purple. **Caryopses** 1.6–1.8 mm, fusiform, light brown. 2*n* = 20.

Muhlenbergia jonesii is endemic to northern California. It grows on open slopes, pumice flats, and in openings in pine forests, at elevations of 1130–2130 m.

45. **Muhlenbergia straminea** Hitchc. [p. 184]
SCREWLEAF MUHLY

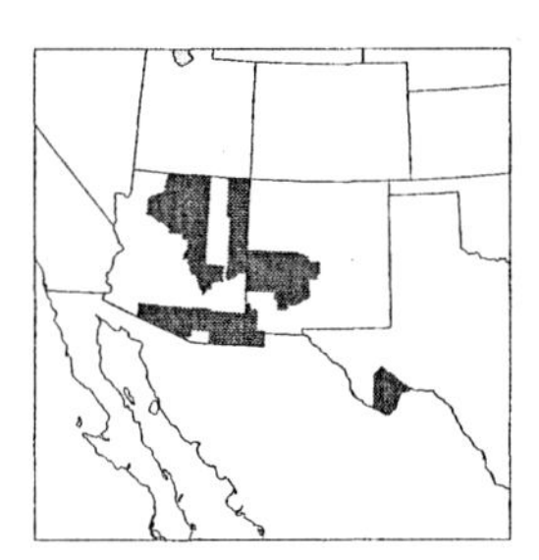

Plants perennial; cespitose. **Culms** 25–70 cm, erect, rounded near the base; **internodes** glabrous. **Sheaths** glabrous, stiff, becoming flattened, ribbonlike or papery, and conspicuously spirally coiled when old; **ligules** (6)10–20 mm, hyaline, acuminate, lacerate; **blades** 7–25 cm long, 1–4 mm wide, flat to involute, scabrous abaxially, spiculate adaxially. **Panicles** 8–25 cm long, 0.5–3 cm wide, not dense; **primary branches** 0.6–8 cm, appressed or diverging up to 30° from the rachises; **pedicels** 0.2–5 mm, scabrous. **Spikelets** 3.5–7 mm, yellowish to pale greenish. **Glumes** (3)3.5–6(7) mm, scabridulous, unawned or awn-tipped; **lower glumes** shorter than the upper glumes, 1-veined; **upper glumes** equaling or exceeding the florets, 3-veined, acuminate to acute, occasionally 3-toothed, awned, awns to 1.5 mm; **lemmas** 3.5–5.5(6) mm, lanceolate, pubescent on the lower ½ of the midveins and margins, hairs to 1 mm, apices scabrous, acuminate, awned, awns 12–27 mm, flexuous; **paleas** 3.5–5.5 mm, lanceolate, pilose between the veins, apices scabridulous, acuminate; **anthers** 2–3.5 mm, purple. **Caryopses** 1.9–2 mm, fusiform, light brown. 2*n* = unknown.

Muhlenbergia straminea grows on rolling, rocky slopes, volcanic tuffs, canyon bottoms, and ridges, usually in open pine forests, at elevations of 1800–2600 m. It is known only from the southwestern United States.

46. **Muhlenbergia montana** (Nutt.) Hitchc. [p. 184]
MOUNTAIN MUHLY

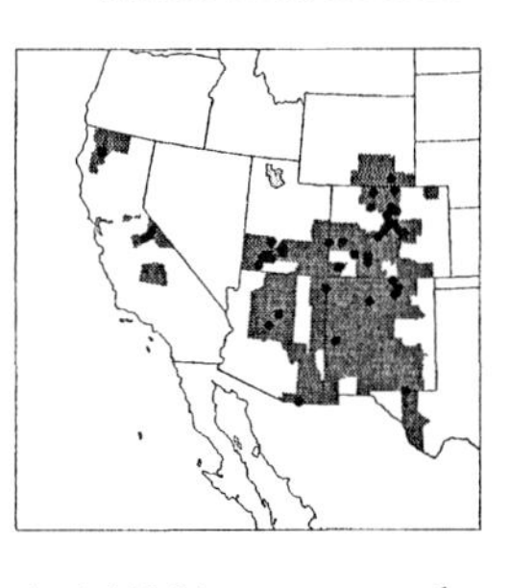

Plants perennial; cespitose, not rhizomatous. **Culms** 10–80 cm, erect, rounded near the base; **internodes** glabrous. **Sheaths** smooth or scabridulous, becoming flattened, papery, and occasionally spirally coiled when old, lower sheaths longer than the internodes; **ligules** 4–14(20) mm, membranous, acute to acuminate; **blades** 6–25 cm long, 1–2.5 mm wide, flat, becoming involute, scabrous abaxially, hirsute adaxially. **Panicles** 4–25 cm long, (1)2–6 cm wide, not dense; **primary branches** 0.5–10 cm, appressed or diverging up to 40° from the rachises; **pedicels** 0.5–6.5 mm, scabrous. **Spikelets** 3–7

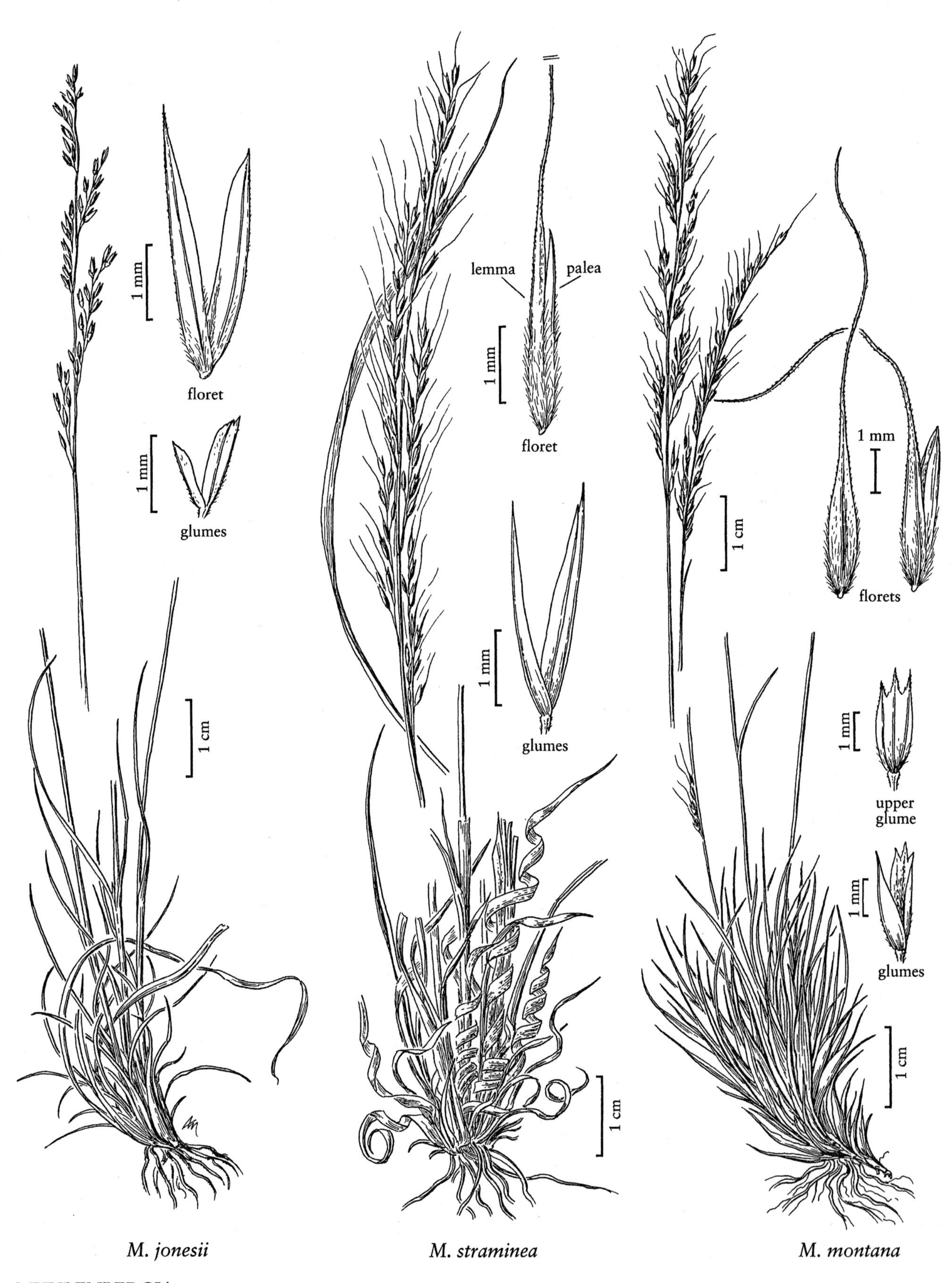

M. jonesii *M. straminea* *M. montana*

mm. **Glumes** subequal, (1)1.5–3.2(4) mm, smooth or scabridulous distally; **lower glumes** 1-veined, sometimes with a less than 1 mm awn; **upper glumes** ⅓–⅔ as long as the lemmas, 3-veined, truncate to acute, 3-toothed, teeth sometimes awned, awns to 1.6 mm; **lemmas** 3–4.5(7) mm, lanceolate, loosely to densely appressed-pubescent on the lower portion of the mid-veins and margins, hairs to 0.8 mm, apices acute to acuminate, awned, awns (2)6–25 mm, flexuous; **paleas** 3–4.5(7) mm, lanceolate, acute to acuminate; **anthers** 1.5–2.3 mm, purplish. **Caryopses** 1.8–2 mm, fusiform, light brown. 2*n* = 20, 40.

Muhlenbergia montana grows on rocky slopes and ridge tops and in dry meadows and open grasslands, at elevations of 1400–3500 m. Its range extends from the western United States to Guatemala. *Muhlenbergia montana* is sometimes difficult to distinguish from *M. filiculmis*, but that species has shorter spikelets and lemma awns and tightly involute or filiform, sharp blades.

47. **Muhlenbergia crispiseta** Hitchc. [p. 186]
MEXICALI MUHLY

Plants annual; tufted. **Culms** 7–16 cm. **Sheaths** longer than the internodes, scabridulous or smooth, margins membranous; **ligules** 1.3–2 mm, membranous, rounded; **blades** 1–5 cm long, 0.7–1.4 mm wide, flat or involute, scabridulous or smooth abaxially, shortly pubescent adaxially. **Panicles** 1.8–4.5 cm long, 1.5–3 cm wide, long-exserted; **primary branches** 1.5–2.8 cm, ascending; **pedicels** 0.4–2 mm, often curved, scabrous. **Spikelets** 1.7–2.2 mm. **Glumes** whitish, mostly smooth, scabrous on the veins; **lower glumes** 1.2–1.6 mm, 1-veined, acute; **upper glumes** 1.6–1.8 mm, wider than the lower glumes, 2- or 3-veined, truncate, 2- or 3-toothed; **lemmas** 1.7–2.2 mm, lanceolate, widest near the middle, whitish with dark green patches, densely hairy on the calluses and lower portion of the lemma bodies, hairs to 0.5 mm, apices glabrous, acuminate, minutely bifid, awned, awns 8–18 mm, sinuous to crisped or curled, olive-green; **paleas** 1.1–1.7 mm, lanceolate, intercostal region loosely pilose on the proximal ⅔, apices acuminate; **anthers** 0.4–0.7 mm, purplish-red. **Caryopses** 0.5–1.1 mm, ellipsoid, brownish. 2*n* = 20.

Muhlenbergia crispiseta grows on rock outcrops, in rocky drainages, and on white tablelands, on soils derived from calcareous parent materials in pine-oak and pinyon-juniper woodlands, at elevations of 1900–2600 m. It is basically a Mexican species, with a disjunct population in Brewster County, Texas.

48. **Muhlenbergia peruviana** (P. Beauv.) Steud. [p. 186]
PERUVIAN MUHLY

Plants annual; tufted. **Culms** 3–27 cm, erect, glabrous. **Sheaths** usually longer than the internodes, smooth or scabridulous; **ligules** 1.5–3 mm, membranous, acute; **blades** 1–5 cm long, 0.6–1.5 mm wide, flat to involute, smooth or scabridulous abaxially, sometimes shortly pubescent adaxially. **Panicles** 2–8 cm long, 0.3–3.4 cm wide, contracted or open; **primary branches** 1–5 cm, diverging up to 80° from the rachises; **pedicels** 0.4–5 mm, smooth or scabrous. **Spikelets** 1.4–4.2 mm. **Glumes** smooth or scabridulous; **lower glumes** 0.8–2.8 mm, narrow to broadly lanceolate, 1-veined, acute, often awn-tipped; **upper glumes** 0.9–3 mm, wider than the lower glumes, lanceolate, (1)2–3-veined, truncate to acute, 2- or 3-toothed; **lemmas** 1.4–4.2 mm, widest near the base, purplish mottled with dark green, hairy on the calluses and lower ⅔ of the lemma bodies, hairs to 0.5 mm, apices acuminate, usually bifid and awned from between the teeth, teeth to 0.5 mm, awns 3–10 mm, flexuous, purplish; **paleas** 1.3–3.8 mm, narrowly lanceolate, acuminate to subacute; **anthers** 0.5–1 mm, purplish to yellowish. **Caryopses** 1–1.6 mm, fusiform, brownish. 2*n* = 30.

Muhlenbergia peruviana grows in open gravelly flats, meadows, rock outcrops, sandy washes, gravelly drainages, rocky slopes, disturbed road cuts, and volcanic flats, in yellow pine forest associations, at elevations of 2000–4600 m. Its primary distribution is to the south of the *Flora* region, extending from the southwestern United States through Mexico to Ecuador, Peru, Bolivia, and Argentina.

As treated here, *Muhlenbergia peruviana* includes what are sometimes identified as *M. pulcherrima* Scribn. *ex* Beal, *M. pusilla* Steud., and *M. peruviana s. s.* There are, however, numerous intermediates among the three extremes represented by these names.

49. **Muhlenbergia emersleyi** Vasey [p. 186]
BULLGRASS

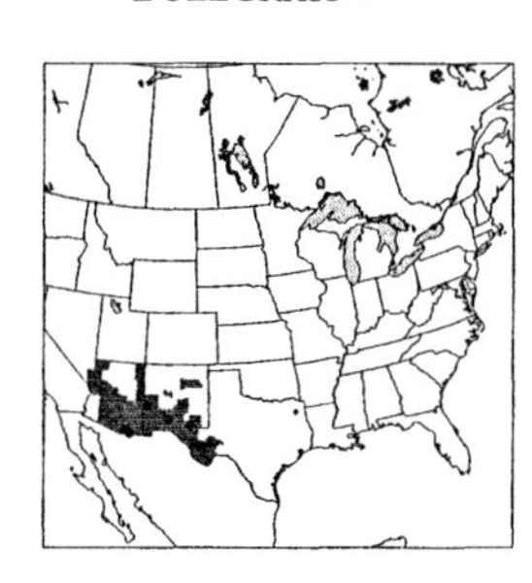

Plants perennial; cespitose, not rhizomatous. **Culms** (50)80–150 cm, stout, erect, not conspicuously branched; **internodes** smooth for most of their length, smooth or scabridulous below the nodes. **Sheaths** shorter or longer than the internodes, glabrous or

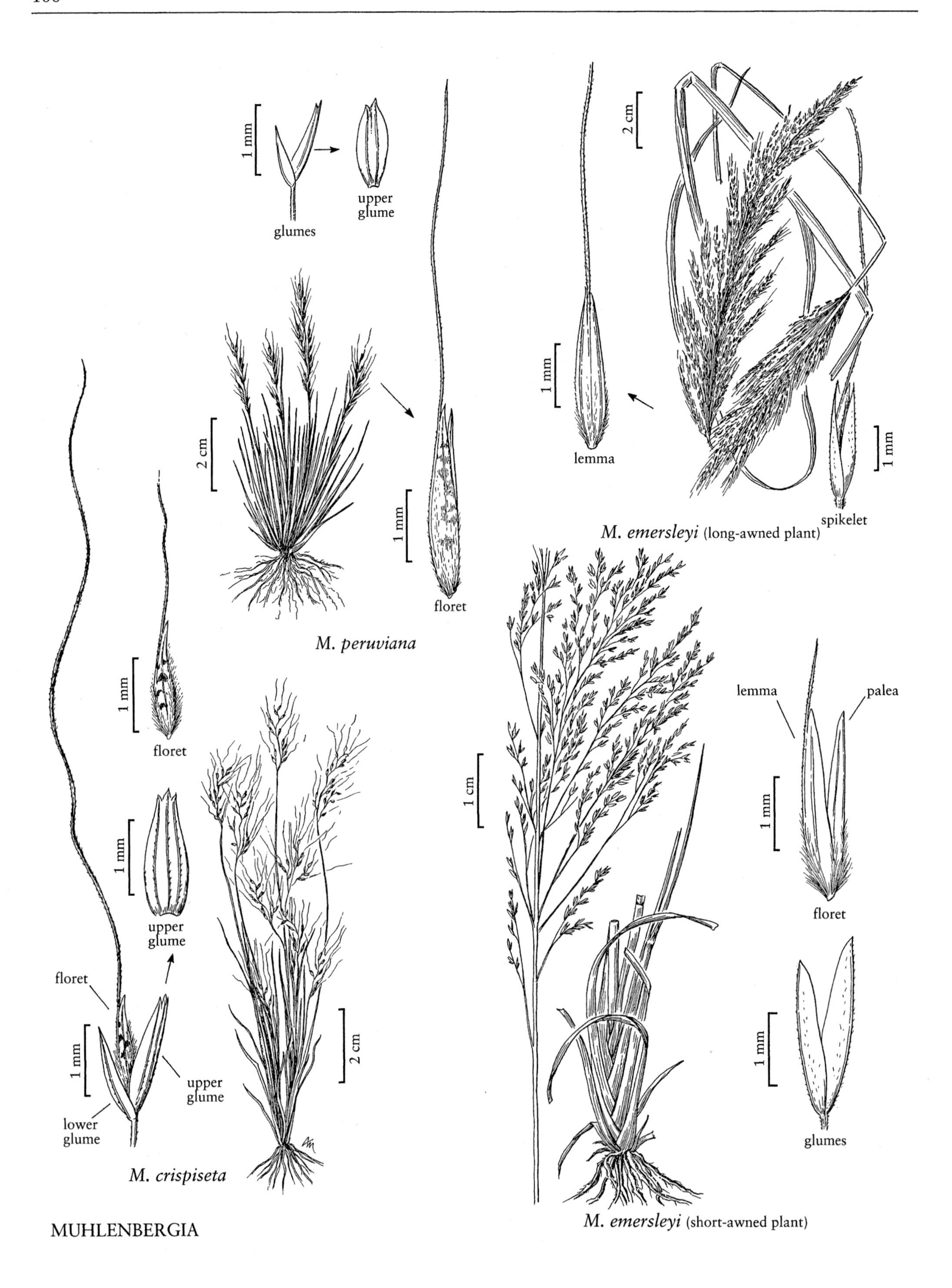

MUHLENBERGIA

puberulent, basal sheaths laterally compressed, usually keeled; **ligules** 10–25 mm, membranous throughout, acuminate, lacerate; **blades** 20–50 cm long, 2–6 mm wide, flat or folded, scabrous abaxially, smooth or scabridulous adaxially. **Panicles** 20–45 cm long, 3–15 cm wide, loosely contracted to open, light purplish to light brownish; **primary branches** 1–17 cm, lax, loosely appressed or diverging up to 70° from the rachises, naked basally; **pedicels** 0.5–3 mm, smooth or scabridulous. **Spikelets** 2.2–3.2 mm. **Glumes** subequal, 2.2–3.2 mm, exceeding the florets, scabridulous to scabrous, faintly 1-veined, acute to obtuse, usually unawned, occasionally awned, awns to 0.2 mm; **lemmas** 2–3 mm, oblong-elliptic, shortly pubescent on the lower ½–¾, apices acute, usually awned, sometimes unawned, awns to 15 mm, flexuous, purplish; **paleas** 1.8–2.9 mm, oblong-elliptic, acute; **anthers** 1.2–1.6 mm, yellowish to purplish. **Caryopses** 1.3–1.6 mm, fusiform, reddish-brown. $2n$ = 24, 26, 28, 40, 42, 46, 60, 64.

Muhlenbergia emersleyi grows on rocky slopes, gravelly washes, canyons, cliffs, and arroyos, often in soils derived from limestone, at elevations of 1200–2500 m, and is also grown as an ornamental. Its range extends from the southwestern United States through Mexico to Panama.

Muhlenbergia emersleyi differs from the closely related *M. longiligula* in its compressed-keeled sheaths, pubescent florets, and membranous ligules.

50. **Muhlenbergia ×involuta** Swallen [p. 189]
CANYON MUHLY

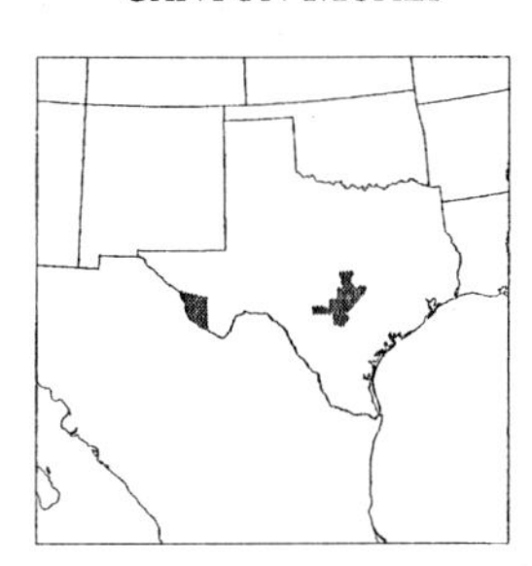

Plants perennial; cespitose, not rhizomatous. **Culms** 60–140 cm, erect; **internodes** puberulent or glabrous for most of their length, puberulent below the nodes. **Sheaths** shorter or longer than the internodes, smooth or scabridulous, tightly imbricate, yellowish-brown, basal sheaths laterally compressed, keeled, not becoming spirally coiled when old; **ligules** 3–12 mm, firm and brown basally, membranous distally, acute; **blades** 10–45 cm long, 1.6–5 mm wide, tightly folded, scabrous abaxially, hirsute adaxially. **Panicles** 18–40 cm long, 1.5–7 cm wide, loosely contracted to open, not dense; **branches** 1–10 cm, ascending or diverging up to 60° from the rachises, stiff, naked basally; **pedicels** 2–8 mm, hirtellous. **Spikelets** 3–4.2 mm, yellowish to purplish. **Glumes** subequal, 2–3 mm, exceeded by the florets, scabridulous or smooth, 1-veined, acute or obtuse, unawned; **lemmas** 3–4.2 mm, lanceolate, glabrous or appressed-pubescent on the lower ¼ of the margins, apices acute to obtuse, usually bifid and awned, teeth to 0.3 mm, awns 0.5–4 mm, straight; **paleas** 3–4.2 mm, lanceolate, glabrous, acute to acuminate; **anthers** 1–1.8 mm, purplish. **Caryopses** not seen. $2n$ = 24.

Muhlenbergia ×involuta grows on rocky, calcareous slopes in openings and along canyons, at elevations of 150–500 m. It has only been found growing naturally in Texas, but it is also available commercially as an ornamental. Swallen (1932) suggested that *M. reverchonii* and *M. lindheimeri* were its parents, but *M. rigida* seems to be another plausible possibility.

51. **Muhlenbergia longiligula** Hitchc. [p. 189]
LONGTONGUE MUHLY

Plants perennial; cespitose, not rhizomatous. **Culms** 60–130 cm, stout, erect; **internodes** mostly smooth, sometimes scabridulous below the nodes. **Sheaths** shorter or longer than the internodes, smooth or scabridulous, not becoming spirally coiled, basal sheaths rounded; **ligules** 10–30 mm, firm and brown basally, membranous distally, acuminate to obtuse; **blades** (10)20–65 cm long, 3–6 mm wide, flat or inrolled at the margins, scabrous abaxially, scabridulous or smooth adaxially. **Panicles** 15–55 cm long, 1–15 cm wide, contracted to open, greenish-tan to purplish; **primary branches** 3–13 cm, narrowly ascending or diverging up to 60° from the rachises, stiff, spikelet-bearing to the base, lower branches with 30–60 spikelets; **pedicels** 0.1–2.5 mm, usually shorter than the spikelets, scabridulous or smooth, strongly divergent. **Spikelets** 2–3.5 mm. **Glumes** subequal, 2–3.5 mm, usually longer than the florets, scabridulous or smooth, 1(2)-veined, acute to acuminate, usually unawned, rarely awned, awns to 0.2 mm; **lemmas** 2–2.9 mm, oblong-elliptic, tan to purplish, smooth or scabridulous, apices acute, often bifid, teeth to 0.2 mm, unawned or awned, awns to 2 mm; **paleas** 2–3 mm, oblong-elliptic, scabridulous or smooth, acute; **anthers** 1–2.1 mm, yellowish to purplish. **Caryopses** 1.1–1.5 mm, fusiform, reddish-brown. $2n$ = 22, 24, 29, 30.

Muhlenbergia longiligula grows on rocky slopes, canyons, and rock outcrops derived from volcanic or calcareous parent materials, at elevations of 1220–2500 m. It is a common species in Arizona and southwestern New Mexico, and extends into northwestern Mexico. It may be confused with *M. emersleyi*, but differs from that species in its rounded basal leaf sheaths, glabrous lemmas, and panicle branches that are spikelet-bearing to the base. It is also similar to *M. lindheimeri*, but differs in its rounded basal sheaths.

52. **Muhlenbergia expansa** (Poir.) Trin. [p. 189]
SAVANNAH HAIRGRASS

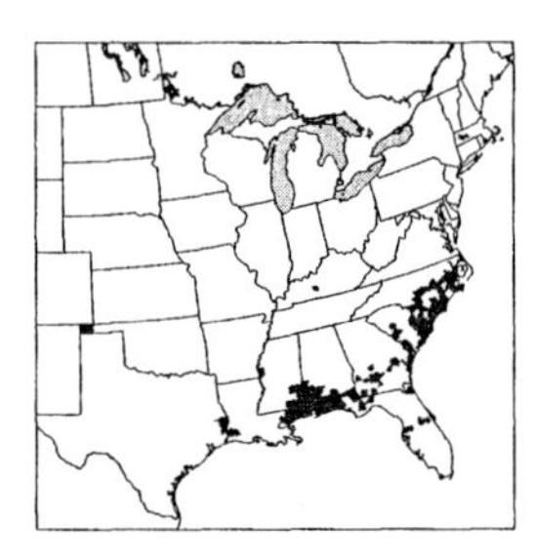

Plants perennial; cespitose, not rhizomatous. **Culms** 60–100 (150) cm, erect from the base, not conspicuously branched; **internodes** mostly glabrous, sometimes puberulent below the nodes. **Sheaths** glabrous or puberulent, rounded basally, becoming fibrous, not flat or spirally coiled, at maturity; **ligules** 1.8–5(10) mm, membranous, firm, strongly decurrent, obtuse; **blades** 20–50(80) cm long, 2–4 mm wide, flat or involute, smooth abaxially, scabrous adaxially. **Panicles** 15–50(60) cm long, 5–30 cm wide, longer than wide, diffuse; **primary branches** 2–20 cm, capillary, spreading 30–100° from the rachises, naked basally, lower branches with 5–20 spikelets; **pedicels** 4–50 mm, longer than the spikelets, capillary, flexible, widely divergent at maturity. **Spikelets** 3–5 mm, often purplish, sometimes brownish or bronze. **Glumes** subequal, 1.5–3.3 mm, shorter than the florets, glabrous; **lower glumes** 1-veined, unawned; **upper glumes** usually 1-veined, rarely 3-veined, acute to acuminate, often erose, sometimes mucronate; **lemmas** 3–5 mm, lanceolate, calluses shortly pubescent, apices acuminate, without setaceous teeth, usually unawned, if, as rarely, awned, awns 1–3 mm, clearly demarcated from the lemma bodies; **paleas** 2–4.5 mm, lanceolate, acuminate, unawned; **anthers** 1.5–2 mm, purple. **Caryopses** 2–2.5 mm, narrowly elliptic, brownish. $2n$ = unknown.

Muhlenbergia expansa grows in perennially moist to wet soils in pitcher plant bogs, pine savannahs, and flatwoods, usually in sandy soils and at elevations of 0–300 m. Its primary range is the coastal plain of the southeastern United States.

53. **Muhlenbergia capillaris** (Lam.) Trin. [p. 191]
HAIRY-AWN MUHLY

Plants perennial; cespitose, not rhizomatous. **Culms** 60–100 (150) cm, erect from the base, not conspicuously branched; **internodes** mostly glabrous, sometimes puberulent below the nodes. **Sheaths** glabrous or puberulent, basal sheaths rounded, often becoming fibrous, but never spirally coiled, at maturity; **ligules** 1.8–5(10) mm, membranous, firm, strongly decurrent, obtuse; **blades** 10–35(80) cm long, 2–4 mm wide, flat or involute, smooth abaxially, scabrous adaxially. **Panicles** 15–50(60) cm long, 5–30(41) cm wide, longer than wide, diffuse; **primary branches** 2–20 cm, capillary, diverging 30–100° from the rachises, naked basally, lower branches with 5–20 spikelets; **pedicels** 10–40(50) mm, longer than the spikelets, capillary, flexible. **Spikelets** 3–5 mm, usually purple, occasionally green, brown, or stramineous. **Glumes** subequal, (0.3)1–1.5(2) mm, usually less than ½ as long as the lemmas, glabrous; **lower glumes** 1-veined, usually unawned, rarely awned, awns 1–3 mm; **upper glumes** 1-veined, rarely 3-veined, acute to acuminate, often erose, usually unawned, rarely awned, awns 1–3(5) mm; **lemmas** 3–5 mm, lanceolate, not shiny, calluses shortly pubescent, apices scabrous, acuminate, sometimes with 2 setaceous teeth, teeth to 1 mm, unawned or awned, awns 2–13(18) mm, clearly demarcated from the lemma bodies; **paleas** 2–4.5 mm, lanceolate, acuminate, usually unawned; **anthers** 1.5–2 mm, purple. **Caryopses** 2–2.5 mm, narrowly elliptic, brownish. $2n$ = unknown.

In the southeastern United States, *Muhlenbergia capillaris* usually grows in rocky or clay soils in open woodlands and savannahs and on calcareous outcrops, at elevations of 0–500 m. In the northeastern states, it is also found on diabase and sandstone outcrops and ridges. Its native range includes the southeastern United States, Bahamas, and possibly various Caribbean islands. It is also grown as an ornamental.

Muhlenbergia capillaris resembles *M. reverchonii* in many respects, but differs it is dull, apically scabrous lemmas.

54. **Muhlenbergia sericea** (Michx.) P.M. Peterson [p. 191]
DUNE HAIRGRASS, PURPLE MUHLY

Plants perennial; cespitose, not rhizomatous. **Culms** 70–140 (160) cm, erect from the base, not conspicuously branched; **internodes** mostly glabrous, sometimes puberulent below the nodes. **Sheaths** glabrous or puberulent, sheaths rounded near the base, rarely becoming fibrous at maturity, not becoming spirally coiled when old; **ligules** 4–8(10) mm, membranous, firm, strongly decurrent, obtuse; **blades** 35–100 cm long, 1–3 mm wide, usually involute, smooth abaxially, scabrous adaxially. **Panicles** 20–60(70) cm long, 15–30(40) cm wide, longer than wide, diffuse; **primary branches** 2–20 cm, capillary, diverging 30–100° from the rachises, naked basally, lower branches with 5–20 spikelets; **pedicels** 4–50 mm, longer than the spikelets, capillary, divergent, stiff or flexible. **Spikelets** 3–5 mm, mostly purplish. **Glumes** subequal, (0.3)1–2 mm (excluding the awns), less than ½ as long as the lemmas, glabrous; **lower glumes** 1-veined, awned, awns 0.5–10 mm; **upper**

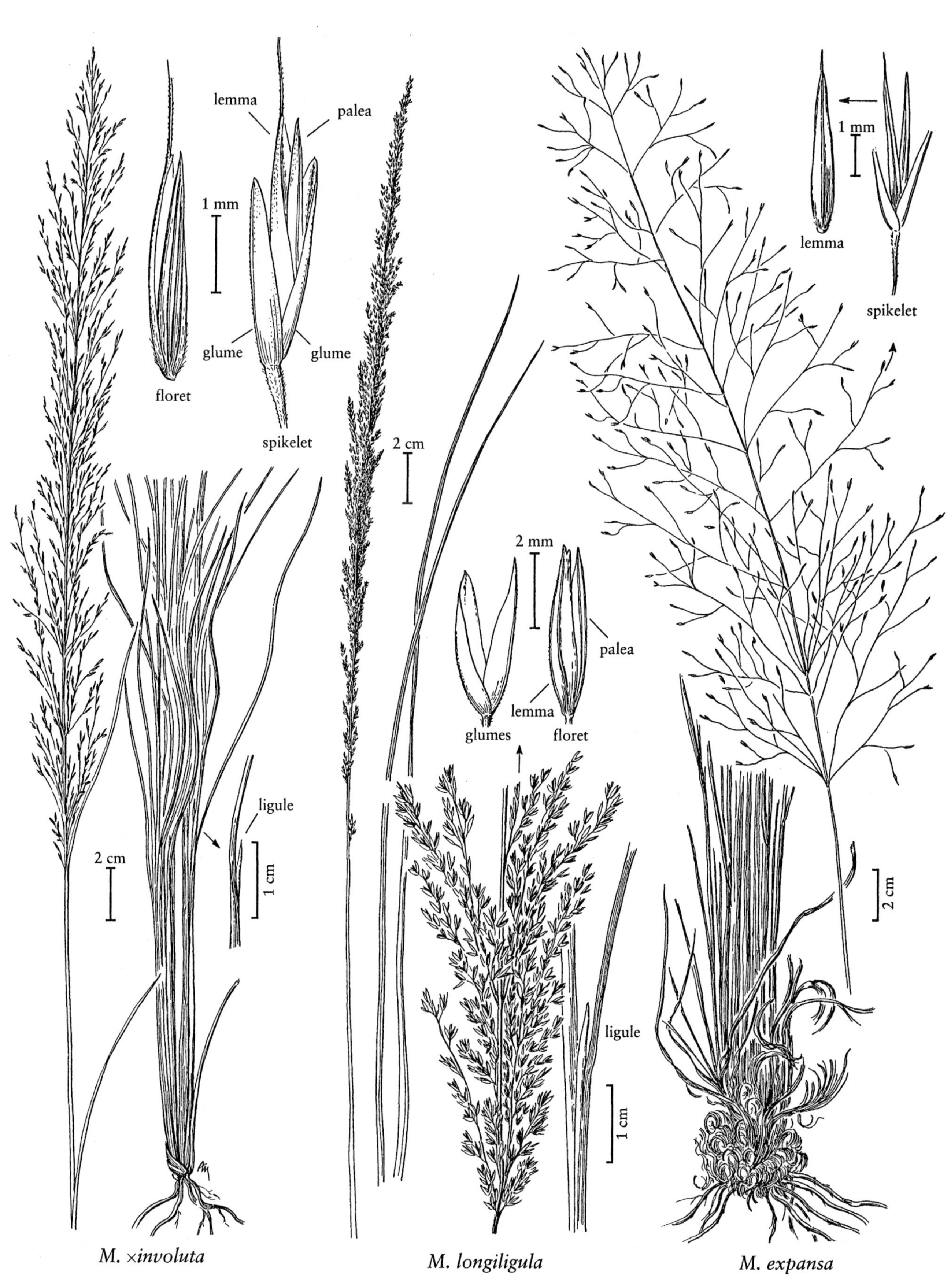

lemma
palea
1 mm
floret
glume
glume
spikelet
2 cm
ligule
1 cm
2 cm
2 mm
palea
lemma
glumes
floret
ligule
1 cm
lemma
1 mm
spikelet
2 cm
M. ×involuta
M. longiligula
M. expansa

glumes 1-veined (rarely 3-veined), acute to acuminate, often erose, awned, awns 2–25 mm; **lemmas** 3–5 mm, lanceolate, calluses shortly pubescent, apices acuminate, with 2 setaceous teeth, teeth 1–5 mm, awned from between the teeth, awns 8–35 mm, clearly demarcated from the lemma bodies; **paleas** 2–4.5 mm, lanceolate, acuminate, veins usually extending into awns to 2 mm; **anthers** 1.5–2 mm, purple. **Caryopses** 2–2.5 mm, narrowly elliptic, brownish. 2*n* = unknown.

Muhlenbergia sericea grows in sandy maritime habitats on the barrier islands and in coastal woodlands of the southeastern United States, at elevations of 0–50 m. It is available as an ornamental, sometimes under the name 'Purple Muhly'.

55. **Muhlenbergia reverchonii** Vasey & Scribn. [p. 191]
SEEP MUHLY

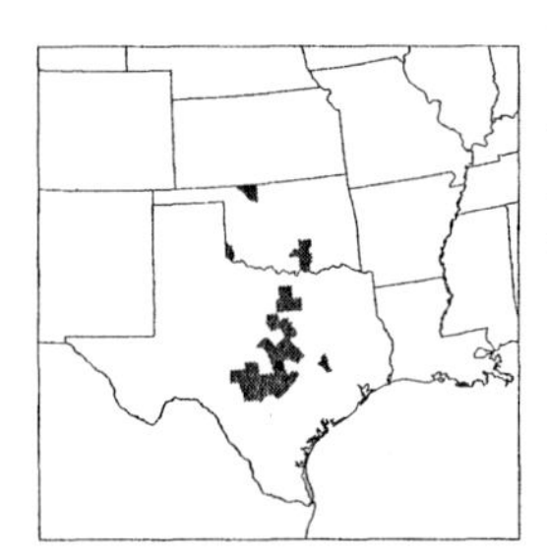

Plants perennial; cespitose, not rhizomatous, typical bunchgrasses in appearance. **Culms** 40–80 cm, stiffly erect from the base, not conspicuously branched; **internodes** glabrous, sometimes puberulent below the nodes. **Sheaths** shorter than the internodes, smooth or scabridulous, basal sheaths rounded, not becoming spirally coiled when old; **ligules** (2)4–7(9) mm, firmer near the base, obtuse, lacerate; **blades** 8–35 cm long, 1–2 mm wide, flat or involute, filiform, glabrous abaxially, densely hirtellous adaxially. **Panicles** 10–20(30) cm long, 4–15 cm wide, about as wide as long, loosely contracted to open but not diffuse; **primary branches** 1.4–10 cm, capillary, diverging up to 80° from the rachises, scabridulous, naked basally, lower branches with 5–15(20) spikelets; **pedicels** 3–25 mm, longer than the spikelets, capillary, stiff, or flexuous. **Spikelets** 3.5–5 mm, stramineous or brownish to purplish. **Glumes** subequal, 1–3 mm, shorter than the florets, hyaline, mostly sparsely hirtellous, apices glabrous, 1-veined (sometimes faintly so), acute, occasionally mucronate, mucros shorter than 0.7 mm; **lemmas** 3.5–5 mm, narrowly lanceolate, calluses hairy, hairs to 0.5 mm, lemma bodies glabrous and smooth, apices scabridulous, acuminate, awned, awns 0.5–4(6) mm, clearly demarcated from the lemma bodies; **paleas** 3.5–5 mm, narrowly lanceolate, mostly glabrous; **anthers** 1.1–2 mm, yellowish to purplish. **Caryopses** 2–2.4 mm, fusiform, brownish. 2*n* = 20.

Muhlenbergia reverchonii grows on calcareous rocky slopes, flats, and limestone rock outcrops, at elevations of 150–650 m. It is restricted to Oklahoma and Texas.

Muhlenbergia reverchonii resembles *M. capillaris* and *M. setifolia* in many respects, but differs from the former in its smooth and shiny lemmas, and from the latter in its wider panicles, spreading panicle branches, acute glumes, and more shortly awned lemmas.

56. **Muhlenbergia rigida** (Kunth) Trin. [p. 193]
PURPLE MUHLY

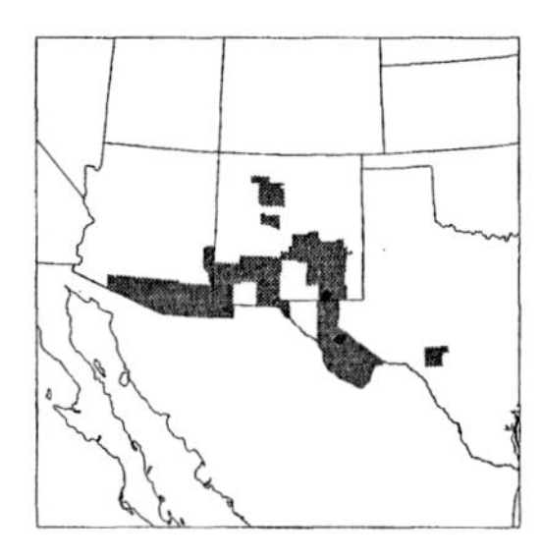

Plants perennial; cespitose, not rhizomatous. **Culms** 40–100 cm, erect, not conspicuously branched; **internodes** mostly smooth, sometimes scabridulous below the nodes. **Sheaths** longer than the internodes, smooth or scabridulous, basal sheaths rounded, not becoming spirally coiled when old; **ligules** (1)3–12(15) mm, firmer basally than distally, obtuse to acute, often lacerate; **blades** 12–35 cm long, 1–3 mm wide, flat or involute, smooth or scabridulous abaxially, scabridulous or hirtellous adaxially. **Panicles** 10–35 cm long, 2–5(12) cm wide, loosely contracted to open, purplish; **primary branches** 0.4–10 cm, lax, capillary, usually appressed to ascending, occasionally diverging up to 80° from the rachises, naked basally; **pedicels** 1–10 mm. **Spikelets** 3.5–5 mm, purplish. **Glumes** equal, 1–1.7(2) mm, exceeded by the florets, usually glabrous, sometimes mostly hirtellous but glabrous distally, 1-veined, obtuse to subacute, unawned; **lemmas** 3.5–5 mm, narrowly lanceolate, purplish, calluses hairy, hairs to 0.5 mm, lemma bodies scabridulous to scabrous, apices acuminate, awned, awns 10–22 mm, clearly demarcated from the lemma bodies, flexuous; **paleas** 3.5–5 mm, narrowly lanceolate, scabridulous, acuminate; **anthers** 1.7–2.3 mm, purplish. **Caryopses** 2–3.5 mm, fusiform, brownish. 2*n* = 40, 44.

Muhlenbergia rigida grows on rocky slopes, ravines, and sandy, gravelly slopes derived from granitic and calcareous substrates, at elevations of 1200–2200 m. It is often a common upland bunchgrass, and is also grown as an ornamental plant.

Muhlenbergia rigida grows in two disjunct areas: the southwestern United States south to Chiapas, Mexico, and in Ecuador, Peru, Bolivia, and Argentina. It differs from *M. setifolia* and *M. reverchonii* in its purplish, scabridulous to scabrous lemmas.

57. **Muhlenbergia elongata** Scribn. *ex* Beal [p. 193]
SYCAMORE MUHLY

Plants perennial; cespitose, not rhizomatous. **Culms** 40–120 cm, erect, not conspicuously branched; **internodes** mostly smooth, sometimes scabridulous below the nodes. **Sheaths** longer than the internodes, smooth or

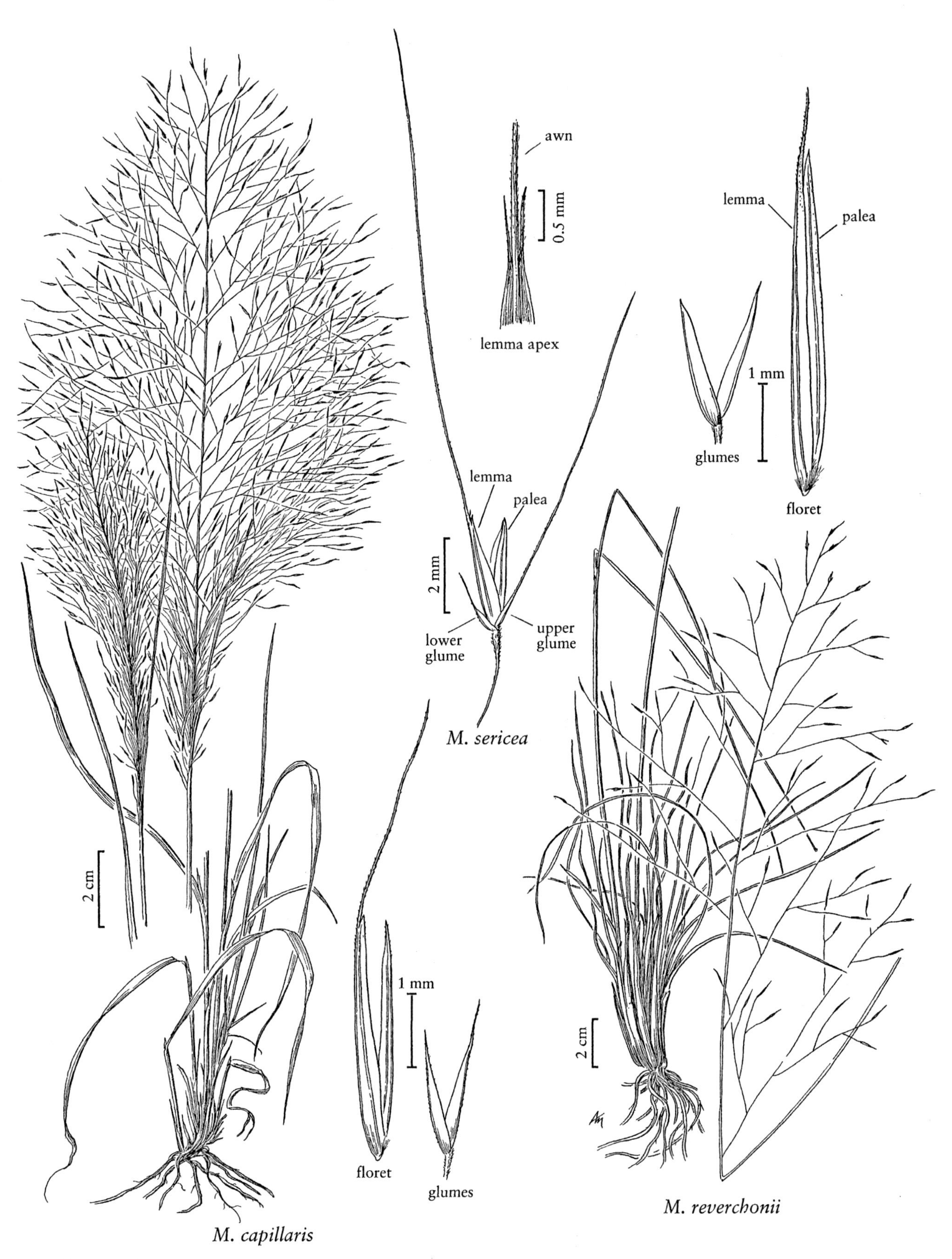

MUHLENBERGIA

scabridulous, rounded towards the base, not becoming spirally coiled when old; **ligules** 2–8 mm, firm below, membranous distally, obtuse to acute; **blades** 15–50 cm long, 0.3–1.5 mm wide, flat to involute, smooth or scabridulous abaxially, hirtellous adaxially. **Panicles** 15–50 cm long, 1–7 cm wide, loosely contracted, not dense; **primary branches** 1–8 cm, appressed or diverging up to 30° from the rachises, naked basally; **pedicels** 1.3–8 mm, hispidulous. **Spikelets** 3–4.2 mm. **Glumes** subequal, 2–3 mm, exceeded by the florets, glabrous below, minutely pubescent or scabridulous distally, 1-veined, acuminate to acute, unawned or awned, awns to 1.6 mm; **lemmas** 2.8–4.2 mm, narrow, lanceolate, calluses hairy, hairs to 1 mm, lemma bodies glabrous and smooth below, scabridulous distally, apices acuminate, awned, awns 8–40 mm, straight to flexuous, demarcation of the awns from the lemma bodies not evident; **paleas** 2.8–4.2 mm, narrowly lanceolate, glabrous, acuminate; **anthers** 1.4–2.2 mm, yellow. **Caryopses** 2–2.2 mm, fusiform, brownish. $2n$ = unknown.

Muhlenbergia elongata grows on rock outcrops, cliffs, canyon walls, and moist rock walls, on rhyolitic and volcanic conglomerates, at elevations of 850–2100 m. It extends south from Arizona into northern Mexico.

58. **Muhlenbergia setifolia** Vasey [p. 193]
CURLYLEAF MUHLY

Plants perennial; cespitose, not rhizomatous. **Culms** 30–80 cm, slightly decumbent basally, erect above, not conspicuously branched, branches, when present, not geniculate and widely divergent; **internodes** mostly glabrous, sometimes hirtellous below the nodes. **Sheaths** shorter than the internodes, glabrous, margins whitish, rounded towards the base, not becoming spirally coiled when old; **ligules** 4–7(10) mm, acuminate, lacerate; **blades** 5–20(25) cm long, 0.2–1.2 mm wide, tightly involute, arcuate, scabrous abaxially, scabrous or hirtellous adaxially. **Panicles** 8–20(25) cm long, 2–5 cm wide, loosely contracted; **primary branches** 0.5–7 cm, capillary, diverging up to 70° from the rachises, naked basally; **pedicels** 3–20 mm. **Spikelets** 3.5–5 mm, stramineous to brown or purplish. **Glumes** subequal, 1.5–2.5 mm, shorter than the florets, thin, hyaline, glabrous; **lower glumes** veinless, truncate or obtuse, often toothed or notched; **upper glumes** 1-veined or veinless, obtuse to acute, often mucronate, mucros shorter than 0.7 mm; **lemmas** 3.5–5 mm, narrowly lanceolate, calluses hairy, hairs to 0.6 mm, lemma bodies glabrous, smooth, shiny, apices acuminate, awned, awns 10–30 mm, clearly demarcated from the lemma bodies, flexuous; **paleas** 3.5–5 mm, narrowly lanceolate, glabrous, acuminate; **anthers** 2–2.6 mm, greenish. **Caryopses** 2.4–3.2 mm, fusiform, brownish. $2n$ = 40.

Muhlenbergia setifolia grows on calcareous rocky slopes, rock outcrops, and in desert grasslands, at elevations of 1000–2250 m. Its range extends from the southwestern United States into northern Mexico. It is similar to *M. reverchonii*, but differs in its narrower panicles, less widespread panicle branches, truncate to obtuse glumes, and longer lemma awns.

59. **Muhlenbergia palmeri** Vasey [p. 195]
PALMER'S MUHLY

Plants perennial; cespitose, not rhizomatous. **Culms** 50–100 cm tall, 2–4 mm thick, erect, rounded near the base, not conspicuously branched, not rooting at the lower nodes; **internodes** mostly glabrous, sometimes puberulent below the nodes. **Sheaths** longer than the internodes, smooth or scabridulous, not becoming spirally coiled when old; **ligules** 1–3 mm, firm to membranous, truncate, ciliolate; **blades** 15–50 cm long, 1–3.5 mm wide, flat to involute, smooth or scabridulous abaxially, scabrous adaxially. **Panicles** 15–35 cm long, 0.5–2(3) cm wide; **primary branches** 0.5–8 cm, appressed; **pedicels** 1–6 mm, hispidulous. **Spikelets** 3–4.3 mm, yellowish-brown to purplish. **Glumes** subequal, (1.7)2–3.1 mm, about ¾ as long as the florets, scabridulous distally, 1-veined, apices acute to acuminate, awned, awns to 1.5 mm; **lemmas** 3–4.3 mm, lanceolate, calluses hairy, hairs to 1.5 mm, lemma bodies scabridulous, apices acuminate, awned, awns 3–10 mm, straight; **paleas** 2.9–4.2 mm, lanceolate, scabridulous, acute to acuminate; **anthers** 1.5–2.5 mm, yellow to purple-tinged. **Caryopses** 2–3 mm, fusiform, brownish. $2n$ = unknown.

Muhlenbergia palmeri grows in rocky drainages and in sandy soil along creeks, at elevations of 1000–2100 m. Its range extends from southern Arizona into northern Mexico.

60. **Muhlenbergia lindheimeri** Hitchc. [p. 195]
LINDHEIMER'S MUHLY

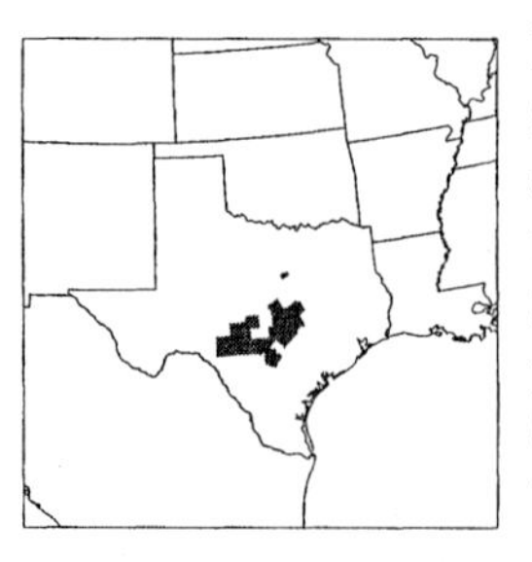

Plants perennial; cespitose, not rhizomatous. **Culms** 50–150 cm, stout, erect, not rooting at the lower nodes; **internodes** mostly glabrous, sometimes puberulent below the nodes. **Sheaths** shorter or longer than the internodes, glabrous, basal

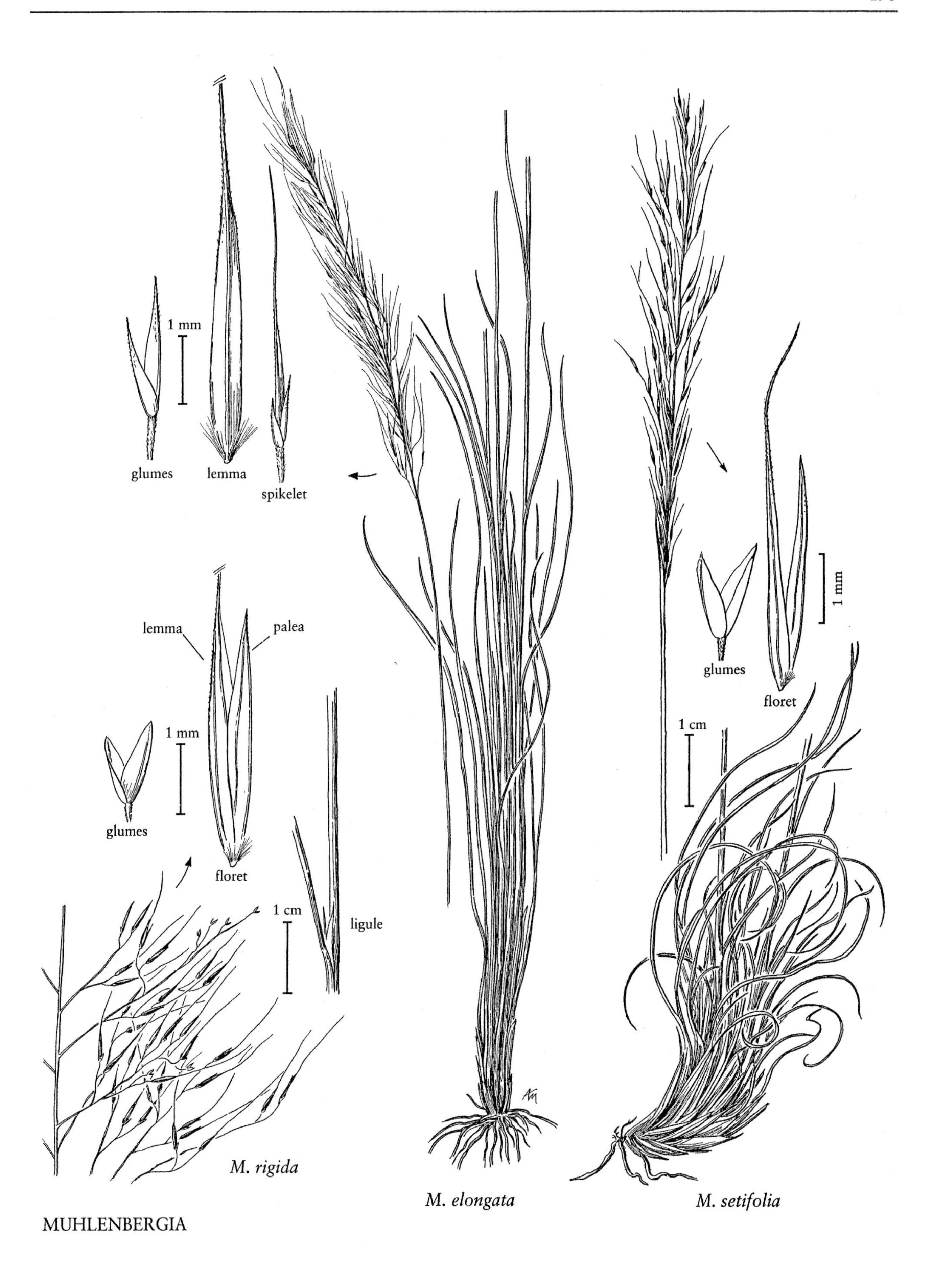

MUHLENBERGIA

sheaths laterally compressed, keeled, not becoming spirally coiled when old; **ligules** 10–35 mm, firm and brown basally, membranous distally, acuminate; **blades** 25–55 cm long, 2–5 mm wide, flat or folded, firm, scabridulous abaxially, scabrous and shortly pubescent adaxially. **Panicles** 15–50 cm long, 0.6–3 cm wide, often purplish-tinged; **primary branches** 0.5–7 cm, appressed or strongly ascending, rarely spreading as much as 20° from the rachises; **pedicels** 0.5–1.2 mm, scabrous. **Spikelets** 2.4–3.5 mm, light grayish. **Glumes** equal, 2–3.5 mm, shorter than or equal to the florets, scabrous or smooth, 1-veined, obtuse to acute, occasionally bifid and the teeth to 0.3 mm, unawned, rarely mucronate, mucros less than 0.2 mm; **lemmas** 2.4–3.5 mm, lanceolate, scabrous or smooth, rarely puberulent near the base, apices obtuse to acute, unawned or awned, awns to 4 mm, straight; **paleas** 2.4–3.5 mm, lanceolate, obtuse; **anthers** 1.1–1.5 mm, purplish. **Caryopses** 1.2–1.6 mm, fusiform, reddish-brown. $2n$ = 20, 26.

Muhlenbergia lindheimeri grows in sandy draws to rocky, calcareous soils, generally in open areas, at elevations of 150–500 m. It is an uncommon species throughout its range, which includes northern Mexico in addition to southern Texas, but it is also grown as an ornamental. It differs from the closely related *M. longiligula* in its compressed-keeled basal sheaths, grayish spikelets, and, when present, bifid glume apices.

61. **Muhlenbergia dubia** E. Fourn. [p. 195]
PINE MUHLY

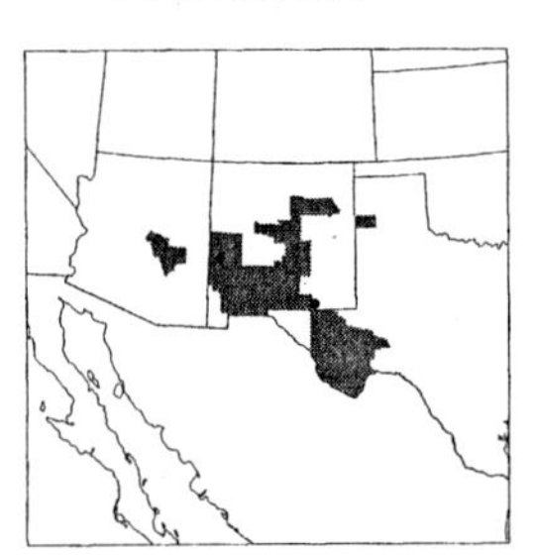

Plants perennial; densely cespitose, not rhizomatous. **Culms** 30–100 cm, erect, rounded near the base, not rooting at the lower nodes; **internodes** glabrous for most of their length, minutely pubescent to hirtellous below the nodes. **Sheaths** longer than the internodes, smooth or scabridulous, not becoming spirally coiled when old; **ligules** 4–10 mm, membranous, firm, acute, lacerate; **blades** 10–60 cm long, 1–2 mm wide, usually involute (occasionally flat), scabrous abaxially, hispidulous adaxially. **Panicles** 10–40 cm long, 0.6–2.4 cm wide, contracted, grayish-green; **primary branches** 0.2–7 cm, diverging up to 40° from the rachises, stiff, spikelet-bearing to the base; **pedicels** 0.1–6 mm, strongly divergent, hispidulous. **Spikelets** 3.8–5 mm, grayish-green. **Glumes** equal, (1.8)2–3 mm, shorter than the florets, glabrous and smooth proximally, scabridulous distally, faintly 1-veined, acute; **lemmas** 3.8–5 mm, narrowly lanceolate, calluses hairy, hairs to 0.5 mm, lemma bodies glabrous and smooth below, scabrous distally, apices acuminate, unawned or awned, awns to 6 mm, straight; **paleas** 3.8–5 mm, narrowly lanceolate, glabrous below, acuminate; **anthers** 1.5–2.2 mm, greenish. **Caryopses** 2.5–3.5 mm, fusiform, brownish. $2n$ = 40, 50.

Muhlenbergia dubia grows on steep slopes, ridge tops, limestone rock outcrops, and along draws, at elevations of 1500–2300 m. Its range extends into northern Mexico. It resembles *M. rigens*, but differs in having looser, contracted (but not spikelike) panicles, longer ligules, olivaceous anthers, and generally longer lemmas.

62. **Muhlenbergia rigens** (Benth.) Hitchc. [p. 197]
DEERGRASS

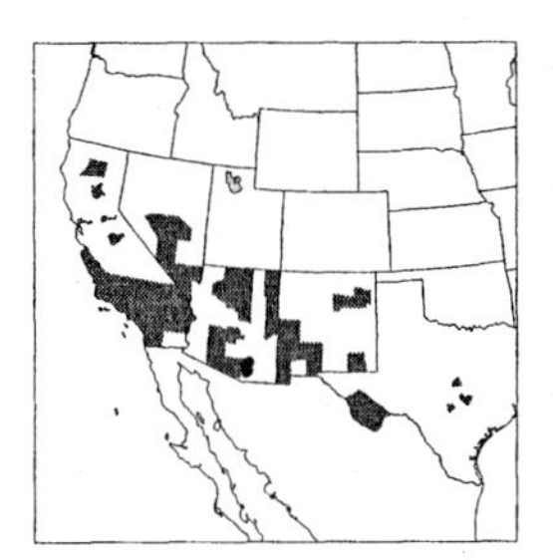

Plants perennial; cespitose, not rhizomatous. **Culms** (35)50–150 cm tall, to 5 mm thick near the base, stiffly erect, not rooting at the lower nodes; **internodes** glabrous. **Sheaths** longer than the internodes, smooth or scabridulous, rounded dorsally, bases somewhat flat and chartaceous; **ligules** 0.5–2(3) mm, firm, truncate, usually ciliolate; **blades** 10–50 cm long, 1.5–6 mm wide, flat or involute, stiff, glabrous abaxially, scabrous adaxially. **Panicles** 15–60 cm long, 0.5–1.2 cm wide, spikelike, dense, grayish-green, often interrupted below; **primary branches** 0.2–4 cm, appressed; **pedicels** 0.2–3 mm, hispidulous, strongly divergent. **Spikelets** 2.4–4 mm. **Glumes** subequal, 1.8–3.2 mm, almost as long as the florets, scabrous to scabridulous, 1-veined, usually acute or obtuse, occasionally acuminate or notched, occasionally mucronate, mucros to 0.6 mm; **lemmas** 2.4–4 mm, lanceolate, pubescent on the calluses, lower ⅛ of the margins, and midveins, hairs to 0.4 mm, apices scabrous, acute or obtuse, unawned, rarely mucronate, mucros to 1 mm; **paleas** 2.3–3.8 mm, lanceolate, acute; **anthers** 1.3–1.8 mm, yellow to purple. **Caryopses** 1.8–2.2 mm, fusiform, brownish. $2n$ = 40.

Muhlenbergia rigens grows in sandy washes, gravelly canyon bottoms, rocky drainages, and moist, sandy slopes, often along small streams, at elevations of 90–2500 m. Its geographic range extends to central Mexico. It is available commercially as an ornamental.

Muhlenbergia rigens is similar to *M. dubia*, but differs in having tighter, spikelike panicles, shorter ligules, yellow or purplish anthers, and shorter lemmas.

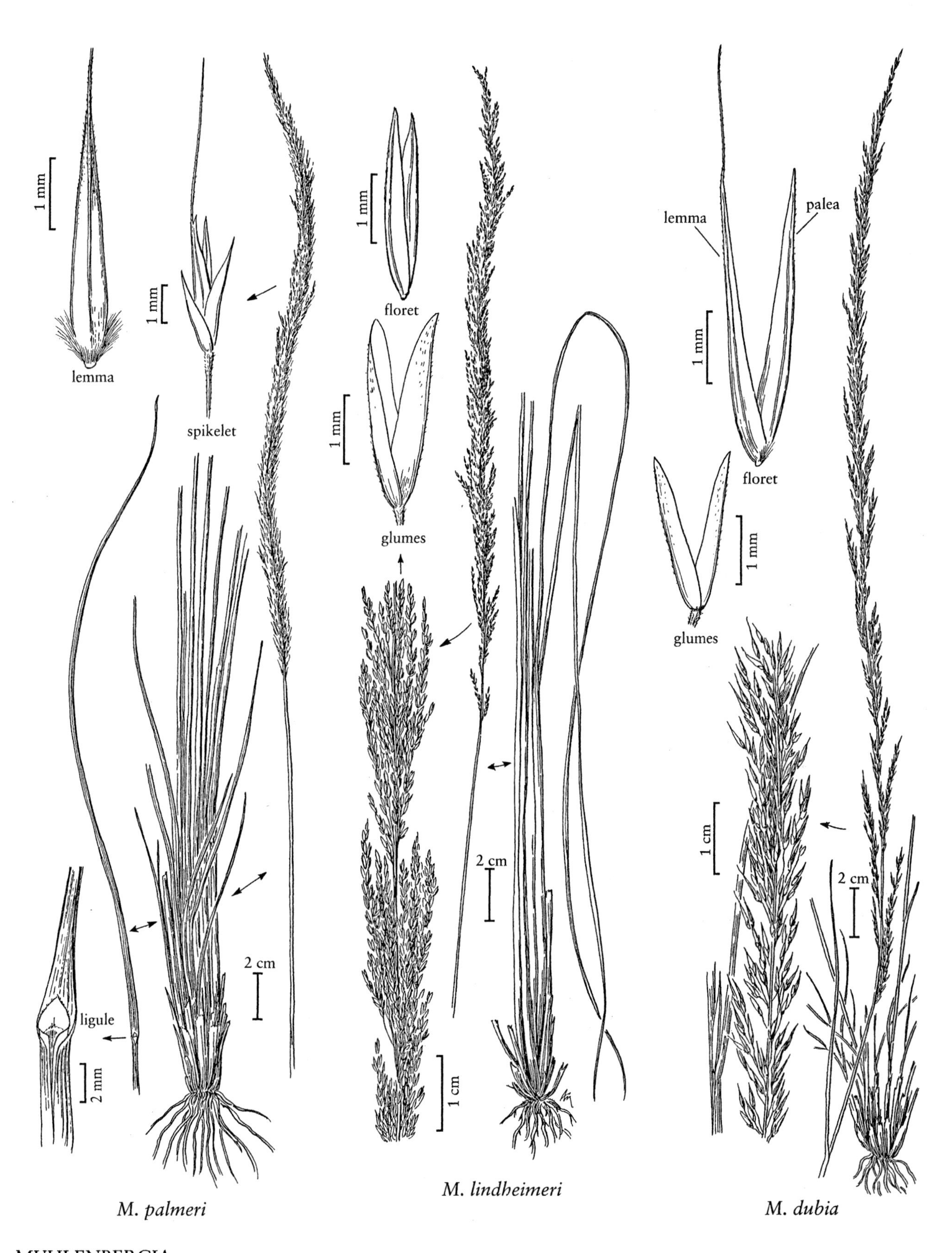

M. palmeri *M. lindheimeri* *M. dubia*

63. Muhlenbergia depauperata Scribn. [p. 197]
SIXWEEKS MUHLY

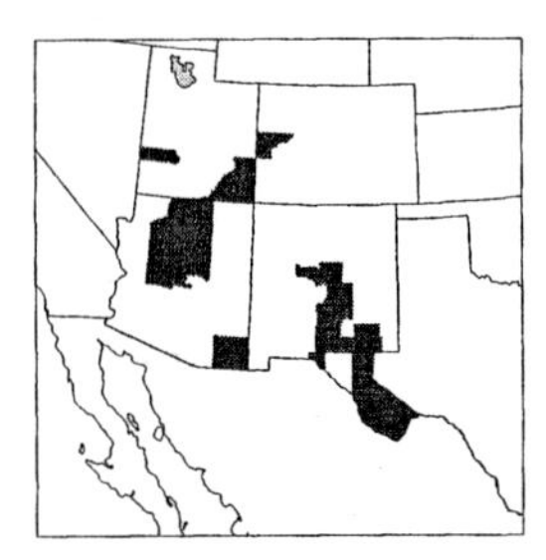

Plants annual; tufted. **Culms** 3–15 cm; **internodes** mostly scabridulous or pubescent, pubescent or strigose below the nodes. **Sheaths** often longer than the internodes, somewhat inflated, smooth or scabrous, keeled, margins scarious, not flattened, papery, or spirally coiled when old; **ligules** 1.4–2.5 mm, membranous, acute, with lateral lobes; **blades** 1–3 cm long, 0.6–1.5 mm wide, flat or involute, scabrous to strigose, midveins and margins thickened, whitish. **Panicles** 2.5–8.5 cm long, 0.5–0.7 cm wide, contracted; **primary branches** 1–2.2 cm, appressed, spikelet-bearing to the base, spikelets borne in subsessile-pedicellate pairs; **longer pedicels** 3–6 mm, scabrous; **disarticulation** beneath the spikelet pairs. **Spikelets** 2.5–5.1 mm, appressed. **Glumes** 2.3–5.1 mm, equaling or exceeding the florets; **lower glumes** 2.3–4 mm, subulate, 2-veined, minutely to deeply bifid, teeth aristate or with awns to 1.3 mm; **upper glumes** 3–5.1 mm, lanceolate, 1-veined, entire, acuminate; **lemmas** 2.5–4.5 mm, narrowly lanceolate, light greenish-brown to purplish, scabrous, appressed-pubescent on the margins and midveins, apices acuminate, awned, awns 6–15 mm, stiff; **paleas** 2.4–3.6 mm, lanceolate, intercostal region appressed-pubescent, apices acuminate; **anthers** 0.5–0.8 mm, purplish to yellowish. **Caryopses** 1.5–2.3 mm, narrowly fusiform, brownish. $2n = 20$.

Muhlenbergia depauperata grows in gravelly flats, rock outcrops, exposed bedrock, and sandy banks, in gramma grassland associations, usually on soils derived from calcareous parent materials, at elevations of 1530–2400 m. Its range extends from the southwestern United States to southern Mexico.

Muhlenbergia depauperata and *M. brevis* share several features with *Lycurus*: spikelets borne in pairs, 2-veined and 2-awned lower glumes, 1-veined and awned upper glumes, acuminate, awned lemmas with short pubescence along the margins, and pubescent paleas.

64. Muhlenbergia brevis C.O. Goodd. [p. 197]
SHORT MUHLY

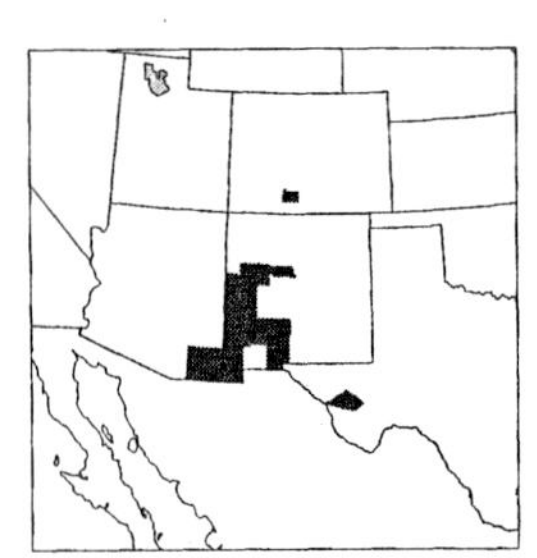

Plants annual; tufted. **Culms** 3–20 cm. **Sheaths** often longer than the internodes, somewhat inflated, smooth or scabrous; **ligules** 1–3 mm, membranous, acute, lacerate, sometimes with lateral lobes; **blades** 1–4.5 cm long, 0.8–2 mm wide, flat to involute, scabrous to strigose, midveins and margins thickened, whitish. **Panicles** 3–11.5 cm long, 0.8–1.8 cm wide, contracted; **primary branches** 1–3.7 cm, closely appressed, spikelets usually in subsessile-pedicellate pairs; **pedicels** 0.2–8 mm, stout, closely appressed, scabrous; **disarticulation** beneath the spikelet pairs. **Spikelets** 2.5–6 mm. **Glumes** to ⅔ as long as the lemmas; **lower glumes** 2–3.5 mm, subulate, 2-veined, minutely to deeply bifid, with 2 aristate teeth or awns to 1.8 mm; **upper glumes** 2.4–4 mm, entire, acuminate to attenuate, 1-veined, awned, awns to 2 mm; **lemmas** 3.5–6 mm, narrowly lanceolate, light greenish-brown to purplish, scabrous, appressed-pubescent on the margins and midveins, apices acuminate, often bifid, awned, awns usually 10–20 mm, stiff; **paleas** 4–6 mm, narrowly lanceolate, intercostal region appressed-pubescent, apices acuminate; **anthers** 0.5–0.9 mm, purplish to yellowish. **Caryopses** 2–2.8 mm, narrowly fusiform, brownish. $2n = 20$.

Muhlenbergia brevis grows on rocky slopes, gravelly flats, and rock outcrops, particularly those derived from calcareous parent materials, at elevations of 1700–2500 m, in gramma grasslands, pinyon-juniper woodlands, and pine-oak woodlands. Its range extends from the southwestern United States to central Mexico.

Like *Muhlenbergia depauperata*, *M. brevis* shares several features with *Lycurus*, notably the paired spikelets with 2-veined and 2-awned lower glumes, 1-veined and awned upper glumes, acuminate, awned lemmas with shortly pubescent margins, and pubescent paleas.

65. Muhlenbergia sinuosa Swallen [p. 199]
MARSHLAND MUHLY

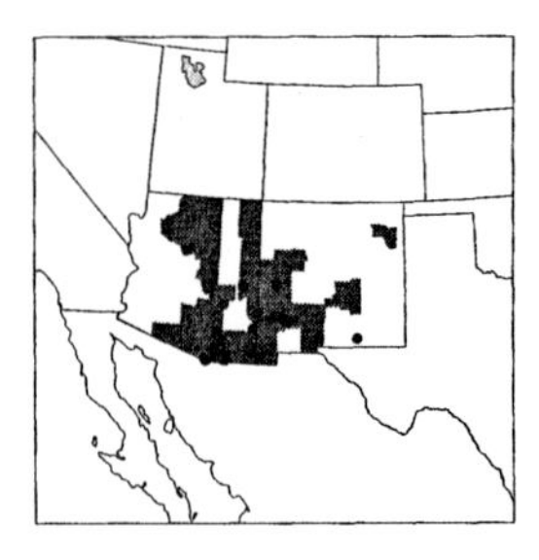

Plants annual; delicate. **Culms** 12–50 cm, erect to geniculate; **internodes** mostly glabrous and smooth or scabridulous, scabridulous or strigulose below the nodes. **Sheaths** usually longer than the internodes, glabrous, smooth or scabridulous; **ligules** 1.5–3.1 mm, hyaline, truncate to obtuse, irregularly toothed to lacerate, with lateral lobes that exceed the central portion; **blades** 2–8.5 cm long, 0.8–2 mm wide, flat, sometimes involute, scabridulous abaxially, shortly pubescent to minutely villous adaxially, midveins prominent abaxially. **Panicles** 10–26 cm long, 2.8–8 cm wide; **primary branches** 2.6–7 cm, often capillary, diverging 25–80° from the rachises; **pedicels** 4–7 mm, usually curved, often through 90° or more. **Spikelets** 1.4–2 mm. **Glumes** equal, 0.7–1.2 mm, usually conspicuously strigulose, particularly near the margins and apices, 1-veined, acute to obtuse, unawned; **lemmas**

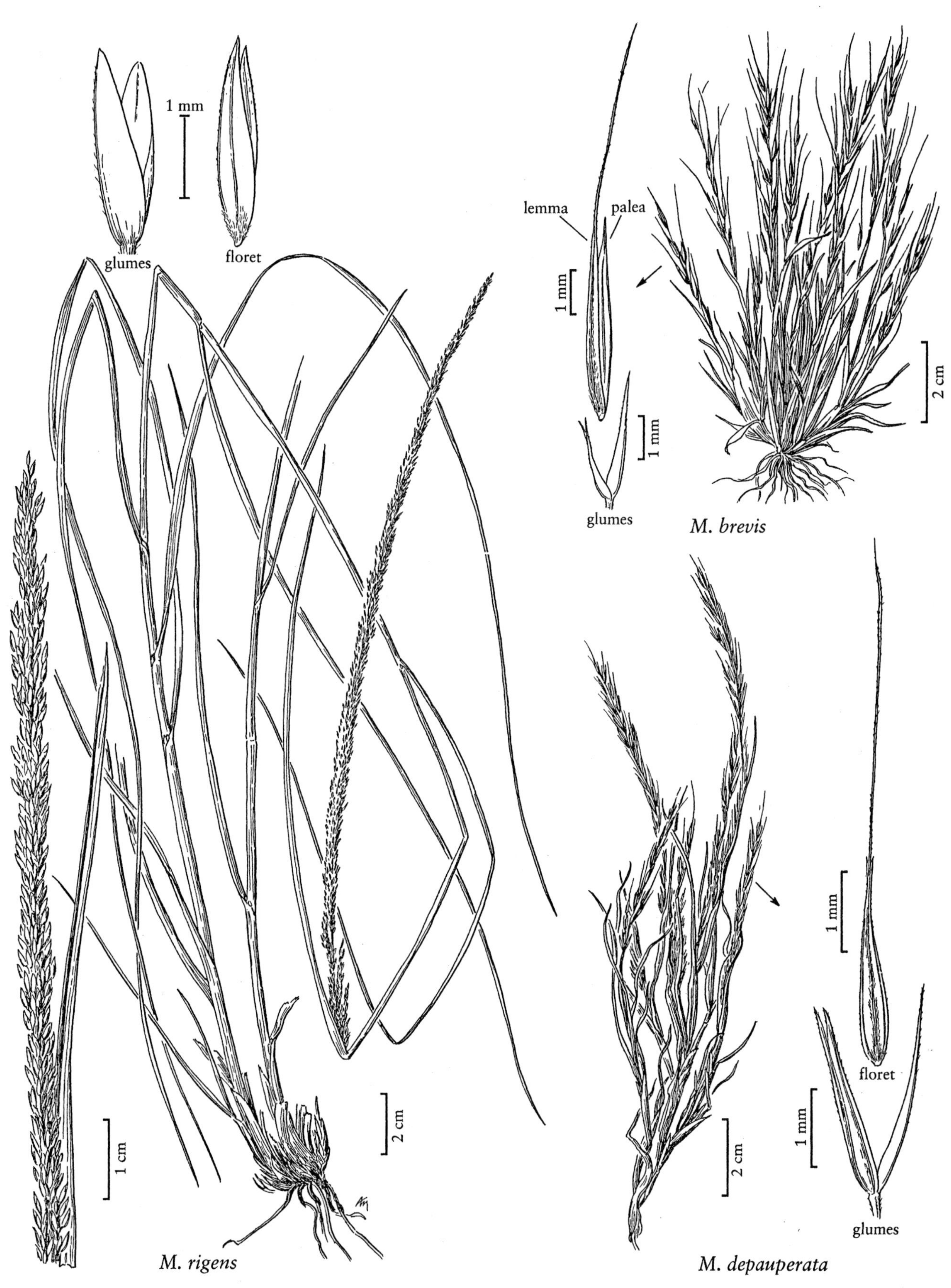

MUHLENBERGIA

1.4–2 mm, oblong-elliptic, greenish, sometimes purplish-tinged, shortly appressed-pubescent on the midveins and margins, apices acute or obtuse, unawned; **paleas** 1.3–1.8 mm, oblong-elliptic, intercostal region sparsely short-pilose or glabrous; **anthers** 0.6–1.2 mm, olivaceous. **Caryopses** 0.8–1.2 mm, fusiform, brownish. 2*n* = 20, 24.

Muhlenbergia sinuosa grows in sandy soil along washes, on open slopes and rocky ledges, and in road-side ditches, at elevations of 1650–2300 m. It is usually found in oak-pine forests, pinyon-juniper woodlands, oak-gramma savannahs, and riverine woodlands. Its range extends from the southwestern United States into northern Mexico.

66. **Muhlenbergia minutissima** (Steud.) Swallen [p. 199]
ANNUAL MUHLY

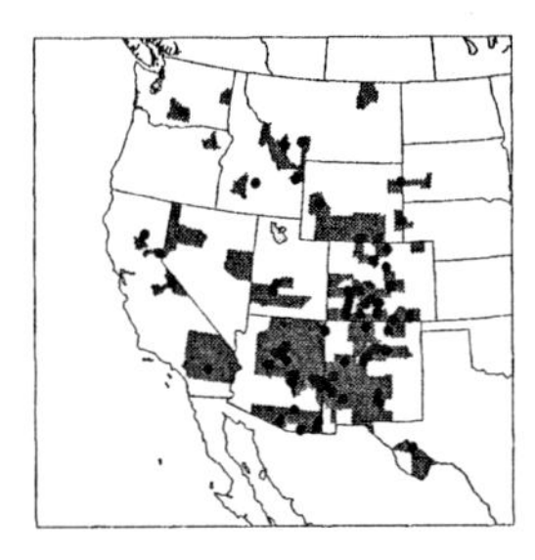

Plants annual. **Culms** 5–40 cm, slender, erect; **internodes** mostly glabrous, scabridulous or smooth, scabridulous or strigulose below the nodes. **Sheaths** shorter or longer than the internodes, smooth or scabridulous; **ligules** 1–2.6 mm, hyaline, truncate to obtuse, sometimes with lateral lobes; **blades** 0.5–4(10) cm long, 0.8–2 mm wide, flat or involute, scabrous abaxially, shortly pubescent adaxially. **Panicles** 5–16.2(21) cm long, 1.5–6.5 cm wide, open; **primary branches** 8–42 mm, often capillary, diverging 25–80° from the rachises; **pedicels** 2–7 mm, straight or curved, but rarely curved as much as 90°. **Spikelets** 0.8–1.5 mm. **Glumes** sparsely strigulose, at least near the apices, 1-veined; **lower glumes** 0.5–0.8 mm, obtuse to acute; **upper glumes** 0.6–0.9 mm, broader than the lower glumes, obtuse; **lemmas** 0.8–1.5 mm, lanceolate, brownish to purplish, glabrous or the midveins and margins appressed-pubescent, apices obtuse to subacute, unawned; **paleas** 0.8–1.4 mm, shortly pubescent or glabrous; **anthers** 0.2–0.7 mm, purplish. **Caryopses** 0.6–0.9 mm, fusiform to elliptic, brownish. 2*n* = 60, 80.

Muhlenbergia minutissima grows in sandy and gravelly drainages, rocky slopes, flats, road cuts, and open sites. It is usually found in yellow pine and oak-pine forests, pinyon-juniper woodlands, thorn-scrub forests, and oak-gramma savannahs, at elevations of 1200–3000 m. Its range extends from the western United States to southern Mexico.

67. **Muhlenbergia texana** Buckley [p. 199]
TEXAS MUHLY

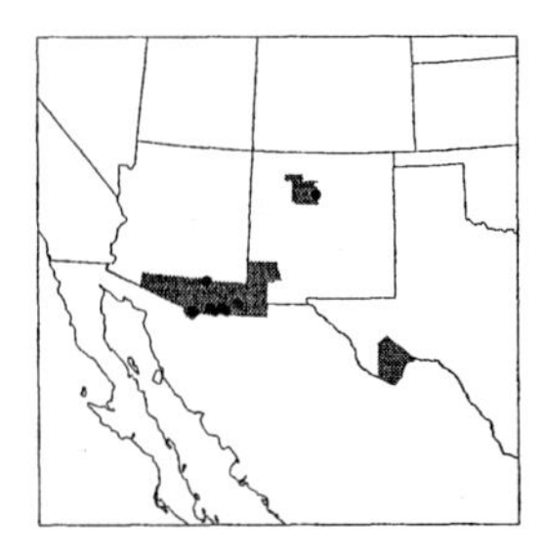

Plants annual; slender. **Culms** 10–35 cm, erect; **internodes** strigulose. **Sheaths** shorter or longer than the internodes, strigulose; **ligules** 0.9–2.5 mm, hyaline, acute to obtuse, irregularly toothed to lacerate; **blades** 1–6(8) cm long, 0.8–2 mm wide, flat or involute, scabrous and sparsely strigulose on both surfaces. **Panicles** 9–21 cm long, 2–7 cm wide; **primary branches** 12–60 mm, occasionally capillary, narrowly ascending or diverging up to 70° from the rachises; **pedicels** 2–7 mm, usually longer than the florets, straight to somewhat curved, rarely bent as much as 90°; **disarticulation** above the glumes. **Spikelets** 1.3–2 mm. **Glumes** sparsely strigulose, particularly on the margins, 1-veined, acute to acuminate; **lower glumes** 0.8–1.2 mm; **upper glumes** 0.9–1.5 mm, bases slightly wider than the lower glumes; **lemmas** 1.3–2 mm, lanceolate, purplish to brown, shortly appressed-pubescent on the lower ½ of the midveins and margins, apices acute to acuminate, awned, awns 0.1–1(2) mm; **paleas** 1.3–2 mm, oblong-elliptic, minutely appressed-pubescent on the lower ½, apices acute; **anthers** 0.4–0.5 mm, purplish. **Caryopses** 0.8–1 mm, fusiform, brownish. 2*n* = 40.

Muhlenbergia texana grows on open slopes, in sandy, gravelly drainages, and on rock outcrops, at elevations of 1200–2750 m. Its range extends from the southwestern United States into northwestern Mexico.

68. **Muhlenbergia eludens** C. Reeder [p. 201]
GRAVELBAR MUHLY

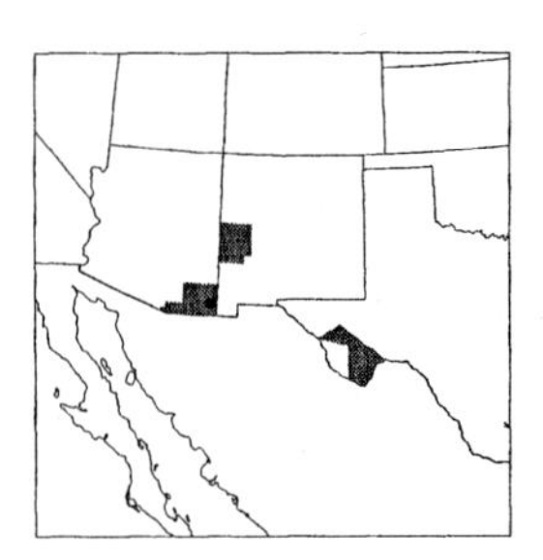

Plants annual. **Culms** 10–40 cm, erect; **internodes** mostly glabrous, strigulose below the nodes. **Sheaths** usually longer than the internodes, scabridulous or puberulent, keeled; **ligules** 1.5–2.5 mm, membranous, acute, erose; **blades** (1)2–5.5(8) cm long, 0.8–2 mm wide, usually involute, occasionally flat, scabridulous abaxially, villous adaxially. **Panicles** 14–24 cm long, 3–7 cm wide, narrowly pyramidal, open; **primary branches** 2–7 cm, ascending, diverging less than 40° from the rachises; **pedicels** 1–2(3) mm, usually shorter than the florets, appressed; **disarticulation** above the glumes. **Spikelets** 1.7–3 mm. **Glumes** subequal, 1.3–1.8 mm, sparsely hirsute, 1-veined; **lower glumes** 1.3–1.6 mm, acuminate, sometimes awned, awns to 0.5 mm; **upper glumes** 1.4–1.8 mm, wider than the lower glumes, acuminate; **lemmas** (1.7)1.9–2.5 mm,

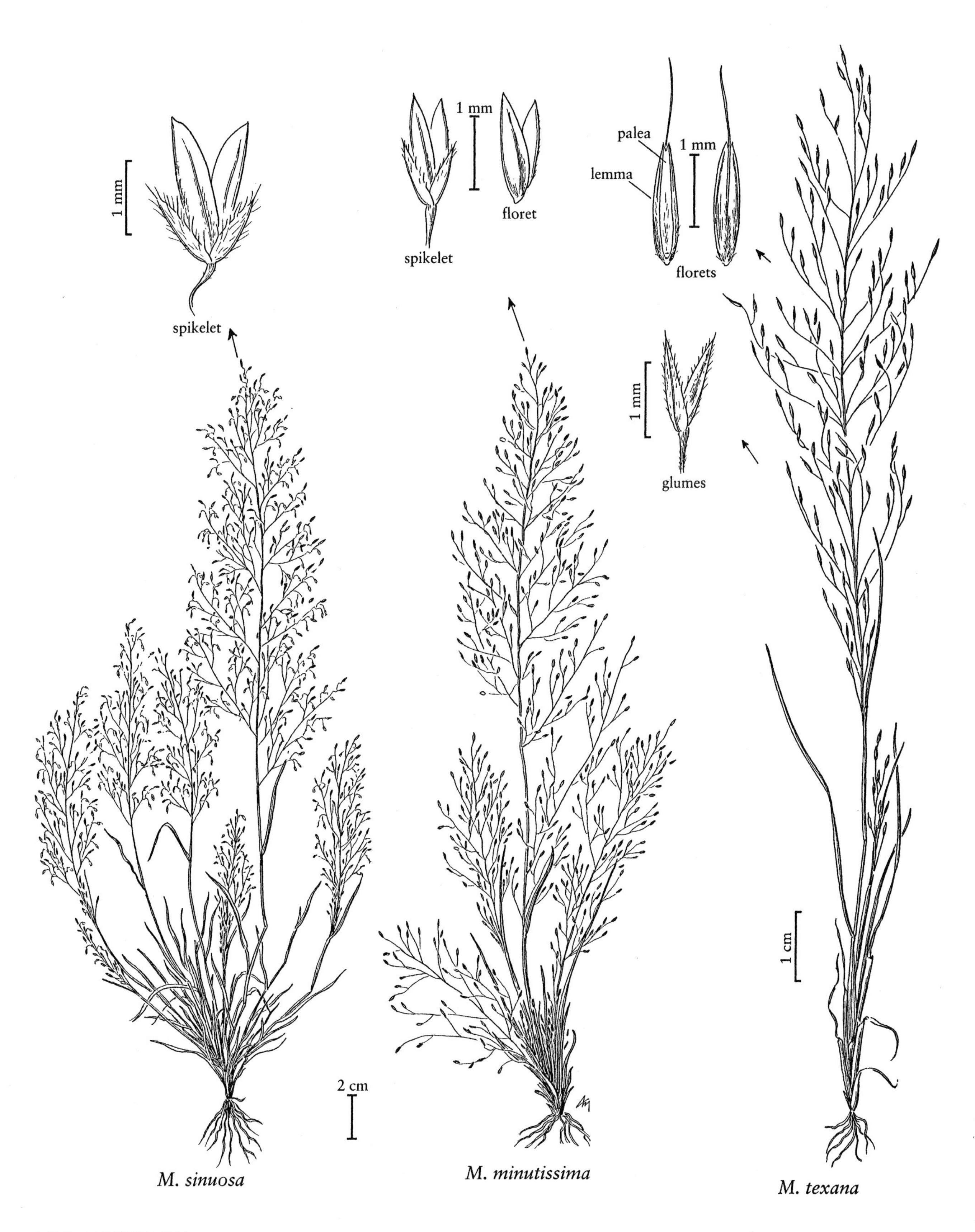
1 mm
spikelet
1 mm
spikelet
floret
palea
1 mm
lemma
florets
1 mm
glumes
2 cm
1 cm
M. sinuosa
M. minutissima
M. texana

lanceolate, purplish or yellowish to light brownish, appressed-pubescent on the midveins and margins, hairs to 0.5 mm, apices awned, awns 1.2–3.5 mm, delicate; **paleas** 1.8–2.4 mm, oblong-elliptic, glabrous; **anthers** 0.4–0.6 mm, purplish. **Caryopses** 1.3–2.3 mm, fusiform to dorsally compressed, light brownish. $2n = 40$.

Muhlenbergia eludens grows in open sandy gullies, washes, rocky slopes, and roadsides. It is found at elevations of 1700–2450 m in the southwestern United States and northwestern Mexico.

69. **Muhlenbergia fragilis** Swallen [p. 201]
DELICATE MUHLY

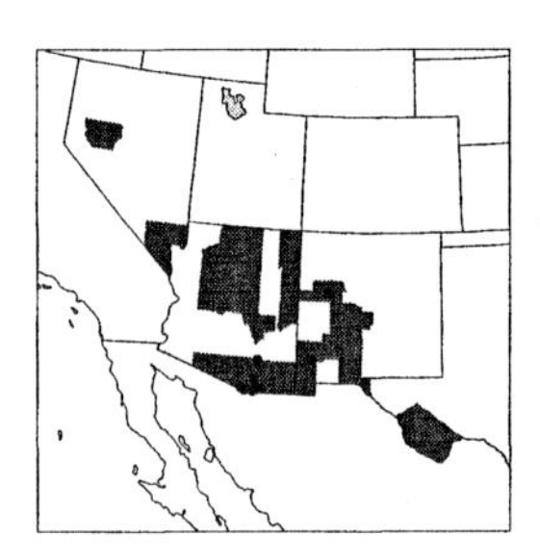

Plants annual. **Culms** 10–38 cm, erect or spreading; **internodes** mostly glabrous, smooth or scabridulous, scabrous or strigulose below the nodes. **Sheaths** often longer than the internodes, scabridulous, margins hyaline; **ligules** 1–3 mm, hyaline, obtuse, irregularly toothed to lacerate, with lateral lobes; **blades** 1–10 cm long, 0.4–2 mm wide, flat, scabrous abaxially, strigulose adaxially, margins and midveins thickened basally, whitish. **Panicles** 10–24 cm long, 3.5–11 cm wide, diffuse; **primary branches** 2.2–6.2 cm long, about 0.1 mm thick, diverging 80–100° from the rachises, straight; **pedicels** 6–10 mm long, about 0.02 mm thick, delicate; **disarticulation** above the glumes. **Spikelets** 1–1.2 mm, appressed to slightly divergent. **Glumes** equal to subequal, 0.5–1 mm, glabrous throughout or obscurely puberulent, hairs about 0.06 mm, 1-veined, obtuse or subacute; **lemmas** 1–1.2 mm, oblong-elliptic, purplish to light brownish, not mottled, glabrous or densely appressed-puberulent on the margins and midveins, apices obtuse, unawned; **paleas** 0.9–1.2 mm, oblong-elliptic; **anthers** 0.3–0.5 mm, purplish. **Caryopses** 0.7–0.9 mm, elliptic, reddish-brown. $2n = 20$.

Muhlenbergia fragilis grows on rocky talus slopes, cliffs, canyon walls, road cuts, and sandy slopes, often over calcareous parent materials, at elevations of 480–2200 m. It is usually found in oak-gramma savannahs, thorn scrub forests, oak-yellow pine forests, and pinyon-juniper woodlands. Its range extends from the southwestern United States to southern Mexico.

Populations may have individual plants with completely glabrous lemmas or may consist entirely of such plants. This morphological variation is not correlated with any distributional or habitat characteristics.

70. **Muhlenbergia ramulosa** (Kunth) Swallen [p. 201]
GREEN MUHLY

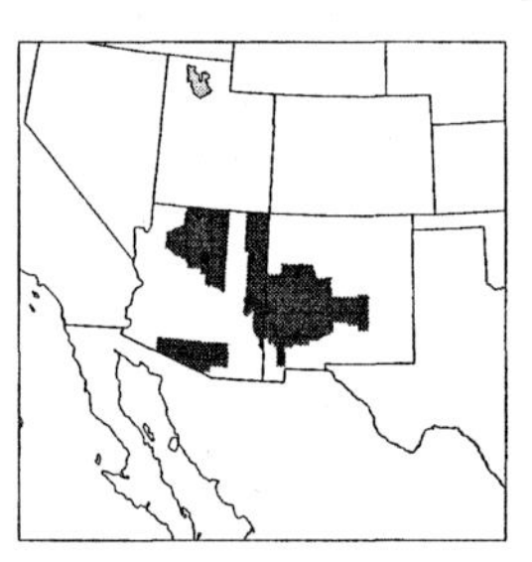

Plants annual. **Culms** (3)5–25 cm, erect, branching, but not rooting, basally; **internodes** glabrous, mostly smooth, sometimes scabridulous below the nodes. **Sheaths** usually shorter than the internodes, glabrous, smooth or scabridulous; **ligules** 0.2–0.5 mm, hyaline, truncate, ciliate, without lateral lobes; **blades** 0.5–3 cm long, 0.8–1.2 mm wide, involute or flat, glabrous abaxially, puberulent adaxially. **Panicles** (1)2–9 cm long, 0.6–2.7 cm wide, open, ovoid or deltoid, spikelets sparse; **primary branches** (0.5)1–3.2 cm long, about 0.1 mm thick, ascending, spreading 20–100° from the rachises; **pedicels** 1–3 mm long, 0.5–1.5 mm thick, stiff. **Spikelets** 0.8–1.3 mm, appressed or divaricate. **Glumes** equal, 0.4–0.7 mm, glabrous, 1-veined, obtuse or subacute, unawned; **lemmas** 0.8–1.3 mm, oval, plump, mottled with greenish-black and greenish-white or ochroleucous patches, glabrous or shortly appressed-pubescent on the margins and midveins, apices acute, unawned; **paleas** 0.7–1.3 mm, oval; **anthers** 0.2–0.3 mm, purplish. **Caryopses** 0.5–1 mm, elliptic, brownish. $2n = 20$.

Muhlenbergia ramulosa grows in open, well-drained areas including slopes, sandy meadows, washes, gravelly road cuts, and rock outcrops in yellow pine-oak forests and in open meadows of pine-fir forests, at elevations of 2100–3500 m. Its range includes the southwestern United States, Mexico, Guatemala, Costa Rica, and Argentina.

17.34 LYCURUS Kunth

Charlotte G. Reeder

Plants perennial; cespitose. **Culms** 10–60 cm, erect to somewhat decumbent, usually branched. **Sheaths** open, compressed-keeled, glabrous, smooth or scabridulous, mostly shorter than the internodes, a 2-veined prophyllum often present; **ligules** hyaline, strongly decurrent, truncate or rounded to elongate and acuminate, sometimes with narrow triangular lobes extending from the edges of the sheath on either side; **blades** folded or flat, rather stiff, with prominent, firm,

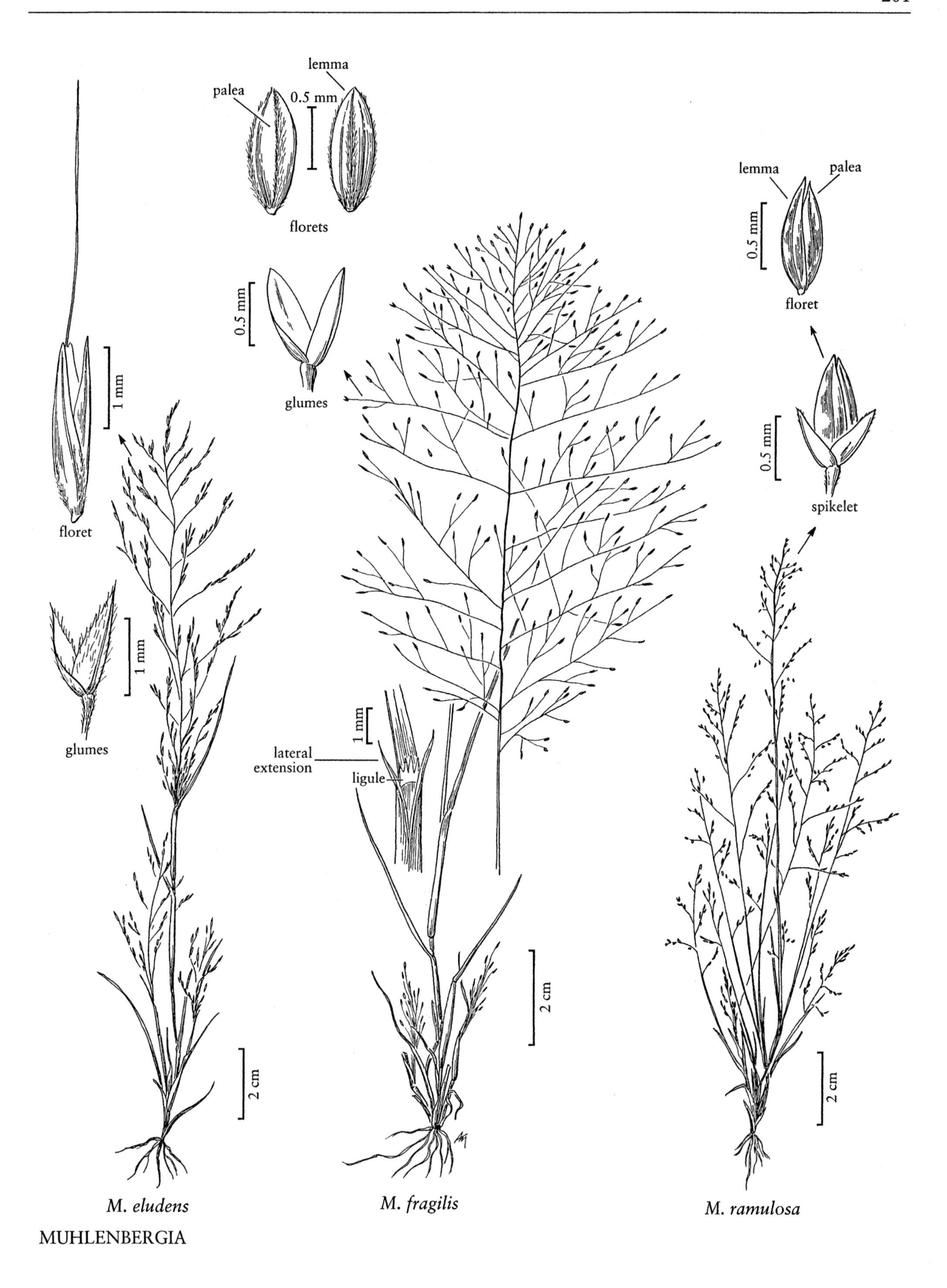

MUHLENBERGIA

scabrous margins, midveins sometimes extending as short mucros or fragile, scabrous, awnlike apices. **Inflorescences** terminal and axillary, dense, bristly, spikelike panicles; **branches** short, fused to the rachis, terminating in a pair of unequally pedicellate spikelets or a pedicellate spikelet and a short secondary branch bearing two spikelets, occasionally in a solitary spikelet, usually the lower spikelet in a pair staminate or sterile and the upper spikelet bisexual, sometimes vice versa, or both spikelets bisexual; **disarticulation** at the fused base of the pedicels or pedicel and branch, paired spikelets falling as a unit, leaving a cuplike tip. **Spikelets** with 1 floret. **Glumes** subequal, awned; **lower glumes** with (1)2(3) awns, usually unequal, awns commonly longer than the body; **upper glumes** 1-veined, with a single flexuous awn that is usually longer than the glume body, rarely a finer second awn present; **lemmas** lanceolate, 3-veined, pubescent on the margins, mostly glabrous over the back, tapering to a scabrous awn that is usually shorter than the lemma body; **paleas** about equal to the lemmas, acute or occasionally the 2 veins extending as very short mucros, pubescent between the veins and on the sides, except for the narrow, glabrous, hyaline margins; **anthers** 3. **Caryopses** fusiform, brownish. x = 10. Name from the Greek *lykos*, 'wolf', and *oura*, 'tail', an allusion to the spikelike inflorescences.

Lycurus is a genus of three species of open rocky slopes and mesas. It is native to two disjunct regions, one extending from Colorado and southern Utah to southern Mexico and Guatemala, the other from Colombia through western South America to west-central Argentina. Two species are native to the *Flora* region. They can only be reliably distinguished by their vegetative characters.

SELECTED REFERENCES **Reeder, C.G.** 1985. The genus *Lycurus* (Gramineae) in North America. Phytologia 57:283–291; **Sánchez, E. and Z.E. Rúgolo de Agrasar.** 1986. Estudio taxonómico sobre el género *Lycurus* (Gramineae). Parodiana 4:267–310.

1. Upper leaves terminating in a fragile, awnlike tip (3)4–7(12) mm long; ligules (2)3–10(12) mm long, elongate, acute or acuminate, sometimes with a small cleft on either side; culms erect .... 1. *L. setosus*
1. Upper leaves acute or with a mucro or bristle 1–3 mm long; ligules 1.5–3 mm long, with evident narrow triangular lobes 1.5–3(4)mm long on the sides; culms erect to ascending, often geniculate .... 2. *L. phleoides*

1. Lycurus setosus (Nutt.) C. Reeder [p. 203]
BRISTLY WOLFSTAIL

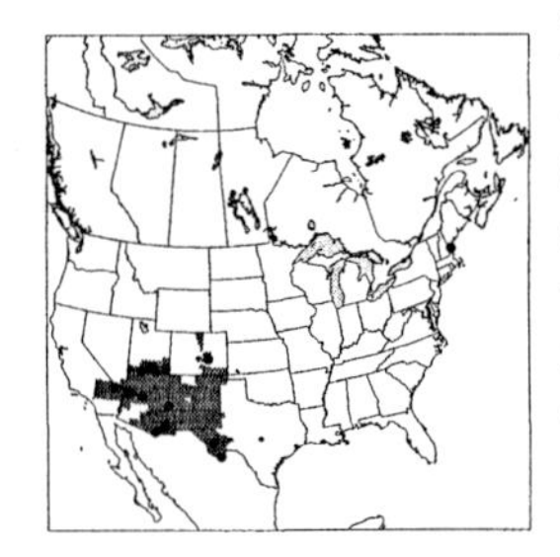

Plants densely cespitose. **Culms** 30–50(60) cm, erect, with several nodes, sparingly branched, scabrous to puberulent at or near the nodes. **Ligules** (2)3–10(12) mm, hyaline, acuminate, sometimes shortly cleft on the sides (in dried specimens the fragile ligules may appear shorter because of the folded tip); **blades** 4–9(13) cm long, 1–2 mm wide, glabrous, smooth or scabridulous abaxially, scabridulous or hispidulous adaxially, with prominent whitish midribs and scabrous margins, midribs extending as fragile, easily broken, awnlike, (3)4–7(12) mm apices. **Panicles** 4–8(10) cm long, (5)7–8 mm wide; **shorter pedicels** 0.8–1(1.5) mm; **longer pedicels** 1–2 mm. **Spikelets** 3–4 mm. **Glumes** 1–1.5(2) mm, scabrous apically; **lower glumes** 2-veined, with (1)2(3) unequal scabridulous awns, shorter awns 1–1.5 mm, longer awns (1)1.5–3(3.5) mm; **upper glumes** 1-veined, with a single, flexuous 2.5–4(5) mm awn; **lemmas** 3–4 mm, tapering to a scabrous 1.5–2(3) mm awn; **anthers** 1.5–2 mm, yellowish. **Caryopses** about 2 mm, brownish. $2n$ = 40.

Lycurus setosus grows on rocky slopes and open mesas, at elevations of 570–3400 m. Its range extends from the southwestern United States to northern Mexico, and, as a disjunct, in Argentina and Bolivia. It was found as an adventive (associated with wool waste) in North Berwick, York County, Maine, in 1902, but it has not been reported from there since. Its flowering time is July–October.

Lycurus setosus is sometimes confused with *Muhlenbergia wrightii*, which has a somewhat similar aspect but is normally found in moist habitats. Also, in *M. wrightii* the first glume is 1-veined with a very short awn, and the lemma is acuminate and unawned or with an awn no more than 1 mm long.

2. **Lycurus phleoides** Kunth [p. 203]
COMMON WOLFSTAIL

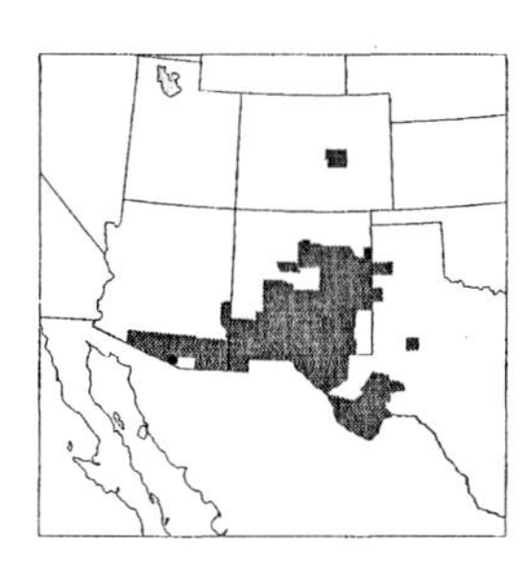

Culms 20–50 cm, erect to ascending, often geniculate. **Ligules** 1.5–3 mm, commonly acute to acuminate, with narrow triangular lobes 1.5–3(4) mm long extending from the sides of the sheaths; **blades** 4–8 cm long, 1–1.5 mm wide, acute or mucronate, midribs sometimes extending up to 3 mm as a short bristle. Similar to *Lycurus setosus* in inflorescence and spikelet characters, except the upper glumes occasionally with a second shorter, more delicate awn. $2n = 40$, ca. 40.

Lycurus phleoides grows on rocky hills and open slopes, at elevations of 670–2600 m. It grows from the southwestern United States to southern Mexico, and in northern South America. It flowers from July–October.

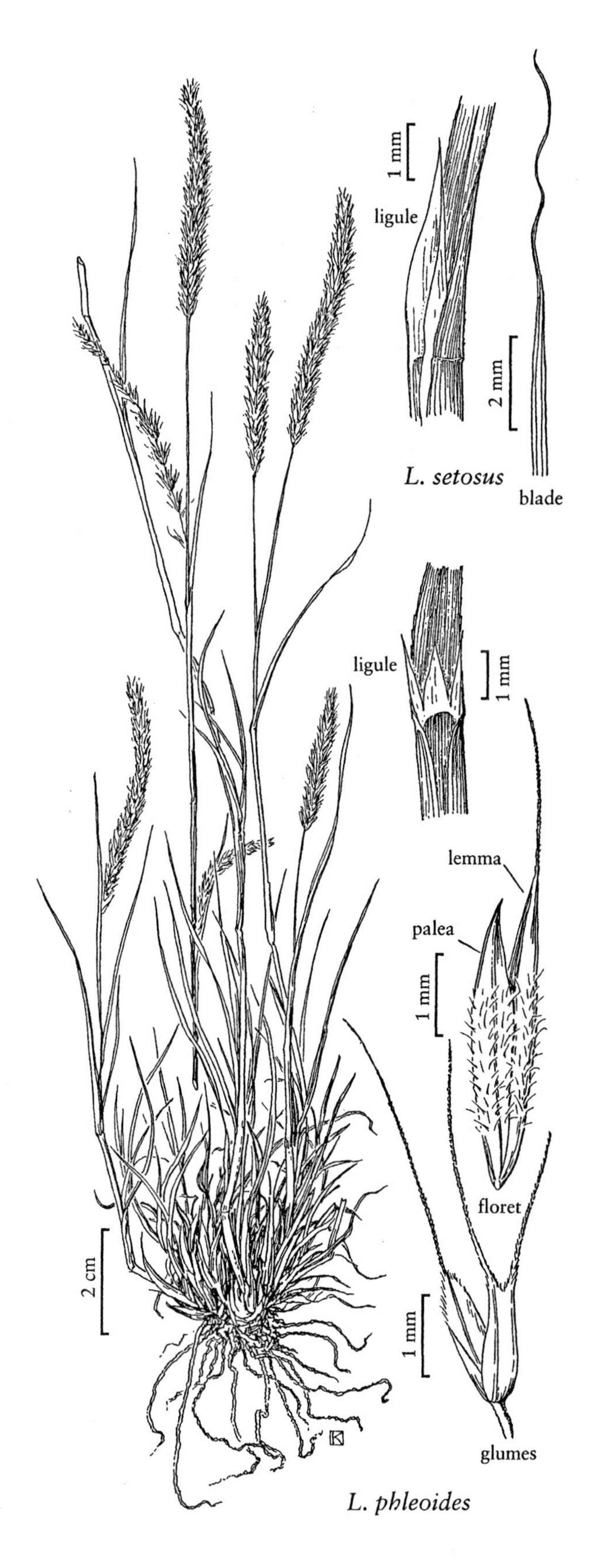

LYCURUS

17.35 CHLORIS Sw.

Mary E. Barkworth

Plants annual or perennial; habit various, rhizomatous, stoloniferous, or cespitose. **Culms** 10–300 cm; **internodes** pith-filled. **Sheaths** strongly keeled, glabrous, scabrous, or pubescent; **ligules** membranous, erose to lacerate or ciliate, occasionally absent; **blades**, particularly those of the basal leaves, often with long, coarse hairs near the base of the adaxial surface and margins. **Inflorescences** terminal, panicles with (1)5–30 spikelike branches, these usually borne digitately, occasionally in 2–several whorls, sometimes with a few isolated branches below the primary whorl(s), all branches usually exceeding the upper leaves; **branches** with spikelets in 2 rows on 1 side of the branch axes. **Spikelets** solitary, sessile to pedicellate, laterally compressed, with 2–3(5) florets, usually only the lowest floret bisexual, rarely the lower 2 florets bisexual, remaining floret(s) sterile or staminate; **florets** laterally compressed or terete, cylindrical to obovoid, awned or unawned, sterile and staminate florets progressively reduced distally if more than 1 present; **disarticulation** usually beneath the lowest floret in the spikelets, all florets falling as a unit, sometimes at the uppermost cauline node, panicles falling intact. **Glumes** unequal, exceeded by the florets, lanceolate, acute to acuminate, usually unawned, occasionally awned, awns to 0.3 mm; **calluses** bearded; **lemmas of bisexual florets** 3-veined, marginal veins pubescent, midveins usually glabrous, sometimes scabrous, usually extending into an awn, sometimes merely mucronate, awns to 37 mm, lemma apices truncate or obtuse, entire or bilobed, lobes, when present, sometimes awn-tipped, awns to 0.6 mm; **paleas** shorter than the lemmas, 2-veined, veins scabrous; **anthers** 3; **lodicules** 2. **Caryopses** ovoid, elliptic, or obovoid. x = (9)10. Named for Chloris, the Greek goddess of flowers.

As interpreted here, *Chloris* is a tropical to subtropical genus of 55–60 species. It is most abundant in the Southern Hemisphere. Of the 18 species treated here, 11 are native, five are species that have been collected in the *Flora* region but seem not to have persisted, one is an introduced species that has become established, and one is known only from cultivation. Many species of *Chloris*, including the native species, provide good forage.

Anderson (1974) interpreted *Chloris* as including *Enteropogon*, *Eustachys*, and *Trichloris*. Support for treating the genus more narrowly can be found in Van den Borre (1994; 2000) and Hilu and Alice (2001), but the limits of many genera in the *Cynodonteae*, including *Chloris*, are not clear.

SELECTED REFERENCES **Anderson, D.E.** 1974. Taxonomy of the genus *Chloris* (Gramineae). Brigham Young Univ. Sci. Bull., Biol. Ser. 19:1–133; **Hilu, K.W.** and **L.A. Alice.** 2001. A phylogeny of the Chloridoideae (Poaceae) based on *matK* sequences. Syst. Bot. 26:386–405; **Hitchcock, A.S.** 1951 [title page 1950]. Manual of the Grasses of the United States, ed. 2, rev. A. Chase. U.S.D.A. Miscellaneous Publication No. 200. U.S. Government Printing Office, Washington, D.C., U.S.A. 1051 pp.; **Van den Borre, A.** 1994. A taxonomy of the Chloridoideae (Poaceae), with special reference to the genus *Eragrostis*. Ph.D. dissertation, Australian National University, Canberra, New South Wales, Australia. 313 pp.; **Van den Borre, A.** and **L. Watson.** 2000. On the classification of the Chloridoideae: Results from morphological and leaf anatomical data analyses. Pp. 180–183 *in* S.W.L. Jacobs and J. Everett (eds.). Grasses: Systematics and Evolution. International Symposium on Grass Systematics and Evolution (3rd:1998). CSIRO Publishing, Collingwood, Victoria, Australia. 408 pp.

1 Panicle branches tightly entangled for most of their length, separable only with difficulty and forming a cylindrical, spikelike inflorescence, the individual branches visibly distinct only at the tip . . . . 1. *C. berroi*

1. Panicle branches sometimes appressed to each other but easily separable, inflorescences evidently composed of multiple spikelike branches borne in 1–several whorls.
   2. Panicles with 1–3 branches.
      3. Lowest (bisexual) lemmas usually glabrous or scabrous, occasionally with a few scattered hairs on the margins.
         4. Lemma of the first sterile floret in each spikelet entire; panicle branches 5–11 cm long, with about 10 spikelets per cm . . . . 9. *C. ventricosa*

4. Lemma of the first sterile floret in each spikelet bilobed for ⅓–½ of its length; panicle branches 4–17 cm long, with 3–7 spikelets per cm on the distal portion ..... 11. *C. divaricata*
3. Lowest (bisexual) lemmas conspicuously hairy on the margins and keels.
5. Lemma of the lowest (bisexual) floret in each spikelet 1.8–2.8 mm long .... 2. *C. ciliata*
5. Lemma of the lowest (bisexual) floret in each spikelet 2.7–4.2 mm long .... 3. *C. canterae*
2. Panicles with 4 or more branches.
6. Spikelets with 3 or more florets (the third often concealed), the lowest (first) floret bisexual, the remainder sterile or staminate, the terminal floret sometimes represented only by a clavate rachilla segment.
7. Second sterile florets ovoid to subspherical, strongly inflated; first sterile florets 0.9–1.3 mm long .... 5. *C. barbata*
7. Second sterile florets widened or not distally, not inflated or inflated only near the apices; first sterile florets 1–3.5 mm long.
8. All sterile or staminate florets awned.
9. Culms to 300 cm tall; panicle branches averaging 10 spikelets per cm; spikelets tawny .... 8. *C. gayana*
9. Culms 30–50 cm tall; panicle branches averaging 6 spikelets per cm; spikelets dark brown to black .... 10. *C. truncata*
8. At least 1 of the sterile or staminate florets in each spikelet unawned.
10. Spikelets barely imbricate, averaging 5–7 per cm; plants annual or short-lived perennials .... 6. *C. pilosa*
10. Spikelets strongly imbricate, averaging 10–14 per cm; plants perennial or annual.
11. Margins of the lowest lemma in each spikelet scabrous, glabrous, or appressed pubescent .... 8. *C. gayana*
11. Margins of the lowest lemma in each spikelet conspicuously hairy, at least distally, the hairs 0.5–3 mm long.
12. Plants annual; third florets, if present, shorter than the subtending rachilla segment .... 7. *C. virgata*
12. Plants perennial; third florets as long as or longer than the subtending rachilla segment.
13. Lowest lemmas 2.7–4.2 mm long .... 3. *C. canterae*
13. Lowest lemmas 1.5–2.8 mm long.
14. Panicles with 2–7 branches; branches 3.5–8 cm long; awns of the first sterile florets 0.9–1.4 mm long .... 2. *C. ciliata*
14. Panicles with (4)8–30 branches; branches (5)8–20 cm long; awns of the first sterile florets 0.8–4 mm long.
15. First sterile florets 1–1.6 mm long .... 4. *C. elata*
15. First sterile florets 2.2–3.2 mm long .... 8. *C. gayana*
6. Spikelets with 2 florets, 1 bisexual and 1 sterile or staminate.
16. Lemmas of the sterile or staminate florets bilobed for at least ⅓ of their length; lateral lobes sometimes awned.
17. Spikelets pectinate, diverging from the branch axes, crowded, averaging 10–14 per cm; plants annual .... 12. *C. pectinata*
17. Spikelets appressed to the branch axes, not crowded, averaging 3–7 per cm; plants perennial .... 11. *C. divaricata*
16. Lemmas of the sterile or staminate florets not bilobed or lobed less than ¼ of their length; lateral lobes, if present, not awned.
18. Lowest (bisexual) lemmas unawned, mucronate, or shortly awned, the awns less than 1.5 mm long.
19. Plants annuals or short-lived perennials, sometimes shortly stoloniferous .... 6. *C. pilosa*
19. Plants perennial, sometimes stoloniferous.
20. Lowest (bisexual) lemmas unawned, sometimes mucronate, the mucros to 1 mm long.

21. Sterile or staminate florets inflated, 1–1.5 mm wide and about equally long . . . . 13. *C. cucullata*
21. Sterile or staminate florets 0.3–0.9 mm wide, usually at least twice as long . . . . 17. *C. submutica*
20. Lowest (bisexual) lemmas always awned, the awns 1–11 mm long.
22. Panicles with 2–9 branches; branches 5–11 cm long . . . . 9. *C. ventricosa*
22. Panicles with 4–28 branches (usually more than 8); branches 2–20 cm long.
23. Panicle branches 10–20, 2–5 cm long; spikelets with only 1 sterile floret . . . . 13. *C. cucullata*
23. Panicle branches 4–28 (usually more than 8), 5–20 cm long, spikelets usually with 2 sterile florets, the lowest often enclosing and concealing the second . . . . 4. *C. elata*
18. Lowest (bisexual) lemmas always awned, the awns 1.5–16 mm long.
24. Lowest lemmas with a conspicuous glabrous or pubescent groove on each side; plants annual or short-lived perennials . . . . 6. *C. pilosa*
24. Lowest lemmas not conspicuously grooved on the sides; plants perennial or annual.
25. Margins of the lowest (bisexual) lemmas conspicuously hairy along most of their length, the hairs strongly divergent, at least some longer than 1 mm.
26. Plants rarely stoloniferous; second (first sterile) florets 1–1.6 mm long . . . . 4. *C. elata*
26. Plants usually stoloniferous; second (first sterile) florets 2.2–3.2 mm long . . . . 8. *C. gayana*
25. Margins of the lowest (bisexual) lemmas glabrous, scabrous, or appressed pubescent, sometimes with a few scattered longer hairs, or with strongly divergent hairs distally but not basally.
27. Panicle branches borne in 2 or more, clearly distinct whorls.
28. Second (first sterile) florets 0.4–0.7 mm long . . . . 18. *C. radiata*
28. Second (first sterile) florets 0.9–2.5 mm long.
29. Panicle branches spikelet-bearing to the base; panicles with 10–16 branches, these usually in several, well separated whorls below a solitary, vertical branch . . . . 14. *C. verticillata*
29. Panicle branches naked on the basal 2–5 cm; panicles with 8–10 branches, these usually digitate, sometimes a second whorl present below the terminal whorl . . . . 16. *C. texensis*
27. Panicle branches usually digitate, in a single terminal cluster, sometimes with a poorly-developed second whorl just below the terminal cluster.
30. Margins of the lowest (bisexual) lemmas with conspicuously longer hairs distally than basally, the distal hairs usually more than 1.5 mm long.
31. Plants annual; third floret, if present, shorter than its subtending rachilla segment . . . . 7. *C. virgata*
31. Plants perennial; third floret, if present, longer than its subtending rachilla segment . . . . 8. *C. gayana*
30. Margins of the lowest (bisexual) lemmas glabrous or with appressed hairs less than 1 mm long, occasionally with a few scattered longer hairs.
32. Panicle branches without spikelets on the basal 2–5 cm . . . . . . 16. *C. texensis*
32. Panicle branches spikelet-bearing to the base or naked for less than 2 cm.
33. Second florets (first sterile florets) 0.1–0.5 mm wide.

34. Plants annual, sometimes rooting at the lower nodes, not stoloniferous; second florets 0.4–0.7 mm long . . . . 18 *C. radiata*
34. Plants perennial, not rooting at the lower nodes but sometimes stoloniferous; second florets 0.7–2.6 mm long.
    35. Culms 10–40 cm; lowest lemmas 1.9–2.7 mm long with awns 1.9–5.2 mm long; second florets 0.9–1.7 mm long . . . . 15. *C. andropogonoides*
    35. Culms to 100 cm; lowest lemmas 2–5.4 mm long, with awns 1–11 mm long; second florets 1–2.6 mm long . . . . 9. *C. ventricosa*

33. Second florets (first sterile florets) 0.5–1 mm wide.
    36. Lemmas of the second florets inconspicuously bilobed; spikelets with (1)2–4 staminate or sterile florets . . . . 8. *C. gayana*
    36. Lemmas of the second florets usually not bilobed; spikelets with 1(2) staminate or sterile florets.
        37. Panicle branches 5–23 cm long; margins of the lowest lemmas appressed pubescent, occasionally sparsely so; second florets awned, the awns 3.1–12.5 mm long; blades without basal hairs . . . . 10. *C. truncata*
        37. Panicle branches 5–11 cm long; margins of the lowest lemmas usually glabrous or scabrous, occasionally sparsely pubescent; second florets awned, the awns 0.5–7.5 mm long; blades with basal hairs up to 3 mm long . . . . 9. *C. ventricosa*

1. **Chloris berroi** Arechav. [p. 209]

Plants perennial; cespitose. **Culms** 15–80 cm. **Sheaths** glabrous; **ligules** ciliate; **blades** 3–15 cm long, 1.5–2 mm wide, glabrous or sparsely pilose near the base. **Panicles** with 2–4 branches, these entangled for most of their length, separable only with difficulty, forming a narrow, cylindrical, spikelike inflorescence, individual branches visibly distinct only at the tips; **branches** 3–12 cm, tightly appressed and adherent, with 9–12 spikelets per cm. **Spikelets** imbricate, with 1 bisexual and 3 sterile florets. **Lower glumes** 1.5–2 mm long, about 0.3 mm wide; **upper glumes** 2.1–2.6 mm long, 0.3–0.6 mm wide; **lowest lemmas** 2.7–3.5 mm, ovate, margins and keels hairy, hairs to 2 mm, awns 2.7–3.4 mm; **second florets** about 1.9 mm, glabrous, awned; **distal florets** unawned. **Caryopses** 1.2–1.8 mm long, 0.5–0.7 mm wide, trigonous. $2n = 40$.

Chloris berroi is native to the Rio de la Plata region of Argentina and Uruguay. It has been cultivated at scattered locations in the United States (Hitchcock 1951), but is not known to be established in the *Flora* region.

2. **Chloris ciliata** Sw. [p. 209]
FRINGED WINDMILL-GRASS

Plants perennial; cespitose. **Culms** 25–60 cm, erect. **Sheaths** glabrous; **ligules** absent or 0.3–0.4 mm, ciliate; **blades** 10–20 cm long, about 5 mm wide, sometimes with long basal hairs, otherwise glabrous or scabrous. **Panicles** digitate, with 2–5(7) evidently distinct branches; **branches** 3.5–6(8) cm, ascending to spreading. **Spikelets** imbricate, with 1 bisexual and 2 sterile florets. **Lower glumes** 1.3–1.7 mm; **upper glumes** 2–2.5 mm; **lowest lemmas** 1.8–2.8 mm long, 0.8–1.1 mm wide, strongly laterally compressed, elliptic, margins and keels conspicuously hairy, hairs 0.5–1.5 mm, apices awned, awns 0.9–1.4 mm; **second florets** 1.3–1.8 mm long, 0.8–1.8 mm wide, widened distally, not inflated, truncate, enclosing the distal florets, awned, awns 0.9–1.4 mm; **distal florets** 0.8–1.1 mm long, 0.9–1.2 mm wide, as long as or longer than the subtending rachilla segments, unawned. **Caryopses** about 1.4 mm long, 0.7 mm wide. $2n = 40$.

Chloris ciliata is a native species of grasslands from the Gulf Coast of Texas, through the Caribbean islands and Mexico to Central America, then, as a disjunct, in Argentina. Argentinean plants differ from northern plants in having long hairs associated with their basal ligules, but no other differences are known. It has been found, as an introduction, in New York.

3. **Chloris canterae** Arechav. [p. 209]
PARAGUAYAN WINDMILL-GRASS

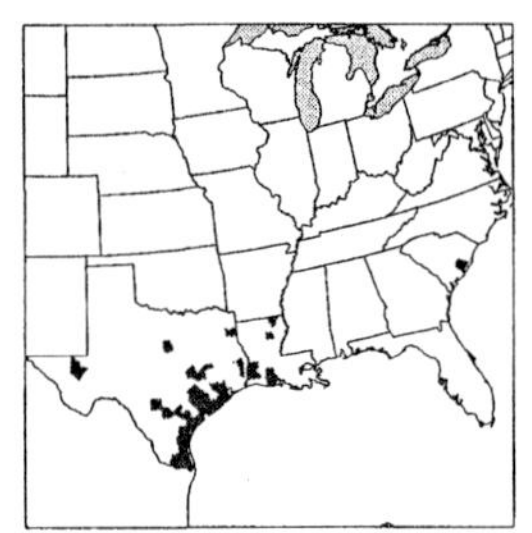

Plants perennial; cespitose. **Culms** to 100 cm, erect. **Sheaths** glabrous; **ligules** about 0.5 mm, membranous, erose; **blades** to 25 cm long, 1–6 mm wide, flat or involute, sometimes appearing filiform, bases with hairs to 7 mm. **Panicles** digitate, with 2–9 evidently separate branches; **branches** 3–14 cm, erect to curving, averaging 11 spikelets per cm. **Spikelets** strongly imbricate, light to medium brown, with 2(3) sterile florets. **Lower glumes** 1.6–2.4 mm; **upper glumes** 2.3–3.8 mm; **lowest lemmas** 2.7–4.2 mm long, 0.6–1.1 mm wide, marginal veins and keels densely and conspicuously hairy, hairs 1.5–3 mm, awns 2.4–5.5 mm; **second florets** 1.1–1.8 mm, about ½ as wide as long, conspicuously widened distally, laterally compressed, glabrous, truncate, awned, awns 1.5–3.5 mm; **distal sterile floret(s)** similar but smaller, longer than the subtending rachilla segments, unawned. **Caryopses** 1.3–2 mm long, 0.8–0.9 mm wide, ovoid-ellipsoid.

Chloris canterae is native to South America. Both of its varieties are found in the coastal plain of Texas and Louisiana. In South America, they are essentially sympatric, but occupy different habitats.

1. Leaves primarily cauline, 2.5–6 mm wide, flat; panicle branches 4–14 cm long . . . var. *canterae*
1. Leaves primarily basal, 1–1.5 mm wide, involute; panicle branches 3–6 cm long . . . . var. *grandiflora*

Chloris canterae Arechav. var. **canterae**

Plants loosely cespitose. **Culms** to 100 cm. **Leaves** primarily cauline; **blades** 2.5–6 mm wide, flat. **Panicles** with 2–9 branches; **branches** (4)5–14 cm. 2*n* = 36.

Chloris canterae var. *canterae* has been collected in Texas and Louisiana. In South America, it grows on moist soils of the *campo* in Paraguay, southern Brazil, and northeastern Argentina.

Chloris canterae var. **grandiflora** (Roseng. & Izag.) D.E. Anderson

Plants densely cespitose. **Culms** 5–30 cm. **Leaves** primarily basal, 1–1.5 mm wide, involute. **Panicles** with 3–5 branches; **branches** 3–6 cm. 2*n* = unknown.

Chloris canterae var. *grandiflora* was collected around woolen mills in southeastern North America in the nineteenth and early twentieth centuries. It has also been found in Texas, but less frequently than var. *canterae*. In South America, it grows in drier, more rocky areas than var. *canterae*.

4. **Chloris elata** Desv. [p. 211]
TALL WINDMILL-GRASS

Plants perennial; usually cespitose, rarely stoloniferous. **Culms** to 135 cm. **Sheaths** mostly glabrous; **ligules** 0.7–1 mm, erose to lacerate; **blades** to 45 cm long, to 15 mm wide, with basal hairs, otherwise usually glabrous, occasionally scabrous. **Panicles** digitate, with (4)8–28 evidently distinct and easily separable branches; **branches** (5)8–20 cm, flexible, usually more or less spreading, sometimes drooping, averaging 12 spikelets per cm. **Spikelets** strongly imbricate, with 1 bisexual and usually 2 sterile florets. **Lower glumes** 1–2.5 mm; **upper glumes** 1.9–3.5 mm; **lowest lemmas** 1.5–2.8 mm long, 0.5–0.9 mm wide, elliptic, strongly laterally compressed, margins conspicuously hairy for most of their length, hairs 1–3 mm, strongly divergent, keels densely appressed pubescent, not or only minutely bilobed, apices awned, awns 1.4–4.8 mm; **second florets** 1–1.6 mm, cylindrical to narrowly turbinate, shortly bilobed, lobes less than ⅕ as long as the lemmas, awned from the sinuses, awns 1.7–4 mm; **third florets** often enclosed by the first sterile florets, 0.5–0.9 mm, turbinate or flabellate, as long as or longer than the subtending rachilla segments, sometimes inflated apically, unawned. **Caryopses** about 1 mm long, about 0.5 mm wide. 2*n* = 72.

The range of *Chloris elata* lies primarily to the south of the *Flora* region, extending from southern Florida and the Caribbean islands to Peru and Argentina.

5. **Chloris barbata** Sw. [p. 211]
SWOLLEN WINDMILL-GRASS

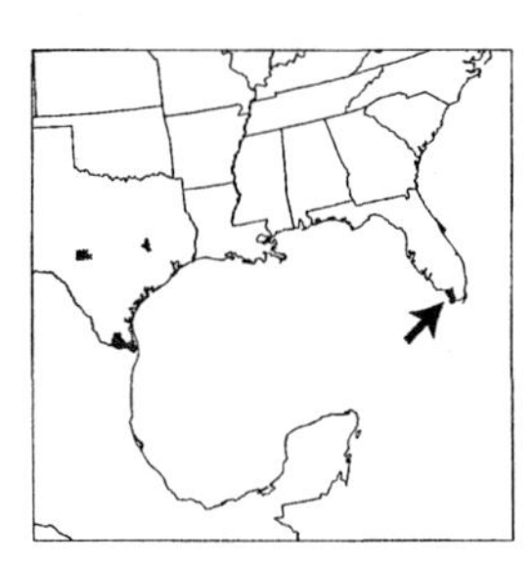

Plants annual. **Culms** 15–95 cm, erect or decumbent and rooting at the lower nodes. **Sheaths** glabrous; **ligules** 0.3–0.5 mm, erose to lacerate; **blades** to 15 cm long, 0.3–0.6 mm wide, with basal hairs, otherwise usually glabrous. **Panicles** digitate, with 7–15 evidently distinct branches; **branches** 3–8 cm, more or less erect, averaging 14 spikelets per cm. **Spikelets** with 1 bisexual and 2(3) sterile florets. **Lower glumes** 1.2–2.1

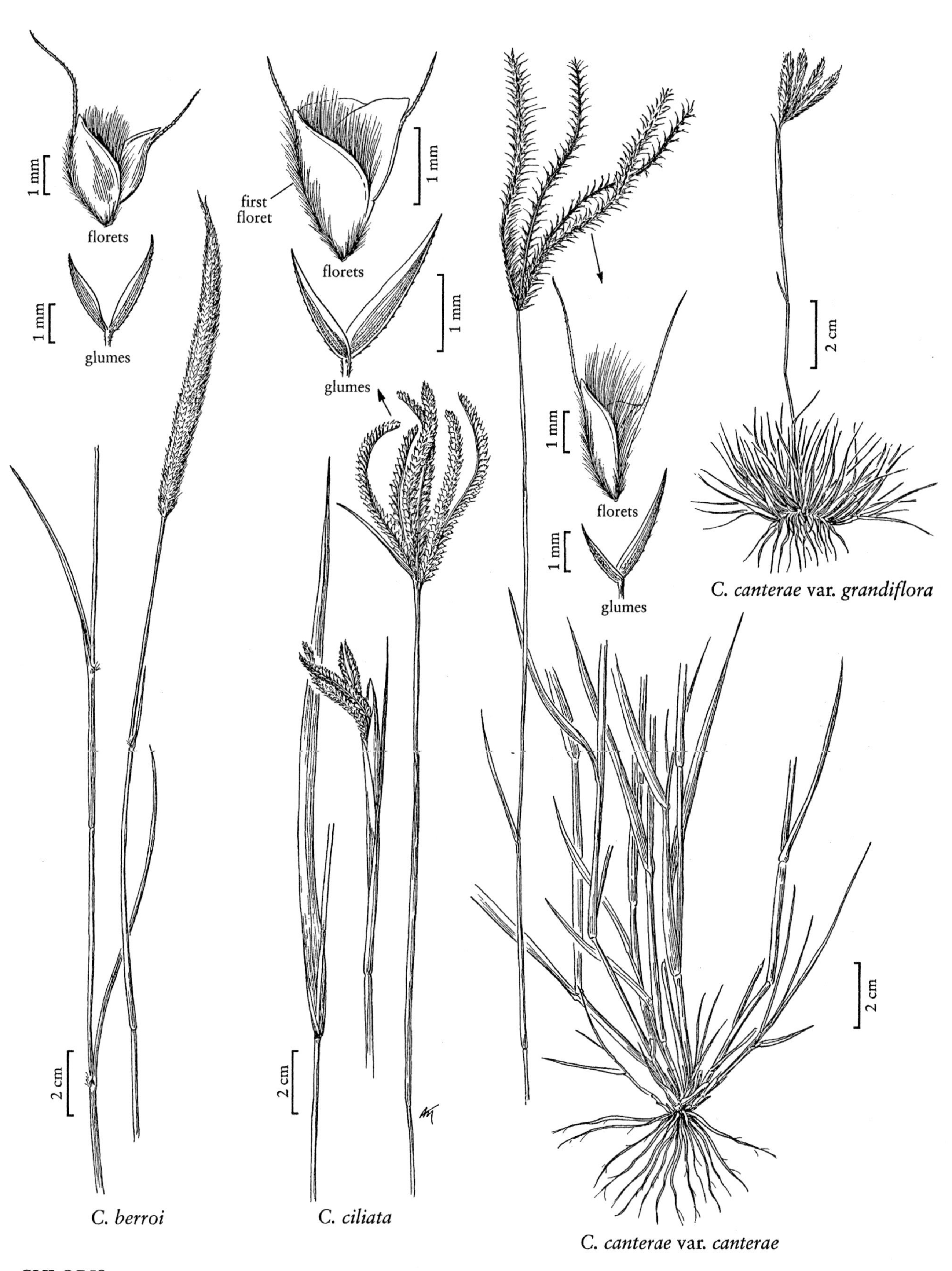
1 mm
florets
1 mm
glumes
first floret
1 mm
florets
1 mm
glumes
1 mm
florets
1 mm
glumes
2 cm
2 cm
2 cm
2 cm
C. canterae var. grandiflora
C. berroi
C. ciliata
C. canterae var. canterae

mm; **upper glumes** 2.3–2.7 mm; **lowest lemmas** 2–2.7 mm, ovate to elliptic, calluses and distal portion of the margins pilose, hairs to 1 mm, keels glabrous or pilose, apices awned, awns 4–7.7 mm; **second florets** 0.9–1.3 mm long, 0.4–0.9 mm wide, slightly to strongly widened distally, inflated, usually glabrous, truncate, awned, awns 5–7 mm; **third florets** obovoid to subspherical, smaller than the first, strongly inflated. **Caryopses** 1.1–1.4 mm. $2n$ = 20, 40, ca. 50.

Chloris barbata grows in subtropical and tropical coastal regions on loams, limestone-derived soils, and along beaches. The main portion of its range lies to the south of the *Flora* region, through the Caribbean and the east coast of Mexico, Central America, and South America. It is a weedy species, often growing in waste areas, but also in cultivated fields.

6. **Chloris pilosa** Schumach. [p. 211]

Plants annual or short-lived perennials; sometimes shortly stoloniferous. **Culms** 30–70(200) cm, erect or somewhat decumbent. **Sheaths** glabrous or sparsely to densely pilose; **blades** to 30 cm long, 2–10 mm wide, with coarse hairs behind the ligule and on the lower portion of the margins. **Panicles** digitate, with 5–9 clearly distinct or easily separable branches; **branches** 3–5 cm, with 5–7 spikelets per cm. **Spikelets** barely imbricate, pale to dark gray, often mottled when mature, with 1 bisexual and (1)2 sterile florets. **Lower glumes** 1.1–1.6 mm; **upper glumes** 1.9–2.3 mm, awned, awns to 0.3 mm; **lowest lemmas** 2.3–3.5 mm, broadly ovate or elliptic, keels gibbous, sides with a conspicuous glabrous or pubescent groove, margins glabrous or appressed pubescent, apices awned, awns to 6 mm; **second florets** 1.5–2.2 mm, widened and inflated distally, mucronate or awned, awns to 3 mm; **distal florets** less than 1 mm, turbinate; **anthers** 0.4–0.5 mm. **Caryopses** 1.3–1.5 mm long, 0.5–0.6 mm wide, trigonous. $2n$ = 20, 30.

Chloris pilosa is native to equatorial Africa, but it is sometimes planted for forage. It has been collected in Kleberg County, Texas, possibly from an experimental forage planting; it is not known to be established in the *Flora* region.

7. **Chloris virgata** Sw. [p. 213]
FEATHER WINDMILL-GRASS, FEATHER FINGERGRASS

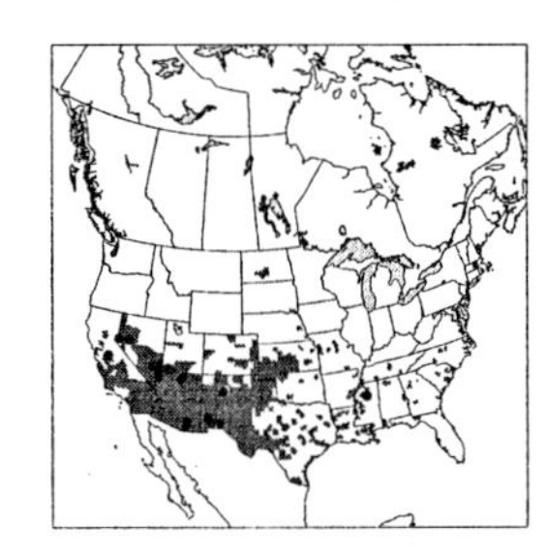

Plants annual; usually tufted, occasionally stoloniferous. **Culms** 10–100+ cm. **Sheaths** usually glabrous; **ligules** to 4 mm, erose or ciliate; **blades** to 30 cm long, to 15 mm wide, basal hairs to 4 mm, otherwise usually glabrous, occasionally pilose. **Panicles** digitate, with 4–20, evidently distinct branches; **branches** 5–10 cm, erect to ascending, averaging 10 spikelets per cm. **Spikelets** strongly imbricate, with 1 bisexual and 1(2) sterile floret(s). **Lower glumes** 1.5–2.5 mm; **upper glumes** 2.5–4.3 mm; **lowest lemmas** 2.5–4.2 mm, keels usually prominently gibbous, glabrous, or conspicuously pilose, sides not grooved, margins glabrous, scabrous or pilose basally, with conspicuously longer hairs distally, hairs longer than 1.5 mm, lemma apices not conspicuously bilobed, awned, awns 2.5–15 mm; **second florets** 1.4–2.9 mm long, 0.4–0.8 mm wide, somewhat widened distally, not inflated, bilobed, lobes less than ⅕ as long as the lemmas, awned from the sinuses, awns 3–9.5 mm; **third florets** greatly reduced, unawned and shorter than the subtending rachilla segment or absent but the rachilla segment present. **Caryopses** 1.5–2 mm long, about 0.5 mm wide, elliptic. $2n$ = 20, 26, 30, 40.

Chloris virgata is a widespread species that grows in many habitats, from tropical to temperate areas with hot summers, including much of the United States. It is a common weed in alfalfa fields of the southwestern United States.

8. **Chloris gayana** Kunth [p. 213]
RHODESGRASS

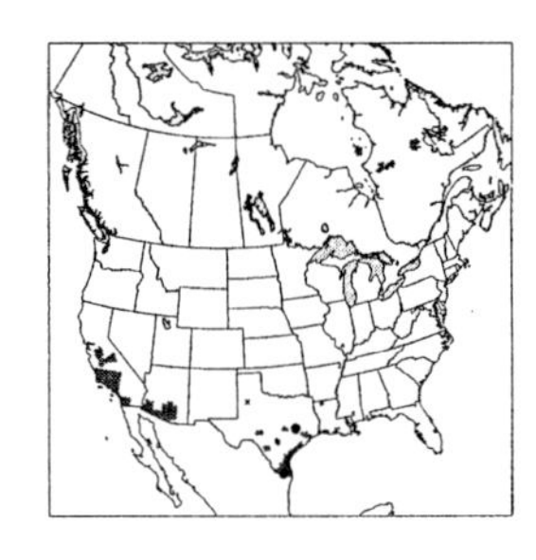

Plants perennial; usually stoloniferous. **Culms** to 300 cm, erect. **Sheaths** glabrous or scabrous, often ciliate apically; **ligules** ciliate; **blades** to 30 cm long, 15 mm wide, scabrous. **Panicles** digitate, with 9–30 evidently distinct branches; **branches** 8–20 cm, usually somewhat divaricate, spikelet-bearing to the base, averaging 10 spikelets per cm. **Spikelets** strongly imbricate, tawny, with 1 bisexual and (1)2–4 usually staminate, sometimes sterile florets. **Lower glumes** 1.4–2.8 mm; **upper glumes** 2.2–3.5 mm; **lowest lemmas** 2.5–4.2 mm long, 0.7–1 mm wide, ovate to obovate or elliptic, somewhat gibbous, sides not grooved, pubescence variable, sides usually glabrous, sometimes

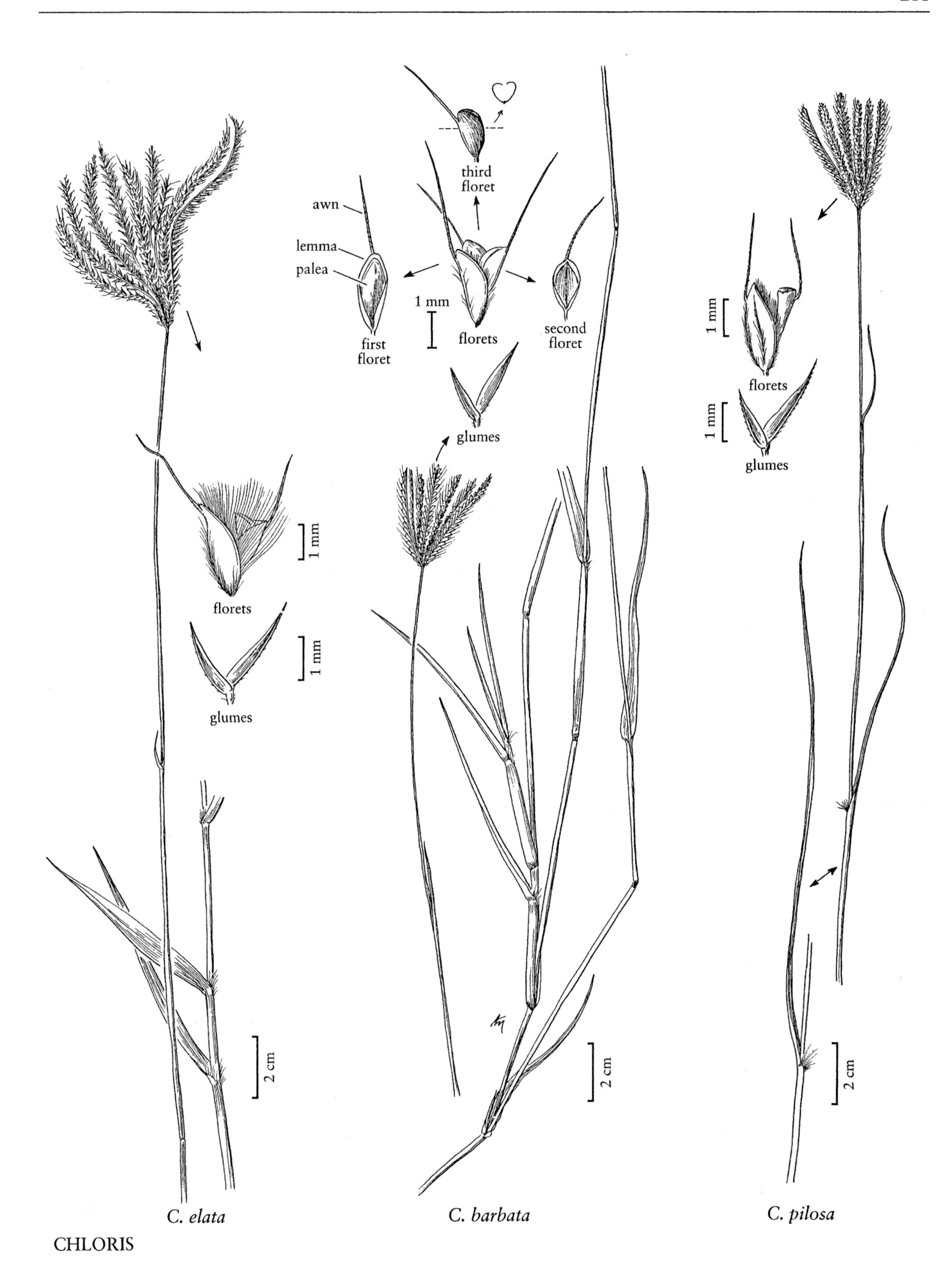
third
floret
awn
lemma
palea
1 mm
first
floret
florets
second
floret
glumes
1 mm
florets
1 mm
glumes
2 cm
2 cm
1 mm
florets
1 mm
glumes
2 cm
C. elata
C. barbata
C. pilosa

CHLORIS

scabrous or appressed pubescent, margins usually glabrous or appressed pubescent on the lower portions, sometimes throughout their length, sometimes with strongly divergent hairs distally, occasionally with strongly divergent hairs their entire length, divergent hairs, when present, 1+ mm, lemma apices inconspicuously bilobed, awned, awns 1.5–6.5 mm; **second florets** staminate or sterile, 2.2–3.2 mm long, 0.3–1 mm wide, similar to the first floret but more cylindrical, not widened distally, inflated, if at all, only near the apices, inconspicuously bilobed, awned, awns 0.8–3.2 mm; **distal florets** progressively smaller, longer than the subtending rachilla segment, awn-tipped or unawned. **Caryopses** 1–1.5 mm long, about 0.5 mm wide. $2n$ = 20, 30, 40.

Chloris gayana grows in warm-temperate to tropical regions throughout the world, including the southern United States. It is cultivated as a meadow grass in irrigated regions of the southwest.

9. **Chloris ventricosa** R. Br. [p. 213]
PLUMP WINDMILL-GRASS

Plants perennial; stoloniferous. **Culms** to 100 cm, erect. **Sheaths** glabrous, scabrous, or partly pilose; **ligules** erose to ciliate; **blades** with basal hairs to 3 mm, otherwise glabrous, scabrous, or sparsely pilose. **Panicles** digitate, with 2–9 evidently distinct branches; **branches** 5–11 cm, spikelet-bearing almost to the base, averaging 10 spikelets per cm. **Spikelets** with 1 bisexual and 1(2) staminate floret(s). **Lower glumes** 1.2–2.3 mm; **upper glumes** 2.5–4.1 mm; **lowest lemmas** 2–5.4 mm long, 0.6–1.2 mm wide, elliptic to obovate, sometimes ventricose, usually glabrous, midveins scabrous, sides not conspicuously grooved, margins inrolled, usually glabrous or scabrous, occasionally with a few scattered hairs, especially distally, apices entire or minutely bilobed, awned, awns 1–11 mm; **second florets** 1–2.6 mm long, 0.3–1 mm wide, cylindrical to narrowly turbinate, apices obtuse to truncate, entire, awned, awns 0.5–7.5 mm; **anthers** 0.7–1.4 mm. **Caryopses** 1.5–2.1 mm long, about 0.4 mm wide, narrowly obovoid to trigonous. $2n$ = unknown.

Chloris ventricosa, an Australian species, has been found near old woolen mills in South Carolina and has been cultivated It is very similar to *C. truncata*, but usually has shorter panicle branches. Other differences include its usually tawny bisexual lemmas and their usually glabrous margins.

10. **Chloris truncata** R. Br. [p. 215]
BLACK WINDMILL-GRASS

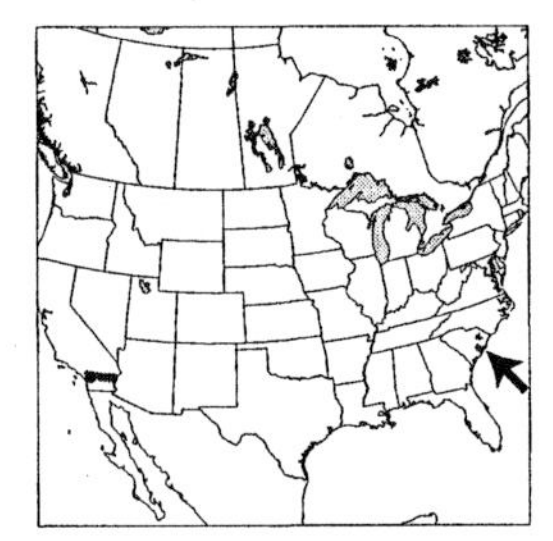

Plants perennial; stoloniferous. **Culms** 30–50 cm. **Sheaths** glabrous; **ligules** short-ciliate; **blades** 9–17 cm long, 0.2–0.3 mm wide, without basal hairs, glabrous, sometimes scabrous. **Panicles** digitate, with 5–13 clearly distinct branches; **branches** 5–23 cm, spikelet-bearing to the base or within 1 cm of the base, averaging 6 spikelets per cm elsewhere. **Spikelets** dark brown to black, with 1 bisexual and 1(2) staminate floret(s). **Lower glumes** 1.4–2.3 mm; **upper glumes** 2.8–4.2 mm; **lowest lemmas** 1.8–4.5 mm long, 0.2–0.7 mm wide, narrowly elliptic, becoming very dark, often almost black at maturity, sides not grooved, mostly glabrous but the margins appressed pubescent, sometimes sparsely so, hairs shorter than 1 mm, apices truncate, awned, awns 3.1–16 mm; **second florets** 1.3–3.5 mm long, 0.5–0.9 mm wide, not inflated, truncate, not or only minutely bilobed, awned, awns 3.1–12.5 mm; **anthers** about 0.6 mm. **Caryopses** 1.7–2.2 mm long, 0.3–0.5 mm wide, ellipsoid to narrowly obovate, trigonous. $2n$ = 40.

Chloris truncata, like the rather similar *C. ventricosa*, is an Australian native that has been found near woolen mills in South Carolina and beside a road near Lake Skinner in Riverside County, California. It usually differs from *C. ventricosa* in having longer panicle branches. Other differences include its very dark, almost black, bisexual lemmas and their usually appressed pubescent margins.

11. **Chloris divaricata** R. Br. [p. 215]
SPREADING WINDMILL-GRASS

Plants perennial; cespitose to shortly stoloniferous. **Culms** 20–50 cm. **Sheaths** glabrous; **ligules** membranous, ciliolate; **blades** to 15 cm long, 1–1.5 mm wide, glabrous or scabrous. **Panicles** digitate, with 3–9 branches; **branches** 4–17 cm, evidently distinct, becoming horizontal, spikelet-bearing to the base or within 0.5 cm of the base, distal portion with 3–7 spikelets per cm. **Spikelets** appressed, with 1 bisexual and 1 sterile floret. **Lower glumes** 0.9–1.8 mm; **upper glumes** 2–2.9 mm; **lowest lemmas** 2.9–4 mm, linear to narrowly lanceolate, mostly glabrous or scabrous, margins glabrous or with a few short, appressed hairs near the apices, 1–3-awned, central awns 7.5–17 mm, lateral lobes unawned or with awns less than 0.4 mm; **second florets** 1.2–1.9 mm,

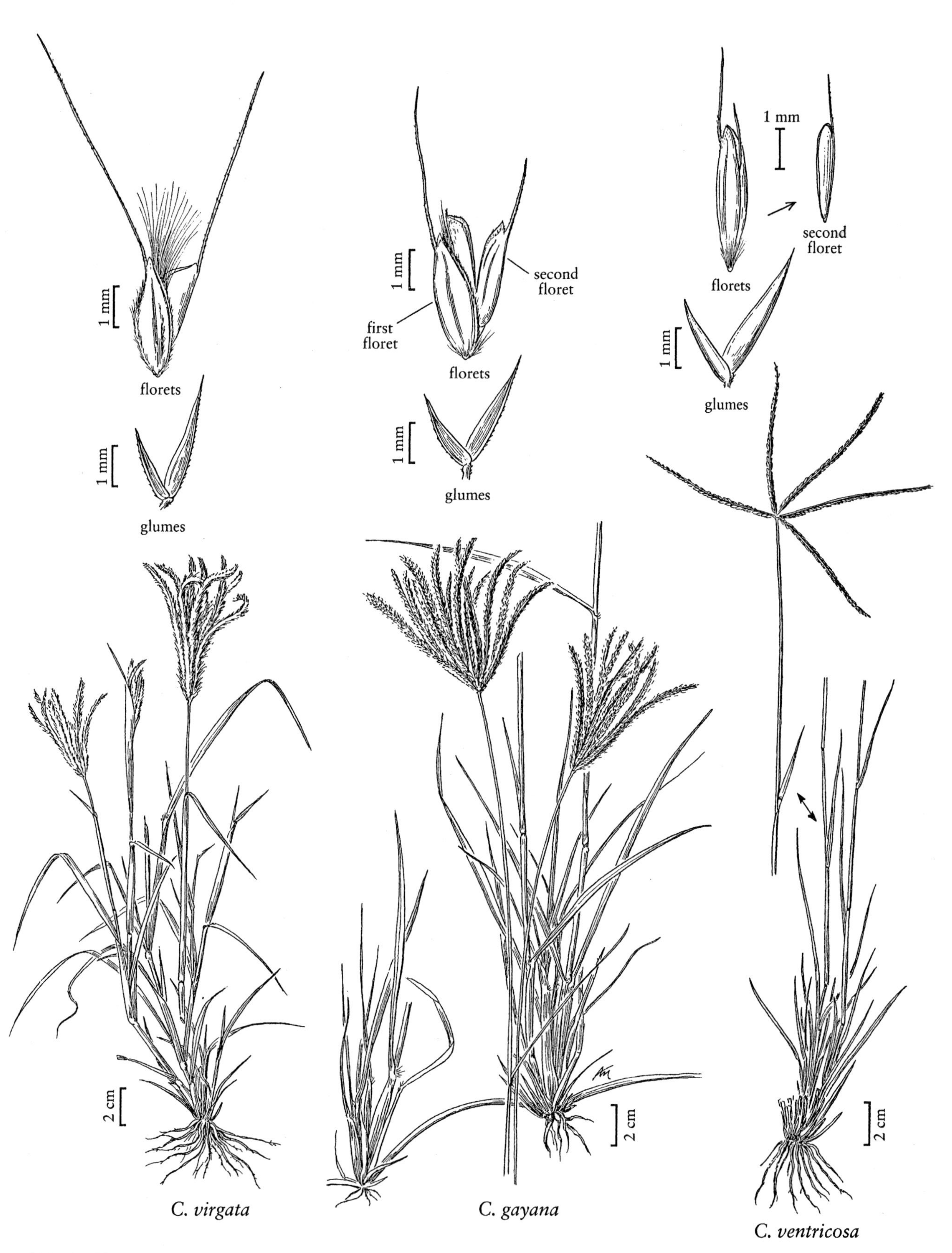
1 mm
florets
1 mm
glumes
1 mm
second floret
first floret
florets
1 mm
glumes
1 mm
second floret
florets
1 mm
glumes
2 cm
2 cm
2 cm
C. virgata
C. gayana
C. ventricosa

CHLORIS

narrowly elliptic, bilobed for ⅓–½ of their length, apices acute, awned from between the lobes, awns 4.5–9.5 mm. **Caryopses** about 2.2 mm long, about 0.4 mm wide, narrowly ellipsoid. $2n$ = unknown.

Chloris divaricata is an Australian species that was collected around woolen mills of South Carolina in the first half of the twentieth century. It has since become established in Texas and New Mexico.

12. **Chloris pectinata** Benth. [p. 215]
COMB WINDMILL-GRASS

Plants annual. **Culms** 20–75 cm, erect, often branched above. **Sheaths** glabrous; **ligules** membranous, ciliate; **blades** to 15 cm long, 2–5 mm wide, sometimes with basal hairs, otherwise glabrous or scabrous. **Panicles** digitate, with 4–13 easily separable or evidently distinct branches; **branches** 5–11 cm, initially erect, becoming divaricate, with 10–14 spikelets per cm. **Spikelets** pectinate, diverging at a wide angle from the branch axes, with 1 bisexual and 1 staminate floret. **Lower glumes** 1.4–2.5 mm; **upper glumes** 2.9–4.3 mm; **lowest lemmas** 3–6.2 mm long, 0.4–0.6 mm wide, linear to narrowly lanceolate, margins glabrous, scabrous, or with hairs less than 0.2 mm, lemma apices bilobed, lobes 0.5–1 mm, sometimes awned, central awns 4–37 mm, awns of lateral lobes, if present, less than 0.6 mm; **second florets** 1.7–2.9 mm long, 0.2–0.3 mm wide, laterally compressed, bilobed, lobes ⅓–½ as long as the lemmas, awned, awns 4–10 mm. **Caryopses** about 2.3 mm long, about 0.3 mm wide, narrowly ellipsoid, trigonous. $2n$ = unknown.

Chloris pectinata is an Australian species that was collected around woolen mills in South Carolina in the first half of the twentieth century.

13. **Chloris cucullata** Bisch. [p. 217]
HOODED WINDMILL-GRASS

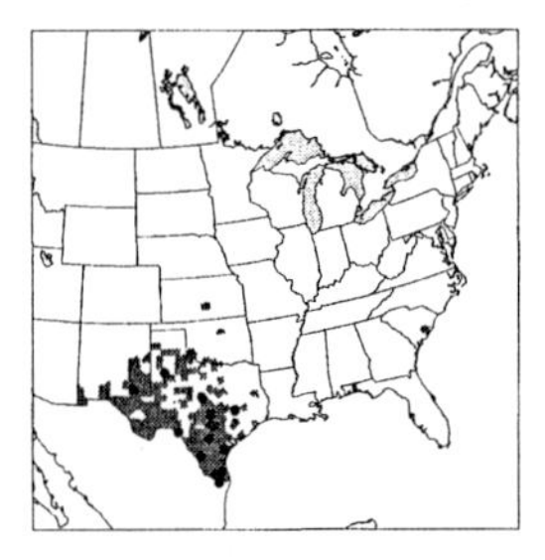

Plants perennial; cespitose. **Culms** 15–60 cm, erect. **Sheaths** glabrous; **ligules** 0.7–1 mm; **blades** to 20 cm long, 2–4 mm wide, without basal hairs, glabrous or scabrous, upper cauline leaves often greatly reduced. **Panicles** with 10–20 branches in several closely-spaced whorls; **branches** 2–5 cm, spreading, with 14–18 spikelets per cm; **disarticulation** beneath the glumes. **Spikelets** with 1 bisexual and 1 sterile floret. **Lower glumes** 0.5–0.7 mm; **upper glumes** 1–1.5 mm; **lowest lemmas** 1.5–2 mm long, 0.7–1 mm wide, broadly elliptic, mostly glabrous but the keels and marginal veins appressed-pilose, obtuse, awned, awns 0.3–1.5 mm; **second florets** 1–1.5 mm long and about equally wide, conspicuously inflated, spherical, with the distal portion of the margins inrolled, not or inconspicuously bilobed, lobes less than ⅕ as long as the lemmas, midveins sometimes excurrent to 1.5 mm. **Caryopses** 0.9–1.2 mm long, about 0.5 mm wide, obovoid. $2n$ = 40.

Chloris cucullata is common along roadsides and in waste areas throughout much of Texas and adjacent portions of New Mexico and Mexico. Records from outside this area probably represent introductions. *Chloris cucullata* hybridizes with both *C. andropogonoides* and *C. verticillata* (see discussion under *C. andropogonoides*).

14. **Chloris verticillata** Nutt. [p. 217]
TUMBLE WINDMILL-GRASS

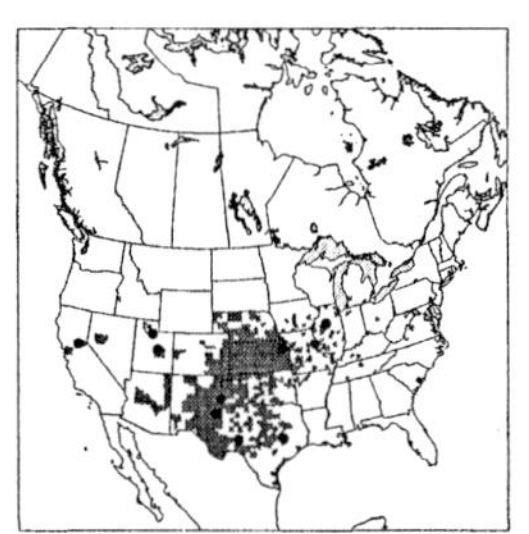

Plants perennial; cespitose. **Culms** 14–40 cm, erect or decumbent, sometimes rooting at the lower nodes. **Sheaths** mostly glabrous, with hairs to 3 mm adjacent to the ligule; **ligules** 0.7–1.3 mm, shortly ciliate; **blades** to 15 cm long, 2–3 mm wide, with basal hairs, otherwise glabrous or scabrous. **Panicles** with 10–16, evidently distinct branches in several well-separate whorls, and a solitary, vertical terminal branch; **branches** 5–15 cm, spikelet-bearing to the base, with 4–7 spikelets per cm; **disarticulation** at the uppermost cauline node, panicles falling intact. **Spikelets** with 1 bisexual and 1 sterile floret. **Lower glumes** 2–3 mm; **upper glumes** 2.8–3.5 mm; **lowest lemmas** 2–3.5 mm long, 1.5–1.9 mm wide, elliptic to lanceolate, keels glabrous or appressed pubescent, sides not conspicuously grooved, margins glabrous or appressed pubescent, acute to obtuse, awned, awns 4.8–9 mm; **second florets** 1.1–2.3 mm, oblong, somewhat inflated, truncate, not or inconspicuously bilobed, lobes less than ⅕ as long as the lemmas, midveins excurrent, forming 3.2–7 mm awns. **Caryopses** 1.3–1.5 mm long, about 0.5 mm wide, elliptic. $2n$ = ca. 28, 40, 63.

Chloris verticillata is a common weed of roadsides, lawns, and waste areas in the central United States. Prior to disruption of the native vegetation, it grew in low areas of the central prairies. It also grows in northern Mexico. *Chloris verticillata* hybridizes with both *C. cucullata* and *C. andropogonoides* (see discussion under *C. andropogonoides*).

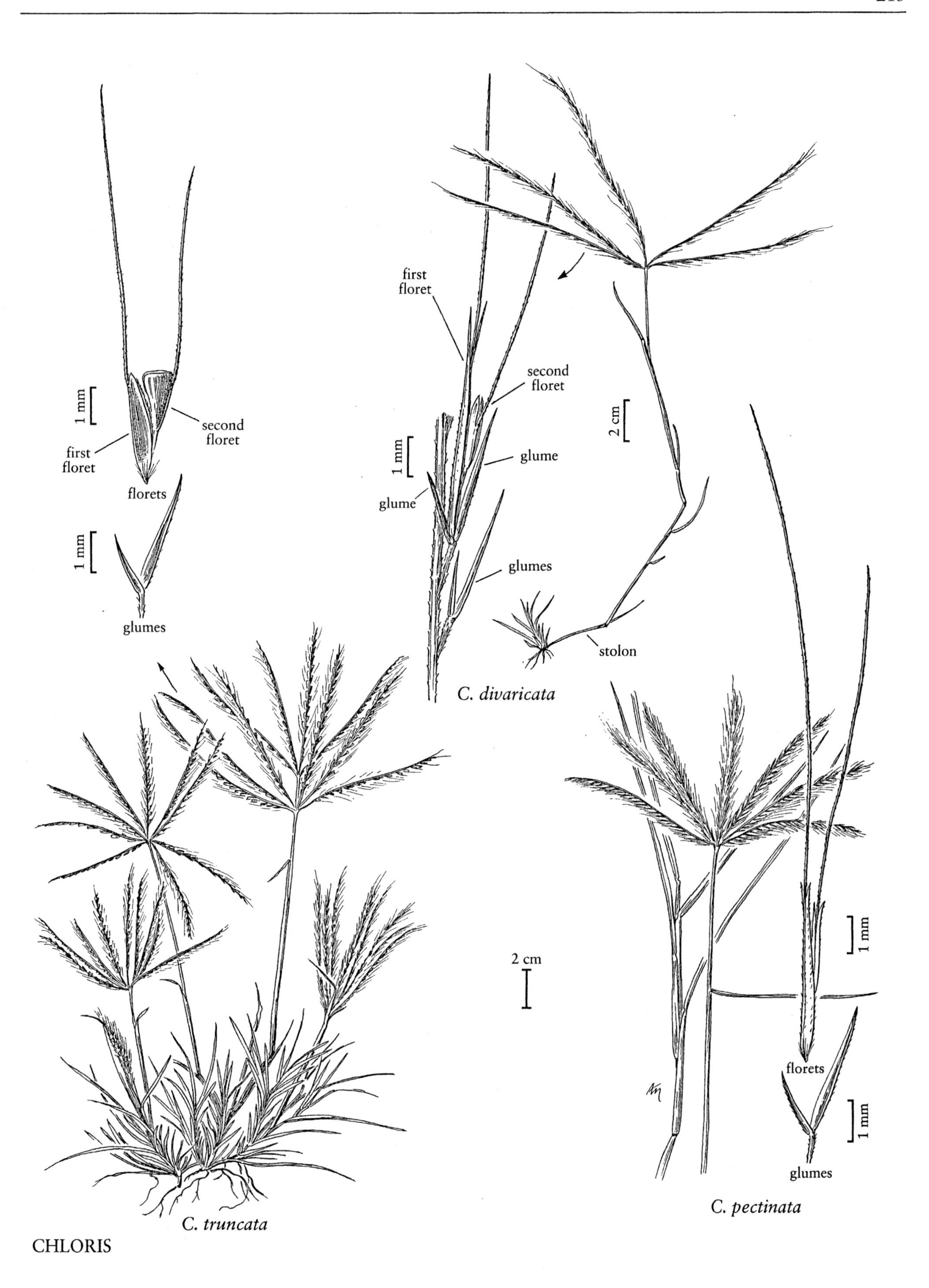
second
floret
first
floret
1 mm
florets
1 mm
glumes
first
floret
second
floret
1 mm
glume
glume
glumes
2 cm
stolon
C. divaricata
C. truncata
2 cm
1 mm
florets
1 mm
glumes
C. pectinata

CHLORIS

15. **Chloris andropogonoides** E. Fourn. [p. 217]
SLIMSPIKE WINDMILL-GRASS

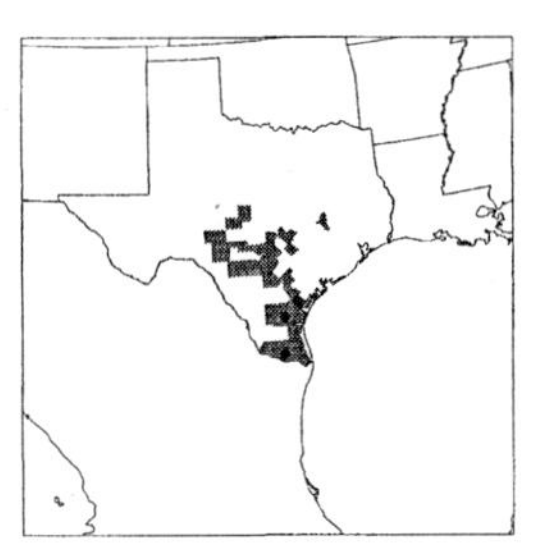

Plants perennial; cespitose to shortly stoloniferous. **Culms** 10–40 cm. **Sheaths** glabrous; **ligules** 0.5–0.8 mm, shortly ciliate; **blades** to 15 cm long, to 1 mm wide, sometimes with basal hairs, mostly glabrous or scabrous. **Panicles** with 6–13, evidently distinct branches, these usually digitate, sometimes with a second, poorly-developed whorl just below the terminal branches; **branches** 4–14 cm, spreading, spikelet-bearing to the base, with 4–7 spikelets per cm; **disarticulation** at the uppermost cauline node, panicles falling intact. **Spikelets** with 1 bisexual and 1 sterile floret. **Lower glumes** 2–2.3 mm; **upper glumes** 3–3.3 mm; **lowest lemmas** 1.9–2.7 mm long, 0.5–0.6 mm wide, lanceolate to elliptic, without conspicuous grooves on the sides, mostly glabrous but the margins and keels appressed pubescent with hairs less than 1 mm, apices acute, awned, awns 1.9–5.2 mm; **second florets** 0.9–1.7 mm, 0.2–0.5 mm wide, narrowly cylindrical, obtuse, bilobed and awned, lobes less than ⅕ as long as the lemmas, awns 2.5–3.5 mm. **Caryopses** 1.3–1.4 mm long, about 0.4 mm wide, ellipsoid. $2n$ = 40.

Chloris andropogonoides grows along grassy roadsides and prairie relicts of the coastal plain of southern Texas and northeastern Mexico.

Hybridization and introgression between *Chloris cucullata*, *Chloris verticillata*, and *Chloris andropogonoides*.

Anderson (1974, pp. 97–103) noted that *Chloris cucullata*, *C. verticillata*, and *C. andropogonoides* are sympatric in southern and central Texas, and often form mixed populations that include many apparent hybrids and introgressants. These plants combine the morphological features of their parents and often have highly irregular meiosis. Diploid counts of about 60 are common in some populations, but seed set is high even in populations with a high level of meiotic irregularity, suggesting apomixis. In some populations, no 'pure' parental plants are found, eliminated either through competition or hybridization. Some of the morphologically-distinct members of such hybrid complexes have been given formal names but, because morphologically-similar hybrids can have different origins, these names do not reflect true taxonomic entities. Among such names are *C. brevispica* Nash, *C. verticillata* var. *aristulata* Torr. & A. Gray, *C. verticillata* var. *intermdia* Vasey, *C. latisquamea* Nash, and *C. subdolichostachya* Müll.-Hal. Plants belonging to such complexes are best named as hybrids between their parents, e.g. "*Chloris verticillata* × *C. andropogonoides*", or as being close to one of the probable parents, e.g., "close to *Chloris andropogonoides* E. Fourn."

16. **Chloris texensis** Nash [p. 219]
TEXAS WINDMILL-GRASS

Plants perennial; cespitose. **Culms** 30–45 cm. **Sheaths** glabrous or sparsely pilose; **ligules** membranous, not or only shortly ciliate; **blades** to 15 cm long, about 4 mm wide, scabrous. **Panicles** with 8–10 clearly separate branches, these usually digitate, occasionally a poorly-developed second whorl present just below the terminal whorl; **branches** to 20 cm, divergent, basal 2–5 cm without spikelets, averaging 3–4 spikelets per cm elsewhere. **Spikelets** with 1 bisexual and 1 sterile floret. **Lower glumes** 2.7–3 mm; **upper glumes** 3.5–3.8 mm; **lowest lemmas** 3.7–4.3 mm long, 0.7–0.8 mm wide, lanceolate to narrowly ovate, sides not conspicuously grooved, margins glabrous or sparsely appressed pubescent distally with hairs shorter than 1 mm, apices acute, awned, awns 7–11 mm; **second florets** 2–2.5 mm long, about 0.5 mm wide, narrowly elliptic, acute, inconspicuously bilobed, awns 4.5–6.5 mm. **Caryopses** about 2.3 mm long, about 0.5 mm wide, ellipsoid, trigonous. $2n$ = unknown.

Chloris texensis appears to be rare. It is endemic to Texas.

17. **Chloris submutica** Kunth [p. 219]
MEXICAN WINDMILL-GRASS

Plants perennial; usually cespitose, occasionally shortly stoloniferous. **Culms** 30–75 cm, erect. **Sheaths** glabrous; **ligules** about 0.5 mm, shortly ciliate; **blades** to 20 cm long, to 5 mm wide, sometimes with long basal hairs, otherwise scabrous. **Panicles** with 5–17, evidently distinct branches in 1–3 closely-spaced whorls; **branches** to 7 cm, usually erect when young, spreading to reflexed at maturity, averaging 12 spikelets per cm. **Spikelets** with 1 bisexual and 1 staminate floret. **Lower glumes** 1.5–3.2 mm; **upper glumes** 2.5–3.4 mm; **lowest lemmas** 2.8–3.7 mm long, 0.6–1.1 mm wide, broadly linear to elliptic, mostly glabrous but the margins appressed pubescent, apices obtuse, not conspicuously bilobed, sometimes shortly mucronate; **second florets** 1.4–2.2 mm long, 0.3–0.9 mm wide, usually at least twice as long as wide, not lobed, unawned, occasionally

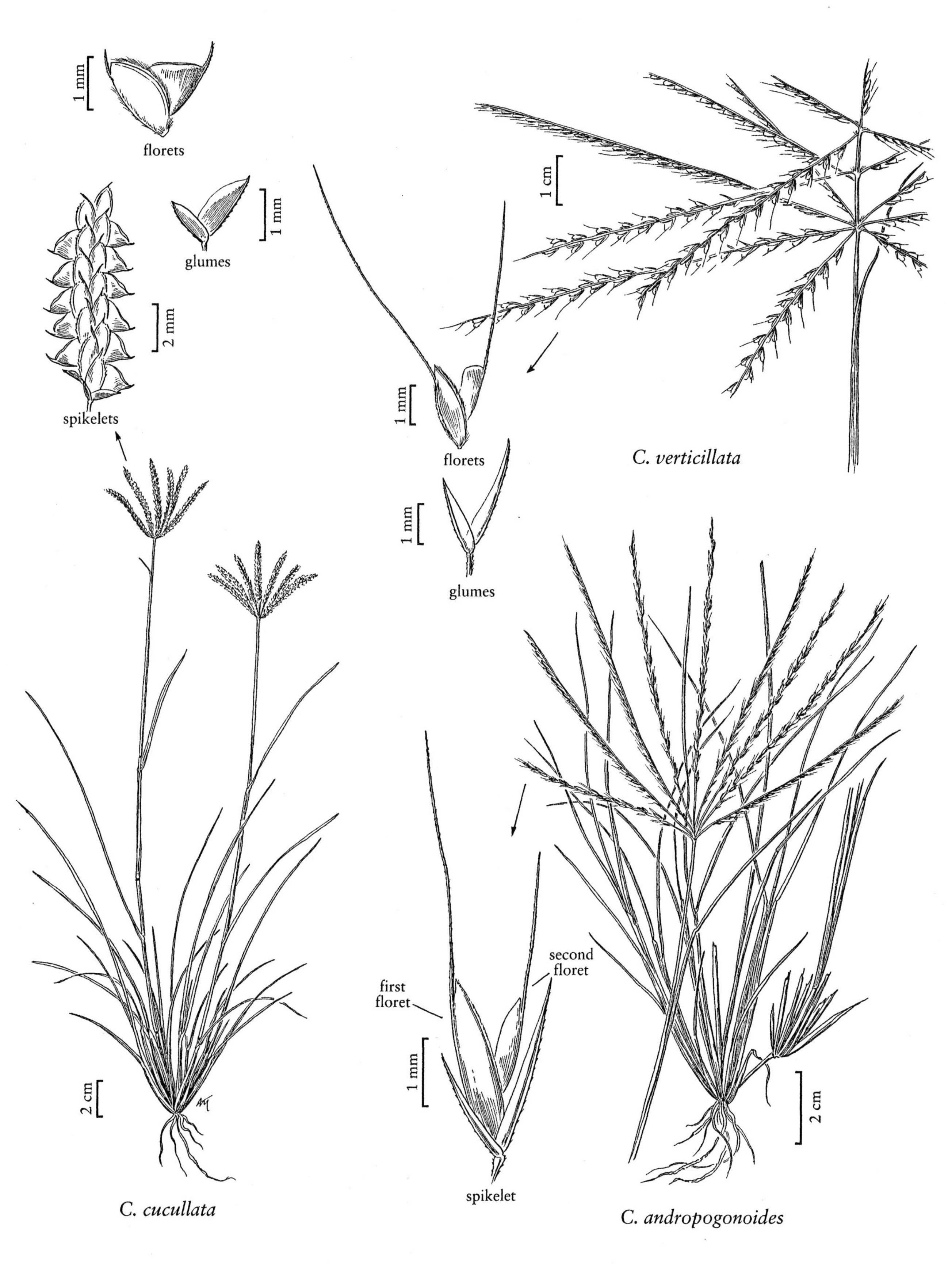
1 mm
florets
1 mm
glumes
2 mm
spikelets
2 cm
C. cucullata
1 cm
1 mm
florets
C. verticillata
1 mm
glumes
first
floret
second
floret
1 mm
spikelet
2 cm
C. andropogonoides

CHLORIS

mucronate. **Caryopses** 1.7–2.3 mm long, 0.5–0.6 mm wide, ellipsoid. $2n=$ ca. 65, 80.

Chloris submutica grows from the southwestern United States through Mexico, Guatemala, and Colombia to Venezuela. In Mexico, it is generally found between 1000–2100 m.

18. **Chloris radiata** (L.) Sw. [p. 219]
RADIATE WINDMILL-GRASS

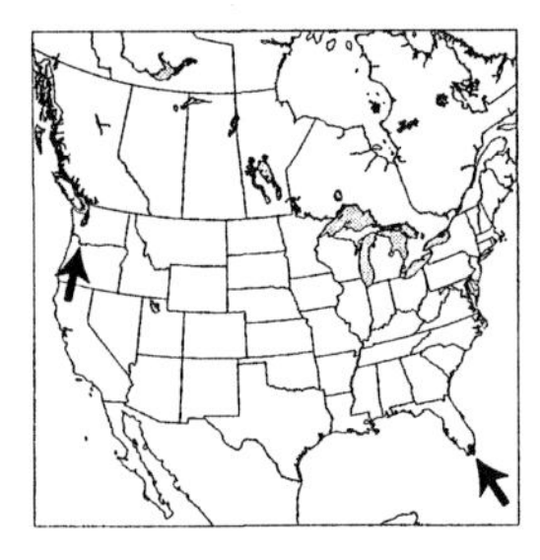

Plants annual; with dense fibrous root growth, not stoloniferous. **Culms** 30–60 cm, erect or decumbent, occasionally rooting at the lower nodes. **Sheaths** usually glabrous, occasionally pilose; **ligules** membranous, shortly ciliate; **blades** 10–30 cm long, to 10 mm wide, sometimes with long basal hairs, usually pilose elsewhere, occasionally glabrous or scabrous. **Panicles** with 5–15, evidently distinct branches in 1–2(3) whorls; **branches** 4.5–8 cm, spikelet-bearing to the base, with 11–15 spikelets per cm distally. **Spikelets** with 1 bisexual and 1 sterile floret. **Lower glumes** 0.7–1.6 mm; **upper glumes** 2–2.7 mm; **lowest lemmas** 2.8–3.3 mm long, 0.4–0.6 mm wide, lanceolate to elliptic, sides not conspicuously grooved, mostly glabrous, margins shortly ciliate distally, hairs less than 1 mm, apices awned, awns 6–13 mm; **second florets** 0.4–0.7 long, about 0.1 mm wide, borne on an equally long or longer rachilla segment, not or inconspicuously bilobed, awned, awns 3–5 mm. **Caryopses** 1.4–1.5 mm long, 0.3–0.4 mm wide, ellipsoidal. $2n = 40$.

Chloris radiata is a weedy species of the eastern Caribbean, Central America, and northern South America. It may be native to Florida, but the record from Linton, Oregon, was from a ballast dump. The species is no longer found in Oregon.

17.36 EUSTACHYS Desv.

Cynthia Aulbach

Plants perennial; cespitose, shortly rhizomatous and often stoloniferous. **Culms** 20–150 cm, erect or decumbent, flattened, glabrous; **internodes** hollow. **Leaves** basal and cauline; **sheaths** open, keeled, strongly compressed, distinctly distichous, equitant; **ligules** to 0.5 mm, scarious, densely short ciliate; **blades** flat or folded, erect to spreading, both surfaces usually glabrous, sometimes scabrous or scabridulous. **Inflorescences** terminal, exceeding the upper leaves, panicles of 1–36 non-disarticulating spikelike branches; **branches** digitately arranged, axes triquetrous, with spikelets in 2 rows on the abaxial sides of the branches. **Spikelets** solitary, diverging strongly from the branch axes, laterally compressed, sessile or subsessile, with 2–3 florets; **lowest florets** bisexual; **second florets** usually reduced to a stipitate empty lemma, occasionally with a palea, staminate if a third floret is present; **third florets,** if present, sterile and usually rudimentary, stipitate; **disarticulation** above the glumes. **Glumes** unequal, 1-veined; **lower glumes** somewhat smaller than the upper glumes, narrow, acuminate; **upper glumes** almost as long as the spikelets, flattened, scarious, glabrous, green, pale, or purplish, veins antrorsely scabrous, apices truncate, bilobed, or bifid, often mucronate; **lowest lemmas** cartilaginous, light to dark brown, scabridulous distally, 3-veined, unawned, mucronate, or with a single awn, awns to 1.2 mm; **paleas** equaling or slightly shorter than the lemmas, glabrous, 2-veined, veins keeled, shortly ciliate or scabridulous; **anthers** 3, deep purple to purple-red. **Caryopses** 1–1.7 mm, trigonous-ellipsoid, glabrous, translucent when fresh, pale to slightly reddish-purple-tinged. $x = 10$. Name from the Greek *eu*, 'truly', and *stachys*, 'spike', probably an allusion to the spikelike panicle branches.

Eustachys, as treated here, is a genus of approximately 12 species, most of which are native to the Western Hemisphere; four are native to the *Flora* region and three have been introduced. It is, in many ways, morphologically similar to *Chloris*, with the placement of a few species being problematic. Molecular data (Alice et al. 2000) support its recognition as a distinct genus, but the relationships between it, *Chloris sensu stricto*, and *Cynodon* are not clear.

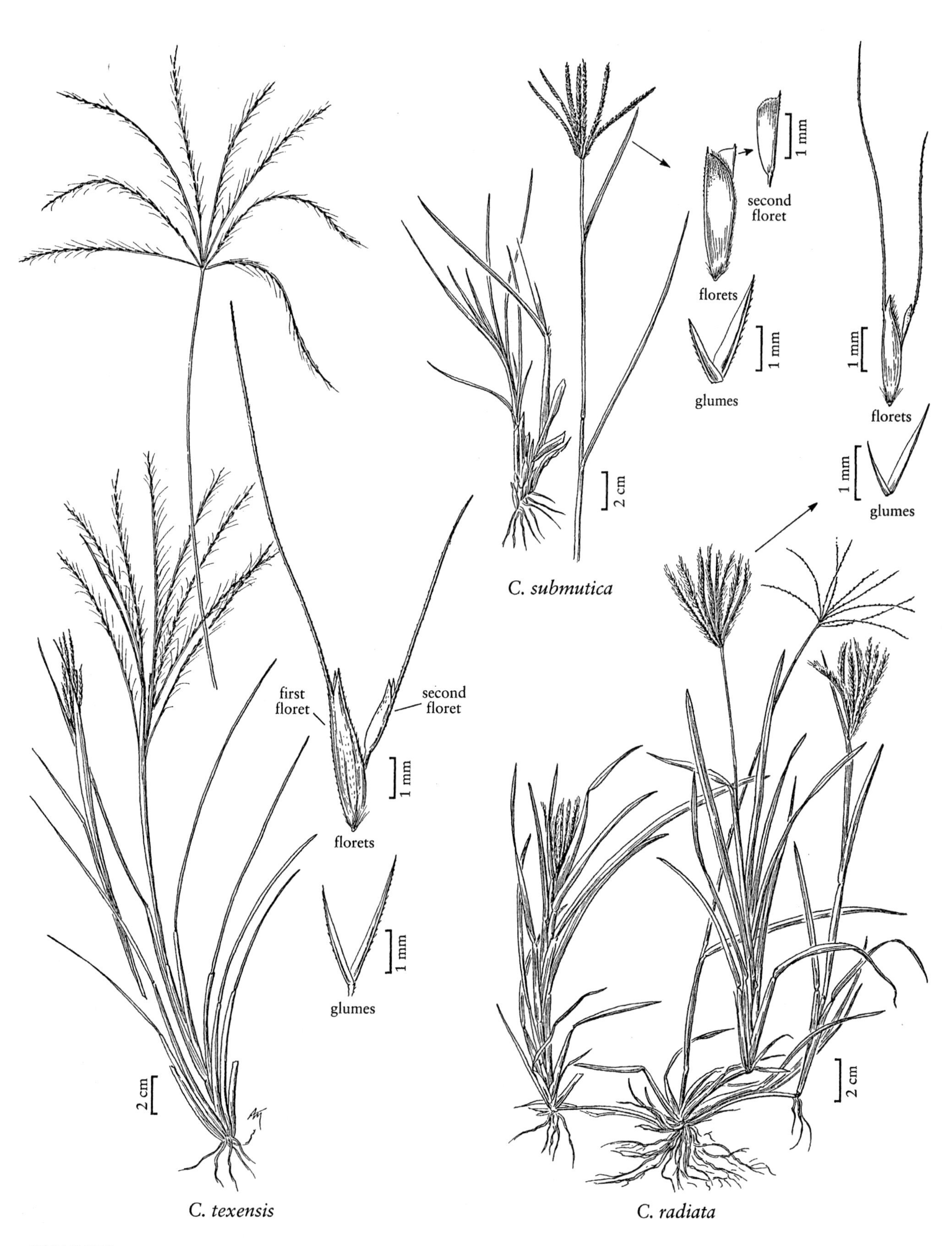
second
floret
1 mm
florets
1 mm
glumes
2 cm
C. submutica
1 mm
florets
1 mm
glumes
first
floret
second
floret
1 mm
florets
1 mm
glumes
2 cm
C. texensis
2 cm
C. radiata

CHLORIS

SELECTED REFERENCES **Alice, L.A., G.G. Borneo,** and **K.W. Hilu.** 2000. Systematics of *Chloris* (*Chloridoideae*; *Poaceae*) and related genera: Evidence from nuclear ITS and chloroplast matK sequences. Amer. J. Bot. 87, Suppl.: 108–109; **Anderson, D.E.** 1974. Taxonomy of the genus *Chloris* (Gramineae). Brigham Young Univ. Sci. Bull., Biol. Ser. 19:1–133.

1. Lateral veins of the lowest lemma in each spikelet usually glabrous, occasionally with a few short, stiff hairs . . . . . 1. *E. glauca*
1. Lateral veins of the lowest lemma in each spikelet pubescent.
   2. Keels of the lowest lemma in each spikelet glabrous.
      3. Spikelets 1.5–2.1 mm long; panicles with 6–15 branches; branches 4–10 cm long, straight, somewhat stiff; leaf blades 5–10 mm wide . . . . . 6. *E. retusa*
      3. Spikelets 2.4–3 mm long; panicles with 10–36 branches; branches 6–15 cm long, flexible; leaf blades 10–15 mm wide . . . . . 7. *E. distichophylla*
   2. Keels of the lowest lemma in each spikelet pubescent.
      4. Spikelets 1.5–2.5 mm long; lowest lemma in each spikelet mucronate.
         5. Lowest lemma in each spikelet dark brown, the lateral veins with appressed hairs shorter than 0.5 mm . . . . . 2. *E. petraea*
         5. Lowest lemma in each spikelet tawny to reddish-brown, the lateral veins with spreading hairs longer than 0.5 mm . . . . . 5. *E. caribaea*
      4. Spikelets 2.6–3.7 mm long; lowest lemma in each spikelet awned, awns 0.4–1.2 mm.
         6. Panicle branches 1–3; awns of lowest lemma in each spikelet 0.4–0.6 mm long; spikelets 3–3.7 mm long . . . . . 3. *E. floridana*
         6. Panicle branches (3)4–9; awns of lowest lemma in each spikelet 0.7–1.2 mm long; spikelets 2.6–3 mm long . . . . . 4. *E. neglecta*

1. **Eustachys glauca** Chapm. [p. 221]
SALTMARSH FINGERGRASS

Culms 60–150 cm, erect. **Leaves** basally disposed, glaucous; **blades** to 30 cm long, 4–15 mm wide, flat or folded, apices obtuse. **Panicles** with 8–24 branches; **branches** 5–15 cm. **Spikelets** 1.5–2(2.4) mm; **florets** 2. **Lower glumes** 1.2–1.4 mm, acute; **upper glumes** 1.3–1.8 mm, narrowly oblong, truncate, awned, awns 0.4–0.8 mm; **calluses** glabrous; **lowest lemmas** 1.6–2.1 mm, narrowly ovate, dark brown at maturity, veins glabrous, occasionally both keels and veins with a few short, stiff hairs, apices acute, unawned, occasionally mucronate; **second lemmas** 1–1.3(1.6) mm, oblong, frequently keeled, acute. **Caryopses** about 1 mm. 2*n* = unknown.

Eustachys glauca grows in the margins of low woods and in ditches and brackish marshes. It is endemic to the southeastern United States.

2. **Eustachys petraea** (Sw.) Desv. [p. 221]
PINEWOODS FINGERGRASS

Culms 20–100 cm, erect, or decumbent and rooting at the nodes. **Blades** (2)5–8(26) cm long, 4–10 mm wide, folded, apices obtuse. **Panicles** with (1)4–6(11) branches; **branches** 2–11.5 cm. **Spikelets** 1.5–2.1 mm; **florets** 2–3. **Lower glumes** 1–1.7 mm, apices acute; **upper glumes** 1.5–1.8 mm, obovate, bilobed, lobes acute or obtuse, awned from between the lobes, awns 0.3–0.9 mm; **calluses** glabrous or with a few hairs, hairs to 0.3 mm; **lowest lemmas** 1.5–2 mm, ovate, dark brown at maturity, lateral veins and keels with appressed, brown hairs, hairs to 0.4 mm, apices mucronate; **second lemmas** about 1 mm, broadly cuneate, occasionally mucronate, apices rounded or truncate. **Caryopses** 1–1.2 mm. 2*n* = 40.

Eustachys petraea grows on dunes and open sandy areas and along roadsides and salt and brackish marshes. Its range extends south from the United States through Mexico to Panama.

first
floret
1 mm
upper
glume
lower
glume
2 cm
2 cm
1 mm
second
lemma
first
lemma
florets
2 cm
palea
1 mm
second
lemma
florets
first
lemma
1 mm
upper
glume
spikelet
lower
glume
E. glauca
E. petraea

3. **Eustachys floridana** Chapm. [p. 223]
FLORIDA FINGERGRASS

Culms 50–100 cm, erect. **Blades** to 30 cm long, to 8.2 mm wide, apices acute. **Panicles** with 1–3 branches; **branches** 5–13 cm. **Spikelets** 3–3.7 mm; **florets** 3. **Lower glumes** 1.8–2 mm, apices acute to obtuse; **upper glumes** 2.2–3.1 mm, oblanceolate, truncate to occasionally bilobed, lobes obtuse, awned from between the lobes, awns 0.5–1 mm; **calluses** with tufts of hairs, hairs to 0.6 mm; **lowest lemmas** 2.9–3.7 mm, ovate, tan to light brown at maturity, lateral veins and keels with appressed, whitish to golden hairs, hairs to 0.7 mm, apices acute, awned, awns 0.4–0.6 mm, arising from just below the apices; **second lemmas** 1.5–2.6 mm, obovate, acute, mucronate. **Caryopses** about 1.7 mm. $2n$ = unknown.

Eustachys floridana grows in dry, sandy woods and old fields. It is endemic to the southeastern United States.

4. **Eustachys neglecta** (Nash) Nash [p. 223]
FOUR-SPIKE FINGERGRASS

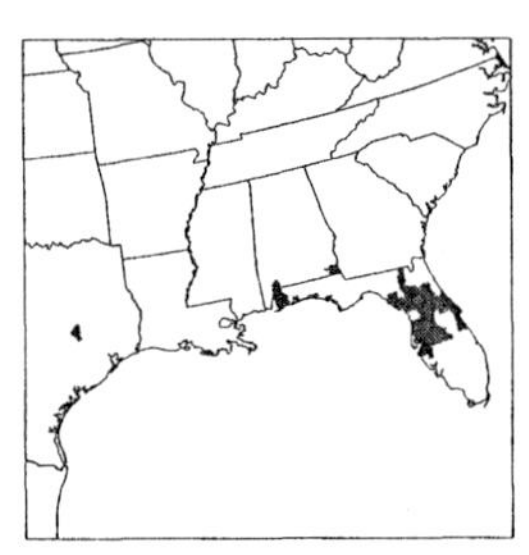

Culms 45–90(120) cm, erect. **Blades** to 21 cm long, to 8 mm wide, acute, occasionally scabridulous basally. **Panicles** with (3)4–9 branches; **branches** 3.3–15.5 cm. **Spikelets** 2.6–3 mm; **florets** 2–3. **Lower glumes** to 2.2 mm, apices obtuse or bifid; **upper glumes** 2.2–2.8 mm, narrowly oblong, apices bilobed, lobes usually acute, occasionally truncate, awned from between the lobes, awns 0.5–1.8 mm; **calluses** with a few hairs, hairs to 0.6 mm; **lowest lemmas** 2.3–3 mm, ovate, medium to dark brown at maturity, lateral veins and keels with slightly spreading, golden to tawny hairs, hairs to 0.7 mm, apices acute, awned, awns 0.7–1.2 mm; **second lemmas** 1.4–2.1 mm, obovate to oblanceolate, apices truncate, awned, awns 0.5–0.8 mm. **Caryopses** about 1.5 mm. $2n$ = unknown.

Eustachys neglecta grows in sandy fields, roadsides, and pinelands. It is endemic to the southeastern United States.

5. **Eustachys caribaea** (Spreng.) Herter [p. 223]
CHICKENFOOT GRASS

Culms 20–70 cm, erect. **Blades** 2–22 cm long, 4–10 mm wide, flat or folded, apices obtuse or mucronate. **Panicles** with 3–10 branches; **branches** 3–9 cm. **Spikelets** 2–2.5 mm; **florets** 2. **Lower glumes** 0.9–1.3 mm, acute; **upper glumes** 1.6–1.8 mm, obovate, apices bilobed, lobes obtuse, awned, awns 0.4–0.7 mm; **calluses** sparsely hairy; **lowest lemmas** 2.1–2.5 mm, ovate, tawny to reddish-brown at maturity, lateral veins and keels with spreading, white, 0.5–1 mm hairs, apices mucronate; **second lemmas** 1.2–1.5 mm, obconic, truncate. **Caryopses** 1.1–1.2 mm. $2n$ = unknown.

Eustachys caribaea has been introduced to the United States from South America. It is now established in North America, growing along roadsides at a few locations within the *Flora* region.

6. **Eustachys retusa** (Lag.) Kunth [p. 223]

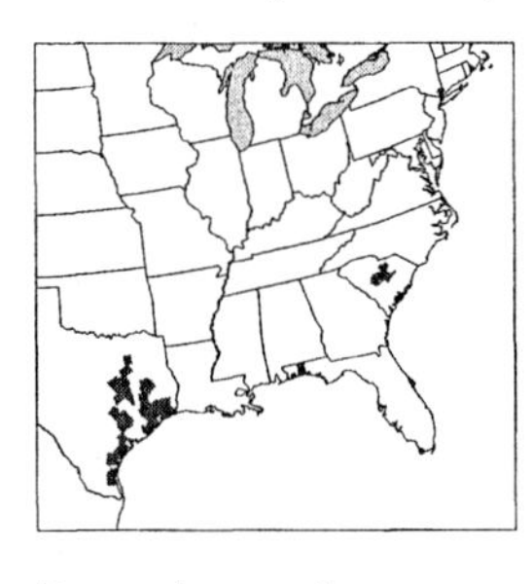

Culms 30–80 cm, erect, or decumbent and rooting at the nodes. **Blades** 6–12 cm long, 5–10 mm wide, usually folded, apices obtuse. **Panicles** with (6)8–15 branches; **branches** 4–10 cm, straight, somewhat stiff. **Spikelets** 1.5–2.1 mm; **florets** 2. **Lower glumes** 1.1–1.3 mm, linear-lanceolate, acute; **upper glumes** 1.3–1.5 mm, oblanceolate, apices truncate or bilobed, mucronate, mucros to 0.4 mm; **calluses** with a few short hairs; **lowest lemmas** 1.7–2 mm, lanceolate to ovate, not strongly keeled, midveins glabrous, lateral veins with widely spreading, white, 1–2 mm hairs, apices acute to mucronate; **second lemmas** 1.1–1.2 mm, broadly cuneate, apices truncate. **Caryopses** about 1 mm. $2n$ = 40.

Eustachys retusa is native to South America, but it is now established along roadsides, sandy fields, and waste areas in the southeastern United States.

7. **Eustachys distichophylla** (Lag.) Nees [p. 223]

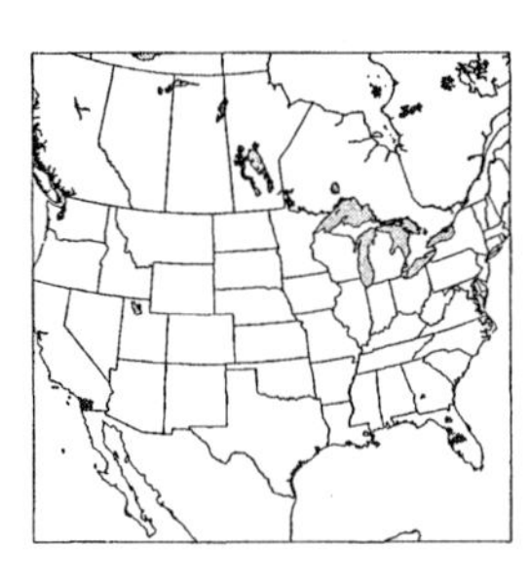

Culms 60–140 cm, erect. **Blades** to 31 cm long, 10–15 mm wide, usually folded, apices obtuse. **Panicles** with 10–36 branches; **branches** 6–17 cm, flexible. **Spikelets** 2.4–3 mm; **florets** 2(3). **Lower glumes** (1.2)1.5–2 mm, lanceolate, apices acute; **upper glumes** 1.6–2.6 mm,

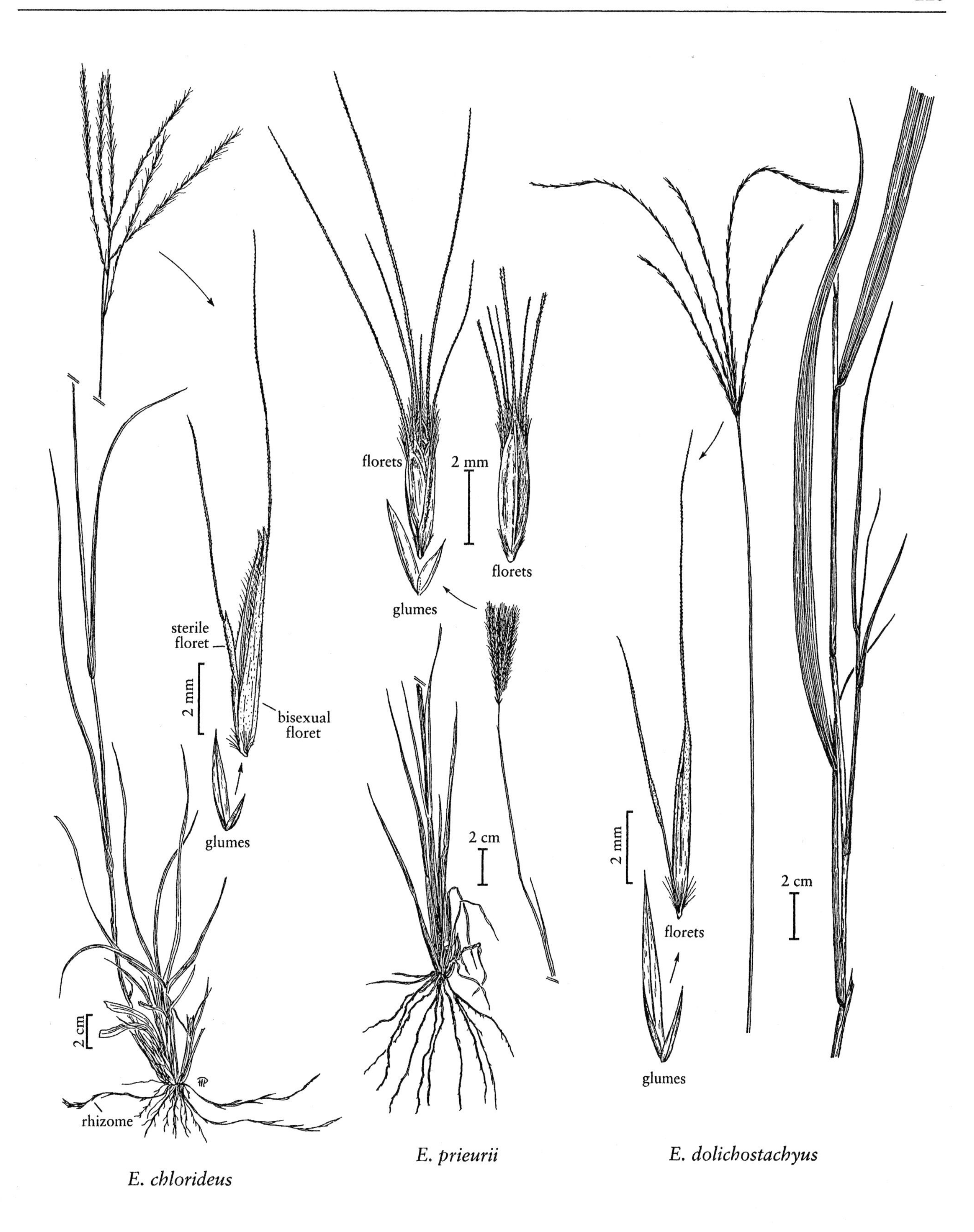

E. chlorideus

E. prieurii

E. dolichostachyus

narrowly oblong, apices truncate or bilobed, awned, awns 0.3–0.6 mm; **calluses** with a few hairs, hairs about 0.3 mm; **lowest lemmas** (2.2)2.5–2.9 mm, lanceolate, not strongly keeled, keels glabrous, lateral veins with strongly spreading, white, 1–1.8 mm hairs, apices acute to mucronate; **second lemmas** 1.2–2 mm, oblanceolate, apices obtuse, third rudimentary lemma occasionally present. **Caryopses** 1.1–1.2 mm. $2n$ = 40.

Eustachys distichophylla is native to South America, but is now established along sandy roadsides, fields, and waste areas in the southern United States.

17.37 ENTEROPOGON Nees

Mary E. Barkworth

Plants annual or perennial; cespitose, sometimes stoloniferous or rhizomatous. **Culms** to 120 cm, not woody. **Sheaths** open; **ligules** membranous, ciliate; **blades** flat. **Inflorescences** terminal, panicles of 1–20 non-disarticulating, spikelike branches, exceeding the upper leaves; **branches** digitately or racemosely arranged, if racemose, the lower nodes usually with more than 1 branch. **Spikelets** solitary, strongly imbricate, appressed to somewhat divergent, dorsally compressed, with 2–6 florets, lowest floret bisexual, elongate, remaining florets progressively reduced, usually the distal florets rudimentary and sterile, occasionally staminate; **disarticulation** beneath the glumes. **Glumes** unequal, subulate to lanceolate, membranous; **upper glumes** much shorter than the lowest florets, acute or shortly awned; **calluses** strigose; **lowest lemmas** stiff, 3-veined, ridged over the midveins, apices acute or bidentate, usually awned from between the teeth, without lateral awns; **paleas** almost as long as the lemmas, 2-keeled and bidentate; **distal lemmas** awned; **lodicules** 2; **anthers** 1–3. **Caryopses** sulcate; **embryos** ¼–⅓(½) as long as the caryopses. x = 10. Name presumably from the Greek *enteron*, 'intestine' and *pogon*, 'beard', although the allusion is unclear.

Enteropogon is a tropical genus of 17 species. Anderson (1974) included it in *Chloris*, but it is now usually recognized as a separate genus. It differs from *Chloris* in its dorsally compressed, indurate lemmas that are conspicuously ridged over the midvein. The caryopses also differ, being dorsally flattened with a shallow ventral groove in *Enteropogon;* in *Chloris*, the caryopses are basically triangular in cross section although there may be a shallow ventral groove. The embryos also tend to be shorter relative to the caryopses in *Enteropogon* than in *Chloris*, but there is some overlap.

There is one species native to the *Flora* region; two others have been found at various locations but have not persisted.

SELECTED REFERENCES **Anderson, D.E.** 1974. Taxonomy of the genus *Chloris* (Gramineae). Brigham Young Univ. Sci. Bull., Biol. Ser. 19:1–133; **Jacobs, S.W.L.** and **J. Highet.** 1988. Re-evaluation of the characters used to distinguish *Enteropogon* from *Chloris* (*Poaceae*). Telopea: 3:217–221; **Lazarides, M.** 1972. A revision of Australian *Chlorideae* (*Gramineae*). Austral. J. Bot. (supp. 5):1–51; **Pohl, R.W.** and **G. Davidse.** 1994. *Enteropogon* Nees. P. 289 *in* G. Davidse, M. Sousa S., and A.O. Chater (eds.). Flora Mesoamericana, vol. 6: Alismataceae a Cyperaceae. Universidad Nacional Autónoma de México, Instituto de Biología: México, D.F., México. 543 pp.

1. Panicles with 3–15 branches racemosely arranged; plants rhizomatous .... 1. *E. chlorideus*
1. Panicles with 1–10 branches in a single, digitate cluster; plants not rhizomatous, sometimes stoloniferous.
    2. Panicle branches 6–11 cm long, erect to slightly diverging; spikelets with 5–6 florets, the distal 4–5 sterile; plants stoloniferous .... 2. *E. prieurii*
    2. Panicle branches 7–25 cm long, divergent to drooping; spikelets with 2 florets, the distal floret sterile; plants not stoloniferous .... 3. *E. dolichostachyus*

1. **Enteropogon chlorideus** (J. Presl) Clayton [p. 226]
BURYSEED UMBRELLAGRASS

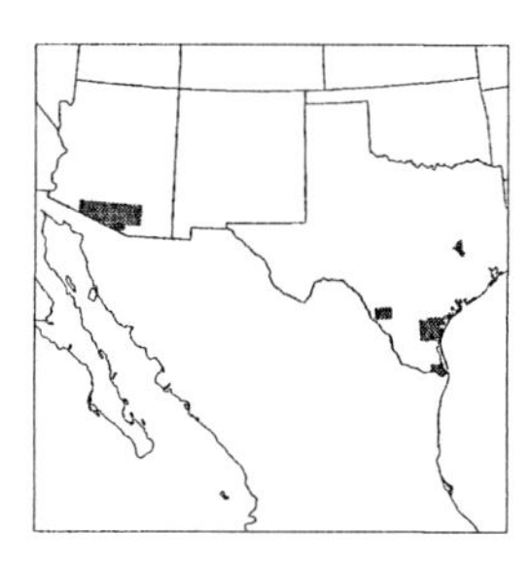

Plants perennial; cespitose and rhizomatous, each rhizome terminating in a cleistogamous spikelet. **Culms** to 100 cm, erect. **Sheaths** sparsely pilose near the ligules; **ligules of lower leaves** with a single prominent tuft of hairs; **ligules of upper leaves** usually glabrous; **blades** to 30 cm long, to 1 cm wide, usually scabrous, occasionally pilose. **Panicles** with 3–10(15) racemosely arranged branches, usually most nodes with more than 1 branch; **branches** 6–10 cm, naked below, with about 4 spikelets per cm distally. **Spikelets** with 1 bisexual and 1 sterile floret. **Lower glumes** 1–2 mm; **upper glumes** 2–3.5 mm; **lower lemmas** 4.5–7.5 mm long, about 1 mm wide, linear to narrowly lanceolate, glabrous or the margins sparsely strigose above, apices acute to acuminate, often bidentate, unawned or awned, awns 6.5–15 mm; **sterile florets** 1.4–3 mm long, to 0.3 mm wide, awns 2–8 mm. **Chasmogamous caryopses** about 4.5 mm long, about 0.8 mm wide; **cleistogamous caryopses** to 4 mm long, about 2.5 mm wide. $2n$ = 40, 80.

Enteropogon chlorideus is native from the southwestern United States through Mexico to Honduras. The spikelet-bearing rhizomes distinguish *Enteropogon* from most other grasses, but they are often missing from herbarium specimens. Seed set is highest in the cleistogamous spikelets.

2. **Enteropogon prieurii** (Kunth) Clayton [p. 226]
PRIEUR'S UMBRELLAGRASS

Plants perennial; stoloniferous. **Culms** to 80 cm. **Sheaths** glabrous, occasionally pilose apically; **ligules** short ciliate to long pilose; **blades** 10–30 cm long, to 5 mm wide, glabrous abaxially, scabrous to pilose adaxially. **Panicles** with 3–7 branches in a single digitate cluster; **branches** 6–11 cm, erect to slightly divergent, with 8–11 spikelets per cm. **Spikelets** with 1 bisexual floret and 4–5 sterile florets. **Lower glumes** 2.1–2.2 mm; **upper glumes** 3.7–4 mm; **lowest lemmas** 3.3–4.7 mm long, 0.4–0.7 mm wide, narrowly elliptic, margins densely strigose distally; **lowest sterile florets** 1.5–2.5 mm, cylindrical, awned, awns 8–17 mm; **distal florets** about 0.3 mm, flabellate. **Caryopses** 2–2.5 mm long, about 0.5 mm wide. $2n$ = unknown.

Enteropogon prieurii is native to the tropics of the Eastern Hemisphere. It was found near wharves in Alabama and North Carolina at the beginning of the twentieth century, but it is not known to be established in the *Flora* region.

3. **Enteropogon dolichostachyus** (Lag.) Keng *ex* Lazarides [p. 226]

Plants perennial; cespitose, neither stoloniferous nor rhizomatous. **Culms** 75–90 cm, branched, ascending or scrambling, sometimes rooting at the lower nodes. **Sheaths** glabrous or pubescent, sometimes hispid on the margins, distal margin hairs stiff, 2–4 mm; **ligules** about 0.3 mm; **blades** 10–17(40) cm long, 2.8–7 mm wide, usually flat, tapering to the apices. **Panicles** with (1)2–4(7) branches in a single digitate cluster; **branches** 7–15(25) cm, divergent to drooping, spikelet-bearing to the base. **Spikelets** with 2 florets. **Glumes** hyaline to thinly membranous, lanceolate, scabrous or smooth, long-acuminate; **lower glumes** 1.5–3 mm; **upper glumes** 4–6.5 mm; **lower lemmas** 3.5–5 mm, scabrous distally, notched, awned, awns 7–15 mm; **distal florets** about 0.8–1.8 mm, sterile, attenuate into the rachilla, awned, awns 3–7 mm. **Caryopses** 2.5–3 mm; **embryos** about 1 mm. $2n$= 20, 40, 60.

Enteropogon dolichostachyus grows from Afghanistan through southeast Asia to Australia. It has been collected from near a woolen mill in South Carolina, but it is not known to be established in the *Flora* region.

17.38 TRICHLORIS E. Fourn. *ex* Benth.

Mary E. Barkworth

Plants perennial; cespitose, sometimes stoloniferous. **Culms** to 150 cm, herbaceous, solid, glabrous. **Sheaths** open, rounded; **ligules** membranous, ciliate, cilia longer than the membranous base, conspicuous tufts of stiff hairs present on either side of the ligules; **blades** linear, flat or folded. **Inflorescences** terminal, panicles of non-disarticulating spikelike branches, exceeding the leaves; **branches** in 1 or more whorl(s), spikelets in 2 rows on the abaxial side of the branches, axes terminating in a functional spikelet. **Spikelets** laterally compressed, with 2–5 florets, lowest 1–2 florets bisexual, distal 1–3 florets progressively reduced and sterile; **disarticulation** above the glumes, all the florets falling as a unit. **Glumes** much shorter than the spikelets, membranous;

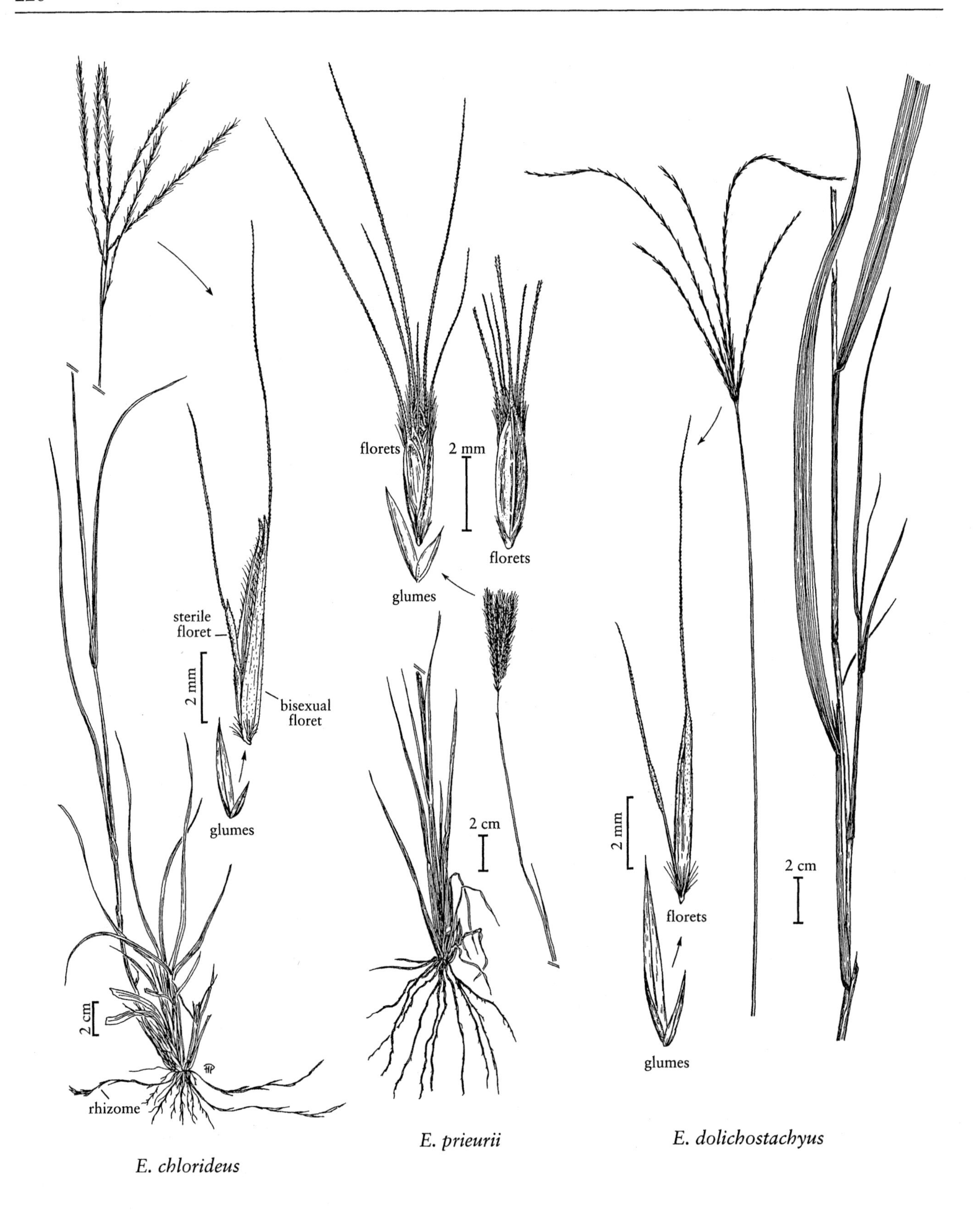

E. chlorideus

E. prieurii

E. dolichostachyus

ENTEROPOGON

lower glumes linear, acuminate; **upper glumes** lanceolate-ovate, awned; **calluses** bearded; **lowest lemmas** linear-lanceolate, 3-veined, veins prolonged into 3 awns; **central awns** 8–12 mm; **lateral awns** 0.5–12 mm; **paleas** 2-keeled, acute; **distal floret(s)** 1–3-awned; **lodicules** 2; **anthers** 2 or 3. **Caryopses** sulcate; **embryos** ½ as long as the caryopses; **hila** punctate. $x = 10$. Name from the Latin *tri*, 'three' and the genus *Chloris*, referring to the 3-awned lemmas and the resemblance to *Chloris*.

Trichloris has two species, both of which are native to the *Flora* region. It differs from *Chloris* in its 3-awned lemmas. Both species of *Trichloris* have a disjunct distribution, populations in North America being widely separated from those in South America.

SELECTED REFERENCE Davidse, G. 1994. *Trichloris* E. Fourn. *ex* Bentham. P. 289 *in* G. Davidse, M. Sousa S., and A.O. Chater (eds.). Flora Mesoamericana, vol. 6: Alismataceae a Cyperaceae. Universidad Nacional Autónoma de México, Instituto de Biología: México, D.F., México. 543 pp.

1. Lowest lemma awns subequal, the central awns 8–12 mm long, equaling or slightly longer than the lateral awns . . . . 1. *T. crinita*
1. Lowest lemma awns unequal, the central awns 8–12 mm long, the lateral awns 0.5–1.5 mm long . . . . 2. *T. pluriflora*

1. **Trichloris crinita** (Lag.) Parodi [p. 229]
FALSE RHODESGRASS

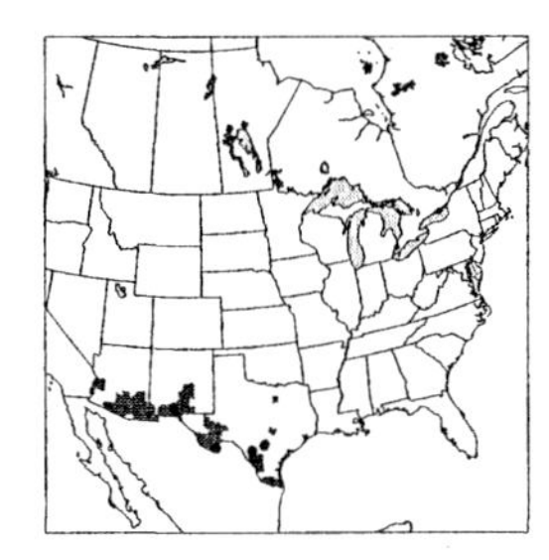

Plants perennial; cespitose, sometimes stoloniferous. **Culms** to 100 cm. **Sheaths** glabrous or sparsely hirsute; **ligules** to 3 mm; **blades** to 20 cm long, 5–10 mm wide, scabrous. **Panicles** with 6–20 branches in several closely-spaced whorls, appearing as a single terminal cluster; **branches** to 15 cm, erect, with 7–9 spikelets per cm. **Spikelets** with 1 bisexual and 1(2) sterile floret(s). **Lower glumes** 0.8–1.1 mm; **upper glumes** 2–2.5 mm, awned, awns to 2 mm; **lowest lemmas** 2.4–3.8 mm long, about 0.5 mm wide, dorsally compressed, narrowly lanceolate to elliptic, scabrous, particularly distally, apices bilobed and 3-awned, central awns 8–12 mm, equaling or slightly longer than the 5–12 mm lateral awns; **first sterile florets** 1–1.5 mm, narrowing to 3 subequal 5–7 mm awns; **second sterile florets**, if present, similar but smaller. **Caryopses** 1.7–2.3 mm, strongly dorsally flattened. $2n = 40$.

Trichloris crinita is a native species that grows in the southwestern United States and northern Mexico, and, as a disjunct, in northern Argentina.

2. **Trichloris pluriflora** (E. Fourn.) Clayton [p. 229]
MULTIFLOWER FALSE RHODESGRASS

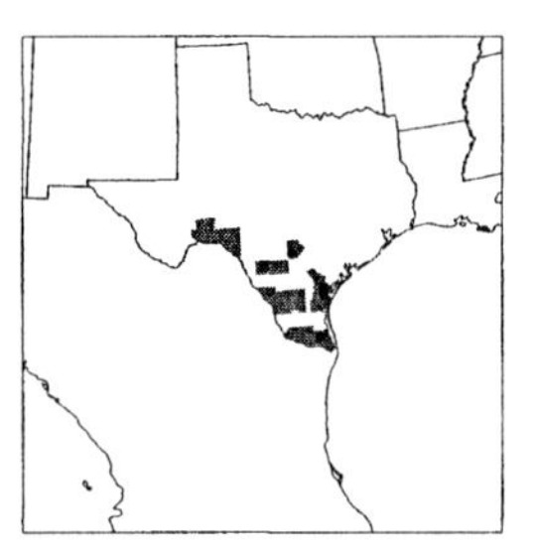

Plants perennial; stoloniferous or cespitose. **Culms** to 150 cm. **Sheaths** glabrous or sparsely hirsute; **ligules** to 3 mm; **blades** to 30 cm long, to 10 mm wide, scabrous or sparsely hirsute. **Panicles** with 7–20 branches in a few, evidently separate whorls; **branches** to 20 cm, ascending, with 7–9 spikelets per cm. **Spikelets** with 1–2 bisexual florets, a third floret with a rudimentary pistil and stamens sometimes present below the (1)2–3 sterile florets. **Lower glumes** 2–3 mm; **upper glumes** 3–5 mm; **lowest lemmas** 3–5 mm, mostly glabrous, margins short-ciliate near the middle, sparsely scabrous distally, apices 3-awned, central awns 8–12 mm, lateral awns 0.5–1.5 mm; **lowest sterile florets** 1.5–3 mm long, about 0.3 mm wide, mostly glabrous, margins sometimes short-ciliate near the middle, apices 3-awned, central awns to 8 mm, lateral awns 0.2–1 mm. **Caryopses** 1.8–2.2 mm, strongly dorsally compressed. $2n = 80$.

Trichloris pluriflora is native from southern Texas to Guatemala and, as a disjunct, from Ecuador to Argentina.

17.39 WILLKOMMIA Hack.

J.K. Wipff

Plants annual or perennial; cespitose or tufted, often stoloniferous. **Culms** 12–80 cm, herbaceous, unbranched distally. **Leaves** involute to flat; **ligules** membranous, ciliate. **Inflorescences** terminal, panicles of several, racemosely arranged, spikelike branches, mostly exceeding the leaves, lower branches sometimes partially included in the upper leaf sheaths at maturity; **branches** not woody, unilateral, with 2 rows of appressed, imbricate, solitary spikelets, terminating in a rudimentary, sterile spikelet, sterile spikelets sometimes consisting of 1 or 2 small scales. **Spikelets** dorsally compressed, with 1 floret; **disarticulation** below the glumes. **Lower glumes** to ⅔ as long as the spikelets, veinless; **upper glumes** equaling the florets, 1-veined; **calluses** acute to pointed; **lemmas** thinly membranous, 3-veined, apices rounded to acute, mucronate, shortly awned, or unawned. **Caryopses** ellipsoid. x = 10. Named for Heinrich Moriz Willkomm (1821–1895), a German botanist and explorer.

Willkommia is a genus of four species, three native to southern tropical Africa and one to the Americas, including the *Flora* region.

1. **Willkommia texana** Hitchc. [p. 229]
WILLKOMMIA

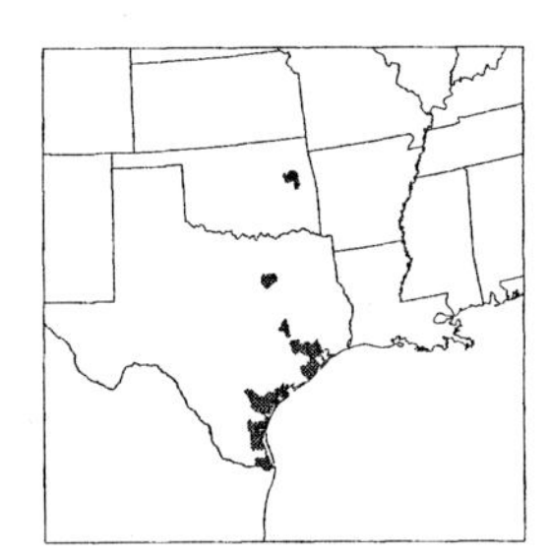

Plants perennial; cespitose or stoloniferous. **Culms** (12)17–65 cm, glabrous; **nodes** 4–6(7), glabrous. **Leaves** cauline; **sheaths** mostly glabrous, upper margins sometimes pubescent; **ligules** 0.3–0.6 mm, truncate; **blades** (1)4–21 cm long, 0.6–3.5 mm wide, involute to flat, glabrous, margins pubescent. **Panicles** (6)7–34(44) cm long, 4–10 mm wide; **branches** 4–13(20), 1–8 cm, appressed, each branch usually terminating in a reduced, sterile spikelet, lowest branches usually partially enclosed in the uppermost sheaths at maturity. **Spikelets** 3.5–5 mm; **disarticulation** usually below the glumes, sometimes the first glume persistent. **Lower glumes** 2–3.5 mm, glabrous; **upper glumes** 3.3–4.8 mm, glabrous and smooth below, scabridulous above; **lemmas** 3.0–4.2 mm, usually appressed pilose between the veins, lateral veins obscure, midveins sometimes excurrent; **paleas** 2.5–3.4 mm, pubescent between the veins; **anthers** 3, 0.3–0.9(1.1) mm, yellow-orange. **Caryopses** 1.4–2 mm. $2n$ = 60.

Willkommia texana grows on clay pans, alkaline flats, and sandy soils, in open or bare areas of Texas and (rarely) Oklahoma and, as a disjunct, in Argentina. The Oklahoma record may represent a recent introduction. North American plants belong to **Willkommia texana** Hitchc. var. **texana**. They differ from the Argentinean variety, **W. texana** var. **stolonifera** Parodi, in being cespitose rather than stoloniferous.

17.40 SCHEDONNARDUS Steud.

Neil Snow

Plants perennial; cespitose. **Culms** 8–55 cm, sometimes geniculate and branched basally, usually curving distally; **internodes** minutely retrorsely pubescent, mostly solid. **Leaves** mostly basal; **sheaths** compressed-keeled, closed, glabrous, margins scarious; **ligules** 1–3.5 mm, membranous, lanceolate; **blades** 1–12 cm long, 0.7–2 mm wide, stiff, usually folded, often spirally twisted, midrib well-developed, margins thick and whitish. **Inflorescences** terminal, panicles of widely spaced, racemosely arranged, spikelike branches, exceeding the upper leaves; **branches** strongly divergent, with distant to slightly imbricate, closely appressed spikelets. **Spikelets** 3–5.5 mm, mostly sessile, compressed laterally, with 1 floret; **florets** bisexual; **disarticulation** at the base of the panicle and above the glumes. **Glumes** lanceolate, unequal, 1-veined; **lemmas** usually exceeding the glumes, 3-veined, unawned or shortly awned; **paleas** subequal to the lemmas; **anthers** 3; **styles** 2. **Caryopses** fusiform. x = 10. Name from the Greek *schedon*, 'near'; Steudel considered *Schedonnardus* to be closely related to *Nardus*.

Schedonnardus is a monotypic North American genus that grows in the prairies and central plains of Canada, the United States, and northwestern Mexico. It has also been found, as a recent introduction, in California and Argentina. It is not known if it is established in California.

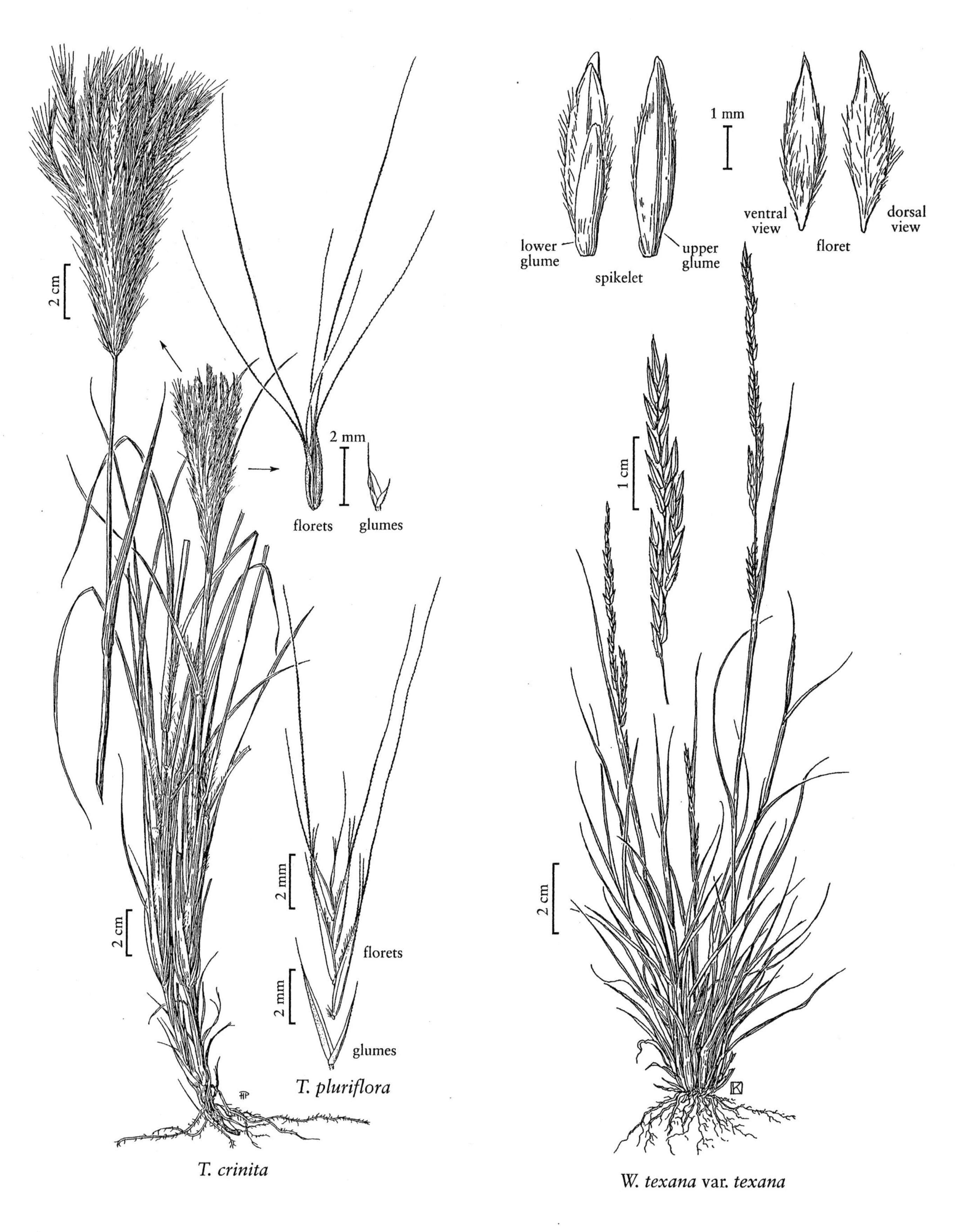

TRICHLORIS

WILLKOMMIA

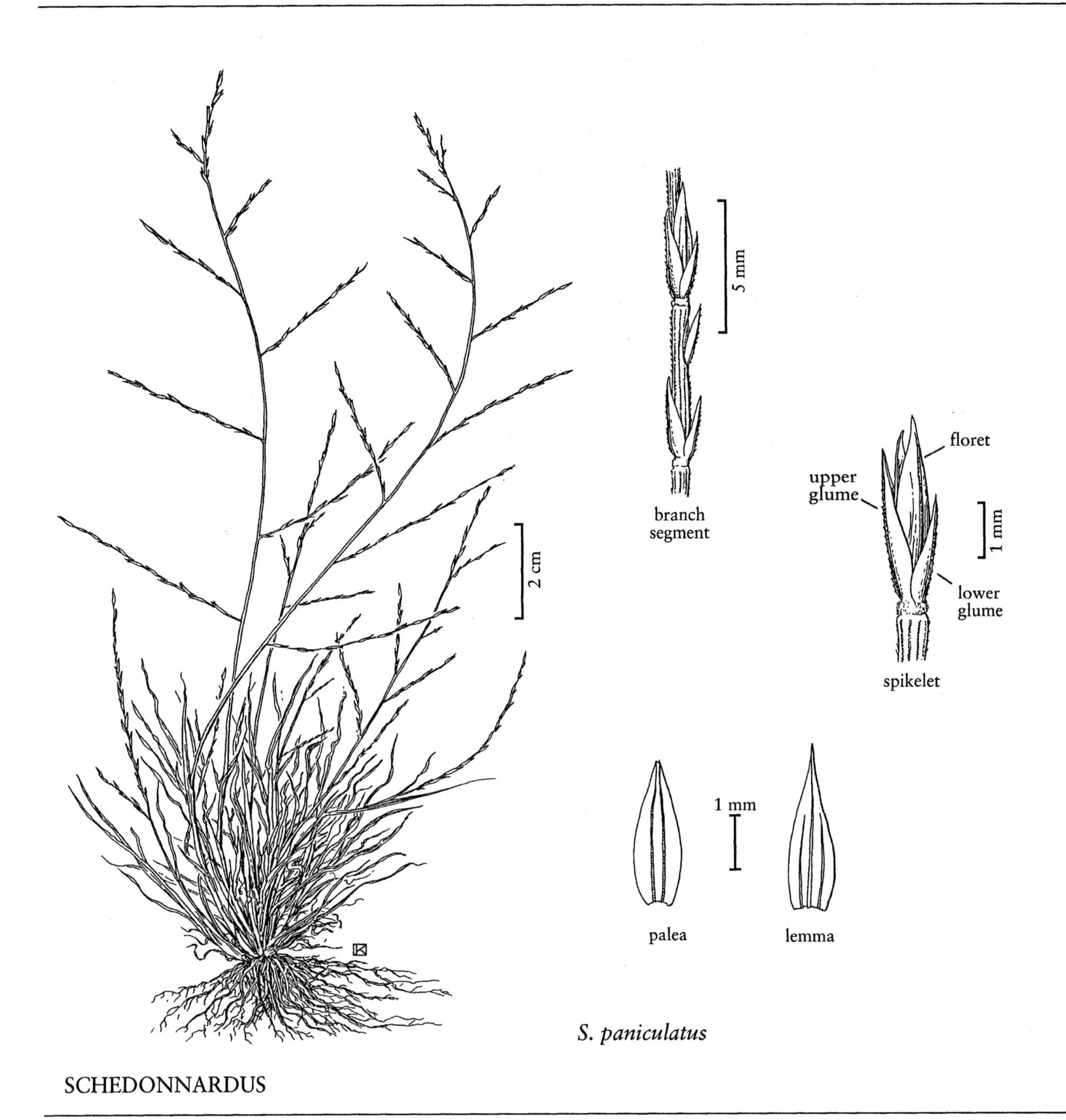

S. paniculatus

SCHEDONNARDUS

1. **Schedonnardus paniculatus** (Nutt.) Trel. [p. 230]
TUMBLEGRASS

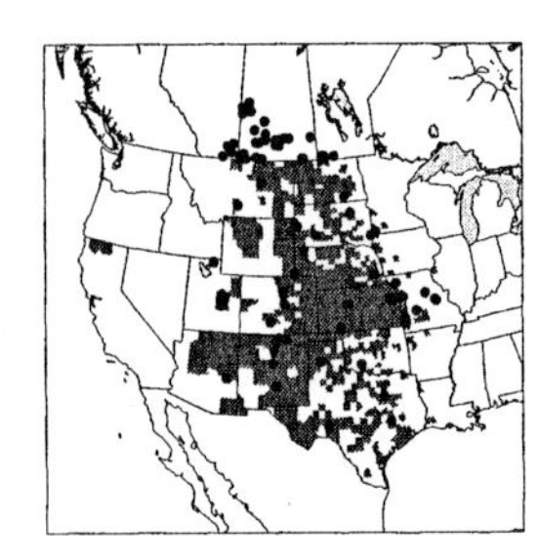

Panicles 5–50 cm, rachises becoming curved; **branches** 2–8(16) cm. **Lower glumes** 1.5–3 mm; **upper glumes** 1.5–4(5.5) mm; **lemmas** 3–5 mm; **anthers** 0.7–1.4 mm. **Caryopses** 2.5–3.5 mm. $2n$ = 20, 30.

Schedonnardus paniculatus is frequently found in disturbed areas. At maturity, the panicle breaks at the base and functions as a tumbleweed for seed dispersal. It is often a conspicuous feature of deserted towns in films of the American West.

17.41 GYMNOPOGON P. Beauv.

James P. Smith, Jr.

Plants usually perennial; often cespitose in appearance, rhizomatous. **Culms** 10–100 cm, erect to decumbent, simple or sparingly branched. **Leaves** cauline, evidently distichous; **sheaths** often strongly overlapping; **auricles** absent; **ligules** 0.1–0.5 mm, membranous, ciliate; **blades** linear to ovate-lanceolate, lacking midribs. **Inflorescences** terminal, panicles of spikelike branches, these subdigitately or racemosely arranged, usually strongly divergent to reflexed, sometimes naked basally, spikelets borne singly. **Spikelets** widely spaced to slightly imbricate, appressed to the branches, shortly pedicellate, laterally compressed, with 1–2(4) florets, only the lowest 1(2) floret(s) bisexual; **rachilla extensions** present, usually with a highly reduced, sterile floret(s); **disarticulation** above the glumes, florets falling together. **Glumes** subequal, usually exceeding the bisexual florets, narrow, acuminate, 1-veined; **lemmas of bisexual florets** 3-veined, midveins prominent, apices minutely bidentate, usually awned from between the teeth, rarely unawned; **anthers** (2)3. x = 10. Name from Greek *gymnos*, 'naked', and *pogon*, 'beard', alluding to the naked prolongation of the rachilla found in many species.

Gymnopogon, a genus of around 15 species, extends from the United States to South America, with one additional species ranging from India to Thailand. Three species are native to the *Flora* region. *Gymnopogon* is most likely to be confused with *Chloris*, but its species differ from most species of *Chloris* in having a more highly reduced, sterile floret at the end of the rachilla extension and in its distichous leaves.

SELECTED REFERENCE Smith, J.P., Jr. 1971. Taxonomic revision of the genus *Gymnopogon* (Gramineae). Iowa State Coll. J. Sci. 45:319–385.

1. Plants with elongate rhizomes; panicle branches naked for at least ⅓ of their length . . . . 1. *G. brevifolius*
1. Plants with short, knotty rhizomes or cespitose with a knotty base; panicle branches naked for less than ⅓ of their length, often spikelet-bearing to the base.
   2. Lemma awns 4–12.2 mm long . . . . 2. *G. ambiguus*
   2. Lemma awns 0–2.2 mm long . . . . 3. *G. chapmanianus*

1. Gymnopogon brevifolius Trin. [p. 233]
SHORTLEAF SKELETONGRASS

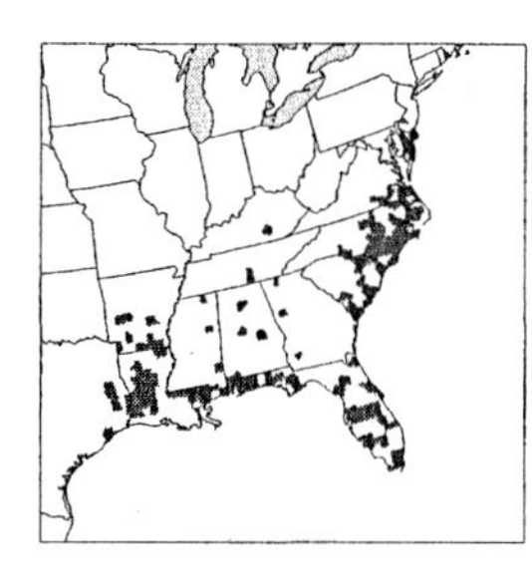

Plants rhizomatous, rhizomes to 9 cm. **Culms** 10–100 cm, erect or decumbent, single or in clumps, simple or sparingly branched. **Sheaths** mostly glabrous, throats pubescent; **collars** mostly glabrous, margins often with hairs; **ligules** about 0.5 mm; **blades** (1)2–8 cm long, 2–8(10) mm wide, glabrous abaxially, glabrous or scabrous adaxially. **Panicles** 10–30 cm; **branches** (6)10–17(20) cm, naked for at least the lower ⅓ of their length, spikelets distant to remote. **Spikelets** with 1(2) florets; **rachilla extensions** naked or with a minute rudimentary floret. **Glumes** (2)3.5–5 mm; **bisexual lemmas** 1.8–3.8 mm, awns 0.8–3 mm; **anthers** 3, 0.8–1 mm. **Caryopses** 1.6–1.9 mm long, 0.3–0.5 mm wide. $2n$ = unknown.

Gymnopogon brevifolius grows in dry to somewhat moist sandy pine woodlands of the southeastern United States, usually in loamy soils. It generally has rather weak, decumbent culms that tend to be obscured by the surrounding vegetation. Plants with stiffer culms tend to be confused with *Gymnopogon ambiguus*, but differ as discussed under *G. ambiguus*. Intermediate plants may be hybrids between the two species; there has been no experimental evaluation of this hypothesis.

2. Gymnopogon ambiguus (Michx.) Britton, Sterns & Poggenb. [p. 233]
BEARDED SKELETONGRASS

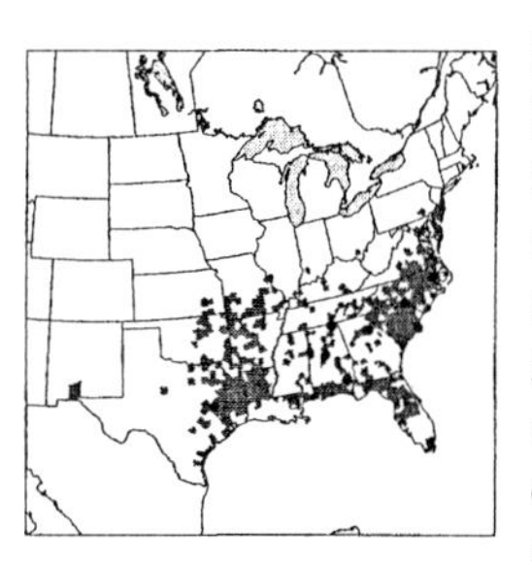

Plants cespitose, with a knotty base of short rhizomes. **Culms** 20–100 cm, suberect to spreading, stiff, simple to sparingly branched. **Sheaths** mostly glabrous, throats sometimes pubescent; **collars** conspicuously pubescent; **ligules** about 0.2 mm; **blades** (1.5)2.5–12 cm long, (2)5–10(18) mm wide, somewhat cordate at the base, mostly glabrous, often pubescent near the basal margins. **Panicles** (6)11.5–30(35) cm; **branches** (3)7–24 cm, stiffly spreading to somewhat reflexed, spikelet-bearing from the base, spikelets remote to slightly imbricate. **Spikelets** with 1(2) florets. **Glumes** 4–7 mm; **calluses** bearded; **bisexual lemmas** 2.5–5(6) mm, awns 4–12.2 mm; **second florets** often reduced to an obliquely inserted 2.4–6.2 mm awn; **anthers** 3, 0.8–1.2 mm. **Caryopses** 2–3 mm long, 0.2–0.5 mm wide. $2n = 40$.

Gymnopogon ambiguus grows in sandy pine woodlands of the southeastern United States, Haiti, and the Dominican Republic. It often grows with *G. brevifolius*, from which it differs in being more robust, having long, wider leaves, longer lemma awns, and, usually, having panicle branches that are spikelet-bearing to the base. Although spikelets of *Gymnopogon ambiguus* usually have only one floret, several plants from Texas have been found in which two florets per spikelet were the norm.

There is an 1853 collection of *G. ambiguus* supposedly from Doña Ana County, New Mexico, but there have been no recent collections from anywhere near there; it is possible that the locality data on the label are incorrect.

3. **Gymnopogon chapmanianus** Hitchc. [p. 233]
CHAPMAN'S SKELETONGRASS

Plants usually perennial; cespitose from a knotty base. **Culms** 20–70 cm, erect to sprawling, simple or sparingly branched from the lower nodes. **Sheaths** glabrous; **ligules** 0.1–0.3 mm; **blades** 1.3–8.5 cm long, 2–8 mm wide, glabrous. **Panicles** 8–23.5 cm; **branches** 2–15 cm, ascending, widely spreading, or reflexed, spikelet-bearing from the base or naked for less than ⅓ of their length. **Spikelets** with (1)2–3(4) florets. **Glumes** to 6 mm, sometimes widely divergent; **lemmas of bisexual florets** 1.5–2.3 mm, unawned or awned, awns 0.7–2.2 mm; **terminal sterile florets** minute, rudimentary, awned, awns not exserted from the spikelets; **anthers** 3, 0.5–0.8 mm. **Caryopses** 1.2–1.5 mm long, 0.3–0.4 mm wide. $2n$ = unknown.

Gymnopogon chapmanianus grows in sandy pine barrens and sites inhabited by dwarf palmetto, *Serenoa repens*. As interpreted here, *G. chapmanianus* includes *G. floridanus* Swallen. Smith (1971) treated the two as distinct species, but he acknowledged that they overlapped morphologically, ecologically, and geographically. Subsequent fieldwork has not supported the recognition of two entities. Smith's most intriguing observation was that only plants fitting the *G. floridanus* end of the morphological range produced mature caryopses. The reproductive biology of *G. chapmanianus* merits examination.

17.42 CTENIUM Panz.

Mary E. Barkworth

Plants usually perennial, sometimes annual; cespitose or rhizomatous. **Culms** 10–150 cm, simple. **Leaves** often aromatic; **ligules** membranous, sometimes ciliate; **blades** flat or convolute, upper blades reduced. **Inflorescences** terminal, panicles of 1–3 strongly pectinate branches, usually exceeding the upper leaves; **branches** usually falcate, if more than 1, digitately arranged, axes crescentic in cross section, with 2 rows of solitary, subsessile spikelets. **Spikelets** strongly divergent, laterally compressed, with 2 well-developed sterile or staminate florets below the single bisexual floret, reduced sterile or staminate florets also present beyond the bisexual floret; **disarticulation** above the glumes. **Glumes** unequal; **lower glumes** shorter than the upper glumes, 1-veined, keeled; **upper glumes** 2–3-veined, awned dorsally; **lemmas** thin, 3-veined, entire or bidentate, awned, awns dorsal, attached just below the lemma apices, or terminal; **lodicules** 2, glabrous; **anthers** 3 in bisexual florets, 2 in staminate florets; **styles** 2. **Caryopses** ellipsoid. $x = 9$. Name from the Greek *ktenion*, 'a little comb', an apt description of the inflorescence branches.

Ctenium is a genus of 17–20 species, native to tropical areas of Africa and the Americas, generally being found in savannah associations. The awned upper glumes and the presence of

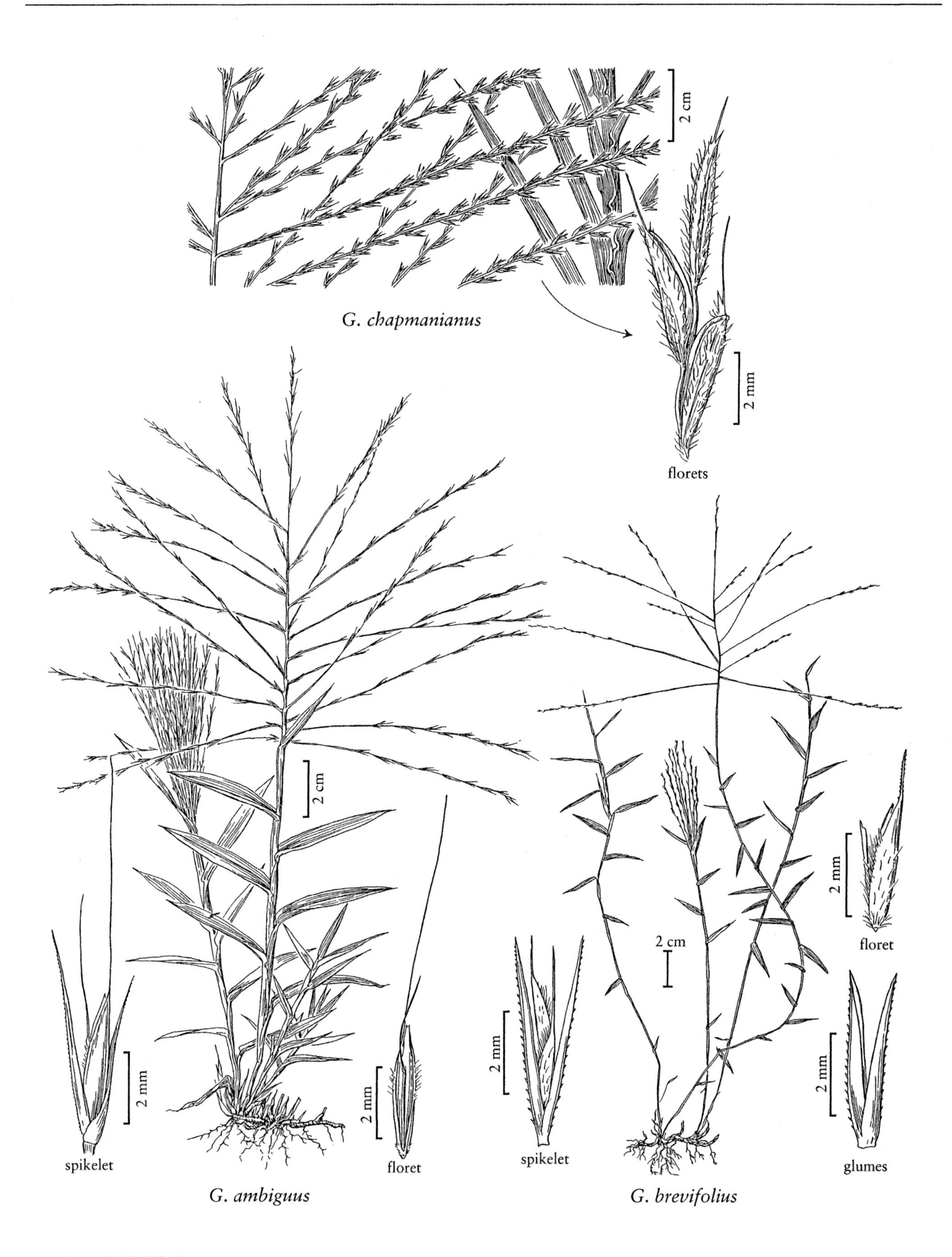
2 cm
G. chapmanianus
2 mm
florets
2 cm
2 mm
spikelet
2 mm
floret
G. ambiguus
2 mm
floret
2 cm
2 mm
spikelet
2 mm
glumes
G. brevifolius

sterile or staminate florets both below and above the fertile floret set it apart from other genera. The aroma of the leaves is described as being like turpentine.

The two native North American species are highly fire-adapted, flourishing in communities that regularly burn on a 1–5 year basis. In the fall, the panicle branches of both species form curves, loops, and corkscrews, which are attractive in floral arrangements.

1. Plants without rhizomes, forming dense tufts; upper glumes with a row of prominent glands on each side of the midvein; awns of the upper glumes strongly diverging at maturity . . . . . . 1. *C. aromaticum*
1. Plants with slender, scaly rhizomes; upper glumes without glands, or the glands inconspicuous; awns of the upper glumes straight to ascending . . . . 2. *C. floridanum*

1. **Ctenium aromaticum** (Walter) Alph. Wood [p. 236]
TOOTHACHE GRASS

Plants perennial; densely cespitose, without rhizomes. **Culms** 1–1.5 m, erect. **Sheaths** persistent, smooth to scabridulous, becoming fibrous at the base at maturity; **ligules** 0.8–2.9 mm; **blades** to 46 cm long, 5.1 mm wide, glabrous abaxially, scabrous adaxially, sometimes also pilose just above the ligules. **Panicles** with 1 branch; **branches** 5–15 cm, curved, axes extending slightly beyond the spikelets, distal spikelets reduced. **Spikelets** 8.5–11 mm. **Lower glumes** 1–2.1 mm; **upper glumes** 4.8–6.1 mm, bidentate, with a row of conspicuous glands on either side of the midveins, awns 3.5–4 mm, strongly divergent, often almost perpendicular to it at maturity; **lemmas of bisexual florets** 4–5 mm, pilose on the lateral veins, awned subapically, awns 3–4 mm, straight or divergent; **distal lemmas** sterile, unawned. $2n$ = unknown.

Ctenium aromaticum is a common species that grows in wet to moist pine flatwoods, savannahs, prairies, pitcher plant bogs, and ecotones between pine uplands and wet streamheads of the southeastern coastal plain. It furnishes fair forage, and the roots are spicy when freshly dug.

2. **Ctenium floridanum** (Hitchc.) Hitchc. [p. 236]
FLORIDA ORANGEGRASS

Plants perennial; rhizomatous, rhizomes scaly, slender. **Culms** 60–100 cm, erect. **Sheaths** smooth or scabridulous, not or only somewhat fibrous at maturity; **ligules** 1–7.9 mm; **blades** to 39 cm long, 5.1 mm wide, glabrous or scabrous. **Panicles** with 1 branch; **branches** 8–15 cm, often twisted. **Spikelets** 7–9.2 mm. **Lower glumes** 1.6–2 mm; **upper glumes** 4.6–5.7 mm, glandless or with a few inconspicuous glands on either side of the midveins, unequally bidentate, awned, awns 3.8–6.2 mm, straight to assending; **lemmas** pubescent on the lateral veins, awns 3–5 mm, straight; **distal lemmas** unawned. $2n$ = unknown.

Ctenium floridanum is an uncommon endemic of Georgia and Florida, where it grows in dry to mesic pine-oak uplands and pine flatwoods. It is also cultivated, the graceful curve of its spikes making it an attractive addition to gardens.

17.43 MICROCHLOA R. Br.

Evangelina Sánchez

Plants perennial (rarely annual); erect and cespitose, or decumbent and mat-forming. **Culms** 5–60 cm; **nodes** glabrous. **Leaves** mostly basal; **ligules** membranous and ciliate or of hairs; **blades** often stiff, convolute. **Inflorescences** terminal, panicles with a solitary (rarely 2), slender, spikelike branch; **rachises** curved or falcate, semi-circular or crescentic in cross section. **Spikelets** solitary, with 1(2) florets, terete to dorsally compressed; **florets** bisexual; **disarticulation** above the glumes. **Glumes** subequal, exceeding the florets, 1-veined; **lower glumes** asymmetric, keeled, keels somewhat twisted; **upper glumes** symmetric, flat, midveins straight; **lemmas** membranous or hyaline, 3-veined, veins hairy; **paleas** with 2 pubescent keels. **Caryopses** 0.9–1.5 mm, glabrous. x = 10, 12. Name from the Greek, *micros*, 'small', and *chloë*, 'grass'.

Microchloa includes three African and one pan-tropical species. The species usually grow in open mesic to xeric habitats in tropical regions, often in shallow, hard soils. One species has become established in the *Flora* region.

The distinctive shape of the rachis links *Microchloa* to *Ctenium*.

1. **Microchloa kunthii** Desv. [p. 236]
KUNTH'S SMALLGRASS

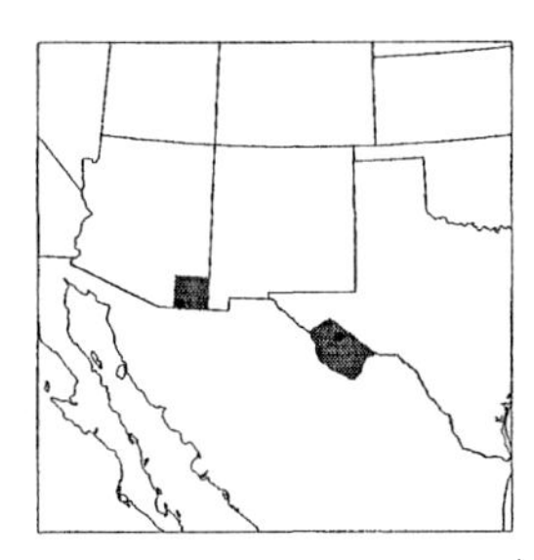

Plants perennial; forming small, dense tufts. **Culms** 5–30 cm. **Sheaths** generally shorter than the internodes; **ligules** 1–1.5 mm, ciliate; **blades** 1–1.5 mm wide, flat or folded, with thick, scabrous, white margins; **cauline blades** 1–2.5 cm; **innovation blades** to 6 cm. **Panicle branches** 6–15 cm; **rachises** ciliate. **Spikelets** 2.5–3.5 mm; **lemmas** 2–2.5 mm, keels pilose, margins densely pilose, hairs 0.2–1 mm. $2n = 24, 40$.

Microchloa kunthii grows on granitic outcrops on rocky slopes. Its range extends southward from Carr Canyon, Huachuca Mountains, Arizona and the Big Bend region of Texas through Mexico to Guatemala.

17.44 CYNODON Rich.

Mary E. Barkworth

Plants perennial; sometimes stoloniferous, sometimes also rhizomatous, often forming dense turf. **Culms** 4–100 cm. **Sheaths** open; **auricles** absent; **ligules** of hairs or membranous; **blades** flat, conduplicate, convolute, or involute, sometimes disarticulating. **Inflorescences** terminal, digitate or subdigitate panicles of spikelike branches; **branches** (1)2–20, 1-sided, with 2 rows of solitary, subsessile, appressed, imbricate spikelets. **Spikelets** laterally compressed, with 1(–3) florets, only the lowest floret functional; **rachilla extension** usually present, sometimes terminating in a reduced floret; **disarticulation** above the glumes. **Glumes** usually shorter than the lemmas, membranous, keeled, usually muticous; **lower glumes** 1-veined; **upper glumes** 1–3-veined, occasionally shortly awned; **lemmas** membranous to cartilaginous, 3-veined, keeled, keels with hairs, occasionally winged, apices mucronate or muticous; **paleas** about as long as the lemmas, 2-keeled; **anthers** 3; **style branches** 2, plumose; **lodicules** 2. $x = 9$. Name from the Greek *kyon*, 'dog', and *odous*, 'tooth', a reference to the sharp, hard scales of the rhizome.

Cynodon is a genus of nine species, all of which are native to tropical regions of the Eastern Hemisphere. Several species are used as lawn and forage grasses in tropical and warm-temperate regions. The most widespread species, *Cynodon dactylon*, is also the most frequently encountered species in the *Flora* region. It is used for lawns, putting greens, and pastures in southern portions of the region, but is generally considered a weed in other parts.

The status of several species in the *Flora* region is unclear. Species other than *C. dactylon* usually grow only under cultivation, but there are scattered records of populations of other species from the southern United States that appear to have become established. Cultivars of *C. aethiopicus* and *C. nlemfuënsis* are used for pasture primarily in tropical Florida. *Cynodon transvaalensis* has had limited commercial distribution as a turf grass.

Many cultivars of *Cynodon* have been developed, some from hybrids between it and other species such as *C. transvaalensis*, *C. aethiopicus*, and *C. nlemfuënsis*. The cultivars may exhibit combinations of features that are not found in the wild species, making it difficult to accommodate them in a key.

SELECTED REFERENCES **Alderson, J.** and **W.C. Sharp.** 1995. Grass Varieties in the United States. CRC Press, Boca Raton, Florida, U.S.A. 296 pp. [previously published by the Soil Conservation Service, U.S. Department of Agriculture, as Agricultural Handbook No. 170, revised 1994]; **Assafa, S., C.M. Taliaferro, M.P. Anderson, B.G. de los Reyes,** and **R.M. Edwards.** 1999. Diversity among *Cynodon* accessions and taxa based

CTENIUM

MICROCHLOA

on DNA amplification fingerprinting. Genome 42:465–474; **Busey, P.** and **S. Boyer.** 2002. Bermudagrass speeds: Can fast greens be green? http://www.floridaturf.com/ballroll.htm; **Caro, J.A.** and **E.A. Sánchez.** 1969. Las especies de *Cynodon* (Gramineae) de la República Argentina. Kurtziana 5:191–252; **de Wet, J.M.J.** and **J.R. Harlan.** 1970. Biosystematics of *Cynodon* L.C. Rich. (Gramineae). Taxon 19:565–569; **Harlan, J.R.** and **J.M.J. de Wet.** 1969. Sources of variation in *Cynodon dactylon* (L.) Pers. Crop Sci. (Madison) 9:774–778; **Harlan, J.R., J.M.J. de Wet, W.W. Huffine,** and **J.R. Deakin.** 1970. A guide to the species of *Cynodon* (Gramineae). Oklahoma Agric. Exp. Sta. Bull. B-673:1–37; **Hitchcock, A.S.** 1951 [title page 1950]. Manual of the Grasses of the United States, ed. 2, rev. A. Chase. U.S.D.A. Miscellaneous Publication No. 200. U.S. Government Printing Office, Washington, D.C., U.S.A. 1051 pp.; **Jones, S.D.** and **G.D. Jones.** 1992. *Cynodon nlemfuënsis*, (Poaceae: Chlorideae) previously unreported in Texas. Phytologia 72:93–95.

1. Lemma keels winged; panicle branches with flattened axes (subg. *Pterolemma*) . . . . 7. *C. incompletus*
1. Lemma keels not winged; panicle branches with triquetrous axes (subg. *Cynodon*).
   2. Glumes 0.1–0.6 mm long . . . . 1. *C. plectostachyus*
   2. Glumes 1.1–2.6 mm long.
      3. Panicles with 1–3(4) branches; culms 5–30 cm tall; blades 1–1.5 mm wide . . . . 2. *C. transvaalensis*
      3. Panicles with (2)4–20 branches; culms 5–100 cm tall; blades (1)2–7 mm wide.
         4. Panicles with 2–6(9) branches in a single whorl; culms 5–40(50) cm tall.
            5. Panicles with (2)4-6(9) branches; anthers dehiscent at maturity . . . . 3. *C. dactylon*
            5. Panicles with 2-4 branches; anthers indehiscent at maturity . . . . 4. *C.* ×*magennisii*
         4. Panicles with 4–20 branches in 1–5 whorls; culms 20–100 cm tall.
            6. Lemma keels glabrous or with a few scattered hairs; panicle branches usually in 2–5 whorls, stiff, frequently red or purple; culms 25–100 cm tall, woody . . . . 5. *C. aethiopicus*
            6. Lemma keels shortly pubescent; panicle branches usually in 1 whorl, lax, usually green; culms 20–60 cm tall, not woody . . . . 6. *C. nlemfuënsis*

1. **Cynodon plectostachyus** (K. Schum.) Pilg. [p. 239]
STARGRASS

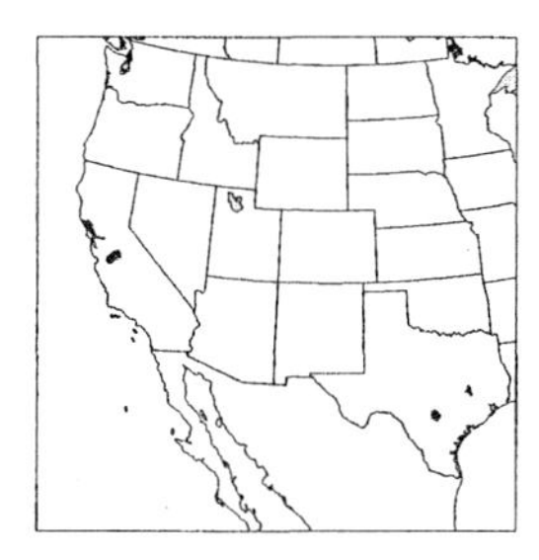

Plants stoloniferous, not rhizomatous; **stolons** to 2 mm thick, arching. **Culms** 60–100 cm tall, 1–4 mm thick, glabrous. **Sheaths** mostly glabrous or sparsely to densely pilose, with long hairs adjacent to the ligules; **ligules** 1–2 mm; **blades** to 30 cm long, 4–8 mm wide, both surfaces scabrous and densely pubescent. **Panicles** with 6–20 subdigitate branches; **branches** 3–10 cm, in (1)2–7 closely spaced whorls, axes triquetrous. **Spikelets** 2.5–3 mm, closely imbricate; **rachillas** prolonged, glabrous, sometimes terminating in a vestigial floret. **Lower glumes** 0.1–0.3 mm; **upper glumes** 0.4–0.6 mm; **lemmas** 2.4–3 mm, keels not winged, keels and margins pubescent, hairs 0.3–0.4 mm; **paleas** stiffly ciliate on the keels. $2n$ = 18, 36.

Cynodon plectostachyus is native to tropical Africa. Its status in the *Flora* region is unclear. The records shown are from non-cultivated plants, but it is not known whether they represent established populations. *Cynodon plectostachyus* is not frost-tolerant.

2. **Cynodon transvaalensis** Burtt Davy [p. 239]
AFRICAN DOGSTOOTH GRASS, FLORIDAGRASS

Plants stoloniferous and rhizomatous; **stolons** slender, prostrate; **rhizomes** slender. **Culms** 5–30 cm tall, to 0.4 mm thick. **Sheaths** glabrous or with scattered hairs; **ligules** to 0.3 mm, membranous and ciliolate; **blades** to 4 cm long, 1–1.5 mm wide, flat or involute and filiform, both surfaces pubescent. **Panicles** with 1–3(4) branches; **branches** 0.7–2.1 cm, in a single whorl, reflexed at maturity, axes triquetrous. **Spikelets** 2–2.7 mm. **Lower glumes** 1.2–1.4 mm; **upper glumes** 1.1–1.3 mm; **lemmas** 2.2–2.7 mm, keels not winged, stiffly and sparsely pubescent, margins glabrous or hispidulous; **paleas** glabrous. $2n$ = 18.

Cynodon transvaalensis is native to southern Africa. Hitchcock (1951, p. 504) reported that it was "coming into cultivation as a lawn grass", but it is no longer sold in the *Flora* region, nor is there any evidence that earlier plantings have led to its establishment. Strains tested in Florida for use in putting greens were unable to withstand the mowing and moisture conditions used to maintain such areas (Busey and Boyer 2002). Strains of the species have, however, been crossed with strains of *C. dactylon* and cultivars developed from these crosses are sometimes used as turf grasses in the southern United States and in similar climates throughout the world.

3. **Cynodon dactylon** (L.) Pers. [p. 239]
BERMUDAGRASS

Plants stoloniferous, usually also rhizomatous. **Culms** 5–40(50) cm, not becoming woody. **Sheaths** glabrous or with scattered hairs; **collars** usually with long hairs, particularly at the margins; **ligules** about 0.5 mm, of hairs; **blades** 1–6(16) cm long, (1)2–4(5) mm wide, flat at maturity, conduplicate or convolute in bud, glabrous or the adaxial surfaces pilose. **Panicles** with (2)4–6(9) branches; **branches** 2–6 cm, in a single whorl, axes triquetrous. **Spikelets** 2–3.2 mm. **Lower glumes** 1.5–2 mm; **upper glumes** 1.4–2.3 mm; **lemmas** 1.9–3.1 mm, keels not winged, pubescent, margins usually less densely pubescent; **anthers** dehiscent at maturity; **paleas** glabrous. $2n$ = 18, 36.

Cynodon dactylon is a variable species, but taxonomists disagree on just how variable. Caro and Sánchez (1969) limited *C. dactylon* to plants with conduplicate leaves, placing those with convolute leaves in a number of other species, such as *C. affinis* Caro & Sánchez and *C. aristiglumis* Caro & Sánchez; de Wet and Harlan (1970) do not mention this character in their study of *Cynodon.* Caro and Sánchez also employed several other characters in the key separating *C. dactylon* from the species with convolute immature leaves, but the overlap between the two sides of the lead is substantial. Pending further study, the broader interpretation, in which *C. dactylon* includes plants with both convolute and conduplicate leaves, has been adopted.

Several varieties of *C. dactylon* have been described, in addition to which numerous cultivars have been developed, some as turf grasses for lawns or putting greens, others as pasture or forage grasses. Their useful range is limited because *C. dactylon* is not cold hardy, going dormant and turning brown when nighttime temperatures fall below freezing or average daytime temperatures are below 10°C.

The most commonly encountered variety, both in the *Flora* region and in other parts of the world, is *C. dactylon* var. *dactylon*, largely because it thrives in severely disturbed, exposed sites; it does not invade natural grasslands or forests. Determining how many other varieties are established in the *Flora* region is almost impossible, because there has been no global study of variation in the species. The presence of numerous cultivars complicates an already difficult problem. The two varieties keyed out below are the only two that grow in the *Flora* region according to de Wet and Harlan (1970), but these authors do not appear to have considered the taxa recognized by Caro and Sánchez (1969). For most purposes, it is probably neither necessary nor feasible to identify the variety of *C. dactylon* encountered.

Cynodon dactylon is considered a weed in many countries and it is true that, once established, it is difficult to eradicate. It does, however, have some redeeming values. It is rich in vitamin C, and its leaves are sometimes used for an herbal tea. It is claimed to have various medicinal properties, but these have not been verified. It is considered a good pasture grass, in addition to which it is sometimes grown as an ornamental and for erosion control on exposed soils.

1. Rhizomes near the surface (sometimes surfacing for a short distance before submerging again), the tips eventually surfacing and, like the lateral buds, producing tillers . . . . var. *dactylon*
1. Rhizomes growing up to 1 m deep, the tips remaining below ground, only the lateral buds producing tillers . . . . var. *aridus*

Cynodon dactylon var. **aridus** J.R. Harlan & de Wet

A cultivar of this variety, 'Giant', has been introduced to the Yuma region of Arizona (Harlan et al. 1970).

Cynodon dactylon (L.) Pers. var. **dactylon**

As noted above, this is by far the most common variety of *Cynodon dactylon.*

4. **Cynodon** ×**magennisii** Hurcombe
MAGENNIS' DOGSTOOTH GRASS

Plants stoloniferous and rhizomatous. **Culms** to 20 cm. **Ligules** of hairs; **blades** 1–1.5 mm wide, pubescent, at least adaxially. **Panicles** with 2–4 branches; **branches** in a single whorl, axes triquetrous. **Spikelets** about 3 mm. **Glumes** about 2 mm, about ⅔ as long as the spikelets; **lemma keels** not winged; **anthers** indehiscent at maturity. $2n$ = 27.

Cynodon ×*magennisii* is a natural triploid hybrid between *C. dactylon* and *C. transvaalensis* (Harlan et al. 1970). Several cultivars have been developed for lawns and golf courses in the southern United States. These exhibit differing mixes of the characteristics of the two parent species.

Like triploid cultivars of *C. dactylon*, *C.* ×*magennisii* fails to produce either pollen or seeds and its anthers remain narrow and indehiscent at maturity. Diploid and tetraploid plants of *C. dactylon* also frequently fail to set seed because they are highly self-sterile, but they produce good pollen and their anthers dehisce at maturity.

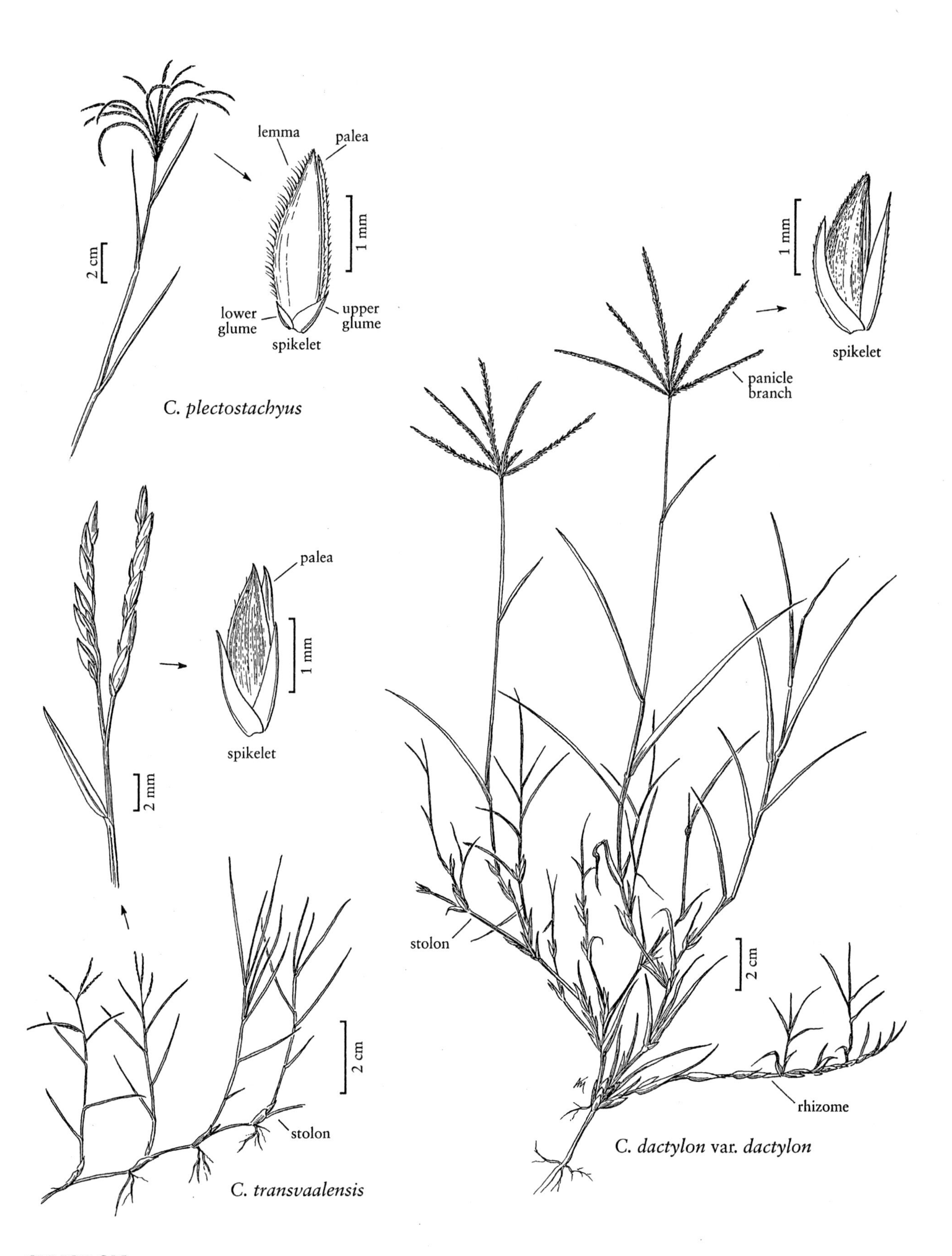
lemma
palea
1 mm
2 cm
lower glume
upper glume
spikelet
C. plectostachyus
palea
1 mm
spikelet
2 mm
2 cm
stolon
C. transvaalensis
1 mm
spikelet
panicle branch
stolon
2 cm
rhizome
C. dactylon var. dactylon

CYNODON

5. **Cynodon aethiopicus** Clayton & J.R. Harlan [p. 241]
ETHIOPIAN DOGSTOOTH GRASS

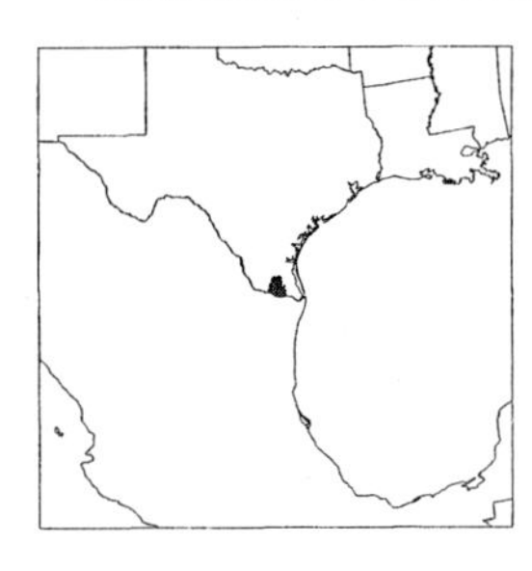

Plants stoloniferous, not rhizomatous; **stolons** stout, woody, lying flat on the ground. **Culms** 25–100 cm tall, 2–6 mm thick, becoming woody. **Sheaths** glabrous; **ligules** about 0.3 mm, membranous, ciliolate; **blades** 3–25 cm long, 3–7 mm wide, glabrous or sparsely pubescent, glaucous. **Panicles** with 4–10(20) branches; **branches** 3.5–7 cm, in (1)2–5 whorls, stiff, usually red or purple, axes triquetrous. **Spikelets** 2–3 mm. **Glumes** equaling to slightly exceeding the florets; **lower glumes** 2–2.2 mm; **upper glumes** 1.7–2.6 mm; **lemmas** 2.1–2.6 mm, keels not winged, glabrous or with a few scattered hairs. $2n$ = 18, 36.

Cynodon aethiopicus is native to the East African rift. It is now established along the canal bank in the Santa Ana National Wildlife Refuge in Texas, and is expected to spread. The cultivar 'McCaleb' has been released as a forage grass for use in Florida.

6. **Cynodon nlemfuënsis** Vanderyst [p. 241]
AFRICAN BERMUDAGRASS

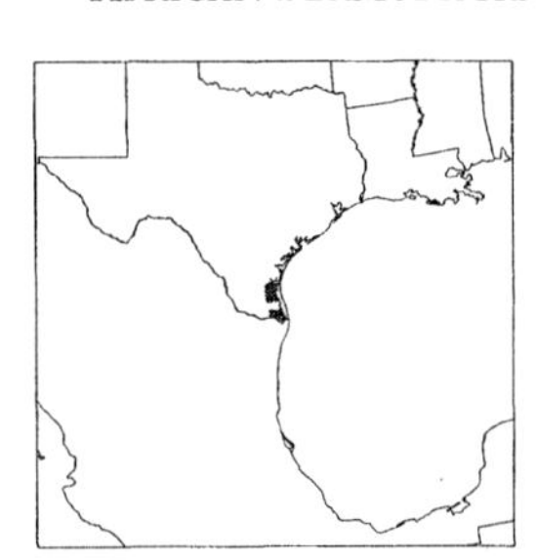

Plants stoloniferous, not rhizomatous; **stolons** stout, woody, usually lying flat on the ground. **Culms** 20–60 cm tall, 1–5 mm thick, not becoming woody. **Sheaths** glabrous; **ligules** about 0.3 mm, membranous, ciliolate; **blades** 5–16 cm long, 2–6 mm wide, abaxial surfaces glabrous or with scattered long hairs, adaxial surfaces with scattered long hairs. **Panicles** with 4–13 branches; **branches** (2)4–7(10) cm, in 1(–3) whorls, lax, usually green, axes triquetrous. **Spikelets** 2–3 mm. **Lower glumes** 1.7–2 mm; **upper glumes** 1.5–2.3(3) mm; **lemmas** 1.9–2.9 mm, keels not winged, shortly pubescent, at least distally; **paleas** glabrous. $2n$ = 18, 36.

Cynodon nlemfuënsis is native to east and central Africa, but it is now established in southern Texas (Jones and Jones 1992), and may be present in other parts of the southern United States. It is similar to *C. dactylon*, but differs in being larger and lacking rhizomes. It is also less hardy, not becoming established where temperatures fall below -4°C. Plants in the *Flora* region belong to **Cynodon nlemfuënsis** var. **nlemfuënsis** which differs from **C. nlemfuënsis** var. **robustus** Clayton & J.R Harlan in having shorter inflorescence branches (2–7(10) cm rather than 6–10 cm) and thinner culms (1–1.5 mm rather than 2–5 mm). Cultivars of *C. nlemfuënsis* include 'Florico', 'Florona', 'Ona', and 'Costa Rica'.

7. **Cynodon incompletus** Nees [p. 241]

Plants stoloniferous, not rhizomatous. **Culms** 5–30 cm. **Sheaths** glabrous; **ligules** membranous; **blades** 1.5–6 cm long, 2–4 mm wide, glabrous or pubescent. **Panicles** with 2–6 branches; **branches** 2–5 cm, in a single whorl, axes flattened. **Spikelets** 2–3 mm, narrowly to broadly ovate. **Glumes** 1.7–2.1 mm, exceeded by the florets; **lemmas** 2.2–2.6 mm, keels winged and pubescent, margins glabrous. $2n$ = 18, 36.

Cynodon incompletus is native to southern Africa. A hybrid between the two varieties identified below, *Cynodon* ×*bradleyi* Stent, is used as a lawn grass in North America (de Wet and Harlan 1970).

1. Blades glabrous or sparsely hirsute; spikelets 2.5–3 mm long, narrowly ovate . . . . var. *incompletus*
1. Blades densely hirsute; spikelets 2–2.5 mm long, broadly ovate . . . . var. *hirsutus*

17.45 **SPARTINA** Schreb.

Mary E. Barkworth

Plants perennial; cespitose from knotty bases or rhizomatous. **Culms** 10–350 cm, erect, terete, solitary or in small to large clumps. **Leaves** mostly cauline; **sheaths** open, smooth, sometimes striate; **ligules** membranous, ciliate, cilia longer than the membranous bases; **blades** flat or involute. **Inflorescences** terminal, usually exceeding the upper leaves, 3–70 cm, panicles of 1–75 spikelike branches attached to an elongate rachis; **branches** racemosely arranged, alternate, opposite, or whorled, appressed to strongly divergent, axes 3-sided, spikelets usually sessile on the 2 lower sides, usually divergent to strongly divergent; **disarticulation** beneath the glumes. **Spikelets** laterally compressed, with 1 floret. **Glumes** unequal, strongly keeled; **lower glumes** shorter than the florets, 1-veined; **upper glumes** usually longer than the florets, 1–6-veined; **lemmas** shorter than the paleas, 1–3-veined, midveins keeled, lateral veins usually obscure;

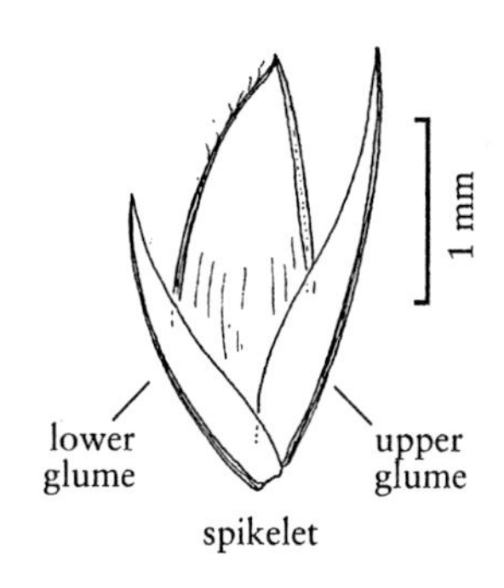

C. *aethiopicus*

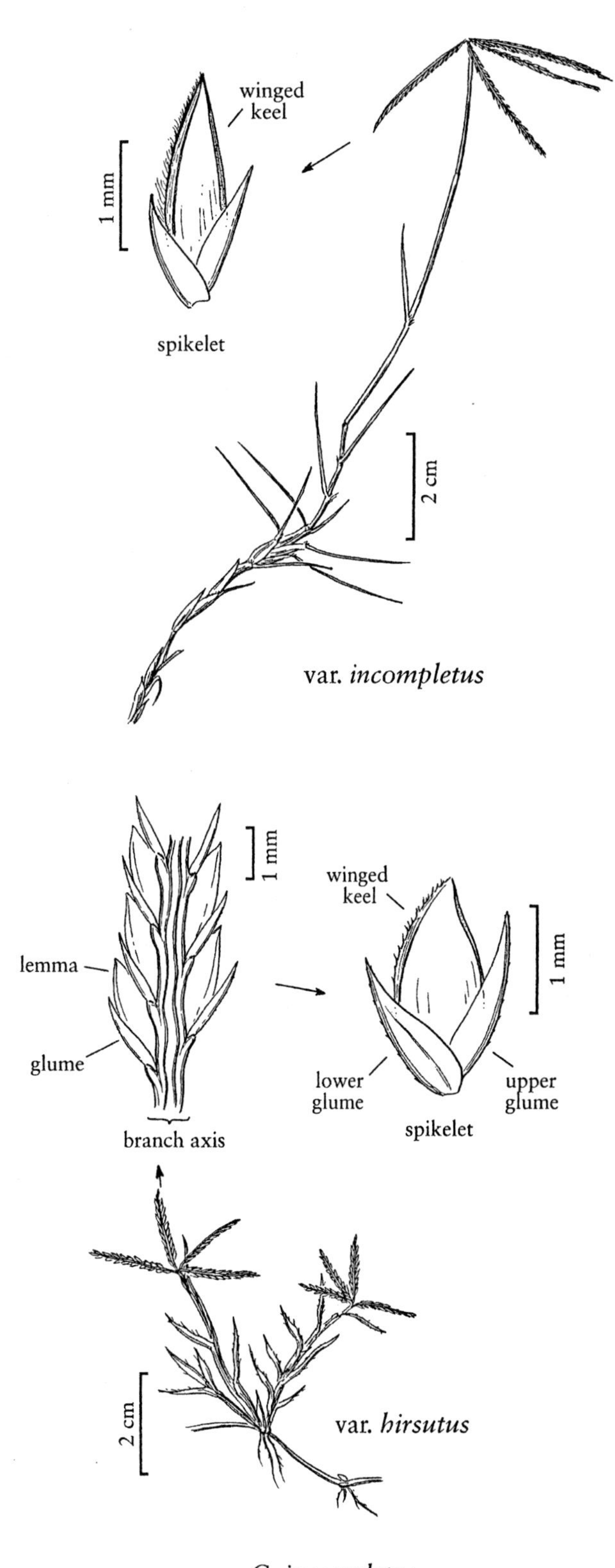

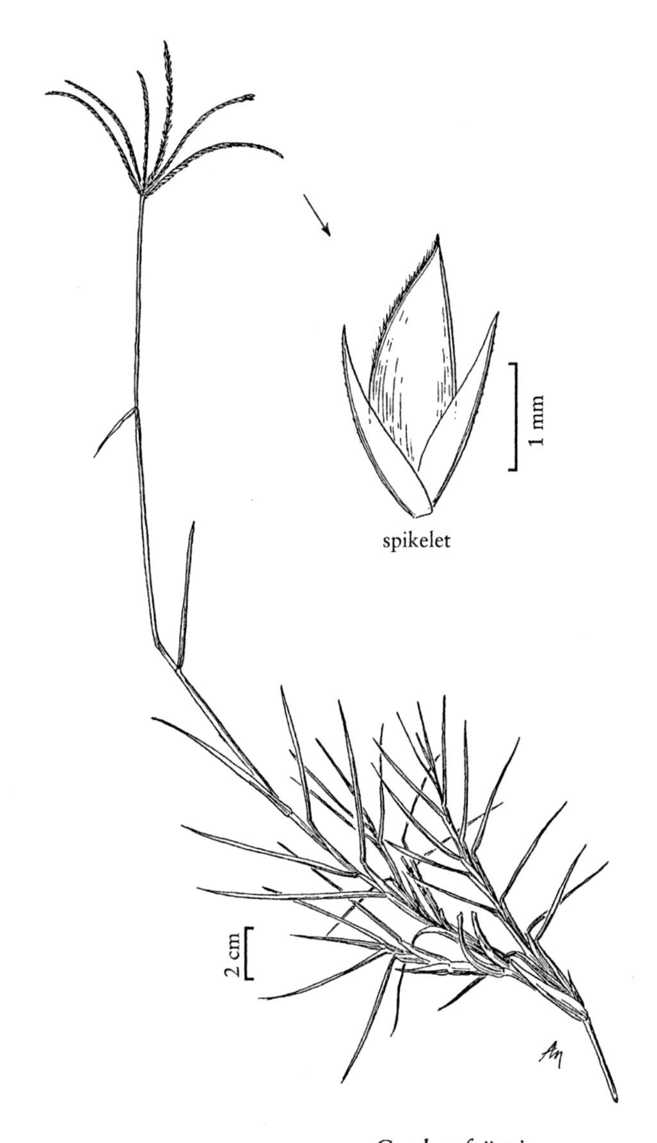

C. *nlemfuënsis*

C. *incompletus*

paleas thin, papery, 2-veined, obscurely keeled; **anthers** 3; **lodicules** absent; **styles** 2, plumose. **Caryopses** rarely produced. $x = 10$. Name from the Greek *spartine*, a cord made from *Spartium junceum*, probably applied to *Spartina* because of the tough leaves.

Spartina is a genus of 15–17 species, most of which grow in moist to wet, saline habitats, both coastal and interior. Reproduction of all the species is almost entirely vegetative.

There are nine native and two introduced species in the *Flora* region, plus three hybrids, one of which is native, the other two being deliberate introductions. One of the introduced species, *S. maritima*, grows in both Europe and at a few locations in Africa; the African populations may also represent introductions.

On the eastern seaboard of North America, the native species of *Spartina* extend as far north as Nova Scotia, but the few species native to the western seaboard do not extend north of California. Two species, *S. alterniflora* and *S. densiflora*, have, however, become established as far north as Washington and now threaten the health of many coastal salt marshes and mud flats (see http://www.spartina.org/ or http://www.wwta.org/environ/environ_topics.htm).

Mobberley (1956), on whose work this treatment is based, described three groups within *Spartina*, but did not give them formal recognition. Most of the species in the *Flora* region belong to his third group, which he characterized as having hard culms, scabrous leaf margins, more or less divergent inflorescence branches, usually closely imbricate spikelets, and hispid keels on their glumes and lemmas. *Spartina alterniflora* and *S. foliosa* belong to Mobberley's second group, the species of which have rather thick, succulent culms, glabrous blades with smooth margins, and less closely imbricate spikelets than in the other two groups. *Spartina spartinae* is the only species of his first group to grow in the *Flora* region. Like other members of the group, it has hard, slender culms, numerous short, closely imbricate inflorescence branches, spikelets that are hispid or villous, at least on the keels of the glumes, and no rhizomes.

SELECTED REFERENCES **Coastal Conservancy.** 2002. Invasive *Spartina* project. http://www.spartina.org/; **Kartesz, J.** and **C.A. Meacham.** 1999. Synthesis of the North American Flora, Version 1.0 (CD-ROM). North Carolina Botanical Garden, Chapel Hill, North Carolina, U.S.A.; **Mobberley, D.G.** 1956. Taxonomy and distribution of the genus *Spartina*. Iowa State Coll. J. Sci. 30:471–574; **Spicher, D.** and **M. Joselyn.** 1985. *Spartina* (Gramineae) in northern California: Distribution and taxonomic notes. Madroño 32:158–167.

1. Leaf blades with smooth or slightly scabrous margins.
    2. Panicles branches 2–8 cm long, usually closely appressed and often twisted, the lower branches evidently less closely imbricate than the upper branches; glumes usually curved; plants of California and Baja California, Mexico . . . . 3. *S. foliosa*
    2. Panicle branches 2–24 cm long, usually loosely appressed or divergent, usually not twisted, lower and upper branches more or less equally imbricate; glumes straight; plants of varied distribution, including California and Baja California, Mexico.
        3. Glumes usually mostly glabrous on the sides, sometimes with appressed hairs; panicles with 3–25 branches . . . . 2. *S. alterniflora*
        3. Glumes usually with appressed hairs on the sides, the margins sometimes glabrous; panicles with 1–12 branches.
            4. Ligules 2–3 mm long; anthers 5–13 mm long, well-filled, dehiscent at maturity . . . . 6. *S. anglica*
            4. Ligules 0.2–1.8 mm long; anthers 3–10 mm long, sometimes poorly filled and indehiscent at maturity.
                5. Ligules 0.2–0.6 mm long; leaf blades 6–12 cm long; anthers 3–6.5 mm long, well-filled, dehiscent at maturity . . . . 4. *S. maritima*
                5. Ligules 1–1.8 mm long; leaf blades 6–30 cm long; anthers 5–10 mm long, poorly filled and indehiscent at maturity . . . . 5. *S. ×townsendii*
1. Leaf blades with strongly scabrous margins.
    6. Panicles smooth in outline, with (6)15–75 tightly appressed panicle branches; branches 0.5–4(7) cm long; plants without rhizomes . . . . 1. *S. spartinae*

6. Panicles not smooth in outline, with 2–67 tightly appressed to strongly divergent branches; branches 1–15 cm; plants with more than 15 panicle branches always strongly rhizomatous, those with less than 16 branches with or without rhizomes.
   7. Plants without rhizomes or the rhizomes short; culms usually clumped; panicle branches 2–16.
      8. Upper glumes 1-veined .... 9. *S. densiflora*
      8. Upper glumes 3–4-veined.
         9. Spikelets 6–9 mm long; culms to 200 cm tall; plants of the southeastern United States .... 7. *S. bakeri*
         9. Spikelets 10–17 mm long; culms to 120 cm tall; plants of the northeastern United States .... 12. *S. ×caespitosa*
   7. Plants with well-developed rhizomes; culms usually solitary, sometimes a few together; panicle branches 3–67.
      10. Rhizomes whitish; upper glumes 1-veined or with all lateral veins on the same side of the keels; panicles with 2–15 branches, the branches 1–9 cm long.
         11. Spikelets 6–11 mm long, ovate to lanceolate; inland plants of western North America, rarely found east of Lake Winnipeg and the Mississippi Valley .... 10. *S. gracilis*
         11. Spikelets 7–17 mm long, linear-lanceolate to ovate-lanceolate; usually coastal, also known from a few inland sites in northeastern North America.
            12. Spikelets 7–12 mm long; blade of the second leaf below the panicles 0.5–4(7) mm wide; plants of disturbed and undisturbed coastal habitats from the Gulf of St. Lawrence to the Gulf of Mexico and, as an introduction, on the west coast of North America .... 11. *S. patens*
            12. Spikelets 10–17 mm long; blade of the second leaf below the panicles 2–7 mm wide; plants of disturbed habitats and artificial wetlands from Maine to Maryland .... 12. *S. ×caespitosa*
      10. Rhizomes light brown to brownish-purple; upper glumes 1-veined or with lateral veins on either side of the keels; panicles with 3–67 branches, the branches 1.5–15 cm long.
         13. Blade of the second leaf below the panicles 2–5(7) mm wide, usually involute even when fresh; panicles with 3–9 branches, the branches 3–9 cm long ..... 12. *S. ×caespitosa*
         13. Blade of the second leaf below the panicles 5–14 mm wide, flat when fresh; panicles with 5–67 branches, the branches 1.5–15 cm long.
            14. Lower glumes ¾ as long as to equaling the adjacent lemmas; upper glumes awned, the awns 3–8 mm, with glabrous, rarely hispid, lateral veins .... 13. *S. pectinata*
            14. Lower glumes less than ½ as long to ⅔ as long as the adjacent lemmas; upper glumes unawned or with awns up to 2 mm long, usually with hispid lateral veins .... 8. *S. cynosuroides*

1. Spartina spartinae (Trin.) Merr. *ex* Hitch. [p. 245]
GULF CORDGRASS

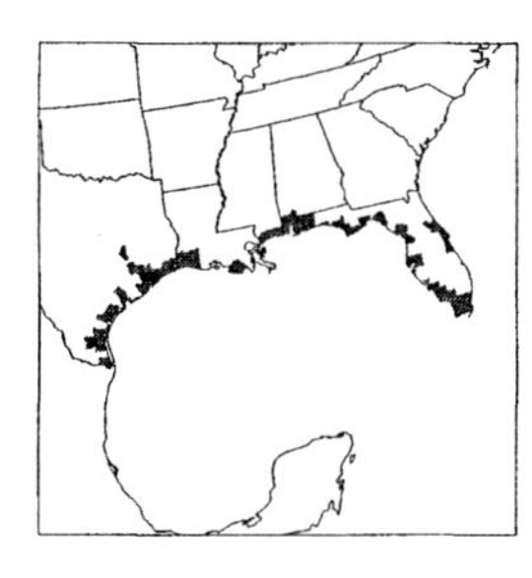

Plants cespitose, not rhizomatous. **Culms** 40–200 cm, in large clumps, hard, usually glabrous, nodes frequently exposed. **Sheaths** mostly glabrous, throat glabrous, sometimes scabrous; **ligules** 1–2 mm; **blades** 1.5–4.5 mm wide, involute when fresh, abaxial surfaces glabrous, adaxial surfaces and margins scabrous. **Panicles** 6–70 cm, smoothly cylindrical in outline, with (6)15–75 branches, internodes shorter than the branches; **branches** 0.5–4(7) cm, lower branches often longer than those above, all branches tightly appressed, closely imbricate, with 10–60 spikelets. **Spikelets** 5–8(10) mm. **Glumes** glabrous or hispidulous, keels hispid; **lower glumes** 2–8 mm, acuminate; **upper glumes** 4–8(10) mm, acuminate to obtuse, keels hispid, lateral veins 1–2, if 2, these on either side of the keel; **lemmas** 5–6 mm, glabrous or hispidulous, keels hispid over the distal ⅔, apices usually acuminate or apiculate, rarely obtuse; **anthers** 3–5 mm, dark red to purple. $2n$ = 28, 42.

Spartina spartinae grows from the Gulf coast through Mexico to Costa Rica in North America and, in South

America, in Paraguay and northern Argentina. In the United States, it grows in sandy beaches, roadsides, ditches, wet meadows, and arid pastures near the coast, the most inland collection being 60 miles from the coast. In other parts of its range it sometimes grows well inland in saline soils where *Pinus palustris* (longleaf pine) is dominant or co-dominant.

2. **Spartina alterniflora** Loisel. [p. 245]
SMOOTH CORDGRASS, SPARTINE ALTERNIFLORE

Plants rhizomatous; **rhizomes** elongate, flaccid, white, scales inflated, not or only slightly imbricate. **Culms** to 250 cm tall, (0.3)5–15(20) mm thick, erect, solitary or in small clumps, succulent, glabrous, having an unpleasant, sulphurous odor when fresh. **Sheaths** mostly glabrous, throat glabrous or minutely pilose, lower sheaths often wrinkled; **ligules** 1–2 mm; **blades** to 60 cm long, 3–25 mm wide, lower blades shorter than those above, usually flat basally, becoming involute distally, abaxial surfaces glabrous, adaxial surfaces glabrous or sparsely pilose, margins usually smooth, sometimes slightly scabrous, apices attenuate. **Panicles** 10–40 cm, with 3–25 branches, often partially enclosed in the uppermost sheath; **branches** 5–15 cm, loosely appressed, not twisted, more or less equally subremote to moderately imbricate throughout the panicle, axes often prolonged beyond the distal spikelets, with 10–30 spikelets. **Spikelets** 8–14 mm, straight, usually divergent, more or less equally imbricate on all the branches. **Glumes** straight, sides usually glabrous, sometimes pilose near the base or appressed pubescent, hairs to 0.3 mm; **lower glumes** 4–10 mm, acute; **upper glumes** 8–14 mm, keels glabrous, lateral veins not present, apices acuminate to obtuse, occasionally apiculate; **lemmas** glabrous or sparsely pilose, apices usually acuminate; **paleas** slightly exceeding the lemmas, thin, papery, apices obtuse or rounded; **anthers** 3–6 mm. $2n = 56, 70$.

Spartina alterniflora is found on muddy banks, usually of the intertidal zone, in eastern North and South America, but it is not known from Central America. In addition, it has become established on the west coast of North America, and in England and southeastern France. It hybridizes with *S. maritima* in Europe, with *S. pectinata* in Massachusetts, and with *S. foliosa* in California.

The rhizomes and scales of *S. alterniflora* have large air spaces, presumably an adaptation to the anaerobic soils of its usual habitat. Decaploid plants tend to be larger than octoploids, but they cannot be reliably distinguished without a chromosome count.

Spartina alterniflora is considered a serious threat to coastal ecosystems in Washington and California. It out-competes many of the native species in these habitats and frequently invades mud flats and channels, converting them to marshlands. Pure *S. alterniflora* grows within the lower elevational marsh zones in its native range but, in San Francisco Bay, its hybrids with *S. foliosa* grow both below and above the range of that species.

3. **Spartina foliosa** Trin. [p. 245]
CALIFORNIA CORDGRASS

Plants occasionally streaked or tinged with purple, rhizomatous; **rhizomes** elongate, flaccid, whitish, scales inflated, not closely imbricate. **Culms** to 150 cm tall, to 10 mm thick, erect, terete, solitary or in small clumps, succulent, glabrous, often with adventitious roots from the lower nodes, having an unpleasant, sulphurous odor when fresh. **Sheaths** mostly glabrous, throats sparsely pilose, lower sheaths sometimes somewhat wrinkled; **ligules** 1–2 mm; **blades** 8–12 mm wide, flat to loosely involute, glabrous, margins usually smooth, sometimes slightly scabrous, apices acuminate. **Panicles** 12–25 cm, with 3–25 branches, smoothly cylindrical, often partially enclosed in the uppermost sheath; **rachises** twisted, glabrous; **branches** 2–8 cm, usually closely appressed and twisted, lower branches noticeably longer and less closely imbricate than the upper branches, all branches with axes rarely extending past the distal spikelets, with 8–30 spikelets. **Spikelets** 8–25 mm, usually appressed, often appearing twisted, those on the lower branches usually less closely imbricate than those on the upper branches. **Glumes** usually curved, sides and keels glabrous, scabrous, or hispid, apices acuminate to obtuse or rounded; **lower glumes** 6–12 mm; **upper glumes** 8–25 mm, 1-veined; **lemmas** glabrous or sparsely appressed pubescent on the sides, keels glabrous, apices obtuse, rounded or lobed; **paleas** slightly exceeding the lemmas, thin, papery, glabrous, apices usually rounded, rarely acuminate; **anthers** 3–6 mm. $2n = 56$.

Spartina foliosa grows in the intertidal zone from northern California to Baja California, Mexico. Populations in San Francisco Bay are threatened by various introduced species of *Spartina*. Of particular concern is *S. alterniflora*, which forms hybrids with *S. foliosa* that have a broader ecological amplitude than either parent. In California, *S. foliosa* is often confused with *S. densiflora*, which is also established in some regions, but *S. foliosa* differs from that species in being rhizomatous and having softer culms and wider leaf blades.

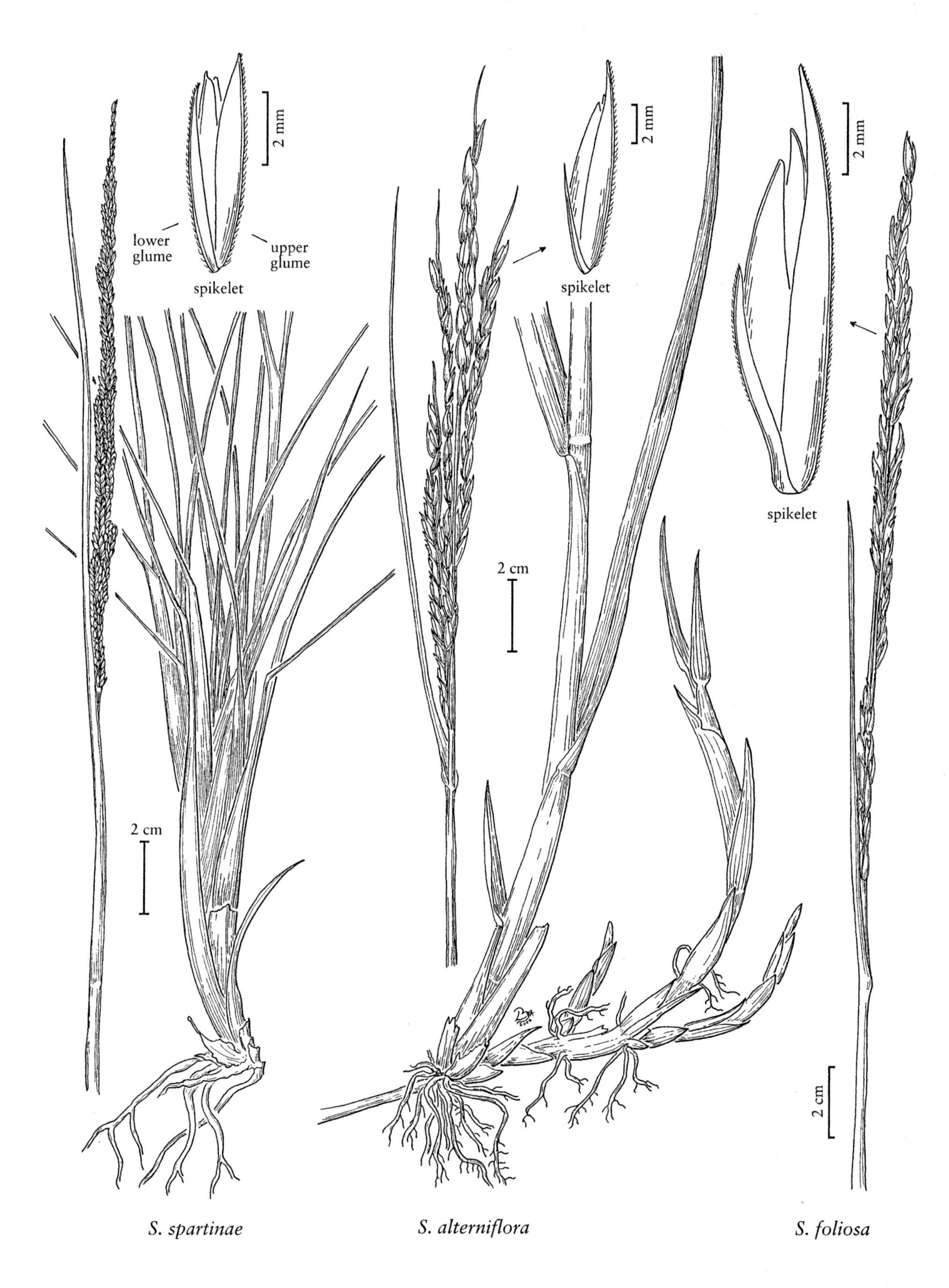

S. spartinae *S. alterniflora* *S. foliosa*

SPARTINA

4. **Spartina maritima** (Curtis) Fernald [p. 248]
Small Cordgrass

Plants rhizomatous; **rhizomes** with scales not inflated, not imbricate. **Culms** to 80 cm, relatively soft, solitary or in small clumps. **Sheaths** mostly glabrous, throat sometimes sparingly pilose, lower sheaths often wrinkled; **ligules** 0.2–0.6 mm; **blades** 6–12 cm long, 5–8 mm wide, loosely involute, disarticulating from the sheaths, abaxial surfaces glabrous, adaxial surfaces glabrous or sparsely pilose, margins smooth. **Panicles** 4–14 cm, with (1)2–3(7) branches; **branches** 2–11 cm, alternate, loosely appressed, not twisted, lower and upper branches more or less equally imbricate, with 5–30 spikelets. **Spikelets** 10–15 mm. **Glumes** straight, mostly appressed pubescent, only the margins glabrous; **lower glumes** 7–10 mm, narrow, acuminate, obtuse, or rounded; **upper glumes** 10–15 mm, acuminate (rarely obtuse); **lemmas** mostly appressed pubescent, margins and basal portion of the keels glabrous, apices acuminate; **anthers** 3–6.5 mm, well-filled, dehiscent at maturity. $2n = 56$.

Spartina maritima is a European species that has been reported as growing in Mississippi (Kartesz and Meacham 1999); the record has not been verified for this treatment. It also grows in Africa, possibly as an introduction.

5. **Spartina ×townsendii** H. Groves & J. Groves [p. 248]
Townsend's Cordgrass

Plants rhizomatous; **rhizomes** whitish, scales not inflated, not closely imbricate. **Culms** to 150 cm, relatively hard, solitary or in small clumps. **Sheaths** mostly glabrous, throats pilose, lower sheaths often wrinkled; **ligules** 1–1.8 mm; **blades** 6–30 cm long, 4–12 mm wide, diverging 20–45° from the culms, flat proximally, involute distally, both surfaces glabrous, margins smooth. **Panicles** 15–25 cm, with 2–10 branches; **branches** 4–24 cm, loosely appressed, with 10–30 spikelets. **Spikelets** 16–22 mm. **Glumes** mostly appressed pubescent, margins glabrous or sparingly hispidulous; **lower glumes** 8–14 mm, linear, acuminate to obtuse; **upper glumes** 16–22 mm, acuminate to obtuse; **lemmas** mostly pubescent, keels glabrous near the base, margins glabrous throughout, apices obtuse to rounded or obscurely lobed; **anthers** 5–10 mm, poorly filled, indehiscent at maturity. $2n = 62$.

Spartina ×townsendii is a sterile hybrid between the European *S. maritima* and the American *S. alterniflora*. It seems to have formed spontaneously at several locations in Europe, often taking over the areas formerly occupied by its progenitors. At some locations it has given rise to the fertile amphiploid *S. anglica*, from which it differes morphologically in its narrower, less divergent upper blades, shorter ligules, shorter, less hairy spikelets, and poorly filled, indehiscent anthers. *Spartina ×townsendii* has been used throughout the world for tideland reclamation because it is easy to establish, but it displaces native species.

6. **Spartina anglica** C.E. Hubb. [p. 248]
English Cordgrass

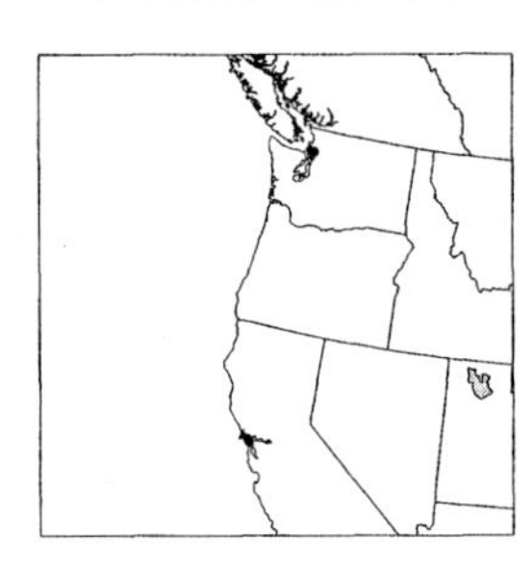

Plants rhizomatous; **rhizomes** elongate, flaccid, thick, whitish, imbricate. **Culms** 30–130 cm, forming large clumps. **Sheaths** glabrous, rounded dorsally; **ligules** 2–3 mm; **blades** 10–46 cm long, 6–15 mm wide, persistent or deciduous, flat or involute, adaxial surfaces ridged, not scabrous, margins smooth or slightly scabrous, sharply pointed, blades of upper leaves strongly divergent. **Panicles** 12–40 cm, with 2–12, more or less equally spaced branches; **branches** 16–25 cm, erect or somewhat divergent, axes pubescent, extending up to 5 cm beyond the spikelets; **disarticulation** at the base of the glumes, spikelets falling intact at maturity. **Spikelets** 14–21 mm long, 2–3 mm wide, narrowly oblong, appressed, closely imbricate. **Glumes** straight, sides appressed pubescent, keels ciliate or hispid, acute; **lower glumes** 10–14 mm, ⅔–⅘ as long as the upper glumes, 1-veined; **upper glumes** exceeding the floret, 3–6-veined; **lemmas** shorter than the upper glumes, shortly appressed pubescent, 1–3-veined, acute; **paleas** a little longer than the lemmas; **anthers** 5–13 mm, well-filled, dehiscent at maturity. $2n = 122–124$.

Spartina anglica is a naturally formed amphidiploid, derived from *Spartina ×townsendii*, that was first recognized as a separate species in 1968. It has been introduced (like *S. ×townsendii*) for reclamation of tidal mudflats. It differs from *Spartina ×townsendii* in its wider and more widely divergent upper blades, longer ligules, longer, more hairy spikelets, and longer, well-filled anthers.

7. **Spartina bakeri** Merr. [p. 248]
Sand Cordgrass

Plants cespitose, bases knotty, not rhizomatous. **Culms** to 200 cm, in large, dense clumps, indurate, often branching from the lower nodes. **Sheaths** smooth to striate, glabrous; **ligules** 0.5–2 mm; **blades** 10–50 cm long, 3–7 mm wide, usually involute, rarely flat, abaxial surfaces glabrous, adaxial surfaces and margins scabrous, apices acuminate. **Panicles** 8–25 cm, usually shallowly sinuous or lobed in outline, with 3–16 branches; **branches** 2–6 cm, usually appressed, moderately imbricate, axes glabrous, sometimes somewhat scabrous on the angles, with 10–30 spikelets. **Spikelets** 6–9 mm. **Glumes** with hispid keels and hispidulous margins, apices acuminate; **lower glumes** 3–6 mm, to ⅔ as long as the upper glumes; **upper glumes** 6–9 mm, hispidulous, 3–4-veined, lateral veins 2–3, prominent, on 1 side of the keel; **lemmas** mostly glabrous, keels hispid, margins glabrous or hispid, apices acute to obtuse, sometimes obscurely lobed; **anthers** about 5 mm, well-filled, dehiscent at maturity. $2n = 42$.

Spartina bakeri grows on sandy maritime beaches and other salt water sites in the southeastern coastal states and on the shores of inland, freshwater lakes in Florida. Its inflorescence is similar to that of *S. patens*, but the branches of *S. patens* usually diverge from the rachises at maturity, whereas those of *S. bakeri* remain appressed. *Spartina bakeri* is distinct from most other species of *Spartina* in North America in forming dense clumps and in being able to grow in freshwater habitats.

8. **Spartina cynosuroides** (L.) Roth [p. 248]
BIG CORDGRASS

Plants strongly rhizomatous; **rhizomes** elongate, purplish-brown or tan, scales closely imbricate. **Culms** 100–350 cm tall, 1–2 cm thick, hard, solitary or few together. **Sheaths** smooth to striate, mostly glabrous, throats often densely pilose, lower sheaths often wrinkled; **ligules** 1–3 mm; **blades** 6–20 mm wide, flat or involute, glabrous on both surfaces, margins strongly scabrous, apices acuminate, second blade below the panicles 5–15 mm wide, usually flat. **Panicles** 15–40 cm, not smooth in outline, with 5–67 branches; **branches** 6–15 cm, usually spreading, with 10–70 spikelets. **Spikelets** 9–14 mm. **Glumes** with hispid keels and hispidulous margins; **lower glumes** 3–7 mm, from less than ½ as long as to ⅔ as long as the adjacent lemmas, linear, acute; **upper glumes** 9–14 mm, usually more than twice as long as the lower glumes, exceeding the florets, mostly glabrous or hispidulous, keels scabrous or hispid, trichomes to 0.3 mm, 2 lateral veins prominent, 1 on each side of the keel, usually hispid, apices unawned or awned, the awns to 2 mm; **lemmas** glabrous or hispidulous, sometimes glabrous proximally and hispidulous distally, apices obtuse to rounded, sometimes shallowly bilobed; **anthers** 4–6 mm, well-filled, dehiscent. $2n = 28, 42$.

Spartina cynosuroides grows in brackish estuaries, tidal lagoons and bays, and in maritime habitats bordering the strand and intertidal zones. It grows primarily on the eastern and Gulf coasts of the United States, but has also been found in Michigan, possibly introduced by shipping. Reports from South Dakota are based on a misidentification.

9. **Spartina densiflora** Brongn. [p. 251]
DENSELY-FLOWERED CORDGRASS

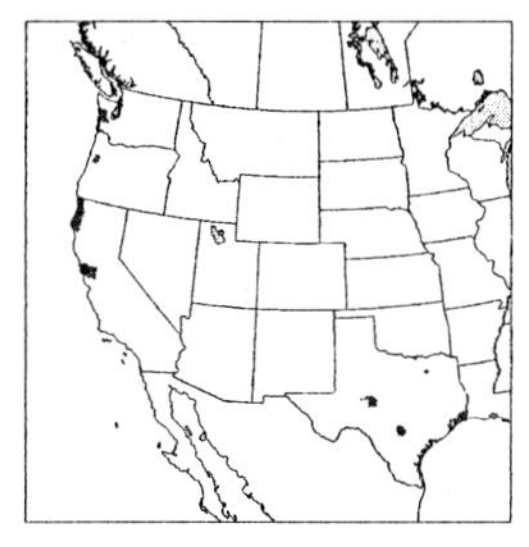

Plants cespitose, rarely rhizomatous; **rhizomes,** when present, short, to 10 mm thick. **Culms** 27–150 cm, forming large clumps, indurate, usually with short extravaginal shoots appressed to the culms. **Sheaths** glabrous, lower sheaths smooth, indurate and shining, upper sheaths dull and somewhat striate; **ligules** 1–2 mm; **blades** 12–43 cm long, 3–8 mm wide, involute when fresh, abaxial surfaces glabrous, adaxial surfaces and margins scabrous, apices acuminate. **Panicles** 10–30 cm long, 4–8 mm wide, sinuous in outline, often twisted, with 2–15 branches; **branches** 1–11 cm long, longer branches narrower than the shorter branches, all branches tightly appressed, moderately imbricate, axes not prolonged beyond the distal spikelets, with 10–30 spikelets. **Spikelets** 8–14 mm. **Glumes** glabrous or sparsely hispidulous, keels hispidulous, margins sparsely hispidulous; **lower glumes** 4–7 mm, usually obtuse; **upper glumes** 8–14 mm, 1-veined, usually acuminate; **lemmas** minutely hispidulous, keels glabrous proximally, hispidulous distally, apices acuminate to obtuse; **paleas** acuminate, keels glabrous basally, hispidulous distally; **anthers** 3–5 mm. $2n$ = unknown.

Spartina densiflora is native to South America, where it grows in coastal marshes and at inland sites. It was introduced to Humboldt Bay, Humboldt County, California, possibly during the nineteenth century. It is now established there and in several locations around San Francisco Bay and in Washington, Oregon, and Texas, as well as the Mediterranean coast of Europe. In California, it has often been mistaken for *S. foliosa*, from which it differs in its indurate culms, narrow, inrolled leaves, and cespitose growth habit and tendency to grow among *Salicornia* in the upper intertidal zone or in open mud.

10. **Spartina gracilis** Trin. [p. 251]
ALKALI CORDGRASS

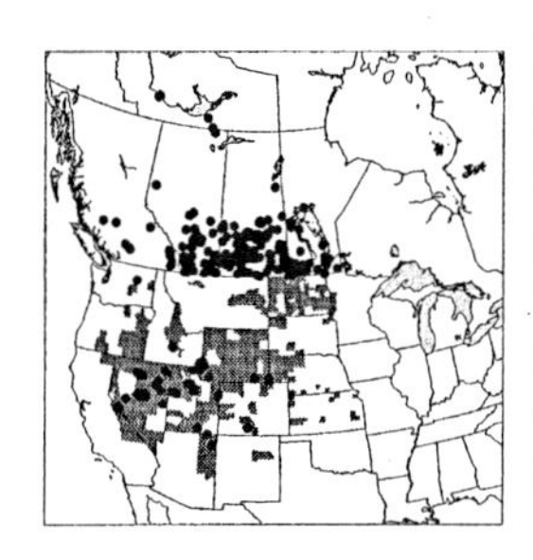

Plants strongly rhizomatous; **rhizomes** elongate, 1.5–5 mm thick, whitish, scales not inflated, closely imbricate. **Culms** 40–100 cm tall, 2–3.5 mm thick, usually solitary, erect, terete, indurate, glabrous. **Sheaths** smooth or striate,

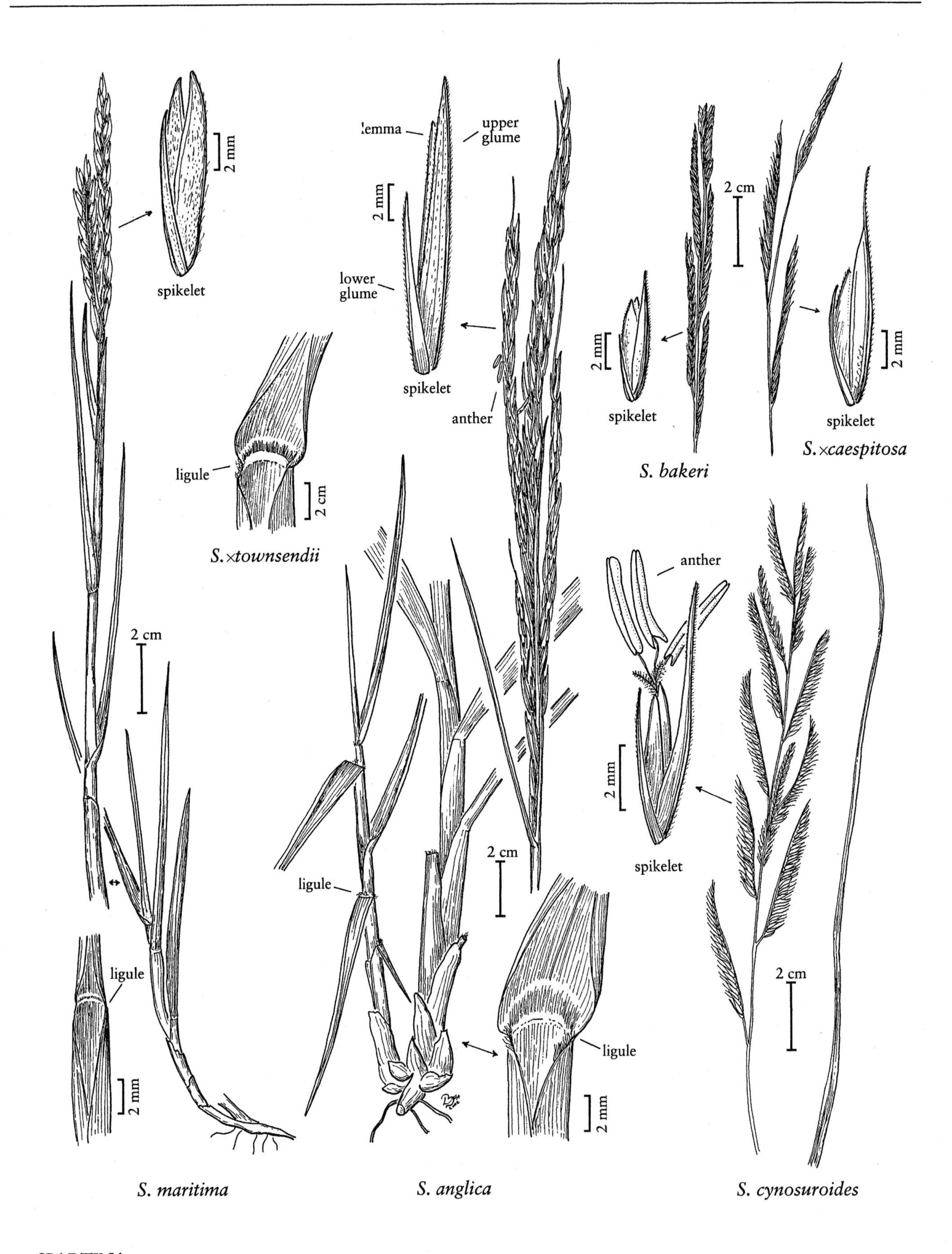

spikelet
2 mm
lemma
upper glume
2 mm
lower glume
spikelet
anther
2 cm
2 mm
spikelet
S. bakeri
2 mm
spikelet
S.×caespitosa
ligule
2 cm
S.×townsendii
anther
2 cm
2 mm
spikelet
ligule
2 cm
ligule
2 mm
ligule
2 mm
2 cm
S. maritima
S. anglica
S. cynosuroides

SPARTINA

mostly or completely glabrous, throats occasionally ciliate; **ligules** 0.5–1 mm; **blades** 6–30 cm long, 2.5–8 mm wide, flat, becoming involute, abaxial surfaces glabrous, adaxial surfaces scabrous, margins scabrous. **Panicles** 8–25 cm, not smooth in outline, with 3–12 branches; **branches** 1.5–8 cm, alternate, differing only slightly in length and spacing within a panicle, usually appressed, rarely spreading, with 10–30 spikelets. **Spikelets** 6–11 mm, ovate to lanceolate. **Glumes** with glabrous or sparingly hispidulous margins, apices acute or mucronate; **lower glumes** 3–7 mm, sides narrow, glabrous or sparsely pubescent, keels glabrous or strigose; **upper glumes** 6–10 mm, usually equaling the florets, keels strigose, hairs 0.2–0.5 mm, lateral veins 2, inconspicuous, both on the same side of the keel; **lemmas** glabrous or sparsely hirsute, keels hirsute, at least distally, hairs 0.3–1 mm, margins sparsely hairy, apices obtuse to rounded, sometimes obscurely lobed; **paleas** sparsely hispid distally, obtuse to slightly rounded; **anthers** 2.5–5 mm, well-filled, dehiscent at maturity. $2n = 42$.

Spartina gracilis is found on the margins of alkaline lakes and along stream margins and river bottoms. Its range extends from the southern portion of the Northwest Territories, Canada, to central Mexico.

11. **Spartina patens** (Aiton) Muhl. [p. 251]
SALTMEADOW CORDGRASS, SPARTINE ÉTALÉE

Plants strongly rhizomatous; **rhizomes** elongate, 1–6 mm thick, whitish, scales not imbricate. **Culms** 15–150 cm tall, 1–6 mm thick, usually solitary, indurate. **Sheaths** glabrous or mostly glabrous, throats occasionally short-pilose; **ligules** about 0.5(1) mm; **blades** 10–50 cm long, 0.5–4(7) mm wide, involute when fresh, abaxial surfaces glabrous, adaxial surfaces scabrous, margins strongly scabrous; blade of the second leaf below the panicles 4–40 cm long, 0.5–3(7) mm wide. **Panicles** 3–15 cm, not smooth in outline even if the branches appressed, with 2–15 branches; **branches** 1–7 cm, alternate, differing only slightly in length and spacing within a panicle, appressed to strongly divergent, with 10–30 spikelets. **Spikelets** 7–12 mm, linear lanceolate to ovate lanceolate. **Glumes** glabrous or sparsely hispidulous on the sides, keels scabrous to hispidulous, trichomes (1)1.5–2.5 mm, apices acuminate; **lower glumes** 3–8 mm, linear; **upper glumes** 7–12 mm, with 2 lateral veins, these on the same side of the keel, usually hispid, apices acuminate, acute, or obtuse; **lemmas** mostly glabrous or sparsely hispidulous, keels hispid distally, apices obtuse, rounded, or obscurely lobed; **anthers** 3–5 mm, well-filled, indehiscent. $2n = 28, 35, 42, 56$.

Spartina patens grows in coastal salt and brackish waters. It is native to the east coast of North and Central America, extending through the Caribbean Islands to the north coast of South America, but is now established at scattered locations on the west coast of Canada and the United States. On the east coast, it is usually one of the dominant components of coastal salt marshes, frequently extending from the dry, sandy beach above the intertidal zone well up into the drier portions of the marshes. The older inland collections are from areas associated with brine deposits or saline soils, but there is some indication that the species' range is increasing inland because of the use of salt to de-ice roads in winter.

The inflorescence of *Spartina patens* is similar to that of *S. bakeri* when young, but its inflorescence branches usually diverge at maturity, whereas those of *S. bakeri* remain appressed.

Spartina patens is probably one of the parents of *S. ×caespitosa*, *S. pectinata* being the other. Unlike *S. ×caespitosa*, *S. patens* grows in both disturbed and undisturbed habitats.

12. **Spartina ×caespitosa** A. A. Eaton [p. 248]
MIXED CORDGRASS

Plants rhizomatous or not; **rhizomes**, when present, thick, usually purplish-brown, scales closely imbricate. **Culms** to 120 cm tall, 1–3 mm thick, indurate, solitary or in small, dense clumps. **Sheaths** mostly glabrous, throats glabrous or short-pilose; **ligules** 0.5–1 mm; **blades** 8–56 cm long, 2–6(7) mm wide, usually involute, abaxial surfaces glabrous, adaxial surfaces glabrous or scabrous, margins strongly scabrous, blade of the second leaf below the panicles 8–56 cm long, 2–5(7) mm wide. **Panicles** 9–20 cm, not smoothly cylindrical, with 3–9 branches; **branches** 3–9 cm, appressed or spreading, with 20–50 spikelets. **Spikelets** 10–17 mm, lanceolate to ovate-lanceolate. **Glumes** glabrous or sparsely hispidulous, keels glabrous, hispid in whole or in part, or ciliate; **lower glumes** 4–9 mm, acuminate or awned; **upper glumes** 10–17 mm, exceeding the florets, keels hispid, lateral veins prominent, 1 on each side of the keel or 2–3 on 1 side of the keel, apices acuminate or awned; **lemmas** glabrous or sparsely hispidulous, apices obtuse, rounded, obscurely lobed, or apiculate; **anthers** 3–6 mm, poorly filled, indehiscent. $2n = 42$.

Spartina ×caespitosa is found in disturbed areas of the drier portions of salt and brackish marshes, at some distance above the intertidal zone. It occurs sporadically along the coast from Maine to Maryland, a region where its putative parents, *S. pectinata* and *S. patens*, are sympatric. None of the populations Mobberley (1956) examined was growing in undisturbed land.

Mobberley's (1956) investigations led him to conclude that the populations of *S.* ×*caespitosa* are polythetic in origin. Part of the evidence for his conclusion was the variability he observed. It is this variability that makes it necessary to bring out the hybrid at several locations in the key. Its distribution is, however, very limited, a fact that may be more useful for identification than any of the morphological characteristics examined.

13. **Spartina pectinata** Link [p. 251]
PRAIRIE CORDGRASS, SPARTINE PECTINÉE

Plants strongly rhizomatous; **rhizomes** elongate, (2)3–8 mm thick, purplish-brown or light brown (drying white), scales closely imbricate. **Culms** to 250 cm tall, 2.5–11 mm thick, solitary or in small clumps, indurate. **Sheaths** mostly glabrous, throats often pilose; **ligules** 1–3 mm; **blades** 20–96 cm long, 5–15 mm wide, flat when fresh, becoming involute when dry, glabrous on both surfaces, margins strongly scabrous, blade of the second leaf below the panicles 32–96 cm long, 5–14 mm wide, usually involute. **Panicles** 10–50 cm, not smooth in outline, with 5–50 branches; **branches** 1.5–15 cm, appressed to somewhat spreading, with 10–80 spikelets. **Spikelets** 10–25 mm. **Glumes** shortly awned, glabrous or sparsely hispidulous; **lower glumes** 5–10 mm, from ¾ as long as to equaling the adjacent lemmas, keels hispid, apices awned; **upper glumes** 10–25 mm (including the awn), exceeding the florets, glabrous or sparsely hispid, keels scabrous to hispid, trichomes about 0.3 mm, lateral veins usually glabrous (rarely hispid), on either side of, and close to, the keels, apices awned, awns 3–8 mm; **lemmas** glabrous, keels pectinate distally, apices bilobed, lobes 0.2–0.9 mm; **anthers** 4–6 mm, well-filled, dehiscent. $2n$ = 42, 70, 84.

Spartina pectinata is native to Canada and the United States, but it has been introduced at scattered locations on other continents. On the Atlantic coast, it grows in marshes, sloughs, and flood plains, being a common constituent of ice-scoured zones of the northeast and growing equally well in salt and fresh water habitats. In western North America, it grows in both wet and dry soils, including dry prairie habitats and along roads and railroads.

Spartina pectinata is thought to be one of the parents of *S.* ×*caespitosa*, the other parent being *S. patens*.

17.46 BOUTELOUA Lag.

J.K. Wipff

Plants annual or perennial; synoecious; habit various, cespitose, stoloniferous, or rhizomatous. **Culms** 1–80 cm. **Leaves** usually mostly basal; **sheaths** open; **ligules** of hairs, membranous, or membranous and ciliate. **Inflorescences** terminal, panicles of 1–80 solitary, spikelike branches, exceeding the upper leaves; **branches** 4–50(75) mm, not woody, 1-sided, usually racemose on elongate rachises, sometimes digitate or subdigitate, with 1–130+ sessile to subsessile spikelets in 2 rows, axes terminating in a spikelet or extending beyond the base of the distal spikelet. **Spikelets** closely imbricate, appressed to pectinate, laterally compressed or terete, with 1–2(3) florets, lowest floret in each spikelet bisexual, distal florets staminate or sterile; **disarticulation** at the base of the branches or above the glumes. **Glumes** unequal or subequal, 1 or both glumes equaled or exceeded by the distal floret, 1-veined, acute or acuminate, sometimes shortly awned; **lower glumes** usually shorter than the lowest floret; **lemmas of lowest florets** entire, bilobed, trilobed, or 4-lobed, 3-veined, veins usually extended into 3 short awns; **paleas of lowest florets** 2-veined, veins sometimes excurrent; **distal floret(s)** staminate or sterile, varying from similar to the lowest floret in shape, size, and venation to sterile and reduced to an awn column with well-developed awns or to a flabellate scale. x = 10. Named for the brothers Claudio (1774–1842) and Esteban (1776–1813) Boutelou y Soldevilla, Spanish botanists.

Bouteloua, a genus of the Western Hemisphere with its center of diversity in Mexico, has about 40 species; all 19 species treated here are native to the *Flora* region. Several of its taxa are important forage grasses, and some are important constituents of the native North

S. densiflora *S. gracilis* *S. patens* *S. pectinata*

SPARTINA

American grasslands. Two that are particularly important in North America are *Bouteloua curtipendula* and *B. gracilis*. These were major constituents of the shortgrass prairie that once covered the drier portions of the Great Plains. Both are excellent forage species. Irrigation has converted much of the area they once occupided to agricultural use, but large areas of *Bouteloua* grasslands remain.

Based on molecular data and morphological similarities in the non-pistillate characters, Columbus (1999) recommended expanding *Bouteloua* to include *Buchloë*, *Opizia*, and *Cathestecum*, plus some other small genera not known from the *Flora* region. The traditional treatment is adopted here, pending corroboration from a wider range of data, both molecular and morphological.

SELECTED REFERENCES **Columbus, J.T.** 1999. An expanded circumscription of *Bouteloua* (Gramineae: Chlorideae): New combinations and names. Aliso 18:61–65; **Esparza Sandoval, S.** and **Y. Herrera-Arrieta.** 1996. Revisión de *Bouteloua barbata* Lagasca (Poaceae: Eragrotideae). Phytologia 80:73–91; **Gould, F.W.** 1979. The genus *Bouteloua* (Poaceae). Ann. Missouri Bot. Gard. 66:348–416; **Griffiths, D.** 1912. The grama grasses: *Bouteloua* and related genera. Contr. U.S. Natl. Herb. 14:343–428; **Reeder, J.R.** and **C.G. Reeder.** 1980. Systematics of *Bouteloua breviseta* and *B. ramosa* (Gramineae). Syst. Bot. 5:312–321; **Reeder, J.R.** and **C.G. Reeder.** 1990. *Bouteloua eludens*: Elusive indeed, but not rare. Desert Pl. 10:19–22, 31–32; **Wipff, J.K.** and **S.D. Jones.** 1996. A new combination in *Bouteloua* (Poaceae). Sida 17:111–114.

1. Panicle branches deciduous, disarticulation occurring at their bases; spikelets usually 1–15 per branch, usually appressed rather than pectinate (subg. *Bouteloua*).
  2. All or most panicle branches with 1 spikelet . . . . . 3. *B. uniflora*
  2. All or most panicle branches with 2–15 spikelets.
    3. First (proximal) spikelet on each branch with 1 floret, the remaining spikelets with 2 florets; plants annual; panicles with 1–15 branches . . . . . 4. *B. aristidoides*
    3. Spikelets all alike or with 2 or more florets; plants annual or perennial; panicles with 1–80 branches.
      4. Central awns of lemmas flanked by 2 membranous lobes at maturity, the lobes 0.5–1.5 mm.
        5. Upper glumes bilobed, awned, the awns arising from between the teeth; inflorescence branch axes with deeply bi- or trifurcate apices; second florets sterile, rudimentary . . . . . 7. *B. rigidiseta*
        5. Upper glumes acute, unawned or awn-tipped; inflorescence branch axes with apices entire; second florets usually staminate.
          6. Base of plants dense, hard and knotty; culms straight, unbranched; panicle branches (15)20–30 mm long; plants rhizomatous . . . . . 9. *B. radicosa*
          6. Base of plants usually not dense, hard, or knotty; culms straight or geniculate, branching; panicle branches 10–20 mm long; plants not rhizomatous . . . . . 8. *B. repens*
      4. Central awns of lemmas, if present, not flanked by membranous lobes or the lobes less than 0.3 mm long.
        7. Upper glumes with hairs, at least over the midveins.
          8. Upper glumes with hairs only over the veins . . . . . 8. *B. repens*
          8. Upper glumes with hairs over the veins and elsewhere.
            9. Panicles 6–10 cm long; branches with 2–6 spikelets . . . . . 5. *B. eludens*
            9. Panicles 2.5–6 cm long; branches with 8–12 spikelets . . . . . 6. *B. chondrosoides*
        7. Upper glumes glabrous, sometimes scabrous.
          10. Second florets sterile, usually rudimentary, usually without paleas; central awns rarely to 7 mm long; panicles with 9–80 branches.
            11. At least some leaf blades more than 2.5 mm wide, flat or folded when dry; ligules 0.3–0.5 mm long; anthers yellow, orange, red, or purple . . . . . 1. *B. curtipendula*
            11. Leaf blades 1–1.5(2.5) mm wide, involute when dry; ligules 1–1.5 mm long; anthers dark purple . . . . . 2. *B. warnockii*
          10. Second florets bisexual, pistillate or staminate, with well-developed paleas; central awns 4–10 mm long; panicles with 2–17 branches.
            12. Base of plants dense, hard and knotty; culms straight, unbranched; panicle branches (15)20–30 mm long; plants rhizomatous . . . . . 9. *B. radicosa*

12. Base of plants usually not dense, hard, or knotty; culms straight or geniculate, branching; panicle branches 10–20 mm long; plants not rhizomatous . . . . 8. *B. repens*

1. Panicle branches persistent; disarticulation above the glumes; spikelets 6–130 or more per branch, pectinate (subg. *Chondrosum*).
    13. Upper glumes of at least some spikelets with papillose-based hairs.
        14. Panicle branches extending beyond the base of the terminal spikelets . . . . 11. *B. hirsuta*
        14. Panicle branches terminating in a spikelet.
            15. Plants tufted annuals or short-lived stoloniferous perennials; panicle branches 4–8, the axes with papillose-based hairs; lowest lemmas 3–4 mm long . . . . 17. *B. parryi*
            15. Plants perennial, often shortly rhizomatous; panicle branches 1–3(6), the axes scabrous, never with papillose-based hairs; lowest lemmas 3.5–6 mm long . . . . 10. *B. gracilis*
    13. Upper glumes glabrous, scabrous, or hairy, but the hairs not papillose-based.
        16. Lower cauline internodes woolly-pubescent . . . . 12. *B. eriopoda*
        16. Lower cauline internodes glabrous or mostly so, sometimes pubescent immediately below the nodes.
            17. Central awns of lemmas not flanked by membranous lobes . . . . 13. *B. trifida*
            17. Central awns of lemmas flanked by 2 membranous lobes.
                18. Lowest paleas in the spikelets awned, awns 1–2 mm long; panicles with 2–20 branches.
                    19. Lowest lemmas glabrous, with awns 3–4 mm long; panicle branches with 6–20 spikelets; plants perennial . . . . 14. *B. kayi*
                    19. Lowest lemmas densely pilose, with awns 0.5–3 mm long; panicle branches with 20–50 spikelets; plants annual or short-lived perennials . . . . . . 15. *B. barbata*
                18. Lowest paleas in the spikelets unawned, but the veins sometimes excurrent for less than 1 mm; panicles with 1–6 branches.
                    20. Plants annual . . . . 16. *B. simplex*
                    20. Plants perennial.
                        21. Culms usually with 2–3 nodes, not woody at the base; caryopses 2.5–3 mm long; lower paleas shallowly bilobed, the veins sometimes excurrent . . . . 10. *B. gracilis*
                        21. Culms usually with 4–5 nodes, somewhat woody at the base; caryopses 1–1.2 mm long; lower paleas acute to acuminate, the veins not excurrent.
                            22. Lower culm internodes with a thick, white, chalky bloom distally; panicle branches stramineous, mostly appressed, usually straight to slightly arcuate; plants rhizomatous, growing on gypsum soils . . . . 18. *B. breviseta*
                            22. Lower culm internodes without a conspicuous bloom; panicle branches dark, mostly ascending to widely divergent, usually becoming arcuate; plants not rhizomatous, growing on limestone soils . . . . 19. *B. ramosa*

Bouteloua Lag. subg. Bouteloua

Panicle branches (1)4–80, with 1–15 spikelets, terminating beyond the distal spikelet in an entire, bifurcate, or trifurcate tip; **spikelets** usually appressed; **disarticulation** at the base of the branches.

1. **Bouteloua curtipendula** (Michx.) Torr. [p. 256]
SIDEOATS GRAMA

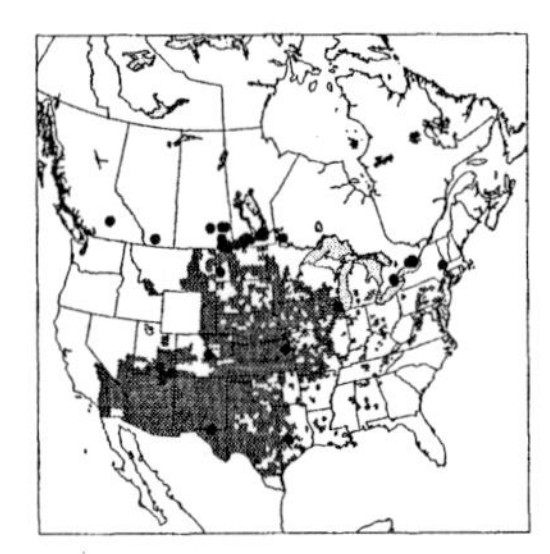

Plants perennial; cespitose or not, with or without rhizomes. **Culms** 8–80 cm, erect or decumbent, solitary or in small to large groups. **Leaves** evenly distributed; **sheaths** mostly glabrous, sometimes with hairs distally; **ligules** 0.3–0.5 mm, membranous, ciliate; **blades** 2–30 cm long, (1.4)2.5–7 mm wide, at least some over 2.5 mm wide, flat or folded when dry, usually smooth abaxially and scabrous adaxially, occasionally pubescent, bases usually with papillose-based hairs on the margins. **Panicles** 13–30 cm, secund, with (12)30–80 reflexed branches; **branches** (5)10–30(40) mm, deciduous, with (1)2–7(15) spikelets, axes terminating 3–5 mm beyond the base of the terminal spikelets, apices entire; **disarticulation** at the base of the branches. **Spikelets** appressed, all alike, with 1 bisexual and 1–2 sterile, rudimentary florets. **Glumes** unequal, glabrous or scabrous; **lower glumes** 2.5–6 mm, ½ or more as long as the upper glumes; **upper glumes** 5.5–8 mm; **lowest lemmas** 3–6.5 mm, glabrous or scabrous-strigose, often minutely rugose, acute or inconspicuously 3-lobed, 3-veined, veins usually extending as short mucros or awns to 6 mm; central mucros or awns not flanked by membranous lobes; **lowest paleas** acute, unawned; **anthers** 1.5–3.5 mm, yellow, orange, red, or purple; **distal floret(s)** 0.4–3.5 mm, sterile, variable, usually a glabrous lemma having a short membranous base, no palea, and 3 unequally-developed awns, central awns 1.5–7 mm. $2n$ = (20), 40, 41–103.

Bouteloua curtipendula is a common, often dominant or co-dominant species in open grasslands and wetlands of the drier portions of the central grasslands of North America. It is highly regarded as a forage species and is also an attractive ornamental. Its range extends from the *Flora* region through Mexico and Central America to western South America.

As the range of chromosome numbers suggests, *B. curtipendula* is an apomictic species. There are three varieties. Two of the three grow in the *Flora* region; the third, *B. curtipendula* var. *tenuis* Gould & Kapadia, is endemic to Mexico.

1. Plants long-rhizomatous; culms solitary or in small clumps . . . . var. *curtipendula*
1. Plants not long-rhizomatous, bases sometimes knotty with short rhizomes; culms in large or small clumps . . . . var. *caespitosa*

Bouteloua curtipendula var. **caespitosa** Gould & Kapadia

Plants cespitose, often with a knotty base, not or shortly rhizomatous. **Culms** in large or small clumps, stiffly erect. **Blades** usually narrow, but at least some over 2.5 mm wide. **Panicles** with 12–80 branches, averaging 2–7 spikelets per branch. **Glumes** and **lemmas** bronze or stramineous to green, or various shades of purple; **anthers** usually yellow or orange, occasionally red or purple. $2n$ = 58–103.

Bouteloua curtipendula var. *caespitosa* grows on loose, sandy or rocky, well drained limestone soils at 200–2500 m in the southwestern United States, Mexico, and South America. It frequently grows, and may hybridize, with *B. warnockii*.

Bouteloua curtipendula (Michx.) Torr. var. **curtipendula**

Plants not cespitose, with long rhizomes. **Culms** solitary or in small clumps. **Blades** 3–7 mm, flat. **Panicles** with 40–70 branches, averaging 3–7 spikelets per branch. **Glumes** and **lemmas** typically purple or purple-tinged; **anthers** red or red-orange, infrequently yellow, orange, or purple. $2n$ = 40, 41–66.

Bouteloua curtipendula var. *curtipendula* is the common variety of *B. curtipendula* in most of the *Flora* region. It grows on rich, loamy, well-drained prairie soils. Its elevational range extends from below 100 m to 2500 m.

2. **Bouteloua warnockii** Gould & Kapadia [p. 256]
WARNOCK'S GRAMA

Plants perennial; cespitose, forming clumps 4–10 cm in diameter, without rhizomes or stolons. **Culms** 20–35(50) cm, stiffly erect. **Leaves** bluish-green, more or less glaucous; **sheaths** mostly glabrous, hairs present distally; **ligules** 1–1.5 mm, of hairs; **blades** 5–15(25) cm long, 1–1.5(2.5) mm wide, stiffly erect or curving, involute when dry; mostly glabrous, ligular area with long and short hairs, bases usually with papillose-based hairs on the margins. **Panicles** 5–13(20) cm, with 9–15(30) branches; **branches** 4–5.5 mm, deciduous, scabrous, with 2–6 spikelets, axes terminating well beyond the terminal spikelets, apices entire; **disarticulation** at the base of the branches. **Spikelets** 5–6.5 mm, with 1 bisexual and 1 sterile floret, appressed, all alike, green, often with a brownish or purplish cast. **Lower glumes** slightly shorter than the upper glumes, both usually exceeded by the lemmas of the lowest florets; **upper glumes** glabrous, sometimes scabrous; **lowest lemmas**

glabrous, acute, 3-awned, awns less than 1 mm, central awns not flanked by 2 membranous lobes; **anthers** 2.2–3.7 mm, dark purple; **second florets** sterile, usually without paleas; **second lemmas** reduced to a glabrous awn column, sometimes moderately well-developed and 3-awned, awns usually not exserted, central awns to 2.5 mm. 2*n* = 21, 22, 23, 24, 25, 28, 38, 40.

Bouteloua warnockii grows on limestone ledges and dry slopes below limestone outcrops. Its range extends from the southwestern United States to the state of Coahuila in northern Mexico. It frequently grows, and may hybridize with, *B. curtipendula* var. *caespitosa*.

3. **Bouteloua uniflora** Vasey [p. 256]
NEALLY'S GRAMA

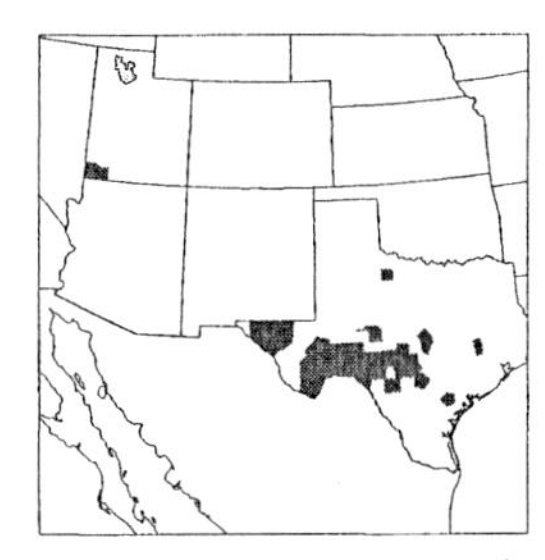

Plants perennial; cespitose, without rhizomes or stolons. **Culms** 20–60 cm, stiffly erect, glabrous. **Sheaths** mostly glabrous, a few long hairs present near the ligules; **ligules** 0.2–0.5 mm, of hairs; **blades** 6–16 cm long, 1–2 mm wide, involute when dry, glabrous, bases usually with papillose-based hairs on the margins. **Panicles** 5–10(14) cm, with 15–70 branches; **branches** 5–9 mm, deciduous, scabrous, with 1 spikelet (lower branches occasionally with 2 spikelets), axes extending 3–4 mm beyond the terminal spikelets, apices entire; **disarticulation** at the base of the branches. **Spikelets** appressed, with 1 bisexual and 0–1 rudimentary florets. **Glumes** acute to slightly cleft and minutely apiculate, midveins usually scabrous; **lower glumes** 2.5–4 mm; **upper glumes** 6.2–8 mm, mostly smooth, midveins usually scabrous; **lowest lemmas** 6–7.5 mm, acute or minutely cleft, glabrous, unawned, sometimes mucronate; **lowest paleas** unawned, glabrous; anthers 2.5–3 mm, bright yellow; **second florets** absent or reduced to 1 or 3 short awns, glabrous. **Caryopses** about 3 mm. 2*n* = 20.

Bouteloua uniflora grows primarily in fertile, rocky, limestone soils of Texas and adjacent Coahuila, Mexico at 300–1000 m. A disjunct collection has been reported from Zion National Park, Utah. Plants in the *Flora* region belong to **Bouteloua uniflora** Vasey var. **uniflora,** which differs from **B. uniflora** var. **coahuilensis** Gould & Kapadia in having taller (40–60 cm, not 20–40 cm) leafy, rather than scapose, culms, longer leaf blades (12–16 cm versus 6–12 cm), and 50–70, rather than 15–40, panicle branches.

4. **Bouteloua aristidoides** (Kunth) Griseb. [p. 258]

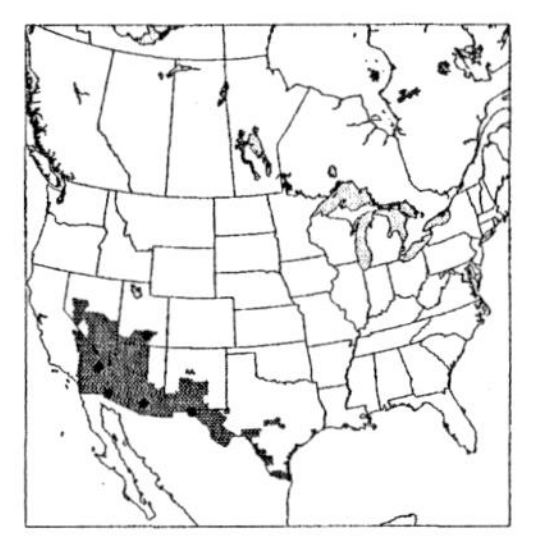

Plants annual; tufted. **Culms** 4–60 cm, outer culms of a tuft decumbent, sometimes geniculate, branched at the lower nodes. **Ligules** 0.2–0.5 mm, membranous, lacerate or ciliate; **blades** 2–5(9) cm long, 0.7–2 mm wide, flat or folded, adaxial surfaces sometimes with papillose-based hairs, margins usually with papillose-based hairs near the ligules. **Panicles** 2.5–10.5 cm, with (1)4–15 branches; **branches** 5–45 mm, deciduous, densely pubescent (at least basally), with 2–10 spikelets per branch, axes extending 2–10 mm beyond the base of the terminal spikelets, apices entire; **disarticulation** at the base of the branches, the break forming a sharp tip. **Spikelets** appressed. **Proximal spikelet on each branch** with 1 floret; **lower glumes** 1.5–3.5 mm, glabrous, narrow to subulate; **upper glumes** 5.5–6.2 mm, densely pubescent, at least on the basal ½; **lemmas** 5.8–6 mm, acuminate, unawned; **lowest paleas** almost as long as the lemmas, bifid, glabrous; **rachillas** prolonged beyond the florets for about 0.5 mm. **Distal spikelets** with 1 bisexual and 1 rudimentary floret, **glumes** unequal, glabrous, minutely scabrous on the keels; narrowly acute or acuminate; **lower glumes** 1.5–2 mm; **upper glumes** 5–6 mm, glabrous or sparsely pubescent basally, often divergent; **lowest lemmas** 6–8 mm, veins pubescent, lateral veins excurrent as short (to 1 mm) awns, acuminate, midvein extended into a setaceous tip or a short awn; **lowest paleas** 5–7 mm, bifid, veins often excurrent as short awns; **anthers** about 2.5 mm, yellow or yellow and red; **distal florets** reduced to a pubescent, 3-awned, awn column, awns 2–7 mm, exserted. **Caryopses** 2.5–3 mm. 2*n* = 40.

There are two varieties, both of which grow in the *Flora* region.

1. Panicle branches with 2–5 spikelets, usually 5–16 mm to the base of the terminal spikelets, axes usually extending an additional 6–10 mm . . . . var. *aristidoides*
1. Panicle branches with 6–10 spikelets, usually 15–35 mm to the base of the terminal spikelets, axes extending an additional 2–5(7) mm . . . . var. *arizonica*

Bouteloua aristidoides (Kunth) Griseb. var. **aristidoides**
NEEDLE GRAMA

Panicle branches 5–16 mm to the base of the terminal spikelets and extending an additional 6–10 mm, with 2–5 spikelets.

Bouteloua aristidoides var. *aristidoides* grows in dry mesas, plains, and washes from near sea level to about 2000 m. It matures rapidly following summer rains, and

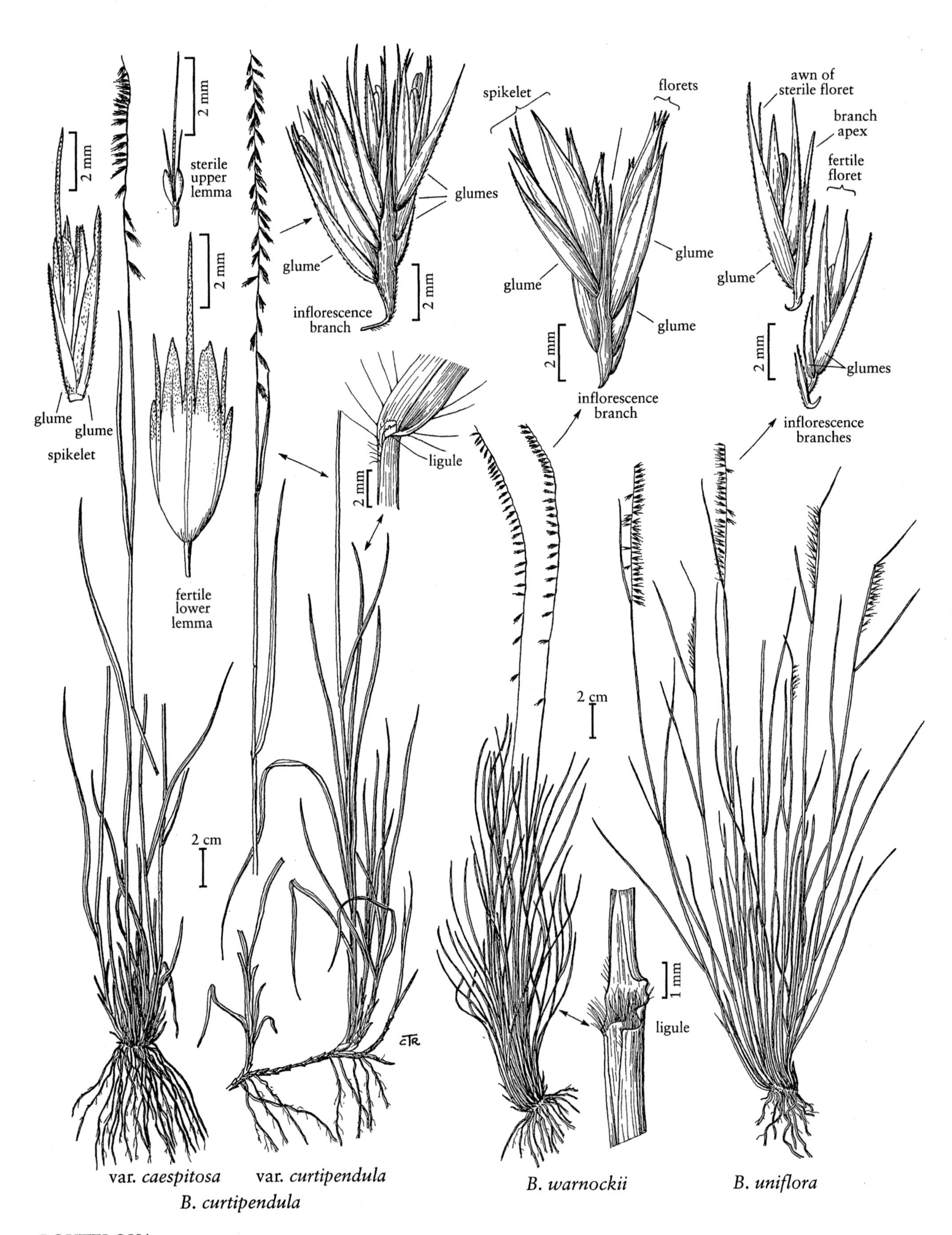
2 mm
glume
glume
spikelet
2 mm
sterile
upper
lemma
2 mm
fertile
lower
lemma
2 cm
glumes
glume
inflorescence
branch
2 mm
ligule
2 mm
spikelet
florets
glume
glume
glume
2 mm
inflorescence
branch
awn of
sterile floret
branch
apex
fertile
floret
glume
2 mm
glumes
inflorescence
branches
2 cm
1 mm
ligule
var. *caespitosa*
var. *curtipendula*
B. curtipendula
B. warnockii
B. uniflora

BOUTELOUA

can be abundant over large areas within its range, which extends from California to western Texas and Mexico.

Bouteloua aristidoides var. **arizonica** M.E. Jones
ARIZONA NEEDLE GRAMA

Panicle branches 15–35 mm to the base of the terminal spikelets and extending an additional 1.5–5(7) mm, with 6–10 spikelets.

Bouteloua aristidoides var. *arizonica* grows in the same kind of habitats as var. *aristidoides*, but only from 500–800 m. It has a more restricted range than *B. aristidoides* var. *aristidoides* (which extends into northern Mexico), being known only from New Mexico, Arizona, and Chihuahua, Mexico. In its extreme form, var. *arizonica* is very different from var. *aristidoides*, but the two varieties do intergrade.

5. **Bouteloua eludens** Griffiths [p. 258]
ELUSIVE GRAMA

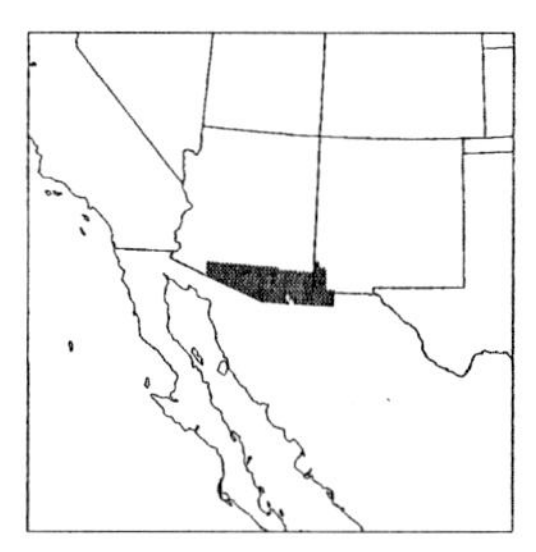

Plants perennial; without rhizomes or stolons. **Culms** 20–60 cm, unbranched. **Leaves** mostly basal; **sheaths** glabrous or sparsely ciliate near the throat, basal sheaths papery, becoming pale; **ligules** to 0.5 mm, of hairs; **blades** to 15 cm long, 1–1.5(3) mm wide, flat, lower leaves variously hispid or scabrous, upper leaves glabrous, with scabrous margins. **Panicles** 6–10 cm, with (8)12–16(20) branches; **branches** 5–11 mm, deciduous, pubescent, with (2)4–6 spikelets, axes extending about 5 mm beyond the terminal spikelets, apices entire; **disarticulation** at the base of the branches. **Spikelets** appressed, all alike, with 1–2 bisexual florets and 1 rudimentary floret. **Glumes** silvery-hispid over and between the veins, at least basally, apices acute, acuminate, or shortly awned; **lower glumes** 5–6 mm; **upper glumes** 6–7 mm, sericeous over the veins and elsewhere, hairs about 0.5 mm, apices mucronate; **lowest florets** bisexual, pistillate, or staminate; **lowest lemmas** 6–7 mm, pubescent between the veins and over the midveins, midveins extending into acuminate or setaceous lobes about the same length as the lateral lobes, not flanked by membranous lobes, lateral veins extending from the lateral lobes for 0.5–2 mm; **lowest paleas** as long as the lemmas, pubescent, acute to acuminate, unawned; **second florets** usually staminate, pistillate, or bisexual (rarely rudimentary); **second lemmas** 8–10 mm; usually well-developed, pubescent, lateral veins extended into 0.5–4 mm awns, midveins extended into a flattened, 1–5 mm awn; **second paleas** as long as the second lemmas, pubescent, usually with 2 short awns; **third florets**, if present, pubescent, variable, resembling the second floret, a 3-awned structure with 2 membranous scales, or a prolongation of the rachilla. **Caryopses** about 5 mm long, about 1.5 mm wide. $2n = 20$.

Bouteloua eludens grows on dry, rocky slopes and rolling desert flats at 1200–1800 m. It is only known from Cochise, Santa Cruz, and eastern Pima counties in Arizona, adjacent portions of New Mexico and Sonora, Mexico. Although its range is small, *B. eludens* is not rare.

Bouteloua eludens resembles *B. chrondrosoides* in having pubescent panicle branches, but *B. eludens* usually has 12–16 branches 5–11 mm long with 2–6 spikelets, whereas *B. chrondrosoides* usually has 3–8 branches 10–15 mm long with 8–12 spikelets per branch. *Bouteloua rigidiseta* is also similar to *B. eludens*, but differs from that species in its glume pubescence and geographic distribution, being only found in Oklahoma, Texas, and northeastern Mexico.

6. **Bouteloua chondrosoides** (Kunth) Benth. *ex* S. Watson [p. 258]
SPRUCETOP GRAMA

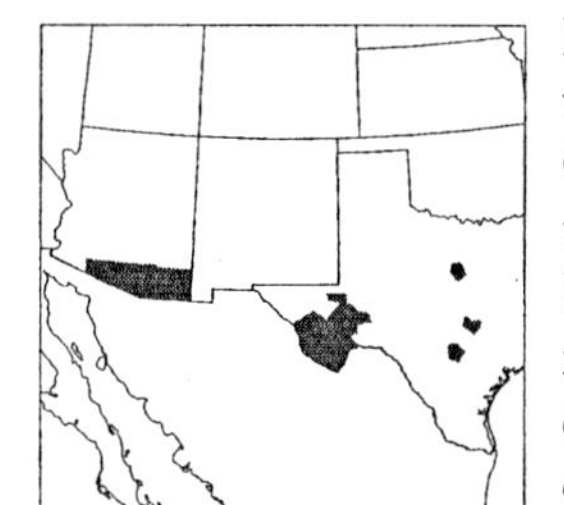

Plants perennial; cespitose, without rhizomes or stolons. **Culms** (10)30–60 cm, erect, unbranched. **Leaves** mostly basal; **sheaths** mostly glabrous, margins often long-ciliate distally; **ligules** 0.3–0.6 mm, of hairs; **blades** 1–10 cm long, 1–2.5(3) mm wide, flat, glaucous, bases with papillose-based hairs on the margins, similar hairs sometimes present on either or both surfaces. **Panicles** 2.5–6 cm, with 3–8(10) branches; **branches** (8)10–15 mm, densely pubescent, with 8–12 spikelets, axes extending to 5 mm beyond the base of the terminal spikelets, apices entire; **disarticulation** at the base of the branches. **Spikelets** appressed, all alike, 7–7.5 mm, with 1 bisexual and 1 rudimentary floret. **Glumes** evidently hairy; **lower glumes** 2.5–4.5 mm; **upper glumes** 4.5–6.5 mm; **lowest lemmas** 4.7–6.2 mm, hairy distally, 3-lobed, lobes unawned or shortly awned; **lowest paleas** 5–7.2 mm, pubescent along the veins and on the margins, bifid, veins excurrent as short awns; **anthers** 2.8–4 mm, yellow; **upper florets** rudimentary, glabrous, 3-awned, awns scabrous, sometimes arising from a short but evident awn column, central awns sometimes with a membranous margin, awns scabrous. **Caryopses** about 2.5 mm long, about 0.9 mm wide. $2n = 20, 22, 40$.

Bouteloua chondrosoides grows on dry, rocky slopes and grassy plateaus at 200–2500 m. Its range extends from southern Arizona and western Texas to Costa Rica. It resembles *B. eludens* in having pubescent panicle branches, but *B. eludens* usually has 12–16 branches 5–11 mm long with 2–6 spikelets whereas *B.*

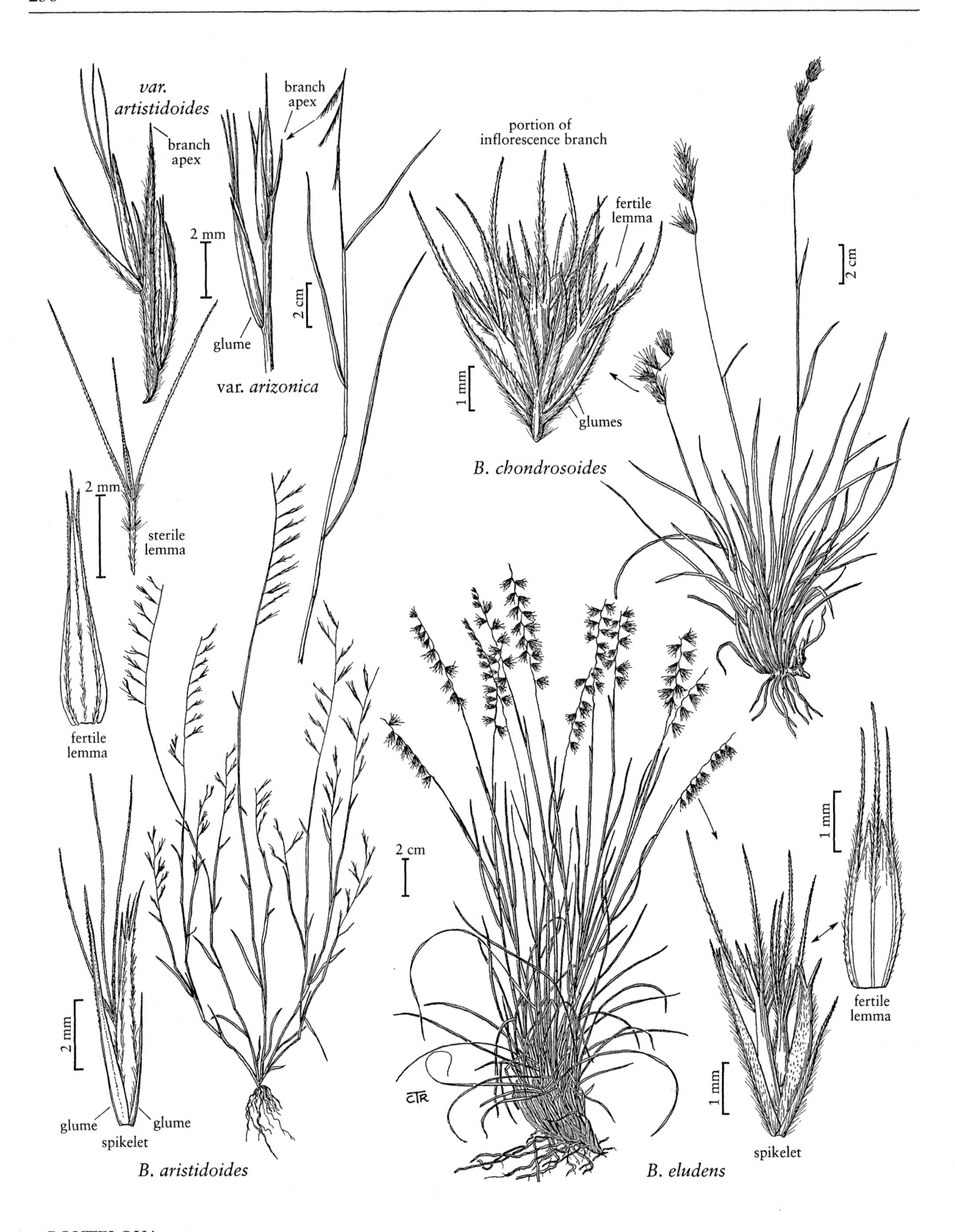
var. artistidoides
branch apex
2 mm
branch apex
glume
2 cm
var. arizonica
portion of inflorescence branch
fertile lemma
1 mm
glumes
B. chondrosoides
2 cm
2 mm
sterile lemma
fertile lemma
2 mm
glume
glume
spikelet
B. aristidoides
2 cm
CTR
1 mm
fertile lemma
1 mm
spikelet
B. eludens

chrondrosoides usually has 3–8 branches 10–15 mm long with 8–12 spikelets per branch.

7. **Bouteloua rigidiseta** (Steud.) Hitchc. [p. 260]
TEXAS GRAMA

Plants perennial; cespitose, without rhizomes or stolons, forming dense, small clumps. **Culms** 10–50 cm, erect, unbranched. **Sheaths** smooth, striate; **ligules** 0.2–0.3 mm, membranous, ciliate; **blades** 4–12(17) cm long, 1–2 mm wide, abaxial surfaces sparsely short pubescent, bases with papillose-based hairs on the margins, similar hairs also present on both surfaces. **Panicles** 3–6 cm, with 6–8 branches; **branches** 8–16 mm, hairy, becoming more sparsely so distally, with 2–6 spikelets, axes terminating beyond the base of the terminal spikelets, apices deeply bi- or trifurcate; **disarticulation** at the base of the branches. **Spikelets** appressed, all alike, with 1 bisexual and 1–2 rudimentary florets. **Glumes** lanceolate, veins pubescent, apices acuminate; **lower glumes** 3–4 mm; **upper glumes** about 6 mm, pubescent over the veins, hairs about 0.7 mm, apices bilobed, awned from the sinuses; **lowest lemmas** 2.5–4 mm, glabrous or sparsely pubescent on the veins, 3-awned, awns wide basally, forming 3 triangular lobes, central awns flanked by 2 membranous 0.5–1.5 mm lobes; **lowest paleas** 4–5 mm, bilobed, veins often excurrent; **second lemmas** glabrous, 3-awned, awns 5–10 mm; **second paleas** 2-lobed, unawned; **third lemmas** similar to the second lemmas but smaller and without paleas. **Caryopses** (2) 3.5–3.7 mm. $2n = 40$.

Bouteloua rigidiseta grows in grassy pastures and openings in woods, usually in clay or sandy clay soils, from near sea level to approximately 700 m. It is both widespread and abundant within its range, which extends from the southern United States to northern Mexico, but has little value as a forage grass. It is one of the earliest flowering warm season grasses. Although similar to *B. eludens*, *B. rigidiseta* differs in its geographic distribution and glume pubescence, so the two taxa are unlikely to be confused in the field.

8. **Bouteloua repens** (Kunth) Scribn. & Merr. [p. 260]
SLENDER GRAMA

Plants perennial; cespitose, usually not dense, hard, or knotty, without rhizomes or stolons. **Culms** 15–65 cm, erect, geniculate, or decumbent, sometimes rooting at the lower nodes, usually branching from the aerial nodes. **Sheaths** glabrous or pubescent; **ligules** 0.2–0.3 mm, membranous, ciliate; **blades** 5–20 cm long, 1–5 mm wide, bases with papillose-based hairs on the margins, both surfaces glabrous or pubescent. **Panicles** 4–14 cm, with (3)7–12 branches; **branches** 10–20 mm, with 2–8 spikelets, extending 4–6 mm beyond the base of the terminal spikelets, apices entire; **disarticulation** at the base of the branches. **Spikelets** appressed, all alike, with 1 bisexual and 1 staminate (rarely rudimentary) floret. **Glumes** glabrous, veins scabrous or strigose; **lower glumes** 4–7 mm; **upper glumes** 4–9 mm, mostly glabrous, sometimes scabrous or strigose over the veins, apices acute, unawned or awn-tipped, awns about 1 mm; **lowest lemmas** 4.5–8 mm, usually glabrous, rarely pubescent basally, 3-awned, awns wide basally, central awns slightly longer than the lateral awns, often flanked by 2 membranous 0.5–1.5 mm lobes; **lowest paleas** 6–8 mm, bilobed, often shortly 2-awned; **anthers** 3–5.5 mm, usually orange or yellow, occasionally red or purple; **second lemmas** 5.5–7 mm, glabrous, 3-awned, central awns 4–10 mm, often flanked by membranous lobes, lateral awns 2–10 mm; **second paleas** 4–7 mm; **anthers** smaller than those of the lowest florets; **rachillas** prolonged beyond the second florets as a short bristle. **Caryopses** 3–4 mm. $2n = 20, 40, 60$.

Bouteloua repens grows in open, usually hilly terrain on many soil types, from sandy ocean shores to montane slopes, reaching elevations of 2500 m. Its native range extends from the southwestern United States through the Caribbean islands, Mexico, and Central America to Colombia and Venezuela.

9. **Bouteloua radicosa** (E. Fourn.) Griffiths [p. 260]
PURPLE GRAMA

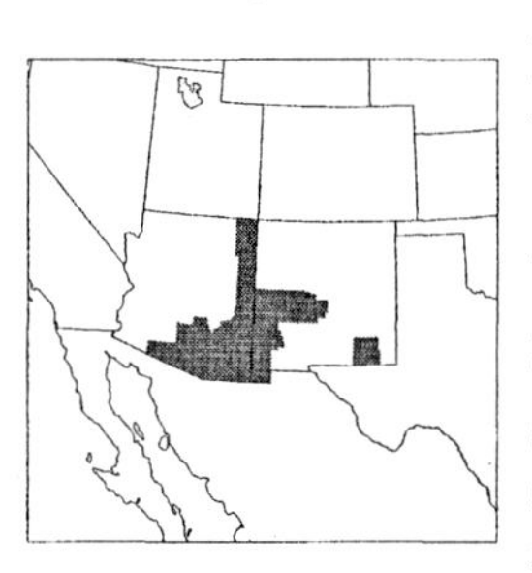

Plants perennial; cespitose, with a dense, hard, knotty base, rhizomatous, rhizomes 2–3 mm thick, with pale cataphylls; **internodes** 4–5 mm. **Culms** (40) 60–80 cm, erect, straight, unbranched. **Sheaths** strongly striate; **ligules** 0.5–1 mm, of hairs; **blades** mostly basal, short and firm, 2–3 mm wide, bases with

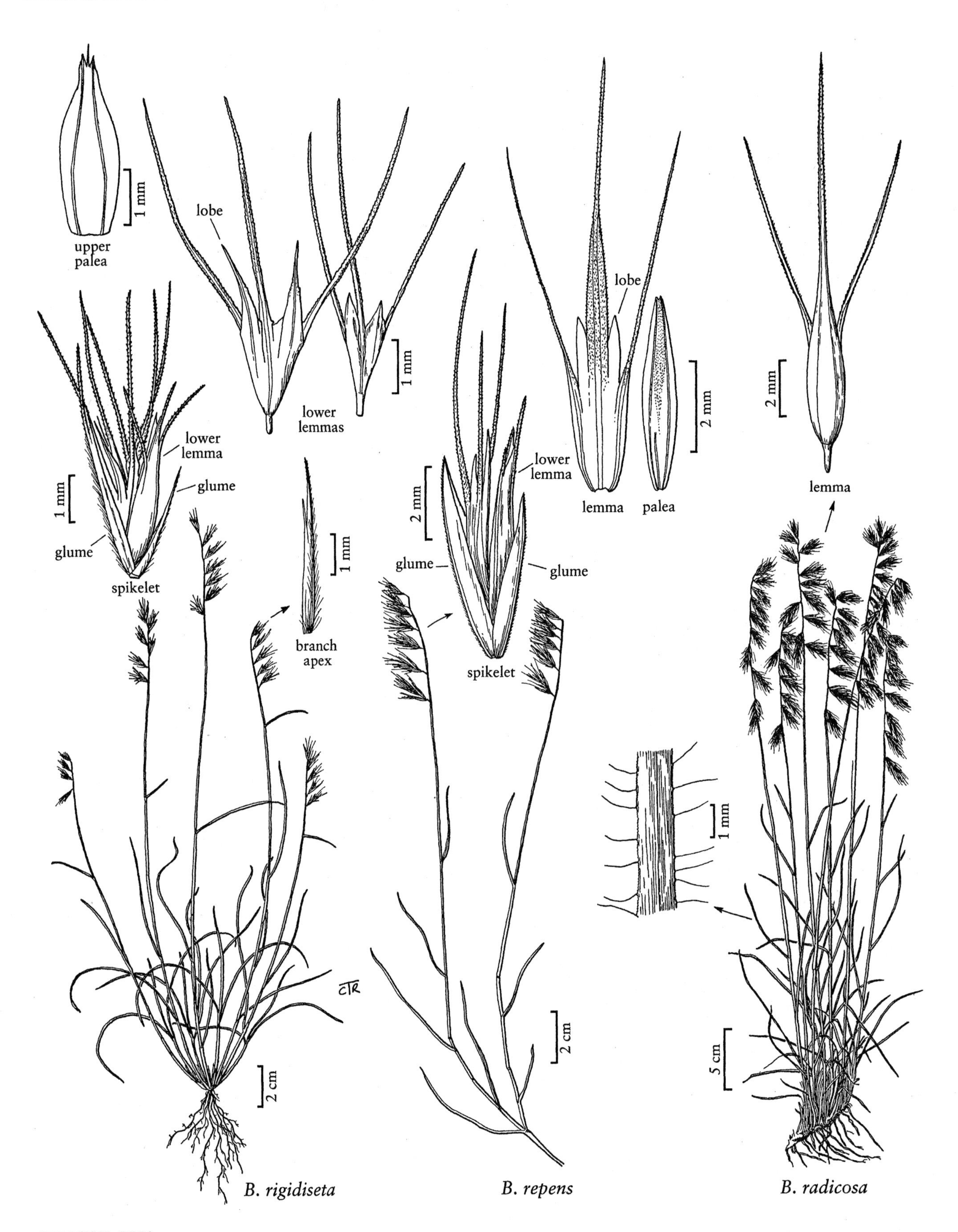
upper palea
1 mm
lobe
lower lemmas
1 mm
lower lemma
glume
1 mm
glume
spikelet
1 mm
branch apex
2 cm
lower lemma
2 mm
glume
glume
spikelet
2 cm
lobe
2 mm
lemma
palea
2 mm
lemma
1 mm
5 cm
B. rigidiseta
B. repens
B. radicosa

papillose-based hairs on the margins. **Panicles** 10–15 cm, usually with 7–12 branches; **branches** (15)20–30 mm, deciduous, with 8–11 spikelets, apices entire; **disarticulation** at the base of the branches. **Spikelets** appressed, all alike, with 2 florets, lowest floret bisexual, upper florets pistillate, bisexual, or staminate. **Glumes** acuminate, glabrous; **lower glumes** about 4 mm; **upper glumes** 5–6 mm; **lowest lemmas** 7–8 mm, smooth, often shortly trilobed, 3-awned, awns extending from the lobes, central awns 2–3 mm, not flanked by membranous lobes, lateral awns about 1 mm; **lower paleas** 6–7 mm, unawned, sometimes mucronate; **upper lemmas** 9–10 mm, central awns 6–8 mm, lateral awns 5–6 mm; **upper paleas** similar to the lower paleas. **Caryopses** 4–5 mm long, 0.75–1 mm wide. 2*n* = 60.

Bouteloua radicosa grows on dry, rocky slopes at 1000–3000 m, from Arizona and southern New Mexico to southern Mexico. It has also become established in Maine, growing in disturbed habtiats, but is not common there.

Bouteloua radicosa frequently grows with *B. repens* at lower elevations, but extends higher than that species. Like *B. repens*, *B. radicosa* exhibits great variation in spikelet and inflorescence characters. Gould (1979) suggested that some of the variation in *B. radicosa* was due to hybridization with *B. repens* in the *Flora* area and **B. williamsii** Swallen in southern Mexico.

Bouteloua subg. Chondrosum (Desv.) A. Gray

Panicle branches 1–20, persistent, with 6–130 spikelets, terminating in a point or spikelet; **spikelets** usually dense and pectinate on the branches; **disarticulation** above the glumes.

10. **Bouteloua gracilis** (Kunth) Lag. *ex* Griffiths [p. 263]
BLUE GRAMA, EYELASH GRASS

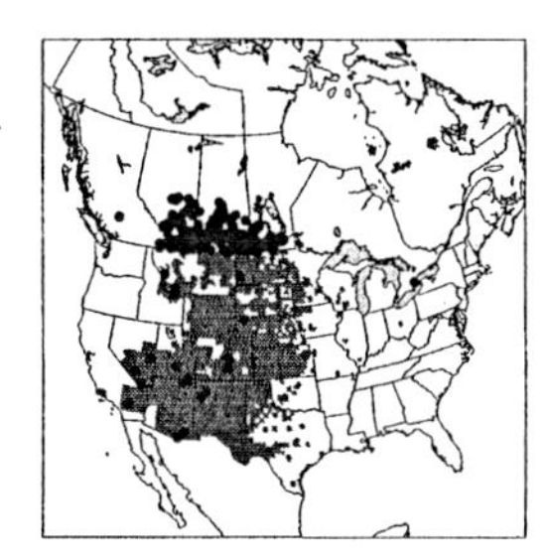

Plants perennial; usually densely cespitose, often with short, stout rhizomes. **Culms** 24–70 cm, not woody basally, erect, geniculate, or decumbent and rooting at the lower nodes, not branched from the aerial nodes; **nodes** usually 2–3, glabrous or puberulent; **lower internodes** glabrous. **Leaves** mainly basal; **sheaths** glabrous or sparsely hirsute; **ligules** 0.1–0.4 mm, of hairs, often with marginal tufts of long hairs; **blades** 2–12(19) cm long, 0.5–2.5 mm wide, flat to involute at maturity, hairs usually present basally. **Panicles** with 1–3(6) branches, these racemose on 2–8.5(12.5) cm rachises or digitate; **branches** 13–50(75) mm, persistent, arcuate, scabrous, without papillose-based hairs, with 40–130 spikelets, terminating in a spikelet; **disarticulation** above the glumes. **Spikelets** pectinate, with 1 bisexual and 1 rudimentary floret. **Glumes** mostly glabrous or scabrous, midveins sometimes with papillose-based hairs; **lower glumes** 1.5–3.5 mm; **upper glumes** 3.5–6 mm; **lowest lemmas** 3.5–6 mm, pubescent at least basally, 5-lobed, central and lateral lobes veined and awned, awns 1–3 mm, central awns flanked by 2 membranous lobes; **lower paleas** about 5 mm, shallowly bilobed, veins excurrent for less than 1 mm; **rachilla segments subtending second florets** with a distal tuft of hairs; **anthers** 1.7–2.9 mm, yellow or purple; **upper florets** sterile, 0.9–3 mm, lobed almost to the base, lobes rounded, 3-awned, awns equal, 1–3 mm. **Caryopses** 2.5–3 mm long, about 0.5 mm wide. 2*n* = 20, 28, 35, 40, 42, 60, 61, 77, 84.

Bouteloua gracilis grows in pure stands in mixed prairie associations and disturbed habitats, usually on rocky or clay soils and mainly at elevations of 300–3000 m. Its native range extends from Canada to central Mexico; records from the eastern portion of the *Flora* represent introductions.

Bouteloua gracilis is an important native forage species and also an attractive ornamental.

11. **Bouteloua hirsuta** Lag. [p. 263]

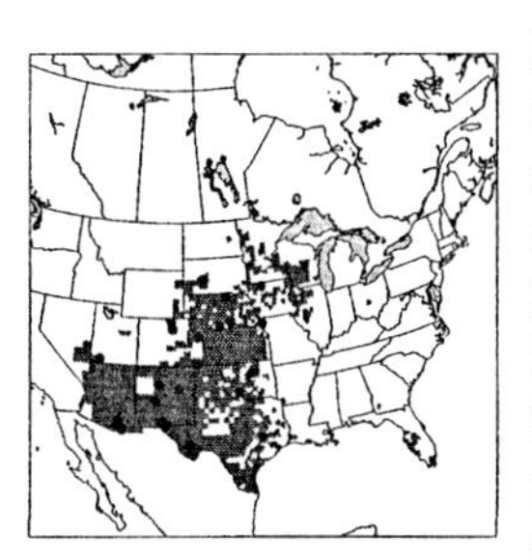

Plants perennial; densely or loosely cespitose, occasionally stoloniferous. **Culms** 15–75 cm, erect or decumbent, sometimes branched basally, sometimes branched aerially; **nodes** 3–6; **internodes** glabrous or sparsely to densely pubescent with papillose-based hairs. **Leaves** basal or mainly cauline; **sheaths** mostly glabrous, finely scabrous, or pubescent, pilose near the ligules; **ligules** 0.2–0.5 mm, of hairs; **blades** 1–30 cm long, 1–2.5 mm wide, flat to involute, papillose-based hairs often present on both surfaces, usually present on the bases of the margins. **Panicles** usually with 0.7–18 cm rachises bearing 1–6 branches, the branches sometimes digitate; **branches** 10–40 mm, persistent, straight, with 20–50 spikelets, axes extending 5–10 mm beyond base of the terminal spikelets; **disarticulation** above the glumes. **Spikelets** pectinate, green to dark purple, with 1 bisexual floret and 1–2 rudimentary florets. **Glumes** acuminate or awn-tipped; **lower glumes** 1.4–3.5 mm; **upper glumes** 3–6 mm, midveins with papillose-based hairs; **lowest lemmas** 2–4.5 mm, pubescent, 1–3-awned, central (or only) awns 0.2–2.5 mm, not flanked by membranous lobes, lateral lobes acuminate, unawned or with awns no longer than the central awn; **lower paleas** ovate, unawned; **anthers** 2–3.4 mm, cream or yellow; **rachilla segments subtending second florets** glabrous or pubescent, sometimes with a distal tuft of hairs; **second lemmas** 0.5–2 mm, bilobed, 3-awned, awns 2–4(6) mm; **third lemmas**, if present, minute, membranous scales, glabrous. **Caryopses** 1.5–2.6 mm. 2*n* = 20, 40, 50, 60; numerous dysploid numbers also reported.

Bouteloua hirsuta is a widespread species, with two subspecies that frequently hybridize in areas of sympatry (Wipff and Jones 1996).

1. Rachilla segments subtending second florets with a distal tuft of hairs; culms erect from the base, usually unbranched . . . . subsp. *pectinata*
1. Rachilla segments subtending second florets without a distal tuft of hairs; culms usually decumbent and branched basally . . . . subsp. *hirsuta*

Bouteloua hirsuta Lag. subsp. **hirsuta**
HAIRY GRAMA

Plants loosely or densely cespitose, sometimes stoloniferous. **Culms** 15–60 cm, usually decumbent and branched basally, sometimes erect, branched or unbranched from the aerial nodes; **nodes** usually 4–6; **internodes** glabrous or sparsely to densely pubescent with papillose-based hairs. **Leaves** basally clustered, sometimes not strongly so; **sheaths** glabrous or pubescent, hairs not papillose-based, sometimes scabrous. **Panicles** with 1–4 branches on 0.7–7.5(9.2) cm rachises or digitate; **branches** 1–4. **Anthers** 2–2.5 mm; **rachilla segments subtending second florets** without a distal tuft of hairs. **Caryopses** 1.4–2 mm. 2*n* = 20, 40, 50, 60; numerous dysploid numbers also reported.

Bouteloua hirsuta subsp. *hirsuta* grows from the open plains to slightly shaded openings in woods and brush on well-drained, often rocky, soils at 50–300 m. It is morphologically, ecologically, and cytologically more variable than subsp. *pectinata*. Its range extends from North Dakota and Minnesota to central Mexico. In the northern portion of its range, it is not densely tufted and the culms are decumbent and branched; in the southwestern United States and northern Mexico, it grows in isolated, dense clumps, with erect, stout, unbranched culms and mostly basal leaves.

Bouteloua hirsuta Lag. subsp. **pectinata** (Feath.) Wipff & S.D. Jones
TALL GRAMA

Plants without rhizomes or stolons. **Culms** 35–75 cm, erect, densely tufted, usually unbranched; **nodes** 3–4; **internodes** glabrous. **Leaves** mostly basal; **sheaths** mostly pubescent. **Panicles** racemose, with (2)3–6 branches on (3)6–18 cm rachises. **Anthers** 2–3.2 mm; **rachilla segments subtending second florets** with a distal tuft of hairs. **Caryopses** 1.5–2.6 mm. 2*n* = 20.

Bouteloua hirsuta subsp. *pectinata* grows in well-drained, relatively undisturbed, calcareous soils, usually on thin-soiled limestone outcrops, at 60–500 m. Its range extends from southern Oklahoma to central Texas. Although restricted in its geographic distribution, where subsp. *pectinata* is sympatric with subsp. *hirsuta*, swarms of morphologically intermediate plants are found (Wipff and Jones 1996).

12. **Bouteloua eriopoda** (Torr.) Torr. [p. 263]
BLACK GRAMA

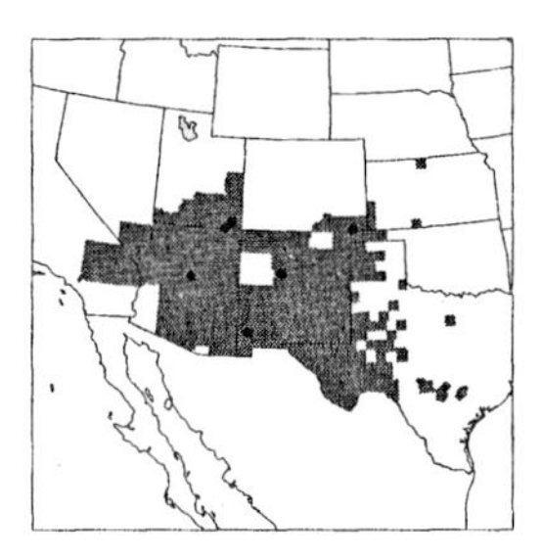

Plants perennial; often shortly rhizomatous, stoloniferous, stolons long, densely woolly-pubescent. **Culms** 20–60(75) cm, wiry, decumbent, rooting at the lower nodes; **lower internodes** densely woolly-pubescent. **Sheaths** mostly glabrous or sparsely pilose, usually pilose near the ligules; **ligules** 0.1–0.4 mm, of hairs; **blades** 2.5–6 cm long, 0.5–2 mm wide, scabrous adaxially, margins with papillose-based hairs basally. **Panicles** (1)2–16 cm, with (1)2–8 branches; **branches** 14–50 mm, persistent, densely woolly-pubescent basally, with 8–18 spikelets, axes terminating in entire, sometimes scarious apices; **disarticulation** above the glumes. **Spikelets** pectinate, with 1 bisexual floret and 1 rudimentary floret. **Glumes** unequal, smooth or scabrous; **lower glumes** 2–4.5 mm; **upper glumes** 4.5–8(9) mm, glabrous, scabrous, or with hairs, hairs to 0.5 mm, not papillose-based; **lower lemmas** 4–7 mm, pubescent basally, glabrous or sparsely puberulent distally, acuminate, central awns 0.5–4 mm, lateral awns absent or shorter than 1 mm; **lower paleas** acuminate, unawned; **anthers** 1.5–3 mm, yellow to orange; **rachilla segment to second florets** about 2 mm, with a distal tuft of hairs; **upper florets** rudimentary, an awn column terminating in 3 awns of 4–9 mm. **Caryopses** 2.5–3 mm. 2*n* = 20, 21, 28.

Bouteloua eriopoda grows on dry plains, foothills, and open forested slopes, often in shrubby habitats, and also in waste ground. It is usually found between 1000–1800 m, but extends to 2500 m. Once a dominant in much of its range, under heavy grazing *B. eriopoda* persists only where protected by shrubs or cacti because it is highly palatable. Its range extends from the southwestern United States to northern Mexico.

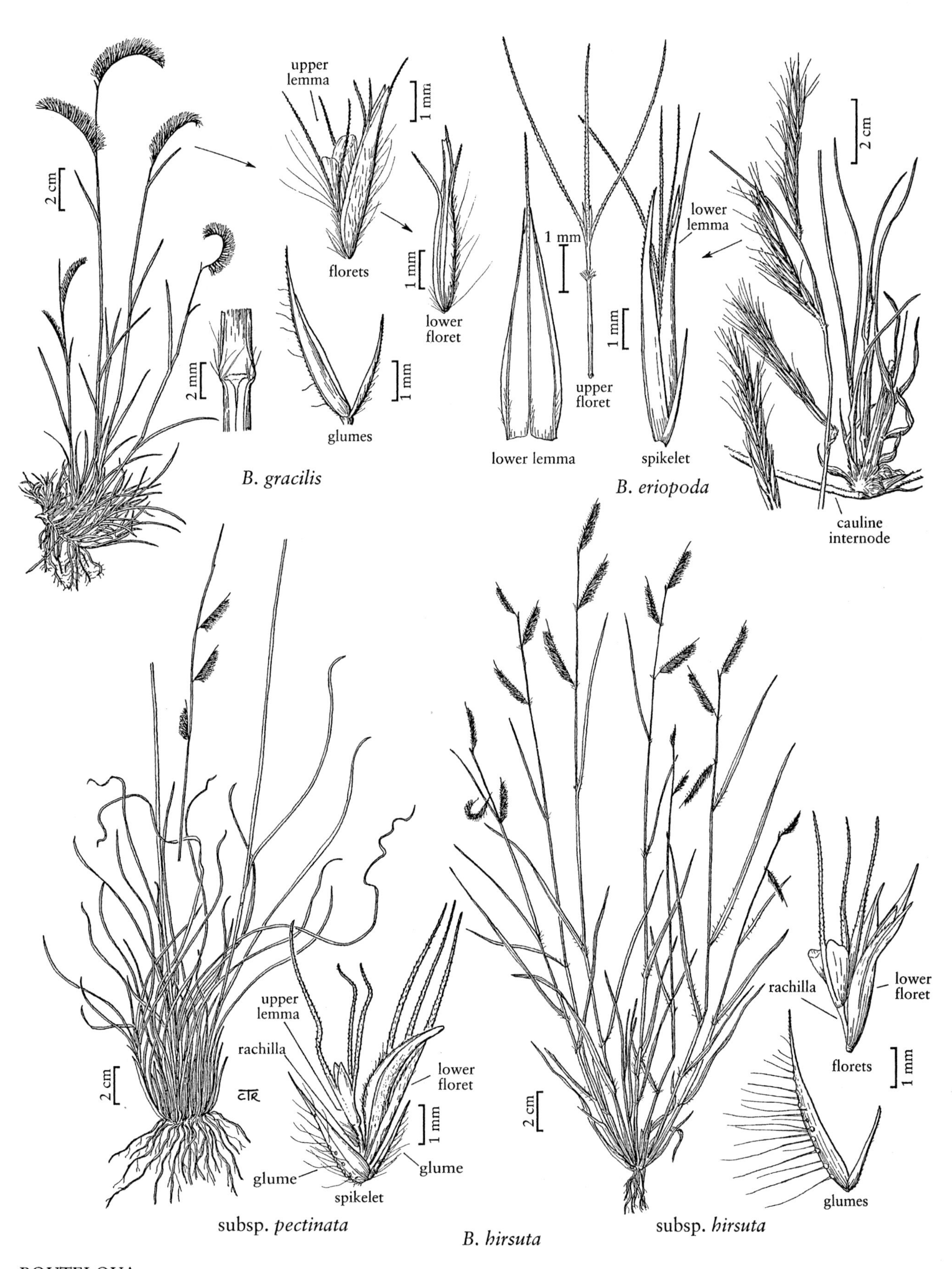
upper
lemma
1 mm
florets
1 mm
lower
floret
2 cm
2 mm
glumes
1 mm
B. gracilis
1 mm
lower lemma
upper
floret
1 mm
lower
lemma
spikelet
B. eriopoda
2 cm
cauline
internode
upper
lemma
rachilla
lower
floret
1 mm
glume
glume
spikelet
2 cm
subsp. pectinata
B. hirsuta
2 cm
rachilla
lower
floret
florets
1 mm
glumes
subsp. hirsuta

13. **Bouteloua trifida** Thurb. *ex* S. Watson [p. 266]
RED GRAMA

Plants perennial; cespitose, older plants occasionally shortly rhizomatous. **Culms** 5–40 cm, slender, wiry, erect or slightly geniculate at the lower nodes; **lower internodes** glabrous, shorter than those above. **Leaves** mostly basal; **sheaths** glabrous, sometimes scabridulous, becoming flattened, persistent; **ligules** 0.2–0.5 mm, of hairs; **blades** 0.7–8 cm long, 0.5–1.5(2) mm wide, scabridulous, margins often with papillose-based hairs basally. **Panicles** 3–9 cm, with 2–7 branches; **branches** 7–25 mm, persistent, spreading, ascending, or appressed, straight to slightly arcuate, with 8–24(32) spikelets, axes terminating in a spikelet; **disarticulation** above the glumes. **Spikelets** appressed to pectinate, reddish-purple; with 1 bisexual floret and 1 rudimentary floret. **Glumes** bilobed; **lower glumes** 1.7–3.4 mm, slightly shorter than the upper glumes, veins excurrent to 0.6 mm; **upper glumes** 1.9–4 mm, glabrous or pubescent, hairs not papillose-based, veins excurrent to 1 mm; **lower lemmas** 1.2–2.2 mm, glabrous, sparsely appressed pubescent along the veins or densely appressed pubescent for much of their length and on the margins, trilobed, lobes veined, tapering into 3 awns, awns 2.2–6.6 mm, central awns not flanked by membranous lobes; **anthers** 0.2–0.4 mm, yellow; **rachilla segments** glabrous; **upper florets** glabrous, of 3 equal awns, awns 2–7 mm. **Caryopses** 0.8–1.5 mm long, 0.3–0.6 mm wide. $2n$ = 20.

Bouteloua trifida grows on dry open plains, shrubby hills, and rocky slopes, at 2200–2500 m. Its range extends from the southwestern United States to central Mexico. It is a drought-resistant species that is sometimes mistaken for *Aristida* because of its delicate, cespitose growth habit and purplish, 3-awned spikelets. Juvenile plants may also be confused with *B. barbata* but that species is annual, with the central awn flanked by two membranous lobes and the lowest paleas 4-lobed and 2-awned.

1. Lower lemmas densely appressed pubescent; awns 2.2–4.5 mm long; anthers 0.2–0.3 mm long . . . . var. *burkii*
1. Lower lemmas glabrous or sparsely appressed pubescent along both sides of the veins; awns (3.2)4–6.6 mm long; anthers 0.3–0.4 mm long . . . . var. *trifida*

Bouteloua trifida var. **burkii** (Scribn. *ex* S. Watson) Vasey *ex* L.H. Dewey

Panicle branches ascending to divergent, rarely appressed. **Lower glumes** 1.7–3.1 mm, veins excurrent for 0.05–0.2 mm, usually not exceeding the apical lobes; **upper glumes** 1.9–3.2 mm, veins excurrent for 0.05–0.2 mm, usually not exceeding the apical teeth; **lower lemmas** conspicuously appressed pubescent across the lower ⅔–⅘ and along most or all of the margins, awns 2.2–4.5 mm; **anthers** 0.2–0.3 mm. **Caryopses** 0.8–1.2 mm long, 0.5–0.6 mm wide, flat or slightly concave adaxially. $2n$ = 20.

Bouteloua trifida var. *burkii* grows in southern New Mexico, southern Texas, and adjacent Mexico.

Bouteloua trifida Thurb. *ex* S. Watson var. **trifida**

Panicle branches appressed to ascending, occasionally divergent. **Lower glumes** 2.2–3.4 mm, midveins excurrent for 0.1–0.6 mm; **upper glumes** 2.7–4 mm, midveins excurrent for 0.2–1 mm; **lower lemmas** glabrous or sparsely appressed pubescent along both sides of the veins, awns (3.2)4–6.6 mm; **anthers** 0.3–0.4 mm. **Caryopses** 1.3–1.5 mm long, 0.4–0.5 mm wide, grooved adaxially. $2n$ = unknown.

Bouteloua trifida var. *trifida* grows in dry plains and rocky slopes, mostly at 300–1500 m, from southern California, Nevada, and Utah to Texas and Mexico.

14. **Bouteloua kayi** Warnock [p. 266]
KAY'S GRAMA

Plants perennial; cespitose, without rhizomes or stolons. **Culms** 10–50 cm, erect; **nodes** glabrous; **internodes** glabrous, scabridulous between the veins. **Leaves** mostly basal; **sheaths** scabridulous or glabrous, sparsely pubescent basally; **ligules** 0.3–0.5 mm, membranous, ciliate; **blades** to 20 cm long, 0.5–1.5 mm wide, involute, scabridulous adaxially. **Panicles** 8–11 cm, with 7–20 branches; **branches** 15–30 mm, persistent, with (6)14–20 pedicellate spikelets, axes terminating in a spikelet; **pedicels** 0.6–0.8 mm; **disarticulation** above the glumes. **Spikelets** 6–8 mm, pectinate, with 1 bisexual and 1 rudimentary floret. **Glumes** subequal, 2.5–4 mm, glabrous, acute or bidentate, midvein sometimes excurrent as a mucro or short awn; **lowest lemmas** 5–7 mm, glabrous, 3-awned, awns 3–4 mm, central awns flanked by 2 membranous, acuminate, 0.4–0.6 mm lobes; **lowest paleas** sometimes reduced to 2 awns, awns 1–2 mm, **anthers** 1.2–1.3 mm, yellow; **second florets** glabrous, reduced to 1–3 awns, awns 3–4 mm. $2n$ = unknown.

Bouteloua kayi is only known from the mountainous limestone terrain along the Rio Grande River in southwestern Brewster County, Texas, at 2200–2500 m. Superficially, it resembles *B. trifida*.

15. **Bouteloua barbata** Lag. [p. 266]

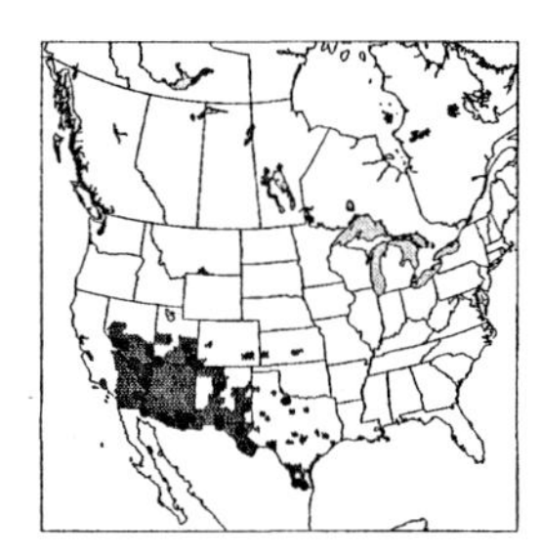

Plants annual or short-lived perennials; tufted, sometimes with stolons. **Culms** 1–75 cm, prostrate, decumbent, or erect, sometimes rooting at the lower nodes; **lower internodes** glabrous. **Leaves** basal or cauline; **sheaths** usually glabrous, except for tufts of long hairs on either side of the collars; **ligules** 0.1–1 mm, membranous, ciliate; **blades** 0.5–10 cm long, 0.7–4 mm wide, adaxial surfaces usually sparsely pubescent with a few papillose-based hairs basally. **Panicles** 0.7–25 cm, with (2)4–9(11) branches; **branches** 10–30 mm, persistent, straight to arcuate, glabrous, scabridulous, or with papillose-based hairs, with 20–55 spikelets, axes terminating in a well-developed spikelet; **disarticulation** above the glumes. **Spikelets** 2.5–5 mm, pectinate, with 1 bisexual and 2 rudimentary florets. **Glumes** unequal, glabrous, sometimes scabridulous, apices sometimes shortly bilobed, acuminate or mucronate; **lower glumes** 0.7–1.5 mm; **upper glumes** 1.5–2.5 mm, glabrous, scabrous, or strigose, hairs not papillose-based; **lowest lemmas** 1.7–4 mm, densely pilose, at least on the margins, 3-awned, awns 0.5–3 mm, central awns flanked by 2 membranous lobes; **lowest paleas** 1.5–4 mm, pubescent on the margins, 4-lobed, 2-awned, awns 1–2 mm; **anthers** 0.4–0.7 mm; **rachilla segments subtending second florets** terminating in a dense tuft of hairs; **second florets** rudimentary, 1.5–4 mm, 2-lobed, lobes rounded, 3-awned, awns 0.5–4 mm; **rachilla segments subtending third florets** with glabrous or puberulent apices; **third florets** rudimentary, flabellate, unawned. **Caryopses** to 1 mm. $2n = 20$.

The range of *Bouteloua barbata* extends from the southwestern United States to southern Mexico. There are three varieties recognized. The two that grow in the *Flora* region are often sympatric, but are usually easily distinguished in the field in this region by their growth habit. According to Gould (1979), in the southern portion of their range the differences between the two varieties are less evident, particularly on herbarium specimens. The third variety, **B. barbata** var. **sonorae** (Griffiths) Gould, is usually stoloniferous; it is known only from the states of Sonora and Sinola, Mexico.

Bouteloua barbata is often confused with juvenile plants of the perennial *B. trifida*, but in *B. barbata* the central awn is flanked by two membranous lobes and the lowest paleas are 4-lobed and 2-awned.

1. Plants annual; culms usually decumbent and geniculate, occasionally rooting at the lower nodes . . . . var. *barbata*
1. Plants short-lived perennials; culms erect from the base . . . . var. *rothrockii*

Bouteloua barbata Lag. var. **barbata**
SIXWEEKS GRAMA

Plants annual; not stoloniferous. **Culms** 1–35 cm, usually decumbent and geniculate, occasionally rooting at the lower nodes. **Ligules** 0.4–1 mm; **blades** 0.5–5(9) cm long, 0.7–3 mm wide. **Panicles** 0.7–9 cm, with (2)4–11 branches; **branches** 10–27 mm, scabrous, with 20–40 spikelets. $2n = 20$.

Bouteloua barbata var. *barbata* grows in loose sands, rocky slopes, and washes, often on disturbed soils, usually at elevations below 2000 m. Its range extends from the southwestern United States to northwestern Mexico.

Bouteloua barbata var. **rothrockii** (Vasey) Gould
ROTHROCK'S GRAMA

Plants short-lived perennials; not stoloniferous. **Culms** 25–60 (75) cm, stiffly erect or slightly geniculate-spreading basally; **ligules** 0.1–0.5 mm; **blades** 6–10 cm long, 1–4 mm wide. **Panicles** (3.5)5–25 cm, with 3–8 branches; **branches** 15–30 mm, scabrous, with 35–50 (55) spikelets. $2n = 40$.

Bouteloua barbata var. *rothrockii* grows on dry slopes and sandy flats, mostly at 750–1700 m. It grows throughout the southwestern United States and Mexico, sometimes covering large areas. It used to be the most important forage grass in southern Arizona and neighboring regions.

Bouteloua barbata var. *rothrockii* resembles *B. parryi* var. *parryi*, but can be easily distinguished from that taxon by the lack of papillose-based hairs on the keels of its upper glumes.

16. **Bouteloua simplex** Lag. [p. 268]
MAT GRAMA

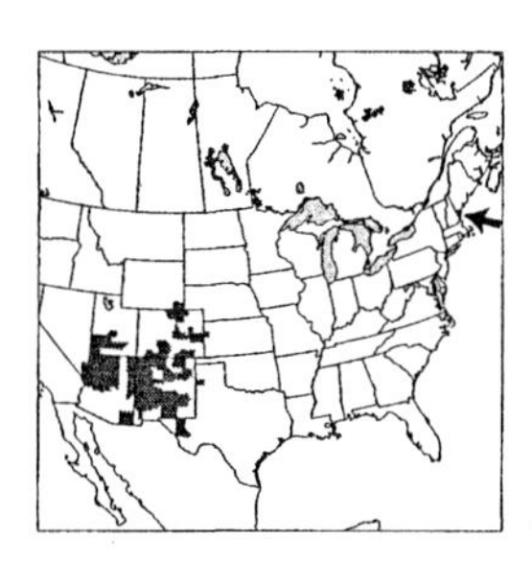

Plants annual. **Culms** 3–35 cm, usually decumbent, occasionally erect, rarely branching; **internodes** glabrous. **Sheaths** smooth, deeply striate; **ligules** 0.1–0.2 mm, of short hairs, sometimes with a few papillose-based hairs on either side; **blades** 2–8 cm long, 0.5–1.5 mm wide, flat to involute, adaxial surfaces mostly glabrous, often pilose basally. **Panicles** usually with only 1 branch (terminating the culm), or with 2–4 branches and subdigitate; **branches** 10–25(40) mm, persistent, straight, arcuate, or circular, with 30–80 spikelets, axes terminating in a reduced spikelet; **disarticulation** above the glumes. **Spikelets** pectinate, with 1 bisexual floret and 1–2 rudimentary florets. **Glumes** glabrous, sometimes scabrous distally, acute or acuminate; **lower glumes** 1.5–2.5 mm; **upper glumes** 3.5–5 mm; **lowest lemmas** 2.5–3.5 mm, pilose over the veins, 3-awned,

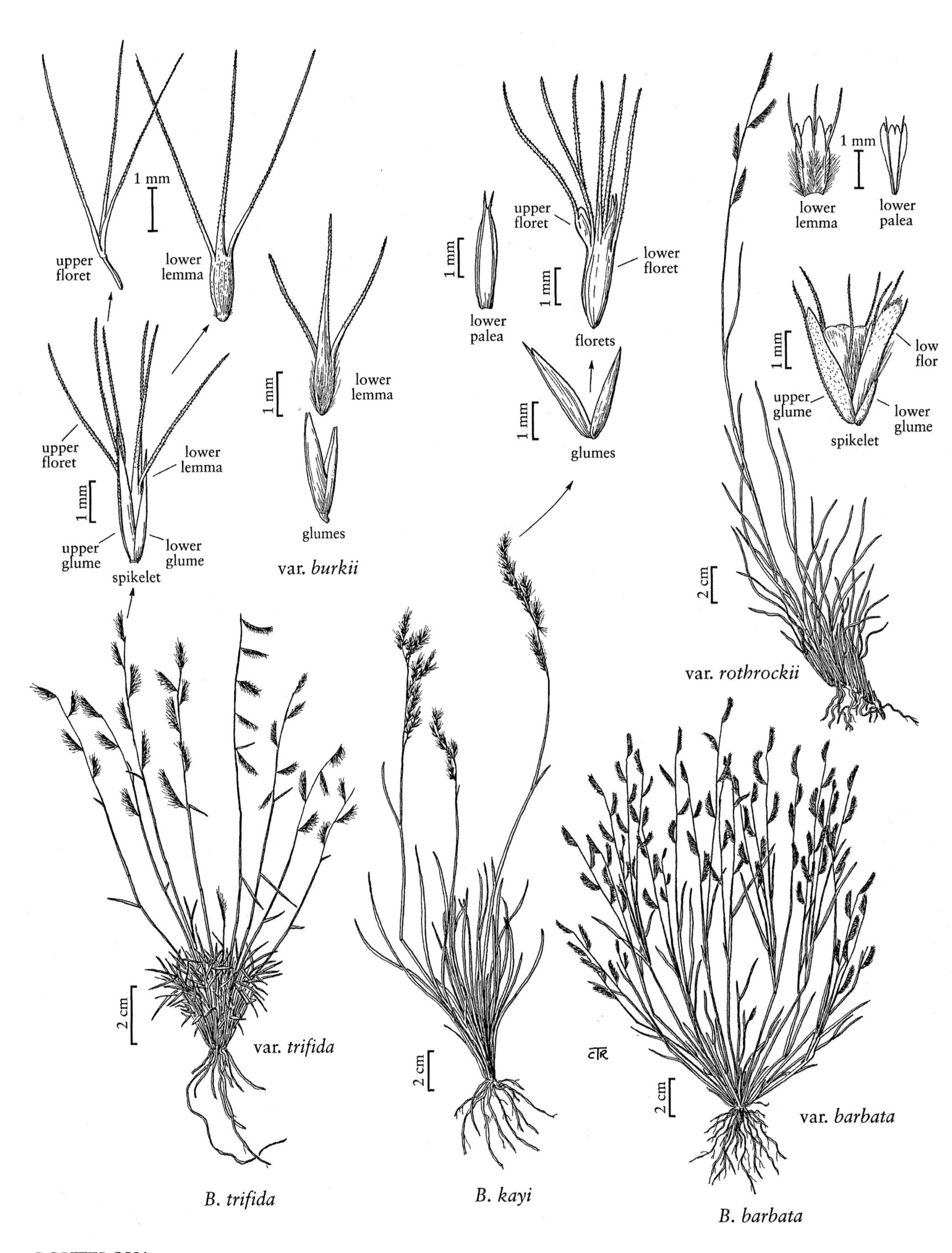
1 mm
upper
floret
lower
lemma
upper
floret
lower
lemma
1 mm
upper
glume
lower
glume
spikelet
1 mm
lower
lemma
1 mm
glumes
var. burkii
upper
floret
lower
floret
1 mm
1 mm
lower
palea
florets
1 mm
glumes
1 mm
lower
lemma
lower
palea
1 mm
low
flor
upper
glume
lower
glume
spikelet
2 cm
var. rothrockii
2 cm
var. trifida
2 cm
2 cm
var. barbata
B. trifida
B. kayi
B. barbata

BOUTELOUA

awns stout and flattened, central awns 1–2 mm, flanked by 2 membranous lobes, lateral awns shorter than the central awns; **lowest paleas** obovate, unawned; **rachilla segments subtending second florets** with densely pubescent apices; **second florets** reduced to an awn column with 3 awns of 5–6 mm; **third florets**, if present, flabellate scales. $2n = 20$.

Bouteloua simplex grows on rocky, open slopes in grassy and open shrub vegetation at 1200–2500 m. Its native range extends from the southwestern United States through Mexico and Central America to western South America. It is adventive in Maine, where it has been grows in disturbed places, but it is not common there.

17. **Bouteloua parryi** (E. Fourn.) Griffiths [p. 268]
PARRY'S GRAMA

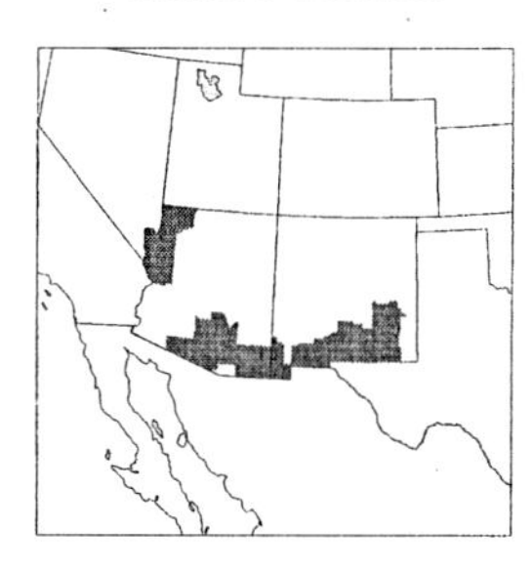

Plants annual or short-lived perennials; tufted, sometimes stoloniferous. **Culms** 20–60 cm, erect or somewhat geniculate at the base. **Leaves** mostly basal; **sheaths** pubescent, usually with tufts of long hairs on either side of the collar; **ligules** 0.1–0.5 mm, of hairs; **blades** 1–3 cm long, 1–2.5 mm wide, margins and usually both surfaces with papillose-based hairs. **Panicles** 2.5–10 cm, with 4–8 branches; **branches** 20–35 mm, persistent, with papillose-based hairs, with 40–65 spikelets, branches terminating in a spikelet; **disarticulation** above the glumes. **Spikelets** pectinate, with 1 bisexual floret and 2 rudimentary florets. **Glumes** unequal; **lower glumes** about 2 mm, glabrous or sparsely pubescent at the base, mucronate; **upper glumes** 3–4 mm, keels with papillose-based hairs, apices bilobed, awned from between the teeth, awns to 0.7 mm; **lowest lemmas** 3–4 mm, pilose or villous proximally, 3-awned, awns 2–3 mm, central awns flanked by 2 membranous lobes; **lowest paleas** about 2.5 mm, 4-lobed, 2-awned; **anthers** 1.8–2 mm, yellow; **rachilla segments subtending second florets** with densely pubescent apices; **second florets** lobed nearly to the base, lobes ovate, awns 2–4 mm, exceeding those of the lowest lemmas, **third florets** minute scales, glabrous, unawned or with a single awn. **Caryopses** 1.3–1.5 mm. $2n = 20$.

Bouteloua parryi grows on sandy slopes and flats at elevations from near sea level to 2000 m. Its range extends from the southwestern United States to central Mexico. Plants in the *Flora* region belong to **B. parryi** (E. Fourn.) Griffiths var. **parryi**, which differs from **B. parryi** var. **gentryi** (Gould) Gould in comprising tufted annuals rather than stoloniferous perennials. *Bouteloua parryi* var. *parryi* resembles *B. barbata* var. *rothrockii*, but differs in the papillose-based hairs on the keels of its upper glumes.

18. **Bouteloua breviseta** Vasey [p. 268]
GYPSUM GRAMA

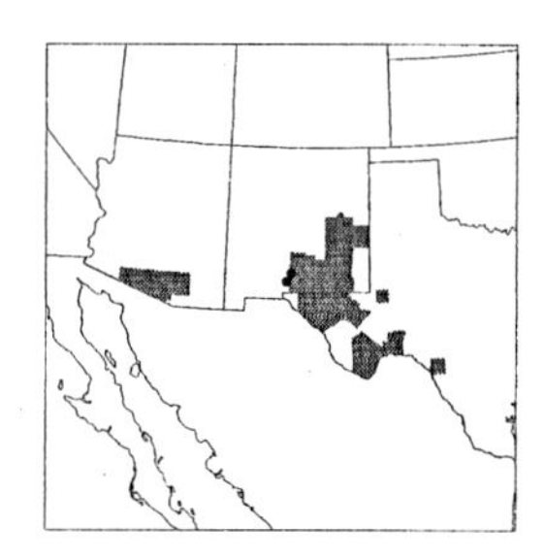

Plants perennial; sometimes cespitose, sometimes rhizomatous, rhizomes 1–3 mm thick, short or elongate, scaly. **Culms** 20–40 cm, erect, somewhat woody at the base, branching at the base and, in late fall, sometimes at the aerial nodes; **nodes** usually 4–5; **internodes** glabrous, distal portions of the lower internodes with a thick, white, chalky bloom. **Ligules** 0.1–0.2 mm, of hairs; **blades** 1–4(7) cm long, 0.5–2 mm wide, flat basally, involute and arcuate to reflexed distally. **Panicles** 2–4 cm, with 1–3(4) branches; **branches** 15–37 mm, persistent, straight to slightly arcuate, mostly appressed, stramineous, with 30–45 spikelets, branches terminating in a reduced, needlelike, 2–5 mm spikelet; **disarticulation** above the glumes. **Spikelets** pectinate, with 1 bisexual floret and 1–2 rudimentary florets. **Glumes** acute to acuminate, glabrous or sparsely short-hairy, hairs not papillose-based; **lower glumes** 2–2.5 mm; **upper glumes** 2–3.5 mm; **lowest lemmas** 2.5–4 mm, sparsely to densely hairy, 3-awned, awns slightly shorter than the lemma bodies, central awns flanked by 2 membranous lobes; **lowest paleas** about 4.5 mm, mostly or completely glabrous, sometimes puberulent distally, acute to acuminate, unawned, veins not excurrent; **second florets** about 4.5 mm, 3-awned, awns 3–5 mm; **rachilla segments subtending second florets** with densely pubescent apices; **third florets**, if present, flabellate scales, 1-awned. **Caryopses** 1–1.2 mm long, about 0.4 mm wide. $2n = 20$.

Bouteloua breviseta is locally abundant on gypsum soils in southeastern New Mexico and the northern portion of the Trans Pecos region in Texas. It also grows in the state of Chihuahua, Mexico. Reeder and Reeder (1980) provide an excellent discussion of *B. breviseta* and *B. ramosa*.

19. **Bouteloua ramosa** Scribn. *ex* Vasey [p. 268]
CHINO GRAMA

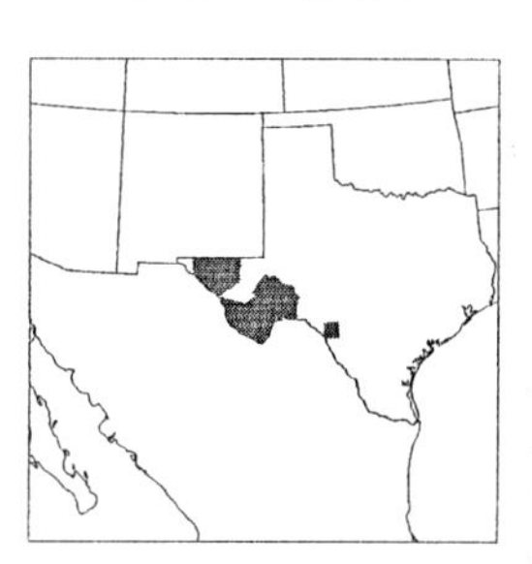

Plants perennial; densely cespitose, bases hard, knotty, without rhizomes or stolons. **Culms** 25–60 cm, numerous, somewhat woody at the base, geniculate, branching profusely from the lower nodes; **nodes** usually 4–5; **lower internodes** glabrous, without a conspicuous, white, chalky bloom. **Ligules** 0.1–0.2 mm, of hairs; **blades** 2–7 cm long, 1–2 mm wide, mostly flat but the tips involute. **Panicles** 1–3(5) cm, with 1–3(4) branches;

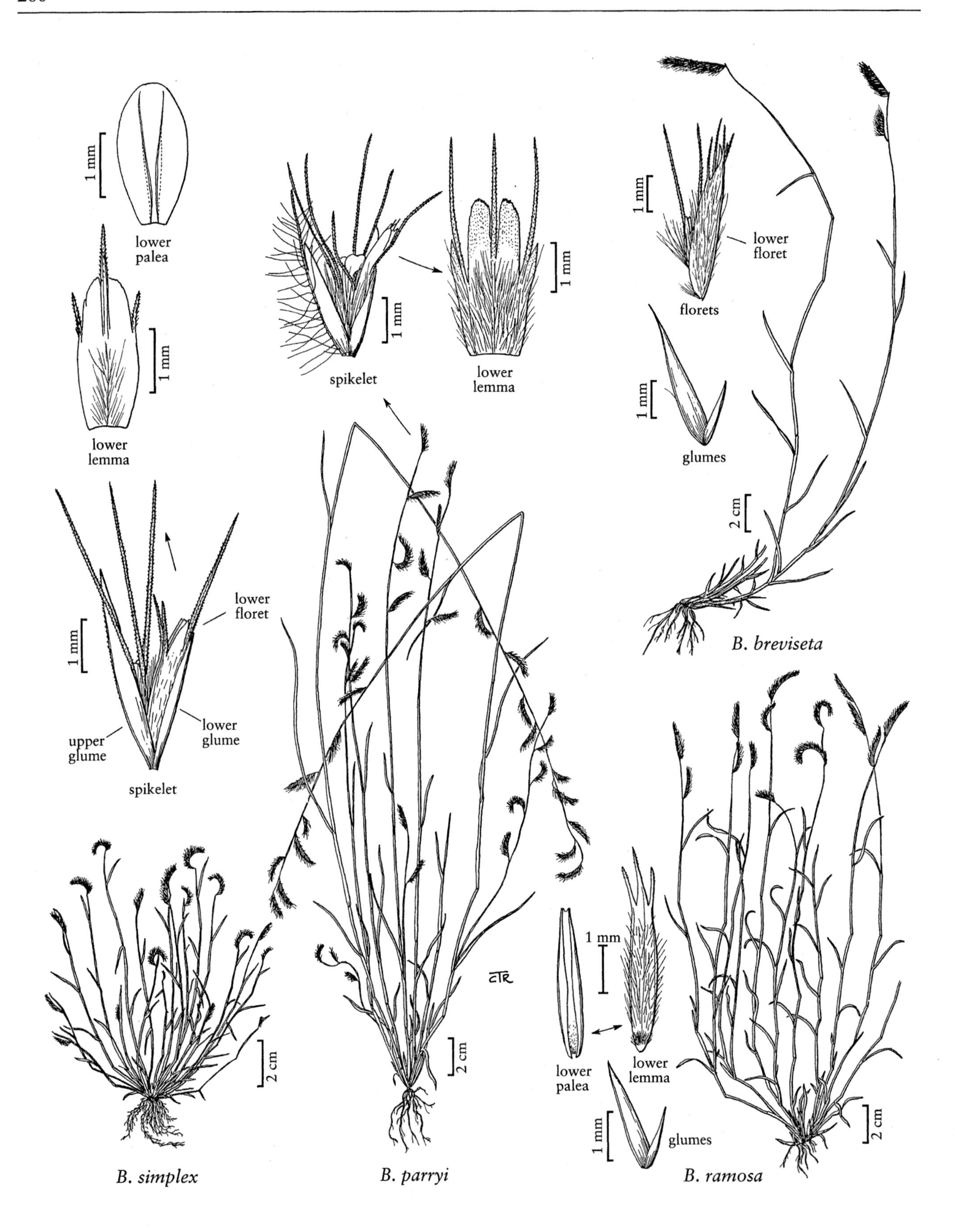
lower
palea
1 mm
lower
lemma
1 mm
spikelet
1 mm
lower
lemma
1 mm
lower
floret
florets
1 mm
glumes
1 mm
2 cm
B. breviseta
lower
floret
1 mm
upper
glume
lower
glume
spikelet
1 mm
lower
palea
lower
lemma
1 mm
glumes
2 cm
2 cm
2 cm
CTR
B. simplex
B. parryi
B. ramosa

branches 10–35 mm, persistent, ascending to widely divergent, becoming arcuate, dark, with 24–45(64) spikelets, branches terminating in a reduced, needlelike, 2–5 mm spikelet; **disarticulation** above the glumes. **Spikelets** with 1 bisexual floret and 1–2 rudimentary florets. **Glumes** acute to acuminate, glabrous or sparsely short-hairy, hairs not papillose-based; **lower glumes** 2–2.5 mm; **upper glumes** 2–3.5 mm; **lowest lemmas** 2.5–4 mm, sparsely to densely hairy, 3-awned, awns slightly shorter than the lemma bodies, central awns flanked by 2 membranous lobes; **lowest paleas** about 4.5 mm, mostly glabrous, sometimes puberulent distally, acute to acuminate, veins not excurrent, unawned; **second florets** about 4.5 mm, 3-awned, awns 3–5 mm; **rachilla segments subtending second florets** with densely pubescent apices; **third florets**, if present, flabellate scales, 1-awned. **Caryopses** 1–1.2 mm long, about 0.4 mm wide. $2n$ = 40.

Bouteloua ramosa is locally common on rocky limestone slopes and flats among shrubs and *Agave lechequilla*. Its range extends from the Trans Pecos region of western Texas to adjacent northern Mexico, particularly the state of Coahuila. Reeder and Reeder (1980) provide an excellent discussion of *B. ramosa* and *B. breviseta*.

17.47 OPIZIA J. Presl

Mary E. Barkworth

Plants annual, perennial, or of indefinite duration; monoecious or dioecious, inflorescences unisexual and dimorphic; stoloniferous and mat-forming. **Culms** 1–15(30) cm, not woody. **Leaves** not clustered, not strongly distichous; **sheaths** open, keeled; **ligules** membranous, not ciliate; **blades** flat. **Inflorescences** terminal, with spikes or spikelike branches on elongate rachises. **Staminate inflorescences** panicles of 1–6 spikelike pectinate branches on elongate rachises, exserted well above the uppermost leaves; **branches** 0.5–2 cm, terminating in a point; **staminate spikelets** glabrous, with 1 floret; **glumes** unequal, much shorter than the the florets; **lemmas** 3-veined, unawned. **Pistillate inflorescences** 1-sided spikes, with 6–12 spikelets, often partially enclosed by the subtending sheaths, sometimes with branches to 6 mm long at the lower nodes; **disarticulation** below the glumes, the spikelets falling intact; **pistillate spikelets** laterally compressed, with 1 bisexual floret and a conspicuously 3-awned rudiment; **lower glumes** rudimentary or absent; **upper glumes** subequal to the lemmas, membranous, flat, not enclosing the florets, unawned; **calluses** blunt, with hairs; **lemmas** coriaceous, keeled, 3-veined, 3-awned, awns emanating from between 4 short, hyaline teeth; **palea keels** adnate to the rachilla basally, widely winged distally. x = unknown. Named for Philippe Maximilian Opiz (1787–1858), a Czech botanist and administrator.

Opizia is a North America genus of two species. Both species are probably native to Mexico, but one of them now grows, possibly from introductions, in Florida and the West Indies.

Columbus (1999) argued that *Opizia* should be included in *Bouteloua*. The traditional treatment is maintained here, pending corroboration of his results from a wider range of data.

SELECTED REFERENCES **Columbus, J.T.** 1999. An expanded circumscription of *Bouteloua* (Gramineae: Chlorideae): New combinations and names. Aliso 18:61–65; **McVaugh, R.** 1983. Flora Novo-Galiciana: A Descriptive Account of the Vascular Plants of Western Mexico, vol. 14: Gramineae (series ed. W.R. Anderson). University of Michigan Press, Ann Arbor, Michigan, U.S.A. 436 pp.

1. **Opizia stolonifera** J. Presl [p. 270]

ACAPULCO GRASS

Culms erect or geniculate. **Sheaths** mostly glabrous, often with a few hairs on either side of the collar; **ligules** 1–1.5 mm; **blades** to 10 cm long, 2-3 mm wide, glabrous abaxially, mostly glabrous or scabrous adaxially, midveins often with a few hairs. **Staminate culms** 5–15(30) cm; **panicles** with 1–6 branches; **branches** 0.5–2 cm; **spikelets** 3–4 mm, glabrous; **anthers** 2–2.5 mm, pale. **Pistillate culms** to 10 cm; **spikes** with 6–12 spikelets (lower nodes sometimes with short branches); **pistillate spikelets** 2.8–4 mm; **glumes** to 3 mm; **lemma bodies** 2–2.5 mm, 3-lobed and 3-awned, awns 3.4–6.8 mm; **palea keels** exceeding the lemmas; **rudiment** 3-awned. $2n$ = unknown.

Opizia stolonifera grows along dry roadsides in Florida. No pistillate plants have been found in the *Flora* region.

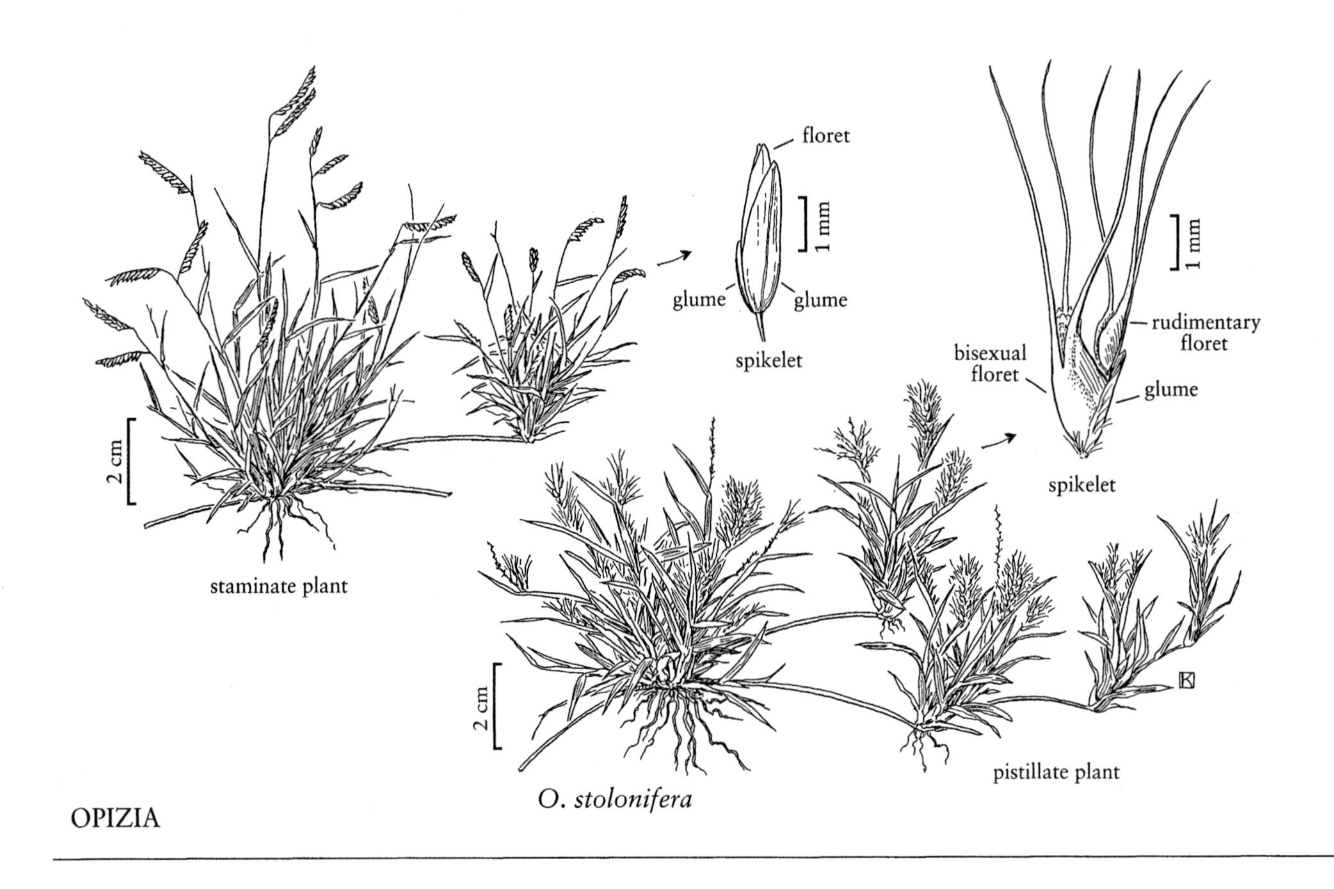

OPIZIA

O. stolonifera

17.48 BUCHLOË Engelm.

Neil Snow

Plants perennial; usually dioecious; strongly stoloniferous, sometimes mat-forming. **Culms** 1–30 cm, erect, solid, mostly unbranched, those of the pistillate inflorescences much shorter than those of the staminate inflorescences; **nodes** mostly glabrous. **Leaves** basally tufted, not clustered or strongly distichous; **sheaths** open, rounded, often sparsely pilose near the collar; **ligules** membranous or of hairs; **blades** usually flat basally, curling when dry, glabrous or sparsely pilose, apices involute. **Staminate inflorescences** terminal, usually exceeding the upper leaves, panicles of 1–3(4) racemosely arranged, unilateral, pectinate branches; **branches** not enclosed at maturity, spikelets densely crowded in 2 rows. **Staminate spikelets** with 2 florets; **glumes** unequal, glabrous, 1- or 2-veined; **lemmas** 3-veined, glabrous, unawned; **anthers** brownish to red or orange. **Pistillate inflorescences** terminal, panicles, partially hidden within bracteate leaf sheaths; **branches** 2–3(4), 2.5–4.5 mm, burlike, with 3–5(7) spikelets; **disarticulation** at the base of the panicle branches. **Pistillate spikelets** with 1 floret, almost completely enclosed by the upper glumes; **lower glumes** irregular and reduced; **branch axes** and **lower portion of upper glumes** globose, white, indurate, terminating in 3 awnlike teeth; **lemmas** firmly membranous, glabrous, 3-veined, unawned or shortly 3-awned. x = 10. Name a contraction of *Bubalochloë*, from the Greek *boubalos*, 'buffalo', and *chloë*, 'grass'.

Buchloë is a monotypic genus of the central plains of North America. It is usually dioecious, infrequently monoecious, or rarely synoecious. On the basis of his molecular studies, Columbus (1999) recommended including it and several other small, usually monoecious or dioecious

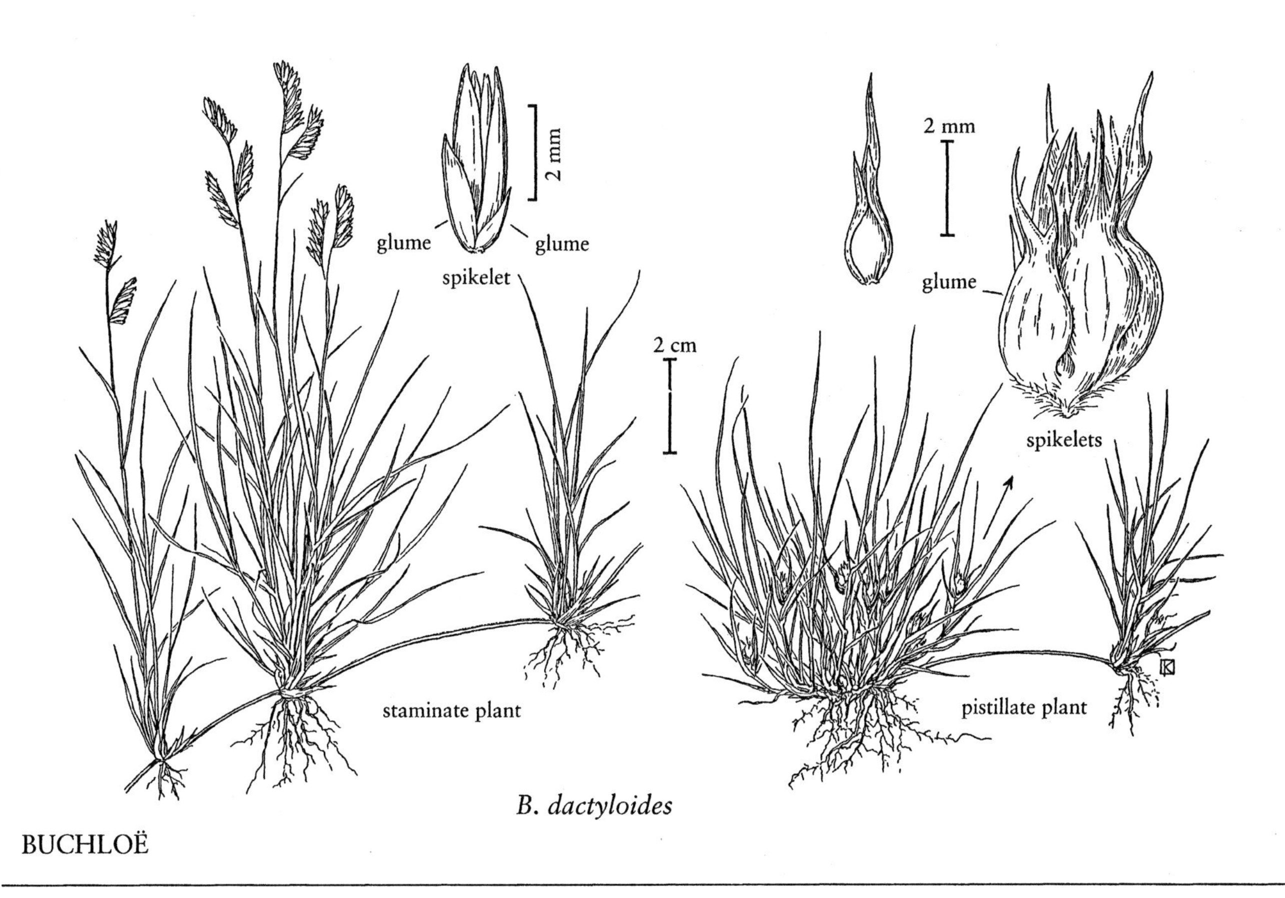

B. dactyloides

BUCHLOË

genera in *Bouteloua*. Morphologically, the segregate genera differ from *Bouteloua* only in their pistillate panicles and spikelets and their reproductive mode, but not in their vegetative and staminate structures. *Buchloë* is maintained here pending corroboration from other studies.

SELECTED REFERENCES **Beetle, A.A.** 1950. Buffalograss–native of the shortgrass plains. Wyoming Agric. Exp. Sta. Bull. 293:1–31; **Columbus, J.T.** 1999. An expanded circumscription of *Bouteloua* (Gramineae: Chlorideae): New combinations and names. Aliso 18:61–65; **Quinn, J.A.** and **J.L. Engel.** 1986. Life-history strategies and sex ratios for a cultivar and a wild population of *Buchloë dactyloides* (Gramineae). Amer. J. Bot. 73:874–881; **Wenger, L.E.** 1940. Inflorescence variations in buffalo grass, *Buchloë dactyloides*. J. Amer. Soc. Agron. 32:274–277.

1. **Buchloë dactyloides** (Nutt.) Engelm. [p. 271]
BUFFALOGRASS

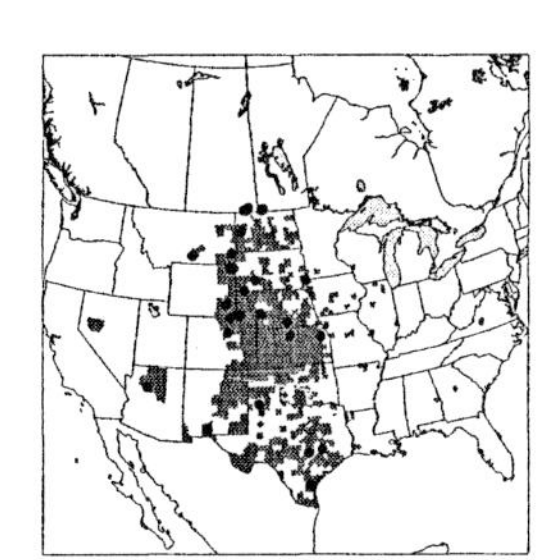

Ligules 0.5–1 mm; **blades** 2–15 cm long, 1-2.5 mm wide. **Staminate spikelets** 4–6 mm long, 1.3–1.8 mm wide; **anthers** 2.5–3 mm. **Pistillate spikelets** to 7 mm long, about 2.5 mm wide. **Caryopses** 2–2.5 mm. $2n$ = 20, 40, 56, 60.

Buchloë dactyloides is a frequent dominant on upland portions of the semi-arid, shortgrass component of the Great Plains, ranging from the southern prairie provinces of Canada through the desert southwest of the United States to much of northern Mexico. Collections from east of the Mississippi River and south of the Ohio River probably represent recent introductions.

Buchloë dactyoides provides valuable forage for livestock and wildlife, and withstands heavy grazing. It may be confused in the southern portion of its range with *Hilaria belangeri*, which consistently has pilose nodes, or in the Big Bend region of Texas with *Cathestecum erectum*, which has three spikelets per node and distinctly awned lemmas.

17.49 CATHESTECUM J. Presl

Plants annual or perennial; monoecious or dioecious; often stoloniferous. **Culms** 5–40 cm, sometimes decumbent. **Leaves** mostly basal; **sheaths** open; **ligules** of hairs; **blades** 0.5–2 mm wide. **Inflorescences** terminal, racemelike panicles of short branches, usually exceeding the upper leaves, sometimes appearing 1-sided; **branches** about 0.5 cm, not appressed to the rachises, with 3 spikelets, bases strongly curved, shortly strigose, axes not extending beyond the distal spikelets; **disarticulation** at the base of the branches. **Spikelets** pedicellate or sessile, lateral spikelets with shorter pedicels than the central spikelet or sessile; **lateral spikelets** with 2 florets, lower florets usually staminate or sterile, rarely pistillate, upper florets much reduced, usually sterile; **central spikelets** pedicellate, with 3 florets, lowest florets bisexual or pistillate (staminate in staminate plants), distal florets staminate or sterile. **Glumes** 2, very unequal, shorter than the spikelets, unawned; **lower glumes** shorter than the upper glumes, those of the central spikelets flabellate, glabrous, veinless; **upper glumes** approximately equal to the adjacent lemmas, 1-veined, acuminate; **lemmas** thinner than the glumes, pilose, 3-veined, 4-lobed, all 3 veins excurrent, forming awns that equal or exceed the lobes, lobes of the sterile florets usually deeper than those of the other florets; **paleas** 2-veined, veins often extending as awns. $x = 10$. Name from the Greek *kathestekos*, 'set fast' or 'stationary'; the allusion is not clear.

Cathestecum is a genus of six species that extends from the southern United States to Guatemala; two are native to the *Flora* region. Columbus (1999) advocated including *Cathestecum* in *Bouteloua*, but the traditional treatment is adopted here.

SELECTED REFERENCES **Beetle, A.A.** 1987. Las Gramineas de México, vol. 2. COTECOCA [Comisión Técnico Consultiva de Coeficientes de Agostadero], México, D.F., México. 344 pp.; **Columbus, J.T.** 1999. An expanded circumscription of *Bouteloua* (Gramineae: Chlorideae): New combinations and names. Aliso 18:61–65; **Pierce, G.J.** 1979. A biosystematic study of *Cathestum* and *Griffithsochloa* (Gramineae). Ph.D. dissertation, University of Wyoming, Laramie, Wyoming, U.S.A. 244 pp.

1. Stolon internodes to 12 cm long, straight or almost so; culms 5–15 cm tall; spikelets frequently reddish or purplish; lateral spikelets of branches with bisexual florets poorly-developed, sterile (rarely staminate) florets; central spikelet on staminate branches with sparsely pilose lemmas . . . . 1. *C. brevifolium*
1. Stolon internodes usually 15 cm or longer, strongly arching; culms 10–30 cm tall; spikelets frequently pale green; lateral spikelets of branches with bisexual florets usually well-developed, usually staminate or sterile; central spikelet on staminate branches with glabrous lemmas . . . . 2. *C. erectum*

1. Cathestecum brevifolium Swallen [p. 275]

Plants perennial; polygamous or dioecious; cespitose, forming dense, small clumps, stoloniferous, stolons thin, internodes to 10(12) cm, straight or only slightly arching. **Culms** 5–15 cm, erect or geniculate. **Leaves** primarily basal; **lower sheaths** densely villous basally, mostly glabrous or sparsely pilose distally, throats densely ciliate; **blades** 1–5 cm long, 1–2 mm wide, stiff, flat to involute, abaxial surfaces glabrous, adaxial surfaces scabrous and pilose, margins scabrous. **Panicles** with 3–10 branches; **branches** divergent, dimorphic, bearing staminate or bisexual spikelets, spikelets usually reddish or purple. **Staminate branches**: all spikelets similar; **lower glumes** to 1 mm, often reduced to a scale, 0–1-veined; **upper glumes** ½–⅔ as long as the spikelets, narrowly trullate to lanceolate, usually glabrous, veins sometimes sparsely pilose, glumes of the lateral spikelets about 2.5 mm, acute or acuminate, those of the central spikelets about 3 mm, minutely lobed and mucronate; **lowest lemmas** to 3 mm, sparsely pilose, with short lobes, mucronate between the lobes; **distal lemmas** to 2.5 mm, similar to the lowest lemmas but with deeper lobes; **anthers** 0.7–2 mm. **Bisexual branches: lateral spikelets** poorly developed, lower glumes to 1 mm, upper glumes to 2.5 mm, veins pilose, florets greatly reduced, sometimes just a cluster of awns, sterile (rarely staminate); **central spikelets** with glumes similar to those of the lateral spikelets; **lowest florets** pistillate, lemmas about 3 mm, scabrous, lobed, lobes about ¼ as long as the lemmas, awned from the sinuses, awns slightly exceeding the lobes; **distal florets** staminate or sterile, about 2.5 mm, deeply lobed, awned from the sinuses, awns exceeding the lobes by 1–3 mm. $2n$ = 20, 40, 60, and 80.

The range of *Cathestecum brevifolium* extends from Ragged Top Mountain, Arizona, to El Salvador and Honduras. The Arizonan plants were originally identified as *C. erectum* but, like many Mexican populations of *C. brevifolium*, they have $2n$ = 60 and are dimorphic, consisting of relatively long-awned pistillate plants and more shortly-awned staminate plants (J.R. Reeder, pers. comm.).

2. **Cathestecum erectum** Vasey & Hack. [p. 275]
False Grama

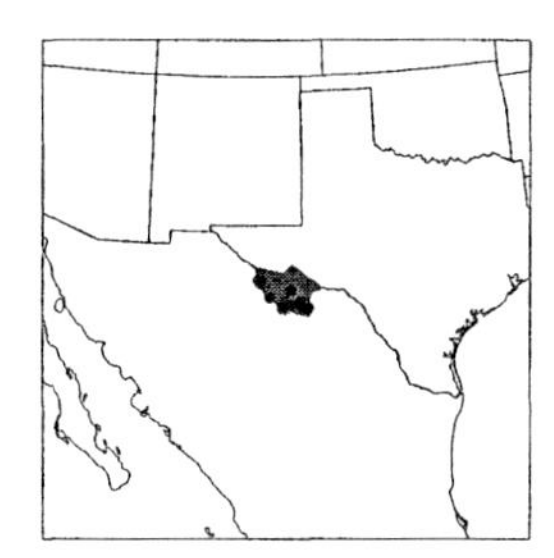

Plants perennial; polygamous; cespitose, forming dense, small clumps, stoloniferous, stolons thin, internodes 15–40 cm, strongly arching. **Culms** 10–30 cm, erect or geniculate, glabrous. **Leaves** primarily basal; **lower sheaths** overlapping, densely villous basally, throats pilose; **upper sheaths** not overlapping, glabrous; **ligules** about 0.3 mm; **blades** 3–12 cm long, 1–1.5 mm wide, involute or flat, glabrous abaxially, scabrous and sparsely pilose adaxially, hairs to 2 mm. **Panicles** with 5–7 branches; **branches** about 5 mm, dimorphic, staminate or bisexual, the 2 forms sometimes on different plants, sometimes mixed within a panicle, spikelets of 1 form all staminate, bisexual form with pistillate (sometimes bisexual) central spikelets and well-developed staminate or sterile lateral spikelets, spikelets frequently pale green. **Staminate branches:** all spikelets similar; **lower glumes** about 1 mm; **upper glumes** glabrous or almost so, those of the lateral spikelets about 3 mm, those of the central spikelets about 4 mm; **lemmas** similar, about 3 mm, glabrous, irregularly lobed, unawned, sometimes mucronate. **Bisexual branches: glumes** villous; **lateral spikelets** with staminate or sterile florets; **lemmas of lateral spikelets** about 3 mm, glabrous or sparsely pubescent, irregularly lobed, awned from the sinuses, awns hispid, equaling or occasionally exceeding the lobes; **anthers** 1.7–2.3 mm; **central spikelets** with the lowest floret pistillate, distal florets staminate or sterile; **lowest lemmas** glabrous or sparsely pubescent, lobed, lobes about ⅓ as long as the lemmas, awned from the sinuses, awns glabrous, subequal to the lobes or the central awns slightly longer; **distal florets** similar to those of the lateral florets, awns 1–2 mm longer than the lobes. $2n$ = 20.

Cathestecum erectum grows on dry hills in the Great Bend region of western Texas and in northern Mexico.

17.50 AEGOPOGON Humb. & Bonpl. *ex* Willd.

Mary E. Barkworth

Plants annual or perennial; synoecious; tufted, cespitose, or sprawling. **Culms** 2–30 cm. **Sheaths** open; **ligules** membranous; **blades** flat. **Inflorescences** terminal, racemelike, 1-sided panicles, usually exceeding the upper leaves; **branches** 0.4–0.8 mm, not appressed to the rachis, with 3 spikelets, bases sharply curved, strigose, axes not extending beyond the distal spikelets; **disarticulation** at the base of the branches. **Spikelets** with 1 floret; **lateral spikelets** pedicellate, staminate or sterile, varying from rudimentary to as large as the central spikelet; **central spikelets** sessile or pedicellate, laterally compressed, bisexual. **Glumes** exceeded by the florets, cuneate, truncate to bilobed, 1-veined, awned from the midveins, sometimes also from the lateral lobes; **lemmas** 3-veined, central veins and sometimes the lateral veins extended into awns, central awns always the longest; **paleas** almost as long as the lemmas, 2-keeled, 2-awned; **lodicules** 2; **anthers** 3; **styles** 2. **Caryopses** about twice as large as the embryos; **hila** punctate. x = 10. Name from the Greek *aix*, 'goat', and *pogon*, 'beard', alluding to the fascicles of the awns of the spikelets.

Aegopogon is an American genus of four species that extends from the southwestern United States to Peru, Bolivia, and northern Argentina. One species is native in the *Flora* region.

SELECTED REFERENCE Beetle, A.A. 1948. The genus *Aegopogon* Humbold. & Bonpl. Univ. Wyoming Publ. 13:17–23.

1. **Aegopogon tenellus** (DC.) Trin. [p. 275]
FRAGILE GRASS

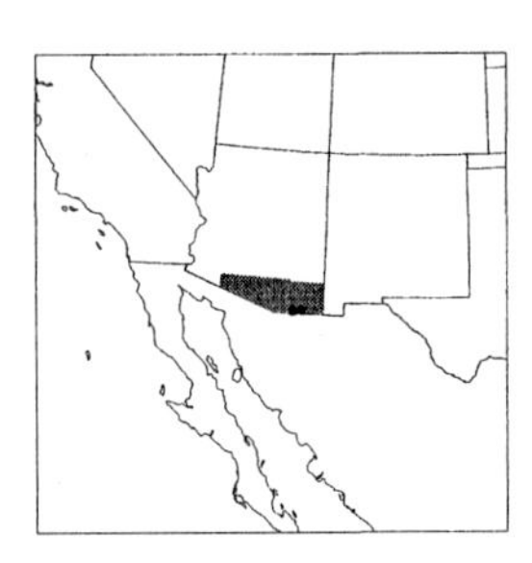

Plants annual; tufted. **Culms** 2–30 cm, bases often prostrate or decumbent, strongly branching. **Sheaths** glabrous or sparsely hirsute; **ligules** 0.7–1.5 mm, lacerate; **blades** 1–7 cm long, 1–2 mm wide, glabrous or puberulent. **Panicles** 2–6 cm. **Lateral spikelets** 1.5–2.3 mm, on 1–1.3 mm pedicels; **central spikelets** 2.5–3.2 mm, on 0.3–0.6 mm pedicels. **Glumes** 1.3–1.8 mm, flabellate, lobes rounded, awns 0.1–0.6 mm; **lemmas** 2.5–3.2 mm, central awns 3–8 mm, lateral awns to 1 mm; **anthers** 0.5–0.8 mm. $2n$ = 20, 60.

Aegopogon tenellus usually grows between 1550–2150 m in shady habitats of moist canyons, but it is sometimes found along roadsides and in other open areas. Its range extends from southern Arizona into northern South America. In some plants, the lateral spikelets are reduced (*Aegopogon tenellus* var. *abortivus* (E. Fourn.) Beetle), but such spikelets (and central spikelets with reduced awns) are also found in plants with normal spikelets, so taxonomic recognition is not warranted.

17.51 HILARIA Kunth

Mary E. Barkworth

Plants perennial or annual; tufted or cespitose, sometimes stoloniferous, perennial species sometimes rhizomatous. **Culms** 5–250 cm, erect or decumbent; **nodes** usually villous or pilose, particularly the upper nodes. **Sheaths** open, glabrous or pilose, lower sheaths often glabrous basally and pilose distally, margins sometimes villous or pilose, upper sheaths often glabrous even if the lower sheaths are pilose; **ligules** 0.5–5 mm, membranous, lacerate or ciliate. **Inflorescences** terminal, spikelike panicles of reduced, disarticulating branches, exceeding the upper leaves; **branches** with 3 spikelets, appressed to the rachises, bases straight, seated in a ciliate, cuplike structure, sometimes with a 0.5–2 mm callus, calluses pilose, axes not extending past the distal florets; **disarticulation** at the base of the branches, leaving the zig-zag rachises. **Lateral spikelets** of each branch shortly pedicellate, with 1–4(5) sterile or staminate florets; **glumes** almost as long as the florets, deeply cleft into 2 or more lobes, with 1 or more dorsal awns; **lemmas** membranous, hyaline. **Central spikelets** sessile, with 1 pistillate or bisexual floret; **glumes** shorter than the florets, rigid, indurate and fused basally, apices with 2 or more lobes; **lemmas** membranous, awned or unawned. x = 9. Named for Auguste François César Prouvençal de St.-Hilaire (1779–1853), a French explorer, botanist, and entomologist.

Hilaria is a genus of 10 species that ranges from the southwestern United States to northern Guatemala, growing primarily in dry grasslands and desert areas. Most of the species are important forage species. The stoloniferous species are important soil binders.

Hilaria is interpreted here as having two groups, the *Hilaria* group and the *Pleuraphis* group [= *Pleuraphis* Torr.]. These are sometimes treated as separate genera but, although they differ consistently in some morphological characters, their overall similarity is striking. One molecular study (Columbus et al. 1998, 2000) has included representatives of both groups. It showed them to be sister taxa; there seems little value in promoting each to generic rank.

In the key and descriptions below, the term "fascicle" refers to a branch and its spikelets. Actual branch lengths are much shorter and harder to measure.

SELECTED REFERENCES **Columbus, J.T., M.S. Kinney, R. Pant, and M.E. Siqueiros Delgado.** 1998. Cladistic parsimony analysis of internal transcribed spacer region (nrDNA) sequences of *Bouteloua* and relatives (Graminese: Chloridoideae). Aliso 17:99–130; **Columbus, J.T., M.S. Kinney, M.E. Siqueiros Delgado, and R. Cerros Tlatilpa.** 2000. Homoplasy, polyphyly, and generic circumscription: The demise of the Boutelouinae (Gramineae: Chloridoideae). [Abstract.] Amer. J. Bot. 87, Suppl.:120; **Sohns, E.R.** 1956. The genus *Hilaria*. J. Wash. Acad. Sci. 46:311–321.

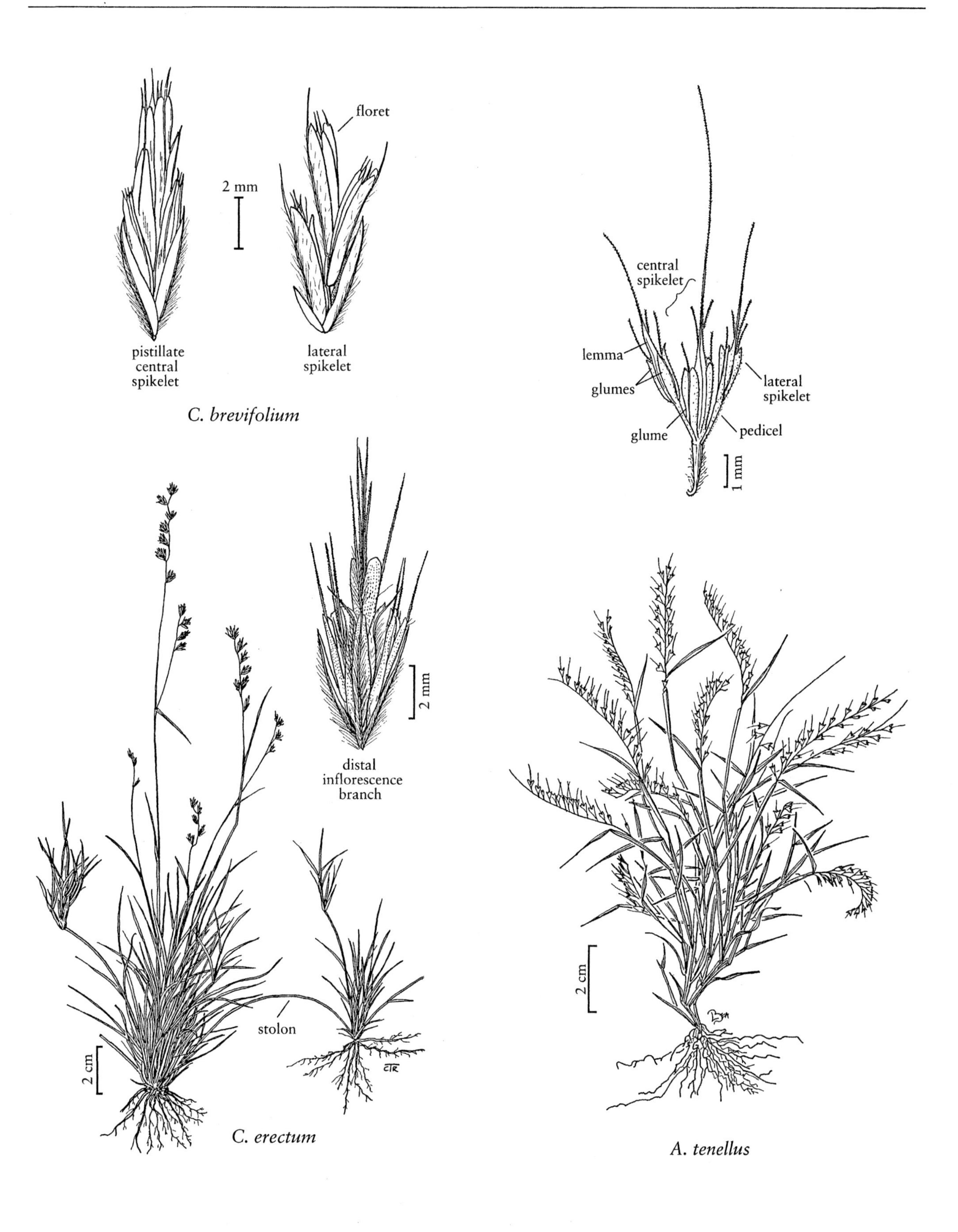

CATHESTECUM

AEGOPOGON

1. Glumes thickened, indurate, and conspicuously fused at the base; central spikelets with 1 pistillate floret (*Hilaria* group).
   2. Glumes pale to purplish, those of the lateral spikelets with dark glands confined to the base or lacking, awned from below midlength . . . . 4. *H. belangeri*
   2. Glumes gray to dark brown, those of the lateral spikelets evenly covered with dark glands, awned from above midlength . . . . 5. *H. swallenii*
1. Glumes papery or membranous throughout, not conspicuously fused at the base; central spikelets with 1 bisexual floret (*Pleuraphis* group).
   3. Glumes of the lateral spikelets flabellate, the awns not exceeding the apical lobes; cauline nodes usually only shortly pubescent, sometimes glabrous . . . . 1. *H. mutica*
   3. Glumes of the lateral spikelets lanceolate or parallel-sided, the awns exceeding the apical lobes; cauline nodes pilose, villous, or glabrous.
      4. Lower cauline internodes tomentose . . . . 2. *H. rigida*
      4. Lower cauline internodes glabrous . . . . 3. *H. jamesii*

1. **Hilaria mutica** (Buckley) Benth. [p. 277]
Tobosagrass

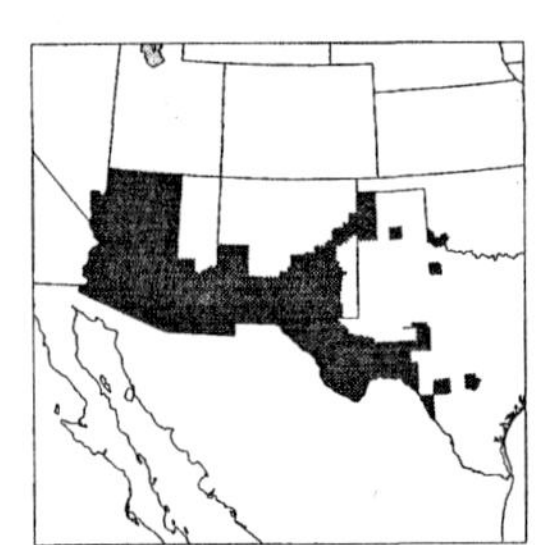

Plants perennial; cespitose, rhizomatous. **Culms** 30–60 cm, erect, geniculate at the middle nodes; **nodes** glabrous or pubescent, hairs to 0.3 mm. **Sheaths** glabrous or sparsely pilose on the margins; **ligules** 0.5–2 mm, lacerate; **blades** 2–15 cm long, 2–4 mm wide, mostly scabrous on both surfaces, with papillose-based hairs behind the ligules. **Panicles** 4–8 cm; **fascicles** 5–8 mm. **Lateral spikelets** with 1 or 2(4) staminate florets; **glumes** not conspicuously fused basally, thin, papery, flabellate, dorsally awned, awns not exceeding the apices, apical lobes rounded, ciliate to finely laciniate, veins not or scarcely excurrent; **anthers** 3, 2.5–3.5 mm. **Central spikelets** with 1 bisexual floret; **glumes** with 1 or more divergent, dorsal awns, apical lobes, ciliate to finely laciniate, veins excurrent; **lemmas** exceeding the glumes, bilobed, mucronate. $2n$ = 36, 54.

Hilaria mutica grows in level upland areas and desert valleys subject to occasional flooding but lacking permanent streams. Its range extends into northern Mexico. Although *H. mutica* has moderate forage value, its palatability is low and it is frequently infected with ergot.

2. **Hilaria rigida** (Thurb.) Benth. *ex* Scribn. [p. 277]
Big Galleta

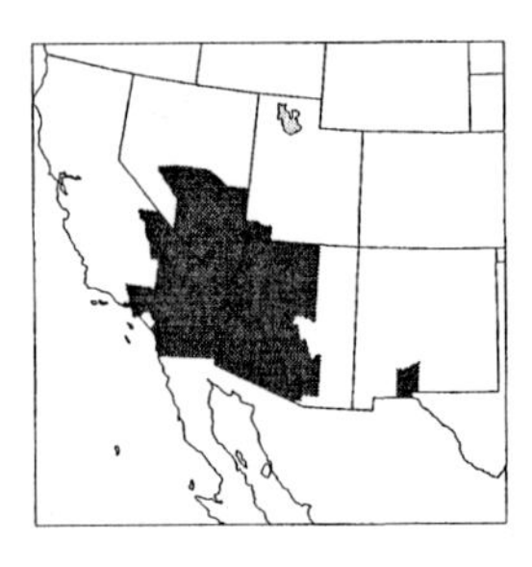

Plants perennial; cespitose, sometimes rhizomatous. **Culms** 35–250 cm, decumbent, much branched above the base, becoming almost woody; **upper nodes** glabrous or villous, hairs to 1.5 mm; **lower internodes** tomentose. **Ligules** 1–2 mm, densely ciliate; **blades** 2–10(16) cm long, 2–5 mm wide, flat basally, involute distally. **Panicles** 4–12 cm; **fascicles** 6–12 mm. **Lateral spikelets** with 2–4 florets, lower 2 florets staminate, other florets (if present) usually sterile; **glumes** thin, membranous, not fused at the base, lanceolate or parallel-sided, 7-veined, awned, awns exceeding the glume apices, apices 2–4-lobed, lobes acute to rounded, long-ciliate, sometimes with 1–3 excurrent veins that form additional slender awns to 1.8 mm; **lower glumes** with dorsal, divergent awns; **upper glumes** with subapical awns; **anthers** 3, 4–4.5 mm. **Central spikelets** equaling or exceeding the lateral spikelets, with 1 stipitate, bisexual floret; **glumes** thin, membranous, narrow, deeply cleft into few–several acuminate, ciliate lobes and slender awns; **lemmas** often exceeding the glumes, thin, ciliate, 2-lobed, midveins excurrent. $2n$ = 18, 36, 54.

Hilaria rigida grows in deserts and open juniper stands, at low elevations, from the southwestern United States to central Mexico. Although almost shrubby, it is very popular with pack horses.

3. **Hilaria jamesii** (Torr.) Benth. [p. 277]
Galleta

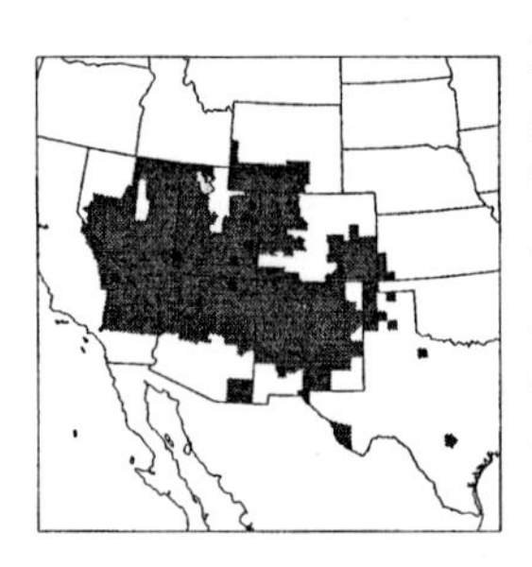

Plants perennial; strongly rhizomatous or stoloniferous. **Culms** 20–65 cm, erect, bases much branched; **nodes** usually pilose or villous, sometimes glabrous; **lower internodes** glabrous. **Sheaths** glabrous, sometimes slightly scabrous; **collars** pilose at the edges; **ligules** 1.5–5 mm, often laciniate; **blades** 2–20 cm long, 2–4 mm wide, involute and curled when dry, sparsely villous behind the ligules, abaxial surfaces scabridulous, adaxial surfaces scabrous. **Panicles** 2–6 cm; **fascicles** 6–8 mm. **Lateral spikelets** with 3 staminate florets; **glumes** thin, membranous, lanceolate or parallel-sided, not conspicuously fused at the base, apices acute to

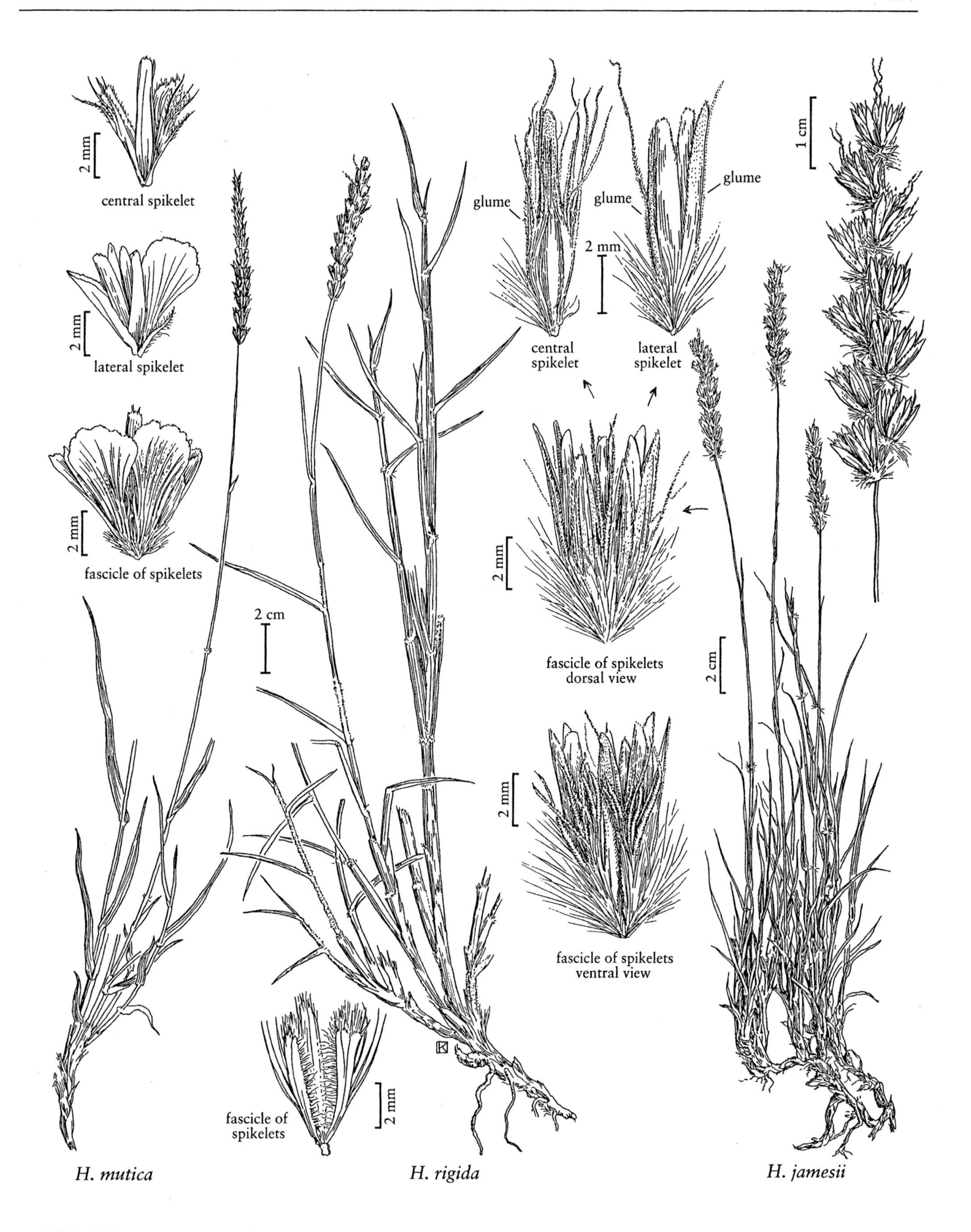

central spikelet
2 mm
lateral spikelet
2 mm
fascicle of spikelets
2 mm
2 cm
fascicle of spikelets
2 mm
H. mutica
H. rigida
glume
glume
glume
2 mm
central spikelet
lateral spikelet
fascicle of spikelets dorsal view
2 mm
fascicle of spikelets ventral view
2 mm
1 cm
2 cm
H. jamesii

HILARIA

rounded, often ciliate, veins rarely excurrent; **lower glumes** dorsally awned, awns exceeding the apices; **anthers** 3, about 5 mm. **Central spikelets** with 1 bisexual floret; **glumes** with excurrent veins forming distinct awns; **lemmas** exceeding the glumes, ciliate, the midveins sometimes excurrent. $2n$ = 18, 36.

Hilaria jamesii is endemic to the southwestern United States, and grows in deserts, canyons, and dry plains. It has medium grazing value but low palatability. It is usually less pubescent than *H. rigida*, the difference being most marked on the lower cauline nodes.

4. **Hilaria belangeri** (Steud.) Nash [p. 279]
CURLY MESQUITE

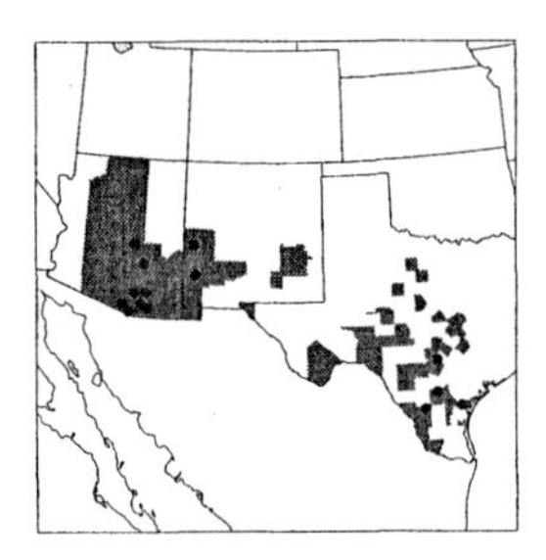

Plants perennial; cespitose, usually stoloniferous. **Culms** 5–35 cm, erect; **nodes** villous. **Sheaths** striate, glabrous; **ligules** 1–3 mm, often lacerate; **blades** 3–15 cm long, 1–3.5 mm wide, adaxial surfaces sparsely pilose, hairs papillose-based, margins sparsely pilose basally, with similar hairs. **Panicles** 2–4 cm; **fascicles** 5–8 mm. **Lateral spikelets** with 2(3) staminate florets, or 1 sterile floret; **glumes** unequal, thick, indurate, and conspicuously fused basally, thinner distally, asymmetrically lobed, scabrous, pale to purplish, bases sometimes spotted with a few dark glands, margins wide, hyaline, awns 1 or more, attached below midlength, equaling or exceeding the central spikelets, antrorsely scabrous; **lower glumes** wider, more deeply lobed, with longer awn(s) than the upper glumes; **anthers** 3, 3–3.7 mm. **Central spikelets** as long as or longer than the lateral spikelets, with 1 pistillate floret; **glumes** terminating in 1 or more antrorsely scabrous awns. $2n$ = 36, 72, 74.

Both varieties of *Hilaria belangeri* are found on mesas and plains within the regions indicated.

1. Plants stoloniferous; blades 3–10 cm long, 1–2 mm wide; ligules about 1–1.5 mm long . . . var. *belangeri*
1. Plants not stoloniferous; blades 3–15 cm long, to 3.5 mm wide; ligules 2.5–3 mm long . . . var. *longifolia*

Hilaria belangeri (Steud.) Nash var. **belangeri**

Plants stoloniferous. **Ligules** 1–1.5 mm; **blades** 3–10 cm long, 1–2 mm wide.

Hilaria belangeri var. *belangeri* grows from Arizona to Texas, and south through Mexico. It was the dominant grass on Texas' shortgrass prairies.

Hilaria belangeri var. **longifolia** (Vasey) Hitchc.

Plants not stoloniferous. **Ligules** 2.5–3 mm; **blades** 3–15 cm long, to 3.5 mm wide.

Hilaria belangeri var. *longifolia* is more restricted than var. *belangeri* in its distribution, growing from Arizona to Texas, and south to northwestern Mexico.

5. **Hilaria swallenii** Cory [p. 279]
SWALLEN'S CURLY MESQUITE

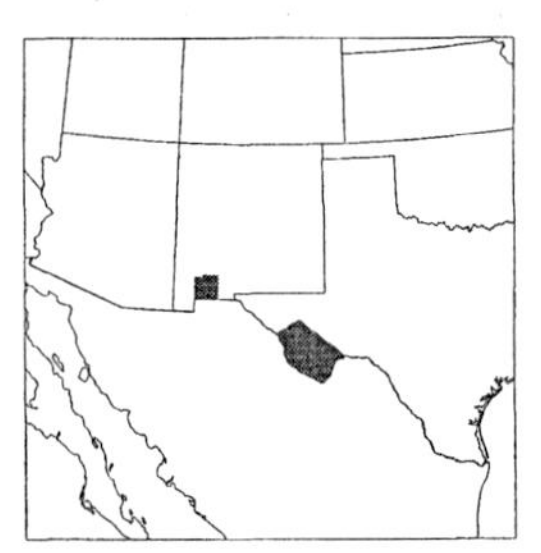

Plants perennial; stoloniferous. **Culms** 10–35 cm, erect; **nodes** villous. **Sheaths** slightly scabrous; **ligules** 2–2.2 mm; **blades** to 8 cm long, 1–2 mm wide, mostly basal, both surfaces scabrous, sometimes also sparsely pilose. **Panicles** 1–4 cm, with 2–8 fascicles; **fascicles** 6.5–8 mm. **Lateral spikelets** with 2 florets, lowest florets usually sterile, distal florets staminate; **glumes** unequal, thick, indurate, and conspicuously fused basally, mostly gray to dark brown, evenly and sparsely to densely spotted with dark glands, awned from above midlength, margins hyaline; **anthers** 3, 3–3.5 mm. **Central spikelets** with 1 pistillate floret; **lemmas** elliptic basally, narrower and parallel-sided distally. $2n$ = 54, 72.

Hilaria swallenii grows on dry plains and rocky mesas in New Mexico, Texas, and northern Mexico. It is considered better forage than *H. belangeri*, but it is less important because it is less common.

17.52 TRAGUS Haller

J.K. Wipff

Plants annual or perennial; cespitose. **Culms** (2)5–65 cm, herbaceous, usually rooting at the lower nodes; **nodes** and **internodes** glabrous. **Leaves** cauline; **sheaths** open, usually shorter than the internodes, mostly glabrous but long-ciliate at the edges of the collar; **ligules** membranous, truncate, ciliate; **blades** usually flat, margins ciliate. **Inflorescences** terminal, exceeding the upper leaves, narrow, cylindrical panicles; **branches** 0.5–5 mm, resembling burs, with 2–5 spikelets; **disarticulation** at the base of the branches. **Spikelets** crowded, attached individually to the branches, with 1 floret; **proximal spikelet(s)** bisexual, larger than the distal spikelet(s); **terminal**

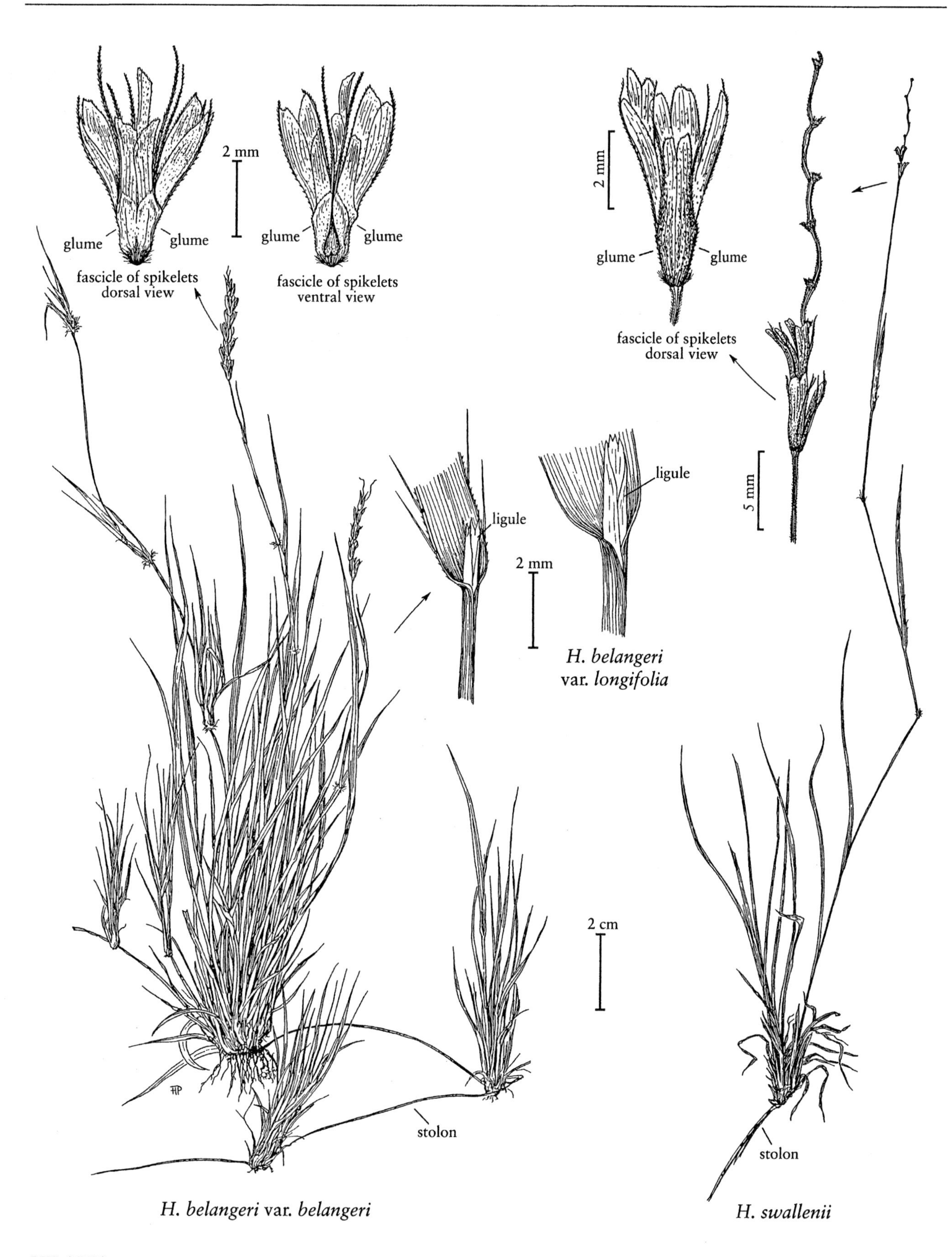

glume
glume
fascicle of spikelets
dorsal view
2 mm
glume
glume
fascicle of spikelets
ventral view
2 mm
glume
glume
fascicle of spikelets
dorsal view
5 mm
ligule
2 mm
ligule
H. belangeri
var. longifolia
2 cm
stolon
stolon
H. belangeri var. belangeri
H. swallenii

spikelets often sterile. **Glumes** unequal; **lower glumes** absent or minute, veinless, membranous; **upper glumes** usually exceeding the florets, 5–7-veined, with 5–7 longitudinal rows of straight or uncinate spinelike projections; **lemmas** 3-veined; **paleas** 2-veined, hyaline, membranous. x = 10. Name from the Greek *tragos*, 'he-goat'.

Tragus has seven species, all of which are native to the tropics and subtropics of the Eastern Hemisphere; four have been introduced into the *Flora* region. The genus is easily recognized by the spinelike projections on the upper glumes. The number of veins in the glume should be determined by examining the adaxial surface, where they appear as green lines.

SELECTED REFERENCE **Anton, A.M.** 1981. The genus *Tragus* (Gramineae). Kew Bull. 36:55–61.

1. Upper glumes with 5 longitudinal rows of spinelike projections, 5-veined.
   2. Proximal internodes of the primary branches not longer than the second internode . . . . . 1. *T. berteronianus*
   2. Proximal internodes of the primary branches 2–3 (or more) times longer than the second internode . . . . . 2. *T. australianus*
1. Upper glumes with (5)6–7 longitudinal rows of spinelike projections, 7-veined.
   3. Panicle branches with 2 (rarely 3) spikelets; proximal spikelets on the branches 3–3.5 mm long . . . . . 3. *T. heptaneuron*
   3. Panicle branches with 3–5 (rarely 2) spikelets; proximal spikelets on the branches 3.8–6.6 mm long . . . . . 4. *T. racemosus*

1. **Tragus berteronianus** Schult. [p. 282]
SPIKE BURGRASS

Plants annual. **Culms** (2)3.5–45 cm. **Ligules** 0.5–1 mm; **blades** (0.5)0.7–8.5 cm long, 1.2–5 mm wide, glabrous. **Panicles** (1)2–13 cm long, (3)4–8 mm wide; **rachises** pubescent; **branches** (0.5)0.7–2.7 mm, pubescent, with 2(3) spikelets, axes occasionally extending past the distal spikelets; **proximal internodes** 0.2–0.6(0.7) mm, shorter than the second internodes. **Proximal spikelets** (1.8)2–4.3 mm; **second spikelets** (0.8)1–3.9 mm, sometimes sterile. **Lower glumes** 0.1–0.6 mm, membranous, minutely pubescent; **upper glumes** 1.8–4.3 mm, minutely pubescent, 5-veined, rarely with 1–2 additional veins adjacent to the midvein; **glume projections** (4)6–14, in 5 rows, (0.2)0.3–1 mm, uncinate; **lemmas** (1.5)1.8–3.1 mm, sparsely pubescent on the back, midveins occasionally excurrent to 0.6 mm; **paleas** (1.3)1.5–2.4 mm; **anthers** 3, 0.4–0.6 mm, yellow, occasionally purple- or green-tinged. **Caryopses** (0.9)1.2–2 mm long, 0.4–0.8 mm wide. $2n$ = 20.

Tragus berteronianus is native to Africa and Asia, and is now established in Arizona, New Mexico, and Texas. It was collected in Maine, Massachusetts, New York, and Virginia in the nineteenth century, and Virginia in 1959.

2. **Tragus australianus** S.T. Blake [p. 282]
AUSTRALIAN BURGRASS

Plants annual. **Culms** 10–45 cm. **Ligules** 0.5–1 mm; **blades** (0.7)3.5–7 cm long, (1.5)2–4 mm wide, surfaces glabrous. **Panicles** (4.5)6–13.5 cm long, 7–9 mm wide; **rachises** pubescent; **branches** 0.7–1.2 mm, pubescent, with 2(3) spikelets, axes rarely extending past the distal spikelets; **proximal internodes** 0.6–1 mm, usually 2–3 (or more) times longer than the second internodes. **Proximal spikelets** 3.1–3.5 mm; **second spikelets** 2.7–3.3 mm. **Lower glumes** absent or to 0.4 mm, glabrous; **upper glumes** 3.1–3.5 mm, minutely pubescent, 5-veined; **glume projections** 7–10, in 5 rows, 0.2–0.8 mm, uncinate; **lemmas** 2.4–2.6 mm, sparsely pubescent on the back, midveins sometimes excurrent to 0.2 mm; **paleas** 2–2.2 mm; **anthers** 3, 0.4–0.5 mm, yellow. **Caryopses** 1.2–1.5 mm long, 0.6 mm wide. $2n$ = unknown.

Tragus australianus is native to Australia, where it becomes established rapidly on disturbed or bare soil after summer rains. In the Western Hemisphere, it is known from Berkeley and Florence counties, South Carolina, and Argentina.

3. **Tragus heptaneuron** Clayton [p. 282]
SEVEN-VEINED BURGRASS

Plants annual. **Culms** 15–41 cm. **Ligules** 0.5–1 mm; **blades** 2.5–7 cm long, 3–5 mm wide, glabrous. **Panicles** 4–9.5 cm long, 7–9 mm wide; **rachises** pubescent; **branches** 0.5–1 mm, pubescent, with 2(3) spikelets, axes not extending past the distal spikelets; **proximal internodes** 0.5–0.8 mm, longer than the second internodes. **Spikelets** bisexual; **proximal spikelets** 3–3.5 mm; **second spikelets** 2.9–3.3 mm. **Lower glumes** absent or to 0.4 mm, glabrous; **upper glumes** 3–3.5 mm, minutely pubescent, 7-veined; **glume projections** 6–9, in 7 rows, 0.1–0.6(1) mm, uncinate; **lemmas** (2)2.3–2.7 mm, sparsely pubescent on the back; **paleas** 2.1–2.3 mm; **anthers** 3, 0.3–0.6 mm, yellow. **Caryopses** (1.2)1.4–1.6 mm long, 0.6 mm wide. $2n$ = unknown.

Tragus heptaneuron is native to tropical Africa, but it has been collected in Florence County, South Carolina.

4. **Tragus racemosus** (L.) All. [p. 282]
STALKED BURGRASS

Plants annual. **Culms** 5–40 cm. **Ligules** 0.5–1.3 mm; **blades** (0.5)1–5.5 cm long, 1.5–4 mm wide, glabrous. **Panicles** (1.5)2–11 cm long, 7–13 mm wide; **rachises** pubescent; **branches** 2.1–4.8 mm, pubescent, with (2)3–5 spikelets, axes extending past the distal spikelets; **proximal internodes** 0.5–1.8 mm, longer than second internodes. **Proximal spikelets** 3.8–6.6 mm; **second spikelets** (2.3)2.9–6.6 mm; **third** and **fourth spikelets** 0.8–4.2 mm; **distal spikelets** sterile. **Lower glumes** 0.7–1.1 mm, glabrous or minutely pubescent; **upper glumes** 3.8–6.6 mm, 7-veined, minutely pubescent; **glume projections** 6–11, in (5)6–7 rows, (0.2)0.6–1.3 mm, usually uncinate; **lemmas** 3.2–4 mm, sparsely pubescent on the back, midveins occasionally excurrent to 4 mm; **paleas** 2.3–3.1 mm; **anthers** 3, 0.6–0.8 mm, yellow. **Caryopses** 1.7–2.3 mm long, 0.6–0.8 mm wide. $2n$ = 40.

Tragus racemosus is native from the Mediterranean region to southwest Asia, but now grows in the United States, primarily in Cochise and Pima Counties, Arizona. It was collected in Maine, Pennsylvania, New York, and North Carolina in the late nineteenth century, but does not appear to be established in these states. Reports for New Mexico and Texas appear to reflect confusion with *T. berteronianus*.

17.53 ZOYSIA Willd.

Sharon J. Anderson

Plants perennial, rhizomatous, mat-forming. **Culms** 5–40 cm tall. **Ligules** to 0.3 mm, of hairs, often with longer hairs at the base of each blade immediately behind the ligule; **blades** usually glabrous abaxially, sometimes with a ciliate callus at the base of each collar, adaxial surfaces glabrous, scabrous, or sparsely pilose, apices often sharply pointed. **Inflorescences** terminal, exceeding the leaves, solitary spikelike racemes (a single spikelet in *Z. minima*), spikelets solitary, shortly pedicellate, laterally appressed to the rachises; **disarticulation** beneath the glumes or not occurring. **Spikelets** laterally compressed, with 1 floret; **florets** bisexual. **Lower glumes** usually absent; **upper glumes** enclosing the floret, chartaceous to coriaceous, awned, awns to 2.5 mm; **lemmas** thin, lanceolate or linear, acute to emarginate, 1-veined; **paleas** thin, rarely present. x = 10. Named for Carl von Zois (1756–1800), a German botanist.

Zoysia has 11 species. They are native to coastal sands between 42°N and 42°S and from Mauritius to Polynesia. Some species also grow in disturbed inland areas. Three species are used as lawn grasses in mesic tropical and subtropical areas, including parts of the *Flora* region. Because lawn grasses are rarely collected, the distribution maps for *Zoysia* show the states in which each species can be successfully grown rather than individual county records.

As is common with commercially important species, several cultivars of *Zoysia* have been

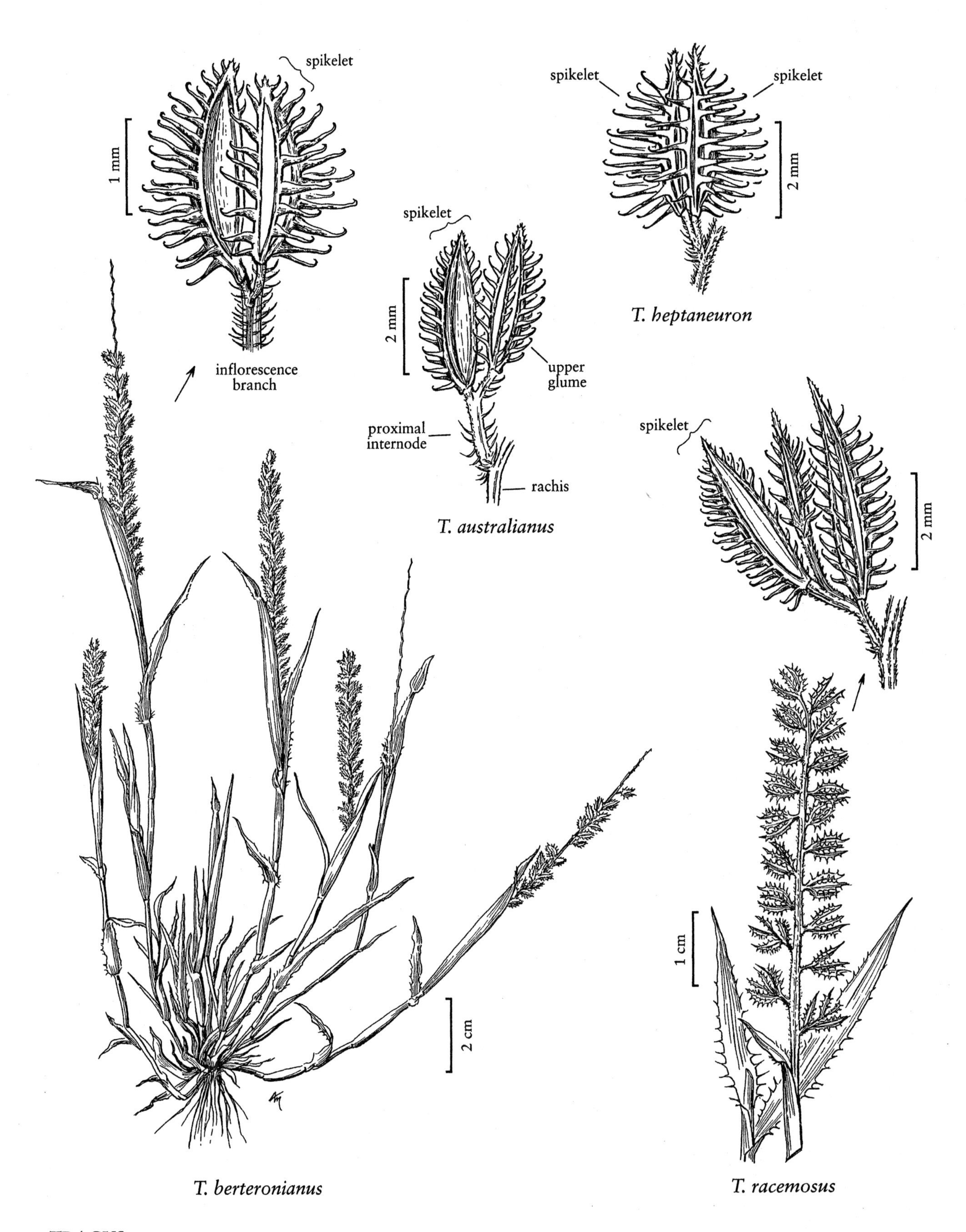
spikelet
1 mm
inflorescence branch
spikelet
2 mm
upper glume
proximal internode
rachis
T. australianus
spikelet
spikelet
2 mm
T. heptaneuron
spikelet
2 mm
1 cm
2 cm
T. berteronianus
T. racemosus

TRAGUS

developed, sometimes from hybrids. Whether or not they are hybrids, cultivars often exceed the normal range of variation for a species in one or more respects. This makes it difficult to account for them in a key. The following key is written for specimens that fall within the normal range of the species concerned. In employing it, caution must be used when interpreting blade widths because blades of *Zoysia* become involute or convolute under drought or salinity stress. In addition, hybridization has resulted in cultivars with vegetative characteristics more like those of one species and reproductive characteristics more like those of another species.

1. Blades to 0.5 mm in diameter; racemes with 3–12 spikelets; peduncles included or extending to 1 cm beyond the sheaths of the flag leaves . . . . 1. *Z. pacifica*
1. Blades 0.5–5 mm wide; racemes with 10–50 spikelets; peduncles extending (0.3)1–6.5 cm beyond the sheaths of the flag leaves.
   2. Pedicels 1.6–3.5 mm long; spikelets ovate, 1–1.4 mm wide; culm internodes 2–10 mm long; blades ascending . . . . 2. *Z. japonica*
   2. Pedicels 0.6–1.6 mm long; spikelets lanceolate, 0.6–1 mm wide; culm internodes 5–40 mm long, all plants with at least some internodes more than 14 mm long; blades patent . . . . 3. *Z. matrella*

1. Zoysia pacifica (Goudswaard) M. Hotta & Kuroki [p. 284]

Mascarenegrass, Korean Velvetgrass

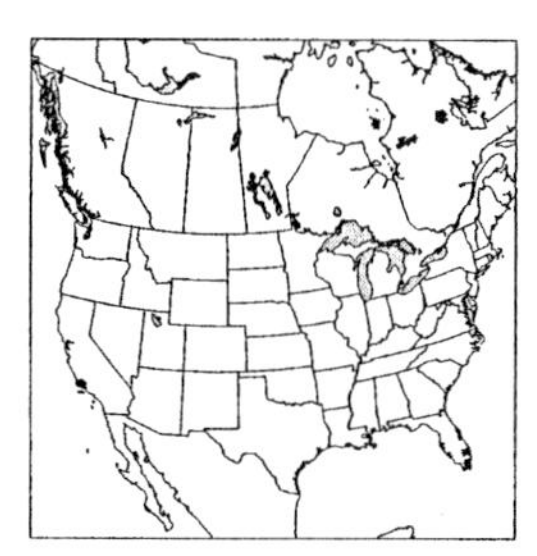

Plants rhizomatous. **Sheaths** glabrous, overlapping; **ligules** 0.07–0.25 mm; **blades** to 3 cm long, to 0.5 mm in diameter, patent, involute to strongly convolute (the margins overlapping), mostly glabrous abaxially, abaxial calluses 0.2–0.6 mm, hairs about 1 mm. **Peduncles** included or extending to 1 cm beyond the sheaths of the flag leaves. **Racemes** 0.4–2 cm, with 3–12 spikelets. **Spikelets** 2.2–2.9 mm long, 0.5–0.8 mm wide, lanceolate to linear, unawned or awned, awns to 0.5 mm. $2n = 40$.

Zoysia pacifica is less cold-tolerant than either *Z. matrella* or *Z. japonica*. It is not a common lawn grass in the *Flora* region, not even in the southern United States. The cultivar 'Cashmere' has many of the characteristics of *Z. pacifica*; it is probably derived from a hybrid between *Z. matrella* and *Z. pacifica*.

2. Zoysia japonica Steud. [p. 284]

Japanese Lawngrass, Korean Lawngrass

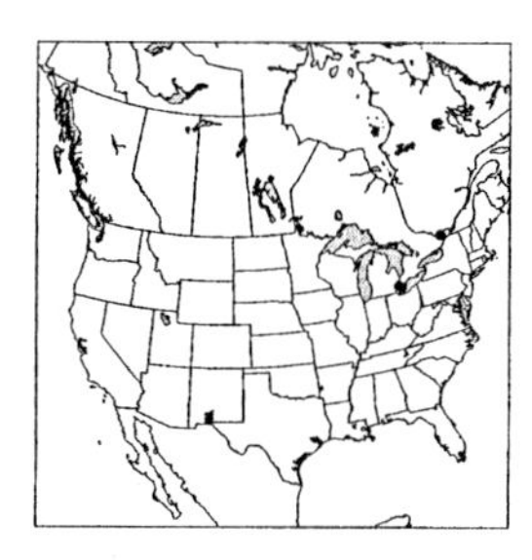

Plants rhizomatous. **Culm internodes** 2–10 mm. **Sheaths** glabrous; **ligules** 0.05–0.25 mm; **blades** to 6.5 cm long, 0.5–5 mm wide, ascending, flat to loosely involute when fully hydrated, involute when stressed, surfaces glabrous or pilose. **Peduncles** exerted, extending 0.8–6.5 cm beyond the sheaths of the flag leaves. **Racemes** 2.5–4.5 cm, with 25–50 spikelets; **pedicels** 1.6–3.5 mm. **Spikelets** 2.5–3.4 mm long, 1–1.4 mm wide, ovate, awned, awns 0.1–1.1 mm. $2n = 40$.

Zoysia japonica was the first species of *Zoysia* introduced into cultivation in the United States, with the introduction of the cultivar 'Meyer' in the 1950s. It is the most cold-tolerant and coarsely textured of the three species that have been introduced to the *Flora* region, and is the only species that is currently available as seed in the United States. The other two species treated here can be established from seed, but there are currently no commercial sources of either one in the United States.

3. Zoysia matrella (L.) Merr. [p. 284]

Manilagrass

Plants rhizomatous. **Culm internodes** 5–40 mm, at least some on each plant longer than 14 mm. **Sheaths** glabrous; **ligules** 0.07–0.25 mm; **blades** to 7 cm long, 0.5–2.5 mm wide, patent, weakly involute when fully hydrated, to strongly involute when stressed, surfaces glabrous or sparsely pilose. **Peduncles** extending (0.3)1–6 cm beyond the sheaths of the flag leaves. **Racemes** 1–4 cm, with 10–40 spikelets; **pedicels** 0.6–1.6 mm. **Spikelets** 2.1–3.2 mm long, 0.6–1 mm wide, lanceolate, unawned or awned, awns to 1 mm. $2n = 20, 40$.

Many of the *Zoysia* lawn grasses grown in the southern and eastern United States are derived from hybrids between *Z. matrella* and *Z. pacifica* or *Z. japonica*, and have retained many of the characteristics of *Z. matrella*. They are used as lawn grasses from Connecticut southwards, and have occasionally been found as escapes in that region.

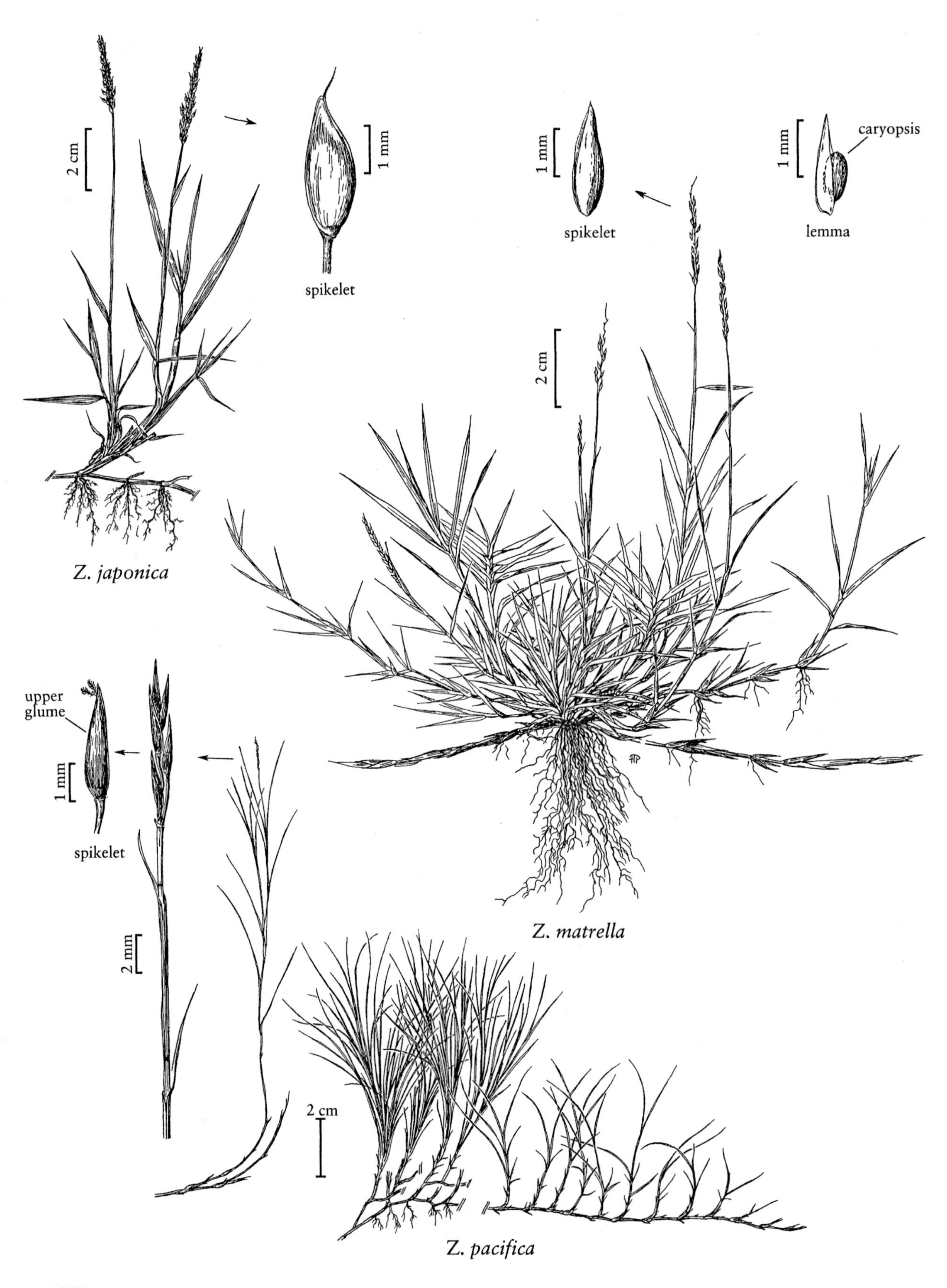
2 cm
1 mm
spikelet
1 mm
spikelet
1 mm
caryopsis
lemma
Z. japonica
2 cm
upper
glume
1 mm
spikelet
2 mm
Z. matrella
2 cm
Z. pacifica

ZOYSIA

18. PAPPOPHOREAE Kunth

John R. Reeder

Plants perennial; rarely annual. **Culms** herbaceous. **Ligules** of hairs; **blade epidermes** with bicellular microhairs. **Inflorescences** terminal, dense, narrow to somewhat open panicles; **disarticulation** above the glumes but not between the florets (except in *Cottea*). **Spikelets** with 3–10 florets, sometimes only the lowest floret bisexual. **Lemmas** 5–13-veined, veins extending into awns, often with intermixed hyaline lobes. **Caryopses** with punctate hila; **embryos** ½ or more as long as the caryopses. $x = 10$.

The tribe *Pappophoreae* includes five genera and approximately 40 species. It is represented in tropical and warm regions around the world.

SELECTED REFERENCES **Reeder, J.R.** 1965. The tribe Orcuttieae and the subtribes of the Pappophoreae (Gramineae). Madroño 18:18–28; **Reeder, J.R. and D.N. Singh.** 1968. Chromosome numbers in the tribe Pappophoreae. Madroño 19:183–187.

1. Spikelets disarticulating above the glumes and between the florets; florets 6–10, the lemmas with both awns and awned teeth, these not forming a pappuslike crown .... 18.03 *Cottea*
1. Spikelets disarticulating above the glumes but not between the florets, these falling as a unit; florets 3–6, the lemmas awned but without awned teeth, the awns forming a pappuslike crown.
   2. Lower glumes 1-veined; lemma awns scabridulous, not plumose .... 18.01 *Pappophorum*
   2. Lower glumes 5–7-veined; lemma awns plumose, not scabridulous .... 18.02 *Enneapogon*

18.01 PAPPOPHORUM Schreb.

John R. Reeder

Plants perennial; cespitose, essentially glabrous throughout. **Culms** 30–130 cm. **Sheaths** open; **ligules** of hairs; **microhairs of blades** with an inflated terminal cell similar in length to the basal cell. **Inflorescences** terminal, narrow, spikelike or somewhat open panicles; **disarticulation** above the glumes but not between the florets, these falling together. **Spikelets** with 3–5 florets, only the lower 1–3 bisexual. **Glumes** subequal, thin, membranous, 1-veined, acute; **lemmas** rounded on the back, hairy at least basally, obscurely (5)7(9)-veined, veins extending into scabridulous awns of unequal length, several additional narrow awnlike lobes usually also present, awns and lobes together forming a pappuslike crown; **paleas** textured like the lemmas, 2-veined, 2-keeled, keels scabrous or hairy; **anthers** 3; **styles** 2. **Caryopses** elliptical, plump, slightly dorsally flattened or nearly terete; **embryos** about ½ as long as the caryopses. $x = 10$. From the Greek *pappos*, 'pappus', and *phoros*, 'bearing', in reference to the pappuslike crown of the lemma.

Pappophorum is an American genus with about eight species. It grows in warm regions of North and South America. Two species are native to the *Flora* region.

SELECTED REFERENCES **Pensiero, J.F.** 1986. Revisión de las especies argentinas del género *Pappophorum* (Gramineae–Eragrostoideae–Pappophoreae). Darwiniana 17:65–87; **Reeder, J.R. and L.J. Toolin.** 1989. Notes on *Pappophorum* (Gramineae: Pappophoreae). Syst. Bot. 14:349–358.

1. Panicles purple-tinged, narrow, but usually with some slightly spreading branches; lemma bodies 3–4 mm, the awns mostly not more than 1.5 times as long, these rarely spreading at right angles .... 1. *P. bicolor*
1. Panicles white or tawny, rarely slightly purplish, tightly contracted; lemma bodies 3–3.2 mm, the awns about twice as long, commonly spreading at right angles when mature .... 2. *P. vaginatum*

1. **Pappophorum bicolor** E. Fourn. [p. 288]
PINK PAPPUSGRASS

Culms 30–80(100) cm. **Sheaths** mostly glabrous, apices with a tuft of hairs on either side; **ligules** about 1 mm; **blades** 10–20(30) cm long, 2–5 mm wide, flat to involute. **Panicles** 12–20 cm, narrow but usually with some slightly spreading branches, pink- or purple-tinged. **Spikelets** with the lower 2 or 3 florets bisexual, distal 1–2 florets sterile. **Glumes** 3–4 mm, thin, glabrous, apices acute or minutely notched and mucronate; **lemmas** somewhat firm, usually faintly 7-veined, with 11–15 awns; **lowest lemma bodies** 3–4 mm, midveins and margins pubescent from the base to about midlength, awns about 1.5 times as long as the lemma bodies; **paleas** subequal to the lemma bodies or slightly longer. **Caryopses** about 2 mm. $2n = 100$.

Pappophorum bicolor grows in open valleys, road right of ways, and grassy plains in Texas and northern Mexico. A report for Arizona (*Mearns 1175*) was apparently due to mixed labels. There have been no subsequent reports of the species from the state.

2. **Pappophorum vaginatum** Buckley [p. 288]
WHIPLASH PAPPUSGRASS

Culms (40)50–100 cm. **Sheaths** mostly glabrous, with a tuft of hairs at the throat; **blades** 10–25 cm long, 2–5 mm wide, flat to involute, adaxial surfaces scabridulous. **Panicles** 10–25 cm, tightly contracted, usually white or tawny, rarely slightly purple-tinged. **Spikelets** with 1(2) bisexual florets and 2 reduced florets. **Glumes** (3)4–4.5 mm, glabrous, acute; **lowest lemma bodies** 3–3.2 mm, midveins and margins pubescent from the base to about midlength, awns about twice as long as the lemma bodies, tending to spread at right angles when mature; **paleas** longer than the lemma bodies. $2n = 60$.

Pappophorum vaginatum grows in similar habitats to *P. bicolor*, the two species sometimes growing together. Its range extends from southern Arizona to Texas and northern Mexico and from Uruguay to Argentina.

18.02 ENNEAPOGON Desv. *ex* P. Beauv.

John R. Reeder

Plants perennial or annual; cespitose, more or less hairy throughout. **Culms** 3–100 cm; **nodes** hairy; **internodes** hollow. **Sheaths** open; **ligules** of hairs; **microhairs of blades** each with an elongated basal cell and an inflated terminal cell. **Inflorescences** terminal, spikelike to somewhat open panicles, bases often included within the uppermost leaf sheath; **disarticulation** above the glumes but not between the florets, florets falling as a unit. **Spikelets** with 3–6 florets, frequently only the lowest floret bisexual, distal florets progressively reduced. **Glumes** subequal, as long as or slightly shorter than the florets (including the awns), more or less pubescent; **lower glumes** 5–7-veined; **lemmas** firm, rounded on the backs, villous below the middle, strongly 9-veined, veins extending into equal, plumose awns 3–5 times as long as the lemma bodies and forming a pappuslike crown; **paleas** longer than the lemmas, entire, thinly membranous, 2-veined, 2-keeled, keels hairy; **anthers** 3, 0.2–1.5 mm; **styles** 2, free to the base, white. $x = 10$. Name from the Greek *ennea*, 'nine', and *pogon*, 'beard', a reference to the nine hairy awns of the lemma.

Enneapogon includes about 28 species. It is found in tropical and warm regions of the world, especially in Africa and Australia. Two species are found in the *Flora* region: one is native, and one is an introduction that has become established in the region.

1. Culms 50–100 cm tall, about 2 mm thick; panicles loosely contracted to somewhat open, up to 3 cm wide at maturity; plants annual . . . . 1. *E. cenchroides*
1. Culms 20–45 cm tall, up to 1 mm thick; panicles usually tightly cylindrical, rarely more than 1 cm wide; plants perennial . . . . 2. *E. desvauxii*

1. **Enneapogon cenchroides** (Licht.) C.E. Hubb. [p. 288]
SOFTFEATHER PAPPUSGRASS

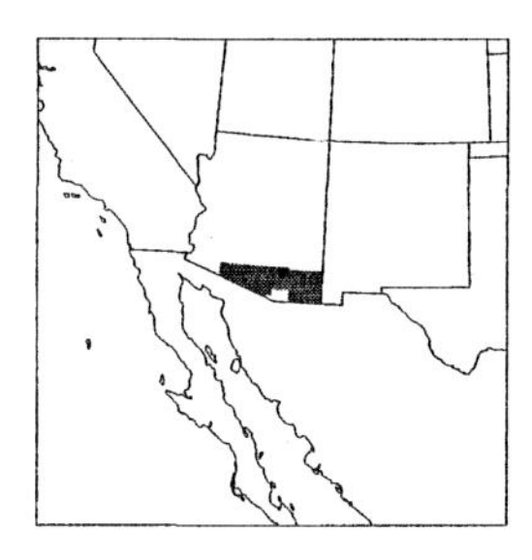

Plants annual. **Culms** 50–80(100) cm, about 2 mm thick, usually rather robust, erect or somewhat geniculate at the base, often branching; **nodes** pubescent. **Sheaths** mostly shorter than the internodes, somewhat loose; **blades** 6–12(20) cm long, 1–7(10) mm wide, flat, becoming involute, apices attenuate. **Panicles** 10–20(30) cm long, to 3 cm wide at the base, loosely contracted to somewhat open. **Spikelets** 3.2–6.8 mm. **Lower glumes** 2.8–5.1 mm; **upper glumes** 3.2–6.8 mm, 3-veined; **lowest lemmas** 1.5–2 mm; **awns** 2.5–5 mm, usually exceeding the glumes; **anthers** 1–1.8 mm. $2n$ = 40.

Enneapogon cenchroides has been introduced and is persisting in the Ajo, Santa Catalina, Tucson, and Galiuro mountains of southern Arizona. Outside the Americas, its range extends from Sudan southward to the Cape Provinces of South Africa, through Arabia to India, and on Ascension Island.

2. **Enneapogon desvauxii** P. Beauv. [p. 288]
NINEAWN PAPPUSGRASS

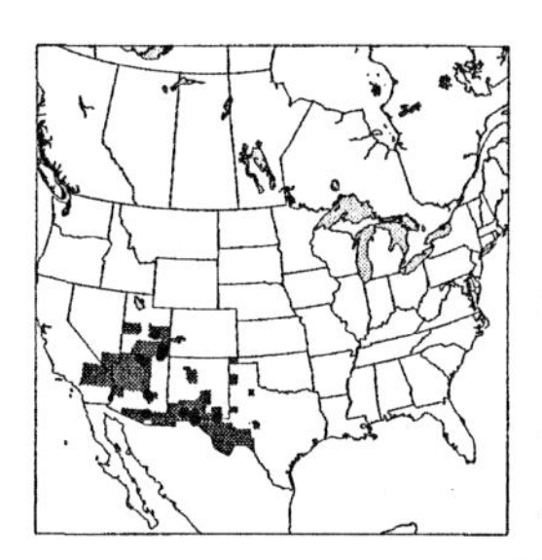

Plants perennial. **Culms** 20–45 cm, about 1 mm thick, ascending to erect from a hard knotty base, often branching; **nodes** pubescent. **Sheaths** usually shorter than the internodes, more or less pubescent; **ligules** about 0.5 mm; **blades** mostly 2–12 cm long, 1–2 mm wide, more or less hairy, soon involute. **Panicles** 2–10 cm, spikelike, grayish-green or lead-colored. **Spikelets** mostly 5–7 mm, usually only the lowest floret bisexual. **Glumes** 3–5 mm, subequal, thin, puberulent; **upper glumes** often 3- or 4-veined; **lowest lemmas** 1.5–2 mm, firm, rounded on the back; **awns** 3–4 mm; **anthers** 0.3–0.5 mm. **Caryopses** 1–1.2 mm, oval, plump; **embryos** subequal to the caryopses. **Cleistogamous spikelets** commonly present in the lower sheaths, their lemmas larger than those of the florets in the aerial panicles, unawned or with awns that are much reduced. $2n$ = 20.

Enneapogon desvauxii grows in open areas of the southwestern United States and in much of Mexico. It also grows in Peru, Bolivia, Argentina, and most of Africa, from which it extends eastward through Arabia and India to China.

18.03 COTTEA Kunth

John R. Reeder

Plants perennial; cespitose, softly pilose throughout. **Culms** 25–70 cm, ascending or erect from hard knotty bases. **Sheaths** pilose, rounded on the backs; **blades** flat, linear; **microhairs of blades** each with an elongated basal cell and an inflated terminal cell. **Inflorescences** terminal, open, rather narrow panicles; **disarticulation** above the glumes and between the florets. **Spikelets** with 6–10 florets, distal florets reduced. **Glumes** subequal, about as long as the lowest lemmas, pilose, 7–13-veined, midveins sometimes prolonged as short awns, apices acuminate or 3-toothed; **lemmas** rounded on the backs, with 9–13 prominent veins, some of these extending into antrorsely barbed awns of various lengths, others into awned teeth, awns and teeth not forming a pappuslike crown; **paleas** slightly longer than the lemmas, 2-veined, 2-keeled, keels hairy; **anthers** 3. **Cleistogamous spikelets**, usually consisting of a single floret, produced in the lower sheaths. x = 10. Named for J.G. Cotta von Cottendorf (1796–1863), a German science patron.

Cottea is a monotypic American genus.

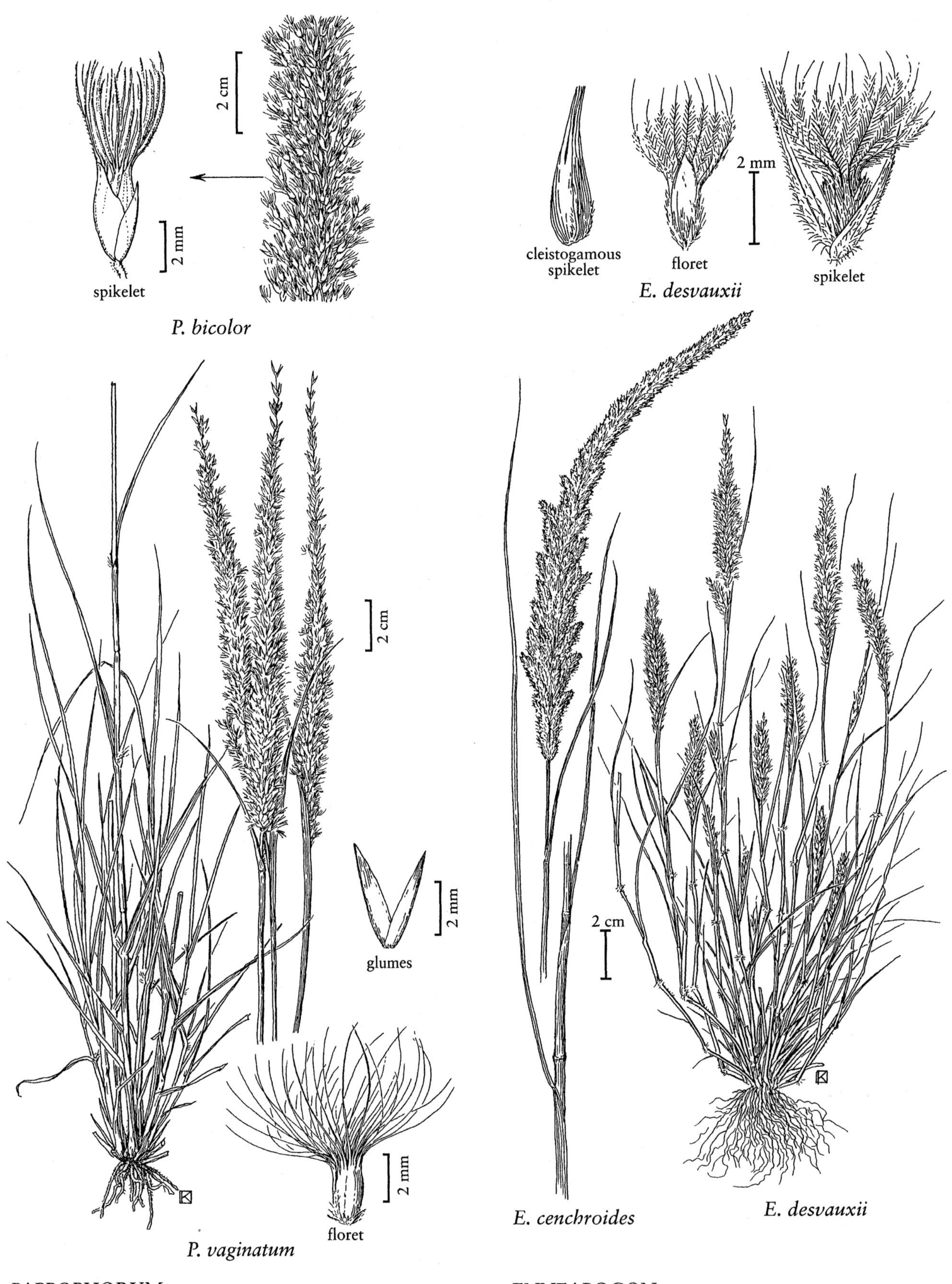

PAPPOPHORUM

ENNEAPOGON

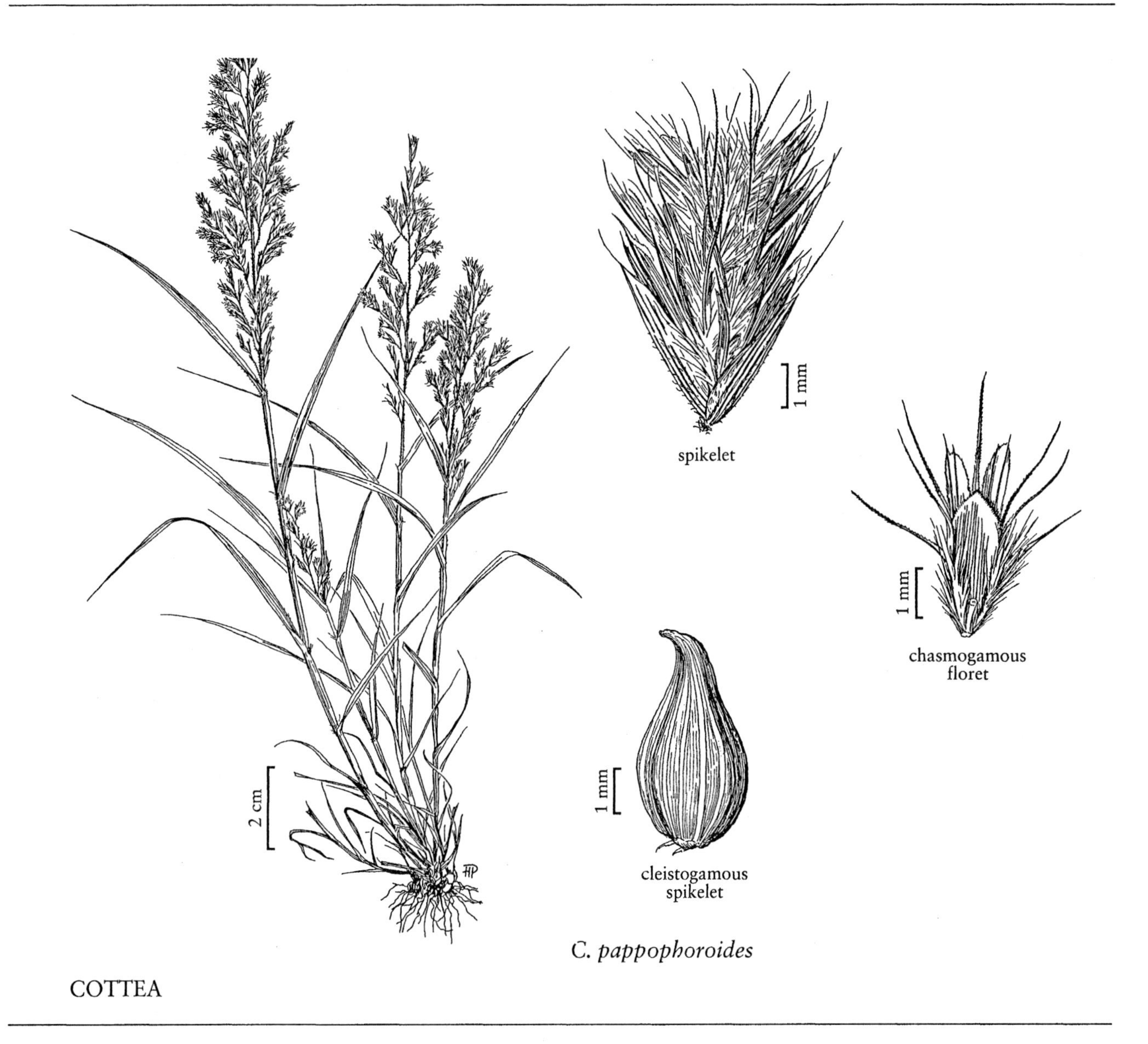

C. pappophoroides

COTTEA

1. **Cottea pappophoroides** Kunth [p. 289]
COTTA GRASS

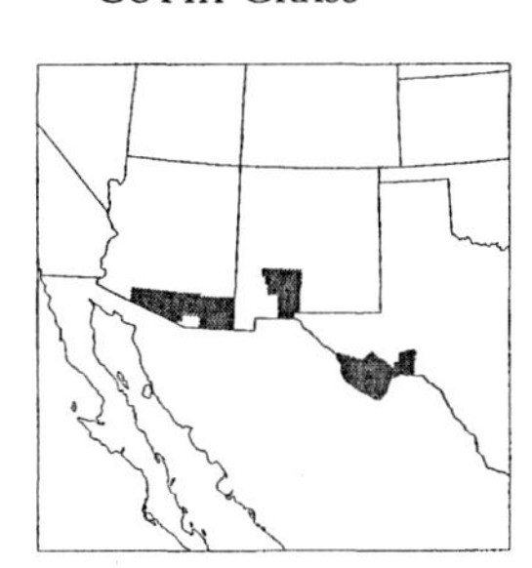

Culms 25–70 cm. **Blades** 5–15 cm long, 4–7 mm wide. **Panicles** 8–15 cm long, 2–6 cm wide, green or purplish; **branches** loosely ascending. **Spikelets** 5–10 mm (including the awns). **Glumes** 4–5 mm; **lemmas** 3–4 mm, conspicuously long-pilose basally, awns and teeth more or less alternating. **Caryopses** about 1.5 mm, plump, elliptical; **embryos** about ½ as long as the caryopses. $2n = 20$.

Cottea pappophoroides grows on open hillsides from Arizona and Texas south to central Mexico, and from Ecuador to Argentina.

19. ORCUTTIEAE Reeder

John R. Reeder

Plants annual; viscid, aromatic. **Culms** with solid, pithy interiors, often with 5 or more nodes. **Leaves** with little or no distinction between sheath and blade; **ligules** absent; **blades** of upper cauline leaves similar in length to those below; **siliceous cells of the leaf epidermis** absent or irregular to dumbbell-shaped; **microhairs of the blades** bicellular, small, sunken, "mushroom-button" shaped. **Inflorescences** terminal, narrow, cylindrical to clavate or capitate, usually dense panicles or spikes, spikelets sessile or shortly pedicellate; **rachillas** tardily disarticulating between the florets. **Spikelets** with 4–25(40) florets. **Glumes** entire, denticulate, toothed, or absent; **lemmas** with 5–17 conspicuous veins; **paleas** subequal to or slightly shorter than the lemmas, well-developed, keels glabrous; **lodicules** absent or 2, obscure, rounded, truncate to slightly emarginate; **anthers** 3. **Caryopses** laterally compressed; **hila** large, punctate, basal; **embryos** ¾ or more as long as the caryopses. $x = 10$.

The tribe *Orcuttieae* includes only three genera and nine species, all of which are restricted to vernal pools and similar habitats in California and Baja California, Mexico.

SELECTED REFERENCES Crampton, B. 1959. The grass genera *Orcuttia* and *Neostapfia*: A study in habitat and morphological specialization. Madroño 15:97–110; Reeder, J.R. 1965. The tribe Orcuttieae and the subtribes of the Pappophoreae (Gramineae). Madroño 18:18–28; Reeder, J.R. 1982. Systematics of the tribe Orcuttieae (Gramineae) and the description of a new segregate genus, *Tuctoria*. Amer. J. Bot. 69:1082–1095.

1. Lemmas deeply cleft into 5 mucronate or awn-tipped teeth, the teeth ⅓ as long as to equaling the lemma bodies; spikelets distichously arranged . . . . 19.01 *Orcuttia*
1. Lemmas entire or denticulate, often with a central mucro; spikelets spirally arranged.
   2. Inflorescences clavate, often partially enclosed at maturity; spikelets laterally compressed, with glumes; lemmas rectangular, not translucent between the veins, the apices mucronate, otherwise entire or denticulate; caryopses without a viscid exudate, the embryo visible through the pericarp . . . . 19.02 *Tuctoria*
   2. Inflorescences cylindrical, usually completely exposed at maturity; spikelets dorsally compressed, without glumes; lemmas flabellate, translucent between the veins, the apices ciliolate; caryopses covered with a viscid exudate, the embryos obscured by the pericarp . . . 19.03 *Neostapfia*

19.01 ORCUTTIA Vasey

John R. Reeder

Plants annual; viscid-aromatic, pilose, sometimes sparsely so, producing long, juvenile, floating basal leaves. **Culms** 3–35 cm, erect, ascending, or decumbent, sometimes becoming prostrate, not breaking apart at the nodes, usually branching only at the lower nodes. **Leaves** without ligules, sometimes with a "collar" line visible at the junction of the sheath and blade, especially when dry; **blades** flat or becoming involute in drying. **Inflorescences** terminal, clavate to capitate spikes, exserted at maturity, spikelets distichously arranged; **disarticulation** tardy, above the glumes and between the florets. **Spikelets** laterally compressed, with 4–40 florets. **Glumes** irregularly 2–5-toothed; **lemmas** deeply cleft and strongly 5-veined, veins terminating in prominent mucronate or awn-tipped teeth ⅓–½ or more as long as the lemma bodies, each tooth with an additional weaker vein on either side of a strong central vein, these extending about halfway to the base of the lemma; **paleas** well-developed, 2-veined; **lodicules** absent; **anthers** 3, white or pinkish, exserted on long, slender, ribbonlike filaments at anthesis; **styles** 2, apical, elongate, filiform, stigmatic for ⅓–½ of their length; **stigmatic hairs** short, often sparse. **Caryopses** slightly compressed laterally, oblong to elliptic; **embryos** ¾ as long as to equaling the caryopses; **epiblast**

absent. $x = 10$, probably. Named for Charles Russell Orcutt (1864–1929), a California botanist.

Orcuttia is a genus of five species, all of which are restricted to vernal pools and similar habitats in California and northern Baja California, Mexico.

1. Lemma teeth unequal, the central tooth the longest.
   2. Lemmas 6–7 mm long, the teeth terminating in awns at least 1 mm long; caryopses 2.3–2.5 mm long . . . . 1. *O. viscida*
   2. Lemmas 4–5 mm long, the teeth sharp-pointed or with awns to 0.5 mm long; caryopses 1.3–1.8 mm long.
      3. Plants sparingly hairy; culms usually prostrate; spikes clavate . . . . 2. *O. californica*
      3. Plants conspicuously hairy, grayish; culms erect or decumbent; spikes somewhat capitate . . . . 3. *O. inaequalis*
1. Lemma teeth essentially equal in length.
   4. Culms usually prostrate; caryopses 1.5–1.8 mm long . . . . 2. *O. californica*
   4. Culms erect, ascending, or decumbent; caryopses 2–3 mm long.
      5. Culms 1–2 mm thick, branching only at the lower nodes; spikes congested, crowded towards the top; leaf blades 3–5 mm wide . . . . 4. *O. pilosa*
      5. Culms 0.5–1 mm thick, often branching from the upper nodes; spikes not congested, even towards the top; leaf blades 1.5–2 mm wide . . . . 5. *O. tenuis*

1. **Orcuttia viscida** (Hoover) Reeder [p. 293]
SACRAMENTO ORCUTTGRASS

Plants pilose, very viscid, strongly aromatic. **Culms** 3–10(15) cm, simple, erect, often spreading in age. **Leaves** usually without a "collar" line; **blades** 2–4 mm wide. **Spikes** 3–5 cm, somewhat congested (less so than those of *O. inaequalis*); **lower** and **upper internodes** 3–7 mm. **Spikelets** with 6–20(30) florets. **Glumes** subequal, 5–6 mm, unequally 3-toothed, teeth as long as the bodies, awn-tipped; **lemmas** 6–7 mm, teeth as long as the lemma bodies, central tooth evidently the longest, awns at least 1 mm; **paleas** at least ¾ as long as the lemmas; **anthers** about 2 mm. **Caryopses** 2.3–2.5 mm, broadly elliptical; **embryos** about as long as the caryopses. $2n = 28$.

Orcuttia viscida grows at elevations below 120 m in Sacramento County, California. Its awn-tipped lemma teeth curve outward at maturity, giving the spikes a distinctive, bristly appearance. It is listed as an endangered species by the U.S. Fish and Wildlife Service.

2. **Orcuttia californica** Vasey [p. 293]
CALIFORNIA ORCUTTGRASS

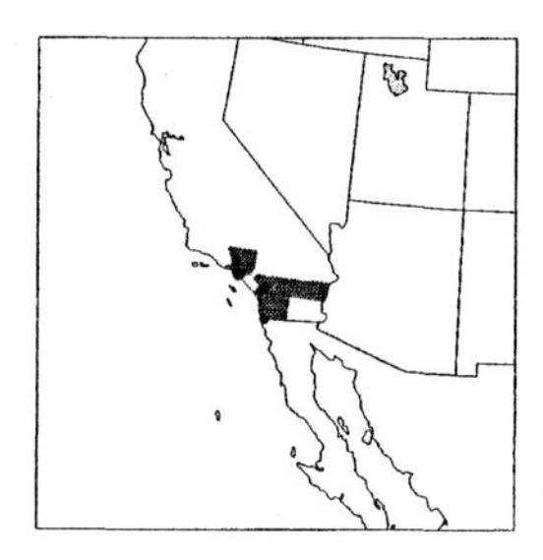

Plants sparsely hairy. **Culms** 5–15(20) cm, simple, sometimes geniculate, usually prostrate and forming mats. **Leaves** with a faint "collar" line usually evident when dry; **blades** 1–2 cm long, 2–3 mm wide. **Spikes** 3–6 cm; **lower internodes** 5–10 mm; **upper internodes** about 1 mm. **Spikelets** with 5–15(25) florets. **Glumes** subequal or the lower glumes a little shorter than the upper, 2–3(4) mm, irregularly toothed; **lemmas** about 5 mm, teeth equal or the central tooth a little longer than the others, awns to 0.5 mm; **paleas** about equal to the lemmas; **anthers** about 2 mm. **Caryopses** 1.5–1.8 mm, narrowly elliptical; **embryos** at least ¾ as long as the caryopses. $2n = 32$.

Orcuttia californica grows at elevations below 625 m in Los Angeles, Riverside, and San Diego counties, California and northern Baja California, Mexico. It is listed as an endangered species by the U.S. Fish and Wildlife Service.

3. **Orcuttia inaequalis** Hoover [p. 293]
SAN JOAQUIN ORCUTTGRASS

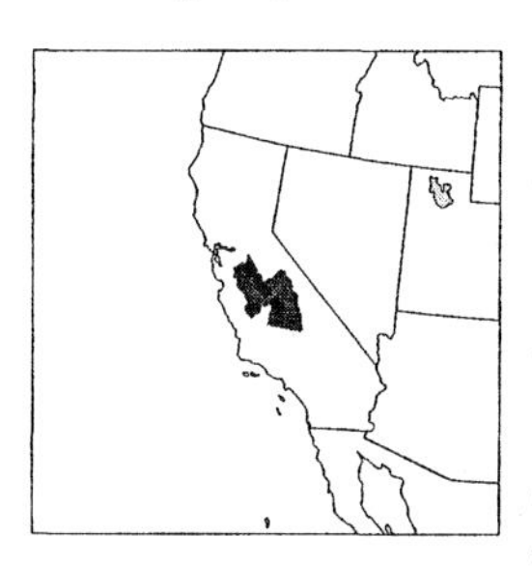

Plants cespitose, conspicuously hairy, grayish. **Culms** 5–15(25) cm, usually ascending to erect, occasionally spreading and forming mats. **Leaves** usually without a "collar" line; **blades** 1–4 cm long, 2–4 mm wide. **Spikes** 2–3.5(5) cm, more or less capitate, usually densely congested; **lower** and **upper internodes** 1–4 mm. **Spikelets** with 4–20(30) florets. **Glumes** subequal, about 3 mm, irregularly toothed; **lemmas** 4–5 mm, teeth about ½ as long as the lemma, central tooth conspicuously longer than the others, teeth sharp, if awn-tipped, awns less than 0.5 mm; **paleas** about equal to the lemmas; **anthers** about 2 mm. **Caryopses** 1.3–1.5 mm, broadly elliptical; **embryos** about as long as the caryopses. $2n = 24$.

Orcuttia inaequalis grows at elevations below 575 m in Fresno, Madera, Merced, Stanislaus, and Tulare counties, California. It is listed as a threatened species by the U.S. Fish and Wildlife Service.

4. **Orcuttia pilosa** Hoover [p. 293]
HAIRY ORCUTTGRASS

Plants cespitose, hairy, usually densely so. **Culms** 5–20(35) cm tall, 1–2 mm thick, simple or branching at the lower nodes, erect or decumbent, sometimes geniculate. **Leaves** with a faint "collar" line usually evident when dry; **blades** 4–6 cm long, 3–5(8) mm wide. **Spikes** to 10 cm; **lower internodes** 5–15 mm; **upper internodes** 1.5–3 mm. **Spikelets** with 10–40 florets. **Glumes** about 3 mm, irregularly 3-toothed; **lemmas** 4–5 mm, teeth about equal and ⅓–½ as long as the lemmas, acute or awn-tipped; **anthers** 2.5–3 mm. **Caryopses** about 2 mm, elliptical; **embryos** ¾ or more as long as the caryopses. $2n = 30$.

Orcuttia pilosa grows at elevations below 150 m in Madera, Merced, Stanislaus, and Tehama counties, California. It is listed as an endangered species by the U.S. Fish and Wildlife Service.

5. **Orcuttia tenuis** Hitchc. [p. 293]
SLENDER ORCUTTGRASS

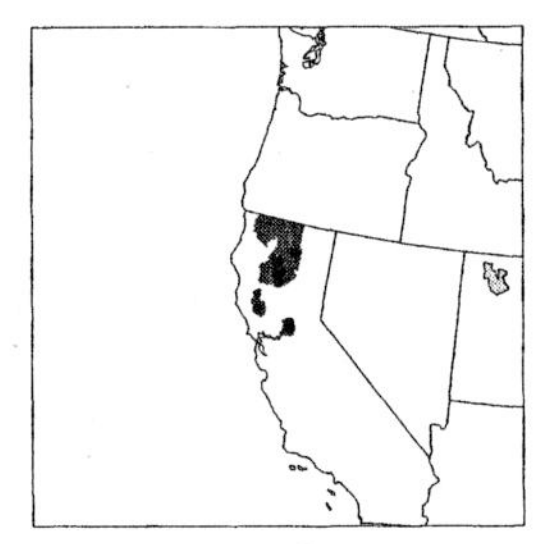

Plants sometimes weakly cespitose, but often with a single main culm branching 2–10 cm above the base, sparsely hairy. **Culms** 5–15(25) cm tall, 0.5–1 mm thick, often strictly erect, but sometimes decumbent when "top heavy" from profuse branching above. **Leaves** usually without a "collar" line; **blades** 1–3 cm long, 1.5–2 mm wide. **Spikes** 5–10 cm, more congested distally than basally; **lower internodes** 5–15 mm; **upper internodes** 2–7 mm. **Spikelets** with 5–20 florets. **Glumes** subequal or the lower glumes a little shorter than the upper, 3–6 mm, with 3–5 teeth to 1 mm; **lemmas** 4.5–6 mm, acute or awn-tipped teeth about equal and ½ as long as the lemma, spreading or slightly recurved; **paleas** slightly shorter than the lemmas; **anthers** about 3 mm. **Caryopses** about 3 mm, narrowly oblong; **embryos** nearly as long as the caryopses. $2n = 26$.

Orcuttia tenuis grows at 25–100 m in Shasta and Tehama counties of the Central Valley of California, with outlying populations in Sacramento County and the lower montane regions of Lake, Shasta, and Siskiyou counties, California. It is listed as a threatened species by the U.S. Fish and Wildlife Service.

19.02 TUCTORIA Reeder

John R. Reeder

Plants annual; viscid-aromatic, more or less hairy throughout, not producing juvenile floating leaves. **Culms** 5–15(30) cm, simple or branching at the upper nodes, erect or ascending, often rather fragile, readily breaking apart at the nodes. **Leaves** without ligules, with little or no distinction between sheath and blade; **blades** flat, becoming involute when dry. **Inflorescences** terminal, clavate spikes, partially included or exserted at maturity, spikelets spirally arranged; **disarticulation** tardy, above the glumes and between the florets. **Spikelets** laterally compressed, with 5–40 florets. **Glumes** irregularly short-toothed or entire; **lemmas** (3)4–7 mm, 11–17-veined, not translucent between the veins, entire or denticulate, usually with a central mucro; **paleas** subequal to or slightly shorter than the lemmas; **lodicules** 2, 0.1–0.5 mm, sometimes fused to the paleas; **anthers** 3, exserted on long, slender, ribbonlike filaments at anthesis; **styles** 2, apical, long, filiform, stigmatic for ⅓–½ of their length; **stigmatic hairs** short, often sparse. **Caryopses** laterally compressed, pyriform to oblong, pericarp not viscid; **embryos** visible through the pericarp, brown, from ¾ as long as to nearly equaling the caryopses; **epiblasts** present. $x = 10$. Name an anagram of *Orcuttia*.

Tuctoria has three species, all of which grow in vernal pools or similar habitats, two in the Central Valley of California and one *T. fragilis* (Swallen) Reeder, in Baja California Sur, Mexico. Both species found in the *Flora* region are endangered by loss of habitat to urbanization and agriculture.

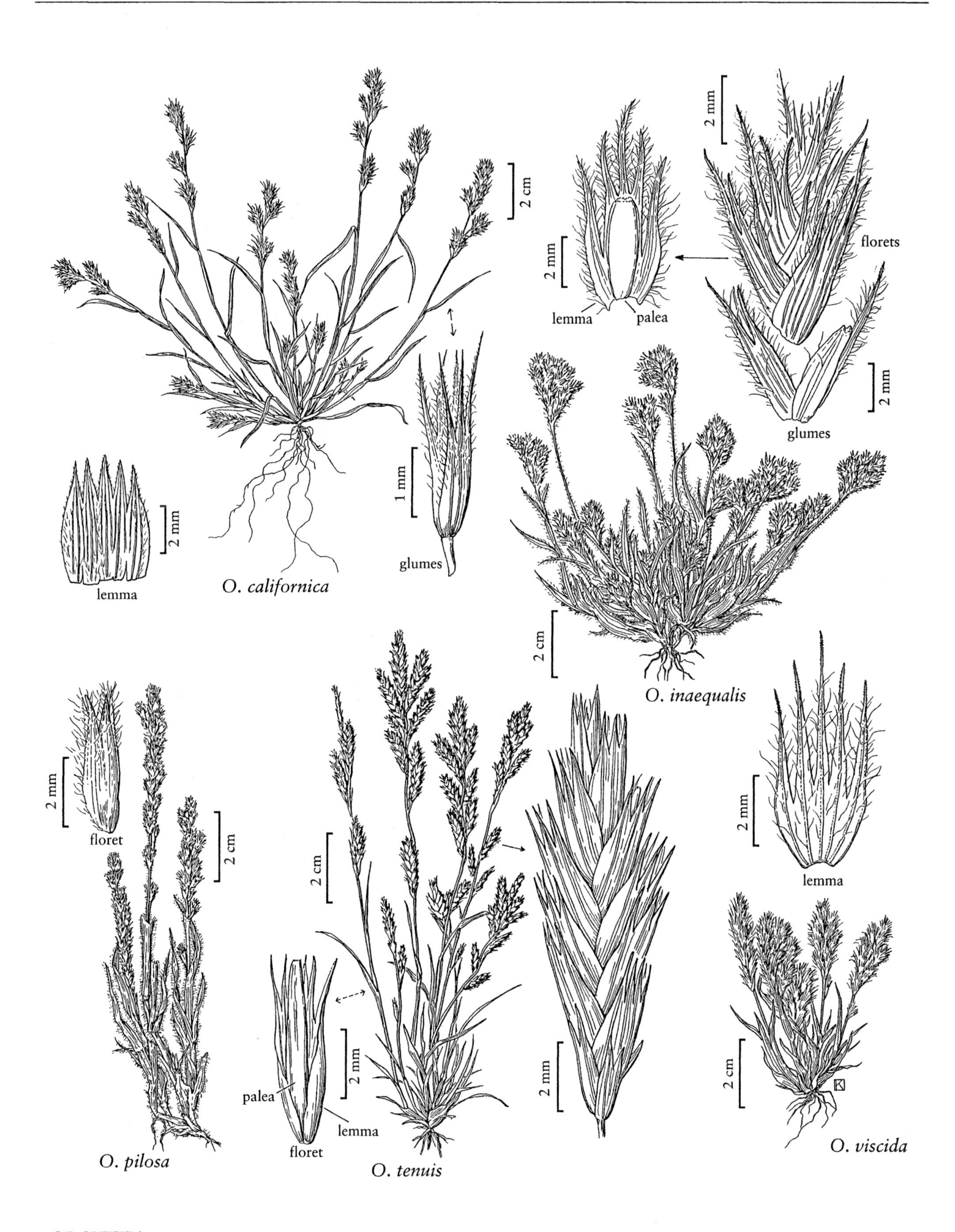

2 cm
2 mm
florets
2 mm
lemma
palea
glumes
2 mm
1 mm
glumes
2 mm
lemma
O. californica
2 cm
O. inaequalis
2 mm
floret
2 cm
2 cm
2 mm
lemma
2 mm
palea
lemma
floret
2 mm
2 cm
O. pilosa
O. tenuis
O. viscida

ORCUTTIA

1. Spikes exserted from the upper leaf sheaths at maturity; lemmas more or less truncate; caryopses about 2 mm long, minutely rugose .... 1. *T. greenei*
1. Spikes partially included in the upper leaf sheaths at maturity; lemmas tapering gradually to a mucronate apex; caryopses about 3 mm long, smooth .... 2. *T. mucronata*

1. **Tuctoria greenei** (Vasey) Reeder [p. 295]
AWNLESS SPIRALGRASS

Culms 5–15(30) cm, erect or decumbent, often geniculate; **nodes** often purplish. **Blades** 1.5–5 cm long, to 5 mm wide, curved outward. **Spikes** 2.5–5(8) cm, exserted at maturity, congested above; **lower internodes** 4–5 mm; **upper internodes** 1–2 mm. **Spikelets** with 5–15(40) florets. **Glumes** subequal, 3–5 mm, prominently many-veined, irregularly dentate at the apices; **lemmas** (3)4–5(6) mm, prominently 9–13-veined, apices truncate and mucronate; **paleas** slightly shorter than the lemmas; **anthers** 3–3.5 mm, whitish; **lodicules** about 0.1 mm, not fused to the paleas. **Caryopses** about 2 mm, slightly compressed laterally, oblong, minutely rugose. $2n = 24$.

Tuctoria greenei grows at elevations below 150 m in the Central Valley of California.

2. **Tuctoria mucronata** (Crampton) Reeder [p. 295]
PRICKLY SPIRALGRASS

Culms to 12 cm, ascending, often decumbent or geniculate at the base; **nodes** usually concealed by the leaves. **Blades** 2–4 cm, involute, usually curved outward, tapering to a fine point. **Spikes** 1.5–6 cm, more or less included in the upper sheaths at maturity, congested throughout; **internodes** 1–2 mm. **Spikelets** with 5–10 florets. **Glumes** subequal (or the lower slightly shorter than the upper), 4–7 mm, mucronate, sometimes with 1 or 2 short lateral teeth; **lemmas** 5–7 mm, prominently 11–15-veined, apices erose or with a few minute teeth, central vein extending into a prominent mucro as much as 1 mm long; **paleas** slightly shorter than the lemmas; **anthers** about 3 mm, yellow, becoming pinkish in drying; **lodicules** about 0.5 mm, fused to the paleas. **Caryopses** about 3 mm, laterally compressed, broadly oblong, smooth. $2n = 40$.

Tuctoria mucronata is known from only two locations in Solano County, California; both locations are at elevations below 10 m.

19.03 NEOSTAPFIA Burtt Davy

John R. Reeder

Plants annual; glabrous or sparsely pilose, not producing juvenile floating leaves. **Culms** 10–30 cm, simple, ascending or decumbent, often geniculate, not breaking apart at the nodes. **Leaves** without ligules, with little or no distinction between sheath and blade. **Inflorescences** terminal, dense cylindrical spikes, usually completely exposed at maturity, spikelets spirally arranged, rachises often extending beyond the spikelets as a naked or scale-covered axis; **disarticulation** below the spikelets. **Spikelets** dorsally compressed, usually with 5 florets. **Glumes** absent; **lemmas** about 5 mm long, flabellate, to 3 mm wide distally, 7–11-veined, translucent between the veins, apices entire, ciliolate; **paleas** slightly shorter and much narrower than the lemmas, hyaline, 2-veined; **lodicules** 2, about 0.2 mm, truncate or slightly emarginate; **anthers** 3, exserted at anthesis, **styles** 2. **Caryopses** laterally compressed, obovoid, pericarp thick and covered with a viscid exudate, obscuring the embryo; **epiblast** present. $x = 10$. Named for Otto Stapf (1857–1933), a botanist at the Royal Botanic Gardens, Kew, England.

Neostapfia is a monotypic genus endemic to the Central Valley of California.

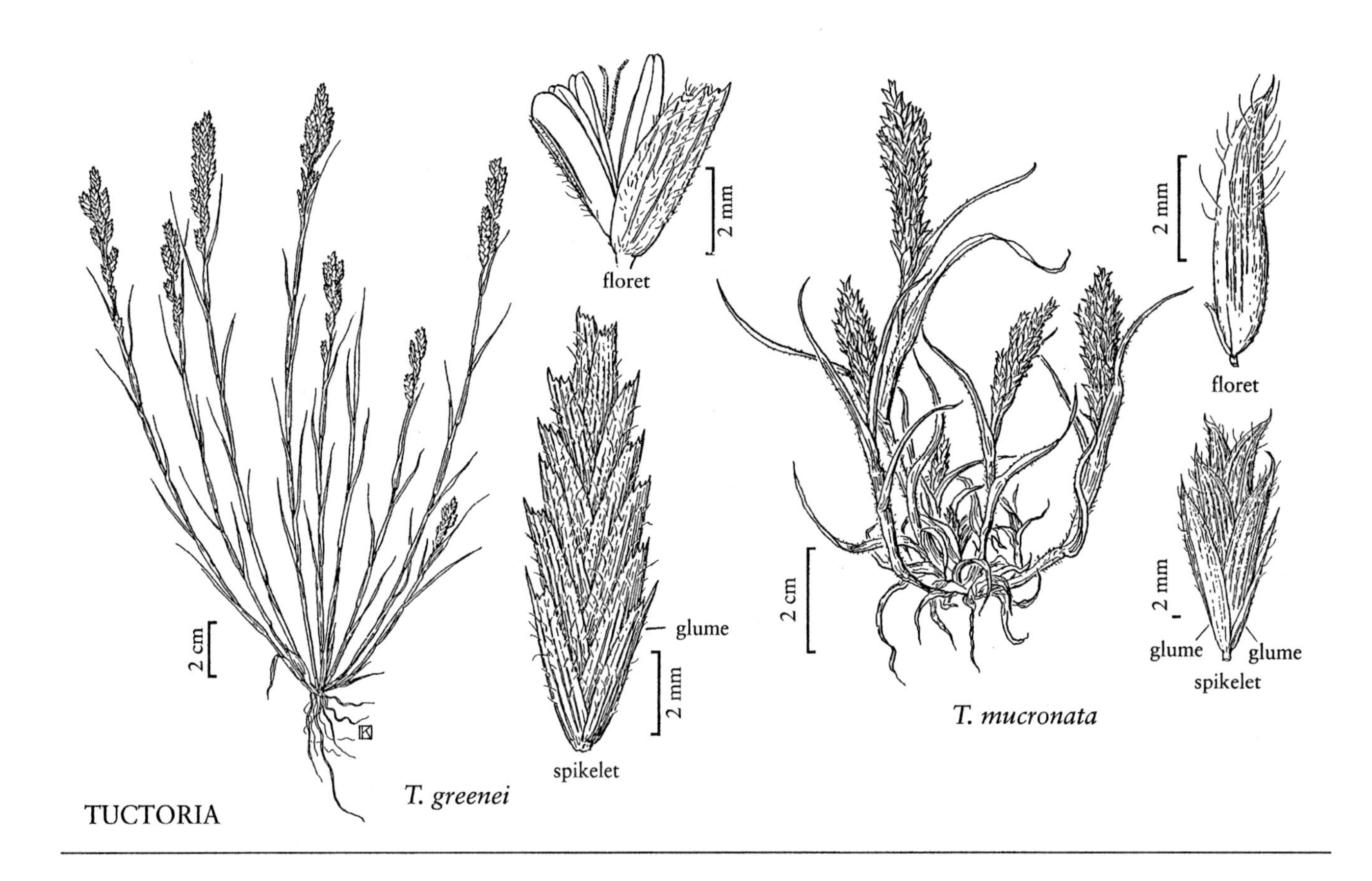

1. **Neostapfia colusana** (Burtt Davy) Burtt Davy [p. 296]
Colusagrass

Plants cespitose, eventually forming rather large clumps, covered with prominent brown viscid glands at maturity. **Culms** 10–30 cm, simple, ascending or decumbent, often geniculate, not breaking apart at the nodes. **Sheaths** loosely enveloping the culms; **blades** 2–5 cm long, 5–12 mm wide. **Spikes** 2–8 cm long, 8–12 mm thick. **Lower lemmas** about 5 mm long, about 3 mm wide, 7–11-veined; **anthers** about 2.5 mm. **Caryopses** about 2.5 mm. $2n = 40$.

Neostapfia colusana grows in vernal pools of Colusa, Merced, Solano, and Stanislaus counties, California, at elevations below 125 m. It is listed by the U.S. Fish and Wildlife Service as a threatened species because of its restricted habitat, much of which has been destroyed. The stout, cylindrical spikes emerging from the sheathing leaves resemble miniature ears of maize. This and the abundant viscid secretion make *N. colusana* a particularly distinctive species.

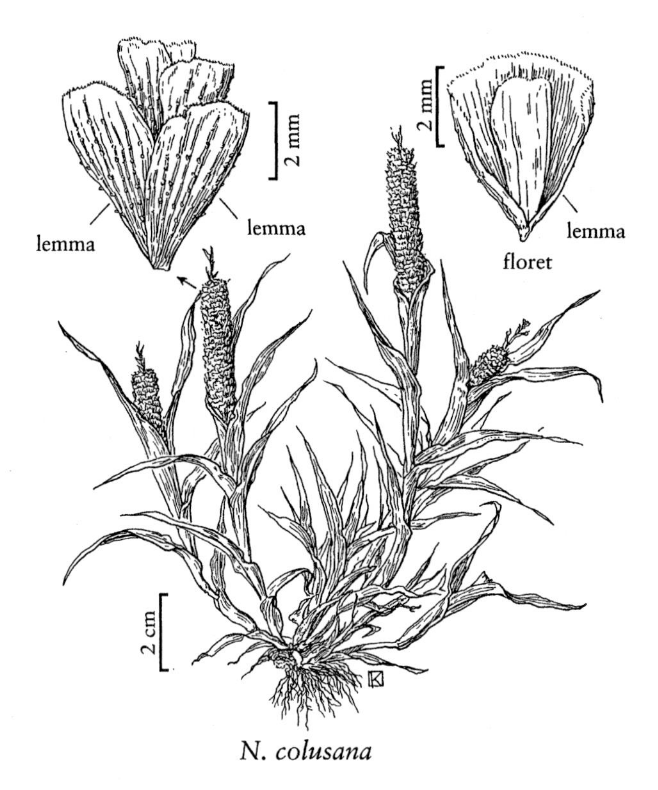

N. colusana

NEOSTAPFIA

7. DANTHONIOIDEAE N.P. Barker & H.P. Linder

Grass Phylogeny Working Group

Plants usually perennial, sometimes annual; when perennial, cespitose, rhizomatous, or stoloniferous. **Culms** usually solid, rarely hollow. **Leaves** distichous; **sheaths** usually open; **abaxial ligules** usually absent; **auricles** usually absent; **adaxial ligules** of hairs or membranous and ciliate; **blades** not pseudopetiolate; **mesophyll** non-radiate; **adaxial palisade layer** absent; **fusoid cells** absent; **arm cells** absent; **kranz anatomy** absent; **midrib** simple, usually with 1 vascular bundle (an arc of bundles in *Cortaderia*); **adaxial bulliform** cells present or not; **stomata** usually with dome-shaped or parallel-sided subsidiary cells (rarely slightly triangular or high dome-shaped); **bicellular microhairs** usually present, distal cell long, narrow; **papillae** usually absent. **Inflorescences** ebracteate (subtending leaf somewhat spatheate in *Urochlaena* Nees), usually paniculate, sometimes racemose or spicate, occasionally a single spikelet; **disarticulation** usually above the glumes and between the florets, sometimes below the glumes or in the culms. **Spikelets** bisexual (sometimes with unisexual florets) or unisexual, with 1–7(20) bisexual or pistillate florets, distal florets in the bisexual spikelets often sterile or staminate; **rachilla extension** present. **Glumes** 2, usually equal, (1)3–7-veined, usually exceeding the distal florets; **florets** laterally compressed; **lemmas** firmly membranous to coriaceous, 3–9-veined, rounded across the back, glabrous or with non-uncinate hairs, these sometimes in tufts or fringes, lemma apices shortly to deeply bilobed, lobes often setaceous, midveins often extended as awns, awns usually geniculate, basal segment often flat and twisted; **paleas** well-developed, sometimes short relative to the lemmas; **lodicules** 2, usually free, usually fleshy, rarely with a membranous apical flap, glabrous or ciliate, often with microhairs, sometimes heavily vascularized; **anthers** 3; **ovaries** usually glabrous, rarely with apical hairs; **haustorial synergids** present, sometimes weakly developed; **styles** 2, bases usually widely separated. **Caryopses** separate from the lemmas and paleas; **hila** punctate or long-linear; **embryos** large or small relative to the caryopses; **endosperm** hard; **starch grains** usually compound; **epiblasts** absent; **scutellar cleft** present; **mesocotyl internode** elongated; **embryonic leaf margins** usually meeting, sometimes overlapping. $x = 6, 7, 9$.

The *Danthonioideae* include only one tribe, the *Danthonieae*, which used to be included in the *Arundinoideae.* Conert (1987) placed *Cortaderia* in a tribe of its own, but its traditional inclusion in the *Danthonieae* is supported by more recent work (Hilu and Esen 1990; Hsiao et al. 1998; Barker et al. 2000; Grass Phylogeny Working Group 2001). The combination of haustorial synergids, ciliate ligules, elongated embryo mesocotyls, and C_3 photosynthesis distinguishes the *Danthonioideae* from other subfamilies of the *Poaceae.*

SELECTED REFERENCES **Barker, N.P., H.P. Linder, and E.H. Harley.** 1999. Sequences of the grass-specific insert in the chloroplast *rpo*C2 gene elucidate generic relationships of the Arundinoideae (Poaceae). Syst. Bot. 23:327–350; **Barker, N.P., C.M. Morton, and H.P. Linder.** 2000. The Danthonieae: Generic composition and relationships. Pp. 221–229 *in* S.W.L. Jacobs and J. Everett (eds.). Grasses: Systematics and Evolution. International Symposium on Grass Systematics and Evolution (3rd:1998). CSIRO Publishing, Collingwood, Victoria, Australia. 408 pp.; **Conert, H.J.** 1987. Current concepts in the systematics of the Arundinoideae. Pp. 239–250 *in* T.R. Soderstrom, K.H. Hilu, C.S. Campbell, and M.E. Barkworth (eds.). Grass Systematics and Evolution. Smithsonian Institution Press, Washington, D.C., U.S.A. 473 pp.; **Grass Phylogeny Working Group.** 2001. Phylogeny and subfamilial classification of the grasses (Poaceae). Ann. Missouri Bot. Gard. 88:373–457; **Hilu, K.W.** and **A. Esen.** 1990. Prolamin and immunological studies in the Poaceae. I. Subfamily Arundinoideae. Pl. Syst. Evol. 173:57–70; **Hsiao, C., S.W.L. Jacobs, N.P. Barker, and N.J. Chatterton.** 1998. A molecular phylogeny of the subfamily Arundinoideae (Poaceae) based on sequences of rDNA (ITS). Austral. Syst. Bot. 11:41–52; **Linder, H.P. and N.P. Barker.** 2000. Biogeography of the Danthonieae. Pp. 231–238 *in* S.W.L. Jacobs and J. Everett (eds.). Grasses: Systematics and Evolution. International Symposium on Grass Systematics and Evolution (3rd:1998). CSIRO Publishing, Collingwood, Victoria, Australia. 408 pp.; **Verboom, G.A., H.P. Linder, and N.P. Barker.** 1994. Haustorial synergids: An important character in the systematics of danthonioid grasses (Arundinoideae: Poaceae). Amer. J. Bot. 81:1601–1610.

20. DANTHONIEAE Zotov

Mary E. Barkworth

See subfamily description.

The *Danthonieae*, the only tribe in the *Danthonioideae*, include approximately 13 genera and 290 species, most of which grow in mesic to xeric, open habitats such as grasslands, heaths, and open woods. It is most abundant in the Southern Hemisphere, with only *Danthonia* being native in the Northern Hemisphere.

Two of the genera recognized here, *Karroochloa* and *Rytidosperma*, are frequently included in *Danthonia*, from which they can be distinguished by the tufts of hairs on their lemmas. It is much more difficult to identify a character, or combination of characters, that will consistently distinguish them from each other. Glume length works in this *Flora* because of the species involved, but it is not generally reliable.

1. Culms 200–700 cm tall; inflorescences plumose, 30–130 cm long . . . . 20.01 *Cortaderia*
1. Culms 2–100 cm tall; inflorescences not plumose, 0.5–12 cm long.
  2. Lemmas with hairs in 1 or more transverse row(s) above the callus and/or at midlength.
    3. Panicles subcapitate; glumes 3.5–7 mm long . . . . 20.04 *Karroochloa*
    3. Panicles narrow; glumes 8–15 mm long . . . . 20.05 *Rytidosperma*
  2. Lemmas glabrous or, if with hairs, the hairs not in transverse rows.
    4. Plants annual . . . . 20.03 *Schismus*
    4. Plants perennial.
      5. Lemma apices entire, acute to acuminate . . . . 20.06 *Tribolium*
      5. Lemma apices bifid, obtuse, acute, or acuminate . . . . 20.02 *Danthonia*

20.01 CORTADERIA Stapf

Kelly W. Allred

Plants perennial; often dioecious or monoecious; cespitose. **Culms** 2–7 m, erect, densely clumped. **Leaves** primarily basal; **sheaths** open, often overlapping, glabrous or hairy; **auricles** absent; **ligules** of hairs; **blades** to 2 m, flat to folded, arching, edges usually sharply serrate. **Inflorescences** terminal, plumose panicles, 30–130 cm, subtended by a long, ciliate bract; **branches** stiff to flexible. **Spikelets** somewhat laterally compressed, usually unisexual, sometimes bisexual, with 2–9 unisexual florets; **disarticulation** above the glumes and below the florets. **Glumes** unequal, nearly as long as the spikelets, hyaline, 1-veined; **calluses** pilose; **lemmas** 3–5(7)-veined, long-acuminate, bifid and awned or entire and mucronate; **lemmas of pistillate** and **bisexual florets** usually long-sericeous; **lemmas of staminate florets** less hairy or glabrous; **lodicules** 2, cuneate and irregularly lobed, ciliate; **paleas** about ½ as long as the lemmas, 2-veined; **anthers** of bisexual florets 3, 1.5–6 mm, those of the pistillate florets smaller or absent. **Caryopses** 1.5–3 mm; **hila** linear, about ½ as long as the caryopses; **embryos** usually shorter than 1 mm. $x = 9$. Name from the Spanish, *cortada,* 'cutting', referring to the sharply serrate blades.

Cortaderia, a genus of about 25 species, is native to South America and New Zealand, with the majority of species being South American. Recent evidence suggests that the species in the two regions represent different lineages, each of which merits generic recognition. The species treated here would remain in *Cortaderia* if this change were made.

Both of the species that are found in North America were originally introduced as ornamental species; both are now considered aggressive weeds in parts of the *Flora* region.

SELECTED REFERENCES **Barker, N.P., C.M. Morton** and **H.P. Linder.** 2000. The Danthonieae: Generic composition and relationships. Pp. 221–229 *in* S.W.L. Jacobs and J. Everett (eds.). Grasses: Systematics and Evolution. International Symposium on Grass Systematics and Evolution (3rd:1998). CSIRO Publishing, Collingwood, Victoria, Australia. 408 pp.; **Connor, H.E.** and **E. Edgar.** 1974. Names and types in *Cortaderia* Stapf (Gramineae). Taxon 23:595–605; **Costas-Lippmann, M.** 1977. More on the weedy "pampas grass" in California. Fremontia 4:25–27; **Hitchcock, A.S.** 1951 [title page 1950]. Manual of the Grasses of the United States, ed. 2, rev. A. Chase. U.S.D.A. Miscellaneous Publication No. 200. U.S. Government Printing Office, Washington, D.C., U.S.A. 1051 pp.; **Linder, H.P.** and **N.P. Barker.** 2000. Biogeography of the Danthonieae. Pp. 231–238 *in* S.W.L. Jacobs and J. Everett (eds.). Grasses: Systematics and Evolution. International Symposium on Grass Systematics and Evolution (3rd:1998). CSIRO Publishing, Collingwood, Victoria, Australia. 408 pp.; **Walsh, N.G.** 1994. *Cortaderia.* Pp. 546–548 *in* N.G. Walsh and T.J. Entwisle. Flora of Victoria, vol. 2: Ferns and Allied Plants, Conifers and Monocotyledons. Inkata Press, Melbourne, Australia. 946 pp.

1. Sheaths hairy; panicles elevated well above the foliage; culms 4–5 times as long as the panicles . . . . . . 1. *C. jubata*
1. Sheaths glabrous or sparsely hairy; panicles elevated only slightly, if at all, above the foliage; culms 2–4 times as long as the panicles . . . . 2. *C. selloana*

1. Cortaderia jubata (Lemoine *ex* Carrière) Stapf [p. 300]
PURPLE PAMPAS GRASS

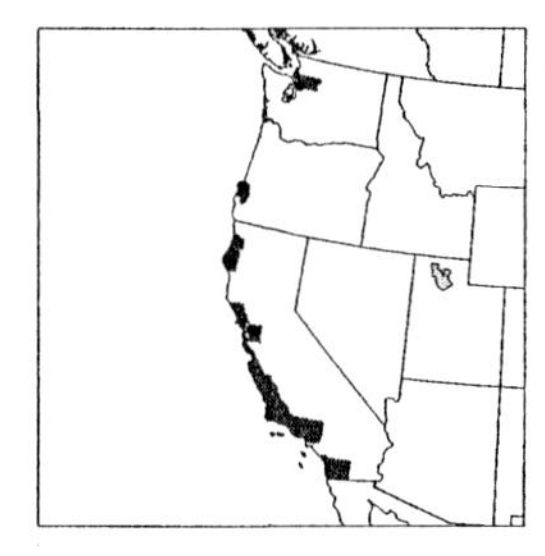

Plants pistillate (in North America). **Culms** 2–7 m, 4–7 times as long as the panicles. **Leaves** primarily basal; **sheaths** hairy, sometimes densely so; **ligules** 1–2 mm; **blades** 1 m long or longer, 2–10 cm wide, mostly flat, often horizontal, dark green, abaxial surfaces hairy near the base. **Panicles** 30–100 cm, elevated well above the basal foliage, deep violet when young. **Spikelets** 14–16 mm, pistillate; **florets** readily disarticulating; **calluses** about 0.6 mm; **lemmas** about 10 mm, long-attenuate to an awn, awns to 1 mm; **paleas** to 4 mm, keels ciliate, apical hairs extending beyond the body of the paleas; **stigmas** usually not exerted. **Caryopses** to 2.5 mm; **embryos** to 1 mm. $2n = 108$.

Cortaderia jubata is found on the west coast of the coterminus United States, growing in disturbed, open ground such as brushy slopes, eroded banks and cliffs, road cuts, cut-over timber areas, and sand dunes. It is native to mountainous areas of Ecuador, Peru, and Bolivia. It was grown in the past as an ornamental because of its attractive panicles, but is now a serious weed in California, reproducing apomictically and invading many open habitats. It was mistakenly called **C. rudiuscula** Stapf by Hitchcock (1951). The florets of *C. rudiuscula* differ from those of *C. jubata* in being longer and narrower, having shorter, less hairy calluses, and in having no hairs that extend beyond the top of the palea. *C. rudiuscula* is not known from North America.

2. Cortaderia selloana (Schult. & Schult. f.) Asch. & Graebn. [p. 300]
PAMPAS GRASS

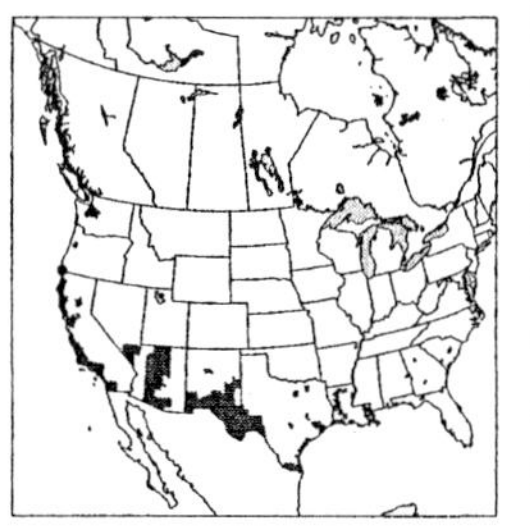

Plants usually dioecious, sometimes monoecious. **Culms** 2–4 m, usually 2–4 times as long as the panicles. **Leaves** primarily basal; **sheaths** mostly glabrous, with a dense tuft of hairs at the collars; **ligules** 1–2 mm; **blades** to 2 m long, 3–8 cm wide, mostly flat, cauline, ascending, arching, bluish-green, abaxial surfaces glabrous basally. **Panicles** 30–130 cm, only slightly, if at all, elevated above the foliage, whitish or pinkish when young. **Spikelets** 15–17 mm; **calluses** to 1 mm, with hairs to 2 mm; **lemmas** long-attenuate to an awn, awns 2.5–5 mm; **paleas** to 4 mm; **stigmas** exerted. **Caryopses** and **florets** not separating easily from the rachilla. $2n = 72$.

Cortaderia selloana is native to central South America. It is cultivated as an ornamental in the warmer parts of North America. It was thought that it would not become a weed problem because most plants sold as ornamentals are unisexual, but it is now considered an aggressive weed in California and Bendigo, Australia. The weedy Australian plants are bisexual (Walsh 1994).

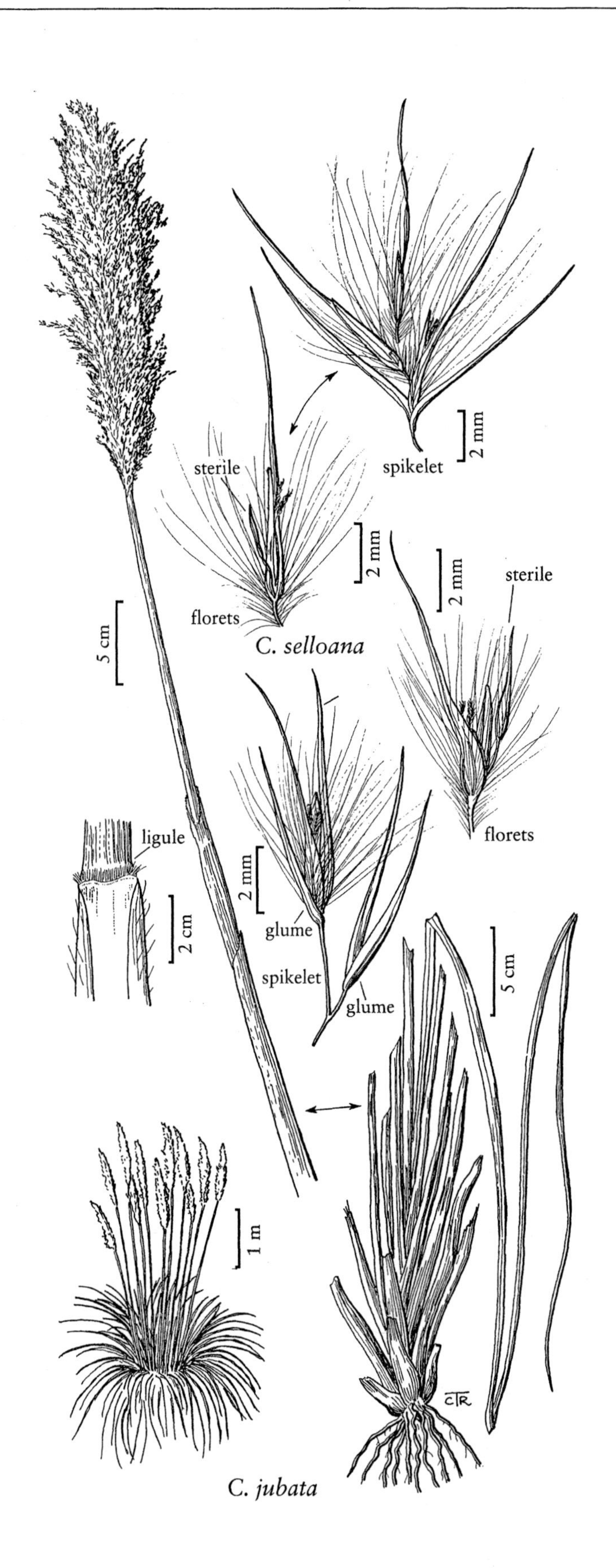
spikelet
2 mm
sterile
florets
2 mm
C. selloana
2 mm
sterile
florets
5 cm
ligule
2 cm
2 mm
glume
spikelet
glume
5 cm
1 m
CTR
C. jubata

CORTADERIA

20.02 DANTHONIA DC.

Stephen J. Darbyshire

Plants perennial; cespitose, sometimes shortly rhizomatous. **Culms** 7–130 cm, erect. **Sheaths** open to the base, with tufts of hairs at the auricle position, sometimes with a line of hairs around the collar; **auricles** absent; **ligules** of hairs; **blades** rolled in the bud, flat or involute when dry. **Inflorescences** terminal; panicles, racemes, or a solitary spikelet, to 12 cm; **rachises, branches,** and **pedicels** scabrous or hirsute. **Spikelets** terete or laterally compressed, with 3–12 florets, terminal floret reduced; **disarticulation** beneath the florets, also at the cauline nodes in some species. **Glumes** subequal or the lower glumes a little longer than the upper glumes, usually exceeding the florets (excluding the awns and lemma teeth), lanceolate, chartaceous, 1–7-veined, keels glabrous or sparsely scabrous; **rachillas** glabrous; **calluses** densely strigose on the sides; **lemma bodies** obscurely (5)7–11-veined, backs glabrous or pilose, margins usually densely pilose proximally, apices with 2 acute to aristate lobes, mucronate or awned between the lobes; **awns,** when present, geniculate and twisted below the geniculation; **paleas** about as long as the lemma bodies, 2-veined, veins scabrous, apices obtuse, sometimes bifid; **lodicules** 2, glabrous or with a few hairs; **anthers** 3, their size depending on whether the florets are chasmogamous or cleistogamous; **ovaries** glabrous. **Caryopses** 1.5–5.5 mm, ovate to obovate, dorsally flattened, brown; **hila** linear, ⅓–¾ as long as the caryopses. **Cleistogenes** usually present in the lower sheaths, with 1(–10) florets, not disarticulating; **rachilla segments** about as long as or longer than the adjacent florets; **lemmas** coriaceous, glabrous or scabrous near the apex, entire, unawned; **paleas** sometimes slightly longer than the lemmas; **anthers** 3, minute; **ovaries** glabrous; **caryopses** more linear than in the aerial florets. $x = 12$. Named for the French botanist Étienne Danthoine, who worked in the early nineteenth century.

Danthonia is interpreted here as a genus of about 20 species that are native in Europe, North Africa, and the Americas. Of the eight species found in the *Flora* region, seven are native and one is an introduction that is now established.

SELECTED REFERENCES **Baeza P., C.M.** 1996. Los géneros *Danthonia* DC. y *Rytidosperma* Steud. (Poaceae) en América–Una revisión. Sendtnera 3:11–93; **Darbyshire, S.J.** and **J. Cayouette.** 1989. The biology of Canadian weeds. 92. *Danthonia spicata* (L.) Beauv. in Roem. & Schult. Canad. J. Pl. Sci. 69:1217–1233; **Dobrenz, A.K.** and **A.A. Beetle.** 1966. Cleistogenes in *Danthonia.* J. Range Managem. 19:292–296; **Dore, W.G.** 1971. *Sieglingia decumbens* (L.) Bernh.–Pulvini of palea. Watsonia 8:297–299; **Dore, W.G.** and **J. McNeill.** 1980. Grasses of Ontario. Research Branch, Agriculture Canada Monograph No. 26. Canadian Government Publishing Centre, Hull, Québec, Canada. 566 pp.; **Linder, H.P.** and **G.A. Verboom.** 1996. Generic limits in the *Rytidosperma* (Danthonieae, Poaceae) complex. Telopea 6:597–627; **Quinn, J.A.** and **D.E. Fairbrothers.** 1971. Habitat ecology and chromosome numbers of natural populations of the *Danthonia sericea* complex. Amer. Midl. Naturalist 85:531–536; **Tzvelev, N.N.** 1976. Zlaki SSSR. Nauka, Leningrad [St. Petersburg], Russia. 788 pp. [In Russian].

NOTE: In the key and descriptions, lemma lengths do not include the apical teeth. Callus characteristics are best seen on the middle to upper florets in the spikelet.

1. Lemmas mucronate, not awned . . . . 1. *D. decumbens*
1. Lemmas not mucronate, with a twisted, geniculate awn.
    2. Calluses of the middle florets from shorter to slightly longer than wide, convex abaxially; lemma bodies 2.5–6 mm long, the backs usually pilose, occasionally glabrous or sparsely pilose.
        3. Awns 10–17 mm; hairs of the lemma margins evidently increasing in length distally, longest hairs 2.5–4 mm long . . . . 2. *D. sericea*
        3. Awns 5–10 mm; hairs of the lemma margins not evidently increasing in length distally, longest hairs 0.5–2 mm long.
            4. Lemma lobes 2–4 mm long, usually ⅔ or more as long as the lemma bodies, aristate; lower inflorescence branches usually flexible, divergent after anthesis; pedicels on the lowest inflorescence branch as long as or longer than the spikelets; leaves not curling at maturity . . . . 3. *D. compressa*

4. Lemma lobes 0.5–2 mm long, less than ⅓ as long as the lemma bodies, acute to aristate; inflorescence branches stiff, appressed to strongly ascending after anthesis; pedicels on the lowest inflorescence branch from shorter than to equaling the spikelets; blades usually becoming curled at maturity . . . . 4. *D. spicata*

2. Calluses of the middle florets longer than wide, concave abaxially; lemma bodies 3–11 mm long, the backs usually glabrous or sparsely pilose (pilose in *D. parryi*).
   5. Lower inflorescence branches (pedicels if the inflorescence racemose) stiff, erect; pedicels from shorter than to as long as the spikelets.
      6. Spikelets 1(–3), if 2–3, the inflorescence a raceme; lemma bodies 5.5–11 mm; mature culms disarticulating at the nodes . . . . 8. *D. unispicata*
      6. Spikelets (4)5–10; lower inflorescence branches usually with 2–3 spikelets; lemma bodies 3–6 mm; mature culms not disarticulating at the nodes . . . . 5. *D. intermedia*
   5. Lower inflorescence branches (pedicels if the inflorescence racemose) flexible, slightly to strongly divergent; pedicels usually as long as or longer than the spikelets (sometimes shorter in *D. parryi*).
      7. Uppermost cauline blades usually strongly divergent or reflexed; inflorescences usually racemose; pedicels usually much longer than the spikelets and usually strongly divergent; lemmas glabrous or sparsely hairy over the back; mature culms disarticulating at the nodes . . . . 7. *D. californica*
      7. Uppermost cauline blades usually erect to ascending; inflorescences usually paniculate; pedicels shorter than to as long as the spikelets; lemmas pilose over the back, at least basally; mature culms not disarticulating at the nodes . . . . 6. *D. parryi*

1. **Danthonia decumbens** (L.) DC. [p. 304]
MOUNTAIN HEATH-GRASS

Culms 8–60 cm, usually erect, sometimes decumbent, not disarticulating. **Sheaths** glabrous or pilose; **blades** 5–15 cm long, 0.5–4 mm wide, usually flat, glabrous or sparsely pilose. **Inflorescences** with up to 15 spikelets; **branches** erect; **lower branches** with 1–3 spikelets. **Spikelets** 6–15 mm; **florets** usually cleistogamous, rarely chasmogamous. **Calluses of middle florets** from as long as to a little longer than wide, convex abaxially; **lemma bodies** 5–6 mm, margins glabrous or pubescent for most of their length, scabrous apically, apices with acute teeth, teeth often scabrous, sometimes scabridulous, mucronate, not awned, from between the teeth; **palea veins** swollen at the base, forming pulvini; **anthers** of the cleistogamous florets 0.2–0.4 mm, those of the chasmogamous florets about 2 mm. **Caryopses** 2.1–2.5 mm long, 1.1–1.8 mm wide. $2n$ = 24, 36, 124.

Danthonia decumbens grows throughout most of Europe, the Caucasus, and northern Turkey, and is now established on the west and east coasts of North America. It grows in heathlands, sandy or rocky meadows, clearings, and sometimes along roadsides. The species is sometimes placed in the monotypic genus *Sieglingia*, as *Sieglingia decumbens* (L.) Bernh.

2. **Danthonia sericea** Nutt. [p. 304]
DOWNY OATGRASS

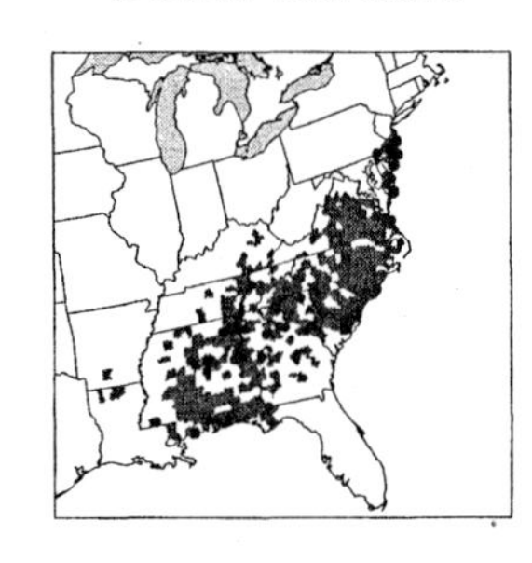

Culms 50–120 cm, not disarticulating. **Sheaths** usually villous, occasionally glabrous; **blades** 10–30 cm long, 2–4 mm wide, pilose or glabrous, usually at least the adaxial surface pilose, uppermost cauline blades erect to ascending. **Inflorescences** with (5)10–25(30) spikelets; **lower branches** erect to ascending, with 2–6 spikelets; **pedicels** on the lowest branch from shorter than to subequal to the spikelets. **Spikelets** 10–20 mm. **Calluses of middle florets** from as long as to a little longer than wide, convex abaxially; **lemma bodies** 4–6 mm, usually pilose over the back (sometimes glabrous), margins densely pilose to beyond midlength, hairs evidently increasing in length distally, longest hairs 2.5–4 mm, longer than apical teeth 2–4.5(5.5) mm, aristate; **awns** 10–17 mm; **anthers** to 2.6 mm. **Caryopses** 1.7–2.4 mm long, 0.8–1.2 mm wide. $2n$ = 36.

Danthonia sericea is restricted to the eastern United States. It grows mostly on sand barrens and in open woods on dry soils. A less common form, with glabrous foliage and lemma backs, is found in bogs, seepage areas, and low moist areas adjacent to lakes and rivers and has been called *D. sericea* var. *epilis* (Scribn.) Gleason. In their study of these ecotypes, Quinn and Fairbrothers (1971) state that "field differences in growth form do not persist under transplant garden and greenhouse conditions." Similar patterns of infraspecific variation are also seen in the leaf and lemma vestiture of *D. spicata* and *D. californica*, but the genetic basis of this variation is probably not taxonomically significant.

3. **Danthonia compressa** Austin [p. 304]
FLATTENED OATGRASS, DANTHONIE COMPRIMÉE

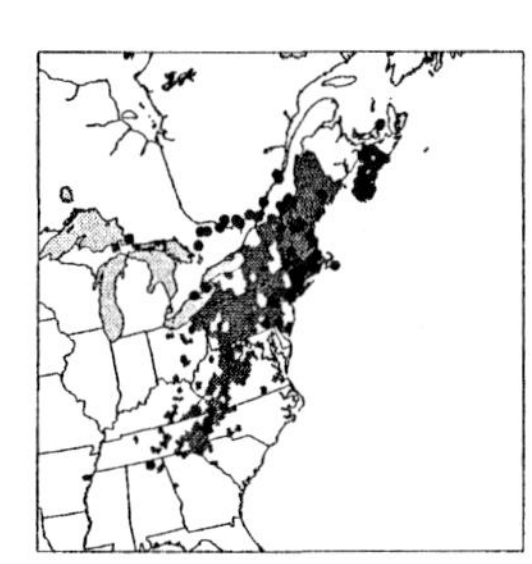

Culms 40–80 cm, disarticulating at the nodes when mature. **Sheaths** glabrous, rarely sparsely pilose, usually reddish above the nodes; **blades** to 30 cm long, 2–4 mm wide, flexible but not curled at maturity, glabrous, sometimes scabrous, uppermost cauline blades erect to ascending. **Inflorescences** with (4)6–17 spikelets; **branches** usually flexible, usually divergent, sometimes strongly so, after anthesis; **lower branches** with 2–3 spikelets; **pedicels** on the lowest branch as long as or longer than the spikelets. **Spikelets** (7)10–16 mm. **Calluses of middle florets** about as long as wide, convex abaxially; **lemma bodies** 2.5–5 mm, pilose over the back, sometimes sparsely so, margins pilose to beyond midlength, distal hairs 0.5–2 mm, apical teeth 2–4 mm, aristate, (½)⅔ or more as long as the lemma bodies; **awns** 6–10 mm; **anthers** to 2.2 mm. **Caryopses** 1.7–2.6 mm long, 0.7–1.1 mm wide. $2n = 36$.

Danthonia compressa grows in open and semi-shaded areas, including meadows, open woods, and woodland openings. Although not a true pioneer species, it may sometimes occur as a weed in perennial crops. It is restricted to eastern North America.

4. **Danthonia spicata** (L.) P. Beauv. *ex* Roem. & Schult. [p. 304]
POVERTY OATGRASS, DANTHONIE À ÉPI

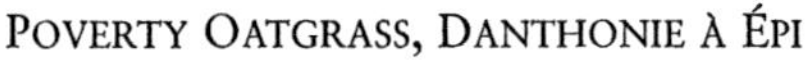

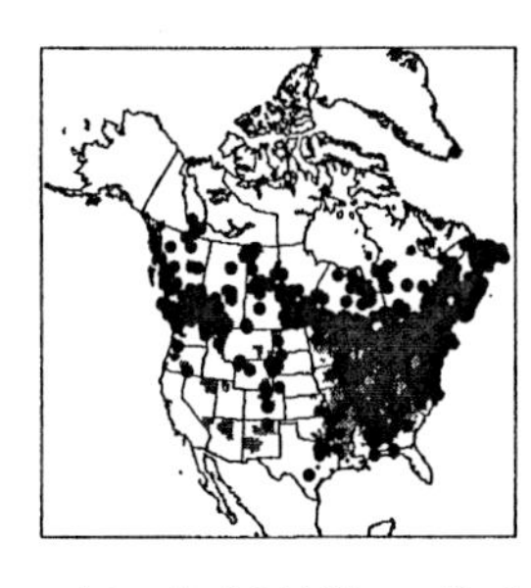

Culms (7)10–70(100) cm, disarticulating at the nodes when mature. **Sheaths** pilose or glabrous; **blades** 6–15(20) cm long, 0.8–3(4) mm wide, usually becoming curled at maturity, glabrous or pilose, uppermost cauline blades erect to ascending. **Inflorescences** with 5–10(18) spikelets; **branches** stiff, appressed to strongly ascending after anthesis; **lower branches** with 1–3 spikelets; **pedicels** on the lowest branch from shorter than to equaling the spikelets. **Spikelets** 7–15 mm. **Calluses of middle florets** about as long as wide, convex abaxially; **lemma bodies** 2.5–5 mm, usually pilose (sometimes glabrous) over the back, margins pilose to about midlength, longest hairs 0.5-2 mm, apical teeth 0.5–2 mm, acute to aristate, less than ⅔ as long as the lemma bodies; **awns** 5–8 mm; **anthers** to 2.5 mm. **Caryopses** 1.5–2(2.3) mm long, 0.7–1 mm wide. $2n = 31, 36$.

Danthonia spicata grows in dry rocky, sandy, or mineral soils, generally in open sunny places. Its range includes most of boreal and temperate North America and extends south into northeastern Mexico.

Phenotypically, *Danthonia spicata* is quite variable, expressing different growth forms under different conditions (Dore and McNeill 1980; Darbyshire and Cayouette 1989). Slow clonal growth, extensive cleistogamy, and limited dispersal contribute to the establishment of morphologically uniform populations, some of which have been given scientific names. For instance, *D. spicata* var. *pinetorum* Piper is sometimes applied to depauperate plants and *D. allenii* Austin misapplied to more robust or 'second growth' plants (Dore and McNeill 1980). Plants of shady or moist habitats often lack the distinctive curled or twisted blades usually found on plants growing in open habits. Such plants, which tend to have smaller spikelets and pilose foliage, have been called *D. spicata* var. *longipila* Scribn. & Merr. The terminal inflorescence is usually primarily cleistogamous, but plants with chasmogamous inflorescences are found throughout the range of the species. Chasmogamous plants differ in having divergent inflorescence branches at anthesis, larger anthers, and well-developed lodicules.

5. **Danthonia intermedia** Vasey [p. 306]
TIMBER OATGRASS, DANTHONIE INTERMÉDIAIRE

Culms 10–50(70) cm, not disarticulating at maturity. **Sheaths** usually glabrous; **blades** 5–10 cm long, 1–3.5 mm wide, glabrous or slightly pilose. **Inflorescences** with (4)5–10 spikelets; **branches** stiff, appressed or strongly ascending; **lower branches** with (1)2–3(5) spikelets; **pedicels** on the lowest branch shorter than the spikelets. **Spikelets** 11–15(19) mm. **Calluses of middle florets** longer than wide, concave abaxially; **lemma bodies** 3–6 mm, glabrous over the back, densely pilose along the margins, teeth 1.5–2.5 mm, acute to acuminate or aristate; **awns** 6.5–8 mm; **anthers** usually tiny, sometimes to 4 mm. **Caryopses** (2)2.3–3 mm long, 0.7–1.1 mm wide. **Cleistogenes** rarely produced. $2n = 36, 98$.

Danthonia intermedia grows in boreal and alpine meadows, open woods, and on rocky slopes and northern plains. Its range extends from Kamchatka, Russia, to North America, south along the cordillera, and east, through boreal and alpine regions, to Quebec

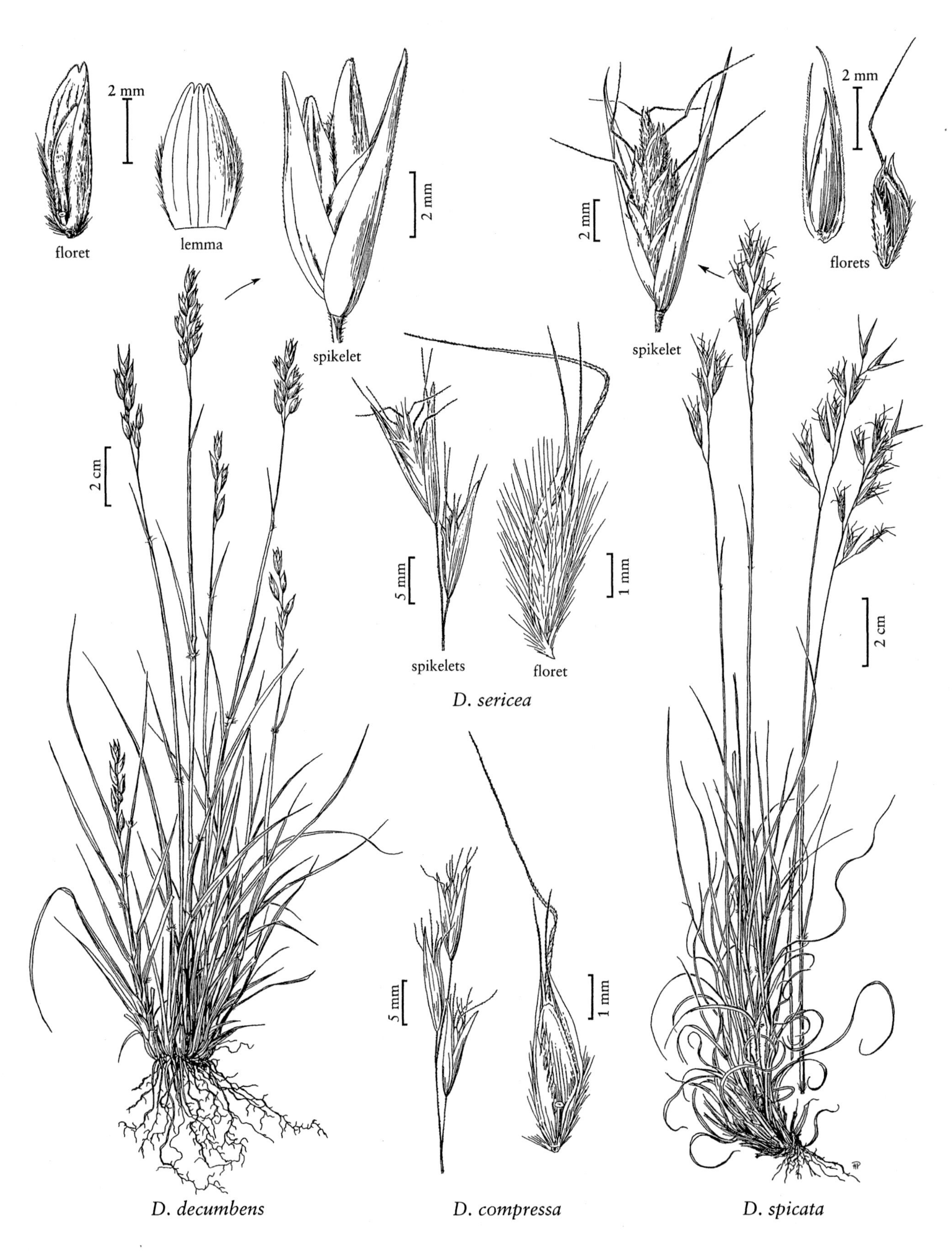
floret
2 mm
lemma
spikelet
2 mm
2 cm
2 mm
spikelet
2 mm
florets
5 mm
spikelets
1 mm
floret
D. sericea
2 cm
5 mm
1 mm
D. decumbens
D. compressa
D. spicata

DANTHONIA

and Newfoundland and Labrador. Its primarily cleistogamous reproduction has probably facilitated its establishment and spread through more boreal and alpine habitats than other members of the genus.

Tzvelev (1976) treats the American plants as *D. intermedia* (Vasey) subsp. *intermedia* and the Russian plants, which have $2n = 18$, as *D. intermedia* subsp. *riabuschinskii* (Kom.) Tzvelev.

6. **Danthonia parryi** Scribn. [p. 306]
PARRY'S OATGRASS

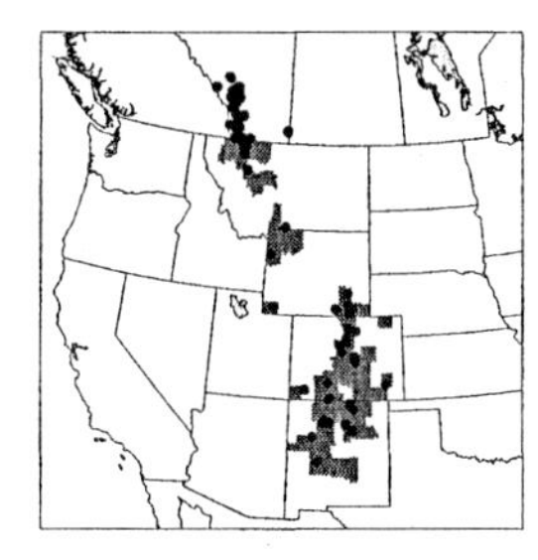

Culms 30–80(100) cm, not disarticulating at maturity. **Sheaths** glabrous or sparsely pubescent; **blades** 15–25 cm long, to 4 mm wide, glabrous or scabrous (rarely pilose), uppermost cauline blades erect or diverging less than 20° from the culm at maturity. **Inflorescences** usually paniculate, sometimes racemose, with (3)4–11 spikelets; **branches** appressed to ascending, somewhat flexible; **pedicels** on the lowest branch from shorter than to as long as the spikelets. **Spikelets** 16–24 mm. **Calluses of middle florets** longer than wide, concave abaxially; **lemma bodies** 5.5–10 mm, backs usually pilose, especially near the base (rarely glabrous), margins pilose, teeth 2.5–8 mm, aristate; **awns** 12–15 mm; **anthers** to 6.5 mm. **Caryopses** rarely produced, 3.5–5.2 mm long, 0.9–1.8 mm wide. $2n = 36$.

Danthonia parryi is endemic to western North America and is often a major component of grasslands on the eastern foothills of the Rocky Mountains. It grows in open grassland, open woods, and rocky slopes, at elevations up to 4000 m. It rarely produces caryopses in the terminal inflorescences. This and its somewhat intermediate morphology have led to speculation that it is derived from hybridization between *D. californica* and *D. intermedia*.

7. **Danthonia californica** Bol. [p. 306]
CALIFORNIA OATGRASS

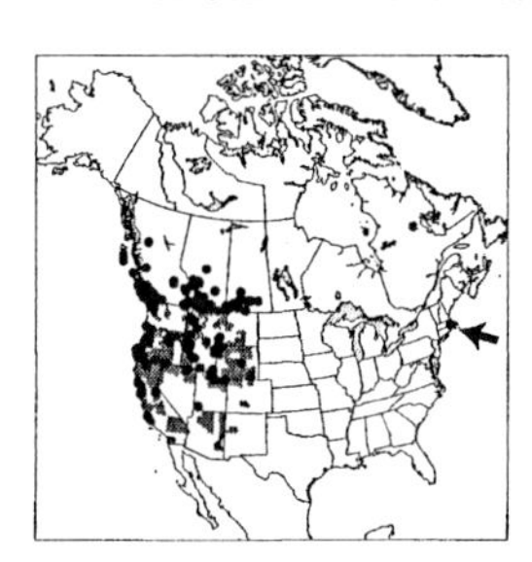

Culms (10)30–130 cm, disarticulating at the nodes at maturity. **Sheaths** glabrous or pilose, upper sheaths usually glabrous or unevenly pilose; **blades** 10–30 cm long, (1)2–5(6) mm wide, flat to rolled or involute, glabrous or pilose, uppermost cauline blades strongly divergent to reflexed at maturity. **Inflorescences** usually racemose, with (2)3–6(10) widely-spreading spikelets; **branches** flexible, strongly divergent to reflexed at maturity, pulvini usually present at the base; **pedicels** on the lowest branch longer than the spikelets, often crinkled. **Spikelets** (10)14–26(30) mm. **Calluses of middle florets** usually longer than wide, concave abaxially; **lemma bodies** 5–10 mm, glabrous or sparsely pilose over the back, margins pubescent (rarely glabrous), apical teeth (2)4–6(7) mm, aristate; **awns** (7)8–12 mm; **anthers** to 4 mm. **Caryopses** 2.5–4.2 mm long, 1.3–1.6 mm wide. $2n = 36$.

Danthonia californica grows in prairies, meadows, and open woods. It has a disjunct distribution, one portion of its range being located in western North America, the other in Chile. An introduced population has been found at Mansfield, Massachusetts.

Plants with pilose foliage have been called *Danthonia californica* var. *americana* (Scribn.) Hitchc. and plants with sparsely pilose lemma backs *D. californica* var. *macounii* Hitchc., but the variation does not appear to be taxonomically significant.

8. **Danthonia unispicata** (Thurb.) Munro *ex* Vasey [p. 306]
ONE-SPIKE OATGRASS

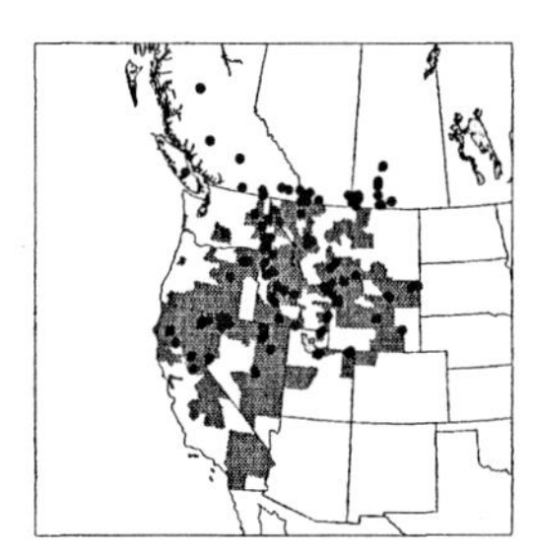

Culms (10)15–30(42) cm, disarticulating at the nodes at maturity. **Sheaths** usually densely pilose, hairs sometimes papillose-based (upper sheaths sometimes glabrous); **blades** 3–8(20) cm long, 1–3 mm wide, both surfaces sparsely to densely pilose, sometimes also scabrous or hirsute (rarely glabrous). **Inflorescences** with 1–2(3) spikelets, if more than 1, racemose; **pedicels** stiff, appressed, shorter than the spikelets. **Spikelets** (8)12–26 mm. **Calluses of middle florets** longer than wide, concave abaxially; **lemma bodies** 5.5–11 mm, glabrous over the back (rarely with a few scattered hairs), margins pilose (rarely glabrous), apical teeth 1.5–7 mm, acute to aristate; **awns** 5.5–13 mm; **anthers** to 3.5 mm. **Caryopses** 2.2–4 mm long, about 1 mm wide. $2n = 36$.

Danthonia unispicata is restricted to western North America, where it grows in prairies and meadows, on rocky slopes, and in dry openings up to timberline in the mountains. It differs from *D. californica* in its shorter stature, usually densely pilose foliage, short, erect pedicels, and the usually erect cauline leaf blades. It is closely related to *D. californica*, and some authors prefer to treat it as *D. californica* var. *unispicata* Thurb.

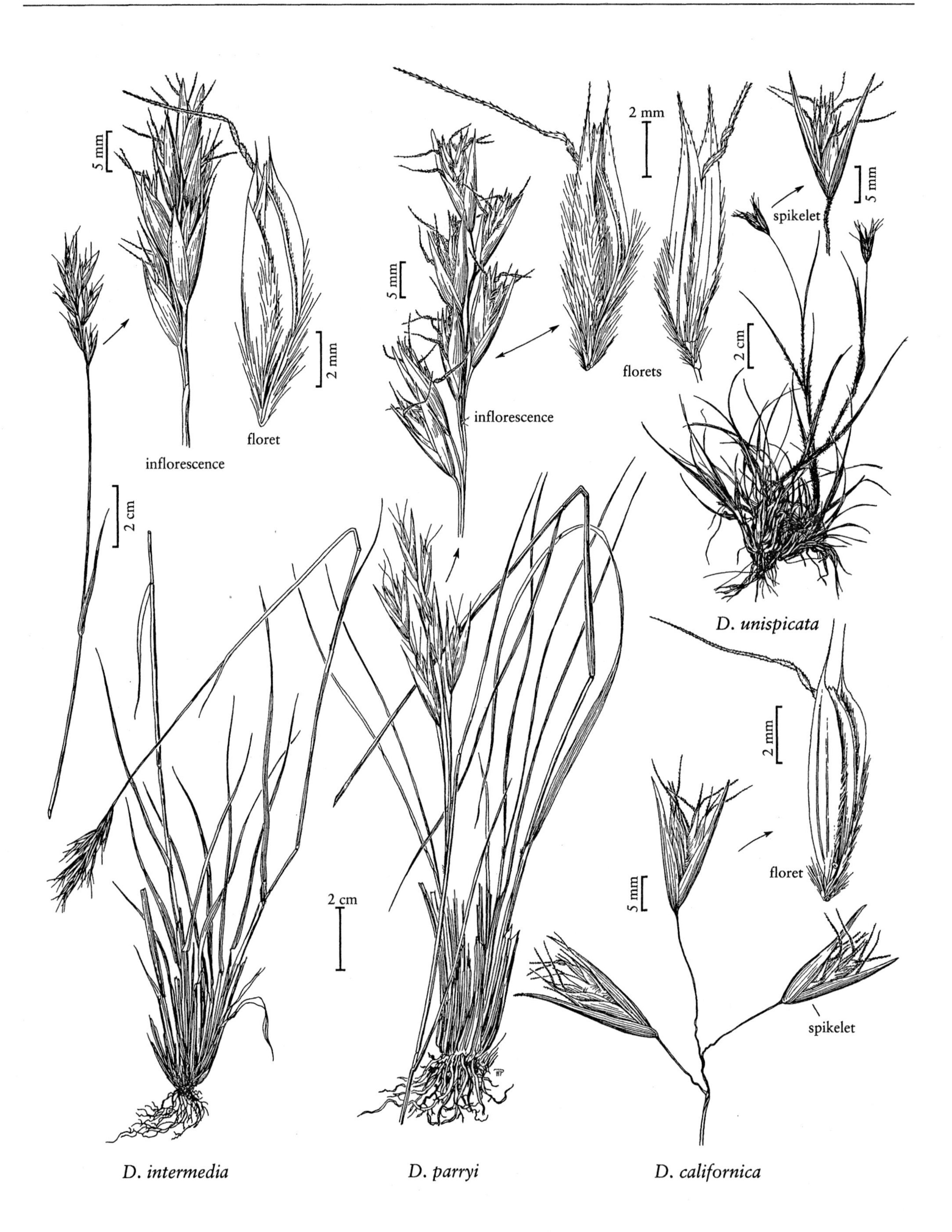
5 mm
2 mm
floret
inflorescence
2 cm
2 mm
5 mm
inflorescence
florets
2 mm
spikelet
5 mm
2 cm
D. unispicata
2 mm
floret
5 mm
spikelet
2 cm
D. intermedia
D. parryi
D. californica

DANTHONIA

20.03 SCHISMUS P. Beauv.

Elizabeth A. Kellogg

Plants annuals or short-lived perennials; tufted. **Culms** 2–30 cm, sometimes decumbent, glabrous. **Sheaths** open, usually shorter than the internodes, with tufts of 1.5–4 mm hairs on the margins of the collars; **auricles** absent; **ligules** membranous, ciliate; **blades** flat or folded, becoming involute on drying. **Inflorescences** terminal, dense panicles, 1–7 cm long, 0.5–2(3) cm wide, branches 1–2 per node; **disarticulation** initially above the glumes, glumes and pedicels sometimes falling together later. **Spikelets** with (4)5–7(10) florets. **Glumes** subequal, exceeding or exceeded by the distal floret, 3–7-veined, margins hyaline; **lemmas** 7–9-veined, margins and intercostal regions usually pubescent, varying to glabrous, margins hyaline, apices bifid or merely notched, sinuses sometimes mucronate, mucros to 1.5 mm; **paleas** spatulate, membranous, 2-veined, 2-keeled; **anthers** 3, 0.2–0.5 mm. **Caryopses** ovoid. $x = 6$. Name from the Greek *schizo*, 'split' referring to the bidentate lemma.

Schismus is a genus of five species that is native to Africa and Asia. Two species are established in the *Flora* region. In using the key and descriptions, the lowest floret in a spikelet should be examined. Succeeding florets tend to have shorter, less acute or acuminate lobes, and a shallower sinus.

SELECTED REFERENCE Conert, H.J. and A.M. Türpe. 1974. Revision der Gattung *Schismus* (Poaceae: Arundinoideae, Danthonieae). Abh. Senckenberg. Naturf. Ges. 532:1–81.

1. Lower glumes equaling or exceeding the distal florets; lemma lobes longer than wide, acute to acuminate; paleas always shorter than the lemmas . . . . 1. *S. arabicus*
1. Lower glumes exceeded by the distal florets; lemma lobes as wide as or wider than long, acute to obtuse; paleas of the lower florets in the spikelets as long as or longer than the lemmas . . . . 2. *S. barbatus*

1. Schismus arabicus Nees [p. 308]
Arabian Schismus

Plants annual. **Culms** (2)6–16 cm. **Ligules** 0.5–1.5 mm, of hairs; **blades** 4–6 cm long, 0.5–2 mm wide, abaxial surfaces glabrous or sparsely pubescent, adaxial surfaces sparsely to densely pubescent. **Panicles** (1)2–3.5 cm. **Spikelets** 5–7 mm. **Lower glumes** 4.2–6.2 mm, equaling or exceeding the distal florets; **upper glumes** 4–6 mm; **lemmas** 1.8–2.6 mm, with dense, spreading pubescence between the veins, lobes longer than wide, acute to acuminate; **paleas** 1.5–2.2 mm, shorter than the lemmas; **anthers** 0.2–0.5 mm. **Caryopses** 0.5–0.8 mm. $2n = 12$.

Schismus arabicus is native to southwestern Asia, but it is now established in the southwestern United States, growing in open and disturbed sites.

2. Schismus barbatus (Loefl. *ex* L.) Thell. [p. 308]
Common Mediterranean Grass

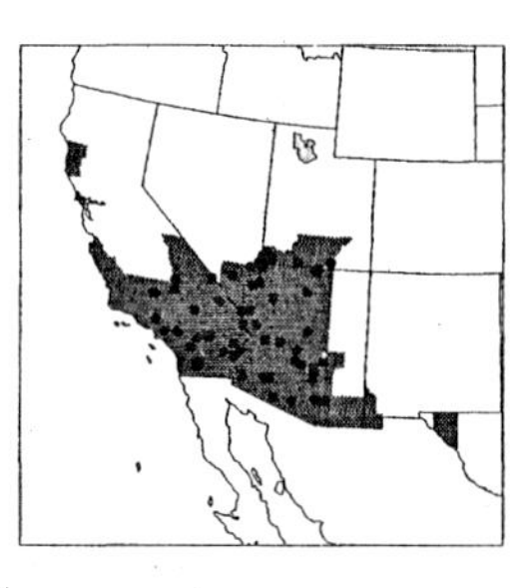

Plants annual. **Culms** 6–27 cm. **Ligules** 0.3–1.1 mm, of hairs; **blades** 3–15 cm long, 0.3–1.5 mm wide, abaxial surfaces glabrous or scabrous, adaxial surfaces scabrous, sparsely long-pubescent near the ligules. **Panicles** 1–6(7.5) cm. **Spikelets** 4.5–7 mm. **Lower glumes** 4–5.2 mm, exceeded by the distal florets; **upper glumes** 4–5.3 mm; **lemmas** 1.5–2(2.5) mm, with sparse, appressed pubescence between the veins, or glabrous and with spreading hairs on the margins, lobes as wide as or wider than long, acute to obtuse; **paleas** 1.7–2.2(2.6) mm, those of the lower florets in the spikelets as long as or longer than the lemmas; **anthers** 0.2–0.4 mm. **Caryopses** 0.6–0.8 mm. $2n = 12$.

Schismus barbatus is native to Eurasia, but it is now established in the southwestern United States. It grows in sandy, disturbed sites along roadsides and fields and in dry riverbeds.

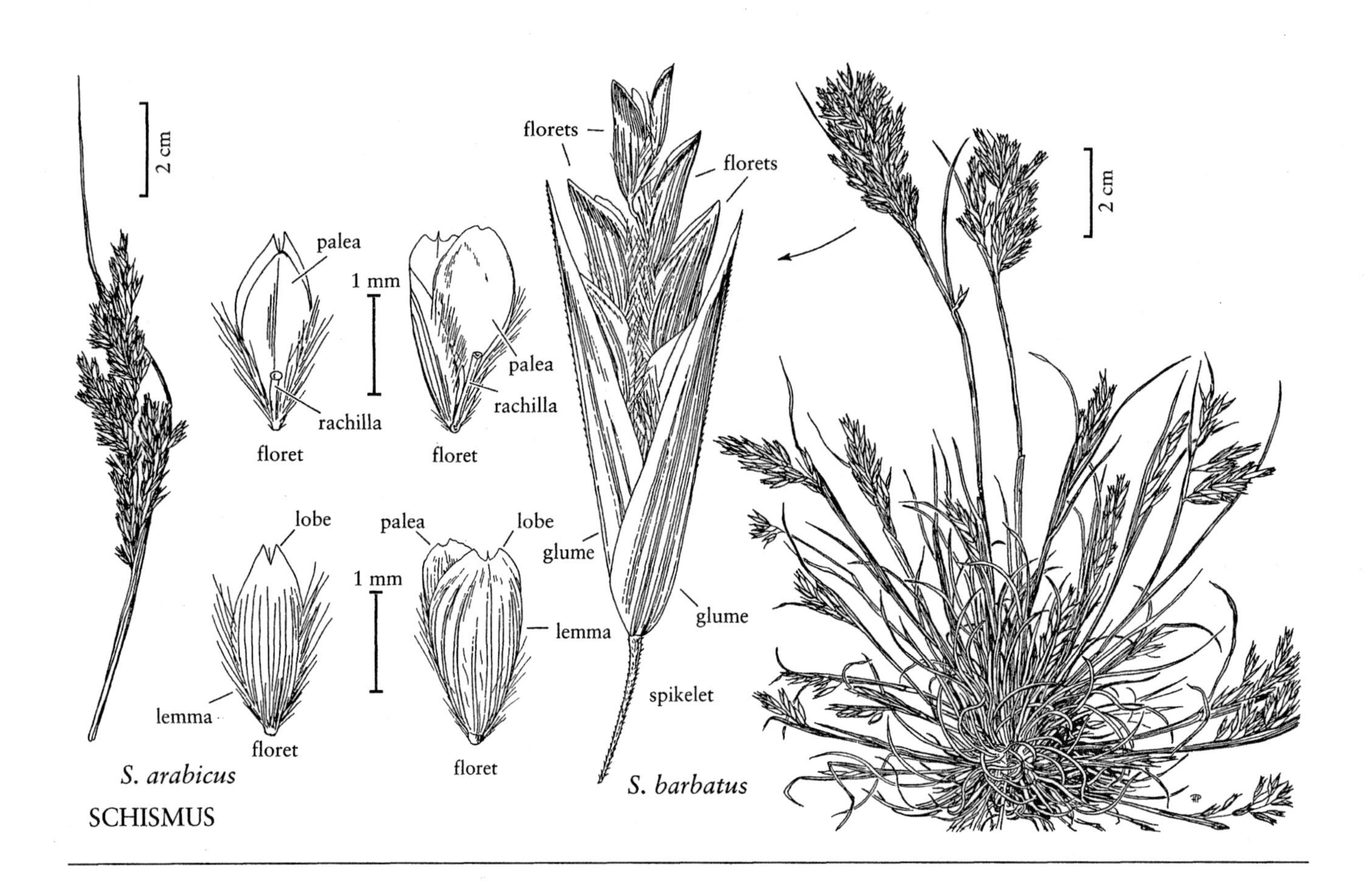

20.04 KARROOCHLOA Conert & Türpe

Stephen J. Darbyshire

Plants annual or perennial; cespitose, stoloniferous or rhizomatous. **Culms** 4–40 cm. **Sheaths** open; **ligules** of hairs; **blades** to 2 mm wide, flat, rolled, or involute. **Inflorescences** terminal, contracted panicles, 0.5–6 cm. **Spikelets** 4–8 mm, laterally compressed, with 3–7 florets; **disarticulation** above the glumes. **Glumes** 3.5–7 mm, subequal, equaling or exceeding the florets, 3–5(7)-veined; **calluses** with lateral tufts of hairs; **lemmas** pubescent and/or with fringes, rows, or tufts of hair, 9-veined, awned from between the 2 apical lobes or teeth, awns twisted and geniculate; **paleas** 2-veined; **lodicules** 2, pubescent; **anthers** 3; **ovaries** glabrous. **Caryopses** about 1 mm; **hila** short. $x = 6$. Name from the Karroo region of South Africa and the Greek *chloa*, 'grass'.

Karroochloa is a southern African genus of four species.

SELECTED REFERENCE **Conert, H.J.** and **A.M. Türpe**. 1969. *Karroochloa*, ein neue Gattung der Gramineen (Poaceae, Arundinoideae, Danthonieae). Senckenberg. Biol. 50:289–318.

1. **Karroochloa purpurea** (L. f.) Conert & Türpe [p. 309]

Plants perennial; densely cespitose, shortly rhizomatous. **Culms** 6–25 cm. **Sheaths** mostly glabrous, sparsely hispid near the throat; **blades** to 4 cm long, to 1 mm wide, rolled, falcate to curled, pilose or hispid. **Panicles** 0.5–2 cm, subcapitate; **pedicels** mostly shorter than or equaling the spikelets. **Spikelets** 4–8 mm, with 3–6 florets. **Glumes** glabrous or scabrous, dark purple fading to brown; **lemma bodies** (1.3)1.5–2 mm, with tufts of hairs on the lower portion and in a transverse row below the sinus, apical teeth 1–2 mm; **awns** (2)3.5–5.5 mm; **anthers** 1.7–2.2 mm. **Caryopses** about 0.8 mm. $2n = 12$.

Karroochloa purpurea was grown in the grass garden of the University of California, Berkeley. There is no evidence that it has become established in the *Flora* region.

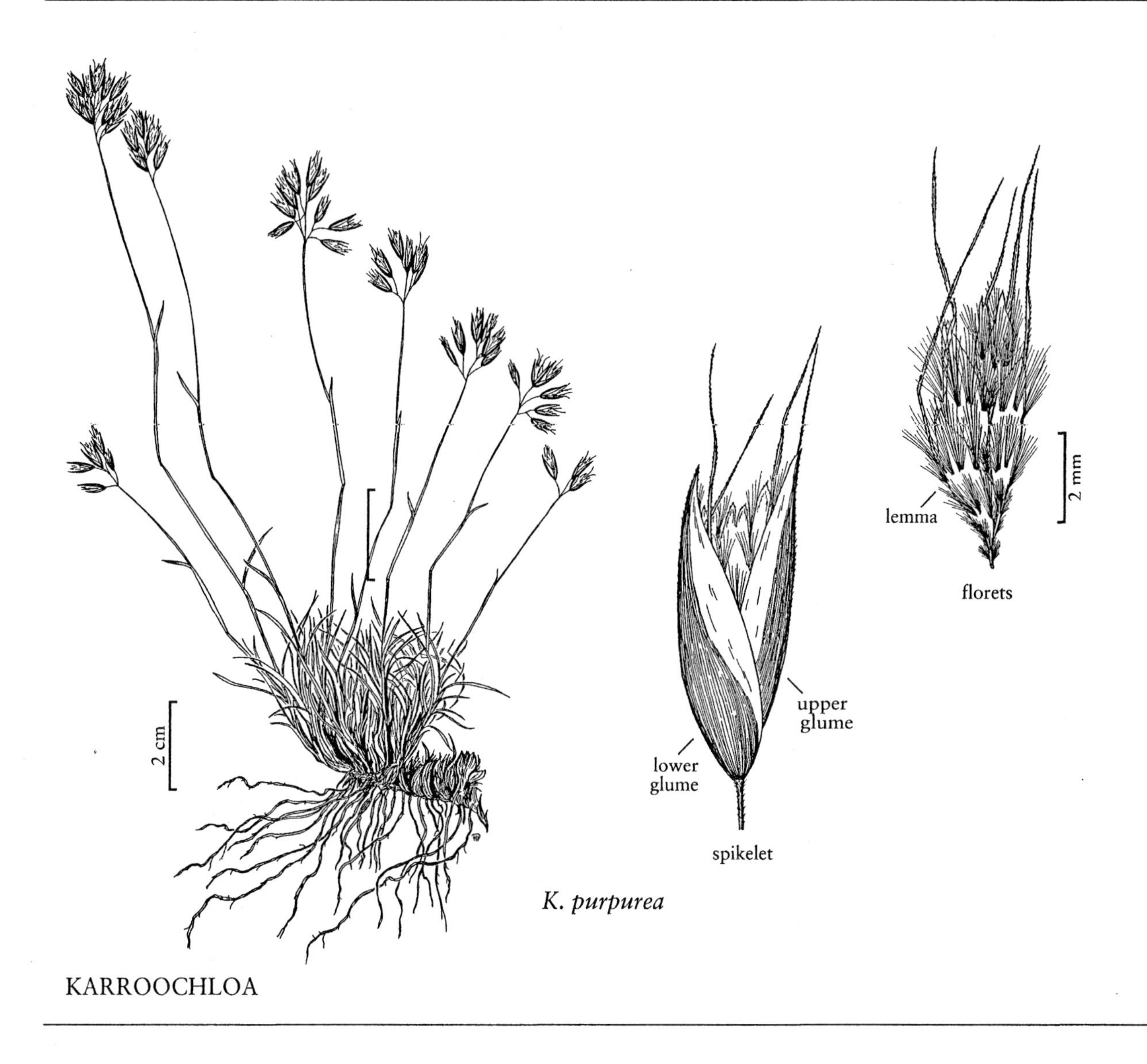

KARROOCHLOA

20.05 RYTIDOSPERMA Steud.

Stephen J. Darbyshire
Henry E. Connor

Plants perennial; cespitose to somewhat spreading, sometimes shortly rhizomatous. **Culms** (1.5)30–90(140) cm. **Sheaths** open, glabrous or hairy, apices with tufts of hair, these sometimes extending across the collar; **ligules** of hairs; **blades** persistent or disarticulating. **Inflorescences** terminal, racemes or panicles. **Spikelets** with 3–10 florets; **florets** bisexual, terminal florets reduced; **disarticulation** above the glumes and between the florets. **Glumes** (2)8–20 mm, subequal or equal, usually exceeding the florets, stiffly membranous; **calluses** with lateral tufts of stiff hairs; **lemmas** ovate to lanceolate, with 2 complete or incomplete transverse rows of tufts of hairs, sometimes reduced to marginal tufts, 5–9-veined, apices bilobed, lobes usually at least as long as the body, acute, acuminate, or aristate, awned from between the lobes, awns longer than the lobes, twisted, usually geniculate; **lodicules** 2, fleshy, with hairs or glabrous. **Caryopses** 1.2–3 mm, obovate to elliptic. **Cleistogenes** absent. $x = 12$. Name from the Greek *rhytidos*, 'wrinkles', and *sperma*, 'seed'.

Rytidosperma, as interpreted here and by Edgar and Connor (2000), is a genus of about 45 species that are native to south and southeastern Asia, Australia, New Zealand, and South America. Linder and Verboom (1996) advocated a narrower interpretation of the genus than Edgar and Connor, but acknowledged that "there is an almost equally strong case for recognizing a single, large genus, *Rytidosperma*" (p. 607). According to their interpretation, all three species treated here would be placed in *Austrodanthonia* H.P. Linder (Linder 1997).

Several species of *Rytidosperma* have been cultivated in research plots or forage trials in North America. The three species treated here have been tried in several states but have escaped cultivation and persisted only in California and Oregon (Weintraub 1953). They have been included in commercial seed mixtures for forage planting in Australia and New Zealand. Other species that have been grown experimentally in both the United States and Canada include *R. caespitosum* (Gaudich.) Connor & Edgar, *R. setaceum* (R. Br.) Connor & Edgar, and *R. tenuius* (Steud.) A. Hansen & P. Sunding. They are not known to have escaped or persisted in North America.

SELECTED REFERENCES **Blumler, M.** 2001. Notes and comments. Fremontia 29:36; **Connor, H.E.** and **E. Edgar**. 1979. *Rytidosperma* Steudel (*Nothodanthonia* Zotov) in New Zealand. New Zealand J. Bot. 17:311–337; **Edgar, E.** and **H.E. Connor**. 2000. Flora of New Zealand, vol. 5. Manaaki Whenua Press, Lincoln, New Zealand. 650 pp.; **Linder, H.P.** 1997. Nomenclatural corrections in the *Rytidosperma* complex (Danthonieae, Poaceae). Telopea 7:269–274; **Linder, H.P.** and **G.A. Verboom**. 1996. Generic limits in the *Rytidosperma* (Danthonieae, Poaceae) complex. Telopea 6:597–627; **Murphy, A.H.** and **R.M. Love**. 1950. Hairy oatgrass, *Danthonia pilosa* R. Br., as a weedy range grass. Bull. Calif. Dep. Agric. 39:118–124; **Myers, W.M.** 1947. Cytology and genetics of forage grasses (concluded). Bot. Rev. 7:369–419; **Vickery, J.W.** 1956. A revision of the Australian species of *Danthonia* DC. Contr. New South Wales Natl. Herb. 2:249–325; **Weintraub, F.C.** 1953. Grasses Introduced into the United States. Agricultural Handbook No. 58. Forest Service, U.S. Department of Agriculture, Washington, D.C., U.S.A. 79 pp.; **Zotov, V.D.** 1963. Synopsis of the grass subfamily Arundinoideae in New Zealand. New Zealand J. Bot. 1:78–136.

1. Upper row of lemma hairs in a more or less continuous row of tufts, the hairs much exceeding the base of the awn; shoots intravaginal, without scaly cataphylls . . . . 2. *R. biannulare*
1. Upper row of lemma hairs in isolated tufts or only at the margins, the hairs not or only just exceeding the base of the awn; some or most shoots extravaginal and with scaly cataphylls.
   2. Callus hairs usually overlapping the lower row of lemma hairs; upper row of lemma hairs often reduced to marginal tufts, sometimes scanty medial tufts also present . . . . 1. *R. penicillatum*
   2. Callus hairs short, rarely reaching the lower row of lemma hairs; upper row of lemma hairs usually with scanty medial tufts . . . . 3. *R. racemosum*

1. **Rytidosperma penicillatum** (Labill.) Connor & Edgar [p. 311]

Hairy Danthonia, Hairy Oatgrass, Poverty Grass

Plants loosely cespitose to somewhat spreading, shortly rhizomatous. **Culms** 30–90 cm, erect, mostly smooth and glabrous, scabrous-pubescent immediately below the inflorescence, branching extravaginal, the new shoots with scaly cataphylls. **Leaves** mostly basal, greatly exceeded by the culms, flag leaf blades usually not reaching the inflorescence; **sheaths** densely hairy or glabrous, with apical tufts of hairs, apical hairs 1–3.5 mm; **ligules** 0.1–1 mm; **blades** to 30 cm long and 5 mm wide, flat, folded, or rolled, pubescent or glabrous. **Inflorescences** 4–10 cm, racemose or paniculate, contracted; **pedicels** much shorter than the spikelets. **Spikelets** 9–15(18) mm, longer than the rachis internodes, with 5–10 florets; **rachilla segments** 0.2–0.5 mm. **Glumes** 8–14(17.5) mm, subequal, lanceolate, sometimes with scattered hairs; **lower glumes** (5)7–9(11)-veined; **upper glumes** 5–7(9)-veined; **calluses** 0.5–1.2 mm, longer than wide, with marginal tufts of hairs usually reaching the lower lemma hairs; **lemma bodies** (2)2.5–4 mm, 9-veined, lower row of hairs continuous or with weak central tufts, hairs of the marginal tufts not or just reaching the upper row of hairs, upper row of hairs composed of 2 marginal tufts, sometimes with 2 additional scanty tufts between, hairs reaching or slightly exceeding the base of the awn; **lobes** 5–13 mm, aristate; **awns** (7)9–16 mm; **paleas** 3–6 mm, exceeding the lemma sinuses, emarginate, intercostal region glabrous or scabrous, margins glabrous or sparsely long-hairy, veins ciliate; **anthers** 0.4–2.5 mm. **Caryopses** 1.8–2.5(3) mm long, 0.8–1.1(1.6) mm wide; **embryos** 0.7–1(1.5) mm; **hila** 0.4–0.5(0.7) mm. $2n$ = unknown.

Rytidosperma penicillatum is endemic to Australia and has been introduced to New Zealand as well as North America. Although considered a poor quality forage, it was introduced and grown experimentally in several states

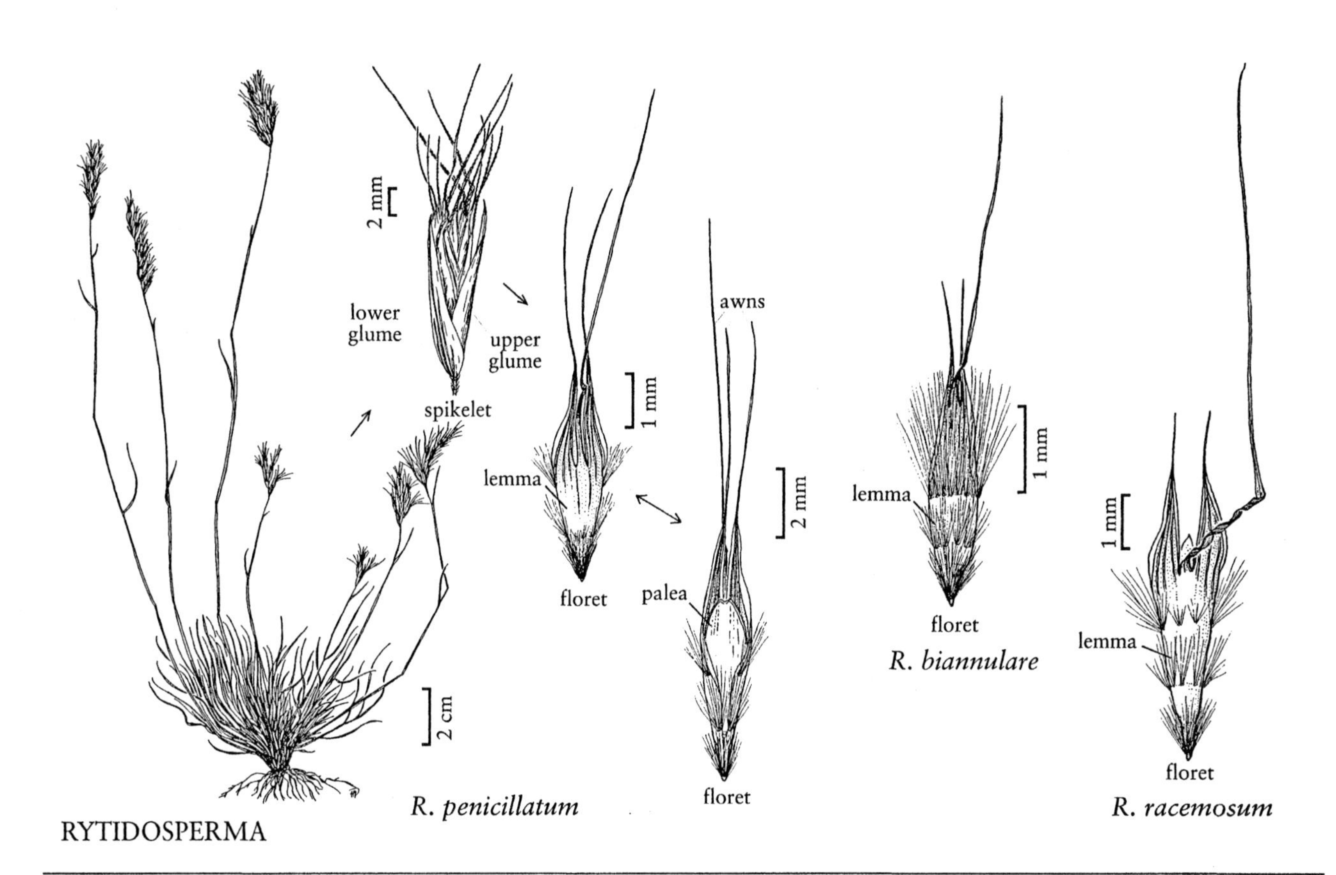

under the name *Danthonia pilosa* R. Br. [= *R. pilosa* (R. Br.) Connor & Edgar]. It has become well-established in northern California and southwestern Oregon, mainly in coastal areas. Since it does well on dry, nutrient depleted soils and competes well with more desirable species, it is considered a troublesome pest.

2. **Rytidosperma biannulare** (Zotov) Connor & Edgar [p. 311]

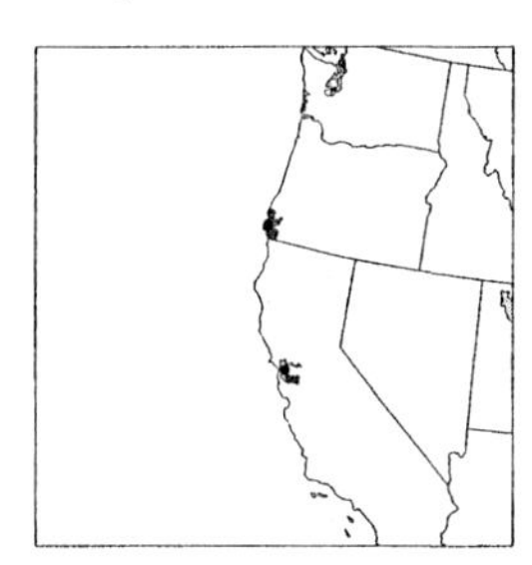

Plants cespitose. **Culms** 30–85 cm, erect, branching intravaginal. **Leaves** mostly basal, exceeded by the culms, flag leaf blades usually reaching or exceeding the inflorescences; **sheaths** mostly glabrous, often purplish distally, with apical tufts of hairs sometimes present, hairs to 5 mm; **ligules** 0.3–0.5(1) mm; **blades** 30–40 cm long, to 5 mm wide, usually involute, margins, apices, and sometimes adaxial surfaces scabrous, young blades sparsely pilose, becoming glabrous at maturity. **Inflorescences** 10–20 cm, narrow, dense panicles; **rachises and pedicels** scabrous. **Spikelets** (7)10–15 mm, longer than the rachis internodes, with 6–7 florets; **rachilla segments** 0.3–0.5 mm. **Glumes** 7.6–11(13.2) mm, subequal, lanceolate, subacute; **lower glumes** 5–7(9)-veined; **upper glumes** 5-veined; **calluses** 0.5–0.7 mm, with marginal tufts of hairs reaching the lower lemma hairs; **lemma bodies** 1.8–2.4(2.8) mm, 7–9-veined, lower and upper rows of hairs dense, hairs of lower rows usually not or only just reaching the upper rows, sometimes ill-defined, hairs of upper rows clearly exceeding the base of the awn but exceeded by the lemma lobes, margins ciliate, other portions of the lemma with short scattered hairs, sometimes glabrous below the upper row of hairs; **lobes** 3.5–5(8.5) mm, aristate; **awns** 6–10(12.5) mm; **paleas** 2.5–4.6 mm, exceeding the lemma sinuses, emarginate, intercostal region sparsely hairy, margins usually long-hairy, veins ciliate; **anthers** 0.8–1.6 mm. **Caryopses** 1.2–1.9 mm long, 0.6–0.8 mm wide; **embryos** 0.5–0.8 mm; **hila** 0.3–0.6 mm. $2n$ = unknown.

Rytidosperma biannulare is endemic to New Zealand. As early as 1905, it was grown experimentally under the name *Danthonia semiannularis* (Labill.) R. Br. in several states. Although frequently mentioned in literature, only a few specimens document its persistence in Oregon and California.

3. **Rytidosperma racemosum** (R. Br.) Connor & Edgar [p. 311]

Plants loosely cespitose, shortly rhizomatous. **Culms** to 90 cm, erect, branching extravaginal, new shoots with scaly cataphylls. **Leaves** mostly basal, exceeded by or as long as the culms, flag leaf blades usually reaching the inflorescences; **sheaths** glabrous or with scattered hairs, becoming brownish, with apical tufts of hairs, hairs to 4 mm; **ligules** 0.2–0.5 mm; **blades** (5)15–25 cm long, to 2 mm wide, flat or involute, glabrous or pubescent. **Inflorescences** 5–15 cm, racemose or with a few branches, narrow; **rachises** and **pedicels** scabrous. **Spikelets** (7)10–13(16) mm, sometimes shorter than the rachis internodes, with 6–7(10) florets; **rachilla segments** 0.1–0.2 mm. **Glumes** (7)8–13(16) mm, subequal, lanceolate, subacute, glabrous; **lower glumes** (5)7-veined; **upper glumes** 5(7)-veined, sometimes with a few hairs; **calluses** (0.6)0.9–1.5(2) mm, with marginal tufts of hairs not or barely reaching the lower lemma hairs; **lemma bodies** 2.5–3.5(4.5) mm, (7)9-veined, lower row of hairs dense, hairs not or just reaching the upper rows, upper row of hairs reaching or slightly exceeding the base of the awn, scanty medial tufts sometimes present; **lobes** 5–10 mm, abruptly aristate; **awns** 11–14 mm; **paleas** 3.5–5 mm, exceeded by the lemma sinuses, emarginate, glabrous or with a few hairs, veins ciliate; **anthers** 0.3–2 mm. **Caryopses** 1.7–2.1(2.5) mm long, 0.8–1.1(1.3) mm wide; **embryos** 0.8–0.9 mm; **hila** 0.4–0.5 mm. $2n = 24$.

Rytidosperma racemosum is endemic to Australia and has been introduced to New Zealand. Grown experimentally in several places in North America, including Berkeley, California, it seems to have become established in only a few places around central California.

20.06 TRIBOLIUM Desv.

Gerrit Davidse

Plants annual or perennial; cespitose, sometimes stoloniferous, occasionally with rhizomes. **Culms** 2–60 cm. **Sheaths** open, throats glabrous or long-ciliate; **ligules** of hairs, or membranous and ciliate, or ciliolate; **blades** flat or rolled, glabrous or villous. **Inflorescences** terminal, sometimes also axillary, panicles, spikes, or racemes. **Spikelets** with 2–5(10) florets, distal florets reduced; **disarticulation** above the glumes and between the florets. **Glumes** exceeding or exceeded by the florets, herbaceous, 3–5-veined, glabrous or scabridulous, sometimes with stiff, papillose-based hairs, glume apices acute or acuminate; **calluses** glabrous; **lemmas** herbaceous, with acute or clavate hairs in marginal rows or variously scattered, lemma apices unlobed, acute to long-acuminate; **paleas** sometimes with tufts of hairs on the margins; **lodicules** 2, obtriangular, glabrous or with bristles, 1–3-veined, lodicule apices at least as thick as the base; **anthers** 3, 0.3–3.1 mm. **Caryopses** 0.7–2.2 mm long, 0.4–1.1 mm wide; **pericarp** poorly separable, dull, smooth or rugulose, glabrous; **embryos** ⅓–½ as long as the caryopses; **hila** punctiform. $x = 6$. Desvaux did not explain the etymology of *Tribolium*. It may be derived from the Greek *trilobos*, a name of various prickley plants (Quattrocchi 1999) or the Latin *tria*, 'three', and *bolus*, a fiery meteor in the form of an arrow, referring to the appearance of the mature spikelets with their three florets and coarsely hairy glumes, which, like the lemmas, may be awned or acuminate (Clifford 1996).

Tribolium is a southern African genus of 10 species. It is unusual in the tribe in having unlobed lemmas, but has the haustorial synergids, bilobed or bi-awned prophylls, and stalked ovaries characteristic of the tribe. Linder and Davidse (1997) suggested that its closest relatives are probably *Schismus*, *Karroochloa*, and *Rytidosperma*.

SELECTED REFERENCES **Clifford, H.T.** 1996. Etymological Dictionary of Grasses, Version 1.0 (CD-ROM). Expert Center for Taxonomic Identification, Amsterdam, The Netherlands; **Linder, H.P.** and **G. Davidse.** 1997. The systematics of *Tribolium* Desv. (Danthonieae: Poaceae). Bot. Jahrb. Syst. 119:445–507; **Quattrocchi, U.** 2000. CRC World Dictionary of Plant Names: Common Names, Scientific Names, Eponyms, Synonyms, and Etymology, vol. 4: R–Z. CRC Press, Boca Raton, Florida, U.S.A. 2896 pp.

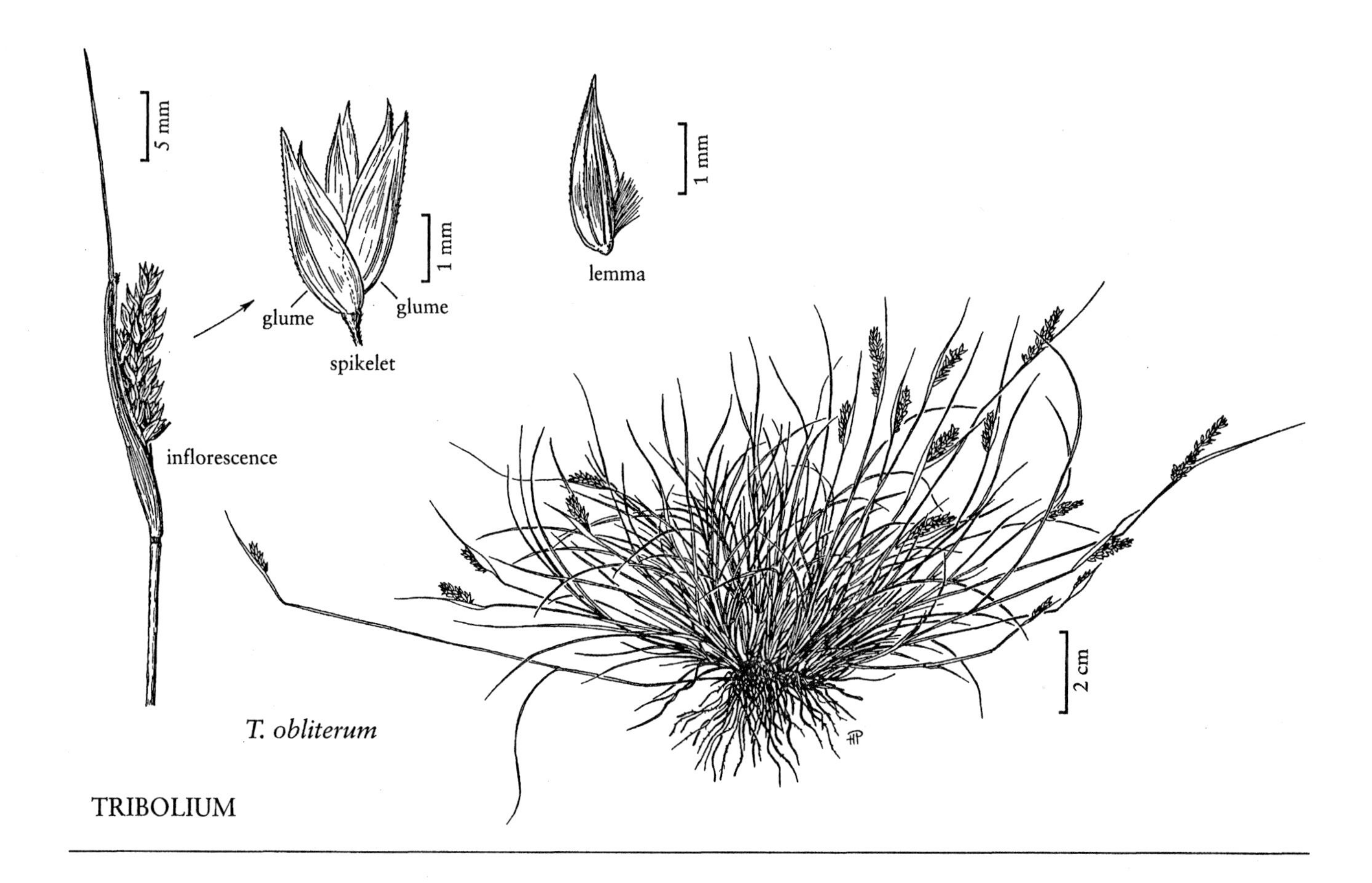

1. **Tribolium obliterum** (Hemsley) Renvoize [p. 313]

Plants perennial; usually cespitose, sometimes stoloniferous, stolons to 30 cm. **Culms** 12–40 cm, decumbent or erect, branching intravaginal. **Sheaths** mostly glabrous, ciliate distally; **ligules** about 0.2 mm, membranous, ciliolate; **blades** 1–8 cm long, 0.5–1 mm wide. **Inflorescences** terminal, panicles, 1–5 cm long, 5–15 mm wide, obovate or cylindrical, with 10–40 spikelets; **branches** scabridulous. **Spikelets** 3.5–4.5 mm, with 5–10 florets. **Glumes** 2.5–3.5 mm, longer than the basal lemmas but exceeded by the distal florets, 5-veined, glabrous or scabridulous over the veins, acute; **lemmas** 2–3.5 mm, with submarginal tufts of hairs in the lower ½, hairs acute, 0.2–0.3 mm, lemma apices acute to acuminate; **paleas** 1.5–2 mm, glabrous between the keels, sometimes with tufts of hairs on the margins; **anthers** 0.3–0.5 mm, pale yellow. **Caryopses** obovate, 0.7–1 mm long, 0.5–0.7 mm wide, plano-convex to concavo-convex, pale brown, smooth; **embryos** 0.4–0.5 mm; **hila** 0.1 mm. $2n$ = 24, 36.

Tribolium obliterum is native to Cape Province, South Africa, where it usually grows in gravelly, well-drained soils at elevations below 600 m. It has been introduced into Australia and St. Helena and was recently discovered in a roadside ditch near Fort Ord, Monterey County, California. It appears to be naturalized there.

8. ARISTIDOIDEAE Caro

Grass Phylogeny Working Group
Kelly W. Allred

Plants annual or perennial; usually cespitose. **Culms** annual, erect, solid or hollow, usually unbranched. **Leaves** distichous; **sheaths** usually open; **auricles** absent; **abaxial ligules** absent or of hairs; **adaxial ligules** membranous and ciliate or of hairs; **blades** without pseudopetioles; **mesophyll cells** radiate or non-radiate; **adaxial palisade layer** absent; **fusoid cells** absent; **arm cells** absent; **kranz anatomy** absent or present, when present, with 1 or 2 parenchyma sheaths; **midribs** simple; **adaxial bulliform cells** present; **stomatal subsidiary cells** dome-shaped or triangular; **bicellular microhairs** present, with long, slender, thin-walled terminal cells. **Inflorescences** terminal, not leafy, usually panicles, sometimes spikes or racemes; **disarticulation** above the glumes. **Spikelets** bisexual, with 1 floret; **rachilla extension** absent. **Glumes** 2, usually longer than the florets, usually acute or acuminate; **florets** terete or laterally compressed, with well-developed calluses; **lemmas** 1- or 3-veined, more or less coriaceous, with a germination flap, lemma margins overlapping at maturity and concealing the paleas, apices evidently 3-awned; **awn bases** often forming a column, lateral awns occasionally reduced or absent; **paleas** less than ½ as long as the lemmas; **lodicules** usually present, 2, free, membranous, glabrous, heavily vascularized; **anthers** 1–3; **ovaries** glabrous; **haustorial synergids** absent; **styles** 2, free to the base but close. **Caryopses** usually fusiform, falling with the lemma and palea attached; **hila** short or long, linear; **endosperm** hard, without lipid; **starch grains** compound; **embryos** small or large relative to the caryopses; **epiblasts** absent; **scutellar cleft** present or absent; **mesocotyl internode** elongated; **embryonic leaf margins** meeting. $x = 11, 12$.

The subfamily *Aristidoideae* includes only one tribe, the *Aristideae*. Other taxonomists have generally included the *Aristidoideae*, with the *Danthonieae* and *Arundineae*, in the *Arundinoideae* (e.g., Watson et al. 1985; Clayton and Renvoize 1986; Kellogg and Campbell 1987), but Esen and Hilu (1991) demonstrated that *Aristida* is clearly distinct from the *Danthonieae* and *Arundineae* in terms of its prolamins. Subsequent work has provided further support for the monophyly of the three tribes, but their position relative to each other and other members of the PACCAD clade is more equivocal.

SELECTED REFERENCES **Caro, J.A.** 1982. Sinopsis taxonómica de las gramíneas argentinas. Dominguezia 4:1–51; **Clayton, W.D.** and **S.A. Renvoize.** 1986. Genera Graminum: Grasses of the World. Kew Bull., Addit. Ser. 13. Her Majesty's Stationery Office, London, England. 389 pp; **Grass Phylogeny Working Group.** 2001. Phylogeny and subfamilial classification of the grasses (Poaceae). Ann. Missouri Bot. Gard. 88:373–457; **Esen, A.** and **K.W. Hilu.** 1991. Electrophoretic and immunological studies of prolamins in the Poaceae. II. Phylogenetic affinities of the Aristideae. Taxon 40:5–17; **Kellogg, E.A.** and **C.S. Campbell.** 1987. Phylogenetic analyses of the Gramineae. Pp. 310–322 *in* T.R. Soderstrom, K.W. Hilu, C.S. Campbell, and M.E. Barkworth (eds.) Grass Systematics and Evolution. Smithsonian Institution Press, Washington, D.C., U.S.A. 473 pp; **Watson, L., H.T. Clifford,** and **M.J. Dallwitz.** 1985. The classification of the Poaceae: Subfamilies and supertribes. Austral. J. Bot. 33:433–484; **Watson, L.** and **M.J. Dallwitz.** 1992. The Grass Genera of the World. C.A.B. International, Wallingford, England. 1038 pp.

21. ARISTIDEAE C.E. Hubb.

Kelly W. Allred

See subfamily description.

The tribe *Aristideae* has three genera and 300–350 species, and is primarily pan-tropical in its distribution. Its members are usually readily recognized by their terete, 3-awned lemmas with overlapping margins. *Aristida*, which has many more species than the other two genera combined, is the only genus found in the Americas.

21.01 ARISTIDA L.

Kelly W. Allred

Plants usually perennial; herbaceous, usually cespitose, occasionally rhizomatous. **Culms** 10–150 cm, not woody, sometimes branched above the base; **internodes** usually pith-filled, sometimes hollow. **Leaves** sometimes predominantly basal, sometimes predominantly cauline; **sheaths** open; **auricles** lacking; **ligules** of hairs or very shortly membranous and long-ciliate, the 2 types generally indistinguishable. **Inflorescences** terminal, usually panicles, sometimes racemes, occasionally spikes; **primary branches** without axillary pulvini and usually appressed to ascending, or with axillary pulvini and ascending to strongly divergent or divaricate. **Spikelets** with 1 floret; **rachillas** not prolonged beyond the florets; **disarticulation** above the glumes. **Glumes** often longer than the florets, thin, usually 1–3-veined, acute to acuminate; **florets** terete or weakly laterally compressed; **calluses** well-developed, hirsute; **lemmas** fusiform, 3-veined, convolute, usually glabrous or scabridulous, usually enclosing the palea at maturity, usually with 3 terminal awns, lateral awns reduced or obsolete in some species, lemma apices sometimes narrowed to a straight or twisted beak below the awns; **awns** ascending to spreading, usually straight, bases sometimes twisted together into a column or the bases of the individual awns coiled, twisted, or otherwise contorted, occasionally disarticulating at maturity; **paleas** shorter than the lemmas, 2-veined, occasionally absent; **anthers** 1 or 3. **Caryopses** fusiform; **hila** linear. x = 11, 12. Name from the Latin *arista*, 'awn'.

Aristida is a tropical to warm-temperate genus of 250–300 species. It grows throughout the world in dry grasslands and savannahs, sandy woodlands, arid deserts, and open, weedy habitats and on rocky slopes and mesas. All 29 species in this treatment are native to the *Flora* region.

The divergent awns aid in wind and animal transportation of the florets and, by holding the florets and the caryopses they contain at an angle to the ground, in establishment. The presence of *Aristida* frequently indicates soil disturbance or abuse. Although generally poor forage grasses and, because of the calluses, potentially harmful to grazing animals, some species of *Aristida* are an important source of spring forage on western rangelands. Quail and small mammals eat small amounts of the seed.

SELECTED REFERENCES **Allred, K.W.** 1984. Morphologic variation and classification of the North American *Aristida purpurea* complex (Gramineae). Brittonia 36:382–395; **Allred, K.W.** 1984, 1985, 1986. Studies in the genus *Aristida* (Gramineae) of the southeastern United States. Rhodora 86:73–77; 87:137–145, 145–155; 88:367–387; **Henrard, J.T.** 1926, 1927, 1928, 1933. A critical revision of the genus *Aristida*. Meded. Rijks-Herb. 54:1–701; 55C:703–747; **Henrard, J.T.** 1929, 1933. A monograph of the genus *Aristida*. Meded. Rijks-Herb. 58:1–325; **Kesler, T.R.** 2000. A taxonomic reevaluation of *Aristida stricta* (Poaceae) using anatomy and morphology. Master's thesis. Florida State University, Tallahassee, Florida, U.S.A. 33 pp.; **Peet, R.K.** 1993. A taxonomic study of *Aristida stricta* and *A. beyrichiana*. Rhodora 95:25–37; **Reeder, J.R.** and **R.S. Felger.** 1989. The *Aristida californica–glabrata* complex (Gramineae). Madroño 36:187–197; **Trent, J.S.** and **K.W. Allred.** 1990. A taxonomic comparison of *Aristida ternipes* and *Aristida hamulosa* (Gramineae). Sida 14:251–261; **Vaughn, J.M.** 1981. Systematics of *Aristida dichotoma*, *basiramea*, and *curtissii* (Poaceae). Master's thesis. University of Oklahoma, Norman, Oklahoma, U.S.A. 38 pp.; **Walters, T.E., D.S. Decker-Walters,** and **D.R. Gordon.** 1994. Restoration considerations for wiregrass (*Aristida stricta*): Allozymic diversity of populations. Conservation Biol. 8:581–585.

NOTE: Lemma lengths are measured from the base of the callus to the divergence of the awns.

1. Lower glumes 3–7-veined.
   2. Awns nearly equal, the lateral awns 8–66 mm long and at least ¾ as long as the central awns . . . . 13. *A. oligantha*
   2. Awns markedly unequal, the lateral awns 1–4 mm long, no more than ½ as long as the central awns, sometimes absent.
      3. Plants annual; inflorescences 5–12 cm long, 2–4 cm wide . . . . 12. *A. ramosissima*
      3. Plants perennial; inflorescences 10–30 cm long, 4–26 cm wide . . . . 7. *A. schiedeana*

1. Lower glumes 1–2(3)-veined.
    4. Central awns spirally coiled at the base.
        5. Lateral awns 1–4 mm long, erect .... 15. *A. dichotoma*
        5. Lateral awns 5–13 mm long, spreading .... 14. *A. basiramea*
    4. Central awns straight to curved, sometimes loosely contorted but not spirally coiled, at the base.
        6. Lateral awns markedly reduced, usually ⅓ or less as long as the central awns.
            7. Panicles 1–6 cm wide, the branches erect-appressed to strongly ascending, without axillary pulvini or the pulvini only weakly developed.
                8. Plants annual; culms often highly branched above the base.
                    9. Awns flattened at the base .... 17. *A. adscensionis*
                    9. Awns terete at the base.
                        10. Lemmas 2.5–10 mm long; central awns curving up to 100° at the base .... 16. *A. longespica*
                        10. Lemmas 8–22 mm long; central awns with a semicircular bend at the base .... 12. *A. ramosissima*
                8. Plants perennial; culms rarely branched above the base.
                    11. Collars densely pilose, the hairs 1–3 mm, often densely tangled and deflexed; blades usually tightly involute, about 0.5 mm in diameter .... 11. *A. gypsophila*
                    11. Collars mostly glabrous or with straight hairs, often with long hairs at the sides; blades usually flat to loosely involute, sometimes tightly involute.
                        12. Lateral awns absent; panicle branches spikelet-bearing to the base; plants of the Florida keys .... 5. *A. floridana*
                        12. Lateral awns usually present, varying from much shorter than to equaling the central awns; panicle branches sometimes naked near the base; plants rarely found east of the Mississippi and not known at all from Florida.
                            13. Primary panicle branches 3–6 cm long; lateral awns (1)8–140 mm long .... 19. *A. purpurea*
                            13. Primary panicle branches 6–16 cm long; lateral awns absent or to 1(3) mm long .... 7. *A. schiedeana*
            7. Panicles 6–45 cm wide, at least the lower branches spreading and having well-developed axillary pulvini.
                14. Lateral awns absent or no more than 3 mm long.
                    15. Central awns often deflexed at a sharp angle when mature; lemma apices often twisted at maturity.
                        16. Blades usually flat, sometimes folded, 1–2 mm wide; plants of juniper, oak, or pine woodlands .... 7. *A. schiedeana*
                        16. Blades usually tightly involute, about 0.5 mm in diameter; plants of thorn-scrub deserts .... 11. *A. gypsophila*
                    15. Central awns usually straight or arcuate; lemma apices not twisted.
                        17. Panicle branches spikelet-bearing from the base; lower glumes longer than the upper glumes .... 5. *A. floridana*
                        17. Panicle branches usually naked at the base; lower glumes about equal to the upper glumes .... 6. *A. ternipes*
                14. Lateral awns 3–23 mm long.
                    18. Anthers 0.8–1 mm long.
                        19. Spikelets usually divergent and the pedicels with axillary pulvini; secondary branches usually absent; primary branches 2–6 cm long; lemma apices with 0–2 twists when mature .... 9. *A. havardii*
                        19. Spikelets usually appressed and the pedicels without axillary pulvini; secondary branches usually well-developed; primary branches 5–13 cm long; lemma apices with 4 or more twists when mature .... 8. *A. divaricata*
                    18. Anthers 1.2–3 mm long.
                        20. Collars glabrous or strigillose; blades with scattered hairs 1.5–3 mm long above the ligule on the adaxial surface; lower glumes about equal to or slightly shorter than the upper glumes .... 6. *A. ternipes*

20. Collars pubescent, with hairs 0.2–0.8 mm long; blades glabrous, sometimes scabridulous, above the ligule on the adaxial surface; lower glumes slightly longer than the upper glumes . . . . 4. *A. patula*

6. Lateral awns well-developed, usually at least ½ as long as the central awns.
21. Blades tightly involute, the adaxial surfaces densely scabrous or densely short-pubescent . . . . 21. *A. stricta*
21. Blades flat or folded and lax, or, if involute, the adaxial surfaces neither densely scabrous nor densely short pubescent.
22. Rachis nodes and leaf sheaths usually lanose or floccose, sheaths occasionally glabrous . . . . 25. *A. lanosa*
22. Rachis nodes glabrous, scabrous, or with straight hairs; leaf sheaths glabrous, pilose, or floccose.
23. Junction of the lemma and awns evident; awns disarticulating at maturity.
24. Plants perennial.
25. Culms 45–100 cm tall; culms unbranched or sparingly branched; blades 12–28 cm long . . . . 18. *A. spiciformis*
25. Culms 10–40 cm tall; culms much branched; blades usually less than 6 cm long . . . . 3. *A. californica*
24. Plants annual.
26. Awns divergent but not arcuate or entwined above the column; cauline internodes pubescent or glabrous . . . . 3. *A. californica*
26. Awns strongly arcuate, often entwined above the column or no column present; cauline internodes glabrous.
27. Glumes 10–17 mm long; lemmas beaked, the beak 2–7 mm long; awns not forming a column; calluses 1–2.5 mm long . . . . . . 1. *A. desmantha*
27. Glumes 20–30 mm long; lemmas not beaked; awns forming a column 8–15 mm long; calluses 3–4 mm long . . . . 2. *A. tuberculosa*
23. Junction of the lemma and awns not evident; awns not disarticulating at maturity.
28. Lemmas terminating in a beak 7–30 mm long; upper glumes awned, the awns 10–12 mm long . . . . 18. *A. spiciformis*
28. Lemmas not beaked or with a beak less than 7 mm long; upper glumes unawned or with an awn to 6 mm long.
29. At least the lower primary panicle branches divergent and with axillary pulvini.
30. Lateral awns about ½ as thick as the central awns . . . . 4. *A. patula*
30. Lateral awns nearly as thick as the central awns.
31. Panicles narrow and contracted above, usually only the lower 1–2 branches spreading and with a pulvinus; lemma apices 0.2–0.3 mm wide . . . . 19. *A. purpurea*
31. Almost all panicle branches spreading and with axillary pulvini; lemma apices 0.1–0.2 mm wide.
32. Anthers 0.8–1 mm long.
33. Spikelets usually divergent and the pedicels with axillary pulvini; secondary branches absent or nearly so; primary branches 2–6 cm long; lemma apices straight or with 1 or 2 twists . . . . 9. *A. havardii*
33. Spikelets usually appressed and the pedicels without axillary pulvini; secondary branches usually well-developed; primary branches 5–13 cm long; lemma apices with 4 or more twists at maturity . . . . . . 8. *A. divaricata*
32. Anthers 1–3 mm long.
34. Base of the blades with scattered hairs 1.5–3 mm long on the adaxial surfaces . . . . 6. *A. ternipes*

34. Base of the blades glabrous or puberulent on the adaxial surface, the hairs, if present, less than 0.5 mm long.
    35. Glumes reddish, the lower glumes often shorter than the upper glumes; awns ascending to divaricate, (8)13–140 mm long; terminal spikelets usually appressed and without axillary pulvini . . . . 19. *A. purpurea*
    35. Glumes brownish, equal or unequal; awns spreading to horizontal, 5–15 mm long; terminal spikelets often spreading from axillary pulvini . . . . 10. *A. pansa*

29. Lower primary panicle branches (pedicels in racemose species) appressed, without axillary pulvini.
    36. Plants with well-developed rhizomes; basal sheaths shredding into threadlike segments at maturity . . . . 22. *A. rhizomophora*
    36. Plants tufted, without rhizomes; basal sheaths not fibrous, not shredding into threadlike segments even when old.
        37. Lower inflorescence nodes with only 1 spikelet; inflorescences spicate or racemose . . . . 23. *A. mohrii*
        37. Lower inflorescence nodes with 2 or more spikelets; inflorescences racemose or paniculate.
            38. Lower glumes usually ⅓–¾ as long as the upper glumes.
                39. Plants annual . . . . 17. *A. adscensionis*
                39. Plants perennial.
                    40. Lemma awns 8–15 mm long; lemmas 5–7 mm long . . . . 29. *A. gyrans*
                    40. Lemma awns (8)15–140 mm long; lemmas 6–16 mm long . . . . 19. *A. purpurea*
            38. Lower glumes usually more than ¾ as long as the upper glumes.
                41. Plants annual.
                    42. Awns flat at the base . . . . 17. *A. adscensionis*
                    42. Awns terete at the base . . . . 16. *A. longespica*
                41. Plants perennial.
                    43. Lemma apices prominently twisted for 3–6 mm; blades usually curled at maturity; leaves forming a basal tuft . . . . 20. *A. arizonica*
                    43. Lemma apices straight or only slightly twisted; blades usually not curled at maturity; leaves variously distributed.
                        44. Lower glumes prominently 2-keeled, (7.5)9–13 mm long; central awns 15–40 mm long . . . . 26. *A. palustris*
                        44. Lower glumes usually 1-keeled, if 2-keeled, 5–9 mm long; central awns 8–25 mm long. [revert to left]

45. Central awns about twice as thick as the lateral awns, divergent to arcuate-reflexed.
    46. All 3 awns divergent to reflexed and contorted at the base; lower rachis nodes usually associated with 2 spikelets (occasionally 1 or 3), 1 pedicellate and 1 sessile . . . . 24. *A. simpliciflora*
    46. Lateral awns usually erect to ascending and not contorted at the base; lower rachis nodes usually associated with more than 2 spikelets, pedicellate to subsessile . . . . . 27. *A. purpurascens*
45. Central awns about the same thickness as the lateral awns, erect to spreading.
    47. Lower glumes 1–4 mm longer than the upper glumes . . . . 27. *A. purpurascens*
    47. Lower glumes from shorter than to 1 mm longer than the upper glumes.

48. Culms usually 3–6 mm thick at the base; primary panicle branches 4–20 cm long; lower glumes 1-veined . . . . 28. *A. condensata*
48. Culms usually 1–4 mm thick at the base; primary panicle branches 1–5 cm long; lower glumes 1–2-veined.
    49. Calluses 0.4–0.8 mm long . . . . 27. *A. purpurascens*
    49. Calluses 1–2 mm long . . . . 29. *A. gyrans*

1. **Aristida desmantha** Trin. & Rupr. [p. 320]
CURLY THREEAWN

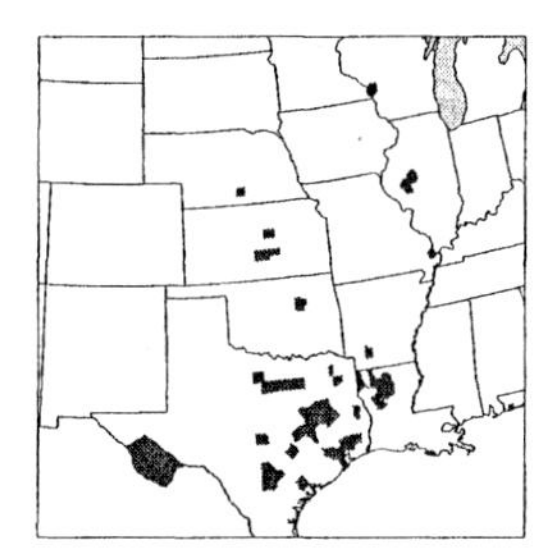

Plants annual. **Culms** (30)45–80 cm, branching at the lower nodes, often diffusely so; **nodes** and **internodes** glabrous. **Leaves** cauline; **sheaths** shorter or slightly longer than the internodes, glabrous or with straight hairs, sometimes pilose-floccose; **collars** glabrous or pilose at the sides; **ligules** about 0.5 mm; **blades** (6)10–20 cm long, 1–2(3) mm wide, yellow-green, turning reddish late in the season, involute to loosely folded, abaxial surfaces smooth or scabrous near the base, sometimes pubescent distally, adaxial surfaces usually glabrous, sometimes scabrous or pubescent, lateral veins about twice as thick as the inner veins. **Inflorescences** paniculate, 10–20 cm long, 2–7 cm wide; **rachis nodes** glabrous; **primary branches** stiffly ascending, with axillary pulvini, with 3–9 spikelets. **Spikelets** in fan-shaped clusters, pedicels with axillary pulvini. **Glumes** 10–17 mm, about equal, light to dark brown, glabrous or rarely sparsely pilose, 1-veined, apices cleft and awned, awns 2–5 mm; **calluses** 1–2.5 mm; **lemmas** 7–10 mm, gray to light brown, narrowing to a 2–5(7) mm beak, junction of the lemma and awns evident; **awns** 20–28 mm, similar in length, curved to strongly arcuate near the base, not forming a column, straight and divergent distally, disarticulating at maturity; **anthers** 3, 1–2 mm, dark purple. **Caryopses** 7–8 mm, smooth, chestnut brown. 2*n* = unknown.

Aristida desmantha grows in sandy fields, dry pine woods, and waste places in the United States. It is generally similar to *A. tuberculosa*, but has shorter glumes, calluses, and awns.

2. **Aristida tuberculosa** Nutt. [p. 320]
SEASIDE THREEAWN

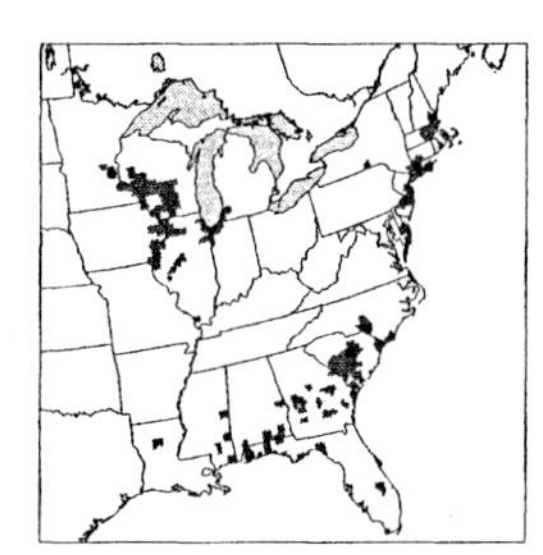

Plants annual. **Culms** (25) 40–100 cm, erect or decumbent to ascending near the base, highly branched above the base; **nodes** and **internodes** glabrous. **Leaves** cauline; **sheaths** usually slightly shorter than the internodes, glabrous or pilose; **collars** often with a line of tangled hairs; **ligules** about 0.5 mm; **blades** 8–25 cm long, 2–4 mm wide, light green, flat to loosely involute, glabrous and smooth abaxially, scabridulous adaxially. **Inflorescences** paniculate, 10–20 cm long, 3–10 cm wide; **rachis nodes** glabrous, scabrous, or strigose; **primary branches** 1–4 cm, stiffly ascending, with axillary pulvini, with 1–4 spikelets per branch. **Spikelets** loosely congested. **Glumes** 20–30 mm, yellowish-brown, 1-veined, apices narrowing to a 5–10 mm awn; **upper glumes** slightly longer than the lower glumes; **calluses** 3–4 mm; **lemmas** 10–14 mm, dark and mottled at maturity, glabrous or occasionally sparsely pubescent, not beaked, junction of the lemma and awns evident; **awns** twisted together basally into a 8–15 mm column, free portions 30–40 mm, those of the central and lateral awns similar in length, strongly curved to arcuate near the base, straight and strongly divergent to reflexed distally, disarticulating at the base of the column at maturity; **anthers** 3, about 2.5 mm, brownish. **Caryopses** 8–10 mm, dark brown. 2*n* = unknown.

Aristida tuberculosa grows in sandy fields, hills, pinelands, and disturbed areas. Along the Atlantic coastal fringe, it grows on maritime dunes; inland it is associated with xeric pine-oak sandhills. It is generally similar to *A. desmantha*, but has longer glumes, calluses, and awns. Like *A. desmantha*, *A. tuberculosa* is restricted to the United States.

3. **Aristida californica** Thurb. [p. 322]
MOJAVE THREEAWN

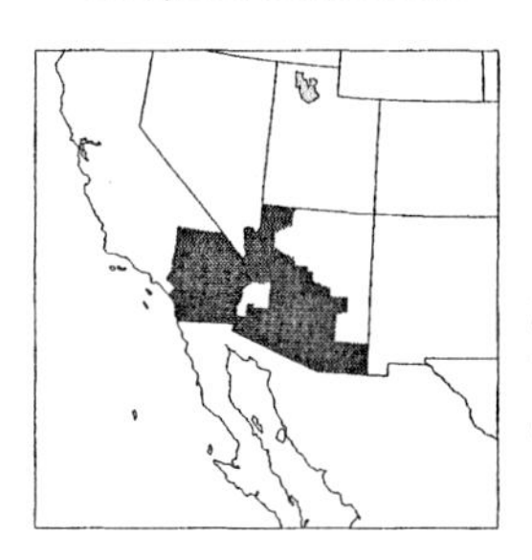

Plants perennial; sometimes flowering the first year. **Culms** 10–40 cm, highly branched above the base in age; **internodes** glabrous or pubescent, sometimes nearly lanose. **Leaves** cauline; **sheaths** shorter than the internodes, glabrous or puberulent; **collars** glabrous or pubescent at the sides; **ligules** 0.5–1 mm; **blades** usually less than 6 cm long, 0.5–1 mm wide, pale green, involute, glabrous or puberulent abaxially. **Inflorescences** paniculate or racemose, 5–10 cm long, 1–2 cm wide, with few spikelets; **rachis nodes** glabrous or with straight hairs; **primary branches** 1–2 cm, appressed, without axillary pulvini. **Spikelets** appressed. **Glumes** unequal, 1–2-veined;

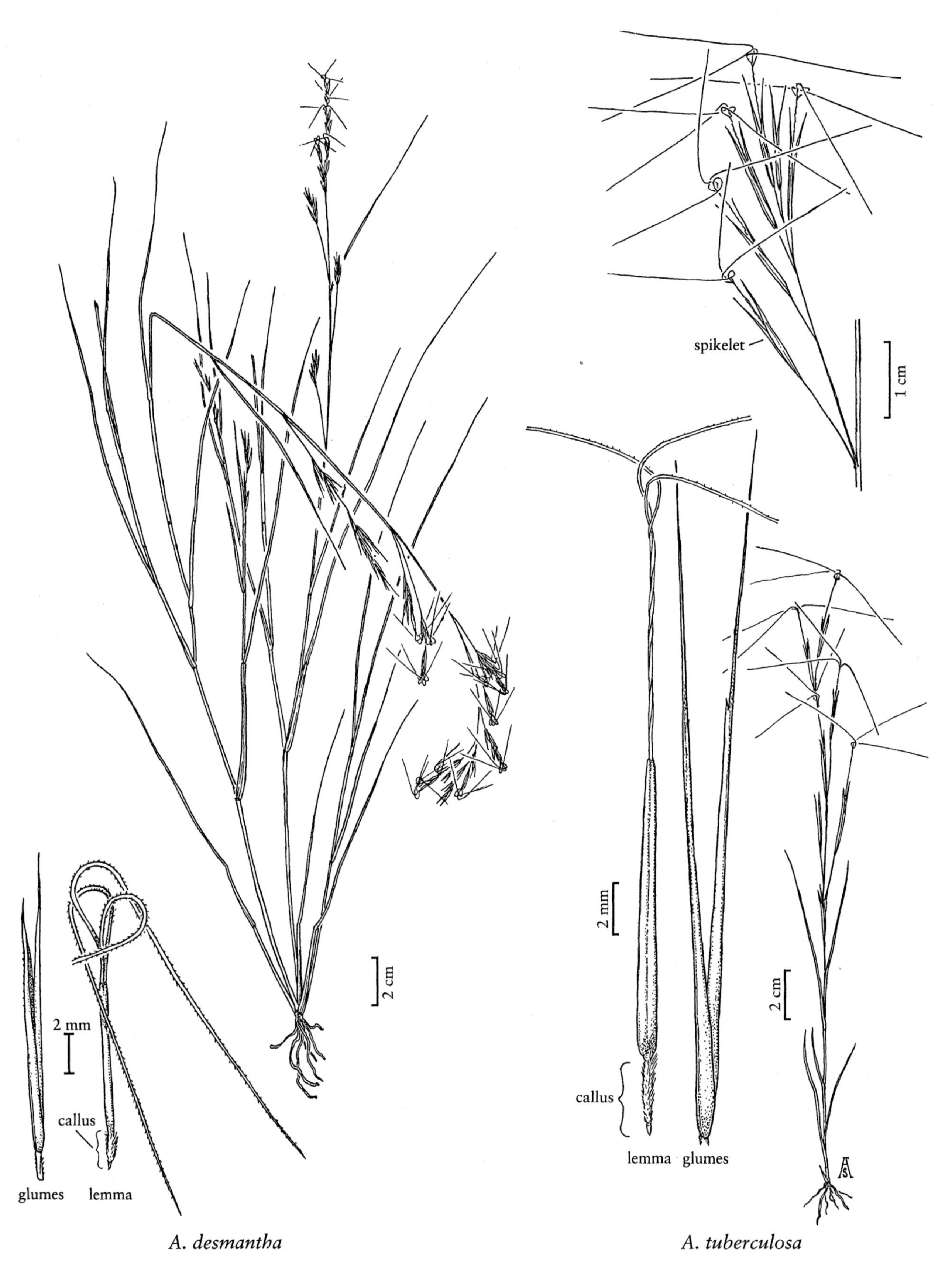

A. desmantha

A. tuberculosa

lower glumes 4–10 mm; **upper glumes** 7–15 mm; **calluses** about 1 mm; **lemmas** 5–7 mm, purple or mottled, junction of the lemma and awns evident; **awns** twisted together basally into a 4–26 mm column, free portions 12–50 mm, those of the central and lateral awns similar in length, curved to arcuate basally, straight and divergent distally, disarticulating at the base of the column at maturity; **anthers** 3, about 2 mm long. $2n$ = 22.

The range of both varieties of *Aristida californica* extends from the southwestern United States into northwestern Mexico.

1. Cauline internodes puberulent to nearly lanose . . . . var. *californica*
1. Cauline internodes glabrous . . . . var. *glabrata*

Aristida californica Thurb. var. **californica**

Cauline internodes puberulent to nearly lanose. **Lower glumes** 6–10 mm; **upper glumes** 11–13 mm; **awns** forming a 7–26 mm column, free portions 25–50 mm.

Aristida californica var. *californica* grows in dry, sandy plains, dunes, and flats of the Sonoran and Mojave deserts at elevations of 0–700 m.

Aristida californica var. **glabrata** Vasey

Cauline internodes glabrous. **Lower glumes** 4–8 mm; **upper glumes** 7–15 mm; **awns** forming a 4–16 mm column, free portions 12–40 mm.

Aristida californica var. *glabrata* grows in sandy to rocky soils of desert grassland and desert thorn-scrub communities in the Sonoran and Mojave deserts at elevations of 500–1400 m, generally to the east of var. *californica*.

4. **Aristida patula** Chapm. *ex* Nash [p. 322]
TALL THREEAWN

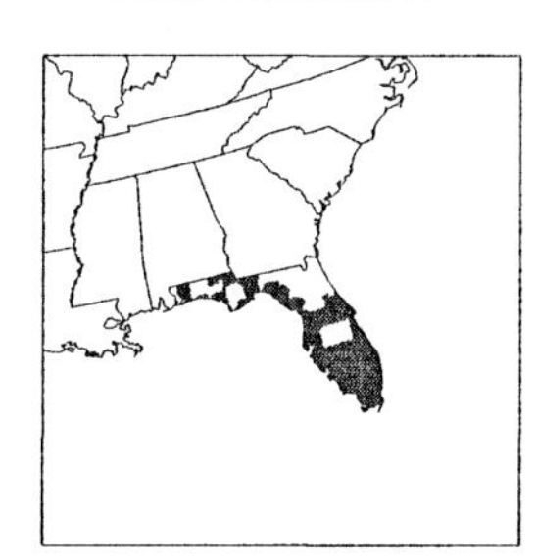

Plants perennial; loosely cespitose, bases knotty, sometimes shortly rhizomatous. **Culms** (60)70–100 cm, stiffly erect, unbranched. **Leaves** cauline; **sheaths** longer than the internodes, mostly glabrous, summit with hairs; **collars** hispid, hairs 0.2–0.8 mm; **ligules** less than 0.5 mm; **blades** 20–55 cm long, 2–4 mm wide, light bluish-green, flat to loosely folded, glabrous abaxially, scabridulous adaxially. **Inflorescences** paniculate, 30–50 cm long, 15–25 cm wide; **rachis nodes** glabrous or with straight hairs shorter than 0.5 mm; **primary branches** 8–22 cm, ascending to divaricate or drooping, with axillary pulvini, basal portion without spikelets. **Spikelets** appressed along the branches. **Glumes** brown to purplish, 1-veined, with a 1–2 mm awn; **lower glumes** 10–13 mm, slightly longer than the upper glumes; **calluses** 0.5–1 mm; **lemmas** 10–12(15) mm, glabrous, light gray to brownish, narrowing to a beak, beak less than 7 mm, not or only slightly twisted, junction with the awns not conspicuous; **awns** unequal, not disarticulating at maturity; **central awns** 20–25 mm, straight; **lateral awns** 3–10 mm, to ½ as long as and about ½ as thick as the central awns, usually divergent; **anthers** 3, about 3 mm, yellow-green. **Caryopses** about 8 mm. $2n$ = unknown.

Aristida patula grows in sandy fields, low pinelands, and roadsides. It is endemic to Florida.

5. **Aristida floridana** (Chapm.) Vasey [p. 322]
FLORIDA THREEAWN

Plants perennial; cespitose. **Culms** 70–100 cm, erect, mostly unbranched. **Leaves** cauline; **sheaths** mostly longer than the internodes, mostly glabrous, summit with hairs; **collars** mostly glabrous, sides usually with straight or wrinkled hairs; **ligules** 0.2–0.3 mm; **blades** 30–55 cm long, 1–2 mm wide, pale yellow-green, loosely involute, lax, glabrous abaxially, with scattered hairs adaxially. **Inflorescences** paniculate, 30–45 cm long, 5–25 cm wide, oblong to ovate; **primary branches** single or paired, ascending and somewhat lax to stiffly divergent, with weakly developed axillary pulvini, spikelet-bearing to the base; **lowermost branches** to 15 cm. **Spikelets** mostly appressed. **Glumes** purplish-tinged, 1-veined; **lower glumes** 10–14 mm; **upper glumes** 8–9 mm; **calluses** 0.5–0.7 mm; **lemmas** 8–12 mm, gray mottled with purple or dark patches, narrowing to a slightly curved and twisted beak, junction with the awns not strongly marked; **awns** not disarticulating at maturity; **central awns** 10–25 mm, falcate; **lateral awns** absent; **anthers** 3, about 1 mm, brown. **Caryopses** 6–7 mm, chestnut-colored. $2n$ = unknown.

Aristida floridana grows in waste places, along roadsides, and on railroad embankments. It is rare in the United States, being known only from Key West and Ramrod Key, Florida. It is more common in the Yucatan Peninsula, Mexico, where it intergrades with *Aristida ternipes*.

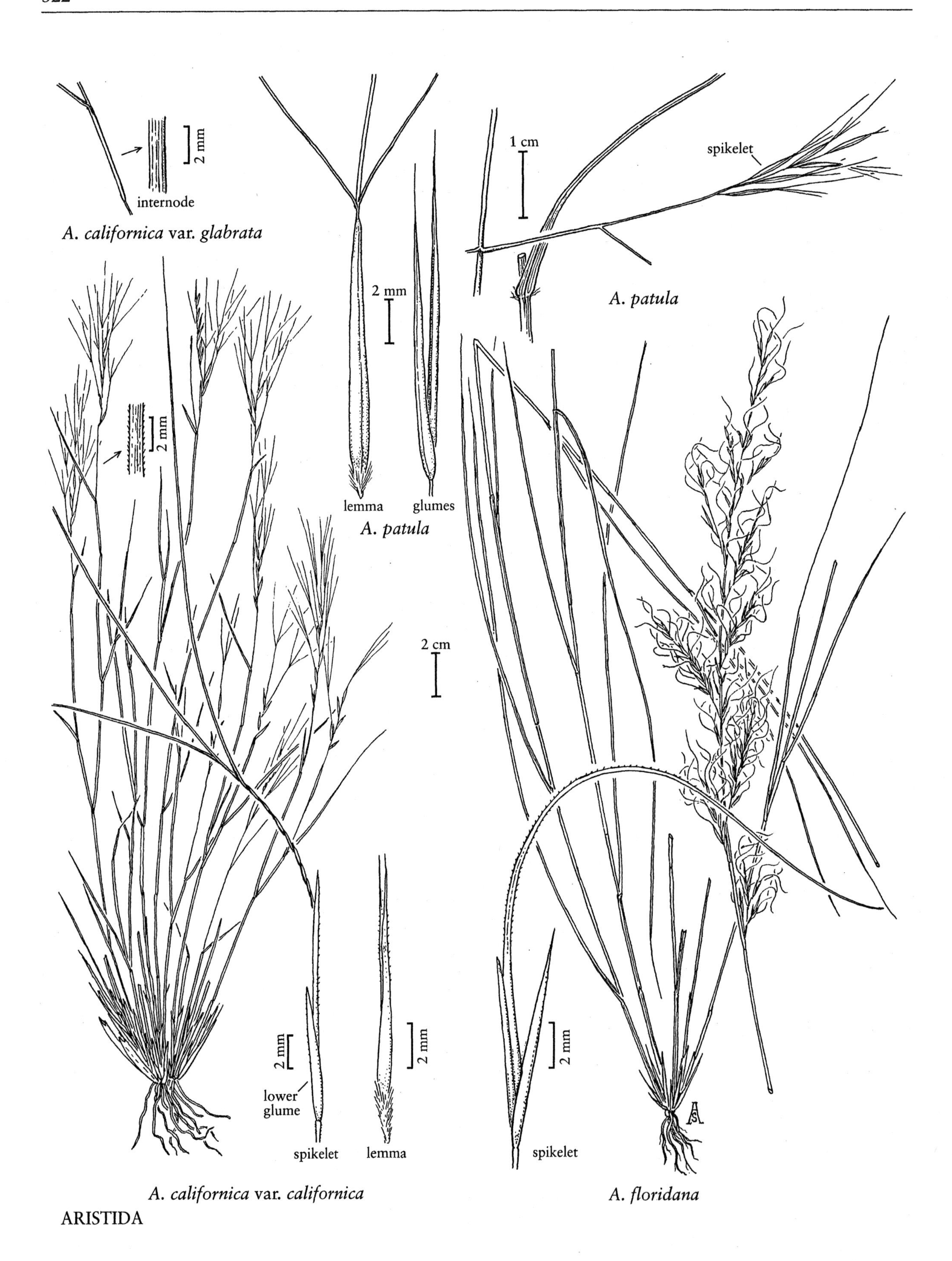
2 mm
internode
A. californica var. glabrata
2 mm
lemma
glumes
A. patula
1 cm
spikelet
A. patula
2 mm
2 cm
2 mm
2 mm
lower glume
spikelet
lemma
2 mm
spikelet
A. californica var. californica
A. floridana

ARISTIDA

6. **Aristida ternipes** Cav. [p. 325]

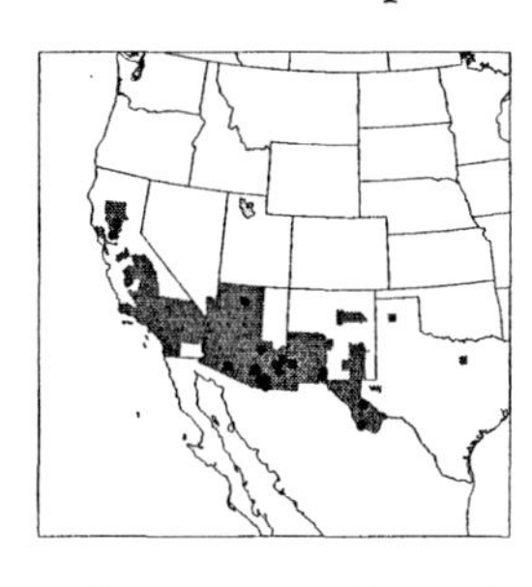

Plants perennial; cespitose. **Culms** 25–120 cm, wiry, erect to sprawling, unbranched. **Leaves** basal and cauline; **sheaths** usually longer than the internodes, glabrous; **collars** glabrous or strigillose; **ligules** less than 0.5 mm; **blades** 5–40 cm long, 1–2.5 mm wide, flat to folded, straight to lax at maturity, adaxial surfaces with scattered, 1.5–3 mm hairs near the ligule. **Inflorescences** paniculate, 15–40 cm long, (8)10–35(45) mm wide; **rachis nodes** glabrous or strigillose; **primary branches** 5–25 cm, remote, stiffly ascending to divaricate, with axillary pulvini, usually naked near the base; **secondary branches** and **pedicels** usually appressed. **Spikelets** usually congested. **Glumes** 9–15 mm, subequal, 1-veined, acuminate; **calluses** 1–1.2 mm; **lemmas** 9–15 mm long, smooth to tuberculate-scabrous, narrowing to slightly keeled, usually not twisted, 0.1–0.2 mm wide apices, junction with the awns not evident; **awns** not disarticulating at maturity, unequal or almost equal; **central awns** 8–25(30) mm, straight to arcuate at the base; **lateral awns** absent or to 0–23 mm; **anthers** 3, 1.2–2.4 mm. **Caryopses** 6–8 mm, light brownish. $2n$ = 22, 24.

1. Lateral awns 2–23 mm long . . . . var. *gentilis*
1. Lateral awns 0–2 mm long . . . . var. *ternipes*

Aristida ternipes var. **gentilis** (Henrard) Allred
HOOK THREEAWN

Awns subequal to unequal, ascending to spreading; **central awns** 10–25(30) mm; **lateral awns** (2)6–23 mm. $2n$ = 44.

Aristida ternipes var. *gentilis* grows on dry slopes and plains and along roadsides from Californica to Texas and south through Mexico to Guatemala.

Aristida ternipes Cav. var. **ternipes**
SPIDERGRASS

Awns markedly unequal; **central awns** 8–25 mm, straight to curving; **lateral awns** absent or to 2 mm, erect. $2n$ = 22, 44.

Aristida ternipes var. *ternipes*, like var. *gentilis*, grows on dry slopes and plains and along roadsides, but its range is somewhat different, extending from Arizona to Texas south through Mexico and Central America to South America.

7. **Aristida schiedeana** Trin. & Rupr. [p. 325]
SINGLE THREEAWN

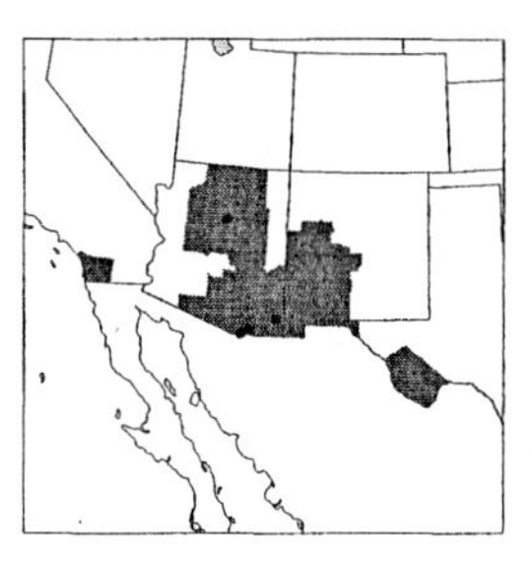

Plants perennial; cespitose. **Culms** 30–120 cm, erect, unbranched. **Leaves** basal and cauline, pale green, sometimes glaucous; **sheaths** longer or shorter than the internodes, glabrous except at the summit; **collars** densely to sparsely pilose or glabrous; **ligules** less than 0.5 mm; **blades** 8–30 cm long, 1–2 mm wide, usually flat, often curled at maturity. **Inflorescences** paniculate, 10–30 cm long, (4)8–26 cm wide; **rachis nodes** with straight hairs, hairs to 0.8 mm; **primary branches** 6–16 cm, abruptly spreading to divaricate, stiff to lax, with axillary pulvini, usually not spikelet-bearing below midlength. **Spikelets** appressed, rarely spreading. **Glumes** 1(3)-veined, brown or purple at maturity, acuminate; **lower glumes** 6–13 mm; **upper glumes** equaling or to 4 mm shorter than the lower glumes; **calluses** 0.8–1.2 mm; **lemmas** 10–15(17) mm, terminating in a strongly twisted, 2–4 mm awnlike beak, junction with the awns not conspicuous; **awns** not disarticulating at maturity; **central awns** 5–12 mm, markedly bent near the base; **lateral awns** absent or to 1(3) mm, erect; **anthers** 1.2–2.2 mm, brownish. **Caryopses** 6–8 mm. $2n$ = 22, 44.

Aristida schiedeana grows on rocky slopes and plains, generally in pinyon-juniper, oak, or ponderosa pine communities. Plants from the southwestern United States and northern Mexico belong to **A. schiedeana** var. **orcuttiana** (Vasey) Allred & Valdés-Reyna, in which the lower glumes are usually glabrous and longer than the upper glumes, and the collar and throat are usually glabrous. **Aristida schiedeana** var. **schiedeana** grows in Mexico, Guatemala, Nicaragua, and Honduras, and has puberulent, equal glumes and pilose collars and throats.

8. **Aristida divaricata** Humb. & Bonpl. *ex* Willd. [p. 325]
POVERTY GRASS

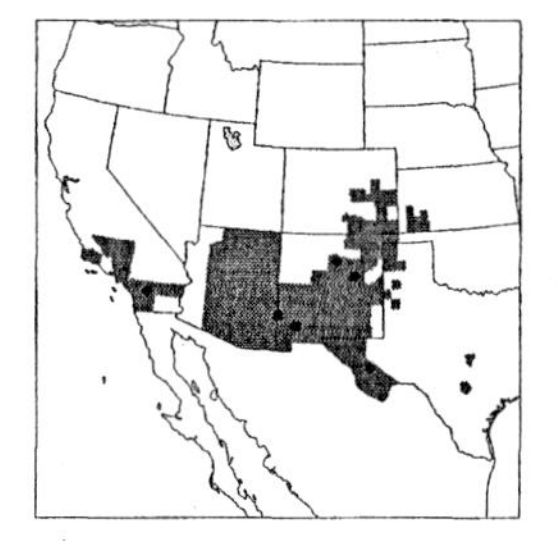

Plants perennial; cespitose. **Culms** 25–70 cm, erect or prostrate, unbranched or sparingly branched. **Leaves** tending to be basal; **sheaths** longer than the internodes, glabrous except at the summit; **collars** densely pilose; **ligules** 0.5–1 mm; **blades** 5–20 cm long, 1–2 mm wide, flat to loosely involute, glabrous. **Inflorescences** paniculate, 10–30 cm long, 6–25 cm wide, peduncles flattened and easily broken; **rachis**

nodes glabrous or with hairs, hairs to 0.5 mm; **primary branches** 5–13 cm, stiffly divaricate to reflexed, with axillary pulvini, usually naked on the basal ½; **secondary branches** usually well-developed. **Spikelets** overlapping, usually appressed, sometimes divergent and the pedicels with axillary pulvini. **Glumes** 8–12 mm, 1-veined, acuminate or shortly awned, awns to 4 mm; **calluses** about 0.5 mm; **lemmas** 8–13 mm long, the terminal 2–3 mm with 4 or more twists when mature, narrowing to 0.1–0.2 mm wide just below the awns, junction with the awns not evident; **awns** (7)10–20 mm, not disarticulating at maturity; **central awns** almost straight to curved at the base, ascending to somewhat divergent distally; **lateral awns** slightly thinner and from much to slightly shorter than the central awns, ascending to divergent; **anthers** 3, 0.8–1 mm. **Caryopses** 8–10 mm, light brown. $2n = 22$.

Aristida divaricata grows on dry hills and plains, especially in pinyon-juniper-grassland zones, from the southwestern United States through Mexico to Guatemala. It occasionally intergrades with *A. havardii*, but that species has lemma beaks that are straight or have only 1–2 twists, shorter primary branches, usually no secondary branches, and pedicels that more frequently have axillary pulvini so the spikelets are more frequently divergent than in *A. divaricata*.

9. **Aristida havardii** Vasey [p. 325]
HAVARD'S THREEAWN

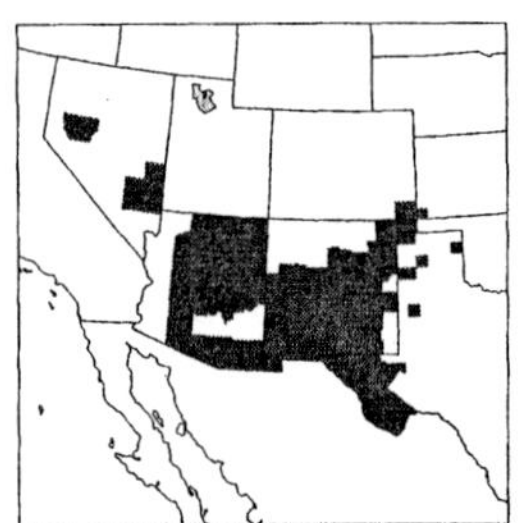

Plants perennial; cespitose. **Culms** 15–40 cm, slender, usually erect, occasionally decumbent, often tightly clustered into hemispheric clumps, unbranched. **Leaves** mostly basal; **sheaths** longer than the internodes, glabrous except at the summit; **collars** densely pilose; **ligules** 0.5–1 mm; **blades** 5–20 cm long, 1–2 mm wide, flat to loosely involute, glabrous. **Inflorescences** paniculate, 8–18 cm long, 4–12 cm wide, peduncles often flattened and easily broken; **rachis nodes** glabrous or with straight, less than 0.3 mm hairs; **primary branches** 2–6 cm, stiffly divaricate to reflexed, with axillary pulvini, usually naked on the lower ½; **secondary branches** usually absent. **Spikelets** usually divergent, pedicels usually with axillary pulvini. **Glumes** 8–12 mm, 1-veined, acuminate or awned, awns to 4 mm; **calluses** about 0.5 mm; **lemmas** 8–13 mm long, glabrous, smooth or scabrous, terminal 2–3 mm straight or with 1–2 twists, narrowing to 0.1–0.2 mm wide, junction with the awns not evident; **awns** (7)10–22 mm, not disarticulating at maturity, from almost straight to somewhat curved basally, ascending to divergent distally; **lateral awns** slightly shorter and thinner than the central awns; **anthers** 3, 0.8–1 mm. **Caryopses** 8–10 mm, light brown. $2n = 22$.

Aristida havardii grows on dry hills and plains in desert grassland to pinyon-juniper zones, and in sandy to rocky ground from the southwestern United States to northern Mexico. It occasionally intergrades with *A. divaricata*, but that species differs in having more twisted lemma beaks, longer primary branches, well-developed secondary branches, and, usually, appressed spikelets.

10. **Aristida pansa** Wooton & Standl. [p. 327]
WOOTON'S THREEAWN

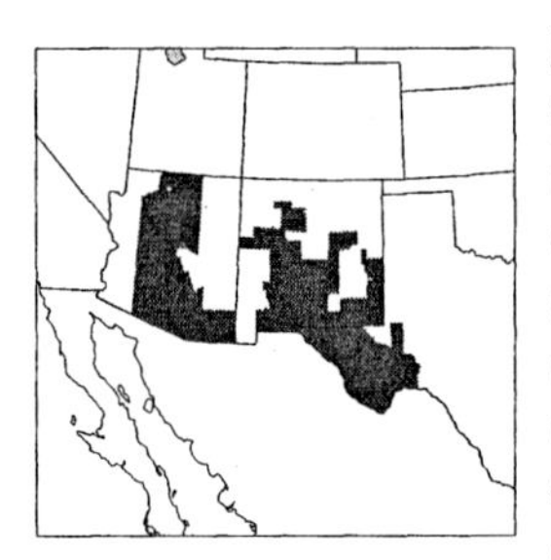

Plants perennial; cespitose. **Culms** 20–60(75) cm, erect, unbranched. **Leaves** basal and cauline; **sheaths** usually longer than the internodes, glabrous except at the summit; **collars** densely pilose, hairs 1–3 mm, cobwebby and tangled, often deflexed; **ligules** less than 0.5 mm; **blades** 4–28 cm long, less than 1 mm wide, usually involute, infrequently flat, usually arcuate, abaxial surfaces glabrous, adaxial surfaces glabrous or puberulent near the base, scabrous or puberulent distally. **Inflorescences** paniculate, 10–20 cm long, 3–10(12) cm wide; **rachis nodes** usually glabrous, sometimes with straight, less than 0.3 mm hairs; **primary branches** 2–11 cm, stiffly ascending to spreading, with axillary pulvini; **secondary branches** and **pedicels** with or without pulvini; **terminal spikelets** often divergent. **Spikelets** clustered on the distal ½ of the branches. **Glumes** equal or subequal, 1-veined, acuminate or awned, awns to 6 mm, brownish; **lower glumes** 5–10 mm; **upper glumes** 6–12 mm; **calluses** 0.5–1 mm; **lemmas** 7–13 mm, terminating in an obscure, narrow beak 1–4 mm long, 0.1–0.2 mm wide, junction with the base of the awns not evident; **awns** 6–15 mm, not disarticulating at maturity, central and lateral awns similar in length and thickness, spreading to horizontal; **anthers** 3, 1–3 mm, brown. **Caryopses** 6–8 mm, tan. $2n$ = unknown.

Aristida pansa grows in desert scrub, commonly in the Chihuahuan Desert of the southwestern United States and Mexico, but its ecological range extends into the lower juniper zones and its geographic range to southern Mexico. It prefers cobbly to sandy, often gypsiferous soil. It is very similar to the single-awned *A. gypsophila*, but it has also been confused with *A. purpurea* var. *perplexa*, which differs in having reddish glumes of unequal length and longer ascending awns.

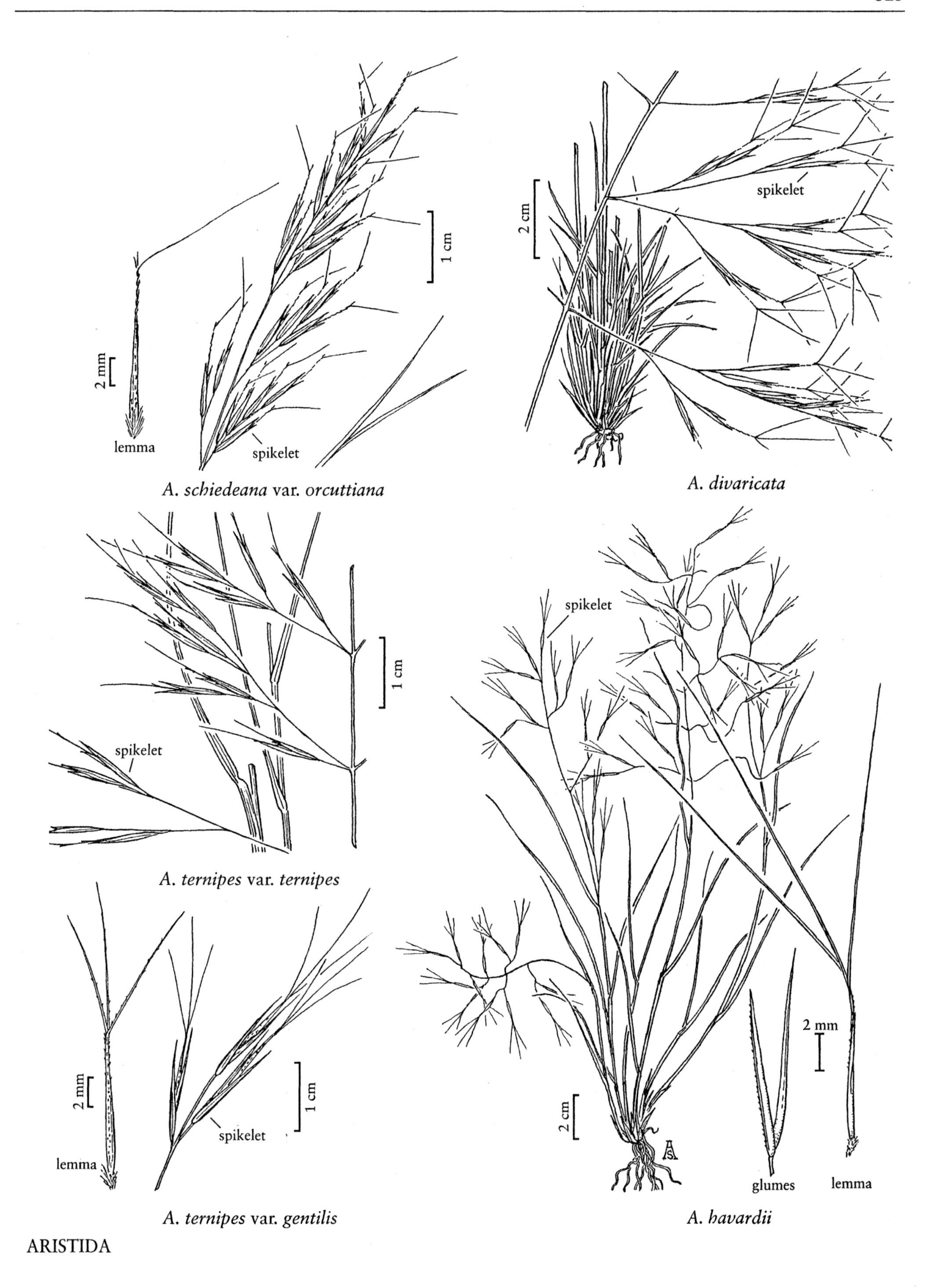

A. schiedeana var. *orcuttiana*

A. divaricata

A. ternipes var. *ternipes*

A. ternipes var. *gentilis*

A. havardii

ARISTIDA

11. **Aristida gypsophila** Beetle [p. 327]
GYPSUM THREEAWN

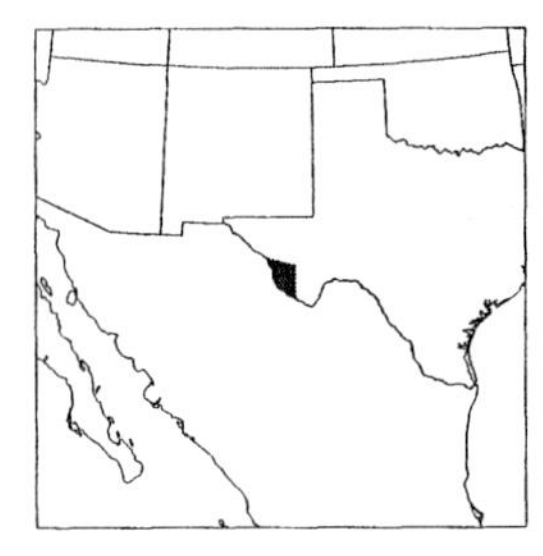

Plants perennial. **Culms** 45–80 cm, erect, usually unbranched. **Leaves** basal and cauline; **sheaths** longer than the internodes, glabrous except at the summit; **collars** densely pilose, hairs 1–3 mm, cobwebby and tangled, often deflexed; **ligules** less than 0.5 mm; **blades** 5–15 cm long, about 0.5 mm wide, usually involute, occasionally loosely folded, glabrous, light green. **Inflorescences** paniculate, 12–20 cm long, 2–8 cm wide; **primary branches** 2–5 cm, erect to horizontal, with or without axillary pulvini, with 1–5 spikelets. **Spikelets** appressed or with axillary pulvini and spreading. **Glumes** 6–10(12) mm, equal or the lower glumes slightly shorter, 1-veined, brownish; **calluses** about 0.5 mm; **lemmas** (6)7–14(16) mm, mostly smooth, mottled, terminating in a 2–4 mm, usually twisted, scabrous beak; **central awns** 5–10 mm, sharply curved at the base, spreading distally; **lateral awns** absent or to 3 mm, erect; **anthers** 3, about 1.5 mm, brown. **Caryopses** 5–8 mm. $2n$ = unknown.

Aristida gypsophila grows on rocky limestone or gypsum hills in thorn-scrub communities of the Chihuahuan Desert, almost always growing in the protection of shrubs. It is very similar to *A. pansa*, which differs in having three well-developed awns and being, usually, shorter in stature. Both species have involute blades with a characteristic tuft of cobwebby hairs at the collar. Plants from the United States have spreading primary branches with axillary pulvini and appressed spikelets. Mexican plants sometimes have primary branches with no axillary pulvini.

12. **Aristida ramosissima** Engelm. *ex* A. Gray [p. 327]
S-CURVE THREEAWN

Plants annual. **Culms** 20–60 cm, wiry. **Leaves** cauline; **sheaths** usually shorter than the internodes, mostly glabrous, sparsely hairy near the summit; **collars** hispidulous, occasionally glabrous; **ligules** 0.2–0.5 mm; **blades** 3–22 cm long, to 2 mm wide, flat to involute, glabrous abaxially, sparsely pilose adaxially, without a prominent midrib but with a thickened vein near each margin, pale to gray-green, sometimes slightly glaucous. **Inflorescences** paniculate or racemose, 5–12 cm long, 2–4 cm wide; **primary branches** barely developed. **Spikelets** appressed to slightly divergent from the axillary pulvini. **Glumes** subequal, apices bifid, awned from the sinuses; **lower glumes** (9)11–16 mm, 3–7-veined, awns 0.5–2(4) mm; **upper glumes** 11–18 mm, 1–3-veined, awns 3–7 mm; **calluses** 0.4–1 mm; **lemmas** (8)15–20(22) mm, dark, occasionally banded or spotted, apices narrowed but not beaklike; **central awns** 12–25 mm, bases with a semicircular bend; **lateral awns** 1–4 mm, erect, occasionally lacking; **anthers** 3, about 3 mm, brown. **Caryopses** 9–11 mm, dark brown to black. $2n$ = unknown.

Aristida ramosissima grows in open, dry, sterile ground, fallow fields, and roadsides. It is restricted to the United States.

13. **Aristida oligantha** Michx. [p. 327]
OLDFIELD THREEAWN

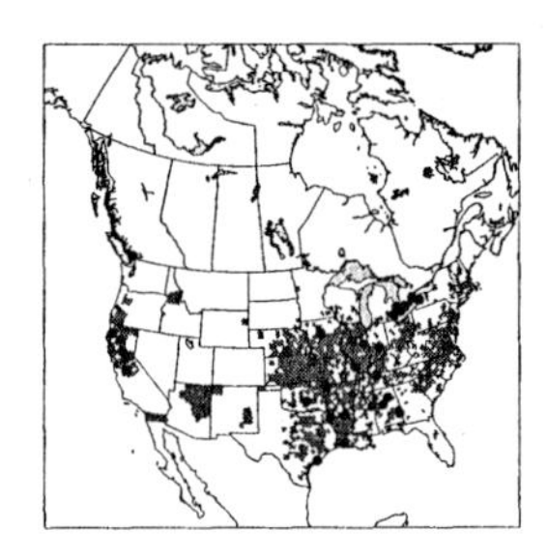

Plants annual. **Culm** 25–55 cm, erect or geniculate at the base, highly branched. **Leaves** cauline; **sheaths** usually shorter than the internodes, lowermost sheaths appressed-pilose basally; **collars** glabrous; **ligules** less than 0.5 mm; **blades** usually 4–12 cm long, 0.5–1.5 mm wide, flat or loosely involute, somewhat lax, glabrous or scabridulous, pale green. **Inflorescences** spicate or racemose, (5)7–20 cm long, 2–4 cm wide; **primary branches** rarely developed. **Spikelets** divergent, pedicels with axillary pulvini. **Glumes** unequal, glabrous, brownish-green with a purple tinge; **lower glumes** (9)12–22(28) mm, 3–7-veined, midvein extended into a 1–13 mm awn between 2 delicate setae; **upper glumes** (7)11–20(24) mm, 1-veined; **calluses** 0.5–2 mm; **lemmas** (9)12–22(23) mm, glabrous, light-colored, often mottled; **awns** (8)12–65(70) mm, subequal, spreading; **anthers** usually 1 and less than 0.5 mm, rarely 3 and 3–4 mm. **Caryopses** 8–14 mm, brown. $2n$ = 22.

Aristida oligantha grows in waste places, dry fields, roadsides, along railroads, and in burned areas, usually in sandy soil. It has been reported from Coahuila, Mexico, but is otherwise unknown outside southern Canada and the United States.

14. **Aristida basiramea** Engelm. *ex* Vasey [p. 329]
FORKTIP THREEAWN, ARISTIDE À RAMEAUX BASILAIRES

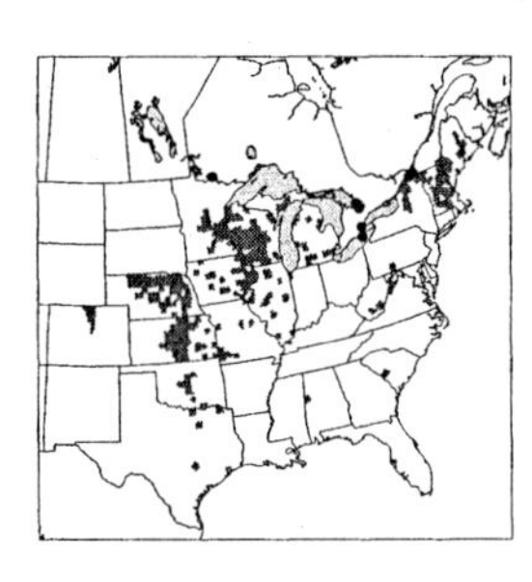

Plants annual. **Culms** 25–45 cm, erect, branching at most nodes. **Leaves** cauline; **sheaths** shorter than the internodes, glabrous or sparsely pilose; **ligules** about 0.3 mm; **blades** 3–8 cm long, 1–1.5 mm wide, flat to folded, becoming involute in age, adaxial surfaces with scattered

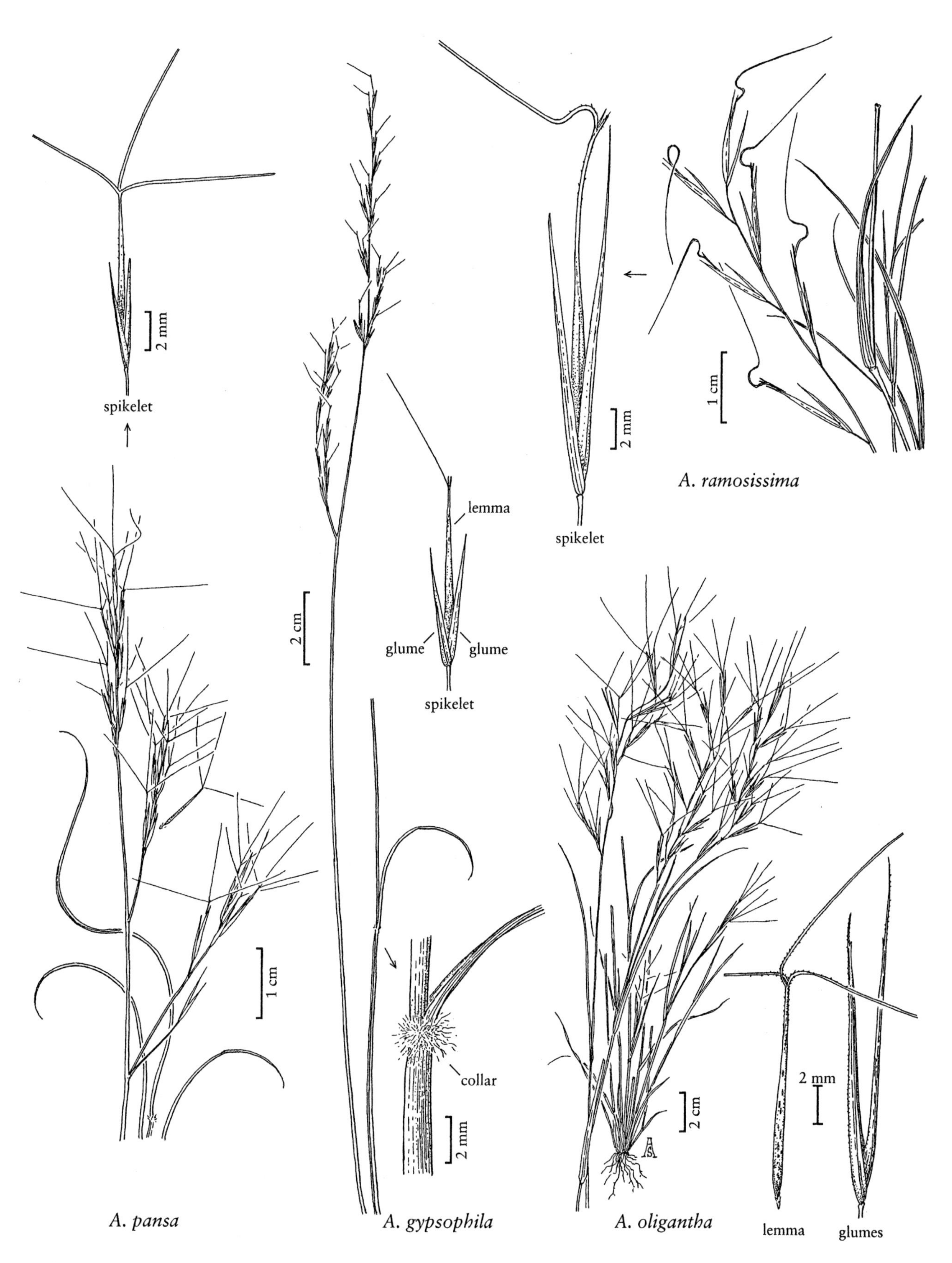
spikelet
2 mm
1 cm
A. pansa
2 cm
lemma
glume
glume
spikelet
collar
2 mm
A. gypsophila
spikelet
2 mm
1 cm
A. ramosissima
2 cm
A. oligantha
2 mm
lemma
glumes

ARISTIDA

pilose hairs, pale green. **Inflorescences** racemose or paniculate, (2)4–10 cm long, 1–2 cm wide, with few (sometimes only 1 or 2) spikelets; **primary branches** weakly developed, to 2 cm, appressed, with 1–3 spikelets. **Spikelets** appressed, only slightly overlapping. **Glumes** 1-veined, acute, awned, awns 1–2 mm, brown to purplish; **upper glumes** 10–12 mm; **lower glumes** 1–2 mm shorter; **calluses** 0.4–0.6 mm; **lemmas** 8–9 mm, light gray, mottled; **awns** erect to divergent; **central awns** 10–15 mm, with 2–3 spiral coils at the base; **lateral awns** 5–10 mm, not coiled but often curved and twisted basally, strongly divergent distally; **anthers** 3, about 3 mm, purplish-brown. **Caryopses** 6–7 mm, light chestnut brown. $2n$ = unknown.

Aristida basiramea grows in open, sandy, often barren ground in southern Ontario and in the United States. It is similar to *A. dichotoma*, differing in its longer lateral awns. Further study may show that the two should be treated as conspecific varieties.

15. **Aristida dichotoma** Michx. [p. 329]
CHURCHMOUSE THREEAWN

Plants annual. **Culms** 15–60 cm, erect or geniculate at the base, branching at most of the nodes. **Leaves** cauline; **sheaths** usually shorter than the internodes, glabrous or sparsely pilose; **collars** glabrous; **ligules** less than 0.5 mm; **blades** 3–10 cm long, 1–2 mm wide, flat to folded basally, involute distally, scabridulous on both surfaces, occasionally sparsely pilose adaxially, light green. **Inflorescences** paniculate or racemose, 2–11 cm long, to 1 cm wide; **nodes** glabrous or strigillose; **primary branches** 1–2 cm, appressed, without axillary pulvini, with 1–2 spikelets. **Spikelets** partly overlapping, often in pairs, 1 spikelet subsessile, the other pedicellate. **Glumes** 1-veined, light gray to dark purplish or brownish; **lower glumes** 3–8(10) mm, from ½ as long as the upper glumes to nearly equaling them; **upper glumes** 4–13 mm; **calluses** 0.3–0.5 mm; **lemmas** 3–11 mm, light gray to purplish, frequently mottled, midveins scabrous, elsewhere glabrous, scabridulous, or sparsely appressed-puberulent, junction with the awns not evident; **central awns** 3–8 mm, coiled at the base, spreading distally; **lateral awns** 1–4 mm, straight, erect; **anthers** 3 and 2–3 mm, or 1 and about 0.25 mm. **Caryopses** light brown. $2n$ = unknown.

Aristida dichotoma grows in sandy fields and clearings, disturbed sites and sterile ground, pine woods, and on granitic outcrops of the United States and southern Ontario. The two varieties have similar ecological preferences and extensive overlap in their ranges, but var. *curtissii* is somewhat more western in its distribution.

Aristida dichotoma is similar to *A. basiramea*, differing in its shorter lateral awns. Further study may show that the two should be treated as conspecific varieties.

1. Glumes unequal; lemmas smooth or scabridulous, 6–11 mm long. . . . . var. *curtissii*
1. Glumes equal or subequal; lemmas sparsely appressed-pubescent, 3–8 mm long . . . . . . var. *dichotoma*

Aristida dichotoma var. curtissii A. Gray

Glumes unequal; **lower glumes** 5–8(10) mm; **upper glumes** 7–13 mm; **lemmas** 6–11 mm, glabrous or scabridulous; **central awns** 4–8 mm; **lateral awns** 2–4 mm. **Caryopses** 4–9 mm.

Aristida dichotoma Michx. var. **dichotoma**

Glumes equal or subequal; **lower glumes** 3–6 mm; **upper glumes** 4–7 mm; **lemmas** 3–8 mm, sparsely appressed-pubescent; **central awns** 3–6 mm; **lateral awns** about 1 mm. **Caryopses** 2–6 mm.

16. **Aristida longespica** Poir. [p. 331]

Plants annual. **Culms** 15–65 cm, erect to spreading, often geniculate-based, sometimes nearly prostrate, usually much-branched. **Leaves** cauline; **sheaths** shorter than the internodes, not disintegrating into threadlike fibers at maturity, glabrous or sparsely pilose, hairs on the throat sometimes to 5 mm; **collars** glabrous; **ligules** about 0.5 mm; **blades** 5–14 cm long, 1–2 mm wide, flat to loosely involute, light green. **Inflorescences** usually paniculate, occasionally racemose or spicate, 6–22 cm long, 1–4(6) cm wide; **nodes** glabrous or with straight hairs, hairs to 0.3 mm; **primary branches** 1–4 cm, appressed to erect, rarely somewhat spreading distally, without axillary pulvini, with 2–5 spikelets per branch. **Spikelets** widely spaced to crowded. **Glumes** 2–11 mm, subequal, 1-veined, acuminate, unawned or awned, awns to 1 mm; **calluses** less than 1 mm; **lemmas** 2.5–10 mm, gray to dark purplish-brown, often horizontally banded or mottled, scabrous-hispid or glabrous, not beaked, apices only slightly narrowed, junction with the awns not evident; **awns** usually unequal, terete and curving up to 100° at the base, erect to reflexed distally, not disarticulating at maturity; **central awns** 1–27 mm; **lateral awns** absent or to 18 mm, shorter than the central awns; **anthers** 1 and 0.2–0.3 mm, or 3 and 3–4 mm. **Caryopses** 3–4 mm, light brown. $2n$ = unknown.

Aristida longespica grows along roadsides and in waste places, sandy fields, and clearings in pine and oak woods of southern Ontario and the eastern and central United States. The two varieties have a similar geographic range and are often found growing together.

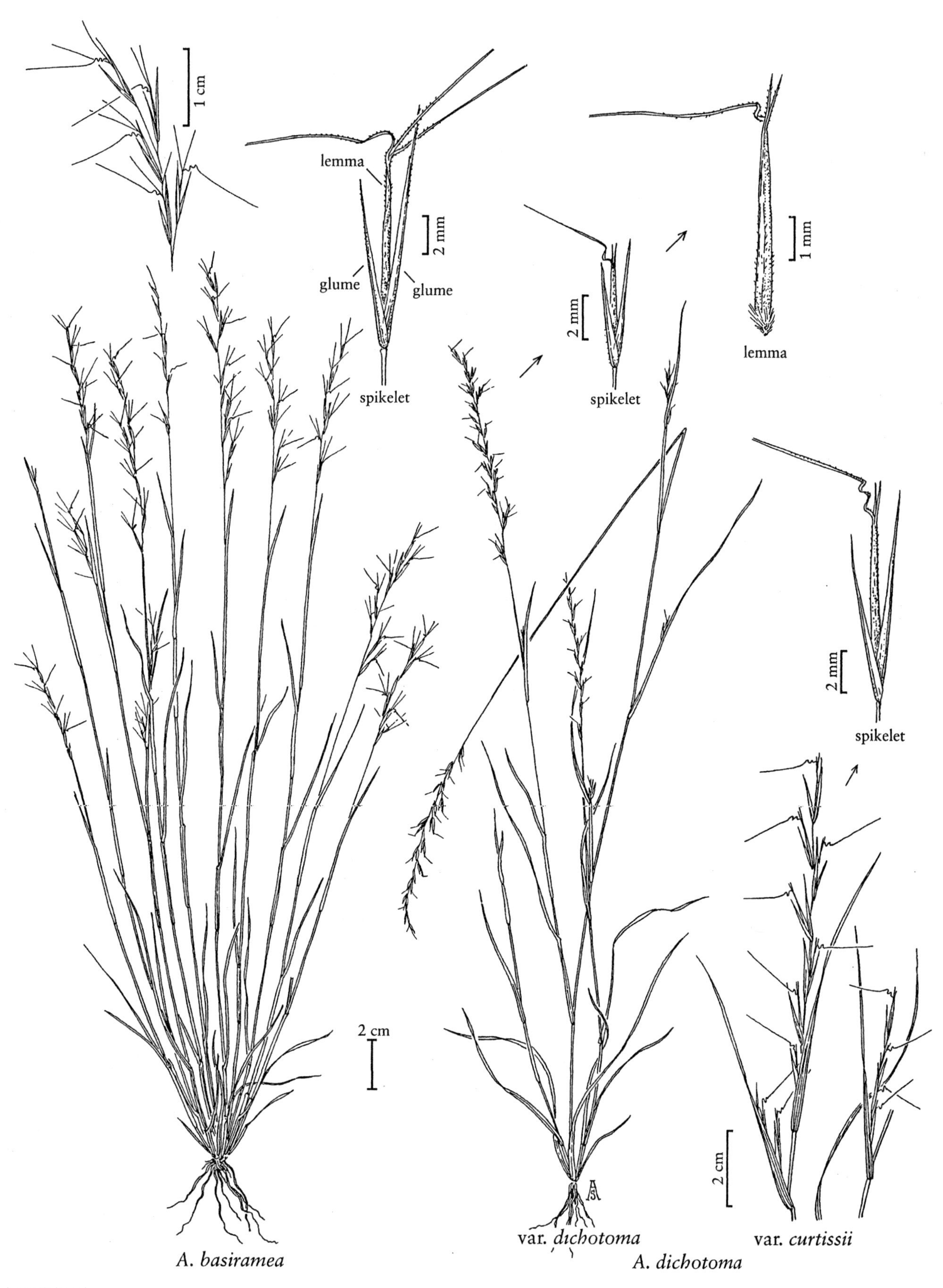
1 cm
lemma
2 mm
glume
glume
spikelet
2 mm
spikelet
1 mm
lemma
2 mm
spikelet
2 cm
2 cm
var. dichotoma
var. curtissii
A. basiramea
A. dichotoma

1. Central awns 8–27 mm long, lateral awns usually 6–18 mm long . . . . var. *geniculata*
1. Central awns 1–14 mm long and/or lateral awns usually 0–5 mm long . . . . var. *longespica*

Aristida longespica var. **geniculata** (Raf.) Fernald
KEARNEY'S THREEAWN

Glumes 4–11 mm; **lemmas** 3.5–10 mm; **central awns** (8)12–27 mm, erect to reflexed; **lateral awns** (1)6–18 mm, erect to horizontal; **anthers** 1 or 3.

Aristida longespica Poir. var. **longespica**
SLIMSPIKE THREEAWN

Glumes 2–8 mm; **lemmas** 2.5–7 mm; **central awns** 1–10(14) mm, erect to reflexed; **lateral awns** absent ot to 5(8) mm, erect to ascending; **anthers** 1, rarely exerted.

17. **Aristida adscensionis** L. [p. 331]
SIXWEEKS THREEAWN

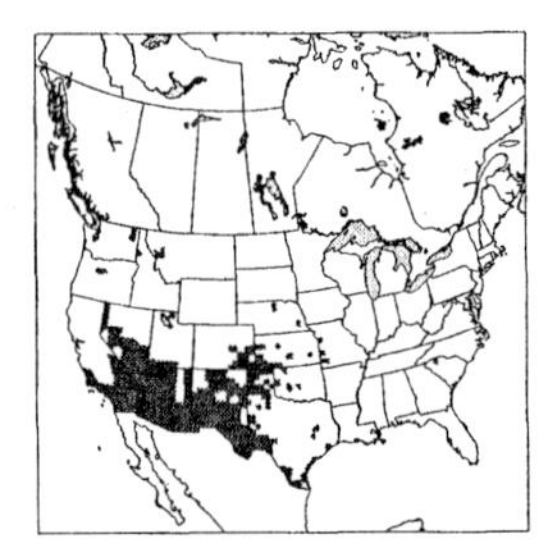

Plants short- to long-lived annuals. **Culms** (3)10–50(80) cm, often highly branched above the base. **Leaves** cauline, glabrous; **sheaths** shorter than the internodes, not disintegrating into threadlike fibers; **ligules** 0.4–1 mm; **blades** 2–14 cm long, 1–2.5 mm wide, flat to involute. **Inflorescences** panicles, 5–15(20) cm long, 0.5–3 cm wide, often interrupted below; **nodes** glabrous or with straight, less than 0.5 mm hairs; **primary branches** 1–4 cm, erect to ascending, without axillary pulvini, with 3–8 spikelets. **Spikelets** crowded. **Glumes** unequal, 1-veined, acuminate; **lower glumes** 4–8 mm; **upper glumes** 6–11 mm; **calluses** 0.5–0.8 mm; **lemmas** 6–9 mm, slightly keeled, midveins scabrous, junction with the awns not evident; **awns** not disarticulating at maturity, flattened and straight to somewhat curved at the base, central rib flanked by equally wide pale wings; **central awns** 7–15(20) mm; **lateral awns** somewhat shorter, occasionally only 1–2 mm; **anthers** 3, 0.3–0.7 mm. $2n$ = 22.

Aristida adscensionis grows in waste ground, along roadsides, and on degraded rangelands and dry hillsides, often in sandy soils. It is associated with woodland, prairie, and desert shrub communities. Its range extends from the United States south through Mexico and Central America to South America.

Because *Aristida adscensionis* is highly variable in height, panicle size, and awn development, several varieties have been described. None are recognized here because most of the variation appears to be environmentally induced.

18. **Aristida spiciformis** Elliott [p. 331]
BOTTLEBRUSH THREEAWN

Plants perennial; cespitose. **Culms** 45–100 cm, unbranched or sparingly branched. **Leaves** cauline; **sheaths** shorter than the internodes, glabrous except for occasional hairs at the summit; **collars** glabrous, or with a few pilose hairs at the sides; **ligules** about 0.5 mm; **blades** 12–28 cm long, 1–3 mm wide, usually folded or loosely involute, light yellow-green, glabrous abaxially, puberulent adaxially. **Inflorescences** dense panicles, (3)8–24 cm long, 2–4 cm wide, somewhat spirally twisted in age; **nodes** glabrous or strigillose, hairs about 0.1 mm; **primary branches** 2–4 cm, tightly appressed, without axillary pulvini, with 6–10 spikelets per branch. **Glumes** lanceolate, 1-veined; **lower glumes** 3.5–4.5 mm, abruptly contracted to a 4–12 mm awn; **upper glumes** 7–10 mm, gradually narrowed to a 10–12 mm awn; **calluses** about 2 mm; **lemmas** 5–6 mm, dark brown or purplish, terminating in a straight or twisted beak 7–30 mm long, about 0.2 mm wide, beak often disarticulating, no obvious zone of articulation developed, junction of the beak and awns sometimes evident; **awns** usually unequal, strongly curved and twisted at the base, straight distally, sometimes disarticulating at maturity; **central awns** (10)20–30 mm; **lateral awns** 10–20(25) mm, at least ½ as long as and evidently thinner than the central awns; **anthers** 3, about 1 mm, brown. **Caryopses** 4–5 mm, including the delicate style column. $2n$ = unknown.

Aristida spiciformis grows in pine savannahs, pine flatwoods, pine-oak sandhills, and oak woods, frequently being associated with *Pinus palustris*. It is a primary fire carrier in these habitats. Its range includes Cuba and Puerto Rico as well as the southeastern United States.

19. **Aristida purpurea** Nutt. [p. 334]

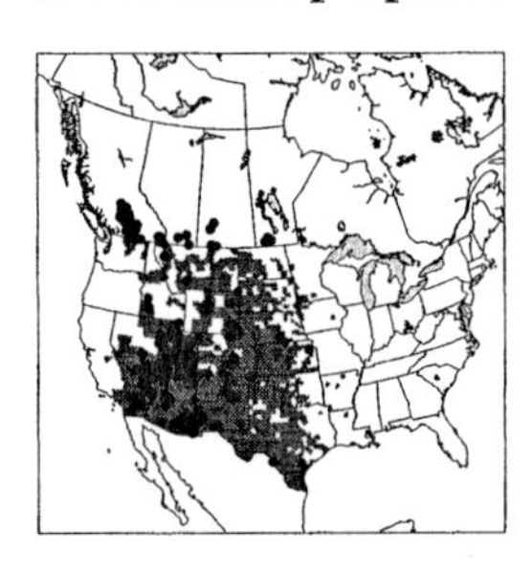

Plants perennial; densely cespitose, without rhizomes. **Culms** 10–100 cm, erect to ascending, usually unbranched. **Leaves** mostly basal or mostly cauline; **sheaths** shorter or longer than the internodes, glabrous, not disintegrating into threadlike fibers at maturity; **collars** glabrous, or sparsely pilose at the sides with straight hairs; **ligules** less than 0.5 mm; **blades** 4–25 cm long, 1–1.5 mm wide, tightly involute to flat, usually glabrous, sometimes scabridulous abaxially, gray-green, lax to curled at maturity. **Inflorescences** usually

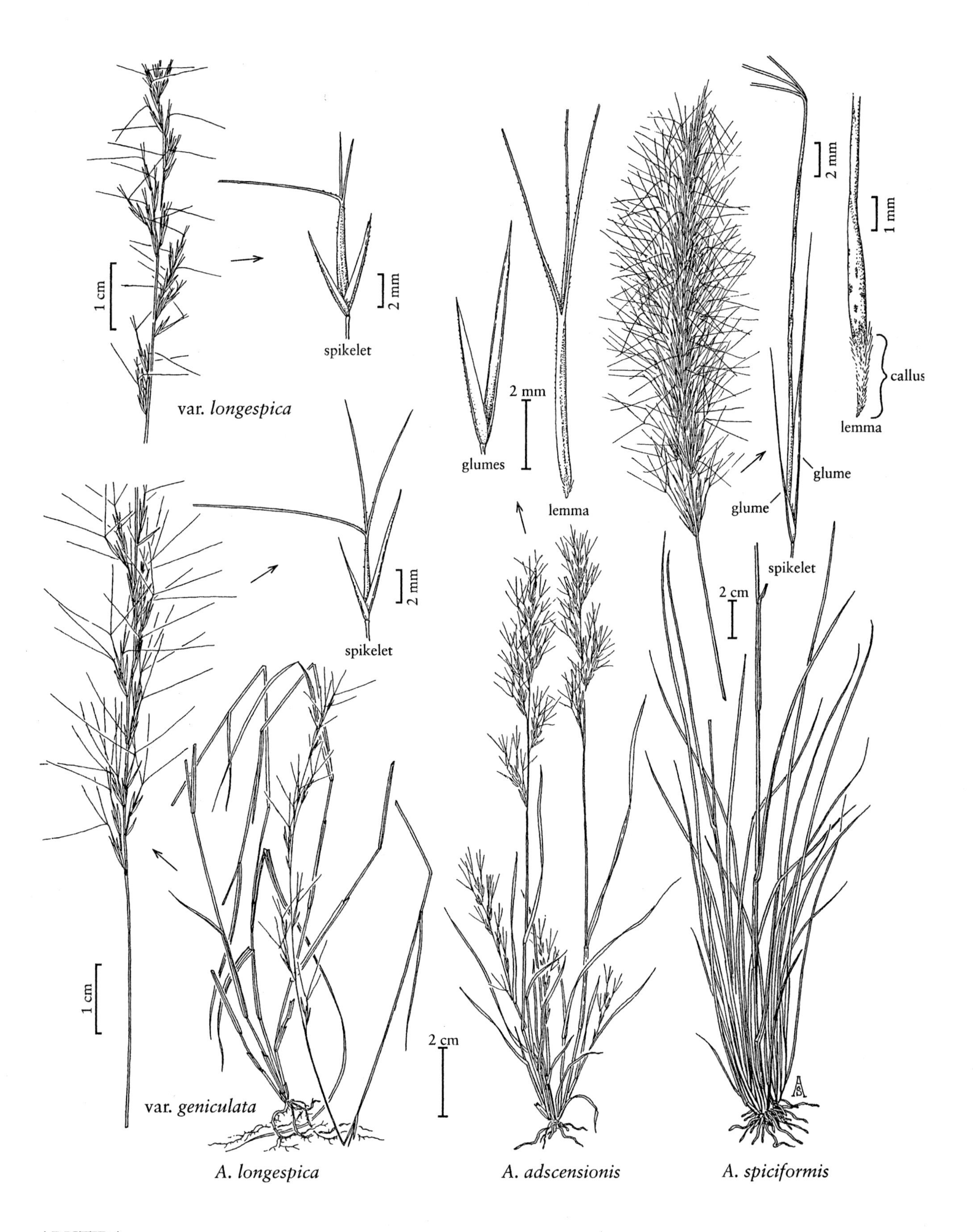

1 cm
spikelet
2 mm
var. *longespica*
2 mm
glumes
lemma
2 mm
1 mm
callus
lemma
glume
glume
spikelet
2 mm
spikelet
2 cm
1 cm
var. *geniculata*
2 cm
A. longespica
A. adscensionis
A. spiciformis

sparingly branched panicles, occasionally racemes, 3–30 cm long, 2–12 cm wide, with 2 or more spikelets per node; **nodes** glabrous or with straight, about 0.5 mm hairs; **primary branches** 3–6 cm, appressed to divaricate, varying sometimes within a panicle, stiff to flexible, bases appressed or abruptly spreading, usually without axillary pulvini. **Spikelets** divergent or appressed, with or without axillary pulvini. **Glumes** usually unequal, lower glumes shorter than the upper glumes, sometimes subequal, light to dark brown or purplish, glabrous, smooth or scabridulous, 1(2)-veined, acuminate, unawned or awned, awns to 1 mm; **lower glumes** 4–12 mm; **upper glumes** 7–25 mm; **calluses** 0.5–1.8 mm; **lemmas** 6–16 mm, glabrous, scabridulous, or tuberculate, whitish to purplish, apices 0.1–0.8 mm wide, not beaked or the beak less than 3 mm, junction with the awns not conspicuous; **awns** (8)13–140 mm, ascending to divaricate, not disarticulating at maturity; **central awns** thicker than the lateral awns; **lateral awns** (8)13–140 mm, usually subequal to the central awns, occasionally less than ⅓ as long as the central awns; **anthers** 3, 0.7–2 mm. **Caryopses** 6–14 mm, tan to chestnut. $2n$ = 22, 44, 66, 88.

Aristida purpurea is composed of several intergrading varieties.

1. Lower or all primary panicle branches stiff, divergent to divaricate from the base, with axillary pulvini; awns 13–30 mm.
    2. Lower glumes ½–⅔ as long as the upper glumes . . . . var. *perplexa*
    2. Lower glumes from ¾ as long as to equaling the upper glumes . . . . var. *parishii*
1. Primary panicle branches appressed or ascending at the base, sometimes drooping distally, without axillary pulvini; awns 8–140 mm.
    3. Awns 35–140 mm long.
        4. Lemmas apices 0.1–0.3 mm wide; awns 0.1–0.2(0.3) mm wide at the base, 35–60 mm long; upper glumes usually shorter than 16 mm long . . . . var. *purpurea*
        4. Lemma apices 0.3–0.8 mm wide; awns 0.2–0.5 mm wide at the base, 40–140 cm long; upper glumes 14–25 mm long . . . var. *longiseta*
    3. Awns 8–35 mm long.
        5. Lemma apices 0.1–0.3 mm wide distally; awns 0.1–0.3 mm wide at the base.
            6. All or most of the panicle branches straight (lower branches sometimes lax); pedicels straight, appressed to ascending . . . . var. *nealleyi*
            6. All or most of the panicle branches and pedicels drooping to sinuous distally . . . . var. *purpurea*
        5. Lemma apices 0.2–0.3 mm wide; awns stout, 0.2–0.3 mm wide at the base.
            7. Mature panicle branches and pedicels flexible, lax or drooping distally . . . var. *purpurea*
            7. Mature panicle branches and pedicels usually stiff, straight.
                8. Panicles usually 3–15 cm long; blades 4–10 cm long . . . . var. *fendleriana*
                8. Panicles usually 15–30 cm long; blades 10–25 cm long.
                    9. Glumes and lemmas reddish or dark-colored at anthesis or earlier (fading to stramineous), usually in marked contrast with the current foliage; panicles dense, the lower nodes with 8–18 spikelets; flowering March to May, after winter rains . . . . . var. *parishii*
                    9. Glumes and lemmas tan to brown (also fading to stramineous), giving the panicle a brownish appearance; old growth gray-green, not in marked contrast with the current foliage; panicles less dense, the lower nodes with 2–10 spikelets. . . . . var. *wrightii*

Aristida purpurea var. **fendleriana** (Steud.) Vasey
FENDLER'S THREEAWN

Culms 10–40 cm. **Leaves** mostly cauline; **blades** 4–10 cm, involute. **Panicles** 3–14(15) cm; **primary branches** mostly appressed, stiff, straight, without axillary pulvini, with few spikelets. **Lower glumes** 5–8 mm; **upper glumes** 10–15 mm; **lemmas** 8–14 mm long, apices 0.2–0.3 wide; **awns** subequal, 18–40 mm long, occasionally slightly longer, 0.2–0.3 mm wide at the base. $2n$ = 22, 44.

Aristida purpurea var. *fendleriana* grows on open slopes, hills, and sandy flats, at low to medium elevations, from the western United States into northern Mexico. It is often confused with var. *longiseta,* having short basal leaves and short panicles, but plants of var. *fendleriana* have narrower lemma apices and thinner, shorter awns than those of var. *longiseta.*

Aristida purpurea var. **longiseta** (Steud.) Vasey
RED THREEAWN

Culms 10–40(50) cm. **Leaves** sometimes mostly basal, sometimes mostly cauline; **blades** 4–16 cm, usually involute. **Panicles** 5–15 cm; **primary branches** appressed or ascending at the base, without axillary pulvini, stout and straight to delicate and drooping distally, usually neither flexible nor tangled. **Lower glumes** 8–12 mm; **upper glumes** (14)16–25 mm; **lemmas** 12–16 mm long, apices 0.3–0.8 mm wide; **awns** subequal, 40–100(140) mm long, 0.2–0.5 mm wide at the base. $2n$ = 22, 44, 66, 88.

Aristida purpurea var. *longiseta* grows on sandy or rocky slopes and plains, and in barren soils of disturbed ground from western Canada to northern Mexico. It is the most variable variety of *Aristida purpurea*, ranging from short plants with basal leaves and short panicles suggestive of var. *fendleriana*, to tall plants with long cauline leaves and long, drooping panicles resembling var. *purpurea*. The length of its glumes, width of its lemma apex, and the length and thickness of its awns distinguish it from all the other varieties. The callus and long, stiff awns are especially troublesome to sheep and cattle.

Aristida purpurea var. **nealleyi** (Vasey) Allred
NEALLEY'S THREEAWN

Culms 20–45 cm. **Blades** 5–15 cm, mostly basal, involute. **Panicles** 8–18(20) cm; **primary branches** and **pedicels** mostly appressed to narrowly ascending, without axillary pulvini, stiff, straight, lower branches occasionally flexible. **Glumes** usually unequal; **lower glumes** 4–7 mm; **upper glumes** (7)8–14 mm; **lemmas** 7–13 mm long, narrowing to about 0.1 mm wide, upper portion sometimes twisted; **awns** 15–22(30) mm long, subequal, about 0.1 mm wide at the base. $2n$ = 22, 44.

Aristida purpurea var. *nealleyi* grows on dry slopes and plains at lower elevations than the other varieties, frequently in desert grassland vegetation. Its range extends from the southwestern United States into Mexico. Although var. *nealleyi* is more distinct than the other varieties, having tight tufts of foliage exceeded by narrow, straw-colored panicles, it grades into var. *purpurea*, and the panicles resemble those of var. *wrightii*. It may also be confused with *A. arizonica*, but differs in having involute, generally straight leaf blades and shorter awns.

Aristida purpurea var. **parishii** (Hitchc.) Allred
PARISH'S THREEAWN

Culms 20–50 cm. **Leaves** mostly cauline; **blades** more than 10 cm, loosely involute to flat. **Panicles** 15–24 cm; **primary branches** stiff, lower branches strongly divergent to divaricate, with axillary pulvini, upper branches appressed to ascending, without axaillary pulvini, lower nodes associated with 8–18 spikelets. **Glumes** red or dark at anthesis, fading to stramineous; **lower glumes** 7–11 mm, ¾ as long as to equaling the upper glumes; **upper glumes** 10–15 mm; **lemmas** 10–13 mm long, narrowing to 0.2–0.3 mm wide near the apex; **awns** subequal, 20–30 mm long, 0.2–0.3 mm wide at the base. $2n$ = unknown.

Aristida purpurea var. *parishii* grows on sandy plains and hills of the southwestern United States and Baja California, Mexico. In many respects it is intermediate between *A. purpurea* and other species of *Aristida* with spreading panicle branches, especially *A. ternipes* var. *gentilis*. Its spikelets are indistinguishable from those of var. *wrightii*, but var. *parishii* frequently has axillary pulvini associated with the lower branches. The two also differ in their phenology: var. *parishii* flowers from March through May in response to winter rains, whereas var. *wrightii* flowers from May through October in response to summer rains.

Aristida purpurea var. **perplexa** Allred & Valdés-Reyna
JORNADA THREEAWN

Culms 30–65 cm. **Blades** 8–20 cm, involute. **Panicles** 8–20 cm; **primary branches** stiff, lower branches diverging or divaricate, with axillary pulvini, upper branches usually strongly divergent, sometimes ascending; **pedicels** often with axillary pulvini; **terminal spikelets** usually appressed. **Glumes** reddish; **lower glumes** (4.5)5–7(7.5) mm, ½–⅔ as long as the upper glumes; **upper glumes** 8–11(12) mm; **lemmas** (8)10–12(13) mm long, narrowing to 0.1–0.2 mm wide; **awns** subequal, (13)18–30 mm long, 0.1–0.2 mm wide. $2n$ = unknown.

Aristida purpurea var. *perplexa* grows in sandy to rocky plains and on mesas in desert grassland and scrub communities, often in calcareous soils, in both the *Flora* region and Mexico. It is sometimes confused with *A. pansa*, which differs in having cobwebby hairs at the collar, equal glumes, and shorter awns.

Aristida purpurea Nutt. var. **purpurea**
PURPLE THREEAWN

Culms 26–60 cm. **Blades** 3–17 cm, basal and cauline, involute. **Panicles** 10–25 cm; **primary branches** appressed at the base, without axillary pulvini, capillary, drooping to sinuous distally; **pedicels** capillary, usually lax to sinuous. **Lower glumes** 4–9 mm; **upper glumes** 7–16 mm; **lemmas** 6–12 mm long, narrowing to 0.1–0.3 mm wide; **awns** subequal, (15)20–60 mm long, 0.1–0.3 mm wide at the base. $2n$ = 22, 44, 66, 88.

Aristida purpurea var. *purpurea* grows in sandy to clay soils, along right of ways, or on dry slopes and mesas. Its range extends from the *Flora* region to Mexico and Cuba. As treated here, var. *purpurea* is, admittedly, a broadly defined taxon, incorporating slender plants with small spikelets that used to be referred to *A. roemeriana* Scheele, but also occasional plants with somewhat flexible branches that are intermediate to var. *wrightii* and var. *nealleyi*.

Aristida purpurea var. **wrightii** (Nash) Allred
WRIGHT'S THREEAWN

Culms 45–100 cm. **Blades** 10–25 cm, involute or flat. **Panicles** (12)14–30 cm; **primary branches** usually erect, without axillary pulvini, stiff, straight, lower nodes associated with 2–10 spikelets. **Glumes** tan to brown, fading to stramineous. **Lower glumes** 5–10 mm; **upper glumes** 9–16 mm; **lemmas** 8–14 mm long, narrowing to

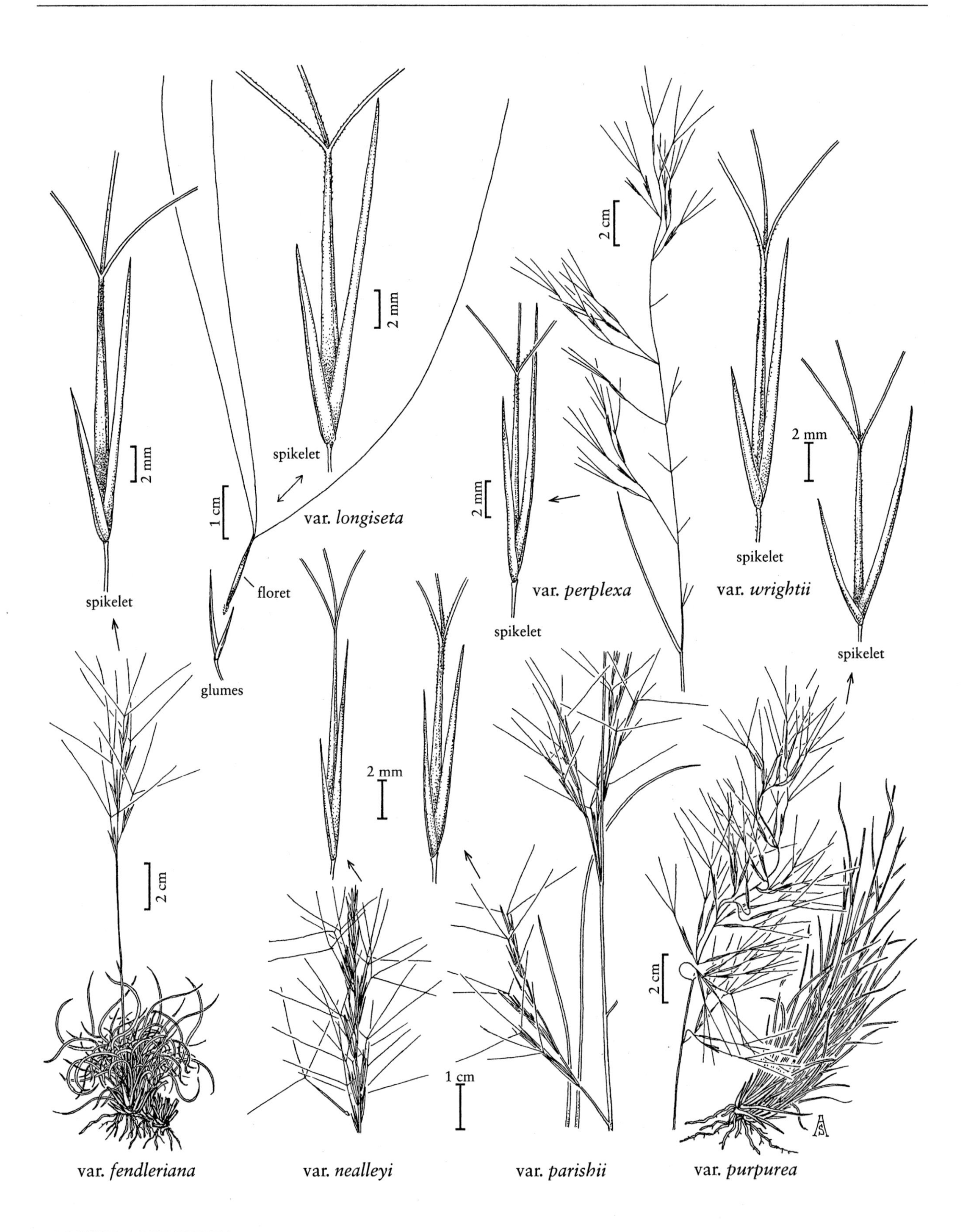

ARISTIDA PURPUREA

0.2–0.3 mm wide; **awns** (8)20–35 mm long, 0.2–0.3 mm wide at the base, lateral awns usually subequal to the central awn, rarely 1–3 mm. $2n$ = 22, 44, 66.

Aristida purpurea var. *wrightii* grows on sandy to gravelly hills and flats from the southwestern United States to southern Mexico. It is the most robust variety of *A. purpurea*, and has dark, stout awns and long panicles. It may be confused with var. *nealleyi*, which has narrower lemmas and awns and a light-colored panicle, but it also intergrades with var. *purpurea* and var. *parishii*. **Aristida purpurea** forma **brownii** (Warnock) Allred & Valdés-Reyna refers to plants with short central awns and lateral awns that are only 1–3 mm long.

20. **Aristida arizonica** Vasey [p. 336]
ARIZONA THREEAWN

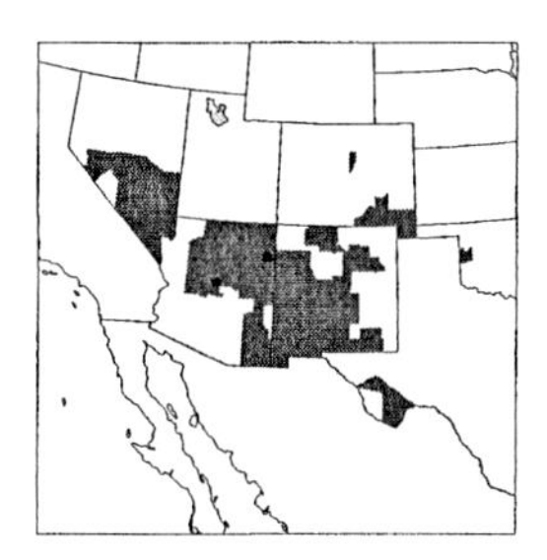

Plants perennial; usually cespitose, occasionally with rhizomes. **Culms** 30–80(100) cm, erect, unbranched. **Leaves** mostly basal; **sheaths** usually longer than the internodes, mostly glabrous, throat sometimes with hairs, not disintegrating into threadlike fibers; **collars** glabrous or with hairs at the sides; **ligules** 0.2–0.4 mm; **blades** 10–25(30) cm long, 1–3 mm wide, usually flat, often curling like wood shavings when mature, glabrous. **Inflorescences** spikelike panicles, 10–25 cm long, 1–3 cm wide; **nodes** glabrous or with straight, about 0.5 mm hairs; **primary branches** 2–6 cm, appressed, without axillary pulvini, with 2–8 spikelets. **Glumes** 10–15(18) mm, brownish, acuminate to awned, awns to 3 mm; **lower glumes** slightly shorter than to equaling the upper glumes, 1–2-veined; **calluses** 1–1.8 mm; **lemmas** 12–18 mm, glabrous, rarely sparsely pilose, terminating in a 3–6 mm twisted column, junction with the awns not conspicuous; **awns** 20–35 mm, straight to curved basally, ascending distally, not disarticulating at maturity; **central awns** 20–35 mm; **lateral awns** slightly shorter than the central awns; **anthers** 3, 1.3–1.9 mm. $2n$ = 22.

Aristida arizonica grows in pine, pine-oak, and pinyon-juniper woodlands from the southwestern United States to southern Mexico. It may be confused with *A. purpurea* var. *nealleyi*, but differs in having flat, curly leaf blades and longer awns.

21. **Aristida stricta** Michx. [p. 336]
WIREGRASS, PINELAND THREEAWN

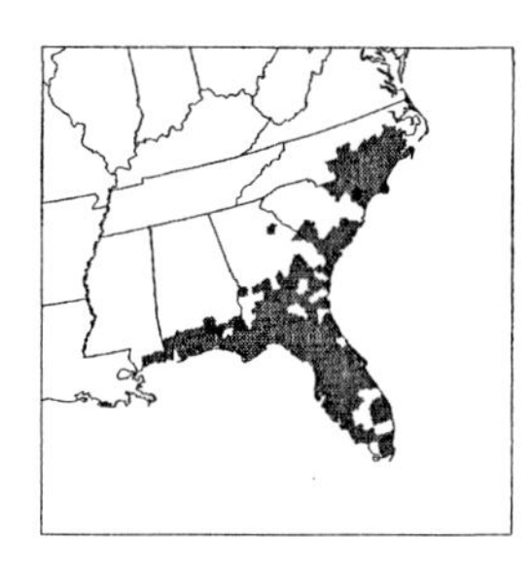

Plants perennial; cespitose, occasionally with rhizomes. **Culms** 60–120 cm, stiffly erect, unbranched. **Leaves** basal or nearly basal; **sheaths** shorter than the internodes, smooth and mostly glabrous abaxially, sometimes with a line of widely spaced, 0.5–1 mm hairs over the midvein, not disintegrating into threadlike fibers at maturity; **collars** glabrous, sometimes with a few conspicuous hairs at the sides; **ligules** 0.1–0.3 mm; **blades** 15–50 cm long, 0.3–1 mm wide, tightly involute, stiff, yellow-green, abaxial surfaces villous on both sides of the midvein, at least on the basal portion, hairs 0.6–1.5 mm, adaxial surfaces densely scabrous or densely short pubescent. **Inflorescences** paniculate, 20–35 cm, 5–8 mm wide; **nodes** glabrous; **primary branches** 2–5 cm, appressed, without axillary pulvini, with 4–12 spikelets, spikelet-bearing to the base. **Spikelets** appressed. **Glumes** subequal, glabrous, light brown or tan, usually 1(2)-veined, bifid and awned, awns 1.5–2.5 mm; **lower glumes** 7–10 mm; **upper glumes** 6–9 mm; **calluses** 0.4–0.6 mm; **lemmas** 6–9 mm, glabrous, light-colored when young, reddish when mature, column 1–2 mm, not twisted, junction with the awns inconspicuous; **awns** (7)10–15(22) mm, subequal, usually horizontally spreading or curving downward, not disarticulating at maturity; **anthers** 3, about 3 mm, reddish-brown. **Caryopses** 4–5 mm, chestnut brown. $2n$ = unknown.

Aristida stricta grows in pine barrens and sandy fields of the coastal plain from Louisiana to North Carolina. Peet (1993) segregated northern populations of *A. stricta* as a separate species, *A. beyrichiana* Trin. & Rupr., based on pubescence patterns of the sheath and blades. Investigations into alloyzyme diversity (Walters et al. 1994), anatomy, morphology, and phenotypic expression (garden transplants) led Kesler (2000) to conclude that such a segregation was not justified; pubescence patterns particularly were inconclusive. Consequently, *A. beyrichiana* is treated here as part of *A. stricta*.

22. **Aristida rhizomophora** Swallen [p. 336]

Plants perennial; cespitose, with well-developed, thick, dark rhizomes. **Culms** 60–100 cm, erect, unbranched. **Leaves** mostly basal; **sheaths** longer than the internodes, glabrous, basal sheaths many-veined, shredding into threadlike segments at maturity; **collars**

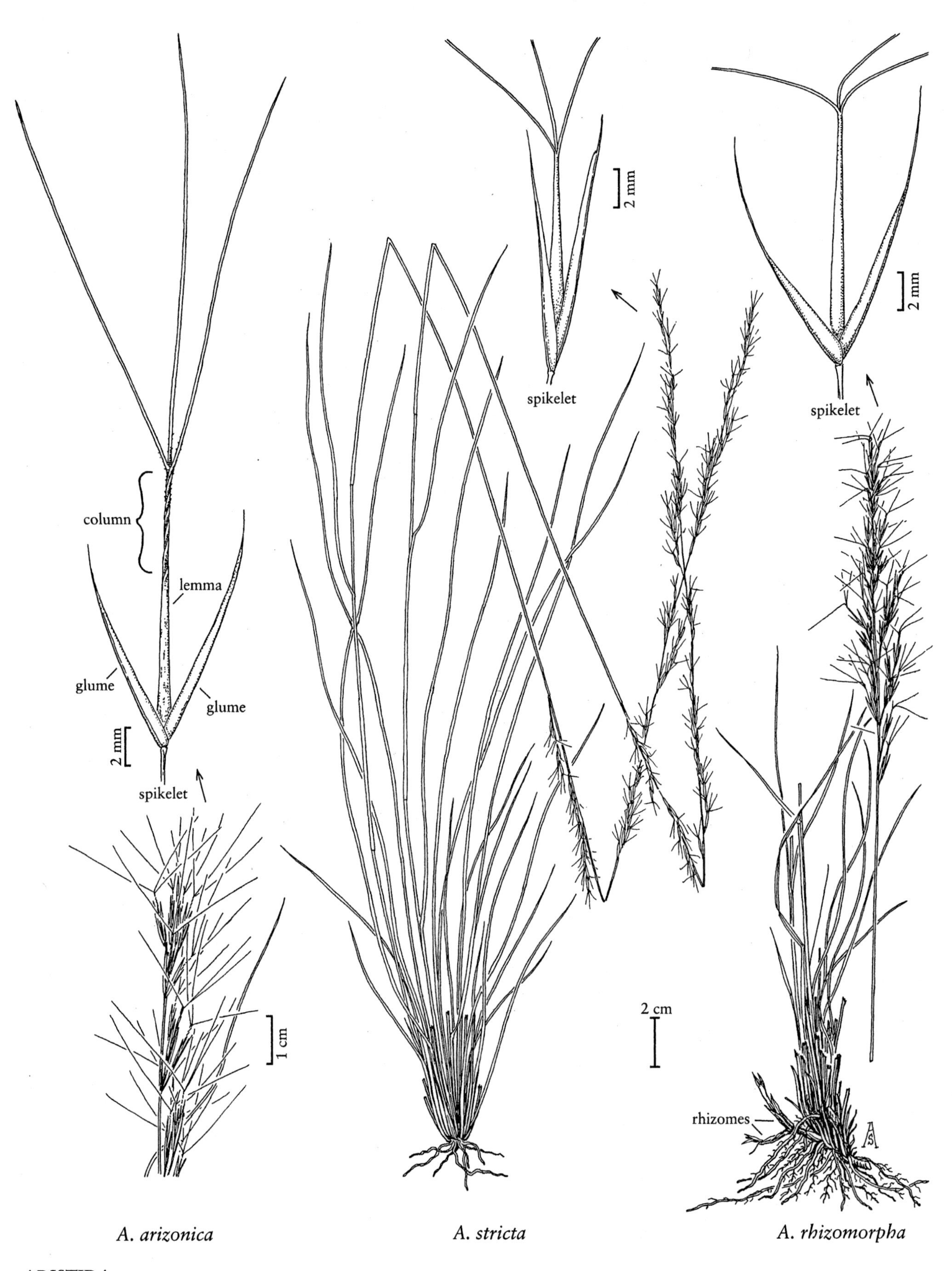

A. arizonica *A. stricta* *A. rhizomorpha*

ARISTIDA

glabrous or sparsely pilose at the corners; **ligules** 0.1–0.2 mm; **blades** 10–55 cm long, 1–3 mm wide, flat to folded, glabrous, pale green to yellow-green, central veins separate and narrow, without a well-defined midrib, lateral veins forming a thickened region on each margin. **Inflorescences** paniculate, (10)20–45 cm long, 2–6 cm wide; **nodes** glabrous; **primary branches** (2)4–15 cm, basal branches appressed, without axillary pulvini, distal branches ascending, occasionally lax or drooping distally. **Spikelets** appressed. **Glumes** unequal, brown to chestnut, 1-veined, awned; **lower glumes** 6–12 mm, awns 2–5 mm; **upper glumes** 13–18 mm, awns 3–6 mm; **calluses** 0.4–0.8 mm; **lemmas** 9–13 mm long, narrowing to a poorly defined beak 1–2 mm long and 0.2–0.3 mm wide, glabrous, tan to brown, junction with the awns not conspicuous; **awns** usually unequal, not disarticulating at maturity; **central awns** 15–30 mm, curved to semicircular at the base, horizontal to reflexed distally; **lateral awns** 13–20 mm, at least ½ as long as the central awns, curved or loosely twisted at the base, straight and strongly divergent distally; **anthers** 3, about 4 mm, yellow. **Caryopses** 6–8 mm, tan to brown. $2n$ = unknown.

Aristida rhizomophora is not well-collected. It is endemic to Florida, where it grows in moist to wet pine flatwoods, and on the borders of ponds and bald-cypress depressions.

23. **Aristida mohrii** Nash [p. 339]
Mohr's Threeawn

Plants perennial; cespitose, bases knotty. **Culms** 55–110 cm, erect, unbranched. **Leaves** cauline; **sheaths** shorter or longer than the internodes, glabrous, not disintegrating at maturity; **collars** glabrous, or sparsely pilose at the sides; **ligules** about 0.2 mm; **blades** 5–25 cm long, 1–2 mm wide, flat or loosely folded, pale green, cauline leaves usually glabrous adaxially, innovation leaves pilose. **Inflorescences** spikelike racemes, 20–45 cm long, 1–2 cm wide; **nodes** glabrous, with only 1 spikelet. **Spikelets** solitary, not overlapping. **Glumes** 9–11 mm, equal or the lower glumes slightly longer than the upper, narrowly oblong, often slightly falcate, tan to brown, 1-veined; **lower glumes** occasionally with 1–2 faint lateral veins, awn-tipped, awns about 0.5 mm; **upper glume** awned, awns 1–2.5 mm; **calluses** 1–1.5 mm; **lemmas** 7–10 mm, brown, lead-colored, or purplish, not beaked, junction with the awns not evident; **awns** equally thick, not disarticulating at maturity; **central awns** 14–20 mm, slightly longer than the lateral awns, strongly curved basally, distal portion reflexed; **lateral awns** horizontal to reflexed; **anthers** 3, about 4.5 mm, purplish. **Caryopses** 4–5 mm, chestnut-colored. $2n$ = unknown.

Aristida mohrii is endemic to the southeastern United States, growing on dry, sandy pinelands and oak barrens, and occasionally in waste places. It is sometimes confused with *A. simpliciflora* because both have reduced, spikelike inflorescences, but *A. simpliciflora* has lateral awns that are only about half as thick as the central awn, and its spikelets are borne in pairs.

24. **Aristida simpliciflora** Chapm. [p. 339]
Southern Threeawn

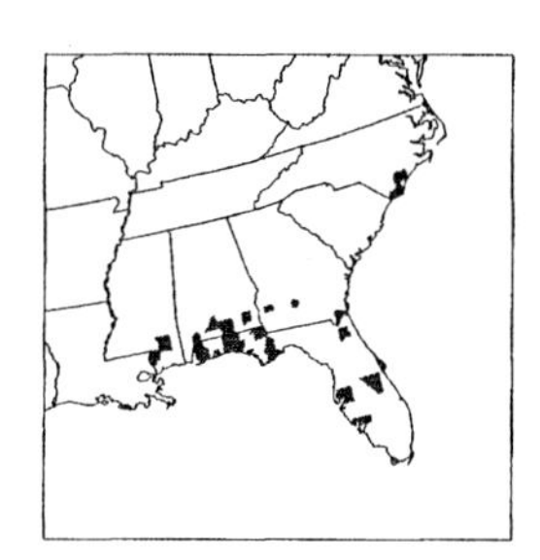

Plants perennial; loosely cespitose. **Culms** 30–80 cm, loosely branched below; **internodes** hollow. **Leaves** cauline, mostly glabrous; **sheaths** shorter than the internodes, remaining intact at maturity; **ligules** about 0.1 mm; **blades** 5–15 cm long, 0.8–1.5 mm wide, usually flat, those of the innovations often sparsely pilose. **Inflorescences** narrowly racemose, 10–30 cm long, 1–2 cm wide, often nodding; **nodes** glabrous; **lower pedicels** appressed. **Spikelets** usually 2(1–3) per node, 1 sessile or short-pedicellate and 1 long-pedicellate. **Glumes** 6–9 mm, subequal, tan to purplish, 1–2-veined, acute to awn-tipped, awns 0.5–1.5 mm; **lower glumes** frequently 2-keeled; **calluses** 0.4–0.6 mm; **lemmas** 5–6 mm, light tan to lead-colored, column not twisted, junction with the awns not conspicuous; **awns** not disarticulating at maturity; **central awns** 10–15 mm, about twice as thick as the lateral awns, reflexed from a semicircular bend; **lateral awns** equal to or slightly shorter than the central awns, divaricate and slightly contorted at the base; **anthers** 3, 2–3 mm, tan to brown. **Caryopses** 4–5 mm, chestnut-colored. $2n$ = unknown.

Aristida simpliciflora grows in wet savannahs, the upper portion of seepage bogs, and the moister portion of ecotones between such bogs and the surrounding dry uplands. It is restricted to the southeastern United States.

Aristida simpliciflora is sometimes confused with *A. mohri* because both have reduced, spikelike inflorescences, but *A. mohri* has lateral awns that are about as thick as the central awn, and its spikelets are solitary.

25. **Aristida lanosa** Muhl. *ex* Elliott [p. 339]
WOOLY THREEAWN

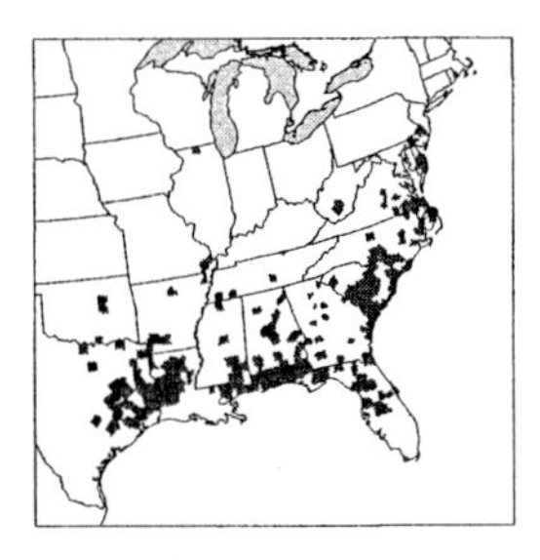

Plants perennial; loosely cespitose. **Culms** 65–150 cm, sometimes thickened at the base, erect, unbranched; **internodes** glabrous; **nodes** concealed. **Leaves** cauline; **sheaths** longer than the internodes, usually lanose-floccose, occasionally glabrate; **ligules** about 0.1 mm; **blades** 10–25(30) cm long, 2–6 mm wide, flat, light green or slightly blue-green, glabrous abaxially. **Inflorescences** paniculate, (25)35–70(82) cm long, (2)3–8(10) cm wide; **rachis nodes** lanose-floccose; **primary branches** 3–12 cm, appressed at the base, without axillary pulvini, ascending to spreading distally, sometimes loose and somewhat flexible, with 4–12 spikelets per branch. **Glumes** usually unequal, 1-veined, brownish-green to dark brown or purplish; **lower glumes** 8.7–18 mm, with a keeled mid-vein; **upper glumes** 8.4–15 mm, awn-tipped, awns to 3 mm; **calluses** 0.5–1 mm; **lemmas** 6.5–10 mm, smooth to scabridulous, mostly dark purplish-mottled, slightly narrowed distally but not beaked, junction with the awns not evident; **central awns** 12–28 mm, curved at the base, often strongly so; **lateral awns** 7–17 mm, at least ½ as long as the central awns; **anthers** 3, about 3 mm, brown. **Caryopses** 5–6 mm, chestnut brown. $2n$ = unknown.

Aristida lanosa is restricted to the eastern United States, where it grows in dry fields, pine-oak woods, and uplands, chiefly in sandy soil. It is sometimes confused with *A. palustris*, but differs in several reproductive, vegetative, and habitat characteristics.

26. **Aristida palustris** (Chapm.) Vasey [p. 339]
LONGLEAF THREEAWN

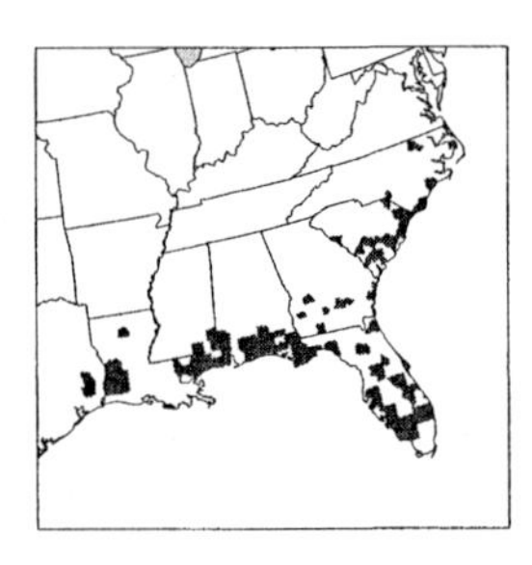

Plants perennial; cespitose, bases hard, knotty. **Culms** 90–150 cm, often thickened basally, stiffly erect, usually unbranched; **internodes** hollow. **Leaves** cauline; **sheaths** usually shorter than the internodes, glabrous, remaining intact at maturity; **collars** glabrous; **ligules** to 0.1 mm; **blades** (8)10–30(35) cm long, 2–4 mm wide, usually flat, occasionally loosely involute, lax, glabrous, light yellow-green to bluish-green when young, drying brownish. **Inflorescences** paniculate, 25–45(55) cm long, 3–6 cm wide; **nodes** glabrous; **primary branches** 2–8 cm, usually single or paired, appressed to erect, occasionally ascending, without axillary pulvini, with (1)2–12 spikelets. **Spikelets** overlapping, appressed. **Glumes** (7.5)9–13.5 mm, subequal, stiff, glabrous or scabridulous, light brown or greenish-brown; **lower glumes** prominently 2-veined, 2-keeled by the development of 1 lateral vein, shortly (1–2 mm) awn-tipped; **upper glumes** 1-veined, shortly (0.5–1 mm) awn-tipped; **calluses** 1–1.4 mm; **lemmas** 6–9 mm, glabrous, 0.3–0.5 mm wide distally, light tan to brown, junction with the awns not evident; **awns** not disarticulating at maturity; **central awns** 15–40 mm, usually strongly curved basally, strongly divergent to horizontal distally; **lateral awns** 8–35 mm, at least ½ as long as the central awns, erect to strongly divergent; **anthers** 3, about 3 mm, purplish. **Caryopses** 4.4–5 mm, chestnut brown. $2n$ = unknown.

Aristida palustris is endemic to the southeastern United States, where it grows in seepage bogs, pitcher plant savannahs, wet pine flatwoods, bald-cypress depressions, and wet prairies. It is a distinctive species of the southeastern coastal plain region that differs from *A. lanosa* in several reproductive, vegetative, and habitat characteristics.

27. **Aristida purpurascens** Poir. [p. 341]

Plants perennial; cespitose, bases knotty, without rhizomes. **Culms** 40–100 cm tall, 1–4 mm thick at the base, erect, branching at the base, shoots becoming thickened and somewhat fan-shaped upwards. **Leaves** cauline; **sheaths** mostly longer than the internode, mostly or completely glabrous, sometimes pilose, particularly along the margins and at the throat, remaining intact at maturity; **collars** glabrous or pilose; **ligules** about 0.2 mm; **blades** 10–25 cm long, 1–3 mm wide, usually flat, usually lax, sometimes sinuous to curling at maturity, glabrous, pale green, drying brownish. **Inflorescences** paniculate, (15)20–55 cm long, 0.5–2(3) cm wide, often nodding; **nodes** glabrous, sometimes scabrous, lower nodes usually associated with more than 2 spikelets; **primary branches** 1–5 cm, tightly appressed to loosely ascending, without axillary pulvini, with 1–8 spikelets. **Spikelets** appressed. **Glumes** 5–10 mm, lower glumes from ¾ as long as to 1–4 mm longer than the upper glumes, glabrous or sparsely appressed-pubescent, 1–2-veined, 1-keeled, tan to purplish, unawned or the awns no longer than 1 mm; **calluses** 0.4–0.8 mm; **lemmas** 4–8 mm, glabrous, mostly light tan or gray, often spotted or banded, beak not twisted, junction with the awns not evident; **awns** 8–25 mm, equal or subequal in length, curved, arcuate, or spirally coiled at the base, not disarticulating at maturity; **central awns** sometimes thicker than the lateral awns, erect to arcuate-reflexed; **lateral awns** straight and erect, ascending, or divergent; **anthers** 1 or 3, 1–1.5 mm,

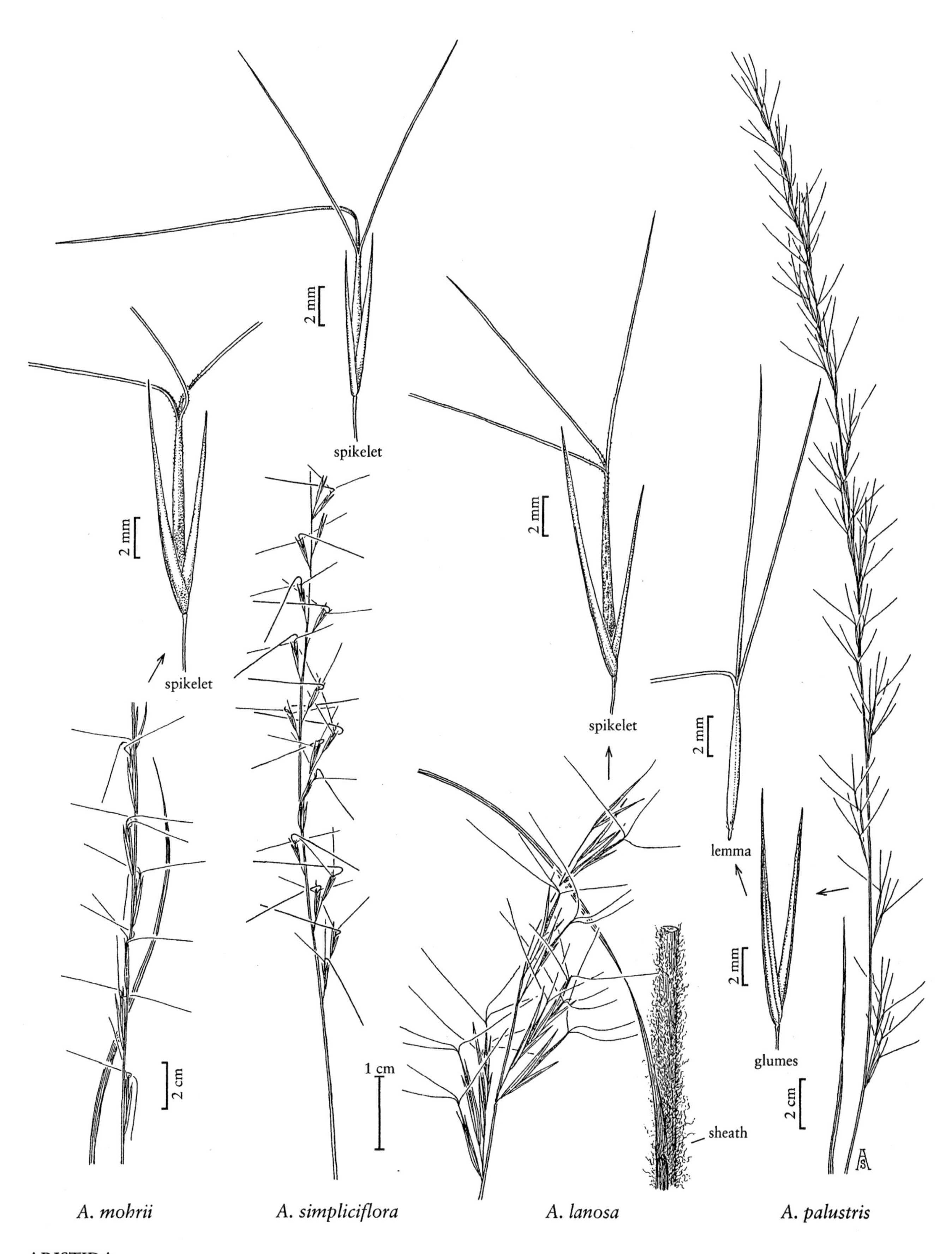

A. mohrii *A. simpliciflora* *A. lanosa* *A. palustris*

ARISTIDA

brown. **Caryopses** 3–5 mm, chestnut brown. 2*n* = unknown.

Aristida purpurascens is composed of three intergrading varieties, as follows:

1. Central awns divaricate to reflexed, about twice as thick at the base as the lateral awns . . . . var. *virgata*
1. Central and lateral awns divergent, all about the same thickness at the base.
   2. Lower glumes usually longer than the upper glumes; awns straight or only slightly contorted at the base; blades 1–3 mm wide, often curling . . . . var. *purpurascens*
   2. Lower glumes shorter than or equal to the upper glumes; awns spirally contorted at the base; blades mostly about 1 mm wide, usually not curling . . . . var. *tenuispica*

Aristida purpurascens Poir. var. **purpurascens**
ARROWFEATHER THREEAWN

Lower sheaths longer than the internodes, usually glabrous, or occasionally with appressed-pilose hairs; **blades** 1–3 mm wide, often curling. **Glumes** unequal; **lower glumes** 6–10 mm, usually longer than the upper glumes; **upper glumes** 5–8 mm; **awns** 15–25 mm, of equal thickness, straight or only slightly contorted, divergent.

Aristida purpurascens var. *purpurascens* grows in waste places, glades, fields, and pine savannahs in sandy or clay soils. Its range extends into northern Mexico.

Aristida purpurascens var. **tenuispica** (Hitchc.) Allred

Lower sheaths shorter or longer than the internodes, glabrous; **blades** usually about 1 mm wide, usually not curling. **Glumes** 8–9 mm, subequal; **awns** 12–18 mm, of equal thickness, spirally contorted at the base, divergent.

Aristida purpurascens var. *tenuispica* grows in pine and oak woods, prairies, and along roadsides, at low elevations and usually in sandy soils. Within the *Flora* region, it grows on the coastal plain from North Carolina to Mississippi. It also grows in Mexico and Honduras.

Aristida purpurascens var. **virgata** (Trin.) Allred

Lower sheaths shorter or longer than the internodes, glabrous. **Glumes** 6–7 mm, equal or the lower glumes slightly longer; **central awns** 13–20 mm, about twice as thick as the lateral awns, divaricate to reflexed at maturity; **lateral awns** 8–13 mm, erect to ascending.

Aristida purpurascens var. *virgata* grows in wet or moist areas such as seepage bogs, sandy pinelands, and wet prairies of the southeastern United States.

28. **Aristida condensata** Chapm. [p. 341]
BIG THREEAWN

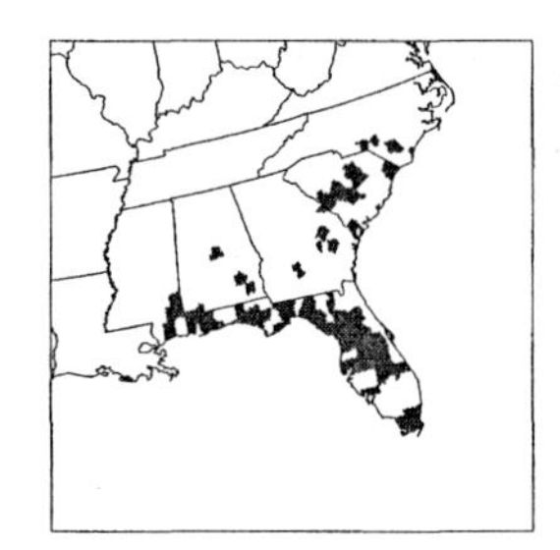

Plants perennial; bases knotty, bleached, not rhizomatous. **Culms** 70–150 cm tall, 3–6 mm thick at the base, erect, rarely branched. **Leaves** mostly cauline; **sheaths** usually longer than the internodes, remaining intact at maturity, glabrous or appressed pilose; **collars** glabrous or pilose at the sides; **ligules** less than 0.5 mm; **blades** (10)15–30 cm long, 1.5–3 mm wide, flat at the base, involute toward the apex, straight to somewhat lax at maturity, glabrous abaxially, yellowish-green when fresh, drying stramineous or darker. **Inflorescences** paniculate, (15)20–55 cm long, 2–4 cm wide; **nodes** glabrous or with short, straight hairs; **primary branches** (4)5–20 cm, appressed to narrowly ascending, without axillary pulvini, naked below, with 5–15 overlapping spikelets distally. **Glumes** 6–10(12) mm, 1-veined, 1-keeled, awns less than 4 mm, brownish; **lower glumes** from ¾ as long as to 1 mm longer than the upper glumes; **calluses** 1–2 mm; **lemmas** 5–8 mm, often reddish-mottled, apices not strongly twisted, junction with the awns not evident; **awns** about equally thick, divergent, spirally contorted at the base but usually not with distinct coils, not disarticulating at maturity; **central awns** 10–15 mm; **lateral awns** 8–13 mm; **anthers** 3, about 2 mm. **Caryopses** 4–5 mm, chestnut-colored. 2*n* = unknown.

Aristida condensata grows on sandy hills, and in pine and oak barrens in the southeastern United States.

29. **Aristida gyrans** Chapm. [p. 341]
CORKSCREW THREEAWN

Plants perennial; tightly cespitose, bases often bleached, without rhizomes. **Culms** 20–65 cm tall, 1–4 mm thick at the base, erect, rarely geniculate at the base, unbranched; **internodes** often in a sequence of 2 short and 1 long. **Leaves** mostly basal; **sheaths** usually shorter than the internodes, glabrous, remaining intact at maturity; **collars** glabrous; **ligules** 0.2–0.3 mm; **blades** (3)5–15 cm long, to 1 mm wide, involute, rarely loosely folded or flat, somewhat stiff and arcuate, bases glabrous abaxially, pale green. **Inflorescences** paniculate or racemose, 10–30 cm long, 1–2 cm wide, slender, lax; **primary branches** 3–5 cm, loosely appressed, without axillary pulvini, with 2–5 spikelets. **Spikelets** appressed. **Glumes** usually unequal, 1-veined, acuminate or awned, awns to 4 mm, tan to

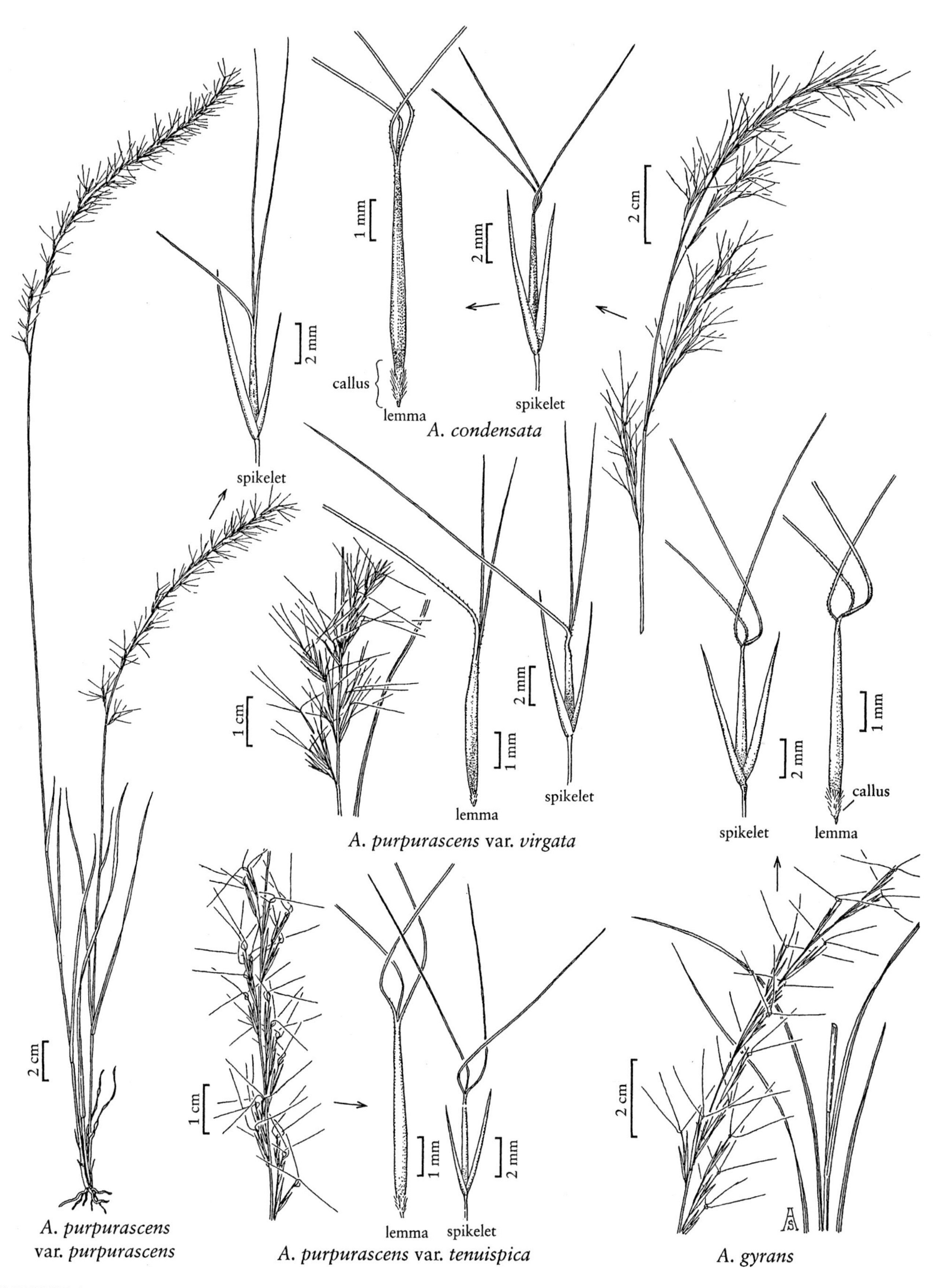
spikelet
2 mm
1 mm
callus
lemma
2 mm
spikelet
A. condensata
2 cm
1 cm
2 mm
1 mm
lemma
spikelet
A. purpurascens var. virgata
2 mm
1 mm
callus
spikelet
lemma
2 cm
1 cm
1 mm
2 mm
lemma
spikelet
2 cm
A. purpurascens
var. purpurascens
A. purpurascens var. tenuispica
A. gyrans

ARISTIDA

dark brownish or purplish; **Lower glumes** 6–9(11) mm; **upper glumes** 9–12 mm; **calluses** 1–2 mm; **lemmas** 5–7 mm, glabrous, brownish, without a column, the junction with the awns not evident; **awns** 8–15 mm, subequal, loosely spirally contorted, but not coiled, just above the base, ascending to spreading distally, not disarticulating at maturity; **anthers** 3, 1–1.5 mm, brownish. **Caryopses** 3–4 mm, somewhat lustrous, chestnut-colored. $2n$ = unknown.

Aristida gyrans is endemic to the southeastern United States, growing in sandy pine woods and oak scrub. It differs from other species in the genus by its combination of narrow blades, unequal glumes, long calluses, and contorted awns.

9. CENTOTHECOIDEAE Soderstr.

Grass Phylogeny Working Group

Plants annual or perennial; rhizomatous or stoloniferous. **Culms** annual, sometimes becoming woody, internodes solid or hollow. **Leaves** distichous; **sheaths** usually open; **auricles** sometimes present; **abaxial ligules** absent or of hairs; **adaxial ligules** membranous, ciliate or not, or of hairs; **blades** often pseudopetiolate; **mesophyll cells** non-radiate; **adaxial palisade layer** often present; **fusoid cells** absent, but fusoidlike cells frequently present as extensions of the outer parenchyma bundle sheath; **arm cells** absent; **kranz anatomy** absent; **midrib** simple; **adaxial bulliform cells** present, large; **stomata** with dome-shaped or triangular subsidiary cells; **bicellular microhairs** present, with long, tapering apical cells; **papillae** absent. **Inflorescences** ebracteate, racemose or paniculate, panicle branches sometimes spikelike; **disarticulation** below the spikelets or the florets, sometimes at the base of the pedicels. **Spikelets** bisexual or unisexual, often laterally compressed, with (1)2–many florets, reduced florets present, distal or basal to the functional florets. **Glumes** shorter than the lemmas; **lemmas** lacking uncinate hairs, usually 5–9-veined, unawned or with single, terminal awns; **paleas** usually well-developed, sometimes short compared to the lemmas; **lodicules** 2 or none, cuneate, usually well-vascularized, varying to not or scarcely vascularized; **stamens** 2; **ovaries** glabrous; **haustorial synergids** presumed absent; **styles** 2, sometimes fused at the base, if free, close. **Hila** basal, punctiform; **endosperm** hard, without lipid; **starch grains** simple; **embryos** small or large relative to the caryopses; **epiblasts** present; **scutellar cleft** present; **mesocotyl internode** present; **embryonic leaf margins** overlapping. x = (11)12.

The subfamily *Centothecoideae* is one of the subfamilies that cannot be characterized by a suite of morphological characteristics, but anatomical, micromorphological, and nucleic acid data all support its recognition. It is most abundant in warm-temperate woodlands and tropical forests. Clayton and Renvoize (1986) suggested that it was an offshoot of the *Arundinoideae*, but molecular data (Hilu et al. 1999; Grass Phylogeny Working Group 2001) argue for a sister group relationship with the *Panicoideae*. The treatment here, in which two tribes are recognized, follows that of the Grass Phylogeny Working Group (2001). Some of the genera, however, are as yet poorly known in terms of the characters used in making such decisions.

SELECTED REFERENCES **Clayton, W.D. and S.A. Renvoize.** 1986. Genera Graminum: Grasses of the World. Kew Bull., Addit. Ser. 13. Her Majesty's Stationery Office, London, England. 389 pp.; **Decker, H.F.** 1964. An anatomic-systematic study of the classical tribe Festuceae (Gramineae). Amer. J. Bot. 51:453–463; **Grass Phylogeny Working Group.** 2001. Phylogeny and subfamilial classification of the grasses (Poaceae). Ann. Missouri Bot. Gard. 88:373–457; **Hilu, K.W., L.A. Alice and J. Liang.** 1999. Phylogeny of the Poaceae inferred from *mat*K sequences. Ann. Missouri Bot. Gard. 86:835–851; **Hilu, K.W. and K. Wright.** 1982. Systematics of Gramineae: A cluster analysis study. Taxon 31:9–36; **Soderstrom, T.R.** 1981.The grass subfamily Centostecoideae. Taxon 30:614–616.

1. Spikelets 4–50 mm long, with 1–15 florets, the lowest florets sometimes sterile, the upper florets bisexual; disarticulation at the base of the florets or the base of the spikelets; leaves not pseudopetiolate; culms 35–150 cm tall; plants not reedlike .... 22. *Centotheceae*
1. Spikelets 1.2–1.8 mm long, with 2(3–4) florets, the lower florets sterile, the upper florets bisexual; disarticulation at the pedicel bases, subsequently below the spikelets; leaves pseudopetiolate; culms 150–400 cm tall; plants reedlike .... 23. *Thysanolaeneae*

22. CENTOTHECEAE Ridl.

J. Gabriel Sánchez-Ken

Plants annual or perennial; cespitose, herbaceous, delicate to sometimes reedlike. **Culms** to 4 m tall, not woody, glabrous; **internodes** solid or hollow. **Sheaths** open; **auricles** sometimes present; **ligules** scarious or membranous, truncate, sometimes ciliolate; **pseudopetiole** present in some genera; **blades** flat, relatively wide, usually with evident cross venation (not in *Chasmanthium*), not disarticulating from the sheaths. **Inflorescences** terminal, paniculate, sometimes with spicate branches. **Spikelets** solitary, pedicellate or subsessile, laterally compressed, with 1–many florets, bisexual, pistillate or staminate, lowest and distal florets often sterile; **rachillas** terminating in a rudimentary floret; **disarticulation** variable, at the base of the pedicels, below the glumes, above the glumes, beneath the florets, between the florets, or some combination of these. **Glumes** sometimes lacking, if present, membranous, shorter than the florets, 2–9-veined, acute to obtuse; **lowest florets** sometimes sterile, with or without paleas; **upper lemmas** membranous, 3–15-veined, unawned or awned; **upper paleas** nearly as long as the lemmas, apices entire or notched; **lodicules** 2, free (rarely fused in some genera), cuneate, truncate or somewhat lobed; **anthers** 1, 2, or 3. **Caryopses** ellipsoid to circular, or trigonous; **embryos** about ⅓ or less as long as the caryopses; **hila** subbasal to basal, punctate. $x = 12$.

The tribe *Centotheceae* has approximately 10 genera and 30 species, most of which grow in tropical forests. *Chasmanthium*, the only member of the tribe represented in the *Flora* region, is also the only genus to extend into temperate regions. Most members of the tribe are easily recognized by the evident cross venation of their wide blades. Unfortunately, *Chasmanthium* is exceptional in this regard, lacking evident cross venation.

SELECTED REFERENCES **Grass Phylogeny Working Group.** 2001. Phylogeny and subfamilial classification of the grasses (Poaceae). Ann. Missouri Bot. Gard. 88:373–457; **Yates, H.O.** 1966. Revision of grasses traditionally referred to *Uniola*, II. *Chasmanthium*. SouthW. Naturalist 11:415–455.

22.01 CHASMANTHIUM Link

J. Gabriel Sánchez-Ken
Lynn G. Clark

Plants perennial; cespitose or loosely colonial, rhizomatous. **Culms** 35–150 cm, simple or branched. **Leaves** cauline; **ligules** membranous, ciliate; **blades** not pseudopetiolate, flat. **Panicles** open or contracted, sometimes becoming racemose distally; **disarticulation** above the glumes and between the florets. **Spikelets** 4–50 mm, laterally compressed, with 2–many florets, lower 1–4 florets sterile. **Glumes** 2, subequal, shorter than the spikelets, glabrous, (2)3–9-veined, acute to acuminate; **lemmas** glabrous, 3–15-veined, compressed-keeled, keels serrate or ciliate, apices acuminate to acute, entire (rarely bifid); **paleas** glabrous, gibbous basally, 2-keeled, keels winged, wings glabrous, scabrous, or pilose; **lodicules** 2, fleshy, cuneate, 2–4-veined, lobed-truncate; **anthers** 1; **ovaries** glabrous; **styles** 2; **style branches** 2, plumose, reddish-purple at anthesis. **Caryopses** 1.9–5 mm, laterally compressed, brown to reddish-black or black. $x = 12$. Name from the Greek *chasma*, 'yawn', and *anthos*, 'flower', presumably for the gaping glumes that expose the grain at maturity.

Chasmanthium, a genus of five species endemic to North America, grows primarily in the southeastern and south-central parts of the United States. It was formerly included in *Uniola*, but it is now recognized as a distinct genus.

SELECTED REFERENCES **Yates, H.O.** 1966a. Morphology and cytology of *Uniola* (Gramineae). SouthW. Naturalist 11:145–189; **Yates, H.O.** 1966b. Revision of grasses traditionally referred to *Uniola*, II. *Chasmanthium*. SouthW. Naturalist 11:451–455.

1. Panicle branches nodding or drooping; pedicels 10–30 mm long; calluses pilose; lower glumes 4.2–9.1 mm long; keels of fertile lemmas winged, the wings scabrous to pilose their full length; caryopses 2.9–5 mm long . . . . 1. *C. latifolium*
1. Panicle branches erect or ascending; pedicels 0.5–2.5(5) mm long; calluses glabrous; lower glumes 1.2–5 mm long; keels of fertile lemmas not winged, scabridulous toward or at the apices; caryopses 1.9–3 mm long.
   2. Spikelets 9.5–24 mm long; fertile lemmas 9–13-veined; caryopses enclosed at maturity; blades 7–16(33) cm long, lanceolate-fusiform; culms leafy for 80% of their height.
      3. Axils of the panicle branches scabrous; fertile florets diverging to 45° from the rachilla; sterile florets (0)1(2); lower glumes 3.1–5 mm long, 7–9-veined; ligules entire . . . . 2. *C. nitidum*
      3. Axils of the panicle branches pilose; fertile florets diverging to 85° from the rachilla; sterile florets 2–4; lower glumes 2.5–2.9 mm long, 2–3-veined; ligules irregularly laciniate . . . 3. *C. ornithorhynchum*
   2. Spikelets 4–10 mm long; fertile lemmas 3–9-veined; caryopses exposed at maturity; blades (8)20–50 cm long, linear-lanceolate; culms leafy for 40–50% of their height.
      4. Collars and sheaths pilose; culms (1)2–3.5 mm thick at the nodes; fertile lemmas 7–9-veined, usually curved or irregularly contorted . . . . 4. *C. sessiliflorum*
      4. Collars and sheaths glabrous; culms to 1 mm thick at the nodes; fertile lemmas 3–7-veined, straight . . . . 5. *C. laxum*

1. **Chasmanthium latifolium** (Michx.) H.O. Yates [p. 347]
BROADLEAF CHASMANTHIUM

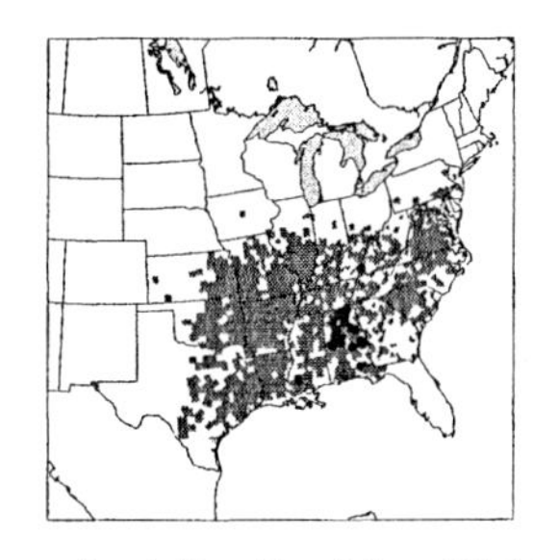

Culms to 150 cm, 2–4 mm thick at the nodes, rarely branched, leafy for 80% of their height. **Sheaths** glabrous; **collars** glabrous; **ligules** 0.7–1 mm, entire; **blades** (7)9–18(22) cm long, (4)10–22 mm wide, lanceolate-fusiform, usually glabrous, sometimes pilose adaxially. **Panicles** (8)10–25(35) cm, open, lax; **branches** nodding or drooping; **axils of panicle branches** sparsely pilose; **pedicels** 10–30 mm. **Spikelets** 15–40(50) mm long, 6–16(20) mm wide, with 6–17(26) florets, lower 1–3 florets sterile, fertile florets diverging to 45°. **Lower glumes** 4.2–9.1 mm, 5–7-veined; **upper glumes** 4.7–8.7 mm, 5–9-veined; **calluses** pilose; **fertile lemmas** 9–12.5 mm, straight, 11–15-veined, keels winged, wings scabrous to pilose their full length; **paleas** 4.6–7.7 mm; **anthers** (0.4)0.6–2.6(3.5) mm, the length varying within a spikelet. **Caryopses** 2.9–5 mm, enclosed, rarely exposed at maturity. $2n=48$.

Chasmanthium latifolium grows along stream and river banks and in rich deciduous woods. It is the most widespread species of the genus, extending further west and east than any of the other four species. The map shows its verifiable range. Yates (1966b) reported seeing one specimen each from New Jersey, New Mexico, and Manitoba, but none of the specimens had clear locality information. In the absence of any other specimens from these regions, the locality data on these three specimens are regarded as probably erroneous.

Flowering in *Chasmanthium latifolium* is sometimes cleistogamous.

2. **Chasmanthium nitidum** (Baldwin) H. O. Yates [p. 347]
SHINY CHASMANTHIUM

Culms 40–120 cm, to 1 mm thick at the nodes, rarely branched, leafy for 80% of their height. **Sheaths** glabrous; **collars** glabrous; **ligules** 0.2–0.3 mm, entire; **blades** 9–16(33) cm long, 4–7 mm wide, lanceolate-fusiform, glabrous adaxially. **Panicles** (9)12–17(29) cm, open, erect; **branches** ascending to divergent; **axils of panicle branches** scabrous; **pedicels** 0.5–2 mm. **Spikelets** 12–24 mm long, (8)9–12(15) mm wide, with (5)7–9(11) florets, lower (0)1(2) florets sterile, fertile florets diverging to 45°. **Lower glumes** 3.1–5 mm, 7–9-veined; **upper glumes** 3–4.6 mm, (5)7–9-veined; **calluses** glabrous; **fertile lemmas** 5.5–8.5 mm, straight, 9–11-veined, keels not winged, scabridulous toward the apices; **paleas** 5–7.5 mm; **anthers** 1.9–2.4 mm, the length invariant within a spikelet. **Caryopses** 2.4–3 mm, enclosed at maturity. $2n=24$.

Chasmanthium nitidum grows along stream and river banks, roadside ditches, and the margins of low, moist woods in the southeastern United States.

3. **Chasmanthium ornithorhynchum** Nees [p. 347]
BIRDBILL CHASMANTHIUM

Culms 35–40(90) cm, 0.8–1 mm thick at the nodes, rarely branched, leafy for 80% of their height. **Sheaths** glabrous; **collars** pilose; **ligules** 0.2–0.3 mm, irregularly laciniate; **blades** 7–9(15) cm long, 3.5–6 mm wide, lanceolate-fusiform, glabrous adaxially. **Panicles** (2.5)5–10.5(12) cm, open, erect; **branches** divergent, sometimes strongly so; **axils of panicle branches** pilose; **pedicels** 0.5–1 mm. **Spikelets** 9.5–12 mm long, 11–17(18) mm wide, with (4)5–10 florets, lower 2–4 florets sterile, fertile florets divergent to 85°. **Lower glumes** 2.5–2.9 mm, 2–3-veined; **upper glumes** 2.6–3.6 mm, 3–5-veined; **calluses** glabrous; **fertile lemmas** 5.8–9.3 mm, straight, (9)11–13-veined, keels not winged, scabrous to scabridulous distally; **paleas** 7.2–9 mm; **anthers** 1.1–1.9 mm, the length invariant within a spikelet. **Caryopses** 2–3 mm, enclosed at maturity. $2n=24$.

Chasmanthium ornithorhynchum grows along stream and river banks in low woods, and on hummocks in swamps. It is most common along the coastal plain from eastern Louisiana to western Florida, but is also found at a few other locations in the southeastern United States.

4. **Chasmanthium sessiliflorum** (Poir.) H.O. Yates [p. 348]
LONGLEAF CHASMANTHIUM

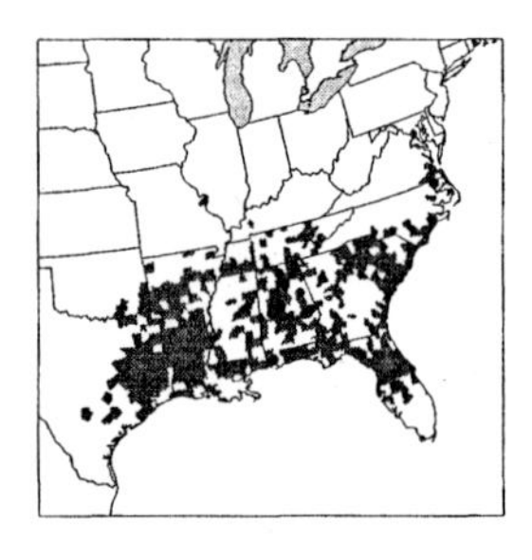

Culms 60–150 cm, (1)2–3.5 mm thick at the nodes, unbranched, leafy for 40% of their height. **Sheaths** pilose; **collars** pilose; **ligules** 0.2–0.3 mm, entire; **blades** (15)20–50 cm long, 4.5–9.5(15) mm wide, linear-lanceolate, sparsely pilose adaxially. **Panicles** (9)20–70 cm, contracted or open, erect; **branches** tightly appressed or ascending to strongly divergent; **axils of panicle branches** glabrous or scabridulous at the edges; **pedicels** 0.3–2.5(5) mm. **Spikelets** 4–10 mm long, 6–9 mm wide, with 4–7(8) florets, lower 1(2) florets sterile, fertile florets divergent to 80°. **Lower glumes** 1.2–2.7 mm, 3–5-veined; **upper glumes** 1.4–2.2 mm, 3–5-veined; **calluses** glabrous; **fertile lemmas** 3.5–5.9 mm, usually curved or irregularly contorted, 7–9-veined, keels not winged, apices scabridulous; **paleas** 2.8–4 mm; **anthers** (0.8)1.3–1.6 mm, varying in length within a spikelet. **Caryopses** 2–2.5 mm, exposed at maturity. $2n=24$.

Chasmanthium sessiliflorum grows in rich woods, meadows, and swamps, especially on the coastal plain. It grows throughout most of the southeastern United States.

5. **Chasmanthium laxum** (L.) H.O. Yates [p. 348]
SLENDER CHASMANTHIUM

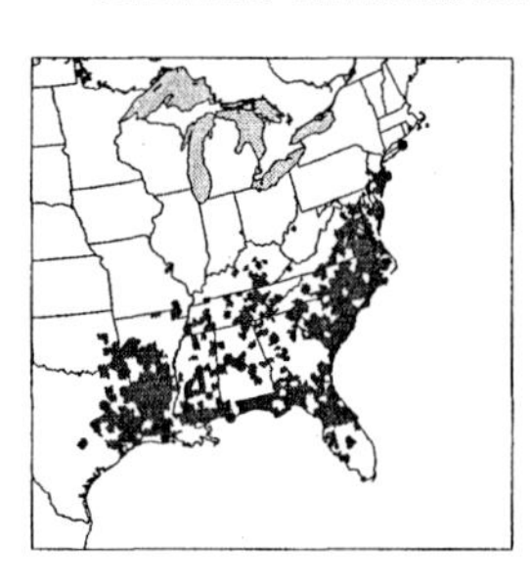

Culms 40–130 cm, to 1 mm thick at the nodes, unbranched, leafy for 50% of their height. **Sheaths** glabrous; **collars** glabrous; **ligules** 0.2–0.4 mm, entire; **blades** (8)15–35(40) cm long, 3–8(11) mm wide, linear-lanceolate, usually glabrous, sometimes sparsely pilose adaxially. **Panicles** (7)12–35(47) cm, contracted, erect; **branches** ascending to appressed; **axils of panicle branches** glabrous; **pedicels** 0.5–2.5 mm. **Spikelets** 4–9 mm long, 2–6 mm wide, with (2)3–5(7) florets, lower 1(2) florets sterile, fertile florets divergent to 45°. **Lower glumes** 1.3–3 mm, (1)3–5-veined; **upper glumes** 1.3–2.5 mm, 3–5-veined; **calluses** glabrous; **fertile lemmas** 2.9–4.5 mm, straight, 3–7-veined, keels not winged, apices scabridulous; **paleas** 2.3–3 mm; **anthers** 1.3–1.5 mm, the length invariant within a spikelet. **Caryopses** 1.9–2.2 mm, exposed at maturity. $2n=24$.

Chasmanthium laxum is almost completely sympatric with *C. sessiliflorum* in the southeastern United States, growing in similar habitats but extending farther into sphagnous stream heads, pine flatwoods, and pine savannahs. Yates (1966b) reported seeing putative, naturally occurring hybrids between *Chasmanthium ornithorhynchum* and *C. laxum* along streams of the outer coastal plain of Mississippi and Louisiana. In general appearance, the hybrids resemble *C. laxum*, their most striking difference being the enlarged, sterile spikelets.

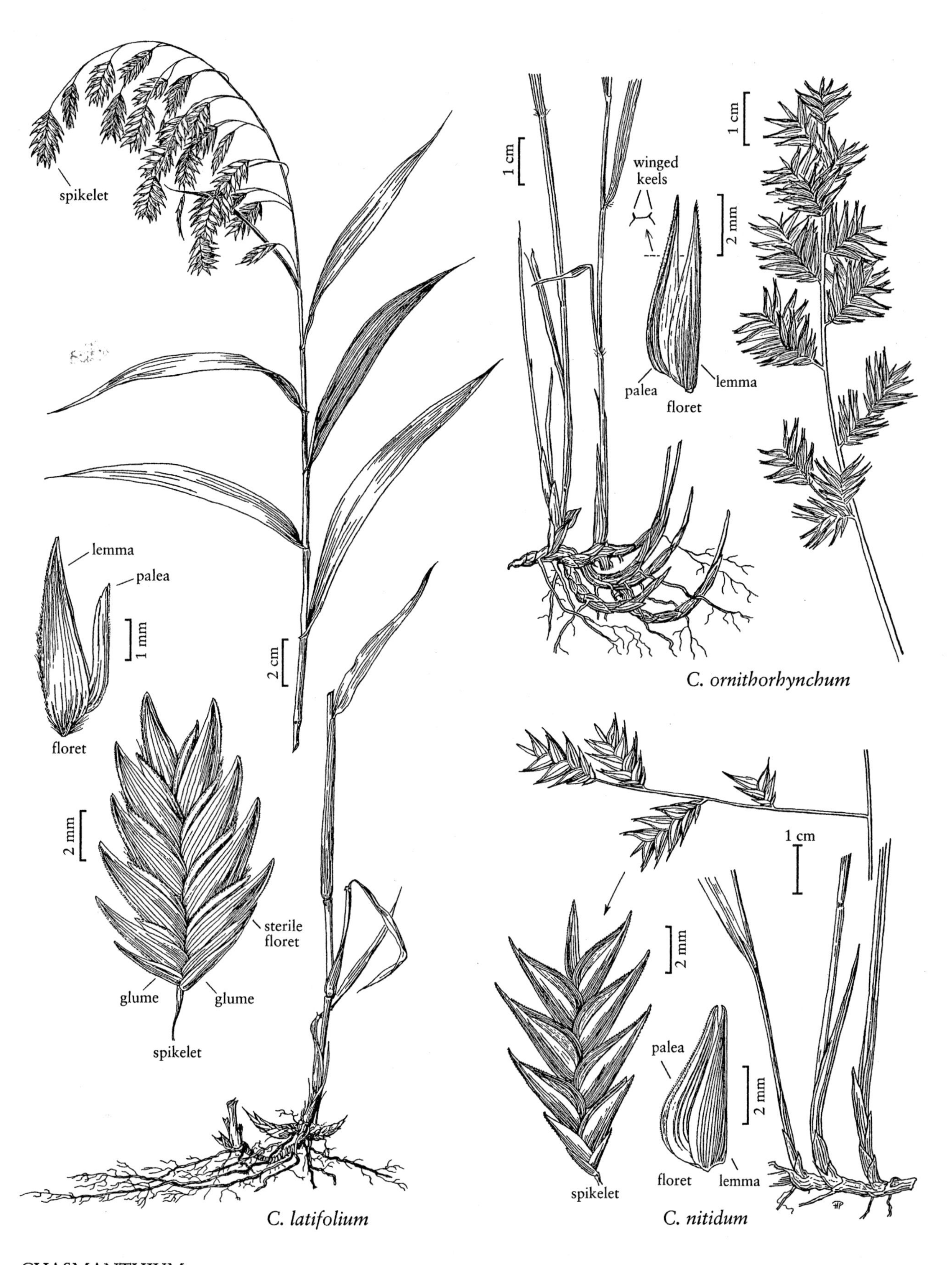

CHASMANTHIUM

CHASMANTHIUM

23. THYSANOLAENEAE C.E. Hubb.

J. Gabriel Sánchez-Ken

Plants perennial; cespitose, shortly rhizomatous, reedlike. **Culms** 1.5–4 m, woody, persistent, glabrous, usually not branched above the base; **internodes** solid. **Sheaths** open; **auricles** sometimes present; **ligules** membranous to somewhat leathery, entire, minutely erose, or ciliolate; **pseudopetioles** poorly developed; **blades** flat, disarticulating from the sheaths when old, cross venation not evident. **Inflorescences** terminal, open diffuse panicles. **Spikelets** solitary, pedicellate, terete, with 2(3–4) florets, lowest floret sterile; **rachillas** prolonged, often terminating in a rudimentary floret; **disarticulation** at the pedicel bases, subsequently below the spikelets. **Glumes** membranous, shorter than the florets, 0–1-veined, obtuse; **lowest florets** equaling the upper florets, sterile; **lowest lemmas** membranous, 1–3-veined; **lowest paleas** not present; **upper florets** bisexual; **upper lemmas** membranous, 3-veined, unawned, marginal veins with papillose-based hairs, hairs 0.8–1.5 mm, strongly diverging at maturity; **upper paleas** about ½ as long as the lemmas, apices notched; **lodicules** 2, free, broadly cuneate, truncate to irregularly lobed; **anthers** 2(3). **Caryopses** nearly spherical to broadly ovoid; **hila** subbasal, punctate; **embryos** large, about ¾ as long as the caryopses. $x = 12$.

The tribe *Thysanolaeneae* is native to tropical Asia and includes only the monotypic genus *Thysanolaena*. It could be included in a more broadly circumscribed *Centotheceae*, but the members of that tribe needs further study before changes to its circumscription are adopted.

23.01 THYSANOLAENA Nees

J. Gabriel Sánchez-Ken

Plants perennial; cespitose, shortly rhizomatous, reedlike. **Culms** 1.5–4 m, woody, persistent, glabrous, usually not branched above the base; **internodes** solid. **Leaves** cauline, distichous; **sheaths** open; **auricles** sometimes present; **ligules** membranous to somewhat leathery, entire, minutely erose, or ciliolate; **pseudopetioles** poorly developed; **blades** flat, disarticulating from the sheaths when old, cross venation not evident. **Inflorescences** terminal, open diffuse panicles. **Spikelets** solitary, pedicellate, terete, with 2(3–4) florets, lowest floret in each spikelet sterile; **rachilla extension** prolonged, often terminating in a rudimentary floret; **disarticulation** at the pedicel bases, subsequently below the spikelets. **Glumes** membranous, shorter than the florets, 0–1-veined, obtuse; **lowest florets** equaling the upper florets, sterile; **lowest lemmas** membranous, 1–3-veined; **lowest paleas** not present; **upper florets** bisexual; **upper lemmas** membranous, 3-veined, unawned, marginal veins with papillose-based hairs, hairs 0.8–1.5 mm, strongly diverging at maturity; **upper paleas** about ½ as long as the lemmas, apices notched; **lodicules** 2, free, broadly cuneate, truncate to irregularly lobed; **anthers** 2(3). **Caryopses** nearly spherical to broadly ovoid; **embryos** about ¾ as long as the caryopses; **hila** subbasal, punctate. $x = 12$. Name from the Greek *thysanos*, 'fringe', and *laina*, 'cloak', a reference to the second lemma.

Thysanolaena is a monotypic genus of tropical Asia, where it grows in open habitats, generally in mountainous areas. Its only species is grown as an ornamental in the *Flora* region.

SELECTED REFERENCE Baaijens, G.J. and J.F. Veldkamp. 1991. *Sporobolus* (Gramineae) in Malesia. Blumea 35:393–458.

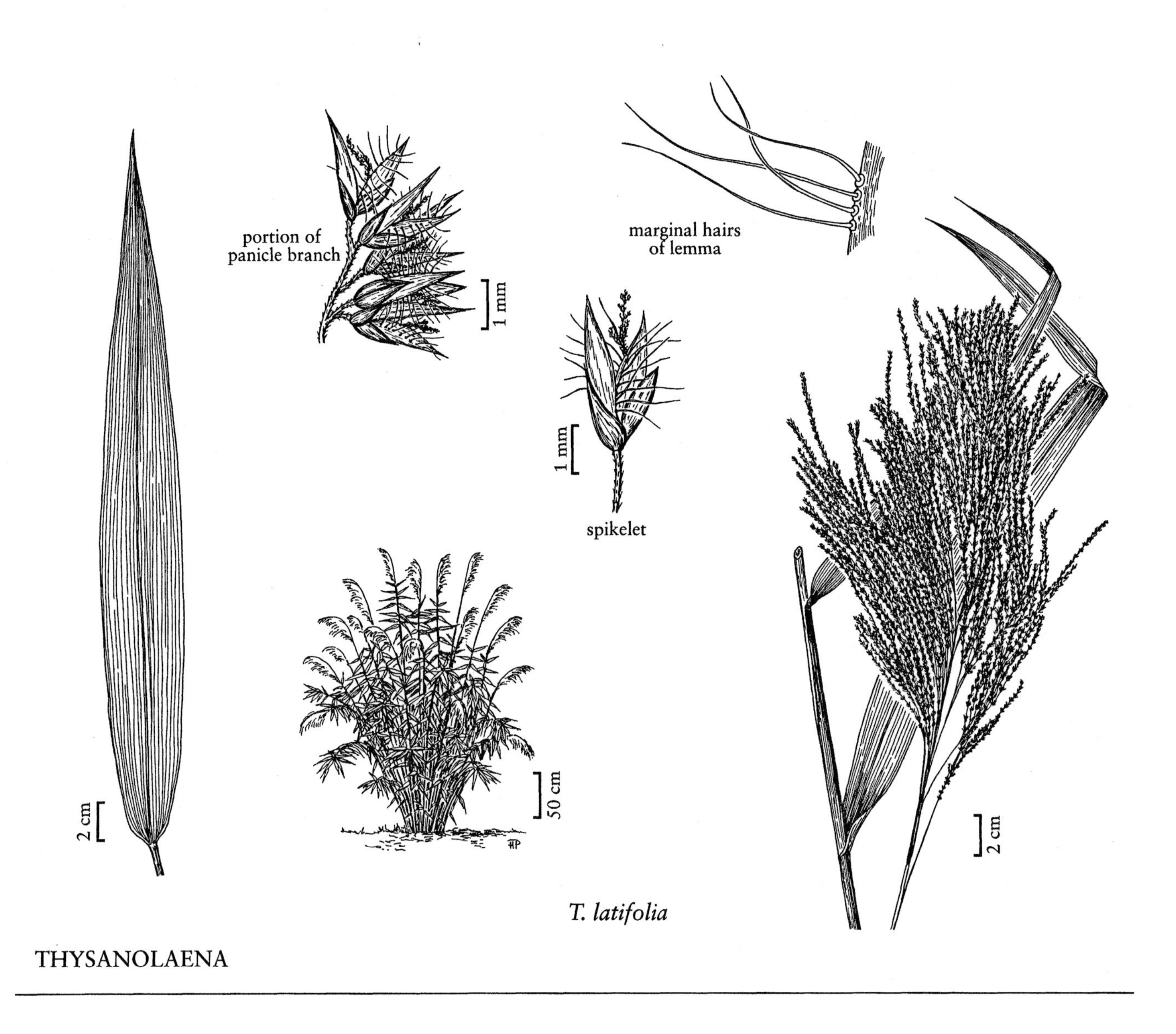

THYSANOLAENA

1. **Thysanolaena latifolia** (Roxb. *ex* Hornem.) Honda [p. 350]
ASIAN BROOMGRASS

Sheaths mostly glabrous, margins often ciliate distally; **collars** pubescent; **blades** 20–65 cm long, 2–7(10) cm wide, flat, usually glabrous or mostly so on both surfaces, occasionally sparsely pubescent abaxially, adaxial surfaces often pubescent immediately distal to the collar and ligule. **Panicles** 17–140 cm; **branches** to 75 cm, often drooping, pulvini pubescent. **Spikelets** 1.2–2.8 mm, frequently abortive. **Lower glumes** 0.5–0.9 mm, broadly ovate; **upper glumes** 0.5–1 mm, broadly elliptic; **lower lemmas** 1.3–2.4 mm, elliptic lanceolate, acute to acuminate; **upper lemmas** 1.2–2.8 mm, lanceolate, acute to acuminate; **anthers** 0.6–1 mm, yellow; **stigmas** purple. **Caryopses** 0.4–0.5 mm. $2n = 24$.

Thysanolaena latifolia is an important pasture species in Asia. It is grown in the United States as an ornamental plant, recommended for frost-free areas with full sunlight or light shade. It is not known to be established in the *Flora* region.

The species has been known as *Thysanolaena maxima* (Roxb.) Kuntze, but Baaijens and Veldkamp (1991) demonstrated that *T. latifolia* has priority at the species level.

10. PANICOIDEAE Link

Grass Phylogeny Working Group

Plants annual or perennial; synoecious, monoecious, or dioecious; primarily herbaceous, habit varied. **Culms** annual, usually solid, sometimes somewhat woody, sometimes decumbent, often branched above the base. **Leaves** distichous; **sheaths** usually open; **auricles** usually absent; **abaxial ligules** usually absent, occasionally present as a line of hairs; **adaxial ligules** membranous, sometimes also ciliate, or of hairs, sometimes absent; **blades** sometimes pseudopetiolate; **mesophyll** radiate or non-radiate; **adaxial palisade layer** absent; **fusoid cells** usually absent; **arm cells** usually absent; **kranz anatomy** absent or present; **midribs** usually simple, rarely complex; **adaxial bulliform cells** present; **stomata** with triangular or dome-shaped subsidiary cells; **bicellular microhairs** usually present, with a long, narrow distal cell; **papillae** absent or present. **Inflorescences** ebracteate (*Paniceae*) or bracteate (most *Andropogoneae*) panicles, racemes, spikes, or complex arrangements of rames (in the *Andropogoneae*), usually bisexual, sometimes unisexual; **disarticulation** usually below the glumes, frequently in the secondary and higher order axes of the inflorescences. **Spikelets** bisexual or unisexual, frequently paired or in triplets, the members of each unit usually with pedicels of different lengths or 1 spikelet sessile. **Glumes** usually 2, equal or unequal, shorter or longer than the adjacent florets, sometimes exceeding the distal florets; **florets** 2(–4), usually dorsally compressed, sometimes terete or laterally compressed; **lower florets** sterile or staminate, frequently reduced to a lemma; **upper florets** usually bisexual; **lemmas** hyaline to coriaceous, lacking uncinate hairs, often terminally awned; **awns** single; **paleas of bisexual florets** well-developed, reduced, or absent; **lodicules** usually 2, sometimes absent, cuneate, free, fleshy, usually glabrous; **anthers** 1–3; **ovaries** usually glabrous; **haustorial synergids** absent; **style branches** 2, free and close or fused at the base. **Caryopses: hila** usually punctate; **endosperm** hard, without lipid; **starch grains** simple; **embryos** large in relation to the caryopses, usually waisted; **epiblasts** usually absent; **scutellar cleft** present; **mesocotyl internode** elongated; **embryonic leaf margins** usually overlapping, rarely just meeting. x = 5, (7), 9, 10, (12), (14).

The subfamily *Panicoideae* is most abundant in tropical and subtropical regions, particularly mesic portions of such regions, but several species grow in temperate regions of the world. Within the *Flora* region, the *Panicoideae* are represented by 59 genera and 364 species. They are most abundant in the eastern United States (Barkworth and Capels 2000). Photosynthesis may be either C_3 or C_4. All three pathways are found in the subfamily, but the PCK and NAD-ME variants appear to have evolved only once, while the NADP-ME pathways seems to have evolved several different times (Giussani et al. 2001).

The *Panicoideae* were first recognized as a distinct unit by Brown (1814), earlier than any of the other subfamilial taxa of the *Poaceae*. Its early recognition is undoubtedly attributable to its distinctive spikelets. Recognition of the tribe *Gynerieae* is recent (Sánchez-Ken and Clark 2001) and its placement in the *Panicoideae*, rather than the *Centothecoideae*, should be regarded as tentative.

Spikelets with two florets are found in many other subfamilies, but rarely do they follow the pattern of the lower floret being sterile or staminate and the upper floret bisexual. Development of unisexual florets within the *Panicoideae* appears to be consistent across the subfamily (LeRoux and Kellogg 1999), but differs from that in the *Ehrhartoideae* (Zaitcheck et al. 2000).

The *Paniceae* and *Andropogoneae* have their conventional interpretation in this *Flora*, so far as the North American taxa are concerned. Molecular studies, however, while strongly supporting the monophyly of the *Andropogoneae*, show the *Paniceae* to be paraphyletic, with two distinct clades. In one of these clades, most taxa have a chromosome base number of x = 9, but some have x = 10, and the taxa are pan-tropical in origin. The taxa in the other clade, with one exception, have a chromosome base number of $x = 10$ and are American in origin. This latter clade is sister to the *Andropogoneae*, which also have a chromosome base number of $x = 10$ (Gómez-Martínez and Culham 2000; Giussanni et al. 2001; Barber et al. 2002).

SELECTED REFERENCES **Barber, J.C., S.A. Aliscioni, L.M. Giussani, J.D. Noll, M.R. Duvall, and E.A. Kellogg.** 2002. Combined analyses of three independent datasets to investigate phylogeny of Poaceae subfamily Panicoideae. [Abstract.] http://www.botany2002.org/; **Barkworth, M.E. and K.M. Capels.** 2000. The Poaceae in North America: A geographic perspective. Pp. 327–346 *in* S.W.L. Jacobs and J. Everett (eds.). Grasses: Systematics and Evolution. International Symposium on Grass Systematics and Evolution (3rd:1998). CSIRO Publishing, Collingwood, Victoria, Australia. 408 pp.; **Brown, R.** 1814. General remarks, geographical and systematical, on the botany of Terra Australis. Pp. 533–613 *in* M. Flinders (ed.). A Voyage to Terra Australis, vol. 2. G. and W. Nicol, London, England. 613 pp.; **Giussani, L.M., J.H. Cota-Sánchez, F.O. Zuloaga, and E.A. Kellogg.** 2001. A molecular phylogeny of the grass subfamily Panicoideae (Poaceae) shows multiple origins of C_4 photosynthesis. Amer. J. Bot. 88:1993–2001; **Gómez-Martínez, R. and A. Culham.** 2000. Phylogeny of the subfamily Panicoideae with emphasis on the tribe Paniceae: Evidence from the *trn*L-F cpDNA region. Pp. 136–140 *in* S.W.L. Jacobs and J. Everett (eds.). Grasses: Systematics and Evolution. International Symposium on Grass Systematics and Evolution (3rd:1998). CSIRO Publishing, Collingwood, Victoria, Australia. 408 pp.; **Grass Phylogeny Working Group.** 2001. Phylogeny and subfamilial classification of the grasses (Poaceae). Ann. Missouri Bot. Gard. 88:373–457; **Hilu, K.W., L.A. Alice and H. Liang.** 1999. Phylogeny of the Poaceae inferred from *mat*K sequences. Ann. Missouri Bot. Gard. 86:835–851; **Kellogg, E.A.** 2000. Molecular and morphological evolution in the Andropogoneae. Pp. 149–158 *in* S.W.L. Jacobs and J. Everett (eds.). Grasses: Systematics and Evolution. International Symposium on Grass Systematics and Evolution (3rd:1998). CSIRO Publishing, Collingwood, Victoria, Australia. 408 pp.; **Le Roux, L.G. and E.A. Kellogg.** 1999. Floral development and the formation of unisexual spikelets in the Andropogoneae (Poaceae). Amer. J. Bot. 86:354–366; **Sánchez-Ken, J.G.and L.G. Clark.** 2001. Gynerieae, a new neotropical tribe of grasses (Poaceae). Novon 11:350–352; **Zaitcheck, B.F., L.G. Le Roux, and E.A. Kellogg.** 2000. Development of male flowers in *Zizania aquatica* (North American wild-rice; Gramineae). Int. J. Pl. Sci. 161:345–351.

1. Blades of leaves on the lower ½ of the culms disarticulating from the sheaths; plants 2–15 m tall, unisexual, without axillary inflorescences; blades with midribs 5–15 mm wide . . . . .24 . *Gynerieae*
1. Blades of most or all cauline leaves remaining attached to the sheaths; plants 0.05–6 m tall, usually bisexual, sometimes with unisexual inflorescences, often with axillary inflorescences; blades with midribs 0.2–5 mm wide.
   2. Glumes usually conspicuously unequal; lower glumes usually greatly exceeded by the upper florets; upper glumes from subequal to longer than the distal florets; lemmas of the upper florets usually coriaceous to indurate; disarticulation usually beneath the glumes, not in the axes of the inflorescence branches . . . . 25. *Paniceae*
   2. Glumes usually subequal, usually exceeding and concealing the florets; lemmas of the upper florets hyaline to membranous; disarticulation frequently in the axes of the inflorescence branches . . . . 26. *Andropogoneae*

24. GYNERIEAE Sánchez-Ken & L.G. Clark

J. Gabriel Sánchez-Ken

Plants perennial; dioecious. **Culms** 2–15 m; **internodes** solid. **Leaves** cauline; **sheaths** longer than the internodes, persistent, distal sheaths strongly overlapping; **blades** leathery, not pseudopetiolate, articulated with the sheaths, midveins conspicuous, those below midculm length disarticulating at maturity. **Inflorescences** terminal, panicles. **Pistillate spikelets** laterally compressed, with 2 florets; **disarticulation** above the glumes and between the florets; **glumes** unequal, exceeding the florets, lower glumes shorter and less firm than the upper glumes; **calluses** linear, glabrous; **lemmas** long silky-pilose distally, apices elongate, narrow, unawned; **style branches** 2, free; **rachilla extension** absent. **Staminate spikelets** with 2–4 florets; **disarticulation** below the terminal florets, glumes and lower floret(s) remaining attached; **glumes** subequal, membranous; **lemmas** membranous, glabrous or sparsely short pilose, (0)1(3)-veined; **anthers** 2. **Caryopses** oblong; **hila** punctate. $x = 11$.

The tribe *Gynerieae* includes only one genus, the neotropical genus *Gynerium*. The position of the tribe in the *Centothecoideae* + *Panicoideae* clade is strongly supported; it is less clear whether it should be included in the *Panicoideae*, as in this treatment, or in the *Centothecoideae*.

SELECTED REFERENCE **Sánchez-Ken, J.G. and L.G. Clark.** 2001. Gynerieae, a new neotropical tribe of grasses (Poaceae). Novon 11:350–352.

24.01 GYNERIUM Willd. *ex* P. Beauv.

J. Gabriel Sánchez-Ken

Plants perennial; dioecious, staminate and pistillate plants similar in gross morphology; rhizomatous. **Sheaths** overlapping, mostly glabrous or the young and bladeless leaves sparsely pilose, margins usually pilose distally; **collars** pilose in young leaves with rudimentary blades; **ligules** pilose; **blades** to 2 m, leathery, glabrous, lower blades disarticulating at maturity, midribs 0.5–1.5 cm wide, sometimes with a line of scattered hairs on either side. **Panicles** large, the pistillate panicles plumose. **Pistillate spikelets** with 2 florets; **disarticulation** above the glumes and between the florets; **glumes** 1–3-veined, tardily disarticulating; **upper glumes** longer and firmer than the lower glumes, exceeding the florets; **lemmas** shortly pilose below, long pilose above, apices elongated and narrowed; **lodicules** 2, membranous, faintly 2–3-veined, sometimes with long hairs, truncate; **staminodes** 2; **styles** 2. **Staminate spikelets** with 2–4 florets; **disarticulation** below the terminal florets; **glumes** 1-veined; **lemmas** hyaline, glabrous, occasionally puberulent below; **lodicules** 2, free, faintly veined, truncate; **anthers** 2; **ovary** abortive. Name from the Greek *gyne*, 'woman', and *erion*, 'wool', a reference to the pistillate lemmas.

Gynerium is a monotypic genus that ranges from Mexico and the West Indies to northeastern Argentina.

1. **Gynerium sagittatum** (Aublet) P. Beauv. [p. 354]
WILDCANE

Culms 2–10(15) m tall, 1–4 cm thick. **Sheaths** distichous, narrowing toward the apices, sometimes with a line of deciduous hairs on the back below the articulation with the blade; **ligules** 0.5–2 mm; **blades** long-attenuate, (0.4)1.5–2 m long, 2–10 cm wide, in a flat, fan-shaped arrangement, margins serrate. **Panicles** 0.5–1.5(2) m. **Pistillate spikelets** 8–11 mm; **lower glumes** 2.5–4 mm, lanceolate, 1(3)-veined; **upper glumes** 7–11 mm, linear-subulate, curved when mature, 3-veined; **lemmas** 4.5–7 mm; **paleas** 1–2 mm, lanceolate. **Staminate spikelets** 3–4 mm; **glumes** (1.5)2–3 mm, hyaline, acute; **lemmas** 3–4 mm, lanceolate, acute to acuminate; **paleas** to 2.5 mm, hyaline; **anthers** 1.5–2 mm, purplish. $2n = 44$.

Gynerium sagittatum is grown as an ornamental in subtropical portions of the *Flora* region. Even when vegetative, it can be identified by its height, the absence of blades on the lower leaves, the strongly distichous, fan-shaped arrangement of the distal leaf blades, and the wide midveins of the blades. It does not flower when grown outdoors in the *Flora* region.

25. PANICEAE R. Br.

Mary E. Barkworth

Plants annual or perennial; habit various. **Culms** 3–800 cm, annual, usually not woody. **Leaves** basal and/or cauline; **sheaths** usually open; **ligules** of hairs or membranous, membranous ligules often ciliate, cilia sometimes longer than the membranous base; **blades** occasionally pseudopetiolate, seldom disarticulating at maturity. **Inflorescences** terminal, sometimes also axillary, occasionally subterranean panicles; **branches** sometimes spikelike and secund, sometimes less than 1 cm; **disarticulation** usually below the glumes, sometimes at the base of the panicle branches, occasionally below the florets. **Spikelets** usually dorsally compressed, varying to terete or laterally compressed, with 2(3) florets, lower florets staminate, sterile, or reduced, upper florets usually bisexual; **calluses** not developed. **Glumes** usually membranous; **lower glumes** usually less than ½ as long as the spikelets, sometimes absent; **upper glumes** usually subequal to the upper florets, occasionally absent; **lower lemmas** similar to the upper glumes in length and texture; **upper lemmas** indurate, coriaceous, or cartilaginous, with a germination flap at the base, margins usually widely separated and involute at maturity, sometimes flat and hyaline; **upper paleas** similar to the upper lemmas in length and texture;

G. sagittatum

GYNERIUM

lodicules short; **anthers** usually 3; **stigmas** usually red. **Caryopses** usually dorsally compressed or terete; **embryos** ½ or more the length of the caryopses. x = 9, 10.

The tribe *Paniceae*, which includes about 100 genera and 2000 species, is primarily tropical in distribution. Within the *Flora* region, it is represented by 27 genera and 262 species, with its greatest representation being in the eastern portion of the contiguous United States (Barkworth and Capels 2000).

The tribe is so morphologically distinct that it was first recognized, in essentially its current sense, by Robert Brown in 1814. Its primary distinguishing features are the unusual spikelet structure combined with the indurate to coriaceous upper florets. Recent molecular studies (Barber et al. 2002; Guissani et al. 2001) show it as comprising two distinct lineages, one of which contains species with a base number of $x = 9$ and the other, species with $x = 10$.

Photosynthesis in the *Paniceae* may follow the C_3 pathway or any of three different C_4 pathways. Most genera are uniform in this regard, but there are some noteworthy exceptions. Guissani et al. (2001) concluded that the C_3 pathway is probably ancestral within the tribe and that two of the three C_4 pathways, NAD-ME and PCK, originated only once within the tribe, whereas the NADP-ME pathway originated independently in several different lineages. Most genera are uniform with respect to their photosyntheticd pathway, but there are some noteworthy exceptions.

The *germination flap* is a small area of soft tissue at the base of the upper lemma through which the primary root of the seedling grows.

SELECTED REFERENCES **Barber, J.C., S.A. Aliscioni, L.M. Giussani, J.D. Noll, M.R. Duvall, and E.A. Kellogg.** 2002. Combined analyses of three independent datasets to investigate phylogeny of Poaceae subfamily Panicoideae. Combined analyses of three independent datasets to investigate phylogeny of Poaceae subfamily Panicoideae. [Abstract.] http://www.botany2002.org/; **Barkworth, M.E. and K.M. Capels.** 2000. The Poaceae in North America: A geographic perspective. Pp. 327–346 *in* S.W.L. Jacobs and J. Everett (eds.). Grasses: Systematics and Evolution. International Symposium on Grass Systematics and Evolution (3rd:1998). CSIRO Publishing, Collingwood, Victoria, Australia. 408 pp.; **Brown, R.** 1814. General remarks, geographical and systematical, on the botany of Terra Australis. Pp. 533–613 *in* M. Flinders (ed.). A Voyage to Terra Australis, vol. 2. G. and W. Nicol, London, England. 613 pp.; **Doust, A.N. and E.A. Kellogg.** 2002. Inflorescence diversification in the panicoid "Bristle Grass" clade (Paniceae, Poaceae): Evidence from molecular phylogenies and developmental morphology. Amer. J. Bot. 89:1203–1222; **Giussani, L.M., J.H. Cota-Sánchez, F.O. Zuloaga, and E.A. Kellogg.** 2001. A molecular phylogeny of the grass subfamily Panicoideae (Poaceae) shows multiple origins of C_4 photosynthesis. Amer. J. Bot. 88:1993–2001; **Gómez-Martínez, R. and A. Culham.** 2000. Phylogeny of the subfamily Panicoideae with emphasis on the tribe Paniceae: Evidence from the *trn*L-F cpDNA region. Pp. 136–140 *in* S.W.L. Jacobs and J. Everett (eds.). Grasses: Systematics and Evolution. International Symposium on Grass Systematics and Evolution (3rd:1998). CSIRO Publishing, Collingwood, Victoria, Australia. 408 pp.

1. Plants developing both subterranean and aerial inflorescences, only the subterranean spikelets setting seed . . . . 25.04 *Amphicarpum*
1. Plants developing only aboveground inflorescences, the spikelets setting seed [*Amphicarpum* is also keyed out here to accommodate situations in which looking for subterranean inflorescences is not permitted or specimens have no underground parts].
   2. Inflorescences spikelike panicles, with the branches partially embedded in the flattened rachises; plants perennial, stoloniferous . . . . 25.22 *Stenotaphrum*
   2. Inflorescences panicles, sometimes spikelike, but the branches not embedded in the rachises or the rachises not flattened; plants annual or perennial, sometimes stoloniferous.
      3. Most spikelets or groups of 2–11 spikelets subtended by 1–many, distinct to more or less connate, stiff bristles or bracts.
         4. Spikelets in groups of 2–11, subtended by 4 flat, narrowly elliptic, coriaceous bracts; terete bristles not present . . . . 25.17 *Anthephora*
         4. Spikelets solitary or in groups, subtended by 1–many stiff, terete bristles, sometimes appearing as an extension of the branch; flat, connate bristles sometimes present distal to the terete bristles.
            5. Bristles falling with the spikelets at maturity; disarticulation at the base of the reduced panicle branches (*fascicles*).
               6. Bristles plumose or antrorsely scabrous, free or fused no more than ½ their length . . . . 25.15 *Pennisetum*
               6. Bristles glabrous, smooth, retrorsely scabrous, or strigose, usually at least some bristles fused for more than ½ their length . . . . 25.16 *Cenchrus*

5. Bristles persistent; disarticulation below the spikelets.
 7. Upper glumes indurate at maturity; lower lemmas somewhat indurate at the base; pedicels subtended by a single bristle . . . . 25.19 *Setariopsis*
 7. Upper glumes membranous to herbaceous at maturity; lower lemmas neither constricted nor indurate at the base; pedicels subtended by 1–many bristles.
  8. Spikelets subtended by 1–many bristles; paleas of the lower florets usually hyaline to membranous at maturity, rarely absent or reduced; paleal veins not keeled . . . . 25.20 *Setaria*
  8. Spikelets subtended by 1 bristle; paleas of the lower florets coriaceous to indurate at maturity, the keels thickened . . . . 25.18 *Ixophorus*

3. All or most spikelets not subtended by stiff bristles, sometimes the terminal spikelet on each branch subtended by a single bristle, and occasionally other spikelets with a single subtending bristle.
 9. Terminal spikelet on each branch subtended by a single bristle; other spikelets occasionally with a single stiff subtending bristle . . . . 25.20 *Setaria*
 9. None of the spikelets subtended by a stiff bristle.
  10. Inflorescences of spikelike branches 1–3.7 cm long, the branch axes extending as a 2.5–4 mm bristle beyond the base of the distal spikelets . . . . 25.21 *Paspalidium*
  10. Inflorescences various but, if of spikelike branches, these terminating in a well-developed or rudimentary spikelet.
   11. Lower glumes or lower lemmas awned, sometimes shortly so (the awn reduced to a point in *Echinochloa colona*).
    12. Upper florets laterally compressed; spikelets also laterally compressed . . . . . 25.12 *Melinis*
    12. Upper florets dorsally compressed; spikelets usually dorsally compressed or terete, sometimes laterally compressed.
     13. Blades linear to linear-lanceolate, usually more than 10 times longer than wide, with prominent midribs; at least the upper leaves, often all leaves, without ligules; ligules usually absent, particularly from the upper leaves, of hairs when present . . . . 25.07 *Echinochloa*
     13. Blades triangular to lanceolate, less than 10 times longer than wide, the midribs not particularly prominent, at least distally; ligules present, of hairs or membranous.
      14. Lower glumes awned, the awns exceeding the florets; upper glumes not ciliate-margined; culms trailing on the ground, frequently rooting and branching at the nodes . . . . 25.06 *Oplismenus*
      14. Lower glumes unawned or shortly awned, the awns exceeded by the florets; upper glumes ciliate-margined; culms erect or decumbent below, sometimes rooting and branching at the lower nodes . . . . 25.03 *Alloteropsis*
   11. Lower glumes and lower lemmas unawned.
    15. Upper florets laterally compressed . . . . 25.12 *Melinis*
    15. Upper florets dorsally compressed or terete.
     16. Upper lemmas and paleas cartilaginous and flexible at maturity; lemma margins flat, hyaline; lower glumes absent or to ¼ the length of the spikelets.
      17. Aerial inflorescences with elongate rachises and glabrous spikelets; spikelets of the aerial panicles rarely setting seed; subterranean spikelets developed, seed-forming . . . . 25.04 *Amphicarpum*
      17. Aerial inflorescences of digitate or subdigitate clusters of spikelike branches with glabrous or pubescent spikelets or with elongate rachises and conspicuously pubescent spikelets; aerial spikelets seed-forming; subterranean spikelets not developed.
       18. Spikelets ellipsoid to obovoid; inflorescences simple panicles with erect to ascending branches on elongate rachises; branches ascending, not conspicuously spikelike . . . 25.02 *Anthenantia*

18. Spikelets lanceoloid to ellipsoid; inflorescences usually panicles with digitate or subdigitate clusters of spikelike branches, sometimes simple panicles with strongly divergent branches . . . . 25.01 *Digitaria*

16. Upper lemmas and paleas chartaceous to indurate and rigid at maturity; lemma margins not hyaline, frequently involute; lower glumes varying from absent to subequal to the spikelets or extending beyond the distal floret.

19. Spikelets subtended by a cuplike callus . . . . 25.14 *Eriochloa*

19. Spikelets not subtended by a cuplike callus.

20. At least the upper leaves, often all leaves, without ligules; ligules, when present, of hairs . . . . 25.07 *Echinochloa*

20. All leaves with ligules, ligules membranous or of hairs.

21. Paleas of the lower florets inflated and indurate at maturity; lower and upper florets standing apart from each other when mature . . . . 25.24 *Steinchisma*

21. Paleas of the lower florets neither inflated nor indurate at maturity; lower and upper florets closely appressed to each other when mature.
[revert to left side]

22. Inflorescences of 1-sided, spikelike primary branches.

23. Spikelets with the lower lemmas and lower glumes (if present) adjacent to the branch axes.

24. Lower glumes absent . . . . 25.25 *Axonopus*

24. Lower glumes present on all or most spikelets.

25. Upper lemmas rugose and verrucose; panicle branches in 2 or more ranks, sometimes verticillate . . . . 25.13 *Urochloa*

25. Upper lemmas smooth; panicle branches in 1 rank . . . . 25.11 *Brachiaria*

23. Spikelets with the upper lemmas and upper glumes (if present) adjacent or appressed to the branch axes.

26. Both glumes absent from all or almost all spikelets, the terminal spikelet on a branch sometimes with upper glumes . . . . 25.27 *Reimarochloa*

26. Upper or both glumes present on all spikelets.

27. Upper lemmas smooth to slightly rugose; lower glumes usually absent . . . . . . 25.26 *Paspalum*

27. Upper lemmas rugose and verrucose; lower glumes present, from ⅓ as long as the spikelets to equaling them . . . . 25.13 *Urochloa*

22. Inflorescences usually panicles with well-developed secondary branches, sometimes spikelike panicles or panicles with spikelike, but not 1-sided, branches.

28. Inflorescences dense, the spikelets concealing at least the distal ½ of the rachises.

29. Upper glumes slightly to strongly saccate, 5–13-veined; panicle branches often fused to the rachises; blades 1.5–22 mm wide; culm internodes hollow . . . . 25.08 *Sacciolepis*

29. Upper glumes not saccate, 3–7-veined; panicle branches not fused to the rachises; blades 12–28 mm wide; culm internodes filled with aerenchyma . . . . 25.23 *Hymenachne*

28. Inflorescences more or less open panicles, the spikelets not concealing the rachises.

30. Lower glumes with saccate bases; glumes and lemmas with woolly pubescent apices; culms weakly lignified, rooting at the nodes . . . . 25.05 *Lasiacis*

30. Lower glumes not saccate basally; glumes and lemmas glabrous or with short, straight hairs, apices sometimes with a tuft of hairs but never woolly pubescent; culms usually not lignified, if lignified, not rooting at the nodes.

31. Lemmas of the upper florets rugose and verrucose; panicle branches usually spikelike and 1-sided, alternate or subopposite, less frequently verticillate . . . . 25.13 *Urochloa*

31. Lemmas of the upper florets usually smooth, if rugose the panicle branches neither verticillate nor 1-sided and spikelike.

32. Plants developing aerial and subterranean panicles; aerial spikelets lanceoloid, often without lower glumes; upper lemmas with flat margins . . . . 25.04 *Amphicarpum*
32. Plants developing aerial, but not subterranean, panicles; spikelets ovoid to ellipsoid or lanceoloid; lower glumes present; upper lemmas with involute margins.
    33. Blades of the basal leaves clearly distinct from the cauline leaves; basal leaves ovate to lanceolate, cauline leaves with longer and narrower blades; basal leaves forming a distinct winter rosette . . . . 25.09 *Dichanthelium*
    33. Blades of the basal and cauline leaves similar, usually linear to lanceolate, varying from filiform to ovate; basal leaves not forming a distinct winter rosette.
        34. Panicles terminating the culms usually appearing in late spring; branches usually developing from the lower and middle cauline nodes in summer, the branches rebranching 1 or more times by fall; upper florets not disarticulating at maturity, plump . . . . 25.09 *Dichanthelium*
        34. Panicles terminating the culms usually appearing after mid-summer; branches usually not developing branches from the lower and middle cauline nodes, when present, rarely rebranched; upper florets disarticulating or not very plump at maturity . . . . 25.10 *Panicum*

25.01 **DIGITARIA** Haller

J.K. Wipff

Plants annual, perennial, or of indefinite duration. **Culms** 5–250 cm, erect or decumbent, branching basally or at aerial culm nodes, when annual or of indefinite duration usually decumbent and rooting at the lower nodes. **Sheaths** open; **ligules** membranous, sometimes ciliate; **blades** usually flat. **Inflorescences** terminal, sometimes also axillary, usually panicles of 1-sided spikelike branches (sometimes only 1 branch) attached digitately or racemosely to a rachis, sometimes simple panicles of solitary, pedicellate spikelets; **spikelike branches**, if present, sometimes with secondary branches, primary branch axes triquetrous, bearing spikelets abaxially, in 2 rows, usually in unequally pedicellate groups of 2–5, occasionally borne singly. **Spikelets** 1.2–8.2 mm, lanceoloid to ellipsoid, dorsally compressed, apices obtuse to acuminate, unawned, with 2 florets; **disarticulation** beneath the glumes. **Lower glumes** absent or to ¼ as long as the spikelets; **upper glumes** usually from ⅙ as long as to equaling the spikelets, occasionally absent, 0–5-veined, usually pubescent; **lower florets** sterile; **lower lemmas** membranous, usually as long as the upper lemmas, usually pubescent, (3)5–7(13)-veined; **lower paleas** absent or reduced; **upper lemmas** mostly stiffly chartaceous to cartilaginous, obscurely veined, with 0.5–1 mm hyaline margins that embrace the upper paleas; **upper paleas** similar to the upper lemmas in texture and size; **lodicules** 3, cuneate; **anthers** 3. **Caryopses** plano-convex; **embryos** ⅕–½ as long as the caryopses; **hila** punctiform to ellipsoid. $x = 9$. Name from the Latin *digitus*, 'finger', a reference to the digitate inflorescence of some species.

Digitaria has approximately 200 species and grows primarily in tropical and warm-temperate regions, often in disturbed, open sites. Some species are grown as cereals; others for forage or as lawn grasses. In North America, the genus is best known for two of its weedy species, *D. sanguinalis* and *D. ciliaris*. There are 29 species known to occur in the *Flora* region; 18 are native to the region.

Most annual species of *Digitaria* will survive several years in regions without a pronounced cold season; such species are described as being of indefinite duration.

SELECTED REFERENCES **Boonbundarl, S.** 1985. A biosystematic study of the *Digitaria leucites* complex in North America. Ph.D. dissertation, Texas A&M University, College Station, Texas, U.S.A. 238 pp.; **Gould, F.W.** 1975. The Grasses of Texas. Texas A&M University Press, College Station, Texas, U.S.A. 653 pp.; **Henrard, J.** 1950. Monograph of the genus *Digitaria.* Universitaire Pers Leiden, Leiden, The Netherlands. 999 pp.; **Kartesz, J.** and **C.A. Meacham.** 1999. Synthesis of the North American Flora, Version 1.0 (CD-ROM). North Carolina Botanical Garden, Chapel Hill, North Carolina, U.S.A.; **Kok, P.D.F., P.J. Robbertse,** and **A.E. van Wyk.** 1989. Systematic study of *Digitaria* section *Digitaria* (Poaceae) in southern Africa. S. African J. Bot. 55:141–153; **Veldkamp, J.F.** 1973. A revision of *Digitaria* in Malesia. Blumea 21:1–80; **Webster, R.D.** 1981. A biosystematic study of the *Digitaria sanguinalis* complex in North America (Poaceae). Ph.D. dissertation, Texas A&M University, College Station, Texas. 156 pp.; **Webster, R.D.** 1983. A revision of the genus *Digitaria* Haller (Paniceae: Poaceae) in Australia. Brunonia 6:131–216; **Webster, R.D.** 1987. Taxonomy of *Digitaria* section *Digitaria* in North America (Poaceae: Paniceae). Sida 12:209–222; **Webster, R.D.** and **S.L. Hatch.** 1981. Taxonomic relationships of Texas specimens of *Digitaria ciliaris* and *Digitaria bicornis* (Poaceae). Sida 9:34–42; **Webster, R.D.** and **S.L. Hatch.** 1990. Taxonomy of *Digitaria* section *Aequiglumae* (Poaceae: Paniceae). Sida 14:145–167; **Wipff, J.K.** and **S.L. Hatch.** 1994. A systematic study of *Digitaria* sect. *Pennatae* (Poaceae: Paniceae) in the New World. Syst. Bot. 19:613–627.

NOTE: The pubescence of the lower lemmas may be mistaken for two white lines between the veins because the individual hairs are not visible, being both tightly packed and closely appressed.

1. Inflorescences simple open panicles, with well-developed primary and secondary branches; branches and pedicels divergent; spikelets solitary.
  2. Upper glumes absent or to 0.6 mm long, veinless . . . . 4. *D. tomentosa*
  2. Upper glumes 1.8–3.8 mm long, 3–7-veined.
    3. Spikelets 3.5–4.6 mm long; upper glumes 5–7-veined . . . . 1. *D. arenicola*
    3. Spikelets 2.2–3.3 mm long; upper glumes 3(5)-veined.
      4. Lower lemmas 7-veined, veins not equally spaced . . . . 2. *D. cognata*
      4. Lower lemmas 5-veined, veins equidistant . . . . 3. *D. pubiflora*
1. Inflorescences panicles of spikelike branches; secondary branches rarely present; spikelets appressed to the branches, in groups of 2--5 on the middle portion of the primary branches.
  5. Spikelets in groups of 3–5 on the middle portions of the primary branches, the longer pedicels in each group often adnate to the branch axes for part of their length.
    6. Upper lemmas pale yellow or gray when immature, light brown to brown when mature, sometimes purple-tinged.
      7. Upper glumes ⅙–⅓ long as the spikelets; sheaths and blades pubescent . . . . 13. *D. serotina*
      7. Upper glumes equaling or almost equaling the spikelets; sheaths and blades usually glabrous.
        8. Upper glumes 5-veined; spikelets elliptic to obovate; plants stoloniferous . . . . 14. *D. longiflora*
        8. Upper glumes 3-veined; spikelets lanceolate; plants not stoloniferous . . . . 15. *D. floridana*
    6. Upper lemmas brown when immature, becoming dark brown when mature.
      9. Primary panicle branches wing-margined, the wings at least ½ as wide as the midribs.
        10. Plants always with axillary panicles in the lower leaf sheaths, these panicles sometimes completely concealed by the sheaths; spikelets 1.7–2.3 mm long . . . . 16. *D. ischaemum*
        10. Plants without axillary panicles; spikelets 1.2–1.7 mm long.
          11. Primary panicle branches, if more than 2, racemose, the terminal branches erect, the other branches usually divergent; upper lemmas light brown to brown at maturity; upper glumes almost as long as the upper lemmas . . . . . . 15. *D. floridana*
          11. Primary panicle branches usually all digitate, sometimes 1 below the others, all the branches erect to ascending; upper lemmas dark brown at maturity; upper glumes ½ as long as to almost equaling the upper lemmas . . . . 17. *D. violascens*
      9. Primary panicle branches not wing-margined or the wings not as wide as the midribs.
        12. Lower lemmas pubescent.
          13. Plants annual, or short-lived perennials, branching at the lower nodes; cauline nodes 3–6 . . . . 7. *D. filiformis*
          13. Plants perennial, not branching at the lower nodes; cauline nodes 1–2.
            14. Upper glumes and lower lemmas with long, glandular-tipped hairs along their margins and intercostal regions . . . . 8. *D. leucocoma*
            14. Upper glumes and lower lemmas glabrous over most of their length, sparsely pubescent near the apices, the hairs short, not glandular-tipped . . . . 6. *D. bakeri*

12. Lower lemmas glabrous.
    15. Upper glumes less than ½ as long as the spikelets . . . . 5. *D. gracillima*
    15. Upper glumes more than ½ as long as the spikelets.
        16. Plants annual, branching at the lower nodes; culm nodes 3–6; upper glumes obtuse . . . . 7. *D. filiformis*
        16. Plants perennial, not branching at the lower nodes; culm nodes 1–2; upper glumes acute . . . . 6. *D. bakeri*

5. Spikelets paired on the middle portions of the primary branches; pedicels not adnate to the branch axes.
   17. Upper lemmas brown when immature, almost always dark brown when mature; primary branches not wing-margined or with wings less than ½ as wide as the midribs.
       18. Spikelets (including pubescence) 1.3–3.1 mm long; lower lemmas (including pubescence) shorter than or no more than 0.5 mm longer than the upper florets; ligules 0.1–1.5 mm long, ciliate; lower lemmas sparsely to densely pubescent, the hairs less than 1 mm long, appressed, not spreading at maturity.
           19. Lower glumes 0.3–1 mm long; plants perennial, with hard, knotty, shortly rhizomatous bases; culms erect, not rooting at the lower nodes; lower lemmas 5-veined, the veins equally spaced . . . . 9. *D. hitchcockii*
           19. Lower glumes absent or to 0.1 mm long; plants annual, or short-lived perennials; culms erect or decumbent, sometimes rooting at the lower nodes; lower lemmas 5–7-veined, the veins unequally spaced, the outer veins closely spaced.
               20. Upper lemmas dark brown at maturity; lower primary panicle branches without secondary branches; upper glumes with clavate to capitate hairs . . . . . . 7. *D. filiformis*
               20. Upper lemmas usually gray, sometimes brown, at maturity; lower primary panicle branches with strongly divergent secondary branches; upper glumes with tapering or parallel-sided hairs . . . . 25. *D. velutina*
       18. Spikelets (including pubescence) 3.7–7.5 mm long; lower lemmas (including pubescence) exceeding the upper florets by 0.8 mm or more; ligules 1–6 mm long, not ciliate; lower lemmas densely pubescent, the hairs 1–6 mm long, usually spreading at maturity.
           21. Terminal pedicels of primary branches 7.4–20 mm long; primary branches usually divergent, sometimes ascending, at maturity, the middle internodes (4.5)6–15 mm long . . . . 10. *D. patens*
           21. Terminal pedicels of primary branches 1.7–6(7) mm long; primary branches appressed to ascending at maturity, the middle internodes 2–6 mm long.
               22. Lower lemmas pubescent between most, sometimes all, of the veins and on the margins . . . . 12. *D. insularis*
               22. Lower lemmas glabrous between the veins, pubescent on the margins, sometimes also on the lateral veins . . . . 11. *D. californica*
   17. Upper lemmas pale yellow, tan, or gray, sometimes purple-tinged, when immature; gray, yellow, tan, light brown, or purple at maturity; primary branches sometimes wing-margined, the margin widths various.
       23. Primary panicle branches not or only narrowly wing-margined, the wings no more than ½ as wide as the midribs.
           24. Spikelets 1.5–2.5 mm long.
               25. Upper glumes glabrous; plants rhizomatous; culms decumbent but usually not rooting at the lower nodes . . . . 18. *D. abyssinica*
               25. Upper glumes shortly villous on the margins, sometimes also elsewhere; plants not rhizomatous; culms erect or decumbent and rooting at the lower nodes . . . . 21. *D. texana*
           24 Spikelets 2.5–3.6 mm long.
               26. Culms usually branching at the aerial nodes, not rooting at the lower nodes; leaf blades 2–2.2 mm wide, flat or folded; upper glumes glabrous . . . 19. *D. pauciflora*
               26. Culms not branching at the aerial nodes, often rooting at the lower nodes; leaf blades 2–7mm wide, flat; upper glumes villous or glabrous.

27. Upper glumes 7–9-veined, glabrous or obscurely pubescent; plants of indefinite duration . . . . 20. *D. simpsonii*
27. Upper glumes (3)5-veined, shortly villous on the margins and sometimes between the margins; plants perennial . . . . 21. *D. texana*

23. Primary branches winged, the wings at least ½ wide as the midribs.
28. Plants perennial, usually stoloniferous, sometimes also rhizomatous.
29. Leaf blades 1–3 mm wide; panicles with 2–4 primary branches, the branches 2–7 cm long . . . . 22. *D. didactyla*
29. Leaf blades 3–13 mm wide; panicles with 2–18 primary branches, the branches 5–25 cm long.
30. Midveins of the lower lemmas scabrous, at least on the distal ½ . . . . . 24. *D. milanjiana*
30. Midveins of the lower lemmas smooth throughout . . . . 23. *D. eriantha*
28. Plants annual or of indefinite duration, usually neither rhizomatous nor stoloniferous.
31. Lateral veins of the lower lemmas scabrous for the distal ⅔ of their length, sometimes scabrous throughout (use 20× magnification); leaf blades usually with papillose-based hairs on both surfaces . . . . 28. *D. sanguinalis*
31. Lateral veins of the lower lemmas smooth throughout or scabrous only on the distal ⅓; leaf blades with or without papillose-based hairs.
32. Lower lemmas of the lower spikelets of each pair 7-veined, the 2 lateral veins on each side crowded together near the margins, the 3 central veins equally spaced.
33. Lower primary panicle branches with strongly divergent secondary branches . . . . 25. *D. velutina*
33. Lower primary panicle branches without secondary branches.
34. Spikelets 2.6–3.7 mm long; spikelets dimorphic with respect to their pubescence; lower lemmas of the upper spikelets of each spikelet pair with marginal hairs that become widely divergent at maturity; lower lemmas of the lower spikelets in each pair glabrous or with hairs that remain appressed at maturity; lowest panicle nodes glabrous or with hairs less than 0.4 mm long . . . . 29. *D. bicornis*
34. Spikelets 1.7–2.8 mm long; spikelets homomorphic with respect to their pubescence; lowest panicle nodes with hairs more than 0.4 mm long.
35. Adaxial surfaces of the blades evenly, sometimes densely, hairy; leaf sheaths usually with scattered hairs; upper glumes ⅓–½ as long as the spikelets . . . . 26. *D. horizontalis*
35. Adaxial surfaces of the blades glabrous or with a few long hairs near the base; leaf sheaths glabrous or with a few long hairs near the base; upper glumes ⅖–⅘ as long as the spikelets . . . . 27. *D. nuda*
32. Lower lemmas of the lower spikelets of each pair 5- or 7-veined, the 2 or 3 lateral veins on each side crowded together near the margins, well-separated from the midvein.
36. Lower glumes absent or no more than 0.1 mm long.
37. Anthers 0.6–1.3 mm long; upper glumes 0.2–1.3 mm long, less than 0.4 times as long as the spikelets . . . . 31. *D. setigera*
37. Anthers 0.3–0.6 mm; upper glumes 1–2.2 mm long, 0.4–0.8 times as long as the spikelets . . . . 27. *D. nuda*
36. Lower glumes 0.1–0.8 mm long.
38. Lower glumes 0.2–0.8 mm long; primary branches glabrous or with hairs shorter than 1 mm; spikelets 2.7–4.1 mm long . . . . 30. *D. ciliaris*
38. Lower glumes 0.1–0.2 mm long; primary branches often with scattered hairs 1–4 mm long; spikelets 1.7–2.8 mm long.

39. Adaxial surfaces of the blades evenly hairy, sometimes densely so; leaf sheaths usually with scattered hairs; upper glumes 0.3–0.5 times as long as the spikelets .... 26. *D. horizontalis*
39. Adaxial surfaces of the blades glabrous or with a few long hairs near the base; leaf sheaths glabrous or with a few long hairs near the base; upper glumes 0.4–0.8 as long as the spikelets .... 27. *D. nuda*

1. Digitaria arenicola (Swallen) Beetle [p. 363]
SAND WITCHGRASS

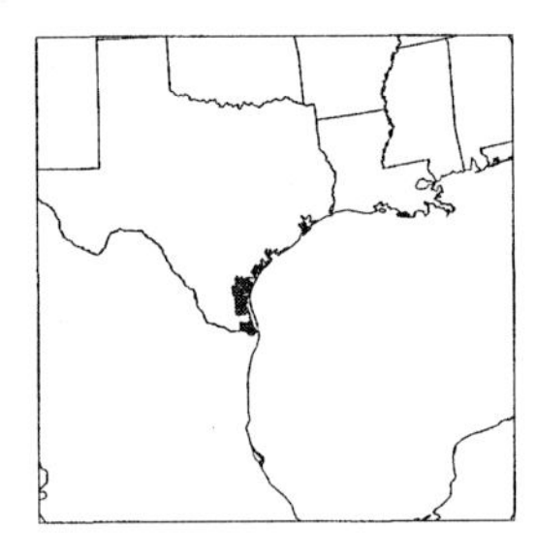

Plants perennial; loosely tufted, with long, creeping rhizomes. **Culms** 20–60 cm, erect; **lower nodes** glabrous or pubescent; **upper nodes** glabrous. **Leaves** mainly cauline; **sheaths** usually glabrous, lower sheaths sometimes pubescent; **ligules** 0.4–1 mm, truncate, entire to lacerate; **blades** 5–11.5 cm long, 3–4.5 mm wide, glabrous, usually flat or folded. **Panicles** simple, 12–24 cm long, 19–40 cm wide, open; **nodes** hispid; **branches** divergent; **lower primary branches** 10–21 cm, with 1–several sterile branches near the base; **pedicels** divergent, spikelets solitary. **Spikelets** 3.5–4.6 mm long, 0.8–1 mm wide, elliptical. **Lower glumes** 0.2–0.5 mm; **upper glumes** 3–3.8 mm, 5–7-veined, densely villous between the veins, hairs white, becoming purple at maturity; **lower lemmas** similar to the upper glumes in size, texture, and pubescence; **upper lemmas** 3–3.7 mm, narrowly acute, dark brown; **anthers** 1.1–1.4 mm. **Caryopses** 1.5–2 mm. $2n$ = 36, 37.

Digitaria arenicola is endemic to deep sands along the coast of Texas, a very restricted habitat and one that is being lost to the development of coastal parks and housing.

2. Digitaria cognata (Schult.) Pilg. [p. 363]
FALL WITCHGRASS

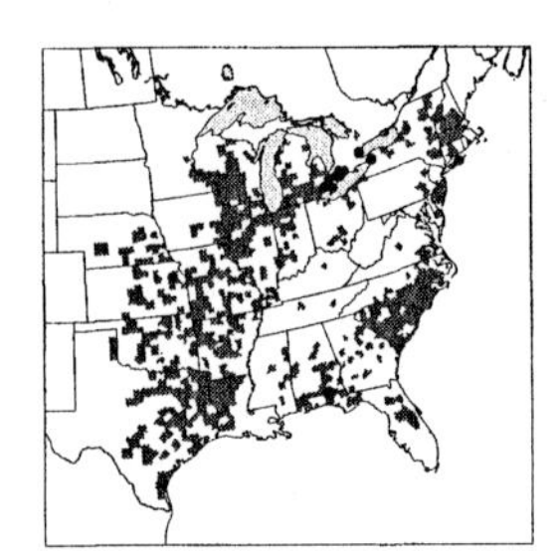

Plants perennial; cespitose, without rhizomes. **Culms** 30–56 cm, erect; **nodes** glabrous. **Leaves** mainly cauline; **sheaths** glabrous or sparsely to densely pubescent, sometimes with papillose-based hairs; **ligules** 0.2–1.5 mm, entire to lacerate; **blades** 2.4–12.6 cm long, 2–5.4 mm wide, glabrous or pubescent. **Panicles** simple, 12.8–27.5 cm long, 16.5–44.5 cm wide, open; **branches** divergent; **lower primary branches** 10.5–24 cm, often with 1–several sterile branches near the base; **pedicels** divergent, spikelets solitary. **Spikelets** 2.2–3.1 mm long, 0.7–1.1 mm wide, obovate or broadly elliptic. **Lower glumes** 0.1–0.8 mm; **upper glumes** 1.8–2.8 mm, 3(5)-veined, glabrous or pubescent between the veins, hairs appearing as a narrow stripe; **lower lemmas** similar to the upper glumes in length, texture, and pubescence, 7-veined, veins unequally spaced, lateral veins closer together than the 3 central veins; **upper lemmas** 1.9–2.9 mm, glabrous, dark brown, narrowly acute; **anthers** 0.5–0.7 mm, yellow or purple. **Caryopses** 1.3–1.6 mm. $2n$ = 36.

Digitaria cognata grows in dry, sandy soils in the eastern portion of the *Flora* region from southern Ontario and Vermont through the United States and thence to southern Mexico.

3. Digitaria pubiflora (Vasey) Wipff [p. 363]
WESTERN WITCHGRASS

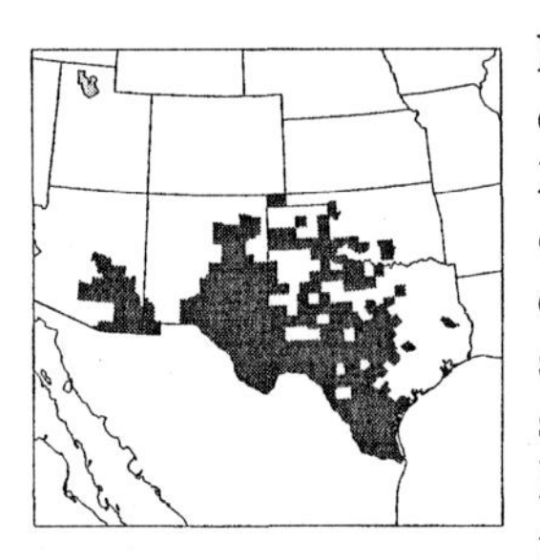

Plants perennial; cespitose, with or without rhizomes. **Culms** 20–70 cm, erect; **nodes** glabrous or pubescent . **Leaves** mainly cauline; **sheaths** glabrous or sparsely to densely pubescent, sometimes with papillose-based hairs; **ligules** 0.5–2.2 mm, entire to lacerate; **blades** 1.3–7.7 cm long, 1.5–4.7 mm wide, glabrous or sparsely to densely pubescent. **Panicles** simple, 4.5–20 cm long, 5.5–31 cm wide, open; **branches** divergent; **lower primary branches** 3.6–17.7 cm, often with 1–several sterile branches near the base; **pedicels** divergent, spikelets solitary. **Spikelets** 2.3–3.3 mm long, 0.6–1 mm wide, narrowly elliptic. **Lower glumes** 0.1–0.4 mm; **upper glumes** 1.8–2.9 mm, 3-veined, densely pubescent between the veins, hairs white, becoming purple at maturity; **lower lemmas** similar to the upper glumes in length, texture, and pubescence, 5-veined, veins equidistant; **upper lemmas** 1.9–2.6 mm, glabrous, dark brown, narrowly acute; **anthers** 0.3–0.5 mm, yellow, red, or purple. **Caryopses** 1.3–1.6 mm. $2n$ = (36), 72.

Digitaria pubiflora grows in dry, sandy or rocky soils from Arizona to central Texas and south to central Mexico.

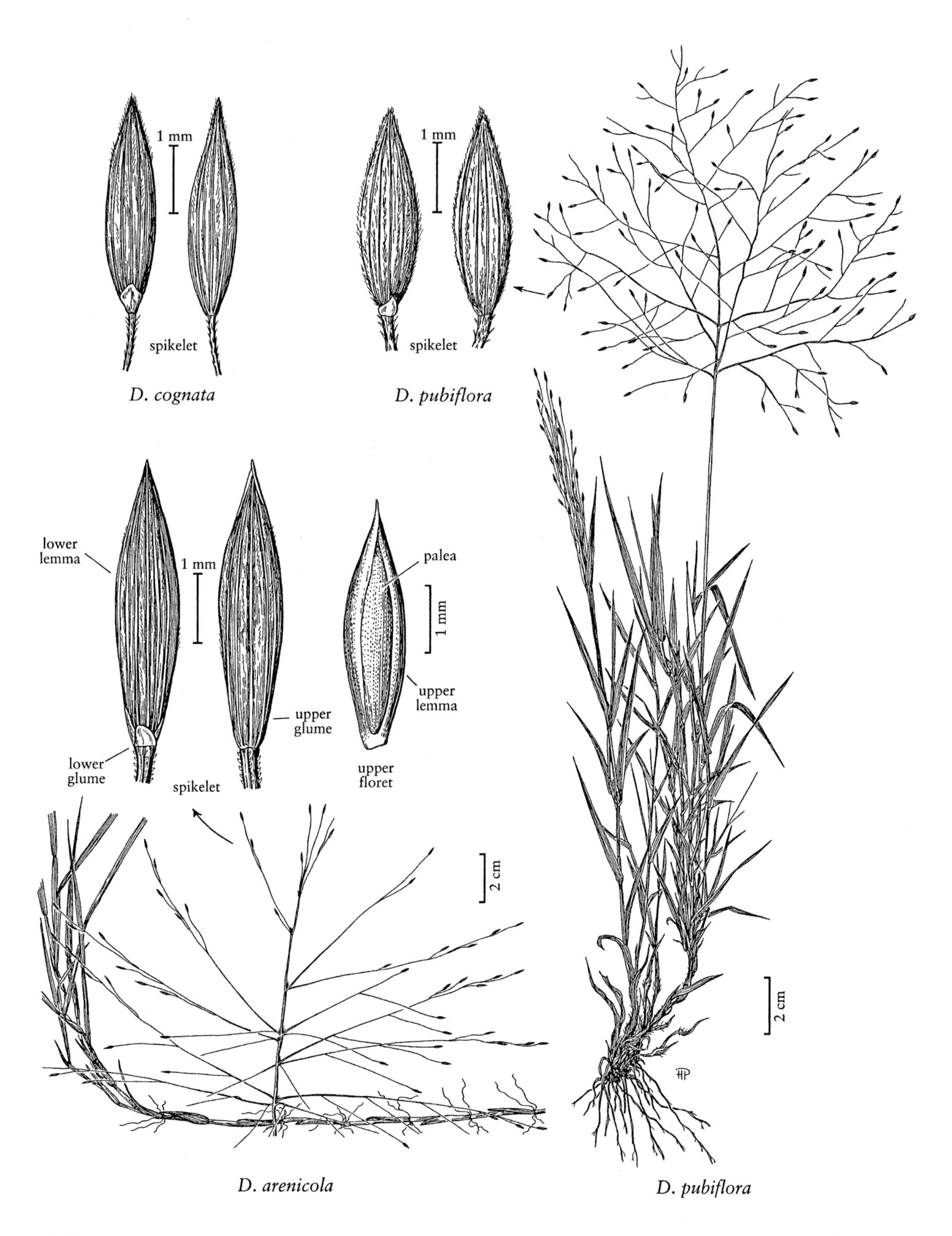
1 mm
spikelet
D. cognata
1 mm
spikelet
D. pubiflora
lower lemma
1 mm
palea
1 mm
upper lemma
upper glume
lower glume
spikelet
upper floret
2 cm
D. arenicola
2 cm
D. pubiflora

DIGITARIA

4. **Digitaria tomentosa** (J. König *ex* Rottler) Henrard [p. 365]

Plants perennial; cespitose, not rhizomatous. **Culms** 40–90 cm, erect, geniculate. **Lower sheaths** densely villous; **upper sheaths** with scattered papillose-based hairs; **ligules** 1.5–3 mm; **blades** 4–7(12) cm long, 3–5 mm wide, villous to nearly glabrous, usually sparsely hairy near the ligules. **Panicles** simple, 10–18 cm, open; **branches** divergent; **pedicels** 0.4–3 mm, divergent, spikelets solitary. **Spikelets** 2–2.5 mm, elliptical, apiculate. **Lower glumes** absent or to 0.2 mm; **upper glumes** absent or to 0.6 mm, veinless; **lower lemmas** as long as the spikelets, 5-veined, margins and intercostal regions more or less pubescent; **upper lemmas** minutely rugose, dark brown. $2n = 36$.

A native of southern India and Ceylon, *Digitaria tomentosa* is a noxious weed that is not known to occur in the *Flora* region. It is included here to help ensure that any introduction is correctly identified.

5. **Digitaria gracillima** (Scribn.) Fernald [p. 365]

Plants perennial; in dense tufts, not rhizomatous. **Culms** 60–100 cm, erect. **Sheaths** villous, particularly the lower sheaths; **ligules** 0.2–0.4, fimbriate; **blades** to 47 cm long, 1–2.1 mm wide, involute, glabrous and smooth abaxially, scabrous adaxially, sometimes with a few long hairs near the base. **Panicles** of 2–3(5) spikelike primary branches on elongate rachises; **primary branches** 8–13(20) cm, narrowly winged, wings no more than ½ as wide as the midribs, bearing spikelets in unequally pedicellate groups of 3–4(5); **secondary branches** rarely present; **longest pedicels** to 5 mm. **Spikelets** 1.7–2.3 mm, elliptical, glabrous. **Lower glumes** absent; **upper glumes** 0.9–1 mm, less than ½ as long as the spikelets, 3-veined, glabrous, broadly rounded to truncate; **lower lemmas** slightly shorter than the spikelets, 5-veined, glabrous; **upper lemmas** smooth, brown when immature, dark brown at maturity. $2n$ = unknown.

Digitaria gracillima is a rare species, endemic to scrub and dry pinelands of peninsular Florida. It was formerly interpreted as including *D. bakeri*, but differs from that species both morphologically and ecologically.

6. **Digitaria bakeri** (Nash) Fernald [p. 365]

Plants perennial; cespitose. **Culms** 40–90 cm, erect, unbranched, glabrous; **nodes** 1–2. **Sheaths** densely hairy, hairs papillose-based; **ligules** 0.8–2.2 mm; **blades** 10–24 cm long, 2–4.5 mm wide, with papillose-based hairs. **Panicles** with 2–3 spikelike primary branches on 4–7 mm rachises; **secondary branches** rarely present; **primary branches** (5)10–22 cm, axes 0.5–0.6 mm wide, not wing-margined, middle portions of the branches bearing spikelets in groups of 3; **pedicels** appressed to the axes. **Spikelets** 2.3–2.4 mm, lanceolate to oblanceolate-elliptic; **lower glumes** absent; **upper glumes** 1.4–1.5 mm, more than ½ as long as the spikelets, 3-veined, truncate to acute, mostly glabrous, apices sparsely hairy; **lower lemmas** about as long as the spikelets, 7-veined, mostly glabrous, apices sparsely hairy; **upper lemmas** 2.3–2.4 mm, dark brown to black, apiculate; **anthers** 1–1.1 mm. $2n$ = unknown.

Digitaria bakeri grows in pastures, particularly horse pastures, from Florida through Mexico to Panama. It is probably more widespread in Florida than the map suggests but, because of its inclusion in *D. gracillima*, little information is available at present.

7. **Digitaria filiformis** (L.) Koeler [p. 367]
SLENDER CRABGRASS

Plants annual, or short-lived perennials; cespitose, not rhizomatous. **Culms** (10)25–150 cm, erect or decumbent, branching, sometimes rooting at the lower nodes; **nodes** 3–6. **Sheaths** keeled, basal sheaths usually with papillose-based hairs, rarely glabrous; **ligules** 0.3–1.5 mm; **blades** 2–18 cm long, 1–6 mm wide, flat or involute, glabrous, scabrous, or pilose. **Panicles** with 2–7 spikelike primary branches, these digitate or the rachises to 1 cm; **longest primary branches** 3–25 cm long, 0.2–0.4 mm wide, axes triquetrous, not wing-margined, bearing spikelets in groups of 2–5 on the lower and middle portions. **Spikelets** 1.3–2.8 mm. **Lower glumes** absent or to 0.1 mm; **upper glumes** 1–2 mm long, from ¾ to almost as long as the spikelets, almost glabrous or sparsely to densely pubescent with clavate to capitate hairs (use 20× magnification), glume apices rounded; **lower lemmas** equaling the spikelets, glabrous or glandular-pubescent, 5–7-veined, veins unequally spaced, outer 3 veins on each side closer to each other than the midvein is to the inner

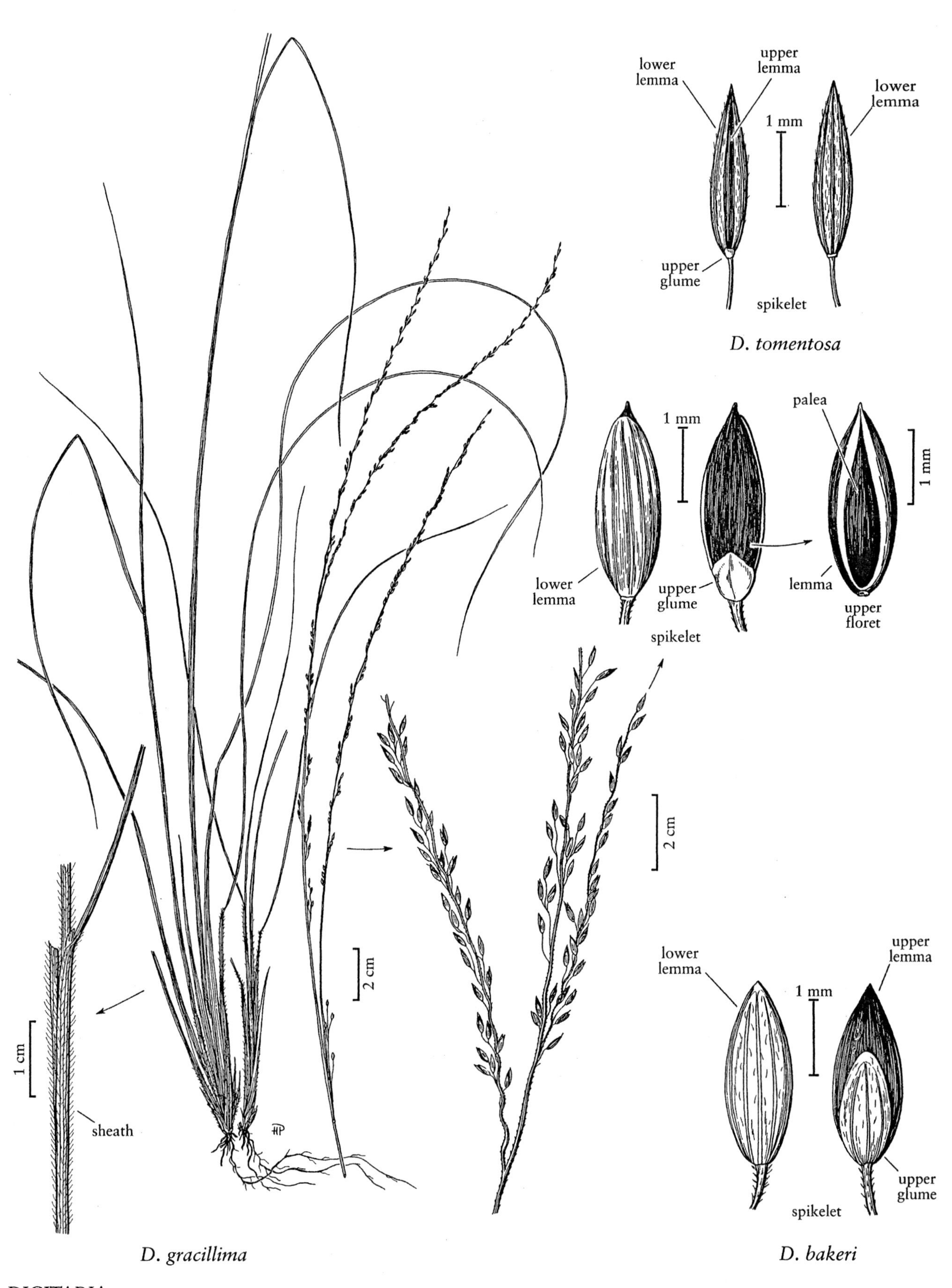
lower lemma
upper lemma
1 mm
lower lemma
upper glume
spikelet
D. tomentosa
palea
1 mm
1 mm
lower lemma
upper glume
lemma
upper floret
spikelet
2 cm
2 cm
1 cm
sheath
D. gracillima
lower lemma
upper lemma
1 mm
upper glume
spikelet
D. bakeri

lateral veins; **upper lemmas** 1.3–2 mm, apiculate, dark brown at maturity; **anthers** 0.3–0.6 mm. $2n$ = 36, 54.

Digitaria filiformis grows throughout the warmer parts of the eastern United States, var. *filifomis* the most widespread of its varieties, extending into Mexico.

1. Lower lemmas glabrous . . . . var. *laeviglumis*
1. Lower lemmas pubescent.
   2. Basal leaf sheaths glabrous; cauline blades about 1 mm wide, folded or involute . . . . var. *dolichophylla*
   2. Basal leaf sheaths with papillose-based hairs; cauline blades 1–6 mm wide, flat.
      3. Spikelets 1.3–1.9 mm long; panicle branches 3–13 cm long; culms 10–80 cm tall . . . . var. *filiformis*
      3. Spikelets 2–2.8 mm long; panicle branches 10–25 cm long; plants 75–150 cm tall . . . . var. *villosa*

Digitaria filiformis var. **dolichophylla** (Henrard) Wipff

Culms 50–115 cm. **Basal leaf sheaths** glabrous; **cauline leaf blades** about 1 mm wide, usually involute. **Spikelets** 1.5–1.6 mm; **lower lemmas** pubescent.

Digitaria filiformis var. *dolichophylla* is an uncommon species of moist pine barrens and open ground in southern Florida.

Digitaria filiformis (L.) Koeler var. **filiformis**

Culms 10–80 cm. **Basal leaf sheaths** with papillose-based hairs; **cauline leaf blades** 1–6 mm wide, flat. **Panicle branches** 3–13 cm. **Spikelets** 1.3–1.9 mm; **lower lemmas** pubescent.

Digitaria filiformis var. *filiformis* is a weed of sandy fields and open, disturbed ground in the southeastern United States and Mexico.

Digitaria filiformis var. **laeviglumis** (Fernald) Wipff

Culms 75–150 cm. **Basal leaf sheaths** with papillose-based hairs or glabrous; **blades** 1–6 mm, flat. **Spikelets** 1.8–2.5 mm; **lower lemmas** glabrous.

Digitaria filiformis var. *laeviglumis* is endemic to sandy soils in New England.

Digitaria filiformis var. **villosa** (Walter) Fernald

Culms 75–150 cm. **Basal leaf sheaths** with papillose-based hairs; **cauline leaf blades** 1–6 mm wide, flat. **Panicle branches** 10–25 cm. **Spikelets** 2–2.8 mm; **lower lemmas** pubescent.

Digitaria filiformis var. *villosa* has essentially the same geographic range as var. *filiformis* and grows in similar habitats. Further study may show that the two varieties should be combined.

8. **Digitaria leucocoma** (Nash) Urb. [p. 367]

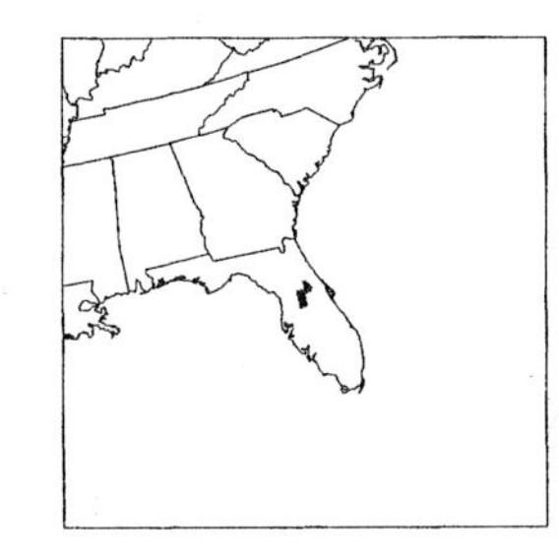

Plants perennial; cespitose. **Culms** to 100 cm, erect, not branching at the lower nodes; **nodes** 1–2. **Sheaths** with appressed hairs, lower sheaths densely hairy, upper sheaths sparingly hairy near the base, otherwise glabrous; **ligules** 2–3 mm; **blades** 10–40 cm long, to 3 mm wide, usually flat, involute when dry. **Panicles** with 2–4 spikelike branches on 4–6 cm rachises; **primary branches** 20–25 cm long, axes triquetrous, not winged; **primary branches** bearing spikelets in unequally pedicellate groups of 3(–5) on the basal ½; **secondary branches** rarely present, longer pedicels often adnate to the branch axes basally. **Spikelets** 2.2–2.5 mm long, 0.8 mm wide, elliptic, acute. **Lower glumes** absent; **upper glumes** 3-veined, margins and intercostal regions with long, glandular-tipped hairs; **lower lemmas** 7-veined, margins and the region between the 2 inner marginal veins with long glandular hairs; **upper lemmas** light to dark brown at maturity, striate, apiculate. $2n$ = unknown.

Digitaria leucocoma is known only from high pinelands near Lake Ella, Lake County, Florida. It has been treated in the past as a synonym of *D. filiformis* var. *villosa*.

9. **Digitaria hitchcockii** (Chase) Stuck. [p. 369]

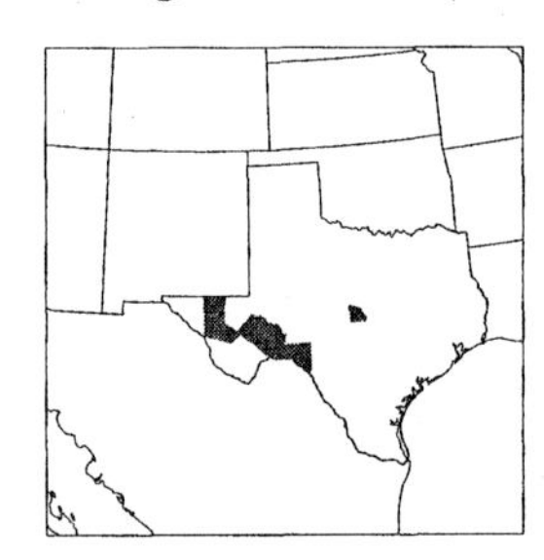

Plants perennial; rhizomatous, rhizomes short, giving the plants hard, knotty, much-branched bases. **Culms** 20–55 cm, erect, sometimes geniculate, not rooting at the lower nodes. **Basal sheaths** tomentose; **culm sheaths** glabrous or variously pubescent (puberulent, ciliate, or sparsely hirsute); **ligules** (0.1)0.5–1(1.5) mm, ciliate; **blades** 2–5.5 cm long, 2–3 mm wide. **Panicles** with 3–6 spikelike primary branches on 6–10(15) cm rachises; **primary branches** 1–6 cm, not or only narrowly winged, bearing spikelets in unequally pedicellate pairs, pedicels not adnate to the branch axes; **shorter pedicels** 1.5–2 mm; **longer pedicels** 3–4 mm. **Spikelets** homomorphic, 2.5–3.1 mm (including pubescence), 2.4–3 mm (excluding pubescence). **Lower glumes** 0.3–1 mm, veinless; **upper glumes** 2.1–3 mm (including pubescence), as long as or exceeding the upper florets by no more than 0.5 mm, 3-veined, densely appressed-pubescent, hairs 0.5–1 mm, white to purple, tapering or parallel-sided, not spreading at maturity; **lower lemmas** 2.3–3.1 mm (including pubescence), as long as or exceeding the upper lemmas by up to 0.5 mm, 5-veined,

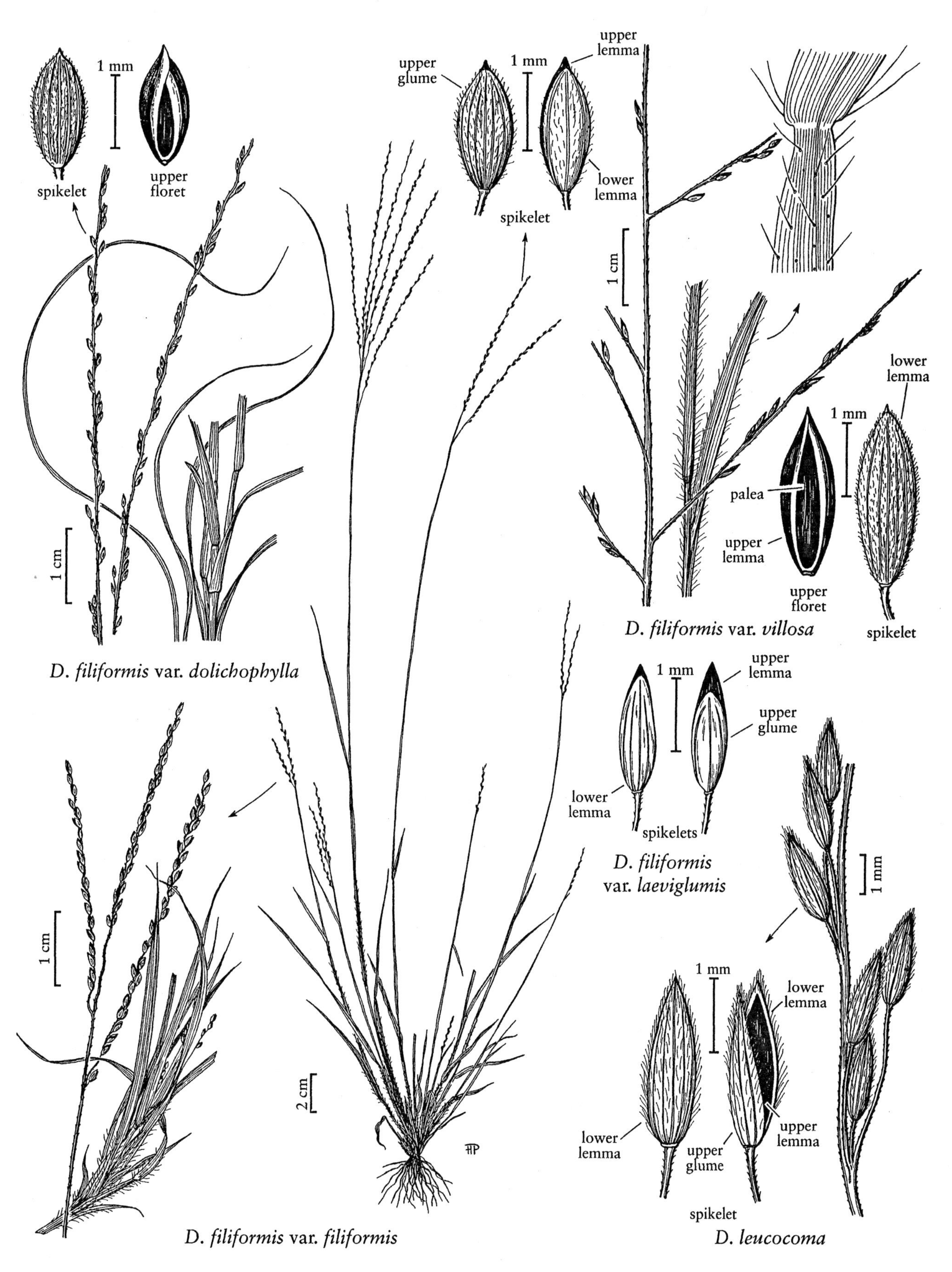
1 mm
spikelet
upper floret
1 cm
D. filiformis var. dolichophylla
upper glume
1 mm
upper lemma
lower lemma
spikelet
1 cm
lower lemma
1 mm
palea
upper lemma
upper floret
spikelet
D. filiformis var. villosa
1 mm
upper lemma
upper glume
lower lemma
spikelets
D. filiformis var. laeviglumis
1 mm
1 cm
2 cm
1 mm
lower lemma
lower lemma
upper glume
upper lemma
spikelet
D. filiformis var. filiformis
D. leucocoma

veins equally spaced, intercostal regions densely appressed-pubescent, hairs 0.5–1 mm, white to purple, tapering or parallel-sided, not spreading at maturity; **upper lemmas** 2.2–2.5 mm, brown when immature, dark brown at maturity. **Caryopses** 2.1–3.4 mm. $2n$ = 36.

Digitaria hitchcockii is an uncommon species of open, dry, gravelly slopes in southwestern Texas and northern Mexico.

10. **Digitaria patens** (Swallen) Henrard [p. 369]

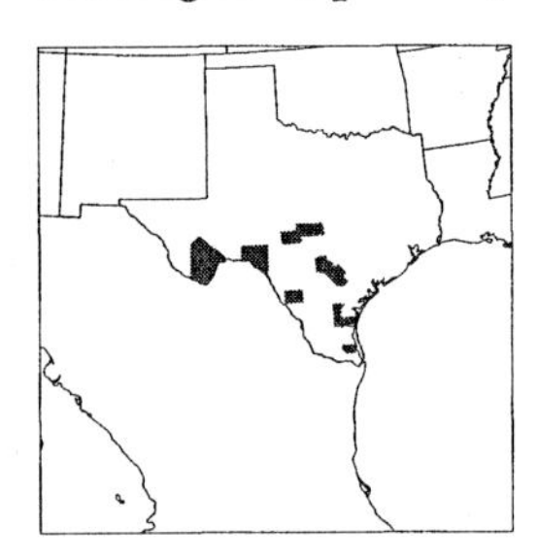

Plants perennial; cespitose, neither rhizomatous nor stoloniferous. **Culms** 40–90 cm, erect, sometimes geniculate, not rooting, at the lower nodes. **Leaves** mainly cauline; **basal sheaths** villous; **upper sheaths** glabrous or sparsely to densely hirsute, hairs papillose-based; **ligules** (1)1.5–4 mm, entire to lacerate; **blades** 5–15 cm long, 1–4 mm wide, glabrous or sparsely pubescent. **Panicles** with 4–10 spikelike primary branches on (4)10–18 cm rachises; **primary branches** 4–10 cm, usually divergent at maturity, varying to ascending, axes not wing-margined, bearing spikelets in unequally pedicellate pairs; **internodes** (4.5)6–15 mm (midbranch); **secondary branches** rarely present; **shorter pedicels** 2–2.5 mm; **longer pedicels** 7–8 mm; **terminal pedicels of primary branches** 7.4–20 mm. **Spikelets** homomorphic, 3.7–5.8 mm (including pubescence), 2.9–4.3 mm (excluding pubescence). **Lower glumes** 0.3–0.5 mm; **upper glumes** 2.4–3.5 mm (excluding pubescence), 3-veined, densely villous, hairs 1.5–4 mm, silvery-white to purple, spreading at maturity; **lower lemmas** 2.8–4.2 mm (excluding pubescence), exceeding the upper lemmas by 0.8–2.2 mm, 5-veined and the veins equally spaced or 7-veined and the lateral veins closer to each other than to the central vein, margins densely villous, hairs 1.5–4 mm, silvery-white to purple, spreading at maturity, apices acuminate; **upper lemmas** 2.6–3.2 mm, lanceolate, brown when immature, dark brown at maturity, acuminate. $2n$ = 72.

Digitaria patens is endemic to southwestern and southern Texas and adjacent Mexico. It grows in well-drained, usually sandy, soils, often in disturbed habitats. Gould (1975) suggested that it might be an octoploid derivative of *D. californica.*

11. **Digitaria californica** (Benth.) Henrard [p. 369]

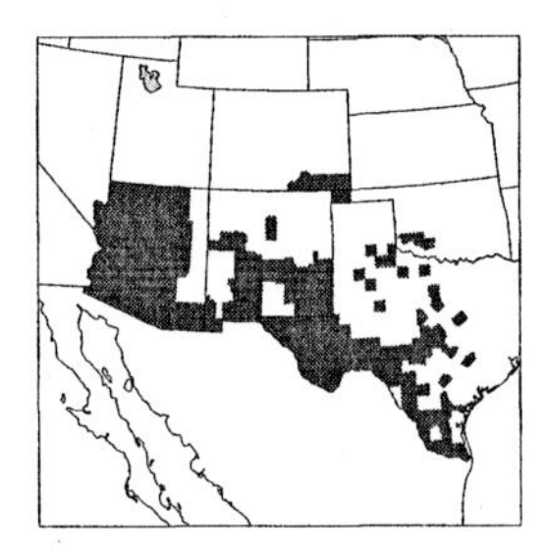

Plants perennial; cespitose, neither rhizomatous nor stoloniferous. **Culms** 40–100 cm, erect, sometimes geniculate, not rooting, at the lower nodes. **Basal sheaths** villous; **upper sheaths** glabrous, densely villous or densely tomentose, or sparsely to densely hairy, with papillose-based hairs; **ligules** (1)1.5–6 mm, entire or lacerate, not ciliate; **blades** 2–12(18) cm long, 2–5(7) mm wide, glabrous or the adaxial surfaces sparsely to densely villous or tomentose. **Panicles** with 4–10 spikelike primary branches on 5–10 cm rachises, rarely with secondary branches; **primary branches** 3–6 cm, appressed to ascending, axes not wing-margined; **internodes** 2–5.5 mm (midbranch), bearing spikelets in unequally pedicellate pairs; **secondary branches** rarely present; **pedicels** not adnate to the branch axes; **shorter pedicels** 0.1–0.3 mm; **longer pedicels** 1–2 mm; **terminal pedicels of branches** 1.7–6(7) mm. **Spikelets** homomorphic, (3.7)4–7.5 mm (including pubescence), 3–5.4 mm (excluding pubescence). **Lower glumes** 0.4–0.6 mm; **upper glumes** 2.5–5.1 mm (excluding pubescence), narrower than the upper florets, 3-veined, densely villous, hairs 1.5–5 mm, silvery-white to purple, widely divergent at maturity; **lower lemmas** 2.7–5 mm (excluding pubescence), pubescence exceeding the upper florets by 2.2–4 mm, 7-veined, veins unequally spaced, only the 3 or 5 central veins visible, margins and outer lateral veins densely pubescent, hairs 1.5–5 mm, silvery-white to purple, widely divergent at maturity, intercostal regions glabrous, apices attenuate (acuminate); **upper lemmas** 2.5–3.4 mm, ovate-lanceolate, brown to dark brown, acuminate. **Caryopses** 1.3–2 mm. $2n$ = 36, 54, 70, 72.

Digitaria californica grows on plains and open ground from Arizona, southern Colorado, and Oklahoma through Mexico and Central America to South America. The name reflects the fact that the first collection was made in Baja California, Mexico. Plants in the *Flora* region belong to **D. californica** (Benth.) Henrard var. **californica.** They differ from those of **D. californica** var. **villosissima** Henrard in having densely villous, rather than densely tomentose, leaves.

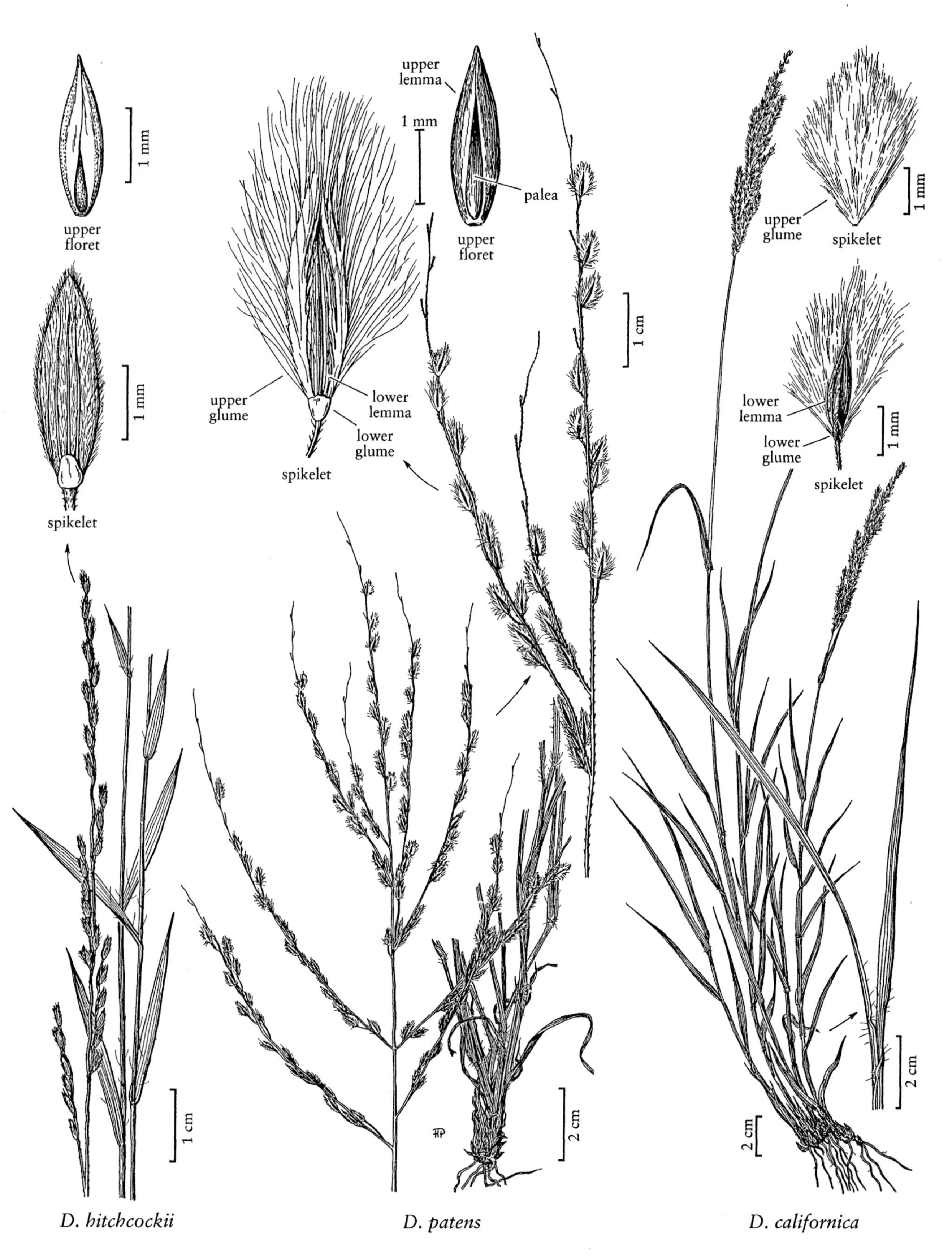
upper floret
1 mm
spikelet
1 mm
1 cm
upper lemma
1 mm
palea
upper floret
upper glume
lower lemma
lower glume
spikelet
1 cm
2 cm
upper glume
spikelet
1 mm
lower lemma
lower glume
spikelet
1 mm
2 cm
2 cm
D. hitchcockii
D. patens
D. californica

DIGITARIA

12. **Digitaria insularis** (L.) Mez *ex* Ekman [p. 371]

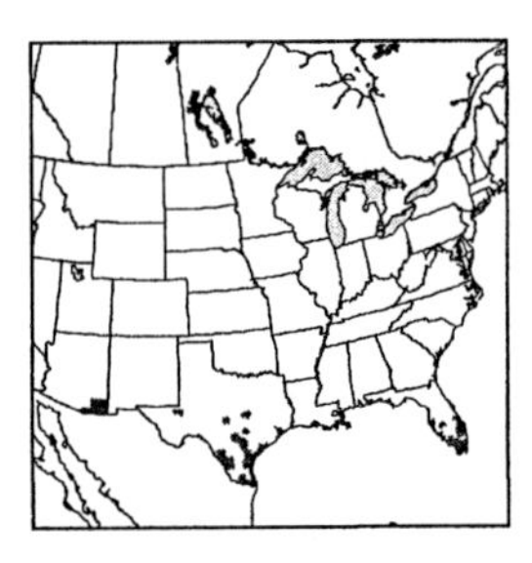

Plants perennial; cespitose, shortly rhizomatous, with knotty bases. **Culms** 80–130 cm, erect, with densely villous cataphylls, branching from the lower and middle nodes. **Sheaths** usually sparsely to densely papillose-hirsute, occasionally glabrous; **ligules** 4–6 mm, usually lacerate, not ciliate; **blades** 20–50 cm long, 10–17 mm wide, lax, smooth or scabridulous abaxially, scabridulous to scabrous adaxially. **Panicles** 20–35 cm long, 2–10 cm wide, with numerous spikelike primary branches; **primary branches** 10–15 cm, appressed to ascending at maturity, axes not wing-margined or with wings less than ½ as wide as the midribs; **internodes** 3–4.5(6) mm (midbranch), bearing spikelets in unequally pedicellate pairs; **secondary branches** rarely present; **pedicels** not adnate to the branches; **shorter pedicels** 0.7–2 mm; **longer pedicels** 2.5–5 mm; **terminal pedicels** 2–5 mm. **Spikelets** 5.5–8.2 mm (including pubescence), 4.2–5.9 mm (excluding pubescence), narrowly ovate, acuminate. **Lower glumes** 0.6–0.8 mm; **upper glumes** 3.5–4.5 mm, 3–5-veined, pubescent on the margins; **lower lemmas** 4.1–5.7 mm (exceeded 1.5–5 mm by pubescence), narrowly ovate, 7-veined, pubescent between most, sometimes all, of the veins and on the margins, veins usually obscured by a dense covering of golden-brown hairs, hairs 3–6 mm, spreading at maturity, intercostal regions on either side of the midvein glabrous or pubescent with shorter, fine, white hairs, sometimes intermixed with the golden-brown hairs; **upper lemmas** 3.2–4.5 mm, narrowly ovate, brown when immature, dark brown at maturity, acuminate; **anthers** 1–1.2 mm. $2n = 36$.

Digitaria insularis grows in low, open ground of the southern United States, and extends to the West Indies, Mexico, and through Central America to Argentina.

13. **Digitaria serotina** (Walter) Michx. [p. 371]

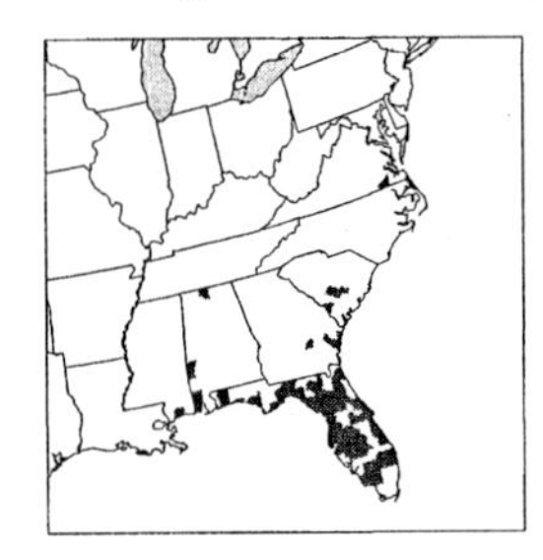

Plants annual; often mat-forming. **Culms** 10–30 cm, decumbent and rooting at the lower nodes. **Sheaths** conspicuously and densely hairy, longer hairs 1.5–2.5 mm, papillose-based, shorter hairs about 0.5 mm, not papillose-based; **ligules** 1.5–2.5 mm; **blades** 2–9 cm long, 3–8 mm wide, conspicuously hairy on both surfaces, longer hairs 1.5–2.5 mm, papillose-based, shorter hairs about 0.5 mm, not papillose-based. **Panicles** with 2–9 spikelike primary branches, digitate or on rachises to 4 cm; **primary branch axes** 3–10 cm, wing-margined, wings wider than the midribs, lower and middle portions bearing spikelets in groups of 3; **secondary branches** rarely present; **shortest pedicels** 0.5–0.8 mm; **midlength pedicels** 1.5–2 mm; **longest pedicels** 3–3.5 mm, adnate to the branch axes basally. **Spikelets** homomorphic, 1.5–1.8 mm, lanceolate. **Lower glumes** absent; **upper glumes** ⅙–⅓ as long as the spikelets, margins and apices with appressed white hairs, hairs about 0.3 mm; **lower lemmas** 7-veined, veins equally spaced, appressed-pubescent between the inner lateral veins and on the margins, hairs 0.3–0.5 mm, minutely verrucose (use 50× magnification); **upper lemmas** yellow or tan at maturity. $2n$ = unknown.

Digitaria serotina is native to the coastal plain of the southeastern United States. It has also been found in Cuba, possibly as an introduction, and on a ballast dump in Philadelphia, Pennsylvania. Its densely hairy sheath and short, densely hairy blades make this one of the more distinctive species of *Digitaria* in the *Flora* region.

14. **Digitaria longiflora** (Retz.) Pers. [p. 371]

Plants of indefinite duration; stoloniferous, stolons long and branching. **Culms** 10–60 cm, occasionally branching from the lower nodes. **Leaves** 3–4, clustered near the base; **sheaths** usually glabrous; **ligules** 0.5–1 mm; **blades** 1.5–4 cm long, 3–5 mm wide, mostly glabrous, bases subcordate and ciliate, with 0.6–1 mm papillose-based hairs. **Panicles** with 2(–4) spikelike primary branches, digitate; **primary branches** 2–5 cm, strongly divergent; **branch axes** about 1 mm wide, wing-margined, wings wider than the central midribs, bearing spikelets in unequally pedicellate groups of 3; **secondary branches** rarely present; **shortest pedicels** about 0.3 mm; **middle pedicels** about 1 mm; **longest pedicels** 1.5–2 mm, adnate to the branch axes basally; **axillary panicles** not present. **Spikelets** 1.2–1.5 mm, elliptic or slightly obovate, acute. **Lower glumes** absent; **upper glumes** equaling or almost equaling the spikelets, 5-veined, minutely pubescent between the veins and on the margins; **lower lemmas** subequal to the upper glumes, 7-veined, usually pubescent on the margins and lateral veins, occasionally glabrous, hairs, if present, 0.2–0.4 mm; **upper lemmas** about 1.2 mm, pale brown or pale gray, becoming light brown at maturity, acute; **anthers** 0.8–0.8 mm. $2n = 18$.

Digitaria longiflora is native to Africa and Asia. It is now established in disturbed areas of Florida, growing on railroad grades and in pastures and lawns.

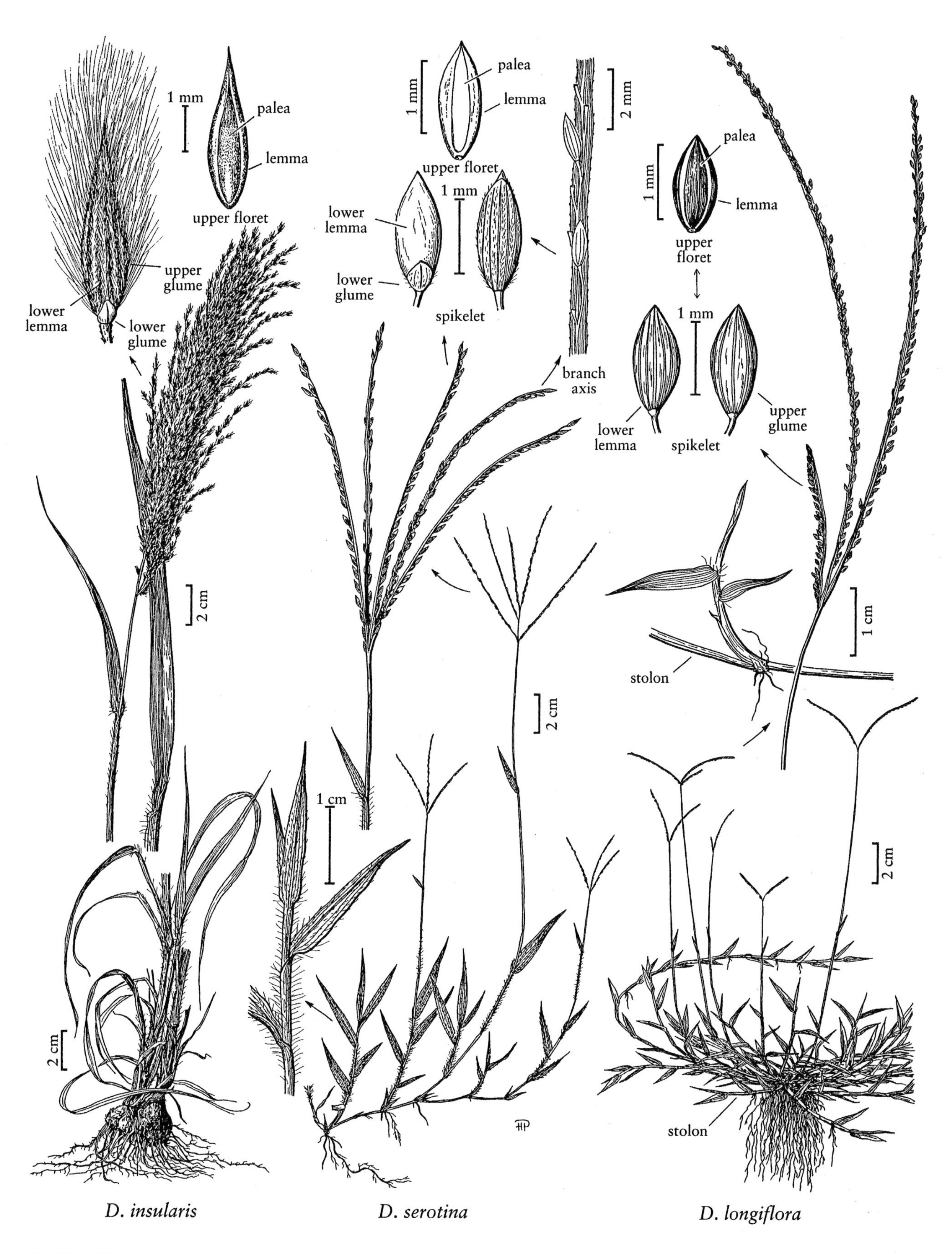

1 mm
palea
lemma
upper floret
upper glume
lower lemma
lower glume
2 cm
2 cm
palea
lemma
1 mm
upper floret
lower lemma
1 mm
lower glume
spikelet
2 mm
branch axis
2 cm
1 cm
palea
1 mm
lemma
upper floret
1 mm
lower lemma
spikelet
upper glume
1 cm
stolon
2 cm
stolon
D. insularis
D. serotina
D. longiflora

DIGITARIA

15. Digitaria floridana Hitchc. [p. 373]

Plants annual or of indefinite duration; not stoloniferous. **Culms** 20–30 cm, decumbent and rooting at the nodes; **nodes** glabrous. **Sheaths** mostly glabrous, throats with papillose-based hairs; **blades** 4–7 cm long, 3–6 mm wide, glabrous. **Panicles** with 2–4 spikelike primary branches, if more than 2, rachises 7–20 mm and the branches racemose; **primary branches** 3–6 cm, terminal branch erect, the other(s) usually divergent, axes wing-margined, wings wider than the midribs, bearing spikelets in unequally pedicellate groups of 3 on the basal and mid-portions; **secondary branches** rarely present; **shortest pedicels** about 0.05 mm; **middle pedicels** about 0.1 mm; **longest pedicels** 0.2–0.3 mm, adnate to the branch axes basally; **axillary panicles** not present. **Spikelets** homomorphic, 1.5–1.7 mm, lanceolate. **Lower glumes** absent; **upper glumes** almost equaling the upper lemmas, conspicuously 3-veined; **lower lemmas** slightly longer than the upper lemmas, 7-veined, veins unequally spaced, all the intercostal regions sparsely hairy, hairs about 0.3 mm; **upper lemmas** light brown when immature, dark brown at maturity. $2n$ = unknown.

Digitaria floridana is a rare species that is known only from sandy pine woods in Hernando County, Florida.

16. Digitaria ischaemum (Schreb.) Muhl. [p. 373]
SMOOTH CRABGRASS, DIGITAIRE ASTRINGENTE

Plants annual or of indefinite duration. **Culms** 20–55(70) cm, decumbent, branching and rooting at the lower nodes; **nodes** 3–4. **Sheaths** glabrous or sparsely pubescent; **ligules** 0.6–2.5 mm; **blades** 1.5–9 cm long, 3–5 mm wide, glabrous, with a few papillose-based hairs basally. **Panicles** terminal and axillary; **terminal panicles** with 2–7 spikelike primary branches, subdigitate or on 0.5–2 cm rachises; **primary branches** 6–15.5 cm, axes wing-margined, wings at least ½ as wide as the midribs, bearing spikelets in groups of 3, lower portions of the longer pedicels adnate to the axes; **secondary branches** rarely present; **axillary inflorescences** always present in some of the lower sheaths, entirely or partially concealed. **Spikelets** 1.7–2.3 mm, homomorphic, narrowly elliptic. **Lower glumes** absent or a veinless, membranous rim; **upper glumes** 1.3–2.3 mm, from ¾ as long as to equaling the upper lemmas, appressed-pubescent; **lower lemmas** 1.7–2.3 mm, 7-veined, veins unequally spaced, smooth, pubescent; **upper lemmas** dark brown at maturity; **anthers** 0.4–0.6 mm. $2n$ = 36.

Digitaria ischaemum is a Eurasian weed that is now common in lawns, gardens, fields, and waste ground in warm-temperate regions throughout the world, including much of the *Flora* region. Larger plants with 5–7 inflorescence branches 8–15 cm long have been called var. *mississippiensis* (Gatt.) Fernald, but they intergrade with more typical plants, and so do not merit taxonomic recognition.

17. Digitaria violascens Link [p. 373]

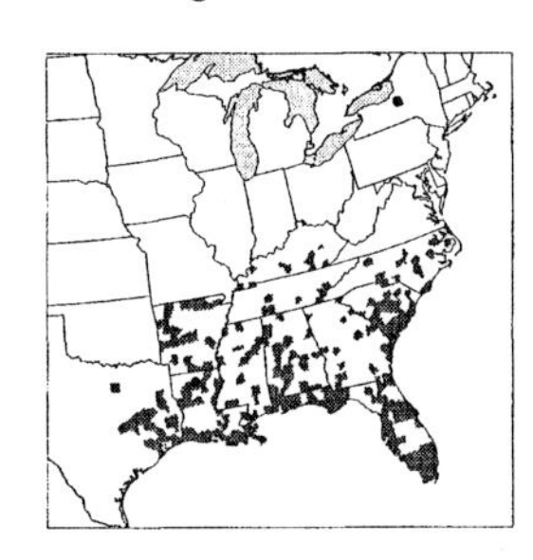

Plants annual or of indefinite duration. **Culms** 15–60 cm, erect, usually not branching from the upper nodes; **nodes** 3–4. **Sheaths** glabrous or sparsely pubescent; **ligules** 0.6–2.5 mm; **blades** 1.5–9 cm long, 3–5 mm wide, glabrous, with papillose-based hairs basally. **Panicles** with 2–7 spikelike primary branches in 1–2 verticils; **primary branches** 3–12 cm, erect to ascending, axes 0.6–1 mm wide, wing-margined, wings at least ½ as wide as the midribs, lower and middle portions of the branches bearing spikelets in groups of 3(4, 5); **secondary branches** rarely present; **axillary inflorescences** absent. **Spikelets** 1.2–1.7 mm, homomorphic, narrowly elliptic. **Lower glumes** absent or a veinless, membranous rim; **upper glumes** 1.2–1.4 mm, ½ as long as to almost equaling the upper lemmas, 3-veined, appressed-pubescent, hairs minutely verrucose; **lower lemmas** 1.2–1.7 mm, 5–7-veined, veins equally spaced, region between the 2 inner lateral veins and the margins appressed-pubescent, hairs 0.3–0.5 mm, smooth or minutely verrucose (use 50× magnification), verrucose hairs most abundant near the lemma bases; **upper lemmas** light brown when immature, dark brown at maturity; **anthers** 0.4–0.6 mm. $2n$ = 36.

Digitaria violascens is a weedy species that is native to tropical regions of the Eastern Hemisphere. It is now established in the *Flora* region, primarily in the southeastern United States, and in Mexico and Central America. It grows in disturbed sites.

18. Digitaria abyssinica (Hochst. *ex* A. Rich.) Stapf [p. 375]

Plants perennial; rhizomatous, mat-forming. **Culms** 5–80 cm, decumbent, occasionally rooting at the lower nodes, branching freely at the base, erect portion 20–40 cm; **nodes** 2–6. **Sheaths of midculm leaves** glabrous or hirsute, with papillose-based hairs; **ligules** 0.8–2.1 mm; **blades** 4–15 cm long, 3–10 mm wide, glabrous or sparsely pubescent with papillose-based hairs. **Panicles**

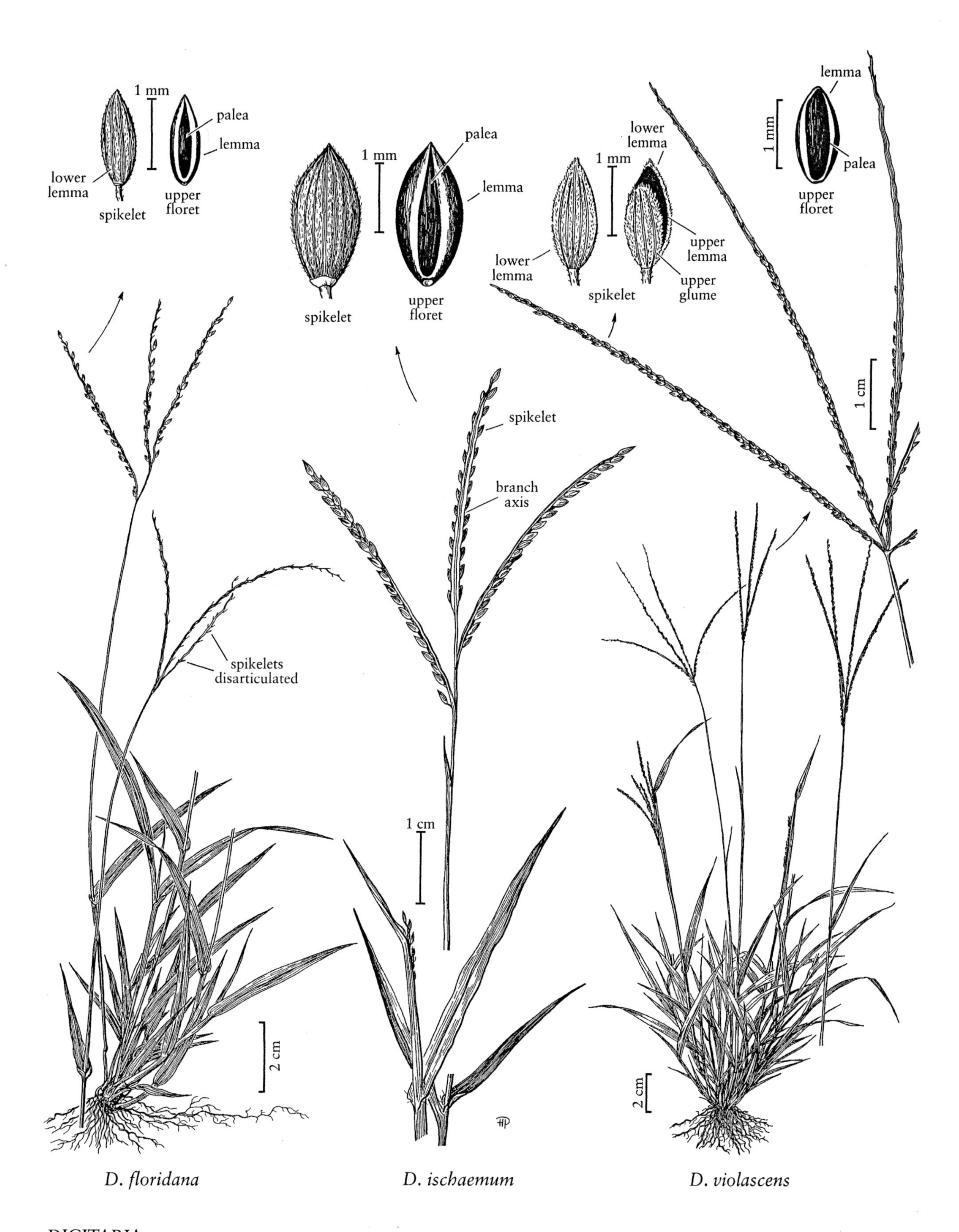

D. floridana *D. ischaemum* *D. violascens*

with 2–25 spikelike primary branches on 1–9 cm rachises; **primary branches** 2–11 cm, axes not winged or narrowly winged, wings less than ½ as wide as the midribs, bearing spikelets in unequally pedicellate pairs; **secondary branches** rarely present; **pedicels** not adnate to the branch axes. **Spikelets** 1.5–2.5 mm long, 0.8–0.95 mm wide, ovate-elliptic to broadly elliptic, usually plump, usually purple-tinged. **Lower glumes** absent or to 0.8 mm and acute; **upper glumes** 1.2–2.4 mm, from 0.8 times as long as to almost equaling the spikelets, glabrous, 3–7-veined, veins usually prominent; **lower lemmas** 1.5–2.5 mm, usually glabrous, occasionally obscurely puberulent on the margins or, very rarely, distinctly pubescent, 7-veined, veins usually prominent; **upper lemmas** light brown, gray, and purple. $2n$ = 36.

Introduced from Africa, *Digitaria abyssinica* is not known to be established in the *Flora* region although it has occasionally been cultivated in the southern United States. It is considered a potentially serious weed threat by the U.S. Department of Agriculture.

19. **Digitaria pauciflora** Hitchc. [p. 375]

Plants perennial; not rhizomatous. **Culms** 50–100 cm, erect to somewhat decumbent, not rooting at the lower nodes, usually branching at the aerial nodes. **Sheaths** grayish-villous; **ligules** 1.5–2 mm; **blades** to 12 cm long, 2–2.2 mm wide, flat or folded, densely grayish-villous on both surfaces. **Panicles** with 2–3 spikelike primary branches, secondary branches not present; **primary branches** 5–11 cm, axes not winged, without spikelets or with widely spaced abortive spikelets on the proximal 1–1.5 cm, bearing spikelets in unequally pedicellate pairs at midlength; **shorter pedicels** about 2 mm; **longer pedicels** about 3 mm, not adnate to the branch axes. **Spikelets** 2.7–3.2 mm, appressed, glabrous. **Lower glumes** minute, rounded, erose; **upper glumes** slightly shorter than the spikelets, 3-veined, glabrous; **lower lemmas** 7-veined, glabrous; **upper lemmas** gray or yellow when immature, becoming purple at maturity. $2n$ = unknown.

Digitaria pauciflora is known only from the type collection, which was collected in pinelands of Dade County, Florida.

20. **Digitaria simpsonii** (Vasey) Fernald [p. 375]

Plants of indefinite duration; not rhizomatous. **Culms** 80–120 cm, erect or decumbent and rooting at the lower nodes, not branching at the aerial nodes. **Sheaths** hirsute, with papillose-based hairs, those of the innovation sheaths compressed-keeled; **ligules** 1–2 mm; **blades** 7–30 cm long, 3–5 mm wide, flat, pilose above and below. **Panicles** with 6–8 spikelike primary branches on 4–6 cm rachises; **primary branches** 8–13 cm, axes triquetrous, narrowly winged, wings less than ½ as wide as the midribs, lower and middle portions of the branches bearing spikelets in appressed, unequally pedicellate pairs; **secondary branches** rarely present; **axillary inflorescences** not present; **pedicels** not adnate to the branch axes. **Spikelets** about 3 mm, elliptic lanceolate, acute. **Lower glumes** absent or minute and hyaline; **upper glumes** 7–9-veined, glabrous or obscurely pubescent; **lower lemmas** 7–9-veined, glabrous or obscurely pubescent; **upper lemmas** elliptic, yellow or gray, becoming purple at maturity, slightly apiculate. $2n$ = unknown.

Digitaria simpsonii is a rare species, known only from sandy fields in Florida.

21. **Digitaria texana** Hitchc. [p. 377]

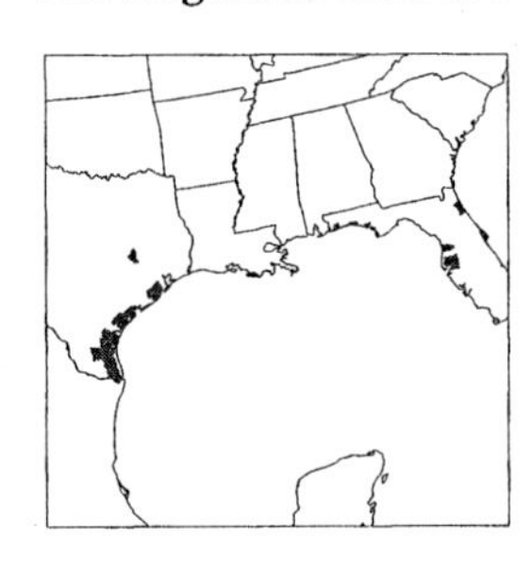

Plants perennial; not rhizomatous. **Culms** 30–80 cm, sometimes erect, usually decumbent and branching and rooting at the lower nodes, not branching at the upper nodes. **Sheaths** of the lower leaves villous, those of the upper leaves sometimes glabrous, those of the flag leaves without axillary panicles; **ligules** 1.5–2 mm; **blades** 10–15 cm long, 2–7 mm wide, hirsute to nearly glabrous. **Panicles** with 5–10 spikelike primary branches on 1–4 cm rachises; **primary branches** 5–10(13) cm, axes triquetrous, narrowly winged, wings less than ½ as wide as the midribs, lower and middle portions of the branches bearing paired spikelets; **secondary branches** rarely present. **Spikelets** 2–3.6 mm, narrowly ovate-oblong, acute. **Lower glumes** absent; **upper glumes** almost as long as the spikelets, 3(5)-veined, shortly villous on the margins and sometimes between the margins; **lower lemmas** similar to the upper glumes; **upper lemmas** gray or yellow, sometimes purple-tinged, becoming purple at maturity. **Caryopses** narrowly oblong. $2n$ = 54.

Digitaria texana grows in sandy oak woods and prairies of southern Texas and Florida.

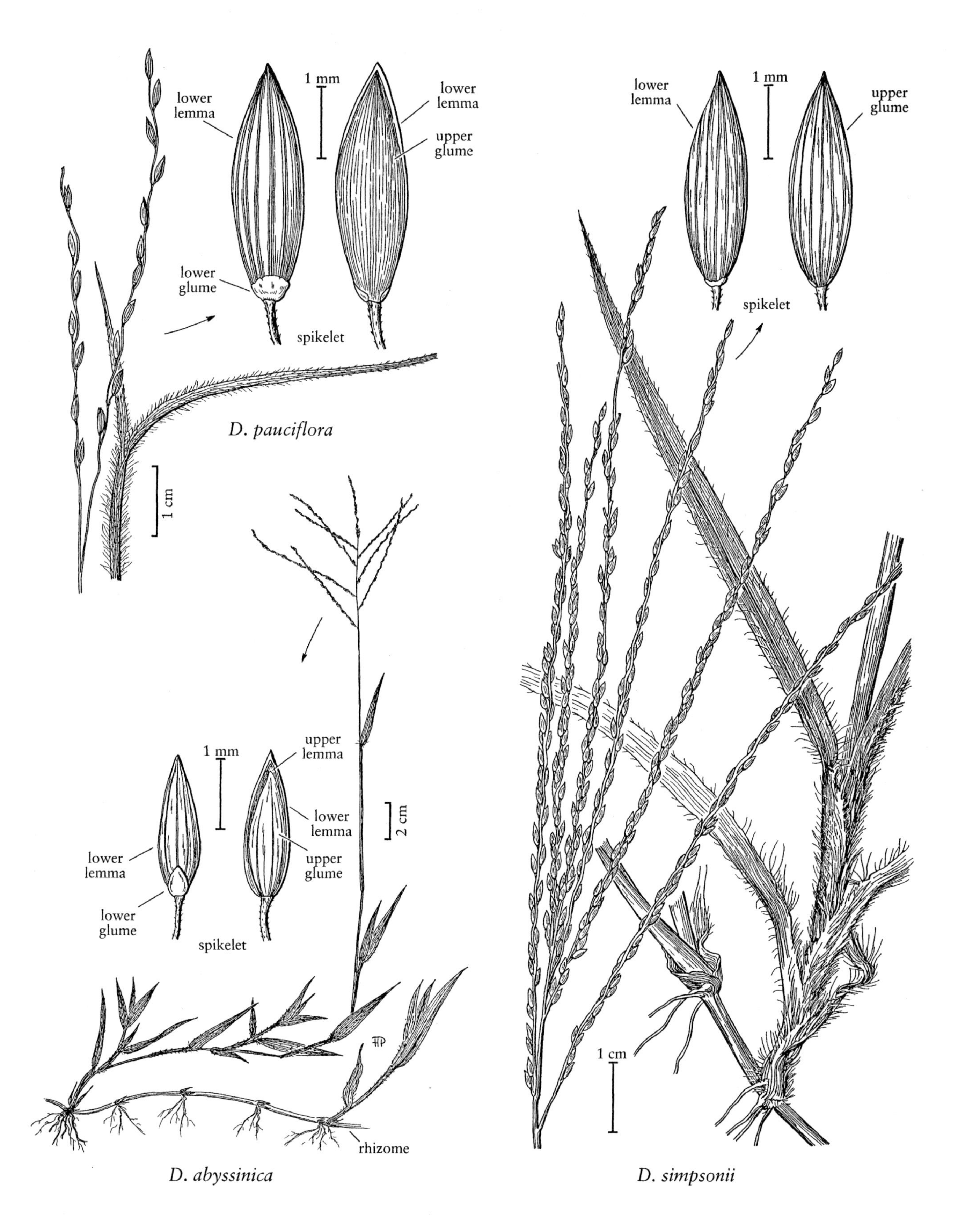
lower
lemma
1 mm
lower
lemma
upper
glume
lower
glume
spikelet
D. pauciflora
1 cm
1 mm
upper
lemma
lower
lemma
upper
glume
lower
lemma
lower
glume
spikelet
2 cm
rhizome
D. abyssinica
lower
lemma
1 mm
upper
glume
spikelet
1 cm
D. simpsonii

22. Digitaria didactyla Willd. [p. 377]
BLUE COUCH

Plants perennial; stoloniferous and rhizomatous, mat-forming. **Culms** 15–40(63) cm, rooting and branching from the lower nodes. **Sheaths** densely to sparsely hairy, with 3–5 mm papillose-based hairs; **ligules** 1–1.5 mm; **blades** 2.5–7 cm long, 1–3 mm wide, flat or folded, usually glabrous, green to bluish-green. **Panicles** with 2–4 spikelike primary branches digitately arranged; **primary branches** 2–7 cm, axes wing-margined, wings at least ½ as wide as the midribs, spikelets somewhat imbricate, in unequally pedicellate pairs; **secondary branches** rarely present; **pedicels** not adnate to the branches; **shorter pedicels** 1–1.5 mm; **longer pedicels** 2–3 mm; **axillary panicles** not present. **Spikelets** homomorphic, 2–2.8 mm long, about 0.8 mm wide. **Lower glumes** to 0.3 mm, triangular; **upper glumes** from ½–¾ as long as the spikelets, 3-veined, pilose on the margins and sometimes between the veins; **upper lemmas** equaling the spikelets, prominently 7-veined, veins equally spaced, margins and sometimes the intercostal regions pilose, hairs 0.3–0.5 mm; **upper lemmas** slightly shorter than the lower lemmas, almost smooth, gray, sometimes purple-tinged, at maturity. 2*n* = unknown.

A native of Africa, *Digitaria didactyla* is often cultivated as a lawn grass in tropical and subtropical regions. It has been grown experimentally in Florida, but is not otherwise known from the *Flora* region.

23. **Digitaria eriantha** Steud. [p. 377]

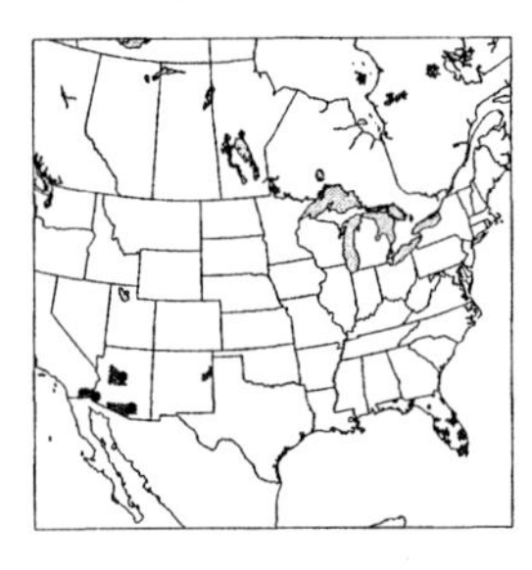

Plants perennial; sometimes stoloniferous, stolons to 6 m, or cespitose, with or without rhizomes, rhizomes, if present, short, giving the plants knotty bases. **Culms** 35–140 cm, erect or decumbent, not rooting at the basal nodes. **Basal sheaths** glabrous or pubescent, often densely so, hairs 4–6 mm, papillose-based; **ligules** (1.8)3–5 mm, erose and ciliate; **blades** 5–40 cm long, 3–6 mm wide, scabridulous, often also papillose-hairy. **Panicles** with 3–15 spikelike primary branches, digitate or with rachises to 3 cm; **primary branches** 5–25 cm, wing-margined, wings wider than the midribs, bearing spikelets in unequally pedicellate pairs; **shorter pedicels** 0.5–1.5 mm; **longer pedicels** 1.5–3 mm. **Spikelets** homomorphic, 2.8–3.5 mm, narrowly lanceolate to narrowly elliptic. **Lower glumes** 0.3–0.5 mm, veinless, acute; **upper glumes** 1.7–1.9 mm, wooly pubescent; **lower lemmas** 2.5–3.5 mm, 7-veined, veins unequally spaced and smooth, occasionally the lateral veins scabridulous over the distal ¼, margins and region between the 2 inner lateral veins appressed-pubescent, with 0.5–1.5 mm hairs; **upper lemmas** gray when immature, becoming brownish at maturity; **anthers** 1.2–1.6 mm, purple. 2*n* = 36.

Digitaria eriantha is an African species that is widely cultivated in warm climates as a pasture grass. Several cultivars have been released for forage and hay use. The appearance of the spikelets varies considerably with the length of the hairs, those of subsp. *eriantha* usually being longer than those of subsp. *pentzii.*

The cultivar, 'Survenola' has been developed from *Digitaria* ×*umfolozi* D.W. Hall, a hybrid between *D. setivalva* Stent [= *D. eriantha* subsp. *eriantha*] and *D. decumbens* Stent [= *D. eriantha* subsp. *pentzii*] and has been released for use in the tropics and on well-fertilized upland soils in Florida. It is described as having much wider leaf blades than any other cultivars that have been released so far (usually 10–13 mm wide, rather than usually less than 8 mm) and glabrous leaf sheaths.

1. Plants cespitose . . . . subsp. *eriantha*
1. Plants stoloniferous . . . . subsp. *pentzii*

Digitaria eriantha Steud. subsp. **eriantha**

Plants cespitose, rhizomatous, rhizomes short. **Basal sheaths** hirsute.

Digitaria eriantha subsp. *eriantha* is seed-producing, but it is not clear whether it is being grown in the *Flora* region.

Digitaria eriantha subsp. **pentzii** (Stent) Kok

Plants stoloniferous (the stolons to 6 m), sterile. **Basal sheaths** glabrous or hirsute. 2*n* = 27 (for *Digitaria decumbens* Stent).

In the Western Hemisphere, *D. eriantha* subsp. *pentzii* has traditionally been referred to as *D. decumbens* Stent. It is widely cultivated as a forage grass throughout the tropics at low and intermediate altitudes (up to 2000 m). It does not set seed; propagation is by sprigging the stolons.

24. **Digitaria milanjiana** (Rendle) Stapf [p. 379]

Plants perennial; rhizomatous and stoloniferous. **Culms** 50–250 cm, erect or decumbent, rooting or not at the lower nodes. **Basal sheaths** glabrous or variously pubescent (pilose, rarely tomentose or with papillose-based hairs); **upper sheaths** glabrous; **ligules** 0.8–2.5 mm; **blades** 6–15(30) cm long, 3.5–8.5(13) mm wide, glabrous adaxially, rarely hirsute, with papillose-based hairs basally, margins scabridulous. **Panicles** with 2–18 spikelike primary branches, these digitate or with rachises to 6 cm; **primary branches** 5–25 cm, axes wing-margined, wings about as wide as the midribs, bearing spikelets in

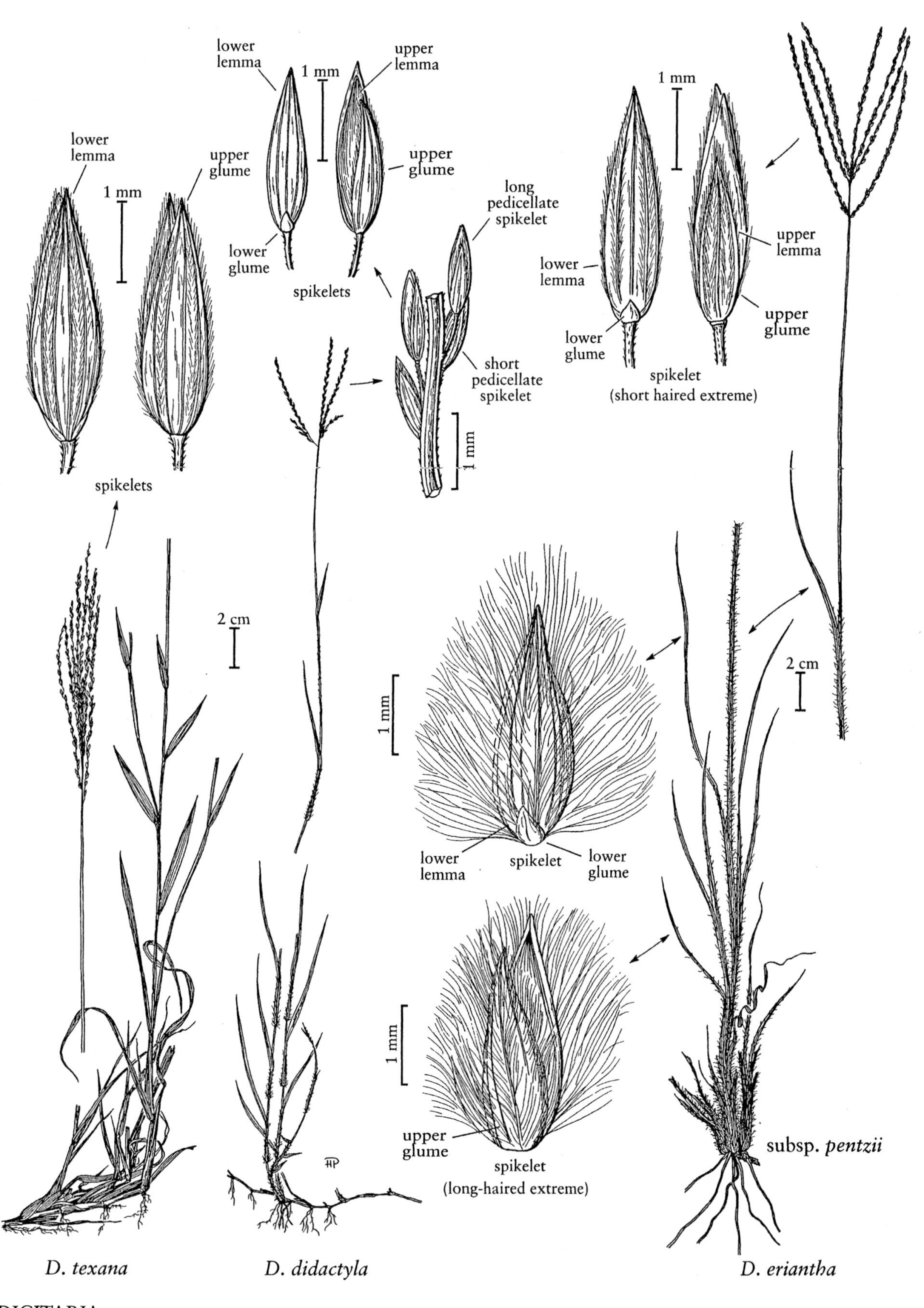
lower
lemma
upper
glume
1 mm
spikelets
lower
lemma
1 mm
upper
lemma
upper
glume
lower
glume
spikelets
long
pedicellate
spikelet
short
pedicellate
spikelet
1 mm
1 mm
upper
lemma
lower
lemma
upper
glume
lower
glume
spikelet
(short haired extreme)
2 cm
1 mm
lower
lemma
spikelet
lower
glume
2 cm
1 mm
upper
glume
spikelet
(long-haired extreme)
subsp. *pentzii*
HP
D. texana
D. didactyla
D. eriantha

DIGITARIA

unequally pedicellate pairs; **secondary branches** rarely present; **shorter pedicels** 0.2–0.3 mm; **longer pedicels** 1–1.5 mm. **Spikelets** homomorphic, 2.5–3.5 mm long, 0.7–0.9 mm wide, lanceolate. **Lower glumes** 0.2–0.5 mm, acute to truncate; **upper glumes** (1.2)1.6–2.3 mm, from ⅗ as long as to almost equaling the spikelets; **lower lemmas** 2.5–3.5 mm, 7-veined, veins unequally spaced, midvein and lateral veins scabrous at least on the distal ½, margins and region between the inner 2 lateral veins with straight, yellowish, 0.6–1 mm hairs; **upper lemmas** gray to tan at maturity. $2n$ = 18, 34, 36, 45, 54, 72(?).

Digitaria milanjiana is native to tropical and subtropical Africa. It has been found as an escape from experimental plantings in Florida.

25. **Digitaria velutina** (Forssk.) P. Beauv. [p. 379]

Plants of indefinite duration; loosely cespitose to straggling. **Culms** 15–80 m, decumbent, rooting and branching at the lower nodes. **Sheaths** pilose, with papillose-based hairs; **ligules** 1.8–2 mm; **blades** 4–15 cm long, 3–10 mm wide, pilose, with papillose-based hairs. **Panicles** with 5–18 spikelike primary branches on 2.5–5 cm rachises, lower branches usually verticillate; **primary branches** 3.5–10 cm long, 0.3–0.5 mm wide, narrowly wing-margined, wings less than ½ as wide as the midribs, bearing spikelets in unequally pedicellate pairs; **secondary branches** often present, often highly divergent; **shorter pedicels** 0.2–0.5 mm; **longer pedicels** 0.8–1.1 mm. **Spikelets** 1.5–2 mm long, about 0.5 mm wide, elliptic-lanceolate. **Lower glumes** absent or to 0.2 mm; **upper glumes** 1.5–1.7 mm, usually to ¾ as long as the spikelets, 3-veined, villous between the veins, hairs tapering or parallel-sided; **lower lemmas** about as long as the spikelets, 7-veined, veins unequally spaced, 2 veins crowded together near each margin, 3 inner veins well-separated, pubescent on the margins and between the inner lateral veins, hairs about 0.2 mm, sometimes sparse, lateral veins smooth throughout or scabridulous only on the distal ⅓; **upper lemmas** 1.5–1.7 mm, usually gray at maturity, sometimes brown; **anthers** about 0.5 mm. $2n$ = 18.

Digitaria velutina is an African species, appearing on the noxious weed list of the U.S. Department of Agriculture. It has been erroneously reported as occurring in Texas (Kartesz and Meacham 1999).

26. **Digitaria horizontalis** Willd. [p. 379]

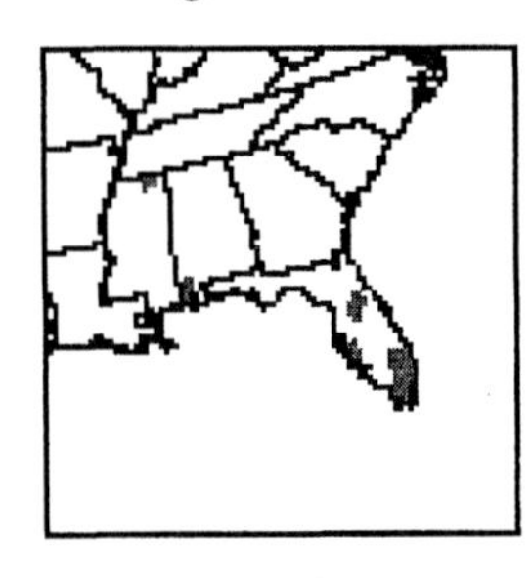

Plants of indefinite duration; sprawling. **Culms** to 1 m, erect portion 20–25 cm, decumbent, rooting and branching freely at the nodes. **Sheaths** usually with scattered papillose-based hairs, hairs more abundant on the lower sheaths; **ligules** 1.5–1.8 mm, erose; **blades** 3–14 cm long, 3–9 mm wide, evenly, often densely pubescent adaxially, hairs papillose-based. **Panicles** with 2–14 spikelike branches attached to 4–15 cm rachises, lower branches whorled, upper branches often paired or solitary; **lower nodes** with hairs more than 0.4 mm; **primary branches** 4–12 cm, axes 0.4–0.7 mm wide, wing-margined, wings at least ½ as wide as the midribs, often with scattered 1–4 mm hairs proximally, bearing spikelets in unequally pedicellate pairs on the proximal and middle portion of the branches; **secondary branches** rarely present; **shorter pedicels** 0.3–0.5 mm; **longer pedicels** 1.3–2 mm. **Spikelets** homomorphic, 2.1–2.4 mm, narrowly ovate. **Lower glumes** 0.1–0.2 mm; **upper glumes** 1–1.1 mm, ⅓–½ as long as the spikelets, 3-veined, margins and apices ciliate; **lower lemmas** about as long as the spikelets, lanceolate, 7-veined, lateral 3 veins on each side unequally or equally spaced, smooth or scabrous over the distal ⅓, lemma margins and the region between the second and third veins densely pubescent, hairs 0.05–0.1 mm, white; **upper lemmas** slightly shorter than the lower lemmas, yellowish or grayish when immature, becoming light brown at maturity, minutely striate. **Caryopses** about 1.8 mm, tan. $2n$ = 36.

Digitaria horizontalis is native to tropical regions of the Americas. It has been found in hammocks and disturbed areas in central and southern Florida and at a few other locations in the southeastern United States, including ballast dumps in Mobile, Alabama. It is probably a recent introduction to the Flora region, even in Florida.

27. **Digitaria nuda** Schumach. [p. 379]

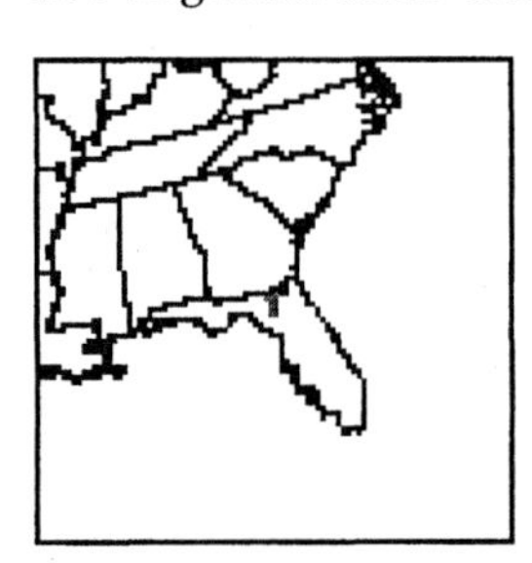

Plants annual or of indefinite duration. **Culms** 20–60 cm, glabrous, decumbent, rooting and branching from the lower nodes, geniculate above. **Sheaths** glabrous or with long hairs near the base; **ligules** 0.8–2.5 mm; **blades** 2–13.5 cm long, 1.5–2.5 mm wide, glabrous on both surfaces or the adaxial surface with a few long hairs near the base. **Panicles** with 3–8 spikelike primary branches, these digitate or with rachises to 2 cm long; **lower panicle nodes** with hairs at least 0.4 mm; **primary branches** 4–15.5(20) cm long, 0.4–0.8 mm wide, axes wing-margined, wings more than ½ as wide as the midribs, proximal portions of the branches often with scattered 1–4 mm hairs, bearing spikelets in unequally pedicellate pairs on the lower and middle portions of the branches; **secondary branches** absent; **pedicels** not adnate to the branches. **Spikelets** homomorphic, 1.7–2.8 mm long, 0.5–0.8 mm wide. **Lower glumes** absent or to 0.2 mm; **upper glumes** 1–2.2 mm, 0.4–0.8 times as long as the spikelets; **lower lemmas** about as long as the spikelets, 7-veined, veins

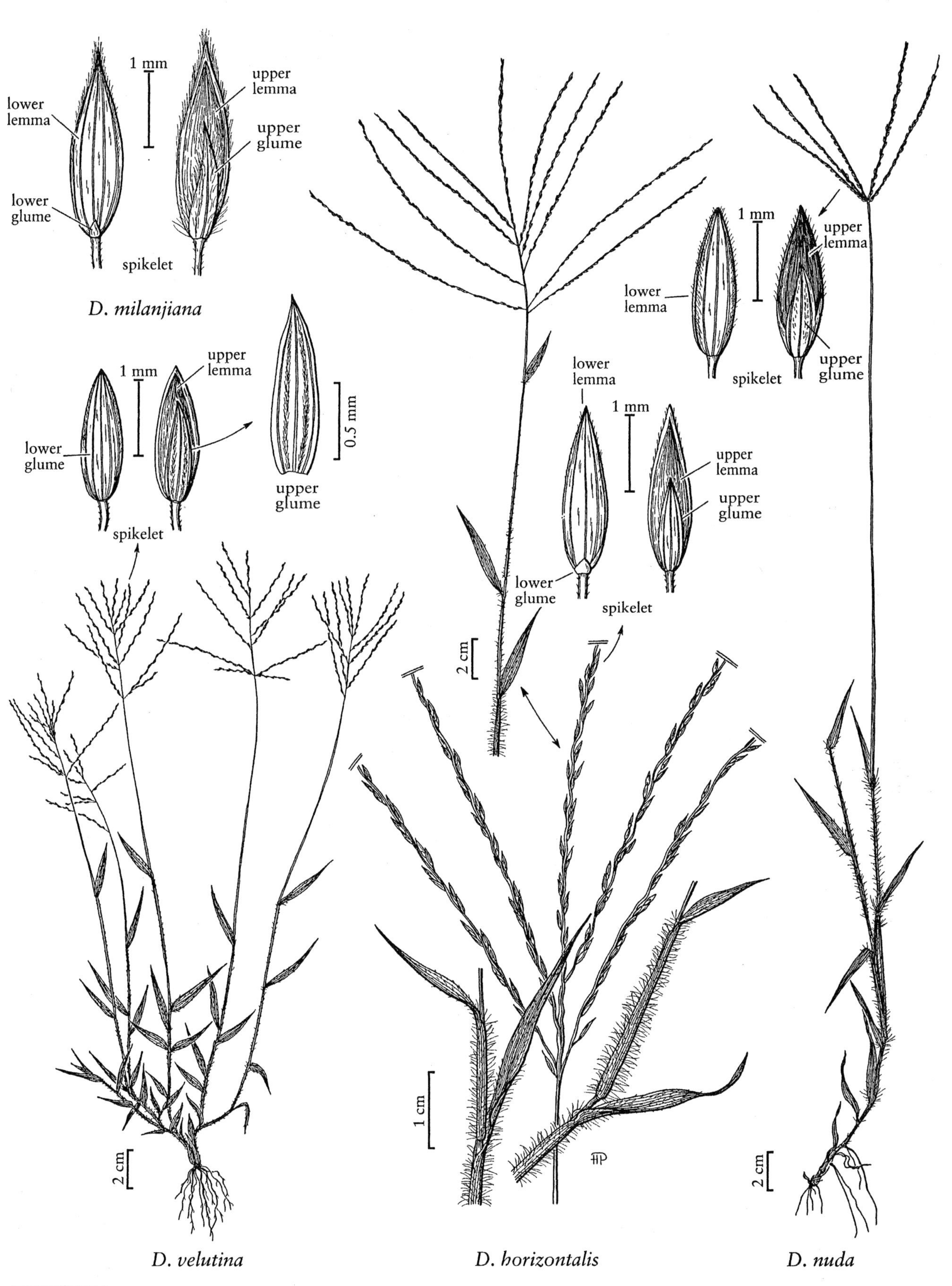
lower lemma
lower glume
1 mm
upper lemma
upper glume
spikelet
D. milanjiana
1 mm
upper lemma
lower glume
0.5 mm
upper glume
spikelet
2 cm
D. velutina
lower lemma
1 mm
upper lemma
upper glume
lower glume
spikelet
2 cm
1 cm
D. horizontalis
1 mm
upper lemma
lower lemma
upper glume
spikelet
2 cm
D. nuda

smooth, lateral veins usually equally spaced, sometimes the inner lateral veins more distant from the other 2, intercostal regions adjacent to the midveins glabrous, those between the lateral veins with 0.5–1 mm hairs, hairs initially appressed, sometimes strongly divergent at maturity; **upper lemmas** yellow to gray when immature, becoming brown at maturity; **anthers** 0.3–0.6 mm. $2n$ = unknown.

Digitaria nuda is an African species that is now established in tropical regions throughout the world, including the Americas. So far as is known, it has only been collected once in the *Flora* region, in Columbia County, Florida.

28. Digitaria sanguinalis (L.) Scop. [p. 381]
Hairy Crabgrass, Digitaire Sanguine

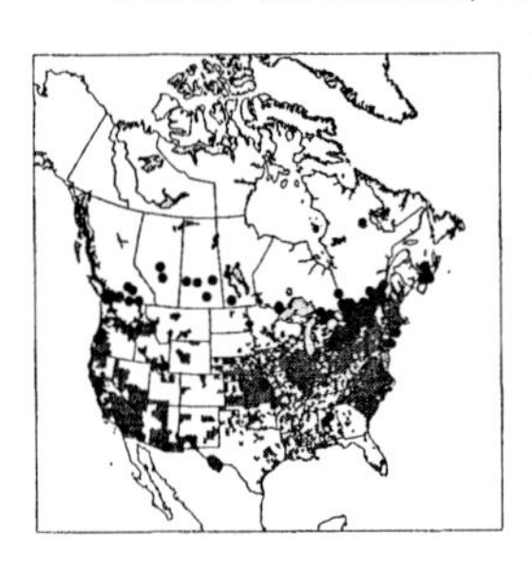

Plants annual. **Culms** 20–70(112), often decumbent and rooting at the lower nodes. **Sheaths** keeled, usually sparsely pubescent with papillose-based hairs; **ligules** 0.5–2.6 mm; **blades** 2–11(14) cm long, 3–8(12) mm wide, usually with papillose-based hairs on both surfaces, sometimes glabrous. **Panicles** with 4–13 spikelike primary branches, these subdigitate or on rachises to 6 cm; **primary branches** 3–30 cm long, 0.7–1.5 mm wide, flattened and winged, wings more than ½ as wide as the midribs, lower and middle portion of the branches bearing spikelets in unequally pedicellate pairs, pedicels not adnate to the branches; **secondary branches** rarely present. **Spikelets** homomorphic, 1.7–3.4 mm long, 0.7–1.1 mm wide. **Lower glumes** 0.2–0.4 mm long, veinless; **upper glumes** 0.9–2 mm, ⅓–½ as long as the spikelets, 3-veined, pubescent on the margins; **lower lemmas** usually exceeded or equaled by the upper florets, sometimes exceeding them but by no more than 0.2 mm, glabrous, 7-veined, lateral (or all) veins scabrous throughout or smooth on the lower ⅓(½) and scabrous distally, 3 middle veins usually widely spaced, remaining veins on each side close together and near the margins; **upper lemmas** 1.7–3 mm, yellow or gray, frequently purple-tinged when immature, often becoming brown at maturity; **anthers** 0.5–0.9 mm. $2n$ = 36, 28, 34, 54.

Digitaria sanguinalis is a weedy Eurasian species that is now found in waste ground of fields, gardens, and lawns throughout much of the world, including the *Flora* region.

29. Digitaria bicornis (Lam.) Roem. & Schult. [p. 381]

Plants of indefinite duration; sometimes stoloniferous. **Culms** with erect portion 10–85 cm, long-decumbent, rooting and branching at the lower nodes. **Sheaths** with papillose-based hairs or the upper sheaths glabrous; **ligules** 1–4 mm; **blades** 3–14 cm long, 2–9 mm wide, mostly glabrous but the adaxial surfaces with papillose-based hairs basally. **Panicles** with (2)3–6 spikelike primary branches, these digitate or a few solitary branches below; **lowest nodes** glabrous or with hairs less than 0.4 mm; **primary branches** 6.5–21 cm long, 0.6–1.3 mm wide, axes winged, wings at least ½ as wide as the midribs, lower and middle portions bearing spikelets in unequally pedicellate pairs, pedicels not adnate to the branches; **secondary branches** absent; **shorter pedicels** about 0.2 mm; **longer pedicels** to 2 mm. **Spikelets** 2.6–3.7 mm, spikelet pairs dimorphic in their pubescence and venation pattern of the lower lemmas. **Lower glumes** absent or to 0.9 mm, deltoid or bifid; **upper glumes** 1.7–2.8 mm, ½–¾ as long as the spikelets, 3-veined; **lower lemmas** 7-veined, veins smooth; **lower lemmas of shortly pedicellate spikelets** with 3 equally spaced, glabrous or shortly pubescent central veins, lemma margins and the region between the 2 lateral veins with appressed or spreading, 0.5–1 mm hairs; **lower lemmas of long-pedicellate spikelets** with unequally spaced veins, midvein well-separated from the 3 lateral veins, lateral veins crowded together near the margins, lemma margins and the region between the 2 inner lateral veins hairy with appressed or strongly divergent, 1–2 mm hairs, sometimes also with longer, glassy yellow hairs; **upper lemmas** of all spikelets usually yellow or gray, sometimes light brown, at maturity; **anthers** 0.5–0.6 mm. $2n$ = 54, 72.

Digitaria bicornis is a common species on the sandy coastal plain of the southeastern United States. Its range extends through Mexico to Costa Rica and northern South America, as well as to the West Indies. The Californian record reflects a 1926 collection; the species is not known to be established in the state.

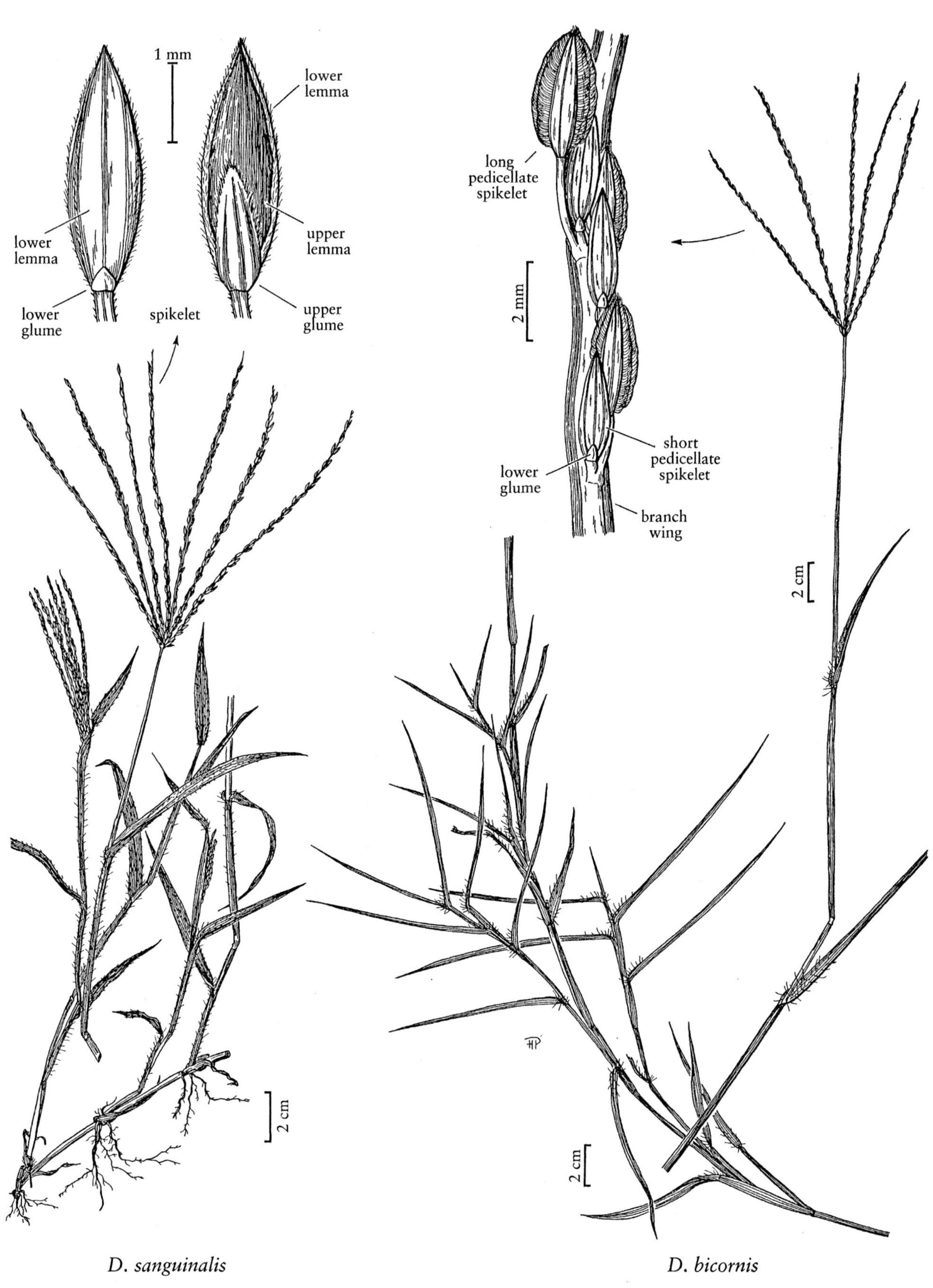

D. sanguinalis

D. bicornis

DIGITARIA

30. Digitaria ciliaris (Retz.) Koeler [p. 383]
SOUTHERN CRABGRASS

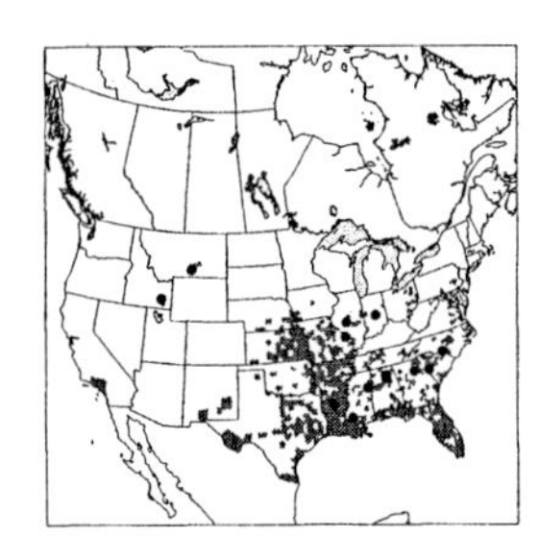

Plants annual or of indefinite duration. **Culms** 10–100 cm long, erect portion 30–60 cm, long-decumbent, rooting and branching at the decumbent nodes, sparingly branched or unbranched from the upper nodes; **nodes** 2–5, glabrous. **Sheaths** with papillose-based hairs; **ligules** 2–3.5 mm, erose; **blades** 1.5–14.4(18.9) cm long, 3–9 mm wide, flat, glabrous, a few scattered papillose-based hairs at the base of the adaxial surfaces (occasionally over the whole adaxial surface), usually also scabrous on both surfaces. **Panicles** with 2–10 spikelike primary branches, these digitate or in 1–3 whorls on rachises to 2 cm; **lowest panicle nodes** with hairs more than 0.4 mm; **primary branches** 3–24 cm long, 0.6–1.2(2) mm wide, glabrous or with less than 1 mm hairs, axes wing-margined, wings at least ½ as wide as the midribs, lower and middle portions of the branches bearing spikelets in unequally pedicellate pairs; **secondary branches** absent; **shorter pedicels** 0.5–1 mm; **longer pedicels** 1.5–4 mm. **Spikelets** (2.7)2.8–4.1 mm long, homomorphic. **Lower glumes** 0.2–0.8 mm, acute; **upper glumes** (1.2)1.5–2.7 mm, about ⅔ to almost as long as the spikelet, 3-veined, margins and apices pilose; **lower lemmas** 2.7–4.1 mm, 7-veined, veins unequally spaced, outer 3 veins crowded together near each margin, well-separated from the midvein, usually smooth, occasionally the lateral veins scabridulous on the distal ⅓, margins and regions between the 2 inner lateral veins hairy, hairs 0.5–1 mm (rarely glabrous), sometimes also with glassy yellow hairs between the 2 inner lateral veins, these more common on the upper spikelets; **upper lemmas** 2.5–4 mm, glabrous, yellow, tan, or gray when immature, becoming brown, often purple-tinged (occasionally completely purple) at maturity; **anthers** 0.6–1 mm. $2n = 54$.

Digitaria ciliaris is a weedy species, found in open, disturbed areas in most warm-temperate to tropical regions, primarily in the eastern United States. It is particularly abundant in the Southeast. So far as is known, the two varieties distinguished in the following key do not differ in any other characters. They are recognized here pending further study.

1. Lower lemmas without glassy yellow hairs . . . var. *ciliaris*
1. Lower lemmas with glassy yellow hairs . . . . var. *chrysoblephara*

Digitaria ciliaris (Retz.) Koeler var. **ciliaris** is the more common of the two varieties in the *Flora* region. **Digitaria ciliaris** var. **chrysoblephara** (Fig. & De Not.) R.R. Stewart is more common in southeast Asia, but it has been found in the northeastern United States.

31. Digitaria setigera Roth [p. 383]

Plants of indefinite duration. **Culms** to 120 cm tall, bases long-decumbent and rooting at the lower nodes. **Sheaths** with papillose-based hairs; **ligules** 2.5–3.5 mm; **blades** 4–28 cm long, 4–12 mm wide, scabrous, usually with some scattered papillose-based hairs on the base of the adaxial surfaces, sometimes with hairs all over. **Panicles** with 3–11 spikelike primary branches in 1–several whorls, rachises to 6 cm; **primary branches** 5–15 cm, axes wing-margined, wings more than ½ as wide as the midribs, lower and middle portions bearing spikelets in unequally pedicellate pairs; **secondary branches** absent; **shorter pedicels** 0.3–0.8 mm; **longer pedicels** 1.7–2.7 mm. **Spikelets** 2.4–3.5 mm, homomorphic, ovate. **Lower glumes** absent or to 0.1 mm; **upper glumes** 0.2–1.3 mm, ⅙–⅓ as long as the spikelets, 1–3-veined, margins and apices with appressed, white hairs about 0.5 mm, truncate or bilobed; **lower lemmas** (5)7-veined, veins smooth or scabrous only over the distal ⅓, unequally spaced, margins and lateral intercostal regions silky-ciliate; **upper lemmas** tan or gray when immature, brown at maturity, acuminate; **anthers** 0.6–1.3 mm. $2n = 70, 72$.

Digitaria setigera is native to southeastern Asia. It is now established in tropical America, growing in disturbed habitats in Florida and Central America, and probably in tropical South America. It has often been confused with *D. sanguinalis*.

Plants in the *Flora* region belong to **Digitaria setigera** Roth var. **setigera**. Unlike plants of **D. setigera** var. **calliblepharata** (Henrard) Veldkamp, they do not have large, glassy hairs on their lower lemmas.

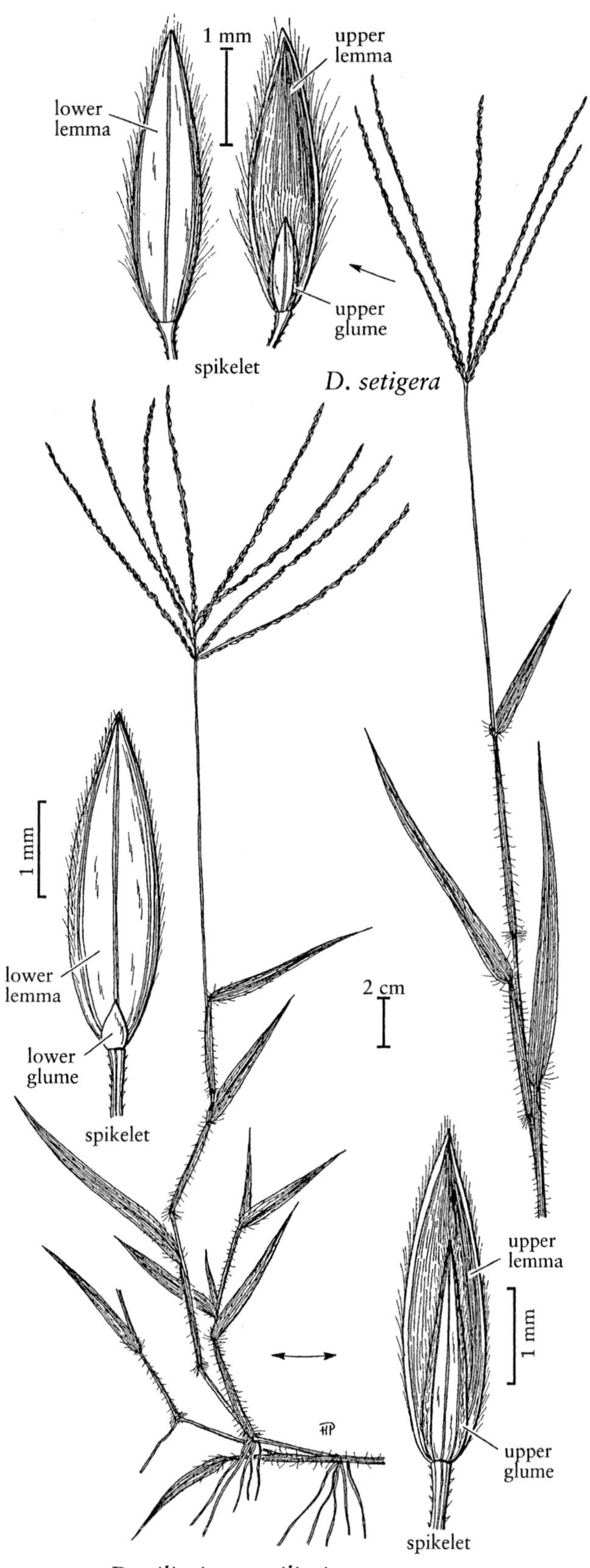
1 mm
upper
lemma
lower
lemma
upper
glume
spikelet
D. setigera
1 mm
lower
lemma
lower
glume
spikelet
2 cm
upper
lemma
1 mm
upper
glume
spikelet
D. ciliaris var. ciliaris

DIGITARIA

25.02 ANTHENANTIA P. Beauv.

J.K. Wipff

Plants perennial; rhizomatous. **Culms** 60–120 cm, stiffly erect, clustered. **Sheaths** open; **ligules** 0.1–0.3 mm, membranous, ciliate; **blades** flat, stiff, upper blades much reduced. **Inflorescences** terminal, simple panicles with elongate rachises; **branches** ascending to erect, not spikelike, lower branches usually with 8 or more spikelets; **disarticulation** beneath the glumes. **Spikelets** 3–4 mm, laterally compressed neither subtended by bristles nor sunken into the rachis, elliptic, obovoid, hirsute, unawned, with 2 florets. **Lower glumes** absent; **upper glumes** as long as the spikelets, obovate, densely hirsute, 5-veined; **lower florets** sterile or staminate; **lower lemmas** densely hirsute, similar to the upper glumes; **lower paleas** present or absent; **upper lemmas** and **upper paleas** cartilaginous, glabrous, dark brown, separating slightly at maturity, exposing the caryopses, lemmas 3-veined, margins flat, hyaline, 0.5–1 mm wide, paleas 2-veined. x = 10. Name from the Greek *anthos*, 'flower', and *enathos*, 'contrary', the spikelet having given Palisot de Beauvois some problems in interpretation.

Anthenantia is a genus of two species, both of which are endemic to the southeastern United States. It is very similar to *Leptocoryphium* Nees, a monotypic genus that extends from Mexico to Argentina, but differs in having laterally, rather than dorsally, compressed spikelets.

Anthenantia is the etymologically correct version of the three alternative spellings used by Palisot de Beauvois (Clayton and Renvoize 1986).

SELECTED REFERENCE **Clayton, W.D.** and **S.A. Renvoize.** 1986. Genera Graminum: Grasses of the World. Kew Bull., Addit. Ser. 13. Her Majesty's Stationery Office, London, England. 389 pp.

1. Junction of the sheath and blade not evident abaxially, the sheath and blades in line with each other; blades 2–5 mm wide, the margins scabrous; leaves 30–60 cm long . . . . 1. *A. rufa*
1. Junction of the sheath and blade evident, the blades not in line with the sheaths; blades 5–10 mm wide, the margins papillose-hispid; leaves mostly less than 30 mm long . . . . 2. *A. villosa*

1. Anthenantia rufa (Elliott) Schult. [p. 386]
PURPLE SILKYSCALE

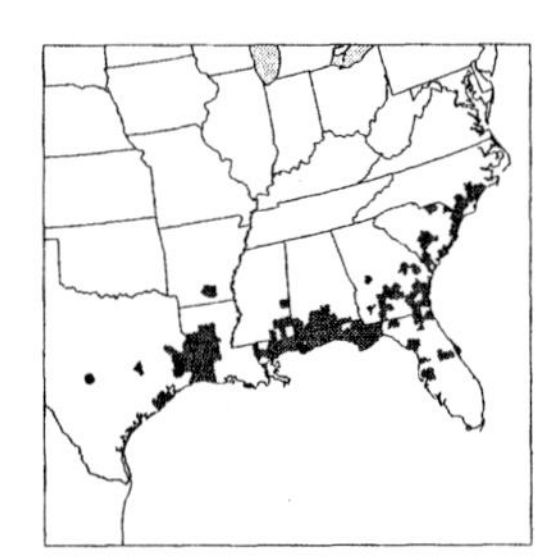

Culms 60–120 cm, from knotty rhizomes. **Leaves** 30–60 cm, junction of the sheath and blade inconspicuous abaxially; **blades** 2–5 mm wide, in line with the sheaths, margins scabrous, otherwise glabrous or the adaxial surfaces inconspicuously pubescent to hirsute. **Panicles** 8–16 cm long, 2–3(5) cm wide; **branches** spreading to erect. **Spikelets** 3–4 mm long, 1.3–1.8 mm wide. **Upper glumes** and **lower lemmas** pubescent, hairs 0.6–1.5 mm, eventually spreading, usually purple-tinged; **anthers** 1–1.5 mm. **Caryopses** 1.2–1.4 mm. $2n$ = 20.

Anthenantia rufa grows in wet pine flatwoods and savannahs, sphagnous streamhead ecotones, and pitcher plant bogs on the southeastern coastal plain from eastern Texas to North Carolina.

2. Anthenantia villosa (Michx.) P. Beauv. [p. 386]
GREEN SILKYSCALE

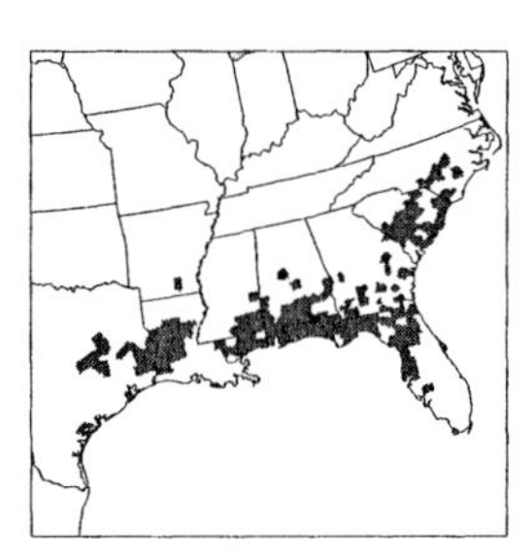

Culms 70–110 cm, erect from scaly rhizomes. **Leaves** mostly shorter than 30 cm, junction of the sheath and blade evident; **blades** 5–10 mm wide, at an angle to the sheaths, surfaces glabrous or the adaxial surfaces hirsute, margins hispid, with papillose-based hairs. **Panicles** 8–16 cm long, 1–3 cm wide; **branches** stiffly to loosely erect. **Spikelets** 3–4 mm long, 0.7–1.5 mm wide. **Upper glumes** and **lower lemmas** pubescent, hairs 0.5–1 mm, appressed or spreading, usually colorless or white; **anthers** 1–1.8 mm. **Caryopses** 1.2–1.4 mm. $2n$ = 20.

Anthenantia villosa grows in dry, usually sandy soil on the southeastern coastal plain from eastern Texas to North Carolina. It usually grows in wetter habitats in southeastern Texas than in other portions of its range.

25.03 ALLOTEROPSIS J. Presl

David W. Hall

Plants annual or perennial; cespitose. **Culms** 15–150 cm, erect or decumbent and rooting at the lower nodes, erect distally; **nodes** glabrous or pubescent. **Leaves** mostly basal; **sheaths** stiffly pubescent; **ligules** of hairs or membranous and ciliate; **blades** linear to lanceolate. **Inflorescences** terminal, panicles of 4–11 digitate or subdigitate, spikelike branches; **branches** naked at the base, spikelets in pairs or triplets, lower lemmas and lower glumes closest to the branch axes; **disarticulation** below the glumes. **Spikelets** 2.5–7 mm, dorsally compressed, with 2 florets. **Glumes** acute to shortly awned; **lower glumes** about ½ as long as the spikelets, 1–2-veined; **upper glumes** equaling the spikelets, 3–5-veined, margins ciliate; **lower florets** staminate; **lower lemmas** papery; **lower paleas** reduced, deeply bifid, 1-veined; **upper florets** bisexual; **upper lemmas** stiffly membranous, margins involute, apices attenuate into a mucro or curved awn; **anthers** 3. **Caryopses** elliptic. $x = 9$. Name from the Greek *allotrios*, 'belonging to another' and *opsis*, 'appearance,' the spikelets and inflorescence resembling those of *Panicum*.

Alloteropsis is a genus of five to eight species that are native to tropical Asia and Australia. One species is now established in the *Flora* region. *Alloteropsis semialata* (R. Br.) Hitchc. is an important forage species in other parts of the world.

1. Alloteropsis cimicina (L.) Stapf [p. 386]
BUGSEED GRASS

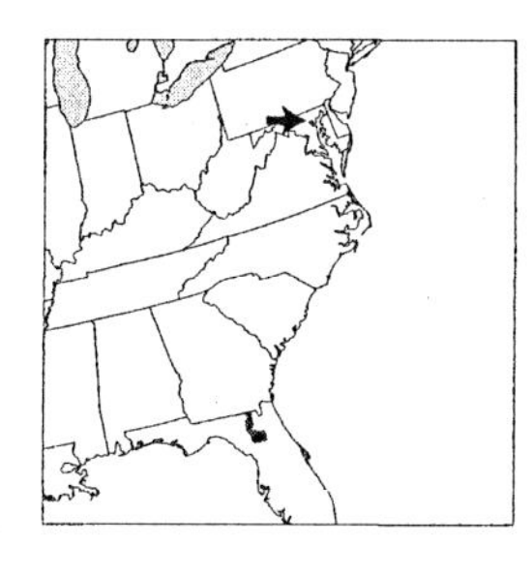

Culms 15–75 cm. **Sheaths** hispid; **ligules** to 2 mm, membranous and ciliate; **blades** to 8 cm long, 15–25 mm wide, flat, cordate, hispid abaxially, glabrous adaxially, margins with papillose-based hairs. **Panicles** with 4–7 digitate branches; **branches** 2–16 cm. **Spikelets** 2.7–3.2 mm, subsessile; **upper lemmas** awned, awns 0.4–1.2 mm. $2n = 36$.

Alloteropsis cimicina is native to southeast Asia but has been collected in Alachua and Columbia counties, Florida, and Baltimore, Maryland. Being a weedy species, it should be sought at other disturbed locations along the east coast of the United States.

25.04 AMPHICARPUM Kunth

J.K. Wipff

Plants annual or perennial; rhizomatous, rhizomes slender, terminating in a reduced panicle of cleistogamous spikelets. **Culms** 30–100 cm, erect or decumbent. **Sheaths** open; **auricles** absent; **ligules** of hairs; **blades** flat. **Inflorescences** subterranean and aerial, only the subterranean inflorescences forming mature caryopses; **subterranean panicles** with 1–5 spikelets; **aerial panicles** terminal, simple, with elongate rachises bearing erect to ascending branches, usually with 15 or more spikelets. **Spikelets** glabrous, unawned, with 2 florets. **Subterranean spikelets** setting seed, with 1 glume; **lower glumes** absent; **upper glumes** and **lower lemmas** similar in size and texture, exceeded by the upper florets; **upper florets** turgid, ellipsoidal; **upper lemmas** mostly indurate, margins thin, flat, apices acuminate; **upper paleas** similar in texture to the lemmas; **anthers** 3; **caryopses** well-developed. **Aerial spikelets** not setting seed, sometimes forming immature caryopses, lanceoloid, dorsally compressed to terete; **glumes** unequal or the lower glumes absent; **upper glumes** and **lower lemmas** similar in size and texture; **upper lemmas** mostly indurate, margins thin, flat, apices acute; **lower florets** staminate or sterile; **upper florets**

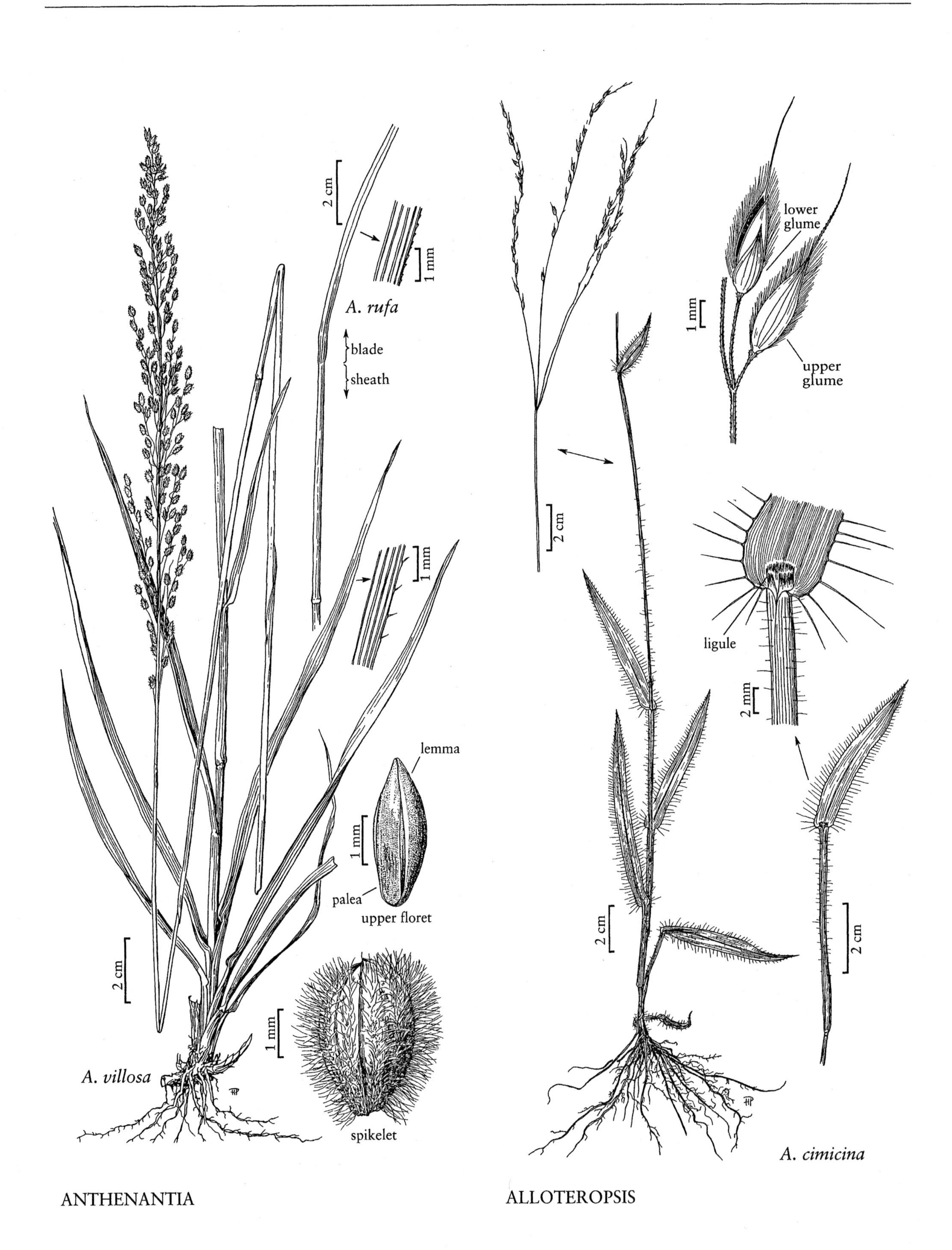

ANTHENANTIA

ALLOTEROPSIS

with pistils but fruit not developed. x = 9. Name from the Greek *amphikarpos*, 'doubly fruit-bearing', a reference to the two kinds of spikelets.

Amphicarpum is a genus of two species, both endemic to the southeastern United States. It differs from all other North American grass genera in its production of subterranean, cleistogamous spikelets. The aerial spikelets occasionally have immature caryopses but, for some unknown reason, these never mature.

1. Leaf blades conspicuously hirsute; plants annual; culms erect . . . . 1. *A. amphicarpon*
1. Leaf blades glabrous or almost glabrous; plants perennial; culms usually decumbent . . . . . 2. *A. muhlenbergianum*

1. Amphicarpum amphicarpon (Pursh) Nash [p. 388]
PURSH'S BLUE MAIDENCANE, HAIRY MAIDENCANE

Plants annual. **Culms** 30–80 cm, erect. **Leaves** mostly basal; **sheaths** hirsute; **blades** 10–15 cm long, 5–15 mm wide, hirsute on both surfaces, margins ciliate. **Subterranean spikelets** 7–8 mm, acuminate. **Aerial panicles** 3–20 cm; **aerial spikelets** 4–5 mm, ellipsoidal. $2n$ = 18.

Amphicarpum amphicarpon grows in sandy pinelands of the eastern United States. It used to be known as *A. purshii* Kunth, but *A. amphicarpon* has priority.

2. Amphicarpum muhlenbergianum (Schult.) Hitchc. [p. 388]
BLUE MAIDENCANE

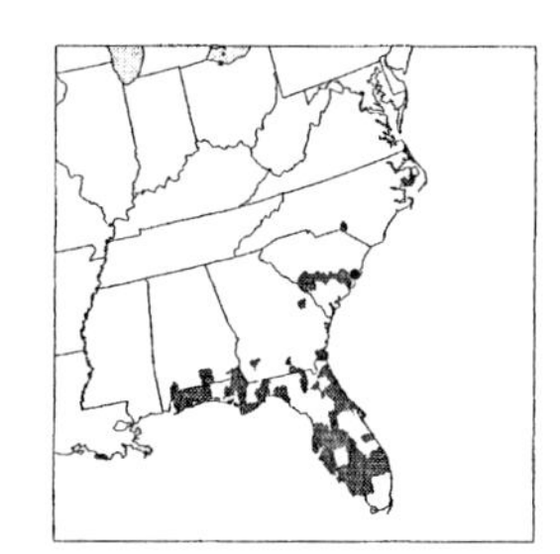

Plants perennial. **Culms** 30–100 cm, usually decumbent, sometimes erect. **Leaves** evenly distributed; **sheaths** usually glabrous, occasionally sparsely hirsute; **blades** to 10(13) cm long, 5–10.5 mm wide, glabrous, margins white. **Subterranean spikelets** 6–9 mm, acuminate. **Aerial panicles** 3–20 cm; **aerial spikelets** 5.5–7 mm, narrowly lanceoloidal. $2n$ = 18.

Amphicarpum muhlenbergianum grows in damp areas, such as dried pond bottoms, ditches, flatwoods, and swampy pinewoods of the southeastern United States.

25.05 LASIACIS (Griseb.) Hitchc.

Gerrit Davidse

Plants perennial (rarely annual); cespitose. **Culms** 0.5–8 m, weakly lignified, erect, arching, climbing, or decumbent, rooting at the nodes. **Sheaths** open; **ligules** membranous, sometimes ciliate; **pseudopetioles** sometimes present; **blades** linear to ovate, bases slightly to strongly asymmetric. **Inflorescences** open or contracted panicles, rachises usually visible, even distally, spikelets attached obliquely to the pedicels; **disarticulation** below the glumes. **Spikelets** subglobose to globose, with 2 florets. **Glumes** membranous, apices lanate pubescent, abruptly apiculate; **lower glumes** ⅓–⅔ as long as the spikelets, 5–13-veined, bases saccate, margins overlapping; **upper glumes** about as long as the upper florets, not saccate, 7–15-veined; **lower florets** sterile or staminate; **lower lemmas** membranous, apices lanate pubescent, abruptly apiculate; **lower paleas** present, sometimes reduced; **upper florets** stipitate, bisexual, appearing to be mucronate or acuminate; **upper lemmas** indurate, usually broadly elliptic to obovate, margins enclosing the edges of the paleas, apices obtuse, somewhat woolly pubescent, usually dark brown at maturity; **upper paleas** similar to the lemmas, but saccate below and gibbous above. **Caryopses** plano-convex, ovoid, or nearly orbicular; **embryo** about ½ as long as the caryopses; **hila** oblong to nearly round. x = 9. Name from the Greek *lasios*, 'woolly', and *akis*, 'point', referring to the tuft of wool at the apex of the spikelets.

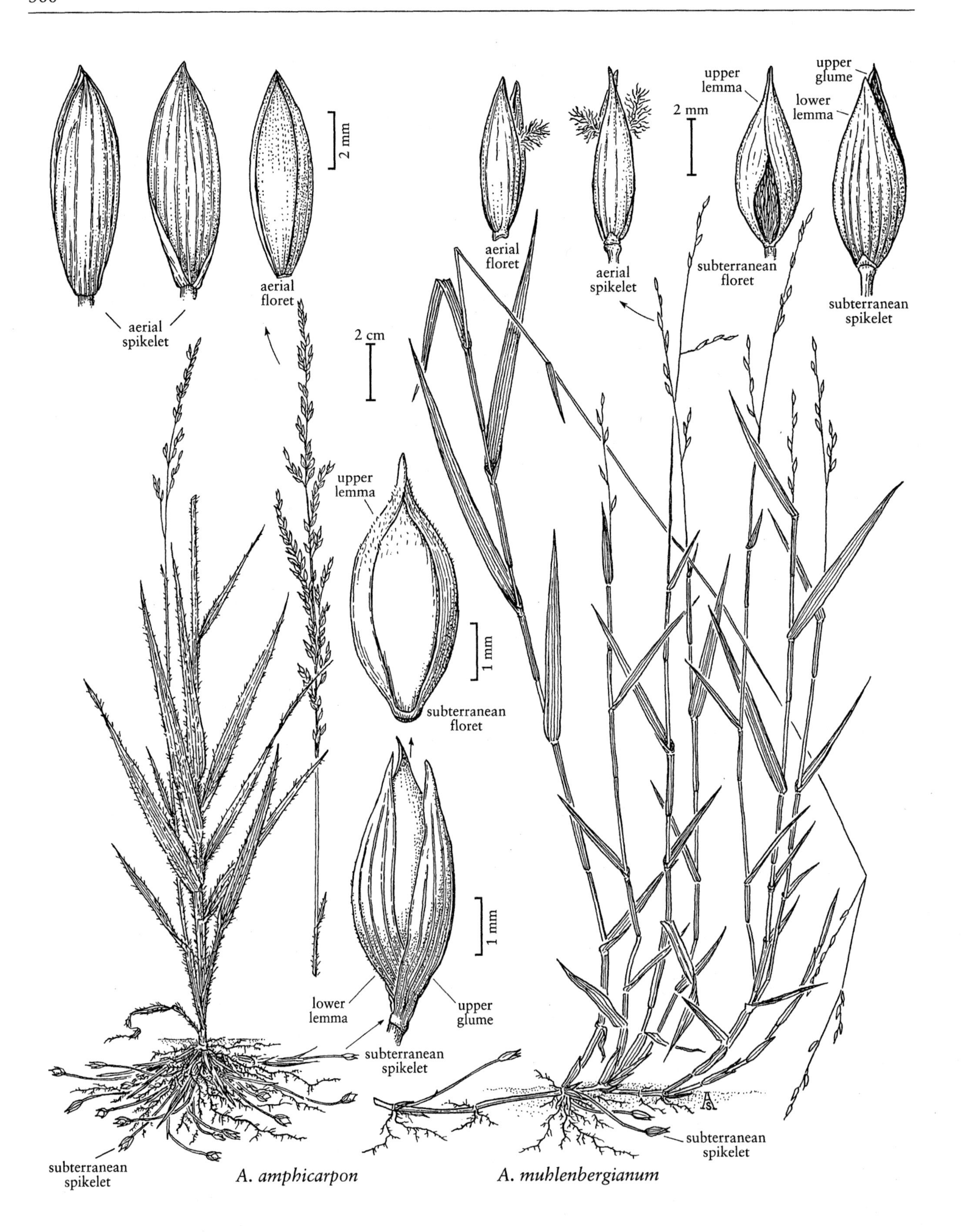

AMPHICARPUM

In *Lasiacis* the upper florets appear to be mucronate or acuminate. The mucro or acuminate apex is formed by the tuft of hairs at the apex of the upper floret.

Lasiacis is a neotropical genus of 16 species that extends from southern Florida to Peru and Argentina. Two species are native to the *Flora* region. The shiny black color of its mature florets and the oil-filled cells of the inner epidermes of the glumes and sterile lemmas distinguish *Lasiacis* from all other grasses. Birds are a common dispersal agent.

SELECTED REFERENCES Davidse, G. 1978. A systematic study of the genus *Lasiacis* (Gramineae: Paniceae). Ann. Missouri Bot. Gard. 65:1133–1254; Davidse, G. and E. Morton. 1973. Bird-mediated fruit dispersal in the tropical grass genus *Lasiacis* (Gramineae: Paniceae). Biotropica 5:162–167.

1. Leaf blades linear-lanceolate to narrowly lanceolate, 3–16 cm long, 3–30 mm wide . . . . 1. *L. divaricata*
1. Leaf blades ovate to broadly lanceolate, 2–16 cm long, 8–56 mm wide . . . . 2. *L. ruscifolia*

1. **Lasiacis divaricata** (L.) Hitchc. [p. 391]

Plants perennial; cespitose. **Culms** (0.5)1–5(7) m long, arching, clambering over brush or bending to the ground if unsupported. **Sheaths** mostly glabrous, margins and throats ciliate; **ligules** 0.8–13 mm, glabrous or ciliate; **blades** (3)5–12(16) cm long, 3–20(30) mm wide, linear-lanceolate to narrowly lanceolate, lower blades often deciduous. **Panicles** 2–12(20) cm; **branches** to 12 cm, with widely-spaced spikelets; **lower branches** usually reflexed; **upper branches** widely spreading. **Spikelets** 3.5–4.5 mm, obovate. **Lower glumes** 1.2–2.5 mm; **lower florets** sterile; **upper florets** 3.4–4 mm long, 1.9–2.4 mm wide, whitish to brown at maturity; **anthers** about 2 mm, white; **stigmas** purple. **Caryopses** whitish. $2n = 36$.

Lasiacis divaricata is a Caribbean species. Its range extends from Florida through the West Indies to Mexico, Panama, and northern Venezuela. In Florida, it usually grows in hammocks, but occasionally in pinelands. The whitish to brown upper florets are unusual in the genus.

Plants in the *Flora* region belong to **Lasiacis divaricata** var. **divaricata**, which differs from the other two varieties in having panicles with fewer spikelets and panicle branches that are usually reflexed.

2. **Lasiacis ruscifolia** (Kunth) Hitchc. [p. 391]

Plants perennial; cespitose. **Culms** 1–8 m long, 5–12 mm thick, hollow, arching or clambering. **Sheaths** glabrous, puberulent, or hispid with papillose-based hairs, margins ciliate; **ligules** 0.2–1 mm, glabrous or ciliate; **blades** 2–16 cm long, 8–56 mm wide, ovate to ovate-lanceolate. **Panicles** 2–22 cm; **lower branches** to 9 cm, divergent. **Spikelets** 2.6–4 mm, globose; **lower glumes** 1.2–2.2 mm, 9–13-veined; **lower florets** sterile; **upper glumes** 11–13-veined; **upper florets** 2.8–3.6 mm long, 2–2.9 mm wide, dark brown to grayish-black at maturity; **upper lemmas** usually with a distinct shelf at the base, from which a sterile projection often arises; **upper paleas** usually deeply concave; **anthers** 1.4–2.3 mm, white; **stigmas** white. **Caryopses** 2–2.5 mm. $2n = 36$.

The range of *Lasiacis ruscifolia* extends from southern Florida to Peru. Plants in the *Flora* region belong to **L. ruscifolia** (Kunth) Hitchc. var. **ruscifolia**, which differs from **L. ruscifolia** var. **velutina** (Swallen) Davidse in being scabrous or puberulent, rather than velutinous, on the panicle branches.

25.06 **OPLISMENUS** P. Beauv.

J.K. Wipff

Plants annual or perennial. **Culms** 10–100 cm, weak, trailing on the ground, branching. **Leaves** cauline; **ligules** membranous and ciliate, or of hairs; **blades** lanceolate. **Inflorescences** terminal, panicles of unilateral branches, spikelets paired (but the first spikelet sometimes reduced), rachises and branches terminating in a spikelet; **branches** 0.1–7 cm, persistent; **disarticulation**

below the glumes. **Spikelets** dorsally compressed, not sunken into the rachis, lacking subtending bristles, with 2 florets. **Lower glumes** awned; **upper glumes** not ciliate on the margins, unawned or with awns shorter than those of the lower glumes, awns of both glumes often becoming viscid; **lower florets** sterile or staminate; **lower lemmas** acute to shortly awned; **lower paleas** present or absent; **upper florets** bisexual; **upper lemmas** papery to leathery, glabrous, smooth, unawned, white or yellow at maturity; **upper paleas** similar to the upper lemmas. $x = 9$. Name from the Greek *hoplismenos*, 'armed', alluding to the awned spikelets.

Oplismenus is a genus of five closely related species that grow in shady, mesic forests of tropical and subtropical regions. One species is native to the *Flora* region. The awns of most species become viscid at maturity, aiding in fruit dispersal (Davidse 1987).

SELECTED REFERENCES **Davey, J.C.** and **W.D. Clayton**. 1977. Some multiple discriminant function studies on *Oplismenus*. Kew Bull. 33:147–157; **Davidse, G.** 1987. Fruit dispersal in Poaceae. Pp. 143–155 *in* T.R. Soderstrom, K.W. Hilu, C.S. Campbell, and M.E. Barkworth (eds.). Grass Systematics and Evolution. Smithsonian Institution Press, Washington, D.C., U.S.A. 473 pp.; **Peterson, P.M., E.E. Terrell, C.A. Davis, H. Scholz**, and **R.J. Soreng**. 1999. *Oplismenus hirtellus* subspecies *undulatifolius*, a new record for North America. Castanea 64:201–202; **Scholz, U.** 1981. Monographie der Gattung *Oplismenus* (Gramineae). Phanerog. Monog. Tomus XIII. J. Cramer, Vaduz, Germany. 217 pp.

1. **Oplismenus hirtellus** (L.) P. Beauv. [p. 391]
BRISTLE BASKETGRASS

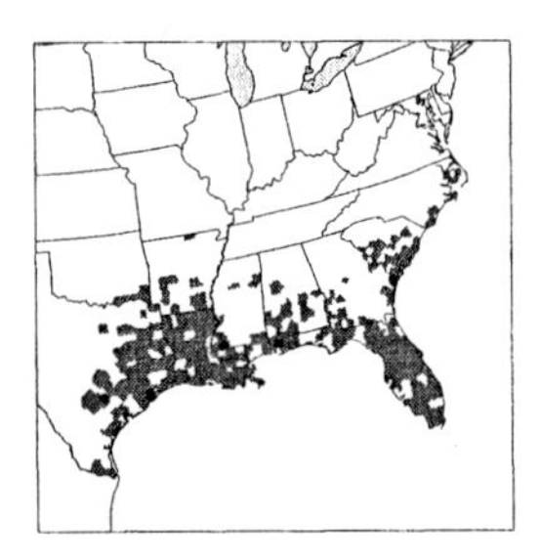

Plants perennial. **Culms** to 90 cm, mostly decumbent and rooting at the nodes, distal 15–35 cm ascending when flowering; **nodes** usually pubescent; **internodes** usually pubescent along 1 side (sometimes glabrous). **Sheaths** conspicuously ciliate on the margins; **ligules** 0.4–1.6(2.3) mm, ciliate; **blades** (0.6)1.3–11.5 cm long, 2–20 mm wide, glabrous, scabrous, or pubescent, margin(s) usually undulating. **Panicles** (1.5)2.5–16.5 cm, with 2–10 primary branches; **branches** 0.1–2.5 cm. **Spikelets** 2.2–3.5(4.5) mm; **calluses** shortly pubescent at the base **awns** usually purplish. **Lower glumes** 1.5–3 mm, scabridulous and/or pubescent, 3–5-veined, awns (1.6)3.2–14.5 mm; **upper glumes** 1.5–2.5 mm, sparsely pubescent, 5–7-veined, awns 0.8–6(10) mm; **lower florets** usually sterile, occasionally staminate; **lower lemmas** 2.2–3.1 mm, sparsely pubescent above, (5)7–9-veined, awns 0.1–1.2 mm; **lower paleas** absent or to 2.3 mm, hyaline; **upper lemmas** (2.1)2.3–3 mm, glabrous, weakly cartilaginous, white to cream-colored; **anthers** 3, 1.3–1.7 mm. **Caryopses** 1.7–1.8 mm long, 0.5–0.9 mm wide, glabrous. $2n$ = 36, (54), 72, (90).

Oplismenus hirtellus grows at scattered locations in the southeastern United States, extending south in subtropical and tropical habitats to Argentina. Scholz (1981) recognized 11 subspecies and two forms within the species, but they overlap, both morphologically and geographically. The key below is included for convenience. It includes the three subspecies attributed to the *Flora* region. In addition, a variegated form cultivated as an ornamental is sold as *Panicum variegatum*.

1. Sheaths and culms noticeably pilose, the hairs 1–3 mm long; lemmas 7-veined . . . . subsp. *undulatifolius*
1. Sheaths and culms glabrous or with a few, scattered hairs less than 1 mm long; lemmas (7)9–11-veined.
   2. Spikelets 2.2–2.7(3.3) mm long; lowest panicle branches 0.1–0.5 cm long . . . . subsp. *setarius*
   2. Spikelets 3–3.4(4.5) mm long; lowest panicle branches 0.5–0.7 cm long . . . . subsp. *fasciculatus*

According to Scholz (1981), *Oplismenus hirtellus* subsp. *fasciculatus* U. Scholz is restricted to southern Louisiana and eastern Mexico; *O. hirtellus* subsp. *setarius* (Lam.) Mez *ex* Ekman is more widely distributed, growing both in the southeastern United States and Mexico. *Oplismenus hirtellus* subsp. *undulatifolius* (Ard.) U. Scholz was found recently in Maryland (Peterson et al. 1999); it is native to the Eastern Hemisphere.

25.07 ECHINOCHLOA P. Beauv.

P.W. Michael

Plants annual or perennial; with or without rhizomes. **Culms** 10–460 cm, prostrate, decumbent or erect, distal portions sometimes floating, sometimes rooting at the lower nodes; **nodes** usually glabrous; **internodes** hollow or solid. **Sheaths** open, compressed; **auricles** absent; **ligules** usually absent but, if present, of hairs; **blades** linear to linear-lanceolate, usually more than 10 times longer than wide, flat, with a prominent midrib. **Inflorescences** terminal, panicles of simple or

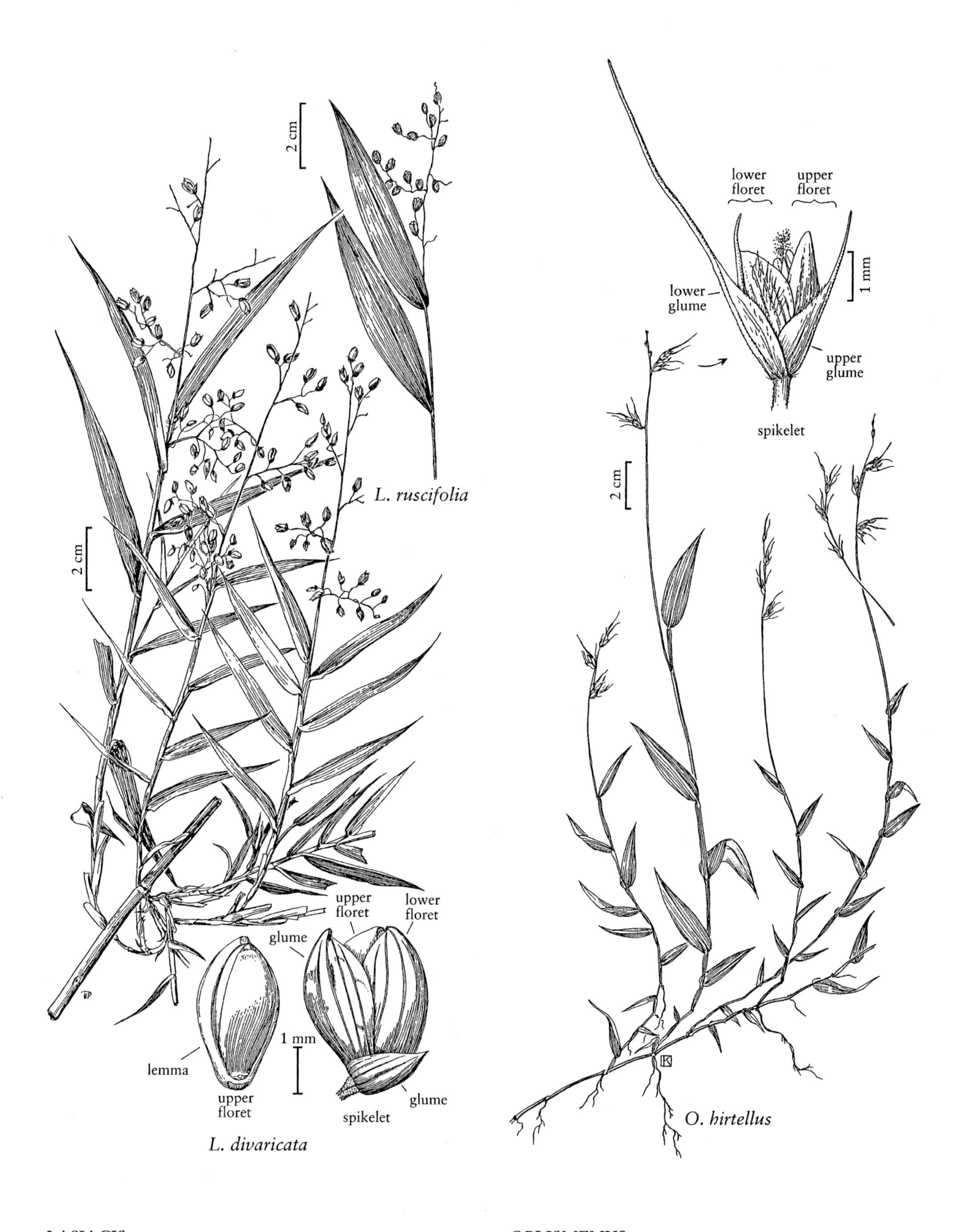

LASIACIS

OPLISMENUS

compound spikelike branches attached to elongate rachises, axes not terminating in a bristle, spikelets subsessile, densely packed on the angular branches; **disarticulation** below the glumes (cultivated taxa not or tardily disarticulating). **Spikelets** plano-convex, with 2(3) florets; **lower florets** sterile or staminate; **upper florets** bisexual, dorsally compressed. **Glumes** membranous; **lower glumes** usually ¼–⅖ as long as the spikelets (varying to more than ½ as long), unawned to minutely awn-tipped; **upper glumes** unawned or shortly awned; **lower lemmas** similar to the upper glumes in length and texture, unawned or awned, awns to 60 mm; **lower paleas** vestigial to well-developed; **upper lemmas** coriaceous, dorsally rounded, mostly smooth, apices short or elongate, firm or membranous, unawned; **upper paleas** free from the lemmas at the apices; **lodicules** absent or minute; **anthers** 3. **Caryopses** ellipsoid, broadly ovoid or spheroid; **embryos** usually 0.7–0.9 times as long as the caryopses. $x = 9$. Name from the Greek *echinos*, 'hedgehog', and *chloa*, 'grass', in reference to the bristly or often awned spikelets.

Echinochloa is a tropical to warm-temperate genus of 40–50 species that are usually associated with wet or damp places. Many of the species are difficult to distinguish because they tend to intergrade. Some of the characters traditionally used for distinguishing taxa, e.g., awn length, are affected by the amount of moisture available; others reflect selection by cultivation, e.g., non-disarticulation in grain taxa, mimicry of rice as weeds of rice fields. There are 13 species in the *Flora* region: five native and one possibly native, four established, two grown as commercial crops, and one in research.

In North America, the most abundant species appears to be the introduced, weedy *Echinochloa crus-galli*, which closely resembles the native *E. muricata*. The confusion between the two species has caused them to be treated as the same species. This confusion is probably reflected in the mapping of both *E. crus-galli* and *E. muricata*. *Echinochloa frumentacea* and *E. esculenta* are grown for grain in India and in China and Japan, respectively, but not in North America. *Echinochloa oryzoides* and *E. oryzicola* are weeds whose success and distribution reflects their adaptation to the periodic inundations of commercial rice fields.

Cytogenetic data suggest that *Echinochloa frumentacea* and *E. esculenta* are domesticated derivatives of *E. crus-galli* and *E. colona*, respectively (Yabuno 1962) and that *E. oryzoides* is very closely related to *E. crus-galli* (Yabuno 1984). Yabuno (1966) suggested that *E. crus-galli* is an allohexaploid produced by natural hybridization between the tetraploid *E. oryzicola* with a not-yet-discovered diploid species of *Echinochloa* and subsequent chromosome doubling. Studies using seed protein electrophoresis and isozyme analyses (Kim et al. 1989; González-Andrés et al. 1996; Asíns et al. 1999), and molecular studies involving RAPD markers and DNA sequences (Hilu 1994; Roy et al. 2000) or PCR-RFLP techniques (Yasuda et al. 2001), will help in clarifying the phylogenetic problems in *Echinochloa*, providing that proper attention is paid to the morphological characterization of the plant materials used and that voucher specimens are preserved.

SELECTED REFERENCES **Asíns, M.J., J.L. Carretero, A. Del Busto, E.A. Carbonell,** and **D. Gomez de Barreda.** 1999. Morphologic and isozyme variation in barnyard grass (*Echinochloa*) weed species. Weed Technol. 13:209–215; **Barrett, S.C.H.** and **D.E. Seaman.** 1980. The weed flora of California rice fields. Aquatic Bot. 9:351–376; **Carretero, J.L.** 1981. El género *Echinochloa* Beauv. en el sudeste de Europa. Anales Jard. Bot. Madrid 38:91–108; **Fishbein, M.** 1995. Noteworthy collections: Arizona. Madroño 42:83; **González-Andrés, F., J.M. Pita,** and **J.M. Ortiz.** Caryopsis isoenzymes of *Echinocloa* [sic] weed species as an aid for taxonomic discrimination. J. Hort. Sci. 71:187–193; **Gould, F.W., M.A. Ali,** and **D.E. Fairbrothers.** 1972. A revision of *Echinochloa* in the United States. Amer. Midl. Naturalist 87:36–59; **Hilu, K.W.** 1994. Evidence from RAPD markers in the evolution of *Echinochloa* millets (Poaceae). Pl. Syst. Evol. 189:247–257; **Hitchcock, A.S.** 1913. Mexican grasses in the United States National Herbarium. Contr. U.S. Natl. Herb. 17^3:181–389; **Jauzein, P.** 1993. Le genre *Echinochloa* en Camargue. Monde Pl. 88, no. 446:1–5; **Kim, K.V., J.H. Kim,** and **I.J. Lee.** 1989. Biochemical identification of *Echinochloa* species collected in Korea. Proc. Conf. Asian-Pacific Weed Sci. Soc. 12th [Proc. II]:519–531; **Kobayashi, H.** and **S. Sakamoto.** 1990. Weed-crop complexes in cereal cultivation. Pp. 67–80 *in* S. Kawano (ed.). Biological Approaches and Evolutionary Trends in Plants. Academic Press, London, England and San Diego, California, U.S.A. 417 pp.; **Michael, P.W.** 1983. Taxonomy and distribution of *Echinochloa* species with a special reference to their occurrence as weeds of rice. Pp. 291–306 *in* International Rice Research Institute. Proc. Conf. Weed Control in Rice (31 Aug.–4 Sept. 1981). International Rice Research Institute, Los Baños, Laguna, Philippines. 422 pp.; **Michael, P.W.** 2001. The taxonomy and distribution of *Echinochloa* species (barnyard grasses) in the Asian-Pacific region, with a review of pertinent biological studies. Proc. Conf. Asian-Pacific Weed Sci. Soc. 18th [Proc. I]:57–66; **Roy, S., J.-P.**

Simon, and **F.-J. Lapointe**. 2000. Determination of the origin of the cold-adapted populations of barnyard grass (*Echinochloa crus-galli*) in eastern North America: A total-evidence approach using RAPD DNA and DNA sequences. Canad. J. Bot. 78:1505–1513; **Vickery, J.W.** 1975. *Echinochloa* Beauv. Pp. 189–211 *in* J.W. Vickery. Flora of New South Wales, No. 19, Gramineae, Part 2 (ed. M.D. Tindale). New South Wales Department of Agriculture, Contributions from the New South Wales National Herbarium Flora Series. National Herbarium of New South Wales, Sydney, New South Wales, Australia. 181 pp.; **Wunderlin, R.P.** 1988. Guide to the Vascular Plants of Florida. University Press of Florida, Gainesville, Florida, U.S.A. 806 pp.; **Yabuno, T.** 1962. Cytotaxonomic studies on the two cultivated species and the wild relatives in the genus *Echinochloa*. Cytologia 27:296–305; **Yabuno, T.** 1966. Biosystematic study of the genus *Echinochloa*. J. Jap. Bot. 19:277–323; **Yabuno, T.** 1984. A biosystematic study on *Echinochloa oryzoides* (Ard.) Fritsch. Cytologia 49:673–678; **Yabuno, T.** and **H. Yamaguchi**. 1996. Hie no hakubutsugaku [A natural history of *Echinochloa*]. Dow Chemicals and Dow Elanco, Tokyo, Japan. 196 pp. [In Japanese]; **Yasuda, K.**, **A. Yano**, **Y. Nakayama**, and **H. Yamaguchi**. 2001. Identification of *Echinochloa oryzicola* and *E. crus-galli* using PCR-RFLP technique. [Abstract.] J. Weed Sci. Technol. 46, Suppl.:204–205. [Text in Japanese, title in Japanese and English].

NOTE: In this treatment, measurements of the spikelets do not include the awns. The color of the caryopses is based on fully ripe caryopses. Chromosome numbers cited in this treatment are only those which are confidently believed to refer to the species described here.

1. Ligules of stiff hairs present on the lower leaves; lower florets staminate; plants perennial.
   2. Plants without scaly rhizomes, sometimes rooting at the lower nodes; lower lemmas usually awned, sometimes merely apiculate; known outside of experimental plantings . . . . 1. *E. polystachya*
   2. Plants with short, scaly rhizomes; lower lemmas unawned, sometimes long-cuspidate; in the *Flora* region, known only from experimental plantings . . . . 2. *E. pyramidalis*
1. Ligules almost always absent from all leaves, the ligule region sometimes pubescent; lower florets sterile or staminate; plants usually annual, sometimes short-lived perennials.
   3. Lower lemmas usually unawned; spikelets, particularly those near the base of the panicles, not disarticulating at maturity; upper lemmas wider and longer than the upper glumes at maturity and, hence, exposed at maturity.
      4. Spikelets always green and pale at maturity, their apices usually obtuse, varying to acute; rachis nodes not or only sparsely hispid with papillose-based hairs; caryopses whitish . . . . 9. *E. frumentacea*
      4. Spikelets purplish to blackish-brown at maturity, their apices obuse to shortly acute; rachis nodes densely hispid with papillose-based hairs; caryopses brownish . . . . 11. *E. esculenta*
   3. Lower lemmas often awned; spikelets disarticulating at maturity; upper lemmas not or scarcely exceeding the upper glumes in length and width at maturity.
      5. Plants essentially obligate weeds of rice, growing in the fields; culms erect, densely tufted; spikelets 3.7–7 mm long; plants resembling rice in their vegetative growth.
         6. Panicles horizontal or drooping at maturity; spikelets broadly ovate to ovate; lower lemmas usually awned; caryopses 1.9–3 mm long, the embryos 70–85% as long as the caryopses . . . . 12. *E. oryzoides*
         6. Panicles erect to slightly drooping; spikelets ovate-elliptical; lower lemmas awned or not; caryopses 1.7–2.6 mm long, the embryos 89–98% as long as the caryopses . . . . . 13. *E. oryzicola*
      5. Plants not obligate weeds of rice, found in summer crops and wet places, and often in rice fields; culms sprawling, decumbent or erect; spikelets 2–5 mm long; plants occasionally resembling rice vegetatively but, if so, the spikelets less than 3 mm long.
         7. Lower florets staminate; anthers of the upper florets 1.2–1.7 mm long . . . . 3. *E. paludigena*
         7. Lower florets sterile; anthers of the upper florets 0.5–1.2 mm long.
            8. Panicle branches 0.7–2(4) cm long, without secondary branches; spikelets 2–3 mm long, unawned . . . . 8. *E. colona*
            8. Panicle branches 1–14 cm long, usually rebranched, the secondary branches often short and inconspicuous; spikelets 2.5–5 mm long, awned or unawned.
               9. Upper lemmas broadly ovate to elliptical, if elliptical, each with a line of minute (need 25× magnification) hairs across the base of the early-withering tips.
                  10. Upper lemmas with rounded or broadly acute coriaceous apices that pass abruptly into a membranous tip, a line of minute hairs present at the base of the tip . . . . 10. *E. crus-galli*
                  10. Upper lemmas with acute or acuminate coriaceous apices that extend into the membranous tip, without hairs at the base of the tip . . . . 4. *E. muricata*

9. Upper lemmas narrowly ovate to elliptical, never with a line of minute hairs across the base of the early-withering, membranous tips.
   11. Spikelets 2.5–3.4 mm long; lower lemmas unawned or with awns 3–10(15) mm long, curved . . . . 6. *E. crus-pavonis*
   11. Spikelets 3–5 mm long; lower lemmas usually with awns 8–25 mm long, typically straight.
      12. Blades 10–35(60) mm wide; sheaths usually hispid and the margins ciliate with prominent papillose-based hairs, sometimes the sheaths only papillose; lower lemmas awned, the awns 8–25(60) mm long; common in the eastern portion of the *Flora* region . . . . 5. *E. walteri*
      12. Blades 5–10 mm wide; sheaths glabrous or with papillose-based hairs; lower lemmas unawned or awned, the awns 8–16(50) mm long; in the *Flora* region, known only from southern Arizona . . . . 7. *E. oplismenoides*

1. **Echinochloa polystachya** (Kunth) Hitchc. [p. 395]
Creeping River Grass

Plants perennial; not rhizomatous. **Culms** 1–2 m tall, to 1 cm thick, erect or decumbent and rooting at the lower nodes, upper portion sometimes floating distally; **nodes** glabrous or antrorsely villous. **Sheaths** mostly glabrous, minutely puberulent, or hispid, hairs papillose-based, throat hispid; **ligules** present on the lower leaves, 1–5 mm, of stiff hairs; **blades** 15–70 cm long, 5–13 mm wide, glabrous. **Panicles** 13–45 cm, erect, rachis nodes hispid, hairs 3–6.5 mm, papillose-based, internodes scabrous; **primary branches** 4–10 cm, subverticillate, ascending, nodes hispid, hairs 2.5–4 mm, papillose-based, internodes scabrous; **secondary branches** short, spikelets subsessile, in clusters. **Spikelets** 4–7 mm, hispid, hairs appressed, disarticulating at maturity. **Lower glumes** at least ½ as long as the spikelets; **lower florets** staminate; **lower lemmas** apiculate or awned, awns to 18 mm; **lower paleas** subequal to the lower lemmas, often purple; **anthers of lower florets** 1.5–3.6 mm, orange; **upper lemmas** 2.5–5 mm, elliptic or narrowly ovate, apices obtuse, with a membranous, soon-withering tip; **anthers of upper florets** shorter than those of the lower florets. **Caryopses** to 3 mm. $2n$ = 54.

Echinochloa polystachya grows in coastal marshes, often in standing water, from Texas, Louisiana, and Florida south through Mexico and the Caribbean islands to Argentina. Two varieties exist. **Echinochloa polystachya** var. **polystachya** has glabrous culms and leaf sheaths; **Echinochloa polystachya** var. **spectabilis** (Nees *ex* Trin.) Mart. Crov. has swollen, pubescent cauline nodes and pubescent leaf sheaths.

2. **Echinochloa pyramidalis** (Lam.) Hitchc. & Chase [p. 395]
Antelope Grass

Plants perennial; with short, scaly rhizomes. **Culms** 1–4.6 m tall, to 2 cm thick, geniculate or long-prostrate and rooting at the lower nodes, often floating distally; **lower** and **upper nodes** glabrous. **Sheaths** mostly glabrous, but usually ciliate at the throat; **ligules** present on the lower leaves, 1–5 mm, of stiff hairs, reduced or absent on the upper leaves; **blades** 8–75 cm long, 5–30 mm wide. **Panicles** 15–40 cm, nodes and internodes scabrous; **primary branches** 2–7.5 cm, solitary to fascicled, erect or ascending, simple or compound, nodes and internodes glabrous or hispid, hairs to 4 mm, papillose-based. **Spikelets** 2.5–4 mm long, 1–1.8 mm wide, disarticulating at maturity, finely pubescent or glabrous, greenish to purple at maturity. **Lower florets** staminate; **lower lemmas** unawned, acute to acuminate or long cuspidate; **anthers of lower florets** 1–1.5 mm; **upper lemmas** apiculate to long cuspidate. **Caryopses** about 2 mm. $2n$ = 54, 72.

Echinochloa pyramidalis is native to Africa, where it is used both as a cereal and a pasture grass. It has been grown experimentally in Gainesville, Florida, but it is not established in North America.

3. **Echinochloa paludigena** Wiegand [p. 397]
Florida Barnyard Grass

Plants annual. **Culms** to 150 cm, erect. **Sheaths** glabrous; **ligules** absent; **blades** 15–60 cm long, 8–20 mm wide, scabrous adaxially. **Panicles** 8–40 cm, erect to slightly drooping, rachis nodes hispid, hairs papillose-based; **primary branches** 2–19 cm, erect to spreading, often widely spaced, longer branches with secondary branching. **Spikelets** 3.3–4.5 mm long, 2.4–2.6 mm wide, disarticulating at maturity, greenish or purplish, scabrous and hispid, hairs to 1 mm, often papillose-based. **Upper glumes** about as long as the spikelets; **lower florets**

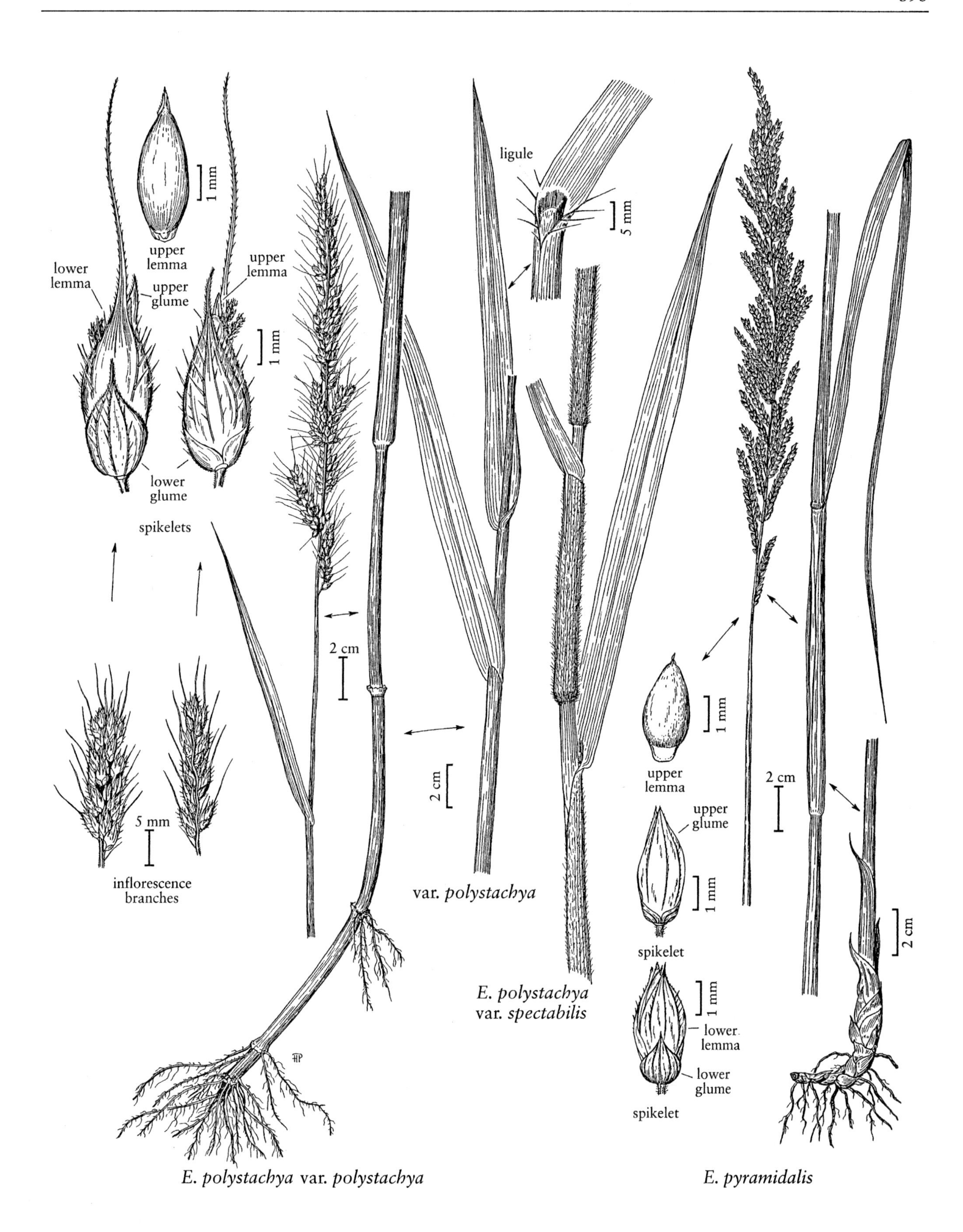

E. polystachya var. *polystachya*

E. pyramidalis

staminate; **lower lemmas** usually awned, awns 1–15 mm, purplish; **lower paleas** well-developed; **upper lemmas** broadly ovate, narrowing abruptly to the acute or acuminate apices; **anthers** of upper florets 1.2–1.7 mm. **Caryopses** 1.5–1.8 mm. $2n$ = unknown.

Echinochloa paludigena is native to swamps, riverbanks, and other wet habitats. Reports from Texas and Louisiana appear to be based on misidentifications; Wunderlin (1988) considers *E. paludigena* as a Florida endemic.

4. **Echinochloa muricata** (P. Beauv.) Fernald [p. 397]
AMERICAN BARNYARD GRASS

Plants annual. **Culms** 80–160 cm, erect or spreading, sometimes rooting at the lowest nodes, often developing short axillary flowering shoots at most upper nodes when mature; **lower nodes** glabrous or puberulent; **upper nodes** glabrous. **Sheaths** glabrous; **ligules** absent; **blades** 1–27 cm long, 0.8–30 mm wide. **Panicles of primary culms** 7–35 cm, rachises and branches glabrous or hispid, hairs to 3 mm, papillose-based; **primary branches** 2–8 cm, usually spreading and rather distant, often with secondary branches. **Spikelets** 2.5–5 mm, disarticulating at maturity, usually purple or streaked with purple, usually hispid, hairs papillose-based. **Upper glumes** about as long as the spikelets; **lower florets** sterile; **lower lemmas** unawned or awned, awns to 16 mm; **lower paleas** well-developed; **upper lemmas** broadly obovoid or orbicular, narrowing to an acute or acuminate coriaceous portion that extends into the membranous tip, boundary between the coriaceous and membranous portions not marked by minute hairs; **anthers** 0.4–1.1 mm. **Caryopses** 1.2–2.5 mm, broadly obovoid or spheroid, yellowish; **embryos** 1.4–2 mm, 80–91% as long as the caryopses. $2n$ = 36.

Echinochloa muricata is native to North America, growing from southern Canada to northern Mexico in moist, often disturbed sites (but not rice fields). It resembles *E. crus-galli* in gross morphology and ecology, but differs consistently by the characters used in the key. The two varieties tend to be distinct, but there is some overlap in both morphology and geography.

1. Spikelets 2.5–3.8 mm long; lower lemmas unawned or awned, the awns to 10 mm long . . . . var. *microstachya*
1. Spikelets 3.5–5 mm long; lower lemmas usually awned, the awns 6–16 mm long . . . . var. *muricata*

Echinochloa muricata var. **microstachya** Wiegand
ÉCHINOCHLOA PIQUANT

Spikelets 2.5–3.8 mm. **Lower glumes** 0.9–1.6 mm; **upper glumes** 2.8–3.8 mm; **lower lemmas** unawned or awned, awns to 10 mm; **anthers** 0.4–0.7 mm.

Echinochloa muricata var. *microstachya* is the common variety in the western part of North America, extending east to the Missouri River and the Texas panhandle.

Echinochloa muricata (P. Beauv.) Fernald var. **muricata**
ÉCHINOCHLOA DE L'OUEST

Spikelets 3.5–5 mm. **Lower glumes** 1–2.6 mm; **upper glumes** 3–5 mm; **lower lemmas** usually awned, awns 6–16 mm, occasionally unawned; **anthers** 0.5–1.1 mm.

Echinochloa muricata var. *muricata* is the common variety in eastern North America.

5. **Echinochloa walteri** (Pursh) A. Heller [p. 399]
COAST BARNYARD GRASS, ÉCHINOCHLOA DE WALTER

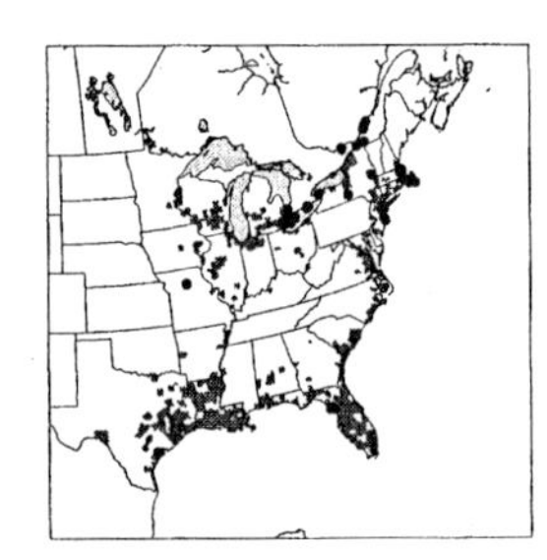

Plants annual. **Culms** (30) 50–200+ cm tall, to 2.5 cm thick; **nodes** pilose or villous, upper nodes usually with sparser and shorter pubescence, occasionally glabrous. **Lower sheaths** usually hispid, hairs papillose-based, sometimes just papillose; **upper sheaths** hispid or glabrous; **ligules** absent; **blades** to 55 cm long, 10–35(60) mm wide, scabrous. **Panicles** 8.5–35 cm, erect to slightly drooping, nodes hispid, hairs 3.5–5 mm, papillose-based, sometimes sparsely so, internodes usually glabrous, sometimes hispid, hairs papillose-based; **primary branches** 1–10 cm, loosely erect, not concealed by the spikelets, nodes usually hispid, hairs papillose-based, sometimes glabrous, internodes scabrous, sometimes also sparsely hispid, hairs papillose-based; **secondary branches** present on the longer primary branches. **Spikelets** 3–5 mm, disarticulating at maturity, scabrous to variously muricate and hairy, hairs usually not papillose-based, margins sometimes with a few papillose-based hairs. **Lower glumes** usually more than ½ as long as the spikelets, abruptly narrowing to a fine, 0.5 mm point; **lower florets** sterile; **lower lemmas** usually awned, awns 8–25(60) mm; **lower paleas** subequal to the lower lemmas; **upper lemmas** 3–5 mm long, about 1.5 mm wide, not or scarcely exceeding the upper glumes, narrowly ovate to elliptical, coriaceous portion subacute, tips acuminate, membranous, without a line of hairs at the base of the tip; **anthers** 0.6–1(1.2) mm. **Caryopses** 1.2–1.8 mm, brownish; **embryos** 52–77% as long as the caryopses. $2n$ = 36.

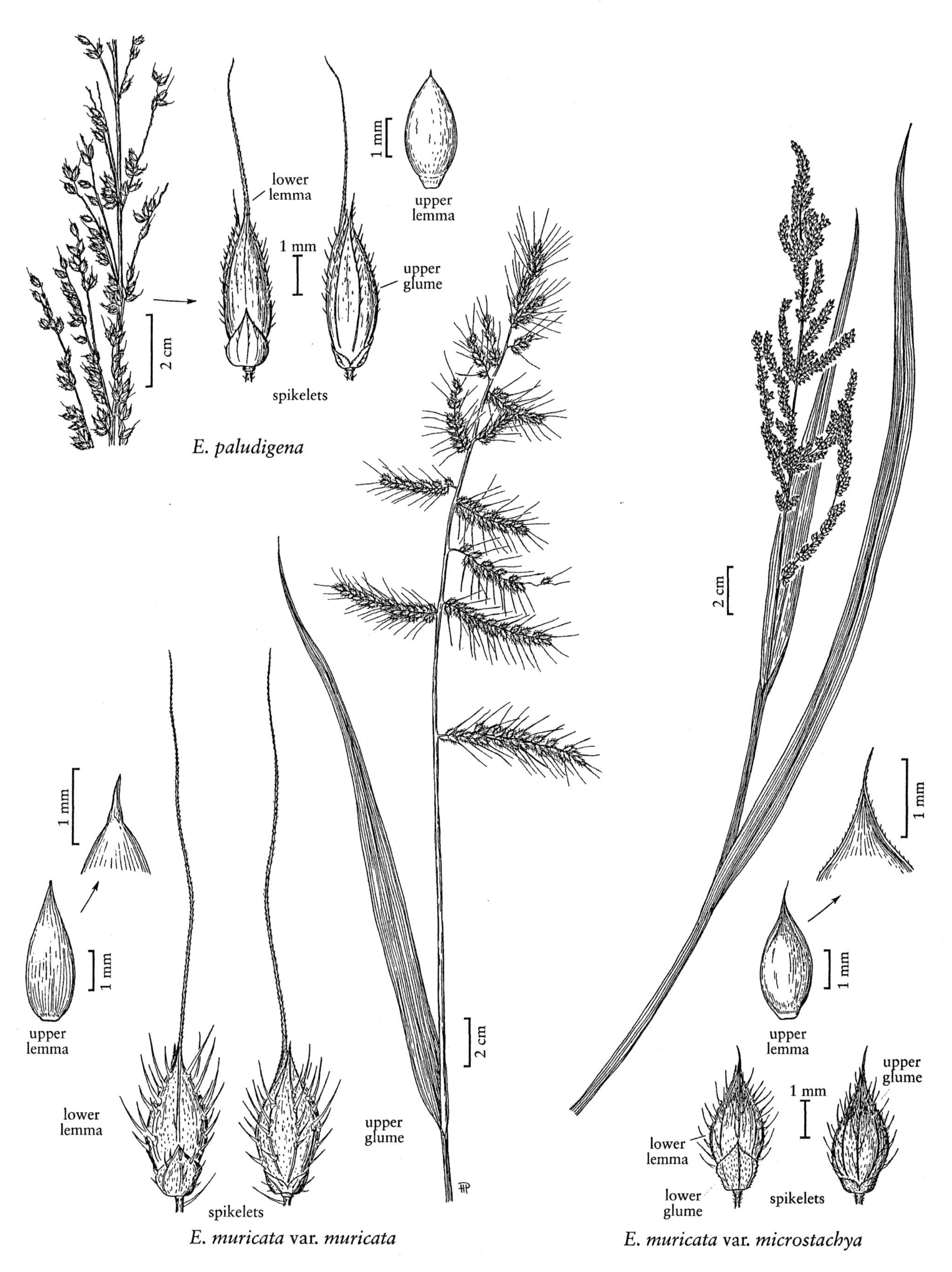
lower
lemma
1 mm
2 cm
spikelets
1 mm
upper
lemma
upper
glume
E. paludigena
2 cm
1 mm
1 mm
upper
lemma
lower
lemma
spikelets
upper
glume
2 cm
E. muricata var. muricata
1 mm
1 mm
upper
lemma
upper
glume
1 mm
lower
lemma
lower
glume
spikelets
E. muricata var. microstachya

Echinochloa walteri grows in wet places, often in shallow water and brackish marshes. It is a native species that extends through Mexico to Guatamala. It is found in both disturbed and undisturbed sites although not in rice fields. Occasional specimens of *E. walteri* with glabrous lower sheaths and short awns can be distinguished from *E. crus-pavonis* by their less dense panicles.

6. **Echinochloa crus-pavonis** (Kunth) Schult. [p. 399]
GULF BARNYARD GRASS

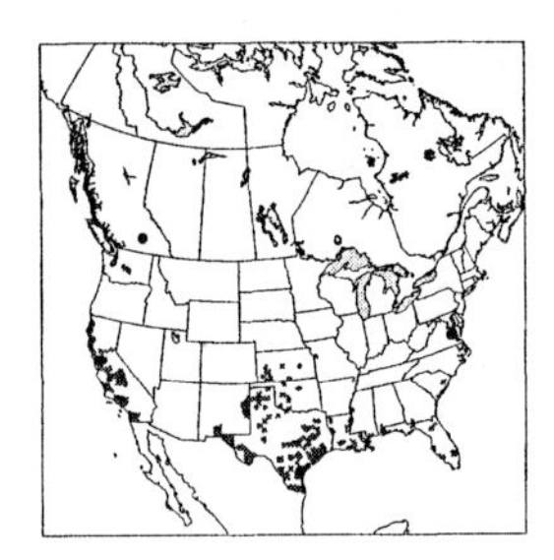

Plants annual or short-lived perennials. **Culms** 30–150 cm; **nodes** glabrous. **Sheaths** glabrous, often purplish; **ligules** absent; **blades** 12–60 cm long, 10–25 mm wide, glabrous. **Panicles** 10–30 cm, erect or drooping, nodes sparsely hispid, hairs papillose-based, internodes glabrous; **primary branches** to 14 cm, nodes sometimes sparsely hispid, hairs papillose-based, internodes usually glabrous; **secondary branches** to 3 cm. **Spikelets** 2.5–3.4 mm long, 1.2–1.4 mm wide, disarticulating at maturity. **Upper glumes** subequal to the spikelets; **lower florets** sterile; **lower lemmas** unawned or awned, awns 3–10(15) mm, curved; **lower paleas** absent, vestigial, or well-developed; **upper lemmas** narrowly elliptic, not or scarcely exceeding the upper glumes, acute or obtuse, with a well-differentiated, early-withering tip, glabrous or pubescent at the base of the tip, hairs not forming a line across the base; **anthers** 0.5–0.7 mm. **Caryopses** 1.2–1.5 mm long, 1–1.3 mm wide; **embryos** 50–70% as long as the caryopses. 2*n* = 36.

Echinochloa crus-pavonis is a native species found in scattered locations from British Columbia to Arizona, east to Florida, and south into South America. It favors marshes and wet places at lower elevations, often being found in the water.

1. Lower paleas more than ½ as long as the lemmas; panicles usually drooping .... var. *crus-pavonis*
1. Lower paleas absent or much less than ½ as long as the lemmas; panicles usually stiffly erect .... var. *macra*

Echinochloa crus-pavonis (Kunth) Schult. var. **crus-pavonis**

Panicles usually drooping. **Lower paleas** ½ or more as long as the lower lemmas.

This is generally the more southern of the two varieties, extending through Mexico and the Caribbean to Bolivia and Argentina. It appears, presumably as an adventive species, as far north as Humboldt County, California.

Echinochloa crus-pavonis var. **macra** (Wiegand) Gould

Panicles usually erect, stiff. **Lower paleas** absent or much less than ½ as long as the lower lemmas.

This variety extends south only as far as northern Mexico. The epithet is frequently spelled '*macera*', but '*macra*' is correct.

7. **Echinochloa oplismenoides** (E. Fourn.) Hitchc. [p. 401]
CHIHUAHUAN BARNYARD GRASS

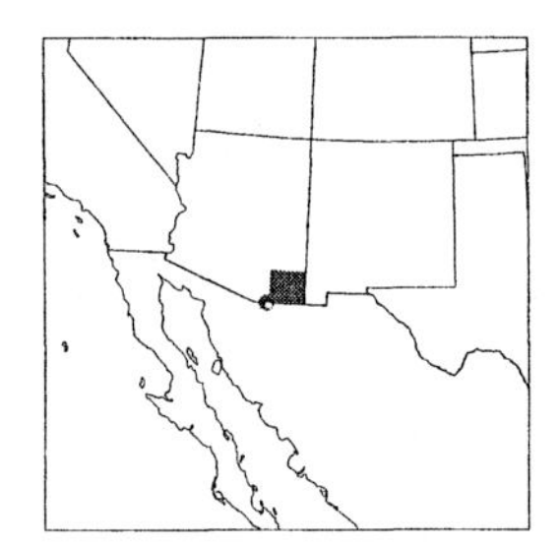

Plants annual. **Culms** to 100 cm, erect, succulent, glabrous, branching from the lower nodes. **Sheaths** glabrous or hispid with papillose-based hairs; **ligules** absent or the ligule region pubescent; **blades** 10–35 cm long, 5–10 mm wide. **Panicles** 15–30 cm, narrow; **primary branches** appressed to ascending, with papillose-based hairs at the base of the spikelets. **Spikelets** 4–5 mm, disarticulating at maturity. **Glumes** with hairs over the veins, glabrous, scabrous, or hispid between the veins; **upper glumes** about equal to the spikelets, muticous or awned, awns to 1 mm; **lower florets** sterile; **lower lemmas** unawned or awned, awns 8–16(50) mm; **lower paleas** absent or hyaline and subequal to the lemmas; **upper lemmas** 4–4.5 mm long, 1.7–1.9 mm wide, elliptic; **anthers** 0.5–0.7 mm, purple. **Caryopses** 2.7-2.9 mm long, 1.7-1.8 mm wide, elliptic in outline, mucronate; **embryos** about 75% as long as the caryopses; **hila** obovate. 2*n* = unknown.

Echinochloa oplismenoides was first found in the United States, in southern Arizona, in 1993 (Fishbein 1995). It was previously known only from Mexico, with a range that extends from northwestern Mexico to Guatemala. The southern Arizonan plants were found near a cattle tank in wet grasslands. Fishbein stated that it was impossible to tell whether they represented a previously overlooked native species or an introduction.

8. **Echinochloa colona** (L.) Link [p. 401]
AWNLESS BARNYARD GRASS

Plants annual; erect or decumbent, cespitose or spreading, rooting from the lower cauline nodes. **Culms** 10–70 cm; **lower nodes** glabrous or hispid, hairs appressed; **upper nodes** glabrous. **Sheaths** glabrous; **ligules** absent, ligule region frequently brown-purple; **blades** 8–22 cm long, 3–6(10) mm wide, mostly glabrous, sometimes hispid, hairs papillose-based on or near the margins. **Panicles**

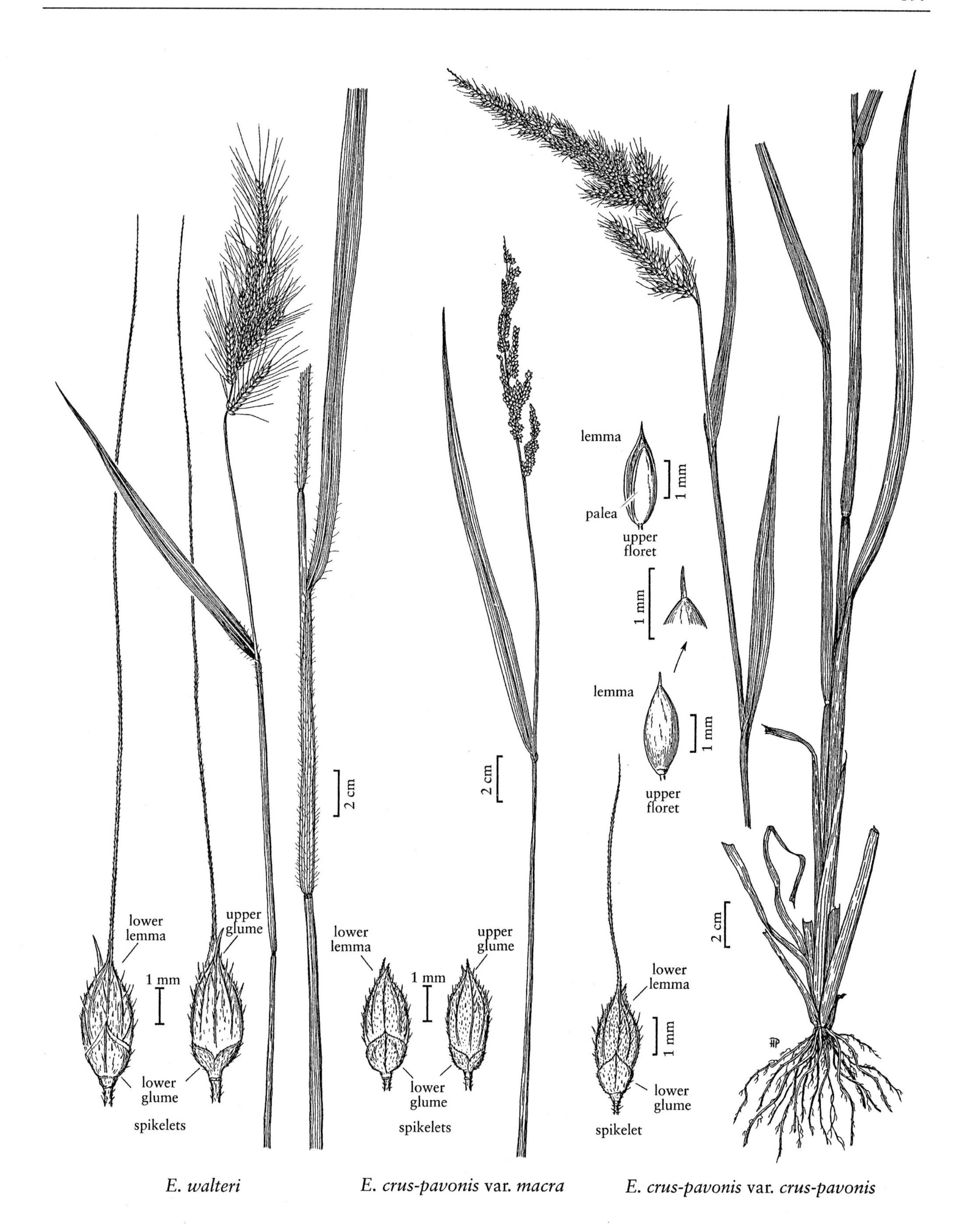

E. walteri *E. crus-pavonis* var. *macra* *E. crus-pavonis* var. *crus-pavonis*

2–12 cm, erect, rachises glabrous or sparsely hispid; **primary branches** 5–10, 0.7–2(4) cm, erect to ascending, spikelike, somewhat distant, without secondary branches, axes glabrous or sparsely hispid, hairs 1.5–2.5 mm, papillose-based. **Spikelets** 2–3 mm, disarticulating at maturity, pubescent to hispid, hairs usually not papillose-based, tips acute to cuspidate. **Lower glumes** about ½ as long as the spikelets; **upper glumes** about as long as the spikelets; **lower florets** usually sterile, occasionally staminate; **lower lemmas** unawned, similar to the upper glumes; **lower paleas** subequal to the lemmas; **upper lemmas** 2.6–2.9 mm, not or scarcely exceeding the upper glumes, elliptic, coriaceous portion rounded distally, passing abruptly into a sharply differentiated, membranous, soon-withering tip; **anthers** 0.7–0.8 mm. **Caryopses** 1.2–1.6 mm, whitish; **embryos** 63–83% as long as the caryopses. $2n$ = 54.

Echinochloa colona is widespread in tropical and subtropical regions. It is adventive and weedy in North America, growing in low-lying, damp to wet, disturbed areas, including rice fields. The unbranched, rather widely-spaced panicle branches make this one of the easier species of *Echinochloa* to recognize.

Hitchcock (1913) considered that '*colonum*' was a non-declining contraction, but dictionaries of Linnaeus' time treated it as a declining adjective. Because Linnaeus was the first to name the species (as "*Panicum colonum*"), it seems best to follow the practice considered correct in his day; hence "*E. colona*".

9. **Echinochloa frumentacea** Link [p. 401]
SIBERIAN MILLET, WHITE PANIC

Plants annual. **Culms** 70–150 cm, erect, glabrous. **Sheaths** glabrous; **ligules** absent; **blades** 8–35 cm long, 3–20(30) mm wide, glabrous. **Panicles** 7–18 cm, erect to slightly drooping at maturity, rachises not or only sparsely hispid, nodes with papillose-based hairs; **branches** numerous, appressed or ascending, spikelike, not or only sparsely hispid, hairs papillose-based; **primary branches** 1.5–4 cm, glabrous or sparsely hispid, hairs to 3 mm, papillose-based; **secondary branches**, if present, usually concealed by the densely packed spikelets; **longer pedicels** 0.2–0.5 mm. **Spikelets** 3–3.5 mm, often with 1 sterile and 2 bisexual florets, not disarticulating at maturity (particularly those near the bases of the panicles), scabrous or short-hispid but without papillose-based hairs, green and pale at maturity, apices usually obtuse, varying to acute. **Upper glumes** narrower and shorter than the upper lemmas; **lower florets** sterile; **lower lemmas** unawned; **lower paleas** subequal to the lower lemmas; **upper lemmas** 2.5–3 mm, ovate to elliptic, coriaceous portion terminating abruptly at the base of the membranous tip; **anthers** 0.8–1 mm. **Caryopses** 1.7–2.2 mm long, 1.6–1.8 mm wide, whitish; **embryos** 66–86% as long as the caryopses. $2n$ = 54.

Echinochloa frumentacea originated in India, and possibly also in Africa. It is grown for grain, fodder, and beer, but not as extensively as in the past. It is found occasionally in the contiguous United States and southern Canada, the primary source being birdseed mixes. It used to be confused with *E. esculenta*, from which it differs in its whitish caryopses and proportionately smaller embryos. Hybrids between *E. frumentacea* and *E. colona* are partially fertile; those with *E. esculenta* are sterile.

10. **Echinochloa crus-galli** (L.) P. Beauv. [p. 403]
BARNYARD GRASS, ÉCHINOCHLOA PIED-DE-COQ

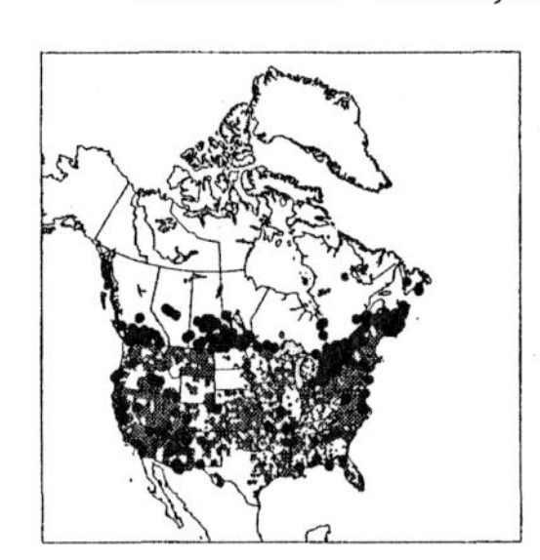

Plants annual. **Culms** 30–200 cm, spreading, decumbent or stiffly erect; **nodes** usually glabrous or the lower nodes puberulent. **Sheaths** glabrous; **ligules** absent, ligule region sometimes pubescent; **blades** to 65 cm long, 5–30 mm wide, usually glabrous, occasionally sparsely hirsute. **Panicles** 5–25 cm, with few–many papillose-based hairs at or below the nodes of the primary axes, hairs sometimes longer than the spikelets; **primary branches** 1.5–10 cm, erect to spreading, longer branches with short, inconspicuous secondary branches, axes scabrous, sometimes also sparsely hispid, hairs to 5 mm, papillose-based. **Spikelets** 2.5–4 mm long, 1.1–2.3 mm wide, disarticulating at maturity. **Upper glumes** about as long as the spikelets; **lower florets** sterile; **lower lemmas** unawned to awned, sometimes varying within a branch, awns to 50 mm; **lower paleas** subequal to the lemmas; **upper lemmas** broadly ovate to elliptical, coriaceous portion rounded distally, passing abruptly into an early-withering, acuminate, membranous tip that is further demarcated from the coriaceous portion by a line of minute hairs (use 25× magnification); **anthers** 0.5–1 mm. **Caryopses** 1.3–2.2 mm long, 1–1.8 mm wide, ovoid or oblong, brownish; **embryos** 59–86% as long as the caryopses. $2n$ = 54.

Echinochloa crus-galli is a Eurasian species that is now widely established in the *Flora* region, where it grows in moist, disturbed sites, including rice fields. Some North American taxonomists have interpreted *E. crus-galli* much more widely; others treat it as here, but recognize several infraspecific taxa based on such characters as trichome length and abundance, and awn length. There are several ecological and physiological ecotypes within the species, but the correlation between these and the species' morphological variation has not been established, so no infraspecific taxa are recognized here.

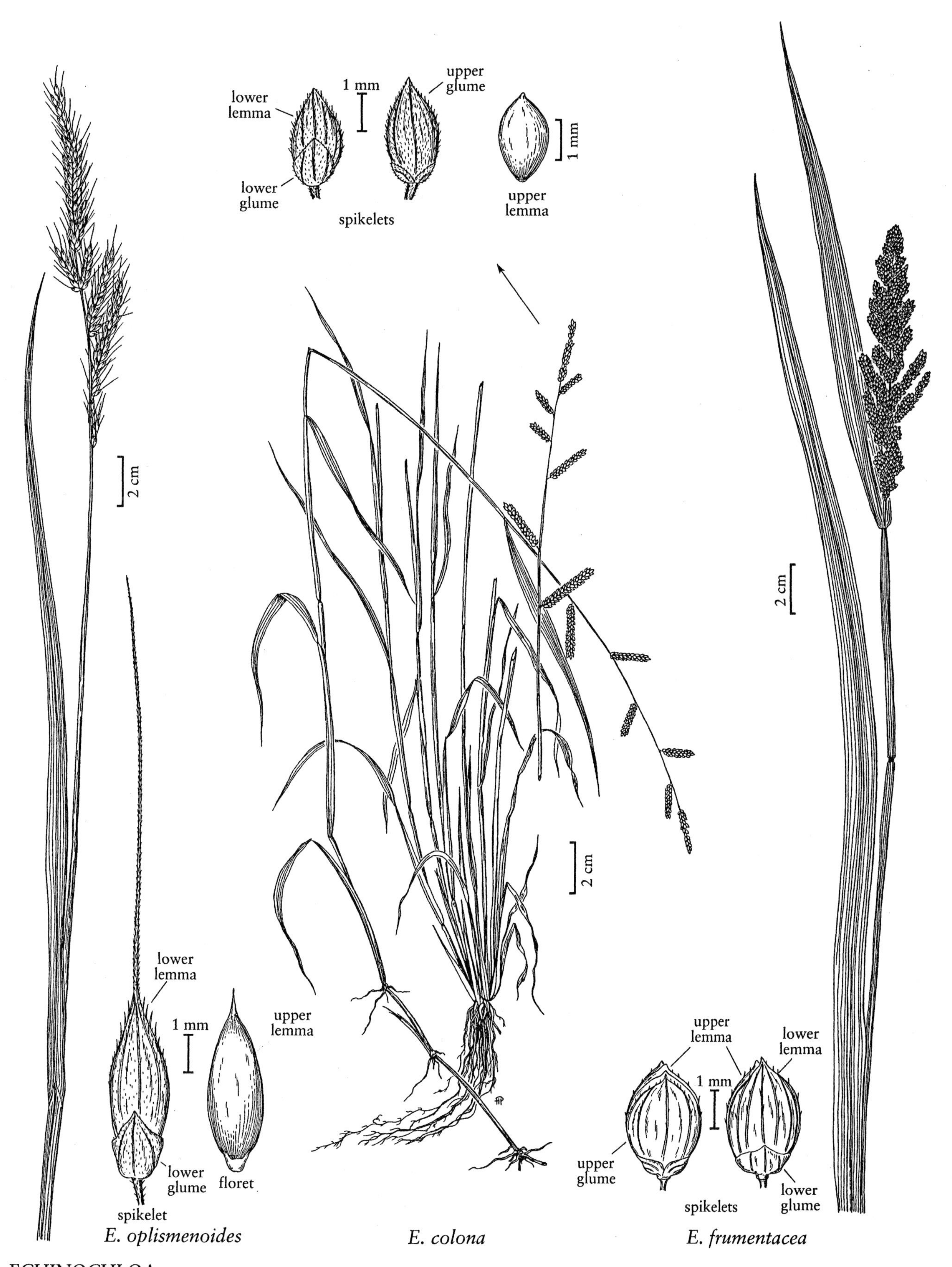

E. oplismenoides *E. colona* *E. frumentacea*

ECHINOCHLOA

11. Echinochloa esculenta (A. Braun) H. Scholtz [p. 403]
JAPANESE MILLET

Plants annual. **Culms** 80–150 cm tall, 4–10 mm thick, glabrous. **Sheaths** glabrous; **ligules** absent, ligule region sometimes pubescent; **blades** 10–50 cm long, 5–25 mm wide. **Panicles** 7–30 cm, dense, rachis nodes densely hispid, hairs papillose-based, internodes scabrous; **primary branches** 2–5 cm, erect or spreading, simple or branched, often incurved at maturity, nodes hispid, hairs papillose-based, internodes usually scabrous; **longer pedicels** 0.5–1 mm. **Spikelets** 3–4 mm long, 2–2.5 mm wide, not or only tardily disarticulating at maturity, obtuse to shortly acute, purplish to blackish-brown at maturity. **Upper glumes** narrower and shorter than the upper lemmas; **lower florets** sterile; **lower lemmas** usually unawned; **lower paleas** shorter and narrower than the lemmas; **upper lemmas** longer and wider than the upper glumes, broadly ovate to ovate-orbicular, shortly apiculate, exposed distally at maturity; **anthers** 1–1.2 mm. **Caryopses** 1.2–2.3 mm, brownish; **embryos** 84–96% as long as the caryopses. 2*n* = 54.

Echinochloa esculenta was derived from *E. crus-galli* in Japan, Korea, and China. It is cultivated for fodder, grain, or birdseed. It has sometimes been included in *E. frumentacea*, from which it differs in its brownish caryopses and longer pedicels. Hybrids between *E. crus-galli* and *E. esculenta* are fully fertile, but those with *E. frumentacea* are sterile.

12. Echinochloa oryzoides (Ard.) Fritsch [p. 403]
EARLY BARNYARD GRASS

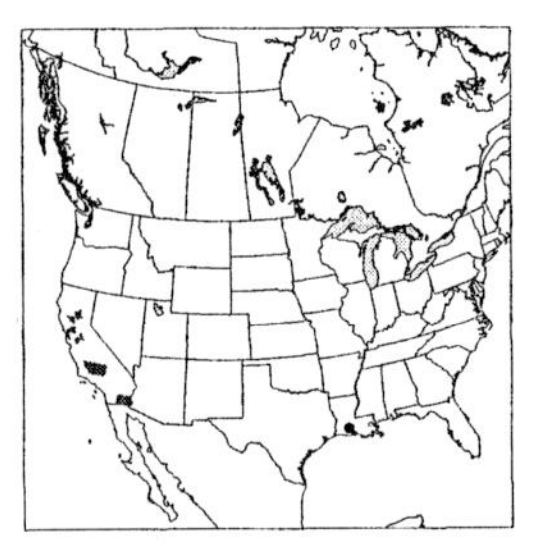

Plants annual. **Culms** 40–120 cm, erect, densely tufted; **nodes** glabrous. **Sheaths** glabrous; **ligules** absent; **blades** lax or drooping, 7–20 cm long, 4–12 mm wide, mostly glabrous. **Panicles** 8–17(25) cm, lax, horizontal to strongly drooping, rachis nodes hispid, hairs to 4 mm, papillose-based, internodes glabrous; **primary branches** to 5 cm, appressed to the rachises, mostly simple, glabrous or sparsely hispid, hairs to 3.5 mm, papillose-based, particularly at the nodes. **Spikelets** 3.7–7 mm long, 1.9–2.4 mm wide, disarticulating at maturity, broadly ovate to ovate. **Lower glumes** usually ¼-⅔ as long as the spikelets, occasionally ½ as long or longer; **upper glumes** subequal to the spikelets; **lower florets** sterile; **lower lemmas** similar in size to the spikelet, usually awned, awns to 5 cm; **lower paleas** well-developed; **upper lemmas** 3.5–4.5 mm, similar in length and width to the upper glumes, broadly elliptic to ovate, with an acute, greenish tip; **anthers** to 0.8 mm. **Caryopses** 1.9–3 mm, light brown or tan; **embryos** 70–85% as long as the caryopses. 2*n* = 54.

Echinochloa oryzoides is a common weed of rice fields throughout the world, growing in the flooded portions of the fields. It was included in *E. oryzicola* by Gould et al. (1972), but it differs in its shorter embryo, lax, strongly drooping panicle, and earlier (June–July) flowering period. This flowering period is also earlier than that of *Oryza*. In addition, *E. oryzoides* is usually conspicuously awned, having longer awns than even the awned variants of *E. oryzicola*, and it is rarely obviously pubescent on the cauline nodes, leaf sheaths, and collars. The earliest known collection of *E. oryzoides* in the United States was made in 1925 (Barrett and Seaman 1980).

13. Echinochloa oryzicola (Vasinger) Vasinger [p. 403]
LATE BARNYARD GRASS

Plants annual. **Culms** 40–150 cm, erect or nearly so, densely tufted; **lower nodes** usually antrorsely scabrous or villous; **upper nodes** glabrous. **Lower sheaths** densely pubescent; **upper sheaths** glabrous or pubescent at the throat, and sometimes on the collar; **ligules** absent; **blades** stiff, ascending, lower blades pubescent, upper blades usually glabrous. **Panicles** 8–20 cm, erect to slightly drooping, rachis nodes hispid ,with papillose-based hairs to 5.6 mm, internodes usually scabrous, sometimes also with a few papillose-based hairs; **primary branches** to 4 cm. **Spikelets** 4–6 mm, ovoid to ellipsoid, disarticulating at maturity. **Lower glumes** usually at least ½ as long as the spikelets; **upper glumes** equaling or exceeding the upper florets; **lower florets** sterile; **lower lemmas** often thickened and somewhat coriaceous, unawned or awned, awns to 1.5 mm; **lower paleas** well-developed; **upper lemmas** broadly ovate to elliptical, coriaceous portion rounded distally, passing abruptly into an early-withering, acuminate, membranus tip that is further demarcated from the coriaceous portion by minute hairs (use 25× magnification); **anthers** 0.9–1.2 mm. **Caryopses** 1.7–2.6 mm, brownish; **embryos** 89–98% as long as the caryopses. 2*n* = 36.

Like *Echinochloa oryzoides*, *E. oryzicola* is an introduced weed of rice fields, where it grows in the flooded portion, with the rice. The two are quite distinct, with *E. oryzicola* flowering after *Oryza* and having a longer embryo and an erect panicle. It is also more likely to have evidently pubescent cauline nodes, leaf sheaths, and collars than *E. oryzoides* and is never conspicuously awned.

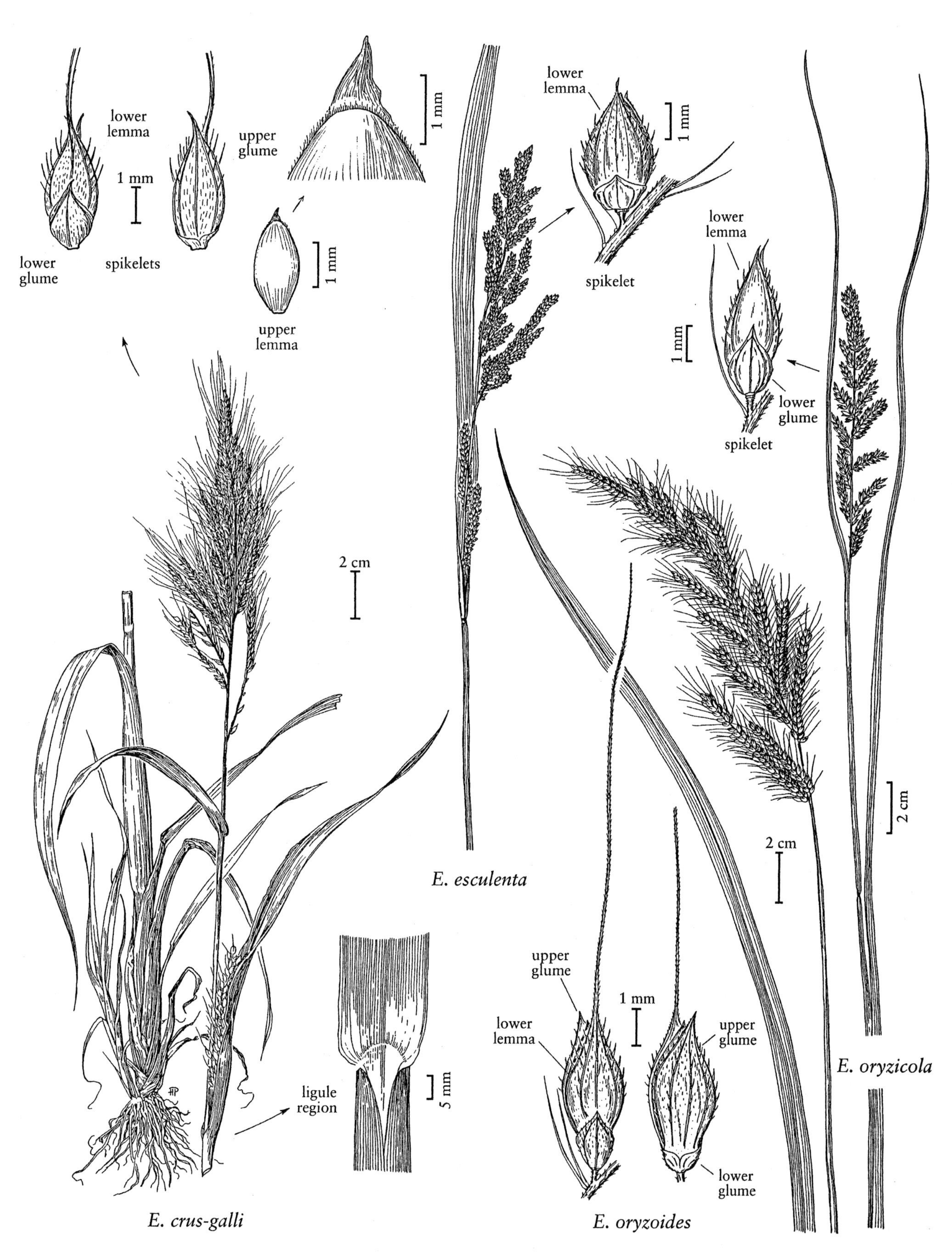

ECHINOCHLOA

25.08 SACCIOLEPIS Nash

J.K. Wipff

Plants annual or perennial; rhizomatous, stoloniferous, or cespitose. **Culms** 5–150 cm, not woody, branched above the base; **internodes** hollow. **Leaves** cauline; **auricles** sometimes present; **ligules** membranous, sometimes ciliate; **blades** flat or rolled, with or without cross venation. **Inflorescences** terminal, usually contracted, dense panicles, distal ½ of the rachises concealed by the spikelets; **branches** fused to the rachises or free and appressed to ascending; **pedicels** with discoid apices; **disarticulation** below the glumes and below the upper florets. **Spikelets** bisexual, with 2 florets, rounded to acute; **rachilla segments** not swollen. **Glumes** unequal, prominently veined, unawned; **lower glumes** 3–7-veined; **upper glumes** as long as or exceeding the upper florets, distinctly saccate or gibbous, 5–13-veined; **lower florets** 0.8–1.9 mm, sterile or staminate, less than ½ as long as the spikelets; **lower lemmas** resembling the upper glumes but not saccate, sometimes with a transverse row of hairs, 5–9-veined, unawned; **lower paleas** present or absent, 0–2-veined; **upper lemmas** subcoriaceous to subindurate, dorsally compressed, glabrous, smooth, margins inrolled or flat, never hyaline, faintly 3–5-veined; **upper paleas** similar to the lemmas, 2-veined; **lodicules** 2, fleshy, glabrous. $x = 9$. Name from the Greek *sakkion*, 'small bag', and *lepis*, 'scale', alluding to the saccate upper glumes.

Sacciolepis is a genus of 30 species. It is represented throughout the tropics and subtropics, primarily in Africa. Two species grow in the *Flora* region. One is native; the other is an introduction that has become established. Most species grow along and in ponds, lakes, streams, ditches, and other moist areas. The prominently multi-veined, saccate upper glumes and contracted panicles distinguish *Sacciolepis* from all other grasses in the *Flora* region.

SELECTED REFERENCES **Judziewicz, E.J.** 1990. A new South American species of *Sacciolepis* (Poaceae: Panicoideae: Paniceae), with a summary of the genus in the New World. Syst. Bot. 15:415–420; **Simon, B.K.** 1972. A revision to the genus *Sacciolepis* (Gramineae) in the "Flora Zambesiaca" area. Kew Bull. 27:387–406.

1. Primary branches fused to the rachises for at least ¾ of their length; lower branches 0.1–0.5 cm long; upper glumes 9-veined; paleas of the lower florets 0.5–1 mm long, to ½ as long as the lower lemmas . . . . 1. *S. indica*
1. Primary branches ascending, free from the rachises; lower branches 0.4–11.5 cm long; upper glumes 11(12)-veined; paleas of the lower florets 2–4 mm long, ¾ to almost as long as the lower lemmas . . . . 2. *S. striata*

1. Sacciolepis indica (L.) Chase [p. 405]
CHASE'S GLENWOODGRASS

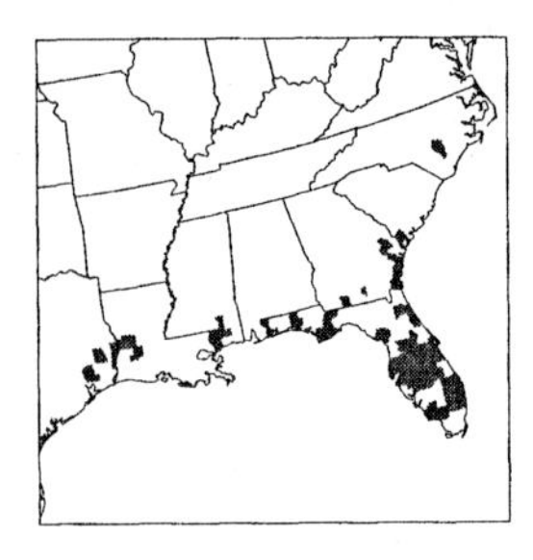

Plants annual; cespitose. **Culms** 5–100 cm, decumbent, spreading, trailing, often rooting at the lower nodes; **nodes** glabrous. **Sheaths** and **collars** glabrous; **ligules** 0.1–0.7 mm long, membranous, truncate; **blades** 1–14.3 cm long, 1.5–5.5 mm wide, glabrous, not cordate at the base. **Panicles** 0.5–9(13) cm long, 4–7 mm wide, contracted; **primary branches** fused to the rachises for at least ¾ of their length; **lower branches** 0.1–0.5 cm; **pedicels** 0.3–1.8 mm. **Spikelets** 2.1–3.3 mm, with or without papillose-based hairs on the upper glumes and lower lemmas, green to dark purple. **Lower glumes** 1.1–1.9 mm, glabrous, 3–5(7)-veined, margins hyaline; **upper glumes** 2–3.3 mm, slightly saccate, glabrous adaxially, 9-veined; **lower florets** sterile (rarely staminate); **lower lemmas** 1.9–3.1 mm, 7–9-veined, veins equidistant; **lower paleas** 0.5–1 mm long, 0.1–0.2 mm wide, ½ or less as long as the lower lemmas, narrow, membranous, white, not veined; **upper lemmas** 1.3–1.6 mm, subcoriaceous, glabrous, shiny, white, with 3–5 obscure veins, acute; **anthers** 3, 0.5–0.8 mm, dark reddish-brown to reddish-purple; **styles** purple. **Caryopses** 1–1.3 mm long, 0.5–0.7 mm wide, glabrous. $2n = 18, 36$.

Sacciolepis indica is native to the Eastern Hemisphere tropics. It is now established in the coastal states of the southeastern United States, where it grows in and along streams, ponds, lakes, ditches, and other moist places. It flowers from late summer to fall.

2. **Sacciolepis striata** (L.) Nash [p. 405]
AMERICAN CUPSCALE

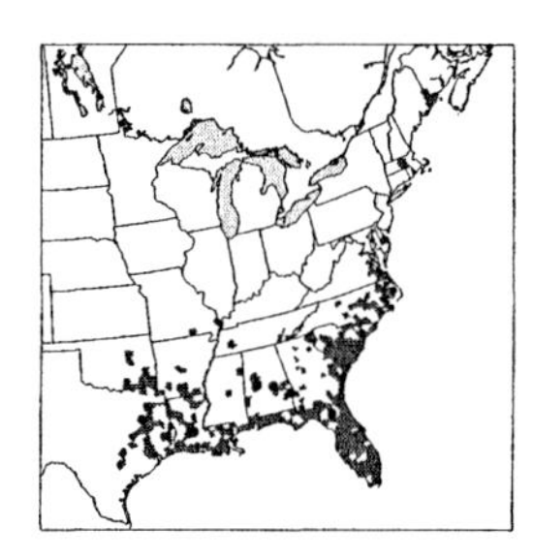

Plants perennial; with or without rhizomes. **Culms** 80–150 cm, erect or decumbent, sprawling, trailing, often rooting at the lower nodes; **nodes** glabrous or pubescent. **Sheaths** mostly glabrous or with papillose-based hairs, margins sometimes ciliate; **collars** pubescent adaxially, usually glabrous abaxially, margins ciliate; **ligules** 0.2–0.7 mm, membranous, ciliate; **blades** (2.3)4–28.5 cm long, 2.9–22 mm wide, glabrous or with papillose-based hairs, bases cordate, margins ciliate basally. **Panicles** 3–29.5 cm long, 0.7–3.1 cm wide, contracted; **branches** ascending to appressed, not fused to the rachises; **lower branches** (0.4)1–11.5 cm; **pedicels** 0.1–5.1 mm. **Spikelets** 2.9–5 mm, green (occasionally mostly purple), tips of the upper glumes and the sterile lemmas dark purple. **Lower glumes** 0.8–1.7 mm, 3–5-veined, margins hyaline; **upper glumes** 2.9–4.8 mm, conspicuously saccate, 11(12)-veined, sparsely puberulent in the distal ⅓, obtuse to acute; **lower florets** staminate; **lower lemmas** 2.8–4.9 mm, 5–7-veined, veins unequally spaced, lateral veins near the margin, margins hyaline and usually overlapping the palea distally; **lower paleas** 2–4 mm long, 0.8–1 mm wide, ¾ to almost equaling the lower lemmas; **anthers** 3, 1.1–1.7 mm, yellow to yellow-red to purple; **upper lemmas** 1.5–2 mm, subcoriaceous, shiny, white, glabrous, with 5 obscure veins, rounded to truncate, margins membranous, not inrolled over the paleas, scabridulous; **upper paleas** similar to the lemmas; **anthers** 3, 0.7–1 mm, yellow; **styles** purple. **Caryopses** 1–1.3 mm long, 0.7–0.9 mm wide, glabrous. $2n = 36$.

Sacciolepis striata is native to the southeastern United States and the West Indies, and from the Guianas to Venezuela and Amapá, Brazil. It grows along and in ponds, lakes, streams, and ditches, and flowers in late summer to fall.

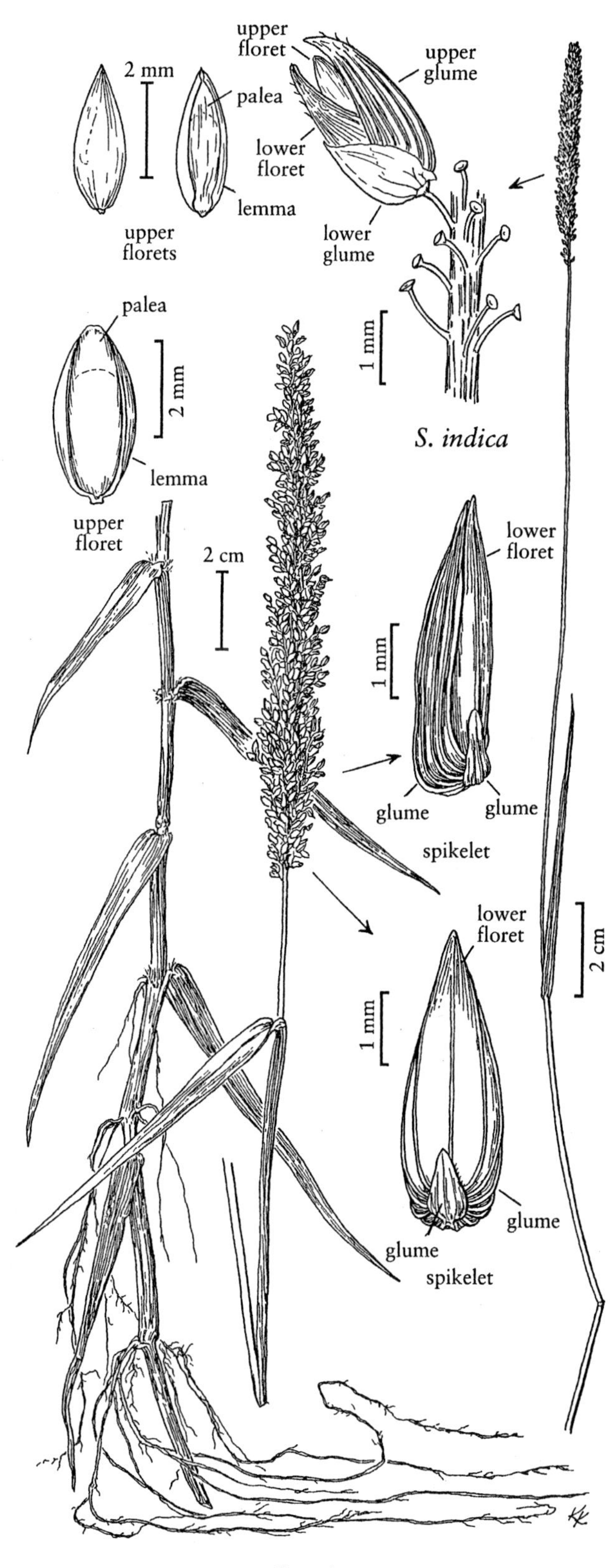

S. striata

SACCIOLEPIS

25.09 DICHANTHELIUM (Hitchc. & Chase) Gould

Robert W. Freckmann
Michel G. Lelong

Plants perennial; cespitose, sometimes rhizomatous, sometimes with hard, cormlike bases, often with basal winter rosettes of leaves having shortly ovate to lanceolate blades, these often sharply distinct from the blades of the cauline leaves. **Culms** 5–150 cm, herbaceous, hollow, usually erect or ascending, rarely sprawling, in the spring often spreading, sometimes decumbent in the fall, usually branching from the mid- or lower culm nodes in summer and fall; **branches** rebranching 1–4 times, terminating in small secondary panicles that are usually partly included in the sheaths. **Cauline leaves** 3–14, usually distinctly longer and narrower than the rosette blades; **ligules** of hairs, membranous, or membranous and ciliate, sometimes absent; **pseudoligules** of 1–5 mm hairs often present at the bases of the blades immediately behind the true ligules; **blades** usually distinctly longer and narrower than those of the basal rosette, cross sections with non-Kranz anatomy; **photosynthesis** C_3. **Inflorescences** panicles, terminal on the culms and branches; **sterile branches** and **bristles** absent; **disarticulation** below the glumes. **Primary panicles** terminating the culms, developing April–June(July), sometimes also in late fall, usually at least partially chasmogamous, often with a lower seed set than the secondary panicles; **secondary panicles** terminating the branches, produced from (May)June to fall, usually partially or totally cleistogamous. **Spikelets** 0.8–5.2 mm, not subtended by bristles, dorsally compressed, surfaces unequally convex, apices unawned. **Glume apices** not or only slightly gaping at maturity; **lower glumes** ⅕–¾ as long as the spikelets, 1–5-veined, truncate, acute, or acuminate; **upper glumes** slightly shorter than the spikelets or exceeding the upper florets by up to 1 mm, 5–11-veined, not saccate, apices rounded to attenuate. **Lower florets** sterile or staminate; **lower lemmas** similar to the upper glumes; **lower paleas** sometimes present, thin, shorter than the lower lemmas; **upper florets** bisexual, sessile, plump, usually apiculate to mucronate, sometimes minutely so, or subacute to (rarely) acute; **upper lemmas** striate, chartaceous-indurate, shiny, usually glabrous, margins involute; **upper paleas** striate; **lodicules** 2; **anthers** 3. **Caryopses** smooth; **pericarp** thin; **endosperm** hard; **hila** round or oval. $x = 9$. Name from the Greek *di*, 'twice' and *anth*, 'flowering', a reference to the two flowering periods.

Dichanthelium is a genus of approximately 72 species, 34 of which are native to the *Flora* region. It is often included in *Panicum*, the two taxa being similar in gross morphology. Recent molecular data reinforce the morphological arguments for recognizing *Dichanthelium* as a distinct genus.

When the branches of *Dichanthelium* develop, in late summer or fall, the culms acquire a very different aspect; comments about the 'fall phase' refer to the appearance of the plant or its culms following this branching. Unless stated otherwise, descriptions and measurements refer to structures of the culms and primary panicles, not those of the branches and secondary panicles. Ligule measurements usually include the hairs of the pseudoligule, if present, because the two are often difficult to distinguish with less than 30× magnification.

SELECTED REFERENCES **Allred, K.W.** and **F.W. Gould**. 1978. Geographic variation in the *Dichanthelium aciculare* complex (Poaceae). Brittonia 30:497–504; **Gould, F.W.** and **C.A. Clark**. 1978. *Dichanthelium* (Poaceae) in the United States and Canada. Ann. Missouri Bot. Gard. 65:1088–1132; **Guissani, L.M., J.H. Cota-Sánchez, F.O. Zuloaga,** and **E.A. Kellogg**. 2001. A molecular phylogeny of the grass subfamily Panicoideae (Poaceae) shows multiple origins of C_4 photosynthesis. Amer. J. Bot. 88:1993–2001; **Hitchcock, A.S.** 1951 [title page 1950]. Manual of the Grasses of the United States, ed. 2, rev. A. Chase. U.S.D.A. Miscellaneous Publication No. 200. U.S. Government Printing Office, Washington, D.C., U.S.A. 1051 pp.; **Hitchcock, A.S.** and **A. Chase**. 1910. The North American species of *Panicum*. Contr. U.S. Natl. Herb. 15:1–396; **LeBlond, R.J.** 2001. Taxonomy of the *Dichotoma* group of *Dichanthelium* (Poaceae). Sida 19:821–837; **Zuloaga, F.O., R.P. Ellis,** and **O. Morrone**. 1993. A Revision of *Panicum* subg. *Dichanthelium* sect. *Dichanthelium* (Poaceae: Panicoideae: Paniceae) in Mesoamerica, the West Indies, and South America. Ann. Missouri Bot. Gard. 80:119–190.

1. Basal leaf blades similar in shape to those of the lower cauline leaves, usually erect to ascending, clustered at the base, sometimes small or vestigial; culms branching from near the base in the fall, with 2–4 leaves, only the upper 2–4 internodes elongated.
   2. Blades soft, 3–12 mm wide, usually ciliate; upper blades less than 20 times as long as wide; fall phase with short panicle-bearing branches, without sterile shoots (sect. *Strigosa*).
      3. Leaf sheaths with retrorse or spreading hairs; upper blades 4–17 cm long, at least ¾ as long as the basal blades; blade margins usually finely short ciliate, the cilia not papillose-based; spikelets with papillose-based hairs . . . . 29. *D. laxiflorum*
      3. Leaf sheaths glabrous or with ascending hairs; upper blades 1.5–6 cm long, less than ¾ as long as the basal blades; blade margins with papillose-based cilia; spikelets glabrous or pubescent, hairs not papillose-based . . . . 30. *D. strigosum*
   2. Blades stiff, 1–5 mm wide, not ciliate; most upper blades at least 20 times as long as wide; fall phase with basal panicles and sterile shoots (sect. *Linearifolia*).
      4. Upper glumes and lower lemmas forming a beak extending 0.2–1 mm beyond the upper florets; spikelets 3.2–4.3 mm long; primary panicles with 7–25 spikelets . . . . 34. *D. depauperatum*
      4. Upper glumes and lower lemmas equaling or exceeding the upper florets by no more than 0.3 mm, not forming a beak; spikelets 2–3.4 mm long; primary panicles with 12–70 spikelets.
         5. Cauline blades 4–8 cm long, all alike; basal blades ascending to spreading . . . . 31 *D. wilcoxianum*
         5. Uppermost cauline blades 10–20 cm long, distinctly longer than the lower blades; basal blades erect to ascending.
            6. Panicles 1–3 cm wide, with ascending branches and appressed pedicels; spikelets turgid, 2.6–3.4 mm long, 1–1.7 mm wide, upper florets obovoid . . . . 32. *D. perlongum*
            6. Panicles 2–6 cm wide, with spreading branches and pedicels; spikelets not turgid, 2–3.2 mm long, 0 8–1.4 mm wide, upper florets ellipsoid . . . . 33. *D. linearifolium*
1. Basal leaf blades usually well-differentiated from the cauline blades, ovate to lanceolate, spreading, forming a rosette, or basal blades absent; culms usually branching from the midculm nodes in the fall, with 3–14 leaves, usually all internodes elongated.
   7. Bases of the culms hard, cormlike; basal rosettes absent; spikelets with papillose-based hairs and attenuate basally (sect. *Pedicellata*).
      8. Culms erect in the spring; cauline leaves 4–7, with thin, glabrous or sparsely hirsute blades that widen distal to the rounded to subcordate bases; lower glumes not encircling the pedicels, subadjacent to the upper glumes . . . . 1. *D. pedicellatum*
      8. Culms decumbent to ascending in the spring; cauline leaves 8–14, with thick, firm, puberulent blades that are parallel-sided distal to the rounded to truncate bases; lower glumes almost to completely encircling the pedicels, attached about 0.2 mm below the upper glumes . . . . 2. *D. nodatum*
   7. Bases of the culms not cormlike; basal rosettes usually present; spikelets not both with papillose-based hairs and attenuate basally.
      9. Blades cordate, thick, with white, cartilaginous margins; spikelets usually spherical to broadly obovoid or broadly ellipsoid, 1–1.8 mm long (sect. *Sphaerocarpa*).
         10. Spikelets 1–1.4 mm long; lower glumes 0.2–0.4 mm long; cauline blades 5–10 mm wide . . . . 23. *D. erectifolium*
         10. Spikelets 1.3–1.8 mm long; lower glumes 0.4–0.8 mm long; cauline blades 5–25 mm wide.
             11. Cauline blades 4–7, 10–25 cm long, 14–25 mm wide, with evident veins; culms nearly erect; panicles less than ½ as wide as long . . . . 24. *D. polyanthes*
             11. Cauline blades 3–4(6), 1.5–10 cm long, 5–14 mm wide, with obscure veins; culms decumbent or ascending; panicles more than ½ as wide as long . . . . . 25. *D. sphaerocarpon*
      9. Blades not cordate or the spikelets not both spherical and less than 1.9 mm long; blade margins usually not white and cartilaginous.
         12. Lower glumes thinner and more weakly veined than the upper glumes, attached about 0.2 mm below the upper glumes, the bases clasping the pedicels; spikelets attenuate basally.

13. Blades 2–7 cm long, about 10 times as long as wide, not or slightly involute, spreading, without raised veins, not longitudinally wrinkled; spikelets obovoid-obpyriform, planoconvex in side view (sect. *Lancearia*) . . . . 26. *D. portoricense*
13. Blades 4–16 cm long, more than 14 times as long as wide, or involute, stiffly erect or ascending, with prominently raised veins, the lower blades usually longitudinally wrinkled; spikelets ellipsoid to obovoid, biconvex in side view (sect. *Angustifolia*).
    14. Culms densely villous; nodes densely bearded; spikelets densely pubescent . . . . 28. *D. consanguineum*
    14. Culms glabrous, puberulent, or pilose with papillose-based hairs; nodes glabrous, puberulent to lightly bearded; spikelets glabrous or pubescent . . . . 27. *D. aciculare*

12. Lower glumes similar in texture and vein prominence to the upper glumes, attached immediately below the upper glumes, the bases not clasping the pedicels; spikelets usually not attenuate basally.
    15. Culms arising from rhizomes 3–5 mm thick, with (5)7–14 cauline blades; sheaths strongly hispid or viscid, mottled with pale spots, constricted at the top (sect. *Clandestina*).
        16. Nodes densely bearded above a viscid glabrous ring, often swollen; blades densely soft pubescent . . . . 10. *D. scoparium*
        16. Nodes glabrous or sparsely pubescent, not swollen; blades glabrous or sparsely pubescent.
            17. Cauline blades 7–15 mm wide, apices involute, long tapering; spikelets glabrous or sparsely puberulent . . . . 8. *D. scabriusculum*
            17. Cauline blades 15–30 mm wide, apices flat, acuminate; spikelets sparsely pubescent . . . . 9. *D. clandestinum*
    15. Culms arising from caudices or from rhizomes to 2 mm thick, with 3–7(9) cauline blades; sheaths not viscid, rarely hispid, not mottled with pale spots or constricted at the top.
        18. Ligules with a membranous base, ciliate distally; culms usually arising from slender rhizomes; lower florets often staminate; cauline blades 5–40 mm wide, often with a cordate base (sect. *Macrocarpa*).
            19. Spikelets ellipsoid, not turgid, with pointed apices; cauline blades 4–6, cordate at the base; sheaths without papillose-based hairs.
                20. Spikelets 2.2–3.2 mm long; ligules about 0.3 mm long; blades 5–25 mm wide; lower floret sterile . . . . 5. *D. commutatum*
                20. Spikelets 2.9–5.2 mm long; ligules 0.4–0.9 mm long; blades 15–40 mm wide; at least some lower florets staminate.
                    21. Nodes glabrous or slightly bearded; spikelets 2.9–3.9 mm long . . . 3. *D. latifolium*
                    21. Nodes densely retrorsely bearded; spikelets 3.8–5.2 mm long . . . . 4. *D. boscii*
            19. Spikelets obovoid, turgid, with rounded apices; cauline blades 3–4, tapered, rounded or truncate to cordate at the base; sheaths with papillose-based hairs.
                22. Blades and spikelets with papillose-based hairs; panicles usually slightly longer than wide, with spreading to ascending branches . . . . . 6. *D. leibergii*
                22. Blades glabrous; spikelets puberulent to almost glabrous; panicles usually more than twice as long as wide, with nearly erect branches . . . . 7. *D. xanthophysum*
        18. Ligules of hairs (except for *D. nudicaule*); culms arising from caudices; lower florets sterile; cauline blades 1–18 mm wide, bases usually tapered, rounded, or truncate at the base, sometimes cordate.
            23. Lower internodes short, upper nodes elongated; flag leaves distant and much reduced; culms rarely branching in the fall; branches, if present, few, developing from basal and subbasal nodes, erect (sect. *Nudicaulia*) . . . . 19. *D. nudicaule*

23. Lower internodes about as long as the upper internodes; flag leaves usually not much reduced; culms branching in the fall; branches often many, developing from mid- or upper culm nodes, often spreading.
    24. Spikelets 2.5–4.3 mm long, usually obovoid, turgid; upper glumes usually with an orange or purple spot at the base, the veins prominent (sect. *Oligosantha*).
        25. Nodes glabrous or sparsely pubescent; abaxial surfaces of the blades glabrous or pubescent, but not velvety pubescent . . . . 11. *D. oligosanthes*
        25. Nodes bearded with spreading to retrorse hairs; abaxial surfaces of the blades softly velvety pubescent.
            26. Spikelets 3.7–4.3 mm long; culms 2–3 mm thick, stiffly erect; ligules 2–5 mm long, without pseudoligules; blades glabrous or sparsely pilose on the adaxial surfaces . . . . 12. *D. ravenelii*
            26. Spikelets 2.5–3.2 mm long; culms usually 1–2 mm thick, erect; ligules 0.5–1 mm long, with the adjacent pseudoligules 1–3 mm long; blades densely velvety pubescent on both surfaces . . . . 13. *D. malacophyllum*
    24. Spikelets 0.8–3 mm long, ellipsoid or obovoid, not turgid; upper glumes lacking an orange or purple spot at the base and the veins not prominent.
        27. Ligules and adjacent pseudoligules 1–5 mm long, or the culms and sheaths with long hairs and also puberulent; spikelets variously pubescent to subglabrous (sect. *Lanuginosa*).
            28. Spikelets 0.8–1.1 mm long, puberulent to subglabrous; culms delicate, 0.3–0.8 mm thick . . . . 16. *D. wrightianum*
            28. Spikelets 1.1–3 mm long, variously pubescent; culms not delicate, usually more than 1 mm thick.
                29. Spikelets 1.1–2.1 mm long; sheaths glabrous or pubescent with hairs no more than 3 mm long . . . . . 14. *D. acuminatum*
                29. Spikelets 1.8–3 mm long; sheaths with hairs to 4 mm long . . . . 15. *D. ovale*
        27. Ligules absent or to 1.8 mm long, without adjacent pseudoligules; culms and at least the upper sheaths glabrous or sparsely pubescent with hairs of 1 length only; spikelets glabrous or pubescent.
            30. Culms (18)40–100 cm tall, rarely delicate, usually more than 1 mm thick; spikelets 1.5–2.7 mm long; blades 3.5–14 cm long, 5–14 mm wide (sect. *Dichanthelium*).
                31. Spikelets glabrous or, if pubescent, either the nodes bearded or the culms weak and prostrate; blade of the flag leaf usually spreading . . . . 17. *D. dichotomum*
                31. Spikelets pubescent; nodes glabrous; culms erect or ascending; blade of the flag leaf erect or ascending . . . . 18. *D. boreale*
            30. Culms 5–40(55) cm tall, delicate, usually less than 1 mm thick; spikelets 1.1–1.7 mm long; longest blades 1.5–6 cm long, 1.5–6 mm wide (sect. *Ensifolia*).
                32. Culms reclining or weakly erect; cauline blades 4–9, usually without prominent white, cartilaginous margins; ligules often more than 1 mm long . . . . 20. *D. ensifolium*
                32. Culms erect, sometimes geniculate basally; cauline blades 3–5, with prominent white, cartilaginous margins; ligules 0.2–0.7 mm long.
                    33. Culms few per clump; the fall phase branching sparingly; cauline blades flat, the bases rounded; blades of the flag leaves much shorter than those of the lower leaves . . . . 21. *D. tenue*

33. Culms many per clump; the fall phase branching extensively; cauline blades often involute, the bases subcordate; blades of the flag leaves only slightly shorter than those of the lower leaves . . . 22. *D. chamaelonche*

Dichanthelium sect. Pedicellata (Hitchc. & Chase) Freckmann & Lelong

Plants usually cespitose, sometimes with knotty rhizomes; **basal rosettes** absent. **Culms** initially erect, ultimately decumbent, often with hard, cormlike bases; **fall phase** freely divaricate-branching before the primary panicles mature. **Cauline leaves** 4–14; **sheaths** glabrous or pubescent, not viscid; **ligules** membranous and ciliate or of hairs; **blades** with papillose-based marginal cilia. **Primary panicles** exserted. **Spikelets** obovoid to obpyriform, with papillose-based hairs, attenuate basally. **Upper florets** with pointed to attenuate, puberulent apices.

1. Dichanthelium pedicellatum (Vasey) Gould [p. 411]
CORM-BASED PANICGRASS

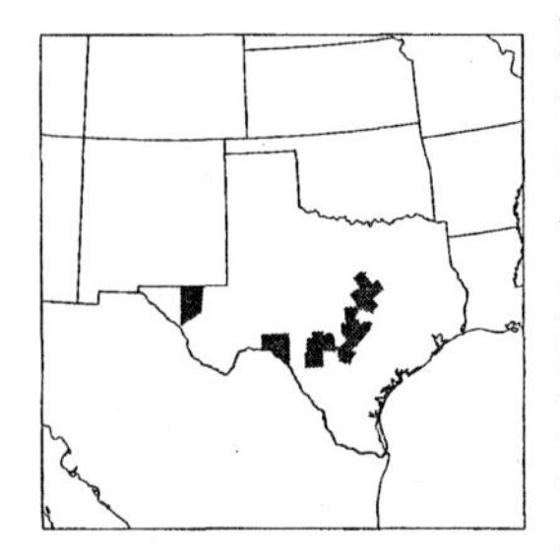

Plants cespitose, not rhizomatous. **Basal rosettes** absent. **Culms** 20–70 cm, initially erect, with hard, cormlike bases; **nodes** puberulent to sparsely hirsute; **internodes** all elongated, puberulent to hirsute; **fall phase** with decumbent culms, developing divaricate branches from the midculm nodes before the primary panicles mature. **Cauline leaves** 4–7; **sheaths** sometimes overlapping, puberulent to papillose-hispid, margins ciliate; **ligules** 0.3–1 mm, membranous and ciliate; **blades** 3–12 cm long, 2–8 mm wide, widening distal to the rounded or subcordate bases, thin, glabrous or sparsely hirsute, margins with papillose-based cilia. **Primary panicles** 3–6 cm long, 2–4 cm wide, exserted; **branches** spreading at maturity; **pedicels** somewhat divergent. **Spikelets** 3.2–4.4 mm long, 1.3–1.6 mm wide, narrowly obovoid-ellipsoid, papillose-hirsute, attenuate to the purplish bases. **Lower glumes** about ½ as long as the spikelets, narrowly triangular, subadjacent to the upper glumes, not encircling the pedicels; **upper glumes** about 0.3 mm shorter than the upper florets; **lower florets** sterile; **upper florets** with pointed, minutely puberulent apices. $2n = 18$.

Dichanthelium pedicellatum grows on limestone outcroppings and in dry, open oak woodlands. Its range extends from Texas into Mexico and Guatemala. Primary panicles develop from late March into June (and sometimes from late August to November) and are open-pollinated; secondary panicles develop from May into fall and are at least partly cleistogamous.

2. Dichanthelium nodatum (Hitchc. & Chase) Gould [p. 411]
SARITA PANICGRASS

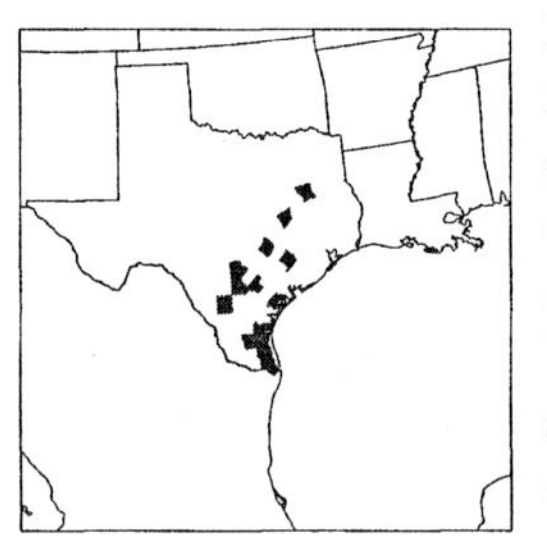

Plants usually cespitose, rarely rhizomatous. **Basal rosettes** absent. **Culms** 20–65 cm, decumbent to ascending even in spring, with hard, cormlike bases; **nodes** puberulent to sparsely pubescent; **internodes** scabrous-puberulent to papillose-hirsute; **fall phase** with geniculate to decumbent culms, developing divaricate branches from the midculm nodes before the primary panicles mature. **Cauline leaves** 8–14; **sheaths** not overlapping, puberulent to papillose-hirsute, margins ciliate; **ligules** 0.1–1 mm, of hairs; **blades** 3–9 cm long, 4–8 mm wide, thick, firm, puberulent, sides parallel above the rounded to truncate bases, margins with papillose-based cilia. **Primary panicles** 3–13 cm long, 2–8 cm wide, exserted; **branches** ascending to divaricate at maturity; **pedicels** appressed. **Spikelets** 3.4–4.4 mm long, 1.3–1.6 mm wide, narrowly obovoid-obpyriform, finely pubescent, hairs papillose-based, bases long, narrow. **Lower glumes** 1.5–2 mm, attached about 0.2 mm below the upper glumes, partly or completely encircling the pedicels; **upper glumes** about 0.3 mm shorter than the upper florets, purplish at the bases; **lower florets** sterile; **upper florets** with pointed, puberulent apices. $2n = 18$ (J. Wipff, pers. comm., 2001).

Dichanthelium nodatum grows in oak savannahs near the Gulf coast from Texas to northeastern Mexico. The primary panicles are produced from April into June (sometimes late August to November) and are at least partly open-pollinated; the secondary panicles are produced from May into fall and are at least partly cleistogamous.

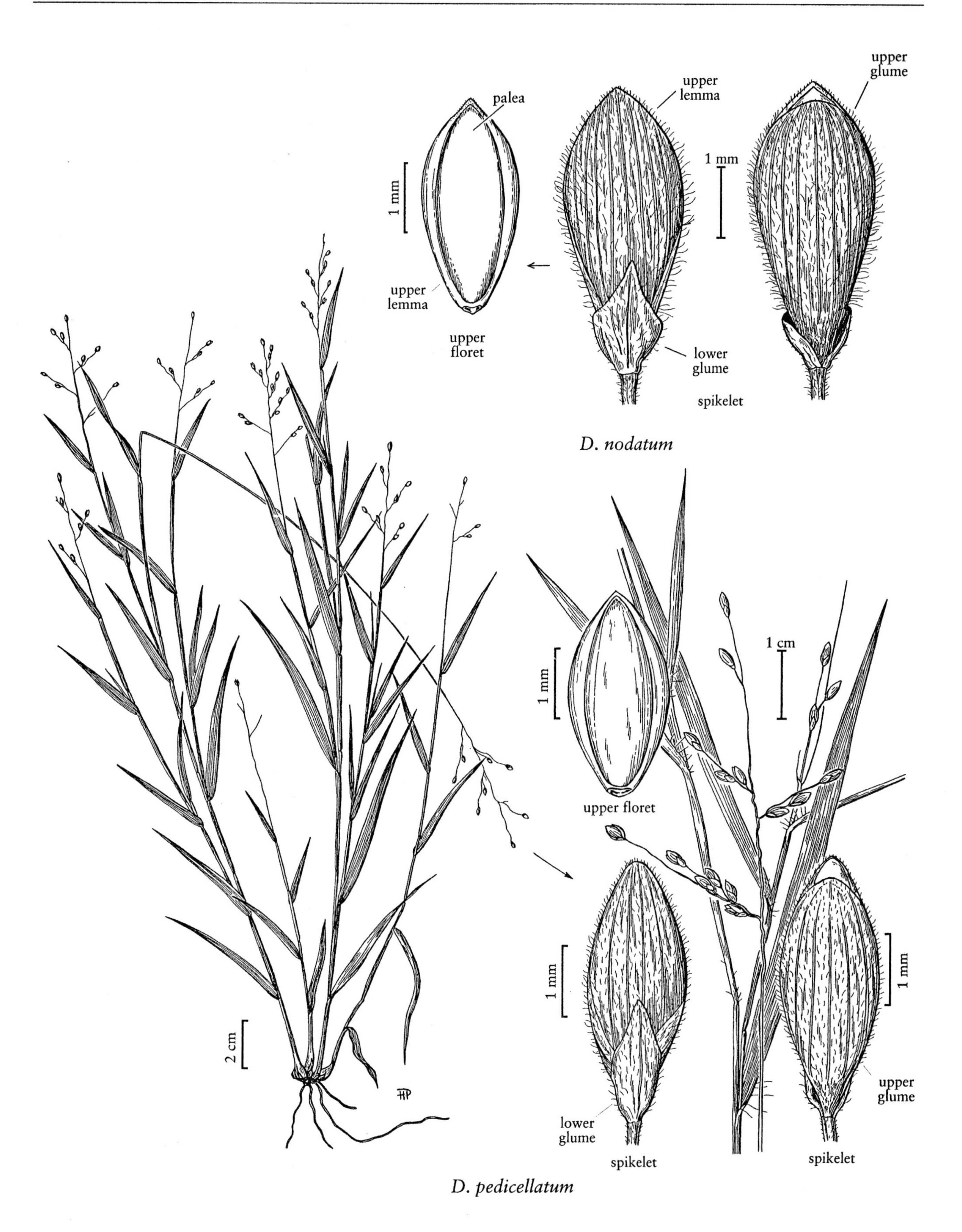

D. nodatum

D. pedicellatum

Dichanthelium sect. Macrocarpa Freckmann & Lelong

Plants cespitose, often with knotty rhizomes, sometimes with caudices. **Basal rosettes** usually well-differentiated. **Culms** 20–110 cm, ascending to erect; **fall phase** sparsely rebranching, not producing dense axillary fascicles. **Cauline leaves** 3–6; **sheaths** glabrous or pubescent, not viscid; **ligules** shortly membranous and ciliate, cilia longer than the membranous portion. **Primary panicles** at least partially exserted. **Spikelets** narrowly ellipsoid to obovoid, pubescent to puberulent, sometimes sparsely so, hairs sometimes papillose-based. **Upper florets** pointed, umbonate, mucronate, or apiculate.

3. **Dichanthelium latifolium** (L.) Harvill [p. 413]
Broadleaved Panicgrass, Panic à Larges Feuilles

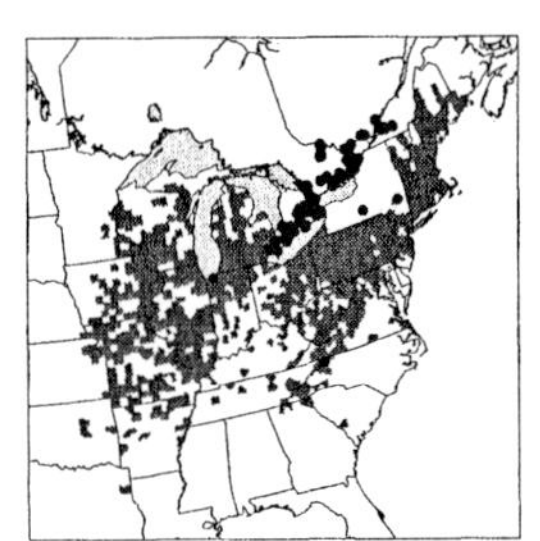

Plants forming small clumps, with knotty rhizomes less than 2 mm thick. **Basal rosettes** well-differentiated; **sheaths** pubescent; **blades** ovate to lanceolate, dark green. **Culms** 45–110 cm, nearly erect; **nodes** glabrous or the lower nodes slightly bearded; **internodes** glabrous or sparsely pubescent; **fall phase** branching from the midculm nodes, branches nearly erect, scarcely rebranching, blades and secondary panicles only slightly reduced. **Cauline leaves** 4–6, often with a transitional leaf above the basal rosette; **sheaths** not overlapping, glabrous or softly villous basally, margins ciliate, collars pubescent; **ligules** 0.4–0.7 mm, membranous, ciliate, cilia longer than the membranous portion; **blades** 3.7–7 times longer than wide, 15–40 mm wide, ovate-lanceolate, glabrous or sparsely pubescent, with 11–13 major veins and 40–120 minor veins, bases cordate-clasping, with papillose-based cilia. **Primary panicles** 7–15 cm long, 4–12 cm wide, 1.5–2 times as long as wide, with 20–80 spikelets, eventually at least partially exserted; **branches** stiff, ascending to spreading. **Spikelets** 2.9–3.9 mm long, 1.6–2 mm wide, ellipsoid, sparsely pubescent. **Lower glumes** ⅓–½ as long as the spikelets, narrowly triangular; **upper glumes** and **lower lemmas** slightly shorter than the spikelets, often red-tinged basally and apically; **lower florets** staminate, anthers exserted prior to those of the upper florets; **upper florets** pointed, apiculate, upper lemmas with a minute fringe of hairs. $2n$ = 18, 36.

Dichanthelium latifolium grows in rich deciduous woods, often in slightly open areas within eastern North America. The primary panicles are open-pollinated and develop in May and June (and sometimes in September and October), the secondary panicles, which are produced from July through September, are rarely open-pollinated.

4. **Dichanthelium boscii** (Poir.) Gould & C.A. Clark [p. 413]
Bosc's Panicgrass, Panic de Bosc

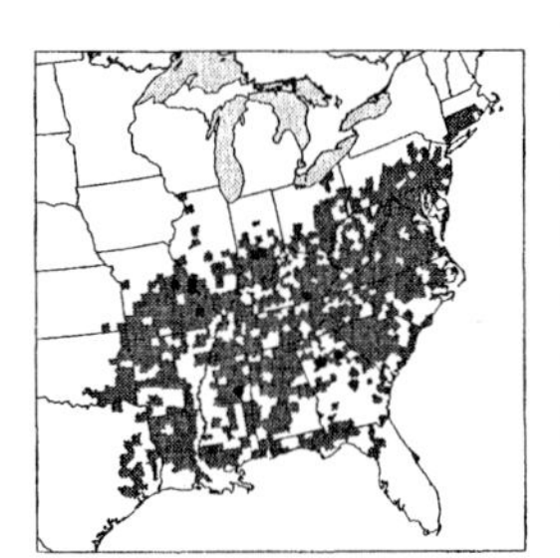

Plants forming small clumps, with knotty rhizomes less than 2 mm thick. **Basal rosettes** well-differentiated; **sheaths** pubescent; **blades** ovate to lanceolate, dark green. **Culms** 25–75 cm, initially erect, often sprawling in the fall, **nodes** densely retrorsely bearded; **internodes** glabrous, or pilose with papillose-based hairs; **fall phase** branching from the midculm nodes, branches nearly erect, sparsely rebranching, blades and secondary panicles only slightly reduced. **Cauline leaves** 4–6, often with a transitional leaf above the basal rosette; **sheaths** not overlapping, bases puberulent to retrorsely pilose, margins ciliate, collars pubescent; **ligules** 0.4–0.9 mm, membranous, ciliate, cilia longer than the membranous portion; **blades** 3–6 times longer than wide, 15–40 mm wide, ovate-lanceolate, glabrous, puberulent, or pilose, with 11–15 major veins and 40–120 minor veins, bases cordate, margins with papillose-based cilia. **Panicles** 4–12 cm long, 4–12 cm wide, about as long as wide when fully expanded, partially included to tardily exserted, with 16–60 spikelets. **Spikelets** 3.8–5.2 mm long, 1.7–2.2 mm wide, narrowly ellipsoid, pubescent or puberulent. **Lower glumes** ⅓–½ as long as the spikelets, narrowly triangular; **upper glumes** shorter than the spikelets; **lower florets** usually staminate; **upper florets** pointed, with a minute tuft of hairs. $2n$ = 18, 36.

Dichanthelium boscii usually grows in semi-open areas in dry oak-hickory woods of the eastern United States. The primary panicles are open-pollinated and are produced from late April through June (and sometimes again in the fall); the secondary panicles are partly open-pollinated, and are produced from July through September.

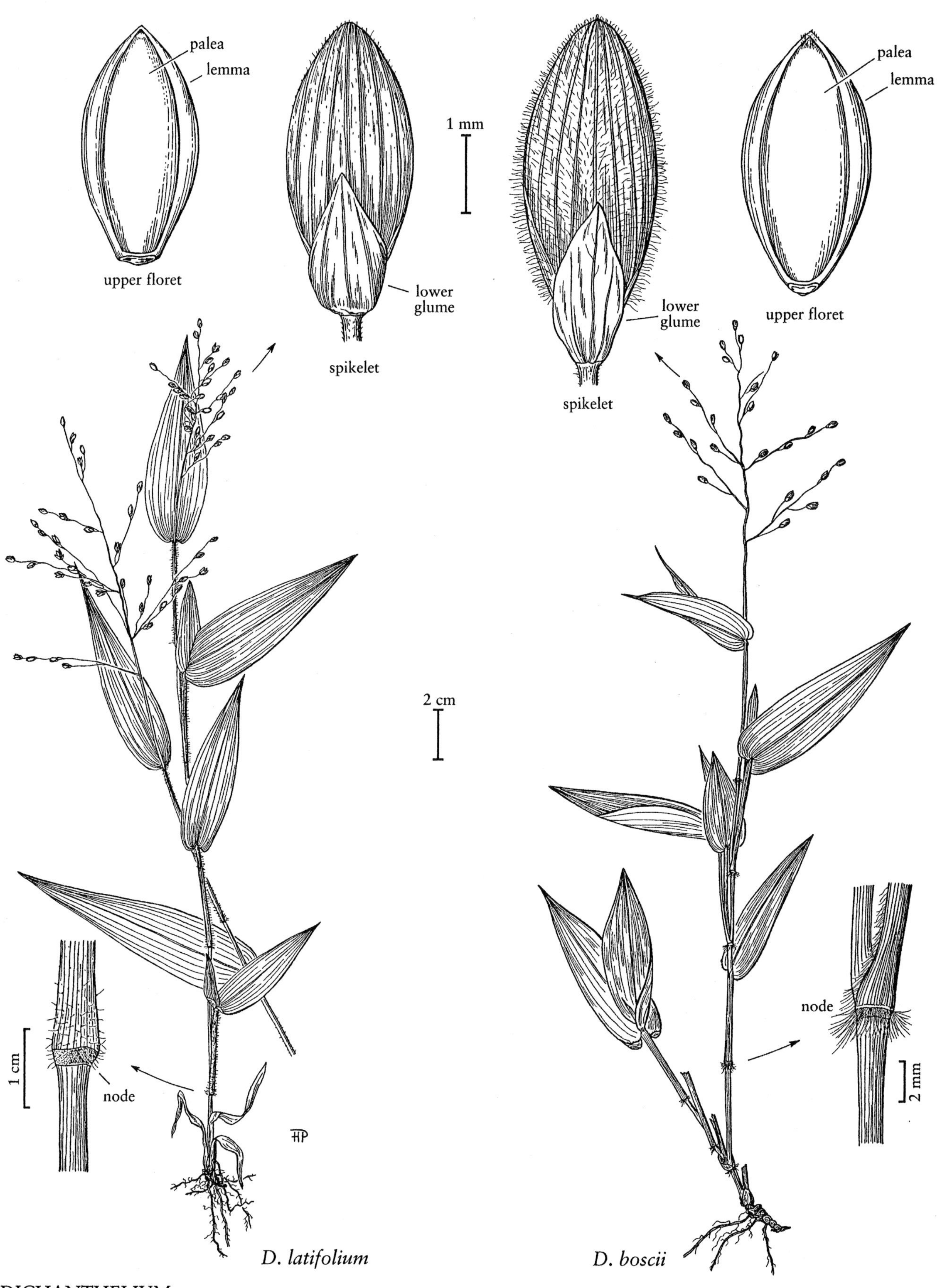
palea
lemma
upper floret
1 mm
lower glume
spikelet
palea
lemma
lower glume
upper floret
spikelet
2 cm
1 cm
node
node
2 mm
HP
D. latifolium
D. boscii

5. Dichanthelium commutatum (Schult.) Gould
[p. 415]
VARIABLE PANICGRASS

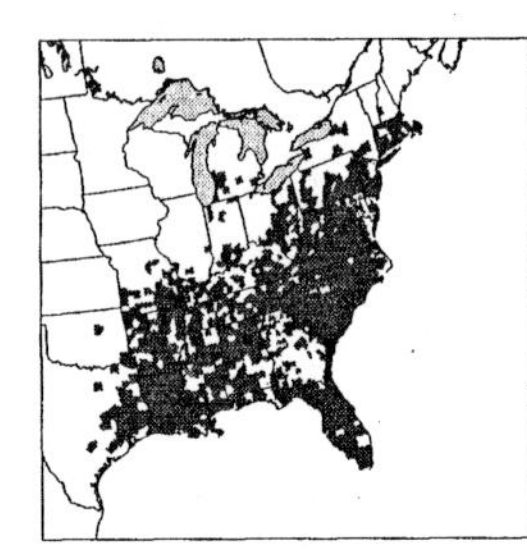

Plants cespitose, with caudices or with rhizomes up to 2 mm thick. **Basal rosettes** well-differentiated; **blades** 1–14 cm long, to 22 mm wide, ovate to lanceolate. **Culms** 20–75 cm, erect or decumbent to sprawling, often purplish; **nodes** and **internodes** glabrous or puberulent to pubescent; **fall phase** initially nearly erect, often sprawling eventually, branches initially erect and apparently dichotomous, later rebranching, blades and secondary panicles smaller than those of the culms. **Cauline leaves** 4–6; **sheaths** not overlapping, often glaucous, purplish, or olivaceous, glabrous or puberulent, margins usually ciliate; **ligules** about 0.3 mm, membranous, ciliate, cilia longer than the membranous portion, rarely with adjacent, about 12 mm hairs; **blades** 5–16 cm long, 5–25 mm wide, linear to ovate-lanceolate, glabrous or puberulent, with 9–13 major veins and 30–80 minor veins, bases cordate-clasping, often asymmetrical, with papillose-based marginal cilia. **Panicles** 5–12 cm long, 3–10 cm wide, open, exserted; **branches** flexuous. **Spikelets** 2.2–3.2 mm long, 1.1–1.3 mm wide, ellipsoid, yellowish-green or purplish, pubescent. **Lower glumes** 0.7–1.8 mm; **upper glumes** and **lower lemmas** equaling or slightly shorter than the spikelets; **lower florets** sterile; **upper florets** often minutely umbonate. $2n = 18$.

Dichanthelium commutatum is fairly common in dry to wet, semi-open woodlands. Its range extends from the eastern United States to South America. The primary panicles are open-pollinated and are produced from April through June; the secondary panicles are primarily cleistogamous and are produced from June through fall.

The four subspecies are fairly distinct in some parts of their ranges, but subsp. *commutatum* intergrades with the other three where they occur together.

1. Culms densely crisp-puberulent; spikelets 2.2–2.7 mm long; cauline blades usually 5–8 cm long, 5–10 mm wide, thick, the bases symmetrical; rosette blades usually less than 3 cm long and to 6 mm wide . . . . subsp. *ashei*
1. Culms usually glabrous or sparsely pubescent; spikelets 2.6–3.2 mm long; cauline blades usually more than 8 cm long and 10 mm wide, thin, bases sometimes asymmetrical; rosette blades large, some more than 4 cm long and 10 mm wide.
   2. Cauline blades nearly linear, 5–14 mm wide, about 10 times as long as wide; spikelets 3–3.2 mm long; lower glumes about ½ as long as the spikelets . . . . subsp. *equilaterale*
   2. Cauline blades ovate-lanceolate, 6–25 mm wide, about 4–8 times as long as wide; spikelets 2.6–3.2 mm long; lower glumes about ¼ as long as the spikelets.
      3. Culms decumbent or sprawling, with loose caudices or rhizomes; blades strongly asymmetric-falcate, often; spikelets 2.9–3.2 mm long; lower lemmas pointed . . . . subsp. *joorii*
      3. Culms more or less erect, with caudices; blades almost symmetrical, green, sometimes glaucous; spikelets 2.6–2.9 mm long; lower lemmas rounded . . . subsp. *commutatum*

Dichanthelium commutatum subsp. **ashei** (T.G. Pearson *ex* Ashe) Freckmann & Lelong

Plants from small caudices. **Culms** slender, wiry, erect, densely crisp-puberulent. **Basal blades** usually shorter than 3 cm, to 6 mm wide. **Cauline blades** 5–8 cm long, 5–10 mm wide, linear-lanceolate, thick, often yellowish-green, bases symmetrical. **Spikelets** 2.2–2.7 mm. **Lower glumes** ⅕–¼ the length of the spikelets; lower lemmas pointed.

Dichanthelium commutatum subsp. *ashei* grows in open woodlands. It sometimes resembles, and may hybridize with, *D. dichotomum.*

Dichanthelium commutatum (Schult.) Gould subsp. **commutatum**

Plants with caudices. **Culms** more or less erect, usually glabrous, sometimes sparsely pubescent or puberulent. **Basal blades** large, usually 8–14 cm long, 7–22 mm wide. **Cauline blades** 3.5–8 times as long as wide, 6–25 mm wide, thin, ovate-lanceolate, green, sometimes glaucous, bases almost symmetrical. **Spikelets** 2.6–2.9 mm. **Lower glumes** about ¼ as long as the spikelets; **lower lemmas** rounded apically.

Dichanthelium commutatum subsp. *commutatum* grows in wet to dry woodlands. Its range extends to South America.

Dichanthelium commutatum subsp. **equilaterale** (Scribn.) Freckmann & Lelong

Plants from loose caudices. **Culms** stiffly erect, glabrous. **Basal blades** large, usually 8–14 cm long, 7–22 mm wide. **Cauline blades** 5–14 mm wide, about 10 times longer than wide, linear, thin, firm, spreading. **Spikelets** 3–3.2 mm; **lower glumes** 1.6–1.8 mm, about ½ as long as the spikelets.

Dichanthelium commutatum subsp. *equilaterale* grows in sandy pine and oak woodlands. Its range extends to southeastern Mexico and Nicaragua.

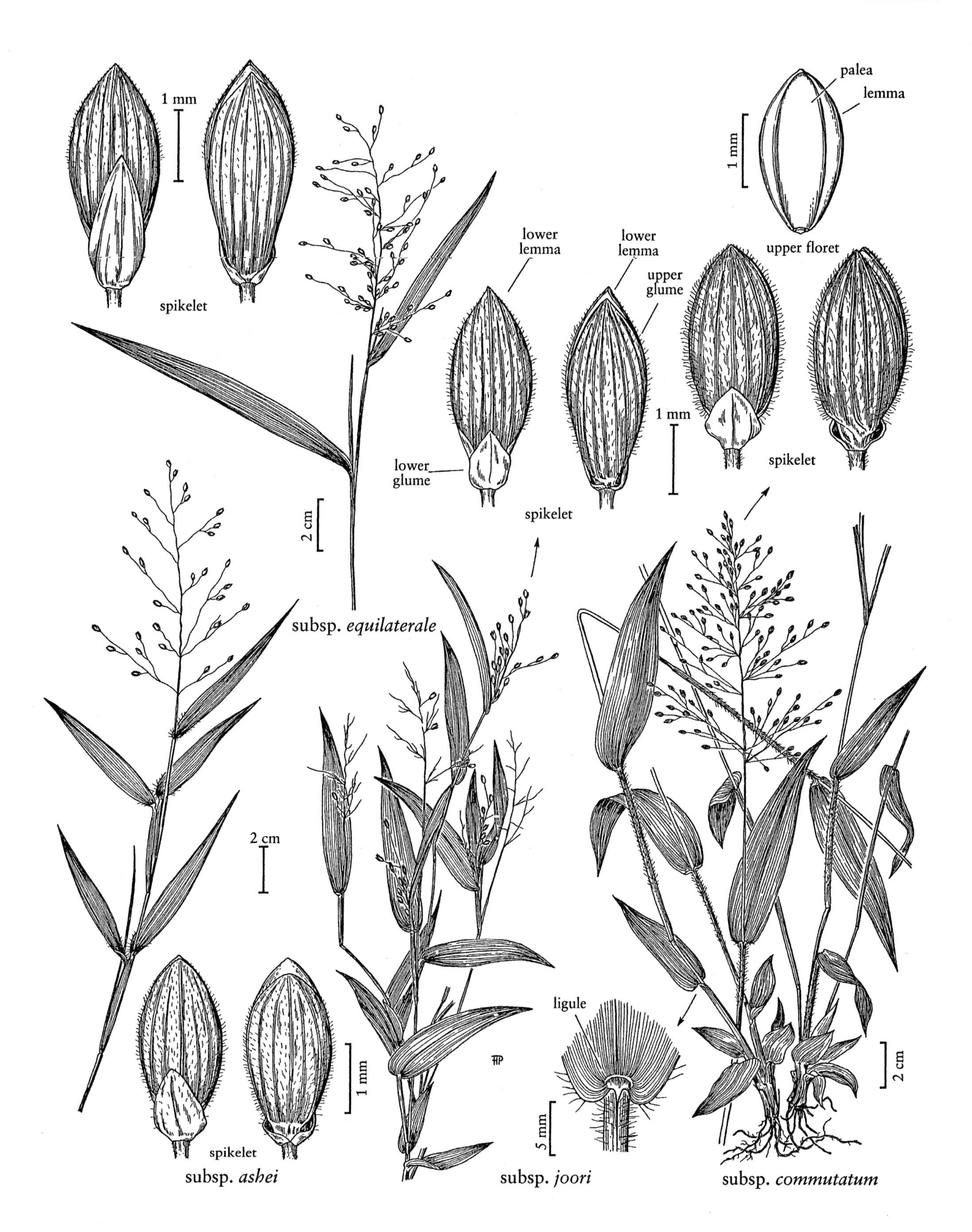

DICHANTHELIUM COMMUTATUM

Dichanthelium commutatum subsp. **joorii** (Vasey) Freckmann & Lelong

Plants from loose caudices or with knotty or loose rhizomes. **Culms** decumbent or sprawling, glabrous, sometimes glaucous, sometimes purplish. **Basal blades** large, usually 8–14 cm long, 7–22 mm wide. **Cauline blades** 8–25 mm wide, 4–8 times longer than wide, thin, ovate-lanceolate, often glaucous, strongly asymmetric-falcate. **Spikelets** 2.9–3.2 mm. **Lower glumes** about ¼ as long as the spikelets; **lower lemmas** pointed.

Dichanthelium commutatum subsp. *joorii* grows in wet woodlands and swamps. Its range extends into Mexico.

6. **Dichanthelium leibergii** (Vasey) Freckmann [p. 417]
LEIBERG'S PANICGRASS

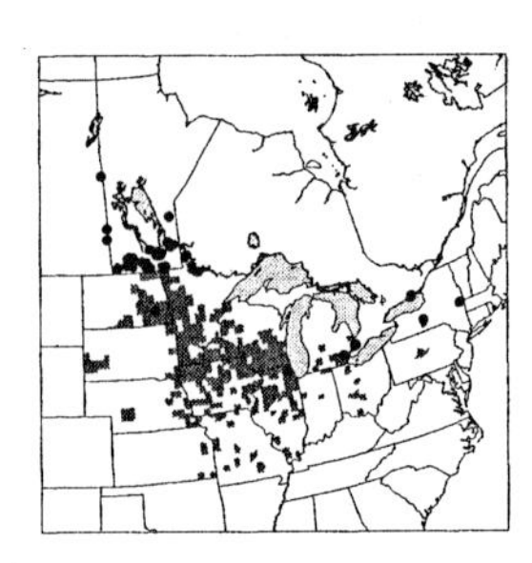

Plants cespitose, with knotty rhizomes no more than 2 mm thick. **Basal rosettes** well-differentiated; **blades** few, small, ovate to lanceolate. **Culms** 24–80 cm, glabrous or puberulent; **nodes** sparsely spreading-pilose; **internodes** mostly elongated, glabrous or puberulent; **fall phase** with a few suberect branches from the lower and midculm nodes, blades slightly reduced, secondary panicles partially exserted. **Cauline leaves** 3–4; **sheaths** not overlapping, with ascending papillose-based hairs; **ligules** 0.3–0.5 mm, membranous, ciliate, cilia longer than the membranous portion; **blades** 5–15 cm long, 7–13 mm wide, ascending to erect, sparsely to densely pubescent with papillose-based hairs, with 9–11 prominent major veins and 25–50 minor veins, bases truncate to cordate, margins with papillose-based cilia. **Panicles** 6–10 cm long, 3–5 cm wide, their length usually less than twice their width, eventually well-exserted, with 20–40 spikelets; **branches** spreading to ascending. **Spikelets** 3.3–3.8 mm long, 1.6–2 mm wide, ellipsoid-obovoid, turgid, pubescent, hairs papillose-based, apices rounded. **Lower glumes** about 1.8 mm, narrowly triangular; **lower florets** staminate; **upper florets** mucronate. $2n = 18$.

Dichanthelium leibergii grows primarily on prairie relics, but is occasionally found in sandy woodlands. It is restricted to the *Flora* region. The primary panicles are produced from mid-May through July, the secondary panicles from late June to September. Sterile putative hybrids with *D. acuminatum* and *D. xanthophysum* are occasionally found.

7. **Dichanthelium xanthophysum** (A. Gray) Freckmann [p. 417]
PALE PANICGRASS, PANIC JAUNÂTRE

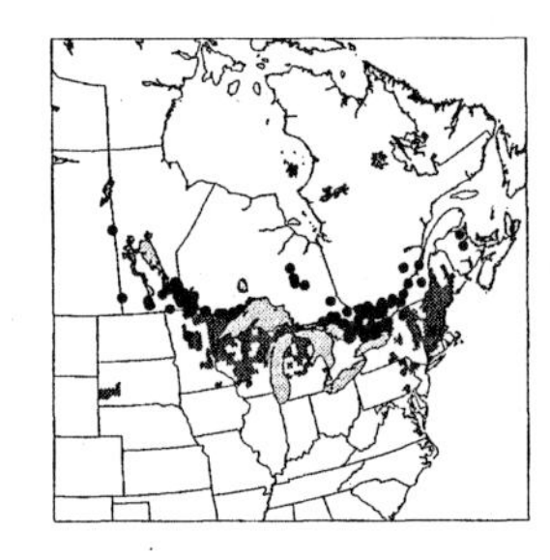

Plants loosely cespitose, with knotty rhizomes to 2 mm thick. **Basal rosettes** often poorly differentiated; **blades** few, grading into the cauline blades. **Culms** 20–55 cm, most forming in the spring, additional culms sometimes produced in the fall; **nodes** glabrous or sparsely ascending-pubescent; **internodes** all elongated, glabrous or puberulent; **fall phase** with a few suberect branches from the lower and midculm nodes, branches not rebranching, blades slightly reduced, secondary panicles partially exserted. **Cauline leaves** 3–4; **lower sheaths** not overlapping, sometimes pubescent; **upper sheaths** overlapping, sparsely to densely pubescent, hairs papillose-based, margins ciliate; **ligules** 0.3–0.5 mm, membranous, ciliate, cilia longer than the membranous bases; **blades** 7–17 cm long, 7–23 mm wide, erect, pale yellow-green to bluish-green, glabrous, with 7–11 prominent major veins and 30–110 minor veins, bases tapered or rounded to truncate, margins with papillose-based cilia. **Panicles** 7–14 cm long, 1–5 cm wide, their length usually more than twice their width, narrowly cylindric, eventually well-exserted, with 9–46 spikelets; **branches** strongly ascending, stiff. **Spikelets** 3.2–4.1 mm long, 1.8–2.2 mm wide, obovoid, turgid, puberulent to subglabrous, with rounded apices. **Lower glumes** 1.7–2.2 mm, narrowly triangular; **lower florets** staminate; **upper florets** longer than the upper glumes, mucronate. $2n = 36$.

Dichanthelium xanthophysum usually grows on sandy or rocky soils in semi-open pine, oak, or aspen woodlands. It extends from eastern Saskatchewan and northeast Montana to Quebec, New England, and West Virginia. Plants from Minnesota and western Quebec approach *D. leibergii* in having cauline blades narrower than 10 mm, and papillose-based hairs. Sterile putative hybrids with *D. leibergii* and *D. boreale* are rare; those with *D. boreale* have been called *Panicum calliphyllum* Ashe.

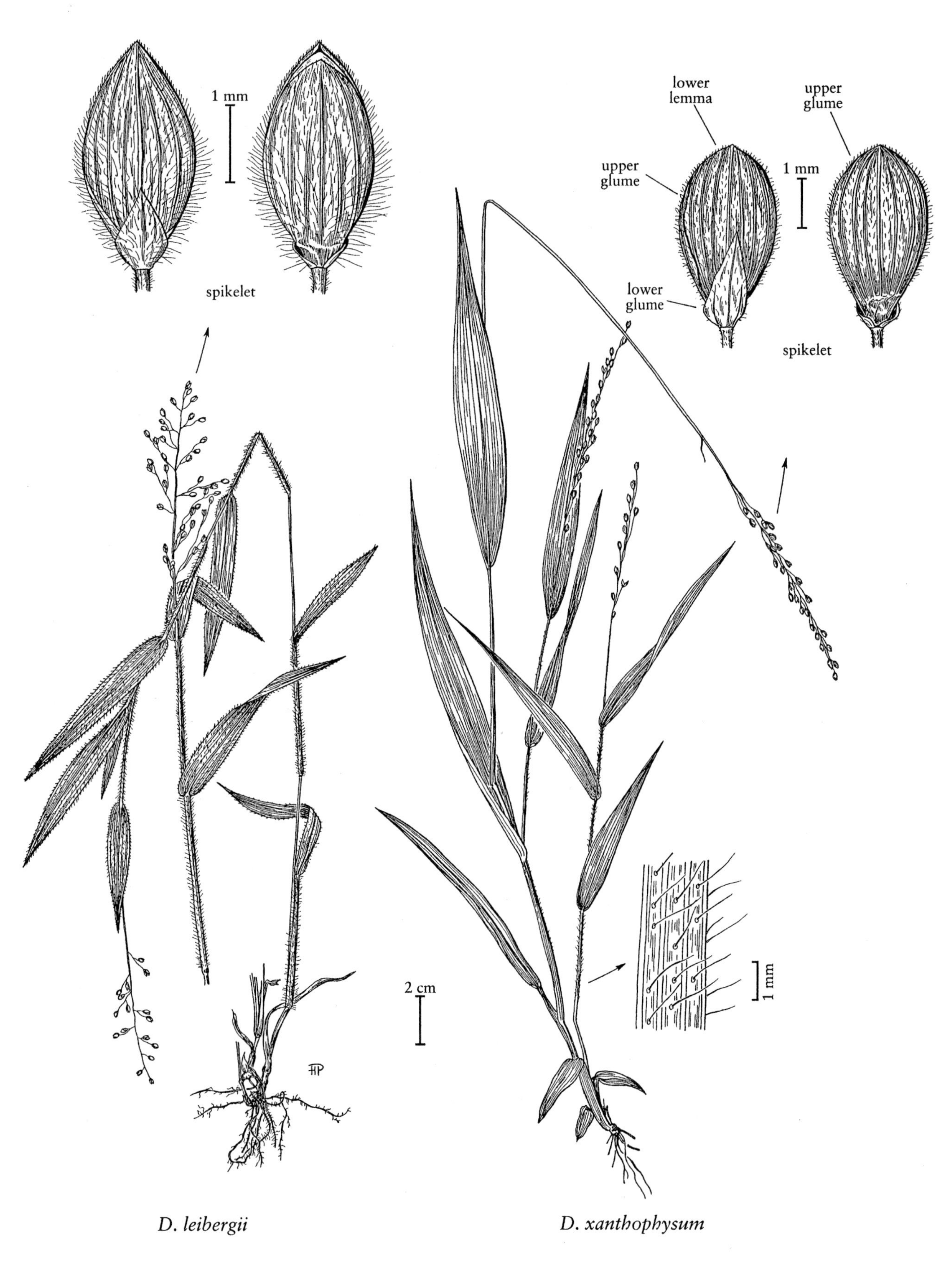

D. leibergii *D. xanthophysum*

Dichanthelium sect. **Clandestina** Freckmann & Lelong

Plants cespitose, with rhizomes 3–5 mm thick. **Basal rosettes** well-differentiated; **blades** 9–25 cm, ovate-lanceolate. **Culms** 50–150 cm; **fall phase** branching from the mid- and upper culm nodes, branches rebranching, often forming dense axillary fascicles, secondary panicles enclosed in the sheaths. **Cauline leaves** 5–14; **sheaths** usually widest near midlength, narrowed to the summit, strongly papillose-hispid or softly pubescent and viscid, summits often mottled with white to yellowish blotches; **ligules** membranous or of hairs. **Primary panicles** exserted. **Spikelets** 2.2–3.6 mm, ellipsoid to ovoid, with prominent veins, pointed. **Upper florets** acute to acuminate, sometimes umbonate or apiculate, apices often with a minute tuft of hairs.

8. **Dichanthelium scabriusculum** (Elliott) Gould & C.A. Clark [p. 420]
TALL-SWAMP PANICGRASS

Plants in large clumps, with rhizomes 3–5 mm thick. **Basal rosettes** well-differentiated; **sheaths** pubescent; **blades** lanceolate. **Culms** 70–150 cm, robust, purplish; **nodes** glabrous or puberulent; **internodes** scabridulous to almost glabrous; **fall phase** branching from the mid- and upper culm nodes, developing numerous, well-separated, dense fascicles of many reduced blades and hidden secondary panicles. **Cauline leaves** 6–14; **sheaths** not overlapping, narrowing above midlength, sparsely to densely papillose-hispid, tops mottled with pale spots, margins ciliate, collars puberulent; **ligules** 0.5–1.2 mm, membranous; **blades** 12–25 cm long, 7–15 mm wide, linear, stiff, ascending to spreading, glabrous or sparsely pubescent, bases subcordate to constricted, margins scabridulous, apices long tapering, involute. **Primary panicles** 10–21 cm long, 6–13 cm wide, eventually well-exserted, with many spikelets; **rachises** and **branches** usually glabrous and mottled. **Spikelets** 2.2–2.8 mm long, 1–1.2 mm wide, ovoid-ellipsoid, often purplish, glabrous, rarely sparsely puberulent. **Lower glumes** 0.5–1 mm, acute; **upper glumes** and **lower florets** exceeding the upper florets, prominently 7–9-veined; **lower florets** sterile; **upper florets** acute to acuminate, with a minute tuft of hairs at the apices. $2n = 18$.

Dichanthelium scabriusculum usually grows in wet, sandy, open sites, including shores, stream banks, swamps, and bogs. It is restricted to the eastern United States. The primary panicles develop from May to July, the secondary panicles, which are usually concealed within the sheaths, from July through November. *Panicum aculeatum* Hitchc. & Chase refers to what appear to be sterile hybrids with *Dichanthelium clandestinum* or robust subspecies of *D. dichotomum*, *P. bennettense* W.V. Br. to hybrids with *D. aciculare*.

9. **Dichanthelium clandestinum** (L.) Gould [p. 420]
DEER-TONGUE GRASS, PANIC CLANDESTIN

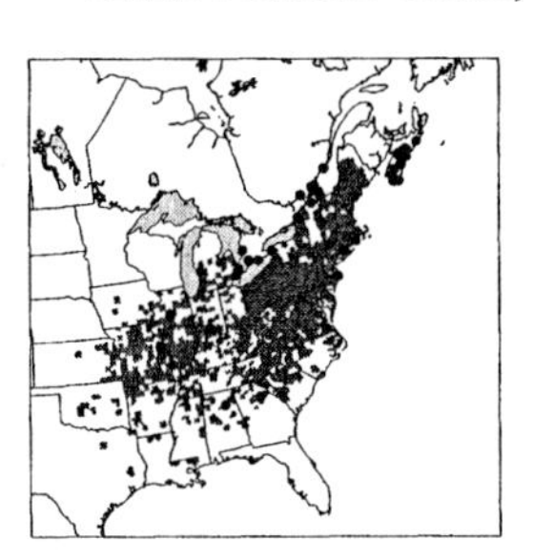

Plants forming large clumps, with rhizomes 3–5 mm thick. **Basal rosettes** well-differentiated; **sheaths** pubescent; **blades** ovate to lanceolate. **Culms** 50–140 cm, stout, pilose with papillose-based hairs to subglabrous; **fall phase** branching from the mid- and upper culm nodes, with a few, nearly erect, elongate branches, sparsely rebranching, sheaths overlapping, concealing the secondary panicles; **nodes** not swollen, glabrous or sparsely pubescent. **Cauline leaves** 5–10; **sheaths** not overlapping, striate-ribbed, narrowing above midlength, hispid to sparsely hirsute, hairs sometimes papillose-based, summits mottled with pale spots, margins ciliate, collars puberulent; **ligules** 0.4–0.9 mm, membranous; **blades** 10–25 cm long, 15–30 mm wide, flat, lanceolate, often rigid, glabrous or sparsely pubescent, with 9–13 major veins and 40–80 minor veins, bases cordate, with papillose-based cilia, apices acuminate. **Primary panicles** 8–16 cm long, 4–12 mm wide, exserted, with many spikelets. **Spikelets** 2.4–3.6 mm long, 1.2–1.5 mm wide, narrowly ellipsoid, sparsely pubescent. **Lower glumes** ⅓–½ as long as the spikelets, narrowly triangular; **upper glumes** and **lower florets** slightly shorter than the spikelets, with 7 or 9 prominent veins; **lower florets** sterile; **upper florets** umbonate, apices with a minute tuft of hairs. $2n = 36$.

Dichanthelium clandestinum usually grows in semi-open areas in damp or sandy woodlands, thickets, or on banks. It is restricted to the eastern part of the *Flora* region. The primary panicles are open-pollinated for a brief period, and produced from late May to early July; the secondary panicles, which are cleistogamous and usually concealed within the sheaths, are produced from July through September.

Panicum recognitum Fernald refers to rare sterile hybrids with *Dichanthelium dichotomum* and perhaps *D. scoparium*; *P. aculeatum* Hitchc. & Chase to putative sterile hybrids with *D. scabriusculum* or *D. dichotomum*.

10. **Dichanthelium scoparium** (Lam.) Gould [p. 420]
VELVETY PANICGRASS

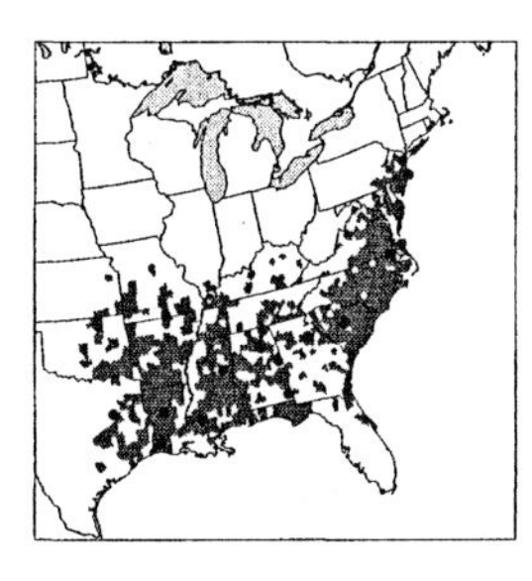

Plants in small clumps, with rhizomes 3–5 mm thick. **Basal rosettes** well-differentiated; **blades** sometimes more than 10 cm, lanceolate. **Culms** 50–150 cm, usually robust, erect; **nodes** often swollen, densely bearded with thin retrorse hairs above a constricted, glabrous, viscid ring; **internodes** grayish-purple, velvety-pubescent; **fall phase** branching from the mid- and upper culm nodes, with long, repeatedly forking and often recurving branches, ultimately with fascicles of reduced blades and included secondary panicles. **Cauline leaves** 7–11; **sheaths** not overlapping, narrowing distally, lustrous, bases sparsely to densely retrorsely villous, hairs papillose-based, summits purplish, with yellowish spots; **collars** densely villous; **ligules** 0.5–2 mm, of hairs; **blades** 9–20 cm long, 9–20 mm wide, thick, densely soft pubescent, bases rounded to subcordate, margins ciliate basally. **Primary panicles** 6–16 cm long, 5–12 cm wide, well-exserted, dense; **rachises** softly pubescent basally; **branches** often mottled with purplish viscid spots, glabrous. **Spikelets** 2.2–2.8 mm long, 1.3–1.5 mm wide, ovoid-ellipsoid, often purplish basally, prominently veined, margins and apices sparsely to densely pubescent, hairs papillose-based. **Lower glumes** 0.6–1.3 mm, subtruncate to acuminate; **lower florets** sterile; **upper florets** minutely apiculate. $2n = 18$.

Dichanthelium scoparium grows in moist, sandy, open, often disturbed areas of the southeastern United States. It is also present in the West Indies. The primary panicles are open-pollinated, produced from May to early August; the secondary panicles are cleistogamous and are produced from July through October.

Panicum glutinoscabrum Fernald may represent rare putative hybrids of *Dichanthelium scoparium* with *D. acuminatum*, and *P. mundum* Fernald, rare hybrids with *D. dichotomum*.

Dichanthelium sect. Oligosantha (Hitchc.) Freckmann & Lelong

Plants usually cespitose, often with caudices, clumps sometimes with only 1 culm. **Basal rosettes** well-differentiated. **Culms** 20–75 cm, with papillose-based hairs; **fall phase** branching from the mid- and upper culm nodes. **Cauline leaves** 4–7; **sheaths** glabrous or hirsute, hairs sometimes papillose-based; **ligules** 0.5–5 mm, of hairs, these often joined at the base into a thickened ring. **Spikelets** 2.5–4.3 mm, ellipsoid to obovoid, turgid. **Upper glumes** usually with an orange or purplish spot at the base, with 7–9 prominent veins. **Upper florets** umbonate, sometimes minutely so.

11. **Dichanthelium oligosanthes** (Schult.) Gould [p. 423]
FEW-FLOWERED PANICGRASS

Plants cespitose, with caudices. **Basal rosettes** well-differentiated; **blades** 2–6 cm, few, ovate to lanceolate. **Culms** 20–75 cm, geniculate basally, stiffly erect distally; **nodes** glabrous or sparsely pubescent; **internodes** often purplish, glabrous, puberulent, or papillose-hirsute; **fall phase** branching from the midculm nodes, branches initially ascending to erect, sometimes developing simultaneously with and overtopping the primary panicles, later rebranching to form short, bushy clumps of blades and small, included secondary panicles. **Cauline leaves** 5–7; **sheaths** not overlapping, glabrous, puberulent, or ascending papillose-hispid, margins ciliate, collars loose, puberulent; **ligules** 1–3 mm, of hairs; **blades** 5–12 cm long, 4–15 mm wide, flat or partly involute, glabrous or pubescent abaxially, with 7–9 major veins only slightly more prominent than the minor veins, bases ciliate, rounded to truncate, margins cartilaginous. **Primary panicles** 5–9 cm long, 3–6 cm wide, partly enclosed to long-exserted, with 6–60 spikelets; **branches** stiff or wiry, puberulent or scabridulous. **Spikelets** 2.7–4.2 mm long, 1.7–2.4 mm wide, ellipsoid to broadly obovoid, turgid, glabrous or sparsely pubescent. **Lower glumes** 1–1.6 mm, acute, similar in texture and vein prominence to the upper glumes; **upper glumes** strongly veined, often orange to purplish at the base; **lower florets** sterile; **upper florets** with minutely umbonate apices. $2n = 18$.

Dichanthelium oligosanthes grows throughout the southern portion of the *Flora* region and extends into northern Mexico. The primary panicles are briefly open-pollinated, then cleistogamous, from late May to early June; the secondary panicles, which are produced from June to November, are cleistogamous. The subspecies intergrade in areas of overlapping range, but they are usually distinct elsewhere.

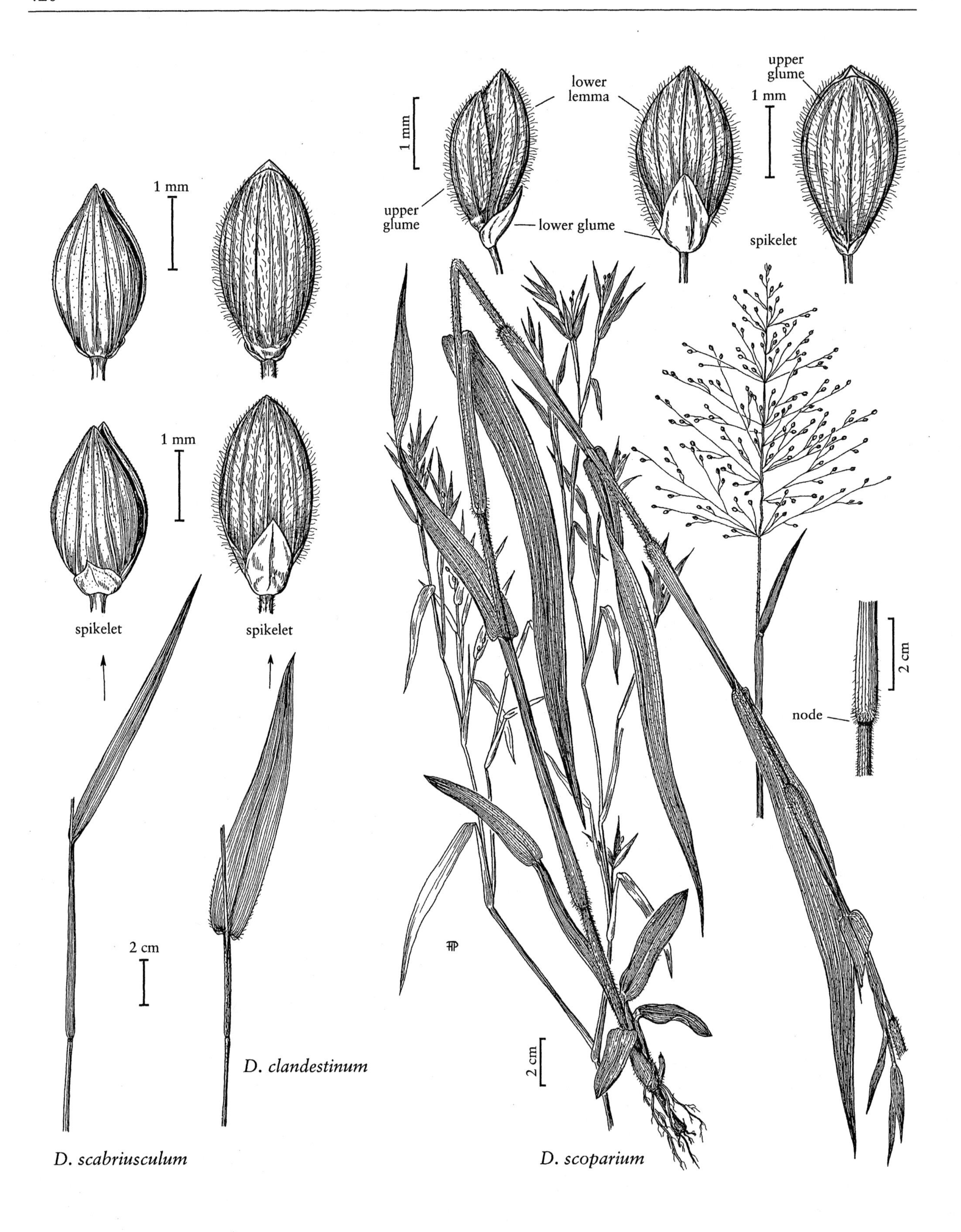
1 mm
1 mm
spikelet
spikelet
2 cm
D. clandestinum
D. scabriusculum
lower lemma
upper glume
1 mm
1 mm
upper glume
lower glume
spikelet
2 cm
node
2 cm
D. scoparium

Specimens of *Dichanthelium oligosanthes* that have few elongated internodes, but those elongated more than usual, are often mistaken for *D. wilcoxianum*. Unlike that species, however, they have turgid spikelets with an orange spot at the base of the lemma, indicating that they belong to *D. oligosanthes*. Such specimens seem to be most common among collections made in the southern and southwestern states during November, February, or March.

Sterile hybrids with *Dichanthelium acuminatum* have often been called *Panicum scoparioides* Ashe. Apparent hybrids with *D. malacophyllum*, *D. ovale*, and *D. acuminatum* subsp. *columbianum* are occasionally found.

1. Spikelets ellipsoid to oblong-obovoid, usually 3.4–4.2 mm long, 1.7–2 mm wide, usually sparsely pubescent; blades usually 4–9 mm wide, more than 10 times longer than wide, often partly involute; ligules 2–3 mm . . . . subsp. *oligosanthes*
1. Spikelets broadly obovoid-ellipsoid, 2.7–3.5 mm long, 2–2.4 mm wide, usually glabrous; blades usually 6–15 mm wide, less than 10 times longer than wide, flat; ligules 1–1.5 mm long . . . . subsp. *scribnerianum*

Dichanthelium oligosanthes (Schult.) Gould subsp. **oligosanthes**

Culms 40–75 cm; **internodes** usually puberulent and also pubescent to pilose or appressed-hispid. **Cauline sheaths** usually puberulent and also pubescent to pilose or appressed-hispid; **ligules** 2–3 mm; **blades** usually 4–9 mm wide, more than 10 times longer than wide, stiff, spreading, usually densely appressed-pubescent abaxially, often involute towards the long-acuminate apices. **Primary panicles** with a few stiff branches; **pedicels** mostly 5–15 mm. **Spikelets** usually 3.4–4.2 mm long, 1.7–2 mm wide, ellipsoid to oblong-obovoid, usually sparsely pubescent. **Upper glumes** often with a faint orange spot at the base.

Dichanthelium oligosanthes subsp. *oligosanthes* grows in dry, open, sandy, oak or pine woodlands. Its range extends from southern Ontario and New Hampshire to the Texas Gulf coast. It has not yet been reported from Mexico.

Dichanthelium oligosanthes subsp. **scribnerianum** (Nash) Freckmann & Lelong
SCRIBNER'S PANICGRASS

Culms 20–50 cm; **internodes** often lustrous, glabrous or sparsely papillose-hispid (rarely puberulent). **Cauline sheaths** often lustrous, glabrous or sparsely papillose-hispid (rarely puberulent); **ligules** 1–1.5 mm; **blades** usually 6–15 mm wide, less than 10 times longer than wide, flat, ascending to spreading, glabrous or sparsely pubescent abaxially, acute. **Primary panicles** denser than in subsp. *oligosanthes*, branches more flexible; **pedicels** mostly shorter than 5 mm. **Spikelets** usually 2.7–3.5 mm long, 2–2.4 mm wide, broadly obovoid-ellipsoid, usually glabrous. **Upper glumes** with a prominent orange to purplish spot at the base.

Dichanthelium oligosanthes subsp. *scribnerianum* grows in sandy or clayey banks and prairies. Its range extends from southern British Columbia to the east coast of the United States, and south into northern Mexico. It is the most widespread of the two varieties.

12. **Dichanthelium ravenelii** (Scribn. & Merr.) Gould [p. 423]
RAVENEL'S PANICGRASS

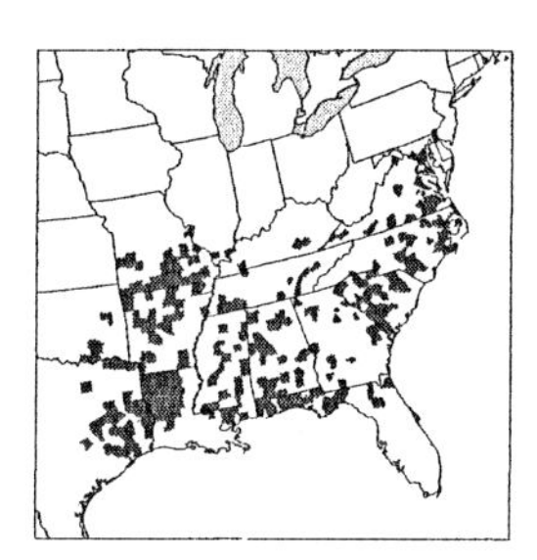

Plants cespitose, with caudices. **Basal rosettes** well-differentiated; **blades** 3–8 cm, ovate to lanceolate. **Culms** 25–75 cm, 2–3 mm thick, erect, purplish; **nodes** densely bearded with spreading to retrorse hairs above a glabrous ring; **internodes** pilose or ascending hirsute, hairs papillose-based, also puberulent; **fall phase** with nearly erect culms, branching from the mid- and upper culm nodes; branches short, ascending, bushy, with several reduced, partly enclosed secondary panicles. **Cauline leaves** 4–6; **sheaths** not overlapping, papillose-hirsute and puberulent; **collars** densely pubescent; **ligules** 2–5 mm, of hairs; **blades** 8–17 cm long, 8–18 mm wide, lanceolate, stiff, thick, abaxial surfaces densely soft-pubescent, velvety, adaxial surfaces glabrous or sparsely pilose, with 9–11 major veins slightly more prominent than the minor veins, bases rounded or subcordate, margins with papillose-based cilia, apices acuminate. **Primary panicles** 5–11 cm, almost as wide as long, shortly exserted, with few spikelets; **rachises** and **branches** scabridulous and finely pubescent, hairs papillose-based. **Spikelets** 3.7–4.3 mm long, 1.6–2.1 mm wide, obovoid, turgid, often shiny, sparsely pustulose-villous. **Lower glumes** 1.8–2.5 mm, loose, strongly veined, acute; **upper glumes** shorter than the spikelets, strongly veined, purplish at the base; **lower florets** sterile; **upper florets** with a minute tuft of hairs around the umbonate apices. $2n = 18$.

Dichanthelium ravenelii grows in dry, sandy woodlands of the southeastern United States. The primary panicles develop from early May through June, and are at least partly open-pollinated. The secondary panicles, which are produced from July through September, are cleistogamous. Putative hybrids with other species are very rare.

13. **Dichanthelium malacophyllum** (Nash) Gould
[p. 423]
SOFT-LEAVED PANICGRASS

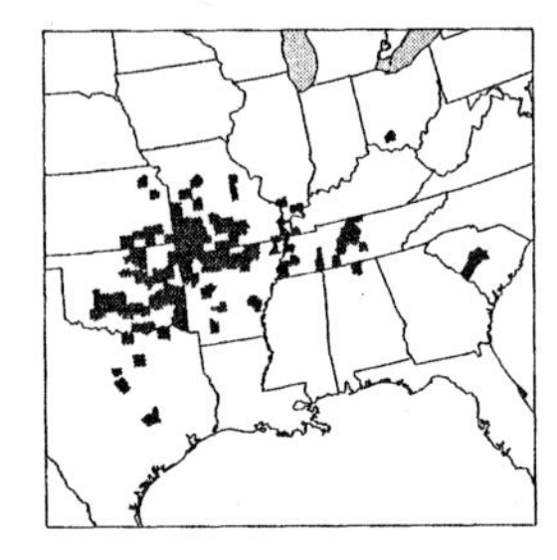

Plants cespitose, with caudices. **Basal rosettes** well-differentiated; **blades** 2–4 cm, ovate to lanceolate. **Culms** 20–70 cm, usually 1–2 mm thick, erect; **nodes** retrorsely bearded; **internodes** puberulent and densely pubescent with soft, spreading to retrorse hairs, hairs papillose-based, papillae small; **fall phase** branching from the mid- and upper culm nodes, ultimately much rebranched, with short, bushy clumps of blades and small, included secondary panicles, this branching beginning before the the primary panicles are exserted. **Cauline leaves** 5–6; **sheaths** not overlapping, pubescence not as dense as on the culms; **collars** puberulent; **ligules** 0.5–1 mm, of hairs, bases of the hairs forming a thickened ring, pseudoligules of 1–3 mm hairs also present; **blades** 5–10 cm long, 6–12 mm wide, lax, both surfaces velvety pubescent, with 9 or 11 major veins, these only slightly more prominent than the minor veins, bases rounded, margins ciliate. **Primary panicles** 3–7 cm long, 2–5 cm wide, tardily and shortly exserted; **rachises** and **branches** densely pubescent. **Spikelets** 2.5–3.2 mm long, 1.5–1.6 mm wide, broadly ellipsoid-obovoid, turgid, with papillose-based hairs, sometimes pilose. **Lower glumes** 1–1.6 mm, strongly veined, acute; **upper glumes** strongly veined, often purplish, especially towards the bases; **lower florets** sterile; **upper florets** minutely umbonate. $2n = 18$.

Dichanthelium malacophyllum usually grows in cedar glades, on dry limestone soils. It is restricted to the United States. The primary panicles are briefly open-pollinated from late May to early June; the secondary panicles, which are produced from June to November, are cleistogamous. The species occasionally intergrades, and perhaps hybridizes, with *D. oligosanthes* and *D. acuminatum*.

Dichanthelium sect. Lanuginosa (Hitchc.) Freckmann & Lelong

Plants cespitose, with caudices. **Basal rosettes** well-differentiated. **Culms** 15–100 cm, erect, ascending, or decumbent; **internodes** glabrous, puberulent, pilose, or densely pubescent or velvety; **fall phase** often much branched. **Cauline leaves** 4–7; **sheaths** glabrous, puberulent, sparsely to densely pubescent, pilose or velvety, often with hairs of two lengths; **ligules** 0.5–3 mm, of hairs, often with an adjacent pseudoligule of 2–5 mm hairs. **Spikelets** 0.8–3 mm, more or less ellipsoid, pubescent to subglabrous, acute to acuminate, upper glumes and lower lemmas not strongly veined. **Upper florets** acute to obtuse, sometimes minutely umbonate or apiculate.

Hybridization, often followed by segregation in autogamous lines, produces a reticulate pattern of intergradation between members of this section. In the descriptions, no distinction is made between the ligules and pseudoligules because of the difficulty of distinguishing the two at less than 30×.

14. **Dichanthelium acuminatum** (Sw.) Gould & C.A. Clark [pp. 427, 428]
HAIRY PANICGRASS, PANIC LAINEUX

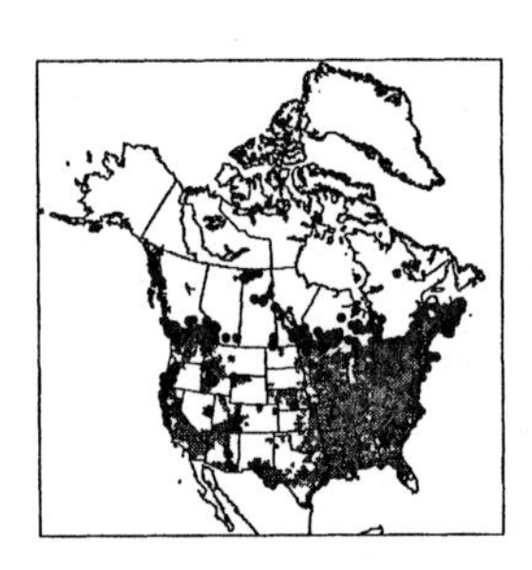

Plants more or less densely cespitose. **Basal rosettes** usually well-differentiated; **blades** ovate to lanceolate. **Culms** 15–100 cm (rarely taller), usually thicker than 1 mm, weak and wiry or relatively stout and rigid, erect, ascending or decumbent; **nodes** occasionally swollen, glabrous or densely pubescent, often with a glabrous or viscid ring below; **internodes** purplish or olive green or grayish-green, to yellowish-green, variously pubescent, with hairs of 2 lengths or glabrous; **fall phase** erect, spreading, or decumbent, usually branching extensively at all but the uppermost nodes, ultimately forming dense fascicles of branchlets with reduced, flat or involute blades and reduced secondary panicles with few spikelets. **Cauline leaves** 4–7; **sheaths** usually shorter than the internodes, glabrous or densely and variously pubescent with hairs shorter than 3 mm, margins ciliate or glabrous; **ligules** and **pseudoligules** 1–5 mm, of hairs; **blades** 2–12 cm long (rarely longer), 2–12 mm wide (rarely wider), firm or lax, spreading to reflexed or stiffly ascending, yellowish-green or grayish-green to olivaceous, densely to sparsely and variously pubescent, margins similar or occasionally whitish-scabridulous, margins often with papillose-based cilia, at least basally, bases rounded or subcordate.

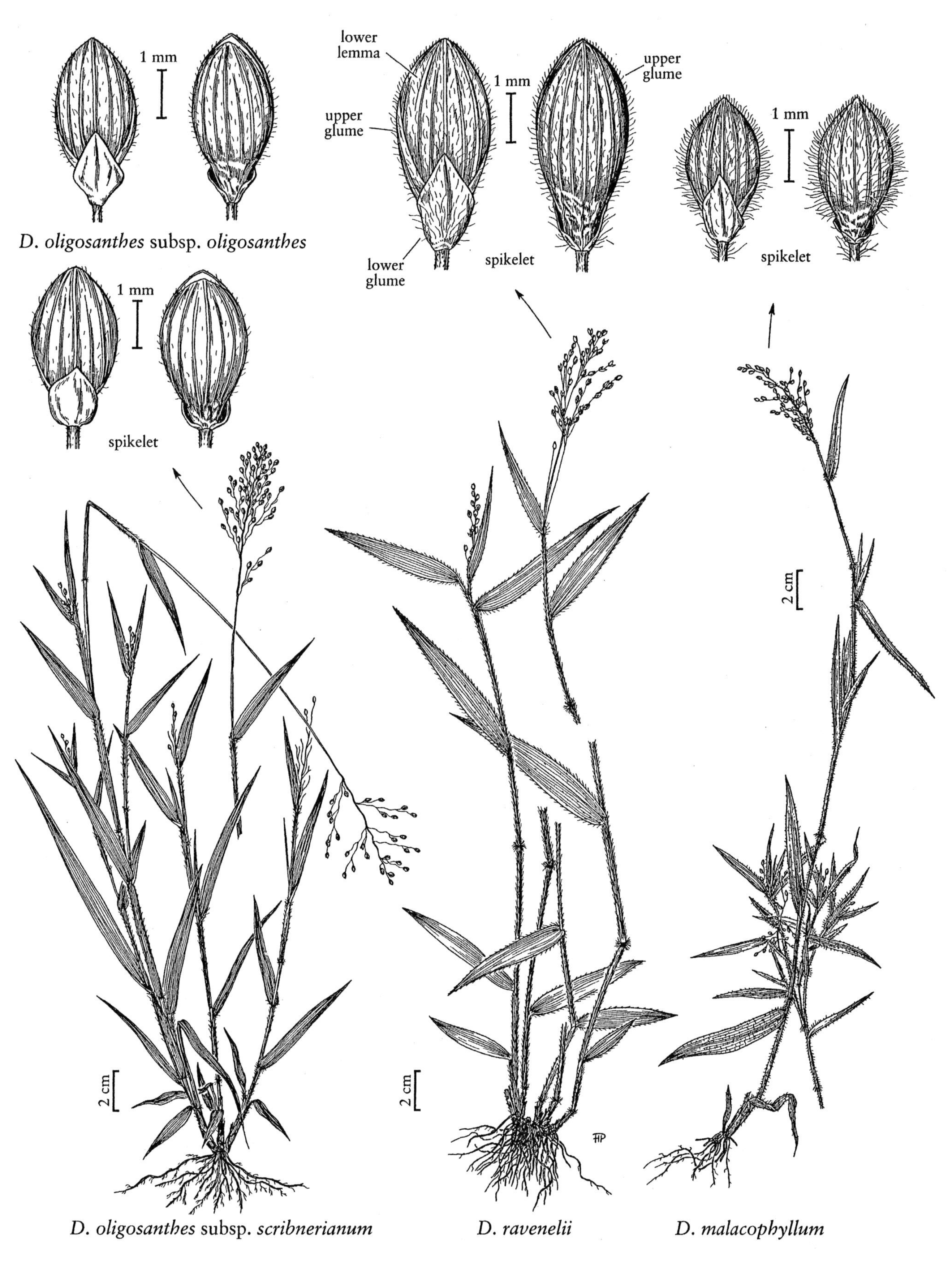

D. oligosanthes subsp. *scribnerianum* *D. ravenelii* *D. malacophyllum*

Primary panicles 3–12 cm, ¼–¾ as wide as long, usually open, well-exserted, rather dense; **rachises** glabrous, puberulent, or more or less densely pilose, at least basally. **Spikelets** 1.1–2.1 mm, obovoid to ellipsoid, yellowish-green to olivaceous or purplish, variously pubescent, obtuse or subacute. **Lower glumes** usually ¼–½ as long as the spikelets, obtuse to acute; **upper glumes** and **lower lemmas** subequal, equaling the upper florets at maturity, or occasionally the upper glumes slightly shorter, not strongly veined; **lower florets** sterile; **upper florets** 1.1–1.7 mm long, 0.6–1 mm wide, ellipsoid, obtuse to acute or minutely umbonate or apiculate. $2n = 18$.

Dichanthelium acuminatum is common and ubiquitous in dry to wet, open, sandy or clayey woods, clearings, bogs, and swamps, or in saline soil near hot springs, growing in much of the *Flora* region and extending into northern South America. It is probably the most polymorphic and troublesome species in the genus. The treatment presented here attempts to delimit the major variants present, but does not fully reflect the intricate reticulate pattern of morphological variation that exists. There is considerable overlap among the nine subspecies recognized and, in addition, there appears to be widespread introgression from other *Dichanthelium* species, such as *D. dichotomum*, *D. sphaerocarpon*, *D. ovale*, and *D. aciculare* into the *D. acuminatum* complex, contributing to the taxonomic difficulties.

1. Lower portion of the culms and lower sheaths usually glabrous or sparsely pubescent.
   2. Primary panicles congested, more than twice as long as wide; spikelets ascending to appressed . . . . subsp. *spretum*
   2. Primary panicles open, less than twice as long as wide; spikelets diverging to ascending.
      3. Blades green or purplish, the margins not conspicuously ciliate at the base; spikelets 1.1–1.5 mm long, usually ellipsoid . . . . subsp. *longiligulatum*
      3. Blades often yellowish-green, the margins usually with long, papillose-based cilia at the base; spikelets 1.3–1.6 mm long, usually obovoid . . . . subsp. *lindheimeri*
1. Lower portion of the culms and lower sheaths densely and variously pubescent or puberulent.
   4. Culms 15–30 cm tall; midculm sheaths nearly as long as the internodes; blades usually 2–6.5 cm long, less than 8 times longer than wide . . . . subsp. *sericeum*
   4. Culms usually 30–100 cm tall; midculm sheaths about ½ as long as the internodes; blades usually 6–12 cm long, more than 8 times longer than wide.
      5. Culms and lower sheaths densely covered with spreading, villous hairs or soft, thin, papillose-based hairs, often with shorter hairs underneath; blades softly pubescent to velvety on the abaxial surfaces.
         6. Primary panicles usually poorly exserted, on peduncles less than 6 cm long; blades suberect, the margins lacking cilia on the distal ½ . . . . subsp. *thermale*
         6. Primary panicles usually well-exserted, on peduncles more than 8 cm long; blades ascending to spreading, the margins ciliate along most of their length . . . . subsp. *acuminatum*
      5. Culms and sheaths pilose with papillose-based hairs to hispid, with mostly ascending hairs, or densely puberulent with a few longer, ascending hairs also present; blades appressed-pubescent or puberulent abaxially, not velvety to the touch.
         7. Sheaths and culms densely puberulent, scattered long hairs often present also . . . . subsp. *columbianum*
         7. Sheaths and culms pilose with papillose-based hairs, the hairs mostly ascending, occasionally with inconspicuous, shorter hairs underneath.
            8. Blades usually 6–12 mm wide, spreading to ascending, the adaxial surfaces nearly glabrous or with hairs shorter than 3 mm long; spikelets 1.5–2 mm . . . . subsp. *fasciculatum*
            8. Blades usually 2–6 mm wide, erect to ascending, spreading or reflexed, the adaxial surfaces glabrous or with hairs 3–6 mm long; spikelets 1.1–1.6 mm long.
               9. Blades erect to ascending, the adaxial surfaces long-pilose; spikelets 1.3–1.6 mm long, usually broadly obovoid . . . . subsp. *implicatum*
               9. Blades ascending, spreading, or reflexed, the adaxial surfaces glabrous or sparsely pubescent; spikelets 1.1–1.5 mm long, usually ellipsoid . . . . subsp. *leucothrix*

Dichanthelium acuminatum (Sw.) Gould & C.A. Clark subsp. **acuminatum** [p. 427]
PANIC LAINEUX

Plants grayish olive green, densely velvety-villous throughout. **Cauline nodes** densely villous, with a glabrous ring below; **fall phase** branching extensively from the midculm nodes, forming conspicuous flabellate fascicles of branches. **Cauline sheaths** densely soft spreading-villous, often with inconspicuous smaller hairs underneath; **midculm sheaths** about ½ as long as the internodes; **blades** 6–12 cm long, to 10 mm wide, ascending to often spreading and slightly incurved, softly pubescent on both surfaces, with papillose-based cilia for most of their length. **Primary panicles** usually well-exserted, on peduncles longer than 8 cm. **Spikelets** 1.6–1.9 mm, broadly ellipsoid or obovoid.

Dichanthelium acuminatum subsp. *acuminatum* grows primarily in moist, open, sandy areas on the Atlantic and Gulf coastal plains. Its range extends through Mexico, the West Indies, and Central America to northern South America.

Dichanthelium acuminatum subsp. **columbianum** (Scribn.) Freckmann & Lelong [p. 427]
PANIC DU DISTRICT DE COLUMBIA

Plants cespitose, pale bluish- or grayish-green. **Culms** erect to ascending, densely puberulent, longer hairs often present also, at least on the lower portion of the culms; **nodes** puberulent; **fall phase** with spreading or decumbent culms, branching early from most nodes, secondary blades not as greatly reduced or as densely crowded as in subspp. *acuminatum, fasciculatum, implicatum*, and *leucothrix*. **Cauline sheaths** pubescent, their pubescence similar to that of the culms but somewhat less dense; **midculm sheaths** about ½ as long as the internodes; **ligules** 1–1.5 mm; **blades** 3–7 cm long, 3–7 mm wide, relatively firm, often ascending, abaxial surfaces densely puberulent to nearly glabrous, adaxial surfaces glabrous or sparsely pilose near the base, margins whitish-scabridulous. **Spikelets** 1.5–1.9 mm, broadly ellipsoid or obovoid, puberulent.

Dichanthelium acuminatum subsp. *columbianum* grows in sandy woods or clearings in the northeastern portion of the species range. It is much less common than the other eastern subspecies of *D. acuminatum*. Occasionally, it resembles the more widespread subsp. *fasciculatum*, subsp. *implicatum*, and subsp. *lindheimeri*.

The culms and sheaths of *Dichanthelium acuminatum* subsp. *columbianum* are always puberulent with very short hairs. This puberulence should not be confused with the slightly longer hairs that develop on the secondary branches of other taxa.

Dichanthelium acuminatum subsp. **fasciculatum** (Torr.) Freckmann & Lelong [p. 427]

Plants yellowish-green to olivaceous or purplish. **Culms** 15–75 cm, suberect, ascending or spreading; **nodes** often with spreading hairs, occasionally with a glabrous ring below. **Cauline sheaths** with ascending to spreading, papillose-based hairs, occasionally with shorter hairs underneath; **midculm sheaths** about ½ as long as the internodes; **blades** 5–12 cm long, 6–12 mm wide, spreading to ascending, bases with papillose-based cilia, abaxial surfaces usually pubescent, adaxial surfaces pilose or glabrous, hairs shorter than 3 mm. **Spikelets** 1.5–2 mm (tending to be longer in the western part of its range), obovoid to ellipsoid.

Dichanthelium acuminatum subsp. *fasciculatum* grows primarily in disturbed areas, open or cut-over woods, thickets, and grasslands, in dry to moist soils, including river banks, lake margins, and marshy areas. It is widespread in temperate North America, growing from Canada to Mexico, but it is somewhat less common in the western part of its range, where it often occurs on moister areas.

Dichanthelium acuminatum subsp. *fasciculatum* includes probably the most widespread, ubiquitous, and variable assemblages of forms in the species. It is not always clearly separable from the other subspecies of *D. acuminatum*, especially subsp. *acuminatum*, subsp. *implicatum*, and subsp. *lindheimeri*. Gene exchange with other *Dichanthelium* species (including *D. dichotomum, D. laxiflorum, D. ovale, D. commutatum*, and *D. boreale*) probably occurs not infrequently.

Dichanthelium acuminatum subsp. **implicatum** (Scribn.) Freckmann & Lelong [p. 427]

Plants densely cespitose. **Culms** seldom over 50 cm, slender, suberect, ascending or spreading; **nodes** more or less densely pubescent; **fall phase** branching extensively from the lower and midculm nodes, with conspicuous, flabellate fascicles of branches and reduced blades. **Cauline sheaths** shorter than the internodes, lower sheaths usually pilose with papillose-based hairs, upper sheaths often short-pubescent; **midculm sheaths** about ½ as long as the internodes; **blades** usually 2–6 mm wide, more than 8 times longer than wide, relatively firm, erect to ascending, often yellowish-green, abaxial surfaces densely pubescent with short papillose-based hairs or short-pubescent with subappressed hairs, adaxial surfaces more or less densely pilose, hairs to 6 mm, conspicuous, erect or ascending, occasionally with shorter hairs underneath. **Spikelets** 1.3–1.6 mm, usually broadly obovoid.

Dichanthelium acuminatum subsp. *implicatum* usually grows in low, moist areas, including open woodlands, meadows, bogs, and cedar and hemlock swamps, and also in drier, sandy areas. Its range extends from south central Canada to the midwestern and northeastern United States. It intergrades occasionally with the more widespread subsp. *fasciculatum.*

Dichanthelium acuminatum subsp. **leucothrix** (Nash) Freckmann & Lelong [p. 427]

Plants cespitose, pale olive green, often purplish-tinged. **Culms** usually 30–100 cm, erect to ascending, sparsely pubescent to almost glabrous, hairs appressed, thin, silvery, papillose-based; **nodes** sparsely pubescent; **fall phase** branching extensively from the lower and midculm nodes, with conspicuous, flabellate fascicles of branches and reduced blades. **Cauline sheaths** shorter than the internodes, sparsely pilose to nearly glabrous, hairs papillose-based, occasionally with shorter soft hairs underneath, margins ciliate; **midculm sheaths** about ½ as long as the internodes; **blades** usually 2–7 cm long, 2–7 mm wide, relatively firm, ascending, spreading, or reflexed, abaxial surfaces densely and softly puberulent, adaxial surfaces glabrous or sparsely appressed-villous, sometimes with a few longer hairs intermixed. **Primary panicles** open, long-exserted, dense. **Spikelets** 1.1–1.5 mm, usually ellipsoid, densely short-pubescent.

Dichanthelium acuminatum subsp. *leucothrix* grows in low, sandy or peaty pine savannahs of the coastal plain. Its range extends through Mexico, the West Indies, and Central America to northern South America. It is closely related, and often sympatric with, the more common, glabrous subsp. *longiligulatum.*

Dichanthelium acuminatum subsp. **lindheimeri** (Nash) Freckmann & Lelong [p. 428]

Culms often yellowish-green, usually glabrous; **nodes** glabrous; **fall phase** usually with stiffly spreading culms with dense fascicles of branches with reduced, often involute blades. **Cauline sheaths** often yellowish-green, usually glabrous or the lowest sheaths sparsely ascending-pubescent; **blades** 4–9 cm long, 4–8 mm wide, stiffly ascending or spreading, often yellowish-green, glabrous on both surfaces or puberulent abaxially, bases rounded, margins faintly whitish-scabridulous, with conspicuous, long, papillose-based cilia at the base. **Primary panicles** 3.5–7 cm, open, less than twice as long as wide. **Spikelets** 1.3–1.6 mm, diverging to ascending, usually obovoid, obtuse.

Dichanthelium acuminatum subsp. *lindheimeri* grows in dry or moist, sandy or clayey, open, often disturbed areas, open woodlands, limestone glades, and roadsides, primarily in the eastern portion of the species range. It intergrades occasionally with the pubescent subsp. *fasciculatum* and subsp. *implicatum.*

Dichanthelium acuminatum subsp. **longiligulatum** (Nash) Freckmann & Lelong [p. 428]

Very similar to subsp. *spretum* vegetatively. **Fall phase** branching profusely from the lower and midculm nodes, producing dense fascicles of reduced branches, blades, and secondary panicles. **Cauline blades** green or purplish. **Primary panicles** 3–8 cm, to ¾ as wide as long, normally expanded; **branches** numerous, slender, ascending, spikelets densely packed. **Spikelets** 1.1–1.5 mm, usually ellipsoid, puberulent.

Dichanthelium acuminatum subsp. *longiligulatum* is common, especially in moist pine savannahs and bogs of the coastal plain; it also grows inland to Tennessee, and in Mexico, the West Indies, Central America, and South America. It is similar to subsp. *leucothrix*, which grows in the same habitat, often at the same sites.

Dichanthelium acuminatum subsp. **sericeum** (Schmoll) Freckmann & Lelong [p. 428]

Plants more or less densely cespitose. **Culms** usually less than 30 cm, stiffly ascending to spreading, densely pubescent. **Midculm sheaths** nearly as long as the internodes; **midculm blades** usually 2–6.5 cm, usually less than 8 times as long as wide. **Primary panicles** usually well-exserted. **Spikelets** mostly 1.6–1.8 mm.

Dichanthelium acuminatum subsp. *sericeum* grows in warm or hot ground around geysers and hot springs in the Rocky Mountains from Banff, Alberta south to Yellowstone National Park and east to Bighorn County, Wyoming.

Dichanthelium acuminatum subsp. **spretum** (Schult.) Freckmann & Lelong [p. 428]

Culms usually glabrous; **nodes** often swollen, glabrous; **fall phase** often with reclining culms, ultimately with fascicles of branches with greatly reduced blades and secondary panicles. **Cauline sheaths** usually glabrous; **blades** 3–9 mm wide, usually firm, ascending to reflexed, puberulent or glabrous abaxially, glabrous adaxially, with sparse papillose-based cilia at the bases. **Primary panicles** 4–12 cm long, ¼–½ as wide as long, usually narrow, congested. **Spikelets** 1.3–1.9 mm, ascending to appressed, usually ellipsoid, usually puberulent (rarely glabrous).

Dichanthelium acuminatum subsp. *spretum* grows in wet to moist, sandy or peaty soil, pine savannahs, and bogs. It is not a common taxon, but is most frequent on the coastal plain and around the Great Lakes. It is very similar to the more common, southern subsp. *longiligulatum.* It also resembles *D. dichotomum* in size and overall habit.

Dichanthelium acuminatum subsp. **thermale** (Bol.) Freckmann & Lelong [p. 428]
GEYSER PANICGRASS

Plants often densely cespitose, densely and softly pubescent throughout, with soft, thin, spreading, papillose-based hairs on the culms and lower sheaths. **Culms** usually over 30 cm. **Midculm sheaths** about ½ as long as the internodes; **blades** at midculm generally 6.5–12 cm long, usually more than 7 times as long as

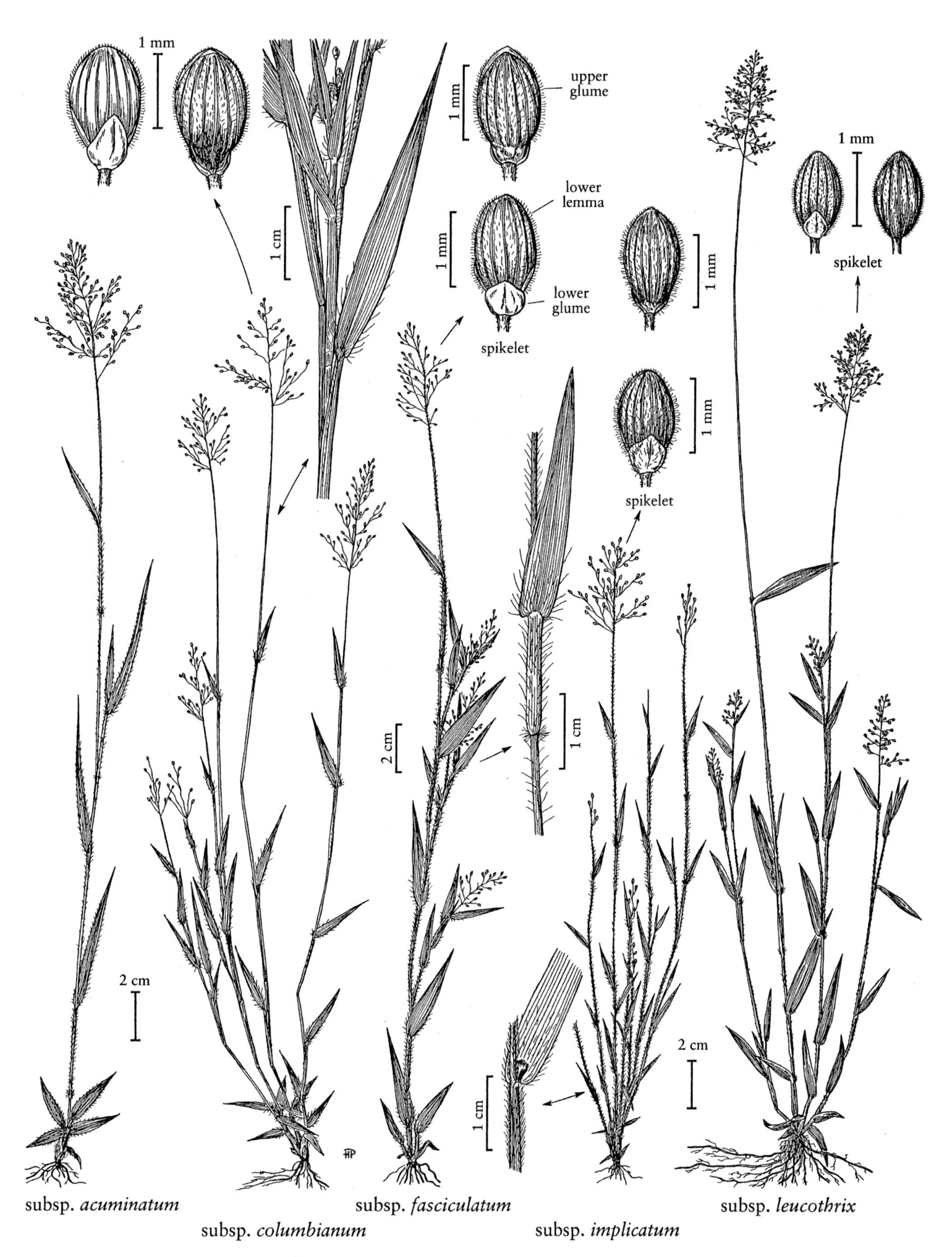

DICHANTHELIUM ACUMINATUM

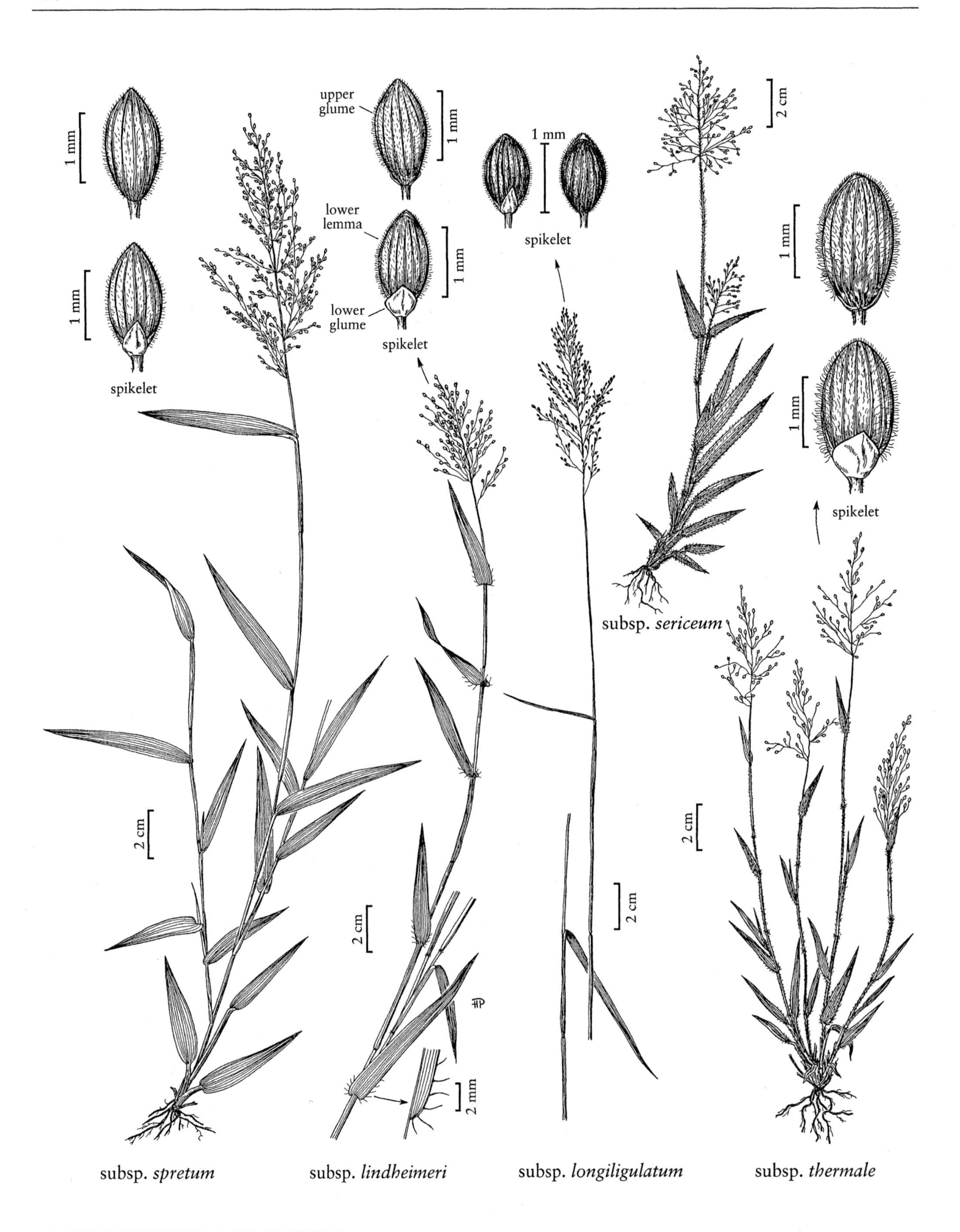

DICHANTHELIUM ACUMINATUM

wide, suberect, softly pubescent on the abaxial surface, without papillose-based cilia on the distal ½. **Primary panicles** usually poorly exserted, peduncles shorter than 6 cm. **Spikelets** mostly 1.8–2 mm.

Dichanthelium acuminatum subsp. *thermale* grows on the mineralized crust of warm, moist soil at the Geysers, Sonoma County, California; it is listed as endangered in that state.

15. **Dichanthelium ovale** (Elliott) Gould & C.A. Clark [p. 431]
STIFF-LEAVED PANICGRASS

Plants cespitose. **Basal rosettes** well-differentiated; **blades** 1–8 cm, lanceolate, often conspicuously ciliate. **Culms** 15–60 cm, usually more than 1 mm thick, not delicate, mostly ascending or spreading, often decumbent; **nodes** densely to sparsely bearded with spreading, retrorse, or appressed hairs; **internodes**, particularly the lower internodes, usually long-hairy with appressed or ascending hairs, occasionally with spreading hairs, occasionally with shorter hairs, rarely nearly glabrous; **fall phase** with decumbent to prostrate culms, branching developing early and forming dense fascicles with erect, slightly reduced blades and greatly reduced secondary panicles. **Cauline leaves** 4–7; **sheaths** shorter than the internodes, pilose, hairs to 4 mm, occasionally with shorter, spreading hairs underneath; **ligules** and **pseudoligules** 1–5 mm, of hairs; **blades** 4–10 cm long, 3–10 mm wide, relatively firm, mostly ascending or spreading, 1 or both surfaces sparsely to densely pubescent with appressed or erect hairs, hairs to 5 mm, bases rounded or slightly narrowed, margins often whitish, ciliate basally, scabridulous elsewhere. **Primary panicles** 3–10 cm long, nearly as wide when fully expanded; **rachises** and **branches** often stiffly ascending or spreading, usually pilose basally. **Spikelets** 1.8–3 mm, ellipsoid or obovoid, densely to sparsely pilose or papillose-pilose, obtuse or slightly acute. **Lower glumes** ⅓–½ as long as the spikelets, often triangular, not strongly veined, usually acute or subacute; **upper glumes** usually slightly shorter than the lower lemmas and upper florets at maturity, not strongly veined; **lower florets** sterile; **upper florets** 1.6–2.5 mm, ellipsoid (slightly less than ½ as wide as long, or wider in subsp. *praecocius*), subacute. $2n = 18$.

Dichanthelium ovale grows in dry, open, sandy or rocky woodland borders, sand barrens, dunes, and dry prairies in southeastern Canada, the eastern United States, the West Indies, Mexico, and Central America. The four subspecies often intergrade, especially subsp. *villosissimum* and subsp. *pseudopubescens* in the southeastern United States, and subsp. *villosissimum* and subsp. *praecocius* in the western part of their range.

The growth form and certain morphological features of *Dichanthelium ovale* resemble those of the widespread *D. laxiflorum*, which usually grows in more mesic habitats. Occasional specimens exhibit traits of *D. acuminatum*, *D. oligosanthes*, and *D. commutatum*.

1. Lower sheaths and lower culm internodes with soft, spreading or retrorse, papillose-based hairs, the longer hairs often longer than 4 mm long; spikelets 1.8–2.5 mm long.
   2. Spikelets 2.1–2.5 mm long; culms usually more than 1 mm thick, stiff; largest blades usually 6–10 mm wide . . . . subsp. *villosissimum*
   2. Spikelets 1.8–2.1 mm long; culms usually less than 1 mm thick, wiry; largest blades usually 2–6 mm wide . . . . subsp. *praecocius*
1. Lower sheaths and lower culm internodes with ascending or appressed, non-papillose-based hairs shorter than 4 mm or nearly glabrous; spikelets 2.1–3 mm long.
   3. Spikelets 2.5–3 mm long; basal blades with long hairs on or near the margins and bases . . . . subsp. *ovale*
   3. Spikelets 2.1–2.6 mm long; basal blades usually without long hairs on or near the margins and bases . . . . subsp. *pseudopubescens*

Dichanthelium ovale (Elliott) Gould & C.A. Clark subsp. **ovale**

Basal blades 3–8 cm, rigid, with long hairs on or near the bases and margins. **Culms** more than 1 mm thick, stiff; **lower internodes** pilose; **upper internodes** short-pilose to nearly glabrous. **Cauline sheaths** with ascending hairs, hairs to 4 mm, not papillose-based: **ligules** 1–4 mm; **blades** 5–12 mm wide, firm, ascending, abaxial surfaces appressed-pubescent, adaxial surfaces nearly glabrous except for the long hairs on or near the scabridulous margins and bases. **Spikelets** 2.5–3 mm, ellipsoid, sparsely to densely pilose. $2n = 18$.

Dichanthelium ovale subsp. *ovale* grows in dry, open, sandy woods, pinelands, and sandhills along the east coast of the United States from New Jersey southwards, extending into the coastal plain from eastern Texas to South Carolina, and in Mexico, Honduras, Guatemala, and Nicaragua. It intergrades somewhat with subsp. *pseudopubescens*. Occasional long-spikelet specimens exhibit morphological characteristics of *D. oligosanthes* and *D. commutatum*.

Dichanthelium ovale subsp. **praecocius** (Hitchc. & Chase) Freckmann & Lelong

Basal blades 1–3 cm, sparsely to densely evenly pilose. **Culms** less than 1 mm thick, wiry; **internodes** with soft, spreading or retrorse papillose-based hairs longer than 4 mm. **Cauline sheaths** with soft, spreading or retrorse hairs, hairs usually longer than 4 mm, papillose-based; **ligules** 3–4 mm; **blades** 2–6 mm wide, both surfaces densely pilose. **Spikelets** 1.8–2.1 mm, obovoid or ellipsoid, pilose with papillose-based hairs.

Dichanthelium ovale subsp. *praecocius* is most common in the midwest and in the tallgrass prairie states. It intergrades with subsp. *villosissimum*, especially in the western parts of the latter's range, and to a lesser extent, with *D. acuminatum* subsp. *fasciculatum* in the northern part of its range.

Dichanthelium ovale subsp. **pseudopubescens** (Nash) Freckmann & Lelong

Basal blades 2–6 cm, evenly pilose. **Culms** more than 1 mm thick, stiff; **lower internodes** sparsely pubescent, with ascending or appressed hairs, hairs shorter than 4 mm, not papillose-based. **Cauline sheaths** with sparse, ascending or appressed hairs, hairs shorter than 4 mm, often with shorter hairs underneath, not papillose-based; **ligules** 1–4 mm; **blades** 3–8 mm wide, both surfaces sparsely appressed-pubescent, margins ciliate basally, scabridulous elsewhere. **Spikelets** 2.1–2.6 mm, ellipsoid or obovoid-ellipsoid, with papillose-based hairs.

Dichanthelium ovale subsp. *pseudopubescens* grows in dry, sandy, open woods, sandhills, and sand dunes, over the same geographic range and in the same habitats as subsp. *villosissimum*, and often intergrades morphologically with that subspecies.

Dichanthelium ovale subsp. **villosissimum** (Nash) Freckmann & Lelong

Basal blades 3–7 cm, evenly long pilose. **Culms** more than 1 mm thick, stiff, often decumbent or prostrate in the fall; **internodes** with soft, spreading or retrorse, papillose-based hairs, hairs longer than 4 mm. **Cauline sheaths** with soft, spreading or retrorse hairs, hairs longer than 4 mm, papillose-based; **ligules** 2–5 mm; **blades** 6–10 mm wide, both surfaces densely pilose, hairs longer than 4 mm, margins short-ciliate basally, scabridulous and faintly whitish elsewhere. **Spikelets** 2.1–2.5 mm, usually ellipsoid, with dense, spreading, papillose-based hairs. **Lower glumes** ⅓–½ as long as the spikelets, usually acute. $2n = 18$.

Dichanthelium ovale subsp. *villosissimum* grows in dry, sandy, open pine and oak woodlands. It and subsp. *pseudopubescens* are the most common and widespread subspecies throughout the eastern United States. The range of subsp. *villosissimum* extends to Mexico, Honduras, Guatemala, and Nicaragua. It grades into the less pubescent subsp. *pseudopubescens*, and occasional specimens with smaller spikelets approach *D. acuminatum* subsp. *acuminatum*, which is usually densely grayish, velvety-pubescent.

16. **Dichanthelium wrightianum** (Scribn.) Freckmann [p. 431]
WRIGHT'S PANICGRASS

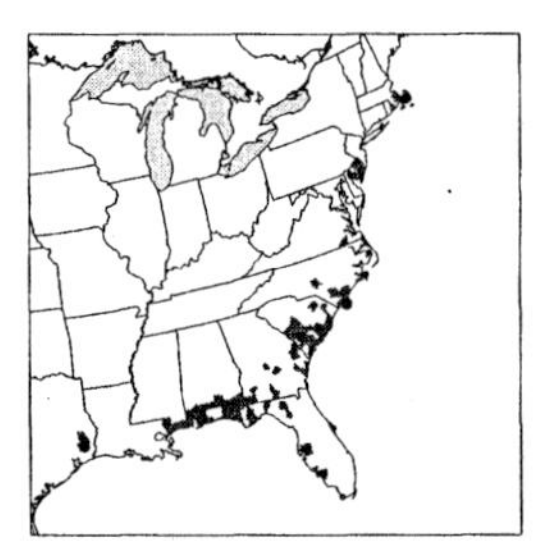

Plants cespitose, with few culms per clump. **Basal rosettes** well-differentiated; **blades** ovate to lanceolate. **Culms** 15–50 cm tall (rarely taller), 0.3–0.8 mm thick, delicate, erect or ascending; **nodes** slightly swollen, often purplish or darker green than the internodes; **internodes** usually puberulent; **fall phase** branching profusely from the lower and midculm nodes, secondary branches and secondary panicles numerous, usually not greatly reduced. **Cauline leaves** 4–7; **sheaths** mostly puberulent or glabrous, margins finely ciliate; **ligules** 1.5–3 mm, of hairs; **blades** 2–4.5 cm long, 2–5 mm wide, ascending or spreading, occasionally involute, finely appressed-pilose adaxially, puberulent abaxially, bases rounded, margins finely whitish-scabridulous. **Primary panicles** 2.5–5.5 cm, ⅓–⅔ as wide as long, well-exserted; **rachises** and **branches** glabrous or sparsely puberulent (at least basally); **ultimate branchlets** and **pedicels** glabrous, somewhat viscid. **Spikelets** 0.8–1.1 mm, ellipsoid to nearly ovoid, often purplish, puberulent or subglabrous, obtuse or subacute. **Lower glumes** ¼–⅓ as long as the spikelets, subacute; **upper glumes** shorter than the lower lemmas; **lower florets** sterile; **upper florets** 0.7–0.9 mm, ellipsoid, subacute. $2n = 18$.

Dichanthelium wrightianum grows in moist, sandy or peaty areas, low pine savannahs, bogs, the margins of ponds, and cypress swamps, in the coastal plain from Massachusetts to Texas and Florida, extending to Cuba, Mexico, Central America, and northern South America.

Occasional specimens of *Dichanthelium wrightianum*, particularly those with subglabrous spikelets, closely resemble *D. chamaelonche*. Others suggest *D. ensifolium*, and a few unusually robust specimens closely approach *D. acuminatum* subsp. *longiligulatum*. All of these taxa often grow together in the same habitats.

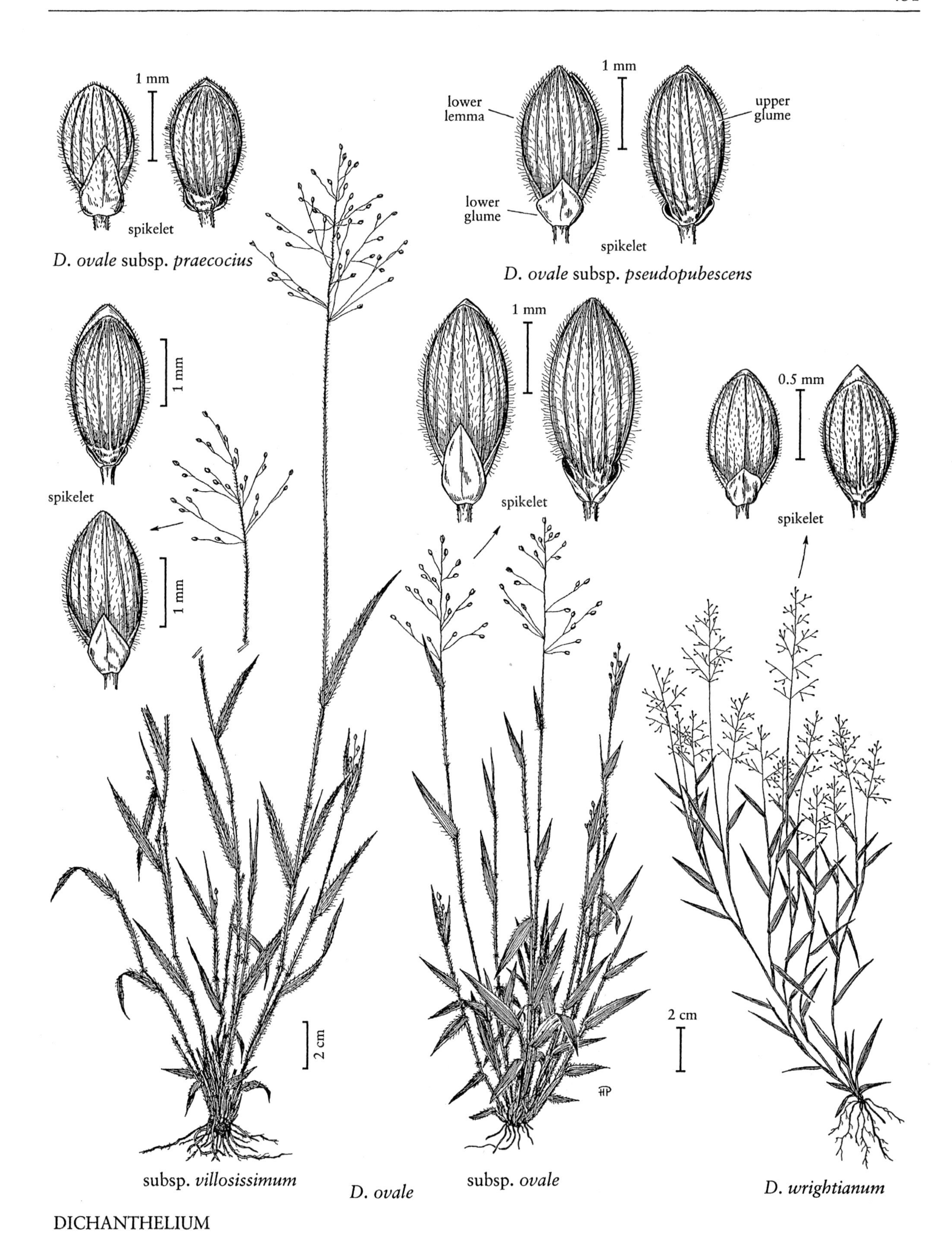
1 mm
spikelet
D. ovale subsp. praecocius
1 mm
lower lemma
upper glume
lower glume
spikelet
D. ovale subsp. pseudopubescens
1 mm
spikelet
1 mm
1 mm
1 mm
spikelet
0.5 mm
spikelet
2 cm
2 cm
HP
subsp. villosissimum
D. ovale
subsp. ovale
D. wrightianum

Dichanthelium (Hitchc. & Chase) Gould sect. Dichanthelium

Plants cespitose, with caudices or knotty crowns. **Basal rosettes** well-differentiated. **Culms** 18–100 cm, decumbent to erect, usually glabrous; **nodes** bearded or glabrous; **fall phase** often much branched and rebranched, with smaller blades and panicles than those of the culms. **Cauline leaves** 3–7; **sheaths** usually glabrous, lower sheaths sometimes pilose; **ligules** 0.2–0.8 mm, of hairs, or absent. **Primary panicles** exserted. **Spikelets** ellipsoid to obovoid, glabrous or pubescent. **Upper florets** acute to obtuse.

Gene exchange between the subspecies of *Dichanthelium dichotomum*, and between *D. dichotomum* and other species in the genus, appears to be rather common.

17. Dichanthelium dichotomum (L.) Gould [p. 435]

FORKED PANICGRASS

Plants in small or large clumps, with knotty crowns. **Basal rosettes** well-differentiated; **blades** ovate to lanceolate. **Culms** 20–100 cm, decumbent to erect, sometimes geniculate; **nodes** usually glabrous, sometimes sparsely pilose or densely bearded with retrorse hairs; **internodes** often purplish or olive green, lowest internodes usually glabrous, varying to sparsely pubescent; **fall phase** usually branching freely, especially from the nodes above the middle, ultimately forming dense, reclining fascicles of divergent branchlets with numerous reduced, thin, often involute blades, secondary panicles often reduced, with few spikelets. **Cauline leaves** 4–7; **sheaths** usually shorter than the internodes, usually glabrous, occasionally the lower sheaths sparsely to densely soft-pubescent, sheaths of the uppermost leaves sometimes with whitish glandular spots between the prominent veins, margins of all sheaths glabrous or ciliate; **ligules** absent or shorter than 1 mm, of hairs; **blades** 3.5–14 cm long, 5–14 mm wide, usually thin, distant, spreading to reflexed or (occasionally) ascending, yellow-green to purplish, usually glabrous on both surfaces or (at least the lower blades) more or less densely and softly pubescent, bases constricted (in narrow-bladed subspecies) or narrowly subcordate (in wide-bladed subspecies), margins glabrous or ciliate basally, glabrous distally, blades of the flag leaves usually spreading. **Primary panicles** 3–12 cm, long-exserted, usually with many spikelets; **branches** wiry, mostly spreading or ascending, usually glabrous, sometimes scabridulous. **Spikelets** 1.5–2.7 mm, usually ellipsoid or obovoid, green or purplish (at least at the base), glabrous or (less commonly) sparsely pubescent or puberulent, often prominently veined, obtuse to acute to beaked. **Lower glumes** usually less than ⅓ as long as the spikelets, obtuse to acute; **upper glumes** usually slightly shorter than or as long as the lower lemmas and upper florets (occasionally extending beyond the floret); **lower florets** sterile; **upper florets** 1.3–2 mm long, usually less than 1 mm wide, ellipsoid, subacute to obtuse.

Dichanthelium dichotomum grows in dry, sandy, clayey, or rocky ground, often in woods, or (more commonly) in moist or wet places, including marshes, bogs, low woods, swamps, and the moist borders of lakes and ponds. Its range extends south from the *Flora* region into the Caribbean. It is a polymorphic and ubiquitous species, with many of its intergrading subspecies exhibiting traits of other widespread and variable species such as *D. commutatum*, *D. laxiflorum*, and *D. sphaerocarpon*, which often grow at the same sites.

1. Lower nodes hairy.
    2. Spikelets 1.5–1.8 mm long, upper floret 0.6–0.8 mm wide . . . . subsp. *microcarpon*
    2. Spikelets 1.8–2.5 mm long; upper floret 0.7–1.0 mm wide.
        3. Spikelets usually glabrous; midculm blades usually 5–7 mm wide . . . . subsp. *dichotomum*
        3. Spikelets pubescent; midculm blades usually 7–14 mm wide.
            4. Lower sheaths and blades glabrous or sparsely pubescent . . . . subsp. *nitidum*
            4. Lower sheaths and blades more or less densely velvety pubescent . . . . subsp. *mattamuskeetense*
1. Lower nodes glabrous.
    5. Larger blades more than 1 cm wide; sheaths often with pale glandular spots between the prominent veins; spikelets 1.9–2.6 mm long, acute to beaked . . . . subsp. *yadkinense*
    5. Larger blades less than 1 cm wide; sheaths without glandular spots; spikelets 1.5–2.3 mm long, obtuse to subacute.

6. Culms weak, ultimately reclining or sprawling, often flattened .... subsp. *lucidum*
6. Culms erect, terete.
   7. Blades usually spreading; spikelets ellipsoid, 1.8–2.3 mm long, rarely purplish at the base....... subsp. *dichotomum*
   7. Blades usually ascending or erect; spikelets broadly ellipsoid or obovoid, 1.5–1.8 mm long, often purplish at the base.. subsp. *roanokense*

Dichanthelium dichotomum (L.) Gould subsp. **dichotomum**

Culms 20–60 cm, usually slender, erect; **nodes** usually glabrous, lowermost nodes sometimes sparsely bearded with soft, retrorse hairs; **internodes** terete, green to purplish, glabrous; **fall phase** branching freely from the midculm nodes, producing dense clusters of reduced, flat to involute blades and reduced secondary panicles. **Cauline sheaths** usually glabrous, lowermost sheaths sometimes sparsely pubescent, margins glabrous or short-ciliate; **blades** usually 3.5–9 cm long, usually 5–7 mm wide (seldom wider), usually spreading, narrowly lanceolate, glabrous on both surfaces, bases constricted. **Primary panicles** well-exserted; **branches** few, flexuous, with fewer spikelets than all the other subspecies apart from subsp. *lucidum*. **Spikelets** 1.8–2.3 mm, ellipsoid, usually glabrous, rarely purplish at the base; **upper florets** 1.7–2 mm long, 0.7–1.0 mm wide. $2n = 18$.

Dichanthelium dichotomum subsp. *dichotomum* usually grows in dry to mesic woods. Its range extends from southern Ontario to Maine and south through Illinois and Missouri to eastern Texas and to the east coast of the United States and central Florida.

Dichanthelium dichotomum subsp. **lucidum** (Ashe) Freckmann & Lelong

Culms occasionally more than 60 cm, very slender, weak; **nodes** usually glabrous; **internodes** often flattened, green, glabrous; **fall phase** with reclining or decumbent culms and numerous axillary branches, branches elongated and widely divergent, not forming fascicles. **Cauline sheaths** usually shorter than the internodes, glabrous; **blades** slightly smaller than those of the other subspecies, ascending or spreading, often lustrous, bright green, glabrous throughout. **Primary panicles** slightly smaller and with fewer spikelets than in the other subspecies (particularly subsp. *dichotomum*, which it closely resembles). **Spikelets** 1.8–2.3 mm, ellipsoid, usually glabrous, obtuse to subacute; **upper florets** 1.7–2 mm. $2n$ = unknown.

Dichanthelium dichotomum subsp. *lucidum* grows in wet woods, the margins of cypress swamps, sphagnum bogs, and other similar, wet habitats. It is primarily a species of the coastal plain, ranging from New Jersey to Florida, southeastern Texas, and up the Mississippi embayment to western Tennessee and, as a disjunct, on the Indiana Dunes of Lake Michigan.

Dichanthelium dichotomum subsp. **mattamuskeetense** (Ashe) Freckmann & Lelong

Plants very similar to subsp. *microcarpon*. **Fall phase** sparingly branched, blades not as greatly reduced as in subsp. *microcarpon*. **Sheaths** and **blades**, particularly those of the lower leaves, more or less densely velvety pubescent. $2n$ = unknown.

Dichanthelium dichotomum subsp. *mattamuskeetense* grows in low, moist, often sandy or peaty, ground and bogs. A relatively uncommon subspecies, it grows on the Atlantic coastal plain from Massachusetts to Florida.

Dichanthelium dichotomum subsp. **microcarpon** (Muhl. *ex* Elliott) Freckmann & Lelong

Culms 30–100 cm, slender, erect or geniculate; **fall phase** freely branching from all nodes, reclining from masses of branchlets and numerous reduced, ciliate blades and secondary panicles; **nodes** conspicuously bearded with retrorse hairs. **Sheaths** usually glabrous, lowermost sheaths sometimes sparsely pubescent, occasionally with whitish spots between the veins, ciliate along the margins; **blades** 5–14 cm long, 5–14 mm wide, thin, spreading to reflexed, glabrous on both surfaces, bases with few to many papillose-based cilia. **Panicles** 5–12 cm, well-exserted, dense. **Spikelets** 1.5–1.8 mm, usually glabrous, rarely slightly pubescent. **Lower glumes** usually less than ¼ as long as the spikelets; **upper glumes** usually shorter than the lower lemmas; **upper florets** 1.3–1.6 mm long, 0.6–0.8 mm wide, subacute. $2n = 18$.

Dichanthelium dichotomum subsp. *microcarpon* grows in wet woods, swamps, and wetland borders. It is a widespread subspecies, extending from southern Michigan to Massachusetts and south to eastern Oklahoma and Texas and throughout the southeast to central Florida.

Dichanthelium dichotomum subsp. **nitidum** (Lam.) Freckmann & Lelong

Plants very similar in most respects to subsp. *microcarpon*. **Fall phase** freely branching from all nodes, reclining from masses of branchlets and numerous reduced, ciliate blades and secondary panicles. **Cauline sheaths** and **blades** usually glabrous, lower sheaths and blades sometimes sparsely pubescent. **Spikelets** 1.8–2.5 mm (rarely longer), puberulent or pubescent. **Lower glumes** less than ⅓ as long as the spikelets, subacute; **upper glumes** and **lower lemmas** subequal; **upper florets** 1.7–2 mm long, 0.7–1.0 mm wide, subobtuse. $2n$ = unknown.

Dichanthelium dichotomum subsp. *nitidum* grows in moist to wet areas, and the borders of swamps. It is primarily a coastal plain taxon, ranging from Virginia to southeastern Texas and Florida.

Dichanthelium dichotomum subsp. *nitidum* is very similar to both subsp. *microcarpon* and subsp. *mattamuskeetense*, and intergrades with each occasionally.

Dichanthelium dichotomum subsp. **roanokense** (Ashe) Freckmann & Lelong

Culms to 100 cm, erect; **nodes** usually glabrous; **internodes** terete, usually glabrous, often slightly glaucous, sometimes olivaceous; **fall phase** with erect or decumbent culms, branching at the mid- and upper culm nodes, with numerous axillary branches, branches elongated and widely divergent, not forming fascicles. **Cauline sheaths** glabrous or the lowest sheaths sparsely pubescent; **blades** usually 5–8 mm wide, stiffly ascending or erect, often olivaceous or purplish abaxially, glabrous or sparsely pubescent basally. **Spikelets** 1.5–1.8 mm (seldom longer), broadly ellipsoid or obovoid, often purplish at the base, glabrous, obtuse to subacute. **Upper florets** 1.4–1.6 mm, broadly ellipsoid. $2n$ = unknown.

Dichanthelium dichotomum subsp. *roanokense* grows in marshes, wet pinelands, wet woods, and the borders of swamps. A relatively uncommon subspecies, it grows on the coastal plain from Delaware to southeastern Texas and in the West Indies.

It is very similar to subsp. *dichotomum* and also exhibits traits of *Dichanthelium sphaerocarpon* and *D. erectifolium*.

Dichanthelium dichotomum subsp. **yadkinense** (Ashe) Freckmann & Lelong

Culms 50–100 cm; **nodes** usually glabrous; **internodes** usually glabrous, often yellowish-green; **fall phase** suberect, sparsely branched from the midculm nodes, blades not as greatly reduced as in the other subspecies. **Cauline sheaths** glabrous, often with pale glandular spots between the prominent veins; **blades** 9–14 cm long, 7–12 mm wide, thin, widest near the middle and tapering to both ends, glabrous on both surfaces. **Spikelets** 1.9–2.6 mm, elliptic to subfusiform, glabrous, apices acute or beaked. **Upper florets** 1.8–2 mm. $2n$ = 18.

Dichanthelium dichotomum subsp. *yadkinense* grows in rich, moist or wet woods. A relatively uncommon subspecies, its range extends from Pennsylvania to Maryland and south through southern Illinois and southeastern Missouri to Georgia and Louisiana, but not to Florida. It exhibits traits of *D. laxiflorum* and *D. commutatum*.

18. **Dichanthelium boreale** (Nash) Freckmann [p. 437]
NORTHERN PANICGRASS, PANIC BOREAL

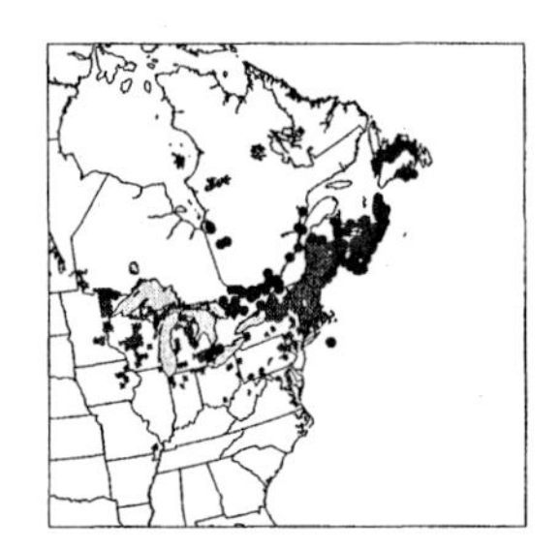

Plants cespitose. **Basal rosettes** well-differentiated; **blades** 2–4 cm, pubescent, reddish. **Culms** 18–75 cm, usually more than 1 mm thick, occasionally delicate, erect or ascending; **nodes** glabrous; **internodes** glabrous; **fall phase** with decumbent culms, branches arising from the lower and midculm nodes, rebranching 2–3 times, with small blades and secondary panicles compared to those on the culms, secondary panicles with 8–10 spikelets, partially included at maturity. **Cauline leaves** 3–5; **sheaths** shorter than the internodes, lower sheaths pubescent, upper sheaths glabrous, margins of all sheaths sparsely ciliate; **ligules** about 0.5 mm, of hairs; **blades** 5–11 cm long, 5–13 mm wide, thin, spreading to erect, usually glabrous, rarely pubescent abaxially, always glabrous adaxially, bases truncate to cordate, ciliate on the margins, blades of the flag leaves erect or ascending. **Primary panicles** 5–11 cm long, 3–8 cm wide, ovoid, long-exserted, with 40–220 spikelets. **Spikelets** 2–2.2 mm long, 0.8–1.3 mm wide, ellipsoid, usually reddish, shortly pubescent, subacute. **Lower glumes** 0.5–1 mm, triangular-ovate; **lower florets** sterile; **upper florets** slightly exceeding the upper glumes and lower lemmas, subacute. $2n$ = 18.

Dichanthelium boreale grows in open woodlands and thickets, wet meadows, and fields. It is restricted to the *Flora* region. The primary panicles are mostly open-pollinated and are produced in May and June; the secondary panicles are predominantly cleistogamous and are produced from mid-June into October.

Dichanthelium boreale occasionally hybridizes with *D. acuminatum* and *D. xanthophysum*, producing a sterile triploid sometimes called *Panicum calliphyllum* Ashe.

Dichanthelium sect. Nudicaulia Freckmann & Lelong

Plants cespitose. **Basal rosettes** somewhat differentiated. **Culms** 20–60 cm, weakly ascending, glabrous; **fall phase** rarely branching from near the base. **Cauline leaves** 3–4; **sheaths** sparsely pilose or glabrous; **ligules** membranous, ciliate; **blades** mostly basal, flat to stiffly involute, ascending to erect, uppermost blades much reduced. **Primary panicles** long-exserted, with few spikelets. **Spikelets** 2.4–3.2 mm, narrowly ellipsoid to ovoid, glabrous. **Upper florets** acute.

Dichanthelium nudicaule is the only member of sect. *Nudicaulia* present in the *Flora* region.

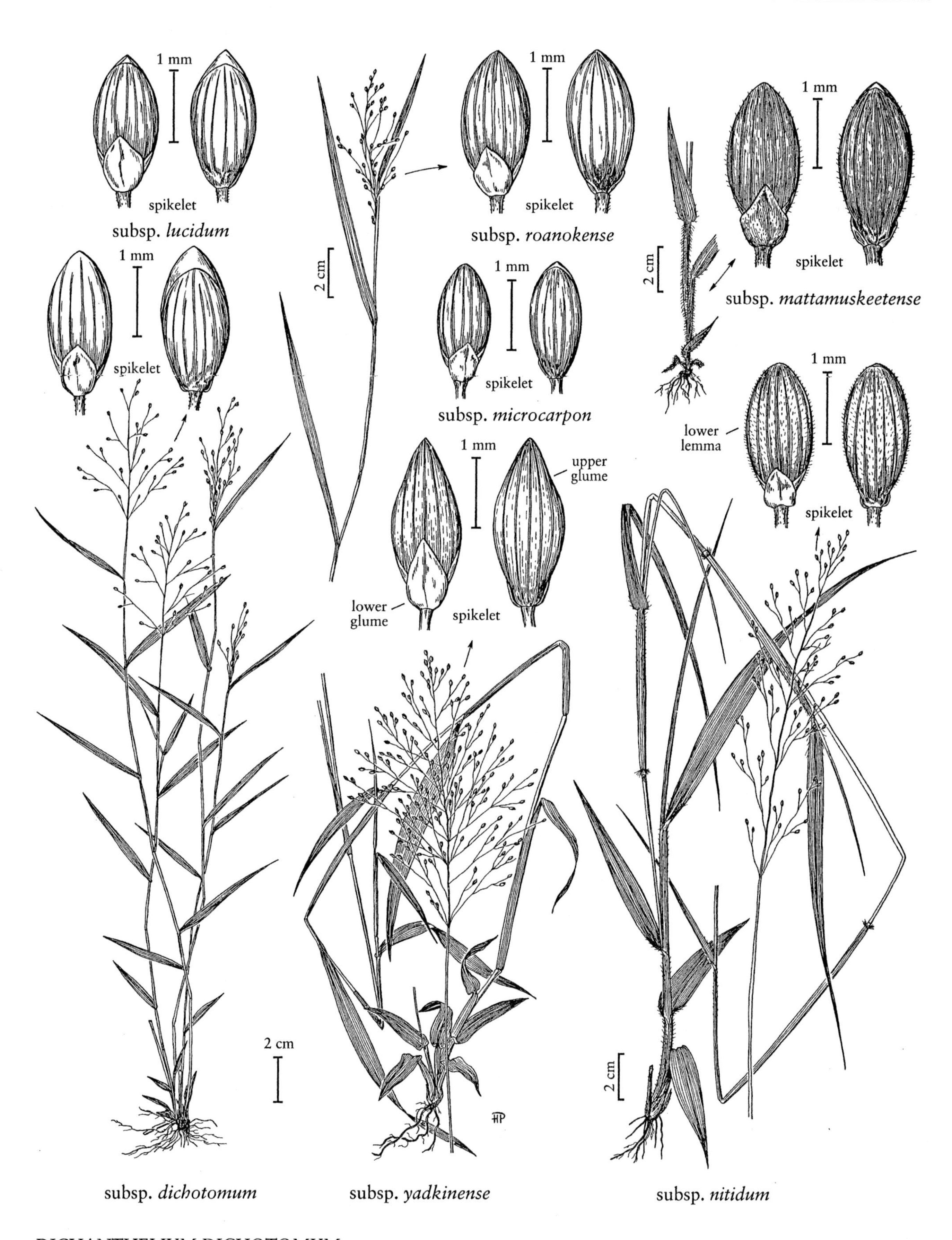

DICHANTHELIUM DICHOTOMUM

19. **Dichanthelium nudicaule** (Vasey) B.F. Hansen & Wunderlin [p. 437]
NAKED-STEMMED PANICGRASS

Plants cespitose, clumps with few culms. **Basal rosettes** somewhat differentiated; **blades** lanceolate. **Culms** 20–60 cm, with caudices, slender, glabrous, weakly ascending, with a tuft of predominantly basal leaves, only the upper 3 internodes elongated; **fall phase** rarely branching, branches, if present, from the basal and subbasal nodes, erect. **Cauline leaves** 3–4; **sheaths** longer than the internodes, lower sheaths sparsely ascending to spreading-pilose, upper sheaths somewhat elongate, striate, glabrous, lustrous; **ligules** usually 0.5–1 mm, membranous, ciliate; **blades** 2–20 cm long, 3–10 mm wide, mostly basal, ascending to erect, widest near midlength, flat to stiffly involute, tapering basally and partly encircling the culm, glabrous, blades of the flag leaves distant from and much smaller than those below. **Primary panicles** 2–7 cm long, almost as wide when expanded, long-exserted, sparse; **branches** few, ascending to spreading, glabrous or scabridulous. **Spikelets** 2.4–3.2 mm long, usually less than 1 mm wide, narrowly ellipsoid to ovoid, often purplish-stained, glabrous. **Lower glumes** less than ⅓ as long as the spikelets, acute; **upper glumes** and **lower lemmas** clearly longer than the upper florets, prominently veined, apices acuminate and usually beaked; **lower florets** sterile; **upper florets** about 2 mm long, about 1 mm wide, ellipsoid, acute. $2n$ = unknown.

Dichanthelium nudicaule is a rare species that grows in wet pine savannas, bogs (including *Sphagnum* mats), and the margins of cypress swamps in eastern Louisiana, southern Mississippi and Alabama, and western Florida. Vegetatively, it exhibits traits of *D. laxiflorum*, but its spikelets resemble those of small plants of *D. scabriusculum*, which are fairly widespread in similar habitats of the Gulf coastal plain. *Dichanthelium nudicaule* is protected by U.S. federal law.

Dichanthelium sect. Ensifolia (Hitchc.) Freckmann & Lelong

Plants cespitose, with caudices. **Basal rosettes** well-differentiated. **Culms** 5–60 cm tall, 0.2–0.8 (1.6) mm thick, erect or reclining, sometimes geniculate at the lower nodes, glabrous or sparsely pubescent near the bases; **fall phase** branching from the lower and midculm nodes. **Cauline leaves** 3–9; **sheaths** glabrous or sparsely pilose, usually ciliate; **ligules** 0.2–1.8 mm, of hairs; **blades** usually 2–5 cm, sometimes with prominent white, cartilaginous margins. **Spikelets** 1.1–1.7 mm, ellipsoid to narrowly obovoid, glabrous or puberulent. **Upper florets** subacute to acute.

20. **Dichanthelium ensifolium** (Baldwin *ex* Elliott) Gould [p. 439]
SWORD-LEAF PANICGRASS

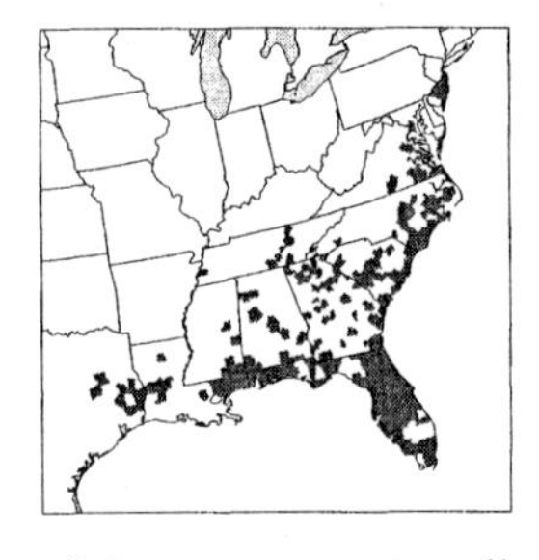

Plants cespitose, with caudices. **Basal rosettes** well-differentiated; **blades** 1.5–6 cm, ovate to lanceolate, soft, glabrous. **Culms** 10–40 cm tall, 0.2–0.8 (1.6) mm thick, weak, erect or reclining; **nodes** usually glabrous, sometimes sparsely bearded; **internodes** usually glabrous, occasionally sparsely pubescent; **fall phase** with spreading culms, sparingly branched, branching mostly from the midculm nodes, occasionally producing small fascicles of leafy branchlets. **Cauline leaves** 4–9; **sheaths** much shorter than the internodes, prominently veined, glabrous or sparsely pilose and ciliate, particularly at the top; **ligules** 0.2–1.8 mm, often more than 1 mm, of hairs, without adjacent pseudoligules; **blades** 1.5–3.5 cm long (seldom longer), 1.5–4 mm wide, all similar in size, thin, spreading or reflexed, abaxial surfaces puberulent, at least apically, or sparsely pilose, adaxial surfaces glabrous or sparsely pilose, at least basally, bases abruptly and strongly constricted, occasionally ciliate, margins entire or faintly scabridulous, rarely white-cartilaginous. **Primary panicles** 1.5–4 cm, nearly as wide as long, long-exserted; **branches** wiry, mostly spreading, minutely scabridulous. **Spikelets** 1.2–1.5 mm, ellipsoid to obovoid, yellow-green to purplish, puberulent or glabrous, subacute or obtuse. **Lower glumes** seldom more than ¼ as long as the spikelets, acute or obtuse; **upper glumes** usually slightly shorter than the lower lemmas and upper florets, not strongly veined; **upper florets** 1.1–1.4 mm long, less than 1 mm wide, ellipsoid, acute. $2n$ = 18.

Dichanthelium ensifolium grows in wet to moist, sandy pinelands, savannahs, and bogs, often on *Sphagnum* mats, primarily on the coastal plain. It extends south into Mesoamerica, and has been reported from Venezuela. Occasional specimens grade towards

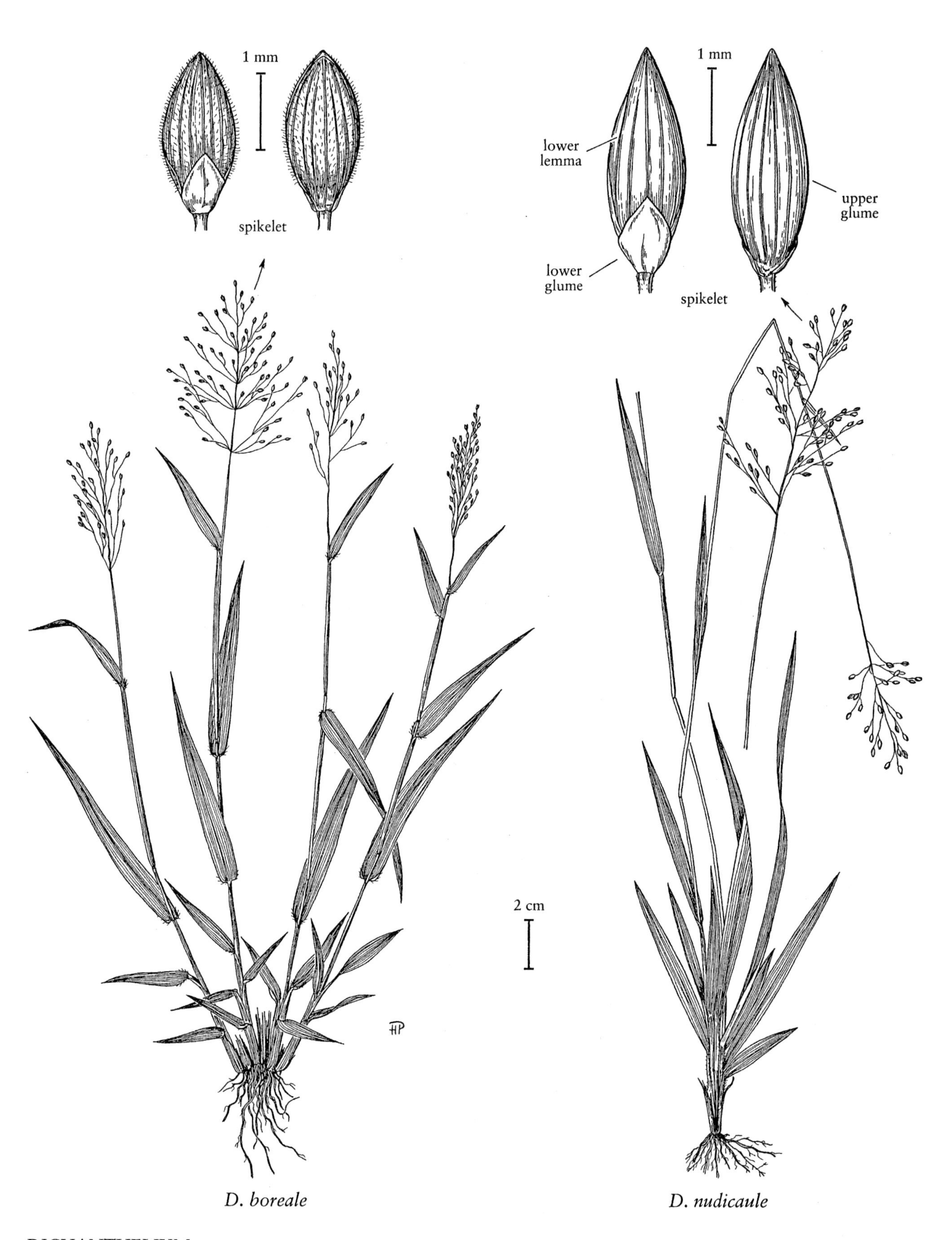

D. boreale

D. nudicaule

the larger *D. tenue*, and are usually found on somewhat drier sites. It also resembles *D. chamaelonche*, but that species is usually more densely cespitose, has slightly smaller, glabrous spikelets, and generally occupies drier, disturbed sites.

The two subspecies are sympatric, often growing together at the same sites.

1. Sheaths sparsely spreading-pilose; ligules usually 1–1.8 mm long; blades sparsely pilose or glabrous on both surfaces . . . . subsp. *curtifolium*
1. Sheaths glabrous; ligules 0.2–1 mm long; blades usually puberulent abaxially, usually glabrous, occasionally pubescent adaxially . . . . . subsp. *ensifolium*

Dichanthelium ensifolium subsp. **curtifolium** (Nash) Freckmann & Lelong

Cauline sheaths sparsely spreading-pilose; **ligules** usually 1–1.8 mm; **blades** sparsely pilose or glabrous on both surfaces.

Dichanthelium ensifolium (Baldwin *ex* Elliott) Gould subsp. **ensifolium**

Cauline sheaths glabrous; **ligules** 0.2–1 mm; **blades** usually puberulent abaxially, glabrous or, occasionally, pubescent adaxially.

21. **Dichanthelium tenue** (Muhl.) Freckmann & Lelong [p. 439]
SLENDER PANICGRASS

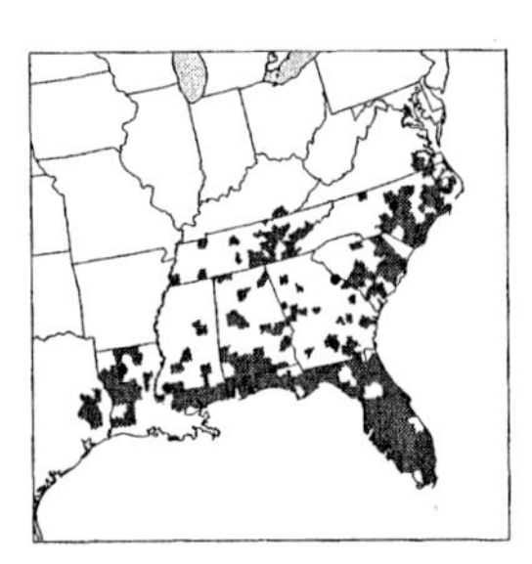

Plants cespitose, with caudices, forming small, often rather dense clumps with few culms. **Basal rosettes** well-differentiated; **blades** 1–5 cm, ovate to lanceolate. **Culms** 15–55 cm tall, 0.2–0.8 mm thick, erect from geniculate bases; **nodes** glabrous; **internodes** mostly glabrous, or the lowest internodes sparsely appressed-pubescent basally; **fall phase** branching sparingly from the lower and midculm nodes. **Cauline leaves** 3–4; **sheaths** much shorter than the internodes, prominently veined, mostly glabrous, margins occasionally ciliate, **ligules** 0.2–0.7 mm, of hairs, without adjacent pseudoligules; **blades** 2–6 cm long, 1.5–6 mm wide, ascending, distant, flat, relatively thick, glabrous on both surfaces or the abaxial surfaces minutely puberulent, bases rounded, margins more or less prominently whitish-scabridulous, blades of the flag leaves much shorter than those of the lower leaves. **Primary panicles** 3–6 cm, nearly as wide as long, long-exserted, dense; **branches** wiry, spreading to ascending, usually scabridulous. **Spikelets** 1.3–1.7 mm long, less than 1 mm wide, ellipsoid, often purplish, densely puberulent, obtuse or subacute. **Lower glumes** usually less than ¼ as long as the spikelets, broadly acute or obtuse; **upper glumes** and **lower lemmas** subequal, or the glumes slightly shorter, exceeded by the upper florets; **lower florets** sterile; **upper florets** 1.3–1.6 mm, ellipsoid, subacute. $2n = 18$.

Dichanthelium tenue grows in moist to dry, sandy woods, savannahs, and disturbed sites. It also grows in Chiapas, Mexico (Zuloaga et al. 1993). It exhibits features of *D. sphaerocarpon* and *D. dichotomum*. It is also closely related to *D. ensifolium*, and occasional specimens are intermediate between them.

22. **Dichanthelium chamaelonche** (Trin.) Freckmann & Lelong [p. 439]
SMALL-SEEDED PANICGRASS

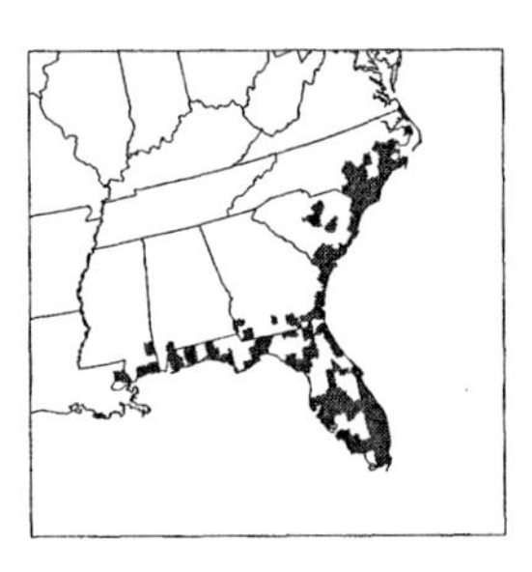

Plants usually densely cespitose, with caudices. **Basal rosettes** well-differentiated; **blades** 1–5 cm, ovate to lanceolate. **Culms** 5–45 cm tall, 0.2–0.8 mm thick, erect, often purplish; **nodes** glabrous or sparsely pubescent; **internodes** often ascending-pubescent below; **fall phase** branching extensively from the basal nodes, usually forming very dense cushions. **Cauline leaves** 3–5; **sheaths** mostly shorter than the internodes, often purplish, glabrous or sparsely pubescent, margins often sparsely ciliate; **ligules** 0.2–0.5 mm, of hairs, without adjacent pseudoligules; **blades** 2–5 cm long (rarely longer), 1–4 mm wide, flat or involute, rather firm, ascending, often purplish, usually glabrous on both surfaces, bases subcordate, often with a few long, stiff cilia, margins narrowly white, cartilaginous, and scabridulous, blades of the flag leaves only slightly shorter than those of the lower leaves. **Primary panicles** 1.5–5 cm (seldom longer), nearly as wide as long, delicate, dense; **branches** numerous, flexuous, spreading, often purplish, glabrous or faintly scabridulous. **Spikelets** 1.1–1.5 mm long, 0.7–1 mm wide, broadly ellipsoid or obovoid, often purple-tinged, glabrous or puberulent, obtuse or subacute. **Lower glumes** approximately ⅓ as long as the spikelets, broadly acute or obtuse; **upper glumes** and **lower lemmas** subequal or the glumes slightly shorter than the lemmas; **lower florets** sterile; **upper florets** 0.9–1.2 mm, ellipsoid, apices exceeding the upper glumes and lower lemmas, subacute.

Dichanthelium chamaelonche grows in low, open, sandy, coastal pine woods, savannahs, and moist depressions in sand dunes. It is restricted to the southeastern United States.

1. Culms 5–20 cm tall, glabrous or puberulent; spikelets 1.3–1.5 mm long, puberulent . . . . . subsp. *breve*
1. Culms 10–45 cm tall, glabrous; spikelets 1.1–1.4 mm long, glabrous . . . . subsp. *chamaelonche*

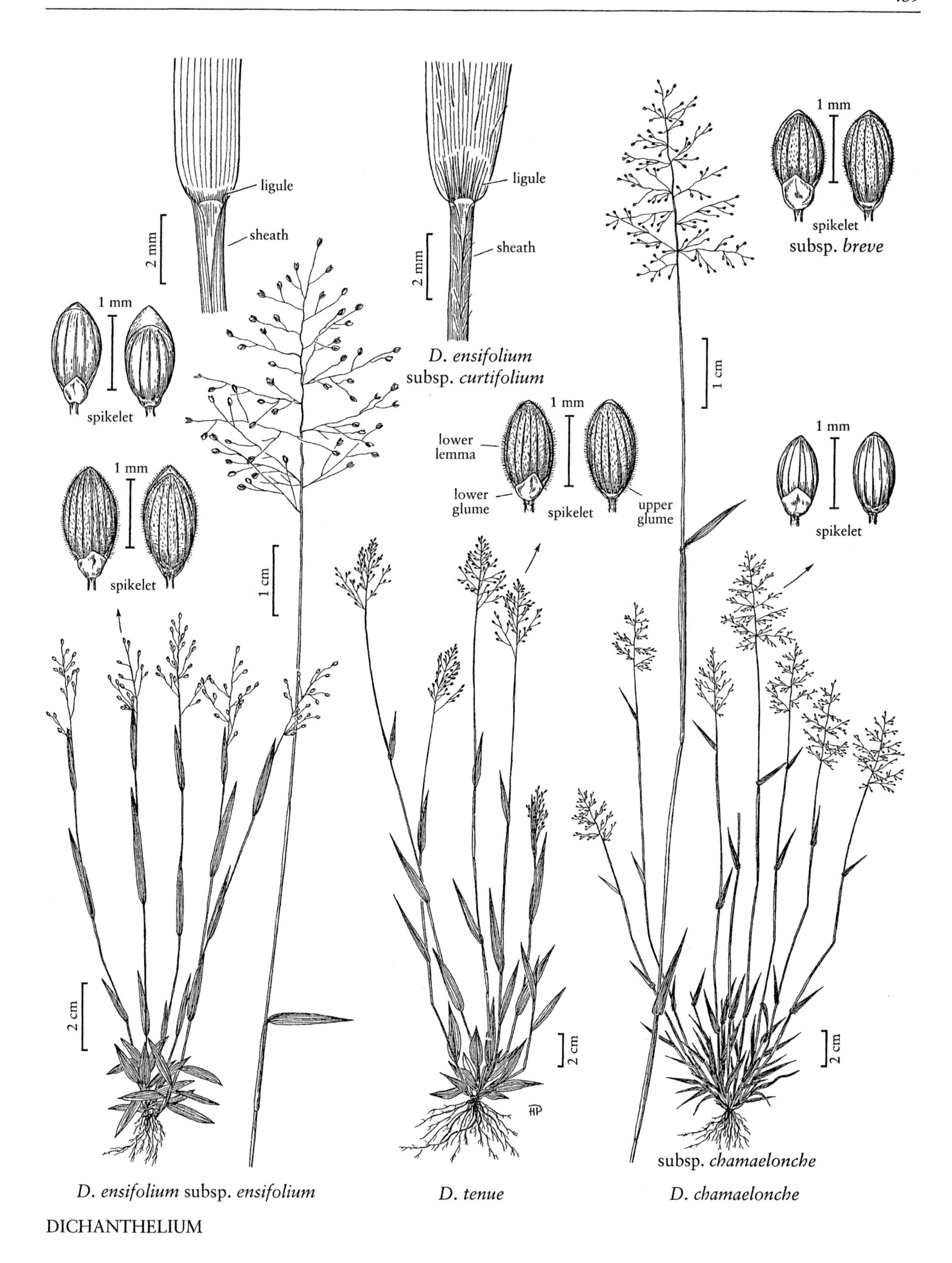
ligule
sheath
2 mm
1 mm
spikelet
1 mm
spikelet
1 cm
2 cm
D. ensifolium subsp. ensifolium
ligule
sheath
2 mm
D. ensifolium
subsp. curtifolium
1 mm
lower lemma
lower glume
spikelet
upper glume
2 cm
D. tenue
1 mm
spikelet
subsp. breve
1 cm
1 mm
spikelet
2 cm
subsp. chamaelonche
D. chamaelonche

Dichanthelium chamaelonche subsp. **breve** (Hitchc. & Chase) Freckmann & Lelong

Culms 5–20 cm, glabrous or puberulent. **Sheaths** puberulent or glabrous; **blades** strongly involute, often arcuate, puberulent or glabrous on both surfaces. **Primary panicles** usually barely exserted above the dense basal tuft of blades. **Spikelets** 1.3–1.5 mm, puberulent. $2n$ = unknown.

Dichanthelium chamaelonche subsp. *breve* grows only in peninsular Florida.

Dichanthelium chamaelonche (Trin.) Freckmann & Lelong subsp. **chamaelonche**

Culms 10–45 cm, glabrous. **Cauline sheaths** glabrous; **blades** flat or involute, glabrous on both surfaces. **Primary panicles** well-exserted above the basal tuft of blades. **Spikelets** 1.1–1.4 mm, glabrous. $2n = 18$.

Dichanthelium chamaelonche subsp. *chamaelonche* grows from North Carolina to Florida and Louisiana.

Dichanthelium sect. Sphaerocarpa (Hitchc. & Chase) Freckmann & Lelong

Plants cespitose, with caudices. **Basal rosettes** well-differentiated; **blades** large, ovate-lanceolate. **Culms** 15–95 cm, usually nearly erect, sometimes spreading to ascending, glabrous or almost glabrous, slightly fleshy or thickened; **fall culms** with sparse branching. **Cauline leaves** 3–7, glabrous or almost glabrous throughout; **ligules** almost obsolete, usually of hairs shorter than 0.8 mm, sometimes with a minute membranous base; **blades** firm, thick, margins cartilaginous, bases cordate, ciliate. **Spikelets** 1–1.8 mm, spherical to broadly obovoid, puberulent or sometimes glabrous. **Upper florets** blunt or minutely umbonate.

Dichanthelium sect. *Sphaerocarpa* extends from the southeastern United States through Central America and Cuba to northern South America. Pairs of species often grow together, with infrequent apparent hybridization.

23. **Dichanthelium erectifolium** (Nash) Gould & C.A. Clark [p. 443]
FLORIDA PANICGRASS, ERECT-LEAF PANICGRASS

Plants cespitose, with few culms. **Basal rosettes** well-differentiated; **blades** numerous, to 15 cm, lowest blades ovate, upper blades lanceolate, grading into the cauline blades. **Culms** 30–75 cm, nearly erect, stiff, slightly fleshy or thickened; **nodes** glabrous, often with a constricted, yellowish ring; **internodes** glabrous; **fall phase** with few, long, suberect branches, sparingly rebranched, branches arising mostly from near the base. **Cauline leaves** 4–7; **sheaths** shorter than the internodes, mostly glabrous, margins ciliate; **ligules** 0.2–0.5 mm; **blades** 5–10 cm long, 5–10 mm wide, stiffly ascending, thick, glabrous, veins evident, bases cordate, with papillose-based cilia, margins whitish, cartilaginous. **Primary panicles** 5–14 cm, ½–⅔ as wide as long, exserted. **Spikelets** 1–1.4 mm, broadly obovoid-spherical, puberulent to subglabrous. **Lower glumes** 0.2–0.4 mm, acute, **upper florets** 0.8–1.1 mm, broadly ellipsoid, minutely umbonate. $2n$ = unknown.

Dichanthelium erectifolium grows in sand and peat in wet pinelands, bogs, and the shores of ponds. Its range extends from the southeastern *Flora* region into the Caribbean.

24. **Dichanthelium polyanthes** (Schult.) Mohlenbr. [p. 443]
MANY-FLOWERED PANICGRASS

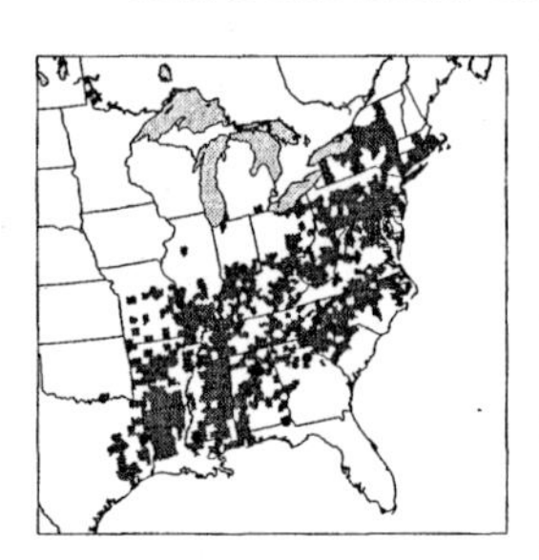

Plants cespitose, with few culms per tuft. **Basal rosettes** well-differentiated; **blades** 3–8 cm long, often to 2 cm wide, ovate-lanceolate. **Culms** 30–95 cm, nearly erect, fairly stout; **nodes** glabrous or puberulent; **internodes** usually glabrous; **fall phase** with few, long-ascending branches, sparingly rebranched, branches arising mostly near the base of the culms. **Cauline leaves** 4–7; **sheaths** shorter than the internodes, mostly glabrous, margins ciliate; **ligules** vestigial; **blades** 10–25 cm long, 14–25 mm wide, thick, firm, often light green, veins evident (some more prominent than others), bases cordate, with papillose-based cilia, margins whitish, cartilaginous. **Primary panicles** 7–20 cm, less than ½ as wide as long, exserted. **Spikelets** 1.3–1.7 mm, broadly ellipsoid-spherical, often purplish at the base, puberulent. **Lower glumes** 0.4–0.7 mm, acute to obtuse, **upper florets** 1.1–1.4 mm, broadly ellipsoid, blunt. 2*n* = 18.

Dichanthelium polyanthes grows in woods, stream banks, and ditches, and is restricted to the eastern United States. It occasionally hybridizes with *D. sphaerocarpon*.

25. **Dichanthelium sphaerocarpon** (Elliott) Gould [p. 443]
ROUND-FRUITED PANICGRASS

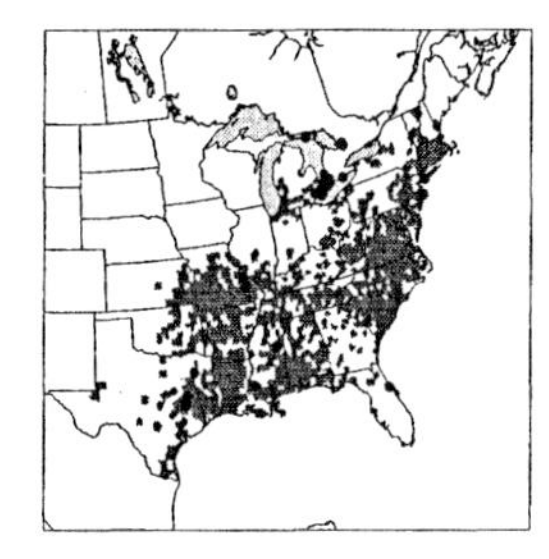

Plants cespitose. **Basal rosettes** well-differentiated; **blades** 2–6 cm long, about 1 cm wide, ovate, the uppermost leaves often resembling the lower cauline blades. **Culms** 15–50 cm, few together, decumbent or ascending, light green, glabrous, slightly fleshy or thickened; **fall phase** branching mostly near the bases, with sparse branching; **nodes** appressed-pubescent or glabrous. **Cauline leaves** 3–4(6); **sheaths** sometimes overlapping near the bases, glabrous, margins ciliate; **ligules** almost obsolete, or of 0.2–0.8 mm hairs from a tiny membranous base; **blades** 1.5–10 cm long, 5–14 mm wide, thick, light green, faintly veined, bases cordate, with papillose-based cilia, margins white, cartilaginous. **Primary panicles** 4–14 cm, more than ½ as wide as long, usually long-exserted. **Spikelets** 1.4–1.8 mm, broadly obovoid-spherical, usually puberulent, sometimes glabrous. **Lower glumes** 0.4–0.8 mm, acute to obtuse, **upper florets** 1.1–1.5 mm, broadly ellipsoid, blunt. 2*n* = 18.

Dichanthelium sphaerocarpon grows in dry, open woods and roadsides. Its range extends from eastern North America to Ecuador and Venezuela. It occasionally hybridizes with other species, including *D. polyanthes*, *D. acuminatum*, and *D. laxiflorum*.

Dichanthelium sect. **Lancearia** (Hitchc.) Freckmann & Lelong

Plants usually densely cespitose, with caudices. **Basal rosettes** usually well-differentiated. **Culms** slender, usually purplish and puberulent; **fall phase** often profusely branched and rebranched. **Cauline leaves** 8–14; **sheaths** usually purplish and puberulent; **ligules** less than 0.5 mm, of hairs; **blades** usually purplish and puberulent. **Primary panicles** exserted. **Spikelets** 1.5–2.6 mm, obovoid-pyriform, planoconvex in side view, bases attenuate. **Lower glumes** thin, weakly veined, attached about 0.2 mm below the upper glumes, clasping at the base. **Upper florets** subacute.

Only one species of sect. *Lancearia*, *Dichanthelium portoricense*, grows in the *Flora* region.

26. **Dichanthelium portoricense** (Desv. *ex* Ham.) B.F. Hansen & Wunderlin [p. 443]
BLUNT-GLUMED PANICGRASS

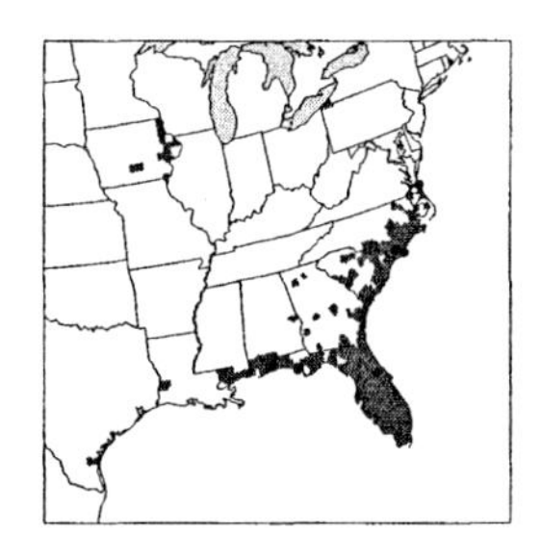

Plants usually densely cespitose. **Basal rosettes** well-differentiated; **blades** 1.5–6 cm, ovate to lanceolate. **Culms** 15–50 cm, slender, wiry; **internodes** olive green to purplish, densely puberulent or glabrous; **fall phase** spreading or decumbent, branching extensively from the lower and midculm nodes, producing numerous congested fascicles of reduced, flat or involute blades and reduced secondary panicles. **Cauline leaves** 4–7; **sheaths** much shorter than the internodes, densely crisp-puberulent, velvety-puberulent, or glabrous, often ciliate along the margins; **ligules** shorter than 0.5 mm; **blades** 2–7 cm long (seldom longer), 2.5–8 mm wide (rarely wider), spreading, firm, flat or slightly involute, without prominently raised veins, not longitudinally wrinkled, densely puberulent or glabrous abaxially, glabrous, sparsely puberulent, or pubescent adaxially, bases subcordate, with papillose-based cilia, margins often whitish and scabridulous. **Primary panicles** 2–7 cm long, ⅔ to nearly as wide as long, with relatively few spikelets, exserted; **branches** flexuous, spreading or reflexed, scabridulous to densely puberulent. **Spikelets** 1.5–2.6 mm, obovoid-pyriform, planoconvex in side view, puberulent, pubescent, or glabrous, attenuate basally, apices usually broadly rounded. **Lower glumes** 0.6–1.4 mm, thin, weakly-veined, attached about 0.2 mm below the upper glumes, clasping at the base; **upper glumes** as long as or slightly shorter than the lower lemmas; **upper florets** 1.4–2 mm, broadly ellipsoid, apices subacute, minutely puberulent. 2*n* = 18.

Dichanthelium portoricense grows in sandy woods, low pinelands, savannahs, and coastal sand dunes, usually in moist places. Its range extends south from the *Flora* region into Mexico, the Caribbean, and Mesoamerica. It is a highly variable species with numerous intergrading forms, some possibly resulting from hybridization with other widespread species in the same region, such as *D. sphaerocarpon* and *D. commutatum*.

1. Spikelets 1.8–2.6 mm long, usually densely pubescent or puberulent (rarely glabrous); cauline blades 4–7 cm long, 3.5–8 mm wide . . . . subsp. *patulum*
1. Spikelets 1.5–2.0 mm long, puberulent to nearly glabrous; cauline blades 2–5 cm long, 2.5–4.5 mm wide . . . . subsp. *portoricense*

Dichanthelium portoricense subsp. **patulum** (Scribn. & Merr.) Freckmann & Lelong

Culms 20–50 cm, often densely puberulent. **Sheaths** puberulent to subglabrous. **Cauline blades** 4–7 cm long, 3.5–8 mm wide. **Primary panicles** 2–7 cm. **Spikelets** 1.8–2.6 mm, usually densely pubescent or puberulent, rarely glabrous.

Dichanthelium portoricense subsp. *patulum* is more common in moist, sandy pinelands and savannahs than subsp. *portoricense*. It also grows in coastal sand dunes, but is less abundant there than subsp. *portoricense*. It is the more variable of the two subspecies, grading into subsp. *portoricense* as well as *D. commutatum*. More robust plants are recognized by some as *Panicum patentifolium* Nash. Occasional specimens, recognized by some as *P. webberianum* Nash, resemble the widespread *D. sphaerocarpon*.

Dichanthelium portoricense (Desv. *ex* Ham.) B.F. Hansen & Wunderlin subsp. **portoricense**

Culms 15–40 cm, glabrous or puberulent. **Sheaths** glabrous or puberulent; **cauline blades** 2–5 cm long, 2.5–4.5 mm wide, usually puberulent abaxially and glabrous adaxially. **Primary panicles** 2–4.5 cm. **Spikelets** 1.5–2.0 mm, puberulent to nearly glabrous.

Dichanthelium portoricense subsp. *portoricense* is more common than subsp. *patulum* in coastal sand dunes. It also grows in sandy pinelands and savannahs. It resembles *D. aciculare* somewhat, but that species usually has ascending-pilose culms, strongly involute or acicular blades, and longer spikelets.

Dichanthelium sect. Angustifolia (Hitchc.) Freckmann & Lelong

Plants grayish-green, densely cespitose, with caudices. **Basal rosettes** sometimes poorly differentiated. **Culms** 15–75 cm, erect; **fall phase** erect or spreading, extensively branched from the mid- and upper culm nodes, secondary panicles and blades of the fascicles much reduced. **Cauline leaves** 3–7; **sheaths** glabrous, pilose, or villous, pubescence sometimes sparse; **ligules** 0.5–2 mm, of hairs, sometimes with a pseudoligule of adjacent longer hairs; **blades** narrow, stiffly ascending to erect, lower blades widest, transitional to the rosette blades, often longitudinally wrinkled, mid-culm blades generally 16–25 times longer than wide, with prominent raised veins, blades of the flag leaves much reduced, often involute. **Primary panicles** usually long-exserted. **Spikelets** ellipsoid to obovoid, narrow to attenuate basally, biconvex in side view. **Lower glumes** thin, obtuse, weakly veined, somewhat remote and clasping at the base; **upper glumes** with 5–9 prominent veins. **Upper florets** blunt to apiculate.

Dichanthelium sect. *Angustifolia* grows from the southeastern United States through Central America and the West Indies to northern South America.

27. **Dichanthelium aciculare** (Desv. *ex* Poir.) Gould & C.A. Clark [p. 445]
Narrow-Leaved Panicgrass

Plants grayish-green, cespitose, with caudices. **Basal rosettes** poorly differentiated; **blades** usually large, ovate to lanceolate, often transitional to the cauline blades. **Culms** 15–75 cm, erect; **nodes** glabrous or sparsely pubescent; **internodes** glabrous or puberulent to pilose basally; **fall phase** with erect to spreading culms, extensively branched from the mid- and upper culm nodes, eventually producing flabellate clusters of reduced, flat or involute blades. **Cauline leaves** 3–7; **sheaths** shorter than the internodes, glabrous or with soft, ascending, papillose-based hairs; **ligules** 0.5–2 mm, of hairs; **lower blades** 4–16 cm long, 3–9 mm wide, stiffly ascending to erect, glabrous or sparsely pilose to pubescent, with prominent raised veins, flat or longitudinally wrinkled, blades of the flag leaves often greatly reduced, often involute. **Primary panicles** 2–10 cm long, 0.5–7 cm wide, open or contracted, well-exserted. **Spikelets** 1.7–3.6 mm long, 1.2–1.8 mm wide, obovoid to ellipsoid, biconvex in side view, glabrous or pubescent, bases narrow to attenuate,

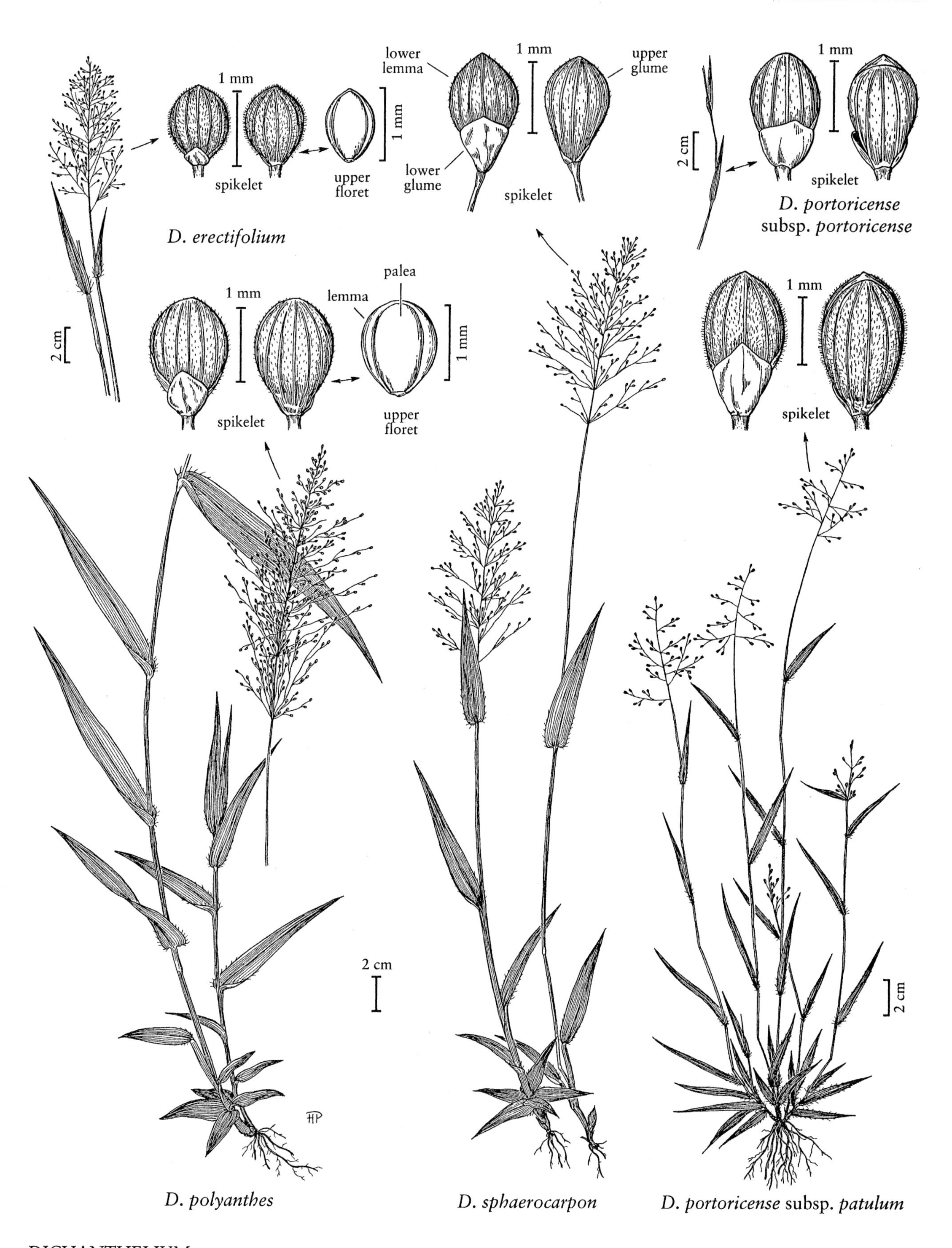
lower lemma
1 mm
upper glume
1 mm
spikelet
1 mm
upper floret
lower glume
spikelet
2 cm
1 mm
spikelet
D. portoricense subsp. portoricense
D. erectifolium
palea
1 mm
lemma
1 mm
spikelet
upper floret
1 mm
2 cm
spikelet
2 cm
2 cm
HP
D. polyanthes
D. sphaerocarpon
D. portoricense subsp. patulum

apices blunt or pointed to beaked. **Lower glumes** thin, weakly veined, about ⅓ as long as the spikelets, attached to 0.5 mm below upper glumes, clasping at the base, broadly triangular to rounded; **upper glumes** with 5–9 prominent veins; **lower florets** sterile; **upper florets** apiculate. $2n = 18$.

Dichanthelium aciculare grows in sandy, open areas in the southeastern United States, the West Indies and the Caribbean, southern Mexico, Central America, and northern South America. It has not been reported from northern Mexico. The primary panicles are open-pollinated (sometimes briefly) and develop from April to June; the secondary panicles are cleistogamous and develop from May into late fall.

The subspecies are often distinct when growing together, perhaps maintained by the predominant autogamy, but they are more difficult to separate over wider geographic areas. Rare, partly fertile putative hybrids with *Dichanthelium consanguineum*, *D. acuminatum*, *D. ovale*, *D. portoricense*, and (possibly) *D. dichotomum* apparently lead to some intergradation with these species.

1. Primary panicles usually contracted; branches appearing 1-sided; culms sparsely pubescent to almost glabrous .... subsp. *neuranthum*
1. Primary panicles not contracted; branches not appearing 1-sided; culms usually pubescent, at least on the lower internodes.
   2. Spikelets 1.7–2.3 mm long, with blunt apices .... subsp. *aciculare*
   2. Spikelets 2.4–3.6 mm long, with pointed or beaked apices.
      3. Spikelets 2.4–3 mm long, not strongly attenuate at the base; lower glumes attached less than 0.2 mm below the upper glumes .... subsp. *angustifolium*
      3. Spikelets 2.9–3.6 mm long, strongly attenuate at the base; lower glumes attached 0.3–0.5 mm below the upper glumes .... subsp. *fusiforme*

Dichanthelium aciculare (Desv. *ex* Poir.) Gould & C.A. Clark subsp. **aciculare**

Plants densely cespitose. **Culms** usually 15–35 cm, usually pubescent, at least on the lower internodes; **nodes** sometimes bearded, usually yellowish; **lower internodes** puberulent to appressed-pilose. **Cauline blades** usually 4–6 cm. **Primary panicles** usually open, branches spreading to ascending, not appearing 1-sided. **Spikelets** 1.7–2.3 mm, obovoid, blunt.

Dichanthelium aciculare subsp. *aciculare* is common in sterile, open sands on the coastal plain. It is restricted to the eastern United States and Mexico.

Dichanthelium aciculare subsp. **angustifolium** (Elliott) Freckmann & Lelong

Plants cespitose. **Culms** usually 35–75 cm, usually pubescent, at least on the lower internodes. **Midculm blades** 6–16 cm, usually glabrous. **Primary panicles** open, branches spreading, not appearing 1-sided. **Spikelets** 2.4–3 mm, narrowly obovoid to ellipsoid, often pointed to beaked. **Lower glumes** attached less than 0.2 mm below the upper glumes.

Dichanthelium aciculare subsp. *angustifolium* grows in open pine woodlands, often in sandy soil with needle duff. It is restricted to the southeastern United States.

Dichanthelium aciculare subsp. **fusiforme** (Hitchc.) Freckmann & Lelong

Resembles subsp. *angustifolium* vegetatively except that the culms are often taller and more slender. **Panicle branches** ascending, not appearing 1-sided. **Spikelets** 2.9–3.6 mm, fusiform, bases strongly attenuate, apices pointed or beaked, **lower glumes** attached 0.3–0.5 mm below the upper glumes.

Dichanthelium aciculare subsp. *fusiforme* grows in sandy pine or oak savannahs. It tends to replace *D. aciculare* subsp. *angustifolium* from southern Florida through Central America and the Antilles.

Dichanthelium aciculare subsp. **neuranthum** (Griseb.) Freckmann & Lelong

Plants cespitose. **Culms** 30–60 cm, stiffly erect, wiry, sparsely pubescent to almost glabrous. **Cauline blades** erect, narrow, often involute. **Primary panicles** usually contracted; **branches** appressed to ascending, appearing 1-sided. **Spikelets** 2–2.8 mm, ellipsoid.

Dichanthelium aciculare subsp. *neuranthum* grows in moist, sandy, open ground and savannahs, primarily on the outer coastal plain and in Cuba and various other Caribbean islands.

28. **Dichanthelium consanguineum** (Kunth) Gould & C.A. Clark [p. 445]
KUNTH'S PANICGRASS

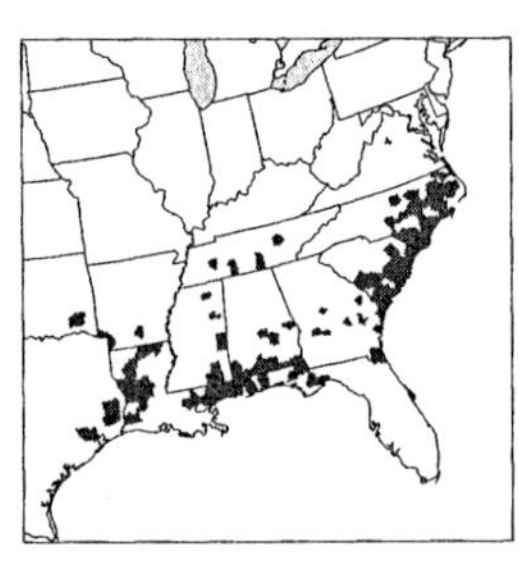

Plants grayish-green, cespitose. **Basal rosettes** poorly differentiated; **blades** 2–8 cm, ovate to lanceolate, grading into the cauline blades. **Culms** 20–55 cm, erect; **nodes** densely bearded; **internodes** densely villous; **fall phase** with spreading culms branching from the lower and midculm nodes, eventually producing flabellate clusters of reduced, flat blades, secondary panicles much reduced. **Cauline leaves** 3–4; **sheaths** shorter than the internodes, pilose with ascending papillose-based hairs to villous; **ligules** 0.5–2 mm, of hairs; **blades** 4–12 cm long, 2–8 mm

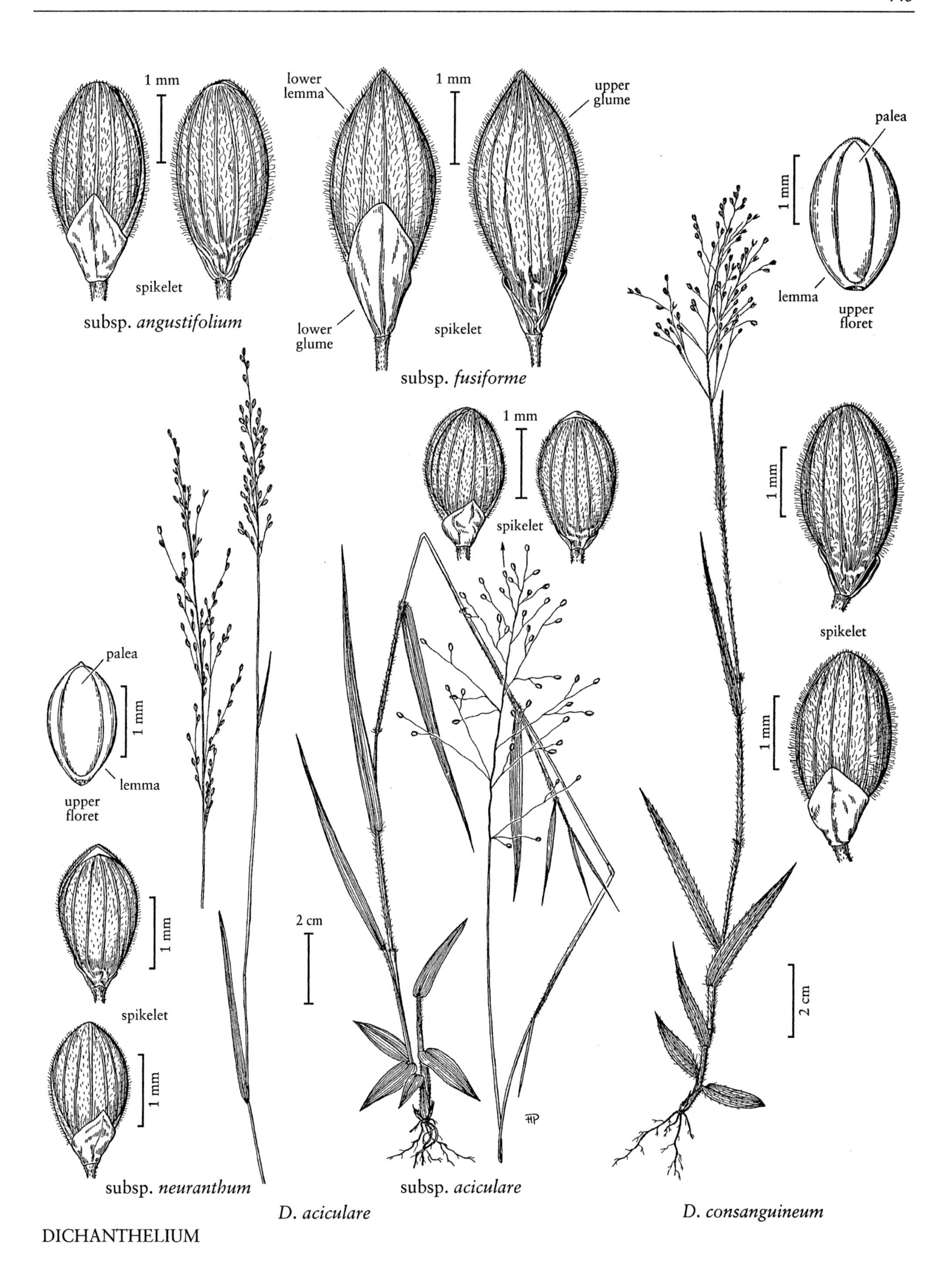
1 mm
spikelet
subsp. angustifolium
lower lemma
1 mm
upper glume
lower glume
spikelet
subsp. fusiforme
palea
1 mm
lemma
upper floret
1 mm
spikelet
palea
1 mm
lemma
upper floret
1 mm
spikelet
1 mm
2 cm
1 mm
spikelet
1 mm
2 cm
subsp. neuranthum
subsp. aciculare
D. aciculare
D. consanguineum

wide, stiffly ascending to erect, often wrinkled along the prominent veins, usually villous on both surfaces, apices involute-pointed, blades of the flag leaves much reduced. **Primary panicles** 3–7 cm long, 1–4 cm wide, well-exserted; **branches** usually ascending, glabrous or puberulent. **Spikelets** 2.3–3 mm long, 1.4–1.8 mm wide, obovoid, biconvex in side view, densely pubescent, attenuate basally. **Lower glumes** about ⅓ as long as the spikelets, attached about 0.2 mm below the upper glumes, clasping at the base, broadly triangular, thinner than the upper glumes, weakly veined; **upper glumes** with 5–9 prominent veins; **lower florets** sterile; **upper florets** broadly ellipsoid, apices blunt, minutely puberulent. $2n = 18$.

Dichanthelium consanguineum grows in sandy woodlands and low, boggy pinelands. It is restricted to the southeastern United States. The primary panicles are open-pollinated and produced from April to June; the secondary panicles are cleistogamous and produced from June into fall. Some specimens of *D. consanguineum* suggest that hybridization occasionally occurs with *D. aciculare* or *D. ovale*.

Dichanthelium sect. **Strigosa** Freckmann & Lelong

Plants densely cespitose, with caudices. **Basal rosettes** poorly differentiated; **blades** scarcely separable from the crowded lower cauline blades. **Culms** 5–55 cm, slender, erect to spreading, lower internodes short, upper 3–5 internodes elongate; **fall phase** usually forming a dense cushion. **Cauline leaves** 2–4; **ligules** membranous, ciliate; **blades** soft, green to yellowish, margins usually ciliate. **Primary panicles** exserted at maturity. **Spikelets** 1.1–2.3 mm, broadly ellipsoid to obovoid, glabrous or pubescent. **Upper florets** subacute or minutely umbonate.

29. **Dichanthelium laxiflorum** (Lam.) Gould [p. 448]
SOFT-TUFTED PANICGRASS

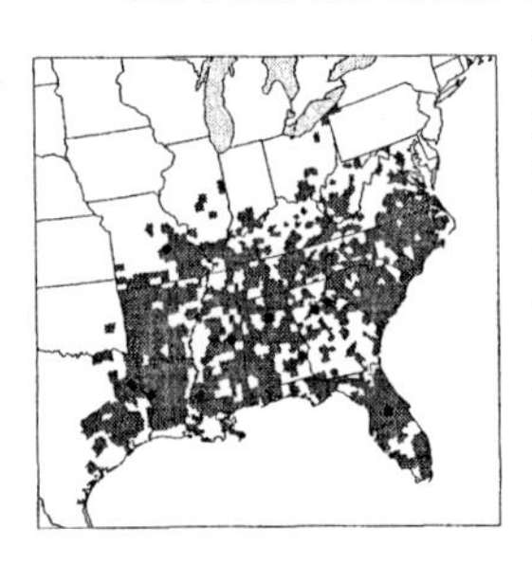

Plants densely cespitose. **Basal rosettes** poorly differentiated; **blades** ovate to lanceolate. **Culms** 15–55 cm, slender, erect or radiating from a large tuft of predominantly basal leaves, lower internodes short, upper 3–5 internodes elongate; **nodes** bearded with soft, spreading or retrorse hairs; **internodes** glabrous; **fall phase** branching extensively from the basal nodes, forming a dense cushion that overwinters. **Cauline leaves** 2–4; **sheaths** usually longer than the internodes, pilose, hairs to 4 mm, retrorse or spreading; **ligules** 0.2–1 mm, at low magnification appearing to be membranous and ciliate, at high magnification evidently of hairs that are coherent at the base; **blades** 4–17 cm long, 4–12 mm wide, lanceolate, at least ¾ as long as the basal blades, spreading to suberect, thin, soft, lax, yellowish-green, nearly glabrous or densely pilose on 1 or both surfaces, margins usually finely short-ciliate, at least on the basal ½, cilia not papillose-based. **Primary panicles** 4–12 cm long, 3–8 cm wide, well-exserted; **secondary panicles** more compact, usually not exserted above the crowded basal leaves; **rachises** and **branches** wiry, spreading or deflexed, often pilose. **Spikelets** 1.7–2.3 mm long, 1–1.2 mm wide, broadly ovate or oblong-obovoid, with papillose-based hairs, obtuse. **Lower glumes** ¼–⅓ as long as the spikelets, broadly deltoid; **upper glumes** and **lower lemmas** subequal, usually fully covering the upper florets; **upper florets** 1.5–1.8 mm long, 1–1.2 mm wide, broadly ellipsoid or obovoid, minutely umbonate. $2n = 18$.

Dichanthelium laxiflorum is a widespread, common species that grows in mesic deciduous woods, and occasionally in drier, more open woodlands. Its range extends south from the *Flora* region into Mexico. The density of the pubescence on the blade surfaces varies greatly.

The primary (spring) panicles are apparently chasmogamous; the secondary panicles are largely cleistogamous and are produced from late spring to winter.

30. **Dichanthelium strigosum** (Muhl. *ex* Elliott) Freckmann [p. 448]
CUSHION-TUFTED PANICGRASS

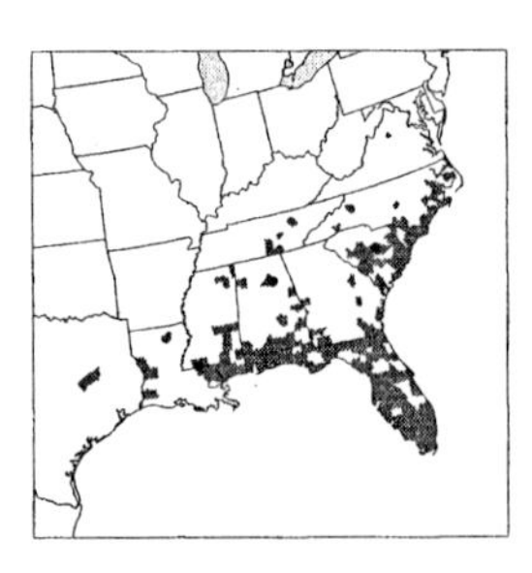

Plants densely cespitose. **Basal rosettes** poorly differentiated; **blades** 1–5 cm, lanceolate, grading into the cauline blades. **Culms** 5–45 cm, slender, erect or spreading; from a dense tuft of predominantly basal leaves, lower internodes short, upper 3–5 internodes elongate; **nodes** glabrous or bearded; **internodes** glabrous or pilose; **fall phase** with spreading culms and branches arising from near the bases forming a dense, flat tuft. **Cauline leaves** 2–4; **lower cauline sheaths** longer than the internodes, mostly glabrous or pilose with ascending hairs, margins finely ciliate; **ligules** 0.2–2 mm, at low magnification appearing to be membranous and ciliate, at high magnification evidently of hairs that are coherent at the base; **blades** 1.5–6 cm long, 3–8 mm wide, lanceolate, glabrous or softly pilose, margins with prominent papillose-based cilia, at least basally. **Primary panicles** short- to long-exserted; **rachises** and **branches** often pilose. **Spikelets** 1.1–2.1 mm, obovoid to broadly ellipsoid, glabrous or pubescent, hairs not papillose-based. **Lower glumes** ⅓–½ as long as the spikelets, acute to obtuse; **upper florets** 0.8–1.7 mm, ellipsoid, subacute.

Dichanthelium strigosum extends from the southeastern *Flora* region south into Mexico, the Caribbean, and into northern South America.

The primary panicles are briefly open-pollinated in April or May; the secondary panicles, which are produced from May through November, are cleistogamous. The three subspecies are mostly sympatric and sometimes grow together, with occasional intergradation.

1. Spikelets pubescent, broadly ellipsoid, 1.6–2.1 mm long; lower glumes about ½ as long as the spikelets; blades glabrous . . . . subsp. *leucoblepharis*
1. Spikelets glabrous, obovoid, 1.1–1.8 mm long; lower glumes about ⅓ as long as the spikelets; blades pilose or glabrous.
   2. Blades pilose; spikelets 1.1–1.6 mm long . . . . subsp. *strigosum*
   2. Blades glabrous or sparsely pilose near the base; spikelets 1.4–1.8 mm long . . . subsp. *glabrescens*

Dichanthelium strigosum subsp. **glabrescens** (Griseb.) Freckmann & Lelong

Culms usually less than 30 cm, glabrous, usually very densely cespitose; **nodes** glabrous. **Cauline blades** mostly glabrous, sometimes sparsely pilose basally. **Spikelets** 1.4–1.8 mm, glabrous, obovoid; **lower glumes** about ⅓ as long as the spikelets. $2n$ = unknown.

Dichanthelium strigosum subsp. *glabrescens* grows in sandy, open pine woods and bogs. Its range extends from Mississippi along the coast to Florida and south through the West Indies.

Dichanthelium strigosum subsp. **leucoblepharis** (Trin.) Freckmann & Lelong

Culms usually 5–30 cm, glabrous or sparsely pubescent; **nodes** glabrous. **Cauline blades** glabrous. **Spikelets** 1.6–2.1 mm, broadly ellipsoid, pubescent; **lower glumes** about ½ as long as the spikelets. $2n$ = unknown.

Dichanthelium strigosum subsp. *leucoblepharis* grows in low, moist, sandy pinelands and bogs. Its range extends from North Carolina along the coastal plain to Florida and eastern Texas and into Mexico.

Dichanthelium strigosum (Muhl. *ex* Elliott) Freckmann subsp. **strigosum**

Culms often 30–45 cm, pilose to subglabrous; **nodes** bearded. **Cauline blades** pilose on both surfaces, those of the upper leaves very reduced. **Spikelets** 1.1–1.6 mm, glabrous, obovoid; **lower glumes** about ⅓ as long as the spikelets. $2n$ = 18.

Dichanthelium strigosum subsp. *strigosum* grows in sandy, low, open pine woods and bogs. It is the most widespread of the three subspecies, extending from southeastern Virginia through the coastal plain to eastern Texas, Florida, Cuba, and the West Indies to Colombia.

Dichanthelium sect. **Linearifolia** Freckmann & Lelong

Plants cespitose, with caudices. **Basal rosettes** poorly differentiated; **blades** narrow, erect or ascending, resembling the lower cauline blades in shape. **Culms** 10–50 cm, erect to spreading or drooping, lower internodes very short, upper 2–4 internodes often much elongated; **fall phase** branching from the basal nodes, usually producing sterile shoots or condensed secondary panicles within about 5 cm of the ground. **Cauline leaves** 2–4; **ligules** 0.5–1 mm, of hairs; **blades** usually erect, stiff, upper blades 1–5 mm wide, 15–60 times as long. **Primary panicles** usually exserted. **Spikelets** narrowly ellipsoid to obovoid, usually pubescent, sometimes glabrous. **Upper florets** subacute to acute or umbonate.

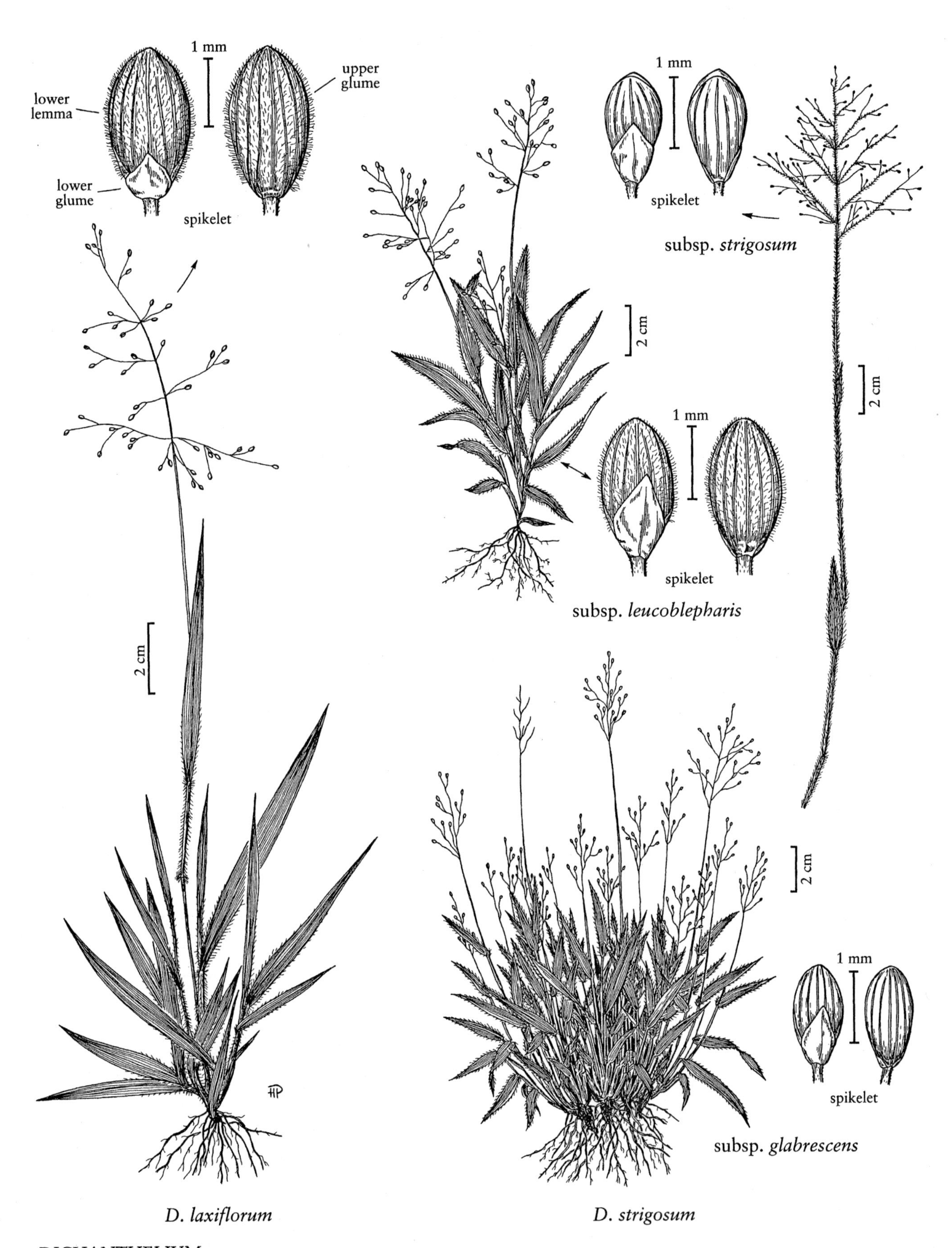

D. laxiflorum *D. strigosum*

31. **Dichanthelium wilcoxianum** (Vasey) Freckmann [p. 451]
WILCOX'S PANICGRASS

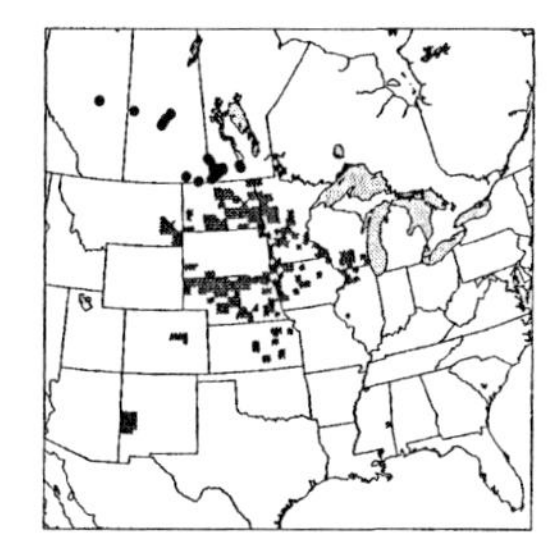

Plants cespitose. **Basal rosettes** poorly differentiated; **sheaths** glabrous; **blades** 2–4 cm, narrow, similar to those of the lower cauline leaves, ascending to spreading. **Culms** 15–35 cm, stiffly erect, all but the upper 2–4 internodes very short; **nodes** glabrous or with weak, reflexed hairs; **internodes** purplish-gray, sparsely pubescent; **fall phase** developing early, forming erect branches from the lower or midculm nodes, each branch terminating in a partially included panicle of 8–16 spikelets, no sterile shoots formed. **Cauline leaves** usually 3; **sheaths** hirsute, hairs papillose-based; **ligules** 0.5–1 mm; **blades** 4–8 cm long, 2–5 mm wide, all alike, stiffly erect, green to grayish-green, flat, not plicate, sparsely pilose. **Primary panicles** 3–5 cm long, 2–4 cm wide, ovoid, open, shortly exserted, with 12–32 spikelets; **branches** short, stiff, spreading; **pedicels** mostly 4–8 mm, spreading. **Spikelets** 2.4–3.2 mm long, 0.7–1.2 mm wide, ellipsoid to obovoid, often reddish throughout, short-pubescent. **Lower glumes** 0.7–1.2 mm, triangular; **upper glumes** and **lower lemmas** about equaling the upper florets; **upper florets** 1.9–2.5 mm, ellipsoid, pointed. $2n$ = 18.

Dichanthelium wilcoxianum grows in dry prairies, especially in sandy or gravelly openings. It is restricted to the *Flora* region. The primary panicles, which are produced from mid-May to early June, are partially open-pollinated; the secondary panicles, which are produced in June, and occasionally also in September, are cleistogamous.

Some specimens of *Dichanthelium oligosanthes* subsp. *scribnerianum* from the southern Great Plains that have prematurely elongating upper internodes resemble *D. wilcoxianum*, but they have greenish spikelets that are 1.7–2.4 mm wide, an orange spot at the base of the glumes, and larger basal rosettes.

32. **Dichanthelium perlongum** (Nash) Freckmann [p. 451]
LONG-STALKED PANICGRASS

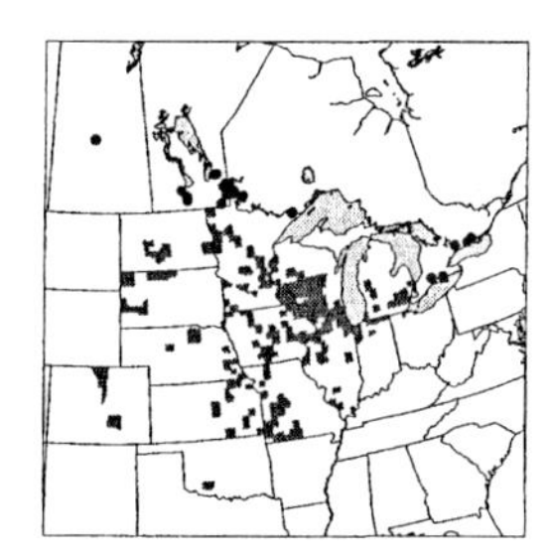

Plants densely cespitose. **Basal rosettes** poorly developed; **sheaths** 2–4 cm; **blades** similar in shape to the lower cauline blades, narrow, ascending. **Culms** 10–50 cm, erect, lower 3–6 internodes telescoped together, forming a slender 2–4 cm column, upper 2 internodes elongated; **nodes** bearded; **internodes** puberulent and pubescent; **fall phase** with sterile branches arising near ground level and foreshortened reproductive branches arising from the higher nodes, secondary panicles small and narrow, enclosed within the sheaths, with 5–10 spikelets. **Cauline leaves** 2–4; **sheaths** longer than the internodes, pilose; **ligules** about 0.5 mm; **blades** 5–20 cm long, 1–3.5 mm wide, stiffly erect, long-tapering, sometimes involute, green or grayish-green, pubescent to pilose, upper 2 or 3 blades much longer than those below. **Primary panicles** 3–8 cm long, 1–3 cm wide, narrowly ellipsoid, long-exserted, with 12–25 spikelets; **branches** ascending; **pedicels** 2–4 mm, appressed. **Spikelets** 2.6–3.4 mm long, 1–1.7 mm wide, ellipsoid-obovoid, turgid, finely pubescent. **Lower glumes** 1–1.4 mm, broadly ovate; **upper glumes** and **lower lemmas** exceeding the upper florets by 0.2–0.3 mm before flowering, slightly pointed at maturity, **upper florets** obovoid, 1.9–2.7 mm, minutely umbonate. $2n$ = 18.

Dichanthelium perlongum grows in dry to mesic prairies, and is restricted to the *Flora* region. It appears to hybridize occasionally with *D. depauperatum* and *D. linearifolium*. The primary panicles are briefly open-pollinated and develop from May to early June; the secondary panicles are cleistogamous and are produced from mid-June through mid-July.

Dichanthelium perlongum is similar to *D. wilcoxianum*, but differs in having only the upper 1 or 2 blades greatly elongated (usually more than 20 times longer than wide), narrow, erect basal blades, and a contracted panicle with ascending branches. *Dichanthelium acuminatum* also may also be confused with *D. perlongum* only if its upper internodes elongate, as tends to be the case after a spring fire, but *D. acuminatum* has less turgid spikelets and hairs in the ligule area that are 3–5 mm long.

33. **Dichanthelium linearifolium** (Scribn.) Gould [p. 451]
LINEAR-LEAVED PANICGRASS, PANIC À FEUILLES LINÉAIRES

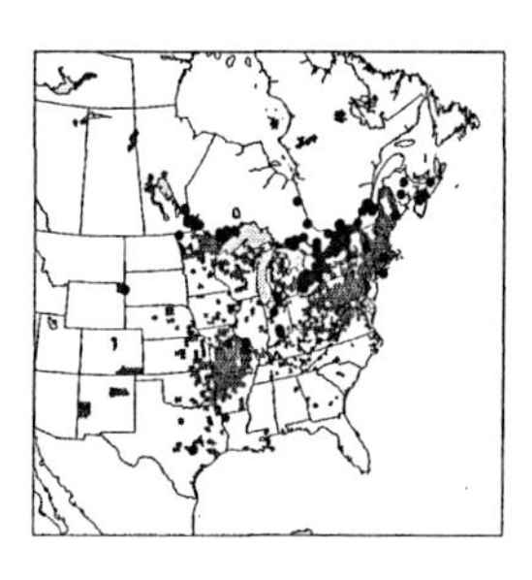

Plants cespitose. **Basal rosettes** poorly differentiated; **blades** similar in shape to the lower cauline blades, narrow, ascending. **Culms** 10–50 cm, very slender, erect to drooping, lower 3–8 internodes telescoped together, less than 2 cm, upper 2 internodes elongated; **nodes** bearded; **internodes** pubescent to almost glabrous; **fall phase** developing a dense mass of erect blades and foreshortened branches arising from the basal nodes, terminating in small, narrow secondary panicles that are enclosed within the sheaths, with 6–15 spikelets. **Cauline leaves** 2–4; **sheaths** longer than the internodes, glabrous or pilose with dense, fine, papillose-based hairs; **ligules** about 0.5 mm; **blades** 5–20 cm long, 2–5 mm wide, stiffly ascending to erect, green to grayish-green, glabrous or densely pilose, apices long-tapering, lower blades shorter than the upper 2 or 3 blades. **Primary panicles** 4–10 cm long, 2–6 cm wide, long-exserted, with 12–70 spikelets; **branches** and **pedicels** spreading. **Spikelets** 2–3.2 mm long, 0.8–1.4 mm wide, ellipsoid, not turgid, sparsely pubescent. **Lower glumes** 0.6–1.1 mm, ovate-triangular; **upper glumes** and **lower lemmas** exceeding the upper florets by about 0.2 mm before flowering, subequal in fruit, slightly pointed at maturity, **upper florets** 1.7–2.3 mm, ovoid-ellipsoid, minutely umbonate. $2n = 18$.

Dichanthelium linearifolium grows in dry, open woodlands, rock outcroppings, and sandy areas. It is restricted to the *Flora* region. The primary panicles are briefly open-pollinated, produced from May to early June; the secondary panicles are cleistogamous, produced from late June through July (rarely in fall). Plants in the northern United States and Canada tend to be shorter and more spreading, subglabrous, and to have spikelets 2–2.6 mm long; they have been called *Panicum werneri* Scribn., but do not merit taxonomic recognition. In the southwestern part of its range, especially in the Ozarks, most plants of *D. linearifolium* are tall, erect, densely pilose, with very elongated blades and spikelets often 2.6–3 mm long; they may hybridize with *D. perlongum*.

34. **Dichanthelium depauperatum** (Muhl.) Gould [p. 451]

STARVED PANICGRASS, PANIC APPAUVRI

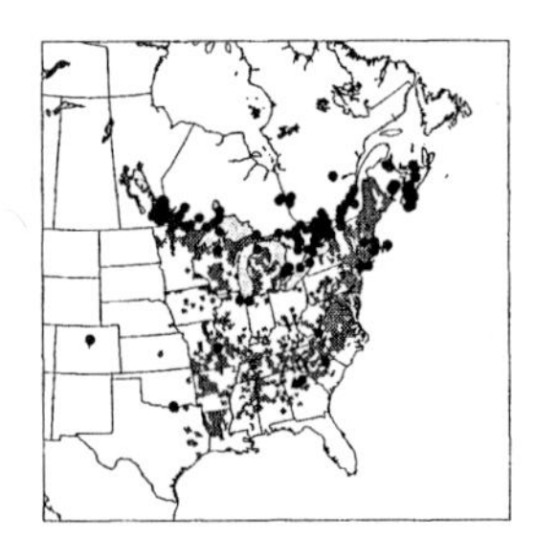

Plants cespitose. **Basal rosettes** poorly differentiated; **blades** similar in shape to the lower cauline blades, narrow, ascending. **Culms** 10–45 cm, erect to spreading, lower 4–10 internodes telescoped together, less than 2 cm, upper 2 internodes elongated; **nodes** bearded; **internodes** pubescent to subglabrous; **fall phase** a dense mass of erect blades and foreshortened branches that arise from the basal culm nodes, about ½ of the branches sterile, others with small, narrow, secondary panicles of 3–7 spikelets that remain enclosed within the sheaths. **Cauline leaves** 2–4; **sheaths** longer than the internodes, glabrous or densely ascending-pilose; **ligules** about 0.5 mm; **blades** 6–15 cm long, 1–4 mm wide, green to grayish-green, sometimes involute, glabrous or densely pilose, apices long-tapering, lower blades small to vestigial, upper 2 or 3 blades longer and stiffly erect. **Primary panicles** 3–6 cm long, 1.5–3 cm wide, usually long-exserted (sometimes contracted and remaining basal), with 7–25 spikelets. **Spikelets** 3.2–4.3 mm long, 1–1.7 mm wide, ellipsoid-pointed, glabrous or finely pubescent. **Lower glumes** 1.2–1.6 mm, narrowly triangular; **upper glumes** and **lower lemmas** exceeding the upper florets by 0.2–1 mm, forming a pointed beak, **upper florets** 1.9–3.1 mm, obovoid, minutely umbonate. $2n = 18$.

Dichanthelium depauperatum grows in dry, open woodlands and open, disturbed areas, especially on sand. It is restricted to the *Flora* region. The primary panicles, which are rarely open-pollinated, are produced from May to early June; the secondary, cleistogamous panicles are produced from late June through July (rarely in fall). The species is linked with *D. perlongum* and *D. linearifolium* by occasional hybrids and hybrid derivatives. In the northern United States and Canada, 80–90% of the plants are glabrous and have been called *Panicum depauperatum* var. *psilophyllum* Fernald, *P. depauperatum* var. *involutum* (Torr.) Alph. Wood, or, if the primary panicles remain near the base, *P. depauperatum* forma *cryptostachys* Fernald; in this treatment, none of these is recognized as a distinct taxonomic entity. The frequency of pilose plants increases southward, where some populations are entirely pilose.

25.10 PANICUM L.

Robert W. Freckmann
Michel G. Lelong

Plants annual or perennial; their habit variable. **Culms** 2–300 cm, herbaceous, sometimes hard and almost woody, or woody, simple or branched, bases sometimes cormlike; **internodes** solid, spongy, or hollow. **Leaves** cauline, basal, or both, basal leaves not forming a winter rosette; **ligules** membranous, usually ciliate; **blades** filiform to ovate, flat to involute, glabrous or pubescent, cross

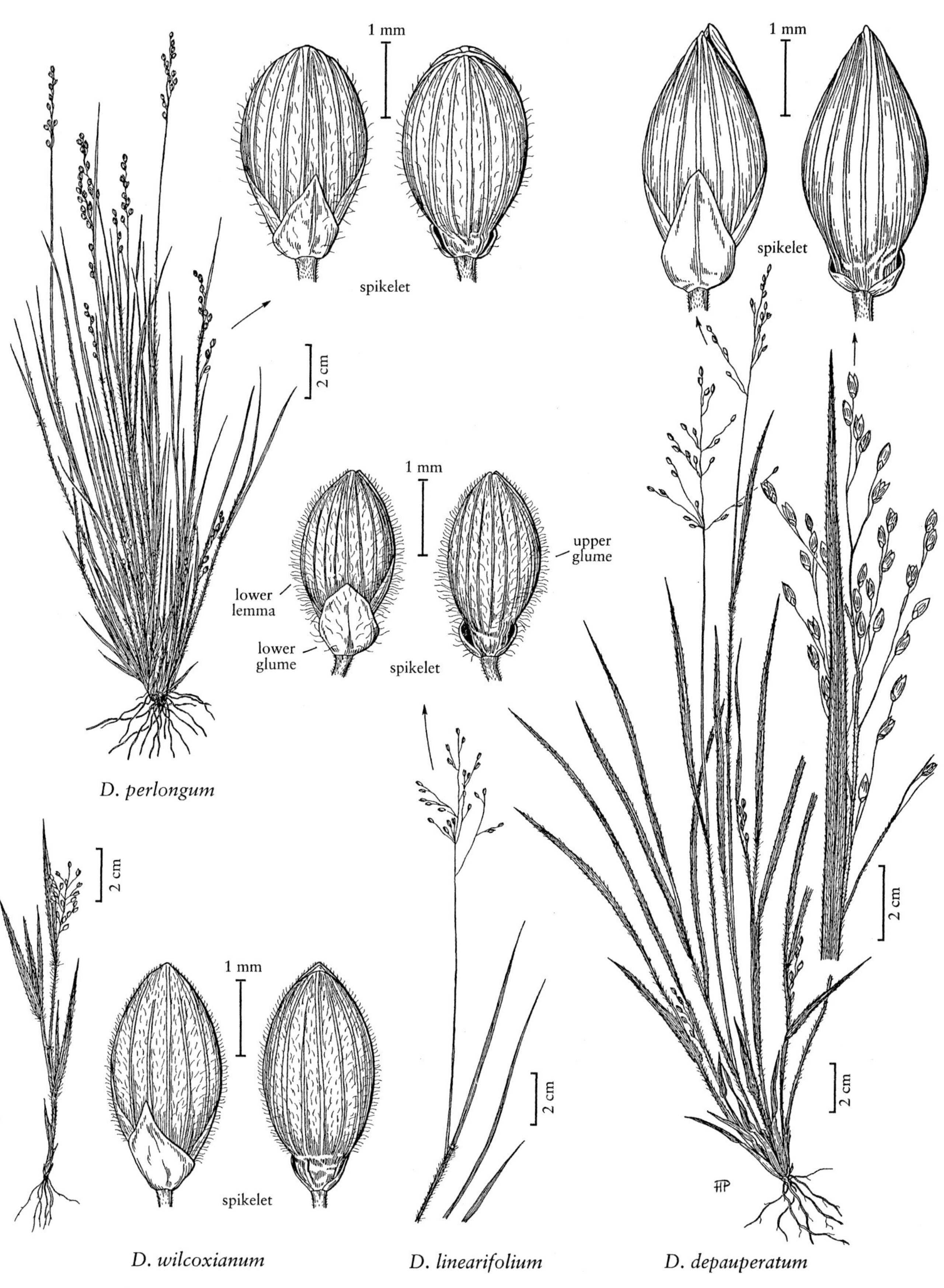
1 mm
spikelet
2 cm
D. perlongum
1 mm
spikelet
2 cm
1 mm
upper glume
lower lemma
lower glume
spikelet
2 cm
1 mm
spikelet
D. wilcoxianum
2 cm
D. linearifolium
2 cm
HP
D. depauperatum

sections with Kranz anatomy and 1 or 2 bundle sheaths or with non-Kranz anatomy; **photosynthesis** C_4 with NAD-me or NADP-me pathways, or, in plants with non-Krantz anatomy, C_3. **Inflorescences** terminal on the culms and branches, often also axillary, terminal panicles typically appearing after midsummer; **sterile branches** and **bristles** absent; **disarticulation** usually below the glumes, sometimes at the base of the upper florets, if at the base of the upper florets, then the florets not very plump at maturity. **Spikelets** 1–8 mm, usually dorsally compressed, sometimes subterete or laterally compressed, unawned. **Glumes** usually unequal, herbaceous, glabrous or pubescent, rarely tuberculate or glandular, apices not or only slightly gaping at maturity; **lower glumes** minute to almost equaling the spikelets, 1–9-veined, truncate, acute, or acuminate; **upper glumes** slightly shorter to much longer than the spikelets, 3–13(15)-veined, bases rarely slightly sulcate, apices rounded to attenuate; **lower florets** sterile or staminate; **lower lemmas** similar to the upper glumes; **lower paleas** absent, or shorter than the lower lemmas and hyaline; **upper florets** bisexual, sessile or stipitate, apices acute, puberulent, or with a tuft of hairs; **upper lemmas** usually more or less rigid and chartaceous-indurate, usually shiny, glabrous or (rarely) pubescent, usually smooth, sometimes verrucose or transversely rugose, margins involute, usually clasping the paleas, rarely with basal wings or lunate scars, apices obtuse, acute, apiculate, or with small green crests; **upper paleas** striate, rarely transversely rugose; **lodicules** 2; **anthers** usually 3. **Caryopses** smooth; **pericarp** thin; **endosperm** hard, without lipid, starch grains simple or compound, or both; **hila** round or oval. $x = 9$ (usually), sometimes 10, with polyploid and dysploid derivatives. Name from the Latin *panis*, 'bread', or *panus*, an ear of millet.

Panicum is a large genus, but just how large is difficult to estimate because its limits are not yet clear. Many taxonomists would treat it as including *Dichanthelium*, *Steinchisma*, and some members of *Urochloa*. Recent work supports some aspects of the treatment presented here, but not all of them. For instance, Guissani et al. (2001) suggest that *Panicum* subg. *Panicum* is a monophyletic group that should have a rank equivalent to *Dichanthelium* and *Steinchisma*. The two other subgenera included here in *Panicum*, subg. *Agrostoidea* and subg. *Phanopyrum*, are not monophyletic, but the relationships of their species to other members of *Panicum sensu lato* are not well enough understood to suggest a better treatment, nor to justify the name changes a differing generic treatment would require.

Most species of *Panicum* are tropical, but many grow in warm, temperate regions. Of the thirty-four species occurring in the *Flora* region, twenty-five are native, seven are established introductions, and two are not established within the region. Within the *Flora* region, *Panicum* is most abundant in the southeastern United States. Many species grow in early seral stages or weedy areas; some grow at forest edges, in prairies, savannahs, deserts, forests, beaches, and in shallow water.

Panicum miliaceum has been grown since prehistory in China and India as a cereal grain, and is a common component of bird seed. Seeds of *P. hirticaule* subsp. *sonorum* have been used for food by the Cocopa tribe of the southwest. Important hay and range species include *P. virgatum*, *P. rigidulum*, *P. bulbosum*, *P. obtusum*, and *P. repens*.

Apomixis, polyploidy, and autogamy have produced numerous microspecies in some groups; hybridization and introgression has resulted in a reticulum of intergrading forms in some complexes. The number of taxa recognized has varied widely over the past century.

SELECTED REFERENCES **Darbyshire, S.J.** and **J. Cayouette.** 1995. Identification of the species in the *Panicum capillare* complex (Poaceae) from eastern Canada and adjacent New York State. Canad. J. Bot. 73:333–348; **Guissani, L.M., J.H. Cota-Sánchez, F.O. Zuloaga,** and **E.A. Kellogg.** 2001. A molecular phylogeny of the grass subfamily Panicoideae (Poaceae) shows multiple origins of C_4 photosynthesis. Amer. J. Bot. 88:1993–2001; **Hitchcock, A.S.** 1951 [title page 1950]. Manual of the Grasses of the United States, ed. 2, rev. A. Chase. U.S.D.A. Miscellaneous Publication No. 200. U.S. Government Printing Office, Washington, D.C., U.S.A. 1051 pp.; **Hitchcock, A.S.** and **A. Chase.** 1910. The North American species of *Panicum*. Contr. U.S. Natl. Herb. 15:1–396; **Reed, C.F.** 1964. A flora of the chrome and manganese ore piles at Canton, in the Port of Baltimore, Maryland and at Newport News, Virginia, with descriptions of genera and species new to the flora of the eastern United

States. Phytologia 10:321–405; **Zuloaga, F.O.** 1987. Systematics of New World Species of *Panicum* (Poaceae: Paniceae). Pp. 287–306 *in* T.R. Soderstrom, K.W. Hilu, C.S. Campbell, and M.E. Barkworth (eds.). Grass Systematics and Evolution. Smithsonian Institution Press, Washington, D.C., U.S.A. 473 pp.; **Zuloaga, F.O.** and **O. Morrone.** 1996. Revisión de las especies Americanas de *Panicum* subgenera *Panicum* sección *Panicum* (Poaceae: Panicoideae: Paniceae). Ann. Missouri Bot. Gard. 83:200–280.

1. Panicle branches 1-sided; spikelets usually subsessile, the longest pedicels usually less than 2 mm long, rarely 3 mm long.
   2. Spikelets 5.5–7 mm long; upper florets less than ⅓ as long as the spikelets (sect. *Phanopyrum*) . . . . . 30. *P. gymnocarpon*
   2. Spikelets 1.6–4.4 mm long; upper florets ⅔ as long as to almost equaling the spikelets.
      3. Lower glumes 5- or 7-veined, about ¾ as long as the spikelets; plants with stolons or shallow rhizomes (sect. *Obtusa*) . . . . . 25. *P. obtusum*
      3. Lower glumes 1- or 3-veined, ½–⅔ as long as the spikelets; plants without stolons, often with rhizomes.
         4. Lower florets staminate; lower paleas subequal to the lower lemmas; upper thin, lemmas flexible, clasping the paleas only at the base (sect. *Hemitoma*) . . . . . 29. *P. hemitomon*
         4. Lower florets sterile; lower paleas no more than ⅔ as long as the lower lemmas; upper lemmas thick, stiff, clasping the paleas throughout their length.
            5. Glumes and lower lemmas without keeled midveins; upper florets with glabrous apices; plants tufted, from knotty rhizomes; panicles with a few spikelets; pedicels with slender hairs near the apices (sect. *Tenera*) . . . . . 24. *P. tenerum*
            5. Glumes and lower lemmas with keeled midveins; upper florets with a tuft of small hairs at the apices; plants often with scaly rhizomes; panicles with many spikelets; pedicels glabrous (sect. *Agrostoidea*).
               6. Plants without conspicuous rhizomes, cespitose; culms and sheaths strongly compressed; spikelets usually 1.6–3.8 mm long, lanceolate, not falcate . . . . . 22. *P. rigidulum*
               6. Plants with conspicuous, stout, short or elongate, scaly rhizomes; culms and sheaths slightly compressed; spikelets 2.3–3.9 mm long, rarely lanceolate, often falcate . . . . . 23. *P. anceps*
1. Panicle branches usually not 1-sided; spikelets not subsessile, the longest pedicels 2–20 mm long.
   7. Upper glumes and lower lemmas warty-tuberculate (sect. *Verrucosa*).
      8. Lower lemmas verrucose with hemispheric warts; spikelets 1.7–2.2 mm long, about 1 mm wide, subacute or obtuse, glabrous; plants of wetlands . . . . . 33. *P. verrucosum*
      8. Lower lemmas tuberculate-hispid; spikelets 3.2–4 mm long, about 1.5 mm wide, acute or acuminate; plants of dry, sandy or clayey areas . . . . . 34. *P. brachyanthum*
   7. Upper glumes and lower lemmas glabrous, villous, or scabridulous, but not warty-tuberculate.
      9. Upper florets faintly to evidently transversally rugose; sheaths keeled; culm bases often cormlike (sect. *Bulbosa*).
         10. Culm bases thickened, cormlike; culms slightly compressed; rhizomes, if present, short and thin; spikelets 2.8–5.4 mm long; lower glumes 1.2–3.5 mm long, ½–⅘ as long as the spikelets . . . . . 26. *P. bulbosum*
         10. Culm bases not cormlike; culms strongly compressed; rhizomes present, long, stout; spikelets 2.5–3.4 mm long; lower glumes usually less than 1.7 mm long, up to ½ as long as the spikelets . . . . . 27. *P. plenum*
      9. Upper florets smooth or striate, rarely inconspicuously rugose; sheaths not keeled; culm bases never cormlike.
         11. Plants with rhizomes about 1 cm thick and with large, pubescent, scalelike leaves; culms hard, almost woody (sect. *Antidotalia*) . . . . . 28. *P. antidotale*
         11. Plants without rhizomes or with rhizomes less than 0.5 cm thick and with small, glabrous, scalelike leaves; culms clearly not woody, except at the base of *P. hirsutum* (subg. *Panicum*).
            12. Glumes, lower lemmas, and upper lemma margins villous, with whitish hairs (sect. *Urvilleana*) . . . . . 21. *P. urvilleanum*

12. Glumes and lemmas usually glabrous, sometimes the lower lemmas sparsely pilose on the margins and near the apices.
    13. Plants perennial, usually with vigorous scaly rhizomes; lower florets staminate (sect. *Repentia*).
        14. Lower glumes 0.5–1.5 mm long, less than ½ as long as the spikelet, 1–5-veined; upper glumes and lower lemmas extending 0.1–0.5 mm beyond the upper florets and scarcely separated (gaping); lower paleas oblong, not hastate-lobed.
            15. Lower glumes subtruncate to broadly acute, faintly veined; upper florets widest at or above the middle, with rounded apices; plants not cespitose, with long, scaly rhizomes . . . . 17. *P. repens*
            15. Lower glumes acute, with evident veins; upper florets widest below the middle, with lightly beaked apices; plants cespitose, with short knotty rhizomes . . . . 18. *P. coloratum*
        14. Lower glumes 1.8–4 mm long, more than ½ as long as the spikelets, with at least 5 veins; upper glumes and lower lemmas extending 0.4–3 mm beyond the upper florets, stiffly separated (gaping); lower paleas hastate-lobed.
            16. Panicles contracted; branches appressed to strongly ascending; plants glabrous throughout . . . . 19. *P. amarum*
            16. Panicles open; branches ascending to spreading; plants often pilose, at least at the base of the leaf blades . . . . 20. *P. virgatum*
    13. Plants annual, or perennials usually without rhizomes, sometimes rooting at the lower nodes; lower florets sterile.
        17. Lower glumes truncate to subacute, ⅛–⅓ as long as the spikelets; sheaths more or less compressed, glabrous or sparsely pubescent; plants slightly succulent or spongy (sect. *Dichotomiflora*).
            18. Plants usually annual, usually terrestrial, rooting at the lower nodes if in water, but not floating; blades 3–25 mm wide . . . . 15. *P. dichotomiflorum*
            18. Plants perennial or of indefinite duration, usually aquatic, sometimes floating, rooting at the lower nodes; blades 2–15 mm wide.
                19. Spikelets 2–2.2 mm long; blades 2–4 mm wide; lower paleas absent; culms succulent . . . . 14. *P. lacustre*
                19. Spikelets 3–4 mm long; blades 5–15 mm wide; lower paleas present; culms spongy . . . . 16. *P. paludosum*
        17. Lower glumes acute to attenuate, usually ⅓–¾ as long as the spikelets; sheaths rounded, usually hirsute or hispid; plants not succulent (sect. *Panicum*).
            20. Spikelets 4–6.5 mm long.
                21. Upper glumes and lower lemmas only slightly exceeding the upper florets; upper florets 2–2.5 mm wide; plants annual; lower paleas truncate to bilobed . . . . 1. *P. miliaceum*
                21. Upper glumes and lower lemmas exceeding the upper florets by 3–4 mm; upper florets 1–1.1 mm wide; plants perennial; lower paleas acute . . . . 8. *P. capillarioides*
            20. Spikelets 1–4.2 mm long.
                22. Plants perennial; panicle branches usually with all or most secondary branches confined to the distal ⅓.
                    23. Lower panicle branches whorled; culms 2–10 mm thick, 50–300 cm tall.
                        24. Sheaths with fragile, prickly hairs causing skin irritation; panicles not breaking at the base and becoming tumbleweeds; lower paleas 1.3–1.7 mm long . . . . 9. *P. hirsutum*

24. Sheaths glabrous or sparsely to densely pubescent but without fragile, prickly hairs; panicles breaking at the base and becoming tumbleweeds; lower paleas 1.4–2.2 mm long . . . . 10. *P. bergii*

23. Lower panicle branches solitary; culms 0.5–10 mm thick, 15–100 cm tall.

25. Blades glabrous and glaucous on the adaxial surface; nodes sericeous or pilose, sometimes almost glabrous . . . . 13. *P. hallii*

25. Blades sparsely to densely hirsute and not glaucous on the adaxial surface; nodes sericeous.

26. Spikelets 2.1–2.9 mm long; culms spreading to weakly ascending; blades spreading, 1–5 mm wide, without a prominent white midrib . . . . 11. *P. diffusum*

26. Spikelets 2.6–3.4 mm long; culms erect to decumbent; blades ascending to erect, 0.5–14 mm wide, with a prominent white midrib . . . . . 12. *P. ghiesbreghtii*

22. Plants annual; panicle branches usually with secondary branches and pedicels attached to the distal ⅔.

27. Blades 2–7 cm long, 5–20 mm wide, lanceolate, 4–6 times longer than wide (sect. *Monticola*, in part) . . . . 31. *P. trichoides*

27. Blades 5–40 cm long, 1–18 mm wide, linear, more than 10 times longer than wide (sect. *Panicum*, in part).

28. Panicles more than 2 times longer than wide at maturity; branches ascending to somewhat divergent; spikelets narrowly ovoid, usually about 3 times longer than wide . . . . 4. *P. flexile*

28. Panicles less than 1.5 times longer than wide at maturity; branches diverging; spikelets variously shaped, less than 3 times longer than wide.

29. Spikelets 2.1–4 mm long, upper glumes and lower lemmas with prominent veins; lower glumes ⅖–¾ as long as the spikelets; lower paleas 0.4–2 mm long, from ⅓ as long as the lower lemmas to equaling them; ligules 0.2–0.4 mm or 1–3.5 mm long.

30. Lower glumes 0.7–1.1 mm long, about ⅖ as long as the spikelets; lower paleas 1–2 mm long; leaf blades 2–8 mm wide, usually completely glabrous, sometimes with a few marginal cilia near the base . . . . 7. *P. psilopodium*

30. Lower glumes 1.2–2.4 mm long, ½–¾ as long as the spikelets; lower paleas 0.2–0.9 mm long; leaf blades 1–30 mm wide, hairs papillose-based.

31. Primary panicle branches appressed to the main axis; culms 2–8 cm long; spikelets 2–2.2 mm long . . . . 6. *P. mohavense*

31. Primary panicle branches divergent; culms 11–110 cm long; spikelets 1.9–4 mm long . . . . 5. *P. hirticaule*

29. Spikelets 1.4–4 mm long, upper glumes and lower lemmas without prominent veins; lower glumes usually less than ½ as long as the spikelets; lower paleas usually small or absent; ligules 0.5–1.5 mm long.

32. Plants mostly glabrous, but the sheaths ciliate on the margins and the blades sometimes sparingly pilose adaxially (sect. *Monticola*, in part) . . . . 32. *P. bisulcatum*

32. Plants mostly hairy, even the sheaths hairy throughout.

33. Panicles usually more than ½ the total height of the plant, breaking at the base of the peduncle at maturity and becoming a tumbleweed; spikelets 1.9–4 mm long; mature upper florets stramineous or nigrescent (sect. *Panicum*, in part) . . . . 2. *P. capillare*

33. Panicles usually less than ½ the total height of the plant, the base of the peduncle usually not breaking at maturity; spikelets 1.4–2.4 mm long; mature upper florets often dark brown . . . . 3. *P. philadelphicum*

Panicum L. subg. Panicum

Plants annual or perennial; usually cespitose. **Culms** usually erect, not compressed. **Sheaths** not keeled; **ligules** of hairs, or membranous and, usually, ciliate; **blades** with vascular bundles separated by 2–6 radially arranged, tabular mesophyll cells and surrounded by a double sheath, cells of the inner sheath thick-walled, cells of the outer sheath with thinner cell walls and usually centripetal chloroplasts; **chloroplasts** with well-developed grana. **Photosynthesis** C_4 NAD-me type. **Panicles** usually pyramidal, lax and diffuse, varying to contracted and condensed; **secondary branches** usually present; **pedicels** divergent to more or less appressed. **Spikelets** ellipsoid to lanceolate, glabrous. **Lower glumes** ⅕–⅘ the length of the spikelets, 1–11-veined; **upper glumes** and **lower lemmas** (5)7–15-veined; **lower florets** usually sterile; **upper florets** smooth, shining; upper paleas with compound or compound and simple papillae towards the apices. $x = 9$.

There are approximately 50 species of *Panicum* subg. *Panicum* in the Western Hemisphere (Zuloaga 1987), 21 of which grow in the *Flora* region.

Panicum L. sect. Panicum

Plants annual or perennial; perennials usually cespitose, sometimes shortly rhizomatous. **Culms** 2–300 cm, erect or decumbent, not succulent, sometimes almost woody at the base, often branching from the lower nodes. **Sheaths** not compressed. **Panicles** usually lax and diffuse; **pedicels** divergent. **Spikelets** ellipsoid to lanceoloid, glabrous. **Lower glumes** (⅓)½–¾ as long as the spikelets, (3)5–7-veined, truncate, obtuse, acute, or acuminate; **lower paleas** present or absent.

Panicum sect. *Panicum* includes approximately 22 species and extends from the southern United States to Argentina. Most species grow in dry, open places, but a few grow in moist sites such as river banks.

1. Panicum miliaceum L. [p. 458]
Broomcorn, Proso Millet, Hog Millet, Panic Millet

Plants annual; sometimes branching from the lower nodes. **Culms** 20–210 cm, stout, not woody; **nodes** puberulent; **internodes** usually with papillose-based hairs, sometimes nearly glabrous, not succulent. **Leaves** numerous; **sheaths** terete, densely pilose, with papillose-based and caducous hairs; **ligules** membranous, ciliate, cilia 1–3 mm; **blades** 15–40 cm long, 7–25 mm wide. **Panicles** 6–20 cm long, 4–11 cm wide, included or shortly exserted at maturity, dense; **branches** stiff, appressed to spreading, spikelets solitary, confined to the distal portions; **pedicels** 1–9 mm, scabrous and sparsely pilose. **Spikelets** 4–6 mm, ovoid, usually glabrous. **Lower glumes** 2.8–3.6 mm, ½–¾ as long as the spikelets, 5–7-veined, veins scabridulous distally, apices attenuate; **upper glumes** 4–5.1 mm, slightly exceeding the upper florets, 11–13(15)-veined, veins scabridulous distally; **lower florets** sterile; **lower lemmas** 4–4.8 mm, slightly exceeding the upper florets, 9–13-veined, veins scabridulous distally; **lower paleas** 1.2–1.6 mm, ½ or less the length of the upper florets, truncate to bilobed; **upper florets** 3–3.8 mm long, 2–2.5 mm wide, smooth or striate, more or less shiny, stramineous to orange, red-brown, or blackish, persisting in the spikelets or disarticulating at maturity. $2n$ = 36, 40, 42, 49, 54, 72.

Panicum miliaceum is native to Asia, where it has been cultivated for thousands of years. In the *Flora* region, it is grown for bird seed and is occasionally planted for game birds. It is also found in corn fields and along roadsides. In Asia, *P. miliaceum* is still grown for fodder and as a cereal, its fast germination and short growth period enabling it to be sown following a spring crop. It also has one of the lowest water requirements of any cereal grain.

1. Mature upper florets blackish, disarticulating at maturity; culms 70–210 cm tall; panicles erect, exserted at maturity, about twice as long as wide; panicle branches ascending to spreading; pulvini well-developed .... subsp. *ruderale*
1. Mature upper florets stramineous to orange, not disarticulating; culms 20–120 cm tall; panicles usually nodding, not fully exserted, more than twice as long as wide; panicle branches ascending to appressed; pulvini almost absent .... subsp. *miliaceum*

Panicum miliaceum L. subsp. **miliaceum**

Culms 20–120 cm. **Panicles** more than twice as long as wide, relatively contracted, usually nodding, not fully exerted; **branches** ascending to appressed; **pulvini** almost absent. **Upper florets** stramineous to orange, not disarticulating at maturity.

Panicum miliaceum subsp. *miliaceum* is the subspecies used in bird seed. It probably rarely persists because of the retention of the upper florets on the plant and, in northern states, poor seed survival over winter.

Panicum miliaceum subsp. **ruderale** (Kitag.) Tzvelev

Culms 70–210 cm. **Panicles** about twice as long as wide, open, erect, exserted; **branches** ascending to spreading; **pulvini** well-developed. **Upper florets** blackish, shiny, disarticulating at maturity.

Panicum miliaceum subsp. *ruderale* is now naturalized over much of the *Flora* region. It may become a major weed, especially in corn fields.

2. **Panicum capillare** L. [p. 458]
WITCHGRASS, PANIC CAPILLAIRE

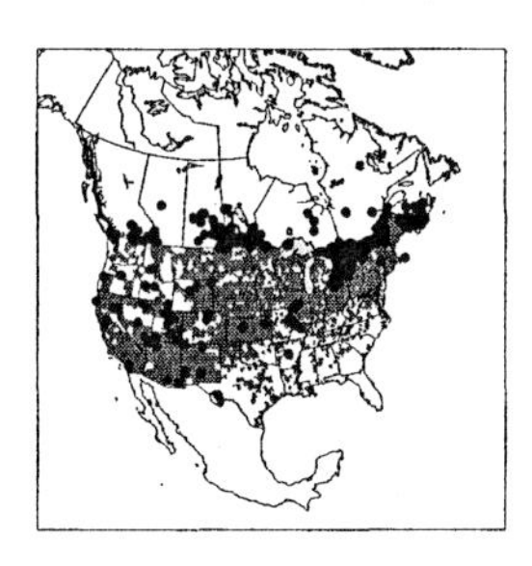

Plants annual; hirsute or hispid, hairs papillose-based, often bluish or purplish. **Culms** 15–130 cm, slender to stout, not woody, erect to decumbent, straight to zigzag, simple to profusely branched; **nodes** sparsely to densely pilose. **Sheaths** rounded, hirsute or hispid, hairs papillose-based; **ligules** membranous, ciliate, cilia 0.5–1.5 mm; **blades** 5–40 cm long, 3–18 mm wide, linear, spreading. **Panicles** 13–50 cm long, 7–24 cm wide, usually more than ½ as long as the plants, included at the base or exserted at maturity, disarticulating at the base of the peduncles at maturity and becoming a tumbleweed; **branches** spreading; **pedicels** 0.5–2.8 mm, scabrous, pilose. **Spikelets** 1.9–4 mm, ellipsoid to lanceoloid, often red-purple, glabrous. **Lower florets** sterile; **lower glumes** ⅓–½ as long as the spikelets, 1–3-veined; **upper glumes** 1.8–3.1 mm, 7–9-veined, midveins scabridulous; **lower lemmas** 1.9–3 mm, extending 0.4–1.1 mm beyond the upper florets, often stiff, straight, prominently veined distally; **upper florets** stramineous or nigrescent, sometimes with a prominent lunate scar at the base, often disarticulating before the glumes, leaving the empty glumes and lower lemmas temporarily persisting on the panicles. $2n = 18$.

Panicum capillare grows in open areas, particularly in disturbed sites such as fields, pastures, roadsides, waste places, ditches, sand, and rock crevices, etc. It grows throughout temperate North America, including northern Mexico. It also grows in Bermuda, the Virgin Islands, and sporadically in South America, and has become naturalized in much of Europe and Asia. It appears to hybridize with *P. philadelphicum*.

1. Upper florets without a lunate scar, usually stramineous; lower paleas absent; pedicels and secondary branches strongly divergent .... subsp. *capillare*
1. Upper florets with a lunate scar at the base, usually nigrescent; lower paleas present; pedicels and secondary branches often appressed, varying to narrowly divergent .... subsp. *hillmanii*

Panicum capillare L. subsp. **capillare**

Culms medium to robust, ascending to erect, rarely delicate or spreading, usually green or red-purple, rarely bluish-green, often branching at the base. **Panicle branches** spreading; **secondary branches** and **pedicels** strongly divergent. **Spikelets** 1.9–4 mm. **Lower paleas** absent; **mature upper florets** about ½ as wide as long, stramineous or tan, sometimes blackish, without a lunate scar at the base.

Panicum capillare subsp. *capillare* is the common subspecies, growing in weedy and dry habitats throughout the range of the species. Plants in the western United States and Canada have spikelets over 2.6 mm long more often than those in the east. Robust plants germinating early in the season and growing on better soils tend to spread more, and have wider, shorter blades and more exserted panicles than plants in the eastern United States and Canada growing under comparable conditions. They are sometimes included in *P. capillare* var. *occidentale* Rydb., but these traits are not well correlated, and several environmental factors apparently affect their expression. Plants in the eastern part of the range with a well-exserted main panicle at anthesis usually arise from seeds germinating relatively late in the season.

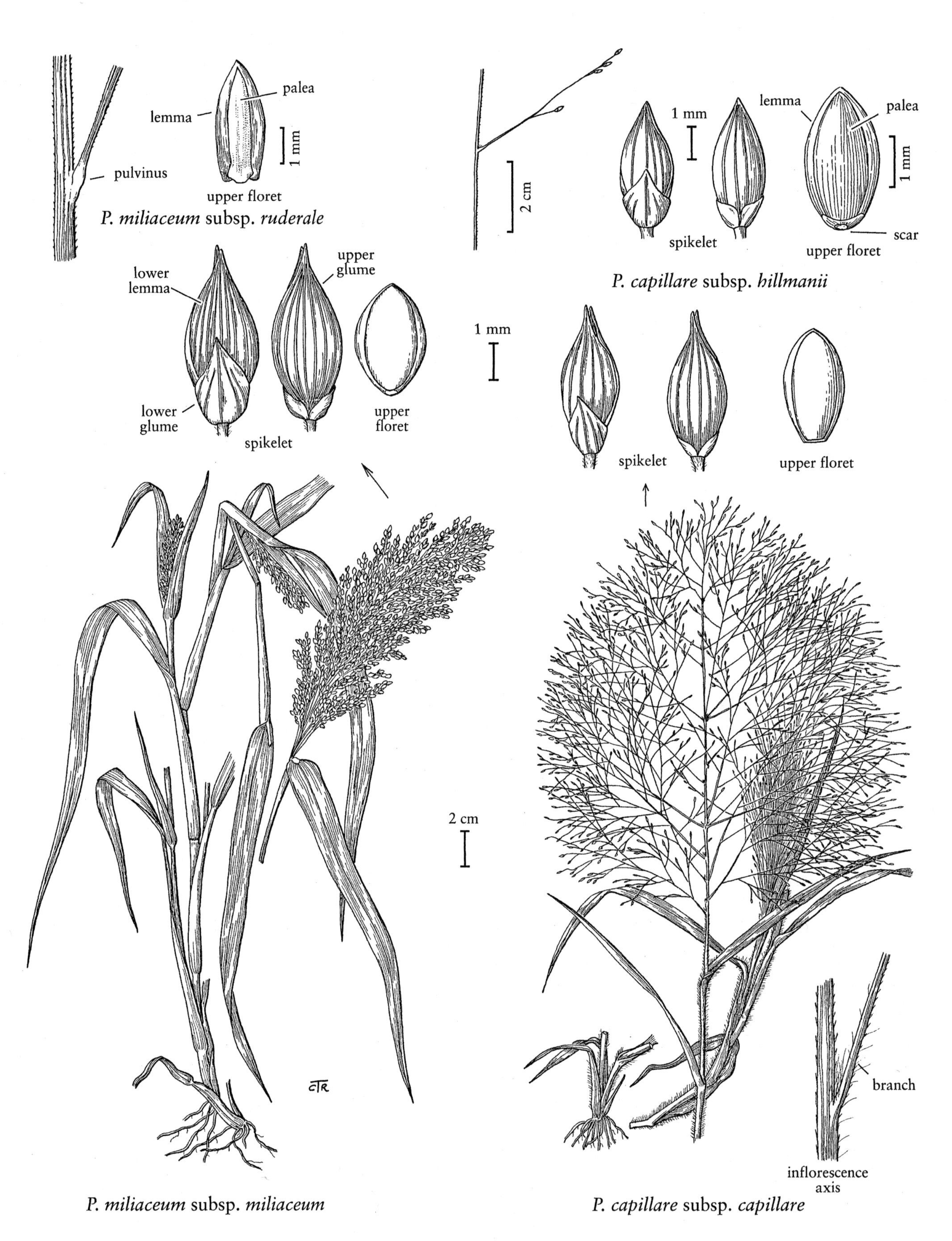
palea
lemma
1 mm
pulvinus
upper floret
P. miliaceum subsp. ruderale
lower lemma
upper glume
lower glume
spikelet
upper floret
2 cm
1 mm
lemma
palea
1 mm
scar
spikelet
upper floret
P. capillare subsp. hillmanii
1 mm
spikelet
upper floret
2 cm
CTR
branch
inflorescence axis
P. miliaceum subsp. miliaceum
P. capillare subsp. capillare

Panicum capillare subsp. **hillmanii** (Chase) Freckmann & Lelong

Culms often stout and stiff, usually bluish-green, usually not branching at the base. **Blades** thick, firm. **Panicle branches** stiff; **secondary branches** and **pedicels** usually appressed, varying to narrowly divergent. **Spikelets** 2.2–3 mm. **Lower paleas** 1–1.8 mm; **mature upper florets** nigrescent, with a prominent lunate scar at the base.

Panicum capillare subsp. *hillmanii* grows in weedy habitats in California, New Mexico, Iowa, Kansas, Oklahoma, and Texas. It may be a southern Great Plains extension of the western plants of subsp. *capillare* that are sometimes called *P. capillare* var. *occidentale* Rydb., but it differs from subsp. *capillare* in more characters than such plants.

3. **Panicum philadelphicum** Bernh. *ex* Trin. [p. 461]
PHILADELPHIA WITCHGRASS

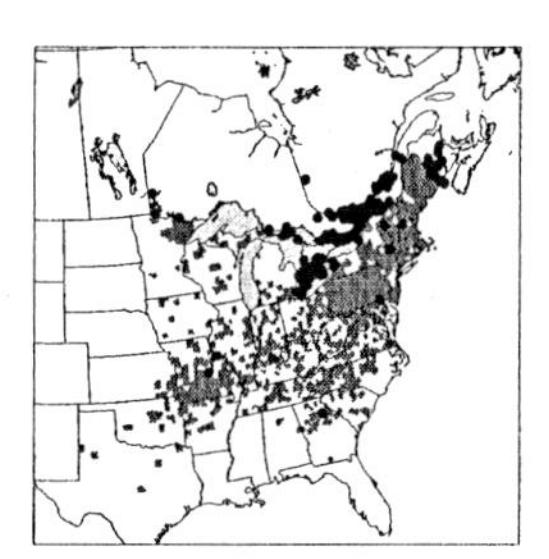

Plants annual; hirsute, hairs papillose-based, usually yellow-green to green, sometimes purplish. **Culms** 8–100 cm tall, about 1 mm thick, erect to decumbent, simple to profusely branched; **nodes** sparsely to densely pilose. **Leaves** often crowded basally; **sheaths** rounded, usually longer than the internodes, hispid, hairs papillose-based, to 5 mm; **ligules** 0.5–1.5 mm; **blades** 3–30 cm long, 2–12 mm wide, linear, ascending to erect, flat, hirsute to sparsely pilose, greenish or purplish, bases truncate to subcordate and ciliate on the margins, apices acute. **Panicles** 7–27 cm long, 4–24 cm wide, ¼–⅓ as long as the plants, diffuse, usually exserted at anthesis, not breaking at the base of the peduncles to become a tumbleweed; **rachises** glabrous or sparsely pilose basally; **primary branches** spreading, secondary branches and pedicels confined to the distal ⅔; **secondary branches** diverging to appressed, with 1–4 spikelets; **pedicels** 3–15 mm, spreading to appressed, scabrous or hirsute; pulvini glabrous or pilose. **Spikelets** 1.4–2.4 mm long, 0.5–0.7 mm wide, usually green, glabrous. **Lower glumes** 0.5–0.9 mm, usually less than ½ as long as the spikelets, 3–4-veined, truncate to acuminate; **upper glumes** 1.6–2 mm, 7-veined, veins not prominent; **lower lemmas** 1.6–1.9 mm, 7–9-veined, veins not prominent; **lower paleas** absent; **lower florets** sterile; **upper florets** 1.5–1.7 mm long, about 0.4 mm wide, often dark brown, sometimes disarticulating, apices minutely papillose. $2n = 18$.

Panicum philadelphicum grows in open areas such as fallow fields, roadside ditches, receding shores, and rock crevices. It is restricted to the eastern part of the *Flora* region. It intergrades with *P. capillare*, possibly as a result of hybridization, especially in the southeastern United States. Seeds germinating on receding shores in late summer often produce tiny plants.

1. Spikelets less than ½ as wide as long; plants purplish . . . . subsp. *lithophilum*
1. Spikelets usually more than ½ as wide as long; plants green or yellow-green.
   2. Spikelets 1.9–2.4 mm long; apices of the upper glumes and lower lemmas straight; secondary branches and pedicels divergent; blades often 6–12 mm wide, those of the flag leaves usually more than ½ as long as the panicles . . . . subsp. *gattingeri*
   2. Spikelets 1.4–2.1 mm long; apices of the upper glumes and lower lemmas curving over the upper florets at maturity; secondary panicle branches and pedicels appressed; blades usually 2–6 mm wide, those of the flag leaves usually less than ½ as long as the panicles . . . . subsp. *philadelphicum*

Panicum philadelphicum subsp. **gattingeri** (Nash) Freckmann & Lelong
PANIC DE GATTINGER

Plants slender to robust. **Culms** to 100 cm, often spreading to decumbent and rooting at the lower nodes, branching freely distally. **Blades** 5–12 mm wide, spreading, those of the flag leaves usually more than ½ as long as the panicles. **Panicles** about ⅓ as long as the plant, exserted; **secondary branches** divergent; **pedicels** divergent, flexible, scabrous. **Spikelets** 1.9–2.4 mm. **Upper glumes** and **lower lemmas** acuminate, straight; **upper florets** about ½ as wide as long, stramineous, not disarticulating.

Panicum philadelphicum subsp. *gattingeri* is commonly found in fields, roadsides, and wet clay on receding shores. This subspecies seems to be more common in the warmer parts of the northeastern United States.

Panicum philadelphicum subsp. **lithophilum** (Swallen) Freckmann & Lelong

Plants slender, sparsely pilose, purplish. **Panicles** with a few spikelets; **pedicels** short, appressed. **Spikelets** 2–2.1 mm, narrow, less than 1 mm wide. **Upper glumes** and **lower lemmas** acuminate; **upper florets** less than ½ as wide as long, blackish at maturity.

Panicum philadelphicum subsp. *lithophilum* is endemic to wet depressions in granitic outcroppings of Georgia and North and South Carolina.

Panicum philadelphicum Bernh. *ex* Trin. subsp. **philadelphicum**
PANIC DE PHILADELPHIE

Plants often slender, pilose, yellowish-green. **Culms** erect or decumbent. **Blades** 2–6 mm wide, often erect, those of the flag leaves usually less than ½ as long as the panicles. **Secondary panicle** branches usually appressed; **pedicels** usually short, appressed. **Spikelets** 1.4–2.1 mm, ovoid-ellipsoid, pale green to slightly reddish. **Upper glumes** and **lower lemmas** hooked over the upper florets; **mature upper florets** more than ½ as wide as long, shiny, blackish, with several pale veins.

Panicum philadelphicum subsp. *philadelphicum* grows in meadows, open woods, sand, and on receding shores.

Plants with decumbent culms, glabrous pulvini, flexuous pedicels without hairs over 0.2 mm long, spikelets 1.4–1.7 mm long, and the mature floret not disarticulating have been called *Panicum tuckermanii* Fernald. They are often fairly distinct on receding lake shores in New England and the Great Lakes area (Darbyshire and Cayoutte 1995), but intergrade with subsp. *philadelphicum* elsewhere.

4. **Panicum flexile** (Gatt.) Scribn. [p. 461]
WIRY WITCHGRASS, PANIC FLEXIBLE

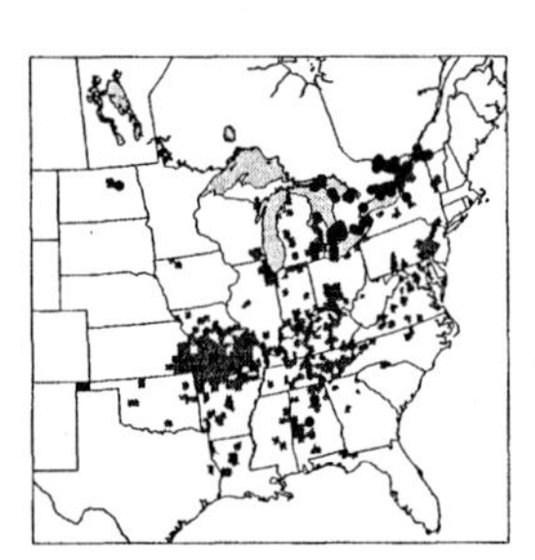

Plants annual; delicate, green or yellow-green. **Culms** 10–75 cm, about 1 mm thick, simple or with erect basal branches; **nodes** densely pilose, hairs ascending; **internodes** glabrous or shortly pubescent distally. **Sheaths** longer than the internodes, green to purplish, hispid, margins sparsely ciliate; **ligules** 0.5–1.5 mm; **blades** 3–32 cm long, 1–7 mm wide, ascending to erect, linear, narrowing basally, flat or the margins involute, surfaces sparsely hirsute or pilose (rarely glabrous), hairs near the base papillose-based, margins prominent, apices acute. **Panicles** 5–45 cm long, 1–6 cm wide, at least ½ as long as the plants and 3 times longer than wide, open; **rachises** glabrous; **primary branches** usually alternate or subopposite, ascending to slightly divergent, secondary branches and pedicels attached to the distal ⅔; **secondary branches** diverging; **pedicels** 0.5–17 mm, ascending to appressed. **Spikelets** 2.5–3.7 mm long, 0.6–1.1 mm wide, narrowly ovoid, glabrous, acute; **lower glumes** 0.8–1.3 mm, ⅓–½ as long as the spikelets, acuminate; **upper glumes** 2.3–3.3 mm, 7–9-veined, exceeding the upper florets by about 0.6 mm; **lower florets** sterile; **lower lemmas** 2.2–2.7 mm, exceeding the upper florets by about 0.6 mm, 7- or 9-veined, apices scabridulous, pointed; **lower paleas** absent; **upper florets** 1.6–1.7 mm long, about 0.6 mm wide, usually smooth, usually pale, sometimes becoming dark at maturity. $2n = 18$.

Panicum flexile grows in fens and other calcareous wetlands, in dry, calcareous or mafic rock barrens, and in open woodlands, especially on limestone derived soils. It is restricted to the *Flora* region.

5. **Panicum hirticaule** J. Presl [p. 463]
ROUGHSTALKED WITCHGRASS

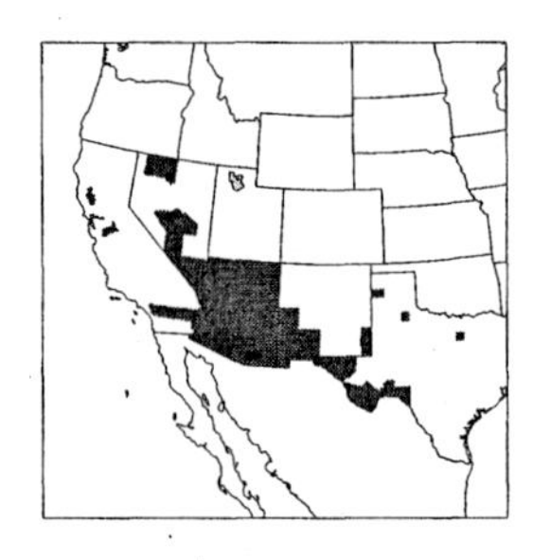

Plants annual; glabrous or hispid, hairs papillose-based. **Culms** 11–110 cm, erect to decumbent; **nodes** shortly hirsute or glabrous. **Sheaths** shorter than the internodes, greenish to purplish, glabrous or with papillose-based hairs, ciliate on 1 margin, glabrous on the other; **collars** hirsute; **ligules** 1.5–3.5 mm, of hairs; **blades** 3–30 cm long, 3–30 mm wide, flat, usually hirsute or sparsely pubescent, hairs papillose-based, sometimes glabrous, bases rounded to cordate-clasping, margins ciliate, cilia papillose-based, apices acute. **Panicles** 9–30 cm long, 5–8 cm wide, erect or nodding, partially included to well-exserted, **rachises** glabrous or sparsely hispid basally; **primary branches** usually alternate to opposite, divergent, secondary branches and pedicels confined to the distal ⅔; **pulvini** inconspicuous; **secondary branches** appressed; **pedicels** 9–27 mm, appressed. **Spikelets** 1.9–4 mm long, 0.8–1 mm wide, ovoid to almost spherical, often reddish-brown, glabrous, veins prominent, scabridulous, apices abruptly acuminate. **Lower glumes** 1.3–2.4 mm, ½–¾ as long as the spikelets, 3–5-veined; **upper glumes** 1.8–3.3 mm, 7–11-veined; **lower florets** sterile; **lower lemmas** similar to the upper glumes, 9-veined; **lower paleas** 0.4–0.9 mm; **upper florets** 1.5–2.4 mm long, 0.4–0.8 mm wide, ellipsoid, smooth or conspicuously papillate, shiny, stramineous, often with a lunate scar at the base.

Panicum hirticaule grows in rocky or sandy soils in waste places, roadsides, ravines, and wet meadows along streams. Its range extends from southeastern California and southwestern Texas southward through Mexico, Central America, Cuba, and Hispaniola to western South America and Argentina.

1. Blades rounded at the base, 3–16 mm wide; lower paleas less than ½ as long as the upper florets; panicles erect . . . . subsp. *hirticaule*
1. Blades cordate, clasping at the base, 4–30 mm wide; lower paleas more than ½ as long as the upper florets; panicles often nodding.
   2. Nodes, sheaths, and blades glabrous or sparsely pilose, hairs papillose-based; culms usually less than 70 cm tall; spikelets 3.2–4. mm long . . . . subsp. *stramineum*

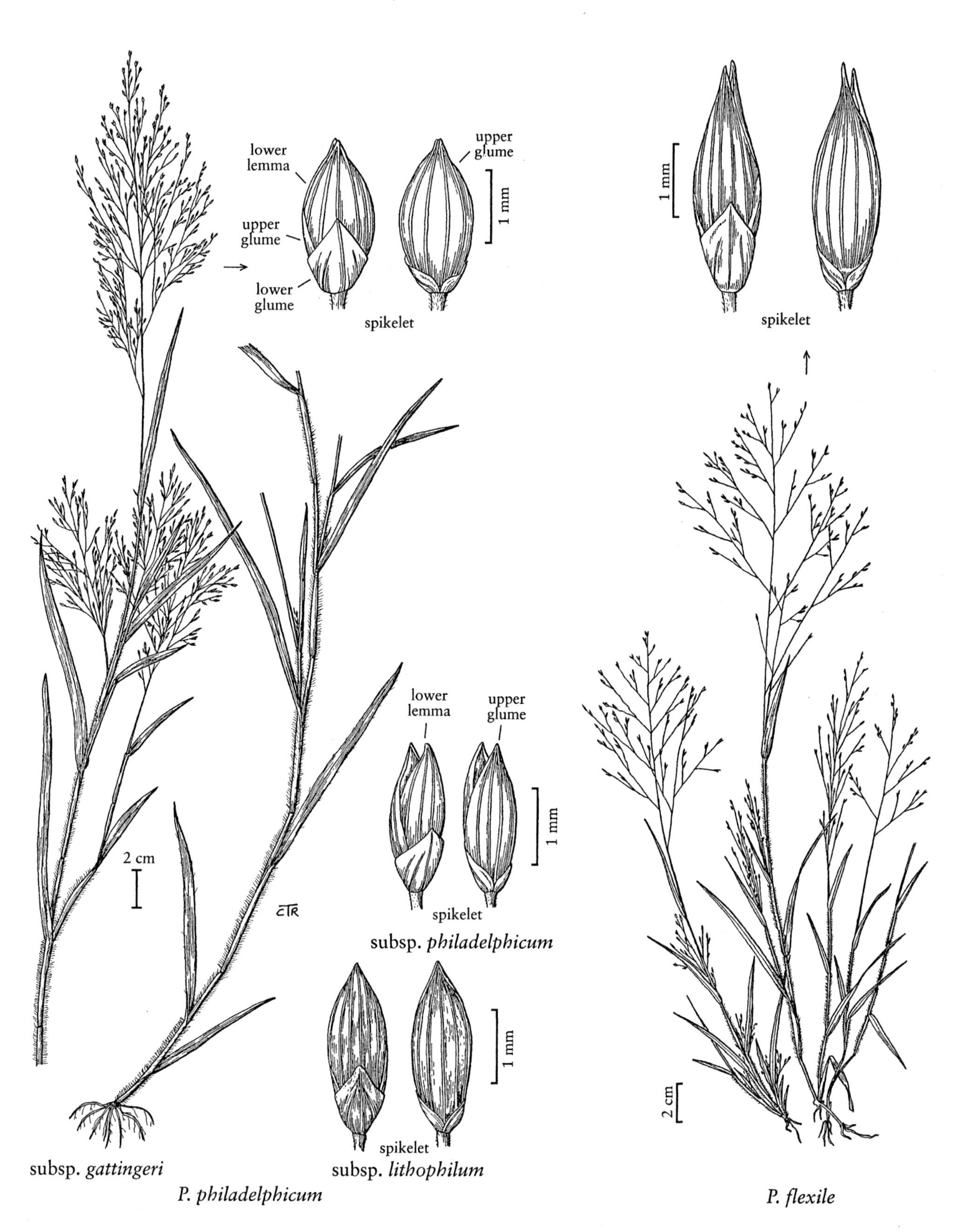

P. philadelphicum

P. flexile

2. Nodes, sheaths, and blades hirsute, hairs papillose-based; culms robust, usually more than 70 cm tall; spikelets 3–3.3 mm long . . . . subsp. *sonorum*

Panicum hirticaule J. Presl subsp. **hirticaule**

Culms 11–70 cm tall, usually simple; **nodes** usually hirsute. **Sheaths** hirsute, hairs papillose-based; **Blades** 3–16 mm wide, rounded basally. **Panicles** erect. **Spikelets** 1.9–3.3 mm. **Lower paleas** less than ½ as long as the upper florets. $2n$ = 18, 36.

Panicum hirticaule subsp. *hirticaule* is the most common of the subspecies, growing throughout the range of the species but occurring more often in arid habitats. It includes *P. alatum* Zuloaga and Morrone, a recently described species that differs from *P. hirticaule* subsp. *hirticaule* by the presence of paired elaiosomes at the base of a slightly stipitate upper floret.

Panicum hirticaule subsp. **sonorum** (Beal) Freckmann & Lelong

Culms 60–100 cm, robust; **nodes** hirsute, hairs papillose-based. **Sheaths** hirsute, hairs papillose-based; **blades** 4–30 mm wide, hirsute, hairs papillose-based, cordate, clasping basally. **Panicles** nodding. **Spikelets** 3–3.3 mm; **lower paleas** more than ½ as long as the upper florets. $2n$ = unknown.

Panicum hirticaule subsp. *sonorum* has been collected only a few times. Its range extends from southern Arizona to Chiapas, Mexico. It may have originated through selection and cultivation. The Cocopa tribe of the extreme lower Colorado River region grow it for the seed, which is used for food.

Panicum hirticaule subsp. **stramineum** (Hitchc. & Chase) Freckmann & Lelong

Culms 20–70 cm, robust, usually freely branching; **nodes** glabrous or sparsely hirsute, hairs papillose-based. **Sheaths** glabrous or sparsely hirsute, hairs papillose-based; **blades** 4–30 mm wide, glabrous or sparsely hirsute, hairs papillose-based, cordate, clasping basally. **Panicles** nodding. **Spikelets** 3.2–4 mm; **lower paleas** more than ½ as long as the upper florets. $2n$ = unknown.

Panicum hirticaule subsp. *stramineum* grows in rich bottomlands in southern Arizona, New Mexico, and western Mexico.

6. **Panicum mohavense** Reeder [p. 463]
MOHAVE WITCHGRASS

Plants annual. **Culms** 2–8 cm, erect-spreading; **nodes** 1–2, hispid; **internodes** pilose, hairs papillose-based. **Sheaths** rounded, much longer than the internodes, with prominent veins, hispid, hairs papillose-based; **ligules** 0.2–0.4 mm, membranous, ciliate; **blades** 1–4 cm long, 1–3 mm wide, flat or involute apically, glabrous basally, margins ciliate, cilia papillose-based. **Panicles** congested, partially included in the sheaths, less than 1.5 times longer than wide; **branches** ascending, narrow; **primary branches** appressed to the main axes, secondary branches and pedicels attached to the distal ⅔; **pedicels** appressed, 1–2 mm. **Spikelets** 2–2.2 mm long, 1–1.3 mm wide, plump-ellipsoid, glabrous. **Lower glumes** 1.2–1.3 mm, acute to attenuate; **upper glumes** and **lower lemmas** 2–2.2 mm, 7–9-veined, apices purplish, acute; **lower florets** sterile; **lower paleas** 0.2–0.4 mm; **upper florets** 1.4–1.8 mm long, about 1 mm wide, broadly ovoid.

Panicum mohavense is known only from arid limestone terraces in Arizona and New Mexico.

7. **Panicum psilopodium** Trin. [p. 463]

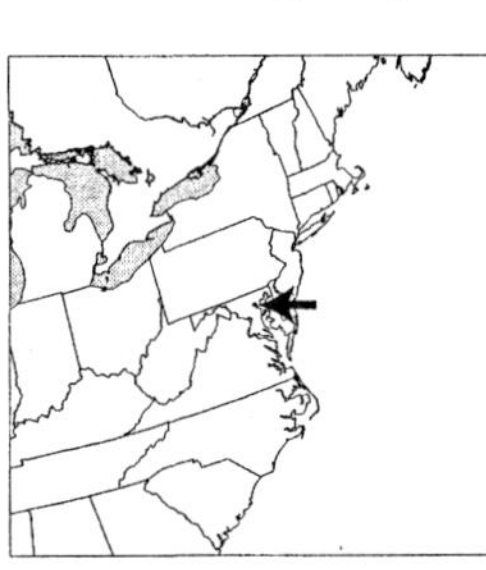

Plants annual; forming small clumps. **Culms** 20–60 cm tall, 0.8–1.2 mm thick, shortly decumbent to geniculate basally, erect distally; **nodes** glabrous; **internodes** glabrous. **Sheaths** shorter or longer than the internodes, rounded, smooth, glabrous; **ligules** about 1 mm; **blades** 5–15 cm long, 2–8 mm wide, flat, linear, glabrous or with a few marginal cilia near the base, bases contracted, apices long-acute. **Panicles** 10–20 cm long, 6–12 cm wide, exserted or partially included; **primary branches** alternate, ascending to strongly divergent, developing secondary branches in the basal ⅓–½; **pedicels** 4–9 mm, ascending. **Spikelets** 2.7–3.2 mm long, 1–1.2 mm wide, ovoid-ellipsoid, green tinged with purple, glaucous, glabrous, acute. **Lower glumes** 0.7–1.1 mm, about ⅓ as long as the spikelets, acute to attenuate; **upper glumes** and **lower lemmas** similar, equaling the spikelets, 11–13-veined, tapering to apiculate apices; **lower paleas** 1–2 mm; **lower florets** sterile; **upper florets** about 2.2 mm, ellipsoid, smooth, shiny, yellow at maturity, apices acute. $2n$ = 54.

Panicum psilopodium is native to eastern Asia. It has been reported from chrome ore piles in Canton,

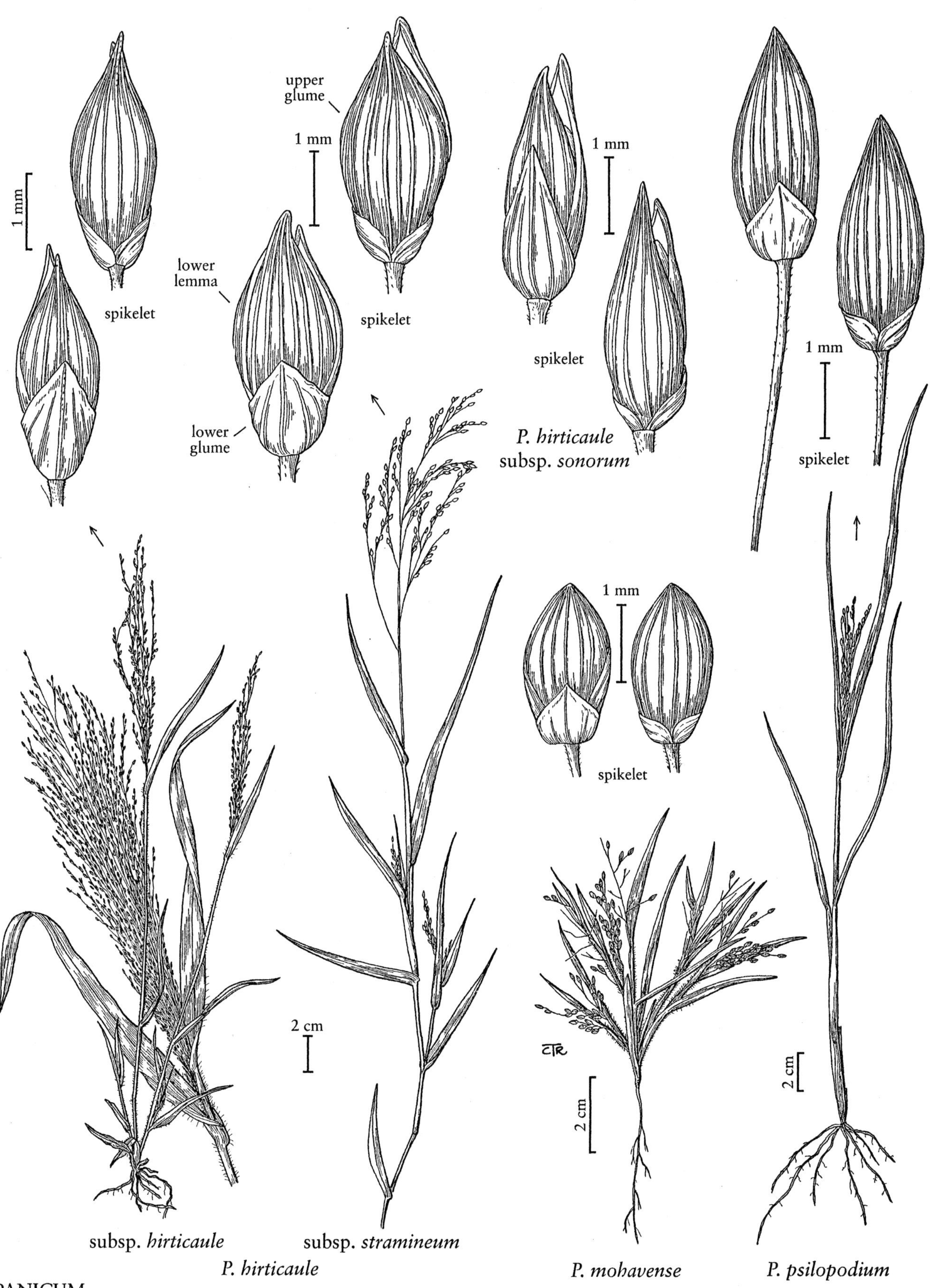

upper glume
1 mm
lower lemma
lower glume
spikelet
1 mm
spikelet
1 mm
spikelet
P. hirticaule subsp. sonorum
1 mm
spikelet
1 mm
spikelet
2 cm
2 cm
2 cm
CTR
subsp. hirticaule
subsp. stramineum
P. hirticaule
P. mohavense
P. psilopodium

PANICUM

Maryland (Reed 1964), but no voucher specimens have been seen. In its native range it grows in open habitats, such as roadsides and waste places.

8. **Panicum capillarioides** Vasey [p. 465]
LONG-BEAKED WITCHGRASS

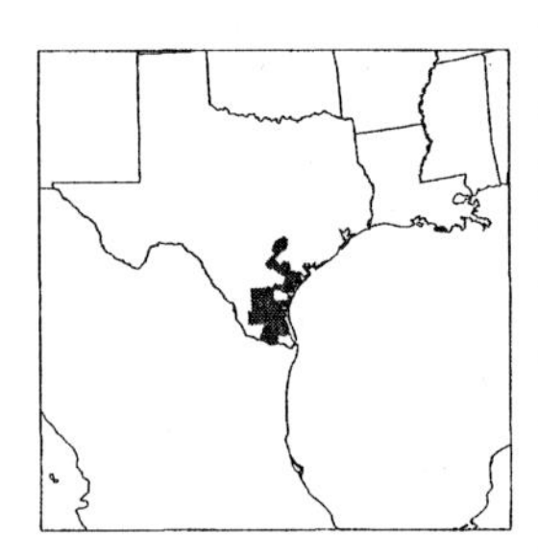

Plants perennial; cespitose from a knotty crown, hirsute, hairs papillose-based or glabrous. **Culms** 30–75 cm tall, 1–2 mm thick, terete to slightly compressed, erect or ascending, stiff, often bent at the nodes, simple or sparingly branched; **nodes** densely pubescent. **Sheaths** shorter than or equaling the internodes, rounded, hirsute, green or tinged with purple, margins ciliate; **ligules** 0.5–1 mm; **blades** 12–30 cm long, 2–12 mm wide, stiffly erect or ascending, flat, pubescent, sometimes sparsely so, hairs papillose-based, bases truncate, apices attenuate. **Panicles** terminal, 15–30 cm long, 10–12(26) cm wide, usually shortly exserted, scarcely overtopping the blades; **rachises** hispid, sometimes glabrous basally; **primary branches** alternate or opposite, divergent, secondary branches divergent, most abundant on the distal ⅓ of the primary branches, with 1–3 spikelets; **pedicels** 2–20 mm, confined to the distal ⅓ of the branches; **pulvini** poorly developed, shortly pilose. **Spikelets** 5–6.5 mm long, 1–1.2 mm wide, glabrous, long-acuminate. **Lower glumes** 2–3 mm, about ½ as long as the spikelets, attached about 0.4 mm below the upper glumes, 5–7-veined, acute to obtuse; **upper glumes** and **lower lemmas** 5–6 mm, exceeding the upper florets by 3–4 mm, 9–13-veined; **lower florets** sterile; **lower paleas** 1.5–2 mm, acute; **upper florets** 1.6–2 mm long, 1–1.1 mm wide, smooth, chestnut brown when mature. $2n = 36$.

Panicum capillarioides grows in sandy grasslands, oak savannahs, and rangelands from southern Texas to northern Mexico.

9. **Panicum hirsutum** Sw. [p. 465]
GIANT WITCHGRASS

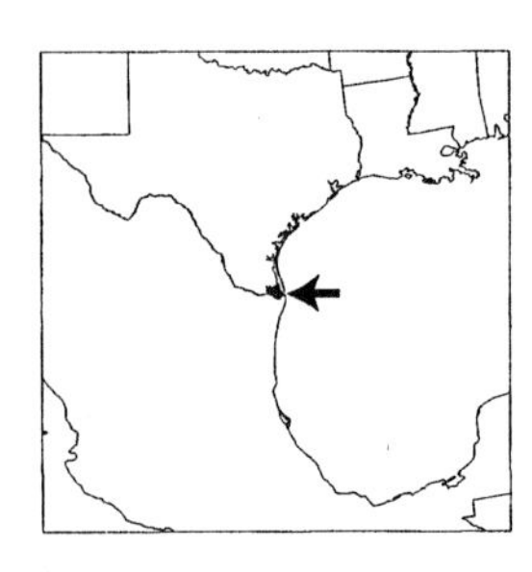

Plants perennial; forming large clumps from short rhizomes. **Culms** 100–300 cm tall, 4–10 mm thick, decumbent, semi-woody at the base, simple or branching from the middle nodes, prophylls prominent, to 15 cm; **nodes** contracted, pilose, sericeous; **internodes** glabrous or with papillose-based hairs below the nodes. **Sheaths** shorter or longer than the internodes, rounded, sparsely hispid, hairs papillose-based, thick, fragile, penetrating and irritating the skin when handled, margins glabrous or ciliate; **collars** more densely pubescent than the sheaths, hairs papillose-based; **ligules** 1.5–2 mm, with longer hairs immediately behind, growing from the base of the blades; **blades** 20–50 cm long, 15–40 mm wide, spreading, flat or with involute margins, bases subcordate to cordate, margins glabrous or sparsely hairy. **Panicles** terminal, 25–45 cm long, 5–15 cm wide, lax, contracted to diffuse, not breaking at the base and becoming tumbleweeds, all or most secondary branches confined to the distal ⅓; **lower branches** whorled; **pedicels** 0.5–2 mm, appressed. **Spikelets** 1.8–2.5 mm long, 0.5–1 mm wide, narrowly ellipsoid, glabrous. **Lower glumes** 0.7–1.4 mm, about ½ as long as the spikelets, 3–5-veined, acute to attenuate; **upper glumes** and **lower lemmas** subequal, about as long as the spikelets, 7–11-veined; **lower florets** sterile; **lower paleas** 1.3–1.7 mm; **upper florets** 1.2–1.6 mm long, 0.5–0.7 mm wide, glabrous, smooth, shiny, chestnut brown to dark brown. $2n = 36$.

Panicum hirsutum grows along river banks or in ditches, often among shrubs in partial shade. Its range extends from southern Texas through eastern Mexico, Central America, Cuba, and the West Indies to Ecuador, Brazil, and Argentina.

10. **Panicum bergii** Arechav. [p. 465]
BERG'S WITCHGRASS

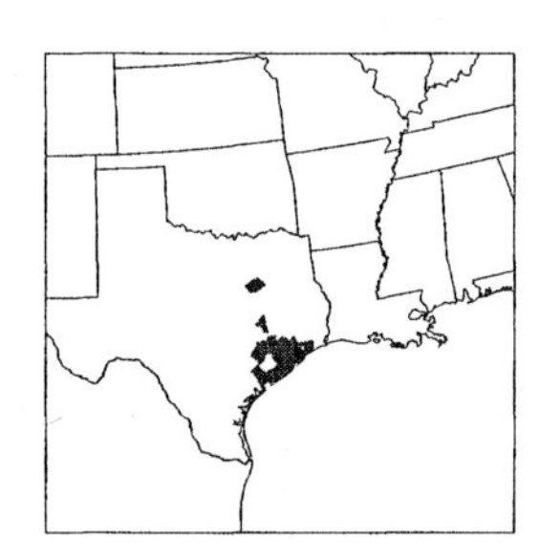

Plants perennial; cespitose, with numerous leaves clustered at the base. **Culms** (10)50–140 cm, stout, stiffly erect, branched from the middle and lower nodes; **lower nodes** sericeous; **lower internodes** sericeous, hairs papillose-based, upper internodes sometimes glabrous. **Sheaths** rounded, glabrous or sparsely to densely hispid, hairs not fragile and prickly, not causing skin irritation, margins ciliate; **ligules** 1–3 mm; **blades** 3–60 cm long, 2–12 mm wide, flat or involute, ascending, adaxial surfaces densely hirsute basally, less densely so elsewhere, bases attenuate, apices acute. **Panicles** (4)15–40 cm long, (3)10–25 cm wide, about ⅓–½ as long as the plants, open, breaking at the base of the peduncles at maturity and dispersed as tumbleweeds, secondary branching mostly confined to the distal ⅓ of the primary branches; **rachises** densely hispid or glabrous; **lower primary branches** in whorls of 4–7, stiffly spreading, naked on the lower ½; **pedicels** 3–20 mm, appressed. **Spikelets** 2–3 mm long, 0.8–1.2 mm wide, glabrous. **Lower glumes** 1–1.6 mm, 5-veined, acuminate; **upper glumes** and **lower lemmas** similar,

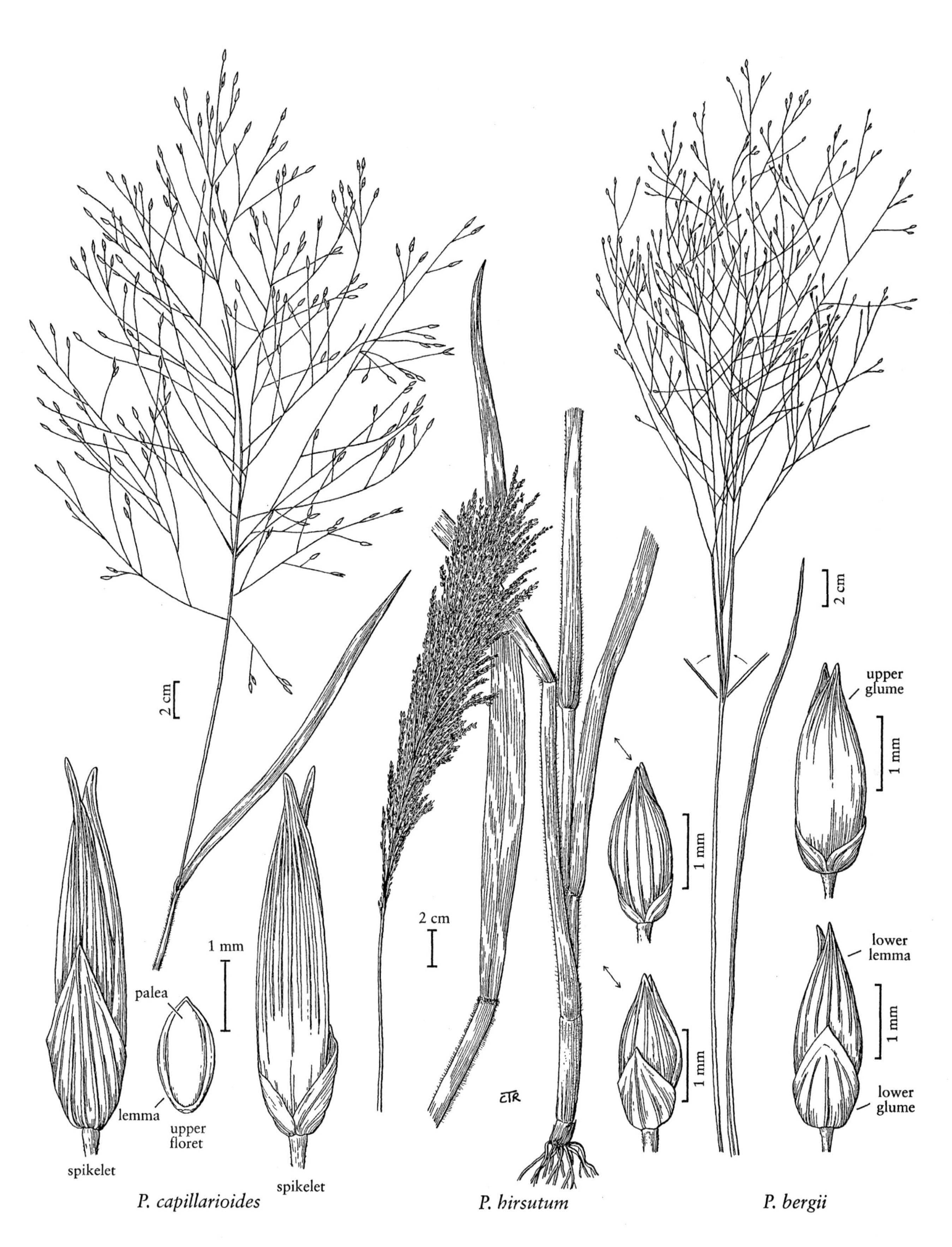
2 cm
spikelet
palea
lemma
upper
floret
1 mm
spikelet
P. capillarioides
2 cm
CTR
P. hirsutum
1 mm
1 mm
2 cm
upper
glume
1 mm
lower
lemma
1 mm
lower
glume
P. bergii

2–2.8 mm, 7–9-veined, exceeding the upper florets by about 0.3 mm; **lower florets** sterile; **lower paleas** 1.4–2.2 mm; **upper florets** 1.5–1.9 mm long, 0.7–1 mm wide, smooth, chestnut brown at maturity. $2n = 36$.

Panicum bergii is an eastern South American species that now grows in southeastern Texas. It occurs in ditches and shallow, and sporadically flooded depressions in grasslands.

11. **Panicum diffusum** Sw. [p. 468]
SPREADING WITCHGRASS

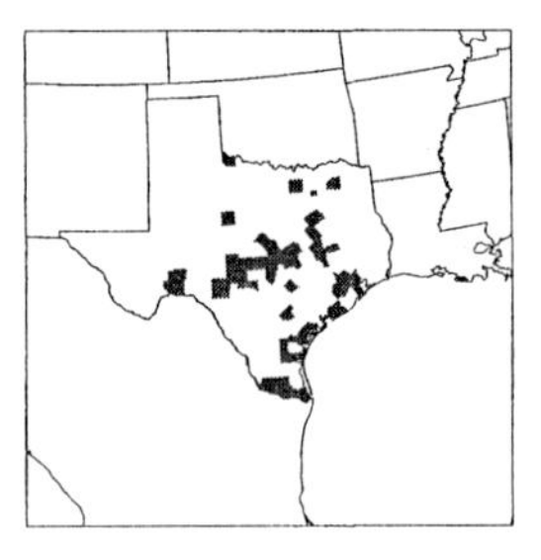

Plants perennial; cespitose and shortly rhizomatous. **Culms** (6)25–100 cm, 0.5–3.5 mm thick, spreading to weakly ascending, usually freely branching; **nodes** pilose, hairs spreading to ascending; **internodes** pilose, with papillose-based hairs, or sparsely hispid. **Sheaths** rounded, glabrous, margins shortly ciliate distally; **ligules** 0.6–4 mm, of hairs; **blades** (3)6–15 cm long, 1–5 mm wide, spreading, abaxial surfaces sparsely hirsute, adaxial surfaces more densely so, hairs papillose-based, midribs prominent and not white, margins involute, bases cuneate, apices subulate. **Terminal panicles** 3–35 cm long, about ½ as wide, shortly exserted; **axillary panicles** smaller, not fully exserted; **rachises** scabridulous; **primary branches** usually solitary, sometimes paired, divergent and widely spaced, secondary branching mostly on the distal ⅓ of the primary branches; **pedicels** 1–4 mm, spreading to appressed, confined to the distal portions of the secondary branches. **Spikelets** 2.1–2.9 mm long, 0.8–1 mm wide, glabrous. **Lower glumes** 1–1.2 mm, to ½ as long as the spikelets, attenuate, 5–9-veined, veins anastomosing apically; **upper glumes** and **lower lemmas** similar, extending about 0.5 mm beyond the upper florets, 11–13-veined; **lower florets** sterile; **lower paleas** 1–1.3 mm; **upper florets** 1.5–1.8 mm long, 0.6–0.8 mm wide, smooth. $2n = 36$.

Panicum diffusum grows along river banks, ditches, and disturbed areas in wet, loamy or clayey soils. Its range extends from Texas to the Caribbean and northern South America.

12. **Panicum ghiesbreghtii** E. Fourn. [p. 468]
GHIESBREGHT'S WITCHGRASS

Plants perennial; cespitose. **Culms** 40–120 cm tall, 2–3 mm thick, decumbent to erect, branching from the base and the middle nodes; **nodes** pilose, hairs spreading; **internodes** hirsute, hairs papillose-based. **Sheaths** usually shorter than the internodes, hirsute, lower sheaths more so than those above, hairs papillose-based; **collars** densely pilose; **ligules** 0.5–4 mm; **blades** 16–55 cm long, 0.5–14 mm wide, erect to ascending, abaxial surfaces hirsute, hairs papillose-based, adaxial surfaces densely pilose, midveins prominent and whitish, bases truncate, margins ciliate basally, apices attenuate. **Terminal panicles** 7–35 cm long, 5–23 cm wide, about ½ as wide as long, shortly exerted or partially included, lax, open; **axillary panicles** smaller, included basally; **primary branches** diverging, lower branches solitary, upper branches solitary to subverticillate; **secondary branching** primarily in the distal ⅓; **pedicels** 1–4 mm, clavate, spreading to appressed. **Spikelets** 2.6–3.4 mm long, 0.9–1.2 mm wide, ovoid, glabrous. **Lower glumes** 1.4–1.7 mm, to ½ as long as the spikelets, acute, 5–7-veined; **upper glumes** and **lower lemmas** similar, exceeding the upper florets by 0.7–0.9 mm, 9–13-veined; **lower florets** sterile; **lower paleas** 0.5–1.3 mm; **upper lorets** 1.7–2.3 mm long, 0.8–1.1 mm wide, smooth, ovoid. $2n$ = unknown.

Panicum ghiesbreghtii grows in low, moist ground, wet thickets, and savannahs, from southern Texas through Mexico, Central America, Cuba, and the West Indies to northern South America.

13. **Panicum hallii** Vasey [p. 468]
HALL'S WITCHGRASS

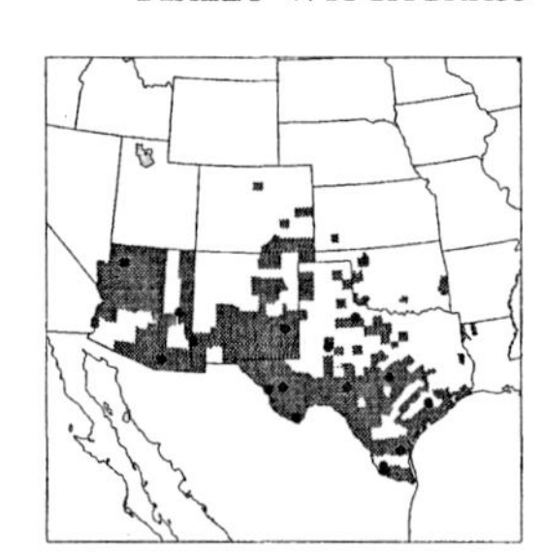

Plants perennial; cespitose. **Culms** 10–100 cm, 2–10 mm thick, erect, simple or sparingly branched basally; **nodes** sericeous, pilose or glabrous; **internodes** usually glaucous. **Leaves** often crowded basally; **sheaths** rounded, glabrous or hirsute, hairs fragile, papillose-based, margins sometimes ciliate distally; **ligules** 0.6–2 mm; **blades** 4–23 cm long, 1–10 mm wide, erect to spreading, flat or sometimes involute (on sterile branches), often curling at maturity, glaucous, abaxial surfaces sometimes with prominent papillae along the midribs, bases rounded or narrowing to the sheaths, margins cartilaginous, ciliate basally, scabridulous elsewhere, apices acute. **Terminal panicles** 7–31 cm long, 3–15 cm wide; rachises glabrous, tending to break at maturity; **branches** usually alternate, slender, stiff, ascending to divergent; **pedicels** 1–15 mm, appressed. **Spikelets** 2.1–4.2 mm long, 0.8–1 mm wide, usually ovoid, glabrous. **Lower glumes** 1.2–2.4 mm, ½–¾ as long as the spikelets, attenuate; **upper glumes** and **lower lemmas** similar, 7–11-veined, acuminate, extending 0.3–1.2 mm beyond the upper florets; **lower florets** sterile; **lower paleas** 0.8–2 mm; **upper florets** 1.5–2.4

mm long, 0.7–1.2 mm wide, ovoid to ellipsoid, smooth, nigrescent. $2n = 18$.

Panicum hallii grows on sandy, gravelly, or rocky land, including roadsides, pastures, rangeland, oak and pine savannahs, chaparral, and moist areas in deserts and on mesas. Its range extends from the southwestern United States to southern Mexico.

1. Spikelets 3–4.2 mm long; panicles usually greatly exceeding the blades, with a few spikelets; blades clustered near the base of the plants, ascending, often curling at maturity . . . subsp. *hallii*
1. Spikelets 2.1–3 mm long; panicles scarcely exceeding the blades, with relatively crowded spikelets; blades not clustered near the base of the plants, lax, spreading, not curled . . . . . . subsp. *filipes*

Panicum hallii subsp. **filipes** (Scribn.) Freckmann & Lelong

Plants often taller than subsp. *hallii*, sparsely pubescent to almost glabrous. **Blades** relatively lax, ascending to spreading, not strongly clustered basally or curling at maturity. **Panicles** scarcely exceeding the blades, with more closely spaced spikelets than in subsp. *hallii*; **main branches** rarely whorled, more crowded. **Spikelets** 2.1–3 mm.

Panicum hallii subsp. *filipes* often grows in moist soil. Its range extends from Arizona, Texas, and Louisiana to southern Mexico.

Panicum hallii Vasey subsp. **hallii**

Plants often depauperate, occasionally large, sparsely to moderately pubescent or hirsute. **Blades** usually erect or ascending, tending to arise from basal clumps and curl at maturity. **Panicles** usually greatly exceeding the blades, with a few spikelets; **main branches** solitary, ascending, well-separated. **Spikelets** 3–4.2 mm.

Panicum hallii subsp. *hallii* usually grows on drier sites than subsp. *filipes*. Its range extends from southern Colorado and Kansas to north central Mexico.

Panicum sect. Dichotomiflora (Hitchc.) Honda

Plants annual or perennial; cespitose, sometimes rhizomatous in the tropics. **Culms** erect to decumbent, compressed, sometimes slightly succulent. **Sheaths** almost glabrous or hispid, hairs papillose-based; **ligules** membranous, ciliate, cilia 2–4 mm. **Panicles** open or contracted; **branches** mostly solitary, stiff, naked basally; **pedicels** short, stiff, subappressed, prominently 3-angled, usually scabrous on the angles, widened and cuplike at the apices. **Spikelets** narrow, ellipsoid to lanceoloid, glabrous, pointed. **Lower glumes** ⅕–⅓ as long as the spikelets, clasping, 0–3-veined, truncate to subacute. **Upper glumes** and **lower lemmas** extending beyond the upper florets, 5–9-veined, veins prominent near the apices; **lower florets** sterile; **lower paleas** long and membranous to vestigial; **upper florets** smooth, shiny, glabrous.

Members of sect. *Dichotomiflora* usually grow in wet, open areas, some as emergents in shallow water. They are often found on disturbed ground. There are about seven species in the Western Hemisphere, but only three grow in the *Flora* region.

14. **Panicum lacustre** Hitchc. & Ekman [p. 471]
CYPRESS-SWAMP PANICUM

Plants perennial; emergent aquatic or terrestrial, rooting at the lower nodes. **Culms** 100–150 cm, erect, succulent, with short innovations; **nodes** glabrous; **internodes** glabrous. **Sheaths** compressed, not keeled, overlapping but narrow, exposing the nodes, bladeless and glabrous or sparsely pilose below the water; **ligules** 1–2 mm, membranous, ciliate; **blades** 1–30 cm long, 2–4 mm wide, narrow, linear, flat or folded, abaxial surfaces sparsely pubescent, adaxial surfaces sparsely pilose. **Panicles** 10–30 cm, open, with many spikelets; **primary branches** fascicled at the base of the panicles, solitary and distant distally; **pedicels** 1–4 mm, sharply 3-angled, appressed. **Spikelets** 2–2.2 mm, glabrous. **Lower glumes** truncate to broadly triangular, ¼ as long as the spikelets, 3-veined; **lower paleas** absent; **upper glumes** and **lower lemmas** equal, slightly exceeding the upper florets, 5- or 7-veined, pointed; **lower florets** sterile; **upper florets** relatively thin, smooth. $2n$ = unknown.

Panicum lacustre grows in shallow water or wet soil at the edge of cypress ponds in the Everglades of southern Florida. It also grows in Cuba.

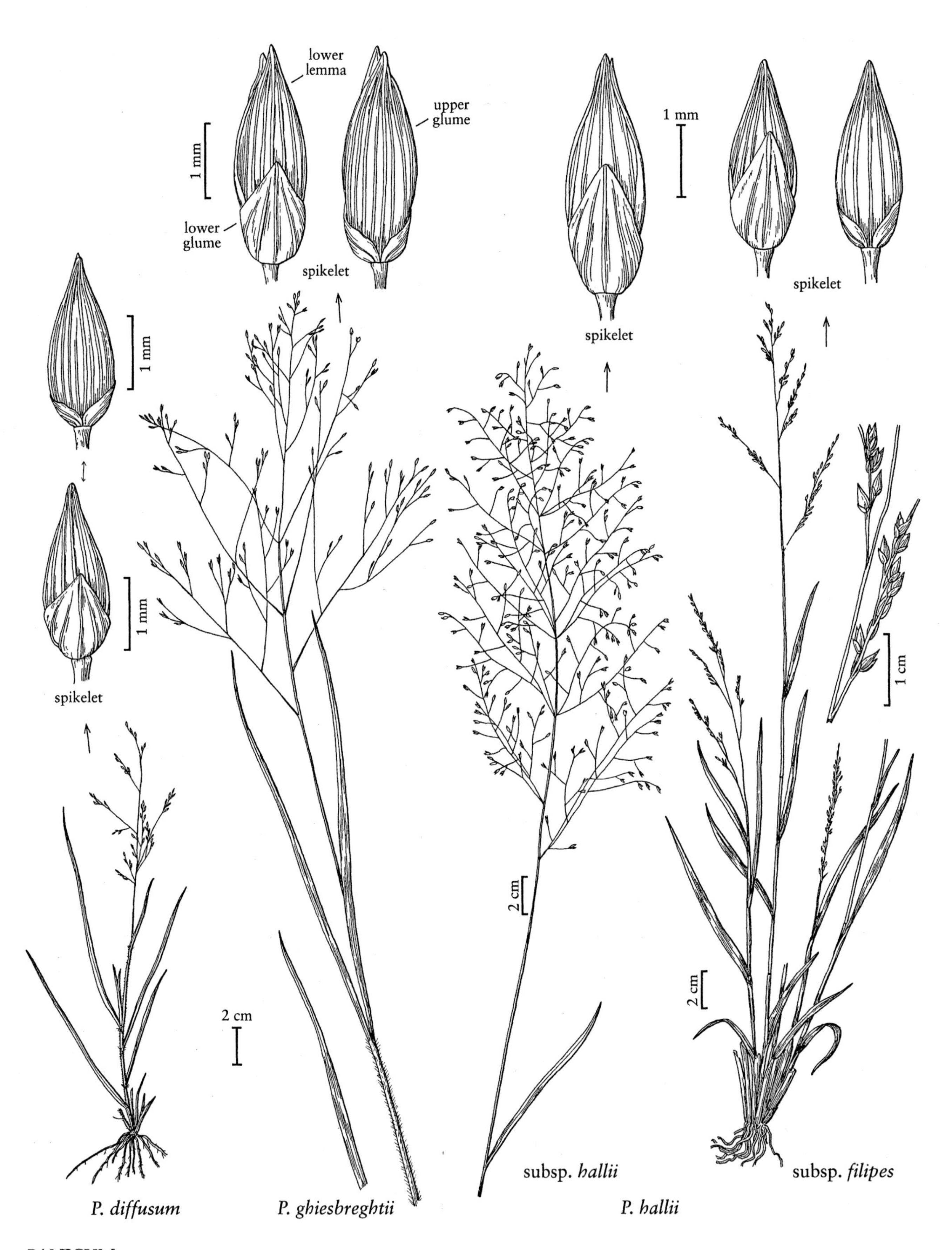
lower
lemma
upper
glume
1 mm
lower
glume
spikelet
1 mm
spikelet
1 mm
1 mm
spikelet
1 mm
spikelet
1 cm
2 cm
2 cm
2 cm
subsp. *hallii*
subsp. *filipes*
P. diffusum
P. ghiesbreghtii
P. hallii

PANICUM

15. **Panicum dichotomiflorum** Michx. [p. 471]
FALL PANICUM, PANIC D'AUTOMNE

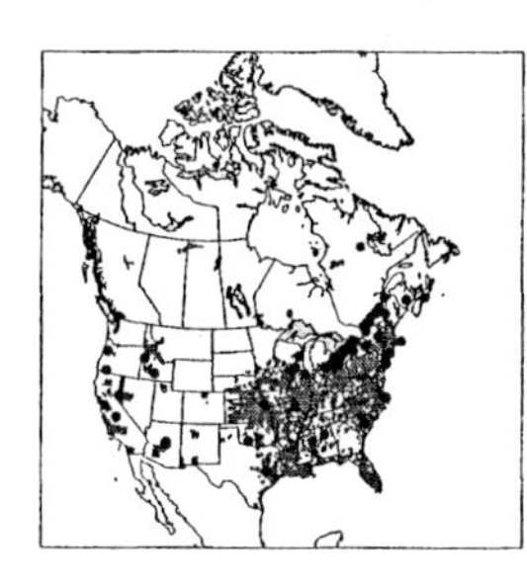

Plants annual or short-lived perennials in the *Flora* region, perennial in the tropics; usually terrestrial, sometimes aquatic but not floating. **Culms** 5–200 cm tall, 0.4–3 mm thick, decumbent to erect, commonly geniculate to ascending, rooting at the lower nodes when in water, simple to divergently branched from the lower and middle nodes, usually succulent, slightly compressed, glabrous; **nodes** usually swollen, sometimes constricted on robust plants, glabrous; **internodes** glabrous, shiny, pale green to purplish. **Sheaths** compressed, inflated, sparsely pubescent near the base, elsewhere mostly glabrous, sparsely pilose, or hispid, hairs sometimes papillose-based, margins or throat ciliate, with papillose-based hairs; **ligules** 0.5–2 mm; **blades** 10–65 cm long, 3–25 mm wide, glabrous or sparsely pilose, often scabrous near the margins, midribs stout, whitish. **Panicles** 4–40 cm, diffuse, lax, with a few spikelets; **branches** to 15 cm, alternate or opposite, occasionally verticillate, ascending to spreading, stiff, scabrous; **pedicels** 1–6 mm, sharply 3-angled, scabrous, expanded to cuplike apices, appressed mostly to the abaxial side of the branches. **Spikelets** 1.8–3.8 mm long, 0.7–1.2 mm wide, ellipsoid to narrowly ovoid, light green to red-purple, glabrous, acute to acuminate. **Lower glumes** 0.6–1.2 mm, ¼–⅓ as long as the spikelets, 0–3-veined, obtuse to acute; **upper glumes** and **lower lemmas** similar, exceeding the upper florets by 0.3–0.6 mm, 7–9-veined; **lower paleas** vestigial to almost as long as the lower lemmas; **lower florets** sterile; **upper florets** 1.4–2.5 mm long, 0.7–1.1 mm wide, narrowly ellipsoid, smooth, shiny, stramineous to nigrescent, with pale veins. $2n$ = 36, 54.

Panicum dichotomiflorum grows in open, often wet, disturbed areas such as cultivated and fallow fields, roadsides, ditches, open stream banks, receding shores, clearings in flood plain woods, and sometimes in shallow water. It is probably native throughout the eastern United States and adjacent Canada, but introduced elsewhere, including in the western United States. Its size and habit may be partly under genetic control, but these features also seem to be strongly affected by moisture levels, soil richness, competition, and the time of germination.

1. Spikelets 1.8–2.2 mm long, widest at the middle, acute; upper glumes and lower lemmas submembranaceous; pedicels often over 3 mm long . . . . subsp. *puritanorum*
1. Spikelets 2.2–3.8 mm long, widest below the middle, acuminate; upper glumes and lower lemmas subcoriaceous; most pedicels less than 3 mm long.
   2. Sheaths glabrous or sparsely pilose, hairs not papillose-based . . . . subsp. *dichotomiflorum*
   2. Sheaths hispid, hairs papillose-based . . . . subsp. *bartowense*

Panicum dichotomiflorum subsp. **bartowense** (Scribn. & Merr.) Freckmann & Lelong

Culms often 100–200 cm, stout, erect, simple or sparingly branched. **Sheaths** loosely overlapping, prominently hispid, hairs papillose-based. **Pedicels** usually less than 3 mm and shorter than the spikelets. **Spikelets** 2.3–2.8 mm, tapered from below the middle to the acuminate apices; **upper glumes** and **lower lemmas** subcoriaceous.

Panicum dichotomiflorum subsp. *bartowense* grows in Florida, Cuba, and the Bahamas. Reports from more northerly areas may represent introductions or misidentifications.

Panicum dichotomiflorum Michx. subsp. **dichotomiflorum**
PANIC D'AUTOMNE DRESSÉ, PANIC À FLEURES DICHOTOM

Culms 5–200 cm. **Sheaths** glabrous or sparsely pilose, not hispid with papillose-based hairs. **Pedicels** usually less than 3 mm and shorter than the spikelets. **Spikelets** 2.3–3.8 mm, tapered from below the middle to the acuminate apices; **upper glumes** and **lower lemmas** subcoriaceous.

Panicum dichotomiflorum subsp. *dichotomiflorum* is the most common of the three subspecies and is found throughout the range of the species. In the past, members of this subspecies have been treated as two different taxa, var. *geniculatum* (Alph. Wood) Fernald and var. *dichotomiflorum*, with more erect, slender plants having fewer long-exserted panicles with slender, ascending branches and less crowded spikelets being placed in var. *dichotomiflorum*. Such plants are more common in the southern part of the subspecies' range, but the traits are poorly correlated and the differences are at least in part affected by photoperiod, nighttime temperatures, and the time of seed germination.

Panicum dichotomiflorum subsp. **puritanorum** (Svenson) Freckmann & Lelong

Culms 30–60 cm, usually slender, sometimes to 2 mm thick. **Sheaths** glabrous or sparsely pilose, hairs not papillose-based. **Pedicels** often over 3 mm, usually longer than the spikelets. **Spikelets** 1.8–2.2 mm, widest at about the middle, acute. **Upper glumes** and **lower lemmas** slightly exceeding the upper florets, submembranacous.

Panicum dichotomiflorum subsp. *puritanorum* has a sporadic distribution on receding shores along the

Atlantic coast from Nova Scotia to Virginia, and around southern Lake Michigan. The small spikelets with thin glumes and thin lower lemmas are probably genetically fixed traits; the commonly delicate habit, however, probably results from late season seed germination following receding water.

16. **Panicum paludosum** Roxb. [p. 471]
AQUATIC PANICUM

Plants perennial; more or less cespitose, rhizomatous or stoloniferous, free-floating or rooting in shallow water. **Culms** 30–150 cm tall, 3–7 mm thick, compressed, spongy, glabrous, decumbent and rooting at the lower nodes in shallow water; **nodes** glabrous; **internodes** glabrous, smooth. **Sheaths** usually shorter than the internodes, not keeled, glabrous or sparsely hispid distally; **ligules** 1–4 mm; **blades** 10–40 cm long, 5–15 mm wide, flat, glabrous, contracted basally, attenuate distally, apices acute. **Panicles** 10–25 cm long, 5–17 cm wide, shortly exserted or included basally; **primary branches** 4–12 cm, ascending to spreading, secondary and higher order branches confined to the distal ⅔; **pedicels** 1–4 mm, sharply 3-angled, ascending to appressed. **Spikelets** 3–4 mm long, 0.8–2 mm wide, lanceolate. **Lower glumes** 0.5–0.9 mm, ⅕–⅓ as long as the spikelets, rounded, glabrous, truncate, weakly 1–3-veined; **upper glumes** and **lemmas** subequal, glabrous, 9–11-veined, veins prominent, apices acute to accuminate; **lower paleas** about ⅔ as long as the lower lemmas; **lower florets** sterile; **upper florets** 2–2.7 mm, narrowly ellipsoid, smooth, shiny, yellowish. $2n = 54$.

Panicum paludosum is an Asian species that grows in shallow water. It has been found in Baltimore, Maryland, but may not be established there.

Panicum sect. Repentia Stapf

Plants perennial; rhizomatous, rhizomes long or short, sometimes with scalelike leaves, sometimes forming a compact, knotty base. **Culms** 20–300 cm, erect, firm, terete, often glaucous. **Sheaths** glabrous or pilose; **ligules** 0.5–6 mm, membranous, ciliate; **blades** linear (sometimes involute), firm. **Panicles** open or contracted. **Spikelets** lanceoloid, glabrous, acute to acuminate. **Lower glumes** about ¼–⅔ as long as the spikelets, 1–7-veined, usually acute or truncate, sometimes acuminate; **upper glumes** and **lower lemmas** unequal, stiffly pointed, upper glumes often exceeding the lower lemmas, the two often separating (gaping) beyond the florets; **lower florets** staminate; **lower paleas** well-developed; **upper florets** smooth, shiny, often pointed.

There are approximately 12 species of *Panicum* sect. *Repentia* in the Western Hemisphere, four of which grow in the *Flora* region. The species generally inhabit wet sites, growing on coastal dunes, sea beaches or along the margins of rivers.

17. **Panicum repens** L. [p. 473]
TORPEDO GRASS

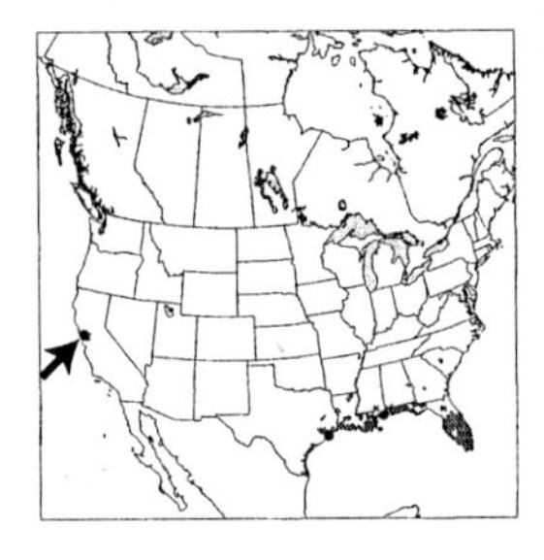

Plants perennial; rhizomatous, forming extensive colonies, rhizomes long, to 5 mm thick, branching, scaly, sharply pointed. **Culms** 20–90 cm tall, 1.8–2.8 mm thick, erect, rigid, simple or branching from the lower and middle nodes; **nodes** glabrous or sparsely hispid; **internodes** glabrous. **Sheaths** generally shorter than the internodes, not keeled, lower nodes glabrous or hispid, hairs papillose-based, particularly near the summits; **ligules** 0.5–1 mm; **blades** 3–25 cm long, 2–8 mm wide, often distichous, flat to slightly involute, firm, adaxial surfaces pilose basally, glabrous or sparsely pubescent abaxially. **Panicles** 3–24 cm long, usually less than 5 cm wide, open; **primary branches** 2–11 cm, alternate, few, stiffly ascending to spreading; **pedicels** 1–6 mm, subappressed. **Spikelets** 2.2–2.8 mm long, 0.8–1.3 mm wide, ellipsoid-ovoid, pale green, acute, upper glumes and lower lemmas sometimes separating (gaping) beyond the florets. **Lower glumes** 0.5–1 mm, ⅕–⅖ as long as the spikelets, glabrous, faintly 1–5-veined, subtruncate to broadly acute; **upper glumes** and **lower lemmas** glabrous, extending 0.1–0.5 mm beyond the upper florets, scarcely separated; **upper glumes** 7–11-veined, shorter than the lower lemmas, acute to short-acuminate; **lower florets** staminate; **lower lemmas** 7–11-veined; **lower paleas** 1.9–2.1 mm, oblong; **upper florets** 1.8–2.7 mm long, 0.7–1.3 mm wide, broadly ellipsoid, broadest at or above the middle, glabrous, shiny, smooth, apices rounded. $2n = 36, 40, 45, 54$.

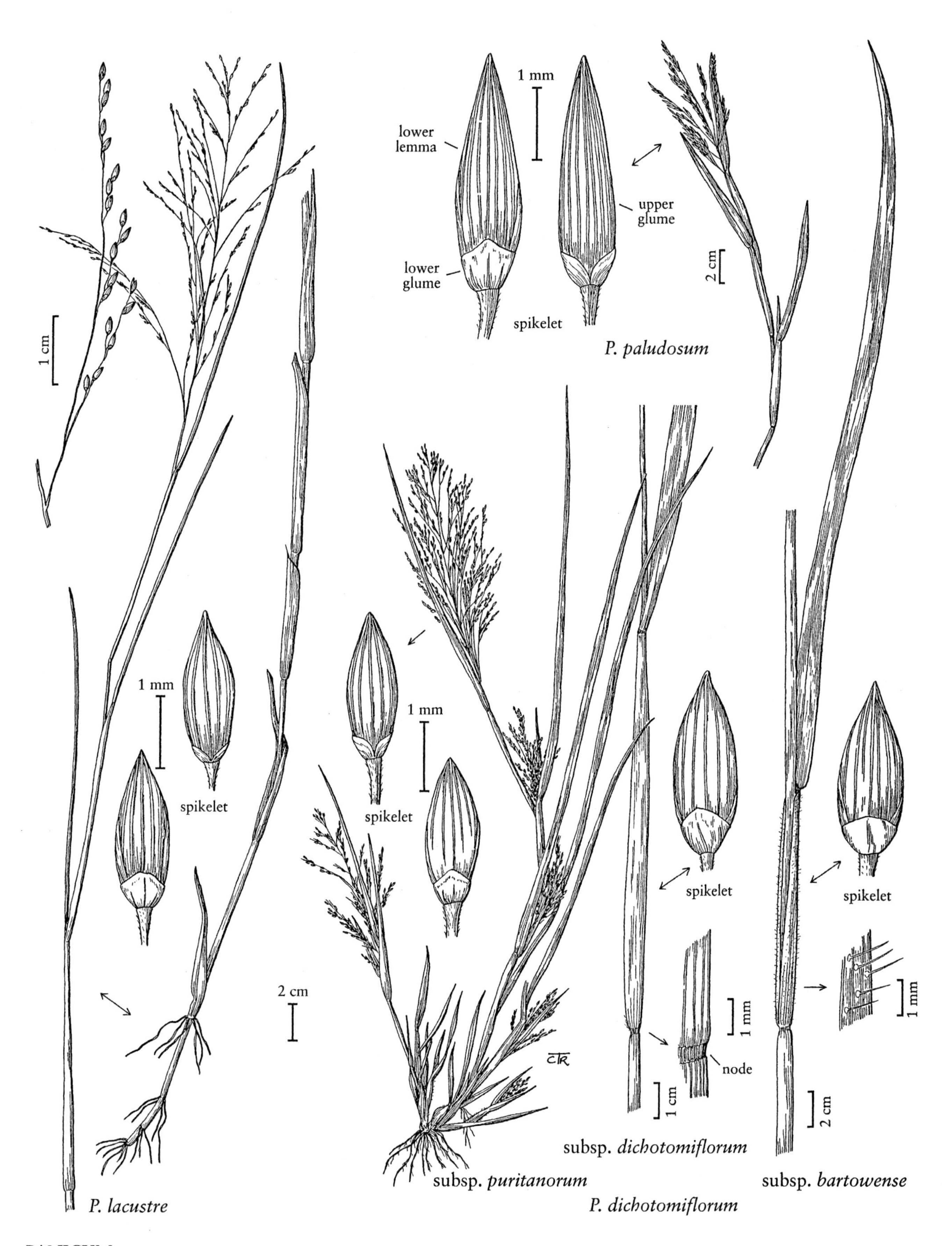
1 mm
lower lemma
upper glume
lower glume
spikelet
2 cm
P. paludosum
1 cm
1 mm
spikelet
1 mm
spikelet
2 cm
spikelet
spikelet
1 mm
1 mm
node
1 cm
2 cm
CTR
subsp. dichotomiflorum
subsp. puritanorum
subsp. bartowense
P. lacustre
P. dichotomiflorum

Panicum repens grows on open, moist, sandy beaches and the shores of lakes and ponds, occasionally extending out into or onto the water. It is mostly, but not exclusively, coastal. It grows on tropical and subtropical coasts throughout the world and may have been introduced to the Americas from elsewhere. Small plants having small, dense panicles of purplish spikelets with longer, subacute lower glumes have been named *Panicum gouinii* E. Fourn., but they intergrade with more typical plants and do not seem to merit taxonomic recognition.

18. **Panicum coloratum** L. [p. 473]
KLEINGRASS

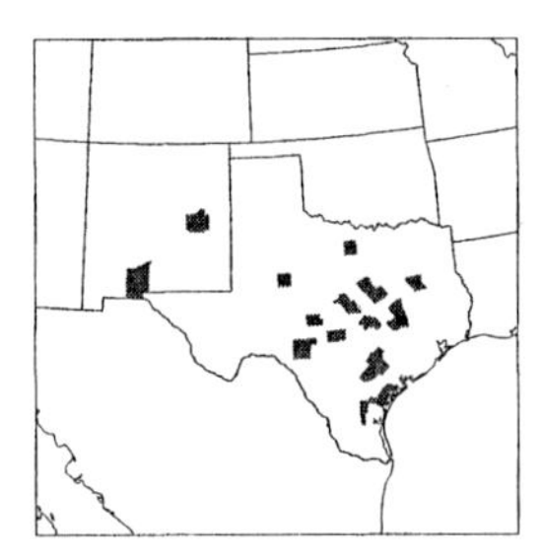

Plants perennial; cespitose, usually with short, knotty rhizomes. **Culms** 50–140 cm tall, 1.5–2.5 mm thick, usually erect, rarely decumbent, firm; **nodes** glabrous or puberulent; **internodes** glabrous. **Sheaths** shorter than the internodes, glabrous or hispid, hairs papillose-based, rounded basally; **ligules** 0.5–2 mm; **blades** 10–30 cm long, 2–8 mm wide, flat, glabrous or sparsely hirsute on 1 or both surfaces. **Panicles** 4–25(40) cm long, 3–14 cm wide, exerted, lax; **primary branches** 3–14 cm, opposite and alternate, ascending, glabrous, branching in the distal ⅔; **pedicels** 1–4 mm, appressed or spreading. **Spikelets** 2.5–3.5 mm long, 1–1.2 mm wide, narrowly ovoid to ellipsoid, glabrous, acute. **Lower glumes** 1–1.5 mm, about ⅓ as long as the spikelets, glabrous, 1–3-veined, acute; **upper glumes** slightly exceeding the lower lemmas, glabrous, acute, scarcely separated from the lower lemmas; **lower florets** staminate; **lower lemmas** similar to the upper glumes; **lower paleas** 2–3 mm, oblong; **upper florets** 2–2.5 mm long, 0.8–1 mm wide, ellipsoid, widest below the middle, glabrous, smooth, shiny, apices lightly beaked. $2n$ = 18, 36, 41, 42, 43, 45, 54, 63 (United States material apparently usually tetraploid, with $2n$ = 36).

Panicum coloratum is an African species that has been widely introduced into tropical and subtropical regions around the world. It is now established in the *Flora* region, growing in open, usually wet ground; it is also occasionally cultivated as a forage grass.

19. **Panicum amarum** Elliott [p. 473]
BITTER BEACHGRASS

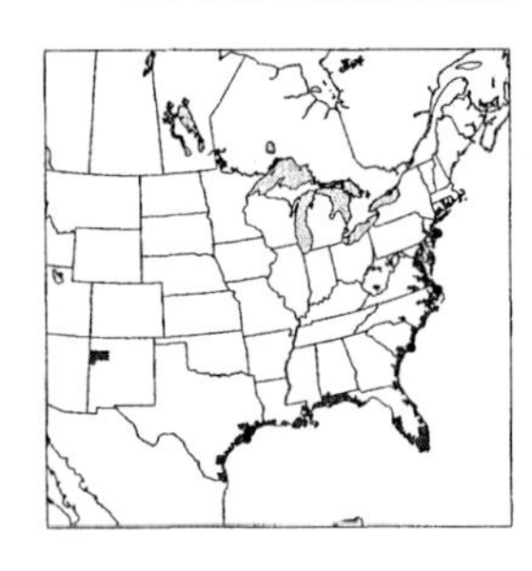

Plants perennial; rhizomatous, rhizomes stout, glabrous and glaucous throughout. **Culms** 20–250 cm tall, 3–10 mm thick, erect or decumbent, simple or branched from the lower nodes; **nodes** glabrous; **internodes** glabrous, glaucous. **Sheaths** shorter or longer than the internodes, not keeled, glabrous; **collars** often glaucous and purplish; **ligules** 1–5 mm; **blades** 7–50 cm long, 2–13 mm wide, erect or ascending, firm, thick, flat basally, more or less involute towards the apices. **Panicles** 10–80 cm long, 2–17 cm wide, contracted, slightly nodding; **primary branches** whorled or opposite, strongly ascending to appressed; **pedicels** 0.5–15 mm, appressed to slightly divergent. **Spikelets** 4–7.7 mm long, 1.5–2 mm wide, narrowly ovoid, glabrous, acuminate; **lower florets** staminate. **Glumes** and **lower lemmas** relatively thick; **lower glumes** 2.8–4 mm, ½–⅘ as long as the spikelets, 3–9-veined, apices of the midveins sometimes scabridulous; **upper glumes** and **lower lemmas** extending 1.5–3 mm beyond the upper florets, apices stiffly gaping; **upper glumes** 3.9–7.6 mm, 5–9-veined; **lower lemmas** slightly shorter than the upper glumes, 7–9-veined, **lower paleas** 3–7 mm, oblong-hastate, folded over the anthers; **lower florets** staminate; **upper florets** 2.4–3.9 mm long, 1–1.8 mm wide, narrowly ovoid to oblong, glabrous, smooth, shiny, lemma margins clasping the paleas only at the base. $2n$ = 36, 54.

Panicum amarum grows in the coastal dunes, wet sandy soils, and the margins of swamps, along the Atlantic Ocean and the Gulf of Mexico from Connecticut to northeastern Mexico. It is also known, as an introduction, from a few inland locations in New Mexico, North Carolina, and West Virginia, as well as in the Bahamas and Cuba.

1. Rhizomes short or ascending; culms often bunched and decumbent, usually more than 120 cm tall; lower glumes with 3–5 less evident veins, the midvein smooth distally; spikelet density high; panicles with 2 or more main branches per node; spikelets 4–5.9 mm long . . . . subsp. *amarulum*
1. Rhizomes horizontally elongate; culms mostly solitary, less than 150 cm tall; lower glumes with 7–9 prominent veins, the midvein scabridulous distally; spikelet density moderate; panicles with 1 or 2 main branches per node; spikelets 4.7–7.7 mm long . . . . subsp. *amarum*

Panicum amarum subsp. **amarulum** (Hitchc. & Chase) Freckmann & Lelong

Plants with short (usually) or ascending rhizomes. **Culms** usually 100–250 cm, robust, decumbent, densely bunched (occasionally with more elongated rhizomes and less bunching in the southern part of its range). **Panicles** usually more than 5 cm wide, spikelet density high; **primary branches** usually 2 or more per node, smooth to moderately scabrous, usually with quaternary branching. **Spikelets** 4–5.9 mm; **lower glumes** 3–5-veined, veins less evident than in subsp. *amarum*, midvein smooth distally.

Panicum amarum subsp. *amarulum* grows in swales behind the first dune and on sandy borders of wet areas.

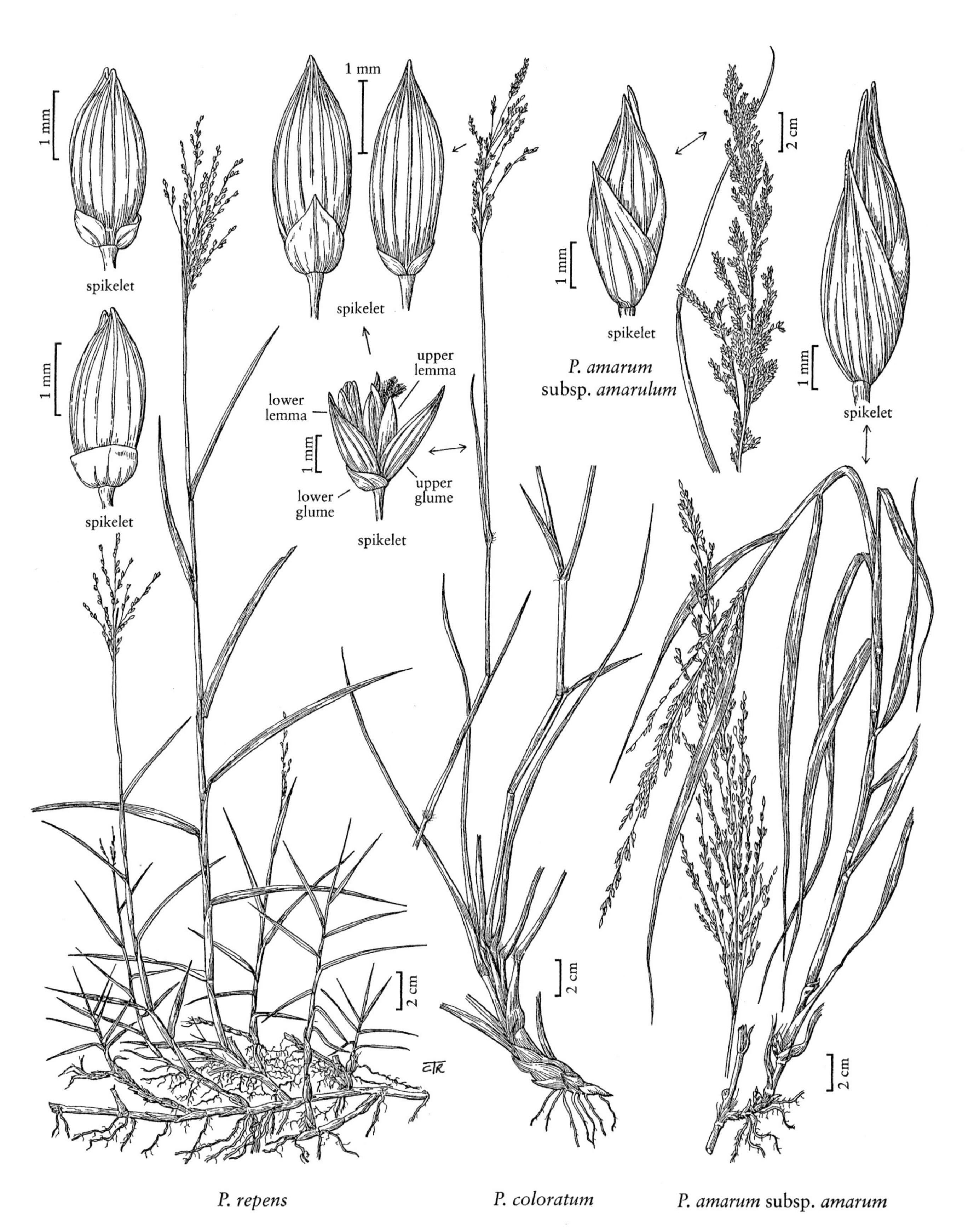

P. repens *P. coloratum* *P. amarum* subsp. *amarum*

It extends as far north as northern New Jersey and extends southward into Mexico. It has been introduced to Massachusetts, West Virginia, Cuba, and the Bahamas.

Panicum amarum subsp. *amarulum* is a fertile tetraploid, and possibly a progenitor of subsp. *amarum*, with which it intergrades in the Gulf region. Plants that intergrade with *P. virgatum* are evident in some coastal areas; they may represent hybrids.

Panicum amarum Elliott subsp. **amarum**

Plants with horizontally elongated rhizomes, internodes mostly over 1 cm, scales not overlapping. **Culms** 20–150 cm, solitary and erect or somewhat bunched and decumbent at the base, often branching from the lower nodes. **Panicles** often less than 5 cm wide, spikelet density less than in subsp. *amarulum*; **primary branches** solitary or paired, often strongly scabrous, with secondary and tertiary branches. **Spikelets** 4.7–7.7 mm. **Lower glumes** with 7–9 prominent veins, midvein scabridulous distally.

Panicum amarum subsp. *amarum* grows on foredunes, where its longer rhizomes probably permit it to respond quickly to being buried under shifting sand, and occasionally in the swales with subsp. *amarulum*. It ranges farther north (into Connecticut) than subsp. *amarulum*, but apparently not into Mexico, Cuba, or the Bahamas. It includes both tetraploids and hexaploids, and is partially sterile. Hybridization with *P. virgatum* may have had some role in its origin.

20. **Panicum virgatum** L. [p. 476]
SWITCHGRASS, PANIC RAIDE

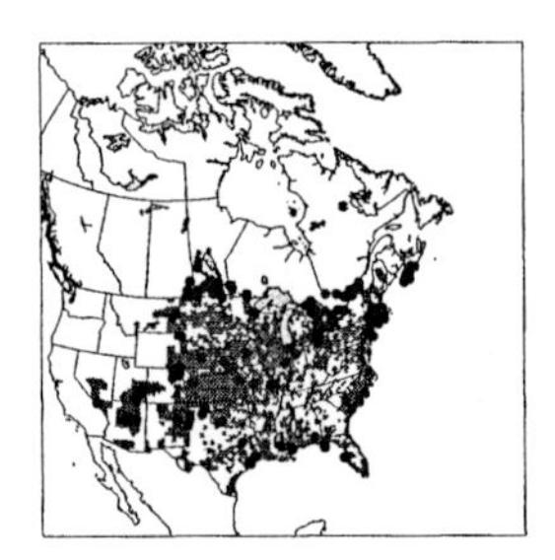

Plants perennial; rhizomatous, rhizomes often loosely interwoven, hard, with closely overlapping scales, sometimes short or forming a knotty crown. **Culms** 40–300 cm tall, 3–5 mm thick, solitary or forming dense clumps, erect or decumbent, usually simple; **nodes** glabrous; **internodes** hard, glabrous or glaucous, green or purplish. **Sheaths** longer than the lower internodes, shorter than those above, glabrous or pilose, especially on the throat, margins usually ciliate; **ligules** 2–6 mm; **blades** 10–60 cm long, 2–15 mm wide, flat, erect, ascending or spreading, glabrous or pubescent, adaxial surfaces sometimes densely pubescent, particularly basally, bases rounded to slightly narrowed, margins scabrous. **Panicles** 10–55 cm long, 4–20 cm wide, exserted, open; **primary branches** thin, straight, solitary to whorled or fascicled, ascending to spreading, scabrous, usually rebranching once; **pedicels** 0.5–20 mm, appressed to spreading. **Spikelets** 2.5–8 mm long, 1.2–2.5 mm wide, narrowly lanceoloid, turgid to slightly laterally compressed, glabrous, acuminate. **Lower glumes** 1.8–3.2 mm, ½–⅘ as long as the spikelets, glabrous, 5–9-veined, acuminate; **upper glumes** and **lower lemmas** extending 0.4–3 mm beyond the upper florets, 7–11-veined, strongly gaping at the apices; **lower florets** staminate; **lower paleas** 3–3.5 mm, ovate-hastate, lateral lobes folded over the anthers before anthesis; **upper florets** 2.3–3 mm long, 0.8–1.1 mm wide, narrowly ovoid, smooth, glabrous, shiny; **upper lemmas** clasping the paleas only at the base. $2n$ = 18, 21, 25, 30, 32, 35, 36, 54–60, 67–72, 74, 77, 90, 108.

Panicum virgatum grows in tallgrass prairies, especially mesic to wet types where it is a major component of the vegetation, and on dry slopes, sand, open oak or pine woodlands, shores, river banks, and brackish marshes. Its range extends, primarily on the eastern side of the Rocky Mountains, from southern Canada through the United States to Mexico, Cuba, Bermuda, and Costa Rica, and, possibly as an introduction, in Argentina. It has also been introduced as a forage grass to other parts of the world.

Panicum virgatum is an important and palatable forage grass, but its abundance in native grasslands decreases with grazing. Several types are planted for range and wildlife habitat improvement. Plants from eastern New Mexico, western Texas, and northern Mexico tend to have larger spikelets (6–8 mm versus 2.5–5.5 mm) and are sometimes called *P. havardii* Vasey.

Tetraploids appear to be the most common ploidy level, especially in the upper midwest and northern plains, with higher ploidy levels being more common southwards, but plants in a small area can range from diploid through duodecaploid, with dysploid derivatives. If morphological markers matched chromosome numbers and ecotypic characters, the species could be considered an aggregate of numerous microspecies. In the absence of such correlations, it must be regarded as simply a wide-ranging, highly variable taxon. Plants identified as *Panicum virgatum* var. *cubense* Griseb. and *P. virgatum* var. *spissum* Linder represent end points of geographic clines.

Panicum virgatum is not always readily separable from *P. amarum*, particularly *P. amarum* subsp. *amarulum*; future work may support their treatment as conspecific taxa.

Panicum sect. **Urvilleana** (Hitchc. & Chase) Pilg.

Plants perennial; rhizomatous, rhizomes stout, horizontal or vertical. **Culms** erect, arising in tufts or solitary. **Ligules** membranous, ciliate. **Panicles** narrow to lax; **branches** ascending to spreading; **secondary branches** and **pedicels** short, crowded, often appressed. **Spikelets** ovoid, densely to sparsely villous, hairs silvery or tawny-white, apices acute. **Lower glumes** about ¾ as long as the spikelets, (5)7–9(11)-veined; **upper glumes** and **lower lemmas** 7–15-veined; **lower florets** staminate; **lower paleas** as large as the lower lemmas; **upper lemmas** villous on the basal portion of the margins.

Panicum sect. *Urvilleana* consists of three species growing on coastal dunes and sand in South America. There is one species in the *Flora* region.

21. **Panicum urvilleanum** Kunth [p. 476]
SILKY PANICGRASS

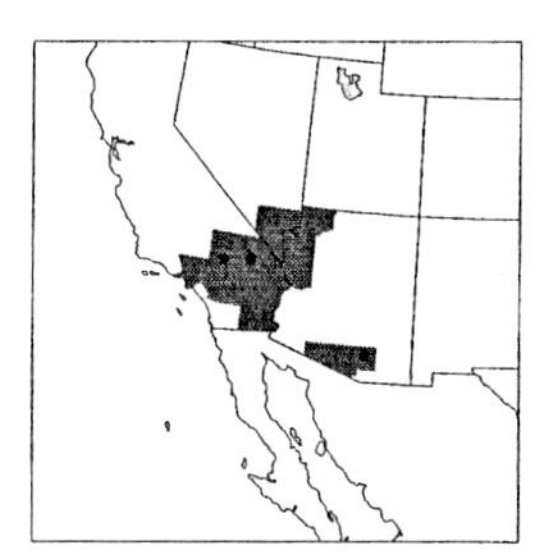

Plants perennial. **Culms** 50–100 cm, erect, solitary or in small tufts from stout, scaly, creeping to vertical rhizomes or stolons, simple or branching at the base; **nodes** densely villous. **Sheaths** densely villous; **ligules** membranous, ciliate, hairs 1.5–2 mm; **blades** 20–60 cm long, 4–10 mm wide, ascending to spreading, strigose to subglabrous, flat basally, tapering to a long, involute point. **Panicles** 20–30 cm long, 3–9 cm wide, narrow, shortly exserted; **branches** slender, ascending; **secondary branches** and **pedicels** 1–4 mm, crowded, ascending to appressed. **Spikelets** 5–7 mm, densely villous, hairs silvery or tawny-white. **Lower glumes** about ¾ the length of the spikelets, 7–11-veined; **upper glumes** and **lower lemmas** 7–15-veined; **lower florets** staminate; **lower paleas** about as long as the lower lemmas; **upper florets** striate, margins of the upper lemmas villous, hairs white; **lodicules** very large. $2n$ = 36.

Panicum urvilleanum grows on desert sand dunes and in creosote bush scrubland in the Mojave and Colorado desert regions of southern California, southern Nevada, and western Arizona. It also grows in Peru, Chile, and Argentina.

Panicum subg. **Agrostoidea** (Nash) Zuloaga

Plants perennial; usually cespitose, often from scaly rhizomes. **Culms** often compressed. **Sheaths** often keeled; **ligules** membranous, papery, often erose, usually ciliate; **blades** with vascular bundles surrounded by a single Kranz sheath with centrifugal chloroplasts and separated by 2–3 isodiametric mesophyll cells; **chloroplasts** without well-developed grana. **Photosynthesis** of the C_4 NADP-me type. **Panicles** contracted to lax, usually with many spikelets; **secondary branches** usually present; **pedicels** short. **Spikelets** ovoid to ellipsoid, glabrous. **Lower glumes** varying in length, 0–3-veined (rarely 7-veined); **upper glumes** and **lower lemmas** usually 3–5-veined (rarely 7–11-veined); **lower florets** usually sterile, occasionally staminate; **lower paleas** absent, small, or large, not thickened; **upper florets** variable. x = 9 or 10.

Panicum subg. *Agrostoidea* is found primarily in warm temperate to tropical regions of the New World, extending from the southern United States through South America. One section is native to India. Species of subg. *Agrostoidea* usually grow in open but mesic places, such as the edges of streams, rivers, ponds, and wet meadows.

Panicum sect. **Agrostoidea** (Nash) C.C. Hsu

Plants perennial; cespitose. **Culms** soft, somewhat compressed. **Ligules** membranous, papery, often erose, with or without cilia. **Panicles** open to contracted, with many, more or less appressed spikelets; **branches** narrow, nearly simple; **pedicels** 0.1–3 mm, crowded, somewhat

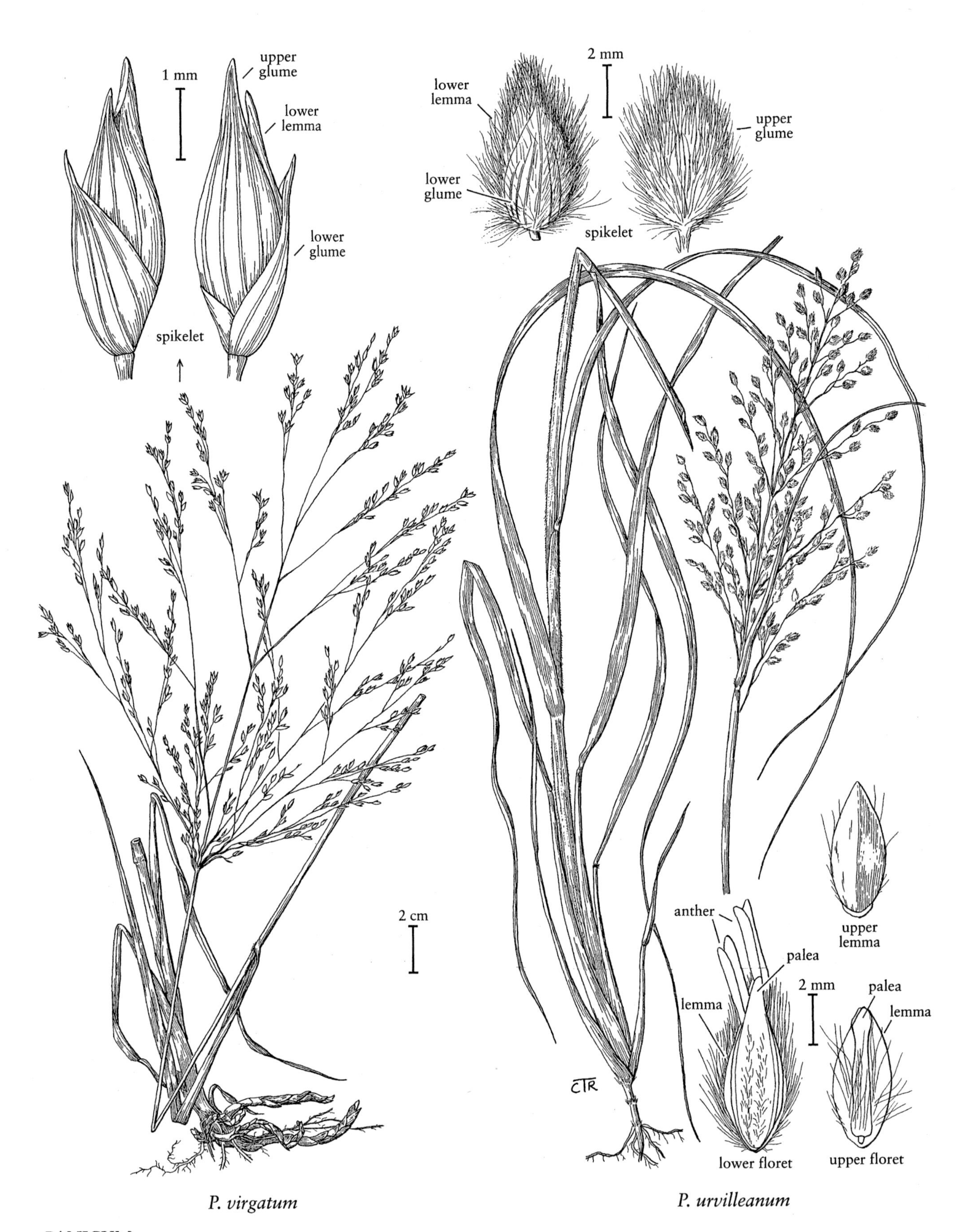

P. virgatum

P. urvilleanum

secund. **Spikelets** long-ellipsoid, acute. **Lower glumes** ⅓–⅔ as long as the spikelets, 3-veined; **upper glumes** and **lower lemmas** 5-veined, strongly keeled along the midveins; **lower florets** sterile; **lower paleas** small; **upper florets** smooth, shiny, with an apical tuft of prickly hairs. $x = 9$.

Panicum sect. *Agrostoidea* includes about four species, one of which is found in the *Flora* region.

22. **Panicum rigidulum** Bosc *ex* Nees [p. 479]
REDTOP PANICUM

Plants perennial; cespitose, not rhizomatous, occasionally purple-tinged throughout, mostly glabrous throughout (except as noted). **Culms** 35–150 cm, stout, compressed. **Sheaths** more or less strongly compressed or keeled, sides usually glabrous or sparsely pubescent distally; **ligules** 0.3–3 mm, membranous, erose or ciliate, cilia often themselves fimbriate; **blades** 8–50 cm long, 2–12 mm wide, flat or folded, both surfaces usually glabrous or scabridulous, or the adaxial surfaces sparsely pilose basally. **Panicles** terminal and axillary, 9–40 cm, ⅓–¾ as wide as long, usually dense; **ultimate branchlets** usually appressed, 1-sided, scabridulous; **pedicels** 0.5–1.5 mm, usually appressed, sometimes with 1–several slender hairs at the apices. **Spikelets** usually 1.6–3.8 mm, usually subsessile, lanceolate, green, purple-tinged, or purple, glabrous. **Lower glumes** ⅖–¾ as long as the spikelets, 3-veined, midveins keeled; **upper glumes** and **lower lemmas** subequal or the glumes slightly longer, often spreading slightly apart at the apices, midveins keeled, usually scabridulous apically; **lower florets** sterile; **lower paleas** to ⅔ as long as the lower lemmas; **upper florets** 1.4–2 mm long, 0.6–0.8 mm wide, ⅖–¾ as long as the spikelets, occasionally stipitate, lustrous, with a tuft of minute, thickish hairs at the apices; **upper lemmas** thick, stiff, clasping the upper paleas throughout their length. $2n = 18$.

Panicum rigidulum grows in swamps, wet woodlands, flood-plain forests, wet pine savannahs, marshy shores of rivers, ponds, and lakes, drainage ditches, and other similar wet to moist places; it is rarely found in dry sites. Its range extends from southern Canada to Mexico, Guatemala, and the Antilles.

1. Sheaths truncate or broadly auriculate; blade bases much narrower than the subtending sheaths . . . . subsp. *abscissum*
1. Sheaths not truncate or broadly auriculate; blade bases about as wide as the subtending sheaths.
    2. Blades usually 5–12 mm wide, flat, mostly glabrous or scabridulous; ligules membranous, 0.3–1 mm long.
        3. Spikelets 1.6–2.5 mm long, usually over 0.6 mm wide, green or purplish-tinged . . . . subsp. *rigidulum*
        3. Spikelets 2.4–3 mm long, usually less than 0.6 mm wide, conspicuouly stipitate, usually purple . . . . subsp. *elongatum*
    2. Blades usually 2–7 mm wide, often folded or involute, usually pilose adaxially, at least near the base; ligules membranous, the cilia usually fimbriate, 0.5–3 mm long.
        4. Spikelets 2–2.7 mm long, green or purplish-stained, often obliquely set on the pedicels . . . . subsp. *pubescens*
        4. Spikelets 2.6–3.8 mm long, usually purple, slender, erect on the pedicels . . . . subsp. *combsii*

Panicum rigidulum subsp. **abscissum** (Swallen) Freckmann & Lelong

Plants yellowish-green. **Culms** compressed, densely cespitose. **Sheaths** strongly keeled, 3–6 mm wide from the keel to the margins, truncate or broadly auriculate, occasionally ciliate distally; **ligules** minute, membranous; **blades** 0.7–1.5 mm wide from the blunt keel to the margins, to 2.5 mm wide overall, thick, curved or flexuous, rigid, glabrous or scabridulous on both surfaces and along the margins, bases much narrower than the subtending sheaths, margins often sparsely ciliate, with long, slender hairs near the base. **Panicles** slender, purplish, long-exserted; **branches** few, appressed or ascending, with relatively few spikelets. **Spikelets** 2.4–2.8(3) mm, purplish, glabrous, obliquely set on the pedicels, mostly abortive. **Lower glumes** 1.6–2 mm, often divergent; **upper glumes** and **lower lemmas** subequal, often spreading apart.

Panicum rigidulum subsp. *abscissum* is endemic to central Florida. It usually grows in marshy, sandy ground, but is occasionally found in dry, sandy sites (e.g., the type specimen, collected near Lake Sebring).

Panicum rigidulum subsp. *abscissum* is very similar vegetatively and reproductively to subsp. *pubescens* and subsp. *combsii*. In addition, its spikelets also suggest those of *P. virgatum*.

Panicum rigidulum subsp. **combsii** (Scribn. & C.R. Ball) Freckmann & Lelong

Plants similar to subsp. *pubescens*, but often shortly rhizomatous, more nearly glabrous, and more deeply and consistently purple throughout. **Ligules** nearly obsolete; **blades** usually 2–7 mm wide, often folded or involute, usually pilose adaxially, at least near the base, bases about equal in width to the subtending sheaths. **Spikelets** 2.6–3.8 mm, usually purple, slender, erect on the pedicels. **Lower glumes** to ¾ as long as the spikelets.

Panicum rigidulum subsp. *combsii* is restricted to the United States, where it grows in the same moist habitats as subsp. *pubescens*, but is much less common. Its long, narrow, purple, often gaping spikelets somewhat resemble those of *P. virgatum*, which often grows in the same habitats.

Panicum rigidulum subsp. **elongatum** (Scribn.) Freckmann & Lelong

Plants similar to subsp. *rigidulum*, but more conspicuously purple-tinged throughout, especially the panicles. **Ligules** 0.3–1 mm, membranous; **blades** usually 5–12 mm wide, flat, mostly glabrous or scabridulous, bases about equal in width to the subtending sheaths. **Panicles** relatively narrow; **branches** few, stiffly ascending. **Spikelets** 2.4–3 mm long, usually less than 0.6 mm wide, conspicuously stipitate, purple, often falcate, subsecund along the branchlets. **Upper florets** stipitate, stipes to 0.4 mm, slender.

Panicum rigidulum subsp. *elongatum* is most common in the piedmont and mountain regions of the eastern United States.

Panicum rigidulum subsp. **pubescens** (Vasey) Freckmann & Lelong

Culms less than 100 cm, usually slender, greatly compressed. **Leaves** mostly basal; **sheaths** narrower than those of subspp. *elongatum* and *rigudulum*, compressed, often pubescent, at least along the sides near the summit or at the throat; **ligules** 0.5–3 mm, membranous, usually fimbriate-ciliate; **blades** usually 2–7 mm wide, often folded or involute, usually pilose adaxially, at least near the base, bases about equal in width to the subtending sheaths. **Panicles** mostly terminal. **Spikelets** 2–2.7 mm, green or slightly purplish, set obliquely on the pedicels. **Lower glumes** about ½ as long as the spikelets; **upper glumes** slightly longer than the lower lemmas, acute; **lower lemmas** acute, the apices diverging slightly from those of the upper glumes.

Panicum rigidulum subsp. *pubescens* grows primarily in moist pine savannahs, bogs, and other similar open, sandy habitats on the Atlantic and Gulf coastal plains of the United States.

Panicum rigidulum Bosc *ex* Nees subsp. **rigidulum**

Culms 40–150 cm, rather robust. **Sheaths** with the throat usually glabrous or sparsely pilose on the sides; **ligules** 0.3–1 mm, membranous; **blades** usually 5–12 mm wide, flat, mostly glabrous or scabridulous, bases about equal in width to the subtending sheaths. **Panicles** terminal and axillary, usually with many spikelets. **Spikelets** 1.6–2.5 mm long, usually over 0.6 mm wide, crowded, green or purplish-tinged, pedicellate, pedicels short, with 1–several slender hairs near the apices.

Panicum rigidulum subsp. *rigidulum* is the most common, most variable, and widest ranging of the five subspecies.

23. **Panicum anceps** Michx. [p. 479]
BEAKED PANICGRASS

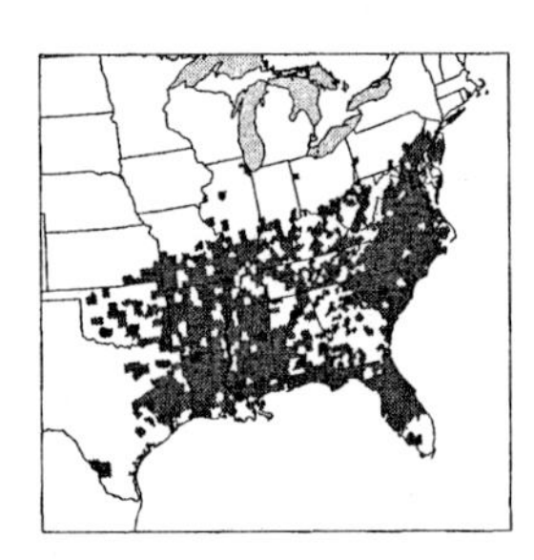

Plants perennial; conspicuously rhizomatous, rhizomes short or elongate, stout, scaly. **Culms** 30–130 cm, terete to slightly compressed. **Sheaths** laterally compressed, glabrous or sparsely to densely pilose or villous, especially at the summit; **ligules** less than 0.5 mm, membranous, erose, often brownish; **blades** 10–50 cm long, 4–12 mm wide, erect, adaxial surfaces pilose at least basally, glabrous or pilose abaxially. **Panicles** 10–40 cm, ¼–⅔ as wide as long, well-exserted at anthesis; **branches** relatively few, stiffly spreading or ascending; **ultimate branchlets** 1-sided; **pedicels** 0.1–3 mm, scabridulous to scabrous, appressed. **Spikelets** 2.3–3.9 mm, narrowly ellipsoid to ovoid, usually subsessile, usually pale to yellowish-green, glabrous, often falcate and gaping at the apices, rarely lanceolate, densely crowded on short, appressed branchlets, set obliquely on short pedicels. **Lower glumes** ⅓–½ as long as the spikelets, 3-veined, keels scabrous, apices acute; **upper glumes** and **lower lemmas** subequal, keeled, beaked, usually gaping at the apices; **lower florets** sterile; **lower paleas** subequal to the lower lemmas; **upper florets** 1.5–2.2 mm long, about 1 mm wide, ⅔–¾ as long as the spikelets, apices with a tuft of minute, thick hairs; **upper lemmas** thick, stiff, clasping the upper paleas throughout their length. $2n$ = 18, 36.

Panicum anceps grows in low, moist, primarily sandy areas, pine savannahs, the borders of flood-plain swamps, mesic woodlands, roadsides, and upland pine-hardwood forests. It is restricted to the United States.

1. Spikelets 2.7–3.9 mm long, often clearly falcate; rhizomes relatively short and stout . . . . subsp. *anceps*
1. Spikelets 2.3–2.8 mm long, not clearly falcate; rhizomes relatively long and slender . . . subsp. *rhizomatum*

Panicum anceps Michx. subsp. **anceps**

Plants rhizomatous, rhizomes relatively short and stout. **Panicles** primarily terminal; **primary branches** usually few, stiffly ascending, with short branchlets and crowded, subsecund spikelets on short pedicels. **Spikelets** 2.7–3.9 mm, often lanceolate, falcate, with prominent green veins, acuminate.

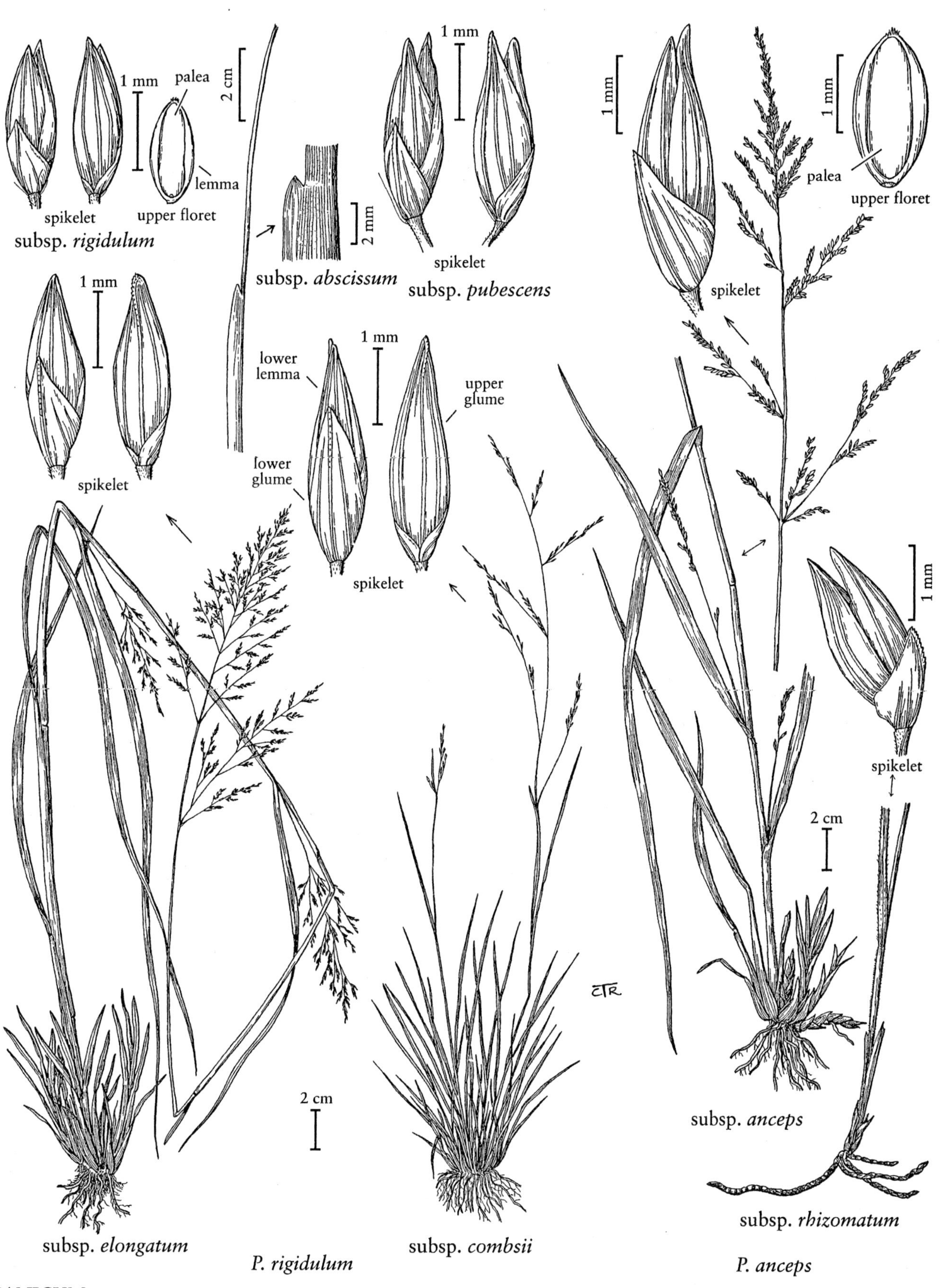

1 mm
palea
lemma
spikelet
upper floret
subsp. rigidulum
2 cm
2 mm
subsp. abscissum
1 mm
spikelet
subsp. pubescens
1 mm
1 mm
palea
upper floret
spikelet
1 mm
spikelet
1 mm
lower lemma
upper glume
lower glume
spikelet
1 mm
spikelet
2 cm
2 cm
subsp. anceps
subsp. rhizomatum
subsp. elongatum
P. rigidulum
subsp. combsii
P. anceps

Panicum anceps subsp. *anceps* is widespread in all physiographic provinces within its range.

Panicum anceps subsp. **rhizomatum** (Hitchc. & Chase) Freckmann & Lelong

Plants rhizomatous, rhizomes relatively long and slender. **Panicles** terminal and axillary; **branches** more numerous, appressed, congested, with dense clusters of spikelets. **Spikelets** 2.3–2.8 mm, not clearly falcate, often ovoid-lanceolate, purplish-stained, acute.

Panicum anceps subsp. *rhizomatum* grows in the Atlantic and Gulf coastal plains.

The small, crowded, often purplish spikelets of this subspecies often closely resemble those of *Panicum rigidulum*.

Panicum sect. Tenera (Hitchc. & Chase) Pilg.

Plants perennial; cespitose. **Culms** wiry, somewhat compressed, erect. **Ligules** less than 0.4 mm, membranous. **Panicles** narrow, with a few spikelets; **branches** appressed; **pedicels** short, with a few long hairs at the apices. **Lower glumes** 1- or 3-veined; **lower florets** sterile; **lower paleas** small; **upper florets** smooth, shiny. $x = 10$.

24. **Panicum tenerum** Beyr. *ex* Trin. [p. 483]
BLUE-JOINT PANICGRASS

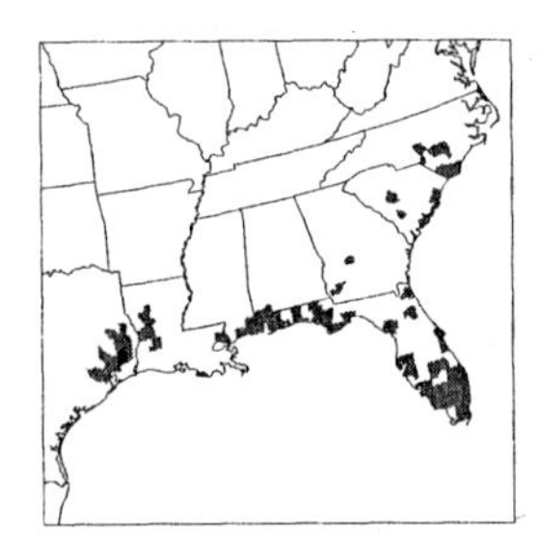

Plants perennial; cespitose, with short, knotted rhizomes. **Culms** 40–100 cm, erect, simple or branching from the lower nodes; **nodes** glabrous; **internodes** glabrous. **Sheaths** shorter than the internodes, usually glabrous, lower sheaths sometimes pilose at the summit, hairs papillose-based; **ligules** 0.1–0.4 mm; **blades** 4–19 cm long, 1.5–4 mm wide, mostly involute at maturity, erect, firm, abaxial surfaces usually glabrous, adaxial surfaces often sparsely pilose, particularly basally. **Panicles** 3–12 cm long, less than 1 cm wide, contracted, with few spikelets; **branches** 1–4 cm, few, ascending-appressed; **ultimate branchlets** 1-sided; **pedicels** 0.5–3 mm, scabridulous, appressed, usually with a few slender hairs at the apices. **Spikelets** 1.8–2.8 mm long, 0.8–1 mm wide, usually subsessile, lanceoloid to narrowly ovoid, green, often purplish-stained, glabrous, acute. **Lower glumes** 0.9–3 mm, ½–⅔ as long as the spikelets, 1–3-veined, not keeled over the midveins, acute or obtuse; **upper glumes** and **lower lemmas** subequal, 5–7-veined, midveins not keeled, acute to short-acuminate, occasionally gaping at the apices; **lower florets** sterile; **lower paleas** about ½–⅔ as long as the lower lemmas; **upper florets** 1.1–1.8 mm long, 0.6–0.8 mm wide, ⅔–¾ as long as the spikelets, lustrous, usually brownish, apices glabrous; **upper lemmas** thick, stiff, clasping the upper paleas throughout their length. $2n = 20$.

Panicum tenerum grows in wet or moist, sandy (often peaty) soil, depressions in pine savannahs, bogs, marshes, pond margins, and interdunal swales. Its range includes the Atlantic and Gulf coastal plains of the United States, the Antilles, Bahamas, and Central America. *Panicum tenerum* exhibits numerous features of the widespread and polymorphic *P. rigidulum*, particularly *P. rigidulum* subsp. *pubescens*.

Panicum sect. Obtusa (Hitchc.) Pilg.

Plants perennial; rhizomatous or stoloniferous. **Culms** wiry; **nodes** swollen, villous (especially on the stolons). **Ligules** membranous, papery. **Panicles** contracted; **branches** spikelike, appressed, 1-sided; **pedicels** 0.1–2.5 mm. **Spikelets** ellipsoid, paired. **Lower glumes** ¾–⅘ as long as the spikelets, 5–7-veined; **upper glumes** and **lower lemmas** 7–11-veined, blunt; **lower florets** staminate; **lower paleas** large; **upper florets** large, with tiny hairs near the bases and apices. $x = 10$.

Panicum sect. *Obtusa* includes only *Panicum obtusum*.

25. **Panicum obtusum** Kunth [p. 483]
VINE MESQUITE

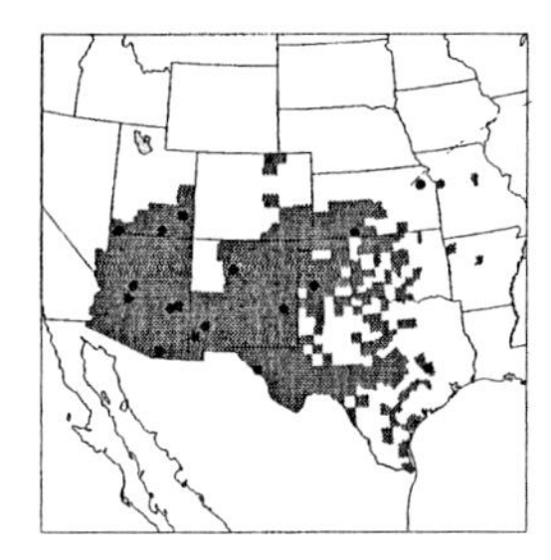

Plants perennial; usually from long slender stolons or shallow rhizomes with swollen, villous nodes. **Culms** 20–80 cm, often in small clumps, compressed, erect or decumbent, glaucous; **lower nodes** pubescent; **upper nodes** glabrous. **Lower sheaths** ascending, pubescent to pilose; **upper sheaths** glabrous; **ligules** 0.2–2 mm, membranous, truncate, irregularly denticulate; **blades** 3–26 cm long, 2–7 mm wide, ascending, firm, glaucous, sparsely pilose near the base, often scabrous on the margins, involute towards the apices. **Panicles** 5–15 cm long, 0.8–1.5 cm wide; **branches** 2–6, spikelike, erect, puberulent, 3-angled; **ultimate branchlets** 1-sided; **pedicels** paired, congested, shorter pedicels 0.1–1 mm, longer pedicels 1.5–2.5 mm. **Spikelets** 2.8–4.4 mm, ellipsoid, terete to slightly laterally compressed, glabrous, obtuse. **Lower glumes** about ¾ as long as the spikelets, 5- or 7-veined; **upper glumes** and **lower lemmas** equaling the spikelets, 5–9-veined; **lower florets** staminate; **lower paleas** 2.5–3.5 mm; **upper florets** puberulent at the bases and apices. $2n$ = 20, 36, 40.

Panicum obtusum grows in seasonally wet sand or gravel, especially on stream banks, ditches, roadsides, wet pastures, and rangeland. Its range extends from the southwestern United States to central Mexico. Flowering is from May through October.

Panicum sect. Bulbosa Zuloaga

Plants perennial; cespitose, rhizomatous. **Culms** slightly compressed, hard, sometimes with a cormlike base. **Sheaths** slightly compressed; **ligules** membranous, erose, ciliate. **Panicles** open, lax, with many spikelets; **pedicels** short, scabrous. **Spikelets** lanceoloid, glabrous. **Lower glumes** up to ⅘ as long as the spikelets, 3–5-veined; **upper glumes** and **lower lemmas** 5–7-veined; lower **florets** sterile or staminate; **lower paleas** large, lanceolate; **upper florets** narrow, transversely rugose, apices pilose. x = 9.

Panicum sect. *Bulbosa* includes approximately three species; two grow in the *Flora* region.

26. **Panicum bulbosum** Kunth [p. 483]
BULB PANICGRASS

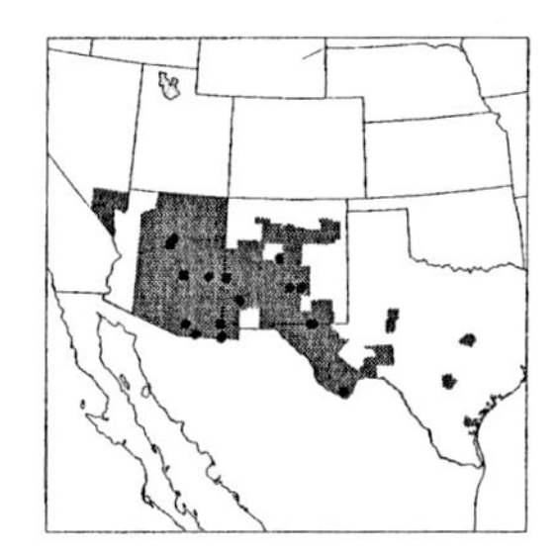

Plants perennial; cespitose, rhizomatous, rhizomes short, thin. **Culms** 50–200 cm tall, 2–3(5) mm thick, with cormlike bases, slightly compressed, erect or geniculate at the lower nodes; **nodes** glabrous or pilose; **internodes** slightly compressed, glabrous. **Sheaths** longer or shorter than the internodes, keeled, glabrous or pilose, hairs papillose-based near the throat; **ligules** 0.5–2 mm, membranous, dissected ciliate; **blades** (6)20–65 cm long, 2–15 mm wide, flat, adaxial surfaces glabrous or densely pubescent, particularly basally, occasionally pubescent on both surfaces, hairs papillose-based, bases subcordate to rounded. **Panicles** 9–50 cm long, 1.5–12 cm wide, open; **branches** opposite and alternate, straight or flexible, strongly ascending to reflexed; **pedicels** 0.5–5 mm, scabridulous, divergent. **Spikelets** 2.8–4.2(5.4) mm long, 1–2 mm wide, ellipsoid or lanceoloid, often purplish, glabrous, acute or obtuse. **Lower glumes** 1.2–3.5 mm, ½–⅘ as long as the spikelets, 3–5-veined; **upper glumes** often longer than the lower lemmas, glabrous, 5–7-veined; **lower florets** sterile or staminate; **lower lemmas** glabrous; **lower paleas** 3–4 mm, sometimes longer than the lower lemmas; **upper florets** 3–4 mm long, 1–1.5 mm wide, equaling or surpassing the lower lemmas, dull, pale, finely transversely rugose, lemma apices puberulent. $2n$ = 36, 54, 70, 72.

Panicum bulbosum grows on gravelly river banks and moist mountain slopes, often in ponderosa pine woodlands, from southern Nevada and Arizona to western Texas and central Mexico. It is an important forage grass and is sometimes cut for hay. Flowering is from July to mid-October. Small plants have been called *P. bulbosum* var. *sciaphilum* (Rupr. *ex* E. Fourn.) Hitchc. & Chase or *P. bulbosum* var. *minor* Vasey, but size and other characters integrade completely.

27. **Panicum plenum** Hitchc. & Chase [p. 483]
CANYON PANICGRASS

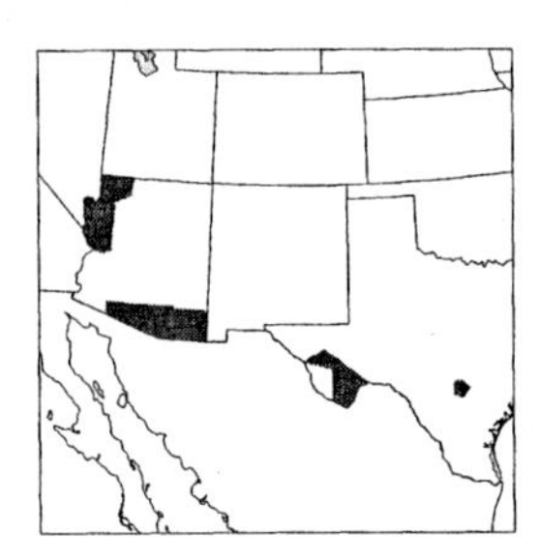

Plants perennial; cespitose, rhizomatous, rhizomes long. **Culms** 50–200 cm, strongly compressed, decumbent-erect, glaucous; **nodes** glabrous. **Sheaths** keeled, glabrous or pubescent near the throat, upper sheaths much shorter than the internodes; **ligules** 0.5–2 mm, membranous, dissected ciliate; **blades** 20–35 cm long, 7–17 mm wide, flat, glabrous on both surfaces or the adaxial surfaces sparsely pilose. **Panicles** 12–50 cm, about ⅔ as wide, open; **primary branches** spreading, often verticillate at the lower nodes; **pedicels** 0.2–4 mm, scabrous, spreading. **Spikelets** 2.5–3.4 mm long, about 1.2 mm wide, ellipsoid, glabrous. **Lower glumes** usually shorter than 1.7 mm, up to ½ as long as the spikelets, 3-veined, subacute; **upper glumes** and **lower lemmas** subequal, scarcely longer than the upper florets, 5-veined; **lower florets** sterile or staminate; **upper florets** 2.9–3 mm long, about 1 mm wide, ellipsoid, glabrous, dull, pale, obscurely transversely rugose, apices minutely pubescent. $2n$ = unknown.

Panicum plenum grows in moist places in canyons, along streams, and on mountain slopes, from Arizona and Texas to central Mexico. It appears to be closely related to *P. bulbosum*. Flowering is from July into October.

Panicum sect. Antidotalia Freckmann & Lelong

Plants perennial; rhizomatous. **Culms** robust, lignified, branching at the middle nodes. **Ligules** membranous, ciliate, cilia longer than the membranous base. **Panicles** pyramidal, more or less lax, with many spikelets. **Spikelets** ellipsoid to ovoid. **Lower glumes** to ½ as long as the spikelets, 1–5-veined; **upper glumes** and **lower lemmas** 7–11-veined; **lower florets** staminate; **upper florets** smooth, shiny. $x = 9$.

In *Panicum* sect. *Antidotalia*, the culms become almost woody at maturity; even a hammer blow fails to flatten them.

28. **Panicum antidotale** Retz. [p. 486]
BLUE PANICGRASS

Plants perennial; cespitose, rhizomatous, rhizomes about 1 cm thick, knotted, pubescent, with large, scalelike leaves. **Culms** 50–300 cm tall, 2–4 mm thick, often compressed, erect or ascending, hard, becoming almost woody; **nodes** swollen, glabrous or pubescent; **internodes** glabrous, glaucous. **Sheaths** not keeled, shorter than or equal to the internodes, glabrous or the lower sheaths at least partially pubescent, hairs papillose-based; **ligules** 0.3–1.5 mm; **blades** 10–60 cm long, 3–20 mm wide, elongate, flat, abaxial surfaces and margins scabrous, adaxial surfaces occasionally pubescent near the base, with prominent, white midveins, bases rounded to narrowed. **Panicles** 10–45 cm, to ½ as wide as long, open or somewhat contracted, with many spikelets; **branches** 4–12 cm, opposite or alternate, ascending to spreading; **pedicels** 0.3–2.5 mm, scabridulous to scabrous, appressed to diverging less than 45° from the branch axes. **Spikelets** 2.4–3.4 mm long, 1–1.3 mm wide, ellipsoid-lanceoloid to narrowly ovoid, often purplish, glabrous, acute. **Lower glumes** 1.4–2.2 mm, ⅓–½ as long as the spikelets, 3–5-veined, obtuse; **upper glumes** and **lower lemmas** subequal, glabrous, 5–9-veined, margins scarious, acute; **lower florets** staminate; **upper florets** 1.8–2.8 mm long, 0.9–1.1 mm wide, smooth, lustrous, acute. $2n = 18, 36$.

Panicum antidotale is native to India. It is grown in the *Flora* region as a forage grass, primarily in the southwestern United States. It is now established in the region, being found in open, disturbed areas and fields.

Panicum subg. Phanopyrum (Raf.) Pilg.

Plants annual or perennial, if perennial, growth habit various. **Culms** erect or decumbent, sometimes succulent or robust. **Ligules** membranous, membranous and ciliate, or lacerate; **blades** with the vascular bundles surrounded by 2 thick-walled sheaths, outer sheaths parenchymatous, chloroplasts few and unspecialized or absent. **Photosynthesis** C_3. **Spikelets** ovoid, ellipsoid, or

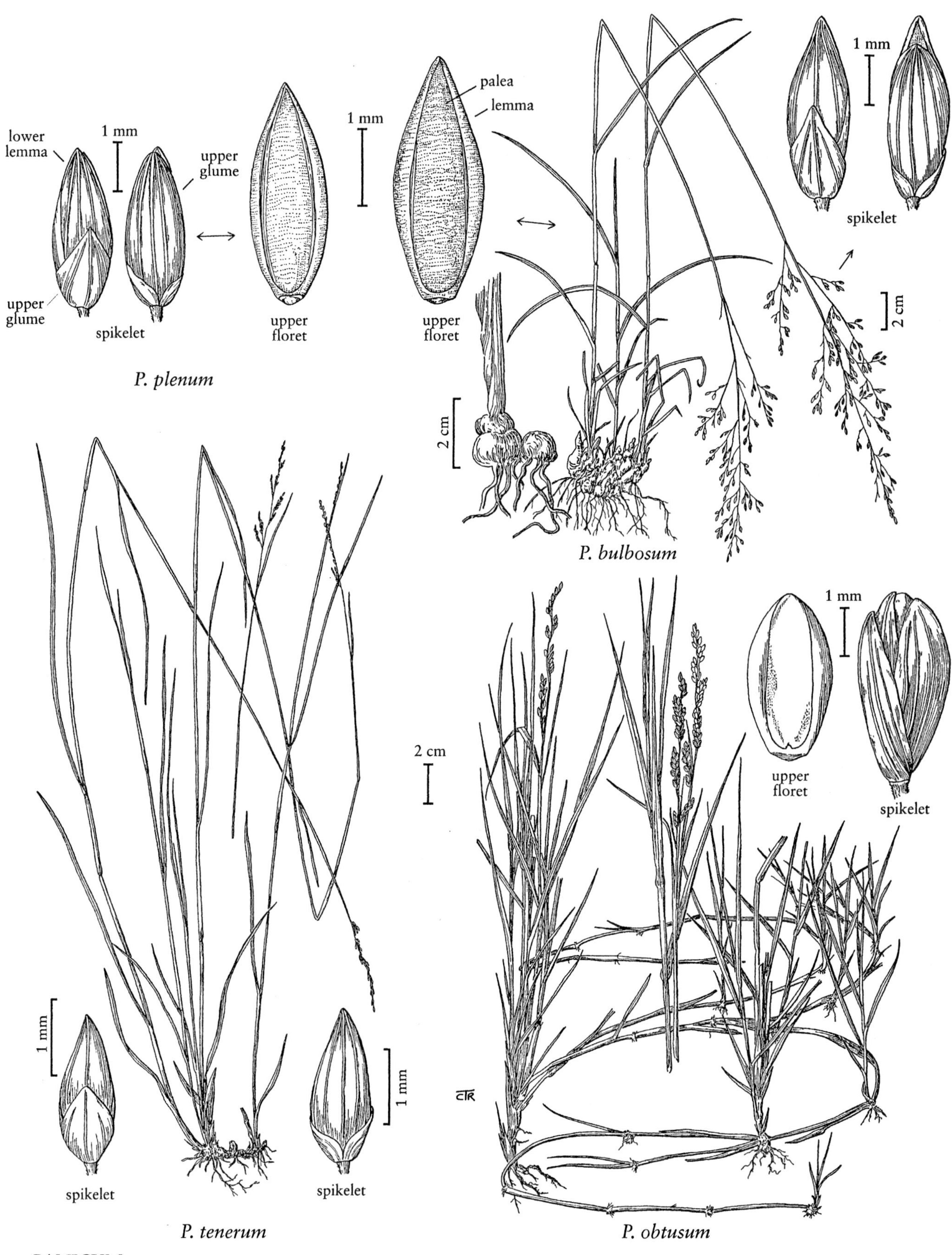
lower lemma
1 mm
upper glume
upper glume
spikelet
upper floret
1 mm
palea
lemma
upper floret
P. plenum
1 mm
spikelet
2 cm
2 cm
P. bulbosum
1 mm
spikelet
1 mm
spikelet
P. tenerum
2 cm
1 mm
upper floret
spikelet
P. obtusum

PANICUM

obovoid, glabrous, pilose, or tuberculate. **Lower glumes** 1/5–4/5 as long as the spikelets, usually 3-veined; **upper glumes** and **lower lemmas** usually 5-veined (rarely 7- or 9-veined); **lower florets** sterile or staminate; **lower paleas** present or absent; **upper florets** sometimes stipitate, indurate to membranous, smooth and shiny, or rugose, or finely pubescent, or with paired glands. $x = 9$ or 10, with some polyploids.

Panicum subg. *Phanopyrum* is most abundant in wet forests of warm temperate and tropical regions. In the Western Hemisphere, its range extends from the southern United States through South America.

Panicum sect. Hemitoma (Hitchc.) Freckmann & Lelong

Plants perennial; aquatic or semi-aquatic, rhizomatous, rhizomes extensive. **Culms** erect or sprawling, often sterile. **Ligules** membranous and ciliate, or lacerate. **Panicles** narrow; **branches** few, spikelike, erect; **ultimate branchlets** 1-sided; **pedicels** less than 2 mm long, appressed. **Spikelets** lanceoloid, laterally compressed, glabrous; **lower florets** staminate; **upper lemmas** thin, flexible, whitish, clasping the paleas only near the base. $x = 9$.

Panicum sect. *Hemitoma* includes only *P. hemitomon*.

29. Panicum hemitomon Schult. [p. 486]

MAIDENCANE

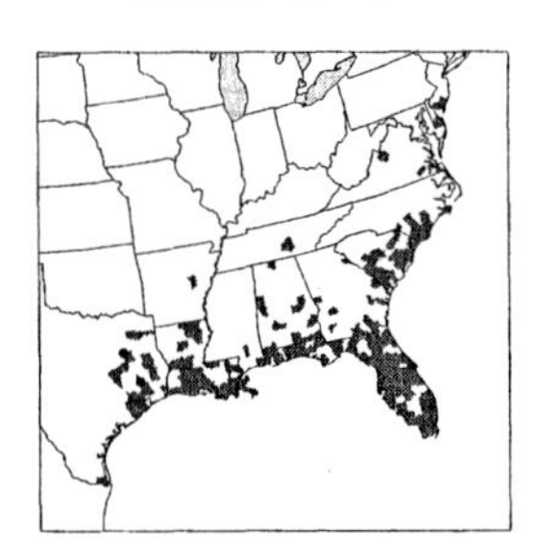

Plants perennial; robust, aquatic or semi-aquatic, forming extensive colonies through spreading rhizomes. **Culms** 50–200 cm, mostly erect and sterile, glabrous, often rooting from the lower nodes if submerged. **Sheaths** usually glabrous, or pilose or hirsute at the lowermost sheath, especially distally; **ligules** shorter than 1 mm; **blades** 8–35 cm long, 5–15 mm wide, ascending or spreading, abaxial surfaces glabrous, adaxial surfaces usually scabridulous or pubescent, bases slightly narrowed, margins scabrous, apices long-tapering. **Panicles** 10–30 cm long, less than 1 cm wide; **branches** mostly short, appressed-ascending, with fascicles of congested spikelets; **ultimate branchlets** 1-sided; **pedicels** 0.2–1.8 mm. **Spikelets** 2–2.8 mm, subsessile, lanceoloid, slightly laterally compressed, glabrous, acute. **Lower glumes** about 1/2 as long as the spikelets, slightly keeled along the midveins, 3-veined, acute; **upper glumes** and **lower lemmas** similar, glumes slightly shorter than the lemmas, faintly keeled on the back, acute; **lower florets** staminate; **lower paleas** subequal to the lower lemmas; **upper florets** 2–2.5 mm, 2/3 to almost as long as the spikelets, narrowly ellipsoid; **upper lemmas** relatively thin, flexible, pale, acuminate, clasping the paleas only at the base. $2n = 36, 40$.

Panicum hemitomon forms extensive, nearly pure stands in water or wet soils such as marshes, swamps, and along the shores of streams, canals, ditches, lakes, and ponds. It is restricted to the United States.

Panicum sect. Phanopyrum Raf.

Plants perennial; stoloniferous. **Culms** thick, decumbent, succulent, rooting profusely at the lower nodes. **Ligules** membranous. **Panicles** open; **branches** spikelike, stiffly ascending, with secund clusters of shortly pedicellate to sessile spikelets. **Spikelets** 6–7 mm, lanceoloid, laterally compressed, glabrous. **Lower glumes** 4/5 as long as the spikelets, 3-veined; **upper glumes** and **lower lemmas** 3-veined; **lower paleas** small, lanceolate; **upper florets** about 1/3 as long as the spikelets, obovoid, stipitate.

Panicum sect. *Phanopyrum* includes only *P. gymnocarpon*.

30. **Panicum gymnocarpon** Elliott [p. 486]

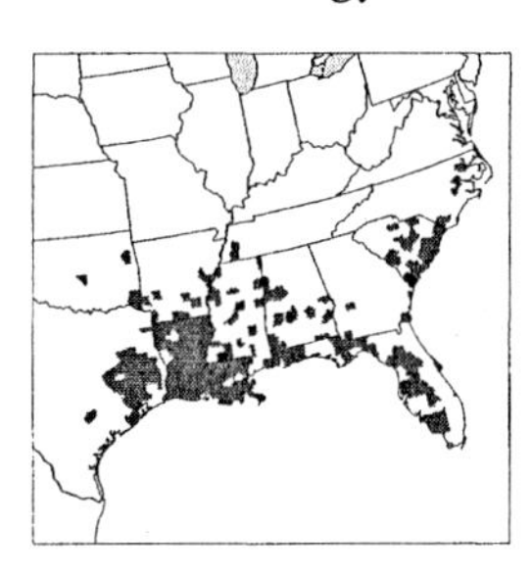

Plants perennial; forming extensive colonies by their long, decumbent, sprawling basal branches and stolons. **Culms** 60–130 cm, thick, glabrous, rooting profusely at the lower nodes; **nodes** glabrous, often with a dark green band. **Sheaths** usually shorter than the internodes, glabrous, prominently veined; **ligules** 0.5–1.5 mm; **blades** 15–40 cm long, 7–25 mm wide, tapering from midlength, flat, both surfaces glabrous, bases subcordate, margins scabrous to smooth, widest at the base, apices acute. **Panicles** 10–40 cm long, 7–20 cm wide, open, with straight, rigid rachises; **branches** whorled, stiffly ascending, with short, appressed, higher order branches; **ultimate branchlets** 1-sided, with solitary spikelets or small clusters of spikelets; **pedicels** 0.1–1.5 mm. **Spikelets** 5.5–7 mm long, about 1 mm wide, narrowly lanceoloid, glabrous. **Glumes** spreading apart at maturity, keeled, prominently veined, scabrous along the midveins; **lower glumes** nearly as long as the lower lemmas; **upper glumes** and **lower lemmas** 3-veined, spreading, greatly exceeding the upper florets, lower lemmas longer than the upper glumes, arcuate; **lower florets** sterile; **lower paleas** thin; **upper florets** 1.9–2.2 mm, less than ⅓ as long as the spikelets, obovoid, lustrous, pale to brownish, acute, often short-stipitate. $2n = 40$.

Panicum gymnocarpon grows in swamps, wet woodlands, and the marshy shores of lakes and streams. It is also found occasionally in shallow water, often in the shade. It is restricted to the United States.

Panicum sect. Monticola Stapf

Plants annual or perennial. **Culms** usually weak, decumbent. **Ligules** membranous and ciliate, or lacerate. **Panicles** usually diffuse, sometimes contracted. **Spikelets** ellipsoid to obovoid, glabrous or pilose, veins evident. **Lower glumes** ⅓–½ as long as the spikelets, 0–3-veined; **upper glumes** and **lower lemmas** 3–5-veined; **lower florets** sterile; **lower paleas** small or absent; **upper florets** rugose, with bicellular microhairs. $x = 9$.

Three species of *Panicum* sect. *Monticola* grow in the Western Hemisphere. Two species grow in the *Flora* region, one of which is native to Asia.

31. **Panicum trichoides** Sw. [p. 489]
SMALL-FLOWERED PANICGRASS

Plants annual. **Culms** 15–100 cm tall, 0.5–1(2) mm thick, sprawling to erect, without cormlike bases, freely branching and rooting from the lower nodes; **nodes** prominent, glabrous or pubescent; **internodes** not succulent, pilose. **Sheaths** shorter than the internodes, rounded, hairs papillose-based; **collars** pilose; **ligules** 0.2–0.5 mm; **blades** 2–7 cm long, 5–20 mm wide, 4–6 times longer than wide, lanceolate, thin, flat, sparsely to densely pilose, hairs papillose-based, bases asymmetrically cordate to subcordate, lower margins ciliate, papillose. **Panicles** 4–24 cm, almost as wide as long, diffuse, partially included or exerted; **primary branches** to 10 cm, alternate, ascending to reflexed, branching in the distal ⅔; **pedicels** 9–20 mm, threadlike. **Spikelets** 1–1.4 mm long, 0.5–0.6 mm wide, not secund, lanceoloid to narrowly ovoid, plano-convex in side view, sparsely pubescent. **Lower glumes** 0.4–0.8 mm, ⅓–½ as long as the spikelets, 1–3-veined, subacute; **upper glumes** 0.8–1.2 mm, arising 0.2 mm above the lower glumes, 3–5-veined; **lower florets** sterile; **lower lemmas** 0.1–0.2 mm longer than the upper glumes, 3–5-veined; **lower paleas** 0.5–0.8 mm, hyaline; **upper florets** 0.8–1.2 mm long, 0.4–0.6 mm wide, finely rugose, lemmas strongly convex. $2n = 18$.

Panicum trichoides grows in moist, often weedy fields, woodlands, and savannahs of Mexico, Central and tropical America, and the Caribbean. It has been found, as a weed, in Brownsville and Austin, Texas, and is probably introduced to the *Flora* region. It has also been introduced into Africa, tropical Asia, and the Pacific islands. In the *Flora* region, it flowers from August through October.

32. **Panicum bisulcatum** Thunb. [p. 489]

Plants annual; loosely tufted, sprawling. **Culms** 30–150 cm tall, 2–4 mm thick, erect or spreading from a geniculate, non-cormous base, not succulent, glabrous throughout. **Sheaths** shorter or longer than the internodes, rounded, often with

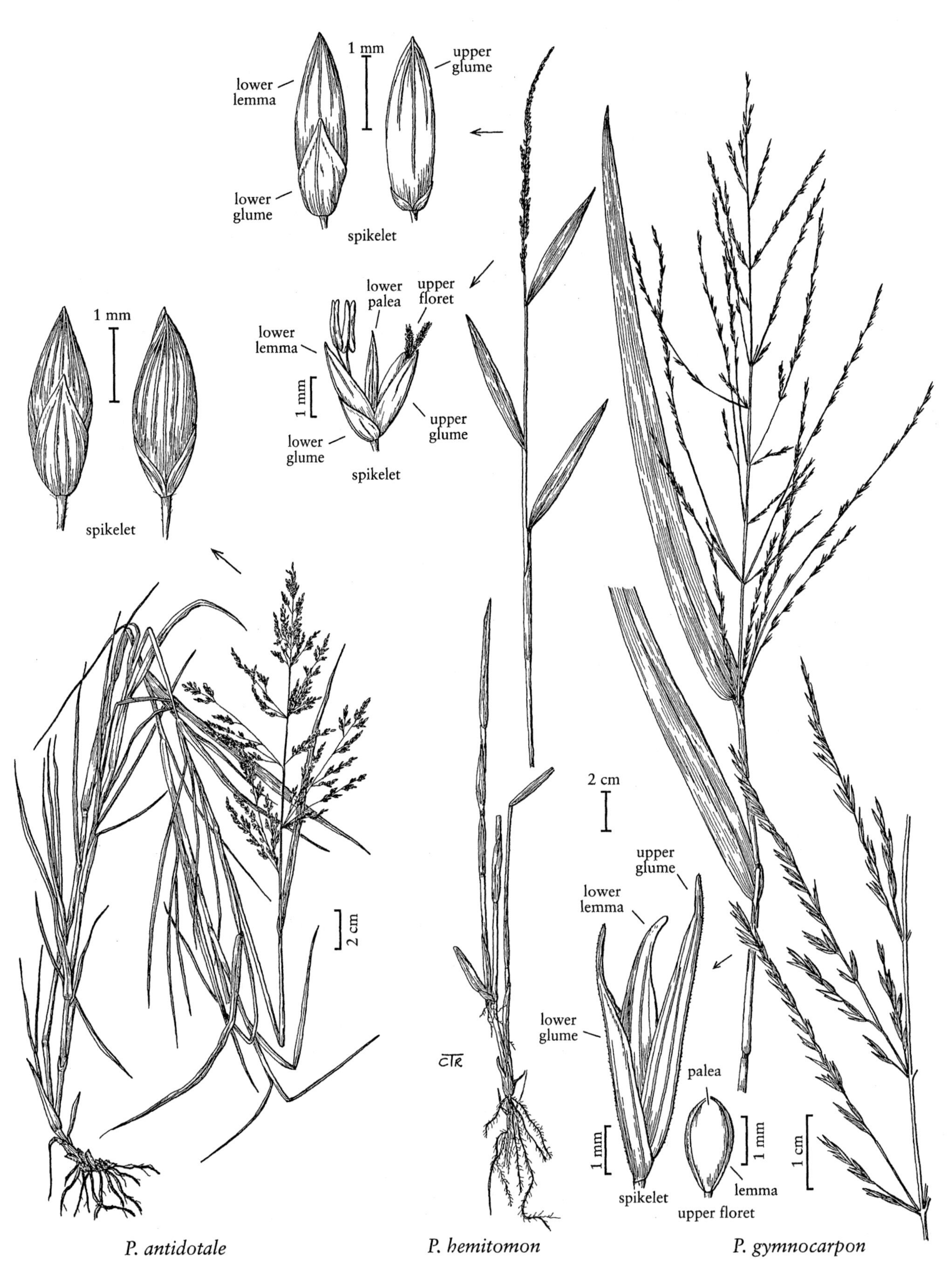

P. antidotale   *P. hemitomon*   *P. gymnocarpon*

PANICUM

minute purple streaks, glabrous, margins shortly ciliate; **ligules** to 0.8 mm; **blades** 5–28 cm long, 4–14 mm wide, linear, more than 10 times longer than wide, thin, flat, glabrous on both surfaces or sparingly pilose adaxially, bases scabridulous near the margins, prominently veined. **Panicles** 12–30 cm long, 9–20 cm wide, usually 1–1.3 times longer than wide, diffuse; **primary branches** 8–15 cm, alternate, divergent, slender, scabridulous, much branched, branches confined to the distal ⅔, secondary branches spreading, spikelets confined to the distal ½ of the branches; **pedicels** 0.5–6 mm. **Spikelets** 1.8–2.7 mm long, 0.8–1 mm wide, ellipsoid, dark green, often purple-tinged, usually glabrous, acute to acuminate. **Lower glumes** ⅓–½ as long as the spikelets, glabrous, deltoid, acute; **upper glumes** and **lower lemmas** subequal, equaling or exceeding the upper florets, smooth, faintly 5-veined, sparsely pilose with short hairs near the margins and apices, acute; **lower florets** sterile; **lower paleas** absent or much shorter than the lower lemmas; **upper florets** 1.5–1.8 mm, ellipsoid, smooth, lustrous, grayish-brown at maturity, apices sparsely puberulent, obtuse to subacute. $2n = 36$.

Panicum bisulcatum is an Asian species that grows in wet, open areas. It has been introduced sporadically, but has rarely become established, on the coastal plain of Georgia and South Carolina. The records from Philadelphia, Pennsylvania are from 1865–1877.

Panicum sect. Verrucosa (Nash) C.C. Hsu

Plants annual. **Culms** weak, sprawling. **Ligules** membranous, ciliate, very short. **Panicles** open, diffuse, lax, with a few spikelets. **Spikelets** warty-tuberculate or verrucose, faintly veined. **Lower glumes** less than ⅕ as long as the spikelets, without veins; **upper glumes** and **lower lemmas** subequal, 5-veined; **lower florets** sterile; **lower paleas** absent; **upper florets** finely longitudinally rugose, papillose and with bicellular microhairs, often with a minute protuberance at the base of the paleas. $x = 9$.

33. **Panicum verrucosum** Muhl. [p. 489]
WARTY PANICGRASS

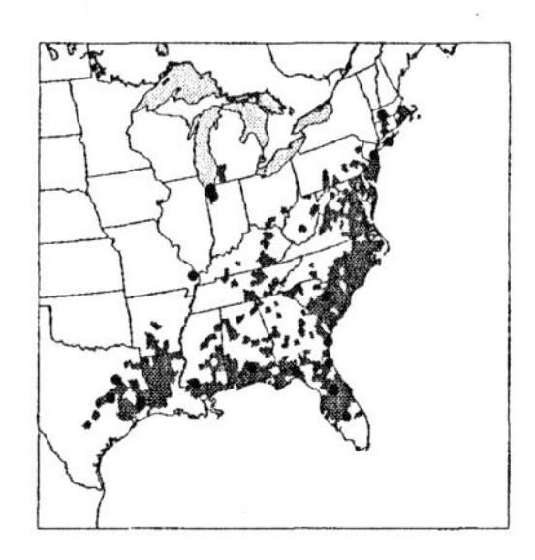

Plants annual; weak, ascending or sprawling. **Culms** 10–150 cm, slender, wiry, erect at first, ultimately decumbent, sprawling, glabrous, often with purple dots and streaks, branching extensively at the base, rooting at the lower nodes. **Sheaths** often shorter than the internodes, loose, glabrous, margins short-ciliate; **ligules** 0.2–0.5 mm, membranous, erose, ciliate; **blades** 5–20 cm long, 3–10 mm wide, thin, flat, glabrous on both surfaces, margins scabridulous, apices long-acuminate. **Panicles** 5–30 cm, nearly as wide as long; **branches** few, capillary, with a few spikelets distally; **pedicels** 0.5–10 mm. **Spikelets** 1.7–2.2 mm long, about 1 mm wide, ellipsoid or obovoid, glabrous, faintly veined, subacute or obtuse at the apices. **Lower glumes** 0.3–0.8 mm, reduced, acute; **upper glumes** and **lower lemmas** subequal or the glumes shorter, distinctly verrucose, with hemispheric warts; **upper florets** 1.6–2 mm long, about 1 mm wide, grayish-brown, dull, minutely papillose, acute. $2n = 36$.

Panicum verrucosum grows primarily in open, moist or wet sandy areas bordering swamps, marshes, or lakes or on roadside ditches; it also grows occasionally in open, drier woodlands. It is restricted to the eastern United States and is mostly, but not exclusively, coastal.

34. **Panicum brachyanthum** Steud. [p. 489]
PRAIRIE PANICGRASS

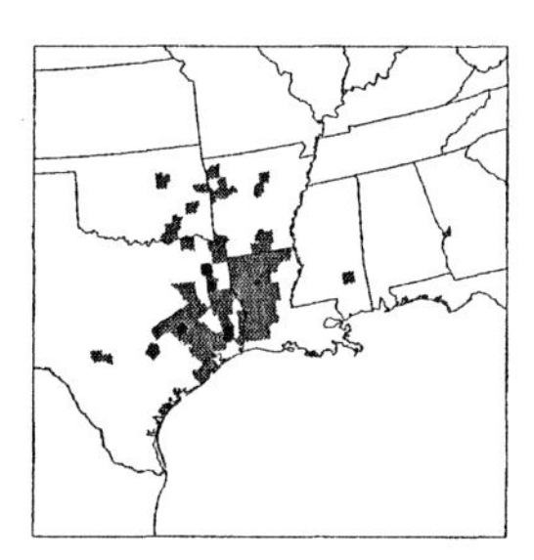

Plants annual; weak, ascending or spreading. **Culms** slender, wiry, glabrous, often with minute purple streaks and dots, ascending from a decumbent base, often branching extensively at the base and rooting at the lower nodes. **Sheaths** usually shorter than the internodes, glabrous, margins short-ciliate; **ligules** usually less than 0.3 mm, membranous, erose, ciliate; **blades** 4–15 cm long (rarely longer), 2–3 mm wide, flat or slightly involute, glabrous on both surfaces, margins scabridulous, especially towards the apices, bases narrowed. **Panicles** 4–17 cm, ½ to nearly as wide as long; **branches** few, capillary, ascending or spreading, scabridulous, with a few spikelets distally; **pedicels** 0.5–10 mm. **Spikelets** 3.2–4 mm long, about 1.5 mm wide, broadly ellipsoid or obovoid, tuberculate, hispid, faintly veined, acute or acuminate at the apices. **Lower glumes** usually less than 1 mm, obtuse or acute; **upper glumes** and **lower lemmas** subequal, distinctly tuberculate, hispid, with stiff hairs arising from wartlike bases; **upper florets** 2.7–3.2 mm long, 1.3–1.6 mm wide, obovoid or ellipsoid, nearly smooth, minutely papillose, or cross-rugulose, subacute to acute. $2n$ = unknown.

Panicum brachyanthum grows in dry, sandy or clayey soils of open areas, remnant prairies, woodland borders, and roadsides and, less commonly, along the margins of bogs and on grassy shores in the western portion of the gulf coast plain. It is restricted to the southern United States. It resembles *P. verrucosum* in its growth habit, but is more restricted in its distribution.

25.11 **BRACHIARIA** (Trin.) Griseb.

J.K. Wipff
Rahmona A. Thompson

Plants annual. **Culms** 10–60 cm, herbaceous, not woody, often creeping. **Leaves** cauline; **sheaths** open, glabrous or pubescent; **ligules** membranous, with a ciliate fringe, fringe longer than the membranous base. **Inflorescences** terminal, secund panicles of 1-sided branches; **branches** erect to ascending, axes triquetrous, terminating in a well-developed spikelet; **secondary branches,** when present, shorter than the primary branches; **disarticulation** below the glumes and beneath the upper florets. **Spikelets** solitary, subsessile, dorsally compressed, unequally convex, in 2 rows, the lower glumes and lemmas appressed or adjacent to the branch axes, with 2 florets; **lower florets** sterile or staminate; **upper florets** stipitate, bisexual, usually glabrous, readily disarticulating, acuminate. **Lower glumes** to 0.5 mm, less than ½ as long as the spikelets, glabrous, adjacent to the branch axes, 0–1-veined; **upper glumes** and **lower lemmas** subequal, villous, 3–5-veined; **upper glumes** subequal to or slightly exceeding the upper florets, not saccate; **lower paleas** present; **anthers** (if present) 3; **upper lemmas** equaling the second glume, glabrous, indurate, smooth, shiny to lustrous, 5- or 7-veined, margins involute, apices round to muticous; **upper paleas** similar to the upper lemmas; **anthers** 3. **Caryopses** ovoid, dorsally compressed. x = 9. Name from the Latin *brachium*, 'forearm', and *-aria*, 'resembling', an allusion to the panicle branches.

Brachiaria, as now interpreted, includes three species, all native to the Eastern Hemisphere. It differs from *Urochloa* in its smooth, rounded, distal floret and from *Panicum* in its secund panicle and stipitate, shiny to lustrous, disarticulating distal floret. Many of the species previously placed in the genus are now placed in *Urochloa*. One species is established in the *Flora* region.

SELECTED REFERENCE **Morrone, O.** and **F.O. Zuloaga.** 1992. Revisión de las especies Sudamericanas nativas e introducisas de los géneros *Brachiaria* y *Urochloa* (Poaceae: Panicoideae: Paniceae). Darwiniana 31:43–109.

1. **Brachiaria eruciformis** (Sm.) Griseb. [p. 491]
SWEET SIGNALGRASS

Plants mat-forming. **Culms** (10)19–60 cm, decumbent, rooting at the lower nodes before geniculately ascending, sometimes branching from the upper nodes; **nodes** pubescent; **internodes** glabrous. **Sheaths** glabrous or pubescent; **ligules** to 1 mm; **blades** 2–6(12) cm long, 3–6 mm wide, pubescent (rarely pilose) on both surfaces, bases subcordate. **Panicles** 4–9 cm long, 0.5–1 cm wide, exserted, with 3–15 erect to appressed branches; **branches** 1–2 cm, hispidulous; **pedicels** 0.1–0.5 mm, pubescent. **Spikelets** (1.6)2–2.6 mm long, 0.8–1 mm wide, ovate. **Lower glumes** 0.3–0.5 mm, to ⅕ the spikelet length; **upper glumes** (1.6)2–2.5 mm; **lower lemmas** longer than the paleas, 5-veined, without cross-venation; **upper lemmas** (1.4)1.7–1.8 mm long, 0.6–0.9 mm wide; **anthers** 0.5–1 mm, reddish. **Caryopses** 1–1.5 mm. $2n$ = 18, 36.

Brachiaria eruciformis is native from the Mediterranean to tropical Africa and India. It tends to be a weedy species in many parts of the world, growing in moist, disturbed sites. It has been grown for evaluation as a forage crop at various experimental stations in the United States. A few of these plantings have resulted in escapes that have persisted persisted for a short time, but the species has not become an established in the *Flora* region.

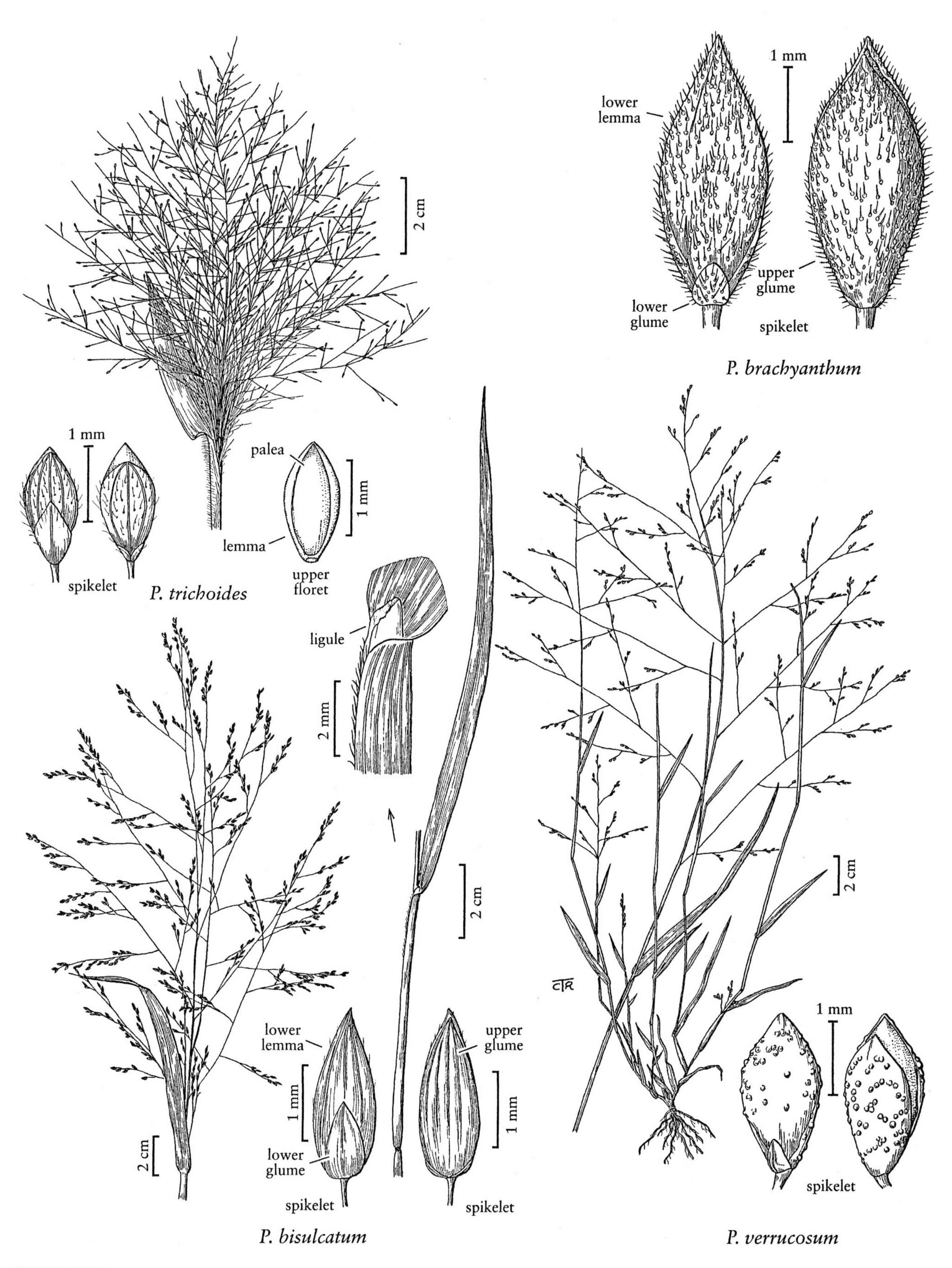
2 cm
1 mm
lower lemma
lower glume
upper glume
spikelet
P. brachyanthum
1 mm
palea
1 mm
lemma
spikelet
P. trichoides
upper floret
ligule
2 mm
2 cm
2 cm
CTR
1 mm
lower lemma
upper glume
1 mm
1 mm
lower glume
spikelet
spikelet
spikelet
2 cm
P. bisulcatum
P. verrucosum

25.12 MELINIS P. Beauv.

J.K. Wipff

Plants annual or perennial; habit various. **Culms** 20–150 cm, erect, decumbent, or prostrate. **Sheaths** open; **ligules** of hairs or membranous and ciliate. **Inflorescences** terminal, simple panicles or panicles of spikelike primary branches, usually with capillary secondary branches and pedicels; **disarticulation** below the glumes, sometimes also below the upper florets, the upper florets then falling first. **Spikelets** with 2 florets. **Lower glumes** present or absent, 0–1-veined, unawned; **upper glumes** equaling or exceeding the florets, sometimes gibbous basally, 5–7-veined, emarginate to bilobed, awned or unawned; **lower florets** staminate or sterile; **lower lemmas** similar to the upper glumes, but not gibbous; **upper florets** bisexual, laterally compressed; **upper lemmas** subcoriaceous, glabrous, smooth, unawned; **upper paleas** resembling the upper lemmas; **lodicules** 2, fleshy or membranous. $x = 9$. Name from the Greek *meline*, 'millet'.

Melinis is an African and western Asian genus of 22 species that grow in savannahs, open grasslands, and disturbed places. Two species have become established in the *Flora* region.

Rhynchelytrum Nees has traditionally been treated as a separate genus, with the number of veins being the diagnostic character. Zizka (1988) showed that this separation was artificial; consequently the older generic name, *Melinis*, is now applied to species that used to be included in *Rhynchelytrum*.

SELECTED REFERENCES **Zizka, G.** 1988. Revision der Melinideae Hitchcock (Poaceae, Panicoideae). Biblioth. Bot. 138:1–149; **Zizka, G.** 1990. Taxonomy of the Melinideae (Poaceae: Panicoideae). Mitt. Inst. Allg. Bot. Hamburg 23:563–572.

1. Glumes and pedicels glabrous, scabridulous; lower florets without paleas .... 1. *M. minutiflora*
1. Glumes, and usually the pedicels, with hairs to 7 mm long; lower florets with paleas .... 2. *M. repens*

1. Melinis minutiflora P. Beauv. [p. 491]
MOLASSES GRASS

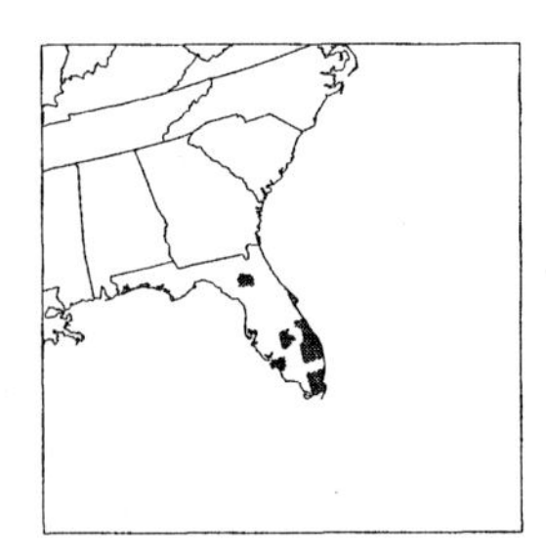

Plants perennial; cespitose; aromatic. **Culms** (50)80–150 cm, branching and sprawling, often becoming matted, usually rooting at the lower nodes; **upper nodes** appressed pubescent; **internodes** glabrous basally, appressed pubescent distally. **Sheaths** densely tomentose, hairs 0.5–5.2 mm, spreading, papillose-based, often sticky and smelling of linseed oil; **ligules** of hairs, 1–2 mm; **blades** 3.5–19 cm long, 4–14 mm wide, flat, pubescent, hairs sometimes papillose-based. **Panicles** (4.5)7–20 cm long, 1–9.5 cm wide, narrowly ovate; **primary branches** to 8 cm; **pedicels** usually shorter than the spikelets, glabrous, scabridulous. **Spikelets** 1.7–2.4 mm, usually purplish; **calluses** glabrous. **Lower glumes** absent or to 0.3 mm, glabrous, scabridulous; **upper glumes** 1.6–2.4 mm, glabrous, unawned, sometimes muticous; **lower florets** sterile; **lower lemmas** bilobed, lobes 0.2–0.7 mm, unawned or awned, awns to 18 mm; **lower paleas** absent; **upper lemmas** 1.4–1.9 mm, glabrous; **upper paleas** 1.5–1.9 mm, usually slightly longer than the upper lemmas; **anthers** 3, 1–1.5 mm, reddish-brown to orange. **Caryopses** 0.9–1.2 mm long, 0.3–0.4 mm wide. $2n = 36$.

Melinis minutiflora is native to Africa, but has been introduced throughout the tropics as a forage crop. It is now regarded as a serious weed in many places. In the *Flora* region, it is only known to be established in southern Florida.

2. Melinis repens (Willd.) Zizka [p. 491]
NATAL GRASS

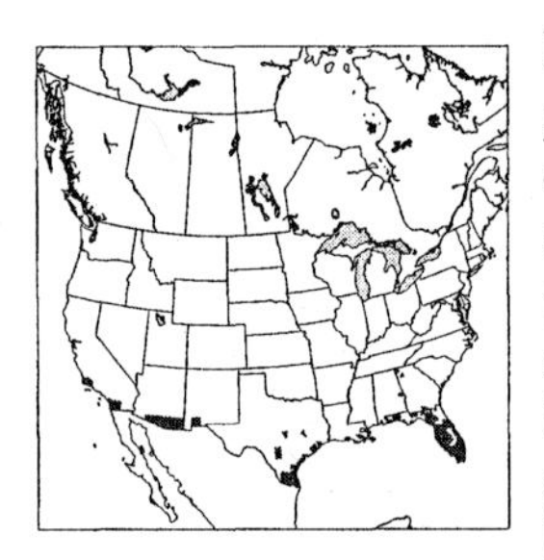

Plants annuals or short-lived perennials. **Culms** (20)40–150 cm, decumbent, usually rooting at the lower nodes; **nodes** pubescent; **internodes** glabrous or with papillose-based hairs, hairs to 4.7 mm. **Sheaths** glabrous or with papillose-based hairs, hairs 0.5–4.7 mm; **ligules** of hairs, 0.7–2.2 mm; **blades** 3.6–27 cm long, 2–9(14) mm wide, flat, glabrous or pubescent, with or without papillose-based hairs. **Panicles** (4)6–22 cm long, (1.5)2.5–12 cm wide; **primary branches** to 11 cm, ends of the primary branches, secondary branches, and pedicels capillary; **pedicels** 0.6–5.3 mm, usually hairy

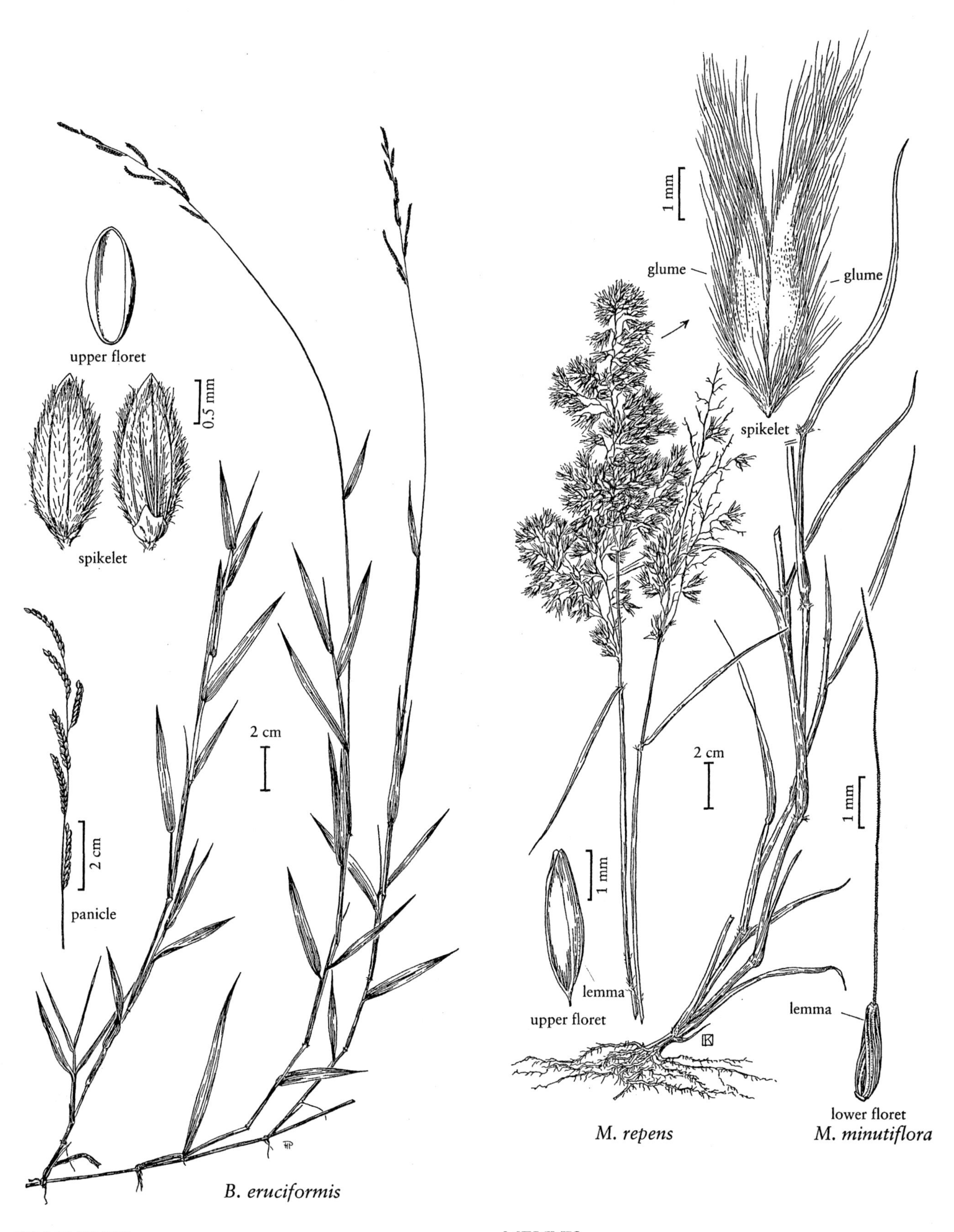

BRACHIARIA

MELINIS

distally, hairs to 6.3 mm. **Spikelets** 2–5.7 mm; **calluses** hairy, hairs to 4 mm. **Lower glumes** 0–1.7 mm, pubescent, sometimes with papillose-based hairs, apices rounded, truncate, or slightly cleft; **upper glumes** (1.9)2.3–4.9 mm, enclosing the upper florets, gibbous basally, densely pubescent, hairs to 7 mm, sometimes papillose-based, varying from white to rose or darkish purple, apices tapering, beaked, glabrous, unawned or awned, awns to 4.1 mm; **lower florets** staminate or sterile; **lower lemmas** 1.9–4.8 mm, unawned or with awns to 4.2 mm; **lower paleas** 0.9–4 mm; **anthers** (0.8)1.5–2.6 mm, orange-brown to orange; **upper lemmas** 1.8–2.7 mm, glabrous; **anthers** 3, 1.2–1.7 mm, orange-brown to orange. **Caryopses** 1.3–1.9 mm long, 0.6–0.9 mm wide. $2n = 36$.

Melinis repens is probably native to Africa and western Asia. It is now established throughout the subtropics, including the southern portion of the *Flora* region. It has been grown as an ornamental, but it is now established and often weedy in warmer portions of the region.

Plants in the *Flora* region belong to **Melinis repens** (Willd.) Zizka subsp. **repens.**

25.13 UROCHLOA P. Beauv.

J.K. Wipff
Rahmona A. Thompson

Plants annual or perennial; usually cespitose, sometimes mat-forming, sometimes stoloniferous. **Culms** 5–500 cm, herbaceous, erect, geniculate, or decumbent and rooting at the lower nodes. **Sheaths** open; **auricles** rarely present; **ligules** apparently of hairs, the basal membranous portion inconspicuous; **blades** ovate-lanceolate to lanceolate, flat. **Inflorescences** terminal or terminal and axillary, usually panicles of spikelike primary branches in 2 or more ranks, rachises not concealed by the spikelets; **primary branches** usually alternate or subopposite, spikelike, and 1-sided, less frequently verticillate, axes flat or triquetrous, usually terminating in a well-developed, rudimentary spikelet; **secondary branches** present or absent, axes flat or triquetrous; **disarticulation** beneath the spikelets. **Spikelets** solitary, paired, or in triplets, subsessile or pedicellate, divergent or appressed, ovoid to ellipsoid, dorsally compressed, in 1–2(4) rows, with 2 florets, lower or upper glumes adjacent to the branch axes. **Glumes** not saccate basally; **lower glumes** usually ⅓–⅔ as long as the spikelets, occasionally equaling the upper florets, (0)1–11-veined; **upper glumes** 5–13-veined; **lower florets** sterile or staminate; **lower lemmas** similar to the upper glumes, 5–9-veined; **lower paleas** if present, usually hyaline, 2-veined; **upper florets** bisexual, sessile, ovoid to ellipsoid, usually plano-convex, usually glabrous, not disarticulating, mucronate or acuminate; **upper lemmas** indurate, transversely rugose and verrucose, 5-veined, margins involute, apices round to mucronate, or aristate; **upper palea** rugose, shiny or lustrous; **lodicules** 2, cuneate, truncate; **anthers** 3. **Caryopses** ovoid to elliptic, dorsally compressed; **embryos** ½–¾ as long as the caryopses; **hila** punctate to linear. $x = 7, 8, 9$, or 10. Name from the Greek *ouros*, 'tail' and *chloa*, 'grass', a reference to the abruptly awned lemmas of some species.

Urochloa is a genus of approximately 100 tropical and subtropical species. There are 19 species found in the *Flora* region. Eight species are established introductions, six are native, three are cultivated as grain or forage crops, and two have been found in the region but are not known to be established.

Urochloa differs from *Brachiaria* in its two or more ranks of panicles and rugose, non-disarticulating, distal florets. The rugose, often mucronate or aristate, distal florets also distinguish it from most species of *Panicum*.

SELECTED REFERENCES **Clayton, W.D.** and **S.A. Renvoize.** 1982. Flora of Tropical East Africa. Gramineae (Part 3). A.A. Balkema, Rotterdam, The Netherlands. 448 pp.; **Davidse, G.** and **R.W. Pohl.** 1994. *Urochloa* P. Beauv. Pp. 331–333 *in* G. Davidse, M. Sousa S., and A.O. Chater (eds.). Flora Mesoamericana, vol. 6: Alismataceae a Cyperaceae. Universidad Nacional Autónoma de México, Instituto de Biología, México, D.F., México. 543 pp.; **Fox, W.E., III** and **S.L. Hatch.** 1996. *Brachiaria eruciformis* and *Urochloa brizantha* (Poaceae: Paniceae) new to Texas. Sida 17:287–288; **Hall, D.W.** 1978. The Grasses of Florida. Ph.D. dissertation, University of Florida, Gainesville, Florida, U.S.A. 498 pp.; **Morrone, O.** and

F.O. Zuloaga. 1992. Revisión de las especies Sudamericanas nativas e introducidas de los géneros *Brachiaria* y *Urochloa* (Poaceae: Panicoideae: Paniceae). Darwiniana 31:43–109; **Morrone, O.** and **F.O. Zuloaga.** 1993. Sinopsis del género *Urochloa* (Poaceae: Panicoideae: Paniceae) para México y America Central. Darwiniana 32:59–75; **Pohl, R.W.** 1980. Flora Costaricensis: Family #15, Gramineae. Fieldiana, Bot., n.s., 4:1–608; **Reed, C.F.** 1964. A flora of the chrome and manganese ore piles at Canton, in the Port of Baltimore, Maryland and at Newport News, Virginia, with descriptions of genera and species new to the flora of the eastern United States. Phytologia 10:321–405; **Sendulsky, T.** 1978. *Brachiaria*: Taxonomy of cultivated and native species in Brazil. Hoehnea 7:99–139; **Veldkamp, J.F.** 1996. *Brachiaria*, *Urochloa* (Gramineae-Paniceae) in Malesia. Blumea 41:413–437; **Webster, R.D.** 1987. The Australian Paniceae (Poaceae). J. Cramer, Berlin and Stuttgart, Germany. 322 pp.; **Wipff, J.K., R.I. Lonard, S.D. Jones,** and **S.L. Hatch.** 1993. The genus *Urochloa* (Poaceae: Paniceae) in Texas, including one previously unreported species for the state. Sida 15:405–413.

1. Branches at the lowest node of the inflorescence verticillate; inflorescences simple, open panicles, the primary branches with well-developed secondary and tertiary branches; plants perennial ...... 18. *U. maxima*
1. Branches at the lowest node of the inflorescence not verticillate; inflorescences usually with spikelike branches, secondary branches absent or inconspicuous, sometimes simple panicles; plants annual or perennial.
    2. Spikelets paired at mid-branch length, sometimes solitary distally.
        3. Axes of the primary panicle branches flat; lower glumes (0)1–3-veined.
            4. Plants annual; culms 10–35 cm tall; spikelets 1.8–2.2 mm long .... 2. *U. reptans*
            4. Plants perennial; culms 20–500 cm tall; spikelets 2.5–5 mm long.
                5. Upper lemmas awned, the awns 0.5–1.2 mm; lower glumes 3-veined, with 1–3 conspicuous, rigid hairs .... 8. *U. mosambicensis*
                5. Upper lemmas unawned; lower glumes 1–3-veined, glabrous .... 1. *U. mutica*
        3. Axes of the primary panicle branches triquetrous; lower glumes 3–7-veined.
            6. Spikelets 4.8–6 mm long; hila linear, about ½ as long as the caryopses .... 3. *U. texana*
            6. Spikelets 2–4.2 mm long; hila punctate.
                7. Branch axes densely hairy, the hairs papillose-based .... 4. *U. arizonica*
                7. Branch axes sometimes densely hairy, with few or no papillose-based hairs.
                    8. Upper glumes with evident cross venation extending from near the bases to the apices; spikelets obovoid; upper glumes and lower lemmas usually glabrous; lower lemmas 7-veined .... 5. *U. fusca*
                    8. Upper glumes without evident cross venation, or the cross venation confined to the distal ½; spikelets ellipsoid; upper glumes and lower lemmas glabrous or pubescent; lower lemmas 5-veined.
                        9. Cauline nodes glabrous; plants 20–120 cm tall; spikelet apices abruptly acuminate .... 6. *U. adspersa*
                        9. Cauline nodes pubescent; plants 10–70 cm tall, spikelets apices broadly acute to acute.
                            10. Upper glumes 5-veined; upper lemmas 1.8–2.1 mm long; glumes separated by an internode about 0.3 mm long .... 11. *U. villosa*
                            10. Upper glumes 7–9-veined; upper lemmas 2.3–3.3 mm long; glumes not separated by a conspicuous internode .... 7. *U. ramosa*
    2. Spikelets solitary at mid-branch length.
        11. Panicle branches triquetrous, 0.2–0.4 mm wide.
            12. Plants perennial; upper glumes (9)11–13-veined; upper lemmas 2.4–2.8 mm long; glumes not separated by a conspicuous internode; lower florets staminate .... 17. *U. ciliatissima*
            12. Plants annual; upper glumes 5-veined; upper lemmas 1.8–2.1 mm long; glumes separated by an internode about 0.3 mm long; lower florets sterile .... 11. *U. villosa*
        11. Panicle branches flat or crescent-shaped in cross section, 0.5–2.5 mm wide.
            13. Upper lemmas awned, the awns 0.3–1.2 mm long, apices rounded.
                14. Plants perennial; lower glumes with 1–3 conspicuous, rigid hairs, the lower glumes ½–¾ as long as the spikelet; lower florets staminate .... 8. *U. mosambicensis*
                14. Plants annual; lower glumes without conspicuous, rigid hairs; the lower glumes ¼–⅓(½) as long as the spikelet; lower florets sterile .... 14. *U. panicoides*
            13. Upper lemmas unawned; apices variable, with or without a mucronate tip.
                15. Spikelets in a single row along the branches; spikelets 4–6 mm long; lower florets staminate; panicle branch axes crescent-shaped in cross section ..... 10. *U. brizantha*

15. Spikelets in 2 rows along the branches; spikelets 2.5–6 mm long; lower florets sterile or staminate; panicle branches flat.
    16. Lower glumes 5–7-veined; glumes scarcely separated, the internode between them shorter than 0.3 mm.
        17. Upper glumes and lower lemmas pubescent, the hairs often long in the distal ⅓ . . . . 9. *U. piligera*
        17. Upper glumes and lower lemmas glabrous.
            18. Plants perennial, stoloniferous; lower florets staminate . . . . 16. *U. arrecta*
            18. Plants annual; lower florets sterile . . . . 15. *U. platyphylla*
    16. Lower glumes (7)9–11-veined; glumes separated by a conspicuous, 0.3–0.5 mm internode.
        19. Lower paleas absent; spikelets usually pubescent, sometimes glabrous . . . . . 9. *U. piligera*
        19. Lower paleas present; spikelets glabrous.
            20. Spikelets 3.3–3.7 mm long; base of blades rounded to subcordate, not clasping the stem . . . . 12. *U. subquadripara*
            20. Spikelets 4–6 mm long; base of blades subcordate to cordate, clasping the stem . . . . 13. *U. plantaginea*

1. **Urochloa mutica** (Forssk.) T.Q. Nguyen [p. 496]
PARAGRASS

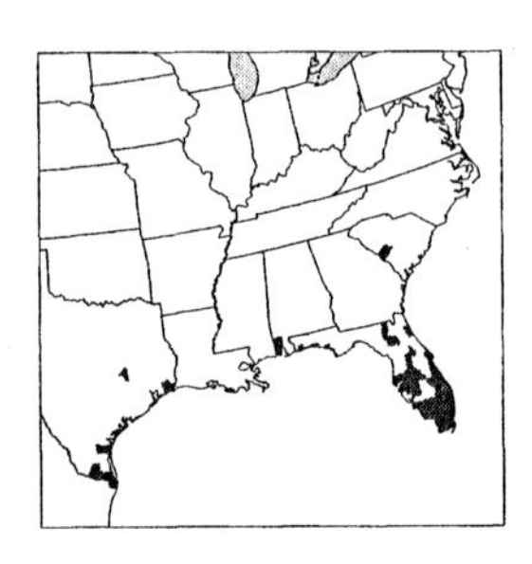

Plants perennial; stoloniferous, straggling. **Culms** to 5 m long, long-decumbent and rooting at the lower nodes, vertical portion 90–200(300) cm; **nodes** villous. **Lower sheaths** with papillose-based hairs, these more dense distally, margins ciliate; **collars** pubescent; **ligules** 1–1.5 mm; **blades** 7.5–35 cm long, 4–20 mm wide, glabrous or sparsely pilose on both surfaces, margins scabrous. **Panicles** 10–25 cm long, 5–10 cm wide, pyramidal, with 10–30 spikelike branches in more than 2 ranks; **primary branches** 2.5–8 cm long, 0.4–0.9 mm wide, ascending to divergent, axils pubescent, axes flat, glabrous or with a few papillose-based hairs, **secondary branches** present or absent; **pedicels** shorter than the spikelets, scabrous, sometimes with hairs. **Spikelets** 2.6–3.5 mm long, 1–1.4 mm wide, mostly in pairs, in 2–4 rows, appressed to the branches, purplish to green. **Glumes** scarcely separate, rachilla internodes short not pronounced; **lower glumes** 0.6–1.1 mm, ⅕–⅓ as long as the spikelets, glabrous, 0–1(3)-veined; **upper glumes** 2.6–3.5 mm, glabrous, 5–(7)-veined, without cross venation; **lower florets** staminate; **lower lemmas** 2.6–3.3 mm, glabrous, 5-veined, without cross venation; **upper lemmas** 2.3–2.8 mm long, 1–1.3 mm wide, apices rounded, mucronate; **anthers** 1–1.5 mm. **Caryopses** 1.8–2 mm. $2n$ = 18, 36.

An African species, *Urochloa mutica* is grown as a forage crop throughout the tropics, but it tends to become weedy. It grows on moist, disturbed soils and is established in the southeastern United States.

2. **Urochloa reptans** (L.) Stapf [p. 496]
SPRAWLING SIGNALGRASS

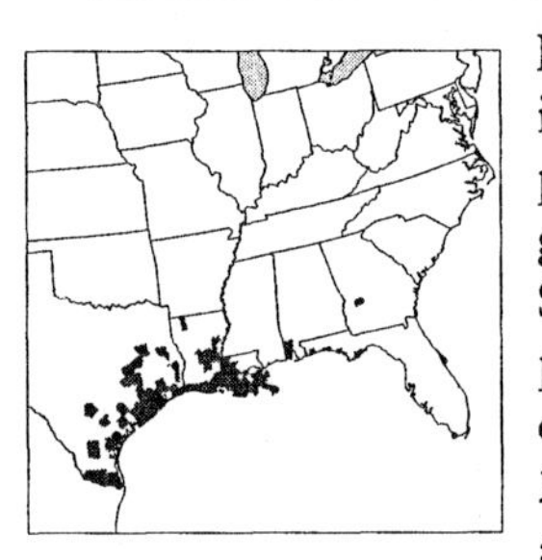

Plants annual; forming sprawling mats. **Culms** 10–35 cm, prostrate to decumbent; **nodes** glabrous or sparsely puberulent. **Sheaths** glabrous or sparsely pubscent, margins densely ciliate; **ligules** 0.5–1 mm; **blades** 2–6 cm long, 4–15 mm wide, adaxial surfaces glabrous or sparsely pubescent with papillose-based hairs, margins ciliate basally. **Panicles** 1.5–6(8) cm long, 4–5 cm wide, ovoid, with 3–16 spikelike branches in more than 2 ranks; **primary branches** 1–4 cm long, 0.2–0.5 mm wide, flat, scabrous; **secondary branches** occasionally present; **pedicels** shorter than the spikelets, scabrous, glabrous or with long hairs distally. **Spikelets** 1.8–2.2 mm long, 0.8–1 mm wide, mostly in pairs, in 2–4 rows, appressed to the branches. **Glumes** scarcely separate, rachilla internodes short, not pronounced; **lower glumes** 0.2–0.6 mm, ⅕–¼ as long as the spikelets, 0–1-veined; **upper glumes** 1.7–2.1 mm, glabrous, 7-veined, without cross venation; **lower florets** sterile or staminate; **lower lemmas** 1.7–2.1 mm, glabrous, 5-veined, without cross venation; **lower paleas** present; **upper lemmas** 1.5–1.8 mm long, 0.8–1 mm wide, apices rounded, mucronate, mucros to about 0.1 mm; **anthers** 0.4–0.6 mm. **Caryopses** 0.8–1.2 mm. $2n$ = 14(18).

Urochloa reptans is widely distributed in tropical and subtropical regions of the world, growing in disturbed habitats. In the *Flora* region, it is found primarily in Texas and Louisiana.

3. Urochloa texana (Buckley) R.D. Webster [p. 496]
TEXAS SIGNALGRASS, TEXAS MILLET

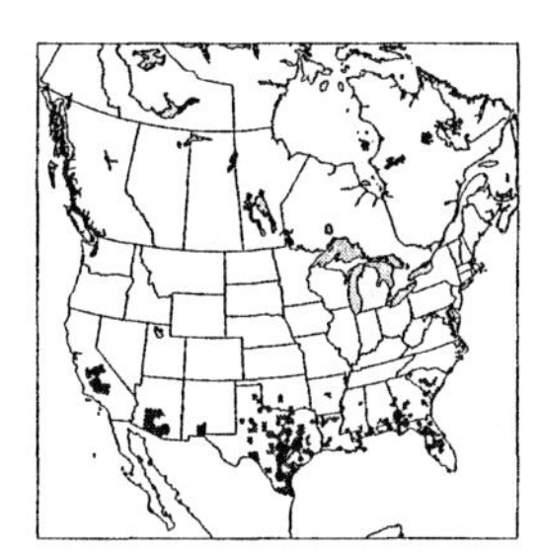

Plants annual. **Culms** (20)40–200 cm, erect or decumbent; **nodes** puberulent; **internodes** pubescent, at least below the nodes. **Sheaths** usually with papillose-based hairs, margins shortly ciliate; **collars** pubescent; **ligules** 1–1.8 mm; **blades** 7–24 cm long, 5–20 mm wide, softly pubescent on both surfaces, margins ciliate basally, scabrous distally. **Panicles** 8–24 cm long, 1–3 cm wide, with spikelike primary branches in more than 2 ranks; **primary branches** 1–8 cm long, 0.3–0.4 mm wide, appressed to ascending, axes triquetrous, pubescent, with some papillose-based hairs; **secondary branches** present on the lower branches, short, appressed; **pedicels** shorter than the spikelets, usually with papillose-based hairs distally. **Spikelets** 4.8–6 mm long, 1.8–2 mm wide, mostly paired, in 2–4 rows, appressed to the branches. **Glumes** scarcely separate, rachilla internodes short, not pronounced; **lower glumes** 2.4–3.2 mm, to ½ as long as the spikelets, 5–7-veined, glabrous or sparsely pubescent distally; **upper glumes** 4–5.5 mm, glabrous or sparsely pilose, 7–9-veined; **lower florets** staminate (sterile); **lower lemmas** 4.7–6 mm, glabrous or sparsely pilose, 5-veined; **lower paleas** present; **upper lemmas** 3.6–4.1 mm long, 1.7–1.9 mm wide, apices acute, beaked; **anthers** (1.6) 2.2–2.7 mm. **Caryopses** 2–3 mm; **hilas** linear, about ½ as long as the caryopses. $2n = 54$.

Urochloa texana grows in sandy, moist soils from the southern United States to northern Mexico. Populations in the United States outside of Texas may represent relatively recent introductions.

4. Urochloa arizonica (Scribn. & Merr.) Morrone & Zuloaga [p. 498]
ARIZONA SIGNALGRASS

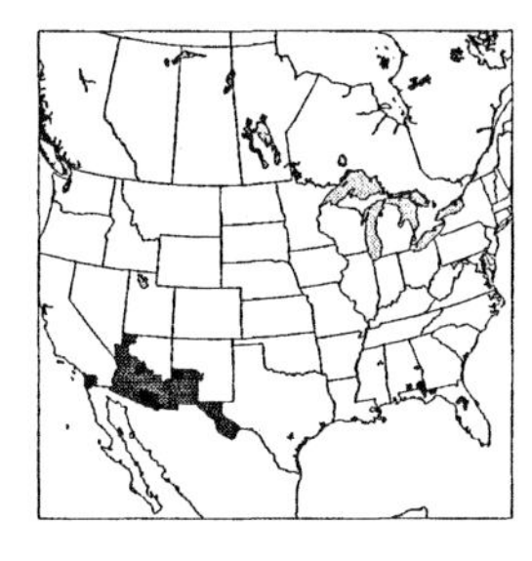

Plants annual. **Culms** 15–65 cm, erect or geniculate, branching from the lower nodes; **nodes** glabrous or hispid. **Sheaths** glabrous or with papillose-based hairs, margins ciliate distally; **ligules** 1–1.6 mm; **blades** 5–15 cm long, 5–12 mm wide, glabrous. **Panicles** 6–20 cm long, 2–5 cm wide, ovoid, with 6–12 spikelike primary branches in more than 2 ranks; **primary branches** 3–7 cm, divergent, axes about 0.4 mm wide, triquetrous, densely pubescent with papillose-based hairs; **secondary branches** short, divergent; **pedicels** shorter than the spikelets, with papillose-based hairs. **Spikelets** 3.2–4 mm long, 1.2–1.6 mm wide, mostly paired, in 2 rows, appressed to the branches. **Glumes** scarcely separate, rachilla internodes short, not pronounced; **lower glumes** 1.5–2 mm, to ½ as long as the spikelets, glabrous, 5-veined, sometimes with evident cross venation near the apices; **upper glumes** 2.5–3.2 mm, glabrous or shortly hirsute, 7-veined, with evident cross venation distally; **lower florets** staminate or sterile; **lower lemmas** 2.5–3.2 mm, glabrous or shortly hirsute, 5-veined, about as long as the spikelet; **lower paleas** present; **upper lemmas** 2.8–3 mm long, 1.2–1.6 mm wide, acute, beaked or mucronate; **anthers** 0.8–1 mm. **Caryopses** 1.5–2 mm; **hila** punctiform. $2n = 36$.

Urochloa arizonica is native to the southwestern United States and northern Mexico, but has been introduced to, and appears to be established in, the southeastern United States. It grows in open, dry areas with rocky or sandy soils.

5. Urochloa fusca (Sw.) B.F. Hansen & Wunderlin [p. 498]
BROWNTOP SIGNALGRASS

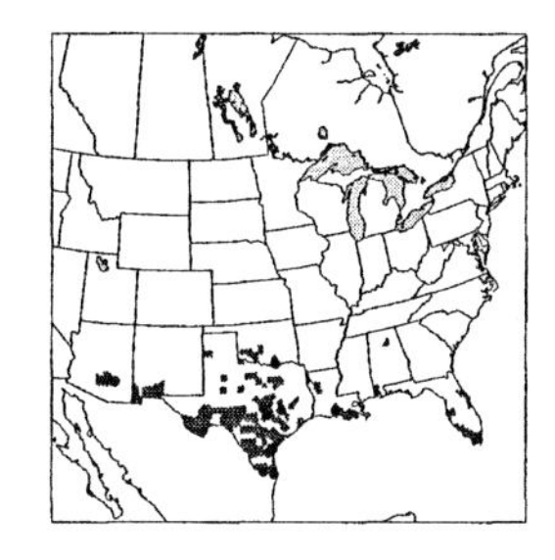

Plants annual; tufted. **Culms** 15–120 cm, geniculate; **nodes** glabrous or shortly pilose. **Sheaths** glabrous or hispid, margins ciliate; **ligules** 1–1.5 mm; **blades** 3–33 cm long, 5–20 mm wide, glabrous or sparsely pilose on both surfaces, margins smooth or scabrous; **collars** pubescent. **Panicles** 5–15 cm long, 2–8 cm wide, simple, with 5–30 spikelike primary branches in more than 2 ranks; **primary branches** 2–10 cm, appressed to divergent, axils glabrous, axes 0.3–0.5 mm wide, triquetrous, scabrous or sparsely pilose; **secondary branches** usually present on the lower primary branches, **pedicels** scabrous and pubescent, shorter than the spikelets. **Spikelets** 2–3.4 mm long, 1.2–1.8 mm wide, obovoid, yellowish to reddish-brown or bronze-colored at maturity, mostly paired, in 2–4 rows, appressed to the branches. **Glumes** scarcely separate, rachilla internodes short, not pronounced; **lower glumes** 1–1.5 mm, at least ⅓ as long as the spikelets, glabrous, (1)3–5-veined; **upper glumes** (2)2.2–3.1 mm, glabrous, 7–9-veined, cross venation evident throughout; **lower florets** usually staminate, sometimes sterile; **lower lemmas** 2–3.1 mm, usually glabrous, 7-veined, cross venation evident throughout; **lower paleas** present; **upper lemmas** 1.8–2.9 mm long, 1.1–1.7 mm wide, apices acute to rounded, mucronate; **anthers** 1–1.6 mm. **Caryopses** 1–1.7 mm; **hila** punctiform. $2n = 18, 36$.

Urochloa fusca grows from the southern United States to Peru, Paraguay, and Argentina, usually in moist, often disturbed areas at low elevations. It

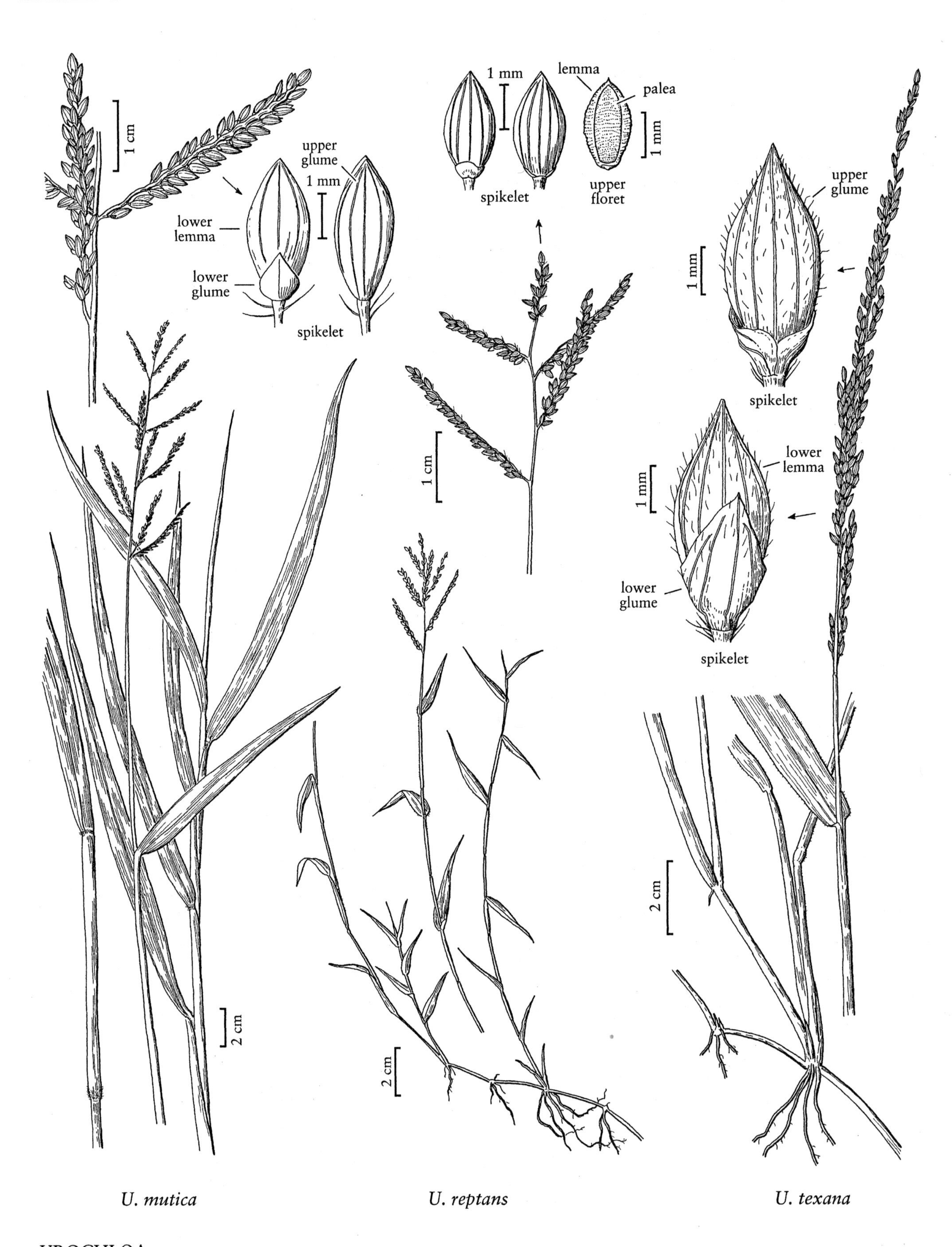
1 cm
upper glume
1 mm
lower lemma
lower glume
spikelet
1 mm
lemma
palea
1 mm
spikelet
upper floret
1 cm
upper glume
1 mm
spikelet
lower lemma
1 mm
lower glume
spikelet
2 cm
2 cm
2 cm
U. mutica
U. reptans
U. texana

UROCHLOA

frequently occurs as a weed, but is occasionally grown for forage and grain.

Plants having smaller, more compact panicles and larger (2.4–3.4 mm), mostly yellowish spikelets have been referred to as *Urochloa fusca* var. *reticulata* (Torr.) B.F. Hansen & Wunderlin. This variety is mainly found in the southwestern United States, but has been introduced into other areas, including Australia. *Urochloa fusca* (Sw.) B.F. Hansen & Wunderlin var. *fusca* has generally larger, more open panicles and smaller (2–2.5 mm), reddish-brown or bronze-colored spikelets. Much intergradation is reported between the two varieties. Further investigation is needed to establish that their recognition is warranted.

6. **Urochloa adspersa** (Trin.) R.D. Webster [p. 498]
DOMINICAN SIGNALGRASS

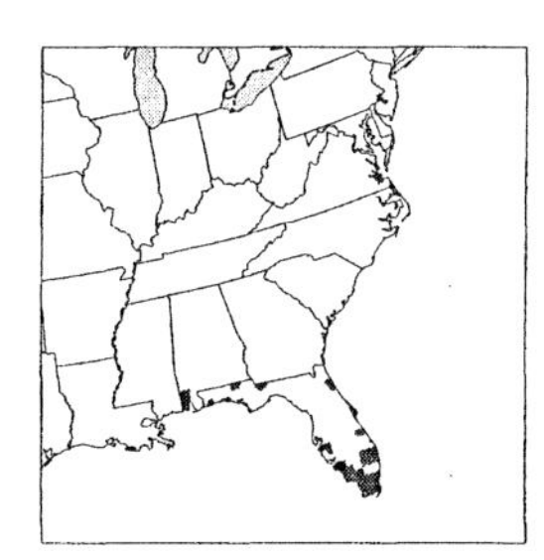

Plants annual. **Culms** 20–120 cm, geniculate or decumbent, usually rooting at the lower nodes; **nodes** glabrous. **Sheaths** glabrous or glabrate, margins ciliate distally; **ligules** 0.5–1 mm; **blades** 2–20 cm long, 7–20 mm wide, glabrous. **Panicles** terminal and axillary, 5–18 cm long, to 1.4 cm wide, with 2–10 spikelike primary branches in more than 2 ranks; **primary branches** 1.5–9 cm, appressed, axes 0.3–0.8 mm wide, triquetrous, scabrous; **secondary branches** present or absent, if present, short, restricted to the lowest panicle branches; **pedicels** scabrous, shorter than the spikelets. **Spikelets** 2.9–3.8 mm long, 1.2–1.4 mm wide, ellipsoid, apices abruptly acuminate, mostly paired, in 2–4 rows, appressed to the branches. **Glumes** scarcely separate, rachilla internodes short, not pronounced; **lower glumes** 1–1.4 mm, glabrous or pubescent, (3)5-veined, ⅓ or less as long as the spikelets; **upper glumes** 2.8–3.7 mm, glabrous or pubescent, 5–7(9)-veined, cross venation not evident or evident only in the distal ½; **lower florets** sterile; **lower lemmas** 2.7–3.6 mm, glabrous or pubescent, 5-veined, usually without cross venation; **upper florets** 2.1–2.9 mm long, 1.3–1.7 mm wide, broadly acute, mucronate; **anthers** 1–1.2 mm. **Caryopses** 1.2–1.8 mm; **hila** punctiform. $2n$ = 54.

Urochloa adspersa grows in southern Florida, the West Indies, and Argentina. It prefers moist, open areas, often on coral limestone. It has also been found on ballast dumps in Mobile, Alabama; Philadelphia, Pennsylvania; and Camden, New Jersey; but it has not persisted at these locations.

7. **Urochloa ramosa** (L.) T.Q. Nguyen [p. 500]
BROWNTOP MILLET

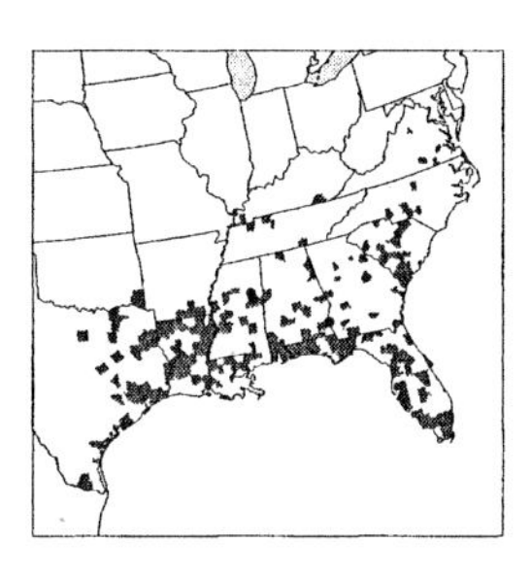

Plants annual; tufted. **Culms** 10–65 cm, decumbent, rooting or not at the lower nodes; **nodes** pubescent. **Sheaths** usually puberulous, sometimes glabrous or sparsely pilose, margins ciliate; **ligules** 0.8–1.7 mm; **blades** 2–25 cm long, 4–14 mm wide, glabrous, margins scabrous. **Panicles** 3–13 cm, simple, with 3–15 spikelike primary branches; **primary branches** 1–8 cm, divergent, axils glabrous, axes 0.4–0.6 mm wide, triquetrous, glabrous, scabrous, or pubescent, with or without some papillose-based hairs; **secondary branches,** if present, confined to the lower branches; **pedicels** shorter than the spikelets, scabrous or pubescent. **Spikelets** 2.5–3.4 mm long, 1.3–2 mm wide, ellipsoid, apices broadly acute to acute, paired, appressed to the branches. **Glumes** scarcely separated, rachilla internode between the glumes not pronounced; **lower glumes** 1–1.5 mm, ⅓–½ as long as the spikelets, glabrous, 3–5-veined; **upper glumes** 2.5–3.4 mm, usually puberulent, sometimes glabrous, margins sometimes somewhat pubescent, 7–9-veined, without evident cross venation; **lower florets** sterile, **lower lemmas** 2.4–3.3 mm, usually puberulent or occasionally glabrous, margins not ciliate, without cross venation, 5-veined; **upper lemmas** 2.3–3.3 mm, acute, mucronate; **anthers** 0.7–1.2 mm. **Caryopses** 1.2–2.3 mm; **hila** punctiform. $2n$ = 36 (usually); also 14, 28, 32, 42, 46, 72.

A weedy species of tropical Africa and Asia, *Urochloa ramosa* has spread throughout the tropics and subtropics, including the southeastern United States. It is considered a weed in the *Flora* area, but it is is cultivated in India as a grain and forage crop; the grain is sometimes used for birdseed.

8. **Urochloa mosambicensis** (Hack.) Dandy [p. 500]
SABI GRASS

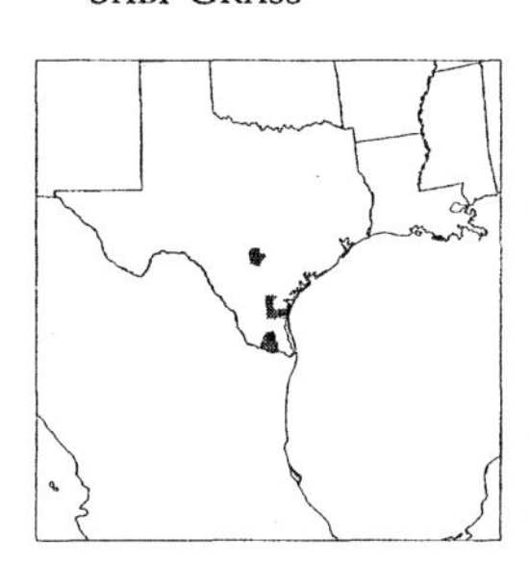

Plants perennial; cespitose, with or without stolons. **Culms** 20–150 cm; **nodes** pubescent; **internodes** with papillose-based hairs. **Sheaths** pubescent, lower sheaths pilose, upper sheaths with papillose-based hairs; **ligules** 1–2 mm; **blades** 3–30 cm long, (1.5) 3–20 mm wide, with scattered papillose-based hairs, margins scabrous. **Panicles** 3–12.5 cm, with 2–6(15) spikelike branches in 2 ranks; **primary branches** 2–10 cm, appressed to ascending, axes 0.8–1.4 mm wide, flat, winged, hispid, hairs not papillose-based; **secondary branches** present;

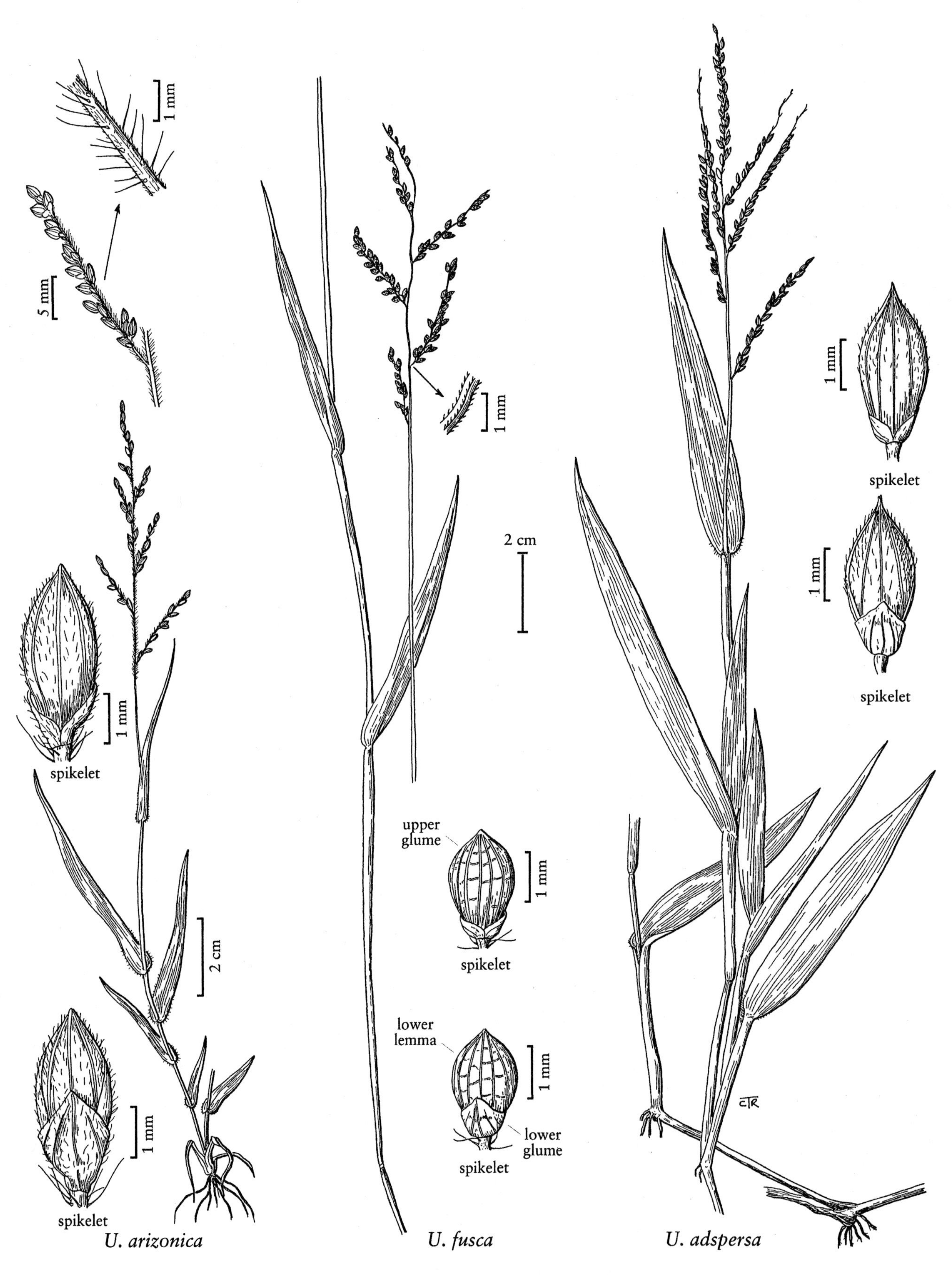
1 mm
5 mm
1 mm
spikelet
1 mm
2 cm
1 mm
spikelet
U. arizonica
1 mm
2 cm
upper
glume
1 mm
spikelet
lower
lemma
1 mm
lower
glume
spikelet
U. fusca
1 mm
spikelet
1 mm
spikelet
CTR
U. adspersa

UROCHLOA

pedicels shorter than the spikelets, scabrous, with 1–3 conspicuous hairs. **Spikelets** (3)4–5 mm long, 1.5–2 mm wide, solitary (rarely paired), appressed to the branch axes, bases glabrous or with a tuft of hairs. **Glumes** scarcely separated, rachilla internode between the glumes not pronounced; **lower glumes** 2.7–3.3 mm, (½)⅔–¾ as long as the spikelets, 3-veined, mostly glabrous but often with 1–3 conspicuous, stiff hairs emanating from the midvein at approximately mid-length; **upper glumes** (3)4–5 mm, glabrous or pubescent, 5-veined; **lower florets** staminate; **lower lemmas** (3)4–5 mm, glabrous or pubescent, 5-veined, with or without a setose fringe along the margins; **lower paleas** present; **upper lemmas** 2.2–2.6 mm, apices rounded, shortly awned, awns 0.5–1.2 mm; **anthers** 1.2–1.5 mm. $2n$ = 30, 42.

Urochloa mosambicensis, native to Africa, has been found in southern Texas (Wipff et al. 1993); it is expected to spread. It is grown for forage and hay in Africa.

9. **Urochloa piligera** (F. Muell. *ex* Benth.) R.D. Webster [p. 500]
WATTLE SIGNALGRASS

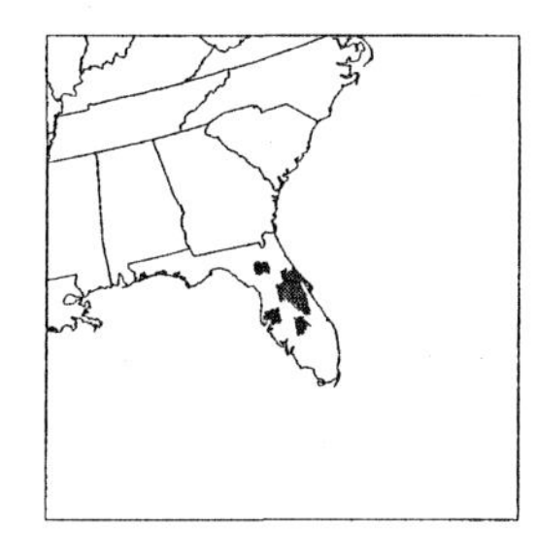

Plants annual. **Culms** 15–60 cm, erect or decumbent, sometimes rooting at the lower nodes; **nodes** glabrous. **Sheaths** glabrous, margins sometimes ciliate; **ligules** 0.7–1.5 mm; **blades** 4–15 cm long, 3–11 mm wide, usually glabrous, sometimes pubescent, margins scabrous. **Panicles** 3–12 cm, with 3–5 spikelike branches in 2 ranks, smooth or scabrous, glabrous or pubescent; **primary branches** 2–6 cm, divergent to reflexed, axils glabrous, axes 1.1–1.6 mm wide, flat, winged, without papillose-based hairs, margins smooth or scabrous; **secondary branches** rarely present; **pedicels** shorter than the spikelets, scabrous. **Spikelets** (3.3)3.8–4.9 mm long, 1.5–1.8 mm wide, ellipsoid, solitary, usually overlapping, appressed to the branch axes, pubescent or sometimes glabrous, in 2 rows, sometimes appearing 1-rowed. **Glumes** separated by 0.3–0.5 mm; **lower glumes** 1.9–2.7 mm, 9–11-veined, glabrous; **upper glumes** 3.2–4.1 mm, 7–9-veined, glabrous or pubescent, margins glabrous or pilose, when pilose, the basal hairs shorter than the distal hairs; **lower florets** sterile; **lower lemmas** 3.1–4.7 mm, resembling the upper glumes in texture and pubescence, 5–7-veined; **lower paleas** absent; **upper lemmas** 2–4 mm, apices recurved, rounded, mucronate; **anthers** 1.5–1.7 mm. $2n$ = unknown.

Urochloa piligera is an Australian species that has been found in Florida. Webster (1987) stated that *U. piligera* has glabrous and pubescent forms that are identical except in the presence or absence of spikelet vestiture. Currently, only the pubescent form has been found in the *Flora* area (Hall 1978). The glabrous form is sometimes confused with *U. subquadripara*. *Urochloa piligera* lacks a lower palea, and has larger and, usually, closely overlapping spikelets; *U. subquadripara* has well-developed paleas and smaller (3.3–3.8 mm), well-separated spikelets.

10. **Urochloa brizantha** (Hochst. *ex* A. Rich.) R.D. Webster [p. 502]
PALISADE SIGNALGRASS

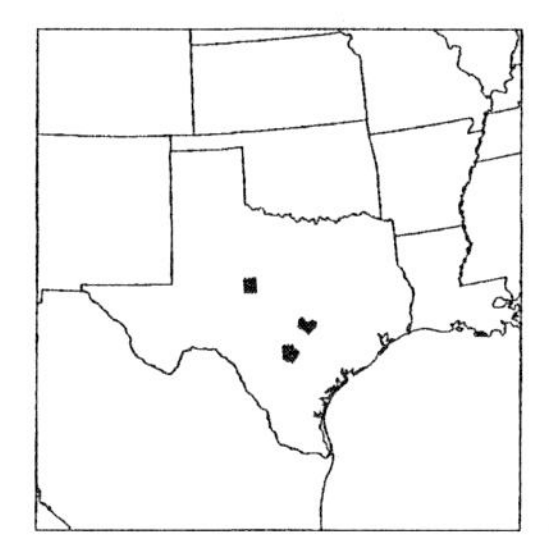

Plants perennial; shortly rhizomatous. **Culms** (30)100–200 cm, erect or geniculately ascending, occasionally branched; **nodes** glabrous. **Sheaths** glabrous or pubescent between the veins, margins glabrous; **collars** glabrous; **ligules** 1–2.2 mm; **blades** 9–40 cm long, 6–20 mm wide, glabrous or hispidulous on both surfaces, margins usually ciliate basally. **Panicles** 3–20 cm long, 2.5–3 cm wide, with 1–7(16) spikelike primary branches in 2 ranks; **rachises** scabrous and pubescent; **primary branches** 4–16(20) cm, ascending to divergent, axils glabrous, axes 0.5–1.2 mm wide, narrowly winged and crescentric (the margins inrolled to produce a crescent-shaped cross section), mostly scabrous, margins ciliate with papillose-based hairs; **secondary branches** absent, **pedicels** shorter than the spikelets, glabrous, scabrous. **Spikelets** 4–6 mm long, 1.8–2.2 mm wide, ovoid to ellipsoid, with 0.3–0.5 mm calluses, solitary, in 1 row (rarely in 2 rows at the base of the lower branches), appressed to the branches. **Glumes** separated by about 0.5 mm; **lower glumes** 1.8–3.3 mm long, about ⅓ as long as the spikelets, 7–11-veined; **upper glumes** 3.6–5.9 mm, 7-veined, glabrous or pubescent, without evident cross venation; **lower florets** staminate; **lower lemmas** 3.8–5.8 mm, glabrous or pubescent, 5-veined; **lower paleas** present; **upper lemmas** 3.3–5.6 mm long, 1.6–2 mm wide, ellipsoid, apices acute, mucronate; **anthers** about 2.2–2.5 mm. $2n$ = 18, 36, 54.

Urochloa brizantha, a native of tropical Africa, was first reported from the *Flora* region in 1993 (Fox and Hatch 1995). It is considered a sporadic introduction in the *Flora* area.

Clayton and Renvoize (1982) report that *Urochloa brizantha* intergrades with **U. decumbens** (Stapf) R.D. Webster, and intermediates are often difficult to separate, although they do not seem to be very common in the wild. The forms widely introduced throughout the tropics as forage were selected from among these intermediates. The selection most commonly used as a forage has the inflorescence characters of *U. brizantha* and the habit features of *U. decumbens*. Davidse and Pohl (1994)

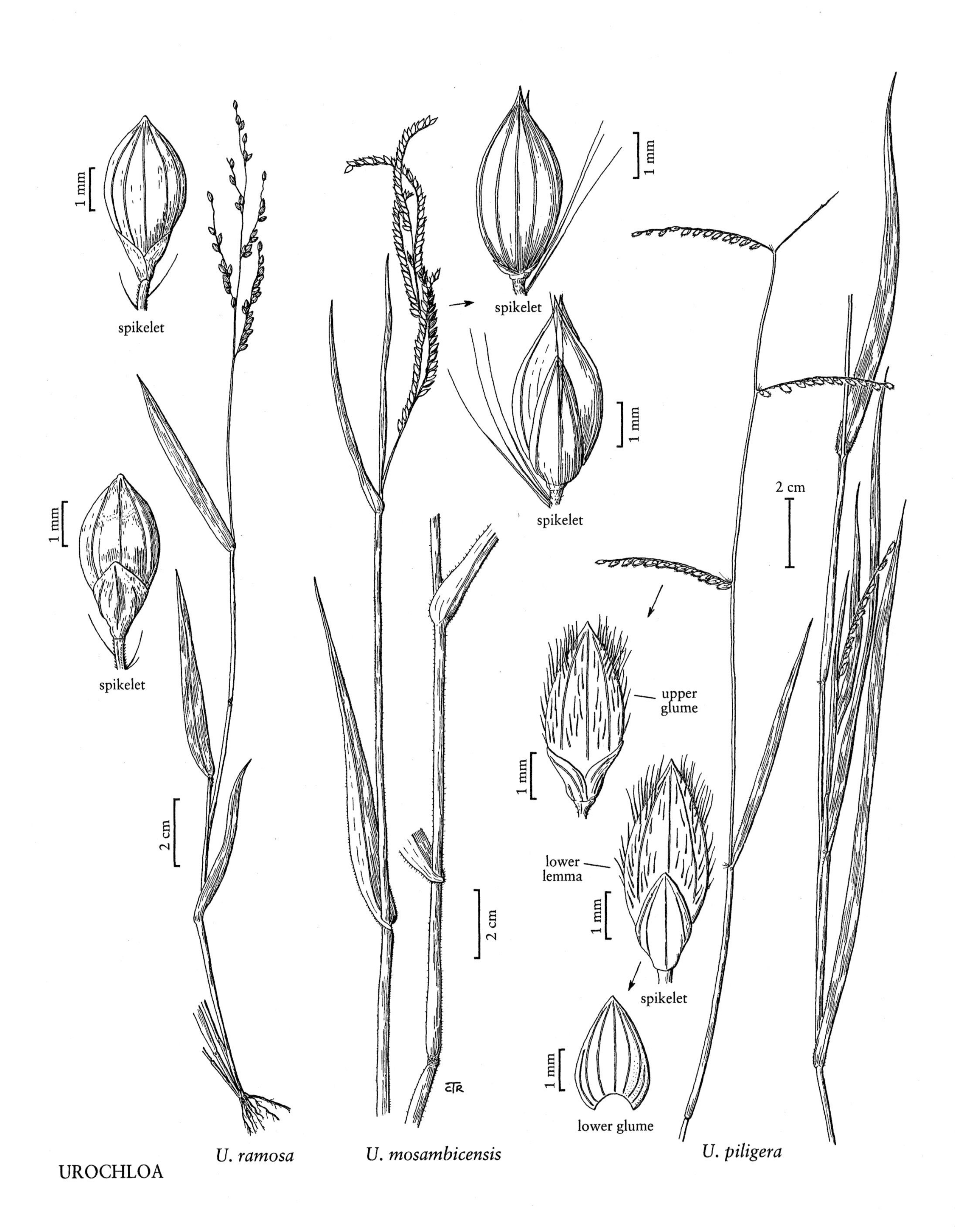
1 mm
spikelet
1 mm
spikelet
2 cm
1 mm
spikelet
1 mm
spikelet
2 cm
CTR
2 cm
upper glume
1 mm
lower lemma
1 mm
spikelet
1 mm
lower glume
U. ramosa
U. mosambicensis
U. piligera

UROCHLOA

reported that the material introduced into Mesoamerica is referable to *U. decumbens*. Although *U. decumbens* has not been reported in the *Flora* area, it is expected in Florida, because it is widely used as forage in the tropics. The two taxa can be distinguished by their panicle branches: *Urochloa decumbens* has flat, ribbonlike panicle branches 1–1.8 mm wide, whereas *U. brizantha* has crescentric panicle branches 0.5–1.2 mm wide.

11. **Urochloa villosa** (Lam.) T.Q. Nguyen [p. 502]
HAIRY SIGNALGRASS

Plants annual; loosely tufted or sprawling. **Culms** 7–50 cm, geniculate, decumbent to prostrate at the base, often much-branched below, usually rooting at the nodes; **nodes** pubescent; **Sheaths** usually pubescent, sometimes densely so, rarely glabrous, margins ciliate; **ligules** 0.3–1.1 mm, **blades** 1–7 cm long, 2–9.3 mm wide; usually densely puberulous on both sides, margins basally ciliate or scabrous. **Panicles** (1.5)4.5–8 cm long, with 4–12 spikelike primary branches in 2 ranks; **primary branches** 0.7–3.3 cm, axes 0.2–0.3 mm wide, triquetrous, pubescent; **secondary branches** rarely present, **pedicels** shorter than the spikelets. **Spikelets** 2–2.8 mm, solitary (or paired), in 1 row, appressed to the branch axes. **Glumes** separated by about 0.3 mm; **lower glumes** 0.7–1.5 mm, ⅓–½ as long as the spikelets, 3-veined; **upper glumes** glabrous or pubescent; 5-veined, **lower florets** sterile; **lower lemmas** similar to the upper glumes; **lower paleas** present; **upper lemmas** 1.8–2.1 mm, granulose to rugulose, acute, mucronate; **anthers** 0.9–1.3 mm. $2n = 36$.

Urochloa villosa is a tropical African and Asian species that Reed (1964) reported collecting from chrome and iron ore piles in Newport News, Virginia in 1959. His voucher specimens, acquired by the Missouri Botanical Garden in 2001, were not available for examination prior to publication of this volume, so the description is from Veldkamp (1996) and Clayton and Renvoize (1982). No additional collections have been reported in the *Flora* area.

12. **Urochloa subquadripara** (Trin.) R.D. Webster [p. 502]
ARMGRASS MILLET

Plants annual or short-lived perennials. **Culms** 10–70 cm, decumbent, branching and rooting at the lower nodes; **nodes** glabrous; **internodes** glabrous or sparsely pilose distally. **Sheaths** mostly pubescent or glabrous, margins ciliate; **ligules** 0.5–1.3 mm; **blades** 2–15(27) cm long, 3–10 (12) mm wide, glabrous or pubescent; margins scabrous, sometimes ciliate basally; bases subcordate, not clasping the stems. **Panicles** 2.5–13(22) cm, with 3–6(9) spikelike primary branches in 2 ranks; **primary branches** 2–7 cm, ascending to reflexed, axes 0.7–1.2 mm wide, flat and narrowly winged, margins scabrous; **secondary branches** absent; **pedicels** shorter than the spikelets, scabrous. **Spikelets** 3.3–3.8 mm long, 1–1.4 mm wide, solitary, appressed to the branches, in 2 rows. **Glumes** separated by 0.3–0.5 mm; **lower glumes** 1.4–1.7 mm, ⅓–½ as long as the spikelets, 9–11-veined, glabrous; **upper glumes** 2.7–3.4 mm, glabrous, 7–9-veined; **lower florets** sterile; **lower lemmas** 2.7–3.4 mm, glabrous, 5-veined, without evident cross venation; **lower paleas** present; **upper lemmas** 2.6–3.4 mm, apices rounded to acute, mucronate; **anthers** 1.1–1.35 mm. **Caryopses** 1.6–2.5 mm. $2n = 36, 54, 72, 84$.

Urochloa subquadripara, native to tropical Asia and Australasia, is established in Florida and, reportedly, Georgia, although no specimens documenting its presence in Georgia have been located. A weedy species that has been introduced into the tropics worldwide, it is reported to have good drought tolerance, and is used as forage in tropical Asia. Its weediness and drought tolerance suggest that it might also become a troublesome weed in some parts of the *Flora* region.

Urochloa subquadripara is similar to **U. distachya** (L.) T.Q. Nguyen, and the two taxa are sometimes treated as one species (e.g., Pohl 1980; Morrone and Zuloaga 1992, 1993). They are maintained here as separate species pending further research. *Urochloa distachya* supposedly has shorter (2.4–3 mm) spikelets and shorter (1.9–2.3 mm) upper florets.

Urochloa subquadripara can also be confused with the glabrous form of *U. piligera*, but it differs in having a well-developed palea and smaller (3.3–3.8 mm), well-separated spikelets.

13. **Urochloa plantaginea** (Link) R.D. Webster [p. 504]
PLANTAIN SIGNALGRASS

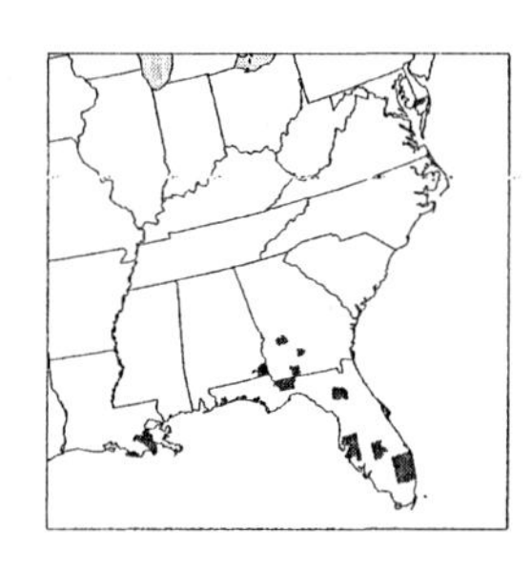

Plants annual. **Culms** 20–100 cm, decumbent, geniculate, branching and rooting at the lower nodes; **nodes** glabrous. **Sheaths** mostly glabrous, except the margins ciliate, with papillose-based hairs; **ligules** 0.5–1.5 mm; **blades** 3–21 cm long, 6–20 mm wide, glabrous, bases subcordate to cordate, clasping the stems, margins sometimes ciliate basally. **Panicles** 6–25 cm long, 2–7 cm wide, with 3–8 spikelike primary branches in 2 ranks; **primary branches** 2–11 cm, axes 1–1.5 mm wide,

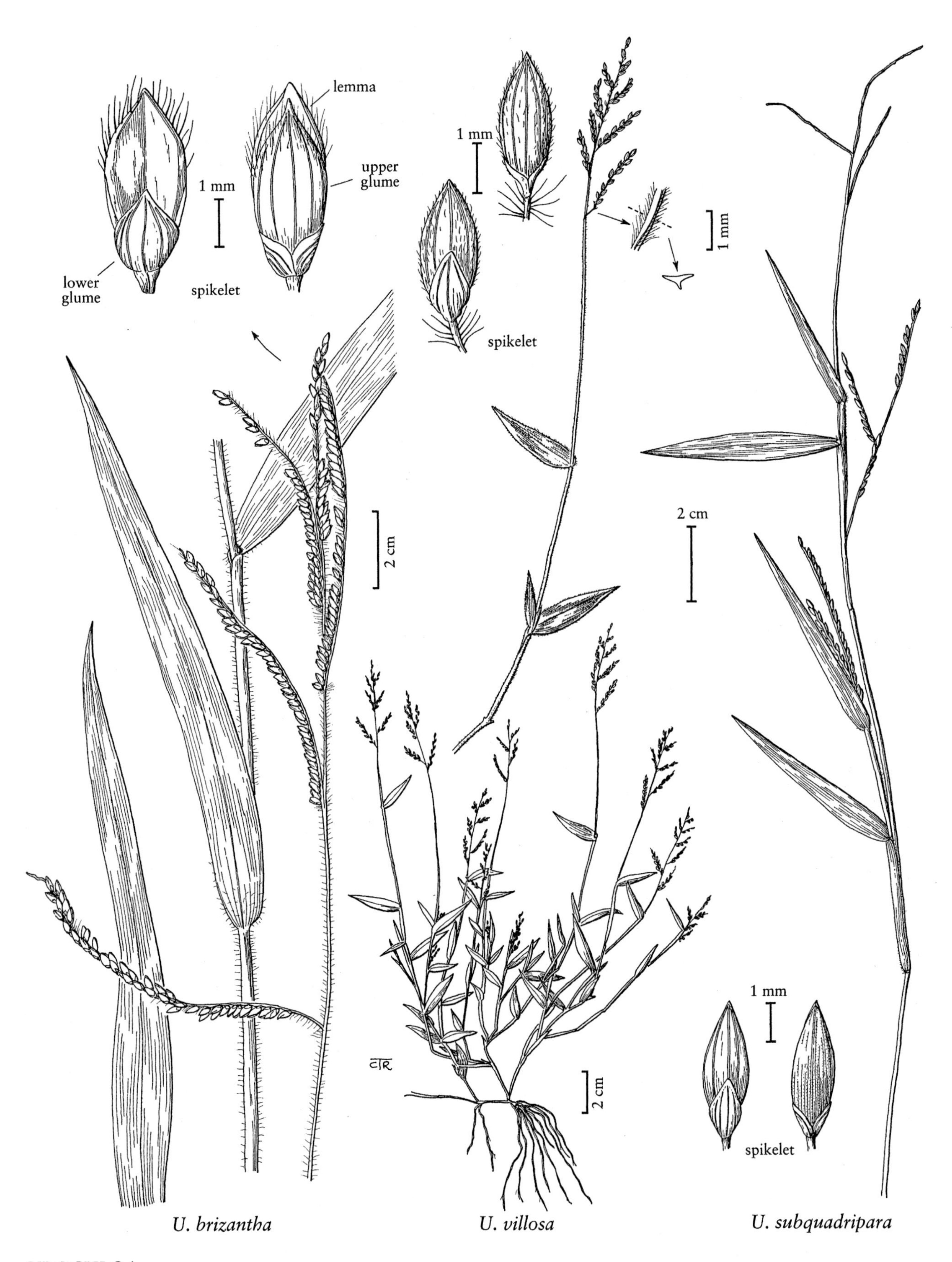
lemma
upper
glume
1 mm
lower
glume
spikelet
2 cm
1 mm
spikelet
1 mm
2 cm
2 cm
1 mm
spikelet
CR
U. brizantha
U. villosa
U. subquadripara

flat, margins scabrous; **secondary branches** absent; **pedicels** shorter than the spikelets, glabrous or scabrous. **Spikelets** (4)4.5–6 mm long, 1.9–2.2 mm wide, solitary, appressed to the branch axes, in 2 rows. **Glumes** separated by an internode of about 0.5 mm; **lower glumes** 1.5–2.5 mm, to ⅓ as long as the spikelets, broadly ovate, glabrous, 9–11-veined; **upper glumes** 3–4.2 mm, glabrous, 7(–9)-veined, without evident cross venation; **lower florets** sterile; **lower lemmas** 3–4.2 mm, glabrous, 5-veined; **lower paleas** present; **upper lemmas** 2.7–3.6 mm long, 1.5–2 mm wide, apices rounded; **anthers** 0.7–1 mm. **Caryopses** 2–2.5 mm. $2n$ = 36, 72.

Urochloa plantaginea, native to western and central Africa, is found from the southeastern United States to Argentina. It is now established in the southeastern United States, growing in loose sand and loam soils. Although considered a weed in the *Flora* area, Sendulsky (1978) stated that it provided good forage.

Hall (1978) reported **Urochloa oligobrachiata** (Pilg.) Kartesz [as *Brachiaria platytaenia* Stapf] from Florida. This report was based on one collection, but the voucher specimen has not been verified. It is similar to *U. plantaginea*, but it differs in having acute to acuminate lower glumes and shortly awned upper lemmas. It is native to western Africa.

14. **Urochloa panicoides** P. Beauv. [p. 504]
LIVERSEED GRASS

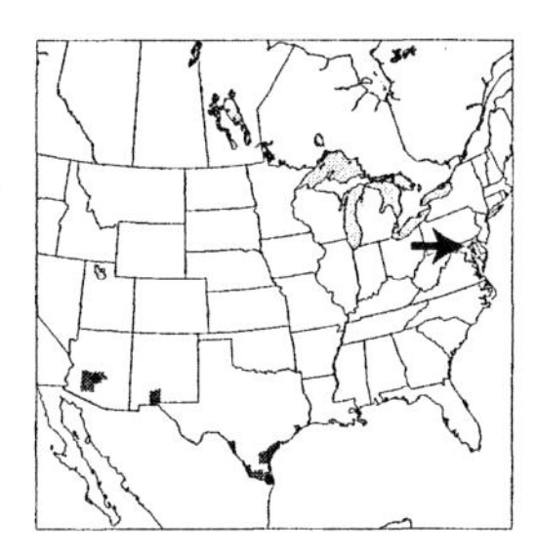

Plants annual. **Culms** (5)10–55(100) cm, erect to decumbent, usually rooting at the lower nodes; **nodes** pubescent; **internodes** glabrous or sparsely pubescent. **Sheaths** hispid, margins ciliate distally; **ligules** 1–1.5 mm; **blades** (2)5–25 cm long, 5–18 mm wide, both surfaces usually with papillose-based hairs, rarely glabrous, bases rounded to subcordate, ciliate, with papillose-based hairs. **Panicles** 3.5–10 cm long, 2–7 (10) cm wide, with 2–7(10) spikelike primary branches in 2 ranks; **primary branches** 1–7 cm, axes about 0.9–1.2 mm wide, flat, scabrous and ciliate with papillose-based hairs; **secondary branches** rarely present; **pedicels** shorter than the spikelets, frequently with 1–5 long hairs. **Spikelets** 2.5–4.5(5.5) mm long, about 1.5–2 mm wide, ellipsoid, solitary, in 2 rows, appressed to the branches. **Glumes** scarcely separated; **lower glumes** 1–1.6 mm, ¼–⅓(½) as long as the spikelets, 3–5-veined, clasping the base of the spikelets, glabrous; **upper glumes** 3.2–4.3 (5) mm, 9–11(13)-veined, glabrous; **lower florets** sterile; **lower lemmas** 3.2–4.3 (5) mm long, glabrous, 5(–7)-veined; **lower paleas** present; **upper lemmas** 2.6–3.5 mm long, about 1.5 mm wide, apices rounded, awned, awns (0.3)0.6–1 mm; **anthers** 0.8–1 mm. **Caryopses** 2–2.5 mm. $2n$ = 30, 36, 48.

Urochloa panicoides is native to Africa, and is considered a noxious weed by the U.S. Department of Agriculture. In the Western Hemisphere, it has been introduced into the southern United States, Mexico, and Argentina. Within the *Flora* region, it has been reported from disturbed sites in the southern and Gulf Coast areas of Texas (Wipff et al. 1993), but it is expected to spread. Populations from New Mexico have been destroyed.

Urochloa panicoides has morphological forms with glabrous, pubescent, or setosely-fringed spikelets. Although the three forms have been recognized taxonomically, Clayton and Renvoize (1982) state that they appear to be of no taxonomic significance. Only the glabrous form is known to occur in the *Flora* area.

15. **Urochloa platyphylla** (Munro *ex* C. Wright) R.D. Webster [p. 504]
BROADLEAF SIGNALGRASS

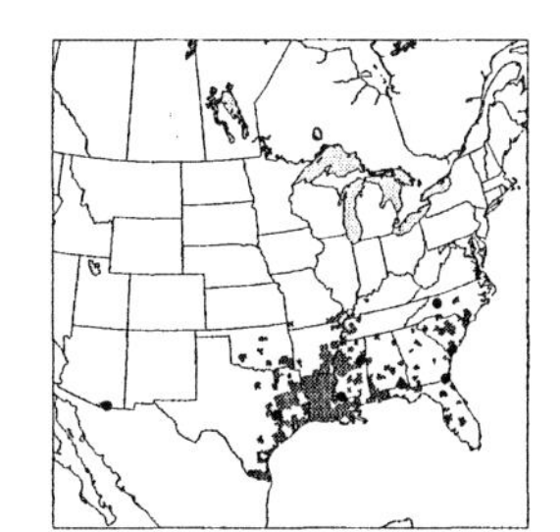

Plants annual. **Culms** 25–100 cm, decumbent, rooting at the lower nodes; **nodes** glabrous. **Sheaths** glabrous or sparsely pilose; **ligules** 0.5–1 mm; **blades** 2.5–17.5 cm long, 3–13 mm wide, glabrous or sparsely pilose, bases subcordate, not clasping the stems, margins ciliate basally, with papillose-based hairs. **Panicles** 6–16 cm long, 2–2.5 cm wide, with 2–8 spikelike primary branches in 2 ranks; **primary branches** 3–8 cm, axils pubescent, axes 1.3–2.5 mm wide, flat, usually glabrous, occasionally pilose dorsally; **secondary branches** rarely present; **pedicels** shorter than the spikelets, scabrous and sparsely pilose. **Spikelets** 3.8–5 mm long, 2–2.5 mm wide, ovoid, bi-convex; solitary, appressed to the branches, in 2 rows. **Glumes** scarcely separated; **lower glumes** 1.2–1.8 mm, to ⅓ as long as the spikelets, obtuse, glabrous, 5(–7)-veined, not clasping the base of the spikelets; **upper glumes** 3.2–4.7 mm, glabrous, 7(–9)-veined; **lower florets** sterile; **lower lemmas** 3.2–4.7 mm, glabrous, 5-veined; **lower paleas** present; **upper lemmas** 2.8–3.4 mm long, 1.8–2.3 mm wide, apices incurved, broadly acute to rounded, mucronulate; **anthers** about 1 mm. **Caryopses** 1.5–2.2 mm. $2n$ = 36.

Urochloa platyphylla is a weedy species found in open, sandy soil in the southeastern United States, West Indies, and South America. Morrone and Zuloaga (1993) considered reports of its occurrence in Mexico as doubtful, possibly being based on misidentified specimens of *Urochloa plantaginea*.

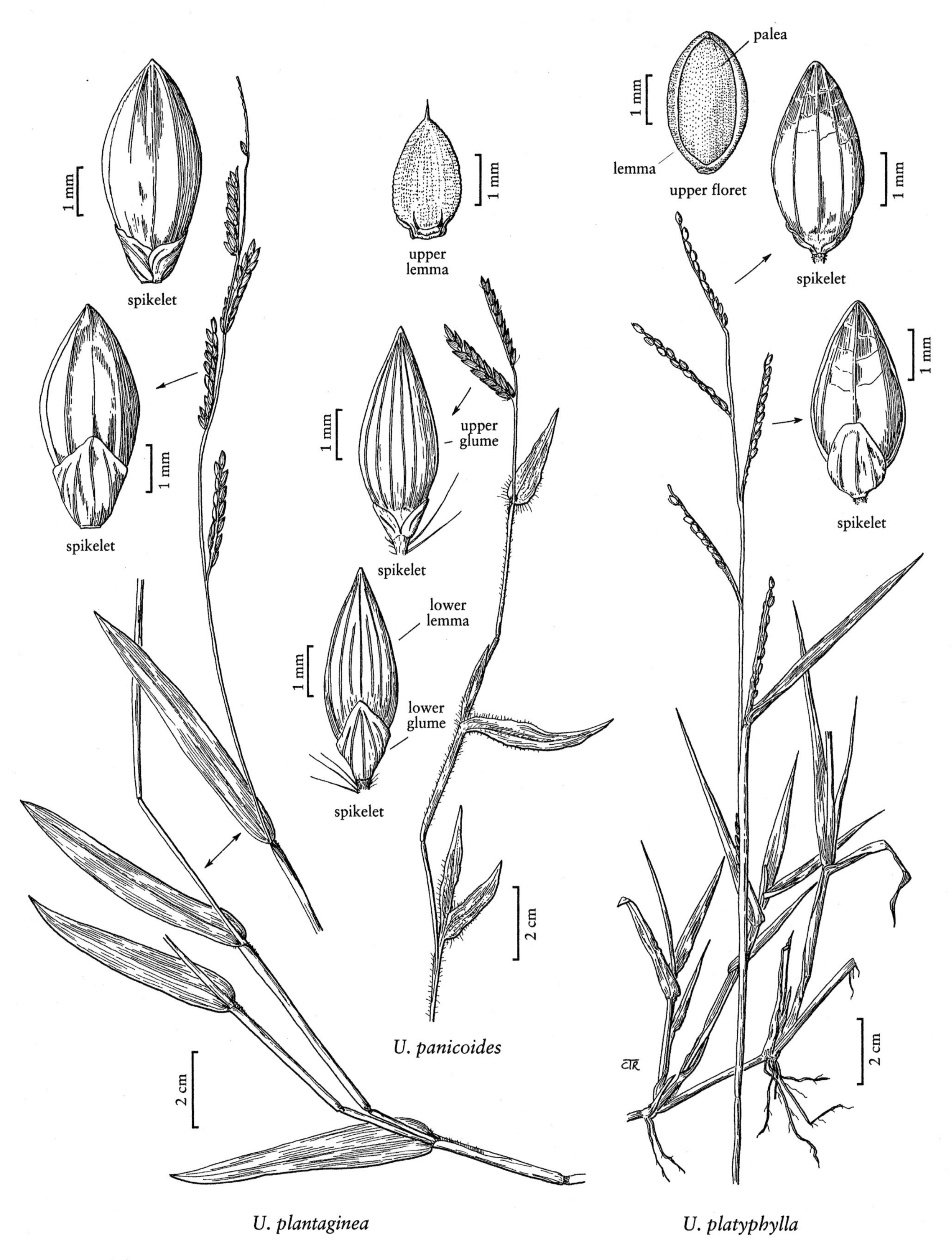

U. plantaginea

U. platyphylla

16. **Urochloa arrecta** (Hack. *ex* T. Durand & Schinz) Morrone & Zuloaga [p. 506]
AFRICAN SIGNALGRASS

Plants perennial; stoloniferous. **Culms** 50–120 cm, branching and rooting at the lower nodes; **nodes** glabrous. **Sheaths** glabrous, margins ciliate; **ligules** about 1 mm; **blades** 5–15 cm long, 7–15 mm wide, glabrous, bases subcordate, margins scabrous. **Panicles** (5)9–18(25) cm long, 3–4 cm wide, with 4–10(15) spikelike primary branches in 2 ranks; **primary branches** (1)2–5(10) cm, axes 0.5–2 mm wide, glabrous, margins scabrous; **secondary branches** rarely present, **pedicels** shorter than the spikelets, mostly scabrous, apices with hairs. **Spikelets** (3)3.3–4.4 mm long, 1.4–1.7 mm wide, ellipsoid, solitary, imbricate, in 2 rows, appressed to the branches. **Glumes** scarcely separated; **lower glumes** 1.5–1.8 mm, glabrous, 5-veined, not clasping the base of the spikelets; **upper glumes** 3.4–4.1 mm, glabrous, 7-veined; **lower florets** staminate; **lower lemmas** 3.4–4.1 mm, glabrous, 5-veined; **upper lemmas** 2.7–3.5 mm long, 1.3–1.6 mm wide, apices rounded, incurved; **anthers** 1.6–1.8 mm. $2n$ = unknown.

Urochloa arrecta is native to Africa, but it has been introduced into Florida and Brazil as a forage grass. It is reported to be established in Collier County, Florida.

17. **Urochloa ciliatissima** (Buckley) R.D. Webster [p. 506]
FRINGED SIGNALGRASS, SANDHILL GRASS

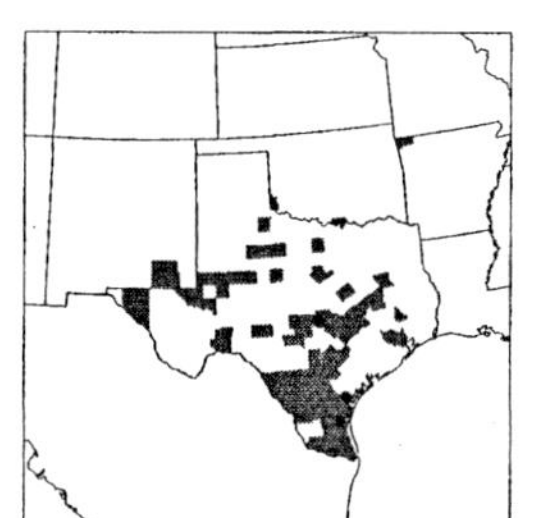

Plants perennial; shortly rhizomatous or with long stolons. **Culms** 10–40 cm, erect to ascending, solitary or in small clumps; **nodes** retrorsely villous; **internodes** glabrous. **Sheaths** glabrous or with papillose-based hairs; **ligules** 0.5–1.5 mm; **blades** 1–7(9) cm long, 2–5 mm wide, glabrous or pilose on both surfaces, margins ciliate basally, with papillose-based hairs. **Panicles** 3–6 cm long, 0.5–1 cm wide, with 2–6 spikelike primary branches in 2 ranks; **rachises** scabridulous; **primary branches** 0.5–2 cm, appressed, axes 0.3–0.4 mm wide, triquetrous, scabridulous, glabrous or puberulent; **secondary branches** rarely present; **pedicels** shorter than the spikelets, scabridulous. **Spikelets** 3–4.5 mm long, 1.5–2 mm wide, plano-convex, solitary, in 2 rows, appressed to the branch axes. **Glumes** scarcely separate, rachilla between the glumes not pronounced; **lower glumes** 2.8–3.2 mm, 5–7-veined, glabrous or with long hairs basally; **upper glumes** 3–4.5 mm, (9)11–13-veined, without cross venation, mostly puberulent, margins pilose-fringed; **lower florets** staminate; **lower lemmas** 3–4.5 mm, 7–9-veined, without cross venation, mostly puberulent, margins pilose-fringed; **lower paleas** present; **upper lemmas** 2.4–2.8 mm long, 1.3–1.4 mm wide, plano-convex, apices broadly acute to rounded, mucronate; **anthers** about 1 mm. **Caryopses** 1.8–3 mm. $2n$ = 36.

Urochloa ciliatissima is endemic to Texas, Oklahoma, and Arkansas, and grows on sandy soils. Reports of its occurrence in Mexico are based on misidentifications (Morrone and Zuloaga 1993).

18. **Urochloa maxima** (Jacq.) R.D. Webster [p. 506]
GUINEA GRASS

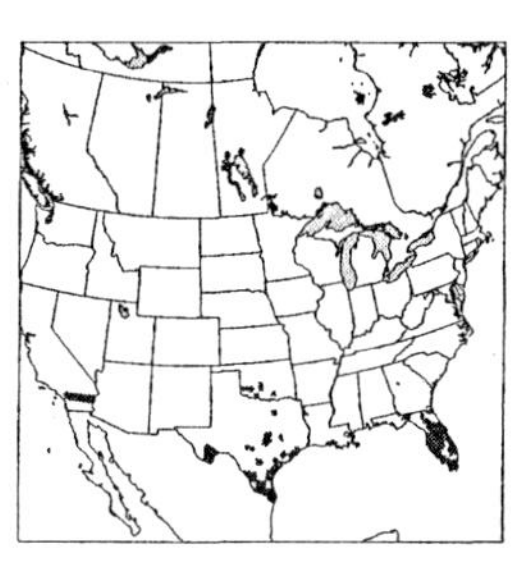

Plants perennial; cespitose, with short, thick rhizomes. **Culms** (60)100–250 cm tall, about 10 mm thick, mostly erect, sometimes geniculate and rooting at the lower nodes; **nodes** pubescent or glabrous. **Sheaths** usually shorter than the internodes, glabrous or pubescent, sometimes with papillose-based hairs, margins sometimes ciliate; **collars** densely pubescent, hairs appressed or divergent; **ligules** 1–3 mm; **blades** (15)30–75(100) cm long, 10–35 mm wide, flat, erect or ascending, glabrous or pubescent, sometimes with appressed papillose-based hairs, margins scabrous, sometimes ciliate basally, midveins conspicuous, sunken, whitish. **Panicles** 20–65 cm, about ⅓ as wide as long, open, rachises smooth or scabrous; **primary branches** usually more than 20, 12–40 cm, axes 0.4–0.6 mm wide, not winged, ascending, those of the lower node(s) verticillate and pilose at the base, upper axils glabrous, lower branches naked basally; **secondary** and **tertiary branches** well-developed; **pedicels** 0.5–1.5 mm, unequal, straight or curved, glabrous or with a single setaceous hair near the apex. **Spikelets** 2.7–3.6 mm long, 0.9–1.1 mm wide, oblong-ellipsoid, usually glabrous (rarely densely covered with papillose-based hairs), solitary, paired (or in triplets), usually appressed to the branch axes. **Glumes** scarcely separate, rachilla between the glumes not pronounced; **lower glumes** 0.8–1.2 mm, 1–3-veined, obtuse or truncate, glabrous; **upper glumes** 2.1–3.5 mm, 5-veined, glabrous; **lower lemmas** 2.1–3.5 mm, subequal, glabrous, 5-veined, without cross venation, acute, muticous or mucronate; **lower florets** staminate; **upper lemmas** 1.9–2.4 mm, ellipsoid, pale, glabrous, apices acute, mucronulate; **anthers** 1.2–2.2 mm. $2n$ = 18, 32, 36, 44, 48.

Urochloa maxima is an important forage grass that is native to Africa. In the *Flora* region, it grows in fields, waste places, stream banks, and hammocks. It is

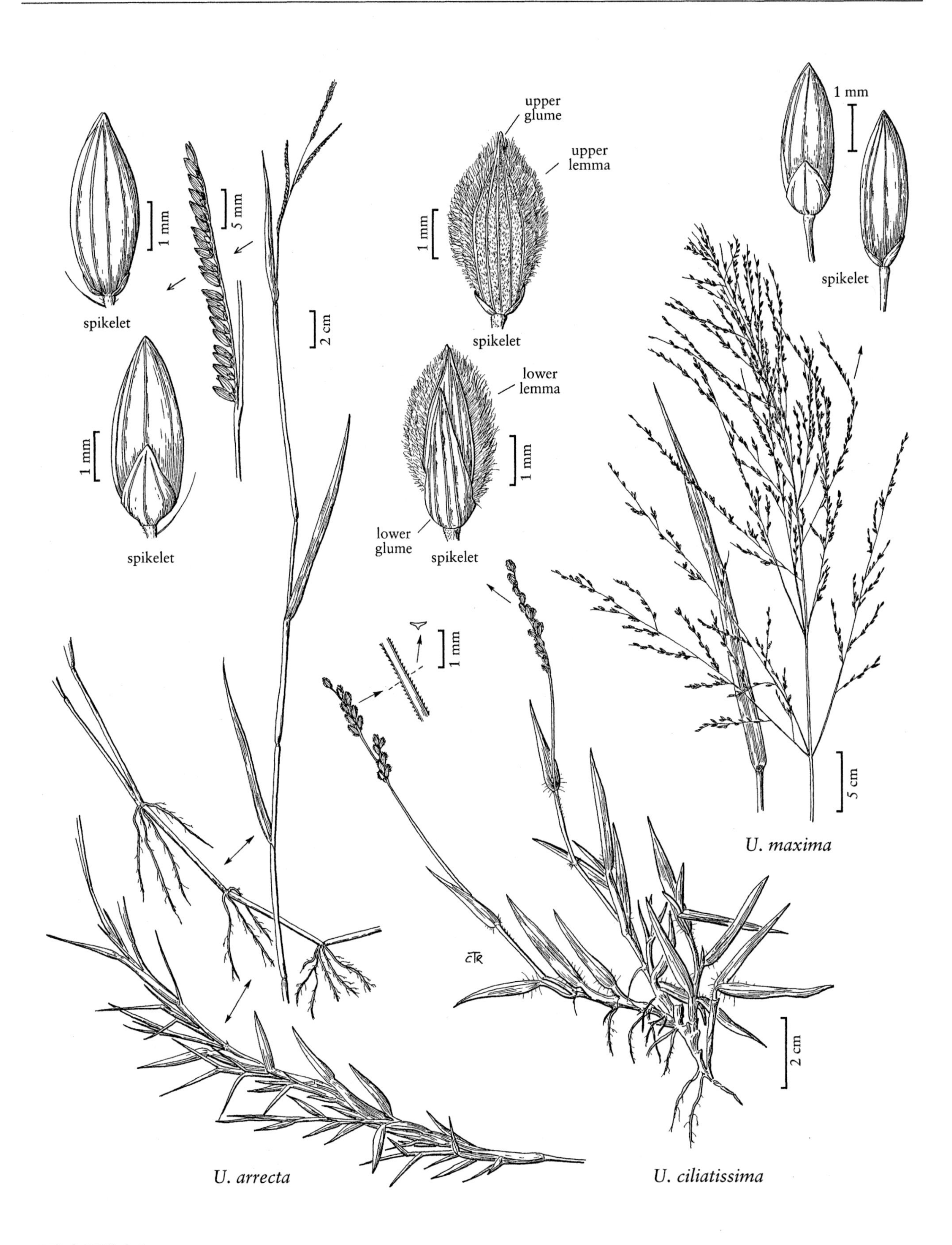
spikelet
1 mm
5 mm
spikelet
1 mm
2 cm
upper glume
upper lemma
1 mm
spikelet
lower lemma
1 mm
lower glume
spikelet
1 mm
1 mm
spikelet
5 cm
U. maxima
2 cm
U. arrecta
U. ciliatissima

cultivated widely as a forage grass at low elevations, especially near the coast, and often escapes.

There are usually two varieties recognized. Only **Urochloa maxima** (Jacq.) R.D. Webster var. **maxima**, which has glabrous spikelets, is known in the *Flora* area. Specimens with densely pubescent spikelets belong to **U. maxima** var. **trichoglumis** (Robyns) R.D. Webster.

25.14 ERIOCHLOA Kunth

Robert B. Shaw
Robert D. Webster
Christine M. Bern

Plants annual or perennial; cespitose, sometimes with short rhizomes or stolons, not producing subterranean spikelets. **Culms** 20–250 cm, erect or decumbent, usually with 2–5 nodes. **Sheaths** open; **auricles** absent; **ligules** membranous, ciliate. **Inflorescences** terminal, panicles of spikelike branches on elongate rachises; **branches** with many pedicellate, loosely appressed spikelets, terminating in a spikelet, without stiff bristles or flat bracts, spikelets in pairs, triplets, or solitary, often solitary distally when in pairs or triplets at the middle of the branches; **pedicels** terminating in a well-developed disk; **disarticulation** below the glume(s). **Spikelets** with 2 florets, lower florets usually sterile, upper florets bisexual. **Lower glumes** typically reduced (sometimes absent) and fused with the glabrous callus to form a cuplike structure; **upper glumes** lanceolate to ovate, glabrous or variously pubescent, 3–9-veined, unawned or awned; **lower lemmas** similar to the upper glumes in length, shape, venation, and pubescence, unawned; **lower paleas** absent to fully developed; **upper lemmas** lanceolate to ovate, indurate, rugose, dull, glabrous, rounded on the back, veins not pronounced, margins involute; **anthers** 3; **lodicules** 2, papery; **styles** with 2 branches, purple, plumose. **Caryopses** not longitudinally grooved; **endosperm** solid. $x = 9$. Name from the Greek *erion*, 'wool', and *chloe*, 'grass', a reference to the usually pubescent pedicels and rachises.

Eriochloa, a genus of 20–30 species, grows in tropical, subtropical, and warm-temperate areas of the world. Eight species of *Eriochloa* are native to the *Flora* region and three are introduced. Of the three introduced species only two, *E. polystachya* and *E. pseudoacrotricha*, have become naturalized.

Only one native species, *Eriochloa sericea*, is abundant enough to be an important forage species. The introduced *E. polystachya* is also used for this purpose.

SELECTED REFERENCES **Shaw, R.B. and F.E. Smeins.** 1981. Some anatomical and morphological characteristics of the North American species of *Eriochloa* (Poaceae: Paniceae). Bot. Gaz. 142:534–544; **Shaw, R.B. and R.D. Webster.** 1987. The genus *Eriochloa* (Poaceae: Paniceae) in North and Central America. Sida 12:165–207.

1. Spikelets solitary at the middle of the branches, sometimes in unequally pedicellate pairs near the base.
   2. Pedicels with more than 12 long (1.5–3 mm) hairs near the apices, densely hirsute or villous below, the hairs mostly about 0.1 mm long, but with some longer hairs interspersed among the short hairs.
      3. Blades 0.5–4 mm wide; spikelets 1.4–1.9 mm wide; plants perennial . . . . 1. *E. sericea*
      3. Blades 5–12 mm wide; spikelets 2–2.5 mm wide; plants annual . . . . 2. *E. villosa*
   2. Pedicels with fewer than 10 long (1.5–3 mm) hairs near the apices; variously hirsute below.
      4. Lower floret of each spikelet with a palea . . . . 3. *E. michauxii*
      4. Lower floret of each spikelet without a palea.
         5. Rachises hairy, the longer hairs 0.1–0.8 mm long; spikelets 3.1–5 mm long, 1.2–1.7 mm wide . . . . 4. *E. contracta*

5. Rachises glabrous or scabrous, not hairy; spikelets 2.7–3.6 mm long, 0.8–1.5 mm wide . . . . 5. *E. fatmensis*

1. Spikelets in unequally pedicellate pairs or triplets at the middle of the branches, sometimes solitary distally.
   6. Adaxial surfaces of the blades velvety to the touch; cauline internodes pubescent to pilose . . . . 6. *E. lemmonii*
   6. Adaxial surfaces of the blades glabrous or hairy, but not velvety to the touch; cauline internodes glabrous or pubescent.
      7. Upper lemmas unawned or the awns shorter than 0.2(0.3) mm.
         8. Plants annual; upper glumes acute to acuminate, often terminating in awnlike apices up to 1.5 mm long; lower florets without paleas.
            9. Longer pedicels of the spikelet pairs or triplets to 1 mm long; upper glumes acute to acuminate, unawned or awned, the awns up to 1.2 mm long . . . . 10. *E. acuminata*
            9. Longer pedicels of the spikelet pairs to 3 mm long; upper glumes always acuminate, awned, the awns 0.5–3.5 mm long . . . . 8. *E. aristata*
         8. Plants perennial; upper glumes acute, unawned; lower floret of each spikelet usually with a palea.
            10. Culms erect, not rooting at the lower nodes; spikelets 3.7–5.7 mm long, 1.3–1.8 mm wide; lower floret of each spikelet always with a palea . . . . 3. *E. michauxii*
            10. Culms decumbent, rooting at the lower nodes; spikelets 3.2–3.9 mm long, 1.1–1.3 mm wide; lower floret of each spikelet with or without a palea . . . . 11. *E. polystachya*
      7. Upper lemmas awned, the awns 0.2–1.5 mm long.
         11. Spikelets 2.7–3.6 mm long, 0.8–1.5 mm wide . . . . 5. *E. fatmensis*
         11. Spikelets 3.6–8.8 mm long, 0.9–1.6 mm wide.
            12. Pedicels uniformly hirsute, the hairs about 0.1 mm long; plants rhizomatous . . . . . 7. *E. punctata*
            12. Pedicels with some hairs 0.5–2.5 mm long, at least distally; plants not rhizomatous.
               13. Plants perennial . . . . 9. *E. pseudoacrotricha*
               13. Plants annual.
                  14. Longer pedicels of the spikelet pairs or triplets to 1 mm long; upper glumes acute to acuminate, unawned or awned, the awns up to 1.2 mm long . . . . 10. *E. acuminata*
                  14. Longer pedicels of the spikelet pairs to 3 mm long; upper glumes always acuminate, awned, the awns 0.5–3.5 mm long . . . . 8. *E. aristata*

1. Eriochloa sericea (Scheele) Munro *ex* Vasey [p. 510]
TEXAS CUPGRASS

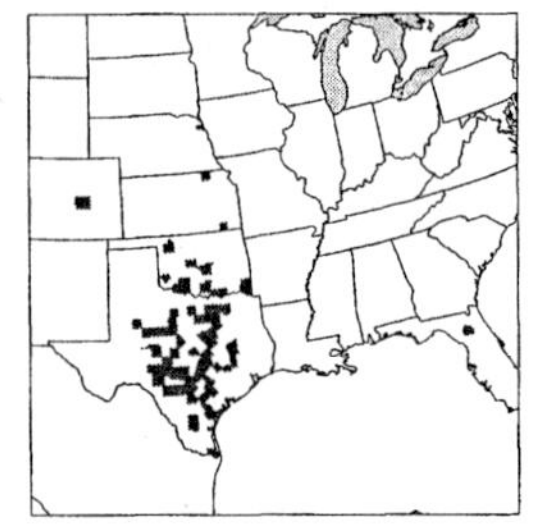

Plants perennial; cespitose, shortly rhizomatous. **Culms** 30–130 cm, erect or decumbent, sometimes rooting at the lower nodes; **internodes** pubescent; **nodes** puberulent to densely pubescent. **Sheaths** 3–14 cm, not overlapping, frequently inflated or spreading from the culm, chartaceous, glabrous or pubescent; **ligules** 0.5–1.5 mm, densely ciliate; **blades** 10–30 cm long, 0.5–4 mm wide, filiform to linear, involute to flat, straight or lax, appressed to divergent, glabrous or with soft pubescence. **Panicles** 4–20 cm long, 0.5–1.5 cm wide, contracted; **rachises** pilose or villous; **branches** (2)4–8(10), 10–35 mm long, 0.4–0.7 mm wide, velutinous, sometimes winged, with 10–20 solitary spikelets; **pedicels** 0.4–0.7 mm, densely hirsute with a mixture of short and long hairs, apices with more than 12 hairs of 1.5–2.5 mm. **Spikelets** 4–5 mm long, 1.4–1.9 mm wide. **Lower glumes** absent; **upper glumes** appressed pubescent, ovate to elliptic, 5–7-veined, acute, unawned; **lower lemmas** 3.8–4.8 mm long, 1.5–1.9 mm wide, indurate, elliptic, pubescent to velutinous, 5–7-veined, acute, unawned; **lower paleas** absent; **anthers** absent; **upper lemmas** 2.7–3.6 mm, elliptic, acute, shortly awned, awns 0.1–0.2 mm; **upper paleas** slightly shorter than the lemmas, indurate, minutely rugose. $2n$ =54.

Eriochloa sericea usually grows on clay or clay-loam soils in prairies, roadsides, or protected areas. It is widespread in the blackland prairie and Edwards Plateau of Texas, but extends into Kansas, Nebraska, and Oklahoma, and onto the coastal prairie and rolling plains of Texas and northern Mexico.

2. **Eriochloa villosa** (Thunb.) Kunth [p. 510]
ÉRIOCHLOÉ VELUE

Plants annual. **Culms** 30–100 cm, erect or decumbent, sometimes rooting at the lower nodes; **nodes** and **internodes** pubescent. **Sheaths** sometimes inflated, glabrous or pubescent; **ligules** 0.5–1 mm; **blades** 10–20 cm long, 5–12 mm wide, flat, adaxial surfaces hairy. **Panicles** 3–16 cm long, 1–3 cm wide; **rachises** villous; **branches** 2–8, 20–70 mm long, 0.8–1.1 mm wide, velutinous, sometimes winged, with 11–24 solitary spikelets (occasionally paired proximally); **pedicels** 0.5–1 mm, densely villous below, often with long hairs intermixed with the short hairs, apices with more than 12 hairs of 1.5–2.5 mm. **Spikelets** 3.9–5.2 mm long, 2–2.5 mm wide, ovate to elliptic. **Lower glumes** occasionally present as a scale; **upper glumes** equaling the lower lemmas, ovate to elliptic, glabrous or pubescent, 7-veined; **lower lemmas** 3.4–5 mm long, 2–2.5 mm wide, 5-veined, acute to apiculate, unawned; **lower paleas** absent; **anthers** absent; **upper lemmas** 3.5–5 mm, ovate to elliptic, acute to apiculate. $2n = 54$.

Eriochloa villosa is a weedy species of eastern Asia that has been found at scattered locations in the *Flora* region.

3. **Eriochloa michauxii** (Poir.) Hitchc. [p. 512]
LONGLEAF CUPGRASS

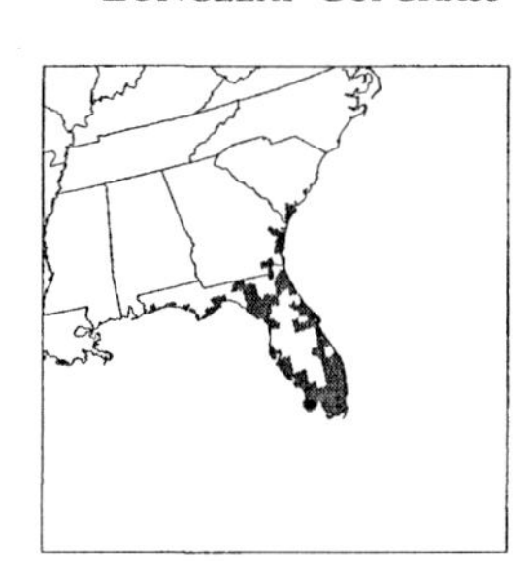

Plants perennial; cespitose and rhizomatous. **Culms** 50–250 cm, erect, not rooting at the lower nodes; **internodes** glabrous; **nodes** 3–7, puberulent to appressed pilose, rarely glabrous. **Sheaths** glabrous; **collars** hairy; **ligules** 0.5–1.5 mm; **blades** 10–60 cm long, 5–15 mm wide, linear, flat to conduplicate, straight or lax, spreading, glabrous adaxially. **Panicles** 10–30 cm long, 2.5–10 cm wide, open or contracted; **rachises** villous or puberulent; **branches** 10–25, 15–60 mm long, 0.5–0.8 mm wide, appressed to spreading, villous to shortly pilose, rarely puberulent or scabrous, not winged, with 16–40 spikelets on the primary branches, spikelets in unequally pedicellate pairs or triplets proximally, solitary distally; **pedicels** 0.3–2.5 mm, variously hirsute below, with fewer than 10 hairs more than 0.5 mm long at the apices. **Spikelets** 3.7–5.7 mm long, 1.3–1.8 mm wide, ovate to elliptic. **Upper glumes** equaling the lower lemmas, ovate, hairy, 5-veined, acute, unawned, rarely mucronate; **lower lemmas** 3.5–5 mm long, 1.3–1.8 mm wide, ovate to elliptic, setose, 5(7)-veined, acute, unawned; **lower paleas** fully developed, as long as or longer than the lemmas, hyaline; **anthers** 3 or absent; **upper lemmas** 3.1–4.6 mm, indurate, elliptic, 5-veined, acute to rounded, mucronate or awned, awns 0.1–0.6 mm; **upper paleas** 3–4.3 mm, indurate, blunt. $2n = 36$.

Eriochloa michauxii is endemic to the southeastern United States. There are two varieties, differing as shown in the following key. Intermediate plants have been collected in Lee and Monroe counties, Florida.

1. Lower florets staminate; blades generally flat, usually 8–15 mm wide . . . . var. *michauxii*
1. Lower floret sterile; blades involute to conduplicate, 5–8 mm wide . . . . var. *simpsonii*

Eriochloa michauxii (Poir.) Hitchc. var. **michauxii** grows in brackish or fresh water marshes, hammocks, and prairies of the southeasten United States, including the whole of Florida.

Eriochloa michauxii var. **simpsonii** (Hitchc.) Hitchc. is a rare variety that grows in low wet areas, roadsides, or on washed sand and shell beaches of southwestern Florida.

4. **Eriochloa contracta** Hitchc. [p. 512]
PRAIRIE CUPGRASS

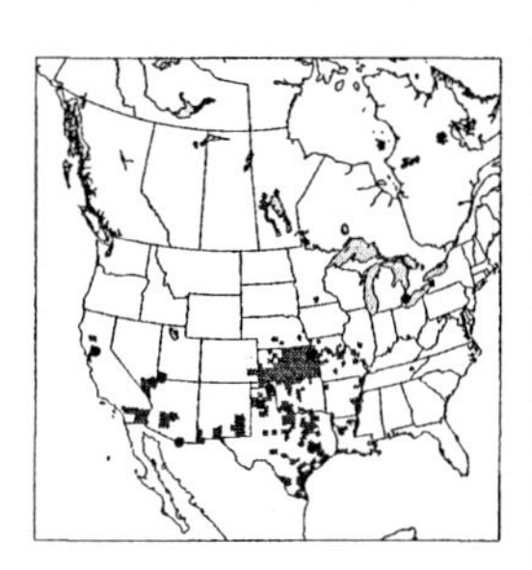

Plants annual; cespitose. **Culms** 20–100 cm, erect or decumbent, sometimes rooting at the lower nodes; **internodes** pilose or pubescent; **nodes** pubescent to puberulent. **Sheaths** sparsely to densely pubescent; **ligules** 0.4–1.1 mm; **blades** 6–12(22) cm long, 2–8 mm wide, linear, flat to conduplicate, straight, appressed to divergent, both surfaces sparsely to densely pubescent with short, evenly spaced hairs. **Panicles** 6–20 cm long, 0.3–1.2 cm wide; **rachises** pilose, longer hairs 0.1–0.8 mm; **branches** 10–20(28), 15–45(60) mm long, 0.2–0.4 mm wide, appressed, pubescent to setose, not winged, with 8–16 mostly solitary spikelets, occasionally paired at the base of the branches; **pedicels** 0.2–1 mm, variously hirsute below, apices with fewer than 10 hairs more than 0.5 mm long. **Spikelets** (3.1)3.5–4.5(5) mm long, 1.2–1.7 mm wide, lanceolate. **Upper glumes** as long as the lower lemmas, with sparsely appressed pubescence on the lower ⅔, scabrous or glabrous distally, 3–9-veined, acuminate and awned, awns 0.4–1 mm; **lower florets** sterile; **lower lemmas** 3–4.3 mm long, 1.2–1.7 mm wide, lanceolate, setose, 3–7-veined, acuminate, unawned or mucronate; **lower paleas** absent; **upper lemmas** 2–2.5 mm, indurate, elliptic, 5–7-veined, acute to rounded and awned, awns 0.4–1.1 mm; **upper paleas** indurate, faintly rugose, blunt. $2n = 36$.

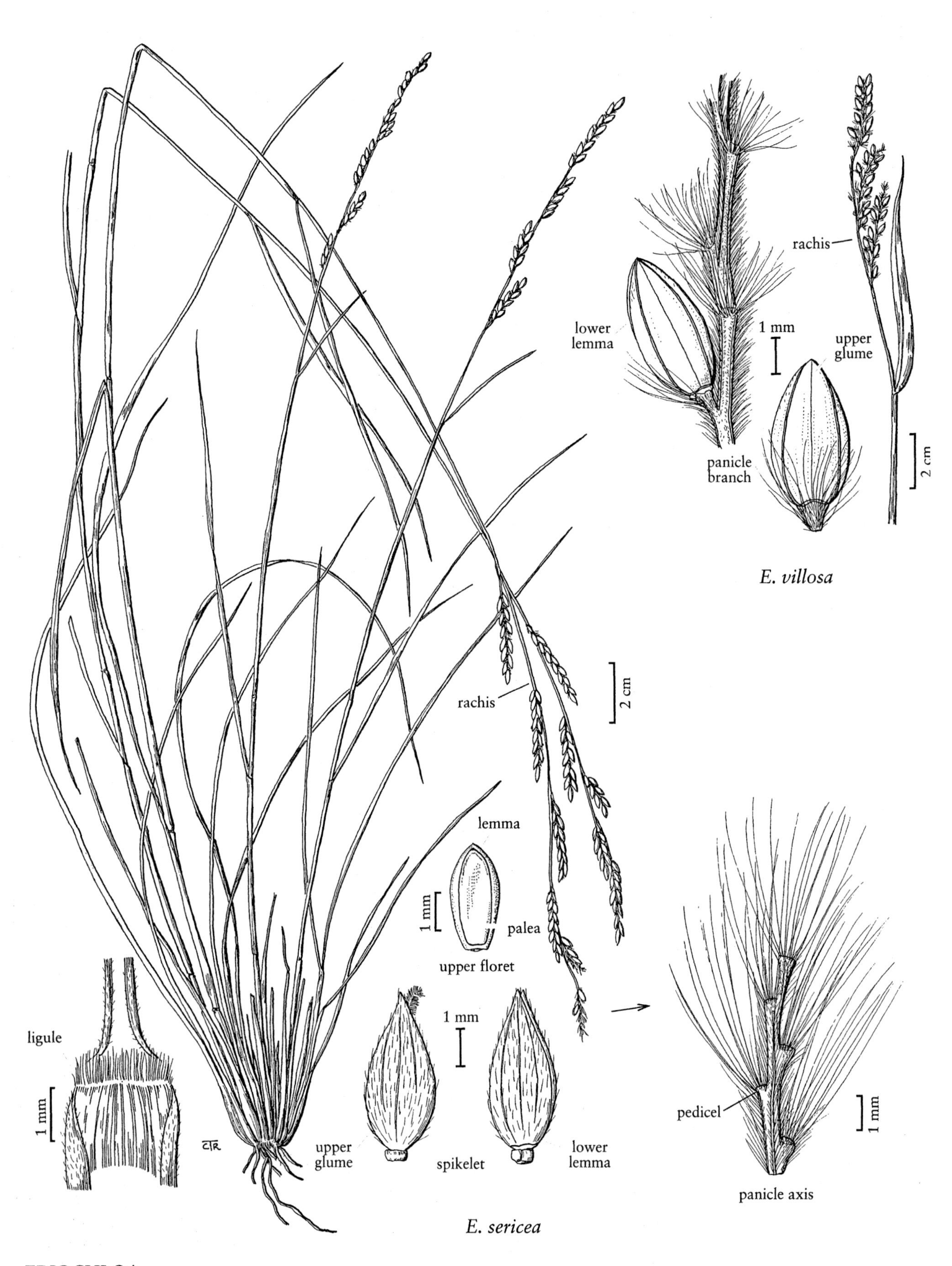
lower lemma
1 mm
panicle branch
rachis
upper glume
2 cm
E. villosa
rachis
2 cm
lemma
1 mm
palea
upper floret
1 mm
upper glume
spikelet
lower lemma
ligule
1 mm
pedicel
1 mm
panicle axis
E. sericea

Eriochloa contracta grows in fields, ditches, and other disturbed areas. It is known only from the United States, being native and common in the central United States, and adventive to the east and southwest. It differs from *E. acuminata* in its tightly contracted, almost cylindrical panicles and longer lemma awns, but intermediate forms can be found. It can also be confused with first-year plants of the perennial *E. punctata*, which have glabrous leaves, narrower and more tapering spikelets, and longer lemma awns.

5. **Eriochloa fatmensis** (Hochst. & Steud.) Clayton [p. 512]

Plants annual. **Culms** 30–120 cm, erect or decumbent, sometimes rooting at the lower nodes; **nodes** and **internodes** glabrous. **Sheaths** smooth, glabrous; **ligules** 0.5–1 mm; **blades** 6–40 cm long, 3–8 mm wide, linear, flat to involute, glabrous adaxially. **Panicles** 6–18 cm long, 0.8–3 cm wide; **rachises** smooth or scabrous; **branches** 3–10, 25–60 mm long, 0.4–0.6 mm wide, sometimes winged, glabrous, with 20–40 spikelets, spikelets solitary or in unequally pedicellate pairs at the middle of the branches; **pedicels** 0.2–0.8 mm, glabrous or hairy below, apices with fewer than 10 hairs to 0.5 mm long. **Spikelets** 2.7–3.6 mm long, 0.8–1.5 mm wide, lanceolate. **Upper glumes** 1.1–1.3 times longer than the lower lemmas, hairy, 5-veined, apices acuminate and acute to awned, awns 0.3–1.5 mm; **lower lemmas** 2.5–3.3 mm long, 0.8–1 mm wide, setose, apices acuminate and mucronate; **lower paleas** absent; **anthers** absent; **upper lemmas** 1.7–2.2 mm, elliptic, acute to rounded, awned, awns 0.2–0.5 mm. $2n$ = unknown.

Eriochloa fatmensis is native to tropical Africa, Arabia, and India, where it usually grows in wet areas or grasslands. It has been found in Tucson, Arizona, and Biloxi, Mississippi, but is probably not established in the *Flora* region.

6. **Eriochloa lemmonii** Vasey & Scribn. [p. 514]
CANYON CUPGRASS

Plants annual; cespitose. **Culms** 20–80 cm, erect or decumbent, sometimes rooting at the lower nodes; **internodes** densely pubescent to pilose; **nodes** pubescent to pilose. **Sheaths** from conspicuously inflated to not inflated, glabrous or pubescent to pilose; **ligules** 0.5–1 mm; **blades** 5–15 cm long, 6–20 mm wide, lanceolate, flat, straight, diverging or ascending, velvety pubescent adaxially. **Panicles** 5–15 cm long, 0.5–4 cm wide, spreading or contracted; **rachises** hairy; **branches** (2)3–8(10), 1–4 cm long, 0.4–0.6 mm wide, appressed or reflexed and spreading, velvety pubescent, not winged, with 10–14 spikelets, spikelets in unequally pedicellate pairs at the middle of the branches, solitary distally; **pedicels** 0.5–1 mm, pilose, apices hairy or glabrous. **Spikelets** 3–4.5(4.9) mm long, 1.2–1.7 mm wide, elliptic. **Upper glumes** equaling the lower lemmas, nearly glabrous or sparsely to densely pilose, elliptic, 5–7-veined, acute, unawned; **lower lemmas** 2.7–4 mm long, 1.2–1.7 mm wide, elliptic, setose to pilose, 5-veined, acute, unawned; **lower paleas** 1–4 mm, hyaline; **anthers** absent or 3; **upper lemmas** 2.3–3.3 mm, elliptic, indurate, dull, rough, occasionally with a few long hairs, acute to rounded, sometimes mucronate; **upper paleas** indurate. $2n$ = 36.

Eriochloa lemmonii, a rare species, grows in canyons and on rocky slopes in Pima County, Arizona, Hidalgo County, New Mexico, and adjacent Mexico. The record from Tennessee reflects an introduction. It is not known if the species has persisted in the region.

Eriochloa lemmonii may hybridize with *E. acuminata*, from which it differs in the frequent presence of lower paleas, raised veins of the upper glumes and lower lemmas, broad, velvety pubescent leaf blades, and blunt spikelets. Reports of *E. lemmonii* from Texas may be based on hybrids between the two species.

7. **Eriochloa punctata** (L.) Desv. *ex* Ham. [p. 514]
LOUISIANA CUPGRASS

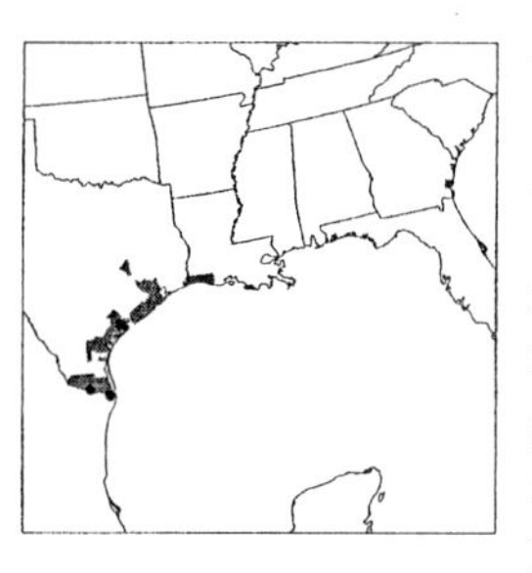

Plants perennial; rhizomatous, often flowering the first year and resembling an annual. **Culms** 30–150 cm, erect or decumbent, not rooting at the lower nodes; **internodes** glabrous; **nodes** 3–10, glabrate. **Sheaths** occasionally inflated, glabrous (rarely puberulent), often purplish at maturity; **ligules** 0.4–1 mm; **blades** 10–50 cm long, (2)4–10(13) mm wide, linear, flat, straight, spreading, glabrous (rarely puberulent) on both surfaces. **Panicles** 9–22 cm long, 1–10 cm wide, contracted, rarely open; **rachises** scabrous to densely pubescent; **branches** (4)8–20(27), 1–6 cm long, 0.3–0.5 mm wide, appressed or divergent, glabrous, not winged, with 28–60 spikelets, spikelets mostly in unequally pedicellate pairs, solitary distally; **pedicels** 0.1–0.7 mm, uniformly pubescent, hairs about 0.1 mm. **Spikelets** (4)4.5–5.7 mm long, 0.9–1.4 mm wide. **Upper glumes** equaling the lower lemmas, lanceolate, sparsely appressed pilose, 5–7-veined,

1 cm
1 cm
1 cm
2 mm
ligule
var. simpsonii
lemma
1 mm
palea
upper floret
upper glume
1 mm
lower lemma
spikelet
2 cm
ligule
5 mm
CTR
var. michauxii
E. michauxii
upper glume
1 mm
spikelet
1 cm
1 mm
E. fatmensis
1 cm
ligule
lemma
palea
1 mm
upper floret
upper glume
1 mm
spikelet
E. contracta

acuminate, sometimes mucronate, mucro shorter than 0.5 mm; **lower lemmas** 4.3–5.5 mm long, 0.9–1.4 mm wide, lanceolate, setose, 5–7-veined, acuminate, unawned or mucronate; **lower paleas** absent; **anthers** absent; **upper lemmas** 2–3.5 mm (excluding the awn), 0.4–0.6 times as long as the lower lemmas, indurate, elliptic, 5-veined, rounded, awned, awns 0.6–1.5 mm; **upper paleas** 0.5–1.2 mm, indurate, blunt. $2n$ = 36.

Eriochloa punctata grows in coastal marshes, along water courses, and in moist swales and ditches of the coastal plain from Texas and Louisiana south through Mexico to Central and South America. It has not been possible to verify the identification of the specimen from Georgia for this treatment. If correct, it suggests that the species may be more widespread than generally thought.

8. **Eriochloa aristata** Vasey [p. 514]
BEARDED CUPGRASS

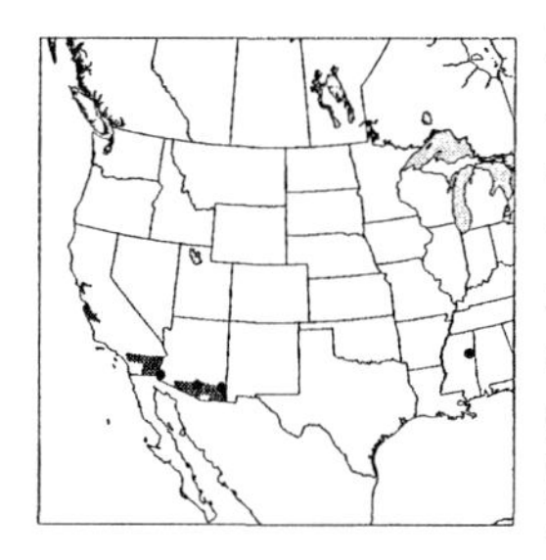

Plants annual; cespitose, not rhizomatous. **Culms** 40–100 cm, erect or decumbent, sometimes rooting at the lower nodes; **internodes** glabrous; **nodes** 3–10, puberulent. **Sheaths** glabrous; **ligules** 0.5–0.8(2) mm; **blades** 6–20 cm long, 6–20 mm wide, linear to lanceolate, flat or folded, straight or lax, glabrous (rarely sparsely pubescent) adaxially. **Panicles** 5–20 cm long, 1–3 cm wide, loosely contracted; **rachises** hairy; **branches** 16–30, 20–35 mm long, 0.3–0.5 mm wide, divergent to spreading, setose, not winged, with 20–35 spikelets in unequally pedicellate pairs; **pedicels** 0.5–3 mm, hairy, with some hairs 0.5–2.5 mm long, at least distally. **Spikelets** 4–8.8 mm long, 1.1–1.6 mm wide, lanceolate. **Upper glumes** 1–1.1 times as long as the lower lemmas, lanceolate, pilose or scabrous above, 5-veined, acuminate and awned, awns 0.5–3.5 mm; **lower florets** sterile; **lower lemmas** 4–8 mm long, 1.1–1.6 mm wide, lanceolate, setose, 3–7-veined, acuminate, mucronate, mucro less than 0.4 mm; **lower paleas** absent; **anthers** absent; **upper lemmas** (2)3–4(6) mm, 0.4–0.6 times as long as the lower lemmas, indurate, elliptic, acute to rounded, 5-veined, awned, awns 0.2–0.8 mm; **upper paleas** indurate, rugose. $2n$ = 36.

Eriochloa aristata is a weed of moist swales, roadsides, and irrigated fields of the southwestern United States. Its range extends through Mexico and Central America to Colombia. There are three specimens from Oktibbeha County, Mississippi, two made in 1890 and one made in 1960. The last collection was from a waste area, which suggests that the species is now established there. The other two were both made in Starkville, where there is an experimental farm. The labels on the two specimens do not state whether the collections were made from experimental plantings or from plants that had escaped from such plantings.

9. **Eriochloa pseudoacrotricha** (Stapf *ex* Thell.) J.M. Black [p. 516]
VERNAL CUPGRASS

Plants perennial; cespitose, not rhizomatous. **Culms** 40–80 cm, erect, not rooting at the lower nodes; **internodes** glabrous; **nodes** hairy. **Sheaths** glabrous, rarely with a few appressed hairs; **collars** glabrous; **ligules** 0.4–1 mm; **blades** 10–20 cm long, 2–4 mm wide, linear, flat, straight, spreading, glabrous or hairy adaxially. **Panicles** 5–15 cm long, 0.3–1 cm wide; **rachises** scabrous; **branches** 2–8, 25–45 mm long, 0.3–0.6 mm wide, divergent to spreading, glabrous, winged, with 2–40 spikelets in unequally pedicellate pairs at the middle of the branches; **pedicels** 1–3 mm, glabrous, apices hairy. **Spikelets** 3.6–5.4 mm long, 1.1–1.4 mm wide, lanceolate. **Upper glumes** 1–1.3 times as long as the lower lemmas, lanceolate, hairy, 5-veined, acuminate, awned, awns 0.5–1.5 mm; **lower florets** sterile; **lower lemmas** 3.5–5.2 mm long, 1.1–1.4 mm wide, lanceolate, setose, 5-veined, mucronate to awned; **lower paleas** absent; **upper lemmas** 2.1–3.7 mm, elliptic, apices rounded, awned, awns 0.4–0.8 mm; **upper paleas** indurate, blunt, not rugose. $2n$ = 36.

Eriochloa pseudoacrotricha is an Australian species that has been introduced into south Texas and Mississippi. It grows in waste areas.

10. **Eriochloa acuminata** (J. Presl) Kunth [p. 516]
SOUTHWESTERN CUPGRASS

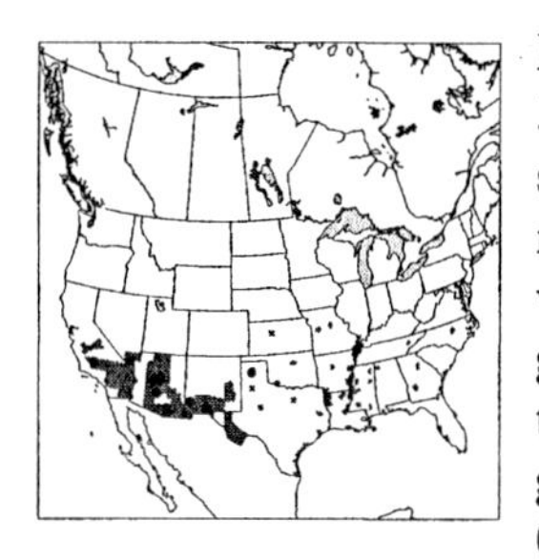

Plants annual; cespitose. **Culms** 30–120 cm, erect or decumbent, sometimes rooting at the lower nodes; **internodes** glabrous or with scattered hairs; **nodes** glabrous or pilose. **Sheaths** sometimes conspicuously inflated, glabrous or pubescent; **ligules** 0.2–1.2 mm; **blades** 5–12(18) cm long, (2)5–12(16) mm wide, linear, flat or folded, straight or lax, glabrous or sparsely pubescent adaxially. **Panicles** 7–16 cm long, 1–6 cm wide, loosely contracted; **rachises** scabrous or hairy; **branches** 5–20, 1–5 cm long, 0.4–0.6 mm wide, appressed to divergent, pubescent, sometimes setose, not winged, with 20–36 spikelets, spikelets mostly in unequally pedicellate pairs, solitary distally; **pedicels** 0.1–1 mm, hairy. **Spikelets** 3.8–5(6) mm long, 1.1–1.4

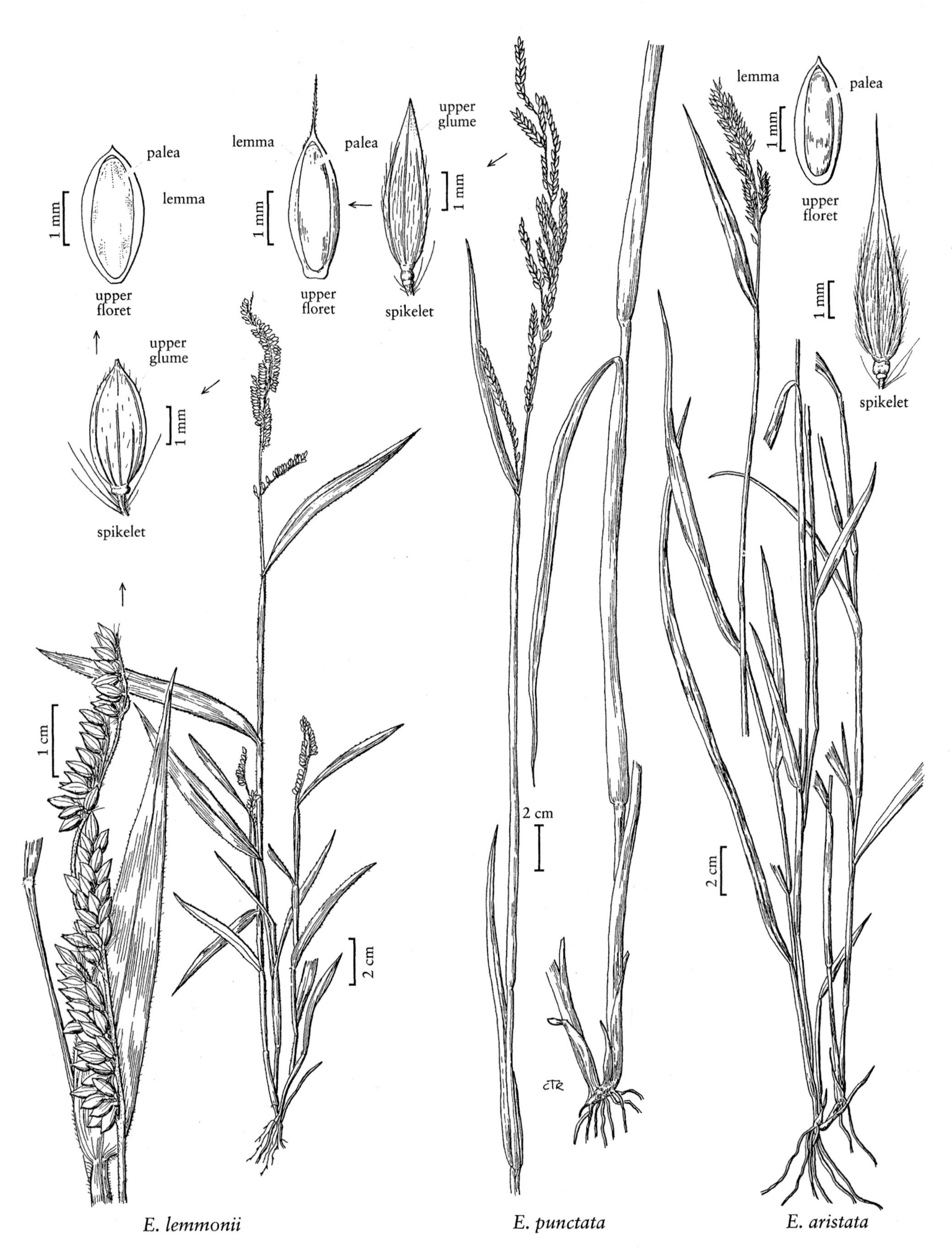
palea
lemma
1 mm
upper floret
upper glume
1 mm
spikelet
1 cm
2 cm
lemma
palea
1 mm
upper floret
upper glume
1 mm
spikelet
2 cm
lemma
palea
1 mm
upper floret
1 mm
spikelet
2 cm
E. lemmonii
E. punctata
E. aristata

ERIOCHLOA

mm wide, lanceolate to ovate. **Lower glumes** absent; **upper glumes** equaling the lower lemmas, lanceolate to ovate, hairy, 5(7)-veined, acuminate to acute, unawned or awned, awns to 1.2 mm; **lower lemmas** 3.6–5 mm long, 1.1–1.4 mm wide, lanceolate to ovate, setose, 5(7)-veined, acuminate to acute, unawned; **lower paleas** absent; **anthers** absent; **upper lemmas** 2.3–3.3 mm, 0.7–0.9 times as long as the lower lemmas, indurate, elliptic, rounded, 5-veined, awned, the awns 0.1–0.3 mm; **upper paleas** indurate, blunt, rugose. $2n = 36$.

Eriochloa acuminata is native to the southern United States and northern Mexico, but has become established outside this region. It may hybridize with *E. lemmonii*, from which it differs in its lack of lower paleas, upper glumes and lower lemmas with level veins, and narrower, glabrous or sparsely pubescent leaf blades.

There are two varieties of *Eriochloa acuminata*, differing as shown in the key below. Both grow in Mexico as well as the United States.

1. Spikelets 4–6 mm long, long-acuminate or tapering to a short awn . . . . var. *acuminata*
1. Spikelets 3.8–4 mm long, acute . . . . var. *minor*

Eriochloa acuminata (J. Presl) Kunth var. **acuminata** generally grows in ditches, fields, right of ways, and other disturbed areas of the southern United States.

Eriochloa acuminata var. **minor** (Vasey) R.B. Shaw is common in irrigated fields, orchards and disturbed areas of the southwestern United States. It is adventive in Maryland.

11. **Eriochloa polystachya** Kunth [p. 516]
CARIBBEAN CUPGRASS

Plants perennial; cespitose, stoloniferous. **Culms** 100–200 cm, decumbent, rooting at the lower nodes; **internodes** glabrous; **nodes** 4–10, densely pilose. **Sheaths** chartaceous to cartilaginous, lower sheaths with papillose-based hairs, upper sheaths glabrous; **collars** hairy; **ligules** 0.6–1.2 mm; **blades** 6–28 cm long, 6–18 mm wide, linear, flat, straight, ascending or drooping, glabrous adaxially. **Panicles** 8–20 cm long, 41–90 mm wide, open; **rachises** sparsely pilose to hirsute; **branches** (5)10–15(18), 2–6 cm long, 0.4–0.6 mm wide, pubescent to setose, not winged, spikelets in unequally pedicellate pairs; **pedicels** 0.5–1 mm, pubescent, apices glabrous. **Spikelets** 3.2–3.9 mm long, 1.1–1.3 mm wide, lanceolate to ovate. **Lower glumes** present as a membranous extension of the calluses; **upper glumes** equaling the lower lemmas, hairy, lanceolate to ovate, 5-veined, acute, unawned; **lower lemmas** 3–3.5 mm long, 1.1–1.3 mm wide, lanceolate to ovate, glabrous or sparsely pubescent, 5-veined, acute, unawned; **lower paleas** fully developed or absent; **anthers** absent or 3; **upper lemmas** 2.2–2.6 mm, indurate, elliptic, rounded, mucronate, mucros less than 0.2 mm; **upper paleas** 2–2.5 mm, indurate. $2n = 36$.

Eriochloa polystachya is native to the West Indies, Costa Rica, Honduras, and South America. It was introduced into the United States as a forage crop and is now established at some locations in Florida and Texas.

25.15 PENNISETUM Rich.

J.K. Wipff

Plants annual or perennial; habit various. **Culms** 3–800 cm, not woody, sometimes branching above the base; **internodes** solid or hollow. **Ligules** membranous and ciliate, or of hairs, rarely completely membranous; **blades** sometimes pseudopetiolate. **Inflorescences** spicate panicles with highly reduced branches termed *fascicles*; **panicles** 1–many per plant, terminal on the culms or on both the culms and the secondary branches, or terminal and axillary, or only axillary, usually completely exposed at maturity; **rachises** usually terete, with (1)5–many fascicles; **fascicle axes** 0.2–7.5(28) mm, with (1)3–130+ bristles and 1–12 spikelets. **Bristles** free or fused at the base, disarticulating with the spikelets at maturity; of 3 kinds, *outer*, *inner*, and *primary*, in some species with all 3 kinds present below each spikelet, in others 1 or more kinds missing from some or all of the spikelets; **outer** (lower) **bristles** antrorsely scabrous, terete; **inner** (upper) **bristles** antrorsely scabrous or long-ciliate, usually flatter and wider than the outer bristles; **primary** (terminal) **bristles** located immediately below the spikelets, solitary, antrorsely scabrous or long-ciliate, often longer than the other bristles associated with the spikelet; **disarticulation** usually at the base of the fascicles, sometimes also beneath the upper florets. **Spikelets** with 2

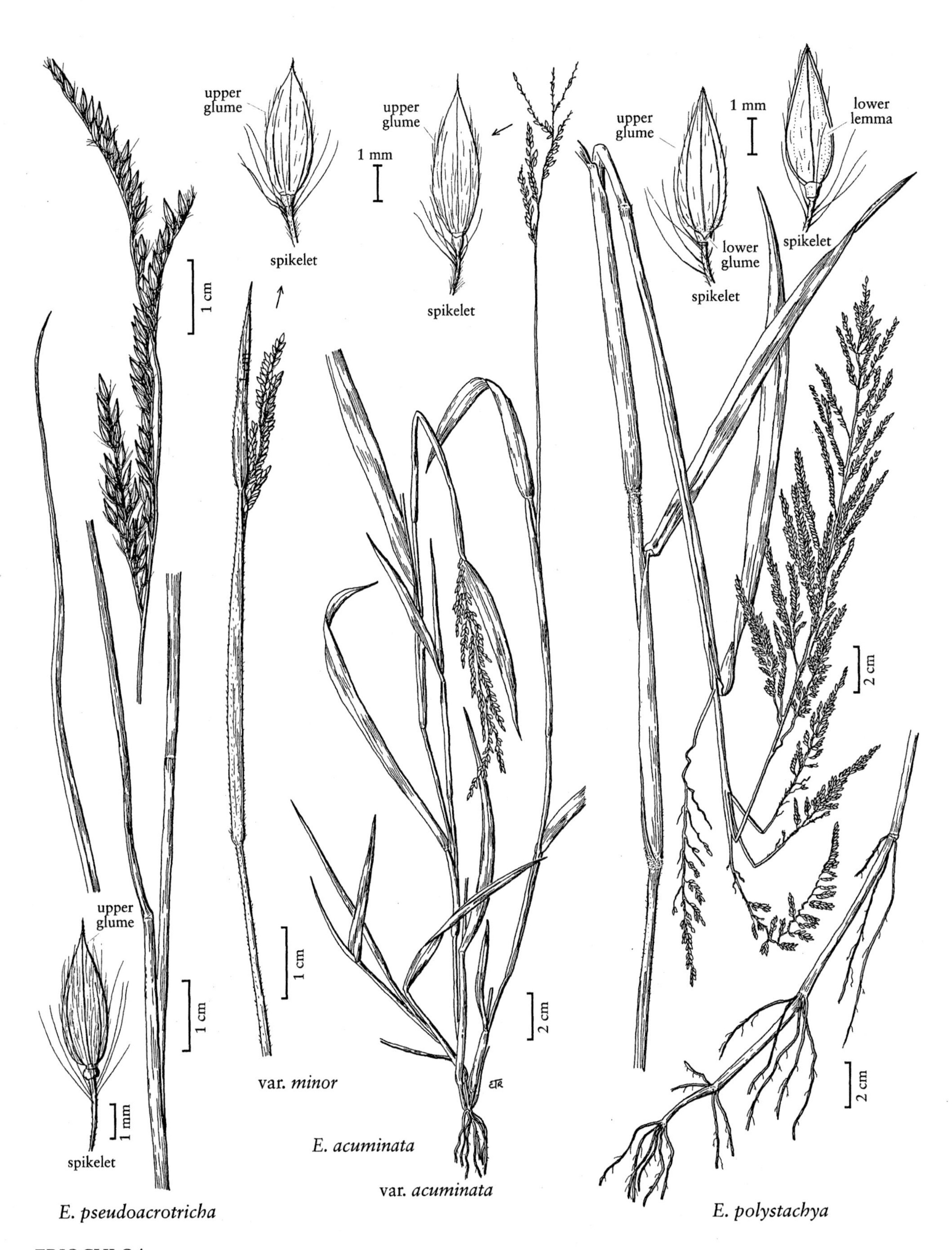
upper glume
spikelet
1 cm
1 mm
upper glume
spikelet
1 mm
1 cm
upper glume
spikelet
1 mm
upper glume
lower glume
spikelet
lower lemma
spikelet
2 cm
1 cm
2 cm
2 cm
1 cm
var. *minor*
E. acuminata
var. *acuminata*
E. pseudoacrotricha
E. polystachya

ERIOCHLOA

florets; **lower glumes** absent or present, 0–5-veined; **upper glumes** longer, 0–11-veined; **lower florets** sterile or staminate; **lower lemmas** usually as long as the spikelets, membranous, 3–15-veined, margins usually glabrous; **lower paleas** present or absent; **upper lemmas** membranous to coriaceous, 5–12-veined; **upper paleas** shorter than the lemmas but similar in texture; **lodicules** 0 or 2, glabrous; **anthers** 3, if present. x = 5, 7, 8, 9 (usually 9). Name from the Latin *penna*, 'feather', and *seta*, 'bristle', an allusion to the plumose bristles of some species.

Pennisetum has 80–130 species, most of which grow in the tropics and subtropics, and occupy a wide range of habitats. Twenty-five species are native to the Western Hemisphere, but none to the *Flora* region. Most of the species treated here are cultivated for food, forage, or as ornamental plants. Many species, including several cultivated species, are weedy. Four are classified as noxious weeds by the U.S. Department of Agriculture. Records known to be based on cultivated plants are not included in the distribution maps but, in many cases, it is not possible to determine whether a record is based on a cultivated plant or an escape.

The placement of the boundary between *Pennisetum* and *Cenchrus* is contentious. As treated here, *Pennisetum* has antrorsely scabrous bristles that are not spiny, fascicle axes that terminate in a bristle, and chromosome base numbers of 5, 7, 8, and 9. *Cenchrus* has retrorsely (rarely antrorsely) scabrous, spiny bristles, fascicle axes that are terminated by a spikelet, and a chromosome base number of 17 (Wipff 2001). In both genera, the bristles are reduced branches (Goebel 1882; Sohns 1955).

SELECTED REFERENCES **Brunken, J.N.** 1977. A systematic study of *Pennisetum* sect. *Pennisetum* (Gramineae). Amer. J. Bot. 64:161–176; **Brunken, J.N.** 1979. Morphometric variation and the classification of *Pennisetum* section *Brevivalvula* (Gramineae) in tropical Africa. Bot. J. Linn. Soc. 79:51–64; **Chase, A.** 1921. The North American species of *Pennisetum*. Contr. U.S. Natl. Herb. 22^4:209–234; **Goebel, K.I.** 1882. Beitrage zur Entwickelungsgeschichte einiger Inflorescenzen. Jahrb. Wiss. Bot. 14:1–39; **Hignight, K.W., E.C. Basha, and M.A. Hussey.** 1991. Cytological and morphological diversity of native apomictic buffelgrass, *Pennisetum ciliare* (L.) Link. Bot. Gaz. 152:214–218; **Schmelzer, G.H.** 1997. Review of *Pennisetum* section *Brevivalvula* (Poaceae). Euphytica 97:1–20; **Sohns, E.R.** 1955. *Cenchrus* and *Pennisetum*: Fascicle morphology. J. Wash. Acad. Sci. 45:135–143; **Wipff, J.K.** 1995. A biosystematic study of selected facultative apomictic species of *Pennisetum* (Poaceae: Paniceae) and their hybrids. Ph.D. dissertation, Texas A&M University, College Station, Texas, U.S.A. 183 pp.; **Wipff, J.K.** 2001. Nomenclatural changes in *Pennisetum* (Poaceae: Paniceae). Sida:19:523–530.

NOTE: Pedicel length is the distance from the base of the primary bristles to the base of the terminal spikelets. Fascicle axis lengths and fascicle densities are measured in the middle of the panicle; spikelet measurements refer to the largest spikelets in the fascicles.

1. Plants stoloniferous; panicles axillary, partially or wholly hidden in the leaf sheaths at maturity, the rachises flattened in cross section, with 1–6 fascicles; spikelets 10–22 mm long, bristles mostly shorter than the spikelet . . . . 1. *P. clandestinum*
1. Plants not stoloniferous; panicles terminal or terminal and axillary, fully exserted at maturity, the rachises terete, with 10–many fascicles; spikelets 2.5–12 mm long, the majority of the bristles as long as or longer than the spikelets.
    2. Fascicles with only 1 bristle and 1 spikelet . . . . 18. *P. petiolare*
    2. Fascicles with 6 or more bristles and 1–12 spikelets.
        3. Most or all bristles scabrous, the primary bristles sometimes sparsely and inconspicuously long-ciliate.
            4. Primary bristles not noticeably longer than all the other bristles in the fascicles.
                5. Terminal panicle erect; fascicles with a stipelike base 1.5–5.6 mm long . . . . 7. *P. alopecuroides*
                5. Terminal panicle drooping; fascicles subsessile, the bases 0.4–0.7 mm long.
                    6. Plants green; most of the bristles only slightly longer than the spikelets; upper glumes (7)9-veined, about as long as the spikelets . . . . 4. *P. nervosum*
                    6. Plants purplish; bristles at least twice as long as the spikelets; upper glumes 1–3-veined, usually about ½ as long as the spikelets . . . . 5. *P. macrostachys*
            4. Primary bristles noticeably longer than all the other bristles in the fascicles.
                7. Panicles dense; rachises with 21–40 fascicles per cm.

8. Rachises pubescent; bristles yellow or purple; leaf blades (4)12–40 mm wide; paleas of lower florets present . . . . . 2. *P. purpureum*
8. Rachises scabrous; bristles white to stramineous; leaf blades 4–12 mm wide; paleas of lower florets absent . . . . . 6. *P. macrourum*

7. Panicles less dense; rachises with 5–16 fascicles per cm.
9. Panicles drooping, terminal and axillary; leaf blades 19–45 mm wide . . . . . 8. *P. latifolium*
9. Panicles erect, all terminal; leaf blades 2–12 mm wide.
10. Plants not rhizomatous; lower part of rachises pubescent . . . . . 7. *P. alopecuroides*
10. Plants rhizomatous; lower part of rachises scabrous . . . . . 14. *P. flaccidum*

3. Bristles, at least the primary bristles, conspicuously long-ciliate.
11. Spikelets 9–12 mm long . . . . . 11. *P. villosum*
11. Spikelets 2.5–7 mm long.
12. Fascicles not disarticulating from the rachises; panicles 4–200 cm long; upper lemmas with pubescent margins; caryopses protruding from the florets at maturity . . . . . 3. *P. glaucum*
12. Fascicles disarticulating from the rachises at maturity; panicles 2–37.5 cm long; upper lemmas with glabrous margins; caryopses concealed by the lemmas and paleas at maturity.
13. Upper florets readily disarticulating at maturity; upper lemmas smooth and shiny, conspicuously different in texture from the lower lemmas.
14. Fascicles with 6–14 long-ciliate inner bristles and 13–30 scabrous outer bristles; fascicle axes 0.2–0.5 mm long; spikelets sessile . . . . . 9. *P. polystachion*
14. Fascicles with 40–90 long-ciliate inner bristles and 10–20 scabrous outer bristles; fascicle axes 1.5–2.5 mm long; spikelets pedicellate, the pedicels 1–3.5 mm long . . . . . 10. *P. pedicellatum*
13. Upper florets not disarticulating at maturity; lower and upper lemmas similar in texture.
15. Lower portion of the rachises glabrous, sometimes scabrous.
16. Inner bristles neither grooved nor fused, even at the base; spikelets 5.2–6.7 mm long, pedicellate, the pedicels 0.1–0.5 mm long . . . . . 14. *P. flaccidum*
16. Inner bristles grooved and fused, at least at the base; spikelets 2.5–5.6 mm long, sessile.
17. Inner bristles fused for up to ¼ their length; many outer bristles exceeding the spikelets; terminal bristles 10.5–23 mm, noticeably longer than the other bristles in the fascicles . . . . . 12. *P. ciliare*
17. Inner bristles fused for ⅓–½ their length; outer bristles not exceeding the spikelets; terminal bristles 2.9–6.5 mm, usually not noticeably longer than other bristles in the fascicles . . . . . 13. *P. setigerum*
15. Lower portion of the rachises pubescent.
18. Plants 200–800 cm tall; midculm leaves (4)12–40 mm wide; panicles golden-yellow or dark purple; rachises with 30–40 fascicles per cm . . . . 2. *P. purpureum*
18. Plants 50–200 cm tall; midculm leaves 2–11 mm wide; panicles white, burgundy, light purple, or pink; rachises with 5–17 fascicles per cm.
19. Midculm leaves 2–3.5 mm wide, convolute or folded, green, the midvein noticeably thickened; lower florets of the spikelets usually sterile, sometimes staminate . . . . . 15. *P. setaceum*
19. Midculm leaves 3–11 mm wide, flat, green or burgundy, the midvein not noticeably thickened; lower florets of the spikelets staminate.
20. Plants shortly rhizomatous; nodes pubescent; panicles erect to slightly arching, white or purple-tinged; leaves green; ligules 1–1.7 mm long; fascicles with 0–24 terete, scabrous outer bristles . . . . . 16. *P. orientale*

20. Plants not rhizomatous; nodes glabrous; panicles conspicuously drooping, burgundy (rarely whitish-green); leaves burgundy (rarely green); ligules 0.5–0.8 mm long; fascicles with 43–68 terete, scabrous outer bristles .... 17. *P. advena*

1. **Pennisetum clandestinum** Hochst. *ex* Chiov. [p. 520]
KIKUYU GRASS

Plants perennial; rhizomatous and stoloniferous. **Culms** 3–45 cm, decumbent, highly branching; **nodes** glabrous. **Sheaths** glabrous or pubescent; **ligules** 1.3–2.2 mm; **blades** 1–15 cm long, 1–6 mm wide, flat or folded, glabrous or pubescent. **Panicles** 2–2.7 cm, axillary, concealed in the sheaths; **rachises** flat, glabrous or scabrous. **Fascicles** 1–6; **axes** to 0.5 mm, with 1–2 spikelets; **outer** and **inner bristles** alike, 6–15, 0.5–10.9 mm; **primary bristles** 10–14 mm, usually not noticeably longer than the other bristles. **Spikelets** 10–22 mm, sessile or pedicellate, pedicels to 0.2 mm; **lower glumes** usually absent, sometimes to 0.5 mm, veinless; **upper glumes** 0–1.3(3.5) mm, veinless; **lower florets** sterile; **lower lemmas** 10–22 mm, 9–13-veined; **lower paleas** usually absent; **upper lemmas** 10–22 mm, 8–12-veined; **upper paleas** 2–7-veined; **anthers** 4.7–7 mm, long-exserted from the florets at anthesis. $2n$ = 36.

Pennisetum clandestinum is native to Africa. It now grows in many parts of the world, often as a forage or lawn grass. The U.S. Department of Agriculture considers it a noxious weed. In parts of the *Flora* region, it is well-established in lawns.

2. **Pennisetum purpureum** Schumach. [p. 520]
ELEPHANT GRASS

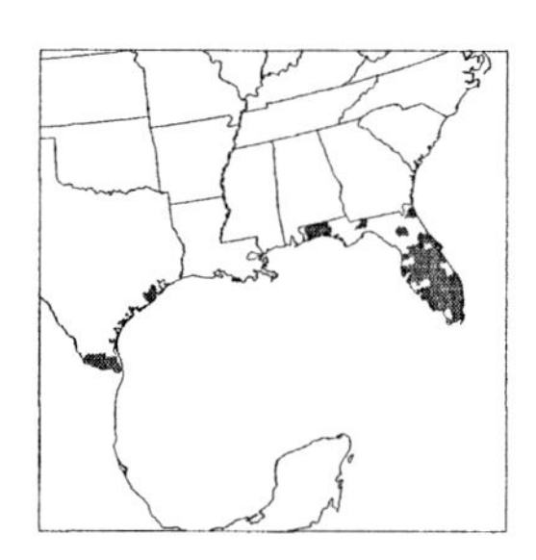

Plants perennial; sometimes rhizomatous. **Culms** 2–8 m, erect, pubescent beneath the panicle; **nodes** glabrous or pubescent. **Sheaths** glabrous or pubescent; **ligules** 1.5–5 mm; **blades** 23–125 cm long, (4)12–40 mm wide, flat, glabrous or pubescent. **Panicles** terminal, 8–30.5 cm long, (10)30–50 mm wide, fully exerted from the leaf sheaths, erect, golden-yellow to dark purple; **rachises** terete, pubescent. **Fascicles** 30–40 per cm, disarticulating at maturity; **fascicle axes** 0.5–1.5 mm, with 1–5 spikelets; **outer bristles** 20–63, 1.5–10.3 mm, yellow or purple, scabrous; **inner bristles** 4–6, 9.1–11.5 mm, yellow or purple, sparsely long-ciliate; **primary bristles** 13–40 mm, noticeably longer than the other bristles, yellow or purple, scabrous. **Spikelets** 5.9–7 mm, pedicellate; **pedicels** of terminal spikelets 0.2–0.4 mm, of other spikelets 1.8–3 mm; **lower glumes** absent or to 0.8 mm; **upper glumes** 0.8–3 mm, 0–1-veined; **lower florets** sterile or staminate; **lower lemmas** 4–5.3 mm, 3–5(6)-veined; **lower paleas** 4–4.7 mm; **anthers** absent or 2.2–3.1 mm, penicillate; **upper lemmas** 4.5–7 mm, subcoriaceous, shiny, 5–7-veined, acuminate; **anthers** 2.7–3.6 mm, penicillate. **Caryopses** 1.8–2.2 mm. $2n$ = 28.

Pennisetum purpureum is native to Africa but now grows in tropical areas throughout the world, frequently becoming naturalized. It is grown as an ornamental in the *Flora* region, and, less commonly, for forage.

3. **Pennisetum glaucum** (L.) R. Br. [p. 522]
PEARL MILLET

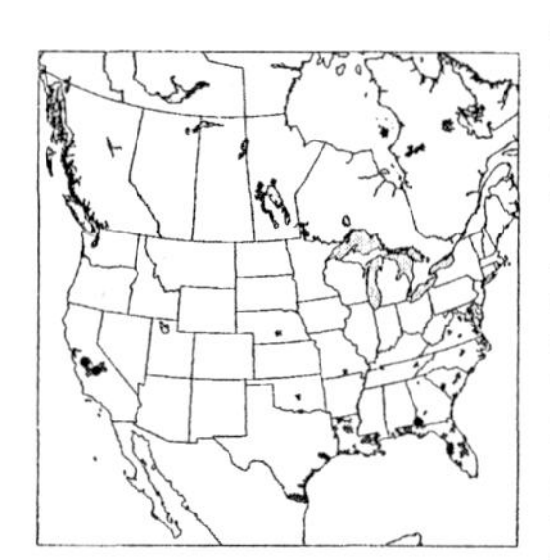

Plants annual. **Culms** 50–300 cm, erect, branching; **nodes** glabrous. **Sheaths** glabrous or pubescent, with or without ciliate margins; **ligules** 2–5 mm; **blades** 15–100 cm long, 7–70 mm wide, flat, glabrous or pubescent. **Panicles** terminal, 4–200 cm long, 2–70 mm wide, fully exerted from the sheaths, erect; **rachises** terete, densely pubescent. **Fascicles** 33–160 per cm; **fascicle axes** 1–28 mm, persistent, with 1–9 spikelets; **outer bristles** 44–131, 0.5–6 mm; **inner bristles** 6–19, 4–6 mm, plumose; **primary bristles** 5.5–6.3, ciliate, sometimes noticeably longer than the other bristles. **Spikelets** 3–7 mm; **pedicels** 0.6–1.8 mm; **lower glumes** absent or to 1.5 mm, veinless; **upper glumes** 0.5–3.5 mm, 3–5-veined; **lower florets** staminate or sterile; **lower lemmas** 1.5–6 mm, glabrous, 3–7-veined, margins ciliate; **lower paleas** vestigial or fully developed, margins ciliate; **anthers** 2.2–2.5 mm, penicillate; **upper florets** coriaceous, shiny; **upper lemmas** 4.3–7 mm, 5–7(9)-veined, margins ciliate; **upper paleas** 3.4–3.9 mm, pubescent, at least near the base, margins ciliate; **anthers** 2–2.2 mm, penicillate. **Caryopses** 2–5.5 mm long, 1.6–3.2 mm wide, protruding from the lemma and palea at maturity. $2n$ = 14.

Pennisetum glaucum, a native of Asia, is cultivated in the United States for grain, forage, and birdseed. It is the most drought tolerant of the tropical cereal crops. Under favorable conditions, 10,000–30,000+ fascicles may be produced. In the *Flora* region, it is used for soil stabilization, partly because it seldom persists for more than 1–2 years.

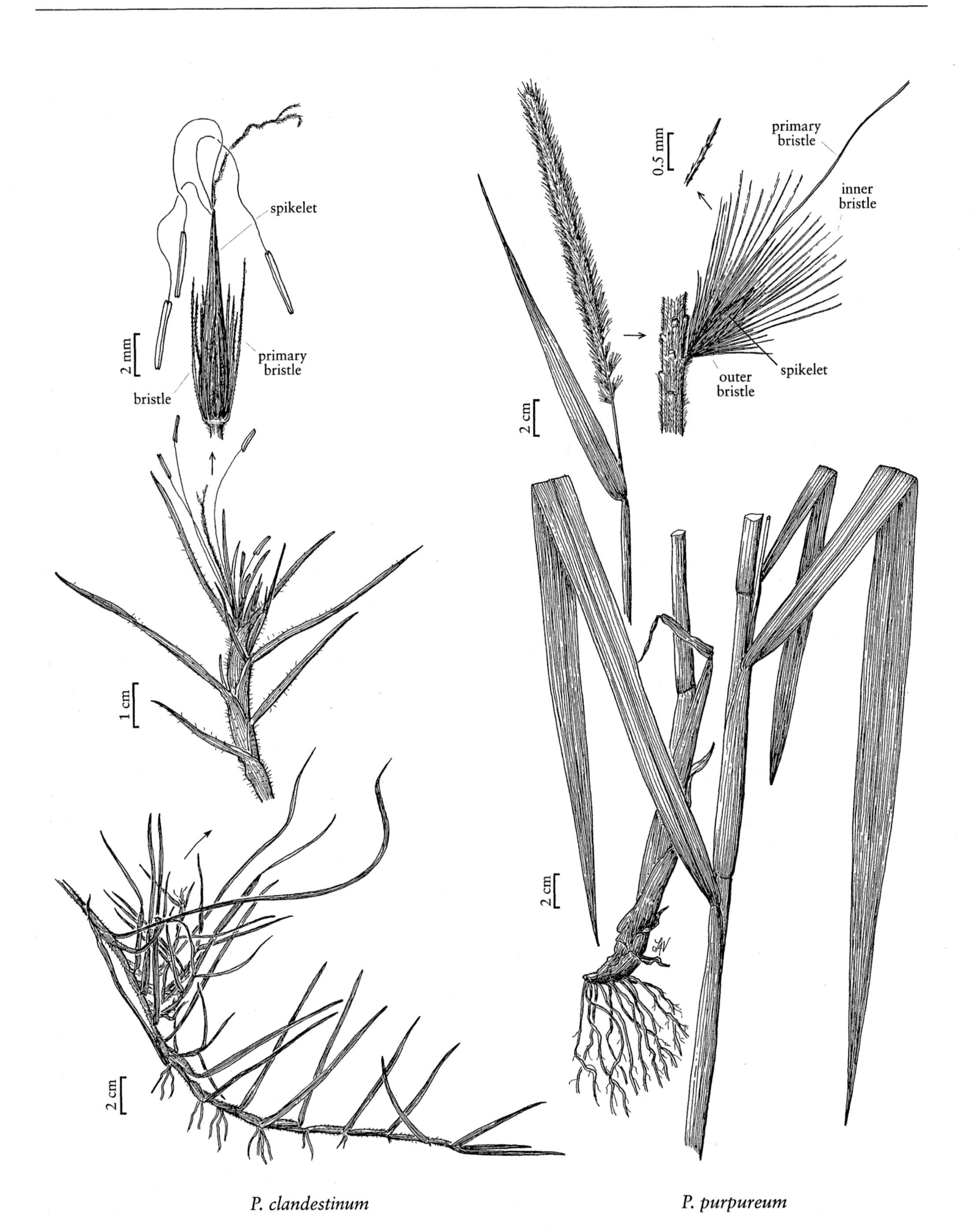

P. clandestinum

P. purpureum

4. **Pennisetum nervosum** (Nees) Trin. [p. 522]
Bentspike Fountaingrass

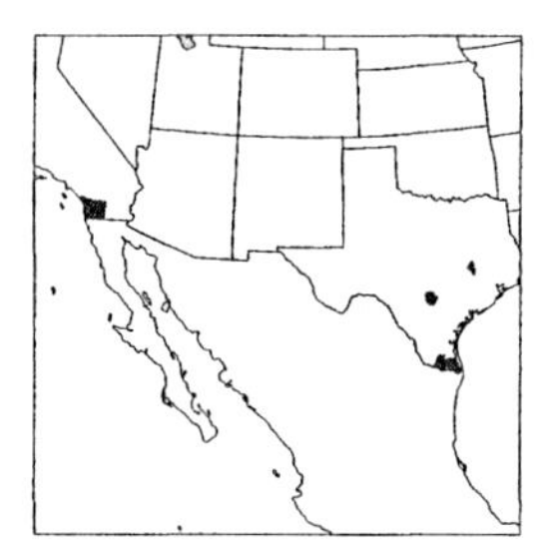

Plants perennial; cespitose. **Culms** 1.5–2(4) m, decumbent, geniculate, branching; **nodes** glabrous. **Leaves** green; **sheaths** glabrous; **ligules** 0.5–1.5 mm; **blades** 23–40 cm long, 7–12 mm wide, flat, glabrous. **Panicles** terminal, 15–22 cm long, 15–21 mm wide, fully exerted from the sheaths, flexible, drooping, green; **rachises** terete, puberulent. **Fascicles** 24–49 per cm; **fascicle axes** 0.4–0.5 mm, with 1 spikelet; **most bristles** only slightly longer than the spikelets; **outer bristles** 27–41, 2.4–8.5 mm; **inner bristles** 3–5, 6.2–11 mm, scabrous; **primary bristles** 7.6–11.8 mm, not noticeably longer than the other bristles, scabrous. **Spikelets** 5.2–6.6 mm, sessile; **lower glumes** 1.7–2.9 mm, 1-veined; **upper glumes** 4.5–6.3 mm, about as long as the spikelets, (7)9-veined; **lower florets** sterile; **lower lemmas** 4.8–6 mm, 7-veined, acuminate to attenuate, midvein excurrent for 0–0.6 mm; **lower paleas** absent; **upper lemmas** 4.9–6 mm, 5-veined, acuminate to attenuate, midvein excurrent for 0–0.6 mm; **anthers** 1.3–1.7 mm. $2n$ = 36.

Pennisetum nervosum is native to South America. It has been introduced into the *Flora* region, being known from populations adjacent to the Rio Grande River in Cameron and Hidalgo counties, Texas, and San Diego County, California.

5. **Pennisetum macrostachys** (Brongn.) Trin. [p. 522]
Pacific Fountaingrass

Plants perennial, or annual in temperate climates; cespitose. **Culms** 100–300 cm, erect, branching; **nodes** glabrous. **Leaves** burgundy; **sheaths** glabrous; **ligules** 0.1–0.3 mm; **blades** 30–53.5 cm long, (15) 18–35 mm wide, flat, glabrous. **Panicles** terminal, (15.5)18–40 cm long, 32–50 mm wide, fully exerted from the sheaths, flexible, drooping, burgundy; **rachises** terete, shortly pubescent. **Fascicles** 17–22 per cm; **fascicle axes** 0.5–0.7 mm, with 1 spikelet; **outer bristles** 21–40, 1.2–22.3 mm, scabrous; **inner bristles** absent; **primary bristles** 20–23 mm, not noticeably longer than the other bristles, scabrous. **Spikelets** 4.4–4.9 mm, sessile or pedicellate, glabrous; **pedicels** to 0.1 mm; **lower glumes** 1.1–1.3 mm, veinless; **upper glumes** 2.1–2.8 mm, usually about ½ as long as the spikelets, 1–3-veined; **lower florets** staminate (sterile); **lower lemmas** 4–4.5 mm, 5-veined; **lower paleas** absent or to 2.6 mm; **anthers** absent or 1.4–1.6 mm; **upper lemmas** 4.3–4.8 mm, 5-veined; **anthers** 1.6–1.8 mm. $2n$ = 68.

Pennisetum macrostachys is native to the South Pacific. It is grown in the *Flora* region as an ornamental species, being sold as 'Burgundy Giant'.

6. **Pennisetum macrourum** Trin. [p. 524]
Waterside Reed

Plants perennial; rhizomatous. **Culms** 60–200 cm, erect; **nodes** pubescent or glabrous. **Sheaths** pubescent or glabrous; **ligules** 1–1.7 mm; **blades** 20–50 cm long, 4–12 mm wide, flat to involute, glabrous, pubescent, or scabrous. **Panicles** terminal, 6–40 cm long, 20–46 mm wide, fully exerted from the sheaths, erect, white to stramineous; **rachises** terete, scabrous. **Fascicles** 21–46 per cm; **fascicle axes** 0.4–0.6(2) mm, with 1(2) spikelet(s); **outer bristles** 15–20, 2.5–8 mm; **inner bristles** 8–10, 4–9 mm, scabrous; **primary bristles** 12.5–20 mm, noticeably longer than the other bristles. **Spikelets** 3–8 mm, sessile or pedicellate, glabrous; **pedicels** to 0.2 mm; **lower glumes** absent or 1.2–1.4 mm, veinless; **upper glumes** 0.8–2 mm, 0–1-veined; **lower florets** sterile; **lower lemmas** 3–7.7 mm, 3–5(7)-veined; **lower paleas** absent; **upper lemmas** 3–6.8 mm, 5-veined; **anthers** 2.6–3.5 mm. $2n$ = 54.

Pennisetum macrourum is native to Africa, where it grows along rivers and lake margins. In the *Flora* region, it is known only from one location in Monterey County, California. Although sometimes recommended as an ornamental grass, the U.S. Department of Agriculture considers it a noxious weed.

7. **Pennisetum alopecuroides** (L.) Spreng. [p. 524]
Foxtail Fountaingrass

Plants perennial; cespitose. **Culms** 30–100 cm, erect; **nodes** glabrous. **Sheaths** glabrous, margins ciliate; **ligules** 0.2–0.5 mm, membranous, ciliate; **blades** (10)30–60 cm long, 2–8(12) mm wide, flat to folded, glabrous, margins ciliate basally. **Panicles** all terminal, 6–20 cm long, 20–53 mm wide, fully exerted from the sheaths, erect, green to brown, deep purple, or stramineous to creamy-white; **rachises** terete, with pubescent hairs. **Fascicles** 9–16 per cm; **fascicle axes** 1.5–5.6 mm, with a stipelike base of 1–5.6 mm and 1(2) spikelet(s); **outer bristles** 13–19, 0.8–15.6 mm; **inner bristles** 7–10, 11.2–30 mm, scabrous; **primary bristles** 26.7–35 mm, scabrous, usually not noticeably longer than the other bristles. **Spikelets** 5.5–8.4 mm, sessile or subsessile, glabrous; **pedicels** to 0.1 mm; **lower glumes** 0.2–1.4 mm, veinless; **upper glumes** 2–4.9 mm, to ½ as long as the spikelet, 1–5-veined, acute to broadly acute; **lower florets** sterile; **lower lemmas** 4.9–8.1 mm, 7–9(10)-veined; **lower paleas** absent; **upper lemmas** 5.2–7.6 mm, 5–7-veined, acuminate; **anthers** 3, 3–4.5 mm. $2n$ = 18.

Pennisetum alopecuroides is native to southeast Asia. It is frequently grown as an ornamental in the *Flora* region.

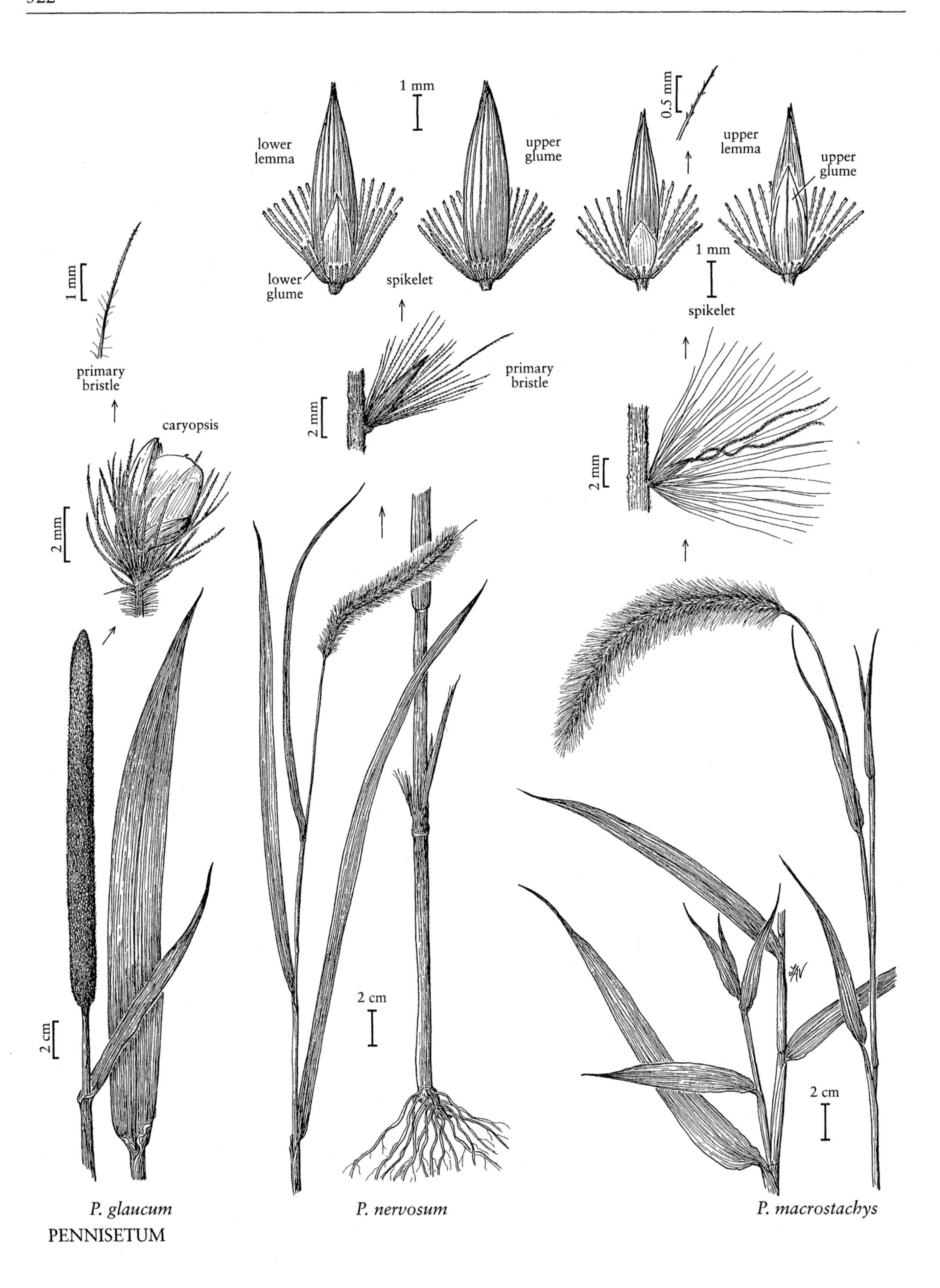
1 mm
lower lemma
upper glume
lower glume
spikelet
0.5 mm
upper lemma
upper glume
1 mm
spikelet
1 mm
primary bristle
primary bristle
2 mm
caryopsis
2 mm
2 mm
2 cm
2 cm
2 cm
P. glaucum
P. nervosum
P. macrostachys

PENNISETUM

8. **Pennisetum latifolium** Spreng. [p. 524]
URUGUAY FOUNTAINGRASS

Plants perennial; cespitose. **Culms** 1.5–3 m, erect; **nodes** pubescent. **Sheaths** pubescent or scabrous, margins ciliate; **ligules** 1.3–2 mm; **blades** 30–75 cm long, 19–45 mm wide, flat, glabrous, pseudopetiolate. **Panicles** terminal on the culms and axillary in the upper leaf sheaths, 2.2–9.5 cm long, 10–15 mm wide, fully exerted from the sheaths, flexible, drooping, green to stramineous; **rachises** terete, scabrous or sparsely pubescent. **Fascicles** 5–13 per cm; **fascicle axes** about 0.3 mm, with 1(2) spikelet(s); **outer bristles** 8–33, 1.1–11 mm, brittle; **inner bristles** absent; **primary bristles** (11.5)15.7–26.5 mm, scabrous, noticeably longer than the other bristles. **Spikelets** 4–7 mm, sessile; **lower glumes** 1.2–2.2 mm, 1(2)-veined, scabridulous and ciliolate distally; **upper glumes** 1.8–3.3 mm, 3-veined, margins ciliolate; **lower florets** sterile; **lower lemmas** 4.8–6.8 mm, 5–8-veined; **lower paleas** absent; **upper lemmas** 4.6–5.5 mm, 5-veined; **upper paleas** 2–3-veined; **anthers** 1.2–1.5 mm. **Caryopses** 2.5–2.8 mm long. $2n$ = 36.

Pennisetum latifolium is native to South America. It is occasionally grown as an ornamental in the *Flora* region.

9. **Pennisetum polystachion** (L.) Schult. [p. 526]
MISSION GRASS

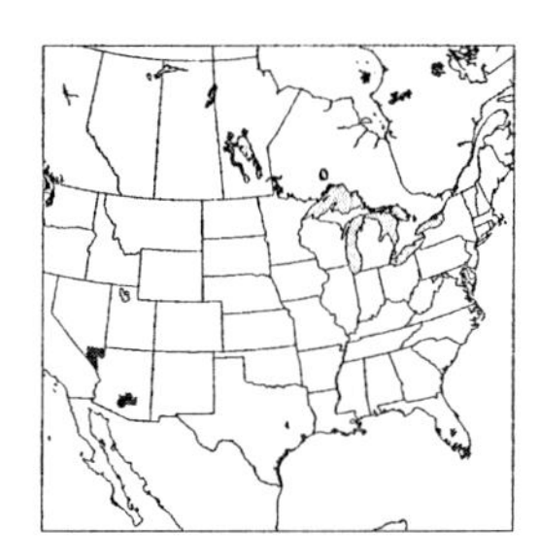

Plants annual or perennial; cespitose from a hard, knotty base. **Culms** 30–200 cm, erect, branching; **nodes** glabrous. **Sheaths** glabrous, margins ciliate; **ligules** 1.5–2.7 mm; **blades** 15–55 cm long, 4–18 mm wide, flat, glabrous or pubescent. **Panicles** terminal, 10–25 cm long, 15–30 mm wide, fully exerted from the sheaths, erect to drooping, white, yellow, light brown, or pink to deep purple; **rachises** terete, scabrous. **Fascicles** 33–45 per cm, disarticulating at maturity; **fascicles axes** 0.2–0.5 mm, with 1 spikelet; **outer bristles** 13–30, 1.3–5 mm, scabrous; **inner bristles** 6–14, 4.3–11.5 mm, long ciliate; **primary bristles** 14–25 mm, long-ciliate, noticeably longer than the other bristles. **Spikelets** 3–4.5 mm, sessile; **lower glumes** absent or to 2 mm, veinless; **upper glumes** 3–4.5 mm, glabrous, 5–7-veined, 3-lobed; **lower florets** sterile or staminate; **lower lemmas** 3–3.9 mm, 5–7-veined, apices lobed; **lower paleas** 2.9–3.7 mm; **anthers** absent or 1.7–2 mm; **upper florets** disarticulating at maturity; **upper lemmas** 1.7–3 mm, coriaceous, shiny, 5-veined, apices ciliate; **anthers** 1.3–2.1 mm. **Caryopses** about 1.7 mm, concealed by the lemma and palea at maturity. $2n$ = 18, 36, 45, 48, 52, 53, 54, 56, 78.

Pennisetum polystachion is a polymorphic, weedy African species that has become established in the tropics and subtropics, including Florida. The U.S. Department of Agriculture considers it a noxious weed. Only **Pennisetum polystachion** subsp. **setosum** (Sw.) Brunken has been found in the *Flora* region. It differs from **P. polystachion** (L.) Schutt. subsp. **polystachion** as indicated:

1. Plants annual, usually profusely branching; fascicles white, pink, red, or deep purple . . . . subsp. *polystachion*
1. Plants perennial, usually sparingly branched; fascicles yellow, light brown, or purplish . . . . subsp. *setosum*

10. **Pennisetum pedicellatum** Trin. [p. 526]
HAIRY FOUNTAINGRASS

Plants usually annual, occasionally perennial; cespitose. **Culms** 40–150 cm, erect, branching; **nodes** glabrous. **Sheaths** glabrous, margins ciliate; **ligules** 1–2 mm; **blades** 6–30 cm long, 4–15 mm wide, flat, pubescent, basal margins ciliate. **Panicles** terminal and axillary, (5)8–19 cm long, 30–35 mm wide, erect, pink to purple; **rachises** terete, puberulent at the base. **Fascicles** 14–15 per cm, disarticulating at maturity; **fascicle axes** 1.5–2.5 mm, with (1)2–5 spikelets; **outer bristles** 10–20, 1.2–2 mm, scabrous; **inner bristles** 40–90, 2.2–14 mm, long ciliate; **primary bristles** 15–25 mm, long-ciliate, noticeably longer than the other bristles. **Spikelets** 3.4–4.8 mm; **pedicels** 0.5–3.5 mm; **lower glumes** 1.2–2.4 mm, 0–1-veined; **upper glumes** 3.4–4.8 mm, glabrous, 5-veined; **lower florets** staminate or sterile; **lower lemmas** 3.1–4 mm, 5(6)-veined; **lower paleas** 2.5–3.5 mm; **anthers** 3, 2.2–2.5 mm; **upper florets** disarticulating at maturity; **upper lemmas** 2–2.7 mm, coriaceous, smooth, shiny, 5-veined, margins glabrous, apices ciliate; **anthers** 1.5–2.5 mm. **Caryopses** about 1.7 mm, concealed by lemma and palea at maturity. $2n$ = 24, 30, 32, 35, 36.

Pennisetum pedicellatum is native to Africa. It now grows in many other areas, including Florida. The U.S. Department of Agriculture considers it a noxious weed.

11. **Pennisetum villosum** R. Br. *ex* Fresen. [p. 526]
FEATHERTOP

Plants perennial; rhizomatous. **Culms** 16–75 cm, erect; **nodes** glabrous. **Sheaths** glabrous, margins ciliate; **ligules** 1–1.3 mm; **blades** 5–40 cm long, 2–4.5 mm wide, flat to folded, glabrous, pubescent, or scabrous, margins ciliate or glabrous basally. **Panicles** terminal, 4–11.5 cm long, 50–75 mm wide, fully exerted from the sheaths, erect, white; **rachises** terete, pubescent (basally). **Fascicles** 7–11

P. macrourum *P. alopecuroides* *P. latifolium*

PENNISETUM

per cm; **fascicle axes** 1.5–2.5 mm, with 1–4 spikelets; **outer bristles** (0)1–8, 1–13.5 mm; **inner bristles** 23–41, 13–50.5 mm, densely plumose; **primary bristles** 40–50 mm, ciliate, usually not noticeably longer than the other bristles. **Spikelets** 9–12 mm, glabrous; **pedicels** 0.1–0.4 mm; **lower glumes** 0.3–1.3 mm, veinless; **upper glumes** 2.5–5.2 mm, 1(3)-veined; **lower florets** staminate or sterile; **lower lemmas** 7.5–10.5 mm, 7–9(10)-veined; **lower paleas** absent or 5.5–8.5 mm; **anthers** absent or 3.8–4.5 mm; **upper lemmas** 9–11 mm, 7-veined, apices scabridulous; **anthers** 3.5–5 mm. **Caryopses** concealed by the lemma and palea at maturity. $2n$ = 45.

Pennisetum villosum is native to Ethiopia, northern Somalia, and the Arabian Peninsula. It is grown as an ornamental in the *Flora* region.

12. **Pennisetum ciliare** (L.) Link [p. 528]
BUFFEL GRASS

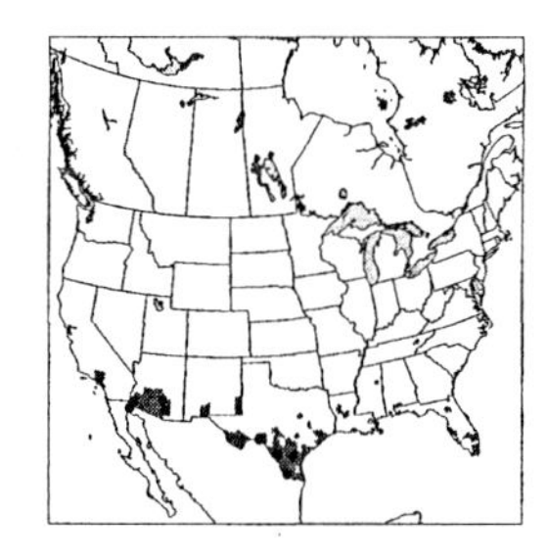

Plants perennial; cespitose from a hard, knotty base, with or without rhizomes. **Culms** 10–150 cm, erect, sometimes branching at the aerial nodes, glabrous, sometimes scabrous beneath the panicle; **nodes** glabrous. **Leaves** green or glaucous; **sheaths** glabrous or pubescent, margins ciliate; **ligules** 0.5–3 mm, membranous, ciliate; **blades** 3–50 cm long, 2–13 mm wide, flat, glabrous or pubescent, margins ciliate or glabrous basally. **Panicles** 2–20 cm long, 4–35 mm wide, fully exerted from the sheaths, erect, green, brown, brown-purple, or dark purple; **rachises** terete, scabrous. **Fascicles** 11–37 per cm, disarticulating at maturity; **fascicle axes** 0.2–1.5 cm, with 1–12 spikelets; **outer bristles** 16–89, 0.3–11.7 mm, many exceeding the spikelets; **inner bristles** 7–20, 3.8–13.8 mm, fused to ¼ of their length, flattened, grooved, ciliate; **primary bristles** 10.5–23 mm, long-ciliate, noticeably longer than the other bristles. **Spikelets** 2.5–5.6 mm, sessile, glabrous; **lower glumes** 1–3 mm, 0–1-veined; **upper glumes** 1.3–3.4 mm, about ½ as long as the spikelet, (0)1–3-veined; **lower florets** staminate or sterile; **lower lemmas** 2.5–5.3 mm, 3–7-veined; **lower paleas** absent or 2.5–5 mm; **anthers** absent or about 1.4 mm; **upper florets** not disarticulating at maturity; **upper lemmas** 2.2–5.4 mm, (3)5(6)-veined, margins glabrous; **anthers** 1.4–2.7 mm. **Caryopses** 1.2–1.9 mm long, 0.4–1 mm wide, concealed by the lemma and palea at maturity. $2n$ = 45.

Pennisetum ciliare is native to Africa, western Asia, and India. It now grows throughout the warmer, drier regions of the world, often as a forage crop, and is established in much of the southeastern United States. It is sometimes included in *Cenchrus*, based solely on the fusion of its bristles.

13. **Pennisetum setigerum** (Vahl) Wipff [p. 528]
BIRDWOOD GRASS

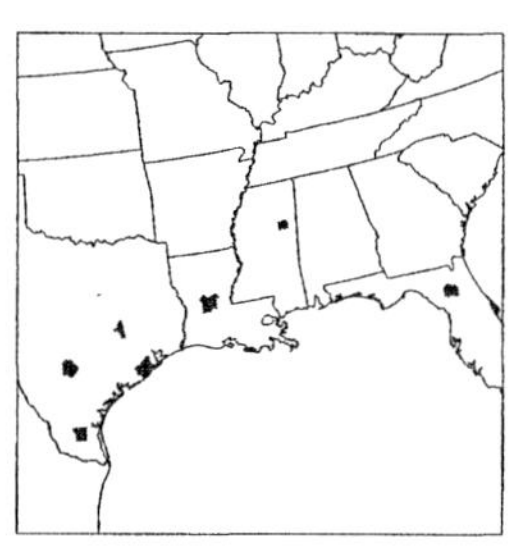

Plants perennial; cespitose from a hard, knotty base, without rhizomes. **Culms** 5–100 cm, erect, sometimes branching, mostly glabrous but sometimes scabrous beneath the panicle; **nodes** glabrous. **Leaves** green; **sheaths** glabrous or pubescent, margins ciliate; **ligules** 0.6–1.2 mm, membranous, ciliate; **blades** 2–45 cm long, 2.5–7 mm wide, flat, glabrous or pubescent, margins ciliate or glabrous basally. **Panicles** 2–13.8 cm long, 4–11 mm wide, erect, green or dark purple; **rachises** terete, scabrous. **Fascicles** 11–24 per cm, disarticulating at maturity; **fascicle axes** 0.2–1.1 mm, with 1–12 spikelets; **outer bristles** 10–62, 0.1–1.8 mm, not exceeding the spikelets; **inner bristles** 6–32, 1.2–5 mm, ciliate, fused for ⅓–½ their length; flattened, grooved; **primary bristles** 2.9–6.5 mm, ciliate, not noticeably longer than the other bristles. **Spikelets** 3.1–5.3 mm, sessile, glabrous; **lower glumes** 1–2.5 mm, 0–1-veined; **upper glumes** 1.5–3.4 mm, (0)1–3-veined, about ½ as long as the spikelet; **lower florets** staminate or sterile; **upper florets** not disarticulating at maturity; **lower lemmas** 2.7–5.3 mm, 3–7-veined; **lower paleas** absent or 2.5–4.5 mm; **anthers** absent or 0.9–3 mm; **upper lemmas** 2.8–5 mm, 3–5-veined; **anthers** 3, 2–3.2 mm. **Caryopses** 1.2–1.8 mm long, 0.4–1 mm wide. $2n$ = 34, 36, 37, 54, 72.

Penniseum setigerum is grown as a forage grass in the southern United States, but is not known to be established in the *Flora* region. It is native to Africa, Arabia, and India. It is sometimes included in *Cenchrus*, based solely on the fusion of its bristles.

14. **Pennisetum flaccidum** Griseb. [p. 528]
HIMALAYAN FOUNTAINGRASS

Plants perennial; rhizomatous. **Culms** 50–200 cm, erect, branching, smooth or scabrous beneath the panicle; **nodes** shortly pubescent or glabrous. **Leaves** green, sometimes glaucous; **sheaths** glabrous, margins ciliate; **ligules** 1–1.5 mm; **blades** 34–44 cm long, 5–10 mm wide, flat, glabrous or pubescent, margins ciliate or glabrous basally. **Panicles** all terminal, 8–17 cm long, 12–28 mm wide, erect, white; **rachises** scabrous below, sometimes puberulent above. **Fascicles** 6–15 per cm; **fascicle axes** 0.6–3.1 mm, with 1–6 spikelets; **outer bristles** 20–60, 0.8–12.3 mm; **inner bristles** 3–7, 5.5–16.5 mm, flattened, neither grooved nor fused, ciliate; **primary bristles** 12.9–22.5 mm, ciliate, noticeably longer than the other bristles. **Spikelets** 5.2–6.7

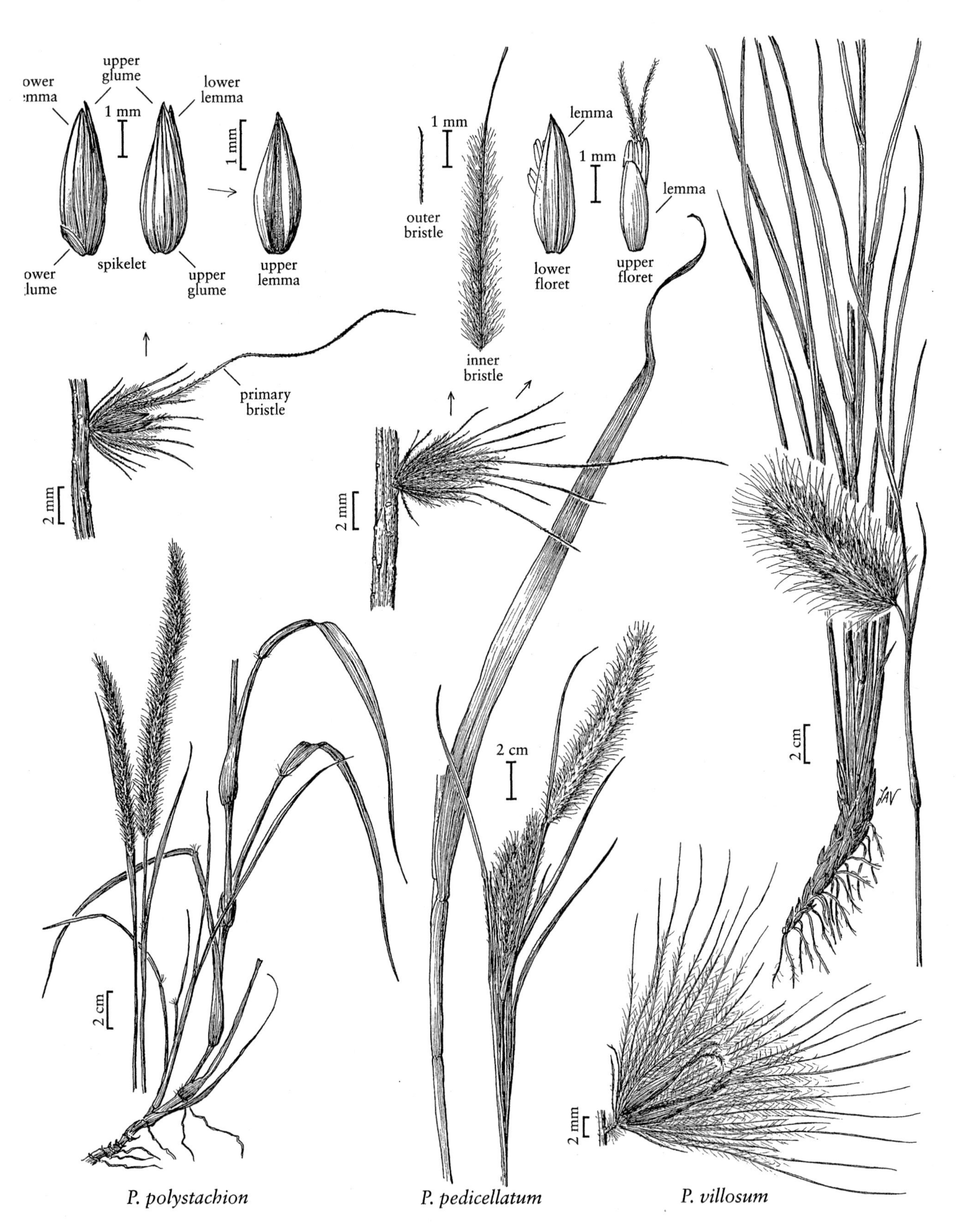

P. polystachion *P. pedicellatum* *P. villosum*

noticeably longer than the other bristles. **Spikelets** 5.2–6.7 mm; **pedicels** 0.1–0.5 mm; **lower glumes** 0.9–1.8 mm, 0–1-veined; **upper glumes** 2.6–4.2 mm, 1–5-veined; **lower florets** staminate; **lower lemmas** 5.1–6.4 mm, 5(6)-veined; **lower paleas** 4.5–5.3 mm; **anthers** 2.8–3.3 mm; **upper florets** not disarticulating at maturity; **upper lemmas** 5–5.8 mm, 5(6)-veined; **upper paleas** with a bifid apex, teeth 0.1–0.2 mm; **anthers** 2.6–3.2 mm. **Caryopses** concealed by the lemma and palea at maturity. $2n$ = 18, 36, 45.

Pennisetum flaccidum is native to central Asia. Although grown primarily as an ornamental, it is reportedly used for forage in the *Flora* region, but only one record, from Brazos County, Texas, is known. It is sometimes sold, incorrectly, as *P. incomptum* Nees *ex* Steud. [= *P. ciliare*].

15. **Pennisetum setaceum** (Forssk.) Chiov. [p. 528]
Tender Fountaingrass

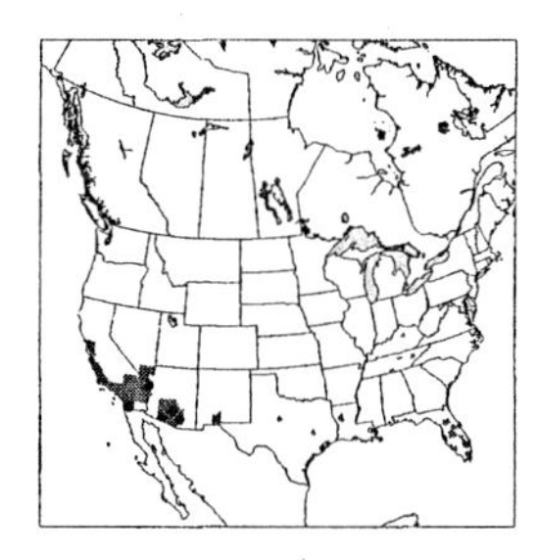

Plants perennial, or annual in temperate climates; cespitose. **Culms** 40–150 cm, erect, pubescent beneath the panicle; **nodes** glabrous. **Leaves** green, sometimes glaucous; **sheaths** glabrous, margins ciliate; **ligules** 0.5–1.1 mm; **blades** 20–65 cm long, 2–3.5 mm wide, convolute or folded, scabrous, midvein noticeably thickened. **Panicles** (6)8–32 cm long, 40–52 mm wide, erect or arching, pink to dark burgundy; **rachises** pubescent. **Fascicles** 8–10 per cm; **fascicle axes** 2.3–4.5 mm, with 1–4 spikelets; **outer bristles** 28–65, 0.9–19 mm; **inner bristles** 8–16, 8–27 mm, ciliate; **primary bristles** 26.5–34.3 mm, ciliate, noticeably longer than the other bristles. **Spikelets** 4.5–7 mm, sessile or pedicellate; **pedicels** to 0.1 mm; **lower glumes** absent or to 0.3 mm, veinless; **upper glumes** 1.2–3.6 mm, (0)1-veined; **lower florets** usually sterile, sometimes staminate; **lower lemmas** 4–6 mm, 3-veined, acuminate, midvein excurrent to 0.7 mm; **lower paleas** usually absent, if present, to 4.4 mm; **anthers** absent or 2.3–2.4 mm; **upper lemmas** 4.5–6.7 mm, attenuate, 5-veined, midvein excurrent to 0.7 mm, margins glabrous; **anthers** 2.1–2.7 mm. $2n$ = 27.

Pennisetum setaceum is a desert grass native to the eastern Mediterranean region. It is a popular ornamental throughout the southern United States, but it is also an invasive weed.

16. **Pennisetum orientale** Willd. *ex* Rich. [p. 530]
White Fountaingrass

Plants perennial; rhizomatous. **Culms** 50–200 cm, erect, pubescent beneath the panicle; **nodes** pubescent. **Leaves** green or glaucous; **sheaths** antrorsely scabridulous, mostly glabrous, margins ciliate; **ligules** 1–1.7 mm; **blades** 25–50 cm long, 3–9 mm wide, flat, antrorsely scabridulous, glabrous or pubescent, margins ciliate or glabrous basally, midvein not noticeably thickened. **Panicles** 11.5–37.3 cm long, 35–50 mm wide, fully exerted from the sheaths, erect to slightly arching, white (sometimes purplish-tinged); **rachises** terete, pubescent. **Fascicles** 5–12 per cm; **fascicle axes** 1–6 mm, with 1–10 spikelets; **outer bristles** 0–24, 0.8–9.6 mm, terete, scabrous; **inner bristles** 6–20, 5.6–17.5 mm, ciliate; **primary bristles** 12.2–23.8 mm, ciliate, noticeably longer than the other bristles. **Spikelets** 5.6–6.7 mm; **pedicels** 0.1–0.2 mm; **lower glumes** 1–2.2 mm, veinless; **upper glumes** 3.1–5.5 mm, 1–3-veined; **lower florets** staminate; **lower lemmas** 5.2–6.5 mm, 4–6-veined; **lower paleas** 3.8–5 mm; **anthers** 2.3–3 mm; **upper florets** not disarticulating at maturity; **upper lemmas** 5.2–6.2 mm, margins glabrous, 5-veined; **upper paleas** bifid, teeth 0.3–0.8 mm; **anthers** 1.9–2.9 mm. **Caryopses** concealed by the lemma and palea at maturity. $2n$ = 18, 27, 36, 45, 54.

Pennisetum orientale is native from North Africa to India. It is grown as an ornamental in the *Flora* region, but has potential as a forage species.

17. **Pennisetum advena** Wipff & Veldkamp [p. 530]
Purple Fountaingrass

Plants perennial, or annual in temperate climates; cespitose. **Culms** 1–1.5 m, erect, sometimes branching above, pubescent beneath the panicle; **nodes** glabrous. **Leaves** burgundy (rarely green); **sheaths** glabrous, margins ciliate; **ligules** 0.5–0.8 mm; **blades** 33–52 cm long, 6–11 mm wide, flat, antrorsely scabridulous, margins ciliate basally, midvein not noticeably thickened. **Panicles** 23–32 cm long, 30–58 mm wide, fully exerted from the sheaths, flexible, drooping, burgundy (rarely pale or whitish-green); **rachises** terete, pubescent. **Fascicles** 10–17 per cm, disarticulating at maturity; **fascicle axes** 1–2 mm, with 1–3 spikelets; **outer bristles** 43–68, 1.2–18.5 mm, terete, scabrous; **inner bristles** 4–10, 11.7–25 mm, long-ciliate; **primary bristles** 21.3–33.6 mm, ciliate, noticeably longer than the other bristles. **Spikelets** 5.3–6.5 mm; **pedicels** 0.1–0.3 mm; **lower glumes** 0.5–1 mm, veinless; **upper glumes** 1.9–3.6 mm, 0–1-veined; **lower florets**

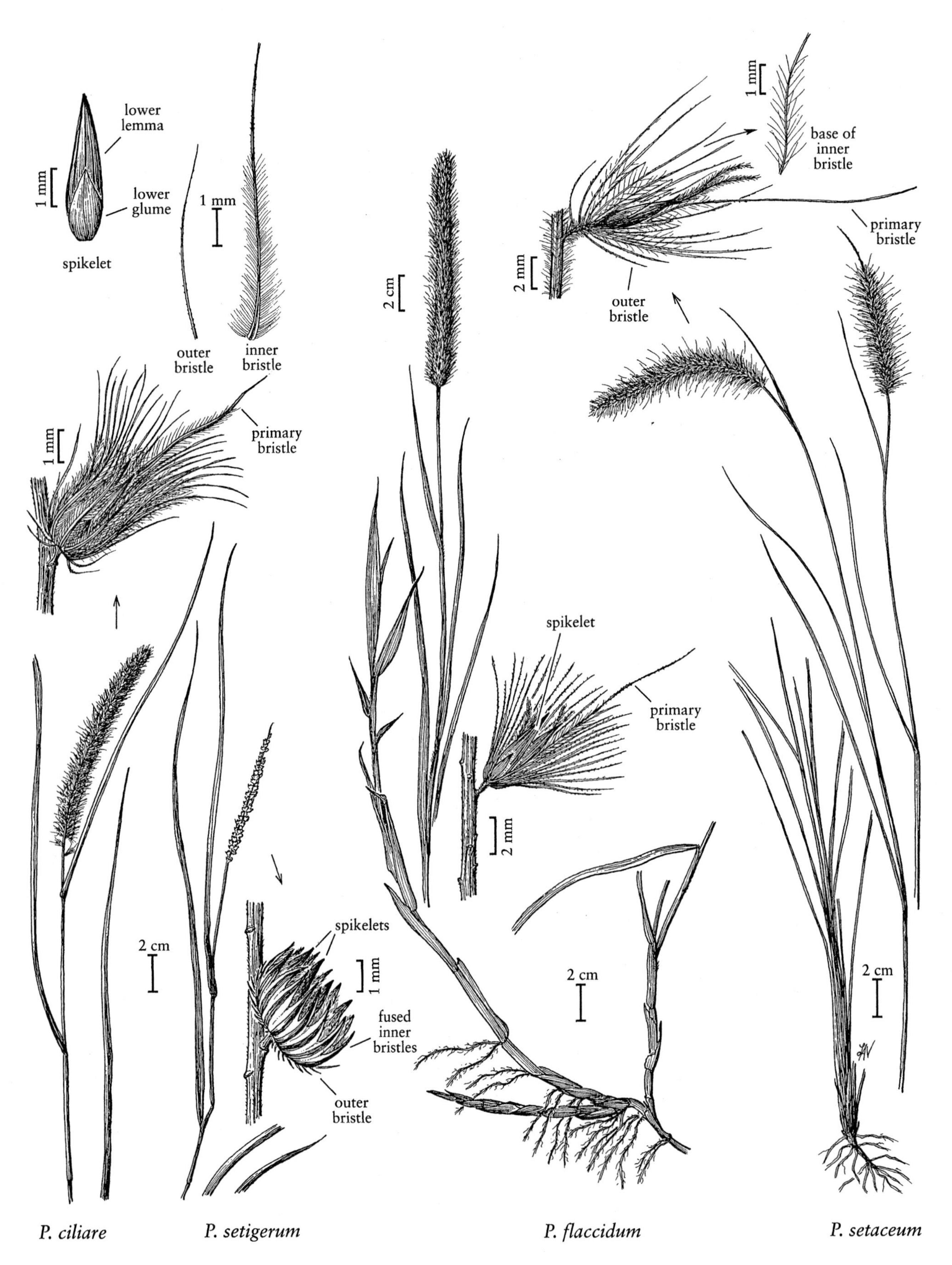
lower
lemma
1 mm
lower
glume
spikelet
1 mm
outer
bristle
inner
bristle
primary
bristle
1 mm
2 cm
spikelets
1 mm
fused
inner
bristles
outer
bristle
2 cm
1 mm
base of
inner
bristle
primary
bristle
2 mm
outer
bristle
spikelet
primary
bristle
2 mm
2 cm
2 cm
P. ciliare
P. setigerum
P. flaccidum
P. setaceum

PENNISETUM

staminate; **lower lemmas** 4.7–6.1 mm, 5(6)-veined; **lower paleas** 4.5–5 mm; **anthers** 2–2.5 mm; **upper florets** not disarticulating at maturity; **upper lemmas** 5.2–6.1 mm, 5-veined; **anthers** 2.5–2.7 mm. **Caryopses** concealed by the lemma and palea at maturity. $2n = 54$.

The origin of *Pennisetum advena* is uncertain. It is frequently cultivated as an ornamental, usually being sold as *P. setaceum* 'Rubrum'.

18. **Pennisetum petiolare** (Hochst.) Chiov. [p. 530]

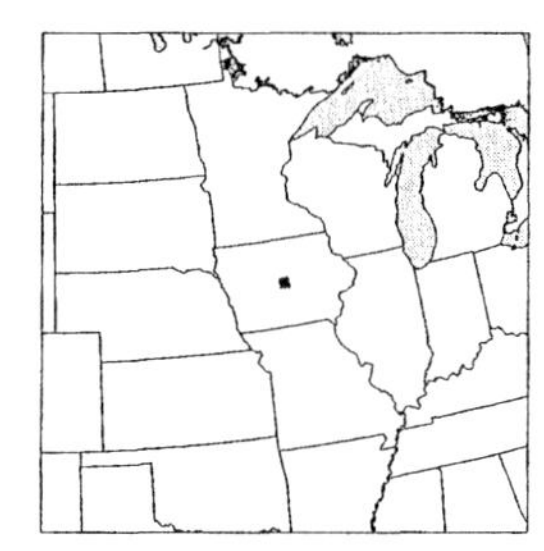

Plants annual; tufted, rooting at the lower nodes. **Culms** 70–200 cm, slender, erect, branching; **nodes** glabrous. **Sheaths** glabrous; **ligules** 0.7–1 mm, of hairs; **blades** 3.5–30 cm long, 8–24 mm wide, flat, sparsely pubescent, abruptly rounded or cordate basally; **lower blades** with a 0.5–13 cm pseudopetiole. **Panicles** terminal and axillary, 3–8 cm, on long peduncles, fully exerted from the sheaths; **rachises** terete, hispid; **disarticulation** beneath the primary bristles, fascicle axes persistent. **Fascicles** 14–24 per cm; **fascicle axes** 0.4–0.6 mm, with 1 spikelet, **outer** and **inner bristles** absent; **primary bristles** 3.8–30 mm, scabrous. **Spikelets** 2.5–3.2 mm, sessile, green to purple. **Glumes** veinless, truncate or emarginate; **lower glumes** 0.2–0.3 mm; **upper glumes** 0.3–0.4 mm; **lower florets** sterile; **lower lemmas** 2.5–3.2 mm, strigulose above, 5–7-veined; **lower paleas** absent; **upper lemmas** 2.3–3 mm, sparsely puberulent, 5-veined; **anthers** 3, 1.5–1.9 mm. $2n$ = unknown.

Pennisetum petiolare is native to northern Africa, where it grows in disturbed habitats. The only collection in the *Flora* region is from Ames, Iowa, where it grew from fallen bird seed. It is not known to be established anywhere in the region.

25.16 CENCHRUS L.

Michael T. Stieber
J.K. Wipff

Plants annual or perennial. **Culms** 5–200 cm, erect or decumbent, usually geniculate; **nodes** and **internodes** usually glabrous. **Sheaths** open, usually glabrous; **ligules** membranous, ciliate, cilia as long as or longer than the basal membrane; **blades** flat or folded, margins cartilaginous, scabridulous. **Inflorescences** terminal, spikelike panicles of highly reduced branches termed *fascicles* ("burs"); **fascicles** consisting of 1–2 series of many, stiff, partially fused, usually retrorsely scabridulous to strigose, sharp bristles surrounding, sometimes almost concealing, 1–4 spikelets; **outer** (lower) **bristles,** if present, in 1 or more whorls, terete or flattened; **inner** (upper) **bristles** usually strongly flattened, fused at least at the base and forming a disk, frequently to more than ½ their length and forming a cupule; **disarticulation** at the base of the fascicles. **Spikelets** sessile, with 2 florets; **lower florets** usually sterile; **upper florets** bisexual. **Lower glumes** ovate, scarious, glabrous, 1-veined, acute to acuminate; **upper glumes** and **lower lemmas** ovate, 3–9-veined; **lower paleas** equaling the lemmas, tawny or purplish; **upper lemmas** and **paleas** subequal, indurate, ovate, obscurely veined, acuminate. **Caryopses** obtrulloid. $x = 17$. Name from the Greek *kengchros*, 'millet'.

Cenchrus has about 16, primarily tropical species, most of which are readily (and painfully) recognized by their spiny fascicles. Most of its species differ from those of *Pennisetum* in having retrorsely scabrous or strigose inner bristles that are fused to well above their bases. The species are generally considered to be undesirable weeds.

Seven species of *Cenchrus* are native to the *Flora* region. The eighth species in this treatment was collected once in Westchester County, New York, but does not appear to have become established in the *Flora* region.

SELECTED REFERENCES Chase, A. 1920. The North American species of *Cenchrus*. Contr. U.S. Natl. Herb. 22[1]:45–77; DeLisle, D.G. 1963. Taxonomy and distribution of the genus *Cenchrus*. Iowa State Coll. J. Sci. 37:259–351; Gayle, E.E. 1892. The spines of *Cenchrus tribuloides* L. Bot. Gaz. 17:126–127; Sohns, E.R. 1955. *Cenchrus* and *Pennisetum*: Fascicle morphology. J. Wash. Acad. Sci. 45:135–143.

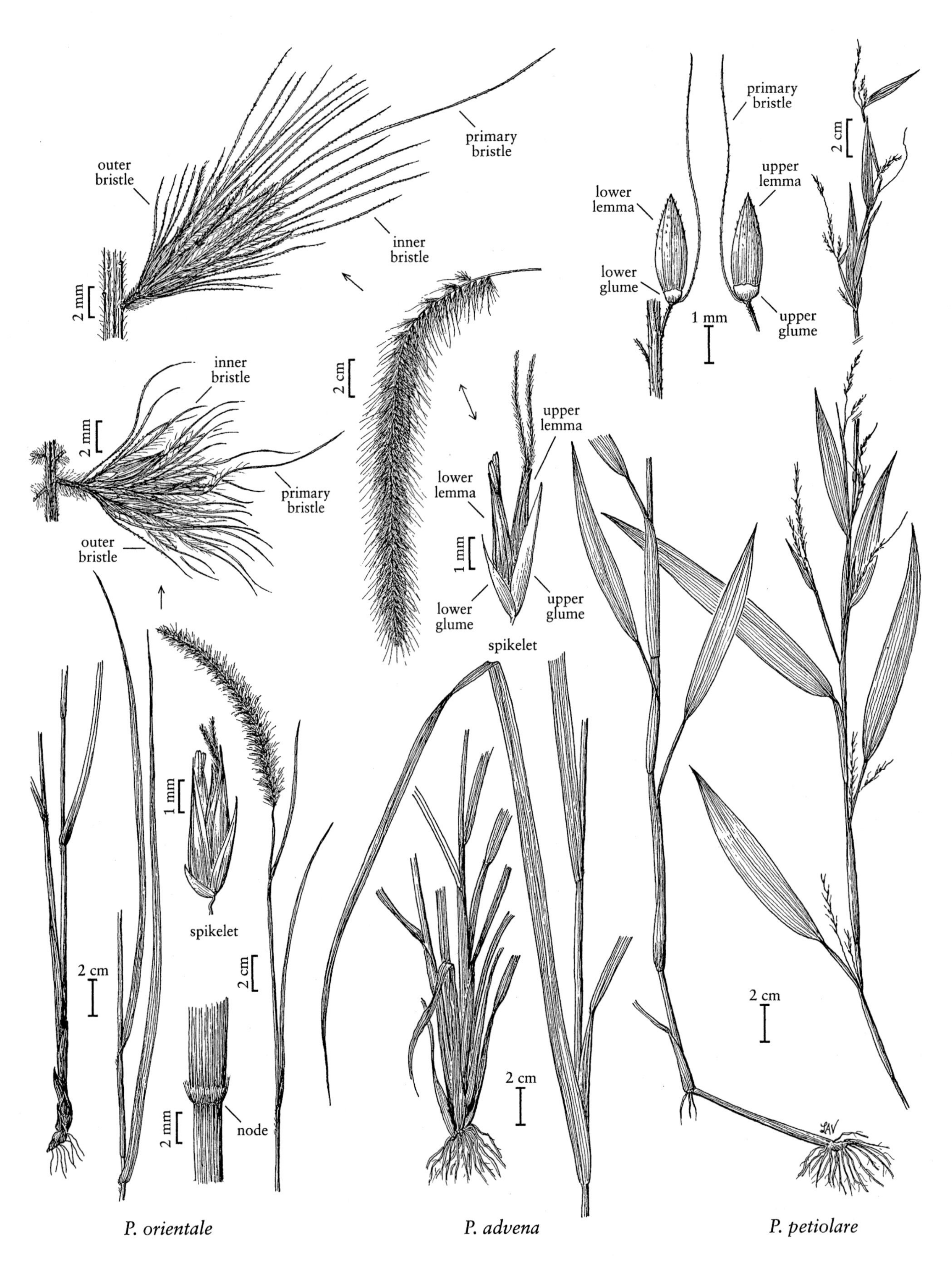
outer
bristle
primary
bristle
inner
bristle
2 mm
inner
bristle
2 mm
primary
bristle
outer
bristle
2 cm
upper
lemma
lower
lemma
1 mm
lower
glume
upper
glume
spikelet
primary
bristle
upper
lemma
lower
lemma
lower
glume
1 mm
upper
glume
2 cm
1 mm
spikelet
2 cm
2 cm
2 mm
node
2 cm
2 cm
P. orientale
P. advena
P. petiolare

PENNISETUM

1. All bristles terete, fused only at the base; fascicles not burlike . . . . 7. *C. myosuroides*
1. Inner bristles flattened, variously fused, forming a shallow disk or distinct cupule; fascicles burlike.
   2. Fascicles having 1 whorl of fused, flattened inner bristles, subtended by 5–25 free, terete, outer bristles.
      3. Inner bristles fused only at the base and forming a shallow disk, their abaxial surfaces with 1–3 grooves . . . . 8. *C. biflorus*
      3. Inner bristles fused for ⅓–½ their length or more, forming a globose cupule, their abaxial surfaces not grooved.
         4. Rachis internodes 0.8–1.7 mm long; the majority of the outer bristles equaling or slightly exceeding the inner, flattened bristles . . . . 1. *C. brownii*
         4. Rachis internodes 2–4 mm long; the majority of the outer bristles about ½ as long as the inner, flattened bristles . . . . 2. *C. echinatus*
   2. Fascicles having more than 1 whorl of flattened inner bristles, these originating at irregular intervals throughout the body of the cupule, sometimes subtended by terete outer bristles.
      5. Plants perennial, long-lived; fascicles not imbricate, usually glabrous; leaf blades 1–3.5 mm wide . . . . 3. *C. gracillimus*
      5. Plants annual or perennial but short-lived; fascicles imbricate, usually pubescent; leaf blades (1)3–14.2 mm wide.
         6. Inner bristles 0.5–0.9(1.4) mm wide at the base; fascicles with 45–75 bristles . . . . 5. *C. longispinus*
         6. Inner bristles 1–3 mm wide at the base; fascicles with 8–43 bristles.
            7. Fascicles densely pubescent, 9–16 mm long, with 1(2) spikelets; inner bristles 4–8 mm long . . . . 6. *C. tribuloides*
            7. Fascicles glabrous or sparsely to moderately pubescent, 5.5–10.2 mm long, with 2–4 spikelets; inner bristles 2–5.8 mm long . . . . 4. *C. spinifex*

1. **Cenchrus brownii** Roem. & Schult. [p. 532]
SLIMBRISTLE SANDBUR, GREEN SANDBUR

Plants annual. **Culms** 25–100 cm, erect or decumbent. **Sheaths** slightly compressed; **ligules** 0.6–1.3 mm; **blades** 6–30 cm long, 0.4–1.1 cm wide, adaxial surfaces glabrous or sparsely pilose. **Panicles** 4–15 cm; **rachis internodes** 0.8–1.7 mm; **fascicles** 5–8 mm long, 2–4.5 mm wide, imbricate, globose, villous at the base, tawny; **outer bristles** 5–25, the majority equaling or slightly exceeding the inner bristles but narrower and terete, arising in a whorl at the base of the fascicles; **inner bristles** 4–10, 2–4 mm long, 0.6–1.8 mm wide at the base, flattened, not grooved, erect or interlocking at maturity, fused for ⅓ their length or more, forming a globose cupule. **Spikelets** 2–3 per fascicle, 3–6 mm. **Lower glumes** 0.5–2.5 mm; **upper glumes** 2.2–4.9 mm, 3–5-veined; **lower lemmas** 3–5.5 mm; **upper florets** 3.6–5.4 mm; **anthers** 0.8–2.3 mm. **Caryopses** 1.9–2.6 mm long, 0.8–1.9 mm wide, ovoid. $2n = 34$.

Cenchrus brownii is native to sandy waste places and forest borders. It occurs infrequently on the coastal plain of the southeastern United States, but is common through the Caribbean, Central America, and the northern coast of South America. It has also been introduced to other parts of the world. The record from Texas may represent an introduction; only one specimen is known from the state.

2. **Cenchrus echinatus** L. [p. 532]
SOUTHERN SANDBUR

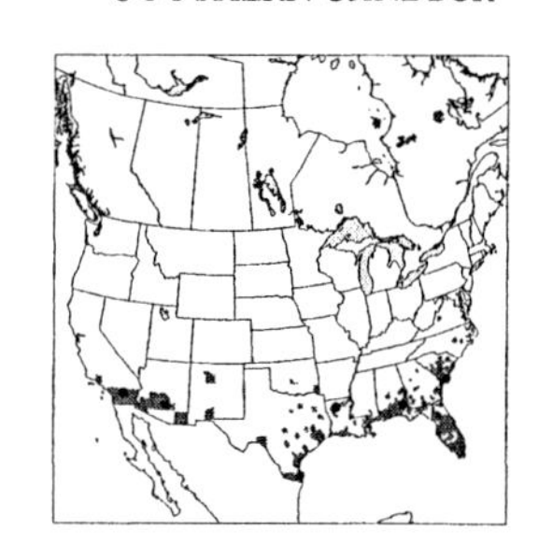

Plants annual. **Culms** 20–100 cm, ascending from a geniculate base. **Sheaths** from shorter than to equaling the internodes, compressed; **ligules** 0.7–1.7 mm; **blades** 4–18(35) cm long, 2–10(14.2) mm wide, adaxial surfaces sparsely pilose, hairs papillose-based. **Panicles** 2.5–12 cm; **rachis internodes** 2–4 mm; **fascicles** 5–10 mm long, 3.5–6(6.3) mm wide, imbricate; **outer bristles** 10–20, terete, the majority no more than ½ as long as the inner bristles; **inner bristles** 2–5 mm long, 0.6–1.5 mm wide, flattened, not grooved, mostly erect, fused for at least ½ their length into a globose cupule, sometimes interlocking at maturity, shortly pubescent, often purple at maturity. **Spikelets** 2–3(4) per fascicle, 4.8–7 mm. **Lower glumes** 1.3–3.4 mm; **upper glumes** 3.8–5.7 mm, 3–7-veined; **lower lemmas** 4.5–6.5 mm; **upper florets** 4.7–7 mm; **anthers** 0.8–2.4 mm. **Caryopses** ovoid, 1.2–3.2 mm long, 1.3–2.2 mm wide. $2n = (34), 68$.

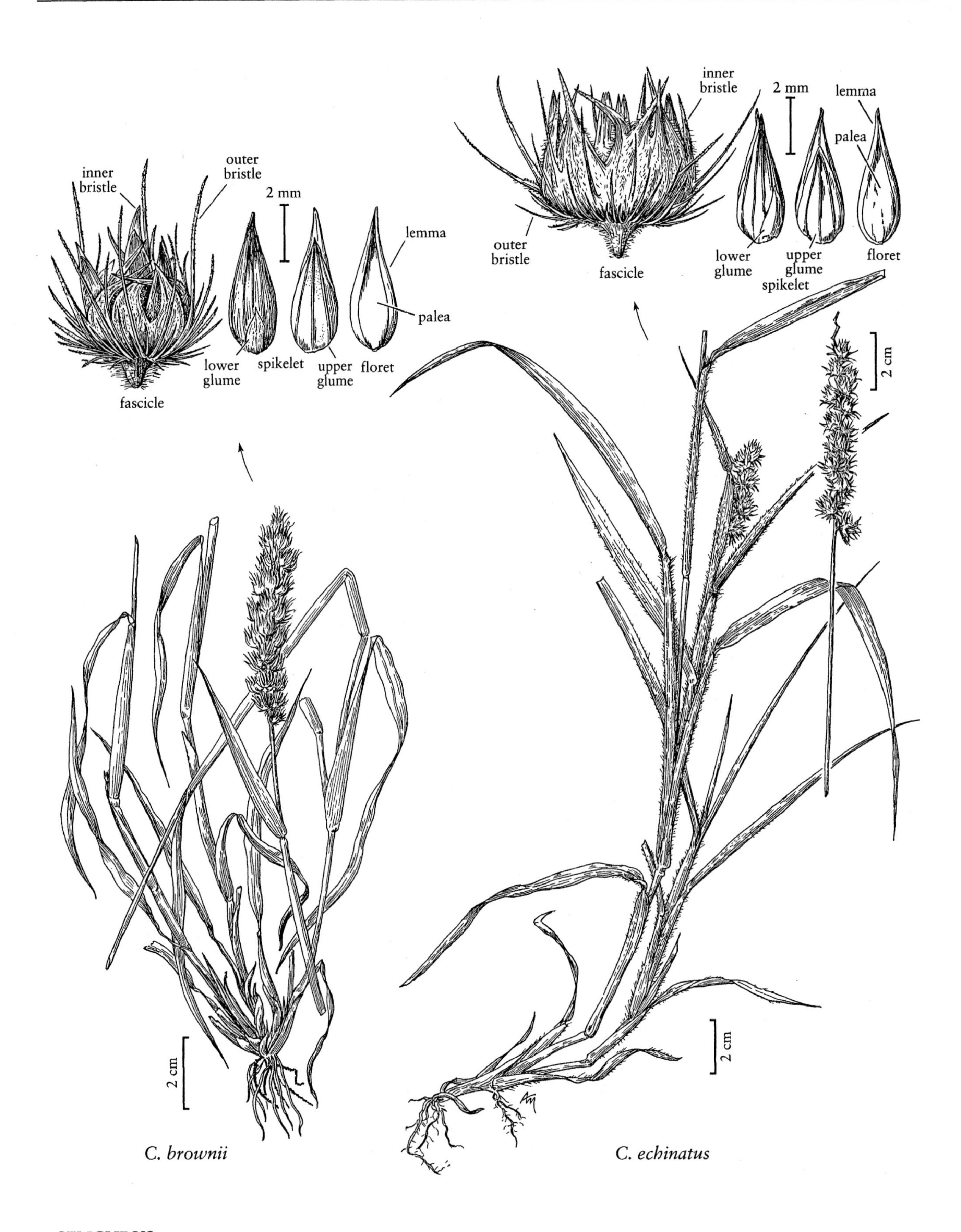
inner
bristle
outer
bristle
2 mm
lemma
palea
lower
glume
spikelet
upper
glume
floret
fascicle
inner
bristle
2 mm
lemma
palea
outer
bristle
fascicle
lower
glume
upper
glume
spikelet
floret
2 cm
2 cm
2 cm
C. brownii
C. echinatus

Cenchrus echinatus grows in disturbed areas throughout the coastal plain and piedmont of the southern United States, Mexico, Central and South America, and, as an unwelcome introduction, elsewhere.

3. **Cenchrus gracillimus** Nash [p. 533]
SLENDER SANDBUR

Plants perennial; sometimes forming dense clumps. **Culms** 20–80 cm, wiry. **Sheaths** shorter than the internodes, keeled, usually glabrous, rarely sparsely pilose; **ligules** 0.2–0.6 mm; **blades** 5–25 cm long, 1–3.5 mm wide, stiff, adaxial surfaces usually glabrous, smooth or scabrous. **Panicles** 2–6(6.8) cm; **rachis internodes** 2–4 mm; **fascicles** 5–13 mm long, 2–4 mm wide, not imbricate, ovoid, glabrous; **outer bristles** sometimes present, flattened; **inner bristles** less than 30, 3.2–6 mm long, 0.2–1 mm wide at the base, in more than 1 whorl, fused for at least ½ their length into a distinct cupule, diverging at irregular intervals from the cupule, somewhat flattened, spreading, purple-tipped at maturity. **Spikelets** 1–3 per fascicle, 4–7 mm. **Lower glumes** 1.4–3.1 mm; **upper glumes** 3.2–5.4 mm, 3–5-veined; **upper lemmas** 4–6 mm, 3–5-veined; **upper florets** 3.9–6.5 mm; **anthers** 0.9–1.9 mm. **Caryopses** 1.8–3 mm long, 1–1.5 mm wide, ovoid-elliptic. $2n = 34$.

Cenchrus gracillimus grows in sandy soils of open pinelands, wet prairies, and river flats of the southeastern United States and the West Indies.

4. **Cenchrus spinifex** Cav. [p. 533]
COASTAL SANDBUR, COMMON SANDBUR

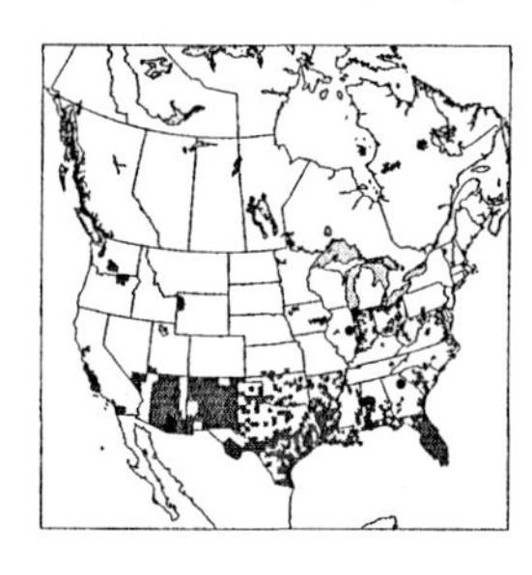

Plants annual or perennial but short-lived; tufted. **Culms** 30–100 cm, geniculate. **Sheaths** compressed, glabrous or sparsely pilose; **ligules** 0.5–1.4 mm; **blades** 3–28 cm long, (1)3–7.2 mm wide, glabrous or sparsely long-pilose adaxially. **Panicles** 3–5(8.5) cm; **fascicles** 5.5–10.2 mm long, 2.5–5 mm wide, imbricate, ovoid to globose, glabrous or sparesely to moderately pubescent; **outer bristles**, when present, mostly flattened; **inner bristles** 8–40 (rarely more), 2–5.8 mm long, 1–2 mm wide, fused at least ½ their length, forming a distinct cupule, the distal portions usually diverging from the cupule at multiple, irregular intervals, sometimes diverging at more or less the same level, ciliate at the base, pubescent, stramineous to mauve or purple, flattened. **Spikelets** 2–4 per fascicle, 3.5–5.9 mm, glabrous. **Lower glumes** 1–3.3 mm; **upper**

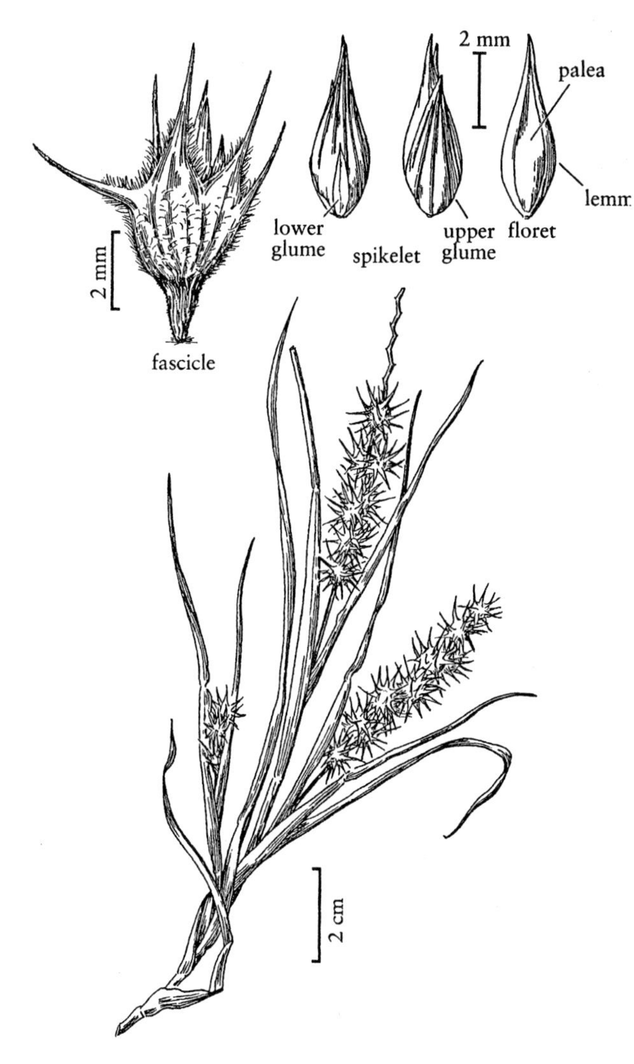

C. spinifex

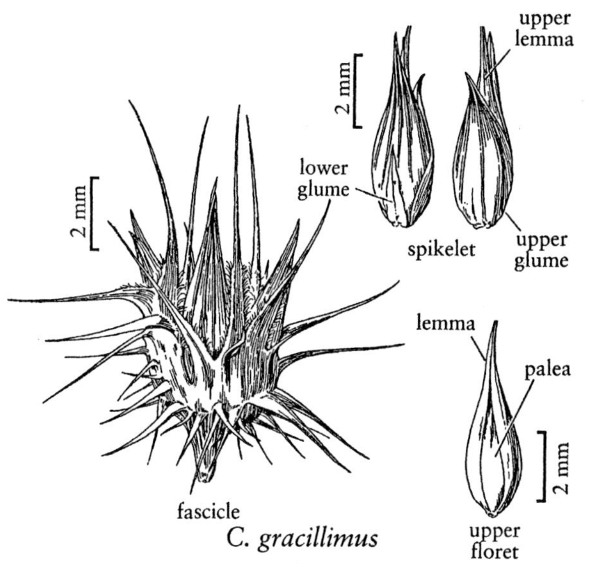

C. gracillimus

CENCHRUS

glumes (2.8)3.5–5 mm, 5–7-veined; **lower florets** sometimes staminate; **lower lemmas** 3–5(5.9) mm, 5–7-veined; **lower paleas** sometimes reduced or absent; **anthers** 1.3–1.6 mm; **upper lemmas** 3.5–5(5.8) mm; **anthers** 0.5–1.2 mm. **Caryopses** about 2.5 mm long, 1–2 mm wide, ovoid. $2n$ = 34 (32).

Cenchrus spinifex is common in sandy woods, fields, and waste places throughout the southern United States and southwards into South America. It may be more widespread than shown in the northern portion of the contiguous United States because it has often been confused with *C. tribuloides*. *Cenchrus spinifex* differs from *C. tribuloides* in its glabrous or less densely pubescent fascicles, narrower inner bristles, and larger number of bristles. It has also been confused with *C. longispinus* but differs in having shorter spikelets, fewer bristles overall, wider inner bristles, and outer bristles that are usually flattened rather than usually terete.

5. **Cenchrus longispinus** (Hack.) Fernald [p. 536]
MAT SANDBUR, LONGSPINE SANDBUR, CENCHRUS À ÉPINES LONGUES

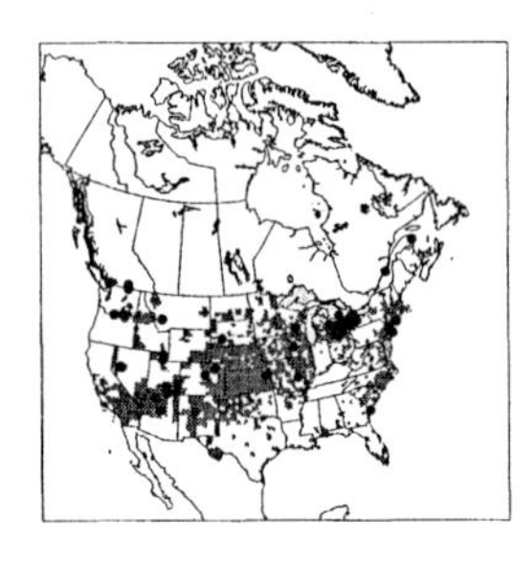

Plants annual; tufted. **Culms** 20–90 cm, sometimes decumbent, often with many branches arising from the base. **Sheaths** strongly compressed-keeled; **ligules** 0.6–1.8 mm; **blades** 4–27 cm long, 1.5–5(7.5) mm wide, adaxial surfaces scabrous or sparsely pilose. **Panicles** 1.5–8(10) cm; **fascicles** 8.3–11.9 mm long, 3.5–6 mm wide, somewhat globose, medium- to short-pubescent; **bristles** 45–75; **outer bristles** numerous, shorter and thinner than the inner bristles, imbricate, mostly terete, reflexed; **inner bristles** 3.5–7 mm long, 0.5–0.9(1.4) mm wide at the base, irregularly placed, fused for ½ their length or more, forming a distinct cupule, the distal portions diverging at irregular intervals from the cupule, often grooved along the margins, purple-tinged. **Spikelets** 2–3(4) per fascicle, (4)5.8–7.8 mm. **Lower glumes** 0.8–3 mm; **upper glumes** 4–6 mm, 3–5-veined; **lower florets** often staminate; **lower lemmas** 4–6.5 mm, 3–7-veined; **anthers** 1.5–2 mm; **upper lemmas** 4–7(7.6) mm; **anthers** 0.7–1 mm, seemingly not well-developed at anthesis. **Caryopses** 2–3.8 mm long, 1.5–2.6 mm wide, ovoid. $2n$ = 34 (38).

Cenchrus longispinus grows in sandy woods, fields, and waste ground in southern Canada and the contiguous United States. Its range extends southwards to Venezuela. It is often confused with *C. spinifex* and *C. tribuloides*; see discussion under those species.

6. **Cenchrus tribuloides** L. [p. 536]
SANDDUNE SANDBUR, DUNE SANDBUR

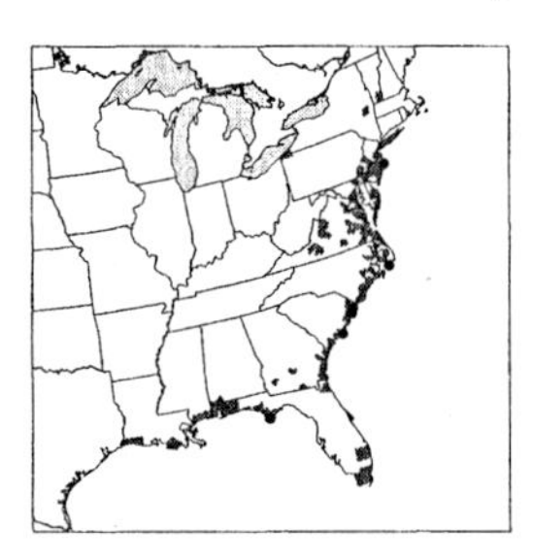

Plants annual. **Culms** 10–70 cm, decumbent, branching and rooting at the lower nodes. **Sheaths** compressed, glabrous or pubescent; **ligules** 1–2.1 mm; **blades** 2–14 cm long, 3–14.2 mm wide. **Panicles** 2–8.2 cm; **fascicles** 9–16 mm long, 4–8 mm wide, imbricate, ovoid, densely pubescent; **bristles** 15–43; **outer bristles** usually present, flattened or terete; **inner bristles** 4–8 mm long, 1.2–3 mm wide, fused for at least ½ their length, forming a distinct cupule, the distal portions diverging at irregular intervals from the cupule, stramineous or purple. **Spikelets** 1(2) per fascicle, 6–8.8 mm. **Lower glumes** 1–4 mm; **upper glumes** 4.9–6.8 mm, 3–7-veined; **lower lemmas** 5.5–7.5 mm, 3–7-veined, enclosing the palea; **upper lemmas** 6–8.7 mm; **anthers** 0.8–2.8 mm. **Caryopses** 2.6–4 mm long, 2.2–3.1 mm wide, ovoid-elliptic. $2n$ = 34.

Cenchrus tribuloides grows in moist, sandy dunes and is restricted to the eastern United States. It differs from *C. spinifex* in its larger spikelets and smaller number of spikelets per fascicle, and from *C. longispinus* in its densely pubescent fascicles, fewer bristles, and wider inner bristles.

7. **Cenchrus myosuroides** Kunth [p. 536]
BIG SANDBUR

Plants perennial. **Culms** 5–200 cm, stout, glaucous. **Sheaths** from shorter than to equaling the internodes; **ligules** 1.5–2(3.4) mm; **blades** 12–40 cm long, 4–13 mm wide, glabrous or sparsely pilose adaxially. **Panicles** 4–23 cm; **fascicles** 3.8–8 mm long, 1.2–2.6 mm wide, composed of several whorls of bristles, not burlike; **bristles** 3–5.8 mm long, 0.2–0.6 mm wide, fused only at the base, not forming a cupule, terete, increasing in size inwards, inner bristles pubescent on the lower ½–⅔. **Spikelets** 1(2–3) per fascicle, 3.8–4.8(5.6) mm. **Lower glumes** 1.5–3 mm; **upper glumes** 3–5 mm, 3–5-veined; **lower lemmas** 3–5.5 mm; **upper lemmas** 3.8–5.4 mm; **anthers** 0.8–2.2 mm. **Caryopses** 1.5–2.6 mm long, 1–1.5 mm wide, ovoid. $2n$ = (54), 70.

Cenchrus myosuroides is a native species that grows mostly along roadsides and in other waste places. Its native range extends through the Caribbean and Central America to northern South America.

8. **Cenchrus biflorus** Roxb. [p. 536]
INDIAN SANDBUR

Plants annual. **Culms** 5–150 cm, erect. **Sheaths** keeled, glabrous, scabrous, or slightly pubescent; **ligules** 1.3–2 mm; **blades** 2–35 cm long, 2–7 mm wide, flat, glabrous or scabrous (sparsely pilose). **Panicles** 2–15 cm; **fascicles** 4–11 mm long, 2–4.5 mm wide; **bristles** 30–60; **outer bristles** numerous, less than ½ as long as the inner bristles, terete; **inner bristles** 2.9–7 mm long, 0.2–1.1 mm wide, flattened, with 1–3 grooves abaxially, fused only at the base, forming a shallow disk, retrorsely scabrous, inner margins long-ciliate. **Spikelets** 1–3(4) per fascicle, 3.5–6 mm long, 1.2–1.9 mm wide; **lower glumes** 0.5–2.5 mm; **upper glumes** 2.5–4.9 mm, 3–5-veined; **lower lemmas** 3.2–5.5 mm, 4–5-veined; **upper lemmas** 3.4–5.9 mm; **anthers** about 1.5 mm. **Caryopses** 2–3.4 mm long, 1–3.5 mm wide, ovoid. $2n = 34$.

Cenchrus biflorus is widely distributed from Africa to India. It was collected once in Westchester County, New York, but has not become established in the *Flora* region.

25.17 ANTHEPHORA Schreb.

Mary E. Barkworth

Plants annual or perennial; tufted or cespitose. **Culms** 15–50 cm, not woody. **Sheaths** open; **ligules** membranous. **Inflorescences** terminal, spikelike panicles, each node supporting a highly reduced branch or *fascicle* of spikelets; **fascicles** imbricate, with a short, thick, basal stipe subtending 4 thick, rigid, coriaceous, many-veined, flat, narrowly elliptic to ovate bracts; **bracts** fused at the base, enclosing 2–11 spikelets, 1–2 of the spikelets sterile or reduced; **disarticulation** beneath the fascicles. **Spikelets** with 2 florets. **Lower glumes** absent; **upper glumes** acicular, 1-veined, awned; **lower florets** sterile, sometimes reduced; **lower lemmas** 7-veined; **lower paleas** subequal to the lower lemmas; **upper florets** bisexual, sterile, or reduced; **upper lemmas** faintly 3-veined. **Caryopses** ellipsoidal. $x = 9$. Name from the Greek *anthe*, 'blossom', and *pherin*, 'to bear', alluding to the superficial resemblance of the inflorescence to a "normal" flower.

Anthephora is a genus of 12 species. Most of the species are native to Africa and Arabia. One species is native to tropical America and has become established in the *Flora* region.

The fascicle bracts are often interpreted as the lower glumes of the spikelets, but the developmental studies needed to evaluate this interpretation have not been conducted.

SELECTED REFERENCE Reeder, J.R. 1960. The systematic position of the grass genus *Anthephora*. Trans. Amer. Microscop. Soc. 79:211–218.

1. **Anthephora hermaphrodita** (L.) Kuntze [p. 538]
OLDFIELD GRASS

Plants annual. **Culms** erect or decumbent, sometimes rooting at the lower nodes, branching from the base. **Sheaths** densely to sparsely pubescent, hairs papillose-based; **ligules** 1.5–3 mm, brownish, entire or dentate, not ciliate; **blades** 4–17 cm long, 2–8 mm wide, flat, pubescent on both surfaces. **Panicles** 4–12 cm long, 5–8 mm wide, with 20–60 fascicles; **fascicles** 5–7.5 mm; **bracts** 4–7 mm, scabrous. **Fertile spikelets** 3.5–4.5 mm, ovoid, scabrid between the veins, acute; **upper lemmas** 3.7–4 mm, glabrous, margins overlapping the edges of the palea. **Caryopses** about 2 mm. $2n = 18$.

Anthephora hermaphrodita is a weedy species, native to maritime beaches, lowland pastures, and disturbed areas from Mexico and the Caribbean Islands to Peru and Brazil. It is now established in Alachua County, Florida, having escaped from plantings at the Experiment Station of the University of Florida, Gainesville.

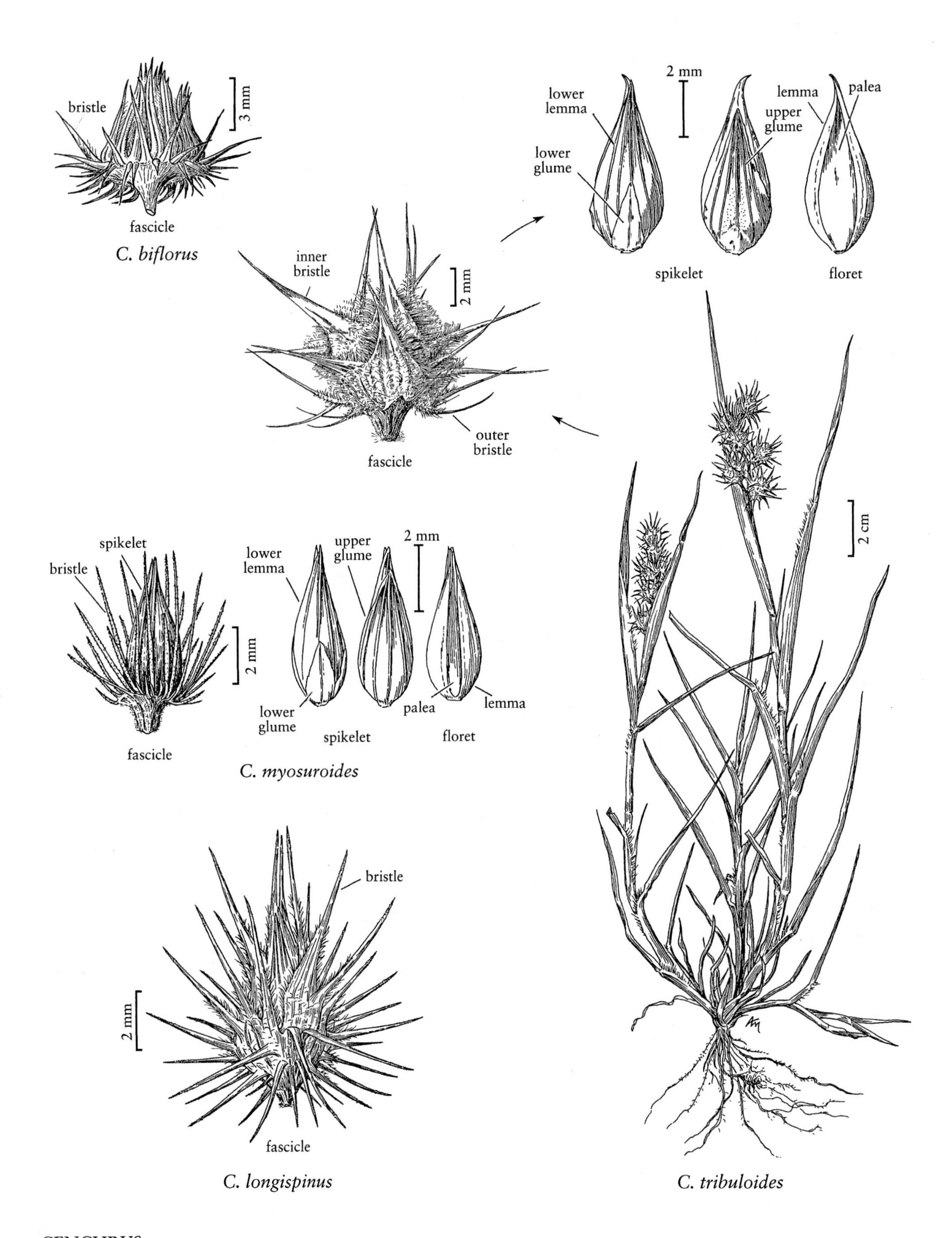
bristle
fascicle
3 mm
C. biflorus
inner bristle
2 mm
outer bristle
fascicle
2 mm
lower lemma
lower glume
upper glume
lemma
palea
spikelet
floret
2 cm
spikelet
bristle
2 mm
fascicle
lower lemma
upper glume
2 mm
lower glume
spikelet
palea
lemma
floret
C. myosuroides
bristle
2 mm
fascicle
C. longispinus
C. tribuloides

25.18 IXOPHORUS Schltdl.

Ken M. Hiser

Plants annuals or short-lived perennials; tufted. **Culms** 15–150 cm tall, 1–10 mm thick, dry to somewhat succulent, longitudinally grooved, glabrous. **Leaves** linear, vernation conduplicate; **sheaths** open, glabrous, compressed laterally, often purple-streaked at the base; **ligules** membranous, long-ciliate; **blades** flat, midvein often white, adaxial surfaces minutely pubescent immediately distal to the ligule, glabrous elsewhere, margins strigillose. **Inflorescences** terminal, open, pyramidal panicles; **rachises** scabridulous, bearing 4–50 alternate, spikelike primary branches; **primary branches** to 7 cm, flexuous, axes scabridulous, bearing spikelets in 2 abaxial rows; **pedicels** shorter than 1 mm, cuplike, each with a single, smooth (occasionally scabridulous), terete bristle; **bristles** 4–12 mm, pale brown to black; **disarticulation** below the glumes. **Spikelets** dorsally compressed, with 2 florets; **lower florets** staminate; **upper florets** pistillate. **Lower glumes** ¼–⅓ as long as the upper glumes, orbicular to triangular, 3-veined; **upper glumes** slightly shorter than the lower lemmas, often purple or green, 11-veined, acute; **lower lemmas** 5-veined, acute; **lower paleas** hyaline, about as long as the upper glumes, accrescent, thinly membranous at anthesis, becoming thicker and stiffer and about 3 times as wide as the lemma in fruit, keels clasping the upper floret at maturity; **anthers** 3, about 2 mm, orange; **upper lemmas** dorsally compressed, indurate, rugose, papillate, bases with a prominent germination flap, margins enclosing the edges of the upper paleas; **upper paleas** flat, indurate, papillate; **stigmas** bright red, plumose. **Caryopses** oblong obtuse, dorsally compressed; **embryos** about ⅓ as long as the caryopses. x = unknown. The origin of the name is obscure.

Ixophorus is native from central Mexico through Central America to northern South America. It is treated here as consisting of a single species, but it has been treated in the past as having as many as three species. These, however, intergrade morphologically with respect to all the characters used to distinguish them (Hiser 2002).

The name is either based on the Greek *ixos*, 'birdlime', "An extremely adhesive viscid substance; hence, anything that ensnares" (Clifford 1966) or the Greek *xiphos*, 'sword', and *phorus*, 'bearing', referring in either case to the bristle. Schlechtendal (1861–1862), in his description of the genus, states that the inflorescence is sticky, but the bristles of living plants, although somewhat lustrous, do not appear to be viscid.

SELECTED REFERENCES **Clifford, H.T.** 1996. Etymological Dictionary of Grasses, Version 1.0 (CD-ROM). Expert Center for Taxonomic Identification, Amsterdam, The Netherlands; **Hiser, K.M.** 2002. Phylogenetic placement, inflorescence development, and taxonomy of the genus *Ixophorus* (Panicoideae: Poaceae). Master's thesis. University of Missouri-St. Louis, St. Louis, Missouri, U.S.A. 88 pp.; **Hitchcock, A.S.** 1919. History of the Mexican grass, *Ixophorus unisetus*. J. Wash. Acad. Sci. 9:546–551; **Schlechtendal, D.F.L.** 1861–1862. Ueber *Setaria* P.B. Linnaea 31:387–509; **Scribner, F.L.** 1897. Studies on American grasses: I. The genus *Ixophorus*. Bull. Div. Agrostol. U.S.D.A. 4[1]:5–7, 40, pl. 1–2.

1. **Ixophorus unisetus** (J. Presl) Schltdl. [p. 538]
TURKEY GRASS, CRANE GRASS

Culms 15–150 cm tall, 1–10 mm thick. **Ligules** 1–1.5 mm; **blades** 5–60 cm long, usually 1–2 cm wide. **Panicles** 10–25 cm, with 4–50 branches. **Spikelets** 3–4 mm.

Ixophorus unisetus has been collected in Kleberg County, Texas, where it was being evaluated for its forage potential. It is not known to be established in the *Flora* region.

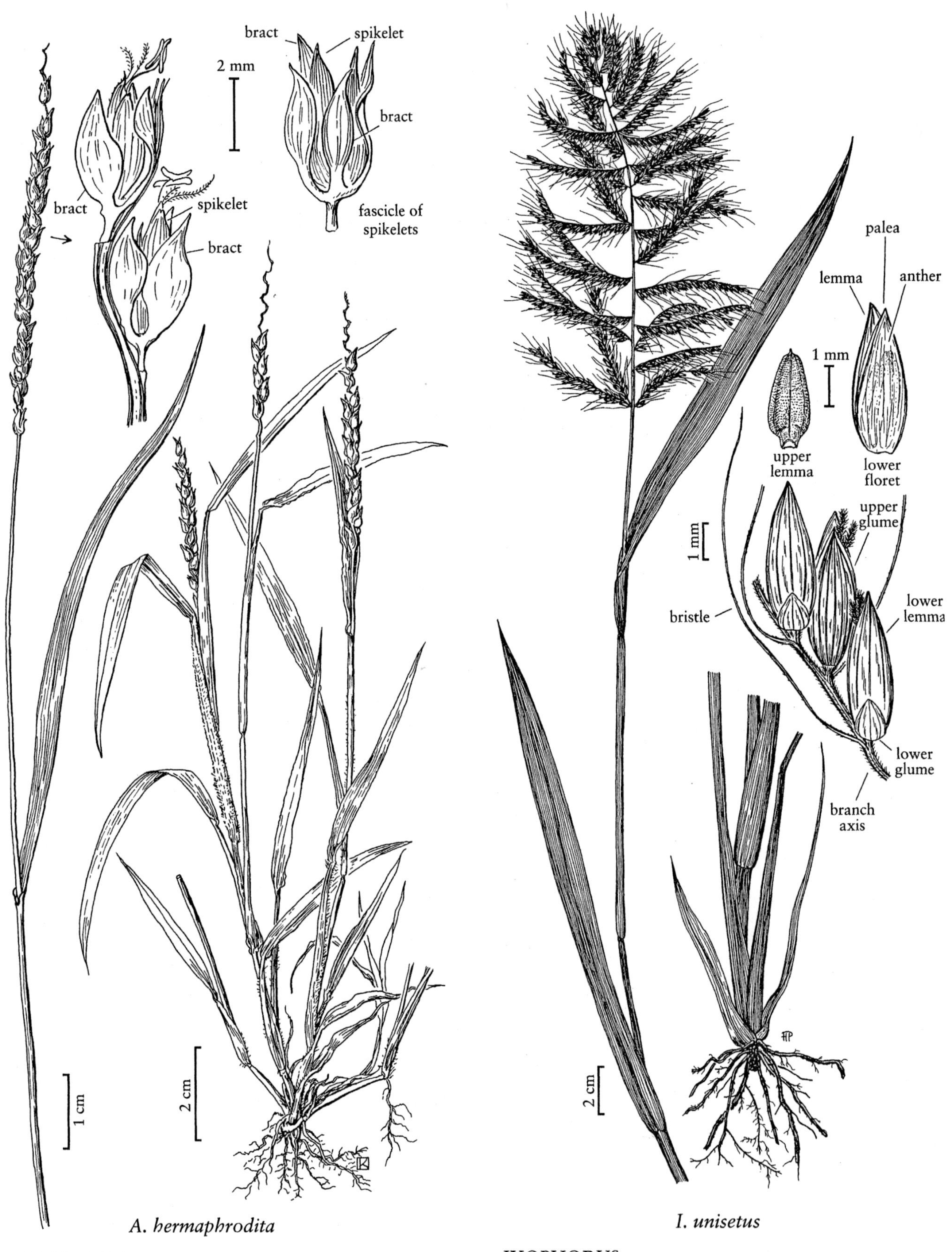

A. hermaphrodita

ANTHEPHORA

I. unisetus

IXOPHORUS

25.19 SETARIOPSIS Scribn.

John R. Reeder

Plants annual. **Culms** 20–80 cm, to about 1 mm thick, solid, branching above the base. **Sheaths** open; **ligules** of hairs; **blades** flat. **Inflorescences** terminal, panicles, 8–23 cm long, 0.8–2 cm wide, with pilose rachises; **branches** 0.5–1.5 cm, spikelets congested, shortly pedicellate, the pedicels subtended by a 3–10 mm, terete bristle; **disarticulation** below the glumes. **Spikelets** dorsally compressed, with 2 florets, lower florets usually sterile, upper florets bisexual. **Lower glumes** about ¼ as long as the spikelets, 5–7-veined, subclasping; **upper glumes** slightly shorter than the spikelets, 11–19-veined, indurate at maturity, constricted at the base, auriculate above the point of constriction; **lower lemmas** longer than the glumes, membranous but somewhat indurate at the base; **lower paleas** usually present, short; **upper lemmas** indurate, finely transversely rugose, apiculate, margins clasping the paleas; **upper paleas** similar to the lemmas in length and texture; **lodicules** 2; **anthers** 3, purple; **ovaries** glabrous; **style branches** 2, free to the base. **Caryopses** ovate, plano-convex; **embryos** about ½ as long as the caryopses. $x = 9$. Name from the grass genus *Setaria* and the Greek *opsis*, 'appearance' or 'likeness'.

Setariopsis includes two species, both of which were thought to be endemic to Mexico until the recent discovery of the following species in Arizona (Reeder 2001).

SELECTED REFERENCES **Pohl, R.W.** 1994. *Setariopsis* Scribner P. 364 *in* G. Davidse, M. Sousa S., and A.O. Chater (eds.). Flora Mesoamericana, vol. 6: Alismataceae a Cyperaceae. Universidad Nacional Autónoma de México, Instituto de Biología, México, D.F., México. 543 pp.; **Reeder, J.R.** 2001. Noteworthy collections: Arizona. Madroño 48:212–213.

1. **Setariopsis auriculata** (E. Fourn.) Scribn. [p. 540]

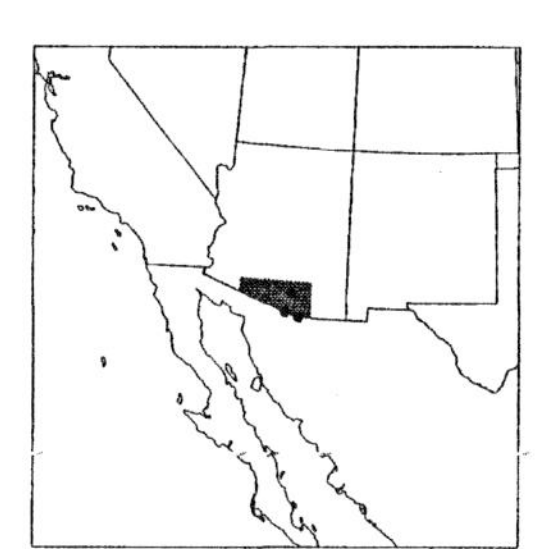

Culms 20–70 cm, weak, erect or ascending, glabrous or with a few, short, appressed hairs on and immediately below the nodes. **Sheaths** mostly glabrous but with soft hairs near the ligule; **ligules** 1–1.5 mm; **blades** to 20 cm long, 5–15 mm wide, glabrous or with a few hairs on the collars. **Panicles** 8–16 cm long, 0.8–1.5 cm wide; **branches** 0.5–3.5 cm; **bristles** 3–10 mm, antrorsely barbed. **Spikelets** 3–3.5 mm, ovate, acute. **Lower glumes** about 1 mm long, 1.5 mm wide, inflated; **upper glumes** 2.7–3.2 mm, ovate, somewhat inflated, abruptly constricted below, margins resembling auricles, apices acute or obtuse; **lower lemmas** ovate-triangular, indurate in the lower ⅓ where the spikelet narrows; **upper lemmas** 2–2.5 mm, finely and transversely rugose, acute and apiculate; **anthers** 0.8–1 mm, purple. $2n = 18$.

Setariopsis auriculata was recently found to be established in Arizona (Reeder 2001), growing in moist, shady habitats. Prior to this discovery, its range was considered to extend from northern Mexico to Nicaragua, Colombia, and Venezuela.

25.20 SETARIA P. Beauv.

James M. Rominger

Plants annual or perennial; cespitose, rarely rhizomatous. **Culms** 10–600 cm, erect or decumbent. **Ligules** membranous and ciliate or of hairs; **blades** flat, folded, or involute, or plicate and petiolate (subg. *Ptychophyllum*). **Inflorescences** terminal, panicles, usually dense and spikelike, occasionally loose and open; **disarticulation** usually below the glumes, spikelets falling intact, bristles persistent. **Spikelets** 1–5 mm, usually lanceoloid-ellipsoid, rarely globose, turgid, subsessile to short pedicellate, in fascicles on short branches or single on a short branch, some or all subtended by 1–several, terete bristles (sterile branchlets). **Lower glumes** membranous, not saccate, less than ½ as long as the spikelets, 1–7-veined; **upper glumes** membranous to herbaceous

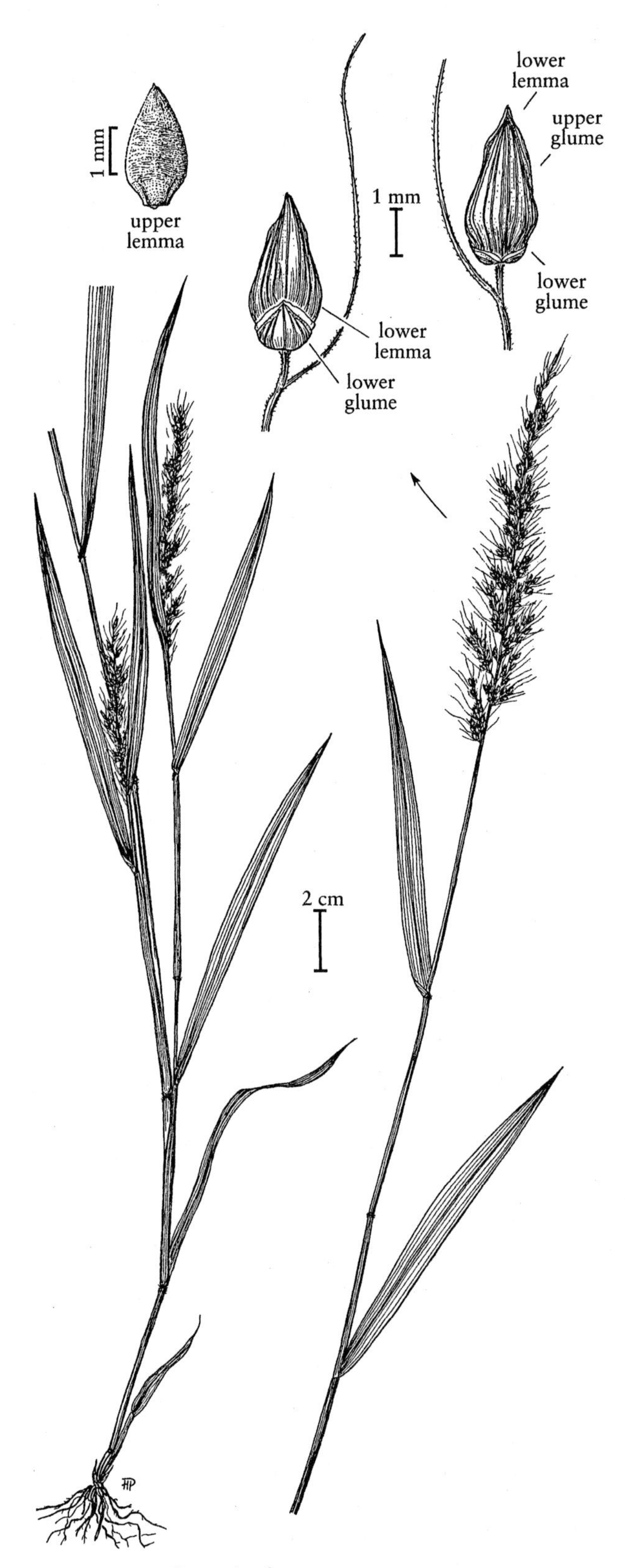

S. auriculata

SETARIOPSIS

at maturity, ½ as long as to nearly equaling the upper lemmas in length, 3–9-veined; **lower florets** staminate or sterile; **lower lemmas** membranous, equaling or rarely exceeding the upper lemmas, rarely absent, not constricted or indurate basally, 5–7-veined; **lower paleas** usually hyaline to membranous at maturity, rarely absent or reduced, veins not keeled; **upper florets** bisexual; **upper lemmas** and **paleas** indurate, transversely rugose, rarely smooth; **anthers** 3, not penicillate; **styles** 2, free or fused basally, white or red. **Caryopses** small, ellipsoid to subglobose, compressed dorsiventrally. $x = 9$. Name from the Latin *seta*, 'bristle' and *aria*, 'possessing'.

Setaria, a genus of about 140 species, grows predominantly in tropical and warm-temperate regions, but it is particularly well-represented in Africa, Asia, and South America. Species from the *Flora* region fall into one of three categories: native to North America, native to South America, or native to the Eastern Hemisphere. There are 27 species in the *Flora* region; fifteen are native, nine are established introductions, one is cultivated, and two are not established or have been collected only at scattered locations. Several species of the native *Setaria macrostachya* complex (*S. macrostachya*, *S. leucopila*, *S. texana*, *S. villosissima*, and *S. scheelei*) provide valuable forage in the southwestern United States. *Setaria italica* has been cultivated for centuries in Asia and Europe, providing food for humans and their livestock. The majority of species in temperate North America are aggressive, exotic annuals which collectively are a major nuisance, particularly in the corn and bean fields of the midwestern states.

SELECTED REFERENCES **Clayton, W.D.** 1979. Notes on *Setaria* (Gramineae). Kew Bull. 33:501–509; **Emery, W.H.P.** 1957. A cyto-taxonomic study of *Setaria macrostachya* (Gramineae) and its relatives in the southwestern United States and Mexico. Bull. Torrey Bot. Club 84:94–105; **Fox, W.E., III** and **S.L. Hatch.** 1999. New combinations in *Setaria* (Poaceae: Paniceae). Sida 18:1037–1047; **Hitchcock, A.S.** 1951 [title page 1950]. Manual of the Grasses of the United States, ed. 2, rev. A. Chase. U.S.D.A. Miscellaneous Publication No. 200. U.S. Government Printing Office, Washington, D.C., U.S.A. 1051 pp.; **Hubbard, F.T.** 1915. A taxonomic study of *Setaria italica* and its immediate allies. Amer. J. Bot. 2:169–198; **Rominger, J.M.** 1962. Taxonomy of *Setaria* (Gramineae) in North America. Illinois Biol. Monogr. 29:1–132.

1. Terminal spikelet of each panicle branch subtended by a single bristle, single bristles occasionally also present below the other spikelets.
  2. Blades not plicate, less than 10 mm wide; bristles present only below the terminal spikelets.
    3. Panicles nodding; spikelets 2-ranked on the branch axes (subg. *Paurochaetium*) . . . . 4. *S. chapmanii*
    3. Panicles erect; spikelets randomly distributed on the branch axes (subg. *Reverchoniae*) . . . . 5. *S. reverchonii*
  2. Blades plicate, more than 10 mm wide; a single bristle sometimes present below the non-terminal spikelets (subg. *Ptychophyllum*).
    4. Plants annual; blades 10–25 mm wide; rachises villous . . . . 1. *S. barbata*
    4. Plants perennial; blades 20–80 mm wide; rachises scabrous or puberulent.
      5. Panicles loosely open, branches lax, 6–10 cm long . . . . 2. *S. palmifolia*
      5. Panicles lanceoloid, branches stiff, 2–5 cm long . . . . 3. *S. megaphylla*
1. All spikelets subtended by 1–several bristles (subg. *Setaria*).
  6. Bristles 4–12 below each spikelet.
    7. Plants annual.
      8. Panicles erect; bristles 3–8 mm long; spikelets 2–3.4 mm long; blades 4–10 mm wide . . . . 27. *S. pumila*
      8. Panicles arching and drooping from near the base; bristles about 10 mm long; spikelets 2.5–3 mm long; blades 10–20 mm wide . . . . 24. *S. faberi*
    7. Plants perennial.
      9. Panicles 3–8(10) cm long, yellow to purple; knotty rhizomes present; native . . . . 25. *S. parviflora*
      9. Panicles 5–25 cm long, usually orange to purple; stout rhizomes present; introduced . . . 26. *S. sphacelata*
  6. Bristles 1–3 (rarely 6) below each spikelet.
    10. Bristles retrorsely scabrous.
      11. Margins of sheaths glabrous; blades strigose on the abaxial surfaces; subtropical . . . . 19. *S. adhaerens*
      11. Margins of sheaths ciliate distally; blades scabrous on the abaxial surfaces; temperate . . . . 20. *S. verticillata*
    10. Bristles antrorsely scabrous.
      12. Plants perennial.

13. Spikelets 2.8–3.2 mm long.
    14. Blades scabrous; plants of Florida and Georgia . . . . 11. *S. macrosperma*
    14. Blades pubescent; plants of Texas and possibly Arizona . . . . 10. *S. villosissima*
13. Spikelets 1.9–2.8(3) mm long.
    15. Panicles 2–6 cm long; spikelets 1.9–2.1 mm long; culms branching at the upper nodes . . . . 6. *S. texana*
    15. Panicles 5–30 cm long; spikelets 2–2.8(3) mm long; culms seldom branching at the upper nodes.
        16. Lower paleas narrow, ½–¾ as long as the lemmas; spikelets elliptical.
            17. Blades usually less than 5 mm wide; panicles 6–15 cm, columnar; bristles ascending . . . . 8. *S. leucopila*
            17. Blades usually more than 5 mm wide; panicles 15–25 cm, tapering to the apex; bristles diverging . . . . 9. *S. scheelei*
        16. Lower paleas broad, subequal to the lemmas in length; spikelets subspherical to ovate-lanceolate.
            18. Panicles dense, cylindrical; spikelets subspherical . . . . 7. *S. macrostachya*
            18. Panicles interrupted, attenuate; spikelets ovate-lanceolate.
                19. Blades 6–12 mm wide; lower glumes ½ as long as the spikelets . . . . 12. *S. setosa*
                19. Blades mostly less than 5 mm wide; lower glumes about ⅓ as long as the spikelets . . . . 13. *S. rariflora*

12. Plants annual.
    20. Upper glumes and lower lemmas with 7 veins, the outer pair of veins not coalescing with the inner 5; lower paleas absent . . . . 15. *S. liebmannii*
    20. Upper glumes and lower lemmas with 5–7 veins, all of which coalesce near the apices; lower paleas present, sometimes reduced or absent.
        21. Upper lemmas smooth and shiny, occasionally obscurely transversely rugose.
            22. Spikelets about 2 mm long, lower paleas equal to the lower lemmas . . . . . . .18. *S. magna*
            22. Spikelets about 3 mm long, lower paleas absent or up to ½ as long as the lower lemmas . . . . 23. *S. italica*
        21. Upper lemmas distinctly transversely rugose, dull.
            23. Upper lemmas coarsely rugose.
                24. Panicles densely spicate; rachises sparsely villous; plants of the southeastern United States . . . . 17. *S. corrugata*
                24. Panicles loosely spicate; rachises scabrous; plants of southern Arizona . . . . 16. *S. arizonica*
            23. Upper lemmas finely rugose.
                25. Panicles verticillate or loosely spicate; rachises visible, scabrous or hispid.
                    26. Panicles verticillate; rachises scabrous; cauline nodes glabrous . . . . 21. *S. verticilliformis*
                    26. Panicles loosely spicate, interrupted; rachises hispid; cauline nodes pubescent . . . . 14. *S. grisebachii*
                25. Panicles densely spicate; rachises not visible, villous.
                    27. Blades softly pilose on the upper surface; spikelets 2.5–3 mm long; panicles nodding from the base . . . . 24. *S. faberi*
                    27. Blades scabrous; spikelets 1.8–2.2 mm long; panicles nodding only from near the apex . . . . 22. *S. viridis*

Setaria subg. Ptychophyllum (A. Braun) Hitchc.

Plants usually robust perennials. **Blades** broad, plicate. **Panicles** usually loose; **bristles** solitary, subtending the terminal spikelet of each ultimate branchlet, occasionally present beneath the other spikelets. **Spikelets** 2.5–5 mm, slender, often acuminate.

Setaria subg. *Ptychophyllum* is primarily an American taxon, but is also found in the Eastern Hemisphere. Three species, all introduced, have been found in the *Flora* region.

1. **Setaria barbata** (Lam.) Kunth [p. 544]
MARY GRASS, CORN GRASS

Plants annual. **Culms** 50–200 cm; **nodes** pubescent. **Sheaths** with ciliate margins distally; **ligules** about 1 mm, ciliate; **blades** 10–25 mm wide, plicate, both surfaces scabrous, adaxial surfaces with parallel rows of papillose-based hairs. **Panicles** to 20 cm, open; **branches** 2–4 cm, axes villous; **bristles** solitary, usually only present below the terminal spikelet on each branch, occasionally below non-terminal spikelets, 5–8 mm, flexible. **Spikelets** 2.5-3 mm. **Lower glumes** about 1 mm, orbicular, 3–5-veined; **upper glumes** about 2 mm, ovate, 7-veined; **lower lemmas** about 2.5 mm, slightly coriaceous, acute; **lower paleas** about equaling the lower lemmas in length and width; **upper lemmas** about 2.3 mm, strongly transversely rugose; **upper paleas** enclosed. $2n$ = 54, 56.

Setaria barbata is an African species that was apparently introduced to the Western Hemisphere from Asia. It is now common throughout the West Indies, but rare in the *Flora* region.

2. **Setaria palmifolia** (J. König) Stapf [p. 544]
PALMGRASS

Plants perennial. **Culms** 1–2 m. **Sheaths** strigose, margins with stiff hairs; **collars** hispid; **ligules** about 2 mm, of hairs; **blades** to 50 cm long, 20–80 mm wide, plicate, tapering at both ends, abaxial surfaces sparsely strigose, adaxial surfaces short pubescent near the base. **Panicles** to 40 cm, open; **branches** 6–10 cm, loosely flexible, axes scabrous; **bristles** solitary, usually present only below the terminal spikelet on each branch, occasionally below non-terminal spikelets, about 5 mm. **Spikelets** 3–4 mm, elliptic, acuminate. **Lower glumes** ½ as long as the spikelets, obtuse, 3–4-veined; **upper glumes** nearly equaling the upper lemmas, 7-veined, acute; **lower lemmas** exceeding the upper lemmas, 5-veined, apices involute; **lower paleas** nearly equaling the lower lemmas in length and width; **upper lemmas** obscurely transversely rugose, yellow, apiculate. $2n$ = 54.

Setaria palmifolia is primarily an Asiatic species. It is a common species in Jamaica, and has been reported from scattered locations around the southern coast of the United States. In the *Flora* region it is occasionally cultivated as an ornamental for the conspicuous, plicate leaves and large panicles. In Southeast Asia the grains are eaten as a substitute for rice and the tender, thickened shoots as a vegetable.

3. **Setaria megaphylla** (Steud.) T. Durand & Schinz [p. 544]
BIGLEAF BRISTLEGRASS

Plants perennial. **Culms** 100–200 cm, nodes villous. **Sheaths** sparsely strigose or glabrous; **ligules** about 2 mm, of hairs; **blades** 40–60 cm long, 20–80 mm wide, strongly plicate, with scattered hairs on each surface. **Panicles** 30–60 cm, lanceoloid; **branches** 2–5 cm, stiff; **bristles** solitary, usually present only below the terminal spikelet on each branch, occasionally below non-terminal spikelets, 1–1.5 cm. **Spikelets** 3–3.5 mm. **Lower glumes** ½ as long as the spikelets, 3-veined; **upper glumes** ⅔ as long as the spikelets, 5–7-veined; **lower lemmas** equaling the upper lemmas, 5-veined; **lower paleas** absent or reduced to a small scale; **upper lemmas** about 3 mm, nearly smooth, shiny. $2n$ = 54.

Setaria megaphylla is a species of tropical Africa and tropical America that has become established in Florida. Hitchcock (1951) stated that **S. poiretiana** (Schult.) Kunth was occasionally cultivated in the United States, but he was referring to *S. megaphylla*.

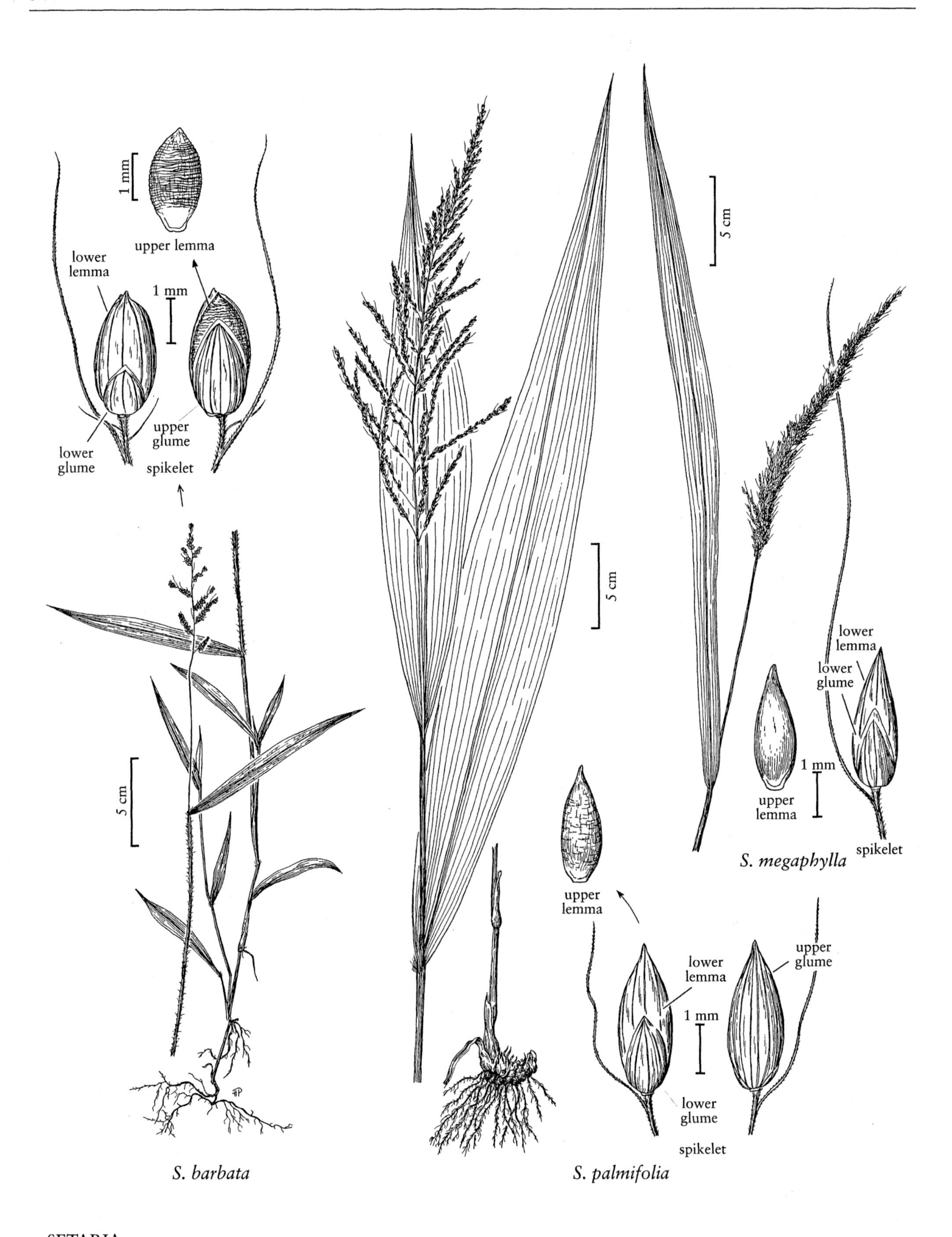
1 mm
upper lemma
lower
lemma
1 mm
upper
glume
lower
glume
spikelet
5 cm
S. barbata
5 cm
5 cm
S. palmifolia
upper
lemma
lower
lemma
1 mm
lower
glume
spikelet
upper
glume
lower
lemma
lower
glume
1 mm
upper
lemma
spikelet
S. megaphylla

Setaria subg. Paurochaetium (Hitchc. & Chase) Rominger

Plants perennial. **Culms** tufted, erect. **Blades** 1–7 mm wide, not plicate. **Panicles** narrow, usually nodding, more or less interrupted; **rachises** scabridulous or scabrous, sometimes also sparsely strigose; **branches** short, appressed; **ultimate branchlets** with 1–several spikelets in 2 ranks, terminal spikelets subtended by a 1–6 mm bristle. **Spikelets** 1.5–2.5 mm, glabrous; **lower florets** usually paleate; **upper lemmas** and **paleas** finely and transversely rugose.

Setaria subg. *Paurochaetium*, as treated by Fox and Hatch (1999), includes seven taxa and extends from southern Florida through the West Indies into the Yucatan region of Mexico and Belize. One species, *S. chapmanii*, grows in the *Flora* region.

Setaria subg. *Paurochaetium* usually differs from subg. *Reverchoniae* in its 2-ranked and smaller spikelets and in the absence of a palea in the lower floret. Unfortunately, *S. chapmanii*, the only representative of subg. *Paurochaetium* in the *Flora* region, is exceptional within the subgenus in lacking a lower palea.

4. **Setaria chapmanii** (Vasey) Pilg. [p. 547]
CHAPMAN'S BRISTLEGRASS

Plants perennial; cespitose. **Culms** 40–100 cm, erect, slender; **nodes** glabrous. **Sheaths** mostly glabrous, margins ciliate distally; **ligules** 0.1–0.4 mm, of stiff hairs; **blades** 15–40 cm long, 2–5 mm wide, those of the basal leaves involute, those of the cauline leaves flat, adaxial surfaces sparsely pilose basally. **Panicles** to 35 cm, nodding, slender, interrupted; **rachises** scabridulous; **branches** 5–20, erect, axes 0.4–3.2 cm, undulating, with 3–12 spikelets in 2 ranks, a single bristle present below the terminal spikelets; **bristles** 3–6 mm. **Spikelets** 1.8–2.2 mm, obovate, turgid. **Lower glumes** 0.6–0.8 mm, about ⅓ as long as the spikelets, 3-veined; **upper glumes** equaling the upper lemmas, 5–7-veined; **lower lemmas** equaling the upper lemmas; **lower paleas** absent; **upper lemmas** finely and transversely rugose; **anthers** 0.9–1.1 mm. $2n$ = unknown.

Setaria chapmanii is native to soils of coral or shell origin in the Florida Keys, the Bahamas, Cuba, and the Yucatan Peninsula, Mexico. The absence of the lower palea makes *S. chapmanii* unusual in subg. *Paurochaetium*.

Setaria subg. Reverchoniae W.E. Fox

Plants perennial. **Culms** tufted, erect. **Blades** to 7 mm wide, not plicate. **Panicles** narrow, erect, more or less interrupted; **rachises** scabrous; **branches** short and appressed; **ultimate branchlets** with 1–several, randomly disposed spikelets, terminal spikelets subtended by a single bristle; **bristles**1–6 mm, glabrous. **Spikelets** 2.1–4.5 mm; **lower florets** usually without paleas; **upper lemmas** and **paleas** finely and transversely rugose.

Setaria subg. *Reverchoniae* is a monotypic subgenus with a range that extends from western New Mexico and southern Oklahoma through western Texas to northeastern Mexico. It usually differs from subg. *Paurochaetium* in the random disposition and larger size of the spikelets and, usually, in the absence of the lower palea.

5. **Setaria reverchonii** (Vasey) Pilg. [p. 547]

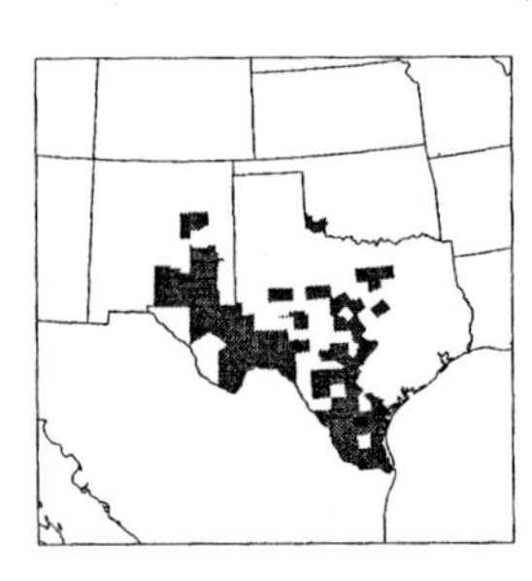

Plants perennial; rhizomatous, rhizomes, short, sometimes knotty. **Culms** 30–90 cm; **nodes** glabrous, strigose, or with appressed hairs. **Sheaths** with papillose-based hairs, sometimes nearly glabrous, margins ciliate distally; **ligules** 1–2 mm, of stiff hairs; **blades** 4–30 cm long, 1–7 mm wide, involute, stiff, scabridulous and narrowed basally. **Panicles** 5–20 cm, erect, slender, interrupted; **rachises** scabrous; **bristles** 2–8 mm. **Spikelets** 2.1–4.5 mm, elliptic to obovate, randomly distributed on the branch axes. **Lower glumes** ½ as long as the spikelets, 5–7-veined; **upper glumes** equaling the upper lemmas, 7–9-veined; **lower lemmas** equaling the upper lemmas; **lower paleas** absent; **upper lemmas** indurate, finely and transversely rugose; **upper paleas** similar to the upper lemmas.

Setaria reverchonii grows in sandy prairies and limestone hills from eastern New Mexico, southwestern Oklahoma, and Texas to northern Mexico.

1. Blades usually more than 15 cm long; spikelets 3.5–4.5 mm long . . . . subsp. *reverchonii*
1. Blades usually less than 15 cm long; spikelets 2.1–3.2 mm long.
   2. Blades 2–4 mm wide; spikelets about 2.5 mm long . . . . subsp. *ramiseta*
   2. Blades 4–7 mm wide; spikelets about 3–3.2 mm long . . . . subsp. *firmula*

Setaria reverchonii subsp. **firmula** (Hitchc. & Chase) W.E. Fox
KNOTGRASS

Sheaths with papillose-based hairs, sometimes nearly glabrous; **blades** 4–10 cm long, 4–7 mm wide. **Spikelets** 3–3.2 mm long, subtending bristles 3–6.4 mm. **Lower glumes** 5–7-veined; **upper glumes** 5–7-veined. $2n$ = 36.

Setaria reverchonii subsp. *firmula* is endemic to the sandy prairies of southeastern Texas.

Setaria reverchonii subsp. **ramiseta** (Scribn.) W.E. Fox
RIO GRANDE BRISTLEGRASS

Sheaths with papillose-based hairs; **blades** 5–15 cm long, 2–4 mm wide, adaxial surfaces with papillose-based hairs. **Spikelets** about 2.5 mm, obovate, turgid, subtending bristles 2–4 mm. **Lower glumes** 5-veined; **upper glumes** 7–9-veined. $2n$ = 36.

Setaria reverchonii subsp. *ramiseta* grows in the sandy plains and prairies of southeastern New Mexico, southern Texas, and northern Mexico.

Setaria reverchonii (Vasey) Pilg. subsp. **reverchonii**
REVERCHON'S BRISTLEGRASS

Sheaths almost glabrous; **blades** 15–30 cm long, 1–3 mm wide. **Spikelets** 3.5–4.5 mm, elliptic, subtending bristles 5–8 mm. **Lower glumes** 5–7-veined; **upper glumes** 7-veined. $2n$ = 36, 72.

Setaria reverchonii subsp. *reverchonii* grows in sandy prairies and limestone hills from southwestern Texas to northern Mexico.

Setaria P. Beauv. subg. Setaria

Plants annual or perennial. **Blades** seldom wider than 20 mm, flat or loosely twisted. **Panicles** usually contracted, spikelike; **branches** short, with 1 or more bristles subtending each spikelet. **Spikelets** obtuse or acutish. **Upper lemmas** transversely rugose, rarely smooth.

Setaria subg. *Setaria* is represented in subtropical and temperate regions throughout the world and is the best represented subgenus in the *Flora* region.

6. **Setaria texana** Emery [p. 549]
TEXAS BRISTLEGRASS

Plants perennial. **Culms** 30–70 cm, wiry, much branched distally. **Sheaths** keeled, margins ciliate distally; **collars** glabrate; **ligules** to 1 mm, densely ciliate; **blades** 5–15 cm long, 2–4 mm wide, flat, scabrous. **Panicles** 2–6 cm, spikelike, basal portion rarely lobed, tapering distally; **rachises** scabrous to puberulent; **bristles** solitary, 3–10 mm. **Spikelets** 1.9–2.1 mm. **Lower glumes** about ½ as long as the spikelets, 3-veined; **upper glumes** about ¾ as long as the spikelets, 5-veined; **lower lemmas** nearly equaling the upper lemmas, 5-veined; **lower paleas** rudimentary to ½ as long as the upper paleas; **upper lemmas** finely and transversely rugose; **upper paleas** narrow. $2n$ = 36.

Setaria texana grows in shaded habitats on sandy loam soils of the Rio Grande plain of south Texas and northeastern Mexico.

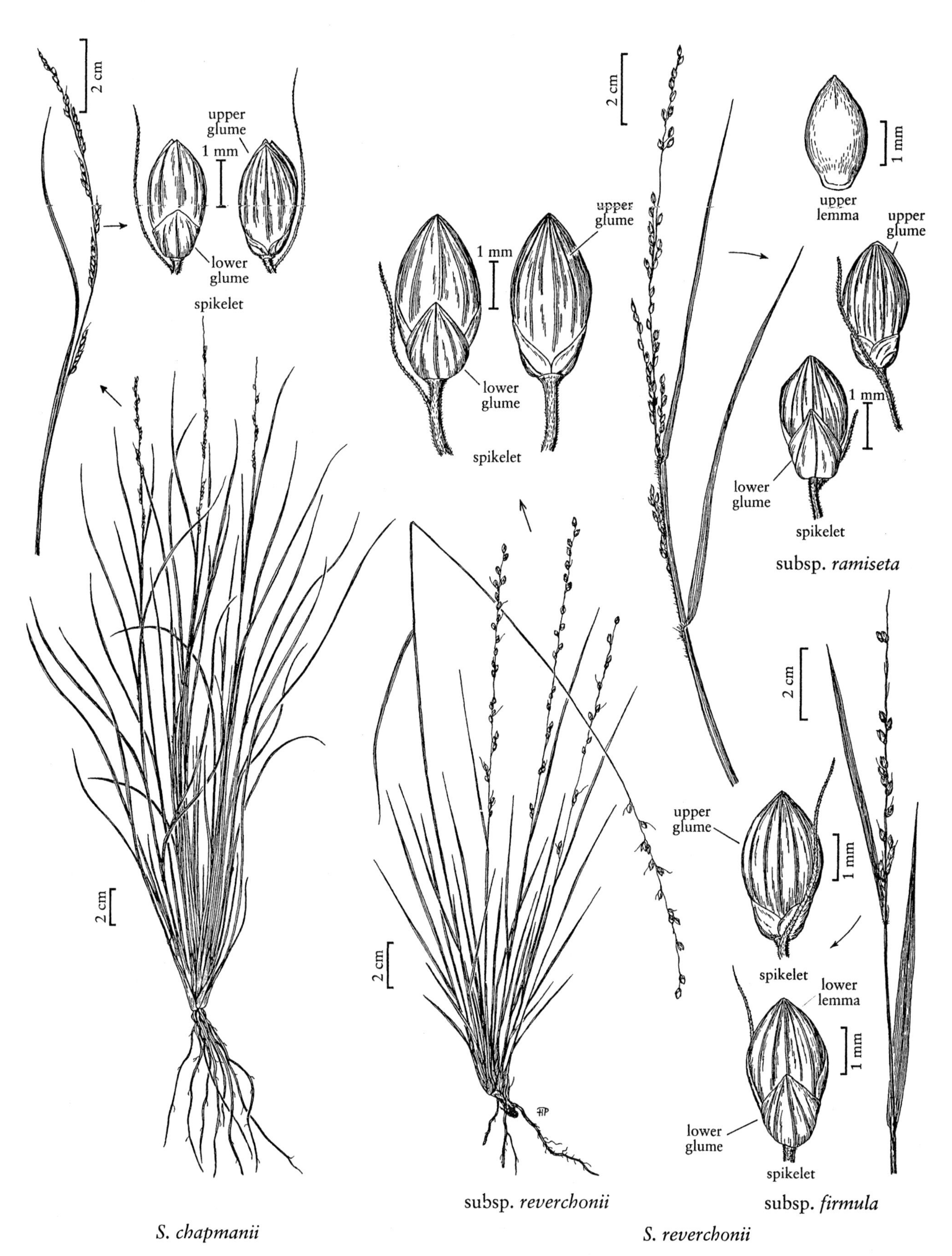
2 cm
upper
glume
1 mm
lower
glume
spikelet
2 cm
2 cm
upper
glume
1 mm
lower
glume
spikelet
upper
lemma
1 mm
upper
glume
1 mm
lower
glume
spikelet
subsp. ramiseta
2 cm
upper
glume
1 mm
spikelet
lower
lemma
1 mm
lower
glume
spikelet
2 cm
subsp. reverchonii
subsp. firmula
S. chapmanii
S. reverchonii

7. **Setaria macrostachya** Kunth [p. 549]
PLAINS BRISTLEGRASS

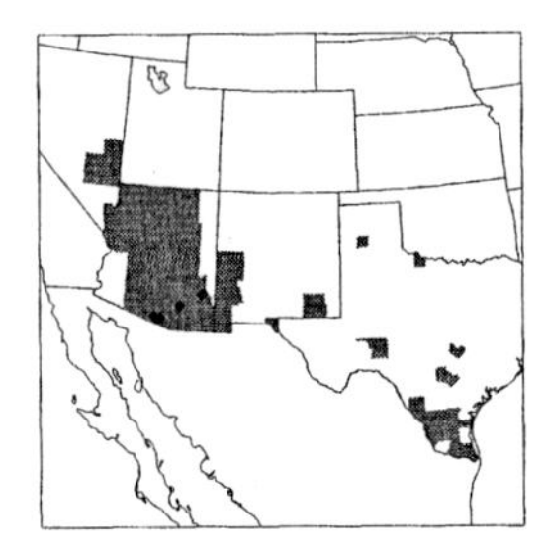

Plants perennial; densely cespitose. **Culms** 60–120 cm, rarely branched distally, scabrous below the nodes and panicles. **Sheaths** keeled, glabrous, usually with a few white hairs at the throat; **ligules** 2–4 mm, densely ciliate; **blades** 15–20 cm long, 7–15 mm wide, flat, adaxial surface scabrous. **Panicles** 10–30 cm long, 1–2 cm wide, uniformly thick from the base to the apex, dense, rarely lobed basally; **rachises** scabrous and loosely pilose; **bristles** usually solitary, 10–20 mm, soft, antrorsely scabrous. **Spikelets** 2–2.3 mm, subspherical. **Lower glumes** ⅓–½ as long as the spikelets, 3–5-veined; **upper glumes** about ¾ as long as the spikelets, 5–7-veined; **lower lemmas** equaling the upper lemmas, 5-veined; **lower paleas** nearly equaling the upper paleas in length and width; **upper lemmas** transversely rugose; **upper paleas** convex, ovate. $2n = 54$.

Setaria macrostachya is abundant in the desert grasslands of the southwestern United States, particularly in southern Arizon and Texas. It extends south through the highlands of central Mexico. It also grows in the West Indies, but is not common there. It is a valuable forage grass in the *Flora* region.

8. **Setaria leucopila** (Scribn. & Merr.) K. Schum. [p. 549]
STREAMBED BRISTLEGRASS

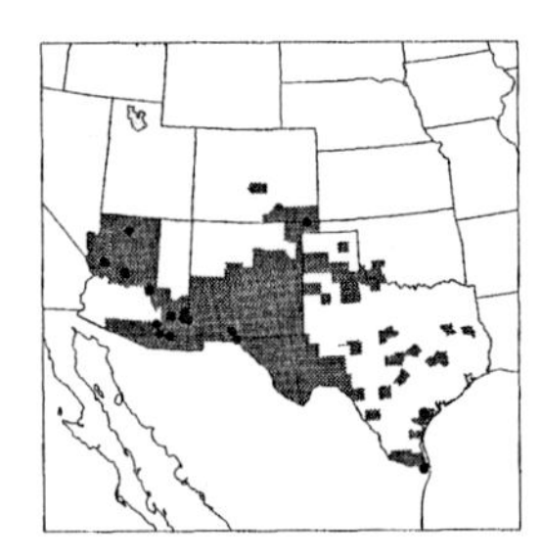

Plants perennial; cespitose. **Culms** 20–100 cm. **Sheaths** compressed, glabrous, margins villous distally; **ligules** 1–2.5 mm, ciliate; **blades** 8–25 cm long, 2–5 mm wide, flat or folded, scabrous on both surfaces. **Panicles** 6–15 cm, tightly spikelike, pale green; **rachises** scabrous or villous; **bristles** usually solitary, 4–15 mm, ascending. **Spikelets** 2.2–2.8(3) mm, elliptical. **Lower glumes** about ½ as long as the spikelets, 3-veined; **upper glumes** from ¾ as long as to equaling the florets, 5-veined; **lower lemmas** equaling the upper lemmas, 5-veined; **lower paleas** ½–¾ as long as the upper paleas, lanceolate; **upper lemmas** apiculate, finely and transversely rugose; **upper paleas** similar. $2n = 54, 68, 72$.

Setaria leucopila grows in the southwestern United States and northern Mexico. It is the most common of the perennial "Plains bristlegrasses."

9. **Setaria scheelei** (Steud.) Hitchc. [p. 549]
SOUTHWESTERN BRISTLEGRASS

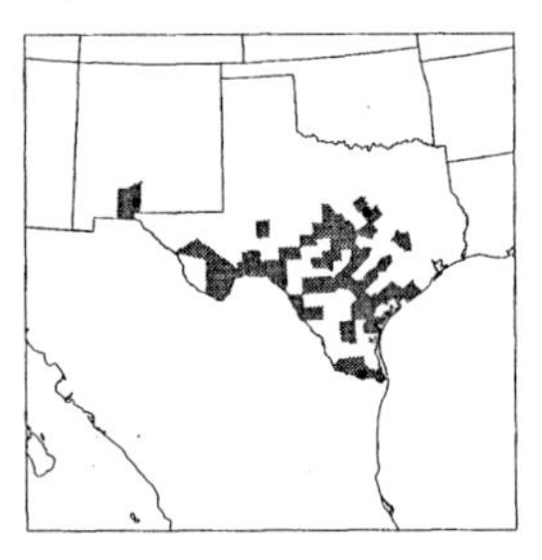

Plants perennial; cespitose. **Culms** 60–120 cm; **nodes** pilose, with appressed hairs. **Sheaths** glabrous or hispid; **ligules** 1–2 mm, hispid; **blades** 15–30 cm long, 6–15 mm wide, flat or folded, scabrous, often pubescent. **Panicles** 15–25 cm, open, tapering from the base; **rachises** pubescent to villous; **lower branches** to 3 cm; **bristles** usually solitary, 10–35 mm, divergent. **Spikelets** 2.2–2.5 mm, elliptical. **Lower glumes** about ½ as long as the spikelets, 3-veined; **upper glumes** from ¾ as long as to equaling the upper florets, 5-veined; **lower lemmas** equaling the upper lemmas, 5-veined; **lower paleas** about ½ as long as the upper paleas, lanceolate; **upper lemmas** finely cross-wrinkled, shortly apiculate; **upper paleas** ovate-lanceolate. $2n = 54$.

Setaria scheelei grows in alluvial soils of canyons and river bottoms of New Mexico and Texas. Within the *Flora* region, it is particularly abundant in the limestone canyons of the Edwards Plateau of central Texas. Its range extends into central Mexico.

10. **Setaria villosissima** (Scribn. & Merr.) K. Schum. [p. 551]
HAIRYLEAF BRISTLEGRASS

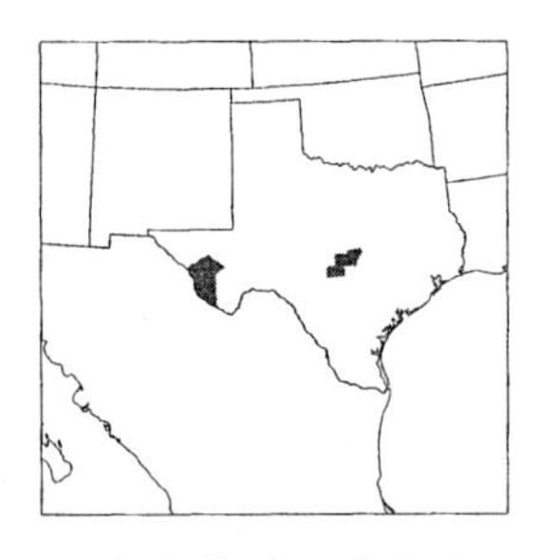

Plants perennial; cespitose. **Culms** 40–100 cm. **Sheaths** villous distally, margins ciliate; **ligules** about 1 mm, densely ciliate, hairs white; **blades** 15–30 cm long, 5–8 mm wide, both surfaces villous. **Panicles** 10–20 cm, loosely spicate; **bristles** usually solitary, 10–20 mm. **Spikelets** 2.8–3 mm. **Lower glumes** about ⅓ as long as the spikelets, broadly ovate, 3-veined; **upper glumes** nearly equaling the spikelets, 5–7-veined; **lower lemmas** equaling the upper lemmas, 5-veined; **lower paleas** about ⅓ as long as the upper paleas, lanceolate; **upper lemmas** finely and transversely undulate-rugose basally, striate and punctate distally; **upper paleas** similar, ovate-lanceolate. $2n = 54$.

Setaria villosissima is a rare species that grows on granitic soils in southwestern Texas and northern Mexico. An Arizona specimen is known, but is not mapped because there are no additional locality data. No other Arizona specimens are known. The villous sheaths and blades and large spikelets of *S. villosissima* aid in its identification.

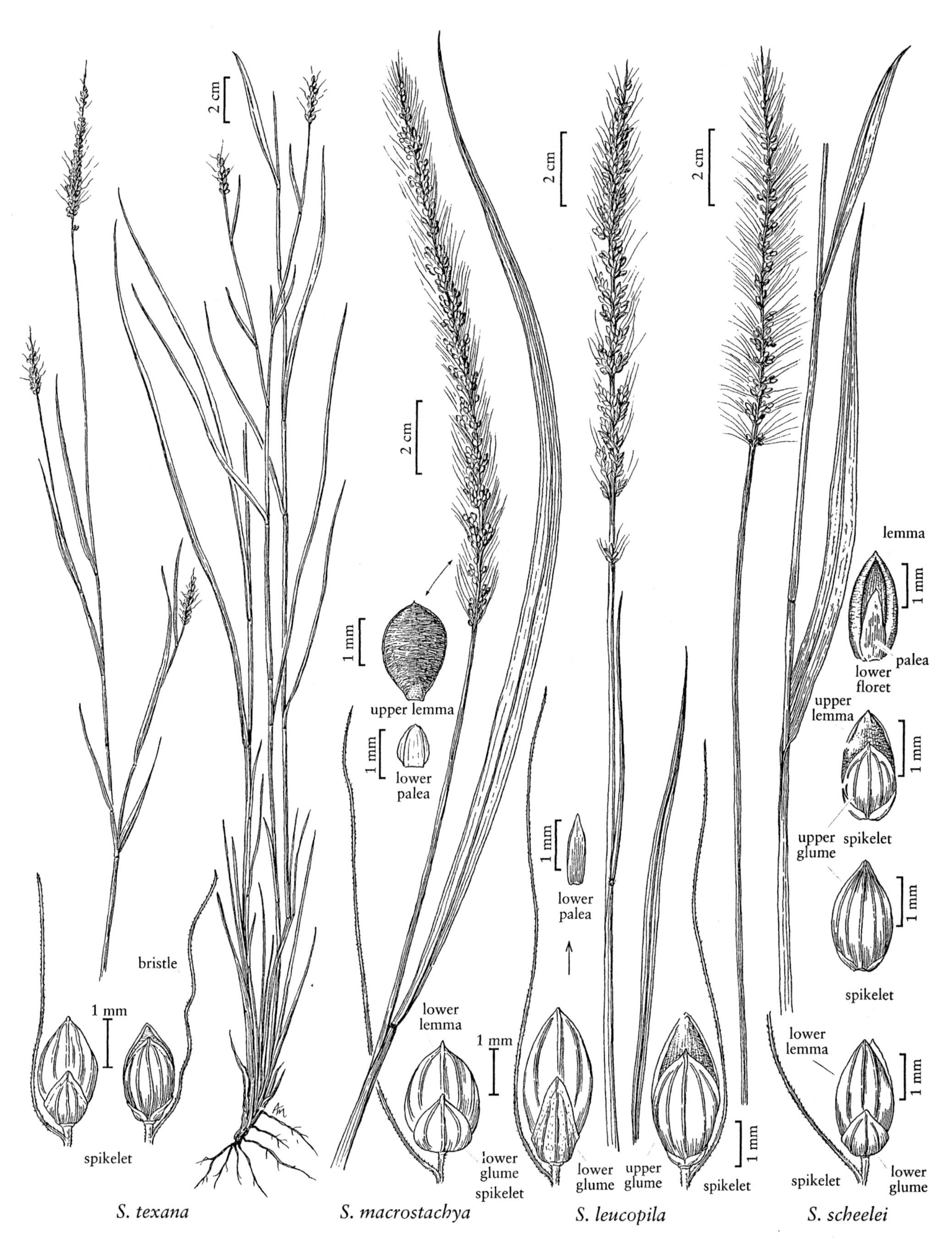
2 cm
2 cm
2 cm
2 cm
2 cm
1 mm
upper lemma
1 mm
lower
palea
lemma
1 mm
palea
lower
floret
upper
lemma
1 mm
upper
glume
spikelet
1 mm
spikelet
1 mm
lower
palea
bristle
1 mm
spikelet
lower
lemma
1 mm
lower
glume
spikelet
lower
glume
upper
glume
1 mm
spikelet
lower
lemma
1 mm
spikelet
lower
glume
S. texana
S. macrostachya
S. leucopila
S. scheelei

11. **Setaria macrosperma** (Scribn. & Merr.) K. Schum. [p. 551]
CORAL BRISTLEGRASS

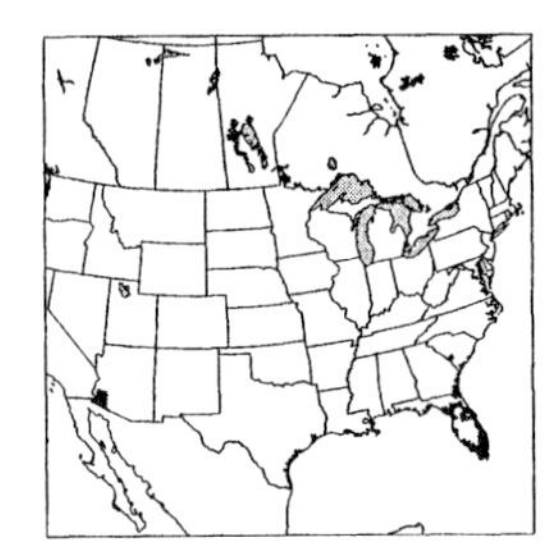

Plants perennial. **Culms** 1–1.5 m. **Sheaths** prominently keeled, margins villous; **ligules** 1–3 mm; **blades** 1–2 cm wide, flat, scabrous. **Panicles** to 25 cm, loosely spicate; **rachises** readily visible, sparsely villous; **bristles** 1(2), 15–30 mm, flexible, antrorsely scabrous. **Spikelets** 3–3.2 mm. **Lower glumes** about ⅓ as long as the spikelets, 3-veined; **upper glumes** about ¾ as long as the spikelets, 5-veined; **lower lemmas** equaling the upper lemmas; **lower paleas** about ½ as long as the upper paleas, hyaline, narrow; **upper lemmas** finely and transversely rugose; **upper paleas** similar to the upper lemmas. $2n$ = unknown.

Setaria macrosperma grows on shell or coral islands, and occasionally in old fields or hammocks. It is most frequent in Florida, but has been collected in both South Carolina and Georgia. It also grows in the Bahamas and Mexico.

12. **Setaria setosa** (Sw.) P. Beauv. [p. 551]
WEST INDIES BRISTLEGRASS

Plants perennial. **Culms** 50–100 cm; **nodes** usually glabrous. **Sheaths** glabrous or finely pubescent, margins ciliate distally; **ligules** of 1 mm hairs; **blades** 15–20 cm long, 6–12 mm wide, flat or folded, often finely pubescent on both surfaces. **Panicles** 15–20 cm, loosely spike-like, interrupted, attenuate; **rachises** often villous; **branches** ascending, lower branches about 2.5 cm; **bristles** usually solitary, less than 10 mm, antrorsely scabrous. **Spikelets** 2–2.5 mm, ovate-lanceolate. **Lower glumes** about ½ as long as the spikelets, 3-veined; **upper glumes** about ⅔ as long as the spikelets, 5–7-veined; **lower lemmas** equaling the upper lemmas; **lower paleas** as long as the upper paleas, broad; **upper lemmas** finely and distinctly transversely rugose. $2n$ = unknown.

Setaria setosa is native to the West Indies and Mexico. It is probably a recent introduction to Florida, but appears to be established there. The specimen from New Jersey was from a ballast dump; the species is not established in that state.

13. **Setaria rariflora** J.C. Mikan *ex* Trin. [p. 551]
BRAZILIAN BRISTLEGRASS

Plants perennial. **Culms** 30–70 cm, usually erect, branching profusely at the base; **nodes** glabrous or sparsely hispid. **Sheaths** keeled, sparsely pubescent, margins ciliate distally; **ligules** to 1 mm, of hairs; **blades** 15–30 cm long, usually less than 5 mm wide, densely pubescent on both surfaces. **Panicles** 5–15 cm, slender, attenuate, interrupted, sparsely flowered; **rachises** pubescent; **branches** mostly shorter than 10 mm, villous; **bristles** usually solitary, 4–7 mm, antrorsely scabrous. **Spikelets** about 2 mm, ovate-lanceolate. **Lower glumes** about ⅓ as long as the spikelets, 3-veined; **upper glumes** about ½ as long as the spikelets, 5–7-veined; **lower lemmas** equaling the upper lemmas; **lower paleas** equaling the upper paleas, broad; **upper lemmas** finely and distinctly transversely rugose; **upper paleas** similar to the upper lemmas. $2n$ = unknown.

Setaria rariflora has its center of distribution in South America. It is probably only recently adventive in North America, where it is known from Florida and the West Indies.

14. **Setaria grisebachii** E. Fourn. [p. 553]
GRISEBACH'S BRISTLEGRASS

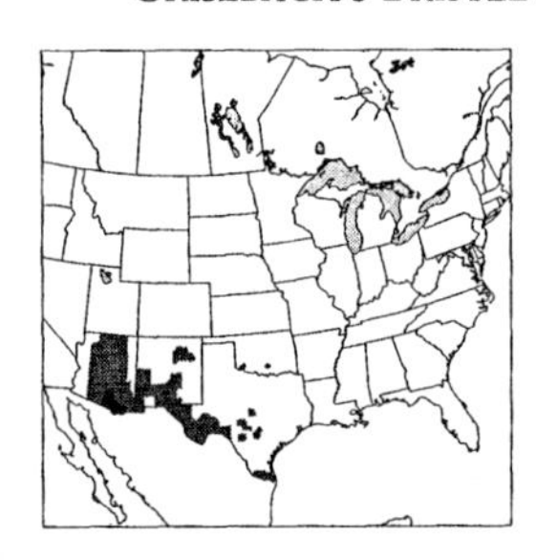

Plants annual. **Culms** 30–100 cm; **nodes** pubescent, hairs appressed. **Sheaths** with ciliate margins; **ligules** ciliate; **blades** to 12(25) cm long, to 10(20) mm wide, flat, hispid on both surfaces. **Panicles** 3–18 cm, loosely spicate, interrupted, often purple; **rachises** hispid; **bristles** 1–3, 5–15 mm, flexible, antrorsely scabrous. **Spikelets** 1.5–2.2 mm. **Lower glumes** about ⅓ as long as the spikelets, distinctly 3-veined, lateral veins coalescing with the central veins below the apices; **upper glumes** nearly equaling the upper lemmas, obtuse, 5-veined; **lower lemmas** equaling the upper lemmas; **lower paleas** about ⅓ as long as the lower lemmas, narrow; **upper lemmas** finely and transversely rugose; **upper paleas** similar to the upper lemmas. $2n$ = unknown.

Setaria grisebachii is the most widespread and abundant native annual species of *Setaria* in the southwestern United States. It grows in open ground and extends along the central highlands of Mexico to Guatemala, usually at elevations of 750–2500 m. The specimens from Maryland were collected on chrome ore piles; the species is not established in the state.

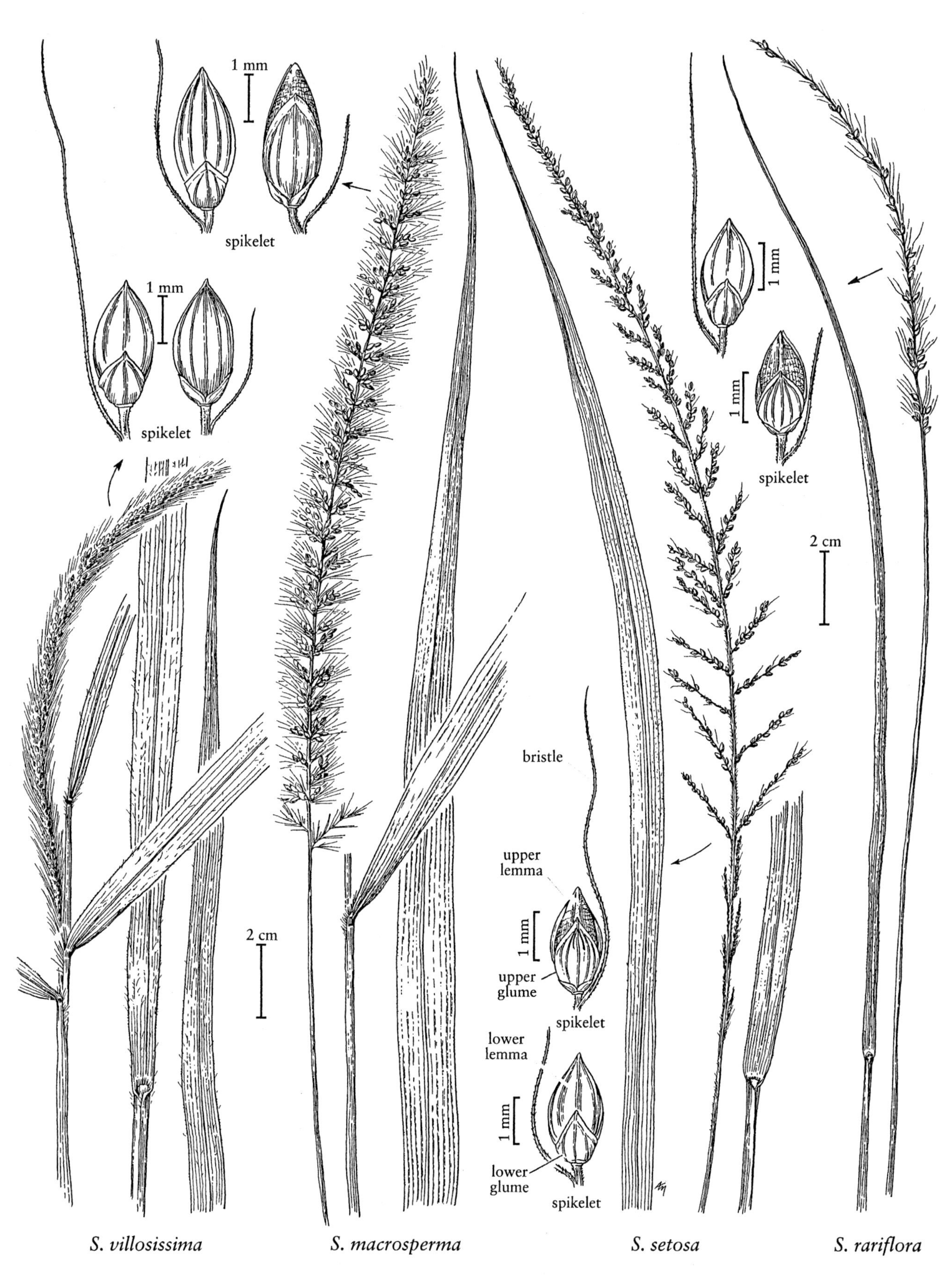
1 mm
spikelet
1 mm
spikelet
2 cm
1 mm
1 mm
spikelet
2 cm
bristle
upper
lemma
1 mm
upper
glume
spikelet
lower
lemma
1 mm
lower
glume
spikelet
S. villosissima
S. macrosperma
S. setosa
S. rariflora

SETARIA

15. **Setaria liebmannii** E. Fourn. [p. 553]
LIEBMANN'S BRISTLEGRASS

Plants annual. **Culms** 30–90 cm. **Sheaths** glabrous, margins ciliate; **ligules** ciliate; **blades** to 20 cm long, 10–20 mm wide, flat, scabrous on both surfaces. **Panicles** 10–25 cm, loosely spicate; **rachises** scabrous; **bristles** solitary, 7–15 mm, slender, antrorsely scabrous. **Spikelets** 2–2.7(3) mm. **Lower glumes** about ⅓ as long as the spikelets, 3-veined; **upper glumes** and **lower lemmas** 7-veined, the 5 central veins coalescing at the apices; **lower paleas** absent; **upper lemmas** gibbous, strongly and coarsely transversely rugose; **upper paleas** similar to the upper lemmas. $2n$ = 18.

Within the *Flora* region, *Setaria liebmannii* is known only from southern Arizona, but it is a common species along the Pacific slope from northern Mexico to Nicaragua, usually growing at elevations below 750 m. The five apically coalescing veins and the additional free pair at the periphery are unique among the *Setaria* species in the *Flora* region.

16. **Setaria arizonica** Rominger [p. 553]
ARIZONA BRISTLEGRASS

Plants annual. **Culms** 25–50 cm; **nodes** pubescent. **Sheaths** glabrous, margins ciliate distally; **ligules** 1–2 mm, ciliate; **blades** 7–15 cm long, 5–8 mm wide, flat, scabrous, abaxial surface conspicuously hispid over the veins with papillose-based hairs, adaxial surface sparsely hispid over the veins; **bristles** solitary, 5–15 mm, flexible. **Panicles** 5–12 cm, loosely spicate; **rachises** scabrous. **Spikelets** 1.8–2 mm. **Lower glumes** about ⅓ as long as the spikelets, 3-veined, lateral veins coalescing with the central vein below the apices; **upper glumes** about ⅔ as long as the upper lemmas, 5-veined, obtuse; **lower paleas** equaling the lower lemmas, broad; **upper lemmas** very strongly and coarsely transversely rugose; **upper paleas** similar to the upper lemmas. $2n$ = unknown.

Setaria arizonica is locally abundant in sandy washes on both sides of the Arizona-Sonora border, southwest of Tucson.

17. **Setaria corrugata** (Elliott) Schult. [p. 553]
COASTAL BRISTLEGRASS

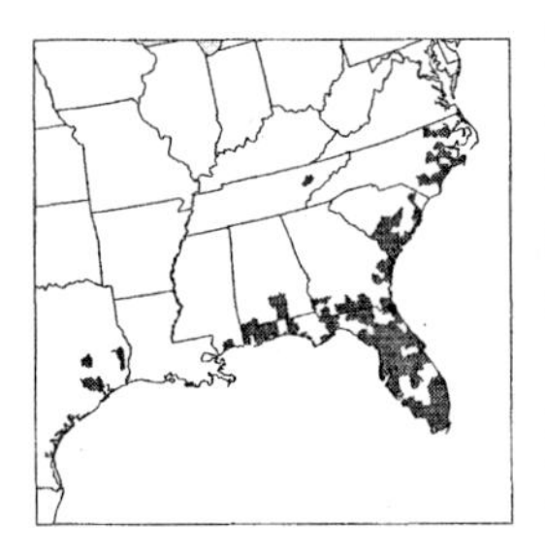

Plants annual. **Culms** to 100 cm; **nodes** hispid, hairs appressed. **Sheaths** glabrous or pilose, margins ciliate distally; **ligules** about 1 mm, ciliate; **blades** 15–30 cm long, 4–7 mm wide, flat, scabrous or pubescent. **Panicles** 3–15 cm, densely spicate; **rachises** rough hispid and sparsely villous; **bristles** 1–3, 5–15 mm, flexible, antrorsely scabrous. **Spikelets** about 2 mm, turgid. **Lower glumes** ⅓–½ as long as the spikelets, 3–5-veined; **upper glumes** about ¾ as long as the upper lemmas, 5–7-veined; **lower lemmas** equaling the upper lemmas; **lower paleas** ¾ as long as the lower lemmas, hyaline; **upper lemmas** very coarsely and transversely rugose; **upper paleas** similar to the upper lemmas. $2n$ = unknown.

Setaria corrugata grows in pinelands and cultivated fields along the southeastern coast of the United States. It is also found in Cuba and the Dominican Republic. Superficially, it resembles *S. viridis*, but is easily distinguished from that species by its coarsely rugose ("corrugated") lower lemmas.

18. **Setaria magna** Griseb. [p. 555]
GIANT BRISTLEGRASS

Plants annual. **Culms** to 6 m tall, 2–3 cm thick at the base. **Sheaths** glabrous, smooth or scabrous, margins villous distally; **ligules** 1–2 mm, ciliate; **blades** to 60 cm long, to 3.5 cm wide, flat. **Panicles** to 50 cm long, to 5 cm wide, densely spikelike; **rachises** densely villous; **bristles** 1 or 2, 10–20 mm, flexible, antrorsely scabrous. **Spikelets** about 2 mm, disarticulating between the lower and upper florets. **Lower glumes** ⅓ as long as the spikelets, 3-veined; **upper glumes** equaling the lower lemmas, 7-veined; **lower florets** often staminate; **lower lemmas** slightly exceeding the upper lemmas; **lower paleas** equaling the lower lemmas, broad, hyaline, minutely pubescent over the veins; **upper lemmas** smooth and shiny brown; **upper paleas** similar to the upper lemmas. $2n$ = 36.

Setaria magna grows in saline marshes along the eastern coast of the United States. There are also disjunct populations in brackish swamps in Arkansas, and in Texas and southeastern New Mexico as well as in Jamaica, Puerto Rico, Bermuda, Mexico, and Costa Rica. It may have been recently introduced to some of these regions, including inland areas of the *Flora* region.

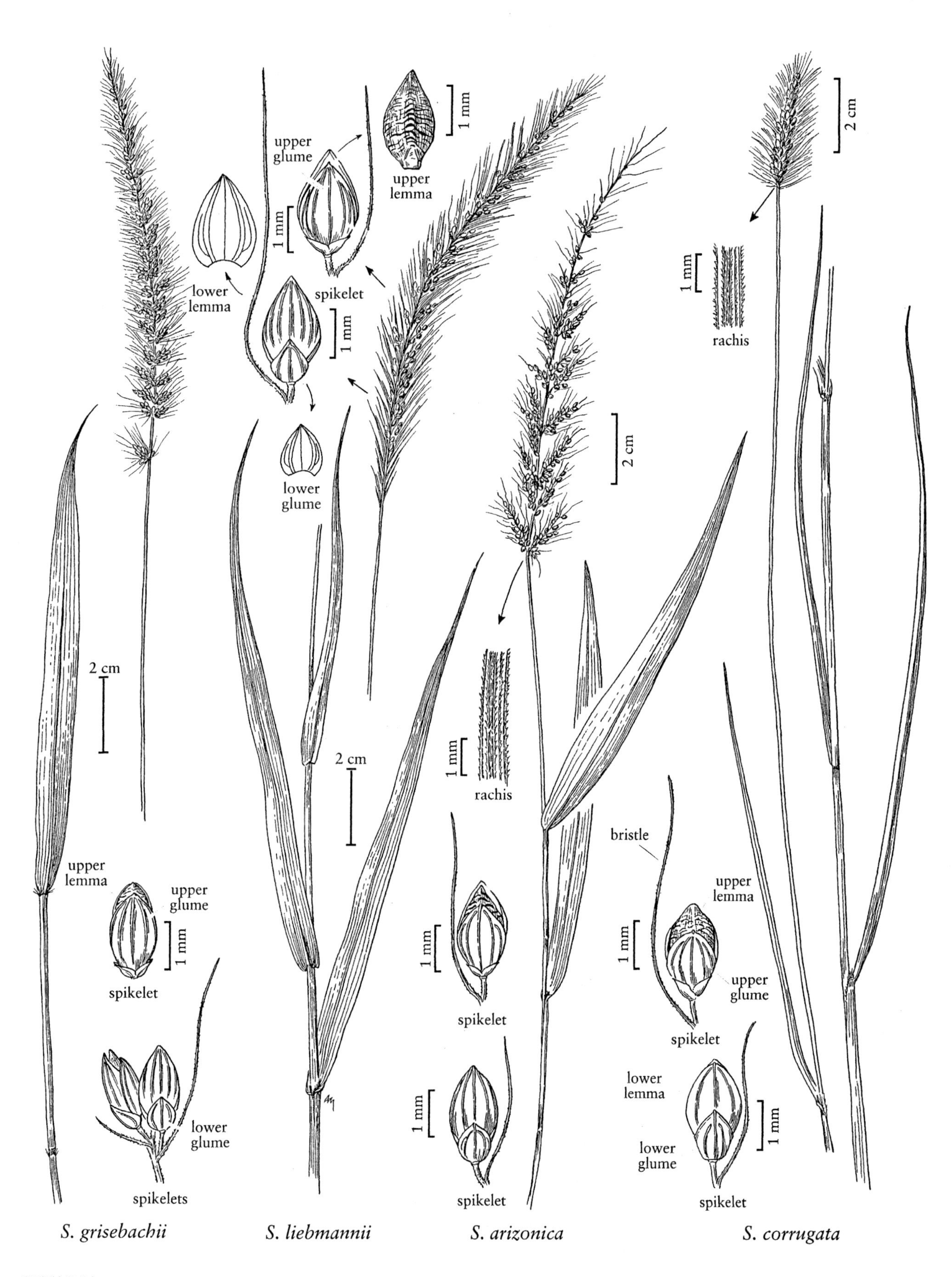
upper
glume
upper
lemma
1 mm
1 mm
lower
lemma
spikelet
1 mm
lower
glume
2 cm
2 cm
upper
lemma
upper
glume
1 mm
spikelet
spikelets
lower
glume
2 cm
1 mm
rachis
1 mm
spikelet
1 mm
spikelet
2 cm
1 mm
rachis
bristle
upper
lemma
1 mm
upper
glume
spikelet
lower
lemma
lower
glume
1 mm
spikelet
S. grisebachii
S. liebmannii
S. arizonica
S. corrugata

19. **Setaria adhaerans** (Forssk.) Chiov. [p. 555]
Bur Bristlegrass, Tropical Barbed Bristlegrass

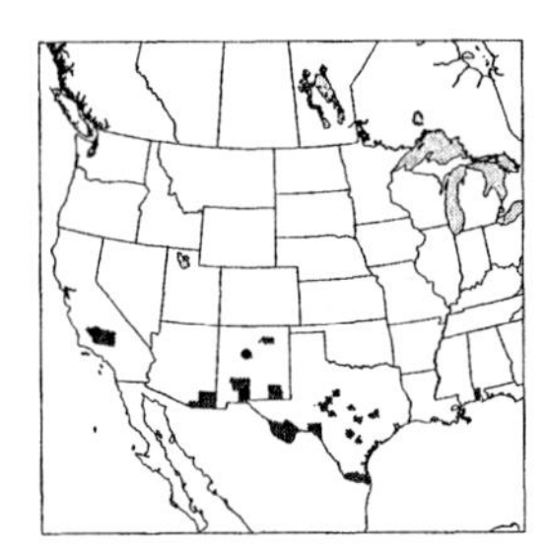

Plants annual. **Culms** 25–60 cm. **Sheaths** glabrous throughout; **ligules** 1–2 mm, of hairs, white; **blades** usually less than 10 cm long, 5–10 mm wide, flat, broad basally, abaxial surfaces conspicuously strigose with papillose-based hairs, tapering abruptly at the apices; **bristles** solitary, about 5 mm, retrorsely scabrous. **Panicles** 2–6 cm, verticillate, green to purple; **rachises** retrorsely rough hispid. **Spikelets** 1.5–2.2 mm. **Lower glumes** about ½ as long as the spikelets, obtuse, 1(3)-veined; **upper glumes** nearly as long as the spikelets, 5–7-veined; **lower lemmas** equaling to slightly exceeding the upper lemmas; **lower paleas** less than ½ as long as the spikelets, scalelike; **upper lemmas** finely and transversely rugose; **upper paleas** similar to the upper lemmas. $2n = 18$.

Setaria adhaerans grows in subtropical regions throughout the world. In North America, it is known from the southern United States, northeastern Mexico, Guatemala, Cuba, and the Bahamas. The Californian record may represent a recent introduction.

Setaria adhaerans resembles the temperate *S. verticillata*, but differs in having shorter panicles, shorter spikelets, glabrous sheath margins, and papillose-based strigose hairs on the blades.

20. **Setaria verticillata** (L.) P. Beauv. [p. 555]
Hooked Bristlegrass, Sétaire Verticillée

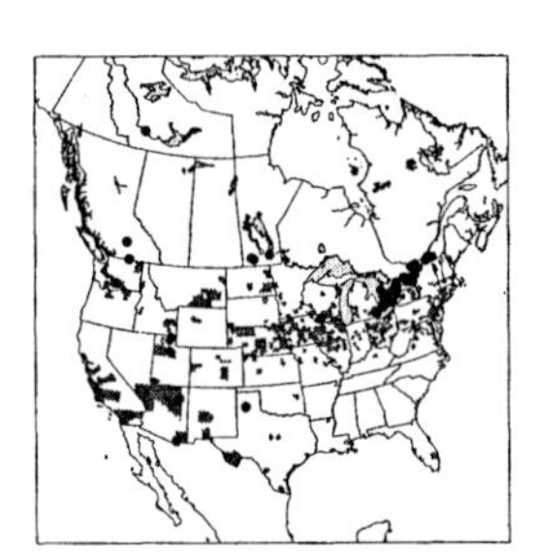

Plants annual. **Culms** 30–100 cm; **nodes** glabrous. **Sheaths** glabrous, margins ciliate distally; **ligules** to 1 mm, densely ciliate; **blades** 5–15 mm wide, flat, abaxial surfaces scabrous. **Panicles** 5–15 cm, tapering to the apices; **rachises** retrorsely rough hispid; **bristles** solitary, 4–7 mm, retrorsely scabrous. **Spikelets** 2–2.3 mm. **Lower glumes** about ⅓ as long as the spikelets, obtuse, 1(3)-veined; **upper glumes** nearly as long as the spikelets; **lower paleas** about ½ as long as the spikelets, broad; **upper lemmas** finely and transversely rugose; **upper paleas** similar to the upper lemmas. $2n = 18, 36, 54, 72, 108$.

Setaria verticillata is a European adventive that is now common throughout the cooler regions of the contiguous United States and in southern Canada. It is an aggressive weed in the vineyards of central California. Reports of *S. carnei* Hitchc. from North America are based on misidentification of this species.

Setaria verticillata resembles the *S. adhaerans* but differs in having longer panicles and spikelets, sheath margins that are ciliate distally, and blades that are scabrous, not hairy. *Setaria verticillata* is a more northern species than *S. adhaerans*, but their ranges overlap in the *Flora* region.

21. **Setaria verticilliformis** Dumort. [p. 555]
Barbed Bristlegrass

Plants annual. **Culms** 30–100 cm; **nodes** glabrous. **Sheaths** mostly glabrous, margins ciliate distally; **ligules** 1–2 mm, of hairs; **blades** 5–15 mm wide, flat, abaxial surfaces scabrous, adaxial surfaces sparsely villous. **Panicles** 5–15 cm, tapering to the apices, branches verticillate; **rachises** antrorsely rough hispid, without villous hairs; **bristles** solitary, 4–7 mm, antrorsely or retrorsely scabrous. **Spikelets** 2–2.3 mm. **Lower glumes** about ⅓ as long as the spikelets, obtuse, 1(3)-veined; **upper glumes** nearly as long as the spikelets; **lower paleas** about ½ as long as the spikelets, broad; **upper lemmas** finely and transversely rugose; **upper paleas** similar to the upper lemmas. $2n = 36$.

Setaria verticilliformis is a European adventive that has been found at scattered, mostly urban, locations in the United States.

22. **Setaria viridis** (L.) P. Beauv. [p. 557]
Green Bristlegrass, Sétaire Verte

Plants annual. **Culms** 20–250 cm; **nodes** glabrous. **Sheaths** glabrous, sometimes scabridulous, margins ciliate distally; **ligules** 1–2 mm, ciliate; **blades** to 20 cm long, 4–25 mm wide, flat, scabrous or smooth, glabrous. **Panicles** 3–20 cm, densely spicate, nodding only from near the apices; **rachises** hispid and villous; **bristles** 1–3, 5–10 mm, antrorsely scabrous, usually green, rarely purple. **Spikelets** 1.8–2.2 mm. **Lower glumes** about ⅓ as long as the spikelets, triangular-ovate, 3-veined; **upper glumes** nearly equaling the upper lemmas, elliptical, 5–6-veined; **lower lemmas** slightly exceeding the upper lemmas, 5-veined; **lower paleas** about ⅓ as long as the lower lemmas, hyaline; **upper lemmas** very finely and transversely rugose, pale green, 5–6-veined; **upper paleas** similar to the upper lemmas. $2n = 18$.

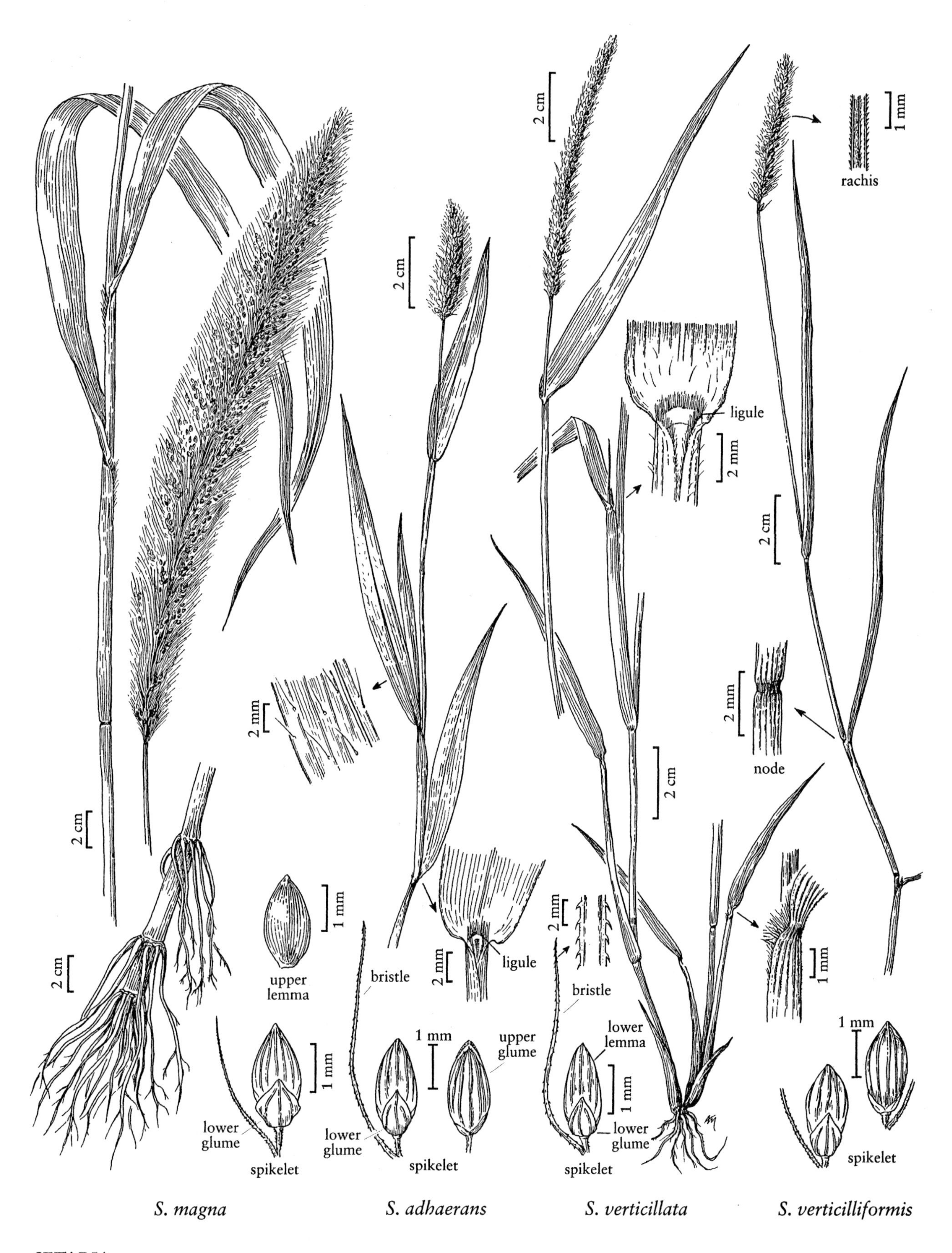

S. magna *S. adhaerans* *S. verticillata* *S. verticilliformis*

Setaria viridis resembles *S. italica* but differs in its shorter spikelets and rugose upper florets, and mode of disarticulation. It is also a more aggressive weed. It is native to Eurasia but is now widespread in warm temperate regions of the world.

1. Culms 100–250 cm tall; blades 10–25 mm wide; panicles 10–20 cm long . . . . var. *major*
1. Culms 20–100 cm tall; blades 4–12 mm wide; panicles 3–8 cm long . . . . var. *viridis*

Setaria viridis var. **major** (Gaudin) Peterm.
GIANT GREEN FOXTAIL

Culms 1–2.5 m; **nodes** 7–12. **Blades** 10–25 mm wide. **Panicles** 10–20 cm, producing around 4,000–6,000 caryopses; **lower branches** about 3 cm, producing a lobed appearance. **Spikelets** morphologically indistinguishable from those of var. *viridis*. $2n = 18$.

Setaria viridis var. *major* is a major adventive weed in corn and bean fields of the midwestern United States, dwarfing var. *viridis* in stature.

Setaria viridis (L.) P. Beauv. var. **viridis**
GREEN FOXTAIL

Culms 20–100 cm; **nodes** 6–7. **Blades** 4–12 mm wide. **Panicles** usually 3–8 cm, producing around 600–800 caryopses. $2n = 18$.

Setaria viridis var. *viridis* is an aggressive adventive weed throughout temperate North America. It is the most common annual representative of *Setaria* in the *Flora* region.

23. **Setaria italica** (L.) P. Beauv. [p. 557]
FOXTAIL MILLET, SÉTAIRE ITALIENNE, SÉTAIRE D'ITALIE, MILLET DES OISEAUX

Plants annual. **Culms** 10–100 cm. **Sheaths** mostly glabrous, margins sparsely ciliate; **ligules** 1–2 mm; **blades** to 20 cm long, 1–3 cm wide, flat, scabrous. **Panicles** 8–30 cm, dense, spikelike, occasionally lobed below; **rachises** hispid to villous; **bristles** 1–3, to 12 mm, tawny or purple. **Spikelets** about 3 mm, disarticulating between the lower and upper florets. **Lower glumes** 3-veined; **upper glumes** 5–7-veined; **lower paleas** absent or ½ as long as the lower lemmas; **upper lemmas** very finely and transversely rugose to smooth and shiny, exposed at maturity. $2n = 18$.

Setaria italica was cultivated in China as early as 2700 B.C. and during the Stone Age in Europe. Nowadays it is grown mostly for hay or as a pasture grass, but it has been used as a substitute for rice in northern China. It is sometimes cultivated in North America, but it is better known as a weed in moist ditches, mostly in the northeastern United States. It is closely related to *S. viridis*, differing in the longer (3 mm) spikelets and smooth, shiny upper florets which readily disarticulate above the lower florets. It exhibits considerable variation in seed and bristle color, bristle length, and panicle shape. Using these characters, Hubbard (1915) recognized several infraspecific taxa; they are not treated here.

24. **Setaria faberi** R.A.W. Herrm. [p. 557]
CHINESE FOXTAIL, SÉTAIRE GÉANTE

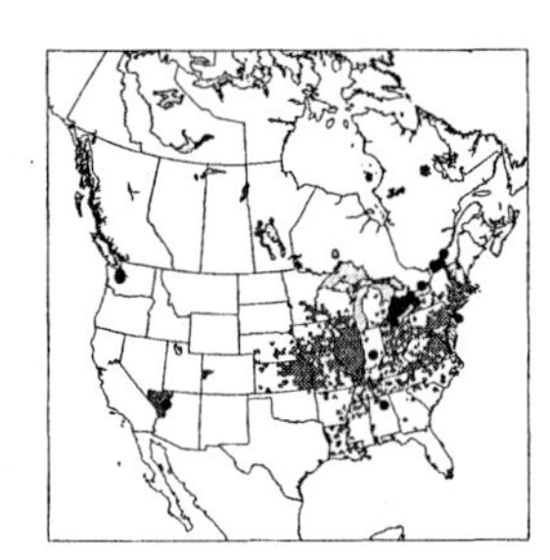

Plants annual. **Culms** 50–200 cm. **Sheaths** glabrous, fringed with white hairs; **ligules** about 2 mm; **blades** 15–30 cm long, 10–20 mm wide, usually with soft hairs on the adaxial surface. **Panicles** 6–20 cm, densely spicate, arching and drooping from near the base; **rachises** densely villous; **bristles** (1)3(6), about 10 mm. **Spikelets** 2.5–3 mm. **Lower glumes** about 1 mm, acute, 3-veined; **upper glumes** about 2.2 mm, obtuse, 5-veined; **lower lemmas** about 2.8 mm, obtuse; **lower paleas** about ⅔ as long as the lower lemmas; **upper lemmas** pale, finely and distinctly transversely rugose; **upper paleas** similar to the upper lemmas. $2n = 36$.

Setaria faberi spread rapidly throughout the North American corn belt after being accidentally introduced from China in the 1920s. It has become a major nuisance in corn and bean fields of the midwestern United States.

25. **Setaria parviflora** (Poir.) Kerguélen [p. 559]
KNOTROOT BRISTLEGRASS

Plants perennial; rhizomatous, rhizomes short, knotty. **Culms** 30–120 cm; **nodes** glabrous. **Sheaths** glabrous; **ligules** shorter than 1 mm, of hairs; **blades** to 25 cm long, 2–8 mm wide, flat, scabrous above. **Panicles** 3–8 (10) cm, of uniform width throughout their length, densely spikelike; **rachises** scabro-hispid; **bristles** 4–12, 2–12 mm, antrorsely barbed, yellow to purple. **Spikelets** 2–2.8 mm, elliptical and turgid. **Lower glumes** about ⅓ as long as the spikelets, 3-veined; **upper glumes** ½–⅔ as long as the spikelets, 5-veined; **lower florets** often staminate; **lower lemmas** occasionally indurate and faintly transversely rugose; **lower paleas** equaling the lower lemmas; **upper lemmas** distinctly transversely rugose, often purple-tipped. $2n = 36, 72$.

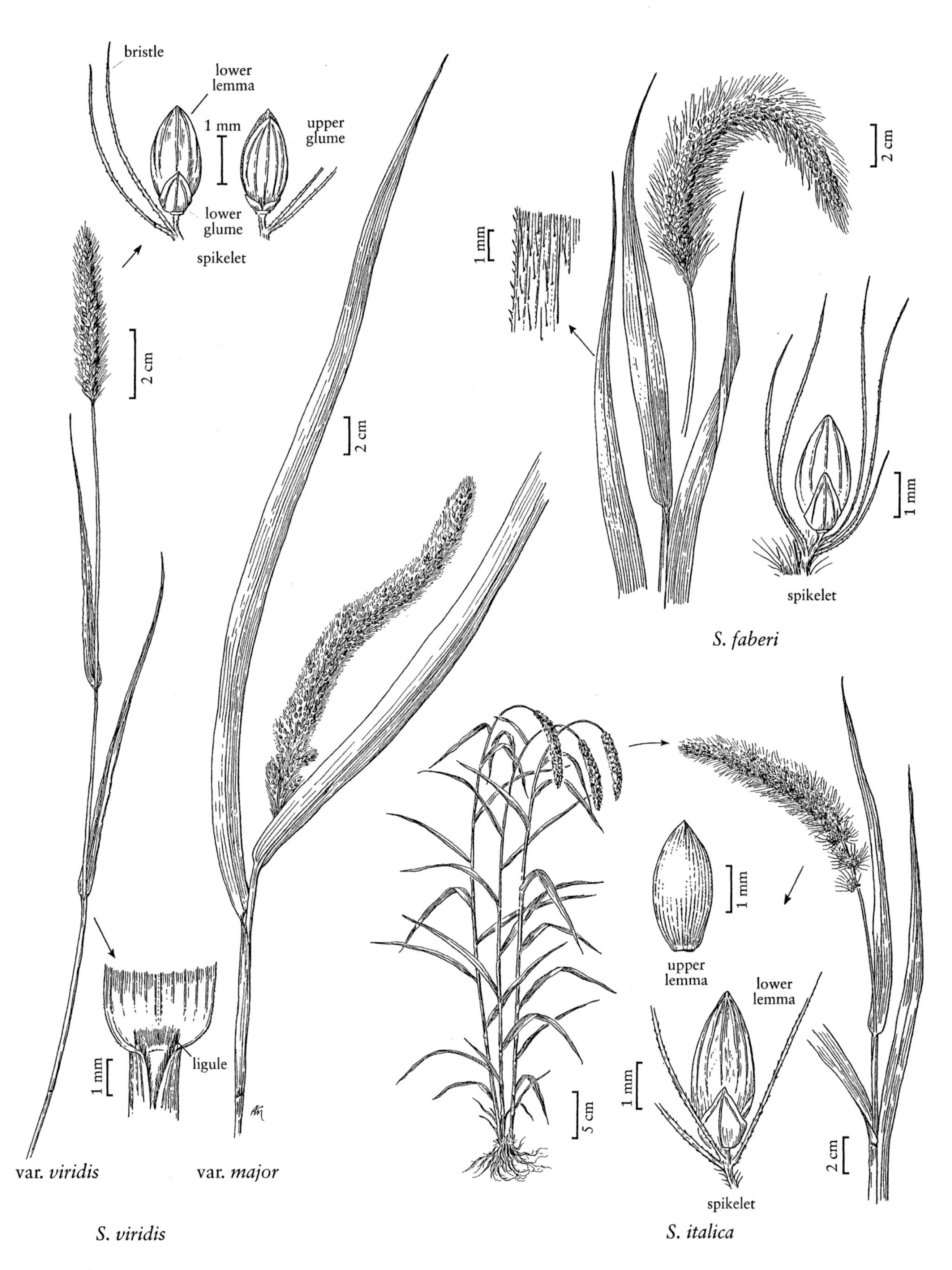
bristle
lower lemma
1 mm
upper glume
lower glume
spikelet
2 cm
2 cm
1 mm
ligule
var. *viridis*
var. *major*
S. viridis
2 cm
1 mm
1 mm
spikelet
S. faberi
1 mm
upper lemma
lower lemma
1 mm
5 cm
spikelet
2 cm
S. italica

SETARIA

Setaria parviflora is a common, native species of moist ground. It is most frequent along the Atlantic and Gulf coasts, but it also grows from the Central Valley of California east through the central United States and southward through Mexico to Central America, as well as in the West Indies. The plant from Oregon was found on a ballast dump; the species is not established in that state.

Setaria parviflora is the most morphologically diverse and widely distributed of the indigenous perennial species of *Setaria*.

26. **Setaria sphacelata** (Schumach.) Stapf & C.E. Hubb. [p. 559]
AFRICAN BRISTLEGRASS

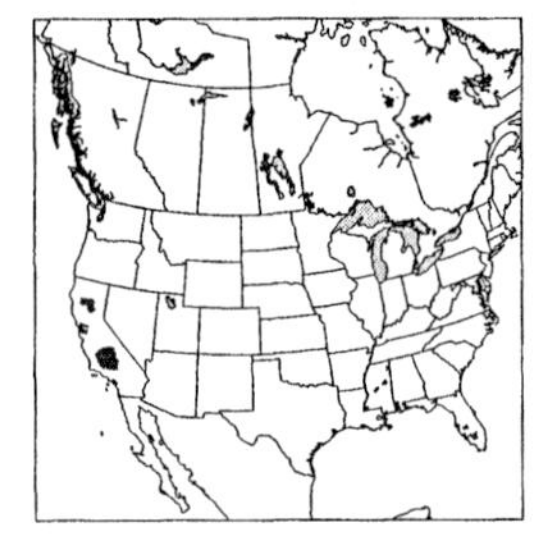

Plants perennial; cespitose, rhizomatous, rhizomes stout. **Culms** 50–150 cm, flattened; **nodes** glabrous. **Sheaths** glabrous; **blades** 15–50 cm long, 4–10 mm wide, flat, rather lax. **Panicles** 5–25 cm long, 4–8 mm thick (excluding the bristles), densely spicate; **bristles** 5 or more, 3–6 mm, usually orange to purple. **Spikelets** 2.5–3 mm, elliptic-oblong. **Lower glumes** about ⅓ as long as the spikelets; **upper florets** staminate; **upper glumes** ½–⅔ as long as the spikelets; **lower lemmas** equaling the upper lemmas; **lower paleas** equaling the upper paleas, broad; **upper lemmas** finely and transversely rugose; **upper paleas** similar to the upper lemmas. $2n = 36, 54$.

Setaria sphacelata is native to tropical Africa, but it has been found at a few scattered locations in the *Flora* region, often near a port. Clayton (1979) recognized five varieties of *Setaria sphacelata*. Those most likely to be introduced into the United States are **Setaria sphacelata** (Schumach.) Stapf & C.E. Hubb. var. **sphacelata** and **S. sphacelata** var. **aurea** (Hochst. *ex* A. Braun) Clayton, with var. *aurea* differing from var. *sphacelata* in having fibrous basal leaf sheaths and upper glumes that are often 3-veined.

27. **Setaria pumila** (Poir.) Roem. & Schult. [p. 559]
YELLOW FOXTAIL, PIGEON GRASS

Plants annual. **Culms** 30–130 cm. **Sheaths** glabrous; **ligules** ciliate; **blades** 4–10 mm wide, loosely twisted, adaxial surfaces with papillose-based hairs basally. **Panicles** 3–15 cm, uniformly thick, erect, densely spicate; **rachises** hispid; **bristles** 4–12, 3–8 mm, antrorsely scabrous. **Spikelets** 2–3.4 mm, strongly turgid. **Lower glumes** about ⅓ as long as the spikelets, 3-veined, acute; **upper glumes** about ½ as long as the spikelets, 5-veined, ovate; **upper florets** often staminate; **lower lemmas** equaling the upper lemmas; **lower paleas** equaling the lower lemmas, broad; **upper lemmas** conspicuously exposed, strongly transversely rugose. $2n = 36, 72$.

1. Spikelets 3–3.4 mm long; bristles yellow . . . subsp. *pumila*
1. Spikelets 2–2.5 mm long; bristles reddish . . . . .subsp. *pallidefusca*

Setaria pumila subsp. **pallidefusca** (Schumach.) B.K. Simon

Blades dark green. **Bristles** reddish. **Spikelets** 2–2.5 mm. $2n = 18, 36, 54, 72$.

Setaria pumila subsp. *pallidefusca* is native to tropical Africa. It is now established as a weed in southeastern Louisiana, but it has also been collected in the past on ballast dumps in Portland, Oregon.

Reports of **S. nigrirostris** (Nees) T. Durand & Schinz in North America have not been verified. It differs from *S. pumila* in being a rhizomatous perennial with spikelets 3.5–5 mm long and bristles with thickened tips (those of *S. pumila* are slender throughout). It is a native of southern Africa.

Setaria pumila (Poir.) Roem. & Schult. subsp. **pumila**
SÉTAIRE GLAUQUE

Blades yellowish-green. **Bristles** yellowish. **Spikelets** 3–3.4 mm. $2n = 36, 72$.

Setaria pumila subsp. *pumila* is a European adventive that has become a common weed in lawns and cultivated fields throughout temperate North America.

25.21 PASPALIDIUM Stapf

Charles M. Allen

Plants annual or perennial. **Culms** to 100 cm, not branching above the base. **Auricles** absent; **ligules** membranous and ciliate or of hairs. **Inflorescences** panicles of racemosely arranged, spikelike branches, branches sometimes highly reduced, each panicle appearing spikelike;

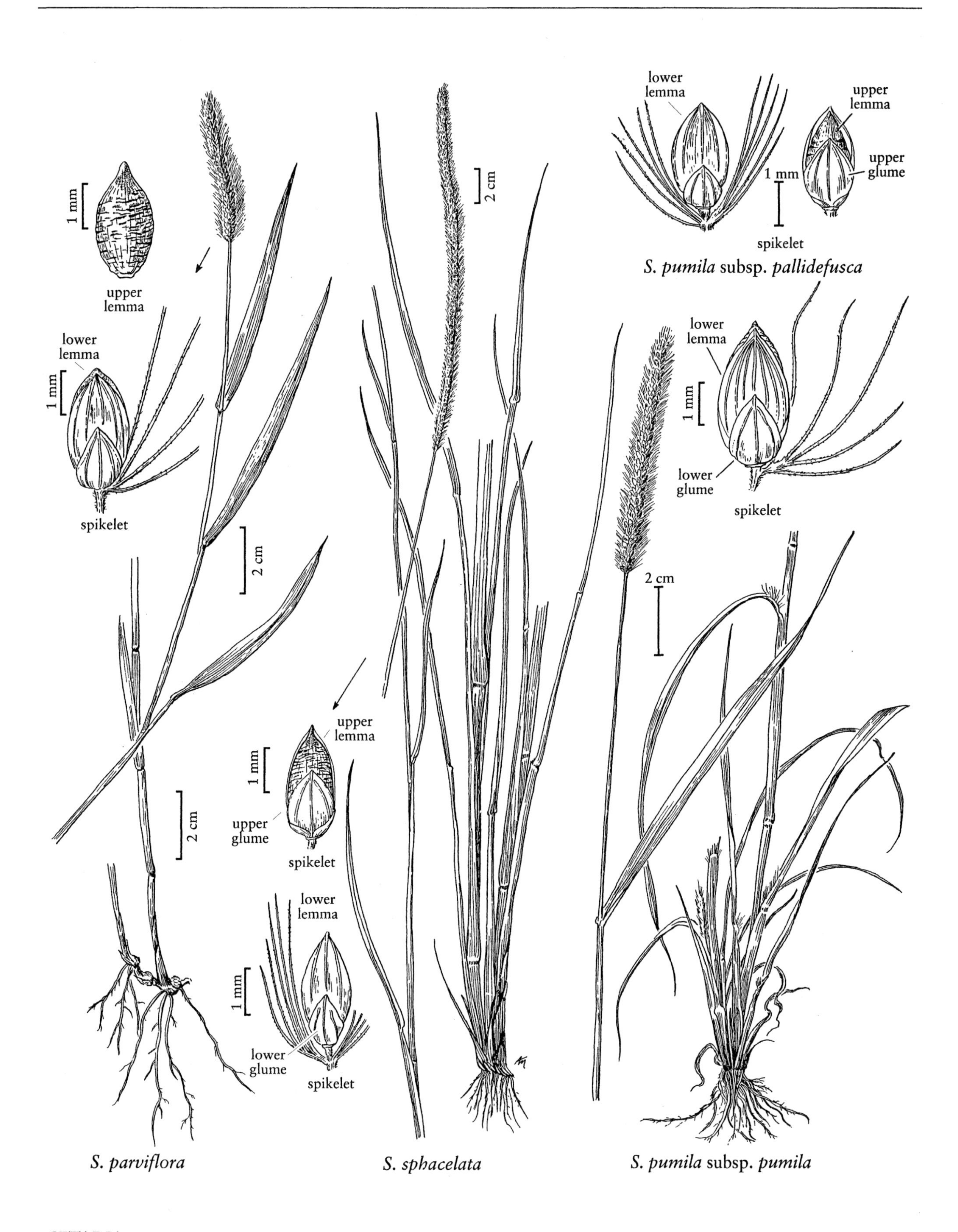
upper
lemma
1 mm
lower
lemma
1 mm
spikelet
2 cm
2 cm
upper
lemma
1 mm
upper
glume
spikelet
lower
lemma
1 mm
lower
glume
spikelet
2 cm
lower
lemma
upper
lemma
1 mm
upper
glume
spikelet
S. pumila subsp. pallidefusca
lower
lemma
1 mm
lower
glume
spikelet
2 cm
S. parviflora
S. sphacelata
S. pumila subsp. pumila

branches 1-sided, terminating in a more or less inconspicuous bristle, bristles 2.5–4 mm; **disarticulation** beneath the spikelets. **Spikelets** subsessile, in 2 rows on 1 side of the branches, lacking subtending bristles, dorsally compressed, with 2 florets, upper glumes and upper florets appressed to the branch axes. **Glumes** membranous; **lower glumes** much shorter than the spikelets; **upper glumes** subequal to the upper florets; **lower florets** sterile or staminate; **lower lemmas** similar to the upper glumes in size and texture; **upper florets** bisexual; **upper lemmas** indurate, rugose, unawned, yellow or brown; **upper paleas** similar to their lemmas; **anthers** 3. $x = 9$. The name is a diminutive of *Paspalum*.

Paspalidium is a genus of approximately 40 species, one of which is native to the *Flora* region. It grows in tropical regions throughout the world. Most of its species have an inflorescence of well-spaced, unilateral, spicate branches and the resemblance to *Paspalum* is evident, but species with closely crowded, highly reduced branches are easily mistaken for species of *Setaria*, the terminal bristle resembling a single, subtending bristle.

1. **Paspalidium geminatum** (Forssk.) Stapf [p. 562]
EGYPTIAN PASPALIDIUM, WATER PASPALIDIUM

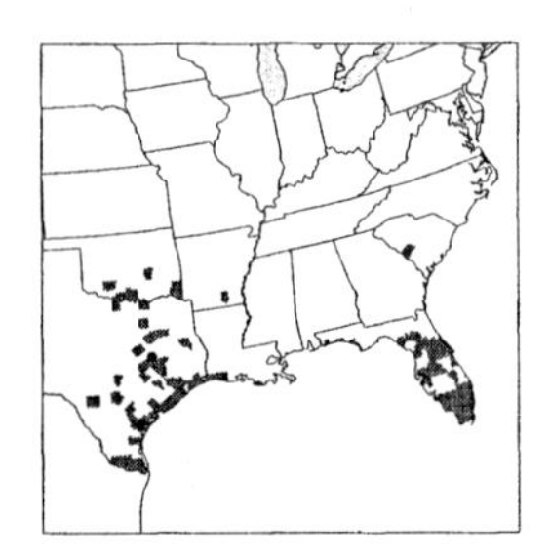

Plants perennial; rhizomatous. **Culms** 25–100 cm, erect. **Leaves** basal and cauline; **sheaths** glabrous, margins scarious, sparsely ciliate distally; **ligules** 0.5–1 mm, of hairs; **blades** flat to conduplicate, glabrous or scabrous. **Panicles** 10–30 cm; **branches** 5–15, 1–3.7 cm, erect, with more than 12 spikelets; **terminal bristles** 2.5–4 mm. **Spikelets** 2.2–3.2 mm, clearly overlapping. **Lower glumes** 0.8–1.2 mm, 1–3-veined, truncate; **upper glumes** and **lower lemmas** 2–2.4 mm, glabrous, 5–7-veined, acuminate; **lower paleas** 2–2.4 mm, scarious; **upper lemmas** and **paleas** 2–2.3 mm, rugose, stramineous to light brown, lemma margins scarious, inrolled, clasping the paleas, lemma apices acuminate; **anthers** 1.2–1.5 mm. **Caryopses** about 1 mm in diameter, spheroidal, slightly flattened, yellow. $2n = 18, 54$.

Paspalidium geminatum grows in moist to wet, fresh to brackish areas. It is native to the southeastern United States, the West Indies, and tropical regions of the Americas.

25.22 STENOTAPHRUM Trin.

Kelly W. Allred

Plants annual or perennial; sometimes rhizomatous or stoloniferous. **Culms** 10–60 cm, usually compressed; **internodes** solid. **Leaves** cauline; **sheaths** shorter than the internodes, compressed; **ligules** membranous and ciliate or of hairs; **blades** flat or folded. **Inflorescences** spikelike panicles; **branches** very short, with fewer than 10 spikelets, appressed to and partially embedded in the flattened, corky rachises; **disarticulation** below the glumes, often with a segment of the branch. **Spikelets** lanceolate to ovate, unawned, lower glumes oriented away from the branch axes. **Glumes** membranous; **lower glumes** scalelike, usually without veins; **upper glumes** 5–7-veined; **lower florets** staminate or sterile, lemmas 3–9-veined; **upper florets** bisexual; **upper lemmas** longer than the glumes, papery to subcoriaceous, 3–5-veined; **upper paleas** generally indurate, 2-veined; **anthers** 3. **Caryopses** lanceolate to ovate, often failing to develop. $x = 9$. Name from the Greek *stenos*, 'narrow', and *taphros*, 'trench', referring to the cavities in the rachis.

Stenotaphrum is a genus of seven species that usually grow on the seashore or near the coast, primarily along the Indian Ocean rim. Three species are endemic to Madagascar, and one species is thought to be native to the *Flora* region.

SELECTED REFERENCES **Busey, P., T.K. Broschat, and B.J. Center.** 1982. Classification of St. Augustinegrass. Crop Sci. (Madison) 22:469–473; **Sauer, J.D.** 1972. Revision of *Stenotaphrum* (Gramineae: Paniceae) with attention to its historical geography. Brittonia 24:202–222.

1. **Stenotaphrum secundatum** (Walter) Kuntze [p. 562]
St. Augustine Grass

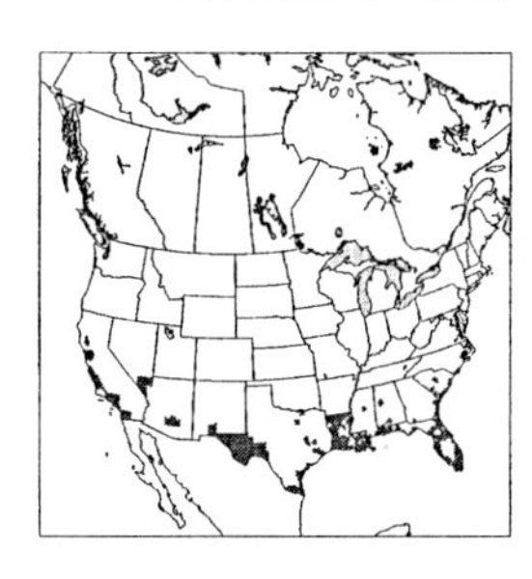

Plants stoloniferous. **Culms** 10–30 cm, decumbent, rooting at the lower nodes, branched above the base, with prominent prophylls. **Sheaths** sparsely pilose, constricted at the summit; **ligules** about 0.5 mm, membranous, ciliate; **blades** 3–15(18) cm long, 4–10 mm wide, thick, flat, glabrous, apices blunt. **Panicles** 4.5–10 cm long, less than 1 cm wide; **rachises** flattened, winged; **branches** 12–20, with 1–5 spikelets. **Spikelets** 3.5–5 mm, partially embedded in 1 side of the branch axes; **lower glumes** about 1 mm, rounded, irregularly toothed; **upper glumes** and **lower lemmas** 3–4 mm, about equal; **upper lemmas** papery, 5-veined, margins weakly clasping the paleas; **anthers** 2–2.5 mm, tan or purple. **Caryopses** about 2 mm, oblong to obovate. $2n$ = 18.

Stenotaphrum secundatum grows on sandy beaches, at the edges of swamps and lagoons, and along inland streams and lakes. It may be native to the southeastern United States, being known from the Carolinas prior to 1800, but it has become naturalized in most tropical and subtropical regions of the world.

Stenotaphrum secundatum is planted for turf in the southern United States and is now established from California to North Carolina and Florida. Numerous cultivars have been developed. Specimens with variegated foliage (often called *S. secundatum* var. *variegatum* Hitchc.) are sometimes used as an ornamental in hanging baskets and greenhouses.

25.23 HYMENACHNE P. Beauv.

Mary E. Barkworth

Plants perennial. **Culms** 50–350 cm, decumbent, often rooting from the lower nodes; **nodes** glabrous; **internodes** glabrous, filled with spongy aerenchyma. **Leaves** evenly distributed; **ligules** membranous. **Inflorescences** narrow, condensed, cylindrical or spikelike panicles; **branches** appressed but not fused to the rachises, densely and evenly short ciliate on the ridges. **Spikelets** dorsally compressed, narrowly lanceolate, with 2 florets. **Glumes** unequal, membranous, with a definite rachilla internode between the 2 glumes; **lower glumes** no more than ½ as long as the spikelet, 1–3-veined; **upper glumes** subequal to or longer than the lower lemmas, membranous, not saccate, 3–7-veined; **lower florets** sterile; **lower lemmas** similar to the upper glumes; **upper florets** bisexual, much shorter than the upper glumes and lower lemmas; **upper lemmas** membranous to leathery, white at maturity, margins clasping the edges of the paleas basally but not distally, apices acute; **paleas** similar in texture to the lemmas; **lodicules** 2; **anthers** 3; **styles** 2, fused. **Caryopses** falling free from the lemmas and paleas. x = 10. Name from the Greek *hymen*, 'membrane', and *achne*, 'chaff', alluding to the membranous upper lemma, an unusual texture for the upper lemma in members of the *Paniceae*.

Hymenachne is a pantropical genus of approximately five species. It is unusual in the *Paniceae* in that all its members grow in aquatic or swampy habitats. It differs from *Sacciolepis* in its aerenchyma-filled internodes, a character that is probably an adaptation to its aquatic habitat. One species is native to the *Flora* region.

SELECTED REFERENCE **Pohl, R.W.** and **N.R. Lersten.** 1975. Stem aerenchyma as a character separating *Hymenachne* and *Sacciolepis* (Gramineae, Panicoideae). Brittonia 27:223–227.

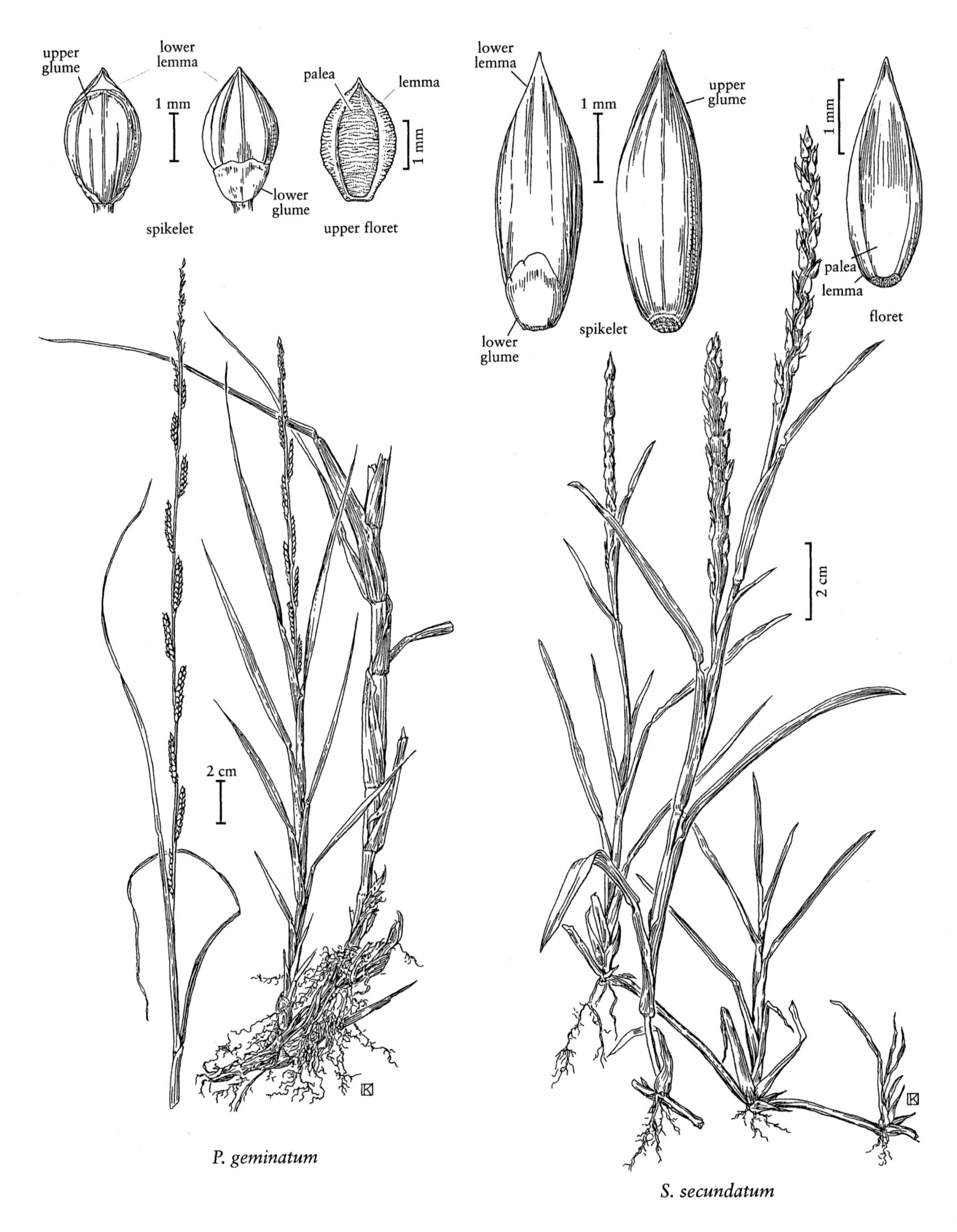

PASPALIDIUM

STENOTAPHRUM

1. **Hymenachne amplexicaulis** (Rudge) Nees [p. 564]
West Indian Marsh Grass

Plants perennial. **Culms** to 3.5 m tall, 1 cm or more thick, decumbent. **Ligules** 1–2.5 mm, brownish; **blades** 15–33 cm long, 12–28 mm wide, lax, flat, glabrous. **Panicles** 10–40 cm long, 0.7–1.2 cm thick, spike-like, dense, sometimes lobed near the base; **basal branches** 1.5–5 cm, strictly erect. **Spikelets** 3.5–5 mm, lanceolate, acuminate. **Lower glumes** 1–1.7 mm, 3-veined; **upper glumes** 2.8–3.9 mm; **lower lemmas** 3.6–4.6 mm, longer than the upper glumes, attenuate to subaristate; **lower paleas** absent; **upper lemmas** 2.5–3.5 mm; **anthers** 1.1–1.2 mm. $2n = 10$.

In the *Flora* region, *Hymenachne amplexicaulis* is known only from low, wet pastures in southern Florida and it is rare even in that state. It is more abundant in the remainder of its range which extends through Mexico to Argentina.

25.24 STEINCHISMA Raf.

Robert W. Freckmann
Michel G. Lelong

Plants perennial; cespitose, rhizomatous, rhizomes short, slender. **Culms** slender, often compressed. **Sheaths** usually keeled; **ligules** minute, membranous, often erose or ciliate; **blades** exhibiting Kranz anatomy, with few organelles in the external sheath and 5–7 isodiametric mesophyll cells between the vascular bundles. **Inflorescences** terminal, open to contracted panicles; **primary branches** few, slender; **pedicels** short, to 1 mm. **Spikelets** ellipsoid or lanceolate, initially somewhat compressed, ultimately expanding greatly. **Glumes** glabrous; **lower glumes** ⅓–½ as long as the spikelets, usually 3(5)-veined, acute; **upper glumes** and **lower lemmas** subequal, 3–5(7)-veined; **lower florets** sterile or staminate, often standing apart from the upper florets at maturity; **lower paleas** longer than the lower lemmas, greatly inflated at maturity, indurate; **upper florets** ovoid or ellipsoid; **upper lemmas** usually dull-colored, minutely papillose, papillae in longitudinal rows, apices acute. $x = 9$ or 10. Name from the Greek *steinos*, 'narrow', and *chasma*, 'yawning', presumably alluding to the gaping glumes and somewhat narrow spikelet when compared to *Panicum* (Clifford 1996).

Steinchisma is a genus of 5–6 species that grow in moist or wet, usually open, sandy areas in warm-temperate and tropical regions of the Western Hemisphere. A single species is native the *Flora* region. It is sometimes included in *Panicum*, but recent studies support its recognition as a separate genus. Photosynthesis in *Steinchisma* is intermediate between C_3 and C_4 plants.

SELECTED REFERENCES **Clifford, H.T.** 1996. Etymological Dictionary of Grasses, Version 1.0 (CD-ROM). Expert Center for Taxonomic Identification, Amsterdam, The Netherlands; **Zuloaga, F.O., O. Morrone, A.S. Vega, and L.M. Giussani.** 1998. Revisión y análisis cladístico de *Steinchisma* (Poaceae: Panicoideae: Paniceae). Ann. Missouri Bot. Gard. 85:631–656.

1. **Steinchisma hians** (Elliott) Nash [p. 564]
Gaping Panicgrass

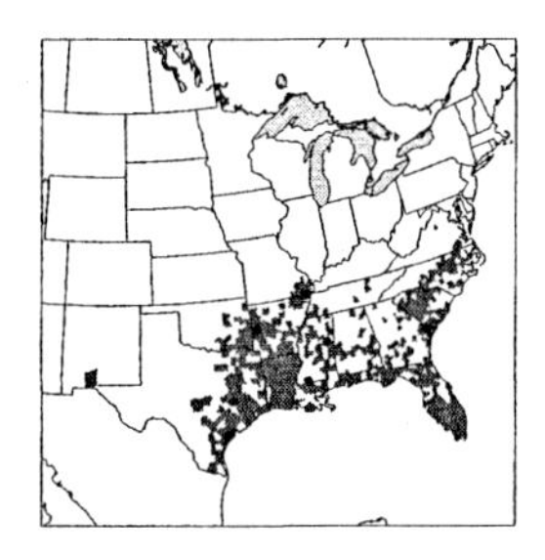

Culms 20–75 cm, often compressed, at least basally, erect to decumbent, glabrous. **Sheaths** usually shorter than the internodes, terete or somewhat compressed, glabrous or sometimes sparsely hispid below the throat, margins scarious or sparsely ciliate at the summit; **ligules** 0.2–0.5 mm, membranous, erose-ciliate; **blades** 6–20 cm long, 2–5 mm wide, relatively long and slender, flat or folded, glabrous abaxially, mostly glabrous adaxially but sparsely pilose basally. **Panicles** 5–20 cm, about ½ as wide as long, delicate, open; **primary branches** flexible, spreading or drooping, with short, crowded secondary branches and pedicels. **Spikelets** 1.8–2.4 mm, often purplish, glabrous. **Lower glumes** acute; **upper glumes** and **lower lemmas** slightly exceeded by the enlarged, indurate, sterile paleas; **upper florets** 1.6–1.9 mm, dull-colored, minutely papillose, acute. $2n = 18, 20$.

Steinchisma hians grows in moist or wet, usually open areas, and in moist pinelands, low woods, and ditches. Its range extends from the southeastern United States, through Mexico and Central America to Colombia, Brazil, and Argentina.

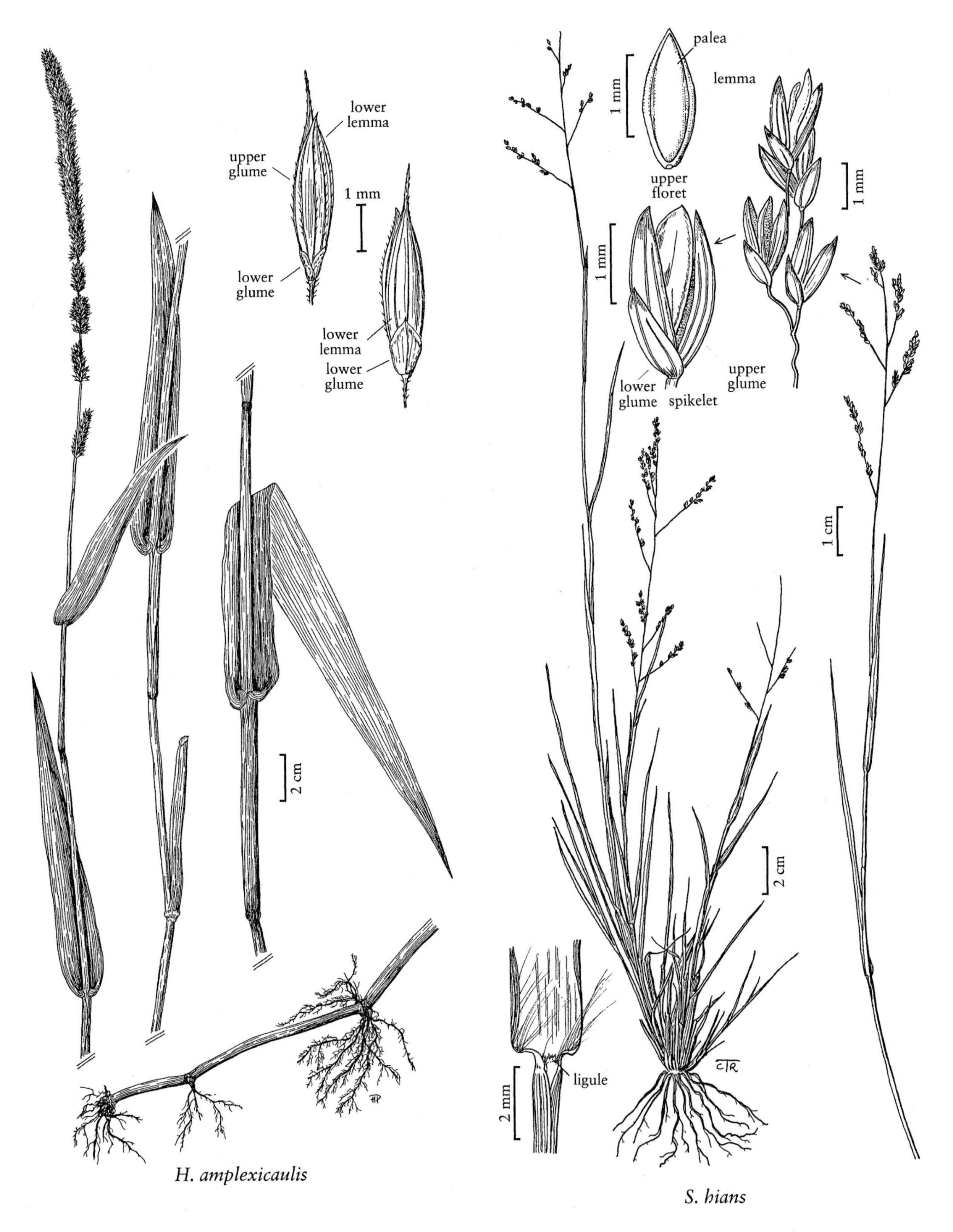

H. amplexicaulis

S. hians

HYMENACHNE

STEINCHISMA

25.25 AXONOPUS P. Beauv.

Mary E. Barkworth

Plants perennial, rarely annual; cespitose, loosely tufted, or mat-forming, sometimes rhizomatous or stoloniferous. **Culms** 7–300 cm, not woody, often decumbent at the base, erect to ascending. **Sheaths** open; **ligules** membranous, truncate, ciliate; **blades** flat or convolute, usually obtuse. **Inflorescences** terminal, sometimes also axillary, panicles of 2–many, digitately, subdigitately, or racemosely arranged spikelike branches; **branches** triquetrous, spikelets subsessile or sessile, solitary, in 2 rows, lower lemmas appressed to the branch axes; **disarticulation** below the glumes. **Spikelets** dorsally compressed, with 2 florets; **lower florets** sterile or staminate; **upper florets** sessile, bisexual. **Lower glumes** absent; **upper glumes** and **lower lemmas** equal, membranous; **lower paleas** absent; **upper lemmas** indurate, usually glabrous, sometimes with an apical tuft of hairs, margins slightly involute, clasping the palea, apices acute to obtuse; **upper paleas** similar to the upper lemmas in texture. **Caryopses** ellipsoid. $x = 10$. Name from the Greek *axon*, 'axis', and *pes*, 'foot'.

Axonopus is a genus of approximately 100 tropical and subtropical species, most of which are native to the Western Hemisphere. Three species are native to the *Flora* region; one additional species has been grown experimentally in Florida.

All the species tend to grow in open habitats, often where the soil is somewhat impermeable and slightly flooded in the rainy season. *Axonopus fissifolius* and *A. compressus* are cultivated for forage in many countries; *A. compressus* is also used as a lawn grass. Both species are inclined to be weedy. The presence of rhizomes or stolons is affected by environmental conditions, with plants growing in crowded conditions, e.g., lawns, rarely producing them.

SELECTED REFERENCE **Black, G.A.** 1963. Grasses of the genus *Axonopus* (a taxonomic treatment) (ed. L.B. Smith). Advancing Frontiers Pl. Sci. 5:1–186.

1. Panicles with 30–100+ branches; lower branches 10–24 cm long; culms (50)100–300 cm tall . . . . . 4. *A. scoparius*
1. Panicles with 2–7 branches; lower branches 1–15 cm long; culms 7–100 cm tall.
   2. Spikelets 3.5–5.5 mm long; upper glumes glabrous; lower lemmas glabrous or sparsely pilose over the veins . . . . . 3. *A. furcatus*
   2. Spikelets 1.6–3.5 mm long; upper glumes and lower lemmas sparsely pilose on the margins or marginal veins.
      3. Upper glumes and lower lemmas extending beyond the upper florets, forming acute to acuminate apices; blades 3–20 mm wide . . . . . 2. *A. compressus*
      3. Upper glumes and lower lemmas not or scarcely extending beyond the upper florets, forming obtuse to subacute apices; blades 1.5–6 mm wide . . . . . 1. *A. fissifolius*

1. **Axonopus fissifolius** (Raddi) Kuhlm. [p. 567]
COMMON CARPETGRASS

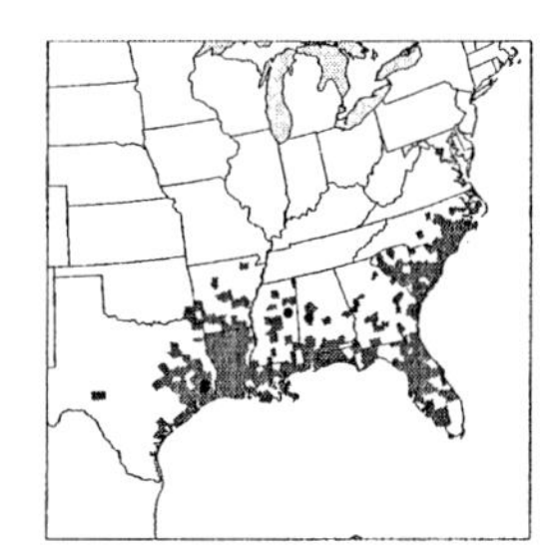

Plants usually cespitose, sometimes stoloniferous, nodes of the stolons often pilose. **Culms** 10–75 cm, erect or depressed-decumbent; **cauline nodes** glabrous or slightly pubescent. **Sheaths** compressed, mostly glabrous, margins ciliate; **ligules** 0.2–0.4 mm; **blades** 4-15 cm long, 1.5–6 mm wide, flat, mostly glabrous, margins with papillose-based cilia. **Panicles** terminal and axillary, 5–11 cm overall, rachises to 3 cm, with 2–7 branches; **branches** 2–9(12) cm, spreading or ascending. **Spikelets** 1.6–2.2(2.8) mm, ovoid or ellipsoid, obtuse to acute. **Upper glumes** and **lower lemmas** scarcely extending beyond the upper florets, 2-veined, margins sparsely pilose, apices obtuse to subacute; **upper lemmas** and **paleas** 1.6–2.1 mm long, 0.5–0.7 mm wide. **Caryopses** 1.5–1.8 mm, gray. $2n = 20, 40, 80, 100$.

Axonopus fissifolius is sometimes used as a lawn or pasture grass, but it is also an invasive weedy species, often growing in moist, disturbed sites. It is native in the southeastern United States and from central Mexico south to Bolivia and Argentina. It has also been introduced into tropical and subtropical regions of the Eastern Hemisphere.

2. **Axonopus compressus** (Sw.) P. Beauv. [p. 567]
Broadleaf Carpetgrass

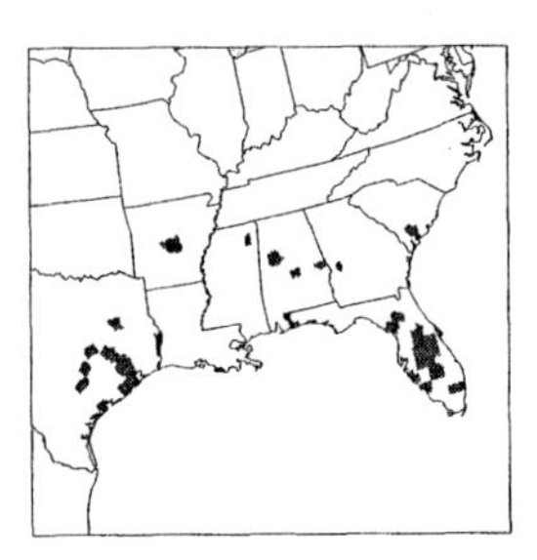

Plants stoloniferous, rarely rhizomatous, rhizomes, when present, 3–5 cm. **Culms** 7–80 cm; **nodes** glabrous or pubescent. **Sheaths** keeled, strongly compressed, pubescent; **ligules** 0.3–0.5 mm; **blades** 3–20 mm wide, glabrous or sparsely pilose, midveins often white and prominent, apices frequently ciliate or pubescent. **Panicles** terminal and axillary, 4-10 cm overall, rachises to 3.5 cm, with 2–5 branches; **branches** 1–13 cm. **Spikelets** 2–3.5 mm, ovoid, ellipsoid, or lanceoloid, acuminate. **Upper glumes** and **lower lemmas** extending beyond the upper florets, 2–5-veined, marginal veins pilose, apices acute to acuminate; **upper lemmas** and **paleas** 1.5–1.8 mm, broadly ellipsoid. **Caryopses** 1.2–1.5 mm, gray. 2*n* = 40, 60, 80.

Axonopus compressus is native from the southeastern United States to Bolivia, Brazil, and Uruguay, and has become established in the Eastern Hemisphere. It is used as a lawn and forage grass but is also weedy, readily growing in moist, disturbed habitats.

3. **Axonopus furcatus** (Flüggé) Hitchc. [p. 567]
Big Carpetgrass

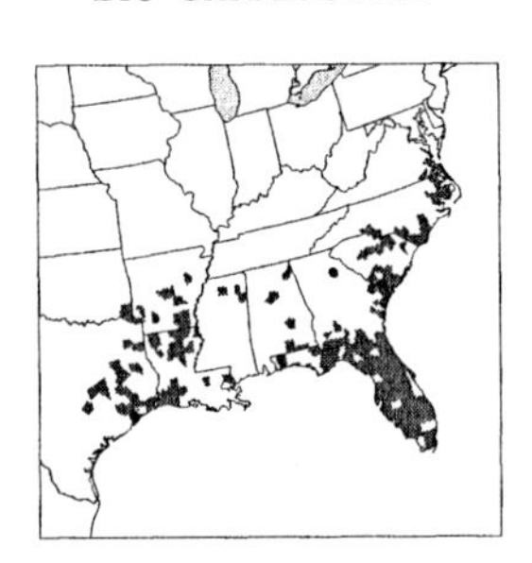

Plants stoloniferous. **Culms** 30–100 cm; **nodes** glabrous or pubescent. **Sheaths** compressed, glabrous or sparsely to densely pilose, hairs appressed; **ligules** 0.3–1 mm; **blades** 3–25 cm long, 2–15 mm wide, margins often with papillose-based hairs near the base, scabrous distally. **Panicles** terminal and axillary, with 2(–4) divergent branches; **branches** 4–15 cm. **Spikelets** 3.5–5.5 mm long, about 1.5 mm wide, sessile or subsessile, ovoid-ellipsoid, acuminate. **Upper glumes** glabrous, 5–7-veined; **lower lemmas** 5–7-veined, glabrous or sparsely pilose over the veins; **upper lemmas** and **paleas** 2.5–3.2 mm, light yellow, obtuse. **Caryopses** 1.8–2.2 mm, obovate, yellow. 2*n* = unknown.

Axonopus furcatus is endemic to the southeastern United States. It grows in moist pine barrens, marshes, river banks, wet ditches, pond margins, and other such damp areas.

4. **Axonopus scoparius** (Humb. & Bonpl. *ex* Flüggé) Kuhlm. [p. 567]
Carpetgrass

Plants rhizomatous, frequently also stoloniferous. **Culms** (50)100–300 cm, to about 1 cm thick, sometimes branching above the base; **nodes** glabrous. **Leaves** primarily cauline; **sheaths** often much wider than the internodes, mostly glabrous but the collars pubescent, lower sheaths compressed; **ligules** 0.5–2.9 mm, ciliolate; **blades** 15–50 cm long, (5)20–35 mm wide, bases usually wider than the sheaths. **Panicles** terminal and axillary, 10–50 cm overall, rachises 2-3.5 cm, with 30–100+ branches; **lower branches** 10–24 cm, frequently fascicled. **Spikelets** 2.1–2.7 mm, ovoid to oblong-ellipsoid, acute or apiculate; **upper glumes** and **lower lemmas** usually 5-veined, sparsely pilose; **upper florets** 0–0.4 mm shorter than the upper glumes and lower lemmas, obtuse to subacute. **Caryopses** usually absent. 2*n* = 20.

Axonopus scoparius is native from southern Mexico to Peru, Bolivia, and Brazil. In Mesoamerica, it rarely sets seed but is grown for forage and often persists after cultivation has ceased. It has been grown experimentally in Florida, but it is not winter hardy even there. Not surprisingly, *A. scoparius* is not established in the *Flora* region.

25.26 PASPALUM L.

Charles M. Allen
David W. Hall

Plants annual or perennial; cespitose, rhizomatous, or stoloniferous. **Culms** 3–400 cm, erect, spreading or prostrate, sometimes trailing for 200+ cm. **Sheaths** open; **auricles** sometimes present; **ligules** membranous. **Inflorescences** terminal, sometimes also axillary, panicles of 1–many spikelike branches, these digitate or racemose on the rachis, spreading to erect, 1 or more branches completely or partially hidden in the sheaths in some species; **branch axes** flattened, usually narrowly to broadly winged, usually terminating in a spikelet, sometimes extending beyond the distal spikelet but never forming a distinct bristle; **disarticulation** below the

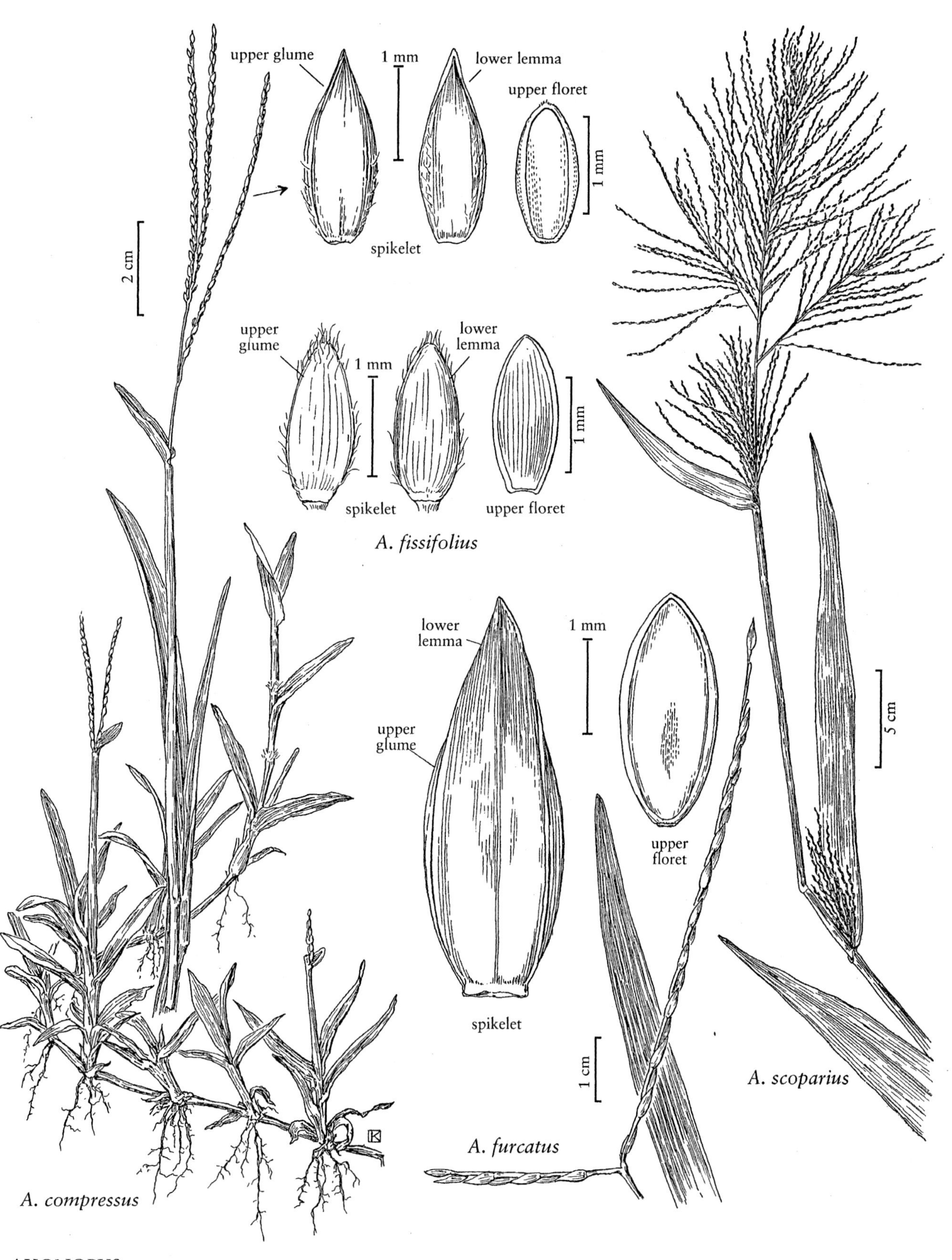

upper glume
1 mm
lower lemma
upper floret
1 mm
spikelet
2 cm
upper glume
1 mm
lower lemma
1 mm
spikelet
upper floret
A. fissifolius
lower lemma
1 mm
upper glume
upper floret
5 cm
spikelet
1 cm
A. scoparius
A. furcatus
A. compressus

AXONOPUS

glumes. **Spikelets** subsessile to shortly pedicellate, plano-convex, rounded to acuminate, dorsally compressed, not subtended by bristles or a ringlike callus, solitary or paired (1 spikelet of the pair reduced in some species), in 2 rows along 1 side of the branches, with 2 florets, first rachilla segment not swollen, upper glumes and upper lemmas adjacent to the branch axes; **lower florets** sterile; **upper florets** sessile or stipitate, bisexual, acute or rounded. **Lower glumes** absent or present only on some spikelets of each branch, without veins or 1-veined, unawned; **upper glumes** and **lower lemmas** subequal, membranous, apices rounded, unawned; **lower paleas** absent or rudimentary; **upper lemmas** convex, indurate, smooth to slightly rugose, stramineous to dark brown, margins scarious, involute, clasping the paleas; **upper paleas** indurate, smooth to slightly rugose, stramineous to dark brown. **Caryopses** orbicular to elliptical, plano-convex or flattened, white, yellow, or brown. $x = 10, 12$. Name from the Greek *paspalos*, a kind of millet.

Paspalum includes 300–400 species, most of which are native to the Western Hemisphere. Forty-three species are found in the *Flora* region; twenty-four are native. *Paspalum scrobiculatum* is grown as a grain in India, and several species are grown as forage plants. There are also many weedy species in the genus. Nineteen of the species growing in the *Flora* region are introduced, and some of them are weedy. Because weeds are under-represented in most herbaria, the distribution maps of such species probably understate their prevalence.

SELECTED REFERENCES **Banks, D.J.** 1966. Taxonomy of *Paspalum setaceum* (Gramineae). Sida 2:269–284; **COTECOCA** (Comisión Técnico Consultiva de Coeficientes de Agostadero). 2000. Las Gramineás de México, vol. 5. Secretaria de Agricultura, Ganaderia y Desarrollo Rural, México, D.F., México. 466 pp.; **Pohl, R.W.** and **G. Davidse.** 1994. *Paspalum* L. Pp. 335–352 *in* G. Davidse, M. Sousa S., and A.O. Chater (eds.). Flora Mesoamericana, vol. 6: Alismataceae a Cyperaceae. Instituto de Biología, Universidad Nacional Autónoma de México, México, D.F., México. 543 pp.

1. Spikelets solitary, not associated with a naked pedicel or rudimentary spikelets.
  2. Panicles with 1–70 branches, if more than 1, the branches racemosely arranged.
    3. Branches 7–70, disarticulating at maturity, the axes extending beyond the distal spikelets . . . . . 1. *P. repens*
    3. Branches 1–6, persistent, terminating in a spikelet.
      4. Upper florets olive to dark brown . . . . . 2. *P. scrobiculatum*
      4. Upper florets pale to stramineous.
        5. Axes of panicle branches 0.6–1.3 mm wide . . . . . 3. *P. laeve*
        5. Axes of panicle branches 1.8–3.3 mm wide.
          6. Spikelets 1.7–2.1 mm long; upper lemmas glabrous throughout . . . . . 4. *P. dissectum*
          6. Spikelets 3.2–4 mm long; upper lemmas with a few short hairs at the apices . . . 5. *P. acuminatum*
  2. Panicles usually composed of a terminal pair of branches, sometimes with 1(–5) additional branches below the terminal pair.
    7. Upper glumes pilose on the margins or shortly pubescent on the back.
      8. Spikelets 1.3–1.9 mm long; upper glumes pilose along the margins . . . . . 6. *P. conjugatum*
      8. Spikelets 2.4–3.2 mm long; upper glumes sparsely short pubescent on the back . . . . . 7. *P. distichum*
    7. Upper glumes glabrous.
      9. Spikelets elliptic, their apices acute to acuminate.
        10. Plants rhizomatous, not appearing cespitose; usually in brackish to salt marsh habitats . . . . . 8. *P. vaginatum*
        10. Plants shortly rhizomatous but appearing cespitose; usually in disturbed inland habitats . . . . . 9. *P. almum*
      9. Spikelets ovate to broadly elliptic, their apices obtuse to broadly acute.
        11. Spikelets 2.5–4 mm long; leaf blades flat or conduplicate . . . . . 10. *P. notatum*
        11. Spikelets 1.9–2.3 mm long; leaf blades flat . . . . . 11. *P. minus*
1. Spikelets paired, if only 1 spikelet functional, a naked pedicel or rudimentary, non-functional spikelet present.
  12. Spikelets 1–1.3 mm long.
    13. Panicle branches 2–6; spikelets elliptic to elliptic-obovate, appressed to the branch axes . . . . . 12. *P. blodgettii*

13. Panicle branches 18–50; spikelets ovate, diverging from the branch axes . . . . 13. *P. paniculatum*
12. Spikelets 1.3–4.1 mm long.
14. Margins of upper glumes and lower lemmas ciliate-lacerate and winged or pilose.
15. Upper glumes and lower lemmas ciliate-lacerate, winged . . . . 14. *P. fimbriatum*
15. Upper glumes and lower lemmas pilose.
16. Panicle branches 2–7; spikelets 2.3–4 mm long . . . . 15. *P. dilatatum*
16. Panicle branches (4)10–30; spikelets 1.8–2.8 mm long . . . . 16. *P. urvillei*
14. Margins of upper glumes and lower lemmas neither ciliate-laceerate nor winged, glabrous or pubescent, if pubescent then the hairs not pilose, often glandular, papillose-based, or wrinkled.
17. Upper florets olive to dark brown.
18. Plants aquatic, the culms decumbent, rooting at the nodes; lower glumes often present . . . . 17. *P. modestum*
18. Plants not aquatic or, if aquatic, the culms erect; lower glumes absent.
19. Panicle branches 10–28 or more.
20. Plants annual; axes of panicle branches broadly winged, wings about as wide as the central portion . . . . 18. *P. boscianum*
20. Plant perennial; axes of panicle branches narrowly winged, wings narrower than the central portion.
21. Axes of panicle branches 1–1.7 mm wide; spikelets 1.8–2.4 mm wide . . . . 19. *P. virgatum*
21. Axes of panicle branches 0.5–1.2 mm wide; spikelets 1.1–1.8 mm wide . . . . 20. *P. conspersum*
19. Branches 1–10(28).
22. Plants annual.
23. Spikelets 1.3–1.8 mm wide, broadly elliptical to orbicular, glabrous; panicles with 1–10(28) branches, the axes 0.7–2.3 mm wide . . . . 18. *P. boscianum*
23. Spikelets 1.7–2.4 mm wide, broadly obovate, shortly pubescent; panicles with 1–5 branches, the axes 0.8–1.3 mm wide . . . . 21. *P. convexum*
22. Plants perennial.
24. Plants cespitose, rhizomes sometimes present but not well-developed; culms 100–200 cm tall, stout; panicle branches ascending, divaricate, or reflexed.
25. Leaf blades 7–18 mm wide . . . . 20. *P. conspersum*
25. Leaf blades 2.5–4 mm wide . . . . 22. *P. plicatulum*
24. Plants not cespitose, rhizomatous; culms 10–150 cm tall, varying in thickness; panicle branches ascending.
26. Rhizomes short, indistinct . . . . 22. *P. plicatulum*
26. Rhizomes long, evident.
27. Plants aquatic; upper florets chestnut brown . . . . 23. *P. wrightii*
27. Plants not aquatic; upper florets dark brown . . . . 24. *P. nicorae*
17. Upper florets white, stramineous, or golden brown.
28. Lower lemmas with well-developed ribs over the veins; upper glumes absent . . . . 25. *P. malacophyllum*
28. Lower lemmas not ribbed over the veins; upper glumes present.
29. Panicles with 15-100 branches.
30. Plants annual; upper glumes and lower lemmas rugose . . . . 26. *P. racemosum*
30. Plants perennial; upper glumes and lower lemmas smooth.
31. Plants rhizomatous, not cespitose; branch axes 0.9–1.2 mm wide; panicle branches often arcuate . . . . 27. *P. intermedium*
31. Plants cespitose, not rhizomatous; branch axes 0.3–0.6 mm wide; panicle branches straight.
32. Panicle branches spreading to reflexed (rarely ascending); leaf blades 10–23 mm wide; axes of panicle branches 0.3–0.4 mm wide . . . . 28. *P. coryphaeum*

32. Panicle branches erect to ascending; leaf blades 4.9–6.1 mm wide; axes of panicle branches 0.5–0.6 mm wide . . . . 29. *P. quadrifarium*

29. Panicles with 1-15 branches.
  33. Spikelet pairs not imbricate; lower glumes usually present . . . . 30. *P. bifidum*
  33. Spikelet pairs imbricate; lower glumes absent or present.
    34. Spikelets 1.3–2.5 mm long.
      35. Upper glumes, usually also the lower lemmas, shortly pubescent.
        36. Lower glumes present . . . . 31. *P. langei*
        36. Lower glumes absent.
          37. Panicles both terminal and axillary, the axillary panicles partially or completely enclosed by the subtending leaf sheath . . . . 32. *P. setaceum*
          37. Panicles all terminal.
            38. Leaf blades involute; culms 80–110 cm tall . . . . 33. *P. laxum*
            38. Leaf blades flat; culms 20–75 cm tall.
              39. Spikelets 1.3–2 mm long, 0.7–1 mm wide, elliptic; upper glumes and lower lemmas 5-veined; culm bases swollen . . . . 34. *P. caespitosum*
              39. Spikelets 2–2.5 mm long, 1.4–1.6 mm wide, ovate; upper glumes and lower lemmas 3-veined; culm bases not swollen . . . . 35. *P. virletii*
      35. Upper glumes and lower lemmas glabrous.
        40. Panicles both terminal and axillary, the axillary panicles partially or completely enclosed by the subtending leaf sheath . . . . 32. *P. setaceum*
        40. Panicles all terminal.
          41. Upper panicle branches erect . . . . 36. *P. monostachyum*
          41. Upper panicle branches spreading to ascending.
            42. Leaf blades mostly involute; plants of sandy or rocky areas, usually on the coast . . . . 37. *P. pleostachyum*
            42. Leaf blades mostly flat; plants of inland areas or, if coastal, then in marshy areas.
              43. Upper glumes and lower lemmas 3-veined.
                44. Leaf blades usually conduplicate, 2.2–8.3 mm wide . . . . 39. *P. praecox*
                44. Leaf blades usually flat, 5–10 mm wide . . . . . 35. *P. virletii*
              43. Upper glumes and lower lemmas 5-veined.
                45. Axes of panicle branches 0.2–0.5 mm wide; ligules 0.2–0.4 mm long . . . . . . 34. *P. caespitosum*
                45. Axes of panicle branches 1.5–2 mm wide; ligules 2.2–4.7 mm long . . . . 38. *P. lividum*
    34. Spikelets 2.5–4.1 mm long.
      46. Upper glumes, and usually lower lemmas, pubescent.
        47. Lower glumes present . . . . 31. *P. langei*
        47. Lower glumes absent.
          48. Leaf blades 2–5 mm wide; upper glumes and lower lemmas abundantly pubescent, most hairs longer than 0.1 mm; spikelets elliptic . . . . 40. *P. hartwegianum*
          48. Leaf blades 4–18 mm wide; upper glumes and lower lemmas glabrous or sparsely pubescent, the hairs shorter than 0.1 mm; spikelets obovate to elliptic . . . . 41. *P. pubiflorum*
      46. Upper glumes, and usually lower lemmas, glabrous.
        49. Upper florets golden brown.

50. Plants not rhizomatous; culms decumbent and rooting at the lower nodes; spikelets 1.3–1.6 mm wide; lower lemmas 5–7-veined; lower glumes often present . . . . 17. *P. modestum*
50. Plants rhizomatous; culms erect, not rooting at the lower nodes; spikelets 1.9–3.1 mm wide; lower lemmas 3-veined.
    51. Panicle branches 1–6; upper glumes 5-veined; leaf blades 3–18 mm wide . . . . 42. *P. floridanum*
    51. Panicle branches 1–3; upper glumes 3-veined; leaf blades 3–4 mm wide . . . . 43. *P. unispicatum*

49. Upper florets stramineous to pale, but not golden brown.
    52. Terminal panicle branches erect.
        53. Blades involute; upper glumes 1-veined . . . 36. *P. monostachyum*
        53. Blades flat; upper glumes 3-veined . . . . 43. *P. unispicatum*
    52. Terminal panicle branches spreading to ascending.
        54. Spikelets 2.2–2.6 mm long.
            55. Spikelets 1.2–1.5 mm wide, elliptic to obovate . . . . 38. *P. lividum*
            55. Spikelets 2–2.8 mm wide, orbicular to suborbicular . . . . 39. *P. praecox*
        54. Spikelets 2.6–4.1 mm long.
            56. Plants decumbent, rooting at the lower nodes, not rhizomatous; spikelets obovate to elliptic . . . . 41. *P. pubiflorum*
            56. Plants rhizomatous, neither decumbent nor rooting at the lower nodes; spikelets orbicular to elliptic.
                57. Spikelets 2.1–3.1 mm long, 2–2.8 mm wide, orbicular to suborbicular; upper glumes 3-veined; leaf blades conduplicate . . . . 39. *P. praecox*
                57. Spikelets 2.9–4.1 mm long, 1.9–3.1 mm wide, suborbicular to elliptic; upper glumes 5-veined; leaf blades flat . . . . 42. *P. floridanum*

1. **Paspalum repens** P.J. Bergius [p. 573]
WATER PASPALUM

Plants annual; aquatic, floating or rhizomatous. **Culms** 4–55 cm, erect; **nodes** pubescent. **Sheaths** glabrous or pubescent; **ligules** 1–4 mm; **blades** 10–40 cm long, 8–22 mm wide, flat, glabrous or sparsely pubescent. **Panicles** terminal, with (7)20–70 racemosely arranged branches; **branches** 1.2–9.5 cm, diverging to spreading, occasionally arcuate, disarticulating at maturity; **branch axes** 0.7–1.5 mm wide, broadly winged, glabrous, margins scabrous, extending beyond the distal spikelet. **Spikelets** 1.1–1.9 mm long, 0.5–0.8 mm wide, solitary, appressed to the branch axes, elliptic, pubescent, white. **Lower glumes** absent; **upper glumes** and **lower lemmas** veinless; **upper florets** white. **Caryopses** 0.8–0.9 mm, translucent, white. $2n = 20$.

Paspalum repens is a native species that grows along the edges of lakes, streams, and roadside ditches in the southeastern United States. Its range extends through tropical America to Peru, Bolivia, and Argentina.

2. **Paspalum scrobiculatum** L. [p. 573]
INDIAN PASPALUM

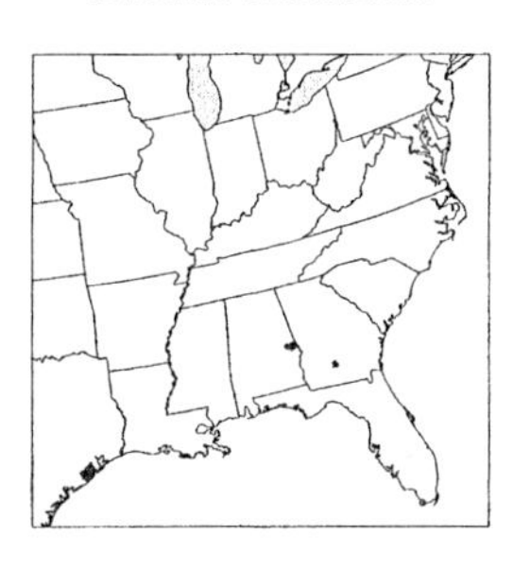

Plants annual. **Culms** 10–150 cm, erect or decumbent; **nodes** glabrous. **Sheaths** glabrous; **ligules** 0.3–1.2 mm, often with a row of hairs behind them; **blades** 5–30 cm long, 2–8(12) mm wide, flat, usually glabrous. **Panicles** terminal, with 1–5 digitately or racemosely arranged branches;

branches 3–10 cm, diverging to spreading, persistent; **branch axes** 1.5–3 mm wide, broadly winged, glabrous, margins scabrous, terminating in a spikelet. **Spikelets** 1.8–3.2 mm long, 2–2.3 mm wide, solitary, diverging from the branch axes, ovate, glabrous, olive green to dark, glossy brown. **Lower glumes** absent; **upper glumes** as long as the lower lemmas, 5–7-veined; **lower lemmas** 3–5-veined; **upper florets** 2.5–3 mm long, 1.4–1.8 mm wide, dark glossy brown. **Caryopses** 1.1–1.5 mm, nearly orbicular. $2n$ = 20, 40, 60, 120.

Paspalum scrobiculatium is native to India. It has been found growing in widely scattered disturbed areas of the southeastern United States, possibly as an escape from cultivation. It is grown as a cereal (*Kodo*) in India.

3. **Paspalum laeve** Michx. [p. 573]
FIELD PASPALUM

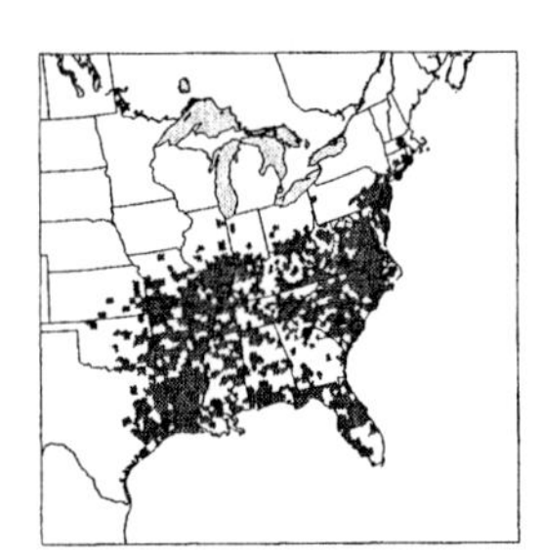

Plants perennial; shortly rhizomatous. **Culms** 40–120 cm, erect; **nodes** glabrous or pubescent. **Sheaths** glabrous or pubescent; **ligules** 1.5–3.8 mm; **blades** to 37 cm long, 2–9.3 mm wide, flat, glabrous or pubescent. **Panicles** terminal, with 1–6 racemosely arranged branches; **branches** 2–10.9 cm, diverging to spreading (rarely erect), persistent; **branch axes** 0.6–1.3 mm wide, glabrous, margins scabrous, terminating in a spikelet. **Spikelets** 2.3–3.3 mm long, 2–2.7 mm wide, solitary, appressed to the branch axes, elliptic to obovate or nearly orbicular, glabrous, stramineous. **Lower glumes** absent; **upper glumes** 3-veined, **lower lemmas** 5-veined; **upper florets** pale to stramineous. **Caryopses** about 2 mm, white to yellow-brown. $2n$ = 20, 58, 70, 80.

Paspalum laeve is restricted to the eastern United States. It grows at the edges of forests and in disturbed areas.

4. **Paspalum dissectum** (L.) L. [p. 574]
MUDBANK PASPALUM

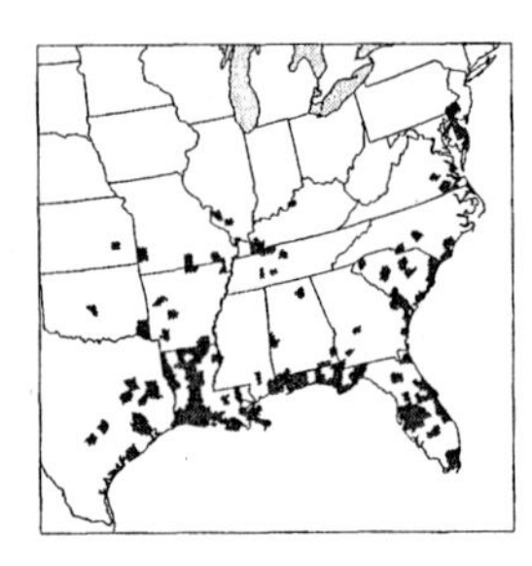

Plants perennial; rhizomatous. **Culms** 10–50 cm, decumbent; **nodes** glabrous or pubescent. **Sheaths** glabrous; **ligules** 2–2.5 mm; **blades** to 12 cm long, 1.3–4.8 mm wide, flat. **Panicles** terminal, with 2–6 racemosely arranged branches; **branches** 1.3–5.3 cm, diverging to erect, often arcuate, persistent; **branch axes** 1.8–3 mm wide, broadly winged, usually conduplicate, glabrous, margins scabrous, terminating in a spikelet. **Spikelets** 1.7–2.1 mm long, 1.1–1.4 mm wide, solitary, appressed to the branch axes, elliptic to ovate, glabrous, stramineous. **Lower glumes** absent; **upper glumes** and **lower lemmas** 5-veined; **upper florets** stramineous, lemmas glabrous throughout. **Caryopses** 1–1.3 mm, white. $2n$ = 40, 60.

Paspalum dissectum grows at the edges of lakes, ponds, rice fields, and wet roadside ditches. It is native to the eastern portion of the contiguous United States and to Cuba.

5. **Paspalum acuminatum** Raddi [p. 574]
BROOK PASPALUM, CANOEGRASS

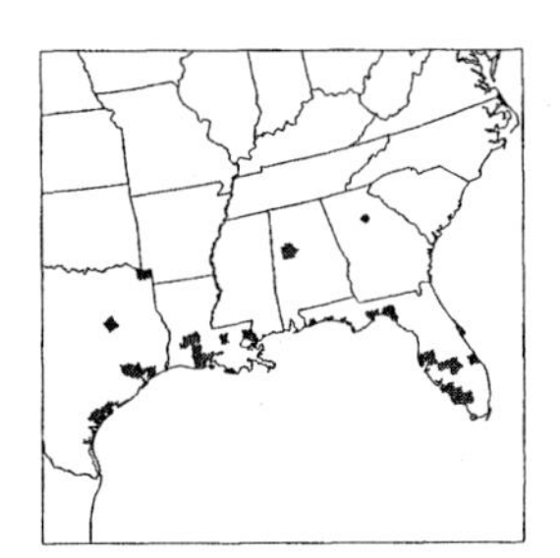

Plants perennial; rhizomatous. **Culms** 30–100 cm, strongly decumbent, upright portion usually not standing more than 20 cm tall, much branched; **nodes** glabrous. **Sheaths** glabrous; **ligules** 1–2.4 mm; **blades** to 7 cm long, 3–6.5 mm wide, flat. **Panicles** terminal, with 2–5 racemosely arranged branches; **branches** 2–6 cm, diverging, persistent; **branch axes** 2–3.3 mm wide, broadly winged, glabrous, margins scabrous, terminating in a spikelet. **Spikelets** 3.2–4 mm long, 1.6–1.7 mm wide, solitary, appressed to the branch axes, elliptic, abruptly pointed, stramineous. **Lower glumes** absent; **upper glumes** and **lower lemmas** glabrous, 5-veined; **upper florets** stramineous, lemmas with a few minute hairs at the apices. **Caryopses** 2–3 mm, white. $2n$ = 40.

Paspalum acuminatum grows at the edges of lakes, ponds, rice fields, and wet roadside ditches. It is native to the Americas, with a range that extends from the southern United States to Argentina.

6. **Paspalum conjugatum** P.J. Bergius [p. 574]
SOUR PASPALUM

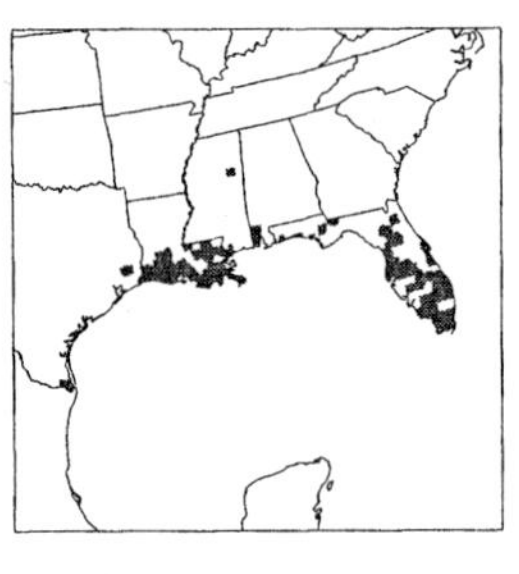

Plants perennial; stoloniferous. **Culms** 15–80 cm, erect; **nodes** glabrous. **Sheaths** glabrous, pubescent distally; **ligules** 0.5–0.8 mm; **blades** 7–23 cm long, 1.5–8 mm wide, flat. **Panicles** terminal, usually composed of a pair of branches, a third branch sometimes present below the terminal pair; **branches** 2.5–12.7 cm, diverging to spreading, often arcuate, persistent; **branch axes** 0.2–0.8 mm wide, glabrous, margins scabrous, terminating in a reduced spikelet. **Spikelets** 1.3–1.9 mm long, 0.8–1.1 mm wide, solitary, appressed to the branch axes, ovate, stramineous. **Lower glumes** absent; **upper glumes** pilose on the margins, veinless or 2–3-veined; **lower lemmas** glabrous, veinless or 2–3-veined; **upper florets** whitish to golden yellow.

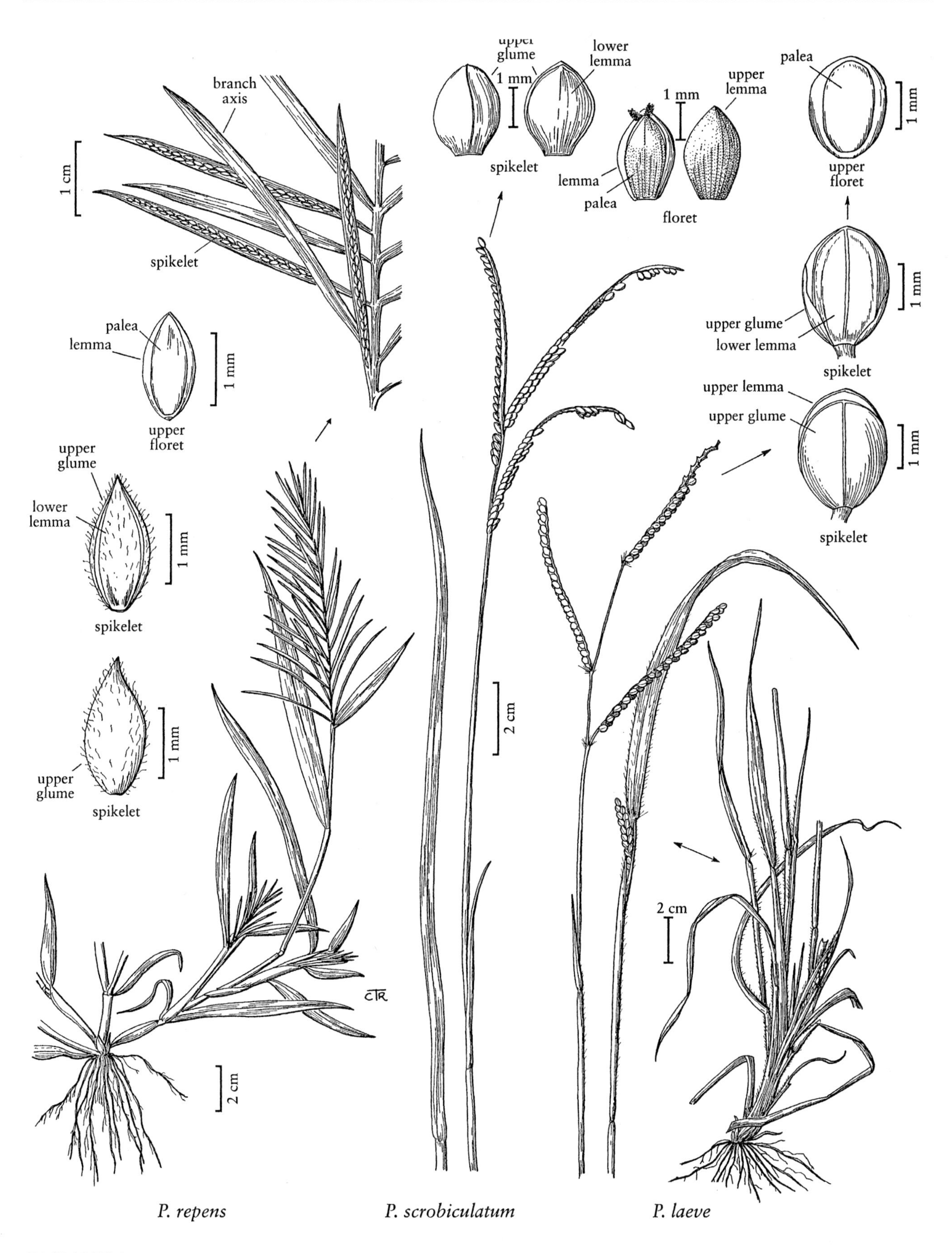
branch
axis
1 cm
spikelet
palea
lemma
1 mm
upper
floret
upper
glume
lower
lemma
1 mm
spikelet
1 mm
upper
glume
spikelet
2 cm
CTR
upper
glume
lower
lemma
1 mm
spikelet
1 mm
upper
lemma
lemma
palea
floret
2 cm
palea
1 mm
upper
floret
1 mm
upper glume
lower lemma
spikelet
upper lemma
upper glume
1 mm
spikelet
2 cm
P. repens
P. scrobiculatum
P. laeve

PASPALUM

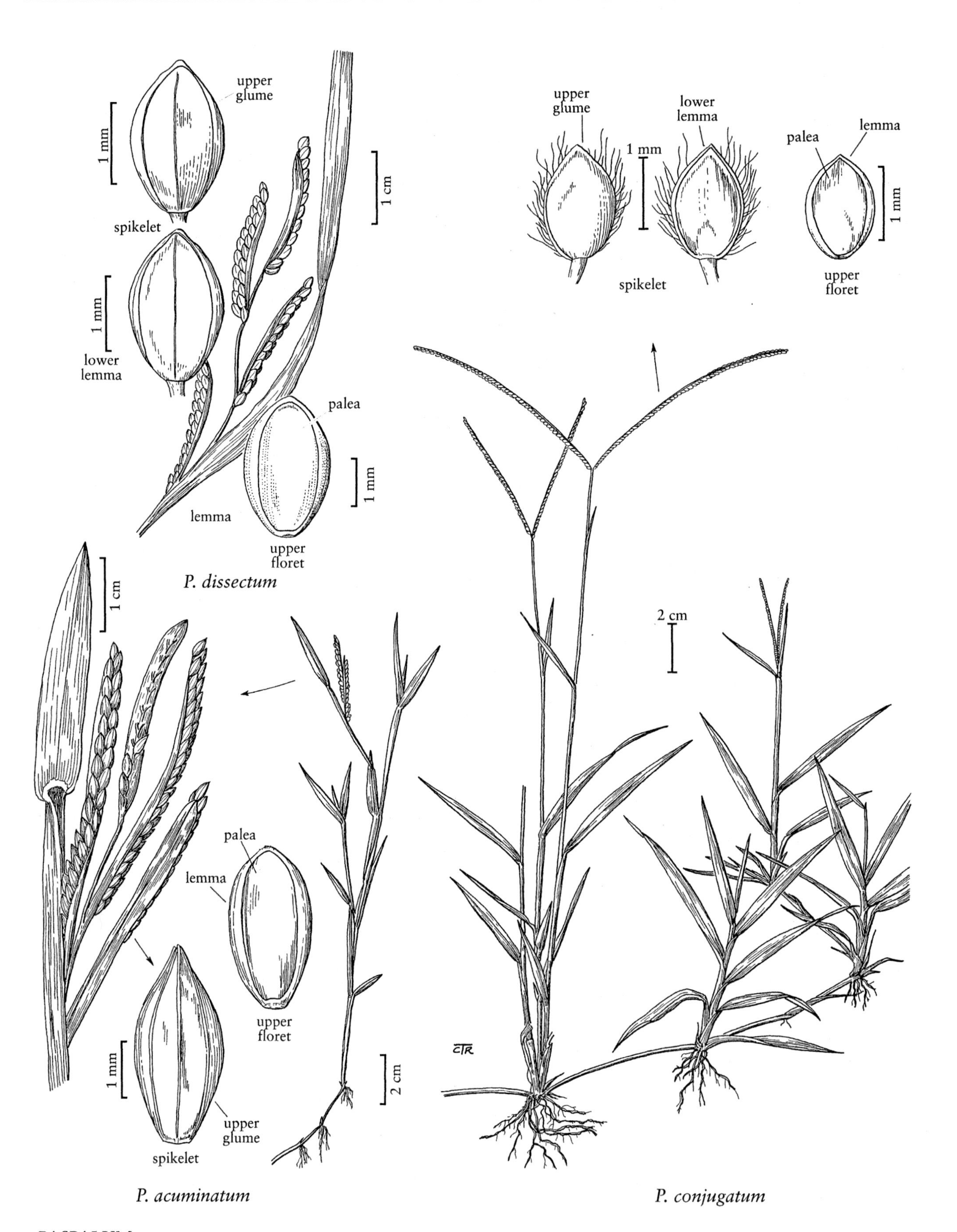
upper
glume
1 mm
spikelet
1 mm
lower
lemma
1 cm
palea
1 mm
lemma
upper
floret
P. dissectum
upper
glume
lower
lemma
1 mm
palea
lemma
1 mm
spikelet
upper
floret
1 cm
palea
lemma
upper
floret
1 mm
upper
glume
spikelet
2 cm
2 cm
CTR
P. acuminatum
P. conjugatum

Caryopses 0.9–1.1 mm, white to yellow. $2n$ = 18, 20, 40, 80.

Pasapalum conjugatum is native to tropical and subtropical regions of both the Western and Eastern hemispheres, including the *Flora* region. It grows in disturbed areas and at the edges of forests, and is sometimes used as a lawn grass.

7. **Paspalum distichum** L. [p. 576]
KNOTGRASS, THOMPSONGRASS

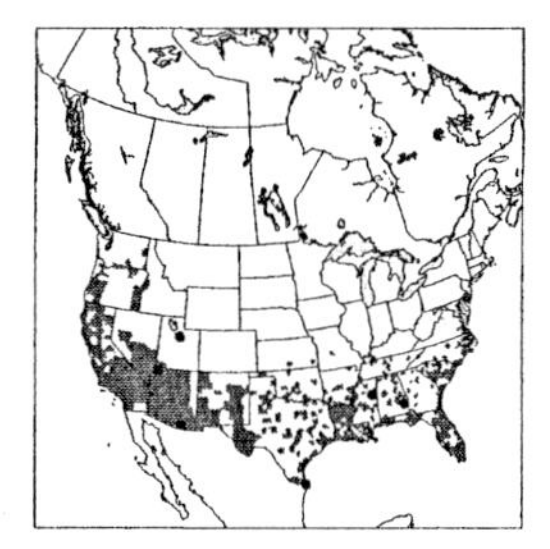

Plants perennial; rhizomatous or cespitose. **Culms** 5–65 cm, erect; **nodes** glabrous. **Sheaths** glabrous, sparsely long pubescent distally; **ligules** 1–2 mm; **blades** to 14 cm long, 1.8–11.5 mm wide, flat or conduplicate, glabrous or pubescent, apices involute. **Panicles** terminal, usually composed of a digitate pair of branches, a third branch sometimes present below; **branches** 1.4–7 cm, diverging, often arcuate; **branch axes** 1.2–2.2 mm wide, winged, glabrous, margins scabrous, terminating in a spikelet. **Spikelets** 2.4–3.2 mm long, 1.1–1.6 mm wide, solitary (rarely paired), appressed to the branch axes, broadly elliptic, stramineous, sometimes partially purple. **Lower glumes** absent or, if present, to 1 mm and triangular; **upper glumes** sparsely and shortly pubescent on the back, 3-veined; **lower lemmas** glabrous, 3-veined; **upper florets** stramineous. **Caryopses** 1.9–2.1 mm, yellow. $2n$ = 20, 30, 40, 48, 60, 61.

Paspalum distichum grows on the edges of lakes, ponds, rice fields, and wet roadside ditches. It is native in warm regions throughout the world, being most abundant in humid areas. In the Western Hemisphere, it grows from the United States to Argentina and Chile.

8. **Paspalum vaginatum** Sw. [p. 576]
SEASHORE PASPALUM

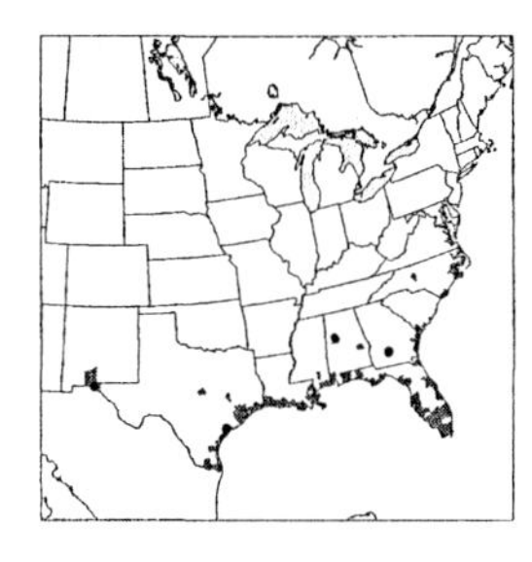

Plants perennial; rhizomatous and/or stoloniferous. **Culms** 10–79 cm, erect; **nodes** glabrous. **Sheaths** glabrous, sparsely long pubescent distally; **ligules** 1–2 mm; **blades** 10–19 cm long, 1.4–8 mm wide, flat or conduplicate, glabrous or pubescent, apices involute. **Panicles** terminal, usually composed of a digitate pair of branches, a third branch sometimes present below; **branches** 1.1–7.9 cm, diverging to erect; **branch axes** 0.4–1.4 mm wide, winged, glabrous, margins scabrous, terminating in a spikelet. **Spikelets** 3–4.5 mm long, 1.1–2 mm wide, solitary, appressed to the branch axes, elliptic-lanceolate, glabrous, light stramineous, apices acute to acuminate. **Lower glumes** absent (rarely present); **upper glumes** and **lower lemmas** glabrous, 3-veined; **upper florets** white. **Caryopses** 2.8–3.1 mm, yellow. $2n$ = 20, 40, 60.

Paspalum vaginatum grows in brackish and salt marshes. It is native to warm, coastal regions around the world, including the Americas. It has been grown for turf and in lawn trials, but is not yet widely used for these purposes.

9. **Paspalum almum** Chase [p. 576]
COMB'S PASPALUM

Plants perennial; cespitose, shortly rhizomatous. **Culms** 10–50 cm, erect. **Sheaths** glabrous or sparsely pubescent; **ligules** 0.5–2 mm; **blades** to 20 cm long, 1.5–3.8 mm wide, flat, pubescent. **Panicles** terminal, usually composed of a digitate pair of branches, 1–5 additional branches sometimes present below; **branches** 1.8–7.1 cm, diverging to erect; **branch axes** 0.8–1.3 mm wide, winged, terminating in a spikelet. **Spikelets** 3–3.6 mm long, 1.3–1.8 mm wide, solitary (rarely paired), appressed to the branch axes, elliptic, glabrous, apices acute to acuminate. **Lower glumes** absent; **upper glumes** and **lower lemmas** glabrous, 5-veined, margins flat; **upper florets** stramineous to golden brown. $2n$ = 12, 24.

Paspalum almum was probably introduced to North America as a forage species. Its native range is Brazil, Paraguay, Uruguay, and eastern Argentina. It has not been reported from Mexico or Central America. In the *Flora* region, it is found along roadsides and in pastures of southeastern Texas and southern Louisiana.

10. **Paspalum notatum** Flüggé [p. 578]
BAHIAGRASS

Plants perennial; rhizomatous. **Culms** 20–110 cm, erect; **nodes** glabrous. **Sheaths** glabrous or pubescent; **ligules** 0.2–0.5 mm; **blades** 5–31 cm long, 2–10 mm wide, flat or conduplicate, glabrous or pubescent. **Panicles** terminal, usually composed of a digitate pair of branches, 1–3 additional branches sometimes present below the terminal pair; **branches** 3–15 cm, diverging to erect; **branch axes** 0.7–1.8 mm wide, narrowly winged, glabrous, margins scabrous, terminating in a spikelet, distal spikelets sometimes reduced. **Spikelets** 2.5–4 mm long, 2–2.8 mm wide, solitary, appressed to the branch axes, broadly elliptic to ovate or obovate, glabrous, light stramineous to white,

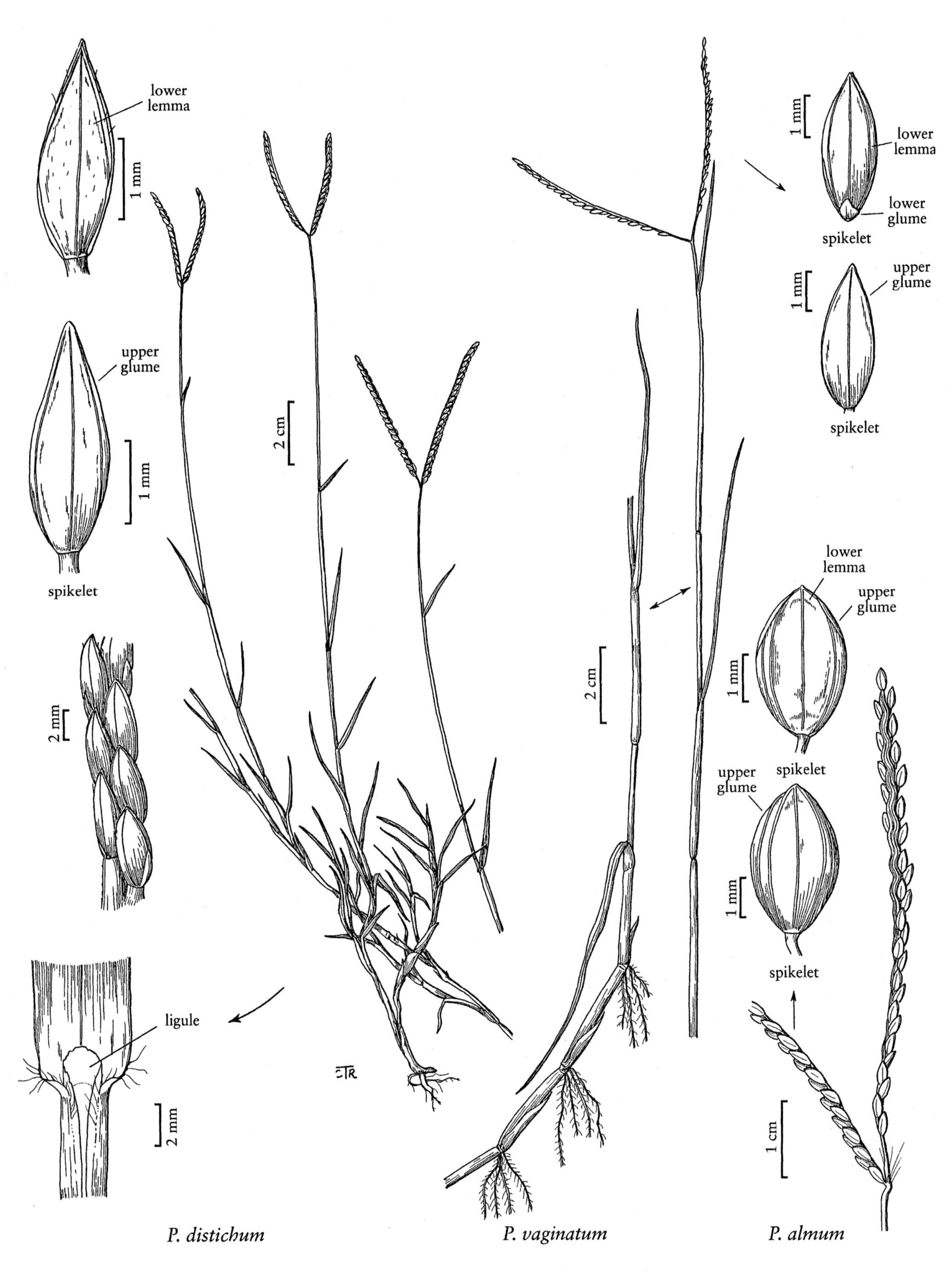
lower
lemma
1 mm
upper
glume
1 mm
spikelet
2 mm
ligule
2 mm
2 cm
ETR
2 cm
1 mm
lower
lemma
lower
glume
spikelet
1 mm
upper
glume
spikelet
lower
lemma
upper
glume
1 mm
spikelet
upper
glume
1 mm
spikelet
1 cm
P. distichum
P. vaginatum
P. almum

apices obtuse to broadly acute. **Lower glumes** absent; **upper glumes** glabrous, 5-veined; **lower lemmas** 5-veined, margins inrolled; **upper florets** light yellow to white. **Caryopses** 2–3 mm, white. 2*n* = 20, 30, 40.

Paspalum notatum is native from Mexico through the Caribbean and Central America to Brazil and northern Argentina. It was introduced to the United States for forage, turf, and erosion control. It is now established, generally being found in disturbed areas and at the edges of forests in the southeastern United States.

Paspalum notatum is sometimes treated as having distinct varieties. They are not recognized here because the variation among them is continuous. A number of cultivars have been developed for use as turf grasses; among these cultivars are 'Common Bahiagrass', 'Pensacola Bahiagrass', and 'Argentine Bahiagrass'.

11. **Paspalum minus** E. Fourn. [p. 578]
MATTED PASPALUM

Plants perennial; shortly rhizomatous. **Culms** 3–60 cm, erect; **nodes** glabrous. **Sheaths** glabrous or pubescent; **ligules** 0.2–0.7 mm; **blades** 8–18 cm long, 2–7.1 mm wide, flat, glabrous or pubescent. **Panicles** terminal, usually composed of a digitate pair of branches, a third branch sometimes present below the terminal pair; **branches** 1.8–6.4 cm, diverging to erect; **branch axes** 0.5–1.3 mm wide, narrowly winged, glabrous, margins scabrous, terminating in a spikelet. **Spikelets** 1.9–2.3 mm long, 1.2–2 mm wide, solitary, appressed to the branch axes, broadly elliptic to ovate to obovate, glabrous, stramineous, apices obtuse. **Lower glumes** absent; **upper glumes** 3-veined, **lower lemmas** faintly 3-veined; **upper florets** stramineous. **Caryopses** 1.8–2.2 mm, white. 2*n* = 20, 40, 50.

Paspalum minus grows in disturbed areas and on the edges of forests. It grows from southern Texas to Florida in the *Flora* region; outside the region, it extends through Mexico and the West Indies to Peru, Bolivia, Brazil, and Paraguay.

12. **Paspalum blodgettii** Chapm. [p. 578]
CORAL PASPALUM

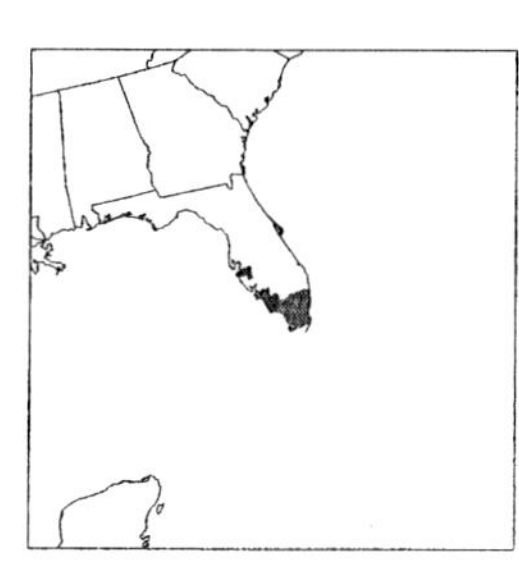

Plants perennial; cespitose, bulbous; scales pubescent. **Culms** 40–100 cm, erect; **nodes** glabrous. **Sheaths** pubescent or glabrous; **ligules** 0.2–0.4 mm; **blades** 5–27 cm long, 1.9–8 mm wide, flat, glabrous, pubescent behind the ligules, margins scabrous, often ciliate basally. **Panicles** terminal, with 2–6 racemosely arranged branches; **branches** 1.5–7.5 cm, diverging to spreading; **branch axes** 0.5–0.8 mm wide, narrowly winged, terminating in a spikelet. **Spikelets** 1–1.3 mm long, 0.7–0.9 mm wide, paired, appressed to the branch axes, elliptic to elliptic-obovate, glandular pubescent, stramineous to light or golden brown. **Lower glumes** absent; **upper glumes** and **lower lemmas** 3-veined; **upper florets** 0.8–1.1 mm, stramineous. **Caryopses** 0.9–1.1 mm in diameter, orbicular, amber. 2*n* = 40.

Paspalum blodgettii grows in hammocks, low pinelands, and along roadsides in southern peninsular Florida, the Bahamas, the Greater Antilles, southeastern Mexico, and Belize.

13. **Paspalum paniculatum** L. [p. 578]
ARROCILLO

Plants perennial; cespitose or rhizomatous. **Culms** to 100 cm, erect; **nodes** pubescent. **Sheaths** pubescent; **ligules** 0.2–0.5 mm; **blades** 12–35 cm long, 10–24 mm wide, flat, scabrous, pubescent near the margins, margins usually undulate. **Panicles** terminal, with 18–50 racemosely arranged branches; **branches** 0.8–8.9 cm, spreading to diverging, often arcuate; **branch axes** 0.2–0.5 mm wide, narrowly winged, scabrous, terminating in a spikelet. **Spikelets** 1.1–1.3 mm long, 0.9–1 mm wide, paired, diverging from the branch axes, ovate, light brown to stramineous. **Lower glumes** absent; **upper glumes** and **lower lemmas** pubescent, 3-veined; **upper florets** 1.1–1.3 mm, stramineous. **Caryopses** 0.7–0.8 mm, light brown. 2*n* = 20, 40, 60.

Paspalum paniculatum is native from Mexico and the West Indies to Argentina. It is now established in Mississippi and southern Florida, growing in disturbed areas.

14. **Paspalum fimbriatum** Kunth [p. 580]
WINGED PASPALUM, PANAMA CROWNGRASS

Plants annual. **Culms** 15–70 cm, erect; **nodes** glabrous or pubescent. **Sheaths** pubescent, sometimes sparsely so; **ligules** 1–1.9 mm; **blades** to 35 cm long, 1.9–16.2 mm wide, flat, sparsely pubescent on both surfaces, margins ciliate basally. **Panicles** terminal, with 2–8 racemosely arranged branches; **branches** 2–6.4 cm, diverging to erect; **branch axes** 0.9–1.6 mm wide, winged, glabrous, margins scabrous, terminating in a spikelet. **Spikelets** 2.5–3.5 mm long, 2.4–3 mm wide,

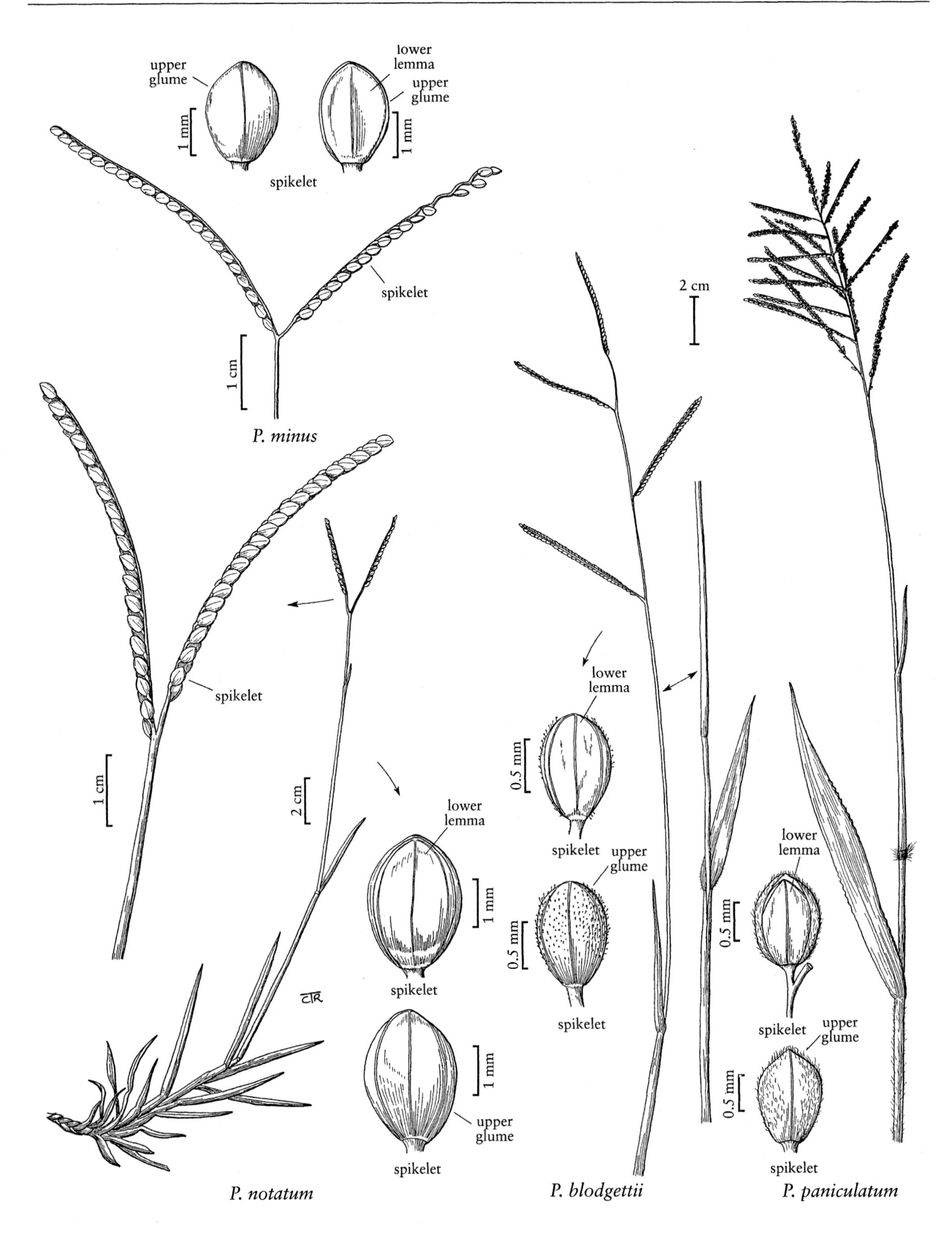
upper glume
lower lemma
upper glume
1 mm
1 mm
spikelet
spikelet
1 cm
P. minus
spikelet
1 cm
2 cm
lower lemma
1 mm
spikelet
1 mm
upper glume
spikelet
CTR
P. notatum
2 cm
lower lemma
0.5 mm
spikelet
upper glume
0.5 mm
spikelet
P. blodgettii
lower lemma
0.5 mm
spikelet
upper glume
0.5 mm
spikelet
P. paniculatum

PASPALUM

paired, appressed to the branch axes, suborbicular, stramineous. **Lower glumes** absent; **upper glumes** and **lower lemmas** 1.9–2.1 mm, ovate, winged, 1-veined, margins ciliate-lacerate; **upper florets** 1.7–1.9 mm, stramineous. **Caryopses** 0.9–1.1 mm, orbicular, white. $2n$ = 20.

Paspalum fimbriatum has probably been introduced into the United States. Its primary range extends from southern Mexico to Colombia, Venezuela, and French Guiana. In the *Flora* region, it grows in disturbed areas of Florida.

15. **Paspalum dilatatum** Poir. [p. 580]
DALLISGRASS

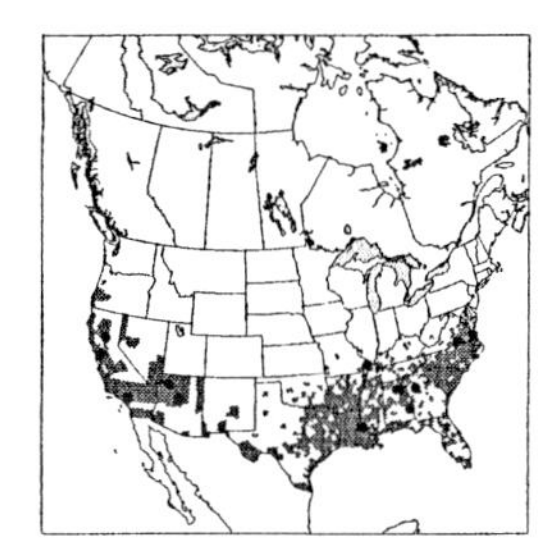

Plants perennial; cespitose, rhizomatous, rhizomes short (less than 1 cm), forming a knotty base. **Culms** 50–175 cm, erect; **nodes** glabrous. **Sheaths** glabrous or pubescent, lower sheaths more frequently pubescent than the upper sheaths; **ligules** 1.5–3.8 mm; **blades** to 35 cm long, 2–16.5 mm wide, flat, mostly glabrous, adaxial surfaces with a few long hairs near the base. **Panicles** terminal, with 2–7 racemosely arranged branches; **branches** 1.5–12 cm, racemose, divergent; **branch axes** 0.7–1.4 mm wide, winged, glabrous, margins scabrous, terminating in a spikelet. **Spikelets** 2.3–4 mm long, 1.7–2.5 mm wide, paired, appressed to the branch axes, ovate, tapering to an acute apex, stramineous (rarely purple). **Lower glumes** absent; **upper glumes** and **lower lemmas** 5–7-veined, margins pilose; **upper florets** stramineous. **Caryopses** 2–2.3 mm, white to brown. $2n$ = 20, 40, 50–63.

Paspalum dilatatum is native to Brazil and Argentina. It is now well established in the *Flora* region, generally as a weed in waste places. It is also used as a turf grass.

16. **Paspalum urvillei** Steud. [p. 580]
VASEYGRASS

Plants perennial; cespitose, with a knotty base composed of very short (less than 1 cm) rhizomes. **Culms** 50–220 cm, erect; **nodes** glabrous or pubescent. **Sheaths** glabrous or pubescent; **ligules** 1–4(7.7) mm; **blades** 12–60 cm long, 2–12 mm wide, flat, mostly glabrous, a few long hairs near the base of the adaxial surface. **Panicles** terminal, with (4)10–30 racemosely arranged branches; **branches** 1.2–11.5 cm, divergent; **branch axes** 0.5–1.1 mm wide, winged, glabrous, margins scabrous, terminating in a spikelet. **Spikelets** 1.8–2.8 mm long, 1.1–1.5 mm wide, paired, appressed to the branch axes, elliptic to slightly obovate, stramineous (rarely purple). **Lower glumes** absent; **upper glumes** and **lower lemmas** 3-veined, margins pilose; **upper florets** stramineous. **Caryopses** 1.2–1.7 mm, white. $2n$ = 40.

Paspalum urvillei has been introduced to the United States from South America. In the *Flora* region it grows in disturbed, moist to wet areas, primarily in the southeastern United States.

17. **Paspalum modestum** Mez [p. 582]
WATER PASPALUM

Plants perennial; usually sprawling, occasionally cespitose. **Culms** 30–110 cm, decumbent and rooting at the lower nodes; **nodes** glabrous. **Sheaths** glabrous; **ligules** 1–2.3 mm; **blades** to 50 cm long, 2–10 mm wide, flat, glabrous or pubescent. **Panicles** terminal, with 2–6(10) racemosely arranged branches; **branches** 3.5–12.5 cm, diverging to erect; **branch axes** 1–2.1 mm wide, glabrous, terminating in a spikelet. **Spikelets** 2.5–3 mm long, 1.3–1.6 mm wide, paired, appressed to the branch axes, elliptic, light brown. **Lower glumes** often present, 0.5–2 mm, brown; **upper glumes** glabrous, 5-veined, margins entire, **lower lemmas** glabrous, 5–7-veined, margins entire; **upper florets** olive, golden brown, or dark brown. **Caryopses** 1.6–1.8 mm, brown. $2n$ = 20, 30, 40.

Paspalum modestum grows in wet roadside ditches and rice fields of Texas and southern Louisiana. It was introduced to the United States from South America. Plants with pale florets may key to *P. lividum*, which differs from *P. modestum* in having shorter ligules.

Until recently, plants belonging to *Paspalum modestum* have been called **P. hydrophilum** Henrard in North America, but experimental studies have shown that the two species are quite distinct and that North American plants belong to *P. modestum*.

18. **Paspalum boscianum** Flüggé [p. 582]
BULL PASPALUM

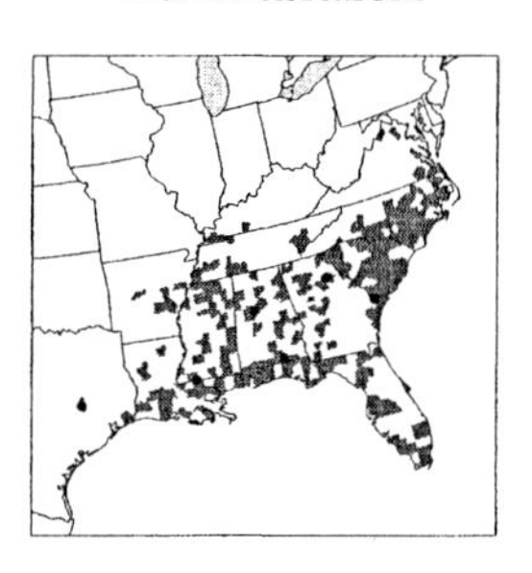

Plants annual. **Culms** 15–96 cm, erect or prostrate, often rooting at the lower nodes; **nodes** glabrous. **Sheaths** glabrous; **ligules** 1–3.2 mm; **blades** to 56 cm long, 2.2–15 mm wide, flat. **Panicles** terminal, with 1–10(28) racemosely arranged branches; **branches** 1.2–8.2 cm, diverging; **branch axes** 0.7–2.3 mm wide, glabrous, broadly

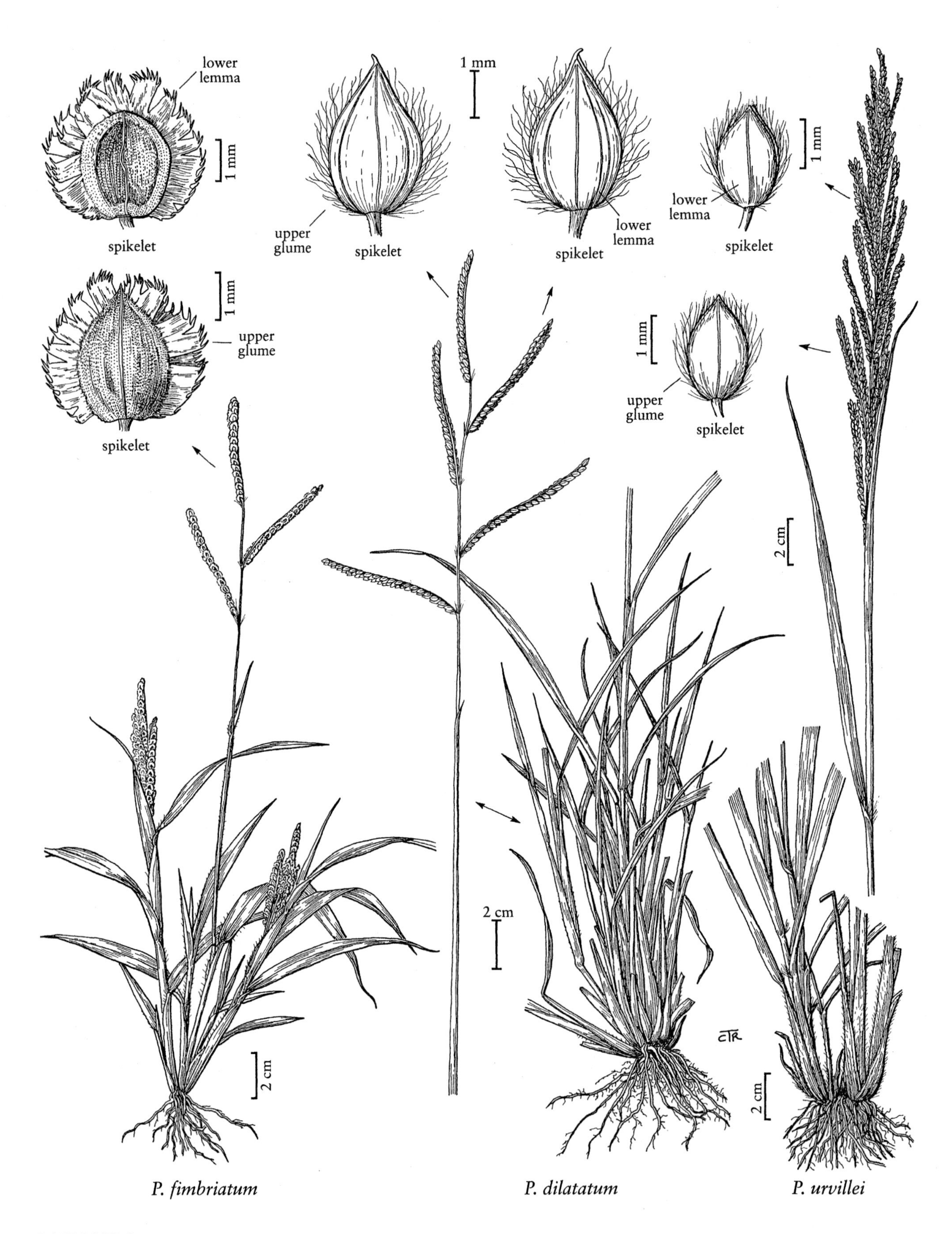

lower lemma
1 mm
spikelet
1 mm
upper glume
spikelet
1 mm
upper glume
spikelet
lower lemma
spikelet
1 mm
lower lemma
spikelet
1 mm
upper glume
spikelet
2 cm
2 cm
2 cm
2 cm
CTR
P. fimbriatum
P. dilatatum
P. urvillei

winged, wings about as wide as the central portion, margins scabrous, terminating in a spikelet. **Spikelets** 2–2.2 mm long, 1.3–1.8 mm wide, paired, appressed to the branch axes, glabrous, broadly elliptic, obovate, or orbicular, light to dark brown. **Lower glumes** absent; **upper glumes** glabrous, 5-veined, margins entire; **lower lemmas** glabrous, 3–5-veined, margins entire; **upper florets** dark glossy brown. **Caryopses** 1.4–1.6 mm, white. 2*n* = 40.

Paspalum boscianum grows in moist to dry, disturbed areas, and at the edges of forests. It is native from the southeastern United States through the West Indies and Mexico to Brazil. The California record came from a weed in a rice field.

19. **Paspalum virgatum** L. [p. 582]
TALQUEZAL

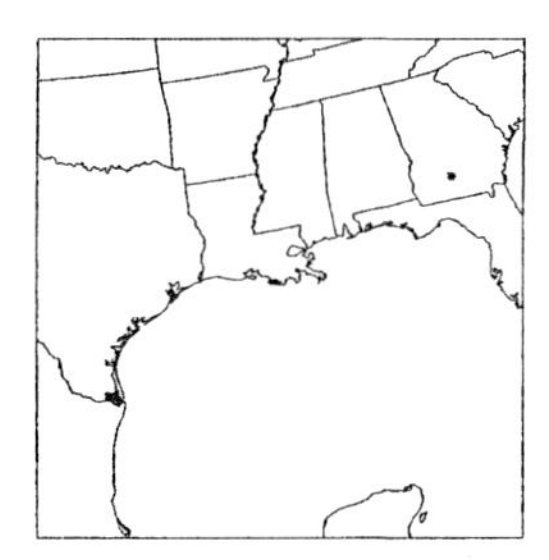

Plants perennial; cespitose. **Culms** 100–200 cm, stout, erect; **nodes** glabrous. **Sheaths** pubescent; **ligules** 1.9–2.2 mm, brown; **blades** 30–90 cm long, 1–3 cm wide, flat, glabrous, pubescent behind the ligules. **Panicles** terminal, with 10–20 racemosely arranged branches; **branches** 3–15 cm, spreading to diverging; **branch axes** 1–1.7 mm wide, winged, wings narrower than the central section, terminating in a spikelet. **Spikelets** 2.2–3.2 mm long, 1.8–2.4 mm wide, paired, appressed to or diverging from the branch axes, obovate, brown. **Lower glumes** absent; **upper glumes** and **lower lemmas** glabrous or variously short pubescent, 5-veined, margins entire; **upper florets** 2.5–2.7 mm, brown. 2*n* = 36, 40, 54, 80.

Paspalum virgatum is native from Mexico to South America. It has been introduced to the southeastern United States, where it grows primarily in disturbed areas and cultivated fields.

20. **Paspalum conspersum** Schrad. [p. 583]
SCATTERED PASPALUM

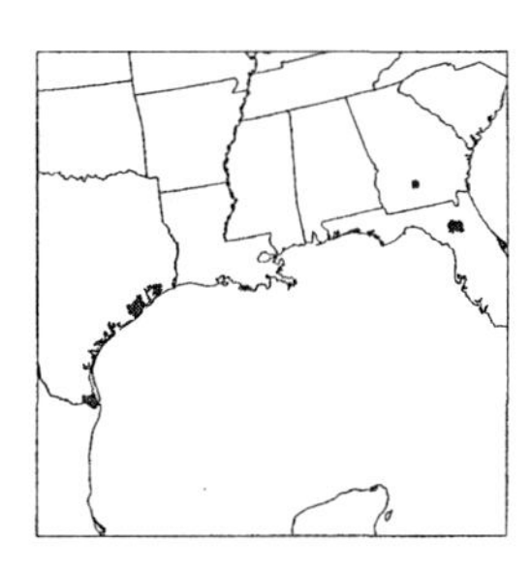

Plants perennial; cespitose. **Culms** 100–200 cm, stout, erect; **nodes** glabrous. **Sheaths** glabrous or sparsely pubescent; **ligules** 1–2 mm, brown; **blades** to 50 cm long, 7–18 mm wide, flat, glabrous or pubescent, margins scabrous, ciliate. **Panicles** terminal, with 4–13 racemosely arranged branches; **branches** 6–11 cm, diverging, divaricate, or reflexed; **branch axes** 0.5–1(1.2) mm wide, winged, wings narrower than the central section, terminating in a spikelet. **Spikelets** 2–2.7(3) mm long, 1.1–1.8 mm wide, paired, appressed to or diverging from the branch axes, elliptic to obovate, pubescent, brown. **Lower glumes** absent; **upper glumes** densely short pubescent, hairs about 0.5 mm; **lower lemmas** glabrous or sparsely short pubescent, margins entire; **upper florets** 1.8–2.2 mm, pubescent, brown. 2*n* = 40, 60.

Paspalum conspersum is native from Mexico to Argentina, but it has been introduced to the southern United States. It is grown for its forage value, and has become established at scattered locations from Texas to Florida, growing along roadsides and in other disturbed areas.

21. **Paspalum convexum** Humb. & Bonpl. *ex* Flüggé [p. 583]
MEXICAN PASPALUM

Plants annual. **Culms** 10–53 cm, erect; **nodes** glabrous. **Sheaths** pubescent or glabrous; **ligules** 2–4.1 mm; **blades** 5–25(80) cm long, 2.9–10.2(12) mm wide, flat. **Panicles** terminal, with 1–5 racemosely arranged branches; **branches** 1.1–5.4(7) cm, divergent; **branch axes** 0.8–1.3 mm wide, not or narrowly winged, glabrous, terminating in a spikelet. **Spikelets** 2.1–2.6 mm long, 1.7–2.4 mm wide, paired, appressed to the branch axes, broadly obovate to suborbicular, shortly pubescent, light to dark brown. **Lower glumes** absent; **upper glumes** and **lower lemmas** shortly pubescent, 5–7-veined, margins entire; **lower paleas** rarely present; **upper florets** dark glossy brown. **Caryopses** 1.3–1.5 mm, white. 2*n* = 30, 32, 40, 60.

Paspalum convexum grows in disturbed areas in the southern United States. It is native from Mexico and the Caribbean Islands to Brazil. It is not considered to have particularly high forage value.

22. **Paspalum plicatulum** Michx. [p. 583]
BROWNSEED PASPALUM

Plants perennial; shortly rhizomatous, often indistinctly so. **Culms** 30–110 cm, stout, erect; **nodes** glabrous. **Sheaths** glabrous; **ligules** 2–3 mm; **blades** to 35 cm long, 2–5.4 mm wide, conduplicate (rarely flat). **Panicles** terminal, with 2–7 racemosely arranged branches; **branches** 1.6–7.1 cm, usually divergent, rarely merely ascending; **branch axes** 0.6–1.1 mm wide, glabrous, terminating in a spikelet. **Spikelets** 2.5–3 mm long, 1.5–2.2 mm wide, paired, appressed to the branch axes,

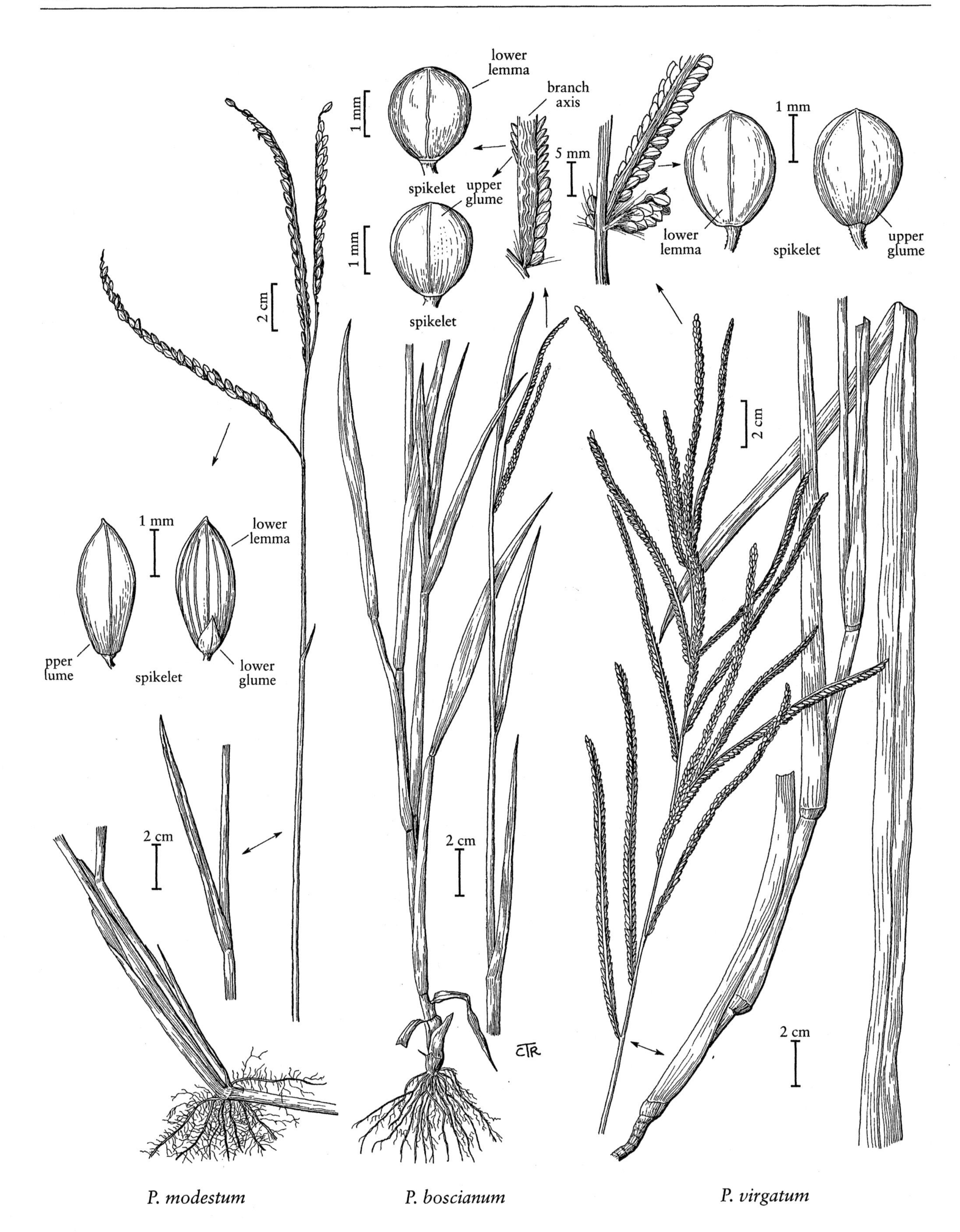

P. modestum *P. boscianum* *P. virgatum*

PASPALUM

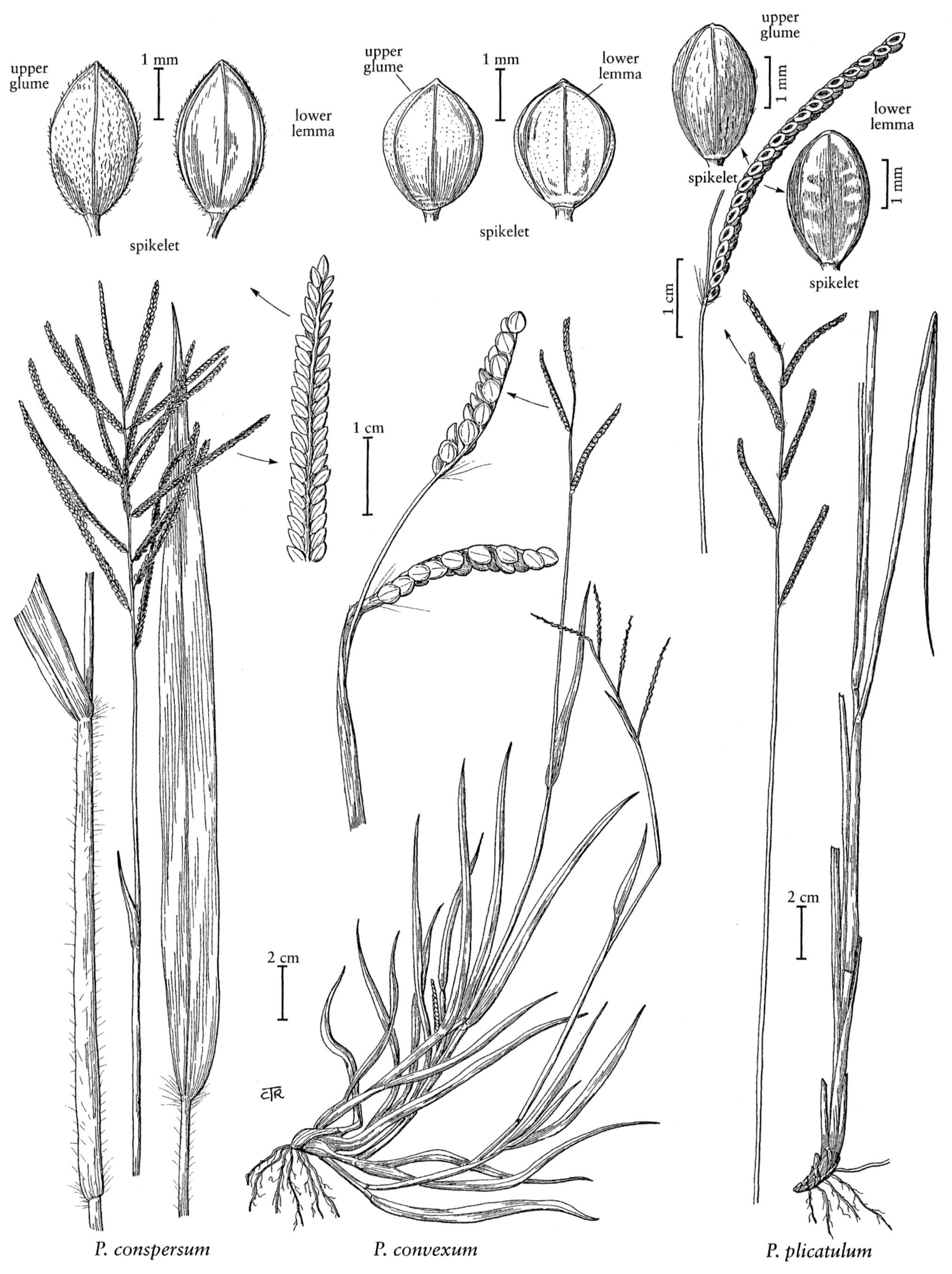

P. conspersum *P. convexum* *P. plicatulum*

PASPALUM

elliptic-ovate, light to dark brown. **Lower glumes** absent; **upper glumes** usually with short, appressed pubescence, rarely glabrous, 5-veined, margins entire; **lower lemmas** with short, appressed pubescence or glabrous, 3-veined, margins entire; **upper florets** dark glossy brown. **Caryopses** 1.4–1.6 mm, brown. $2n$ = 20, 40, 60.

Paspalum plicatulum grows in prairies, along forest margins, and in disturbed areas. Its range extends from the southeastern United States through the Caribbean and Mexico to Bolivia, Paraguay, and Argentina.

23. **Paspalum wrightii** Hitchc. & Chase [p. 585]
WRIGHT'S PASPALUM

Plants perennial; aquatic to semi-aquatic, conspicuously rhizomatous or stoloniferous. **Culms** 80–150 cm, erect; **nodes** glabrous. **Sheaths** glabrous or sparsely pubescent; **ligules** 1–3 mm; **blades** to 35 cm long, 2–7 mm wide, flat, glabrous above, pubescent below, especially basally. **Panicles** terminal, with 5–8 racemosely arranged branches; **branches** 3.5–11 cm, divergent to erect; **branch axes** 0.6–1.1 mm wide, glabrous, terminating in a spikelet. **Spikelets** 2.2–2.7 mm long, 1–1.4 mm wide, paired, appressed to or divergent from the branch axes, elliptic, glabrous, light brown. **Lower glumes** absent; **upper glumes** glabrous, 3-veined; **lower lemmas** glabrous, 5-veined; **upper florets** glossy chestnut brown. $2n$ = unknown.

The range of *Paspalum wrightii* extends from Cuba and Campeche, Mexico, to Bolivia, Paraguay, and Argentina. It is now established in the *Flora* region, growing along wet, roadside ditches, primarily on the Gulf coast of Texas.

24. **Paspalum nicorae** Parodi [p. 585]
BRUNSWICKGRASS

Plants perennial; rhizomatous, rhizomes 5–25 cm, conspicuous. **Culms** 10–70 cm, erect to ascending; **nodes** glabrous. **Sheaths** glabrous, pubescent apically; **ligules** 1.2–1.5 mm; **blades** 6–20 cm long, 4–5 mm wide, flat to conduplicate, glabrous or pubescent. **Panicles** terminal, with 2–5 racemosely arranged branches; **branches** 1.4–5.2 cm, divergent; **branch axes** about 0.8 mm wide, glabrous, terminating in a spikelet. **Spikelets** 2.3–2.7 mm long, 1.4–1.8 mm wide, paired, appressed to or divergent from the branch axes, elliptic, dark brown. **Lower glumes** absent; **upper glumes** shortly pubescent, 5-veined, margins entire; **lower lemmas** transversely rugose at maturity, glabrous, 5-veined, margins entire; **upper florets** dark glossy brown. **Caryopses** about 1.8 mm long, 1.4 mm wide, ellipsoidal. $2n$ = 40.

Paspalum nicorae is native to Brazil, Uruguay, and Argentina. It was introduced to the United States for use in pastures and as a cover crop in waterways. It is now established in the southeastern United States, growing as a weed in pastures, turf, and other disturbed areas.

25. **Paspalum malacophyllum** Trin. [p. 585]
RIBBED PASPALUM

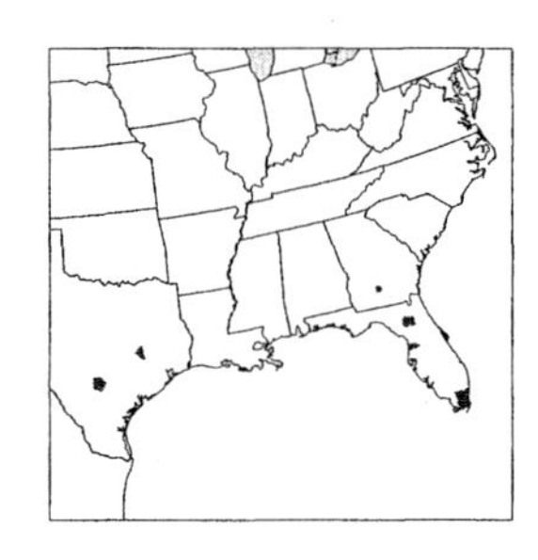

Plants perennial; cespitose, sometimes with short rhizomes. **Culms** 90–200 cm, erect; **nodes** sunken, glabrous or pubescent, brown. **Sheaths** pubescent; **ligules** 4–5 mm, membranous, brown, acute; **blades** 12–40 cm long, 8–35 mm wide, flat or conduplicate, pubescent below, glabrous above, distinctly pubescent basally. **Panicles** terminal, with 8–25 racemosely arranged branches; **branches** 1–8 cm, divergent to erect; **branch axes** 1–1.2 mm wide, margins scabrous, terminating in a spikelet; **pedicels** 0.2–0.4 and 0.5–1.2 mm long, flattened, scabrous. **Spikelets** 1.8–2 mm, paired, appressed to or divergent from the branch axes, oblong-elliptic, white to stramineous. **Glumes** absent; **lower lemmas** glabrous, ribbed over the veins, sulcate between, 5-veined, margins entire; **upper lemmas** as long as the lower ones, longitudinally papillose-striate, glabrous, pale-colored. **Upper florets** white to stramineous. $2n$ = 40, 60.

Paspalum malacophyllum is native from Mexico to Bolivia and Argentina. It was introduced to the southern United States for forage and soil conservation, and is now established in the southeastern United States, growing in disturbed sites at scattered locations.

26. **Paspalum racemosum** Lam. [p. 587]
PERUVIAN PASPALUM

Plants annual; cespitose or rhizomatous. **Culms** 40–90 cm, erect; **nodes** purple. **Sheaths** glabrous; **ligules** 0.1–0.3 mm; **blades** 4–13 cm long, 10–22 mm wide, flat, glabrous. **Panicles** terminal, with 40–75 racemosely arranged branches; **branches** 1–2.5 cm, divergent to erect; **branch axes** 1–1.5 mm wide, terminating in a pedicellate spikelet. **Spikelets** 2.5–2.9 mm long, 0.8–1.2

branch
axis
1 mm
lower
lemma
1 mm
spikelet
upper
glume
1 mm
spikelet
2 cm
CTR
upper
glume
1 mm
lower
lemma
spikelet
upper
lemma
1 mm
lower
lemma
spikelet
5 mm
2 cm
2 cm
P. wrightii
P. nicorae
P. malacophyllum

PASPALUM

mm wide, paired, appressed to or divergent from the branch axes, linear-elliptic, pubescent, stramineous or purplish. **Lower glumes** absent; **upper glumes** and **lower lemmas** rugose, shortly ciliate; **lower lemmas** lacking ribs over the veins; **upper florets** 1.3–1.6 mm, stramineous, oblong elliptic, pale, shiny. **Caryopses** white. $2n$ = unknown.

Paspalum racemosum is native to Colombia, Ecuador, and Peru. Within the *Flora* region, it is known from disturbed sites at a few widely scattered locations.

27. **Paspalum intermedium** Munro *ex* Morong & Britton [p. 587]
INTERMEDIATE PASPALUM

Plants perennial; shortly rhizomatous. **Culms** to 200 cm, erect; **nodes** glabrous. **Sheaths** glabrous; **ligules** 2–3 mm; **blades** to 57 cm long, 2–3 cm wide, flat, glabrous below, appressed pubescent above. **Panicles** terminal, with 60–100 racemosely arranged branches; **branches** 1–13 cm, divergent to spreading, often arcuate; **branch axes** 0.9–1.2 mm wide, winged, margins scabrous, long pubescent. **Spikelets** 2–2.4 mm long, 0.9–1.2 mm wide, paired, divergent to spreading from the branch axes, elliptic to ovate, glabrous or pubescent, stramineous, sometimes partially purple. **Lower glumes** absent; **upper glumes** smooth, 3-veined, margins entire, sparsely short-pubescent, at least distally; **lower lemmas** smooth, lacking ribs over the veins, 3-veined, margins entire, glabrous or shortly pubescent; **upper florets** stramineous to white. **Caryopses** 1.5–1.7 mm, golden brown. $2n$ = 20, 40.

Paspalum intermedium is an introduced roadside weed in the *Flora* region. It is found in Mexico and South America, but not in Central America (Pohl and Davidse 1994).

28. **Paspalum coryphaeum** Trin. [p. 587]
EMPEROR PASPALUM

Plants perennial; cespitose, not rhizomatous. **Culms** 65–400 cm, erect; **nodes** pilose. **Sheaths** papillose-hirsute (upper sheaths sometimes glabrous); **ligules** 1–4.5 mm; **blades** 30–50 cm long, 10–23 mm wide, flat, with long hairs behind the ligules, otherwise glabrous or puberulent adaxially. **Panicles** terminal, with (6)15–44 racemosely arranged branches; **branches** 5–13 cm, straight, spreading to reflexed, rarely merely divergent; **branch axes** 0.3–0.4 mm wide, narrowly winged, glabrous, margins scabrous, pubescent, terminating in a spikelet. **Spikelets** 2–2.5 mm long, 1.8–1.9 mm wide, paired, divergent to spreading from the branch axes, elliptic, brown to stramineous, often purple-tinged. **Lower glumes** usually absent, if present, to 0.9 mm, triangular; **upper glumes** smooth, papillose-hirsute, 3-veined; **lower lemmas** smooth, papillose-hirsute or glabrous, 3-veined; **upper florets** white. $2n$ = 20, 40, 60.

Paspalum coryphaeum is native from Costa Rica and the Caribbean south to northern South America. In the *Flora* region, it grows in disturbed habitats at scattered southeastern locations.

29. **Paspalum quadrifarium** Lam. [p. 589]
PAJA MANSE, PAJA COLORADA, TUSSOCK PASPALUM

Plants perennial; cespitose. **Culms** (50)100–180 cm, erect; **nodes** pubescent. **Sheaths** pubescent, margins extending into auricles; **ligules** 1–6.3 mm; **blades** 15–62 cm long, 4.9–6.1 mm wide, involute to flat, glabrous. **Panicles** terminal, with 15–44 racemosely arranged branches; **branches** 1.2–8.5 cm, straight, erect to ascending, lower branches longer than those above; **branch axes** 0.5–0.6 mm wide, narrowly winged, glabrous, margins scabrous, pubescent, terminating in a spikelet. **Spikelets** 2–2.5(3) mm long, 0.9–1.3 mm wide, paired, divergent to spreading from the branch axes, elliptic, brown to stramineous, often purple-tinged. **Lower glumes** usually absent, if present, to 0.9 mm, triangular; **upper glumes** shortly pubescent, 3-veined, purple-spotted, margins entire; **lower lemmas** glabrous or pubescent, lacking ribs over the veins, 3-veined, margins entire; **upper florets** 2.2–2.5 mm, white. $2n$ = 20, 30, 40.

Paspalum quadrifarium is native to Uruguay, Paraguay, Brazil, and Argentina. It is grown as an ornamental in Florida, but has also become established in disturbed habitats of the southeastern United States. It is considered a noxious weed in New South Wales, Australia.

30. **Paspalum bifidum** (Bertol.) Nash [p. 589]
PITCHFORK PASPALUM

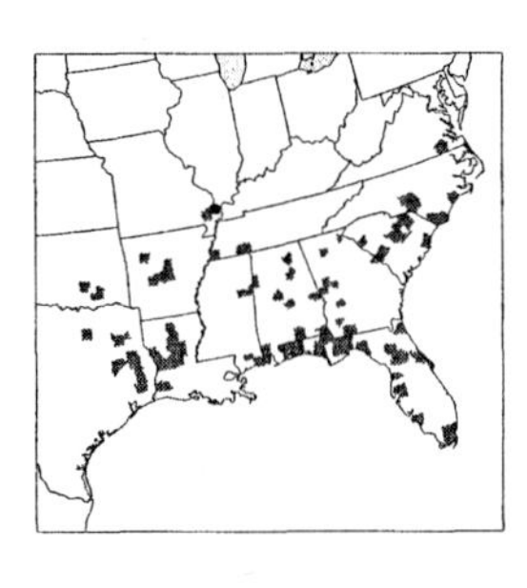

Plants perennial; rhizomatous. **Culms** 60–140 cm, erect; **nodes** glabrous. **Sheaths** pubescent; **ligules** 2–4 mm; **blades** to 37 cm long, 2.2–11 mm wide, flat. **Panicles** terminal, with 2–5 racemosely arranged branches; **branches** 3.7–13 cm, divergent to erect; **branch axes** 0.2–0.8

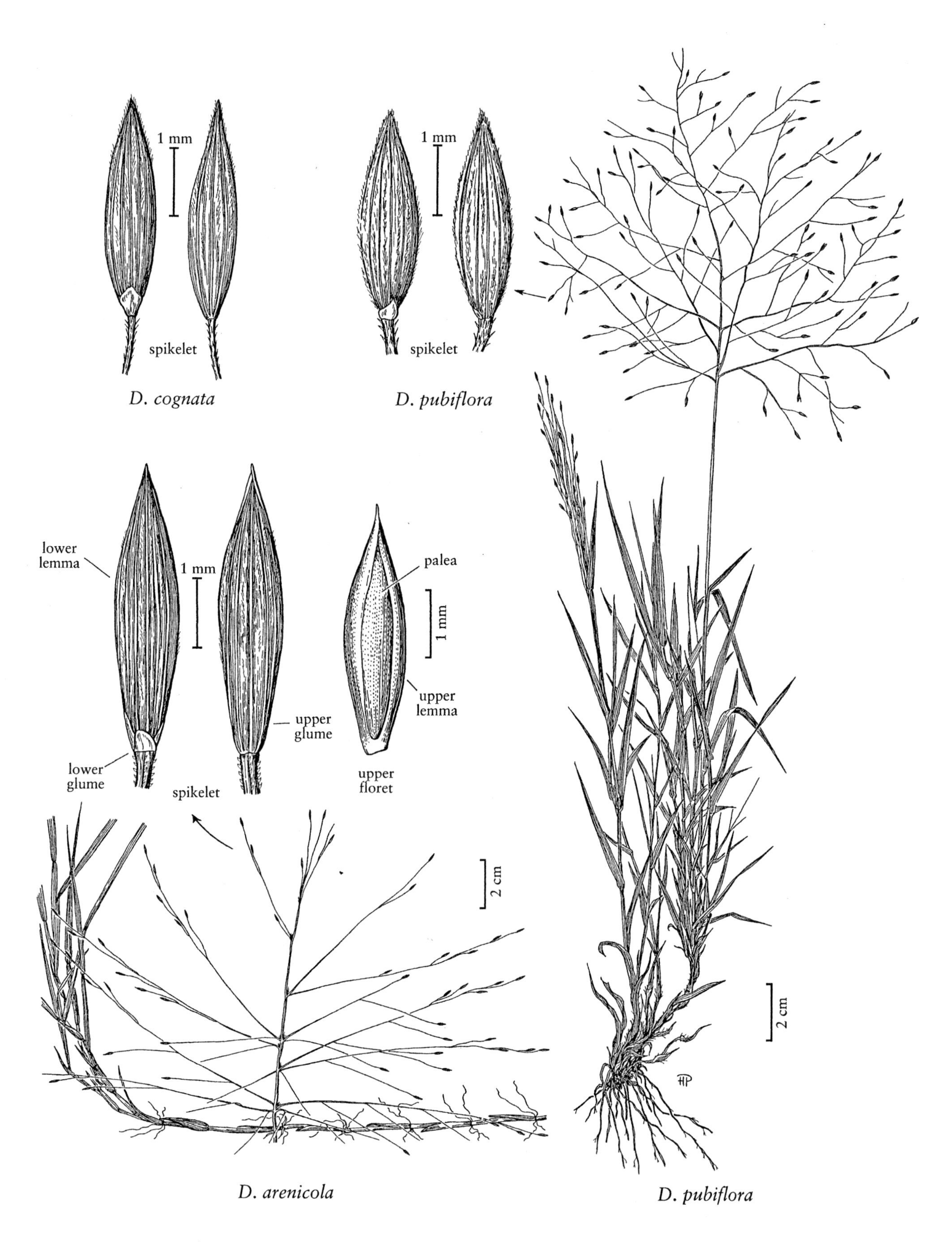

1 mm
spikelet
D. cognata
1 mm
spikelet
D. pubiflora
lower lemma
1 mm
palea
1 mm
upper lemma
upper glume
lower glume
spikelet
upper floret
2 cm
2 cm
D. arenicola
D. pubiflora

mm wide, glabrous, margins scabrous, terminating in a spikelet. **Spikelets** 3.1–4 mm long, 2–2.5 mm wide, paired, not imbricate, appressed to the branch axes, elliptic to obovate, yellow-brown. **Lower glumes** present or absent; **upper glumes** glabrous or sparsely pubescent basally, (6)7-veined, margins entire; **lower lemmas** glabrous or sparsely pubescent basally, lacking ribs over the veins, 5-veined, margins entire; **upper florets** white. **Caryopses** 2.6–2.9 mm, purple. 2*n* = unknown.

Paspalum bifidum is restricted to the southeastern United States. It grows at the edges of forests in longleaf pine-oak-grass ecosystems, usually in dry to mesic loamy sandy soils. It grows vigorously following fire.

31. **Paspalum langei** (E. Fourn.) Nash [p. 589]
RUSTYSEED PASPALUM

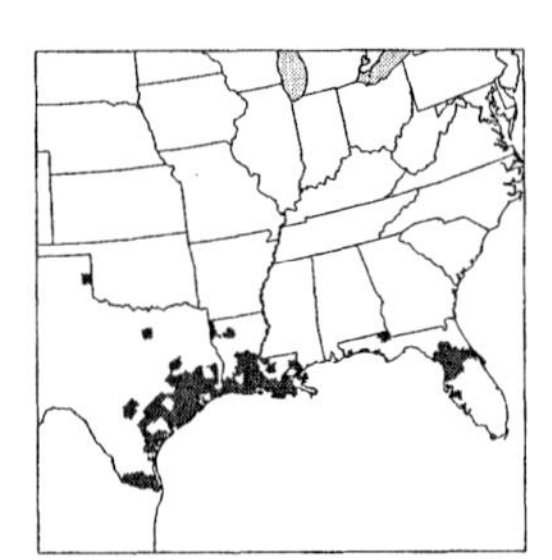

Plants perennial; cespitose. **Culms** 23–125 cm, erect; **nodes** glabrous or pubescent. **Sheaths** glabrous or pubescent; **ligules** 0.6–1.9 mm; **blades** to 38 cm long, 4–18 mm wide, flat, glabrous or pubescent, dark green. **Panicles** terminal, with 1–3(4) racemosely arranged branches; **branches** 2.3–13.4 cm, erect to divergent, terminating in a spikelet; **branch axes** 0.2–1 mm wide, glabrous, margins scabrous. **Spikelets** 2.1–3.3 mm long, 1.3–1.6 mm wide, paired, imbricate, appressed to the branch axes, elliptic to obovate, stramineous to brown. **Lower glumes** 0.4–1.2(1.8) mm, stramineous to brown; **upper glumes** with papillose-based short pubescence, 3- or 5-veined, margins entire, **lower lemmas** with papillose-based short pubescence, lacking ribs over the veins, 3-veined, margins entire; **upper florets** light stramineous. **Caryopses** 1.3–1.5 mm, light to dark brown. 2*n* = 40, 60.

Paspalum langei is native from Texas to Florida, and extends through Mexico to Venezuela and the Antilles. It grows at the edges of moist woods and in disturbed areas.

32. **Paspalum setaceum** Michx. [pp. 591, 593, 595]

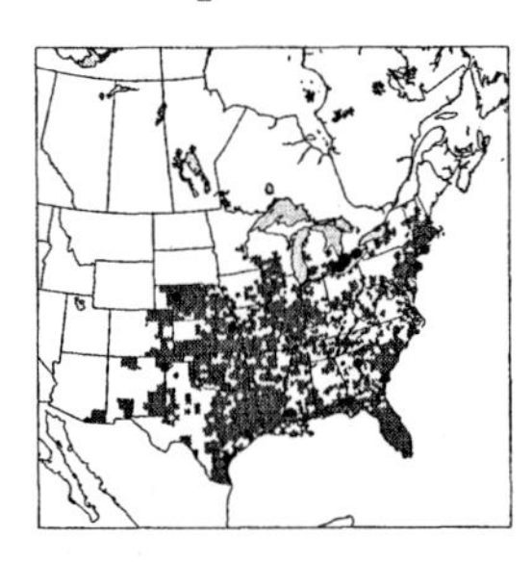

Plants perennial; cespitose or shortly rhizomatous. **Culms** 25–110 cm, erect, spreading, or prostrate; **nodes** glabrous or pubescent. **Sheaths** glabrous or pubescent; **ligules** 0.2–0.5 mm; **blades** flat, glabrous or pubescent. **Panicles** terminal and axillary, with 1–6 racemosely arranged branches, axillary panicles partially or completely enclosed by the subtending leaf sheath; **branches** 2–12(17) cm, ascending to spreading, often arcuate, terminating in a spikelet; **branch axes** 0.2–1.2 mm wide, glabrous, sometimes scabrous. **Spikelets** 1.4–2.6 mm long, paired, imbricate, appressed to the branch axes, elliptic to obovate to ovate to orbicular, stramineous or brown. **Lower glumes** absent; **upper glumes** and **lower lemmas** glabrous or shortly glandular-pubescent, 3-veined, margins entire; **lower lemmas** lacking ribs over the veins; **upper florets** stramineous. **Caryopses** elliptic to suborbicular, white. 2*n* = 20.

Paspalum setaceum is a variable species that grows east of the Rocky Mountains in the contiguous United States and Mexico. The following treatment summarizes the major patterns of variation within the species. Some specimens will be hard to place, particularly old herbarium specimens that have lost their color. Nine varieties grow in the *Flora* region.

1. Leaf blades conspicuously basal, recurved, 3–10 mm wide; lower lemmas without evident midveins.
   2. Leaf blades yellowish-green, usually glabrous; spikelets usually glabrous, not spotted . . . . var. *longepedunculatum*
   2. Leaf blades grayish-green, hirsute; spikelets short pubescent, often spotted . . . . var. *villosissimum*
1. Leaf blades more evenly distributed, lax to straight, 1.5–20 mm wide; lower lemmas with or without evident midveins.
   3. Leaf blades glabrous (or almost so) on the surfaces, sometimes ciliate on the margins.
      4. Leaf blades 2.4–6.1 mm wide, stiff; spikelets 2–2.6 mm long . . . . var. *rigidifolium*
      4. Leaf blades 3–18 mm wide, lax to somewhat stiff but, if somewhat stiff, more than 6 mm wide; spikelets 1.7–2.4 mm long.
         5. Lower lemmas without evident midveins; blades yellowish-green to dark green . . . . var. *stramineum*
         5. Lower lemmas with evident midveins; blades dark green to purplish . . . . var. *ciliatifolium*
   3. Leaf blades evidently hirsute on the surfaces as well as on the margins.
      6. Plants widely spreading to prostrate.
         7. Leaf blades grayish-green; lower lemmas without evident midveins . . . . var. *psammophilum*
         7. Leaf blades yellowish-green; lower lemmas with or without evident midveins . . . . var. *supinum*
      6. Plants erect to spreading.
         8. Lower lemmas usually with evident midveins; spikelets 1.8–2.5 mm long, usually glabrous, light green to green . . . . var. *muhlenbergii*

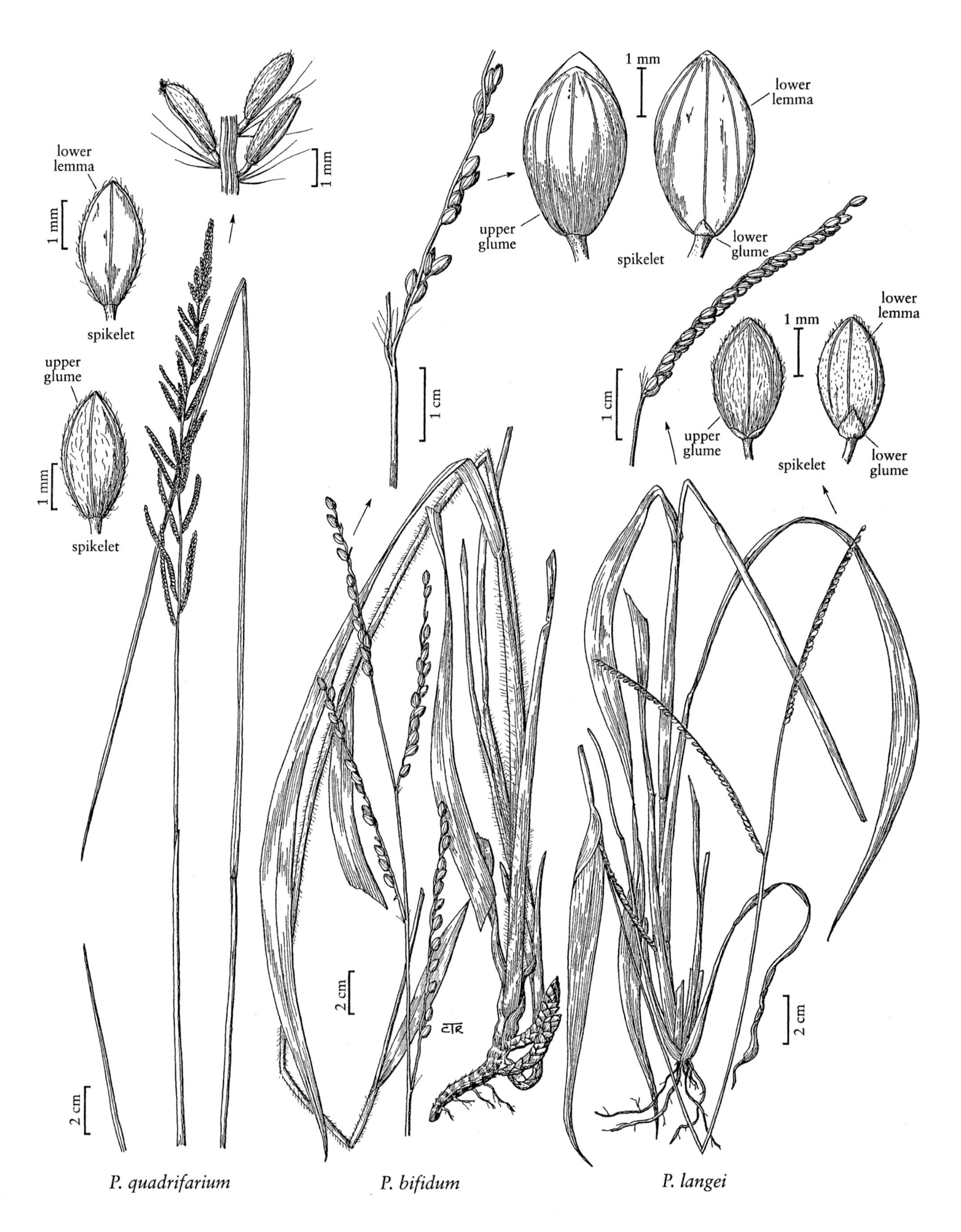
lower
lemma
1 mm
spikelet
upper
glume
1 mm
spikelet
1 mm
1 cm
1 mm
upper
glume
spikelet
lower
lemma
lower
glume
lower
lemma
1 mm
1 cm
upper
glume
spikelet
lower
glume
2 cm
2 cm
2 cm
P. quadrifarium
P. bifidum
P. langei

PASPALUM

8. Lower lemmas usually without evident midveins; spikelets 1.4–2.4 mm long, usually pubescent, pale yellow to light green.
   9. Leaves grayish-green; blades 1.5–7 mm wide, always conspicuously hirsute .... var. *setaceum*
   9. Leaves yellowish-green to dark green; blades 3.3–13.5 mm wide, almost glabrous or conspicuously hirsute .... var. *stramineum*

Paspalum setaceum var. **ciliatifolium** (Michx.) Vasey [p. 591]
FRINGELEAF PASPALUM

Plants erect to spreading. **Leaves** mostly cauline; **blades** 2–32 cm long, 3–18 mm wide, lax to somewhat stiff, glabrous or with a few hairs along the midrib (rarely shortly pilose), dark green to purple, margins scabrous, ciliate. **Panicle branches** 2–11.9 cm; **branch axes** 0.6–1.2 mm wide. **Spikelets** 1.7–2 mm long, 1.2–1.5 mm wide, elliptic to obovate, pubescent or occasionally glabrous; **lower lemmas** with an evident midvein; **upper florets** 1.7–2 mm.

Paspalum setaceum var. *ciliatifolium* is the most variable and widespread of the nine varieties of *P. setaceum*. It usually grows in sandy soil in open areas, including disturbed areas, of prairies and forest margins. Its range extends from Louisiana and the eastern United States to Panama, the West Indies, and Bermuda.

Paspalum setaceum var. **longepedunculatum** (Leconte) Alph. Wood [p. 591]
BARESTEM PASPALUM

Plants erect. **Leaves** mostly basal; **blades** to 15 cm long, 3–10 mm wide, glabrous, usually recurved, yellow-green, margins ciliate. **Panicle branches** 2.4–9.2 cm long, ascending or nodding, arcuate; **branch axes** 0.2–0.6 mm wide. **Spikelets** 1.4–1.8 mm long, 0.9–1.3 mm wide, elliptic to obovate or suborbicular, glabrous or with scattered glandular hairs, not spotted; **lower lemmas** without an evident midvein; **upper florets** 1.4–1.8 mm.

Paspalum setaceum var. *longepedunculatum* grows on open ground, usually in moist areas such as along ditches and roadsides, as well as in flatwoods. It is found primarily in the coastal plain of the southeastern United States, but has also been found in Ohio, Kentucky, and Tennessee. It is similar to var. *villosissimum*, differing in its glabrous leaves, more delicate habit, and more poorly-developed rhizomes. Both varieties grow in peninsular Florida, but var. *longepedunculatum* also grows along the coast as far west as the Mississippi delta and as far north as southern North Carolina.

Paspalum setaceum var. **muhlenbergii** (Nash) D.J. Banks [p. 591]
HURRAHGRASS

Plants mostly erect. **Leaves** mostly cauline; **blades** to 25 cm long, 2–10.2 mm wide, lax to straight, surfaces and margins evenly hirsute, hairs 1.5–5.5 mm, light to dark green. **Panicle branches** 1.9–14.3 cm; **branch axes** 0.5–1 mm wide. **Spikelets** 1.8–2.5 mm long, 1.5–2 mm wide, oval to suborbicular, glabrous or with a few hairs, light green to green; **lower lemmas** usually with an evident midvein; **upper florets** 1.8–2.3 mm.

Paspalum setaceum var. *muhlenbergii* is endemic to the *Flora* region, extending from southern Ontario to the Gulf coast of Texas and northern Florida. It grows in disturbed areas and on the margins of forests. It resembles var. *supinum*, differing in its erect habit and, usually, in its spikelet shape and presence of a midvein on the lower lemma.

Paspalum setaceum var. **psammophilum** (Nash) D.J. Banks [p. 591]
SAND PASPALUM

Plants spreading to prostrate. **Leaves** mostly cauline; **blades** to 16 cm long, 2.8–8.3 mm wide, lax to straight, grayish-green, surfaces and margins evidently hirsute, with soft, short hairs. **Panicle branches** 3.1–5.9 cm; **branch axes** 0.7–1 mm wide. **Spikelets** 1.8–2.2 mm long, 1.6–1.8 mm wide, suborbicular to orbicular, pubescent; **lower lemmas** without an evident midvein; **upper florets** 1.8–2.1 mm.

Paspalum setaceum var. *psammophilum* grows in sandy, maritime habitats and, inland, along sandy roadsides and in dry fields, from Massachusetts to the District of Columbia. The combination of its spreading to prostrate habit and densely puberulent foliage distinguishes it from other varieties of *P. setaceum*.

Paspalum setaceum var. **rigidifolium** (Nash) D.J. Banks [p. 593]
STIFF PASPALUM

Plants erect to spreading. **Leaves** mostly cauline; **blades** to 30 cm long, 2.4–6.1 mm wide, conspicuously rigid, stiff, pubescent adaxially, glabrous or sparsely pubescent abaxially, margins sometimes ciliate. **Panicle branches** 4.8–11.3 cm; **branch axes** 0.7–1.1 mm wide. **Spikelets** 2–2.6 mm long, 1.6–1.8 mm wide, obovate to ovate, pubescent, sometimes sparsely so; **lower lemmas** with or without an evident midvein; **upper florets** 2–2.4 mm.

Paspalum setaceum var. *rigidifolium* grows on hammocks, sand barrens, high pinelands, and flatwoods of Georgia, Florida, and Cuba.

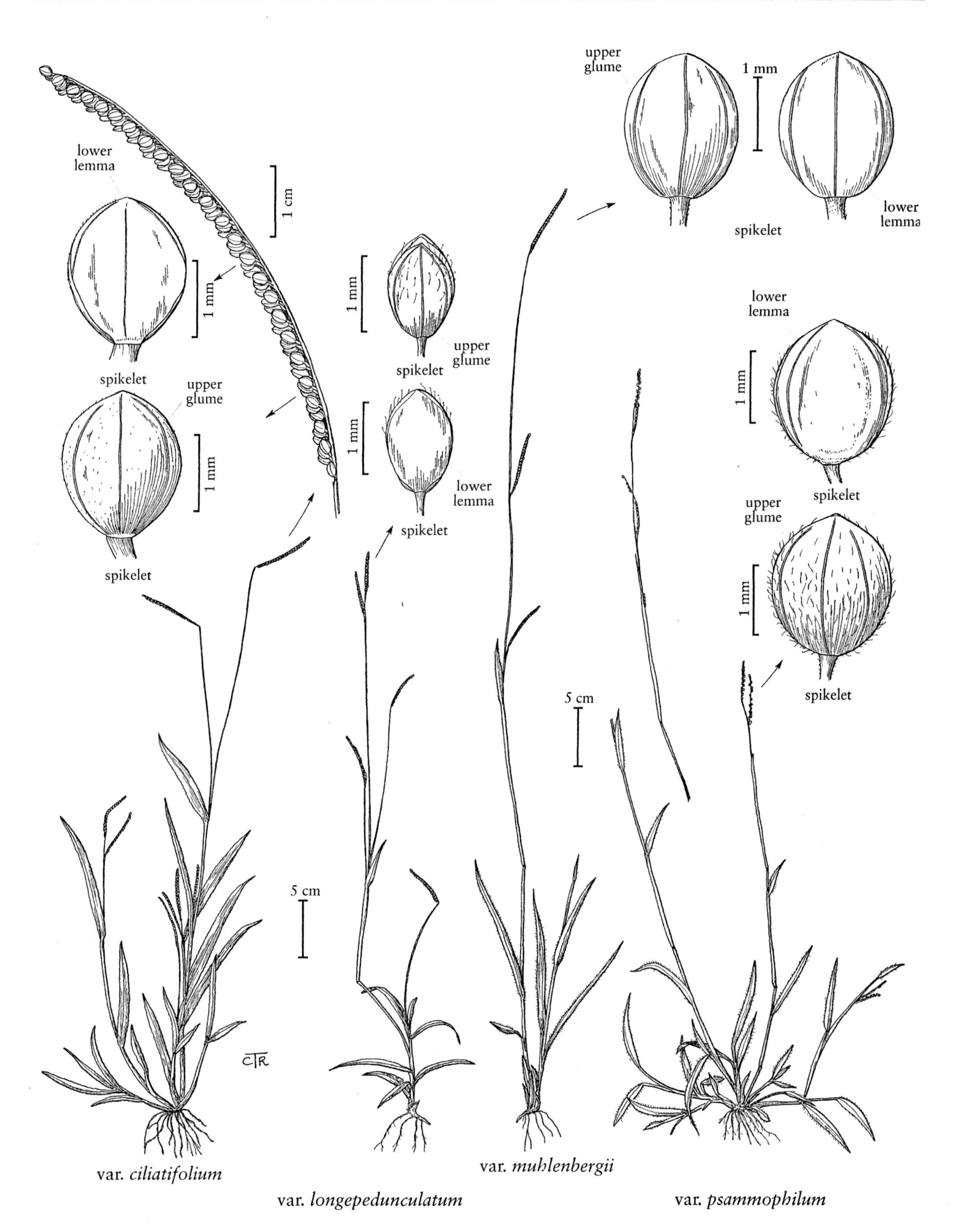

PASPALUM SETACEUM

Paspalum setaceum Michx. var. **setaceum** [p. 593]
THIN PASPALUM

Plants erect. **Leaves** mostly cauline; **blades** to 22 cm long, 1.5–7 mm wide, lax to straight, conspicuously hirsute, usually with long stiff hairs and short soft hairs, grayish-green, margins hirsute. **Panicle branches** 2–11.2 cm; **branch axes** 0.3–0.9 mm wide. **Spikelets** 1.4–1.9 mm long, 1.1–1.6 mm wide, elliptic, obovate, orbicular, or suborbicular, pubescent to nearly glabrous; **lower lemmas** without an evident midvein; **upper lemmas** 1.3–2 mm.

Paspalum setaceum var. *setaceum* grows in open areas and sandy soils, often at the edges of forests, primarily on the southeastern coastal plain of the United States, from southern New England to eastern Mexico, but extending inland to western Virginia, Missouri, and Arkansas. It also grows in Cuba.

Paspalum setaceum var. **stramineum** (Nash) D.J. Banks [p. 593]
YELLOW SAND PASPALUM

Plants erect to spreading. **Leaves** mostly cauline; **blades** to 30 cm long, 3.3–13.5 mm wide, lax to somewhat stiff, glabrous or with a few hairs along the midrib, sometimes pubescent, yellow-green to dark green, margins scabrous, ciliate. **Panicle branches** 4–12 cm; **branch axes** 0.6–1.1 mm wide. **Spikelets** 1.7–2.4 mm long, 1.5–2.1 mm wide, obovate to suborbicular, pubescent or occasionally glabrous; **lower lemmas** without an evident midvein; **upper florets** 1.7–2.1 mm.

Paspalum setaceum var. *stramineum* grows at the edges of forests and in disturbed areas with sandy soil. Its range extends from the central plains and eastern United States to Mexico, Bermuda, and the West Indies.

Paspalum setaceum var. **supinum** (Bosc *ex* Poir.) Trin. [p. 593]
SUPINE THIN PASPALUM

Plants usually spreading. **Leaves** mostly cauline; **blades** to 31 cm long, 2.2–19 mm wide, lax to straight, long pubescent, hairs 1.5–4 mm, yellow-green, margins ciliate, with long stiff hairs. **Panicle branches** 2.2–9.9 cm; **branch axes** 0.6–1.5 mm wide. **Spikelets** 1.7–2.1 mm long, 1.2–1.6 mm wide, elliptic to obovate (rarely suborbicular), glabrous or pubescent; **lower lemmas** with or without an evident midvein; **upper florets** 1.7–2.1 mm.

Paspalum setaceum var. *supinum* grows at the edges of forests and in disturbed areas. Within the *Flora* region, its range extends from Texas, Arkansas, and Louisana to South Carolina, Georgia, and Florida.

Paspalum setaceum var. *supinum* resembles var. *muhlenbergii*, differing in its spreading habit and, usually, in its spikelet shape and lack of a midvein on the lower lemma.

Paspalum setaceum var. **villosissimum** (Nash) D.J. Banks [p. 595]
HAIRY PASPALUM

Plants erect. **Leaves** conspicuously basal; **blades** to 15 cm long, 3–10 mm wide, recurved, hirsute, usually with long stiff hairs and short soft hairs, grayish-green, margins ciliate, with long stiff hairs. **Panicle branches** 2–11.2 cm; **branch axes** 0.3–0.9 mm wide. **Spikelets** 1.4–1.9 mm long, 1.1–1.6 mm wide, elliptic, obovate, orbicular, or suborbicular, short pubescent to nearly glabrous, often purple-spotted; **lower lemmas** without an evident midvein; **upper lemmas** 1.3–2 mm.

Paspalum setaceum var. *villosissimum* grows in sandy fields and flatwoods of Florida and Cuba. It resembles var. *longepedunculatum*, differing in its pubescent leaves, more robust habit, and more developed rhizomes.

33. **Paspalum laxum** Lam. [p. 595]
COCONUT PASPALUM

Plants perennial; cespitose to short rhizomatous. **Culms** 80–110 cm, erect; **nodes** glabrous. **Sheaths** glabrous, sparsely pubescent apically; **ligules** 1–2.9 mm; **blades** 9–41 cm long, 3–7 mm wide, mostly involute, pubescent above, glabrous below. **Panicles** terminal, with 1–5(10) racemosely arranged branches; **branches** 1.9–11.4 cm, erect to divergent, terminating in a spikelet; **branch axes** 0.4–0.7 mm wide, very narrowly winged, scabrous. **Spikelets** 1.6–2.2 mm long, 1.1–1.3 mm wide, paired, imbricate, appressed to the branch axes, elliptic-obovate to ovate. **Lower glumes** absent; **upper glumes** shortly pubescent, 5-veined, margins entire; **lower lemmas** glabrous or shortly pubescent, lacking ribs over the veins, 3-veined, margins entire; **upper florets** 1.4–2 mm, white to stramineous. $2n = 60$.

Paspalum laxum grows in hammocks and along roads, often in sandy or limestone soils. It used to be common in coconut groves, hence the English-language name. It grows in southern Florida, the Antilles, and Belize.

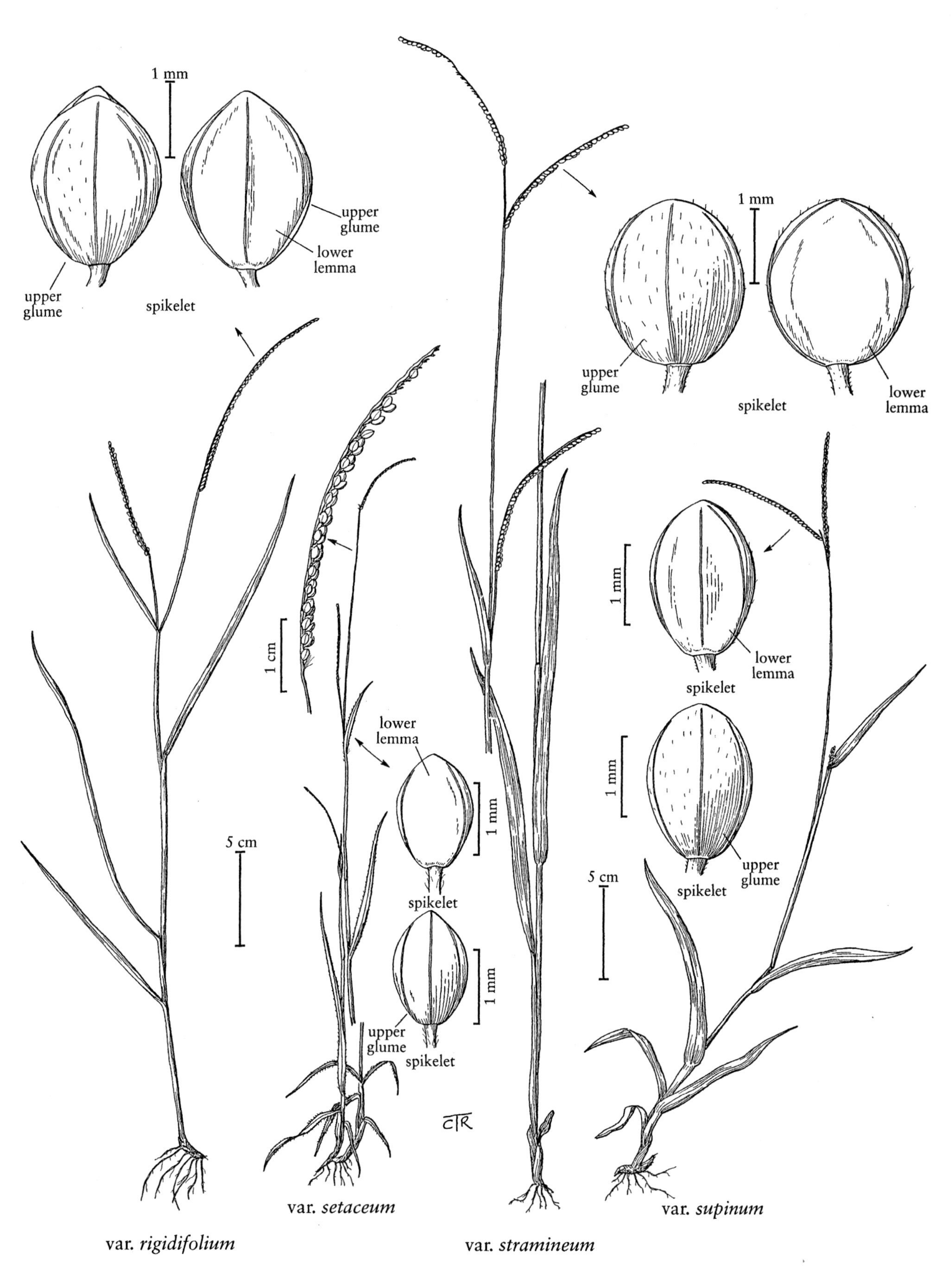

PASPALUM SETACEUM

34. **Paspalum caespitosum** Flüggé [p. 595]
BLUE PASPALUM

Plants perennial; cespitose. **Culms** 20–60 cm, erect, base swollen, bulblike; **cataphylls** pubescent; **nodes** sparsely pubescent or glabrous. **Sheaths** pubescent or glabrous; **ligules** 0.2–0.4 mm; **blades** to 25 cm long, 1.9–6.2 mm wide, flat, glabrous, pubescent behind the ligules, margins scabrous, often ciliate basally. **Panicles** terminal, with 2–5(8) racemosely arranged branches; **branches** 0.9–4.4 cm, divergent to spreading, terminating in a spikelet; **branch axes** 0.2–0.5 mm wide, narrowly winged. **Spikelets** 1.3–2 mm long, 0.7–1 mm wide, paired (rarely appearing solitary as a result of aborted spikelets), imbricate, appressed to the branch axes, elliptic. **Lower glumes** absent; **upper glumes** and **lower lemmas** glabrous or sparsely and shortly pubescent basally or around the margins, 5-veined, margins entire; **lower lemmas** lacking ribs over the veins; **upper florets** 1.3–1.8 mm, stramineous to golden brown. **Caryopses** 1.2–1.4 mm, ellipsoid, amber. $2n = 40$.

Paspalum caespitosum grows in hammocks and sandy pinelands. It is native in southern Alabama, Florida, the West Indies, Mexico, and Central America.

35. **Paspalum virletii** E. Fourn. [p. 596]
VIRLET'S PASPALUM

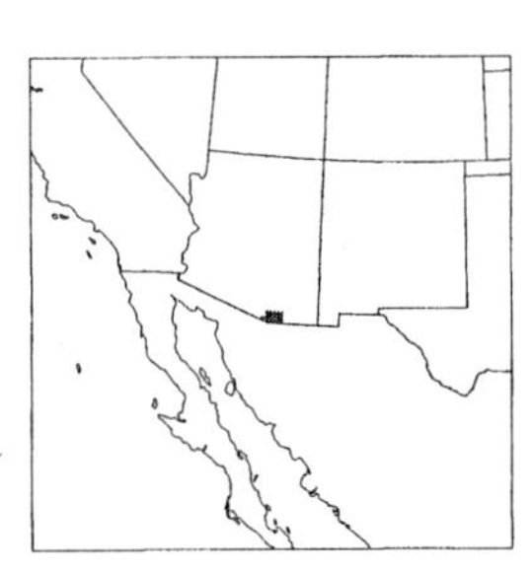

Plants perennial; cespitose. **Culms** 40–75 cm, erect, not swollen at the base; **nodes** pubescent. **Sheaths** pubescent; **blades** to 15 cm long, 5–10 mm wide, flat, pubescent. **Panicles** terminal, with 3–8 racemosely arranged branches; **branches** 2–7 cm, spreading, terminating in a spikelet; **branch axes** narrow, sparsely pubescent. **Spikelets** 2–2.5 mm long, 1.4–1.6 mm wide, paired, imbricate, appressed to the branch axes, ovate. **Lower glumes** absent; **upper glumes** shortly pubescent, 3-veined, margins entire; **lower lemmas** glabrous, lacking ribs over the veins, 3-veined, margins entire; **upper florets** pale to stramineous or golden brown. $2n$ = unknown.

Paspalum virletii grows in dry, sandy soils in disturbed habits. It is known only from Arizona, where it is considered a rare species, and from Mexico, where it also appears to be either rare or poorly collected (COTECOCA 2000).

36. **Paspalum monostachyum** Vasey [p. 596]
GULFDUNE PASPALUM

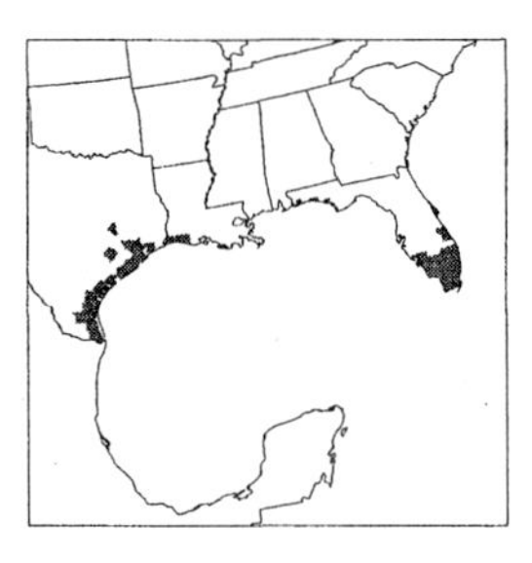

Plants perennial; rhizomatous. **Culms** 60–120 cm, erect; **nodes** glabrous. **Sheaths** glabrous; **ligules** 0.5–3 mm; **blades** to 50 cm long, 0.2–2(8) mm wide, involute (rarely flat), glabrous, pubescent behind the ligules. **Panicles** terminal, with 1–3 racemosely arranged branches; **branches** 5.6–23.3 cm, erect (rarely divergent), terminating in a spikelet; **branch axes** 0.5–1.2 mm wide, glabrous, margins scabrous to pubescent. **Spikelets** 2.3–3.7 mm long, 1.3–1.9 mm wide, paired, imbricate, appressed to the branch axes, elliptic to narrowly ovate, glabrous, stramineous (rarely partially purple). **Lower glumes** usually absent; **upper glumes** glabrous, 1-veined, margins entire; **lower lemmas** glabrous, lacking ribs over the veins, 3-veined, margins entire; **upper florets** stramineous. **Caryopses** 2–2.4 mm, yellow to golden brown. $2n$ = unknown.

Paspalum monostachyum grows in sand and muck soils on coastal sand dunes, wet prairie, marshes, and disturbed habitats of the southern coastal plain from Florida to eastern Mexico.

37. **Paspalum pleostachyum** Döll [p. 596]
TROPICAL PASPALUM

Plants perennial; cespitose. **Culms** 80–110 cm, erect; **nodes** glabrous. **Sheaths** glabrous, sparsely pubescent apically; **ligules** 1–2.9 mm; **blades** to 62 cm long, 3–7 mm wide, mostly involute, pubescent above, glabrous below. **Panicles** terminal, with 3–15 racemosely arranged branches; **branches** 5.2–12.5 cm, divergent to spreading, terminating in a spikelet; **branch axes** 0.3–0.8 mm wide, very narrowly winged, scabrous. **Spikelets** 2.2–2.5 mm long, 1–1.3 mm wide, paired, appressed to the branch axes, elliptic to obovate, white to light stramineous. **Lower glumes** absent; **upper glumes** glabrous, 5-veined; **lower lemmas** glabrous, 3-veined; **upper florets** 1.4–2 mm, white to stramineous. **Caryopses** 1–1.6 mm, suborbicular, brown. $2n$ = unknown.

Paspalum pleostachyum grows in sandy soil or rocky areas in Florida, the West Indies, and from northern South America to Brazil. It is usually found along the coast.

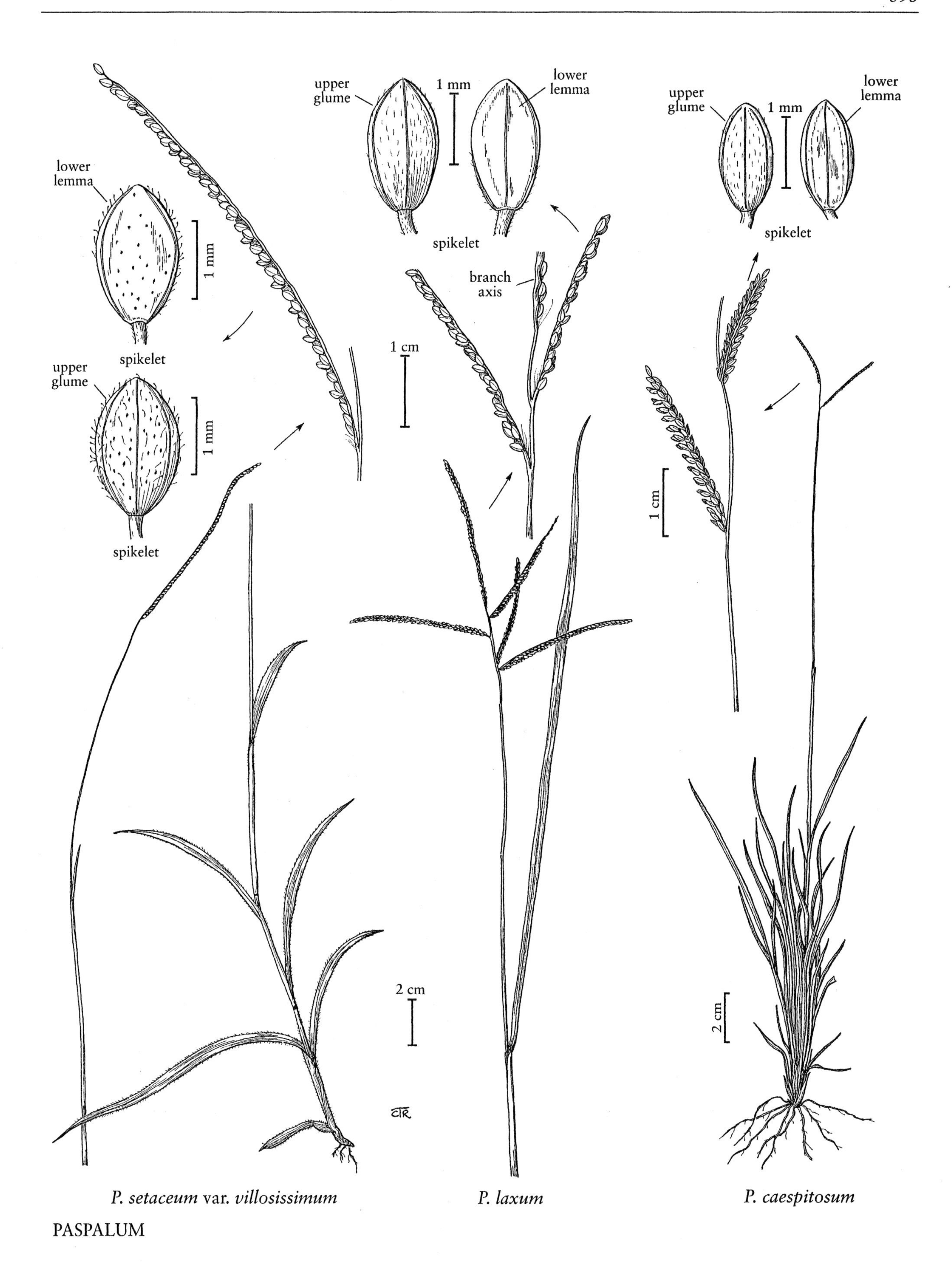

P. setaceum var. *villosissimum*   *P. laxum*   *P. caespitosum*

PASPALUM

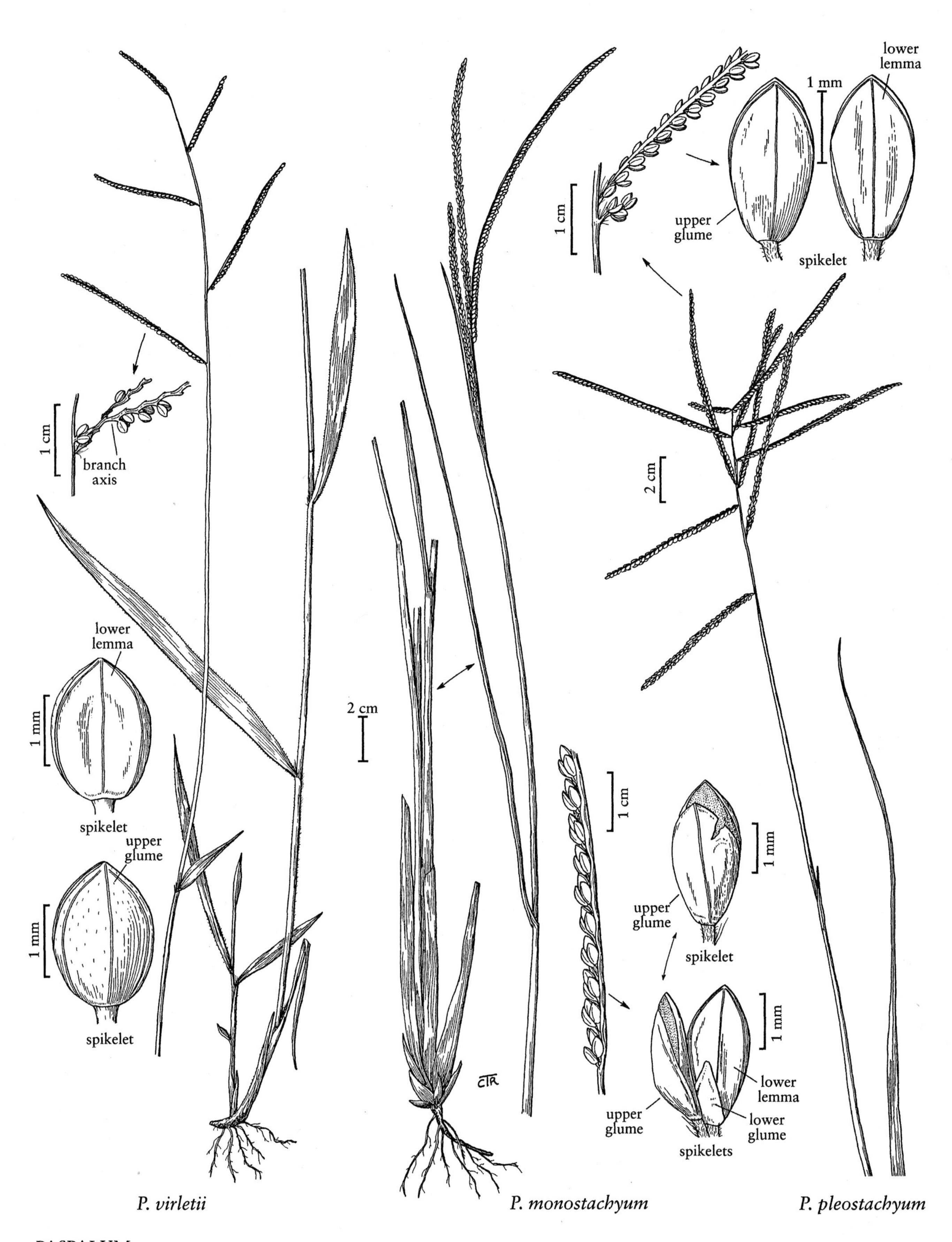

1 cm
branch
axis
lower
lemma
1 mm
spikelet
upper
glume
1 mm
spikelet
2 cm
CTR
1 cm
upper
glume
lower
lemma
1 mm
spikelet
2 cm
1 cm
1 mm
upper
glume
spikelet
1 mm
lower
lemma
upper
glume
lower
glume
spikelets
P. virletii
P. monostachyum
P. pleostachyum

38. **Paspalum lividum** Trin. *ex* Schltdl. [p. 598]
LONGTOM

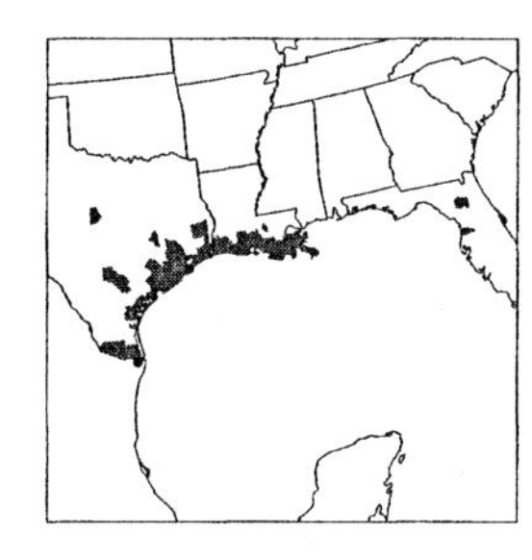

Plants perennial; decumbent or cespitose. **Culms** 30–97 cm, erect; **nodes** glabrous. **Sheaths** glabrous or pubescent; **ligules** 2.2–4.7 mm; **blades** to 38 cm long, 2.3–6.2 mm wide, flat, glabrous or pubescent. **Panicles** terminal, with 3–11 racemosely arranged branches; **branches** 1.5–4 cm, divergent, occasionally arcuate, terminating in a spikelet; **branch axes** 1.5–2 mm wide, broadly winged, glabrous or sparsely pubescent, margins scabrous, usually slightly conduplicate, occasionally purple. **Spikelets** 2.2–2.6 mm long, 1.2–1.5 mm wide, paired, imbricate, appressed to divergent from the branch axes, elliptic to obovate, stramineous (rarely purple-spotted), margins scabrous apically. **Lower glumes** absent; **upper glumes** and **lower lemmas** glabrous, 5-veined, margins entire; **lower lemmas** lacking ribs over the veins; **upper florets** white to pale. **Caryopses** 2–2.2 mm, brown. $2n$ = 40, 60.

Paspalum lividum grows in fresh and brackish marshes and ditches. It is native from the Gulf coast of the United States southward through Mexico and Central America to Cuba and Argentina. Plants of *P. modestum* with pale upper florets may be mistaken for *P. lividum*, but will have ligules that are only 1–2.3 mm long.

Zuloaga and Morrone regard *Paspalum lividum* as a synonym of *P. denticulatum* Trin. (http://mobot.mobot.org/W3T/Search/nwgc.html December 9, 2002).

39. **Paspalum praecox** Walter [p. 598]
EARLY PASPALUM

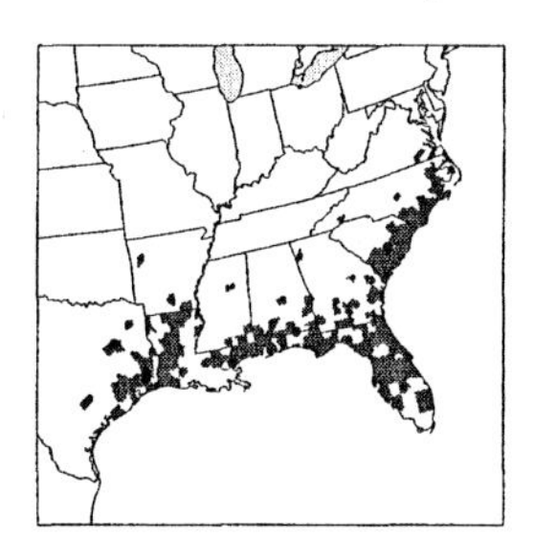

Plants perennial; shortly rhizomatous. **Culms** 5–160 cm, erect, not rooting at the lower nodes; **nodes** glabrous. **Sheaths** densely pubescent, occasionally glabrous; **ligules** 1–2.2 mm; **blades** to 55 cm long, 2.2–8.3 mm wide, conduplicate (occasionally flat), glabrous below, pubescent above. **Panicles** terminal, with 2–10 racemosely arranged branches; **branches** 0.8–10.3 cm, divergent to spreading, often arcuate, terminating in a spikelet; **branch axes** 0.8–2 mm wide, narrowly winged, glabrous, margins scabrous. **Spikelets** 2.1–3.1 mm long, 2–2.8 mm wide, paired, imbricate, appressed to divergent from the branch axes, orbicular to suborbicular, stramineous. **Lower glumes** absent; **upper glumes** and **lower lemmas** glabrous, 3-veined, margins entire; **upper florets** white to light yellow. **Caryopses** 1.9–2.1 mm, brown. $2n$ = 20, 40.

Paspalum praecox grows in pitcher plant bogs, wet pine flatwoods, wet savannahs, prairies, and wet streamhead ecotones. It is restricted to the United States, growing predominantly on the southeastern coastal plain.

40. **Paspalum hartwegianum** E. Fourn. [p. 598]
HARTWEG'S PASPALUM

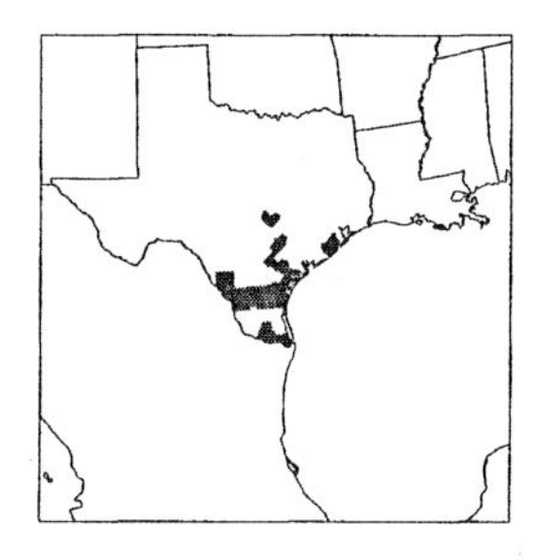

Plants perennial; decumbent or cespitose. **Culms** 50–120 cm, erect; **nodes** glabrous. **Sheaths** glabrous, sparsely pubescent apically; **ligules** 2–5 mm; **blades** to 21 cm long, 2–5 mm wide, flat, glabrous, pubescent behind the ligules, margins ciliate basally. **Panicles** terminal, with 4–9 racemosely arranged branches; **branches** 2–6.5 cm, divergent to erect, terminating in a spikelet; **branch axes** 1.2–1.5 mm wide, winged, glabrous, margins scabrous. **Spikelets** 2.8–3 mm long, 1.5–1.6 mm wide, paired, imbricate, appressed to divergent from the branch axes, elliptic, stramineous. **Lower glumes** absent; **upper glumes** and **lower lemmas** abundantly pubescent, hairs longer than 0.1 mm, 3-veined, margins entire; **lower lemmas** lacking ribs over the veins; **upper florets** 2.5–2.7 mm, white to stramineous. $2n$ = 60.

Paspalum hartwegianum grows in wet prairies, ditches, and swales from southern Texas through Mexico and Central America to Paraguay and Argentina.

41. **Paspalum pubiflorum** Rupr. *ex* E. Fourn. [p. 600]
HAIRYSEED PASPALUM

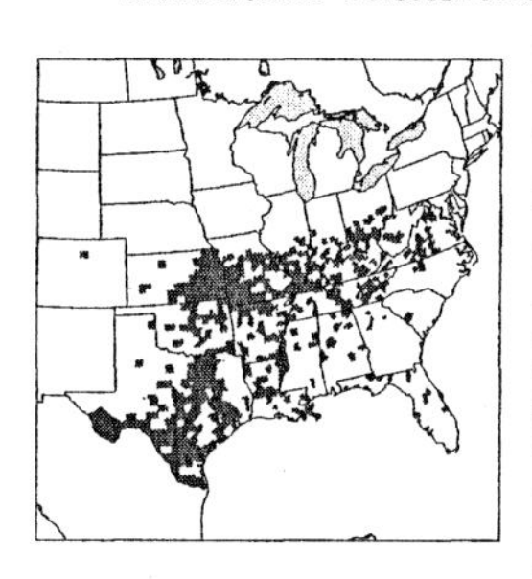

Plants perennial; usually decumbent, rooting at the nodes. **Culms** 30–130 cm, decumbent; **nodes** glabrous or pubescent. **Sheaths** glabrous or pubescent; **ligules** 1–3.2 mm; **blades** to 31 cm long, 4–18 mm wide, flat, glabrous, with a few hairs behind the ligules. **Panicles** terminal, with 2–7 racemosely arranged branches; **branches** 2.2–7.9 cm, divergent to spreading, terminating in a spikelet; **branch axes** 1.1–2.3 mm wide, narrowly winged, glabrous, margins scabrous. **Spikelets** 2.8–3.6 mm long, 1.5–2 mm wide, paired, imbricate, appressed to divergent from the branch axes, elliptic to obovate, pubescent or glabrous, light brown to stramineous. **Lower glumes** absent; **upper glumes** and **lower lemmas** glabrous or sparsely pubescent, hairs shorter than 0.1 mm, 3-veined, margins entire; **lower lemmas** lacking ribs over the veins; **upper florets**

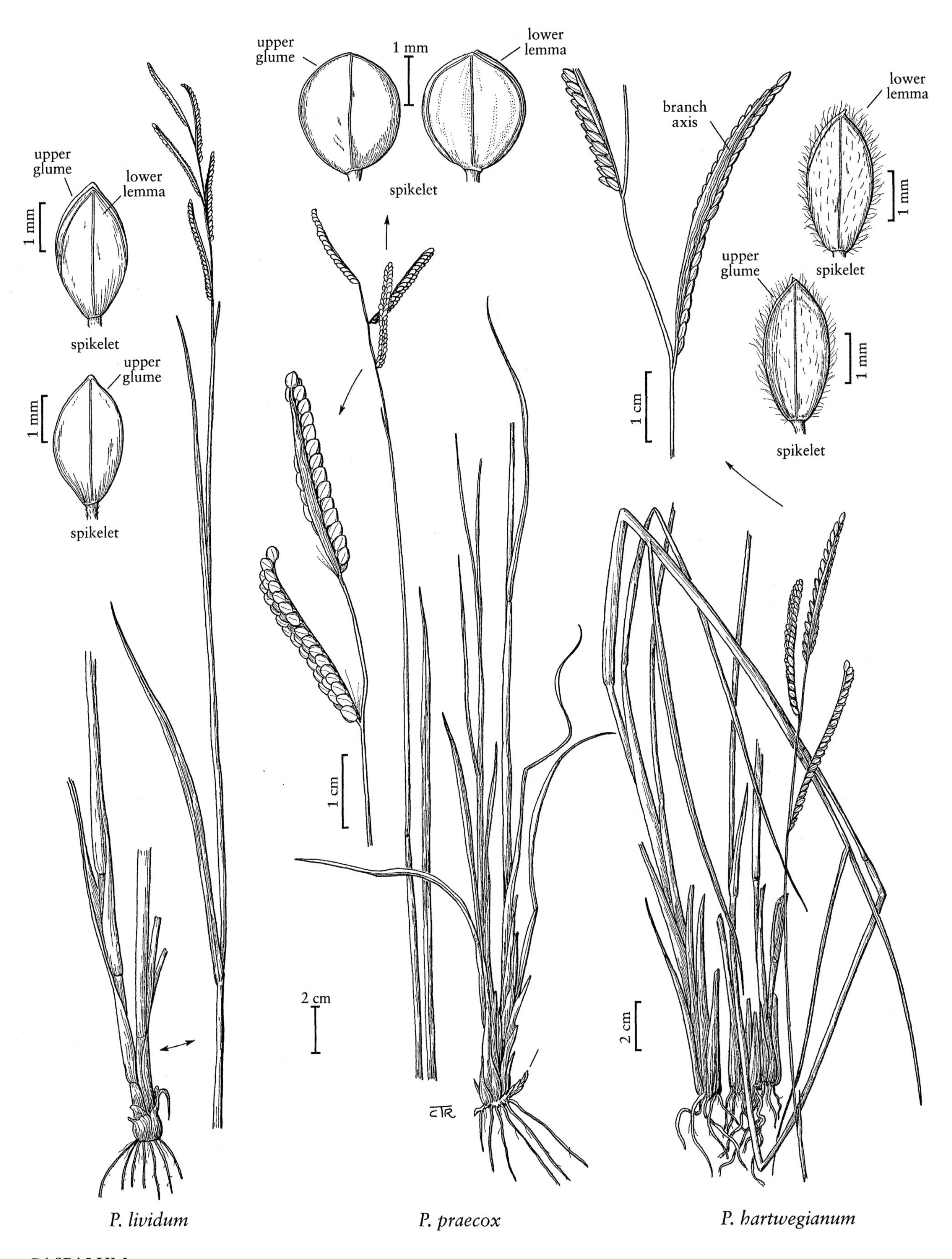
upper glume
lower lemma
1 mm
spikelet
upper glume
1 mm
spikelet
upper glume
1 mm
lower lemma
spikelet
1 cm
2 cm
CTR
branch axis
lower lemma
1 mm
spikelet
upper glume
1 mm
spikelet
1 cm
2 cm
P. lividum
P. praecox
P. hartwegianum

stramineous. **Caryopses** 1.8–2 mm, golden brown or white. $2n$ = 60, ca. 64.

Paspalum pubiflorum grows on the edges of forests and in disturbed areas. It is native to the southeastern United States, Mexico, and Cuba.

42. **Paspalum floridanum** Michx. [p. 600]
FLORIDA PASPALUM

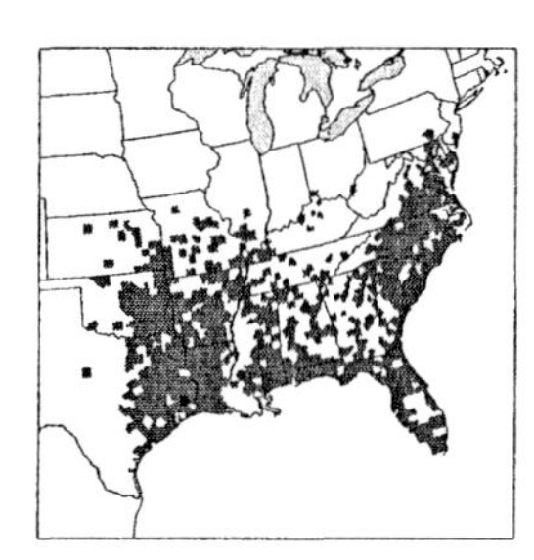

Plants perennial; rhizomatous. **Culms** 80–210 cm, erect; **nodes** glabrous or pubescent. **Sheaths** glabrous or pubescent; **ligules** 1.2–3.3 mm; **blades** to 52 cm long, 3–18 mm wide, flat, glabrous or pubescent, usually densely pubescent behind the ligules. **Panicles** terminal, with 1–6 racemosely arranged branches; **branches** 3–17.9 cm, divergent to erect, terminating in a spikelet; **branch axes** 0.3–1.8 mm wide, glabrous, the margins scabrous. **Spikelets** 2.9–4.1 mm long, 1.9–3.1 mm wide, paired, imbricate, appressed to the branch axes, elliptic to suborbicular to orbicular, glabrous, stramineous. **Lower glumes** absent; **upper glumes** glabrous, 5-veined, margins entire; **lower lemmas** glabrous, lacking ribs over the veins, 3-veined, margins entire; **upper florets** golden brown. **Caryopses** 2.8 mm, amber. $2n$ = 120, 140, ca. 160–170.

Paspalum floridanum grows along the edges of forests, flatwoods, and pinewoods and in open areas. It is a frequent component of dry-mesic soils in longleaf pine-oak-grass ecosystems, and is restricted to the eastern United States.

43. **Paspalum unispicatum** (Scribn. & Merr.) Nash [p. 600]
ONE-SPIKE PASPALUM

Plants perennial; rhizomatous, not rooting at the lower nodes. **Culms** 50–80 cm, erect. **Sheaths** glabrous, pubescent apically, margins scarious; **ligules** 1–2 mm, membranous, lacerate; **blades** 3–4 mm wide, flat, glabrous, pubescent behind the ligules, margins papillose-ciliate. **Inflorescence** terminal, erect, a spicate raceme 7–15 cm long, or a panicle with 1–2 subterminal spicate branches that are wholly or partially enclosed in the upper sheath, often arcuate; **branches** terminating in a spikelet. **Spikelets** 2.7–3 mm, paired, imbricate, obovate, stramineous. **Lower glumes** absent, or 1–2.3 mm; **upper glumes** and **lower lemmas** glabrous, 3-veined, margins entire; **lower lemma** lacking ribs over the veins; **lower florets** often staminate; **lower paleas** 2.5–2.9 mm, membranous; **upper florets** 2.3–2.9 mm, white, stramineous, or golden brown. $2n$ = 40.

Paspalum unispicatum grows in sandy soil in the coastal plain of Texas and extends southward through Mexico and Central America to Cuba and Paraguay, Uruguay, and Argentina. It has not been reported from Brazil.

25.27 **REIMAROCHLOA** Hitchc.

Plants annual; sometimes stoloniferous. **Culms** 10–100 cm, erect to ascending, branching above the base. **Leaves** mostly cauline; **ligules** of hairs; **blades** linear. **Inflorescences** numerous, terminal and axillary, subdigitate or racemose panicles of spikelike branches, spikelets borne singly in 2 rows on the abaxial sides of the branches; **disarticulation** at the base of the spikelets. **Spikelets** dorsally compressed, with 2 florets, upper floret appressed to the branch axes. **Glumes** mostly absent, upper glume sometimes present on the terminal spikelet of a branch; **lower florets** sterile; **upper florets** bisexual; **lower lemmas** subequal to the upper lemmas; **upper lemmas** membranous to coriaceous, margins narrow; **upper paleas** similar in texture to the lemmas, their bases enclosed by the lemmas; **anthers** 1 or 2. x = unknown. Named for J.A.H. Reimarus (1729–1814), a German botanist and professor of natural history and physics at Hamburg, and the Greek *chloa*, 'grass'.

Reimarochloa is a genus of three species, all of which grow in damp habitats. The range of the genus extends from the southern United States to Argentina. One species is native to the *Flora* region.

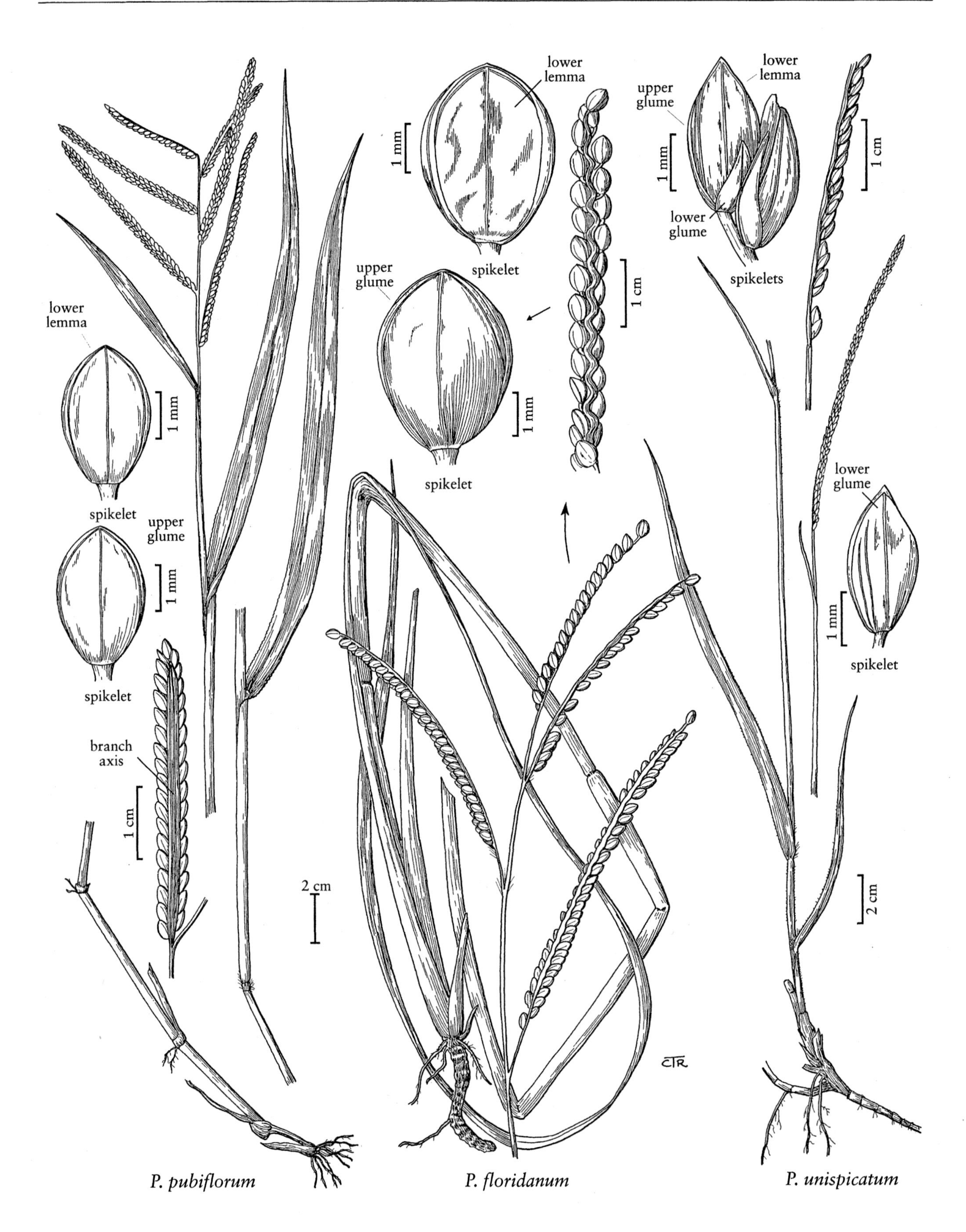
lower
lemma
1 mm
spikelet
upper
glume
1 mm
spikelet
branch
axis
1 cm
2 cm
lower
lemma
1 mm
spikelet
upper
glume
1 mm
spikelet
1 cm
CTR
lower
lemma
upper
glume
1 mm
lower
glume
spikelets
1 cm
lower
glume
1 mm
spikelet
2 cm
P. pubiflorum
P. floridanum
P. unispicatum

1. **Reimarochloa oligostachya** (Munro *ex* Benth.) Hitchc. [p. 601]
FLORIDA REIMARGRASS

Plants stoloniferous. **Culms** 20–40 cm or longer, compressed, decumbent, rooting at the lower nodes. **Sheaths** loose, lower sheaths pubescent, upper sheaths glabrous; **blades** to 13 cm long, 2–4 mm wide, glabrous, scabrous, or pubescent. **Panicle branches** (1)2–3(4), 2.5–8 cm. **Spikelets** 3.8–5.2 mm long, 1–1.5 mm wide, glabrous; **upper lemmas** thinly coriaceous. $2n$ = unknown.

Reimarochloa oligostachya grows in water or wet soil of hammocks, riverbanks, ditches, and disturbed areas. While not common in the *Flora* region, it grows in peninsular Florida, Mobile County, Alabama, and Cuba. The Alabama record is probably an introduction.

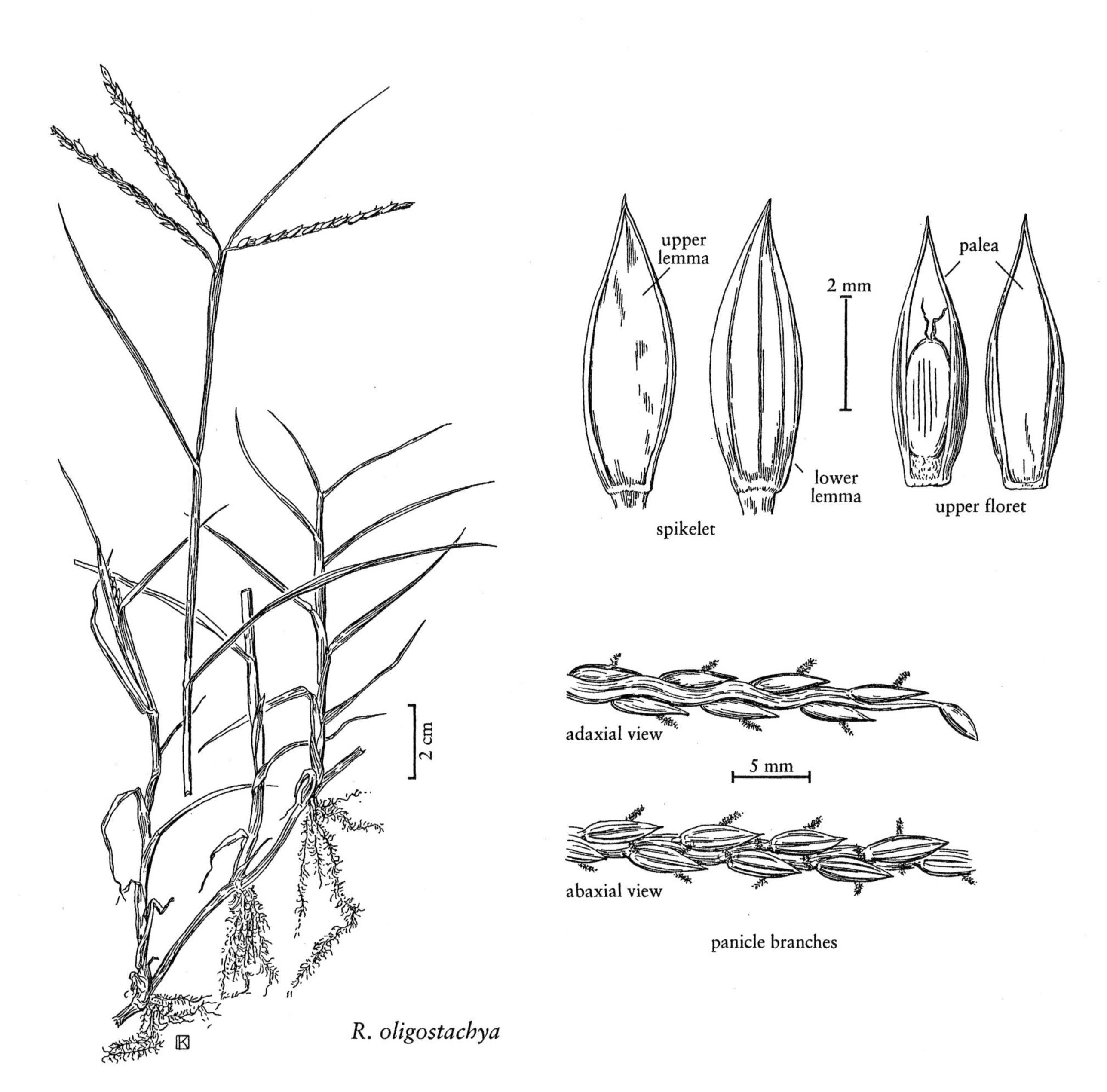

REIMAROCHLOA

26. ANDROPOGONEAE Dumort.

Mary E. Barkworth

Plants usually perennial. **Culms** 7–600 cm, annual, not woody, often reddish or purple, particularly at the nodes, often branched above the base. **Sheaths** open; **ligules** usually scarious to membranous, ciliate or not; **blades** mostly well-developed, leaves subtending an inflorescence or an inflorescence unit often with reduced blades. **Photosynthetic pathway** NADP-ME; **bundle sheaths** single. **Inflorescences** terminal, frequently on both the culms and their branches, sometimes also axillary, usually of 1–many spikelike branches, these in digitate clusters of 1–13+ on a peduncle or attached, directly or indirectly, to elongate rachises, often partially to almost completely enclosed by the subtending leaf sheath at maturity, in some taxa axillary inflorescences composed of multiple-stalked pedunculate clusters of inflorescence branches subtended by a modified leaf; **disarticulation** usually in the branch axes beneath the sessile florets, the dispersal unit being a sessile floret, the internode to the next sessile floret, the pedicel, and the pedicellate spikelet (branches with disarticulating axes are termed *rames* in the following accounts), sometimes beneath the glumes, the branch axes remaining intact. **Spikelets** in unequally pedicellate pairs, sessile-pedicellate pairs, or triplets, or apparently solitary and sessile, pedicellate spikelets and sometimes the pedicels reduced or absent, triplets usually with 1 sessile and 2 pedicellate spikelets, terminal spikelet units on the branches often with 2 pedicellate spikelets even if the others have only 1 (all spikelet units with 2 sessile and 1 pedicellate spikelets in *Polytrias*). **Spikelet pairs** or **triplets** *homogamous* (spikelets in the unit sexually alike) or *heterogamous* (spikelets in the unit sexually dissimilar); **spikelets of unequally pedicellate pairs** usually homogamous and homomorphic; **spikelets in sessile-pedicellate pairs** or **triplets** usually heterogamous and heteromorphic; **sessile spikelets** usually bisexual; **pedicellate spikelets** usually smaller than the sessile spikelets, often staminate or sterile, sometimes absent. **Spikelets** usually with 2 florets (1 in *Polytrias*). **Glumes** exceeding and usually concealing the florets (excluding the awns), rounded or dorsally compressed, usually tougher than the lemmas; **lower florets** in bisexual or pistillate spikelets sterile or staminate, often reduced to a hyaline scale; **upper florets** bisexual or pistillate, lemmas often hyaline, sometimes with an awn that exceeds the glumes; **lodicules** cuneate; **anthers** usually 3. **Pedicels** free or fused to the rachis internodes. **Pedicellate spikelets** variable, sometimes similar to the sessile spikelets, sometimes differing in sexuality and shape, sometimes missing. x = usually 9 or 10, or possibly 5 with 9 and 10 reflecting ancient polyploidy.

The tribe *Andropogoneae* includes about 87 genera and 1060 species, of which 31 genera and 102 species have been found in the *Flora* region; some of these have not become established. The tribe is common in tropical and subtropical regions, particularly in areas with significant summer rains, such as the central plains of North America. Two of the grasses that used to dominate the prairies of central North America, *Andropogon gerardii* and *Schizachyrium scoparium* (Big and Little Bluestem, respectively), are member of the *Andropogoneae*. The reddish-purplish coloration that characterizes the culms and leaves of many *Andropogoneae* gives a striking aspect to grasslands (and lawns) dominated by its members.

Members of the *Andropogoneae* differ from those of *Paniceae* in the reduced lemmas and paleas of their florets and, usually, in their paired, unequally pedicellate spikelets, disarticulating inflorescence branches (*rames*), and the manner in which these branches are aggregated into inflorescences. Unequally pedicellate spikelet pairs are found in many other tribes, but they are more common, and the pedicels more strikingly unequal in length, in the *Andropogoneae*. Recent molecular work supports recognition of the tribe with one modification of its traditional limits, the incorporation of *Arundinella* and *Tristachya* (Kellogg 2000). There is less agreement on the tribe's internal structure and its relationship to the *Paniceae* (Clayton and Renvoize 1986; Kellogg 2000; Spangler 2000; Guissani et al. 2001).

Inflorescence Structures

Describing inflorescence structures in the *Andropogoneae* is not simple. There is a basic pattern, but its many modifications have resulted in great structural diversity. The following paragraphs provide an overview of this diversity and explain the words and phrases used in describing it. Diagrammatic representations of many of the structures mentioned are presented on pages 604 and 605.

Spikelets

Members of the *Andropogoneae,* like those of the *Paniceae*, generally have two florets per spikelet, the lower floret usually being reduced in size and sterile or staminate, and the upper floret bisexual (p. 604). Despite this similarity, spikelets of the two tribes are easy to distinguish. In the *Paniceae*, the lowest glume is usually much shorter than the floret, and the upper florets usually have lemmas that are thicker and tougher than the glumes and lower lemmas. In the *Andropogoneae*, the glumes usually exceed and enclose both florets, and are thicker and tougher than the lemmas. The florets of the *Andropogoneae* contrast strongly with the glumes, having hyaline or thinly membranous lemma bodies and hyaline paleas, or, in many cases, no palea. They are almost always completely concealed by the glumes, except that the upper floret often has an awn that projects beyond the glumes.

In some *Andropogoneae*, the glumes are merely thickly membranous, but most genera have coriaceous or indurate glumes. The lower glumes are sometimes tougher and larger than the upper glumes, and may even conceal the upper glumes as, for example, in *Heteropogon* (p. 681). In such genera, the lower glumes may be mistaken for lemmas. In dioecious species, or monoecious species with strongly differentiated staminate and pistillate spikelets, the staminate spikelets usually have softer glumes than the pistillate spikelets.

Spikelet Units

The basic element of the inflorescence structure in the *Andropogoneae* is the *spikelet unit.* These units usually consist of pairs of spikelets, one sessile and one pedicellate (e.g., *Saccharum bengalense*, p. 615), but they may consist of a pair of unequally pedicellate spikelets (e.g., *Miscanthus sacchariflorus*, p. 619) or of three spikelets (*e.g.*, *Chysopogon fulvus*, p. 635). If there are three spikelets in the unit, one is usually sessile and the other two pedicellate, but a few genera, such as *Polytrias*, have two sessile spikelets and one pedicellate spikelet.

Unequally pedicellate spikelet pairs or triplets are found in other tribes, but in the *Andropogoneae* they usually differ in size, shape, and sexuality. Spikelet units with spikelets that differ in their sexuality are described as *heterogamous*; those with sexually similar spikelets are said to be *homogamous*. Spikelet units with morphologically dissimilar spikelets are *heteromorphic* (e.g., *Andropogon longiberbis*, p. 663); those with morphologically similar spikelets are *homomorphic* (e.g., *Chrysopogon zizanioides*, p. 636). In most *Andropogoneae*, the spikelet units are heterogamous and heteromorphic. The sessile spikelets usually contain a bisexual or pistillate floret, and often exhibit features such as awns and calluses that are related to seed dispersal and establishment (Peart 1984); the pedicellate spikelets are usually staminate, sterile, vestigial, or even absent. In some genera the situation is reversed, the pedicellate spikelets

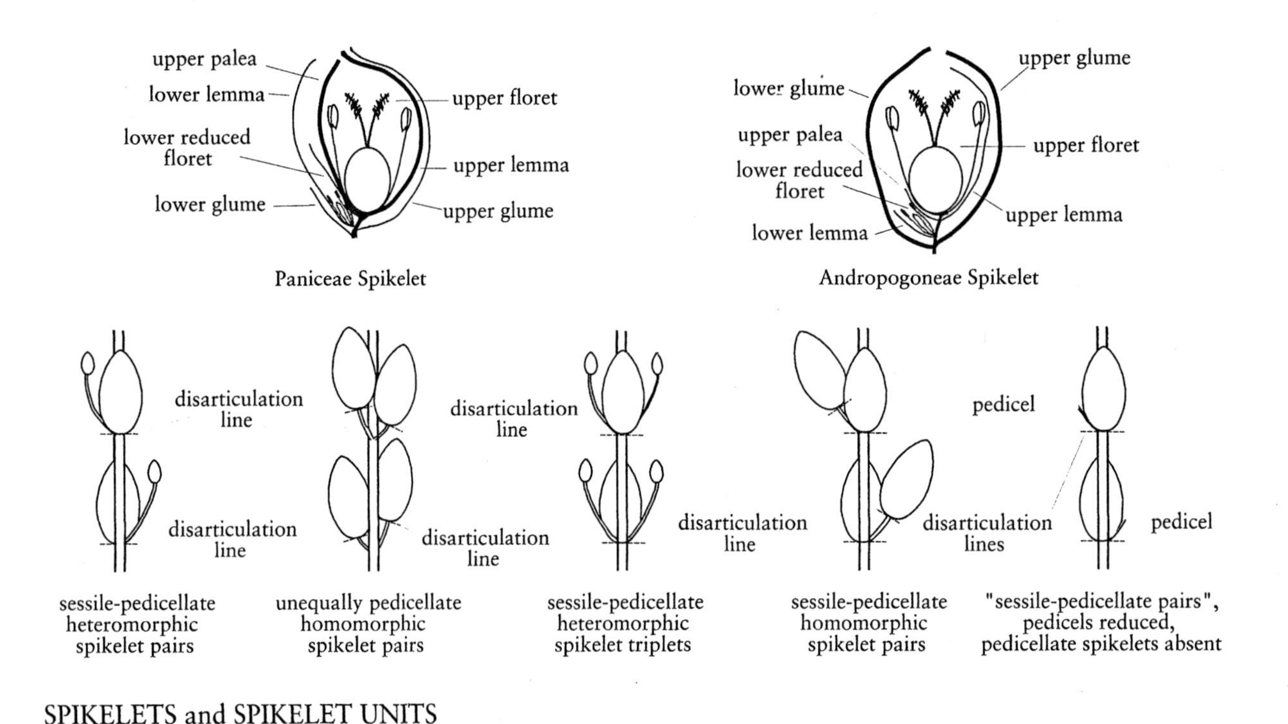

SPIKELETS and SPIKELET UNITS

being bisexual or pistillate, and the sessile spikelets staminate or sterile. Sterile and staminate spikelets are sometimes morphologically similar to the pistillate or bisexual spikelets, but usually lack the features associated with seed dispersal and establishment.

A few genera have no staminate or sterile spikelets, merely empty pedicels associated with the bisexual sessile spikelets, as in *Sorghastrum* (p. 632) or even, as in *Arthraxon* (p. 679), with only a stump where the pedicel and its spikelet would be.

Inflorescence Structure

Further complexity is introduced to the *Andropogoneae* inflorescence structure by the manner in which the spikelet units are aggregated and the mode of disarticulation. Three patterns can be identified. The simplest pattern consists of inflorescences similar to those common in other tribes, in which neither the rachis nor the inflorescence branches break up at maturity. Genera with such inflorescences [e.g., *Miscanthus* (p. 619) and *Imperata* (p. 622)] have unequally pedicellate spikelets, and disarticulation is below the glumes. Such inflorescences are, however, in the minority within the *Andropogoneae*.

A more common situation is for the spikelets to be in sessile-pedicellate pairs and disarticulation to be in the branch axes, immediately below the attachment of the sessile spikelets. The resulting dispersal unit consists of the spikelet pair plus the internode that extends from the sessile spikelet to the next most distal sessile spikelet. These disarticulating inflorescence branches, termed *rames* in this *Flora*, form the basic unit of the typical *Andropogoneae* inflorescence. In other publications, the rames are often called *racemes*, a word that is restricted in this *Flora* to an entire infloresence, not just an inflorescence branch.

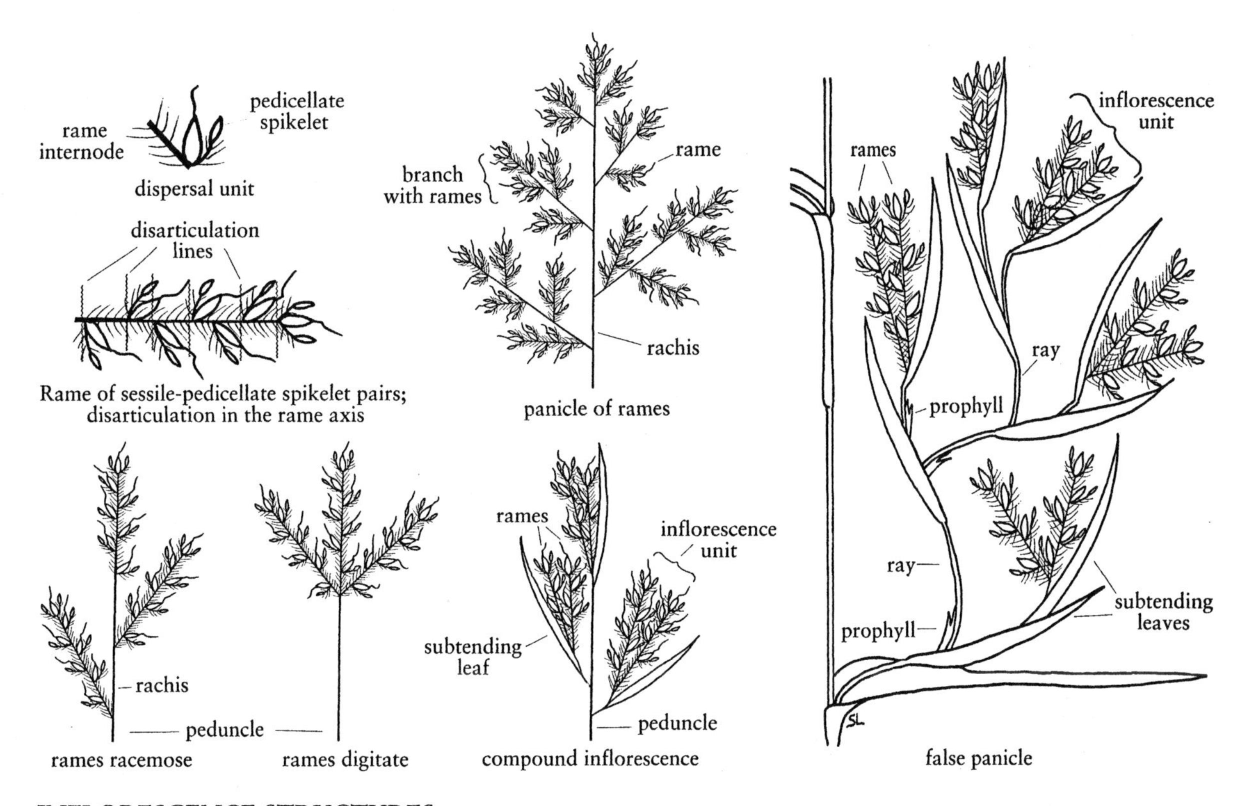

INFLORESCENCE STRUCTURES

Rames are usually composed of several spikelet units, but sometimes of only one. The spikelets may be evenly distributed, or the base of the rame axis may be naked. Individual plants may bear few to many rames, and the rames themselves may be aggregated in a wide array of primary and secondary arrangements; they may also be branched.

One or more rames may be borne on a single stalk. If this stalk is attached to a rachis, the unit formed by the stalk and its rame(s) constitutes an inflorescence branch. Such a pattern is seen, for example, in *Sorghum halepense* (p. 629) and *Bothriochloa bladhii* (p. 647). A more common situation is for one or more rames to be attached digitately to a common stalk, the peduncle. This peduncle may terminate a culm (as in *Dichanthium annulatum* [p. 638] or *Elionurus* [p. 686]) or be axillary to a subtending leaf (as in *Andropogon hallii* [p. 654] and *Hemarthria altissima* [p. 686]). Each peduncle and its associated rame(s) constitutes an *inflorescence unit.*

False panicles represent a further level of complexity. In these, the inflorescence units terminate *rays*, each of which has a *prophyll*, a 2-veined structure, in its axil. Several rays may develop within the axil of a single leaf sheath, and rays may themselves give rise to subtending leaves with multiple rays in their axils. The result is a complex, tiered inflorescence in which only the ultimate units are easily described. Such inflorescences are found, for example, in *Andropogon glomeratus* (p. 663) and *Cymbopogon citratus* (p. 667). Fortunately, identification of the *Andropogoneae* does not require analyzing false panicles, merely their ultimate inflorescence units.

In another inflorescence pattern, the rame axes are thick and the pedicels are either closely appressed or even fused to the rame axes. In these genera, the pedicellate spikelets are often highly reduced or absent. Pistillate rames of *Tripsacum* (p. 697) and wild taxa of *Zea* (p. 700) represent an extreme example of this pattern. In these genera, the sessile spikelets are completely embedded in the rame axes, the lower glumes being indurate and completely concealing the florets. Less extreme examples are seen in *Coelorachis* (p. 689) and *Hackelochloa* (p. 694).

SELECTED REFERENCES **Clayton, W.D.** 1969. A Revision of the Genus *Hyparrhenia*. Kew Bull., Addit. Ser. 2. Her Majesty's Stationery Office, London, England. 196 pp.; **Clayton, W.D.** 1987. Andropogoneae. Pp. 307–309 *in* T.R. Soderstrom, K.W. Hilu, C.S. Campbell, and M.E. Barkworth (eds.). Grass Systematics and Evolution. Smithsonian Institution Press, Washington, D.C., U.S.A. 473 pp.; **Clayton, W.D., G. Davidse, F. Gould, M. Lazarides,** and **T.R. Soderstrom.** 1994. A Revised Handbook to the Flora of Ceylon, vol. 8 (ed. M.D. Dassanayake). Amerind Publishing Co., New Delhi, India. 458 pp.; **Clayton, W.D.** and **S.A. Renvoize.** 1986. Genera Graminum: Grasses of the World. Kew Bull., Addit. Ser. 13. Her Majesty's Stationery Office, London, England. 389 pp.; **Guissani, L.M., J.H. Cota-Sánchez, F.O. Zuloaga,** and **E.A. Kellogg.** 2001. A molecular phylogeny of the grass subfamily Panicoideae (Poaceae) shows multiple origins of C_4 photosynthesis. Amer. J. Bot. 88:1993–2001; **Kellogg, E.A.** 2000. Molecular and morphological evolution in the Andropogoneae. Pp. 149–158 *in* S.W.L. Jacobs and J. Everett (eds.). Grasses: Systematics and Evolution. International Symposium on Grass Systematics and Evolution (3rd:1998). CSIRO Publishing, Collingwood, Victoria, Australia. 408 pp.; **Peart, M.H.** 1984. The effects of morphology, orientation, and position of grass diaspores on seedling survival. J. Ecol. 69:425–436; **Pohl, R.W.** 1968. How to Know the Grasses, rev. ed. Wm. C. Brown Co., Dubuque, Iowa, U.S.A. 244 pp.; **Spangler. R.E.** 2000. Andropogoneae systematics and generic limits in *Sorghum*. Pp. 167–170 *in* S.W.L. Jacobs and J. Everett (eds.). Grasses: Systematics and Evolution. International Symposium on Grass Systematics and Evolution (3rd:1998). CSIRO Publishing, Collingwood, Victoria, Australia. 408 pp.

1. Leaves smelling of lemon oil or citronella, the sheaths without glandular depressions on the keel; plants perennial, not reaching reproductive maturity in the *Flora* region when grown outdoors . . . . . 26.16 *Cymbopogon*
1. Leaves usually not aromatic or, if aromatic and smelling of citronella, the sheaths with glandular depressions along the keel and plants annual; plants reaching reproductive maturity in the *Flora* region.
    2. All spikelets unisexual, the pistillate and staminate spikelets in separate inflorescences or the pistillate spikelets below the staminate spikelets in the same inflorescence.
        3. Pistillate spikelets completely concealed within a hard, globose, beadlike structure (a modified leaf sheath) from which the staminate rames protrude . . . . . 26.30 *Coix*
        3. Pistillate spikelets exposed or enclosed by 1 or more subtending leaf sheaths and a hyaline prophyll; staminate spikelets either distal on the same branch or in a separate inflorescence on the same plant.
            4. Staminate and pistillate spikelets in the same inflorescence and on the same branch, the staminate spikelets distal to the pistillate spikelets . . . . . 26.28 *Tripsacum*
            4. Staminate and pistillate inflorescences usually separate; staminate inflorescences terminal on the culms and branches; pistillate inflorescences terminal on axillary peduncles, sometimes aggregated in false panicles . . . . . 26.29 *Zea*
    2. Some spikelets bisexual (usually the sessile or more shortly pedicellate spikelet of each spikelet pair or triplet).
        5. Spikelets apparently solitary and sessile, the pedicellate spikelets absent; pedicels absent or present.
            6. Culms decumbent, scrambling; leaf blades ovate to ovate-lanceolate; pedicels absent or shorter than 3 mm . . . . . 26.18 *Arthraxon*
            6. Culms erect; leaf blades lanceolate to linear-lanceolate; pedicels always present, usually longer than 3 mm.
                7. Inflorescences terminal and axillary, composed of digitate clusters of 1–13 rames on a common peduncle; peduncles subtended by, and often partially included in, a modified leaf . . . . . 26.15 *Andropogon*
                7. Inflorescences terminal, with elongate rachises and brancheswith several to many rames; peduncles and branches not subtended by a modified leaf . . . . . 26.09 *Sorghastrum*
        5. Spikelets in sessile-pedicellate or unequally pedicellate pairs or triplets, the pedicellate spikelets often smaller than the sessile spikelets, sometimes rudimentary.
            8. Pedicels strongly appressed or fused to the thick rame axes, or the rames with only 1 spikelet unit, this a triplet with 2 unequally pedicellate spikelets; bisexual spikelets usually unawned; inflorescences of rames.
                9. Lower glumes of the sessile spikelets rugose, pitted, tuberculate, or alveolate or the keels winged or with spinelike projections at the base.
                    10. Keels of the lower glumes with spinelike projections on the base, sometimes winged distally, the surface between the keels smooth; spikelets unawned . . . . 26.25 *Eremochloa*
                    10. Keels of the lower glumes winged throughout or not winged, the surface between the keels rough, rugose, pitted, tuberculate, or alveolate; spikelets unawned or awned.

11. Sessile spikelets awned . . . . 26.13 *Ischaemum*
11. Sessile spikelets unawned.
12. Plants perennial; sessile spikelets ovate, the lower glumes smooth, rugose, or pitted . . . . 26.24 *Coelorachis*
12. Plants annual; sessile spikelets hemispherical, the lower glumes alveolate . . . . 26.27 *Hackelochloa*
9. Lower glumes of the sessile spikelets smooth or scabrous, not sculptured, the keels without spinelike projections.
13. Inflorescences false panicles; individual rames to 1 cm long, with 1 spikelet unit; spikelet units composed of 1 sessile and 2 unequally pedicellate and dissimilar spikelets . . . . 26.14 *Apluda*
13. Inflorescences usually solitary rames, sometimes with 2 rames in a digitate cluster; individual rames 2–15 cm long, with more than 1 spikelet unit; spikelet units composed of sessile pedicellate pairs, the pedicellate spikelets often rudimentary or absent.
14. Pedicels appressed, but not fused, to the rame axes.
15. Pedicellate spikelets 1–3 mm long . . . . 26.24 *Coelorachis*
15. Pedicellate spikelets 4–8 mm long . . . . 26.22 *Elionurus*
14. Pedicels at least partially fused to the rame axes.
16. Plants perennial; sheaths mostly glabrous, sparsely ciliate basally . . . 26.23 *Hemarthria*
16. Plants annual; sheaths with stiff, papillose-based hairs 1–3 mm long . . . . 26.26 *Rottboellia*
8. Pedicels free; rame or branch internodes slender, sometimes thickened distally; bisexual spikelets usually awned; inflorescences of rames with the spikelets in sessile-pedicellate pairs or of non-disarticulating branches with the spikelets in unequally pedicellate pairs.
17. All spikelet units homogamous, frequently also homomorphic.
18. Terminal inflorescences a single rame or a digitate or subdigitate cluster of rames.
19. Terminal inflorescences a digitate or subdigitate cluster of (1)2–6 rames; rames 3–7 cm long . . . . 26.06 *Microstegium*
19. Terminal inflorescences solitary rames; rames 2–3 cm long . . . . 26.05 *Polytrias*
18. Terminal inflorescences with elongated rachises.
20. Spikelets in unequally pedicellate pairs; disarticulation below the glumes, the branches remaining intact at maturity.
21. Spikelets usually awned; inflorescence branches usually 7–35 cm long . . . . 26.03 *Miscanthus*
21. Spikelets unawned; inflorescence branches 1–7 cm long . . . . 26.04 *Imperata*
20. Spikelets in sessile-pedicellate pairs or triplets; disarticulation in the rames, below the sessile spikelets.
22. Culms to 100 cm tall, often decumbent and straggling; terminal inflorescences composed of 2–6 subdigitately to racemosely arranged rames . . . . 26.06 *Microstegium*
22. Culms 40–600 cm tall, erect; terminal inflorescences panicles, with more than 6 primary branches; branches usuallywith 2 or more rames.
23. Panicle branches alternate, with multiple rames; rames with more than 5 spikelet units . . . . 26.02 *Saccharum*
23. Panicle branches subverticillate, with 1–3 rames; rames with 2–5 spikelet units . . . . 26.01 *Spodiopogon*
17. All or most spikelet units heterogamous, usually also heteromorphic, sometimes the proximal units on the rames or racemes homomorphic and homogamous.
24. Terminal inflorescences with elongated rachises.
25. Rame internodes and pedicels with a translucent median line . . . . 26.12 *Bothriochloa*
25. Rame internodes and pedicels without a translucent median line.
26. Sessile spikelets terete or laterally compressed, calluses usually sharp, sometimes blunt . . . . 26.10 *Chrysopogon*

26. Sessile spikelets dorsally compressed, calluses usually blunt, sometimes sharp . . . . . 26.08 *Sorghum*

24. Terminal inflorescences and individual inflorescence units without elongated rachises, composed of 1–13 rames, or a raceme in which disarticulation occurs below the pedicellate spikelets.
    27. Inflorescences composed of rames, disarticulation being in the rame axes; rame internodes and pedicels with a translucent median groove . . . . . 26.12 *Bothriochloa*
    27. Inflorescences composed of rames, with disarticulation in the axes or a spikelike raceme (occasionally of 2 subdigitate spikelike branches) with disarticulation below the pedicellate spiklets; inflorescence internodes and pedicels without a translucent median groove.
        28. All spikelet units in the inflorescence heterogamous.
            29. Disarticulation occurring below the pedicellate spikelets, not in the inflorescence axes; pedicellate spikelets bisexual and awned, awns (4)6–15 cm long, pilose on the column, the hairs 1–2 mm long; sessile or subsessile spikelets staminate or sterile, unawned . . . . . 26.07 *Trachypogon*
            29. Disarticulation occurring below the sessile spikelets, in the inflorescence axes; pedicellate spikelets staminate, sterile, rudimentary, or absent, unawned or with awns to 6 mm long; sessile spikelets bisexual or pistillate, awned.
                30. Rames usually solitary on the peduncles, occasionally 2; rame internodes cupulate or fimbriate distally; lower glumes of the sessile spikelets veined between the keels . . . . . 26.17 *Schizachyrium*
                30. Rames usually 2–13 on the peduncles, occasionally solitary; rame internodes neither fimbriate nor cupulate distally; lower glumes of the sessile spikelets usually without veins between the keels . . . . . 26.15 *Andropogon*
        28. Basal spikelet units on each rame homomorphic and homogamous, sterile or staminate, unawned.
            31. Awns 5–15 cm long; rames with 3–10 homogamous spikelet units . . . . . 26.20 *Heteropogon*
            31. Awns 1–5(19) cm long; rames with 1–2 homogamous spikelet units.
                32. Inflorescences terminal on the culms, axillary inflorescences not present or few in number . . . . . 26.11 *Dichanthium*
                32. Inflorescences terminal and axillary, axillary inflorescences numerous.
                    33. Homogamous spikelets distinctive, forming an involucre around the rame bases . . . . . 26.21 *Themeda*
                    33. Homogamous spikelets not distinctive, not forming an involucre around the rame bases . . . . . 26.19 *Hyparrhenia*

26.01 SPODIOPOGON Trin.

Mary E. Barkworth

Plants usually perennial; sometimes rhizomatous. **Culms** 40–150 cm, erect, simple or branching. **Leaves** not aromatic; **ligules** membranous; **blades** lanceolate to broadly linear, sometimes pseudopetiolate. **Inflorescences** terminal, open or contracted panicles, with evident rachiseswith numerous subverticellate branches that terminate in 1–3 short rames; **rames** with slender internodes and 2–5 sessile-pedicellate homogamous spikelet pairs; **disarticulation** in the rames, below the sessile spikelets. **Spikelets** usually lanceolate. **Glumes** equal, chartaceous, often pilose, scarcely keeled, with several raised veins, acute; **calluses** glabrous or densely hairy; **lower florets** usually staminate, unawned; **upper florets** bisexual; **upper lemmas** bilobed, with a geniculate awn; **anthers** 3. x = 10. **Pedicels** slender, not fused to the rames axes. Name from the Greek *spodios*, 'ash-colored' or 'gray', and *pogon*, 'beard', a reference to the spikelet hairs.

Spodiopogon is a genus of 10–15 species, most of which grow in subtropical regions of the Eastern Hemisphere, although *Spodiopogon sibiricus* extends north to Irkutsk, Russia. One species is cultivated in the *Flora* region.

1. **Spodiopogon sibiricus** Trin. [p. 610]
SILVER SPIKE

Plants rhizomatous. **Culms** 90–150 cm tall, 2–4 mm thick. **Basal leaves** bladeless or with reduced blades; **cauline sheaths** mostly glabrous, but pilose at the collar; **ligules of cauline leaves** 2–3 mm, ciliate on the erose margin; **cauline blades** to 35 cm long, 8–20 mm wide, pilose on both surfaces, margins ciliate near the base, cilia papillose-based. **Panicles** 12–20 cm long, 2–4 cm wide, shortly exserted; **rachises** glabrous, smooth; **primary branches** 2–6 cm; **rames** 2–3.5 cm. **Spikelets** 4.5–5.5 mm. **Lower glumes** pilose throughout, 5–9-veined; **upper glumes** of the sessile spikelets pilose on the margins, those of the pedicellate spikelets pilose throughout; **callus hairs** about ¼ as long as the spikelets; **awns** 0.7–1.2 cm; **anthers** of the sessile spikelets about 2 mm, those of the pedicellate spikelets about 3 mm. $2n$ = 40, 42.

Spodiopogon sibiricus is native to the grasslands of the montane regions that extend from central China to northeastern Siberia. It is grown as an ornamental in Canada and the contiguous United States.

26.02 SACCHARUM L.

Robert D. Webster

Plants perennial; cespitose, often with a knotty crown, sometimes rhizomatous, rhizomes usually short but elongate in some species, rarely stoloniferous. **Culms** 0.8–6 m, erect. **Leaves** cauline, not aromatic; **sheaths** usually glabrous, sometimes ciliate at the throats; **ligules** membranous, ciliate; **blades** flat, lax, smooth, usually glabrous. **Inflorescences** terminal, large, often plumose, fully exserted panicles with evident rachises and numerous, ascending to appressed branches terminating in multiple rames, branches alternate, sometimes naked below; **rames** with numerous sessile-pedicellate spikelet pairs and a terminal triad of 1 sessile and 2 pedicellate spikelets, internodes slender, without a translucent median groove; **disarticulation** beneath the pedicellate spikelets and in the rames beneath the sessile spikelets, sessile spikelets falling with the adjacent internode and pedicel. **Spikelet pairs** homogamous and homomorphic, or almost so, not embedded in the rame axes, dorsally compressed. **Sessile spikelets: calluses** truncate, usually with silky hairs; **glumes** subequal, chartaceous to coriaceous, glabrous or villous, 2-keeled, veins not raised; **lower florets** sterile; **lower lemmas** hyaline or membranous; **lower paleas** absent or vestigial, entire; **upper florets** bisexual; **upper lemmas** entire or bidentate, muticous or awned; **lodicules** 2, truncate; **anthers** 2 or 3. **Pedicels** neither appressed nor fused to the rame axes.

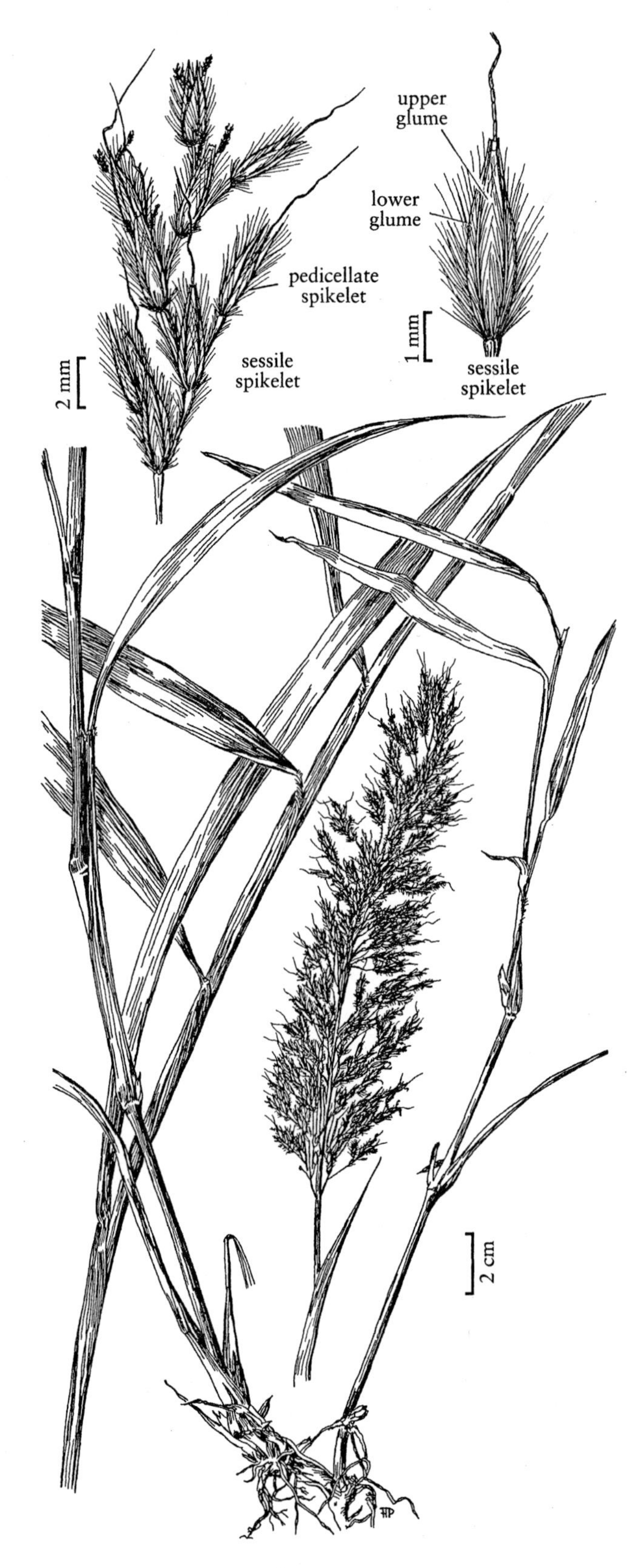

S. sibiricus

SPODIOPOGON

Pedicellate spikelets well developed, from slightly shorter than to equaling the sessile spikelets. x = 10. Name from the Latin *saccharum*, 'sugar', a reference to the sweet juice.

Saccharum is a genus of 35–40 species that grow throughout the tropics and subtropics. Nine species can be found in the *Flora* region; five are native, two are grown as ornamentals, one is grown for agriculture, and one for research. Some species of *Saccharum* hybridize naturally with other, presumably closely related, genera such as *Miscanthus*, *Imperata*, and *Sorghum*. Species with awned lemmas are sometimes placed in a separate genus, *Erianthus*. The most familiar species of *Saccharum* is *S. officinarum*, sugar cane.

SELECTED REFERENCE Webster, R.D. and R.B. Shaw. 1995. Taxonomy of the North American species of *Saccharum* (Poaceae: Andropogoneae). Sida 16:551–580.

1. Spikelets unawned, or with awns less than 5 mm long; anthers 3.
   2. Spikelets with visible awns, the awns 2–5 mm long . . . . 6. *S. ravennae*
   2. Spikelets unawned, or the awns concealed by the glumes.
      3. Lower glumes of sessile spikelets pubescent . . . . 9. *S. bengalense*
      3. Lower glumes of sessile spikelets mostly glabrous, sometimes ciliate distally.
         4. Culms clumped, 2–5 cm thick; rhizomes short; blades 20–60 mm wide . . . . 8. *S. officinarum*
         4. Culms solitary or few together, 0.6–2 cm thick; rhizomes elongate; blades 10–25 mm wide . . . . 7. *S. spontaneum*
1. Spikelets awned, the awns 10–26 mm long; anthers 2.
   5. Awns spirally coiled at the base.
      6. Callus hairs 3–7 mm long, equal to or shorter than the spikelets, white to brown; rachises glabrous or sparsely pilose . . . . 3. *S. brevibarbe*
      6. Callus hairs 9–14 mm long, exceeding the spikelets, silvery or tinged with purple; rachises densely pubescent . . . . 2. *S. alopecuroides*
   5. Awns straight to curved at the base.
      7. Callus hairs longer than the spikelets; lowest panicle nodes densely pilose . . . . 1. *S. giganteum*
      7. Callus hairs absent or no more than equaling the spikelets; lowest panicle nodes glabrous or sparsely pilose.
         8. Calluses glabrous or with hairs to 2 mm long and exceeded by the spikelets; panicles 1–2.5 cm wide . . . . 5. *S. baldwinii*
         8. Callus hairs 3–7 mm long, often equaling the spikelets; panicles 3–10 cm wide.
            9. Awns flat basally; lower lemmas of sessile spikelets not or indistinctly veined; upper lemmas 0.9–1 times as long as the lower lemmas . . . . 3. *S. brevibarbe*
            9. Awns terete basally; lower lemmas of the sessile spikelets typically 3-veined; upper lemmas 0.7–0.8 times as long as the lower lemmas . . . . 4. *S. coarctatum*

1. Saccharum giganteum (Walter) Pers. [p. 613]
SUGARCANE PLUMEGRASS

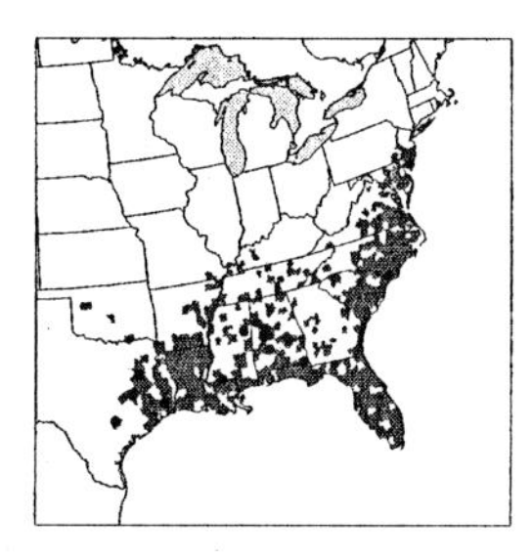

Plants rhizomatous. **Culms** 1–2.5 m; **nodes** sericeous, hairs to 5 mm. **Sheaths** glabrate or glabrous; **auricles** absent; **ligules** 2–6 mm; **blades** usually 35–70 cm long, 8–30 mm wide, adaxial surfaces glabrous or pilose. **Peduncles** 40–80 cm, pilose; **panicles** 6–15 cm wide, oblong or lanceolate; **rachises** 15–30 cm, pilose; **lowest nodes** densely pilose; **primary branches** 2–13 cm, ascending or appressed to the rachises; **rame internodes** 2–5.5 mm, pilose. **Sessile spikelets** 4.2–6 mm long, 0.8–1.1 mm wide, straw-colored. **Callus hairs** (7)15–20(25) mm, longer than the spikelets, straw-colored or brown; **glumes** usually glabrous; **lower glumes** smooth, indistinctly 5-veined; **lower lemmas** 3–5 mm, without veins; **upper lemmas** 2.5–3.5 mm, 1-veined, entire; **awns** 12–26 mm, straight or curved, terete basally; **lodicule veins** sometimes extending into hairlike projections; **anthers** 2. **Pedicels** 2.5–5 mm, pilose. **Pedicellate spikelets** similar to the sessile spikelets, except frequently pilose. $2n$ = 30, 60, 90.

Saccharum giganteum grows in wet soils of bogs, swales, and swamps. Its range extends from the eastern and southeastern United States to Central America. It is a polymorphic, primarily chasmogamous species that intergrades morphologically with the primarily cleistogamous **S. trinii** (Hack.) Renvoize in Central America. The combination of long callus hairs and straight awns distinguishes it from all other species of *Saccharum* in the *Flora* region.

2. Saccharum alopecuroides (L.) Nutt. [p. 613]
SILVER PLUMEGRASS

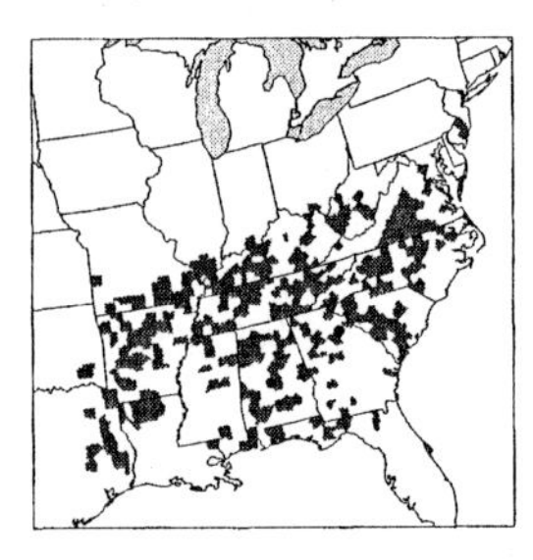

Plants rhizomatous. **Culms** 1–2.5 m; **nodes** hairy, occasionally glabrate, hairs 7–12 mm. **Sheaths** mostly glabrous, ciliate distally; **auricles** absent; **ligules** 1–3 mm; **blades** 30–60 cm long, 14–28 mm wide, glabrous at maturity. **Peduncles** 40–60 cm, pilose; **panicles** 3–10 cm wide, oblong to lanceolate; **rachises** 15–34 cm, densely pilose; **lowest nodes** glabrous or sparsely pilose; **primary branches** 3–12 cm, appressed; **rame internodes** 3–5 mm, pilose. **Sessile spikelets** 6–7 mm long, 1.1–1.4 mm wide, straw-colored; **callus hairs** 9–14 mm, exceeding the spikelets, silvery or purple-tinged; **lower glumes** 5-veined, smooth; **upper glumes** 3–5-veined; **lower lemmas** 4.8–5.6 mm, without veins or 1-veined; **upper lemmas** 4–4.6 mm, 1-veined, bifid, teeth 1.8–2 mm, ciliate; **awns** 14–20 mm, flattened and spirally coiled at the base; **lodicule veins** not extending into hairlike projections; **anthers** 2. **Pedicels** 2.5–4 mm, pilose. **Pedicellate spikelets** similar to the sessile spikelets, except frequently pilose. $2n$ = 30.

Saccharum alopecuroides grows in damp woods, open areas, and field margins. It is restricted to the southeastern United States. It is rare or non-existent on the sandy coastal plain, and there are few specimens from southern Florida and the higher elevations of the Appalachian Mountains. The combination of long rhizomes, long silvery callus hairs, and spirally coiled awns distinguish *S. alopecuroides* from all other species in the region.

3. Saccharum brevibarbe (Michx.) Pers. [p. 613]
SHORTBEARD PLUMEGRASS

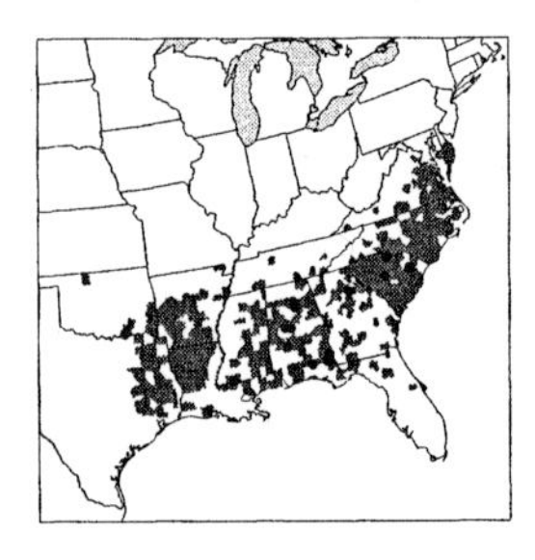

Plants rhizomatous. **Culms** 0.8–2.5 m; **nodes** glabrous or pubescent. **Sheaths** not ciliate; **auricles** absent; **ligules** 1–2 mm; **blades** usually 40–60 cm long, 7–25 mm wide, glabrous. **Peduncles** 45–75 cm, usually glabrous, occasionally pubescent or minutely pilose; **panicles** 3–10 cm wide, linear or oblong; **rachises** (10)30–50 cm, glabrous or sparsely pilose; **lowest nodes** glabrous or sparsely pilose; **primary branches** 7–14 cm, appressed; **rame internodes** 4–6 mm, with hairs. **Sessile spikelets** 6.5–10.5 mm long, 1.2–1.5 mm wide, purple or straw-colored. **Callus hairs** 3–7 mm, from shorter than to equaling the spikelets, white to straw-colored or brown; **lower glumes** 5-veined, smooth basally, scabrous distally; **lower lemmas** 5.5–8 mm, not or indistinctly veined, initially entire, sometimes becoming bifid, teeth 2–2.5 mm; **upper lemmas** 5.5–8 mm, 0.9–1 times as long as the lower lemmas, 3-veined, entire or bifid; **awns** 10–22 mm, always flattened below, sometimes spirally coiled; **lodicule veins** sometimes extending as hairlike projections; **anthers** 2. **Pedicels** 3–4 mm, with hairs. **Pedicellate spikelets** similar to the sessile spikelets. $2n$ = 60.

Saccharum brevibarbe grows only in the southeastern United States.

1. Awns 15–22 mm long, straight or sinuous at the base; upper lemmas of the sessile spikelets entire at maturity . . . . var. *brevibarbe*
1. Awns 10–18 mm long, spirally coiled at the base, usually with 2–4 coils; upper lemmas of the sessile spikelets bifid at maturity, teeth about 2–2.5 mm long . . . . var. *contortum*

Saccharum brevibarbe (Michx.) Pers. var. **brevibarbe**

Upper lemmas of sessile spikelets entire at maturity; **awns** 15–22 mm, straight or sinuous throughout.

Saccharum brevibarbe var. *brevibarbe* grows in the southeastern coastal states and is common in central and southern Arkansas, eastern Oklahoma, the piney woods region of eastern Texas, and northern Louisiana.

Saccharum brevibarbe var. **contortum** (Baldwin) R.D. Webster

Upper lemmas of sessile spikelets initially entire, becoming bifid at maturity, teeth about 2 mm; **awns** 10–18 mm, spirally coiled basally.

Saccharum brevibarbe var. *contortum* grows in moist, sandy pinelands and open ground of the coastal plain, from Maryland to Florida and inland to Tennessee and Oklahoma. Initially, the awns in var. *contortum* are not coiled and the lemmas are entire but, as the spirals develop, they tear the lemmas, creating the bifid apices.

4. Saccharum coarctatum (Fernald) R.D. Webster [p. 613]
COMPRESSED PLUMEGRASS

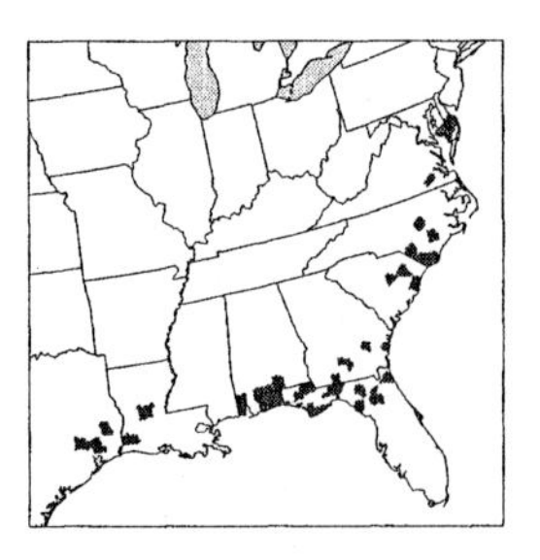

Plants cespitose, not or shortly rhizomatous. **Culms** 1–2.5 m; **nodes** with 1–3 mm hairs. **Sheaths** glabrous; **auricles** 0.3–3 mm; **ligules** 1–2 mm; **blades** 15–40 cm long, 7–12 mm wide. **Peduncles** 35–45 cm, glabrous; **panicles** 3–7 cm wide, linear to oblong; **rachises** 13–35 cm, glabrous or sparsely pilose; **primary branches** 5–12 cm, appressed; **rame internodes** 3–6 mm, with hairs. **Sessile spikelets** 6–8 mm long, 0.9–1.2 mm wide, brown. **Callus hairs** 3–5 mm, from shorter than to equaling the spikelets, white or straw-colored; **lower glumes** smooth

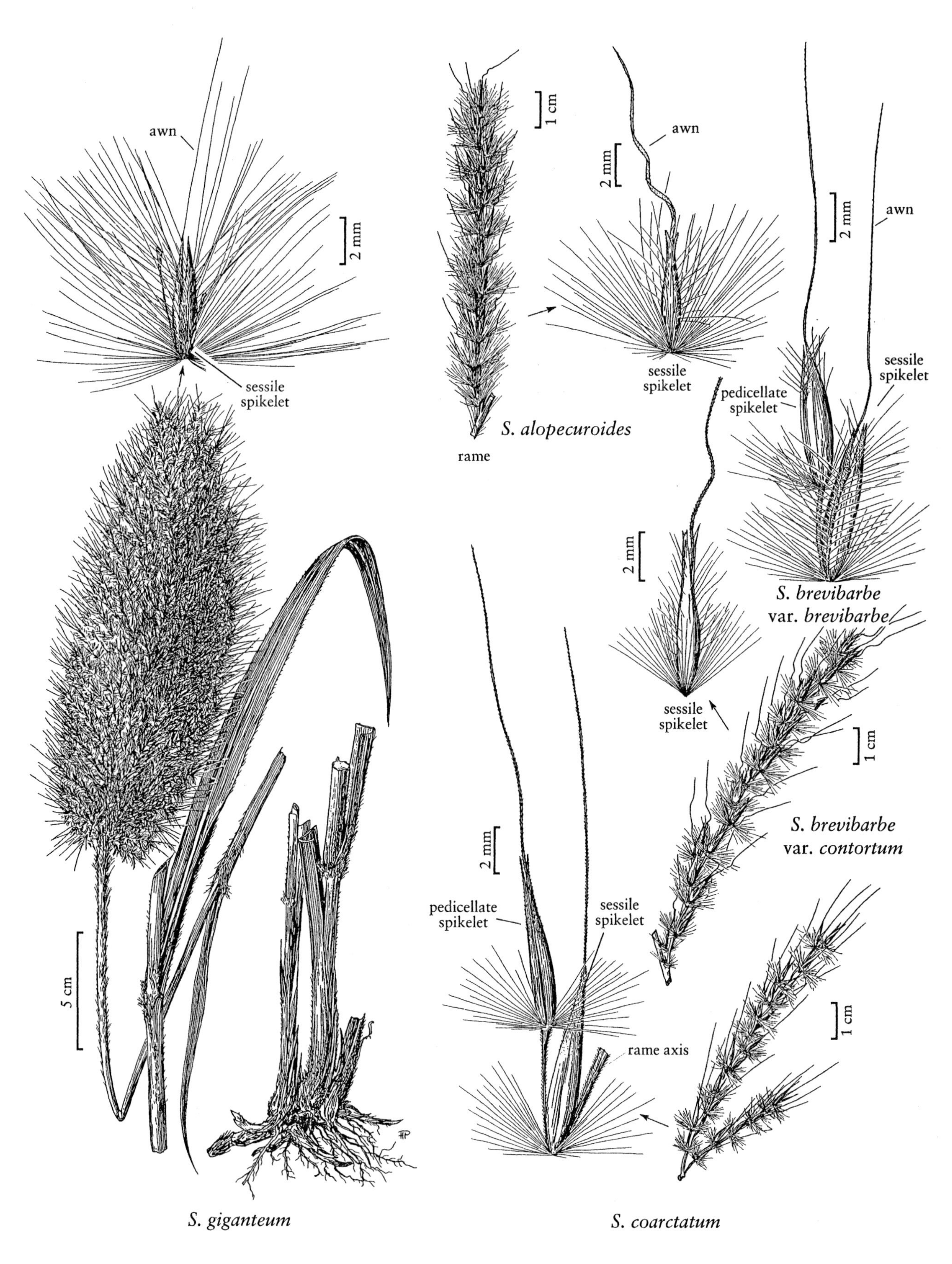
awn
2 mm
sessile spikelet
5 cm
S. giganteum
1 cm
rame
S. alopecuroides
awn
2 mm
sessile spikelet
2 mm
awn
sessile spikelet
pedicellate spikelet
S. brevibarbe var. brevibarbe
2 mm
sessile spikelet
1 cm
S. brevibarbe var. contortum
2 mm
pedicellate spikelet
sessile spikelet
rame axis
1 cm
S. coarctatum

SACCHARUM

or scabrous, 5-veined; **lower lemmas** 5.8–7.5 mm, usually 3-veined; **upper lemmas** 4–5.5 mm, 0.7–0.8 times as long as the lower lemmas, 3-veined, entire; **awns** 16–26 mm, terete and straight to curving basally; **lodicule veins** extending into hairlike projections to 0.6 mm long; **anthers** 2. **Pedicels** 3–5 mm, sparsely and shortly pilose. **Pedicellate spikelets** similar to the sessile spikelets. 2*n* = 60.

Saccharum coarctatum is common in wet, peaty or sandy soils of swales, pond margins, and meadows of the coastal plain of the southeastern United States. It is unusual in having lodicule veins that extend into hairlike projections up to 0.6 mm long.

5. **Saccharum baldwinii** Spreng. [p. 615]
NARROW PLUMEGRASS

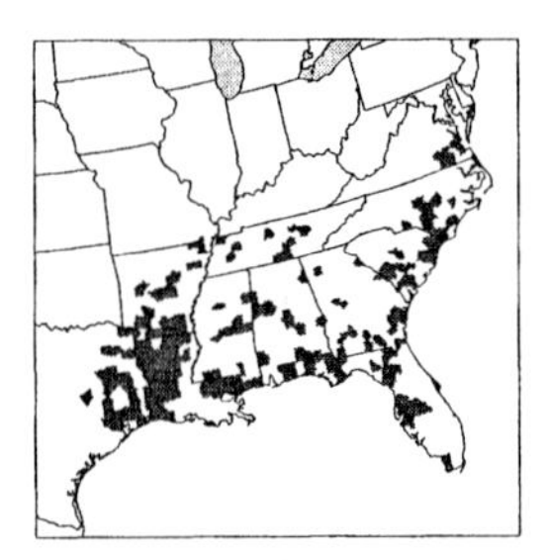

Plants cespitose, rarely stoloniferous. **Culms** 0.9–1.8 m; **nodes** glabrous or with hairs to 0.5 mm. **Sheaths** glabrous; **ligules** 1–3 mm, with lateral lobes; **blades** 18–60 cm long, 5–12 mm wide, glabrous. **Peduncles** 30–40 cm, glabrous; **panicles** 1–2.5 cm wide, linear; **lowest nodes** glabrous or sparsely pilose; **rachises** 10–35 cm, glabrous or sparsely pubescent; **primary branches** 6–18 cm, appressed; **rame internodes** 3–5 mm, glabrous. **Sessile spikelets** 7–10 mm long, 1.1–1.5 mm wide, brown. **Callus hairs** absent or to 2 mm, shorter than the spikelets, straw-colored; **lower glumes** scabrous, 5-veined; **lower lemmas** 6–8 mm, 2-veined; **upper lemmas** 0.9–1 times as long as the lower lemmas, 3-veined, entire; **awns** 17–24 mm, terete, straight or curved at the base; **lodicule veins** extending into hairlike projections; **anthers** 2. **Pedicels** 3–5 mm, glabrous. **Pedicellate spikelets** similar to the sessile spikelets. 2*n* = 30.

Saccharum baldwinii commonly grows in sandy, shaded river and stream bottoms. It occurs throughout the southeastern United States, but it is not as common as other members of the genus, and is rare or completely absent from higher elevations of the Appalachian Mountains.

6. **Saccharum ravennae** (L.) L. [p. 615]
RAVENNAGRASS

Plants cespitose. **Culms** 2–4 m, glabrous; **nodes** glabrous. **Sheaths** glabrous; **auricles** absent; **ligules** 0.6–1.1 mm; **blades** 50–100 cm long, 5–14 mm wide, glabrous. **Peduncles** 40–80 cm, glabrous; **panicles** lanceolate; **rachises** 30–70 cm, glabrous; **primary branches** 6–20 cm, appressed or spreading; **rame internodes** 1–2 mm, with hairs. **Sessile spikelets** 4–6 mm long, 0.7–0.9 mm wide, straw-colored. **Callus hairs** 4–6 mm, subequal to the spikelets, white; **lower glumes** smooth, 4–5-veined; **upper glumes** 3-veined; **lower lemmas** 3–5 mm, 1-veined; **upper lemmas** subequal to the lower lemmas, without veins, entire; **awns** 2–5 mm, flat, straight or curved at the base; **lodicule veins** not extending into hairlike projections; **anthers** 3. **Pedicels** 1–3 mm, pubescent. **Pedicellate spikelets** similar to the sessile spikelets. 2*n* = 20.

Saccharum ravennae is native to southern Europe and western Asia. It is grown as an ornamental in the *Flora* region, occasionally escaping and persisting.

7. **Saccharum spontaneum** L. [p. 615]
WILD SUGARCANE

Plants with long rhizomes. **Culms** 2–4 m tall, 0.6–2 cm thick, solitary or few together. **Sheaths** usually glabrous; **ligules** 1.5–3 mm; **blades** 50–100 cm long, 10–25 mm wide, usually glabrous, markedly hirsute above the ligules. **Peduncles** pilose; **panicles** 40–70 cm, narrowly oblong to widely ovate, **rachises** 25–50 cm, densely pilose; **primary branches** 2.5–7 cm. **Sessile spikelets** 3.5–7 mm. **Callus hairs** to 12 mm; **glumes** glabrous over the back, ciliate toward the tip; **lower lemmas** about 3 mm; **upper lemmas** subequal to the lower lemmas, entire; **awns** absent; **anthers** 3. **Pedicels** 1.5–3 mm, ciliate. **Pedicellate spikelets** similar to the sessile spikelets. 2*n* = 20, 24–30, 32, 36, 38, 40, 48–60, 64, 69.

Saccharum spontaneum is a weedy species, native to tropical Africa and Asia, that is now established in Mesoamerica but not, so far as is known, in the *Flora* region. It is listed as a noxious weed by the U.S. Department of Agriculture, but it is grown in breeding programs as a source of potentially useful genes for *S. officinarum* (sugar cane), with which it readily hybridizes. Because of the potential economic damage of uncontrolled hybridization between *S. spontaneum* and *S. officinarum*, the U.S. Department of Agriculture should be notified of plants found growing outside a controlled planting.

8. **Saccharum officinarum** L. [p. 615]
SUGARCANE

Plants with short rhizomes. **Culms** 3–6 m tall, 2–5 cm thick, clumped, glabrous throughout or nearly so, lower internodes swollen. **Sheaths** sometimes ciliate at the collar margins; **auricles** present; **ligules** 2–3 mm; **blades** 70–150 cm long, 20–60 mm wide, usually glabrous, occasionally with hairs on the adaxial surfaces. **Peduncles** 20–80 cm, glabrous; **panicles** 50–100 cm long, to 20 cm wide, lanceolate; **rachises** 30–80 cm,

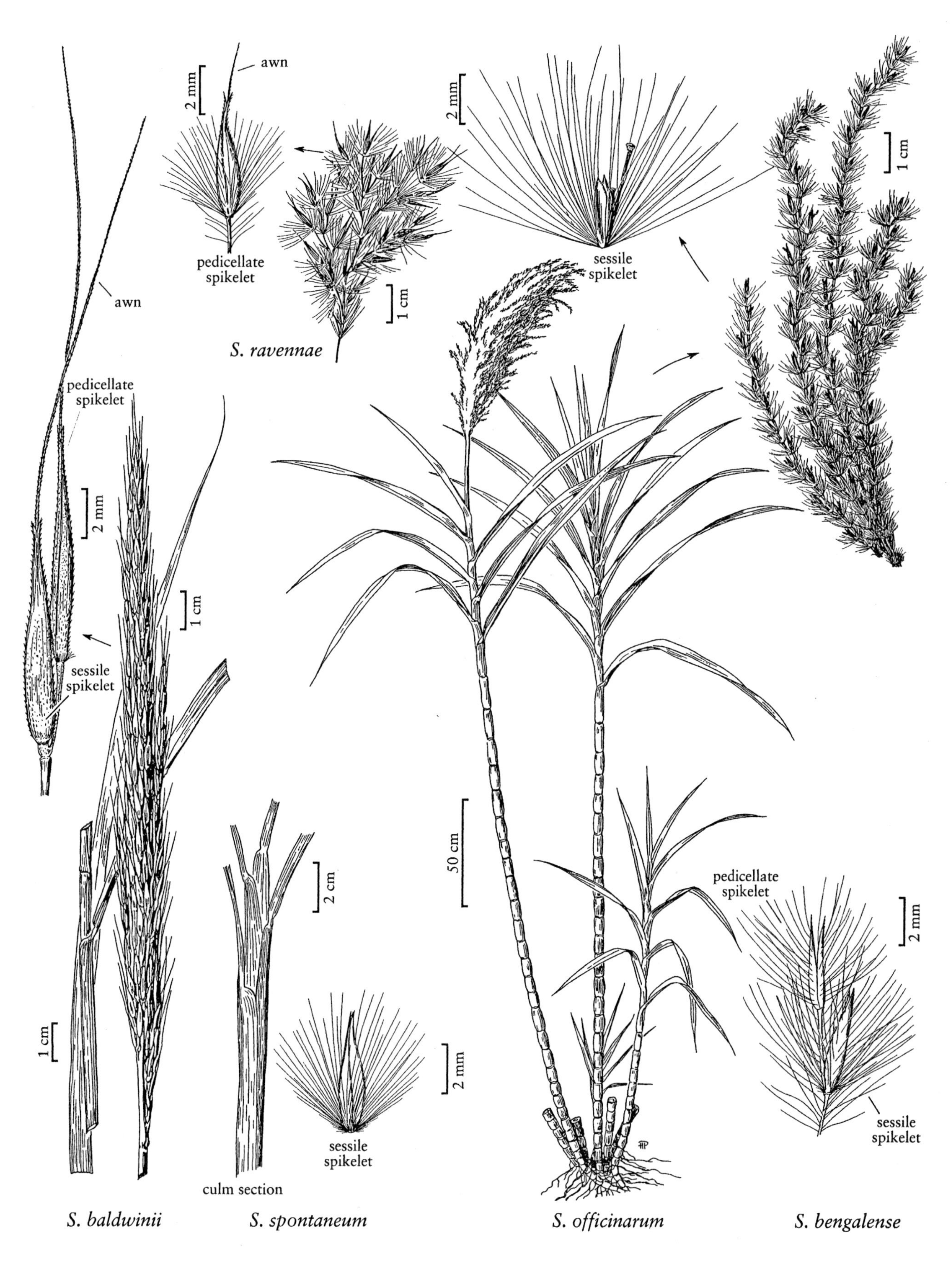
awn
2 mm
pedicellate
spikelet
1 cm
S. ravennae
2 mm
sessile
spikelet
1 cm
awn
pedicellate
spikelet
2 mm
1 cm
sessile
spikelet
50 cm
2 cm
pedicellate
spikelet
2 mm
1 cm
2 mm
sessile
spikelet
culm section
sessile
spikelet
S. baldwinii
S. spontaneum
S. officinarum
S. bengalense

SACCHARUM

glabrous; **primary branches** 10–25 cm, appressed to spreading; **rame internodes** 3–6 mm, glabrous. **Sessile spikelets** 3–5 mm long, 0.8–0.9 mm wide, white to gray. **Callus hairs** 6–10 mm, exceeding the spikelets, white; **lower glumes** glabrous, 2–4-veined; **upper glumes** 3-veined; **lower lemmas** 3–4.5 mm, 2–3-veined; **upper lemmas** without veins, entire; **awns** absent; **lodicule veins** not extending into hairlike projections; **anthers** 3. **Pedicels** 2–5 mm, glabrous. **Pedicellate spikelets** similar to the sessile spikelets. $2n$ = 80.

Saccharum officinarum is native to tropical Asia and the Pacific islands. It is cultivated for sugar production in various parts of the world, including Texas, Louisiana, and Florida. It is also becoming popular as an ornamental plant for gardens in warmer parts of the contiguous United States, and appears to be established in some parts of the southeastern United States. A number of different, clonally propagated color forms are available. It hybridizes with *S. spontaneum* (see discussion above).

9. **Saccharum bengalense** Retz. [p. 615]
TALL CANE

Plants cespitose, not rhizomatous. **Culms** to 5 m, glabrous. **Blades** to 2 m long, 3–25 mm wide, flat or channeled, glaucous and scabrous. **Panicles** 20–90 cm, compact; **primary branches** 2–5 cm, considerably shorter than the supporting branches; **rame internodes** hirsute, hairs to 7 mm. **Sessile spikelets** 4–6 mm long, somewhat heteromorphic. **Sessile spikelets: callus hairs** to 2.5 mm, white to gray; **glumes** equal; **lower glumes** membranous, pubescent; **upper glumes** glabrous; **lower lemmas** oblong-elliptic, pubescent; **upper lemmas** oblong-elliptic, ciliate on the margins, acute to shortly awned; **awns** about 1.3 mm, not visible beyond the glumes; **anthers** 3. **Pedicels** shorter than the sessile spikelet. **Pedicellate spikelets** pilose on the glumes, hairs 4–9 mm. $2n$ = 20, 22, 40, 60.

Saccharum bengalense is native from Iran to northern India. It is sometimes cultivated as an ornamental in the *Flora* region.

26.03 MISCANTHUS Andersson

Mary E. Barkworth

Plants perennial; cespitose, sometimes rhizomatous. **Culms** 40–400 cm, erect. **Leaves** not aromatic; **sheaths** open; **ligules** membranous, truncate, ciliate; **blades** flat. **Inflorescences** terminal, ovoid or corymbose panicles, with elongate rachises and numerous ascending, spikelike branches; **branches** usually more than 10 cm long, with unequally pedicellate spikelet pairs, spikelets homogamous and homomorphic; **disarticulation** below the glumes. **Calluses** short, blunt, pilose, with fine hairs, hairs often exceeding the spikelets. **Glumes** membranous to coriaceous; **lower glumes** broadly convex to weakly 2-keeled, without raised veins; **lower florets** sterile; **upper florets** bisexual; **upper lemmas** entire and unawned or bidentate and awned from the sinuses; **anthers** 2 or 3. **Pedicels** free. x = 19. Name from the Greek *mischos*, 'pedicel', and *anthos*, 'flower', both spikelets ("flowers") being pedicellate.

Miscanthus is a genus of approximately 25 species. Most of the species are native to southeast Asia; a few extend into Africa. Some species hybridize with *Saccharum*, from which *Miscanthus* differs in its non-disarticulating branches and unequally pedicellate, rather than sessile-pedicellate, spikelets.

The five species found in the *Flora* region are all grown as ornamentals because of their large, plumose panicles and striking growth habit. They flower in late summer to fall. The differing chromosome numbers within *Miscanthus sinensis* and *M. sacchariflorus* are associated with morphological differences in Japan, but it is not known if this is true for cultivated plants.

SELECTED REFERENCES **Adati, S.** and **I. Shiotani**. 1962. The cytotaxonomy of the genus *Miscanthus* and its phylogenetic status. Bull. Fac. Agric. Mie Univ. 25:1–24; **Edgar, E.** and **H.E. Connor**. 2000. Flora of New Zealand, vol. 5. Manaaki Whenua Press, Lincoln, New Zealand. 650 pp.; **Koyama, T.** 1987. Grasses of Japan and Its Neighboring Regions: An Identification Manual. Kodansha, Ltd., Tokyo, Japan. 370 pp.; **Scott, T.** 1995. Whispering grasses. Gardens Ill. 15:56–65.

1. Callus hairs 2–4 times as long as the spikelets.
   2. Spikelets 4–6 mm long; upper lemmas unawned or the awns not exceeding the glumes . . . . . 4. *M. sacchariflorus*
   2. Spikelets 2–2.8 mm long; upper lemmas awned, the awns 9–13 mm long, exceeding the glumes . . . . . 3. *M. nepalensis*
1. Callus hairs from shorter than to twice as long as the spikelets.
   3. Culms few together or solitary; basal leaves with reduced blades, only the cauline leaves with long blades; panicles loose, with 2–5 branches . . . . . 5. *M. oligostachyus*
   3. Culms densely tufted, forming large clumps; many basal leaves with long blades; panicles usually with more than 15 branches.
      4. Spikelets 3.5–7 mm long; blades 6–20 mm wide; rachises ⅓–⅔ as long as the panicles . . . . . 2. *M. sinensis*
      4. Spikelets 3–3.5 mm long; blades 15–40 mm wide; rachises ¾–⅘ as long as the panicles . . . 1. *M. floridulus*

1. **Miscanthus floridulus** (Labill.) Warb. *ex* K. Schum. & Lauterb. [p. 619]
GIANT CHINESE SILVERGRASS

Plants cespitose, forming large clumps. **Culms** 1.5–4 m tall, 8–16 mm thick below. **Leaves** crowded at the base; **sheaths** glabrous or sparsely pubescent, margins glabrous or ciliate; **ligules** 1–3 mm; **blades** 30–80 cm long, 15–40 mm wide, adaxial surfaces pubescent near the bases, glabrous elsewhere, midveins whitish, conspicuous both ab- and adaxially. **Panicles** 30–50 cm long, 10–20 cm wide, exserted, dense, ovoid-ellipsoid, white, usually with more than 15 branches; **rachises** 25–40 cm, hispid-pubescent, ¾–⅘ as long as the panicles; **branches** 10–25 cm long, 8–10 mm wide, often branched at the base; **internodes** 3–5 mm, glabrous. **Shorter pedicels** 1–1.5 mm; **longer pedicels** 2.5–3.5 mm, becoming somewhat recurved. **Spikelets** 3–3.5 mm, lanceolate to lance-ovate; **callus hairs** 4–6 mm, to twice as long as the spikelets, white. **Lower glumes** glabrous or puberulent distally; **awns of upper lemmas** 5–15 mm, weakly geniculate. $2n$ = 36, 38, 57.

Miscanthus floridulus is the most widespread species of *Miscanthus* in southeast Asia. The culms are used for arrow-shafts in Papua New Guinea and as support and drying racks for climbing vegetables and tobacco in the Philippines. In North America it is grown as an ornamental. The blades of the lower leaves tend to fall off in late summer, leaving the culms naked at the base. It is tolerant of wind and salt spray.

2. **Miscanthus sinensis** Andersson [p. 619]
EULALIA

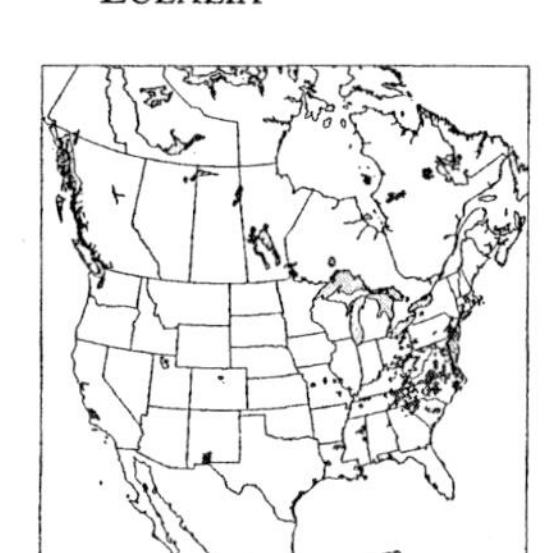

Plants cespitose, forming large clumps, with short, thick rhizomes. **Culms** 60–200 cm tall, 3–7 mm thick below. **Leaves** predominantly basal; **sheaths** mostly glabrous, throats pilose; **ligules** 1–2 mm; **blades** 20–70 cm long, 6–20 mm wide, midveins conspicuous abaxially, 1–2 mm wide, whitish. **Panicles** 15–25 cm long, 8–28 cm wide, dense to loose, usually with more than 15 branches; **rachises** 6–15 cm, ⅓–⅔ as long as the inflorescences; **branches** 8–15(30) cm long, about 10 mm wide, sometimes branched at the base; **internodes** 4–8 mm, glabrous. **Shorter pedicels** 1.5–2.5 mm; **longer pedicels** 3.5–6 mm, slightly recurved at maturity. **Spikelets** 3.5–7 mm, lanceolate to lance-ovate; **callus hairs** 6–12 mm, to twice as long as the spikelets, white, stramineous to reddish. **Glumes** subequal; **lower glumes** 3-veined, ciliolate on the margins; **upper glumes** 1-veined; **awns of upper lemmas** 6–12 mm, geniculate below. $2n$ = 38, 40, and dysploids from 35–42.

Miscanthus sinensis is native to southeastern Asia. It is frequently cultivated in the United States and southern Canada, and is now established in some parts of the United States. Approximately 40 forms and cultivars are available, some having white-striped leaves, others differently colored callus hairs and, consequently, differently colored panicles.

3. **Miscanthus nepalensis** (Trin.) Hack. [p. 619]
HIMALAYA FAIRYGRASS

Plants cespitose, shortly rhizomatous. **Culms** 40–80(150) cm. **Sheaths** more or less keeled, with scattered hairs, particularly below the collar; **ligules** 2–3.5 mm, obtuse, lacerate and shortly pubescent; **blades** 20–60 cm long, 4–10 mm wide, stiff, flat or folded, abaxial surfaces with scattered fine hairs, adaxial surfaces glabrous. **Panicles** 10–20 cm, flabellate, golden brown, with more than 15 branches; **rachises** about ½ as long as the panicles; **branches** 3.5-10.5 cm, ascending. **Shorter pedicels** 1.5-2 mm; **longer pedicels** 2.5-3.5 mm. **Spikelets** 2–2.8 mm; **callus hairs** 3–4 times longer than the spikelets, golden brown. **Lower glumes** hairy on the lower margins, hairs to 3 times longer than the glumes; **awns of upper lemmas** 9–13 mm, exceeding the glumes, flexuous to weakly geniculate. $2n$ = 40.

Miscanthus nepalensis is native from Pakistan through the Himalayas to Myanmar. It is cultivated occasionally in the *Flora* region. Edgar and Conner (2000) report that, in New Zealand, *M. nepalensis* has escaped cultivation and is spreading.

4. **Miscanthus sacchariflorus** (Maxim.) Benth. [p. 619]
Amur Silvergrass, Miscanthus

Plants rhizomatous, rhizomes 3–6 mm wide. **Culms** 60–250 cm tall, 5–8 mm thick below; **nodes** pilose. **Leaves** evenly distributed; **ligules** 0.5–1 mm; **blades** 20–80 cm long, 0.5–3 cm wide, adaxial surfaces densely pilose basally, midribs prominent, whitish. **Panicles** 15–40 cm long, 8–16 cm wide, white to yellowish-brown, usually with more than 15 branches; **rachises** 4–10 cm; **nodes** pilose; **branches** 10–35 cm long, about 10 mm wide, sometimes branching at the base. **Shorter pedicels** 1.5–3 mm; **longer pedicels** 3–7 mm, strongly curved at maturity. **Spikelets** 4–6 mm; **callus hairs** 2–4 times as long as the spikelets, copious, white. **Lower glumes** 2-keeled above, margins densely pilose distally, hairs to 15 mm; **upper glumes** 4–5 mm, 3-veined, margins ciliate distally; **awns of upper lemmas** absent or short, not exceeding the glumes. $2n$ = 38, 57, 64, 76, 95.

Miscanthus sacchariflorus is native to the margins of rivers or marshes in temperate to north-temperate regions of eastern Asia, and appears to require cold and humidity for optimum growth. It has escaped from cultivation in various parts of the *Flora* region. It combines a large, plumose panicle with recurving leaves that turn orange in the fall.

5. **Miscanthus oligostachyus** Stapf [p. 620]
Small Japanese Silvergrass

Plants cespitose, rhizomatous. **Culms** 80–150 cm tall, 2–3 mm thick below, few together or solitary; **nodes** finely pubescent. **Sheaths** mostly glabrous, pilose near the summits; **ligules** 2–3 mm, rounded; **blades** well-developed only on the cauline sheaths, 8–35 cm long, 6–25 mm wide, adaxial surfaces densely pilose basally. **Panicles** long-exserted, loose, with 2–5 erect to suberect branches; **branches** 7–15 cm, densely pilose, with white or purplish-white hairs. **Shorter pedicels** 1.5–2 mm; **longer pedicels** 5–6 mm, sulcate on 1 side. **Spikelets** 6–8 mm; **callus hairs** from ½ as long as to equaling the spikelets, silky, white. **Lower glumes** 6–8 mm, sparsely pilose, 2-keeled above, 2-toothed, teeth densely white-ciliate; **upper glumes** equaling the lower glumes, 3–5-veined; **awns of upper lemmas** (4)8–15 mm, twisted at the bases; **anthers** 2.5–3 mm. $2n$ = 38.

Miscanthus oligostachyus is a native of Japanese and Korean forests that is sold as an ornamental species in the United States. It does best in regions with cool summers. Koyama (1987) recognized three subspecies of *M. oligostachyus*; they have not been evaluated for this treatment.

26.04 IMPERATA Cirillo

Mark L. Gabel

Plants perennial; strongly rhizomatous. **Culms** 10–150(217) cm, mostly erect and unbranched, usually with 3–4 nodes. **Leaves** not aromatic; **sheaths** open, ciliate at the margins of the collars; **ligules** membranous; **blades** of the basal leaves linear to lanceolate, sometimes ciliate basally, those of the cauline leaves reduced. **Inflorescences** terminal, cylindrical to conical panicles with an evident rachis; **rachises** often with numerous long hairs; **inflorescence branches** 1–7 cm, usually shorter than the rachises, with spikelets in unequally pedicellate pairs; **disarticulation** below the glumes. **Spikelets** homogamous and homomorphic, unawned; **calluses** very short, hairy, hairs 7–16 mm. **Glumes** equal to subequal, membranous, 3–9-veined, with hairs longer than the florets over at least the lower ½; **lower florets** reduced to hyaline or membranous lemmas; **upper florets** bisexual, lemmas, if present, hyaline, unawned; **anthers** 1–2, yellow to brown; **stigmas** elongate, purple to brown; **styles** connate or free. **Pedicels** not fused to the branch axes, terminating in cuplike tips. **Caryopses** ovate to obovate, light to dark brown. x = 10. Named after Ferrante Imperato (1550–1625) of Naples, an apothecary and author of a folio work on natural history.

Imperata has nine species and is widely distributed in warm regions of both hemispheres. Its economic importance is primarily negative, as both *I. cylindrica* and *I. brasiliensis* are weedy (Gabel 1989), but new shoots of both species are used for hay or grazing. *Imperata* is thought to be closely related to *Miscanthus*. One species is native to the *Flora* region, and two have been introduced.

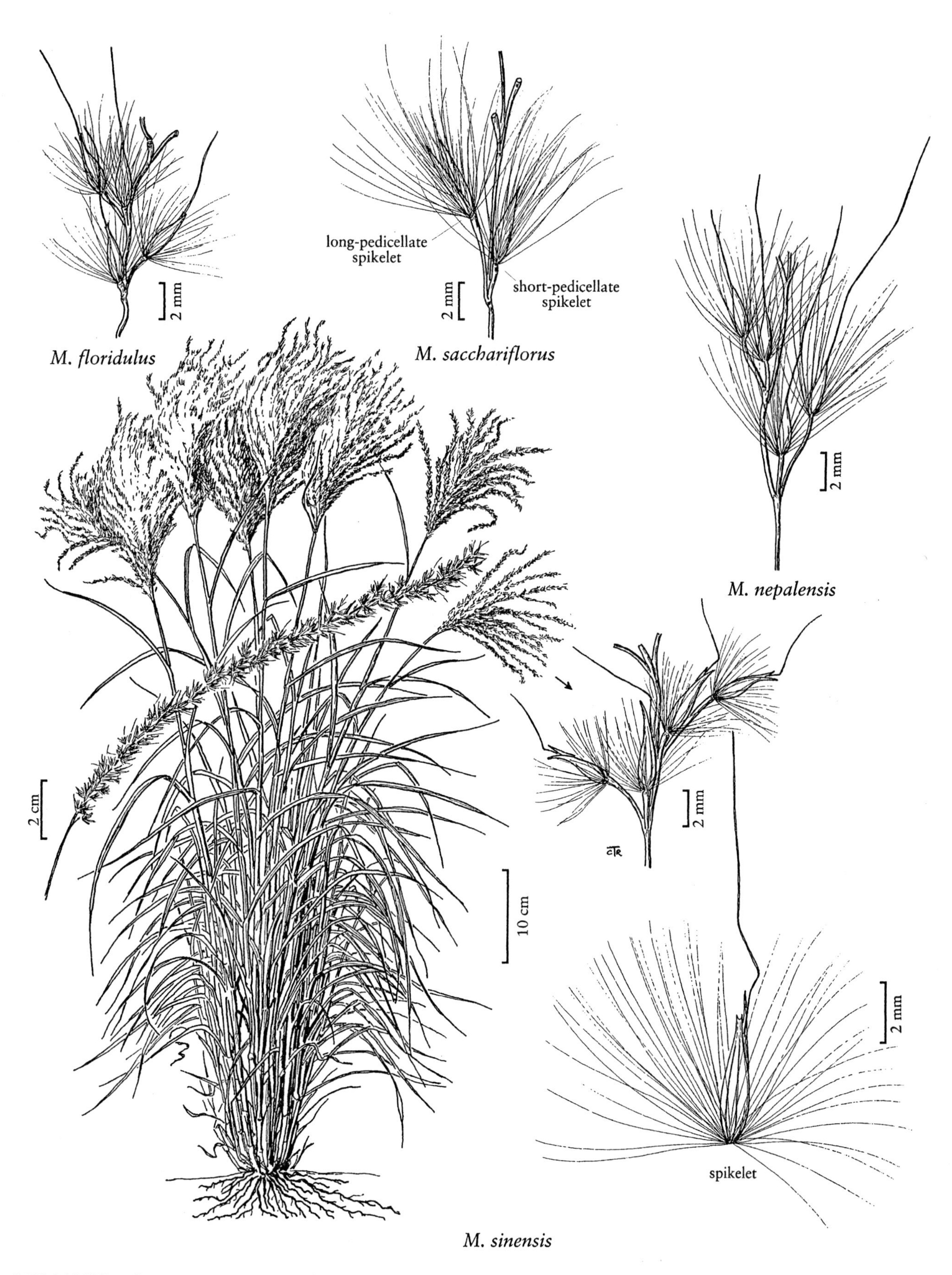
2 mm
M. floridulus
long-pedicellate spikelet
short-pedicellate spikelet
2 mm
M. sacchariflorus
2 mm
M. nepalensis
2 cm
10 cm
2 mm
2 mm
spikelet
M. sinensis

MISCANTHUS

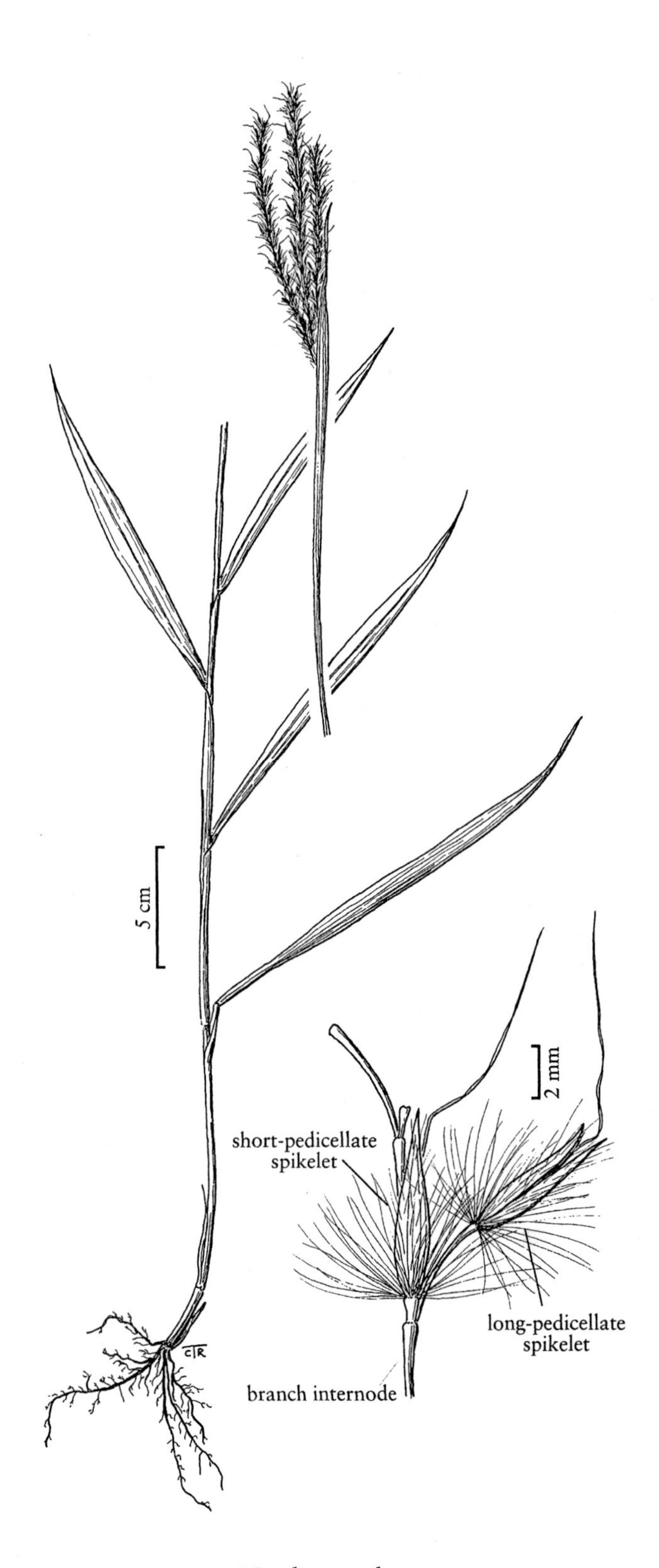

M. oligostachyus

SELECTED REFERENCES **Gabel, M.L.** 1989. Federal noxious weed identification bulletin. U.S. Dept. of Agriculture, Animal & Plant Health Inspection Service, Plant Protection & Quarantine Bull. 28:1–10; **Hall, D.W.** 1978. The Grasses of Florida. Ph.D. dissertation, University of Florida, Gainesville, Florida, U.S.A. 498 pp.

1. Stamens 2, filaments not dilated at the base . . . . 3. *I. cylindrica*
1. Stamens 1, filaments dilated at the base.
   2. Panicles 7.5–14(17) cm long; lower branches 1–3.5 cm long, appressed; upper florets usually without lemmas; southeastern United States . . . . 1. *I. brasiliensis*
   2. Panicles 16–34 cm long; lower branches 2–5 cm long, divergent; both florets with lemmas; southwestern United States . . . . 2. *I. brevifolia*

1. **Imperata brasiliensis** Trin. [p. 622]
BRAZILIAN BLADYGRASS, BRAZILIAN SATINTAIL

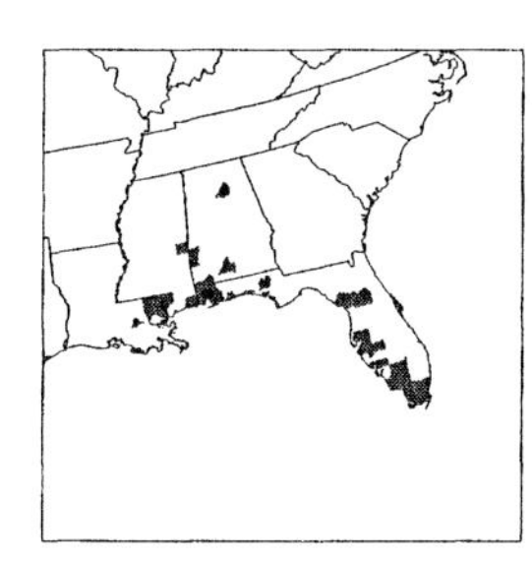

Culms 22–98 cm. **Ligules** 0.5–1.7 mm; **blades** 3–13(19) mm wide, linear-lanceolate. **Panicles** 7.5–14(17) cm; **lower branches** 1–3.5 cm, appressed. **Callus hairs** 7–13 mm; **glumes** 2.4–4.5 mm; **lower lemmas** 1–3.4 mm long, 0.5–1.1 mm wide; **upper lemmas** usually absent, if present, about 1 mm long, 0.3 mm wide; **stamens** 1, bases of the filaments dilated; **anthers** 1.4–2.8 mm; **styles** 1.1–4.7 mm; **stigmas** 2.4–6.7 mm. 2*n* = unknown.

The current range of *Imperata brasiliensis* includes South America and Central America, Mexico, and Cuba. It is now thought to be established in the southeastern United States, although it is considered to be eliminated from Florida (Hall 1978); collections of *Imperata* made there since 1970 having proved to be *I. cylindrica*. The two species differ in the number of their stamens and the frequent absence of the lower lemma in *I. brasiliensis*.

Imperata brasiliensis is listed as a noxious weed by the U.S. Department of Agriculture. Burning stimulates its flowering; consequently many specimens have burned leaves.

2. **Imperata brevifolia** Vasey [p. 622]
SATINTAIL

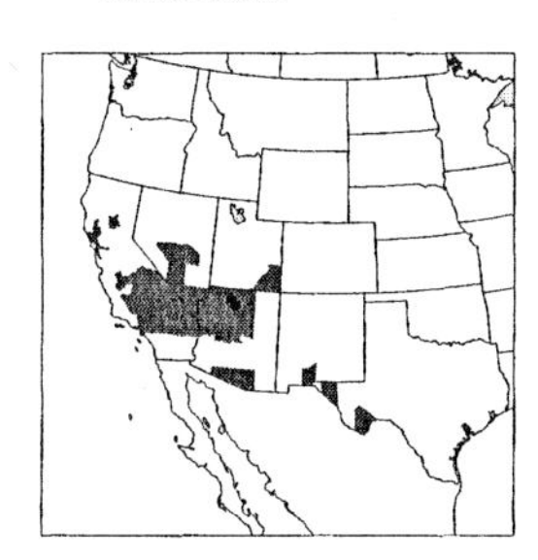

Culms 51–129 cm. **Ligules** 0.7–2.9 mm; **blades** 7–14 mm wide, linear to lanceolate, abaxial surfaces smooth, adaxial surfaces sometimes densely pilose basally, otherwise scabrous. **Panicles** 16–34 cm, dense; **lower branches** 2–5 cm, divergent. **Callus hairs** 8–12 mm; **glumes** 2.7–4.1 mm; **lower lemmas** 2.5–3.9 mm, membranous, glumelike; **upper lemmas** 1.4–2.4 mm, completely surrounding the ovary; **stamens** 1, filaments dilated at the base; **anthers** 1.3–2.3 mm, yellow to orange; **styles** 0.9–2.4 mm; **stigmas** 2.1–4 mm, purple to brown. 2*n* = 20.

Once known from wet or moist sites in the southwestern deserts from southern California, Nevada, and Utah to western Texas, *Imperata brevifolia* is currently known only from populations in Grand Canyon National Park. It was last collected outside the park in the early 1970s at a site that is now under Lake Powell. Most collections were made before 1945, in sites that are now used for housing or agriculture.

Imperata brevifolia is listed as a noxious weed by the state of California. The reason for the listing is not clear; it may stem from confusion of this native species with one of the introduced weedy species.

3. **Imperata cylindrica** (L.) Raeusch. [p. 622]
COGONGRASS, BLADYGRASS

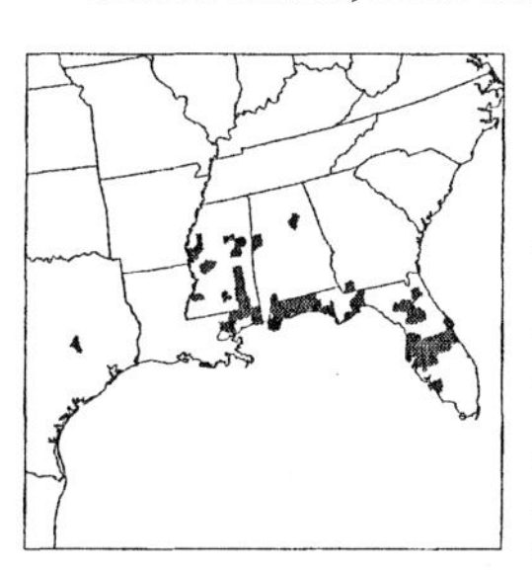

Culms (10)30–95(217) cm. **Ligules** 0.2–3.5 mm; **blades** to 150 cm long, (1)3–11(28) mm wide, linear-lanceolate, bases narrowed to the broad midrib, often with hairs on the margins. **Panicles** 5.7–22.3(52) cm, narrowly cylindrical; **lower branches** 1–3.2(7) cm, appressed. **Callus hairs** 9–16 mm; **glumes** 2.6–5.5 mm; **lower lemmas** 1.4–4.5 mm; **upper lemmas** (0.7)1.3–2.3(3.4) mm; **stamens** 2, filaments not dilated at the base; **anthers** (1.5)2.2–4.2 mm, orange to brown; **styles** 0.5–3.4 mm; **stigmas** 2.8–5.2(8.3) mm, purple to brown. 2*n* = 20, 40, 60.

Imperata cylindrica is the most variable species in the genus. Several varieties have been recognized but, although there are statistically significant differences between plants from different regions, identification to variety without knowledge of a plant's geographic origin is risky. All North American plants examined have had 2*n* = 20.

Imperata cylindrica is one of the world's 10 worst weeds, and is listed as a noxious weed by the U.S. Department of Agriculture. It was introduced to

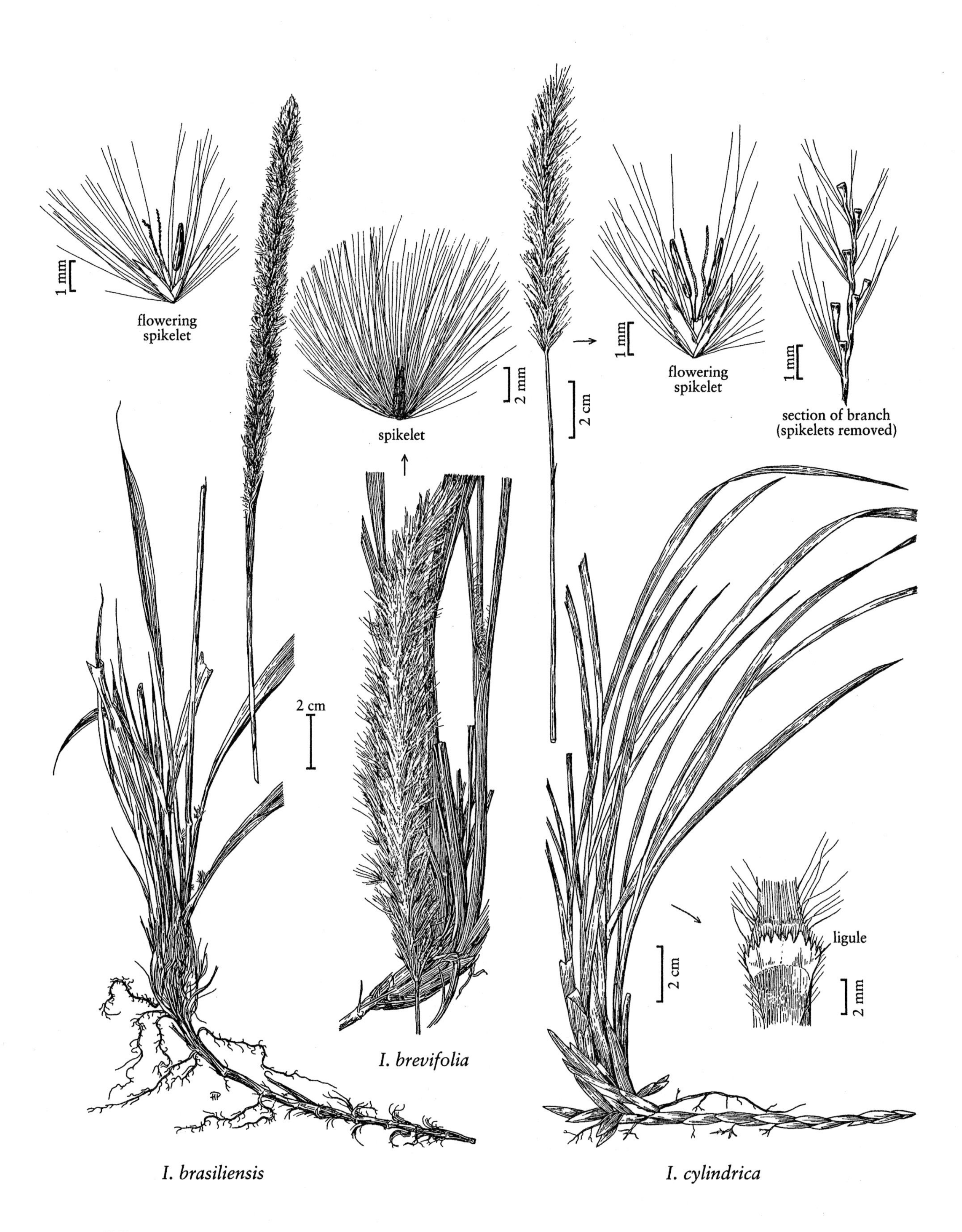

1 mm
flowering spikelet
spikelet
2 mm
2 cm
1 mm
flowering spikelet
1 mm
section of branch (spikelets removed)
2 cm
I. brevifolia
ligule
2 cm
2 mm
I. brasiliensis
I. cylindrica

IMPERATA

Alabama by 1912, and has spread considerably through the southeastern United States since then. The cultivar 'Red Baron' is diminutive and non-weedy, but individual shoots may revert to the aggressive form. Such reversion is particularly common in plants grown from tissue culture.

26.05 POLYTRIAS Hack.

Mary E. Barkworth

Plants perennial; stoloniferous. **Culms** 10–40 cm, often decumbent and rooting at the lower nodes. **Leaves** not aromatic; **ligules** membranous, ciliate or fimbriate. **Inflorescences** terminal, solitary rames, spikelets in homomorphic sessile-pedicellate triplets of 2 sessile spikelets and 1 pedicellate spikelet; **internodes** without a median translucent line; **disarticulation** in the rames below the sessile spikelets, sometimes also beneath the pedicellate spikelets. **Spikelets** dorsally compressed, with 1 floret; **sessile spikelets** bisexual; **pedicellate spikelets** bisexual, unisexual, or sterile. **Glumes** equal, oblong, truncate, membranous; **lower glumes** with the margins incurved over the upper glumes; **upper glumes** keeled; **florets** bisexual; **lemmas** hyaline, bifid almost to the base, awned from the cleft; **awns** twisted, geniculate; **anthers** 3. **Pedicels** not fused to the rame axes. $x = 10$. Name from the Greek *polys*, 'many', and *trias*, 'in threes', a reference to the large number and unusual arrangement of the spikelet triads.

Polytrias is a monotypic genus of the Asian tropics that has become naturalized in Africa and the Western Hemisphere. It is unusual within the *Andropogoneae* in having only one floret, rather than two, in its spikelets.

1. **Polytrias amaura** (Büse) Kuntze [p. 625]
JAVA GRASS

Plants highly stoloniferous. **Culms** 10–40 cm, decumbent, rooting at the lower nodes, erect portions 10–20 cm; **nodes** pubescent; **internodes** glabrous. **Leaves** cauline, often purplish; **sheaths** keeled, pubescent basally and sometimes sparsely so distally, margins ciliate; **ligules** 0.2–0.5 mm, truncate; **blades** 0.5–7 cm long, 1–7 mm wide, flat, pubescent. **Rames** 2–3 cm; **internodes** 2–3 mm, flat, ciliate on the edges and distally. **Sessile spikelets** 3–4 mm, ovate, pilose, brown or yellow-brown; **calluses** blunt; **glumes** concealing the floret; **lower glumes** 2–3 mm; **lemmas** about 1 mm; **awns** 4–12 mm, exserted, geniculate, twisted below the bend, brown. **Pedicels** 4–4.5 mm, slender, free of the rame axes. **Pedicellate spikelets** similar to the sessile spikelets or somewhat smaller, sometimes staminate. **Caryopses** 1.5–1.8 mm. $2n = 20$.

Polytrias amaura is native to southeastern Asia. It used as a lawn grass in tropical and subtropical regions, including Florida. It gives a purplish cast to a lawn.

26.06 MICROSTEGIUM Nees

John W. Thieret

Plants annual or perennial; straggling. **Culms** to 100 cm, often decumbent. **Leaves** not aromatic; **ligules** membranous; **blades** narrowly-elliptic to lanceolate, often pseudopetiolate. **Inflorescences** terminal, subdigitate to racemose clusters of 1–few rames; **rame internodes** slender, without a translucent longitudinal groove; **disarticulation** in the rames beneath the sessile spikelets, and below the pedicellate spikelets. **Spikelets** in homogamous, homomorphic, sessile-pedicellate pairs, with 1 or 2 florets. **Lower glumes** herbaceous to cartilaginous, longitudinally grooved, margins inflexed, 4–6-veined, usually keeled; **upper glumes** 3-veined, mucronate or shortly awned; **lower florets** absent, or reduced and sterile; **upper florets** bisexual; **upper lemmas** usually awned; **anthers** (2)3. $x = 10$. **Pedicels** not fused to the rame axes. Name from the Greek *micros*, 'small', and *stege*, 'cover', possibly alluding to small glumes.

Microstegium is a genus of approximately 15 species, most of which are native to southeastern Asia; one is established in the *Flora* region.

SELECTED REFERENCES **Fairbrothers, D.E. and J.R. Gray.** 1972. *Microstegium vimineum* (Trin.) A. Camus (Gramineae) in the United States. Bull. Torrey Bot. Club 99:97–100; **Hunt, D.M. and R.E. Zaremba.** 1992. The northeastward spread of *Microstegium vimineum* (Poaceae) into New York and adjacent states. Rhodora 94:167–170; **Mehrhoff, L.J.** 2000. Perennial *Microstegium vimineum* (Poaceae): An apparent misidentification. J. Torrey Bot. Soc. 127:251–254.

1. **Microstegium vimineum** (Trin.) A. Camus [p. 625]
NEPALESE BROWNTOP

Plants annual. **Culms** 40–100 cm tall, 1–1.5 mm thick, freely branching, lower portions prostrate, rooting at the nodes, terminal portions and flowering branches erect; **nodes** glabrous. **Sheaths** shorter than the internodes, mostly glabrous or sparsely pubescent above, margins ciliate, becoming pilose at the throat; **ligules** 0.5–0.8 mm, truncate; **blades** 3–10 cm long, 8–15 mm wide, glabrous or sparsely pubescent, bases cuneate, midveins white, apices attenuate, acute. **Cleistogamous inflorescences** concealed in the upper sheaths; **chasmogamous inflorescence** exserted, of (1)2–4(6) racemose to subdigitate, erect to ascending rames; **rames** 3–7 cm, glaucous-green; **internodes** 3.5–5 mm, gradually widened above, ciliate. **Spikelets** 3.7–6.5 mm. **Lower glumes** 2-keeled, subtruncate to shallowly 2-toothed; **upper glumes** acute; upper lemmas usually awned, awns 2–5(15) mm, often concealed by the glumes; **anthers** 3, 0.7–1 mm. **Pedicels** 3–4 mm. $2n = 40$.

Microstegium vimineum was introduced to Tennessee from Asia around 1919 and is now established in much of the eastern United States. Although often associated with forested and wetland areas, it also does well in many disturbed areas. In suitable habitats it quickly spreads by rooting from its prostrate culms, forming dense, monospecific stands. It is sometimes confused with **Leersia virginica** Willd., but differs from that species in its glabrous cauline nodes and the presence of hairs at the summit of the leaf sheaths. In addition, *M. vimineum* flowers in late September and October and is clearly a member of the *Andropogoneae,* whereas *L. viriginica* flowers in June through July and is a member of the *Oryzeae.*

26.07 TRACHYPOGON Nees

Kelly W. Allred

Plants annual or perennial; cespitose or shortly rhizomatous. **Culms** 30–200 cm, unbranched; **internodes** semi-solid. **Leaves** cauline, not aromatic; **sheaths** shorter than the internodes, rounded; **ligules** membranous; **blades** flat to involute. **Inflorescences** terminal, solitary racemes of heterogamous subsessile-pedicellate spikelets pairs (rarely of 2 digitate spikelike branches), axes slender, without a translucent median groove; **disarticulation** beneath the pedicellate spikelets. **Subsessile spikelets** staminate or sterile, without a callus and unawned, otherwise similar to the pedicellate spikelets. **Pedicels** slender, not fused to the rames axes. **Pedicellate spikelets** bisexual; **calluses** sharp, strigose; **glumes** firm, enclosing the florets; **lower glumes** several-veined, encircling the upper glumes; **upper glumes** 3-veined; **lower florets** sterile; **upper florets** bisexual, lemmas firm but hyaline at the base, tapering to an awn; **awns** (4)6–15 cm, twisted, pubescent to plumose; **paleas** absent; **anthers** 3. $x = 10$. Name from the Greek *trachys*, 'rough', and *pogon*, 'beard', referring to the plumose awn of the bisexual florets.

Trachypogon is a tropical or warm-temperate genus that is native to Africa and tropical to subtropical America. Estimates of the number of species included range from one to ten. One species, *Trachypogon secundus*, is native to the *Flora* region, but some taxonomists (e.g., Dávila 1994) include it in **T. plumosus** (Humb. & Bonpl. *ex* Willd.) Nees and others (e.g., Judziewicz 1990) include it, *T. plumosus*, and various other taxa in **T. spicatus** (L. f.) Kuntze. The traditional treatment and nomenclature for North American plants is retained here, pending formal study of the taxa involved.

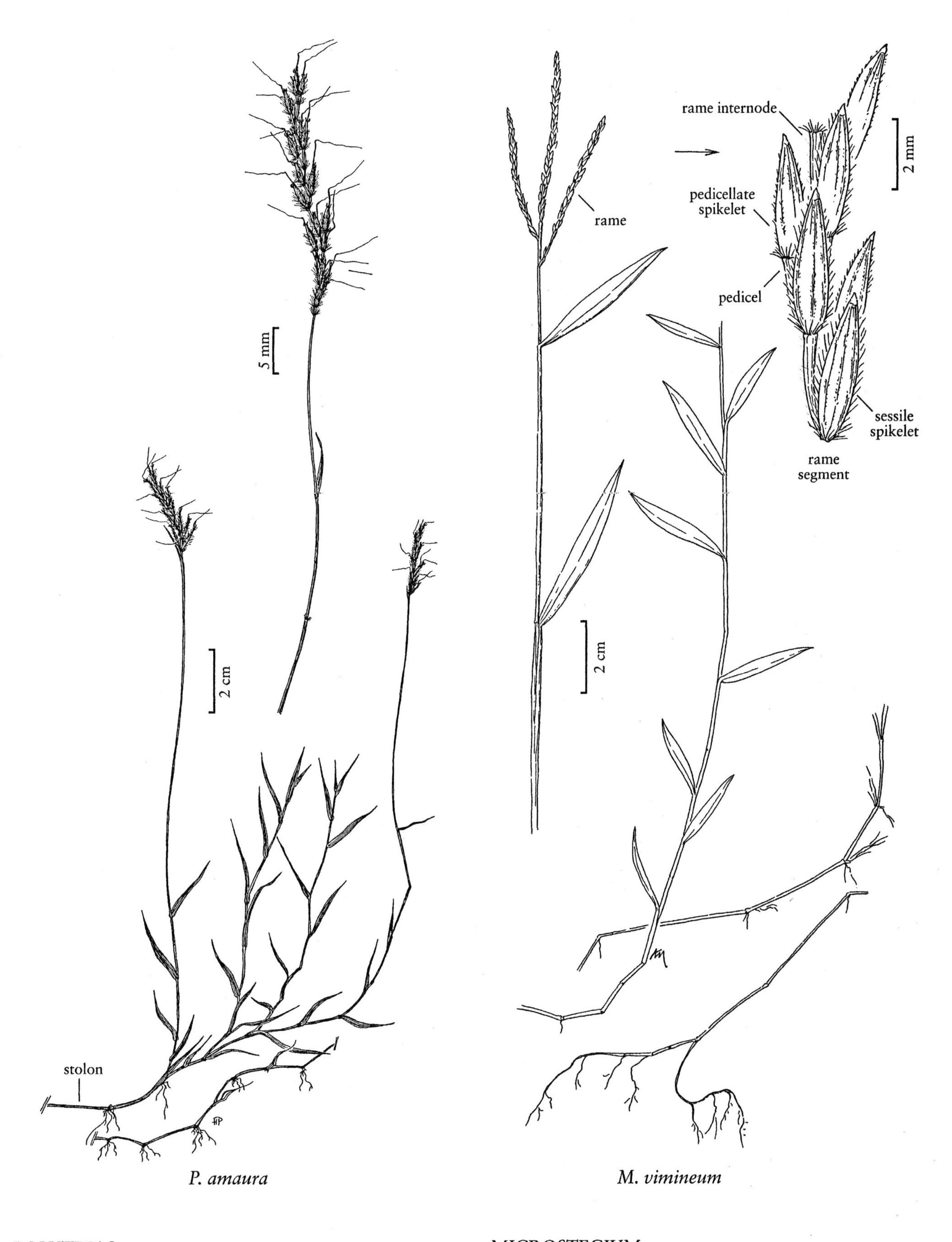

P. amaura

M. vimineum

POLYTRIAS

MICROSTEGIUM

SELECTED REFERENCES **Dávila, P.D.** 1994. *Trachypogon* Nees. Pp. 380–381 *in* G. Davidse, M. Sousa S., and A.O. Chater (eds.). Flora Mesoamericana, vol. 6: Alismataceae a Cyperaceae. Universidad Nacional Autónoma de México, Instituto de Biología, México, D.F., México. 543 pp.; **Judziewicz, E.J.** 1990. Flora of the Guianas: 187. Poaceae (Gramineae). Series A: Phanerogams, Fascicle 8 (series ed. A.R.A. Görts-Van Rijn). Koeltz Scientific Books, Koenigstein, Germany. 727 pp.

1. **Trachypogon secundus** (J. Presl) Scribn. [p. 627]
CRINKLE-AWN

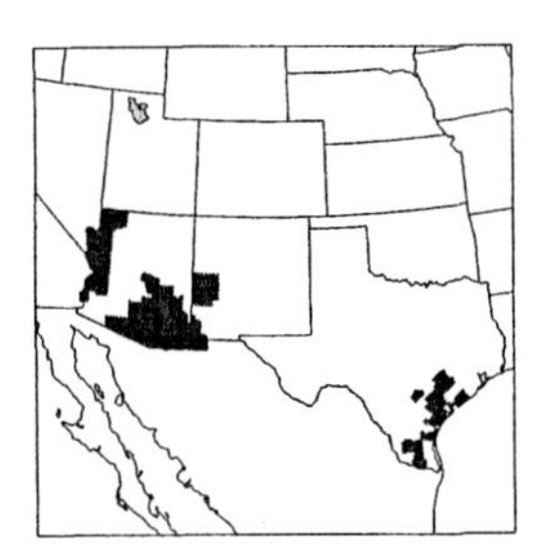

Plants perennial. **Culms** 60–120 cm, erect; **nodes** appressed-hirsute. **Sheaths** sparsely appressed-pilose; **ligules** 2–5 mm, stiff, acute; **blades** usually 12–35 cm long, 3–8 mm wide, with a broad midrib. **Racemes** 10–18 cm, the internodes glabrous. **Pedicellate spikelets** 6–8 mm; **glumes** pilose; **awns** 4–6 cm, pilose below, with 1–2 mm hairs, nearly glabrous distally; **anthers** 4–5 mm, orange. $2n = 20$.

Trachypogon secundus is found in sandy prairies, woodlands, rocky hills, and canyons, in well-drained soils at 500–2000m. Statements about is range are difficult to make because of disagreement as to whether northern plants, such as those found in the *Flora* region, belong to the same species as those found elsewhere.

Trachypogon secundus resembles *Heteropogon*, but differs in the longer, non-disarticulating inflorescence and shorter, pale awns. It rates as fairly good fodder when green, but is seldom abundant enough to be an important forage grass.

26.08 SORGHUM Moench

Mary E. Barkworth

Plants annual or perennial. **Culms** 50–500+ cm; **internodes** solid. **Leaves** not aromatic, basal and cauline; **auricles** absent; **ligules** membranous and ciliate or of hairs; **blades** usually flat. **Inflorescences** terminal, panicles with evident rachises; **primary branches** whorled, compound, the ultimate units rames; **rames** with most spikelets in heterogamous sessile-pedicellate spikelet pairs, terminal spikelet unit on each rame usually a triplet of 1 sessile and 2 pedicellate spikelets, rame axes without a translucent median line; **disarticulation** in the rames below the sessile spikelets, sometimes also below the pedicellate spikelets (cultivated taxa not or only tardily disarticulating). **Sessile spikelets** dorsally compressed, **calluses** blunt or pointed; **lower glumes** dorsally compressed and rounded basally, 2-keeled or winged distally, 5–15-veined, usually unawned; **upper glumes** 2-keeled, sometimes awned; **lower florets** reduced to hyaline lemmas; **upper florets** pistillate or bisexual, lemmas hyaline, sometimes awned. **Pedicels** slender, neither appressed nor fused to the rame axes. **Pedicellate spikelets** staminate or sterile, well-developed, often subequal to the sessile spikelets in size. $x = 10$. Name from the Italian word for the plant, *sorgho*.

Most of the approximately 25 species of *Sorghum* are native to tropical and subtropical regions of the Eastern Hemisphere, but one is native to Mexico. Two have been introduced into the *Flora* region. Some species are grown as forage, although they produce cyanogenic compounds. *Sorghum bicolor* is widely cultivated, being used as a grain, for syrup, and as a flavoring for beer.

Spangler (2000) found, using *ndh*F data, that *Sorghum* is polyphyletic, forming two distinct clades. The two species treated here were in the same clade. He found *Microstegium* and *Miscanthus* to be more closely related to *Sorghum* than *Sorghastrum*.

SELECTED REFERENCES **Clayton, W.D.** and **S.A Renvoize.** 1982. Flora of Tropical East Africa. Gramineae (Part 3). A.A. Balkema, Rotterdam, The Netherlands. 448 pp.; **de Wet, J.M.J.** 1978. Systematics and evolution of *Sorghum* sect. *Sorghum* (Gramineae). Amer. J. Bot. 65:477–484; **Dillon, S.L., P.K. Lawrence,** and **R.J. Henry.** 2001. The use of ribosomal ITS to determine phylogenetic relationships within *Sorghum*. Pl. Syst. Evol. 230:97–110; **Harlan, J.R.** and **J.M.J. de Wet.** 1972. Sources of variation in *Cynodon dactylon* (L.) Pers. Crop Sci. 12:172–176; **Harlan, J.R., J.M.J. de Wet,** and **A.B.L. Stemler** (eds.). 1976. Origins of African Plant Domestication. Mouton Press, The Hague, The Netherlands. 498 pp.; **Spangler. R.E.** 2000. Andropogoneae systematics and generic limits in *Sorghum*. Pp. 167–170 *in* S.W.L. Jacobs and J. Everett (eds.). Grasses: Systematics and Evolution. International Symposium on Grass Systematics and Evolution (3rd:1998). CSIRO Publishing, Collingwood, Victoria, Australia. 408 pp.

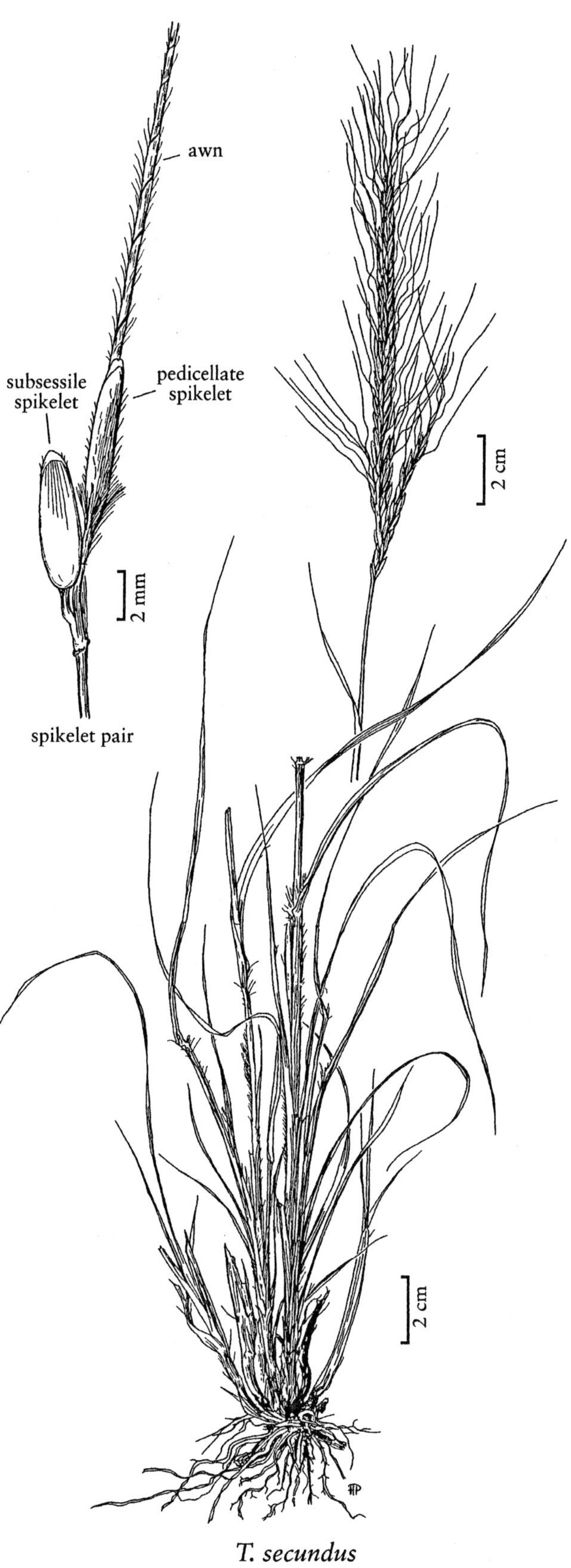

T. secundus

TRACHYPOGON

1. Plants perennial, rhizomatous; spikelets disarticulating at maturity; caryopses not exposed at maturity . . . . 1. *S. halepense*
1. Plants usually annual, sometimes short-lived perennials; spikelets either not disarticulating or doing so tardily; caryopses often exposed at maturity . . . . 2. *S. bicolor*

1. **Sorghum halepense** (L.) Pers. [p. 629]
JOHNSON GRASS

Plants perennial; rhizomatous. **Culms** 50–200 cm tall, 0.4–2 cm thick; **nodes** appressed pubescent; **internodes** glabrous. **Ligules** 2–6 mm, membranous, conspicuously ciliate; **blades** 10–90 cm long, 8–40 mm wide. **Panicles** 10–50 cm long, 5–25 cm wide, primary branches compound, terminating in rames of 1–5 spikelet pairs; **disarticulation** usually beneath the sessile spikelets, sometimes also beneath the pedicellate spikelets. **Sessile spikelets** bisexual, 3.8–6.5 mm long, 1.5–2.3 mm wide; **calluses** blunt; **glumes** indurate, shiny, appressed pubescent; **upper lemmas** unawned, or with a geniculate, twisted awn to 13 mm; **anthers** 1.9–2.7 mm. **Pedicels** 1.8–3.3 mm. **Pedicellate spikelets** staminate, 3.6–5.6 mm; **glumes** membranous to coriaceous, unawned. **Caryopses** not exposed at maturity. $2n = 20$, 40; several dysploid counts also reported.

Sorghum halepense is native to the Mediterranean region. It is sometimes grown for forage in North America, but it is considered a serious weed in warmer parts of the United States. It hybridizes readily with *S. bicolor*, and derivatives of such hybrids are widespread. The annual **Sorghum** ×**almum** Parodi, which has wider (2–2.8 mm) sessile spikelets with more veins in the lower glumes (13–15 versus 10–13) than *S. halepense*, is one such derivative.

2. **Sorghum bicolor** (L.) Moench [p. 629]
SORGHUM

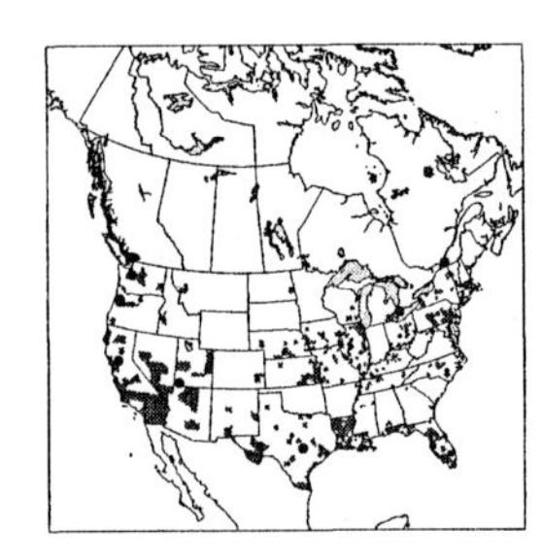

Plants annual or short-lived perennials; often tillering, without rhizomes. **Culms** 50–500+ cm tall, 1–5 cm thick, sometimes branching above the base; **nodes** glabrous or appressed pubescent; **internodes** glabrous. **Ligules** 1–4 mm; **blades** 5–100 cm long, 5–100 mm wide, sometimes glabrous. **Panicles** 5–60 cm long, 3–30 cm wide, open or contracted, primary branches compound, terminating in rames with 2–7 spikelet pairs; **disarticulation** usually not occurring or tardy. **Sessile spikelets** bisexual, 3–9 mm, lanceolate to ovate; **calluses** blunt; **glumes** coriaceous to membranous, glabrous, densely hirsute, or pubescent, keels usually winged; **upper lemmas** unawned or with a geniculate, twisted, 5–30 mm awn; **anthers** 2–2.8 mm. **Pedicels** 1–2.6 mm. **Pedicellate spikelets** 3–6 mm, usually shorter than the sessile spikelets, staminate or sterile. **Caryopses** often exposed at maturity. $2n = 20$, 40.

Sorghum bicolor was domesticated in Africa 3000 years ago, reached northwestern India before 2500 B.C., and became an important crop in China after the Mongolian conquest. It was introduced to the Western Hemisphere in the early sixteenth century, and is now an important crop in the United States and Mexico. Numerous cultivated strains exist, some of which have been formally named. They are all interfertile with each other and with other wild species of *Sorghum*.

The treatment presented here is based on de Wet (1978) and is somewhat artificial. *Sorghum bicolor* subsp. *arundinaceum* is the wild progenitor of the cultivated strains, all of which are treated as *S. bicolor* subsp. *bicolor*. These strains tend to lose their distinguishing characteristics if left to themselves. They will also hybridize with subsp. *arundinaceum*, and these hybrids can backcross to either parent, resulting in plants that may strongly resemble one parent while having some characteristics of the other. All such hybrids and backcrosses are treated here as *S. bicolor* subsp. ×*drummondii*.

1. Inflorescences branches remaining intact at maturity; caryopses exposed at maturity; sessile spikelets 3–9 mm long, elliptic to oblong . . . subsp. *bicolor*
1. Inflorescences branches rames, disarticulating at maturity, sometimes tardily; caryopses not exposed at maturity; sessile spikelets 5–8 mm long, lanceolate to elliptic.
   2. Rames readily disarticulating . . . subsp. *arundinaceum*
   2. Rames disarticulating tardily . . . subsp. ×*drummondii*

Sorghum bicolor subsp. **arundinaceum** (Desv.) de Wet & J.R. Harlan

Plants annual or weakly biennial. **Culms** to 4 m, slender to stout. **Rames** readily disarticulating at maturity, with 1–5 nodes. **Sessile spikelets** 5–8 mm, lanceolate to elliptic. **Caryopses** not exposed at maturity.

Sorghum bicolor subsp. *arundinaceum* is native to, and most common, in Africa, but some strains have been introduced into the Western Hemisphere.

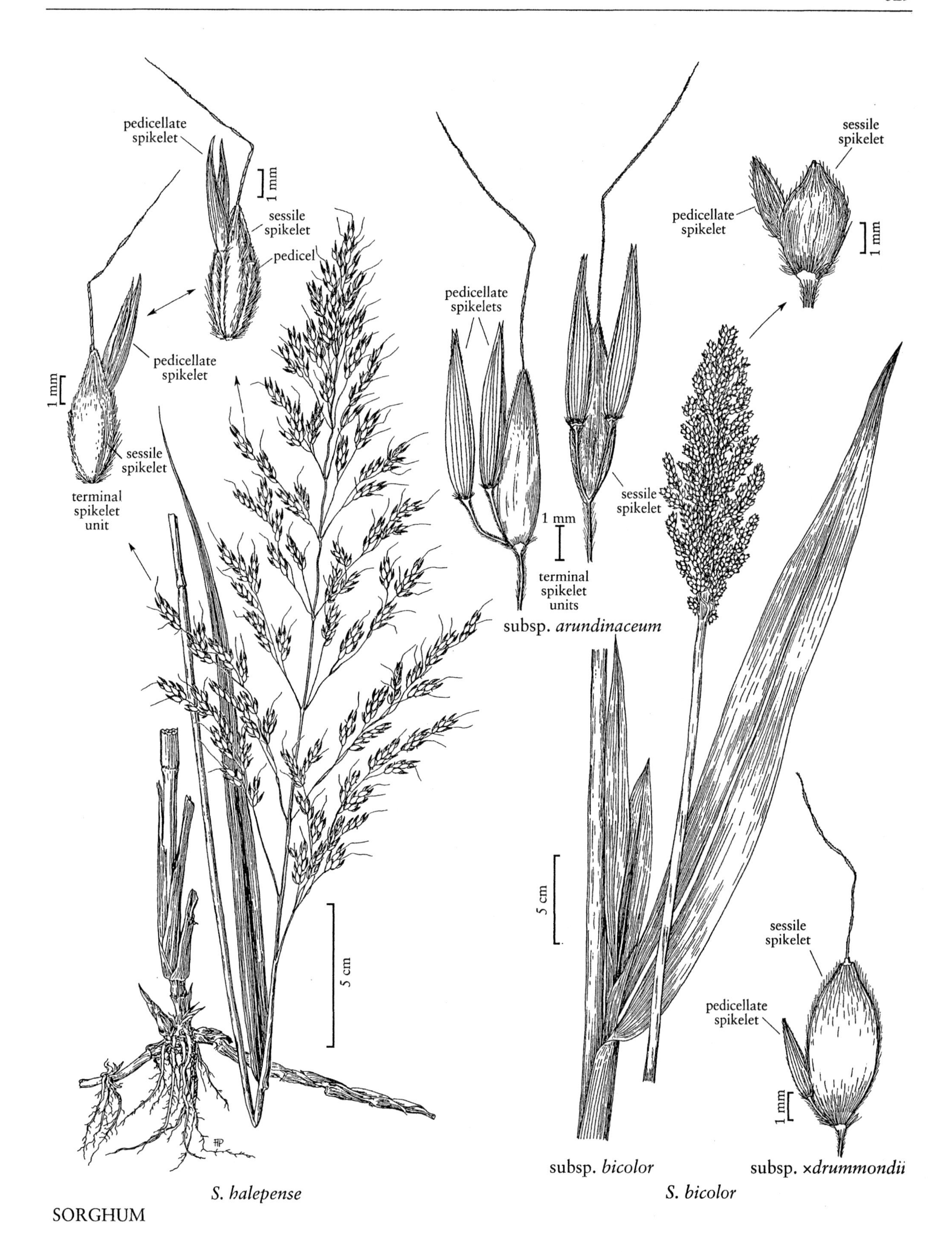
pedicellate
spikelet
1 mm
sessile
spikelet
pedicel
pedicellate
spikelet
1 mm
sessile
spikelet
terminal
spikelet
unit
5 cm
pedicellate
spikelets
1 mm
terminal
spikelet
units
sessile
spikelet
subsp. arundinaceum
sessile
spikelet
pedicellate
spikelet
1 mm
5 cm
sessile
spikelet
pedicellate
spikelet
1 mm
subsp. bicolor
subsp. ×drummondii
S. halepense
S. bicolor

SORGHUM

Sorghum bicolor (L.) Moench subsp. **bicolor**
SORGHUM, BROOMCORN, SORGO

Plants annual. **Culms** to 5 m or more, stout, frequently tillering. **Inflorescence branches** remaining intact at maturity, with 1–5 nodes. **Sessile spikelets** 3–9 mm long, 2–5 mm wide, elliptic to oblong. **Caryopses** exposed at maturity.

All the cultivated sorghums are placed in *Sorghum bicolor* subsp. *bicolor*. 'Grain sorghums' have short panicles and panicle branches, 'broomcorns' have elongate panicles and panicle branches, and 'sweet sorghums' or 'sorgo' produce an abundance of sweet juice in their stems. For a more detailed treatment, see Harlan and de Wet (1972).

Sorghum bicolor subsp. ×**drummondii** (Steud.) de Wet
CHICKEN CORN, SUDANGRASS

Plants annual. **Culms** to 4 m, relatively stout. **Rames** usually tardily disarticulating, mostly with 3–5 nodes. **Sessile spikelets** 5–6 mm, lanceolate to elliptic. **Caryopses** not exposed at maturity.

The hybrids treated here as *Sorgum bicolor* subsp. ×*drummondii* are most common in the Eastern Hemisphere, but a few are cultivated in the United States. Among these are the plants known as 'chicken corn' and 'Sudangrass' [= *S. sudanense* (Piper) Stapf] (de Wet 1978).

26.09 SORGHASTRUM Nash

Patricia D. Dávila Aranda
Stephan L. Hatch

Plants annual or perennial; cespitose, sometimes rhizomatous. **Culms** 50–300+ cm, erect, nodding or clambering, unbranched; **nodes** densely pubescent, particularly in young plants. **Leaves** not aromatic; **ligules** membranous, glabrous or pubescent; **blades** flat, involute, or folded. **Inflorescences** terminal, secund or equilateral panicles with evident rachises and numerous branches, not subtended by modified leaves; **branches** capillary, rebranching, with many rames, not subtended by modified leaves; **disarticulation** in the rames, beneath the sessile spikelets. **Spikelets** sessile, subtending a hairy pedicel (2 pedicels in the terminal spikelet units), dorsally compressed. **Calluses** blunt or sharp; **glumes** coriaceous; **lower glumes** pubescent, 5–9-veined, acute; **upper glumes** slightly longer, usually glabrous, 5-veined, truncate; **lower florets** reduced to hyaline lemmas; **upper florets** bisexual, lemmas hyaline, bifid, awned from the sinuses; **awns** usually once- or twice-geniculate, often spirally twisted, shortly strigose, brownish; **anthers** 3; **ovaries** glabrous. **Caryopses** flattened. **Pedicels** 3–6.5 mm, slender, not fused to the rame axes; **pedicellate spikelets** absent. $x = 10$. Name from *Sorghum* and the Latin suffix *astrum*, 'a poor imitation of', alluding to its similarity to *Sorghum*.

Sorghastrum includes about 18 species. Most are native to tropical or subtropical America, two are African, and four are native to the *Flora* region. Absence of the pedicellate spikelet, while confusing at first, makes *Sorghastrum* a readily recognizable genus. Its species range from sea level to approximately 3000 m, and can be found in a wide range of habitats. Two species, neither of which occur in the *Flora* region, are considered good forage.

SELECTED REFERENCES **Dávila, P.D.** 1988. Systematic revision of the genus *Sorghastrum* (Poaceae: Andropogoneae). Ph.D. dissertation, Iowa State University, Ames, Iowa, U.S.A. 333 pp.; **Hall, D.W.** 1982. *Sorghastrum* (Poaceae) in Florida. Sida 9:302–308; **Sorrie, B.A.** and **S.W. Leonard.** 1999. Noteworthy records of Mississippi vascular plants. Sida 18:889–908.

1. Awns 10–22(30) mm long, once-geniculate; plants rhizomatous . . . . 3. *S. nutans*
1. Awns 21–40 mm long, twice-geniculate; plants not rhizomatous.
   2. Pedicels sharply curved to recurved; panicles secund; sessile spikelets 0.8–1.2 mm wide . . . . . . 2. *S. secundum*
   2. Pedicels flexuous; panicles not secund; sessile spikelets 1.1–1.8 mm wide . . . . 1. *S. elliottii*

1. **Sorghastrum elliottii** (C. Mohr) Nash [p. 632]
SLENDER INDIANGRASS

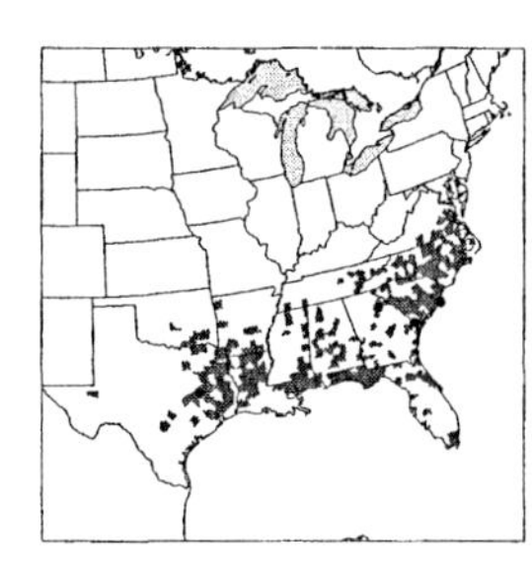

Plants not rhizomatous. **Culms** 70–190 cm tall, 1.2–2.4 mm thick; **internodes** glabrous. **Sheaths** mostly glabrous, throats pubescent; **ligules** 2–5 mm, decurrent, ciliate; **blades** 20–55 cm long, 2.5–5.5(8) mm wide, mostly glabrous. **Panicles** 10–35 cm, open, arching, dark purple; **rachises** 0.3–0.8 mm thick 1–2 mm above the lowest node; **branches** capillary, flexible, longest branches 19.5–34.5 cm. **Spikelets** 6–7.5 mm long, 1.1–1.4 mm wide, dark chestnut brown at maturity. **Calluses** 1–1.3 mm, blunt; **lower glumes** 5.5–7.3 mm, glabrous, 5-veined; **upper glumes** 6.2–7.5 mm; **awns** 25–40 mm, 5 times longer than the spikelets, twice-geniculate; **anthers** 2–3 mm. **Caryopses** 2–2.5 mm. **Pedicels** 3–6.5 mm, flexuous. $2n = 20$.

Sorghastrum elliottii usually grows in dry, open woods on sandy terraces of the lowlands in the southeastern United States, often over a clay subsoil. Plants with straight panicles and sessile spikelets that are 1.3–1.8 mm wide are sometimes called *S. apalachicolense* D.W. Hall, but the variation appears to be continuous and such plants are included here as *S. elliottii*.

2. **Sorghastrum secundum** (Elliott) Nash [p. 632]
LOPSIDED INDIANGRASS

Plants not rhizomatous. **Culms** 90–180 cm tall, 1.5–3 mm wide; **internodes** glabrous or pubescent beneath the nodes. **Sheaths** usually glabrous, occasionally pubescent in young plants; **ligules** 2.5–4(5.7) mm; **blades** 20–50 cm long, (1.8)3–6 mm wide, scabrous, particularly on the adaxial surfaces. **Panicles** 15–40 cm, straight to slightly arching, secund, somewhat open; **nodes** glabrous or almost so; **branches** erect or nearly so. **Spikelets** 6–8 mm long, 0.8–1.2 mm wide, lanceolate, dark brown to golden brown at maturity. **Calluses** 1–1.2 mm, blunt, densely bearded; **lower glumes** 6–7.5 mm, pubescent, truncate, 7–9-veined; **upper glumes** 6.5–8 mm, glabrous, acuminate, 5-veined; **awns** 30–40 mm, 4–5 times longer than the spikelets, twice-geniculate, dark brown; **anthers** 2.5–4.5 mm. **Caryopses** 2–3 mm. **Pedicels** 4–7.5 mm, pubescent, sharply curved to recurved. $2n = 20$.

Sorghastrum secundum grows in woodlands, sandy soils, and occasionally at the edges of marshes, at elevations below 1000 m. Its native range extends north and west from Florida to the Appalachian Mountains; other records probably reflect introductions. The mountains may have effectively prevented its further spread to the northwest.

Sorghastrum secundum is easily confused with plants of *S. elliottii* that are not at anthesis, because both species have straight to slightly arching panicles with ascending branches. However, the rachis nodes of *S. secundum* are glabrous or almost glabrous.

3. **Sorghastrum nutans** (L.) Nash [p. 632]
INDIANGRASS, FAUX-SORGHO PENCHÉ

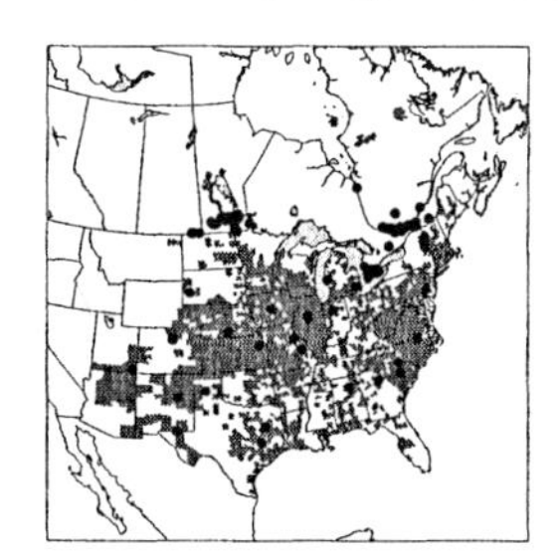

Plants rhizomatous, rhizomes short, stout, scaly. **Culms** 50–240 cm tall, 1.5–4.5 mm thick, erect; **internodes** glabrous. **Sheaths** glabrous or sparsely hispid; **ligules** 2–6 mm, usually with thick, pointed auricles; **blades** 10–70 cm long, 1–4 mm wide, usually glabrous. **Panicles** 20–75 cm, loosely contracted, yellowish to brownish; **branches** often flexible. **Spikelets** 5–8.7 mm. **Calluses** blunt, villous; **lower glumes** 5–8 mm, pubescent, 7–9-veined; **upper glumes** 5–8 mm, 5-veined; **awns** 10–22(30) mm, about 2–3 times longer than the spikelets, once-geniculate; **anthers** (2)3–5 mm. **Caryopses** 2–3 mm. **Pedicels** 3–6 mm, flexible. $2n = 20, 40, 80$.

Sorghastrum nutans grows in a wide range of habitats, from prairies to woodlands, savannahs, and scrubland vegetation. It is native from Canada to Mexico and was one of the four principal grasses of the tallgrass prairie that occupied the central United States prior to agricultural development of the region. It is frequently used for forage, for erosion control on slopes and along highways, and in restoration work. It is an attractive plant and can be used to advantage in flower arrangements. It grows readily from seed if adequate moisture is available. There are several cultivars on the market.

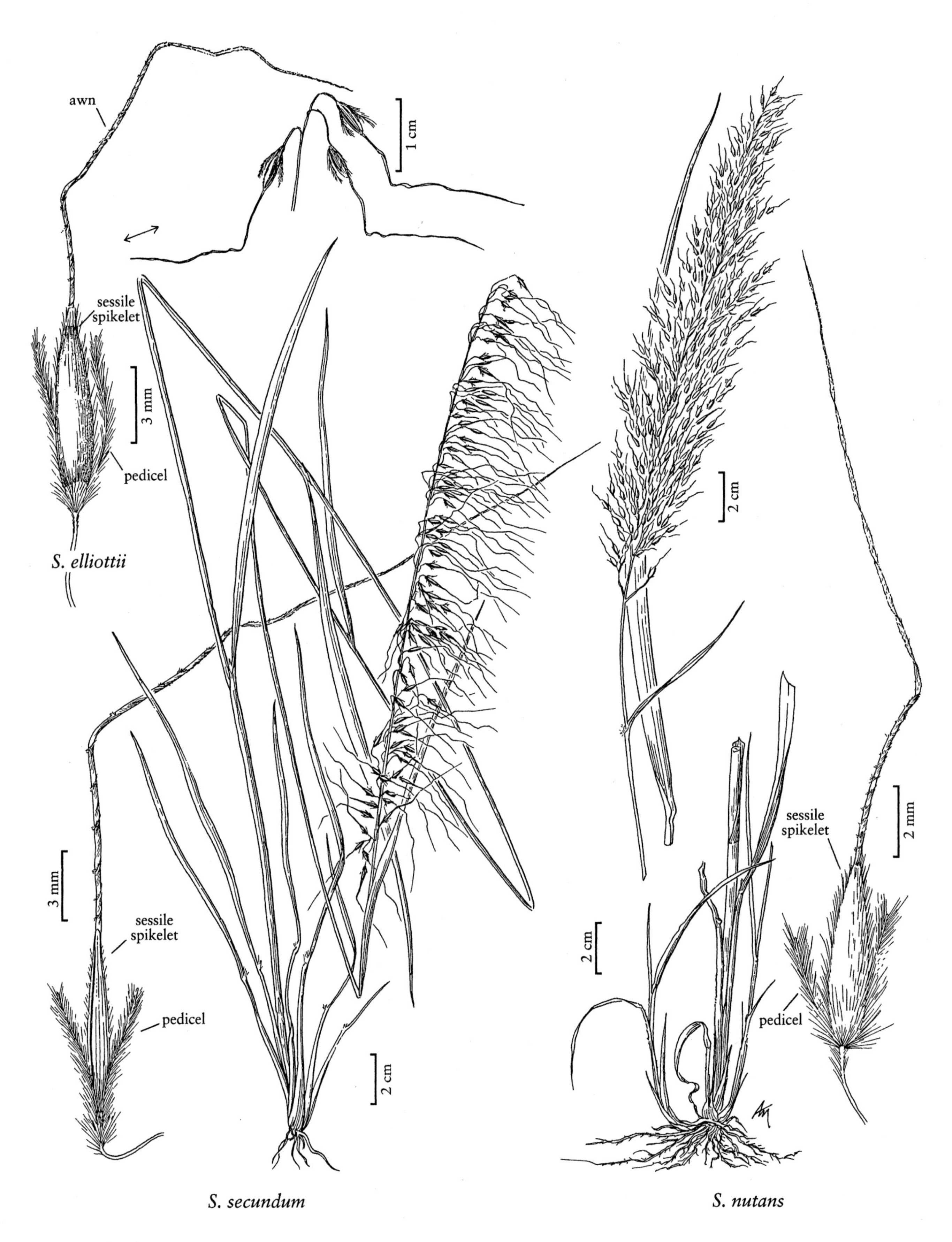
awn
1 cm
sessile spikelet
3 mm
pedicel
S. elliottii
2 cm
3 mm
sessile spikelet
pedicel
2 cm
2 cm
2 mm
sessile spikelet
pedicel
S. secundum
S. nutans

26.10 CHRYSOPOGON Trin.

David W. Hall
John W. Thieret

Plants annual or perennial; if perennial, sometimes cespitose, sometimes rhizomatous or stoloniferous. **Culms** 15–300 cm, erect, sometimes decumbent. **Leaves** not aromatic; mostly basal; **auricles** absent; **ligules** shortly membranous and ciliolate to ciliate or of hairs; **blades** often rough and glaucous. **Inflorescences** terminal panicles with elongate rachises and numerous branches, branches often naked for a considerable distance before terminating in a rame; **rames** often with only a single heterogamous triplet of 1 sessile and 2 pedicellate spikelets, sometimes with 1(–3) heterogamous sessile-pedicellate spikelet pairs below the terminal triplet, internodes without a translucent median groove; **disarticulation** oblique, below the sessile spikelets. **Sessile spikelets** terete or laterally compressed; **calluses** usually sharp, setose, hairs white or yellow to brown; **glumes** leathery to stiff, involute or folded and keeled above; **lower glumes** rounded or laterally compressed; **lower florets** sterile; **upper florets** bisexual, unawned or awned. **Pedicels** slender, not fused to the rame axes, without a translucent groove. **Pedicellate spikelets** dorsally compressed or absent, if present, lower florets sterile and unawned, upper florets sterile or staminate, awned or unawned. x = 5 or 10. Name from the Greek *chrysos*, 'golden', and *pogon*, 'beard', an allusion to the yellow, bearded callus.

Chrysopogon is a tropical and subtropical genus of 26 species. All but one species are native to the Eastern Hemisphere tropics, the majority to India. *Chrysopogon pauciflorus* is native to Florida and Cuba; four species have been introduced in the *Flora* region.

Some species of *Chrysopogon* are ecologically or economically important outside the region, *C. zizanioides* being used for controlling soil erosion, *C. fulvus* for forage, and *C. aciculatus* for lawns. *Chrysopogon aciculatus*, however, is also an aggressive weed whose sharp calluses can pierce the stomach of grazing animals and get in the feet of soft-footed animals.

SELECTED REFERENCES **National Research Council**. 1993. Vetiver Grass: A Thin Green Line Against Erosion. National Academy Press, Washington, D.C., U.S.A. 171 pp.; **Veldkamp, J.F.** 1999. A revision of *Chrysopogon* Trin. including *Vetiveria* Bory (Gramineae) in Thailand and Malesia with notes on some other species from Africa and Australia. Austrobaileya 5:503–533.

1. Upper lemmas of sessile spikelets awned, the awns 2–16 cm long.
   2. Plants annual; pedicellate spikelets 7.2–15 mm long . . . . 1. *C. pauciflorus*
   2. Plants perennial; pedicellate spikelets 2.5–8 mm long . . . . 2. *C. fulvus*
1. Upper lemmas of sessile spikelets unawned or the awns no more than 8 mm long.
   3. Calluses of the sessile spikelets 0.6–0.8 mm long, rounded, with white hairs; plants not stoloniferous . . . . 4. *C. zizanioides*
   3. Calluses of the sessile spikelets 3–6.4 mm long, sharp, with golden-yellow hairs; plants extensively stoloniferous . . . . 3. *C. aciculatus*

1. **Chrysopogon pauciflorus** (Chapm.) Benth. *ex* Vasey
FLORIDA RHAPHIS [p. 635]

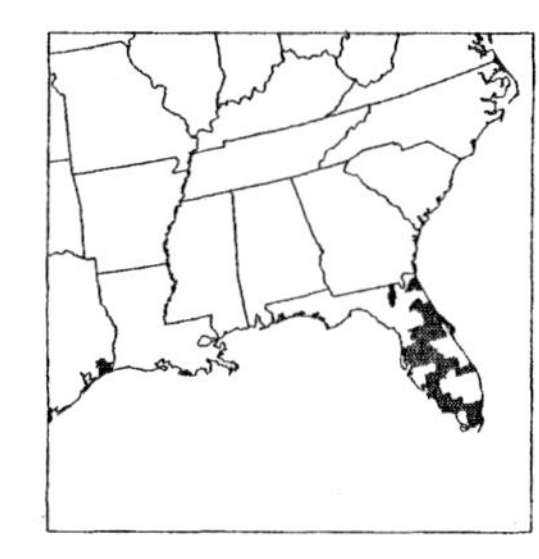

Plants annual. **Culms** 60–110 cm, erect or somewhat decumbent. **Sheaths** glabrous; **ligules** membranous, ciliolate; **blades** to 31 cm long, 4–10 mm wide, flat or folded, mostly or completely glabrous, adaxial surfaces sometimes with scattered pubescence at the base. **Panicles** 20–30 cm, open; **branches** 5–8 cm, capillary, strongly divergent; **rames** usually a triplet of spikelets. **Sessile spikelets** 8.1–10 mm; **calluses** about 7 mm, sharp; **glumes** smooth below, scabrous distally; **lower glumes** shortly awned or mucronate from the sinuses of the minutely bilobed apices; **upper lemmas** awned, awns 10.6–16 cm, geniculate, twisted below. **Pedicellate spikelets** 7.2–15 mm. $2n$ = unknown.

Chrysopogon pauciflorus is native, but infrequently encountered, in the southeastern United States, primarily in Florida; it also occurs in Cuba. It grows in flatwoods, abandoned fields, pinelands, marsh edges, and various disturbed sites.

2. **Chrysopogon fulvus** (Spreng.) Chiov. [p. 635]
GOLDEN BEARDGRASS

Plants perennial; cespitose, not stoloniferous. **Culms** 20–80(120) cm, geniculately ascending. **Leaves** mostly basal; **sheaths** glabrous; **ligules** 0.2–0.5 mm, membranous, ciliolate; **blades** 2–30 cm long, 2–3(9) mm wide, mostly glabrous or puberulous adaxially, bases sometimes with hispid hairs. **Panicles** 4–8(16) cm long, 1.5–3 cm wide, ovate, with many branches; **branches** 3–7 cm, sharply ascending, capillary, naked basal portions 2–6 cm, puberulous, terminating in a rame; **rames** with a triplet of spikelets. **Sessile spikelets** 3.5–5.2(8) mm (including the callus); **calluses** 0.7–1.5 mm, sharp, setose, hairs 1.5–1.9 mm, golden; **lower glumes** laterally compressed, smooth, hispidulous distally, acute; **upper glumes** with a dorsal fringe of hairs, awns 4.1–5.3(10) mm; **upper lemmas** awned, awns 2–3 cm, slightly geniculate, column twisted, puberulous, hairs 0.2–0.4 mm. **Pedicels** 1–2.5 mm, setose on the edges, hairs 3–4.9 mm. **Pedicellate spikelets** 2.5–8 mm; **lower glumes** muticous or awned, awns to 0.7 cm. $2n$ = 40.

Chrysopogon fulvus is native from southern India to Thailand, where it is considered a good forage grass. It was grown at the experiment station in Gainesville, Florida, and subsequently found in adjacent flatwoods as an escape.

3. **Chrysopogon aciculatus** (Retz.) Trin. [p. 635]
MACKIE'S PEST, LOVEGRASS

Plants perennial; extensively stoloniferous, with numerous sterile, leafy shoots. **Culms** 15–50 cm, often decumbent at the base, otherwise ascending or erect. **Sheaths** entirely or mostly glabrous, sometimes ciliate on the upper margins; **ligules** 0.1–0.3 mm, membranous, ciliolate; **blades** 1.5–11(23) cm long, 3–7 mm wide, adaxial surfaces mostly glabrous, or with a few papillose-based hairs near the base. **Panicles** 3–10 cm long, 1–3 cm wide, with many branches; **branches** 1.5–3.5 cm, stiffly ascending or appressed, naked lower portions 1.3–2 cm, terminating in a rame; **rames** 5–15 mm, with 1(–4) spikelet pairs. **Sessile spikelets** 7.5–9 mm (including the callus); **calluses** 3–6.4 mm, sharp, setose, hairs 0.4–1.1 mm, golden; **lower glumes** smooth on the lower portion, scabrous distally, acute or shortly bilobed; **upper glumes** mucronate, mucros 0.5–1.3 mm; **upper lemmas** awned, awns 4–8 mm, exerted, more or less straight. **Pedicels** 2–4 mm, mostly glabrous, hispidulous distally. **Pedicellate spikelets** 4.4–7.1 mm, staminate; **glumes** acute to acuminate; **anthers** 1.5–2.7 mm. $2n$ = 20.

Chrysopogon aciculatus is native to tropical Asia, Australia, and Polynesia. In the contiguous United States, it is known only from controlled plantings at the experiment station in Gainesville, Florida. It is a vigorous colonizer of bare ground that can withstand heavy grazing and trampling, and is difficult to eradicate once established. The sharp calluses are injurious to grazing animals. The U.S. Department of Agriculture considers *C. aciculatus* a noxious weed, and should be informed if the species is found growing in other than a controlled planting.

4. **Chrysopogon zizanioides** (L.) Roberty [p. 636]
VETIVER, KHUS-KHUS, KHAS-KHAS

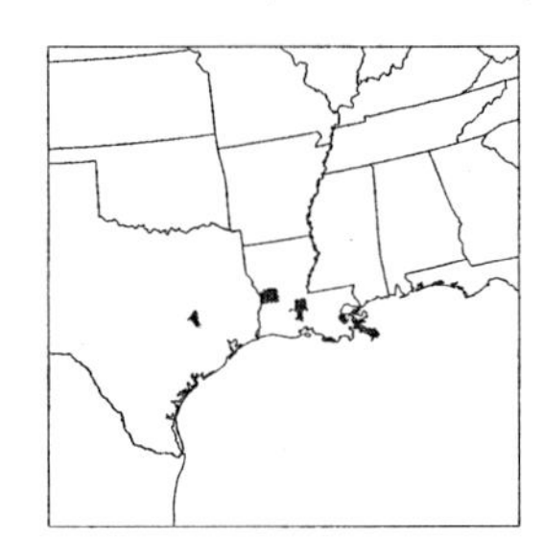

Plants perennial; cespitose, not stoloniferous. **Culms** 1–3 m, not branched, not woody. **Sheaths** glabrous, keeled; **Ligules** 0.3–1.5 mm, of hairs; **blades** 23–140 cm long, 2.5–13 mm wide, flat or folded, mostly glabrous but the adaxial surfaces usually pilose basally. **Panicles** 16–33 cm long, 2.5–9 cm wide, with many branches; **branches** 5.5–12 cm, ascending, naked basal portions 1–4 cm, terminating in a rame; **rames** to 10 cm, with 5–13 spikelet pairs and a terminal triplet. **Sessile spikelets** 3.8–6 mm (including the callus); **calluses** 0.6–0.8 mm, rounded, laterally ciliate basally, hairs 0.1–1.4 mm, white; **lower glumes** scabrous or setulose to spinulose distally, particularly on the veins, acute to acuminate; **upper glumes** setulose distally, particularly on the veins, without a dorsal fringe of hairs, muticous; **upper lemmas** muticous to awned, awns to 2(4.5) mm, straight. **Pedicels** 2.2–4.3 mm. **Pedicellate spikelets** 2.8–4.6 mm, staminate; **glumes** muticous; **anthers** 1.6–2 mm. $2n$ = 20.

Chrysopogon zizanioides, which used to be included in *Vetiveria*, is native to river banks and flood plains in the south Asian tropics and subtropics, but it has been deliberately established in the warmer areas of the United States. It grows in a variety of soils, from heavy clays to dune sand, and will tolerate windy coastal conditions.

Hedges of *Chrysopogon zizanioides* can control soil erosion or restore eroded land. Once established, they are effective even in desert areas subject to flash flooding. The deep root system reaches water far below the surface and prevents the plants from being washed away while the dense, aboveground growth traps silt and sediment. Because *C. zizanioides* does not spread vegetatively and many cultigens have low or no seed production, contour hedges can be planted around cultivated fields or engi-

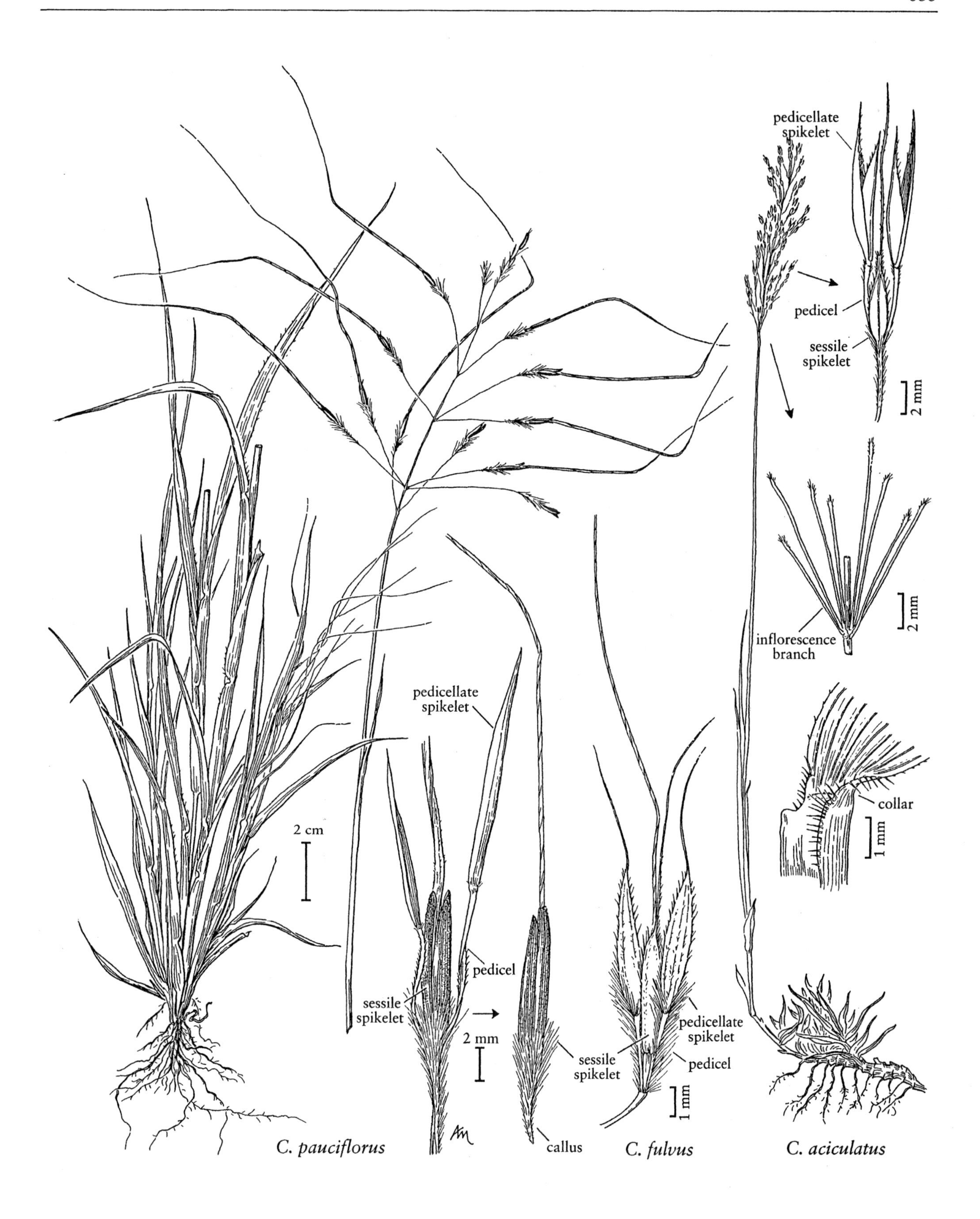
pedicellate spikelet
pedicel
sessile spikelet
2 mm
2 mm
inflorescence branch
collar
1 mm
2 cm
pedicellate spikelet
pedicel
sessile spikelet
2 mm
callus
sessile spikelet
pedicellate spikelet
pedicel
1 mm
C. pauciflorus
C. fulvus
C. aciculatus

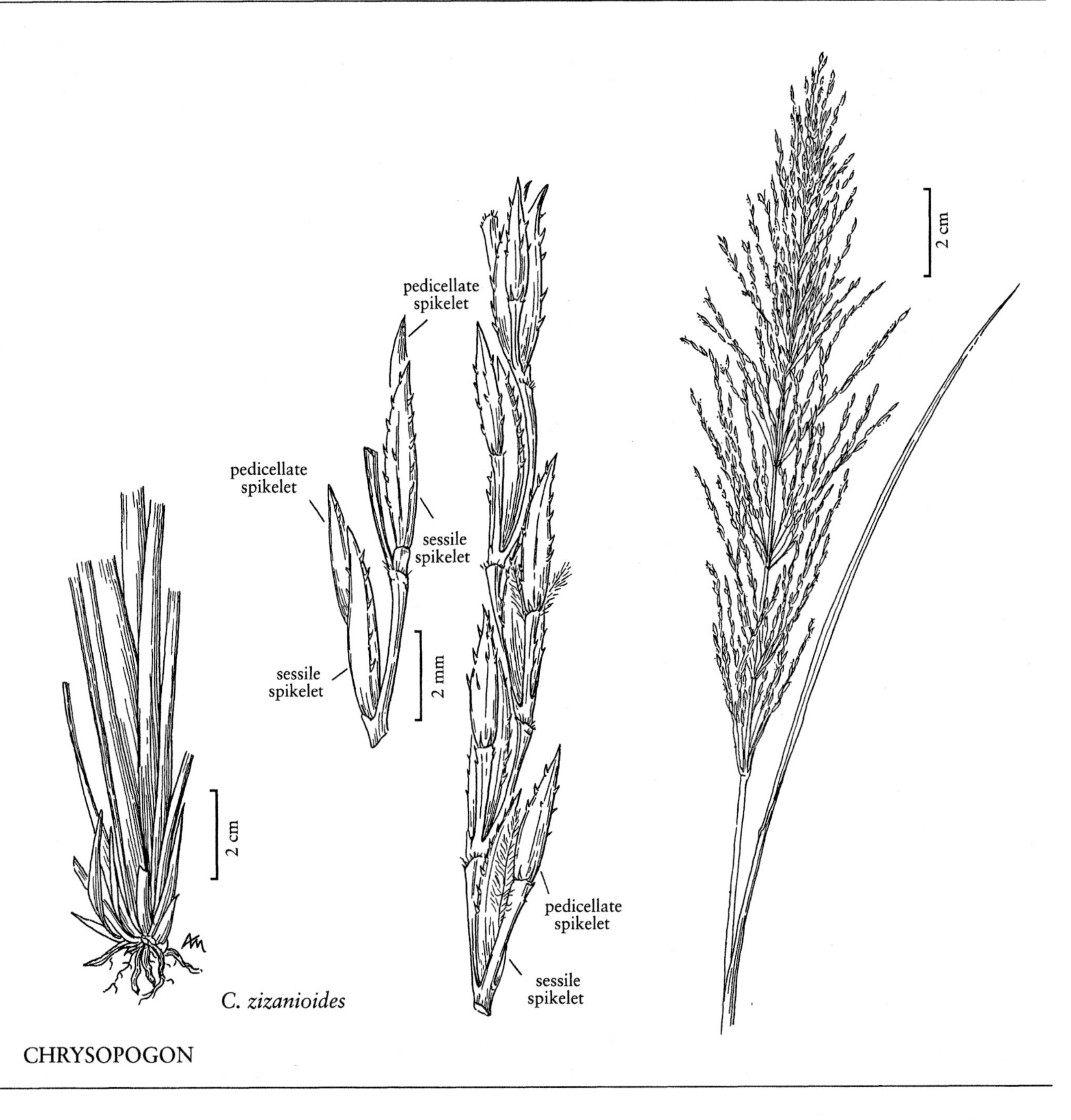

CHRYSOPOGON

hedges can be planted around cultivated fields or engineering structures without fear of invasion.

Essential oils from the aromatic roots are sometimes used as perfume, and numerous biocidal effects are reported. For current information on uses of *Chrysopogon zizanioides*, see http://www.vetiver.org/.

26.11 DICHANTHIUM Willemet

Mary E. Barkworth

Plants annual or perennial; cespitose, sometimes with extensive creeping stolons. **Culms** 15–200 cm. **Leaves** usually not aromatic; **ligules** membranous, sometimes ciliate; **blades** 2-4 mm wide. **Inflorescences** terminal, sometimes also axillary but the axillary inflorescences not numerous; **peduncles** with 1–many rames in digitate or subdigitate clusters; **rames** sometimes naked basally, axes terete to slightly flattened, without a translucent, longitudinal groove, bearing 1–many sessile-pedicellate spikelet pairs and a terminal triplet of 1 sessile and 2 pedicellate spikelets, basal pair(s) homomorphic and homogamous, staminate or sterile, unawned, persistent, distal spikelet pairs homomorphic but heterogamous, sessile spikelets bisexual and awned, pedicellate spikelets staminate or sterile and unawned; **disarticulation** in the rames, beneath the bisexual sessile spikelets. **Sessile spikelets** often imbricate, dorsally compressed, with blunt calluses; **lower glumes** chartaceous to cartilaginous, broadly convex to slightly concave, sometimes pitted; **lower florets** reduced, sterile; **upper florets** sterile or staminate and unawned in the homogamous pairs, bisexual and awned in the heterogamous pairs; **awns** 1–3.5 cm, usually glabrous; **anthers** (2)3. **Pedicels** free of the rame axes, terete to somewhat flattened, slender, not grooved. **Pedicellate spikelets** sterile or staminate. $x = 10$. Name from the Greek *dicha*, 'in two', as in two separate things, and *anthos*, 'flower', a reference to the presence of homogamous and heterogamous spikelets.

Dichanthium, a genus of 20 species, grows in habitats ranging from subdeserts to marshlands in tropical Asia and Australia. It is frequently found in disturbed areas, and some species are considered to provide good forage. Three species have been introduced to the *Flora* region, one of which is sometimes used as a lawn grass.

1. Lower glume of the sessile spikelets with a subapical arch of hairs; pedicellate spikelets usually sterile . . . . 1. *D. sericeum*
1. Lower glume of the sessile spikelets without a subapical arch of hairs; pedicellate spikelets usually staminate.
   2. Rame bases pilose; lower glume of the sessile spikelets more or less obovate . . . . 2. *D. aristatum*
   2. Rame bases glabrous; lower glume of the sessile spikelets elliptic to oblong . . . . 3. *D. annulatum*

1. **Dichanthium sericeum** (R. Br.) A. Camus [p. 638]
QUEENSLAND BLUEGRASS

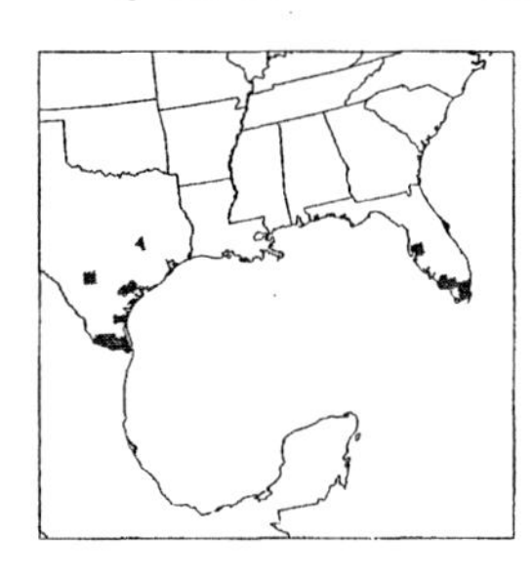

Plants annual or perennial; tufted or cespitose. **Culms** 50–120 cm; **nodes** densely pilose, hairs about 2 mm. **Sheaths** with scattered papillose-based hairs; **ligules** 1–2 mm; **blades** 5–25 cm long, 2–5 mm wide. **Rames** 1–7, 3–7 cm, subdigitate, often glaucous and white-villous, spikelet-bearing to the base, basal spikelet pairs consisting only of glumes; **internodes** pubescent, hairs immediately below the nodes to 1.5 mm. **Sessile bisexual spikelets** 2.5–4.5 mm long, 1–1.4 mm wide; **lower glumes** 5–10-veined, with 0.7–1.5 mm hairs on the basal ½ and about 3 mm papillose-based hairs on the distal portion of the keels and in a transverse subapical arch; **awns** 2–3.5 cm, twice-geniculate. **Pedicellate spikelets** about 3 mm, usually sterile. $2n = 20$.

Dichanthium sericeum is an Australian species. There are two subspecies: **D. sericeum** (R. Br.) A. Camus subsp. **sericeum** is a perennial with sessile spikelets 4–4.5 mm long and to 1–1.4 mm wide, 9–10-veined lower glumes, and rames more than 4 cm long; **D. sericeum** subsp. **humilius** (J.M. Black) B.K. Simon is an annual, with sessile spikelets up to 4 mm long and about 1 mm wide, 5–7-veined lower glumes, and rames less than 4 cm long. *Dichanthium sericeum* subsp. *sericeum* is established in Texas and Florida.

2. Dichanthium aristatum (Poir.) C.E. Hubb. [p. 638]
AWNED DICHANTHIUM

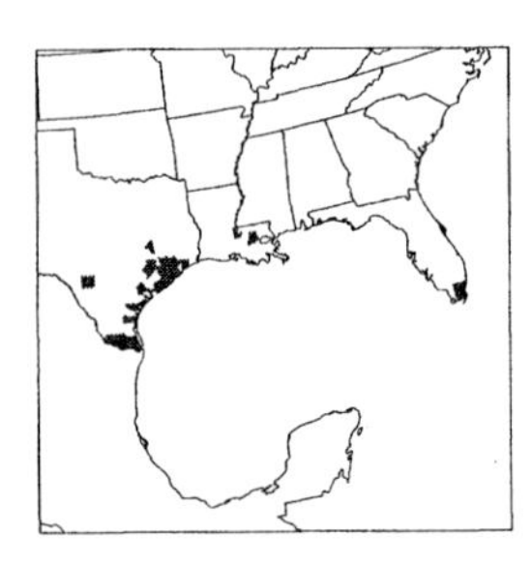

Plants perennial; stoloniferous, stolons often 2 m or longer. **Culms** 70–100 cm, decumbent, erect portions generally about 35 cm, pubescent beneath the inflorescences; **nodes** glabrous or densely short pubescent. **Sheaths** glabrous; **ligules** 1–1.3 mm; **blades** 6–25 cm long, 3–6 mm wide, glabrous or hispid. **Rames** (2)3–5(8), 4–7 cm, subdigitate, erect to divergent, bases pilose, without spikelets; **internodes** pilose. **Sessile spikelets** 4–5 mm; **lower glumes** more or less obovate, often involute, margins ciliate basally, keels winged distally, apices obtuse; **awns** 1.5–2.5 cm, twice-geniculate. **Pedicellate spikelets** 4–5 mm, usually staminate. $2n = 20$.

Dichanthium aristatum was introduced to the Americas from southern Asia. It is sometimes used as a lawn grass in Texas, Louisiana, and Florida.

3. Dichanthium annulatum (Forssk.) Stapf [p. 638]
RINGED DICHANTHIUM

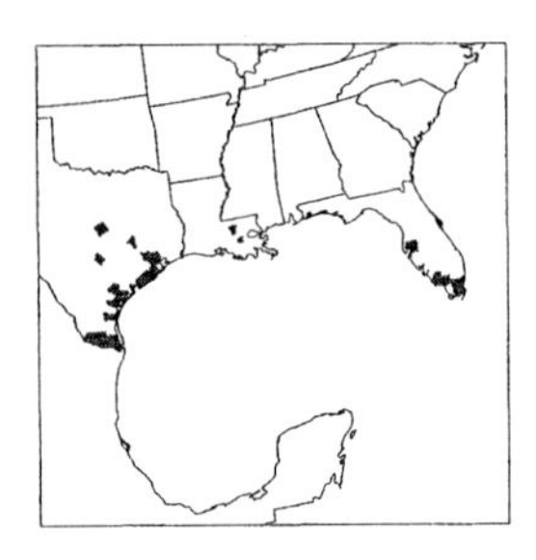

Plants perennial; stoloniferous. **Culms** to 100 cm, decumbent, erect portions generally to 60 cm, often branched above the bases, glabrous beneath the inflorescences; **nodes** glabrous or short-pubescent. **Sheaths** glabrous; **ligules** 1–1.8 mm, truncate; **blades** 3–30 cm long, 2–7 mm wide, scabrous, sparsely pilose, hairs sometimes papillose-based. **Rames** 2–9, 2.5–7 cm, subdigitate, erect to ascending, bases without spikelets, glabrous, internodes ciliate on the margins. **Sessile spikelets** 2.5–5 mm long, 1–1.5 mm wide; **lower glumes** elliptic or oblong, sparsely pubescent below, apices obtuse, irregularly 2–3-toothed, 5–9-veined; **upper glumes** 3-veined; **awns** 1.3–2.2 cm, twice-geniculate. **Pedicellate spikelets** 2.5–5 mm, usually staminate. $2n = 20, 40$.

Dichanthium annulatum is native to southeastern Asia and is a highly esteemed forage grass, especially in India. It is now established at scattered locations in Texas, Louisiana, and Florida.

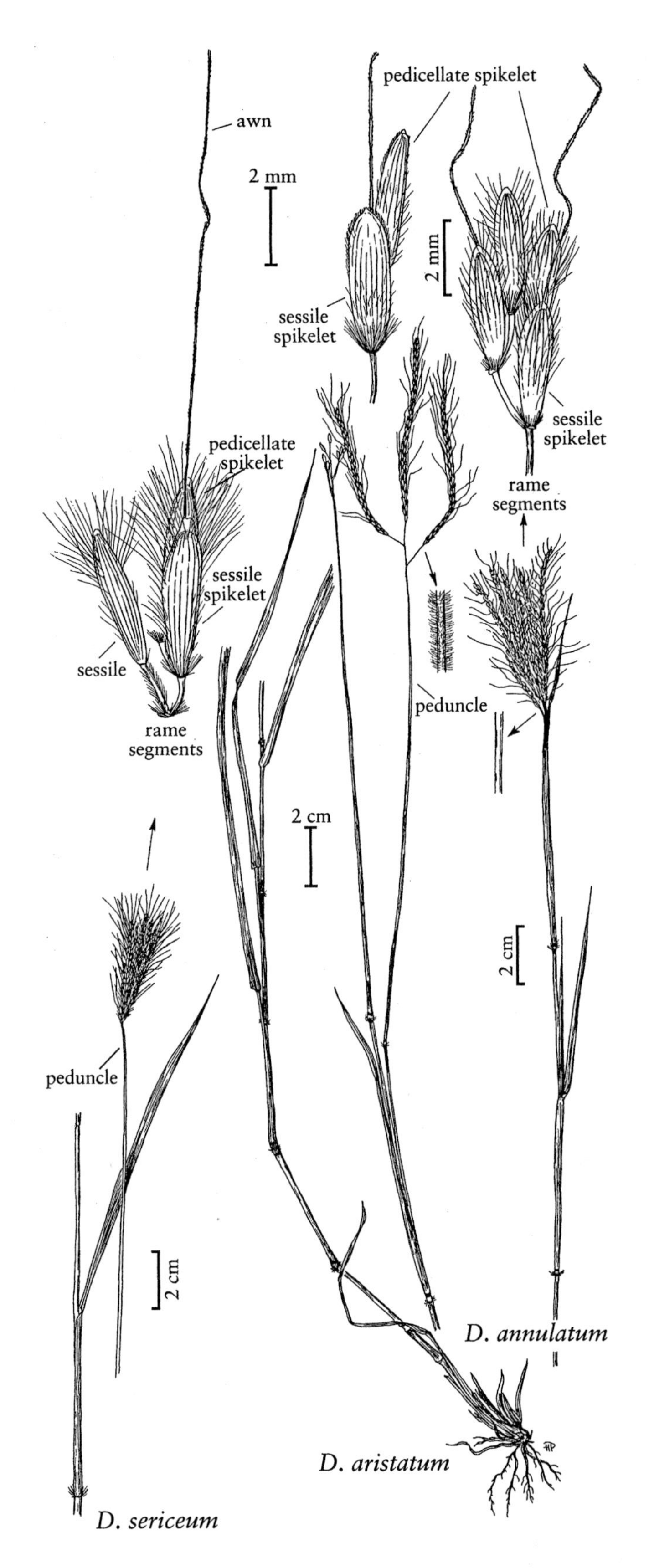

DICHANTHIUM

26.12 BOTHRIOCHLOA Kuntze

Kelly W. Allred

Plants perennial; cespitose or stoloniferous. **Culms** 30–250 cm, with pithy internodes. **Leaves** basal or cauline, not aromatic; **sheaths** open; **auricles** absent; **ligules** membranous, sometimes also ciliate; **blades** usually flat, convolute in the bud. **Inflorescences** terminal, panicles of subdigitate to racemosely arranged branches, each branch with (1)2–many rames, branches not subtended by modified leaves; **rames** with spikelets in heterogamous sessile-pedicellate pairs, internodes with a translucent, longitudinal groove, often villous on the margins; **disarticulation** in the rames, beneath the sessile spikelets. **Spikelets** dorsally compressed; **sessile spikelets** with 2 florets; **lower glumes** rounded, several-veined, sometimes with a dorsal pit, margins clasping the upper glume; **upper glumes** somewhat keeled, 3-veined; **lower florets** hyaline scales, unawned; **upper florets** bisexual; **upper lemmas** with a midvein that usually extends into a twisted, geniculate awn, occasionally unawned; **anthers** 3. **Pedicels** similar to the internodes. **Pedicellate spikelets** reduced or well-developed, sterile or staminate, unawned. **Caryopses** lanceolate to oblong, somewhat flattened; **hila** punctate, basal; **embryos** about ½ as long as the caryopses. $x = 10$. Name from the Greek *bothros*, 'trench' or 'pit', and *chloë*, 'grass', alluding either to the groove in the pedicels or to the pit in the lower glumes of some species.

Bothriochloa is a genus of about 35 species that grow in tropical to warm-temperate regions. Nine are native to the *Flora* region; three Eastern Hemisphere species have been introduced into the southern United States for forage and range rehabilitation. Most species provide fair forage in summer and fall. Polyploidy has been an important mechanism of speciation in the genus.

SELECTED REFERENCES **Allred, K.W.** 1983. Systematics of the *Bothriochloa saccharoides* complex (Poaceae: Andropogoneae). Syst. Bot. 8:168–184; **de Wet, J.M.J.** 1968. Biosystematics of the *Bothriochloa barbinodis* complex (Gramineae). Amer. J. Bot. 55:1246–1250; **de Wet, J.M.J.** and **J.R. Harlan.** 1970. *Bothriochloa intermedia*–a taxonomic dilemma. Taxon 19:339–340; **Vega, A.S.** 2000. Revisión taxonómica de las especies americanas del género *Bothriochloa* (Poaceae: Panicoideae: Andropogoneae). Darwiniana 38:127–186.

1. Pedicellate spikelets about as long as the sessile spikelets.
   2. Sessile spikelets 5.5–7 mm long . . . . 1. *B. wrightii*
   2. Sessile spikelets 3–4.5 mm long.
      3. Rachises longer than the branches . . . . 10. *B. bladhii*
      3. Rachises shorter than the branches.
         4. Lower glumes of the sessile spikelets with a dorsal pit . . . . 12. *B. pertusa*
         4. Lower glumes of the sessile spikelets without a dorsal pit . . . . 11. *B. ischaemum*
1. Pedicellate spikelets much shorter than the sessile spikelets.
   5. Sessile spikelets 2.5–4.5 mm long; awns absent or less than 17 mm long.
      6. Sessile spikelets unawned or with awns less than 6 mm long . . . . 4. *B. exaristata*
      6. Sessile spikelets with awns 8–17 mm long.
         7. Panicles reddish when mature; hairs below the sessile spikelets about ¼ as long as the spikelets, sparse, not obscuring the spikelets . . . . 10. *B. bladhii*
         7. Panicles silvery-white or light tan; hairs below the sessile spikelets at least ½ as long as the spikelets, copious, at least somewhat obscuring the spikelets.
            8. Panicles 9–20 cm long; sessile spikelets narrowly ovate to lanceolate; glumes acute; leaves evenly distributed on the culms; culms 2–4 mm thick . . . . 3. *B. longipaniculata*
            8. Panicles 4–12(14) cm long; sessile spikelets ovate; glumes blunt; leaves often clustered at the base of the culms; culms usually less than 2 mm thick . . . . 2. *B. laguroides*
   5. Sessile spikelets 4.5–8.5 mm long; awns 18–35 mm long.
      9. Rachises 5–20 cm long, with numerous branches.
         10. Panicles of the larger shoots 14–25 cm long; culms 130–250 cm tall, 2–4 mm thick, stiffly erect, little-branched distally, glaucous below the nodes; nodes with spreading hairs, the hairs 2–6 mm long . . . . 5. *B. alta*

10. Panicles of the larger shoots 5–14(20) cm long; culms usually 60–120 cm tall, usually less than 2 mm thick, tending to be bent at the base and often branched at maturity, not glaucous below the nodes; nodes with ascending hairs less than 3 mm long . . . . 6. *B. barbinodis*
9. Rachises usually less than 5 cm long, with 2–9 branches.
 11. Cauline nodes densely pubescent, the hairs 3–7 mm long, white, spreading . . . . 7. *B. springfieldii*
 11. Cauline nodes glabrous or puberulent, the hairs always less than 3 mm long, usually off-white and ascending.
  12. Lower branches of the inflorescences rebranched; sessile spikelets 4.5–6.5 mm long; lower glumes sparsely hairy near the base; leaves primarily cauline, the blades 2–5 mm wide . . . . 8. *B. hybrida*
  12. Lower branches of the inflorescences simple, not rebranched; sessile spikelets 5–8 mm long; lower glumes glabrous; leaves primarily basal, the blades usually less than 2 mm wide . . . . 9. *B. edwardsiana*

1. **Bothriochloa wrightii** (Hack.) Henrard [p. 641]
WRIGHT'S BLUESTEM

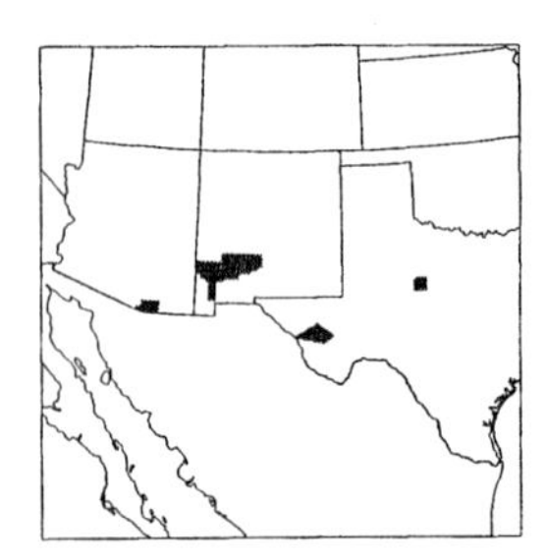

Culms to 70 cm, erect, sparingly branched; **nodes** glabrous or hirsute, hairs about 1 mm. **Leaves** cauline, glaucous; **ligules** 1–2 mm; **blades** 15–25 cm long, 3–7 mm wide, glabrous. **Panicles** 5–6 cm, oblong to fan-shaped; **rachises** 1–3 cm, with 4–5 branches; **branches** 4–6 cm, lacking axillary pulvini, with 1 rame; **rame internodes** with stiff, 1–3 mm marginal hairs. **Sessile spikelets** 5.5–7 mm, lanceolate-elliptic; **lower glumes** glabrous, usually without a dorsal pit; **awns** 10–15 mm, twisted, once-geniculate; **anthers** about 3 mm. **Pedicellate spikelets** staminate, subequal to the sessile spikelets. $2n = 120$.

Bothriochloa wrightii grows in rocky grasslands and shrubby slopes of the pine-oak woodlands of southern Arizona, New Mexico, Texas, and northern Mexico, at 1200–1800 m. It was last collected in the United States in 1930. It differs from *B. barbinodis* in its glaucous foliage, short, fan-shaped panicles, and large, pedicellate spikelets.

2. **Bothriochloa laguroides** (DC.) Herter [p. 641]
SILVER BLUESTEM

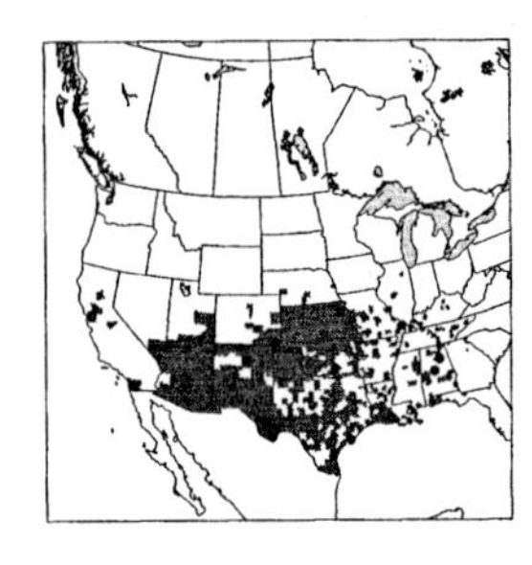

Culms 35–115(130) cm tall, usually less than 2 mm thick, erect or geniculate at the base, branched at maturity; **nodes** shortly hirsute, pilose with erect hairs, or glabrous. **Leaves** usually basal (sometimes cauline on robust plants), usually glaucous; **ligules** 1–3 mm; **blades** 5–25 cm long, 2–7 mm wide, flat to folded, mostly glabrous. **Panicles** 4–12(14) cm, narrowly oblong or lanceolate, silvery-white or light tan; **rachises** 4–8 cm, with more than 10 branches; **branches** 1–5.5 cm, erect-appressed, rarely with axillary pulvini, lower branches shorter than the rachises, usually with more than 1 rame; **rame internodes** with a groove wider than the margins, margins copiously hairy, hairs 3–9 mm, at least somewhat obscuring the spikelets. **Sessile spikelets** 2.5–4.5 mm, ovate, somewhat glaucous, apices blunt; **lower glumes** glabrous or hirtellous, rarely with a dorsal pit; **awns** 8–16 mm; **anthers** 0.6–1.4 mm. **Pedicellate spikelets** 1.5–2.5(3.5) mm, shorter than the sessile spikelets, sterile. $2n = 60$.

Bothriochloa laguroides grows in well-drained soils of grasslands, prairies, roadsides, river bottoms, and woodlands, often on limestone, usually at 20–2100 m. Plants from the United States and northern Mexico belong to **B. laguroides** subsp. **torreyana** (Steud.) Allred & Gould, which differs from **B. laguroides** (DC.) Herter subsp. **laguroides** in its glabrous, or almost glabrous, nodes, long internode hairs, and pilose throat region. Occasional plants are found with spreading branches and axillary pulvini; they do not merit formal recognition. *Bothriochloa laguroides* subsp. *torreyana* is used in landscaping. It does well on rocky slopes and sandy banks.

Bothriochloa laguroides has been confused with *B. saccharoides* (Sw.) Rydb., a more southern species that differs from *B. laguroides* in having pilose leaves, a narrow central groove in the internodes and pedicels, and panicle branches with axillary pulvini.

3. **Bothriochloa longipaniculata** (Gould) Allred & Gould [p. 641]
LONGSPIKE SILVER BLUESTEM

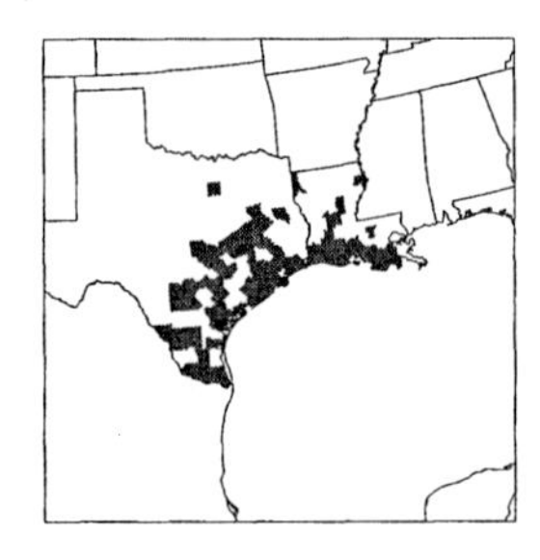

Culms 60–150(200) cm tall, 2–4 mm thick, robust; **nodes** glabrous or shortly hirsute. **Leaves** cauline, evenly distributed, glabrous, dark green; **ligules** 2.5–3 mm; **blades** 12–20 cm long, (3)4–7 mm wide, flat to folded. **Panicles** 9–20 cm, narrowly lanceolate, silvery-white or light tan; **rachises** 7–15

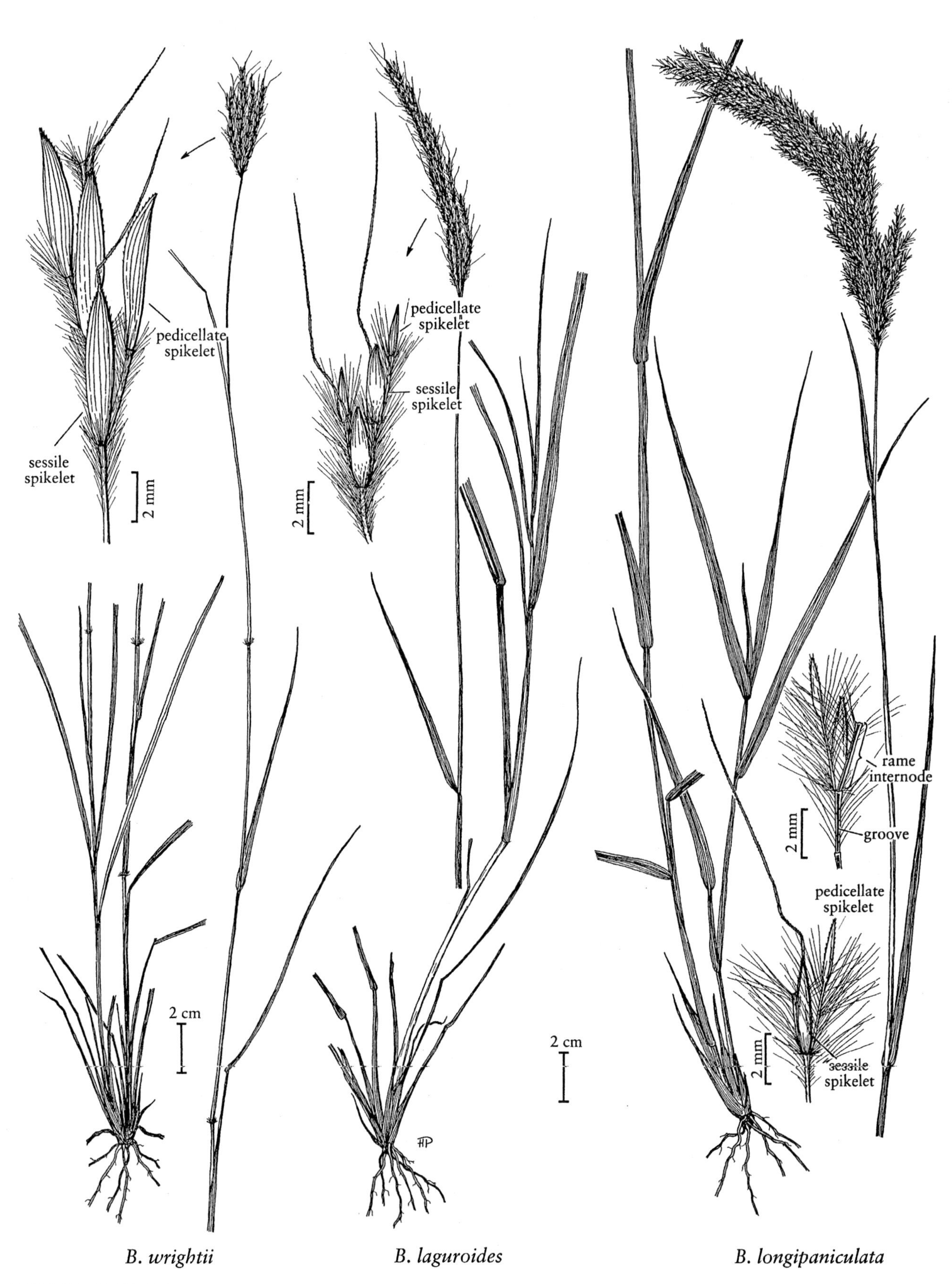

B. wrightii *B. laguroides* *B. longipaniculata*

BOTHRIOCHLOA

cm, with numerous branches; **branches** 3–5 cm, shorter than the rachises, erect, without axillary pulvini, with multiple rames; **rame internodes** with a membranous groove wider than the margins, margins copiously hairy, hairs 3–8 mm, at least somewhat obscuring the spikelets. **Sessile spikelets** (3)3.5–4.5 mm, narrowly ovate to lanceolate, shiny green, apices acute; **lower glumes** hirtellous on the lower ½, hairs shorter than 0.8 mm, lacking a dorsal pit; **awns** 9–14 mm; **anthers** 1–2 mm. **Pedicellate spikelets** 1.8–2.8 mm, sterile. $2n = 120$.

Bothriochloa longipaniculata grows at 2–200 m, along roadsides and in fields, open woodlands, disturbed ground, and swales of the Gulf coastal prairie, often in heavy clay soil. Its range extends from southern Texas and Louisiana to northeastern Mexico and possibly Panama.

4. **Bothriochloa exaristata** (Nash) Henrard [p. 643]
AWNLESS BLUESTEM

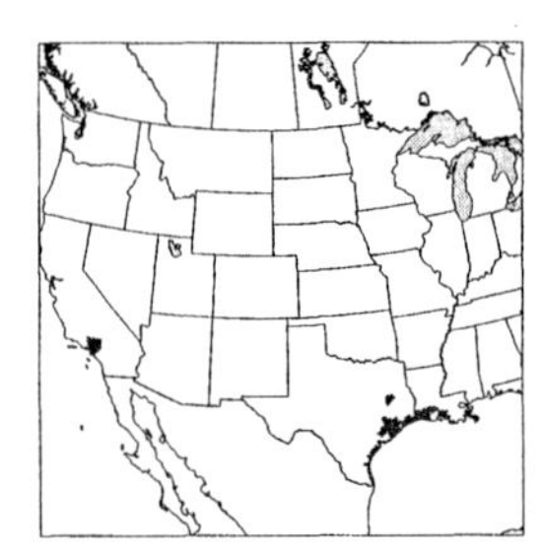

Culms 40–150 cm, erect; **nodes** glabrous, uppermost node often concealed within the sheaths. **Leaves** cauline, mostly glabrous; **sheaths** with a white, powdery bloom; **ligules** 1–2.2 mm; **blades** 10–20 cm long, 3–6(8) mm wide, flat to folded. **Panicles** 4.5–15 cm, lanceolate; **rachises** with numerous branches; **branches** shorter than the rachises, erect-appressed, lacking axillary pulvini; **rame internodes** with a central groove about as wide as the margins, margins densely villous, hairs 4–6 mm, obscuring the spikelets. **Sessile spikelets** 2.5–4 mm long, 0.6–0.8 mm wide, narrowly ovate; **lower glumes** glabrous or sparsely short-pilose, lacking a dorsal pit; **awns** absent or to 6 mm; **anthers** 0.5–1.5 mm. **Pedicellate spikelets** shorter than the sessile spikelets, sterile. $2n = 60$.

Bothriochloa exaristata grows in heavy soils of fields and roadsides of the Gulf coastal prairie, at 2–150 m, as well as in coastal areas of southern Brazil and adjacent Argentina, and inland along the Rio Pilcomayo to Paraguay. It has been reported from Los Angeles County, California. When growing in dense grassland thickets, *B. exaristata* has rather spindly basal growth, but branches abundantly from the middle and upper nodes.

5. **Bothriochloa alta** (Hitchc.) Henrard [p. 643]
TALL BLUESTEM

Culms 1.3–2.5 m tall, 2–4 mm wide, stiffly erect, not or only sparingly branched; **nodes** hirsute, hairs 2–6 mm, stiff, spreading, tan; **internodes** glaucous below the nodes. **Leaves** cauline; **ligules** 1–3 mm; **blades** 20–30 cm long, 4–10 mm wide, glabrous or sparsely pilose near the base. **Panicles** 14–25 cm long on the larger shoots, 3–6 cm wide when pressed, oblong, dense; **rachises** 10–20 cm, with numerous branches, rachises and branches kinked and wavy at the base from being compressed in the sheath; **branches** 2–8 cm, much shorter than the rachises, erect to appressed, with multiple rames; **rame internodes** villous on the margins, with 5–8 mm distal hairs. **Sessile spikelets** 4.5–6 mm, ovate; **lower glumes** shortly pilose, with or without a dorsal pit; **awns** 18–22 mm; **anthers** about 1 mm, often remaining in the floret, light brown. **Pedicellate spikelets** 3.8–4.4 mm. $2n = 120$.

Bothriochloa alta grows along roads, drainage ways, and gravelly slopes in the desert grasslands of the southwestern United States, at 600–1200 m, and extends south to Bolivia and Argentina. It is not a common species in the *Flora* region. It often grows with and is mistaken for *B. barbinodis*, but differs from that species in having longer culms, panicles, and nodal hairs, and $2n = 120$. Plants in the southwestern United States have larger spikelets and more hairy panicles than those of central Mexico, where the species was originally described.

6. **Bothriochloa barbinodis** (Lag.) Herter [p. 643]
CANE BLUESTEM

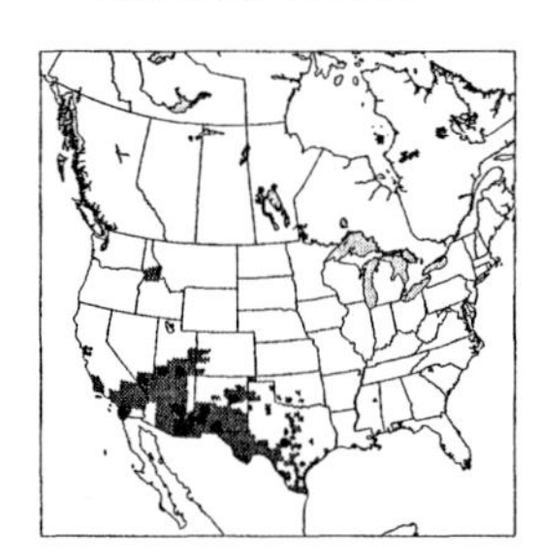

Culms 60–120 cm tall, rarely more than 2 mm thick, erect, geniculate at the base, often branched at maturity, not glaucous below the nodes; **nodes** hirsute, hairs 3–4 mm, mostly erect to ascending, tan or off-white. **Leaves** cauline; **ligules** 1–2 mm, often erose; **blades** 20–30 cm long, 2–7 mm wide, not glaucous, glabrous or sparingly pilose near the throat. **Panicles** 5–14(20) cm on the larger shoots, oblong to somewhat fan-shaped, silvery-white; **rachises** 5–10 cm, straight, exserted or partially included in the sheath, with numerous branches; **branches** 4–9 cm, erect, with several rames; **rame internodes** with a membranous groove wider than the margins, margins densely pilose, longest hairs 3–7 mm, concentrated distally. **Sessile spikelets** 4.5–7.3 mm; **lower glumes** short pilose, with

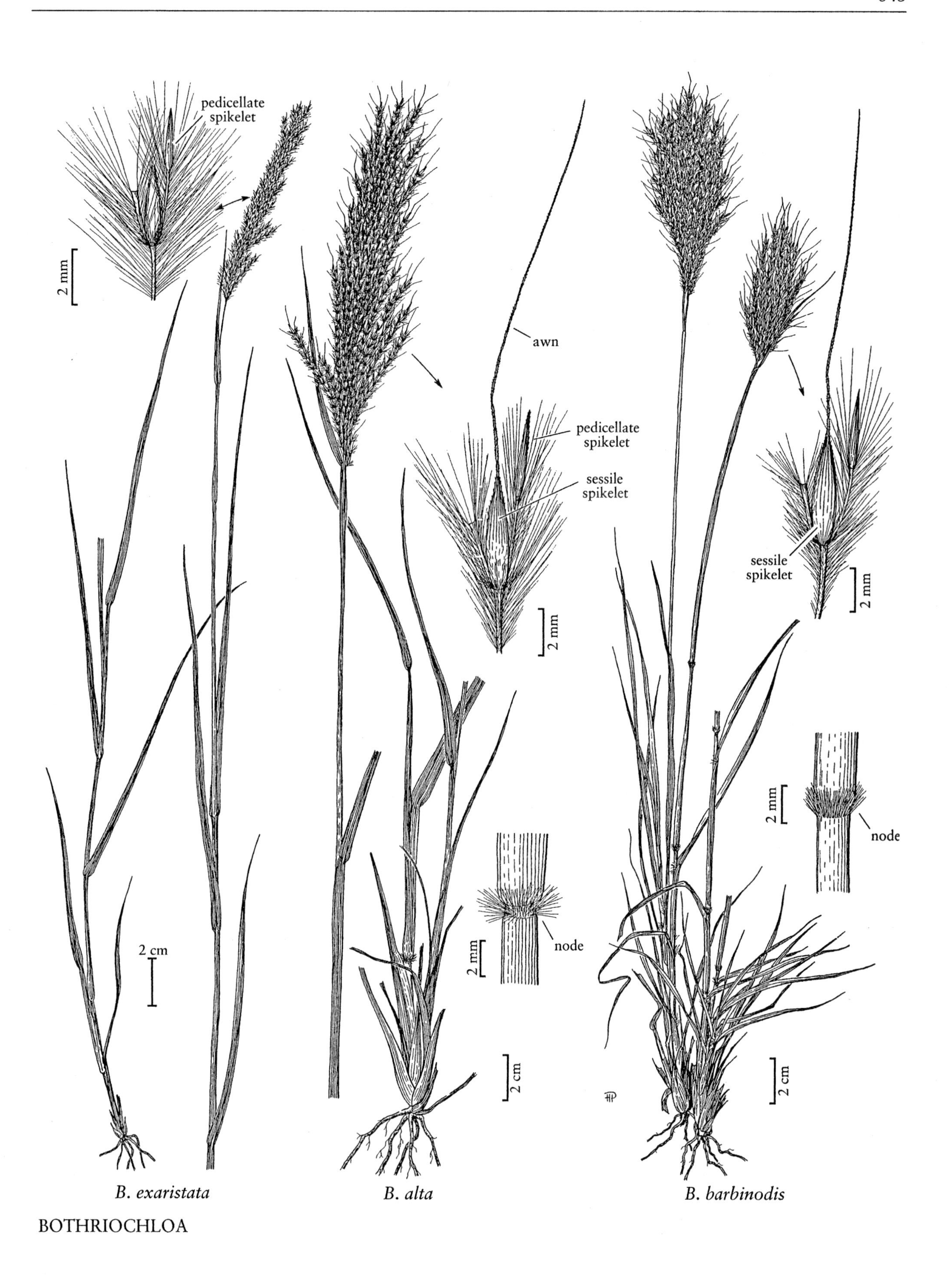

BOTHRIOCHLOA

or without a dorsal pit; **awns** 20–35 mm; **anthers** 0.5–1 mm, often remaining within the spikelet. **Pedicellate spikelets** 3–4 mm, narrowly lanceolate, sterile. 2*n* = 180.

Bothriochloa barbinodis is a common species, at 500–1200 m, along roadsides, drainage ways, and gravelly slopes in desert grasslands, from the southwestern United States through Mexico and Central America to Bolivia and Argentina, and has been found in the Hawaiian Islands. Plants with a pit on the back of their lower glumes occur sporadically; they do not differ in any other respect from those without pits. The species is sometimes used as an ornamental. It is tolerant of coastal conditions and will grow as far north as Vancouver, British Columbia.

Bothriochloa barbinodis has been confused with three other species in the *Flora* region. It differs from *B. wrightii* in not having glaucous foliage, and in having oblong to merely somewhat fan-shaped panicles with pedicellate spikelets that are definitely shorter than the sessile spikelets; from *B. alta* in having shorter culms, panicles, and nodal hairs; and from *B. springfieldii* in having taller culms, wider leaves, shorter nodal hairs, and more, less hairy panicles branches.

7. **Bothriochloa springfieldii** (Gould) Parodi [p. 645]
SPRINGFIELD BLUESTEM

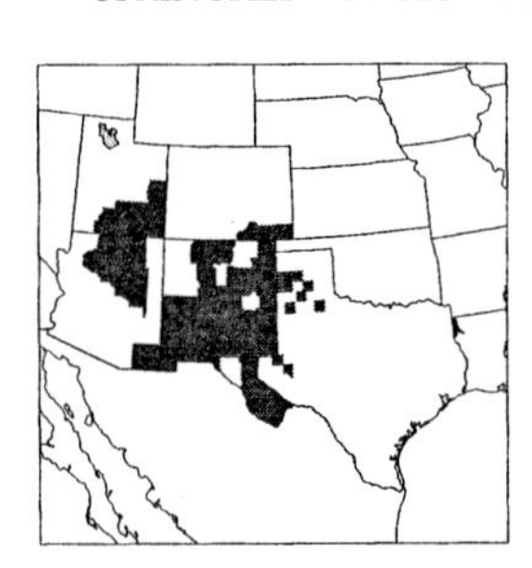

Culms 30–80 cm, erect, unbranched; **nodes** prominently bearded, hairs 3–7 mm, spreading, silvery-white. **Leaves** mostly basal; **ligules** 1–2.5 mm; **blades** 5–30 cm long, 2–3(5) mm wide, flat to folded, glabrous or sparsely hispid adaxially, pilose near the throat. **Panicles** 4–9 cm, oblong to fan-shaped; **rachises** 1–5 cm, with 2–9 branches; **branches** 4–8 cm, longer than the rachises, with 1(2) rames; **rame internodes** with a membranous groove wider than the margins, margins densely white-villous, hairs 5–10 mm, obscuring the sessile spikelets. **Sessile spikelets** 5.5–8.5 mm, lanceolate; **lower glumes** densely short-pilose on the lower ½, sometimes with a dorsal pit; **awns** 18–26 mm; **anthers** 1–1.5 mm. **Pedicellate spikelets** 3.5–5.5 mm, sterile. 2*n* = 120.

Bothriochloa springfieldii grows in rocky uplands, ravines, plains, sandy areas, and roadsides, from southern Utah to western Texas and Mexico at 900–2500 m. and, as a disjunct in northwest Louisiana. It differs from *B. barbinodis* in its less robust habit, narrower blades, longer nodal hairs, and fewer, more hairy panicle branches, and from *B. edwardsiana* in its pubescent nodes and wider, non-ciliate leaf blades.

8. **Bothriochloa hybrida** (Gould) Gould [p. 645]
HYBRID BLUESTEM

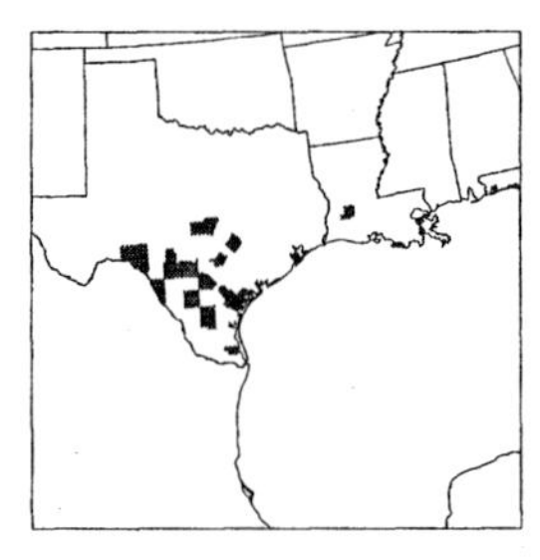

Culms 30–80 cm, stiffly erect, moderately branched above the base; **nodes** glabrous or puberulent. **Leaves** primarily cauline; **sheaths** glabrous, green, sometimes glaucous; **ligules** 1–2 mm; **blades** 5–25 cm long, 2–5 mm wide, flat to folded, usually ciliate, with long hairs near the base and some hairs on the adaxial surface. **Panicles** 5–12 cm, lanceolate; **rachises** usually shorter than 5 cm; **branches** 3–8, without axillary pulvini, lower branches longer than the rachises; at least the lower branches rebranched and with multiple rames; **rame internodes** with 5–7 mm marginal hairs. **Sessile spikelets** 4.5–6.5 mm, narrowly ovate; **lower glumes** 4.5–5.7(6.5) mm, sparsely hairy near the base, with a dorsal pit above the middle; **awns** 18–25 mm; **anthers** 0.5–1 mm. **Pedicellate spikelets** 2.2–3.6 mm, sterile. 2*n* = 120.

Bothriochloa hybrida grows in open grasslands, rangeland pastures, disturbed ground, and roadsides, often on calcareous soil, usually at 50–500 m. Its range extends from southern Texas and Louisiana to central Mexico. It resembles *B. edwardsiana* in some respects, but the latter species has a less robust habit, more predominantly basal foliage, and narrower leaf blades.

9. **Bothriochloa edwardsiana** (Gould) Parodi [p. 645]
MERRILL'S BLUESTEM

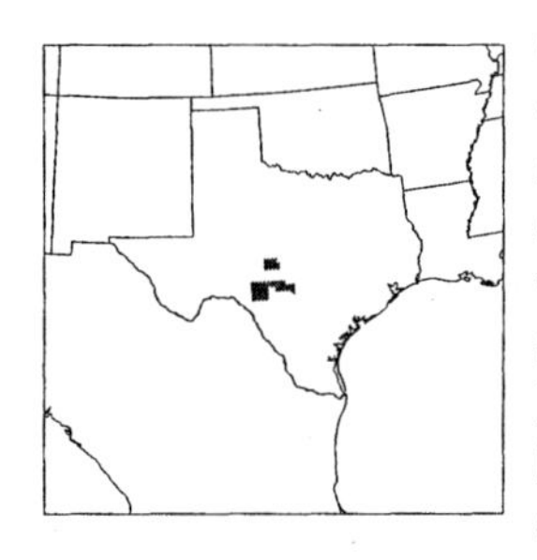

Culms 35–65 cm, slender, stiffly erect, rarely geniculate; **lower nodes** shortly hairy, hairs shorter than 3 mm, usually off-white and ascending; **upper nodes** glabrous or glabrate. **Leaves** mostly basal, glaucous; **ligules** 1–1.5 mm; **blades** 10–25 cm long, 1–2(3.5) mm wide, flat to rolled, with 3–7 mm hairs below the middle. **Panicles** 6–12 cm, loose, fan-shaped; **rachises** shorter than 5 cm, with 3–6 branches; **branches** longer than the rachises, not rebranched, with 1 rame; **rame internodes** with 3–5 mm marginal hairs. **Sessile spikelets** 5–8 mm, lanceolate; **lower glumes** 5.5–7 mm, glabrous, shiny, with a deep dorsal pit, tapering to a narrow, slightly bifid apex; **awns** 20–28 mm; **anthers** 0.5–1 mm. **Pedicellate spikelets** 2.5–3.5 mm, sterile. 2*n* = 60.

Bothriochloa edwardsiana grows in the rocky plains and prairies of the Edwards Plateau of Texas, on calcareous soil, at 300–600 m. It also grows in northern Mexico and Uruguay. It resembles *B. hybrida* in some respects, but that species has a more robust habit, predominantly cauline foliage, and wider leaf blades.

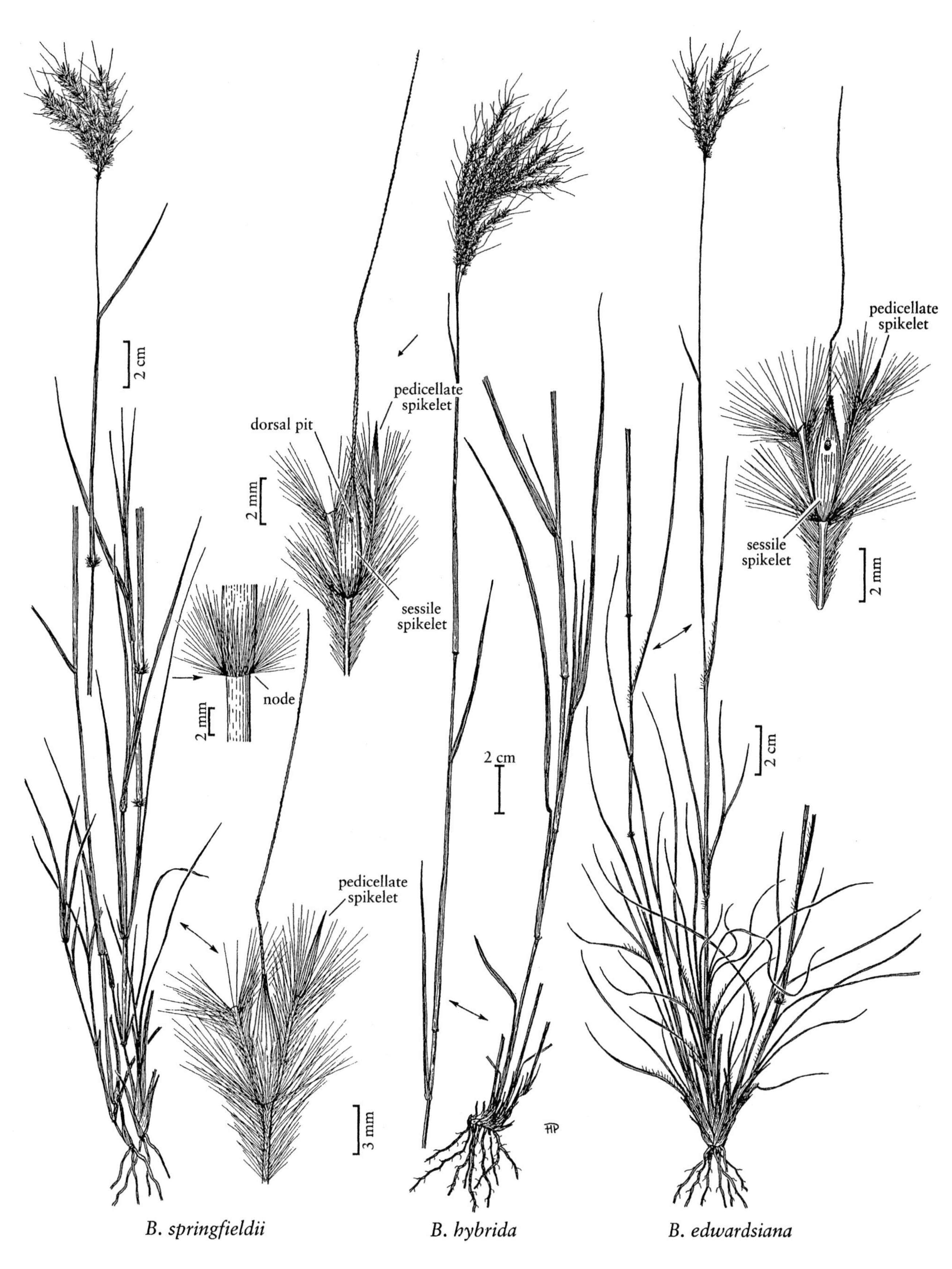

B. springfieldii *B. hybrida* *B. edwardsiana*

10. **Bothriochloa bladhii** (Retz.) S.T. Blake [p. 647]
AUSTRALIAN BLUESTEM

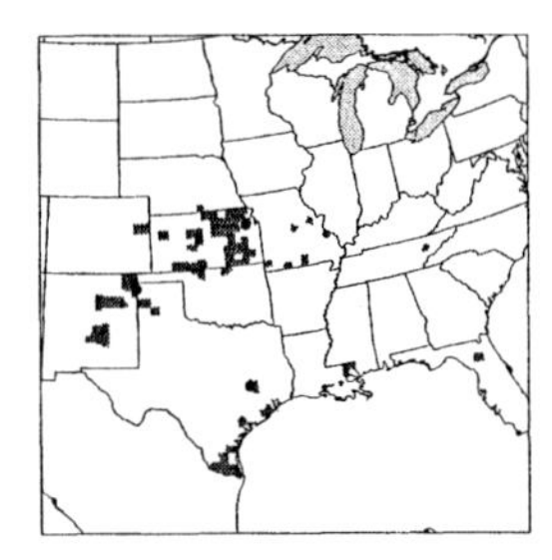

Culms 40–90(150) cm, usually erect; **nodes** glabrous or short hispid, with mostly appressed, less than 2 mm hairs. **Leaves** cauline; **ligules** 0.5–1.5 mm; **blades** (10)20–35(40) cm long, 1–4.5(5.5) mm wide, mostly glabrous. **Panicles** 5–15(24) cm, elliptic to lanceolate, reddish at maturity; **rachises** 6–12(20) cm, with numerous branches; **branches** 3–7 cm, shorter than the rachises, erect to spreading during anthesis, with axillary pulvini, lower branches with multiple rames; **rame internodes** with darkened grooves, with sparse, about 1 mm marginal hairs. **Sessile spikelets** 3.5–4 mm, oblong-ovate; **lower glumes** glabrous or scabrous, with or without a dorsal pit; **awns** 10–17 mm, twisted, geniculate; **anthers** 1–2 mm. **Pedicellate spikelets** about the same size and shape as the sessile spikelets, or about ½ their size, staminate or sterile. $2n$ = 40, 60, 80.

Bothriochloa bladhii grows along roadsides and in rangeland pastures, waste ground, and open disturbed areas, at 150–1800 m. It is native to subtropical Asia and Africa and was introduced to the *Flora* region as a forage grass. It is now established in the southern and central United States. A similar species, *B. decipiens* (Hack.) C.E. Hubb., has been grown at some experiment stations in the United States. It is not known to be established in North America. *Bothriochloa decipiens* differs from *B. bladhii* in having longer (4.7–5.3 mm) sessile spikelets and a single anther.

The Eastern Hemisphere species of *Bothriochloa* are thought to be closely related to *Capillipedium* and *Dichanthium*, largely because *B. bladhii* hybridizes with those genera as well as with *B. ischaemum*.

11. **Bothriochloa ischaemum** (L.) Keng [p. 647]

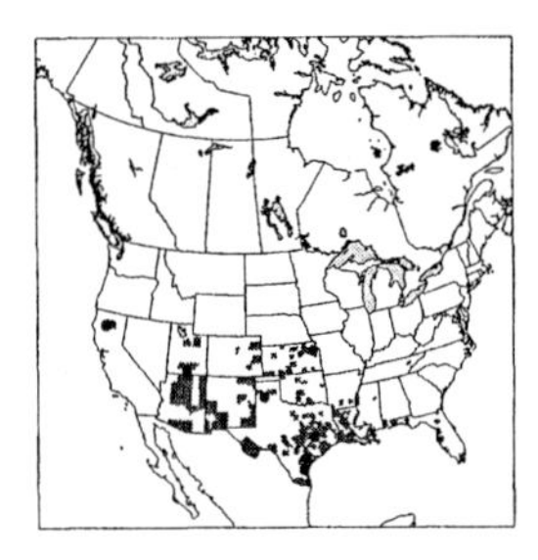

Plants usually cespitose, occasionally stoloniferous or almost rhizomatous under close grazing or cutting. **Culms** 30–80(95) cm, stiffly erect; **nodes** glabrous or short hirsute. **Leaves** tending to be basal; **ligules** 0.5–1.5 mm; **blades** 5–25 cm long, 2–4.5 mm wide, flat to folded, glabrous or with long, scattered hairs at the base of the blade. **Panicles** 5–10 cm, fan-shaped, silvery reddish-purple; **rachises** 0.5–2 cm, with (1)2–8 branches; **branches** 3–9 cm, longer than the rachises, erect to somewhat spreading from the axillary pulvini, usually with only 1 rame; **rame internodes** with a central groove narrower than the margins, margins ciliate, with 1–3 mm hairs. **Sessile spikelets** 3–4.5 mm, narrowly ovate; **lower glumes** hirsute below, with about 1 mm hairs, lacking a dorsal pit; **awns** 9–17 mm, twisted, geniculate; **anthers** 1–2 mm. **Pedicellate spikelets** about as long as the sessile spikelets, but usually narrower, sterile or staminate. $2n$ = 40, 50, 60.

Bothriochloa ischaemum grows along roadsides and in waste ground and rangeland pastures, at 50–1200 m. It is native to southern Europe and Asia. It was introduced to the United States for erosion control along right of ways and for livestock forage in the southwest. It is now established in the region and has spread along roadsides into other central and southern states. There are two variants that are sometimes recognized as varieties, plants with glabrous nodes being called *B. ischaemum* var. *ischaemum* and plants with pubescent nodes being called *B. ischaemum* var. *songarica* (Rupr. *ex* Fisch. & C.A. Mey.) Celarier & J.R. Harlan. The varieties are not recognized here.

12. **Bothriochloa pertusa** (L.) A. Camus [p. 647]
PITTED BLUESTEM

Plants cespitose or stoloniferous. **Culms** to 100 cm, often decumbent or stoloniferous, freely branching; **nodes** bearded. **Leaves** mostly basal, green, sometimes glaucous; **sheaths** glabrous, keeled; **ligules** 0.7–1.5 mm; **blades** 3–15 cm long, 3–4 mm wide, flat, margins and ligule regions hairy. **Panicles** 3–5 cm, fan-shaped, often purplish; **rachises** 0.2–2 cm, with 3–8 branches; **branches** 3–4.5 cm, longer than the rachises, usually with 1 rame; **rame internodes** with villous margins, with 1–3 mm hairs. **Sessile spikelets** 3–4 mm, lanceolate; **callus hairs** about 1 mm; **lower glumes** sparsely hirtellous, with a prominent dorsal pit near the middle; **awns** 10–17 mm; **anthers** 1–1.8 mm, yellow. **Pedicellate spikelets** the same size as the sessile spikelets, sterile, pitted or not, occasionally with 2 pits. $2n$ = 40, 60.

Bothriochloa pertusa is native to the Eastern Hemisphere, and was introduced to the southern United States as a warm-season pasture grass. It now grows in disturbed, moist, grassy places and pastures in the region, at elevations of 2–200 m. It has not persisted at all locations shown on the map.

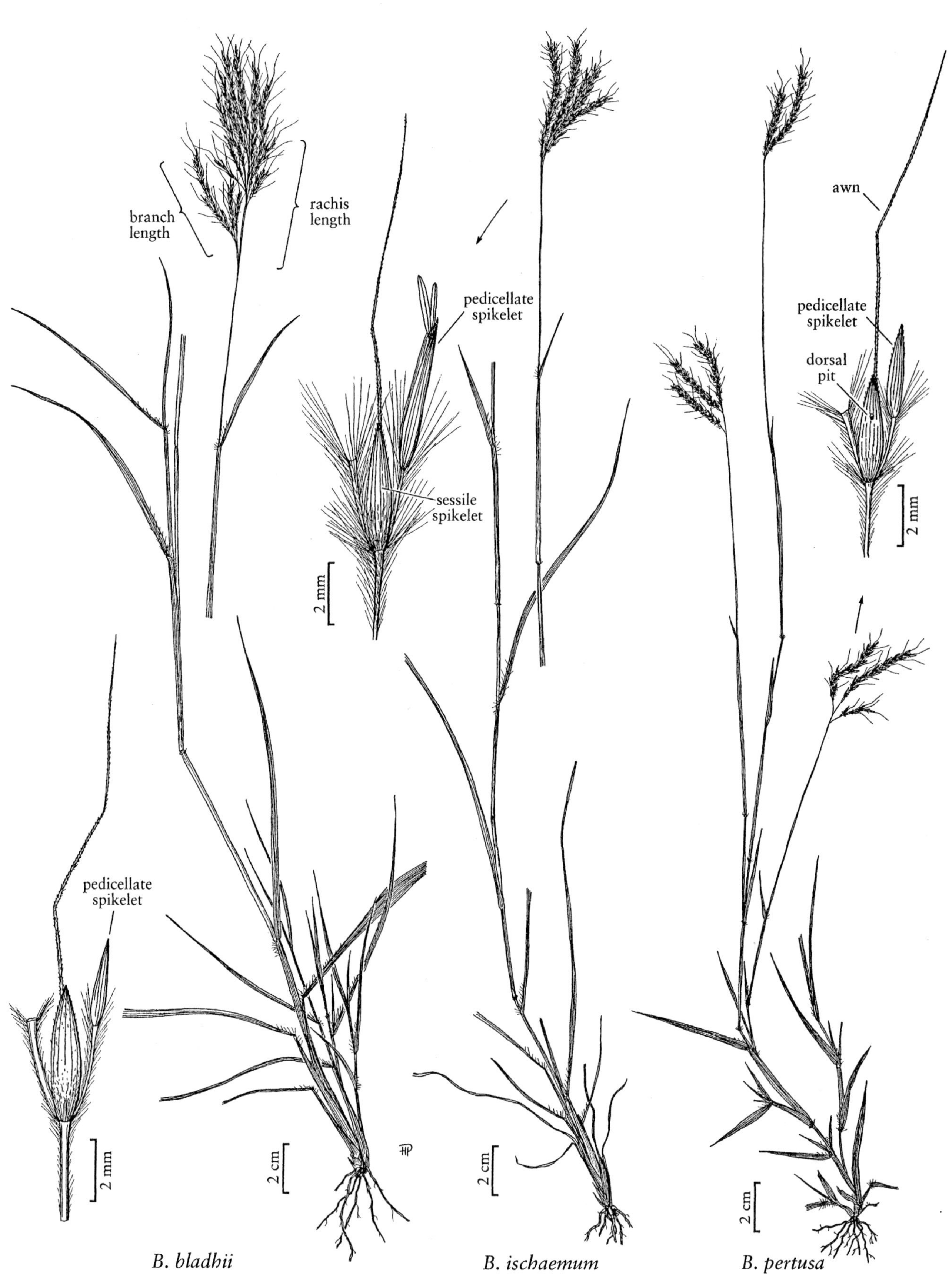

BOTHRIOCHLOA

26.13 ISCHAEMUM L.

Mary E. Barkworth

Plants annual or perennial. **Culms** 10–350 cm, often decumbent, sometimes branched above the base. **Leaves** not aromatic; **sheaths** open; **ligules** membranous, glabrous or ciliate, sides often higher than the middle. **Inflorescences** terminal, sometimes also axillary; **inflorescence units** with (1)2–many rames on a common peduncle; **rames** secund, ascending, members of a cluster sometimes so closely appressed as to appear as one; **internodes** stoutly linear to clavate. **Spikelets** in homogamous or heterogamous sessile-pedicellate or unequally pedicellate pairs; **disarticulation** in the rames, below both the sessile and pedicellate spikelets. **Sessile spikelets** dorsally compressed; **glumes** subequal; **lower glumes** 2-keeled, keels sometimes winged; **upper glumes** keeled, sometimes awned; **lower florets** staminate; **upper florets** bisexual, lemmas usually bifid and awned from the sinus. **Pedicels** fused to the rame axes, clavate or inflated, sometimes as wide as the spikelets. **Pedicellate spikelets** morphologically and sexually similar to the sessile spikelets or staminate and reduced. x = 9, 10. Name from the Greek *ischion*, 'hip' or 'hip-joint socket'.

Ischaemum includes approximately 65 species, all of which are native to tropical regions of the Eastern Hemisphere. None of the species is known to be established in North America. The genus is included in this treatment because two of its species are considered serious weed threats by the U.S. Department of Agriculture.

SELECTED REFERENCES **Koyama, T.** 1987. Grasses of Japan and Its Neighboring Regions: An Identification Manual. Kodansha, Ltd., Tokyo, Japan. 370 pp.; **Reed, C.F.** 1964. A flora of the chrome and manganese ore piles at Canton, in the Port of Baltimore, Maryland and at Newport News, Virginia, with descriptions of genera and species new to the flora of the eastern United States. Phytologia 10:321–405.

1. Lower glumes of the sessile spikelets not winged, rugose, with 4–5 ridges; plants annuals or short-lived perennials . . . . 1. *I. rugosum*
1. Lower glumes of the sessile spikelets winged on the keels, not rugose; plants perennial . . . . 2. *I. indicum*

1. Ischaemum rugosum Salisb. [p. 650]
RIBBED MURAINAGRASS

Plants annual, or short-lived perennials. **Culms** 30–130 cm, erect or geniculate at the base, simple to strongly branched below; **nodes** antrorsely pilose; **internodes** glabrous. **Leaves** cauline; **sheaths** mostly glabrous or sparsely pilose, margins ciliate distally; **collars** villous; **ligules** 1–5.5 mm; **blades** 8–20 cm long, 7–15 mm wide, flat, usually pilose on both surfaces, sometimes with papillose-based hairs, occasionally glabrous. **Inflorescence units** with 2 rames; **rames** 3–8 cm long, 3–4 mm wide; **internodes** 2.5–3.5 mm, slightly clavate distally. **Sessile spikelets** 3–5 mm long, 1.8–2.2 mm wide; **calluses** shortly pubescent; **lower glumes** coriaceous, yellowish, not winged, coarsely transversely rugose with 4–5 ridges on the proximal ⅔–⅘, thickly chartaceous distally and tapering to the apices; **upper glumes** thickly chartaceous, ciliate, awned, awns 1.5–2 cm, geniculate below the middle; **anthers** about 2 mm. **Pedicellate spikelets** varying from 0.5 mm to equaling the sessile spikelets. $2n$ = 18, 20, 44.

Ischaemum rugosum is native to southern Asia, and is now established in moist, tropical habitats around the world, including Mexico. It has been found in southern Texas and on chrome ore piles in Canton, Maryland, but is thought to have been eliminated from both areas. The U.S. Department of Agriculture considers it a noxious weed; plants found growing within the continental United States should be promptly reported to that agency.

2. Ischaemum indicum (Houtt.) Merr. [p. 650]
INDIAN MURAINAGRASS

Plants perennial, forming loose clumps. **Culms** 20–90 cm, decumbent, often rooting from the lower nodes; **nodes** antrorsely pilose; **internodes** glabrous. **Leaves** crowded towards the base; **sheaths** open, sparsely to densely pilose; **ligules** about 1 mm, membranous; **blades** 4–15 cm long, 3–8 mm wide, pilose on both surfaces. **Inflorescence units** with 2 rames; **rames** 3–5 cm long, 4–5 mm wide; **internodes** 3–4 mm long, 0.5–0.7 mm wide, slightly triquetrous. **Sessile spikelets** 4–5 mm long, 1.2–1.5

mm wide; **calluses** strigose; **lower glumes** coriaceous below, thickly membranous above, rough but not rugose, 5–7-veined, 2-keeled, keels narrowly winged; **upper glumes** 1-keeled, keels winged distally, awned, awns 1–1.5 cm; **anthers** 2–2.5 mm. **Pedicels** 2.5–3.5 mm, ciliate on the edges. **Pedicellate spikelets** slightly smaller than the sessile spikelets. 2*n* = 36.

Reed (1964) reported finding *Ischaemum indicum* growing on chrome ore piles in Canton, Maryland. This reported introduction has not been verified.

26.14 APLUDA L.

Mary E. Barkworth

Plants perennial; often scrambling. **Culms** to 3 m, decumbent. **Leaves** not aromatic; **sheaths** open; **ligules** membranous; **blades** linear, often pseudopetiolate. **Inflorescences** false panicles, individual inflorescence units with solitary rames; **rames** to 1 cm, often enclosed by the subtending leaf sheath, with 1 sessile and 2 unequally pedicellate spikelets; **disarticulation** at the base of the sessile spikelets, sometimes also at the base of the pedicellate spikelets. **Sessile spikelets** laterally compressed, with a large, bulbous callus; **lower glumes** coriaceous, without keels or wings, smooth, bidentate; **upper glumes** unawned; **upper lemmas** awned or unawned. **Pedicels** flat, wide, adjacent to each other, appressed but not fused to the rame axes. **Pedicellate spikelets** usually unequal, unawned, 1 staminate or bisexual and as large as the sessile spikelet, the other sterile and usually smaller. *x* = 10. Name from the Latin *apluda*, 'chaff', a reference to the appearance of the inflorescence after the spikelets have fallen.

Apluda is treated here as consisting of a single weedy species that is native to tropical Asia and Australia, where it grows primarily in thickets and forest margins. It is not known to be established in the *Flora* region.

SELECTED REFERENCES **Clayton, W.D.** 1994. *Apluda*. Pp. 35–37 *in* W.D. Clayton, G. Davidse, F. Gould, M. Lazarides, and T.R. Soderstrom. A Revised Handbook to the Flora of Ceylon, vol. 8 (ed. M.D. Dassanayake). Amerind Publishing Co., New Delhi, India. 458 pp.; **Reed, C.F.** 1964. A flora of the chrome and manganese ore piles at Canton, in the Port of Baltimore, Maryland and at Newport News, Virginia, with descriptions of genera and species new to the flora of the eastern United States. Phytologia 10:321–405.

1. **Apluda mutica** L. [p. 650]
MAURITIAN GRASS

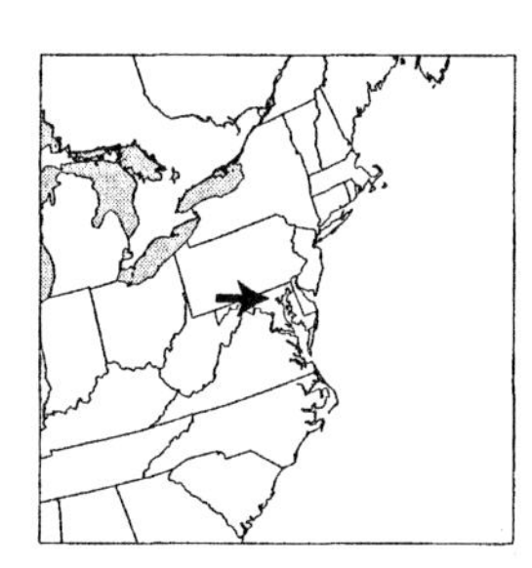

Culms to 3 m, decumbent and rooting from the lower nodes. **Blades** 5–25 cm long, 2–10 mm wide, flat, attenuate distally. **Inflorescences** 3–40 cm; **rames** to 10 mm long, **subtending sheaths** 3.5–10 mm, narrowly ovate in side view. **Sessile spikelets** 2–6 mm; **lower glumes** narrowly elliptic-lanceolate; **upper lemmas** unawned or awned; **awns** 4–12 mm. **Pedicellate spikelets** broadly lanceolate; **larger spikelets** 2–5 mm; **smaller spikelets** 2–4 mm. 2*n* = 20, 40.

Reed (1964) reported finding *Apluda mutica* on chrome ore piles in Canton, Maryland (a temporary unloading ground for ores). The report has not been verified because his voucher specimens were not accessible at the time of writing. They were acquired by the Missouri Botanical Garden in 2001.

26.15 ANDROPOGON L.

Christopher S. Campbell

Plants perennial; usually cespitose, sometimes rhizomatous. **Culms** 20–310 cm, erect, much-branched distally. **Leaves** not aromatic; **ligules** membranous, sometimes ciliate; **blades** linear, flat, folded, or convolute. **Inflorescences** terminal and axillary or a false panicle; **inflorescence units** 1–600+ per culm; **peduncles** initially concealed by the subtending leaf sheaths, sometimes exserted beyond the sheaths at maturity, with (1)2–5(13) rames; **rames** not reflexed at maturity,

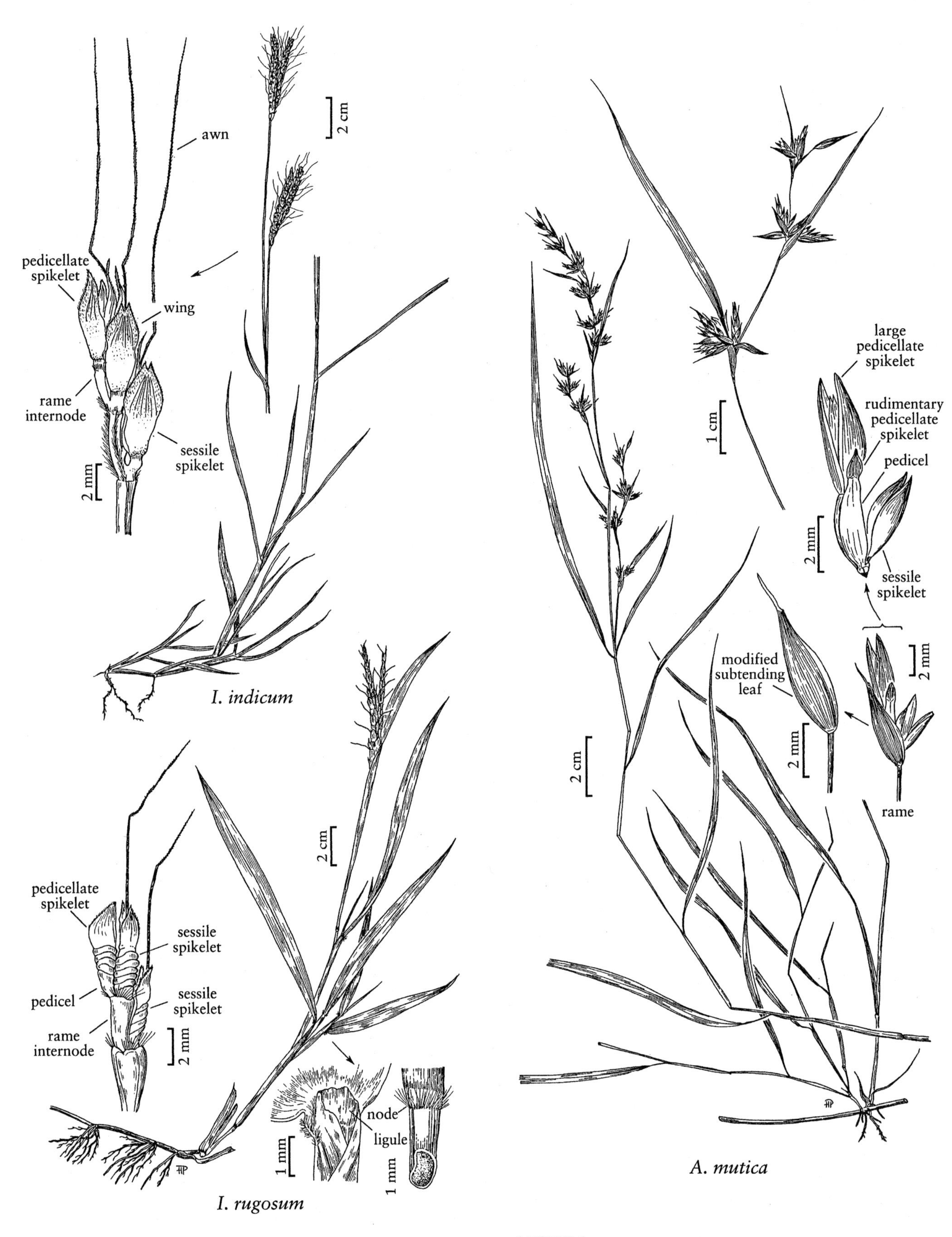

ISCHAEMUM

APLUDA

axes slender, terete to flattened, not longitudinally grooved, usually conspicuously pubescent, with spikelets in heterogamous sessile-pedicellate pairs (the terminal spikelets sometimes in triplets of 1 sessile and 2 pedicellate spikelets), apices of the internodes neither cupulate nor fimbriate; **disarticulation** in the rames, below the sessile spikelets. **Sessile spikelets** bisexual, awned, with short, blunt calluses; **lower glumes** 2-keeled, flat or concave, usually not veined between the keels, sometimes 2–9-veined; **anthers** 1, 3(2). **Pedicels** usually longer than 3 mm, similar to the rame internodes in shape, length, and pubescence color, not fused to the rame axes. **Pedicellate spikelets** usually vestigial or absent, sometimes well-developed and staminate. x = 10. Name from the Greek *andro*, 'man', and *pogon*, 'beard', referring to the pubescent pedicels of the staminate spikelets.

Andropogon is a cosmopolitan genus of tropical and temperate zones, comprising approximately 120 species. Thirteen species are native to the *Flora* region. *Andropogon bicornis* has been found in the region but is not known to be established. All but *A. hallii* grow in the southeastern United States.

Several taxa are ecologically important in North America. *Andropogon gerardii* is one of the most important native grasses in North America, being one of the dominant species in the tallgrass prairies that used to cover the center of the continent. Many varieties of *A. glomeratus* and *A. virginicus* aggressively colonize abandoned fields, cutover timberlands, and roadsides. Some species are used in restoration and landscaping.

Species of *Andropogon* with solitary rames are easily confused with *Schizachyrium* but, in *Andropogon*, the lower glumes of the sessile spikelets are flat or concave and the rame internodes are not cupulate, whereas *Schizachyrium* has convex glumes and rame internodes with strongly cupulate apices. Successful identification of species in *Andropogon* sect. *Leptopogon* (numbers 3–14) requires mature, complete specimens and careful field study (Campbell 1983, 1986).

SELECTED REFERENCES **Barnes, P.W.** 1986. Variation in the big bluestem (*Andropogon gerardii*)-sand bluestem (*Andropogon hallii*) complex along a local dune/meadow gradient in the Nebraska sandhills. Amer. J. Bot. 73:172–184; **Campbell, C.S.** 1983. Systematics of the *Andropogon virginicus* complex (Gramineae). J. Arnold Arbor. 64:171–254; **Campbell, C.S.** 1986. Phylogenetic reconstructions and two new varieties in the *Andropogon virginicus* complex (Poaceae: Andropogoneae). Syst. Bot. 11:280–292; **Clayton, W.D.** 1964. Studies in the Gramineae: V. New species of *Andropogon*. Kew Bull. 17:465–470.

1. Pedicellate spikelets usually well-developed, (3.5)6–12 mm long, usually staminate; sessile spikelets 5–12 mm long (sect. *Andropogon*).
    2. Sessile spikelets with awns 8–25 mm long; ligules 0.4–2.5 mm long; hairs of the rame internodes 2.2–4.2 mm long, sparse to dense; rhizomes sometimes present, the internodes usually less than 2 cm .... 1. *A. gerardii*
    2. Sessile spikelets unawned or with awns less than 11 mm long; ligules (0.9)2.5–4.5 mm long; hairs of the rame internodes 3.7–6.6 mm long, usually dense; rhizomes always present, the internodes often more than 2 cm long .... 2. *A. hallii*
1. Pedicellate spikelets usually vestigial or absent, those of the terminal spikelet units occasionally well-developed and staminate; sessile spikelets 2.6–8.4 mm long (sect. *Leptopogon*).
    3. Peduncles with solitary rames; plants of southern Florida .... 3. *A. gracilis*
    3. Peduncles with (1)2–13 rames; plants of varied distribution, including southern Florida.
        4. Rames not or scarcely exserted at maturity; peduncles mostly less than 15 mm long at maturity.
            5. Culms 30–140 (usually about 80) cm tall; blades 0.8–5 (usually about 2.5) mm wide; inflorescence units 2–31 per culm .... 9. *A. gyrans*
            5. Culms 20–250 (usually more than 90) cm tall; blades 1.7–9.5 (usually more than 3) mm wide; inflorescence units 3–600 per culm.
                6. Blades pubescent, most hairs appressed; callus hairs 1.5–5 mm long .... 13. *A. longiberbis*
                6. Blades glabrous or with spreading (rarely appressed) hairs; callus hairs 1–3 mm long.

7. Blades 11–52 cm long; sheaths smooth, rarely somewhat scabrous; ligules 0.2–1 mm long; keels of the lower glumes usually smooth below midlength, scabrous distally . . . . 12. *A. virginicus*
7. Blades 13–109 cm long; sheaths usually scabrous; ligules 0.6–2.2 mm long; keels of the lower glumes sometimes scabrous below midlength . . . . 14. *A. glomeratus*

4. Rames sometimes exserted above their subtending sheaths at maturity; 1 or more peduncles more than 15 mm long at maturity.
8. Anthers 3.
9. Sessile spikelets 4.5–8.4 mm long; pedicellate spikelets 1.5–3.6 mm long, sterile; plants common and widespread in the southeastern United States . . . . 4. *A. ternarius*
9. Sessile spikelets 3–4 mm long; pedicellate spikelets mostly vestigial or absent, those of the terminal spikelet units well-developed and staminate; in the *Flora* region, known only from southern Florida . . . . 5. *A. bicornis*
8. Anthers 1 (rarely 3).
10. Peduncles with 2–13 rames . . . . 8. *A. liebmannii*
10. Peduncles usually with 2 (infrequently up to 4) rames or (in *A. gyrans* var. *gyrans* and *A. virginicus* var. *virginicus*), 2–5 (infrequently up to 7) rames.
11. Culms 30–120(140) (usually less than 100) cm tall; blades 0.8–5 (usually less than 3) mm wide; inflorescence units 2–31 per culm.
12. Peduncleswith 2–5 rames; anthers 0.6–1.7 mm long; sessile spikelets (3)4.1–4.4(5.7) mm long . . . . 9. *A. gyrans*
12. Peduncleswith 2 rames; anthers 1.2–2 mm long; sessile spikelets (4)4.8–5(5.5) mm long . . . . 10. *A. tracyi*
11. Culms (20)90–310 (usually more than 100) cm tall; blades 1.7–9.5 (usually more than 3) mm wide; inflorescence units 5–210 per culm.
13. Upper portion of the plants open, the branches conspicuously arching . . . . 11. *A. brachystachyus*
13. Upper portion of the plants dense, the branches usually straight and erect to ascending.
14. Rame internodes usually densely and uniformly pubescent over their entire length; anthers 1.3–3.5 mm long; sessile spikelets (3.8)4–6.1 mm long.
15. Blades 15–35 cm long, often more or less pubescent; sheaths smooth, very rarely somewhat scabrous; anthers 2–3.5 mm long; inflorescence units 5–45 per culm . . . . 6. *A. arctatus*
15. Blades 32–61 cm long, usually glabrous; sheaths often scabrous; anthers 1.3–2 mm long; inflorescence units usually at least 50 (9–210) per culm . . . . 7. *A. floridanus*
14. Rame internodes sparsely pubescent basally, more densely pubescent distally; anthers 0.5–1.5 mm long; sessile spikelets 2.6–4(5).
16. Blades 11–52 cm long; sheaths smooth, rarely somewhat scabrous; ligules 0.2–1 mm long; keels of the lower glumes usually smooth below midlength, scabrous distally . . . . 12. *A. virginicus*
16. Blades 13–109 cm long; sheaths usually scabrous, sometimes smooth; ligules 0.6–2.2 mm long; keels of the lower glumes sometimes scabrous below midlength . . . . 14. *A. glomeratus*

Andropogon L. sect. Andropogon

Inflorescences usually only terminal, axillary inflorescences absent or few. **Lower glumes** thinly coriaceous, with 3–9 veins between the keels; **keels** often winged. **Pedicellate spikelets** usually well-developed, staminate.

1. **Andropogon gerardii** Vitman [p. 654]
BIG BLUESTEM, BARBON DE GERARD

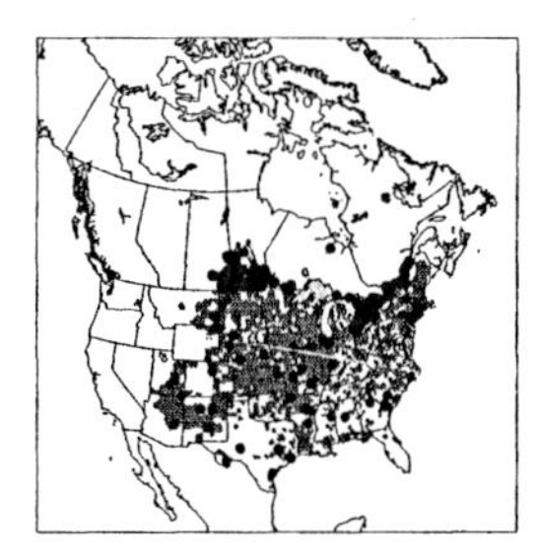

Plants often forming large clumps, rhizomes, if present, with internodes shorter than 2 cm. **Culms** 1–3 m, often glaucous. **Sheaths** glabrous or pilose; **ligules** 0.4–2.5 mm; **blades** 5–50 cm long, (2)5–10 mm wide, usually pilose adaxially, at least near the collar. **Inflorescence units** usually only terminal; **peduncles** with 2–6(10) rames; **rames** 5–11 cm, exserted at maturity, usually purplish, sometimes yellowish; **internodes** sparsely to densely pubescent, hairs 2.2–4.2 mm, usually white, rarely yellowish. **Sessile spikelets** 5–11 mm, scabrous; **awns** 8–25 mm; **anthers** 3, 2.5–4.5 mm. **Pedicellate spikelets** 3.5–12 mm, usually well-developed and staminate. $2n$ = 20, 40, 60 (usually), 70, 80.

Andropogon gerardii grows in prairies, meadows, and generally dry soils. It is a widespread species, extending from southern Canada to Mexico, and was once dominant over much of its range. It is frequently planted for erosion control, restoration, or as an ornamental; the records from Washington and central Montana reflect such plantings. It hybridizes with *A. hallii*, the two sometimes being treated as conspecific subspecies.

2. **Andropogon hallii** Hack. [p. 654]
SAND BLUESTEM

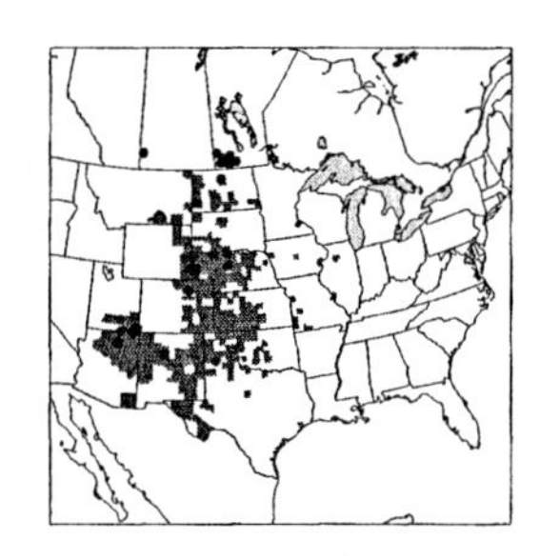

Plants strongly rhizomatous, rhizome internodes often longer than 2 cm. **Culms** (40)60–150(200) cm, strongly glaucous. **Ligules** (0.9)2.5–4.5 mm, ciliate; **blades** 3–40(51) cm long, (1.5)2–10 mm wide, often pilose, at least near the collar. **Inflorescence units** usually only terminal; **peduncles** with 2–7 rames; **rames** 4–7(9) cm, exerted at maturity; **internodes** usually densely pubescent, hairs 3.7–6.6 mm, often strongly yellowish. **Sessile spikelets** (5)6.5–12 mm; **lower glumes** often ciliate; **awns** absent or to 11 mm; **anthers** 3, (2.3)4–6 mm. **Pedicellate spikelets** 3.5–12 mm, usually well-developed and staminate. $2n$ = 60 (usually), 70, 100.

Andropogon hallii grows on sandhills and in sandy soil. Its range extends through the central plains into northern Mexico. It is similar to *A. gerardii*, differing primarily in its rhizomatous habit, more densely pubescent rames and pedicels, and greater drought tolerance. *Andropogon hallii* and *A. gerardii* are sympatric in some locations. The two species can hybridize and are sometimes treated as conspecific subspecies.

Andropogon sect. Leptopogon Stapf

Inflorescences false panicles; **inflorescence units** numerous. **Lower glumes** membranous; **keels** not winged, intercostal region usually not veined or with inconspicuous veins. **Pedicellate spikelets** usually vestigial or absent.

3. **Andropogon gracilis** Spreng. [p. 654]
WIRE BLUESTEM

Plants densely cespitose. **Culms** 20–60 cm, wiry, glabrous. **Sheaths** smooth; **ligules** to 1.4 mm; **blades** to 45 cm long, to 4 mm wide, involute and filiform, or folded. **Inflorescence units** 3–50+ per culm; **peduncles** 2–13.2 cm, with 1 rame; **rames** 2–4 cm, usually long-exserted at maturity; **internodes** densely pubescent, hairs to 8 mm. **Sessile spikelets** 4–6 mm; **lower glumes** scabrous in the distal ½; **awns** 11–20 mm. **Pedicellate spikelets** reduced to an awned or unawned glume, sterile. $2n$ = 40.

Andropogon gracilis grows on oölite in openings and rocky margins of pine woodlands of southern Florida and the West Indies. Although not uncommon, it is frequently overlooked. It has sometimes been placed in *Schizachyrium* because of its solitary rames.

4. **Andropogon ternarius** Michx. [p. 656]
SPLIT BLUESTEM

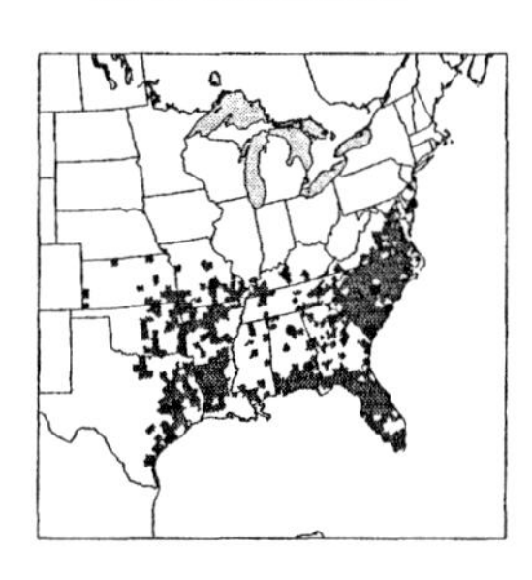

Plants cespitose. **Culms** 70–150 cm. **Sheaths** smooth or scabrous, sometimes pilose; **ligules** 0.4–1.5 mm, ciliate; **blades** 1–3 mm wide, pubescent or glabrous and glaucous. **Inflorescence units** 2–30+ per culm; **peduncles** usually 5–20 mm, with (1)2 rames; **rames** 3–4 cm, exerted at maturity, terminating in a sessile-pedicellate spikelet pair; **internodes** sparsely to densely villous, hairs from as long as to twice as long as the sessile spikelets. **Sessile spikelets** 4.5–8.4 mm; callus

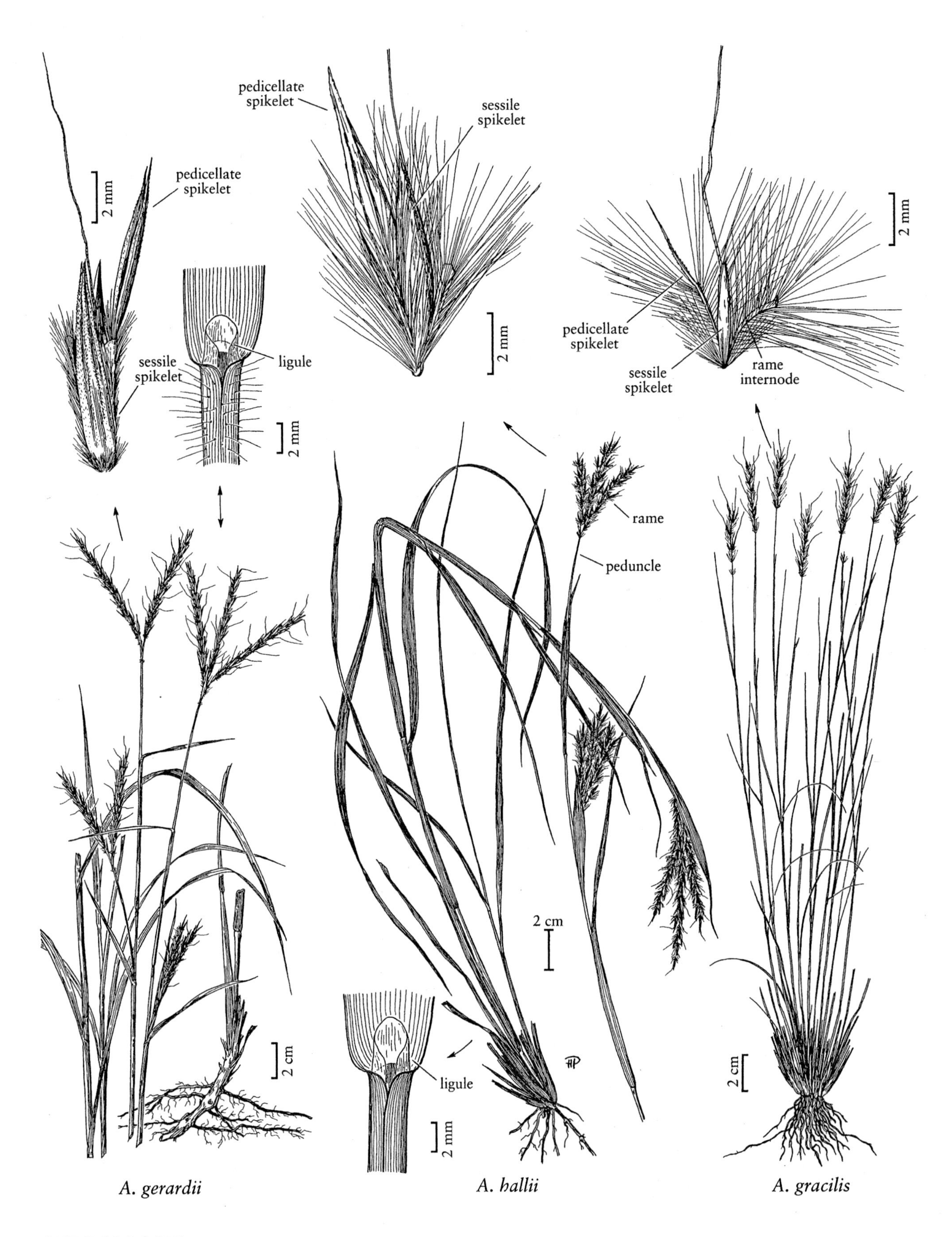
pedicellate spikelet
sessile spikelet
ligule
2 mm
2 mm
2 cm
pedicellate spikelet
sessile spikelet
2 mm
rame
peduncle
2 cm
ligule
2 mm
pedicellate spikelet
sessile spikelet
rame internode
2 mm
2 cm
A. gerardii
A. hallii
A. gracilis

hairs to 8 mm; **awns** 10–25 mm; **anthers** 3, 1.2–2.3 mm. **Pedicellate spikelets** 1.5–3.6 mm, sterile. 2*n* = 40, 60.

Andropogon ternarius grows in the southeastern United States and northern Mexico. It is planted as an ornamental and for erosion control on slopes in poor and sandy soils, and is tolerant of coastal conditions.

Andropogon ternarius is similar to *A. arctatus* but differs in its possession of three anthers and usually in its longer spikelets, both sessile and pedicellate.

1. Rames densely villous, with hairs about twice as long as the sessile spikelets and more or less obscuring them; lower glumes of the sessile spikelets sometimes scabrous, without conspicuous veins between the keels . . . . . . var. *cabanisii*
1. Rames sparsely villous, with hairs about as long as the sessile spikelets, but not obscuring them; lower glumes of the sessile spikelets scabrous, often conspicuously 2-veined between the keels . . . . var. *ternarius*

Andropogon ternarius var. **cabanisii** (Hack.) Fernald & Griscom

Rames densely villous; **internode hairs** about twice as long as and more or less obscuring the sessile spikelets. **Lower glumes of sessile spikelets** glabrous, sometimes scabrous, not conspicuously veined between the keels.

Andropogon ternarius var. *cabanisii* grows in dry pine woods and scrublands of peninsular Florida.

Andropogon ternarius Michx. var. **ternarius**

Rames sparsely villous; **internode hairs** about as long as, but not obscuring, the sessile spikelets. **Lower glumes of sessile spikelets** scabrous, often with 2 conspicuous veins between the keels.

Andropogon ternarius var. *ternarius* grows in dry, sandy woods, fields, openings, and roadsides of the southeastern United States and Mexico.

5. **Andropogon bicornis** L. [p. 656]
BARBAS DE INDIO

Plants densely cespitose, upper portion dense, obovate to obpyramidal. **Culms** 60–250 cm; **internodes** not glaucous. **Sheaths** smooth; **ligules** 0.6–1 mm; **blades** 20–70 cm long, 2–7 mm wide, usually glabrous or scabrous on the margins. **Inflorescence units** 50–500; **subtending sheaths** 2.5–4.5 cm long, 2–3 mm wide; **peduncles** 20–70 mm, with 2(3) rames; **rames** 2–4 cm, exerted at maturity; **internodes** filiform, densely and evenly pubescent, hairs 3–9 mm. **Sessile spikelets** 3–4 mm; unawned; **callus hairs** 0.5–1 mm; **keels of lower glumes** scabrous above the midpoint; **anthers** 3, 1–1.4 mm. **Pedicellate spikelets** mostly vestigial or absent, 1–2 of those in the terminal units on each rame 3–5 mm and staminate. 2*n* = 60, 120.

Andropogon bicornis is a widespread species of the Western Hemisphere tropics. It was collected in the early 1960s in Dade County, Florida, near the track of a major hurricane, but may not be established in the *Flora* region.

6. **Andropogon arctatus** Chapm. [p. 658]
PINEWOODS BLUESTEM

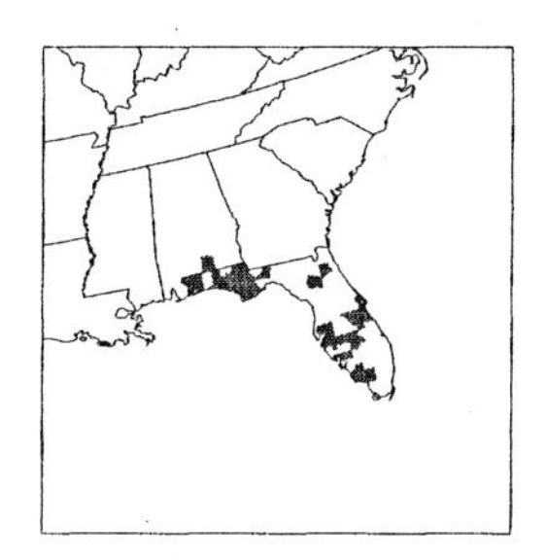

Plants cespitose or somewhat rhizomatous, upper portion dense, oblong to ovate. **Culms** 90–170 cm; **internodes** occasionally somewhat glaucous just below the nodes; **branches** straight, erect to ascending. **Sheaths** smooth, rarely somewhat scabrous; **ligules** 0.3–0.9 mm, sometimes ciliate, cilia to 0.5 mm; **blades** 15–35 cm long, 3–8 mm wide, glabrous or densely pubescent, hairs spreading. **Inflorescence units** 5–45 per culm; **subtending sheaths** (3.3)3.8–6.8(9) cm long, (2.5)3.2–4(5) mm wide; **peduncles** (9)26–66(115) mm, with 2(4) rames; **rames** (2.2)2.6–4.3(5.3) cm, usually exserted at maturity, pubescence either evenly distributed or more dense distally within each internode. **Sessile spikelets** (4.3)4.9–5.4(6.1) mm; **callus hairs** 1.5–2.5 mm; **keels of lower glumes** scabrous from below the midpoint; **awns** 5–16 mm; **anthers** 1(3), 2–3.5 mm, red. **Pedicellate spikelets** vestigial or absent. 2*n* = 20.

Andropogon arctatus grows in flatwoods, bogs, and scrublands of southern Alabama and Florida. Its flowering appears to be stimulated by fire but, unlike other members of sect. *Leptopogon* in the *Flora* region, the effect lasts only one or two years, the plants then remaining vegetative until the next fire occurs. It is similar to *A. ternarius*, but differs in its long, usually solitary anther and shorter spikelets.

7. **Andropogon floridanus** Scribn. [p. 658]
FLORIDA BLUESTEM

Plants cespitose, usually densely obpyramidal to oblanceolate above. **Culms** 70–210 cm; **internodes** occasionally somewhat glaucous just below the node; **branches** straight, mostly erect to ascending. **Sheaths** often scabrous, somteimes smooth; **ligules** 0.4–1.2 mm, ciliate, cilia 0.2–1.3 mm; **blades** 32–61 cm long, 2.9–5 mm wide, glabrous, rarely sparsely pubescent. **Inflorescence units**

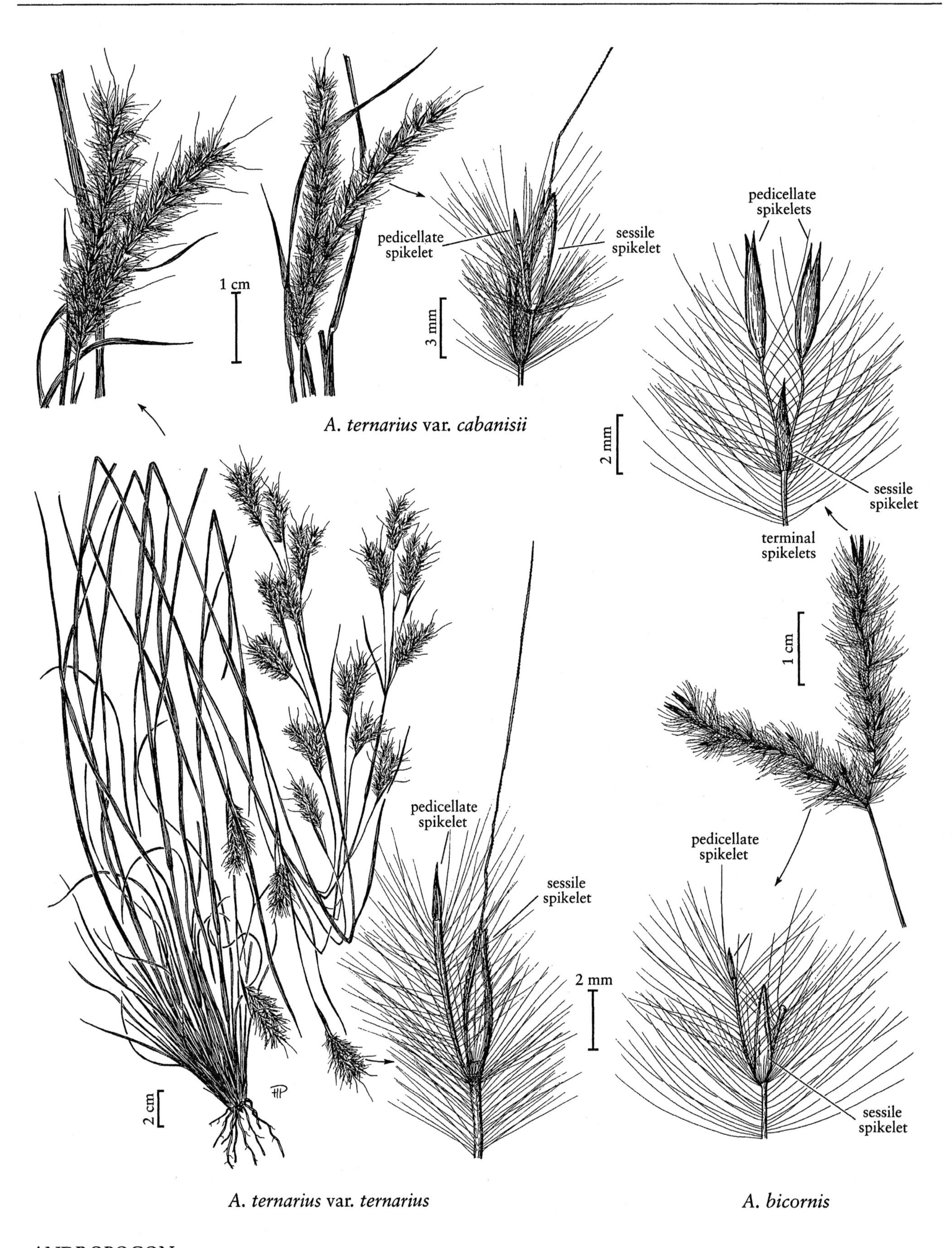
1 cm
pedicellate spikelet
sessile spikelet
3 mm
A. ternarius var. cabanisii
pedicellate spikelets
2 mm
sessile spikelet
terminal spikelets
1 cm
pedicellate spikelet
sessile spikelet
2 mm
pedicellate spikelet
sessile spikelet
2 cm
A. ternarius var. ternarius
A. bicornis

ANDROPOGON

(9)50–210 per culm; **subtending sheaths** (3)4–5.9(7) cm long, (1.5)2–2.7(3.6) mm wide; **peduncles** (10)19–48(93) mm, with 2(4) rames; **rames** (2)2.5–3.7(4.5) cm, usually exserted at maturity, **internodes** evenly pubescent. **Sessile spikelets** (3.8)4.4–4.8(5.5) mm; **callus hairs** 1–3 mm; **keels of lower glumes** glabrous below midlength; **awns** 5–15 mm; **anthers** 1(3), 1.3–2 mm, usually yellow (sometimes purple). **Pedicellate spikelets** vestigial or absent. $2n = 20$.

Andropogon floridanus grows on sandy soils in southeastern Georgia and Florida, being most abundant in *Pinus clausa* scrublands. It usually occurs in small stands, but stands of about a hundred individuals have been observed.

8. **Andropogon liebmannii** Hack. [p. 658]
LIEBMANN'S BLUESTEM, MOHR'S BLUESTEM

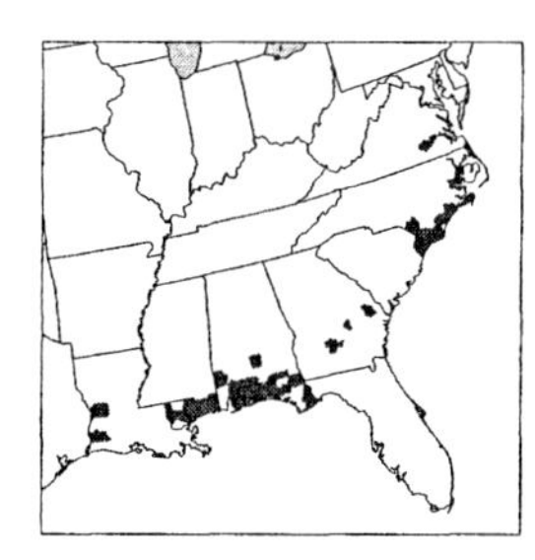

Plants cespitose, cylindrical to oblong above. **Culms** 20–170 cm; **internodes** not glaucous; **branches** mostly erect, straight. **Sheaths** smooth; **ligules** 0.7–1.2 mm, sometimes ciliate, cilia to 0.4 mm; **blades** 3–35 cm long, 2.5–7.5 mm wide, sparsely to densely pubescent with spreading, shaggy hairs. **Inflorescence units** 7–50 per culm; **subtending sheaths** (4)4.9–7.4(10) cm long, (3)4.2–6.1(10.1) mm wide; **peduncles** (10)24–68(130) mm, at least some extending beyond the subtending sheaths at maturity, with 2–13 rames; **rames** (2)2.4–4(5) cm, usually exserted at maturity, pubescence increasing in density distally within each internode. **Sessile spikelets** (3)4–4.5(6.9) mm; **keels of lower glumes** scabrous above (and sometimes below) the midpoint; **awns** 17–24 mm; **anthers** 1, 0.7–1.4 mm, yellow. **Pedicellate spikelets** vestigial or absent. $2n = 20$.

Andropogon liebmannii has two varieties. **Andropogon liebmannii** var. **pungensis** (Ashe) C.S. Campb., the variety found in the *Flora* region, differs from **A. liebmannii** Hack. var. **liebmannii**, which grows in Mexico, in having culms that are usually more than 80 cm tall, leaves that are more than 15 cm long, and sessile spikelets that are more than 4.2 mm long; in var. *pungensis* the culms are usually less than 90 cm tall, the leaves less than 15 cm long, and the sessile spikelets less than 4.2 mm long.

Andropogon liebmannii var. *pungensis* grows along the coastal plain of the southeastern United states in bog, swamp, savannahs, and flatwoods. It used to be treated as a species, which was known as *A. mohrii*. The English name "Mohr's Bluestem" reflects this treatment.

9. **Andropogon gyrans** Ashe [p. 660]

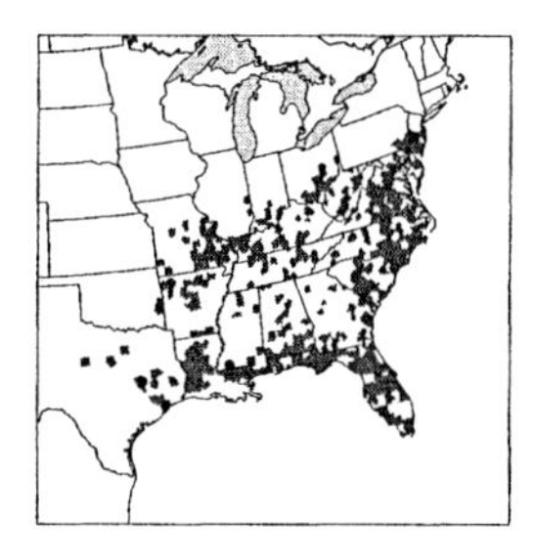

Plants cespitose, cylindrical to ovate above. **Culms** 30–100 (140) cm; **internodes** usually glaucous; **branches** mostly erect, straight. **Sheaths** smooth; **ligules** 0.3–1.5 mm, sometimes ciliate, cilia to 0.7 mm; **blades** 6–48 cm long, 0.8–5 mm wide, glabrous or densely pubescent with spreading hairs. **Inflorescence units** 2–31 per culm; **subtending sheaths** (2.6)4.1–4.5(13.5) cm long, (1.5)2.7–4.7(8) mm wide; **peduncles** (1)5–31(195) mm, with 2–5 rames; **rames** (1.5)2.8–4.2(6) cm, exserted or not at maturity, pubescence increasing in density distally within each internode. **Sessile spikelets** (3)3.9–4.7(5.7) mm; **callus hairs** 1–5 mm; **keels of lower glumes** scabrous only beyond midlength; **awns** 8–24 mm; **anthers** 1, 0.6–1.4(1.7) mm, yellow or purple. **Pedicellate spikelets** vestigial or absent. $2n = 20$.

Andropogon gyrans extends from the southeastern United States to the Caribbean and Central America.

1. Ligules 0.3–1.1 mm long; rames usually hidden within the more or less overlapping and inflated upper sheaths at maturity; plants usually of well-drained soils .... var. *gyrans*
1. Ligules 0.8–1.5 mm long; rames usually exposed at maturity; plants of wet habitats ...... var. *stenophyllus*

Andropogon gyrans Ashe var. **gyrans**
ELLIOTT'S BEARDGRASS

Ligules 0.3–0.8(1.1) mm. **Inflorescence units** usually with 2–5 rames; **rames** usually concealed at maturity.

Andropogon gyrans var. *gyrans* generally grows in dry, sandy soil of roadsides, embankments, fields, and pine or oak woods, occasionally in moister soil. Its range extends south from the United States to the Caribbean and Central America. Plants from Florida and Mississippi do not have inflated sheaths.

Andropogon gyrans var. **stenophyllus** (Hack.) C.S. Campb.

Ligules (0.8)1.1–1.5 mm. **Inflorescence units** usually with 2 rames; **rames** usually exposed at maturity.

Andropogon gyrans var. *stenophyllus* grows in ditches, bogs, savannahs, and pond margins of the coastal plain, from eastern Texas to North Carolina.

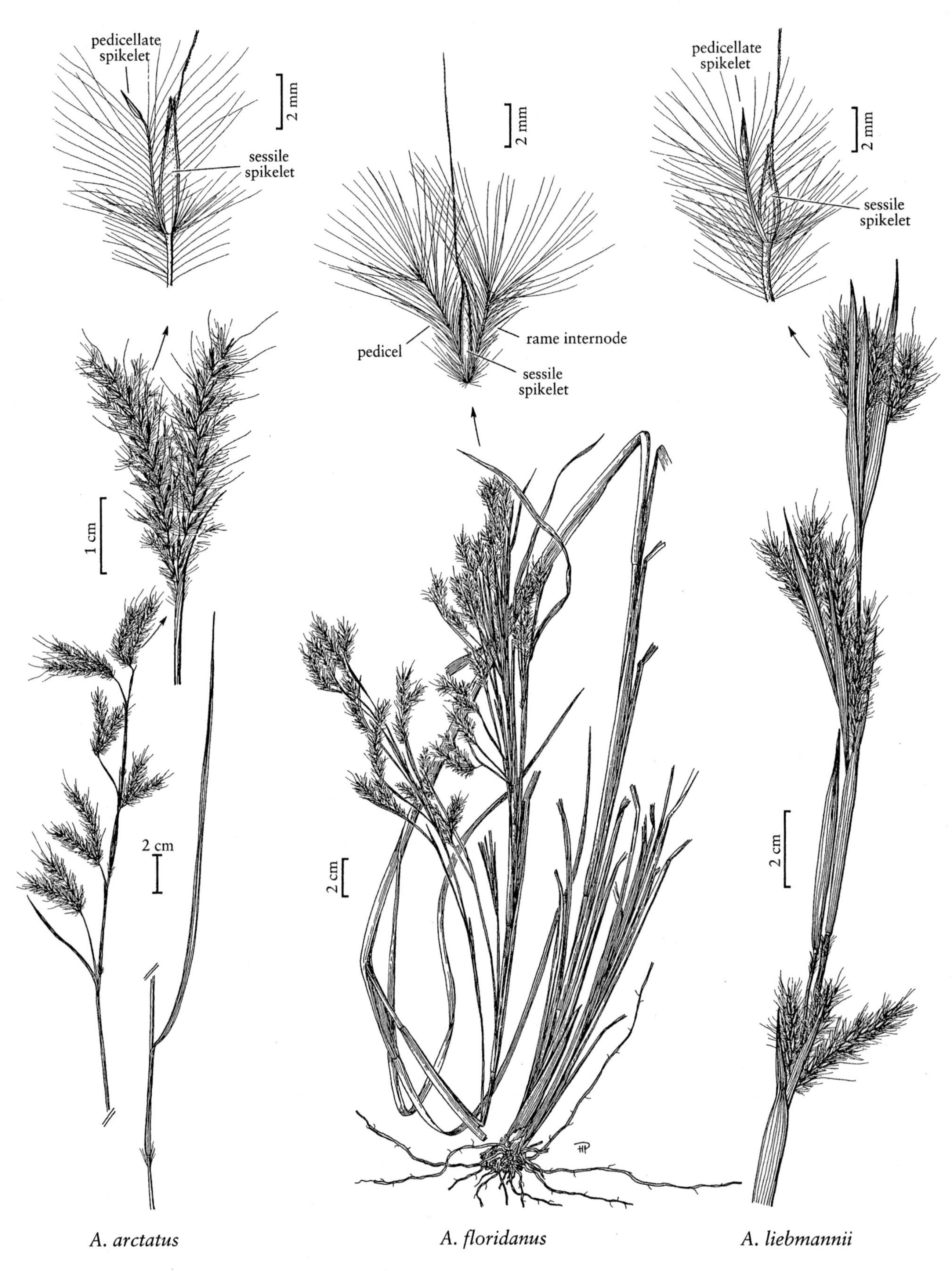

A. arctatus *A. floridanus* *A. liebmannii*

10. **Andropogon tracyi** Nash [p. 660]
TRACY'S BLUESTEM

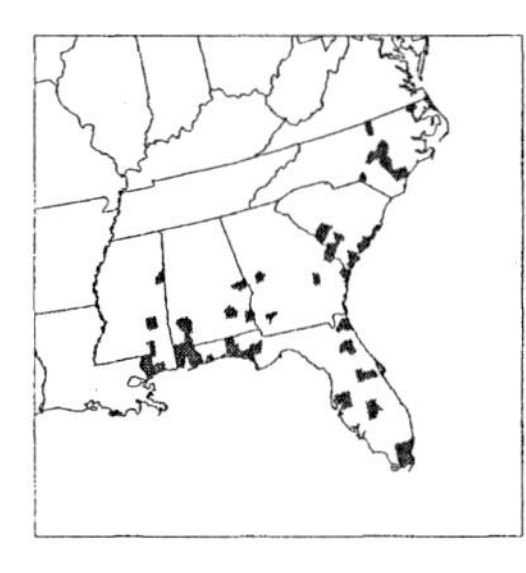

Plants cespitose, upper portion dense, cylindrical. **Culms** 50–120 cm; **internodes** not glaucous; **branches** mostly erect, straight. **Sheaths** smooth; **ligules** 0.2–0.5 mm, ciliate, cilia 0.2–0.8 mm; **blades** 10–22 cm long, 1.2–2.6 mm wide, glabrous or sparsely pubescent, with spreading hairs. **Inflorescence units** 3–11 per culm; **subtending sheaths** (2.8)4.1–5.8(7.2) cm long, (3)4–4.7(5.8) mm wide; **peduncles** (9)14–31(65) mm, with 2 rames; **rames** (1.5)2.4–3.6(4.2) cm, usually exserted at maturity, pubescence increasing in density distally within each internode. **Sessile spikelets** (4)4.8–5(5.5) mm; **callus hairs** 1.5–3.5 mm; **keels of lower glumes** scabrous only above the midpoint; **awns** 11–23 mm; **anthers** 1, 1.2–2 mm, yellow. **Pedicellate spikelets** vestigial or absent. $2n = 20$.

Andropogon tracyi grows on sandhills, sandy pinelands, and scrublands of the southeastern United States. It resembles *A. longiberbis*, but usually differs in having sparsely pubescent blades and a more slender appearance.

11. **Andropogon brachystachyus** Chapm. [p. 662]
SHORTSPIKE BLUESTEM

Plants cespitose, open and ovate to obpyramidal above. **Culms** 1.1–3.1 m; **internodes** not glaucous; **branches** arching. **Sheaths** smooth; **ligules** 0.2–0.5 mm, ciliate, cilia 0.6–1.5 mm; **blades** 21–54 cm long, 2.3–6 mm wide, glabrous or sparsely pubescent, with spreading hairs. **Inflorescence units** 12–190 per culm; **subtending sheaths** (2.1)2.4–3.5(4.1) cm long, (2.3)2.6–3(3.8) mm wide; **peduncles** (13)20–31(43) mm, with 2(3) rames; **rames** (1.2)1.5–2.1(2.6) cm, usually exserted at maturity, pubescence increasing in density distally within each internode. **Sessile spikelets** (4.1)4.4–4.6(5) mm; **callus hairs** 1–1.5 mm; **keels of lower glumes** scabrous only above the midpoint; **awns** 2–11 mm; **anthers** 1, 1.4–2.4 mm, red. **Pedicellate spikelets** vestigial or absent. $2n = 20$.

Andropogon brachystachyus grows in sandy, often seasonally wet soils of flatwoods, savannahs, pond margins, and scrublands of the southeastern United States. It sometimes forms large populations, but does not invade disturbed sites as do some morphologically similar forms of *A. virginicus* var. *virginicus*.

12. **Andropogon virginicus** L. [p. 662]
BROOMSEDGE BLUESTEM

Plants cespitose, dense and cylindrical to obpyramidal above. **Culms** 40–210 cm; **internodes** glaucous or not; **branches** erect to ascending, usually straight, sometimes arching. **Sheaths** usually smooth, rarely somewhat scabrous; **ligules** 0.2–1 mm, ciliate, cilia 0.2–1.3 mm; **blades** 11–52 cm long, 1.7–6.5 mm wide, smooth and glabrous or sparsely to densely pubescent with spreading hairs. **Inflorescence units** 6–195 per culm; **subtending sheaths** (2.1)3.1–4.6(6.7) cm long, (1.7)3–3.8(5.6) mm wide; **peduncles** usually (1)4–6(30) mm, with 2–7 rames; **rames** (0.5)1.7–2.8(4.4) cm, sometimes exerted at maturity, pubescence sparse basally and increasing in density distally within each internode. **Sessile spikelets** (2.6)3.5–3.8(4.7) mm; **callus hairs** 1–3 mm; **keels of lower glume** usually smooth below midlength, scabrous distally; **awns** 6–21 mm; **anthers** 1(3), 0.6–1.5 mm, yellow or purple. **Pedicellate spikelets** vestigial to absent. $2n = 20$.

Andropogon virginicus is native from the southeastern United States to northern South America, but has become established outide its native range in California, Hawaii, Japan, and Australia. Three varieties are recognized, two of which contain morphologically distinct variants. *Andropogon virginicus* hybridizes with *A. glomeratus* and *A. longiberbis* (Campbell 1986).

1. Leaves bluish-green, more or less strongly glaucous . . . . var. *glaucus*
1. Leaves green, sometimes somewhat glaucous.
   2. Sheaths subtending the inflorescence units (1.7)2.4–3.1(4) mm wide; inflorescences units usually with 2 rames; rames (1.3)1.5–2.3(3) cm long; peduncles (1) 4–9 (30) mm long . . . . var. *decipiens*
   2. Sheaths subtending the inflorescences units (2.2)3.3–4.4(5.6) mm wide; inflorescence units with 2–5(7) rames; rames (0.5)1.9–3.3(4.4) cm long; peduncles (2)3–6(12) mm long . . . . var. *virginicus*

Andropogon virginicus var. **decipiens** C.S. Campb.

Culms 70–170 cm. **Leaves** green, sometimes slightly glaucous. **Inflorescence units** usually with 2 rames; **subtending sheaths** (1.7)2.4–3.1(4) mm wide; **peduncles** (1) 4–9 (30) mm; **rames** (1.3)1.5–2.3(3) cm.

Andropogon virginicus var. *decipiens* grows in flatwoods, scrublands, and disturbed sites, such as roadsides and cleared timberlands, of the southeastern coastal plain.

2 mm
sessile spikelet
rame internode
pedicel
2 cm
2 mm
pedicel
sessile spikelet
rame internode
2 cm
blade
2 cm
A. gyrans var. stenophyllus
A. gyrans var. gyrans
A. tracyi

ANDROPOGON

Andropogon virginicus var. **glaucus** Hack.

Culms 60–180 cm. **Leaves** bluish-green, more or less strongly glaucous. **Inflorescence units** with 2(3) rames; **subtending sheaths** (2.7)3.1–3.8(5.5) mm wide; **peduncles** (2)3–4(10) mm; **rames** (1.4)1.7–3(4) cm, not exserted.

Andropogon virginicus var. *glaucus* grows on moist or dry soils of the coastal plain, from southern New Jersey to eastern Texas. Plants growing on sandy, well-drained soils differ from those on poorly drained slopes in being glabrous (rather than pubescent) beneath the subtending sheaths of the inflorescence units, and in tending to have shorter rames.

Andropogon virginicus L. var. **virginicus**

Culms 40–210 cm; **internodes** green. **Leaves** green or slightly glaucous, glabrous or pubescent, at least on the margins near the collar. **Inflorescence units** with 2–5(7) rames; **subtending sheaths** (2.2)3.3–4.4(5.6) mm wide; **peduncles** (2)3–6(12) mm; **rames** (0.5)1.9–3.3(4.4) cm, not exserted.

Andropogon virginicus var. *virginicus* is the widespread and weedy variety of *A. virginicus* that grows as a native species from the central plains through Mexico and Central America to Colombia and, as a naturalized species, in California, Hawaii, Japan, and Australia.

Plants colonizing openings in mature vegetation created by disturbance have green culms and green, pubescent leaves. Those growing in poorly drained soils of pond margins, swales, and cutover flatwoods have glaucous culms and glabrous, green to somewhat glaucous leaves. Glaucous plants of *A. virginicus* var. *virginicus* differ from those of var. *decipiens* in having no exposed rames and, often, wider sheaths subtending the inflorescence units.

13. **Andropogon longiberbis** Hack. [p. 663]
HAIRY BLUESTEM

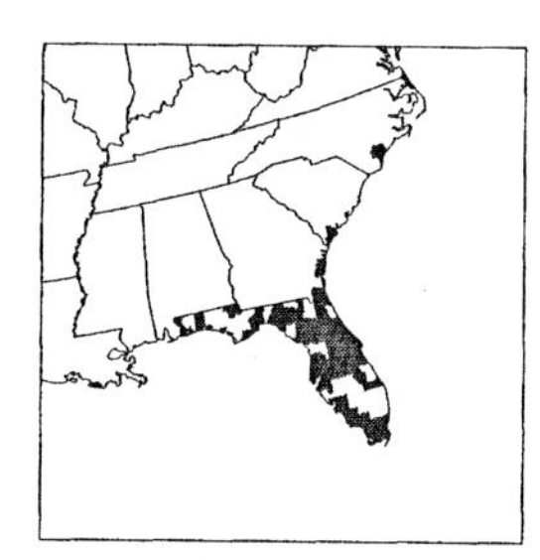

Plants cespitose; cylindrical to oblong and more or less open in the upper portion. **Culms** 50–100(150) cm; **internodes** green, sometimes somewhat glaucous just below the node; **branches** mostly erect, straight. **Sheaths** not scabrous; **ligules** 0.2–0.6 mm, ciliate, cilia 0.3–0.6 mm; **blades** 11–50 cm long, 2–5.5 mm wide, sparsely to densely pubescent, most hairs appressed. **Inflorescence units** 7–97 (usually about 45) per culm; **subtending sheaths** (2.5)3–4.5(6) cm long, (2.5)3.2–4.1(5.5) mm wide; **peduncles** (1)3–4(13) mm, with 2(3) rames; **rames** (1.3)1.8–2.6(4) cm, not exserted at maturity, pubescence increasing in density distally within each internode. **Sessile spikelets** (3.5)4.1–4.5(5) mm; **callus hairs** 1.5–5 mm; **keels of lower glumes** scabrous only above the midpoint; **awns** 10–21 mm; **anthers** 1, 0.9–1.6 mm, yellow. **Pedicellate spikelets** vestigial or absent. $2n = 20$.

Andropogon longiberbis grows in sandy or rocky soils of roadsides, dunes, sandhills, pinelands, and fields, from the southeastern United States to the Bahamas. It resembles *A. tracyi*, but usually differs in having more densely pubescent blades and a less slender appearance.

Andropogon longiberbis appears to hybridize with both *A. virginicus* var. *virginicus* and *A. glomeratus* var. *pumilus*.

14. **Andropogon glomeratus** (Walter) Britton, Sterns & Poggenb. [p. 663]
BUSHY BLUESTEM, BUSHY BEARDGRASS

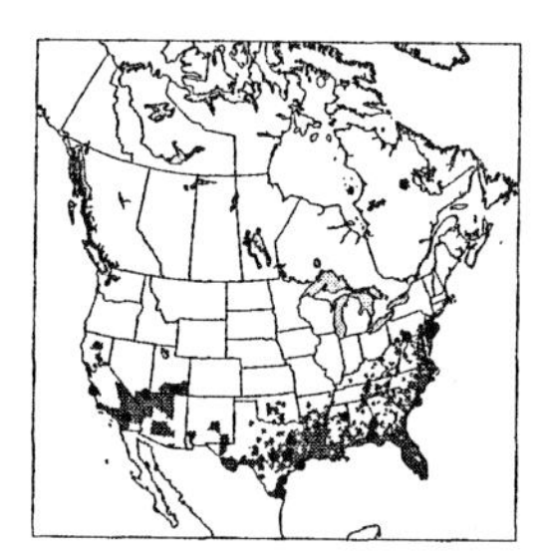

Plants cespitose, upper portion dense, oblong to oblanceolate or obpyramidal. **Culms** 20–250 cm; **internodes** green, sometimes glaucous; **branches** mostly erect, straight. **Sheaths** usually scabrous, sometimes smooth; **ligules** 0.6–2.2 mm, sometimes ciliate, cilia to 0.9 mm; **blades** 13–109 cm long, 2.9–9.5 mm wide, glabrous or sparsely to densely pubescent, hairs usually spreading, rarely appressed. **Inflorescence units** 10–600 per culm; **subtending sheaths** (2.0)2.9–4.4(6.5) cm long, (1.5)2.3–3.4(4.4) mm wide; **peduncles** (1)6–14(60) mm, with 2(4) rames; **rames** (1)1.7–2.5(3.5) cm, exserted or not at maturity, pubescence sparse basally and increasing in density distally within each internode. **Sessile spikelets** 3–5 mm; **callus hairs** 1–2.5 mm; **keels of lower glumes** sometimes scabrous below midlength, usually scabrous distally; **awns** 6–19 mm; **anthers** 1(3), 0.5–1.5 mm, yellow, red, or purple. **Pedicellate spikelets** vestigial or absent, sterile. $2n = 20$.

Andropogon glomeratus hybridizes with both *A. longiberbis* and *A. virginicus*. Some of its varieties are morphologically similar to the latter species.

1. Blades glaucous, glabrous, and smooth . . . var. *glaucopsis*
1. Blades green, often pubescent or scabrous.
   2. Sheaths subtending the inflorescence units 1.5–3 mm wide; leaf sheaths usually smooth; ligules ciliate, the cilia 0.2–0.9 mm long . . . . var. *pumilus*
   2. Sheaths subtending the inflorescence units (1.5)2.3–3.4(4.4) mm wide; leaf sheaths often scabrous; ligules, when ciliate, with the cilia no more than 0.5 mm long.
      3. Keels of the lower glumes scabrous below and beyond midlength . . . . var. *scabriglumis*

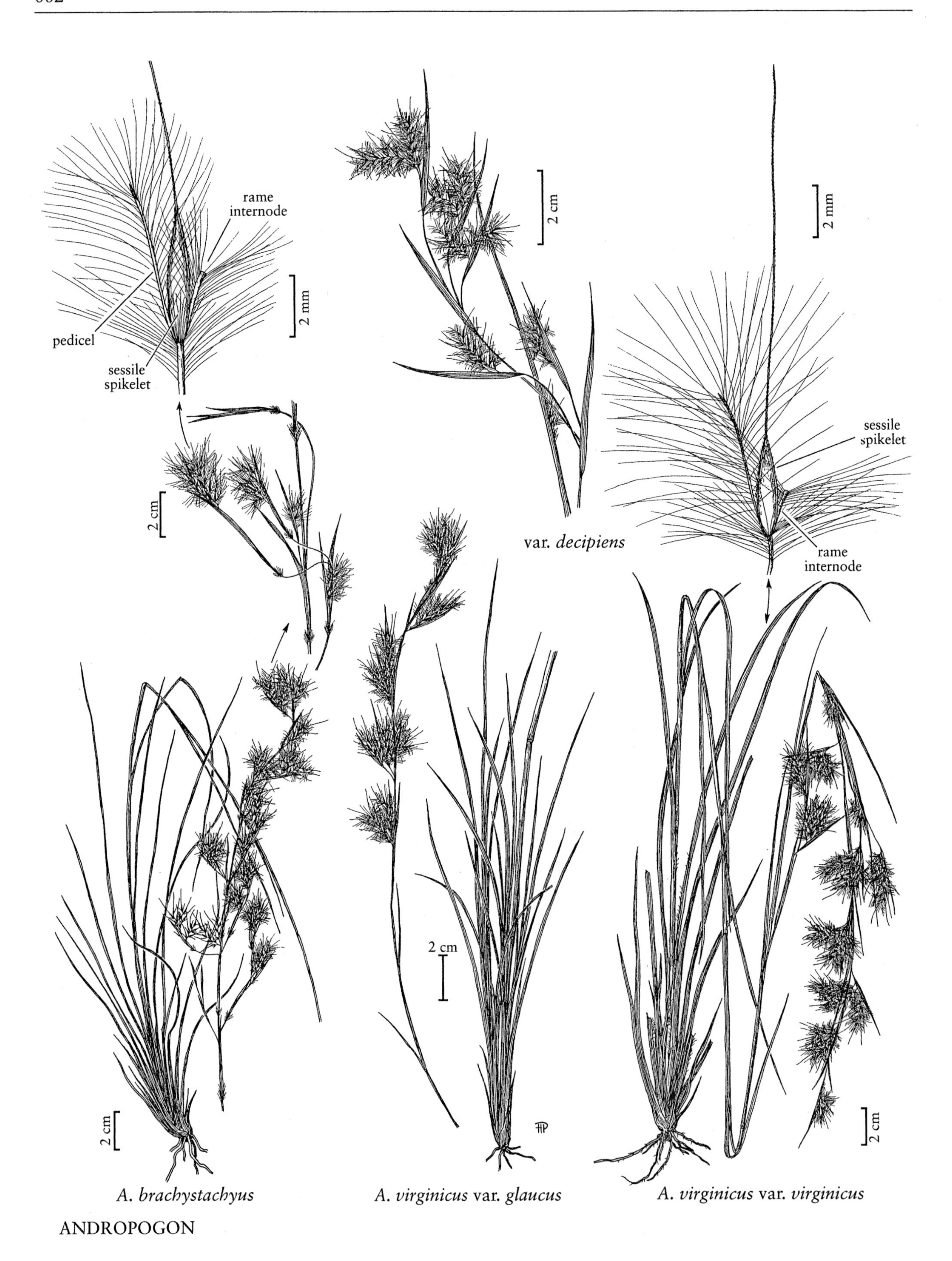
rame
internode
pedicel
sessile
spikelet
2 mm
2 cm
2 cm
var. *decipiens*
2 mm
sessile
spikelet
rame
internode
2 cm
2 cm
2 cm
A. brachystachyus
A. virginicus var. *glaucus*
A. virginicus var. *virginicus*

ANDROPOGON

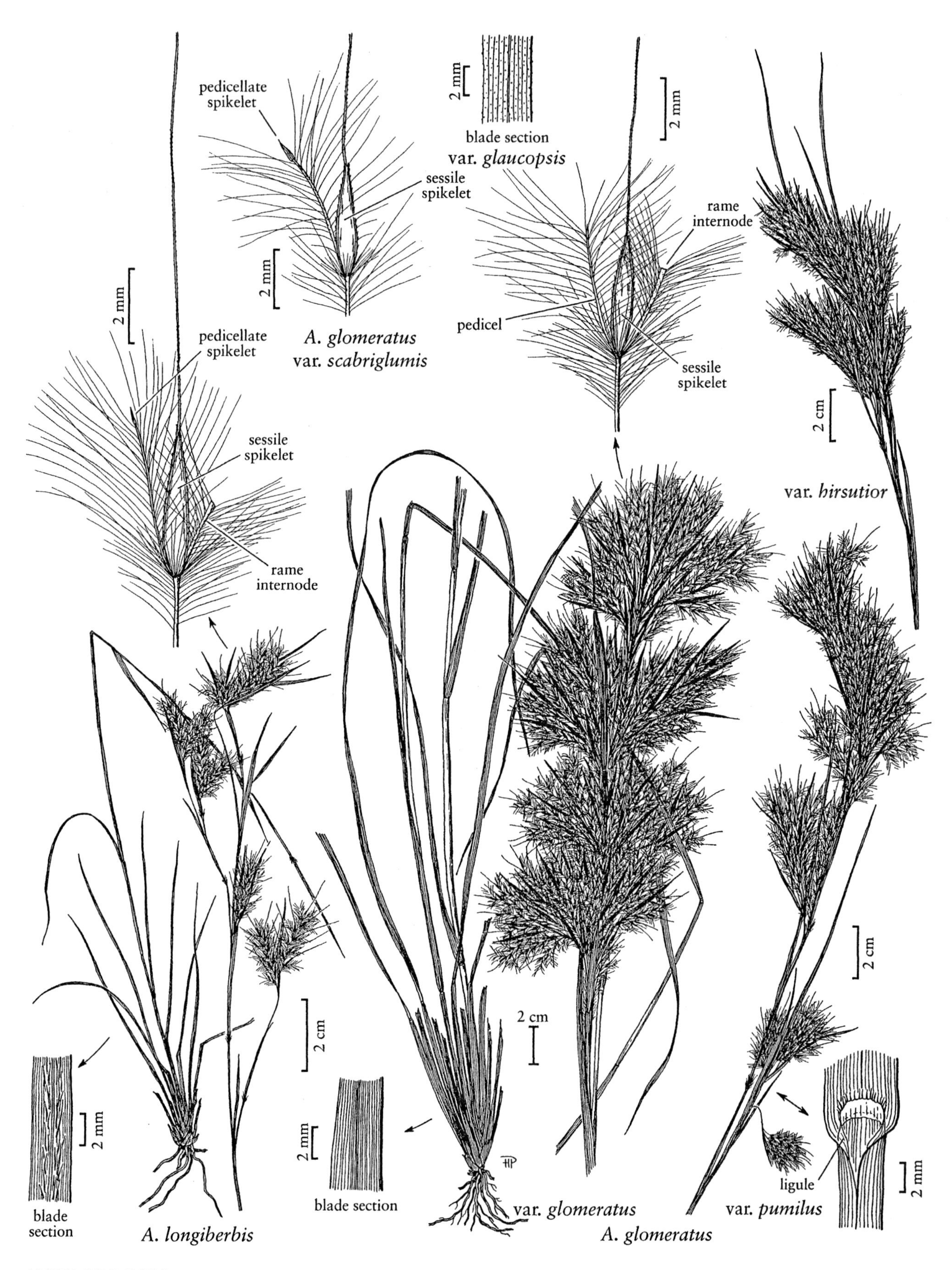
pedicellate spikelet
sessile spikelet
2 mm
A. glomeratus var. scabriglumis
2 mm
blade section
var. glaucopsis
2 mm
rame internode
pedicel
sessile spikelet
2 cm
var. hirsutior
2 mm
pedicellate spikelet
sessile spikelet
rame internode
2 cm
2 mm
blade section
A. longiberbis
2 mm
blade section
2 cm
var. glomeratus
A. glomeratus
2 cm
ligule
var. pumilus
2 mm

ANDROPOGON

3. Keels of the lower glumes usually smooth below midlength, scabrous distally.
   4. Upper portion of the plants oblong to obpyramidal; mature peduncles (4)11–35(60) mm long; anthers eventually falling .... var. *glomeratus*
   4. Upper portion of the plants cylindrical to oblong; mature peduncles 2–5 (8) mm long; withered remnants of anthers retained within the spikelets .... var. *hirsutior*

Andropogon glomeratus var. **glaucopsis** (Elliott) C. Mohr

Plants usually oblong in the upper portion, sometimes cylindrical. **Culms** 1.2–2.2 m; **internodes** glaucous. **Sheaths** (1.3)2.3–3.1(4.7) mm wide, usually smooth; **ligules** ciliate or not, cilia to 0.2 mm; **blades** glaucous, glabrous, smooth. **Subtending sheaths of inflorescence units** usually 2–2.5 mm wide; **peduncles** shorter than 10 mm; **rames** usually shorter than 2 cm, not exerted at maturity. **Keels of lower glumes** usually scabrous beyond midlength.

Andropogon glomeratus var. *glaucopsis* grows in flatwoods, bogs, ditches, swamps, pond margins, and swales of the southeastern coastal plain.

Andropogon glomeratus (Walter) Britton, Sterns & Poggenb. var. **glomeratus**

Plants oblong to pyramidal in the upper portion. **Culms** 60–160 cm; **internodes** not glaucous. **Sheaths** (2)2.5–3.4(4.7) mm wide, scabrous; **ligules** sometimes ciliate, cilia to 0.3 mm; **blades** green, usually scabrous. **Subtending sheaths of inflorescence units** usually 2.5–3.4 mm wide; **peduncles** (4)11–35(60) mm; **rames** usually 2.1–2.9 cm, exerted. **Keels of lower glumes** usually smooth below midlength, scabrous distally; **anthers** eventually deciduous.

Andropogon glomeratus var. *glomeratus* grows in bogs, swamps, savannahs, flatwoods, and ditches of the southeastern United States.

Andropogon glomeratus var. **hirsutior** (Hack.) C. Mohr

Plants usually oblong in the upper portion, sometimes cylindrical. **Culms** 1–2 m; **internodes** not glaucous. **Sheaths** scabrous; **ligules** ciliate or not, cilia to 0.3 mm; **blades** green, usually pubescent. **Subtending sheaths of inflorescence units** (2)2.4–3.1(4) mm wide; **peduncles** 2–5(8) mm; **rames** usually 1.7–2.8 cm, not exserted at maturity. **Keels of lower glumes** usually smooth below midlength, scabrous distally; **anthers** often retained within the spikelets.

Andropogon glomeratus var. *hirsutior* grows in ditches, swales, bogs, flatwoods, and savannahs of the southeastern coastal plain, often forming very large populations in cleared, low ground.

Andropogon glomeratus var. **pumilus** (Vasey) L.H. Dewey

Plants oblanceolate to obpyramidal in the upper portion. **Culms** to 2.5 m, but as short as 20 cm in poor soils; **internodes** not glaucous. **Leaf sheaths** usually smooth; **ligules** ciliate, cilia 0.2–0.9 mm; **blades** green, smooth or pubescent. **Subtending sheaths of inflorescence units** (2)2.9–4.3(5.2) cm long, 1.5–3 mm wide; **peduncles** (2)8–15(40) mm; **rames** 1.3–3 cm, exerted. **Keels of lower glumes** usually smooth below midlength, scabrous distally; **anthers** often retained within the spikelets.

Andropogon glomeratus var. *pumilus* is weedy and grows in disturbed, wet or moist sites. It is abundant and widespread, extending from the southern United States through Central America to northern South America.

Andropogon glomeratus var. **scabriglumis** C.S. Campb.

Plants oblanceolate to obpyramidal in the upper portion. **Culms** 80–150 cm; **internodes** not glaucous. **Sheaths** usually scabrous; **ligules** ciliate, cilia 0.2–0.5 mm; **blades** green, smooth or pubescent. **Subtending sheaths of inflorescence units** (2.3)2.9–4.5(6.3) cm long, (1.5)2.3–3.3(4.4) mm wide; **peduncles** (2)5–10(16) mm; **rames** (1.7)1.9–2.3(2.8) cm, exerted. **Keels of lower glumes** usually scabrous below and above the midpoint.

Andropogon glomeratus var. *scabriglumis* grows in moist soils of seepage slopes and the edges of springs, from California to New Mexico and southward into Mexico.

26.16 CYMBOPOGON Spreng.

Mary E. Barkworth

Plants usually perennial; cespitose. **Culms** 15–300 cm. **Leaves** aromatic, smelling of lemon oil or citronella; **sheaths** open, not strongly keeled except near the summit; **ligules** membranous; **blades** usually glabrous or mostly so, with long filiform apices. **Inflorescences** terminal and axillary, false panicles; **peduncles** often enclosed in the subtending leaf sheaths at maturity, with 2 rames; **rames** with 4–7 heterogamous spikelet pairs, axes slender, without a median groove, lower rame of each pair with 1 homogamous spikelet pair at the base, its pedicel swollen and more or less fused to the adjacent internodes, upper rames with short, sterile, flattened bases that are usually deflexed at maturity, without homogamous spikelet units. **Heterogamous spikelet units: sessile spikelets** dorsally compressed, with 2 florets; **lower glumes** chartaceous, concave or flat, 2-keeled, with or without intercostal veins, often streaked with oil glands; **upper florets** with a short, glabrous awn (rarely unawned); **pedicels** linear, free from the rame axes; **pedicellate spikelets** well-developed. $x = 10$. Name from the Greek *kymbe*, 'boat', and *pogon*, 'beard', referring to the boat-shaped leaf sheaths subtending the usually hairy rames (Clifford 1996).

Cymbopogon comprises 55 species, and is native to the tropics and subtropics of the Eastern Hemisphere. It is cultivated in southern Florida and California, sometimes persisting for a considerable period. Plants grown outdoors in the *Flora* region generally remain vegetative, but can usually be identified to genus by their lemony aroma. *Heteropogon melanocarpus* also smells lemony when fresh but, unlike the species of *Cymbopogon* that have been grown in the United States, it has a row of glandular depressions over the well-developed keels of the lower leaf sheaths.

Identification of any grass to species in the absence of reproductive parts is difficult. The key for use on vegetative plants of the three species of *Cymbopogon* reported from the *Flora* region should be used only as a last resort.

Cymbopogon citratus and *C. nardus* are cultivated commercially for their oils (lemon oil and citronella oil, respectively), which are used in cooking and perfume, and as an insect repellent; *C. citratus* and *C. jwarancusa* are also grown for their medicinal value.

SELECTED REFERENCES **Clifford, H.T.** 1996. Etymological Dictionary of Grasses, Version 1.0 (CD-ROM). Expert Center for Taxonomic Identification, Amsterdam, The Netherlands; **Soenarko, S.** 1977. The genus *Cymbopogon* Spreng. (Gramineae). Reinwardtia 9:225–375.

REPRODUCTIVE KEY

1. Pedicels pilose on the margins, glabrous dorsally . . . . 2. *C. nardus*
1. Pedicels pilose on the margins and the dorsal surface.
   2. Lower glumes of the sessile spikelets shallowly concave distally; the keels not winged; ligules 2–6 mm long; blades whitish . . . . 1. *C. jwarancusa*
   2. Lower glumes of the sessile spikelets flat above, the keels narrowly winged; ligules 0.5–2 mm long; blades green . . . . 3. *C. citratus*

VEGETATIVE KEY

1. Ligules 0.5–2 mm long, truncate, the nodes not swollen . . . . 3. *C. citratus*
1. Ligules 2–6 mm long, truncate to acute, the nodes usually swollen.
   2. Basal sheaths purplish-red; blades 3–16 mm wide . . . . 2. *C. nardus*
   2. Basal sheaths whitish; blades 1.5–4 mm wide . . . . 1. *C. jwarancusa*

1. **Cymbopogon jwarancusa** (Jones) Schult. [p. 667]
IWARANCUSA GRASS

Plants perennial. **Culms** to 150 cm, erect or geniculate; **nodes** often swollen. **Basal sheaths** glabrous, smooth, whitish-green; **ligules** 2–6 mm, truncate to acute; **blades** to 30 cm long, 1.5–4 mm wide, whitish. **Inflorescences** 15–40 cm, erect; **rames** 1322 mm; **internodes** and **pedicels** densely pilose on the margins and dorsal surface. **Sessile spikelets of heterogamous pairs** 4.5–5.5 mm; **lower glumes** lanceolate, shallowly concave distally, sharply keeled, keels not winged; **upper lemmas** awned, awns 7–10 mm. **Pedicellate spikelets** about 6 mm. $2n = 20$.

Cymbopogon jwarancusa is native to Asia, where it is grown for perfume and as a medicine for fevers. It is grown as an ornamental in the United States, and may persist for a considerable time after planting in the warmest parts of the *Flora* region.

2. **Cymbopogon nardus** (L.) Rendle [p. 667]
CITRONELLA GRASS, NARDGRASS

Plants perennial. **Culms** to 250 cm; **nodes** often swollen. **Basal sheaths** glabrous, smooth, purplish-red; **ligules** 3–6 mm, acute; **blades** to 100 cm long, 3–16 mm wide, surfaces smooth or scabrous. **Inflorescences** to 100 cm, linear, interrupted; **rames** 10–17 mm; **internodes** and **pedicels** pilose on the margins, glabrous dorsally. **Sessile spikelets of heterogamous pairs** 3–4.5(6) mm; **lower glumes** with narrowly winged keels; **upper lemmas** unawned or awned, awns 6–10 mm. $2n$ = 20, 40, 60.

Cymbopogon nardus has been cultivated in the United States, but the variety involved is not known. **Cymbopogon nardus** (L.) Rendle var. **nardus**, which is native to Sri Lanka, is the common citronella grass. It differs from **C. nardus** var. **confertiflorus** (Steud.) Stapf *ex* Bor, which is native to both Indian and Sri Lanka, in having unawned spikelets and $2n$ = 20, rather than awned spikelets and $2n$ = 40, 60. Both varieties have been widely introduced beyond their native range.

3. **Cymbopogon citratus** (DC.) Stapf [p. 667]
LEMON GRASS

Plants perennial. **Culms** to 200 cm, flexuous; **nodes** not swollen. **Basal sheaths** closely overlapping, gaping at maturity, forming somewhat flattened fans, glabrous, strongly glaucous; **ligules** 0.5–2 mm, truncate; **blades** to 90 cm long, 6.5–15 mm wide. **Inflorescences** to 60 cm, nodding; **rames** 10–25 mm; **internodes** and **pedicels** pilose on the margins and dorsal surface. **Sessile spikelets of heterogamous pairs** 5–6 mm; **lower glumes** shallowly concave below, flat distally, keels narrowly winged; **upper lemmas** entire or bidentate, unawned or with a 1–2 mm awn. **Pedicellate spikelets** 4–4.5 mm, unawned. $2n$ = 40, 60.

Cymbopogon citratus is now known only in cultivation, even in Asia. Young shoots are used as a spice, and the oils are extracted for lemon oil. It has been grown in Florida.

26.17 SCHIZACHYRIUM Nees

J.K. Wipff

Plants annual or perennial; cespitose or rhizomatous, sometimes both cespitose and shortly rhizomatous. **Culms** 7–210 cm, branched above the bases, often purplish near the nodes. **Leaves** not aromatic, **sheaths** open; **auricles** usually absent; **ligules** membranous; **blades** flat, folded, or involute, those of the uppermost leaves often greatly reduced. **Inflorescences** axillary and terminal, of 1, rarely 2, rames, peduncles subtended by a modified leaf; **rames** not reflexed, with spikelets in heterogamous sessile-pedicellate spikelet pairs, internodes more or less flattened, filiform to clavate, without a median groove, apices cupulate or fimbriate; **disarticulation** in the rame axes, below the sessile spikelets. **Spikelets** somewhat dorsiventrally compressed. **Sessile spikelets** with 2 florets; **glumes** exceeding the florets, lanceolate to linear, membranous; **lower glumes** enclosing the upper glumes, convex, weakly 2-keeled, with several (sometimes inconspicuous) intercostal veins; **lower florets** reduced to hyaline lemmas; **upper florets** bisexual, lemmas hyaline, bilobed or bifid to ⅞ of their length (rarely entire), awned from the sinuses; **anthers** 3. **Pedicels** free of the rame axes, usually pubescent. **Pedicellate spikelets** usually shorter than to as long as the sessile spikelets, occasionally longer, sterile or staminate, with 1 floret, often disarticulating as the rame matures; **lemmas** present in staminate spikelets, hyaline, unawned or with a straight awn of less than 10 mm. x = 10. Name from the Greek *schizo*, 'split', and *achyron*, 'chaff', referring to the divided lemma.

Schizachyrium is a genus of approximately 60 species that are native to tropical and subtropical regions of the world; nine are native to the *Flora* region. In North America, the best known species is *S. scoparium*, which was one of the major constituents of the grasslands that used to cover the central plains. Hitchcock (1951) included both *Schizachyrium* and *Bothriochloa* in *Andropogon*. Most species of *Schizachyrium* differ from species of the other two genera in having only one rame per peduncle, but *S. spadiceum* has two. More reliable, but less conspicuous distinguishing features of *Schizachyrium* are the cupulate tips of the rame

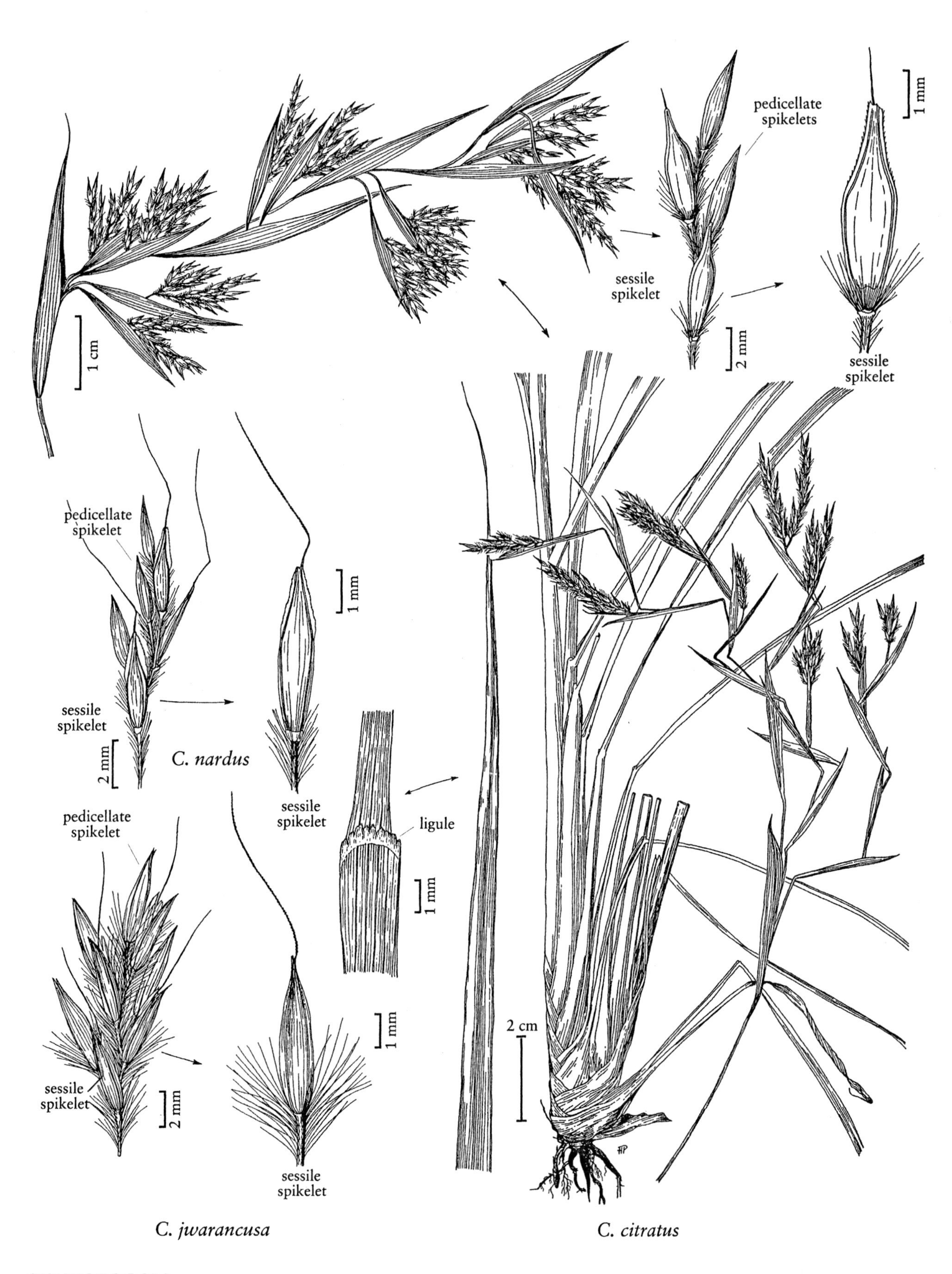
pedicellate
spikelets
1 mm
sessile
spikelet
2 mm
sessile
spikelet
1 cm
pedicellate
spikelet
1 mm
sessile
spikelet
2 mm
C. nardus
sessile
spikelet
ligule
1 mm
pedicellate
spikelet
1 mm
2 cm
sessile
spikelet
2 mm
sessile
spikelet
C. jwarancusa
C. citratus

CYMBOPOGON

internodes, the convex lower glumes, and the presence of veins between the keels of the lower glumes. A few species of *Andropogon* have solitary rames, but they do not have these other features.

SELECTED REFERENCES **Bruner, J.L.** 1987. Systematics of the *Schizachyrium scoparium* (Poaceae) complex in North America. Ph.D. dissertation, Ohio State University, Columbus, Ohio, U.S.A. 167 pp.; **Gandhi, K.N.** 1989. A biosystematic study of the *Schizachyrium scoparium* complex. Ph.D. dissertation, Texas A&M University, College Station, Texas, U.S.A. 188 pp.; **Grelen, H.E.** 1974. Pinehills bluestem, *Andropogon scoparius* Anderss. *ex* Hack., an anomaly of the *A. scoparius* complex. Amer. Midl. Naturalist 91:438–444; **Hatch, S.L.** 1975. A biosystematic study of the *Schizachyrium cirratum–Schizachyrium sanguineum* complex (Poaceae). Ph.D. dissertation, Texas A&M University, College Station, Texas, U.S.A. 112 pp.; **Hitchcock, A.S.** 1951 [title page 1950]. Manual of the Grasses of the United States, ed. 2, rev. A. Chase. U.S.D.A. Miscellaneous Publication No. 200. U.S. Government Printing Office, Washington, D.C., U.S.A. 1051 pp.; **Wipff, J.K.** 1996. Nomenclatural combinations in *Schizachyrium* (Poaceae: Andropogoneae). Phytologia 80:35–39.

1. Peduncles with 2 rames . . . . 1. *S. spadiceum*
1. Peduncles with only 1 rame.
   2. Leaf blades 0.5–2 mm wide, with a longitudinal stripe of white, spongy tissue (formed of bulliform cells) on their adaxial surfaces; plants cespitose; pedicellate spikelets about as long as the sessile spikelets . . . . 6. *S. tenerum*
   2. Leaf blades (1)1.5–9 mm wide, without a longitudinal stripe of white, spongy tissue on their adaxial surfaces; plants cespitose or rhizomatous; pedicellate spikelets equal to or smaller than the sessile spikelets.
      3. Plants rooting and branching at the lower nodes and at aerial nodes in contact with the soil; leaf collars usually elongate and narrow; plants of sandy coastal habitats.
         4. Ligules 0.5–1 mm long, pedicellate spikelets 4.5–8.5 mm long . . . . 4. *S. maritimum*
         4. Ligules 1.5–2 mm long, pedicellate spikelets 1.5–5 mm long . . . . 5. *S. littorale*
      3. Plants not rooting or branching at the lower nodes; leaf collars neither elongate nor particularly narrow; plants of varied habitats.
         5. Pedicel bases 0.2–0.5 mm wide, gradually widening to 0.3–1 mm distally, straight, often somewhat stiff, not tending to curve outward; rames appearing linear.
            6. Pedicellate spikelets 6–8 mm long, about as long as the sessile spikelets, usually staminate, sometimes sterile, unawned . . . . 9. *S. cirratum*
            6. Pedicellate spikelets 0.7–10 mm long, usually shorter than the sessile spikelets, sterile, unawned or awned, the awns up to 6 mm long.
               7. Upper lemmas cleft for ⅔–⅞ of their length; lower glumes glabrous or pubescent . . . . 8. *S. sanguineum*
               7. Upper lemmas cleft for up to ½ of their length; lower glumes glabrous . . . . 2. *S. scoparium*
         5. Pedicel bases 0.1–0.2 mm wide, flaring above midlength to about 0.5 mm wide, tending to curve outward; rames appearing somewhat open.
            8. Upper lemmas indurate at the base, cleft ¾–⅞ of their length; leaf blades 2.5–10 cm long; pedicellate spikelets 0.5–2 mm long; plants cespitose; known only from peninsular Florida . . . . 7. *S. niveum*
            8. Upper lemmas membranous at the base, cleft for up to ½ of their length; leaf blades 7–105 cm long; pedicellate spikelets 0.7–10 mm long; plants cespitose or not; widespread, including Florida.
               9. Plants cespitose, not or shortly rhizomatous . . . . 2. *S. scoparium*
               9. Plants not cespitose, strongly rhizomatous.
                  10. Pedicellate spikelets awned, awns to 4 mm; leaf blades usually 3.5–9 mm wide; culms usually 1–3 mm thick; plants of sandy soils . . . . 2. *S. scoparium*
                  10. Pedicellate spikelets unawned or the awns less than 1 mm; leaf blades 1–3 mm wide; culms usually less than 1 mm thick; plants of oölitic soil . . . . 3. *S. rhizomatum*

1. **Schizachyrium spadiceum** (Swallen) Wipff [p. 669]
HONEY BLUESTEM

Plants cespitose. **Culms** 60–95 cm, slender, erect, glabrous. **Leaves** glaucous; **sheaths** compressed, scabridulous, glabrous or almost so; **ligules** 1–1.5 mm, truncate, erose-ciliate; **blades** 10–25 cm long, 2–2.5 mm wide, flat, scabrous, young blades ciliate basally. **Peduncles** 5–9 cm, mostly erect, often included in the subtending leaves, with 2 rames; **rames** 3.5–5 cm, enclosed or exerted at maturity; **internodes** 4–6.3 mm, ciliate proximally, densely villous on the distal ½–⅔, hairs 4–7 mm. **Sessile spikelets** 7–8 mm; **calluses** 0.2–0.5 mm, hairs 1–2 mm; **lower glumes** glabrous, keels scabridulous distally; **awns** 14.5–17.5 mm, once-geniculate. **Pedicels** 5–6 mm, hairs 5–7 mm. **Pedicellate spikelets** 0.8–4 mm, sterile, unawned. $2n$ = unknown.

Schizachyrium spadiceum was once thought to be a Mexican endemic, but it is now known from limestone slopes in Brewster County, Texas.

2. **Schizachyrium scoparium** (Michx.) Nash [p. 671]

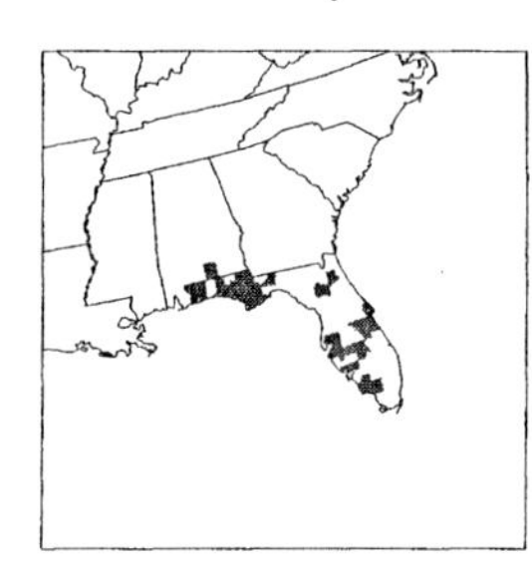

Plants cespitose or rhizomatous, green to purplish, sometimes glaucous. **Culms** 7–210 cm tall, usually 1–3 mm thick, not rooting or branching at the lower nodes. **Sheaths** rounded or keeled, glabrous or pubescent, sometimes glaucous; **ligules** 0.5–2 mm, collars neither elongate nor narrowed; **blades** 7–105 cm long, 1.5–9 mm wide, without a longitudinal stripe of white, spongy tissue. **Peduncles** 0.8–10 cm; **rames** 2.5–8 cm, partially to completely exserted, usually somewhat open; **internodes** 3–7 mm, usually arcuate at maturity, ciliate on at least the distal ½ (sometimes throughout), hairs 1.5–6 mm. **Sessile spikelets** 3–11 mm; **calluses** 0.5–1(2) mm, hairs 0.3–4 mm; **lower glumes** glabrous; **upper lemmas** membranous throughout, cleft to ½ their length; **awns** 2.5–17 mm. **Pedicels** 3–7.5 mm long, 0.1–0.2 mm wide at the base, flaring above midlength to 0.3–0.5 mm, straight or curving outwards. **Pedicellate spikelets** 0.7–10 mm, sometimes shorter than the sessile spikelets, sterile or staminate, unawned or awned, awns to 4 mm, when sterile, the lemma usually absent. $2n$ = 40.

Schizachyrium scoparium is a widespread grassland species extending from Canada to Mexico. It is one of the principal grasses in the tallgrass prairies that used to dominate the central plains of North America. It exhibits considerable variation, much of it clinal. The following

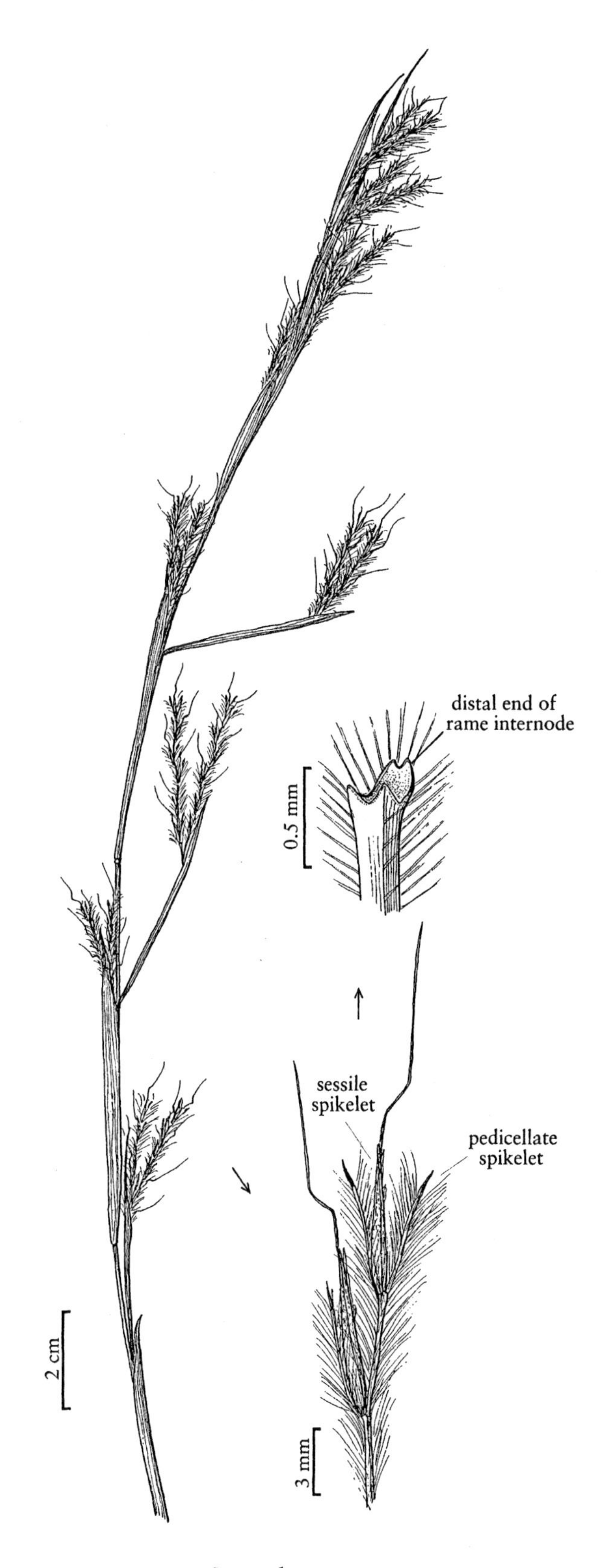

SCHIZACHYRIUM

varieties are recognized because they are morphologically, ecologically, and geographically distinctive.

1. Plants not cespitose, strongly rhizomatous; pedicellate spikelets sterile .... var. *stoloniferum*
1. Plants usually cespitose, not or shortly rhizomatous; pedicellate spikelets staminate or sterile.
   2. Pedicellate spikelets of the proximal spikelet units on each rame staminate, 5–10 mm long, with a lemma, pedicellate spikelets of the distal units usually smaller (1–4 mm) and sterile; sheaths and blades densely tomentose to glabrate .... var. *divergens*
   2. Most pedicellate spikelets sterile, 1–6 mm long, without a lemma; sheaths and blades usually glabrous, occasionally pubescent .... var. *scoparium*

Schizachyrium scoparium var. **divergens** (Hack.) Gould
PINEHILL BLUESTEM

Plants cespitose. **Culms** 7–180 cm. **Sheaths** densely tomentose initially, sometimes glabrate, margins usually pilose distally; **blades** 12–30 cm long, 3.5–5 mm wide; initially densely tomentose, becoming glabrous. **Rames** 3–5 cm, with 7–12 spikelets, usually partially to wholly exserted, sometimes appearing linear; **internodes** 3.5–6 mm, often sparsely pubescent, hairs 1.5–3 mm. **Sessile spikelets** 6–10 mm; **calluses** about 0.3 mm, hairs to 2 mm; **awns** 9–15 mm. **Pedicels** 3.5–6.5 mm, curving out at maturity. **Pedicellate spikelets** on the proximal portion of the rames 5–10 mm, mostly staminate, with lemmas, distal spikelets often smaller (1–4 mm) and sterile, unawned or awned, awns to 2 mm.

Schizachyrium scoparium var. *divergens* is common in the south central pinelands of the United States. The pubescence of the leaves varies across its range, western plants having longer and more villous leaves than those in the east and, towards Mississippi, the pubescence is confined to the sheaths. *Schizachyrium scoparium* var. *divergens* intergrades with var. *scoparium*.

Grelen (1974) found that plants of *S. scoparium* var. *divergens* from western Louisiana, Oklahoma, and Arkansas produced mostly staminate pedicellate spikelets, plants from southeastern Louisiana and southeastern Mississippi produced mostly sterile pedicellate spikelets, and plants from western Mississippi varied in this character.

Schizachyrium scoparium (Michx.) Nash var. **scoparium**
LITTLE BLUESTEM, BROOM BEARDGRASS, BROOM BLUESTEM, SCHIZACHYRIUM À BALAIS

Plants usually cespitose, sometimes producing short rhizomes. **Culms** 30–210 cm. **Sheaths** usually glabrous, keeled; **blades** 9–45 cm long, 1.5–9 mm wide, flat, usually glabrous, occasionally pubescent. **Peduncles** to 10 cm; **rames** 2–8 cm, with 6–13 spikelets, exserted. **Sessile spikelets** 6–11 mm; **calluses** about 0.5 mm, hairs to 2.5 mm, **awns** 2.5–17 mm; **Pedicels** 3–7.5 mm, straight or curving out at maturity. **Pedicellate spikelets** usually 1–6 mm, sterile, without lemmas, occasionally staminate and with a lemma, unawned or awned, awns to 4 mm.

Schizachyrium scoparium var. *scoparium* grows in a variety of soils and in open habitats. It was once a dominant component of the prairie grasslands that extended through the central plains of North America and into Mexico, but it has largely been replaced by fields of maize, wheat, sorghum, sunflowers, and field mustard. It is the most variable of the varieties recognized within *S. scoparium*, with morphological features that vary independently and continuously across its range, coming together in distinctive combinations in some regions. Some of these phases have been named as varieties, or even species, but they have proven to be untenable taxonomic entities when plants from throughout the range of the species are considered.

Schizachyrium scoparium var. **stoloniferum** (Nash) Wipff
CREEPING BLUESTEM

Plants not cespitose, with long, scaly rhizomes. **Culms** 58–210 cm. **Sheaths** usually pubescent near the collars; **blades** 10–39 cm long, 3.5–9 mm wide, pubescent near the collars. **Rames** 2–6.5 cm, with 6–14 spikelets, usually partially to fully exserted; **internodes** pubescent, hairs to 4.5 mm. **Sessile spikelets** 5–10 mm; **calluses** with hairs to 2.5 mm; **awns** 6–14 mm. **Pedicels** 3.5–5 mm, curving out at maturity. **Pedicellate spikelets** 0.75–4 mm, sterile, awned, awns 1–3 mm.

Schizachyrium scoparium var. *stoloniferum* grows in sandy soils of woodland openings and roadsides from southern Alabama and Georgia south to the Everglades. Northern populations consist of widely spaced, weak culms growing in rather bare sand; southern populations consist of dense, vigorous stands with taller, more robust culms growing primarily along roadsides, possibly spread by grading equipment. Some clones, particularly in the south, are largely sterile.

3. **Schizachyrium rhizomatum** (Swallen) Gould [p. 673]
FLORIDA LITTLE BLUESTEM

Plants with short, scaly rhizomes. **Culms** 50–90 cm tall, usually less than 1 mm thick, not rooting or branching at the lower nodes, usually glabrous. **Ligules** about 0.5 mm; **blades** 9.5–25 cm long, 1–3 mm wide, usually folded, without a longitudinal stripe of white, spongy tissue. **Peduncles**

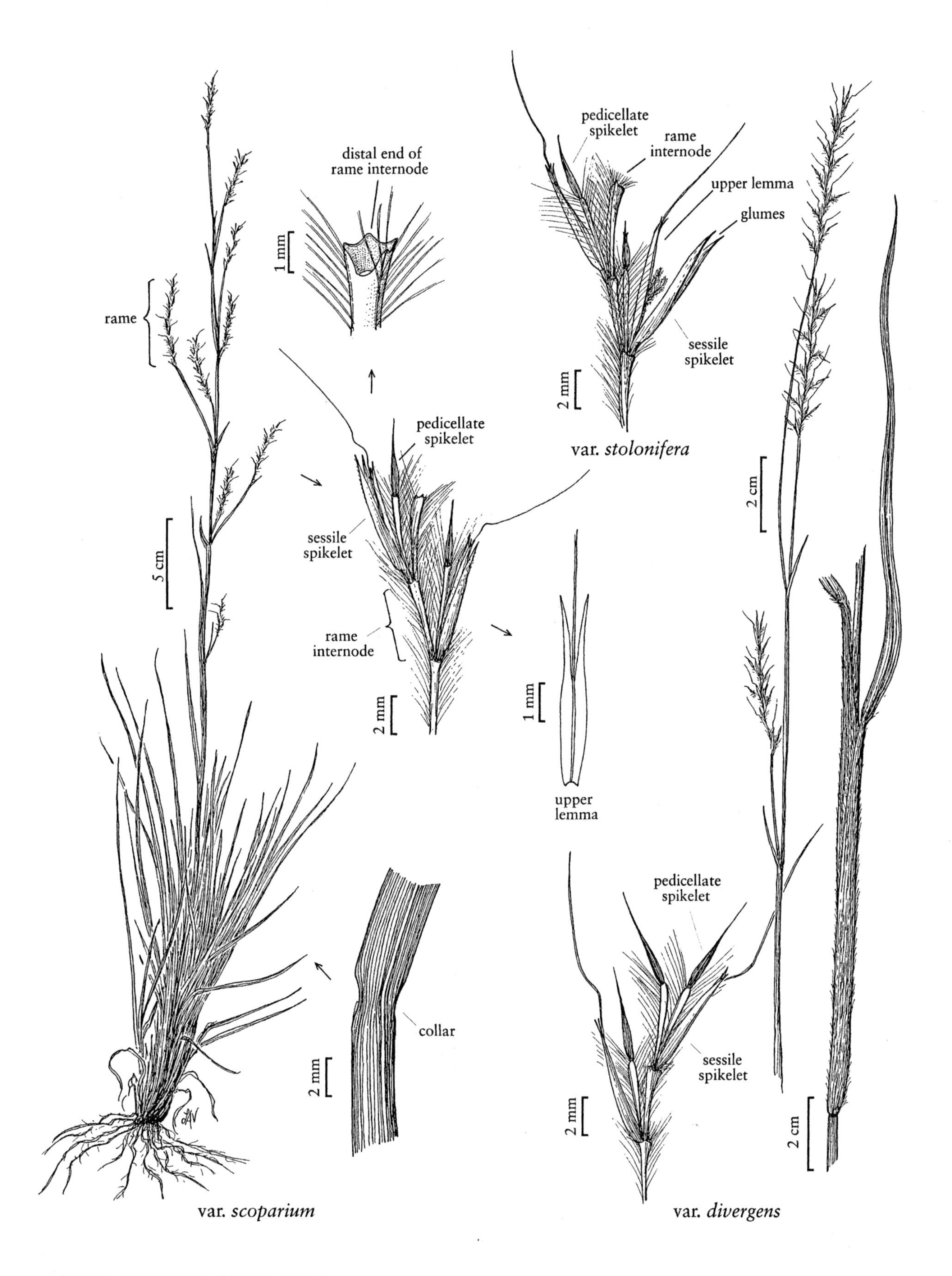

SCHIZACHYRIUM SCOPARIUM

3–7 cm; **rames** 2–5.5 cm, with 5–14 spikelets, partially to fully exserted, collars neither elongate nor particularly narrow. **Sessile spikelets** 4–7.5 mm; **calluses** sparsely pubescent, hairs to 1.5 mm; **awns** 2.5–10 mm; **upper lemmas** membranous throughout, apices cleft for about ¼ of their length. **Pedicels** 3.5–5 mm, ciliate, hairs to 2.3 mm, pedicel bases 0.1–0.2 mm wide, flaring above midlength to about 0.5 mm wide, tending to curve outward, rames appearing somewhat open. **Pedicellate spikelets** 2.5–5.5 mm, unawned or with awns to 1 mm.

Schizachyrium rhizomatum grows in open glades and on the margins of pine woodlands and is endemic to Florida. It is restricted to thin, oölitic soils that are often saturated with water, and forms sparse stands, occasionally mixed with *Andropogon gracilis*, in the Florida Keys.

4. **Schizachyrium maritimum** (Chapm.) Nash [p. 673]
GULF BLUESTEM

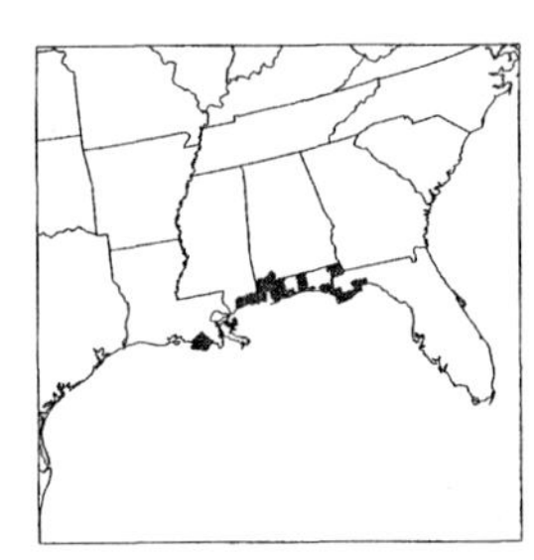

Plants often appearing rhizomatous. **Culms** 35–80 cm, solitary, decumbent, branching at the lower nodes, often rooting from nodes in contact with the soil. **Leaves** glaucous throughout; **sheaths** shorter than the internodes, keeled; **collars** constricted, elongate; **ligules** 0.5–1 mm; **blades** 11–142 cm long, 3.5–5.5 mm wide, folded, without a longitudinal stripe of white, spongy tissue. **Peduncles** 1–6 cm; **subtending leaf sheaths** 3.2–6.6 cm long, 3–6.5 mm wide; **rames** 2.5–6.5 cm, flexuous, usually partially exserted, appearing somewhat open; **internodes** 4–5.5 mm, straight, pubescent for ½–¾ of their length, hairs 2.5–6 mm. **Sessile spikelets** 9–11 mm; **calluses** 0.3–0.5 mm; hairs to 1 mm; **awns** 8–13 mm. **Pedicels** 5–7 mm, as conspicuously villous as the rachis. **Pedicellate spikelets** 4.5–8.5 mm, staminate, unawned or awned, awns to 3.5 mm. $2n = 40$.

Schizachyrium maritimum is endemic to the southeastern United States, growing in sandy areas, usually at the ocean waterline but also along roads in low, dune areas, from Louisiana to the Florida panhandle.

The plants often appear rhizomatous because the lower, decumbent portions of the culms are frequently covered by sand. It is an effective sand binder and can withstand frequent inundation by sea water, the constricted collar permitting the blades to sway freely when subjected to wind or wave action.

5. **Schizachyrium littorale** (Nash) E.P. Bicknell [p. 675]
SHORE BLUESTEM, DUNE BLUESTEM

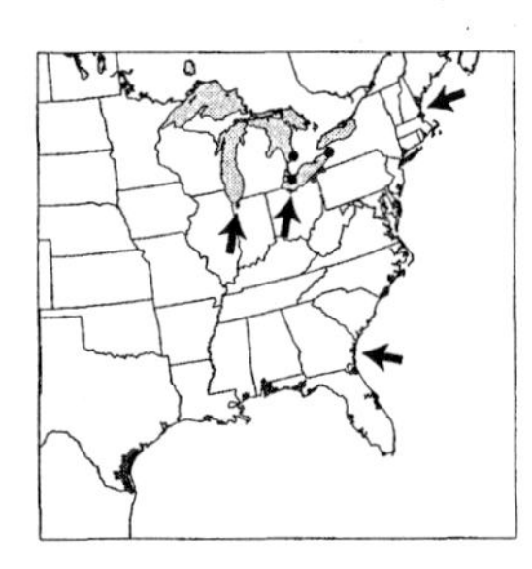

Plants cespitose, sometimes appearing rhizomatous, glaucous. **Culms** 39–160 cm, branching at the lower nodes, often rooting from nodes in contact with the soil; **lower internodes** usually shortened and compressed. **Leaves** glaucous; **collars** usually constricted, elongate; **auricles** flexible, yellow; **ligules** 1.5–2 mm; **blades** 10–30 cm long, 3.5–6.5 mm wide, without a longitudinal stripe of white, spongy tissue. **Peduncles** 0.5–5 mm; **rames** 3–9 cm, with 13–19 spikelets, arcuate at maturity; **internodes** 4–6 mm, densely villous, hairs 3–7.5 mm. **Sessile spikelets** 6–10 mm; **calluses** 0.2–0.5 mm, glabrous; **lower glumes** glabrous; **awns** 9–20 mm. **Pedicels** 5–7 mm, hairy distally, hairs 5–7 mm. **Pedicellate spikelets** 1.5–5 mm, often staminate, unawned or awned, awns to 3.5 mm. $2n = 40$.

Schizachyrium littorale is restricted to shifting, coastal sand dunes of the Gulf, Atlantic, and Great Lakes coasts of the United States. It often appears rhizomatous because the lower nodes are frequently covered by sand.

6. **Schizachyrium tenerum** Nees [p. 675]
SLENDER BLUESTEM

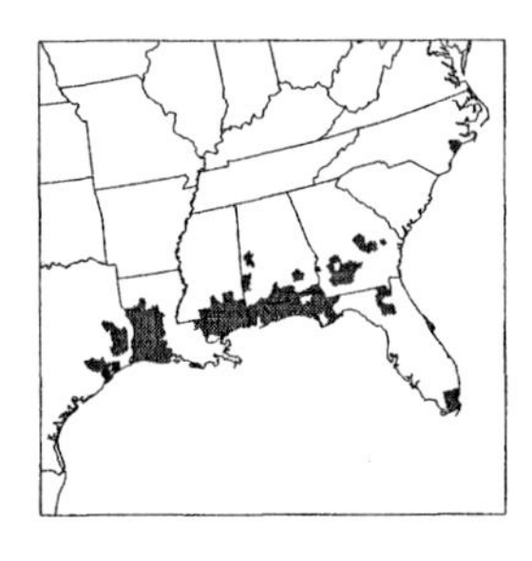

Plants cespitose. **Culms** 60–100 cm, sometimes reclining or decumbent, glabrous. **Collars** not elongate, about as wide as the blade; **ligules** to 0.5 mm, ciliolate; **blades** 5–15 cm long, 0.5–2 mm wide, involute or flat, glabrous or sparsely hairy basally, with a wide central zone of bulliform cells evident on the adaxial surfaces as a longitudinal stripe of white, spongy tissue. **Rames** 2–6 cm, eventually long-exserted; **internodes** 2–4 mm, straight, glabrous. **Sessile spikelets** 3.5–4.5 mm; **calluses** 0.5–1 mm, hairs to 1.2 mm; **lower glumes** glabrous; **upper lemmas** acute, entire; **awns** 6–10 mm. **Pedicels** 3–5 mm, glabrous. **Pedicellate spikelets** usually as long as or slightly longer than the sessile spikelets, sterile, unawned. $2n = 60$.

Schizachyrium tenerum is an uncommon species in the southeastern United States, where it grows on sandy soils in pine forest openings and coastal prairies. Its range extends through Central America into South America.

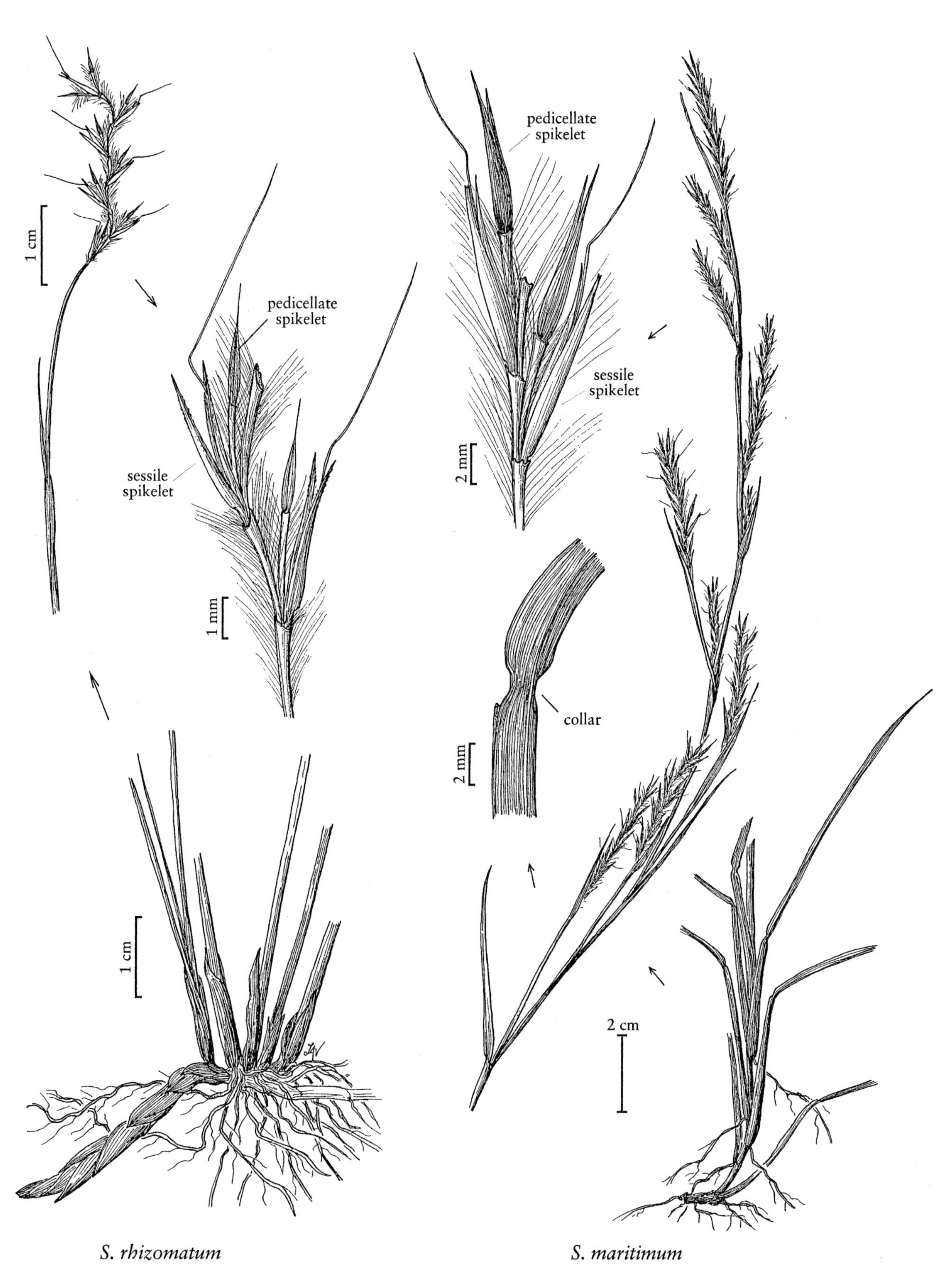

S. rhizomatum *S. maritimum*

SCHIZACHYRIUM

7. **Schizachyrium niveum** (Swallen) Gould [p. 676]
PINESCRUB BLUESTEM

Plants cespitose. **Culms** 49–90 cm, not rooting or branching at the lower nodes. **Leaves** usually completely glabrous; **sheaths** keeled; **ligules** 0.5–1 mm; **blades** 2.5–10 cm long, (1)2–4 mm wide, flat, without a longitudinal stripe of white, spongy tissue. **Peduncles** 2–4.6 cm; **subtending leaf sheaths** 2.5–4 cm long, 1.5–3.5 mm wide; **rames** 2.5–4.5 cm, somewhat open and usually partially exserted, varying from included to completely exserted; **internodes** 3–7 mm, straight, densely villous for their full length, hairs 0.5–2.5 mm, silvery-white. **Sessile spikelets** 5–6.5 mm; **calluses** with 0.5–1 mm hairs; **lemmas** slightly indurate at the base (unique among the species treated here in this respect), cleft for ¾–⅞ of their length; **awns** 10.5–15 mm. **Pedicels** 5–6.5 mm long, 0.1–0.2 mm wide at the base, flaring beyond midlength to about 0.5 mm, densely villous. **Pedicellate spikelets** 0.5–2 mm, sterile, unawned or awned, awns 1–2 mm. $2n$ = 40.

Schizachyium niveum is an endangered, rare species known only from central peninsular Florida, where it occurs in openings and sandhills of *Ceratiola*-pine-oak woodlands. It has been reported from south central Georgia, but Bruner (1987) found no evidence for the report. Of the two recent collections in Florida, he relocated one, in an area favored by real estate developers.

8. **Schizachyrium sanguineum** (Retz.) Alston [p. 676]

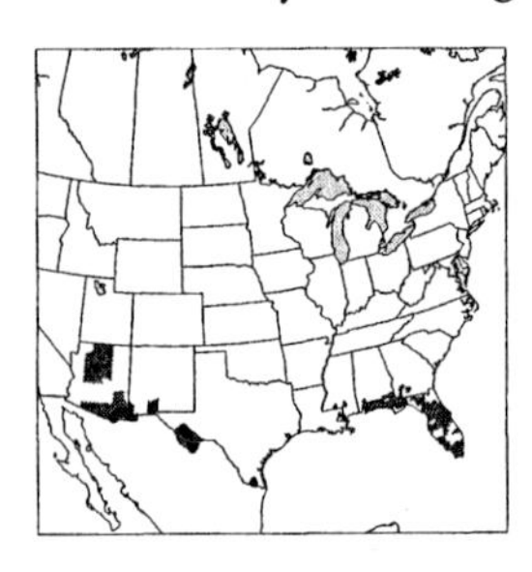

Plants cespitose. **Culms** 40–120 cm, erect, not rooting or branching at the lower nodes, glabrous. **Sheaths** glabrous, rounded; **ligules** 0.7–2 mm; **blades** 7–20 cm long, 1–6 mm wide, usually with long, papillose-based hairs basally, glabrous elsewhere, sometimes scabrous, without a longitudinal stripe of white, spongy tissue. **Peduncles** 4–6 cm; **rames** 4–15 cm, not open, usually almost fully exserted at maturity; **internodes** 4–6 mm, straight, from mostly glabrous with a tuft of hairs at the base to densely hirsute all over. **Sessile spikelets** 5–9 mm; **calluses** 0.5–1 mm, hairs to 2 mm; **lower glumes** glabrous or densely pubescent; **upper lemmas** cleft for (⅔)¾–⅞ of their length; **awns** 15–25 mm. **Pedicels** 3–6 mm long, 0.3–0.5 mm wide at the base, gradually widening to about 0.6–0.8 mm at the top, straight. **Pedicellate spikelets** 3–5 mm, usually evidently shorter than the sessile spikelets, sterile or staminate, awned, awns 0.3–6 mm.

Schizachyrium sanguineum extends from the southern United States to Chile, Paraguay, and Uruguay.

1. Lower glumes of the sessile spikelets glabrous or scabrous; pedicels ciliate on 1 edge . . . var. *sanguineum*
1. Lower glumes of the sessile spikelets pubescent to hirsute; pedicels ciliate on both edges . . . var. *hirtiflorum*

Schizachyrium sanguineum var. **hirtiflorum** (Nees) S.L. Hatch
HAIRY CRIMSON BLUESTEM

Culms 40–120 cm, branching at the upper nodes, glaucous. **Ligules** 1–2 mm; **blades** 10–20 cm long, 1.5–5 mm wide. **Rames** 4–10(12) cm; **internodes** scabrous, glabrous or hirsute. **Sessile spikelets** 5–9 mm; **lower glumes** sparsely to densely hirsute on the back; **awns** 15–25 mm. **Pedicels** arcuate at maturity, ciliate on both edges distally. **Pedicellate spikelets** 3–5 mm, staminate or sterile, awns 0.3–5 mm. $2n$ = 40, 60, 70, 100.

In the *Flora* region, *Schizachyrium sanguineum* var. *hirtiflorum* grows on rocky slopes and well-drained soils from Arizona to southwestern Texas and Florida, and is considered a good forage species. Its range extends through Central America to Chile, Paraguay, and Uruguay.

Schizachyrium sanguineum (Retz.) Alston var. **sanguineum**
CRIMSON BLUESTEM

Culms 60–120 cm. **Ligules** 0.7–1.5 mm; **blades** usually 1–6 mm wide. **Rames** 5–8 cm; **internodes** 4–6 mm, glabrous except for a tuft of hairs at the base. **Sessile spikelets: glumes** glabrous or scabrous; **awns** 15–25 mm. **Pedicels** straight, ciliate on 1 edge distally. **Pedicellate spikelets** about 3 mm, sterile, awns 3–6 mm. $2n$ = 40, 50, 60, 70, 80.

Schizachyrium sanguineum var. *sanguineum* grows in tropical and subtropical regions of America, Africa, and Asia. Within the *Flora* region, it is known only from the pine woods of Alabama and Florida.

9. **Schizachyrium cirratum** (Hack.) Wooton & Standl. [p. 676]
TEXAS SCHIZACHYRIUM, TEXAS BEARDGRASS

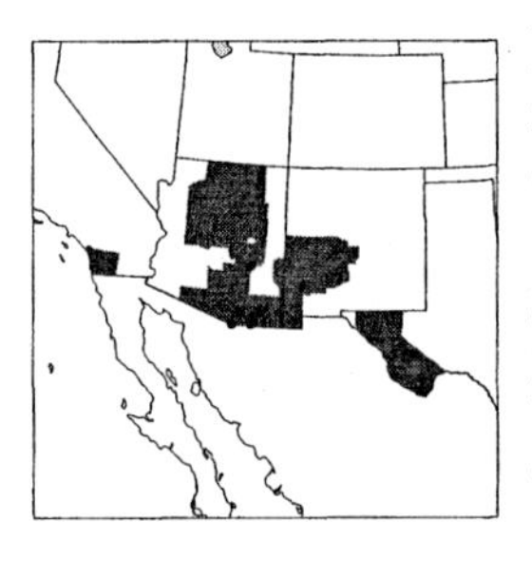

Plants cespitose or shortly rhizomatous. **Culms** 31–75 cm, often decumbent, not rooting or branching at the lower nodes, glabrous, glaucous, sometimes purplish. **Ligules** 1–2.5 mm; **blades** 6–17 cm long, 2–4 mm wide, glabrous, without a longitudinal stripe of white, spongy tissue. **Rames** 4–6 cm, usually exerted, straight, often somewhat stiff, not flexuous, appearing linear;

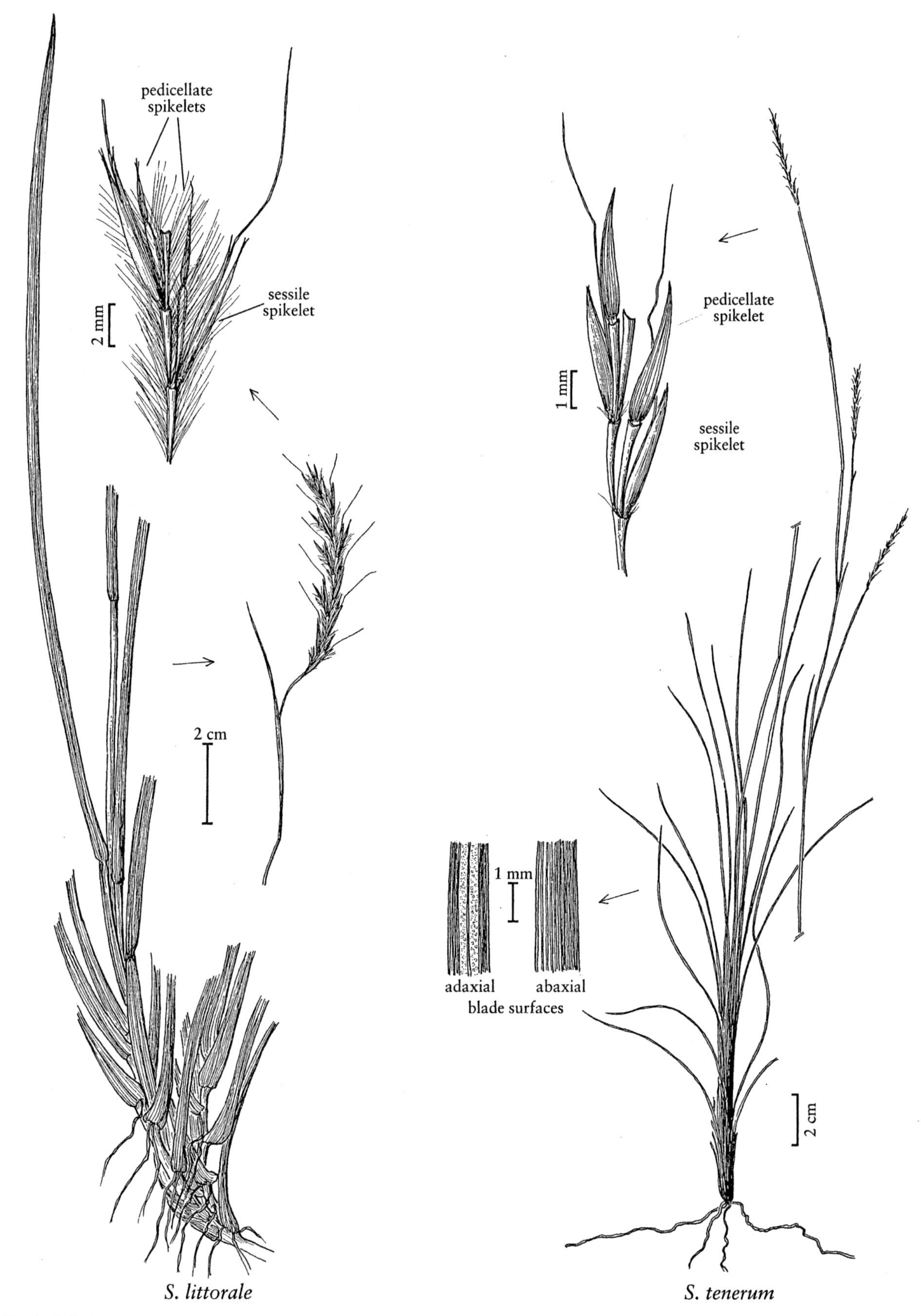

S. littorale

S. tenerum

SCHIZACHYRIUM

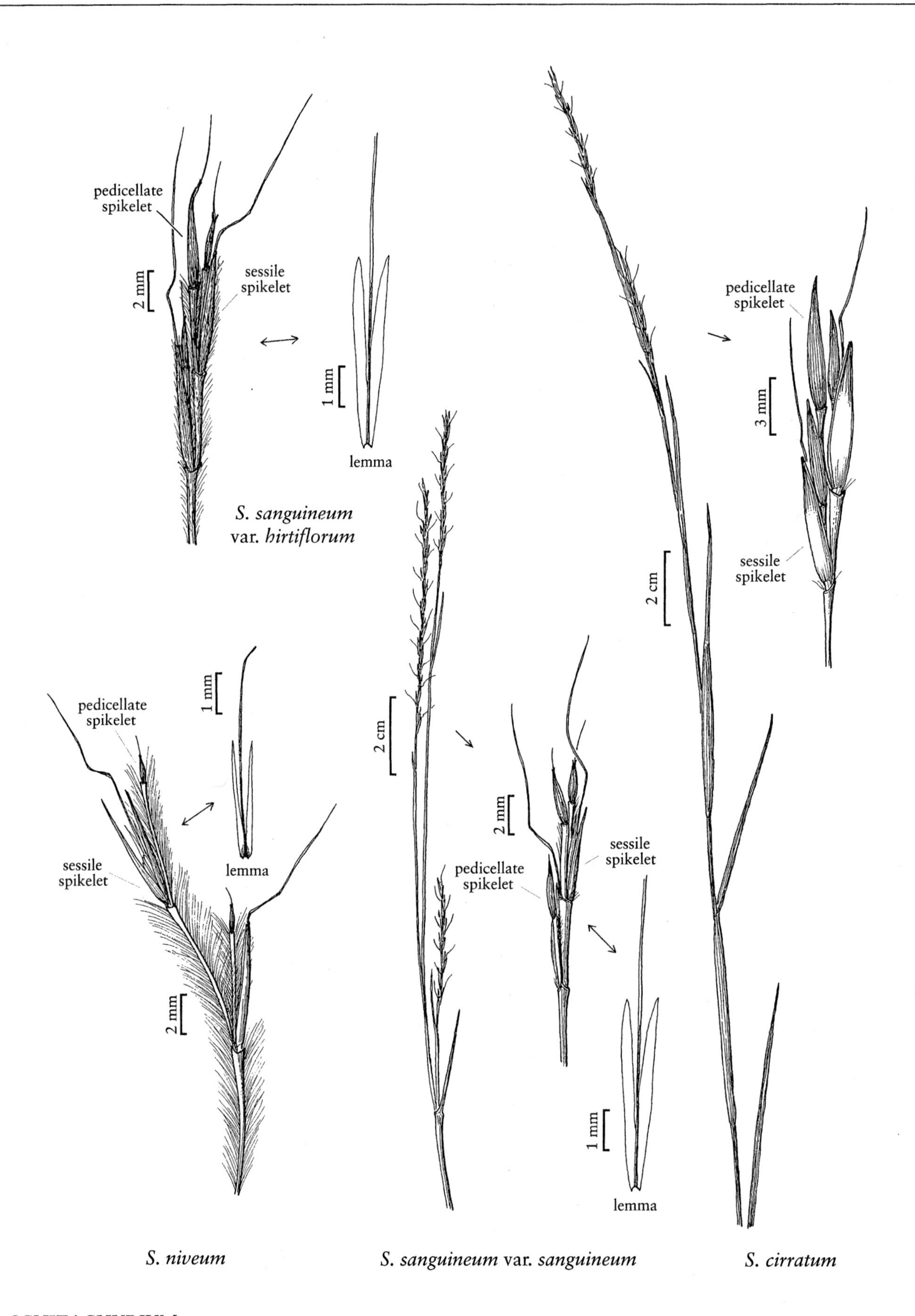

S. niveum *S. sanguineum* var. *sanguineum* *S. cirratum*

internodes straight, with a tuft of hairs near the base, elsewhere glabrous or ciliate on the margins. **Sessile spikelets** 8–10 mm; **calluses** 0.3–0.6 mm, hairs 0.5–1.2 mm; **glumes** glabrous or scabrous; **awns** 13–24 mm. **Pedicels** 3.5–5 mm long, 0.2–0.5 mm wide at the base, widening to 0.5–1 mm, straight, with a tuft of hairs at the base, distal ½ usually ciliate on 1 side, sometimes on both sides. **Pedicellate spikelets** 6–8 mm, about as long as the sessile spikelets, usually staminate, sometimes sterile, unawned. $2n = 20$ (for var. *cirratum*).

Schizachyrium cirratum grows on rocky slopes, mostly at elevations of 5000 feet or higher, from southern California to western Texas into Mexico, and is known from South America. It is an excellent forage grass. Plants in the *Flora* region differ from those in central Mexico in being essentially non-rhizomatous and in having glabrous rame axes and pedicels that are ciliate only on the distal half.

26.18 ARTHRAXON P. Beauv.

John W. Thieret

Plants annual or perennial; scrambling. **Culms** 0.5–2 m, ascending to decumbent, often rooting at the nodes, branched. **Leaves** not aromatic; **sheaths** open, at least the outer margins pubescent, usually with papillose-based hairs; **ligules** membranous, fimbriate or ciliate; **blades** ovate to ovate-lanceolate. **Inflorescences** terminal and axillary, panicles of subdigitate, often flabellate, clusters of rames; **rame internodes** not sulcate; **disarticulation** in the rames, beneath the sessile spikelets. **Spikelets** in heteromorphic sessile-pedicellate pairs or appearing solitary and sessile, pedicels greatly reduced and lacking spikelets. **Sessile spikelets** bisexual, with 2 florets; **calluses** absent or blunt; **glumes** equal or subequal; **lower florets** sterile, reduced to an unawned lemma; **upper florets** bisexual, awned (rarely unawned); **anthers** 2 or 3. **Pedicels** 0.2-3 mm, not thickened, not fused to the rame axes. **Pedicellate spikelets** absent or rudimentary. $x = 9, 10$. Name from the Greek *arthron*, 'segment', and *axon*, 'axis', referring to the jointed inflorescence axes.

Arthraxon is a genus of seven species that are native to tropical and subtropical regions of the Eastern Hemisphere; one species is established in the *Flora* region.

SELECTED REFERENCE Welzen, P.C. van. 1981. A taxonomic revision of *Arthraxon* Beauv. (Gramineae). Blumea 27:255–300.

1. **Arthraxon hispidus** (Thunb.) Makino [p. 679]
JOINTHEAD, SMALL CARPETGRASS

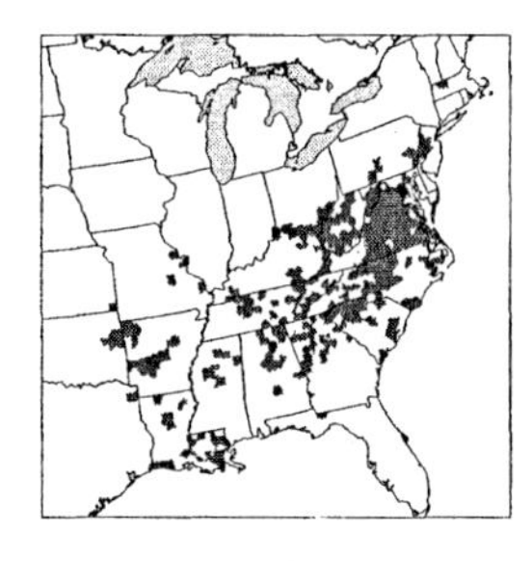

Plants annual. **Culms** 0.5–1(2) m, weak, often decumbent and rooting at the lower nodes; **nodes** hispid. **Leaves** cauline; **sheaths** usually shorter than the internodes; **ligules** 0.4–3.5 mm, ciliate; **lower blades** 1–7.5 cm long, 4–20 mm wide, cordate-clasping, ovate to ovate-lanceolate, flat, margins ciliate (sometimes sparingly so), surfaces usually glabrous, abaxial surface rarely hispidulous; **upper blades** greatly reduced. **Panicles** 1.3–7 cm, flabellate or contracted, with 12–20 rames; **rames** 1–6(11) cm. **Sessile spikelets: glumes** 3–5.5 mm, lanceolate; **lower glumes** several-veined; **upper glumes** 1- or 3-veined; **awns** 0.3–9 mm, included or exserted, usually twisted below, sometimes geniculate at midlength; **anthers** usually 2, 0.5–0.7 mm. **Pedicels** absent or to 2 mm. **Pedicellate spikelets** absent. $2n = 36$.

Arthraxon hispidus is native to Asia, but is naturalized and spreading along roadsides, shores, ditches and in low woods and fields of the eastern United States. It is also naturalized in Mexico, Central America, and the West Indies. Plants in the *Flora* region belong to **A. hispidus** (Thunb.) Makino var. **hispidus**, the most widespread and variable of the four varieties. **Arthraxon castratus** (Griff.) V. Naray. *ex* Bor reported from Puerto Rico, differs from *A. hispidus* in having pilose lemma margins, a palea in its second floret, and three anthers.

26.19 HYPARRHENIA Andersson *ex* E. Fourn.

Mary E. Barkworth

Plants annual or perennial; cespitose, often with short rhizomes. **Culms** 30–350(400) cm, usually erect, much branched above the bases. **Leaves** not aromatic; **ligules** membranous, not ciliate; **blades** usually flat or folded. **Inflorescences** false panicles with numerous inflorescence units; **peduncles** with 2 rames in digitate clusters; **rames** with naked, often deflexed bases, axes without a translucent median groove; **disarticulation** in the rames, beneath the bisexual spikelets. **Spikelets** in sessile-pedicellate pairs, basal 1–2 pairs on each rame homogamous, morphologically similar to the heterogamous pairs, staminate or sterile, unawned, not forming an involucre, tardily deciduous, remaining pairs heterogamous. **Heterogamous spikelet units: sessile spikelets** dorsally compressed or subterete; **calluses** blunt to sharp, strigose; **glumes** equal, pubescent; **lower glumes** coriaceous, rounded, without keels, truncate to slightly bilobed; **upper glumes** narrower, shallowly keeled; **lower florets** sterile, reduced; **upper florets** bisexual, awned from between the teeth of the bifid lemma; **awns** usually present, to 3.5(19) cm, pubescent on the lower portion. **Caryopses** oblong, subterete. **Pedicels** slender, not adnate to the rame axes. **Pedicellate spikelets** usually slightly longer than the sessile spikelets, staminate or sterile, usually unawned, lower glumes sometimes aristulate. $x = 10, 15$. Name from the Greek *hypo*, 'under', and *arrhen*, 'masculine', referring to the pair of staminate spikelets at the base of the rames of some species.

Hyparrhenia is a genus of approximately 55 mostly African species. Two have been introduced into the *Flora* region, but only one is known to be established. Clayton (1969) provides a detailed discussion of the structure of the inflorescence.

SELECTED REFERENCE Clayton, W.D. 1969. A Revision of the Genus *Hyparrhenia*. Kew Bull., Addit. Ser. 2. Her Majesty's Stationery Office, London, England. 196 pp.

1. Spikelets with whitish to dark yellow hairs . . . . 1. *H. hirta*
1. Spikelets with reddish hairs . . . . 2. *H. rufa*

1. Hyparrhenia hirta (L.) Stapf [p. 679]
THATCHING GRASS

Plants perennial; cespitose but with short rhizomes. **Culms** 30–100 cm. **Sheaths** glabrous; **blades** 2–40 cm long, 1–3(4) mm wide. **Peduncles** 5–10 cm; **rames** 1–3.5(4) cm, 1 almost sessile, the other with a 5–10 mm base, both with 8–14 heterogamous spikelet pairs. **Glumes** of all spikelets densely pubescent, hairs to 0.3 mm, white to dark yellow. **Sessile spikelets of homogamous pairs** 4.9–5.6 mm; **sessile spikelets of heterogamous pairs** 4–4.5 mm; **lemmas** awned, awns 1–3.5 cm. **Pedicellate spikelets** 4.8–6.5 mm. $2n = 30, 44, 45$.

Hyparrhenia hirta is native to southern Africa, where it grows on stony soils and is sometimes used for thatching. It has been cultivated in Texas and Florida, but is not currently known to be established in the *Flora* region. A report of its occurrence in Los Angeles County, California, has not been verified.

2. Hyparrhenia rufa (Nees) Stapf [p. 679]
JARAGUA GRASS

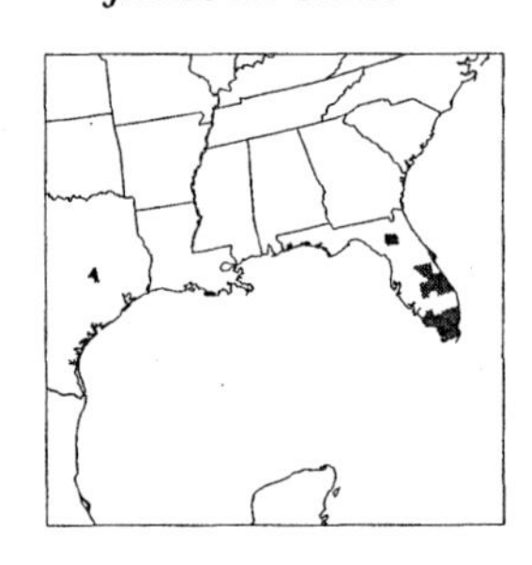

Plants usually perennial; cespitose but with short rhizomes. **Culms** 30–350 cm. **Sheaths** glabrous; **blades** 30–60 cm long, 2–8 mm wide. **Peduncles** 0.7–7 cm; **rames** 1.5–2.5 cm, 1 almost sessile, the other with a 6–10 mm stalk, both with 7–14 heterogamous spikelet pairs. **Glumes** of all spikelets moderately densely pubescent, hairs reddish. **Sessile spikelets of homogamous pairs** 3–5.5 mm, **sessile spikelets of heterogamous pairs** 3.2–4.2 mm; **lemmas** awned, awns 2–3 cm. **Pedicellate spikelets** 3–5 mm. $2n = 30, 36, 40$.

Hyparrhenia rufa is native to the Eastern Hemisphere tropics, but is now established in tropical America. It grows in ditches, pastures, swamps, and pine flatwoods, and along roadsides, in the southeastern United States.

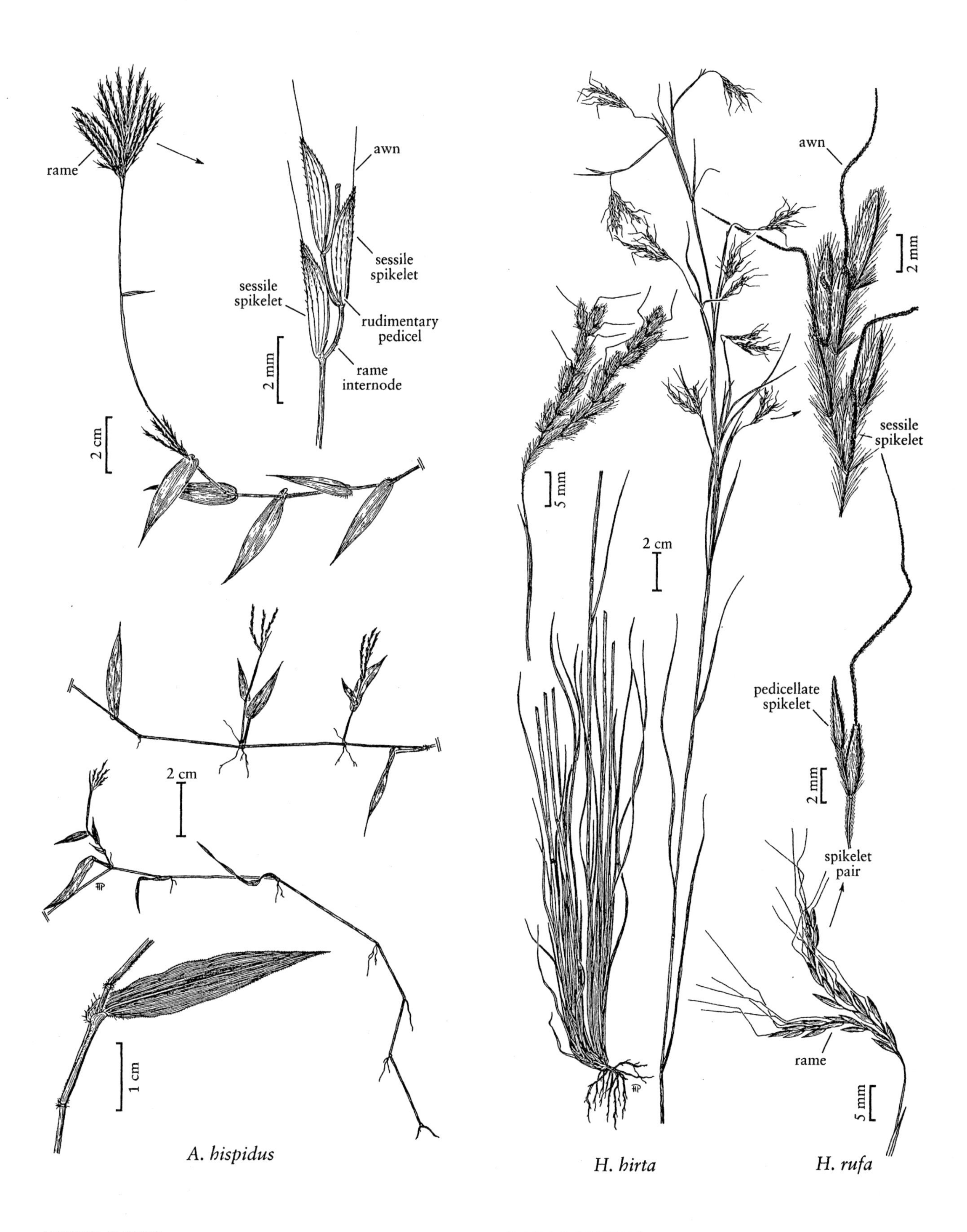

ARTHRAXON

HYPARRHENIA

26.20 HETEROPOGON Pers.

Mary E. Barkworth

Plants annual or perennial; cespitose. **Culms** 20–200 cm, simple or branched. **Leaves** sometimes aromatic and smelling of lemon oil or citronella; **sheaths** keeled, sometimes with a row of glandular depressions on the keel; **ligules** membranous, glabrous or ciliate. **Inflorescences** terminal and axillary; **peduncles** usually with 1 rame, sometimes with several in a digitate cluster; **rames** with 3–10 homogamous, unawned, sessile-pedicellate spikelet pairs on the lower ¼–⅔ and heterogamous, awned, sessile-pedicellate spikelet pairs distally, axes slender, without a translucent median groove; **disarticulation** in the rames, beneath the sessile spikelets of the heterogamous spikelet pairs, sometimes also below their pedicellate spikelets. **Homogamous spikelet units** sterile or staminate; **calluses** poorly developed; **glumes** membranous, many-veined, keels winged above. **Heterogamous spikelet units: sessile spikelets** bisexual, terete; **calluses** 1.5–3 mm, sharp, antrorsely strigose, hairs golden brown; **glumes** coriaceous, pubescent, concealing the florets; **lower glumes** enclosing the upper glumes, obscurely 5–9-veined; **upper glumes** sulcate, 3-veined; **lower florets** sterile, reduced to a hyaline lemma; **upper florets** bisexual, lemmas with conspicuous, geniculate awns; **awns** 5–15 cm, with hairs. **Caryopses** lanceolate, sulcate on 1 side. **Pedicels** short, free of the rame axes, not grooved; **pedicellate spikelets** sterile or staminate, larger than the sessile spikelets; **calluses** long, glabrous, functioning as pedicels; **glumes** membranous, many-veined, keels winged above. x = 10, 11. Name from the Greek *heteros*, 'different', and *pogon*, 'beard', alluding to the difference between the calluses of the spikelets in the heterogamous pairs.

Heteropogon is a pantropical genus of eight to ten species. Two species grow in the *Flora* region; probably both are introduced. Many grow well on poor soils.

1. Glumes of the pedicellate spikelets of the heterogamous spikelet units without glandular pits; plants perennial . . . . 1. *H. contortus*
1. Glumes of the pedicellate spikelets of the heterogamous spikelet units with a row of glandular pits along the midvein; plants annual . . . . 2. *H. melanocarpus*

1. **Heteropogon contortus** (L.) P. Beauv. *ex* Roem. & Schult. [p. 681]
Tanglehead

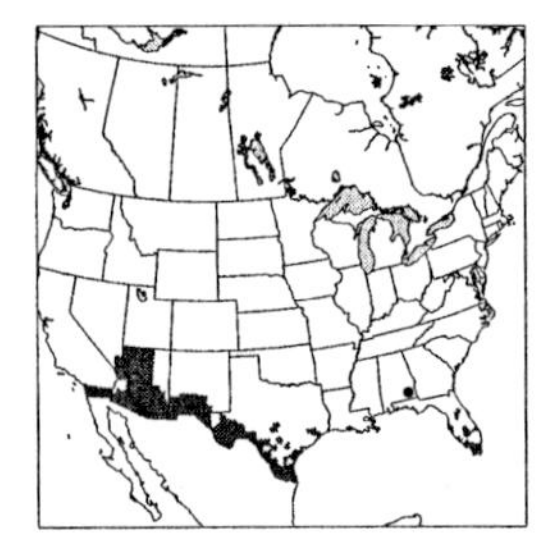

Plants perennial. **Culms** 20–150 cm, erect. **Sheaths** smooth, reddish; **ligules** 0.5–0.8 mm, cilia 0.2–0.5 mm; **blades** 10–15 cm long, 2–7 mm wide, flat or folded, glabrous or pubescent. **Rames** 3–7 cm, secund, with 12–22, brown to reddish-brown, sessile-pedicellate spikelet pairs. **Homogamous spikelets** 6–10 mm. **Heterogamous spikelets: sessile spikelets** 5–10 mm, brown, awned; **calluses** 1.8–2 mm, strigose; **awns** 6–10 cm; **pedicellate spikelets** 6–10 mm, unawned; **glumes** ovate-lanceolate, glabrous or with papillose-based hairs distally, without glandular pits, greenish to purplish-brown, becoming stramineous when dry. $2n$ = 40, 50, 60.

Heteropogon contortus grows on rocky hills and canyons in the southern United States into Mexico, and worldwide in subtropical and tropical areas, occupying a variety of different habitats, including disturbed habitats. It is probably native to the eastern hemisphere but is now found in tropical and subtropical areas throughout the world.

Heteropogon contortus is a valuable forage grass if continuously grazed so as to prevent the calluses from developing. It is also considered a weed, being able to establish itself in newly disturbed and poor soils.

2. **Heteropogon melanocarpus** (Elliott) Benth. [p. 681]
Sweet Tanglehead

Plants annual. **Culms** 50–200 cm, often with prop roots, freely branching above the base. **Sheaths** glabrous, with a row of glandular depressions along the keel; **ligules** 2–4 mm, erose to lacerate, glabrous; **blades** 30–50 cm long, 3–12 mm wide, usually folded, abaxial surfaces with dark glandular depressions along the keel, adaxial surfaces with scattered papillose-based hairs near the base, scabrous elsewhere. **Rames** 2.5–6.5 cm. **Homogamous spikelets** 10–14 mm, green; **lower glumes** glabrous, unawned. **Heterogamous spikelets: sessile spikelets** 8–11.5 mm, dark brown, awned; **calluses** about 3 mm; **awns** 10–15 cm; **pedicellate spikelets** 16–21 mm, unawned; **lower glumes** scabrous or sparsely ciliate distally, midveins glandular, pitted. $2n = 20$.

Heteropogon melanocarpus is probably native to the Eastern Hemisphere, but is now found in tropical regions throughout the world. It grows in pine woods, fields, and disturbed areas of the southern United States. When fresh, plants of *H. melanocarpus* smell like citronella oil.

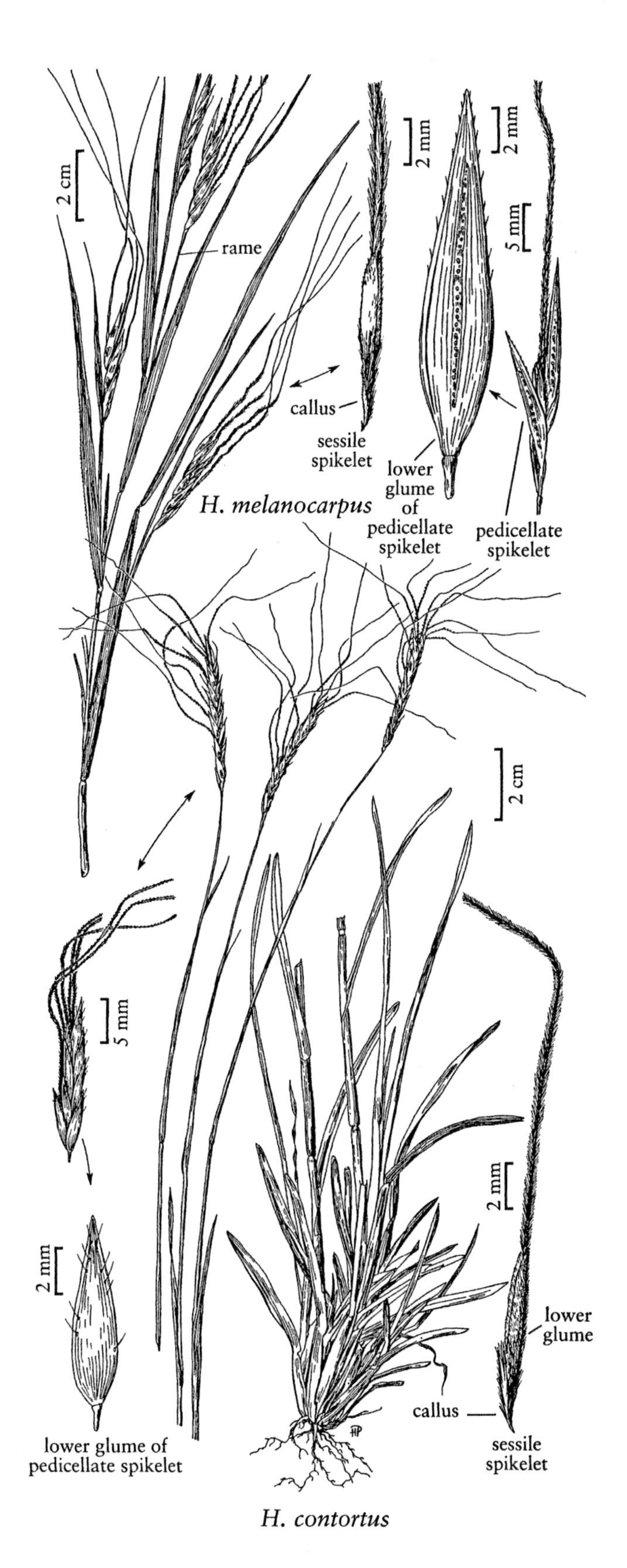

HETEROPOGON

26.21 THEMEDA Forssk.

Mary E. Barkworth

Plants annual or perennial; usually cespitose, rarely stoloniferous. **Culms** 30–310 cm, erect. **Leaves** not aromatic; **sheaths** open; **ligules** membranous, sometimes ciliate. **Inflorescences** numerous, terminal and axillary, false panicles; **peduncles** shorter than the subtending sheaths, with 1–8 rames; **rames** spikelet-bearing to the base, axes slender, without a longitudinal translucent groove, with 2 large, solitary, homogamous sessile-pedicellate spikelet pairs at the base and 1–4 smaller, heterogamous sessile-pedicellate spikelet pairs distally, terminal or only unit sometimes a triplet with 1 sessile and 2 pedicellate spikelets; **disarticulation** in the rames below the sessile spikelets of the heterogamous spikelet units, occasionally beneath the homogamous spikelets. **Homogamous spikelet pairs** distinctive, forming an involucre around the rame bases, separated by internodes less than ½ as long as the spikelets; **spikelets** subequal, strongly compressed dorsally, staminate or sterile, unawned; **lower glumes** membranous, 2-keeled. **Heterogamous spikelet pairs: sessile spikelets** subterete or dorsally compressed, awned; **calluses** bearded, usually sharp; **glumes** coriaceous; **lower glumes** wrapped around and concealing the upper glumes, obscurely veined but not keeled, truncate; **upper glumes** sulcate, with thin margins; **lower florets** highly reduced, sterile; **upper florets** bisexual, upper lemmas usually terminating in a geniculate awn; **awns** usually present, to 9 cm, puberulent to pubescent, sometimes absent. **Pedicels** slender, not sulcate, not fused to the rame axes; **pedicellate spikelets** similar to the homogamous spikelets except narrower, staminate or sterile, and unawned. **Caryopses** narrowly ovate or linear, subterete or channeled on 1 side. $x = 5, 10$. Name from the Arabic *thaemed*, a depression where water collects after rain and later evaporates, referring to the habitat of some species of this genus.

Themeda is a genus of approximately 18 species, all of which are native to tropical and subtropical regions of the Eastern Hemisphere, primarily southeast Asia. One species has been established in the *Flora* region, one is grown as an ornamental, and a third was introduced but is not known to be established.

SELECTED REFERENCES Reed, C.F. 1964. A flora of the chrome and manganese ore piles at Canton, in the Port of Baltimore, Maryland and at Newport News, Virginia, with descriptions of genera and species new to the flora of the eastern United States. Phytologia 10:321–405; Simon, B.K. 1992. *Themeda*. Pp. 1227–1228 *in* J.R. Wheeler (ed.). Flora of the Kimberley Region. Department of Conservation and Land Management, Western Australian Herbarium, Como, Western Australia, Australia. 1327 pp.; Towne, E.G. and I. Barnard. 2000. *Themeda quadrivalvis* (Poaceae: Andropogoneae) in Kansas: An exotic plant introduced from birdseed. Sida 19:201–203.

1. Awns 7–9 cm long; calluses 3–4 mm long .... 1. *T. arguens*
1. Awns 2.5–7 cm long; calluses 0.5–4 mm long.
   2. Plants annual; homogamous spikelets 4–7 mm long; sessile spikelets of heterogamous pairs4–6 mm long .... 2. *T. quadrivalvis*
   2. Plants perennial; homogamous spikelets 6–14 mm long; sessile spikelets of heterogamous pairs 6–14 mm long .... 3. *T. triandra*

1. Themeda arguens (L.) Hack. [p. 683]
CHRISTMAS GRASS

Plants annual. **Culms** 1–3 m, erect, or geniculate and ascending. **Sheaths** glabrous, not flattened at the base. **False panicles** 20–100 cm; **sheaths subtending the rame clusters** 35–40 mm, pilose, hairs papillose-based; **rames** 3–5 per cluster, with 2–3 heterogamous spikelets. **Homogamous spikelets** 8–10 mm; **lower glumes** glabrous, acuminate. **Heterogamous sessile spikelets** 9–11 mm; **calluses** 3–4 mm; **lower glumes** hispid and tuberculate-scabrous, dark brown; **awns** 7–9 cm; **pedicellate spikelets** 8–10 mm. $2n = 20$.

Themeda arguens is native to northern Australia and southeastern Asia. Reed (1964) reported finding it on ore piles in Maryland and Virginia.

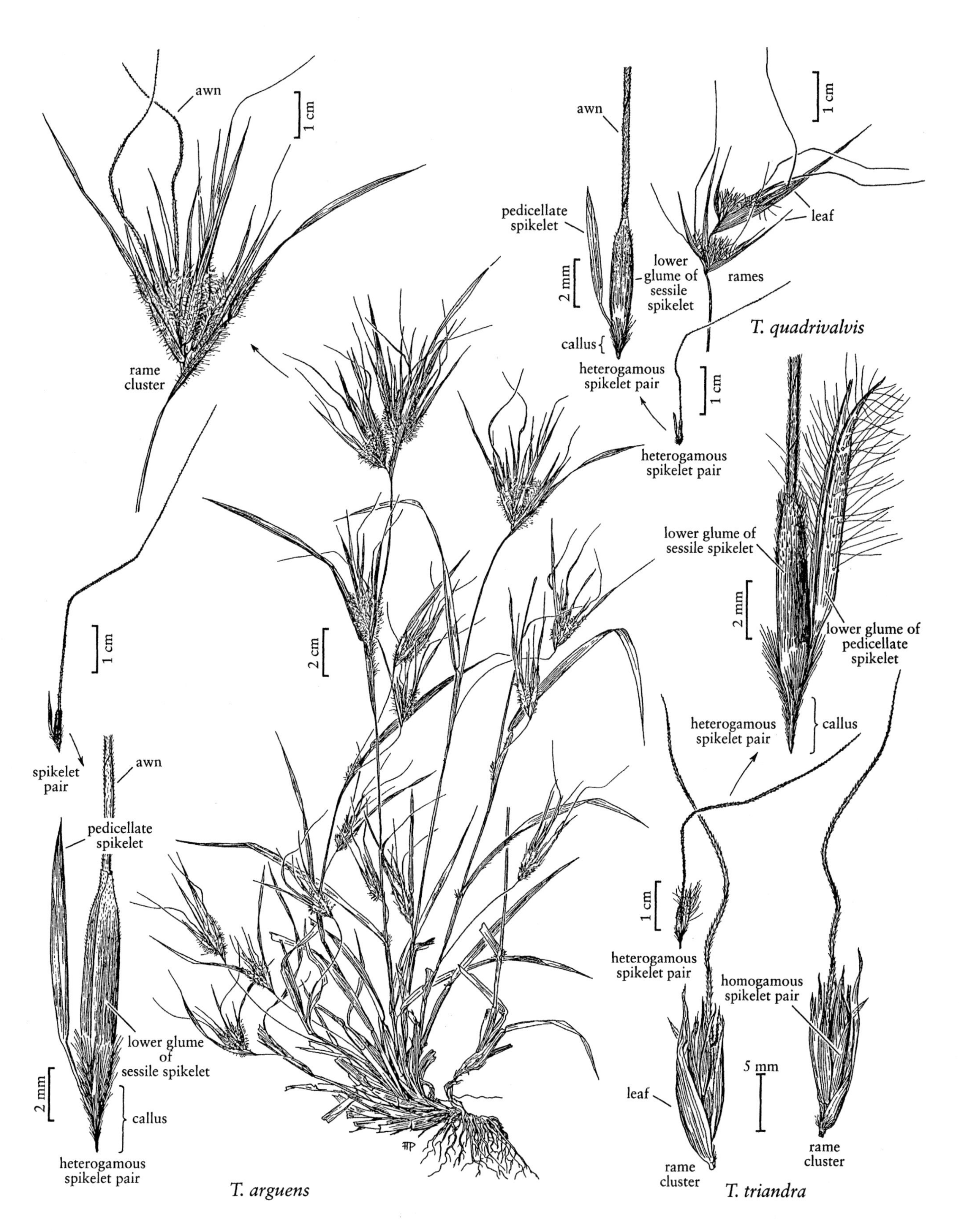
awn
1 cm
rame cluster
1 cm
spikelet pair
awn
pedicellate spikelet
lower glume of sessile spikelet
2 mm
callus
heterogamous spikelet pair
2 cm
T. arguens
awn
pedicellate spikelet
2 mm
lower glume of sessile spikelet
callus
heterogamous spikelet pair
1 cm
heterogamous spikelet pair
1 cm
leaf
rames
T. quadrivalvis
lower glume of sessile spikelet
2 mm
lower glume of pedicellate spikelet
heterogamous spikelet pair
callus
1 cm
heterogamous spikelet pair
homogamous spikelet pair
5 mm
leaf
rame cluster
rame cluster
T. triandra

THEMEDA

2. **Themeda quadrivalvis** (L.) Kuntze [p. 683]
KANGAROO GRASS

Plants annual. **Culms** to 2 m, glabrous. **Sheaths** glabrous, flattened at the base; **ligules** 1–2.5 mm; **blades** to 60 cm long, 1–6 mm wide, usually folded. **False panicles** to 130 cm; **sheaths subtending the rame clusters** 17–50 mm, distal sheaths shorter and more strongly keeled, margins tuberculate; **rames** 1–3 per cluster, 8–10 mm, with 1–2 heterogamous spikelet pairs. **Homogamous spikelets** 4–7 mm; **lower glumes** many-veined, hairy distally, hairs papillose-based; **upper glumes** subequal to the lower glumes, 3-veined. **Sessile heterogamous spikelets** 4–6 mm; **calluses** 0.5–3 mm; **lower glumes** glabrous or sparsely hirsute; **upper glumes** 4.5–5.5 mm; **awns** 4–5 cm; **pedicellate spikelets** 4.5–5.5 mm, sterile. $2n = 18$.

A native of Malaysia, *Themeda quadrivalvis* has been found at scattered locations in the contiguous United States. Towne and Ballard (2002) reported that it is a common contaminant of the thistle seed sold for bird feeders. Most of the seeds are sterile, but a few are not. So far as is known, the species is not established in the *Flora* region.

3. **Themeda triandra** Forssk. [p. 683]
ROOIGRAS

Plants perennial. **Culms** 30–300 cm, erect. **Blades** to 30 cm long, 1–8 mm wide. **False panicles** 20–50 cm; **sheaths subtending the rame clusters** 1.5–3.5 cm, reddish, glabrous or pilose, hairs papillose-based; **rames** 15–20 mm, with 1 heterogamous spikelet pair. **Homogamous spikelets** 6–14 mm, glabrous or with papillose-based hairs. **Sessile heterogamous spikelets** 6–14 mm; **calluses** 2–4 mm, with rufous hairs; **lower glumes** mostly smooth, glabrous, apices appressed-pubescent; **awns** 2.5–7 cm; **pedicellate spikelets** 6–14 mm, glabrous or with papillose-based hairs. $2n = 20, 21, 22, 24, 40, 60, 70, 80$.

Themeda triandra is native to India, Korea, China, and Japan. Many infraspecific taxa have been described in *T. triandra* and in *T. australis*, a species that is sometimes included in *T. triandra. Themeda triandra* subsp. *japonica* (Willd.) T. Koyama is sold as an ornamental in the *Flora* region but, in the absence of a treatment that covers both species throughout their range, it is impossible to state how it differs from other members of the complex.

26.22 ELIONURUS Humb. & Bonpl. *ex* Willd.

Mary E. Barkworth

Plants perennial, occasionally annual; cespitose, sometimes with short rhizomes. **Culms** 10–150 cm, erect, sometimes branching above the base. **Leaves** sometimes aromatic; **sheaths** without glandular pits; **ligules** shortly membranous and densely ciliate or of hairs; **blades** involute, flat, or folded. **Inflorescences** terminal, sometimes also axillary, composed of solitary, flexuous rames; **rame internodes** columnar to clavate, apices strongly oblique, not hollowed or rimmed; **disarticulation** in the rames, below the sessile spikelets. **Spikelets** in sessile-pedicellate pairs. **Sessile spikelets** dorsally compressed; **calluses** blunt, sometimes resembling a short pedicel; **lower glumes** enclosing the upper glumes, subcoriaceous, 2-keeled, keels prominently ciliate, intercarinal surface smooth, apices cuspidate to bilobed, rarely entire; **lower florets** reduced, sterile; **upper florets** bisexual, unawned. **Pedicels** stout, appressed but not fused to the rame axes, pubescent or ciliate on the angles. **Pedicellate spikelets** 3–8 mm, about equal to the sessile spikelets, staminate, muticous to awn-tipped. $x = 5$. Name from the Greek *eluein*, 'mouse', and *oura*, 'tail', alluding to the narrowly cylindrical inflorescence.

Elionurus has 15 species. Most of the species are native to tropical Africa and America; one species is Australian. Several of the species are considered important elements of native pastures, but neither of the two species native to the *Flora* region is ever sufficiently abundant to be important in this regard.

1. Lower glumes densely pilose; pedicels pilose dorsally; culms antrorsely hirsute below the nodes . . . . 1. *E. barbiculmis*
1. Lower glumes glabrous or nearly so; pedicels ciliate on the angles, usually glabrous elsewhere; culms glabrous throughout . . . . 2. *E. tripsacoides*

1. **Elionurus barbiculmis** Hack. [p. 686]
WOOLYSPIKE BALSAMSCALE

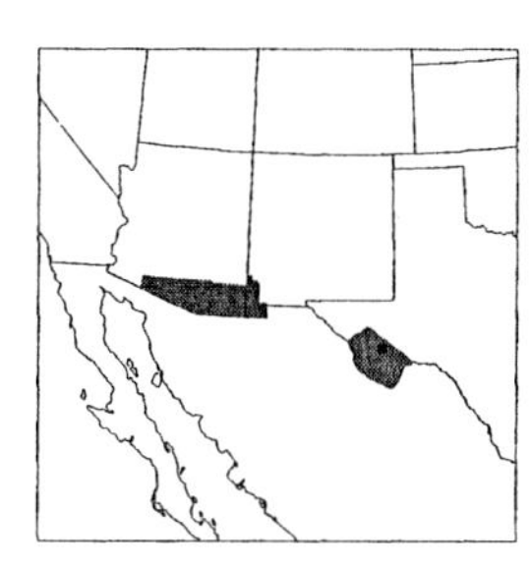

Plants cespitose. **Culms** 40–60 cm, erect, usually unbranched, densely antrorsely hirsute beneath the nodes. **Sheaths** mostly glabrous, often ciliate on the margins, particularly at the throat; **ligules** with 1–2 mm hairs; **blades** 15–30 cm long, 1–2(4) mm wide, usually involute, abaxial surfaces with scattered long hairs adjacent to the margins, adaxial surfaces usually densely pilose. **Rames** 5–10 cm, internodes densely villous. **Sessile spikelets** 4.5–8 mm; **calluses** about 0.5 mm, hirsute; **lower glumes** densely hirsute, acuminate, bifid, teeth 1.5–2.5 mm; **pedicels** densely pilose dorsally. **Pedicellate spikelets** with densely pilose lower glumes. $2n = 20$.

Elionurus barbiculmis grows on mesas, rocky slopes, hills, and in canyons, usually above 1200 m. Its range extends from southern Arizona and southwestern Texas into northern Mexico.

2. **Elionurus tripsacoides** Humb. & Bonpl. *ex* Willd. [p. 686]
PAN-AMERICAN BALSAMSCALE

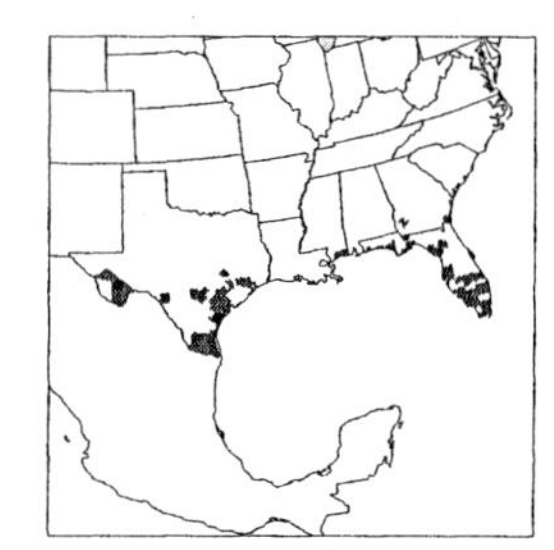

Plants cespitose; with short, knotty, rhizomatous bases. **Culms** 60–120 cm, glabrous throughout. **Sheaths** glabrous or pilose on the margins; **ligules** 0.5–1 mm, membranous, shortly ciliate; **blades** 16–30 cm long, 2–4 mm wide, adaxial surfaces with hairs to 5 mm basally, glabrous distally, margins ciliate near the bases. **Rames** 6–15 cm; **internodes** pilose. **Sessile spikelets** 6–8 mm; **calluses** about 1 mm, antrorsely hirsute; **lower glumes** usually mostly glabrous, rarely sparsely pilose dorsally, keels ciliate distally, apices acuminate, bidentate, teeth about 0.5 mm; **pedicels** hispid on the margins, usually glabrous elsewhere. **Pedicellate spikelets** similar to the sessile spikelets. $2n = 20$.

Elionurus tripsacoides grows in moist pine woods and low prairies around southern Texas and the Gulf coast to Georgia, and south through Mexico and Central America to Argentina.

26.23 HEMARTHRIA R. Br.

Charles M. Allen

Plants perennial. **Culms** to 150 cm, erect or decumbent, rooting at the nodes, usually branched above the bases. **Leaves** not aromatic; **sheaths** mostly glabrous, sometimes ciliate near the base; **ligules** membranous, ciliate; **blades** usually linear-lanceolate, sometimes linear. **Inflorescences** terminal and axillary, with 1(2) flattened rames borne on a common peduncle, spikelets partially embedded in the rame axes; **disarticulation** in the rames, usually oblique and often tardy. **Spikelets** in heterogamous sessile-pedicellate pairs, dorsally compressed. **Sessile spikelets** with 2 florets; **calluses** blunt; **lower glumes** coriaceous, smooth; **upper glumes** equaling the lower glumes, chartaceous to membranous, sometimes partially adnate to the rame axes, sometimes awned; **lower florets** reduced to hyaline lemmas; **upper florets** bisexual, lemmas unawned. **Pedicels** thick, fused to the rame axes. **Pedicellate spikelets** morphologically similar to the sessile spikelets, staminate or sterile. $x = 9, 10$. Name from the Greek *hemi*, 'half', and *arthron*, 'segment', because of the tardily disarticulating rames.

Hemarthria is a genus of 12 species, native to the tropics and subtropics of the Eastern Hemisphere, and possibly to the Western Hemisphere. All the species grow in or near water. One species has been introduced into the *Flora* region.

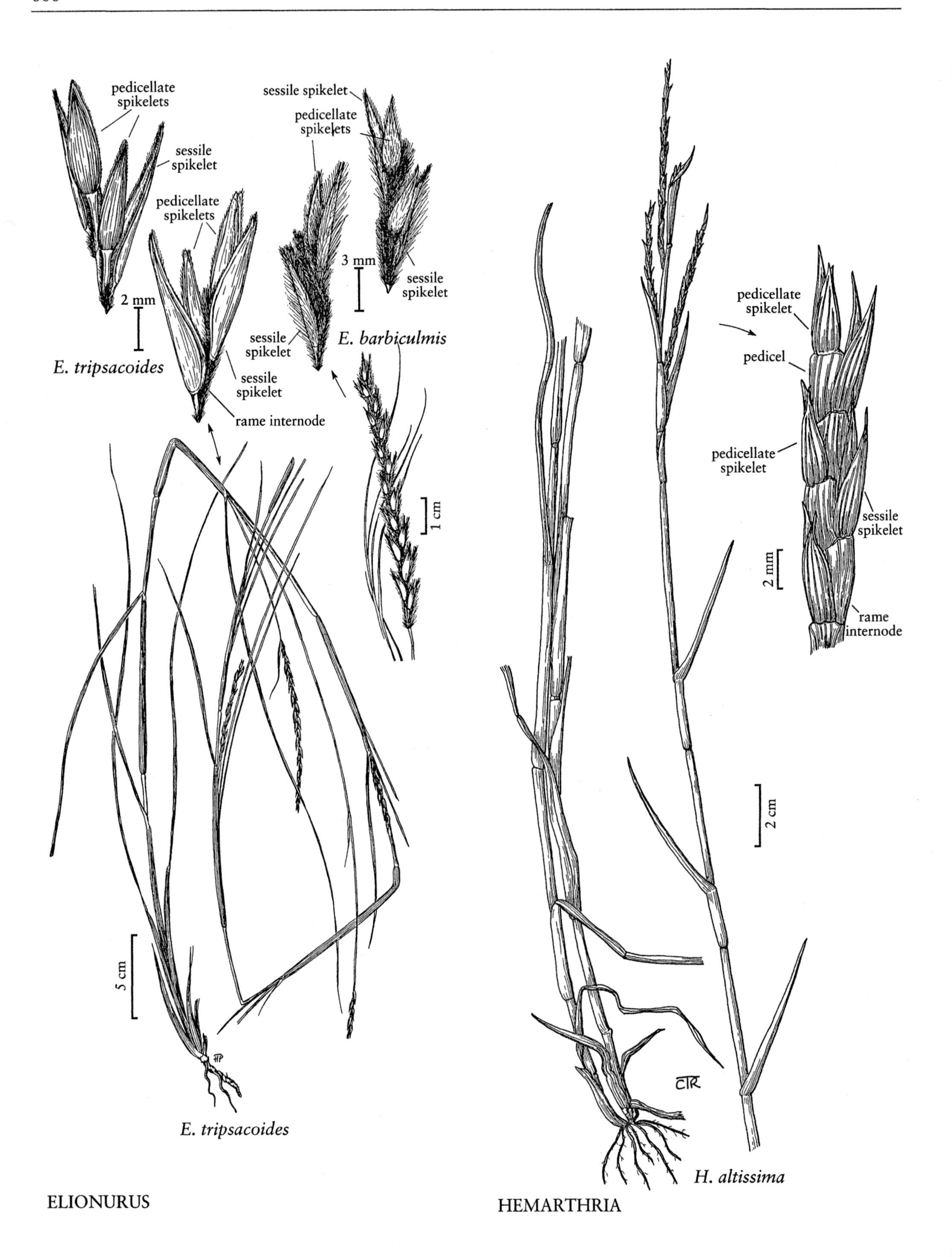
pedicellate spikelets
sessile spikelet
pedicellate spikelets
2 mm
E. tripsacoides
sessile spikelet
rame internode
sessile spikelet
pedicellate spikelets
3 mm
sessile spikelet
sessile spikelet
E. barbiculmis
1 cm
5 cm
E. tripsacoides
ELIONURUS
pedicellate spikelet
pedicel
pedicellate spikelet
sessile spikelet
2 mm
rame internode
2 cm
H. altissima
HEMARTHRIA

1. **Hemarthria altissima** (Poir.) Stapf & C.E. Hubb. [p. 686]

Plants perennial; rhizomatous and/or stoloniferous. **Culms** 30–150 cm, erect to ascending, flattened. **Leaves** basal and cauline; **sheaths** mostly glabrous, margins sparsely ciliate basally, scabrous distally; **ligules** 0.2–1 mm; **blades** flat to conduplicate, glabrous, margins ciliate basally. **Rames** 2–10 cm, erect. **Sessile spikelets: lower glumes** 4–5 mm, 10–15-veined, lateral veins distinct, margins scarious, apices acute; **upper glumes** smooth, hyaline to membranous, acute; **anthers** 3. **Pedicels** 4–5 mm. **Pedicellate spikelets** 4–7 mm, acuminate. $2n$ = 20, 36.

Hemarthria altissima grows in tropical and subtropical regions throughout the world, including southern Texas and Florida. It is considered native to the Mediterranean region. Although an excellent forage grass, it is not sufficiently abundant in the *Flora* region to be important in this regard.

26.24 COELORACHIS Brongn.

Charles M. Allen

Plants perennial; cespitose or rhizomatous. **Culms** 60–400 cm, erect. **Leaves** not aromatic; basal and cauline; **sheaths** open, glabrous, margins scarious; **auricles** lacking; **ligules** membranous, ciliate; **blades** flat to conduplicate, glabrous or sparsely pubescent, margins scarious, sometimes scabrous. **Inflorescences** terminal and axillary, composed of a solitary, pedunculate rame; **rames** stout; **disarticulation** in the rames, below the sessile spikelets. **Spikelets** dorsally compressed, in heterogamous sessile-pedicellate pairs. **Sessile spikelets** embedded in the rame axes, ovate, with 2 florets, unawned; **lower glumes** indurate, smooth, rugose, or pitted, 7–11-veined, not keeled; **upper glumes** coriaceous, keeled, 1-veined; **lower florets** sterile; **upper florets** bisexual, unawned; **anthers** 3. **Pedicels** short, thick, appressed or partly fused to the side of the rame axes. **Pedicellate spikelets** 1–3 mm, usually reduced. **Caryopses** ellipsoid to broadly ellipsoid, yellow. x = 9. Name from the Greek *koilos*, 'hollow', and *rachis*, 'axis', in reference to the axes of the inflorescence, which are concave.

Coelorachis is a tropical genus of approximately 20 species; four are native to the southeastern United States. Most species tend to favor damp soils. Veldkamp et al. (1986) recommended combining *Coelorachis* and *Hackelochloa* with some other small genera in *Mnesithea* Kunth, but these two seem to be sufficiently distinct to be maintained until more data are available.

SELECTED REFERENCE Veldkamp, J.F., R. de Koning, and M.S.M. Sosef. 1986. Generic delimitation of *Rottboellia* and related genera (Gramineae). Blumea 31:281–307.

1. Culms and sheaths terete; lower glumes of the sessile spikelets with circular pits on the sides, the central region initially smooth, usually developing rectangular pits at maturity, occasionally remaining smooth . . . . 1. *C. cylindrica*
1. Culms and sheaths compressed-keeled; lower glumes of the sessile spikelets transversely rugose, rectangular-pitted, or smooth.
   2. Lower glumes of the sessile spikelets rectangular-pitted . . . . 2. *C. tessellata*
   2. Lower glumes of the sessile spikelets transversely rugose or smooth.
      3. Lower glumes of the sessile spikelets distinctly transversely rugose; rachises distinctly indented below the sessile spikelets . . . . 3. *C. rugosa*
      3. Lower glumes of the sessile spikelets smooth to slightly transversely rugose; rachises not, or only slightly, indented below the sessile spikelets . . . . 4. *C. tuberculosa*

1. **Coelorachis cylindrica** (Michx.) Nash [p. 689]
CAROLINA JOINTGRASS

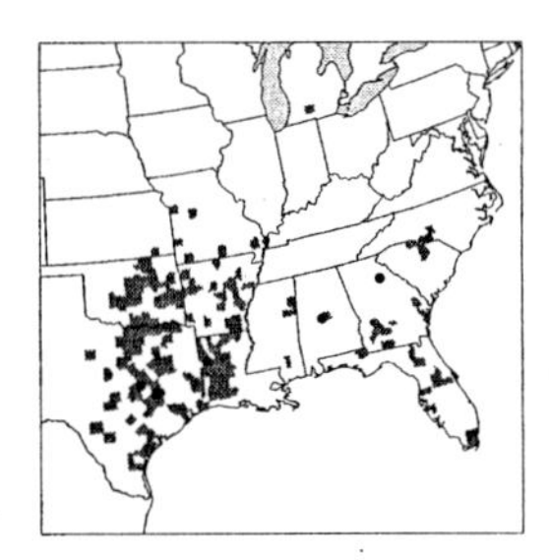

Plants shortly rhizomatous. **Culms** 60–120 cm, terete. **Sheaths** terete; **ligules** 0.2–0.8 mm. **Rames** 6.5–12.5 cm, often purple. **Sessile spikelets** 5–5.5 mm; **lower glumes** with circular pits on the sides, the central region initially smooth, usually developing rectangular pits at maturity, occasionally remaining smooth; **upper lemmas** and **paleas** 4–4.5 mm. **Pedicellate spikelets** 1–2 mm. **Caryopses** about 2.2 mm. $2n = 18$.

Coelorachis cylindrica is native to the southeastern United States, where it grows in tallgrass prairies, the edges of forests, and roadsides. The specimen from Michigan was found in an old field, in association with many native species. Its source is unknown.

2. **Coelorachis tessellata** (Steud.) Nash [p. 689]
PITTED JOINTGRASS

Plants cespitose. **Culms** 80–120 cm, compressed-keeled. **Sheaths** compressed-keeled; **blades** to 41 cm long, to 7.8 mm wide, folded to flat, scabrous above. **Rames** 4.5–7(12) cm. **Sessile spikelets** 3.9–6.2 mm long, 2.1–2.4 mm wide; **lower glumes** with rectangular pits, keels narrowly winged distally. **Pedicellate spikelets** 2.3–2.7 mm, reduced to scales. $2n$ = unknown.

Coelorachis tessellata is endemic to the southern coastal plain of the United States, extending from Louisiana to northern Florida, although it is rare in Florida. It grows in bogs and moist pine woods, especially flatwoods.

3. **Coelorachis rugosa** (Nutt.) Nash [p. 689]
WRINKLED JOINTGRASS

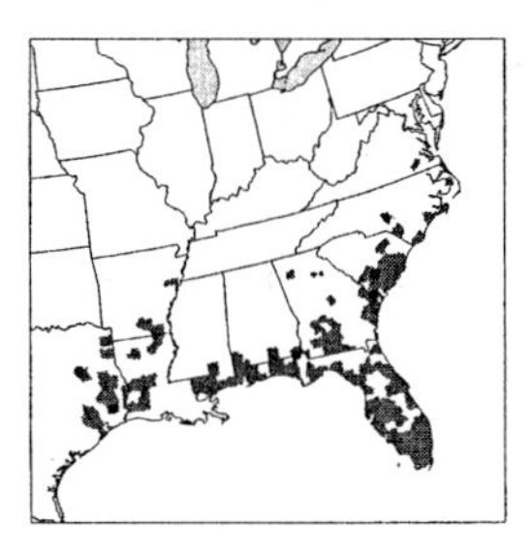

Plants cespitose. **Culms** 60–120 cm, compressed-keeled. **Sheaths** compressed-keeled; **ligules** 0.5–1 mm. **Rames** 3–9.5 cm; **rachises** distinctly indented adjacent to the sessile spikelets. **Sessile spikelets** 3–4 mm; **lower glumes** transversely rugose; **upper lemmas** and **paleas** 2–3 mm. **Pedicellate spikelets** 1–3 mm. **Caryopses** about 2 mm, broadly ellipsoid. $2n$ = unknown.

Coelorachis rugosa is endemic to the southeastern United States. It grows in moist to wet areas in prairies, bogs, and pine woods, especially flatwoods and savannahs.

4. **Coelorachis tuberculosa** (Nash) Nash [p. 689]
SMOOTH JOINTGRASS

Plants cespitose. **Culms** 60–120 cm, compressed-keeled. **Sheaths** compressed-keeled, glabrous; **blades** to 31 cm long, to 7.8 mm wide, folded or flat, glabrous. **Rames** 4–8 cm; **rachises** not or only slightly indented adjacent to the sessile spikelets. **Sessile spikelets** 3.3–4.3 mm long, 1.3–2 mm wide; **lower glumes** smooth or sparsely and shallowly transversely rugose, keels narrowly winged. **Pedicellate spikelets** 1.9–2.6 mm, reduced to scales. $2n$ = unknown.

Coelorachis tuberculosa is an uncommon species, endemic to the southeastern United States. It grows in moist to wet areas such as bogs and pine woods, especially flatwoods and savannahs.

26.25 **EREMOCHLOA** Büse

John W. Thieret

Plants perennial; cespitose, sometimes stoloniferous. **Culms** 10–70 cm, herbaceous, erect to decumbent, basal branching extravaginal. **Leaves** not aromatic; **sheaths** open; **auricles** absent; **ligules** membranous, truncate; **blades** flaccid, linear to lanceolate. **Inflorescences** terminal, of long-pedunculate, solitary, 1-sided rames (axillary rames occasionally present) with more than 1 spikelet unit; **rame internodes** clavate, glabrous; **disarticulation** in the rames. **Spikelets** in heteromorphic sessile-pedicellate pairs, pedicellate spikelets absent or rudimentary. **Sessile spikelets** imbricate, not embedded in the rame axes, dorsally compressed, with 2 florets, unawned; **calluses** truncate; **glumes** exceeding the florets, differing in shape; **lower glumes** 4–9-veined, 2-keeled, keels with spinelike projections and often winged distally, smooth between the

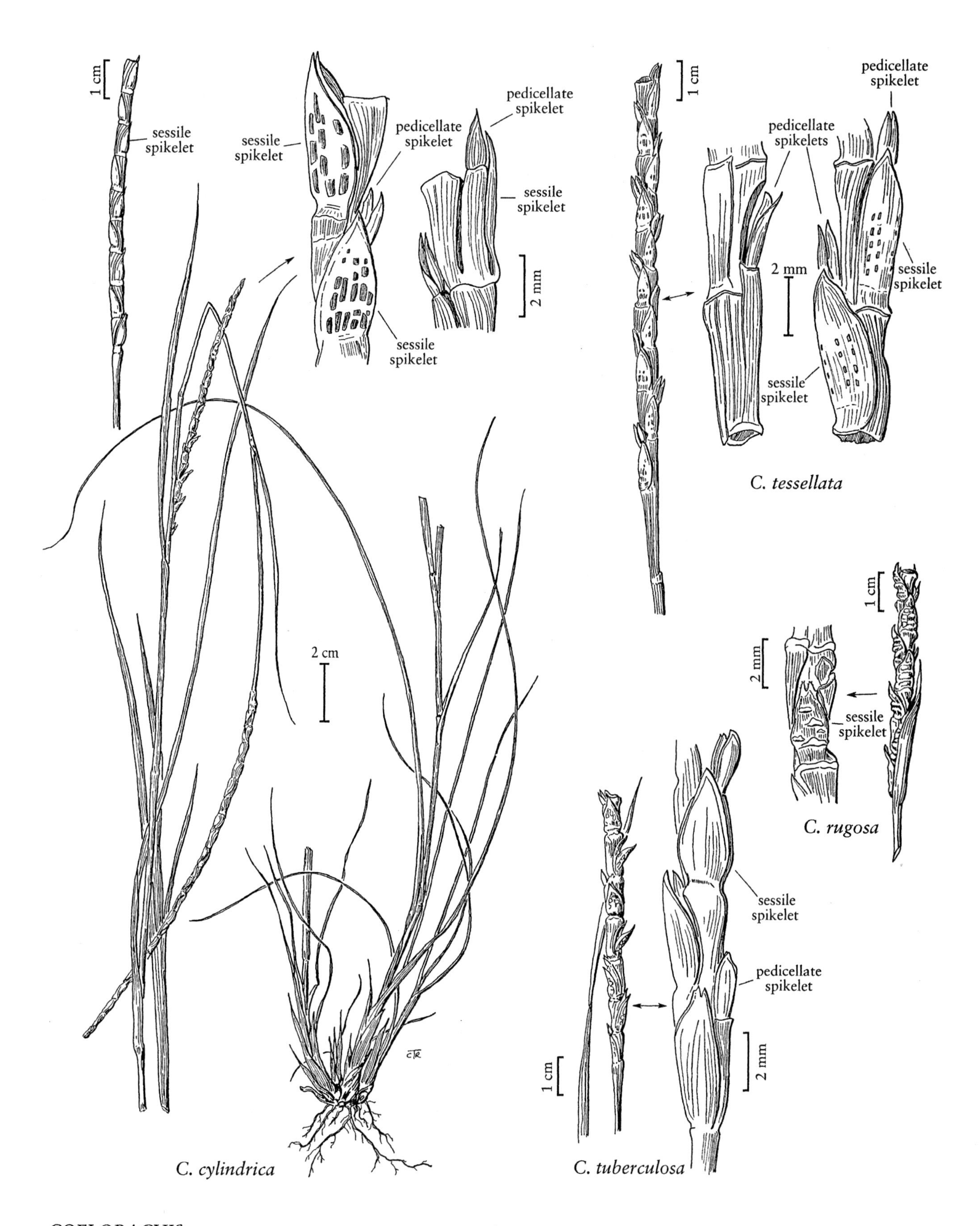

COELORACHIS

keels, margins folded inward; **upper glumes** often shorter than the lower glumes, 3–5-veined, keels entire; **lower florets** staminate; **lower paleas** present; **upper florets** bisexual or pistillate; **lemmas** and **paleas** hyaline, unawned; **anthers** 3; **style branches** 2, red, free to the base. **Pedicels** closely appressed but not fused to the rame axes, flattened, thick, widening above the bases, glabrous. **Pedicellate spikelets** usually absent or rudimentary, occasionally well-developed. x = 9. Name from the Greek *eremos*, 'solitary', and *chloa*, 'grass', a reference to the solitary rame.

Eremochloa is a genus of 11 species that are native to Asia and Australia. One species is naturalized in the southeastern United States; another was found once in California but is not established in the *Flora* region.

SELECTED REFERENCE **Buitenhuis, A.G.** and **J.F. Veldkamp.** 2001. Revision of *Eremochloa* (Gramineae–Andropogoneae–Rottboellinae). Blumea 46:399–420.

1. Keels of the lower glumes of the sessile spikelets winged distally, with 1–several 0.2–0.3 mm hooklike spines at the base . . . . 1. *E. ophiuroides*
1. Keels of the lower glumes of the sessile spikelets not winged, spine-bearing throughout, the basal spines 1–3 mm . . . . 2. *E. ciliaris*

1. **Eremochloa ophiuroides** (Munro) Hack. [p. 692]
CENTIPEDE GRASS

Plants mat-forming, stoloniferous, stolons to 150 cm, often branched, with well-developed leaves, and (usually) axillary fascicles of closely imbricate leaves. **Culms** 10–35 cm, unbranched. **Sheaths** mostly glabrous, margins sometimes pilose, keeled; **leaves** mostly basal, **blades of basal leaves** 0.5–15 cm long, 1–5 mm wide, glabrous or pilose, with papillose-based hairs near the base, margins glabrous or pectinate near the base; **blades of upper leaves** reduced to obsolete. **Rames** 1–3, 3–12 cm, straight; **internodes** 2–2.3 mm. **Sessile spikelets** (2.2)3–4 mm long, (1.1)1.8–2.2 mm wide, elliptic; **calluses** sparsely pubescent; **glumes** glabrous; **lower glumes** 5–7-veined, obtuse to truncate, often notched, keels with 1–several, 0.2–0.3 mm hooklike spines near the base, winged distally; **upper glumes** 3-veined, elliptic, acute; **anthers of lower florets** about 0.3 mm; **anthers of upper florets** 1.5–1.7 mm. **Pedicels** 2.8–3.5 mm long, about 0.5 mm wide at midlength. **Pedicellate spikelets** absent or to 3.4 mm, occasionally well-developed. **Caryopses** 1.5–2 mm, purple to reddish-brown or brown. $2n$ = 18.

Eremochloa ophiuroides, an east Asian species, was introduced into the southeastern United States as a lawn grass about 1920. It is now established along roadsides and in woods, fallow fields, and dunes in the region. It flowers from spring to fall, and sporadically at other times. The common name refers to the appearance of the leafy stolons.

2. **Eremochloa ciliaris** (L.) Merr. [p. 692]

Plants cespitose, sometimes shortly stoloniferous. **Culms** 30–70 cm, compressed, sometimes branching from the upper nodes. **Leaves** mostly basal; **sheaths** keeled, strongly distichous and imbricate, often loose, basal sheaths pubescent below; **ligules** to 0.6 mm, membranous; **blades** to 25 cm long, to 6 mm wide, flat or folded, basal blades glabrous. **Rames** 4–7 cm, straight to falcate; **internodes** 2–3 mm, clavate, shortly pubescent. **Sessile spikelets** 4–5 mm, ovoid-oblong; **calluses** pubescent; **lower glumes** elliptic, obscurely 7–9-veined, keeled, keels with conspicuous spines, basal spines to 3 mm, those near the apices to 0.3 mm; **upper glumes** 3–5-veined, keeled below; **lower florets** staminate; **anthers** about 2 mm, yellow; **upper florets** bisexual. **Pedicels** to 3 mm. **Pedicellate spikelets** not differentiated. $2n$ = 36.

Eremochloa ciliaris is native to southeast Asia. It was collected in San Francisco in the nineteenth century, but has not been reported since from the *Flora* region.

26.26 ROTTBOELLIA L. f.

J.K. Wipff

Plants annual; cespitose. **Culms** 30–300+ cm, glabrous or sparsely pubescent below the nodes, branching above the bases. **Leaves** mostly cauline, not aromatic; **sheaths** sometimes with papillose-based hairs; **auricles** absent; **ligules** membranous, ciliate; **blades** flat. **Inflorescences** terminal and axillary, solitary rames with more than 1 spikelet unit, spikelets partially embedded in the rame axes; **disarticulation** in the rame axes. **Spikelets** heterogamous, in sessile-pedicellate pairs, dorsally compressed, unawned. **Sessile spikelets** with 2 florets; **lower glumes** coriaceous, smooth or scabridulous, not pitted, 2-keeled, narrowly winged above; **upper glumes** coriaceous, 1-keeled, winged; **lower florets** staminate or sterile; **upper florets** bisexual; **lemmas** and **paleas** hyaline; **anthers** 3; **ovaries** glabrous. **Pedicels** thick, fused to the rame axes. **Pedicellate spikelets** sterile or staminate; **glumes** herbaceous. **Caryopses** with a hard endosperm. x = 9, 10. Named for Christen Friis Rottboell (1727–1797), a Danish botanist.

Rottboellia is a genus of five species, all native to tropical Africa and Asia, represented in all tropical regions of the world by *R. cochinchinensis*, the species that has been introduced into the *Flora* region.

SELECTED REFERENCES **Hall, D.W. and D.T. Patterson.** 1992. Itchgrass–stop the trains? Weed Technol. 6:239–241; **Veldkamp, J.F., R. de Koning, and M.S.M. Sosef.** 1986. Generic delimitation of *Rottboellia* and related genera (Gramineae). Blumea 31:281–307; **Wipff, J.K. and B.S. Rector.** 1993. *Rottboellia cochinchinensis* (Poaceae: Andropogoneae) new to Texas. Sida 15:419–424.

1. **Rottboellia cochinchinensis** (Lour.) Clayton [p. 692]
ITCHGRASS

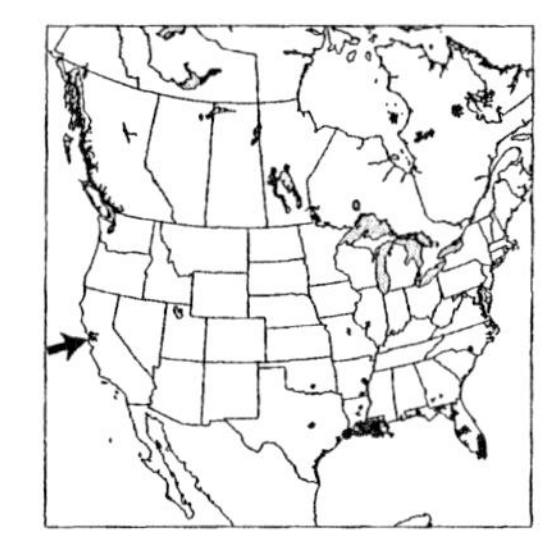

Culms developing prop roots from the lower nodes. **Sheaths** with 1–3 mm, stiff, papillose-based hairs; **ligules** 1–1.3 mm, light brown; **blades** (9)20–50(60) cm long, 10–20(25) mm wide, glabrous abaxially, adaxial surfaces mostly sparsely pubescent, densely pubescent behind the ligules, hairs papillose-based. **Rames** (3)6–15 cm long, 2–4 mm wide; **internodes** 6–12 mm. **Sessile spikelets: lower glumes** 3.5–7 mm long, 1.4–2 mm wide, convex to flat, 11–13(15)-veined, apices bifid; **upper glumes** 5–6.2 mm, navicular, almost completely enclosing the florets, (13)15–17-veined; **anthers of lower florets** 2.2–2.3 mm; **anthers of upper florets** 1.2–2.1 mm; **stigmas** purple. **Pedicels** 3–6.5 mm long, 1.5–2.2 mm wide, flat. **Pedicellate spikelets** 3–8 mm, sterile. **Caryopses** 3–4 mm long, 2–2.2 mm wide. $2n$ = 20, 40.

Rottboellia cochinchinensis is a native of southeast Asia. The species is considered one of the world's worst weeds, and is classified as a noxious weed by the U.S. Department of Agriculture. It is established in the southeastern United States, and has been reported from scattered locations elsewhere in the contiguous United States. In Africa, it is controlled by tillage or by fire, followed by shallow, then deep, plowing (Hall and Patterson 1992). 'Itchgrass' aptly describes the effects of the hairs on the skin.

26.27 HACKELOCHLOA Kuntze

John W. Thieret

Plants annual; cespitose. **Culms** 20–120 cm, erect to decumbent, often rooting at the lower nodes, branching above the bases. **Leaves** not aromatic; **sheaths** open; **auricles** absent; **ligules** membranous, ciliate. **Inflorescences** terminal and axillary, solitary, 2-sided rames, these sometimes fascicled and partially enclosed in subtending leaf sheaths at maturity; **disarticulation** in the rames, beneath the sessile spikelets. **Spikelets** in heterogamous sessile-pedicellate pairs.

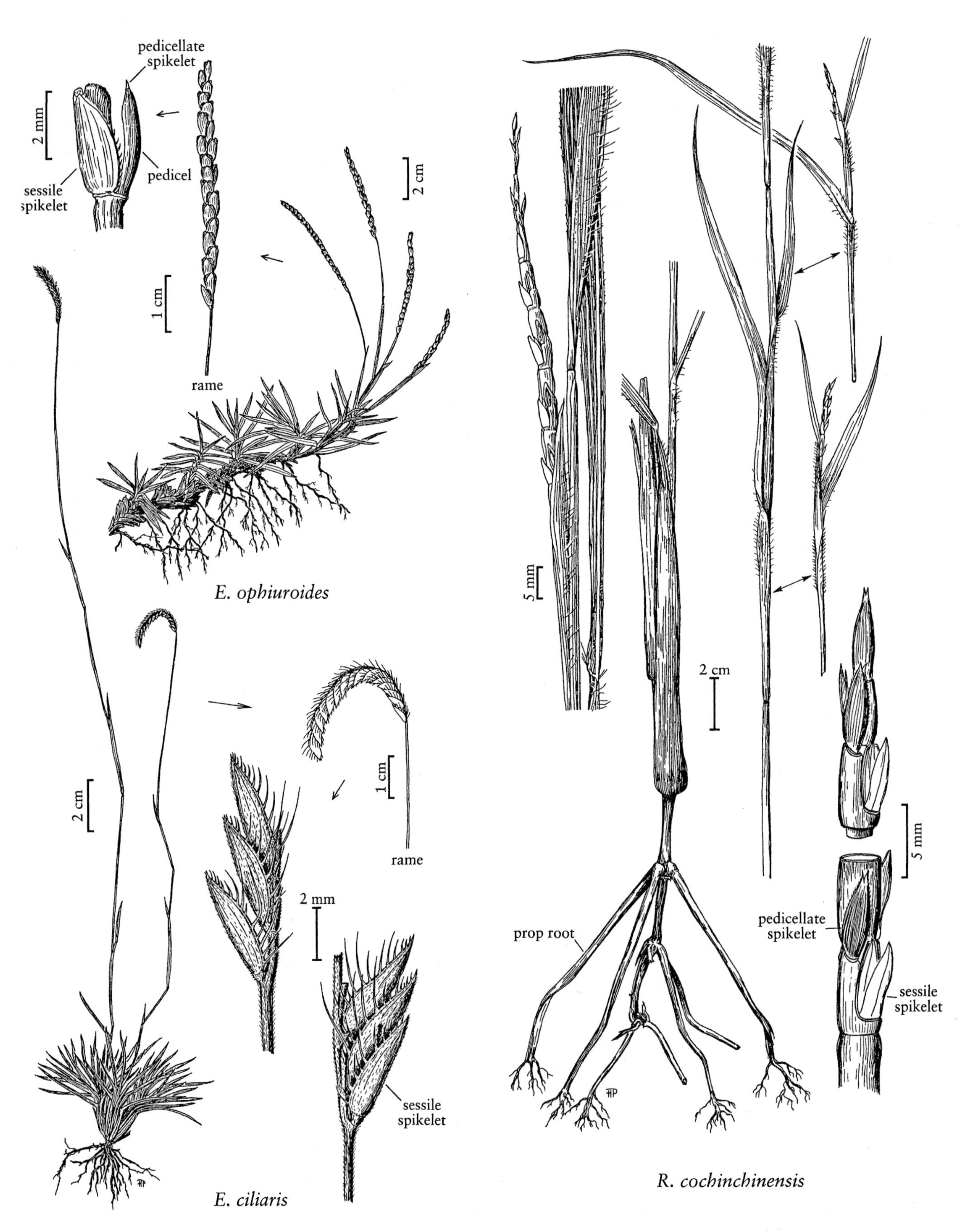

EREMOCHLOA

ROTTBOELLIA

Sessile spikelets hemispherical, partly embedded in the rame axes; **lower glumes** as long as the spikelets, indurate, alveolate, indistinctly 7–11-veined, not keeled, margins involute; **upper glumes** chartaceous, 3-veined, usually adherent to the rame axes; **lower florets** sterile; **upper florets** bisexual; **anthers** 3. **Pedicels** adnate to the rame axes, concealed by the sessile spikelets. **Pedicellate spikelets** as long as or longer than the sessile spikelets, ovate; **lower glumes** dorsally compressed, 5–9-veined; **upper glumes** laterally compressed, 5–7-veined; **lower florets** sterile; **upper florets** staminate; **anthers** 3. x = 7 (probably). Named for Eduard Hack. (1850–1926), an Austrian agrostologist, and the Greek *chloa*, 'grass'.

Hackelochloa is treated here as a monospecific genus that is widely distributed in warm regions of the world, often as a weed. Veldkamp et al. (1986) combined it with *Coelorachis* Brongn., *Heteropholis* C.E. Hubb., *Ratzeburgia* Kunth, and *Rottboellia formosa* R. Br. in *Mnesithea* Kunth. The traditional treatment for *Hackelochloa* is retained here.

SELECTED REFERENCES **Allred, K.W.** 1993. A Field Guide to the Grasses of New Mexico. New Mexico Agricultural Experiment Station, Department of Agricultural Communications. New Mexico State University, Las Cruces, New Mexico, U.S.A. 258 pp.; **Hitchcock, A.S.** 1951 [title page 1950]. Manual of the Grasses of the United States, ed. 2, rev. A. Chase. U.S.D.A. Miscellaneous Publication No. 200. U.S. Government Printing Office, Washington, D.C., U.S.A. 1051 pp.; **Veldkamp, J.F., R. de Koning, and M.S.M. Sosef.** 1986. Generic delimitation of *Rottboellia* and related genera (Gramineae). Blumea 31:281–307.

1. **Hackelochloa granularis** (L.) Kuntze [p. 694]
PITSCALE GRASS

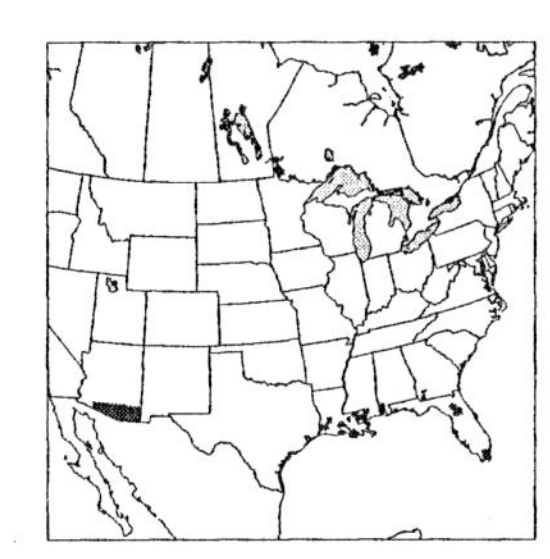

Culms 20–120 cm, glabrous or hispid, with papillose-based hairs. **Leaves** sparsely to densely hispid throughout, hairs papillose-based; **sheaths** shorter than the internodes; **ligules** 2–3 mm; **blades** 3–20 cm long, 6–13 mm wide, subcordate. **Rames** (2)7–27 mm. **Sessile spikelets** 1–1.3 mm; **anthers** 0.3–0.4 mm. **Pedicellate spikelets** 1.6–2.2 mm; **glumes** chartaceous; **lower glumes** winged on 1 keel; **upper glumes** with the midvein narrowly winged; **anthers** 1–1.2 mm. **Caryopses** 0.7–0.9 mm, elliptic to nearly orbicular in outline, brown to yellow-brown. $2n$ = 14.

Hackelochloa granularis is a native of the Eastern Hemisphere that has become established in cultivated land, roadsides, and weedy areas of the southern region of the United States. Its range extends south through Mexico and Central and South America. Hitchcock (1951) reported it from New Mexico, but it is not established there (Allred 1993).

26.28 TRIPSACUM L.

Mary E. Barkworth

Plants perennial; monoecious, staminate and pistillate spikelets evidently distinct, located in the same inflorescences, pistillate spikelets below the staminate spikelets. **Culms** 0.7–5 m. **Leaves** not aromatic; **sheaths** open; **ligules** membranous, erose to ciliate. **Inflorescences** terminal and axillary, panicles of 1–several subdigitate to racemose rames; **rames** with pistillate spikelets proximally and staminate spikelets distally. **Disarticulation** in the rames, beneath the pistillate spikelets and at the base of the staminate portions. **Pistillate spikelets** exposed, solitary, embedded in the indurate rame axes; **lower glumes** coriaceous, closing the hollows in the rachises and concealing the florets; **upper glumes** similar but smaller; **lower florets** sterile; **upper florets** pistillate; **lemmas** and **paleas** hyaline, unawned; **styles** 2, not fused. **Staminate spikelets** paired, both sessile or both subsessile, or 1 sessile and the other pedicellate; **glumes** coriaceous, chartaceous, or membranous; **lemmas** and **paleas** hyaline, unawned. **Pedicels** (when present) not fused to the rame axes. x = 9. The origin of the name is unknown.

Tripsacum is a genus of 12 species, all of which are native to tropical and subtropical regions of the Western Hemisphere; three are native to the *Flora* region. They are good forage grasses, but are rarely sufficiently abundant in the *Flora* region to be important in this regard. The genus is of interest to plant breeders because of its relationship to *Zea*.

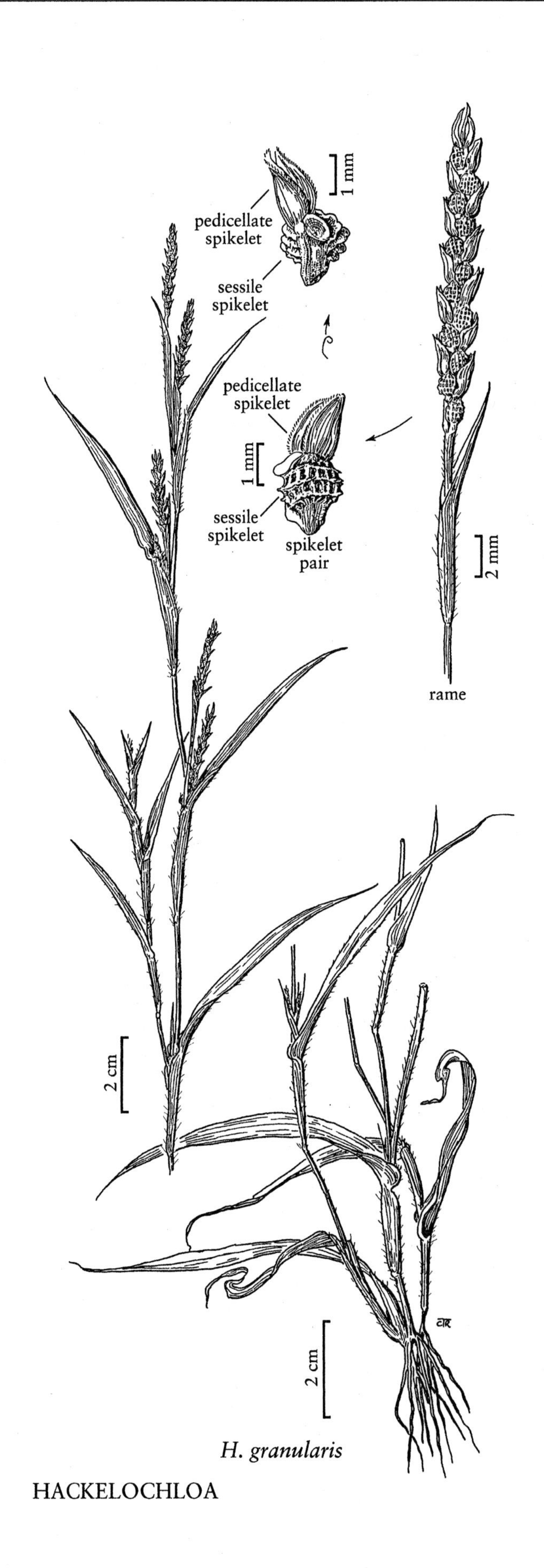

H. granularis

HACKELOCHLOA

Measurements of the pistillate spikelets are based on measurements of the lower glumes of the sessile spikelets, the remainder of the spikelet being concealed between the rachis and the lower glume.

SELECTED REFERENCES **Gray, J.R.** 1974. The genus *Tripsacum* L. (Gramineae): Taxonomy and chemosystematics. Ph.D. dissertation, University of Illinois at Urbana-Champaign, Urbana, Illinois, U.S.A. 191 pp.; **de Wet, J.M.J., D.E. Brink, and C.E. Cohen.** 1983. Systematics of *Tripsacum* section *Fasciculata* (Gramineae). Amer. J. Bot. 70:1139–1146; **de Wet, J.M.J., J.R. Harlan, and D.E. Brink.** 1982. Systematics of *Tripsacum dactyloides.* Amer. J. Bot. 69:1251–1257; **Li, Y.G., C.L. Dewald, and V.A. Sokolov.** 2000. Sectional delineation of sexual *Tripsacum dactyloides–T. maizar* allotriploids. Ann. Bot. (London), n.s., 85:845–850.

1. Staminate spikelets in sessile-pedicellate pairs, the pedicels almost flat to plano-convex in cross section, 2–5 mm long, less than 0.3 mm wide; glumes usually membranous (sect. *Fasciculata*) . . . . 1. *T. lanceolatum*
1. Staminate spikelets sessile, subsessile, or in sessile-pedicellate pairs, the pedicels triangular in cross section, to 2 mm long and about 0.5–0.8 mm wide; glumes somewhat coriaceous (sect. *Tripsacum*).
   2. Blades 9–35(45) mm wide, flat; culms 1–2(4) m tall . . . . 2. *T. dactyloides*
   2. Blades 1–7(15) mm wide, involute or folded; culms to 1 m tall . . . . 3. *T. floridanum*

Tripsacum sect. Fasciculata Hitchc.

Staminate spikelets in sessile-pedicellate pairs, the pedicels slender and relatively flexible.

1. Tripsacum lanceolatum Rupr. *ex* E. Fourn. [p. 697]
MEXICAN GAMAGRASS

Plants rhizomatous. **Culms** 1–2 m tall, 2–4 mm thick. **Lower sheaths** hispid; **upper sheaths** essentially glabrous; **ligules** erose, not ciliate; **blades** to 100 cm long, 8–30 mm wide, glabrous or slightly pubescent. **Terminal inflorescences** with 4–7(10) rames. **Pistillate spikelets** 2–3 mm wide, beadlike in appearance. **Staminate spikelets** in sessile-pedicellate pairs; **glumes** 5–10 mm long, 1.5–2 mm wide, usually membranous, acute; **pedicels** 2–5 mm long, less than 0.3 mm wide, almost flat to plano-convex in cross section, flexible. $2n = 72$.

Tripsacum lanceolatum grows in moist soil (often in canyon bottoms) of mountains from southeastern Arizona and southwestern New Mexico through Mexico to Guatemala. It has not been found in New Mexico since the 1800s.

Tripsacum L. sect. Tripsacum

Staminate spikelets in sessile pairs or 1 member of each pair on a short, stout pedicel.

2. Tripsacum dactyloides (L.) L. [p. 697]
EASTERN GAMAGRASS

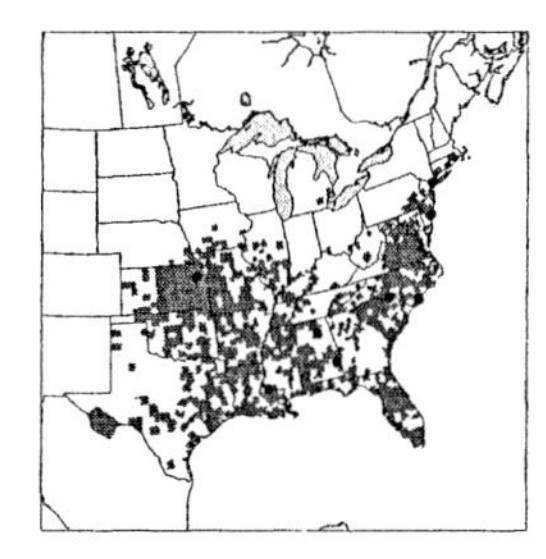

Plants with short, knotty rhizomes. **Culms** 1–2(4) m tall, 3–5 mm thick, clumped. **Sheaths** usually glabrous, occasionally slightly pilose; **ligules** ciliate; **blades** 30–75(120) cm long, 9–35(45) mm wide, flat, usually glabrous, tapering to attenuate apices. **Terminal inflorescences** erect, with (1)2–3(6) rames; **rames** 12–25 cm. **Pistillate spikelets** 6–8 mm long, 3–5.5 mm wide. **Staminate spikelets** all sessile or subsessile; **glumes** 5–12 mm, coriaceous, blunt, acute, or bifid; **pedicels**, when present, about 1 mm long, 0.5–0.8 mm wide, triangular in cross section, rigid. $2n = 36, 54, 72$.

Tripsacum dactyloides grows in water courses and limestone outcrops from the central and eastern United States through Mexico to northern South America. Plants from the United States and northern Mexico belong to **Tripsacum dactyloides** var. **dactyloides.** They differ from those of the other two varieties in their erect stems and sessile staminate spikelets. Narrow-bladed plants of *T. dactyloides* from Texas resemble *T. floridanum*, but on transplanting to favorable conditions develop the wider blades characteristic of *T. dactyloides*. The two species can hybridize; the hybrids are partially sterile.

Growing *Tripsacum dactyloides* for forage has proven practical only in South America. It is also used as an ornamental grass, the chief attraction being its foliage.

3. **Tripsacum floridanum** Porter *ex* Vasey [p. 697]
FLORIDA GAMAGRASS

Plants with short, thick rhizomes. **Culms** to 1 m tall, to 2 mm thick, usually solitary or in small clumps. **Sheaths** glabrous; **blades** to 60 cm long, 1–7(15) mm wide, involute or folded, glabrous. **Terminal inflorescences** erect, with 1–2 rames. **Pistillate spikelets** 3.5–4.5 mm wide. **Staminate spikelets** sessile-pedicellate; **spikelets** 5–7 mm; **glumes** coriaceous, acute; **pedicels** to 2 mm long, to 0.5 mm wide, triangular in cross section. $2n = 36$.

Tripsacum floridanum grows along roadsides and in pine woods, often in wet soils, of Florida and Cuba. It is grown as an ornamental, but it reseeds rather too readily under some conditions. Reports of *T. floridanum* from Texas are based on narrow-bladed specimens of *T. dactyloides*.

26.29 ZEA L.

Hugh H. Iltis

Plants annual or perennial; monoecious, inflorescences unisexual or bisexual with the pistillate spikelets basal and the staminate spikelets distal. **Culms** (0.2)0.5–6 m tall, 1–5 cm thick, solitary or several to many together, monopodial, often branching (branches frequently highly reduced and hidden within the subtending leaf sheath), usually succulent when young, becoming woody with age; **lower nodes** with prop roots; **internodes** pith-filled. **Leaves** not aromatic, cauline, distichous; **sheaths** open; **auricles** sometimes present; **ligules** membranous, shortly ciliate; **blades** 2–12 cm wide, flat. **Pistillate or partially pistillate inflorescences** terminal on axillary branches; **staminate inflorescences** (*tassels*) paniculate, of 1-many branches or rames, sometimes with secondary and tertiary branching. WILD TAXA: **Pistillate inflorescences** solitary, distichous rames (*ears*), these often tightly clustered in false panicles, each usually wholly or partially enclosed by a thin prophyll and an equally thin bladeless leaf sheath; **rames** composed of 5–15 spikelets in 2 ranks; **disarticulation** in the rame axes, dispersal units (*fruitcases*) consisting of an indurate, shiny rame segment and its embedded spikelet. **Pistillate spikelets** solitary, sessile, with 1 floret; **pedicels** and **pedicellate spikelets** suppressed; **lower glumes** exceeding the floret, indurate on the central, exposed portion, hyaline on the margins, concealing the caryopses at maturity. DOMESTICATED TAXON: **Pistillate inflorescences** solitary, polystichous spikes (*ears*) terminating reduced branches, each spike surrounded by several to many, often bladeless leaf sheaths and a prophyll (*husks*), with 60–1000+ spikelets in 8–24 rows, neither spikes nor spikelets disarticulating at maturity. **Pistillate spikelets** in subsessile pairs, each spikelet with 1 functional floret; **glumes** shorter than the spikelets, indurate basally, hyaline distally; **lower florets** suppressed. ALL TAXA: **lemmas** and **paleas** hyaline, unawned; **lodicules** absent; **ovaries** glabrous; **styles** (*silks*) 2, appearing solitary by being fused except at the very tip, filamentous, sides stigmatic. **Caryopses** subspherical to dorsally compressed; **hila** round; **embryos** about ⅔ as long as the caryopses. WILD TAXA: **Staminate panicles** terminal on the culms and primary branches, sometimes also on the secondary branches and pistillate inflorescences; **rames** distichous, similar in thickness and structure, axes

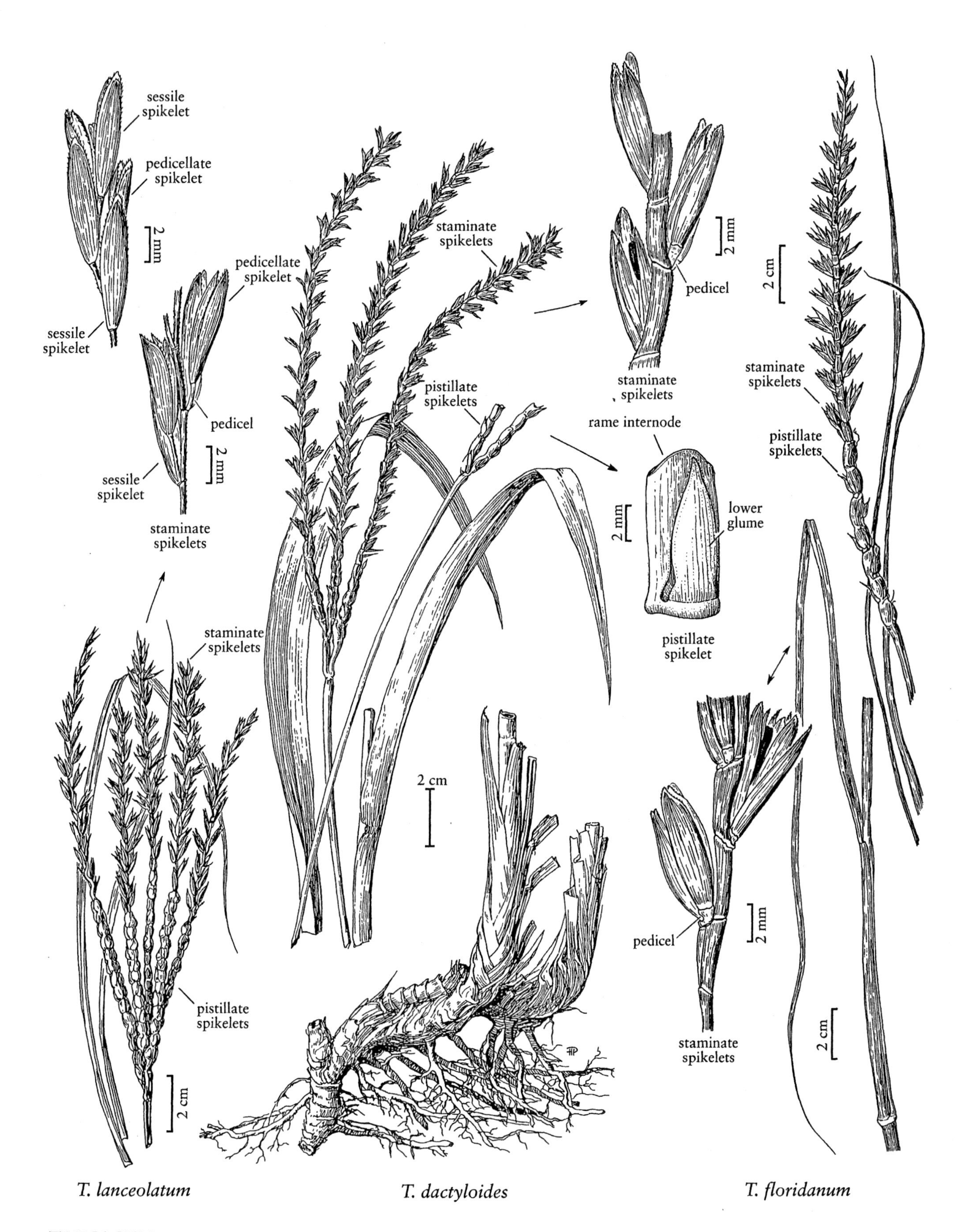
sessile spikelet
pedicellate spikelet
2 mm
sessile spikelet
pedicellate spikelet
pedicel
2 mm
sessile spikelet
staminate spikelets
staminate spikelets
pistillate spikelets
2 cm
staminate spikelets
pistillate spikelets
2 cm
2 mm
pedicel
staminate spikelets
rame internode
2 mm
lower glume
pistillate spikelet
2 cm
staminate spikelets
pistillate spikelets
pedicel
2 mm
staminate spikelets
2 cm
T. lanceolatum
T. dactyloides
T. floridanum

TRIPSACUM

disarticulating below the sessile spikelets after pollination, abscission layers evident. DOMESTICATED TAXON: **Staminate panicles** terminal on the culms, central axes always much thicker than the lateral branches and irregularly polystichous, lateral branches distichous to more or less polystichous, not disarticulating, without abscission layers below the sessile spikelets. ALL TAXA: **Staminate spikelets** in sessile-pedicellate pairs, each with 2 staminate florets; **glumes** membranous to chartaceous, stiff to flexible, sometimes with a pair of winged keels, 5–14(28)-veined, acute; **lemmas** and **paleas** hyaline; **lodicules** 2; **anthers** 3. $x = 10$. Name from the Greek *zea* or *zeia*, a kind of grain.

Zea is an American genus of five species, four of which are native to montane Mexico and Central America. The fifth species, *Z. nicaraguensis* H.H. Iltis & B.F. Benz, is said to have been ubiquitous at one time in coastal Pacific Nicaragua, but is now known from only four or five small populations near sea level in seasonally flooded savannahs and riverine forests inland from the Bay of Fonseca, Nicaragua.

The often weedy, wild taxa, known as 'teosinte', are used in plant breeding as well as in developmental and evolutionary studies. The genus has also been the focus of physiological and genetic research, mostly involving the domesticated taxon, *Zea mays* subsp. *mays*. Examples of such work include Barbara McClintock's Nobel prize-winning discovery that genes can "jump" from one chromosome to another, and recent work on the evolution of tassel morphology (e.g., Westerbergh and Doebley 2002).

Zea mays subsp. *mays*, the most widespread taxon in the genus, was first domesticated about 7,000 years ago and soon became widely planted in the Americas. It is now grown in all warmer parts of the world and is the world's third most important crop plant. No other American grass has such agricultural importance.

In the *Flora* region, *Zea mays* subsp. *mays* is widely grown commercially; *Z. luxurians* is sometimes grown for forage; while *Z. diploperennis* and *Z. perennis*, and the other subspecies of *Z. mays*, are almost completely confined to research plantings.

SELECTED REFERENCES **Doebley, J.F.** 1990. Molecular systematics of *Zea* (Gramineae). Maydica 35:143–150; **Doebley, J.F. and H.H. Iltis.** 1980. Taxonomy of *Zea* (Gramineae). I. A subgeneric classification with a key to taxa. Amer. J. Bot. 67:982–993; **Doebley, J.F., A. Stec, and C. Justus.** 1995. *Teosinte branched1* and the origin of maize: Evidence for epistatis and the evolution of dominance. Genetics 141:333–346; **Eubanks, M.W.** 2001. The mysterious origin of maize. Econ. Bot. 55:492–514; **Hitchcock, A.S.** 1951 [title page 1950]. Manual of the Grasses of the United States, ed. 2, rev. A. Chase. U.S.D.A. Miscellaneous Publication No. 200. U.S. Government Printing Office, Washington, D.C., U.S.A. 1051 pp.; **Iltis, H.H.** 2000. Homeotic sexual translocations and the origin of maize (*Zea mays*, Poaceae): A new look at an old problem. Econ. Bot. 54:7–42; **Iltis, H.H.** and **J.F. Doebley.** 1980. Taxonomy of *Zea* (Gramineae). II. Subspecific categories in the *Zea mays* complex and a generic synopsis. Amer. J. Bot. 67:996–1004; **McVaugh, R.** 1983. Flora Novo-Galiciana: A Descriptive Account of the Vascular Plants of Western Mexico, vol. 14: Gramineae (series ed. W.R. Anderson). University of Michigan Press, Ann Arbor, Michigan, U.S.A. 436 pp.; **Westerbergh A.** and **J. F. Doebley.** 2002. Morphological traits defining species differences in wild relatives of maize are controlled by multiple quantitative trait loci. Evolution 56:273–283.

1. Pistillate inflorescences terete, with 2+ rows of paired spikelets, each spikelet with a functional floret, hence the spikelets in 4+ rank; staminate spikelets of wild taxa only slightly imbricate; lower glumes of the staminate spikelets flexible and translucent, loosely enclosing the upper glumes before anthesis, rounded on the back, the lateral veins not more prominent than those between, never forming winged keels; plants annual (sect. *Zea*) .... 4. *Z. mays*
1. Pistillate inflorescences somewhat flattened, with 2 rows of solitary spikelets, hence the spikelets appearing 2-ranked; staminate inflorescences with densely imbricate spikelets; lower glumes of the staminate spikelets stiff, not translucent, strongly enclosing the upper glumes before anthesis, with more or less flat backs, the lateral veins evidently more prominent than those between, keeled and winged distally; plants annual or perennial (sect. *Luxuriantes*).
    2. Plants annual; lower glumes of the staminate spikelets narrowly winged; pistillate inflorescence units 1–many per node, usually enclosed by their subtending leaf sheaths, occasionally with 1–2 rames on naked peduncles that exceed the subtending leaf sheaths .... 3. *Z. luxurians*

2. Plants perennial, rhizomatous; lower glumes of the staminate spikelets strongly winged; pistillate inflorescence units 1–4(5) per node, almost always exceeding the subtending leaf sheaths.
   3. Rhizomes to 15 cm long, with internodes 0.2–0.6 cm long, often forming scaly, tuberous short shoots; culms to 3.5 m tall and 3 cm thick; leaf blades to 40 cm long and 4–5.5 cm wide . . . . 1. *Z. diploperennis*
   3. Rhizomes to 40 cm long or more, with internodes 1–6 cm long, lacking tuberlike shoots; culms to 2.5 m tall and 1.5–2 cm thick; leaf blades often to 65(80) cm long and 2–4.5 cm wide . . . . 2. *Z. perennis*

1. **Zea diploperennis** H.H. Iltis, Doebley & R. Guzmán [p. 700]

Plants perennial; rhizomatous, rhizomes to 15 cm, internodes 0.2–0.6 cm, often forming scaly, tuberous short shoots 1–2 cm thick. **Culms** 1–3.5 m tall, (1)2–3 cm thick, solitary or in large clumps. **Blades** usually to 40 cm long, 4–5.5 cm wide, linear-lanceolate. **Pistillate peduncles** (1)2–4(5) per node, 5–25(52) cm, the 3–5 longer peduncles extending far beyond the subtending leaf sheaths, each peduncle with 1(2) pistillate rames; **pistillate rames** 5–10 cm long, 4–5 mm thick, distichous, with 5–10 solitary spikelets, frequently not enclosed in a leaf sheath; **fruitcases** trapezoidal in side view, 6–9 mm on the long side, 2.5–4.5 mm on the short side, 4–5 mm in diameter. **Caryopses** concealed by the lower glumes. **Terminal staminate panicles** 6–18 cm, with 2–15 branches; **branches** 6–15 cm, erect to divergent, internodes 2–6 mm; **spikelets** 8.5–11.5 mm long, about 3 mm wide, densely imbricate; **lower glumes** flat dorsally, stiff, not translucent, margins tightly enclosing the upper glumes, the 2 principal sublateral veins prominently keeled and apically winged. $2n = 20$.

Zea diploperennis, although locally abundant, is rare in the wild, being known only from a few populations in the Sierra de Manantlán, Jalisco, Mexico. It grows at elevations of 1400–2400 m, sometimes forming large clones or extensive colonies in old maize fields and on the edges of oak-pine cloud forests. It is grown for genetic research and plant breeding in many countries and occasionally as an ornamental plant in warmer parts of the contiguous United States. It hybridizes infrequently with *Z. mays* subsp. *mays* in its native range.

2. **Zea perennis** (Hitchc.) Reeves & Mangelsd. [p. 700]
PERENNIAL TEOSINTE

Plants perennial; rhizomatous, rhizomes to 40 cm or longer, internodes 1–6 cm, not forming tuberous short shoots. **Culms** 1–2.5 m tall, 1.5–2 cm thick, loosely clumped, usually branched above. **Blades** 20–65(80) cm long, 2–4.5 cm wide, linear. **Pistillate peduncles** 1–2(3) per node, 10–25 cm, at least 1 and sometimes 2 extending far beyond the terminal leaf sheaths; **pistillate rames** 4–8 cm long, 4–6 mm thick, distichous, with 5–10 solitary spikelets, distal portions often staminate; **fruitcases** trapezoidal in side view, 6–9 mm on the long side, 2.5–4.5 mm on the short side, 4–5 mm in diameter. **Caryopses** concealed by the lower glumes. **Terminal staminate panicles** 12–20 cm, with 2–8 branches; **branches** 9–15 cm, erect to nodding, internodes 2.4–6.2 mm; **spikelets** 8.5–11 mm long, 2–2.5(3.2) mm wide, densely imbricate; **lower glumes** flat dorsally, stiff, not translucent, margins tightly enclosing the upper glumes, lateral veins prominent, strongly winged distally. $2n = 40$.

Zea perennis is parapatric to *Z. diploperennis*, being native to the northern base of the Volcán de Colima, Jalisco, Mexico, at elevations of 1520–2200 m. It is rare, although locally abundant, in and around maize fields and orchards in former open oak and pine forests and savannahs. *Zea perennis* crosses infrequently with *Z. mays* subsp. *mays*. The hybrids, being triploid, are sterile. It has also been cultivated at research stations in the United States for many years and Hitchcock (1951) reported that it was established at James Island, South Carolina. It is not known if the population has persisted.

3. **Zea luxurians** (Durieu & Asch.) R.M. Bird [p. 700]
GUATEMALA TEOSINTE, FLORIDA TEOSINTE

Plants annual. **Culms** 1–3(4) m tall, 1–4 cm thick, unbranched in dense stands, abundantly branched in open areas. **Blades** 20–80 cm long, 3–8 cm wide, glabrous. **Pistillate inflorescences** 1–many per node, usually in dense, sheathed, axillary clusters; **peduncles** usually (0)1–8 cm, slender and not exceeding the leaf sheaths, occasionally 1(2) peduncles as long as 23 cm and exceeding the leaf sheaths; **pistillate rames** distichous, 6–9 cm, subtended by a sheath, with 5–9 solitary spikelets; **fruitcases** trapezoidal in side view, 7–11.5 mm on the long side, 3.7–6.5 mm on the short side, 3–5 mm in diameter. **Caryopses** concealed. **Terminal staminate panicles** 12–24 cm, with (4)10–28 stiffly ascending branches; **branches** 7–16(21) cm, internodes 3–6 mm; **pedicels** 3–5 mm; **spikelets** 4.6–12 mm, densely imbricate; **lower glumes** flat dorsally, stiff, not translucent, margins tightly enclosing the upper glumes, (9)12–20(28)-veined, the 2 sublateral veins prominent, keeled, ciliate, narrowly winged distally. $2n = 20$.

Zea luxurians is endemic to Central America, growing from Guatemala to Honduras, at elevations of 600–1200 m, and may extend into Oaxaca, Mexico. It was frequently grown for forage about a century ago, and is

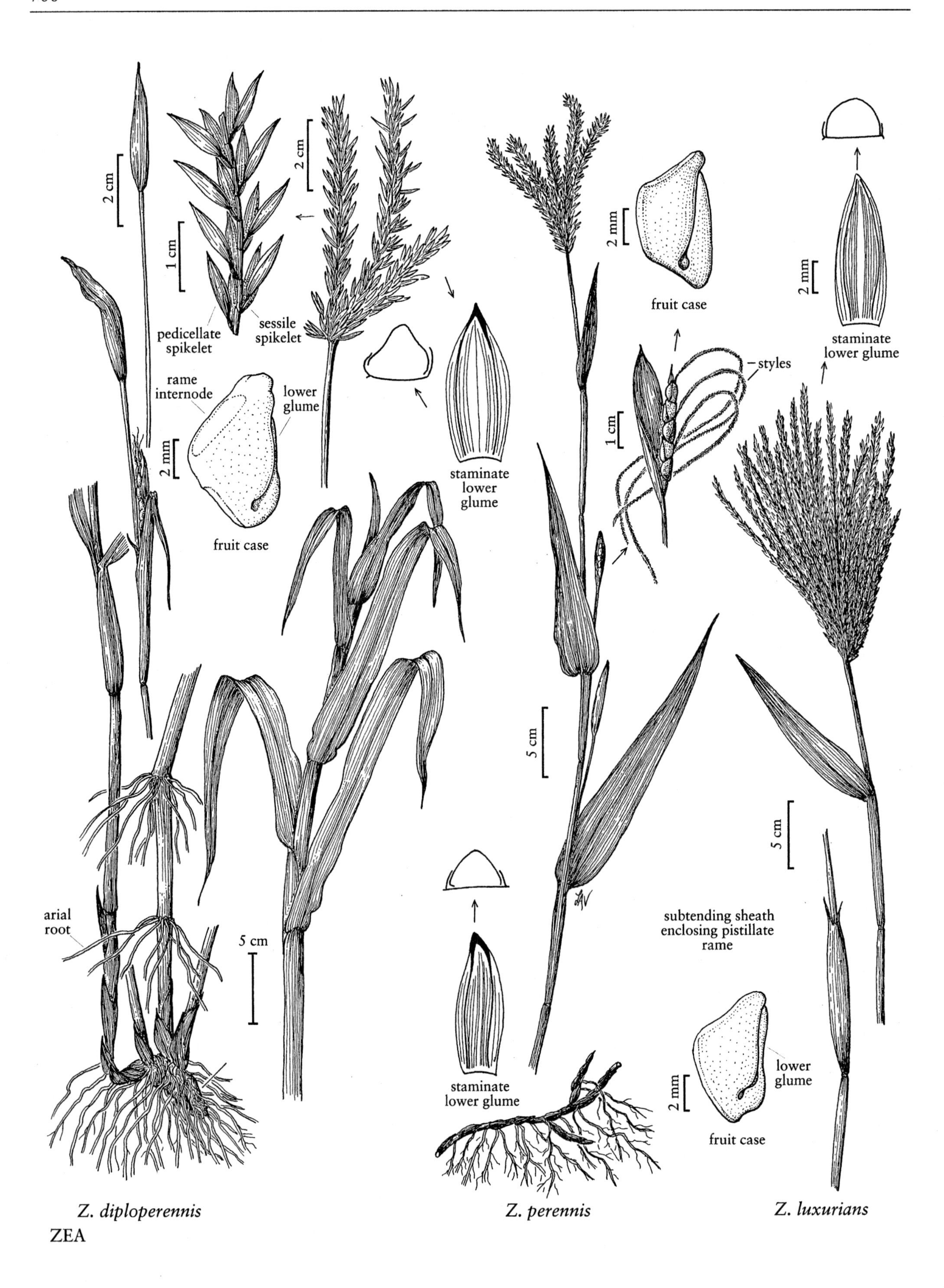
2 cm
1 cm
2 cm
pedicellate spikelet
sessile spikelet
rame internode
lower glume
2 mm
fruit case
staminate lower glume
arial root
5 cm
2 mm
fruit case
styles
1 cm
5 cm
staminate lower glume
2 mm
staminate lower glume
5 cm
subtending sheath enclosing pistillate rame
lower glume
2 mm
fruit case
Z. diploperennis
Z. perennis
Z. luxurians

ZEA

still sometimes grown for this purpose in the southern United States. It hardly ever tillers in the wild, but forms as many as 50 tillers in favorable agricultural settings and longer day lengths than in its native range. Although it can hybridize with *Z. mays* subsp. *mays*, *Z. luxurians* rarely does so in the wild.

4. **Zea mays** L. [p. 702]

Plants annual. **Culms** (0.5)1–3(6) m tall, (0.5)1–5 cm thick. **Blades** mostly 30–90 cm long, 2.5–12 cm wide. **Pistillate inflorescences** rames or spikes, usually shortly pedunculate (sometimes sessile), solitary, 4–30(40) cm long, (0.5)1–10 cm thick, with 2 or more rows of paired spikelets, hence the spikelets 4 or more ranked, rarely terminating in an unbranched staminate inflorescence. **Caryopses** concealed in fruitcases (wild taxa) or exposed (domesticated taxon); **fruitcases of wild taxa** distichous, triangular in side view; **domesticated taxon** without fruitcases, glumes reduced and shallow or collapsed and embedded in the rachis. **Staminate panicles** 10–25+ cm, with 1–60(235) branches, internodes 1.5–8.2 mm; **spikelets** 9–14 mm long, 2.5–5 mm wide; **lower glumes** rounded dorsally, flexible, translucent, papery, loosely enclosing the upper glumes, the 2 lateral veins subequal to the others, not winged. $2n = 20$.

Of the five subspecies of *Zea mays*, only the domesticated subspecies, *Z. mays* subsp. *mays*, is widely grown outside of research programs. Three wild subspecies are treated here, albeit briefly, because of their importance as genetic resources for *Z. mays* subsp. *mays*.

1. Pistillate inflorescences cylindrical spikes, 2–5(10) cm thick, with 8–24+ rows of spikelets pairs, each inflorescence tightly and permanently enclosed by several leaf sheaths and a large prophyll, not disarticulating at maturity; caryopses 60–1000+, not concealed by the glumes; staminate panicle branches not disarticulating below the sessile spikelets, lacking abscission layers; central axis of the staminate panicles polystichous, much thicker than the lateral branches; obligate domesticate .... subsp. *mays*
1. Pistillate inflorescences cylindrical, distichous rames, less than 1 cm thick, with 2 rows of spikelet pairs, each rame usually enclosed by a single leaf sheath and a prophyll, disarticulating at maturity into fruitcases; caryopses 4–15, each one concealed within a fruitcase; staminate panicles composed of rames that disarticulate below the sessile spikelets and have evident abscission layers; central axis of the staminate panicles similar in width to the rames; in the *Flora* region, wild taxa are known only from research plantings.
   2. Staminate spikelets (6.6)7.5–10.5 mm long; fruitcases 6–10 mm long, 4–6 mm wide; staminate panicles with 1–35+ ascending to divergent, rather stiff branches ..... subsp. *mexicana*
   2. Staminate spikelets 4.6–7.2(7.9) mm long; fruitcases 5–8 mm long, 3–5 mm wide; staminate panicles usually with 10–100(235) divergent to nodding branches.
      3. Leaves pubescent; staminate panicles with (2)10–100(235) branches ... subsp. *parviglumis*
      3. Leaves glabrous or almost so; staminate panicles usually with fewer than 40 branches .... subsp. *huehuetenangensis*

Zea mays subsp. **huehuetenangensis** (H.H. Iltis & Doebley) Doebley
HUEHUETENANGO TEOSINTE

Zea mays subsp. *huehuetenangensis* is morphologically similar to subsp. *parviglumis* (see below), but it often grows more than 5 m tall, and has essentially glabrous leaves, and smaller staminate panicles with fewer (less than 40), firmer branches, and different ecological, phenological, and molecular characteristics.

It is endemic to the Province of Huehuetenango, Guatemala, where it grows as a common weed on the edges of, and in, maize fields, and in seasonally moist oak cloud and tropical deciduous forests, at elevations from 900–1650 m. In its native range, it commonly hybridizes with *Z. mays* subsp. *mays*, both subspecies flowering from mid-December to mid-January, at the end of the wet season.

Zea mays L. subsp. **mays**
CORN, INDIAN CORN, MAIZE, MAÏS

Culms (1)2–4(6) m tall, (1)2–5 cm thick. **Blades** 50–90 cm long, 3–12 cm wide. **Pistillate inflorescences** spikes, 15–25(40) cm long, 2–5(10) cm thick, cylindrical, tightly and permanently enclosed in several to many leaf sheaths and a large prophyll, with 8–24 or more rows of paired spikelets on a thickened, strongly vascularized, tough rachis (*cob*), not disarticulating at maturity; **fruitcases** not developed, rachis internodes fused into the extra-vascular cylinder, glumes reduced, shallow. **Caryopses** 60–1000+, exposed and naked. **Staminate panicles** with a polystichous central axis and non-disarticulating branches; **central axes** usually much denser and thicker than the usually distichous lateral branches, these lacking abscission layers. $2n = 20$.

Zea mays subsp. *mays* is the familiar domesticated corn (or maize), from which around 400 indigenous races and many different kinds of cultivars have been developed. It is an obligate cultigen, unable to persist

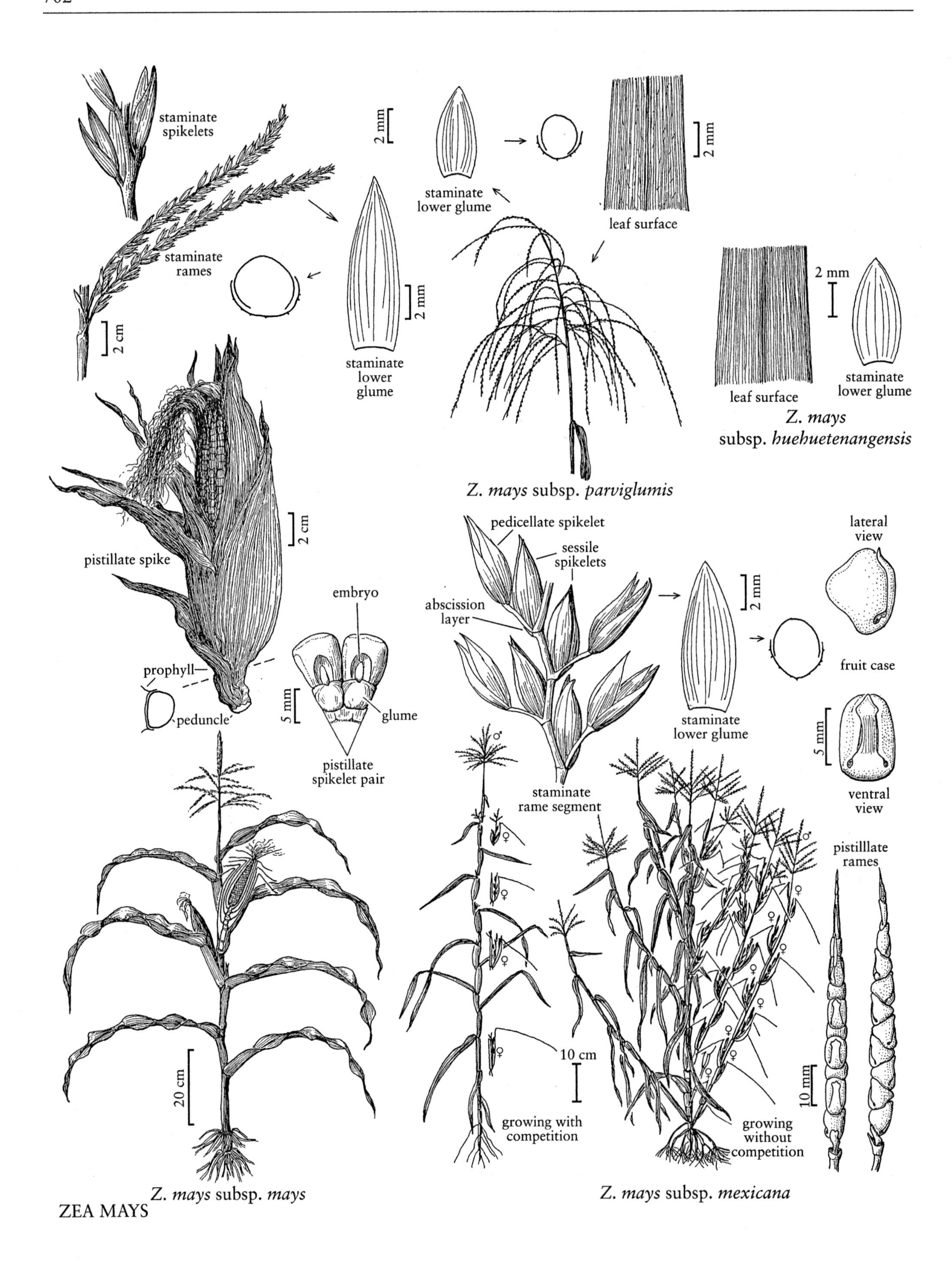
staminate spikelets
staminate rames
2 cm
staminate lower glume
2 mm
staminate lower glume
2 mm
leaf surface
2 mm
Z. mays subsp. parviglumis
leaf surface
2 mm
staminate lower glume
Z. mays subsp. huehuetenangensis
pistillate spike
2 cm
prophyll
peduncle
embryo
5 mm
glume
pistillate spikelet pair
pedicellate spikelet
sessile spikelets
abscission layer
staminate rame segment
2 mm
staminate lower glume
lateral view
fruit case
5 mm
ventral view
pistilllate rames
10 mm
20 cm
10 cm
growing with competition
growing without competition
Z. mays subsp. mays
Z. mays subsp. mexicana

ZEA MAYS

outside of cultivation because the caryopses are permanently attached to the rachis and enclosed by the subtending leaf sheaths. Supersweet cultivars have a double recessive gene that delays the conversion of sugar to starch; flint corns have unusually hard endosperm; and waxy cultivars have endosperm with an unusually high level of proteins and oils. Popcorns have a core of soft, relatively moist endosperm surrounded by hard endosperm. The grains "pop" when heat causes the moisture of the inner endosperm to vaporize.

Zea mays subsp. **mexicana** (Schrad.) H.H. Iltis
CHALCO TEOSINTE, CENTRAL-PLATEAU TEOSINTE, NOBOGAME TEOSINTE

Culms (0.5)1–3 m tall, (0.5)1–2(3) cm thick, unbranched when growing in corn fields to branched and with staminate inflorescences terminating the many branches when growing in the open. **Blades** 30–85 cm long, 3–8 cm wide. **Pistillate inflorescences** 5–8(10) cm long, 0.6–0.8 cm thick, distichous, with 2 rows of spikelets embedded in the rachis; **rachises** disarticulating at maturity; **fruitcases** (6)9–12(15), 6–10 mm long, 4–6 mm wide, triangular in side view, with pointed apices. **Caryopses** concealed in the fruitcases. **Staminate panicles** with 1–35+ ascending to divergent, somewhat stiff, disarticulating branches; **central axes** of staminate panicles as slender as the lateral branches. **Staminate spikelets** (6.6)7.5–10.5 mm. $2n = 20$.

Zea mays subsp. *mexicana* is a weedy taxon, native to upland Mexico. It is most abundant in the Meseta Central of the Mexican neo-volcanic plateau, at elevations of 1700–2500 m. It grows almost entirely in and around cornfields, and readily hybridizes with subsp. *mays*. The long-day tolerant 'Northern Teosinte' is the result of a series of backcrosses between such hybrids and the northernmost population of subsp. *mexicana*, 'Nobogame Teosinte'.

Zea mays subsp. **parviglumis** H.H. Iltis & Doebley
BALSAS TEOSINTE, GUERRERO TEOSINTE

Culms (0.5)2–4 m, unbranched or branched above the middle, thinner than in subsp. *mexicana*. **Leaves** pubescent. **Fruitcases** 5–8 mm long, 3–5 mm wide. **Caryopses** concealed. **Staminate panicles** with (2)10–100(235) slender, often drooping branches; **spikelets** 4.6–7.2 mm, distant. $2n = 20$.

Zea mays subsp. *parviglumis*, which has the smallest fruitcases of all the wild taxa, is endemic to the Pacific slope of southern Mexico, from Oaxaca to Jalisco, being most abundant in the Balsas River drainage. It grows in highly seasonal, sunny thorn scrub, and open tropical deciduous forests and savannahs, at elevations of (450)600–1400(1950) m. One of its higher elevation populations appears to be the ancestor of subsp. *mays*. In the southern United States, *Z. mays* subsp. *parviglumis* is grown as part of breeding programs. In its native habitat, it tends to be seasonally isolated from subsp. *mays*, flowering a few weeks later, but the two sometimes form abundant hybrids in local areas.

26.30 COIX L.

John W. Thieret

Plants annual or perennial; monoecious, pistillate and staminate spikelets on separate rames in the same inflorescence. **Culms** to 3 m, erect, creeping, or floating, branched; **internodes** solid. **Leaves** not aromatic; **ligules** membranous. **Inflorescences** axillary, of 2(3) rames, 1 pistillate, the other(s) staminate, pistillate rames completely enclosed in indurate, globose to cylindric, modified leaf sheaths, termed *involucres*, from which the staminate rames protrude. **Pistillate rames** each with 3 spikelets, 1 sessile and pistillate, the other 2 pedicellate and rudimentary; **sessile spikelets** somewhat dorsally compressed; **glumes** coriaceous, beaked; **stigmas** protruding from the involucres. **Caryopses** more or less globose. **Staminate rames** flexible, exserted from the involucre; **spikelets** in pairs or triplets, 1 sessile, the other(s) pedicellate, reduced, or absent; **lower glumes** chartaceous, with 15 or more veins, 2-keeled, keels winged above; **upper glumes** similar, with 1 keel; **lower florets** sometimes sterile; **upper florets** staminate; **stamens** 0 or 3; **lodicules** 2. **Pedicels** not fused to the rame axes. $x = 5$. Name from the Greek *koix*, a palm.

Coix is a genus of about five species, one of which has been introduced into the *Flora* region. All the species are native to tropical Asia, where *C. lacryma-jobi* and, to a lesser extent, *C. gigantea* are harvested for food.

SELECTED REFERENCES **Jain, S.K.** and **D.K. Banerjee.** 1974. Preliminary observations on the ethnobotany of the genus *Coix*. Econ. Bot. 28:38–42; **Watt, G.** 1904. *Coix* spp. or Job's tears: A review of all available information. Agric. Ledger 13:189–229.

1. Coix lacryma-jobi L. [p. 704]
JOB'S-TEARS

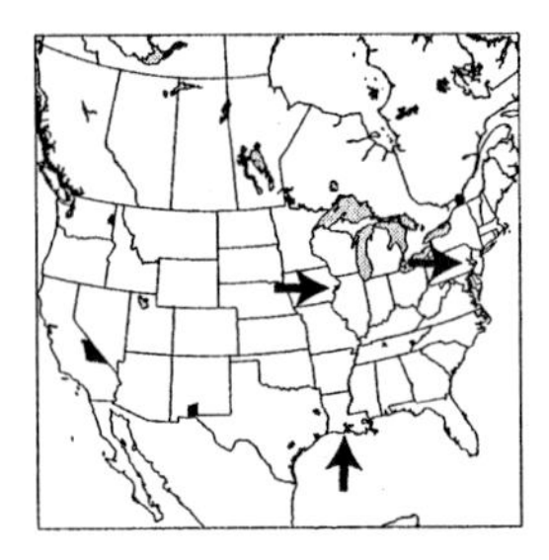

Plants annual or perennial. **Culms** to 3 m. **Leaves** mostly cauline, evidently distichous; **blades** to 75 cm long, 1.5–6 cm wide. **Involucres** usually 8–12 mm, varying in color. **Lower glumes of functional pistillate spikelets** 6–10 mm, hyaline below, 5–7-veined, with a 1-3 mm coriaceous beak. **Staminate rames** 10–35 mm, with 3–25 spikelet pairs, disarticulating at maturity; **spikelets** 5–9 mm, dorsally compressed; **glumes** exceeding the florets, with 15+ veins; **lower glumes** elliptic to obovate, somewhat asymmetrical, margins folded inward, apices obtuse; **upper glumes** lanceolate to narrowly elliptic, keels often winged, apices acute; **upper lemmas** 5–8 mm, hyaline, elliptic to ovate, 3-veined; **upper paleas** similar but 2-veined; **anthers** 3–6 mm. $2n = 20$.

Coix lacryma-jobi is a tall, maize-like plant. In North America, it is usually grown as an ornamental, but it has become established at scattered locations in the *Flora* region. The involucres, which can be used as beads, may be white, blue, pink, straw, gray, brown, or black, with the color being distributed evenly, irregularly, or in stripes. Cultivars with easily removed involucres are grown for food and beverage, especially in Asia.

COIX

GEOGRAPHIC BIBLIOGRAPHY

The Geographic Bibliography shows written sources of information and Web sites used in developing the geographic database from which the maps in this volume were generated as well those placed on the Web site (http://herbarium.usu.edu/webmanual/). Herbaria and individuals that provided files or information for the database are acknowledged individually in the Preface (see Data Contributors).

Book titles are listed in full; the names of journals are abbreviated according to Lawrence et al. (1968) and its supplement (Bridson and Smith 1991).

Aiken, S.G., L.L. Consaul, and M.J. Dallwitz. 1995 onwards. Poaceae of the Canadian Arctic Archipelago: Descriptions, illustrations, identification, and information retrieval. http://www.mun.ca/biology/delta/arcticf/poa/index.htm.

Aiken, S.G., W.G. Dore, L.P. Lefkovitch, and K.C. Armstrong. 1989. *Calamagrostis epigejos* (Poaceae) in North America, especially Ontario. Canad. J. Bot. 67:3205–3218.

Alex, J.F. 1987. Quackgrass: Origin, distribution, description and taxonomy. Pp. 1–16 *in* Quackgrass Action Committee Workshop Technical Proceedings: March 10 and 11, Westin Hotel, Winnipeg, Manitoba. Monsanto Canada Ltd.

Albee, B.J., L.M. Shultz, and S. Goodrich. 1988. Atlas of the Vascular Plants of Utah. Utah Museum of Natural History Occasional Publication No. 7. Utah Museum of Natural History, Salt Lake City, Utah. 670 pp.

Alford, M.H. 1999. The vascular flora of Amite County, Mississippi. M.S. thesis. Duke University, Raleigh, North Carolina. 176 pp.

Alinot, S.F. 1973. The Vascular Flora of Glen Helen, Clifton Gorge, and John Bryan State Park. Ohio Biological Survey, Biological Notes No. 5. Ohio State University, Columbus, Ohio. 49 pp.

Allen, C.M. 1992. Grasses of Louisiana. 2nd ed. Cajun Prairie Habitat Preservation Society, Eunice, Louisiana. 320 pp.

Allen, C.M., C.S. Reid, and C.H. Doffitt. 1999. *Bouteloua rigidiseta* (Poaceae) new to Louisiana. Sida 18:1285.

Allen, L. and M.L. Curto. 1996. Noteworthy collections. Madroño 43:337–338.

Allred, K.W. 1997. A Field Guide to the Grasses of New Mexico. 2nd ed. New Mexico Agricultural Experiment Station, Department of Agricultural Communications, New Mexico State University, Las Cruces, New Mexico. 258 pp.

Amoroso, J.L. and W.S. Judd. 1995. A floristic study of the Cedar Key Scrub State Reserve, Levy County, Florida. Castanea 60:210–232.

Anderson, D.E. 1961. Taxonomy and distribution of the genus *Phalaris*. Iowa State Coll. J. Sci. 36:1–96.

Anderson, L.C. 1995. Noteworthy plants from north Florida: VI. Sida 16:581–587.

Anderson, L.C. 2000. Noteworthy plants from north Florida: VII. Sida 19:211–216.

Anderson, L.C. and D.W. Hall. 1993. *Luziola bahiensis* (Poaceae): new to Florida. Sida 15:619–622.

Andreas, B.K. 1989. The vascular flora of the glaciated Allegheny Plateau region of Ohio. Bull. Ohio Biol. Surv., n.s., 8, no. 1:1–191.

Angelo, R. and D.E. Boufford. 1998. Atlas of the flora of New England: Poaceae. Rhodora 100:101–233. [see also http://neatlas.huh.harvard.edu/].

Anonymous. [Date unknown]. Camp Dodge—grasses, sedges, rushes [unpublished manuscript]. 2 pp.

Arizona Rare Plant Committee. 2001. Arizona Rare Plant Field Guide. Arizona Rare Plant Committee, Arizona [loose-leaf format, unpaginated].

Banks, D.L. and **S. Boyd.** 1998. Noteworthy collections. Madroño 45:85–86.

Barkley, T.M. (ed). 1977. Atlas of the Flora of the Great Plains. Iowa State University Press, Ames, Iowa. 600 pp.

Barrett, S.C.H. and **D.E. Seaman.** 1980. The weed flora of Californian rice fields. Aquatic Bot. 9:351–376.

Basinger, M.A. and **P. Rogertson.** 1996. Vascular flora and ecological survey of an old-growth forest remnant in the Ozark Hills of southern Illinois. Phytologia 80:352–367.

Beatley, J.C. 1969. Vascular plants of the Nevada Test Site, Nellis Air Force Range, and Ash Meadows. University of California Laboratory of Nuclear Medicine and Radiation Biology, Los Angeles, California. 122 pp.

Beatley, J.C. 1971. Vascular plants of Ash Meadows, Nevada. University of California Laboratory of Nuclear Medicine and Radiation Biology, Los Angeles, California. 59 pp.

Beatley, J.C. 1973. Additions to vascular plants of the Nevada Test Site, Nellis Air Force Range, and Ash Meadows. University of California Laboratory of Nuclear Medicine and Radiation Biology, Los Angeles, California. 19 pp.

Beatley, J.C. 1973. Check list of vascular plants of the Nevada Test Site and central-southern Nevada. Department of Biological Sciences, University of Cincinnati, Cincinnati, Ohio. 42 pp.

Best, C., J.T. Howell, W. Knight, I. Knight, and **M. Wells.** 1996. A Flora of Sonoma County: Manual of the Flowering Plants and Ferns of Sonoma County, California. California Native Plant Society, Sacramento, California. 347 pp.

Best, K.F., J.D. Banting, and **G.G. Bowes.** 1978. The biology of Canadian weeds. 31. *Hordeum jubatum* L. Canad. J. Pl. Sci. 58:699–708.

Bittner, R.T. and **D.J. Gibson.** 1998. Microhabitat relations of the rare reed bent grass, *Calamagrostis porteri* subsp. *insperata* (Poaceae), with implications for its conservation. Ann. Missouri Bot. Gard. 85:69–80.

Blondeau, M. and **J. Cayouette.** 2002. La Flore Vasculaire de la Baie Wakeham et du Havre Douglas, Nord-du-Québec, Détroit d'Hudson [unpublished manuscript].

Borowski, M. and **W.C. Holmes.** 1996. *Phyllostachys aurea* Riv. (Gramineae: Bambuseae) in Texas. Phytologia 80:30–34.

Bough, M., J.C. Colosi, and **P.B. Cavers.** 1985. The major weedy biotypes of proso millet (*Panicum miliaceum*) in Canada. Canad. J. Bot. 64:1188–1198.

Bowcutt, F. 1996. A floristic study of Delta Meadows River Park, Sacramento County, California. Madroño 43:417–431.

Boyd, S. 1998. Noteworthy collections. Madroño 45:326–328.

Boyle, W.S. 1945. A cyto-taxonomic study of the North American species of *Melica*. Madroño 8:1–26.

Braun, E.L. 1967. The Monocotyledoneae: Cat-tails to Orchids. The Vascular Flora of Ohio, Volume 1. Ohio State University Press, Columbus, Ohio. 464 pp.

Brind'amour, M. and **V. Lavoie.** 1985. Addition à la flore vasculaire des marais intertidaux du Saint-Laurent (Québec): *Spartina* ×*caespitosa* A.A. Eaton. Naturaliste Canad. 112:431–432.

Brown, L.E. 1993. The deletion of *Sporobolus heterolepis* (Poaceae) from the Texas and Louisiana floras, and the addition of *Sporobolus silveanus* to the Oklahoma flora. Phytologia 74:371–381.

Brown, L.E. and **I.S. Elsik.** 2002. Notes on the flora of Texas with additions and other significant records: II. Sida 20:437–444.

Brown, L.E. and **S.J. Marcus.** 1998. Notes on the flora of Texas with additions and other significant records. Sida 18:315–324.

Brown, L.E. and **J. Schultz.** 1991. *Arthraxon hispidus* (Poaceae), new to Texas. Phytologia 71:379–381.

Bruner, J.L. 1987. Systematics of the *Schizachyrium scoparium* (Poaceae) complex in North America. Ph.D. dissertation. Ohio State University, Columbus, Ohio. 167 pp.

Bryson, C.T. 1993. *Sacciolepis indica* (Poaceae) new to Mississippi. Sida 15:555.

Calder, J.A. and **R.L. Taylor.** 1968. Flora of the Queen Charlotte Islands, Part I: Systematics of the Vascular Plants. Canada Department of Agriculture Monograph No. 4, Part 1. Research Branch, Canada Department of Agriculture, Ottawa, Ontario. 659 pp.

CalFlora. 2000. Information on California plants for conservation, research, and education. http://www.calflora.org/.

Callihan, R.H. 1987. Eradication technology of matgrass (*Nardus strictus* L.) in the Clearwater National Forest. Department of Plant, Soil, and Entomological Sciences, University of Idaho, Moscow, Idaho. 51 pp.

Callihan, R.H. and **T.W. Miller.** 1998. Noxious weed identification guide. http://www.oneplan.org/Crop/noxWeeds/nxWeed00.htm.

Callihan, R.H. and **D. Pavek.** 1988. May, 1988 *Milium vernale* survey in Idaho County, Idaho. Department of Plant, Soil, and Entomological Sciences, University of Idaho, Moscow, Idaho. 6 pp.

Campbell, J.J.N. 1992. Atlas of the Kentucky Flora: First Release of Data on Ferns, Gymnosperms, and Monocots. Julian J.N. Campbell, Lexington, Kentucky. 296 pp.

Caplow, F. 2002. New species of *Spartina* in Washington. BEN [Botanical Electronic News] #286. http://www.ou.edu/cas/botany-micro/ben/.

Carter, J.L. 1962. The vascular flora of Cherokee County. Proc. Iowa Acad. Sci. 69:60–70.

Catling, P.M., D.S. Erskine, and R.B. MacLaren. 1985. The Plants of Prince Edward Island, with New Records, Nomenclatural Changes, and Corrections and Deletions. Research Branch, Agriculture Canada Publication No. 1798. Canadian Government Publishing Centre, Ottawa, Ontario. 272 pp.

Catling, P.M. and S.M. McKay. 1980. Halophytic plants in southern Ontario. Canad. Field-Naturalist 94:248–258.

Catling, P.M. and S.M. McKay. 1981. A review of the occurrence of halophytes in the eastern Great Lakes region. Michigan Bot. 20:167–179.

Catling, P.M., A.A. Reznicek, and J.L. Riley. 1977. Some new and interesting grass records from southern Ontario. Canad. Field-Naturalist 91:350–359.

Catling, P.M., A. Sinclair, and D. Cuddy. 2001. Vascular plants of a successional alvar burn 100 days after a severe fire and their mechanisms of re-establishment. Canad. Field-Naturalist 115:214–222.

Cayouette, J. 1987. La Flore vasculaire de la région du Lac Chavigny (58° 12' N, 75° 08' O[W]), Nouveau-Québec. Provancheria 20:1–51.

Cayouette, J. and S.J. Darbyshire. 1987. La répartition de *Danthonia intermedia* dans l'est du Canada. Naturaliste Canad. 114:217–220.

Cayouette, R. 1972. Additions à la flore adventice du Québec. Naturaliste Canad. 99:135–136.

Chester, E.W. 1996. Rare and noteworthy vascular plants from the Fort Campbell Military Reservation, Kentucky and Tennessee. Sida 17:269–274.

Chester, E.W., B.E. Wofford, H.R. DeSelm, and A.M. Evans. 1993. Atlas of Tennessee Vascular Plants. Pteridophytes, Gymnosperms, Angioisperms: Monocots, Vol. 1. Austin Peay State University Miscellaneous Publication No. 9. The Center for Field Biology, Austin Peay State University, Clarksville, Tennessee. 118 pp. [see also http://www.bio.utk.edu/botany/herbarium/vascular/vascular.html].

Chester, T.J. 2003. Flora of Torrey Pines State Reserve. http://tchester.org/plants/floras/coast/torrey_pines.html.

Clark, C. 1990. Vascular plants of the underdeveloped areas of California State Polytechnic University, Pomona. Crossosoma 16:1–7.

Clark, D.A. 1996. A Floristic Survey of the Mesa de Maya Region, Las Animas County, Colorado. Natural History Inventory of Colorado No. 17. University of Colorado Museum, Boulder, Colorado. 44 pp.

Clark, D.A. and T. Hogan. 2000. Noteworthy collections. Madroño 47:142–144.

Clay, K. 1995. Noteworthy collections. Castanea 60:84–85.

Clokey, I.W. 1951. Flora of the Charleston Mountains, Clark County, Nevada. University of California Publications in Botany No. 24. 274 pp.

Cochrane, T.S., M.M. Rice, and W.E. Rice. 1984. The flora of Rock County, Wisconsin: Supplement I. Michigan Bot. 23:121–133.

Cody, W.J. 1988. Plants of Riding Mountain National Park, Manitoba. Research Branch, Agriculture Canada Publication 1818/E. Canadian Government Publishing Centre, Ottawa, Ontario. 319 pp.

Cody, W.J. 1996. Flora of the Yukon Territory. NRC [National Resource Council of Canada], Ottawa, Ontario. 643 pp.

Cody, W.J., S.J. Darbyshire, and C.E. Kennedy. 1990. A bluegrass, *Poa pseudoabbreviata* Roshev., new to the flora of Canada, and some additional records from Alaska. Canad. Field-Naturalist 104:589–591.

Cody, W.J., C.E. Kennedy, and B. Bennett. 1998. New records of vascular plants in the Yukon Territory. Canad. Field-Naturalist 112:289–328.

Cody, W.J., C.E. Kennedy, and B. Bennett. 2000. New records of vascular plants in the Yukon Territory II. Canad. Field-Naturalist 114:417–443.

Cody, W.J., C.E. Kennedy, and B. Bennett. 2001. New records of vascular plants in the Yukon Territory III. Canad. Field-Naturalist 115:301–322.

Cody, W.J., K.L. MacInnes, J. Cayouette, and S.J. Darbyshire. 2000. Alien and invasive native vascular plants along the Norman Wells pipeline, District of Mackenzie, Northwest Territories. Canad. Field-Naturalist 114:126–137.

Cope, E.A. 1994. Further notes on beachgrasses (*Ammophila*) in northeastern North America. Newslett. New York Fl. Assoc. 5:5–7.

Crampton, B. 1955. *Scribneria* in California. Leafl. W. Bot. 7:219–220.

Crouch, V.E. and M.S. Golden. 1997. Floristics of a bottomland forest and adjacent uplands near the Tombigbee River, Choctaw County, Alabama. Castanea 62:219–238.

Curto, M.L. 1993. Grasses of San Luis Obispo County, California [unpublished manuscript]. 79 pp.

Curto, M.L. 1998. A new *Stipa* (Poaceae: Stipeae) from Idaho and Nevada. Madroño 45:57–63.

Curto, M.L. and L. Allen. 1992. California Native Plant Society grass walk: Cuyamaca Rancho State Park [brochure]. 16 pp.

Cusick, A.W. and G.M. Silberhorn. 1977. The vascular plants of unglaciated Ohio. Bull. Ohio Biol. Surv., n.s., 5, no. 4:1–157.

Cutler, H.C. and E. Anderson. 1941. A preliminary survey of the genus *Tripsacum*. Ann. Missouri Bot. Gard. 28:249–269.

Darbyshire, S.J. 1996. Tall wheatgrass (*Elytrigia elongata* (Host) Nevski; Poaceae) new to Nova Scotia [unpublished manuscript]. 6 pp.

Darbyshire, S.J. and J. Cayouette. 1989. The biology of Canadian weeds. 92. *Danthonia spicata* (L.) Beauv. *in* Roem. & Schult. Canad. J. Pl. Sci. 69:1217–1233.

Darbyshire, S.J. and **J. Cayouette**. 1995. Identification of the species in the *Panicum capillare* complex (Poaceae) from eastern Canada and adjacent New York State. Canad. J. Bot. 73:333–348.

Davis, J.I. 1990. *Puccinellia howellii* (Poaceae), a new species from California. Madroño 37:55–58.

Deam, C.C. 1940. Flora of Indiana. Department of Conservation, Division of Forestry, State of Indiana, Indianapolis, Indiana. 1236 pp.

DeSelm, H.R. 1975. *Schizachyrium stoloniferum* Nash var. *wolfei* DeSelm. Sida 6:114–115.

Deshaye, J. and **J. Cayouette**. 1988. La flore vasculaire des îles et de la presqu'île de Manitounuk, Baie d'Hudson: Structure phytogéographique et interprétation bioclimatique. Provancheria 21:1–74.

Dewey, S. 1997. Medusahead [*Taeniatherum caput-medusae*] 1995 status survey [unpublished report consisting of annotated maps]. 4 pp.

Diamond, A.R., Jr. and **J.D. Freeman**. 1993. Vascular flora of Conecuh Co., Alabama. Sida 15:623–638.

Diamond, A.R., Jr., M. Woods, J.A. Hall, and **B.H. Martin.** 2002. The vascular flora of the Pike County Pocosin Nature Preserve. Southeastern Naturalist 1:45–54.

Diggs, G.M., Jr., B.L. Lipscomb, and **R.H. O'Kennon.** 1999. Shinners & Mahler's Illustrated Flora of North Central Texas. Sida Botanical Miscellaney No. 16. Botanical Research Institute of Texas and Austin College, Fort Worth, Texas. 1626 pp.

Dignard, N. 2000. Additions récentes à la flore de l'île d'Anticosti. Ludoviciana 29:69–72.

Dix, W.L. 1945. Will the stowaway, *Molinia caerulea*, become naturalized? Bartonia 23:41–42.

Dore, W.G. and **C.J. Marchant.** 1968. Observations on the hybrid cord-grass, *Spartina* ×*caespitosa* in the maritime provinces. Canad. Field-Naturalist 82:181–184.

Dore, W.G. and **J. McNeill.** 1980. Grasses of Ontario. Research Branch, Agriculture Canada Monograph No. 26. Canadian Government Publishing Centre, Hull, Québec. 568 pp.

Douglas, B.J., A.G. Thomas, I.N. Morrison, and **M.G. Maw.** 1985. The biology of Canadian weeds. 70. *Setaria viridis* (L.) Beauv. Canad. J. Pl. Sci. 65:669–690.

Douglas, G.W., D.V. Meidinger, and **J.L. Penny.** 2002. Rare Native Plants of British Columbia. 2nd ed. Conservation Data Centre, Victoria, British Columbia. 358 pp.

Douglas, G.W., D.V. Meidinger, and **J.J. Pojar.** 2002. Illustrated Flora of British Columbia, Vol. 8 [revision, in part, of The Vascular Plants of British Columbia, Part 4 (1994)]. Ministry of Sustainable Resource Management, Victoria, British Columbia. 457 pp.

Douglas, G.W., G.B. Straley, and **D.V. Meldinger** (eds). 1994. The Vascular Plants of British Columbia. Part 4—Monocotyledons. Research Branch, Ministry of Forests, Victoria, British Columbia. 257 pp.

Douglas, G.W. and **R.J. Taylor.** 1970. Contributions to the flora of Washington. Rhodora 72:496–501.

Doyon, D., C.J. Bouchard, and **R. Néron.** 1986. Répartition géographique et importance dans les cultures de quatre adventices du Québec: *Abutilon theophrasti*, *Amaranthus powellii*, *Acalypha rhomboïdea*, et *Panicum dichotomiflorum.* Naturaliste Canad. 113:115–123.

Doyon, D., C.J. Bouchard, and **R. Néron.** 1988. Extension de la répartition géographique de *Setaria faberii* au Québec. Naturaliste Canad. 115:125–129.

Doyon, D. and **W.G. Dore.** 1967. Notes on the distribution of two grasses, *Sporobolus neglectus* and *Leersia virginica*, in Québec. Canad. Field-Naturalist 81:30–32.

Drew, M.B., L.K. Kirkman, and **A.K. Gholson, Jr.** 1988. The vascular flora of Ichauway, Baker County, Georgia: A remnant longleaf pine/wiregrass ecosystem. Castanea 63:1–24.

Dubé, M. 1983. Addition de *Festuca gigantea* à la flore du Canada. Ludoviciana 14:213–215.

Easterly, N.W. 1951. The flora of Iowa County. Proc. Iowa Acad. Sci. 58:71–95.

Eddy, T.L. 1983. A vascular flora of Green Lake County, Wisconsin. M.S. thesis. University of Wisconsin-Oshkosh, Oshkosh, Wisconsin. 130 pp.

Eddy, T.L. and **N.A. Harriman.** 1992. *Muhlenbergia richardsonis* in Wisconsin. Michigan Bot. 31:39–40.

Eilers, L.J. 1974. The flora of Brush Creek Canyon State Preserve. Proc. Iowa Acad. Sci. 81:150–157.

Eilers, L.J. and **D.M. Roosa.** 1994. The Vascular Plants of Iowa. University of Iowa Press, Iowa City, Iowa. 304 pp.

Emily de Camp Herbarium. 1999. Plant Communities of the Emily de Camp Herbarium, Island Beach State Park. Emily de Camp Herbarium, Seaside Park, New Jersey. 18 pp.

Ertter, B. 1997. Annotated Checklist of the East Bay Flora: Native and Naturalized Vascular Plants of Alameda and Contra Costa Counties, California. California Native Plant Society, East Bay Chapter, Berkeley, California. 114 pp.

Ertter, B. 1997. Grasses of Mt. Diablo [unpublished manuscript]. 28 pp.

Fassett, N.C. 1951. Grasses of Wisconsin: The Taxonomy, Ecology, and Distribution of the Gramineae Growing in the State Without Cultivation. University of Wisconsin Press, Madison, Wisconsin. 173 pp.

Fay, M.J. 1951. The flora of Cedar County, Iowa. Proc. Iowa Acad. Sci. 58:107–131.

Fay, M.J. and **R.F. Thorne.** 1953. Additions to the flora of Cedar County, Iowa. Proc. Iowa Acad. Sci. 60:122–130.

Ferlatte, W.J. 1974. A Flora of the Trinity Alps of Northern California. University of California Press, Berkeley, California. 206 pp.

Fishbein, M. 1995. Noteworthy collections. Madroño 42:83.

Fleming, G.P. and **C. Ludwig.** 1996. Noteworthy collections. Castanea 61:89–95.

Fleming, K.M., J.R. Singhurst, and **W.C. Holmes.** 2002. Vascular flora of Big Lake Bottom Wildlife Management Area, Anderson County, Texas. Sida 20:355–372.

Fleming, P. and **R. Kanal.** 1995. Annotated checklist of vascular plants of Rock Creek Park, National Park Service, Washington, D.C. Castanea 60:283–316.

Fleurbec. 1985. Plantes Sauvages du Bord de la Mer: Guide d'Identification Fleurbec. Groupe Fleurbec, Saint-Augustin, Québec. 286 pp.

Flora of Texas Consortium. 2000. Herbarium Specimen Browser. http://www.csdl.tamu.edu/FLORA/tracy/main1.html.

Fox, W.E., III and **S.L. Hatch.** 1996. *Brachiaria eruciformis* and *Urochloa brizantha* (Poaceae: Paniceae) new to Texas. Sida 17:287–288.

Franklin, J.F. and **C. Wiberg.** 1979. Goat Marsh Research Natural Area. Federal Research Natural Areas in Oregon and Washington: A Guidebook for Scientists and Educators, Supplement No. 10. U.S. Department of Agriculture–Forest Service, Pacific Northwest Forest and Range Experiment Station, Corvallis, Oregon. 19 pp.

Freckmann, R.W. 1972. Grasses of Central Wisconsin. Reports on the Fauna and Flora of Wisconsin, Report No. 6. Museum of Natural History, Stevens Point, Wisconsin. 81 pp.

Freeman, C.C., R.L. McGregor, and **C.A. Morse.** 1998. Vascular plants new to Kansas. Sida 18:593–604.

Freese, E.L. [date unknown]. Vascular flora checklist, Cedar Hills sand prairie, Black Hawk County, Iowa [unpublished manuscript]. 3 pp.

Frelich, L. 1979. Vascular Plants of Newport State Park, Wisconsin. Department of Natural Resources Research Report No. 100. Department of Natural Resources, Madison, Wisconsin. 34 pp.

Friends of Mount Revelstoke and Glacier. 1996. Vascular Plant Checklist: Mount Revelstoke and Glacier National Parks. Natural History Handbook No. 1. Friends of Mount Revelstoke and Glacier, Revelstoke, British Columbia. 19 pp.

Gandhi, K.N. 1989. A biosystematic study of the *Schizachyrium scoparium* complex. Ph.D. dissertation. Texas A&M University, College Station, Texas. 188 pp.

Garlitz, R. 1998. Rare and interesting grass records from the northeastern Lower Peninsula. Michigan Bot. 28:67–71.

Garlitz, R. and **D. Garlitz.** 1986. *Bouteloua gracilis*, a new grass to Michigan. Michigan Bot. 25:123–124.

Gervias, C., R. Trahan, D. Moreno, and **A.-M. Drolet.** 1993. Le *Phragmites australis* au Québec: Distribution géographique, nombres chromosomiques et reproduction. Canad. J. Bot. 71:1386–1393.

Gillett, G.W., J.T. Howell, and **H. Leschke.** 1995. A Flora of Lassen Volcanic National Park, California, rev. Vernon H. Oswald, David W. Showers, and Mary Ann Showers. Revised ed. California Native Plant Society, Sacramento, California. 216 pp.

Gilman, A.V. 1993. Four recent additions to the vascular flora of Vermont. Maine Naturalist 1:31–32.

Gould, F.W. 1969. Taxonomy of the *Bouteloua repens* complex. Brittonia 21:261–274.

Gould, F.W. 1979. The genus *Bouteloua* (Poaceae). Ann. Missouri Bot. Gard. 66:348–416.

Gould, F.W., M.A. Ali, and **D.E. Fairbrothers.** 1972. A revision of *Echinochloa* in the United States. Amer. Midl. Naturalist 87:36–59.

Gould, F.W. and **Z.J. Kapadia.** 1964. Biosystematic studies in the *Bouteloua curtipendula* complex. II. Taxonomy. Brittonia 16:182–207.

Grant, M.L. 1950. Dickinson County flora: A preliminary check-list of the vascular plants of Dickinson County, Iowa, based largely on the herbarium of the Iowa Lakeside Laboratory. Proc. Iowa Acad. Sci. 57:91–129.

Grant, M.L. 1953. Additions to and notes on the flora of Dickinson County, Iowa. Proc. Iowa Acad. Sci. 60:131–140.

Gray, J.R. 1974. The genus *Tripsacum* L. (Gramineae): Taxonomy and chemosystematics. Ph.D. dissertation. University of Illinois at Urbana-Champaign, Urbana, Illinois. 191 pp.

Griffin, J.R. 1990. Flora of Hastings Reservation, Carmel Valley, California. 3rd ed. Hastings Natural History Reservation, University of California-Berkeley, Berkeley, California. 92 pp.

Guala, G.F. 1988. *Poa bulbosa* L. (Poaceae) in Michigan. Michigan Bot. 27:13–14.

Guldner, L.F. 1960. The Vascular Plants of Scott and Muscatine Counties, with Some Reference to Adjoining Areas of Surrounding Counties in Iowa and to Rock Island and Whiteside Counties in Illinois. Davenport Public Museum Publications in Botany No. 1. Davenport Public Museum, Davenport, Iowa. 228 pp.

Hall, D.W. 1982. *Sorghastrum* (Poaceae) in Florida. Sida 9:302–308.

Hall, H.M. 1902. A Botanical Survey of San Jacinto Mountain. University of California Publications in Botany No. 1. 140 pp.

Hallsten, G.P., Q.D. Skinner, and **A.A. Beetle.** 1987. Grasses of Wyoming. 3rd ed. Research Journal No. 202. Agricultural Experiment Station, University of Wyoming, Laramie, Wyoming. 432 pp.

Hansen, B.F. and **R.P. Wunderlin.** 1996. Grasses of Florida: A Checklist of the Poaceae of Florida Along with Their Distribution by County. Institute for Systematic Botany, University of South Florida, Tampa, Florida. 133 pp. [see also http://www. plantatlas.usf.edu/].

Harris, S.K. 1975. The Flora of Essex County, Massachusetts. Peabody Museum, Salem, Massachusetts. 81 pp.

Hartley, T.G. 1966. The Flora of the "Driftless Area". University of Iowa Studies in Natural History, Vol. XXI, No. 1. University of Iowa, Iowa City, Iowa. 174 pp.

Harvey, L.H. 1948. *Eragrostis* in North and Middle America. Ph.D. dissertation. University of Michigan, Ann Arbor, Michigan. 270 pp.

Harvill, A.M., Jr., T.R. Bradley, C.E. Stevens, T.F. Wieboldt, D.M.E. Ware, D.W. Ogle, G.W. Ramsey, and G.P. Fleming. 1992. Atlas of the Virginia Flora. 3rd ed. Virginia Botanical Associates, Burkeville, Virginia. 144 pp.

Hatch, S.L. 1975. A biosystematic study of the *Schizachyrium cirratum–Schizachyrium sanguineum* complex (Poaceae). Ph.D. dissertation. Texas A&M University, College Station, Texas. 112 pp.

Hatch, S.L., W.E. Fox, III, and J.E. Dawson, III. 1998. *Triraphis mollis* (Poaceae: Arundineae), a species reported new to the United States. Sida 18:365–368.

Hatch, S.L., D.J. Rosen, J.A. Thomas, and J.E. Dawson, III. 1998. *Luziola peruviana* (Poaceae: Oryzeae) previously unreported from Texas and a key to Texas species. Sida 18:611–614.

Hawkins, T.K. and E.L. Richards. 1995. A floristic study of two bogs on Crowley's Ridge in Greene County, Arkansas. Castanea 60:233–244.

Hay, S.G. 1989. La migration récente de plantes halophytes dans la région Montréalaise. Quatre-Temps 13:7–11.

Hayden, A. 1943. A botanical survey in the Iowa lake region of Clay and Palo Alto Counties. Iowa State Coll. J. Sci. 17:277–416.

Hays, J. 1995. A floristic survey of Falls Hollow sandstone glades, Pulaski County, Missouri. Phytologia 78:264–276.

Hämet-Ahti, L. 1965. Vascular plants of Wells Gray Provincial Park and its vicinity, in eastern British Columbia. Ann. Bot. Fenn. 2:138–164.

Heidel, B. 1996. Noteworthy collections. Madroño 43:436–440.

Heise, K.L. and A.M. Merenlender. 1999. Flora of a vernal pool complex in the Mayacmas Mountains of southeastern Mendocino County, California. Madroño 46:38–45.

Hellquist, C.E. and G.E. Crow. 1997. The bryophyte and vascular flora of Little Dollar Lake peatland, Mackinac County, Michigan. Rhodora 99:195–222.

Henry, L.K. 1978. Vascular Flora of Bedford County, Pennsylvania: An Annotated Checklist. Carnegie Museum of Natural History, Pittsburgh, Pennsylvania. 29 pp.

Herrera-Arrieta, Y. 1998. A revision of the *Muhlenbergia montana* (Nutt.) Hitchc. complex (Poaceae: Chloridoideae). Brittonia 50:23–50.

Hinds, H. 2000. Flora of New Brunswick. 2nd ed. Department of Biology, University of New Brunswick, Fredericton, New Brunswick. 695 pp.

Hodgdon, A.R., G.E. Crow, and F.L. Steele. 1979. Grasses of New Hampshire: I. Tribes Poeae (Festuceae) and Triticeae (Hordeae). New Hampshire Agricultural Experiment Station Bulletin No. 512. New Hampshire Agricultural Experiment Station, University of New Hampshire, Durham, New Hampshire. 53 pp.

Holiday, S. 2000. A floristic study of Tsegi Canyon, Arizona. Madroño 47:29–42.

Horton, D. 1998. Swamp White Oak Savanna Preserve (Muscatine Co., Iowa) plant list (by family) [unpublished manuscript]. 2 pp.

Hough, M.Y. 1983. New Jersey Wild Plants. Harmony Press, Harmony, New Jersey. 414 pp.

Hounsell, R.W. and E.C. Smith. 1966. Contributions of the flora of Nova Scotia. VIII. Distribution of arctic-alpine and boreal disjuncts. Rhodora 68:409–419.

Howell, J.T. 1970. Marin Flora: Manual of the Flowering Plants and Ferns of Marin County, California. 2nd ed. University of California Press, Berkeley, California. 366 pp.

Howell, J.T. 1979. A reconsideration of *Trisetum projectum* (Gramineae). Wasmann J. Biol. 37:21–23.

Howell, J.T., G.H. True, and C. Best. 1981. Notes on Marin County plants 1970–1980. Four Seasons 6:3–6.

Hrusa, F., B. Ertter, A. Sanders, G. Leppig, and E. Dean. 2002. Catalogue of non-native vascular plants occurring spontaneously in California beyond those addressed in *The Jepson Manual*—Part I. Madroño 49:61–98.

Hunt, D.M. and R.E. Zaremba. 1992. The northeastward spread of *Microstegium vimineum* (Poaceae) into New York and adjacent states. Rhodora 94:167–170.

Illinois Natural History Survey. 2002. Database of grasses in the Illinois Natural History Survey herbarium (ILLS). http://www.inhs.uiuc.edu/cbd/collections/ plants.html.

Iverson, L.R., D. Ketzner, and J. Karnes. 2002. Illinois Plant Information Network (ILPIN) database. http://www. fs.fed.us/ne/delaware/ilpin/ilpin.html.

Jacobson, A.L., F.C. Weinmann, and P.F. Zika. 2001. Noteworthy collections. Madroño 48:213–214.

John, T. 1993. City of Rocks vascular flora [unpublished manuscript]. 2 pp.

Johnson, A.E. and A. Blyth. 1988. Rediscovery of *Calamovilfa curtissii* (Gramineae) in the Florida panhandle. Sida 13:137–140.

Johnson-Groh, C.L. and D.R. Farrar. 1985. Flora and phytogeographical history of Ledges State Park, Boone County, Iowa. Proc. Iowa Acad. Sci. 92:137–143.

Johnson-Groh, C.L., D.Q. Lewis, and J.F. Shearer. 1987. Vegetation communities and flora of Dolliver State Park, Webster County, Iowa. Proc. Iowa Acad. Sci. 94:84–88.

Jones, S.B., Jr. and **N.C. Coile.** 1988. The Distribution of the Vascular Flora of Georgia. University of Georgia, Athens, Georgia. 230 pp.

Jones, S.D. and **G. Jones.** 1992. *Cynodon nlemfuënsis* (Poaceae: Chlorideae) previously unreported in Texas. Phytologia 72:93–95.

Jones, S.D. and **J.K. Wipff.** 1992. *Eustachys retusa* (Poaceae), the first report in Florida and a key to *Eustachys* in Florida. Phytologia 73:274–276.

Judziewicz, E.J. and **R.G. Koch.** 1993. Flora and vegetation of the Apostle Islands National Lakeshore and Madeline Island, Ashland and Bayfield Counties, Wisconsin. Michigan Bot. 32:43–193.

Junak, S., T. Ayers, R. Scott, D. Wilken, and **D. Young.** 1995. A Flora of Santa Cruz Island. Santa Barbara Botanic Garden, Santa Barbara, California. 397 pp.

Kartesz, J. and **H.N. Mozingo.** 1982. Preliminary checklist of grasses for Nevada [unpublished manuscript]. 31 pp.

Kearney, T.H. and **R.H. Peebles.** 1951. Arizona Flora. University of California Press, Berkeley and Los Angeles, California. 1032 pp.

Keck, D.D. 1949. *Poa* of western North America [unpublished typewritten manuscript; manuscript and annotations on file at the United States National Herbarium]. 288 pp.

Kennedy, C.E., C.A. Scott Smith, and **D.A. Cooley.** 2001. Observations of change in the cover of polargrass, *Arctagrostis latifolia*, and arctic lupine, *Lupinus arcticus*, in upland tundra on Herschel Island, Yukon Territory. Canad. Field-Naturalist 115:323–328.

Kiger, R.W. 1971. *Arthraxon hispidus* (Gramineae) in the United States: Taxonomic and floristic status. Rhodora 73:39–46.

Krings, A. 2002. Additions to the flora of Nags Head Woods (Dare County, North Carolina) and the Outer Banks of North Carolina. Sida 20:839–843.

L'abrecque, J. 2002. La situation de la muhlenbergie ténue variété ténue (*Muhlenbergia tenuiflora* var. *tenuiflora*) au Québec [unpublished manuscript].

Labrecque, J. and **G. Lavoie.** 2002. Les Plantes Vasculaires Menacées ou Vulnérables du Québec. Ministère de l'Environnement, Direction du Patrimoine Écologique et du Développement Durable, Québec City, Québec. 118 pp.

Lackschewitz, K. 1991. Vascular Plants of West-Central Montana—Identification Guidebook. General Technical Report INT-277. U.S. Department of Agriculture–Forest Service, Intermountain Research Station, Ogden, Utah. 648 pp.

Lammers, T.G. 1980. The vascular flora of Starr's Cave State Preserve. Proc. Iowa Acad. Sci. 87:148–158.

Lammers, T.G. 1983. The vascular flora of Des Moines County, Iowa. Proc. Iowa Acad. Sci. 90:55–71.

Landry, G.P. 1996. Noteworthy collections. Castanea 61:197.

Lange, K.I. 1998. Flora of Sauk County and Caledonia Township, Columbia County, South Central Wisconsin. Technical Bulletin No. 190. Department of Natural Resources, Madison, Wisconsin. 169 pp.

Lawrence, D.L. and **J.T. Romo.** 1995. Tree and shrub communities of wooded draws near the Matador Research Station in southern Saskatchewan. Canad. Field-Naturalist 108:397–409.

Layser, E.F., Jr. 1980. Flora of Pend Oreille County, Washington. Washington State University Cooperative Extension, Pullman, Washington. 146 pp.

Lelong, M.G. 1977. Annotated list of vascular plants in Mobile, Alabama. Sida 7:118–146.

Lelong, M.G. 1988. Noteworthy monocots of Mobile and Baldwin Counties, Alabama. Sida 13:101–113.

Lesica, P. 1985. Checklist of the Vascular Plants of Glacier National Park. Proceedings of the Montana Academy of Sciences Monograph No. 4. Montana Academy of Sciences, Missoula, Montana. 42 pp.

Lesica, P. 1994. Noteworthy collections. Madroño 41:231.

Lesica, P. 1998. Noteworthy collections. Madroño 45:328–330.

Lewis, M.E. 1971. Flora and major plant communities of the Ruby–East Humboldt Mountains, with special emphasis on Lamoille Canyon. Report to the U.S. Forest Service, Region 4, Humboldt National Forest, Nevada. 62 pp.

Lloyd, R.M. and **R.S. Mitchell.** 1965. Plants of the White Mountains, California and Nevada. Revised ed. University of California White Mountain Research Station, Berkeley, California. 61 pp.

Lomer, F. 1996. Introduced bog plants, Vancouver, British Columbia. BEN [Botanical Electronic News] #128. http://www.ou.edu/cas/botany-micro/ben/.

Lomer, F. 1996. Six new introduced species in British Columbia. BEN [Botanical Electronic News] #128. http://www.ou.edu/cas/botany-micro/ben/.

Lomer, F. 2001. Ephemeral introductions of vascular plants around Vancouver, British Columbia (Part 2). BEN [Botanical Electronic News] #270. http://www.ou.edu/cas/botany-micro/ben/.

Lonard, R.I. 1993. Guide to the Grasses of the Lower Rio Grande Valley, Texas. University of Texas-Pan American, Edinburg, Texas. 240 pp.

MacDonald, J. 1996. A survey of the flora of Monroe County, Mississippi. M.S. thesis. Mississippi State University, Mississippi State, Mississippi. 162 pp.

MacRoberts, B.R. and **M.H. MacRoberts.** 1993. Vascular flora of sandstone outcrop communities in western Louisiana, with notes on rare and noteworthy species. Phytologia 75:463–480.

MacRoberts, B.R. and **M.H. MacRoberts.** 1995. Floristics of xeric sandhills in northwestern Louisiana. Phytologia 79:123–131.

MacRoberts, B.R. and **M.H. MacRoberts**. 1995. Noteworthy vascular plant collections on the Kisatchie National Forest, Louisiana. Phytologia 78:291–313.

MacRoberts, B.R. and **M.H. MacRoberts**. 1995. Vascular flora of two calcareous prairie remnants on the Kisatchie National Forest. Phytologia 78:18–27.

MacRoberts, B.R. and **M.H. MacRoberts**. 1996. The floristics of calcareous prairies on the Kisatchie National Forest, Louisiana. Phytologia 81:35–43.

MacRoberts, B.R. and **M.H. MacRoberts**. 1996. Floristics of xeric sandhills in east Texas. Phytologia 80:1–7.

MacRoberts, B.R. and **M.H. MacRoberts**. 1997. Floristics of beech-hardwood forest in east Texas. Phytologia 82:20–29.

MacRoberts, B.R. and **M.H. MacRoberts**. 1998. Floristics of muck bogs in east central Texas. Phytologia 85:61–73.

MacRoberts, B.R. and **M.H. MacRoberts**. 1998. Floristics of wetland pine savannas in the Big Thicket National Preserve, southeast Texas. Phytologia 85:40–50.

MacRoberts, B.R. and **M.H. MacRoberts**. 1998. Noteworthy vascular plant collections on the Angelina and Sabine National Forests, Texas. Phytologia 84:1–27.

MacRoberts, B.R., M.H. MacRoberts, and **L.E. Brown**. 2002. Annotated checklist of the vascular flora of the Hickory Creek unit of the Big Thicket National Preserve, Tyler County, Texas. Sida 20:781–795.

Magee, D.M. 1993. Manual of the Vascular Flora of New England and Adjacent New York. Normandeau Associates, Inc., Bedford, New Hampshire. 49 pp.

Maley, A. 1994. A Floristic Survey of the Black Forest of the Colorado Front Range. Natural History Inventory of Colorado No. 14. University of Colorado Museum, Boulder, Colorado. 31 pp.

Mansfield, D.H. 1995. Vascular Flora of Steens Mountain, Oregon. Journal of the Idaho Academy of Science, Vol. 31, No. 2. Idaho Academy of Science, Moscow, Idaho. 88 pp.

Matthews, M.A. 1997. An Illustrated Field Key to the Flowering Plants of Monterey County and Ferns, Fern Allies, and Conifers. California Native Plant Society, Sacramento, California. 401 pp.

Maze, J. and **K.A. Robson**. 1996. A new species of *Achnatherum* (*Oryzopsis*) from Oregon. Madroño 43:393–403.

McClain, W.E. and **J.E. Ebinger**. 2002. A comparison of the vegetation of three limestone glades in Calhoun County, Illinois. Southeastern Naturalist 1:179–188.

McClintock, E., P. Reeberg, and **W. Knight**. 1990. A Flora of the San Bruno Mountains, San Mateo County, California. Special Publication No. 8. California Native Plant Society, Sacramento, California. 223 pp.

McNeill, J. 1981. *Apera*, silky-bent or windgrass, an important weed genus recently discovered in Ontario, Canada. Canad. J. Pl. Sci. 61:479–485.

Meeks, D.N. 1984. A floristic study of northern Tippah County, Mississippi. M.A. thesis. Mississippi State University, Missippi State, Mississippi. 69 pp.

Michener-Foote, J. and **T. Hogan**. 1999. The Flora and Vegetation of the Needle Mountains, San Juan Range, Southwestern Colorado. Natural History Inventory of Colorado No. 18. University of Colorado Museum, Boulder, Colorado. 39 pp.

Mitchell, R.S. and **G.C. Tucker**. 1997. Revised Checklist of New York State Plants. Contributions to a Flora of New York State, Checklist IV. New York State Museum Bulletin No. 490. State Education Department, University of the State of New York, Albany, New York. 400 pp.

Mitchell, W.W. and **H.J. Hodgson**. 1965. The status of hybridization between *Agropyron sericeum* and *Elymus sibiricus* in Alaska. Canad. J. Bot. 43:855–859.

Moe, L.M. and **E.C. Twisselmann**. 1995. A Key to Vascular Plant Species of Kern County, California and A Flora of Kern County, California. California Native Plant Society, Sacramento, California. 620 pp. [Flora section reprinted from Wasmann J. Biol. 25:1–395 (1967)].

Mohlenbrock, R.H. and **D.M. Ladd**. 1978. Distribution of Illinois Vascular Plants. Southern Illinois University Press, Carbondale, Illinois. 282 pp.

Montana Natural History Program. 1996. Animal and plant species of Beaverhead County. http://nris.state.mt.us/index.html.

Montana State University. 1997. Checklist of the grasses of Montana. http://gemini.oscs.montana.edu/~mlavin/herb/grass.htm.

Morris, M.W. 1987. The vascular flora of Grenada County, Mississippi. M.S. thesis. Mississippi State University, Missippi State, Mississippi. 123 pp.

Morris, M.W. 1997. Contributions to the flora and ecology of the northern longleaf pine belt in Rankin County, Mississippi. Sida 17:615–626.

Morrone, O., A.S. Vega, and **F.O. Zuloaga**. 1996. Revisión de las especies del género *Paspalum* L. (Poaceae: Panicoideae: Paniceae), grupo Dissecta (*s. str.*). Candollea 51:103–138.

Morrone, O. and **F.O. Zuloaga**. 1992. Revisión de las especies sudamericanas nativas e introducidas de los géneros *Brachiaria* y *Urochloa* (Poaceae: Panicoideae: Paniceae). Darwiniana 31:43–109.

Morrone, O. and **F.O. Zuloaga**. 1993. Synopsis del género *Urochloa* (Poaceae: Panicoideae: Paniceae) para México y América Central. Darwiniana 32:59–75.

Morton, J.K. and **J.M. Venn**. 1984. The Flora of Manitoulin Island and the Adjacent Islands of Lake Huron, Georgian Bay and the North Channel. 2nd revised ed. Department of Biology, University of Waterloo, Waterloo, Ontario. 106 pp.

Moseley, R.K. 1996. Vascular flora of subalpine parks in the Coeur d'Alene River drainage, northern Idaho. Madroño 43:479–492.

Moss, E.H. 1983. Flora of Alberta: A Manual of Flowering Plants, Conifers, Ferns and Fern Allies Found Growing Without Cultivation in the Province of Alberta, Canada, rev. John G. Packer. Revised ed. University of Toronto Press, Toronto, Ontario. 687 pp.

Mount Allison University. 2002. Native flora database [area near Sackville, New Brunswick]. http://www.mta.ca/~rthompso/nativeflora/floraindex.html.

Munro, M. 1998. Roland's Flora of Nova Scotia, Vol. 2. 3rd ed. Nova Scotia Museum and Nimbus Publishing, Province of Nova Scotia. 1297 pp.

Musselman, L.J., T.S. Cochrane, W.E. Rice, and M.M. Rice. 1971. The flora of Rock County, Wisconsin. Michigan Bot. 10:145–205.

Naczi, R.F.C., R.L. Jones, F.J. Metzmeier, M.A. Gorton, and **T.J. Weckman.** 2002. Native flowering plant species new or otherwise significant in Kentucky. Sida 20:397–402.

Naumann, T. 1996. Plant List, Including Scientific and Common Names, Dinosaur National Monument. Dinosaur Nature Association, Vernal, Utah. 20 pp.

Negrete, I.G., A.D. Nelson, J.R. Goetze, L. Macke, T. Wilburn, and **A. Day.** 1999. A checklist for the vascular plants of Padre Island National Seashore. Sida 18:1227–1245.

Nelson, J.B. and **K.B. Kelly.** 1997. Noteworthy collections. Castanea 62:283–287.

Nesom, G.L. and **L.E. Brown.** 1998. Annotated checklist of the vascular plants of Walker, Montgomery, and San Jacinto Counties, east Texas. Phytologia 84:98–106.

New York Flora Association. 1990. Preliminary Vouchered Atlas of New York State Flora. New York State Museum Institute, Albany, New York. 496 pp.

New York Flora Association. 1990–1998. Miscellaneous articles and species lists. Newslett. New York Fl. Assoc. 1–9.

Neyland, R., B.J. Hoffman, M. Mayfield, and **L.E. Urbatsch.** 2000. A vascular flora survey of Calcasieu Parish, Louisiana. Sida 19:361–386.

Niemann, D.A. and **R.Q. Landers, Jr.** 1974. Forest communities in Woodman Hollow State Preserve, Iowa. Proc. Iowa Acad. Sci. 81:176–184.

Northam, F.E. 1995. Range extension of southwestern cupgrass (*Eriochloa acuminata*) into Kansas. Trans. Kansas Acad. Sci. 98:68–71.

Northam, F.E. and **R.H. Callihan.** 1992. Morphology and phenology of interrupted windgrass in northern Idaho. J. Idaho Acad. Sci. 28:15–19.

Northam, F.E., R.H. Callihan, and **R.R. Old.** 1989. *Sporobolus vaginiflorus* (Torrey *ex* Gray) Wood: Biology and pest implications of an alien grass recorded in Idaho. J. Idaho Acad. Sci. 25:49–55.

Northam, F.E., R.H. Callihan, and **R.R. Old.** 1991. Range extensions of four introduced grasses in Idaho. J. Idaho Acad. Sci. 27:19–21.

Northam, F.E., R.R. Old, and **R.H. Callihan.** 1993. Little lovegrass (*Eragrostis minor*) distribution in Idaho and Washington. Weed Technol. 7:771–775.

Ohio Department of Natural Resources. 2002. Database of county distributions for Ohio's rare plants. http://www.dnr.state.oh.us/dnap/heritage/corange.html.

Oldham, M.J. 2001. Hardgrass (*Sclerochloa dura*), new to British Columbia. BEN [Botanical Electronic News] #269. http://www.ou.edu/cas/botany-micro/ben/.

Oldham, M.J. and **S.J. Darbyshire.** 1993. The adventive grasses, *Apera interrupta* and *Deschampsia danthonoides*, new to Maine. Maine Naturalist 1:231–232.

Oldham, M.J., S.J. Darbyshire, D. McLeod, D.A. Sutherland, D. Tiedje, and **J.M. Bowles.** 1995. New and noteworthy Ontario grass (Poaceae) records. Michigan Bot. 34:105–132.

Oswald, V. and **L. Ahart.** 1994. Flora of Butte County, California. California Native Plant Society, Sacramento, California. 348 pp.

Ownbey, G.B. and **T. Morley.** 1991. Vascular Plants of Minnesota: A Checklist and Atlas. University of Minnesota, Minneapolis, Minnesota. 307 pp.

Parker, D.S. and **J.M. Stucky.** 1995. Southwestern cupgrass (*Eriochloa acuminata* (Presl) Kunth (Poaceae: Paniceae) in North Carolina [unpublished manuscript]. 16 pp.

Peck, J.H., L.J. Eilers, and **D.M. Roosa.** 1978. The vascular plants of Fremont County, Iowa. Iowa Bird Life 48:3–24.

Peck, J.H., B.W. Haglan, L.J. Eilers, D.M. Roosa, and **D. vander Zee.** 1984. Checklist of the vascular flora of Lyon and Sioux Counties, Iowa. Proc. Iowa Acad. Sci. 91:92–97.

Peck, J.H., T.G. Lammers, B.W. Haglan, D.M. Roosa, and **L.J. Eilers.** 1981. A checklist of the vascular flora of Lee County, Iowa. Proc. Iowa Acad. Sci. 88:159–171.

Peck, J.H., D.M. Roosa, and **L.J. Eilers.** 1980. A checklist of the vascular flora of Allamakee County, Iowa. Proc. Iowa Acad. Sci. 87:62–75.

Peet, R.K. 1993. A taxonomic study of *Aristida stricta* and *A. beyrichiana*. Rhodora 95:25–37.

Perkins, B.E. and **T.S. Patrick.** 1980. Status report on Tennessee populations of *Calamovilfa arcuata*. University of Tennessee, Knoxville, Tennessee. 13 pp.

Peterson, P.M. 1986. A flora of the Cottonwood Mountains, Death Valley National Monument, California. Wasmann J. Biol. 44:73–126.

Peterson, P.M., E.E. Terrell, E.C. Uebel, C. Davis, H. Scholz, and **R.J. Soreng.** 1999. *Oplismenus hirtellus* subspecies *undulatifolius*, a new record for North America. Castanea 64:201–202.

Plunkett, G.M. and **G.W. Hall.** 1995. The vascular flora and vegetation of western Isle of Wight County, Virginia. Castanea 60:30–59.

Pohl, R.W. 1959. Introduced weedy grasses in Iowa. Proc. Iowa Acad. Sci. 66:160–162.

Pohl, R.W. 1966. The grasses of Iowa. Iowa State Coll. J. Sci. 40:341–566.

Poole, J.P. 1978. An addition to the flora of the Gaspé Peninsula. Rhodora 80:154.

Popovich, S.J. and D. Henderson. 1994. Noteworthy collections. Madroño 41:149–150.

Popovich, S.J., W.D. Shepperd, D.W. Reichert, and M.A. Cone. 1993. Flora of the Fraser Experimental Forest, Colorado. General Technical Report RM-233. U.S. Department of Agriculture–Forest Service, Rocky Mountain Forest and Range Experiment Station, Fort Collins, Colorado. 62 pp.

Porsild, A.E. and W.J. Cody. 1980. Vascular Plants of the Continental Northwest Territories, Canada. National Museum of Natural Sciences, National Museum of Canada, Ottawa, Ontario. 667 pp.

Powell, A.M. 1994. Grasses of the Trans-Pecos and Adjacent Areas. University of Texas Press, Austin, Texas. 377 pp.

Provance, M.C. and A.C. Sanders. 2000. Noteworthy collections. Madroño 47:139–141.

Radford, A.E., H.E. Ahles, and C.R. Bell. 1965. Atlas of the Vascular Flora of the Carolinas. North Carolina Agricultural Experiment Station Technical Bulletin No. 165. North Carolina Agricultural Experiment Station, University of North Carolina-Raleigh, Raleigh, North Carolina. 208 pp.

Raven, P.H., H.J. Thompson, and B.A. Prigge. 1986. Flora of the Santa Monica Mountains, California. 2nd ed. Southern California Botanists, Special Publication No. 2. University of California-Los Angeles, Los Angeles, California. 181 pp.

Rawinski, T.J., M.N. Rasmussen, and S.C. Rooney. 1989. Discovery of *Sporobolous asper* (Poaceae) in Maine. Rhodora 91:220–221.

Read, J.C. and B.J. Simpson. 1992. Documented chromosome numbers 1992: 3. Documentation and notes on the distribution of *Melica montezumae*. Sida 15:151–152.

Redman, D.E. 1995. Distribution and habitat types for Nepal microstegium [*Microstegium vimineum* (Trin.) Camus] in Maryland and the District of Columbia. Castanea 60:270–275.

Redman, D.E. 1995. Noteworthy collections. Castanea 60:82–84.

Reed, C.F. 1964. A flora of the chrome and manganese ore piles at Canton, in the Port of Baltimore, Maryland and at Newport News, Virginia, with descriptions of genera and species new to the flora of the eastern United States. Phytologia 10:321–405.

Reeder, J.R. 1991. A new species of *Panicum* (Gramineae) from Arizona. Phytologia 71:300–303.

Reeder, J.R. 1995. *Stipa tenuissima* (Gramineae) in Arizona—a comedy of errors. Madroño 41:328–329.

Reeder, J.R. 2001. Noteworthy collections. Madroño 48:212–213.

Reeder, J.R. and C.G. Reeder. 1990. *Bouteloua eludens*: Elusive indeed, but not rare. Desert Pl. 10:19–22, 31.

Reeder, J.R. and C.G. Reeder. 1980. Systematics of *Bouteloua breviseta* and *B. ramosa* (Gramineae). Syst. Bot. 5:312–321.

Rhoads, A.F. and W.M. Klein, Jr. 1993. The Vascular Flora of Pennsylvania: Annotated Checklist and Atlas. American Philosophical Society, Philadelphia, Pennsylvania. 636 pp.

Rice, P. 1999. INVADERS database for early detection and tracking of invasive alien plants and weedy natives. http://invader.dbs.umt.edu/.

Richards, C.D., F. Hyland, and L.M. Eastman. 1983. Check-List of the Vascular Plants of Maine. 2nd revised ed. Bulletin of the Josselyn Botanical Society No. 11. Lincoln Press, Sanford, Maine. 73 pp.

Riefner, R.E., Jr. and D.R. Pryor. 1996. New locations and interpretations of vernal pools in southern California. Phytologia 80:296–327.

Riggins, R. 1977. A biosystematic study of the *Sporobolus asper* complex (Gramineae). Iowa State J. Res. 51:287–321.

Riley, J.L. 1979. Some new and interesting vascular plant records from northern Ontario. Canad. Field-Naturalist 93:355–362.

Riley, J.L. and S.M. McKay. 1980. The Vegetation and Phytogeography of Coastal Southwestern James Bay. Life Sciences Contributions, Royal Ontario Museum No. 124. Royal Ontario Museum, Toronto, Ontario. 81 pp.

Rill, K.D. 1983. A vascular flora of Winnebago County, Wisconsin. Trans. Wisconsin Acad. Sci. 71:155–180.

Roalson, E.H. and K.W. Allred. 1998. A floristic study in the Diamond Creek drainage area, Gila National Forest, New Mexico. Aliso 17:47–62.

Roberts, F.M., Jr. 1989. A Checklist of the Vascular Plants of Orange County, California. Museum of Systematic Biology Research Series No. 6. University of California-Irvine, Irvine, California. 58 pp.

Rogers, B.S. and A. Tiehm. 1979. Vascular plants of the Sheldon National Wildlife Refuge, with special reference to possible threatened and endangered species. U.S. Department of the Interior–Fish and Wildlife Service, Region 1, Portland, Oregon. 87 pp.

Roosa, D.M., L.J. Eilers, and S. Zaber. 1991. An annotated checklist of the vascular plant flora of Guthrie County, Iowa. J. Iowa Acad. Sci 98:14–30.

Rosen, D.J., S.D. Jones, and J.K. Wipff. 2001. *Phyllostachys bambusoides* (Poaceae: Bambuseae) previously unreported from Louisiana. Sida 19:731–734.

Rouleau, E. and G. Lamoureux. 1992. Atlas of the Vascular Plants of the Island of Newfoundland and of the Islands of Saint-Pierre et Miquelon. Fleurbec, Québec, Québec. 777 pp.

Rousseau, C. 1968. Histoire, habitat et distribution de 220 plantes introduites au Québec. Naturaliste Canad. 95:49–169.

Rousseau, C. 1974. Géographie Floristique du Québec-Labrador: Distribution des Principales Espèces Vascularies. Les Presses de l'Université Laval, Québec, Québec. 615 pp.

Rubtzoff, P. 1961. Notes on fresh-water marsh and aquatic plants in California. Leafl. W. Bot. 9:165–180.

Russell, N.H. 1956. A checklist of the vascular flora of Poweshiek County, Iowa. Proc. Iowa Acad. Sci. 63:161–176.

Saichuk, J.K., C.M. Allen, and **W.D. Reese.** 2000. *Alopecurus myosuroides* and *Sclerochloa dura* (Poaceae) new to Louisiana. Sida 19:411–412.

Sanders, A.C. 1996. Noteworthy collections. Madroño 43:524–532.

Scheffer, T.H. 1945. The introduction of *Spartina alternifolia* to Washington with oyster culture. Leafl. W. Bot. 4:163–164.

Scott, R.W. 1995. The Alpine Flora of the Rocky Mountains, Vol. 1: The Middle Rockies. University of Utah Press, Salt Lake City, Utah. 901 pp.

Seabloom, E.W. and **A.M. Wiedemann.** 1994. Distribution and effects of *Ammophila breviligulata* Fern. (American beachgrass) on the foredunes of the Washington coast. J. Coastal Res. 10:178–188.

Seagrist, R.V. and **K.J. Taylor.** 1998. Alpine vascular flora of Buffalo Peaks, Mosquite Range, Colorado, USA. Madroño 45:319–325.

Seagrist, R.V. and **K.J. Taylor.** 1999. Alpine vascular flora of Hasley Basin, Elk Mountains, Colorado, USA. Madroño 45:310–318.

Sharma, M.P. and **W.H. Vanden Born.** 1978. The biology of Canadian weeds. 27. *Avena fatua* L. Canad. J. Pl. Sci. 58:141–157.

Simmons, M.P., D.M.E. Ware, and **W.J. Hayden.** 1995. The vascular flora of the Potomac River watershed of King George County, Virginia. Castanea 60:179–200.

Simpson, M.G., S.C. McMillan, and **B.L. Stone.** 1995. Checklist of the Vascular Plants of San Diego County. San Diego State University Herbarium Press, San Diego, CA. 80 pp. [see also http://www.sdnhm.org/research/botany/sdplants/index.html].

SMASCH Project. 2001. Specimen MAnagement System for California Herbaria database [University of California, Berkeley (UC) and Jepson (JEPS) herbaria]. http://ww.mip.berkeley.edu/www_apps/ smasch/.

Smith, C.F. 1976. A Flora of the Santa Barbara Region, California. Santa Barbara Museum of Natural History, Santa Barbara, California. 331 pp.

Smith, E.B. (ed). 1988. An Atlas and Annotated List of the Vascular Plants of Arkansas. Edwin B. Smith, Fayetteville, Arkansas. 489 pp.

Smith, G.L. and **C.R. Wheeler.** 1990–1991. A flora of the vascular plants of Mendocino County, California. Wasmann J. Biol. 48/49:63–81.

Snow, N. 1992–1994. The vascular flora of southeastern Yellowstone National Park and the headwaters region of the Yellowstone River. Wasmann J. Biol. 50:52–95.

Sorrie, B.A. 1987. Notes on the rare flora of Massachusetts. Rhodora 89:113–196.

Sorrie, B.A. 1998. Noteworthy collections. Castanea 63:496–500.

Sorrie, B.A. and **P.W. Dunwiddie.** 1990. *Amphicarpum purshii* (Poaceae), a genus and species new to New England. Rhodora 92:105–107.

Sorrie, B.A. and **S.W. Leonard.** 1999. Noteworthy records of Mississippi vascular plants. Sida 18:889–908.

Sorrie, B.A. and **P. Somers.** 1999. The Vascular Plants of Massachusetts: A County Checklist. Natural Heritage and Endangered Species Program, Massachusetts Division of Fisheries and Wildlife, Westborough, Massachusetts. 189 pp.

Sorrie, B.A., B. Van Eerden, and **M.J. Russo.** 1997. Noteworthy plants from Fort Bragg and Camp MacKall, North Carolina. Castanea 62:239–259.

Soza, V. 2000. Noteworthy collections. Madroño 47:141–142.

Sparks, L.H., R. Del Moral, A.F. Watson, and **A.R. Kruckeberg.** 1976. The distribution of vascular plant species on Sergief Island, southeast Alaska. Syesis 10:5–9.

Spellenberg, R., D. Anderson, and **R. Brozka.** 1993. Noteworthy colletions. Madroño 40:136–138.

Spicher, D. and **M. Josselyn.** 1985. *Spartina* (Gramineae) in northern California: Distribution and taxonomic notes. Madroño 32:158–167.

Spribille, T. 2002. Noteworthy collections. Madroño 49:55–58.

Stalter, R. and **E. Lamont.** 1996. Noteworthy collections. Castanea 61:396–397.

Stalter, R. and **J. Tamory.** 1999. The vascular flora of Biscayne National Park, Florida. Sida 18:1207–1226.

Steel, M.G., P.B. Cavers, and **S.M. Lee.** 1983. The biology of Canadian weeds. 59. *Setaria glauca* (L.) Beauv. and *S. verticillata* (L.) Beauv. Canad.J. Pl. Sci. 63:711–725.

Steury, B.W. 2000. Noteworthy collections. Castanea 65:168.

Stevenson, G.A. 1965. Notes on the more recently adventive flora of the Brandon area, Manitoba. Canad. Field-Naturalist 79:174–177.

Stewart, H. and **R.J. Hebda.** 2000. Grasses of the Columbia Basin of British Columbia. Research Branch, Ministry of Forests, British Columbia Working Paper No. 45. Ministry of Forests Research Program, Victoria, British Columbia. 228 pp. [see also http://livinglandscapes.bc.ca/cb_grasses/index_grasses.html].

Stickney, P.F. 1961. Range of rough fescue (*Festuca scabrella* Torr.) in Montana. Proc. Montana Acad. Sci. 20:12–17.

Stiles, B.J. and **C.L. Howel.** 1998. Floristic survey of Rabun County, Georgia, part II. Castanea 63:154–160.

Stuart, J.D., T. Worley, and **A.C. Buell.** 1996. Plant associations of Castle Crags State Park, Shasta County, California. Madroño 43:273–291.

Sundell, E., R.D. Thomas, C. Amason, and **C.H. Doffitt.** 2002. Noteworthy vascular plants from Arkansas: II. Sida:409–418.

Sundell, E., R.D. Thomas, C. Amason, R.L. Stuckey, and **J. Logan.** 1991. Noteworthy vascular plants from Arkansas. Sida 18:877–887.

Talbot, S.S., B.A. Yurtsev, D.F. Murray, G.W. Argus, C. Bay, and **A. Elvebakk.** 1999. Atlas of rare endemic vascular plants of the Arctic. Conservation of Arctic Flora and Fauna (CAFF) Technical Report No. 3. U.S. Department of the Interior–Fish and Wildlife Service, Anchorage, Alaska. 73 pp.

Taylor, R.J. and **C.E. Taylor.** 1980. Status report on *Calamovilfa arcuata*. University of Tennessee, Knoxville, Tennessee. 27 pp.

Taylor, R.J. and **C.E. Taylor.** 1987. Additions to the vascular flora of Oklahoma—IV. Sida 12:233–237.

Terrell, E.E. and **J.L. Reveal.** 1996. Noteworthy collections. Castanea 61:95–96.

Terrell, E.E., J.L. Reveal, R.W. Spjut, R.F. Whitcomb, J.H. Kirkbride, and **M.T. Cimino.** 2000. Annotated List of the Flora of the Beltsville Agricultural Research Center, Beltsville, Maryland. U.S. Department of Agriculture–Agricultural Research Center, Beltsville, Maryland. 89 pp.

Texas A&M University. 1996. Bioinformatics Working Group herbarium specimen browser [*Bouteloua* only]. http://www.csdl.tamu.edu/FLORA/.

Thomas, R.D. 2002. *Cynosurus echinatus* (Poaceae) new to Texas. Sida 20:837.

Thorne, R.F. 1955. The flora of Johnson County, Iowa. Proc. Iowa Acad. Sci. 62:155–196.

Thorne, R.F., B.A. Prigge, and **J. Henrickson.** 1981. A flora of the higher ranges and the Kelso Dunes of the eastern Mojave Desert in California. Aliso 10:71–186.

Titus, J.H., S. Moore, M. Arnot, and **P.J. Titus.** 1998. Inventory of the vascular flora of the blast zone, Mount St. Helens, Washington. Madroño 45:146–161.

Towne, E.G. 2002. Vascular plants of Konza Prairie Biological Station: An annotated checklist of species in a Kansas tallgrass prairie. Sida 20:269–294.

Towne, E.G. and **I. Barnard.** 2000. *Themeda quadrivalvis* (Poaceae: Andropogoneae) in Kansas: An exotic plant introduced from birdseed. Sida 19:201–203.

Tucker, G.C. 1996. The genera of Pooideae (Gramineae) in the southeastern United States. Harvard Pap. Bot. 9:11–90.

University of British Columbia. 2002. Database of grasses in the University of British Columbia herbarium. http://www.botany.ubc.ca/herbarium/.

University of Oklahoma. 2002. Atlas of the flora of Oklahoma. http://geo.ou.edu/Botanical/.

University of South Carolina. 2001. South Carolina plant atlas. http://cricket.biol.sc.edu/herb/.

Urban, K.A. 1971. Common Plants of Craters of the Moon National Monument. Craters of the Moon Natural History Association, Arco, Idaho. 30 pp.

U.S. Department of Agriculture–Animal and Plant Health Inspection Service. 1990. Weed alert! Be on the lookout for Serrated Tussock, a Federal noxious weed (*Nassella trichotoma*). U.S. Department of Agriculture–Animal and Plant Health Inspection Service [USDA–APHIS], in conjunction with the Illinois Department of Agriculture. 3 pp.

U.S. Department of Agriculture–Forest Service, Rocky Mountain Research Station. 1999. Manitou Experimental Forest: Species list of principal plants. http://lamar.colostate.edu/ ~rwu4451/ manitou/.

Vanderhorst, J.P. 1993. Flora of the Flat Tops, White River Plateau, and vicinity in northwestern Colorado. M.S. thesis. University of Wyoming, Laramie, Wyoming. 129 pp.

Vega, A.S. 2000. Revisión taxonómica de las especies americanas del género *Bothriochloa* (Poaceae: Panicoideae: Andropogoneae). Darwiniana 38:127–186.

Vermont Botanical and Bird Club. 1973. Check List of Vermont Plants, Including All Vascular Plants Growing Without Cultivation. Vermont Botanical and Bird Club, Burlington, Vermont. 90 pp.

Villamil, C.B. 1969. El género *Monanthochloë* (Gramineae): Estudios morfológicos y taxonómicos con especial referencia a la especia Argentina. Kurtziana 5:369–391.

Voss, E.G. 1972. Michigan Flora: A Guide to the Identification and Occurrence of the Native and Naturalized Seed-Plants of the State. Part I, Gymnosperms and Monocots. Cranbrook Institute of Science, Bloomfield Hills, Michigan. 488 pp.

Wagenknecht, B.L. 1954. The flora of Washington County, Iowa. Proc. Iowa Acad. Sci. 61:184–204.

Ward, G.H. 1948. A flora of Chelan County, Washington. M.S. thesis. State College of Washington [Washington State University], Pullman, Washington. 179 pp.

Warwick, S.I. 1979. The biology of Canadian weeds. 37. *Poa annua* L. Canad. J. Pl. Sci. 59:1053–1066.

Warwick, S.I. and **S.G. Aiken.** 1986. Electrophoretic evidence for the recognition of two species in annual wild rice (*Zizania*, Poaceae). Syst. Bot. 11:464–473.

Warwick, S.I. and **L.D. Black.** 1983. The biology of Canadian weeds. 61. *Sorghum halepense* (L.) Pers. Canad. J. Pl. Sci. 63:997–1014.

Warwick, S.I., L.D. Black, and **B.F. Zilkey.** 1985. The biology of Canadian weeds. 72. *Apera spica-venti* (L.) Beauv. Canad. J. Pl. Sci. 63:997–1014.

Watson, W.C. 1989. The vascular flora of Pilot Knob State Preserve. J. Iowa Acad. Sci 96:6–13.

Weber, W.A. 1984. A new genus of grasses from the western oil shales. Phytologia 55:1–2.

Weber, W.A. 1995. Checklist of Vascular Plants of Boulder County, Colorado. Natural History Inventory of Colorado No. 16. University of Colorado Museum, Boulder, Colorado. 68 pp.

Werner, P.A. and **R. Rioux.** 1977. The biology of Canadian weeds. 24. *Agropyron repens* (L.) Beauv. Canad. J. Pl. Sci. 57:905–919.

Whitney, K.D. 1996. Noteworthy collections. Madroño 43:336–337.

Williams, A.H. 1997. Range expansion northward in Illinois and into Wisconsin of *Tridens flavus* (Poaceae). Rhodora 99:344–351.

Wilson, B.L. 1992. Checklist of the vascular flora of Page County, Iowa. J. Iowa Acad. Sci 99:22–33.

Winstead, R. 1990. A taxonomic and ecological survey of the plant communities of Attala County, Mississippi. M.S. thesis. Mississippi State University, Mississippi State, Mississippi. 347 pp.

Wipff, J.K. and **S.L. Hatch.** 1992. *Eustachys caribaea* (Poaceae: Chlorideae) in Texas. Sida 15:160–161.

Wipff, J.K. and **S.D. Jones.** 1994. *Melica subulata* (Poaceae: Meliceae): The first report for Colorado. Sida 16:210–211.

Wipff, J.K., S.D. Jones, and **C.T. Bryson.** 1994. *Eustachys glauca* and *E. caribaea* (Poaceae: Chlorideae): The first reports for Mississippi. Sida 16:211.

Wipff, J.K., R.I. Lonard, S.D. Jones, and **S.L. Hatch.** 1993. The genus *Urochloa* (Poaceae: Paniceae) in Texas, including one previously unreported species for the state. Sida 15:405–413.

Wolden, B.O. 1956. The flora of Emmet County, Iowa. Proc. Iowa Acad. Sci. 63:118–156.

Yadon, V. 1995. Checklist of the vascular plants of Monterey County, California, revised to conform with the Jepson Manual [unpublished manuscript]. 7 pp.

Yatskievych, G. 1999. Steyermark's Flora of Missouri, Vol. 1. Revised ed. Missouri Department of Convervation, Jefferson City, Missouri. 991 pp.

Zebryk, T.M. 1998. Noteworthy collections. Castanea 63:78–79.

Zika, P.F. 1989. Noteworthy collections. Madroño 36:207.

Zika, P.F. 1990. Range expansion of some grasses in Vermont. Rhodora 92:80–89.

Zika, P.F. 2000. Noteworthy collections. Madroño 47:214–216.

Zika, P.F. 2000. Two more weeds in Maine. Rhodora 102:208–209.

Zika, P.F. and **E.J. Marshall.** 1991. Contributions to the flora of the Lake Champlain Valley, New York and Vermont, III. Bull. Torrey Bot. Club 118:58–61.

Zika, P.F., R.J. Stern, and **H.E. Ahles.** 1983. Contributions to the flora of the Lake Champlain Valley, New York and Vermont. Bull. Torrey Bot. Club 110:366–369.

Zika, P.F. and **B.L. Wilson.** 1998. Noteworthy collections. Madroño 45:86–87.

Zobel, D.B. and **C.R. Wasem.** 1979. Pyramid Lake Research Natural Area. Federal Research Natural Areas in Oregon and Washington: A Guidebook for Scientists and Educators, Supplement No. 8. U.S. Department of Agriculture–Forest Service, Pacific Northwest Forest and Range Experiment Station, Corvallis, Oregon. 17 pp.

GENERAL BIBLIOGRAPHY

The General Bibliography shows the complete citation for all of the "Selected References" in the present volume. The notes in square brackets refer to treatment(s) in which a reference is cited, but many of them are of more general application. Book titles are listed in full; the names of journals are abbreviated according to Lawrence et al. (1968) and its supplement (Bridson and Smith 1991).

Adati, S. and **I. Shiotani**. 1962. The cytotaxonomy of the genus *Miscanthus* and its phylogenetic status. Bull. Fac. Agric. Mie Univ. 25:1–24. **[Miscanthus]**

Alderson, J. and **W.C. Sharp**. 1995. Grass Varieties in the United States. CRC Press, Boca Raton, Florida, U.S.A. 296 pp. [previously published by the Soil Conservation Service, U.S. Department of Agriculture, as Agricultural Handbook No. 170, revised 1994]. **[Cynodon]**

Alice, L.A., G.G. Borneo, and **K.W. Hilu**. 2000. Systematics of *Chloris* (*Chloridoideae*; *Poaceae*) and related genera: Evidence from nuclear ITS and chloroplast matK sequences. Amer. J. Bot. 87:108–109. **[Eustachys]**

Allred, K.W. 1983. Systematics of the *Bothriochloa saccharoides* complex (Poaceae: Andropogoneae). Syst. Bot. 8:168–184. **[Bothriochloa]**

Allred, K.W. 1984. Morphologic variation and classification of the North American *Aristida purpurea* complex (Gramineae). Brittonia 36:382–395. **[Aristida]**

Allred, K.W. 1984, 1985, 1986. Studies in the genus *Aristida* (Gramineae) of the southeastern United States. Rhodora 86:73–77; 87:137–145, 145–155; 88:367–387. **[Aristida]**

Allred, K.W. 1993. A Field Guide to the Grasses of New Mexico. New Mexico Agricultural Experiment Station, Department of Agricultural Communications. New Mexico State University, Las Cruces, New Mexico, U.S.A. 258 pp. **[Hackelochloa]**

Allred, K.W. and **F.W. Gould**. 1978. Geographic variation in the *Dichanthelium aciculare* complex (Poaceae). Brittonia 30:497–504. **[Dichanthelium]**

Anderson, D.E. 1974. Taxonomy of the genus *Chloris* (Gramineae). Brigham Young Univ. Sci. Bull., Biol. Ser. 19:1–133. **[Chloris; Enteropogon; Eustachys]**

Angiosperm Phylogeny Working Group. 1998. An ordinal classification for the families of flowering plants. Ann. Missouri Bot. Gard. 85:531–553. [Introduction].

Anton, A.M. 1981. The genus *Tragus* (Gramineae). Kew Bull. 36:55–61. **[Tragus]**

Anton, A.M. and **A.T. Hunziker**. 1978. El género *Munroa* (Poaceae): Sinopsis morfológica y taxonómica. Bol. Acad. Nac. Ci. 52:229–252. **[Munroa]**

Asíns, M.J., J.L. Carretero, A. Del Busto, E.A. Carbonell, and **D. Gomez de Barreda**. 1999. Morphologic and isozyme variation in barnyard grass (*Echinochloa*) weed species. Weed Technol. 13:209–215. **[Echinochloa]**

Assafa, S., C.M. Taliaferro, M.P. Anderson, B.G. de los Reyes, and **R.M. Edwards**. 1999. Diversity among *Cynodon* accessions and taxa based on DNA amplification fingerprinting. Genome 42:465–474. **[Cynodon]**

Baaijens, G.J. and **J.F. Veldkamp**. 1991. *Sporobolus* (Gramineae) in Malesia. Blumea 35:393–458. **[Sporobolus; Thysanolaena]**

Baeza P., C.M. 1996. Los géneros *Danthonia* DC. y *Rytidosperma* Steud. (Poaceae) en América–Una revisión. Sendtnera 3:11–93. **[Danthonia]**

Banks, D.J. 1966. Taxonomy of *Paspalum setaceum* (Gramineae). Sida 2:269–284. **[Paspalum]**

Barber, J.C., S.A. Aliscioni, L.M. Giussani, J.D. Noll, M.R. Duvall, and **E.A. Kellogg**. 2002. Combined analyses of three independent datasets to investigate phylogeny of Poaceae subfamily Panicoideae. [Abstract.] http://www.botany2002.org/. **[Paniceae; Panicoideae]**

Barker, N.P., H.P. Linder, and **E.H. Harley**. 1998. Sequences of the grass-specific insert in the chloroplast *rpo*C2 gene elucidate generic relationships of the Arundinoideae (Poaceae). Syst. Bot. 23:327–350. **[Arundinoideae]**

Barker, N.P., H.P. Linder, and **E.H. Harley**. 1995. Polyphyly of Arundinoideae (Poaceae): Evidence from *rbcL* sequence data. Syst. Bot. 20:423–435. **[Arundinoideae; Chloridoideae]**

Barker, N.P., H.P. Linder, and **E.H. Harley**. 1999. Sequences of the grass-specific insert in the chloroplast *rpo*C2 gene elucidate generic relationships of the Arundinoideae (Poaceae). Syst. Bot. 23:327–350. **[Danthonioideae]**

Barker, N.P., C.M. Morton, and H.P. Linder. 2000. The Danthonieae: Generic composition and relationships. Pp. 221–229 *in* S.W.L. Jacobs and J. Everett (eds.). Grasses: Systematics and Evolution. International Symposium on Grass Systematics and Evolution (3rd:1998). CSIRO Publishing, Collingwood, Victoria, Australia. 408 pp. [**Cortaderia; Danthonioideae**]

Barkworth, M.E. and K.M. Capels. 2000. The Poaceae in North America: A geographic perspective. Pp. 327–346 *in* S.W.L. Jacobs and J. Everett (eds.). Grasses: Systematics and Evolution. International Symposium on Grass Systematics and Evolution (3rd:1998). CSIRO Publishing, Collingwood, Victoria, Australia. 408 pp. [Introduction; **Chloridoideae; Paniceae; Panicoideae**]

Barnes, P.W. 1986. Variation in the big bluestem (*Andropogon gerardii*)-sand bluestem (*Andropogon hallii*) complex along a local dune/meadow gradient in the Nebraska sandhills. Amer. J. Bot. 73:172–184. [**Andropogon**]

Barrett, S.C.H. and D.E. Seaman. 1980. The weed flora of California rice fields. Aquatic Bot. 9:351–376. [**Echinochloa**]

Beetle, A.A. 1943. The North American variations of *Distichlis spicata.* Bull. Torrey Bot. Club 70:638–650. [**Distichlis**]

Beetle, A.A. 1948. The genus *Aegopogon* Humbold. & Bonpl. Univ. Wyoming Publ. 13:17–23. [**Aegopogon**]

Beetle, A.A. 1950. Buffalograss–native of the shortgrass plains. Wyoming Agric. Exp. Sta. Bull. 293:1–31. [**Buchloë**]

Beetle, A.A. 1987. Las Gramineas de México, vol. 2. COTECOCA [Comisión Técnico Consultiva de Coeficientes de Agostadero], México, D.F., México. 344 pp. [**Cathesticum**]

Black, G.A. 1963. Grasses of the genus *Axonopus* (a taxonomic treatment) (ed. L.B. Smith). Advancing Frontiers Pl. Sci. 5:1–186. [**Axonopus**]

Black, J.M. 1978. Gramineae. Pp. 88–249 *in* J.M. Black. Flora of South Australia, Part I, ed. 3 (rev. J.P. Jessop). D.J. Woolman, Government Printer, Adelaide, South Australia, Australia. 466 pp. [**Dactyloctenium**]

Blumler, M. 2001. Notes and comments. Fremontia 29:36. [**Rytidosperma**]

Boonbundarl, S. 1985. A biosystematic study of the *Digitaria leucites* complex in North America. Ph.D. dissertation, Texas A&M University, College Station, Texas, U.S.A. 238 pp. [**Digitaria**]

Bridson, G.D.R. and E.R. Smith (eds.). 1991. B–P–H/S: Botanico–Periodicum–Huntianum/Supplementum. Hunt Institute for Botanical Documentation, Carnegie Mellon University, Pittsburgh, Pennsylvania, U.S.A. 1068 pp. [Introduction; General Bibliography; Geographic Bibliography]

Brown, R. 1814. General remarks, geographical and systematical, on the botany of Terra Australis. Pp. 533–613 *in* M. Flinders (ed.). A Voyage to Terra Australis, vol. 2. G. and W. Nicol, London, England. 613 pp. [**Paniceae; Panicoideae**]

Brummitt, R.K. and C.E. Powell (eds.). 1992. Authors of Plant Names: A list of authors of scientific names of plants, with recommended standard forms of their names, including abbreviations. Royal Botanic Gardens, Kew, England. 732 pp. [Introduction; Names and Synonyms]

Bruner, J.L. 1987. Systematics of the *Schizachyrium scoparium* (Poaceae) complex in North America. Ph.D. dissertation, Ohio State University, Columbus, Ohio, U.S.A. 167 pp. [**Schizachyrium**]

Brunken, J.N. 1977. A systematic study of *Pennisetum* sect. *Pennisetum* (Gramineae). Amer. J. Bot. 64:161–176. [**Pennisetum**]

Brunken, J.N. 1979. Morphometric variation and the classification of *Pennisetum* section *Brevivalvula* (Gramineae) in tropical Africa. Bot. J. Linn. Soc. 79:51–64. [**Pennisetum**]

Buitenhuis, A.G. and J.F. Veldkamp. 2001. Revision of *Eremochloa* (Gramineae–Andropogoneae–Rottboellinae). Blumea 46:399–420. [**Eremochloa**]

Burbidge, N.T. 1953. The genus *Triodia* R. Br. (Gramineae). Austral. J. Bot. 1:121–184. [**Tridens**]

Busey, P. and S. Boyer. 2002. Bermudagrass speeds: Can fast greens be green? http://www.floridaturf.com/ballroll.htm/. [**Cynodon**]

Busey, P., T.K. Broschat, and B.J. Center. 1982. Classification of St. Augustinegrass. Crop Sci. (Madison) 22:469–473. [**Stenotaphrum**]

Campbell, C.S. 1983. Systematics of the *Andropogon virginicus* complex (Gramineae). J. Arnold Arbor. 64:171–254. [**Andropogon**]

Campbell, C.S. 1985. The subfamilies and tribes of Gramineae (Poaceae) in the southeastern United States. J. Arnold Arbor. 66:123–199. [**Chloridoideae; Cynodonteae**]

Campbell, C.S. 1986. Phylogenetic reconstructions and two new varieties in the *Andropogon virginicus* complex (Poaceae: Andropogoneae). Syst. Bot. 11:280–292. [**Andropogon**]

Caro, J.A. 1981. Rehabilitación del género *Dasyochloa* (Gramineae). Dominguezia 2:1–17. [**Dasyochloa**]

Caro, J.A. 1982. Sinopsis taxonómica de las gramíneas argentinas. Dominguezia 4:1–51. [**Aristidoideae**]

Caro, J.A. and E.A. Sánchez. 1969. Las especies de *Cynodon* (Gramineae) de la República Argentina. Kurtziana 5:191–252. [**Cynodon**]

Carretero, J.L. 1981. El género *Echinochloa* Beauv. en el sudeste de Europa. Anales Jard. Bot. Madrid 38:91–108. [**Echinochloa**]

Chase, A. 1920. The North American species of *Cenchrus.* Contr. U.S. Natl. Herb. 22^1:45–77. [**Cenchrus**]

Chase, A. 1921. The North American species of *Pennisetum.* Contr. U.S. Natl. Herb. 22^4:209–234. [**Pennisetum**]

Clark, L.G., W. Zhang, and J.F. Wendel. 1995. A phylogeny of the grass family (Poaceae) based on *ndhF* sequence data. Syst. Bot. 20:436–460. [**Arundinoideae; Cynodonteae**]

Clayton, W.D. 1964. Studies in the Gramineae: V. New species of *Andropogon*. Kew Bull. 17:465–470. **[Andropogon]**

Clayton, W.D. 1967. Studies in the Gramineae. XIV. Kew Bull. 21:111–117. **[Phragmites]**

Clayton, W.D. 1969. A Revision of the Genus *Hyparrhenia*. Kew Bull., Addit. Ser. 2. Her Majesty's Stationery Office, London, England. 196 pp. **[Andropogoneae; Hyparrhenia]**

Clayton, W.D. 1970. Flora of Tropical East Africa, Gramineae (Part 1). Crown Agents for Oversea Governments and Administrations, London, England. 176 pp. **[Phragmites]**

Clayton, W.D. 1979. Notes on *Setaria* (Gramineae). Kew Bull. 33:501–509. **[Setaria]**

Clayton, W.D. 1987. Andropogoneae. Pp. 307–309 *in* T.R. Soderstrom, K.W. Hilu, C.S. Campbell, and M.E. Barkworth (eds.). Grass Systematics and Evolution. Smithsonian Institution Press, Washington, D.C., U.S.A. 473 pp. **[Andropogoneae]**

Clayton, W.D. 1994. *Apluda*. Pp. 35–37 *in* W.D. Clayton, G. Davidse, F. Gould, M. Lazarides, and T.R. Soderstrom. A Revised Handbook to the Flora of Ceylon, vol. 8 (ed. M.D. Dassanayake). Amerind Publishing Co., New Delhi, India. 458 pp. **[Apluda]**

Clayton, W.D., G. Davidse, F. Gould, M. Lazarides, and **T.R. Soderstrom.** 1994. A Revised Handbook to the Flora of Ceylon, vol. 8 (ed. M.D. Dassanayake). Amerind Publishing Co., New Delhi, India. 458 pp. **[Andropogoneae]**

Clayton, W.D., S.M. Phillips, and **S.A. Renvoize.** 1974. Flora of Tropical East Africa. Gramineae (Part 2) (ed. R.M. Pohill). Whitefriars Press, Ltd., London, England. 373 pp. **[Dactyloctenium]**

Clayton, W.D. and **S.A Renvoize.** 1982. Flora of Tropical East Africa. Gramineae (Part 3). A.A. Balkema, Rotterdam, The Netherlands. 448 pp. **[Sorghum; Urochloa]**

Clayton, W.D. and **S.A. Renvoize.** 1986. Genera Graminum: Grasses of the World. Kew Bull., Addit. Ser. 13. Her Majesty's Stationery Office, London, England. 389 pp. [Introduction; **Andropogoneae; Anthenantia; Aristidoideae; Arundinoideae; Centothecoideae; Cynodonteae]**

Clifford, H.T. 1996. Etymological Dictionary of Grasses, Version 1.0 (CD-ROM). Expert Center for Taxonomic Identification, Amsterdam, The Netherlands. **[Acrachne; Cymbopogon; Eragrostis; Ixophorus; Steinchisma; Tribolium]**

Coastal Conservancy. 2002. Invasive *Spartina* project. http://www.spartina.org/. **[Spartina]**

Columbus, J.T. 1999. An expanded circumscription of *Bouteloua* (Gramineae: Chlorideae): New combinations and names. Aliso 18:61–65. **[Bouteloua; Buchloë; Cathesticum; Opizia]**

Columbus, J.T., M.S. Kinney, R. Pant, and **M.E. Siqueiros Delgado.** 1998. Cladistic parsimony analysis of internal transcribed spacer region (nrDNA) sequences of *Bouteloua* and relatives (Gramineae: Chloridoideae). Aliso 17:99–130. **[Cynodonteae; Hilaria]**

Columbus, J.T., M.S. Kinney, M.E. Siqueiros Delgado, and **R. Cerros Tlatilpa.** 2000. Homoplasy, polyphyly, and generic circumscription: The demise of the Boutelouinae (Gramineae: Chloridoideae). [Abstract.] Amer. J. Bot. 87:120. **[Hilaria]**

Conert, H.J. 1987. Current concepts in the systematics of the Arundinoideae. Pp. 239–250 *in* T.R. Soderstrom, K.W. Hilu, C.S. Campbell, and M.E. Barkworth (eds.). Grass Systematics and Evolution. Smithsonian Institution Press, Washington, D.C., U.S.A. 473 pp. **[Arundinoideae; Danthonioideae]**

Connor, H.E. and **E. Edgar.** 1974. Names and types in *Cortaderia* Stapf (Gramineae). Taxon 23:595–605. **[Cortaderia]**

Connor, H.E. and **E. Edgar.** 1979. *Rytidosperma* Steudel (*Nothodanthonia* Zotov) in New Zealand. New Zealand J. Bot. 17:311–337. **[Rytidosperma]**

Conert, H.J. and **A.M. Türpe.** 1969. *Karroochloa*, ein neue gattung der Gramineen (Poaceae, Arundinoideae, Danthonieae). Senckenberg. Biol. 50:289–318. **[Karroochloa]**

Conert, H.J. and **A.M. Türpe.** 1974. Revision der Gattung *Schismus* (Poaceae: Arundinoideae, Danthonieae). Abh. Senckenberg. Naturf. Ges. 532:1–81. **[Schismus]**

Costas-Lippmann, M. 1977. More on the weedy "pampas grass" in California. Fremontia 4:25–27. **[Cortaderia]**

COTECOCA (Comisión Técnico Consultiva de Coeficientes de Agostadero). 2000. Las Gramíneas de México, vol. 5. Secretaria de Agricultura, Ganaderia y Desarrollo Rural, México, D.F., México. 466 pp. **[Paspalum]**

Crampton, B. 1959. The grass genera *Orcuttia* and *Neostapfia*: A study in habitat and morphological specialization. Madroño 15:97–110. **[Orcuttieae]**

Cronquist, A. 1981. An Integrated System of Classification of Flowering Plants. Columbia University Press, New York. 1262 pp. [Introduction]

Dahlgren, R.M.T. and **H.T. Clifford.** 1982. The Monocotyledons: A Comparative Study. Botanical Systematics: An Occasional Series of Monographs (series ed. V.H. Heywood). Academic Press, London and New York. 378 pp. [Introduction]

Darbyshire, S.J. and **J. Cayouette.** 1989. The biology of Canadian weeds. 92. *Danthonia spicata* (L.) Beauv. in Roem. & Schult. Canad. J. Pl. Sci. 69:1217–1233. **[Danthonia]**

Darbyshire, S.J. and **J. Cayouette.** 1995. Identification of the species in the *Panicum capillare* complex (Poaceae) from eastern Canada and adjacent New York State. Canad. J. Bot. 73:333–348. **[Panicum]**

Davey, J.C. and W.D. Clayton. 1977. Some multiple discriminant function studies on *Oplismenus*. Kew Bull. 33:147–157. [**Oplismenus**]

Davidse, G. 1978. A systematic study of the genus *Lasiacis* (Gramineae: Paniceae). Ann. Missouri Bot. Gard. 65:1133–1254. [**Lasiacis**]

Davidse, G. 1987. Fruit dispersal in Poaceae. Pp. 143–155 *in* T.R. Soderstrom, K.W. Hilu, C.S. Campbell, and M.E. Barkworth (eds.). Grass Systematics and Evolution. Smithsonian Institution Press, Washington, D.C., U.S.A. 473 pp. [**Oplismenus**]

Davidse, G. 1994. *Trichloris* E. Fourn. *ex* Bentham. P. 289 *in* G. Davidse, M. Sousa S., and A.O. Chater (eds.). Flora Mesoamericana, vol. 6: Alismataceae a Cyperaceae. Universidad Nacional Autónoma de México, Instituto de Biología, México, D.F., México. 543 pp. [**Trichloris**]

Davidse, G. and E. Morton. 1973. Bird-mediated fruit dispersal in the tropical grass genus *Lasiacis* (Gramineae: Paniceae). Biotropica 5:162–167. [**Lasiacis**]

Davidse, G. and R.W. Pohl. 1994. *Urochloa* P. Beauv. Pp. 331–333 *in* G. Davidse, M. Sousa S., and A.O. Chater (eds.). Flora Mesoamericana, vol. 6: Alismataceae a Cyperaceae. Universidad Nacional Autónoma de México, Instituto de Biología, México, D.F., México. 543 pp. [**Urochloa**]

Dávila, P.D. 1988. Systematic revision of the genus *Sorghastrum* (Poaceae: Andropogoneae). Ph.D. dissertation, Iowa State University, Ames, Iowa, U.S.A. 333 pp. [**Sorghastrum**]

Dávila, P.D. 1994. *Trachypogon* Nees. Pp. 380–381 *in* G. Davidse, M. Sousa S., and A.O. Chater (eds.). Flora Mesoamericana, vol. 6: Alismataceae a Cyperaceae. Universidad Nacional Autónoma de México, Instituto de Biología, México, D.F., México. 543 pp. [**Trachypogon**]

Decker, H.F. 1964. An anatomic-systematic study of the classical tribe Festuceae (Gramineae). Amer. J. Bot. 51:453–463. [**Centothecoideae**]

DeLisle, D.G. 1963. Taxonomy and distribution of the genus *Cenchrus*. Iowa State Coll. J. Sci. 37:259–351. [**Cenchrus**]

de Wet, J.M.J. 1968. Biosystematics of the *Bothriochloa barbinodis* complex (Gramineae). Amer. J. Bot. 55:1246–1250. [**Bothriochloa**]

de Wet, J.M.J. 1978. Systematics and evolution of *Sorghum* sect. *Sorghum* (Gramineae). Amer. J. Bot. 65:477–484. [**Sorghum**]

de Wet, J.M.J., D.E. Brink, and C.E. Cohen. 1983. Systematics of *Tripsacum* section *Fasciculata* (Gramineae). Amer. J. Bot. 70:1139–1146. [**Tripsacum**]

de Wet, J.M.J. and J.R. Harlan. 1970a. *Bothriochloa intermedia*–a taxonomic dilemma. Taxon 19:339–340. [**Bothriochloa**]

de Wet, J.M.J. and J.R. Harlan. 1970b. Biosystematics of *Cynodon* L.C. Rich. (Gramineae). Taxon 19:565–569. [**Cynodon**]

de Wet, J.M.J., J.R. Harlan, and D.E. Brink. 1982. Systematics of *Tripsacum dactyloides*. Amer. J. Bot. 69:1251–1257. [**Tripsacum**]

Dix, W.L. 1945. Will the stowaway, *Molinia caerulea*, become naturalized? Bartonia 23:41–42. [**Molinia**]

Dobrenz, A.K. and A.A. Beetle. 1966. Cleistogenes in *Danthonia*. J. Range Managem. 19:292–296. [**Danthonia**]

Doebley, J.F. 1990. Molecular systematics of *Zea* (Gramineae). Maydica 35:143–150. [**Zea**]

Doebley, J.F. and H.H. Iltis. 1980. Taxonomy of *Zea* (Gramineae). I. A subgeneric classification with a key to taxa. Amer. J. Bot. 67:982–993. [**Zea**]

Doebley, J.F., A. Stec, and C. Justus. 1995. *Teosinte branched1* and the origin of maize: Evidence for epistatis and the evolution of dominance. Genetics 141:333-346. [**Zea**]

Dore, W.G. 1971. *Sieglingia decumbens* (L.) Bernh.–Pulvini of palea. Watsonia 8:297–299. [**Danthonia**]

Dore, W.G. and J. McNeill. 1980. Grasses of Ontario. Research Branch, Agriculture Canada Monograph No. 26. Canadian Government Publishing Centre, Hull, Québec, Canada. 566 pp. [**Danthonia**]

Douglas, G.W., D. Meldinger, and J. Pojar. 2002. Illustrated Flora of British Columbia, vol. 8. British Columbia Ministry of Sustainable Resource Management and British Columbia Ministry of Forests, Victoria, British Columbia, Canada. 457 pp. [**Muhlenbergia**]

Edgar, E. and H.E. Connor. 2000. Flora of New Zealand, vol. 5. Manaaki Whenua Press, Lincoln, New Zealand. 650 pp. [**Miscanthus; Rytidosperma**]

Emery, W.H.P. 1957. A cyto-taxonomic study of *Setaria macrostachya* (Gramineae) and its relatives in the southwestern United States and Mexico. Bull. Torrey Bot. Club 84:94–105. [**Setaria**]

Esen, A. and K.W. Hilu. 1991. Electrophoretic and immunological studies of prolamins in the Poaceae. II. Phylogenetic affinities of the Aristideae. Taxon 40:5–17. [**Aristidoideae**]

Esparza Sandoval, S. and Y. Herrera-Arrieta. 1996. Revisión de *Bouteloua barbata* Lagasca (Poaceae: Eragrotideae). Phytologia 80:73–91. [**Bouteloua**]

Fairbrothers, D.E. and J.R. Gray. 1972. *Microstegium vimineum* (Trin.) A. Camus (Gramineae) in the United States. Bull. Torrey Bot. Club 99:97–100. [**Microstegium**]

Fishbein, M. 1995. Noteworthy collections: Arizona. Madroño 42:83. [**Echinochloa**]

Fox, W.E., III and S.L. Hatch. 1996. *Brachiaria eruciformis* and *Urochloa brizantha* (Poaceae: Paniceae) new to Texas. Sida 17:287–288. [**Urochloa**]

Fox, W.E., III and S.L. Hatch. 1999. New combinations in *Setaria* (Poaceae: Paniceae). Sida 18:1037–1047. [**Setaria**]

Gabel, M.L. 1989. Federal noxious weed identification bulletin. U.S. Dept. of Agriculture, Animal & Plant Health Inspection Service, Plant Protection & Quarantine Bull. 28:1–10. **[Imperata]**

Gandhi, K.N. 1989. A biosystematic study of the *Schizachyrium scoparium* complex. Ph.D. dissertation, Texas A&M University, College Station, Texas, U.S.A. 188 pp. **[Schizachyrium]**

Gayle, E.E. 1892. The spines of *Cenchrus tribuloides* L. Bot. Gaz. 17:126–127. **[Cenchrus]**

Goebel, K.I. 1882. Beitrage zur Entwickelungsgeschichte einiger Inflorescenzen. Jahrb. Wiss. Bot. 14:1–39. **[Pennisetum]**

Gómez-Martínez, R. and **A. Culham.** 2000. Phylogeny of the subfamily Panicoideae with emphasis on the tribe Paniceae: Evidence from the *trn*L-F cpDNA region. Pp. 136–140 *in* S.W.L. Jacobs and J. Everett (eds.). Grasses: Systematics and Evolution. International Symposium on Grass Systematics and Evolution (3rd:1998). CSIRO Publishing, Collingwood, Victoria, Australia. 408 pp. **[Paniceae; Panicoideae]**

Gómez-Sánchez, M., P. Dávila-Aranda, and **J. Valdés-Reyna.** Estudio anatómica de *Swallenia* (Poaceae: Eragroistideae: Monanthochloinae), un género monotípico de Norte América. Madroño 48:152–161 (2001) [publication date 2002]. **[Swallenia]**

González-Andrés, F., J.M. Pita, and **J.M. Ortiz.** Caryopsis isoenzymes of *Echinocloa* [*sic*] weed species as an aid for taxonomic discrimination. J. Hort. Sci. 71:187–193. **[Echinochloa]**

Gould, F.W. 1975. The Grasses of Texas. Texas A&M University Press, College Station, Texas, U.S.A. 653 pp. **[Digitaria; Tridens]**

Gould, F.W. 1979. The genus *Bouteloua* (Poaceae). Ann. Missouri Bot. Gard. 66:348–416. **[Bouteloua]**

Gould, F.W., M.A. Ali, and **D.E. Fairbrothers.** 1972. A revision of *Echinochloa* in the United States. Amer. Midl. Naturalist 87:36–59. **[Echinochloa]**

Gould, F.W. and **C.A. Clark.** 1978. *Dichanthelium* (Poaceae) in the United States and Canada. Ann. Missouri Bot. Gard. 65:1088–1132. **[Dichanthelium]**

Gould, F.W. and **R.B. Shaw.** 1983. Grass Systematics, ed. 2. Texas A&M University Press, College Station, Texas, U.S.A. 397 pp. **[Chloridoideae]**

Grass Phylogeny Working Group. 2000. A phylogeny of the grass family (Poaceae), as inferred from eight character sets. Pp. 3–7 *in* S.W.L. Jacobs and J. Everett (eds.). Grasses: Systematics and Evolution. International Symposium on Grass Systematics and Evolution (3rd:1998). CSIRO Publishing, Collingwood, Victoria, Australia. 408 pp. [**Arundinoideae;** PACCAD Grasses]

Grass Phylogeny Working Group. 2001. Phylogeny and subfamilial classification of the grasses (Poaceae). Ann. Missouri Bot. Gard. 88:373–457. [Introduction; **Aristidoideae; Arundinoideae; Centotheceae; Centothecoideae; Chloridoideae; Danthonioideae;** PACCAD Grasses; **Panicoideae; Uniola**]

Gray, J.R. 1974. The genus *Tripsacum* L. (Gramineae): Taxonomy and chemosystematics. Ph.D. dissertation, University of Illinois at Urbana-Champaign, Urbana, Illinois, U.S.A. 191 pp. **[Tripsacum]**

Grelen, H.E. 1974. Pinehills bluestem, *Andropogon scoparius* Anderss. *ex* Hack., an anomaly of the *A. scoparius* complex. Amer. Midl. Naturalist 91:438–444. **[Schizachyrium]**

Greuter, W., J. McNeill, F.R. Barrie, H.M. Burdet, V. Demoulin, T.S. Filgueiras, D.H. Nicolson, P.C. Silva, J.E. Skog, P. Trehane, and **N.J. Turland** (eds.). 2000. International Code of Botanical Nomenclature (Saint Louis Code): Adopted by the Sixteenth International Botanical Congress, St. Louis, Missouri, July–August 1999. Regnum Vegetabile [series], vol. 138. Koeltz Scientific Books, Königstein, Germany. 474 pp. [Introduction; Names & Synonyms]

Griffiths, D. 1912. The grama grasses: *Bouteloua* and related genera. Contr. U.S. Natl. Herb. 14:343–428. **[Bouteloua]**

Giussani, L.M, J.H. Cota-Sánchez, F.O. Zuloaga, and **E.A. Kellogg.** 2000. A molecular phylogeny of the subfamily *Panicoideae* (*Poaceae*) using *ndhF* sequences. Amer. J. Bot. 87(suppl.):129. **[Dichanthelium; Panicum]**

Guissani, L.M., J.H. Cota-Sánchez, F.O. Zuloaga, and **E.A. Kellogg.** 2001. A molecular phylogeny of the grass subfamily Panicoideae (Poaceae) shows multiple origins of C_4 photosynthesis. Amer. J. Bot. 88:1993–2001. **[Andropogoneae; Paniceae; Panicoideae]**

Hall, D.W. 1978. The Grasses of Florida. Ph.D. dissertation, University of Florida, Gainesville, Florida, U.S.A. 498 pp. **[Imperata; Urochloa]**

Hall, D.W. and **D.T. Patterson.** 1992. Itchgrass–stop the trains? Weed Technol. 6:239–241. **[Rottboellia]**

Hammel, B.E. and **J.R. Reeder.** 1979. The genus *Crypsis* (Gramineae) in the United States. Syst. Bot. 4:267–280. **[Crypsis]**

Harlan, J.R. and **J.M.J. de Wet.** 1969. Sources of variation in *Cynodon dactylon* (L.) Pers. Crop Sci. (Madison) 9:774–778. **[Cynodon]**

Harlan, J.R. and **J.M.J. de Wet.** 1972. Sources of variation in *Cynodon dactylon* (L.) Pers. Crop Sci. 12:172–176. **[Sorghum]**

Harlan, J.R., J.M.J. de Wet, W.W. Huffine, and **J.R. Deakin.** 1970. A guide to the species of *Cynodon* (Gramineae). Oklahoma Agric. Exp. Sta. Bull. B-673:1–37. **[Cynodon]**

Harlan, J.R., J.M.J. de Wet, and **A.B.L. Stemler** (eds.). 1976. Origins of African Plant Domestication. Mouton Press, The Hague, The Netherlands. 498 pp. **[Sorghum]**

Harvey, L.H. 1948. *Eragrostis* in North and Middle America. Ph.D. dissertation, University of Michigan, Ann Arbor, Michigan, U.S.A. 269 pp. **[Eragrostis]**

Harvey, L.H. 1975. *Eragrostis*. Pp. 177–201 *in* F.W. Gould. The Grasses of Texas. Texas A&M University Press, College Station, Texas, U.S.A. 635 pp. **[Eragrostis]**

Hatch, S.L. 1975. A biosystematic study of the *Schizachyrium cirratum–Schizachyrium sanguineum* complex (Poaceae). Ph.D. dissertation, Texas A&M University, College Station, Texas, U.S.A. 112 pp. [Schizachyrium]

Henrard, J.T. 1926, 1927, 1928, 1933. A critical revision of the genus *Aristida*. Meded. Rijks-Herb. 54:1–701; 55C:703–747. [Aristida]

Henrard, J.T. 1929, 1933. A monograph of the genus *Aristida*. Meded. Rijks-Herb. 58:1–325. [Aristida]

Henrard, J. 1950. Monograph of the genus *Digitaria*. Universitaire Pers Leiden, Leiden, The Netherlands. 999 pp. [Digitaria]

Henry, M.A. 1979. A rare grass on the Eureka dunes. Fremontia 7:3–6. [Swallenia]

Herrera-Arrieta, Y. 1998. A revision of the *Muhlenbergia montana* (Nutt.) Hitchc. complex (Poaceae: Chloridoideae). Brittonia 50:23–50. [Muhlenbergia]

Hignight, K.W., E.C. Bashaw, and M.A. Hussey. 1991. Cytological and morphological diversity of native apomictic buffelgrass, *Pennisetum ciliare* (L.) Link. Bot. Gaz. 152:214–218. [Pennisetum]

Hilu, K.W. 1980. Noteworthy collections: *Eleusine tristachya*. Madroño 27:177–178. [Eleusine]

Hilu, K.W. 1994. Evidence from RAPD markers in the evolution of *Echinochloa* millets (Poaceae). Pl. Syst. Evol. 189:247–257. [Echinochloa]

Hilu, K.W. and L.A. Alice. 2001. A phylogeny of the Chloridoideae (Poaceae) based on *matK* sequences. Syst. Bot. 26:386–405. [Chloridoideae; Chloris; Cynodonteae]

Hilu, K.W., L.A. Alice, and H. Liang. 1999. Phylogeny of the Poaceae inferred from *matK* sequences. Ann. Missouri Bot. Gard. 86:835–851. [Centothecoideae; Chloridoideae; Panicoideae]

Hilu, K.W. and A. Esen. 1990. Prolamin and immunological studies in the Poaceae. I. Subfamily Arundinoideae. Pl. Syst. Evol. 173:57–70. [Arundinoideae; Danthonioideae]

Hilu, K.W. and A. Esen. 1993. Prolamin and immunological studies in the Poacae: III. Subfamily Chloridoideae. Amer. J. Bot. 80:104–113. [Chloridoideae]

Hilu, K.W. and J.L. Johnson. 1997. Systematics of *Eleusine* Gaertn. (Poaceae, Chloridoideae): Chloroplast DNA and total evidence. Ann. Missouri Bot. Gard. 84:841–847. [Eleusine]

Hilu, K.W. and K. Wright. 1982. Systematics of Gramineae: A cluster analysis study. Taxon 31:9–36. [Centothecoideae; Chloridoideae]

Hiser, K.M. 2002. Phylogenetic placement, inflorescence development, and taxonomy of the genus *Ixophorus* (Panicoideae: Poaceae). Master's thesis. University of Missouri-St. Louis, St. Louis, Missouri, U.S.A. 88 pp. [Ixophorus]

Hitchcock, A.S. 1913. Mexican grasses in the United States National Herbarium. Contr. U.S. Natl. Herb. 17[3]:181–389. [Echinochloa]

Hitchcock, A.S. 1919. History of the Mexican grass, *Ixophorus unisetus*. J. Wash. Acad. Sci. 9:545-551. [Ixophorus]

Hitchcock, A.S. 1935a. Manual of the Grasses of the United States. U.S. Government Printing Office, Washington, D.C., U.S.A. 1040 pp. [Introduction]

Hitchcock, A.S. 1935b. *Muhlenbergia* Schreb. Pp. 431–476 *in* M.A. Howe, H.A. Gleason, and J.H. Barnhart (eds.). North American Flora, vol. 17, part 6. New York Botanical Garden, New York, New York, U.S.A. 64 pp. [Muhlenbergia]

Hitchcock, A.S. 1951 [title page 1950]. Manual of the Grasses of the United States, ed. 2, rev. A. Chase. U.S.D.A. Miscellaneous Publication No. 200. U.S. Government Printing Office, Washington, D.C., U.S.A. 1051 pp. [Introduction; Chloris; Cortaderia; Cynodon; Dichanthelium; Hackelochloa; Muhlenbergia; Panicum; Schizachyrium; Setaria; Tridens; Zea]

Hitchcock, A.S. and A. Chase. 1910. The North American species of *Panicum*. Contr. U.S. Natl. Herb. 15:1–396. [Dichanthelium; Panicum]

Hocking, P.C., C.M. Finlayson, and A.J. Chick. 1983. The biology of Australian weeds. 12. *Phragmites australis* Trin. *ex* Steud. J. Austral. Inst. Agric. Sci. 49:123–132. [Phragmites]

Hsaio, C., S.W.L. Jacobs, N.P. Barker, and N.J. Chatterton. 1998. A molecular phylogeny of the subfamily Arundinoideae (Poaceae) based on sequences of rDNA (ITS). Austral. Syst. Bot. 11:41–52. [Arundinoideae; Danthonioideae]

Hsaio, C., S.W.L. Jacobs, N.J. Chatterton, and K.H. Asay. 1999. A molecular phylogeny of the grass family (Poaceae) based on the sequences of nuclear ribosomal DNA (ITS). Austral. Sys. Bot. 11:667–688. [Chloridoideae]

Hubbard, F.T. 1915. A taxonomic study of *Setaria italica* and its immediate allies. Amer. J. Bot. 2:169–198. [Setaria]

Hunt, D.M. and R.E. Zaremba. 1992. The northeastward spread of *Microstegium vimineum* (Poaceae) into New York and adjacent states. Rhodora 94:167–170. [Microstegium]

Hunziker, A.T. and A.M. Anton. 1979. A synoptical revision of *Blepharidachne* (Poaceae). Brittonia 31:446–453. [Blepharidachne]

Iltis, H.H. 2000. Homeotic sexual translocations and the origin of maize (*Zea mays*, Poaceae): A new look at an old problem. Econ. Bot. 54:7–42. [Zea]

Iltis, H.H. and J.F. Doebley. 1980. Taxonomy of *Zea* (Gramineae). II. Subspecific categories in the *Zea mays* complex and a generic synopsis. Amer. J. Bot. 67:996–1004. [Zea]

Jacobs, S.W.L. 1987. Systematics of the Chloridoid grasses. Pp. 871–903 *in* T.R. Soderstrom, K.W. Hilu, C.S. Campbell, and M.E. Barkworth (eds.) Grass Systematics and Evolution. Smithsonian Institution Press, Washington, D.C., U.S.A. 473 pp. **[Chloridoideae]**

Jacobs, S.W.L. and S.M. Hastings. 1993. *Dactyloctenium*. Pp. 527–529 *in* G.J. Harden (ed.). Flora of New South Wales, vol. 4. New South Wales University Press, Kensington, New South Wales, Australia. 775 pp. **[Dactyloctenium]**

Jacobs, S.W.L. and J. Highet. 1988. Re-evaluation of the characters used to distinguish *Enteropogon* from *Chloris* (*Poaceae*). Telopea 3:217–221. **[Enteropogon]**

Jain, S.K. and D.K. Banerjee. 1974. Preliminary observations on the ethnobotany of the genus *Coix*. Econ. Bot. 28:38–42. **[Coix]**

Jauzein, P. 1993. Le genre *Echinochloa* en Camargue. Monde Pl. 88, no. 446: 1–5. **[Echinochloa]**

Jones, S.D. and G.D. Jones. 1992. *Cynodon nlemfuënsis*, (Poaceae: Chlorideae) previously unreported in Texas. Phytologia 72:93–95. **[Cynodon]**

Judd, W.S., C.S. Campbell, E.A. Kellogg, P.F. Stevens, and **M.J. Donoghue.** 2002. Plant Systematics: A Phylogenetic Approach, ed. 2. Sinauer Associates, Sunderland, Massachusetts, U.S.A. 576 pp. [Introduction]

Judziewicz, E.J. 1990. A new South American species of *Sacciolepis* (Poaceae: Panicoideae: Paniceae), with a summary of the genus in the New World. Syst. Bot. 15:415–420. **[Sacciolepis]**

Judziewicz, E.J. 1990. Flora of the Guianas: 187. Poaceae (Gramineae). Series A: Phanerogams, Fascicle 8 (series ed. A.R.A. Görts-Van Rijn). Koeltz Scientific Books, Koenigstein, Germany. 727 pp. **[Trachypogon]**

Kartesz, J. and C.A. Meacham. 1999. Synthesis of the North American Flora, Version 1.0 (CD-ROM). North Carolina Botanical Garden, Chapel Hill, North Carolina, U.S.A. **[Digitaria; Muhlenbergia; Spartina]**

Kearney, T.H. and R.H. Peebles. 1951. Arizona Flora. University of California Press, Berkeley and Los Angeles, California, U.S.A. 1032 pp. **[Muhlenbergia]**

Kellogg, E.A. 2000. Molecular and morphological evolution in the Andropogoneae. Pp. 149–158 *in* S.W.L. Jacobs and J. Everett (eds.). Grasses: Systematics and Evolution. International Symposium on Grass Systematics and Evolution (3rd:1998). CSIRO Publishing, Collingwood, Victoria, Australia. 408 pp. **[Andropogoneae; Panicoideae]**

Kellogg, E.A. and C.S. Campbell. 1987. Phylogenetic analyses of the Gramineae. Pp. 310–322 *in* T.R. Soderstrom, K.W. Hilu, C.S. Campbell, and M.E. Barkworth (eds.) Grass Systematics and Evolution. Smithsonian Institution Press, Washington, D.C., U.S.A. 473 pp. **[Aristidoideae; Arundinoideae; Chloridoideae]**

Kesler, T.R. 2000. A taxonomic reevaluation of *Aristida stricta* (Poaceae) using anatomy and morphology. Master's thesis. Florida State University, Tallahassee, Florida, U.S.A. 33 pp. **[Aristida]**

Kim, K.V., J.H. Kim, and I.J. Lee. 1989. Biochemical identification of *Echinochloa* species collected in Korea. Proc. Conf. Asian-Pacific Weed Sci. Soc. 12th [Proc. II]:519–531. **[Echinochloa]**

Kobayashi, H. and S. Sakamoto. 1990. Weed-crop complexes in cereal cultivation. Pp. 67–80 *in* S. Kawano (ed.). Biological Approaches and Evolutionary Trends in Plants. Academic Press, London, England and San Diego, California, U.S.A. 417 pp. **[Echinochloa]**

Koch, S.D. 1974. The *Eragrostis pectinacea–pilosa* complex in North and Central America. Illinois Biol. Monogr. 48:1–74. **[Eragrostis]**

Koekemoer, M. 1991. *Dactyloctenium* Willd. Pp. 99–101 *in* G.E. Gibbs Russell, L. Watson, M. Koekemoer, L. Smook, N.P. Barker, H.M. Anderson, and M.J. Dallwitz. Grasses of Southern Africa (ed. O.A. Leistner). National Botanic Gardens, Botanical Research Institute, Pretoria, Republic of South Africa. 437 pp. **[Dactyloctenium]**

Kok, P.D.F., P.J. Robbertse, and A.E. van Wyk. 1989. Systematic study of *Digitaria* section *Digitaria* (Poaceae) in southern Africa. S. African J. Bot. 55:141–153. **[Digitaria]**

Koyama, T. 1987. Grasses of Japan and Its Neighboring Regions: An Identification Manual. Kodansha, Ltd., Tokyo, Japan. 370 pp. **[Hakonechloa; Ischaemum; Miscanthus; Phragmites]**

Lægaard, S. and P.M. Peterson. 2001. Flora of Ecuador 68 (ed. G. Harling & L. Andersson). 214(2). Gramineae (part 2): Subfam. Chloridoideae. Botanical Institute, University of Göteborg, Göteborg, Sweden and Section for Botany, Riksmuseum, Stockholm, Sweden. 131 pp. **[Sporobolus]**

Lawrence, G.H.M., A.F.G. Buchheim, G.S. Daniels, and H. Dolezal (eds.). 1968. B–P–H: Botanico–Periodicum–Huntianum. Hunt Botanical Library, Pittsburgh, Pennsylvania, U.S.A. 1063 pp. [Introduction; General Bibliography; Geographic Bibliography]

Lazarides, M. 1972. A revision of Australian *Chlorideae* (*Gramineae*). Austral. J. Bot. (supp. 5):1–51. **[Enteropogon]**

LeBlond, R.J. 2001. Taxonomy of the *Dichotoma* group of *Dichanthelium* (Poaceae). Sida 19:821–837. **[Dichanthelium]**

Le Roux, L.G. and E.A. Kellogg. 1999. Floral development and the formation of unisexual spikelets in the Andropogoneae (Poaceae). Amer. J. Bot. 86:354–366. **[Panicoideae]**

Linder, H.P. 1997. Nomenclatural corrections in the *Rytidosperma* complex (Danthonieae, Poaceae). Telopea 7:269–274. **[Rytidosperma]**

Linder, H.P. and N.P. Barker. 2000. Biogeography of the Danthonieae. Pp. 231–238 *in* S.W.L. Jacobs and J. Everett (eds.). Grasses: Systematics and Evolution. International Symposium on Grass Systematics and Evolution (3rd:1998). CSIRO Publishing, Collingwood, Victoria, Australia. 408 pp. **[Cortaderia; Danthonioideae]**

Linder, H.P. and G. Davidse. 1997. The systematics of *Tribolium* Desv. (Danthonieae: Poaceae). Bot. Jahrb. Syst. 119:445–507. **[Tribolium]**

Linder, H.P. and G.A. Verboom. 1996. Generic limits in the *Rytidosperma* (Danthonieae, Poaceae) complex. Telopea 6:597–627. **[Danthonia; Rytidosperma]**

Linder, H.P., G.A. Verboom, and N.P. Barker. 1997. Phylogeny and evolution in the *Crinipes* group of grasses (*Arundinoideae: Poaceae*). Kew Bull. 52:91–110. **[Arundineae; Arundinoideae]**

McGregor, R.L. 1990. Seed dormancy and germination in the annual cleistogamous species of *Sporobolus* (Poaceae). Trans. Kansas Acad. Sci. 93:8–11. **[Sporobolus]**

McGregor, R.L., T.M. Barkley, R.E. Brooks, and E.K. Schofield. 1986. Flora of the Great Plains. Contribution No. 84-135-B, Division of Biology, Kansas Agricultural Experiment Station, Kansas State University. Contribution No. 1254, Department of Botany, North Dakota Agricultural Experiment Station, North Dakota State University. University Press of Kansas, Lawrence, Kansas, U.S.A. 1392 pp. **[Muhlenbergia]**

McVaugh, R. 1983. Flora Novo-Galiciana: A Descriptive Account of the Vascular Plants of Western Mexico, vol. 14: Gramineae (series ed. W.R. Anderson). University of Michigan Press, Ann Arbor, Michigan, U.S.A. 436 pp. **[Opizia; Zea]**

Mehrhoff, L.J. 2000. Perennial *Microstegium vimineum* (Poaceae): An apparent misidentification. J. Torrey Bot. Soc. 127:251–254. **[Microstegium]**

Michael, P.W. 1983. Taxonomy and distribution of *Echinochloa* species with a special reference to their occurrence as weeds of rice. Pp. 291–306 *in* International Rice Research Institute. Proc. Conf. Weed Control in Rice (31 Aug.–4 Sept. 1981). International Rice Research Institute, Los Baños, Laguna, Philippines. 422 pp. **[Echinochloa]**

Michael, P.W. 2001. The taxonomy and distribution of *Echinochloa* species (barnyard grasses) in the Asian-Pacific region, with a review of pertinent biological studies. Proc. Conf. Asian-Pacific Weed Sci. Soc. 18th [Proc. I]:57–66. **[Echinochloa]**

Michelangeli, F.A., J.I. Davis, and D.W. Stevenson. 2003. Phylogenetic relationships among Poaceae and related families as inferred from morphology, inversions in the plastid genome, and sequence data from the mitochondrial and plastid genomes. Amer. J. Bot. 90:93–106. [Introduction]

Mobberley, D.G. 1956. Taxonomy and distribution of the genus *Spartina*. Iowa State Coll. J. Sci. 30:471–574. **[Spartina]**

Morden, C.W. and S.L. Hatch. 1984. Cleistogamy in *Muhlenbergia cuspidata* (Poaceae). Sida 10:254–255. **[Muhlenbergia]**

Morden, C.W. and S.L. Hatch. 1996. Morphological variation and synopsis of the *Muhlenbergia repens* complex (Poaceae). Sida 17:349–365. **[Muhlenbergia]**

Morrone, O. and F.O. Zuloaga. 1992. Revisión de las especies Sudamericanas nativas e introducidas de los géneros *Brachiaria* y *Urochloa* (Poaceae: Panicoideae: Paniceae). Darwiniana 31:43–109. **[Brachiaria; Urochloa]**

Morrone, O. and F.O. Zuloaga. 1993. Sinopsis del género *Urochloa* (Poaceae: Panicoideae: Paniceae) para México y America Central. Darwiniana 32:59–75. **[Urochloa]**

Murphy, A.H. and R.M. Love. 1950. Hairy oatgrass, *Danthonia pilosa* R. Br., as a weedy range grass. Bull. Calif. Dep. Agric. 39:118–124. **[Rytidosperma]**

Myers, W.M. 1947. Cytology and genetics of forage grasses (concluded). Bot. Rev. 7:369–419. **[Rytidosperma]**

National Research Council. 1993. Vetiver Grass: A Thin Green Line Against Erosion. National Academy Press, Washington, D.C., U.S.A. 171 pp. **[Chrysopogon]**

Nicora, E.G. 1995. Los géneros *Diplachne* y *Leptochloa* (Gramineae, Eragrosteae) de la Argentina y países limítrofes. Darwiniana 33:233–256. **[Leptochloa]**

Parodi, L.R. 1934. Contribución al estudio de las gramíneas del género *Munroa*. Revista Mus. La Plata 34:171–193. **[Munroa]**

Pavlik, B.M. and M.G. Barbour. 1988. Demographic monitoring of endemic sand dune plants, Eureka Valley, California. Biol. Conservation 46:217–242. **[Swallenia]**

Peart, M.H. 1984. The effects of morphology, orientation, and position of grass diaspores on seedling survival. J. Ecol. 69:425–436. **[Andropogoneae]**

Peet, R.K. 1993. A taxonomic study of *Aristida stricta* and *A. beyrichiana*. Rhodora 95:25–37. **[Aristida]**

Pensiero, J.F. 1986. Revisión de las especies argentinas del género *Pappophorum* (Gramineae–Eragrostoideae–Pappophoreae). Darwiniana 17:65–87. **[Pappophorum]**

Peterson, P.M. 2000. Systematics of the Muhlenbergiinae (Chloridoideae: Eragrostideae). Pp. 195–212 *in* S.W.L. Jacobs and J. Everett (eds.). Grasses: Systematics and Evolution. International Symposium on Grass Systematics and Evolution (3rd:1998). CSIRO Publishing, Collingwood, Victoria, Australia. 408 pp. **[Muhlenbergia]**

Peterson, P.M. and C.R. Annable. 1990. A revision of *Blepharoneuron* (Poaceae: Eragrostideae). Syst. Bot. 15:515–525. **[Blepharoneuron]**

Peterson, P.M. and **C.R. Annable.** 1991. Systematics of the annual species of *Muhlenbergia* (Poaceae-Eragrostideae). Syst. Bot. Monogr. 31:1–109. **[Muhlenbergia]**

Peterson, P.M. and **Y. Herrera-Arrieta.** 2001. A leaf blade anatomical survey of *Muhlenbergia* (Poaceae: Muhlenbergiinae). Sida 19:469–506. **[Muhlenbergia]**

Peterson, P.M., R.J. Soreng, G. Davidse, T.S. Filgueras, F.O. Zuloaga, and **E. Judziewicz.** 2001. Catalogue of New World Grasses (Poaceae): II. Subfamily Chloridoideae. Contr. U.S. Nat. Herb. 41:1–255. **[Cynodonteae]**

Peterson, P.M., E.E. Terrell, C.A. Davis, H. Scholz, and **R.J. Soreng.** 1999. *Oplismenus hirtellus* subspecies *undulatifolius*, a new record for North America. Castanea 64:201–202. **[Oplismenus]**

Peterson, P.M., R.D. Webster, and **J. Valdés-Reyna.** 1995. Subtribal classification of the New World Eragrostideae (Poaceae: Chloridoideae). Sida 16:529–544. **[Cynodonteae]**

Peterson, P.M., R.D. Webster and **J. Valdés-Reyna.** 1997. Genera of New World Eragrostideae (Poaceae: Chloridoideae). Smithsonian Contr. Bot. 87:1–50. **[Sporobolus]**

Phillips, S.M. 1972. A survey of the genus *Eleusine* Gaertner (Gramineae) in Africa. Kew Bull. 27:251–70. **[Eleusine]**

Phillips, S.M. 1995. Flora of Ethiopia and Eritrea, vol. 7 (I. Hedberg and S. Edwards, eds.). National Herbarium, Biology Department, Science Faculty, Addis Ababa University, Addis Ababa, Ethiopia and Department of Systematic Botany, Uppsala University, Uppsala, Sweden. 420 pp. **[Acrachne]**

Pierce, G.J. 1979. A biosystematic study of *Cathestum* and *Griffithsochloa* (Gramineae). Ph.D. dissertation, University of Wyoming, Laramie, Wyoming, U.S.A. 244 pp. **[Cathesticum]**

Pohl, R.W. 1968. How to Know the Grasses, rev. ed. Wm. C. Brown Co., Dubuque, Iowa, U.S.A. 244 pp. **[Andropogoneae]**

Pohl, R.W. 1980. Flora Costaricensis: Family #15, Gramineae. Fieldiana, Bot., n.s., 4:1–608. **[Urochloa]**

Pohl, R.W. 1994. *Setariopsis* Scribner. P. 364 *in* G. Davidse, M. Sousa S., and A.O. Chater (eds.). Flora Mesoamericana, vol. 6: Alismataceae a Cyperaceae. Universidad Nacional Autónoma de México, Instituto de Biología, México, D.F., México. 543 pp. **[Setariopsis]**

Pohl, R.W. and **G. Davidse.** 1994a. *Enteropogon* Nees. P. 289 *in* G. Davidse, M. Sousa S., and A.O. Chater (eds.). Flora Mesoamericana, vol. 6: Alismataceae a Cyperaceae. Universidad Nacional Autónoma de México, Instituto de Biología, México, D.F., México. 543 pp. **[Enteropogon]**

Pohl, R.W. and **G. Davidse.** 1994b. *Paspalum* L. Pp. 335–352 *in* G. Davidse, M. Sousa S., and A.O. Chater (eds.). Flora Mesoamericana, vol. 6: Alismataceae a Cyperaceae. Universidad Nacional Autónoma de México, Instituto de Biología, México, D.F., México. 543 pp. **[Paspalum]**

Pohl, R.W. and **N.R. Lersten.** 1975. Stem aerenchyma as a character separating *Hymenachne* and *Sacciolepis* (Gramineae, Panicoideae). Brittonia 27:223–227. **[Hymenachne]**

Prat, H. 1936. La systématique des Graminées. Ann. Sci. Nat., Bot., Ser. 10, 18:165–258. **[Chloridoideae]**

Quattrocchi, U. 2000. CRC World Dictionary of Plant Names: Common Names, Scientific Names, Eponyms, Synonyms, and Etymology, vol. 4: R–Z. CRC Press, Boca Raton, Florida, U.S.A. 2896 pp. **[Tribolium]**

Quinn, J.A. and **J.L. Engel.** 1986. Life-history strategies and sex ratios for a cultivar and a wild population of *Buchloë dactyloides* (Gramineae). Amer. J. Bot. 73:874–881. **[Buchloë]**

Quinn, J.A. and **D.E. Fairbrothers.** 1971. Habitat ecology and chromosome numbers of natural populations of the *Danthonia sericea* complex. Amer. Midl. Naturalist 85:531–536. **[Danthonia]**

Reed, C.F. 1964. A flora of the chrome and manganese ore piles at Canton, in the Port of Baltimore, Maryland and at Newport News, Virginia, with descriptions of genera and species new to the flora of the eastern United States. Phytologia 10:321–405. **[Apluda; Dinebra; Ischaemum; Panicum; Themeda; Urochloa]**

Reeder, C.G. 1985. The genus *Lycurus* (Gramineae) in North America. Phytologia 57:283–291. **[Lycurus]**

Reeder, J.R. 1960. The systematic position of the grass genus *Anthephora*. Trans. Amer. Microscop. Soc. 79:211–218. **[Anthephora]**

Reeder, J.R. 1965. The tribe Orcuttieae and the subtribes of the Pappophoreae (Gramineae). Madroño 18:18–28. **[Orcuttieae; Pappophoreae]**

Reeder, J.R. 1976. Systematic position of *Redfieldia* (Gramineae). Madroño 23:434–438. **[Redfieldia]**

Reeder, J.R. 1982. Systematics of the tribe Orcuttieae (Gramineae) and the description of a new segregate genus, *Tuctoria*. Amer. J. Bot. 69:1082–1095. **[Orcuttieae]**

Reeder, J.R. 2001 [publication date 2002]. Noteworthy collections: Arizona. Madroño 48:212–213. **[Setariopsis]**

Reeder, J.R. and **R.S. Felger.** 1989. The *Aristida californica–glabrata* complex (Gramineae). Madroño 36:187–197. **[Aristida]**

Reeder, J.R. and **C.G. Reeder.** 1980. Systematics of *Bouteloua breviseta* and *B. ramosa* (Gramineae). Syst. Bot. 5:312–321. **[Bouteloua]**

Reeder, J.R. and **C.G. Reeder.** 1990. *Bouteloua eludens*: Elusive indeed, but not rare. Desert Pl. 10:19–22, 31–32. **[Bouteloua]**

Reeder, J.R. and **D.N. Singh.** 1968. Chromosome numbers in the tribe Pappophoreae. Madroño 19:183–187. **[Pappophoreae]**

Reeder, J.R. and L.J. Toolin. 1987. *Scleropogon* (Gramineae), a monotypic genus with disjunct distribution. Phytologia 62:267–275. [Scleropogon]

Reeder, J.R. and L.J. Toolin. 1989. Notes on *Pappophorum* (Gramineae: Pappophoreae). Syst. Bot. 14:349–358. [Pappophorum]

Riggins, R. 1977. A biosystematic study of the *Sporobolus asper* complex (Gramineae). Iowa State J. Res. 51:287–321. [Sporobolus]

Rominger, J.M. 1962. Taxonomy of *Setaria* (Gramineae) in North America. Illinois Biol. Monogr. 29:1–132. [Setaria]

Roy, S., J.-P. Simon, and F.-J. Lapointe. 2000. Determination of the origin of the cold-adapted populations of barnyard grass (*Echinochloa crus-galli*) in eastern North America: A total-evidence approach using RAPD DNA and DNA sequences. Canad. J. Bot. 78:1505–1513. [Echinochloa]

Saltonstall, K. 2002. Cryptic invasion by a non-native genotype of the common reed, *Phragmites australis*, into North America. Proc. Natl. Acad. Sci. U.S.A. 99:2445–2449. [Phragmites]

Sánchez, E. 1979. Anatomía foliar de las especies y variedades argentinas de los géneros *Tridens* Roem. et Schult. y *Erioneuron* Nash (Gramineae–Eragrostoideae–Eragrosteae). Darwiniana 22:159–175. [Erioneuron]

Sánchez, E. 1983. *Dasyochloa* Willdenow *ex* Rydberg (Poaceae). Lilloa 36:131–8. [Dasyochloa]

Sánchez, E. 1984. Estudios anatómicos en el género *Munroa* (Poaceae, Chloridoideae, Eragrostideae). Darwiniana 25:43–57. [Munroa]

Sánchez, E. and Z.E. Rúgolo de Agrasar. 1986. Estudio taxonómico sobre el género *Lycurus* (Gramineae). Parodiana 4:267–310. [Lycurus]

Sánchez-Ken, J.G. and L.G. Clark. 2001. Gynerieae, a new neotropical tribe of grasses (Poaceae). Novon 11:350–352. [Gynerieae; Panicoideae]

Sánchez Vega, I. and S.D. Koch. 1988. Estudio biosistemático de *Eragrostis mexicana*, *E. neomexicana*, *E. orcuttiana*, y *E. virescens* (Gramineae: Chloridoideae). Bol. Soc. Bot. 48:95–112. [Eragrostis]

Sauer, J.D. 1972. Revision of *Stenotaphrum* (Gramineae: Paniceae) with attention to its historical geography. Brittonia 24:202–222. [Stenotaphrum]

Savile, D.B.O. 1979. Fungi as aids in higher plant classification. Bot. Rev. 45:377–503. [Chloridoideae]

Schlechtendal, D.F.L. 1861–1862. Ueber *Setaria* P.B. Linnaea 31:387–509. [Ixophorus]

Schmelzer, G.H. 1997. Review of *Pennisetum* section *Brevivalvula* (Poaceae). Euphytica 97:1–20. [Pennisetum]

Scholz, U. 1981. Monographie der Gattung *Oplismenus* (Gramineae). Phanerog. Monog. Tomus XIII. J. Cramer, Vaduz, Germany. 217 pp. [Oplismenus]

Scott, T. 1995. Whispering grasses. Gardens Ill. 15:56–65. [Miscanthus]

Scribner, F.L. 1897. Studies on American grasses: I. The Genus *Ixophorus*. Bull. Div. Agrostol. U.S.D.A. 4^1:5-7. [Ixophorus]

Sendulsky, T. 1978. *Brachiara*: Taxonomy of cultivated and native species in Brazil. Hoehnea 7:99–139. [Urochloa]

Shaw, R.B. and F.E. Smeins. 1981. Some anatomical and morphological characteristics of the North American species of *Eriochloa* (Poaceae: Paniceae). Bot. Gaz. 142:534–544. [Eriochloa]

Shaw, R.B. and R.D. Webster. 1987. The genus *Eriochloa* (Poaceae: Paniceae) in North and Central America. Sida 12:165–207. [Eriochloa]

Simon, B.K. 1972. A revision to the genus *Sacciolepis* (Gramineae) in the "Flora Zambesiaca" area. Kew Bull. 27:387–406. [Sacciolepis]

Simon, B.K. 1992. *Themeda*. Pp. 1227–1228 *in* J.R. Wheeler (ed.). Flora of the Kimberley Region. Department of Conservation and Land Management, Western Australian Herbarium, Como, Western Australia, Australia. 1327 pp. [Themeda]

Simon, B.K. and S.W.L. Jacobs. 1999. Revision of the genus *Sporobolus* (Poaceae, Chloridoideae) in Australia. Austral. J. Bot. 12:375–448. [Sporobolus]

Smith, J.P., Jr. 1971. Taxonomic revision of the genus *Gymnopogon* (Gramineae). Iowa State Coll. J. Sci. 45:319–385. [Gymnopogon]

Snow, N. 1997. Phylogeny and systematics of *Leptochloa* P. Beauv. *sensu lato* (Poaceae, Chloridoideae). Ph.D. dissertation, Washington University, St. Louis, Missouri, U.S.A. 506 pp. [Leptochloa]

Snow, N. 1998. Nomenclatural changes in *Leptochloa* P. Beauvois *sensu lato* (Poaceae, Chloridoideae). Novon 8:77–80. [Leptochloa]

Soderstrom, T.R. 1967. Taxonomic study of subgenus *Podosemum* and section *Epicampes* of *Muhlenbergia* (Gramineae). Contr. U.S. Natl. Herb. 34:75–189. [Muhlenbergia]

Soderstrom, T.R. 1981.The grass subfamily Centostecoideae [*sic*]. Taxon 30:614–616. [Centothecoideae]

Soderstrom, T.R. and H.F. Decker. 1965. *Allolepis*: A new segregate of *Distichlis* (Gramineae). Madroño 18:33–64. [Allolepis]

Soderstrom, T.R., K.W. Hilu, C.S. Campbell, and M.E. Barkworth (eds.). Grass Systematics and Evolution. Smithsonian Institution Press, Washington, D.C., U.S.A. 473 pp. [Introduction]

Soenarko, S. 1977. The genus *Cymbopogon* Sprengel (Gramineae). Reinwardtia 9:225–375. [Cymbopogon]

Sohns, E.R. 1955. *Cenchrus* and *Pennisetum*: Fascicle morphology. J. Wash. Acad. Sci. 45:135–143. [Cenchrus; Pennisetum]

Sohns, E.R. 1956. The genus *Hilaria*. J. Wash. Acad. Sci. 46:311–321. [Hilaria]

Soreng, R.J. and **J.I. Davis.** 1998. Phylogenetics and character evolution in the grass family (Poaceae): Simultaneous analysis of morphological and chloroplast DNA restriction site character sets. Bot. Rev. (Lancaster) 64:1–85. [**Chloridoideae**]

Spangler, R.E. 2000. Andropogoneae systematics and generic limits in *Sorghum*. Pp. 167–170 *in* S.W.L. Jacobs and J. Everett (eds.). Grasses: Systematics and Evolution. International Symposium on Grass Systematics and Evolution (3rd:1998). CSIRO Publishing, Collingwood, Victoria, Australia. 408 pp. [**Andropogoneae; Sorghum**]

Spicher, D. and **M. Joselyn.** 1985. *Spartina* (Gramineae) in northern California: Distribution and taxonomic notes. Madroño 32:158–167. [**Spartina**]

Stafleu, F.A. and **R.S. Cowan.** 1976–1988. Taxonomic Literature: A Selective Guide to Botanical Publications and Collections with Dates, Commentaries and Types. 7 vols, ed. 2. Regnum Vegetabile [series], vols. 94, 98, 105, 110, 112, 115–116. Bohn, Scheltema and Holkema, Utrecht, The Netherlands. [Introduction]

Stafleu, F.A. and **E.A. Mennega.** 1992+. Taxonomic Literature: A Selective Guide to Botanical Publications and Collections with Dates, Commentaries and Types. Supplement. 6+ vols. Regnum Vegetabile [series], vols. 125, 130, 132, 134–135, 137+. Koeltz Scientific Books, Königstein, Germany. [Introduction]

Swallen, J.R. 1932. Five new grasses from Texas. Amer. J. Bot. 19:436–442. [**Muhlenbergia**]

Tateoka, T. 1961. A biosystematic study of *Tridens* (Gramineae). Amer. J. Bot. 48: 565–573. [**Tridens**]

Thieret, J.W. 1956. Synopsis of the genus *Calamovilfa* (Gramineae). Castanea 31:145–152. [**Calamovilfa**]

Trent, J.S. and **K.W. Allred.** 1990. A taxonomic comparison of *Aristida ternipes* and *Aristida hamulosa* (Gramineae). Sida 14:251–261. [**Aristida**]

Tzvelev, N.N. 1976. Zlaki SSSR. Nauka, Leningrad [St. Petersburg], Russia. 788 pp. [In Russian]. [**Aeluropus; Danthonia**]

Valdés-Reyna, J. and **S.L. Hatch.** 1997. A revision of *Erioneuron* and *Dasyochloa* (Poaceae: Eragrostideae). Sida 17:645–666. [Dasyochloa] [**Erioneuron**]

Van den Borre, A. 1994. A taxonomy of the Chloridoideae (Poaceae), with special reference to the genus *Eragrostis*. Ph.D. dissertation, Australian National University, Canberra, New South Wales, Australia. 313 pp. [**Chloris; Cynodonteae**]

Van den Borre, A. and **L. Watson.** 1997. On the classification of the Chloridoideae (Poaceae). Austral. Sys. Bot. 10:491–531. [**Chloridoideae; Cynodonteae**]

Van den Borre, A. and **L. Watson.** 2000. On the classification of the Chloridoideae: Results from morphological and leaf anatomical data analyses. Pp. 180–183 *in* S.W.L. Jacobs and J. Everett (eds.). Grasses: Systematics and Evolution. International Symposium on Grass Systematics and Evolution (3rd:1998). CSIRO Publishing, Collingwood, Victoria, Australia. 408 pp. [**Chloris; Chloridoideae; Eragrostis**]

Vaughn, J.M. 1981. Systematics of *Aristida dichotoma*, *basiramea*, and *curtissii* (Poaceae). Master's thesis. University of Oklahoma, Norman, Oklahoma, U.S.A. 38 pp. [**Aristida**]

Veldkamp, J.F. 1973. A revision of *Digitaria* in Malesia. Blumea 21:1–80. [**Digitaria**]

Veldkamp, J.F. 1996. *Brachiaria*, *Urochloa* (Gramineae-Paniceae) in Malesia. Blumea 41:413–437. [**Urochloa**]

Veldkamp, J.F. 1999. A revision of *Chrysopogon* Trin. including *Vetiveria* Bory (Gramineae) in Thailand and Malesia with notes on some other species from Africa and Australia. Austrobaileya 5:503–533. [**Chrysopogon**]

Veldkamp, J.F., R. de Koning, and **M.S.M. Sosef.** 1986. Generic delimitation of *Rottboellia* and related genera (Gramineae). Blumea 31:281–307. [**Coelorachis; Hackelochloa; Rottboellia**]

Verboom, G.A., H.P. Linder, and **N.P. Barker.** 1994. Haustorial synergids: An important character in the systematics of danthonioid grasses (Arundinoideae: Poaceae). Amer. J. Bot. 81:1601–1610. [**Danthonioideae**]

Vickery, J.W. 1956. A revision of the Australian species of *Danthonia* DC. Contr. New South Wales Natl. Herb. 2:249–325. [**Rytidosperma**]

Vickery, J.W. 1975. *Echinochloa* Beauv. Pp. 189–211 *in* J.W. Vickery. Flora of New South Wales, No. 19, Gramineae, Part 2 (ed. M.D. Tindale). New South Wales Department of Agriculture, Contributions from the New South Wales National Herbarium Flora Series. National Herbarium of New South Wales, Sydney, New South Wales, Australia. 181 pp. [**Echinochloa**]

Villamil, C.B. 1969. El género *Monanthochloë* (Gramineae). Estudios morfológicos y taxonómicos con especial referencia a la especie argentina. Kurtziana 5:369–391. [**Monanthochloë**]

Walsh, N.G. 1994. *Cortaderia*. Pp. 546–548 *in* N.G. Walsh and T.J. Entwisle. Flora of Victoria, vol. 2: Ferns and Allied Plants, Conifers and Monocotyledons. Inkata Press, Melbourne, Australia. 946 pp. [**Cortaderia**]

Walters, T.E., D.S. Decker-Walters, and **D.R. Gordon.** 1994. Restoration considerations for wiregrass (*Aristida stricta*): Allozymic diversity of populations. Conservation Biol. 8:581–585. [**Aristida**]

Watson, L., H.T. Clifford, and **M.J. Dallwitz.** 1985. The classification of the Poaceae: Subfamilies and supertribes. Austral. J. Bot. 33:433–484. [**Aristidoideae; Arundinoideae**]

Watson, L. and **M.J. Dallwitz.** 1992. The Grass Genera of the World. C.A.B. International, Wallingford, England. 1038 pp. [**Aristidoideae; Arundinoideae**]

Watt, G. 1904. *Coix* spp. or Job's tears: A review of all available information. Agric. Ledger 13:189–229. [**Coix**]

Weakley, A.S. and **P.M. Peterson.** 1998. Taxonomy of the *Sporobolus floridanus* complex (Poaceae: Sporobolinae). Sida 18:247–270. [**Sporobolus**]

Webster, R.D. 1981. A biosystematic study of the *Digitaria sanguinalis* complex in North America (Poaceae). Ph.D. dissertation, Texas A&M University, College Station, Texas, U.S.A. 156 pp. [**Digitaria**]

Webster, R.D. 1983. A revision of the genus *Digitaria* Haller (Paniceae: Poaceae) in Australia. Brunonia 6:131–216. [**Digitaria**]

Webster, R.D. 1987a. Taxonomy of *Digitaria* section *Digitaria* in North America (Poaceae: Paniceae). Sida 12:209–222. [**Digitaria**]

Webster, R.D. 1987b. The Australian Paniceae (Poaceae). J. Cramer, Berlin and Stuttgart, Germany. 322 pp. [**Urochloa**]

Webster, R.D. and **S.L. Hatch.** 1981. Taxonomic relationships of Texas specimens of *Digitaria ciliaris* and *Digitaria bicornis* (Poaceae). Sida 9:34–42. [**Digitaria**]

Webster, R.D. and **S.L. Hatch.** 1990. Taxonomy of *Digitaria* section *Aequiglumae* (Poaceae: Paniceae). Sida 14:145–167. [**Digitaria**]

Webster, R.D. and **R.B. Shaw.** 1995. Taxonomy of the North American species of *Saccharum* (Poaceae: Andropogoneae). Sida 16:551–580. [**Saccharum**]

Weintraub, F.C. 1953. Grasses Introduced into the United States. Agricultural Handbook No. 58. Forest Service, U.S. Department of Agriculture, Washington, D.C., U.S.A. 79 pp. [**Rytidosperma**]

Welsh, S.L., N.D. Atwood, S. Goodrich, and **L.C. Higgins** (eds.). 1993. A Utah Flora, ed. 2, revised. Monte L. Bean Life Science Museum, Brigham Young University, Provo, Utah, U.S.A. 986 pp. [**Muhlenbergia**]

Welzen, P.C. van. 1981. A taxonomic revision of *Arthraxon* Beauv. (Gramineae). Blumea 27:255–300. [**Arthraxon**]

Wendel, J.F. 2000. Genome evolution in polyploids. Pl. Molec. Biol. 42:225–249. [Introduction]

Wenger, L.E. 1940. Inflorescence variations in buffalo grass, *Buchloë dactyloides*. J. Amer. Soc. Agron. 32:274–277. [**Buchloë**]

Westerbergh, A. and **J. F. Doebley.** 2002. Morphological traits defining species differences in wild relatives of maize are controlled by multiple quantitative trait loci. Evolution 56:273–283. [**Zea**]

Wipff, J.K. 1995. A biosystematic study of selected facultative apomictic species of *Pennisetum* (Poaceae: Paniceae) and their hybrids. Ph.D. dissertation, Texas A&M University, College Station, Texas, U.S.A. 183 pp. [**Pennisetum**]

Wipff, J.K. 1996. Nomenclatural combinations in *Schizachyrium* (Poaceae: Andropogoneae). Phytologia 80:35–39. [**Schizachyrium**]

Wipff, J.K. 2001. Nomenclatural changes in *Pennisetum* (Poaceae: Paniceae). Sida:19:523–530. [**Pennisetum**]

Wipff, J.K. and **S.L. Hatch.** 1994. A systematic study of *Digitaria* sect. *Pennatae* (Poaceae: Paniceae) in the New World. Syst. Bot. 19:613–627. [**Digitaria**]

Wipff, J.K. and **S.D. Jones.** 1996. A new combination in *Bouteloua* (Poaceae). Sida 17:111–114. [**Bouteloua**]

Wipff, J.K., R.I. Lonard, S.D. Jones, and **S.L. Hatch.** 1993. The genus *Urochloa* (Poaceae: Paniceae) in Texas, including one previously unreported species for the state. Sida 15:405–413. [**Urochloa**]

Wipff, J.K. and **B.S. Rector.** 1993. *Rottboellia cochinchinensis* (Poaceae: Andropogoneae) new to Texas. Sida 15:419–424. [**Rottboellia**]

Wolf, N.M. 1776. Genera Plantarum. [publisher unknown, Danzig, Germany]. 177 pp. [**Eragrostis**]

Wunderlin, R.P. 1988. Guide to the Vascular Plants of Florida. University Press of Florida, Gainesville, Florida, U.S.A. 806 pp. [**Echinochloa**]

Yabuno, T. 1962. Cytotaxonomic studies on the two cultivated species and the wild relatives in the genus *Echinochloa*. Cytologia 27:296–305. [**Echinochloa**]

Yabuno, T. 1966. Biosystematic study of the genus *Echinochloa*. J. Jap. Bot. 19:277–323. [**Echinochloa**]

Yabuno, T. 1984. A biosystematic study on *Echinochloa oryzoides* (Ard.) Fritsch. Cytologia 49:673–678. [**Echinochloa**]

Yabuno, T. and **H. Yamaguchi.** 1996. Hie no hakubutsugaku [A natural history of *Echinochloa*]. Dow Chemicals and Dow Elanco, Tokyo, Japan. 196 pp. [In Japanese]. [**Echinochloa**]

Yasuda, K., A. Yano, Y. Nakayama, and **H. Yamaguchi.** 2001. Identification of *Echinochloa oryzicola* and *E. crus-galli* using PCR-RFLP technique. [Abstract]. J. Weed Sci. Technol. 46, Suppl.:204–205. [Text in Japanese, title in Japanese and English]. [**Echinochloa**]

Yates, H.O. 1966a. Morphology and cytology of *Uniola* (Gramineae). SouthW. Naturalist 11:145–189. [**Chasmanthium; Uniola**]

Yates, H.O. 1966b. Revision of grasses traditionally referred to *Uniola*, I. *Uniola* and *Leptochloöpsis*. SouthW. Naturalist 11:372–394. [**Uniola**]

Yates, H.O. 1966c. Revision of grasses traditionally referred to *Uniola*, II. *Chasmanthium*. SouthW. Naturalist 11:415–455. [**Centotheceae; Chasmanthium; Uniola**]

Zaitcheck, B.F., L.G. Le Roux, and **E.A. Kellogg.** 2000. Development of male flowers in *Zizania aquatica* (North American wild-rice; Gramineae). Int. J. Pl. Sci. 161:345–351. [**Panicoideae**]

Zizka, G. 1988. Revision der Melinideae Hitchcock (Poaceae, Panicoideae). Biblioth. Bot. 138:1–149. [**Melinis**]

Zizka, G. 1990. Taxonomy of the Melinideae (Poaceae: Panicoideae). Mitt. Inst. Allg. Bot. Hamburg 23:563–572. [**Melinis**]

Zotov, V.D. 1963. Synopsis of the grass subfamily Arundinoideae in New Zealand. New Zealand J. Bot. 1:78–136. [**Rytidosperma**]

Zuloaga, F.O., R.P. Ellis, and **O. Morrone.** 1993. A Revision of *Panicum* subg. *Dichanthelium* sect. *Dichanthelium* (Poaceae: Panicoideae: Paniceae) in Mesoamerica, the West Indies, and South America. Ann. Missouri Bot. Gard. 80:119–190. [**Dichanthelium**]

Zuloaga, F.O. and **O. Morrone.** 1996. Revisión de las especies Americanas de *Panicum* subgenera *Panicum* sección *Panicum* (Poaceae: Panicoideae: Paniceae). Ann. Missouri Bot. Gard. 83:200–280. [**Panicum**]

Zuloaga, F.O., O. Morrone, A.S. Vega, and **L.M. Giussani.** 1998. Revisión y análisis cladístico de *Steinchisma* (Poaceae: Panicoideae: Paniceae). Ann. Missouri Bot. Gard. 85:631–656. [**Steinchisma**]

Names and Synonyms

The list of names that follows contains all the scientific names of taxa that are described in this volume plus the names of the genera, tribes, and subfamilies that will be treated in volume 24. Names followed by "[24]" belong to taxa that will be treated in volume 24. The infrageneric taxa that will be treated in that volume are not listed.

The listing also shows synonyms of the taxa mentioned in this volume as well as some of those to be treated in volume 24. The listing of synonyms is not complete, containing only names found in current floras, herbarium databases, and miscellaneous other publications.

Accepted names are in **boldface**, followed by the name(s) of their author(s) and the page number on which the taxon is described, if the taxon is treated in this volume. The names of the authors are abbreviated according to Brummitt and Powell (1992). Some of the accepted names have been inappropriately applied to taxa in the *Flora* region. These names are followed by the statement: "within the *Flora* region, misapplied to . . .".

Synonyms are in *italics*, and are followed by the name used in this volume and the page on which that taxon is treated. Names in standard typeface fall into two categories. Some are names that violate the International Code of Botanical Nomenclature (Greuter et al. 2000) in some way and therefore cannot be used; others are suspected by the author of the treatment concerned to apply to a hybrid for which no appropriate scientific name has been published.

Infraspecific names are listed alphabetically, based on the *infraspecific epithet*; the authors of tautonymic subspecies and varieties (subspecies and varieties in which the specific and infraspecific epithet are the same) are ignored as far as the alphabetic listing is concerned. In general, names of forms are listed only if they are mentioned in the text.

NOMENCLATURAL INDEX

Names of accepted taxa are listed in **boldface**; names of other taxa are in *italics*; vernacular names are in SMALL CAPS. Page numbers in **boldface** refer to the primary treatment; page numbers in *italics* refer to the illustration; other page references are in regular type.

Infraspecific names are listed alphabetically, based on the *infraspecific epithet*; the rank (subspecies, variety, or form) is ignored. The authors of tautonymic infraspecific taxa (subspecies, varieties, and forms in which the specific and infraspecific epithet are the same) are ignored as far as the alphabetic listing is concerned.

If the spellings are the same, vernacular names are listed after all the scientific names for that genus.

Numerical Listing of the Subfamilies, Tribes, and Genera

The list below shows the subfamilies, tribes, and genera of this volume in the order in which they are treated and the pages on which they are treated. For the tribes and genera, the numbers on the left show the relative position of the genera within the volume, and correspond to the numbers shown in the page headers.

Alphabetical Listing of the Subfamilies, Tribes, and Genera

The list below shows the subfamilies, tribes, and genera treated in this volume in alphabetical order, followed by the page number on which they are treated.

Political Map of North America North of Mexico

Canadian Provinces

Alta.	*Alberta*	N.S.	*Nova Scotia*
B.C.	*British Columbia*	Ont.	*Ontario*
Man.	*Manitoba*	P.E.I.	*Prince Edward Island*
N.B.	*New Brunswick*	Que.	*Quebec*
Nfld.	*Newfoundland (incl. Labrador)*	Sask.	*Saskatchewan*
N.W.T.	*Northwest Territories*	Yukon	

United States

Ala.	*Alabama*	Mont.	*Montana*
Alaska		Nebr.	*Nebraska*
Ariz.	*Arizona*	Nev.	*Nevada*
Ark.	*Arkansas*	N.H.	*New Hampshire*
Calif.	*California*	N.J.	*New Jersey*
Colo.	*Colorado*	N. Mex.	*New Mexico*
Conn.	*Connecticut*	N.Y.	*New York*
Del.	*Delaware*	N.C.	*North Carolina*
D.C.	*District of Columbia*	N. Dak.	*North Dakota*
Fla.	*Florida*	Ohio	
Ga.	*Georgia*	Okla.	*Oklahoma*
Idaho		Oreg.	*Oregon*
Ill.	*Illinois*	Pa.	*Pennsylvania*
Ind.	*Indiana*	R.I.	*Rhode Island*
Iowa		S.C.	*South Carolina*
Kans.	*Kansas*	S. Dak.	*South Dakota*
Ky.	*Kentucky*	Tenn.	*Tennessee*
La.	*Louisiana*	Tex.	*Texas*
Maine		Utah	
Md.	*Maryland*	Vt.	*Vermont*
Mass.	*Massachusetts*	Va.	*Virginia*
Mich.	*Michigan*	Wash	*Washington*
Minn.	*Minnesota*	W. Va.	*West Virginia*
Miss.	*Mississippi*	Wis.	*Wisconsin*
Mo.	*Missouri*	Wyo.	*Wyoming*